Machine Learning

A Bayesian and Optimization Perspective

Machine Learning

A Bayesian and Optimization Perspective

2nd Edition

Sergios Theodoridis
Department of Informatics and Telecommunications
National and Kapodistrian University of Athens
Athens, Greece

Shenzhen Research Institute of Big Data
The Chinese University of Hong Kong
Shenzhen, China

ELSEVIER
ACADEMIC PRESS
An imprint of Elsevier

Academic Press is an imprint of Elsevier
125 London Wall, London EC2Y 5AS, United Kingdom
525 B Street, Suite 1650, San Diego, CA 92101, United States
50 Hampshire Street, 5th Floor, Cambridge, MA 02139, United States
The Boulevard, Langford Lane, Kidlington, Oxford OX5 1GB, United Kingdom

Notices

Knowledge and best practice in this field are constantly changing. As new research and experience broaden our understanding, changes in research methods, professional practices, or medical treatment may become necessary.

Practitioners and researchers must always rely on their own experience and knowledge in evaluating and using any information, methods, compounds, or experiments described herein. In using such information or methods they should be mindful of their own safety and the safety of others, including parties for whom they have a professional responsibility.

To the fullest extent of the law, neither the Publisher nor the authors, contributors, or editors, assume any liability for any injury and/or damage to persons or property as a matter of products liability, negligence or otherwise, or from any use or operation of any methods, products, instructions, or ideas contained in the material herein.

Library of Congress Cataloging-in-Publication Data
A catalog record for this book is available from the Library of Congress

British Library Cataloguing-in-Publication Data
A catalogue record for this book is available from the British Library

ISBN: 978-0-12-818803-3

For information on all Academic Press publications
visit our website at https://www.elsevier.com/books-and-journals

Publisher: Mara Conner
Acquisitions Editor: Tim Pitts
Editorial Project Manager: Charlotte Rowley
Production Project Manager: Paul Prasad Chandramohan
Designer: Greg Harris

Typeset by VTeX

Printed in Great Britain
Last digit is the print number: 10 9 8 7 6 5 4 3 2 1

Στο Δεσποινὰκι
For Everything
All These Years

Contents

About the Author

Sergios Theodoridis is professor of machine learning and signal processing with the National and Kapodistrian University of Athens, Athens, Greece and with the Chinese University of Hong Kong, Shenzhen, China. He has received a number of prestigious awards, including the 2014 IEEE Signal Processing Magazine Best Paper Award, the 2009 IEEE Computational Intelligence Society Transactions on Neural Networks Outstanding Paper Award, the 2017 European Association for Signal Processing (EURASIP) Athanasios Papoulis Award, the 2014 IEEE Signal Processing Society Education Award, and the 2014 EURASIP Meritorious Service Award. He has served as president of EURASIP and vice president for the IEEE Signal Processing Society. He is a Fellow of EURASIP and a Life Fellow of IEEE. He is the coauthor of the book Pattern Recognition, 4th edition, Academic Press, 2009 and of the book Introduction to Pattern Recognition: A MATLAB Approach, Academic Press, 2010.

Preface

Machine learning is a name that is gaining popularity as an umbrella and evolution for methods that have been studied and developed for many decades in different scientific communities and under different names, such as statistical learning, statistical signal processing, pattern recognition, adaptive signal processing, image processing and analysis, system identification and control, data mining and information retrieval, computer vision, and computational learning. The name "machine learning" indicates what all these disciplines have in common, that is, to *learn from data*, and then *make predictions*. What one tries to learn from data is their underlying structure and regularities, via the development of a *model*, which can then be used to provide predictions.

To this end, a number of diverse approaches have been developed, ranging from optimization of cost functions, whose goal is to optimize the deviation between what one observes from data and what the model predicts, to probabilistic models that attempt to model the statistical properties of the observed data.

The goal of this book is to approach the machine learning discipline in a unifying context, by presenting major paths and approaches that have been followed over the years, without giving preference to a specific one. It is the author's belief that all of them are valuable to the newcomer who wants to learn the secrets of this topic, from the applications as well as from the pedagogic point of view. As the title of the book indicates, the emphasis is on the processing and analysis front of machine learning and not on topics concerning the theory of learning itself and related performance bounds. In other words, the focus is on methods and algorithms closer to the application level.

The book is the outgrowth of more than three decades of the author's experience in research and teaching various related courses. The book is written in such a way that individual (or pairs of) chapters are as self-contained as possible. So, one can select and combine chapters according to the focus he/she wants to give to the course he/she teaches, or to the topics he/she wants to grasp in a first reading. Some guidelines on how one can use the book for different courses are provided in the introductory chapter.

Each chapter grows by starting from the basics and evolving to embrace more recent advances. Some of the topics had to be split into two chapters, such as sparsity-aware learning, Bayesian learning, probabilistic graphical models, and Monte Carlo methods. The book addresses the needs of advanced graduate, postgraduate, and research students as well as of practicing scientists and engineers whose interests lie *beyond black-box* approaches. Also, the book can serve the needs of short courses on specific topics, e.g., sparse modeling, Bayesian learning, probabilistic graphical models, neural networks and deep learning.

Second Edition

The first edition of the book, published in 2015, covered advances in the machine learning area up to 2013–2014. These years coincide with the start of a real booming in research activity in the field of *deep learning* that really reshaped our related knowledge and revolutionized the field of machine learning. The main emphasis of the current edition was to, basically, rewrite Chapter 18. The chapter now covers a review of the field, starting from the early days of the perceptron and the perceptron rule, until the most recent advances, including convolutional neural networks (CNNs), recurrent neural networks (RNNs), adversarial examples, generative adversarial networks (GANs), and capsule networks.

Also, the second edition covers in a more extended and detailed way nonparametric Bayesian methods, such as Chinese restaurant processes (CRPs) and Indian buffet processes (IBPs). It is the author's belief that Bayesian methods will gain in importance in the years to come. Of course, only time can tell whether this will happen or not. However, the author's feeling is that uncertainty is going to be a major part of the future models and Bayesian techniques can be, at least in principle, a reasonable start. Concerning the other chapters, besides the (omnipresent!) typos that have been corrected, changes have been included here and there to make the text easier to read, thanks to suggestions by students, colleagues, and reviewers; I am deeply indebted to all of them.

Most of the chapters include *MATLAB®* exercises, and the related code is freely available from the book's companion website. Furthermore, in the second edition, all the computer exercises are also given in *Python* together with the corresponding code, which are also freely available via the website of the book. Finally, some of the computer exercises in Chapter 18 that are related to deep learning, and which are closer to practical applications, are given in *Tensorflow*.

The *solutions manual* as well *lecture slides* are available from the book's website for instructors.

In the second edition, all *appendices* have been moved to the website associated with the book, and they are freely downloadable. This was done in an effort to save space in a book that is already more than 1100 pages. Also, some sections dedicated to methods that were present in various chapters in the first edition, which I felt do not constitute basic knowledge and current mainstream research topics, while they were new and "fashionable" in 2015, have been moved, and they can be downloaded from the companion website of the book.

Instructor site URL:

http://textbooks.elsevier.com/web/Manuals.aspx?isbn=9780128188033

Companion Site URL:

https://www.elsevier.com/books-and-journals/book-companion/9780128188033

Acknowledgments

Writing a book is an effort on top of everything else that must keep running in parallel. Thus, writing is basically an early morning, after five, and over the weekends and holidays activity. It is a big effort that requires dedication and persistence. This would not be possible without the support of a number of people—people who helped in the simulations, in the making of the figures, in reading chapters, and in discussing various issues concerning all aspects, from proofs to the structure and the layout of the book.

First, I would like to express my gratitude to my mentor, friend, and colleague Nicholas Kalouptsidis, for this long-lasting and fruitful collaboration.

The cooperation with Kostas Slavakis over the more recent years has been a major source of inspiration and learning and has played a decisive role for me in writing this book.

I am indebted to the members of my group, and in particular to Yannis Kopsinis, Pantelis Bouboulis, Simos Chouvardas, Kostas Themelis, George Papageorgiou, Charis Georgiou, Christos Chatzichristos, and Emanuel Morante. They were next to me the whole time, especially during the difficult final stages of the completion of the manuscript. My colleagues Aggelos Pikrakis, Kostas Koutroumbas, Dimitris Kosmopoulos, George Giannakopoulos, and Spyros Evaggelatos gave a lot of their time for discussions, helping in the simulations and reading chapters.

Without my two sabbaticals during the spring semesters of 2011 and 2012, I doubt I would have ever finished this book. Special thanks go to all my colleagues in the Department of Informatics and Telecommunications of the National and Kapodistrian University of Athens.

During my sabbatical in 2011, I was honored to be a holder of an Excellence Chair in Carlos III University of Madrid and spent the time with the group of Anibal Figuieras-Vidal. I am indebted to Anibal for his invitation and all the fruitful discussions and the bottles of excellent red Spanish wine we had together. Special thanks go to Jerónimo Arenas-García and Antonio Artés-Rodríguez, who have also introduced me to aspects of traditional Spanish culture.

During my sabbatical in 2012, I was honored to be an Otto Mønsted Guest Professor at the Technical University of Denmark with the group of Lars Kai Hansen. I am indebted to him for the invitation and our enjoyable and insightful discussions, as well as his constructive comments on chapters of the book and the visits to the Danish museums on weekends. Also, special thanks go to Morten Mørup and the late Jan Larsen for the fruitful discussions.

The excellent research environment of the Shenzhen Research Institute of Big Data of the Chinese University of Hong Kong ignited the spark and gave me the time to complete the second edition of the book. I am deeply indebted to Tom Luo who offered me this opportunity and also introducing me to the secrets of Chinese cooking.

A number of colleagues were kind enough to read and review chapters and parts of the book and come back with valuable comments and criticism. My sincere thanks go to Tulay Adali, Kostas Berberidis, Jim Bezdek, Soterios Chatzis, Gustavo Camps-Valls, Rama Chellappa, Taylan Cemgil and his students, Petar Djuric, Paulo Diniz, Yannis Emiris, Mario Figuieredo, Georgios Giannakis, Mark Girolami, Dimitris Gunopoulos, Alexandros Katsioris, Evaggelos Karkaletsis, Dimitris Katselis, Athanasios Liavas, Eleftherios Kofidis, Elias Koutsoupias, Alexandros Makris, Dimitirs Manatakis,

Elias Manolakos, Petros Maragos, Francisco Palmieri, Jean-Christophe Pesquet, Bhaskar Rao, George Retsinas, Ali Sayed, Nicolas Sidiropoulos, Paris Smaragdis, Isao Yamada, Feng Yin, and Zhilin Zhang.

Finally, I would like to thank Tim Pitts, the Editor at Academic Press, for all his help.

Notation

I have made an effort to keep a consistent mathematical notation throughout the book. Although every symbol is defined in the text prior to its use, it may be convenient for the reader to have the list of major symbols summarized together. The list is presented below:

- Vectors are denoted with **boldface** letters, such as $\boldsymbol{x}$.
- Matrices are denoted with capital letters, such as A.
- The determinant of a matrix is denoted as $\det\{A\}$, and sometimes as $|A|$.
- A diagonal matrix with elements $a_1, a_2, \ldots, a_l$ in its diagonal is denoted as $A = \text{diag}\{a_1, a_2, \ldots, a_l\}$.
- The identity matrix is denoted as I.
- The trace of a matrix is denoted as $\text{trace}\{A\}$.
- Random variables are denoted with roman fonts, such as x, and their corresponding values with *mathmode* letters, such as x.
- Similarly, random vectors are denoted with roman **boldface**, such as $\mathbf{x}$, and the corresponding values as $\boldsymbol{x}$. The same is true for random matrices, denoted as X and their values as X.
- Probability values for discrete random variables are denoted by capital P, and probability density functions (PDFs), for continuous random variables, are denoted by lower case p.
- The vectors are assumed to be column-vectors. In other words,

$$\boldsymbol{x} = \begin{bmatrix} x_1 \\ x_2 \\ \vdots \\ x_l \end{bmatrix}, \text{ or } \boldsymbol{x} = \begin{bmatrix} x(1) \\ x(2) \\ \vdots \\ x(l) \end{bmatrix}.$$

  That is, the ith element of a vector can be represented either with a subscript, x_i, or as $x(i)$.
- Matrices are written as

$$X = \begin{bmatrix} x_{11} & x_{12} & \ldots & x_{1l} \\ \vdots & \vdots & \ddots & \vdots \\ x_{l1} & x_{l2} & \ldots & x_{ll} \end{bmatrix}, \text{ or } X = \begin{bmatrix} X(1,1) & X(1,2) & \ldots & X(1,l) \\ \vdots & \vdots & \ddots & \vdots \\ X(l,1) & X(l,2) & \ldots & X(l,l) \end{bmatrix}.$$

- Transposition of a vector is denoted as $\boldsymbol{x}^T$ and the Hermitian transposition as $\boldsymbol{x}^H$.
- Complex conjugation of a complex number is denoted as x^* and also $\sqrt{-1} := j$. The symbol ":=" denotes definition.
- The sets of real, complex, integer, and natural numbers are denoted as $\mathbb{R}$, $\mathbb{C}$, $\mathbb{Z}$, and $\mathbb{N}$, respectively.
- Sequences of numbers (vectors) are denoted as x_n ($\boldsymbol{x}_n$) or $x(n)$ ($\boldsymbol{x}(n)$) depending on the context.
- Functions are denoted with lower case letters, e.g., f, or in terms of their arguments, e.g., $f(x)$ or sometimes as $f(\cdot)$, if no specific argument is used, to indicate a function of a single argument, or $f(\cdot, \cdot)$ for a function of two arguments and so on.

CHAPTER

INTRODUCTION

1

CONTENTS

1.1 THE HISTORICAL CONTEXT

During the period that covers, roughly, the last 250 years, humankind has lived and experienced three *transforming* revolutions, which have been powered by technology and science. The *first industrial* revolution was based on the use of water and steam and its origins are traced to the end of the 18th century, when the first organized factories appeared in England. The *second industrial* revolution was powered by the use of electricity and mass production, and its "birth" is traced back to around the turn of the 20th century. The *third industrial* revolution was fueled by the use of electronics, information technology, and the adoption of automation in production. Its origins coincide with the end of the Second World War.

Although difficult for humans, including historians, to put a stamp on the age in which they themselves live, more and more people are claiming that the *fourth industrial* revolution has already started and is fast transforming everything that we know and learned to live with so far. The fourth industrial revolution builds upon the third one and is powered by the *fusion* of a number of technologies, e.g.,

Machine Learning. https://doi.org/10.1016/B978-0-12-818803-3.00010-6

computers and communications (internet), and it is characterized by the convergence of the *physical*, *digital*, and *biological* spheres.

The terms artificial intelligence (AI) and machine learning are used and spread more and more to denote the type of automation technology that is used in the production (industry), in the distribution of goods (commerce), in the service sector, and in our economic transactions (e.g., banking). Moreover, these technologies affect and shape the way we socialize and interact as humans via social networks, and the way we entertain ourselves, involving games and cultural products such as music and movies.

A distinct qualitative difference of the fourth, compared to the previous industrial revolutions, is that, before, it was the manual skills of humans that were gradually replaced by "machines." In the one that we are currently experiencing, mental skills are also replaced by "machines." We now have automatic answering software that runs on computers, less people are serving us in banks, and many jobs in the service sector have been taken over by computers and related software platforms. Soon, we are going to have cars without drivers and drones for deliveries. At the same time, new jobs, needs, and opportunities appear and are created. The labor market is fast changing and new competences and skills are and will be required in the future (see, e.g., [22,23]).

At the center of this historical happening, as one of the key enabling technologies, lies a discipline that deals with data and whose goal is to extract information and related knowledge that is hidden in it, in order to make predictions and, subsequently, take decisions. That is, the goal of this discipline is to *learn* from data. This is analogous to what humans do in order to reach decisions. Learning through the senses, personal experience, and the knowledge that propagates from generation to generation is at the heart of human intelligence. Also, at the center of any scientific field lies the development of models (often called theories) in order to explain the available experimental evidence. In other words, *data* comprise a major source of *learning*.

1.2 ARTIFICIAL INTELLIGENCE AND MACHINE LEARNING

The title of the book refers to *machine learning*, although the term *artificial intelligence* is used more and more, especially in the media but also by some experts, to refer to any type of algorithms and methods that perform tasks that traditionally required human intelligence. Being aware that definitions of terms can never be exact and there is always some "vagueness" around their respective meanings, I will still attempt to clarify what I mean by machine learning and in which aspects this term means something different from AI. No doubt, there may be different views on this.

Although the term machine learning was popularized fairly recently, as a scientific field it is an old one, whose roots go back to statistics, computer science, information theory, signal processing, and automatic control. Examples of some related names from the past are statistical learning, pattern recognition, adaptive signal processing, system identification, image analysis, and speech recognition. What all these disciplines have in common is that they process data, develop models that are *data-adaptive*, and subsequently make predictions that can lead to decisions. Most of the basic theories and algorithmic tools that are used today had already been developed and known before the dawn of this century. With a "small" yet important difference: the available data, as well as the computer power prior to 2000, were not enough to use some of the more elaborate and complex models that had been developed. The terrain started changing after 2000, in particular around 2010. Large data sets were gradually created and the computer power became affordable to allow the use of more complex models. In turn, more and more applications adopted such algorithmic techniques. "Learning from data" became the new trend and the term machine learning prevailed as an umbrella for such techniques.

Moreover, the big difference was made with the use and "rediscovery" of what is today known as *deep neural networks*. These models offered impressive predictive accuracies that had never been achieved by previous models. In turn, these successes paved the way for the adoption of such models in a wide range of applications and also ignited intense research, and new versions and models have been proposed. These days, another term that is catching up is "data science," indicating the emphasis on how one can develop robust machine learning and computational techniques that deal efficiently with large-scale data.

However, the main rationale, which runs the spine of all the methods that come under the machine learning umbrella, remains the same and it has been around for many decades. The main concept is to estimate a set of parameters that describe the model, using the available data and, in the sequel, to make *predictions* based on *low-level information and signals*. One may easily argue that there is not much *intelligence* built in such approaches. No doubt, deep neural networks involve much more "intelligence" than their predecessors. They have the potential to *optimize* the *representation* of their low-level input information to the computer.

The term "representation" refers to the way in which related information that is hidden in the input data is quantified/coded so that it can be subsequently processed by a computer. In the more technical jargon, each piece of such information is known as a feature (see also Section 1.5.1). As discussed in detail in Chapter 18, where neural networks (NNs) are defined and presented in detail, what makes these models distinctly different from other data learning methods is their *multilayer* structure. This allows for the "building" up of a *hierarchy* of representations of the input information at various abstraction levels. Every layer builds upon the previous one and the higher in hierarchy, the more abstract the obtained representation is. This structure offers to neural networks a significant performance advantage over alternative models, which restrict themselves to a single representation layer. Furthermore, this single-level representation was rather hand-crafted and designed by the users, in contrast to the deep networks that "learn" the representation layers from the input data via the use of optimality criteria.

Yet, in spite of the previously stated successes, I share the view that we are still very far from what an intelligent machine should be. For example, once trained (estimating the parameters) on one data set, which has been developed for a specific task, it is not easy for such models to *generalize* to other tasks. Although, as we are going to see in Chapter 18, advances in this direction have been made, we are still very far from what human intelligence can achieve. When a child sees one cat, readily recognizes another one, even if this other cat has a different color or if it turns around. Current machine learning systems need thousands of images with cats, in order to be trained to "recognize" one in an image. If a human learns to ride a bike, it is very easy to transfer this knowledge and learn to ride a motorbike or even to drive a car. Humans can easily transfer knowledge from one task to another, without forgetting the previous one. In contrast, current machine learning systems lack such a generalization power and tend to forget the previous task once they are trained to learn a new one. This is also an open field of research, where advances have also been reported.

Furthermore, machine learning systems that employ deep networks can even achieve superhuman prediction accuracies on data similar to those with which they have been trained. This is a significant achievement, not to be underestimated, since such techniques can efficiently be used for dedicated jobs; for example, to recognize faces, to recognize the presence of various objects in photographs, and also to annotate images and produce text that is related to the content of the image. They can recognize speech, translate text from one language to another, detect which music piece is currently playing in the bar, and whether the piece belongs to the jazz or to the rock musical genre. At the same time, they can be fooled by carefully constructed examples, known as adversarial examples, in a way that no human would be fooled to produce a wrong prediction (see Chapter 18).

Concerning AI, the term "artificial intelligence" was first coined by John McCarthy in 1956 when he organized the first dedicated conference (see, e.g., [20] for a short history). The concept at that time, which still remains a goal, was whether one can build an intelligent machine, realized on software and hardware, that can possess *human-like* intelligence. In contrast to the field of machine learning, the concept for AI was not to focus on low-level information processing with emphasis on predictions, but on the high-level *cognitive* capabilities of humans to *reason* and *think*. No doubt, we are still very far from this original goal. Predictions are, indeed, part of intelligence. Yet, intelligence is much more than that. Predictions are associated with what we call *inductive* reasoning. Yet what really differentiates human from the animals intelligence is the power of the human mind to form *concepts* and *create conjectures* for explaining data and more general the World in which we live. *Explanations* comprise a high-level facet of our intelligence and constitute the basis for scientific theories and the creation of our civilization. They are assertions concerning the "why" 's and the "how" 's related to a task, e.g., [5,6,11].

To talk about AI, at least as it was conceived by pioneers such as Alan Turing [16], systems should have built-in capabilities for *reasoning* and giving *meaning*, e.g., in language processing, to be able to infer *causality*, to model efficient representations of *uncertainty*, and, also, to pursue long-term goals [8]. Possibly, towards achieving these challenging goals, we may have to understand and implement notions from the theory of mind, and also build machines that implement self-awareness. The former psychological term refers to the understanding that others have their own beliefs and intentions that justify their decisions. The latter refers to what we call *consciousness*. As a last point, recall that human intelligence is closely related to feelings and emotions. As a matter of fact, the latter seem to play an important part in the *creative* mental power of humans (e.g., [3,4,17]). Thus, in this more theoretical perspective AI still remains a *vision for the future*.

The previous discussion should not be taken as an attempt to get involved with philosophical theories concerning the nature of human intelligence and AI. These topics comprise a field in itself, for more than 60 years, which is much beyond the scope of this book. My aim was to make the newcomer in the field aware of some views and concerns that are currently being discussed.

In the more practical front, for the early years, the term AI was used to refer to techniques built around knowledge-based systems that sought to hard-code knowledge in terms of formal languages, e.g., [13]. Computer "reasoning" was implemented via a set of logical inference rules. In spite of the early successes, such methods seem to have reached a limit, see, e.g., [7]. It was the alternative path of machine learning, via learning from data, that gave a real push into the field. These days, the term AI is used as an umbrella to cover all methods and algorithmic approaches that are related to the machine intelligence discipline, with machine learning and knowledge-based techniques being parts of it.

1.3 ALGORITHMS CAN LEARN WHAT IS HIDDEN IN THE DATA

It has already been emphasized that data lie at the heart of machine learning systems. Data are the *beginning*. It is the information hidden in the data, in the form of underlying regularities, correlations, or structure, which a machine learning system tries to "learn." Thus, irrespective of how intelligent a software algorithm is designed to be, it cannot learn more than what the data which it has been trained on allow.

Collecting the data and building the data set on which an "intelligent" system is going to be trained is highly critical. Building data sets that address human needs and developing systems that are going to make decisions on issues, where humans and their lives are involved, requires special attention, and above all, responsibility. This is not an easy issue and good intentions are not enough. We are all products of the societies in which we live, which means that our beliefs, to a large extent, are formed by the prevailing social stereotypes concerning, e.g., gender, racial, ethnic, religious, cultural, class-related, and political views. Most importantly, most of these beliefs take place and exist at a subconscious level. Thus, sampling "typical" cases to form data sets may have a strong flavor of *subjectivity* and introduce biases. A system trained on such data can affect lives, and it may take time for this to be found out. Furthermore, our world is fast changing and these changes should continuously be reflected in the systems that make decisions on our behalf. Outsourcing our lives to computers should be done cautiously and above all in an ethical framework, which is much wider and general than the set of the existing legal rules.

Of course, although this puts a burden on the shoulders of the individuals, governments, and companies that develop data sets and "intelligent" systems, it cannot be left to their good will. On the one hand, a specialized legal framework that guides the *design*, *implementation*, and *use* of such platforms and systems is required to protect our ethical standards and social values. Of course, this does not concern only the data that are collected but also the overall system that is built. Any system that replaces humans should be (a) transparent, (b) fair, and (c) accurate. Not that the humans act, necessarily, according to what the previous three terms mean. However, humans can, also, reason and discuss, we have feelings and emotions, and we do not just perform predictions.

On the other hand, this may be the time when we can develop and build more "objective" systems; that is, to go beyond human subjectivity. However, such "objectivity" should be based on science, rules, criteria, and principles, which are not yet here. As Michael Jordan [8] puts it, the development of such systems will require perspectives from the *social sciences* and *humanities*. Currently, such systems are built following an ad hoc rather than a principled way. Karl Popper [12], one of the most influential philosophers of science, stressed that all knowledge creation is theory-laden. Observations are never free of an underlying theory or explanation. Even if one believes that the process begins with observations, the act of observing requires a *point of view* (see, also, [1,6]).

If I take the liberty to make a bit of a science fiction (something trendy these days), when AI, in the context of its original conception, is realized, then data sampling and creation of data sets for training could be taken care of by specialized algorithms. Maybe such algorithms will be based on scientific principles that in the meantime will have been developed. After all, this may be the time of dawn for the emergence of a new scientific/engineering field that integrates in a principle way data-focused disciplines. To this end, another statement of Karl Popper may have to be implemented, i.e., that of *falsification*, yet in a slightly abused interpretation. An emphasis on building intelligent systems should be directed on criticism and experimentations for finding evidence that *refutes* the principles, which were employed for their development. Systems can only be used if they survive the falsification test.

1.4 TYPICAL APPLICATIONS OF MACHINE LEARNING

It is hard to find a discipline in which machine learning techniques for "learning" from data have not been applied. Yet, there are some areas, which can be considered as typical applications, maybe due to their economic and social impact. Examples of such applications are summarized below.

SPEECH RECOGNITION

Speech is the primary means of communication among humans. Language and speech comprise major attributes that differentiate humans from animals. Speech recognition has been one of the main research topics whose roots date back to the early 1960s. The goal of speech recognition is to develop methods and algorithms that enable the recognition and the subsequent representation in a computer of spoken language. This is an interdisciplinary field involving signal processing, machine learning, linguistics, and computer science.

Examples of speech recognition tasks and related systems that have been developed over the years range from the simplest isolated word recognition, where the speaker has to wait between utterances, to the more advanced continuous speech recognizers. In the latter, the user can speak almost naturally, and concurrently the computer can determine the content. Speaker recognition is another topic, where the system can identify the speaker. Such systems are used, for example, for security purposes.

Speech recognition embraces a wide spectrum of applications. Some typical cases where speech recognizers have been used include automatic call processing in telephone networks, query-based information systems, data entry, voice dictation, robotics, as well as assistive technologies for people with special needs, e.g., blind people.

COMPUTER VISION

This is a discipline that has been inspired by the human visual system. Typical tasks that are addressed within the computer vision community include the automatic extraction of edges from images, representation of objects as compositions of smaller structures, object detection and recognition, optical flow, motion estimation, inference of shape from various cues, such as shading and texture, and three-dimensional reconstruction of scenes from multiple images. Image morphing, that is, changing one image to another through a seamless transition, and image stitching, i.e., creating a panoramic image from a number of images, are also topics in the computer vision research. More recently, there is more and more interaction between the field of computer vision and that of graphics.

MULTIMODAL DATA

Both speech recognition and computer vision process information that originates from single *modalities*. However, humans perceive the natural world in a *multimodal way*, via their multiple senses, e.g., vision, hearing, and touch. There is complementary information in each one of the involved modalities that the human brain exploits in order to understand and perceive the surrounding world.

Inspired by that, *multimedia* or multimodal understanding, via *cross-media integration*, has given birth to a related field whose goal is to improve the performance in the various scientific tasks that arise in problems that deal with multiple modalities. An example of modality blending is to combine together image/video, speech/audio, and text. A related summary that also touches issues concerning the mental processes of human *sensation*, *perception*, and *cognition* is presented in [9].

NATURAL LANGUAGE PROCESSING

This is the discipline that studies the processing of a language using computers. An example of a natural language processing (NLP) task is that of SPAM detection. Currently, the NLP field is an area of intense research with typical topics being the development of automatic translation algorithms and software, sentiment analysis, text summarization, and authorship identification. Speech recognition has a strong affinity with NLP and, strictly speaking, could be considered as a special subtopic of it. Two case studies related to NLP are treated in the book, one in Chapter 11 concerning authorship identification and one in Chapter 18 related to neural machine translation (NMT).

ROBOTICS

Robots are used to perform tasks in the manufacturing industry, e.g., in an assembly line for car production, or by space agencies to move objects in the space. More recently, the so-called social robots are built to interact with people in their social environment. For example, social robots are used to benefit hospitalized children [10].

Robots have been used in situations that are difficult or dangerous for humans, such as bomb detonation and work in difficult and hazardous environments, e.g., places of high heat, deep oceans, and areas of high radiation. Robots have also been developed for teaching.

Robotics is an interdisciplinary field that, besides machine learning, includes disciplines such as mechanical engineering, electronic engineering, computer science, computer vision, and speech recognition.

AUTONOMOUS CARS

An autonomous or *self-driving* car is a vehicle that can move around with no or little human intervention. Most of us have used self-driving trains in airports. However, these operate in a very well-controlled environment. Autonomous cars are designed to operate in the city streets and in motorways. This field is also of interdisciplinary nature, where areas such as radar, lidar, computer vision, automatic control, sensor networks, and machine learning meet together. It is anticipated that the use of self-driving cars will reduce the number of accidents, since, statistically, most of the accidents occur because of human errors, due to alcohol, high speed, stress, fatigue, etc.

There are various levels of automation that one can implement. At level 0, which is the category in which most of the cars currently operate, the driver has the control and the automated built-in system may issue warnings. The higher the level, the more autonomy is present. For example, at level 4, the driver would be first notified whether conditions are safe, and then the driver can decide to switch the vehicle into the autonomous driving mode. At the highest level, level 5, the autonomous driving requires absolutely no human intervention [21].

Besides the aforementioned examples of notable machine learning applications, machine learning has been applied in a wide range of other areas, such as healthcare, bioinformatics, business, finance, education, law, and manufacturing.

CHALLENGES FOR THE FUTURE

In spite of the impressive advances that have been achieved in machine learning, there are a number of challenges for the foreseeable future, besides the long-term ones that were mentioned before, while presenting AI. In the Berkeley report [18], the following list of challenges are summarized:

- Designing systems that learn *continually* by interacting with a *dynamic environment*, while making decisions that are *timely*, *robust*, and *secure*.
- Designing systems that enable personalized applications and services, yet do not compromise users' *privacy and security*.
- Designing systems that can train on data sets owned by different organizations without compromising their *confidentiality*, and in the process provide AI capabilities that span the boundaries of potentially competing organizations.
- Developing domain-specific architectures and software systems to address the performance needs of future applications, including custom chips, edge-cloud systems to efficiently process data at the edge, and techniques for abstracting and sampling data.

Besides the above more technology-oriented challenges, important social challenges do exist. The new technologies are influencing our daily lives more and more. In principle, they offer the potential, much more than ever before, to manipulate and shape beliefs, views, interests, entertainment, customs, and culture, independent of the societies. Moreover, they offer the potential for accessing personal data that in the sequel can be exploited for various reasons, such as economic, political, or other malicious purposes. As M. Schaake, a member of the European Parliament, puts it, "When algorithms affect human rights, public values or public decision-making, we need oversight and transparency." However, what was said before should not mobilize technophobic reactions. On the contrary, human civilization has advanced because of leaps in science and technology. All that is needed is social sensitivity, awareness of the possible dangers, and a related legal "shielding."

Putting it in simple words, as Henri Bergson said [2], history is not deterministic. History is a *creative* evolution.

1.5 MACHINE LEARNING: MAJOR DIRECTIONS

As has already been stated before, machine learning is the scientific field whose goal is to develop methods and algorithms that "learn" from data; that is, to extract information that "resides" in the data, which can subsequently be used by the computer to perform a task. To this end, the starting point is an available data set. Depending on the type of information that one needs to acquire, in the context of a specific task, different types of machine learning have been developed. They are described below.

1.5.1 SUPERVISED LEARNING

Supervised learning refers to the type of machine learning where all the available data have been *labeled*. In other words, data are represented in pairs of *observations*, e.g., $(y_n, \boldsymbol{x}_n)$, $n = 1, 2, \ldots, N$, where each $\boldsymbol{x}_n$ is a vector or, in general, a set of variables. The variables in $\boldsymbol{x}_n$ are called the input variables, also known as the *independent* variables or *features*, and the respective vector is known as the *feature vector*. The variables y_n are known as the output or *dependent* or *target* or *label* variables.

In some cases, y_n can also be a vector. The goal of learning is to obtain/estimate a functional mapping to, given the value of the input variables, predict the value of the respective output one. Two "pillars" of supervised learning are the classification and the regression tasks.

Classification

The goal in classification is to assign a *pattern* to one of a set of possible classes, whose number is considered to be known. For example, in X-ray mammography, we are given an image where a region indicates the existence of a tumor. The goal of a computer-aided diagnosis system is to predict whether this tumor corresponds to the *benign* or the *malignant* class. Optical character recognition (OCR) systems are also built around a classification system, in which the image corresponding to each letter of the alphabet has to be recognized and assigned to one of the 26 (for the Latin alphabet) classes; see Example 18.3 for a related case study. Another example is the prediction of the authorship of a given text. Given a text written by an unknown author, the goal of a classification system is to predict the author among a number of authors (classes); this application is treated in Section 11.15. The receiver in a digital communications system can also be viewed as a classification system. Upon receiving the transmitted data, which have been contaminated by noise and also by other transformations imposed by the transmission channel (Chapter 4), the receiver has to reach a decision on the value of the originally transmitted symbols. For example, in a binary transmitted sequence, the original symbols belong either to the $+1$ or to the -1 class. This task is known as *channel equalization.*

The first step in designing any machine learning task is to decide how to represent each pattern in the computer. This is achieved during the preprocessing stage; one has to "encode" related information that resides in the raw data (e.g., image pixels or strings of words) in an efficient and information-rich way. This is usually done by transforming the raw data into a new space and representing each pattern by a vector, $\boldsymbol{x} \in \mathbb{R}^l$. This comprises the feature vector and its l elements the corresponding feature values. In this way, each pattern becomes a single point in an l-dimensional space, known as the *feature space* or the *input space*. We refer to this transformation of the raw data as the *feature generation* or *feature extraction* stage. One starts with generating some large value, K, of possible features and eventually selects the l most informative ones via an optimizing procedure known as the *feature selection* stage. As we will see in Section 18.12, in the context of convolutional neural networks, the previous two stages are merged together and the features are obtained and optimized in a combined way, together with the estimation of the functional mapping, which was mentioned before.

Having decided upon the input space in which the data are represented, one has to train a classifier; that is, a predictor. This is achieved by first selecting a set of N data points/samples/examples, whose class is known, and this comprises the *training set.* This is the set of observation pairs, $(y_n, \boldsymbol{x}_n)$, $n = 1, 2, \ldots, N$, where y_n is the (output) variable denoting the class in which $\boldsymbol{x}_n$ belongs, and it is known as the corresponding *class label*; the class labels take values over a *discrete* set, e.g., $\{1, 2, \ldots, M\}$, for an M-class classification task. For example, for a two-class classification task, $y_n \in \{-1, +1\}$ or $y_n \in \{0, +1\}$. To keep our discussion simple, let us focus on the two-class case. Based on the training data, one then designs a function, f, which is used to predict the output label, given the input feature vector, $\boldsymbol{x}$. In general, we may need to design a set of such functions.

Once the function, f, has been designed, the system is ready to make predictions. Given a pattern whose class is unknown, we obtain the corresponding feature vector, $\boldsymbol{x}$, from the raw data. Depending on the value of $f(\boldsymbol{x})$, the pattern is classified in one of the two classes. For example, if the labels take the values ± 1, then the predicted label is obtained as $\hat{y} = \text{sgn}\{f(\boldsymbol{x})\}$. This operation defines the

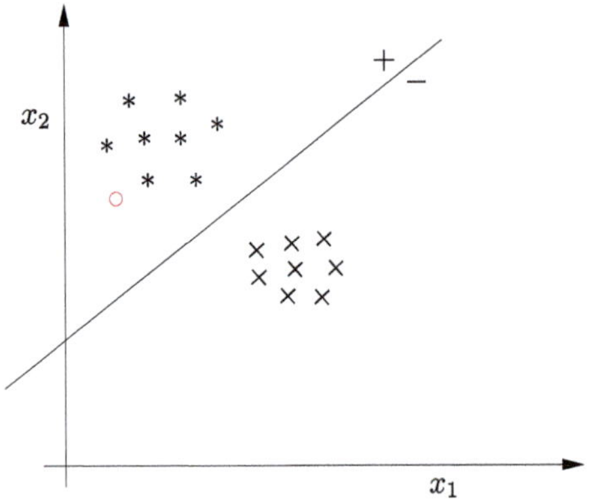

FIGURE 1.1

The classifier (linear in this simple case) has been designed in order to separate the training data into two classes. The graph (straight line) of the linear function, $f(\boldsymbol{x}) = 0$, has on its positive side the points coming from one class and on its negative side those of the other. The "red" point, whose class is unknown, is classified in the same class as the "star" points, since it lies on the positive side of the line.

classifier. If the function f is linear (nonlinear) we say that the respective classification task is linear (nonlinear) or, in a slight abuse of terminology, that the classifier is a linear (nonlinear) one.

Fig. 1.1 illustrates the classification task. Initially, we are given the set of points, each one representing a pattern in the two-dimensional space (two features used, x_1, x_2). Stars belong to one class, and the crosses to the other, in a two-class classification task. These are the training points, which are used to obtain a classifier. For our very simple case, this is achieved via a linear function,

$$f(\boldsymbol{x}) = \theta_1 x_1 + \theta_2 x_2 + \theta_0, \tag{1.1}$$

whose graph, for all the points such that $f(\boldsymbol{x}) = 0$, is the straight line shown in the figure. The values of the parameters $\theta_1, \theta_2, \theta_0$ are obtained via an estimation method based on the training set. This phase, where a classifier is *estimated*, is also known as the *training* or *learning* phase.

Once a classifier has been "learned," we are ready to perform predictions, that is, to predict the class label of a pattern $\boldsymbol{x}$. For example, we are given the point denoted by the red circle, whose class is unknown to us. According to the classification system that has been designed, this belongs to the same class as the points denoted by stars, which all belong to, say, class $+1$. Indeed, every point on one side of the straight line will give a positive value, $f(\boldsymbol{x}) > 0$, and all the points on its other side will give a negative value, $f(\boldsymbol{x}) < 0$. The predicted label, $\hat{y}$, for the point denoted with the red circle will then be $\hat{y} = \text{sgn}\{f(\boldsymbol{x})\} > 0$, and it is classified in class $+1$, to which the star points belong.

Our discussion to present the classification task was based on features that take numeric values. Classification tasks where the features are of categorical type do exist and are of major importance, too.

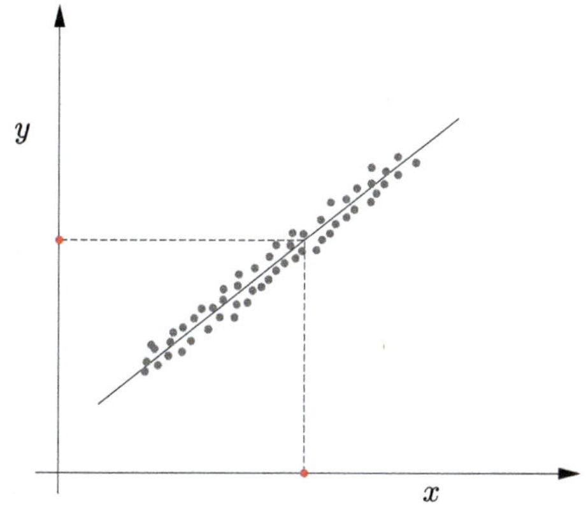

FIGURE 1.2

In a regression task, once a function (linear in this case) f has been designed, for its graph to fit the available training data set, given a new (red) point, x, the prediction of the associated output (red) value is given by $y = f(x)$.

Regression

Regression shares to a large extent the feature generation/selection stage, as described before; however, now the output variable, y, is *not* discrete, but it takes values in an interval in the real axis or in a region in the complex numbers' plane. Generalizations to vector-valued outputs are also possible. Our focus here is on real variables. The regression task is basically a function (curve/surface) fitting problem.

We are given a set of training samples, $(y_n, \boldsymbol{x}_n)$, $y_n \in \mathbb{R}$, $\boldsymbol{x}_n \in \mathbb{R}^l$, $n = 1, 2, \ldots, N$, and the task is to estimate a function f, whose graph fits the data. Once we have found such a function, when a new sample $\boldsymbol{x}$, outside the training set, arrives, we can predict its output value. This is shown in Fig. 1.2. The training data in this case are the gray points. Once the function fitting task has been completed, given a new point x (red), we are ready to predict its output value as $\hat{y} = f(\boldsymbol{x})$. In the simple case of the figure, the function f is linear and thus its graph is a straight line.

The regression task is a generic one that embraces a number of problems. For example, in financial applications, one can predict tomorrow's stock market prices given current market conditions and other related information. Each piece of information is a measured value of a corresponding feature. Signal and image restoration come under this common umbrella of tasks. Signal and image denoising can also be seen as a special type of the regression task. Deblurring of a blurred image can also be treated as regression (see Chapter 4).

1.6 UNSUPERVISED AND SEMISUPERVISED LEARNING

The goal of supervised learning is to establish a functional relationship between the input and output variables. To this end, labeled data are used, which comprise the set of output–input pairs, on which the learning of the unknown mapping is performed.

In the antipode of supervised learning lies *unsupervised learning*, where only input variables are provided. No output or label information is available. The aim of unsupervised learning is to unravel the structure that underlies the given set of data. This is an important part in data learning methods. Unsupervised learning comes under a number of facets.

One of the most important types of unsupervised learning is that of *clustering*. The goal of any clustering task is to unravel the way in which the points in a data set are grouped assuming that such a group structure exists. As an example, given a set of newspaper articles, one may want to group them together according to how similar their content is. As a matter of fact, at the heart of any clustering algorithm lies the concept of similarity, since patterns that belong to the same group (cluster) are assumed to be more similar than patterns that belong to different clusters.

One of the most classical clustering schemes, the so-called k-means clustering, is presented and discussed in Section 12.6.1. However, clustering is not a main topic of this book, and the interested reader may look at more specialized references (e.g., [14,15]).

Another type of unsupervised learning is *dimensionality* reduction. The goal is also to reveal a particular structure of the data, which is of a different nature than that of the groupings. For example, although the data may be represented in a high-dimensional space, they may lie around a lower-dimensional subspace or a manifold. Such methods are very important in machine learning for compressed representations or computational reduction reasons. Dimensionality reduction methods are treated in detail in Chapter 19.

Probability distribution estimation can also be considered as a special case of unsupervised learning. Probabilistic modeling is treated extensively in Chapters 12, 13, 15, and 16.

More recently, unsupervised learning is used for data generation. The so-called generative adversarial networks (GANs) comprise a new way of dealing with this old topic, by employing game theoretic arguments, and they are treated in Chapter 18.

Semisupervised lies in between supervised and unsupervised learning. In semisupervised learning, there are labeled data but not enough to get a good estimate of the output–input dependence. The existence of a number of unlabeled patterns can assist the task, since it can reveal additional structure of the input data that can be efficiently utilized. Semisupervised learning is treated in, e.g., [14].

Finally, another type of learning, which is increasingly gaining in importance, is the so-called *reinforcement learning* (RL). This is also an old field, with origins in automatic control. At the heart of this type of learning lies a set of rules and the goal is to learn sequences of actions that will lead an *agent* to achieve its goal or to maximize its objective function. For example, if the agent is a robot, the goal may be to move from point A to point B. Intuitively, RL attempts to learn actions by *trial and error*. In contrast to supervised learning, optimal actions are not learned from labels but from what is known as a *reward*. This scalar value informs the system whether the outcome of whatever it did was right or wrong. Taking actions that maximize the reward is the goal of RL.

Reinforcement learning is beyond the scope of this book and the interested reader may consult, e.g., [19].

1.7 STRUCTURE AND A ROAD MAP OF THE BOOK

In the discussion above, we saw that seemingly different applications, e.g., authorship identification and channel equalization, as well as financial prediction and image deblurring, can be treated in a unified framework. Many of the techniques that have been developed for machine learning are no different than techniques used in statistical signal processing or adaptive signal processing. Filtering comes under the general framework of regression (Chapter 4), and "adaptive filtering" is the same as "online learning" in machine learning. As a matter of fact, as will be explained in more detail, this book can serve the needs of more than one advanced graduate or postgraduate courses.

Over the years, a large number of techniques have been developed, in the context of different applications. Most of these techniques belong to one of two schools of thought. In one of them, the involved parameters that define an unknown function, for example, $\theta_1, \theta_2, \theta_0$ in Eq. (1.1), are treated as random variables. Bayesian learning builds upon this rationale. Bayesian methods learn distributions that describe the randomness of the involved parameters/variables. According to the other school, parameters are treated as nonrandom variables. They correspond to a fixed, yet unknown value. We will refer to such parameters as deterministic. This term is justified by the fact that, in contrast to random variables, if the value of a nonrandom variable is known, then its value can be "predicted" exactly. Learning methods that build around deterministic variables focus on optimization techniques to obtain estimates of the corresponding values. In some cases, the term "frequentist" is used to describe the latter type of techniques (see Chapter 12).

Each of the two previous schools of thought has its pros and cons, and I firmly believe that there is always more than one road that leads to the "truth." Each can solve some problems more efficiently than the other. Maybe in a few years, the scene will be more clear and more definite conclusions can be drawn. Or it may turn out, as in life, that the "truth" is somewhere in the middle.

In any case, every newcomer to the field has to learn the basics and the classics. That is why, in this book, all major directions and methods will be discussed, in an equally balanced manner, to the greatest extent possible. Of course, the author, being human, could not avoid giving some more emphasis to the techniques with which he is most familiar through his own research. This is healthy, since writing a book is a means of sharing the author's expertise and point of view with the readers. This is why I strongly believe that a new book does not serve to replace previous ones, but to complement previously published points of view.

Chapter 2 is an introduction to probability and statistics. Random processes are also discussed. Readers who are familiar with such concepts can bypass this chapter. On the other hand, one can focus on different parts of this chapter. Readers who would like to focus on statistical signal processing/adaptive processing can focus more on the random processes part. Those who would like to follow a probabilistic machine learning point of view would find the part presenting the various distributions more important. In any case, the multivariate normal (Gaussian) distribution is a must for those who are not yet familiar with it.

Chapter 3 is an overview of the parameter estimation task. This is a chapter that presents an overview of the book and defines the main concepts that run across its pages. This chapter has also been written to stand alone as an introduction to machine learning. Although it is my feeling that all of it should be read and taught, depending on the focus of the course and taking into account the omnipresent time limitations, one can focus more on the parts of her or his interest. Least-squares and ridge regression are discussed alongside the maximum likelihood method and the presentation of the basic notion of the Bayesian approach. In any case, the parts dealing with the definition of the inverse problems, the bias–variance tradeoff, and the concepts of generalization and regularization are a must.

Chapter 4 is dedicated to the mean-square error (MSE) linear estimation. For those following a statistical signal processing course, the whole chapter is important. The rest of the readers can bypass the parts related to complex-valued processing and also the part dealing with computational complexity issues, since this is only of importance if the input data are random processes. Bypassing this part will not affect reading later parts of the chapter that deal with the MSE estimation of linear models, the Gauss–Markov theorem, and the Kalman filtering.

Chapter 5 introduces the stochastic gradient descent family of algorithms. The first part, dealing with the stochastic approximation method, is a must for every reader. The rest of the chapter, which deals with the least-mean-squares (LMS) algorithm and its offsprings, is more appropriate for readers who are interested in a statistical signal processing course, since these families are suited for tracking time-varying environments. This may not be the first priority for readers who are interested in classification and machine learning tasks with data whose statistical properties are not time-varying.

Chapter 6 is dedicated to the least-squares (LS) method, which is of interest to all readers in machine learning and signal processing. The latter part, dealing with the total least-squares method, can be bypassed in a first reading. Emphasis is also put on ridge regression and its geometric interpretation. Ridge regression is important to the newcomer, since he/she becomes familiar with the concept of regularization; this is an important aspect in any machine learning task, tied directly with the generalization performance of the designed predictor.

I have decided to compress the part dealing with fast LS algorithms, which are appropriate when the input is a random process/signal that imposes a special structure on the involved covariance matrices, into a discussion section. It is the author's feeling that this is of no greater interest than it was a decade or two ago. Also, the main idea, that of a highly structured covariance matrix that lies behind the fast algorithms, is discussed in some detail in Chapter 4, in the context of Levinson's algorithm and its lattice and lattice-ladder by-products.

Chapter 7 is a must for any machine learning course. Important classical concepts, including classification in the context of the Bayesian decision theory, nearest neighbor classifiers, logistic regression, Fisher's discriminant analysis and decision trees are discussed. Courses on statistical signal processing can also accommodate the first part of the chapter dealing with the classical Bayesian decision theory.

The aforementioned six chapters comprise the part of the book that deals with more or less classical topics. The rest of the chapters deal with more advanced techniques and can fit with any course dealing with machine learning or statistical/adaptive signal processing, depending on the focus, the time constraints, and the background of the audience.

Chapter 8 deals with convexity, a topic that is receiving more and more attention recently. The chapter presents the basic definitions concerning convex sets and functions and the notion of projection. These are important tools used in a number of recently developed algorithms. Also, the classical projections onto convex sets (POCS) algorithm and the set-theoretic approach to online learning are discussed as an alternative to gradient descent-based schemes. Then, the task of optimization of nonsmooth convex loss functions is introduced, and the family of proximal mapping, alternating direction method of multipliers (ADMM), and forward backward-splitting methods are presented. This is a chapter that can be used when the emphasis of the course is on optimization. Employing nonsmooth loss functions and/or nonsmooth regularization terms, in place of the squared error and its ridge regression relative, is a trend of high research and practical interest.

Chapters 9 and 10 deal with sparse modeling. The first of the two chapters introduces the main concepts and ideas and the second deals with algorithms for batch as well for as online learning scenarios. Also, in the second chapter, a case study in the context of time-frequency analysis is discussed. Depending on time constraints, the main concepts behind sparse modeling and compressed sensing can be taught in a related course. These two chapters can also be used as a specialized course on sparsity on a postgraduate level.

Chapter 11 deals with learning in reproducing kernel Hilbert spaces and nonlinear techniques. The first part of the chapter is a must for any course with an emphasis on classification. Support vector regression and support vector machines are treated in detail. Moreover, a course on statistical signal processing with an emphasis on nonlinear modeling can also include material and concepts from this chapter. A case study dealing with authorship identification is discussed at the end of this chapter.

Chapters 12 and 13 deal with Bayesian learning. Thus, both chapters can be the backbone of a course on machine learning and statistical signal processing that intends to emphasize Bayesian methods. The former of the chapters deals with the basic principles and it is an introduction to the expectation-maximization (EM) algorithm. The use of this celebrated algorithm is demonstrated in the context of two classical applications: linear regression and Gaussian mixture modeling for probability density function estimation. The second chapter deals with approximate inference techniques, and one can use parts of it, depending on the time constraints and the background of the audience. Sparse Bayesian learning and the relevance vector machine (RVM) framework are introduced. At the end of this chapter, nonparametric Bayesian techniques such as the Chinese restaurant process (CRP), the Indian buffet process (IBP), and Gaussian processes are discussed. Finally, a case study concerning hyperspectral image unmixing is presented. Both chapters, in their full length, can be used as a specialized course on Bayesian learning.

Chapters 14 and 17 deal with Monte Carlo sampling methods. The latter chapter deals with particle filtering. Both chapters, together with the two previous ones that deal with Bayesian learning, can be combined in a course whose emphasis is on statistical methods of machine learning/statistical signal processing.

Chapters 15 and 16 deal with probabilistic graphical models. The former chapter introduces the main concepts and definitions, and at the end it introduces the message passing algorithm for chains and trees. This chapter is a must for any course whose emphasis is on probabilistic graphical models. The latter of the two chapters deals with message passing algorithms on junction trees and then with approximate inference techniques. Dynamic graphical models and hidden Markov models (HMMs) are introduced. The Baum–Welch and Viterbi schemes are derived as special cases of message passaging algorithms by treating the HMM as a special instance of a junction tree.

Chapter 18 deals with neural networks and deep learning. In the second edition, this chapter has been basically rewritten to accommodate advances in this topic that have taken place after the first edition was published. This chapter is also a must in any course with an emphasis on classification. The chapter starts from the early days of the perceptron algorithm and perceptron rule and moves on to the most recent advances in deep learning. The feed-forward multilayer architecture is introduced and a number of stochastic gradient-type algorithmic variants, for training networks, are presented. Different types of nonlinearities and cost functions are discussed and their interplay with respect to the vanishing/exploding phenomenon in the gradient propagation are presented. Regularization and the dropout method are discussed. Convolutional neural networks (CNNs) and recurrent neural networks (RNNs) are reviewed in some detail. The notions of attention mechanism and adversarial examples are presented. Deep belief networks, GANs, and variational autoencoders are considered in some detail. Capsule networks are introduced and, at the end, a discussion on transfer learning and multitask learning is provided. The chapter concludes with a case study related to neural machine translation (NMT).

Chapter 19 is on dimensionality reduction techniques and latent variable modeling. The methods of principal component analysis (PCA), canonical correlations analysis (CCA), and independent component analysis (ICA) are introduced. The probabilistic approach to latent variable modeling is discussed, and the probabilistic PCA (PPCA) is presented. Then, the focus turns to dictionary learning and to robust PCA. Nonlinear dimensionality reduction techniques such as kernel PCA are discussed, along with classical manifold learning methods: local linear embedding (LLE) and isometric mapping (ISOMAP). Finally, a case study in the context of functional magnetic resonance imaging (fMRI) data analysis, based on ICA, is presented.

Each chapter starts with the basics and moves on to cover more recent advances in the corresponding topic. This is also true for the whole book and the first six chapters cover more classical material.

In summary, we provide the following suggestions for different courses, depending on the emphasis that the instructor wants to place on various topics.

- Machine learning with emphasis on classification:
  - Main chapters: 3, 7, 11, and 18.
  - Secondary chapters: 12 and 13, and possibly the first part of 6.
- Statistical signal processing:
  - Main chapters: 3, 4, 6, and 12.
  - Secondary chapters: 5 (first part) and 13–17.
- Machine learning with emphasis on Bayesian techniques:
  - Main chapters: 3 and 12–14.
  - Secondary chapters: 7, 15, and 16, and possibly the first part of 6.
- Adaptive signal processing:
  - Main chapters: 3–6.
  - Secondary chapters: 8, 9, 10, 11, 14, and 17.

I believe that the above suggestions of following various combinations of chapters is possible, since the book has been written in such a way as to make individual chapters as self-contained as possible.

At the end of most of the chapters, there are computer exercises, mainly based on the various examples given in the text. The exercises are given in MATLAB® and the respective code is available on the book's website. Moreover, all exercises are provided, together with respective codes, in Python and are also available on the book's website. Some of the exercises in Chapter 18 are in the context of TensorFlow.

The solutions manual as well as all the figures of the book are available on the book's website.

REFERENCES

[1] R. Bajcsy, Active perception, Proceedings of the IEEE 76 (8) (1988) 966–1005.
[2] H. Bergson, Creative Evolution, McMillan, London, 1922.
[3] A. Damasio, Descartes' Error: Emotion, Reason, and the Human Brain, Penguin, 2005 (paperback reprint).
[4] S. Dehaene, H. Lau, S. Kouider, What is consciousness, and could machines have it?, Science 358 (6362) (2017) 486–492.
[5] D. Deutsch, The Fabric of Reality, Penguin, 1998.
[6] D. Deutsch, The Beginning of Infinity: Explanations That Transform the World, Allen Lane, 2011.
[7] H. Dreyfus, What Computers Still Can't Do, M.I.T. Press, 1992.
[8] M. Jordan, Artificial intelligence: the revolution hasn't happened yet, https://medium.com/@mijordan3/artificial-intelligence-the-revolution-hasnt-happened-yet-5e1d5812e1e7, 2018.

[9] P. Maragos, P. Gros, A. Katsamanis, G. Papandreou, Cross-modal integration for performance improving in multimedia: a review, in: P. Maragos, A. Potamianos, P. Gros (Eds.), Multimodal Processing and Interaction: Audio, Video, Text, Springer, 2008.
[10] MIT News, Study: social robots can benefit hospitalized children, http://news.mit.edu/2019/social-robots-benefit-sick-children-0626.
[11] J. Pearl, D. Mackenzie, The Book of Why: The New Science of Cause and Effect, Basic Books, 2018.
[12] K. Popper, The Logic of Scientific Discovery, Routledge Classics, 2002.
[13] S. Russell, P. Norvig, Artificial Intelligence, third ed., Prentice Hall, 2010.
[14] S. Theodoridis, K. Koutroumbas, Pattern Recognition, fourth ed., Academic Press, Amsterdam, 2009.
[15] S. Theodoridis, A. Pikrakis, K. Koutroumbas, D. Cavouras, Introduction to Pattern Recognition: A MATLAB Approach, Academic Press, Amsterdam, 2010.
[16] A. Turing, Computing machinery intelligence, MIND 49 (1950) 433–460.
[17] N. Schwarz, I. Skurnik, Feeling and thinking: implications for problem solving, in: J.E. Davidson, R. Sternberg (Eds.), The Psychology of Problem Solving, Cambridge University Press, 2003, pp. 263–292.
[18] I. Stoica, et al., A Berkeley View of Systems Challenges for AI, Technical Report No. UCB/EECS-2017-159, 2017.
[19] R.C. Sutton, A.G. Barto, Reinforcement Learning: An Introduction, MIT Press, 2018.
[20] The history of artificial intelligence, University of Washington, https://courses.cs.washington.edu/courses/csep590/06au/projects/history-ai.pdf.
[21] USA Department of Transportation Report, https://www.nhtsa.gov/sites/nhtsa.dot.gov/files/documents/13069a_ads2.0-0906179a_tag.pdf.
[22] The Changing Nature of Work, World Bank Report, 2019, http://documents.worldbank.org/curated/en/816281518818814423/pdf/2019-WDR-Report.pdf.
[23] World Economic Forum, The future of jobs, http://www3.weforum.org/docs/WEF_Future_of_Jobs_2018.pdf, 2018.

CHAPTER

PROBABILITY AND STOCHASTIC PROCESSES

2

CONTENTS

Machine Learning. https://doi.org/10.1016/B978-0-12-818803-3.00011-8

2.1 INTRODUCTION

The goal of this chapter is to provide the basic definitions and properties related to probability theory and stochastic processes. It is assumed that the reader has attended a basic course on probability and statistics prior to reading this book. So, the aim is to help the reader refresh her/his memory and to establish a common language and a commonly understood notation.

Besides probability and random variables, random processes are briefly reviewed and some basic theorems are stated. A number of key probability distributions that will be used later on in a number of chapters are presented. Finally, at the end of the chapter, basic definitions and properties related to information theory and stochastic convergence are summarized.

The reader who is familiar with all these notions can bypass this chapter.

2.2 PROBABILITY AND RANDOM VARIABLES

A random variable, x, is a variable whose variations are due to chance/randomness. A random variable can be considered as a function, which assigns a value to the outcome of an experiment. For example, in a coin tossing experiment, the corresponding random variable, x, can assume the values $x_1 = 0$ if the result of the experiment is "heads" and $x_2 = 1$ if the result is "tails."

We will denote a random variable with a lower case roman, such as x, and the values it takes once an experiment has been performed, with mathmode italics, such as x.

A random variable is described in terms of a set of *probabilities* if its values are of a discrete nature, or in terms of a *probability density function* (PDF) if its values lie anywhere within an interval of the real axis (noncountably infinite set). For a more formal treatment and discussion, see [4,6].

2.2.1 PROBABILITY

Although the words "probability" and "probable" are quite common in our everyday vocabulary, the mathematical definition of probability is not a straightforward one, and there are a number of different definitions that have been proposed over the years. Needless to say, whatever definition is adopted, the end result is that the properties and rules which are derived remain the same. Two of the most commonly used definitions are the following

Relative Frequency Definition

The probability $P(A)$ of an event A is the limit

$$P(A) = \lim_{n \longrightarrow \infty} \frac{n_A}{n}, \tag{2.1}$$

where n is the number of total trials and n_A the number of times event A occurred. The problem with this definition is that in practice in any physical experiment, the numbers n_A and n can be large, yet they are always finite. Thus, the limit can only be used as a *hypothesis* and not as something that can be attained experimentally. In practice, often, we use

$$P(A) \approx \frac{n_A}{n} \tag{2.2}$$

for large values of n. However, this has to be used with caution, especially when the probability of an event is very small.

Axiomatic Definition

This definition of probability is traced back to 1933 to the work of Andrey Kolmogorov, who found a close connection between probability theory and the mathematical theory of sets and functions of a real variable, in the context of measure theory, as noted in [5].

The probability $P(A)$ of an event is a nonnegative number assigned to this event, or

$$P(A) \geq 0. \tag{2.3}$$

The probability of an event C which is certain to occur is equal to one, i.e.,

$$P(C) = 1. \tag{2.4}$$

If two events A and B are mutually exclusive (they cannot occur simultaneously), then the probability of occurrence of either A *or* B (denoted as $A \cup B$) is given by

$$P(A \cup B) = P(A) + P(B). \tag{2.5}$$

It turns out that these three defining properties, which can be considered as the respective *axioms*, suffice to develop the rest of the theory. For example, it can be shown that the probability of an impossible event is equal to zero, e.g., [6].

The previous two approaches for defining probability are not the only ones. Another interpretation, which is in line with the way we are going to use the notion of probability in a number of places in this book in the context of *Bayesian learning*, has been given by Cox [2]. There, probability was seen as a measure of *uncertainty* concerning an event. Take, for example, the uncertainty whether the Minoan civilization was destroyed as a consequence of the earthquake that happened close to the island of Santorini. This is obviously not an event whose probability can be tested with repeated trials. However, putting together historical as well as scientific evidence, we can quantify our expression of uncertainty concerning such a conjecture. Also, we can modify the degree of our uncertainty once more historical evidence comes to light due to new archeological findings. Assigning numerical values to represent degrees of belief, Cox developed a set of axioms encoding common sense properties of such beliefs, and he came to a set of rules equivalent to the ones we are going to review soon; see also [4].

The origins of probability theory are traced back to the middle 17th century in the works of Pierre Fermat (1601–1665), Blaise Pascal (1623–1662), and Christian Huygens (1629–1695). The concepts of probability and the mean value of a random variable can be found there. The original motivation for developing the theory seems not to be related to any purpose for "serving society"; the purpose was to serve the needs of gambling and games of chance!

2.2.2 DISCRETE RANDOM VARIABLES

A discrete random variable x can take any value from a finite or *countably* infinite set $\mathcal{X}$. The probability of the event, "$\mathrm{x} = x \in \mathcal{X}$," is denoted as

$$P(\mathrm{x} = x) \quad \text{or simply } P(x). \tag{2.6}$$

The function P is known as the *probability mass function* (PMF). Being a probability of events, it has to satisfy the first axiom, so $P(x) \geq 0$. Assuming that no two values in $\mathcal{X}$ can occur simultaneously and that after any experiment a single value will always occur, the second and third axioms combined give

$$\sum_{x \in \mathcal{X}} P(x) = 1. \tag{2.7}$$

The set $\mathcal{X}$ is also known as the *sample* or *state space*.

Joint and Conditional Probabilities

The *joint probability* of two events, A, B, is the probability that both events occur simultaneously, and it is denoted as $P(A, B)$. Let us now consider two random variables, x, y, with sample spaces $\mathcal{X} = \{x_1, \ldots, x_{n_x}\}$ and $\mathcal{Y} = \{y_1, \ldots, y_{n_y}\}$, respectively. Let us adopt the relative frequency definition and assume that we carry out n experiments and that each one of the values in $\mathcal{X}$ occurred $n_1^x, \ldots, n_{n_x}^x$ times and each one of the values in $\mathcal{Y}$ occurred $n_1^y, \ldots, n_{n_y}^y$ times. Then,

$$P(x_i) \approx \frac{n_i^x}{n}, \; i = 1, 2, \ldots, n_x, \quad \text{and} \quad P(y_j) \approx \frac{n_j^y}{n}, \; j = 1, 2, \ldots, n_y.$$

Let us denote by n_{ij} the number of times the values x_i and y_j occurred simultaneously. Then, $P(x_i, y_j) \approx \frac{n_{ij}}{n}$. Simple reasoning dictates that the total number, n_i^x, that value x_i occurred is equal to

$$n_i^x = \sum_{j=1}^{n_y} n_{ij}. \tag{2.8}$$

Dividing both sides in the above by n, the following *sum rule* readily results.

$$\boxed{P(x) = \sum_{y \in \mathcal{Y}} P(x, y): \quad \text{sum rule.}} \tag{2.9}$$

The *conditional probability* of an event A, *given* another event B, is denoted as $P(A|B)$, and it is defined as

$$\boxed{P(A|B) := \frac{P(A, B)}{P(B)}: \quad \text{conditional probability,}} \tag{2.10}$$

provided $P(B) \neq 0$. It can be shown that this is indeed a probability, in the sense that it respects all three axioms [6]. We can better grasp its physical meaning if the relative frequency definition is adopted. Let n_{AB} be the number of times that both events occurred simultaneously, and let n_B be the number of times event B occurred, out of n experiments. Then we have

$$P(A|B) = \frac{n_{AB}}{n} \frac{n}{n_B} = \frac{n_{AB}}{n_B}. \tag{2.11}$$

In other words, the conditional probability of an event A, given another event B, is the relative frequency that A occurred, not with respect to the total number of experiments performed, but relative to the times event B occurred.

Viewed differently and adopting similar notation in terms of random variables, in conformity with Eq. (2.9), the definition of the conditional probability is also known as the *product rule* of probability, written as

$$\boxed{P(x, y) = P(x|y)P(y): \quad \text{product rule.}} \tag{2.12}$$

To differentiate from the joint and conditional probabilities, probabilities $P(x)$ and $P(y)$ are known as *marginal probabilities*. The product rule is generalized in a straightforward way to l random variables, i.e.,

$$P(x_1, x_2, \dots, x_l) = P(x_l|x_{l-1}, \dots, x_1)P(x_{l-1}, \dots, x_1),$$

which recursively leads to the product

$$P(x_1, x_2, \dots, x_l) = P(x_l|x_{l-1}, \dots, x_1)P(x_{l-1}|x_{l-2}, \dots, x_1) \dots P(x_1).$$

Statistical independence: Two random variables are said to be statistically independent *if and only if* their joint probability is equal to the product of the respective marginal probabilities, i.e.,

$$P(x, y) = P(x)P(y). \tag{2.13}$$

Bayes Theorem

The Bayes theorem is a direct consequence of the product rule and the symmetry property of the joint probability, $P(x, y) = P(y, x)$, and it is stated as

$$\boxed{P(y|x) = \frac{P(x|y)P(y)}{P(x)}: \quad \text{Bayes theorem,}} \tag{2.14}$$

where the marginal, $P(x)$, can be written as

$$P(x) = \sum_{y \in \mathcal{Y}} P(x, y) = \sum_{y \in \mathcal{Y}} P(x|y)P(y),$$

and it can be considered as the normalizing constant of the numerator on the right-hand side in Eq. (2.14), which guarantees that summing up $P(y|x)$ with respect to all possible values of $y \in \mathcal{Y}$ results in one.

The Bayes theorem plays a central role in machine learning, and it will be the basis for developing Bayesian techniques for estimating the values of unknown parameters.

2.2.3 CONTINUOUS RANDOM VARIABLES

So far, we have focused on discrete random variables. Our interest now turns to the extension of the notion of probability to random variables which take values on the real axis, $\mathbb{R}$.

The starting point is to compute the probability of a random variable, x, to lie in an interval, $x_1 < \mathrm{x} \leq x_2$. Note that the two events, $\mathrm{x} \leq x_1$ and $x_1 < \mathrm{x} \leq x_2$, are mutually exclusive. Thus, we can write that

$$P(\mathrm{x} \leq x_1) + P(x_1 < \mathrm{x} \leq x_2) = P(\mathrm{x} \leq x_2). \tag{2.15}$$

We define the *cumulative distribution function* (CDF) of x as

$$\boxed{F_{\mathrm{x}}(x) := P(\mathrm{x} \leq x): \quad \text{cumulative distribution function.}} \tag{2.16}$$

Then, Eq. (2.15) can be written as

$$P(x_1 < \mathrm{x} \leq x_2) = F_{\mathrm{x}}(x_2) - F_{\mathrm{x}}(x_1). \tag{2.17}$$

Note that F_{x} is a monotonically increasing function. Furthermore, if it is continuous, the random variable x is said to be of a *continuous* type. Assuming that it is also differentiable, we can define the *probability density function* (PDF) of x as

$$\boxed{p_{\mathrm{x}}(x) := \frac{dF_{\mathrm{x}}(x)}{dx}: \quad \text{probability density function,}} \tag{2.18}$$

which then leads to

$$P(x_1 < \mathrm{x} \leq x_2) = \int_{x_1}^{x_2} p_{\mathrm{x}}(x)dx. \tag{2.19}$$

Also,

$$F_{\mathrm{x}}(x) = \int_{-\infty}^{x} p_{\mathrm{x}}(z)dz. \tag{2.20}$$

Using familiar arguments from calculus, the PDF can be interpreted as

$$\Delta P(x < \mathrm{x} \leq x + \Delta x) \approx p_{\mathrm{x}}(x)\Delta x, \tag{2.21}$$

which justifies its name as a "density" function, being the probability (ΔP) of x lying in a small interval Δx, divided by the length of this interval. Note that as $\Delta x \longrightarrow 0$ this probability tends to zero. Thus, the probability of a continuous random variable taking any single value is zero. Moreover, since $P(-\infty < \mathrm{x} < +\infty) = 1$, we have

$$\int_{-\infty}^{+\infty} p_{\mathrm{x}}(x)dx = 1. \tag{2.22}$$

Usually, in order to simplify notation, the subscript x is dropped and we write $p(x)$, unless it is necessary for avoiding possible confusion. Note, also, that we have adopted the lower case "p" to denote a PDF and the capital "P" to denote a probability.

All previously stated rules concerning probabilities are readily carried out for the case of PDFs, in the following way:

$$p(x|y) = \frac{p(x,y)}{p(y)}, \quad p(x) = \int_{-\infty}^{+\infty} p(x,y)\,dy. \tag{2.23}$$

2.2.4 MEAN AND VARIANCE

Two of the most common and useful quantities associated with any random variable are the respective mean value and variance. The mean value (or sometimes called expected value) is denoted as

$$\boxed{\mathbb{E}[\mathrm{x}] := \int_{-\infty}^{+\infty} x p(x)\,dx: \quad \text{mean value,}} \tag{2.24}$$

where for discrete random variables the integration is replaced by summation $\left(\mathbb{E}[\mathrm{x}] = \sum_{x \in \mathcal{X}} x P(x)\right)$.

The variance is denoted as σ_{x}^2 and it is defined as

$$\boxed{\sigma_{\mathrm{x}}^2 := \int_{-\infty}^{+\infty} (x - \mathbb{E}[\mathrm{x}])^2 p(x)\,dx: \quad \text{variance,}} \tag{2.25}$$

where integration is replaced by summation for discrete variables. The variance is a measure of the spread of the values of the random variable around its mean value.

The definition of the mean value is generalized for any function $f(\mathrm{x})$, i.e.,

$$\mathbb{E}[f(\mathrm{x})] := \int_{-\infty}^{+\infty} f(x)p(x)dx. \tag{2.26}$$

It is readily shown that the mean value with respect to two random variables, y, x, can be written as the product

$$\mathbb{E}_{\mathrm{x,y}}[f(\mathrm{x},\mathrm{y})] = \mathbb{E}_{\mathrm{x}}\left[\mathbb{E}_{\mathrm{y}|\mathrm{x}}[f(\mathrm{x},\mathrm{y})]\right], \tag{2.27}$$

where $\mathbb{E}_{\mathrm{y}|x}$ denotes the mean value with respect to $p(y|x)$. This is a direct consequence of the definition of the mean value and the product rule of probabilities.

Given two random variables x, y, their *covariance* is defined as

$$\text{cov}(\mathrm{x}, \mathrm{y}) := \mathbb{E}\big[(\mathrm{x} - \mathbb{E}[\mathrm{x}])(\mathrm{y} - \mathbb{E}[\mathrm{y}])\big], \tag{2.28}$$

and their *correlation* as

$$r_{\mathrm{xy}} := \mathbb{E}[\mathrm{xy}] = \text{cov}(\mathrm{x}, \mathrm{y}) + \mathbb{E}[\mathrm{x}]\,\mathbb{E}[\mathrm{y}]. \tag{2.29}$$

A *random vector* is a collection of random variables, $\mathbf{x} = [\mathrm{x}_1, \ldots, \mathrm{x}_l]^T$, and $p(\mathbf{x})$ is the joint PDF (probability mass for discrete variables),

$$p(\mathbf{x}) = p(\mathrm{x}_1, \ldots, \mathrm{x}_l). \tag{2.30}$$

The *covariance matrix* of a random vector $\mathbf{x}$ is defined as

$$\boxed{\text{Cov}(\mathbf{x}) := \mathbb{E}\left[(\mathbf{x} - \mathbb{E}[\mathbf{x}])(\mathbf{x} - \mathbb{E}[\mathbf{x}])^T\right]: \quad \text{covariance matrix,}} \tag{2.31}$$

or

$$\text{Cov}(\mathbf{x}) = \begin{bmatrix} \text{cov}(\mathrm{x}_1, \mathrm{x}_1) & \ldots & \text{cov}(\mathrm{x}_1, \mathrm{x}_l) \\ \vdots & \ddots & \vdots \\ \text{cov}(\mathrm{x}_l, \mathrm{x}_1) & \ldots & \text{cov}(\mathrm{x}_l, \mathrm{x}_l) \end{bmatrix}. \tag{2.32}$$

Another symbol that will be used to denote the covariance matrix is $\Sigma_{\mathbf{x}}$. Similarly, the *correlation matrix* of a random vector $\mathbf{x}$ is defined as

$$\boxed{R_{\mathbf{x}} := \mathbb{E}\left[\mathbf{x}\mathbf{x}^T\right]: \quad \text{correlation matrix,}} \tag{2.33}$$

or

$$\begin{aligned} R_{\mathbf{x}} &= \begin{bmatrix} \mathbb{E}[\mathrm{x}_1, \mathrm{x}_1] & \ldots & \mathbb{E}[\mathrm{x}_1, \mathrm{x}_l] \\ \vdots & \ddots & \vdots \\ \mathbb{E}[\mathrm{x}_l, \mathrm{x}_1] & \ldots & \mathbb{E}[\mathrm{x}_l, \mathrm{x}_l] \end{bmatrix} \\ &= \text{Cov}(\mathbf{x}) + \mathbb{E}[\mathbf{x}]\,\mathbb{E}[\mathbf{x}^T]. \end{aligned} \tag{2.34}$$

Often, subscripts are dropped in other to simplify notation, and the corresponding symbols, e.g., r, Σ and R are used instead, unless it is necessary to avoid confusion when different random variables are involved.[1] Both the covariance and correlation matrices have a very rich structure, which will be exploited in various parts of this book to lead to computational savings whenever they are present in calculations. For the time being, observe that both are symmetric and positive semidefinite. The

[1] Note that in the subsequent chapters, to avoid having bold letters in subscripts, which can be cumbersome when more that one vector variables are involved, the notation has been slightly relaxed. For example, R_x is used in place of $R_{\mathbf{x}}$. For the sake of uniformity, the same applies to the rest of the variables corresponding to correlations, variances and covariance matrices.

symmetry, $\Sigma = \Sigma^T$, is readily deduced from the definition. An $l \times l$ symmetric matrix A is called *positive semidefinite* if

$$\mathbf{y}^T A \mathbf{y} \geq 0, \quad \forall \mathbf{y} \in \mathbb{R}^l. \tag{2.35}$$

If the inequality is a strict one, the matrix is said to be *positive definite*. For the covariance matrix, we have

$$\mathbf{y}^T \mathbb{E}\left[(\mathbf{x} - \mathbb{E}[\mathbf{x}])(\mathbf{x} - \mathbb{E}[\mathbf{x}])^T\right] \mathbf{y} = \mathbb{E}\left[\left(\mathbf{y}^T(\mathbf{x} - \mathbb{E}[\mathbf{x}])\right)^2\right] \geq 0,$$

and the claim has been proved.

Complex Random Variables

A complex random variable, $\mathrm{z} \in \mathbb{C}$, is a sum

$$\mathrm{z} = \mathrm{x} + j\mathrm{y}, \tag{2.36}$$

where x, y are real random variables and $j := \sqrt{-1}$. Note that for complex random variables, the PDF *cannot* be defined since inequalities of the form $\mathrm{x} + j\mathrm{y} \leq x + jy$ have no meaning. When we write $p(z)$, we mean the joint PDF of the real and imaginary parts, expressed as

$$p(z) := p(x, y). \tag{2.37}$$

For complex random variables, the notions of mean and covariance are defined as

$$\mathbb{E}[\mathrm{z}] := \mathbb{E}[\mathrm{x}] + j\,\mathbb{E}[\mathrm{y}], \tag{2.38}$$

and

$$\mathrm{cov}(\mathrm{z}_1, \mathrm{z}_2) := \mathbb{E}\left[(\mathrm{z}_1 - \mathbb{E}[\mathrm{z}_1])(\mathrm{z}_2 - \mathbb{E}[\mathrm{z}_2])^*\right], \tag{2.39}$$

where "$*$" denotes complex conjugation. The latter definition leads to the variance of a complex variable,

$$\sigma_z^2 = \mathbb{E}\left[|\mathrm{z} - \mathbb{E}[\mathrm{z}]|^2\right] = \mathbb{E}\left[|\mathrm{z}|^2\right] - |\mathbb{E}[\mathrm{z}]|^2. \tag{2.40}$$

Similarly, for complex random vectors, $\mathbf{z} = \mathbf{x} + j\mathbf{y} \in \mathbb{C}^l$, we have

$$p(\mathbf{z}) := p(x_1, \ldots, x_l, y_1, \ldots, y_l), \tag{2.41}$$

where $x_i, y_i, i = 1, 2, \ldots, l$, are the components of the involved real vectors, respectively. The covariance and correlation matrices are similarly defined as

$$\mathrm{Cov}(\mathbf{z}) := \mathbb{E}\left[(\mathbf{z} - \mathbb{E}[\mathbf{z}])(\mathbf{z} - \mathbb{E}[\mathbf{z}])^H\right], \tag{2.42}$$

where "H" denotes the Hermitian (transposition and conjugation) operation.

For the rest of the chapter, we are going to deal mainly with real random variables. Whenever needed, differences with the case of complex variables will be stated.

2.2.5 TRANSFORMATION OF RANDOM VARIABLES

Let $\mathbf{x}$ and $\mathbf{y}$ be two random vectors, which are related via the vector transform,

$$\mathbf{y} = f(\mathbf{x}), \tag{2.43}$$

where $f : \mathbb{R}^l \longmapsto \mathbb{R}^l$ is an *invertible* transform. That is, given y, $x = f^{-1}(y)$ can be uniquely obtained. We are given the joint PDF, $p_{\mathbf{x}}(x)$, of $\mathbf{x}$ and the task is to obtain the joint PDF, $p_{\mathbf{y}}(y)$, of $\mathbf{y}$.

The Jacobian matrix of the transformation is defined as

$$J(\mathbf{y}; \mathbf{x}) := \frac{\partial(y_1, y_2, \ldots, y_l)}{\partial(x_1, x_2, \ldots, x_l)} := \begin{bmatrix} \frac{\partial y_1}{\partial x_1} & \cdots & \frac{\partial y_1}{\partial x_l} \\ \vdots & \ddots & \vdots \\ \frac{\partial y_l}{\partial x_1} & \cdots & \frac{\partial y_l}{\partial x_l} \end{bmatrix}. \tag{2.44}$$

Then, it can be shown (e.g., [6]) that

$$\boxed{p_{\mathbf{y}}(y) = \left.\frac{p_{\mathbf{x}}(x)}{|\det(J(\mathbf{y}; \mathbf{x}))|}\right|_{x=f^{-1}(y)},} \tag{2.45}$$

where $|\det(\cdot)|$ denotes the absolute value of the determinant of a matrix. For real random variables, as in $\mathrm{y} = f(\mathrm{x})$, Eq. (2.45) simplifies to

$$p_{\mathrm{y}}(y) = \left.\frac{p_{\mathrm{x}}(x)}{\left|\frac{dy}{dx}\right|}\right|_{x=f^{-1}(y)}. \tag{2.46}$$

The latter can be graphically understood from Fig. 2.1. The following two events have equal probabilities:

$$P(x < \mathrm{x} \leq x + \Delta x) = P(y + \Delta y < \mathrm{y} \leq y), \quad \Delta x > 0, \ \Delta y < 0.$$

Hence, by the definition of a PDF we have

$$p_{\mathrm{y}}(y)|\Delta y| = p_{\mathrm{x}}(x)|\Delta x|, \tag{2.47}$$

which leads to Eq. (2.46).

Example 2.1. Let us consider two random vectors that are related via the linear transform

$$\mathbf{y} = A\mathbf{x}, \tag{2.48}$$

where A is invertible. Compute the joint PDF of $\mathbf{y}$ in terms of $p_{\mathbf{x}}(x)$.

The Jacobian of the transformation is easily computed and given by

$$J(\mathbf{y}; \mathbf{x}) = \begin{bmatrix} a_{11} & \ldots & a_{1l} \\ a_{21} & \ldots & a_{2l} \\ \vdots & \ddots & \vdots \\ a_{l1} & \ldots & a_{ll} \end{bmatrix} = A.$$

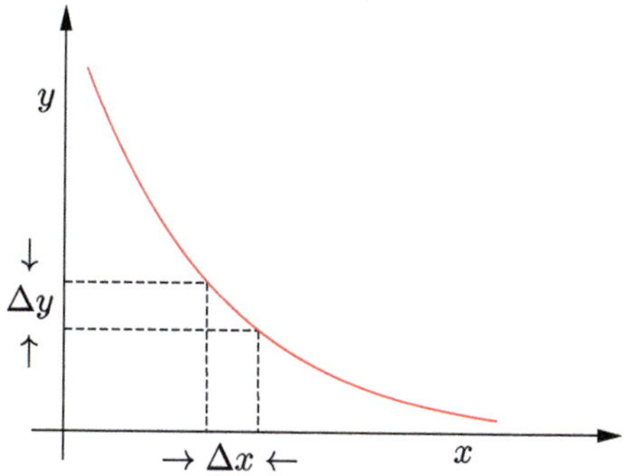

FIGURE 2.1

Note that by the definition of a PDF, $p_{\mathrm{y}}(y)|\Delta y| = p_{\mathrm{x}}(x)|\Delta x|$.

Hence,

$$p_{\mathbf{y}}(\mathbf{y}) = \frac{p_{\mathbf{x}}(A^{-1}\mathbf{y})}{|\det(A)|}. \tag{2.49}$$

2.3 EXAMPLES OF DISTRIBUTIONS

In this section, some notable examples of distributions are provided. These are popular for modeling the random nature of variables met in a wide range of applications, and they will be used later in this book.

2.3.1 DISCRETE VARIABLES

The Bernoulli Distribution

A random variable is said to be distributed according to a Bernoulli distribution if it is binary, $\mathcal{X} = \{0, 1\}$, with

$$P(\mathrm{x} = 1) = p, \quad P(\mathrm{x} = 0) = 1 - p.$$

In a more compact way, we write $\mathrm{x} \sim \mathrm{Bern}(x|p)$, where

$$\boxed{P(x) = \mathrm{Bern}(x|p) := p^x(1-p)^{1-x}.} \tag{2.50}$$

Its mean value is equal to

$$\mathbb{E}[\mathrm{x}] = 1p + 0(1-p) = p \tag{2.51}$$

and its variance is equal to

$$\sigma_{\mathrm{x}}^2 = (1-p)^2 p + p^2(1-p) = p(1-p). \tag{2.52}$$

The Binomial Distribution

A random variable x is said to follow a binomial distribution with parameters n, p, and we write $\mathrm{x} \sim \mathrm{Bin}(x|n, p)$, if $\mathcal{X} = \{0, 1, \dots, n\}$ and

$$\boxed{P(\mathrm{x} = k) := \mathrm{Bin}(k|n, p) = \binom{n}{k} p^k (1-p)^{n-k}, \quad k = 0, 1, \dots, n,} \tag{2.53}$$

where by definition

$$\binom{n}{k} := \frac{n!}{(n-k)!k!}. \tag{2.54}$$

For example, this distribution models the times that heads occurs in n successive trials, where $P(\mathrm{Heads}) = p$. The binomial is a generalization of the Bernoulli distribution, which results if in Eq. (2.53) we set $n = 1$. The mean and variance of the binomial distribution are (Problem 2.1)

$$\mathbb{E}[\mathrm{x}] = np \tag{2.55}$$

and

$$\sigma_{\mathrm{x}}^2 = np(1-p). \tag{2.56}$$

Fig. 2.2A shows the probability $P(k)$ as a function of k for $p = 0.4$ and $n = 9$. Fig. 2.2B shows the respective cumulative distribution. Observe that the latter has a staircase form, as is always the case for discrete variables.

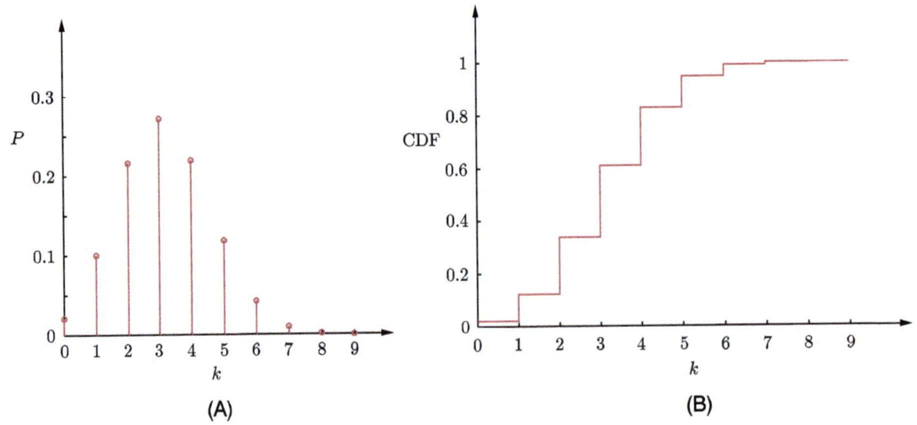

FIGURE 2.2

(A) The probability mass function (PMF) for the binomial distribution for $p = 0.4$ and $n = 9$. (B) The respective cumulative distribution function (CDF). Since the random variable is discrete, the CDF has a staircase-like graph.

The Multinomial Distribution

This is a generalization of the binomial distribution if the outcome of each experiment is not binary but can take one out of K possible values. For example, instead of tossing a coin, a die with K sides is thrown. Each one of the possible K outcomes has probability $P_1, P_2, \ldots, P_K$, respectively, to occur, and we denote

$$\boldsymbol{P} = [P_1, P_2, \ldots, P_K]^T.$$

After n experiments, assume that $x_1, x_2, \ldots, x_K$ times sides $\mathrm{x} = 1, \mathrm{x} = 2, \ldots, \mathrm{x} = K$ occurred, respectively. We say that the random (discrete) vector,

$$\mathbf{x} = [\mathrm{x}_1, \mathrm{x}_2, \ldots, \mathrm{x}_K]^T, \tag{2.57}$$

follows a multinomial distribution, $\mathbf{x} \sim \mathrm{Mult}(\boldsymbol{x}|n, \boldsymbol{P})$, if

$$\boxed{P(\boldsymbol{x}) = \mathrm{Mult}(\boldsymbol{x}|n, \boldsymbol{P}) := \binom{n}{x_1, x_2, \ldots, x_K} \prod_{k=1}^{K} P_k^{x_k},} \tag{2.58}$$

where

$$\binom{n}{x_1, x_2, \ldots, x_K} := \frac{n!}{x_1! x_2! \ldots x_K!}.$$

Note that the variables $\mathrm{x}_1, \ldots, \mathrm{x}_K$ are subject to the constraint

$$\sum_{k=1}^{K} \mathrm{x}_k = n,$$

and also

$$\sum_{k=1}^{K} P_K = 1.$$

The mean value, the variances, and the covariances are given by

$$\mathbb{E}[\mathbf{x}] = n\boldsymbol{P},\ \sigma_{\mathrm{x}_k}^2 = nP_k(1 - P_k),\ k = 1, 2, \ldots, K,\ \mathrm{cov}(\mathrm{x}_i, \mathrm{x}_j) = -nP_iP_j,\ i \neq j. \tag{2.59}$$

The special case of the multinomial, where only one experiment, $n = 1$, is performed, is known as the *categorical* distribution. The latter can be considered as the generalization of the Bernoulli distribution.

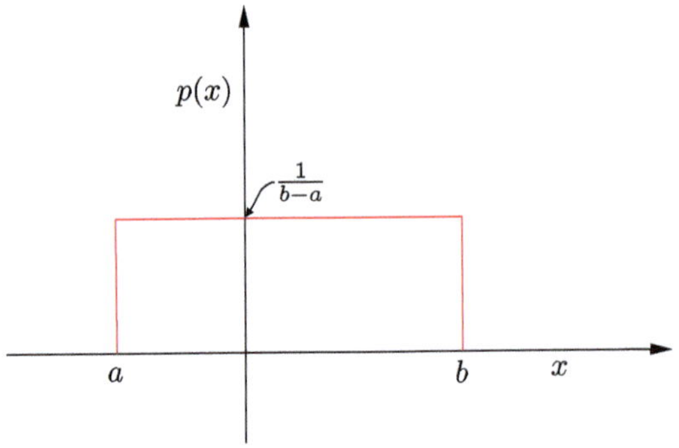

FIGURE 2.3

The PDF of a uniform distribution $\mathcal{U}(a, b)$.

2.3.2 CONTINUOUS VARIABLES

The Uniform Distribution

A random variable x is said to follow a *uniform* distribution in an interval $[a, b]$, and we write $\mathrm{x} \sim \mathcal{U}(a, b)$, with $a > -\infty$ and $b < +\infty$, if

$$\boxed{p(x) = \begin{cases} \frac{1}{b-a}, & \text{if } a \leq x \leq b, \\ 0, & \text{otherwise.} \end{cases}} \tag{2.60}$$

Fig. 2.3 shows the respective graph. The mean value is equal to

$$\mathbb{E}[\mathrm{x}] = \frac{a+b}{2}, \tag{2.61}$$

and the variance is given by (Problem 2.2)

$$\sigma_{\mathrm{x}}^2 = \frac{1}{12}(b-a)^2. \tag{2.62}$$

The Gaussian Distribution

The Gaussian or normal distribution is one among the most widely used distributions in all scientific disciplines. We say that a random variable x is *Gaussian* or *normal* with parameters μ and σ^2, and we write $\mathrm{x} \sim \mathcal{N}(\mu, \sigma^2)$ or $\mathcal{N}(x|\mu, \sigma^2)$, if

$$p(x) = \frac{1}{\sqrt{2\pi}\sigma} \exp\left(-\frac{(x-\mu)^2}{2\sigma^2}\right). \tag{2.63}$$

It can be shown that the corresponding mean and variance are

$$\mathbb{E}[\mathrm{x}] = \mu \quad \text{and} \quad \sigma_{\mathrm{x}}^2 = \sigma^2. \tag{2.64}$$

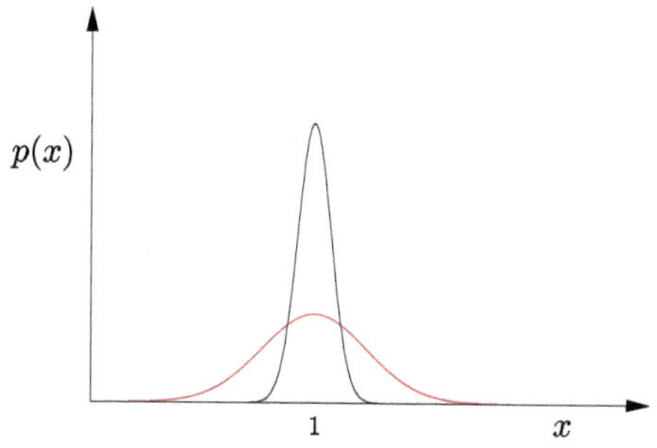

FIGURE 2.4

The graphs of two Gaussian PDFs for $\mu = 1$ and $\sigma^2 = 0.1$ (red) and $\sigma^2 = 0.01$ (gray).

Indeed, by the definition of the mean value, we have

$$\begin{aligned} \mathbb{E}[x] &= \frac{1}{\sqrt{2\pi}\sigma} \int_{-\infty}^{+\infty} x \exp\left(-\frac{(x-\mu)^2}{2\sigma^2}\right) dx \\ &= \frac{1}{\sqrt{2\pi}\sigma} \int_{-\infty}^{+\infty} (y+\mu) \exp\left(-\frac{y^2}{2\sigma^2}\right) dy. \end{aligned} \tag{2.65}$$

Due to the symmetry of the exponential function, performing the integration involving y gives zero and the only surviving term is due to μ. Taking into account that a PDF integrates to one, we obtain the result.

To derive the variance, from the definition of the Gaussian PDF, we have

$$\int_{-\infty}^{+\infty} \exp\left(-\frac{(x-\mu)^2}{2\sigma^2}\right) dx = \sqrt{2\pi}\sigma. \tag{2.66}$$

Taking the derivative of both sides with respect to σ, we obtain

$$\int_{-\infty}^{+\infty} \frac{(x-\mu)^2}{\sigma^3} \exp\left(-\frac{(x-\mu)^2}{2\sigma^2}\right) dx = \sqrt{2\pi} \tag{2.67}$$

or

$$\sigma_x^2 := \frac{1}{\sqrt{2\pi}\sigma} \int_{-\infty}^{+\infty} (x-\mu)^2 \exp\left(-\frac{(x-\mu)^2}{2\sigma^2}\right) dx = \sigma^2, \tag{2.68}$$

which proves the claim.

Fig. 2.4 shows the graph for two cases, $\mathcal{N}(x|1, 0.1)$ and $\mathcal{N}(x|1, 0.01)$. Both curves are symmetrically placed around the mean value $\mu = 1$. Observe that the smaller the variance is, the sharper around the mean value the PDF becomes.

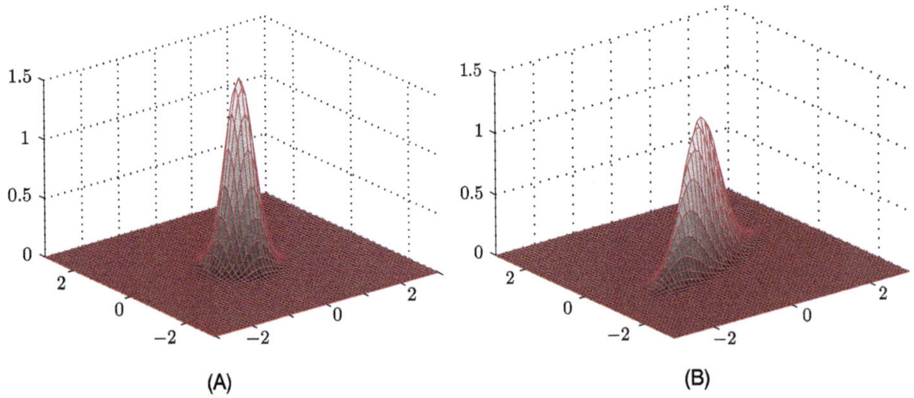

FIGURE 2.5

The graphs of two-dimensional Gaussian PDFs for $\boldsymbol{\mu} = \mathbf{0}$ and different covariance matrices. (A) The covariance matrix is diagonal with equal elements along the diagonal. (B) The corresponding covariance matrix is nondiagonal.

The generalization of the Gaussian to vector variables, $\mathbf{x} \in \mathbb{R}^l$, results in the so-called *multivariate Gaussian* or *normal* distribution, $\mathbf{x} \sim \mathcal{N}(\boldsymbol{x}|\boldsymbol{\mu}, \Sigma)$ with parameters $\boldsymbol{\mu}$ and Σ, which is defined as

$$p(\boldsymbol{x}) = \frac{1}{(2\pi)^{l/2}|\Sigma|^{1/2}} \exp\left(-\frac{1}{2}(\boldsymbol{x}-\boldsymbol{\mu})^T \Sigma^{-1} (\boldsymbol{x}-\boldsymbol{\mu})\right): \quad \text{Gaussian PDF,} \tag{2.69}$$

where $|\cdot|$ denotes the determinant of a matrix. It can be shown (Problem 2.3) that the respective mean values and the covariance matrix are given by

$$\mathbb{E}[\mathbf{x}] = \boldsymbol{\mu} \quad \text{and} \quad \Sigma_{\mathbf{x}} = \Sigma. \tag{2.70}$$

Fig. 2.5 shows the two-dimensional normal PDF for two cases. Both share the same mean vector, $\boldsymbol{\mu} = \mathbf{0}$, but they have different covariance matrices,

$$\Sigma_1 = \begin{bmatrix} 0.1 & 0.0 \\ 0.0 & 0.1 \end{bmatrix}, \quad \Sigma_2 = \begin{bmatrix} 0.1 & 0.01 \\ 0.01 & 0.2 \end{bmatrix}. \tag{2.71}$$

Fig. 2.6 shows the corresponding isovalue contours for equal probability density values. In Fig. 2.6A, the contours are circles, corresponding to the symmetric PDF in Fig. 2.5A with covariance matrix Σ_1. The one shown in Fig. 2.6B corresponds to the PDF in Fig. 2.5B associated with Σ_2. Observe that, in general, the isovalue curves are ellipses/hyperellipsoids. They are centered at the mean vector, and the orientation of the major axis as well their exact shape is controlled by the eigenstructure of the associated covariance matrix. Indeed, all points $\boldsymbol{x} \in \mathbb{R}^l$, which score the same probability density value, obey

$$(\boldsymbol{x}-\boldsymbol{\mu})^T \Sigma^{-1} (\boldsymbol{x}-\boldsymbol{\mu}) = \text{constant} = c. \tag{2.72}$$

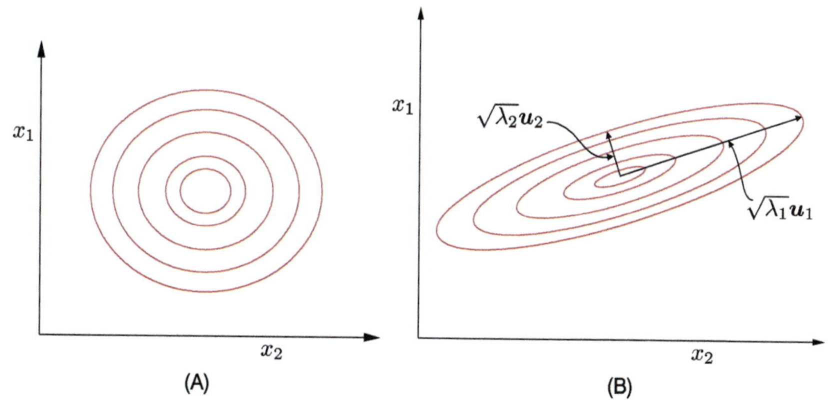

FIGURE 2.6

The isovalue contours for the two Gaussians of Fig. 2.5. The contours for the Gaussian in Fig. 2.5A are circles, while those corresponding to Fig. 2.5B are ellipses. The major and minor axes of the ellipse are determined by the eigenvectors/eigenvalues of the respective covariance matrix, and they are proportional to $\sqrt{\lambda_1 c}$ and $\sqrt{\lambda_2 c}$, respectively. In the figure, they are shown for the case of $c = 1$. For the case of the diagonal matrix, with equal elements along the diagonal, all eigenvalues are equal, and the ellipse becomes a circle.

We know that the covariance matrix besides being positive definite is also *symmetric*, $\Sigma = \Sigma^T$. Thus, its eigenvalues are real and the corresponding eigenvectors can be chosen to form an orthonormal basis (Appendix A.2), which leads to its diagonalization,

$$\Sigma = U \Lambda U^T, \tag{2.73}$$

with

$$U := [\boldsymbol{u}_1, \ldots, \boldsymbol{u}_l], \tag{2.74}$$

where $\boldsymbol{u}_i$, $i = 1, 2, \ldots, l$, are the orthonormal eigenvectors, and

$$\Lambda := \text{diag}\{\lambda_1, \ldots, \lambda_l\} \tag{2.75}$$

comprise the respective eigenvalues. We assume that Σ is invertible, hence all eigenvalues are positive (being positive definite it has positive eigenvalues, Appendix A.2). Due to the orthonormality of the eigenvectors, matrix U is orthogonal as expressed in $UU^T = U^T U = I$. Thus, Eq. (2.72) can now be written as

$$\boldsymbol{y}^T \Lambda^{-1} \boldsymbol{y} = c, \tag{2.76}$$

where we have used the linear transformation

$$\boldsymbol{y} := U^T (\boldsymbol{x} - \boldsymbol{\mu}), \tag{2.77}$$

which corresponds to a rotation of the axes by U and a translation of the origin to $\boldsymbol{\mu}$. Eq. (2.76) can be written as

$$\frac{y_1^2}{\lambda_1}+\cdots+\frac{y_l^2}{\lambda_l}=c, \tag{2.78}$$

where it can be readily observed that it is an equation describing a (hyper)ellipsoid in the $\mathbb{R}^l$. From Eq. (2.77), it is easily seen that it is centered at $\boldsymbol{\mu}$ and that the major axes of the ellipsoid are parallel to $\boldsymbol{u}_1,\ldots,\boldsymbol{u}_l$ (plug in place of $\boldsymbol{x}$ the standard basis vectors, $[1,0,\ldots,0]^T$, etc.). The sizes of the respective axes are controlled by the corresponding eigenvalues. This is shown in Fig. 2.6B. For the special case of a diagonal covariance with equal elements across the diagonal, all eigenvalues are equal to the value of the common diagonal element and the ellipsoid becomes a (hyper)sphere (circle) as shown in Fig. 2.6A.

The Gaussian PDF has a number of nice properties, which we are going to discover as we move on in this book. For the time being, note that if the covariance matrix is diagonal,

$$\Sigma=\text{diag}\{\sigma_1^2,\ldots,\sigma_l^2\},$$

that is, when the covariance of all the elements $\text{cov}(\mathrm{x}_i,\mathrm{x}_j)=0,\ i,j=1,2,\ldots,l$, then the random variables comprising $\mathbf{x}$ are statistically *independent*. In general, this is not true. Uncorrelated variables are not necessarily independent; independence is a much stronger condition. This is true, however, if they follow a multivariate Gaussian. Indeed, if the covariance matrix is diagonal, then the multivariate Gaussian is written as

$$p(\boldsymbol{x})=\prod_{i=1}^{l}\frac{1}{\sqrt{2\pi}\sigma_i}\exp\left(-\frac{(x_i-\mu_i)^2}{2\sigma_i^2}\right). \tag{2.79}$$

In other words,

$$p(\boldsymbol{x})=\prod_{i=1}^{l}p(x_i), \tag{2.80}$$

which is the condition for statistical independence.

The Central Limit Theorem

This is one of the most fundamental theorems in probability theory and statistics and it partly explains the popularity of the Gaussian distribution. Consider N mutually *independent* random variables, each following its own distribution with mean values μ_i and variances $\sigma_i^2,\ i=1,2,\ldots,N$. Define a new random variable as their sum,

$$\mathrm{x}=\sum_{i=1}^{N}\mathrm{x}_i. \tag{2.81}$$

Then the mean and variance of the new variable are given by

$$\mu=\sum_{i=1}^{N}\mu_i \text{ and } \sigma^2=\sum_{i=1}^{N}\sigma_i^2. \tag{2.82}$$

It can be shown (e.g., [4,6]) that as $N \longrightarrow \infty$ the distribution of the normalized variable

$$\mathrm{z} = \frac{\mathrm{x} - \mu}{\sigma} \tag{2.83}$$

tends to the *standard* normal distribution, and for the corresponding PDF we have

$$p(z) \xrightarrow[N \to \infty]{} \mathcal{N}(z|0, 1). \tag{2.84}$$

In practice, even summing up a relatively small number, N, of random variables, one can obtain a good approximation to a Gaussian. For example, if the individual PDFs are smooth enough and each random variable is *independent and identically distributed* (i.i.d.), a number N between 5 and 10 can be sufficient. The term i.i.d. will be used a lot in this book. The term implies that successive samples of a random variable are drawn independently from the same distribution that describes the respective variable.

The Exponential Distribution

We say that a random variable follows an exponential distribution with parameter $\lambda > 0$, if

$$p(x) = \begin{cases} \lambda \exp(-\lambda x), & \text{if } x \geq 0, \\ 0, & \text{otherwise.} \end{cases} \tag{2.85}$$

The distribution has been used, for example, to model the time between arrivals of telephone calls or of a bus at a bus stop. The mean and variance can be easily computed by following simple integration rules, and they are

$$\mathbb{E}[\mathrm{x}] = \frac{1}{\lambda}, \quad \sigma_{\mathrm{x}}^2 = \frac{1}{\lambda^2}. \tag{2.86}$$

The Beta Distribution

We say that a random variable, $\mathrm{x} \in [0, 1]$, follows a beta distribution with positive parameters, a, b, and we write, $\mathrm{x} \sim \text{Beta}(x|a, b,)$, if

$$p(x) = \begin{cases} \frac{1}{B(a,b)} x^{a-1}(1-x)^{b-1}, & \text{if } 0 \leq x \leq 1, \\ 0, & \text{otherwise,} \end{cases} \tag{2.87}$$

where $B(a, b)$ is the beta function, defined as

$$B(a, b) := \int_0^1 x^{a-1}(1-x)^{b-1}\, dx. \tag{2.88}$$

The mean and variance of the beta distribution are given by (Problem 2.4)

$$\mathbb{E}[\mathrm{x}] = \frac{a}{a+b}, \quad \sigma_{\mathrm{x}}^2 = \frac{ab}{(a+b)^2(a+b+1)}. \tag{2.89}$$

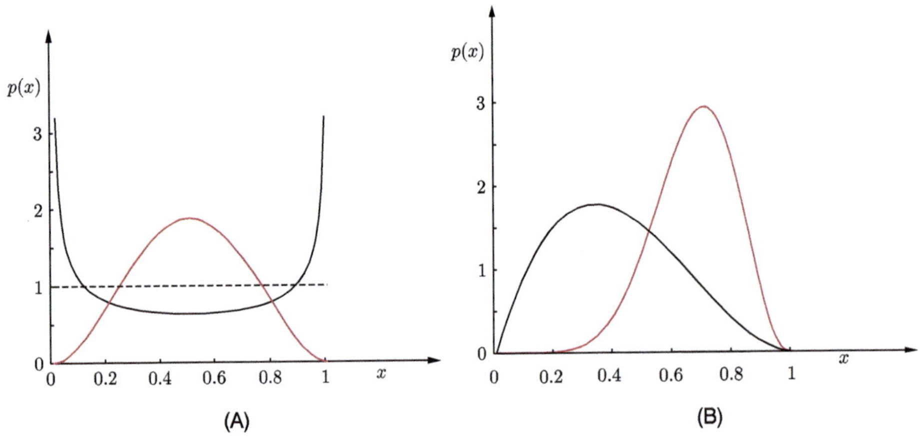

FIGURE 2.7

The graphs of the PDFs of the beta distribution for different values of the parameters. (A) The dotted line corresponds to $a = 1, b = 1$, the gray line to $a = 0.5, b = 0.5$, and the red one to $a = 3, b = 3$. (B) The gray line corresponds to $a = 2, b = 3$, and the red one to $a = 8, b = 4$. For values $a = b$, the shape is symmetric around $1/2$. For $a < 1, b < 1$, it is convex. For $a > 1, b > 1$, it is zero at $x = 0$ and $x = 1$. For $a = 1 = b$, it becomes the uniform distribution. If $a < 1$, $p(x) \longrightarrow \infty, x \longrightarrow 0$ and if $b < 1$, $p(x) \longrightarrow \infty, x \longrightarrow 1$.

Moreover, it can be shown (Problem 2.5) that

$$B(a, b) = \frac{\Gamma(a)\Gamma(b)}{\Gamma(a+b)}, \tag{2.90}$$

where Γ is the gamma function defined as

$$\Gamma(a) = \int_0^\infty x^{a-1} e^{-x}\, dx. \tag{2.91}$$

The beta distribution is very flexible and one can achieve various shapes by changing the parameters a, b. For example, if $a = b = 1$, the uniform distribution results. If $a = b$, the PDF has a symmetric graph around $1/2$. If $a > 1, b > 1$, then $p(x) \longrightarrow 0$ both at $x = 0$ and $x = 1$. If $a < 1$ and $b < 1$, it is convex with a unique minimum. If $a < 1$, it tends to ∞ as $x \longrightarrow 0$, and if $b < 1$, it tends to ∞ for $x \longrightarrow 1$. Figs. 2.7A and B show the graph of the beta distribution for different values of the parameters.

The Gamma Distribution

A random variable follows the gamma distribution with positive parameters a, b, and we write $\mathrm{x} \sim \mathrm{Gamma}(x|a, b)$, if

$$p(x) = \begin{cases} \dfrac{b^a}{\Gamma(a)} x^{a-1} e^{-bx}, & x > 0, \\ 0, & \text{otherwise}. \end{cases} \tag{2.92}$$

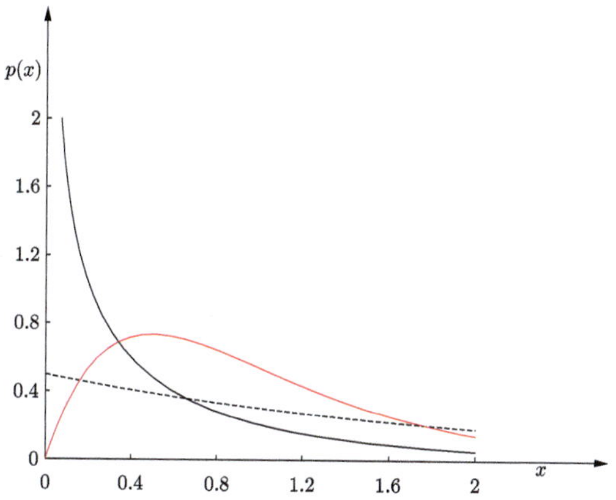

FIGURE 2.8

The PDF of the gamma distribution takes different shapes for the various values of the following parameters: $a = 0.5, b = 1$ (full line gray), $a = 2, b = 0.5$ (red), $a = 1, b = 2$ (dotted).

The mean and variance are given by

$$\mathbb{E}[\mathrm{x}] = \frac{a}{b}, \quad \sigma_{\mathrm{x}}^2 = \frac{a}{b^2}. \tag{2.93}$$

The gamma distribution also takes various shapes by varying the parameters. For $a < 1$, it is strictly decreasing and $p(x) \longrightarrow \infty$ as $x \longrightarrow 0$ and $p(x) \longrightarrow 0$ as $x \longrightarrow \infty$. Fig. 2.8 shows the resulting graphs for various values of the parameters.

Remarks 2.1.

- Setting in the gamma distribution a to be an integer (usually $a = 2$), the *Erlang* distribution results. This distribution is used to model waiting times in queueing systems.
- The *chi-squared* is also a special case of the gamma distribution, and it is obtained if we set $b = 1/2$ and $a = \nu/2$. The chi-squared distribution results if we sum up ν squared normal variables.

The Dirichlet Distribution

The Dirichlet distribution can be considered as the multivariate generalization of the beta distribution. Let $\mathbf{x} = [\mathrm{x}_1, \ldots, \mathrm{x}_K]^T$ be a random vector, with components such as

$$0 \leq \mathrm{x}_k \leq 1, \quad k = 1, 2, \ldots, K, \quad \text{and} \quad \sum_{k=1}^{K} \mathrm{x}_k = 1. \tag{2.94}$$

In other words, the random variables lie on $(K-1)$-dimensional *simplex*, Fig. 2.9. We say that the random vector $\mathbf{x}$ follows a Dirichlet distribution with (positive) parameters $\boldsymbol{a} = [a_1, \ldots, a_K]^T$, and we

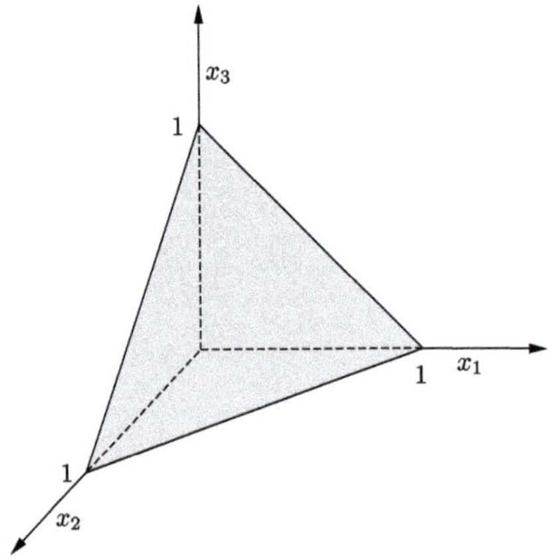

FIGURE 2.9

The two-dimensional simplex in $\mathbb{R}^3$.

write $\mathbf{x} \sim \text{Dir}(\boldsymbol{x}|\boldsymbol{a})$, if

$$\boxed{p(\boldsymbol{x}) = \text{Dir}(\boldsymbol{x}|\boldsymbol{a}) := \frac{\Gamma(\bar{a})}{\Gamma(a_1)\ldots\Gamma(a_K)} \prod_{k=1}^{K} x_k^{a_k-1},} \tag{2.95}$$

where

$$\bar{a} = \sum_{k=1}^{K} a_k. \tag{2.96}$$

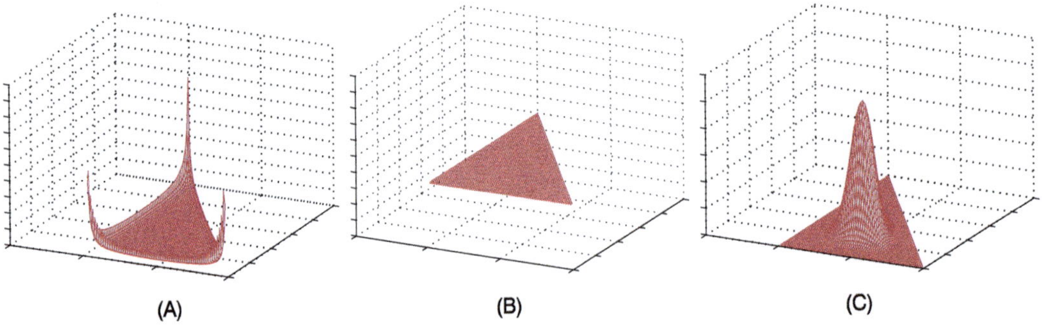

FIGURE 2.10

The Dirichlet distribution over the two-dimensional simplex for (A) (0.1,0.1,0.1), (B) (1,1,1), and (C) (10,10,10).

The mean, variance, and covariances of the involved random variables are given by (Problem 2.7), i.e.,

$$\mathbb{E}[\mathbf{x}] = \frac{1}{\bar{a}}\boldsymbol{a}, \quad \sigma_{\mathrm{x}_k}^2 = \frac{a_k(\bar{a}-a_k)}{\bar{a}^2(\bar{a}+1)}, \quad \mathrm{cov}(\mathrm{x}_i, \mathrm{x}_j) = -\frac{a_i a_j}{\bar{a}^2(\bar{a}+1)}, \; i \neq j. \tag{2.97}$$

Fig. 2.10 shows the graph of the Dirichlet distribution for different values of the parameters, over the respective two-dimensional simplex.

2.4 STOCHASTIC PROCESSES

The notion of a random variable has been introduced to describe the result of a random experiment whose outcome is a single value, such as heads or tails in a coin tossing experiment, or a value between one and six when throwing the die in a backgammon game.

In this section, the notion of a *stochastic process* is introduced to describe random experiments where the outcome of each experiment is a function or a sequence; in other words, the outcome of each experiment is an infinite number of values. In this book, we are only going to be concerned with stochastic processes associated with sequences. Thus, the result of a random experiment is a *sequence*, u_n (or sometimes denoted as $u(n)$), $n \in \mathbb{Z}$, where $\mathbb{Z}$ is the set of integers. Usually, n is interpreted as a time index, and u_n is called a *time series*, or in signal processing jargon, a *discrete-time signal*. In contrast, if the outcome is a function, $u(t)$, it is called a *continuous-time* signal. We are going to adopt the time interpretation of the free variable n for the rest of the chapter, without harming generality.

When discussing random variables, we used the notation x to denote the random variable, which assumes a value, x, from the sample space once an experiment is performed. Similarly, we are going to use u_n to denote the specific sequence resulting from a single experiment and the roman font, u_n, to denote the corresponding *discrete-time* random process, that is, the rule that assigns a specific sequence as the outcome of an experiment. A stochastic process can be considered as a family or *ensemble* of sequences. The individual sequences are known as *sample sequences* or simply as *realizations*.

For our notational convention, in general, we are going to reserve different symbols for processes and random variables. We have already used the symbol u and not x; this is only for pedagogical reasons, just to make sure that the reader readily recognizes when the focus is on random variables and when it is on random processes. In signal processing jargon, a stochastic process is also known as a *random signal*. Fig. 2.11 illustrates the fact that the outcome of an experiment involving a stochastic process is a sequence of values.

Note that fixing the time to a specific value, $n = n_0$, makes u_{n_0} a random variable. Indeed, for each random experiment we perform, a single value results at time instant n_0. From this perspective, a random process can be considered the collection of infinite random variables, $\{\mathrm{u}_n, n \in \mathbb{Z}\}$. So, is there a need to study a stochastic process separate from random variables/vectors? The answer is yes, and the reason is that we are going to allow certain time dependencies among the random variables, corresponding to different time instants, and study the respective effect on the time evolution of the random process. Stochastic processes will be considered in Chapter 5, where the underlying time dependencies will be exploited for computational simplifications, and in Chapter 13 in the context of Gaussian processes.

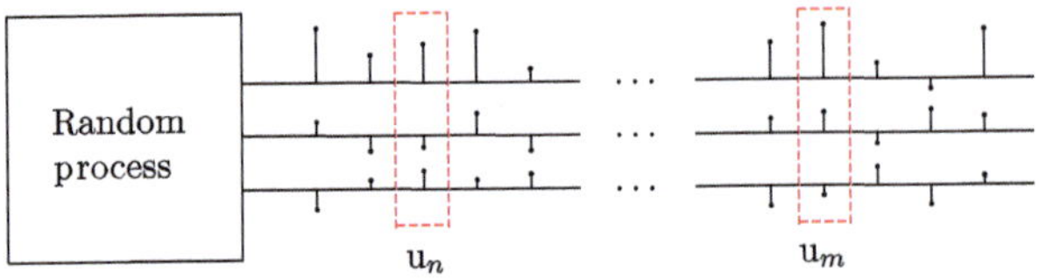

FIGURE 2.11

The outcome of each experiment, associated with a *discrete-time* stochastic process, is a *sequence* of values. For each one of the realizations, the corresponding values obtained at any instant (e.g., n or m) comprise the outcomes of a corresponding random variable, u_n or u_m, respectively.

2.4.1 FIRST- AND SECOND-ORDER STATISTICS

For a stochastic process to be fully described, one must know the joint PDFs (PMFs for discrete-valued random variables)

$$p(u_n, u_m, \ldots, u_r; n, m, \ldots, r), \tag{2.98}$$

for all possible combinations of random variables, $\mathrm{u}_n, \mathrm{u}_m, \ldots, \mathrm{u}_r$. Note that, in order to emphasize it, we have explicitly denoted the dependence of the joint PDFs on the involved time instants. However, from now on, this will be suppressed for notational convenience. Most often, in practice, and certainly in this book, the emphasis is on computing first- and second-order statistics only, based on $p(u_n)$ and $p(u_n, u_m)$. To this end, the following quantities are of particular interest.

Mean at time n:

$$\mu_n := \mathbb{E}[\mathrm{u}_n] = \int_{-\infty}^{+\infty} u_n p(u_n) du_n. \tag{2.99}$$

Autocovariance at time instants n, m:

$$\mathrm{cov}(n, m) := \mathbb{E}\left[\left(\mathrm{u}_n - \mathbb{E}[\mathrm{u}_n]\right)\left(\mathrm{u}_m - \mathbb{E}[\mathrm{u}_m]\right)\right]. \tag{2.100}$$

Autocorrelation at time instants n, m:

$$r(n, m) := \mathbb{E}\left[\mathrm{u}_n \mathrm{u}_m\right]. \tag{2.101}$$

Note that for notational simplicity, subscripts have been dropped from the respective symbols, e.g., $r(n, m)$ is used instead of the more formal notation, $r_{\mathrm{u}}(n, m)$. We refer to these mean values as *ensemble averages* to stress that they convey statistical information over the ensemble of sequences that comprise the process.

The respective definitions for complex stochastic processes are

$$\mathrm{cov}(n, m) = \mathbb{E}\left[\left(\mathrm{u}_n - \mathbb{E}[\mathrm{u}_n]\right)\left(\mathrm{u}_m - \mathbb{E}[\mathrm{u}_m]\right)^*\right], \tag{2.102}$$

and

$$r(n, m) = \mathbb{E}\left[\mathrm{u}_n \mathrm{u}_m^*\right]. \tag{2.103}$$

2.4.2 STATIONARITY AND ERGODICITY

Definition 2.1 (*Strict-sense stationarity*). A stochastic process u_n is said to be *strict-sense stationary* (SSS) if its statistical properties are invariant to a shift of the origin, or if $\forall k \in \mathbb{Z}$

$$p(u_n, u_m, \ldots, u_r) = p(u_{n-k}, u_{m-k}, \ldots, u_{r-k}), \tag{2.104}$$

and for *any* possible combination of time instants, $n, m, \ldots, r \in \mathbb{Z}$.

In other words, the stochastic processes u_n and u_{n-k} are described by the same joint PDFs of all orders. A weaker version of stationarity is that of the mth-order stationarity, where joint PDFs involving up to m variables are invariant to the choice of the origin. For example, for a second-order ($m=2$) stationary process, we have $p(u_n) = p(u_{n-k})$ and $p(u_n, u_r) = p(u_{n-k}, u_{r-k})$, $\forall n, r, k \in \mathbb{Z}$.

Definition 2.2 (*Wide-sense stationarity*). A stochastic process u_n is said to be *wide-sense stationary* (WSS) if the mean value is constant over all time instants and the autocorrelation/autocovariance sequences depend on the difference of the involved time indices, or

$$\mu_n = \mu, \quad \text{and} \quad r(n, n-k) = r(k). \tag{2.105}$$

Note that WSS is a weaker version of the second-order stationarity; in the latter case, all possible second-order statistics are independent of the time origin. In the former, we only require the autocorrelation (autocovariance) and the mean value to be independent of the time origin. The reason we focus on these two quantities (statistics) is that they are of major importance in the study of linear systems and in the mean-square estimation, as we will see in Chapter 4.

Obviously, an SSS process is also WSS but, in general, not the other way around. For WSS processes, the autocorrelation becomes a *sequence* with a *single* time index as the free parameter; thus its value, which measures a relation of the random variables at two time instants, depends *solely on how much these time instants differ*, and not on their specific values.

From our basic statistics course, we know that given a random variable x, its mean value can be approximated by the sample mean. Carrying out N successive independent experiments, let $x_n, n = 1, 2, \ldots, N$, be the obtained values, known as *observations*. The *sample mean* is defined as

$$\hat{\mu}_N := \frac{1}{N} \sum_{n=1}^{N} x_n. \tag{2.106}$$

For large enough values of N, we expect the sample mean to be close to the true mean value, $\mathbb{E}[\mathrm{x}]$. In a more formal way, this is guaranteed by the fact that $\hat{\mu}_N$ is associated with an *unbiased* and *consistent* estimator. We will discuss such issues in Chapter 3; however, we can refresh our memory at this point. Every time we repeat the N random experiments, different samples result and hence a different estimate $\hat{\mu}_N$ is computed. Thus, the values of the estimates define a new random variable, $\hat{\mathrm{\mu}}_n$, known as the estimator. This is unbiased, because it can easily be shown that

$$\mathbb{E}[\hat{\mathrm{\mu}}_N] = \mathbb{E}[\mathrm{x}], \tag{2.107}$$

and it is consistent because its variance tends to zero as $N \longrightarrow +\infty$ (Problem 2.8). These two properties guarantee that, with high probability, for large values of N, $\hat{\mu}_N$ will be close to the true mean value.

To apply the concept of sample mean approximation to random processes, one must have at her/his disposal a number of N realizations, and compute the sample mean at different time instants "across the process," using *different* realizations, representing the ensemble of sequences. Similarly, sample mean arguments can be used to approximate the autocovariance/autocorrelation sequences. However, this is a costly operation, since now each experiment results in an infinite number of values (a sequence of values). Moreover, it is common in practical applications that only one realization is available to the user.

To this end, we will now define a special type of stochastic processes, where the sample mean operation can be significantly simplified.

Definition 2.3 (*Ergodicity*). A stochastic process is said to be *ergodic* if the complete statistics can be determined by any one of the realizations.

In other words, if a process is ergodic, every single realization carries identical statistical information and it can describe the entire random process. Since from a single sequence only one set of PDFs can be obtained, we conclude that *every ergodic process is necessarily stationary.* A nonstationary process has infinite sets of PDFs, depending upon the choice of the origin. For example, there is only one mean value that can result from a single realization and be obtained as a (time) average over the values of the sequence. Hence, the mean value of a stochastic process that is ergodic must be constant for all time instants, or independent of the time origin. The same is true for all higher-order statistics.

A special type of ergodicity is that of the *second-order* ergodicity. This means that only statistics up to a second order can be obtained from a single realization. Second-order ergodic processes are necessarily WSS. For second-order ergodic processes, the following are true:

$$\mathbb{E}[\mathrm{u}_n] = \mu = \lim_{N\to\infty} \hat{\mu}_N, \tag{2.108}$$

where

$$\hat{\mu}_N := \frac{1}{2N+1} \sum_{n=-N}^{N} u_n.$$

Also,

$$\mathrm{cov}(k) = \lim_{N\to\infty} \frac{1}{2N+1} \sum_{n=-N}^{N} (u_n - \mu)(u_{n-k} - \mu), \tag{2.109}$$

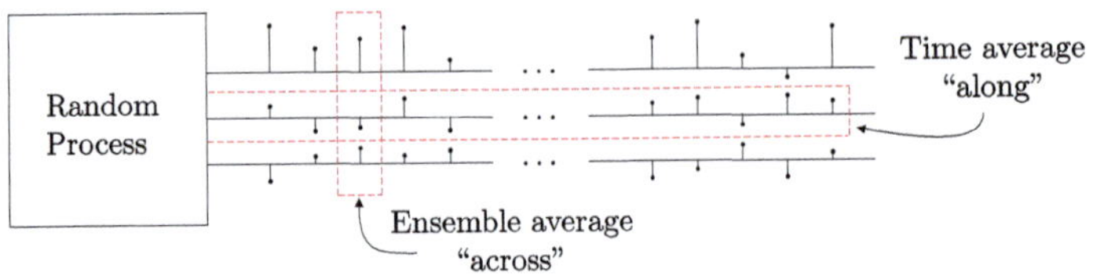

FIGURE 2.12

For ergodic processes, the common mean value, for all time instants (ensemble averaging "across" the process), is computed as the time average "along" the process.

where both limits are in the mean-square sense; that is,

$$\lim_{N\to\infty} \mathbb{E}\left[|\hat{\mu}_N - \mu|^2\right] = 0,$$

and similarly for the autocovariance. Note that often, ergodicity is only required to be assumed for the computation of the mean and covariance and not for all possible second-order statistics. In this case, we talk about *mean-ergodic* and *covariance-ergodic* processes.

In summary, when ergodic processes are involved, ensemble averages "across the process" can be obtained as time averages "along the process"; see Fig. 2.12.

In practice, when only a finite number of samples from a realization is available, then the mean and covariance are approximated as the respective sample means.

An issue is to establish conditions under which a process is mean-ergodic or covariance-ergodic. Such conditions do exist, and the interested reader can find such information in more specialized books [6]. It turns out that the condition for mean-ergodicity relies on second-order statistics and the condition for covariance-ergodicity on fourth-order statistics.

It is very common in statistics as well as in machine learning and signal processing to subtract the mean value from the data during the *preprocessing stage*. In such a case, we say that the data are *centered*. The resulting new process has now *zero mean* value, and the covariance and autocorrelation sequences coincide. From now on, we will assume that the mean is known (or computed as a sample mean) and then subtracted. Such a treatment simplifies the analysis without harming generality.

Example 2.2. The goal of this example is to construct a process that is WSS yet not ergodic. Let a WSS process, u_n,

$$\mathbb{E}[\mathrm{u}_n] = \mu,$$

and

$$\mathbb{E}[\mathrm{u}_n \mathrm{u}_{n-k}] = r_{\mathrm{u}}(k).$$

Define the process

$$\mathrm{v}_n := \mathrm{a}\mathrm{u}_n, \tag{2.110}$$

where a is a random variable taking values in $\{0, 1\}$, with probabilities $P(0) = P(1) = 0.5$. Moreover, a and u_n are statistically independent. Then, we have

$$\mathbb{E}[\mathrm{v}_n] = \mathbb{E}[\mathrm{a}\mathrm{u}_n] = \mathbb{E}[\mathrm{a}]\,\mathbb{E}[\mathrm{u}_n] = 0.5\mu, \tag{2.111}$$

and

$$\mathbb{E}[\mathrm{v}_n \mathrm{v}_{n-k}] = \mathbb{E}[\mathrm{a}^2]\,\mathbb{E}[\mathrm{u}_n \mathrm{u}_{n-k}] = 0.5 r_{\mathrm{u}}(k). \tag{2.112}$$

Thus, v_n is WSS. However, it is not ergodic. Indeed, some of the realizations will be equal to zero (when $a = 0$), and the mean value and autocorrelation, which will result from them as time averages, will be zero, which is different from the ensemble averages.

2.4.3 POWER SPECTRAL DENSITY

The Fourier transform is an indispensable tool for representing in a compact way, in the frequency domain, the variations that a function/sequence undergoes in terms of its free variable (e.g., time). Stochastic processes are inherently related to time. The question that is now raised is whether stochastic processes can be described in terms of a Fourier transform. The answer is affirmative, and the vehicle to achieve this is via the autocorrelation sequence for processes that are at least WSS. Prior to providing the necessary definitions, it is useful to summarize some common properties of the autocorrelation sequence.

Properties of the Autocorrelation Sequence

Let u_n be a WSS process. Its autocorrelation sequence has the following properties, which are given for the more general complex-valued case:

- *Property I.*

$$r(k) = r^*(-k), \quad \forall k \in \mathbb{Z}. \tag{2.113}$$

  This property is a direct consequence of the invariance with respect to the choice of the origin. Indeed,

$$r(k) = \mathbb{E}[\mathrm{u}_n \mathrm{u}_{n-k}^*] = \mathbb{E}[\mathrm{u}_{n+k}\mathrm{u}_n^*] = r^*(-k).$$

- *Property II.*

$$r(0) = \mathbb{E}\left[|\mathrm{u}_n|^2\right]. \tag{2.114}$$

  That is, the value of the autocorrelation at $k = 0$ is equal to the mean-square of the magnitude of the respective random variables. Interpreting the square of the magnitude of a variable as its energy, $r(0)$ can be interpreted as the corresponding (average) power.
- *Property III.*

$$r(0) \geq |r(k)|, \quad \forall k \neq 0. \tag{2.115}$$

  The proof is provided in Problem 2.9. In other words, the correlation of the variables, corresponding to two different time instants, cannot be larger (in magnitude) than $r(0)$. As we will see in Chapter 4, this property is essentially the Cauchy–Schwarz inequality for the inner products (see also Appendix of Chapter 8).
- *Property IV.* The autocorrelation sequence of a stochastic process is *positive definite*. That is,

$$\boxed{\sum_{n=1}^{N}\sum_{m=1}^{N} a_n a_m^* r(n,m) \geq 0, \quad \forall a_n \in \mathbb{C},\ n = 1, 2, \ldots, N,\ \forall N \in \mathbb{Z}.} \tag{2.116}$$

  Proof. The proof is easily obtained by the definition of the autocorrelation,

$$0 \leq \mathbb{E}\left[\left|\sum_{n=1}^{N} a_n \mathrm{u}_n\right|^2\right] = \sum_{n=1}^{N}\sum_{m=1}^{N} a_n a_m^* \, \mathbb{E}\left[\mathrm{u}_n \mathrm{u}_m^*\right], \tag{2.117}$$

which proves the claim. Note that strictly speaking, we should say that it is semipositive definite. However, the "positive definite" name is the one that has survived in the literature. This property will be useful when introducing *Gaussian processes* in Chapter 13. □

- *Property V.* Let u_n and v_n be two WSS processes. Define the new process

$$z_n = u_n + v_n.$$

Then,

$$r_z(k) = r_u(k) + r_v(k) + r_{uv}(k) + r_{vu}(k), \tag{2.118}$$

where the *cross-correlation* between two jointly WSS stochastic processes is defined as

$$\boxed{r_{uv}(k) := \mathbb{E}[u_n v^*_{n-k}],\ k \in \mathbb{Z}: \quad \text{cross-correlation.}} \tag{2.119}$$

The proof is a direct consequence of the definition. Note that if the two processes are *uncorrelated*, as when $r_{uv}(k) = r_{vu}(k) = 0$, then

$$r_z(k) = r_u(k) + r_v(k).$$

Obviously, this is also true if the processes u_n and v_n are independent and of zero mean value, since then $\mathbb{E}[u_n v^*_{n-k}] = \mathbb{E}[u_n]\mathbb{E}[v^*_{n-k}] = 0$. It should be stressed here that uncorrelatedness is a weaker condition and it *does not* necessarily imply independence; the opposite is true for zero mean values.

- *Property VI.*

$$r_{uv}(k) = r^*_{vu}(-k). \tag{2.120}$$

The proof is similar to that of Property I.

- *Property VII.*

$$r_u(0)r_v(0) \geq |r_{uv}(k)|^2, \quad \forall k \in \mathbb{Z}. \tag{2.121}$$

The proof is also given in Problem 2.9.

Power Spectral Density

Definition 2.4. Given a WSS stochastic process u_n, its *power spectral density* (PSD) (or simply the *power spectrum*) is defined as the Fourier transform of its autocorrelation sequence,

$$\boxed{S(\omega) := \sum_{k=-\infty}^{\infty} r(k) \exp(-j\omega k): \quad \text{power spectral density.}} \tag{2.122}$$

Using the Fourier transform properties, we can recover the autocorrelation sequence via the *inverse Fourier transform* in the following manner:

$$\boxed{r(k) = \frac{1}{2\pi} \int_{-\pi}^{+\pi} S(\omega) \exp(j\omega k)\, d\omega.} \tag{2.123}$$

Due to the properties of the autocorrelation sequence, the PSD has some interesting and useful properties, from a practical point of view.

Properties of the PSD

- The PSD of a WSS stochastic process is a *real* and *nonnegative* function of ω. Indeed, we have

$$\begin{aligned} S(\omega) &= \sum_{k=-\infty}^{+\infty} r(k)\exp(-j\omega k) \\ &= r(0) + \sum_{k=-\infty}^{-1} r(k)\exp(-j\omega k) + \sum_{k=1}^{\infty} r(k)\exp(-j\omega k) \\ &= r(0) + \sum_{k=1}^{+\infty} r^*(k)\exp(j\omega k) + \sum_{k=1}^{\infty} r(k)\exp(-j\omega k) \\ &= r(0) + 2\sum_{k=1}^{+\infty} \text{Real}\{r(k)\exp(-j\omega k)\}, \end{aligned} \tag{2.124}$$

which proves the claim that PSD is a real number.[2] In the proof, Property I of the autocorrelation sequence has been used. We defer the proof concerning the nonnegativity to the end of this section.

- The area under the graph of $S(\omega)$ is proportional to the power of the stochastic process, as expressed by

$$\mathbb{E}\left[|\mathrm{u}_n|^2\right] = r(0) = \frac{1}{2\pi}\int_{-\pi}^{+\pi} S(\omega)d\omega, \tag{2.125}$$

which is obtained from Eq. (2.123) if we set $k=0$. We will come to the physical meaning of this property very soon.

Transmission Through a Linear System

One of the most important tasks in signal processing and systems theory is the *linear filtering operation* on an *input* time series (*signal*) to generate another *output* sequence. The block diagram of the filtering operation is shown in Fig. 2.13. From the linear system theory and signal processing basics, it is established that for a class of linear systems known as *linear time-invariant*, the input–output relation is given via the elegant *convolution* between the input sequence and the *impulse response* of the filter,

$$\boxed{\mathrm{d}_n = w_n * \mathrm{u}_n := \sum_{i=-\infty}^{+\infty} w_i^* \mathrm{u}_{n-i}: \quad \text{convolution sum,}} \tag{2.126}$$

where $\ldots, w_0, w_1, w_2, \ldots$ are the parameters comprising the impulse response describing the filter [8]. In case the impulse response is of finite duration, for example, $w_0, w_1, \ldots, w_{l-1}$, and the rest of the

[2] Recall that if $z = a + jb$ is a complex number, its real part $\text{Real}\{z\} = a = \frac{1}{2}(z + z^*)$.

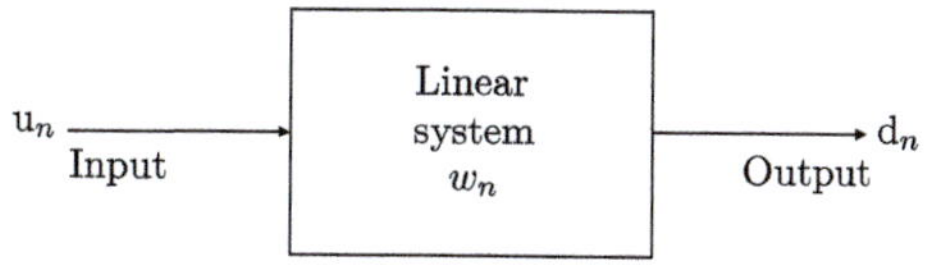

FIGURE 2.13

The linear system (filter) is excited by the input sequence (signal) u_n and provides the output sequence (signal) d_n.

values are zero, then the convolution can be written as

$$\mathrm{d}_n = \sum_{i=0}^{l-1} w_i^* \mathrm{u}_{n-i} = \boldsymbol{w}^H \mathbf{u}_n, \tag{2.127}$$

where

$$\boldsymbol{w} := [w_0, w_1, \ldots, w_{l-1}]^T, \tag{2.128}$$

and

$$\mathbf{u}_n := [\mathrm{u}_n, \mathrm{u}_{n-1}, \ldots, \mathrm{u}_{n-l+1}]^T \in \mathbb{R}^l. \tag{2.129}$$

The latter is known as the *input vector* of order l and at time n. It is interesting to note that this is a random vector. However, its elements are part of the stochastic process at *successive* time instants. This gives the respective autocorrelation matrix certain properties and a rich structure, which will be studied and exploited in Chapter 4. As a matter of fact, this is the reason that we used different symbols to denote processes and general random vectors; thus, the reader can readily remember that when dealing with a process, the elements of the involved random vectors have this *extra structure*. Moreover, observe from Eq. (2.126) that if the impulse response of the system is zero for negative values of the time index n, this guarantees *causality*. That is, the output depends only on the values of the input at the current and previous time instants, and there is no dependence on future values. As a matter of fact, this is also a necessary condition for causality; that is, if the system is causal, then its impulse response is zero for negative time instants [8].

Theorem 2.1. *The PSD of the output* d_n *of a linear time-invariant system, when it is excited by a WSS stochastic process* u_n*, is given by*

$$\boxed{S_{\mathrm{d}}(\omega) = |W(\omega)|^2 S_{\mathrm{u}}(\omega),} \tag{2.130}$$

where

$$W(\omega) := \sum_{n=-\infty}^{+\infty} w_n \exp(-j\omega n). \tag{2.131}$$

Proof. First, it is shown (Problem 2.10) that

$$\boxed{r_{\mathrm{d}}(k) = r_{\mathrm{u}}(k) * w_k * w_{-k}^*.} \tag{2.132}$$

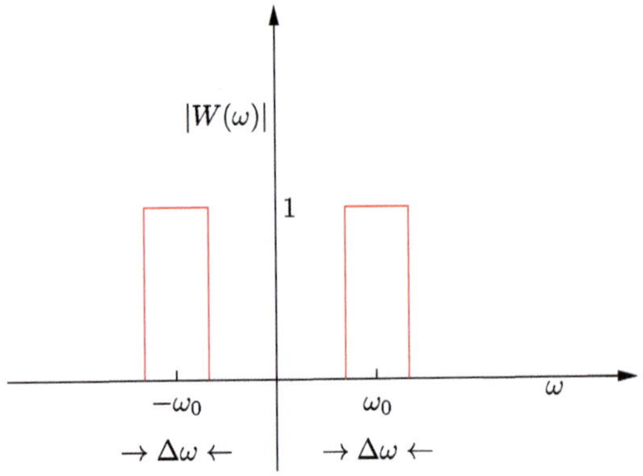

FIGURE 2.14

An ideal bandpass filter. The output contains frequencies only in the range of $|\omega - \omega_o| < \Delta\omega/2$.

Then, taking the Fourier transform of both sides, we obtain Eq. (2.130). To this end, we used the well-known properties of the Fourier transform,

$$r_{\mathrm{u}}(k) * w_k \longmapsto S_{\mathrm{u}}(\omega)W(\omega), \quad \text{and} \quad w^*_{-k} \longmapsto W^*(\omega). \qquad \square$$

Physical Interpretation of the PSD

We are now ready to justify why the Fourier transform of the autocorrelation sequence was given the specific name of "power spectral density." We restrict our discussion to real processes, although similar arguments hold true for the more general complex case. Fig. 2.14 shows the magnitude of the Fourier transform of the impulse response of a very special linear system. The Fourier transform is unity for any frequency in the range $|\omega - \omega_o| \leq \frac{\Delta\omega}{2}$ and zero otherwise. Such a system is known as *bandpass filter*. We assume that $\Delta\omega$ is very small. Then, using Eq. (2.130) and assuming that within the intervals $|\omega - \omega_o| \leq \frac{\Delta\omega}{2}$, $S_{\mathrm{u}}(\omega) \approx S_{\mathrm{u}}(\omega_o)$, we have

$$S_{\mathrm{d}}(\omega) = \begin{cases} S_{\mathrm{u}}(\omega_o), & \text{if } |\omega - \omega_o| \leq \frac{\Delta\omega}{2}, \\ 0, & \text{otherwise.} \end{cases} \tag{2.133}$$

Hence,

$$\Delta P := \mathbb{E}\left[|\mathrm{d}_n|^2\right] = r_{\mathrm{d}}(0) = \frac{1}{2\pi}\int_{-\infty}^{+\infty} S_{\mathrm{d}}(\omega)d\omega \approx S_{\mathrm{u}}(\omega_o)\frac{\Delta\omega}{\pi}, \tag{2.134}$$

due to the symmetry of the PSD ($S_{\mathrm{u}}(\omega) = S_{\mathrm{u}}(-\omega)$). Hence,

$$\frac{1}{\pi}S_{\mathrm{u}}(\omega_o) = \frac{\Delta P}{\Delta\omega}. \tag{2.135}$$

In other words, the value $S_{\mathrm{u}}(\omega_o)$ can be interpreted as the power density (power per frequency interval) in the frequency (spectrum) domain.

Moreover, this also establishes what was said before that the PSD is a *nonnegative* real function for any value of $\omega \in [-\pi, +\pi]$ (the PSD, being the Fourier transform of a sequence, is periodic with period 2π, e.g., [8]).

Remarks 2.2.

- Note that for any WSS stochastic process, there is only one autocorrelation sequence that describes it. However, the converse is not true. A single autocorrelation sequence can correspond to more than one WSS process. Recall that the autocorrelation is the mean value of the product of random variables. However, many random variables can have the same mean value.
- We have shown that the Fourier transform, $S(\omega)$, of an autocorrelation sequence, $r(k)$, is nonnegative. Moreover, if a sequence $r(k)$ has a nonnegative Fourier transform, then it is positive definite and we can *always* construct a WSS process that has $r(k)$ as its autocorrelation sequence (e.g., [6, pages 410, 421]). Thus, the *necessary and sufficient condition for a sequence to be an autocorrelation sequence is the nonnegativity of its Fourier transform.*

Example 2.3 (White noise sequence). A stochastic process η_n is said to be *white noise* if the mean and its autocorrelation sequence satisfy

$$\boxed{\mathbb{E}[\eta_n] = 0 \quad \text{and} \quad r(k) = \begin{cases} \sigma_\eta^2, & \text{if } k = 0, \\ 0, & \text{if } k \neq 0. \end{cases} \quad : \quad \text{white noise,}} \tag{2.136}$$

where σ_η^2 is its variance. In other words, all variables at different time instants are uncorrelated. If, in addition, they are independent, we say that it is *strictly white noise*. It is readily seen that its PSD is given by

$$S_\eta(\omega) = \sigma_\eta^2. \tag{2.137}$$

That is, it is constant, and this is the reason it is called white noise, analogous to white light, whose spectrum is equally spread over all wavelengths.

2.4.4 AUTOREGRESSIVE MODELS

We have just seen an example of a stochastic process, namely, white noise. We now turn our attention to generating WSS processes via appropriate modeling. In this way, we will introduce controlled correlation among the variables, corresponding to the various time instants. We focus on the real data case, to simplify the discussion.

Autoregressive processes are among the most popular and widely used models. An autoregressive process of order l, denoted as AR(l), is defined via the following *difference equation*:

$$\boxed{\mathrm{u}_n + a_1 \mathrm{u}_{n-1} + \cdots + a_l \mathrm{u}_{n-l} = \eta_n : \quad \text{autoregressive process,}} \tag{2.138}$$

where η_n is a white noise process with variance σ_η^2.

As is always the case with any difference equation, one starts from some initial conditions and then generates samples recursively by plugging into the model the input sequence samples. The input samples here correspond to a white noise sequence and the initial conditions are set equal to zero, $u_{-1} = \dots u_{-l} = 0$.

There is no need to mobilize mathematics to see that such a process is not stationary. Indeed, time instant $n = 0$ is distinctly different from all the rest, since it is the time in which initial conditions are applied. However, the effects of the initial conditions tend asymptotically to zero if all the roots of the corresponding characteristic polynomial,

$$z^l + a_1 z^{l-1} + \dots + a_l = 0,$$

have magnitude *less than unity* (the solution of the corresponding homogeneous equation, without input, tends to zero) [7]. Then, it can be shown that asymptotically, the AR(l) becomes WSS. This is the assumption that is usually adopted in practice, which will be the case for the rest of this section. Note that the mean value of the process is zero (try it).

The goal now becomes to compute the corresponding autocorrelation sequence, $r(k), k \in \mathbb{Z}$. Multiplying both sides in Eq. (2.138) with $\mathrm{u}_{n-k}, k > 0$, and taking the expectation, we obtain

$$\sum_{i=0}^{l} a_i \, \mathbb{E}[\mathrm{u}_{n-i}\mathrm{u}_{n-k}] = \mathbb{E}[\eta_n \mathrm{u}_{n-k}], \quad k > 0,$$

where $a_0 := 1$, or

$$\sum_{i=0}^{l} a_i r(k-i) = 0. \tag{2.139}$$

We have used the fact that $\mathbb{E}[\eta_n \mathrm{u}_{n-k}]$, $k > 0$, is zero. Indeed, u_{n-k} depends recursively on η_{n-k}, $\eta_{n-k-1} \dots$, which are all uncorrelated to η_n, since this is a white noise process. Note that Eq. (2.139) is a difference equation, which can be solved provided we have the initial conditions. To this end, multiply Eq. (2.138) by u_n and take expectations, which results in

$$\sum_{i=0}^{l} a_i r(i) = \sigma_\eta^2, \tag{2.140}$$

since u_n recursively depends on η_n, which contributes the σ_η^2 term, and $\eta_{n-1}, \dots$, which result to zeros. Combining Eqs. (2.140)–(2.139) the following *linear* system of equations results:

$$\begin{bmatrix} r(0) & r(1) & \dots & r(l) \\ r(1) & r(0) & \dots & r(l-1) \\ \vdots & \vdots & \vdots & \vdots \\ r(l) & r(l-1) & \dots & r(0) \end{bmatrix} \begin{bmatrix} 1 \\ a_1 \\ \vdots \\ a_l \end{bmatrix} = \begin{bmatrix} \sigma_\eta^2 \\ 0 \\ \vdots \\ 0 \end{bmatrix}. \tag{2.141}$$

These are known as the *Yule–Walker equations*, whose solution results in the values $r(0), \dots, r(l)$, which are then used as the initial conditions to solve the difference equation in (2.139) and obtain $r(k)$, $\forall k \in \mathbb{Z}$.

Observe the special structure of the matrix in the linear system. This type of matrix is known as *Toeplitz*, and this is the property that will be exploited to efficiently solve such systems, which result when the autocorrelation matrix of a WSS process is involved; see Chapter 4.

Besides the autoregressive models, other types of stochastic models have been suggested and used. The *autoregressive-moving average* (ARMA) model of order (l, m) is defined by the difference equation

$$\mathrm{u}_n + a_1 \mathrm{u}_{n-1} + \ldots + a_l \mathrm{u}_{n-l} = b_1 \eta_n + \ldots + b_m \eta_{n-m}, \tag{2.142}$$

and the *moving average* model of order m, denoted as MA(m), is defined as

$$\mathrm{u}_n = b_1 \eta_n + \cdots + b_m \eta_{n-m}. \tag{2.143}$$

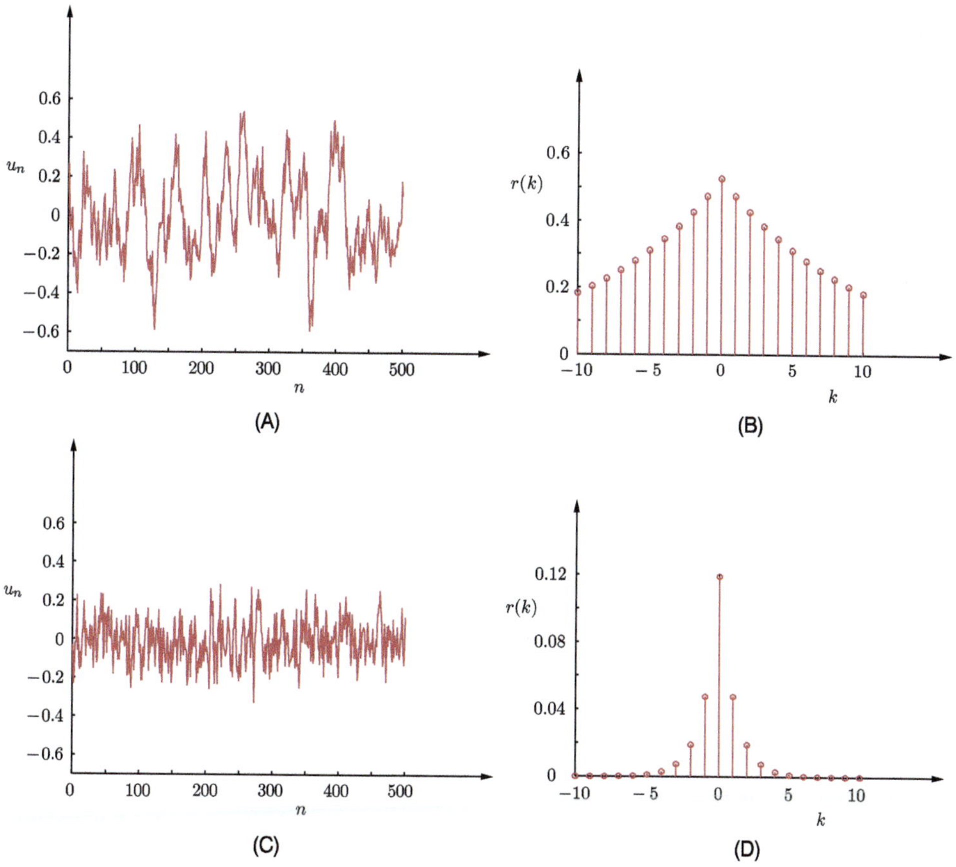

FIGURE 2.15

(A) The time evolution of a realization of the AR(1) with $a = -0.9$ and (B) the respective autocorrelation sequence. (C) The time evolution of a realization of the AR(1) with $a = -0.4$ and (D) the corresponding autocorrelation sequence.

Note that the AR(l) and the MA(m) models can be considered as special cases of the ARMA(l, m). For a more theoretical treatment of the topic, see [1].

Example 2.4. Consider the AR(1) process,

$$\mathrm{u}_n + a\mathrm{u}_{n-1} = \eta_n.$$

Following the general methodology explained before, we have

$$r(k) + ar(k-1) = 0, \quad k = 1, 2, \ldots,$$
$$r(0) + ar(1) = \sigma_\eta^2.$$

Taking the first equation for $k = 1$ together with the second one readily results in

$$r(0) = \frac{\sigma_\eta^2}{1-a^2}.$$

Plugging this value into the difference equation, we recursively obtain

$$r(k) = (-a)^{|k|}\frac{\sigma_\eta^2}{1-a^2}, \quad k = 0, \pm 1, \pm 2, \ldots, \tag{2.144}$$

where we used the property $r(k) = r(-k)$. Observe that if $|a| > 1$, $r(0) < 0$, which is meaningless. Also, $|a| < 1$ guarantees that the magnitude of the root of the characteristic polynomial ($z_* = -a$) is smaller than one. Moreover, $|a| < 1$ guarantees that $r(k) \longrightarrow 0$ as $k \longrightarrow \infty$. This is in line with common sense, since variables that are far away must be uncorrelated.

Fig. 2.15 shows the time evolution of two AR(1) processes (after the processes have converged to be stationary) together with the respective autocorrelation sequences, for two cases, corresponding to $a = -0.9$ and $a = -0.4$. Observe that the larger the magnitude of a, the smoother the realization becomes and time variations are *slower*. This is natural, since nearby samples are highly correlated and so, on average, they tend to have similar values. The opposite is true for small values of a. For comparison purposes, Fig. 2.16A is the case of $a = 0$, which corresponds to a white noise. Fig. 2.16B shows the PSDs corresponding to the two cases of Fig. 2.15. Observe that the faster the autocorrelation approaches zero, the more spread out the PSD is, and vice versa.

2.5 INFORMATION THEORY

So far in this chapter, we have looked at some basic definitions and properties concerning probability theory and stochastic processes. In the same vein, we will now focus on the basic definitions and notions related to *information theory*. Although information theory was originally developed in the context of communications and coding disciplines, its application and use has now been adopted in a wide range of areas, including machine learning. Notions from information theory are used for establishing cost functions for optimization in parameter estimation problems, and concepts from information theory are employed to estimate unknown probability distributions in the context of constrained optimization tasks. We will discuss such methods later in this book.

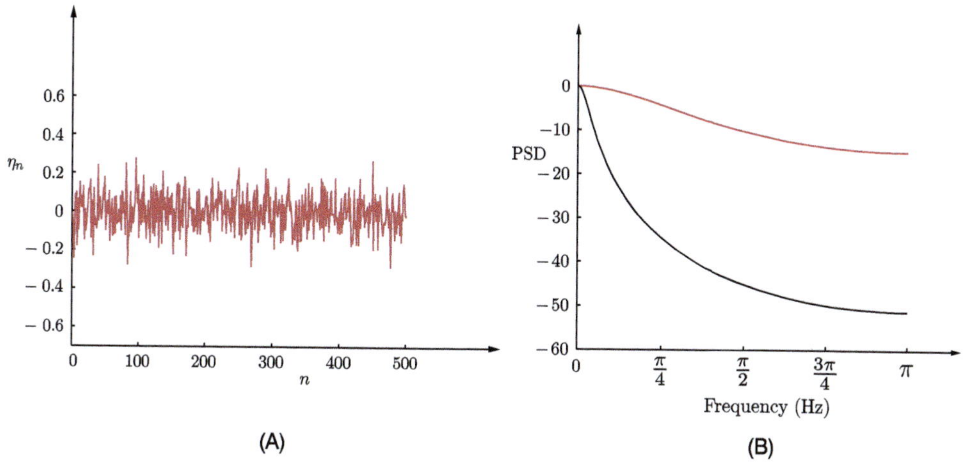

FIGURE 2.16

(A) The time evolution of a realization from a white noise process. (B) The PSDs in dBs, for the two AR(1) sequences of Fig. 2.15. The red one corresponds to $a = -0.4$ and the gray one to $a = -0.9$. The smaller the magnitude of a, the closer the process is to a white noise, and its PSD tends to increase the power with which high frequencies participate. Since the PSD is the Fourier transform of the autocorrelation sequence, observe that the broader a sequence is in time, the narrower its Fourier transform becomes, and vice versa.

The father of information theory is *Claude Elwood Shannon* (1916–2001), an American mathematician and electrical engineer. He founded information theory with the landmark paper "A mathematical theory of communication," published in the Bell System Technical Journal in 1948. However, he is also credited with founding digital circuit design theory in 1937, when, as a 21-year-old Master's degree student at the Massachusetts Institute of Technology (MIT), he wrote his thesis demonstrating that electrical applications of Boolean algebra could construct and resolve any logical, numerical relationship. So he is also credited as a father of digital computers. Shannon, while working for the national defense during the Second World War, contributed to the field of cryptography, converting it from an art to a rigorous scientific field.

As is the case for probability, the notion of information is part of our everyday vocabulary. In this context, an event carries information if it is either unknown to us, or if the probability of its occurrence is very low and, in spite of that, it happens. For example, if one tells us that the sun shines bright during summer days in the Sahara desert, we could consider such a statement rather dull and useless. On the contrary, if somebody gives us news about snow in the Sahara during summer, that statement carries a lot of information and can possibly ignite a discussion concerning climate change.

Thus, trying to formalize the notion of information from a mathematical point of view, it is reasonable to define it in terms of the negative logarithm of the probability of an event. If the event is certain to occur, it carries zero information content; however, if its probability of occurrence is low, then its information content has a large positive value.

2.5.1 DISCRETE RANDOM VARIABLES

Information

Given a discrete random variable x, which takes values in the set $\mathcal{X}$, the *information* associated with any value $x \in \mathcal{X}$ is denoted as $I(x)$ and it is defined as

$$\boxed{I(x) = -\log P(x): \quad \text{information associated with } \mathrm{x} = x \in \mathcal{X}.} \tag{2.145}$$

Any base for the logarithm can be used. If the natural logarithm is chosen, information is measured in terms of *nats* (natural units). If the base 2 logarithm is employed, information is measured in terms of *bits* (binary digits). Employing the logarithmic function to define information is also in line with common sense reasoning that the information content of two statistically independent events should be the sum of the information conveyed by each one of them individually; $I(x, y) = -\log P(x, y) = -\log P(x) - \log P(y)$.

Example 2.5. We are given a binary random variable $\mathrm{x} \in \mathcal{X} = \{0, 1\}$, and we assume that $P(1) = P(0) = 0.5$. We can consider this random variable as a source that generates and emits two possible values. The information content of each one of the two equiprobable events is

$$I(0) = I(1) = -\log_2 0.5 = 1 \text{ bit}.$$

Let us now consider another source of random events, which generates *code words* comprising k binary variables together. The output of this source can be seen as a random vector with binary-valued elements, $\mathbf{x} = [\mathrm{x}_1, \ldots, \mathrm{x}_k]^T$. The corresponding probability space, $\mathcal{X}$, comprises $K = 2^k$ elements. If all possible values have the same probability, $1/K$, then the information content of each possible event is equal to

$$I(\boldsymbol{x}_i) = -\log_2 \frac{1}{K} = k \text{ bits}.$$

We observe that in the case where the number of possible events is larger, the information content of each individual one (assuming equiprobable events) becomes larger. This is also in line with common sense reasoning, since if the source can emit a large number of (equiprobable) events, then the occurrence of any one of them carries more information than a source that can only emit a few possible events.

Mutual and Conditional Information

Besides marginal probabilities, we have already been introduced to the concept of conditional probability. This leads to the definition of mutual information.

Given two discrete random variables, $\mathrm{x} \in \mathcal{X}$ and $\mathrm{y} \in \mathcal{Y}$, the information content provided by the occurrence of the event $\mathrm{y} = y$ about the event $\mathrm{x} = x$ is measured by the *mutual information*, denoted as $I(x; y)$ and defined by

$$\boxed{I(x; y) := \log \frac{P(x|y)}{P(x)}: \quad \text{mutual information.}} \tag{2.146}$$

Note that if the two variables are statistically independent, then their mutual information is zero; this is most reasonable, since observing y says nothing about x. On the contrary, if by observing y it is certain that x will occur, as when $P(x|y) = 1$, then the mutual information becomes $I(x; y) = I(x)$, which is again in line with common reasoning. Mobilizing our now familiar product rule, we can see that

$$I(x; y) = I(y; x).$$

The *conditional information* of x given y is defined as

$$\boxed{I(x|y) = -\log P(x|y): \quad \text{conditional information.}} \tag{2.147}$$

It is straightforward to show that

$$I(x; y) = I(x) - I(x|y). \tag{2.148}$$

Example 2.6. In a communications channel, the source transmits binary symbols, x, with probability $P(0) = P(1) = 1/2$. The channel is noisy, so the received symbols y may have changed polarity, due to noise, with the following probabilities:

$$\begin{aligned} P(\mathrm{y} = 0|\mathrm{x} = 0) &= 1 - p, \\ P(\mathrm{y} = 1|\mathrm{x} = 0) &= p, \\ P(\mathrm{y} = 1|\mathrm{x} = 1) &= 1 - q, \\ P(\mathrm{y} = 0|\mathrm{x} = 1) &= q. \end{aligned}$$

This example illustrates in its simplest form the effect of a *communications channel*. Transmitted bits are hit by noise and what the receiver receives is the noisy (possibly wrong) information. The task of the receiver is to decide, upon reception of a sequence of symbols, which was the originally transmitted one.

The goal of our example is to determine the mutual information about the occurrence of $\mathrm{x} = 0$ and $\mathrm{x} = 1$ once $\mathrm{y} = 0$ has been observed. To this end, we first need to compute the marginal probabilities,

$$P(\mathrm{y} = 0) = P(\mathrm{y} = 0|\mathrm{x} = 0)P(\mathrm{x} = 0) + P(\mathrm{y} = 0|\mathrm{x} = 1)P(\mathrm{x} = 1) = \frac{1}{2}(1 - p + q),$$

and similarly,

$$P(\mathrm{y} = 1) = \frac{1}{2}(1 - q + p).$$

Thus, the mutual information is

$$\begin{aligned} I(0; 0) &= \log_2 \frac{P(\mathrm{x} = 0|\mathrm{y} = 0)}{P(\mathrm{x} = 0)} = \log_2 \frac{P(\mathrm{y} = 0|\mathrm{x} = 0)}{P(\mathrm{y} = 0)} \\ &= \log_2 \frac{2(1 - p)}{1 - p + q}, \end{aligned}$$

and

$$I(1; 0) = \log_2 \frac{2q}{1 - p + q}.$$

Let us now consider that $p = q = 0$. Then $I(0; 0) = 1$ bit, which is equal to $I(\mathrm{x} = 0)$, since the output specifies the input with certainty. If on the other hand $p = q = 1/2$, then $I(0; 0) = 0$ bits, since the noise can randomly change polarity with equal probability. If now $p = q = 1/4$, then $I(0; 0) = \log_2 \frac{3}{2} = 0.587$ bits and $I(1; 0) = -1$ bit. Observe that the mutual information can take negative values, too.

Entropy and Average Mutual Information

Given a discrete random variable $\mathrm{x} \in \mathcal{X}$, its *entropy* is defined as the average information over all possible outcomes,

$$H(\mathrm{x}) := -\sum_{x \in \mathcal{X}} P(x) \log P(x) : \quad \text{entropy of x.} \tag{2.149}$$

Note that if $P(x) = 0$, $P(x) \log P(x) = 0$, by taking into consideration that $\lim_{x \to 0} x \log x = 0$.

In a similar way, the *average mutual information* between two random variables, x, y, is defined as

$$\begin{aligned} I(\mathrm{x}; \mathrm{y}) &:= \sum_{x \in \mathcal{X}} \sum_{y \in \mathcal{Y}} P(x, y) I(x; y) \\ &= \sum_{x \in \mathcal{X}} \sum_{y \in \mathcal{Y}} P(x, y) \log \frac{P(x|y)}{P(x)} \\ &= \sum_{x \in \mathcal{X}} \sum_{y \in \mathcal{Y}} P(x, y) \log \frac{P(x|y)P(y)}{P(x)P(y)} \end{aligned}$$

or

$$I(\mathrm{x}; \mathrm{y}) = \sum_{x \in \mathcal{X}} \sum_{y \in \mathcal{Y}} P(x, y) \log \frac{P(x, y)}{P(x)P(y)} : \quad \text{average mutual information.} \tag{2.150}$$

It can be shown that

$$I(\mathrm{x}; \mathrm{y}) \geq 0,$$

and it is zero if x and y are statistically independent (Problem 2.12).

In comparison, the *conditional entropy* of x given y is defined as

$$H(\mathrm{x}|\mathrm{y}) := -\sum_{x \in \mathcal{X}} \sum_{y \in \mathcal{Y}} P(x, y) \log P(x|y) : \quad \text{conditional entropy.} \tag{2.151}$$

It is readily shown, by taking into account the probability product rule, that

$$I(\mathrm{x}; \mathrm{y}) = H(\mathrm{x}) - H(\mathrm{x}|\mathrm{y}). \tag{2.152}$$

Lemma 2.1. *The entropy of a random variable $\mathrm{x} \in \mathcal{X}$ takes its maximum value if all possible values $x \in \mathcal{X}$ are equiprobable.*

Proof. The proof is given in Problem 2.14. □

In other words, the entropy can be considered as a measure of randomness of a source that emits symbols randomly. The maximum value is associated with the maximum uncertainty of what is going to be emitted, since the maximum value occurs if all symbols are equiprobable. The smallest value of the entropy is equal to zero, which corresponds to the case where all events have zero probability with the exception of one, whose probability to occur is equal to one.

Example 2.7. Consider a binary source that transmits the values 1 or 0 with probabilities p and $1-p$, respectively. Then the entropy of the associated random variable is

$$H(\mathrm{x}) = -p\log_2 p - (1-p)\log_2(1-p).$$

Fig. 2.17 shows the graph for various values of $p \in [0, 1]$. Observe that the maximum value occurs for $p = 1/2$.

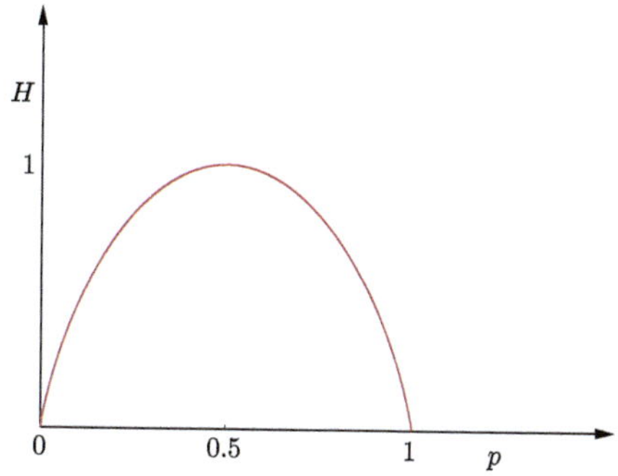

FIGURE 2.17

The maximum value of the entropy for a binary random variable occurs if the two possible events have equal probability, $p = 1/2$.

2.5.2 CONTINUOUS RANDOM VARIABLES

All the definitions given before can be generalized to the case of continuous random variables. However, this generalization must be made with caution. Recall that the probability of occurrence of any single value of a random variable that takes values in an interval in the real axis is zero. Hence, the corresponding information content is infinite.

To define the entropy of a continuous variable x, we first *discretize* it and form the corresponding discrete variable x_Δ, i.e.,

$$\mathrm{x}_\Delta := n\Delta, \text{ if } (n-1)\Delta < \mathrm{x} \leq n\Delta, \tag{2.153}$$

where $\Delta > 0$. Then,

$$P(\mathrm{x}_\Delta = n\Delta) = P(n\Delta - \Delta < \mathrm{x} \leq n\Delta) = \int_{(n-1)\Delta}^{n\Delta} p(x)\,dx = \Delta\bar{p}(n\Delta), \tag{2.154}$$

where $\bar{p}(n\Delta)$ is a number between the maximum and the minimum value of $p(x), x \in (n\Delta - \Delta, n\Delta]$ (such a number exists by the mean value theorem of calculus). Then we can write

$$H(\mathrm{x}_\Delta) = -\sum_{n=-\infty}^{+\infty} \Delta\bar{p}(n\Delta)\log\left(\Delta\bar{p}(n\Delta)\right), \tag{2.155}$$

and since

$$\sum_{n=-\infty}^{+\infty} \Delta\bar{p}(n\Delta) = \int_{-\infty}^{+\infty} p(x)\,dx = 1,$$

we obtain

$$H(\mathrm{x}_\Delta) = -\log\Delta - \sum_{n=-\infty}^{+\infty} \Delta\bar{p}(n\Delta)\log\left(\bar{p}(n\Delta)\right). \tag{2.156}$$

Note that $\mathrm{x}_\Delta \longrightarrow \mathrm{x}$ as $\Delta \longrightarrow 0$. However, if we take the limit in Eq. (2.156), then $-\log\Delta$ goes to infinity. This is the crucial difference compared to the discrete variables.

The entropy for a continuous random variable x is defined as the limit

$$H(\mathrm{x}) := \lim_{\Delta\to 0}\left(H(\mathrm{x}_\Delta) + \log\Delta\right),$$

or

$$\boxed{H(\mathrm{x}) = -\int_{-\infty}^{+\infty} p(x)\log p(x)\,dx : \quad \text{entropy.}} \tag{2.157}$$

This is the reason that the entropy of a continuous variable is also called *differential entropy*.

Note that the entropy is still a measure of randomness (uncertainty) of the distribution describing x. This is demonstrated via the following example.

Example 2.8. We are given a random variable $\mathrm{x} \in [a, b]$. Of all the possible PDFs that can describe this variable, find the one that maximizes the entropy.

This task translates to the following constrained optimization task:

$$\text{maximize with respect to } p: \quad H = -\int_a^b p(x)\ln p(x)dx,$$

$$\text{subject to:} \quad \int_a^b p(x)dx = 1.$$

The constraint guarantees that the function to result is indeed a PDF. Using calculus of variations to perform the optimization (Problem 2.15), it turns out that

$$p(x) = \begin{cases} \frac{1}{b-a}, & \text{if } x \in [a, b], \\ 0, & \text{otherwise.} \end{cases}$$

In other words, the result is the uniform distribution, which is indeed the most random one since it gives no preference to any particular subinterval of $[a, b]$.

We will come to this method of estimating PDFs in Section 12.8.1. This elegant method for estimating PDFs comes from Jaynes [3,4], and it is known as the *maximum entropy method*. In its more general form, more constraints are involved to fit the needs of the specific problem.

Average Mutual Information and Conditional Information

Given two continuous random variables, the average mutual information is defined as

$$I(\mathrm{x}; \mathrm{y}) := \int_{-\infty}^{+\infty} \int_{-\infty}^{+\infty} p(x, y) \log \frac{p(x, y)}{p(x)p(y)} \, dx \, dy \tag{2.158}$$

and the conditional entropy of x, given y,

$$H(\mathrm{x}|\mathrm{y}) := -\int_{-\infty}^{+\infty} \int_{-\infty}^{+\infty} p(x, y) \log p(x|y) \, dx \, dy. \tag{2.159}$$

Using standard arguments and the product rule, it is easy to show that

$$I(\mathrm{x}; \mathrm{y}) = H(\mathrm{x}) - H(\mathrm{x}|\mathrm{y}) = H(\mathrm{y}) - H(\mathrm{y}|\mathrm{x}). \tag{2.160}$$

Relative Entropy or Kullback–Leibler Divergence

The *relative entropy* or *Kullback–Leibler divergence* is a quantity that has been developed within the context of information theory for measuring similarity between two PDFs. It is widely used in machine learning optimization tasks when PDFs are involved; see Chapter 12. Given two PDFs, p and q, their Kullback–Leibler divergence, denoted as $\mathrm{KL}(p||q)$, is defined as

$$\mathrm{KL}(p||q) := \int_{-\infty}^{+\infty} p(x) \log \frac{p(x)}{q(x)} \, dx : \quad \text{Kullback–Leibler divergence.} \tag{2.161}$$

Note that

$$I(\mathrm{x}; \mathrm{y}) = \mathrm{KL}\big(p(x, y)||p(x)p(y)\big).$$

The Kullback–Leibler divergence is *not* symmetric, i.e., $\mathrm{KL}(p||q) \neq \mathrm{KL}(q||p)$, and it can be shown that it is a nonnegative quantity (the proof is similar to the proof that the mutual information is nonnegative; see Problem 12.7 of Chapter 12). Moreover, it is zero if and only if $p = q$.

Note that all we have said concerning entropy and mutual information is readily generalized to the case of random vectors.

2.6 STOCHASTIC CONVERGENCE

We will close this memory refreshing tour of the theory of probability and related concepts with some definitions concerning convergence of sequences of random variables.

Consider a sequence of random variables

$$\mathrm{x}_0, \mathrm{x}_1, \dots, \mathrm{x}_n \dots.$$

We can consider this sequence as a discrete-time stochastic process. Due to the randomness, a realization of this process, as shown by

$$x_0, x_1, \dots, x_n \dots,$$

may converge or may not. Thus, the notion of convergence of random variables has to be treated carefully, and different interpretations have been developed.

Recall from our basic calculus that a sequence of numbers, x_n, converges to a value x if $\forall \epsilon > 0$ there exists a number $n(\epsilon)$ such that

$$|x_n - x| < \epsilon, \quad \forall n \geq n(\epsilon). \tag{2.162}$$

CONVERGENCE EVERYWHERE

We say that a random sequence *converges everywhere* if every realization, x_n, of the random process converges to a value x, according to the definition given in Eq. (2.162). Note that every realization converges to a different value, which itself can be considered as the outcome of a random variable x, and we write

$$\mathrm{x}_n \xrightarrow[n\to\infty]{} \mathrm{x}. \tag{2.163}$$

It is common to denote a realization (outcome) of a random process as $\mathrm{x}_n(\zeta)$, where ζ denotes a specific experiment.

CONVERGENCE ALMOST EVERYWHERE

A weaker version of convergence, compared to the previous one, is the *convergence almost everywhere*. Consider the set of outcomes ζ, such that

$$\lim \mathrm{x}_n(\zeta) = \mathrm{x}(\zeta), \quad n \longrightarrow \infty.$$

We say that the sequence x_n converges almost everywhere if

$$P(\mathrm{x}_n \longrightarrow \mathrm{x}) = 1, \quad n \longrightarrow \infty. \tag{2.164}$$

Note that $\{\mathrm{x}_n \longrightarrow \mathrm{x}\}$ denotes the event comprising *all* the outcomes such that $\lim \mathrm{x}_n(\zeta) = \mathrm{x}(\zeta)$. The difference with the convergence everywhere is that now it is allowed to a finite or countably infinite number of realizations (that is, to a set of zero probability) not to converge. Often, this type of convergence is referred to as *almost sure* convergence or convergence *with probability 1*.

CONVERGENCE IN THE MEAN-SQUARE SENSE

We say that a random sequence x_n converges to the random variable x in the *mean-square sense* if

$$\mathbb{E}\left[|\mathrm{x}_n - \mathrm{x}|^2\right] \longrightarrow 0, \quad n \longrightarrow \infty. \tag{2.165}$$

CONVERGENCE IN PROBABILITY

Given a random sequence x_n, a random variable x, and a nonnegative number ϵ, then $\{|\mathrm{x}_n - \mathrm{x}| > \epsilon\}$ is an event. We define the new sequence of numbers, $P(\{|\mathrm{x}_n - \mathrm{x}| > \epsilon\})$. We say that x_n converges to x *in probability* if the constructed sequence of numbers tends to zero,

$$P(\{|\mathrm{x}_n - \mathrm{x}| > \epsilon\}) \longrightarrow 0, \ n \longrightarrow \infty, \ \forall \epsilon > 0. \tag{2.166}$$

CONVERGENCE IN DISTRIBUTION

Given a random sequence x_n and a random variable x, let $F_n(x)$ and $F(x)$ be the CDFs, respectively. We say that x_n converges to x *in distribution* if

$$F_n(x) \longrightarrow F(x), \quad n \longrightarrow \infty, \tag{2.167}$$

for *every* point x of continuity of $F(x)$.

It can be shown that if a random sequence converges either almost everywhere or in the mean-square sense, then it necessarily converges in probability, and if it converges in probability, then it necessarily converges in distribution. The converse arguments are not necessarily true. In other words, the weakest version of convergence is that of convergence in distribution.

PROBLEMS

2.1 Derive the mean and variance for the binomial distribution.

2.2 Derive the mean and variance for the uniform distribution.

2.3 Derive the mean and covariance matrix of the multivariate Gaussian.

2.4 Show that the mean and variance of the beta distribution with parameters a and b are given by

$$\mathbb{E}[\mathrm{x}] = \frac{a}{a+b}$$

and

$$\sigma_{\mathrm{x}}^2 = \frac{ab}{(a+b)^2(a+b+1)}.$$

Hint: Use the property $\Gamma(a+1) = a\Gamma(a)$.

2.5 Show that the normalizing constant in the beta distribution with parameters a, b is given by

$$\frac{\Gamma(a+b)}{\Gamma(a)\Gamma(b)}.$$

2.6 Show that the mean and variance of the gamma PDF

$$\text{Gamma}(x|a, b) = \frac{b^a}{\Gamma(a)} x^{a-1} e^{-bx}, \quad a, b, x > 0,$$

are given by

$$\mathbb{E}[\mathrm{x}] = \frac{a}{b},$$
$$\sigma_{\mathrm{x}}^2 = \frac{a}{b^2}.$$

2.7 Show that the mean and variance of a Dirichlet PDF with K variables $\mathrm{x}_k, k = 1, 2, \ldots, K$, and parameters $a_k, k = 1, 2, \ldots, K$, are given by

$$\mathbb{E}[\mathrm{x}_k] = \frac{a_k}{\overline{a}}, \quad k = 1, 2, \ldots, K,$$
$$\sigma_{\mathrm{x}_k}^2 = \frac{a_k(\overline{a} - a_k)}{\overline{a}^2(1 + \overline{a})}, \quad k = 1, 2, \ldots, K,$$
$$\mathrm{cov}[\mathrm{x}_i \mathrm{x}_j] = -\frac{a_i a_j}{\overline{a}^2(1 + \overline{a})}, \quad i \neq j,$$

where $\overline{a} = \sum_{k=1}^{K} a_k$.

2.8 Show that the sample mean, using N i.i.d. drawn samples, is an unbiased estimator with variance that tends to zero asymptotically, as $N \longrightarrow \infty$.

2.9 Show that for WSS processes

$$r(0) \geq |r(k)|, \quad \forall k \in \mathbb{Z},$$

and that for jointly WSS processes

$$r_{\mathrm{u}}(0) r_{\mathrm{v}}(0) \geq |r_{\mathrm{uv}}(k)|^2, \quad \forall k \in \mathbb{Z}.$$

2.10 Show that the autocorrelation of the output of a linear system, with impulse response $w_n, n \in \mathbb{Z}$, is related to the autocorrelation of the input WSS process via

$$r_{\mathrm{d}}(k) = r_{\mathrm{u}}(k) * w_k * w_{-k}^*.$$

2.11 Show that

$$\ln x \leq x - 1.$$

2.12 Show that

$$I(\mathrm{x}; \mathrm{y}) \geq 0.$$

Hint: Use the inequality of Problem 2.11.

2.13 Show that if $a_i, b_i, i = 1, 2, \ldots, M$, are positive numbers such that

$$\sum_{i=1}^{M} a_i = 1 \quad \text{and} \quad \sum_{i=1}^{M} b_i \leq 1,$$

then

$$-\sum_{i=1}^{M} a_i \ln a_i \leq -\sum_{i=1}^{M} a_i \ln b_i .$$

2.14 Show that the maximum value of the entropy of a random variable occurs if all possible outcomes are equiprobable.

2.15 Show that from all the PDFs that describe a random variable in an interval $[a, b]$, the uniform one maximizes the entropy.

REFERENCES

[1] P.J. Brockwell, R.A. Davis, Time Series: Theory and Methods, second ed., Springer, New York, 1991.
[2] R.T. Cox, Probability, frequency and reasonable expectation, Am. J. Phys. 14 (1) (1946) 1–13.
[3] E.T. Jaynes, Information theory and statistical mechanics, Phys. Rev. 106 (4) (1957) 620–630.
[4] E.T. Jaynes, Probability Theory: The Logic of Science, Cambridge University Press, Cambridge, 2003.
[5] A.N. Kolmogorov, Foundations of the Theory of Probability, second ed., Chelsea Publishing Company, New York, 1956.
[6] A. Papoulis, S.U. Pillai, Probability, Random Variables and Stochastic Processes, fourth ed., McGraw Hill, New York, 2002.
[7] M.B. Priestly, Spectral Analysis and Time Series, Academic Press, New York, 1981.
[8] J. Proakis, D. Manolakis, Digital Signal Processing, second ed., MacMillan, New York, 1992.

CHAPTER

LEARNING IN PARAMETRIC MODELING: BASIC CONCEPTS AND DIRECTIONS

3

CONTENTS

3.1 INTRODUCTION

Parametric modeling is a theme that runs across the spine of this book. A number of chapters focus on different aspects of this important problem. This chapter provides basic definitions and concepts related to the task of learning when parametric models are mobilized to describe the available data.

Machine Learning. https://doi.org/10.1016/B978-0-12-818803-3.00012-X

As has already been pointed out in the introductory chapter, a large class of machine learning problems ends up as being equivalent to a function estimation/approximation task. The function is "learned" during the learning/training phase by digging in the information that resides in the available training data set. This function relates the so-called input variables to the output variable(s). Once this functional relationship is established, one can in turn exploit it to *predict* the value(s) of the output(s), based on measurements from the respective input variables; these predictions can then be used to proceed to the decision making phase.

In parametric modeling, the aforementioned functional dependence that relates the input to the output is defined via a set of parameters, whose number is *fixed* and a-priori known. The values of the parameters are unknown and have to be estimated based on the available input–output observations. In contrast to the parametric, there are the so-called *nonparametric* methods. In such methods, parameters may still be involved to establish the input–output relationship, yet their number is *not* fixed; it depends on the size of the data set and it grows with the number of observations. Nonparametric methods will also be treated in this book (e.g., Chapters 11 and 13). However, the emphasis in the current chapter lies on parametric models.

There are two possible paths to deal with the uncertainty imposed by the unknown values of the involved parameters. According to the first one, parameters are treated as *deterministic* nonrandom variables. The task of learning is to obtain estimates of their unknown values. For each one of the parameters a single value estimate is obtained. The other approach has a stronger statistical flavor. The unknown parameters are treated as *random variables* and the task of learning is to *infer* the associated probability distributions. Once the distributions have been learned/inferred, one can use them to make predictions. Both approaches are introduced in the current chapter and are treated in more detail later on in various chapters of the book.

Two of the major machine learning tasks, namely, regression and classification, are presented and the main directions in dealing with these problems are exposed. Various issues that are related to the parameter estimation task, such as estimator efficiency, bias–variance dilemma, overfitting, and the curse of dimensionality, are introduced and discussed. The chapter can also be considered as a road map to the rest of the book. However, instead of just presenting the main ideas and directions in a rather "dry" way, we chose to deal and work with the involved tasks by adopting simple models and techniques, so that the reader gets a better feeling of the topic. An effort was made to pay more attention to the scientific notions than to algebraic manipulations and mathematical details, which will, unavoidably, be used to a larger extent while "embroidering" the chapters to follow.

The least-squares (LS), maximum likelihood (ML), regularization, and Bayesian inference techniques are presented and discussed. An effort has been made to assist the reader to grasp an informative view of the big picture conveyed by the book. Thus, this chapter could also be used as an overview introduction to the parametric modeling task in the realm of machine learning.

3.2 PARAMETER ESTIMATION: THE DETERMINISTIC POINT OF VIEW

The task of estimating the value of an unknown parameter vector, $\boldsymbol{\theta}$, has been at the center of interest in a number of application areas. For example, in the early years at university, one of the very first subjects any student has to study is the so-called curve fitting problem. Given a set of data points, one must find a curve or a surface that "fits" the data. The usual path to follow is to *adopt* a functional form, such as a linear function or a quadratic one, and try to estimate the associated unknown parameters so that the graph of the function "passes through" the data and follows their deployment in space as close as possible. Figs. 3.1A and B are two such examples. The data lie in the $\mathbb{R}^2$ space and are given to us as a set of points (y_n, x_n), $n = 1, 2, \dots, N$. The adopted functional form for the curve corresponding to Fig. 3.1A is

$$y = f_{\boldsymbol{\theta}}(x) = \theta_0 + \theta_1 x, \tag{3.1}$$

and for the case of Fig. 3.1B it is

$$y = f_{\boldsymbol{\theta}}(x) = \theta_0 + \theta_1 x + \theta_2 x^2. \tag{3.2}$$

The unknown parameter vectors are $\boldsymbol{\theta} = [\theta_0, \theta_1]^T$ and $\boldsymbol{\theta} = [\theta_0, \theta_1, \theta_2]^T$, respectively. In both cases, the parameter values, which define the curves drawn by the red lines, provide a much better fit compared to the values associated with the black ones. In both cases, the task comprises two steps: (a) first adopt a specific *parametric functional* form which we reckon to be more appropriate for the data at hand and (b) estimate the values of the unknown parameters in order to obtain a "good" fit.

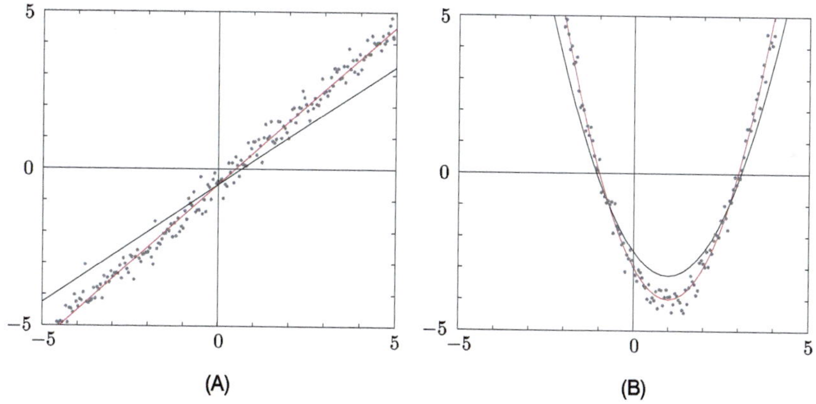

FIGURE 3.1

(A) Fitting a linear function. (B) Fitting a quadratic function. The red lines are the optimized ones.

In the more general and formal setting, the task can be defined as follows. Given a set of data points, $(y_n, \boldsymbol{x}_n)$, $y_n \in \mathbb{R}$, $\boldsymbol{x}_n \in \mathbb{R}^l$, $n = 1, 2, \dots, N$, and a parametric[1] set of functions,

$$\mathcal{F} := \left\{ f_{\boldsymbol{\theta}}(\cdot) : \boldsymbol{\theta} \in \mathcal{A} \subseteq \mathbb{R}^K \right\}, \tag{3.3}$$

find a function in $\mathcal{F}$, which will be denoted as $f(\cdot) := f_{\boldsymbol{\theta}_*}(\cdot)$, such that given a value of $\boldsymbol{x} \in \mathbb{R}^l$, $f(\boldsymbol{x})$ best approximates the corresponding value $y \in \mathbb{R}$. The set $\mathcal{A}$ is a constraint set in case we wish to constrain the unknown K parameters to lie within a specific region in $\mathbb{R}^K$. Constraining the parameters to

[1] Recall from the Notations section that we use the symbol $f(\cdot)$ to denote a function of a single argument; $f(\boldsymbol{x})$ denotes the value that the function $f(\cdot)$ has at $\boldsymbol{x}$.

be within a subset of the parameter space $\mathbb{R}^K$ is almost omnipresent in machine learning. We will deal with constrained optimization later on in this chapter (Section 3.8). We start our discussion by considering y to be a real variable, $y \in \mathbb{R}$, and as we move on and understand better the various "secrets," we will allow it to move to higher-dimensional spaces. The value $\boldsymbol{\theta}_*$ is the value that results from an estimation procedure. The values of $\boldsymbol{\theta}_*$ that define the red curves in Figs. 3.1A and B are

$$\boldsymbol{\theta}_* = [-0.5, 1]^T, \quad \boldsymbol{\theta}_* = [-3, -2, 1]^T, \tag{3.4}$$

respectively.

To reach a decision with respect to the choice of $\mathcal{F}$ is not an easy task. For the case of the data in Fig. 3.1, we were a bit "lucky." First, the data live in the two-dimensional space, where we have the luxury of visualization. Second, the data were scattered along curves whose shape is pretty familiar to us; hence, a simple inspection suggested the proper family of functions, for each one of the two cases. Obviously, real life is hardly as generous as that and in the majority of practical applications, the data reside in high-dimensional spaces and/or the shape of the surface (hypersurface, for spaces of dimensionality higher than three) can be quite complex. Hence, the choice of $\mathcal{F}$, which dictates the functional form (e.g., linear, quadratic, etc.), is not easy. In practice, one has to use as much a priori information as possible concerning the physical mechanism that underlies the generation of the data, and most often use different families of functions and finally keep the one that results in the best performance, according to a chosen criterion.

Having adopted a parametric family of functions, $\mathcal{F}$, one has to get estimates for the set of the unknown parameters. To this end, a measure of fitness has to be adopted. The more classical approach is to adopt a *loss* function, which quantifies the deviation/error between the measured value of y and the predicted one using the corresponding measurements $\boldsymbol{x}$, as in $f_{\boldsymbol{\theta}}(\boldsymbol{x})$. In a more formal way, we adopt a *nonnegative* (loss) function,

$$\mathcal{L}(\cdot, \cdot) : \mathbb{R} \times \mathbb{R} \longmapsto [0, \infty),$$

and compute $\boldsymbol{\theta}_*$ so as to minimize the total loss, or as we say the *cost*, over all the data points, or

$$f(\cdot) := f_{\boldsymbol{\theta}_*}(\cdot) : \ \boldsymbol{\theta}_* = \arg\min_{\boldsymbol{\theta} \in \mathcal{A}} J(\boldsymbol{\theta}), \tag{3.5}$$

where

$$J(\boldsymbol{\theta}) := \sum_{n=1}^{N} \mathcal{L}\big(y_n, f_{\boldsymbol{\theta}}(\boldsymbol{x}_n)\big), \tag{3.6}$$

assuming that a minimum exists. Note that, in general, there may be more than one optimal value $\boldsymbol{\theta}_*$, depending on the shape of $J(\boldsymbol{\theta})$.

As the book evolves, we are going to see different loss functions and different parametric families of functions. For the sake of simplicity, for the rest of this chapter we will adhere to the squared error loss function,

$$\mathcal{L}\big(y, f_{\boldsymbol{\theta}}(\boldsymbol{x})\big) = \big(y - f_{\boldsymbol{\theta}}(\boldsymbol{x})\big)^2,$$

and to the linear class of functions.

The squared error loss function is credited to the great mathematician Carl Frederich Gauss, who proposed the fundamentals of the least-squares (LS) method in 1795 at the age of 18. However, it was Adrien-Marie Legendre who first published the method in 1805, working independently. Gauss published it in 1809. The strength of the method was demonstrated when it was used to predict the location of the asteroid Ceres. Since then, the squared error loss function has "haunted" all scientific fields, and even if it is not used directly, it is, most often, used as the standard against which the performance of more modern alternatives are compared. This success is due to some nice properties that this loss criterion has, which will be explored as we move on in this book.

The combined choice of linearity with the squared error loss function turns out to simplify the algebra and hence becomes very pedagogic for introducing the newcomer to the various "secrets" that underlie the area of parameter estimation. Moreover, understanding linearity is very important. Treating nonlinear tasks, most often, turns out to finally resort to a linear problem. Take, for example, the nonlinear model in Eq. (3.2) and consider the transformation

$$\mathbb{R} \ni x \longmapsto \boldsymbol{\phi}(x) := \begin{bmatrix} x \\ x^2 \end{bmatrix} \in \mathbb{R}^2. \tag{3.7}$$

Then Eq. (3.2) becomes

$$y = \theta_0 + \theta_1 \phi_1(x) + \theta_2 \phi_2(x). \tag{3.8}$$

That is, the model is now linear with respect to the components $\phi_k(x)$, $k = 1, 2$, of the two-dimensional image $\boldsymbol{\phi}(x)$ of x. As a matter of fact, this simple trick is at the heart of a number of nonlinear methods that will be treated later on in the book. No doubt, the procedure can be generalized to any number K of functions $\phi_k(\boldsymbol{x})$, $k = 1, 2, \ldots, K$, and besides monomials, other types of nonlinear functions can be used, such as exponentials, splines, and wavelets. In spite of the nonlinear nature of the input–output dependence modeling, we still consider this model to be linear, because it retains its linearity with respect to the involved unknown parameters θ_k, $k = 1, 2, \ldots, K$. Although for the rest of the chapter we will adhere to linear functions, in order to keep our discussion simple, everything that will be said applies to nonlinear ones as well. All that is needed is to replace $\boldsymbol{x}$ with $\boldsymbol{\phi}(\boldsymbol{x}) := [\phi_1(\boldsymbol{x}), \ldots, \phi_K(\boldsymbol{x})]^T \in \mathbb{R}^K$.

In the sequel, we will present two examples in order to demonstrate the use of parametric modeling. These examples are generic and can represent a wide class of problems.

3.3 LINEAR REGRESSION

In statistics, the term *regression* was coined to define the task of modeling the relationship of a *dependent* random variable y, which is considered to be the response of a system, when this is activated by a set of random variables $\mathrm{x}_1, \mathrm{x}_2, \ldots, \mathrm{x}_l$, which will be represented as the components of an equivalent random vector $\mathbf{x}$. The relationship is modeled via an additive disturbance or noise term η. The block diagram of the process, which relates the involved variables, is given in Fig. 3.2. The noise variable η is an *unobserved* random variable. The goal of the regression task is to estimate the parameter vector $\boldsymbol{\theta}$, given a set of measurements/observations $(y_n, \boldsymbol{x}_n)$, $n = 1, 2, \ldots, N$, that we have at our disposal. This is also known as the *training data set*. The dependent variable is usually known as the *output* variable and the vector $\mathbf{x}$ as the *input* vector or the *regressor*. If we model the system as a linear combiner, the dependence relationship is written as

$$\mathrm{y} = \theta_0 + \theta_1 \mathrm{x}_1 + \cdots + \theta_l \mathrm{x}_l + \eta = \theta_0 + \boldsymbol{\theta}^T \mathbf{x} + \eta. \tag{3.9}$$

The parameter θ_0 is known as the *bias* or the *intercept*. Usually, this term is absorbed by the parameter vector $\boldsymbol{\theta}$ with a simultaneous increase in the dimension of $\mathbf{x}$ by adding the constant 1 as its last element. Indeed, we can write

$$\theta_0 + \boldsymbol{\theta}^T \mathbf{x} + \eta = [\boldsymbol{\theta}^T, \theta_0] \begin{bmatrix} \mathbf{x} \\ 1 \end{bmatrix} + \eta.$$

From now on, the regression model will be written as

$$\mathrm{y} = \boldsymbol{\theta}^T \mathbf{x} + \eta, \tag{3.10}$$

and, unless otherwise stated, this notation means that the bias term has been absorbed by $\boldsymbol{\theta}$ and $\mathbf{x}$ has been extended by adding 1 as an extra component. Because the noise variable is unobserved, we need a model to be able to *predict* the output value of y, given an observed value, $\boldsymbol{x}$, of the random vector $\mathbf{x}$.

In linear regression, given an estimate $\hat{\boldsymbol{\theta}}$ of $\boldsymbol{\theta}$, we adopt the following prediction model:

$$\hat{y} = \hat{\theta}_0 + \hat{\theta}_1 x_1 + \cdots + \hat{\theta}_l x_l := \hat{\boldsymbol{\theta}}^T \boldsymbol{x}. \tag{3.11}$$

Using the squared error loss function, the estimate $\hat{\boldsymbol{\theta}}$ is set equal to $\boldsymbol{\theta}_*$, which minimizes the square difference between $\hat{y}_n$ and y_n over the set of the available observations; that is, one minimizes, with respect to $\boldsymbol{\theta}$, the cost function

$$J(\boldsymbol{\theta}) = \sum_{n=1}^{N} (y_n - \boldsymbol{\theta}^T \boldsymbol{x}_n)^2. \tag{3.12}$$

We start our discussion by considering no constraints; hence, we set $\mathcal{A} = \mathbb{R}^K$, and we are going to search for solutions that lie anywhere in $\mathbb{R}^K$. Taking the derivative (gradient) with respect to $\boldsymbol{\theta}$ (see Appendix A) and equating to the zero vector, $\mathbf{0}$, we obtain (Problem 3.1)

$$\left(\sum_{n=1}^{N} \boldsymbol{x}_n \boldsymbol{x}_n^T \right) \hat{\boldsymbol{\theta}} = \sum_{n=1}^{N} y_n \boldsymbol{x}_n. \tag{3.13}$$

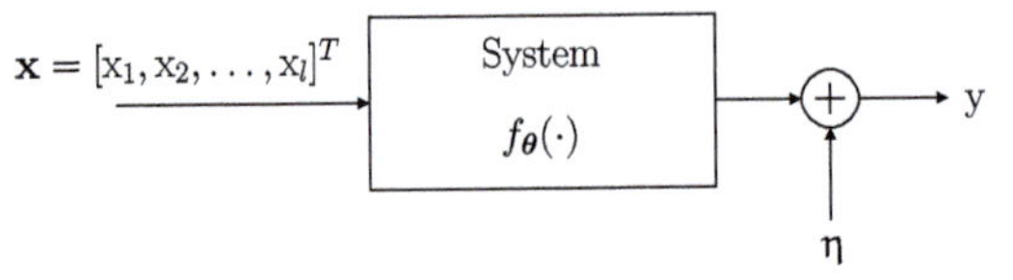

FIGURE 3.2

Block diagram showing the input–output relation in a regression model.

Note that the sum on the left-hand side is an $(l+1) \times (l+1)$ matrix, being the sum of N outer vector products, i.e., $\boldsymbol{x}_n \boldsymbol{x}_n^T$. As we know from linear algebra, we need at least $N = l+1$ observation vectors to guarantee that the matrix is invertible, assuming of course linear independence among the vectors (see, e.g., [35]).

For those who are not (yet) very familiar with working with vectors, note that Eq. (3.13) is just a generalization of the scalar case. For example, in a "scalar" world, the input–output pairs would comprise scalars (y_n, x_n) and the unknown parameter would also be a scalar, θ. The cost function would be $\sum_{n=1}^{N}(y_n - \theta x_n)^2$. Taking the derivative and setting it equal to zero leads to

$$\left(\sum_{n=1}^{N} x_n^2\right)\hat{\theta} = \sum_{n=1}^{N} y_n x_n, \text{ or } \hat{\theta} = \frac{\sum_{n=1}^{N} y_n x_n}{\sum_{n}^{N} x_n^2}.$$

A more popular way to write Eq. (3.13) is via the so-called input matrix X, defined as the $N \times (l+1)$ matrix which has as rows the (extended) regressor vectors $\boldsymbol{x}_n^T$, $n = 1, 2, \ldots, N$, expressed as

$$X := \begin{bmatrix} \boldsymbol{x}_1^T \\ \boldsymbol{x}_2^T \\ \vdots \\ \boldsymbol{x}_N^T \end{bmatrix} = \begin{bmatrix} x_{11} & \ldots & x_{1l} & 1 \\ x_{21} & \ldots & x_{2l} & 1 \\ \vdots & \ddots & \vdots & \\ x_{N1} & \ldots & x_{Nl} & 1 \end{bmatrix}. \tag{3.14}$$

Then it is straightforward to see that Eq. (3.13) can be written as

$$(X^T X)\hat{\boldsymbol{\theta}} = X^T \boldsymbol{y}, \tag{3.15}$$

where

$$\boldsymbol{y} := [y_1, y_2, \ldots, y_N]^T. \tag{3.16}$$

Indeed,

$$X^T X = [\boldsymbol{x}_1, \ldots, \boldsymbol{x}_N] \begin{bmatrix} \boldsymbol{x}_1^T \\ \vdots \\ \boldsymbol{x}_N^T \end{bmatrix} = \sum_{n=1}^{N} \boldsymbol{x}_n \boldsymbol{x}_n^T,$$

and similarly,

$$X^T \boldsymbol{y} = [\boldsymbol{x}_1, \ldots, \boldsymbol{x}_N] \begin{bmatrix} y_1 \\ \vdots \\ y_N \end{bmatrix} = \sum_{n=1}^{N} y_n \boldsymbol{x}_n.$$

Thus, finally, the LS estimate is given by

$$\boxed{\hat{\boldsymbol{\theta}} = (X^T X)^{-1} X^T \boldsymbol{y}: \quad \text{the LS estimate},} \tag{3.17}$$

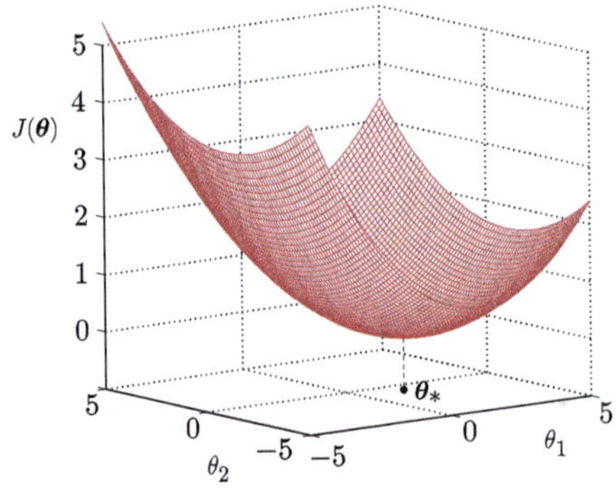

FIGURE 3.3

The squared error loss function has a unique minimum at the point $\boldsymbol{\theta}_*$.

assuming, of course, that $(X^T X)^{-1}$ exists.

In other words, the obtained estimate of the parameter vector is given by a *linear set of equations*. This is a major advantage of the squared error loss function, when applied to a linear model. Moreover, this solution is *unique*, provided that the $(l + 1) \times (l + 1)$ matrix $X^T X$ is invertible. The uniqueness is due to the parabolic shape of the graph of the sum of squared errors cost function. This is illustrated in Fig. 3.3 for the two-dimensional space. It is readily observed that the graph has a unique minimum. This is a consequence of the fact the sum of squared errors cost function is a strictly convex one. Issues related to convexity of loss functions will be treated in more detail in Chapter 8.

Example 3.1. Consider the system that is described by the following model:

$$\mathrm{y} = \theta_0 + \theta_1 \mathrm{x}_1 + \theta_2 \mathrm{x}_2 + \eta := [0.25, -0.25, 0.25] \begin{bmatrix} \mathrm{x}_1 \\ \mathrm{x}_2 \\ 1 \end{bmatrix} + \eta, \tag{3.18}$$

where η is a Gaussian random variable of zero mean and variance $\sigma^2 = 1$. Observe that the generated data are spread, due to the noise, around the plane that is defined as

$$f(\boldsymbol{x}) = \theta_0 + \theta_1 x_1 + \theta_2 x_2, \tag{3.19}$$

in the two-dimensional space (see Fig. 3.4C).

The random variables x_1 and x_2 are assumed to be mutually independent and uniformly distributed over the interval [0, 10]. Furthermore, both variables are independent of the noise variable η. We generate $N = 50$ i.i.d. points[2] for each of the three random variables, i.e., x_1, x_2, η. For each triplet, we use Eq. (3.18) to generate the corresponding value, y, of y. In this way, the points $(y_n, \boldsymbol{x}_n)$, $n = 1, 2, \ldots, 50$, are generated, where each observation $\boldsymbol{x}_n$ lies in $\mathbb{R}^2$. These are used as the training points to obtain the

2 Independent and identically distributed samples (see also Section 2.3.2, following Eq. (2.84)).

LS estimates of the parameters of the linear model

$$\hat{y} = \hat{\theta}_0 + \hat{\theta}_1 x_1 + \hat{\theta}_2 x_2. \tag{3.20}$$

Then we repeat the experiments with $\sigma^2 = 10$. Note that, in general, Eq. (3.20) defines a different plane than the original (3.19).

The values of the LS optimal estimates are obtained by solving a 3×3 linear system of equations. They are

$$\text{(a) } \hat{\theta}_0 = 0.028, \ \hat{\theta}_1 = 0.226, \ \hat{\theta}_2 = -0.224,$$
$$\text{(b) } \hat{\theta}_0 = 0.914, \ \hat{\theta}_1 = 0.325, \ \hat{\theta}_2 = -0.477,$$

for the two cases, respectively. Figs. 3.4A and B show the recovered planes. Observe that in the case of Fig. 3.4A, corresponding to a noise variable of small variance, the obtained plane follows the data points much closer than that of Fig. 3.4B.

Remarks 3.1.

- The set of points $(\hat{y}_n, x_{n1}, \dots, x_{nl})$, $n = 1, 2, \dots, N$, lie on a *hyperplane* in the $\mathbb{R}^{l+1}$ space. Equivalently, they lie on a hyperplane that crosses the origin and, thus, it is a linear subspace in the extended space $\mathbb{R}^{l+2}$ when one absorbs θ_0 in $\boldsymbol{\theta}$, as explained previously.
- Note that the prediction model in Eq. (3.11) could still be used, even if the true system structure does not obey the linear model in Eq. (3.9). For example, the true dependence between y and **x** may be a nonlinear one. In such a case, the predictions of y based on the model in Eq. (3.11) may, however, not be satisfactory. It all depends on the deviation of our adopted model from the true structure of the system that generates the data.
- The prediction performance of the model also depends on the statistical properties of the noise variable. This is an important issue. We will see later on that, depending on the statistical properties of the noise variable, some loss functions and methods may be more suitable than others.
- The two previous remarks suggest that in order to quantify the performance of an estimator some related criteria are necessary. In Section 3.9, we will present some theoretical touches that shed light on certain aspects related to the performance of an estimator.

3.4 CLASSIFICATION

Classification is the task of predicting the class to which an object, known as *pattern*, belongs. The pattern is assumed to belong to one and only one among a number of a priori known classes. Each pattern is *uniquely* represented by a set of values, known as *features*. One of the early stages in designing a classification system is to select an appropriate set of feature variables. These should "encode" as much class-discriminatory information so that, by measuring their value for a given pattern, we are able to predict, with high enough probability, the class of the pattern. Selecting the appropriate set of features for each problem is not an easy task, and it comprises one of the most important areas within the field of *pattern recognition* (see, e.g., [12,37]). Having selected, say, l feature (random) variables, $\mathrm{x}_1, \mathrm{x}_2, \dots, \mathrm{x}_l$, we stack them as the components of the so-called *feature vector* $\mathbf{x} \in \mathbb{R}^l$. The goal is to

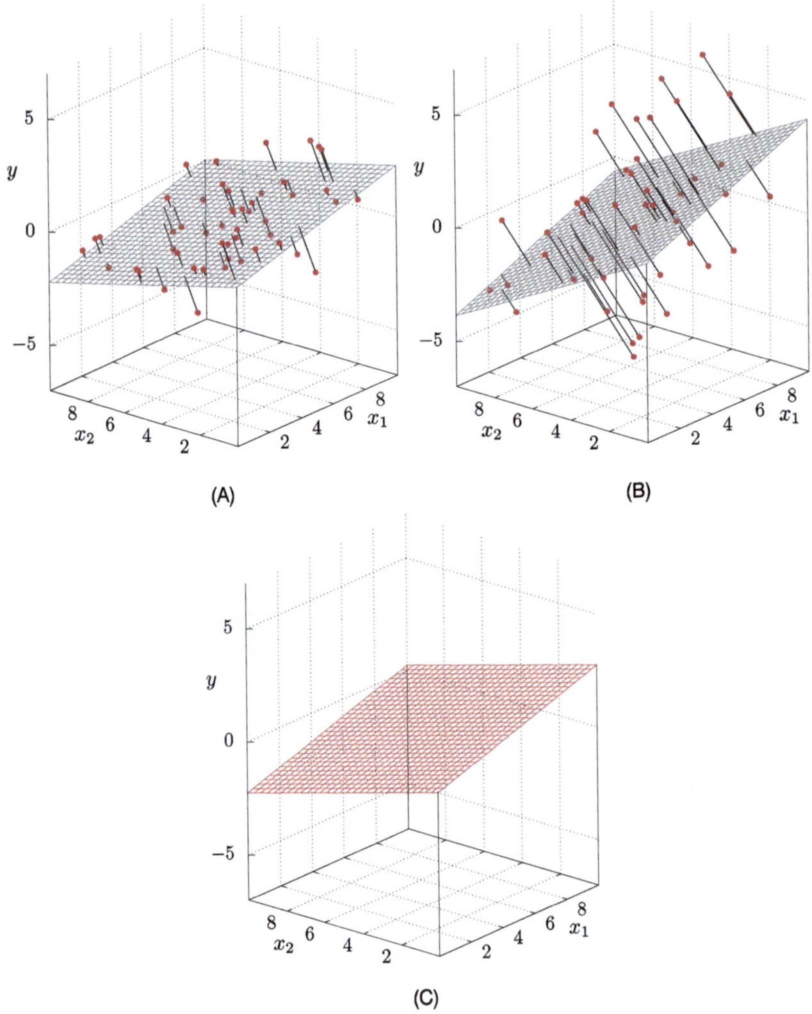

FIGURE 3.4

Fitting a plane using the LS method for (A) a low variance and (B) a high variance noise case. In (C), the true plane that corresponds to the true coefficients is shown for comparison reasons. Note that when the noise variance in the regression model is smaller, a better fit to the data set is obtained.

design a *classifier*, such as a function[3] $f(\boldsymbol{x})$, or equivalently a *decision surface*, $f(\boldsymbol{x}) = 0$, in $\mathbb{R}^l$, so that given the values in a feature vector $\boldsymbol{x}$, which corresponds to a pattern, we will be able to *predict* the class to which the pattern belongs.

[3] In the more general case, a set of functions.

To formulate the task in mathematical terms, each class is represented by the *class label* variable y. For the simple two-class classification task, this can take either of two values, depending on the class (e.g., $1, -1$ or $1, 0$, etc.). Then, given the value of $\boldsymbol{x}$, corresponding to a specific pattern, its class label is predicted according to the rule

$$\hat{y} = \phi(f(\boldsymbol{x})),$$

where $\phi(\cdot)$ is a nonlinear function that indicates on which side of the decision surface $f(\boldsymbol{x}) = 0$ lies $\boldsymbol{x}$. For example, if the class labels are ± 1, the nonlinear function is chosen to be the sign function, or $\phi(\cdot) = \text{sgn}(\cdot)$. It is now clear that what we have said so far in the previous section can be transferred here and the task becomes that of estimating a function $f(\cdot)$, based on a set of training points $(y_n, \boldsymbol{x}_n) \in D \times \mathbb{R}^l$, $n = 1, 2, \ldots, N$, where D denotes the discrete set in which y lies. Function $f(\cdot)$ is selected so as to belong in a specific parametric class of functions, $\mathcal{F}$, and the goal is, once more, to estimate the parameters so that the deviation between the true class labels, y_n, and the predicted ones, $\hat{y}_n$, is minimum according to a preselected criterion. So, is the classification any different from the regression task?

The answer to the previous question is that although they may share some similarities, they are different. Note that in a classification task, the dependent variables are of a *discrete* nature, in contrast to the regression, where they lie in an interval. This suggests that, in general, different techniques have to be adopted to optimize the parameters. For example, the most obvious choice for a criterion in a classification task is the probability of error. However, in a number of cases, one could attack both tasks using the same type of loss functions, as we will do in this section; even if such an approach is adopted, in spite of the similarities in their mathematical formalism, the goals of the two tasks remain different.

In the regression task, the function $f(\cdot)$ has to "explain" the data generation mechanism. The corresponding surface in the $(y, \boldsymbol{x})$ space $\mathbb{R}^{l+1}$ should develop so as to follow the spread of the data in the space as closely as possible. In contrast, in classification, the goal is to place the corresponding surface $f(\boldsymbol{x}) = 0$ in $\mathbb{R}^l$, so as to separate the data that belong to different classes as much as possible. The goal of a classifier is to *partition* the space where the feature vectors lie into regions and associate each region with a class. Fig. 3.5 illustrates two cases of classification tasks. The first one is an example of two linearly separable classes, where a straight line can separate the two classes; the second case is an example of two nonlinearly separable classes, where the use of a linear classifier would have failed to separate the two classes.

Let us now make what we have said so far more concrete. We are given a set of training patterns, $\boldsymbol{x}_n \in \mathbb{R}^l$, $n = 1, 2, \ldots, N$, that belong to either of two classes, say, ω_1 and ω_2. The goal is to design a hyperplane

$$\begin{aligned} f(\boldsymbol{x}) &= \theta_0 + \theta_1 x_1 + \cdots + \theta_l x_l \\ &= \boldsymbol{\theta}^T \boldsymbol{x} = 0, \end{aligned}$$

where we have absorbed the bias θ_0 in $\boldsymbol{\theta}$ and we have extended the dimension of $\boldsymbol{x}$, as explained before. Our aim is to place this hyperplane in between the two classes. Obviously, any point lying on this hyperplane scores a zero, $f(\boldsymbol{x}) = 0$, and the points lying on either side of the hyperplane score either a positive ($f(\boldsymbol{x}) > 0$) or a negative value ($f(\boldsymbol{x}) < 0$), depending on which side of the hyperplane they lie. We should therefore train our classifier so that the points from one class score a positive value and the points of the other a negative one. This can be done, for example, by labeling all points from

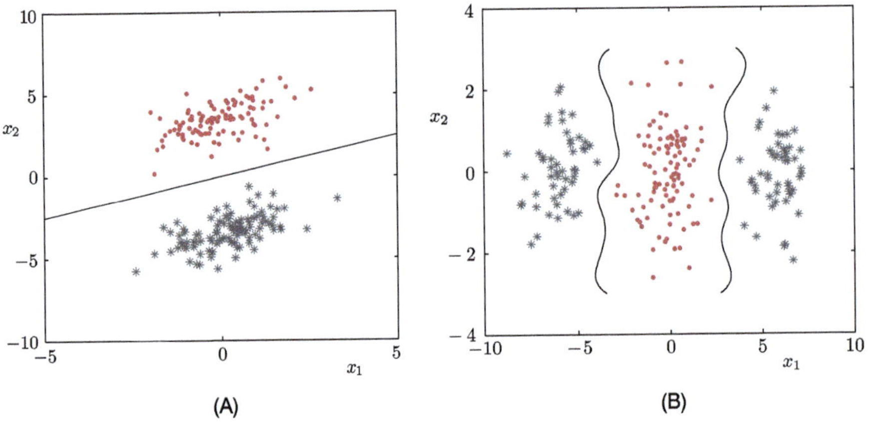

FIGURE 3.5

Examples of two-class classification tasks. (A) A linearly separable classification task. (B) A nonlinearly separable classification task. The goal of a classifier is to partition the space into regions and associate each region with a class.

class ω_1 with $y_n = 1, \forall n : \boldsymbol{x}_n \in \omega_1$, and all points from class ω_2 with $y_n = -1, \forall n : \boldsymbol{x}_n \in \omega_2$. Then the squared error is mobilized to compute $\boldsymbol{\theta}$ so as to minimize the cost, i.e.,

$$J(\boldsymbol{\theta}) = \sum_{n=1}^{N} \left(y_n - \boldsymbol{\theta}^T \boldsymbol{x}_n \right)^2 .$$

The solution is exactly the same as Eq. (3.13). Fig. 3.6 shows the resulting LS classifiers for two cases of data. Observe that in the case of Fig. 3.6B, the resulting classifier cannot classify correctly all the data points. Our desire to place all data which originate from one class on one side and the rest on the other cannot be satisfied. All that the LS classifier can do is to place the hyperplane so that the sum of squared errors between the desired (true) values of the labels y_n and the predicted outputs $\boldsymbol{\theta}^T \boldsymbol{x}_n$ is minimal. It is mainly for cases such as overlapping classes, which are usually encountered in practice, where one has to look for an alternative to the squared error criteria and methods, in order to serve better the needs and the goals of the classification task. For example, a reasonable optimality criterion would be to minimize the probability of error, that is, the percentage of points for which the true labels y_n and the ones predicted by the classifier, $\hat{y}_n$, are different. Chapter 7 presents methods and loss functions appropriate for the classification task. In Chapter 11 support vector machines are discussed and in Chapter 18 neural networks and deep learning methods are presented. These are currently among the most powerful techniques for classification problems.

GENERATIVE VERSUS DISCRIMINATIVE LEARNING

The path that we have taken in order to introduce the classification task was to consider a functional dependence between the output variable (label) y and the input variables (features) **x**. The involved

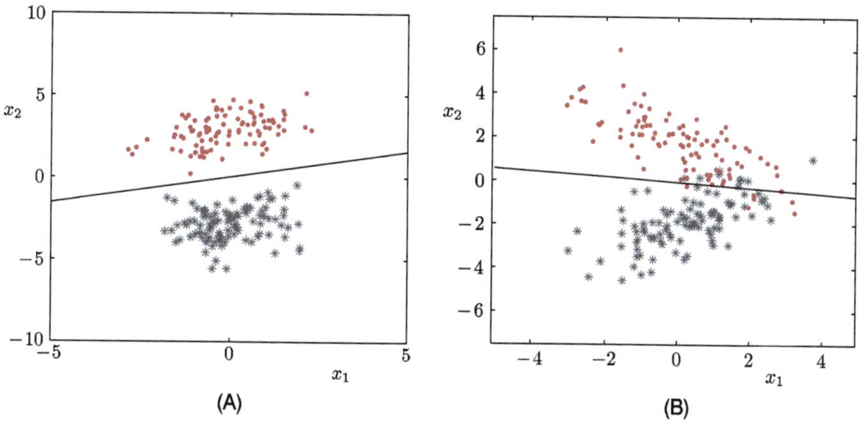

FIGURE 3.6

Design of a linear classifier, $\theta_0 + \theta_1 x_1 + \theta_2 x_2 = 0$, based on the squared error loss function. (A) The case of two linearly separable classes. (B) The case of nonseparable classes. In the latter case, the classifier cannot separate fully the two classes. All it can do is to place the separating (decision) line so as to minimize the deviation between the true labels and the predicted output values in the LS sense.

parameters were optimized with respect to a cost function. This path of modeling is also known as *discriminative learning*. We were not concerned with the statistical nature of the dependence that ties these two sets of variables together. In a more general setting, the term discriminative learning is also used to cover methods that model directly the posterior probability of a class, represented by its label y, given the feature vector $\boldsymbol{x}$, as in $P(y|\boldsymbol{x})$. The common characteristic of all these methods is that they bypass the need of modeling the input data distribution explicitly. From a statistical point of view, discriminative learning is justified as follows.

Using the product rule for probabilities, the joint distribution between the input data and their respective labels can be written as

$$p(y, \boldsymbol{x}) = P(y|\boldsymbol{x})p(\boldsymbol{x}).$$

In discriminative learning, only the first of the two terms in the product is considered; a functional form is adopted and parameterized appropriately as $P(y|\boldsymbol{x};\boldsymbol{\theta})$. Parameters are then estimated by optimizing a cost function. The distribution of the input data is ignored. Such an approach has the advantage that simpler models can be used, especially if the input data are described by a joint probability density function (PDF) of a complex form. The disadvantage is that the input data distribution is ignored, although it can carry important information, which could be exploited to the benefit of the overall performance.

In contrast, the alternative path, known as *generative learning*, exploits the input data distribution. Once more, employing the product rule, we have

$$p(y, \boldsymbol{x}) = p(\boldsymbol{x}|y)P(y),$$

where $P(y)$ is the probability concerning the classes and $p(\boldsymbol{x}|y)$ is the distribution of the input given the class label. For such an approach, we end up with one distribution per class, which has to be learned. In parametric modeling, a set of parameters is associated with each one of these conditional distributions. Once the input–output joint distribution has been learned, the prediction of the class label of an unknown pattern, $\boldsymbol{x}$, is performed based on the a posteriori probability, i.e.,

$$P(y|\boldsymbol{x}) = \frac{p(y, \boldsymbol{x})}{p(\boldsymbol{x})} = \frac{p(y, \boldsymbol{x})}{\sum_y p(y, \boldsymbol{x})}.$$

We will return to these issues in more detail in Chapter 7.

3.5 BIASED VERSUS UNBIASED ESTIMATION

In supervised learning, we are given a set of training points, $(y_n, \boldsymbol{x}_n)$, $n = 1, 2, \ldots, N$, and we return an estimate of the unknown parameter vector, say, $\hat{\boldsymbol{\theta}}$. However, the training points themselves are random variables. If we are given another set of N observations of the *same* random variables, these are going to be different, and obviously the resulting estimate will also be different. In other words, by changing our training data, different estimates will result. Hence, we can assume that the resulting estimate, of a fixed yet unknown parameter, is itself a random variable. This, in turn, poses questions on how good an estimator is. Undoubtedly, each time the obtained estimate is optimal with respect to the adopted loss function *and* the specific training set used. However, who guarantees that the resulting estimates are "close" to the true value, assuming that there is one? In this section, we will try to address this task and to illuminate some related theoretical aspects. Note that we have already used the term *estimator* in place of the term *estimate*. Let us elaborate a bit on their difference, before presenting more details.

An estimate, such as $\hat{\boldsymbol{\theta}}$, has a specific value, which is the result of a function acting on a set of observations on which our chosen estimate depends (see Eq. (3.17)). In general, we could generalize Eq. (3.17) and write

$$\hat{\boldsymbol{\theta}} = g(\boldsymbol{y}, X).$$

However, once we allow the set of observations to change randomly, and the estimate becomes itself a random variable, we write the previous equation in terms of the corresponding random variables, i.e.,

$$\hat{\boldsymbol{\uptheta}} = g(\mathbf{y}, \mathrm{X}),$$

and we refer to this functional dependence as the *estimator* of the unknown vector $\boldsymbol{\theta}$.

In order to simplify the analysis and focus on the insight behind the methods, we will assume that our parameter space is that of real numbers, $\mathbb{R}$. We will also assume that the model (i.e., the set of functions $\mathcal{F}$) which we have adopted for modeling our data is the correct one and that the (unknown to us) value of the associated true parameter is equal to[4] θ_o. Let $\hat{\uptheta}$ denote the random variable of the

[4] Not to be confused with the intercept; the subscript here is "o" and not "0."

associated estimator. Adopting the squared error loss function to quantify deviations, a reasonable criterion to measure the performance of an estimator is the *mean-square error* (MSE),

$$\text{MSE} = \mathbb{E}\left[(\hat{\theta} - \theta_o)^2\right], \tag{3.21}$$

where the mean $\mathbb{E}$ is taken over *all* possible training data sets of size N. If the MSE is small, then we expect that, on average, the resulting estimates are close to the true value. Note that although θ_o is not known, studying the way in which the MSE depends on various terms will still help us to learn how to proceed in practice and unravel possible paths that one can follow to obtain good estimators.

Indeed, this simple and "natural" looking criterion hides some interesting surprises for us. Let us insert the mean value $\mathbb{E}[\hat{\theta}]$ of $\hat{\theta}$ in Eq. (3.21). We get

$$\begin{aligned} \text{MSE} &= \mathbb{E}\left[\left\{\left(\hat{\theta} - \mathbb{E}[\hat{\theta}]\right) + \left(\mathbb{E}[\hat{\theta}] - \theta_o\right)\right\}^2\right] \\ &= \underbrace{\mathbb{E}\left[\left(\hat{\theta} - \mathbb{E}[\hat{\theta}]\right)^2\right]}_{\text{Variance}} + \underbrace{\left(\mathbb{E}[\hat{\theta}] - \theta_o\right)^2}_{\text{Bias}^2}, \end{aligned} \tag{3.22}$$

where, for the second equality, we have taken into account that the mean value of the product of the two involved terms turns out to be zero, as can be readily seen. What Eq. (3.22) suggests is that the MSE consists of two terms. The first one is the variance around the mean value and the second one is due to the bias, that is, the deviation of the mean value of the estimator from the true one.

3.5.1 BIASED OR UNBIASED ESTIMATION?

One may naively think that choosing an estimator that is *unbiased*, as is $\mathbb{E}[\hat{\theta}] = \theta_o$, such that the second term in Eq. (3.22) becomes zero, is a reasonable choice. Adopting an unbiased estimator may also be appealing from the following point of view. Assume that we have L different training sets, each comprising N points. Let us denote each data set by $\mathcal{D}_i$, $i = 1, 2, \dots, L$. For each one, an estimate $\hat{\theta}_i$, $i = 1, 2, \dots, L$, will result. Then, we form a new estimator by taking the average value,

$$\hat{\theta}^{(L)} := \frac{1}{L}\sum_{i=1}^{L} \hat{\theta}_i.$$

This is also an unbiased estimator, because

$$\mathbb{E}[\hat{\theta}^{(L)}] = \frac{1}{L}\sum_{i=1}^{L} \mathbb{E}[\hat{\theta}_i] = \theta_o.$$

Moreover, assuming that the involved estimators are mutually uncorrelated, i.e.,

$$\mathbb{E}\left[(\hat{\theta}_i - \theta_o)(\hat{\theta}_j - \theta_o)\right] = 0,$$

and of the same variance, σ^2, the variance of the new estimator is now much smaller (Problem 3.2), i.e.,

$$\sigma^2_{\hat{\theta}^{(L)}} = \mathbb{E}\left[\left(\hat{\theta}^{(L)} - \theta_o\right)^2\right] = \frac{\sigma^2}{L}.$$

Hence, by averaging a large number of such unbiased estimators, we expect to get an estimate close to the true value. However, in practice, data are a commodity that is not always abundant. As a matter of fact, very often the opposite is true and one has to be very careful about how to exploit them. In such cases, where one cannot afford to obtain and average a large number of estimators, an unbiased estimator may not necessarily be the best choice. Going back to Eq. (3.22), there is no reason to suggest that by making the second term equal to zero, the MSE (which, after all, is the quantity of interest to us) becomes minimal. Indeed, let us look at Eq. (3.22) from a slightly different point of view. Instead of computing the MSE for a given estimator, let us replace $\hat{\theta}$ with θ in Eq. (3.22) and compute an estimator that will minimize the MSE with respect to θ directly. In this case, focusing on unbiased estimators, or $\mathbb{E}[\theta] = \theta_o$, introduces a constraint to the task of minimizing the MSE, and it is well known that an unconstrained minimization problem always results in loss function values that are less than or equal to any value generated by a constrained counterpart, i.e.,

$$\min_{\theta} \text{MSE}(\theta) \leq \min_{\theta\colon \mathbb{E}[\theta]=\theta_o} \text{MSE}(\theta), \tag{3.23}$$

where the dependence of the MSE on the estimator θ in Eq. (3.23) is explicitly denoted.

Let us denote by $\hat{\theta}_{\text{MVU}}$ a solution of the task $\min_{\theta\colon \mathbb{E}[\theta]=\theta_o} \text{MSE}(\theta)$. It can be readily verified by Eq. (3.22) that $\hat{\theta}_{\text{MVU}}$ is an unbiased estimator of minimum variance. Such an estimator is known as the *minimum variance unbiased* (MVU) estimator and we assume that such an estimator exists. An MVU does not always exist (see [20], Problem 3.3). Moreover, if it exists it is unique (Problem 3.4). Motivated by Eq. (3.23), our next goal is to search for a *biased* estimator, which results, hopefully, in a smaller MSE. Let us denote this estimator as $\hat{\theta}_b$. For the sake of illustration, and in order to limit our search for $\hat{\theta}_b$, we consider here only $\hat{\theta}_b$s that are scalar multiples of $\hat{\theta}_{\text{MVU}}$, so that

$$\hat{\theta}_b = (1+\alpha)\hat{\theta}_{\text{MVU}}, \tag{3.24}$$

where $\alpha \in \mathbb{R}$ is a free parameter. Note that $\mathbb{E}[\hat{\theta}_b] = (1+\alpha)\theta_o$. By substituting Eq. (3.24) into Eq. (3.22) and after some simple algebra we obtain

$$\text{MSE}(\hat{\theta}_b) = (1+\alpha)^2 \text{MSE}(\hat{\theta}_{\text{MVU}}) + \alpha^2\theta_o^2. \tag{3.25}$$

In order to get $\text{MSE}(\hat{\theta}_b) < \text{MSE}(\hat{\theta}_{\text{MVU}})$, α must be in the range (Problem 3.5) of

$$-\frac{2\text{MSE}(\hat{\theta}_{\text{MVU}})}{\text{MSE}(\hat{\theta}_{\text{MVU}}) + \theta_o^2} < \alpha < 0. \tag{3.26}$$

It is easy to verify that this range implies that $|1+\alpha| < 1$. Hence, $|\hat{\theta}_b| = |(1+\alpha)\hat{\theta}_{\text{MVU}}| < |\hat{\theta}_{\text{MVU}}|$. We can go a step further and try to compute the optimum value of α, which corresponds to the minimum

MSE. By taking the derivative of $\text{MSE}(\hat{\theta}_b)$ in Eq. (3.25) with respect to α, it turns out (Problem 3.6) that this occurs for

$$\alpha_* = -\frac{\text{MSE}(\hat{\theta}_{\text{MVU}})}{\text{MSE}(\hat{\theta}_{\text{MVU}}) + \theta_o^2} = -\frac{1}{1 + \frac{\theta_o^2}{\text{MSE}(\hat{\theta}_{\text{MVU}})}}. \tag{3.27}$$

Therefore, we have found a way to obtain the optimum estimator among those in the set $\{\hat{\theta}_b = (1 + \alpha)\hat{\theta}_{\text{MVU}} : \alpha \in \mathbb{R}\}$, which results in the minimum MSE. This is true, but as many nice things in life, this is not, in general, realizable. The optimal value for α is given in terms of the unknown θ_o! However, Eq. (3.27) is useful in a number of other ways. First, there are cases where the MSE is proportional to θ_o^2; hence, this formula can be used. Also, for certain cases, it can be used to provide useful bounds [19]. Moreover, as far as we are concerned in this book, it says something very important. If we want to do better than the MVU, then, looking at the text following Eq. (3.26), a possible way is to *shrink* the norm of the MVU estimator. Shrinking the norm is a way of introducing bias into an estimator. We will discuss ways to achieve this in Section 3.8 and later on in Chapters 6 and 9.

Note that what we have said so far is readily generalized to parameter vectors. An unbiased parameter vector estimator satisfies

$$\mathbb{E}[\hat{\boldsymbol{\theta}}] = \boldsymbol{\theta}_o,$$

and the MSE around the true value $\boldsymbol{\theta}_o$ is defined as

$$\text{MSE} = \mathbb{E}\left[(\hat{\boldsymbol{\theta}} - \boldsymbol{\theta}_o)^T(\hat{\boldsymbol{\theta}} - \boldsymbol{\theta}_o)\right] = \sum_{i=1}^{l} \mathbb{E}\left[(\hat{\theta}_i - \theta_{oi})^2\right].$$

Looking carefully at the previous definition reveals that the MSE for a parameter vector is the sum of the MSEs of the components $\hat{\theta}_i$, $i = 1, 2, \ldots, l$, around the corresponding true values θ_{oi}.

3.6 THE CRAMÉR–RAO LOWER BOUND

In the previous sections, we saw how one can improve the performance of the MVU estimator, provided that this exists and it is known. However, how can one know that an unbiased estimator that has been obtained is also of minimum variance? The goal of this section is to introduce a criterion that can provide such information.

The *Cramér–Rao lower bound* [9,31] is an elegant theorem and one of the most well-known techniques used in statistics. It provides a lower bound on the variance of *any* unbiased estimator. This is very important because (a) it offers the means to assert whether an unbiased estimator has minimum variance, which, of course, in this case coincides with the corresponding MSE in Eq. (3.22), (b) if this is not the case, it can be used to indicate how far away the performance of an unbiased estimator is from the optimal one, and finally (c) it provides the designer with a tool to know the best possible performance that can be achieved by an unbiased estimator. Because our main purpose here is to focus on the insight and physical interpretation of the method, we will deal with the simple case where our unknown parameter is a real number. The general form of the theorem, involving vectors, is given in Appendix B.

We are looking for a bound of the variance of an unbiased estimator, whose randomness is due to the randomness of the training data, as we change from one set to another. Thus, it does not come as a surprise that the bound involves the joint PDF of the data, parameterized in terms of the unknown parameter θ. Let $\mathcal{X} = \{\boldsymbol{x}_1, \boldsymbol{x}_2, \ldots, \boldsymbol{x}_N\}$ denote the set of N observations, corresponding to a random vector[5] $\mathbf{x}$ that depends on the unknown parameter. Also, let the respective joint PDF of the observations be denoted as $p(\mathcal{X}; \theta)$.

Theorem 3.1. *It is assumed that the joint PDF satisfies the following* regularity *condition:*

$$\mathbb{E}\left[\frac{\partial \ln p(\mathcal{X}; \theta)}{\partial \theta}\right] = 0, \quad \forall \theta. \tag{3.28}$$

This regularity condition is a weak one and holds for most of the cases in practice (Problem 3.7). Then, the variance of any unbiased estimator, $\hat{\theta}$, must satisfy the following inequality:

$$\boxed{\sigma_{\hat{\theta}}^2 \geq \frac{1}{I(\theta)} : \quad \textit{Cramér–Rao lower bound,}} \tag{3.29}$$

where

$$I(\theta) := -\mathbb{E}\left[\frac{\partial^2 \ln p(\mathcal{X}; \theta)}{\partial \theta^2}\right]. \tag{3.30}$$

Moreover, the necessary and sufficient condition for obtaining an unbiased estimator that achieves the bound is the existence of a function $g(\cdot)$ such that for all possible values of θ,

$$\frac{\partial \ln p(\mathcal{X}; \theta)}{\partial \theta} = I(\theta)\left(g(\mathcal{X}) - \theta\right). \tag{3.31}$$

The MVU estimate is then given by

$$\hat{\theta} = g(\mathcal{X}) := g(\boldsymbol{x}_1, \boldsymbol{x}_2, \ldots, \boldsymbol{x}_N), \tag{3.32}$$

and the variance of the respective estimator is equal to $1/I(\theta)$.

When an MVU estimator attains the Cramér–Rao bound, we say that it is *efficient*. All the expectations before are taken with respect to $p(\mathcal{X}; \theta)$. The interested reader may find more on the topic in more specialized books on statistics [20,27,36].

Example 3.2. Let us consider the simplified version of the linear regression model in Eq. (3.10), where the regressor is real-valued and the bias term is zero,

$$y_n = \theta x + \eta_n, \tag{3.33}$$

[5] Note that here $\mathbf{x}$ is treated as a random quantity in a general setting, and not necessarily in the context of the regression/classification tasks.

where we have explicitly denoted the dependence on n, which runs over the number of available observations. Note that in order to further simplify the discussion, we have assumed that our N observations are the result of different realizations of the noise variable only, and that we have kept the value of the input, x, constant, which can be considered to be equal to one, without harming generality; that is, our task is reduced to that of estimating a parameter from its noisy measurements. Thus, for this case, the observations are the scalar outputs y_n, $n = 1, 2, \dots, N$, which we consider to be the components of a vector $\boldsymbol{y} \in \mathbb{R}^N$. We further assume that η_n are samples of a white Gaussian noise (Section 2.4) with zero mean and variance equal to σ_η^2; that is, successive samples are i.i.d. drawn and, hence, they are mutually uncorrelated ($\mathbb{E}[\eta_i \eta_j] = 0$, $i \neq j$). Then, the joint PDF of the output observations is given by

$$p(\boldsymbol{y}; \theta) = \prod_{n=1}^{N} \frac{1}{\sqrt{2\pi\sigma_\eta^2}} \exp\left(-\frac{(y_n - \theta)^2}{2\sigma_\eta^2}\right), \tag{3.34}$$

or

$$\ln p(\boldsymbol{y}; \theta) = -\frac{N}{2}\ln(2\pi\sigma_\eta^2) - \frac{1}{2\sigma_\eta^2}\sum_{n=1}^{N}(y_n - \theta)^2. \tag{3.35}$$

We will derive the corresponding Cramér–Rao bound. Taking the derivative of the logarithm with respect to θ we have

$$\frac{\partial \ln p(\boldsymbol{y}; \theta)}{\partial \theta} = \frac{1}{\sigma_\eta^2}\sum_{n=1}^{N}(y_n - \theta) = \frac{N}{\sigma_\eta^2}(\bar{y} - \theta), \tag{3.36}$$

where

$$\bar{y} := \frac{1}{N}\sum_{n=1}^{N} y_n,$$

that is, the sample mean of the observations. The second derivative, as required by the theorem, is given by

$$\frac{\partial^2 \ln p(\boldsymbol{y}; \theta)}{\partial \theta^2} = -\frac{N}{\sigma_\eta^2},$$

and hence,

$$I(\theta) = \frac{N}{\sigma_\eta^2}. \tag{3.37}$$

Eq. (3.36) is in the form of Eq. (3.31), with $g(\boldsymbol{y}) = \bar{y}$; thus, an efficient estimator can be obtained and the lower bound of the variance of any unbiased estimator, for our data model of Eq. (3.33), is

$$\sigma_{\hat{\theta}}^2 \geq \frac{\sigma_\eta^2}{N}. \tag{3.38}$$

We can easily verify that the corresponding estimator $\bar{y}$ is indeed an unbiased one under the adopted model of Eq. (3.33),

$$\mathbb{E}[\bar{y}] = \frac{1}{N}\sum_{n=1}^{N}\mathbb{E}[y_n] = \frac{1}{N}\sum_{n=1}^{N}\mathbb{E}[\theta + \eta_n] = \theta.$$

Moreover, the previous formula, combined with Eq. (3.36), also establishes the regularity condition, as required by the Cramér–Rao theorem.

The bound in Eq. (3.38) is a very natural result. The Cramér–Rao lower bound depends on the variance of the noise source. The higher this is, and therefore the higher the uncertainty of each observation with respect to the value of the true parameter is, the higher the minimum variance of an estimator is expected to be. On the other hand, as the number of observations increases and more "information" is disclosed to us, the uncertainty decreases and we expect the variance of our estimator to decrease.

Having obtained the lower bound for our task, let us return our attention to the LS estimator for the specific regression model of Eq. (3.33). This results from Eq. (3.13) by setting $x_n = 1$, and a simple inspection shows that the LS estimate is nothing but the sample mean, $\bar{y}$, of the observations. Furthermore, the variance of the corresponding estimator is given by

$$\begin{aligned}
\sigma_{\bar{y}}^2 &= \mathbb{E}\left[(\bar{y}-\theta)^2\right] = \mathbb{E}\left[\frac{1}{N^2}\left(\sum_{n=1}^{N}(y_n-\theta)\right)^2\right] \\
&= \frac{1}{N^2}\mathbb{E}\left[\left(\sum_{n=1}^{N}\eta_n\right)^2\right] = \frac{1}{N^2}\mathbb{E}\left[\sum_{i=1}^{N}\eta_i\sum_{j=1}^{N}\eta_j\right] \\
&= \frac{1}{N^2}\sum_{i=1}^{N}\sum_{j=1}^{N}\mathbb{E}[\eta_i\eta_j] = \frac{\sigma_\eta^2}{N},
\end{aligned}$$

which coincides with our previous finding via the use of the Cramér–Rao theorem. In other words, for this particular task and having assumed that the noise is white Gaussian, the LS estimator $\bar{y}$ is an MVU estimator and it attains the Cramér–Rao bound. However, if the input is not fixed, but it also varies from experiment to experiment and the training data become (y_n, x_n), then the LS estimator attains the Cramér–Rao bound only asymptotically, for large values of N (Problem 3.8). Moreover, it has to be pointed out that if the assumptions for the noise being Gaussian *and* white are not valid, then the LS estimator is not efficient anymore.

It turns out that this result, which has been obtained for the real axis case, is also true for the general regression model given in Eq. (3.10) (Problem 3.9). We will return to the properties of the LS estimator in more detail in Chapter 6.

Remarks 3.2.

- The Cramér–Rao bound is not the only one that is available in the literature. For example, the Bhattacharyya bound makes use of higher-order derivatives of the PDF. It turns out that in cases where an efficient estimator does not exist, the Bhattacharyya bound is tighter compared to the Cramér–Rao one with respect to the variance of the MVU estimator [27]. Other bounds also exist [21]; however, the Cramér–Rao bound is the easiest to determine.

3.7 SUFFICIENT STATISTIC

If an efficient estimator does not exist, this does not necessarily mean that the MVU estimator cannot be determined. It may exist, but it will not be an efficient one, in the sense that it does not satisfy the Cramér–Rao bound. In such cases, the notion of *sufficient statistic* and the Rao–Blackwell theorem come into the picture.[6] Note that such techniques are beyond the focus of this book and they are mentioned here in order to provide a more complete picture of the topic. In our context, concerning the needs of the book, we will refer to and use the sufficient statistic notion when dealing with the exponential family of distributions in Chapter 12. The section can be bypassed in a first reading.

The notion of sufficient statistic is due to Sir Ronald Aylmer Fisher (1890–1962). Fisher was an English statistician and biologist who made a number of fundamental contributions that laid out many of the foundations of modern statistics. Besides statistics, he made important contributions in genetics.

In short, given a random vector **x**, whose distribution depends on a parameter θ, a sufficient statistic for the unknown parameter is a function,

$$T(\mathcal{X}) := T(\boldsymbol{x}_1, \boldsymbol{x}_2, \ldots, \boldsymbol{x}_N),$$

of the respective observations, which contains *all* information about θ. From a mathematical point of view, a statistic $T(\mathcal{X})$ is said to be sufficient for the parameter θ if the conditional joint PDF

$$p\left(\mathcal{X}|T(\mathcal{X});\theta\right)$$

does not depend on θ. In such a case, it becomes apparent that $T(\mathcal{X})$ must provide *all* information about θ, which is contained in the set $\mathcal{X}$. Once $T(\mathcal{X})$ is known, $\mathcal{X}$ is no longer needed, because no further information can be extracted from it; this justifies the name of "sufficient statistic." The concept of sufficient statistic is also generalized to parameter vectors $\boldsymbol{\theta}$. In such a case, the sufficient statistic may be a *set* of functions, called a *jointly sufficient statistic*. Typically, there are as many functions as there are parameters; in a slight abuse of notation, we will still write $T(\mathcal{X})$ to denote this set (vector of) functions.

A very important theorem, which facilitates the search for a sufficient statistic in practice, is the following [27].

Theorem 3.2 (Factorization theorem). *A statistic $T(\mathcal{X})$ is sufficient if and only if the respective joint PDF can be factored as*

$$p(\mathcal{X};\boldsymbol{\theta}) = h(\mathcal{X})g\left(T(\mathcal{X}),\boldsymbol{\theta}\right).$$

That is, the joint PDF is factored into two parts: one part that depends only on the statistic and the parameters and a second part that is independent of the parameters. The theorem is also known as the Fisher–Neyman factorization theorem.

[6] It must be pointed out that the use of the sufficient statistic in statistics extends much beyond the search for MVU estimators.

Once a sufficient statistic has been found and under certain conditions related to the statistic, the Rao–Blackwell theorem determines the MVU estimator by taking the expectation conditioned on $T(\mathcal{X})$. A by-product of this theorem is that if an unbiased estimator is expressed *solely* in terms of the sufficient statistic, then it is necessarily the unique MVU estimator [23]. The interested reader can obtain more on these issues from [20,21,27].

Example 3.3. Let x be a Gaussian random variable, $\mathcal{N}(\mu, \sigma^2)$, and let the set of i.i.d. observations be $\mathcal{X} = \{x_1, x_2, \ldots, x_N\}$. Assume the mean value, μ, to be the unknown parameter. Show that

$$S_\mu = \frac{1}{N}\sum_{n=1}^{N} x_n$$

is a sufficient statistic for the parameter μ.

The joint PDF is given by

$$p(\mathcal{X}; \mu) = \frac{1}{(2\pi\sigma^2)^{\frac{N}{2}}} \exp\left(-\frac{1}{2\sigma^2}\sum_{n=1}^{N}(x_n - \mu)^2\right).$$

Plugging the obvious identity

$$\sum_{n=1}^{N}(x_n - \mu)^2 = \sum_{n=1}^{N}(x_n - S_\mu)^2 + N(S_\mu - \mu)^2$$

into the joint PDF, we obtain

$$p(\mathcal{X}; \mu) = \frac{1}{(2\pi\sigma^2)^{\frac{N}{2}}} \exp\left(-\frac{1}{2\sigma^2}\sum_{n=1}^{N}(x_n - S_\mu)^2\right) \exp\left(-\frac{N}{2\sigma^2}(S_\mu - \mu)^2\right),$$

which, according to the factorization theorem, proves the claim.

In a similar way, one can prove (Problem 3.10) that if the unknown parameter is the variance σ^2, then $\bar{S}_{\sigma^2} := \frac{1}{N}\sum_{n=1}^{N}(x_n - \mu)^2$ is a sufficient statistic, and if both μ and σ^2 are unknown, then a sufficient statistic is the set (S_μ, S_{σ^2}), where

$$S_{\sigma^2} = \frac{1}{N}\sum_{n=1}^{N}(x_n - S_\mu)^2.$$

In other words, in this case, all information concerning the unknown set of parameters that can be possibly extracted from the available N observations can be fully recovered by considering only the sum of the observations and the sum of their squares.

3.8 REGULARIZATION

We have already seen that the LS estimator is an MVU estimator, under the assumptions of linearity of the regression model and in the presence of a Gaussian white noise source. We also know that one can improve the performance by shrinking the norm of the MVU estimator. There are various ways to achieve this goal, and they will be discussed later on in this book. In this section, we focus on one possibility. Moreover, we will see that trying to keep the norm of the solution small serves important needs in the context of machine learning.

Regularization is a mathematical tool to impose a priori information on the structure of the solution, which comes as the outcome of an optimization task. Regularization was first suggested by the great Russian mathematician Andrey Nikolayevich Tychonoff (sometimes spelled Tikhonov) for the solution of integral equations. Sometimes it is also referred as Tychonoff–Phillips regularization, to honor Thomas Phillips as well, who developed the method independently [29,39].

In the context of our task and in order to shrink the norm of the parameter vector estimate, we can reformulate the sum of squared errors minimization task, given in Eq. (3.12), as

$$\text{minimize} \quad J(\boldsymbol{\theta}) = \sum_{n=1}^{N} \left(y_n - \boldsymbol{\theta}^T \boldsymbol{x}_n \right)^2, \tag{3.39}$$

$$\text{subject to} \quad \|\boldsymbol{\theta}\|^2 \leq \rho, \tag{3.40}$$

where $\|\cdot\|$ stands for the Euclidean norm of a vector. In this way, we do not allow the LS criterion to be completely "free" to reach a solution, but we *limit* the space in which we search for it. Obviously, using different values of ρ, we can achieve different levels of shrinkage. As we have already discussed, the optimal value of ρ cannot be analytically obtained, and one has to experiment in order to select an estimator that results in a good performance. For the squared error loss function and the constraint used before, the optimization task can equivalently be written as [6,8]

$$\boxed{\text{minimize} \quad L(\boldsymbol{\theta}, \lambda) = \sum_{n=1}^{N} \left(y_n - \boldsymbol{\theta}^T \boldsymbol{x}_n \right)^2 + \lambda \|\boldsymbol{\theta}\|^2 : \quad \text{ridge regression.}} \tag{3.41}$$

It turns out that for specific choices of $\lambda \geq 0$ and ρ, the two tasks are equivalent. Note that this new cost function, $L(\boldsymbol{\theta}, \lambda)$, involves one term that measures the model *misfit* and a second one that quantifies the *size* of the norm of the parameter vector. It is straightforward to see that taking the gradient of L in Eq. (3.41) with respect to $\boldsymbol{\theta}$ and equating to zero, we obtain the *regularized version* of the LS solution for the linear regression task of Eq. (3.13),

$$\left(\sum_{n=1}^{N} \boldsymbol{x}_n \boldsymbol{x}_n^T + \lambda I \right) \hat{\boldsymbol{\theta}} = \sum_{n=1}^{N} y_n \boldsymbol{x}_n, \tag{3.42}$$

where I is the identity matrix of appropriate dimensions. The presence of λ biases the new solution away from that which would have been obtained from the unregularized LS formulation. The task is also known as *ridge regression*. Ridge regression attempts to reduce the norm of the estimated vector and *at the same time* tries to keep the sum of squared errors small; in order to achieve this *combined* goal, the vector components, θ_i, are modified in such a way so that the contribution in the misfit measuring term, from the less informative directions in the input space, is minimized. In other words, those of the components that are associated with less informative directions will be pushed to smaller values to keep the norm small and at the same time have minimum influence on the misfit measuring term. We will return to this in more detail in Chapter 6. Ridge regression was first introduced in [18].

It has to be emphasized that in practice, the bias parameter θ_0 is left out from the norm in the regularization term; penalization of the bias would make the procedure dependent on the origin chosen for y. Indeed, it is easily checked that adding a constant term to each one of the output values, y_n, in the cost function would not result in just a shift of the predictions by the same constant, if the bias term is included in the norm. Hence, usually, ridge regression is formulated as

$$\text{minimize} \quad L(\boldsymbol{\theta}, \lambda) = \sum_{n=1}^{N} \left(y_n - \theta_0 - \sum_{i=1}^{l} \theta_i x_{ni} \right)^2 + \lambda \sum_{i=1}^{l} |\theta_i|^2. \tag{3.43}$$

It turns out (Problem 3.11) that minimizing Eq. (3.43) with respect to θ_i, $i = 0, 1, \ldots, l$, is equivalent to minimizing Eq. (3.41) using *centered* data and neglecting the intercept. That is, one solves the task

$$\text{minimize} \quad L(\boldsymbol{\theta}, \lambda) = \sum_{n=1}^{N} \left((y_n - \bar{y}) - \sum_{i=1}^{l} \theta_i (x_{ni} - \bar{x}_i) \right)^2 + \lambda \sum_{i=1}^{l} |\theta_i|^2, \tag{3.44}$$

and the estimate of θ_0 in Eq. (3.43) is given in terms of the obtained estimates $\hat{\theta}_i$, i.e.,

$$\hat{\theta}_0 = \bar{y} - \sum_{i=1}^{l} \hat{\theta}_i \bar{x}_i,$$

where

$$\bar{y} = \frac{1}{N} \sum_{n=1}^{N} y_n \quad \text{and} \quad \bar{x}_i = \frac{1}{N} \sum_{n=1}^{N} x_{ni}, \ i = 1, 2, \ldots, l.$$

In other words, $\hat{\theta}_0$ compensates for the differences between the sample means of the output and input variables. Note that similar arguments hold true if the Euclidean norm, used in Eq. (3.42) as a regularizer, is replaced by other norms, such as ℓ_1 or in general ℓ_p, $p > 1$, norms (Chapter 9).

From a different viewpoint, reducing the norm can be considered as an attempt to "simplify" the structure of the estimator, because a smaller number of components of the regressor now have an important say. This viewpoint becomes more clear if one considers nonlinear models, as discussed in Section 3.2. In this case, the existence of the norm of the respective parameter vector in Eq. (3.41) forces the model to get rid of the less important terms in the nonlinear expansion, $\sum_{k=1}^{K} \theta_k \phi_k(\boldsymbol{x})$, and effectively pushes K to lower values.

Although in the current context the complexity issue emerges in a rather disguised form, one can make it a major player in the game by choosing to use different functions and norms for the regularization term; as we will see next, there are many reasons that justify such choices.

INVERSE PROBLEMS: ILL-CONDITIONING AND OVERFITTING

Most tasks in machine learning belong to the so-called *inverse problems*. The latter term encompasses all the problems where one has to infer/predict/estimate the values of a model based on a set of available output/input observations (training data). In a less mathematical terminology, in inverse problems one has to unravel unknown causes from known effects, or in other words, to reverse the cause–effect relations. Inverse problems are typically *ill-posed*, as opposed to the *well-posed* ones. Well-posed problems are characterized by (a) the existence of a solution, (b) the uniqueness of the solution, and (c) the stability of the solution. The latter condition is usually violated in machine learning problems. This means that the obtained solution may be very sensitive to changes of the training set. *Ill-conditioning* is another term used to describe this sensitivity. The reason for this behavior is that the model used to describe the data can be complex, in the sense that the number of the unknown free parameters is large with respect to the number of data points. The "face" with which this problem manifests itself in machine learning is known as *overfitting*. This means that during training, the estimated parameters of the unknown model learn too much about the idiosyncrasies of the specific training data set, and the model performs badly when it deals with data sets other than that used for training. As a matter of fact, the MSE criterion discussed in Section 3.5 attempts to quantify exactly this data dependence of the task, that is, the mean deviation of the obtained estimates from the true value, by changing the training sets.

When the number of training samples is small with respect to the number of unknown parameters, the available information is not enough to "reveal" a sufficiently good model that fits the data, and it can be misleading due to the presence of the noise and possible outliers. Regularization is an elegant and efficient tool to cope with the complexity of the model, that is, to make it less complex, more smooth. There are different ways to achieve this. One way is by constraining the norm of the unknown vector, as ridge regression does. When dealing with more complex, compared to linear, models, one can use constraints on the smoothness of the involved nonlinear function, for example by involving derivatives of the model function in the regularization term. Also, regularization can help when the adopted model and the number of training points are such that no solution is possible. For example, in the LS linear regression task of Eq. (3.13), if the number N of training points is less than the dimension of the regressors $\boldsymbol{x}_n$, then the $(l+1) \times (l+1)$ matrix, $\bar{\Sigma} = \sum_n \boldsymbol{x}_n \boldsymbol{x}_n^T$, is not invertible. Indeed, each term in the summation is the outer product of a vector with itself and hence it is a matrix of rank one. Thus, as we know from linear algebra, we need at least $l+1$ linearly independent terms of such matrices to guarantee that the sum is of full rank, and hence invertible. However, in ridge regression, this can be bypassed, because the presence of λI in Eq. (3.42) guarantees that the left-hand matrix is invertible. Furthermore, the presence of λI can also help when $\bar{\Sigma}$ is invertible but it is ill-conditioned. Usually in such cases, the resulting LS solution has a very large norm and, thus, it is meaningless. Regularization helps to replace the original ill-conditioned problem with a "nearby" one, which is well-conditioned and whose solution approximates the target one.

Another example where regularization can help to obtain a solution, or even a unique solution to an otherwise unsolvable problem, is when the model's order is large compared to the number of data, although we know that it is sparse. That is, only a very small percentage of the model's parameters are nonzero. For such a task, a standard LS linear regression approach has no solution. However, regularizing the sum of squared errors cost function using the ℓ_1 norm of the parameter vector can lead to a unique solution; the ℓ_1 norm of a vector comprises the sum of the absolute values of its components. This problem will be considered in Chapters 9 and 10.

Regularization is closely related to the task of using priors in Bayesian learning, as we will discuss in Section 3.11. Finally, note that regularization is not a panacea for facing the problem of overfitting. As a matter of fact, selecting the right set of functions $\mathcal{F}$ in Eq. (3.3) is the first crucial step. The issue of the complexity of an estimator and the consequences on its "average" performance, as measured over all possible data sets, is discussed in Section 3.9.

Example 3.4. The goal of this example is to demonstrate that the estimator obtained via ridge regression can score a better MSE performance compared to the unconstrained LS solution. Let us consider, once again, the scalar model exposed in Example 3.2, and assume that the data are generated according to

$$y_n = \theta_o + \eta_n, \quad n = 1, 2, \ldots, N,$$

where, for simplicity, we have assumed that the regressors $x_n \equiv 1$ and η_n, $n = 1, 2, \ldots, N$, are i.i.d. zero mean Gaussian noise samples of variance σ_η^2.

We have already seen in Example 3.2 that the solution to the LS parameter estimation task is the sample mean $\hat{\theta}_{\text{MVU}} = \frac{1}{N}\sum_{n=1}^N y_n$. We have also shown that this solution scores an MSE of σ_η^2/N and under the Gaussian assumption for the noise it achieves the Cramér–Rao bound. The question now is whether a biased estimator, $\hat{\theta}_b$, which corresponds to the solution of the associated ridge regression task, can achieve an MSE lower than $\text{MSE}(\hat{\theta}_{\text{MVU}})$.

It can be readily verified that Eq. (3.42), adapted to the needs of the current linear regression scenario, results in

$$\hat{\theta}_b(\lambda) = \frac{1}{N+\lambda}\sum_{n=1}^{N} y_n = \frac{N}{N+\lambda}\hat{\theta}_{\text{MVU}},$$

where we have explicitly expressed the dependence of the estimate $\hat{\theta}_b$ on the regularization parameter λ. Note that for the associated estimator we have $\mathbb{E}[\hat{\theta}_b(\lambda)] = \frac{N}{N+\lambda}\theta_o$.

A simple inspection of the previous relation takes us back to the discussion related to Eq. (3.24). Indeed, by following a sequence of steps similar to those in Section 3.5.1, one can verify (see Problem 3.12) that the minimum value of $\text{MSE}(\hat{\theta}_b)$ is

$$\text{MSE}(\hat{\theta}_b(\lambda_*)) = \frac{\frac{\sigma_\eta^2}{N}}{1 + \frac{\sigma_\eta^2}{N\theta_o^2}} < \frac{\sigma_\eta^2}{N} = \text{MSE}(\hat{\theta}_{\text{MVU}}), \tag{3.45}$$

attained at $\lambda_* = \sigma_\eta^2/\theta_o^2$. The answer to the question whether the ridge regression estimate offers an improvement to the MSE performance is therefore positive in the current context. As a matter of fact, there *always* exists a $\lambda > 0$ such that the ridge regression estimate, which solves the general task of Eq. (3.41), achieves an MSE lower than the one corresponding to the MVU estimate [5, Section 8.4].

We will now demonstrate the previous theoretical findings via some simulations. To this end, the true value of the model was chosen to be $\theta_o = 10^{-2}$. The noise was Gaussian of zero mean value and variance $\sigma_\eta^2 = 0.1$. The number of i.i.d. generated samples was $N = 100$. Note that this is quite large, compared to the single parameter we have to estimate. The previous values imply that $\theta_o^2 < \sigma_\eta^2/N$.

Table 3.1 Attained values of the MSE for ridge regression and different values of the regularization parameter.

λ	$\mathrm{MSE}(\hat{\theta}_b(\lambda))$
0.1	9.99082×10^{-4}
1.0	9.79790×10^{-4}
100.0	2.74811×10^{-4}
$\lambda_* = 10^3$	9.09671×10^{-5}

The attained MSE for the unconstrained LS estimate was $\mathrm{MSE}(\hat{\theta}_{MVU}) = 1.00108 \times 10^{-3}$.

Then it can be shown that for any value of $\lambda > 0$, we can obtain a value for $\mathrm{MSE}(\hat{\theta}_b(\lambda))$ which is smaller than that of $\mathrm{MSE}(\hat{\theta}_{\mathrm{MVU}})$ (see Problem 3.12). This is verified by the values shown in Table 3.1. To compute the MSE values in the table, the expectation operation in the definition in Eq. (3.21) was approximated by the respective sample mean. To this end, the experiment was repeated L times and the MSE was computed as

$$\mathrm{MSE} \approx \frac{1}{L}\sum_{i=1}^{L}(\hat{\theta}_i - \theta_o)^2.$$

To get accurate results, we perform $L = 10^6$ trials. The corresponding MSE value for the unconstrained LS task is equal to $\mathrm{MSE}(\hat{\theta}_{\mathrm{MVU}}) = 1.00108 \times 10^{-3}$. Observe that substantial improvements can be attained when using regularization, in spite of the relatively large number of training data.

However, the percentage of performance improvement depends heavily on the specific values that define the model, as Eq. (3.45) suggests. For example, if $\theta_o = 0.1$, the obtained values from the experiments were $\mathrm{MSE}(\hat{\theta}_{\mathrm{MVU}}) = 1.00061 \times 10^{-3}$ and $\mathrm{MSE}(\hat{\theta}_b(\lambda_*)) = 9.99578 \times 10^{-4}$. The theoretical ones, as computed from Eq. (3.45), are 1×10^{-3} and 9.99001×10^{-4}, respectively. The improvement obtained by using the ridge regression is now rather insignificant.

3.9 THE BIAS–VARIANCE DILEMMA

This section goes one step beyond Section 3.5. There, the MSE criterion was used to quantify the performance with respect to the unknown parameter. Such a setting was useful to help us understand some trends and also better digest the notions of "biased" versus "unbiased" estimation. Here, although the criterion will be the same, it will be used in a more general setting. To this end, we shift our interest from the unknown parameter to the dependent variable and our goal becomes obtaining an estimator of the value y, given a measurement of the regressor vector, $\mathbf{x} = \boldsymbol{x}$. Let us first consider the more general form of regression,

$$\mathrm{y} = g(\mathbf{x}) + \eta, \tag{3.46}$$

where, for simplicity and without loss of generality, once more we have assumed that the dependent variable takes values in the real axis, $\mathrm{y} \in \mathbb{R}$. The first question to be addressed is whether there exists an estimator that guarantees minimum MSE performance.

3.9.1 MEAN-SQUARE ERROR ESTIMATION

Our goal is estimating an unknown (nonlinear in general) function $g(\boldsymbol{x})$. This problem can be cast in the context of the more general estimation task setting.

Consider the *jointly distributed* random variables $\mathrm{y}, \mathbf{x}$. Then, given any observation, $\mathbf{x} = \boldsymbol{x} \in \mathbb{R}^l$, the task is to obtain a function $\hat{y} := \hat{g}(\boldsymbol{x}) \in \mathbb{R}$, such that

$$\hat{g}(\boldsymbol{x}) = \arg\min_{f:\mathbb{R}^l \to \mathbb{R}} \mathbb{E}\left[(\mathrm{y} - f(\boldsymbol{x}))^2\right], \tag{3.47}$$

where the expectation is taken with respect to the conditional probability of y given the value of $\boldsymbol{x}$, or in other words, $p(y|\boldsymbol{x})$.

We will show that the optimal estimator is the mean value of y, or

$$\boxed{\hat{g}(\boldsymbol{x}) = \mathbb{E}\left[\mathrm{y}|\boldsymbol{x}\right] := \int_{-\infty}^{+\infty} y p(y|\boldsymbol{x})\, dy : \quad \text{optimal MSE estimator.}} \tag{3.48}$$

Proof. We have

$$\begin{aligned}
\mathbb{E}\left[(\mathrm{y} - f(\boldsymbol{x}))^2\right] &= \mathbb{E}\left[(\mathrm{y} - \mathbb{E}[\mathrm{y}|\boldsymbol{x}] + \mathbb{E}[\mathrm{y}|\boldsymbol{x}] - f(\boldsymbol{x}))^2\right] \\
&= \mathbb{E}\left[(\mathrm{y} - \mathbb{E}[\mathrm{y}|\boldsymbol{x}])^2\right] + \mathbb{E}\left[(\mathbb{E}[\mathrm{y}|\boldsymbol{x}] - f(\boldsymbol{x}))^2\right] \\
&\quad + 2\mathbb{E}\left[(\mathrm{y} - \mathbb{E}[\mathrm{y}|\boldsymbol{x}])(\mathbb{E}[\mathrm{y}|\boldsymbol{x}] - f(\boldsymbol{x}))\right],
\end{aligned}$$

where the dependence of the expectation on $\boldsymbol{x}$ has been omitted for notational convenience. It is readily seen that the last (product) term on the right-hand side is zero, hence, we are left with the following:

$$\mathbb{E}\left[(\mathrm{y} - f(\boldsymbol{x}))^2\right] = \mathbb{E}\left[(\mathrm{y} - \mathbb{E}[\mathrm{y}|\boldsymbol{x}])^2\right] + (\mathbb{E}[\mathrm{y}|\boldsymbol{x}] - f(\boldsymbol{x}))^2, \tag{3.49}$$

where we have taken into account that, for fixed $\boldsymbol{x}$, the terms $\mathbb{E}[y|\boldsymbol{x}]$ and $f(\boldsymbol{x})$ are not random variables. From Eq. (3.49) we finally obtain our claim,

$$\mathbb{E}\left[(\mathrm{y} - f(\boldsymbol{x}))^2\right] \geq \mathbb{E}\left[(\mathrm{y} - \mathbb{E}[\mathrm{y}|\boldsymbol{x}])^2\right]. \tag{3.50}$$

□

Note that this is a very elegant result. The optimal estimate, in the MSE sense, of the value of the unknown function at a point $\boldsymbol{x}$ is given as $\hat{g}(\boldsymbol{x}) = \mathbb{E}[\mathrm{y}|\boldsymbol{x}]$. Sometimes, the latter is also known as the *regression of* y *conditioned on* $\mathbf{x} = \boldsymbol{x}$. This is, in general, a nonlinear function. It can be shown that if $(\mathrm{y}, \mathbf{x})$ take values in $\mathbb{R} \times \mathbb{R}^l$ and are jointly Gaussian, then the optimal MSE estimator $\mathbb{E}[\mathrm{y}|\boldsymbol{x}]$ is a linear (affine) function of $\boldsymbol{x}$.

The previous results generalize to the case where $\mathbf{y}$ is a random vector that takes values in $\mathbb{R}^k$. The optimal MSE estimate, given the values of $\mathbf{x} = \boldsymbol{x}$, is equal to

$$\hat{g}(\boldsymbol{x}) = \mathbb{E}[\mathbf{y}|\boldsymbol{x}],$$

where now $\hat{g}(\boldsymbol{x}) \in \mathbb{R}^k$ (Problem 3.15). Moreover, if $(\mathbf{y}, \mathbf{x})$ are jointly Gaussian random vectors, the MSE optimal estimator is also an affine function of $\boldsymbol{x}$ (Problem 3.16).

The findings of this subsection can be fully justified by physical reasoning. Assume, for simplicity, that the noise source in Eq. (3.46) is of zero mean. Then, for a fixed value $\mathbf{x} = \boldsymbol{x}$, we have $\mathbb{E}[\mathrm{y}|\boldsymbol{x}] = g(\boldsymbol{x})$ and the respective MSE is equal to

$$\text{MSE} = \mathbb{E}\left[(\mathrm{y} - \mathbb{E}[\mathrm{y}|\boldsymbol{x}])^2\right] = \sigma_\eta^2. \tag{3.51}$$

No other function of $\boldsymbol{x}$ can do better, because the optimal one achieves an MSE equal to the noise variance, which is irreducible; it represents the intrinsic uncertainty of the system. As Eq. (3.49) suggests, any other function $f(\boldsymbol{x})$ will result in an MSE larger by the factor $(\mathbb{E}[\mathrm{y}|\boldsymbol{x}] - f(\boldsymbol{x}))^2$, which corresponds to the deviation from the optimal one.

3.9.2 BIAS–VARIANCE TRADEOFF

We have just seen that the optimal estimate, in the MSE sense, of the dependent variable in a regression task is given by the conditional expectation $\mathbb{E}[\mathrm{y}|\boldsymbol{x}]$. In practice, any estimator is computed based on a specific training data set, say, $\mathcal{D}$. Let us make the dependence on the training set explicit and express the estimate as a function of $\boldsymbol{x}$ parameterized on $\mathcal{D}$, or $f(\boldsymbol{x};\mathcal{D})$. A reasonable measure to quantify the performance of an estimator is its mean-square deviation from the optimal one, expressed by $\mathbb{E}_{\mathcal{D}}[(f(\boldsymbol{x};\mathcal{D}) - \mathbb{E}[\mathrm{y}|\boldsymbol{x}])^2]$, where the mean is taken with respect to all possible training sets, because each one results in a different estimate. Following a similar path as for Eq. (3.22), we obtain

$$\mathbb{E}_{\mathcal{D}}\left[(f(\boldsymbol{x};\mathcal{D}) - \mathbb{E}[\mathrm{y}|\boldsymbol{x}])^2\right] = \underbrace{\mathbb{E}_{\mathcal{D}}\left[(f(\boldsymbol{x};\mathcal{D}) - \mathbb{E}_{\mathcal{D}}[f(\boldsymbol{x};\mathcal{D})])^2\right]}_{\text{Variance}} + \underbrace{\left(\mathbb{E}_{\mathcal{D}}[f(\boldsymbol{x};\mathcal{D})] - \mathbb{E}[\mathrm{y}|\boldsymbol{x}]\right)^2}_{\text{Bias}^2}. \tag{3.52}$$

As was the case for the MSE parameter estimation task when changing from one training set to another, the mean-square deviation from the optimal estimate comprises two terms. The first one is the variance of the estimator around its own mean value and the second one is the squared difference of the mean from the optimal estimate, that is, the bias. It turns out that one *cannot* make *both* terms small simultaneously. For a fixed number of training points N in the data sets $\mathcal{D}$, trying to minimize the variance term results in an increase of the bias term and vice versa. This is because in order to reduce the bias term, one has to increase the complexity (more free parameters) of the adopted estimator $f(\cdot;\mathcal{D})$. This, in turn, results in higher variance as we change the training sets. This is a manifestation of the overfitting issue that we have already discussed. The only way to reduce both terms simultaneously is to increase the number of the training data points N, and at the same time increase the complexity of the model *carefully*, so as to achieve the aforementioned goal. If one increases the number of training points and at the same time increases the model complexity excessively, the overall MSE may increase. This is known as the *bias–variance dilemma* or *tradeoff*. This is an issue that is omnipresent in any estimation task. Usually, we refer to it as *Occam's razor* rule.

Occam was a logician and a nominalist scholastic medieval philosopher who expressed the following law of parsimony: "Plurality must never be posited without necessity." The great physicist Paul Dirac expressed the same statement from an esthetics point of view, which underlies mathematical theories: "A theory with a mathematical beauty is more likely to be correct than an ugly one that fits the data." In our context of model selection, it is understood that one has to select the simplest model that can "explain" the data. Although this is not a scientifically proven result, it underlies the rationale behind a number of developed model selection techniques [1,32,33,40] and [37, Chapter 5], which trade off complexity with accuracy.

Let us now try to find the MSE, given $\boldsymbol{x}$, by considering all possible sets $\mathcal{D}$. To this end, note that the left-hand side of Eq. (3.52) is the mean with respect to $\mathcal{D}$ of the second term in Eq. (3.49), if $\mathcal{D}$ is brought explicitly in the notation. It is easy to see that, by reconsidering Eq. (3.49) and taking the expectation on both y and $\mathcal{D}$, given the value of $\mathbf{x} = \boldsymbol{x}$, the resulting MSE becomes (try it, following similar arguments as for Eq. (3.52))

$$\begin{aligned}\text{MSE}(\boldsymbol{x}) &= \mathbb{E}_{\mathrm{y}|\boldsymbol{x}}\,\mathbb{E}_{\mathcal{D}}\left[\left(\mathrm{y} - f(\boldsymbol{x};\mathcal{D})\right)^2\right] \\ &= \sigma_\eta^2 + \mathbb{E}_{\mathcal{D}}\left[\left(f(\boldsymbol{x};\mathcal{D}) - \mathbb{E}_{\mathcal{D}}\left[f(\boldsymbol{x};\mathcal{D})\right]\right)^2\right] \\ &\quad + \left(\mathbb{E}_{\mathcal{D}}\left[f(\boldsymbol{x};\mathcal{D})\right] - \mathbb{E}[\mathrm{y}|\boldsymbol{x}]\right)^2, \end{aligned} \tag{3.53}$$

where Eq. (3.51) has been used and the product rule, as stated in Chapter 2, has been exploited. In the sequel, one can take the mean over $\boldsymbol{x}$. In other words, this is the prediction MSE over all possible inputs, averaged over all possible training sets. The resulting MSE is also known as the *test* or *generalization error* and it is a measure of the performance of the respective adopted model. Note that the generalization error in Eq. (3.53) involves averaging over (theoretically) all possible training data sets of certain size N. In contrast, the so-called *training error* is computed over a single data set, the one used for the training, and this results in an overoptimistic estimate of the error. We will come back to this important issue in Section 3.13.

Example 3.5. Let us consider a simplistic, yet pedagogic, example that demonstrates this tradeoff between bias and variance. We are given a set of training points that are generated according to a regression model of the form

$$\mathrm{y} = g(\mathrm{x}) + \eta. \tag{3.54}$$

The graph of $g(x)$ is shown in Fig. 3.7. The function $g(x)$ is a fifth-order polynomial. Training sets are generated as follows. For each set, the x-axis is sampled at N equidistant points x_n, $n = 1, \dots, N$, in the interval $[-1, 1]$. Then, each training set $\mathcal{D}_i$ is created as

$$\mathcal{D}_i = \left\{(g(x_n) + \eta_{n,i}, x_n) : n = 1, 2, \dots, N\right\}, \quad i = 1, 2, \dots,$$

where $\eta_{n,i}$ denotes different noise samples, drawn i.i.d. from a white noise process. In other words, all training points have the same x-coordinate but different y-coordinates due to the different values of the noise. The gray dots in the (x, y)-plane in Fig. 3.7 correspond to one realization of a training set, for the case of $N = 10$. For comparison, the set of noiseless points, $(g(x_n), x_n)$, $n = 1, 2, \ldots, 10$, is shown as red dots on the graph of $g(x)$.

First, we are going to be very naive, and choose a fixed linear model to fit the data,

$$\hat{y} = f_1(x) = \theta_0 + \theta_1 x,$$

where the values θ_1 and θ_0 have been chosen arbitrarily, irrespective of the training data. The graph of this straight line is shown in Fig. 3.7. Because no training was involved and the model parameters are fixed, there is no variation as one changes the training sets and $\mathbb{E}_{\mathcal{D}}[f_1(x)] = f_1(x)$, with the variance term being equal to zero. On the other hand, the square of the bias, which is equal to $(f_1(x) - \mathbb{E}[\mathrm{y}|x])^2$, is expected to be large because the choice of the model was arbitrary, without paying attention to the training data.

In the sequel, we go to the other "extreme." We choose a relatively complex class of functions, such as a high-order (10th-order) polynomial f_2. The dependence of the resulting estimates on the respective set $\mathcal{D}$, which is used for training, is explicitly denoted as $f_2(\cdot; \mathcal{D})$. Note that for each one of the sets, the corresponding graph of the resulting optimal model is expected to go through the training points, i.e.,

$$f_2(x_n; \mathcal{D}_i) = g(x_n) + \eta_{n,i}, \quad n = 1, 2, \ldots, N.$$

This is shown in Fig. 3.7 for one curve. Note that, in general, this is always the case if the order of the polynomial (model) is large and the number of parameters to be estimated is larger than the number of

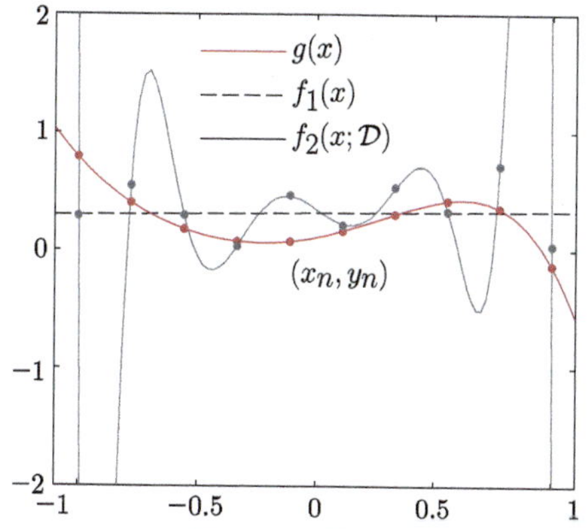

FIGURE 3.7

The observed data are denoted as gray dots. These are the result of adding noise to the red points, which lie on the red curve associated with the unknown $g(\cdot)$. Fitting the data by the fixed polynomial $f_1(x)$ results in high bias; observe that most of the data points lie outside the straight line. On the other hand, the variance of the "estimator" will be zero. In contrast, fitting a high-degree polynomial $f_2(x; \mathcal{D})$ results in low bias, because the corresponding curve goes through all the data points; however, the respective variance will be high.

training points. For the example shown in the figure, we are given 10 training points and 11 parameters have to be estimated, for a 10th-order polynomial (including the bias). One can obtain a perfect fit on the training points by solving a linear system comprising 10 equations ($N = 10$) with 11 unknowns. This is the reason that when the order of the model is large, a *zero* (or very small, in practice) error can be achieved on the training data set. We will come back to this issue in Section 3.13.

For this high-order polynomial fitting setup of the experiment, the following reasoning holds true. The bias term at each point x_n, $n = 1, 2, \ldots, N$, is zero, because

$$\mathbb{E}_{\mathcal{D}}[f_2(x_n; \mathcal{D})] = \mathbb{E}_{\mathcal{D}}[g(x_n) + \eta] = g(x_n) = \mathbb{E}_{\mathcal{D}}[\mathrm{y}|x_n].$$

On the other hand, the variance term at the points x_n, $n = 1, 2, \ldots, N$, is expected to be large, because

$$\mathbb{E}_{\mathcal{D}}\left[(f_2(x_n; \mathcal{D}) - g(x_n))^2\right] = \mathbb{E}_{\mathcal{D}}\left[(g(x_n) + \eta - g(x_n))^2\right] = \sigma_\eta^2.$$

Assuming that the functions f_2 and g are continuous and smooth enough and the points x_n are sampled densely enough to cover the interval of interest in the real axis, we expect similar behavior at all points $x \neq x_n$.

Example 3.6. This is a more realistic example that builds upon the previous one. The data are generated as before, via the regression model in Eq. (3.54) using the fifth-order polynomial for g. The number of training points is equal to $N = 10$ and 1000 training sets $\mathcal{D}_i$, $i = 1, 2 \ldots, 1000$, are generated. Two sets of experiments have been run.

The first one attempts to fit in the noisy data a high-order polynomial of degree equal to 10 (as in the previous example) and the second one adopts a second-order polynomial. For each one of the two setups, the experiment is repeated 1000 times, each time with a different data set $\mathcal{D}_i$. Figs. 3.8A and C show 10 (for visibility reasons) out of the 1000 resulting curves for the high- and low-order polynomials, respectively. The substantially higher variance for the case of the high-order polynomial is readily noticed. Figs. 3.8B and D show the corresponding curves which result from averaging over the 1000 performed experiments, together with the graph of the "unknown" (original g) function. The high-order polynomial results in an excellent fit of very low bias. The opposite is true for the case of the second-order polynomial.

Thus, in summary, for a fixed number of training points, the more complex the prediction model is (larger number of parameters), the larger the variance becomes, as we change from one training set to another. On the other hand, the more complex the model is, the smaller the bias gets; that is, the average model by training over different data sets gets closer to the optimal MSE one. The reader may find more information on the bias–variance dilemma task in [16].

3.10 MAXIMUM LIKELIHOOD METHOD

So far, we have approached the estimation problem as an optimization task around a set of training examples, without paying any attention to the underlying statistics that generates these points. We only used statistics in order to check under which conditions the estimators were efficient. However, the optimization step did not involve any statistical information. For the rest of the chapter, we are going to

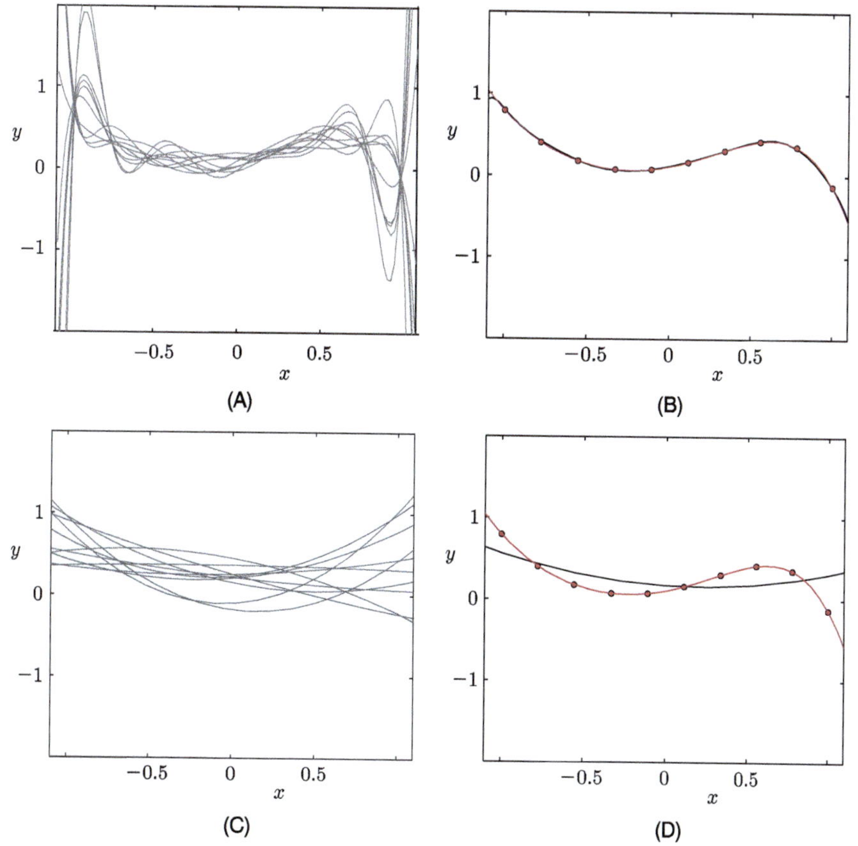

FIGURE 3.8

(A) Ten of the resulting curves from fitting a 10th-order polynomial. (B) The corresponding average over 1000 different experiments. The red curve represents the unknown polynomial. The dots indicate the points that give birth to the training data, as described in the text. (C), (D) The results from fitting a second-order polynomial. Observe the bias–variance tradeoff as a function of the complexity of the fitted model.

involve statistics more and more. In this section, the ML method is introduced. It is not an exaggeration to say that ML and LS are two of the major pillars on which parameter estimation is based and new methods are inspired from. The ML method was suggested by Sir Ronald Aylmer Fisher.

Once more, we will first formulate the method in a general setting, independent of the regression/classification tasks. We are given a set of, say, N observations, $\mathcal{X} = \{\boldsymbol{x}_1, \boldsymbol{x}_2, \ldots, \boldsymbol{x}_N\}$, drawn from a probability distribution. We assume that the joint PDF of these N observations is of a known parametric functional type, denoted as $p(\mathcal{X}; \boldsymbol{\theta})$, where the parameter vector $\boldsymbol{\theta} \in \mathbb{R}^K$ is unknown and the task is to estimate its value. This joint PDF is known as the *likelihood function* of $\boldsymbol{\theta}$ with respect to the given set of observations $\mathcal{X}$. According to the ML method, the estimate is provided by

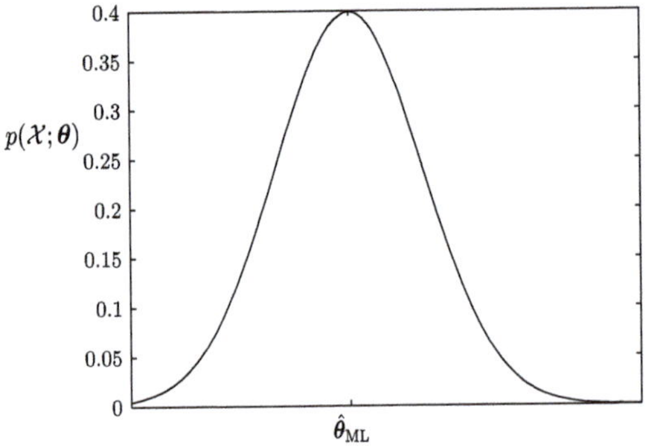

FIGURE 3.9

According to the maximum likelihood method, we assume that, given the set of observations, the estimate of the unknown parameter is the value that maximizes the corresponding likelihood function.

$$\boxed{\hat{\boldsymbol{\theta}}_{\text{ML}} := \arg\max_{\boldsymbol{\theta}\in\mathcal{A}\subset\mathbb{R}^K} p(\mathcal{X};\boldsymbol{\theta}): \quad \text{maximum likelihood estimate.}} \tag{3.55}$$

For simplicity, we will assume that the constraint set, $\mathcal{A}$, coincides with $\mathbb{R}^K$, i.e., $\mathcal{A} = \mathbb{R}^K$, and that the parameterized family $\{p(\mathcal{X};\boldsymbol{\theta}) : \boldsymbol{\theta} \in \mathbb{R}^K\}$ enjoys a unique minimizer with respect to the parameter $\boldsymbol{\theta}$. This is illustrated in Fig. 3.9. In other words, given the set of observations $\mathcal{X} = \{\boldsymbol{x}_1, \boldsymbol{x}_2, \ldots, \boldsymbol{x}_N\}$, one selects the unknown parameter vector so as to make this joint event the most likely one to happen.

Because the logarithmic function $\ln(\cdot)$ is monotone and increasing, one can instead search for the maximum of the *log-likelihood function*,

$$\left.\frac{\partial \ln p(\mathcal{X};\boldsymbol{\theta})}{\partial \boldsymbol{\theta}}\right|_{\boldsymbol{\theta}=\hat{\boldsymbol{\theta}}_{\text{ML}}} = \mathbf{0}. \tag{3.56}$$

Assuming the observations to be i.i.d., the ML estimator has some very attractive properties, namely:

- The ML estimator is asymptotically unbiased; that is, assuming that the model of the PDF which we have adopted is correct and there exists a true parameter $\boldsymbol{\theta}_o$, we have

$$\lim_{N\to\infty} \mathbb{E}[\hat{\boldsymbol{\theta}}_{\text{ML}}] = \boldsymbol{\theta}_o. \tag{3.57}$$

- The ML estimate is asymptotically *consistent* so that given any value of $\epsilon > 0$,

$$\lim_{N\to\infty} \text{Prob}\left\{\left|\hat{\boldsymbol{\theta}}_{\text{ML}} - \boldsymbol{\theta}_o\right| > \epsilon\right\} = 0, \tag{3.58}$$

  that is, for large values of N, we expect the ML estimate to be very close to the true value with high probability.
- The ML estimator is asymptotically efficient; that is, it achieves the Cramér–Rao lower bound.
- If there exists a sufficient statistic $T(\mathcal{X})$ for an unknown parameter, then only $T(\mathcal{X})$ suffices to express the respective ML estimate (Problem 3.20).

- Moreover, assuming that an efficient estimator does exist, this estimator is optimal in the ML sense (Problem 3.21).

Example 3.7. Let $\boldsymbol{x}_1, \boldsymbol{x}_2, \ldots, \boldsymbol{x}_N$ be the observation vectors stemming from a multivariate normal distribution with known covariance matrix and unknown mean (Chapter 2), that is,

$$p(\boldsymbol{x}_n; \boldsymbol{\mu}) = \frac{1}{(2\pi)^{l/2}|\Sigma|^{1/2}} \exp\left(-\frac{1}{2}(\boldsymbol{x}_n - \boldsymbol{\mu})^T \Sigma^{-1}(\boldsymbol{x}_n - \boldsymbol{\mu})\right).$$

Assume that the observations are mutually independent. Obtain the ML estimate of the unknown mean vector.

For the N statistically independent observations, the joint log-likelihood function is given by

$$L(\boldsymbol{\mu}) = \ln \prod_{n=1}^{N} p(\boldsymbol{x}_n; \boldsymbol{\mu}) = -\frac{N}{2} \ln\left((2\pi)^l |\Sigma|\right) - \frac{1}{2}\sum_{n=1}^{N}(\boldsymbol{x}_n - \boldsymbol{\mu})^T \Sigma^{-1}(\boldsymbol{x}_n - \boldsymbol{\mu}).$$

Taking the gradient with respect to $\boldsymbol{\mu}$, we obtain[7]

$$\frac{\partial L(\boldsymbol{\mu})}{\partial \boldsymbol{\mu}} := \begin{bmatrix} \frac{\partial L}{\partial \mu_1} \\ \frac{\partial L}{\partial \mu_2} \\ \vdots \\ \frac{\partial L}{\partial \mu_l} \end{bmatrix} = \sum_{n=1}^{N} \Sigma^{-1}(\boldsymbol{x}_n - \boldsymbol{\mu}),$$

and equating to $\mathbf{0}$ leads to

$$\hat{\boldsymbol{\mu}}_{\text{ML}} = \frac{1}{N}\sum_{n=1}^{N} \boldsymbol{x}_n.$$

In other words, for Gaussian distributed data, the ML estimate of the mean is the sample mean. Moreover, note that the ML estimate is expressed in terms of its sufficient statistic (see Section 3.7).

3.10.1 LINEAR REGRESSION: THE NONWHITE GAUSSIAN NOISE CASE

Consider the linear regression model

$$\mathrm{y} = \boldsymbol{\theta}^T \mathbf{x} + \eta.$$

We are given N training data points $(y_n, \boldsymbol{x}_n)$, $n = 1, 2, \ldots, N$. The corresponding (unobserved) noise samples η_n, $n = 1, \ldots, N$, are assumed to follow a jointly Gaussian distribution with zero mean and

---

[7] Recall from matrix algebra that $\frac{\partial(\boldsymbol{x}^T\boldsymbol{b})}{\partial \boldsymbol{x}} = \boldsymbol{b}$ and $\frac{\partial(\boldsymbol{x}^T A\boldsymbol{x})}{\partial \boldsymbol{x}} = 2A\boldsymbol{x}$ if A is symmetric (Appendix A).

covariance matrix equal to Σ_η. That is, the corresponding random vector of all the noise samples stacked together, $\eta = [\eta_1, \ldots, \eta_N]^T$, follows the multivariate Gaussian distribution

$$p(\eta) = \frac{1}{(2\pi)^{1/N}|\Sigma_\eta|^{1/2}} \exp\left(-\frac{1}{2}\eta^T \Sigma_\eta^{-1}\eta\right).$$

Our goal is to obtain the ML estimate of the parameters, $\boldsymbol{\theta}$.

Replacing η with $\boldsymbol{y} - X\boldsymbol{\theta}$, and taking the logarithm, the joint log-likelihood function of $\boldsymbol{\theta}$ with respect to the training set is given by

$$L(\boldsymbol{\theta}) = -\frac{N}{2}\ln(2\pi) - \frac{1}{2}\ln|\Sigma_\eta| - \frac{1}{2}(\boldsymbol{y} - X\boldsymbol{\theta})^T \Sigma_\eta^{-1}(\boldsymbol{y} - X\boldsymbol{\theta}), \tag{3.59}$$

where $\boldsymbol{y} := [y_1, y_2, \ldots, y_N]^T$, and $X := [\boldsymbol{x}_1, \ldots, \boldsymbol{x}_N]^T$ stands for the input matrix. Taking the gradient with respect to $\boldsymbol{\theta}$, we get

$$\frac{\partial L(\boldsymbol{\theta})}{\partial \boldsymbol{\theta}} = X^T \Sigma_\eta^{-1}(\boldsymbol{y} - X\boldsymbol{\theta}), \tag{3.60}$$

and equating to the zero vector, we obtain

$$\hat{\boldsymbol{\theta}}_{\mathrm{ML}} = \left(X^T \Sigma_\eta^{-1} X\right)^{-1} X^T \Sigma_\eta^{-1}\boldsymbol{y}. \tag{3.61}$$

Remarks 3.3.

- Compare Eq. (3.61) with the LS solution given in Eq. (3.17). They are different, unless the covariance matrix of the successive noise samples, Σ_η, is diagonal and of the form $\sigma_\eta^2 I$, that is, if the noise is Gaussian as well as white. In this case, the LS and ML solutions coincide. However, if the noise sequence is nonwhite, the two estimates differ. Moreover, it can be shown (Problem 3.9) that *in this case of colored Gaussian noise, the ML estimate is an efficient one and it attains the Cramér–Rao bound, even if N is finite.*

3.11 BAYESIAN INFERENCE

In our discussion so far, we have assumed that the parameters associated with the functional form of the adopted model are deterministic constants whose values are unknown to us. In this section, we will follow a different rationale. The unknown parameters will be treated as random variables. Hence, whenever our goal is to estimate their values, this is conceived as an effort to estimate the values of a *specific* realization that corresponds to the observed data. A more detailed discussion concerning the Bayesian inference rationale is provided in Chapter 12. As the name Bayesian suggests, the heart of the method beats around the celebrated Bayes theorem. Given two jointly distributed random vectors, say, $\mathbf{x}$, $\boldsymbol{\theta}$, the Bayes theorem states that

$$p(\boldsymbol{x}, \boldsymbol{\theta}) = p(\boldsymbol{x}|\boldsymbol{\theta})p(\boldsymbol{\theta}) = p(\boldsymbol{\theta}|\boldsymbol{x})p(\boldsymbol{x}). \tag{3.62}$$

David Bayes (1702–1761) was an English mathematician and a Presbyterian minister who first developed the basics of the theory. However, it was Pierre-Simon Laplace (1749–1827), the great French mathematician, who further developed and popularized it.

Assume that $\mathbf{x}$, $\boldsymbol{\theta}$ are two statistically dependent random vectors. Let $\mathcal{X} = \{\boldsymbol{x}_n \in \mathbb{R}^l, n = 1, 2, \ldots, N\}$ be the set of observations resulting from N successive experiments. Then the Bayes theorem gives

$$p(\boldsymbol{\theta}|\mathcal{X}) = \frac{p(\mathcal{X}|\boldsymbol{\theta})p(\boldsymbol{\theta})}{p(\mathcal{X})} = \frac{p(\mathcal{X}|\boldsymbol{\theta})p(\boldsymbol{\theta})}{\int p(\mathcal{X}|\boldsymbol{\theta})p(\boldsymbol{\theta})\,d\boldsymbol{\theta}}. \tag{3.63}$$

Obviously, if the observations are i.i.d., then we can write

$$p(\mathcal{X}|\boldsymbol{\theta}) = \prod_{n=1}^{N} p(\boldsymbol{x}_n|\boldsymbol{\theta}).$$

In the previous formulas, $p(\boldsymbol{\theta})$ is the a priori or prior PDF concerning the statistical distribution of $\boldsymbol{\theta}$, and $p(\boldsymbol{\theta}|\mathcal{X})$ is the conditional or a posteriori or posterior PDF, formed after the set of N observations has been obtained. The prior probability density, $p(\boldsymbol{\theta})$, can be considered as a constraint that *encapsulates our prior knowledge* about $\boldsymbol{\theta}$. Undoubtedly, our uncertainty about $\boldsymbol{\theta}$ is modified after the observations have been received, because more information is now disclosed to us. If the adopted assumptions about the underlying models are sensible, we expect the posterior PDF to be a more accurate one to describe the statistical nature of $\boldsymbol{\theta}$. We will refer to the process of approximating the PDF of a random quantity based on a set of training data as *inference*, to differentiate it from the process of estimation, which returns a single value for each parameter/variable. So, according to the inference approach, one attempts to draw conclusions about the nature of the randomness that underlies the variables of interest. This information can in turn be used to make predictions and to take decisions.

We will exploit Eq. (3.63) in two ways. The first refers to our familiar goal of obtaining an estimate of the parameter vector $\boldsymbol{\theta}$, which "controls" the model that describes the generation mechanism of our observations, $\boldsymbol{x}_1, \boldsymbol{x}_2, \ldots, \boldsymbol{x}_N$. Because $\mathbf{x}$ and $\boldsymbol{\theta}$ are two statistically dependent random vectors, we know from Section 3.9 that the MSE optimal estimate of the value of $\boldsymbol{\theta}$, given $\mathcal{X}$, is

$$\hat{\boldsymbol{\theta}} = \mathbb{E}[\boldsymbol{\theta}|\mathcal{X}] = \int \boldsymbol{\theta}\, p(\boldsymbol{\theta}|\mathcal{X})\,d\boldsymbol{\theta}. \tag{3.64}$$

Another direction along which one can exploit the Bayes theorem, in the context of statistical inference, is to obtain an estimate of the PDF of $\mathbf{x}$ given the observations $\mathcal{X}$. This can be done by *marginalizing* over a distribution, using the equation

$$p(\boldsymbol{x}|\mathcal{X}) = \int p(\boldsymbol{x}|\boldsymbol{\theta})p(\boldsymbol{\theta}|\mathcal{X})\,d\boldsymbol{\theta}, \tag{3.65}$$

where the conditional independence of $\mathbf{x}$ on $\mathcal{X}$, given the value $\boldsymbol{\theta} = \boldsymbol{\theta}$, expressed as $p(\boldsymbol{x}|\mathcal{X}, \boldsymbol{\theta}) = p(\boldsymbol{x}|\boldsymbol{\theta})$, has been used. Indeed, if the value $\boldsymbol{\theta}$ is given, then the conditional $p(\boldsymbol{x}|\boldsymbol{\theta})$ is fully defined and does not depend on $\mathcal{X}$. The dependence of $\mathbf{x}$ on $\mathcal{X}$ is through $\boldsymbol{\theta}$, if the latter is unknown. Eq. (3.65) provides an estimate of the unknown PDF, by exploiting the information that resides in the obtained observations and in the adopted functional dependence on the parameters $\boldsymbol{\theta}$. Note that, in contrast to what we did in the case of the ML method, where we used the observations to obtain an estimate of the parameter vector, here we assume the parameters to be random variables, provide our prior knowledge about $\boldsymbol{\theta}$ via $p(\boldsymbol{\theta})$, and integrate the joint PDF, $p(\boldsymbol{x}, \boldsymbol{\theta}|\mathcal{X})$, over $\boldsymbol{\theta}$.

Once $p(\boldsymbol{x}|\mathcal{X})$ is available, it can be used for prediction. Assuming that we have obtained the observations $\boldsymbol{x}_1, \ldots, \boldsymbol{x}_N$, our estimate about the next value, $\boldsymbol{x}_{N+1}$, can be determined via $p(\boldsymbol{x}_{N+1}|\mathcal{X})$. Obviously, the form of $p(\boldsymbol{x}|\mathcal{X})$ is, in general, changing as new observations are obtained, because each time an observation becomes available, part of our uncertainty about the underlying randomness is removed.

Example 3.8. Consider the simplified linear regression task of Eq. (3.33) and assume $x = 1$. As we have already said, this problem is that of estimating the value of a constant buried in noise. Our methodology will follow the Bayesian philosophy. Assume that the noise samples are i.i.d. drawn from a Gaussian PDF of zero mean and variance σ_η^2. However, we impose our a priori knowledge concerning the unknown θ via the prior distribution

$$p(\theta) = \mathcal{N}(\theta_0, \sigma_0^2). \tag{3.66}$$

That is, we assume that we know that the values of θ lie around θ_0, and σ_0^2 quantifies our degree of uncertainty about this prior knowledge. Our goals are first to obtain the posterior PDF, given the set of observations $\boldsymbol{y} = [y_1, \ldots, y_N]^T$, and then to obtain $\mathbb{E}[\theta|\boldsymbol{y}]$, according to Eqs. (3.63) and (3.64) after adapting them to our current notational needs. We have

$$\begin{aligned} p(\theta|\boldsymbol{y}) &= \frac{p(\boldsymbol{y}|\theta)p(\theta)}{p(\boldsymbol{y})} = \frac{1}{p(\boldsymbol{y})}\left(\prod_{n=1}^{N} p(y_n|\theta)\right)p(\theta) \\ &= \frac{1}{p(\boldsymbol{y})}\left(\prod_{n=1}^{N}\frac{1}{\sqrt{2\pi}\sigma_\eta}\exp\left(-\frac{(y_n-\theta)^2}{2\sigma_\eta^2}\right)\right) \\ &\quad \times \frac{1}{\sqrt{2\pi}\sigma_0}\exp\left(-\frac{(\theta-\theta_0)^2}{2\sigma_0^2}\right). \end{aligned} \tag{3.67}$$

After some algebraic manipulations of Eq. (3.67) (Problem 3.25), one ends up with the following:

$$p(\theta|\boldsymbol{y}) = \frac{1}{\sqrt{2\pi}\sigma_N}\exp\left(-\frac{(\theta-\bar{\theta}_N)^2}{2\sigma_N^2}\right), \tag{3.68}$$

where

$$\bar{\theta}_N = \frac{N\sigma_0^2\bar{y}_N + \sigma_\eta^2\theta_0}{N\sigma_0^2 + \sigma_\eta^2}, \tag{3.69}$$

with $\bar{y}_N = \frac{1}{N}\sum_{n=1}^{N} y_n$ being the sample mean of the observations and

$$\sigma_N^2 = \frac{\sigma_\eta^2 \sigma_0^2}{N\sigma_0^2 + \sigma_\eta^2}. \tag{3.70}$$

In other words, if the prior and conditional PDFs are Gaussians, then the posterior is also Gaussian. Moreover, the mean and the variance of the posterior are given by Eqs. (3.69) and (3.70), respectively.

Observe that as the number of observations increases, $\bar{\theta}_N$ tends to the sample mean of the observations; recall that the latter is the estimate that results from the ML method. Also, note that the variance keeps decreasing as the number of observations increases, which is in line with common sense, because more observations mean less uncertainty. Fig. 3.10 illustrates the previous discussion. Data samples, y_n, were generated using a Gaussian pseudorandom generator with the mean equal to $\theta = 1$ and variance equal to $\sigma_\eta^2 = 0.1$. So the true value of our constant is equal to 1. We used a Gaussian prior PDF with mean value equal to $\theta_0 = 2$ and variance $\sigma_0^2 = 6$. We observe that as N increases, the posterior PDF gets narrower and its mean tends to the true value of 1.

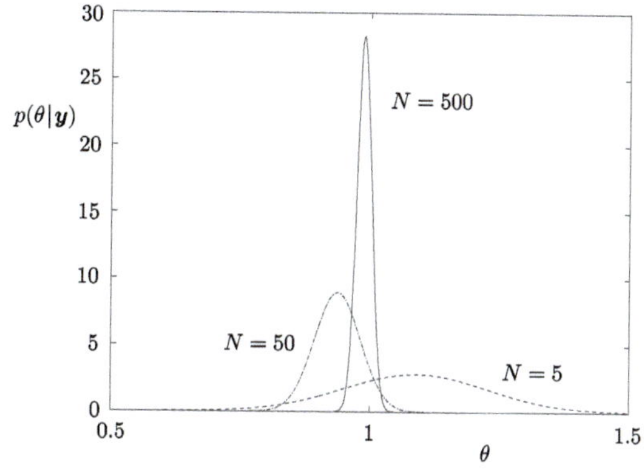

FIGURE 3.10

In the Bayesian inference approach, note that as the number of observations increases, our uncertainty about the true value of the unknown parameter is reduced and the mean of the posterior PDF tends to the true value and the variance tends to zero.

It should be pointed out that in this example case, both the ML and LS estimates become identical, or

$$\hat{\theta} = \frac{1}{N}\sum_{n=1}^{N} y_n = \bar{y}_N.$$

This will also be the case for the mean value in Eq. (3.69) if we set σ_0^2 very large, as might happen if we have no confidence in our initial estimate of θ_0 and we assign a very large value to σ_0^2. In effect, this is equivalent to using no prior information.

Let us now investigate what happens if our prior knowledge about θ_0 is "embedded" in the LS criterion in the form of a constraint. This can be done by modifying the constraint in Eq. (3.40) such that

$$(\theta - \theta_0)^2 \leq \rho, \tag{3.71}$$

which leads to the equivalent minimization of the following Lagrangian:

$$\text{minimize} \quad L(\theta, \lambda) = \sum_{n=1}^{N} (y_n - \theta)^2 + \lambda \left((\theta - \theta_0)^2 - \rho \right). \tag{3.72}$$

Taking the derivative with respect to θ and equating to zero, we obtain

$$\hat{\theta} = \frac{N \bar{y}_N + \lambda \theta_0}{N + \lambda},$$

which, for $\lambda = \sigma_\eta^2 / \sigma_0^2$, becomes identical to Eq. (3.69). The world is small after all! This has happened only because we used Gaussians both for the conditional and for the prior PDFs. For different forms of PDFs, this would not be the case. However, this example shows that a close relationship ties priors and constraints. They both attempt to impose prior information. Each method, in its own unique way, is associated with the respective pros and cons. In Chapters 12 and 13, where a more extended treatment of the Bayesian inference task is provided, we will see that the very essence of regularization, which is a means against overfitting, lies at the heart of the Bayesian approach.

Finally, one may wonder if the Bayesian inference has offered us any more information, compared to the deterministic parameter estimation path. After all, when the aim is to obtain a specific value for the unknown parameter, taking the mean of the Gaussian posterior comes to the same solution which results from the regularized LS approach. Well, even for this simple case, the Bayesian inference readily provides a piece of extra information; this is an estimate of the variance around the mean, which is very valuable in order to assess our trust of the recovered estimate. Of course, all these are valid provided that the adopted PDFs offer a good description of the statistical nature of the process at hand [24].

Finally, it can be shown (Problem 3.26) that the previously obtained results can be generalized for the more general linear regression model, of nonwhite Gaussian noise, considered in Section 3.10, which is modeled as

$$\boldsymbol{y} = X\boldsymbol{\theta} + \boldsymbol{\eta}.$$

It turns out that the posterior PDF is also Gaussian with mean value equal to

$$\mathbb{E}[\boldsymbol{\theta}|\boldsymbol{y}] = \boldsymbol{\theta}_0 + \left(\Sigma_0^{-1} + X^T \Sigma_\eta^{-1} X \right)^{-1} X^T \Sigma_\eta^{-1} (\boldsymbol{y} - X\boldsymbol{\theta}_0) \tag{3.73}$$

and covariance matrix

$$\Sigma_{\theta|y} = \left(\Sigma_0^{-1} + X^T \Sigma_\eta^{-1} X \right)^{-1}. \tag{3.74}$$

3.11.1 THE MAXIMUM A POSTERIORI PROBABILITY ESTIMATION METHOD

The maximum a posteriori probability estimation technique, usually denoted as MAP, is based on the Bayesian theorem, but it does not go as far as the Bayesian philosophy allows. The goal becomes that of obtaining an estimate which maximizes Eq. (3.63); in other words,

$$\boxed{\hat{\boldsymbol{\theta}}_{\text{MAP}} = \arg\max_{\boldsymbol{\theta}} p(\boldsymbol{\theta}|\mathcal{X}) : \quad \text{MAP estimate},} \tag{3.75}$$

and because $p(\mathcal{X})$ is independent of $\boldsymbol{\theta}$, this leads to

$$\begin{aligned} \hat{\boldsymbol{\theta}}_{\text{MAP}} &= \arg\max_{\boldsymbol{\theta}} p(\mathcal{X}|\boldsymbol{\theta})p(\boldsymbol{\theta}) \\ &= \arg\max_{\boldsymbol{\theta}} \{\ln p(\mathcal{X}|\boldsymbol{\theta}) + \ln p(\boldsymbol{\theta})\}. \end{aligned} \tag{3.76}$$

If we consider Example 3.8, it is a matter of simple exercise to obtain the MAP estimate and show that

$$\hat{\theta}_{\text{MAP}} = \frac{N\bar{y}_N + \frac{\sigma_\eta^2}{\sigma_0^2}\theta_0}{N + \frac{\sigma_\eta^2}{\sigma_0^2}} = \bar{\theta}_N. \tag{3.77}$$

Note that for this case, the MAP estimate coincides with the regularized LS solution for $\lambda = \sigma_\eta^2/\sigma_0^2$. Once more, we verify that adopting a prior PDF for the unknown parameter acts as a regularizer which embeds into the problem the available prior information.

Remarks 3.4.

- Observe that for the case of Example 3.8, all three estimators, namely, ML, MAP, and the Bayesian (taking the mean), result *asymptotically*, as N increases, in the same estimate. This is a more general result and it is true for other PDFs as well as for the case of parameter vectors. As the number of observations increases, our uncertainty is reduced and $p(\mathcal{X}|\boldsymbol{\theta})$, $p(\boldsymbol{\theta}|\mathcal{X})$ peak sharply around a value of $\boldsymbol{\theta}$. This forces all the methods to result in similar estimates. However, the obtained estimates are different for finite values of N. More recently, as we will see in Chapters 12 and 13, Bayesian methods have become very popular, and they seem to be the preferred choice for a number of practical problems.
- Choosing the prior PDF in the Bayesian methods is not an innocent task. In Example 3.8, we chose the conditional PDF (likelihood function) as well as the prior PDF to be Gaussians. We saw that the posterior PDF was also Gaussian. The advantage of such a choice was that we could come to closed form solutions. This is not always the case, and then the computation of the posterior PDF needs sampling methods or other approximate techniques. We will come to that in Chapters 12 and 14. However, the family of Gaussians is not the only one with this nice property of leading to closed form solutions. In probability theory, if the posterior is of the same form as the prior, we say that $p(\boldsymbol{\theta})$ is a *conjugate prior* of the likelihood function $p(\mathcal{X}|\boldsymbol{\theta})$ and then the involved integrations can be carried out in closed form (see, e.g., [15,30] and Chapter 12). Hence, the Gaussian PDF is a conjugate of itself.

- Just for the sake of pedagogical purposes, it is useful to recapitulate some of the nice properties that the Gaussian PDF possesses. We have met the following properties in various sections and problems in the book so far: (a) it is a conjugate of itself; (b) if two random variables (vectors) are jointly Gaussian, then their marginal PDFs are also Gaussian and the posterior PDF of one with respect to the other is also Gaussian; (c) moreover, the linear combination of jointly Gaussian variables turns out to be Gaussian; (d) as a by-product, it turns out that the sum of statistically independent Gaussian random variables is also a Gaussian one; and finally (e) the central limit theorem states that the sum of a large number of independent random variables tends to be Gaussian, as the number of the summands increases.

3.12 CURSE OF DIMENSIONALITY

In a number of places in this chapter, we mentioned the need of having a large number of training points. In Section 3.9.2, while discussing the bias–variance tradeoff, it was stated that in order to end up with a low overall MSE, the complexity (number of parameters) of the model should be small enough with respect to the number of training points. In Section 3.8, overfitting was discussed and it was pointed out that if the number of training points is small with respect to the number of parameters, overfitting occurs.

The question that is now raised is how big a data set should be, in order to be more relaxed concerning the performance of the designed predictor. The answer to the previous question depends largely on the dimensionality of the input space. It turns out that the larger the dimension of the input space is, the more data points are needed. This is related to the so-called *curse of dimensionality*, a term coined for the first time in [4].

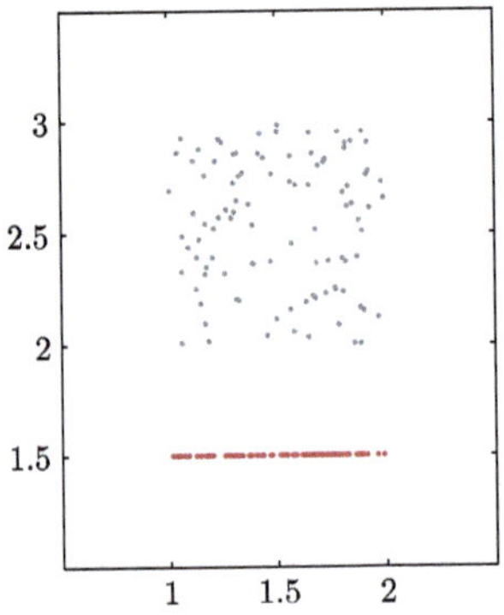

FIGURE 3.11

A simple experiment which demonstrates the curse of dimensionality. A number of 100 points are generated randomly, drawn from a uniform distribution, in order to fill the one-dimensional segment of length equal to one ($[1, 2] \times \{1.5\}$) (red points), and the two-dimensional rectangular region of unit area $[1, 2] \times [2, 3]$ (gray points). Observe that although the number of points in both cases is the same, the rectangular region is more sparsely populated than the densely populated line segment.

Let us assume that we are given the same number of points, N, thrown randomly in a unit cube (hypercube) in two different spaces, one being of low and the other of very high dimension. Then, the average distance of the points in the latter case will be much larger than that in the low-dimensional space case. As a matter of fact, the average distance shows a dependence that is analogous to the exponential term ($N^{-1/l}$), where l is the dimensionality of the space [14,37]. For example, the average distance between two out of 10^{10} points in the two-dimensional space is 10^{-5}, and in the 40-dimensional space it is equal to 1.83. Fig. 3.11 shows two cases, each one consisting of 100 points. The red points lie on a (one-dimensional) line segment of length equal to one and were generated according to the uniform distribution. Gray points cover a (two-dimensional) square region of unit area, which were also generated by a two-dimensional uniform distribution. Observe that the square area is more sparsely populated compared to the line segment. This is the general trend and high-dimensional spaces are sparsely populated; thus, many more data points are needed in order to fill in the space with enough data. Fitting a model in a parameter space, one must have enough data covering sufficiently well all regions in the space, in order to be able to learn well enough the input–output functional dependence (Problem 3.13).

There are various ways to cope with the curse of dimensionality and try to exploit the available data set in the best possible way. A popular direction is to resort to suboptimal solutions by projecting the input/feature vectors in a lower-dimensional subspace or manifold. Very often, such an approach leads to small performance losses, because the original training data, although they are generated in a high-dimensional space, may in fact "live" in a lower-dimensional subspace or manifold, due to physical dependencies that restrict the number of free parameters. Take as an example a case where the data are three-dimensional vectors, but they lie around a straight line, which is a one-dimensional linear manifold (affine set or subspace if it crosses the origin) or around a circle (one-dimensional nonlinear manifold) embedded in the three-dimensional space. In this case, the true number of free parameters is equal to one; this is because one free parameter suffices to describe the location of a point on a circle or on a straight line. The true number of free parameters is also known as the *intrinsic dimensionality* of the problem. The challenge, now, becomes that of learning the subspace/manifold onto which to project. These issues will be considered in more detail in Chapter 19.

Finally, it has to be noted that the dimensionality of the input space is not always the crucial issue. In pattern recognition, it has been shown that the critical factor is the so-called *VC dimension* of a classifier. In a number of classifiers, such as (generalized) linear classifiers or neural networks (to be considered in Chapter 18), the VC dimension is directly related to the dimensionality of the input space. However, one can design classifiers, such as the support vector machines (Chapter 11), whose performance is not directly related to the input space and they can be efficiently designed in spaces of very high (or even infinite) dimensionality [37,40].

3.13 VALIDATION

From previous sections, we already know that what is a "good" estimate according to one set of training points is not necessarily a good one for other data sets. This is an important aspect in any machine learning task; the performance of a method may vary with the random choice of the training set. A major phase, in any machine learning task, is to quantify/predict the performance that the designed (prediction) model is expected to exhibit in practice. It will not come as a surprise that "measuring" the

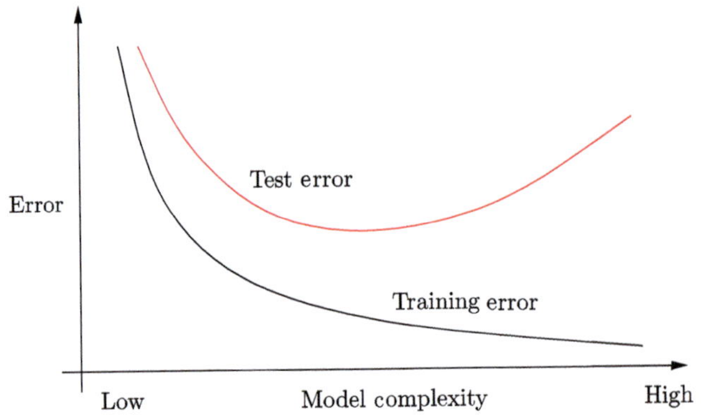

FIGURE 3.12

The training error tends to zero as the model complexity increases; for complex enough models with a large number of free parameters, a perfect fit to the training data is possible. However, the test error initially decreases, because more complex models "learn" the data better, up to a certain point. After that point of complexity, the test error increases.

performance against the training data set would lead to an "optimistic" value of the performance index, because this is computed on the same set on which the estimate was optimized; this trend has been known since the early 1930s [22]. For example, if the model is complex enough, with a large number of free parameters, the training error may even become zero, since a perfect fit to the data can be achieved. What is more meaningful and fair is to look for the so-called *generalization* performance of an estimator, that is, its average performance computed over *different* data sets which *did not* participate in the training (see the last paragraph of Section 3.9.2). The error associated with this average performance is known as the test error or the generalization error.[8]

Fig. 3.12 shows a typical performance that is expected to result in practice. The error measured on the (single) training data set is shown together with the (average) test error as the model complexity varies. If one tries to fit a complex model, with respect to the size of the available training set, then the error measured on the training set will be overoptimistic. On the contrary, the true error, as this is represented by the test error, takes large values; in the case where the performance index is the MSE, this is mainly contributed by the variance term (Section 3.9.2). On the other hand, if the model is too simple the test error will also attain large values; for the MSE case, this time the contribution is mainly due to the bias term. The idea is to have a model complexity that corresponds to the minimum of the respective curve. As a matter of fact, this is the point that various model selection techniques try to predict.

For some simple cases and under certain assumptions concerning the underlying models, we are able to have analytical formulas that quantify the average performance as we change data sets. However, in practice, this is hardly ever the case, and one must have a way to test the performance of an

[8] Note that some authors use the term generalization error to denote the difference between the test and the training errors. Another term for this difference is generalization gap.

obtained classifier/predictor using different data sets. This process is known as *validation*, and there are a number of alternatives that one can resort to.

Assuming that enough data are at the designer's disposal, one can split the data into one part, to be used for training, and another part for testing the performance. For example, the probability of error is computed over the test data set for the case of a classifier, or the MSE for the case of a regression task; other measures of fit can also be used. If this path is taken, one has to make sure that both the size of the training set and the size of the test set are large enough with respect to the model complexity; a large test data set is required in order to provide a statistically sound result on the test error. Especially if different methods are compared, the smaller the difference in their comparative performance is expected to be, the larger the size of the test set must be, in order to guarantee reliable conclusions [37, Chapter 10].

CROSS-VALIDATION

In practice, very often the size of the available data set is not sufficient and one cannot afford to "lose" part of it from the training set for the sake of testing. *Cross-validation* is a very common technique that is usually employed. Cross-validation has been rediscovered a number of times; however, to our knowledge, the first published description can be traced back to [25]. According to this method, the data set is split into K roughly equal-sized parts. We repeat training K times, each time selecting one (different each time) part of the data for testing and the remaining $K-1$ parts for training. This gives us the advantage of testing with a part of the data that has not been involved in the training, so it can be considered independent, and at the same time using, eventually, all the data, both for training and testing. Once we finish, we can (a) combine the obtained K estimates by averaging or via another more advanced way and (b) combine the errors from the test sets to get a better estimate of the test error that our estimator is expected to exhibit in real-life applications. This method is known as K-fold cross-validation. An extreme case is when we use $K = N$, so that each time one sample is left for testing. This is sometimes referred to as the *leave-one-out* (LOO) cross-validation method. The price one pays for K-fold cross-validation is the complexity of training K times. In practice, the value of K depends very much on the application, but typical values are of the order of 5 to 10.

The cross-validation estimator of the test error is very nearly unbiased. The reason for the slight bias is that the training set in cross-validation is slightly smaller than the actual data set. The effect of this bias will be conservative in the sense that the estimated fit will be slightly biased in the direction suggesting a poorer fit. In practice, this bias is rarely a concern, especially in the LOO case, where each time only one sample is left out. The variance, however, of the cross-validation estimator can be large, and this has to be taken into account when comparing different methods. In [13], the use of *bootstrap* techniques is suggested in order to reduce the variance of the obtained error predictions by the cross-validation method.

Moreover, besides complexity and high variance, cross-validation schemes are not beyond criticisms. Unfortunately, the overlap among the training sets introduces unknowable dependencies between runs, making the use of formal statistical tests difficult [11]. All this discussion reveals that the validation task is far from innocent. Ideally, one should have at her/his disposal large data sets and divide them in several *nonoverlapping* training sets, of whatever size is appropriate, along with separate test sets (or a single one) that are (is) large enough. More on different validation schemes and their properties can be found in, e.g., [3,12,17,37] and an insightful related discussion provided in [26].

3.14 EXPECTED LOSS AND EMPIRICAL RISK FUNCTIONS

What was said before in our discussion concerning the generalization and the training set-based performance of an estimator can be given a more formal statement via the notion of *expected loss*. Sometimes the expected loss is referred to as the *risk* function. Adopting a loss function, $\mathcal{L}(\cdot,\cdot)$, in order to quantify the deviation between the predicted value, $\hat{y} = f(\boldsymbol{x})$, and the respective true one, y, the corresponding expected loss is defined as

$$J(f) := \mathbb{E}\left[\mathcal{L}\left(\mathrm{y}, f(\mathbf{x})\right)\right], \tag{3.78}$$

or more explicitly

$$\boxed{J(f) = \int \cdots \int \mathcal{L}\big(y, f(\boldsymbol{x})\big) p(y, \boldsymbol{x}) dy d\boldsymbol{x}: \quad \text{expected loss function,}} \tag{3.79}$$

where the integration is replaced by summation whenever the respective variables are discrete. As a matter of fact, this is the ideal cost function one would like to optimize with respect to $f(\cdot)$, in order to get the optimal estimator over *all* possible values of the input–output pairs. However, such an optimization would in general be a very hard task, even if one knew the functional form of the joint distribution. Thus, in practice, one has to be content with two approximations. First, the functions to be searched are constrained within a certain family $\mathcal{F}$ (in this chapter, we focused on parametrically described families of functions). Second, because the joint distribution is either unknown and/or the integration may not be analytically tractable, the expected loss is approximated by the so-called *empirical risk* version, defined as

$$\boxed{J_N(f) = \frac{1}{N} \sum_{n=1}^{N} \mathcal{L}\big(y_n, f(\boldsymbol{x}_n)\big): \quad \text{empirical risk function.}} \tag{3.80}$$

As an example, the MSE function, discussed earlier, is the expected loss associated with the squared error loss function and the sum of squared errors cost is the respective empirical version. For large enough values of N and provided that the family of functions is restricted enough,[9] we expect the outcome from optimizing J_N to be close to that which would be obtained by optimizing J (see, e.g., [40]).

From the validation point of view, given any prediction function $f(\cdot)$, what we called test error corresponds to the corresponding value of J in Eq. (3.79) and what we called the training error corresponds to that of J_N in Eq. (3.80).

We can now take the discussion a little further, and we will reveal some more secrets concerning the accuracy–complexity tradeoff in machine learning. Let f_* be the function that optimizes the expected loss,

$$f_* := \arg\min_f J(f), \tag{3.81}$$

[9] That is, the family of functions is not very large. To keep the discussion simple, take the example of the quadratic class of functions. This is larger than that of the linear ones, because the latter is a special case (subset) of the former.

and let $f_{\mathcal{F}}$ be the optimum after *constraining* the task within the family of functions $\mathcal{F}$,

$$f_{\mathcal{F}} := \arg\min_{f \in \mathcal{F}} J(f). \tag{3.82}$$

Let us also define

$$f_N := \arg\min_{f \in \mathcal{F}} J_N(f). \tag{3.83}$$

Then, we can readily write that

$$\boxed{\begin{aligned}\mathbb{E}\left[J(f_N) - J(f_*)\right] = & \underbrace{\mathbb{E}\left[J(f_{\mathcal{F}}) - J(f_*)\right]}_{\text{approximation error}} + \\ & \underbrace{\mathbb{E}\left[J(f_N) - J(f_{\mathcal{F}})\right]}_{\text{estimation error}}.\end{aligned}} \tag{3.84}$$

The *approximation error* measures the deviation in the test error if instead of the overall optimal function one uses the optimal obtained within a certain family of functions. The *estimation error* measures the deviation due to optimizing the empirical risk instead of the expected loss. If one chooses the family of functions to be very large, then it is expected that the approximation error will be small, because there is a high probability f_* will be close to one of the members of the family. However, the estimation error is expected to be large, because for a fixed number of data points N, fitting a complex function is likely to lead to overfitting. For example, if the family of functions is the class of polynomials of a very large order, a very large number of parameters are to be estimated and overfitting will occur. The opposite is true if the class of functions is a small one. In parametric modeling, complexity of a family of functions is related to the number of free parameters. However, this is not the whole story. As a matter of fact, complexity is really measured by the so-called *capacity* of the associated set of functions. The VC dimension mentioned in Section 3.12 is directly related to the capacity of the family of the considered classifiers. More concerning the theoretical treatment of these issues can be obtained from [10,40,41].

LEARNABILITY

In a general setting, an issue of fundamental importance is whether the expected loss in Eq. (3.79) can be minimized to within an *arbitrary* precision based on only a finite set of observations, $(y_n, \boldsymbol{x}_n)$, $n = 1, 2, \ldots, N$, as $N \longrightarrow \infty$. The issue here is not an algorithmic one, that is, how efficiently this can be done. The *learnability* issue, as it is called, refers to whether it is statistically possible to employ the empirical risk function in Eq. (3.80) in place of the expected cost function.

It has been established that for the supervised classification and regression problems, a task is learnable *if and only if* the empirical cost, $J_N(f)$, converges *uniformly* to the expected loss function for *all* $f \in \mathcal{F}$ (see, e.g., [2,7]). However, this is not necessarily the case for other learning tasks. As a matter of fact, it can be shown that there exist tasks that cannot be learned via the empirical risk function. Instead, they can be learned via alternative mechanisms (see, e.g., [34]). For such cases, the notion of the *algorithmic stability* replaces that of uniform convergence.

3.15 NONPARAMETRIC MODELING AND ESTIMATION

The focus of this chapter was on the task of parameter estimation and on techniques that spring from the idea of parametric functional modeling of an input–output dependence. However, as said in the beginning of the chapter, besides parametric modeling, an alternative philosophy that runs across the field of statistical estimation is that of *nonparametric modeling*. This alternative path presents itself with two faces.

In its most classical version, the estimation task involves no parameters. Typical examples of such methods are the histogram approximation of an unknown distribution, the closely related Parzen windows method, [28], and the k-nearest neighbor density estimation (see, e.g., [37,38]). The latter approach is related to one of the most widely known and used methods for classification, known as the k-nearest neighbor classification rule, which is discussed in Chapter 7. For the sake of completeness, the basic rationale behind such methods is provided in the Additional Material part that is related to the current chapter and can be downloaded from the site of the book.

The other path of nonparametric modeling is when parameters pop in, yet their number is not fixed and a priori selected, but it grows with the number of training examples. We will treat such models in the context of reproducing kernel Hilbert spaces (RKHSs) in Chapter 11. There, instead of parameterizing the family of functions, in which one constrains the search for finding the prediction model, the candidate solution is constrained to lie within a specific *functional space*. Nonparametric models in the context of Bayesian learning are also treated in Chapter 13.

PROBLEMS

3.1 Prove the least squares optimal solution for the linear regression case given in Eq. (3.13).

3.2 Let $\hat{\boldsymbol{\theta}}_i$, $i = 1, 2, \ldots, m$, be unbiased estimators of a parameter vector $\boldsymbol{\theta}$, so that $\mathbb{E}[\hat{\boldsymbol{\theta}}_i] = \boldsymbol{\theta}$, $i = 1, \ldots, m$. Moreover, assume that the respective estimators are uncorrelated to each other and that all have the same (total) variance, $\sigma^2 = \mathbb{E}[(\boldsymbol{\theta}_i - \boldsymbol{\theta})^T(\boldsymbol{\theta}_i - \boldsymbol{\theta})]$. Show that by averaging the estimates, e.g.,

$$\hat{\boldsymbol{\theta}} = \frac{1}{m}\sum_{i=1}^{m} \hat{\boldsymbol{\theta}}_i,$$

the new estimator has total variance $\sigma_c^2 := \mathbb{E}[(\hat{\boldsymbol{\theta}} - \boldsymbol{\theta})^T(\hat{\boldsymbol{\theta}} - \boldsymbol{\theta})] = \frac{1}{m}\sigma^2$.

3.3 Let a random variable x being described by a uniform PDF in the interval $[0, \frac{1}{\theta}]$, $\theta > 0$. Assume a function[10] g, which defines an estimator $\hat{\theta} := g(\mathrm{x})$ of θ. Then, for such an estimator to be unbiased, the following must hold:

$$\int_0^{\frac{1}{\theta}} g(x)\,dx = 1.$$

However, show that such a function g does not exist.

[10] To avoid any confusion, let g be Lebesgue integrable on intervals of $\mathbb{R}$.

3.4 A family $\{p(\mathcal{D};\boldsymbol{\theta}):\boldsymbol{\theta}\in\mathcal{A}\}$ is called *complete* if, for any vector function $\boldsymbol{h}(\mathcal{D})$ such that $\mathbb{E}_{\mathcal{D}}[\boldsymbol{h}(\mathcal{D})]=\boldsymbol{0}, \forall\boldsymbol{\theta}$, we have $\boldsymbol{h}=\boldsymbol{0}$.

Show that if $\{p(\mathcal{D};\boldsymbol{\theta}):\boldsymbol{\theta}\in\mathcal{A}\}$ is complete and there exists an MVU estimator, then this estimator is unique.

3.5 Let $\hat{\theta}_u$ be an unbiased estimator, so that $\mathbb{E}[\hat{\theta}_u]=\theta_o$. Define a biased one by $\hat{\theta}_b=(1+\alpha)\hat{\theta}_u$. Show that the range of α where the MSE of $\hat{\theta}_b$ is smaller than that of $\hat{\theta}_u$ is

$$-2<-\frac{2\text{MSE}(\hat{\theta}_u)}{\text{MSE}(\hat{\theta}_u)+\theta_o^2}<\alpha<0.$$

3.6 Show that for the setting of Problem 3.5, the optimal value of α is equal to

$$\alpha_*=-\frac{1}{1+\frac{\theta_o^2}{\text{var}(\hat{\theta}_u)}},$$

where, of course, the variance of the unbiased estimator is equal to the corresponding MSE.

3.7 Show that the regularity condition for the Cramér–Rao bound holds true if the order of integration and differentiation can be interchanged.

3.8 Derive the Cramér–Rao bound for the LS estimator when the training data result from the linear model

$$y_n=\theta x_n+\eta_n,\quad n=1,2,\ldots,$$

where x_n are i.i.d. samples of a zero mean random variable with variance σ_x^2 and η_n are i.i.d. noise samples drawn from a Gaussian with zero mean and variance σ_η^2. Assume also that x and η are independent. Then, show that the LS estimator achieves the Cramér–Rao bound only asymptotically.

3.9 Let us consider the regression model

$$y_n=\boldsymbol{\theta}^T\boldsymbol{x}_n+\eta_n,\quad n=1,2,\ldots,N,$$

where the noise samples $\boldsymbol{\eta}=[\eta_1,\ldots,\eta_N]^T$ come from a zero mean Gaussian random vector, with covariance matrix Σ_η. If $X=[\boldsymbol{x}_1,\ldots,\boldsymbol{x}_N]^T$ stands for the input matrix and $\boldsymbol{y}=[y_1,\ldots,y_N]^T$, then show that

$$\hat{\boldsymbol{\theta}}=\left(X^T\Sigma_\eta^{-1}X\right)^{-1}X^T\Sigma_\eta^{-1}\boldsymbol{y},$$

is an efficient estimate.

Note that the previous estimate coincides with the ML one. Moreover, bear in mind that in the case where $\Sigma_\eta=\sigma^2 I$, the ML estimate becomes equal to the LS one.

3.10 Assume a set of i.i.d. $\mathcal{X}=\{x_1,x_2,\ldots,x_N\}$ samples of a random variable, with mean μ and variance σ^2. Define also the quantities

$$S_\mu:=\frac{1}{N}\sum_{n=1}^{N}x_n,\quad S_{\sigma^2}:=\frac{1}{N}\sum_{n=1}^{N}(x_n-S_\mu)^2,$$

$$\bar{S}_{\sigma^2} := \frac{1}{N}\sum_{n=1}^{N}(x_n - \mu)^2.$$

Show that if μ is considered to be known, a sufficient statistic for σ^2 is $\bar{S}_{\sigma^2}$. Moreover, in the case where both (μ, σ^2) are unknown, a sufficient statistic is the pair (S_μ, S_{σ^2}).

3.11 Show that solving the task

$$\text{minimize} \quad L(\boldsymbol{\theta}, \lambda) = \sum_{n=1}^{N}\left(y_n - \theta_0 - \sum_{i=1}^{l}\theta_i x_{ni}\right)^2 + \lambda\sum_{i=1}^{l}|\theta_i|^2$$

is equivalent to minimizing

$$\text{minimize} \quad L(\boldsymbol{\theta}, \lambda) = \sum_{n=1}^{N}\left((y_n - \bar{y}) - \sum_{i=1}^{l}\theta_i (x_{ni} - \bar{x}_i)\right)^2 + \lambda\sum_{i=1}^{l}|\theta_i|^2,$$

and the estimate of θ_0 is given by

$$\hat{\theta}_0 = \bar{y} - \sum_{i=1}^{l}\hat{\theta}_i \bar{x}_i.$$

3.12 This problem refers to Example 3.4, where a linear regression task with a real-valued unknown parameter θ_o is considered. Show that $\text{MSE}(\hat{\theta}_b(\lambda)) < \text{MSE}(\hat{\theta}_{\text{MVU}})$ or the ridge regression estimate shows a lower MSE performance than the one for the MVU estimate, if

$$\begin{cases} \lambda \in (0, \infty), & \theta_o^2 \leq \frac{\sigma_\eta^2}{N}, \\ \lambda \in \left(0, \dfrac{2\sigma_\eta^2}{\theta_o^2 - \dfrac{\sigma_\eta^2}{N}}\right), & \theta_o^2 > \dfrac{\sigma_\eta^2}{N}. \end{cases}$$

Moreover, the minimum MSE performance for the ridge regression estimate is attained at $\lambda_* = \sigma_\eta^2/\theta_o^2$.

3.13 Consider, once more, the same regression model as that of Problem 3.9, but with $\Sigma_\eta = I_N$. Compute the MSE of the predictions $\mathbb{E}[(\mathrm{y} - \hat{\mathrm{y}})^2]$, where y is the true response and $\hat{\mathrm{y}}$ is the predicted value, given a test point $\boldsymbol{x}$ and using the LS estimator

$$\hat{\boldsymbol{\theta}} = \left(X^T X\right)^{-1} X^T \mathbf{y}.$$

The LS estimator has been obtained via a set of N measurements, collected in the input matrix X and $\mathbf{y}$, where the notation has been introduced previously in this chapter. The expectation $\mathbb{E}[\cdot]$ is taken with respect to y, the training data $\mathcal{D}$, and the test points $\mathbf{x}$. Observe the dependence of the MSE on the dimensionality of the space.

Hint: Consider, first, the MSE, given the value of a test point $\boldsymbol{x}$, and then take the average over all the test points.

3.14 Assume that the model that generates the data is

$$y_n = A \sin\left(\frac{2\pi}{N}kn + \phi\right) + \eta_n,$$

where $A > 0$, and $k \in \{1, 2, \dots, N-1\}$. Assume that η_n are i.i.d. samples from a Gaussian noise of variance σ_η^2. Show that there is no unbiased estimator for the phase ϕ based on N measurement points, $y_n, n = 0, 1, \dots N-1$, that attains the Cramér–Rao bound.

3.15 Show that if $(\mathbf{y}, \mathbf{x})$ are two jointly distributed random vectors, with values in $\mathbb{R}^k \times \mathbb{R}^l$, then the MSE optimal estimator of $\mathbf{y}$ given the value $\mathbf{x} = \boldsymbol{x}$ is the regression of $\mathbf{y}$ conditioned on $\boldsymbol{x}$, or $\mathbb{E}[\mathbf{y}|\boldsymbol{x}]$.

3.16 Assume that $\mathbf{x}, \mathbf{y}$ are jointly Gaussian random vectors, with covariance matrix

$$\Sigma := \mathbb{E}\left[\begin{bmatrix}\mathbf{x} - \boldsymbol{\mu}_x \\ \mathbf{y} - \boldsymbol{\mu}_y\end{bmatrix}\left[(\mathbf{x} - \boldsymbol{\mu}_x)^T, (\mathbf{y} - \boldsymbol{\mu}_y)^T\right]\right] = \begin{bmatrix}\Sigma_x & \Sigma_{xy} \\ \Sigma_{yx} & \Sigma_y\end{bmatrix}.$$

Assuming also that the matrices Σ_x and $\bar{\Sigma} := \Sigma_y - \Sigma_{yx}\Sigma_x^{-1}\Sigma_{xy}$ are nonsingular, show that the optimal MSE estimator $\mathbb{E}[\mathbf{y}|\boldsymbol{x}]$ takes the following form:

$$\mathbb{E}[\mathbf{y}|\boldsymbol{x}] = \mathbb{E}[\mathbf{y}] + \Sigma_{yx}\Sigma_x^{-1}(\boldsymbol{x} - \boldsymbol{\mu}_x).$$

Note that $\mathbb{E}[\mathbf{y}|\boldsymbol{x}]$ is an affine function of $\boldsymbol{x}$. In other words, for the case where $\mathbf{x}$ and $\mathbf{y}$ are jointly Gaussian, the optimal estimator of $\mathbf{y}$, in the MSE sense, which is in general a nonlinear function, becomes an affine function of $\boldsymbol{x}$.

In the special case where x, y are scalar random variables, we have

$$\mathbb{E}[\mathrm{y}|x] = \mu_y + \frac{\alpha\sigma_y}{\sigma_x}(x - \mu_x),$$

where α stands for the *correlation coefficient*, defined as

$$\alpha := \frac{\mathbb{E}\left[(\mathrm{x} - \mu_x)(\mathrm{y} - \mu_y)\right]}{\sigma_x\sigma_y},$$

with $|\alpha| \leq 1$. Note, also, that the previous assumption on the nonsingularity of Σ_x and $\bar{\Sigma}$ translates, in this special case, to $\sigma_x \neq 0 \neq \sigma_y$.

Hint: Use the matrix inversion lemma from Appendix A, in terms of the Schur complement $\bar{\Sigma}$ of Σ_x in Σ and the fact that $\det(\Sigma) = \det(\Sigma_y)\det(\bar{\Sigma})$.

3.17 Assume a number l of jointly Gaussian random variables $\{\mathrm{x}_1, \mathrm{x}_2, \dots, \mathrm{x}_l\}$ and a nonsingular matrix $A \in \mathbb{R}^{l\times l}$. If $\mathbf{x} := [\mathrm{x}_1, \mathrm{x}_2, \dots, \mathrm{x}_l]^T$, then show that the components of the vector $\mathbf{y}$, obtained by $\mathbf{y} = A\mathbf{x}$, are also jointly Gaussian random variables.

A direct consequence of this result is that any linear combination of jointly Gaussian variables is also Gaussian.

3.18 Let $\mathbf{x}$ be a vector of jointly Gaussian random variables of covariance matrix Σ_x. Consider the general linear regression model

$$\mathbf{y} = \Theta\mathbf{x} + \boldsymbol{\eta},$$

where $\Theta \in \mathbb{R}^{k\times l}$ is a parameter matrix and $\boldsymbol{\eta}$ is the noise vector which is considered to be Gaussian, with zero mean, and with covariance matrix Σ_η, independent of $\mathbf{x}$. Then show that $\mathbf{y}$ and $\mathbf{x}$ are jointly Gaussian, with the covariance matrix given by

$$\Sigma = \begin{bmatrix} \Theta\Sigma_x\Theta^T + \Sigma_\eta & \Theta\Sigma_x \\ \Sigma_x\Theta^T & \Sigma_x \end{bmatrix}.$$

3.19 Show that a linear combination of Gaussian independent variables is also Gaussian.

3.20 Show that if a sufficient statistic $T(\mathcal{X})$ for a parameter estimation problem exists, then $T(\mathcal{X})$ suffices to express the respective ML estimate.

3.21 Show that if an efficient estimator exists, then it is also optimal in the ML sense.

3.22 Let the observations resulting from an experiment be x_n, $n = 1, 2, \ldots, N$. Assume that they are independent and that they originate from a Gaussian PDF $\mathcal{N}(\mu, \sigma^2)$. Both the mean and the variance are unknown. Prove that the ML estimates of these quantities are given by

$$\hat{\mu}_{\text{ML}} = \frac{1}{N}\sum_{n=1}^{N} x_n, \quad \hat{\sigma}^2_{\text{ML}} = \frac{1}{N}\sum_{n=1}^{N}(x_n - \hat{\mu}_{\text{ML}})^2.$$

3.23 Let the observations $x_n, n = 1, 2, \ldots, N$, come from the uniform distribution

$$p(x;\theta) = \begin{cases} \frac{1}{\theta}, & 0 \le x \le \theta, \\ 0, & \text{otherwise.} \end{cases}$$

Obtain the ML estimate of θ.

3.24 Obtain the ML estimate of the parameter $\lambda > 0$ of the exponential distribution

$$p(x) = \begin{cases} \lambda \exp(-\lambda x), & x \ge 0, \\ 0, & x < 0, \end{cases}$$

based on a set of measurements x_n, $n = 1, 2, \ldots, N$.

3.25 Assume a $\mu \sim \mathcal{N}(\mu_0, \sigma_0^2)$ and a stochastic process $\{x_n\}_{n=-\infty}^{\infty}$, consisting of i.i.d. random variables, such that $p(x_n|\mu) = \mathcal{N}(\mu, \sigma^2)$. Consider N observations so that $\mathcal{X} := \{x_1, x_2, \ldots, x_N\}$, and prove that the posterior $p(x|\mathcal{X})$, of any $x = x_{n_0}$ conditioned on $\mathcal{X}$, turns out to be Gaussian with mean μ_N and variance σ_N^2, where

$$\mu_N := \frac{N\sigma_0^2\bar{x} + \sigma^2\mu_0}{N\sigma_0^2 + \sigma_\eta^2}, \quad \sigma_N^2 := \frac{\sigma^2\sigma_0^2}{N\sigma_0^2 + \sigma_\eta^2}.$$

3.26 Show that for the linear regression model,

$$y = X\theta + \eta,$$

the a posteriori probability $p(\theta|y)$ is a Gaussian one if the prior distribution probability is given by $p(\theta) = N(\theta_0, \Sigma_0)$, and the noise samples follow the multivariate Gaussian distribution $p(\eta) = N(\mathbf{0}, \Sigma_\eta)$. Compute the mean vector and the covariance matrix of the posterior distribution.

3.27 Assume that x_n, $n = 1, 2, \ldots, N$, are i.i.d. observations from a Gaussian $\mathcal{N}(\mu, \sigma^2)$. Obtain the MAP estimate of μ if the prior follows the exponential distribution

$$p(\mu) = \lambda \exp(-\lambda\mu), \quad \lambda > 0, \mu \geq 0.$$

MATLAB® EXERCISES

3.28 Write a MATLAB program to reproduce the results and figures of Example 3.1. Play with the value of the noise variance.

3.29 Write a MATLAB program to reproduce the results of Example 3.6. Play with the number of training points, the degrees of the involved polynomials, and the noise variance in the regression model.

REFERENCES

[1] H. Akaike, A new look at the statistical model identification, IEEE Trans. Autom. Control 19 (6) (1970) 716–723.
[2] N. Alon, S. Ben-David, N. Cesa-Bianci, D. Haussler, Scale-sensitive dimensions, uniform convergence and learnability, J. Assoc. Comput. Mach. 44 (4) (1997) 615–631.
[3] S. Arlot, A. Celisse, A survey of cross-validation procedures for model selection, Stat. Surv. 4 (2010) 40–79.
[4] R.E. Bellman, Dynamic Programming, Princeton University Press, Princeton, 1957.
[5] A. Ben-Israel, T.N.E. Greville, Generalized Inverses: Theory and Applications, second ed., Springer-Verlag, New York, 2003.
[6] D. Bertsekas, A. Nedic, O. Ozdaglar, Convex Analysis and Optimization, Athena Scientific, Belmont, MA, 2003.
[7] A. Blumer, A. Ehrenfeucht, D. Haussler, W. Warmuth, Learnability and the Vapnik-Chernovenkis dimension, J. Assoc. Comput. Mach. 36 (4) (1989) 929–965.
[8] S. Boyd, L. Vandenberghe, Convex Optimization, Cambridge University Press, Cambridge, 2004.
[9] H. Cramer, Mathematical Methods of Statistics, Princeton University Press, Princeton, 1946.
[10] L. Devroy, L. Györfi, G. Lugosi, A Probabilistic Theory of Pattern Recognition, Springer, New York, 1991.
[11] T.G. Dietterich, Approximate statistical tests for comparing supervised classification learning algorithms, Neural Comput. 10 (1998) 1895–1923.
[12] R. Duda, P. Hart, D. Stork, Pattern Classification, second ed., Wiley, New York, 2000.
[13] A. Efron, R. Tibshirani, Improvements on cross-validation: the 632+ bootstrap method, J. Am. Stat. Assoc. 92 (438) (1997) 548–560.
[14] J.H. Friedman, Regularized discriminant analysis, J. Am. Stat. Assoc. 84 (1989) 165–175.
[15] A. Gelman, J.B. Carlin, H.S. Stern, D.B. Rubin, Bayesian Data Analysis, second ed., CRC Press, Boca Raton, FL, 2003.
[16] S. Geman, E. Bienenstock, R. Doursat, Neural networks and the bias-variance dilemma, Neural Comput. 4 (1992) 1–58.
[17] T. Hastie, R. Tibshirani, J. Friedman, The Elements of Statistical Learning, second ed., Springer, New York, 2009.
[18] A.E. Hoerl, R.W. Kennard, Ridge regression: biased estimation for nonorthogonal problems, Technometrics 12 (1) (1970) 55–67.
[19] S. Kay, Y. Eldar, Rethinking biased estimation, IEEE Signal Process. Mag. 25 (6) (2008) 133–136.

[20] S. Kay, Statistical Signal Processing, Prentice Hall, Upper Saddle River, NJ, 1993.
[21] M. Kendall, A. Stuart, The Advanced Theory of Statistics, vol. 2, MacMillan, New York, 1979.
[22] S.C. Larson, The shrinkage of the coefficient of multiple correlation, J. Educ. Psychol. 22 (1931) 45–55.
[23] E.L. Lehmann, H. Scheffe, Completeness, similar regions, and unbiased estimation: Part II, Sankhyā 15 (3) (1955) 219–236.
[24] D. McKay, Probable networks and plausible predictions—a review of practical Bayesian methods for supervised neural networks, Netw. Comput. Neural Syst. 6 (1995) 469–505.
[25] F. Mosteller, J.W. Tukey, Data analysis, including statistics, in: Handbook of Social Psychology, Addison-Wesley, Reading, MA, 1954.
[26] R.M. Neal, Assessing relevance determination methods using DELVE, in: C.M. Bishop (Ed.), Neural Networks and Machine Learning, Springer-Verlag, New York, 1998, pp. 97–129.
[27] A. Papoulis, S.U. Pillai, Probability, Random Variables, and Stochastic Processes, fourth ed., McGraw Hill, New York, NY, 2002.
[28] E. Parzen, On the estimation of a probability density function and mode, Ann. Math. Stat. 33 (1962) 1065–1076.
[29] D.L. Phillips, A technique for the numerical solution of certain integral equations of the first kind, J. Assoc. Comput. Mach. 9 (1962) 84–97.
[30] H. Raiffa, R. Schlaifer, Applied Statistical Decision Theory, Division of Research, Graduate School of Business Administration, Harvard University, Boston, 1961.
[31] R.C. Rao, Information and the accuracy attainable in the estimation of statistical parameters, Bull. Calcutta Math. Soc. 37 (1945) 81–89.
[32] J. Rissanen, A universal prior for integers and estimation by minimum description length, Ann. Stat. 11 (2) (1983) 416–431.
[33] G. Schwartz, Estimating the dimension of the model, Ann. Stat. 6 (1978) 461–464.
[34] S. Shalev-Shwartz, O. Shamir, N. Srebro, K. Shridharan, Learnability, stability and uniform convergence, J. Mach. Learn. Res. (JMLR) 11 (2010) 2635–2670.
[35] G. Strang, Introduction to Linear Algebra, fifth ed., Wellesley-Cambridge Press and SIAM, 2016.
[36] J. Shao, Mathematical Statistics, Springer, New York, 1998.
[37] S. Theodoridis, K. Koutroumbas, Pattern Recognition, fourth ed., Academic Press, New York, 2009.
[38] S. Theodoridis, A. Pikrakis, K. Koutroumbas, D. Cavouras, An Introduction to Pattern Recognition: A MATLAB Approach, Academic Press, New York, 2010.
[39] A.N. Tychonoff, V.Y. Arsenin, Solution of Ill-Posed Problems, Winston & Sons, Washington, 1977.
[40] V.N. Vapnik, The Nature of Statistical Learning Theory, Springer-Verlag, New York, 1995.
[41] V.N. Vapnik, Statistical Learning Theory, John Wiley & Sons, New York, 1998.

CHAPTER

MEAN-SQUARE ERROR LINEAR ESTIMATION

4

CONTENTS

4.1 INTRODUCTION

Mean-square error (MSE) linear estimation is a topic of fundamental importance for parameter estimation in statistical learning. Besides historical reasons, which take us back to the pioneering works of Kolmogorov, Wiener, and Kalman, who laid the foundations of the optimal estimation field, understanding MSE estimation is a must, prior to studying more recent techniques. One always has to grasp the basics and learn the classics prior to getting involved with new "adventures." Many of the concepts to be discussed in this chapter are also used in the next chapters.

Machine Learning. https://doi.org/10.1016/B978-0-12-818803-3.00013-1

Optimizing via a loss function that builds around the square of the error has a number of advantages, such as a single optimal value, which can be obtained via the solution of a linear set of equations; this is a very attractive feature in practice. Moreover, due to the relative simplicity of the resulting equations, the newcomer in the field can get a better feeling of the various notions associated with optimal parameter estimation. The elegant geometric interpretation of the MSE solution, via the orthogonality theorem, is presented and discussed. In the chapter, emphasis is also given to computational complexity issues while solving for the optimal solution. The essence behind these techniques remains exactly the same as that inspiring a number of computationally efficient schemes for online learning, to be discussed later in this book.

The development of the chapter is around real-valued variables, something that will be true for most of the book. However, complex-valued signals are particularly useful in a number of areas, with communications being a typical example, and the generalization from the real to the complex domain may not always be trivial. Although in most of the cases the difference lies in changing matrix transpositions by Hermitian ones, this is not the whole story. This is the reason that we chose to deal with complex-valued data in separate sections, whenever the differences from the real data are not trivial and some subtle issues are involved.

4.2 MEAN-SQUARE ERROR LINEAR ESTIMATION: THE NORMAL EQUATIONS

The general estimation task has been introduced in Chapter 3. There, it was stated that given two dependent random vectors, $\mathbf{y}$ and $\mathbf{x}$, the goal of the estimation task is to obtain a function, g, so as, given a value $\boldsymbol{x}$ of $\mathbf{x}$, to be able to predict (estimate), in some optimal sense, the corresponding value $\boldsymbol{y}$ of $\mathbf{y}$, or $\hat{\boldsymbol{y}} = g(\boldsymbol{x})$. The MSE estimation was also presented in Chapter 3 and it was shown that the optimal MSE estimate of $\mathbf{y}$ given the value $\mathbf{x} = \boldsymbol{x}$ is

$$\hat{\boldsymbol{y}} = \mathbb{E}[\mathbf{y}|\boldsymbol{x}].$$

In general, this is a nonlinear function. We now turn our attention to the case where g is *constrained* to be a linear function. For simplicity and in order to pay more attention to the concepts, we will restrict our discussion to the case of scalar dependent (output) variables. The more general case will be discussed later on.

Let $(\mathrm{y}, \mathbf{x}) \in \mathbb{R} \times \mathbb{R}^l$ be two jointly distributed random entities of *zero mean values*. In case the mean values are not zero, they are subtracted. Our goal is to obtain an estimate of $\boldsymbol{\theta} \in \mathbb{R}^l$ in the linear estimator model,

$$\hat{\mathrm{y}} = \boldsymbol{\theta}^T \mathbf{x}, \tag{4.1}$$

so that

$$J(\boldsymbol{\theta}) = \mathbb{E}\left[(\mathrm{y} - \hat{\mathrm{y}})^2\right], \tag{4.2}$$

is minimum, or

$$\boldsymbol{\theta}_* := \arg\min_{\boldsymbol{\theta}} J(\boldsymbol{\theta}). \tag{4.3}$$

In other words, the optimal estimator is chosen so as to minimize the variance of the error random variable

$$\mathrm{e} = \mathrm{y} - \hat{\mathrm{y}}. \tag{4.4}$$

Minimizing the cost function $J(\boldsymbol{\theta})$ is equivalent to setting its gradient with respect to $\boldsymbol{\theta}$ equal to zero (see Appendix A),

$$\begin{aligned}\nabla J(\boldsymbol{\theta}) &= \nabla \mathbb{E}\left[\left(\mathrm{y} - \boldsymbol{\theta}^T \mathbf{x}\right)\left(\mathrm{y} - \mathbf{x}^T \boldsymbol{\theta}\right)\right] \\ &= \nabla \left\{\mathbb{E}[\mathrm{y}^2] - 2\boldsymbol{\theta}^T \mathbb{E}[\mathbf{x}\mathrm{y}] + \boldsymbol{\theta}^T \mathbb{E}[\mathbf{x}\mathbf{x}^T]\boldsymbol{\theta}\right\} \\ &= -2\boldsymbol{p} + 2\Sigma_x \boldsymbol{\theta} = \mathbf{0}\end{aligned}$$

or

$$\boxed{\Sigma_x \boldsymbol{\theta}_* = \boldsymbol{p}: \quad \text{normal equations},} \tag{4.5}$$

where the input–output cross-correlation vector $\boldsymbol{p}$ is given by[1]

$$\boldsymbol{p} = \left[\mathbb{E}[\mathrm{x}_1 \mathrm{y}], \ldots, \mathbb{E}[\mathrm{x}_l \mathrm{y}]\right]^T = \mathbb{E}[\mathbf{x}\mathrm{y}], \tag{4.6}$$

and the respective covariance matrix is given by

$$\Sigma_x = \mathbb{E}\left[\mathbf{x}\mathbf{x}^T\right].$$

Thus, the weights of the optimal linear estimator are obtained via a linear system of equations, provided that the covariance matrix is *positive definite* and hence it can be inverted (Appendix A). Moreover, in this case, the solution is *unique*. On the contrary, if Σ_x is singular and hence cannot be inverted, there are infinitely many solutions (Problem 4.1).

4.2.1 THE COST FUNCTION SURFACE

Elaborating on the cost function $J(\boldsymbol{\theta})$, as defined in (4.2), we get

$$J(\boldsymbol{\theta}) = \sigma_y^2 - 2\boldsymbol{\theta}^T \boldsymbol{p} + \boldsymbol{\theta}^T \Sigma_x \boldsymbol{\theta}. \tag{4.7}$$

Adding and subtracting the term $\boldsymbol{\theta}_*^T \Sigma_x \boldsymbol{\theta}_*$ and taking into account the definition of $\boldsymbol{\theta}_*$ from (4.5), it is readily seen that

$$\boxed{J(\boldsymbol{\theta}) = J(\boldsymbol{\theta}_*) + (\boldsymbol{\theta} - \boldsymbol{\theta}_*)^T \Sigma_x (\boldsymbol{\theta} - \boldsymbol{\theta}_*),} \tag{4.8}$$

where

$$J(\boldsymbol{\theta}_*) = \sigma_y^2 - \boldsymbol{p}^T \Sigma_x^{-1} \boldsymbol{p} = \sigma_y^2 - \boldsymbol{\theta}_*^T \Sigma_x \boldsymbol{\theta}_* = \sigma_y^2 - \boldsymbol{p}^T \boldsymbol{\theta}_* \tag{4.9}$$

is the minimum achieved at the optimal solution. From (4.8) and (4.9), the following remarks can be made.

---

[1] The cross-correlation vector is often denoted as $\boldsymbol{r}_{xy}$. Here we will use $\boldsymbol{p}$, in order to simplify the notation.

Remarks 4.1.

- The cost at the optimal value $\boldsymbol{\theta}_*$ is always less than the variance $\mathbb{E}[\mathrm{y}^2]$ of the output variable. This is guaranteed by the positive definite nature of Σ_x or Σ_x^{-1}, which makes the second term on the right-hand side in (4.9) always positive, unless $\boldsymbol{p} = \mathbf{0}$; however, the cross-correlation vector will only be zero if $\mathbf{x}$ and y are uncorrelated. Well, in this case, one cannot say anything (make any prediction) about y by observing samples of $\mathbf{x}$, at least as far as the MSE criterion is concerned, which turns out to involve information residing up to the second-order statistics. In this case, the variance of the error, which coincides with $J(\boldsymbol{\theta}_*)$, will be equal to the variance σ_{y}^2; the latter is a measure of the "intrinsic" uncertainty of y around its (zero) mean value. On the contrary, if the input–output variables are correlated, then observing $\mathbf{x}$ removes part of the uncertainty associated with y.
- For any value $\boldsymbol{\theta}$ other than the optimal $\boldsymbol{\theta}_*$, the error variance increases as (4.8) suggests, due to the positive definite nature of Σ_x. Fig. 4.1 shows the cost function (MSE) surface defined by $J(\boldsymbol{\theta})$ in (4.8). The corresponding isovalue contours are shown in Fig. 4.2. In general, they are ellipses, whose axes are determined by the eigenstructure of Σ_x. For $\Sigma_x = \sigma^2 I$, where all eigenvalues are equal to σ^2, the contours are circles (Problem 4.3).

4.3 A GEOMETRIC VIEWPOINT: ORTHOGONALITY CONDITION

A very intuitive view of what we have said so far comes from the geometric interpretation of the random variables. The reader can easily check out that the set of random variables is a *linear space* over the field of real (and complex) numbers. Indeed, if x and y are any two random variables then $\mathrm{x} + \mathrm{y}$, as well as $\alpha\mathrm{x}$, are also random variables for every $\alpha \in \mathbb{R}$.[2] We can now equip this linear space with an inner product operation, which also implies a norm and makes it an *inner product space*. The

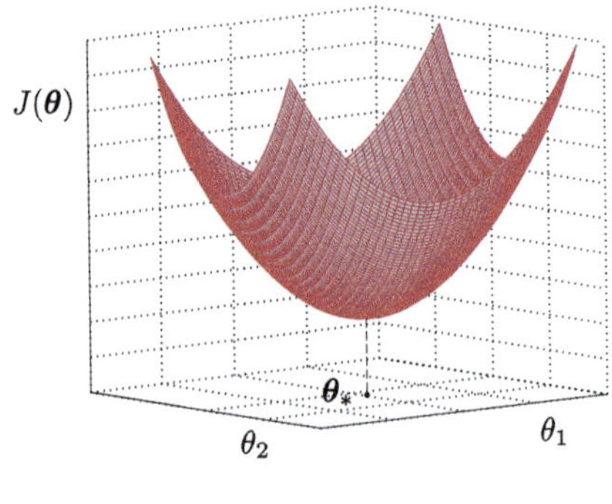

FIGURE 4.1

The MSE cost function has the form of a (hyper)paraboloid.

2 These operations also satisfy all the properties required for a set to be a linear space, including associativity, commutativity, and so on (see [47] and the appendix associated with Chapter 8).

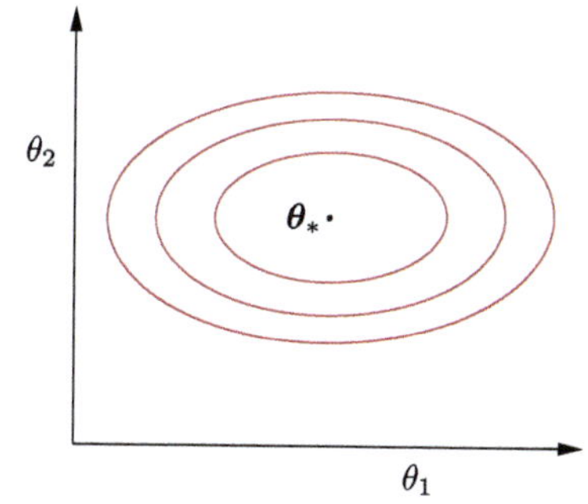

FIGURE 4.2

The isovalue contours for the cost function surface corresponding to Fig. 4.1. They are ellipses; the major and the minor axes of each ellipse are determined by the maximum and minimum eigenvalues, $\lambda_{\max}$ and $\lambda_{\min}$, of the covariance matrix, Σ, of the input random variables. The largest the ratio $\frac{\lambda_{\max}}{\lambda_{\min}}$ is, the more elongated the ellipses are. The ellipses become circles if the covariance matrix has the special form of $\sigma^2 I$. That is, all variables are mutually uncorrelated and they have the same variance. By varying Σ, different shapes of the ellipses and different orientations result.

reader can easily check that the mean value operation has all the properties required for an operation to be called an inner product. Indeed, for any subset of random variables,

- $\mathbb{E}[\mathrm{x}\mathrm{y}] = \mathbb{E}[\mathrm{y}\mathrm{x}]$,
- $\mathbb{E}[(\alpha_1 \mathrm{x}_1 + \alpha_2 \mathrm{x}_2)\mathrm{y}] = \alpha_1 \mathbb{E}[\mathrm{x}_1 \mathrm{y}] + \alpha_2 \mathbb{E}[\mathrm{x}_2 \mathrm{y}]$,
- $\mathbb{E}[\mathrm{x}^2] \geq 0$, with equality if and only if $\mathrm{x} = 0$.

Thus, the norm induced by this inner product,

$$\|\mathrm{x}\| := \sqrt{\mathbb{E}[\mathrm{x}^2]},$$

coincides with the respective standard deviation (assuming $\mathbb{E}[\mathrm{x}] = 0$). From now on, given two uncorrelated random variables x, y, or $\mathbb{E}[\mathrm{x}\mathrm{y}] = 0$, we can call them *orthogonal*, because their inner product is zero. We are now free to apply to our task of interest the orthogonality theorem, which is known to us from our familiar finite-dimensional (Euclidean) linear (vector) spaces.

Let us now rewrite (4.1) as

$$\hat{\mathrm{y}} = \theta_1 \mathrm{x}_1 + \cdots + \theta_l \mathrm{x}_l.$$

Thus, the random variable, $\hat{\mathrm{y}}$, which is now interpreted as a point in a vector space, results as a linear combination of l elements in this space. Thus, the variable $\hat{\mathrm{y}}$ will necessarily lie in the subspace spanned by these points. In contrast, the true variable, y, will not lie, in general, in this subspace. Because our goal is to obtain a $\hat{\mathrm{y}}$ that is a good approximation of y, we have to seek the specific linear combination that makes the norm of the error, $\mathrm{e} = \mathrm{y} - \hat{\mathrm{y}}$, minimum. This specific linear combination corresponds to the *orthogonal* projection of y onto the subspace spanned by the points $\mathrm{x}_1, \mathrm{x}_2, \ldots, \mathrm{x}_l$. This is equivalent

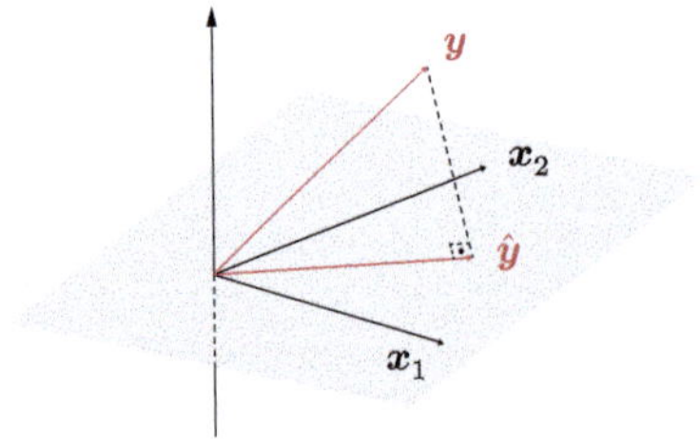

FIGURE 4.3

Projecting y on the subspace spanned by x_1, x_2 (shaded plane) guarantees that the deviation between y and ŷ corresponds to the minimum MSE.

to requiring

$$\boxed{\mathbb{E}[e x_k] = 0, \quad k = 1, \ldots, l: \quad \text{orthogonality condition.}} \tag{4.10}$$

The error variable being orthogonal to every point x_k, $k = 1, 2, \ldots, l$, will be orthogonal to the respective subspace. This is illustrated in Fig. 4.3. Such a choice guarantees that the resulting error will have the minimum norm; by the definition of the norm, this corresponds to the minimum MSE, i.e., to the minimum $\mathbb{E}[e^2]$.

The set of equations in (4.10) can now be written as

$$\mathbb{E}\left[\left(y - \sum_{i=1}^{l} \theta_i x_i\right) x_k\right] = 0, \quad k = 1, 2, \ldots, l,$$

or

$$\sum_{i=1}^{l} \mathbb{E}[x_i x_k]\theta_i = \mathbb{E}[x_k y], \quad k = 1, 2, \ldots, l, \tag{4.11}$$

which leads to the linear set of equations in (4.5).

This is the reason that this elegant set of equations is known as *normal equations*. Another name is *Wiener–Hopf equations*. Strictly speaking, the Wiener–Hopf equations were first derived for continuous-time processes in the context of the causal estimation task [49,50]; for a discussion see [16,44].

Nobert Wiener was a mathematician and philosopher. He was awarded a PhD at Harvard at the age of 17 in mathematical logic. During the Second World War, he laid the foundations of linear estimation theory in a classified work, independently of Kolmogorov. Later on, Wiener was involved in pioneering work embracing automation, artificial intelligence, and cognitive science. Being a pacifist, he was regarded with suspicion during the Cold War years.

The other pillar on which linear estimation theory is based is the pioneering work of Andrey Nikolaevich Kolmogorov (1903–1987) [24], who developed his theory independent of Wiener. Kolmogorov's contributions cover a wide range of topics in mathematics, including probability, computational complexity, and topology. He is the father of the modern axiomatic foundation of the notion of probability (see Chapter 2).

Remarks 4.2.

- So far, in our theoretical findings, we have assumed that $\mathbf{x}$ and y are jointly distributed (correlated) variables. If, in addition, we assume that they are linearly related according to the linear regression model,

$$\mathrm{y} = \boldsymbol{\theta}_o^T \mathbf{x} + \eta, \quad \boldsymbol{\theta}_o \in \mathbb{R}^k, \tag{4.12}$$

  where η is a zero mean noise variable independent of $\mathbf{x}$, then, if the dimension k of the true system $\boldsymbol{\theta}_o$ is equal to the number of parameters, l, adopted for the model, so that $k = l$, then it turns out that (Problem 4.4)

$$\boldsymbol{\theta}_* = \boldsymbol{\theta}_o,$$

  and the optimal MSE is equal to the variance of the noise, σ_η^2.
- *Undermodeling*. If $k > l$, then the order of the model is less than that of the true system, which relates y and $\mathbf{x}$ in (4.12); this is known as *undermodeling*. It is easy to show that if the variables comprising $\mathbf{x}$ are uncorrelated, then (Problem 4.5)

$$\boldsymbol{\theta}_* = \boldsymbol{\theta}_o^1,$$

  where

$$\boldsymbol{\theta}_o := \begin{bmatrix} \boldsymbol{\theta}_o^1 \\ \boldsymbol{\theta}_o^2 \end{bmatrix}, \quad \boldsymbol{\theta}_o^1 \in \mathbb{R}^l, \quad \boldsymbol{\theta}_o^2 \in \mathbb{R}^{k-l}.$$

  In other words, the MSE optimal estimator identifies the first l components of $\boldsymbol{\theta}_o$.

4.4 EXTENSION TO COMPLEX-VALUED VARIABLES

Everything that has been said so far can be extended to complex-valued signals. However, there are a few subtle points involved and this is the reason that we chose to treat this case separately. Complex-valued variables are very common in a number of applications, as for example in communications, e.g., [41].

Given two real-valued variables, (x, y), one can either consider them as a vector quantity in the two-dimensional space, $[x, y]^T$, or describe them as a complex variable, $z = x + jy$, where $j^2 := -1$. Adopting the latter approach offers the luxury of exploiting the operations available in the field $\mathbb{C}$ of complex numbers, i.e., multiplication and division. The existence of such operations greatly facilitates the algebraic manipulations. Recall that such operations are not defined in vector spaces.[3]

[3] Multiplication and division can also be defined for groups of four variables, (x, ϕ, z, y), known as quaternions; the related algebra was introduced by Hamilton in 1843. The real and complex numbers as well as quaternions are all special cases of the so-called Clifford algebras [39].

Let us assume that we are given a complex-valued (output) random variable

$$\mathrm{y} := \mathrm{y}_r + j\mathrm{y}_i \tag{4.13}$$

and a complex-valued (input) random vector

$$\mathbf{x} = \mathbf{x}_r + j\mathbf{x}_i. \tag{4.14}$$

The quantities $\mathrm{y}_r, \mathrm{y}_i, \mathbf{x}_r$, and $\mathbf{x}_i$ are real-valued random variables/vectors. The goal is to compute a linear estimator defined by a complex-valued parameter vector $\boldsymbol{\theta} = \boldsymbol{\theta}_r + j\boldsymbol{\theta}_i \in \mathbb{C}^l$, so as to minimize the respective MSE,

$$\mathbb{E}\left[|\mathrm{e}|^2\right] := \mathbb{E}\left[\mathrm{e}\mathrm{e}^*\right] = \mathbb{E}\left[|\mathrm{y} - \boldsymbol{\theta}^H\mathbf{x}|^2\right]. \tag{4.15}$$

Looking at (4.15), it is readily observed that in the case of complex variables the inner product operation between two complex-valued random variables should be defined as $\mathbb{E}[\mathrm{x}\mathrm{y}^*]$, so as to guarantee that the implied norm by the inner product, $\|\mathrm{x}\| = \sqrt{\mathbb{E}[\mathrm{x}\mathrm{x}^*]}$, is a valid quantity. Applying the orthogonality condition as before, we rederive the normal equations as in (4.11),

$$\Sigma_x \boldsymbol{\theta}_* = \boldsymbol{p}, \tag{4.16}$$

where now the covariance matrix and cross-correlation vector are given by

$$\Sigma_x = \mathbb{E}\left[\mathbf{x}\mathbf{x}^H\right], \tag{4.17}$$

$$\boldsymbol{p} = \mathbb{E}\left[\mathbf{x}\mathrm{y}^*\right]. \tag{4.18}$$

Note that (4.16)–(4.18) can alternatively be obtained by minimizing (4.15) (Problem 4.6). Moreover, the counterpart of (4.9) is given by

$$J(\boldsymbol{\theta}_*) = \sigma_y^2 - \boldsymbol{p}^H \Sigma_x^{-1} \boldsymbol{p} = \sigma_y^2 - \boldsymbol{p}^H \boldsymbol{\theta}_*. \tag{4.19}$$

Using the definitions in (4.13) and (4.14), the cost in (4.15) is written as

$$\begin{aligned} J(\boldsymbol{\theta}) &= \mathbb{E}[|\mathrm{e}|^2] = \mathbb{E}[|\mathrm{y} - \hat{\mathrm{y}}|^2] \\ &= \mathbb{E}[|\mathrm{y}_r - \hat{\mathrm{y}}_r|^2] + \mathbb{E}[|\mathrm{y}_i - \hat{\mathrm{y}}_i|^2], \end{aligned} \tag{4.20}$$

where

$$\boxed{\hat{\mathrm{y}} := \hat{\mathrm{y}}_r + j\hat{\mathrm{y}}_i = \boldsymbol{\theta}^H\mathbf{x}: \quad \text{complex linear estimator},} \tag{4.21}$$

or

$$\begin{aligned} \hat{\mathrm{y}} &= (\boldsymbol{\theta}_r^T - j\boldsymbol{\theta}_i^T)(\mathbf{x}_r + j\mathbf{x}_i) \\ &= (\boldsymbol{\theta}_r^T\mathbf{x}_r + \boldsymbol{\theta}_i^T\mathbf{x}_i) + j(\boldsymbol{\theta}_r^T\mathbf{x}_i - \boldsymbol{\theta}_i^T\mathbf{x}_r). \end{aligned} \tag{4.22}$$

Eq. (4.22) reveals the true flavor behind the complex notation; that is, its *multichannel* nature. In multichannel estimation, we are given more than one set of input variables, namely, $\mathbf{x}_r$ and $\mathbf{x}_i$, and we want to generate, jointly, more than one output variable, namely, $\hat{y}_r$ and $\hat{y}_i$. Eq. (4.22) can equivalently be written as

$$\begin{bmatrix} \hat{y}_r \\ \hat{y}_i \end{bmatrix} = \Theta \begin{bmatrix} \mathbf{x}_r \\ \mathbf{x}_i \end{bmatrix}, \tag{4.23}$$

where

$$\Theta := \begin{bmatrix} \boldsymbol{\theta}_r^T & \boldsymbol{\theta}_i^T \\ -\boldsymbol{\theta}_i^T & \boldsymbol{\theta}_r^T \end{bmatrix}. \tag{4.24}$$

Multichannel estimation can be generalized to more than two outputs and to more than two input sets of variables. We will come back to the more general multichannel estimation task toward the end of this chapter.

Looking at (4.23), we observe that starting from the direct generalization of the linear estimation task for real-valued signals, which led to the adoption of $\hat{y} = \boldsymbol{\theta}^H \mathbf{x}$, resulted in a matrix, Θ, of a *very special structure*.

4.4.1 WIDELY LINEAR COMPLEX-VALUED ESTIMATION

Let us define the linear two-channel estimation task starting from the definition of a linear operation in vector spaces. The task is to generate a vector output, $\hat{\mathbf{y}} = [\hat{y}_r, \hat{y}_i]^T \in \mathbb{R}^2$, from the input vector variables, $\mathbf{x} = [\mathbf{x}_r^T, \mathbf{x}_i^T]^T \in \mathbb{R}^{2l}$, via the linear operation

$$\hat{\mathbf{y}} = \begin{bmatrix} \hat{y}_r \\ \hat{y}_i \end{bmatrix} = \Theta \begin{bmatrix} \mathbf{x}_r \\ \mathbf{x}_i \end{bmatrix}, \tag{4.25}$$

where

$$\Theta := \begin{bmatrix} \boldsymbol{\theta}_{11}^T & \boldsymbol{\theta}_{12}^T \\ \boldsymbol{\theta}_{21}^T & \boldsymbol{\theta}_{22}^T \end{bmatrix}, \tag{4.26}$$

and compute the matrix Θ so as to minimize the total error variance

$$\Theta_* := \arg\min_{\Theta} \left\{ \mathbb{E}\left[(y_r - \hat{y}_r)^2 \right] + \mathbb{E}\left[(y_i - \hat{y}_i)^2 \right] \right\}. \tag{4.27}$$

Note that (4.27) can equivalently be written as

$$\Theta_* := \arg\min_{\Theta} \left\{ \mathbb{E}\left[\mathbf{e}^T \mathbf{e} \right] \right\} = \arg\min_{\Theta} \left\{ \text{trace}\left\{ \mathbb{E}\left[\mathbf{e}\mathbf{e}^T \right] \right\} \right\},$$

where

$$\mathbf{e} := \mathbf{y} - \hat{\mathbf{y}}.$$

Minimizing (4.27) is equivalent to minimizing the two terms individually; in other words, treating each channel separately (Problem 4.7). Thus, the task can be tackled by solving two sets of normal equations, namely,

$$\Sigma_\varepsilon \begin{bmatrix} \boldsymbol{\theta}_{11} \\ \boldsymbol{\theta}_{12} \end{bmatrix} = \boldsymbol{p}_r, \quad \Sigma_\varepsilon \begin{bmatrix} \boldsymbol{\theta}_{21} \\ \boldsymbol{\theta}_{22} \end{bmatrix} = \boldsymbol{p}_i, \tag{4.28}$$

where

$$\begin{aligned} \Sigma_\varepsilon &:= \mathbb{E}\left[\begin{bmatrix} \mathbf{x}_r \\ \mathbf{x}_i \end{bmatrix} \begin{bmatrix} \mathbf{x}_r^T, & \mathbf{x}_i^T \end{bmatrix}\right] \\ &= \begin{bmatrix} \mathbb{E}[\mathbf{x}_r\mathbf{x}_r^T] & \mathbb{E}[\mathbf{x}_r\mathbf{x}_i^T] \\ \mathbb{E}[\mathbf{x}_i\mathbf{x}_r^T] & \mathbb{E}[\mathbf{x}_i\mathbf{x}_i^T] \end{bmatrix} := \begin{bmatrix} \Sigma_r & \Sigma_{ri} \\ \Sigma_{ir} & \Sigma_i \end{bmatrix} \end{aligned} \tag{4.29}$$

and

$$\boldsymbol{p}_r := \mathbb{E}\begin{bmatrix} \mathbf{x}_r \mathrm{y}_r \\ \mathbf{x}_i \mathrm{y}_r \end{bmatrix}, \quad \boldsymbol{p}_i := \mathbb{E}\begin{bmatrix} \mathbf{x}_r \mathrm{y}_i \\ \mathbf{x}_i \mathrm{y}_i \end{bmatrix}. \tag{4.30}$$

The obvious question that is now raised is whether we can tackle this more general task of the two-channel linear estimation task by employing complex-valued arithmetic. The answer is in the affirmative. Let us define

$$\boldsymbol{\theta} := \boldsymbol{\theta}_r + j\boldsymbol{\theta}_i, \quad \boldsymbol{v} := \boldsymbol{v}_r + j\boldsymbol{v}_i, \tag{4.31}$$

and

$$\mathbf{x} = \mathbf{x}_r + j\mathbf{x}_i.$$

Then we define

$$\boldsymbol{\theta}_r := \frac{1}{2}(\boldsymbol{\theta}_{11} + \boldsymbol{\theta}_{22}), \quad \boldsymbol{\theta}_i := \frac{1}{2}(\boldsymbol{\theta}_{12} - \boldsymbol{\theta}_{21}) \tag{4.32}$$

and

$$\boldsymbol{v}_r := \frac{1}{2}(\boldsymbol{\theta}_{11} - \boldsymbol{\theta}_{22}), \quad \boldsymbol{v}_i := -\frac{1}{2}(\boldsymbol{\theta}_{12} + \boldsymbol{\theta}_{21}). \tag{4.33}$$

Under the previous definitions, it is a matter of simple algebra (Problem 4.8) to prove that the set of equations in (4.25) is equivalent to

$$\boxed{\hat{\mathrm{y}} := \hat{\mathrm{y}}_r + j\hat{\mathrm{y}}_i = \boldsymbol{\theta}^H\mathbf{x} + \boldsymbol{v}^H\mathbf{x}^* : \quad \text{widely linear complex estimator.}} \tag{4.34}$$

To distinguish from (4.21), this is known as *widely linear* complex-valued estimator. Note that in (4.34), $\mathbf{x}$ and its complex conjugate $\mathbf{x}^*$ are *simultaneously* used in order to cover all possible solutions, as those are dictated by the vector space description, which led to the formulation in (4.25).

Circularity Conditions

We now turn our attention to investigating conditions under which the widely linear formulation in (4.34) breaks down to (4.21); that is, the conditions for which the optimal widely linear estimator turns out to have $\boldsymbol{v} = \mathbf{0}$.

Let

$$\boldsymbol{\varphi} := \begin{bmatrix} \boldsymbol{\theta} \\ \boldsymbol{v} \end{bmatrix} \quad \text{and} \quad \tilde{\mathbf{x}} := \begin{bmatrix} \mathbf{x} \\ \mathbf{x}^* \end{bmatrix}. \tag{4.35}$$

Then the widely linear estimator is written as

$$\hat{\mathrm{y}} = \boldsymbol{\varphi}^H \tilde{\mathbf{x}}.$$

Adopting the orthogonality condition in its complex formulation

$$\mathbb{E}\left[\tilde{\mathbf{x}}\mathrm{e}^*\right] = \mathbb{E}\left[\tilde{\mathbf{x}}\left(\mathrm{y} - \hat{\mathrm{y}}\right)^*\right] = \mathbf{0},$$

we obtain the following set of normal equations for the optimal $\boldsymbol{\varphi}_*$:

$$\mathbb{E}\left[\tilde{\mathbf{x}}\tilde{\mathbf{x}}^H\right]\boldsymbol{\varphi}_* = \mathbb{E}\left[\tilde{\mathbf{x}}\tilde{\mathbf{x}}^H\right]\begin{bmatrix} \boldsymbol{\theta}_* \\ \boldsymbol{v}_* \end{bmatrix} = \begin{bmatrix} \mathbb{E}[\mathbf{x}\mathrm{y}^*] \\ \mathbb{E}[\mathbf{x}^*\mathrm{y}^*] \end{bmatrix},$$

or

$$\begin{bmatrix} \Sigma_x & P_x \\ P_x^* & \Sigma_x^* \end{bmatrix}\begin{bmatrix} \boldsymbol{\theta}_* \\ \boldsymbol{v}_* \end{bmatrix} = \begin{bmatrix} \boldsymbol{p} \\ \boldsymbol{q}^* \end{bmatrix}, \tag{4.36}$$

where Σ_x and $\boldsymbol{p}$ have been defined in (4.17) and (4.18), respectively, and

$$P_x := \mathbb{E}[\mathbf{x}\mathbf{x}^T], \quad \boldsymbol{q} := \mathbb{E}[\mathbf{x}\mathrm{y}]. \tag{4.37}$$

The matrix P_x is known as the *pseudocovariance/autocorrelation* matrix of $\mathbf{x}$. Note that (4.36) is the equivalent of (4.28); to obtain the widely linear estimator, one needs to solve one set of complex-valued equations whose number is double compared to that of the linear (complex) formulation.

Assume now that

$$\boxed{P_x = O \text{ and } \boldsymbol{q} = \mathbf{0}: \quad \text{circularity conditions.}} \tag{4.38}$$

We say that in this case, the input–output variables are *jointly circular* and the input variables in $\mathbf{x}$ obey the (second-order) *circular* condition. It is readily observed that under the previous circularity assumptions, (4.36) leads to $\boldsymbol{v}_* = \mathbf{0}$ and the optimal $\boldsymbol{\theta}_*$ is given by the set of normal equations (4.16)–(4.18), which govern the more restricted linear case. Thus, adopting the linear formulation leads to optimality only under certain conditions, which do not always hold true in practice; a typical example of such variables, which do not respect circularity, are met in fMRI imaging (see [1] and the references therein). It can be shown that the MSE achieved by a widely linear estimator is always less than or equal to that obtained via a linear one (Problem 4.9).

The notions of circularity and of the widely linear estimation were treated in a series of fundamental papers [35,36]. A stronger condition for circularity is based on the PDF of a complex random variable: A random variable x is circular (or strictly circular) if x and $\mathrm{x}e^{j\phi}$ are distributed according to the same PDF; that is, the PDF is *rotationally invariant* [35]. Fig. 4.4A shows the scatter plot of points generated by a circularly distributed variable, and Fig. 4.4B corresponds to a noncircular one. Strict circularity implies the second-order circularity, but the converse is not always true. For more on complex random variables, the interested reader may consult [3,37]. In [28], it is pointed out that the full second-order

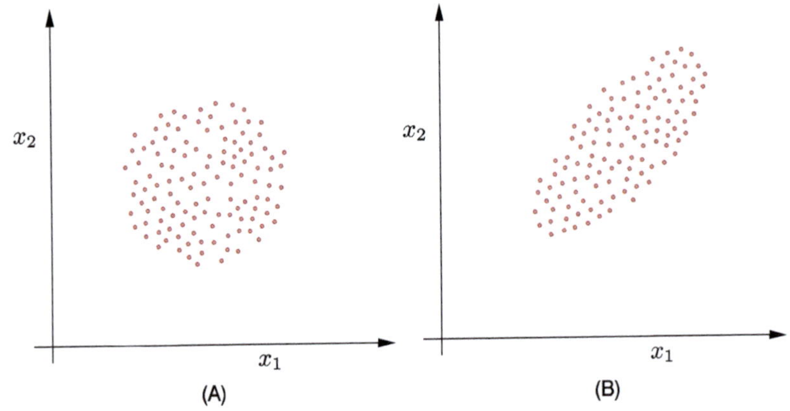

FIGURE 4.4

Scatter plots of points corresponding to (A) a circular process and (B) a noncircular one, in the two-dimensional space.

statistics of the error, without doubling the dimension, can be achieved if instead of the MSE one employs the Gaussian entropy criterion.

Finally, note that substituting in (4.29) the second-order circularity conditions, given in (4.38), one obtains (Problem 4.10)

$$\Sigma_r = \Sigma_i, \quad \Sigma_{ri} = -\Sigma_{ir}, \quad \mathbb{E}[\mathbf{x}_r \mathrm{y}_r] = \mathbb{E}[\mathbf{x}_i \mathrm{y}_i], \quad \mathbb{E}[\mathbf{x}_i \mathrm{y}_r] = -\mathbb{E}[\mathbf{x}_r \mathrm{y}_i], \tag{4.39}$$

which then implies that $\boldsymbol{\theta}_{11} = \boldsymbol{\theta}_{22}$ and $\boldsymbol{\theta}_{12} = -\boldsymbol{\theta}_{21}$; in this case, (4.33) verifies that $\boldsymbol{v} = \mathbf{0}$ and that the optimal in the MSE sense solution has the special structure of (4.23) and (4.24).

4.4.2 OPTIMIZING WITH RESPECT TO COMPLEX-VALUED VARIABLES: WIRTINGER CALCULUS

So far, in order to derive the estimates of the parameters, for both the linear as well as the widely linear estimators, the orthogonality condition was mobilized. For the complex linear estimation case, the normal equations were derived in Problem 4.6, by direct minimization of the cost function in (4.20). Those who got involved with solving the problem have experienced a procedure that was more cumbersome compared to the real-valued linear estimation. This is because one has to use the real and imaginary parts of all the involved complex variables and express the cost function in terms of the equivalent real-valued quantities *only*; then the required gradients for the optimization have to be performed. Recall that any complex function $f : \mathbb{C} \rightarrow \mathbb{R}$ is not differentiable with respect to its complex argument, because the Cauchy–Riemann conditions are violated (Problem 4.11). The previously stated procedure of splitting the involved variables into their real and imaginary parts can become cumbersome with respect to algebraic manipulations. Wirtinger calculus provides an equivalent formulation that is based on simple rules and principles, which bear a great resemblance to the rules of standard complex differentiation.

Let $f : \mathbb{C} \longmapsto \mathbb{C}$ be a complex function defined on $\mathbb{C}$. Obviously, such a function can be regarded as either defined on $\mathbb{R}^2$ or $\mathbb{C}$ (i.e., $f(z) = f(x + jy) = f(x, y)$). Furthermore, it may be regarded as either complex-valued, $f(x, y) = f_r(x, y) + jf_i(x, y)$, or as vector-valued $f(x, y) = (f_r(x, y), f_i(x, y))$. We say that f is *differentiable in the real sense* if both f_r and f_i are differentiable. Wirtinger's calculus considers the complex structure of f and the real derivatives are described using an equivalent formulation that greatly simplifies calculations; moreover, this formulation bears a surprising similarity to the complex derivatives.

Definition 4.1. The *Wirtinger derivative* or *W-derivative* of a complex function f at a point $z_0 \in \mathbb{C}$ is defined as

$$\boxed{\frac{\partial f}{\partial z}(z_0) = \frac{1}{2}\left(\frac{\partial f_r}{\partial x}(z_0) + \frac{\partial f_i}{\partial y}(z_0)\right) + \frac{j}{2}\left(\frac{\partial f_i}{\partial x}(z_0) - \frac{\partial f_r}{\partial y}(z_0)\right) : \text{ W-derivative.}}$$

The *conjugate Wirtinger's derivative* or *CW-derivative* of f at z_0 is defined as

$$\boxed{\frac{\partial f}{\partial z^*}(z_0) = \frac{1}{2}\left(\frac{\partial f_r}{\partial x}(z_0) - \frac{\partial f_i}{\partial y}(z_0)\right) + \frac{j}{2}\left(\frac{\partial f_i}{\partial x}(z_0) + \frac{\partial f_r}{\partial y}(z_0)\right) : \text{ CW-derivative.}}$$

For some of the properties and the related proofs regarding Wirtinger's derivatives, see Appendix A.3. An important property for us is that if f is real-valued (i.e., $\mathbb{C} \longmapsto \mathbb{R}$) and z_0 is a (local) optimal point of f, it turns out that

$$\boxed{\frac{\partial f}{\partial z}(z_0) = \frac{\partial f}{\partial z^*}(z_0) = 0 : \quad \text{optimality conditions.}} \tag{4.40}$$

In order to apply Wirtinger's derivatives, the following simple *tricks* are adopted:

- express function f in terms of z *and* z^*;
- to compute the W-derivative apply the usual differentiation rule, treating z^* as a constant;
- to compute the CW-derivative apply the usual differentiation rule, treating z as a constant.

It should be emphasized that all these statements must be regarded as useful computational tricks rather than rigorous mathematical rules. Analogous definitions and properties carry on for complex vectors z, and the W-gradient and CW-gradients

$$\nabla_z f(z_0), \quad \nabla_{z^*} f(z_0)$$

result from the respective definitions if partial derivatives are replaced by partial gradients, ∇_x, ∇_y.

Although Wirtinger's calculus has been known since 1927 [51], its use in applications has a rather recent history [7]. Its revival was ignited by the widely linear filtering concept [27]. The interested reader may obtain more on this issue from [2,25,30]. Extensions of Wirtinger's derivative to general Hilbert (infinite-dimensional) spaces was done more recently in [6] and to the subgradient notion in [46].

Application in linear estimation. The cost function in this case is

$$J(\boldsymbol{\theta}, \boldsymbol{\theta}^*) = \mathbb{E}\left[|\mathrm{y} - \boldsymbol{\theta}^H \mathbf{x}|^2\right] = \mathbb{E}\left[\left(\mathrm{y} - \boldsymbol{\theta}^H \mathbf{x}\right)\left(\mathrm{y}^* - \boldsymbol{\theta}^T \mathbf{x}^*\right)\right].$$

Thus, treating $\boldsymbol{\theta}$ as a constant, the optimal occurs at

$$\nabla_{\boldsymbol{\theta}^*} J = \mathbb{E}\left[\mathbf{x}\mathrm{e}^*\right] = \mathbf{0},$$

which is the orthogonality condition leading to the normal equations (4.16)–(4.18).

Application in widely linear estimation. The cost function is now (see notation in (4.35))

$$J(\boldsymbol{\varphi}, \boldsymbol{\varphi}^*) = \mathbb{E}\left[\left(\mathrm{y} - \boldsymbol{\varphi}^H \tilde{\mathbf{x}}\right)\left(\mathrm{y}^* - \boldsymbol{\varphi}^T \tilde{\mathbf{x}}^*\right)\right],$$

and treating $\boldsymbol{\varphi}$ as a constant,

$$\nabla_{\boldsymbol{\varphi}^*} J = \mathbb{E}\left[\tilde{\mathbf{x}}\mathrm{e}^*\right] = \mathbb{E}\begin{bmatrix}\mathbf{x}\mathrm{e}^* \\ \mathbf{x}^*\mathrm{e}^*\end{bmatrix} = \mathbf{0},$$

which leads to the set derived in (4.36).

Wirtinger's calculus will prove very useful in subsequent chapters for deriving gradient operations in the context of online/adaptive estimation in Euclidean as well as in reproducing kernel Hilbert spaces.

4.5 LINEAR FILTERING

Linear statistical filtering is an instance of the general estimation task, when the notion of time evolution needs to be taken into consideration and estimates are obtained at each time instant. There are three major types of problems that emerge:

- *Filtering*, where the estimate at time instant n is based on all previously received (measured) input information *up to and including* the current time index, n.
- *Smoothing*, where data over a time interval $[0, N]$ are first collected and an estimate is obtained at each time instant $n \leq N$, using *all* the available information in the interval $[0, N]$.
- *Prediction*, where estimates at times $n + \tau, \ \tau > 0$, are to be obtained based on the information up to and including time instant n.

To fit in the above definitions more with what has been said so far in the chapter, take for example a time-varying case, where the output variable at time instant n is y_n and its value depends on observations included in the corresponding input vector $\mathbf{x}_n$. In filtering, the latter can include measurements received only at time instants $n, n-1, \ldots, 0$. This restriction in the index set is directly related to *causality*. In contrast, in smoothing, future time instants are included in addition to the past, i.e., $\ldots, n+2, n+1, n, n-1, \ldots$.

Most of the effort in this book will be spent on filtering whenever time information enters into the picture. The reason is that this is the most commonly encountered task and, also, the techniques used for smoothing and prediction are similar in nature to that of filtering, with usually minor modifications.

In signal processing, the term filtering is usually used in a more specific context, and it refers to the operation of a *filter*, which acts on an input random process/signal (u_n), to transform it into another one (d_n); see Section 2.4.3. Note that we have switched into the notation, introduced in Chapter 2, used to denote random processes. We prefer to keep different notation for processes and random variables,

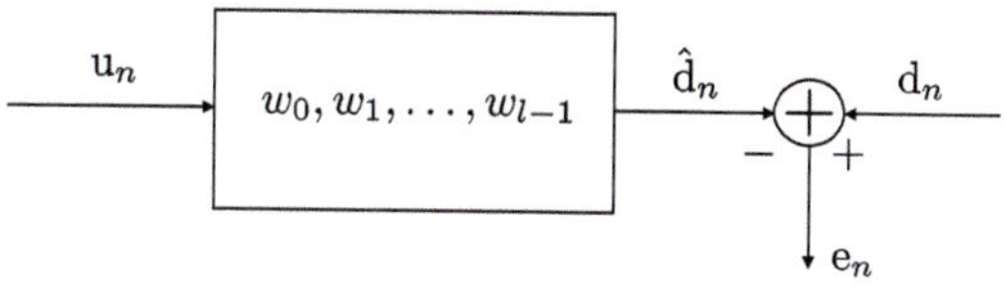

FIGURE 4.5

In statistical filtering, the impulse response coefficients are estimated so as to minimize the error between the output and the desired response processes. In MSE linear filtering, the cost function is $\mathbb{E}[\mathrm{e}_n^2]$.

because in the case of random processes, the filtering task obtains a special structure and properties, as we will soon see. Moreover, although the mathematical formulation of the involved equations, for both cases, may end up to be the same, we feel that it is good for the reader to keep in mind that there is a different underlying mechanism for generating the data.

The task in statistical linear filtering is to compute the coefficients (impulse response) of the filter so that the output process of the filter, $\hat{\mathrm{d}}_n$, when the filter is excited by the input random process, u_n, is as close as possible to a *desired* response process, d_n. In other words, the goal is to minimize, in some sense, the corresponding error process (see Fig. 4.5). Assuming that the unknown filter is of a finite impulse response (FIR) (see Section 2.4.3 for related definitions), denoted as $w_0, w_1, \ldots, w_{l-1}$, the output $\hat{\mathrm{d}}_n$ of the filter is given as

$$\boxed{\hat{\mathrm{d}}_n = \sum_{i=0}^{l-1} w_i \mathrm{u}_{n-i} = \boldsymbol{w}^T \mathbf{u}_n : \quad \text{convolution sum,}} \tag{4.41}$$

where

$$\boldsymbol{w} = [w_0, w_1, \ldots, w_{l-1}]^T \quad \text{and} \quad \mathbf{u}_n = [\mathrm{u}_n, \mathrm{u}_{n-1}, \ldots, \mathrm{u}_{n-l+1}]^T. \tag{4.42}$$

Fig. 4.6 illustrates the convolution operation of the linear filter, when its input is excited by a realization u_n of the input processes to provide in the output the signal/sequence $\hat{d}_n$.

Alternatively, (4.41) can be viewed as a *linear* estimator function; given the jointly distributed variables, at time instant n, $(\mathrm{d}_n, \mathbf{u}_n)$, (4.41) provides the estimator, $\hat{\mathrm{d}}_n$, given the values of $\mathbf{u}_n$. In order to obtain the coefficients, $\boldsymbol{w}$, the MSE criterion will be adopted. Furthermore, we will assume that:

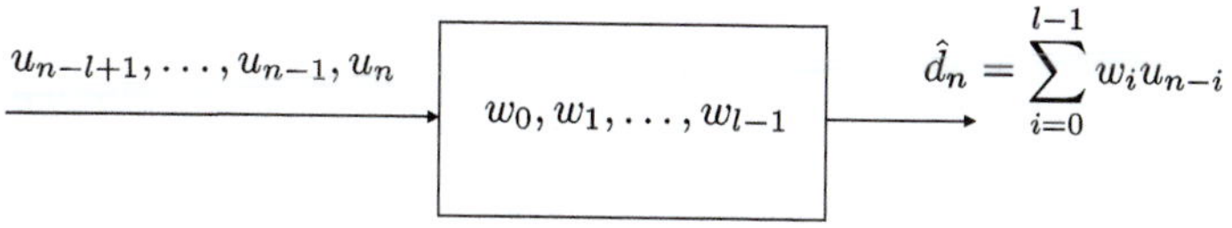

FIGURE 4.6

The linear filter is excited by a realization of an input process. The output signal is the convolution between the input sequence and the filter's impulse response.

- The processes u_n, d_n are *wide-sense stationary* real random processes.
- Their mean values are equal to zero, in other words, $\mathbb{E}[\mathrm{u}_n] = \mathbb{E}[\mathrm{d}_n] = 0$, $\forall n$. If this is not the case, we can subtract the respective mean values from the processes, u_n and d_n, during a preprocessing stage. Due to this assumption, the autocorrelation and covariance matrices of $\mathbf{u}_n$ coincide, so that

$$R_u = \Sigma_u.$$

The normal equations in (4.5) now take the form

$$\Sigma_u \boldsymbol{w} = \boldsymbol{p},$$

where

$$\boldsymbol{p} = \left[\mathbb{E}[\mathrm{u}_n \mathrm{d}_n], \dots, \mathbb{E}[\mathrm{u}_{n-l+1}\mathrm{d}_n]\right]^T,$$

and the respective covariance/autocorrelation matrix, of order l, of the input process is given by

$$\Sigma_u := \mathbb{E}\left[\mathbf{u}_n \mathbf{u}_n^T\right] = \begin{bmatrix} r(0) & r(1) & \dots & r(l-1) \\ r(1) & r(0) & \dots & r(l-2) \\ \vdots & & \ddots & \\ r(l-1) & r(l-2) & \dots & r(0) \end{bmatrix}, \tag{4.43}$$

where $r(k)$ is the autocorrelation sequence of the input process. Because we have assumed that the involved processes are wide-sense stationary, we have

$$r(n, n-k) := \mathbb{E}[\mathrm{u}_n \mathrm{u}_{n-k}] = r(k).$$

Also, recall that, for real wide-sense stationary processes, the autocorrelation sequence is symmetric, or $r(k) = r(-k)$ (Section 2.4.3). Observe that in this case, where the input vector results from a random process, the covariance matrix has a special structure, which will be exploited later on to derive efficient schemes for the solution of the normal equations.

For the complex linear filtering case, the only differences are:

- the output is given as $\hat{\mathrm{d}}_n = \boldsymbol{w}^H \mathbf{u}_n$,
- $\boldsymbol{p} = \mathbb{E}[\mathbf{u}_n \mathrm{d}_n^*]$,
- $\Sigma_u = \mathbb{E}[\mathbf{u}_n \mathbf{u}_n^H]$,
- $r(-k) = r^*(k)$.

4.6 MSE LINEAR FILTERING: A FREQUENCY DOMAIN POINT OF VIEW

Let us now turn our attention to the more general case, and assume that our filter is of *infinite impulse response* (IIR). Then (4.41) becomes

$$\hat{\mathrm{d}}_n = \sum_{i=-\infty}^{+\infty} w_i \mathrm{u}_{n-i}. \tag{4.44}$$

Moreover, we have allowed the filter to be *noncausal*.[4] Following similar arguments as those used to prove the MSE optimality of $\mathbb{E}[y|\boldsymbol{x}]$ in Section 3.9.1, it turns out that the optimal filter coefficients must satisfy the following condition (Problem 4.12):

$$\boxed{\mathbb{E}\left[\left(\mathrm{d}_n - \sum_{i=-\infty}^{+\infty} w_i \mathrm{u}_{n-i}\right)\mathrm{u}_{n-j}\right] = 0, \quad j \in \mathbb{Z}.} \tag{4.45}$$

Observe that this is a generalization (involving an infinite number of terms) of the orthogonality condition stated in (4.10). A rearrangement of the terms in (4.45) results in

$$\sum_{i=-\infty}^{+\infty} w_i \, \mathbb{E}[\mathrm{u}_{n-i}\mathrm{u}_{n-j}] = \mathbb{E}[\mathrm{d}_n \mathrm{u}_{n-j}], \; j \in \mathbb{Z}, \tag{4.46}$$

and finally to

$$\sum_{i=-\infty}^{+\infty} w_i r(j-i) = r_{du}(j), \; j \in \mathbb{Z}, \tag{4.47}$$

where $r_{du}(j)$ denotes the cross-correlation sequence between the processes d_n and u_n. Eq. (4.47) can be considered as the generalization of (4.5) to the case of random processes. The problem now is how one can solve (4.47) that involves infinite many parameters. The way out is to cross into the frequency domain. Eq. (4.47) can be seen as the convolution of the unknown sequence with the autocorrelation sequence of the input process, which gives rise to the cross-correlation sequence. However, we know that convolution of two sequences corresponds to the product of the respective Fourier transforms (e.g., [42] and Section 2.4.2). Thus, we can now write that

$$\boxed{W(\omega)S_u(\omega) = S_{du}(\omega),} \tag{4.48}$$

where $W(\omega)$ is the Fourier transform of the sequence of the unknown parameters and $S_u(\omega)$ is the *power spectral density* of the input process, defined in Section 2.4.3. In analogy, the Fourier transform $S_{du}(\omega)$ of the cross-correlation sequence is known as the *cross-spectral density*. If the latter two quantities are available, then once $W(\omega)$ has been computed, the unknown parameters can be obtained via the inverse Fourier transform.

DECONVOLUTION: IMAGE DEBLURRING

We will now consider an important application in order to demonstrate the power of MSE linear estimation. Image deblurring is a typical *deconvolution* task. An image is degraded due to its transmission via a nonideal system; the task of deconvolution is to optimally recover (in the MSE sense in our case)

---

[4] A system is called *causal* if the output $\hat{\mathrm{d}}_n$ depends *only* on input values u_m, $m \leq n$. A necessary and sufficient condition for causality is that the impulse response is zero for negative time instants, meaning that $w_n = 0$, $n < 0$. This can easily be checked out; try it.

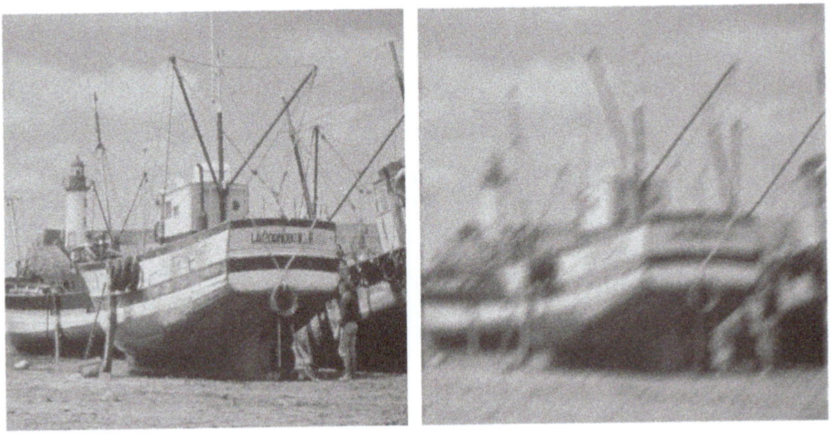

(A) (B)

FIGURE 4.7

(A) The original image and (B) its blurred and noisy version.

the original undegraded image. Fig. 4.7A shows the original image and 4.7B a blurred version (e.g., taken by a nonsteady camera) with some small additive noise.

At this point, it is interesting to recall that deconvolution is a process that our human brain performs all the time. The human (and not only) vision system is one of the most complex and highly developed biological systems that has been formed over millions years of a continuous evolution process. Any raw image that falls on the retina of the eye is *severely blurred.* Thus, one of the main early processing activities of our visual system is to deblur it (see, e.g., [29] and the references therein for a related discussion).

Before we proceed any further, the following assumptions are adopted:

- The image is a *wide-sense stationary* two-dimensional random process. Two-dimensional random processes are also known as *random fields* (see Chapter 15).
- The image is of an infinite extent; this can be justified for the case of large images. This assumption will grant us the "permission" to use (4.48). The fact that an image is a two-dimensional process does not change anything in the theoretical analysis; the only difference is that now the Fourier transforms involve two frequency variables, ω_1, ω_2, one for each of the two dimensions.

A gray image is represented as a two-dimensional array. To stay close to the notation used so far, let $d(n,m),\ n,m \in \mathbb{Z}$, be the original undegraded image (which for us is now the desired response), and $u(n,m),\ n,m \in \mathbb{Z}$, be the degraded one, obtained as

$$u(n,m) = \sum_{i=-\infty}^{+\infty} \sum_{j=-\infty}^{+\infty} h(i,j)d(n-i,m-j) + \eta(n,m), \tag{4.49}$$

where $\eta(n,m)$ is the realization of a noise field, which is assumed to be zero mean and independent of the input (undegraded) image. The sequence $h(i,j)$ is the *point spread sequence* (impulse response) of the system (e.g., camera). We will assume that this is known and it has, somehow, been measured.[5]

Our task now is to estimate a two-dimensional filter, $w(n, m)$, which is applied to the degraded image to optimally reconstruct (in the MSE sense) the original undegraded image. In the current context, Eq. (4.48) is written as

$$W(\omega_1, \omega_2) S_u(\omega_1, \omega_2) = S_{du}(\omega_1, \omega_2).$$

Following similar arguments as those used to derive Eq. (2.130) of Chapter 2, it is shown that (Problem 4.13)

$$S_{du}(\omega_1, \omega_2) = H^*(\omega_1, \omega_2) S_d(\omega_1, \omega_2) \tag{4.50}$$

and

$$S_u(\omega_1, \omega_2) = |H(\omega_1, \omega_2)|^2 S_d(\omega_1, \omega_2) + S_\eta(\omega_1, \omega_2), \tag{4.51}$$

where "*" denotes complex conjugation and S_η is the power spectral density of the noise field. Thus, we finally obtain

$$\boxed{W(\omega_1, \omega_2) = \frac{1}{H(\omega_1, \omega_2)} \frac{|H(\omega_1, \omega_2)|^2}{|H(\omega_1, \omega_2)|^2 + \frac{S_\eta(\omega_1,\omega_2)}{S_d(\omega_1,\omega_2)}}.} \tag{4.52}$$

Once $W(\omega_1, \omega_2)$ has been computed, the unknown parameters could be obtained via an inverse (two-dimensional) Fourier transform. The deblurred image then results as

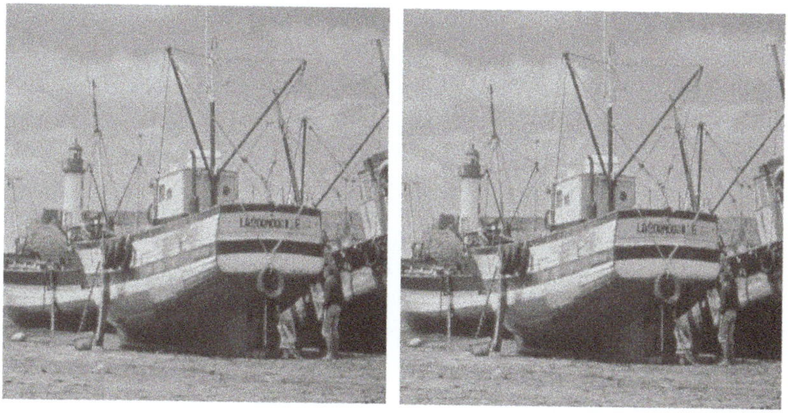

(A) (B)

FIGURE 4.8

(A) The original image and (B) the deblurred one for $C = 2.3 \times 10^{-6}$. Observe that in spite of the simplicity of the method, the reconstruction is pretty good. The differences become more obvious to the eye when the images are enlarged.

[5] Note that this is not always the case.

$$\hat{d}(n,m) = \sum_{i=-\infty}^{+\infty} \sum_{j=-\infty}^{+\infty} w(i,j)u(n-i,m-j). \tag{4.53}$$

In practice, because we are not really interested in obtaining the weights of the deconvolution filter, we implement (4.53) in the frequency domain

$$\hat{D}(\omega_1,\omega_2) = W(\omega_1,\omega_2)U(\omega_1,\omega_2),$$

and then obtain the inverse Fourier transform. Thus all processing is efficiently performed in the frequency domain. Software packages to perform Fourier transforms (via the fast Fourier transform [FFT]) of an image array are "omnipresent" on the internet.

Another important issue is that in practice we do not know $S_d(\omega_1,\omega_2)$. An approximation which is usually adopted and renders sensible results can be made by assuming that $\frac{S_\eta(\omega_1,\omega_2)}{S_d(\omega_1,\omega_2)}$ is a constant, C, and trying different values of it. Fig. 4.8 shows the deblurred image for $C = 2.3 \times 10^{-6}$. The quality of the end result depends a lot on the choice of this value (MATLAB® exercise 4.25). Other, more advanced, techniques have also been proposed. For example, one can get a better estimate of $S_d(\omega_1,\omega_2)$ by using information from $S_\eta(\omega_1,\omega_2)$ and $S_u(\omega_1,\omega_2)$. The interested reader can obtain more on the image deconvolution/restoration task from, e.g., [14,34].

4.7 SOME TYPICAL APPLICATIONS

Optimal linear estimation/filtering has been applied in a wide range of diverse applications of statistical learning, such as regression modeling, communications, control, biomedical signal processing, seismic signal processing, and image processing. In the sequel, we present some typical applications in order for the reader to grasp the main rationale of how the previously stated theory can find its way in solving practical problems. In all cases, wide-sense stationarity of the involved random processes is assumed.

4.7.1 INTERFERENCE CANCELATION

In interference cancelation, we have access to a mixture of two signals expressed as $d_n = y_n + s_n$. Ideally, we would like to remove one of them, say, y_n. We will consider them as realizations of respective

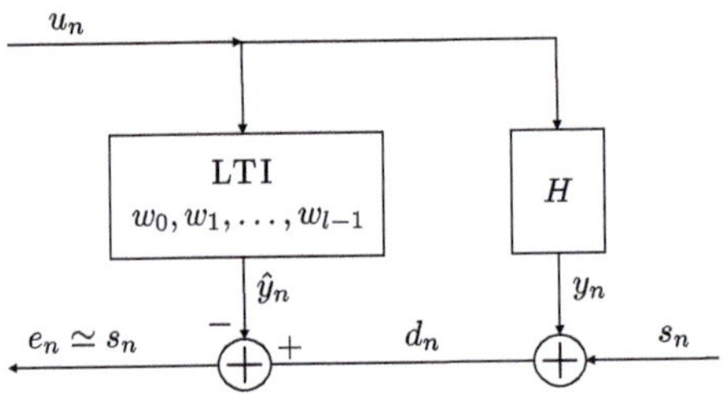

FIGURE 4.9

A basic block diagram illustrating the interference cancelation task.

random processes/signals, or d_n, y_n, and s_n. To achieve this goal, the only available information is another signal, say, u_n, that is statistically related to the unwanted signal, y_n. For example, y_n may be a filtered version of u_n. This is illustrated in Fig. 4.9, where the corresponding realizations of the involved random processes are shown.

Process y_n is the output of an unknown system H, whose input is excited by u_n. The task is to model H by obtaining estimates of its impulse response (assuming that it is linear time-invariant and of known order). Then the output of the model will be an approximation of y_n when this is activated by the same input, u_n. We will use d_n as the desired response process. The optimal estimates of $w_0, \ldots, w_{l-1}$ (assuming the order of the unknown system H to be l) are provided by the normal equations

$$\Sigma_u \boldsymbol{w}_* = \boldsymbol{p}.$$

However,

$$\begin{aligned}\boldsymbol{p} = \mathbb{E}[\mathbf{u}_n \mathrm{d}_n] &= \mathbb{E}\left[\mathbf{u}_n (\mathrm{y}_n + \mathrm{s}_n)\right] \\ &= \mathbb{E}\left[\mathbf{u}_n \mathrm{y}_n\right], \end{aligned} \tag{4.54}$$

because the respective input vector $\mathbf{u}_n$ and s_n are considered statistically independent. That is, the previous formulation of the problem leads to the same normal equations as when the desired response was the signal y_n, which we want to remove! Hence, the output of our model will be an approximation (in the MSE sense), $\hat{\mathrm{y}}_n$, of y_n, and if subtracted from d_n the resulting (error) signal e_n will be an approximation to s_n. How good this approximation is depends on whether l is a good "estimate" of the true order of H. The cross-correlation in the right-hand side of (4.54) can be approximated by computing the respective sample mean values, in particular over periods where s_n is absent. In practical systems, online/adaptive versions of this implementation are usually employed, as we will see in Chapter 5.

Interference cancelation schemes have been widely used in many systems such as noise cancelation, echo cancelation in telephone networks, and video conferencing, and in biomedical applications, for example, in order to cancel the maternal interference in a fetal electrocardiograph.

Fig. 4.10 illustrates the echo cancelation task in a video conference application. The same setup applies to the hands-free telephone service in a car. The *far-end* speech signal is considered to be a realization u_n of a random process u_n; through the loudspeakers, it is broadcasted in room A (car) and it is reflected in the interior of the room. Part of it is absorbed and part of it enters the microphone; this is denoted as y_n. The equivalent response of the room (reflections) on u_n can be represented by a filter, H, as in Fig. 4.9. Signal y_n returns back and the speaker in location B listens to her or his own voice, together with the *near-end* speech signal, s_n, of the speaker in A. In certain cases, this feedback path from the loudspeakers to the microphone can cause instabilities, giving rise to a "howling" sound effect. The goal of the echo canceler is to optimally remove y_n.

4.7.2 SYSTEM IDENTIFICATION

System identification is similar in nature to the interference cancelation task. Note that in Fig. 4.9, one basically models the unknown system. However, the focus there was on replicating the output y_n and not on the system's impulse response.

In system identification, the aim is to model the impulse response of an unknown plant. To this end, we have access to its input signal as well as to a *noisy* version of its output. The task is to design a

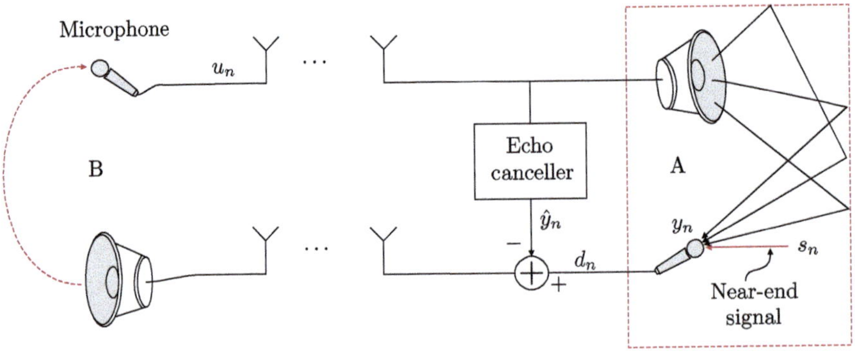

FIGURE 4.10

The echo canceler is optimally designed to remove the part of the far-end signal, u_n, that interferes with the near-end signal, s_n.

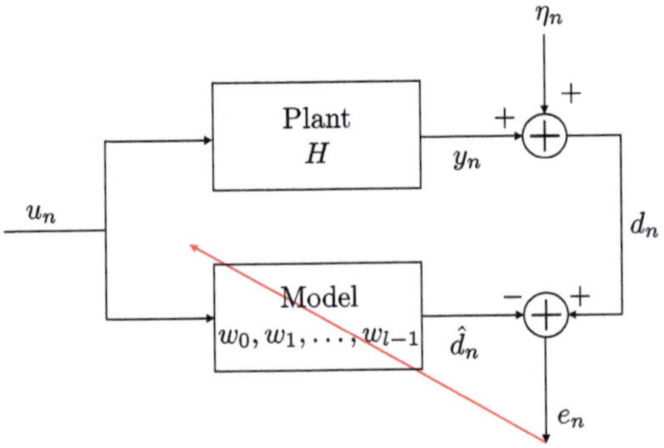

FIGURE 4.11

In system identification, the impulse response of the model is optimally estimated so that the output is close, in the MSE sense, to that of the unknown plant. The red line indicates that the error is used for the optimal estimation of the unknown parameters of the filter.

model whose impulse response approximates that of the unknown plant. To achieve this, we optimally design a linear filter whose input is the same signal as the one that activates the plant and its desired response is the noisy output of the plant (see Fig. 4.11). The associated normal equations are

$$\Sigma_u \boldsymbol{w}_* = \mathbb{E}[\mathbf{u}_n \mathrm{d}_n] = \mathbb{E}[\mathbf{u}_n \mathrm{y}_n] + 0,$$

assuming the noise η_n is statistically independent of $\mathbf{u}_n$. Thus, once more, the resulting normal equations are as if we had provided the model with a desired response equal to the noiseless output of the unknown plant, expressed as $\mathrm{d}_n = \mathrm{y}_n$. Hence, the impulse response of the model is estimated so that its output is close, in the MSE sense, to the true (noiseless) output of the unknown plant. System identification is of major importance in a number of applications. In control, it is used for driving the as-

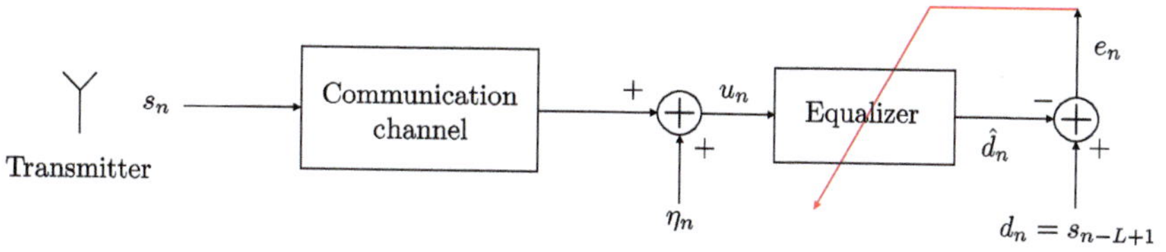

FIGURE 4.12

The task of an equalizer is to optimally recover the originally transmitted information sequence s_n, delayed by L time lags.

sociated controllers. In data communications, it is used for estimating the transmission channel in order to build up maximum likelihood estimators of the transmitted data. In many practical systems, adaptive versions of the system identification scheme are implemented, as we will discuss in following chapters.

4.7.3 DECONVOLUTION: CHANNEL EQUALIZATION

Note that in the cancelation task the goal was to "remove" the (filtered version of the) input signal (u_n) to the unknown system H. In system identification, the focus was on the (unknown) system itself. In *deconvolution*, the emphasis is on the input of the unknown system. That is, our goal now is to recover, in the MSE optimal sense, a (delayed) input signal, $d_n = s_{n-L+1}$, where L is the delay in units of the sampling period T. The task is also called *inverse system identification*. The term *equalization* or *channel equalization* is used in communications. The deconvolution task was introduced in the context of image deblurring in Section 4.6. There, the required information about the *unknown* input process was obtained via an approximation. In the current framework, this can be approached via the transmission of a training sequence.

The goal of an *equalizer* is to recover the transmitted information symbols, by mitigating the so-called *intersymbol interference* (ISI) that any (imperfect) dispersive communication channel imposes on the transmitted signal; besides ISI, additive noise is also present in the transmitted information bits (see Example 4.2). Equalizers are "omnipresent" in these days; in our mobile phones, in our modems, etc. Fig. 4.12 presents the basic scheme for an equalizer. The equalizer is trained so that its output is as close as possible to the transmitted data bits delayed by some time lag L; the delay is used in order to account for the overall delay imposed by the channel equalizer system. Deconvolution/channel equalization is at the heart of a number of applications besides communications, such as acoustics, optics, seismic signal processing, and control. The channel equalization task will also be discussed in the next chapter in the context of online learning via the decision feedback equalization mode of operation.

Example 4.1 (Noise cancelation). The noise cancelation application is illustrated in Fig. 4.13. The signal of interest is a realization of a process s_n, which is contaminated by the noise process $v_1(n)$. For example, s_n may be the speech signal of the pilot in the cockpit and $v_1(n)$ the aircraft noise at the location of the microphone. We assume that $v_1(n)$ is an AR process of order one, expressed as

$$v_1(n) = a_1 v_1(n-1) + \eta_n.$$

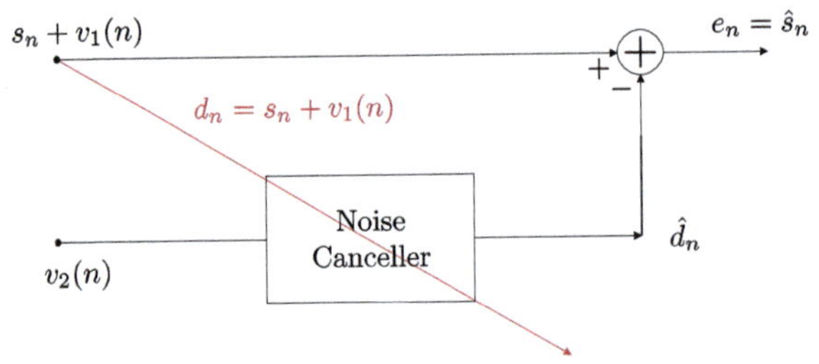

FIGURE 4.13

A block diagram for a noise canceler. The signals shown are realizations of the corresponding random variables that are used in the text. Using as desired response the contaminated signal, the output of the optimal filter is an estimate of the noise component.

The random signal $\mathrm{v}_2(n)$ is a noise sequence[6] which is related to $\mathrm{v}_1(n)$, but it is statistically independent of s_n. For example, it may be the noise picked up from another microphone positioned at a nearby location. This is also assumed to be an AR process of the first order,

$$\mathrm{v}_2(n) = a_2 \mathrm{v}_2(n-1) + \eta_n.$$

Note that both $\mathrm{v}_1(n)$ and $\mathrm{v}_2(n)$ are generated by the same noise source, η_n, which is assumed to be white of variance σ_η^2. For example, in an aircraft it can be assumed that the noise at different points is due to a "common" source, especially for nearby locations.

The goal of the example is to compute estimates of the weights of the noise canceler, in order to optimally remove (in the MSE sense) the noise $\mathrm{v}_1(n)$ from the mixture $\mathrm{s}_n + \mathrm{v}_1(n)$. Assume the canceler to be of order two.

The input to the canceler is $\mathrm{v}_2(n)$ and as desired response the mixture signal, $\mathrm{d}_n = \mathrm{s}_n + \mathrm{v}_1(n)$, will be used. To establish the normal equations, we need to compute the covariance matrix, Σ_2, of $\mathrm{v}_2(n)$ and the cross-correlation vector, $\boldsymbol{p}_2$, between the input random vector, $\mathbf{v}_2(n)$, and d_n.

Because $\mathrm{v}_2(n)$ is an AR process of the first order, recall from Section 2.4.4 that the autocorrelation sequence is given by

$$r_2(k) = \frac{a_2^k \sigma_\eta^2}{1 - a_2^2}, \quad k = 0, 1, \ldots. \tag{4.55}$$

Hence,

$$\Sigma_2 = \begin{bmatrix} r_2(0) & r_2(1) \\ r_2(1) & r_2(0) \end{bmatrix} = \begin{bmatrix} \frac{\sigma_\eta^2}{1-a_2^2} & \frac{a_2\sigma_\eta^2}{1-a_2^2} \\ \frac{a_2\sigma_\eta^2}{1-a_2^2} & \frac{\sigma_\eta^2}{1-a_2^2} \end{bmatrix}.$$

6 We use the index n in parenthesis to unclutter notation due to the presence of a second subscript.

Next, we are going to compute the cross-correlation vector. We have

$$\begin{aligned} p_2(0) :&= \mathbb{E}[\mathrm{v}_2(n)\mathrm{d}_n] = \mathbb{E}[\mathrm{v}_2(n)\,(\mathrm{s}_n + \mathrm{v}_1(n))] \\ &= \mathbb{E}[\mathrm{v}_2(n)\mathrm{v}_1(n)] + 0 = \mathbb{E}\left[(a_2\mathrm{v}_2(n-1) + \eta_n)\,(a_1\mathrm{v}_1(n-1) + \eta_n)\right] \\ &= a_2 a_1 p_2(0) + \sigma_\eta^2, \end{aligned}$$

or

$$p_2(0) = \frac{\sigma_\eta^2}{1 - a_2 a_1}. \tag{4.56}$$

We used the fact that $\mathbb{E}[\mathrm{v}_2(n-1)\eta_n] = \mathbb{E}[\mathrm{v}_1(n-1)\eta_n] = 0$, because $\mathrm{v}_2(n-1)$ and $\mathrm{v}_1(n-1)$ depend recursively on previous values, i.e., $\eta(n-1)$, $\eta(n-2)$, ..., and also that η_n is a white noise sequence, hence the respective correlation values are zero. Also, due to stationarity, $\mathbb{E}[\mathrm{v}_2(n)\mathrm{v}_1(n)] = \mathbb{E}[\mathrm{v}_2(n-1)\mathrm{v}_1\,(n-1)]$.

For the other value of the cross-correlation vector we have

$$\begin{aligned} p_2(1) &= \mathbb{E}[\mathrm{v}_2(n-1)\mathrm{d}_n] = \mathbb{E}[\mathrm{v}_2(n-1)\,(\mathrm{s}_n + \mathrm{v}_1(n))] \\ &= \mathbb{E}[\mathrm{v}_2(n-1)\mathrm{v}_1(n)] + 0 = \mathbb{E}\left[\mathrm{v}_2(n-1)\,(a_1\mathrm{v}_1(n-1) + \eta_n)\right] \\ &= a_1 p_2(0) = \frac{a_1\sigma_\eta^2}{1 - a_1 a_2}. \end{aligned}$$

In general, it is easy to show that

$$p_2(k) = \frac{a_1^k \sigma_\eta^2}{1 - a_2 a_1}, \quad k = 0, 1, \ldots. \tag{4.57}$$

Recall that because the processes are real-valued, the covariance matrix is symmetric, meaning $r_2(k) = r_2(-k)$. Also, for (4.55) to make sense (recall that $r_2(0) > 0$), $|a_2| < 1$. The same holds true for $|a_1|$, following similar arguments for the autocorrelation process of $\mathrm{v}_1(n)$.

Thus, the optimal weights of the noise canceler are given by the following set of normal equations:

$$\begin{bmatrix} \dfrac{\sigma_\eta^2}{1-a_2^2} & \dfrac{a_2\sigma_\eta^2}{1-a_2^2} \\ \dfrac{a_2\sigma_\eta^2}{1-a_2^2} & \dfrac{\sigma_\eta^2}{1-a_2^2} \end{bmatrix} \boldsymbol{w} = \begin{bmatrix} \dfrac{\sigma_\eta^2}{1-a_1a_2} \\ \dfrac{a_1\sigma_\eta^2}{1-a_1a_2} \end{bmatrix}.$$

Note that the canceler optimally "removes" from the mixture, $\mathrm{s}_n + \mathrm{v}_1(n)$, the component that is correlated to the input, $\mathrm{v}_2(n)$; observe that $\mathrm{v}_1(n)$ basically acts as the desired response.

Fig. 4.14A shows a realization of the signal $d_n = s_n + v_1(n)$, where $s_n = \cos(\omega_0 n)$ with $\omega_0 = 2 * 10^{-3} * \pi$., $a_1 = 0.8$, and $\sigma_\eta^2 = 0.05$. Fig. 4.14B is the respective realization of the signal $\mathrm{s}_n + \mathrm{v}_1(n) -$

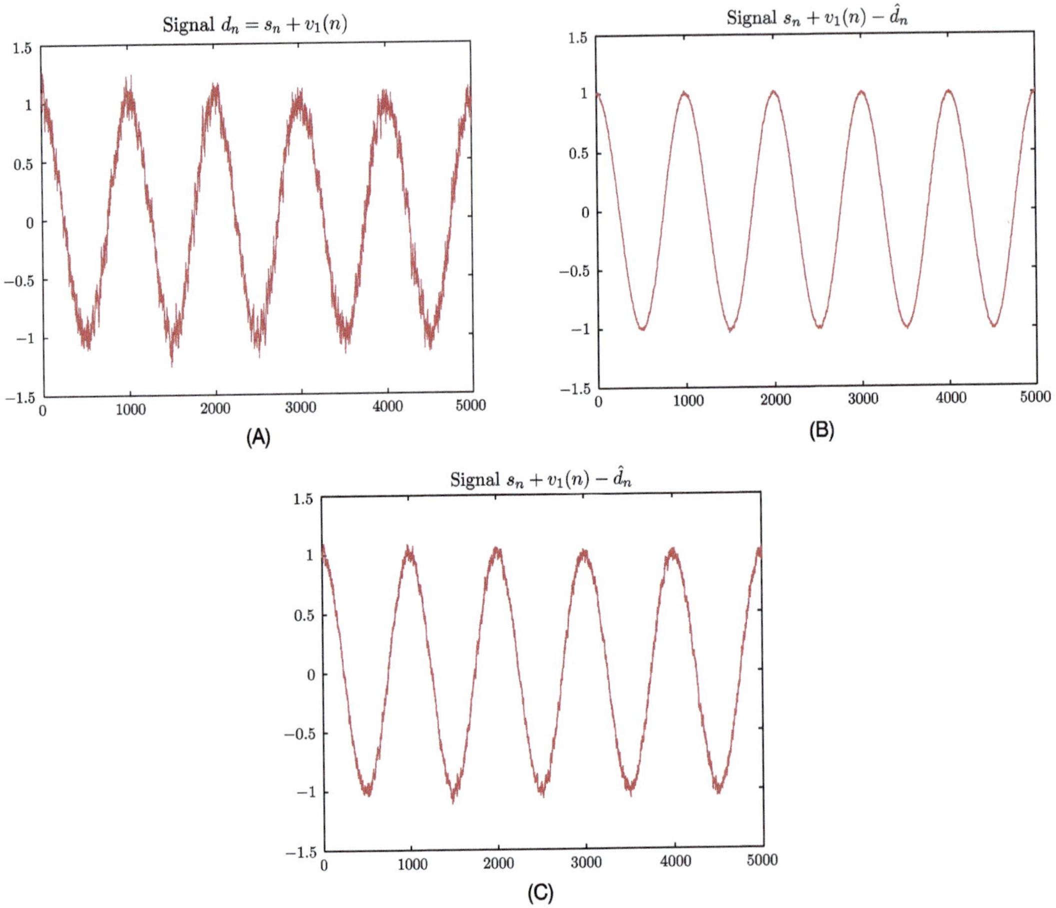

FIGURE 4.14

(A) The noisy sinusoid signal of Example 4.1. (B) The denoised signal for strongly correlated noise sources, v_1 and v_2. (C) The obtained denoised signal for less correlated noise sources.

$\hat{d}(n)$ for $a_2 = 0.75$. The corresponding weights for the canceler are $\boldsymbol{w}_* = [1, 0.125]^T$. Fig. 4.14C corresponds to $a_2 = 0.5$. Observe that the higher the cross-correlation between $v_1(n)$ and $v_2(n)$, the better the obtained result becomes.

Example 4.2 (Channel equalization). Consider the channel equalization setup in Fig. 4.12, where the output of the channel, which is sensed by the receiver, is given by

$$u_n = 0.5 s_n + s_{n-1} + \eta_n. \tag{4.58}$$

The goal is to design an equalizer comprising three taps, $\boldsymbol{w} = [w_0, w_1, w_2]^T$, so that

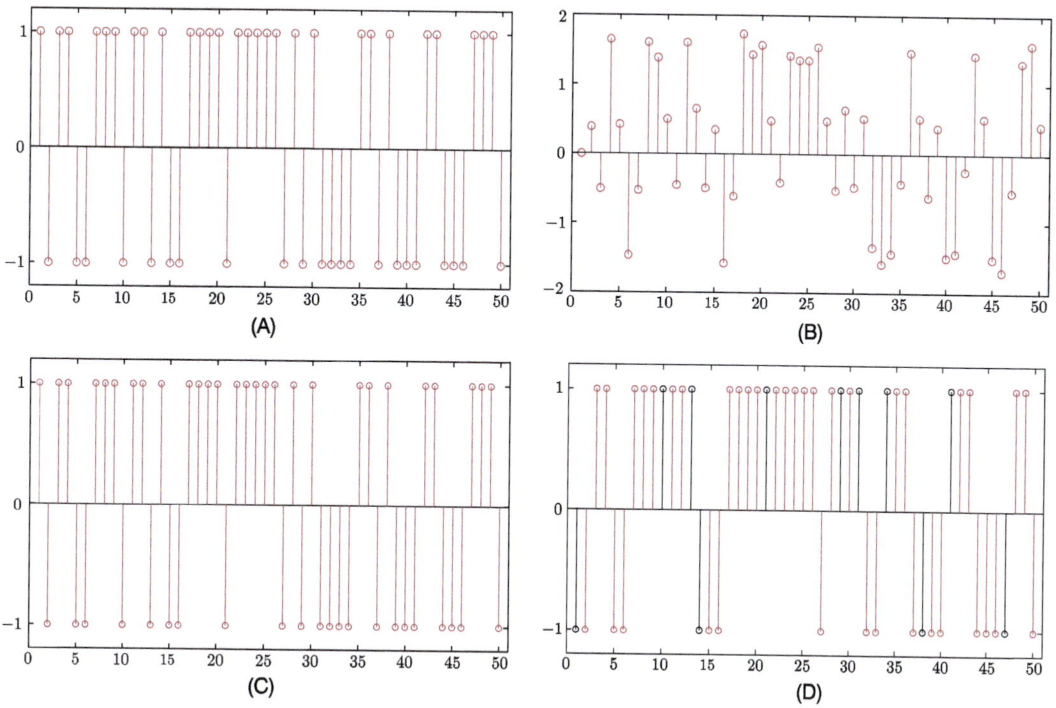

FIGURE 4.15

(A) A realization of the information sequence comprising equiprobable, randomly generated, ±1 samples of Example 4.2. (B) The received at the receiver-end corresponding sequence. (C) The sequence at the output of the equalizer for a low channel noise case. The original sequence is fully recovered with no errors. (D) The output of the equalizer for high channel noise. The samples in gray are in error and of opposite polarity compared to the originally transmitted samples.

$$\hat{\mathrm{d}}_n = \boldsymbol{w}^T \mathbf{u}_n,$$

and estimate the unknown taps using as a desired response sequence $\mathrm{d}_n = \mathrm{s}_{n-1}$. We are given that $\mathbb{E}[\mathrm{s}_n] = \mathbb{E}[\eta_n] = 0$ and

$$\Sigma_s = \sigma_s^2 I, \quad \Sigma_\eta = \sigma_\eta^2 I.$$

Note that for the desired response we have used a delay $L = 1$. In order to better understand the reason that a delay is used and without going into many details (for the more experienced reader, note that the channel is nonminimum phase, e.g., [41]), observe that at time n, most of the contribution to u_n in (4.58) comes from the symbol s_{n-1}, which is weighted by one, while the sample s_n is weighted by 0.5; hence, it is most natural from an intuitive point of view, at time n, having received u_n, to try to obtain an estimate for s_{n-1}. This justifies the use of the delay.

Fig. 4.15A shows a realization of the input information sequence s_n. It consists of randomly generated, equiprobable ±1 samples. The effect of the channel is (a) to combine successive information samples together (ISI) and (b) to add noise; the purpose of the equalizer is to optimally remove both of them. Fig. 4.15B shows the respective realization sequence of u_n, which is received at the receiver's front end. Observe that, by looking at it, one cannot recognize in it the original sequence; the noise together with the ISI has really changed its "look."

Following a similar procedure as in the previous example, we obtain (Problem 4.14)

$$\Sigma_u = \begin{bmatrix} 1.25\sigma_s^2 + \sigma_\eta^2 & 0.5\sigma_s^2 & 0 \\ 0.5\sigma_s^2 & 1.25\sigma_s^2 + \sigma_\eta^2 & 0.5\sigma_s^2 \\ 0 & 0.5\sigma_s^2 & 1.25\sigma_s^2 + \sigma_\eta^2 \end{bmatrix}, \quad \boldsymbol{p} = \begin{bmatrix} \sigma_s^2 \\ 0.5\sigma_s^2 \\ 0 \end{bmatrix}.$$

Solving the normal equations,

$$\Sigma_u \boldsymbol{w}_* = \boldsymbol{p},$$

for $\sigma_s^2 = 1$ and $\sigma_\eta^2 = 0.01$, results in

$$\boldsymbol{w}_* = [0.7462, 0.1195, -0.0474]^T.$$

Fig. 4.15C shows the recovered sequence by the equalizer ($\boldsymbol{w}_*^T \boldsymbol{u}_n$), after appropriate thresholding. It is exactly the same as the transmitted one: no errors. Fig. 4.15D shows the recovered sequence for the case where the variance of the noise was increased to $\sigma_\eta^2 = 1$. The corresponding MSE optimal equalizer is equal to

$$\boldsymbol{w}_* = [0.4132, 0.1369, -0.0304]^T.$$

This time, the sequence reconstructed by the equalizer has errors with respect to the transmitted one (gray lines).

A slightly alternative formulation for obtaining Σ_u, instead of computing each one of its elements individually, is the following. Verify that the input vector to the equalizer (with tree taps) at time n is given by

$$\begin{aligned} \mathbf{u}_n &= \begin{bmatrix} 0.5 & 1 & 0 & 0 \\ 0 & 0.5 & 1 & 0 \\ 0 & 0 & 0.5 & 1 \end{bmatrix} \begin{bmatrix} s_n \\ s_{n-1} \\ s_{n-2} \\ s_{n-3} \end{bmatrix} + \begin{bmatrix} \eta_n \\ \eta_{n-1} \\ \eta_{n-2} \\ \eta_{n-3} \end{bmatrix} \\ &:= H\mathbf{s}_n + \boldsymbol{\eta}_n, \end{aligned} \tag{4.59}$$

which results in

$$\begin{aligned} \Sigma_{\mathrm{u}} = \mathbb{E}\left[\mathbf{u}_n \mathbf{u}_n^T\right] &= H\sigma_s^2 H^T + \Sigma_\eta \\ &= \sigma_s^2 H H^T + \sigma_\eta^2 I. \end{aligned}$$

The reader can easily verify that this is the same as before. Note, however, that (4.59) reminds us of the linear regression model. Moreover, note the special structure of the matrix H. Such matrices are also known as *convolution* matrices. This structure is imposed by the fact that the elements of $\mathbf{u}_n$ are time-shifted versions of the first element, because the input vector corresponds to a random process. This is exactly the property that will be exploited next to derive efficient schemes for the solution of the normal equations.

4.8 ALGORITHMIC ASPECTS: THE LEVINSON AND LATTICE-LADDER ALGORITHMS

The goal of this section is to present algorithmic schemes for the efficient solution of the normal equations in (4.16). The filtering case where the input and output entities are random processes will be considered. In this case, we have already pointed out that the input covariance matrix has a special structure. The main concepts to be presented here have a generality that goes beyond the specific form of the normal equations. A vast literature concerning efficient (fast) algorithms for the least-squares task as well as a number of its online/adaptive versions have their roots to the schemes to be presented here. At the heart of all these schemes lies the specific structure of the input vector, whose elements are *time-shifted versions* of its first element, u_n.

Recall from linear algebra that in order to solve a general linear system of l equations with l unknowns, one requires $O(l^3)$ operations (multiplications and additions (MADS)). Exploiting the rich structure of the autocorrelation/covariance matrix, associated with random processes, an algorithm with $O(l^2)$ operations will be derived. The more general complex-valued case will be considered.

The autocorrelation/covariance matrix (for zero mean processes) of the input random vector has been defined in (4.17). That is, it is Hermitian as well as semipositive definite. From now on, we will assume that it is positive definite. The covariance matrix in $\mathbb{C}^{m\times m}$, associated with a complex wide-sense stationary process, is given by

$$\Sigma_m = \begin{bmatrix} r(0) & r(1) & \cdots & r(m-1) \\ r(-1) & r(0) & \cdots & r(m-2) \\ \vdots & \vdots & \ddots & \vdots \\ r(-m+1) & r(-m+2) & \cdots & r(0) \end{bmatrix}$$

$$= \begin{bmatrix} r(0) & r(1) & \cdots & r(m-1) \\ r^*(1) & r(0) & \cdots & r(m-2) \\ \vdots & \vdots & \ddots & \vdots \\ r^*(m-1) & r^*(m-2) & \cdots & r(0) \end{bmatrix},$$

where the property

$$r(i) := \mathbb{E}[\mathrm{u}_n \mathrm{u}_{n-i}^*] = \mathbb{E}\left[\left(\mathrm{u}_{n-i}\mathrm{u}_n^*\right)^*\right] := r^*(-i)$$

has been used. We have relaxed the notational dependence of Σ on u and we have instead explicitly indicated the order of the matrix, because this will be a very useful index from now on.

We will follow a recursive approach, and our aim will be to express the optimal filter solution of order m, denoted from now on as $\boldsymbol{w}_m$, in terms of the optimal one, $\boldsymbol{w}_{m-1}$, of order $m-1$.

The covariance matrix of a wide-sense stationary process is a *Toeplitz* matrix; all the elements along *any* of its diagonals are equal. This property together with its Hermitian nature gives rise to the following *nested* structure:

$$\Sigma_m = \begin{bmatrix} \Sigma_{m-1} & J_{m-1}\boldsymbol{r}_{m-1} \\ \boldsymbol{r}_{m-1}^H J_{m-1} & r(0) \end{bmatrix} \tag{4.60}$$

$$= \begin{bmatrix} r(0) & \boldsymbol{r}_{m-1}^T \\ \boldsymbol{r}_{m-1}^* & \Sigma_{m-1} \end{bmatrix}, \tag{4.61}$$

where

$$\boldsymbol{r}_{m-1} := \begin{bmatrix} r(1) \\ r(2) \\ \vdots \\ r(m-1) \end{bmatrix} \tag{4.62}$$

and J_{m-1} is the antidiagonal matrix of dimension $(m-1) \times (m-1)$, defined as

$$J_{m-1} := \begin{bmatrix} 0 & 0 & \cdots & 1 \\ 0 & 0 & \iddots & 0 \\ 0 & 1 & \cdots & 0 \\ 1 & 0 & \cdots & 0 \end{bmatrix}.$$

Note that right-multiplication of any matrix by J_{m-1} reverses the order of its columns, while multiplying it from the left reverses the order of the rows as follows:

$$\boldsymbol{r}_{m-1}^H J_{m-1} = \left[r^*(m-1) \;\; r^*(m-2) \;\; \cdots \;\; r^*(1) \right],$$

and

$$J_{m-1}\boldsymbol{r}_{m-1} = \left[r(m-1) \;\; r(m-2) \;\; \cdots \;\; r(1) \right]^T.$$

Applying the matrix inversion lemma from Appendix A.1 for the upper partition in (4.60), we obtain

$$\Sigma_m^{-1} = \begin{bmatrix} \Sigma_{m-1}^{-1} & \boldsymbol{0} \\ \boldsymbol{0}^T & 0 \end{bmatrix} + \begin{bmatrix} -\Sigma_{m-1}^{-1} J_{m-1}\boldsymbol{r}_{m-1} \\ 1 \end{bmatrix} \frac{1}{\alpha_{m-1}^b} \left[-\boldsymbol{r}_{m-1}^H J_{m-1} \Sigma_{m-1}^{-1} \;\; 1 \right], \tag{4.63}$$

where for this case the so-called *Schur complement* is the scalar

$$\alpha_{m-1}^b = r(0) - \boldsymbol{r}_{m-1}^H J_{m-1} \Sigma_{m-1}^{-1} J_{m-1} \boldsymbol{r}_{m-1}. \tag{4.64}$$

The cross-correlation vector of order m, $\boldsymbol{p}_m$, admits the following partition:

$$\boldsymbol{p}_m = \begin{bmatrix} \mathbb{E}[\mathrm{u}_n \mathrm{d}_n^*] \\ \vdots \\ \mathbb{E}[\mathrm{u}_{n-m+2} \mathrm{d}_n^*] \\ \mathbb{E}[\mathrm{u}_{n-m+1} \mathrm{d}_n^*] \end{bmatrix} = \begin{bmatrix} \boldsymbol{p}_{m-1} \\ p_{m-1} \end{bmatrix}, \quad \text{where} \quad p_{m-1} := \mathbb{E}[\mathrm{u}_{n-m+1} \mathrm{d}_n^*]. \tag{4.65}$$

Combining (4.63) and (4.65), the following elegant relation results:

$$\boldsymbol{w}_m := \Sigma_m^{-1} \boldsymbol{p}_m = \begin{bmatrix} \boldsymbol{w}_{m-1} \\ 0 \end{bmatrix} + \begin{bmatrix} -\boldsymbol{b}_{m-1} \\ 1 \end{bmatrix} k_{m-1}^w, \tag{4.66}$$

where

$$\boldsymbol{w}_{m-1} = \Sigma_{m-1}^{-1} \boldsymbol{p}_{m-1}, \quad \boldsymbol{b}_{m-1} := \Sigma_{m-1}^{-1} J_{m-1} \boldsymbol{r}_{m-1}$$

and

$$k_{m-1}^w := \frac{p_{m-1} - \boldsymbol{r}_{m-1}^H J_{m-1} \boldsymbol{w}_{m-1}}{\alpha_{m-1}^b}. \tag{4.67}$$

Eq. (4.66) is an order recursion that relates the optimal solution $\boldsymbol{w}_m$ with $\boldsymbol{w}_{m-1}$. In order to obtain a complete recursive scheme, all one needs is a recursion for updating $\boldsymbol{b}_m$.

FORWARD AND BACKWARD MSE OPTIMAL PREDICTORS

Backward prediction: The vector $\boldsymbol{b}_m = \Sigma_m^{-1} J_m \boldsymbol{r}_m$ has an interesting physical interpretation: it is the MSE optimal backward predictor of order m. That is, it is the linear filter which optimally estimates/predicts the value of u_{n-m} given the values of $\mathrm{u}_{n-m+1}, \mathrm{u}_{n-m+2}, \ldots, \mathrm{u}_n$. Thus, in order to design the optimal backward predictor of order m, the desired response must be

$$\mathrm{d}_n = \mathrm{u}_{n-m},$$

and from the respective normal equations we get

$$\boldsymbol{b}_m = \Sigma_m^{-1} \begin{bmatrix} \mathbb{E}[\mathrm{u}_n \mathrm{u}_{n-m}^*] \\ \mathbb{E}[\mathrm{u}_{n-1} \mathrm{u}_{n-m}^*] \\ \vdots \\ \mathbb{E}[\mathrm{u}_{n-m+1} \mathrm{u}_{n-m}^*] \end{bmatrix} = \Sigma_m^{-1} J_m \boldsymbol{r}_m. \tag{4.68}$$

Hence, the MSE optimal backward predictor coincides with $\boldsymbol{b}_m$, i.e.,

$$\boxed{\boldsymbol{b}_m = \Sigma_m^{-1} J_m \boldsymbol{r}_m : \quad \text{MSE optimal backward predictor.}}$$

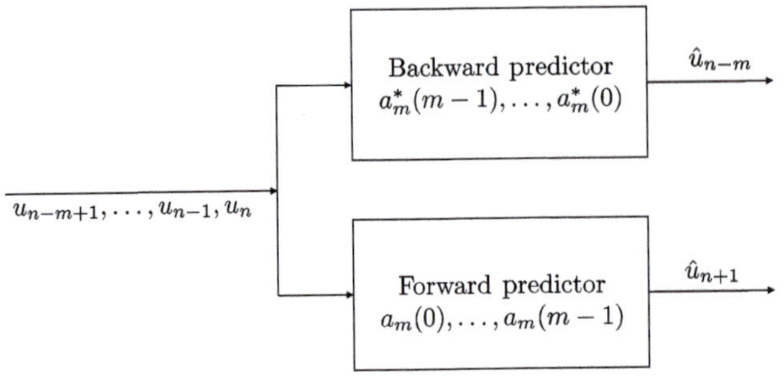

FIGURE 4.16

The impulse response of the backward predictor is the conjugate reverse of that of the forward predictor.

Moreover, the corresponding minimum MSE, adapting (4.19) to our current needs, is equal to

$$J(\boldsymbol{b}_m) = r(0) - \boldsymbol{r}_m^H J_m \Sigma_m^{-1} J_m \boldsymbol{r}_m = \alpha_m^b.$$

That is, the Schur complement in (4.64) is equal to the respective optimal MSE!

Forward prediction: The goal of the forward prediction task is to predict the value u_{n+1}, given the values $\mathrm{u}_n, \mathrm{u}_{n-1}, \ldots, \mathrm{u}_{n-m+1}$. Thus, the MSE optimal forward predictor of order m, $\boldsymbol{a}_m$, is obtained by selecting the desired response $\mathrm{d}_n = \mathrm{u}_{n+1}$, and the respective normal equations become

$$\boldsymbol{a}_m = \Sigma_m^{-1} \begin{bmatrix} \mathbb{E}[\mathrm{u}_n \mathrm{u}_{n+1}^*] \\ \mathbb{E}[\mathrm{u}_{n-1} \mathrm{u}_{n+1}^*] \\ \vdots \\ \mathbb{E}[\mathrm{u}_{n-m+1} \mathrm{u}_{n+1}^*] \end{bmatrix} = \Sigma_m^{-1} \begin{bmatrix} r^*(1) \\ r^*(2) \\ \vdots \\ r^*(m) \end{bmatrix} \tag{4.69}$$

or

$$\boxed{\boldsymbol{a}_m = \Sigma_m^{-1} \boldsymbol{r}_m^* : \quad \text{MSE optimal forward predictor.}} \tag{4.70}$$

From (4.70), it is not difficult to show (Problem 4.16) that (recall that $J_m J_m = I_m$)

$$\boldsymbol{a}_m = J_m \boldsymbol{b}_m^* \Rightarrow \boldsymbol{b}_m = J_m \boldsymbol{a}_m^* \tag{4.71}$$

and that the optimal MSE for the forward prediction, $J(\boldsymbol{a}_m) := \alpha_m^f$, is equal to that for the backward prediction, i.e.,

$$J(\boldsymbol{a}_m) = \alpha_m^f = \alpha_m^b = J(\boldsymbol{b}_m).$$

Fig. 4.16 depicts the two prediction tasks. In other words, the optimal forward predictor is the conjugate reverse of the backward one, so that

$$\boldsymbol{a}_m := \begin{bmatrix} a_m(0) \\ \vdots \\ a_m(m-1) \end{bmatrix} = J_m \boldsymbol{b}_m^* := \begin{bmatrix} b_m^*(m-1) \\ \vdots \\ b_m^*(0) \end{bmatrix}.$$

This property is due to the stationarity of the involved process. Because the statistical properties only depend on the difference of the time instants, forward and backward predictions are not much different; in both cases, given a set of samples, $u_{n-m+1}, \dots, u_n$, we predict *one* sample ahead in the future (u_{n+1} in the forward prediction) or *one* sample back in the past (u_{n-m} in the backward prediction).

Having established the relationship between $\boldsymbol{a}_m$ and $\boldsymbol{b}_m$ in (4.71), we are ready to complete the missing step in (4.66); that is, to complete an order recursive step for the update of $\boldsymbol{b}_m$. Since (4.66) holds true for any desired response d_n, it also applies for the special case where the optimal filter to be designed is the forward predictor $\boldsymbol{a}_m$; in this case, $\mathrm{d}_n = \mathrm{u}_{n+1}$. Replacing in (4.66) $\boldsymbol{w}_m$ ($\boldsymbol{w}_{m-1}$) with $\boldsymbol{a}_m$ ($\boldsymbol{a}_{m-1}$) results in

$$\boldsymbol{a}_m = \begin{bmatrix} \boldsymbol{a}_{m-1} \\ 0 \end{bmatrix} + \begin{bmatrix} -J_{m-1}\boldsymbol{a}_{m-1}^* \\ 1 \end{bmatrix} k_{m-1}, \tag{4.72}$$

where (4.71) has been used and

$$k_{m-1} = \frac{r^*(m) - \boldsymbol{r}_{m-1}^H J_{m-1} \boldsymbol{a}_{m-1}}{\alpha_{m-1}^b}. \tag{4.73}$$

Combining (4.66), (4.67), (4.71), (4.72), and (4.73), the following algorithm, known as *Levinson's* algorithm, for the solution of the normal equations results.

Algorithm 4.1 (Levinson's algorithm).

- Input
  - $r(0), r(1), \dots, r(l)$
  - $p_k = \mathbb{E}[\mathrm{u}_{n-k}\mathrm{d}_n^*]$, $k = 0, 1, \dots, l-1$
- Initialize
  - $w_1 = \frac{p_0}{r(0)}$, $a_1 = \frac{r^*(1)}{r(0)}$, $\alpha_1^b = r(0) - \frac{|r(1)|^2}{r(0)}$
  - $k_1^w = \frac{p_1 - r^*(1)w_1}{\alpha_1^b}$, $k_1 = \frac{r^*(2) - r^*(1)a_1}{\alpha_1^b}$
- **For** $m = 2, \dots, l-1$, **Do**
  - $\boldsymbol{w}_m = \begin{bmatrix} \boldsymbol{w}_{m-1} \\ 0 \end{bmatrix} + \begin{bmatrix} -J_{m-1}\boldsymbol{a}_{m-1}^* \\ 1 \end{bmatrix} k_{m-1}^w$
  - $\boldsymbol{a}_m = \begin{bmatrix} \boldsymbol{a}_{m-1} \\ 0 \end{bmatrix} + \begin{bmatrix} -J_{m-1}\boldsymbol{a}_{m-1}^* \\ 1 \end{bmatrix} k_{m-1}$
  - $\alpha_m^b = \alpha_{m-1}^b(1 - |k_{m-1}|^2)$
  - $k_m^w = \frac{p_m - \boldsymbol{r}_m^H J_m \boldsymbol{w}_m}{\alpha_m^b}$

- $k_m = \frac{r^*(m+1) - \boldsymbol{r}_m^H J_m \boldsymbol{a}_m}{\alpha_m^b}$

- **End For**

Note that the update for α_m^b is a direct consequence of its definition in (4.64) and (4.72) (Problem 4.17). Also note that $\alpha_m^b \geq 0$ implies that $|k_m| \leq 1$.

Remarks 4.3.

- The complexity per order recursion is $4m$ MADS, hence for a system with l equations this amounts to $2l^2$ MADS. This computational saving is substantial compared to the $O(l^3)$ MADS, required by adopting a general purpose scheme. The previous very elegant scheme was proposed in 1947 by Levinson [26]. A formulation of the algorithm was also independently proposed by Durbin [12]; the algorithm is sometimes called the *Levinson–Durbin* algorithm. In [11], it was shown that Levinson's algorithm is redundant in its prediction part and the *split Levinson* algorithm was developed, whose recursions evolve around symmetric vector quantities leading to further computational savings.

4.8.1 THE LATTICE-LADDER SCHEME

So far, we have been involved with the so-called *transversal* implementation of a linear time-invariant FIR filter; in other words, the output is expressed as a convolution between the impulse response and the input of the linear structure. Levinson's algorithm provided a computationally efficient scheme for obtaining the MSE optimal estimate $\boldsymbol{w}_*$. We now turn our attention to an equivalent implementation of the corresponding linear filter, which comes as a direct consequence of Levinson's algorithm.

Define the error signals associated with the mth-order optimal forward and backward predictors at time instant n as

$$e_m^f(n) := \mathrm{u}_n - \boldsymbol{a}_m^H \mathbf{u}_m(n-1), \tag{4.74}$$

where $\mathbf{u}_m(n)$ is the input random vector of the mth-order filter, and the order of the filter has been explicitly brought into the notation.[7] The backward error is given by

$$\begin{aligned} e_m^b(n) &:= \mathrm{u}_{n-m} - \boldsymbol{b}_m^H \mathbf{u}_m(n) \\ &= \mathrm{u}_{n-m} - \boldsymbol{a}_m^T J_m \mathbf{u}_m(n). \end{aligned} \tag{4.75}$$

Employing (4.75), the order recursion in (4.72), and the partitioning of $\mathbf{u}_m(n)$, which is represented by

$$\mathbf{u}_m(n) = [\mathbf{u}_{m-1}^T(n), \mathrm{u}_{n-m+1}]^T = [\mathrm{u}_n, \mathbf{u}_{m-1}^T(n-1)]^T, \tag{4.76}$$

in (4.74), we readily obtain

$$e_m^f(n) = e_{m-1}^f(n) - e_{m-1}^b(n-1)k_{m-1}^*, \quad m = 1, 2, \ldots, l, \tag{4.77}$$

$$e_m^b(n) = e_{m-1}^b(n-1) - e_{m-1}^f(n)k_{m-1}, \quad m = 1, 2, \ldots, l, \tag{4.78}$$

[7] The time index is now given in parentheses, to avoid having double subscripts.

with $e_0^f(n) = e_0^b(n) = u_n$ and $k_0 = \frac{r^*(1)}{r(0)}$. This pair of recursions is known as *lattice* recursions. Let us focus a bit more on this set of equations.

Orthogonality of the Optimal Backward Errors

From the vector space interpretation of random signals, it is apparent that $e_m^b(n)$ lies in the subspace spanned by $u_{n-m}, \ldots, u_n$, and we can write

$$e_m^b(n) \in \text{span}\{u(n-m), \ldots, u(n)\}.$$

Moreover, because $e_m^b(n)$ is the error associated with the MSE optimal backward predictor, $e_m^b(n) \perp \text{span}\{u(n-m+1), \ldots, u(n)\}$. However, the latter subspace is the one where $e_{m-k}^b(n),\ k = 1, 2, \ldots, m$, lie. Hence, for $m = 1, 2, \ldots, l-1$, we can write

$$\boxed{e_m^b(n) \perp e_k^b(n),\ k < m: \quad \text{orthogonality of the backward errors.}}$$

Moreover, it is obvious that

$$\text{span}\{e_0^b(n), e_1^b(n), \ldots, e_{l-1}^b(n)\} = \text{span}\{u_n, u_{n-1}, \ldots, u_{n-l+1}\}.$$

Hence, the normalized vectors

$$\boxed{\tilde{e}_m^b(n) := \frac{e_m^b(n)}{||e_m^b(n)||}, \quad m = 0, 1, \ldots, l-1: \quad \text{orthonormal basis}}$$

form an *orthonormal* basis in $\text{span}\{u_n, u_{n-1}, \ldots, u_{n-l+1}\}$ (see Fig. 4.17). As a matter of fact, the pair of (4.77) and (4.78) comprises a *Gram–Schmidt* orthogonalizer [47].

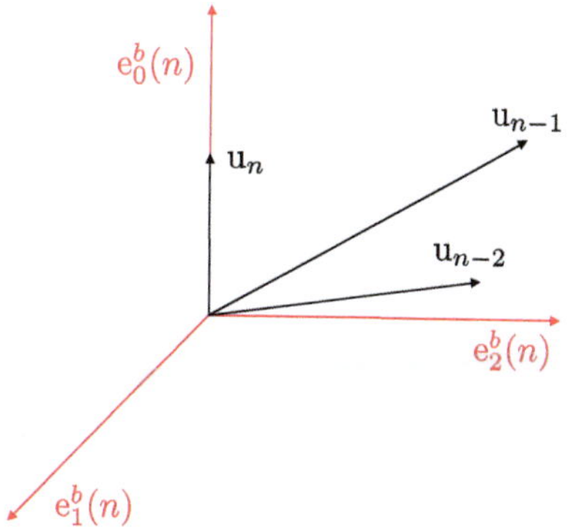

FIGURE 4.17

The optimal backward errors form an orthogonal basis in the respective input random signal space.

Let us now express $\hat{\mathrm{d}}_n$, or the projection of d_n in span$\{\mathrm{u}_n, ..., \mathrm{u}_{n-l+1}\}$, in terms of the new set of orthogonal vectors,

$$\hat{\mathrm{d}}_n = \sum_{m=0}^{l-1} h_m \mathrm{e}_m^b(n), \tag{4.79}$$

where the coefficients h_m are given by

$$\begin{aligned} h_m &= \langle \hat{\mathrm{d}}_n, \frac{\mathrm{e}_m^b(n)}{||\mathrm{e}_m^b(n)||^2} \rangle = \frac{\mathbb{E}[\hat{\mathrm{d}}_n \mathrm{e}_m^{b*}(n)]}{||\mathrm{e}_m^b(n)||^2} = \frac{\mathbb{E}[(\mathrm{d}_n - \mathrm{e}_n)\mathrm{e}_m^{b*}(n)]}{||\mathrm{e}_m^b(n)||^2} \\ &= \frac{\mathbb{E}[\mathrm{d}_n \mathrm{e}_m^{b*}(n)]}{||\mathrm{e}_m^b(n)||^2}, \end{aligned} \tag{4.80}$$

where the orthogonality of the error e_n, with the subspace spanned by the backward errors, has been taken into account. From (4.67) and (4.80), and taking into account the respective definitions of the involved quantities, we readily obtain

$$h_m = k_m^{w*}.$$

That is, the coefficients k_m^w, $m = 0, 1, \ldots, l-1$, in Levinson's algorithm are the parameters in the expansion of $\hat{\mathrm{d}}_n$ in terms of the orthogonal basis. Combining (4.77)–(4.79) the *lattice-ladder* scheme of Fig. 4.18 results, whose output is the MSE approximation $\hat{\mathrm{d}}_n$ of d_n.

Remarks 4.4.

- The lattice-ladder scheme is a highly efficient, *modular* structure. It comprises a sequence of successive similar stages. To increase the order of the filter, it suffices to add an extra stage, which is a highly desirable property in VLSI implementations. Moreover, lattice-ladder schemes enjoy a higher robustness, compared to Levinson's algorithm, with respect to numerical inaccuracies.

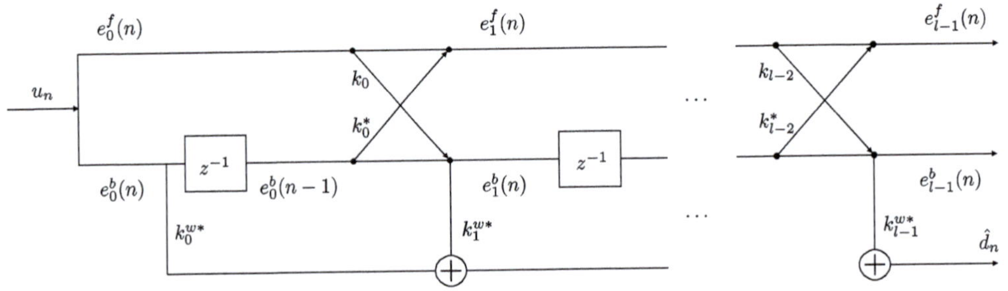

FIGURE 4.18

The lattice-ladder structure. In contrast to the transversal implementation in terms of $\boldsymbol{w}_m$, the parameterization is now in terms of k_m, k_m^w, $m = 0, 1, \ldots, l-1$, $k_0 = \frac{r^*(1)}{r(0)}$, and $k_0^w = \frac{p_0^*}{r(0)}$. Note the resulting highly modular structure.

- *Cholesky factorization.* The orthogonality property of the optimal MSE backward errors leads to another interpretation of the involved parameters. From the definition in (4.75), we get

$$\mathbf{e}_l^b(n) := \begin{bmatrix} \mathrm{e}_0^b(n) \\ \mathrm{e}_1^b(n) \\ \vdots \\ \mathrm{e}_{l-1}^b(n) \end{bmatrix} = U^H \begin{bmatrix} \mathrm{u}_n \\ \mathrm{u}_{n-1} \\ \vdots \\ \mathrm{u}_{n-l+1} \end{bmatrix} = U^H \mathbf{u}_l(n), \tag{4.81}$$

where

$$U^H := \begin{bmatrix} 1 & 0 & 0 & \cdots & 0 \\ -a_1(0) & 1 & 0 & \cdots & 0 \\ \vdots & \vdots & \vdots & \vdots & 0 \\ -a_l(l-1) & -a_l(l-2) & \cdots & \cdots & 1 \end{bmatrix}$$

and

$$\boldsymbol{a}_m := [a_m(0), a_m(1), \ldots, a_m(m-1)]^T, \; m = 1, 2, \ldots, l.$$

Due to the orthogonality of the involved backward errors,

$$\mathbb{E}[\mathbf{e}_l^b(n)\mathbf{e}_l^{bH}(n)] = U^H \Sigma_l U = D,$$

where

$$D := \operatorname{diag}\left\{\alpha_0^b, \alpha_1^b, \ldots, \alpha_{l-1}^b\right\},$$

or

$$\Sigma_l^{-1} = U D^{-1} U^H = (U D^{-1/2})(U D^{-1/2})^H.$$

That is, the prediction error powers and the weights of the optimal forward predictor provide the *Cholesky factorization* of the inverse covariance matrix.
- *The Schur algorithm.* In a parallel processing environment, the inner products involved in Levinson's algorithm pose a bottleneck in the flow of the algorithm. Note that the updates for $\boldsymbol{w}_m$ and $\boldsymbol{a}_m$ can be performed fully in parallel. Schur's algorithm [45] is an alternative scheme that overcomes the bottleneck, and in a multiprocessor environment the complexity can go down to $O(l)$. The parameters involved in Schur's algorithm perform a Cholesky factorization of Σ_l (e.g., [21,22]).
- Note that all these algorithmic schemes for the efficient solution of the normal equations owe their existence to the rich structure that the (autocorrelation) covariance matrix and the cross-correlation vector acquire when the involved jointly distributed random entities are random processes; their *time sequential nature* imposes such a structure. The derivation of the Levinson and lattice-ladder schemes reveal the flavor of the type of techniques that can be (and have extensively been) used to derive computational schemes for the online/adaptive versions and the related least-squares error loss function, to be discussed in Chapter 6. There, the algorithms may be computationally more involved, but the essence behind them is the same as for those used in the current section.

4.9 MEAN-SQUARE ERROR ESTIMATION OF LINEAR MODELS

We now turn our attention to the case where the underlying model that relates the input–output variables is a linear one. To prevent confusion with what was treated in the previous sections, it must be stressed that, so far, we have been concerned with the linear estimation task. At no point in this stage of our discussion has the generation model of the data been brought in (with the exception in the comment of Remarks 4.2). We just adopted a linear estimator and obtained the MSE solution for it. The focus was on the solution and its properties. The emphasis here is on cases where the input–output variables are related via a linear data generation model.

Let us assume that we are given two jointly distributed random vectors, $\mathbf{y}$ and $\boldsymbol{\theta}$, which are related according to the following linear model,

$$\mathbf{y} = X\boldsymbol{\theta} + \boldsymbol{\eta}, \tag{4.82}$$

where $\boldsymbol{\eta}$ denotes the set of the involved noise variables. Note that such a model covers the case of our familiar regression task, where the unknown parameters $\boldsymbol{\theta}$ are considered random, which is in line with the Bayesian philosophy, as discussed in Chapter 3. Once more, we assume zero mean vectors; otherwise, the respective mean values are subtracted. The dimensions of $\mathbf{y}$ ($\boldsymbol{\eta}$) and $\boldsymbol{\theta}$ may not necessarily be the same; to be in line with the notation used in Chapter 3, let $\mathbf{y}, \boldsymbol{\eta} \in \mathbb{R}^N$ and $\boldsymbol{\theta} \in \mathbb{R}^l$. Hence, X is an $N \times l$ matrix. Note that the matrix X is considered to be deterministic and not random.

Assume the covariance matrices of our zero mean variables,

$$\Sigma_\theta = \mathbb{E}[\boldsymbol{\theta}\boldsymbol{\theta}^T], \quad \Sigma_\eta = \mathbb{E}[\boldsymbol{\eta}\boldsymbol{\eta}^T],$$

are known. The goal is to compute a matrix H of dimension $l \times N$ so that the linear estimator

$$\hat{\boldsymbol{\theta}} = H\mathbf{y} \tag{4.83}$$

minimizes the MSE cost

$$J(H) := \mathbb{E}\left[(\boldsymbol{\theta} - \hat{\boldsymbol{\theta}})^T(\boldsymbol{\theta} - \hat{\boldsymbol{\theta}})\right] = \sum_{i=1}^{l} \mathbb{E}\left[(\theta_i - \hat{\theta}_i)^2\right]. \tag{4.84}$$

Note that this is a *multichannel* estimation task and it is equivalent to solving l optimization tasks, one for each component, θ_i, of $\boldsymbol{\theta}$. Defining the error vector as

$$\boldsymbol{\epsilon} := \boldsymbol{\theta} - \hat{\boldsymbol{\theta}},$$

the cost function is equal to the trace of the corresponding *error covariance matrix*, so that

$$J(H) := \text{trace}\left\{\mathbb{E}\left[\boldsymbol{\epsilon}\boldsymbol{\epsilon}^T\right]\right\}.$$

Focusing on the ith component in (4.83), we write

$$\hat{\theta}_i = \boldsymbol{h}_i^T\mathbf{y}, \quad i = 1, 2, \ldots, l, \tag{4.85}$$

where $\boldsymbol{h}_i^T$ is the ith row of H and its optimal estimate is given by

$$\boldsymbol{h}_{*,i} := \arg\min_{\boldsymbol{h}_i} \mathbb{E}\left[(\theta_i - \hat{\theta}_i)^2\right] = \mathbb{E}\left[(\theta_i - \boldsymbol{h}_i^T \mathbf{y})^2\right]. \tag{4.86}$$

Minimizing (4.86) is exactly the same task as that of the linear estimation considered in the previous section (with $\mathbf{y}$ in place of $\mathbf{x}$ and θ_i in place of y); hence,

$$\Sigma_y \boldsymbol{h}_{*,i} = \boldsymbol{p}_i, \quad i = 1, 2, \dots, l,$$

where

$$\Sigma_y = \mathbb{E}[\mathbf{y}\mathbf{y}^T] \quad \text{and} \quad \boldsymbol{p}_i = \mathbb{E}[\mathbf{y}\theta_i], \quad i = 1, 2, .., l,$$

or

$$\boldsymbol{h}_{*,i}^T = \boldsymbol{p}_i^T \Sigma_y^{-1}, \quad i = 1, 2, \dots, l,$$

and finally,

$$H_* = \Sigma_{y\theta} \Sigma_y^{-1}, \quad \hat{\boldsymbol{\theta}} = \Sigma_{y\theta} \Sigma_y^{-1} \mathbf{y}, \tag{4.87}$$

where

$$\Sigma_{y\theta} := \begin{bmatrix} \boldsymbol{p}_1^T \\ \boldsymbol{p}_2^T \\ \vdots \\ \boldsymbol{p}_l^T \end{bmatrix} = \mathbb{E}[\boldsymbol{\theta}\mathbf{y}^T] \tag{4.88}$$

is an $l \times N$ cross-correlation matrix. All that is now required is to compute Σ_y and $\Sigma_{y\theta}$. To this end,

$$\begin{aligned} \Sigma_y = \mathbb{E}\left[\mathbf{y}\mathbf{y}^T\right] &= \mathbb{E}\left[(X\boldsymbol{\theta} + \boldsymbol{\eta})\left(\boldsymbol{\theta}^T X^T + \boldsymbol{\eta}^T\right)\right] \\ &= X \Sigma_\theta X^T + \Sigma_\eta, \end{aligned} \tag{4.89}$$

where the independence of the zero mean vectors $\boldsymbol{\theta}$ and $\boldsymbol{\eta}$ has been used. Similarly,

$$\Sigma_{y\theta} = \mathbb{E}\left[\boldsymbol{\theta}\mathbf{y}^T\right] = \mathbb{E}\left[\boldsymbol{\theta}\left(\boldsymbol{\theta}^T X^T + \boldsymbol{\eta}^T\right)\right] = \Sigma_\theta X^T, \tag{4.90}$$

and combining (4.87), (4.89), and (4.90), we obtain

$$\hat{\boldsymbol{\theta}} = \Sigma_\theta X^T \left(\Sigma_\eta + X \Sigma_\theta X^T\right)^{-1} \mathbf{y}. \tag{4.91}$$

Employing from Appendix A.1 the matrix identity

$$\left(A^{-1} + B^T C^{-1} B\right)^{-1} B^T C^{-1} = A B^T \left(B A B^T + C\right)^{-1}$$

in (4.91) we obtain

$$\boxed{\hat{\boldsymbol{\theta}} = (\Sigma_\theta^{-1} + X^T \Sigma_\eta^{-1} X)^{-1} X^T \Sigma_\eta^{-1} \mathbf{y} : \quad \text{MSE linear estimator.}} \tag{4.92}$$

In case of complex-valued variables, the only difference is that transposition is replaced by Hermitian transposition.

Remarks 4.5.

- Recall from Chapter 3 that the optimal MSE estimator of $\boldsymbol{\theta}$ given the values of $\mathbf{y}$ is provided by

  $$\mathbb{E}[\boldsymbol{\theta}|\mathbf{y}].$$

  However, as shown in Problem 3.16, if $\boldsymbol{\theta}$ and $\mathbf{y}$ are jointly Gaussian vectors, then the optimal estimator is linear (affine for nonzero mean variables), and it coincides with the MSE linear estimator of (4.92).
- If we allow nonzero mean values, then instead of (4.83) the affine model should be adopted,

  $$\hat{\boldsymbol{\theta}} = H\mathbf{y} + \boldsymbol{\mu}.$$

  Then

  $$\mathbb{E}[\hat{\boldsymbol{\theta}}] = H\,\mathbb{E}\left[\mathbf{y}\right] + \boldsymbol{\mu} \Rightarrow \boldsymbol{\mu} = \mathbb{E}[\hat{\boldsymbol{\theta}}] - H\,\mathbb{E}[\mathbf{y}].$$

  Hence,

  $$\hat{\boldsymbol{\theta}} = \mathbb{E}[\hat{\boldsymbol{\theta}}] + H\big(\mathbf{y} - \mathbb{E}[\mathbf{y}]\big),$$

  and finally,

  $$\hat{\boldsymbol{\theta}} - \mathbb{E}[\hat{\boldsymbol{\theta}}] = H\big(\mathbf{y} - \mathbb{E}[\mathbf{y}]\big),$$

  which justifies our approach to subtract the means and work with zero mean value variables. For nonzero mean values, the analogue of (4.92) is

  $$\hat{\boldsymbol{\theta}} = \mathbb{E}[\hat{\boldsymbol{\theta}}] + \left(\Sigma_\theta^{-1} + X^T \Sigma_\eta^{-1} X\right)^{-1} X^T \Sigma_\eta^{-1} \big(\mathbf{y} - \mathbb{E}[\mathbf{y}]\big). \tag{4.93}$$

  Note that for zero mean noise $\boldsymbol{\eta}$, $\mathbb{E}[\mathbf{y}] = X\,\mathbb{E}[\boldsymbol{\theta}]$.
- Compare (4.93) with (3.73) for the Bayesian inference approach. They are identical, provided that the covariance matrix of the prior (Gaussian) PDF is equal to Σ_θ and $\boldsymbol{\theta}_0 = \mathbb{E}[\hat{\boldsymbol{\theta}}]$ for a zero mean noise variable.

4.9.1 THE GAUSS–MARKOV THEOREM

We now turn our attention to the case where $\boldsymbol{\theta}$ in the regression model is considered to be an (unknown) constant, instead of a random vector. Thus, the linear model is now written as

$$\mathbf{y} = X\boldsymbol{\theta} + \boldsymbol{\eta}, \tag{4.94}$$

and the randomness of $\mathbf{y}$ is solely due to $\boldsymbol{\eta}$, which is assumed to be zero mean with covariance matrix Σ_η. The goal is to design an *unbiased linear* estimator of $\boldsymbol{\theta}$, which minimizes the MSE,

$$\hat{\boldsymbol{\theta}} = H\mathbf{y}, \tag{4.95}$$

and select H such that

$$\begin{aligned} &\text{minimize} \quad \text{trace}\left\{\mathbb{E}\left[(\boldsymbol{\theta} - \hat{\boldsymbol{\theta}})(\boldsymbol{\theta} - \hat{\boldsymbol{\theta}})^T\right]\right\} \\ &\text{subject to} \quad \mathbb{E}[\hat{\boldsymbol{\theta}}] = \boldsymbol{\theta}. \end{aligned} \tag{4.96}$$

From (4.94) and (4.95), we get

$$\mathbb{E}[\hat{\boldsymbol{\theta}}] = H\,\mathbb{E}[\mathbf{y}] = H\,\mathbb{E}\left[(X\boldsymbol{\theta} + \boldsymbol{\eta})\right] = HX\boldsymbol{\theta},$$

which implies that the unbiased constraint is equivalent to

$$HX = I. \tag{4.97}$$

Employing (4.95), the error vector becomes

$$\boldsymbol{\epsilon} = \boldsymbol{\theta} - \hat{\boldsymbol{\theta}} = \boldsymbol{\theta} - H\mathbf{y} = \boldsymbol{\theta} - H(X\boldsymbol{\theta} + \boldsymbol{\eta}) = -H\boldsymbol{\eta}. \tag{4.98}$$

Hence, the constrained minimization in (4.96) can now be written as

$$\begin{aligned} &H_* = \arg\min_H \text{trace}\{H\Sigma_\eta H^T\}, \\ &\text{s.t.} \quad HX = I. \end{aligned} \tag{4.99}$$

Solving (4.99) results in (Problem 4.18)

$$H_* = (X^T\Sigma_\eta^{-1}X)^{-1}X^T\Sigma_\eta^{-1}, \tag{4.100}$$

and the associated minimum MSE is

$$J(H_*) := \text{MSE}(H_*) = \text{trace}\left\{(X^T\Sigma_\eta^{-1}X)^{-1}\right\}. \tag{4.101}$$

The reader can verify that

$$J(H) \geq J(H_*)$$

for any other linear unbiased estimator (Problem 4.19).

The previous result is known as the *Gauss–Markov* theorem. The optimal MSE linear unbiased estimator is given by

$$\boxed{\hat{\boldsymbol{\theta}} = (X^T\Sigma_\eta^{-1}X)^{-1}X^T\Sigma_\eta^{-1}\mathbf{y}: \quad \text{BLUE.}} \tag{4.102}$$

It is also known as the *best linear unbiased estimator* (BLUE), or the *minimum variance unbiased linear estimator*. For complex-valued variables, the transposition is simply replaced by the Hermitian one.

Remarks 4.6.

- For the BLUE to exist, $X^T \Sigma_\eta^{-1} X$ must be invertible. This is guaranteed if Σ_η is positive definite and the $N \times l$ matrix X, $N \geq l$, is full rank (Problem 4.20).
- Observe that the BLUE coincides with the maximum likelihood estimator (Chapter 3) if $\boldsymbol{\eta}$ follows a multivariate Gaussian distribution; recall that under this assumption, the Cramér–Rao bound is achieved. If this is not the case, there may be another unbiased estimator (nonlinear), which results in lower MSE. Recall also from Chapter 3 that there may be a biased estimator that results in lower MSE; see [13,38] and the references therein for a related discussion.

Example 4.3 (Channel identification). The task is illustrated in Fig. 4.11. Assume that we have access to a set of input–output observations, u_n and d_n, $n = 0, 1, 2, \ldots, N-1$. Moreover, we are given that the impulse response of the system comprises l taps, it is zero mean, and its covariance matrix is Σ_w. Also, the second-order statistics of the zero mean noise are known, and we are given its covariance matrix, Σ_η. Then, assuming that the plant starts from zero initial conditions, we can adopt the following model relating the involved random variables (in line with the model in (4.82)):

$$\mathbf{d} := \begin{bmatrix} \mathrm{d}_0 \\ \mathrm{d}_1 \\ \vdots \\ \mathrm{d}_{l-1} \\ \vdots \\ \mathrm{d}_{N-1} \end{bmatrix} = U \begin{bmatrix} \mathrm{w}_0 \\ \mathrm{w}_1 \\ \vdots \\ \mathrm{w}_{l-1} \end{bmatrix} + \begin{bmatrix} \eta_0 \\ \eta_1 \\ \vdots \\ \eta_{l-1} \\ \vdots \\ \eta_{N-1} \end{bmatrix}, \tag{4.103}$$

where

$$U := \begin{bmatrix} u_0 & 0 & 0 & \cdots & 0 \\ u_1 & u_0 & 0 & \cdots & 0 \\ \cdots & \cdots & \cdots & \cdots & \cdots \\ u_{l-1} & u_{l-2} & \cdots & \cdots & u_0 \\ \cdots & \cdots & \cdots & \cdots & \cdots \\ u_{N-1} & \cdots & \cdots & \cdots & u_{N-l} \end{bmatrix}.$$

Note that U is treated deterministically. Then, recalling (4.92) and plugging in the set of obtained measurements, the following estimate results:

$$\hat{\boldsymbol{w}} = (\Sigma_w^{-1} + U^T \Sigma_\eta^{-1} U) U^T \Sigma_\eta^{-1} \boldsymbol{d}. \tag{4.104}$$

4.9.2 CONSTRAINED LINEAR ESTIMATION: THE BEAMFORMING CASE

We have already dealt with a constrained linear estimation task in Section 4.9.1 in our effort to obtain an unbiased estimator of a fixed-value parameter vector. In the current section, we will see that the pro-

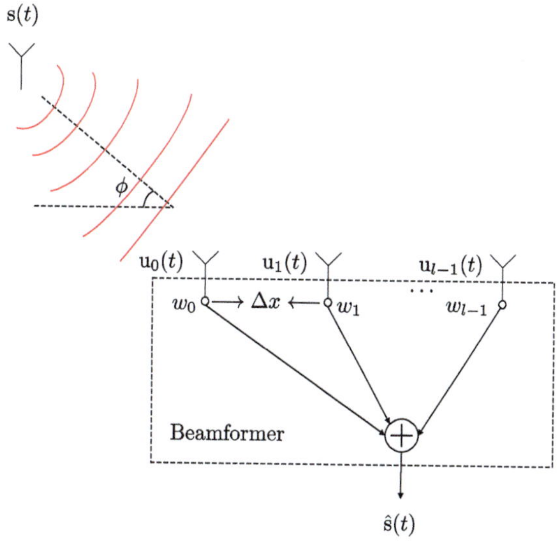

FIGURE 4.19

The task of the beamformer is to obtain estimates of the weights $w_0, \ldots, w_{l-1}$, so as to minimize the effect of noise and at the same time impose a constraint that, in the absence of noise, would leave signals impinging the array from the desired angle, ϕ, unaffected.

cedure developed there is readily applicable for cases where the unknown parameter vector is required to respect certain linear constraints.

We will demonstrate such a constrained task in the context of *beamforming*. Fig. 4.19 illustrates the basic block diagram of the beamforming task. A beamformer comprises a set of antenna elements. We consider the case where the antenna elements are uniformly spaced along a straight line. The goal is to linearly combine the signals received by the individual antenna elements, so as to:

- turn the main beam of the array to a specific direction in space, and
- optimally reduce the noise.

The first goal imposes a constraint to the designer, which will guarantee that the gain of the array is high for the specific desired direction; for the second goal, we will adopt MSE arguments.

In a more formal way, assume that the transmitter is far enough away, so as to guarantee that the wavefronts that the array "sees" are planar. Let $\mathrm{s}(t)$ be the information random process transmitted at a carrier frequency, ω_c. Then the modulated signal is

$$\mathrm{r}(t) = \mathrm{s}(t)e^{j\omega_c t}.$$

If Δx is the distance between successive elements of the array, then a wavefront that arrives at time t_0 at the first element will reach the ith element delayed by

$$\Delta_{t_i} = t_i - t_0 = i\frac{\Delta x \cos\phi}{c}, \quad i = 0, 1, \ldots, l-1,$$

where c is the speed of propagation, ϕ is the angle formed by the array and the direction propagation of the wavefronts, and l is the number of array elements. We know from our basic electromagnetic courses that

$$c = \frac{\omega_c \lambda}{2\pi},$$

where λ is the respective wavelength. Taking a snapshot at time t, the signal received from direction ϕ at the ith element will be

$$\begin{aligned} \mathrm{r}_i(t) &= \mathrm{s}(t - \Delta t_i) e^{j\omega_c (t - i\frac{2\pi \Delta x \cos\phi}{\omega_c \lambda})} \\ &\simeq \mathrm{s}(t) e^{j\omega_c t} e^{-2\pi j \frac{i \Delta x \cos\phi}{\lambda}}, \quad i = 0, 1, \ldots, l-1, \end{aligned}$$

where we have assumed a relatively low time signal variation. After converting the received signals in the baseband (multiplying by $e^{-j\omega_c t}$), the vector of the received signals (one per array element) at time t can be written in the following linear regression-type formulation:

$$\mathbf{u}(t) := \begin{bmatrix} \mathrm{u}_0(t) \\ \mathrm{u}_1(t) \\ \vdots \\ \mathrm{u}_{l-1}(t) \end{bmatrix} = \boldsymbol{x}\mathrm{s}(t) + \boldsymbol{\eta}(t), \tag{4.105}$$

where

$$\boldsymbol{x} := \begin{bmatrix} 1 \\ e^{-2\pi j \frac{\Delta x \cos\phi}{\lambda}} \\ \vdots \\ e^{-2\pi j \frac{(l-1)\Delta x \cos\phi}{\lambda}} \end{bmatrix},$$

and the vector $\boldsymbol{\eta}(t)$ contains the additive noise plus any other interference due to signals coming from directions other than ϕ, so that

$$\boldsymbol{\eta}(t) = [\eta_0(t), \ldots, \eta_{l-1}(t)]^T,$$

and it is assumed to be of zero mean; $\boldsymbol{x}$ is also known as the *steering vector*. The output of the beamformer, acting on the input vector signal, will be

$$\hat{\mathrm{s}}(t) = \boldsymbol{w}^H \mathbf{u}(t),$$

where the Hermitian transposition has to be used, because now the involved signals are complex-valued.

We will first impose the constraint. Ideally, in the absence of noise, one would like to recover signals that impinge on the array from the desired direction, ϕ, exactly. Thus, $\boldsymbol{w}$ should satisfy the following constraint:

$$\boldsymbol{w}^H \boldsymbol{x} = 1, \tag{4.106}$$

which guarantees that $\hat{\mathrm{s}}(t) = \mathrm{s}(t)$ in the absence of noise. Note that (4.106) is an instance of (4.97) if we consider $\boldsymbol{w}^H$ and $\boldsymbol{x}$ in place of H and X, respectively. To account for the noise, we require the MSE

$$\mathbb{E}\left[|\mathrm{s}(t) - \hat{\mathrm{s}}(t)|^2\right] = \mathbb{E}\left[|\mathrm{s}(t) - \boldsymbol{w}^H \mathbf{u}(t)|^2\right]$$

to be minimized. However,

$$\mathrm{s}(t) - \boldsymbol{w}^H \mathbf{u}(t) = \mathrm{s}(t) - \boldsymbol{w}^H (\boldsymbol{x}\mathrm{s}(t) + \boldsymbol{\eta}(t)) = -\boldsymbol{w}^H \boldsymbol{\eta}(t).$$

Hence, the optimal $\boldsymbol{w}_*$ results from the following constrained task:

$$\begin{aligned} \boldsymbol{w}_* &:= \arg\min_{w} (\boldsymbol{w}^H \Sigma_\eta \boldsymbol{w}) \\ \text{s.t.} \quad & \boldsymbol{w}^H \boldsymbol{x} = 1, \end{aligned} \tag{4.107}$$

which is an instance of (4.99) and the solution is given by (4.100); adapting it to the current notation and to its complex-valued formulation, we get

$$\boxed{\boldsymbol{w}_*^H = \frac{\boldsymbol{x}^H \Sigma_\eta^{-1}}{\boldsymbol{x}^H \Sigma_\eta^{-1} \boldsymbol{x}}} \tag{4.108}$$

and

$$\hat{\mathrm{s}}(t) = \boldsymbol{w}_*^H \mathbf{u}(t) = \frac{\boldsymbol{x}^H \Sigma_\eta^{-1} \mathbf{u}(t)}{\boldsymbol{x}^H \Sigma_\eta^{-1} \boldsymbol{x}}. \tag{4.109}$$

The minimum MSE is equal to

$$\mathrm{MSE}(\boldsymbol{w}_*) = \frac{1}{\boldsymbol{x}^H \Sigma_\eta^{-1} \boldsymbol{x}}. \tag{4.110}$$

An alternative formulation for the cost function in order to estimate the weights of the beamformer, which is often met in practice, builds upon the goal to minimize the output power, subject to the same constraint as before,

$$\begin{aligned} \boldsymbol{w}_* &:= \arg\min_{\boldsymbol{w}} \mathbb{E}\left[|\boldsymbol{w}^H \mathbf{u}(t)|^2\right], \\ \text{s.t.} \quad & \boldsymbol{w}^H \boldsymbol{x} = 1, \end{aligned}$$

or equivalently

$$\begin{aligned} \boldsymbol{w}_* &:= \arg\min_{w} \boldsymbol{w}^H \Sigma_u \boldsymbol{w}, \\ \text{s.t.} \quad & \boldsymbol{w}^H \boldsymbol{x} = 1. \end{aligned} \tag{4.111}$$

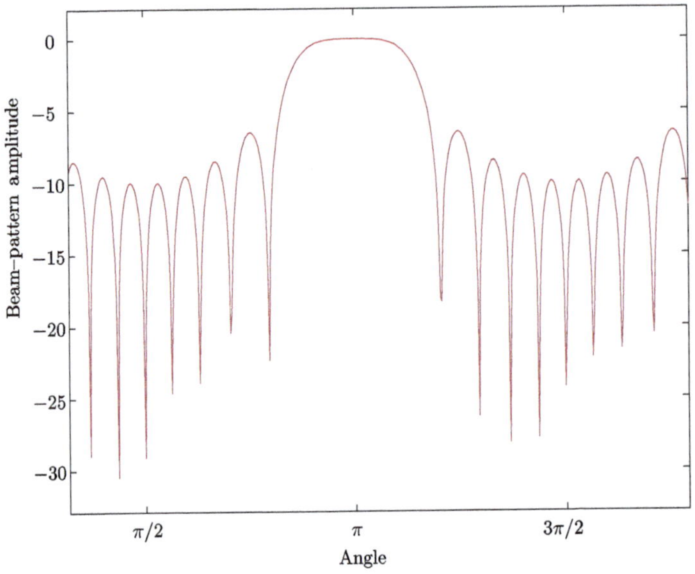

FIGURE 4.20

The amplitude beam pattern, in dBs, as a function of the angle ϕ with respect to the planar array.

This time, the beamformer is pushed to reduce its output signal, which, due to the presence of the constraint, is equivalent to optimally minimizing the contributions originating from the noise as well as from all other interference sources impinging on the array from different, to ϕ, directions. The resulting solution of (4.111) is obviously the same as (4.109) and (4.110) if one replaces Σ_η with Σ_u.

This type of linearly constraint task is known as *linearly constrained minimum variance* (LMV) or *Capon* beamforming or *minimum variance distortionless response (MVDR) beamforming*. For a concise introduction to beamforming, see, e.g., [48].

Widely linear versions for the beamforming task have also been proposed (e.g., [10,32]) (Problem 4.21).

Fig. 4.20 shows the resulting beam pattern as a function of the angle ϕ. The desired angle for designing the optimal set of weights in (4.108) is $\phi = \pi$. The number of antenna elements is $l = 10$, the spacing has been chosen as $\frac{\Delta x}{\lambda} = 0.5$, and the noise covariance matrix is chosen as $0.1I$. The beam pattern amplitude is in dBs, meaning the vertical axis shows $20\log_{10}(|\boldsymbol{w}_*^H \boldsymbol{x}(\phi)|)$. Thus, any signal arriving from directions ϕ not close to $\phi = \pi$ will be absorbed. The main beam can become sharper if more elements are used.

4.10 TIME-VARYING STATISTICS: KALMAN FILTERING

So far, our discussion about the linear estimation task has been limited to stationary environments, where the statistical properties of the involved random variables are assumed to be invariant with time.

However, very often in practice this is not the case, and the statistical properties may be different at different time instants. As a matter of fact, a large effort in the subsequent chapters will be devoted to studying the estimation task under time-varying environments.

Rudolf Kalman is the third scientist, the other two being Wiener and Kolmogorov, whose significant contributions laid the foundations of estimation theory. Kalman was Hungarian-born and emigrated to the United States. He is the father of what is today known as system theory based on the state-space formulation, as opposed to the more limited input–output description of systems.

In 1960, in two seminal papers, Kalman proposed the celebrated Kalman filter, which exploits the state-space formulation in order to accommodate in an elegant way time-varying dynamics [18,19]. We will derive the basic recursions of the Kalman filter in the general context of two jointly distributed random vectors $\mathbf{y}$, $\mathbf{x}$. The task is to estimate the values of $\mathbf{x}$ given observations of $\mathbf{y}$. Let $\mathbf{y}$ and $\mathbf{x}$ be linearly related via the following set of recursions:

$$\mathbf{x}_n = F_n \mathbf{x}_{n-1} + \boldsymbol{\eta}_n, \quad n \geq 0: \quad \text{state equation,} \tag{4.112}$$

$$\mathbf{y}_n = H_n \mathbf{x}_n + \mathbf{v}_n, \quad n \geq 0: \quad \text{output equation,} \tag{4.113}$$

where $\boldsymbol{\eta}_n, \mathbf{x}_n \in \mathbb{R}^l$, $\mathbf{v}_n, \mathbf{y}_n \in \mathbb{R}^k$. The vector $\mathbf{x}_n$ is known as the *state* of the system at time n and $\mathbf{y}_n$ is the output, which is the vector that can be observed (measured); $\boldsymbol{\eta}_n$ and $\mathbf{v}_n$ are the noise vectors, known as *process* noise and *measurement* noise, respectively. Matrices F_n and H_n are of appropriate dimensions and they are assumed to be known. Observe that the so-called *state equation* provides the information related to the time-varying dynamics of the corresponding system. It turns out that a large number of real-world tasks can be brought into the form of (4.112) and (4.113). The model is known as the *state-space* model for $\mathbf{y}_n$. In order to derive the time-varying estimator, $\hat{\mathbf{x}}_n$, given the measured values of $\mathbf{y}_n$, the following assumptions will be adopted:

- $\mathbb{E}[\boldsymbol{\eta}_n \boldsymbol{\eta}_n^T] = Q_n$, $\mathbb{E}[\boldsymbol{\eta}_n \boldsymbol{\eta}_m^T] = O$, $n \neq m$,
- $\mathbb{E}[\mathbf{v}_n \mathbf{v}_n^T] = R_n$, $\mathbb{E}[\mathbf{v}_n \mathbf{v}_m^T] = O$, $n \neq m$,
- $\mathbb{E}[\boldsymbol{\eta}_n \mathbf{v}_m^T] = O$, $\forall n, m$,
- $\mathbb{E}[\boldsymbol{\eta}_n] = \mathbb{E}[\mathbf{v}_n] = \mathbf{0}$, $\forall n$,

where O denotes a matrix with zero elements. That is, $\boldsymbol{\eta}_n$, $\mathbf{v}_n$ are uncorrelated; moreover, noise vectors at different time instants are also considered uncorrelated. Versions where some of these conditions are relaxed are also available. The respective covariance matrices, Q_n and R_n, are assumed to be known.

The development of the time-varying estimation task evolves around two types of estimators for the state variables:

- The first one is denoted as

$$\hat{\mathbf{x}}_{n|n-1},$$

  and it is based on all information that has been received up to and including time instant $n-1$; in other words, the obtained observations of $\mathbf{y}_0, \mathbf{y}_1, \ldots, \mathbf{y}_{n-1}$. This is known as the a priori or *prior* estimator.
- The second estimator at time n is known as the *posterior* one, it is denoted as

$$\hat{\mathbf{x}}_{n|n},$$

and it is computed by updating $\hat{\mathbf{x}}_{n|n-1}$ after $\mathbf{y}_n$ has been observed.

For the development of the algorithm, assume that at time $n-1$ all required information is available; that is, the value of the posterior estimator as well the respective error covariance matrix

$$\hat{\mathbf{x}}_{n-1|n-1}, \quad P_{n-1|n-1} := \mathbb{E}\left[\mathbf{e}_{n-1|n-1}\mathbf{e}_{n-1|n-1}^T\right],$$

where

$$\mathbf{e}_{n-1|n-1} := \mathbf{x}_{n-1} - \hat{\mathbf{x}}_{n-1|n-1}.$$

Step 1: Using $\hat{\mathbf{x}}_{n-1|n-1}$, predict $\hat{\mathbf{x}}_{n|n-1}$ using the state equation; that is,

$$\hat{\mathbf{x}}_{n|n-1} = F_n\hat{\mathbf{x}}_{n-1|n-1}. \tag{4.114}$$

In other words, ignore the contribution from the noise. This is natural, because prediction cannot involve the unobserved variables.
Step 2: Obtain the respective error covariance matrix,

$$P_{n|n-1} = \mathbb{E}\left[(\mathbf{x}_n - \hat{\mathbf{x}}_{n|n-1})(\mathbf{x}_n - \hat{\mathbf{x}}_{n|n-1})^T\right]. \tag{4.115}$$

However,

$$\begin{aligned}\mathbf{e}_{n|n-1} := \mathbf{x}_n - \hat{\mathbf{x}}_{n|n-1} &= F_n\mathbf{x}_{n-1} + \boldsymbol{\eta}_n - F_n\hat{\mathbf{x}}_{n-1|n-1} \\ &= F_n\mathbf{e}_{n-1|n-1} + \boldsymbol{\eta}_n.\end{aligned} \tag{4.116}$$

Combining (4.115) and (4.116), it is straightforward to see that

$$P_{n|n-1} = F_nP_{n-1|n-1}F_n^T + Q_n. \tag{4.117}$$

Step 3: Update $\hat{\mathbf{x}}_{n|n-1}$. To this end, adopt the following recursion:

$$\hat{\mathbf{x}}_{n|n} = \hat{\mathbf{x}}_{n|n-1} + K_n\mathbf{e}_n, \tag{4.118}$$

where

$$\mathbf{e}_n := \mathbf{y}_n - H_n\hat{\mathbf{x}}_{n|n-1}. \tag{4.119}$$

This time update recursion, once the observations for $\mathbf{y}_n$ have been received, has a form that we will meet over and over again in this book. The "new" (posterior) estimate is equal to the "old" (prior) one, which is based on *the past history, plus a correction term*; the latter is proportional to the error $\mathbf{e}_n$ in predicting the newly arrived observations vector and its prediction based on the "old" estimate. Matrix K_n, known as the *Kalman gain*, controls the amount of correction and its value is computed so as to minimize the MSE; in other words,

$$J(K_n) := \mathbb{E}\left[\mathbf{e}_{n|n}^T\mathbf{e}_{n|n}\right] = \text{trace}\left\{P_{n|n}\right\}, \tag{4.120}$$

where

$$P_{n|n} = \mathbb{E}\left[\mathbf{e}_{n|n}\mathbf{e}_{n|n}^T\right] \tag{4.121}$$

and

$$\mathbf{e}_{n|n} := \mathbf{x}_n - \hat{\mathbf{x}}_{n|n}.$$

It can be shown that the optimal Kalman gain is equal to (Problem 4.22)

$$K_n = P_{n|n-1} H_n^T S_n^{-1}, \tag{4.122}$$

where

$$S_n = R_n + H_n P_{n|n-1} H_n^T. \tag{4.123}$$

Step 4: The final recursion that is now needed in order to complete the scheme is that for the update of $P_{n|n}$. Combining the definitions in (4.119) and (4.121) with (4.118), we obtain the following result (Problem 4.23):

$$P_{n|n} = P_{n|n-1} - K_n H_n P_{n|n-1}. \tag{4.124}$$

The algorithm has now been derived. All that is now needed is to select the initial conditions, which are chosen such that

$$\hat{\boldsymbol{x}}_{1|0} = \mathbb{E}[\mathbf{x}_1] \tag{4.125}$$

$$P_{1|0} = \mathbb{E}\left[(\mathbf{x}_1 - \hat{\boldsymbol{x}}_{1|0})(\mathbf{x}_1 - \hat{\boldsymbol{x}}_{1|0})^T\right] = \Pi_0, \tag{4.126}$$

for some initial guess Π_0. The Kalman algorithm is summarized in Algorithm 4.2.

Algorithm 4.2 (Kalman filtering).

- Input: $F_n,\ H_n,\ Q_n,\ R_n,\ \boldsymbol{y}_n,\ n = 1, 2, \ldots$
- Initialization:
  - $\hat{\boldsymbol{x}}_{1|0} = \mathbb{E}[\mathbf{x}_1]$
  - $P_{1|0} = \Pi_0$
- **For** $n = 1, 2, \ldots,$ **Do**
  - $S_n = R_n + H_n P_{n|n-1} H_n^T$
  - $K_n = P_{n|n-1} H_n^T S_n^{-1}$
  - $\hat{\boldsymbol{x}}_{n|n} = \hat{\boldsymbol{x}}_{n|n-1} + K_n(\boldsymbol{y}_n - H_n \hat{\boldsymbol{x}}_{n|n-1})$
  - $P_{n|n} = P_{n|n-1} - K_n H_n P_{n|n-1}$
  - $\hat{\boldsymbol{x}}_{n+1|n} = F_{n+1} \hat{\boldsymbol{x}}_{n|n}$
  - $P_{n+1|n} = F_{n+1} P_{n|n} F_{n+1}^T + Q_{n+1}$
- **End For**

For complex-valued variables, transposition is replaced by the Hermitian operation.

Remarks 4.7.

- Besides the previously derived basic scheme, there are a number of variants. Although, in theory, they are all equivalent, their practical implementation may lead to different performance. Observe that $P_{n|n}$ is computed as the difference of two positive definite matrices; this may lead to a $P_{n|n}$ that is not positive definite, due to numerical errors. This can cause the algorithm to diverge. A popular alternative is the so-called *information filtering* scheme, which propagates the inverse state-error covariance matrices, $P_{n|n}^{-1}$ and $P_{n|n-1}^{-1}$ [20]. In contrast, the scheme in Algorithm 4.2 is known as the *covariance* Kalman algorithm (Problem 4.24).

To cope with the numerical stability issues, a family of algorithms propagates the factors of $P_{n|n}$ (or $P_{n|n}^{-1}$), resulting from the respective Cholesky factorization [5,40].

- There are different approaches to arrive at the Kalman filtering recursions. An alternative derivation is based on the orthogonality principle applied to the so-called *innovations process* associated with the observation sequence, so that

$$\boldsymbol{\epsilon}(n) = \mathbf{y}_n - \hat{\mathbf{y}}_{n|1:n-1},$$

where $\hat{\mathbf{y}}_{n|1:n-1}$ is the prediction based on the past observations history [17]. In Chapter 17, we are going to rederive the Kalman recursions looking at it as a Bayesian network.

- Kalman filtering is a generalization of the optimal MSE linear filtering. It can be shown that when the involved processes are stationary, Kalman filter converges in its steady state to our familiar normal equations [31].
- *Extended Kalman filters*. In (4.112) and (4.113), both the state and the output equations have a linear dependence on the state vector $\mathbf{x}_n$. Kalman filtering, in a more general formulation, can be cast as

$$\mathbf{x}_n = \boldsymbol{f}_n(\mathbf{x}_{n-1}) + \boldsymbol{\eta}_n,$$

$$\mathbf{y}_n = \boldsymbol{h}_n(\mathbf{x}_n) + \mathbf{v}_n,$$

where $\boldsymbol{f}_n$ and $\boldsymbol{h}_n$ are nonlinear vector functions. In the extended Kalman filtering (EKF), the idea is to *linearize* the functions $\boldsymbol{h}_n$ and $\boldsymbol{f}_n$, at each time instant, via their Taylor series expansions, and keep the linear term only, so that

$$F_n = \left.\frac{\partial \boldsymbol{f}_n(\boldsymbol{x}_n)}{\partial \boldsymbol{x}_n}\right|_{\boldsymbol{x}_n = \hat{\boldsymbol{x}}_{n-1|n-1}},$$

$$H_n = \left.\frac{\partial \boldsymbol{h}_n(\boldsymbol{x}_n)}{\partial \boldsymbol{x}_n}\right|_{\boldsymbol{x}_n = \hat{\boldsymbol{x}}_{n|n-1}},$$

and then proceed by using the updates derived for the linear case.

By its very definition, the EKF is suboptimal and often in practice one may face divergence of the algorithm; in general, it must be stated that its practical implementation needs to be carried out with care. Having said that, it must be pointed out that it is heavily used in a number of practical systems.

Unscented Kalman filters represent an alternative way to cope with the nonlinearity, and the main idea springs forth from probabilistic arguments. A set of points are deterministically selected from a Gaussian approximation of $p(\boldsymbol{x}_n|\boldsymbol{y}_1, \ldots, \boldsymbol{y}_n)$; these points are propagated through the nonlinearities, and estimates of the mean values and covariances are obtained [15]. Particle filtering, to be discussed in Chapter 17, is another powerful and popular approach to deal with nonlinear state-space models via probabilistic arguments.

More recently, extensions of Kalman filtering in reproducing kernel Hilbert spaces offers an alternative approach to deal with nonlinearities [52].

A number of Kalman filtering versions for distributed learning (Chapter 5) have appeared in, e.g., [9,23,33,43]. In the latter of the references, subspace learning methods are utilized in the prediction stage associated with the state variables.

- The literature on Kalman filtering is huge, especially when applications are concerned. The interested reader may consult more specialized texts, for example [4,8,17] and the references therein.

Example 4.4 (Autoregressive process estimation). Let us consider an AR process (Chapter 2) of order l, represented as

$$\mathrm{x}_n = -\sum_{i=1}^{l} a_i \mathrm{x}_{n-i} + \eta_n, \tag{4.127}$$

where η_n is a white noise sequence of variance σ_η^2. Our task is to obtain an estimate $\hat{x}_n$ of x_n, having observed a noisy version of it, y_n. The corresponding random variables are related as

$$\mathrm{y}_n = \mathrm{x}_n + \mathrm{v}_n. \tag{4.128}$$

To this end, the Kalman filtering formulation will be used. Note that the MSE linear estimation, presented in Section 4.9, cannot be used here. As we have already discussed in Chapter 2, an AR process is asymptotically stationary; for finite-time samples, the initial conditions at time $n = 0$ are "remembered" by the process and the respective second-order statistics are time dependent, hence it is a nonstationary process. However, Kalman filtering is specially suited for such cases.

Let us rewrite (4.127) and (4.128) as

$$\begin{bmatrix} \mathrm{x}_n \\ \mathrm{x}_{n-1} \\ \mathrm{x}_{n-2} \\ \vdots \\ \mathrm{x}_{n-l+1} \end{bmatrix} = \begin{bmatrix} -a_1 & -a_2 & \cdots & -a_{l-1} & -a_l \\ 1 & 0 & \cdots & 0 & 0 \\ 0 & 1 & \cdots & 0 & 0 \\ 0 & 0 & \cdots & 1 & 0 \end{bmatrix} \begin{bmatrix} \mathrm{x}_{n-1} \\ \mathrm{x}_{n-2} \\ \mathrm{x}_{n-3} \\ \vdots \\ \mathrm{x}_{n-l} \end{bmatrix} + \begin{bmatrix} \eta_n \\ 0 \\ \vdots \\ 0 \end{bmatrix},$$

$$\mathrm{y}_n = \begin{bmatrix} 1 & 0 & \cdots & 0 \end{bmatrix} \begin{bmatrix} \mathrm{x}_n \\ \vdots \\ \mathrm{x}_{n-l+1} \end{bmatrix} + \mathrm{v}_n$$

or

$$\mathbf{x}_n = F\mathbf{x}_{n-1} + \boldsymbol{\eta}, \tag{4.129}$$

$$\mathrm{y}_n = H\mathbf{x}_n + \mathrm{v}_n, \tag{4.130}$$

where the definitions of $F_n := F$ and $H_n := H$ are obvious and

$$Q_n = \begin{bmatrix} \sigma_n^2 & 0 & \cdots & 0 \\ 0 & 0 & \cdots & 0 \\ 0 & 0 & \cdots & 0 \end{bmatrix}, \quad R_n = \sigma_{\mathrm{v}}^2 \quad \text{(scalar)}.$$

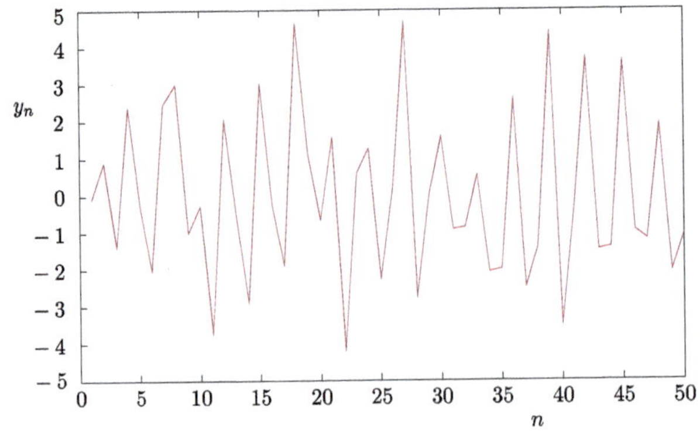

(A)

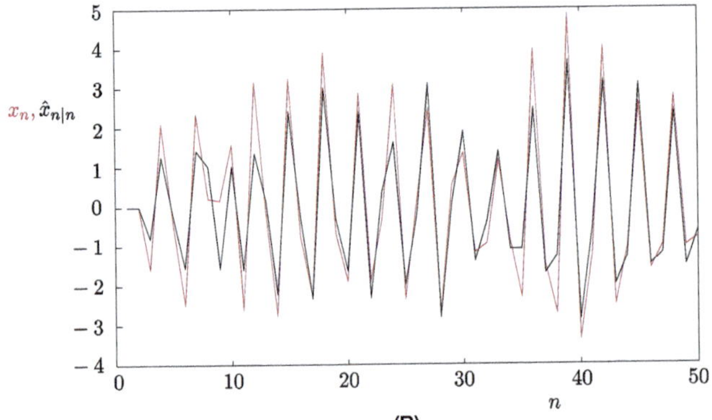

(B)

FIGURE 4.21

(A) A realization of the observation sequence, y_n, which is used by the Kalman filter to obtain the predictions of the state variable. (B) The AR process (state variable) in red together with the predicted by the Kalman filter sequence (gray), for Example 4.4. The Kalman filter has removed the effect of the noise v_n.

Fig. 4.21A shows the values of a specific realization y_n, and Fig. 4.21B shows the corresponding realization of the AR(2) (red) together with the predicted Kalman filter sequence $\hat{x}_n$. Observe that the match is very good. For the generation of the AR process we used $l = 2$, $\alpha_1 = 0.95$, $\alpha_2 = 0.9$, $\sigma_\eta^2 = 0.5$. For the Kalman filter output noise, $\sigma_v^2 = 1$.

PROBLEMS

4.1 Show that the set of equations

$$\Sigma\boldsymbol{\theta} = \boldsymbol{p}$$

has a unique solution if $\Sigma > 0$ and infinitely many if Σ is singular.

4.2 Show that the set of equations

$$\Sigma\boldsymbol{\theta} = \boldsymbol{p}$$

always has a solution.

4.3 Show that the shape of the isovalue contours of the graph of $J(\boldsymbol{\theta})$ (MSE),

$$J(\boldsymbol{\theta}) = J(\boldsymbol{\theta}_*) + (\boldsymbol{\theta} - \boldsymbol{\theta}_*)^T \Sigma (\boldsymbol{\theta} - \boldsymbol{\theta}_*),$$

are ellipses whose axes depend on the eigenstructure of Σ.
Hint: Assume that Σ has discrete eigenvalues.

4.4 Prove that if the true relation between the input $\mathbf{x}$ and the true output y is linear, meaning

$$\mathrm{y} = \boldsymbol{\theta}_o^T \mathbf{x} + \mathrm{v}_n, \quad \boldsymbol{\theta}_o \in \mathbb{R}^l,$$

where v is independent of $\mathbf{x}$, then the optimal MSE estimate $\boldsymbol{\theta}_*$ satisfies

$$\boldsymbol{\theta}_* = \boldsymbol{\theta}_o.$$

4.5 Show that if

$$\mathrm{y} = \boldsymbol{\theta}_o^T \mathbf{x} + \mathrm{v}, \quad \boldsymbol{\theta}_o \in \mathbb{R}^k,$$

where v is independent of $\mathbf{x}$, then the optimal MSE $\boldsymbol{\theta}_* \in \mathbb{R}^l$, $l < k$, is equal to the top l components of $\boldsymbol{\theta}_o$ if the components of $\mathbf{x}$ are uncorrelated.

4.6 Derive the normal equations by minimizing the cost in (4.15).
Hint: Express the cost $\boldsymbol{\theta}$ in terms of the real part $\boldsymbol{\theta}_r$ and its imaginary part $\boldsymbol{\theta}_i$, and optimize with respect to $\boldsymbol{\theta}_r, \boldsymbol{\theta}_i$.

4.7 Consider the multichannel filtering task

$$\hat{\mathbf{y}} = \begin{bmatrix} \hat{\mathrm{y}}_r \\ \hat{\mathrm{y}}_i \end{bmatrix} = \Theta \begin{bmatrix} \mathbf{x}_r \\ \mathbf{x}_i \end{bmatrix}.$$

Estimate Θ so as to minimize the error norm

$$\mathbb{E}\left[\|\mathbf{y} - \hat{\mathbf{y}}\|^2\right].$$

4.8 Show that (4.34) is the same as (4.25).

4.9 Show that the MSE achieved by a linear complex-valued estimator is always larger than that obtained by a widely linear one. Equality is achieved only under the circularity conditions.

4.10 Show that under the second-order circularity assumption, the conditions in (4.39) hold true.

4.11 Show that if

$$f : \mathbb{C} \longrightarrow \mathbb{R},$$

then the Cauchy–Riemann conditions are violated.

4.12 Derive the optimality condition in (4.45).

4.13 Show Eqs. (4.50) and (4.51).

4.14 Derive the normal equations for Example 4.2.

4.15 The input to the channel is a white noise sequence s_n of variance σ_s^2. The output of the channel is the AR process

$$y_n = a_1 y_{n-1} + s_n. \tag{4.131}$$

The channel also adds white noise η_n of variance σ_η^2. Design an optimal equalizer of order two, which at its output recovers an approximation of s_{n-L}. Sometimes, this equalization task is also known as whitening, because in this case the action of the equalizer is to "whiten" the AR process.

4.16 Show that the forward and backward MSE optimal predictors are conjugate reverse of each other.

4.17 Show that the MSE prediction errors ($\alpha_m^f = \alpha_m^b$) are updated according to the recursion

$$\alpha_m^b = \alpha_{m-1}^b (1 - |\kappa_{m-1}|^2).$$

4.18 Derive the BLUE for the Gauss–Markov theorem.

4.19 Show that the MSE (which in this case coincides with the variance of the estimator) of any linear unbiased estimator is higher than that associated with the BLUE.

4.20 Show that if Σ_η is positive definite, then $X^T \Sigma_\eta^{-1} X$ is also positive definite if X is full rank.

4.21 Derive an MSE optimal linearly constrained widely linear beamformer.

4.22 Prove that the Kalman gain that minimizes the error covariance matrix

$$P_{n|n} = \mathbb{E}\left[(\mathbf{x}_n - \hat{\mathbf{x}}_{n|n})(\mathbf{x}_n - \hat{\mathbf{x}}_{n|n})^T\right]$$

is given by

$$K_n = P_{n|n-1} H_n^T \left(R_n + H_n P_{n|n-1} H_n^T\right)^{-1}.$$

Hint: Use the following formulas:

$$\frac{\partial \operatorname{trace}\{AB\}}{\partial A} = B^T \ (AB \text{ a square matrix}),$$
$$\frac{\partial \operatorname{trace}\{ACA^T\}}{\partial A} = 2AC, \ (C = C^T).$$

4.23 Show that in Kalman filtering, the prior and posterior error covariance matrices are related as

$$P_{n|n} = P_{n|n-1} - K_n H_n P_{n|n-1}.$$

4.24 Derive the Kalman algorithm in terms of the inverse state-error covariance matrices, $P_{n|n}^{-1}$. In statistics, the inverse error covariance matrix is related to Fisher's information matrix; hence, the name of the scheme.

MATLAB® EXERCISES

4.25 Consider the image deblurring task described in Section 4.6.

- Download the "boat" image from Waterloo's Image repository.[8] Alternatively, you may use any grayscale image of your choice. You can load the image into MATLAB®'s memory using the "imread" function (also, you may want to apply the function "im2double" to get an array consisting of doubles).
- Create a blurring point spread function (PSF) using MATLAB®'s command "fspecial." For example, you can write

```
PSF = fspecial('motion',20,45);
```

  The blurring effect is produced using the "imfilter" function

```
J = imfilter(I,PSF,'conv', 'circ');
```

  where I is the original image.
- Add some white Gaussian noise to the image using MATLAB®'s function "imnoise," as follows:

```
J = imnoise(J, 'gaussian', noise_mean, noise_var);
```

  Use a small value of noise variance, such as 10^{-6}.
- To perform the deblurring, you need to employ the "deconvwnr" function. For example, if J is the array that contains the blurred image (with the noise) and PSF is the point spread function that produced the blurring, then the command

```
K = deconvwnr(J, PSF, C);
```

  returns the deblurred image K, provided that the choice of C is reasonable. As a first attempt, select $C = 10^{-4}$. Use various values for C of your choice. Comment on the results.

4.26 Consider the noise cancelation task described in Example 4.1. Write the necessary code to solve the problem using MATLAB® according to the following steps:

(a) Create 5000 data samples of the signal $s_n = \cos(\omega_0 n)$, for $\omega_0 = 2 \times 10^{-3}\pi$.

(b) Create 5000 data samples of the AR process $v_1(n) = a_1 \cdot v_1(n-1) + \eta_n$ (initializing at zero), where η_n represents zero mean Gaussian noise with variance $\sigma_\eta^2 = 0.0025$ and $a_1 = 0.8$.

(c) Add the two sequences (i.e., $d_n = s_n + v_1(n)$) and plot the result. This represents the contaminated signal.

(d) Create 5000 data samples of the AR process $v_2(n) = a_2 v_2(n-1) + \eta_n$ (initializing at zero), where η_n represents the same noise sequence and $a_2 = 0.75$.

(e) Solve for the optimum (in the MSE sense) $\boldsymbol{w} = [w_0, w_1]^T$. Create the sequence of the restored signal $\hat{s}_n = d_n - w_0 v_2(n) - w_1 v_2(n-1)$ and plot the result.

(f) Repeat steps (b)–(e) using $a_2 = 0.9, 0.8, 0.7, 0.6, 0.5, 0.3$. Comment on the results.

---

[8] http://links.uwaterloo.ca/.

(g) Repeat steps (b)–(e) using $\sigma_\eta^2 = 0.01, 0.05, 0.1, 0.2, 0.5$, for $a_2 = 0.9, 0.8, 0.7, 0.6, 0.5, 0.3$. Comment on the results.

4.27 Consider the channel equalization task described in Example 4.2. Write the necessary code to solve the problem using MATLAB® according to the following steps:

(a) Create a signal s_n consisting of 50 equiprobable ± 1 samples. Plot the result using MATLAB®'s function "stem."

(b) Create the sequence $u_n = 0.5s_n + s_{n-1} + \eta_n$, where η_n denotes zero mean Gaussian noise with $\sigma_\eta^2 = 0.01$. Plot the result with "stem."

(c) Find the optimal $\boldsymbol{w}_* = [w_0, w_1, w_2]^T$, solving the normal equations.

(d) Construct the sequence of the reconstructed signal $\hat{s}_n = \text{sgn}(w_0 u_n + w_1 u_{n-1} + w_2 u_{n-2})$. Plot the result with "stem" using red color for the correctly reconstructed values (i.e., those that satisfy $s_n = \hat{s}_n$) and black color for errors.

(e) Repeat steps (b)–(d) using different noise levels for σ_η^2. Comment on the results.

4.28 Consider the autoregressive process estimation task described in Example 4.4. Write the necessary code to solve the problem using MATLAB® according to the following steps:

(a) Create 500 samples of the AR sequence $x_n = -a_1 x_{n-1} - a_2 x_{n-2} + \eta_n$ (initializing at zeros), where $a_1 = 0.2$, $a_2 = 0.1$, and η_n denotes zero mean Gaussian noise with $\sigma_n^2 = 0.5$.

(b) Create the sequence $y_n = x_n + v_n$, where v_n denotes zero mean Gaussian noise with $\sigma_v^2 = 1$.

(c) Implement the Kalman filtering algorithm as described in Algorithm 4.2, using y_n as inputs and the matrices F, H, Q, R as described in Example 4.4. To initialize the algorithm, you can use $\hat{\boldsymbol{x}}_{1|0} = [0, 0]^T$ and $P_{1|0} = 0.1 \cdot I_2$. Plot the predicted values $\hat{x}_n$ (i.e., $\hat{x}_{n|n}$) versus the original sequence x_n. Play with the values of the different parameters and comment on the obtained results.

REFERENCES

[1] T. Adali, V.D. Calhoun, Complex ICA of brain imaging data, IEEE Signal Process. Mag. 24 (5) (2007) 136–139.
[2] T. Adali, H. Li, Complex-valued adaptive signal processing, in: T. Adali, S. Haykin (Eds.), Adaptive Signal Processing: Next Generation Solutions, John Wiley, 2010.
[3] T. Adali, P. Schreier, Optimization and estimation of complex-valued signals: theory and applications in filtering and blind source separation, IEEE Signal Process. Mag. 31 (5) (2014) 112–128.
[4] B.D.O. Anderson, J.B. Moore, Optimal Filtering, Prentice Hall, Englewood Cliffs, NJ, 1979.
[5] G.J. Bierman, Factorization Methods for Discrete Sequential Estimation, Academic Press, New York, 1977.
[6] P. Bouboulis, S. Theodoridis, Extension of Wirtinger's calculus to reproducing kernel Hilbert spaces and the complex kernel LMS, IEEE Trans. Signal Process. 53 (3) (2011) 964–978.
[7] D.H. Brandwood, A complex gradient operator and its application in adaptive array theory, IEEE Proc. 130 (1) (1983) 11–16.
[8] R.G. Brown, P.V.C. Hwang, Introduction to Random Signals and Applied Kalman Filtering, second ed., John Wiley Sons, Inc., 1992.
[9] F.S. Cattivelli, A.H. Sayed, Diffusion strategies for distributed Kalman filtering and smoothing, IEEE Trans. Automat. Control 55 (9) (2010) 2069–2084.
[10] P. Chevalier, J.P. Delmas, A. Oukaci, Optimal widely linear MVDR beamforming for noncircular signals, in: Proceedings of the IEEE International Conference on Acoustics, Speech and Signal Processing, ICASSP, 2009, pp. 3573–3576.
[11] P. Delsarte, Y. Genin, The split Levinson algorithm, IEEE Trans. Signal Process. 34 (1986) 470–478.
[12] J. Dourbin, The fitting of time series models, Rev. Int. Stat. Inst. 28 (1960) 233–244.
[13] Y.C. Eldar, Minimax, MSE estimation of deterministic parameters with noise covariance uncertainties, IEEE Trans. Signal Process. 54 (2006) 138–145.

[14] R.C. Gonzalez, R.E. Woods, Digital Image Processing, Addison-Wesley, 1993.
[15] S. Julier, A skewed approach to filtering, Proc. SPIE 3373 (1998) 271–282.
[16] T. Kailath, An innovations approach to least-squares estimation: Part 1. Linear filtering in additive white noise, IEEE Trans. Automat. Control AC-13 (1968) 646–655.
[17] T. Kailath, A.H. Sayed, B. Hassibi, Linear Estimation, Prentice Hall, Englewood Cliffs, 2000.
[18] R.E. Kalman, A new approach to linear filtering and prediction problems, Trans. ASME J. Basic Eng. 82 (1960) 34–45.
[19] R.E. Kalman, R.S. Bucy, New results in linear filtering and prediction theory, Trans. ASME J. Basic Eng. 83 (1961) 95–107.
[20] P.G. Kaminski, A.E. Bryson, S.F. Schmidt, Discrete square root filtering: a survey, IEEE Trans. Autom. Control 16 (1971) 727–735.
[21] N. Kalouptsidis, S. Theodoridis, Parallel implementation of efficient LS algorithms for filtering and prediction, IEEE Trans. Acoust. Speech Signal Process. 35 (1987) 1565–1569.
[22] N. Kalouptsidis, S. Theodoridis (Eds.), Adaptive System Identification and Signal Processing Algorithms, Prentice Hall, 1993.
[23] U.A. Khan, J. Moura, Distributing the Kalman filter for large-scale systems, IEEE Trans. Signal Process. 56 (10) (2008) 4919–4935.
[24] A.N. Kolmogorov, Stationary sequences in Hilbert spaces, Bull. Math. Univ. Moscow 2 (1941) (in Russian).
[25] K. Kreutz-Delgado, The complex gradient operator and the $\mathbb{CR}$-calculus, http://citeseerx.ist.psu.edu/viewdoc/download?doi=10.1.1.86.6515&rep=rep1&type=pdf, 2006.
[26] N. Levinson, The Wiener error criterion in filter design and prediction, J. Math. Phys. 25 (1947) 261–278.
[27] H. Li, T. Adali, Optimization in the complex domain for nonlinear adaptive filtering, in: Proceedings, 33rd Asilomar Conference on Signals, Systems and Computers, Pacific Grove, CA, 2006, pp. 263–267.
[28] X.-L. Li, T. Adali, Complex-valued linear and widely linear filtering using MSE and Gaussian entropy, IEEE Trans. Signal Process. 60 (2012) 5672–5684.
[29] D.J.C. MacKay, Information Theory, Inference, and Learning Algorithms, Cambridge University Press, 2003.
[30] D. Mandic, V.S.L. Guh, Complex Valued Nonlinear Adaptive Filters, John Wiley, 2009.
[31] J.M. Mendel, Lessons in Digital Estimation Theory, Prentice Hall, Englewood Cliffs, NJ, 1995.
[32] T. McWhorter, P. Schreier, Widely linear beamforming, in: Proceedings 37th Asilomar Conference on Signals, Systems, Computers, Pacific Grove, CA, 1993, p. 759.
[33] P.V. Overschee, B.D. Moor, Subspace Identification for Linear Systems: Theory, Implementation, Applications, Kluwer Academic Publishers, 1996.
[34] M. Petrou, C. Petrou, Image Processing: The Fundamentals, second ed., John Wiley, 2010.
[35] B. Picinbono, On circularity, IEEE Trans. Signal Process. 42 (12) (1994) 3473–3482.
[36] B. Picinbono, P. Chevalier, Widely linear estimation with complex data, IEEE Trans. Signal Process. 43 (8) (1995) 2030–2033.
[37] B. Picinbono, Random Signals and Systems, Prentice Hall, 1993.
[38] T. Piotrowski, I. Yamada, MV-PURE estimator: minimum-variance pseudo-unbiased reduced-rank estimator for linearly constrained ill-conditioned inverse problems, IEEE Trans. Signal Process. 56 (2008) 3408–3423.
[39] I.R. Porteous, Clifford Algebras and Classical Groups, Cambridge University Press, 1995.
[40] J.E. Potter, New statistical formulas, in: Space Guidance Analysis Memo, No 40, Instrumentation Laboratory, MIT, 1963.
[41] J. Proakis, Digital Communications, second ed., McGraw Hill, 1989.
[42] J.G. Proakis, D.G. Manolakis, Digital Signal Processing: Principles, Algorithms and Applications, second ed., MacMillan, 1992.
[43] O.-S. Reza, Distributed Kalman filtering for sensor networks, in: Proceedings IEEE Conference on Decision and Control, 2007, pp. 5492–5498.
[44] A.H. Sayed, Fundamentals of Adaptive Filtering, John Wiley, 2003.
[45] J. Schur, Über Potenzreihen, die im Innern des Einheitskreises beschränkt sind, J. Reine Angew. Math. 147 (1917) 205–232.
[46] K. Slavakis, P. Bouboulis, S. Theodoridis, Adaptive learning in complex reproducing kernel Hilbert spaces employing Wirtinger's subgradients, IEEE Trans. Neural Netw. Learn. Syst. 23 (3) (2012) 425–438.
[47] G. Strang, Linear Algebra and Its Applications, fourth ed., Hartcourt Brace Jovanovich, 2005.
[48] M. Viberg, Introduction to array processing, in: R. Chellappa, S. Theodoridis (Eds.), Academic Library in Signal Processing, vol. 3, Academic Press, 2014, pp. 463–499.
[49] N. Wiener, E. Hopf, Über eine klasse singulärer integralgleichungen, S.B. Preuss. Akad. Wiss. (1931) 696–706.
[50] N. Wiener, Extrapolation, Interpolation and Smoothing of Stationary Time Series, MIT Press, Cambridge, MA, 1949.
[51] W. Wirtinger, Zur formalen Theorie der Funktionen von mehr komplexen Veränderlichen, Math. Ann. 97 (1927) 357–375.
[52] P. Zhu, B. Chen, J.C. Principe, Learning nonlinear generative models of time series with a Kalman filter in RKHS, IEEE Trans. Signal Process. 62 (1) (2014) 141–155.

CHAPTER

ONLINE LEARNING: THE STOCHASTIC GRADIENT DESCENT FAMILY OF ALGORITHMS

5

CONTENTS

Machine Learning. https://doi.org/10.1016/B978-0-12-818803-3.00014-3

5.1 INTRODUCTION

In Chapter 4, we introduced the notion of mean-square error (MSE) optimal linear estimation and stated the normal equations for computing the paramneters/coefficients of the optimal estimator/filter. A prerequisite for the normal equations is the knowledge of the second-order statistics of the involved processes/variables, so that the covariance matrix of the input and the input–output cross-correlation vector can be obtained. However, most often in practice, all the designer has at her/his disposal is a set of training samples; thus, the covariance matrix and the cross-correlation vector have to be estimated somehow. More importantly, in a number of practical applications, the underlying statistics may be time-varying. We discussed this scenario while introducing Kalman filtering. The path taken there was to adopt a state-space representation and assume that the time dynamics of the model were known. However, although Kalman filtering is an elegant tool, it does not scale well in high-dimensional spaces due to the involved matrix operations and inversions.

The focus of this chapter is to introduce *online learning* techniques for estimating the unknown parameter vector. These are time-iterative schemes, which update the available estimate every time a measurement set (input–output pair of observations) is acquired. Thus, in contrast to the so-called *batch processing* methods, which process the whole block of data as a single entity, online algorithms operate on a single data point at a time; therefore, such schemes do not require the training data set to be known and stored in advance. Online algorithmic schemes learn the underlying statistics from the data in a *time-iterative* fashion. Hence, one does not have to provide further statistical information. Another characteristic of the algorithmic family, to be developed and studied in this chapter, is its computational simplicity. The required complexity for updating the estimate of the unknown parameter vector is linear with respect to the number of the unknown parameters. This is one of the major reasons that have made such schemes very popular in a number of practical applications; besides complexity, we will discuss other reasons that have contributed to their popularity. A discussion concerning batch versus online algorithms is provided in Section 8.12.

The fact that such learning algorithms work in a time-iterative mode gives them the agility to learn and *track* slow time variations of the statistics of the involved processes/variables; this is the reason these algorithms are also called *time-adaptive* or simply *adaptive*, because they can adapt to the needs of a changing environment. Online/time-adaptive algorithms have been used extensively since the early 1960s in a wide range of applications, including signal processing, control, and communications. More recently, the philosophy behind such schemes has been gaining in popularity in the context of applications where data reside in large datasets, with a massive number of samples; for such tasks, storing all the data points in the memory may not be possible, and they have to be considered one at a time. Moreover, the complexity of batch processing techniques can amount to prohibitive levels, for today's technology. The current trend is to refer to such applications as *big data* problems.

In this chapter, we focus on a very popular class of online/adaptive algorithms that springs from the classical gradient descent method for optimization. Although our emphasis will be on the squared error loss function, the same rationale can also be adopted for other (differentiable) loss functions. The case of nondifferentiable loss functions will be treated in Chapter 8. A number of variants of the stochastic gradient rationale, in the context of deep neural networks, is given in Chapter 18. The online processing rationale will be a recurrent theme in this book.

5.2 THE STEEPEST DESCENT METHOD

Our starting point is the method of *gradient descent*, one of the most widely used methods for iterative minimization of a differentiable cost function, $J(\boldsymbol{\theta})$, $\boldsymbol{\theta} \in \mathbb{R}^l$. As does any other iterative technique, the method starts from an initial estimate, $\boldsymbol{\theta}^{(0)}$, and generates a sequence, $\boldsymbol{\theta}^{(i)}$, $i = 1, 2, \ldots$, such that

$$\boldsymbol{\theta}^{(i)} = \boldsymbol{\theta}^{(i-1)} + \mu_i \Delta\boldsymbol{\theta}^{(i)}, \quad i > 0, \tag{5.1}$$

where $\mu_i > 0$. All the schemes for the iterative minimization of a cost function, which we will deal with in this book, have the general form of (5.1). Their differences are in the way that μ_i and $\Delta\boldsymbol{\theta}^{(i)}$ are chosen; the latter vector is known as the *update direction* or the *search direction*. The sequence μ_i is known as the *step size* or the *step length* at the ith iteration; note that the values of μ_i may either be constant or change at each iteration. In the gradient descent method, the choice of $\Delta\boldsymbol{\theta}^{(i)}$ is made to guarantee that

$$J(\boldsymbol{\theta}^{(i)}) < J((\boldsymbol{\theta}^{(i-1)}),$$

except at a minimizer, $\boldsymbol{\theta}_*$.

Assume that at the $(i-1)$th iteration step the value $\boldsymbol{\theta}^{(i-1)}$ has been obtained. Then, for sufficiently small μ_i and mobilizing a first-order Taylor expansion around $\boldsymbol{\theta}^{(i-1)}$, we can write

$$J(\boldsymbol{\theta}^{(i)}) = J(\boldsymbol{\theta}^{(i-1)} + \mu_i \Delta\boldsymbol{\theta}^{(i)}) \approx J(\boldsymbol{\theta}^{(i-1)}) + \mu_i \nabla J(\boldsymbol{\theta}^{(i-1)})^T \Delta\boldsymbol{\theta}^{(i)}.$$

Looking carefully at the above approximation, what we basically do is to *linearize* locally the cost function with respect to $\Delta\boldsymbol{\theta}^{(i)}$. Selecting the search direction so that

$$\nabla J(\boldsymbol{\theta}^{(i-1)})^T \Delta\boldsymbol{\theta}^{(i)} < 0, \tag{5.2}$$

this guarantees that $J(\boldsymbol{\theta}^{(i-1)} + \mu_i \Delta\boldsymbol{\theta}^{(i)}) < J(\boldsymbol{\theta}^{(i-1)})$. For such a choice, $\Delta\boldsymbol{\theta}^{(i)}$ and $\nabla J(\boldsymbol{\theta}^{(i-1)})$ must form an *obtuse* angle. Fig. 5.1 shows the graph of a cost function in the two-dimensional case, $\boldsymbol{\theta} \in \mathbb{R}^2$, and Fig. 5.2 shows the respective isovalue contours in the two-dimensional plane. Note that, in general, the contours can have any shape and are not necessarily ellipses; it all depends on the functional form of $J(\boldsymbol{\theta})$. However, because $J(\boldsymbol{\theta})$ has been assumed differentiable, the contours must be smooth and accept at any point a (unique) tangent plane, as this is defined by the respective gradient. Furthermore, recall from basic calculus that the gradient vector, $\nabla J(\boldsymbol{\theta})$, is perpendicular to the plane (line) tangent to the corresponding isovalue contour at the point $\boldsymbol{\theta}$ (Problem 5.1). The geometry is illustrated in Fig. 5.3; to facilitate the drawing and unclutter notation, we have removed the iteration index i. Note that by selecting the search direction which forms an obtuse angle with the gradient, it places $\boldsymbol{\theta}^{(i-1)} + \mu_i \Delta\boldsymbol{\theta}^{(i)}$ at a point on a contour which corresponds to a lower value of $J(\boldsymbol{\theta})$. Two issues are now raised: (a) to choose the best search direction along which to move and (b) to compute how far along this direction one can go. Even without much mathematics, it is obvious from Fig. 5.3 that if $\mu_i ||\Delta\boldsymbol{\theta}^{(i)}||$ is too large, then the new point can be placed on a contour corresponding to a larger value than that of the current

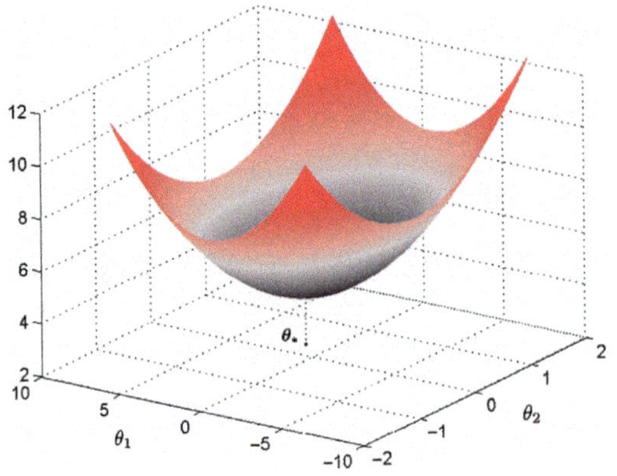

FIGURE 5.1

A cost function in the two-dimensional parameter space.

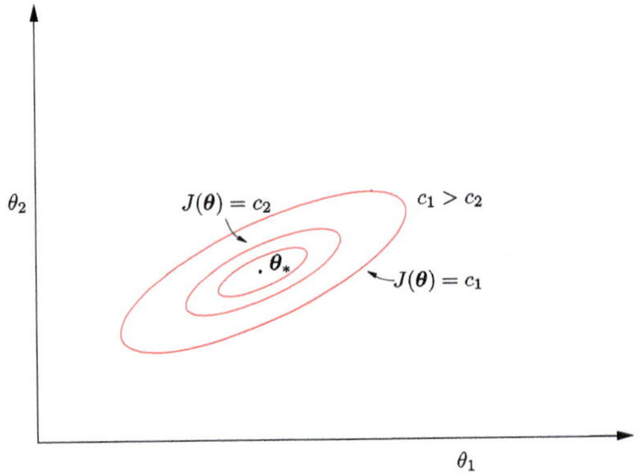

FIGURE 5.2

The corresponding isovalue curves of the cost function in Fig. 5.1, in the two-dimensional plane. All the points $\boldsymbol{\theta}$ lying on the same (isovalue) ellipse score the same value for the cost $J(\boldsymbol{\theta})$. Note that as we move away from the optimal value, $\boldsymbol{\theta}_*$, the values of c increase.

contour; after all, the first-order Taylor expansion holds approximately true for small deviations from $\boldsymbol{\theta}^{(i-1)}$.

To address the first of the two issues, let us assume $\mu_i = 1$ and search for all vectors, $\boldsymbol{z}$, with unit Euclidean norm and having their origin at $\boldsymbol{\theta}^{(i-1)}$. Then it does not take long to see that for all possible directions, the one that gives the most negative value of the inner product, $\nabla J\left(\boldsymbol{\theta}^{(i-1)}\right)^T \boldsymbol{z}$, is that of the negative gradient

$$\boldsymbol{z} = -\frac{\nabla J\left(\boldsymbol{\theta}^{(i-1)}\right)}{\left\|\nabla J\left(\boldsymbol{\theta}^{(i-1)}\right)\right\|}.$$

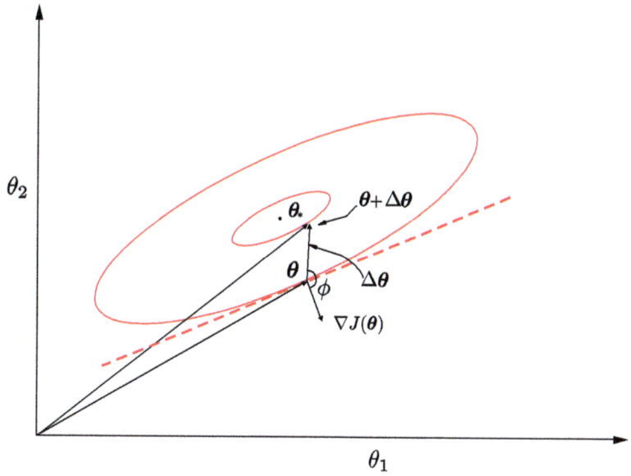

FIGURE 5.3

The gradient vector at a point $\boldsymbol{\theta}$ is perpendicular to the tangent plane (dotted line) at the isovalue curve crossing $\boldsymbol{\theta}$. The descent direction forms an obtuse angle, ϕ, with the gradient vector.

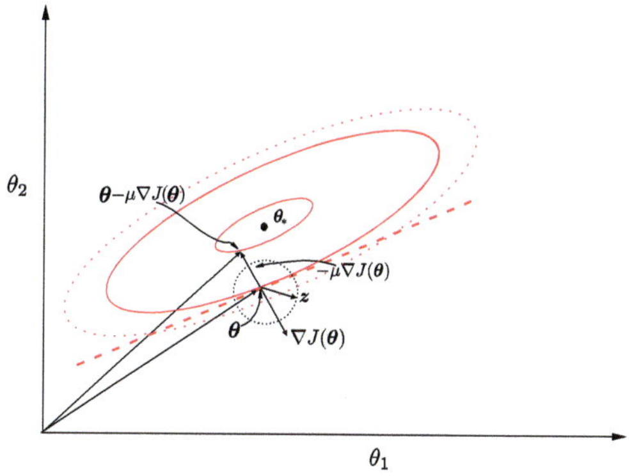

FIGURE 5.4

From all the descent directions of unit Euclidean norm (dotted circle centered at $\boldsymbol{\theta}^{(i-1)}$, which for notational simplicity is shown as $\boldsymbol{\theta}$), the negative gradient one leads to the maximum decrease of the cost function.

This is illustrated in Fig. 5.4. Center the unit Euclidean norm ball at $\boldsymbol{\theta}^{(i-1)}$. Then from all the unit norm vectors having their origin at $\boldsymbol{\theta}^{(i-1)}$, choose the one pointing to the negative gradient direction. Thus, for all unit Euclidean norm vectors, the steepest descent direction coincides with the (negative) *gradient descent* direction, and the corresponding update recursion becomes

$$\boxed{\boldsymbol{\theta}^{(i)} = \boldsymbol{\theta}^{(i-1)} - \mu_i \nabla J\left(\boldsymbol{\theta}^{(i-1)}\right): \quad \text{gradient descent scheme.}} \tag{5.3}$$

Note that we still have to address the second point, concerning the choice of μ_i. The choice must be made in such a way to guarantee convergence of the minimizing sequence. We will come to this issue soon.

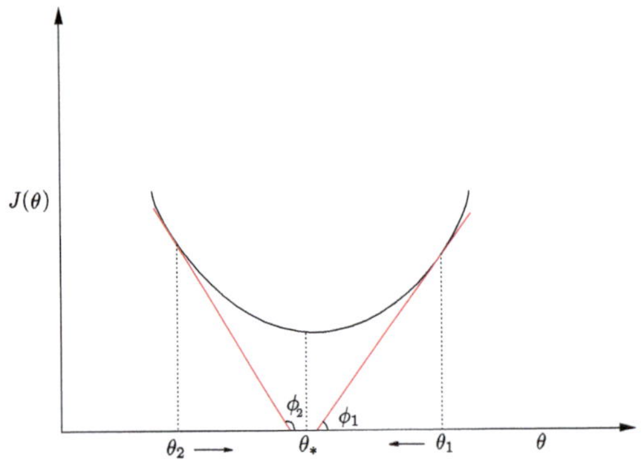

FIGURE 5.5

Once the algorithm is at θ_1, the gradient descent will move the point to the left, towards the minimum. The opposite is true for point θ_2.

Iteration (5.3) is illustrated in Fig. 5.5 for the one-dimensional case. If at the current iteration the algorithm has "landed" at θ_1, then the derivative of $J(\boldsymbol{\theta})$ at this point is positive (the tangent of an acute angle, ϕ_1), and this will force the update to move to the left towards the minimum. The scenario is different if the current estimate was θ_2. The derivative is negative (the tangent of an obtuse angle, ϕ_2) and this will push the update to the right, toward, again, the minimum. Note, however, that it is important how far to the left or to the right one has to move. A large move from, say, θ_1 to the left may land the update on the other side of the optimal value. In such a case, the algorithm may oscillate around the minimum and never converge. A major effort in this chapter will be devoted in providing theoretical frameworks for establishing bounds for the values of the step size that guarantee convergence.

The gradient descent method exhibits approximately *linear convergence*; that is, the error between $\boldsymbol{\theta}^{(i)}$ and the true minimum converges to zero asymptotically in the form of a *geometric series*. However, the convergence rate depends heavily on the eigenvalue spread of the Hessian matrix of $J(\boldsymbol{\theta})$. The dependence of the convergence rate on the eigenvalues will be unraveled in Section 5.3. For large eigenvalue spreads, the rate of convergence can become extremely slow. On the other hand, the great advantage of the method lies in its low computational requirements.

Finally, it has to be pointed out that we arrived at the scheme in Eq. (5.3) by searching all directions via the unit Euclidean norm. However, there is nothing "sacred" around Euclidean norms. One can employ other norms, such as the ℓ_1 and quadratic $\boldsymbol{z}^T P \boldsymbol{z}$ norms, where P is a positive definite matrix. Under such choices, one will end up with alternative update iterations (see, e.g., [23]). We will return to this point in Chapter 6, when dealing with Newton's iterative and coordinate descent minimization schemes.

5.3 APPLICATION TO THE MEAN-SQUARE ERROR COST FUNCTION

Let us apply the gradient descent scheme to derive an iterative algorithm to minimize our familiar cost function

$$J(\boldsymbol{\theta}) = \mathbb{E}\left[\left(\mathrm{y} - \boldsymbol{\theta}^T \mathbf{x}\right)^2\right]$$
$$= \sigma_y^2 - 2\boldsymbol{\theta}^T \boldsymbol{p} + \boldsymbol{\theta}^T \Sigma_x \boldsymbol{\theta}, \tag{5.4}$$

where $\Sigma_x = \mathbb{E}[\mathbf{x}\mathbf{x}^T]$ is the input covariance matrix and $\boldsymbol{p} = \mathbb{E}[\mathbf{x}\mathrm{y}]$ is the input–output cross-correlation vector (Chapter 4). The respective cost function gradient with respect to $\boldsymbol{\theta}$ is easily seen to be (e.g., Appendix A)

$$\nabla J(\boldsymbol{\theta}) = 2\Sigma_x \boldsymbol{\theta} - 2\boldsymbol{p}. \tag{5.5}$$

In this chapter, we will also adhere to zero mean jointly distributed input–output random variables, except if otherwise stated. Thus, the covariance and correlation matrices coincide. If this is not the case, the covariance in (5.5) is replaced by the correlation matrix. The treatment is focused on real data and we will point out differences with the complex-valued data case whenever needed.

Employing (5.5), the update recursion in (5.3) becomes

$$\boldsymbol{\theta}^{(i)} = \boldsymbol{\theta}^{(i-1)} - \mu\left(\Sigma_x \boldsymbol{\theta}^{(i-1)} - \boldsymbol{p}\right)$$
$$= \boldsymbol{\theta}^{(i-1)} + \mu\left(\boldsymbol{p} - \Sigma_x \boldsymbol{\theta}^{(i-1)}\right), \tag{5.6}$$

where the step size has been considered constant and it has also absorbed the factor 2. The more general case of iteration dependent values of the step size will be discussed soon after. Our goal now becomes that of searching for all values of μ that guarantee convergence. To this end, define

$$\boldsymbol{c}^{(i)} := \boldsymbol{\theta}^{(i)} - \boldsymbol{\theta}_*, \tag{5.7}$$

where $\boldsymbol{\theta}_*$ is the (unique) optimal MSE solution that results by solving the respective normal equations (Chapter 4),

$$\Sigma_x \boldsymbol{\theta}_* = \boldsymbol{p}. \tag{5.8}$$

Subtracting $\boldsymbol{\theta}_*$ from both sides of (5.6) and plugging in (5.7), we obtain

$$\boldsymbol{c}^{(i)} = \boldsymbol{c}^{(i-1)} + \mu\left(\boldsymbol{p} - \Sigma_x \boldsymbol{c}^{(i-1)} - \Sigma_x \boldsymbol{\theta}_*\right)$$
$$= \boldsymbol{c}^{(i-1)} - \mu \Sigma_x \boldsymbol{c}^{(i-1)} = (I - \mu \Sigma_x)\, \boldsymbol{c}^{(i-1)}. \tag{5.9}$$

Recall that Σ_x is a symmetric positive definite matrix (Chapter 2); hence all its eigenvalues are positive, and moreover (Appendix A.2) it can be written as

$$\Sigma_x = Q \Lambda Q^T, \tag{5.10}$$

where

$$\Lambda := \text{diag}\{\lambda_1, \dots, \lambda_l\} \quad \text{and} \quad Q := [\boldsymbol{q}_1, \boldsymbol{q}_2, \dots, \boldsymbol{q}_l],$$

with $\lambda_j, \boldsymbol{q}_j,\ j = 1, 2, \ldots, l$, being the eigenvalues and the respective normalized (*orthogonal*) eigenvectors of the covariance matrix,[1] represented by

$$\boldsymbol{q}_k^T \boldsymbol{q}_j = \delta_{kj}, \quad k, j = 1, 2, \ldots, l \Longrightarrow Q^T = Q^{-1}.$$

That is, the matrix Q is orthogonal. Plugging the factorization of Σ_x into (5.9), we obtain

$$\boldsymbol{c}^{(i)} = Q\,(I - \mu\Lambda)\,Q^T \boldsymbol{c}^{(i-1)},$$

or

$$\boldsymbol{v}^{(i)} = (I - \mu\Lambda)\,\boldsymbol{v}^{(i-1)}, \tag{5.11}$$

where

$$\boldsymbol{v}^{(i)} := Q^T \boldsymbol{c}^{(i)}, \quad i = 1, 2, \ldots. \tag{5.12}$$

The previously used "trick" is a standard one and its aim is to "decouple" the various components of $\boldsymbol{\theta}^{(i)}$ in (5.6). Indeed, each one of the components $v^{(i)}(j),\ j = 1, 2, \ldots, l$, of $\boldsymbol{v}^{(i)}$ follows an iteration path, which is independent of the rest of the components; in other words,

$$\begin{aligned} v^{(i)}(j) &= (1 - \mu\lambda_j) v^{(i-1)}(j) = (1 - \mu\lambda_j)^2 v^{(i-2)}(j) \\ &= \cdots = (1 - \mu\lambda_j)^i v^{(0)}(j), \end{aligned} \tag{5.13}$$

where $v^{(0)}(j)$ is the jth component of $\boldsymbol{v}^{(0)}$, corresponding to the initial vector. It is readily seen that if

$$|1 - \mu\lambda_j| < 1 \Longleftrightarrow -1 < 1 - \mu\lambda_j < 1, \quad j = 1, 2, \ldots, l, \tag{5.14}$$

the geometric series tends to zero and

$$\boldsymbol{v}^{(i)} \longrightarrow \boldsymbol{0} \Longrightarrow Q^T(\boldsymbol{\theta}^{(i)} - \boldsymbol{\theta}_*) \longrightarrow \boldsymbol{0} \Longrightarrow \boldsymbol{\theta}^{(i)} \longrightarrow \boldsymbol{\theta}_*. \tag{5.15}$$

Note that (5.14) is equivalent to

$$\boxed{0 < \mu < 2/\lambda_{\max} : \quad \text{condition for convergence,}} \tag{5.16}$$

where $\lambda_{\max}$ denotes the maximum eigenvalue of Σ_x.

Time constant: Fig. 5.6 shows a typical sketch of the evolution of $v^{(i)}(j)$ as a function of the iteration steps for the case $0 < 1 - \mu\lambda_j < 1$. Assume that the envelope, denoted by the red line, is (approximately) of an exponential form, $f(t) = \exp(-t/\tau_j)$. Plug into $f(t)$, as the values corresponding at the time instants $t = iT$ and $t = (i-1)T$, the values of $v^{(i)}(j),\ v^{(i-1)}(j)$ from (5.13); then the *time constant* results as

$$\tau_j = \frac{-1}{\ln(1 - \mu\lambda_j)},$$

[1] In contrast to other chapters, we denote eigenvectors with $\boldsymbol{q}$ and not as $\boldsymbol{u}$, since at some places the latter is used to denote the input random vector.

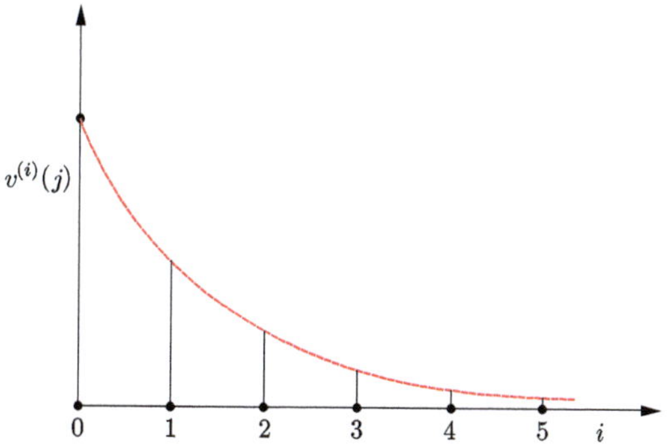

FIGURE 5.6

Convergence curve for one of the components of the transformed error vector. Note that the curve is of an approximate exponentially decreasing type.

assuming that the sampling time between two successive iterations is $T = 1$. For small values of μ, we can write

$$\tau_j \approx \frac{1}{\mu \lambda_j}, \quad \text{for } \mu \ll 1.$$

That is, the slowest rate of convergence is associated with the component that corresponds to the smallest eigenvalue. However, this is only true for small enough values of μ. For the more general case, this may not be true. Recall that the rate of convergence depends on the value of the term $1 - \mu\lambda_j$. This is also known as the jth *mode*. Its value depends not only on λ_j but also on μ. Let us consider as an example the case of μ taking a value very close to the maximum allowable one, $\mu \simeq 2/\lambda_{\max}$. Then the mode corresponding to the maximum eigenvalue will have an absolute value very close to one. On the other hand, the time constant of the mode corresponding to the minimum eigenvalue will be controlled by the value of $|1 - 2\lambda_{\min}/\lambda_{\max}|$, which can be much smaller than one. In such a case, the mode corresponding to the maximum eigenvalue exhibits slower convergence.

To obtain the optimum value for the step size, one has to select its value in such a way that the resulting maximum absolute mode value is minimized. This is a min/max task,

$$\begin{aligned} \mu_o = &\quad \arg\min_\mu \max_j |1 - \mu\lambda_j|, \\ \text{s.t.} &\quad |1 - \mu\lambda_j| < 1, \;\; j = 1, 2, \ldots, l. \end{aligned}$$

The task can be solved easily graphically. Fig. 5.7 shows the absolute values of the modes (corresponding to the maximum, minimum, and an intermediate eigenvalue). The (absolute) values of the modes initially decrease as μ increases, and then they start increasing. Observe that the optimal value results at the point where the curves for the maximum and minimum eigenvalues intersect. Indeed, this corresponds to the minimum-maximum value. Moving μ away from μ_o, the maximum mode value increases; increasing μ_o, the mode corresponding to the maximum eigenvalue becomes larger, and decreasing it, the mode corresponding to the minimum eigenvalue is increased. At the intersection we have

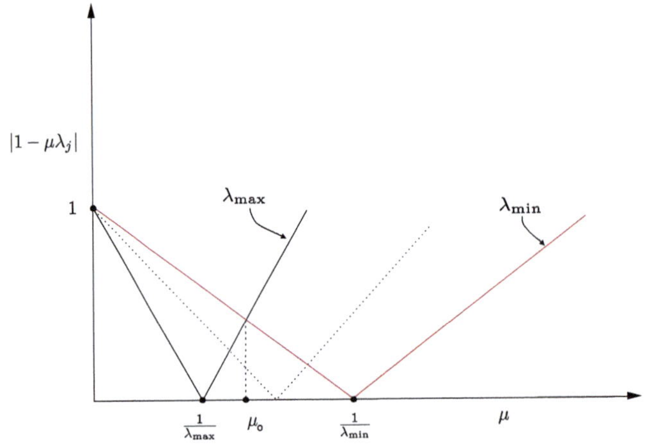

FIGURE 5.7

For each mode, increasing the value of the step size, the time constant starts decreasing and then after a point starts increasing. The full black line corresponds to the maximum eigenvalue, the red one to the minimum, and the dotted curve to an intermediate eigenvalue. The overall optimal, μ_o, corresponds to the value where the red and full black curves intersect.

$$1 - \mu_o \lambda_{\min} = -(1 - \mu_o \lambda_{\max}),$$

which results in

$$\mu_o = \frac{2}{\lambda_{\max} + \lambda_{\min}}. \tag{5.17}$$

At the optimal value, μ_o, there are two slowest modes; one corresponding to $\lambda_{\min}$ (i.e., $1 - \mu_o \lambda_{\min}$) and another one corresponding to $\lambda_{\max}$ (i.e., $1 - \mu_o \lambda_{\max}$). They have equal magnitudes but opposite signs, and they are given by

$$\pm \frac{\rho - 1}{\rho + 1},$$

where

$$\rho := \frac{\lambda_{\max}}{\lambda_{\min}}.$$

In other words, the *convergence rate depends on the eigenvalues spread of the covariance matrix.*

Parameter error vector convergence: From the definitions in (5.7) and (5.12), we get

$$\begin{aligned}
\boldsymbol{\theta}^{(i)} &= \boldsymbol{\theta}_* + Q\boldsymbol{v}^{(i)} \\
&= \boldsymbol{\theta}_* + [\boldsymbol{q}_1, \ldots, \boldsymbol{q}_l][v^{(i)}(1), \ldots, v^{(i)}(l)]^T \\
&= \boldsymbol{\theta}_* + \sum_{k=1}^{l} \boldsymbol{q}_k v^{(i)}(k),
\end{aligned} \tag{5.18}$$

or

$$\theta^{(i)}(j) = \theta_*(j) + \sum_{k=1}^{l} q_k(j) v^{(0)}(k)(1 - \mu\lambda_k)^i, \ j = 1, 2, \dots l. \tag{5.19}$$

In other words, the components of $\boldsymbol{\theta}^{(i)}$ converge to the respective components of the optimum vector $\boldsymbol{\theta}_*$ as a weighted average of powers of $1 - \mu\lambda_k$, i.e., $(1 - \mu\lambda_k)^i$. Computing the respective time constant in closed form is not possible; however, we can state lower and upper bounds. The lower bound corresponds to the time constant of the fastest converging mode and the upper bound to the slowest of the modes. For small values of $\mu \ll 1$, we can write

$$\frac{1}{\mu\lambda_{\max}} \le \tau \le \frac{1}{\mu\lambda_{\min}}. \tag{5.20}$$

The learning curve: We now turn our focus to the mean-square error (MSE). Recall from (4.8) that

$$J\left(\boldsymbol{\theta}^{(i)}\right) = J\left(\boldsymbol{\theta}_*\right) + \left(\boldsymbol{\theta}^{(i)} - \boldsymbol{\theta}_*\right)^T \Sigma_x \left(\boldsymbol{\theta}^{(i)} - \boldsymbol{\theta}_*\right), \tag{5.21}$$

or, mobilizing (5.18) and (5.10) and taking into consideration the orthonormality of the eigenvectors, we obtain

$$J\left(\boldsymbol{\theta}^{(i)}\right) = J\left(\boldsymbol{\theta}_*\right) + \sum_{j=1}^{l} \lambda_j |v^{(i)}(j)|^2 \Longrightarrow$$

$$J\left(\boldsymbol{\theta}^{(i)}\right) = J\left(\boldsymbol{\theta}_*\right) + \sum_{j=1}^{l} \lambda_j (1 - \mu\lambda_j)^{2i} |v^{(0)}(j)|^2, \tag{5.22}$$

which converges to the minimum value $J(\boldsymbol{\theta}_*)$ asymptotically. Moreover, observe that this convergence is monotonic, because $\lambda_j(1 - \mu\lambda_j)^2$ is positive. Following similar arguments as before, the respective time constants for each one of the modes are now

$$\tau_j^{\text{mse}} = \frac{-1}{2\ln(1 - \mu\lambda_j)} \approx \frac{1}{2\mu\lambda_j}. \tag{5.23}$$

Example 5.1. The aim of the example is to demonstrate what we have said so far, concerning the convergence issues of the gradient descent scheme in (5.6). The cross-correlation vector was chosen to be

$$\boldsymbol{p} = [0.05, 0.03]^T,$$

and we consider two different covariance matrices,

$$\Sigma_1 = \begin{bmatrix} 1 & 0 \\ 0 & 0.1 \end{bmatrix}, \quad \Sigma_2 = \begin{bmatrix} 1 & 0 \\ 0 & 1 \end{bmatrix}.$$

Note that for the case of Σ_2, both eigenvalues are equal to 1, and for Σ_1 they are $\lambda_1 = 1$ and $\lambda_2 = 0.1$ (for diagonal matrices the eigenvalues are equal to the diagonal elements of the matrix).

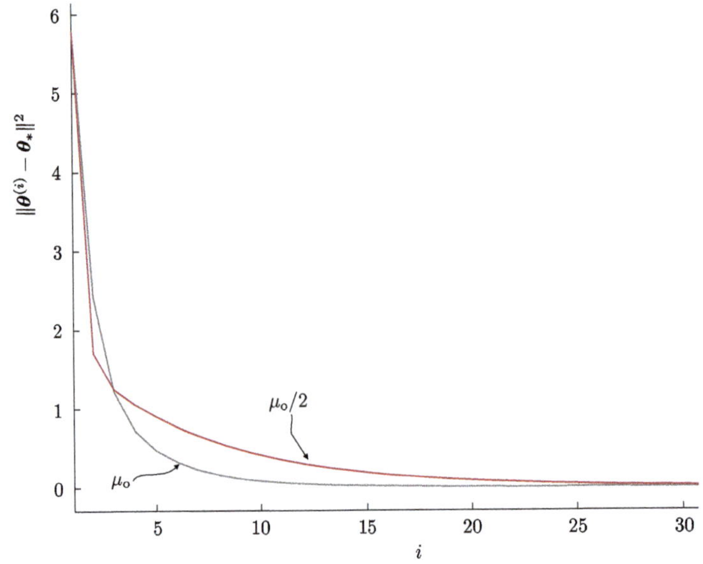

FIGURE 5.8

The black curve corresponds to the optimal value $\mu = \mu_o$ and the gray one to $\mu = \mu_o/2$, for the case of an input covariance matrix with unequal eigenvalues.

Fig. 5.8 shows the error curves for two values of μ for the case of Σ_1; the gray one corresponds to the optimum value ($\mu_o = 1.81$) and the red one to $\mu = \mu_o/2 = 0.9$. Observe the faster convergence towards zero that is achieved by the optimal value. Note that it may happen, as is the case in Fig. 5.8, that initially the convergence for some $\mu \neq \mu_o$ will be faster compared to μ_o. What the theory guarantees is that, eventually, the curve corresponding to the optimal will tend to zero faster than for any other value of μ. Fig. 5.9 shows the respective trajectories of the successive estimates in the two-dimensional space, together with the isovalue curves; the latter are ellipses, as we can readily deduce if we look carefully at the form of the quadratic cost function written as in (5.21). Compare the zig-zag path, which corresponds to the larger value of $\mu = 1.81$, with the smoother one, obtained for the smaller step size $\mu = 0.9$.

For comparison reasons, to demonstrate the dependence of the convergence speed on the eigenvalues spread, Fig. 5.10 shows the error curves using the same step size, $\mu = 1.81$, for both cases, Σ_1 and Σ_2. Observe that large eigenvalues spread of the input covariance matrix slows down the convergence rate. Note that if the eigenvalues of the covariance matrix are equal to, say, λ, the isovalue curves are circles; the optimal step size in this case is $\mu = 1/\lambda$ and convergence is achieved in only one step (Fig. 5.11).

TIME-VARYING STEP SIZES

The previous analysis cannot be carried out for the case of an iteration dependent step size. It can be shown (Problem 5.2) that in this case, the gradient descent algorithm converges if

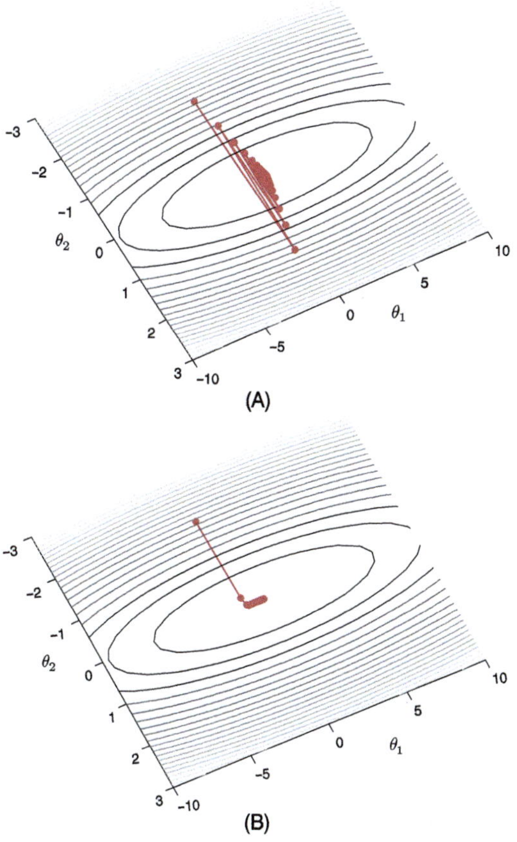

FIGURE 5.9

The trajectories of the successive estimates (dots) obtained by the gradient descent algorithm for (A) the larger value of $\mu = 1.81$ and (B) the smaller value of $\mu = 0.9$. In (B), the trajectory toward the minimum is smooth. In contrast, in (A), the trajectory consists of zig-zags.

- $\mu_i \longrightarrow 0, \text{ as } i \longrightarrow \infty,$
- $\sum_{i=1}^{\infty} \mu_i = \infty.$

A typical example of sequences which comply with both conditions are those that satisfy the following:

$$\sum_{i=1}^{\infty} \mu_i^2 < \infty, \quad \sum_{i=1}^{\infty} \mu_i = \infty, \tag{5.24}$$

as, for example, the sequence

$$\mu_i = \frac{1}{i}.$$

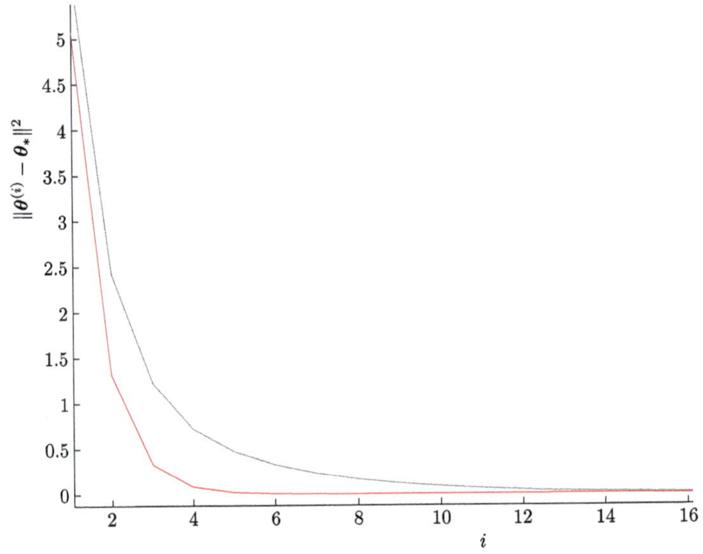

FIGURE 5.10

For the same value of $\mu = 1.81$, the error curves for the case of unequal eigenvalues ($\lambda_1 = 1$ and $\lambda_2 = 0.1$) (gray) and for equal eigenvalues ($\lambda_1 = \lambda_2 = 1$). For the latter case, the isovalue curves are circles; if the optimal value $\mu_o = 1$ is used, the algorithm converges in one step. This is demonstrated in Fig. 5.11.

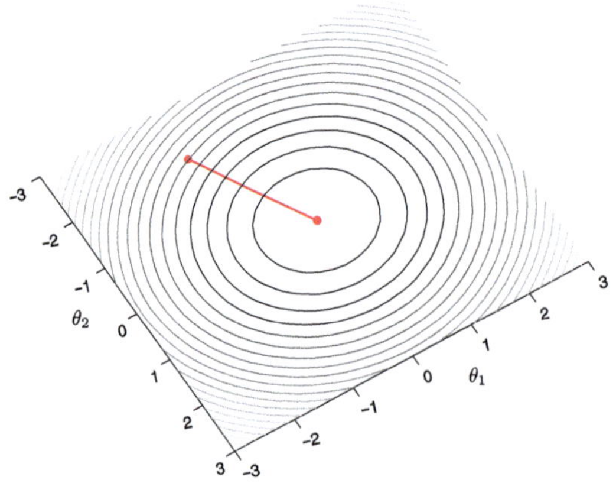

FIGURE 5.11

When the eigenvalues of the covariance matrix are all equal to a value, λ, the use of the optimal $\mu_o = 1/\lambda$ achieves convergence in one step.

Note that the two (sufficient) conditions require that the sequence tends to zero, yet its infinite sum diverges. We will meet this pair of conditions in various parts of this book. The previous conditions state that the step size has to become smaller and smaller as iterations progress, but this should not take place in a very aggressive manner, so that the algorithm is left to be active for a sufficient number of iterations to learn the solution. If the step size tends to zero very fast, then updates are practically frozen after a few iterations, without the algorithm having acquired enough information to get close to the solution.

5.3.1 THE COMPLEX-VALUED CASE

In Section 4.4.2 we stated that a function $f : \mathbb{C}^l \longmapsto \mathbb{R}$ is not differentiable with respect to its complex argument. To deal with such cases, the Wirtinger calculus was introduced. In this section, we use this mathematically convenient tool to derive the corresponding steepest descent direction.

To this end, we again employ a first-order Taylor series approximation [22]. Let

$$\boldsymbol{\theta} = \boldsymbol{\theta}_r + j\boldsymbol{\theta}_i.$$

Then the cost function

$$J(\boldsymbol{\theta}) : \mathbb{C}^l \longmapsto [0, +\infty)$$

is approximated as

$$\begin{aligned} J(\boldsymbol{\theta} + \Delta\boldsymbol{\theta}) &= J(\boldsymbol{\theta}_r + \Delta\boldsymbol{\theta}_r, \boldsymbol{\theta}_i + \Delta\boldsymbol{\theta}_i) \\ &= J(\boldsymbol{\theta}_r, \boldsymbol{\theta}_i) + \Delta\boldsymbol{\theta}_r^T \nabla_r J(\boldsymbol{\theta}_r, \boldsymbol{\theta}_i) + \Delta\boldsymbol{\theta}_i^T \nabla_i J(\boldsymbol{\theta}_r, \boldsymbol{\theta}_i), \end{aligned} \tag{5.25}$$

where ∇_r (∇_i) denotes the gradient with respect to $\boldsymbol{\theta}_r$ $(\boldsymbol{\theta}_i)$. Taking into account that

$$\Delta\boldsymbol{\theta}_r = \frac{\Delta\boldsymbol{\theta} + \Delta\boldsymbol{\theta}^*}{2}, \quad \Delta\boldsymbol{\theta}_i = \frac{\Delta\boldsymbol{\theta} - \Delta\boldsymbol{\theta}^*}{2j},$$

it is easy to show (Problem 5.3) that

$$J(\boldsymbol{\theta} + \Delta\boldsymbol{\theta}) = J(\boldsymbol{\theta}) + \operatorname{Re}\{\Delta\boldsymbol{\theta}^H \nabla_{\boldsymbol{\theta}^*} J(\boldsymbol{\theta})\}, \tag{5.26}$$

where $\nabla_{\boldsymbol{\theta}^*} J(\boldsymbol{\theta})$ is the CW-derivative, defined in Section 4.4.2 as

$$\nabla_{\boldsymbol{\theta}^*} J(\boldsymbol{\theta}) = \frac{1}{2}\big(\nabla_r J(\boldsymbol{\theta}) + j\nabla_i J(\boldsymbol{\theta})\big).$$

Looking carefully at (5.26), it is straightforward to observe that the direction

$$\Delta\boldsymbol{\theta} = -\mu \nabla_{\boldsymbol{\theta}^*} J(\boldsymbol{\theta})$$

makes the updated cost equal to

$$J(\boldsymbol{\theta} + \Delta\boldsymbol{\theta}) = J(\boldsymbol{\theta}) - \mu \big\|\nabla_{\boldsymbol{\theta}^*} J(\boldsymbol{\theta})\big\|^2,$$

which guarantees that $J(\boldsymbol{\theta} + \Delta\boldsymbol{\theta}) < J(\boldsymbol{\theta})$; it is straightforward to see, by taking into account the definition of an inner product, that the above search direction is the one of the largest decrease. Thus, the counterpart of (5.3) becomes

$$\boxed{\boldsymbol{\theta}^{(i)} = \boldsymbol{\theta}^{(i-1)} - \mu_i \nabla_{\boldsymbol{\theta}^*} J\left(\boldsymbol{\theta}^{(i-1)}\right): \quad \text{complex gradient descent scheme.}} \tag{5.27}$$

For the MSE cost function and for the linear estimation model, we get

$$\begin{aligned} J(\boldsymbol{\theta}) &= \mathbb{E}\left[\left(\mathrm{y} - \boldsymbol{\theta}^H \mathbf{x}\right)\left(\mathrm{y} - \boldsymbol{\theta}^H \mathbf{x}\right)^*\right] \\ &= \sigma_y^2 + \boldsymbol{\theta}^H \Sigma_x \boldsymbol{\theta} - \boldsymbol{\theta}^H \boldsymbol{p} - \boldsymbol{p}^H \boldsymbol{\theta}, \end{aligned}$$

and taking the gradient with respect to $\boldsymbol{\theta}^*$, by treating $\boldsymbol{\theta}$ as a constant (Section 4.4.2), we obtain

$$\nabla_{\boldsymbol{\theta}^*} J(\boldsymbol{\theta}) = \Sigma_x \boldsymbol{\theta} - \boldsymbol{p},$$

and the respective gradient descent iteration is the same as in (5.6).

5.4 STOCHASTIC APPROXIMATION

Solving for the normal equations (Eq. (5.8)) as well as using the gradient descent iterative scheme (for the case of the MSE), one has to have access to the second-order statistics of the involved variables. However, in most of the cases, this is not known and it has to be approximated using a set of observations. In this section, we turn our attention to algorithms that can learn the statistics iteratively via the training set. The origins of such techniques are traced back to 1951, when Robbins and Monro introduced the method of *stochastic approximation* [79], or the *Robbins–Monro algorithm.*

Let us consider the case of a function that is defined in terms of the expected value of another one, namely,

$$f(\boldsymbol{\theta}) = \mathbb{E}\left[\phi(\boldsymbol{\theta}, \boldsymbol{\eta})\right], \quad \boldsymbol{\theta} \in \mathbb{R}^l,$$

where $\boldsymbol{\eta}$ is a random vector of unknown distribution. The goal is to compute a root of $f(\boldsymbol{\theta})$. If the distribution was known, the expectation could be computed, at least in principle, and one could use any root-finding algorithm to compute the roots. The problem emerges when the statistics are unknown; hence the exact form of $f(\boldsymbol{\theta})$ is not known. All one has at her/his disposal is a sequence of i.i.d. observations $\boldsymbol{\eta}_0, \boldsymbol{\eta}_1, \ldots$. Robbins and Monro proved that the following algorithm,[2]

$$\boxed{\boldsymbol{\theta}_n = \boldsymbol{\theta}_{n-1} - \mu_n \phi(\boldsymbol{\theta}_{n-1}, \boldsymbol{\eta}_n): \quad \text{Robbins–Monro scheme,}} \tag{5.28}$$

starting from an arbitrary initial condition, $\boldsymbol{\theta}_{-1}$, converges[3] to a root of $f(\boldsymbol{\theta})$, under some general conditions and provided that (Problem 5.4)

[2] The original paper dealt with scalar variables only and the method was later extended to more general cases; see [96] for a related discussion.

[3] Convergence here is meant to be in probability (see Section 2.6).

$$\boxed{\sum_n \mu_n^2 < \infty, \quad \sum_n \mu_n \longrightarrow \infty: \quad \text{convergence conditions.}} \tag{5.29}$$

In other words, in the iteration (5.28), we get rid of the expectation operation and use the value of $\phi(\cdot, \cdot)$, which is computed using the current observations and the currently available estimate. That is, the algorithm learns both the statistics as well as the root; two in one! The same comments made for the convergence conditions, met in the iteration dependent step size case in Section 5.3, are valid here as well.

In the context of optimizing a general differentiable cost function of the form

$$J(\boldsymbol{\theta}) = \mathbb{E}\left[\mathcal{L}(\boldsymbol{\theta}, \mathrm{y}, \mathbf{x})\right], \tag{5.30}$$

the Robbins–Monro scheme can be mobilized to find a root of the respected gradient, i.e.,

$$\nabla J(\boldsymbol{\theta}) = \mathbb{E}\left[\nabla \mathcal{L}(\boldsymbol{\theta}, \mathrm{y}, \mathbf{x})\right],$$

where the expectation is with respect to the pair $(\mathrm{y}, \mathbf{x})$. As we have seen in Chapter 3, such cost functions are also known as the *expected risk* or the *expected loss* in machine learning terminology. Given the sequence of observations $(y_n, \boldsymbol{x}_n)$, $n = 0, 1, \ldots$, the recursion in (5.28) now becomes

$$\boxed{\boldsymbol{\theta}_n = \boldsymbol{\theta}_{n-1} - \mu_n \nabla \mathcal{L}(\boldsymbol{\theta}_{n-1}, y_n, \boldsymbol{x}_n).} \tag{5.31}$$

Let us now assume, for simplicity, that the expected risk has a unique minimum, $\boldsymbol{\theta}_*$. Then, according to the Robbins–Monro theorem and using an appropriate sequence μ_n, $\boldsymbol{\theta}_n$ will converge to $\boldsymbol{\theta}_*$. However, although this information is important, it is not by itself enough. In practice, one has to cease iterations after a *finite* number of steps. Hence, one has to know something more concerning the rate of convergence of such a scheme. To this end, two quantities are of interest, namely, the mean and the covariance matrix of the estimator at iteration n, i.e.,

$$\mathbb{E}[\boldsymbol{\theta}_n], \ \mathrm{Cov}(\boldsymbol{\theta}_n).$$

It can be shown (see [67]) that if $\mu_n = \mathcal{O}(1/n)$[4] and assuming that iterations have brought the estimate close to the optimal value, then

$$\boxed{\mathbb{E}[\boldsymbol{\theta}_n] = \boldsymbol{\theta}_* + \frac{1}{n}\boldsymbol{c}} \tag{5.32}$$

and

$$\boxed{\mathrm{Cov}(\boldsymbol{\theta}_n) = \frac{1}{n}V + \mathcal{O}(1/n^2),} \tag{5.33}$$

where $\boldsymbol{c}$ and V are constants that depend on the form of the expected risk. The above formulas have also been derived under some further assumptions concerning the eigenvalues of the Hessian matrix of the expected risk.[5] It is worth pointing out here that, in general, the convergence analysis of even

4 The symbol $\mathcal{O}$ denotes order of magnitude.

5 The proof is a bit technical and the interested reader can look at the provided reference.

simple algorithms is a mathematically tough task, and it is common to carry it out under a number of assumptions. What is important to keep from (5.32) and (5.33) is that both the mean and the variances of the components follow an $\mathcal{O}(1/n)$ pattern (in (5.33), $\mathcal{O}(1/n)$ is the prevailing pattern, since the $\mathcal{O}(1/n^2)$ dependence goes to zero much faster). Furthermore, these formulas indicate that the parameter vector estimate *fluctuates* around the optimal value. Indeed, in contrast to Eq. (5.15), where $\boldsymbol{\theta}^{(i)}$ converges to the optimal value, here it is the corresponding expected value that converges. The spread around the expected value is controlled by the respective covariance matrix.

The previously reported fluctuation depends on the choice of the sequence μ_n, being smaller for smaller values of the step size sequence. However, μ_n cannot be made to decrease very fast due to the two convergence conditions, as discussed before. This is the price one pays for using the *noisy version of the gradient* and it is the reason that such schemes suffer from relatively slow convergence rates. However, this does not mean that such schemes are, necessarily, the poor relatives of other more "elaborate" algorithms. As we will discuss in Chapter 8, their low complexity requirements make this algorithmic family to be the preferred choice in a number of practical applications.

APPLICATION TO THE MSE LINEAR ESTIMATION

Let us apply the Robbins–Monro algorithm to solve for the optimal MSE linear estimator if the covariance matrix and the cross-correlation vector are unknown. We know that the solution corresponds to the root of the gradient of the cost function, which can be written in the form (recall the orthogonality theorem from Chapter 3 and Eq. (5.8))

$$\Sigma_x \boldsymbol{\theta} - \boldsymbol{p} = \mathbb{E}\left[\mathbf{x}(\mathbf{x}^T \boldsymbol{\theta} - \mathrm{y})\right] = \mathbf{0}.$$

Given the training sequence of observations, $(y_n, \boldsymbol{x}_n)$, which are assumed to be i.i.d. drawn from the joint distribution of $(\mathrm{y}, \mathbf{x})$, the Robbins–Monro algorithm becomes

$$\boldsymbol{\theta}_n = \boldsymbol{\theta}_{n-1} + \mu_n \boldsymbol{x}_n \left(y_n - \boldsymbol{x}_n^T \boldsymbol{\theta}_{n-1}\right), \tag{5.34}$$

which converges to the optimal MSE solution provided that the two conditions in (5.29) are satisfied. Compare (5.34) with (5.6). Taking into account the definitions, $\Sigma_x = \mathbb{E}[\mathbf{x}\mathbf{x}^T]$, $\boldsymbol{p} = \mathbb{E}[\mathbf{x}\mathrm{y}]$, the former equation results from the latter one by dropping out the expectation operations and using an iteration dependent step size. Observe that the iterations in (5.34) coincide with *time updates*; time has now explicitly entered into the scene. This prompts us to start thinking about modifying such schemes appropriately to track time-varying environments. Algorithms such as the one in (5.34), which result from the generic gradient descent formulation by replacing the expectation by the respective instantaneous observations, are also known as *stochastic gradient descent schemes*.

Remarks 5.1.

- All the algorithms to be derived next can also be applied to nonlinear estimation/filtering tasks of the form

$$\hat{\mathrm{y}} = \sum_{k=1}^{l} \theta_k \phi_k(\mathbf{x}) = \boldsymbol{\theta}^T \boldsymbol{\phi},$$

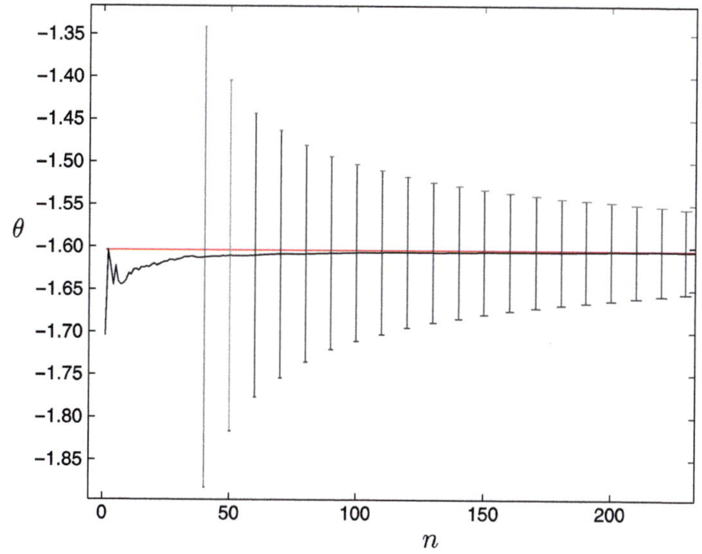

FIGURE 5.12

The red line corresponds to the true value of the unknown parameter. The black curve corresponds to the average over 200 realizations of the experiment. Observe that the mean value converges to the true value. The bars correspond to the respective standard deviation, which keeps decreasing as n grows.

and the place of $\mathbf{x}$ is taken by $\boldsymbol{\phi}$, where

$$\boldsymbol{\phi} = [\phi_1(\mathbf{x}), \ldots, \phi_l(\mathbf{x})]^T.$$

Example 5.2. The aim of this example is to demonstrate the pair of Eqs. (5.32) and (5.33), which characterize the convergence properties of the stochastic gradient scheme.

Data samples were first generated according to the regression model

$$y_n = \boldsymbol{\theta}^T \boldsymbol{x}_n + \eta_n,$$

where $\boldsymbol{\theta} \in \mathbb{R}^2$ was randomly chosen and then fixed. The elements of $\boldsymbol{x}_n$ were i.i.d. generated via a normal distribution $\mathcal{N}(0, 1)$ and η_n are samples of a white noise sequence with variance equal to $\sigma^2 = 0.1$. Then the observations $(y_n, \boldsymbol{x}_n)$ were used in the recursive scheme in (5.34) to obtain an estimate of $\boldsymbol{\theta}$. The experiment was repeated 200 times and the mean and variance of the obtained estimates were computed for each iteration step. Fig. 5.12 shows the resulting curve for one of the parameters (the trend for the other one being similar). Observe that the mean values of the estimates tend to the true value, corresponding to the red line, and the standard deviation keeps decreasing as n grows. The step size was chosen equal to $\mu_n = 1/n$.

5.5 THE LEAST-MEAN-SQUARES ADAPTIVE ALGORITHM

The stochastic gradient descent algorithm in (5.34) converges to the optimal MSE solution provided that μ_n satisfies the two convergence conditions. Once the algorithm has converged, it "locks" at the obtained solution. In a case where the statistics of the involved variables/processes and/or the unknown parameters start changing, the algorithm cannot track the changes. Note that if such changes occur, the error term

$$e_n = y_n - \boldsymbol{\theta}_{n-1}^T \boldsymbol{x}_n$$

will attain larger values; however, because μ_n is very small, the increased value of the error will not lead to corresponding changes of the estimate at time n. This can be overcome if one sets the value of μ_n to a preselected *fixed* value, μ. The resulting algorithm is the celebrated least-mean-squares (LMS) algorithm [102].

Algorithm 5.1 (The LMS algorithm).

- Initialize
  - $\boldsymbol{\theta}_{-1} = \mathbf{0} \in \mathbb{R}^l$; other values can also be used.
  - Select the value of μ.
- **For** $n = 0, 1, \ldots,$ **Do**
  - $e_n = y_n - \boldsymbol{\theta}_{n-1}^T \boldsymbol{x}_n$
  - $\boldsymbol{\theta}_n = \boldsymbol{\theta}_{n-1} + \mu e_n \boldsymbol{x}_n$
- **End For**

In case the input is a time series,[6] u_n, the initialization also involves the samples, $u_{-1}, \ldots, u_{-l+1} = 0$, to form the input vectors, $\boldsymbol{u}_n, n = 0, 1, \ldots, l-2$. The complexity of the algorithm amounts to $2l$ multiplications/additions (MADs) per time update. We have assumed that observations start arriving at time instant $n = 0$, to be in line with most references treating the LMS.

Let us now comment on this simple structure. Assume that the algorithm has converged close to the solution; then the error term is expected to take small values and thus the updates will remain close to the solution. If the statistics and/or the system parameters now start changing, the error values are expected to increase. Given that μ has a constant value, the algorithm has now the "agility" to update the estimates in an attempt to "push" the error to lower values. This small variation of the iterative scheme has important implications. The resulting algorithm is no more a member of the Robbins–Monro stochastic approximation family. Thus, one has to study its convergence conditions as well as its performance properties. Moreover, since the algorithm now has the potential to track changes in the values of the underlying parameters, as well as the statistics of the involved processes/variables, one has to study its performance in nonstationary environments; this is associated to what is known as the *tracking* performance of the algorithm, and it will be treated at the end of the chapter.

[6] Recall our adopted notation from Chapter 2; in this case we use $\boldsymbol{u}_n$ in place of $\boldsymbol{x}_n$.

5.5.1 CONVERGENCE AND STEADY-STATE PERFORMANCE OF THE LMS IN STATIONARY ENVIRONMENTS

The goal of this subsection is to study the performance of the LMS in stationary environments, that is, to answer the following questions. (a) Does the scheme converge and under which conditions? And if it converges, where does it converge? Although we introduced the scheme having in mind nonstationary environments, we still have to know how it behaves under stationarity; after all, the environment can change very slowly, and it can be considered "locally" stationary.

The convergence properties of the LMS, as well as of any other online/adaptive algorithm, are related to its *transient* characteristics; that is, the period from the initial estimate until the algorithm reaches a "steady-state" mode of operation. In general, analyzing the transient performance of an online algorithm is a formidable task indeed. This is also true even for the very simple structure of the LMS summarized in Algorithm 5.1. The LMS update recursions are equivalent to a time-varying, nonlinear (Problem 5.5), and stochastic in nature estimator. Many papers, some of them of high scientific insight and mathematical skill, have been produced. However, with the exception of a few rare and special cases, the analysis involves approximations. Our goal in this book is not to treat this topic in detail. Our focus will be restricted to the most "primitive" of the techniques, which are easier for the reader to follow compared with more advanced and mathematically elegant theories; after all, even this primitive approach provides results that turn out to be in agreement with what one experiences in practice.

Convergence of the Parameter Error Vector

Define

$$\boldsymbol{c}_n := \boldsymbol{\theta}_n - \boldsymbol{\theta}_*,$$

where $\boldsymbol{\theta}_*$ is the optimal solution resulting from the normal equations. The LMS update recursion can now be written as

$$\boldsymbol{c}_n = \boldsymbol{c}_{n-1} + \mu \boldsymbol{x}_n (y_n - \boldsymbol{\theta}_{n-1}^T \boldsymbol{x}_n + \boldsymbol{\theta}_*^T \boldsymbol{x}_n - \boldsymbol{\theta}_*^T \boldsymbol{x}_n).$$

Because we are going to study the statistical properties of the obtained estimates, we have to switch our notation from that referring to observations to the one involving the respective random variables. Then we can write

$$\begin{aligned}
\mathbf{c}_n &= \mathbf{c}_{n-1} + \mu \mathbf{x}(\mathrm{y} - \boldsymbol{\theta}_{n-1}^T \mathbf{x} + \boldsymbol{\theta}_*^T \mathbf{x} - \boldsymbol{\theta}_*^T \mathbf{x}) \\
&= \mathbf{c}_{n-1} - \mu \mathbf{x}\mathbf{x}^T \mathbf{c}_{n-1} + \mu \mathbf{x} \mathrm{e}_* \\
&= (I - \mu \mathbf{x}\mathbf{x}^T)\mathbf{c}_{n-1} + \mu \mathbf{x} \mathrm{e}_*, \qquad (5.35)
\end{aligned}$$

where

$$\mathrm{e}_* = \mathrm{y} - \boldsymbol{\theta}_*^T \mathbf{x} \qquad (5.36)$$

is the error random variable associated with the optimal $\boldsymbol{\theta}_*$. Compare (5.35) with (5.9). They look similar, yet they are very different. First, the latter of the two involves the expected value, Σ_x, in place of the respective variables. Moreover in (5.35), there is a second term that acts as an external input to the difference stochastic equation. From (5.35) we obtain

$$\mathbb{E}[\mathbf{c}_n] = \mathbb{E}\left[(I - \mu \mathbf{x}\mathbf{x}^T)\mathbf{c}_{n-1}\right] + \mu\,\mathbb{E}[\mathbf{x}\mathrm{e}_*]. \tag{5.37}$$

To proceed, it is time to introduce assumptions.

Assumption 1. The involved random variables are jointly linked via the regression model,

$$\mathrm{y} = \boldsymbol{\theta}_o^T \mathbf{x} + \eta, \tag{5.38}$$

where η is the noise variable with variance σ_η^2 and it is assumed to be independent of $\mathbf{x}$. Moreover, successive samples η_n, which generate the data, are assumed to be i.i.d. We have seen in Remarks 4.2 and Problem 4.4 that in this case, $\boldsymbol{\theta}_* = \boldsymbol{\theta}_o$ and $\sigma_{e_*}^2 = \sigma_\eta^2$. Also, due to the orthogonality condition, $\mathbb{E}[\mathbf{x}\mathrm{e}_*] = \mathbf{0}$. In addition, a stronger condition will be adopted, and e_* and $\mathbf{x}$ will be assumed to be *statistically independent*. This is justified by the fact that under the above model, $e_{*,n} = \eta_n$, and the noise sequence has been assumed to be independent of the input.

Assumption 2. (Independence assumption) Assume that $\mathbf{c}_{n-1}$ is statistically independent of both $\mathbf{x}$ and e_*. No doubt this is a strong assumption, but one we will adopt to simplify computations. Sometimes there is a tendency to "justify" this assumption by resorting to some special cases, which we will not do. If one is not happy with the assumption, he/she has to look for more recent methods, based on more rigorous mathematical analysis; of course, this does not mean that such methods are free of assumptions.

I. *Convergence in the mean*: Having adopted the previous assumptions, (5.37) becomes

$$\begin{aligned}\mathbb{E}[\mathbf{c}_n] &= \mathbb{E}\left[\left(I - \mu \mathbf{x}\mathbf{x}^T\right)\mathbf{c}_{n-1}\right] \\ &= (I - \mu \Sigma_x)\,\mathbb{E}[\mathbf{c}_{n-1}].\end{aligned} \tag{5.39}$$

Following similar arguments as in Section 5.3, we obtain

$$\mathbb{E}[\mathbf{v}_n] = (I - \mu \Lambda)\,\mathbb{E}[\mathbf{v}_{n-1}],$$

where $\Sigma_x = Q \Lambda Q^T$ and $\mathbf{v}_n = Q^T \mathbf{c}_n$. The last equation leads to

$$\mathbb{E}[\boldsymbol{\theta}_n] \longrightarrow \boldsymbol{\theta}_* \quad \text{as } n \longrightarrow \infty,$$

provided that

$$0 < \mu < \frac{2}{\lambda_{\max}}.$$

In other words, in a stationary environment, the LMS converges to the optimal MSE solution *in the mean*. Thus, by fixing the value of the step size to be a constant, we lose something; the obtained estimates, even after convergence, hover around the optimal solution. The obvious task to be considered next is to study the respective covariance matrix.

II. *Error vector covariance matrix*: From (5.39), applying it recursively and assuming for the initial condition that $\mathbb{E}[\mathbf{c}_{-1}] = \mathbf{0}$, we have $\mathbb{E}[\mathbf{c}_n] = \mathbf{0}$. In any case, borrowing the same arguments used in establishing the convergence in the mean, it turns out that the latter is approximately true for large enough values of n, irrespective of the initialization. Thus, from (5.35) we get

$$\begin{aligned}\Sigma_{c,n} := \mathbb{E}[\mathbf{c}_n\mathbf{c}_n^T] = \Sigma_{c,n-1} - \mu\,\mathbb{E}[\mathbf{x}\mathbf{x}^T\mathbf{c}_{n-1}\mathbf{c}_{n-1}^T] \\ - \mu\,\mathbb{E}[\mathbf{c}_{n-1}\mathbf{c}_{n-1}^T\mathbf{x}\mathbf{x}^T] + \mu^2\,\mathbb{E}[\mathrm{e}_*^2\mathbf{x}\mathbf{x}^T] \\ + \mu^2\,\mathbb{E}[\mathbf{x}\mathbf{x}^T\mathbf{c}_{n-1}\mathbf{c}_{n-1}^T\mathbf{x}\mathbf{x}^T],\end{aligned} \tag{5.40}$$

where we have used the independence between e_* and $\mathbf{c}_{n-1}$ and the fact that e_* is orthogonal to $\mathbf{x}$ in order to set to zero some of the terms. Taking into consideration the adopted independence assumptions and assuming that the input vector follows a *Gaussian distribution*, (5.40) becomes

$$\begin{aligned}\Sigma_{c,n} &= \Sigma_{c,n-1} - \mu\Sigma_x\Sigma_{c,n-1} - \mu\Sigma_{c,n-1}\Sigma_x \\ &\quad + 2\mu^2\Sigma_x\Sigma_{c,n-1}\Sigma_x + \mu^2\Sigma_x\mathrm{trace}\{\Sigma_x\Sigma_{c,n-1}\} \\ &\quad + \mu^2\sigma_\eta^2\Sigma_x,\end{aligned} \tag{5.41}$$

where the Gaussian assumption has been exploited to express the term involving fourth-order moments (e.g., [74]) as

$$\mathbb{E}[\mathbf{x}\mathbf{x}^T\Sigma_{c,n-1}\mathbf{x}\mathbf{x}^T] = 2\Sigma_x\Sigma_{c,n-1}\Sigma_x + \Sigma_x\mathrm{trace}\{\Sigma_x\Sigma_{c,n-1}\}.$$

Mobilizing the definition of $\mathbf{v}_n = Q^T\mathbf{c}_n$, (5.41) results in (Problem 5.6)

$$\begin{aligned}\Sigma_{v,n} &= Q^T\Sigma_{c,n}Q = \Sigma_{v,n-1} - \mu\Lambda\Sigma_{v,n-1} - \mu\Sigma_{v,n-1}\Lambda \\ &\quad + 2\mu^2\Lambda\Sigma_{v,n-1}\Lambda + \mu^2\Lambda\mathrm{trace}\{\Lambda\Sigma_{v,n-1}\} + \mu^2\sigma_\eta^2\Lambda.\end{aligned} \tag{5.42}$$

Note that our interest lies at the diagonal elements of $\Sigma_{v,n}$, since these correspond to the variances of the respective elements of $\boldsymbol{\theta}_n - \boldsymbol{\theta}_*$, and correspondingly to $\mathbf{v}_n - \boldsymbol{v}_*$. Collecting all the *diagonal* elements in a vector, $\boldsymbol{s}_n$, a close inspection of the diagonal elements of $\Sigma_{v,n}$ in (5.42) can persuade the reader that the following difference equation is true:

$$\boldsymbol{s}_n = (I - 2\mu\Lambda + 2\mu^2\Lambda^2 + \mu^2\boldsymbol{\lambda}\boldsymbol{\lambda}^T)\boldsymbol{s}_{n-1} + \mu^2\sigma_\eta^2\boldsymbol{\lambda}, \tag{5.43}$$

where

$$\boldsymbol{\lambda} := [\lambda_1, \lambda_2, \ldots, \lambda_l]^T.$$

It is well known from linear system theory that the difference equation in (5.43) is stable if the eigenvalues of the matrix,

$$\begin{aligned}A &:= I - 2\mu\Lambda + 2\mu^2\Lambda^2 + \mu^2\boldsymbol{\lambda}\boldsymbol{\lambda}^T \\ &= (I - \mu\Lambda)^2 + \mu^2\Lambda^2 + \mu^2\boldsymbol{\lambda}\boldsymbol{\lambda}^T,\end{aligned} \tag{5.44}$$

have magnitude less than one. This can be guaranteed if the step size μ is chosen such as (Problem 5.7)

$$0 < \mu < \frac{2}{\sum_{i=1}^{l} \lambda_i},$$

or

$$\boxed{\mu < \frac{2}{\text{trace}\{\Sigma_x\}}.} \tag{5.45}$$

The last condition guarantees that the variances remain *bounded*. Recall the number of assumptions made. Thus, to be on the safe side, μ must be selected so that it is not close to this upper bound.

III. *Excess mean-square error*: We know that the minimum MSE is achieved at $\boldsymbol{\theta}_*$. Any other weight vector results in higher values of the MSE. We have already said that in the steady state, the estimates obtained via the LMS fluctuate randomly around $\boldsymbol{\theta}_*$; thus, the MSE will be larger than the minimum $J_{\min}$. This "extra" error power, denoted as J_{exc}, is known as the *excess* MSE. Also, the ratio

$$\mathcal{M} := \frac{J_{\text{exc}}}{J_{\min}}$$

is known as the *misadjustment*. No doubt, we should seek for the relationship of $\mathcal{M}$ with μ and in practice we would like to adjust μ accordingly, in order to get a value of $\mathcal{M}$ as small as possible. Unfortunately, we will soon see that there is a tradeoff in achieving that. Making $\mathcal{M}$ small, the convergence speed becomes slower and vice versa; there is no free lunch!

By the respective definitions, we have[7]

$$\begin{aligned} \mathrm{e}_n &= \mathrm{y}_n - \boldsymbol{\theta}_{n-1}^T \mathbf{x} \\ &= \mathrm{e}_{*,n} - \mathbf{c}_{n-1}^T \mathbf{x}, \end{aligned}$$

or

$$\mathrm{e}_n^2 = \mathrm{e}_{*,n}^2 + \mathbf{c}_{n-1}^T \mathbf{x}\mathbf{x}^T \mathbf{c}_{n-1} - 2\mathrm{e}_{*,n} \mathbf{c}_{n-1}^T \mathbf{x}. \tag{5.46}$$

Taking the expectation on both sides and exploiting the assumed independence between $\mathbf{c}_{n-1}$ and $\mathbf{x}$ and $\mathrm{e}_{*,n}$, as well as the orthogonality between $\mathrm{e}_{*,n}$ and $\mathbf{x}$, we get

$$\begin{aligned} J_n := \mathbb{E}[\mathrm{e}_n^2] &= J_{\min} + \mathbb{E}\left[\mathbf{c}_{n-1}^T \mathbf{x}\mathbf{x}^T \mathbf{c}_{n-1}\right] \\ &= J_{\min} + \mathbb{E}\left[\text{trace}\{\mathbf{c}_{n-1}^T \mathbf{x}\mathbf{x}^T \mathbf{c}_{n-1}\}\right] \\ &= J_{\min} + \text{trace}\{\Sigma_x \Sigma_{c,n-1}\}, \end{aligned} \tag{5.47}$$

where the property $\text{trace}\{AB\} = \text{trace}\{BA\}$ has been used. Thus, we can finally write

$$\boxed{J_{\text{exc},n} = \text{trace}\{\Sigma_x \Sigma_{c,n-1}\}: \quad \text{excess MSE at time instant } n.} \tag{5.48}$$

[7] The time index n is explicitly used for e, e_*, y, since the formula to be derived is also valid for time-varying environments, and it is going to be used later on for the time-varying statistics case.

Let us now elaborate on it a bit more. Taking into account that $QQ^T = I$, we get

$$\begin{aligned} J_{\text{exc},n} &= \text{trace}\{QQ^T \Sigma_x QQ^T \Sigma_{c,n-1} QQ^T\} \\ &= \text{trace}\{Q\Lambda \Sigma_{v,n-1} Q^T\} = \text{trace}\{\Lambda \Sigma_{v,n-1}\} \\ &= \sum_{i=1}^{l} \lambda_i \left[\Sigma_{v,n-1}\right]_{ii} = \boldsymbol{\lambda}^T \boldsymbol{s}_{n-1}, \end{aligned} \tag{5.49}$$

where $\boldsymbol{s}_n$ is the vector of the diagonal elements of $\Sigma_{v,n}$ and obeys the difference equation in (5.43). Assuming that μ has been chosen so that convergence is guaranteed, then for large values of n the *steady state* has been reached. In a more formal way, an online algorithm has reached the steady state if

$$\mathbb{E}[\boldsymbol{\theta}_n] = \mathbb{E}[\boldsymbol{\theta}_{n-1}] = \text{Constant}, \tag{5.50}$$

$$\Sigma_{\theta,n} = \Sigma_{\theta,n-1} = \text{Constant}. \tag{5.51}$$

Thus, in steady state, we assume in Eq. (5.43) that $\boldsymbol{s}_n = \boldsymbol{s}_{n-1}$; if this is exploited in (5.49) this leads to (Problem 5.10)

$$J_{\text{exc},\infty} := \lim_{n\to\infty} J_{\text{exc},n} \simeq \frac{\mu\sigma_\eta^2 \text{trace}\{\Sigma_x\}}{2 - \mu\text{trace}\{\Sigma_x\}}, \tag{5.52}$$

and for the misadjustment (since under our assumptions $J_{\min} = \sigma_\eta^2$),

$$\mathcal{M} \simeq \frac{\mu\text{trace}\{\Sigma_x\}}{2 - \mu\text{trace}\{\Sigma_x\}},$$

which, for small values of μ, leads to

$$\boxed{J_{\text{exc},\infty} \simeq \frac{1}{2}\mu\sigma_\eta^2 \text{trace}\{\Sigma_x\}: \quad \text{excess MSE}} \tag{5.53}$$

and

$$\boxed{\mathcal{M} \simeq \frac{1}{2}\mu\text{trace}\{\Sigma_x\}: \quad \text{misadjustment.}} \tag{5.54}$$

That is, the smaller the value of μ, the smaller the excess MSE.

IV. *Time constant.* Note that the transient behavior of the LMS is described by the difference equation in (5.43) and its convergence rate (the speed with which it forgets the initial conditions until it settles to its steady-state operation) depends on the eigenvalues of A in (5.44). To simplify the formulas, assume μ to be small enough so that A is approximated as $(I - \mu\Lambda)^2$. Following similar arguments as those used for the gradient descent in Section 5.3, we can write

$$\tau_j^{LMS} \simeq \frac{1}{2\mu\lambda_j}.$$

That is, the time constant for each one of the modes is inversely proportional to μ. Hence, the slower the rate of convergence (small values of μ), the lower the misadjustment and vice versa. Viewing it differently, the more time the algorithm spends on learning, prior to reaching the steady state, the smaller is its deviation from the optimal.

5.5.2 CUMULATIVE LOSS BOUNDS

In the previously reported analysis method for the LMS performance, there is an underlying assumption that the training samples are generated by a linear model. The focus of the analysis was to investigate how well our algorithm estimates the unknown model once steady state has been reached. This path of analysis is very popular and well suited for a number of tasks, such as system identification.

An alternative route for studying the performance of an algorithm, shedding light from a different angle, is via the so-called *cumulative loss*. Recall that the main goal in machine learning is *prediction*; hence, measuring the prediction accuracy of an algorithm, given a set of observations, becomes the main goal. However, this performance index should be measured against the generalization ability of the algorithm, as pointed out in Chapter 3. In practice, this can be done in different ways, such as via the leave-one-out method (Section 3.13).

For the case of the squared error loss function, the cumulative loss over N observation samples is defined as

$$\boxed{\mathcal{L}_{\text{cum}} = \sum_{n=0}^{N-1} (y_n - \hat{y}_n)^2 = \sum_{n=0}^{N-1} \left(y_n - \boldsymbol{\theta}_{n-1}^T \boldsymbol{x}_n\right)^2 : \quad \text{cumulative loss.}} \tag{5.55}$$

Note that $\boldsymbol{\theta}_{n-1}$ has been estimated based on observations up to and including the time instant $n-1$. So, the training pair $(y_n, \boldsymbol{x}_n)$ can be considered as a test sample, not involved in the training, for measuring the error. It must be pointed out, however, that the cumulative loss is *not* a direct measure of the generalization performance associated with the finally obtained parameter vector. Such a measure should involve $\boldsymbol{\theta}_{N-1}$, tested against a number of test samples.

The goal of the family of methods which build around the cumulative loss is to derive corresponding upper bounds. Our aim here is to outline the essence behind such approaches, without resorting to proofs; these comprise a series of bounds and can become a bit technical. The interested reader can consult the related references. We will return to the cumulative loss and related bounds in the context of the so-called *regret analysis* in Chapter 8.

In the context of the LMS, the following theorem has been proved in [21].

Theorem 5.1. *Let $C = \max_n ||\boldsymbol{x}_n||$, $\mu = \beta/C^2$, $0 < \beta < 2$. Then the set of predictions, $\hat{y}_0, \ldots, \hat{y}_{N-1}$, generated by Algorithm 5.1 satisfies the following bound:*

$$\sum_{n=0}^{N-1} (y_n - \hat{y}_n)^2 \leq \inf_{\boldsymbol{\theta}} \left\{ \frac{C^2 ||\boldsymbol{\theta}||^2}{2\beta(1-\beta)c} + \frac{\mathcal{L}(\boldsymbol{\theta}, S)}{(2-\beta)^2 c(1-c)} \right\}, \tag{5.56}$$

where $0 < c < 1$ and

$$\mathcal{L}(\boldsymbol{\theta}, S) = \sum_{n=0}^{N-1} \left(y_n - \boldsymbol{\theta}^T \boldsymbol{x}_n\right)^2, \tag{5.57}$$

where $S = \{(y_n, \boldsymbol{x}_n), \quad n = 0, 1, \ldots, N-1\}$.

One can then tune β optimally to minimize the upper bound. Note that this is a *worst-case* scenario. The tuning is achieved by restricting the set of linear functions, so that $||\boldsymbol{\theta}|| \leq \Theta$. Let

$$L_{\Theta}(S) = \min_{||\boldsymbol{\theta}|| \leq \Theta} \mathcal{L}(\boldsymbol{\theta}, S), \tag{5.58}$$

and also assume that there is an upper bound L, such that

$$|L_{\Theta}(S)| \leq L. \tag{5.59}$$

Then it can be shown [21] that

$$\sum_{n=0}^{N-1} (y_n - \hat{y}_n)^2 \leq L_{\Theta}(S) + 2\Theta C\sqrt{L} + (\Theta C)^2. \tag{5.60}$$

Note that the previous analysis has been carried out without invoking any probabilistic arguments. Alternative bounds are derived by mobilizing different assumptions [21]. Bounds of this kind, involving similar assumptions, are frequently encountered for the analysis of various algorithms. We will meet such examples later in this book.

Remarks 5.2.

- The analysis method presented in this section, concerning the transient and steady-state performance of the LMS, can be considered as the most primitive and goes back to the early work of Widrow and Hoff [102]. Another popular path for analyzing stochastic algorithms in an averaging sense is the so-called *averaging method*, which operates under the assumption of small values of the step size μ [56]. Another approach that builds around the assumption of small step sizes, $\mu \simeq 0$, is the so-called ordinary-differential-equation approach (ODE) [57]. The difference update equation is "transformed" to a differential equation, which paves the way for using arguments from the Lyapunov stability theory. An alternative elegant theoretical tool, which can be used as a vehicle for analyzing the transient, the steady state, and the tracking performance of adaptive schemes is the so-called *energy conservation* method, developed in [82] and later on extended in [3,105]. More on the performance analysis of the LMS as well as of other online schemes can be found in more specialized books and papers [4,13,47,62,83,91,92,96,103].
- *H^{∞} optimality of the LMS*. It may come as a surprise that the LMS algorithm, a very simple structure, has survived the time and is one of the most popular and widely used schemes in practical applications. The reason is that, besides its low complexity, it enjoys the luxury of robustness. An alternative optimization flavor of the LMS algorithm has been given via the theory of H^{∞} optimization for estimation.

  Assume that our data obey the regression model of (5.38), where now no assumption is made on the nature of η. Given the sequence of output observation samples, $y_0, y_1, \ldots, y_{N-1}$, the goal is to obtain estimates, $\hat{s}_{n|n-1}$, of s_n, generated as $s_n = \boldsymbol{\theta}^T \boldsymbol{x}_n$, based on the training set up to and including time $n-1$ (causality), such that

$$\frac{\sum_{n=0}^{N-1} |\hat{s}_{n|n-1} - s_n|^2}{\mu^{-1}||\boldsymbol{\theta}||^2 + \sum_{n=0}^{N-1} |\eta_n|^2} < \gamma^2. \tag{5.61}$$

The numerator is the total squared estimation error. The denominator involves two terms. One is the noise/disturbance energy, and the other is the norm of the unknown parameter vector. Assuming that one starts iterations from $\boldsymbol{\theta}_{-1} = \mathbf{0}$, this term measures the energy of the disturbance from the initial guess. It turns out that the LMS scheme is the one that minimizes the following cost:

$$\gamma_{\text{opt}}^2 = \inf_{\{\hat{s}_{n|n-1}\}} \sup_{\{\boldsymbol{\theta}, \eta_n\}} \left(\frac{\sum_{n=0}^{\infty} |\hat{s}_{n|n-1} - s_n|^2}{\mu^{-1}||\boldsymbol{\theta}||^2 + \sum_{n=0}^{\infty} |\eta_n|^2} \right).$$

Moreover, it turns out that the optimum corresponds to $\gamma_{\text{opt}}^2 = 1$ [46,83]. Note that, basically, the LMS optimizes a worst-case scenario. It makes the estimation error minimum under the worst (maximum) disturbance circumstances. This type of optimality explains the robust performance of the LMS under "nonideal" environments, which are often met in practice, where a number of modeling assumptions are not valid. Such deviations from the model can be accommodated in η_n, which then "loses" its i.i.d., white, Gaussian, or any other mathematically attractive property.
Finally, it is interesting to point out the similarity of the bound in (5.61) to that in (5.56). Indeed, in the former we make the following substitutions:

$$\hat{s}_{n|n-1} = \hat{y}_n, \quad s_n = y_n, \quad \eta_n = y_n - \boldsymbol{\theta}^T \boldsymbol{x}_n.$$

Ignoring the values of the constants, the involved quantities are the same. After all, H^∞ is about maximizing the worst-case scenario [55].

5.6 THE AFFINE PROJECTION ALGORITHM

As will soon be verified in the simulations section, a major drawback of the basic LMS scheme is its fairly slow convergence speed. In an attempt to improve upon it, a number of variants have been proposed over the years. The *affine projection algorithm* (APA) belongs to the so-called *data reusing* family, where, at each time instant, past data are reused. Such a rationale helps the algorithm to "learn faster" and improve the convergence speed. However, besides the increased complexity, the faster convergence speed is achieved at the *expense of an increased misadjustment level.*

The APA was proposed originally in [48] and later on in [72]. Let the currently available estimate be $\boldsymbol{\theta}_{n-1}$. According to APA, the updated estimate, $\boldsymbol{\theta}$, must satisfy the following constraints:

$$\boldsymbol{x}_{n-i}^T \boldsymbol{\theta} = y_{n-i}, \quad i = 0, 1, \ldots, q-1.$$

In other words, we *force* the parameter vector $\boldsymbol{\theta}$ to provide at its output the desired response samples, for the q most recent time instants, where q is a user-defined parameter. At the same time, APA requires $\boldsymbol{\theta}$ to be as close as possible, in the Euclidean norm sense, to the current estimate, $\boldsymbol{\theta}_{n-1}$. Thus, APA, at each time instant, solves the following constrained optimization task:

$$\boldsymbol{\theta}_n = \arg\min_{\boldsymbol{\theta}} ||\boldsymbol{\theta} - \boldsymbol{\theta}_{n-1}||^2$$

$$\text{s.t.} \quad \boldsymbol{x}_{n-i}^T \boldsymbol{\theta} = y_{n-i}, \quad i = 0, 1, \ldots, q-1. \tag{5.62}$$

If one defines the $q \times l$ matrix

$$X_n = \begin{bmatrix} \boldsymbol{x}_n^T \\ \vdots \\ \boldsymbol{x}_{n-q+1}^T \end{bmatrix},$$

then the set of constraints can be compactly written as

$$X_n \boldsymbol{\theta} = \boldsymbol{y}_n,$$

where

$$\boldsymbol{y}_n = [y_n \dots y_{n-q+1}]^T.$$

Using Lagrange multipliers in (5.62) (Appendix C) results in (Problem 5.11)

$$\boldsymbol{\theta}_n = \boldsymbol{\theta}_{n-1} + X_n^T \left(X_n X_n^T \right)^{-1} \boldsymbol{e}_n, \tag{5.63}$$

$$\boldsymbol{e}_n = \boldsymbol{y}_n - X_n \boldsymbol{\theta}_{n-1}, \tag{5.64}$$

provided that $X_n X_n^T$ is invertible. The resulting scheme is summarized in Algorithm 5.2.

Algorithm 5.2 (The affine projection algorithm).

- Initialize
  - $\boldsymbol{x}_{-1} = \dots = \boldsymbol{x}_{-q+1} = \boldsymbol{0}$, $y_{-1} \dots y_{-q+1} = 0$
  - $\boldsymbol{\theta}_{-1} = \boldsymbol{0} \in \mathbb{R}^l$ (or any other value).
  - Choose $0 < \mu < 2$ and δ to be small.
- **For** $n = 0, 1, \dots,$ **Do**
  - $\boldsymbol{e}_n = \boldsymbol{y}_n - X_n \boldsymbol{\theta}_{n-1}$
  - $\boldsymbol{\theta}_n = \boldsymbol{\theta}_{n-1} + \mu X_n^T \left(\delta I + X_n X_n^T \right)^{-1} \boldsymbol{e}_n$
- **End For**

When the input is a time series, the corresponding input vector, denoted as $\boldsymbol{u}_n$, is initialized by setting to zero all required samples with negative time index, $u_{-1}, u_{-2}, \dots$. Note that in the algorithm, a parameter, δ, of a small value has also been used to prevent numerical problems in the associated matrix inversion. Also, a step size μ has been introduced, which controls the size of the update and whose presence will be justified soon. The complexity of APA is increased compared with that of the LMS, due to the involved matrix inversion and matrix operations, requiring $O(q^3)$ MADs. Fast versions of the APA, which exploit the special structure of $X_n X_n^T$, for the case where the involved input–output variables are realizations of stochastic processes, have also been developed (see [42,43]).

The convergence analysis of the APA is more involved than that of the LMS. It turns out that provided that $0 < \mu < 2$, stability of the algorithm is guaranteed. The misadjustment is approximately given by [2,34,83]

$$\boxed{\mathcal{M} \simeq \frac{\mu q \sigma_\eta^2}{2-\mu} \mathbb{E}\left[\frac{1}{||\mathbf{x}_n||^2}\right] \text{trace}\{\Sigma_x\}: \quad \text{misadjustment for the APA.}}$$

In words, the misadjustment increases as the parameter q increases; that is, as the number of the reused past data samples increases.

GEOMETRIC INTERPRETATION OF APA

Let us look at the optimization task in (5.62), associated with APA. Each one of the q constraints defines a hyperplane in the l-dimensional space. Hence, since $\boldsymbol{\theta}_n$ is constrained to lie on all these hyperplanes, it will lie on their *intersection*. Provided that $\boldsymbol{x}_{n-i},\ i=0,\dots,q-1$, are linearly independent, these hyperplanes share a nonempty intersection, which is an *affine set* of dimension $l-q$. An affine set is the translation of a linear subspace (i.e., a plane crossing the origin) by a constant vector; that is, it defines a plane in a general position. Thus, $\boldsymbol{\theta}_n$ can lie anywhere in this affine set. From the infinite number of points lying in this set, APA selects the one that lies closest, in the Euclidean distance sense, to $\boldsymbol{\theta}_{n-1}$. In other words, $\boldsymbol{\theta}_n$ is the *projection* of $\boldsymbol{\theta}_{n-1}$ on the affine set defined by the intersection of the q hyperplanes. Recall from geometry that the projection $P_H(\boldsymbol{a})$ of a point $\boldsymbol{a}$ on a linear subspace/affine set H is the point in H whose distance from $\boldsymbol{a}$ is minimum. Fig. 5.13 illustrates the geometry for the case of $q=2$; this special case of APA is also known as the binormalized data reusing LMS [7].

In the ideal noiseless case, the unknown parameter vector would lie in the intersection of all the hyperplanes defined by $(y_n, \boldsymbol{x}_n),\ n=0,1,\dots,q-1$, and this is the information that APA tries to exploit to speed up convergence. However, this is also its drawback. In any practical system, noise is present; thus forcing the updates to lie in the intersection of these hyperplanes is not necessarily good, since their position in space is also determined by the noise. As a matter of fact, the reason that μ is introduced is to account for such cases. An alternative technique, which exploits projections, and at the same time replaces hyperplanes by hyperslabs (whose width depends on the noise variance) to account for the noise, will be treated in Chapter 8. In addition, there, no matrix inversion will be required.

ORTHOGONAL PROJECTIONS

Projections and projection matrices/operators play a crucial part in machine learning, signal processing, and optimization in general; after all, a projection corresponds to a minimization task when the loss is interpreted as a "distance." Let A be an $l \times k,\ k<l$, matrix with column vectors, $\boldsymbol{a}_i,\ i=1,\dots,k$, and $\boldsymbol{x}$ an l-dimensional vector. The orthogonal projection of $\boldsymbol{x}$ on the subspace spanned by the columns of A (assumed to be linearly independent) is given by (Appendix A)

$$\boxed{P_{\{\boldsymbol{a}_i\}}(\boldsymbol{x}) = A(A^T A)^{-1} A^T \boldsymbol{x},} \tag{5.65}$$

where in complex spaces the transpose operation is replaced by the Hermitian one. One can easily check that $P^{\perp}_{\{\boldsymbol{a}_i\}}(\boldsymbol{x}) := (I - P_{\{\boldsymbol{a}_i\}})\boldsymbol{x}$ is orthogonal to $P_{\{\boldsymbol{a}_i\}}(\boldsymbol{x})$ and

$$\boxed{\boldsymbol{x} = P_{\{\boldsymbol{a}_i\}}(\boldsymbol{x}) + P^{\perp}_{\{\boldsymbol{a}_i\}}(\boldsymbol{x}).}$$

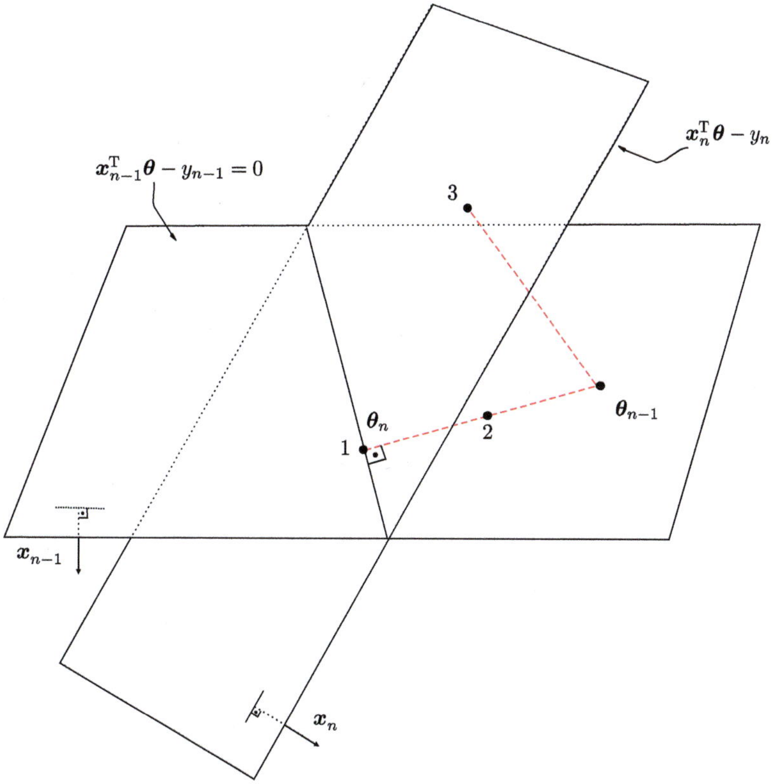

FIGURE 5.13

The geometry associated with the APA, for $q = 2$ and $l = 3$. The intersection of the two hyperplanes is a straight line (affine set of dimension $3 - 2 = 1$); θ_n is the projection of θ_{n-1} on this line (point 1) for $\mu = 1$ and $\delta = 0$. Point 2 corresponds to the case $\mu < 1$. Point 3 is the projection of θ_n on the hyperplane defined by (y_n, x_n). This is the case for $q = 1$. The latter case corresponds to the normalized LMS (NLMS) of Section 5.6.1.

When A has orthonormal columns, we obtain the (familiar from geometry) expansion

$$P_{\{a_i\}}(x) = AA^{\mathrm{T}}x = \sum_{i=1}^{k}(a_i^{\mathrm{T}}x)a_i.$$

Thus, the factor $(A^{\mathrm{T}}A)^{-1}$, for the general case, accounts for the lack of orthonormality of the columns of A. The matrix $A(A^TA)^{-1}A^T$ is known as the respective *projection matrix* and $I - A(A^TA)^{-1}A^T$ as the projection matrix on the respective *orthogonal complement* space.

The simplest case occurs when $k = 1$; then the projection of x onto a_1 is equal to

$$P_{\{a_1\}}(x) = \frac{a_1 a_1^T}{||a_1||^2}x,$$

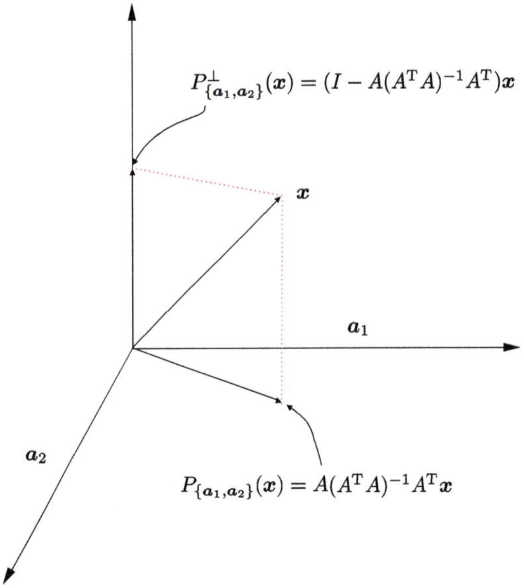

FIGURE 5.14

Geometry indicating the orthogonal projection operation on the subspace spanned by $\boldsymbol{a}_1, \boldsymbol{a}_2$, using the projection matrix. Note that $A = [\boldsymbol{a}_1, \boldsymbol{a}_2]$.

and the corresponding projection matrices are given by

$$P_{\{\boldsymbol{a}_1\}} = \frac{\boldsymbol{a}_1\boldsymbol{a}_1^T}{||\boldsymbol{a}_1||^2}, \quad P^{\perp}_{\{\boldsymbol{a}_1\}} = I - \frac{\boldsymbol{a}_1\boldsymbol{a}_1^T}{||\boldsymbol{a}_1||^2}.$$

Fig. 5.14 illustrates the geometry.

Let us apply the previously reported linear algebra results in the case of the APA of (5.63) and (5.64), and rewrite them as

$$\begin{aligned}\boldsymbol{\theta}_n &= \left(I - X_n^T(X_nX_n^T)^{-1}X_n\right)\boldsymbol{\theta}_{n-1} \\ &\quad + X_n^T(X_nX_n^T)^{-1}\boldsymbol{y}_n.\end{aligned}$$

The first term on the right-hand side is the projection $P^{\perp}_{\{\boldsymbol{x}_n,\ldots,\boldsymbol{x}_{n-q+1}\}}(\boldsymbol{\theta}_{n-1})$. This is the most natural. By the definition of the respective affine set, as the intersection of the hyperplanes

$$\boldsymbol{x}_{n-i}^T\boldsymbol{\theta} - y_{n-i} = 0, \quad i = 0, \ldots, q-1,$$

each vector $\boldsymbol{x}_{n-i}$ is *orthogonal* to the respective hyperplane (Fig. 5.13) (Problem 5.12). Hence, projecting $\boldsymbol{\theta}_{n-1}$ on the intersection of all these hyperplanes is equivalent to projecting on an affine set, which is *orthogonal* to all $\boldsymbol{x}_n, \ldots, \boldsymbol{x}_{n-q+1}$. Note that the matrix $X_n^T(X_nX_n^T)^{-1}X_n$ is the projection matrix

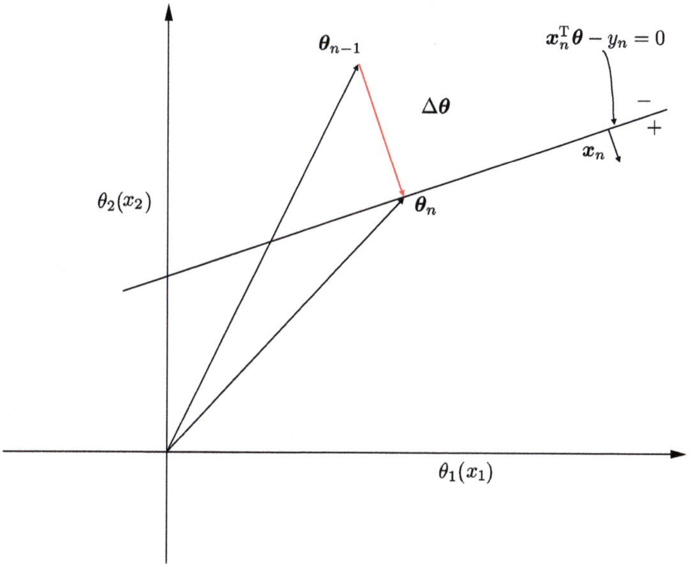

FIGURE 5.15

$\Delta\boldsymbol{\theta}$ is equal to the (signed) distance of $\boldsymbol{\theta}_{n-1}$ from the plane times the unit vector $\frac{\boldsymbol{x}_n}{||\boldsymbol{x}_n||}$; $\boldsymbol{x}_n$ should be at the origin, but it is drawn so that it is clearly shown that it is perpendicular to the line segment.

that projects on the subspace spanned by $\boldsymbol{x}_n, \ldots, \boldsymbol{x}_{n-q+1}$. The second term accounts for the fact that the affine set on which we project does not include the origin, but it is translated to another point in the space, whose direction is determined by $\boldsymbol{y}_n$. Fig. 5.15 illustrates the case for $l = 2$ and $q = 1$. Because $\boldsymbol{\theta}_{n-1}$ does not lie on the line (plane) $\boldsymbol{x}_n^T\boldsymbol{\theta} - y_n = 0$, whose direction is defined by $\boldsymbol{x}_n$, we know from geometry (and it is easily checked) that its distance from this line is $s = \frac{|\boldsymbol{x}_n^T\boldsymbol{\theta}_{n-1} - y_n|}{||\boldsymbol{x}_n||}$. Also, $\boldsymbol{\theta}_{n-1}$ lies on the negative side of the straight line, so that $\boldsymbol{x}_n^T\boldsymbol{\theta}_{n-1} - y_n < 0$. Hence, taking into account the directions of the involved vectors, it turns out that

$$\Delta\boldsymbol{\theta} = \frac{y_n - \boldsymbol{x}_n^T\boldsymbol{\theta}_{n-1}}{||\boldsymbol{x}_n||}\frac{\boldsymbol{x}_n}{||\boldsymbol{x}_n||}.$$

Thus, for this specific case, the correction

$$\boldsymbol{\theta}_n = \boldsymbol{\theta}_{n-1} + \Delta\boldsymbol{\theta}$$

coincides with the recursion of the APA.

5.6.1 THE NORMALIZED LMS

The NLMS is a special case of the APA corresponding to $q = 1$ (see Fig. 5.13). We treat it separately due to its popularity, and it is summarized in Algorithm 5.3.

Algorithm 5.3 (The normalized LMS).

- Initialization
  - $\boldsymbol{\theta}_{-1} = \mathbf{0} \in \mathbb{R}^l$, or any other value.
  - Choose $0 < \mu < 2$, and δ a small value.
- **For** $n = 0, 1, 2, \ldots,$ **Do**
  - $e_n = y_n - \boldsymbol{\theta}_{n-1}^T \boldsymbol{x}_n$
  - $\boldsymbol{\theta}_n = \boldsymbol{\theta}_{n-1} + \frac{\mu}{\delta + \boldsymbol{x}_n^T \boldsymbol{x}_n} \boldsymbol{x}_n e_n$
- **End For**

The complexity of the NLMS is $3l$ MADs. Stability of the NLMS is guaranteed if

$$0 < \mu < 2.$$

One can look at the NLMS as an LMS whose step size is left to vary with the iterations, as represented by

$$\mu_n = \frac{\mu}{\delta + \boldsymbol{x}_n^T \boldsymbol{x}_n},$$

which turns out to have a beneficial effect on the convergence speed, compared with the LMS. More on the performance analysis of the NLMS can be found in, e.g., [15,83,90].

Remarks 5.3.

- To deal with sparse models, the so-called *proportionate* NLMS and related versions of the other algorithms have been derived [11,35]. The idea is to use a separate step size for each one of the parameters. This gives the freedom to the coefficients which correspond to small (or zero) values of the model to adapt at a different rate than the rest, and this has a significant effect on the convergence performance of the algorithm. Such schemes can be considered as the ancestors of the more theoretically elegant sparsity promoting online algorithms, to be treated in Chapter 10.
- Another trend that has received attention more recently is to appropriately (convexly) combine the outputs of two (or more) learning structures. This has the effect of decreasing the sensitivity of the learning algorithms to choices of parameters such as the step size, or to the dimensionality of the problem (size of the filter). The two (or more) algorithms run independently and the mixing parameters of the outputs are learned during the training. In general, this approach turns out to be more robust in the choice of the involved user-defined parameters [8,81].
- In some papers, the user-defined parameter, δ, in the NLMS and the APA is suggested to be given a very small positive value. Often, the explanation for this option is that the use of δ is to avoid division by zero. However, in practice, there are cases, such as the echo cancelation task, where δ needs to be set to quite large values (even larger than 1) to attain good performance. It is fairly recent that the importance of δ was emphasized and a formula for its proper tuning was proposed both for the case of NLMS and APA and their proportionate counterparts [12,73]. There, it is indicated that without the proper setup of this parameter, the performance of these algorithms may be significantly affected, and they may not even converge.

5.7 THE COMPLEX-VALUED CASE

In Section 4.4, when

$$y_n \in \mathbb{C} \quad \text{and} \quad \mathbf{x}_n \in \mathbb{C}^l,$$

the *widely linear* formulation of the estimation task was introduced to deal with complex-valued data that do not obey the circularity conditions. The output of a widely linear estimator is given by

$$\hat{y}_n = \boldsymbol{\varphi}^H \tilde{\mathbf{x}}_n,$$

where

$$\boldsymbol{\varphi} := \begin{bmatrix} \boldsymbol{\theta} \\ \boldsymbol{v} \end{bmatrix} \quad \text{and} \quad \tilde{\mathbf{x}}_n = \begin{bmatrix} \mathbf{x}_n \\ \mathbf{x}_n^* \end{bmatrix},$$

with $\boldsymbol{\theta}, \boldsymbol{v} \in \mathbb{C}^l$. The MSE cost function is

$$J(\boldsymbol{\varphi}) = \mathbb{E}\left[\left|y_n - \boldsymbol{\varphi}^H \tilde{\mathbf{x}}_n\right|^2\right],$$

and following standard arguments analogous to Section 5.3.1, the minimum with respect to $\boldsymbol{\varphi}$ is given by the root of

$$\Sigma_{\tilde{x}} \boldsymbol{\varphi} - \begin{bmatrix} \boldsymbol{p} \\ \boldsymbol{q}^* \end{bmatrix} = \mathbf{0} \text{ or } \mathbb{E}\left[\tilde{\mathbf{x}}_n \tilde{\mathbf{x}}_n^H\right] \boldsymbol{\varphi} = \mathbb{E}\left[\tilde{\mathbf{x}}_n y_n^*\right].$$

THE WIDELY LINEAR LMS

Employing the Robbins–Monro scheme and fixing the value of μ_n to be a constant, we obtain

$$\boldsymbol{\varphi}_n = \boldsymbol{\varphi}_{n-1} + \mu \tilde{\boldsymbol{x}}_n e_n^*,$$

$$e_n = y_n - \boldsymbol{\varphi}_{n-1}^H \tilde{\boldsymbol{x}}_n.$$

Breaking $\boldsymbol{\varphi}_n$ and $\tilde{\boldsymbol{x}}_n$ into their components, the widely linear LMS results.

Algorithm 5.4 (The widely linear LMS).

- Initialize
  - $\boldsymbol{\theta}_{-1} = \mathbf{0}, \ \boldsymbol{v}_{-1} = \mathbf{0}$
  - Choose μ.
- **For** $n = 0, 1, \ldots,$ **Do**
  - $e_n = y_n - \boldsymbol{\theta}_{n-1}^H \boldsymbol{x}_n - \boldsymbol{v}_{n-1}^H \boldsymbol{x}_n^*$
  - $\boldsymbol{\theta}_n = \boldsymbol{\theta}_{n-1} + \mu \boldsymbol{x}_n e_n^*$
  - $\boldsymbol{v}_n = \boldsymbol{v}_{n-1} + \mu \boldsymbol{x}_n^* e_n^*$
- **End For**

Note that whatever has been said for the stability conditions concerning the LMS is applied here as well, if Σ_x is replaced by $\Sigma_{\tilde{x}}$. For circularly symmetric variables, we set $\boldsymbol{v}_n = \mathbf{0}$ and the *complex linear LMS* results.

THE WIDELY LINEAR APA

Let $\boldsymbol{\varphi}_n$ and $\tilde{\boldsymbol{x}}_n$ be defined as before. The widely linear APA results; it is given in Algorithm 5.5 (Problem 5.13).

Algorithm 5.5 (The widely linear APA).

- Initialize
  - $\boldsymbol{\varphi}_{-1} = \boldsymbol{0}$
  - Choose μ.
- **For** $n = 0, 1, 2, \ldots,$ **Do**
  - $\boldsymbol{e}_n^* = \boldsymbol{y}_n^* - \tilde{X}_n \boldsymbol{\varphi}_{n-1}$
  - $\boldsymbol{\varphi}_n = \boldsymbol{\varphi}_{n-1} + \mu \tilde{X}_n^H (\delta I + \tilde{X}_n \tilde{X}_n^H)^{-1} \boldsymbol{e}_n^*$
- **End For**

Note that

$$\tilde{X}_n = \begin{bmatrix} \tilde{\boldsymbol{x}}_n^H \\ \vdots \\ \tilde{\boldsymbol{x}}_{n-q+1}^H \end{bmatrix} = \begin{bmatrix} \boldsymbol{x}_n^H, \ \boldsymbol{x}_n^T \\ \vdots \\ \boldsymbol{x}_{n-q+1}^H, \ \boldsymbol{x}_{n-q+1}^T \end{bmatrix},$$

and $\boldsymbol{\varphi}_n \in \mathbb{C}^{2l}$. For circular variables/processes, the complex linear APA results by setting

$$\tilde{X}_n = X_n = \begin{bmatrix} \boldsymbol{x}_n^H \\ \vdots \\ \boldsymbol{x}_{n-q+1}^H \end{bmatrix}$$

and

$$\boldsymbol{\varphi}_n = \boldsymbol{\theta}_n \in \mathbb{C}^l.$$

5.8 RELATIVES OF THE LMS

In addition to the three basic stochastic gradient descent schemes that were previously reviewed, a number of variants have been proposed over the years, in an effort to either improve performance or reduce complexity. Some notable examples are described below.

THE SIGN-ERROR LMS

The update recursion for this algorithm becomes (e.g., [17,30,63])

$$\boldsymbol{\theta}_n = \boldsymbol{\theta}_{n-1} + \mu \operatorname{csgn}[e_n^*] \boldsymbol{x}_n,$$

where the complex sign of a complex number, $z = x + jy$, is defined as

$$\text{csgn}(z) = \text{sgn}(x) + j\,\text{sgn}(y).$$

If in addition μ is chosen to be a power of two, then the recursion becomes *multiplication-free*, and l multiplications are only needed for the computation of the error. It turns out that the algorithm minimizes, in the stochastic approximation sense, the following cost function:

$$J(\boldsymbol{\theta}) = \mathbb{E}\left[|\mathrm{y} - \boldsymbol{\theta}^H \mathbf{x}|\right],$$

and stability is guaranteed for sufficiently small values of μ [83].

THE LEAST-MEAN-FOURTH (LMF) ALGORITHM

The scheme minimizes the following cost function:

$$J(\boldsymbol{\theta}) = \mathbb{E}\left[|\mathrm{y} - \boldsymbol{\theta}^H \mathbf{x}|^4\right],$$

and the corresponding update recursion is given by

$$\boldsymbol{\theta}_n = \boldsymbol{\theta}_{n-1} + \mu |e_n|^2 \boldsymbol{x}_n e_n^*.$$

It has been shown [101] that minimization of the fourth power of the error may lead to an adaptive scheme with better compromise between convergence rate and excess MSE than the LMS if the noise source is sub-Gaussian. In a sub-Gaussian distribution, the tails of the PDF graph are decaying at a faster rate compared to the Gaussian one. The LMF could be seen as a version of the LMS with time-varying step size, equal to $\mu|e_n|^2$. Hence, when the error is large, the step size increases, which helps the LMF to converge faster. On the other hand, when the error is small, the equivalent step size is reduced, leading to lower values of the excess MSE. This idea turns out to work well when the noise is sub-Gaussian. However, the LMF tends to become unstable in the presence of outliers. This is also understood, because for very large error values the equivalent step size becomes very large, leading to instability. This is, for example, the case when the noise follows a distribution with high tails; that is, when the corresponding PDF curve decays at slower rates compared to the Gaussian, such as in the case of super-Gaussian distributions. Results concerning the analysis of the LMF can be found in [69,70,83]. Robustness and stability of the algorithm are guaranteed for sufficiently small values of μ [83].

TRANSFORM-DOMAIN LMS

We have already commented that the convergence speed of the LMS heavily depends on the condition number $\left(\frac{\lambda_{\max}}{\lambda_{\min}}\right)$ of the covariance matrix; this will soon be demonstrated in the examples below.

Transform-domain techniques exploit the decorrelation properties of certain transforms, such as the DFT and DCT, in order to decorrelate the input variables. When the input comprises a stochastic process, we say that such transforms "*prewhiten*" the input process. Moreover, in the case where the involved variables are part of a random process, one can exploit the time-shifting property and employ *block processing techniques*; by processing a block of data samples per time instant, one can exploit efficient implementations of certain transforms, such as the fast Fourier transform (FFT), to reduce the overall complexity. Such schemes are appropriate for applications where long filters are involved. For example, in some applications, such as echo cancelation, filter orders of a few hundred taps are commonly encountered [10,39,53,66,68].

Let T be a unitary transform in the complex domain, represented by $TT^H = T^H T = I$. Define

$$\hat{\boldsymbol{x}}_n = T^H \boldsymbol{x}_n, \tag{5.66}$$

and apply the transform matrix T^H on both sides of the respective recursion in Algorithm 5.4, for the widely linear LMS, to obtain

$$\hat{\boldsymbol{\theta}}_n = \hat{\boldsymbol{\theta}}_{n-1} + \mu \hat{\boldsymbol{x}}_n e_n^*, \tag{5.67}$$

where

$$\hat{\boldsymbol{\theta}}_n = T^H \boldsymbol{\theta}_n. \tag{5.68}$$

Note that

$$\begin{aligned} e_n &= y_n - \boldsymbol{\theta}_{n-1}^H \boldsymbol{x}_n = y_n - \boldsymbol{\theta}_{n-1}^H T T^H \boldsymbol{x}_n \\ &= y_n - \hat{\boldsymbol{\theta}}_{n-1}^H \hat{\boldsymbol{x}}_n. \end{aligned}$$

Hence, the error term is not affected. Note that until now, we have not achieved much. Indeed,

$$\Sigma_{\hat{x}} = T^H \Sigma_x T$$

is a similarity transformation that does not affect the condition number of the matrix Σ_x; it is known from linear algebra that both matrices share the same eigenvalues (Problem 5.14). Let us now choose $T = Q$, where Q is the unitary matrix comprising the orthonormal eigenvectors of Σ_x. For this case, $\Sigma_{\hat{x}} = \Lambda$, which is the diagonal matrix with entries the eigenvalues of Σ_x. In the sequel, we modify the transform-domain LMS in (5.67), so that we use a different step size per component, according to the following scenario:

$$\hat{\boldsymbol{\theta}}_n = \hat{\boldsymbol{\theta}}_{n-1} + \mu \Lambda^{-1} \hat{\boldsymbol{x}}_n e_n^*, \tag{5.69}$$

or

$$\bar{\boldsymbol{\theta}}_n = \bar{\boldsymbol{\theta}}_{n-1} + \mu \bar{\boldsymbol{x}}_n e_n^*, \tag{5.70}$$

where

$$\bar{\boldsymbol{\theta}}_n := \Lambda^{1/2} \hat{\boldsymbol{\theta}}_n$$

and

$$\bar{\boldsymbol{x}}_n := \Lambda^{-1/2} \hat{\boldsymbol{x}}_n.$$

Note that the error is not affected, because $\bar{\boldsymbol{\theta}}_n^H \bar{\boldsymbol{x}}_n = \hat{\boldsymbol{\theta}}_n^H \hat{\boldsymbol{x}}_n$. We have now achieved our original goal, since

$$\Sigma_{\bar{x}} = \Lambda^{-1/2} \Sigma_{\hat{x}} \Lambda^{-1/2} = \Lambda^{-1/2} \Lambda \Lambda^{-1/2} = I.$$

That is, the condition number of $\Sigma_{\bar{x}}$ is equal to 1. In practice, this technique is difficult to apply, due to the complexity associated with the eigendecomposition task; more importantly, Σ_x must be known, which in adaptive implementations is not the case. In practice, we resort to unitary transforms, T, that approximately whiten the input, such as the DFT and DCT. Then (5.69) is replaced by

$$\hat{\boldsymbol{\theta}}_n = \hat{\boldsymbol{\theta}}_{n-1} + \mu D^{-1} \hat{\boldsymbol{x}}_n e_n^*,$$

where D is the diagonal matrix with entries the variances of the respective components of $\hat{\boldsymbol{x}}_n$, or

$$[D]_{ii} = \mathbb{E}\left[(\hat{\mathrm{x}}_n(i))^2 \right] = \sigma_i^2, \quad i = 1, 2, \ldots, l,$$

where $\hat{\mathrm{x}}_n(i)$ is the ith entry of $\hat{\mathbf{x}}_n$, which has been assumed to be a zero mean vector; this is justified from the fact that if $\Sigma_{\hat{x}}$ is truly diagonal and equal to Λ, the eigenvalues correspond to the variances of the respective elements. The entries σ_i^2 are estimated in a time-adaptive fashion, and the scheme given in Algorithm 5.6 results.

Algorithm 5.6 (The transform-domain LMS).

- Initialization
  - $\hat{\boldsymbol{\theta}}_{-1} = \mathbf{0}$; or any other value.
  - $\sigma_{-1}^2(i) = \delta, \ i = 1, 2, \ldots, l$; δ a small value.
  - Choose μ, and $0 \ll \beta < 1$.

- **For** $n = 0, 1, 2, \ldots,$ **Do**
  - $\hat{\boldsymbol{x}}_n = T^H \boldsymbol{x}_n$
  - $e_n = y_n - \hat{\boldsymbol{\theta}}_{n-1}^H \hat{\boldsymbol{x}}_n$
  - $\hat{\boldsymbol{\theta}}_n = \hat{\boldsymbol{\theta}}_{n-1} + \mu D^{-1} \hat{\boldsymbol{x}}_n e_n^*$
  - **For** $i = 1, 2, \ldots, l$, **Do**
    - $\ast$ $\sigma_i^2(n) = \beta \sigma_i^2(n-1) + (1-\beta) |\hat{\mathrm{x}}_n(i)|^2$
  - **End For**
  - $D = \mathrm{diag}\{\sigma_i^2(n)\}$
- **End For**

Subband adaptive filters is a related family of algorithms where whitening is achieved via block data processing and the use of a multirate filter band. Such schemes were first proposed in [41,53,54] and have successfully been used in applications such as echo cancelation [45,53].

Another approach that has been adopted to improve the convergence rate, for the case of system identification applications, is to select the input excitation signal to be of a specific type. For example, in [5] it is pointed out that the excitation signal that optimizes the convergence speed of the NLMS algorithm is a deterministic perfect periodic sequence (PPSEQ) with period equal to the impulse response of the system. Such sequences have been used in [24,25] to derive versions of the LMS which not only converge fast but are also of very low computational complexity, requiring only one multiplication, one addition, and one subtraction per update.

5.9 SIMULATION EXAMPLES

Example 5.3. The goal of this example is to demonstrate the sensitivity of the convergence rate of the LMS to the eigenvalues spread of the input covariance matrix. To this end, two experiments were conducted, in the context of a regression/system identification task. Data were generated according to our familiar model

$$y_n = \boldsymbol{\theta}_o^T \boldsymbol{x}_n + \eta_n.$$

The (unknown) parameters $\boldsymbol{\theta}_o \in \mathbb{R}^{10}$ were randomly chosen from $\mathcal{N}(0, 1)$ and then frozen. In the first experiment, the input vectors were formed by a white noise sequence with samples i.i.d. drawn from $\mathcal{N}(0, 1)$. Thus, the input covariance matrix was diagonal with all the elements being equal to the corresponding noise variance (Section 2.4.3). The noise samples η_n were also i.i.d. drawn from a Gaussian with zero mean and variance $\sigma^2 = 0.01$. In the second experiment, the input vectors were formed by an AR(1) process with coefficient equal to $a_1 = 0.85$ and the corresponding white noise

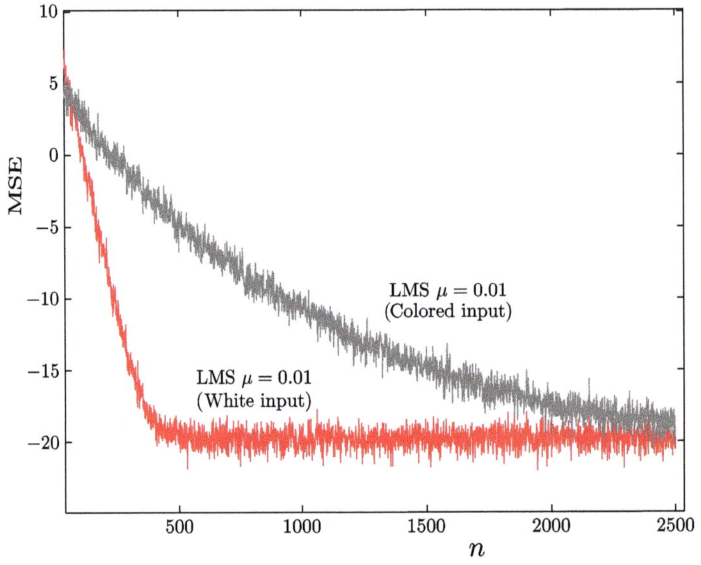

FIGURE 5.16

Observe that for the same step size, the convergence of the LMS is faster when the input is white. The two curves are the result of averaging 100 independent experimental realizations.

excitation was of variance equal to 1 (Section 2.4.4). Thus, the input covariance matrix is no more diagonal and the eigenvalues are not equal. The LMS was run on both cases with the same step size $\mu = 0.01$. Fig. 5.16 summarizes the results. The vertical axis (denoted as MSE) shows the squared error, e_n^2, in dBs ($10\log_{10} e_n^2$) and the horizontal axis shows the time instants (iterations) n. Note that both curves level out at the same error floor. However, the convergence rate for the case of the white noise input is significantly higher. The curves shown in the figure are the result of averaging 100 independent experimental realizations.

It must be emphasized that when comparing convergence performance of different algorithms, either all algorithms should converge to the same error floor and compare the respective convergence rates, or all algorithms should have the same convergence rate and compare respective error floors.

Example 5.4. In this example, the dependence of the LMS on the choice of the step size is demonstrated. The unknown parameters $\boldsymbol{\theta}_o \in \mathbb{R}^{10}$ and the data were exactly the same as in the white noise case of Example 5.3.

The LMS was run using the generated samples, with two different step sizes, namely, $\mu = 0.01$ and $\mu = 0.075$. The obtained averaged (over 100 realizations) curves are shown in Fig. 5.17. Observe that the larger the step size, for the same set of observation samples, the faster the convergence becomes albeit at the expense of a higher error floor (misadjustment), in accordance to what was discussed in Section 5.5.1.

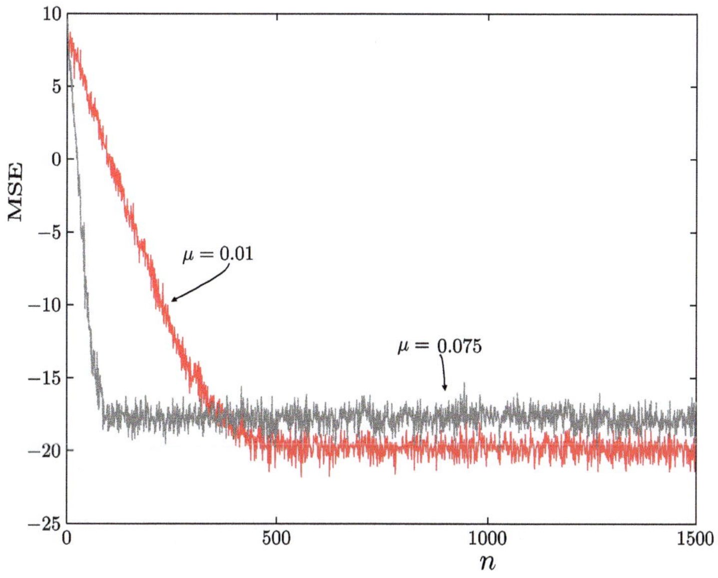

FIGURE 5.17

For the same input, the larger the step size for the LMS, the faster the convergence becomes, albeit at the expense of higher error floor (MSE in dBs).

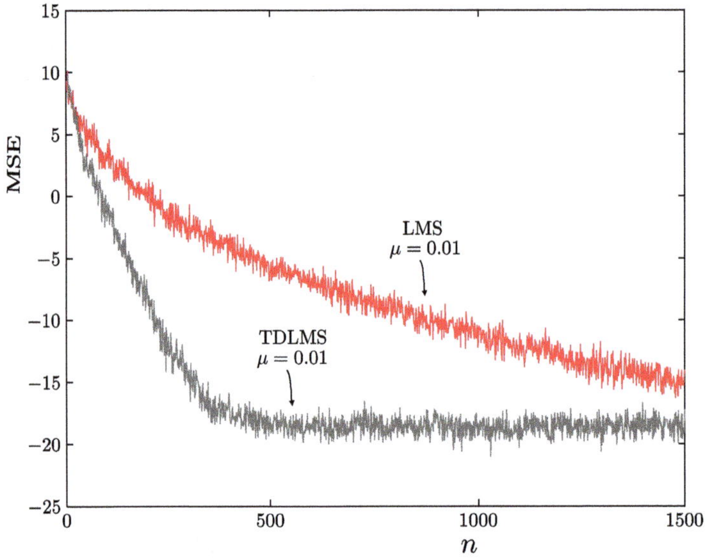

FIGURE 5.18

Observe that for the same step size, the convergence of the transform-domain LMS is significantly faster compared to the LMS, for similar error floors. The higher the eigenvalues spread of the input covariance matrix is, the more the obtained performance improvement becomes (MSE in dBs).

Example 5.5 (LMS versus transform-domain LMS). In this example, the stage of the experimental setup is exactly the same as that considered in Example 5.3 for the AR(1) case. The goal is to compare the LMS and the transform-domain LMS. Fig. 5.18 shows the obtained averaged error curves. The step size was the same as the one used in Example 5.3, $\mu = 0.01$; hence the curve for the LMS is the same as the corresponding one appearing in Fig. 5.16. Observe the significantly faster convergence achieved by the transform-domain LMS, due to its (approximate) whitening effect on the input.

Example 5.6. The experimental setup is similar to that of Example 5.4, with the only exception that the unknown parameter vector was of higher dimension, $\boldsymbol{\theta}_o \in \mathbb{R}^{60}$, so that the differences in the performance of the algorithms is more clear. The goal is to compare the LMS, the NLMS, and the APA. The step size of the LMS was chosen equal to $\mu = 0.025$ and for the NLMS $\mu = 0.35$ and $\delta = 0.001$, so that both algorithms had similar convergence rates. The step size for the APA was chosen equal to $\mu = 0.1$, so that $q = 10$ will settle at the same error floor as that of the NLMS. For the APA, we also chose $\delta = 0.001$. The results are shown in Fig. 5.19.

Observe the lower error floor, for the same convergence rate, obtained by the NLMS compared to the LMS and the improved performance obtained by the APA for $q = 10$. Increasing the number of past data samples (re)used in APA to $q = 30$, we can see the improved convergence rate that is obtained, although at the expense of higher error floor, as predicted by the theoretical results reported in Section 5.6.

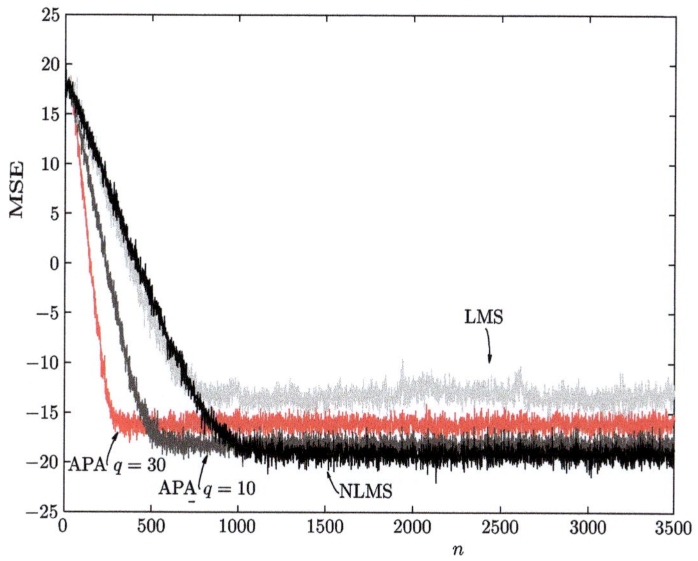

FIGURE 5.19

For the same step size, the NLMS converges at the same rate to a lower error floor compared to the LMS. For the APA, increasing q improves the convergence, at the expense of higher error floors (MSE in dBs).

5.10 ADAPTIVE DECISION FEEDBACK EQUALIZATION

The task of channel equalization was introduced in Fig. 4.12. The input to the equalizer is a stochastic process (random signal), which, according to the notational convention introduced in Section 2.4, will be denoted as u_n. Note that upon receiving the noisy and distorted by the (communications) channel sample, u_n, one has to obtain an estimate of the originally transmitted information sequence, s_n, delayed by L time lags, which accounts for the various delays imposed by the overall transmission system involved. Thus, at time n, the equalizer decides for $\hat{s}_{n-L+1}$. Ideally, if one knew the true values of the initially transmitted information sequence up to and including time instant $n-L$, represented by $s_{n-L}, s_{n-L-1}, s_{n-L-2}\ldots$, it could only be beneficial to use this information, together with the received sequence, u_n, to recover an estimate of $\hat{s}_{n-L+1}$. This idea is explored in the *decision feedback equalizer* (DFE). The equalizer's output, for the complex-valued data case, is now written as

$$\begin{aligned}\hat{d}_n &= \sum_{i=0}^{L-1} w_i^{f*} u_{n-i} + \sum_{i=0}^{l-1} w_i^{b*} s_{n-L-i} \\ &= \boldsymbol{w}^H \boldsymbol{u}_{e,n}, \end{aligned} \tag{5.71}$$

where

$$\boldsymbol{w} := \begin{bmatrix} \boldsymbol{w}^f \\ \boldsymbol{w}^b \end{bmatrix} \in \mathbb{C}^{L+l}, \quad \boldsymbol{u}_{e,n} := \begin{bmatrix} \boldsymbol{u}_n \\ \boldsymbol{s}_n \end{bmatrix} \in \mathbb{C}^{L+l},$$

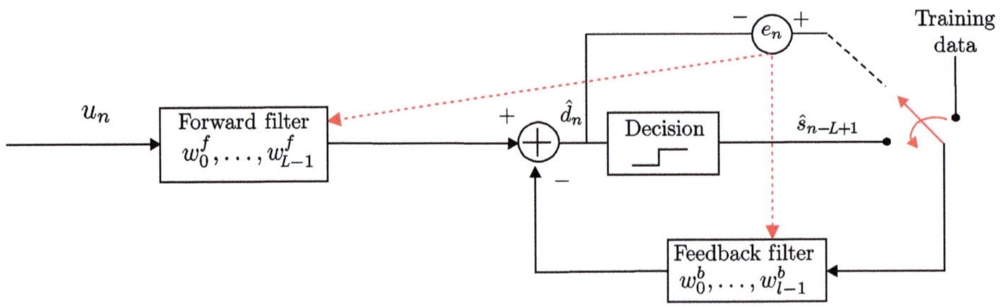

FIGURE 5.20

The forward part of the DFE acts on the received samples, while the backward part acts on the training data/decisions, depending on the mode of operation.

and $\boldsymbol{s}_n := [s_{n-L}, \ldots, s_{n-L-l+1}]^T$. The desired response at time n is

$$d_n = s_{n-L+1}.$$

In practice, after the initial training period, the information samples s_{n-L-i} are replaced by their computed estimates, $\hat{s}_{n-L-i}$, $i = 0, 1, \ldots, l-1$, which are available from decisions taken in previous time instants. It is said that the equalizer operates in the *decision-directed* mode. The basic DFE structure is shown in Fig. 5.20. Note that during the training period, the parameter vector, $\boldsymbol{w}$, is trained so as to minimize the power of the error,

$$e_n = d_n - \hat{d}_n = s_{n-L+1} - \hat{d}_n.$$

Once all the available training samples have been used, training carries on using the estimates $\hat{s}_{n-L+1}$. For example, for a binary information sequence $s_n \in \{1, -1\}$, the decision concerning the estimate at time n is obtained by passing $\hat{d}_n$ through a threshold device and $\hat{s}_{n-L+1}$ is obtained. Note that DFE is one of the early examples of *semisupervised* learning, where training data are not enough, and the estimates are used for training [94]. In this way, assuming that at the end of the training phase $\hat{s}_{n-L+1} = s_{n-L+1}$, and also that time variations are slow, so as to guarantee that $\hat{d}_n \simeq d_n$, we expect, with good enough probability, that $\hat{s}_{n-L+1}$ will remain equal to s_{n-L+1}, so that the equalizer can track the changes. More on DFEs and their error performance can be found in [77].

Any of the adaptive schemes treated so far can be used in a DFE scenario by replacing in the input vector, $\boldsymbol{u}_e$, the term $\boldsymbol{s}_n$ by $\hat{\boldsymbol{s}}_n$, when operating in the decision-directed mode. Note that adaptive algorithms in the context of the equalization task were employed first in [44,78]. A version of the DFE, operating in the frequency domain, has been proposed for the first time in [14].

Thus the LMS recursion for the linear DFE, in its complex-valued formulation, becomes

$$\hat{d}_n = \boldsymbol{w}_{n-1}^H \boldsymbol{u}_{e,n}$$
$$d_n = s_{n-L+1}; \quad \text{in the training mode, or}$$
$$d_n = T\left[\hat{d}_n\right]; \quad \text{in the decision directed mode,}$$

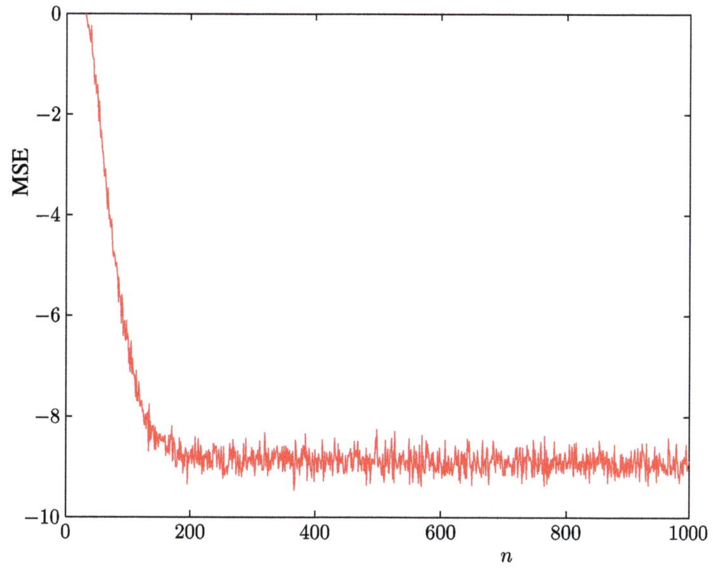

FIGURE 5.21

The MSE in dBs for the DFE of Example 5.7. After the time instant $n = 250$, the LMS is trained with the decisions $\hat{s}_{n-L+1}$.

$$e_n = d_n - \hat{d}_n,$$
$$\boldsymbol{w}_n = \boldsymbol{w}_{n-1} + \mu \boldsymbol{u}_{e,n} e_n^*,$$

where $T[\cdot]$ denotes the thresholding operation.

Example 5.7. Let us consider a communication system where the input information sequence comprises a stream of randomly generated symbols $s_n = \pm 1$, with equal probability. This sequence is sent to a channel with impulse response,

$$\boldsymbol{h} = [0.04, -0.05, 0.07, -0.21, 0.72, 0.36, 0.21, 0.03, 0.07]^T.$$

The output of the channel is contaminated by white Gaussian noise at the SNR = 11 dB level. A DFE is used with the length of the feed-forward section being equal to $L = 21$ and the length of the feedback part equal to $l = 10$. The DFE was trained with 250 symbols; then, it was switched on to the decision-directed mode and it was run for 10,000 iterations. At each iteration, the decision (sgn $(\hat{d}_n)$) was compared with the true transmitted symbol s_{n-L+1}. The error rate (total number of errors over the corresponding number of transmitted symbols) was approximately 1%. Fig. 5.21 shows the averaged MSE curve as a function of the number of iterations. For the LMS, we used $\mu = 0.025$.

5.11 THE LINEARLY CONSTRAINED LMS

The task of linearly constrained MSE estimation was introduced in Section 4.9.2. Here, we turn our attention to its online stochastic gradient counterpart. The discussion will be carried out within the notational convention used for linear filtering involving processes, since a typical application is that of beamforming; this is also the reason that we will involve the complex-valued formulation. However, everything to be said carries on to the more general linear estimation task.

In Section 4.9.2, the goal was to minimize, either the noise variance or the output of the filter (beamformer), subject to a set of constraints, leading to the costs $\boldsymbol{w}^H \Sigma_\eta \boldsymbol{w}$ or $\boldsymbol{w}^H \Sigma_u \boldsymbol{w}$, respectively. In a more general setting, we will require the output to be "close" to a desired response random signal d_n. For the beamforming setting, this corresponds to a desired (training) signal; that is, besides the desired direction, which is provided via the constraints, we are also given a desired signal sequence. Moreover, we will assume that our solution has to satisfy more than one constraint. The task now becomes

$$\begin{aligned} \boldsymbol{w}_* &= \arg\min_{\boldsymbol{w}} \mathbb{E}\left[\left|\mathrm{d}_n - \boldsymbol{w}^H \mathbf{u}_n\right|^2\right] \\ \text{s.t.} \quad & \boldsymbol{w}^H \boldsymbol{c}_i = g_i, \quad i = 1, 2, \ldots, m, \end{aligned} \tag{5.72}$$

for some $g_i \in \mathbb{R}$. The corresponding Lagrangian is

$$\begin{aligned} L(\boldsymbol{w}) &= \sigma_d^2 + \boldsymbol{w}^H \mathbb{E}[\mathbf{u}_n \mathbf{u}_n^H]\boldsymbol{w} - \boldsymbol{w}^H \mathbb{E}[\mathbf{u}_n \mathrm{d}_n^*] - \mathbb{E}[\mathbf{u}_n^H \mathrm{d}_n]\boldsymbol{w} \\ &\quad + \sum_{i=1}^m \lambda_i \left(\boldsymbol{w}^H \boldsymbol{c}_i - g_i\right), \end{aligned}$$

where λ_i, $i = 1, 2, \ldots, m$, are the corresponding Lagrange multipliers. Taking the gradient with respect to $\boldsymbol{w}^*$ and considering $\boldsymbol{w}$ to be a constant, we get

$$\nabla_{w^*} L(\boldsymbol{w}) = \mathbb{E}[\mathbf{u}_n \mathbf{u}_n^H]\boldsymbol{w} - \mathbb{E}[\mathbf{u}_n \mathrm{d}_n^*] + \sum_{i=1}^m \lambda_i \boldsymbol{c}_i.$$

Applying the Robbins–Monro scheme to find the root, we get

$$\boldsymbol{w}_n = \boldsymbol{w}_{n-1} + \mu \boldsymbol{u}_n e_n^* - \mu C \boldsymbol{\lambda}_n, \tag{5.73}$$

where we have used a constant step size μ,

$$\hat{d}_n = \boldsymbol{w}_{n-1}^H \boldsymbol{u}_n,$$

and

$$e_n = d_n - \hat{d}_n,$$

and we have allowed $\boldsymbol{\lambda}$ to be time dependent; C is defined as the matrix having as columns the vectors $\boldsymbol{c}_i$, $i = 1, 2, \ldots, m$.

Plugging (5.73) into the constraints in (5.72), which can be compactly written as $C^H \boldsymbol{w} = \boldsymbol{g}$ (recall $g_i \in \mathbb{R}$), we readily obtain

$$-\mu \boldsymbol{\lambda}_n = (C^H C)^{-1} \boldsymbol{g} - (C^H C)^{-1} C^H \left(\boldsymbol{w}_{n-1} + \mu \boldsymbol{u}_n e_n^* \right),$$

and the update recursion becomes

$$\boldsymbol{w}_n = \left(I - C(C^H C)^{-1} C^H \right) \left(\boldsymbol{w}_{n-1} + \mu \boldsymbol{u}_n e_n^* \right) + C \left(C^H C \right)^{-1} \boldsymbol{g}.$$

Note that $\left(I - C(C^H C)^{-1} C^H \right)$ is the orthogonal projection matrix on the intersection (affine set) of the hyperplanes defined by the constraints (recall that $C(C^H C)^{-1} C^H$ is the respective projection matrix on the subspace spanned by $\boldsymbol{c}_i$, $i = 1, 2, \ldots, m$). In the case the goal is to minimize the output, one has to set in e_n the desired response $d_n = 0$.

The constrained LMS was first treated in [40]. Besides the previously used constraints, other constraints can also be used; for example, for the constrained NLMS, one additionally demands

$$\boldsymbol{w}_n^H \boldsymbol{u}_n = d_n$$

(see, e.g., [6]).

5.12 TRACKING PERFORMANCE OF THE LMS IN NONSTATIONARY ENVIRONMENTS

We have already considered the convergence behavior of the LMS and made related comments for the other algorithms that have been discussed. As stated before, convergence is a *transient* phenomenon; that is, it concerns the period from the initial kick-off point to the steady state. The steady state was also discussed for stationary environments, that is, environments in which the unknown parameter vector as well as the underlying statistics of the involved random variables/processes remain unchanged.

We now turn our focus to cases where the true (yet unknown) parameter vector/system undergoes changes. Thus, this affects the output observations and consequently their statistics. Note that the statistics of the input can also change. However, we are not going to consider such cases, as the analysis can become quite involved. Our goal is to study the *tracking* performance of the LMS, that is, the ability of the algorithm to track changes of the unknown parameter vector. Note that tracking is a *steady-state* phenomenon. In other words, we assume that enough time has elapsed so that the influence of the initial conditions has been forgotten and has no effect, any more, on the algorithm. Tracking agility and convergence speed are two *different* properties of an algorithm. An algorithm may converge fast, but it *may not* necessarily have a good tracking performance, and vice versa. We will see such cases later on.

The setting of our discussion will be similar to that of Section 5.5.1. In conformity with the discussion there, we consider the real-data case; similar results hold true for the complex-valued linear estimation scenario. However, in contrast to the adopted model in (5.38), a *time-varying* model is adopted here, using the following assumptions.

Assumption 1. The output observations are generated according to the model

$$\mathrm{y}_n = \boldsymbol{\theta}_{o,n-1}^T \mathbf{x} + \eta, \tag{5.74}$$

which is in line with the prediction model used in LMS, at time n. That is, the unknown set of parameters is a time-varying one. The statistical properties of the input vector $\mathbf{x}$, as well as the noise variable η, are assumed to be time independent, and this is the reason we have not used the time index; equivalently, in the case where the input is a random process, u_n, it is assumed to be *stationary*. Moreover, the input variables are assumed to be independent of the zero mean noise variable, η. Furthermore, successive samples of η are i.i.d. (white noise sequence) of variance σ_η^2. So far, we have not gone much beyond Assumption 1, stated in Section 5.5.1.

Assumption 2. The time-varying model follows a random walk variation, represented by

$$\boldsymbol{\theta}_{o,n} = \boldsymbol{\theta}_{o,n-1} + \boldsymbol{\omega}. \tag{5.75}$$

The random vector $\boldsymbol{\omega}$ is assumed to be zero mean with covariance matrix

$$\mathbb{E}\left[\boldsymbol{\omega}\boldsymbol{\omega}^T\right] = \Sigma_\omega.$$

Note that the variance of a random walk grows unbounded with time; this is readily shown by applying (5.75) recursively.

A variant of this model would be more sensible to use,

$$\boldsymbol{\theta}_{o,n} = a\boldsymbol{\theta}_{o,n-1} + \boldsymbol{\omega},$$

with $|a| < 1$ [106]. However, the analysis gets more involved, so we will stick with the model in (5.75). After all, our goal here is to highlight and have a first touch on the notion of tracking and to get an idea of its effects on the misadjustment in steady state.

Assumption 3. As in Section 5.5.1, we assume that $\mathbf{c}_{n-1} := \boldsymbol{\theta}_{n-1} - \boldsymbol{\theta}_{o,n-1}$ is independent of $\mathbf{x}$ and η. This time, we will also assume independence of $\mathbf{c}_n$ and $\boldsymbol{\omega}$.

Recall from (5.48) that the excess MSE at time n is given by

$$J_{\text{exc},n} = \text{trace}\{\Sigma_x \Sigma_{c,n-1}\}.$$

Thus, our goal now is to compute $\Sigma_{c,n-1}$ for the time-varying model case. It is straightforward to see that the counterpart of (5.35) now becomes

$$\mathbf{c}_n = \left(I - \mu\mathbf{x}\mathbf{x}^T\right)\mathbf{c}_{n-1} + \mu\mathbf{x}\eta - \boldsymbol{\omega}. \tag{5.76}$$

Adopting the previously stated three assumptions, as well as the *Gaussian* assumption for $\mathbf{x}$, and following exactly the same steps used for (5.41), we end up with

$$\begin{aligned}\Sigma_{c,n} &= \Sigma_{c,n-1} - \mu\Sigma_x\Sigma_{c,n-1} - \mu\Sigma_{c,n-1}\Sigma_x + 2\mu^2\Sigma_x\Sigma_{c,n-1}\Sigma_x \\ &\quad + \mu^2\Sigma_x\text{trace}\{\Sigma_x\Sigma_{c,n-1}\} + \mu^2\sigma_\eta^2\Sigma_x + \Sigma_\omega.\end{aligned} \tag{5.77}$$

Note that if complex data are involved, the only difference is that the fourth term on the right-hand side is not multiplied by 2 (Problem 5.15). This equation governs the propagation of $\Sigma_{c,n}$, which in turn can provide the excess MSE error.

A more convenient form results for small values of μ, where the fourth and fifth terms on the right-hand side can be neglected, being small with respect to $\mu\Sigma_x\Sigma_{c,n-1}$. Moreover, at the steady state, $\Sigma_{c,n} = \Sigma_{c,n-1} := \Sigma_c$, and taking the trace on both sides, we end up with (recall trace$\{A + B\} =$

trace{A} + trace{B} and trace{AB} = trace{BA})

$$J_{\text{exc}} = \text{trace}\{\Sigma_x \Sigma_c\} = \frac{1}{2}\left(\mu\sigma_\eta^2 \text{trace}\{\Sigma_x\} + \frac{1}{\mu}\text{trace}\{\Sigma_\omega\}\right). \tag{5.78}$$

Note that this is exactly the same approximation as the one resulting from the more sound theory of energy conservation [83].

Compare (5.78) with (5.53); in the current setting, there is another term associated with the noise, which drifts the model around its mean. Thus, (a) the excess MSE is contributed by the inability of the LMS to obtain the optimum value exactly, and (b) there is an extra term measuring its "inertia" to track the changes of the model fast enough. This is the most important outcome of the current discussion. In time-varying environments, the misadjustment increases. Moreover, looking at (5.78) and at the effect of μ, it is observed that small values of μ have a beneficial effect on the first term, but they increase the contribution of the second one. The opposite is true for relatively big values of μ. This is natural. Small step sizes give the algorithm the chance to learn better under stationary environments, but the algorithm cannot track the changes fast enough. Thus, the choice of μ should be the outcome of a tradeoff. Minimizing the excess error in (5.78), it is easily shown to result in

$$\mu_{\text{opt}} = \sqrt{\frac{\text{trace}\{\Sigma_\omega\}}{\sigma_\eta^2 \text{trace}\{\Sigma_x\}}}.$$

Note, however, that such choices for μ are of theoretical importance only. In practice, the time variation of the system can hardly correspond to that of the adopted model; the latter, due to the complexity of the analysis, was chosen to be a simple one in an effort to simplify the mathematical manipulations. Moreover, for the sake of simplifying the analysis, a number of assumptions were adopted. In practice, μ is chosen more according to the user's practical experience, after experimentation, than based on the theory. The theory, however, has pointed out the tradeoff between the speed of convergence and that of tracking.

More mathematically rigorous analysis of the performance of online/adaptive schemes in nonstationary environments can be obtained from [16,38,47,61,70,83]. Simulation results demonstrating the tracking performance of the LMS compared with other algorithms are given in Example 6.3 in Chapter 6, where the recursive least-squares (RLS) algorithm is presented.

5.13 DISTRIBUTED LEARNING: THE DISTRIBUTED LMS

The focus of our attention is now turned toward a problem that has been of an increasing importance over the last decade or so. There is a growing number of applications where data are received/reside in different sensors/databases, which are spatially distributed. However, all this spatially distributed information has to be exploited towards achieving a common goal, i.e., to perform a *common estimation/inference* task. We refer to such tasks as *distributed* or *decentralized* learning. At the heart of this problem lies the concept of *cooperation*, which is another name for the process of exchanging learning experience/information in order to reach a common goal/decision. Human societies have survived because of cooperation (and have disappeared due to lack of cooperation).

Distributed learning is common in many biological systems, where no individual/agent is in charge, yet the group exhibits a high degree of intelligence (we humans refer to it as instinct). Look, for example, at the way birds fly in formation and bees swarm in a new hive.

Besides sociology and biology, science and engineering have used the concept of distributed learning; *wireless sensor networks* (WSNs) are a typical example. WSNs were originally suggested as spatially distributed autonomous sensors to monitor physical and environmental conditions, such as pressure, temperature, and sound, and to cooperatively pass their data to a central unit. Although WSNs were originally motivated for military applications, today they are targeted at a diverse number of applications, such as traffic control, homeland security and surveillance, health care, and environmental modeling. Each sensor node is equipped with an onboard processor, in order to perform locally some simple processing and transmit the required and *partially* processed data. Sensors/nodes are characterized by low processing, memory, and communication capabilities due to low energy and bandwidth constraints [1,104].

Other typical examples of distributed learning applications are the modeling and study of the way individuals are linked in social networks, modeling pathways defined over complex power grids, cognitive radio systems, and pattern recognition; the common characteristic of all these applications is that data are partially processed in each individual node/agent, and the processed information is passed over to the network under a *certain protocol*.

The obvious question that the unfamiliar reader may ask is why not use only one node/agent and the locally residing information to perform the inference task. The answer, of course, is that we can come with better estimates/results by exploiting the available data/information across the whole network. This brings us to the notion of *consensus*.

According to the American Heritage dictionary, "consensus" is defined as "an opinion or position reached by a group as a whole"; that is, consensus is the process that guarantees an "accepted agreement" within the group. The term "accepted agreement" is not uniquely defined. In some cases, this may refer to a *unanimous* decision; in other cases it refers to a *majority* rule. In some cases, all the opinions of all agents are equally weighted whereas in others, different weights are imposed, based on some relative significance measures. However, in all cases, the essence of any consensus-based process is the trust that a "better" decision is reached when compared to the process of each agent/person acting individually.

In this section, we focus on the task of *parameter estimation*. Each individual agent has access to partial information via a "local" acquisition process of data. Although each agent has access to a different set of data, they all share a common goal, i.e., to estimate the *same* unknown set of parameters. This task will be achieved in a collaborative manner. However, different cooperation scenarios can be adopted.

5.13.1 COOPERATION STRATEGIES

In distributed learning, each individual agent is represented as a node in a graph. Edges between nodes indicate that the respective agents can exchange information. Undirected edges indicate that information can be exchanged in both directions, while directed edges indicate the allowed direction of information flow.[8]

[8] More rigorous definitions on graphs are given in Chapter 15.

Centralized Networks

Under this scenario of cooperation, nodes communicate their measurements to a central *fusion* unit for processing. The obtained estimate can be communicated back to each one of the nodes. Fig. 5.22

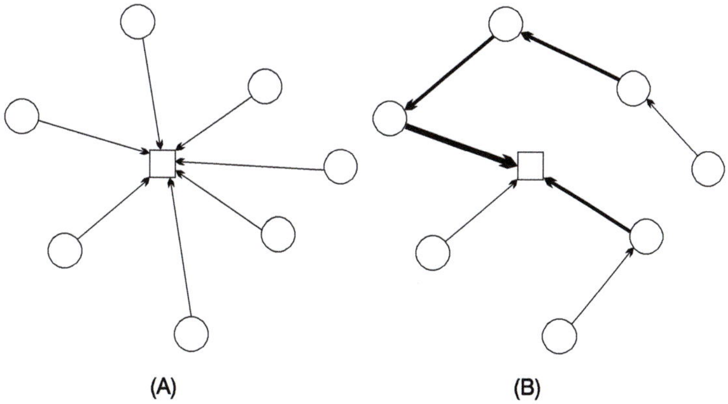

FIGURE 5.22

The square indicates the fusion center. (A) All nodes communicate directly to the fusion center. (B) Some nodes are connected directly to the fusion center. Others communicate their own data to a neighboring node, and so on, until the information reaches the fusion center. The bolder a connection is drawn, the higher the amount of data transmitted via the corresponding link.

illustrates the topology. In Fig. 5.22A all nodes are connected directly to the fusion center, indicated by a square. In Fig. 5.22B, some of the nodes can be linked directly to the fusion center, while others communicate their measurements to a linked neighbor, which then passes the received as well as the locally available observations/measurements either to a neighboring node or to the fusion center. The major advantage in this cooperation strategy is that the fusion center can compute optimal estimates, since it has access to all the available information. However, the optimality is obtained under a number of drawbacks, such as demand for increased communication costs and delays, especially for large networks. In addition, when the fusion center breaks down, the whole network collapses. Moreover, in certain applications, privacy issues are involved. For example, when data concern medical records, the nodes do not wish to send the available (training) data, but it is preferably to communicate certain locally obtained processed information. To overcome the drawbacks of the centralized processing scenario, different distributed processing schemes have been proposed.

Decentralized Networks

Under this scenario, there is no central fusion center. Processing is performed *locally* at each node, employing the locally received measurements, and in the sequel, each node communicates the locally obtained estimates to its neighbors, that is, to the nodes it is linked with. These links are denoted as edges in the respective graph. There are different decentralized schemes.

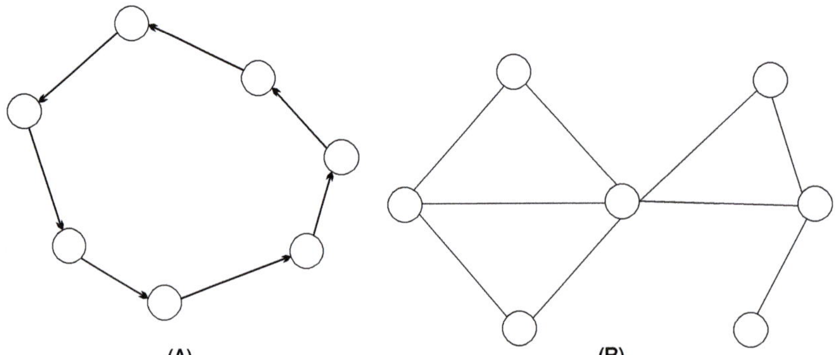

FIGURE 5.23

(A) The incremental or ring topology. The information flow follows a cyclic path. (B) Topology corresponding to a diffusion strategy. Each node communicates information to the nodes with which it shares an edge.

- *Incremental/ring networks*: These require the existence of a *cyclic path*[9] following the edges through the network. Starting from a node, such a cycle has to visit every node at least once, and then return to the first one. Such a topology implements an *iterative* computational scheme. At each iteration, every node performs its data acquisition and processing locally and communicates the required information to its neighbor in the cyclic path. It has been shown that incremental schemes achieve global performance (e.g., [58]). The main disadvantage of this mode of cooperation is that cycling information around at each iteration is a problem in large networks. It is also important to stress at this point that the construction and maintenance of a cyclic graph, visiting each node, is an NP-hard task [52]. Moreover, the whole network collapses if one node malfunctions. The corresponding graph topology is shown in Fig. 5.23A.
- *Ad hoc networks*: According to this philosophy of cooperation, nodes perform locally data acquisition as well as processing, at each iteration. However, the constraint of a cyclic path is removed. Each node communicates information to its neighboring nodes with which it shares an edge; in this way, information is *diffused* across the whole network. An advantage of such schemes is that operation does not cease if some nodes are malfunctioning. Also, the topology of the network need not be fixed. The price one pays for such "extras" is that the final obtained performance, after convergence, is inferior to those obtained by its incremental counterpart and by the centralized processing. This is natural, since at each iteration every node has access to only a limited amount of information. Fig. 5.23B illustrates an example of the topology for ad hoc networks.

Besides the previous schemes, a number of variants exists. For example, the neighbors of each node may change probabilistically, which introduces randomness in the way information is diffused at each iteration [33,60].

[9] Consider a set of nodes $x_1, \ldots, x_k$ in a graph such that there is an edge connecting (x_{i-1}, x_i), $i = 2, \ldots k$. The set of edges connecting the k nodes is a path.

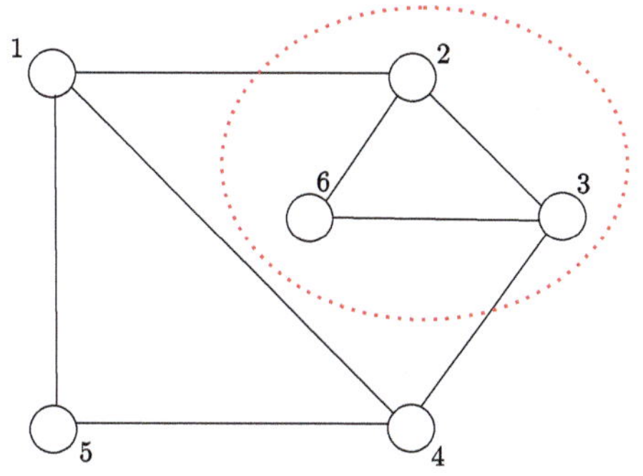

FIGURE 5.24

A graph corresponding to a network operating under a diffusion strategy. The red dotted line encircles the nodes comprising the neighborhood of node $k = 6$.

In this section, our focus will be on diffusion schemes. Our aim is to provide the reader with a sample of basic techniques around the LMS scheme and not to cover distributed learning in general, which is a field on its own with a long history; see [9,19,29,51,76,95,107] for sample references from classical to more recent contributions to the field. Besides distributed inference, a number of related aspects concerning the topology of the network and learning over graphs are attracting a lot of attention in the context of the emerging field of *complex networks*; see, [18,37,87,89,100] and the references therein.

5.13.2 THE DIFFUSION LMS

Let us consider a network of K agents/nodes. Each node exchanges information with the nodes in its neighborhood. Given a node k in a graph, let $\mathcal{N}_k$ be the set of nodes with which this node shares an edge; moreover, node k is also included in $\mathcal{N}_k$. This comprises the neighborhood set of k. We will denote the cardinality of this set as n_k. Fig. 5.24 shows a graph with six nodes. For example, the neighborhood of node $k = 6$ is $\mathcal{N}_6 = \{2, 3, 6\}$, with cardinality $n_6 = 3$. The cardinality of $\mathcal{N}_k$ is also known as the *degree* of node k. On the contrary, nodes $k = 6$ and $k = 5$ are not neighbors, because they are not directly linked via an edge. For the needs of the section, we assume that the graph is a *strongly connected* one; that is, there is at least one path of edges that connects any pair of nodes.

Each node in the network has access to a local data acquisition process, which provides the pair of training data[10] $(y_k(n),\ \boldsymbol{x}_k(n))$, $k = 1, 2, \ldots, K$, $n = 0, 1, \ldots,$ which are i.i.d. observations drawn from the respected stochastic zero mean jointly distributed variables, $\mathrm{y}_k, \mathbf{x}_k$, $k = 1, \ldots, K$. We further assume that, in all cases, the pairs of the input–output variables are associated with a common to all (unknown) parameter vector $\boldsymbol{\theta}_o$. For example, in every node, the data are assumed to be generated by a corresponding regression model

$$\mathrm{y}_k = \boldsymbol{\theta}_o^T \mathbf{x}_k + \eta_k, \quad k = 1, 2, \ldots, K, \tag{5.79}$$

10 The time index is used in parentheses, to unclutter notation due to the presence of the node index k.

where $\mathbf{x}_k$ and the zero mean noise variable η_k obey, in general, *different* statistical properties in each node. We will discuss such applications soon.

Treating each node individually, the MSE optimal solution, which minimizes the *local* cost function,

$$J_k(\boldsymbol{\theta}) = \mathbb{E}\left[\left|\mathrm{y}_k - \boldsymbol{\theta}^T \mathbf{x}_k\right|^2\right],$$

will be given by the respective normal equations, involving the respective covariance matrix and cross-correlation vector; in other words,

$$\Sigma_{x_k}\boldsymbol{\theta}_* = \boldsymbol{p}_k. \tag{5.80}$$

Recall that for the case of a regression model, $\boldsymbol{\theta}_* = \boldsymbol{\theta}_o$ and the same solution results from all nodes. Undoubtedly, if the statistics Σ_{x_k}, $\boldsymbol{p}_k$, $k = 1, 2, \ldots, K$, were known, we could stop here. However, we already know that this is not the case and in practice they have to be estimated. Alternatively, one has to resort to iterative techniques to learn the statistics as well as the unknown parameters. This is where one has to consider all nodes, to benefit from all the observations, which are distributed across the network. Thus, a more natural criterion to adopt is

$$J(\boldsymbol{\theta}) = \sum_{k=1}^{K} J_k(\boldsymbol{\theta}) = \sum_{k=1}^{K} \mathbb{E}\left[\left|\mathrm{y}_k - \boldsymbol{\theta}^T \mathbf{x}_k\right|^2\right]. \tag{5.81}$$

Using the standard arguments, which we have employed a number of times so far, it is readily seen that the (common) estimate of the unknown $\boldsymbol{\theta}_o$ will be provided as a solution of

$$\left(\sum_{k=1}^{K} \Sigma_{x_k}\right)\boldsymbol{\theta}_* = \sum_{k=1}^{K} \boldsymbol{p}_k.$$

Let us use the global cost in (5.81) as our kick-off point to apply a gradient descent optimization scheme,

$$\boldsymbol{\theta}^{(i)} = \boldsymbol{\theta}^{(i-1)} + \mu \sum_{k=1}^{K} \left(\boldsymbol{p}_k - \Sigma_{x_k}\boldsymbol{\theta}^{(i-1)}\right), \tag{5.82}$$

from which a corresponding stochastic gradient scheme results, by replacing expectations with instantaneous observations and associating iteration steps with time updates, so that

$$\boldsymbol{\theta}_n = \boldsymbol{\theta}_{n-1} + \mu \sum_{k=1}^{K} \boldsymbol{x}_k(n) e_k(n),$$
$$e_k(n) = y_k(n) - \boldsymbol{\theta}_{n-1}^T \boldsymbol{x}_k(n),$$

and a constant step size has been used, adopting the rationale behind the classical LMS formulation. Such an LMS-type recursion is perfect for a centralized scenario, where all data are transmitted to a fusion center. This is one of the extremes, having at its opposite end the scenario with nodes acting individually without cooperation. However, there is an intermediate path, which will lead us to the distributed diffusion mode of operation.

Instead of trying to minimize (5.81), let us select a specific node k and construct a local cost as the weighted aggregate in $\mathcal{N}_k$, represented by

$$J_k^{loc}(\boldsymbol{\theta}) = \sum_{m \in \mathcal{N}_k} c_{mk} J_m(\boldsymbol{\theta}), \quad k = 1, 2, \ldots, K, \tag{5.83}$$

so that

$$\sum_{k=1}^{K} c_{mk} = 1, \quad c_{mk} \geq 0, \quad \text{and } c_{mk} = 0 \quad \text{if } m \notin \mathcal{N}_k, \quad m = 1, 2, \ldots, K. \tag{5.84}$$

Let C be the $K \times K$ matrix with entries $[C]_{mk} = c_{mk}$. Then the summation condition in (5.84) can be written as

$$C\mathbf{1} = \mathbf{1}, \tag{5.85}$$

where $\mathbf{1}$ is the vector with all its entries being equal to 1. That is, all the entries across a row are summing to 1. Such matrices are known as *right stochastic* matrices. In contrast, a matrix is said to be left stochastic if

$$C^T\mathbf{1} = \mathbf{1}.$$

Also, a matrix that is both left and right stochastic is known as *doubly stochastic* (Problem 5.16). Note that due to this matrix constraint, we still have

$$\sum_{k=1}^{K} J_k^{loc}(\boldsymbol{\theta}) = \sum_{k=1}^{K} \sum_{m \in \mathcal{N}_k} c_{mk} J_m(\boldsymbol{\theta}) = \sum_{k=1}^{K} \sum_{m=1}^{K} c_{mk} J_m(\boldsymbol{\theta})$$
$$= \sum_{m=1}^{K} J_m(\boldsymbol{\theta}) = J(\boldsymbol{\theta}).$$

That is, summing all local costs, the global one results.

Let us focus on minimizing (5.83). The gradient descent scheme results in

$$\boldsymbol{\theta}_k^{(i)} = \boldsymbol{\theta}_k^{(i-1)} + \mu_k \sum_{m \in \mathcal{N}_k} c_{mk} \left(\boldsymbol{p}_m - \Sigma_{x_m} \boldsymbol{\theta}_k^{(i-1)} \right).$$

However, since nodes in the neighborhood exchange information, they could also share their current estimates. This is justified by the fact that the ultimate goal is to reach a common estimate; thus, exchanging current estimates could be used for the benefit of the algorithmic process to achieve this goal. To this end, we will modify the cost in (5.83) by regularizing it, leading to

$$\tilde{J}_k^{loc}(\boldsymbol{\theta}) = \sum_{m \in \mathcal{N}_k} c_{mk} J_m(\boldsymbol{\theta}) + \lambda ||\boldsymbol{\theta} - \tilde{\boldsymbol{\theta}}||^2, \tag{5.86}$$

where $\tilde{\boldsymbol{\theta}}$ encodes information with respect to the unknown vector, which is obtained by the neighboring nodes and $\lambda > 0$. Applying the gradient descent on (5.86) (and absorbing the factor "2," which comes from the exponents, into the step size), we obtain

$$\boldsymbol{\theta}_k^{(i)} = \boldsymbol{\theta}_k^{(i-1)} + \mu_k \sum_{m\in\mathcal{N}_k} c_{mk}\left(\boldsymbol{p}_m - \Sigma_{x_m}\boldsymbol{\theta}_k^{(i-1)}\right) + \mu_k\lambda\left(\tilde{\boldsymbol{\theta}} - \boldsymbol{\theta}_k^{(i-1)}\right), \tag{5.87}$$

which can be broken down into the following two steps:

$$\text{Step 1:} \quad \boldsymbol{\psi}_k^{(i)} = \boldsymbol{\theta}_k^{(i-1)} + \mu_k \sum_{m\in\mathcal{N}_k} c_{mk}\left(\boldsymbol{p}_m - \Sigma_{x_m}\boldsymbol{\theta}_k^{(i-1)}\right),$$

$$\text{Step 2:} \quad \boldsymbol{\theta}_k^{(i)} = \boldsymbol{\psi}_k^{(i)} + \mu_k\lambda\left(\tilde{\boldsymbol{\theta}} - \boldsymbol{\theta}_k^{(i-1)}\right).$$

Step 2 can be slightly modified and replace $\boldsymbol{\theta}_k^{(i-1)}$ by $\boldsymbol{\psi}_k^{(i)}$, since this encodes more recent information, and we obtain

$$\boldsymbol{\theta}_k^{(i)} = \boldsymbol{\psi}_k^{(i)} + \mu_k\lambda\left(\tilde{\boldsymbol{\theta}} - \boldsymbol{\psi}_k^{(i)}\right).$$

Furthermore, a reasonable choice of $\tilde{\boldsymbol{\theta}}$, at each iteration step, would be

$$\tilde{\boldsymbol{\theta}} = \tilde{\boldsymbol{\theta}}^{(i)} := \sum_{m\in\mathcal{N}_{k\setminus k}} b_{mk}\boldsymbol{\psi}_m^{(i)},$$

where

$$\sum_{m\in\mathcal{N}_{k\setminus k}} b_{mk} = 1, \quad b_{mk} \geq 0,$$

and $\mathcal{N}_{k\setminus k}$ denotes the elements in $\mathcal{N}_k$ excluding k. In other words, at each iteration, we update $\boldsymbol{\theta}_k$ so as to move it toward the descent direction of the local cost and at the same time we constrain it to stay close to the convex combination of the rest of the updates, which are obtained during the computations in step 1 from *all* the nodes in its neighborhood. Thus, we end up with the following recursions.

Diffusion gradient descent

$$\text{Step 1:} \quad \boldsymbol{\psi}_k^{(i)} = \boldsymbol{\theta}_k^{(i-1)} + \mu_k \sum_{m\in\mathcal{N}_k} c_{mk}\left(\boldsymbol{p}_m - \Sigma_{x_m}\boldsymbol{\theta}_k^{(i-1)}\right), \tag{5.91}$$

$$\text{Step 2:} \quad \boldsymbol{\theta}_k^{(i)} = \sum_{m\in\mathcal{N}_k} a_{mk}\boldsymbol{\psi}_m^{(i)}, \tag{5.92}$$

where we set

$$a_{kk} = 1 - \mu_k\lambda \quad \text{and} \quad a_{mk} = \mu_k\lambda b_{mk}, \tag{5.93}$$

which leads to

$$\sum_{m\in\mathcal{N}_k} a_{mk} = 1, \quad a_{mk} \geq 0, \tag{5.94}$$

for small enough values of $\mu_k\lambda$. Note that by setting $a_{mk}=0$, $m \notin \mathcal{N}_k$ and defining A to be the matrix with entries $[A]_{mk}=a_{mk}$, we can write

$$\sum_{m=1}^{K} a_{mk}=1 \ \Rightarrow\ A^T\mathbf{1}=\mathbf{1}, \tag{5.95}$$

that is, A is a left stochastic matrix. It is important to stress here that, irrespective of our derivation before, *any* left stochastic matrix A can be used in (5.92).

A slightly different path to arrive at (5.87) is via the interpretation of the gradient descent scheme as a minimizer of a regularized linearization of the cost function around the currently available estimate. The regularizer used is $||\boldsymbol{\theta}-\boldsymbol{\theta}^{(i-1)}||^2$ and it tries to keep the new update as close as possible to the currently available estimate. In the context of the distributed learning, instead of $\boldsymbol{\theta}^{(i-1)}$ we can use a convex combination of the available estimates obtained in the neighborhood [26,27,84].

We are now ready to state the first version of the diffusion LMS (DiLMS), by replacing in (5.91) and (5.92) expectations with instantaneous observations and interpreting iterations as time updates.

Algorithm 5.7 (The adapt-then-combine diffusion LMS).

- Initialize
  - **For** $k=1,2,\ldots,K$, **Do**
    - $\boldsymbol{\theta}_k(-1)=\mathbf{0}\in\mathbb{R}^l$; or any other value.
  - **End For**
  - Select μ_k, $k=1,2,\ldots,K$; a small positive number.
  - Select C: $C\mathbf{1}=\mathbf{1}$
  - Select A: $A^T\mathbf{1}=\mathbf{1}$
- **For** $n=0,1,\ldots,$ **Do**
  - **For** $k=1,2,\ldots,K$, **Do**
    - **For** $m\in\mathcal{N}_k$, **Do**
      - $e_{k,m}(n)=y_m(n)-\boldsymbol{\theta}_k^T(n-1)\boldsymbol{x}_m(n)$; For complex-valued data, change $T\to H$.
    - **End For**
    - $\boldsymbol{\psi}_k(n)=\boldsymbol{\theta}_k(n-1)+\mu_k\sum_{m\in\mathcal{N}_k} c_{mk}\boldsymbol{x}_m(n)e_{k,m}(n)$; For complex-valued data, $e_{k,m}(n)\to e^*_{k,m}(n)$.
  - **End For**
  - **For** $k=1,2,\ldots,K$
    - $\boldsymbol{\theta}_k(n)=\sum_{m\in\mathcal{N}_k} a_{mk}\boldsymbol{\psi}_m(n)$
  - **End For**
- **End For**

The following comments are in order:

- This form of DiLMS is known as *adapt-then-combine* (ATC) DiLMS since the first step refers to the update and the combination step follows.

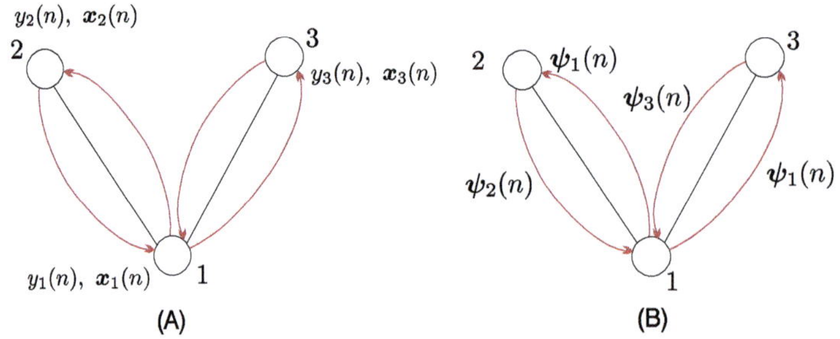

FIGURE 5.25

Adapt-then-combine. (A) In step 1, adaptation is carried out after the exchange of the received observations. (B) In step 2, the nodes exchange their locally computed estimates to obtain the updated one.

- In the special case of $C = I$, the adaptation step becomes

$$\boldsymbol{\psi}_k(n) = \boldsymbol{\theta}_k(n-1) + \mu \boldsymbol{x}_k(n) e_k(n),$$

  and nodes *need not exchange* their observations.
- The adaptation rationale is illustrated in Fig. 5.25. At time n, all three neighbors exchange the received data. In case the input vector corresponds to a realization of a random signal, $\mathrm{u}_k(n)$, the exchange of information comprises two values $(y_k(n), u_k(n))$ in each direction for each one of the links. In the more general case, where the input is a random vector of jointly distributed variables, all l variables have to be exchanged. After this message passing, adaptation takes place, as shown in Fig. 5.25A. Then, the nodes exchange their obtained estimates, $\boldsymbol{\psi}_k(n),\ k = 1, 2, 3$, across the links (Fig. 5.25B).

A different scheme results if one reverses the order of the two steps and performs first the combination and then the adaptation.

Algorithm 5.8 (The combine-then-adapt diffusion LMS).

- Initialization
  - **For** $k = 1, 2, \ldots, K$, **Do**
    - * $\boldsymbol{\theta}_k(-1) = \mathbf{0} \in \mathbb{R}^l$; or any other value.
  - **End For**
  - Select C: $\quad C\mathbf{1} = \mathbf{1}$
  - Select A: $\quad A^T\mathbf{1} = \mathbf{1}$
  - Select $\mu_k,\ k = 1, 2, \ldots, K$; a small value.
- **For** $n = 0, 1, 2, \ldots,$ **Do**
  - **For** $k = 1, 2, \ldots, K$, **Do**
    - * $\boldsymbol{\psi}_k(n-1) = \sum_{m \in \mathcal{N}_k} a_{mk} \boldsymbol{\theta}_m(n-1)$
  - **End For**

- **For** $k = 1, 2, \ldots, K$, **Do**
  - **For** $m \in \mathcal{N}_k$, **Do**
    - $e_{k,m}(n) = y_m(n) - \boldsymbol{\psi}_k^T(n-1)\boldsymbol{x}_m(n)$; For complex-valued data, change $T \rightarrow H$.
  - **End For**
  - $\boldsymbol{\theta}_k(n) = \boldsymbol{\psi}_k(n-1) + \mu_k \sum_{m \in \mathcal{N}_k} \boldsymbol{x}_m(n) e_{k,m}(n)$; For complex-valued data, $e_{k,m}(n) \rightarrow e_{k,m}^*(n)$.
- **End For**
- **End For**

The rationale of this adaptation scheme is the reverse of that illustrated in Fig. 5.25, where the phase in 5.25B precedes that of 5.25A. In the case $C = I$, there is no input–output data information exchange and the parameter update for the k node becomes

$$\boldsymbol{\theta}_k(n) = \boldsymbol{\psi}_k(n-1) + \mu_k \boldsymbol{x}_k(n) e_k(n).$$

Remarks 5.4.

- One of the early reports on the DiLMS can be found in [59]. In [80,93], versions of the algorithm for diminishing step sizes are presented and its convergence properties are analyzed. Besides the DiLMS, a version for incremental distributed cooperation has been proposed in [58]. For a related review, see [84,86,87].
- So far, nothing has been said about the choice of the matrices C (A). There are a number of possibilities. Two popular choices are the following.

  Averaging rule:

$$c_{mk} = \begin{cases} \frac{1}{n_k}, & \text{if } k = m \text{ or if nodes } k \text{ and } m \text{ are neighbors}, \\ 0, & \text{otherwise}, \end{cases}$$

  and the respective matrix is left stochastic.

  Metropolis rule:

$$c_{mk} = \begin{cases} \frac{1}{\max\{n_k, n_m\}}, & \text{if } k \neq m \text{ and } k, m \text{ are neighbors}, \\ 1 - \sum_{i \in \mathcal{N}_k \setminus k} c_{ik}, & m = k, \\ 0, & \text{otherwise}, \end{cases}$$

  which makes the respective matrix to be doubly stochastic.
- Distributed LMS-based algorithms for the case where different nodes estimate different, yet overlapping, parameter vectors have also been derived [20,75].

5.13.3 CONVERGENCE AND STEADY-STATE PERFORMANCE: SOME HIGHLIGHTS

In this subsection, we will summarize some findings concerning the performance analysis of the DiLMS. We will not give proofs. The proofs follow similar steps as for the standard LMS, with slightly more involved algebra. The interested reader can obtain proofs by looking at the original papers as well as in [84].

- The gradient descent scheme in (5.91), (5.92) is guaranteed to converge, meaning

$$\boldsymbol{\theta}_k^{(i)} \xrightarrow[i\to\infty]{} \boldsymbol{\theta}_*,$$

provided that

$$\mu_k \leq \frac{2}{\lambda_{\max}\{\Sigma_k^{loc}\}},$$

where

$$\Sigma_k^{loc} = \sum_{m\in\mathcal{N}_k} c_{mk}\Sigma_{x_m}. \tag{5.96}$$

This corresponds to the condition in (5.16).
- If one assumes that C is doubly stochastic, it can be shown that the convergence rate to the solution for the distributed case is higher than that corresponding to the noncooperative scenario, when each node operates individually, using the same step size, $\mu_k = \mu$, for all cases and provided this common value guarantees convergence. In other words, cooperation *improves the convergence speed*. This is in line with the general comments made in the beginning of the section.
- Assume that in the model in (5.79), the involved noise sequences are both spatially and temporally white, as represented by

$$\mathbb{E}\left[\eta_k(n)\eta_k(n-r)\right] = \sigma_k^2\delta_r, \qquad \delta_r = \begin{cases} 1, & r=0,\\ 0, & r\neq 0,\end{cases}$$

$$\mathbb{E}\left[\eta_k(n)\eta_m(r)\right] = \sigma_k^2\delta_{km}\delta_{nr}, \qquad \delta_{km},\delta_{nr} = \begin{cases} 1, & k=m,\ n=r,\\ 0, & \text{otherwise}.\end{cases}$$

Also, the noise sequences are independent of the input vectors,

$$\mathbb{E}\left[\mathbf{x}_m(n)\eta_k(n-r)\right] = \mathbf{0}, \quad k,m=1,2,\dots,K, \forall r,$$

and finally, the independence assumption is mobilized among the input vectors, spatially as well as temporally, namely,

$$\mathbb{E}\left[\mathbf{x}_k(n)\mathbf{x}_m^T(n-r)\right] = O, \text{ if } k\neq m, \text{ and } \forall r.$$

Under the previous assumptions, which correspond to the assumptions adopted when studying the performance of the LMS, the following hold true for the DiLMS.
Convergence in the mean: Provided that

$$\mu_k < \frac{2}{\lambda_{\max}\{\Sigma_k^{\text{loc}}\}}, \tag{5.97}$$

we have

$$\mathbb{E}\left[\boldsymbol{\theta}_k(n)\right] \xrightarrow[n\to\infty]{} \boldsymbol{\theta}_*, \quad k=1,2,\dots,K.$$

It is important to state here that the stability condition in (5.97) depends on C and not on A.

- If in addition to the previous assumption, C is chosen to be doubly stochastic, then the convergence in the mean, in any node under the distributed scenario, is *faster* than that obtained if the node is operating individually without cooperation, provided $\mu_k = \mu$ is the same and it is chosen so as to guarantee convergence.
- *Misadjustment*: under the assumptions of C and A being doubly stochastic, the following are true:
  - The average misadjustment over all nodes in the steady state for the ATC strategy is always smaller than that of the combine-then-adapt one.
  - The average misadjustment over all the nodes of the network in the distributed operation is always lower than that obtained if nodes are adapted individually, without cooperation, by using the same $\mu_k = \mu$ in all cases. That is, cooperation does not only improve convergence speed but it also *improves the steady-state performance*.

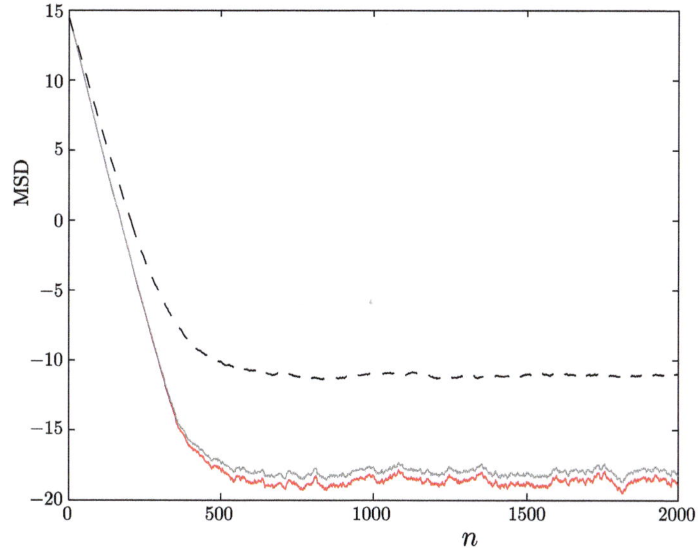

FIGURE 5.26

Average (over all the nodes) error convergence curves (MSD) for the LMS in noncooperative mode of operation (dotted line) and for the case of the DiLMS in the ATC mode (red line) and the CTA mode (gray line). The step size μ was the same in all three cases. Cooperation among nodes significantly improves performance. For the case of the DiLMS, the ATC version results in slightly better performance compared to that of the CTA (MSD in dBs).

Example 5.8. In this example, a network of $L = 10$ nodes is considered. The nodes were randomly connected with a total number of 32 connections; it was checked the resulting network was strongly connected. In each node, data are generated according to a regression model, using the same vector $\boldsymbol{\theta}_o \in \mathbb{R}^{30}$. The latter was randomly generated by $\mathcal{N}(0, 1)$. The input vectors, $\boldsymbol{x}_k$ in (5.79), were i.i.d. generated according to $\mathcal{N}(0, 1)$ and the noise level was different for each node, varying from 20 to 25 dBs.

Three experiments were carried out. The first involved the distributed LMS in its adapt-then-combine (ATC) form and the second one the combine-then-adapt (CTA) version. In the third experiment, the LMS algorithm was run independently for each node, without cooperation. In all cases, the step size was chosen equal to $\mu = 0.01$. Fig. 5.26 shows the average (over all nodes) $\text{MSD}(n) : \frac{1}{K}\sum_{k=1}^{K} ||\boldsymbol{\theta}_k(n) - \boldsymbol{\theta}_o||^2$ obtained for each one of the experiments. It is readily seen that cooperation improves the performance significantly, both in terms of convergence as well as in steady-state error floor. Moreover, as stated in Section 5.13.3, the ATC performs slightly better than the CTA version.

5.13.4 CONSENSUS-BASED DISTRIBUTED SCHEMES

An alternative path for deriving an LMS version for distributed networks was followed in [64,88]. Recall that, so far, in our discussion in deriving the DiLMS, we required the update at each node to be close to a convex combination of the available estimates in the respective neighborhood. Now we will demand such a requirement to become very strict. Although we are not going to get involved with details, since this would require to divert quite a lot from the material and the algorithmic tools which have been presented so far, let us state the task in the context of the linear MSE estimation.

To bring (5.81) in a distributed learning context, let us modify it by allowing different parameter vectors for each node k, so that

$$J(\boldsymbol{\theta}_1, \ldots, \boldsymbol{\theta}_K) = \sum_{k=1}^{K} \mathbb{E}\left[|\mathrm{y}_k - \boldsymbol{\theta}_k^T \mathbf{x}_k|^2\right].$$

Then the task is cast according to the following constrained optimization problem:

$$\{\hat{\boldsymbol{\theta}}_k, k = 1, \ldots, K\} = \arg \min_{\{\boldsymbol{\theta}_k,\ k=1,\ldots,K\}} J(\boldsymbol{\theta}_1, \ldots, \boldsymbol{\theta}_K)$$

$$\text{s.t.} \quad \boldsymbol{\theta}_k = \boldsymbol{\theta}_m, \quad k = 1, 2, \ldots, K, \ m \in \mathcal{N}_k.$$

In other words, one demands *equality* of the estimates within a neighborhood. As a consequence, these constraints lead to network-wise equality, since the graph that represents the network has been assumed to be connected. The optimization is carried out iteratively by employing stochastic approximation arguments and building on the alternating direction method of multipliers (ADMM) (Chapter 8) [19]. The algorithm, besides updating the vector estimates, has to update the associated Lagrange multipliers as well.

In addition to the previously reported ADMM-based scheme, a number of variants known as *consensus-based algorithms* have been employed in several studies [19,33,49,50,71]. A formulation around which a number of stochastic gradient consensus-based algorithms evolve is the following [33, 49,50]:

$$\boldsymbol{\theta}_k(n) = \boldsymbol{\theta}_k(n-1) + \mu_k(n)\left[\boldsymbol{x}_k(n)e_k(n) + \lambda \sum_{m \in \mathcal{N}_{k \setminus k}} \left(\boldsymbol{\theta}_k(n-1) - \boldsymbol{\theta}_m(n-1)\right)\right], \tag{5.98}$$

where

$$e_k(n) := y_k(n) - \boldsymbol{\theta}_k^T(n-1)\boldsymbol{x}_k(n)$$

and for some $\lambda > 0$. Observe the form in (5.98); the term in brackets on the right-hand side is a regularizer whose goal is to enforce equality among the estimates within the neighborhood of node k. Several alternatives to Eq. (5.98) have been proposed. For example, in [49] a different step size is employed for the consensus summation on the right-hand side of (5.98). In [99], the following formulation is provided:

$$\boldsymbol{\theta}_k(n) = \boldsymbol{\theta}_k(n-1) + \mu_k(n)\left[\boldsymbol{x}_k(n)e_k(n) + \sum_{m\in\mathcal{N}_k\setminus k} b_{m,k}\left(\boldsymbol{\theta}_k(n-1) - \boldsymbol{\theta}_m(n-1)\right)\right], \tag{5.99}$$

where $b_{m,k}$ stands for some nonnegative coefficients. If one defines the weights,

$$a_{m,k} := \begin{cases} 1 - \sum_{m\in\mathcal{N}_k\setminus k} \mu_k(n)b_{m,k}, & m = k, \\ \mu_k(n)b_{m,k}, & m \in \mathcal{N}_k \setminus k, \\ 0, & \text{otherwise}, \end{cases} \tag{5.100}$$

recursion (5.99) can be equivalently written as

$$\boxed{\boldsymbol{\theta}_k(n) = \sum_{m\in\mathcal{N}_k} a_{m,k}\boldsymbol{\theta}_m(n-1) + \mu_k(n)\boldsymbol{x}_k(n)e_k(n).} \tag{5.101}$$

The update rule in (5.101) is also referred to as *consensus strategy* (see, e.g., [99]). Note that the step size is considered to be time-varying. In particular, in [19,71], a diminishing step size is employed, within the stochastic gradient rationale, which has to satisfy the familiar pair of conditions in order to guarantee convergence to a consensus value over all the nodes,

$$\sum_{n=0}^{\infty} \mu_k(n) = \infty, \quad \sum_{n=0}^{\infty} \mu_k^2(n) < \infty. \tag{5.102}$$

Observe the update recursion in (5.101). It is readily seen that the update $\boldsymbol{\theta}_k(n)$ involves only the error $e_k(n)$ of the corresponding node. In contrast, looking carefully at the corresponding update recursions in both Algorithms 5.7 and 5.8, $\boldsymbol{\theta}_k(n)$ is updated according to the average error within the neighborhood. This is an important difference.

The theoretical properties of the consensus recursion (5.101), which employs a *constant step size*, and a comparative analysis against the diffusion schemes have been presented in [86,99]. There, it has been shown that the diffusion schemes outperform the consensus-based ones, in the sense that (a) they converge faster, (b) they reach a lower steady-state mean-square deviation error floor, and (c) their mean-square stability is insensitive to the choice of the combination weights.

5.14 A CASE STUDY: TARGET LOCALIZATION

Consider a network consisting of K nodes, whose goal is to estimate and track the location of a specific target. The location of the unknown target, say, $\boldsymbol{\theta}_o$, is assumed to belong to the two-dimensional space.

The position of each node is denoted by $\boldsymbol{\theta}_k = [\theta_{k1}, \theta_{k2}]^T$, and the true distance between node k and the unknown target is equal to

$$r_k = \|\boldsymbol{\theta}_o - \boldsymbol{\theta}_k\|. \tag{5.103}$$

The vector whose direction points from node k to the unknown source is given by

$$\boldsymbol{g}_k = \frac{\boldsymbol{\theta}_o - \boldsymbol{\theta}_k}{\|\boldsymbol{\theta}_o - \boldsymbol{\theta}_k\|}. \tag{5.104}$$

Obviously, the distance can be rewritten in terms of the direction vector as

$$r_k = \boldsymbol{g}_k^T(\boldsymbol{\theta}_o - \boldsymbol{\theta}_k). \tag{5.105}$$

It is reasonable to assume that each node k "senses" the distance and the direction vectors via noisy observations. For example, such a noisy information can be inferred from the strength of the received signal or other related information. Following a similar rationale as in [84,98], the noisy distance can be modeled as

$$\hat{r}_k(n) = r_k + v_k(n), \tag{5.106}$$

where n stands for the discrete-time instance and $v_k(n)$ for the additive noise term. The noise in the direction vector is a consequence of two effects: (a) a deviation occurring along the perpendicular direction to $\boldsymbol{g}_k$ and (b) a deviation that takes place along the parallel direction of $\boldsymbol{g}_k$. All in one, the noisy direction vector (see Fig. 5.27) occurring at time instance n can be written as

$$\hat{\boldsymbol{g}}_k(n) = \boldsymbol{g}_k + v_k^{\perp}(n)\boldsymbol{g}_k^{\perp} + v_k^{\parallel}(n)\boldsymbol{g}_k, \tag{5.107}$$

where $v_k^{\perp}(n)$ is the noise corrupting the unit norm perpendicular direction vector $\boldsymbol{g}_k^{\perp}$ and $v_k^{\parallel}(n)$ is the noise occurring at the parallel direction vector. Taking into consideration the noisy terms, (5.106) is written as

$$\hat{r}_k(n) = \hat{\boldsymbol{g}}_k^T(n)(\boldsymbol{\theta}_o - \boldsymbol{\theta}_k) + \eta_k(n), \tag{5.108}$$

where

$$\eta_k(n) = v_k(n) - v_k^{\perp}(n)\boldsymbol{g}_k^{\perp T}(\boldsymbol{\theta}_o - \boldsymbol{\theta}_k) - v_k^{\parallel}(n)\boldsymbol{g}_k^T(\boldsymbol{\theta}_o - \boldsymbol{\theta}_k). \tag{5.109}$$

Eq. (5.109) can be further simplified if one recalls that by construction $\boldsymbol{g}_k^{\perp T}(\boldsymbol{\theta}_o - \boldsymbol{\theta}_k) = 0$. Moreover, typically, the contribution of $v_k^{\perp}(n)$ is assumed to be significantly larger than the contribution of $v_k^{\parallel}(n)$. Henceforth, taking into consideration these two arguments, (5.109) can be simplified to

$$\eta_k(n) \approx v_k(n). \tag{5.110}$$

If one defines $y_k(n) := \hat{r}_k(n) + \hat{\boldsymbol{g}}_k^T(n)\boldsymbol{\theta}_k$ and combines (5.108) with (5.110) the following model results:

$$y_k(n) \approx \boldsymbol{\theta}_o^T \hat{\boldsymbol{g}}_k(n) + v_k(n). \tag{5.111}$$

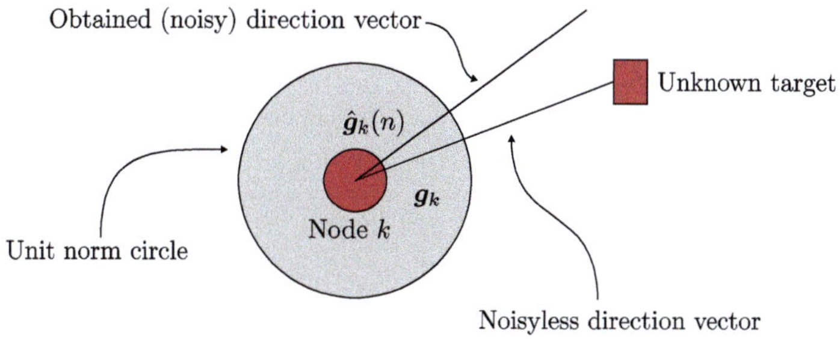

FIGURE 5.27

Illustration of a node, the target source, and the direction vectors.

Eq. (5.111) is a linear regression model. Using the available estimates, for each time instant, one has access to $y_k(n)$, $\hat{\boldsymbol{g}}_k(n)$ and any form of distributed algorithm can be adopted in order to obtain a better estimate of $\boldsymbol{\theta}_o$.

Indeed, it has been verified that the information exchange and fusion enhances significantly the ability of the nodes to estimate and track the target source. The nodes can possibly represent fish schools which seek a nutrition source, bee swarms which search for their hive, or bacteria seeking nutritive sources [28,84,85,97].

Some other typical applications of distributed learning are social networks [36], radio resource allocation [32], and network cartography [65].

5.15 SOME CONCLUDING REMARKS: CONSENSUS MATRIX

In our treatment of the DiLMS, we used the combination matrices A (C), which we assumed to be left (right) stochastic. Also, in the performance-related section, we pointed out that some of the reported results hold true if these matrices are, in addition, doubly stochastic. Although it was not needed in this chapter, in the general distributed processing theory, a matrix of significant importance is the so-called *consensus matrix*. A matrix $A \in \mathbb{R}^{K \times K}$ is said to be a consensus matrix if, in addition to being doubly stochastic, as represented by

$$A\mathbf{1} = \mathbf{1}, \quad A^T\mathbf{1} = \mathbf{1},$$

it also satisfies the property

$$\left|\lambda_i\left\{A^T - \frac{1}{K}\mathbf{1}\mathbf{1}^T\right\}\right| < 1, \quad i = 1, 2, \ldots, K.$$

In other words, all eigenvalues of the matrix

$$A^T - \frac{1}{K}\mathbf{1}\mathbf{1}^T$$

have magnitude strictly less than one. To demonstrate its usefulness, we will state a fundamental theorem in distributed learning.

Theorem 5.2. *Consider a network consisting of K nodes, each one of them having access to a state vector* $\boldsymbol{x}_k$. *Consider the recursion*

$$\boxed{\boldsymbol{\theta}_k^{(i)} = \sum_{m \in \mathcal{N}_k} a_{mk} \boldsymbol{\theta}_m^{(i-1)}, \quad k = 1, 2, \dots, K, \ i > 0 : \textit{consensus iteration},}$$

with

$$\boldsymbol{\theta}_k^{(0)} = \boldsymbol{x}_k, \quad k = 1, 2, \dots, K.$$

Define $A \in \mathbb{R}^{K \times K}$ *to be the matrix with entries*

$$[A]_{mk} = a_{mk}, \quad m, k = 1, 2, \dots, K,$$

where $a_{mk} \geq 0$, $a_{mk} = 0$ *if* $m \notin \mathcal{N}_k$. *If A is a consensus matrix, then [31]*

$$\boldsymbol{\theta}_k^{(i)} \longrightarrow \frac{1}{K} \sum_{k=1}^{K} \boldsymbol{x}_k.$$

The opposite is also true. If convergence is always *guaranteed, then A is a consensus matrix.*

In other words, this theorem states that updating each node by convexly combining, with appropriate weights, the current estimates in its neighborhood, the network converges to the average value in a consensus rationale (Problem 5.17).

PROBLEMS

5.1 Show that the gradient vector is perpendicular to the tangent at a point of an isovalue curve.

5.2 Prove that if

$$\sum_{i=1}^{\infty} \mu_i^2 < \infty, \quad \sum_{i=1}^{\infty} \mu_i = \infty,$$

the gradient descent scheme, for the MSE cost function and for the iteration dependent step size case, converges to the optimal solution.

5.3 Derive the steepest gradient descent direction for the complex-valued case.

5.4 Let θ, x be two jointly distributed random variables. Let also the function (regression)

$$f(\theta) = \mathbb{E}[\mathrm{x}|\theta].$$

assumed to be an increasing one. Show that under the conditions in (5.29), the recursion

$$\theta_n = \theta_{n-1} - \mu_n x_n$$

converges in probability to the root of $f(\theta)$.

5.5 Show that the LMS algorithm is a nonlinear estimator.

5.6 Show Eq. (5.42).

5.7 Derive the bound in (5.45).
Hint: Use the well-known property from linear algebra that the eigenvalues of a matrix $A \in \mathbb{R}^{l\times l}$ satisfy the following bound:

$$\max_{1\le i\le l} |\lambda_i| \le \max_{1\le i\le l} \sum_{j=1}^{l} |a_{ij}| := \|A\|_1.$$

5.8 *Gershgorin circle theorem.* Let A be an $l \times l$ matrix, with entries $a_{ij}, i, j = 1, 2, \ldots, l$. Let $R_i := \sum_{\substack{j=1\\ j\neq i}}^{l} |a_{ij}|$ be the sum of absolute values of the nondiagonal entries in row i. Show that if λ is an eigenvalue of A, then there exists at least one row i such that the following is true:

$$|\lambda - a_{ii}| \le R_i.$$

The last bound defines a circle which contains the eigenvalue λ.

5.9 Apply the Gershgorin circle theorem to prove the bound in (5.45).

5.10 Derive the misadjustment formula given in (5.52).

5.11 Derive the APA iteration scheme.

5.12 Consider the hyperplane that comprises all the vectors $\boldsymbol{\theta}$ such as

$$\boldsymbol{x}_n^T\boldsymbol{\theta} - y_n = 0,$$

for a pair $(y_n, \boldsymbol{x}_n)$. Show that $\boldsymbol{x}_n$ is perpendicular to the hyperplane.

5.13 Derive the recursions for the widely linear APA.

5.14 Show that a similarity transformation of a square matrix via a unitary matrix does not affect the eigenvalues.

5.15 Show that if $\mathbf{x} \in \mathbb{R}^l$ is a Gaussian random vector, then

$$F := \mathbb{E}[\mathbf{x}\mathbf{x}^{\mathrm{T}} S \mathbf{x}\mathbf{x}^{\mathrm{T}}] = \Sigma_x \mathrm{trace}\{S\Sigma_x\} + 2\Sigma_x S \Sigma_x,$$

and if $\mathbf{x} \in \mathbb{C}^l$,

$$F := \mathbb{E}[\mathbf{x}\mathbf{x}^{\mathrm{H}} S \mathbf{x}\mathbf{x}^{\mathrm{H}}] = \Sigma_x \mathrm{trace}\{S\Sigma_x\} + \Sigma_x S \Sigma_x.$$

5.16 Show that if an $l \times l$ matrix C is right stochastic, then all its eigenvalues satisfy

$$|\lambda_i| \le 1, \quad i = 1, 2, \ldots, l.$$

The same holds true for left and doubly stochastic matrices.

5.17 Prove Theorem 5.2.

MATLAB® EXERCISES

5.18 Consider the MSE cost function in (5.4). Set the cross-correlation equal to $\boldsymbol{p} = [0.05, 0.03]^T$. Also, consider two covariance matrices,

$$\Sigma_1 = \begin{bmatrix} 1 & 0 \\ 0 & 0.1 \end{bmatrix}, \quad \Sigma_2 = \begin{bmatrix} 1 & 0 \\ 0 & 1 \end{bmatrix}.$$

Compute the corresponding optimal solutions, $\boldsymbol{\theta}_{(*,1)} = \Sigma_1^{-1}\boldsymbol{p}$, $\boldsymbol{\theta}_{(*,2)} = \Sigma_2^{-1}\boldsymbol{p}$. Apply the gradient descent scheme of (5.6) to estimate $\boldsymbol{\theta}_{(*,2)}$; set the step size equal to (a) its optimal value μ_o according to (5.17) and (b) equal to $\mu_o/2$. For these two choices for the step size, plot the error $\|\boldsymbol{\theta}^{(i)} - \boldsymbol{\theta}_{(*,2)}\|^2$ at each iteration step i. Compare the convergence speeds of these two curves towards zero. Moreover, in the two-dimensional space, plot the coefficients of the successive estimates, $\boldsymbol{\theta}^{(i)}$, for both step sizes, together with the isovalue contours of the cost function. What do you observe regarding the trajectory towards the minimum?
Apply (5.6) for the estimation of $\boldsymbol{\theta}_{(*,1)}$ employing Σ_1^{-1} and $\boldsymbol{p}$. Use as step size μ_o of the previous experiment. Plot, in the same figure, the previously computed error curve $\|\boldsymbol{\theta}^{(i)} - \boldsymbol{\theta}_{(*,2)}\|^2$ together with the error curve $\|\boldsymbol{\theta}^{(i)} - \boldsymbol{\theta}_{(*,1)}\|^2$. Compare the convergence speeds.
Now set the step size equal to the optimum value associated with Σ_1. Again, in the two-dimensional space, plot the values of the successive estimates and the isovalue contours of the cost function. Compare the number of steps needed for convergence, with the ones needed in the previous experiment. Play with different covariance matrices and step sizes.

5.19 Consider the linear regression model

$$y_n = \boldsymbol{x}_n^T \boldsymbol{\theta}_o + \eta_n,$$

where $\boldsymbol{\theta}_o \in \mathbb{R}^2$. Generate the coefficients of the unknown vector $\boldsymbol{\theta}_o$ randomly according to the normalized Gaussian distribution, $\mathcal{N}(0, 1)$. The noise is assumed to be white Gaussian with variance 0.1. The samples of the input vector are i.i.d. generated via the normalized Gaussian. Apply the Robbins–Monro algorithm in (5.34) for the optimal MSE linear estimation with a step size equal to $\mu_n = 1/n$. Run 1000 independent experiments and plot the mean value of the first coefficient of the 1000 produced estimates, at each iteration step. Also, plot the horizontal line crossing the true value of the first coefficient of the unknown vector. Furthermore, plot the standard deviation of the obtained estimate, every 30 iteration steps. Comment on the results. Play with different rules of diminishing step sizes and comment on the results.

5.20 Generate data according to the regression model

$$y_n = \boldsymbol{x}_n^T \boldsymbol{\theta}_o + \eta_n,$$

where $\boldsymbol{\theta}_o \in \mathbb{R}^{10}$, and whose elements are randomly obtained using the Gaussian distribution $\mathcal{N}(0, 1)$. The noise samples are also i.i.d. generated from $\mathcal{N}(0, 0.01)$.

Generate the input samples as part of two processes: (a) a white noise sequence, i.i.d. generated via $\mathcal{N}(0, 1)$, and (b) an autoregressive AR(1) process with $a_1 = 0.85$ and the corresponding white noise excitation of variance equal to 1. For these two choices of the input, run the LMS algorithm on the generated training set $(y_n, \boldsymbol{x}_n)$, $n = 0, 1, \ldots$, to estimate $\boldsymbol{\theta}_o$. Use a step size equal to $\mu = 0.01$. Run 100 independent experiments and plot the average error per iteration in dBs, using $10\log_{10}(e_n^2)$, with $e_n^2 = (y_n - \boldsymbol{\theta}_{n-1}^T \boldsymbol{x}_n)^2$. What do you observe regarding the convergence speed of the algorithm for the two cases? Repeat the experiment with different values of the AR coefficient a_1 and different values of the step size. Observe how the learning curve changes with the different values of the step size and/or the value of the AR coefficient. Choose, also, a relatively large value for the step size and make the LMS algorithm to diverge. Comment and justify theoretically the obtained results concerning convergence speed and the error floor at the steady state after convergence.

5.21 Use the data set generated from the AR(1) process of the previous exercise. Employ the transform-domain LMS (Algorithm 5.6) with step size equal to 0.01. Also, set $\delta = 0.01$ and $\beta = 0.5$. Moreover, employ the DCT transform. As in the previous exercise, run 100 independent experiments and plot the average error per iteration. Compare the results with those of the LMS with the same step size.
Hint: Compute the DCT transformation matrix using the *dctmtx* MATLAB® function.

5.22 Generate the same experimental setup as in Exercise 5.20, with the difference that $\boldsymbol{\theta}_o \in \mathbb{R}^{60}$. For the LMS algorithm set $\mu = 0.025$ and for the NLMS (Algorithm 5.3) $\mu = 0.35$ and $\delta = 0.001$. Employ also the APA (Algorithm 5.2) with parameters $\mu = 0.1$, $\delta = 0.001$, and $q = 10, 30$. Plot in the same figure the error learning curves of all these algorithms, as in the previous exercises. How does the choice of q affect the behavior of the APA, in terms of both convergence speed and the error floor at which it settles after convergence? Play with different values of q and of the step size μ.

5.23 Consider the decision feedback equalizer described in Section 5.10.
- **(a)** Generate a set of 1000 random ± 1 values (BPSK) (i.e., s_n). Direct this sequence into a linear channel with impulse response $\boldsymbol{h} = [0.04, -0.05, 0.07, -0.21, 0.72, 0.36, 0.21, 0.03, 0.07]^T$ and add to the output 11 dB white Gaussian noise. Denote the output as u_n.
- **(b)** Design the adaptive DFE using $L = 21$, $l = 10$, and $\mu = 0.025$ following the training mode only. Perform a set of 500 experiments feeding the DFE with different random sequences from the ones described in step (a). Plot the MSE (averaged over the 500 experiments). Observe that around $n = 250$ the algorithm achieves convergence.
- **(c)** Design the adaptive decision feedback equalizer using the parameters of step (b). Feed the equalizer with a series of 10,000 random values generated as in step (a). After the 250th data sample, change the DFE to decision-directed mode. Count the percentage of the errors performed by the equalizer from the 251th to the 10,000th sample.
- **(d)** Repeat steps (a) to (c) changing the level of the white Gaussian noise added to the BPSK values to 15, 12, 10 dB. Then, for each case, change the delay to $L = 5$. Comment on the results.

5.24 Develop the MATLAB® code for the two forms of the DiLMS, ATC and CTA, and reproduce the results of Example 5.8. Play with the choice of the various parameters. Make sure that the resulting network is strongly connected.

REFERENCES

[1] I.F. Akyildiz, W. Su, Y. Sankarasubramaniam, E. Cayirci, A survey on sensor networks, IEEE Commun. Mag. 40 (8) (2002) 102–114.

[2] S.J.M. Almeida, J.C.M. Bermudez, N.J. Bershad, M.H. Costa, A statistical analysis of the affine projection algorithm for unity step size and autoregressive inputs, IEEE Trans. Circuits Syst. I 52 (7) (2005) 1394–1405.

[3] T.Y. Al-Naffouri, A.H. Sayed, Transient analysis of data-normalized adaptive filters, IEEE Trans. Signal Process. 51 (3) (2003) 639–652.

[4] S. Amari, Theory of adaptive pattern classifiers, IEEE Trans. Electron. Comput. 16 (3) (1967) 299–307.

[5] C. Antweiler, M. Dörbecker, Perfect sequence excitation of the NLMS algorithm and its application to acoustic echo control, Ann. Telecommun. 49 (7–8) (1994) 386–397.

[6] J.A. Appolinario, S. Werner, P.S.R. Diniz, T.I. Laakso, Constrained normalized adaptive filtering for CDMA mobile communications, in: Proceedings, EUSIPCO, Rhodes, Greece, 1998.

[7] J.A. Appolinario, M.L.R. de Campos, P.S.R. Diniz, The binormalized data-reusing LMS algorithm, IEEE Trans. Signal Process. 48 (2000) 3235–3242.

[8] J. Arenas-Garcia, A.R. Figueiras-Vidal, A.H. Sayed, Mean-square performance of a convex combination of two adaptive filters, IEEE Trans. Signal Process. 54 (3) (2006) 1078–1090.

[9] S. Barbarossa, G. Scutari, Bio-inspired sensor network design: distributed decisions through self-synchronization, IEEE Signal Process. Mag. 24 (3) (2007) 26–35.

[10] J. Benesty, T. Gänsler, D.R. Morgan, M.M. Sondhi, S.L. Gay, Advances in Network and Acoustic Echo Cancellation, Springer Verlag, Berlin, 2001.

[11] J. Benesty, S.L. Gay, An improved PNLMS algorithm, in: IEEE International Conference on Acoustics, Speech, and Signal Processing, ICASSP, vol. 2, 2002.

[12] J. Benesty, C. Paleologu, S. Ciochina, On regularization in adaptive filtering, IEEE Trans. Audio Speech Lang. Process. 19 (6) (2011) 1734–1742.

[13] A. Benvenniste, M. Metivier, P. Piouret, Adaptive Algorithms and Stochastic Approximations, Springer-Verlag, NY, 1987.

[14] K. Berberidis, P. Karaivazoglou, An efficient block adaptive DFE implemented in the frequency domain, IEEE Trans. Signal Process. 50 (9) (2002) 2273–2285.

[15] N.J. Bershad, Analysis of the normalized LMS with Gaussian inputs, IEEE Trans. Acoust. Speech Signal Process. 34 (4) (1986) 793–806.

[16] N.J. Bershad, O.M. Macchi, Adaptive recovery of a chirped sinusoid in noise. Part 2: Performance of the LMS algorithm, IEEE Trans. Signal Process. 39 (1991) 595–602.

[17] J.C.M. Bermudez, N.J. Bershad, A nonlinear analytical model for the quantized LMS algorithm: the arbitrary step size case, IEEE Trans. Signal Process. 44 (1996) 1175–1183.

[18] A. Bertrand, M. Moonen, Seeing the bigger picture, IEEE Signal Process. Mag. 30 (3) (2013) 71–82.

[19] D.P. Bertsekas, J.N. Tsitsiklis, Parallel and Distributed Computations: Numerical Methods, Athena Scientific, Belmont, MA, 1997.

[20] N. Bogdanovic, J. Plata-Chaves, K. Berberidis, Distributed incremental-based LMS for node-specific adaptive parameter estimation, IEEE Trans. Signal Process. 62 (20) (2014) 5382–75397.

[21] N. Cesa-Bianchi, P.M. Long, M.K. Warmuth, Worst case quadratic loss bounds for prediction using linear functions and gradient descent, IEEE Trans. Neural Netw. 7 (3) (1996) 604–619.

[22] P. Bouboulis, S. Theodoridis, Extension of Wirtinger's calculus to reproducing kernel Hilbert spaces and the complex kernel LMS, IEEE Trans. Signal Process. 53 (3) (2011) 964–978.

[23] S. Boyd, L. Vandenberghe, Convex Optimization, Cambridge University Press, 2004.

[24] A. Carini, Efficient NLMS and RLS algorithms for perfect and imperfect periodic sequences, IEEE Trans. Signal Process. 58 (4) (2010) 2048–2059.

[25] A. Carini, G.L. Sicuranza, V.J. Mathews, Efficient adaptive identification of linear-in-the-parameters nonlinear filters using periodic input sequences, Signal Process. 93 (5) (2013) 1210–1220.

[26] F.S. Cattivelli, A.H. Sayed, Diffusion LMS strategies for distributed estimation, IEEE Trans. Signal Process. 58 (3) (2010) 1035–1048.

[27] F.S. Cattivelli, Distributed Collaborative Processing Over Adaptive Networks, PhD Thesis, University of California, LA, 2010.

[28] J. Chen, A.H. Sayed, Bio-inspired cooperative optimization with application to bacteria mobility, in: IEEE International Conference on Acoustics, Speech and Signal Processing, ICASSP, 2011, pp. 5788–5791.

[29] S. Chouvardas, Y. Kopsinis, S. Theodoridis, Sparsity-aware distributed learning, in: S. Cui, A. Hero, J. Moura, Z.Q. Luo (Eds.), Big Data Over Networks, Cambridge University Press, 2014.

[30] T.A.C.M. Claasen, W.F.G. Mecklenbrauker, Comparison of the convergence of two algorithms for adaptive FIR digital filters, IEEE Trans. Acoust. Speech Signal Process. 29 (1981) 670–678.
[31] M.H. DeGroot, Reaching a consensus, J. Am. Stat. Assoc. 69 (345) (1974) 118–121.
[32] P. Di Lorenzo, S. Barbarossa, Swarming algorithms for distributed radio resource allocation, IEEE Signal Process. Mag. 30 (3) (2013) 144–154.
[33] A.G. Dimakis, S. Kar, J.M.F. Moura, M.G. Rabbat, A. Scaglione, Gossip algorithms for distributed signal processing, Proc. IEEE 98 (11) (2010) 1847–1864.
[34] P.S.R. Diniz, Adaptive Filtering: Algorithms and Practical Implementation, fourth ed., Springer, 2013.
[35] D.L. Duttweiler, Proportionate NLMS adaptation in echo cancelers, IEEE Trans. Audio Speech Lang. Process. 8 (2000) 508–518.
[36] C. Chamley, A. Scaglione, L. Li, Models for the diffusion of belief in social networks, IEEE Signal Process. Mag. 30 (3) (2013) 16–28.
[37] C. Eksin, P. Molavi, A. Ribeiro, A. Jadbabaie, Learning in network games with incomplete information, IEEE Signal Process. Mag. 30 (3) (2013) 30–42.
[38] D.C. Farden, Tracking properties of adaptive signal processing algorithms, IEEE Trans. Acoust. Speech Signal Process. 29 (1981) 439–446.
[39] E.R. Ferrara, Fast implementations of LMS adaptive filters, IEEE Trans. Acoust. Speech Signal Process. 28 (1980) 474–475.
[40] O.L. Frost III, An algorithm for linearly constrained adaptive array processing, Proc. IEEE 60 (1972) 926–935.
[41] I. Furukawa, A design of canceller of broadband acoustic echo, in: Proceedings, International Teleconference Symposium, 1984, pp. 1–8.
[42] S.L. Gay, S. Tavathia, The fast affine projection algorithm, in: Proceedings International Conference on Acoustics, Speech and Signal Processing, ICASSP, 1995, pp. 3023–3026.
[43] S.L. Gay, J. Benesty, Acoustical Signal Processing for Telecommunications, Kluwer, 2000.
[44] A. Gersho, Adaptive equalization of highly dispersive channels for data transmission, Bell Syst. Tech. J. 48 (1969) 55–70.
[45] A. Gilloire, M. Vetterli, Adaptive filtering in subbands with critical sampling: analysis, experiments and applications to acoustic echo cancellation, IEEE Trans. Signal Process. 40 (1992) 1862–1875.
[46] B. Hassibi, A.H. Sayed, T. Kailath, H^∞ optimality of the LMS algorithm, IEEE Trans. Signal Process. 44 (2) (1996) 267–280.
[47] S. Haykin, Adaptive Filter Theory, fourth ed., Pentice Hall, 2002.
[48] T. Hinamoto, S. Maekawa, Extended theory of learning identification, IEEE Trans. 95 (10) (1975) 227–234 (in Japanese).
[49] S. Kar, J. Moura, Convergence rate analysis of distributed gossip (linear parameter) estimation: fundamental limits and tradeoffs, IEEE J. Sel. Top. Signal Process. 5 (4) (2011) 674–690.
[50] S. Kar, J. Moura, K. Ramanan, Distributed parameter estimation in sensor networks: nonlinear observation models and imperfect communication, IEEE Trans. Inf. Theory 58 (6) (2012) 3575–3605.
[51] S. Kar, J.M.F. Moura, Consensus + innovations distributed inference over networks, IEEE Signal Process. Mag. 30 (3) (2013) 99–109.
[52] R.M. Karp, Reducibility among combinational problems, in: R.E. Miller, J.W. Thatcher (Eds.), Complexity of Computer Computations, Plenum Press, NY, 1972, pp. 85–104.
[53] W. Kellermann, Kompensation akustischer echos in frequenzteilbandern, in: Aachener Kolloquim, Aachen, Germany, 1984, pp. 322–325.
[54] W. Kellermann, Analysis and design of multirate systems for cancellation of acoustical echos, in: Proceedings, IEEE International Conference on Acoustics, Speech and Signal Processing, New York, 1988, pp. 2570–2573.
[55] J. Kivinen, M.K. Warmuth, B. Hassibi, The p-norm generalization of the LMS algorithms for filtering, IEEE Trans. Signal Process. 54 (3) (2006) 1782–1793.
[56] H.J. Kushner, G.G. Yin, Stochastic Approximation Algorithms and Applications, Springer, New York, 1997.
[57] L. Ljung, System Identification: Theory for the User, Prentice Hall, Englewood Cliffs, NJ, 1987.
[58] C.G. Lopes, A.H. Sayed, Incremental adaptive strategies over distributed networks, IEEE Trans. Signal Process. 55 (8) (2007) 4064–4077.
[59] C.G. Lopes, A.H. Sayed, Diffusion least-mean-squares over adaptive networks: formulation and performance analysis, IEEE Trans. Signal Process. 56 (7) (2008) 3122–3136.
[60] C. Lopes, A.H. Sayed, Diffusion adaptive networks with changing topologies, in: Proceedings International Conference on Acoustics, Speech and Signal Processing, CASSP, Las Vegas, April 2008, pp. 3285–3288.

[61] O.M. Macci, N.J. Bershad, Adaptive recovery of chirped sinusoid in noise. Part 1: Performance of the RLS algorithm, IEEE Trans. Signal Process. 39 (1991) 583–594.
[62] O. Macchi, Adaptive Processing: The Least-Mean-Squares Approach With Applications in Transmission, Wiley, New York, 1995.
[63] V.J. Mathews, S.H. Cho, Improved convergence analysis of stochastic gradient adaptive filters using the sign algorithm, IEEE Trans. Acoust. Speech Signal Process. 35 (1987) 450–454.
[64] G. Mateos, I.D. Schizas, G.B. Giannakis, Performance analysis of the consensus-based distributed LMS algorithm, EURASIP J. Adv. Signal Process. (2009) 981030, https://doi.org/10.1155/2009/981030.
[65] G. Mateos, K. Rajawat, Dynamic network cartography, IEEE Signal Process. Mag. 30 (3) (2013) 129–143.
[66] R. Merched, A. Sayed, An embedding approach to frequency-domain and subband filtering, IEEE Trans. Signal Process. 48 (9) (2000) 2607–2619.
[67] N. Murata, A statistical study on online learning, in: D. Saad (Ed.), Online Learning and Neural Networks, Cambridge University Press, UK, 1998, pp. 63–92.
[68] S.S. Narayan, A.M. Peterson, Frequency domain LMS algorithm, Proc. IEEE 69 (1) (1981) 124–126.
[69] V.H. Nascimento, J.C.M. Bermudez, Probability of divergence for the least mean fourth algorithm, IEEE Trans. Signal Process. 54 (2006) 1376–1385.
[70] V.H. Nascimento, M.T.M. Silva, Adaptive filters, in: R. Chellappa, S. Theodoridis (Eds.), Signal Process, E-Ref. 1, 2014, pp. 619–747.
[71] A. Nedic, A. Ozdaglar, Distributed subgradient methods for multi-agent optimization, IEEE Trans. Autom. Control 54 (1) (2009) 48–61.
[72] K. Ozeki, T. Umeda, An adaptive filtering algorithm using an orthogonal projection to an affine subspace and its properties, IEICE Trans. 67-A (5) (1984) 126–132 (in Japanese).
[73] C. Paleologu, J. Benesty, S. Ciochina, Regularization of the affine projection algorithm, IEEE Trans. Circuits Syst. II, Express Briefs 58 (6) (2011) 366–370.
[74] A. Papoulis, S.U. Pillai, Probability, Random Variables and Stochastic Processes, fourth ed., McGraw Hill, 2002.
[75] J. Plata-Chaves, N. Bogdanovic, K. Berberidis, Distributed diffusion-based LMS for node-specific adaptive parameter estimation, arXiv:1408.3354, 2014.
[76] J.B. Predd, S.R. Kulkarni, H.V. Poor, Distributed learning in wireless sensor networks, IEEE Signal Process. Mag. 23 (4) (2006) 56–69.
[77] J. Proakis, Digital Communications, fourth ed., McGraw Hill, New York, 2000.
[78] J. Proakis, J.H. Miller, Adaptive receiver for digital signalling through channels with intersymbol interference, IEEE Trans. Inf. Theory 15 (1969) 484–497.
[79] H. Robbins, S. Monro, A stochastic approximation method, Ann. Math. Stat. 22 (1951) 400–407.
[80] S.S. Ram, A. Nedich, V.V. Veeravalli, Distributed stochastic subgradient projection algorithms for convex optimization, J. Optim. Theory Appl. 147 (3) (2010) 516–545.
[81] M. Martinez-Ramon, J. Arenas-Garcia, A. Navia-Vazquez, A.R. Figueiras-Vidal, An adaptive combination of adaptive filters for plant identification, in: Proceedings the 14th International Conference on Digital Signal Processing, DSP, 2002, pp. 1195–1198.
[82] A.H. Sayed, M. Rupp, Error energy bounds for adaptive gradient algorithms, IEEE Trans. Signal Process. 44 (8) (1996) 1982–1989.
[83] A.H. Sayed, Fundamentals of Adaptive Filtering, John Wiley, 2003.
[84] A.H. Sayed, Diffusion adaptation over networks, in: R. Chellappa, S. Theodoridis (Eds.), Academic Press Library in Signal Processing, vol. 3, Academic Press, 2014, pp. 323–454.
[85] A.H. Sayed, S.-Y. Tu, X. Zhao, Z.J. Towfic, Diffusion strategies for adaptation and learning over networks, IEEE Signal Process. Mag. 30 (3) (2013) 155–171.
[86] A.H. Sayed, Adaptive networks, Proc. IEEE 102 (4) (2014) 460–497.
[87] A.H. Sayed, Adaptation, learning, and optimization over networks, Found. Trends Mach. Learn. 7 (4–5) (2014) 311–801.
[88] I.D. Schizas, G. Mateos, G.B. Giannakis, Distributed LMS for consensus-based in-network adaptive processing, IEEE Trans. Signal Process. 57 (6) (2009) 2365–2382.
[89] D.I. Shuman, S.K. Narang, A. Ortega, P. Vandergheyrst, The emerging field of signal processing on graphs, IEEE Signal Process. Mag. 30 (3) (2013) 83–98.
[90] D.T. Slock, On the convergence behavior of the LMS and normalized LMS algorithms, IEEE Trans. Signal Process. 40 (9) (1993) 2811–2825.

[91] V. Solo, X. Kong, Adaptive Signal Processing Algorithms: Stability and Performance, Prentice Hall, Upper Saddle River, NJ, 1995.
[92] V. Solo, The stability of LMS, IEEE Trans. Signal Process. 45 (12) (1997) 3017–3026.
[93] S.S. Stankovic, M.S. Stankovic, D.M. Stipanovic, Decentralized parameter estimation by consensus based stochastic approximation, IEEE Trans. Autom. Control 56 (3) (2011) 531–543.
[94] S. Theodoridis, K. Koutroumbas, Pattern Recognition, fourth ed., Academic Press, 2009.
[95] J.N. Tsitsiklis, Problems in Decentralized Decision Making and Computation, PhD Thesis, MIT, 1984.
[96] Y.Z. Tsypkin, Adaptation and Learning in Automatic Systems, Academic Press, New York, 1971.
[97] S.-Y. Tu, A.H. Sayed, Foraging behavior of fish schools via diffusion adaptation, in: Proceedings Cognitive Information Processing, CIP, 2010, pp. 63–68.
[98] S.-Y. Tu, A.H. Sayed, Mobile adaptive networks, IEEE J. Sel. Top. Signal Process. 5 (4) (2011) 649–664.
[99] S.-Y. Tu, A.H. Sayed, Diffusion strategies outperform consensus strategies for distributed estimation over adaptive networks, IEEE Trans. Signal Process. 60 (12) (2012) 6217–6234.
[100] K. Vikram, V.H. Poor, Social learning and Bayesian games in multiagent signal processing, IEEE Signal Process. Mag. 30 (3) (2013) 43–57.
[101] E. Walach, B. Widrow, The least mean fourth (LMF) adaptive algorithm and its family, IEEE Trans. Inf. Theory 30 (2) (1984) 275–283.
[102] B. Widrow, M.E. Hoff, Adaptive switching circuits, in: IRE Part 4, IRE WESCON Convention Record, 1960, pp. 96–104.
[103] B. Widrow, S.D. Stearns, Adaptive Signal Processing, Prentice Hall, Englewood Cliffs, 1985.
[104] J.-J. Xiao, A. Ribeiro, Z.-Q. Luo, G.B. Giannakis, Distributed compression-estimation using wireless networks, IEEE Signal Process. Mag. 23 (4) (2006) 27–741.
[105] N.R. Yousef, A.H. Sayed, A unified approach to the steady-state and tracking analysis of adaptive filters, IEEE Trans. Signal Process. 49 (2) (2001) 314–324.
[106] N.R. Yousef, A.H. Sayed, Ability of adaptive filters to track carrier offsets and random channel nonstationarities, IEEE Trans. Signal Process. 50 (7) (2002) 1533–1544.
[107] F. Zhao, J. Lin, L. Guibas, J. Reich, Collaborative signal and information processing, Proc. IEEE 91 (8) (2003) 1199–1209.

CHAPTER

THE LEAST-SQUARES FAMILY

6

CONTENTS

6.1 INTRODUCTION

The squared error loss function was at the center of our attention in the previous two chapters. The sum of squared errors cost was introduced in Chapter 3, followed by the mean-square error (MSE) version, treated in Chapter 4. The stochastic gradient descent technique was employed in Chapter 5 to

Machine Learning. https://doi.org/10.1016/B978-0-12-818803-3.00015-5

help us bypass the need to perform expectations for obtaining the second-order statistics of the data, as required by the MSE formulation.

In this chapter, we return to the original formulation of the sum of error squares, and our goal is to look more closely at the resulting family of algorithms and their properties. An emphasis is given on the geometric interpretation of the least-squares (LS) method as well as on some of the most important statistical properties of the resulting solution. The singular value decomposition (SVD) of a matrix is introduced for a first time in this book. Its geometric orthogonalizing properties are discussed and its connection with what is known as dimensionality reduction is established; the latter topic is extensively treated in Chapter 19. Also, a major part of the chapter is dedicated to the recursive LS (RLS) algorithm, which is an online scheme that solves the LS optimization task. The spine of the RLS scheme comprises an efficient update of the inverse (sample) covariance matrix of the input data, whose rationale can also be adopted in the context of different learning methods for developing related online schemes; this is one of the reasons we pay special tribute to the RLS algorithm. The other reason is its popularity in a large number of signal processing/machine learning tasks, due to some attractive properties that this scheme enjoys. Two major optimization schemes are introduced, namely, Newton's method and the coordinate descent method, and their use in solving the LS task is discussed. The bridge between the RLS scheme and Newton's optimization method is established. Finally, at the end of the chapter, a more general formulation of the LS task, known as the total least-squares (TLS) method, is also presented.

6.2 LEAST-SQUARES LINEAR REGRESSION: A GEOMETRIC PERSPECTIVE

The focus of this section is to outline the geometric properties of the LS method. This provides an alternative view on the respective minimization method and helps in its understanding, by revealing a physical structure that is associated with the obtained solution. Geometry is very important when dealing with concepts related to the dimensionality reduction task.

We begin with our familiar linear regression model. Given a set of observations,

$$y_n = \boldsymbol{\theta}^T \boldsymbol{x}_n + \eta_n, \quad n = 1, 2, \ldots, N, \; y_n \in \mathbb{R}, \; \boldsymbol{x}_n \in \mathbb{R}^l, \; \boldsymbol{\theta} \in \mathbb{R}^l,$$

where η_n denotes the (unobserved) values of a *zero* mean noise source, the task is to obtain an estimate of the unknown parameter vector, $\boldsymbol{\theta}$, so that

$$\hat{\boldsymbol{\theta}}_{\text{LS}} = \arg\min_{\boldsymbol{\theta}} \sum_{n=1}^{N} (y_n - \boldsymbol{\theta}^T \boldsymbol{x}_n)^2. \tag{6.1}$$

Our stage of discussion is that of real numbers, and we will point out differences with the complex number case whenever needed. Moreover, we assume that our data have been centered around their sample means; alternatively, the intercept, θ_0, can be absorbed in $\boldsymbol{\theta}$ with a corresponding increase in the dimensionality of $\boldsymbol{x}_n$. Define

$$\boldsymbol{y} = \begin{bmatrix} y_1 \\ \vdots \\ y_N \end{bmatrix} \in \mathbb{R}^N, \quad X := \begin{bmatrix} \boldsymbol{x}_1^T \\ \vdots \\ \boldsymbol{x}_N^T \end{bmatrix} \in \mathbb{R}^{N \times l}. \tag{6.2}$$

Eq. (6.1) can be recast as

$$\hat{\boldsymbol{\theta}}_{\text{LS}} = \arg\min_{\boldsymbol{\theta}} \|\boldsymbol{e}\|^2,$$

where

$$\boldsymbol{e} := \boldsymbol{y} - X\boldsymbol{\theta}$$

and $\|\cdot\|$ denotes the Euclidean norm, which measures the "distance" between the respective vectors in $\mathbb{R}^N$, i.e., $\boldsymbol{y}$ and $X\boldsymbol{\theta}$. Indeed, the nth component of the vector $\boldsymbol{e}$ is equal to $e_n = y_n - \boldsymbol{x}_n^T\boldsymbol{\theta}$, which, due to the symmetry of the inner product, is equal to $y_n - \boldsymbol{\theta}^T\boldsymbol{x}_n$; furthermore, the square Euclidean norm of the vector is the sum of the squares of its components, which makes the square norm of $\boldsymbol{e}$ identical to the sum of squared errors cost in Eq. (6.1).

Let us now denote as $\boldsymbol{x}_1^c, \dots, \boldsymbol{x}_l^c \in \mathbb{R}^N$ the columns of X, i.e.,

$$X = [\boldsymbol{x}_1^c, \dots, \boldsymbol{x}_l^c].$$

Then the matrix-vector product above can be written as the linear combination of the columns of matrix X, i.e.,

$$\hat{\boldsymbol{y}} := X\boldsymbol{\theta} = \sum_{i=1}^{l} \theta_i \boldsymbol{x}_i^c,$$

and

$$\boldsymbol{e} = \boldsymbol{y} - \hat{\boldsymbol{y}}.$$

Note the $\boldsymbol{e}$ can be viewed as the *error vector* between the vector of the output observations, $\boldsymbol{y}$, and $\hat{\boldsymbol{y}}$; the latter is the prediction of $\boldsymbol{y}$ based on the input observations, stacked in X, and given a value for $\boldsymbol{\theta}$. Obviously, the N-dimensional vector $\hat{\boldsymbol{y}}$, being a linear combination of the columns of X, lies in the span$\{\boldsymbol{x}_1^c, \dots, \boldsymbol{x}_l^c\}$. By definition, the latter is the subspace of $\mathbb{R}^l$ that is generated by all possible linear combinations of the l columns of X (see Appendix A). Thus, naturally, our task now becomes that of selecting $\boldsymbol{\theta}$ so that the error vector between $\boldsymbol{y}$ and $\hat{\boldsymbol{y}}$ has minimum norm. In general, the observations vector, $\boldsymbol{y}$, does not lie in the subspace spanned by the columns of X, due to the existence of the noise.

According to the Pythagorean theorem of orthogonality for Euclidean spaces, the minimum norm error is obtained if $\hat{\boldsymbol{y}}$ is chosen as the *orthogonal projection* of $\boldsymbol{y}$ onto the span $\{\boldsymbol{x}_1^c, \dots, \boldsymbol{x}_l^c\}$. Recalling the concept of orthogonal projections (Appendix A and Section 5.6, Eq. (5.65)), the orthogonal projection of $\boldsymbol{y}$ onto the subspace spanned by the columns of X is given by

$$\boxed{\hat{\boldsymbol{y}} = X(X^TX)^{-1}X^T\boldsymbol{y}: \quad \text{LS estimate,}} \tag{6.3}$$

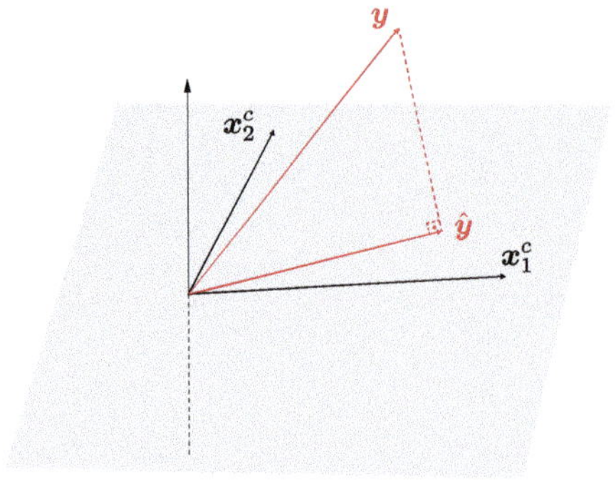

FIGURE 6.1

In the figure, $\boldsymbol{y}$ lies outside the two-dimensional (shaded) plane that is defined by the two column vectors, $\boldsymbol{x}_1^c$ and $\boldsymbol{x}_2^c$, i.e., span$\{\boldsymbol{x}_1^c, \boldsymbol{x}_2^c\}$. From all the points on this plane, the one that is closest to $\boldsymbol{y}$, in the minimum error norm sense, is its respective orthogonal projection, i.e., the point $\hat{\boldsymbol{y}}$. The parameter vector $\boldsymbol{\theta}$ associated with this orthogonal projection coincides with the LS estimate.

assuming that $X^T X$ is invertible. Recalling the definition of $\hat{\boldsymbol{y}}$, the above corresponds to the LS estimate for the unknown set of parameters, as we know it form Chapter 3, i.e.,

$$\hat{\boldsymbol{\theta}} = (X^T X)^{-1} X^T \boldsymbol{y}.$$

The geometry is illustrated in Fig. 6.1.

It is common to describe the LS solution in terms of the *Moore–Penrose pseudoinverse* of X, which for a tall matrix[1] is defined as

$$\boxed{X^{\dagger} := (X^T X)^{-1} X^T : \quad \text{pseudoinverse of a tall matrix } X,} \tag{6.4}$$

and hence we can write

$$\hat{\boldsymbol{\theta}}_{\text{LS}} = X^{\dagger} \boldsymbol{y}. \tag{6.5}$$

Thus, we have rederived Eq. (3.17) of Chapter 3, this time via geometric arguments. Note that the pseudoinverse is a generalization of the notion of the inverse of a square matrix. Indeed, if X is square, then it is readily seen that the pseudoinverse coincides with X^{-1}. For complex-valued data, the only difference is that transposition is replaced by the Hermitian one.

[1] A matrix, e.g., $X \in \mathbb{R}^{N \times l}$, is called *tall*, if $N > l$. If $N < l$ it is known as *fat*. If $l = N$, it is a *square* matrix.

6.3 STATISTICAL PROPERTIES OF THE LS ESTIMATOR

Some of the statistical properties of the LS estimator were touched on in Chapter 3, for the special case of a random real parameter. Here, we will look at this issue in a more general setting. Assume that there exists a true (yet unknown) parameter/weight vector $\boldsymbol{\theta}_o$ that generates the output (dependent) random variables (stacked in a random vector $\mathbf{y} \in \mathbb{R}^N$), according to the model

$$\mathbf{y} = X\boldsymbol{\theta}_o + \boldsymbol{\eta},$$

where $\boldsymbol{\eta}$ is a *zero* mean noise vector. Observe that we have assumed that X is fixed and not random; that is, the randomness underlying the output variables $\mathbf{y}$ is due solely to the noise. Under the previously stated assumptions, the following properties hold.

THE LS ESTIMATOR IS UNBIASED

The LS estimator for the parameters is given by

$$\begin{aligned} \hat{\boldsymbol{\theta}}_{\text{LS}} &= (X^T X)^{-1} X^T \mathbf{y}, \\ &= (X^T X)^{-1} X^T (X\boldsymbol{\theta}_o + \boldsymbol{\eta}) = \boldsymbol{\theta}_o + (X^T X)^{-1} X^T \boldsymbol{\eta}, \end{aligned} \tag{6.6}$$

or

$$\mathbb{E}[\hat{\boldsymbol{\theta}}_{\text{LS}}] = \boldsymbol{\theta}_o + (X^T X)^{-1} X^T \mathbb{E}[\boldsymbol{\eta}] = \boldsymbol{\theta}_o,$$

which proves the claim.

COVARIANCE MATRIX OF THE LS ESTIMATOR

Let, in addition to the previously adopted assumptions,

$$\mathbb{E}[\boldsymbol{\eta}\boldsymbol{\eta}^T] = \sigma_\eta^2 I,$$

that is, the source generating the noise samples is white. By the definition of the covariance matrix, we get

$$\Sigma_{\hat{\theta}_{\text{LS}}} = \mathbb{E}\left[(\hat{\boldsymbol{\theta}}_{\text{LS}} - \boldsymbol{\theta}_o)(\hat{\boldsymbol{\theta}}_{\text{LS}} - \boldsymbol{\theta}_o)^T\right],$$

and substituting $\hat{\boldsymbol{\theta}}_{\text{LS}} - \boldsymbol{\theta}_o$ from (6.6), we obtain

$$\begin{aligned} \Sigma_{\hat{\theta}_{\text{LS}}} &= \mathbb{E}\left[(X^T X)^{-1} X^T \boldsymbol{\eta}\boldsymbol{\eta}^T X (X^T X)^{-1}\right] \\ &= (X^T X)^{-1} X^T \mathbb{E}[\boldsymbol{\eta}\boldsymbol{\eta}^T] X (X^T X)^{-1} \\ &= \sigma_\eta^2 (X^T X)^{-1}. \end{aligned} \tag{6.7}$$

Note that, for large values of N, we can write

$$X^T X = \sum_{n=1}^{N} \boldsymbol{x}_n \boldsymbol{x}_n^T \approx N \Sigma_x,$$

where Σ_x is the covariance matrix of our (zero mean) input variables, i.e.,

$$\Sigma_x := \mathbb{E}[\mathbf{x}_n \mathbf{x}_n^T] \approx \frac{1}{N}\sum_{n=1}^{N} \boldsymbol{x}_n \boldsymbol{x}_n^T.$$

Thus, for large values of N, we can write

$$\boxed{\Sigma_{\hat{\theta}_{LS}} \approx \frac{\sigma_\eta^2}{N}\Sigma_x^{-1}.} \tag{6.8}$$

In other words, under the adopted assumptions, the LS estimator is not only unbiased, but its covariance matrix *tends asymptotically to zero*. That is, with high probability, the estimate $\hat{\boldsymbol{\theta}}_{\text{LS}}$, which is obtained via a large number of measurements, will be close to the true value $\boldsymbol{\theta}_o$. Viewing it slightly differently, note that the LS solution tends to the MSE solution, which was discussed in Chapter 4. Indeed, for the case of centered data,

$$\lim_{N\to\infty} \frac{1}{N}\sum_{n=1}^{N} \boldsymbol{x}_n \boldsymbol{x}_n^T = \Sigma_x,$$

and

$$\lim_{N\to\infty} \frac{1}{N}\sum_{n=1}^{N} \boldsymbol{x}_n y_n = \mathbb{E}[\mathbf{x}\mathrm{y}] = \boldsymbol{p}.$$

Moreover, we know that for the linear regression modeling case, the normal equations, $\Sigma_x \boldsymbol{\theta} = \boldsymbol{p}$, result in the solution $\boldsymbol{\theta} = \boldsymbol{\theta}_o$ (Remarks 4.2).

THE LS ESTIMATOR IS BLUE IN THE PRESENCE OF WHITE NOISE

The notion of the best linear unbiased estimator (BLUE) was introduced in Section 4.9.1 in the context of the *Gauss–Markov* theorem. Let $\hat{\boldsymbol{\theta}}$ denote any other *linear* unbiased estimator, under the assumption that

$$\mathbb{E}[\boldsymbol{\eta}\boldsymbol{\eta}^T] = \sigma_\eta^2 I.$$

Then, due to the linearity assumption, the estimator will have a linear dependence on the output random variables that are observed, i.e.,

$$\hat{\boldsymbol{\theta}} = H\mathbf{y}, \quad H \in \mathbb{R}^{l\times N}.$$

It will be shown that the variance of such an estimator can never become smaller than that of the LS one, i.e.,

$$\mathbb{E}\left[(\hat{\boldsymbol{\theta}} - \boldsymbol{\theta}_o)^T(\hat{\boldsymbol{\theta}} - \boldsymbol{\theta}_o)\right] \geq \mathbb{E}\left[(\hat{\boldsymbol{\theta}}_{LS} - \boldsymbol{\theta}_o)^T(\hat{\boldsymbol{\theta}}_{LS} - \boldsymbol{\theta}_o)\right]. \tag{6.9}$$

Indeed, from the respective definitions we have

$$\hat{\boldsymbol{\theta}} = H(X\boldsymbol{\theta}_o + \boldsymbol{\eta}) = HX\boldsymbol{\theta}_o + H\boldsymbol{\eta}. \tag{6.10}$$

However, because $\hat{\boldsymbol{\theta}}$ has been assumed unbiased, (6.10) implies that $HX = I$ and

$$\hat{\boldsymbol{\theta}} - \boldsymbol{\theta}_o = H\boldsymbol{\eta}.$$

Thus,

$$\begin{aligned} \Sigma_{\hat{\theta}} &:= \mathbb{E}\left[(\hat{\boldsymbol{\theta}} - \boldsymbol{\theta}_o)(\hat{\boldsymbol{\theta}} - \boldsymbol{\theta}_o)^T\right] \\ &= \sigma_\eta^2 H H^T. \end{aligned}$$

However, taking into account that $HX = I$, it is easily checked (try it) that

$$\sigma_\eta^2 H H^T = \sigma_\eta^2 (H - X^\dagger)(H - X^\dagger)^T + \sigma_\eta^2 (X^T X)^{-1},$$

where $X^\dagger$ is the respective pseudoinverse matrix, defined in (6.4).

Because $\sigma_n^2 (H - X^\dagger)(H - X^\dagger)^T$ is a positive semidefinite matrix (Appendix A), its trace is nonnegative (Problem 6.1) and thus we have

$$\text{trace}\{\sigma_n^2 H H^T\} \geq \text{trace}\{\sigma_n^2 (X^T X)^{-1}\},$$

and recalling (6.7), we have proved that

$$\text{trace}\{\Sigma_{\hat{\theta}}\} \geq \text{trace}\{\Sigma_{\hat{\theta}_{\text{LS}}}\}. \tag{6.11}$$

However, recalling from linear algebra the property of the trace (Appendix A), we have

$$\text{trace}\{\Sigma_{\hat{\theta}}\} = \text{trace}\left\{\mathbb{E}\left[(\hat{\boldsymbol{\theta}} - \boldsymbol{\theta}_o)(\hat{\boldsymbol{\theta}} - \boldsymbol{\theta}_o)^T\right]\right\} = \mathbb{E}\left[(\hat{\boldsymbol{\theta}} - \boldsymbol{\theta}_o)^T(\hat{\boldsymbol{\theta}} - \boldsymbol{\theta}_o)\right],$$

and similarly for $\Sigma_{\hat{\theta}_{\text{LS}}}$. Hence, Eq. (6.11) above leads directly to (6.9). Moreover, equality holds only if

$$H = X^\dagger = (X^T X)^{-1} X^T.$$

Note that this result could have been obtained directly from (4.102) by setting $\Sigma_\eta = \sigma_\eta^2 I$. This also emphasizes the fact that if the noise is not white, then the LS parameter estimator is *no more* BLUE.

THE LS ESTIMATOR ACHIEVES THE CRAMÉR–RAO BOUND FOR WHITE GAUSSIAN NOISE

The concept of the Cramér–Rao lower bound was introduced in Chapter 3. There, it was shown that, under the *white Gaussian noise* assumption, the LS estimator of a real number was *efficient*; that is, it achieves the Cramér–Rao bound. Moreover, in Problem 3.9, it was shown that if $\boldsymbol{\eta}$ is zero mean Gaussian noise with covariance matrix Σ_η, then the efficient estimator is given by

$$\hat{\boldsymbol{\theta}} = (X^T \Sigma_\eta^{-1} X)^{-1} X^T \Sigma_\eta^{-1} \mathbf{y},$$

which for $\Sigma_\eta = \sigma_\eta^2 I$ coincides with the LS estimator. In other words, under the white Gaussian noise assumption, the LS estimator becomes a *minimum variance unbiased estimator* (MVUE). This is a strong result. No other unbiased estimator (*not necessarily linear*) will do better than the LS one. Note that this result holds true not only asymptotically, but also for a finite number of samples N. If one wishes to decrease further the mean-square error, then *biased* estimators, as produced via regularization, have to be considered; this has already been discussed in Chapter 3; see also [16,50] and the references therein.

ASYMPTOTIC DISTRIBUTION OF THE LS ESTIMATOR

We have already seen that the LS estimator is unbiased and that its covariance matrix is approximately (for large values of N) given by (6.8). Thus, as $N \longrightarrow \infty$, the variance around the true value, $\boldsymbol{\theta}_o$, is becoming increasingly small. Furthermore, there is a stronger result, which provides the distribution of the LS estimator for large values of N. Under some general assumptions, such as independence of successive observation vectors and that the white noise source is independent of the input, and mobilizing the central limit theorem, it can be shown (Problem 6.2) that

$$\boxed{\sqrt{N}(\hat{\boldsymbol{\theta}}_{LS} - \boldsymbol{\theta}_o) \longrightarrow \mathcal{N}(\mathbf{0}, \sigma_\eta^2 \Sigma_x^{-1}),} \tag{6.12}$$

where the limit is meant to be in *distribution* (see Section 2.6). Alternatively, we can write

$$\hat{\boldsymbol{\theta}}_{LS} \sim \mathcal{N}\left(\boldsymbol{\theta}_o, \frac{\sigma_\eta^2}{N} \Sigma_x^{-1}\right).$$

In other words, the LS parameter estimator is asymptotically distributed according to the normal distribution.

6.4 ORTHOGONALIZING THE COLUMN SPACE OF THE INPUT MATRIX: THE SVD METHOD

The *singular value decomposition* (SVD) of a matrix is among the most powerful tools in linear algebra. As a matter of fact, it will be the tool that we are going to use as a starting point to deal with dimensionality reduction in Chapter 19. Due to its importance in machine learning, we present the basic theory here and exploit it to shed light on our LS estimation task from a different angle. We start by considering the general case, and then we tailor the theory to our specific needs.

Let X be an $m \times l$ matrix and allow its rank r not to be necessarily full (Appendix A), i.e.,

$$r \leq \min\{m, l\}.$$

Then there exist *orthogonal* matrices,[2] U and V, of dimensions $m \times m$ and $l \times l$, respectively, so that

[2] Recall that a square matrix U is called orthogonal if $U^T U = UU^T = I$. For complex-valued square matrices, if $U^H U = UU^H = I$, U is called unitary.

$$\boxed{X = U \begin{bmatrix} D & O \\ O & O \end{bmatrix} V^T : \text{ singular value decomposition of } X,} \tag{6.13}$$

where D is an $r \times r$ *diagonal* matrix[3] with elements $\sigma_i = \sqrt{\lambda_i}$, known as the *singular values* of X, where λ_i, $i = 1, 2, \ldots, r$, are the *nonzero* eigenvalues of XX^T; matrices denoted as O comprise zero elements and are of appropriate dimensions.

Taking into account the zero elements in the diagonal matrix, (6.13) can be rewritten as

$$X = U_r D V_r^T = \sum_{i=1}^{r} \sigma_i \boldsymbol{u}_i \boldsymbol{v}_i^T, \tag{6.14}$$

where

$$U_r := [\boldsymbol{u}_1, \ldots, \boldsymbol{u}_r] \in \mathbb{R}^{m \times r}, \quad V_r := [\boldsymbol{v}_1, \ldots, \boldsymbol{v}_r] \in \mathbb{R}^{l \times r}. \tag{6.15}$$

Eq. (6.14) provides a matrix factorization of X in terms of U_r, V_r, and D. We will make use of this factorization in Chapter 19, when dealing with dimensionality reduction techniques. Fig. 6.2 offers a schematic illustration of (6.14).

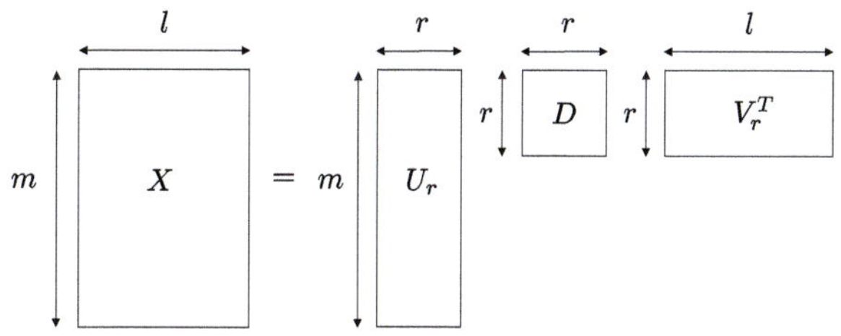

FIGURE 6.2

The $m \times l$ matrix X, of rank $r \leq \min\{m, l\}$, factorizes in terms of the matrices $U_r \in \mathbb{R}^{m \times r}$, $V_r \in \mathbb{R}^{l \times r}$ and the $r \times r$ diagonal matrix D.

It turns out that $\boldsymbol{u}_i \in \mathbb{R}^m$, $i = 1, 2, \ldots, r$, known as *left singular vectors*, are the normalized eigenvectors corresponding to the nonzero eigenvalues of XX^T, and $\boldsymbol{v}_i \in \mathbb{R}^l$, $i = 1, 2, \ldots, r$, are the normalized eigenvectors associated with the nonzero eigenvalues of $X^T X$, and they are known as *right singular vectors*. Note that both XX^T and $X^T X$ share the same eigenvalues (Problem 6.3).

Proof. By the respective definitions (Appendix A), we have

$$XX^T \boldsymbol{u}_i = \lambda_i \boldsymbol{u}_i, \quad i = 1, 2, \ldots, r, \tag{6.16}$$

[3] Usually it is denoted as Σ, but here we avoid the notation so as not to confuse it with the covariance matrix Σ; D reminds us of its diagonal structure.

and

$$X^T X \boldsymbol{v}_i = \lambda_i \boldsymbol{v}_i, \quad i = 1, 2, \ldots, r. \tag{6.17}$$

Moreover, because XX^T and $X^T X$ are symmetric matrices, it is known from linear algebra that their eigenvalues are real[4] and the respective eigenvectors are orthogonal, which can then be normalized to unit norm to become orthonormal (Problem 6.4). It is a matter of simple algebra (Problem 6.5) to show from (6.16) and (6.17) that

$$\boldsymbol{u}_i = \frac{1}{\sigma_i} X \boldsymbol{v}_i, \quad i = 1, 2, \ldots, r. \tag{6.18}$$

Thus, we can write

$$\sum_{i=1}^{r} \sigma_i \boldsymbol{u}_i \boldsymbol{v}_i^T = X \sum_{i=1}^{r} \boldsymbol{v}_i \boldsymbol{v}_i^T = X \sum_{i=1}^{l} \boldsymbol{v}_i \boldsymbol{v}_i^T = X V V^T,$$

where we used the fact that for eigenvectors corresponding to $\sigma_i = 0$ ($\lambda_i = 0$), $i = r+1, \ldots, l$, $X\boldsymbol{v}_i = \boldsymbol{0}$. However, due to the orthonormality of $\boldsymbol{v}_i$, $i = 1, 2, \ldots, l$, $VV^T = I$ and the claim in (6.14) has been proved. □

PSEUDOINVERSE MATRIX AND SVD

Let us now elaborate on the SVD expansion and investigate its geometric implications. By the definition of the pseudoinverse, $X^\dagger$, and assuming the $N \times l$ $(N > l)$ data matrix to be full column rank $(r = l)$, employing (6.14) in (6.5) we get (Problem 6.6)

$$\begin{aligned} \hat{\boldsymbol{y}} &= X \hat{\boldsymbol{\theta}}_{\text{LS}} = X (X^T X)^{-1} X^T \boldsymbol{y} \\ &= U_l U_l^T \boldsymbol{y} = [\boldsymbol{u}_1, \ldots, \boldsymbol{u}_l] \begin{bmatrix} \boldsymbol{u}_1^T \boldsymbol{y} \\ \vdots \\ \boldsymbol{u}_l^T \boldsymbol{y} \end{bmatrix}, \end{aligned}$$

or

$$\boxed{\hat{\boldsymbol{y}} = \sum_{i=1}^{l} (\boldsymbol{u}_i^T \boldsymbol{y}) \boldsymbol{u}_i : \quad \text{LS estimate in terms of an orthonormal basis.}} \tag{6.19}$$

The latter represents the *projection* of $\boldsymbol{y}$ onto the column space of X, i.e., span$\{\boldsymbol{x}_1^c, \ldots, \boldsymbol{x}_l^c\}$ using a corresponding *orthonormal* basis, $\{\boldsymbol{u}_1, \ldots, \boldsymbol{u}_l\}$, to describe the subspace (see Fig. 6.3). Note that each $\boldsymbol{u}_i$, $i = 1, 2, \ldots, l$, lies in the space spanned by the columns of X as suggested from Eq. (6.18). In other words, the SVD of matrix X provides the orthonormal basis that describes the respective column space.

[4] This is also true for complex matrices, XX^H, $X^H X$.

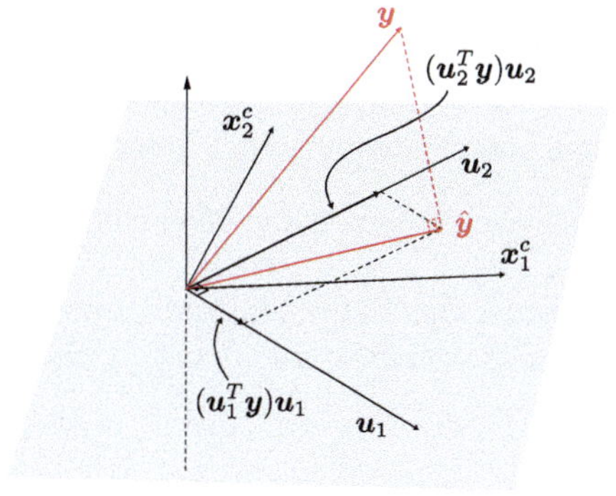

FIGURE 6.3

The eigenvectors u_1, u_2, lie in the column space of X, i.e., in the (shaded) plane span$\{x_1^c, x_2^c\}$, and form an orthonormal basis. Because $\hat{y}$ is the projection of y onto this subspace, it can be expressed as a linear combination of the two vectors of the orthonormal basis. The respective weights for the linear combination are $u_1^T y$ and $u_2^T y$.

We can further use the previous results to express the pseudoinverse in terms of the eigenvalues/eigenvectors of $X^T X$ ($X X^T$). It is easily shown that we can write

$$X^\dagger = (X^T X)^{-1} X^T = V_l D^{-1} U_l^T = \sum_{i=1}^{l} \frac{1}{\sigma_i} \boldsymbol{v}_i \boldsymbol{u}_i^T .$$

As a matter of fact, this is in line with the more general definition of a pseudoinverse in linear algebra, including matrices that are not full rank (i.e., $X^T X$ is not invertible), namely,

$$\boxed{X^\dagger := V_r D^{-1} U_r^T = \sum_{i=1}^{r} \frac{1}{\sigma_i} \boldsymbol{v}_i \boldsymbol{u}_i^T : \quad \text{pseudoinverse of a matrix of rank } r.} \tag{6.20}$$

In the case of matrices with $N < l$, and assuming that the rank of X is equal to N, it is readily verified that the previous generalized definition of the pseudoinverse is equivalent to

$$\boxed{X^\dagger = X^T (X X^T)^{-1} : \quad \text{pseudoinverse of a fat matrix } X.} \tag{6.21}$$

Note that a system with N equations and $l > N$ unknowns,

$$X\boldsymbol{\theta} = \boldsymbol{y},$$

has infinite solutions. Such systems are known as *underdetermined*, to be contrasted with the *overdetermined* systems for which $N > l$. It can be shown that for underdetermined systems, the solution $\boldsymbol{\theta} = X^\dagger \boldsymbol{y}$ is the one with the *minimum Euclidean norm*. We will consider the case of such systems of equations in more detail in Chapter 9, in the context of sparse models.

Remarks 6.1.

Here we summarize some important properties from linear algebra that are related to the SVD decomposition of the matrix. We will make use of these properties in various parts of the book later on.

- Computing the pseudoinverse using the SVD is numerically more robust than the direct method via the inversion of $(X^T X)^{-1}$.
- *k rank matrix approximation*: The best rank $k < r \leq \min(m, l)$ approximation matrix, $\hat{X} \in \mathbb{R}^{m \times l}$, of $X \in \mathbb{R}^{m \times l}$ in the Frobenius, $\|\cdot\|_F$, as well as in the spectral, $\|\cdot\|_2$, norms sense is given by (e.g., [26])

$$\hat{X} = \sum_{i=1}^{k} \sigma_i \boldsymbol{u}_i \boldsymbol{v}_i^T, \tag{6.22}$$

with the previously stated norms defined as (Problem 6.9)

$$\boxed{\|X\|_F := \sqrt{\sum_i \sum_j |X(i,j)|^2} = \sqrt{\sum_{i=1}^{r} \sigma_i^2} : \quad \text{Frobenius norm of } X,} \tag{6.23}$$

and

$$\boxed{\|X\|_2 := \sigma_1 : \quad \text{spectral norm of } X,} \tag{6.24}$$

where $\sigma_1 \geq \sigma_2 \geq \ldots \geq \sigma_r > 0$ are the singular values of X. In other words, $\hat{X}$ in (6.22) minimizes the error matrix norms,

$$\|X - \hat{X}\|_F \quad \text{and} \quad \|X - \hat{X}\|_2.$$

Moreover, it turns out that the approximation error is given by (Problems 6.10 and 6.11)

$$\|X - \hat{X}\|_F = \sqrt{\sum_{i=k+1}^{r} \sigma_i^2}, \quad \|X - \hat{X}\|_2 = \sigma_{k+1}.$$

This is also known as the *Eckart–Young–Mirsky* theorem.
- *Null and range spaces of X*: Let the rank of an $m \times l$ matrix X be equal to $r \leq \min\{m, l\}$. Then the following easily shown properties hold (Problem 6.13). The null space of X, $\mathcal{N}(X)$, defined as

$$\mathcal{N}(X) := \{\boldsymbol{x} \in \mathbb{R}^l : X\boldsymbol{x} = \boldsymbol{0}\}, \tag{6.25}$$

is also expressed as

$$\mathcal{N}(X) = \text{span}\{\boldsymbol{v}_{r+1}, \ldots, \boldsymbol{v}_l\}. \tag{6.26}$$

Furthermore, the range space of X, $\mathcal{R}(X)$, defined as

$$\mathcal{R}(X) := \{\boldsymbol{x} \in \mathbb{R}^l : \exists\, \boldsymbol{a} \text{ such as } X\boldsymbol{a} = \boldsymbol{x}\}, \tag{6.27}$$

is expressed as

$$\mathcal{R}(X) = \text{span}\{\boldsymbol{u}_1, \dots, \boldsymbol{u}_r\}. \tag{6.28}$$

- Everything that has been said before transfers to complex-valued data, trivially, by replacing transposition with the Hermitian one.

6.5 RIDGE REGRESSION: A GEOMETRIC POINT OF VIEW

In this section, we shed light on the ridge regression task from a different perspective. Instead of a dry optimization task, we are going to look at it by mobilizing statistical geometric arguments that the SVD decomposition offers to us.

Ridge regression was introduced in Chapter 3 as a means to impose bias on the LS solution and also as a major path to cope with overfitting and ill-conditioning problems. In ridge regression, the minimizer results as

$$\hat{\boldsymbol{\theta}}_R = \arg\min_{\boldsymbol{\theta}} \left\{ \|\boldsymbol{y} - X\boldsymbol{\theta}\|^2 + \lambda\|\boldsymbol{\theta}\|^2 \right\},$$

where $\lambda > 0$ is a user-defined parameter that controls the importance of the regularizing term. Taking the gradient with respect to $\boldsymbol{\theta}$ and equating to zero results in

$$\hat{\boldsymbol{\theta}}_R = (X^T X + \lambda I)^{-1} X^T \boldsymbol{y}. \tag{6.29}$$

Looking at (6.29), we readily observe (a) its "stabilizing" effect from the numerical point of view, when $X^T X$ is ill-conditioned and its inversion poses problems, and (b) its biasing effect on the (unbiased) LS solution. Note that ridge regression provides a solution even if $X^T X$ is not invertible, as is the case when $N < l$. Let us now employ the SVD expansion of (6.14) in (6.29). Assuming a full column rank matrix X, we obtain (Problem 6.14)

$$\hat{\boldsymbol{y}} = X\hat{\boldsymbol{\theta}}_R = U_l D(D^2 + \lambda I)^{-1} D U_l^T \boldsymbol{y},$$

or

$$\boxed{\hat{\boldsymbol{y}} = \sum_{i=1}^{l} \frac{\sigma_i^2}{\lambda + \sigma_i^2} (\boldsymbol{u}_i^T \boldsymbol{y})\boldsymbol{u}_i : \quad \text{ridge regression shrinks the weights.}} \tag{6.30}$$

Comparing (6.30) and (6.19), we observe that the components of the projection of $\boldsymbol{y}$ onto the $\text{span}\{\boldsymbol{u}_1, \dots, \boldsymbol{u}_l\}$ ($\text{span}\{\boldsymbol{x}_1^c, \dots, \boldsymbol{x}_l^c\}$) are *shrunk* with respect to their LS counterpart. Moreover, the shrinking level depends on the singular values σ_i; the smaller the value of σ_i, the higher the shrinking of the corresponding component. Let us now turn our attention to the investigation of the geometric interpretation of this algebraic finding. This small diversion will also provide more insight in the interpretation of $\boldsymbol{v}_i$ and $\boldsymbol{u}_i$, $i = 1, 2, \dots, l$, which appear in the SVD method.

Recall that $X^T X$ is a scaled version of the sample covariance matrix for centered regressors. Also, by the definition of the $\boldsymbol{v}_i$s, we have

$$(X^T X)\boldsymbol{v}_i = \sigma_i^2 \boldsymbol{v}_i, \quad i = 1, 2, \dots, l,$$

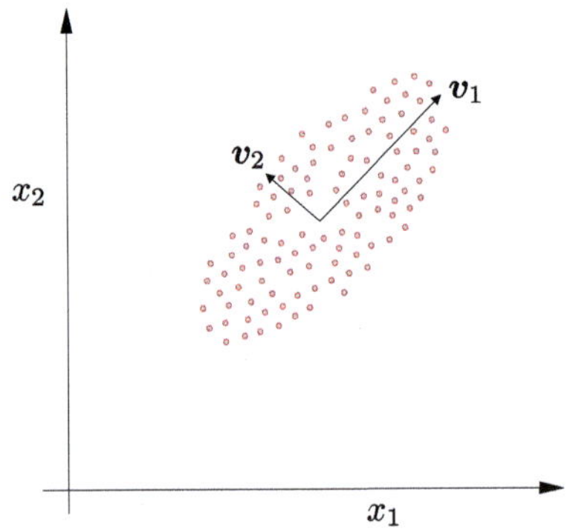

FIGURE 6.4

The singular vector v_1, which is associated with the singular value $\sigma_1 > \sigma_2$, points to the direction where most of the (variance) activity in the data space takes place. The variance in the direction of v_2 is smaller.

and in a compact form,

$$(X^T X)V_l = V_l \operatorname{diag}\{\sigma_1^2, \ldots, \sigma_l^2\} \Rightarrow$$
$$(X^T X) = V_l D^2 V_l^T = \sum_{i=1}^{l} \sigma_i^2 \boldsymbol{v}_i \boldsymbol{v}_i^T, \tag{6.31}$$

where the orthogonality property of V_l has been used for the inversion. Note that in (6.31), the (scaled) sample covariance matrix is written as a sum of rank one matrices, $\boldsymbol{v}_i \boldsymbol{v}_i^T$, each one weighted by the square of the respective singular value, σ_i^2. We are now close to revealing the physical/geometric meaning of the singular values. To this end, define

$$\boldsymbol{q}_j := X\boldsymbol{v}_j = \begin{bmatrix} \boldsymbol{x}_1^T \boldsymbol{v}_j \\ \vdots \\ \boldsymbol{x}_N^T \boldsymbol{v}_j \end{bmatrix} \in \mathbb{R}^N, \quad j = 1, 2, \ldots, l. \tag{6.32}$$

Note that $\boldsymbol{q}_j$ is a vector in the column space of X. Moreover, the respective squared norm of $\boldsymbol{q}_j$ is given by

$$\sum_{n=1}^{N} q_j^2(n) = \boldsymbol{q}_j^T \boldsymbol{q}_j = \boldsymbol{v}_j^T X^T X \boldsymbol{v}_j = \boldsymbol{v}_j^T \left(\sum_{i=1}^{l} \sigma_i^2 \boldsymbol{v}_i \boldsymbol{v}_i^T \right) \boldsymbol{v}_j = \sigma_j^2,$$

due to the orthonormality of the $\boldsymbol{v}_j$s. That is, σ_j^2 is equal to the (scaled) sample variance of the elements of $\boldsymbol{q}_j$. However, by the definition in (6.32), this is the sample variance of the projections of the input vectors (regressors), $\boldsymbol{x}_n$, $n = 1, 2, \ldots, N$, along the direction $\boldsymbol{v}_j$. The larger the value of σ_j, the larger the spread of the (input) data along the respective direction. This is shown in Fig. 6.4, where $\sigma_1 \gg \sigma_2$. From the variance point of view, $\boldsymbol{v}_1$ is the more *informative* direction, compared to $\boldsymbol{v}_2$. It is the direction where most of the activity takes place. This observation is at the heart of *dimensionality reduction*, which will be treated in more detail in Chapter 19. Moreover, from (6.18), we obtain

$$\boldsymbol{q}_j = X\boldsymbol{v}_j = \sigma_j \boldsymbol{u}_j. \tag{6.33}$$

In other words, $\boldsymbol{u}_j$ points in the direction of $\boldsymbol{q}_j$. Thus, (6.30) suggests that while projecting $\boldsymbol{y}$ onto the column space of X, the directions $\boldsymbol{u}_j$ associated with larger values of variance are weighted more heavily than the rest. *Ridge regression respects and assigns higher weights to the more informative directions, where most of the data activity takes place.* Alternatively, the less important directions, those associated with small data variance, are shrunk the most.

One final comment concerning ridge regression is that the ridge solutions are not invariant under scaling of the input variables. This becomes obvious by looking at the respective equations. Thus, in practice, often the input variables are standardized to unit variances.

PRINCIPAL COMPONENTS REGRESSION

We have just seen that the effect of the ridge regression is to enforce a shrinking rule on the parameters, which decreases the contribution of the less important of the components $\boldsymbol{u}_i$ in the respective summation. This can be considered as a soft shrinkage rule. An alternative path is to adopt a hard thresholding rule and keep only the m most significant directions, known as the *principal axes* or *directions*, and forget the rest by setting the respective weights equal to zero. Equivalently, we can write

$$\hat{\boldsymbol{y}} = \sum_{i=1}^{m} \hat{\theta}_i \boldsymbol{u}_i, \tag{6.34}$$

where

$$\hat{\theta}_i = \boldsymbol{u}_i^T \boldsymbol{y}, \quad i = 1, 2, \dots, m. \tag{6.35}$$

Furthermore, employing (6.18) we have

$$\hat{\boldsymbol{y}} = \sum_{i=1}^{m} \frac{\hat{\theta}_i}{\sigma_i} X\boldsymbol{v}_i, \tag{6.36}$$

or equivalently, the weights for the expansion of the solution in terms of the input data can be expressed as

$$\boldsymbol{\theta} = \sum_{i=1}^{m} \frac{\hat{\theta}_i}{\sigma_i} \boldsymbol{v}_i. \tag{6.37}$$

In other words, the prediction $\hat{\boldsymbol{y}}$ is performed in a subspace of the column space of X, which is spanned by the m principal axes, that is, the subspace where most of the data activity takes place.

6.6 THE RECURSIVE LEAST-SQUARES ALGORITHM

In previous chapters, we discussed the need for developing recursive algorithms that update the estimates every time a new pair of input–output observations is received. Solving the LS problem using a general purpose solver would amount to $\mathcal{O}(l^3)$ multiplications and additions (MADS), due to the involved matrix inversion. Also, $\mathcal{O}(Nl^2)$ operations are required to compute the (scaled) sample covariance matrix $X^T X$. In this section, the special structure of $X^T X$ will be taken into account in order to obtain a computationally efficient online scheme for the solution of the LS task. Moreover, when dealing with time-recursive techniques, one can also care for time variations of the statistical properties of the involved data. We will allow for such applications, and the sum of squared errors cost will be slightly modified in order to accommodate time-varying environments.

For the needs of the section, our notation will be slightly "enriched" and we will use explicitly the time index, n. Also, to be consistent with the online schemes discussed in Chapter 5, we will assume that the time starts at $n = 0$ and the received observations are $(y_n, \boldsymbol{x}_n)$, $n = 0, 1, 2, \ldots$. To this end, let us denote the input matrix at time n as

$$X_n^T = [\boldsymbol{x}_0, \boldsymbol{x}_1, \ldots, \boldsymbol{x}_n].$$

Moreover, the cost function in Eq. (6.1) is modified to involve a *forgetting factor*, $0 < \beta \leq 1$. The purpose of its presence is to help the cost function slowly forget past data samples by weighting heavier the more recent observations. This will equip the algorithm with the ability to track changes that occur in the underlying data statistics. Moreover, since we are interested in time-recursive solutions, starting from time $n = 0$, we are forced to introduce regularization. During the initial period, corresponding to time instants $n < l - 1$, the corresponding system of equations will be underdetermined and $X_n^T X_n$ is not invertible. Indeed, we have

$$X_n^T X_n = \sum_{i=0}^{n} \boldsymbol{x}_i \boldsymbol{x}_i^T.$$

In other words, $X_n^T X_n$ is the sum of rank one matrices. Hence, for $n < l - 1$ its rank is necessarily less than l, and it cannot be inverted (see Appendix A). For larger values of n, it can become full rank, provided that at least l of the input vectors are linearly independent, which is usually assumed to be the case. The previous arguments lead to the following modifications of the "conventional" least-squares, known as the *exponentially weighted sum of squared errors* cost function, minimized by

$$\boldsymbol{\theta}_n = \arg\min_{\boldsymbol{\theta}} \left(\sum_{i=0}^{n} \beta^{n-i} (y_i - \boldsymbol{\theta}^T \boldsymbol{x}_i)^2 + \lambda \beta^{n+1} \|\boldsymbol{\theta}\|^2 \right), \tag{6.38}$$

where β is a user-defined parameter very close to unity, for example, $\beta = 0.999$. In this way, the more recent samples are weighted heavier than the older ones. Note that the regularizing parameter has been made time-varying. This is because for large values of n, no regularization is required. Indeed, for $n > l$, matrix $X_n^T X_n$ becomes, in general, invertible. Moreover, recall from Chapter 3 that the use of regularization also takes precautions for overfitting. However, for very large values of $n \gg l$, this is not a problem, and one wishes to get rid of the imposed bias. The parameter $\lambda > 0$ is also a user-defined variable and its choice will be discussed later on.

Minimizing (6.38) results in

$$\Phi_n \boldsymbol{\theta}_n = \boldsymbol{p}_n, \tag{6.39}$$

where

$$\Phi_n = \sum_{i=0}^{n} \beta^{n-i} \boldsymbol{x}_i \boldsymbol{x}_i^T + \lambda \beta^{n+1} I \tag{6.40}$$

and

$$\boldsymbol{p}_n = \sum_{i=0}^{n} \beta^{n-i} \boldsymbol{x}_i y_i, \tag{6.41}$$

which for $\beta = 1$ coincides with the ridge regression.

TIME-ITERATIVE COMPUTATIONS

By the respective definitions, we have

$$\Phi_n = \beta \Phi_{n-1} + \boldsymbol{x}_n \boldsymbol{x}_n^T, \tag{6.42}$$

and

$$\boldsymbol{p}_n = \beta \boldsymbol{p}_{n-1} + \boldsymbol{x}_n y_n. \tag{6.43}$$

Recall Woodbury's matrix inversion formula (Appendix A.1),

$$(A + BD^{-1}C)^{-1} = A^{-1} - A^{-1}B(D + CA^{-1}B)^{-1}CA^{-1}.$$

Plugging it in (6.42), after the appropriate inversion and substitutions we obtain

$$\Phi_n^{-1} = \beta^{-1}\Phi_{n-1}^{-1} - \beta^{-1}\boldsymbol{k}_n \boldsymbol{x}_n^T \Phi_{n-1}^{-1}, \tag{6.44}$$

$$\boldsymbol{k}_n = \frac{\beta^{-1}\Phi_{n-1}^{-1}\boldsymbol{x}_n}{1 + \beta^{-1}\boldsymbol{x}_n^T \Phi_{n-1}^{-1}\boldsymbol{x}_n}. \tag{6.45}$$

The term $\boldsymbol{k}_n$ is known as the *Kalman gain*. For notational convenience, define

$$P_n = \Phi_n^{-1}.$$

Also, rearranging the terms in (6.45), we get

$$\boldsymbol{k}_n = \left(\beta^{-1}P_{n-1} - \beta^{-1}\boldsymbol{k}_n \boldsymbol{x}_n^T P_{n-1}\right)\boldsymbol{x}_n,$$

and taking into account (6.44) results in

$$\boldsymbol{k}_n = P_n \boldsymbol{x}_n. \tag{6.46}$$

TIME UPDATING OF THE PARAMETERS

From (6.39) and (6.43)–(6.45) we obtain

$$\begin{aligned}\boldsymbol{\theta}_n &= \left(\beta^{-1}P_{n-1} - \beta^{-1}\boldsymbol{k}_n\boldsymbol{x}_n^T P_{n-1}\right)\beta\boldsymbol{p}_{n-1} + P_n\boldsymbol{x}_n y_n \\ &= \boldsymbol{\theta}_{n-1} - \boldsymbol{k}_n\boldsymbol{x}_n^T\boldsymbol{\theta}_{n-1} + \boldsymbol{k}_n y_n,\end{aligned}$$

and finally,

$$\boldsymbol{\theta}_n = \boldsymbol{\theta}_{n-1} + \boldsymbol{k}_n e_n, \tag{6.47}$$

where,

$$e_n := y_n - \boldsymbol{\theta}_{n-1}^T\boldsymbol{x}_n. \tag{6.48}$$

The derived algorithm is summarized in Algorithm 6.1.

Note that the basic recursive update of the vector of the parameters follows the same rationale as the LMS and the gradient descent schemes that were discussed in Chapter 5. The updated estimate of the parameters $\boldsymbol{\theta}_n$ at time n equals that of the previous time instant, $n-1$, plus a correction term that is proportional to the error e_n. As a matter of fact, this is the *generic scheme* that we are going to meet for all the recursive algorithms in this book, including those used for the very "trendy" case of neural networks. The main difference from algorithm to algorithm lies in how one computes the multiplicative factor for the error. In the case of the LMS, this factor was equal to $\mu\boldsymbol{x}_n$. For the case of the RLS, this factor is $\boldsymbol{k}_n$. As we shall soon see, the RLS algorithm is closely related to an alternative to the gradient descent family of optimization algorithms, known as Newton's iterative optimization of cost functions.

Algorithm 6.1 (The RLS algorithm).

- Initialize
  - $\boldsymbol{\theta}_{-1} = \mathbf{0}$; any other value is also possible.
  - $P_{-1} = \lambda^{-1}I$; $\lambda > 0$ a user-defined variable.
  - Select β; close to 1.
- **For** $n = 0, 1, \ldots,$ **Do**
  - $e_n = y_n - \boldsymbol{\theta}_{n-1}^T\boldsymbol{x}_n$
  - $\boldsymbol{z}_n = P_{n-1}\boldsymbol{x}_n$
  - $\boldsymbol{k}_n = \frac{\boldsymbol{z}_n}{\beta + \boldsymbol{x}_n^T\boldsymbol{z}_n}$
  - $\boldsymbol{\theta}_n = \boldsymbol{\theta}_{n-1} + \boldsymbol{k}_n e_n$
  - $P_n = \beta^{-1}P_{n-1} - \beta^{-1}\boldsymbol{k}_n\boldsymbol{z}_n^T$
- **End For**

Remarks 6.2.

- The complexity of the RLS algorithm is of the order $\mathcal{O}(l^2)$ per iteration, due to the matrix-product operations. That is, there is an order of magnitude difference compared to the LMS and the other schemes that were discussed in Chapter 5. In other words, RLS does not scale well with dimensionality.

- The RLS algorithm shares similar numerical behavior with the Kalman filter, which was discussed in Section 4.10; P_n may lose its positive definite and symmetric nature, which then leads the algorithm to divergence. To remedy such a tendency, symmetry preserving versions of the RLS algorithm have been derived; see [65,68]. Note that the use of $\beta < 1$ has a beneficial effect on the error propagation [30,34]. In [58], it is shown that for $\beta = 1$ the error propagation mechanism is of a random walk type, and hence the algorithm is unstable. In [5], it is pointed out that due to numerical errors the term $\frac{1}{\beta + \boldsymbol{x}_n^T P_{n-1} \boldsymbol{x}_n}$ may become negative, leading to divergence. The numerical performance of the RLS becomes a more serious concern in implementations using limited precision, such as fixed point arithmetic. Compared to the LMS, RLS would require the use of higher precision implementations; otherwise, divergence may occur after a few iteration steps. This adds further to its computational disadvantage compared to the LMS.
- The choice of λ in the initialization step has been considered in [46]. The related theoretical analysis suggests that λ has a direct influence on the convergence speed, and it should be chosen so as to be a small positive for high signal-to-noise (SNR) ratios and a large positive constant for low SNRs.
- In [56], it has been shown that the RLS algorithm can be obtained as a special case of the Kalman filter.
- The main advantage of the RLS is that it converges to the steady state much faster than the LMS and the rest of the members of the LMS family. This can be justified by the fact that the RLS can been seen as an offspring of Newton's iterative optimization method.
- Distributed versions of the RLS have been proposed in [8,39,40].

6.7 NEWTON'S ITERATIVE MINIMIZATION METHOD

The gradient descent formulation was presented in Chapter 5. It was noted that it exhibits a linear convergence rate and a heavy dependence on the condition number of the Hessian matrix associated with the cost function. Newton's method is a way to overcome this dependence on the condition number and at the same time improve upon the rate of convergence toward the solution.

In Section 5.2, a first-order Taylor expansion was used around the current value $J\left(\boldsymbol{\theta}^{(i-1)}\right)$. Let us now consider a second-order Taylor expansion (assume $\mu_i = 1$) that involves second-order derivatives,

$$\begin{aligned} J\left(\boldsymbol{\theta}^{(i-1)} + \Delta\boldsymbol{\theta}^{(i)}\right) &= J\left(\boldsymbol{\theta}^{(i-1)}\right) + \left(\nabla J\left(\boldsymbol{\theta}^{(i-1)}\right)\right)^T \Delta\boldsymbol{\theta}^{(i)} \\ &\quad + \frac{1}{2}\left(\Delta\boldsymbol{\theta}^{(i)}\right)^T \nabla^2 J\left(\boldsymbol{\theta}^{(i-1)}\right)\Delta\boldsymbol{\theta}^{(i)}. \end{aligned}$$

Recall that the second derivative, $\nabla^2 J$, being the derivative of the gradient vector, is an $l \times l$ matrix (see Appendix A). Assuming that $\nabla^2 J\left(\boldsymbol{\theta}^{(i-1)}\right)$ to be positive definite (this is always the case if $J(\boldsymbol{\theta})$ is a strictly convex function[5]), the above turns out to be a convex quadratic function with respect to the

[5] See Chapter 8 for related definitions.

step $\Delta\boldsymbol{\theta}^{(i)}$; the latter is computed so as to *minimize* the above second-order approximation. Due to the convexity, there is a single minimum, which results by equating the corresponding gradient to $\mathbf{0}$, which results in

$$\Delta\boldsymbol{\theta}^{(i)} = -\left(\nabla^2 J\left(\boldsymbol{\theta}^{(i-1)}\right)\right)^{-1} \nabla J\left(\boldsymbol{\theta}^{(i-1)}\right). \tag{6.49}$$

Note that this is indeed a descent direction, because

$$\nabla^T J\left(\boldsymbol{\theta}^{(i-1)}\right)\Delta\boldsymbol{\theta}^{(i)} = -\nabla^T J\left(\boldsymbol{\theta}^{(i-1)}\right)\left(\nabla^2 J\left(\boldsymbol{\theta}^{(i-1)}\right)\right)^{-1} \nabla J\left(\boldsymbol{\theta}^{(i-1)}\right) < 0,$$

due to the positive definite nature of the Hessian[6]; equality to zero is achieved only at a minimum, where the gradient becomes zero. Thus, the iterative scheme takes the following form:

$$\boxed{\boldsymbol{\theta}^{(i)} = \boldsymbol{\theta}^{(i-1)} - \mu_i \left(\nabla^2 J\left(\boldsymbol{\theta}^{(i-1)}\right)\right)^{-1} \nabla J\left(\boldsymbol{\theta}^{(i-1)}\right): \quad \text{Newton's iterative scheme.}} \tag{6.50}$$

Fig. 6.5 illustrates the method. Note that if the cost function is quadratic, then the minimum is achieved in one iteration!

As is apparent from the derived recursion, the difference of Newton's schemes with the gradient descent family of algorithms lies in the step size. This is no more a scalar, i.e., μ_i. The step size involves the inverse of a matrix; that is, the second derivative of the cost function with respect to the parameters' vector. Here lies the power and at the same time the drawback of this type of algorithms. The use of the second derivative provides extra information concerning the local shape of the cost (see below) and as a consequence leads to faster convergence. At the same time, it increases substantially

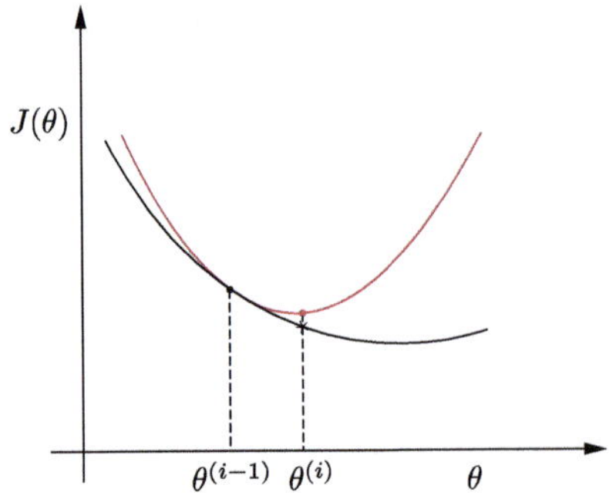

FIGURE 6.5

According to Newton's method, a local quadratic approximation of the cost function is considered (red curve), and the correction pushes the new estimate toward the minimum of this approximation. If the cost function is quadratic, then convergence can be achieved in one step.

6 Recall from Appendix A that if A is positive definite, then $\boldsymbol{x}^T A\boldsymbol{x} > 0, \forall\, \boldsymbol{x}$.

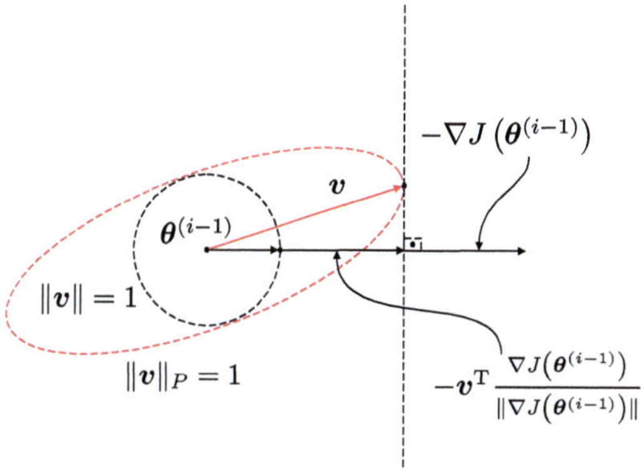

FIGURE 6.6

The graphs of the unit Euclidean (black circle) and quadratic (red ellipse) norms centered at $\theta^{(i-1)}$ are shown. In both cases, the goal is to move as far as possible in the direction of $-\nabla J(\theta^{(i-1)})$, while remaining at the ellipse (circle). The result is different for the two cases. The Euclidean norm corresponds to the gradient descent and the quadratic norm to Newton's method.

the computational complexity. Also, matrix inversions need always a careful treatment. The associated matrix may become ill-conditioned, its determinant gets small values, and in such cases numerical stability issues have to be carefully considered.

Observe that in the case of Newton's algorithm, the correction direction is not that of 180° with respect to $\nabla J(\theta^{(i-1)})$, as is the case for the gradient descent method. An alternative point of view is to look at (6.50) as the steepest descent direction under the following norm (see Section 5.2):

$$\|v\|_P = (v^T P v)^{1/2},$$

where P is a symmetric positive definite matrix. For our case, we set

$$P = \nabla^2 J\left(\theta^{(i-1)}\right).$$

Then searching for the respective normalized steepest descent direction, i.e.,

$$v = \arg\min_{z} z^T \nabla J\left(\theta^{(i-1)}\right)$$
$$\text{s.t.} \quad \|z\|_P^2 = 1,$$

results in the normalized vector pointing in the same direction as the one in (6.49) (Problem 6.15). For $P = I$, the gradient descent algorithm results. The geometry is illustrated in Fig. 6.6. Note that Newton's direction accounts for the *local shape* of the cost function.

The convergence rate for Newton's method is, in general, high and it becomes quadratic close to the solution. Assuming θ_* to be the minimum, quadratic convergence means that at each iteration i, the deviation from the optimum value follows the following pattern:

$$\boxed{\ln \ln \frac{1}{||\boldsymbol{\theta}^{(i)} - \boldsymbol{\theta}_*||^2} \propto i : \quad \text{quadratic convergence rate.}} \tag{6.51}$$

In contrast, for the linear convergence, the iterations approach the optimal according to

$$\boxed{\ln \frac{1}{||\boldsymbol{\theta}^{(i)} - \boldsymbol{\theta}_*||^2} \propto i : \quad \text{linear convergence rate.}} \tag{6.52}$$

Furthermore, the presence of the Hessian in the correction term remedies, to a large extent, the influence of the condition number of the Hessian matrix on the convergence [6] (Problem 6.16).

6.7.1 RLS AND NEWTON'S METHOD

The RLS algorithm can be rederived following Newton's iterative scheme applied to the MSE and adopting *stochastic approximation* arguments. Let

$$J(\boldsymbol{\theta}) = \frac{1}{2}\mathbb{E}\left[(\mathrm{y} - \boldsymbol{\theta}^T \mathbf{x})^2\right] = \frac{1}{2}\sigma_y^2 + \frac{1}{2}\boldsymbol{\theta}^T \Sigma_x \boldsymbol{\theta} - \boldsymbol{\theta}^T \boldsymbol{p},$$

or

$$-\nabla J(\boldsymbol{\theta}) = \boldsymbol{p} - \Sigma_x \boldsymbol{\theta} = \mathbb{E}[\mathbf{x}\mathrm{y}] - \mathbb{E}[\mathbf{x}\mathbf{x}^T]\boldsymbol{\theta} = \mathbb{E}[\mathbf{x}(\mathrm{y} - \mathbf{x}^T\boldsymbol{\theta})] = \mathbb{E}[\mathbf{x}\mathrm{e}],$$

and

$$\nabla^2 J(\boldsymbol{\theta}) = \Sigma_x.$$

Newton's iteration becomes

$$\boldsymbol{\theta}^{(i)} = \boldsymbol{\theta}^{(i-1)} + \mu_i \Sigma_x^{-1} \mathbb{E}[\mathbf{x}\mathrm{e}].$$

Following stochastic approximation arguments and replacing iteration steps with time updates and expectations with observations, we obtain

$$\boldsymbol{\theta}_n = \boldsymbol{\theta}_{n-1} + \mu_n \Sigma_x^{-1} \boldsymbol{x}_n e_n.$$

Let us now adopt the approximation,

$$\Sigma_x \simeq \frac{1}{n+1}\Phi_n = \left(\frac{1}{n+1}\lambda\beta^{n+1} I + \frac{1}{n+1}\sum_{i=0}^{n} \beta^{n-i} \boldsymbol{x}_i \boldsymbol{x}_i^T\right),$$

and set

$$\mu_n = \frac{1}{n+1}.$$

Then

$$\boldsymbol{\theta}_n = \boldsymbol{\theta}_{n-1} + \boldsymbol{k}_n e_n,$$

with

$$k_n = P_n x_n,$$

where

$$P_n = \left(\sum_{i=0}^{n} \beta^{n-i} x_i x_i^T + \lambda \beta^{n+1} I \right)^{-1},$$

which then, by using similar steps as for (6.42)–(6.44), leads to the RLS scheme.

Note that this point of view explains the fast converging properties of the RLS and its relative insensitivity to the condition number of the covariance matrix.

Remarks 6.3.

- In Section 5.5.2, it was pointed out that the LMS is optimal with respect to a min/max robustness criterion. However, this is not true for the RLS. It turns out that while LMS exhibits the best worst-case performance, the RLS is expected to have better performance on average [23].

6.8 STEADY-STATE PERFORMANCE OF THE RLS

Compared to the stochastic gradient techniques, which were considered in Chapter 5, we do not have to worry whether RLS converges and where it converges. The RLS computes the *exact* solution of the minimization task in (6.38) in an iterative way. Asymptotically and for $\beta = 1$, $\lambda = 0$ RLS solves the MSE optimization task. However, we have to consider its steady-state performance for $\beta \neq 1$. Even for the stationary case, $\beta \neq 1$ results in an excess mean-square error. Moreover, it is important to get a feeling of its tracking performance in time-varying environments. To this end, we adopt the same setting as the one followed in Section 5.12. We will not provide all the details of the proof, because this follows similar steps as in the LMS case. We will point out where differences arise and state the results. For the detailed derivation, the interested reader may consult [15,48,57]; in the latter one, the energy conservation theory is employed.

As in Chapter 5, we adopt the following models:

$$\mathrm{y}_n = \boldsymbol{\theta}_{o,n-1}^T \mathbf{x}_n + \eta_n \tag{6.53}$$

and

$$\boldsymbol{\theta}_{o,n} = \boldsymbol{\theta}_{o,n-1} + \boldsymbol{\omega}_n, \tag{6.54}$$

with

$$\mathbb{E}[\boldsymbol{\omega}_n \boldsymbol{\omega}_n^T] = \Sigma_\omega.$$

Hence, taking into account (6.53), (6.54), and the RLS iteration involving the respective random variables, we get

$$\boldsymbol{\theta}_n - \boldsymbol{\theta}_{o,n} = \boldsymbol{\theta}_{n-1} + \mathbf{k}_n \mathrm{e}_n - \boldsymbol{\theta}_{o,n-1} - \boldsymbol{\omega}_n,$$

Table 6.1 The steady-state excess MSE, for small values of μ and β.

Algorithm	Excess MSE, J_{exc}, at steady state
LMS	$\frac{1}{2}\mu\sigma_\eta^2\text{trace}\{\Sigma_x\} + \frac{1}{2}\mu^{-1}\text{trace}\{\Sigma_\omega\}$
APA	$\frac{1}{2}\mu\sigma_\eta^2\text{trace}\{\Sigma_x\}\mathbb{E}\left[\frac{q}{\Vert x\Vert^2}\right] + \frac{1}{2}\mu^{-1}\text{trace}\{\Sigma_x\}\text{trace}\{\Sigma_\omega\}$
RLS	$\frac{1}{2}(1-\beta)\sigma_\eta^2 l + \frac{1}{2}(1-\beta)^{-1}\text{trace}\{\Sigma_\omega\Sigma_x\}$

For $q=1$, the normalized LMS results. Under a Gaussian input assumption and for long system orders, l, in the APA, $\mathbb{E}\left[\frac{q}{\Vert x\Vert}\right] \simeq \frac{q}{\sigma_x^2(l-2)}$ [11].

or

$$\mathbf{c}_n := \boldsymbol{\theta}_n - \boldsymbol{\theta}_{o,n} = \mathbf{c}_{n-1} + P_n\mathbf{x}_n e_n - \boldsymbol{\omega}_n$$
$$= (I - P_n\mathbf{x}_n\mathbf{x}_n^T)\mathbf{c}_{n-1} + P_n\mathbf{x}_n\eta_n - \boldsymbol{\omega}_n,$$

which is the counterpart of (5.76). Note that the time indices for the input and the noise variables can be omitted, because their statistics is assumed to be time-invariant.

We adopt the same assumptions as in Section 5.12. In addition, we assume that P_n is changing slowly compared to $\mathbf{c}_n$. Hence, every time P_n appears inside an expectation, it is substituted by its mean $\mathbb{E}[P_n]$, i.e.,

$$\mathbb{E}[P_n] = \mathbb{E}[\Phi_n^{-1}],$$

where

$$\Phi_n = \lambda\beta^{n+1}I + \sum_{i=0}^{n}\beta^{n-i}\mathbf{x}_i\mathbf{x}_i^T$$

and

$$\mathbb{E}[\Phi_n] = \lambda\beta^{n+1}I + \frac{1-\beta^{n+1}}{1-\beta}\Sigma_x.$$

Assuming $\beta \simeq 1$, the variance at the steady state of Φ_n can be considered small and we can adopt the following approximation:

$$\mathbb{E}[P_n] \simeq [\mathbb{E}[\Phi_n]]^{-1} = \left[\beta^{n+1}\lambda I + \frac{1-\beta^{n+1}}{1-\beta}\Sigma_x\right]^{-1}.$$

Based on all the previously stated assumptions, repeating carefully the same steps as in Section 5.12, we end up with the result shown in Table 6.1, which holds for small values of β. For comparison reasons, the excess MSE is shown together with the values obtained for the LMS as well as the APA algorithms. In stationary environments, one simply sets $\Sigma_\omega = 0$.

According to Table 6.1, the following remarks are in order.

Remarks 6.4.

- For stationary environments, the performance of the RLS is independent of Σ_x. Of course, if one knows that the environment is stationary, then ideally $\beta = 1$ should be the choice. Yet recall that for $\beta = 1$, the algorithm has stability problems.
- Note that for small μ and $\beta \simeq 1$, there is an "equivalence" of $\mu \simeq 1 - \beta$, for the two parameters in the LMS and RLS. That is, larger values of μ are beneficial to the tracking performance of LMS, while smaller values of β are required for faster tracking; this is expected because the algorithm forgets the past.
- It is clear from Table 6.1 that an algorithm may converge to the steady state quickly, but it may not necessarily track fast. It all depends on the specific scenario. For example, under the modeling assumptions associated with Table 6.1, the optimal value μ_{opt} for the LMS (Section 5.12) is given by

$$\mu_{\text{opt}} = \sqrt{\frac{\text{trace}\{\Sigma_\omega\}}{\sigma_\eta^2 \text{trace}\{\Sigma_x\}}},$$

which corresponds to

$$J_{\min}^{LMS} = \sqrt{\sigma_\eta^2 \text{trace}\{\Sigma_x\}\text{trace}\{\Sigma_\omega\}}.$$

Optimizing with respect to β for the RLS, it is easily shown that

$$\beta_{\text{opt}} = 1 - \sqrt{\frac{\text{trace}\{\Sigma_\omega \Sigma_x\}}{\sigma_\eta^2 l}},$$

$$J_{\min}^{\text{RLS}} = \sqrt{\sigma_\eta^2 l \text{trace}\{\Sigma_\omega \Sigma_x\}}.$$

Hence, the ratio

$$\boxed{\frac{J_{\min}^{\text{LMS}}}{J_{\min}^{\text{RLS}}} = \sqrt{\frac{\text{trace}\{\Sigma_x\}\text{trace}\{\Sigma_\omega\}}{l\text{trace}\{\Sigma_\omega \Sigma_x\}}}}$$

depends on Σ_ω and Σ_x. *Sometimes LMS tracks better, yet in other problems RLS is the winner.* Having said that, it must be pointed out that the RLS *always* converges faster, and the difference in the rate, compared to the LMS, increases with the condition number of the input covariance matrix.

6.9 COMPLEX-VALUED DATA: THE WIDELY LINEAR RLS

Following similar arguments as in Section 5.7, let

$$\boldsymbol{\varphi} = \begin{bmatrix} \boldsymbol{\theta} \\ \boldsymbol{v} \end{bmatrix}, \quad \tilde{\boldsymbol{x}}_n = \begin{bmatrix} \boldsymbol{x}_n \\ \boldsymbol{x}_n^* \end{bmatrix},$$

with

$$\hat{y}_n = \boldsymbol{\varphi}^H \tilde{\boldsymbol{x}}_n.$$

The associated regularized cost becomes

$$J(\boldsymbol{\varphi}) = \sum_{i=0}^{n} \beta^{n-i}(y_n - \boldsymbol{\varphi}^H \tilde{\boldsymbol{x}}_n)(y_n - \boldsymbol{\varphi}^H \tilde{\boldsymbol{x}}_n)^* + \lambda \beta^{n+1} \boldsymbol{\varphi}^H \boldsymbol{\varphi},$$

or

$$J(\boldsymbol{\varphi}) = \sum_{i=0}^{n} \beta^{n-i}|y_n|^2 + \sum_{i=0}^{n} \beta^{n-i} \boldsymbol{\varphi}^H \tilde{\boldsymbol{x}}_n \tilde{\boldsymbol{x}}_n^H \boldsymbol{\varphi}$$
$$- \sum_{i=0}^{n} \beta^{n-i} y_n \tilde{\boldsymbol{x}}_n^H \boldsymbol{\varphi} - \sum_{i=0}^{n} \beta^{n-i} \boldsymbol{\varphi}^H \tilde{\boldsymbol{x}}_n y_n^* + \lambda \beta^{n+1} \boldsymbol{\varphi}^H \boldsymbol{\varphi}.$$

Taking the gradient with respect to $\boldsymbol{\varphi}^*$ and equating to zero, we obtain

$$\boxed{\tilde{\Phi}_n \boldsymbol{\varphi}_n = \tilde{\boldsymbol{p}}_n : \qquad \text{widely linear LS estimate,}} \tag{6.55}$$

where

$$\tilde{\Phi}_n = \beta^{n+1}\lambda I + \sum_{i=0}^{n} \beta^{n-i} \tilde{\boldsymbol{x}}_n \tilde{\boldsymbol{x}}_n^H, \tag{6.56}$$

$$\tilde{\boldsymbol{p}}_n = \sum_{i=0}^{n} \beta^{n-i} \tilde{\boldsymbol{x}}_n y_n^*. \tag{6.57}$$

Following similar steps as for the real-valued RLS, Algorithm 6.2 results, where $\tilde{P}_n := \tilde{\Phi}^{-1}$.

Algorithm 6.2 (The widely linear RLS algorithm).

- Initialize
  - $\boldsymbol{\varphi}_0 = \boldsymbol{0}$
  - $\tilde{P}_{-1} = \lambda^{-1} I$
  - Select β
- **For** $n = 0, 1, 2, \ldots,$ **Do**
  - $e_n = y_n - \boldsymbol{\varphi}_{n-1}^H \tilde{\boldsymbol{x}}_n$
  - $z_n = \tilde{P}_{n-1} \tilde{\boldsymbol{x}}_n$
  - $\boldsymbol{k}_n = \frac{z_n}{\beta + \tilde{\boldsymbol{x}}_n^H z_n}$
  - $\boldsymbol{\varphi}_n = \boldsymbol{\varphi}_{n-1} + \boldsymbol{k}_n e_n^*$
  - $\tilde{P}_n = \beta^{-1} \tilde{P}_{n-1} - \beta^{-1} \boldsymbol{k}_n \boldsymbol{z}_n^H$
- **End For**

Setting $\boldsymbol{v}_n = 0$ and replacing $\tilde{\boldsymbol{x}}_n$ with $\boldsymbol{x}_n$ and $\boldsymbol{\varphi}_n$ with $\boldsymbol{\theta}_n$, the linear complex-valued RLS results.

6.10 COMPUTATIONAL ASPECTS OF THE LS SOLUTION

The literature concerning the efficient solution of the LS method as well as the computationally efficient implementation of the RLS is huge. In this section, we will only highlight some of the basic directions that have been followed over the years. Most of the available software packages implement such efficient schemes.

A major direction in developing various algorithmic schemes was to cope with the numerical stability issues, as we have already discussed in Remarks 6.2. The main concern is to guarantee the symmetry and positive definiteness of Φ_n. The path followed toward this end is to work with the square root factors of Φ_n.

CHOLESKY FACTORIZATION

It is known from linear algebra that every positive definite symmetric matrix, such as Φ_n, accepts the following factorization:

$$\Phi_n = L_n L_n^T,$$

where L_n is lower triangular with positive entries along its diagonal. Moreover, this factorization is unique.

Concerning the LS task, one focuses on updating the factor L_n, instead of Φ_n, in order to improve numerical stability. Computation of the Cholesky factors can be achieved via a modified version of the Gauss elimination scheme [22].

QR FACTORIZATION

A better option for computing square factors of a matrix, from a numerical stability point of view, is via the QR decomposition method. To simplify the discussion, let us consider $\beta = 1$ and $\lambda = 0$ (no regularization). Then the positive definite (sample) covariance matrix can be factored as

$$\Phi_n = U_n^T U_n.$$

From linear algebra [22], we know that the $(n+1) \times l$ matrix U_n can be written as a product

$$U_n = Q_n R_n,$$

where Q_n is an $(n+1) \times (n+1)$ orthogonal matrix and R_n is an $(n+1) \times l$ upper triangular matrix. Note that R_n is related to the Cholesky factor L_n^T. It turns out that working with the QR factors of U_n is preferable, with respect to numerical stability, to working on the Cholesky factorization of Φ_n. QR factorization can be achieved via different paths:

- *Gram–Schmidt* orthogonalization of the input matrix columns. We have seen this path in Chapter 4 while discussing the lattice-ladder algorithm for solving the normal equations for the filtering case. Under the time shift property of the input signal, lattice-ladder-type algorithms have also been developed for the LS filtering task [31,32].
- *Givens rotations*: This has also been a popular line [10,41,52,54].

- *Householder reflections*: This line has been followed in [53,55]. The use of Householder reflections leads to a particularly robust scheme from a numerical point of view. Moreover, the scheme presents a high degree of parallelism, which can be exploited appropriately in a parallel processing environment.

A selection of related to QR factorization review papers is given in [2].

FAST RLS VERSIONS

Another line of intense activity, especially in the 1980s, was that of exploiting the special structure associated with the filtering task; that is, the input to the filter comprises the samples from a realization of a random signal/process. Abiding by our adopted notational convention, the input vector will now be denoted as $\boldsymbol{u}$ instead of $\boldsymbol{x}$. Also, for the needs of the discussion we will bring into the notation the order of the filter, m. In this case, the input vectors (regressors) at two successive time instants share all but two of their components. Indeed, for an mth-order system, we have

$$\boldsymbol{u}_{m,n} = \begin{bmatrix} u_n \\ \vdots \\ u_{n-m+1} \end{bmatrix}, \quad \boldsymbol{u}_{m,n-1} = \begin{bmatrix} u_{n-1} \\ \vdots \\ u_{n-m} \end{bmatrix},$$

and we can partition the input vector as

$$\boldsymbol{u}_{m,n} = \left[u_n \ \boldsymbol{u}_{m-1,n-1}\right]^T = \left[\boldsymbol{u}_{m-1,n}, \ u_{n-m+1}\right]^T.$$

This property is also known as *time shift* structure. Such a partition of the input vector leads to

$$\Phi_{m,n} = \begin{bmatrix} \sum_{i=0}^{n} u_i^2 & \sum_{i=0}^{n} \boldsymbol{u}_{m-1,i-1}^T u_i \\ \sum_{i=0}^{n} \boldsymbol{u}_{m-1,i-1} u_i & \Phi_{m-1,n-1} \end{bmatrix}$$
$$= \begin{bmatrix} \Phi_{m-1,n} & \sum_{i=0}^{n} \boldsymbol{u}_{m-1,i} u_{i-m+1} \\ \sum_{i=0}^{n} \boldsymbol{u}_{m-1,i}^T u_{i-m+1} & \sum_{i=0}^{n} u_{i-m+1}^2 \end{bmatrix}, \quad m = 2, 3, \ldots, l, \qquad (6.58)$$

where for complex variables transposition is replaced by the Hermitian one. Compare (6.58) with (4.60). The two partitions look alike, yet they are different. Matrix $\Phi_{m,n}$ is no longer Toeplitz. Its low partition is given in terms of $\Phi_{m-1,n-1}$. Such matrices are known as *near-to-Toeplitz*. All that is needed is to "correct" $\Phi_{m-1,n-1}$ back to $\Phi_{m-1,n}$ subtracting a rank one matrix, i.e.,

$$\Phi_{m-1,n-1} = \Phi_{m-1,n} - \boldsymbol{u}_{m-1,n} \boldsymbol{u}_{m-1,n}^T.$$

It turns out that such corrections, although they may slightly complicate the derivation, can still lead to computational efficient order recursive schemes, via the application of the matrix inversion lemma, as was the case in the MSE of Section 4.8. Such schemes have their origin in the pioneering PhD thesis of Martin Morf at Stanford [42]. Levinson-type, Schur-type, split-Levinson-type, and lattice-ladder algorithms have been derived for the LS case [3,27,28,43,44,60,61]. Some of the schemes noted previously under the QR factorization exploit the time shift structure of the input signal.

Besides the order recursive schemes, a number of fixed-order fast RLS-type schemes have been developed following the work in [33]. Recall from the definition of the Kalman gain in (6.45) that for an lth-order system we have

$$\begin{aligned} \boldsymbol{k}_{l+1,n} = \Phi_{l+1,n}^{-1}\boldsymbol{u}_{l+1,n} &= \begin{bmatrix} * & * \\ * & \Phi_{l,n-1} \end{bmatrix}^{-1} \begin{bmatrix} * \\ \boldsymbol{u}_{l,n-1} \end{bmatrix} \\ &= \begin{bmatrix} \Phi_{l,n} & * \\ * & * \end{bmatrix}^{-1} \begin{bmatrix} \boldsymbol{u}_{l,n} \\ * \end{bmatrix}, \end{aligned}$$

where $*$ denotes any value of the element. Without going into detail, the low partition can relate the Kalman gain of order l and time $n-1$ to the Kalman gain of order $l+1$ and time n (step up). Then the upper partition can be used to obtain the time update Kalman gain at order l and time n (step down). Such a procedure bypasses the need for matrix operations leading to $\mathcal{O}(l)$ RLS-type algorithms [7,9] with complexity $7l$ per time update. However, these versions turned out to be numerically unstable. Numerically stabilized versions, at only a small extra computational cost, were proposed in [5,58]. All the aforementioned schemes have also been developed for solving the (regularized) exponentially weighted LS cost function.

Besides this line, variants that obtain approximate solutions have been derived in an attempt to reduce complexity; these schemes use an approximation of the covariance or inverse covariance matrix [14,38]. The *fast Newton transversal filter* (FNTF) algorithm [45] approximates the inverse covariance matrix by a banded matrix of width p. Such a modeling has a specific physical interpretation. A banded inverse covariance matrix corresponds to an AR process of order p. Hence, if the input signal can sufficiently be modeled by an AR model, FNTF obtains a least-squares performance. Moreover, this performance is obtained at $\mathcal{O}(p)$ instead of $\mathcal{O}(l)$ computational cost. This can be very effective in applications where $p \ll l$. This is the case, for example, in audio conferencing, where the input signal is speech. Speech can efficiently be modeled by an AR of the order of 15, yet the filter order can be of a few hundred taps [49]. FNTF bridges the gap between LMS ($p=1$) and (fast) RLS ($p=l$). Moreover, FNTF builds upon the structure of the stabilized fast RLS. More recently, the banded inverse covariance matrix approximation has been successfully applied in spectral analysis [21].

More on the efficient LS schemes can be found in [15,17,20,24,29,57].

6.11 THE COORDINATE AND CYCLIC COORDINATE DESCENT METHODS

So far, we have discussed the gradient descent and Newton's method for optimization. We will conclude the discussion with a third method, which can also be seen as a member of the steepest descent family of methods. Instead of the Euclidean and quadratic norms, let us consider the following minimization task for obtaining the normalized descent direction:

$$\boldsymbol{v} = \arg\min_{\boldsymbol{z}} \boldsymbol{z}^T \nabla J, \tag{6.59}$$

$$\text{s.t.} \quad ||z||_1 = 1, \tag{6.60}$$

where $||\cdot||_1$ denotes the ℓ_1 norm, defined as

$$||z||_1 := \sum_{i=1}^{l} |z_i|.$$

Most of Chapter 9 is dedicated to this norm and its properties. Observe that this is not differentiable. Solving the minimization task (Problem 6.17) results in

$$\boldsymbol{v} = -\operatorname{sgn}\left((\nabla J)_k\right) \boldsymbol{e}_k,$$

where $\boldsymbol{e}_k$ is the direction of the coordinate corresponding to the component $(\nabla J)_k$ with the largest absolute value, i.e.,

$$|(\nabla J)_k| > |(\nabla J)_j|, \quad j \neq k,$$

and sgn$(\cdot)$ is the sign function. The geometry is illustrated in Fig. 6.7. In other words, the descent direction is along a single basis vector; that is, each time only a *single component* of $\boldsymbol{\theta}$ is updated. It is the component that corresponds to the directional derivative, $\left(\nabla J(\boldsymbol{\theta}^{(i-1)})\right)_k$, with the largest increase and the update rule becomes

$$\theta_k^{(i)} = \theta_k^{(i-1)} - \mu_i \frac{\partial J\left(\boldsymbol{\theta}^{(i-1)}\right)}{\partial \theta_k} : \quad \text{coordinate descent scheme,} \tag{6.61}$$

$$\theta_j^{(i)} = \theta_j^{(i-1)}, \quad j = 1, 2, \ldots, l, \ j \neq k. \tag{6.62}$$

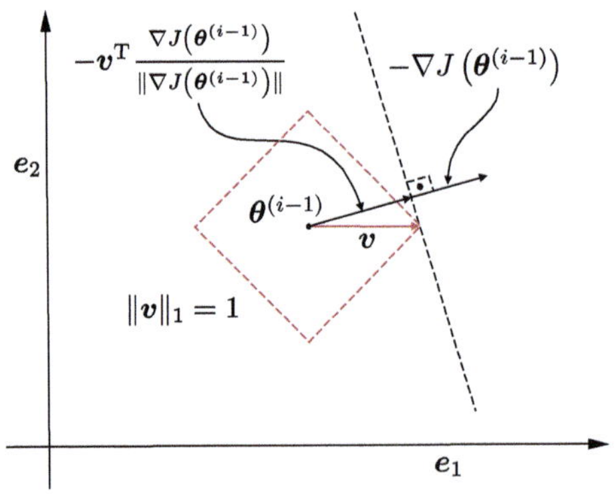

FIGURE 6.7

The unit norm $||\cdot||_1$ ball centered at $\boldsymbol{\theta}^{(i-1)}$ is a rhombus (in $\mathbb{R}^2$). The direction $\boldsymbol{e}_1$ is the one corresponding to the largest component of ∇J. Recall that the components of the vector ∇J are the respective directional derivatives.

Because only one component is updated at each iteration, this greatly simplifies the update mechanism. The method is known as *coordinate descent.*

Based on this rationale, a number of variants of the basic coordinate descent have been proposed. The *cyclic coordinate descent* (CCD) method in its simplest form entails a *cyclic* update with respect to one coordinate per iteration cycle; that is, at the ith iteration the following minimization is solved:

$$\theta_k^{(i)} := \arg\min_{\theta} J(\theta_1^{(i)}, \dots, \theta_{k-1}^{(i)}, \theta, \theta_{k+1}^{(i-1)}, \dots, \theta_l^{(i-1)}).$$

In words, all components but θ_k are assumed constant; those components $\theta_j,\ j < k$, are fixed to their updated values, $\theta_j^{(i)},\ j = 1, 2, \dots, k-1$, and the rest, $\theta_j,\ j = k+1, \dots, l$, to the available estimates, $\theta_j^{(i-1)}$, from the previous iteration. The nice feature of such a technique is that a simple closed form solution for the minimizer may be obtained. A revival of such techniques has happened in the context of sparse learning models (Chapter 10) [18,67]. Convergence issues of CCD have been considered in [36,62]. CCD algorithms for the LS task have also been considered in [66] and the references therein. Besides the basic CCD scheme, variants are also available, using different scenarios for the choice of the direction to be updated each time, in order to improve convergence, ranging from a random choice to a change of the coordinate systems, which is known as an *adaptive coordinate descent* scheme [35].

6.12 SIMULATION EXAMPLES

In this section, simulation examples are presented concerning the convergence and tracking performance of the RLS compared to algorithms of the gradient descent family, which have been derived in Chapter 5.

Example 6.1. The focus of this example is to demonstrate the comparative performance, with respect to the convergence rate of the RLS, NLMS, and APA algorithms, which have been discussed in Chapter 5. To this end, we generate data according to the regression model

$$y_n = \boldsymbol{\theta}_o^T \boldsymbol{x}_n + \eta_n,$$

where $\boldsymbol{\theta}_o \in \mathbb{R}^{200}$. Its elements are generated randomly according to the normalized Gaussian. The noise samples are i.i.d. generated via the zero mean Gaussian with variance equal to $\sigma_\eta^2 = 0.01$. The elements of the input vector are also i.i.d. generated via the normalized Gaussian. Using the generated samples $(y_n, \boldsymbol{x}_n),\ n = 0, 1, \dots,$ as the training sequence for all three previously stated algorithms, the convergence curves of Fig. 6.8 are obtained. The curves show the squared error in dBs $(10\log_{10}(e_n^2))$, averaged over 100 different realizations of the experiments, as a function of the time index n. The following parameters were used for the involved algorithms. (a) For the NLMS, we used $\mu = 1.2$ and $\delta = 0.001$; (b) for the APA, we used $\mu = 0.2$, $\delta = 0.001$, and $q = 30$; and (c) for the RLS, we used $\beta = 1$ and $\lambda = 0.1$. The parameters for the NLMS algorithm and the APA were chosen so that both algorithms converge to the same error floor. The improved performance of the APA concerning the convergence rate compared to the NLMS is readily seen. However, both algorithms fall short when compared to the RLS. Note that the RLS converges to lower error floor, because no forgetting factor

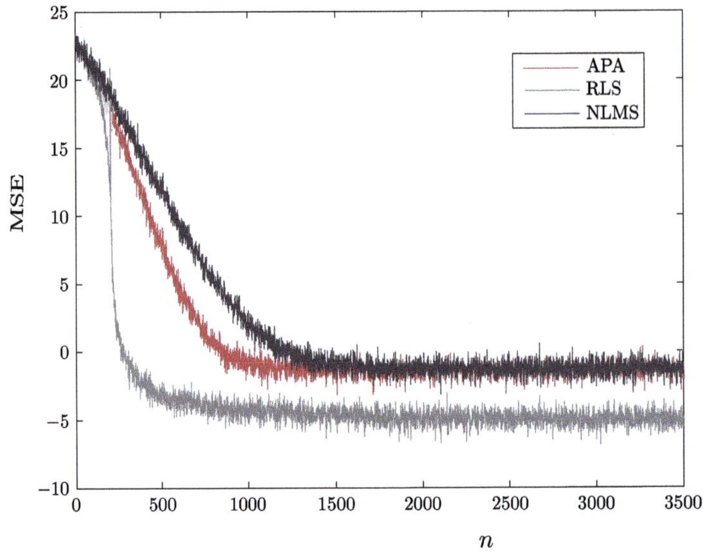

FIGURE 6.8

MSE curves as a function of the number of iterations for NLMS, APA, and RLS. The RLS converges faster and at lower error floor.

was used. To be consistent, a forgetting factor $\beta < 1$ should have been used in order for this algorithm to settle at the same error floor as the other two algorithms; this would have a beneficial effect on the convergence rate. However, having chosen $\beta = 1$, it is demonstrated that the RLS can converge really fast, even to lower error floors. On the other hand, this improved performance is obtained at substantial higher complexity. In case the input vector is part of a random process, and the special time shift structure can be exploited, as discussed in Section 6.10, the lower-complexity versions are at the disposal of the designer. A further comparative performance example, including another family of online algorithms, will be given in Chapter 8.

However, it has to be stressed that this notable advantage (between RLS- and LMS-type schemes) in convergence speed from the initial conditions to steady state may not be the case concerning the tracking performance, when the algorithms have to track time-varying environments. This is demonstrated next.

Example 6.2. This example focuses on the comparative tracking performance of the RLS and NLMS. Our goal is to demonstrate some cases where the RLS fails to do as well as the NLMS. Of course, it must be kept in mind that according to the theory, the comparative performance is very much dependent on the specific application.

For the needs of our example, let us mobilize the time-varying model of the parameters given in (6.54) in its more practical version and generate the data according to the following linear system:

$$y_n = \boldsymbol{x}_n^T \boldsymbol{\theta}_{o,n-1} + \eta_n, \tag{6.63}$$

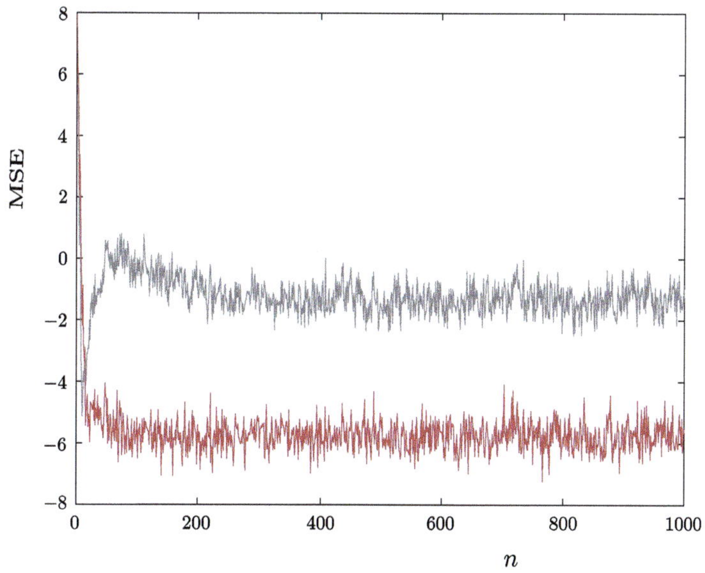

FIGURE 6.9

For a fast time-varying parameter model, the RLS (gray) fails to track it, in spite of its very fast initial convergence, compared to the NLMS (red).

where

$$\boldsymbol{\theta}_{o,n} = \alpha \boldsymbol{\theta}_{o,n-1} + \boldsymbol{\omega}_n,$$

with $\boldsymbol{\theta}_{o,n} \in \mathbb{R}^5$. It turns out that such a time-varying model is closely related (for the right choice of the involved parameters) to what is known in communications as a Rayleigh fading channel, if the parameters comprising $\boldsymbol{\theta}_{o,n}$ are thought to represent the impulse response of such a channel [57]. Rayleigh fading channels are very common and can adequately model a number of transmission channels in wireless communications. Playing with the parameter α and the variance of the corresponding noise source, $\boldsymbol{\omega}$, one can achieve fast or slow time-varying scenarios. In our case, we chose $\alpha = 0.97$ and the noise followed a Gaussian distribution of zero mean and covariance matrix $\Sigma_\omega = 0.1I$.

Concerning the data generation, the input samples were generated i.i.d. from a Gaussian $\mathcal{N}(0, 1)$, and the noise was also Gaussian of zero mean value and variance equal to $\sigma_\eta^2 = 0.01$. Initialization of the time-varying model ($\boldsymbol{\theta}_{o,0}$) was randomly done by drawing samples from $\mathcal{N}(0, 1)$.

Fig. 6.9 shows the obtained MSE curve as a function of the iterations for the NLMS and the RLS. For the RLS, the forgetting factor was set equal to $\beta = 0.995$ and for the NLMS, $\mu = 0.5$ and $\delta = 0.001$. Such a choice resulted in the best performance, for both algorithms, after extensive experimentation. The curves are the result of averaging out 200 independent runs.

Fig. 6.10 shows the resulting curves for medium and slow time-varying channels, corresponding to $\Sigma_\omega = 0.01I$ and $\Sigma_\omega = 0.001I$, respectively.

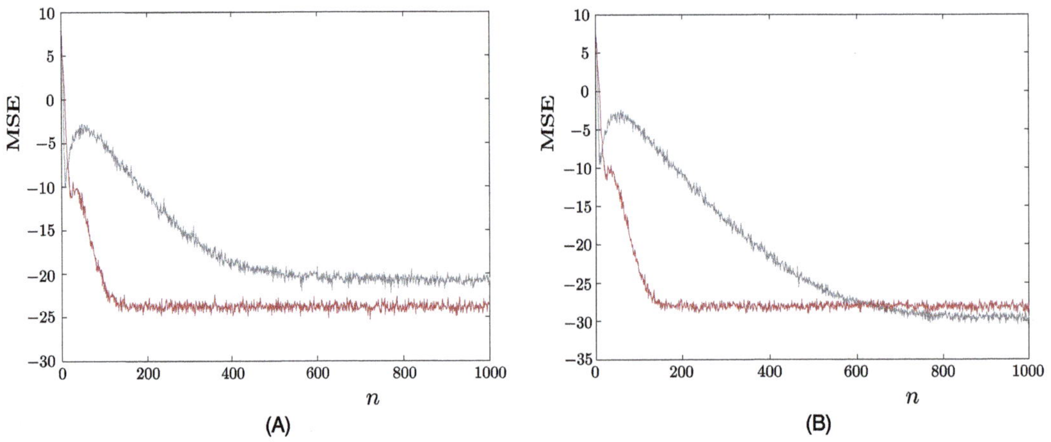

FIGURE 6.10

MSE curves as a function of iteration for (A) a medium and (B) a slow time-varying parameter model. The red curve corresponds to the NLMS and the gray one to the RLS.

6.13 TOTAL LEAST-SQUARES

In this section, the LS task will be formulated from a different perspective. Assume zero mean (centered) data and our familiar linear regression model, employing the observed samples,

$$\boldsymbol{y} = X\boldsymbol{\theta} + \boldsymbol{\eta},$$

as in Section 6.2. We have seen that the LS task is equivalent to (orthogonally) projecting $\boldsymbol{y}$ onto the $\text{span}\{\boldsymbol{x}_1^c, \dots, \boldsymbol{x}_l^c\}$ of the columns of X, hence making the error

$$\boldsymbol{e} = \boldsymbol{y} - \hat{\boldsymbol{y}}$$

orthogonal to the column space of X. Equivalently, this can be written as

$$\begin{aligned} &\text{minimize} \quad \|\boldsymbol{e}\|^2, \\ &\text{s.t.} \quad \boldsymbol{y} - \boldsymbol{e} \in \mathcal{R}(X), \end{aligned} \tag{6.64}$$

where $\mathcal{R}(X)$ is the range space of X (see Remarks 6.1 for the respective definition). Moreover, once $\hat{\boldsymbol{\theta}}_{\text{LS}}$ has been obtained, we can write

$$\hat{\boldsymbol{y}} = X\hat{\boldsymbol{\theta}}_{\text{LS}} = \boldsymbol{y} - \boldsymbol{e}$$

or

$$[X \vdots \boldsymbol{y} - \boldsymbol{e}] \begin{bmatrix} \hat{\boldsymbol{\theta}}_{\text{LS}} \\ -1 \end{bmatrix} = \boldsymbol{0}, \tag{6.65}$$

where $[X \vdots \boldsymbol{y} - \boldsymbol{e}]$ is the matrix that results after extending X by an extra column, $\boldsymbol{y} - \boldsymbol{e}$. Thus, all the points $(y_n - e_n, \boldsymbol{x}_n) \in \mathbb{R}^{l+1}$, $n = 1, 2, \ldots, N$, lie on the same hyperplane, crossing the origin, as shown in Fig. 6.11. In other words, in order to fit a hyperplane to the data, the LS method applies a correction e_n, $n = 1, 2, \ldots, N$, to the *output samples*. Thus, we have silently assumed that the regressors have been obtained via exact measurements and that the noise affects *only* the output observations.

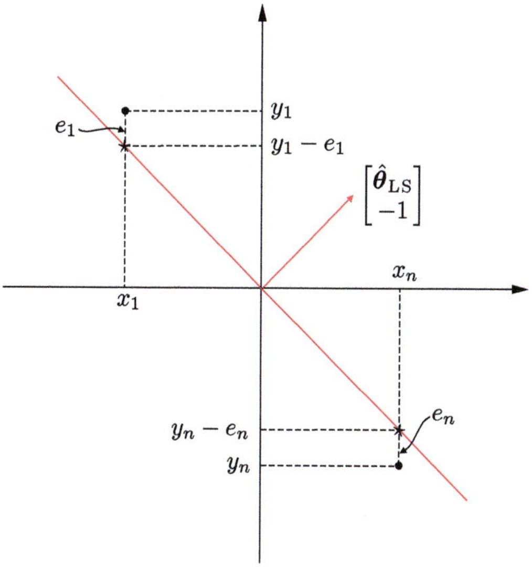

FIGURE 6.11

According to the LS method, *only* the output points y_n are corrected to $y_n - e_n$, so that the pairs $(y_n - e_n, x_n)$ lie on a hyperplane, crossing the origin for centered data. If the data are not centered, it crosses the centroid, $(\bar{y}, \bar{x})$.

In this section, the more general case will be considered, where we allow both the input (regressors) and the output variables to be perturbed by (unobserved) noise samples. Such a treatment has a long history, dating back to the 19th century [1]. The method remained in obscurity until it was revived 50 years later for two-dimensional models by Deming [13]; it is sometimes known as *Deming regression* (see also [19] for a historical overview). Such models are also known as *errors-in-variables* regression models.

Our kick-off point is the formulation in (6.64). Let $\boldsymbol{e}$ be the correction vector to be applied on $\boldsymbol{y}$ and E the correction matrix to be applied on X. The method of *total least-squares* (TLS) computes the unknown parameter vector by solving the following optimization task:

$$\begin{aligned} \text{minimize} \quad & \left\| [E \vdots \boldsymbol{e}] \right\|_F, \\ \text{s.t.} \quad & \boldsymbol{y} - \boldsymbol{e} \in \mathcal{R}(X - E). \end{aligned} \tag{6.66}$$

Recall (Remarks 6.1) that the Frobenius norm of a matrix is defined as the square root of the sum of squares of all its entries, and it is the direct generalization of the Euclidean norm defined for vectors.

Let us first focus on solving the task in (6.66), and we will comment on its geometric interpretation later on. The set of constraints in (6.66) can equivalently be written as

$$(X - E)\boldsymbol{\theta} = \boldsymbol{y} - \boldsymbol{e}. \tag{6.67}$$

Define

$$F := X - E, \tag{6.68}$$

and let $\boldsymbol{f}_i^T \in \mathbb{R}^l$, $i = 1, 2, \dots, N$, be the rows of F, i.e.,

$$F^T = [\boldsymbol{f}_1, \dots, \boldsymbol{f}_N],$$

and $\boldsymbol{f}_i^c \in \mathbb{R}^N$, $i = 1, 2, \dots, l$, the respective columns, i.e.,

$$F = [\boldsymbol{f}_1^c, \dots, \boldsymbol{f}_l^c].$$

Let also

$$\boldsymbol{g} := \boldsymbol{y} - \boldsymbol{e}. \tag{6.69}$$

Hence, (6.67) can be written in terms of the columns of F, i.e.,

$$\boxed{\theta_1 \boldsymbol{f}_1^c + \dots + \theta_l \boldsymbol{f}_l^c - \boldsymbol{g} = \boldsymbol{0}.} \tag{6.70}$$

Eq. (6.70) implies that the $l+1$ vectors, $\boldsymbol{f}_1^c, \dots, \boldsymbol{f}_l^c, \boldsymbol{g} \in \mathbb{R}^N$, are linearly dependent, which in turn dictates that

$$\boxed{\text{rank}\{[F \vdots \boldsymbol{g}]\} \le l.} \tag{6.71}$$

There is a subtle point here. The opposite is not necessarily true; that is, (6.71) does not necessarily imply (6.70). If the rank$\{F\} < l$, there is, in general, no $\boldsymbol{\theta}$ to satisfy (6.70). This can easily be verified, for example, by considering the extreme case where $\boldsymbol{f}_1^c = \boldsymbol{f}_2^c = \dots = \boldsymbol{f}_l^c$. Keeping that in mind, we need to impose some extra assumptions.

Assumptions:

1. The $N \times l$ matrix X is full rank. This implies that all its singular values are nonzero, and we can write (recall (6.14))

$$X = \sum_{i=1}^{l} \sigma_i \boldsymbol{u}_i \boldsymbol{v}_i^T,$$

where we have assumed that

$$\sigma_1 \ge \sigma_2 \ge \dots \ge \sigma_l > 0. \tag{6.72}$$

2. The $(N \times (l+1))$ matrix $[X \vdots y]$ is also full rank; hence

$$[X \vdots \boldsymbol{y}] = \sum_{i=1}^{l+1} \bar{\sigma}_i \bar{\boldsymbol{u}}_i \bar{\boldsymbol{v}}_i^T$$

with

$$\bar{\sigma}_1 \geq \bar{\sigma}_2 \geq \ldots \geq \bar{\sigma}_{l+1} > 0. \tag{6.73}$$

3. Assume that

$$\bar{\sigma}_{l+1} < \sigma_l.$$

As we will see soon, this guarantees the existence of a unique solution. If this condition is not valid, solutions can still exist; however, this corresponds to a degenerate case. Such solutions have been the subject of study in the related literature [37,64]. We will not deal with such cases here. Note that, in general, it can be shown that $\bar{\sigma}_{l+1} \leq \sigma_l$ [26]. Thus, our assumption demands *strict* inequality.

4. Assume that

$$\bar{\sigma}_l > \bar{\sigma}_{l+1}.$$

This condition will also be used in order to guarantee uniqueness of the solution.

We are now ready to solve the following optimization task:

$$\begin{aligned} &\underset{F,\boldsymbol{g}}{\text{minimize}} \quad \left\| [X \vdots \boldsymbol{y}] - [F \vdots \boldsymbol{g}] \right\|_F^2, \\ &\text{s.t.} \quad \text{rank}\{[F \vdots \boldsymbol{g}]\} = l. \end{aligned} \tag{6.74}$$

In words, compute the best, in the Frobenius norm sense, rank l approximation, $[F \vdots \boldsymbol{g}]$, to the (rank $l+1$) matrix $[X \vdots \boldsymbol{y}]$. We know from Remarks 6.1 that

$$[F \vdots \boldsymbol{g}] = \sum_{i=1}^{l} \bar{\sigma}_i \bar{\boldsymbol{u}}_i \bar{\boldsymbol{v}}_i^T, \tag{6.75}$$

and consequently

$$[E \vdots \boldsymbol{e}] = \bar{\sigma}_{l+1} \bar{\boldsymbol{u}}_{l+1} \bar{\boldsymbol{v}}_{l+1}^T, \tag{6.76}$$

with the corresponding Frobenius and spectral norms of the error matrix being equal to

$$\left\| E \vdots \boldsymbol{e} \right\|_F = \bar{\sigma}_{l+1} = \left\| E \vdots \boldsymbol{e} \right\|_2. \tag{6.77}$$

Note that the above choice is unique, because $\bar{\sigma}_{l+1} < \bar{\sigma}_l$.

So far, we have uniquely solved the task in (6.74). However, we still have to recover the estimate $\hat{\boldsymbol{\theta}}_{\text{TLS}}$, which will satisfy (6.70). In general, the existence of a unique vector cannot be guaranteed from the F and $\boldsymbol{g}$ given in (6.75). Uniqueness is imposed by assumption (3), which guarantees that the rank of F is equal to l.

Indeed, assume that the rank of F, k, is less than l, i.e., $k < l$. Let the best (in the Frobenius/spectral norm sense) rank k approximation of X be X_k and $X - X_k = E_k$. We know from Remarks 6.1 that

$$\|E_k\|_F = \sqrt{\sum_{i=k+1}^{l} \sigma_i^2} \geq \sigma_l.$$

Also, because E_k is the perturbation (error) associated with the best approximation, we have

$$\|E\|_F \geq \|E_k\|_F$$

or

$$\|E\|_F \geq \sigma_l.$$

However, from (6.77) we have

$$\bar{\sigma}_{l+1} = \|E \vdots \boldsymbol{e}\|_F \geq \|E\|_F \geq \sigma_l,$$

which violates assumption (3). Thus, $\text{rank}\{F\} = l$. Hence, there is a *unique* $\hat{\boldsymbol{\theta}}_{\text{TLS}}$, such that

$$[F \vdots \boldsymbol{g}] \begin{bmatrix} \hat{\boldsymbol{\theta}}_{\text{TLS}} \\ -1 \end{bmatrix} = \mathbf{0}. \tag{6.78}$$

In other words, $[\hat{\boldsymbol{\theta}}_{\text{TLS}}^T, -1]^T$ belongs to the null space of

$$\begin{bmatrix} F \vdots \boldsymbol{g} \end{bmatrix}, \tag{6.79}$$

which is a rank-deficient matrix; hence, its null space is of dimension one, and it is easily checked that it is spanned by $\bar{\boldsymbol{v}}_{l+1}$, leading to

$$\begin{bmatrix} \hat{\boldsymbol{\theta}}_{\text{TLS}} \\ -1 \end{bmatrix} = \frac{-1}{\bar{v}_{l+1}(l+1)} \bar{\boldsymbol{v}}_{l+1},$$

where $\bar{v}_{l+1}(l+1)$ is the last component of $\bar{\boldsymbol{v}}_{l+1}$. Moreover, it can be shown that (Problem 6.18)

$$\boxed{\hat{\boldsymbol{\theta}}_{\text{TLS}} = (X^T X - \bar{\sigma}_{l+1}^2 I)^{-1} X^T \boldsymbol{y} : \quad \text{total least-squares estimate.}} \tag{6.80}$$

Note that Assumption (3) guarantees that $X^T X - \bar{\sigma}_{l+1}^2 I$ is positive definite (think of why this is so).

GEOMETRIC INTERPRETATION OF THE TOTAL LEAST-SQUARES METHOD

From (6.67) and the definition of F in terms of its rows, $\boldsymbol{f}_1^T, \ldots, \boldsymbol{f}_N^T$, we get

$$\boldsymbol{f}_n^T \hat{\boldsymbol{\theta}}_{\text{TLS}} - g_n = 0, \quad n = 1, 2, \ldots, N, \tag{6.81}$$

or

$$\hat{\boldsymbol{\theta}}_{\text{TLS}}^T(\boldsymbol{x}_n - \boldsymbol{e}_n) - (y_n - e_n) = 0. \tag{6.82}$$

In words, both the regressors $\boldsymbol{x}_n$ and the outputs y_n are corrected in order for the points $(y_n - e_n, \boldsymbol{x}_n - \boldsymbol{e}_n)$, $n = 1, 2, \ldots, N$, to lie on a hyperplane in $\mathbb{R}^{l+1}$. Also, once such a hyperplane is computed and it is unique, it has an interesting interpretation. It is the hyperplane that *minimizes the total square distance* of all the training points $(y_n, \boldsymbol{x}_n)$ from it. Moreover, the corrected points $(y_n - e_n, \boldsymbol{x}_n - \boldsymbol{e}_n) = (g_n, \boldsymbol{f}_n)$, $n = 1, 2, \ldots, N$, are the *orthogonal projections* of the respective $(y_n, \boldsymbol{x}_n)$ training points onto this hyperplane. This is shown in Fig. 6.12.

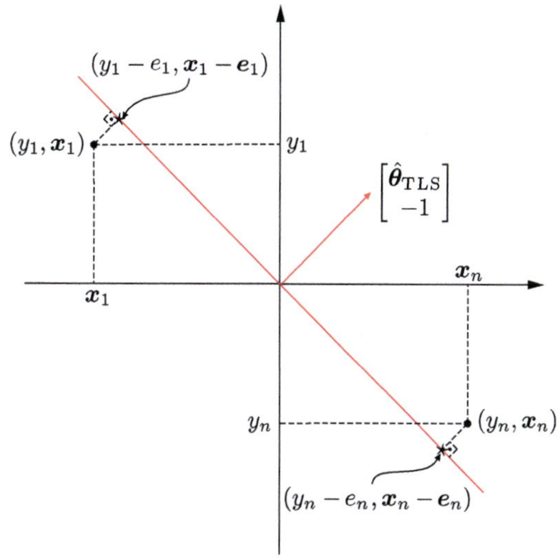

FIGURE 6.12

The total least-squares method corrects both the values of the output variable as well as the input vector so that the points, after the correction, lie on a hyperplane. The corrected points are the orthogonal projections of $(y_n, \ \boldsymbol{x}_n)$ on the respective hyperplane; for centered data, this crosses the origin. For noncentered data, it crosses the centroid $(\bar{y}, \bar{\boldsymbol{x}})$.

To prove the previous two claims, it suffices to show (Problem 6.19) that the direction of the hyperplane that minimizes the total distance from a set of points $(y_n, \boldsymbol{x}_n)$, $n = 1, 2, \ldots, N$, is that defined by $\bar{\boldsymbol{v}}_{l+1}$; the latter is the eigenvector associated with the smallest singular value of $[X \vdots \boldsymbol{y}]$, assuming $\bar{\sigma}_l > \bar{\sigma}_{l+1}$.

To see that $(g_n, \boldsymbol{f}_n)$ is the orthogonal projection of $(y_n, \boldsymbol{x}_n)$ on this hyperplane, recall that our task minimizes the following Frobenius norm:

$$||X - F \vdots \boldsymbol{y} - \boldsymbol{g}||_F^2 = \sum_{n=1}^{N} \left((y_n - g_n)^2 + ||\boldsymbol{x}_n - \boldsymbol{f}_n||^2 \right).$$

However, each one of the terms in the above summation is the Euclidean distance between the points $(y_n, \boldsymbol{x}_n)$ and $(g_n, \boldsymbol{f}_n)$, which is minimized if the latter is the orthogonal projection of the former on the hyperplane.

Remarks 6.5.

- The hyperplane defined by the TLS solution,

$$\left[\hat{\boldsymbol{\theta}}_{\text{TLS}}^T, -1\right]^T,$$

minimizes the total distance of all the points $(y_n, \boldsymbol{x}_n)$ from it. We know from geometry that the squared distance of each point from this hyperplane is given by

$$\frac{|\hat{\boldsymbol{\theta}}_{\text{TLS}}^T \boldsymbol{x}_n - y_n|^2}{||\boldsymbol{\theta}_{\text{TLS}}^T||^2 + 1}.$$

Thus, $\hat{\boldsymbol{\theta}}_{\text{TLS}}$ minimizes the following ratio:

$$\hat{\boldsymbol{\theta}}_{\text{TLS}} = \arg\min_{\boldsymbol{\theta}} \frac{||X\boldsymbol{\theta} - \boldsymbol{y}||^2}{||\boldsymbol{\theta}||^2 + 1}.$$

This is basically a normalized (weighted) version of the LS cost. Looking at it more carefully, TLS promotes vectors of larger norm. This could be seen as a "deregularizing" tendency of the TLS. From a numerical point of view, this can also be verified by (6.80). The matrix to be inverted for the TLS solution is more ill-conditioned than its LS counterpart. Robustness of the TLS can be improved via the use of regularization. Furthermore, extensions of TLS that employ other cost functions, in order to address the presence of outliers, have also been proposed.
- The TLS method has also been extended to deal with the more general case where $\boldsymbol{y}$ and $\boldsymbol{\theta}$ become matrices. For further reading, the interested reader can look at [37,64] and the references therein. A distributed algorithm for solving the TLS task in ad hoc sensor networks has been proposed in [4]. A recursive scheme for the efficient solution of the TLS task has appeared in [12].
- TLS has widely been used in a number of applications, such as computer vision [47], system identification [59], speech and image processing [25,51], and spectral analysis [63].

Example 6.3. To demonstrate the potential of the TLS to improve upon the performance of the LS estimator, in this example, we use noise not only in the input but also in the output samples. To this end, we generate randomly an input matrix, $X \in \mathbb{R}^{150\times 90}$, filling it with elements according to the normalized Gaussian, $\mathcal{N}(0, 1)$. In the sequel, we generate the vector $\boldsymbol{\theta}_o \in \mathbb{R}^{90}$ by randomly drawing samples also from the normalized Gaussian. The output vector is formed as

$$y = X\theta_o.$$

Then we generate a noise vector $\eta \in \mathbb{R}^{150}$, filling it with elements randomly drawn from $\mathcal{N}(0, 0.01)$, and form

$$\tilde{y} = y + \eta.$$

A noisy version of the input matrix is obtained as

$$\tilde{X} = X + E,$$

where E is filled randomly with elements by drawing samples from $\mathcal{N}(0, 0.2)$.

Using the generated $\tilde{y}$, X, $\tilde{X}$ and pretending that we do not know θ_o, the following three estimates are obtained for its value.

- Using the LS estimator (6.5) together with X and $\tilde{y}$, the average (over 10 different realizations) Euclidean distance of the obtained estimate $\hat{\theta}$ from the true one is equal to $\|\hat{\theta} - \theta_o\| = 0.0125$.
- Using the LS estimator (6.5) together with $\tilde{X}$ and $\tilde{y}$, the average (over 10 different realizations) Euclidean distance of the obtained estimate $\hat{\theta}$ from the true one is equal to $\|\hat{\theta} - \theta_o\| = 0.4272$.
- Using the TLS estimator (6.80) together with $\tilde{X}$ and $\tilde{y}$, the average (over 10 different realizations) Euclidean distance of the obtained estimate $\hat{\theta}$ from the true one is equal to $\|\hat{\theta} - \theta_o\| = 0.2652$.

Observe that using noisy input data, the LS estimator resulted in higher error compared to the TLS one. Note, however, that the successful application of the TLS presupposes that the assumptions that led to the TLS estimator are valid.

PROBLEMS

6.1 Show that if $A \in \mathbb{C}^{m\times m}$ is positive semidefinite, its trace is nonnegative.

6.2 Show that under (a) the independence assumption of successive observation vectors and (b) the presence of white noise independent of the input, the LS estimator is asymptotically distributed according to the normal distribution, i.e.,

$$\sqrt{N}(\boldsymbol{\theta} - \boldsymbol{\theta}_0) \longrightarrow \mathcal{N}(\mathbf{0}, \sigma_\eta^2 \Sigma_x^{-1}),$$

where σ_η^2 is the noise variance and Σ_x the covariance matrix of the input observation vectors, assuming that it is invertible.

6.3 Let $X \in \mathbb{C}^{m\times l}$. Then show that the two matrices

$$XX^H \text{ and } X^H X$$

have the same nonzero eigenvalues.

6.4 Show that if $X \in \mathbb{C}^{m\times l}$, then the eigenvalues of XX^H $(X^H X)$ are real and nonnegative. Moreover, show that if $\lambda_i \neq \lambda_j$, $\boldsymbol{v}_i \perp \boldsymbol{v}_j$.

6.5 Let $X \in \mathbb{C}^{m \times l}$. Show that if $\boldsymbol{v}_i$ is the normalized eigenvector of $X^H X$, corresponding to $\lambda_i \neq 0$, then the corresponding normalized eigenvector $\boldsymbol{u}_i$ of XX^H is given by

$$\boldsymbol{u}_i = \frac{1}{\sqrt{\lambda_i}} X \boldsymbol{v}_i.$$

6.6 Show Eq. (6.19).

6.7 Show that the right singular vectors, $\boldsymbol{v}_1, \dots, \boldsymbol{v}_r$, corresponding to the r singular values of a rank r matrix X solve the following iterative optimization task: compute $\boldsymbol{v}_k$, $k = 2, 3, \dots, r$, such that

$$\text{minimize} \quad \frac{1}{2}||X\boldsymbol{v}||^2, \tag{6.83}$$

$$\text{subject to} \quad ||\boldsymbol{v}||^2 = 1, \tag{6.84}$$

$$\boldsymbol{v} \perp \{\boldsymbol{v}_1, \dots, \boldsymbol{v}_{k-1}\}, \; k \neq 1, \tag{6.85}$$

where $|| \cdot ||$ denotes the Euclidean norm.

6.8 Show that projecting the rows of X onto the k rank subspace, $V_k = \text{span}\{\boldsymbol{v}_1, \dots, \boldsymbol{v}_k\}$, results in the largest variance, compared to any other k-dimensional subspace, Z_k.

6.9 Show that the squared Frobenius norm is equal to the sum of the squared singular values.

6.10 Show that the best k rank approximation of a matrix X or rank $r > k$, in the Frobenius norm sense, is given by

$$\hat{X} = \sum_{i=1}^{k} \sigma_i \boldsymbol{u}_i \boldsymbol{v}_i^T,$$

where σ_i are the singular values and $\boldsymbol{v}_i, \boldsymbol{u}_i$, $i = 1, 2, \dots, r$, are the right and left singular vectors of X, respectively. Then show that the approximation error is given by

$$\sqrt{\sum_{i=k+1}^{r} \sigma_i^2}.$$

6.11 Show that $\hat{X}$, as given in Problem 6.10, also minimizes the spectral norm and that

$$||X - \hat{X}||_2 = \sigma_{k+1}.$$

6.12 Show that the Frobenius and spectral norms are unaffected by multiplication with orthogonal matrices, i.e.,

$$\|X\|_F = \|QXU\|_F$$

and

$$\|X\|_2 = \|QXU\|_2,$$

if $QQ^T = UU^T = I$.

6.13 Show that the null and range spaces of an $m \times l$ matrix X of rank r are given by

$$\mathcal{N}(X) = \text{span}\{\boldsymbol{v}_{r+1}, \ldots, \boldsymbol{v}_l\},$$
$$\mathcal{R}(X) = \text{span}\{\boldsymbol{u}_1, \ldots, \boldsymbol{u}_r\},$$

where

$$X = [\boldsymbol{u}_1, \ldots, \boldsymbol{u}_m] \begin{bmatrix} D & O \\ O & O \end{bmatrix} \begin{bmatrix} \boldsymbol{v}_1^T \\ \vdots \\ \boldsymbol{v}_l^T \end{bmatrix}.$$

6.14 Show that for the ridge regression

$$\hat{\boldsymbol{y}} = \sum_{i=1}^{l} \frac{\sigma_i^2}{\lambda + \sigma_i^2} (\boldsymbol{u}_i^T \boldsymbol{y}) \boldsymbol{u}_i.$$

6.15 Show that the normalized steepest descent direction of $J(\boldsymbol{\theta})$ at a point $\boldsymbol{\theta}_0$ for the quadratic norm $\|v\|_P$ is given by

$$\boldsymbol{v} = -\frac{P^{-1}\nabla J(\boldsymbol{\theta}_0)}{||P^{-1}\nabla J(\boldsymbol{\theta}_0)||_P}.$$

6.16 Explain why the convergence of Newton's iterative minimization method is relatively insensitive to the Hessian matrix.
Hint: Let P be a positive definite matrix. Define a change of variables,

$$\tilde{\boldsymbol{\theta}} = P^{\frac{1}{2}} \boldsymbol{\theta},$$

and carry out gradient descent minimization based on the new variable.

6.17 Show that the steepest descent direction $\boldsymbol{v}$ of $J(\boldsymbol{\theta})$ at a point $\boldsymbol{\theta}_0$, constrained to

$$\|\boldsymbol{v}\|_1 = 1,$$

is given by $\boldsymbol{e}_k$, where $\boldsymbol{e}_k$ is the standard basis vector in the direction k, such that

$$|(\nabla J(\boldsymbol{\theta}_0))_k| > |(\nabla J(\boldsymbol{\theta}_0))_j|, \quad k \neq j.$$

6.18 Show that the TLS solution is given by

$$\hat{\boldsymbol{\theta}} = \left(X^T X - \bar{\sigma}_{l+1}^2 I\right)^{-1} X^T \boldsymbol{y},$$

where $\bar{\sigma}_{l+1}$ is the smallest singular value of $\left[X \vdots \boldsymbol{y}\right]$.

6.19 Given a set of centered data points, $(y_n, \boldsymbol{x}_n) \in \mathbb{R}^{l+1}$, derive a hyperplane

$$\boldsymbol{a}^T \boldsymbol{x} + y = 0,$$

which crosses the origin, such that the total square distance of all the points from it is minimum.

MATLAB® EXERCISES

6.20 Consider the regression model

$$y_n = \boldsymbol{\theta}_o^T \boldsymbol{x}_n + \eta_n,$$

where $\boldsymbol{\theta}_o \in \mathbb{R}^{200}$ ($l = 200$) and the coefficients of the unknown vector are obtained randomly via the Gaussian distribution $\mathcal{N}(0, 1)$. The noise samples are also i.i.d., according to a Gaussian of zero mean and variance $\sigma_\eta^2 = 0.01$. The input sequence is a white noise one, i.i.d. generated via the Gaussian $\mathcal{N}(0, 1)$.

Using as training data the samples $(y_n, \boldsymbol{x}_n) \in \mathbb{R} \times \mathbb{R}^{200}$, $n = 1, 2, \ldots$, run the APA (Algorithm 5.2), the NLMS algorithm (Algorithm 5.3), and the RLS algorithm (Algorithm 6.1) to estimate the unknown $\boldsymbol{\theta}_o$.

For the APA, choose $\mu = 0.2$, $\delta = 0.001$, and $q = 30$. Furthermore, in the NLMS set $\mu = 1.2$ and $\delta = 0.001$. Finally, for the RLS set the forgetting factor β equal to 1. Run 100 independent experiments and plot the average error per iteration in dBs, i.e., $10\log_{10}(e_n^2)$, where $e_n^2 = (y_n - \boldsymbol{x}_n^T \boldsymbol{\theta}_{n-1})^2$. Compare the performance of the algorithms.

Keep playing with different parameters and study their effect on the convergence speed and the error floor in which the algorithms converge.

6.21 Consider the linear system

$$y_n = \boldsymbol{x}_n^T \boldsymbol{\theta}_{o,n-1} + \eta_n, \tag{6.86}$$

where $l = 5$ and the unknown vector is time-varying. Generate the unknown vector with respect to the following model:

$$\boldsymbol{\theta}_{o,n} = \alpha \boldsymbol{\theta}_{o,n-1} + \boldsymbol{\omega}_n,$$

where $\alpha = 0.97$ and the coefficients of $\boldsymbol{\omega}_n$ are i.i.d. drawn from the Gaussian distribution, with zero mean and variance equal to 0.1. Generate the initial value $\boldsymbol{\theta}_{o,0}$ with respect to $\mathcal{N}(0, 1)$.

The noise samples are i.i.d., having zero mean and variance equal to 0.001. Furthermore, generate the input samples so that they follow the Gaussian distribution $\mathcal{N}(0, 1)$. Compare the performance of the NLMS and RLS algorithms. For the NLMS, set $\mu = 0.5$ and $\delta = 0.001$. For the RLS, set the forgetting factor β equal to 0.995. Run 200 independent experiments and plot the average error per iteration in dBs, i.e., $10\log_{10}(e_n^2)$, with $e_n^2 = (y_n - \boldsymbol{x}_n^T \boldsymbol{\theta}_{n-1})^2$. Compare the performance of the algorithms.

Keep the same parameters, but set the variance associated with $\boldsymbol{\omega}_n$ equal to 0.01, 0.001. Play with different values of the parameters and the variance of the noise $\boldsymbol{\omega}$.

6.22 Generate an 150×90 matrix X, the entries of which follow the Gaussian distribution $\mathcal{N}(0, 1)$. Generate the vector $\boldsymbol{\theta}_o \in \mathbb{R}^{90}$. The coefficients of this vector are i.i.d. obtained, also, via the Gaussian $\mathcal{N}(0, 1)$. Compute the vector $\boldsymbol{y} = X\boldsymbol{\theta}_o$. Add a 90×1 noise vector, $\boldsymbol{\eta}$, to $\boldsymbol{y}$ in order to generate $\tilde{\boldsymbol{y}} = \boldsymbol{y} + \boldsymbol{\eta}$. The elements of $\boldsymbol{\eta}$ are generated via the Gaussian $\mathcal{N}(0, 0.01)$. In the sequel, add a 150×90 noise matrix, E, so as to produce $\tilde{X} = X + E$; the elements of E are generated according to the Gaussian $\mathcal{N}(0, 0.2)$. Compute the LS estimate via (6.5) by employing (a) the true input matrix X and the noisy output $\tilde{\boldsymbol{y}}$; and (b) the noisy input matrix $\tilde{X}$ and the noisy output $\tilde{\boldsymbol{y}}$.

In the sequel, compute the TLS estimate via (6.80) using the noisy input matrix $\tilde{X}$ and the noisy output $\tilde{\boldsymbol{y}}$.

Repeat the experiments a number of times and compute the average Euclidean distances between the obtained estimates for the previous three cases and the true parameter vector $\boldsymbol{\theta}_o$.

Play with different noise levels and comment on the results.

REFERENCES

[1] R.J. Adcock, Note on the method of least-squares, Analyst 4 (6) (1877) 183–184.

[2] J.A. Apolinario Jr. (Ed.), QRD-RLS Adaptive Filtering, Springer, New York, 2009.

[3] K. Berberidis, S. Theodoridis, Efficient symmetric algorithms for the modified covariance method for autoregressive spectral analysis, IEEE Trans. Signal Process. 41 (1993) 43.

[4] A. Bertrand, M. Moonen, Consensus-based distributed total least-squares estimation in ad hoc wireless sensor networks, IEEE Trans. Signal Process. 59 (5) (2011) 2320–2330.

[5] J.L. Botto, G.V. Moustakides, Stabilizing the fast Kalman algorithms, IEEE Trans. Acoust. Speech Signal Process. 37 (1989) 1344–1348.

[6] S. Boyd, L. Vandenberghe, Convex Optimization, Cambridge University Press, 2004.

[7] G. Carayannis, D. Manolakis, N. Kalouptsidis, A fast sequential algorithm for least-squares filtering and prediction, IEEE Trans. Acoust. Speech Signal Process. 31 (1983) 1394–1402.

[8] F.S. Cattivelli, C.G. Lopes, A.H. Sayed, Diffusion recursive least-squares for distributed estimation over adaptive networks, IEEE Trans. Signal Process. 56 (5) (2008) 1865–1877.

[9] J.M. Cioffi, T. Kailath, Fast recursive-least-squares transversal filters for adaptive filtering, IEEE Trans. Acoust. Speech Signal Process. 32 (1984) 304–337.

[10] J.M. Cioffi, The fast adaptive ROTOR's RLS algorithm, IEEE Trans. Acoust. Speech Signal Process. 38 (1990) 631–653.

[11] M.H. Costa, J.C.M. Bermudez, An improved model for the normalized LMS algorithm with Gaussian inputs and large number of coefficients, in: Proceedings, IEEE Conference in Acoustics Speech and Signal Processing, 2002, pp. 1385–1388.

[12] C.E. Davila, An efficient recursive total least-squares algorithm for FIR adaptive filtering, IEEE Trans. Signal Process. 42 (1994) 268–280.

[13] W.E. Deming, Statistical Adjustment of Data, J. Wiley and Sons, 1943.

[14] P.S.R. Diniz, M.L.R. De Campos, A. Antoniou, Analysis of LMS-Newton adaptive filtering algorithms with variable convergence factor, IEEE Trans. Signal Process. 43 (1995) 617–627.

[15] P.S.R. Diniz, Adaptive Filtering: Algorithms and Practical Implementation, third ed., Springer, 2008.

[16] Y.C. Eldar, Minimax MSE estimation of deterministic parameters with noise covariance uncertainties, IEEE Trans. Signal Process. 54 (2006) 138–145.

[17] B. Farhang-Boroujeny, Adaptive Filters: Theory and Applications, J. Wiley, NY, 1999.

[18] J. Friedman, T. Hastie, H. Hofling, R. Tibshirani, Pathwise coordinate optimization, Ann. Appl. Stat. 1 (2007) 302–332.

[19] J.W. Gillard, A Historical Review of Linear Regression With Errors in Both Variables, Technical Report, University of Cardiff, School of Mathematics, 2006.

[20] G. Glentis, K. Berberidis, S. Theodoridis, Efficient least-squares adaptive algorithms for FIR transversal filtering, IEEE Signal Process. Mag. 16 (1999) 13–42.

[21] G.O. Glentis, A. Jakobsson, Superfast approximative implementation of the IAA spectral estimate, IEEE Trans. Signal Process. 60 (1) (2012) 472–478.

[22] G.H. Golub, C.F. Van Loan, Matrix Computations, The Johns Hopkins University Press, 1983.

[23] B. Hassibi, A.H. Sayed, T. Kailath, H^∞ optimality of the LMS algorithm, IEEE Trans. Signal Process. 44 (1996) 267–280.

[24] S. Haykin, Adaptive Filter Theory, fourth ed., Prentice Hall, NJ, 2002.

[25] K. Hermus, W. Verhelst, P. Lemmerling, P. Wambacq, S. Van Huffel, Perceptual audio modeling with exponentially damped sinusoids, Signal Process. 85 (1) (2005) 163–176.

[26] R.A. Horn, C.R. Johnson, Matrix Analysis, second ed., Cambridge University Press, 2013.

[27] N. Kalouptsidis, G. Carayannis, D. Manolakis, E. Koukoutsis, Efficient recursive in order least-squares FIR filtering and prediction, IEEE Trans. Acoust. Speech Signal Process. 33 (1985) 1175–1187.

[28] N. Kalouptsidis, S. Theodoridis, Parallel implementation of efficient LS algorithms for filtering and prediction, IEEE Trans. Acoust. Speech Signal Process. 35 (1987) 1565–1569.

[29] N. Kalouptsidis, S. Theodoridis, Adaptive System Identification and Signal Processing Algorithms, Prentice Hall, 1993.
[30] A.P. Liavas, P.A. Regalia, On the numerical stability and accuracy of the conventional recursive least-squares algorithm, IEEE Trans. Signal Process. 47 (1999) 88–96.
[31] F. Ling, D. Manolakis, J.G. Proakis, Numerically robust least-squares lattice-ladder algorithms with direct updating of the reflection coefficients, IEEE Trans. Acoust. Speech Signal Process. 34 (1986) 837–845.
[32] D.L. Lee, M. Morf, B. Friedlander, Recursive least-squares ladder estimation algorithms, IEEE Trans. Acoust. Speech Signal Process. 29 (1981) 627–641.
[33] L. Ljung, M. Morf, D. Falconer, Fast calculation of gain matrices for recursive estimation schemes, Int. J. Control 27 (1984) 304–337.
[34] S. Ljung, L. Ljung, Error propagation properties of recursive least-squares adaptation algorithms, Automatica 21 (1985) 157–167.
[35] I. Loshchilov, M. Schoenauer, M. Sebag, Adaptive coordinate descent, in: Proceedings Genetic and Evolutionary Computation Conference, GECCO, ACM Press, 2011, pp. 885–892.
[36] Z. Luo, P. Tseng, On the convergence of the coordinate descent method for convex differentiable minimization, J. Optim. Theory Appl. 72 (1992) 7–35.
[37] I. Markovsky, S. Van Huffel, Overview of total least-squares methods, Signal Process. 87 (10) (2007) 2283–2302.
[38] D.F. Marshall, W.K. Jenkins, A fast quasi-Newton adaptive filtering algorithm, IEEE Trans. Signal Process. 40 (1993) 1652–1662.
[39] G. Mateos, I. Schizas, G.B. Giannakis, Distributed recursive least-squares for consensus-based in-network adaptive estimation, IEEE Trans. Signal Process. 57 (11) (2009) 4583–4588.
[40] G. Mateos, G.B. Giannakis, Distributed recursive least-squares: stability and performance analysis, IEEE Trans. Signal Process. 60 (7) (2012) 3740–3754.
[41] J.G. McWhirter, Recursive least-squares minimization using a systolic array, Proc. SPIE Real Time Signal Process. VI 431 (1983) 105–112.
[42] M. Morf, Fast Algorithms for Multivariable Systems, PhD Thesis, Stanford University, Stanford, CA, 1974.
[43] M. Morf, T. Kailath, Square-root algorithms for least-squares estimation, IEEE Trans. Autom. Control 20 (1975) 487–497.
[44] M. Morf, B. Dickinson, T. Kailath, A. Vieira, Efficient solution of covariance equations for linear prediction, IEEE Trans. Acoust. Speech Signal Process. 25 (1977) 429–433.
[45] G.V. Moustakides, S. Theodoridis, Fast Newton transversal filters: a new class of adaptive estimation algorithms, IEEE Trans. Signal Process. 39 (1991) 2184–2193.
[46] G.V. Moustakides, Study of the transient phase of the forgetting factor RLS, IEEE Trans. Signal Process. 45 (1997) 2468–2476.
[47] M. Mühlich, R. Mester, The role of total least-squares in motion analysis, in: H. Burkhardt (Ed.), Proceedings of the 5th European Conference on Computer Vision, Springer-Verlag, 1998, pp. 305–321.
[48] V.H. Nascimento, M.T.M. Silva, Adaptive filters, in: R. Chellappa, S. Theodoridis (Eds.), Signal Process, E-Ref. 1, 2014, pp. 619–747.
[49] T. Petillon, A. Gilloire, S. Theodoridis, Fast Newton transversal filters: an efficient way for echo cancellation in mobile radio communications, IEEE Trans. Signal Process. 42 (1994) 509–517.
[50] T. Piotrowski, I. Yamada, MV-PURE estimator: minimum-variance pseudo-unbiased reduced-rank estimator for linearly constrained ill-conditioned inverse problems, IEEE Trans. Signal Process. 56 (2008) 3408–3423.
[51] A. Pruessner, D. O'Leary, Blind deconvolution using a regularized structured total least norm algorithm, SIAM J. Matrix Anal. Appl. 24 (4) (2003) 1018–1037.
[52] P.A. Regalia, Numerical stability properties of a QR-based fast least-squares algorithm, IEEE Trans. Signal Process. 41 (1993) 2096–2109.
[53] A.A. Rondogiannis, S. Theodoridis, On inverse factorization adaptive least-squares algorithms, Signal Process. 52 (1997) 35–47.
[54] A.A. Rondogiannis, S. Theodoridis, New fast QR decomposition least-squares adaptive algorithms, IEEE Trans. Signal Process. 46 (1998) 2113–2121.
[55] A. Rontogiannis, S. Theodoridis, Householder-based RLS algorithms, in: J.A. Apolonario Jr. (Ed.), QRD-RLS Adaptive Filtering, Springer, 2009.
[56] A.H. Sayed, T. Kailath, A state space approach to adaptive RLS filtering, IEEE Signal Process. Mag. 11 (1994) 18–60.
[57] A.H. Sayed, Fundamentals of Adaptive Filtering, J. Wiley Interscience, 2003.

[58] D.T.M. Slock, R. Kailath, Numerically stable fast transversal filters for recursive least-squares adaptive filtering, IEEE Trans. Signal Process. 39 (1991) 92–114.
[59] T. Söderström, Errors-in-variables methods in system identification, Automatica 43 (6) (2007) 939–958.
[60] S. Theodoridis, Pipeline architecture for block adaptive LS FIR filtering and prediction, IEEE Trans. Acoust. Speech Signal Process. 38 (1990) 81–90.
[61] S. Theodoridis, A. Liavas, Highly concurrent algorithm for the solution of ρ-Toeplitz system of equations, Signal Process. 24 (1991) 165–176.
[62] P. Tseng, Convergence of a block coordinate descent method for nondifferentiable minimization, J. Optim. Theory Appl. 109 (2001) 475–494.
[63] D. Tufts, R. Kumaresan, Estimation of frequencies of multiple sinusoids: making linear prediction perform like maximum likelihood, Proc. IEEE 70 (9) (1982) 975–989.
[64] S. Van Huffel, J. Vandewalle, The Total-Least-Squares Problem: Computational Aspects and Analysis, SIAM, Philadelphia, 1991.
[65] M.H. Verhaegen, Round-off error propagation in four generally-applicable, recursive, least-squares estimation schemes, Automatica 25 (1989) 437–444.
[66] G.P. White, Y.V. Zakharov, J. Liu, Low complexity RLS algorithms using dichotomous coordinate descent iterations, IEEE Trans. Signal Process. 56 (2008) 3150–3161.
[67] T.T. Wu, K. Lange, Coordinate descent algorithms for lasso penalized regression, Ann. Appl. Stat. 2 (2008) 224–244.
[68] B. Yang, A note on the error propagation analysis of recursive least-squares algorithms, IEEE Trans. Signal Process. 42 (1994) 3523–3525.

CHAPTER

CLASSIFICATION: A TOUR OF THE CLASSICS

7

CONTENTS

7.1 INTRODUCTION

The classification task was introduced in Chapter 3. There, it was pointed out that, in principle, one could employ the same loss functions as those used for regression in order to optimize the design of a classifier; however, for most cases in practice, this is not the most reasonable way to attack such problems. This is because in classification the output random variable, y, is of a *discrete nature*; hence, different measures than those used for the regression task are more appropriate for quantifying performance quality.

Machine Learning. https://doi.org/10.1016/B978-0-12-818803-3.00016-7

The goal of this chapter is to present a number of widely used loss functions and methods. Most of the techniques covered are conceptually simple and constitute the basic pillars on which classification is built. Besides their pedagogical importance, these techniques are still in use in a number of practical applications and often form the basis for the development of more advanced methods, to be covered later in the book.

The classical Bayesian classification rule, the notion of minimum distance classifiers, the logistic regression loss function, Fisher's linear discriminant, classification trees, and the method of combining classifiers, including the powerful technique of boosting, are discussed. The perceptron rule, although it boasts to be among the most basic classification rules, will be treated in Chapter 18 and it will be used as the starting point for introducing neural networks and deep learning techniques. Support vector machines are treated in the framework of reproducing kernel Hilbert spaces in Chapter 11.

In a nutshell, this chapter can be considered a beginner's tour of the task of designing classifiers.

7.2 BAYESIAN CLASSIFICATION

In Chapter 3, a linear classifier was designed via the least-squares (LS) method. However, the squared error criterion cannot serve well the needs of the classification task. In Chapters 3 and 6, we have proved that the LS estimator is an efficient one only if the conditional distribution of the output variable y, given the feature values $\boldsymbol{x}$, follows a Gaussian distribution of a special type. However, in classification, the dependent variable is discrete, hence it is not Gaussian; thus, the use of the squared error criterion cannot be justified, in general. We will return to this issue in Section 7.10 (Remarks 7.7), when the squared error is discussed against other loss functions used in classification.

In this section, the classification task will be approached via a different path, inspired by the *Bayesian decision theory*. In spite of its conceptual simplicity, which ties very well with common sense, Bayesian classification possesses a strong optimality flavor with respect to the probability of error, that is, the probability of wrong decisions/class predictions that a classifier commits.

Bayesian classification rule: Given a set of M classes, ω_i, $i = 1, 2, \ldots, M$, and the respective *posterior probabilities* $P(\omega_i|\boldsymbol{x})$, classify an unknown feature vector, $\boldsymbol{x}$, according to the following rule[1]:

$$\boxed{\text{Assign } \boldsymbol{x} \text{ to } \omega_i = \arg\max_{\omega_j} P(\omega_j|\boldsymbol{x}), \quad j = 1, 2, \ldots, M.} \tag{7.1}$$

In words, the unknown pattern, represented by $\boldsymbol{x}$, is assigned to the class for which the posterior probability becomes maximum.

Note that prior to receiving any observation, our uncertainty concerning the classes is expressed via the prior class probabilities, denoted by $P(\omega_i)$, $i = 1, 2, \ldots, M$. Once the observation $\boldsymbol{x}$ has been obtained, this extra information removes part of our original uncertainty, and the related statistical information is now provided by the posterior probabilities, which are then used for the classification.

[1] Recall that probability values for discrete random variables are denoted by capital P, and PDFs, for continuous random variables, by lower case p.

Employing in (7.1) the Bayes theorem,

$$P(\omega_j|\boldsymbol{x}) = \frac{p(\boldsymbol{x}|\omega_j)P(\omega_j)}{p(\boldsymbol{x})}, \quad j = 1, 2, \ldots, M, \tag{7.2}$$

where $p(\boldsymbol{x}|\omega_j)$ are the respective conditional PDFs, the Bayesian classification rule becomes

$$\text{Assign } \boldsymbol{x} \text{ to } \omega_i = \arg\max_{\omega_j} p(\boldsymbol{x}|\omega_j)P(\omega_j), \quad j = 1, 2 \ldots, M. \tag{7.3}$$

Note that the probability denstity function (PDF) of the data, $p(\boldsymbol{x})$, in the denominator of (7.2) does not enter in the maximization task, because it is a positive quantity independent of the classes ω_j; hence, it does not affect the maximization. In other words, the classifier depends on the a priori class probabilities and the respective conditional PDFs. Also, recall that

$$p(\boldsymbol{x}|\omega_j)P(\omega_j) = p(\omega_j, \boldsymbol{x}) := p(y, \boldsymbol{x}),$$

where in the current context, the output variable, y, denotes the label associated with the corresponding class ω_i. The last equation verifies what was said in Chapter 3: the Bayesian classifier is a generative modeling technique.

We now turn our attention to how one can obtain estimates of the involved quantities. Recall that in practice, all one has at one's disposal is a set of training data, from which estimates of the prior probabilities as well as the conditional PDFs must be obtained. Let us assume that we are given a set of training points, $(y_n, \boldsymbol{x}_n) \in D \times \mathbb{R}^l$, $n = 1, 2, \ldots, N$, where D is the discrete set of class labels, and consider the general task comprising M classes. Assume that each class ω_i, $i = 1, 2, \ldots, M$, is represented by N_i points in the training set, with $\sum_{i=1}^{M} N_i = N$. Then the a priori probabilities can be approximated by

$$P(\omega_i) \approx \frac{N_i}{N}, \quad i = 1, 2, \ldots, M. \tag{7.4}$$

For the conditional PDFs, $p(\boldsymbol{x}|\omega_i)$, $i = 1, 2 \ldots, M$, any method for estimating PDFs can be mobilized. For example, one can assume a known parametric form for each one of the conditionals and adopt the maximum likelihood (ML) method, discussed in Section 3.10, or the maximum a posteriori (MAP) estimator, discussed in Section 3.11.1, in order to obtain estimates of the parameters using the training data from each one of the classes. Another alternative is to resort to nonparametric histogram-like techniques, such as Parzen windows and the k-nearest neighbor density estimation techniques, as discussed in Section 3.15. Other methods for PDF estimation can also be employed, such as mixture modeling, to be discussed in Chapter 12. The interested reader may also consult [38,39].

THE BAYESIAN CLASSIFIER MINIMIZES THE MISCLASSIFICATION ERROR

In Section 3.4, it was pointed out that the goal of designing a classifier is to partition the space in which the feature vectors lie into regions, and associate each one of the regions to one and only one class. For a two-class task (the generalization to more classes is straightforward), let $\mathcal{R}_1, \mathcal{R}_2$ be the two regions in $\mathbb{R}^l$, where we decide in favor of class ω_1 and ω_2, respectively. The probability of classification error is given by

$$P_e = P(\boldsymbol{x} \in \mathcal{R}_1, \boldsymbol{x} \in \omega_2) + P(\boldsymbol{x} \in \mathcal{R}_2, \boldsymbol{x} \in \omega_1). \tag{7.5}$$

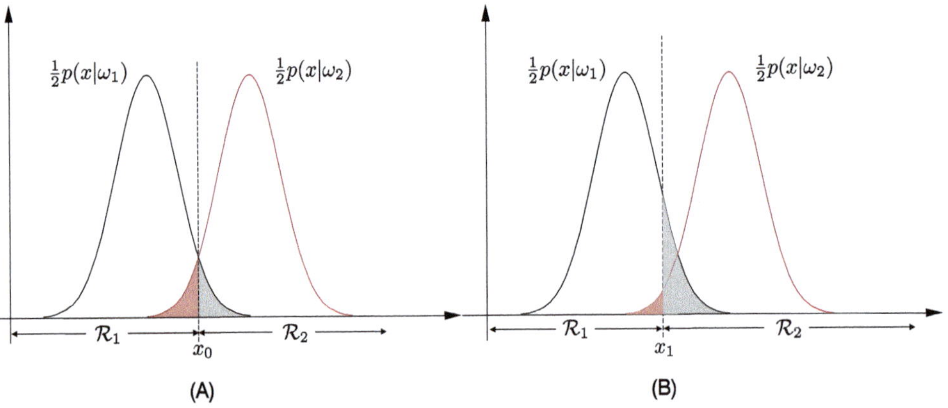

FIGURE 7.1

(A) The classification error probability for partitioning the feature space, according to the Bayesian optimal classifier, is equal to the area of the shaded region. (B) Moving the threshold value away from the value corresponding to the optimal Bayes rule increases the probability of error, as is indicated by the increase of the area of the corresponding shaded region.

That is, it is equal to the probability of the feature vector to belong to class ω_1 (ω_2) *and* to lie in the "wrong" region $\mathcal{R}_2$ ($\mathcal{R}_1$) in the feature space.

Eq. (7.5) can be written as

$$\boxed{P_e = P(\omega_2)\int_{\mathcal{R}_1} p(\boldsymbol{x}|\omega_2)d\boldsymbol{x} + P(\omega_1)\int_{\mathcal{R}_2} p(\boldsymbol{x}|\omega_1)d\boldsymbol{x}: \quad \text{probability of error.}} \tag{7.6}$$

It turns out that the Bayesian classifier, as defined in (7.3), minimizes P_e with respect to $\mathcal{R}_1$ and $\mathcal{R}_2$ [11,38]. This is also true for the general case of M classes (Problem 7.1).

Fig. 7.1A demonstrates geometrically the optimality of the Bayesian classifier for the two-class one-dimensional case and assuming equiprobable classes ($P(\omega_1) = P(\omega_2) = 1/2$). The region $\mathcal{R}_1$, to the left of the threshold value x_0, corresponds to $p(x|\omega_1) > p(x|\omega_2)$, and the opposite is true for region $\mathcal{R}_2$. The probability of error is equal to the area of the shaded region, which is equal to the sum of the two integrals in (7.6). In Fig. 7.1B, the threshold has been moved away from the optimal Bayesian value, and as a result the probability of error, given by the total area of the corresponding shaded region, increases.

7.2.1 AVERAGE RISK

Because in classification the dependent variable (label), y, is of a discrete nature, the classification error probability may seem like the most natural cost function to be optimized. However, this is not always true. In certain applications, not all errors are of the same importance. For example, in a medical diagnosis system, committing an error by predicting the class of a finding in an X-ray image as being "malignant" while its true class is "normal" is less significant than an error the other way around. In the former case, the wrong diagnosis will be revealed in the next set of medical tests. However, the opposite may have unwanted consequences. For such cases, one uses an alternative to the probability of error cost function that puts relative weights on the errors according to their importance. This cost function is known as the *average risk* and it results in a rule that resembles that of the Bayesian classifier, yet it is slightly modified due to the presence of the weights.

Let us start from the simpler two-class case. The *risk* or *loss* associated with each one of the two classes is defined as

$$r_1 = \lambda_{11} \int_{\mathcal{R}_1} p(\boldsymbol{x}|\omega_1)d\boldsymbol{x} + \lambda_{12} \underbrace{\int_{\mathcal{R}_2} p(\boldsymbol{x}|\omega_1)d\boldsymbol{x}}_{\text{error}}, \tag{7.7}$$

$$r_2 = \lambda_{21} \underbrace{\int_{\mathcal{R}_1} p(\boldsymbol{x}|\omega_2)d\boldsymbol{x}}_{\text{error}} + \lambda_{22} \int_{\mathcal{R}_2} p(\boldsymbol{x}|\omega_2)d\boldsymbol{x}. \tag{7.8}$$

Typically, $\lambda_{11} = \lambda_{22} = 0$, since they correspond to correct decisions. The *average* risk, to be minimized, is given by

$$r = P(\omega_1)r_1 + P(\omega_2)r_2.$$

Then, following similar arguments as before for the optimal Bayes classifier, the optimal average risk classifier rule becomes

$$\boxed{\text{Assign } \boldsymbol{x} \text{ to } \omega_1\ (\omega_2) \text{ if}: \lambda_{12} P(\omega_1|\boldsymbol{x}) > (<)\ \lambda_{21} P(\omega_2|\boldsymbol{x}).} \tag{7.9}$$

Equivalently, we can write

$$\text{Assign } \boldsymbol{x} \text{ to } \omega_1\ (\omega_2) \text{ if}: \underbrace{\lambda_{12} P(\omega_1)}_{P'(\omega_1)} p(\boldsymbol{x}|\omega_1) > (<)\ \underbrace{\lambda_{21} P(\omega_2)}_{P'(\omega_2)} p(\boldsymbol{x}|\omega_2). \tag{7.10}$$

Note that if λ_{12} is *large*, compared to λ_{21}, this means that class ω_1 is more "important." Looking at it from a slightly different view, one can interpret the use of the weights as a way to *increase* the prior probability for class ω_1 with respect to that of the class ω_2, i.e.,

$$\frac{P'(\omega_1)}{P'(\omega_2)} > \frac{P(\omega_1)}{P(\omega_2)}.$$

For the M-class problem, the *risk* (*loss*) associated with class ω_k is defined as

$$r_k = \sum_{i=1}^{M} \lambda_{ki} \int_{\mathcal{R}_i} p(\boldsymbol{x}|\omega_k)\, d\boldsymbol{x}, \tag{7.11}$$

where $\lambda_{kk} = 0$ and λ_{ki} is the weight that controls the significance of committing an error by assigning a pattern from class ω_k to class ω_i. The *average risk* is given by

$$r = \sum_{k=1}^{M} P(\omega_k) r_k = \sum_{i=1}^{M} \int_{\mathcal{R}_i} \left(\sum_{k=1}^{M} \lambda_{ki} P(\omega_k) p(\boldsymbol{x}|\omega_k) \right) d\boldsymbol{x}, \tag{7.12}$$

which is minimized if we partition the input space by selecting each $\mathcal{R}_i$ (where we decide in favor of class ω_i) so that each one of the M integrals in the summation becomes minimum; this is achieved if we adopt the rule

$$\text{Assign } \boldsymbol{x} \text{ to } \omega_i : \sum_{k=1}^{M} \lambda_{ki} P(\omega_k) p(\boldsymbol{x}|\omega_k) < \sum_{k=1}^{M} \lambda_{kj} P(\omega_k) p(\boldsymbol{x}|\omega_k), \quad \forall j \neq i,$$

or equivalently,

$$\boxed{\text{Assign } \boldsymbol{x} \text{ to } \omega_i : \sum_{k=1}^{M} \lambda_{ki} P(\omega_k|\boldsymbol{x}) < \sum_{k=1}^{M} \lambda_{kj} P(\omega_k|\boldsymbol{x}), \quad \forall j \neq i.} \tag{7.13}$$

It is common to consider the weights λ_{ij} as defining an $M \times M$ matrix

$$L := [\lambda_{ij}], \quad i, j = 1, 2, \ldots, M, \tag{7.14}$$

which is known as the *loss* matrix. Note that if we set $\lambda_{ki} = 1$, $k = 1, 2, \ldots, M, i = 1, 2, \ldots, M, k \neq i$, then we obtain the Bayes rule (verify it).

Remarks 7.1.

- *The reject option*: Bayesian classification relies on the maximum value of the posterior probabilities, $P(\omega_i|\boldsymbol{x})$, $i = 1, 2, \ldots, M$. However, often in practice, it may happen that for some value $\boldsymbol{x}$, the maximum value is comparable to the values the posterior obtains for other classes. For example, in a two-class task, it may turn out that $P(\omega_1|\boldsymbol{x}) = 0.51$ and $P(\omega_2|\boldsymbol{x}) = 0.49$. If this happens, it may be more sensible not to make a decision for this particular pattern, $\boldsymbol{x}$. This is known as the reject option. If such a decision scenario is adopted, a user-defined threshold value θ is chosen and classification is carried out only if the maximum posterior is larger than this threshold value, so that $P(\omega_i|\boldsymbol{x}) > \theta$. Otherwise, no decision is taken. Similar arguments can be adopted for the average risk classification.

Example 7.1. In a two-class, one-dimensional classification task, the data in the two classes are distributed according to the following two Gaussians:

$$p(x|\omega_1) = \frac{1}{\sqrt{2\pi}} \exp\left(-\frac{x^2}{2}\right),$$

and

$$p(x|\omega_2) = \frac{1}{\sqrt{2\pi}} \exp\left(-\frac{(x-1)^2}{2}\right).$$

The problem is more sensitive with respect to errors committed on patterns from class ω_1, which is expressed via the following loss matrix:

$$L = \begin{bmatrix} 0 & 1 \\ 0.5 & 0 \end{bmatrix}.$$

In other words, $\lambda_{12} = 1$ and $\lambda_{21} = 0.5$. The two classes are considered equiprobable. Derive the threshold value x_r, which partitions the feature space $\mathbb{R}$ into the two regions $\mathcal{R}_1, \mathcal{R}_2$ in which we decide in favor of class ω_1 and ω_2, respectively. What is the value of the threshold when the Bayesian classifier is used instead?

Solution: According to the average risk rule, the region for which we decide in favor of class ω_1 is given by

$$\mathcal{R}_1 : \lambda_{12}\frac{1}{2}p(x|\omega_1) > \lambda_{21}\frac{1}{2}p(x|\omega_2),$$

and the respective threshold value x_r is computed by the equation

$$\exp\left(-\frac{x_r^2}{2}\right) = 0.5\exp\left(-\frac{(x_r-1)^2}{2}\right),$$

which, after taking the logarithm and solving the respective equation, trivially results in

$$x_r = \frac{1}{2}(1 - 2\ln 0.5).$$

The threshold for the Bayesian classifier results if we set $\lambda_{21} = 1$, which gives

$$x_{\mathrm{B}} = \frac{1}{2}.$$

The geometry is shown in Fig. 7.2. In other words, the use of the average risk moves the threshold to the right of the value corresponding to the Bayesian classifier; that is, it enlarges the region in which we decide in favor of the more significant class, ω_1. Note that this would also be the case if the two classes were not equiprobable, as shown by $P(\omega_1) > P(\omega_2)$ (for our example, $P(\omega_1) = 2P(\omega_2)$).

7.3 DECISION (HYPER)SURFACES

The goal of any classifier is to partition the feature space into regions. The partition is achieved via points in $\mathbb{R}$, curves in $\mathbb{R}^2$, surfaces in $\mathbb{R}^3$, and hypersurfaces in $\mathbb{R}^l$. Any hypersurface S is expressed in terms of a function

$$g : \mathbb{R}^l \longmapsto \mathbb{R},$$

and it comprises all the points such that

$$S = \left\{ \boldsymbol{x} \in \mathbb{R}^l : \quad g(\boldsymbol{x}) = 0 \right\}.$$

Recall that all points lying on one side of this hypersurface score $g(\boldsymbol{x}) > 0$ and all the points on the other side score $g(\boldsymbol{x}) < 0$. The resulting (hyper)surfaces are knows as *decision (hyper)surfaces*, for obvious reasons. Take as an example the case of the two-class Bayesian classifier. The respective decision hypersurface is (implicitly) formed by

$$g(\boldsymbol{x}) := P(\omega_1|\boldsymbol{x}) - P(\omega_2|\boldsymbol{x}) = 0. \tag{7.15}$$

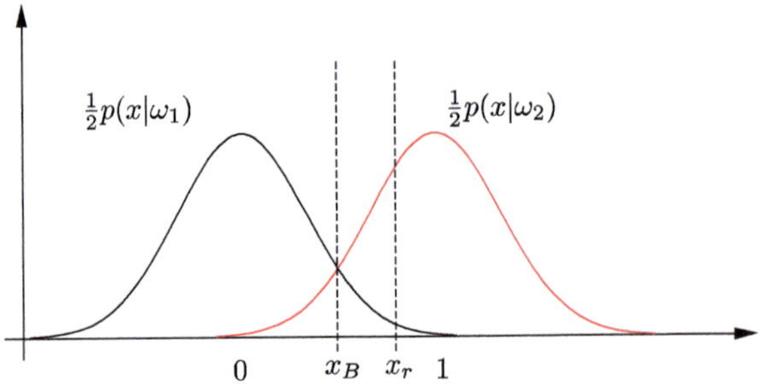

FIGURE 7.2

The class distributions and the resulting threshold values for the two cases of Example 7.1. Note that minimizing the average risk enlarges the region in which we decide in favor of the most sensitive class, ω_1.

Indeed, we decide in favor of class ω_1 (region $\mathcal{R}_1$) if $\boldsymbol{x}$ falls on the positive side of the hypersurface defined in (7.15), and in favor of ω_2 for the points falling on the negative side (region $\mathcal{R}_2$). This is illustrated in Fig. 7.3. At this point, recall the reject option from Remarks 7.1. Points where no decision is taken are those that lie close to the decision hypersurface.

Once we move away from the Bayesian concept of designing classifiers (as we will soon see, and this will be done for a number of reasons), different families of functions for selecting $g(\boldsymbol{x})$ can be adopted and the specific form will be obtained via different optimization criteria, which are not necessarily related to the probability of error/average risk.

In the sequel, we focus on investigating the form that the decision hypersurfaces take for the special case of the Bayesian classifier and where the data in the classes are distributed according to the Gaussian PDF. This can provide further insight into the way a classifier partitions the feature space and it will also lead to some useful implementations of the Bayesian classifier, under certain scenarios. For simplicity, the focus will be on two-class classification tasks, but the results are trivially generalized to the more general M-class case.

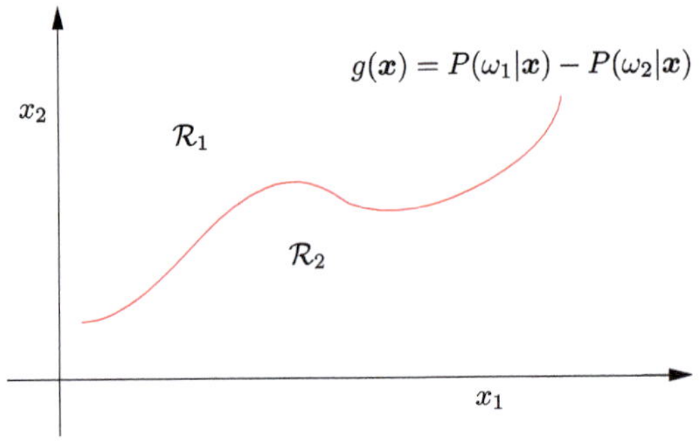

FIGURE 7.3

The Bayesian classifier implicitly forms hypersurfaces defined by $g(\boldsymbol{x}) = P(\omega_1|\boldsymbol{x}) - P(\omega_2|\boldsymbol{x}) = 0$.

7.3.1 THE GAUSSIAN DISTRIBUTION CASE

Assume that the data in each class are distributed according to the Gaussian PDF, so that

$$p(\boldsymbol{x}|\omega_i) = \frac{1}{(2\pi)^{l/2}|\Sigma_i|^{1/2}} \exp\left(-\frac{1}{2}(\boldsymbol{x}-\boldsymbol{\mu}_i)^T \Sigma_i^{-1}(\boldsymbol{x}-\boldsymbol{\mu}_i)\right), \quad i = 1, 2, \ldots, M. \tag{7.16}$$

Because the logarithmic function is a monotonically increasing one, it does not affect the point where the maximum of a function occurs. Thus, taking into account the exponential form of the Gaussian, the computations can be facilitated if the Bayesian rule is expressed in terms of the following functions:

$$g_i(\boldsymbol{x}) := \ln\left(p(\boldsymbol{x}|\omega_i)P(\omega_i)\right) = \ln p(\boldsymbol{x}|\omega_i) + \ln P(\omega_i), \quad i = 1, 2, \ldots, M, \tag{7.17}$$

and search for the class for which the respective function scores the maximum value. Such functions are also known as *discriminant functions*.

Let us now focus on the two-class classification task. The decision hypersurface, associated with the Bayesian classifier, can be expressed as

$$g(\boldsymbol{x}) = g_1(\boldsymbol{x}) - g_2(\boldsymbol{x}) = 0, \tag{7.18}$$

which, after plugging into (7.17) the specific forms of the Gaussian conditionals, and after a bit of trivial algebra, becomes

$$\begin{aligned} g(\boldsymbol{x}) = & \underbrace{\frac{1}{2}\left(\boldsymbol{x}^T \Sigma_2^{-1}\boldsymbol{x} - \boldsymbol{x}^T \Sigma_1^{-1}\boldsymbol{x}\right)}_{\text{quadratic terms}} \\ & \underbrace{+\boldsymbol{\mu}_1^T \Sigma_1^{-1}\boldsymbol{x} - \boldsymbol{\mu}_2^T \Sigma_2^{-1}\boldsymbol{x}}_{\text{linear terms}} \\ & \underbrace{-\frac{1}{2}\boldsymbol{\mu}_1^T \Sigma_1^{-1}\boldsymbol{\mu}_1 + \frac{1}{2}\boldsymbol{\mu}_2^T \Sigma_2^{-1}\boldsymbol{\mu}_2 + \ln\frac{P(\omega_1)}{P(\omega_2)} + \frac{1}{2}\ln\frac{|\Sigma_2|}{|\Sigma_1|}}_{\text{constant terms}} = 0. \end{aligned} \tag{7.19}$$

This is of a quadratic nature; hence the corresponding (hyper)surfaces are *(hyper)quadrics*, including (hyper)ellipsoids, (hyper)parabolas, and hyperbolas. Fig. 7.4 shows two examples, in the two-dimensional space, corresponding to $P(\omega_1) = P(\omega_2)$, and

$$\text{(a)} \quad \boldsymbol{\mu}_1 = [0, 0]^T, \ \boldsymbol{\mu}_2 = [4, 0]^T, \ \Sigma_1 = \begin{bmatrix} 0.3 & 0.0 \\ 0.0 & 0.35 \end{bmatrix}, \ \Sigma_2 = \begin{bmatrix} 1.2 & 0.0 \\ 0.0 & 1.85 \end{bmatrix},$$

and

$$\text{(b)} \quad \boldsymbol{\mu}_1 = [0, 0]^T, \ \boldsymbol{\mu}_2 = [3.2, 0]^T, \ \Sigma_1 = \begin{bmatrix} 0.1 & 0.0 \\ 0.0 & 0.75 \end{bmatrix}, \ \Sigma_2 = \begin{bmatrix} 0.75 & 0.0 \\ 0.0 & 0.1 \end{bmatrix},$$

respectively. In Fig. 7.4A, the resulting curve for scenario (a) is an ellipse, and in Fig. 7.4B, the corresponding curve for scenario (b) is a hyperbola.

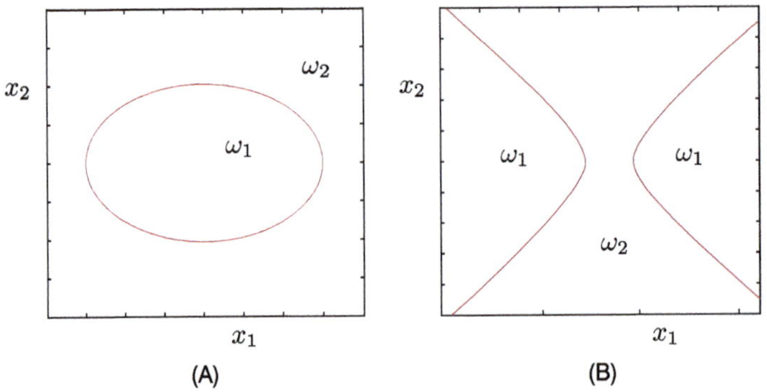

FIGURE 7.4

The Bayesian classifier for the case of Gaussian distributed classes partitions the feature space via quadrics. (A) The case of an ellipse and (B) the case of a hyperbola.

Looking carefully at (7.19), it is readily noticed that once the covariance matrices for the two classes become equal, the quadratic terms cancel out and the discriminant function becomes linear; thus, the corresponding hypersurface is a *hyperplane*. That is, under the previous assumptions, the optimal Bayesian classifier becomes a *linear* classifier, which after some straightforward algebraic manipulations (try it) can be written as

$$g(\boldsymbol{x}) = \boldsymbol{\theta}^T(\boldsymbol{x} - \boldsymbol{x}_0) = 0, \tag{7.20}$$

$$\boldsymbol{\theta} := \Sigma^{-1}(\boldsymbol{\mu}_1 - \boldsymbol{\mu}_2), \tag{7.21}$$

$$\boldsymbol{x}_0 := \frac{1}{2}(\boldsymbol{\mu}_1 + \boldsymbol{\mu}_2) - \ln \frac{P(\omega_1)}{P(\omega_2)} \frac{\boldsymbol{\mu}_1 - \boldsymbol{\mu}_2}{||\boldsymbol{\mu}_1 - \boldsymbol{\mu}_2||^2_{\Sigma^{-1}}}, \tag{7.22}$$

where Σ is common to the two-class covariance matrix and

$$||\boldsymbol{\mu}_1 - \boldsymbol{\mu}_2||_{\Sigma^{-1}} := \sqrt{(\boldsymbol{\mu}_1 - \boldsymbol{\mu}_2)^T \Sigma^{-1}(\boldsymbol{\mu}_1 - \boldsymbol{\mu}_2)}$$

is the Σ^{-1} norm of the vector $(\boldsymbol{\mu}_1 - \boldsymbol{\mu}_2)$; alternatively, this is also known as the *Mahalanobis distance* between $\boldsymbol{\mu}_1$ and $\boldsymbol{\mu}_2$. The Mahalanobis distance is a generalization of the Euclidean distance; note that for $\Sigma = I$ it becomes the Euclidean distance.

Fig. 7.5 shows three cases for the two-dimensional space. The full black line corresponds to the case of equiprobable classes with a covariance matrix of the special form, $\Sigma = \sigma^2 I$. The corresponding decision hyperplane, according to Eq. (7.20), is now written as

$$g(\boldsymbol{x}) = (\boldsymbol{\mu}_1 - \boldsymbol{\mu}_2)^T(\boldsymbol{x} - \boldsymbol{x}_0) = 0. \tag{7.23}$$

The separating line (hyperplane) crosses the middle point of the line segment joining the mean value points, $\boldsymbol{\mu}_1$ and $\boldsymbol{\mu}_2$ ($\boldsymbol{x}_0 = \frac{1}{2}(\boldsymbol{\mu}_1 + \boldsymbol{\mu}_2)$). Also, it is perpendicular to this segment, defined by the vector $\boldsymbol{\mu}_1 - \boldsymbol{\mu}_2$, as is readily verified by the above hyperplane definition. Indeed, for *any* point $\boldsymbol{x}$ on the

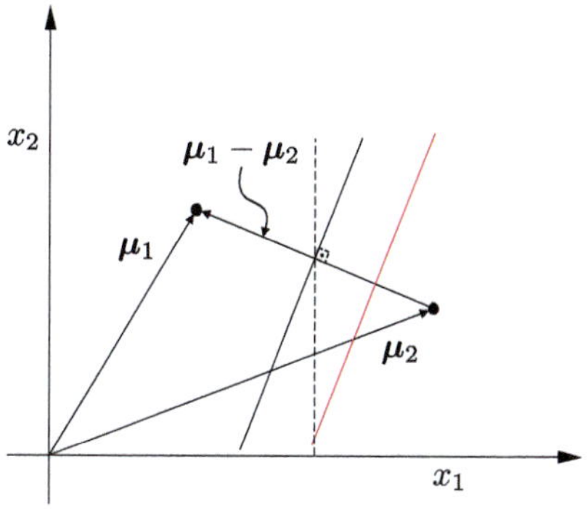

FIGURE 7.5

The full black line corresponds to the Bayesian classifier for two equiprobable Gaussian classes that share a common covariance matrix of the specific form $\Sigma = \sigma^2 I$; the line bisects the segment joining the two mean values (minimum Euclidean distance classifier). The red one is for the same case but for $P(\omega_1) > P(\omega_2)$. The dotted line is the optimal classifier for equiprobable classes and a common covariance of a more general form, different from $\sigma^2 I$ (minimum Mahalanobis distance classifier).

decision curve (full black line in Fig. 7.5), since the middle point $\boldsymbol{x}_0$ also lies on this curve, the corresponding vector $\boldsymbol{x} - \boldsymbol{x}_0$ is parallel to this line. Hence, the inner product in Eq. (7.23) being equal to zero means that the decision line is perpendicular to $\boldsymbol{\mu}_1 - \boldsymbol{\mu}_2$. The red line corresponds to the case where $P(\omega_1) > P(\omega_2)$. It gets closer to the mean value point of class ω_2, thus enlarging the region where one decides in favor of the more probable class. Note that in this case, the logarithm of the ratio in Eq. (7.22) is positive. Finally, the dotted line corresponds to the equiprobable case with the common covariance matrix being of a more general form, $\Sigma \neq \sigma^2 I$. The separating hyperplane crosses $\boldsymbol{x}_0$ but it is rotated in order to be perpendicular to the vector $\Sigma^{-1}(\boldsymbol{\mu}_1 - \boldsymbol{\mu}_2)$, according to (7.20)–(7.21). For each one of the three cases, an unknown point is classified according to the side of the respective hyperplane on which it lies.

What was said before for the two-class task is generalized to the more general M-class problem; the separating hypersurfaces of two contiguous regions $\mathcal{R}_i, \mathcal{R}_j$ associated with classes ω_i, ω_j obey the same arguments as the ones adopted before. For example, assuming that all covariance matrices are the same, the regions are partitioned via hyperplanes, as illustrated in Fig. 7.6. Moreover, each region $\mathcal{R}_i$, $i = 1, 2, \ldots, M$, is convex (Problem 7.2); in other words, joining any two points within $\mathcal{R}_i$, all the points lying on the respective segment lie in $\mathcal{R}_i$ as well.

Two special cases are of particular interest, leading to a simple classification rule. The rule will be expressed for the general M-class problem.

Minimum Distance Classifiers

There are two special cases, where the optimal Bayesian classifier becomes very simple to compute and it also has a strong geometric flavor.

- *Minimum Euclidean distance classifier*: Under the assumptions of (a) Gaussian distributed data in each one of the classes, i.e., Eq. (7.16), (b) equiprobable classes, and (c) common covariance matrix

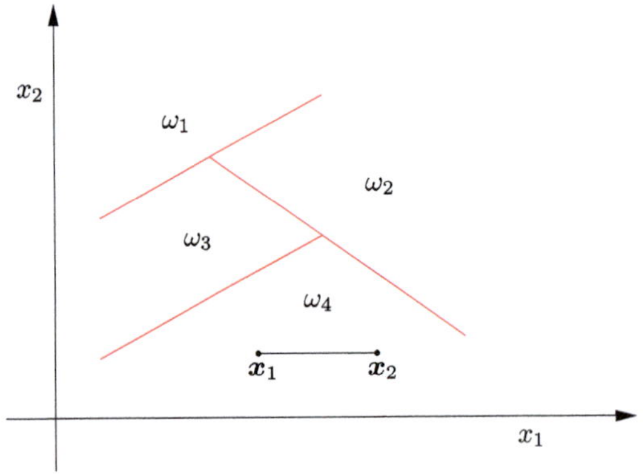

FIGURE 7.6

When data are distributed according to the Gaussian distribution and they share the same covariance matrix in all classes, the feature space is partitioned via hyperplanes, which form polyhedral regions. Note that each region is associated with one class and it is convex.

in all classes of the special form $\Sigma = \sigma^2 I$ (individual features are independent and share a common variance), the Bayesian classification rule becomes equivalent to

$$\text{Assign } \boldsymbol{x} \text{ to class } \omega_i : i = \arg\min_j (\boldsymbol{x} - \boldsymbol{\mu}_j)^T (\boldsymbol{x} - \boldsymbol{\mu}_j), \quad j = 1, 2, \ldots M. \tag{7.24}$$

This is a direct consequence of the Bayesian rule under the adopted assumptions. In other words, the Euclidean distance of $\boldsymbol{x}$ is computed from the mean values of all classes and it is assigned to the class for which this distance becomes smaller.

For the case of the two classes, this classification rule corresponds to the full black line of Fig. 7.5. Indeed, recalling our geometry basics, any point that lies on the left side of this hyperplane that *bisects* the segment $\boldsymbol{\mu}_1 - \boldsymbol{\mu}_2$ is closer to $\boldsymbol{\mu}_1$ than to $\boldsymbol{\mu}_2$, in the Euclidean distance sense. The opposite is true for any point lying on the right of the hyperplane.

- *Minimum Mahalanobis distance classifier*: Under the previously adopted assumptions, but with the covariance matrix being of the more general form $\Sigma \neq \sigma^2 I$, the rule becomes

$$\text{Assign } \boldsymbol{x} \text{ to class } \omega_i : i = \arg\min_j (\boldsymbol{x} - \boldsymbol{\mu}_j)^T \Sigma^{-1} (\boldsymbol{x} - \boldsymbol{\mu}_j), \quad j = 1, 2, \ldots M. \tag{7.25}$$

Thus, instead of looking for the minimum Euclidean distance, one searches for the minimum Mahalanobis distance; the latter is a weighted form of the Euclidean distance, in order to account for the shape of the underlying Gaussian distributions [38]. For the two-class case, this rule corresponds to the dotted line of Fig. 7.5.

Remarks 7.2.

- In statistics, adopting the Gaussian assumption for the data distribution is sometimes called *linear discriminant analysis* (LDA) or *quadratic discriminant analysis* (QDA), depending on the adopted assumptions with respect to the underlying covariance matrices, which will lead to linear or quadratic discriminant functions, respectively. In practice, the ML method is usually employed in order to obtain estimates of the unknown parameters, namely, the mean values and the covariance matrices. Recall from Example 3.7 of Chapter 3 that the ML estimate of the mean value of a Gaussian PDF, obtained via N observations, $\boldsymbol{x}_n$, $n = 1, 2, \ldots, N$, is equal to

$$\hat{\boldsymbol{\mu}}_{ML} = \frac{1}{N}\sum_{n=1}^{N} \boldsymbol{x}_n.$$

Moreover, the ML estimate of the covariance matrix of a Gaussian distribution, using N observations, is given by (Problem 7.4)

$$\hat{\Sigma}_{ML} = \frac{1}{N}\sum_{n=1}^{N}(\boldsymbol{x}_n - \hat{\boldsymbol{\mu}}_{ML})(\boldsymbol{x}_n - \hat{\boldsymbol{\mu}}_{ML})^T. \tag{7.26}$$

This corresponds to a biased estimator of the covariance matrix. An unbiased estimator results if (Problem 7.5)

$$\hat{\Sigma} = \frac{1}{N-1}\sum_{n=1}^{N}(\boldsymbol{x}_n - \hat{\boldsymbol{\mu}}_{ML})(\boldsymbol{x}_n - \hat{\boldsymbol{\mu}}_{ML})^T.$$

Note that the number of parameters to be estimated in the covariance matrix is $O(l^2/2)$, taking into account its symmetry.

Example 7.2. Consider a two-class classification task in the two-dimensional space, with $P(\omega_1) = P(\omega_2) = 1/2$. Generate 100 points, 50 from each class. The data from each class, ω_i, $i = 1, 2$, stem from a corresponding Gaussian, $\mathcal{N}(\boldsymbol{\mu}_i, \Sigma_i)$, where

$$\boldsymbol{\mu}_1 = [0, -2]^T, \quad \boldsymbol{\mu}_2 = [0, 2]^T,$$

and (a)

$$\Sigma_1 = \Sigma_2 = \begin{bmatrix} 1.2 & 0.4 \\ 0.4 & 1.2 \end{bmatrix},$$

or (b)

$$\Sigma_1 = \begin{bmatrix} 1.2 & 0.4 \\ 0.4 & 1.2 \end{bmatrix}, \quad \Sigma_2 = \begin{bmatrix} 1 & -0.4 \\ -0.4 & 1 \end{bmatrix}.$$

Fig. 7.7 shows the decision curves formed by the Bayesian classifier. Observe that in the case of Fig. 7.7A, the classifier turns out to be a linear one, while for the case of Fig. 7.7B, it is nonlinear of a parabola shape.

Example 7.3. In a two-class classification task, the data in each one of the two equiprobable classes are distributed according to the Gaussian distribution, with mean values $\boldsymbol{\mu}_1 = [0, 0]^T$ and $\boldsymbol{\mu}_2 = [3, 3]^T$,

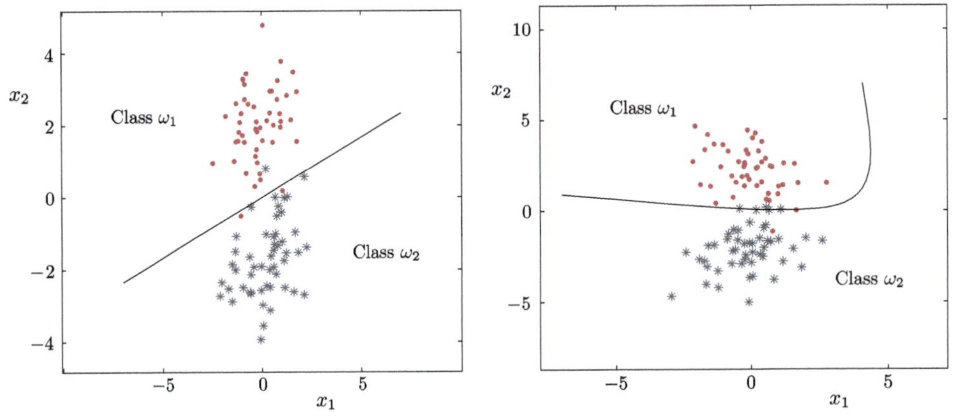

FIGURE 7.7

If the data in the feature space follow a Gaussian distribution in each one of the classes, then (A) if all the covariance matrices are equal, the Bayesian classifier is a hyperplane; (B) otherwise, it is a quadric hypersurface.

respectively, sharing a common covariance matrix

$$\Sigma = \begin{bmatrix} 1.1 & 0.3 \\ 0.3 & 1.9 \end{bmatrix}.$$

Use the Bayesian classifier to classify the point $\boldsymbol{x} = [1.0, 2.2]^T$ into one of the two classes.

Because the classes are equiprobable, are distributed according to the Gaussian distribution, and share the same covariance matrix, the Bayesian classifier is equivalent to the minimum Mahalanobis distance classifier. The (square) Mahalanobis distance of the point $\boldsymbol{x}$ from the mean value of class ω_1 is

$$d_1^2 = [1.0, 2.2] \begin{bmatrix} 0.95 & -0.15 \\ -0.15 & 0.55 \end{bmatrix} \begin{bmatrix} 1.0 \\ 2.2 \end{bmatrix} = 2.95,$$

where the matrix in the middle on the left-hand side is the inverse of the covariance matrix. Similarly for class ω_2, we obtain

$$d_2^2 = [-2.0, -0.8] \begin{bmatrix} 0.95 & -0.15 \\ -0.15 & 0.55 \end{bmatrix} \begin{bmatrix} -2.0 \\ -0.8 \end{bmatrix} = 3.67.$$

Hence, the pattern is assigned to class ω_1, because its distance from $\boldsymbol{\mu}_1$ is smaller compared to that from $\boldsymbol{\mu}_2$. Verify that if the Euclidean distance were used instead, the pattern would be assigned to class ω_2.

7.4 THE NAIVE BAYES CLASSIFIER

We have already seen that in the case the covariance matrix is to be estimated, the number of unknown parameters is of the order of $\mathcal{O}(l^2/2)$. For high-dimensional spaces, besides the fact that this estimation task is a formidable one, it also requires a large number of data points, in order to obtain statistically good estimates and avoid overfitting, as discussed in Chapter 3. In such cases, one has to be content with suboptimal solutions. Indeed, adopting an optimal method while using bad estimates of the involved parameters can lead to a bad overall performance.

The *naive Bayes classifier* is a typical and popular example of a suboptimal classifier. The basic assumption is that the components (features) in the feature vector are statistically independent; hence, the joint PDF can be written as a product of l marginals,

$$p(\boldsymbol{x}|\omega_i) = \prod_{k=1}^{l} p(x_k|\omega_i), \quad i = 1, 2, \ldots, M.$$

Having adopted the Gaussian assumption, each one of the marginals is described by two parameters, the mean and the variance; this leads to a total of $2l$ unknown parameters to be estimated per class. This is a substantial saving compared to the $O(l^2/2)$ number of parameters. It turns out that this simplistic assumption can end up with better results compared to the optimal Bayes classifier when the size of the data samples is limited.

Although the naive Bayes classifier was introduced in the context of Gaussian distributed data, its use is also justified for the more general case. In Chapter 3, we discussed the curse of dimensionality issue and it was stressed that high-dimensional spaces are sparsely populated. In other words, for a fixed finite number of data points, N, within a cube of fixed size for each dimension, the larger the dimension of the space is, the larger the average distance between any two points becomes. Hence, in order to get good estimates of a set of parameters in large spaces, an increased number of data is required. Roughly speaking, if N data points are needed in order to get a good enough estimate of a PDF in the real axis, N^l data points would be needed for similar accuracy in an l-dimensional space. Thus, by assuming the features to be mutually independent, one will end up estimating l one-dimensional PDFs, hence substantially reducing the need for data.

The independence assumption is a common one in a number of machine learning and statistics tasks. As we will see in Chapter 15, one can adopt more "mild" independence assumptions that lie in between the two extremes, which are full independence and full dependence.

7.5 THE NEAREST NEIGHBOR RULE

Although the Bayesian rule provides the optimal solution with respect to the classification error probability, its application requires the estimation of the respective conditional PDFs; this is not an easy task once the dimensionality of the feature space assumes relatively large values. This paves the way for considering alternative classification rules, which becomes our focus from now on.

The *k-nearest neighbor* (k-NN) rule is a typical nonparametric classifier and it is one of the most popular and well-known classifiers. In spite of its simplicity, it is still in use and stands next to more elaborate schemes.

Consider N training points, $(y_n, \boldsymbol{x}_n)$, $n = 1, 2, \ldots, N$, for an M-class classification task. At the heart of the method lies a parameter k, which is a user-defined parameter. Once k is selected, then given a pattern $\boldsymbol{x}$, assign it to the class in which the majority of its k nearest (according to a metric, e.g., Euclidean or Mahalanobis distance) neighbors, among the training points, belong. The parameter k should not be a multiple of M, in order to avoid ties. The simplest form of this rule is to assign the pattern to the class in which its nearest neighbor belongs, meaning $k = 1$.

It turns out that this conceptually simple rule tends to the Bayesian classifier if (a) $N \longrightarrow \infty$, (b) $k \longrightarrow \infty$, and (c) $k/N \longrightarrow 0$. In practical terms, these conditions mean that N and k must be large, yet k must be relatively small with respect to N. More specifically, it can be shown that the classification errors P_{NN} and $P_{k\text{NN}}$ satisfy, asymptotically, the following bounds [9]:

$$P_{\text{B}} \leq P_{\text{NN}} \leq 2P_{\text{B}} \tag{7.27}$$

for the $k = 1$ NN rule, and

$$P_{\text{B}} \leq P_{k\text{NN}} \leq P_{\text{B}} + \sqrt{\frac{2P_{\text{NN}}}{k}} \tag{7.28}$$

for the more general k-NN version; P_{B} is the error corresponding to the optimal Bayesian classifier. These two formulas are quite interesting. Take for example (7.27). It says that the simple NN rule will never give an error larger than twice the optimal one. If, for example, $P_{\text{B}} = 0.01$, then $P_{\text{NN}} \leq 0.02$. This is not bad for such a simple classifier. All this says is that if one has an easy task (as indicated by the very low value of P_{B}), the NN rule can also do a good job. This, of course, is not the case if the problem is not an easy one and larger error values are involved. The bound in (7.28) says that for large values of k (provided, of course, N is large enough), the performance of the k-NN tends to that of the optimal classifier. In practice, one has to make sure that k does not get values close to N, but remains a relatively small fraction of it.

One may wonder how a performance close to the optimal classifier can be obtained, even in theory and asymptotically, because the Bayesian classifier exploits the statistical information for the data distribution while the k-NN does not take into account such information. The reason is that if N is a very large value (hence the space is densely populated with points) and k is a relatively small number, with respect to N, then the nearest neighbors will be located very close to $\boldsymbol{x}$. Then, due to the *continuity* of the involved PDFs, the values of their posterior probabilities will be close to $P(\omega_i|\boldsymbol{x})$, $i = 1, 2, \ldots, M$. Furthermore, for large enough k, the majority of the neighbors must come from the class that scores the maximum value of the posterior probability given $\boldsymbol{x}$.

A major drawback of the k-NN rule is that every time a new pattern is considered, its distance from all the training points has to be computed, then selecting the k closest to it points. To this end, various searching techniques have been suggested over the years. The interested reader may consult [38] for a related discussion.

Remarks 7.3.

- The use of the k-NN concept can also be adopted in the context of the regression task. Given an observation $\boldsymbol{x}$, one searches for its k closest input vectors in the training set, denoted as $\boldsymbol{x}_{(1)}, \ldots, \boldsymbol{x}_{(k)}$, and computes an estimate of the output value $\hat{y}$ as an average of the respective outputs in the training

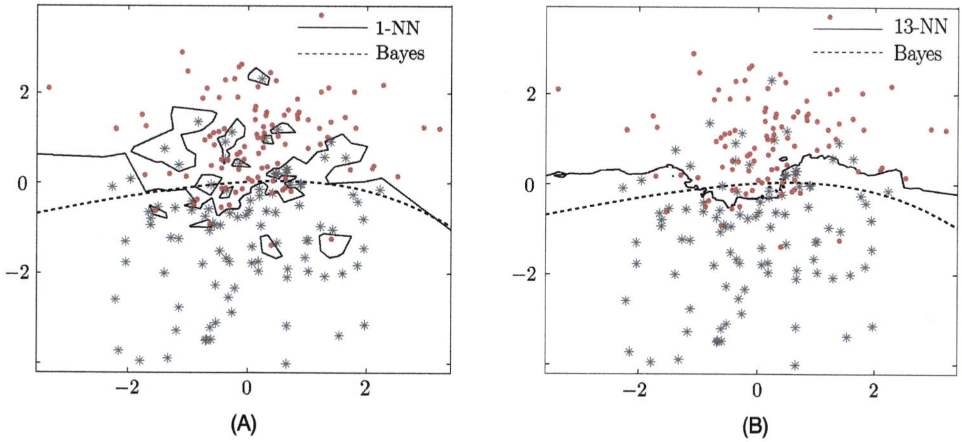

FIGURE 7.8

A two-class classification task. The dotted curve corresponds to the optimal Bayesian classifier. The full line curves correspond to the (A) 1-NN and (B) 13-NN classifiers. Observe that the 13-NN is closer to the Bayesian one.

set, represented by

$$\hat{y} = \frac{1}{k}\sum_{i=1}^{k} y_{(i)}.$$

Example 7.4. An example that illustrates the decision curves for a two-class classification task in the two-dimensional space, obtained by the Bayesian, the 1-NN, and the 13-NN classifier, is given in Fig. 7.8. A number of $N = 100$ data are generated for each class by Gaussian distributions. The decision curve of the Bayes classifier has the form of a parabola, while the 1-NN classifier exhibits a highly nonlinear nature. The 13-NN rule forms a decision line close to the Bayesian one.

7.6 LOGISTIC REGRESSION

In Bayesian classification, the assignment of a pattern in a class is performed based on the posterior probabilities, $P(\omega_i|\boldsymbol{x})$. The posteriors are estimated via the respective conditional PDFs, which is not, in general, an easy task. The goal in this section is to model the posterior probabilities directly, via the *logistic regression* method. This name has been established in the statistics community, although the model refers to classification and not to regression. This is a typical example of the discriminative modeling approach, where the distribution of data is not taken into account.

The two-class case: The starting point is to model the ratio of the posteriors as

$$\boxed{\ln\frac{P(\omega_1|\boldsymbol{x})}{P(\omega_2|\boldsymbol{x})} = \boldsymbol{\theta}^T\boldsymbol{x}: \quad \text{two-class logistic regression,}} \tag{7.29}$$

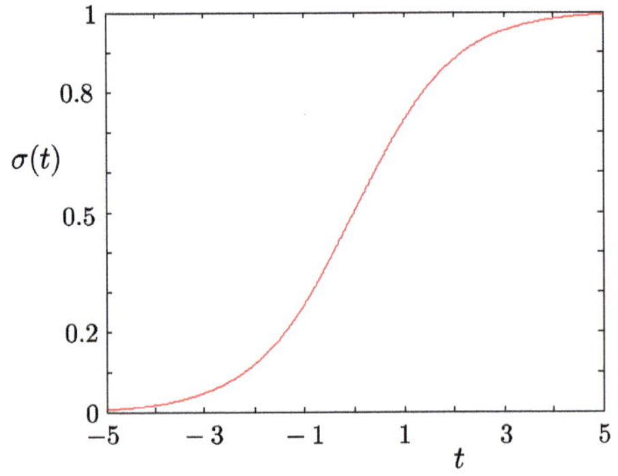

FIGURE 7.9

The sigmoid link function.

where the constant term, θ_0, has been absorbed in $\boldsymbol{\theta}$. Taking into account that

$$P(\omega_1|\boldsymbol{x}) + P(\omega_2|\boldsymbol{x}) = 1$$

and defining

$$t := \boldsymbol{\theta}^T \boldsymbol{x},$$

it is readily seen that the model in (7.29) is equivalent to

$$P(\omega_1|\boldsymbol{x}) = \sigma(t), \tag{7.30}$$

$$\sigma(t) := \frac{1}{1 + \exp(-t)}, \tag{7.31}$$

and

$$P(\omega_2|\boldsymbol{x}) = 1 - P(\omega_1|\boldsymbol{x}) = \frac{\exp(-t)}{1 + \exp(-t)}. \tag{7.32}$$

The function $\sigma(t)$ is known as the *logistic sigmoid* or *sigmoid link* function and it is shown in Fig. 7.9.

Although it may sound a bit "mystical" as to how one thought of such a model, it suffices to look more carefully at (7.17)–(7.18) to demystify it. Let us assume that the data in a two-class task follow Gaussian distributions with $\Sigma_1 = \Sigma_2 \equiv \Sigma$. Under such assumptions, and taking into account the Bayes theorem, one can write

$$\ln \frac{P(\omega_1|\boldsymbol{x})}{P(\omega_2|\boldsymbol{x})} = \ln \frac{p(\boldsymbol{x}|\omega_1)P(\omega_1)}{p(\boldsymbol{x}|\omega_2)P(\omega_2)} \tag{7.33}$$

$$= \ln p(\boldsymbol{x}|\omega_1) + \ln P(\omega_1) - \big(\ln p(\boldsymbol{x}|\omega_2) + \ln P(\omega_2)\big) \tag{7.34}$$

$$= g(\boldsymbol{x}). \tag{7.35}$$

Furthermore, we know that under the previous assumptions, $g(\boldsymbol{x})$ is given by Eqs. (7.20)–(7.22); hence, we can write

$$\ln \frac{P(\omega_1|\boldsymbol{x})}{P(\omega_2|\boldsymbol{x})} = (\boldsymbol{\mu}_1 - \boldsymbol{\mu}_2)^T \Sigma^{-1}\boldsymbol{x} + \text{constants}, \tag{7.36}$$

where "constants" refers to all terms that do not depend on $\boldsymbol{x}$. In other words, when the distributions that describe the data are Gaussians with a common covariance matrix, then the log ratio of the posteriors is a *linear* function. Thus, in logistic regression, all we do is to go one step ahead and adopt such a linear model, irrespective of the underlying data distributions.

Moreover, even if the data are distributed according to Gaussians, it may still be preferable to adopt the logistic regression formulation instead of that in (7.36). In the latter formulation, the covariance matrix has to be estimated, amounting to $\mathcal{O}(l^2/2)$ parameters. The logistic regression formulation only involves $l+1$ parameters. That is, once we know about the linear dependence of the log ratio on $\boldsymbol{x}$, we can use this a priori information to simplify the model. Of course, assuming that the Gaussian assumption is valid, if one can obtain good estimates of the covariance matrix, employing this extra information can lead to more efficient estimates, in the sense of lower variance. The issue is treated in [12]. This is natural, because more information concerning the distribution of the data is exploited. In practice, it turns out that using the logistic regression is, in general, a safer bet compared to the linear discriminant analysis (LDA).

The parameter vector $\boldsymbol{\theta}$ is estimated via the ML method applied on the set of training samples, $(y_n, \boldsymbol{x}_n)$, $n = 1, 2, \ldots, N$, $y_n \in \{0, 1\}$. The likelihood function can be written as

$$P(y_1, \ldots, y_N; \boldsymbol{\theta}) = \prod_{n=1}^{N} \left(\sigma(\boldsymbol{\theta}^T \boldsymbol{x}_n)\right)^{y_n} \left(1 - \sigma(\boldsymbol{\theta}^T \boldsymbol{x}_n)\right)^{1-y_n}. \tag{7.37}$$

Indeed, if $\boldsymbol{x}_n$ originates from class ω_1, then $y_n = 1$ and the corresponding probability is given by $\sigma(\boldsymbol{\theta}^T \boldsymbol{x}_n)$. On the other hand, if $\boldsymbol{x}_n$ comes from ω_2, then $y_n = 0$ and the respective probability is given by $1 - \sigma(\boldsymbol{\theta}^T \boldsymbol{x}_n)$. Assuming independence among the successive observations, the likelihood is given as the product of the respective probabilities.

Usually, we consider the negative log-likelihood given by

$$L(\boldsymbol{\theta}) = -\sum_{n=1}^{N} \big(y_n \ln s_n + (1 - y_n) \ln(1 - s_n)\big), \tag{7.38}$$

where

$$s_n := \sigma(\boldsymbol{\theta}^T \boldsymbol{x}_n). \tag{7.39}$$

The log-likelihood cost function in (7.38) is also known as the *cross-entropy* error. Minimization of $L(\boldsymbol{\theta})$ with respect to $\boldsymbol{\theta}$ is carried out iteratively by any iterative minimization scheme, such as the gradient descent or Newton's method. Both schemes need the computation of the respective gradient, which in turn is based on the derivative of the sigmoid link function (Problem 7.6)

$$\frac{d\sigma(t)}{dt} = \sigma(t)\big(1 - \sigma(t)\big). \tag{7.40}$$

The gradient is given by (Problem 7.7)

$$\begin{aligned}\nabla L(\boldsymbol{\theta}) &= \sum_{n=1}^{N} (s_n - y_n)\boldsymbol{x}_n \\ &= X^T(\boldsymbol{s} - \boldsymbol{y}),\end{aligned} \tag{7.41}$$

where

$$X^T = [\boldsymbol{x}_1, \ldots, \boldsymbol{x}_N], \quad \boldsymbol{s} := [s_1, \ldots, s_N]^T, \quad \boldsymbol{y} = [y_1, \ldots, y_N]^T.$$

The Hessian matrix is given by (Problem 7.8)

$$\begin{aligned}\nabla^2 L(\boldsymbol{\theta}) &= \sum_{n=1}^{N} s_n(1 - s_n)\boldsymbol{x}_n\boldsymbol{x}_n^T \\ &= X^T R X,\end{aligned} \tag{7.42}$$

where

$$R := \text{diag}\{s_1(1 - s_1), \ldots, s_N(1 - s_N)\}. \tag{7.43}$$

Note that because $0 < s_n < 1$, by definition of the sigmoid link function, matrix R is positive definite (see Appendix A); hence, the Hessian matrix is also positive definite (Problem 7.9). This is a necessary and sufficient condition for convexity.[2] Thus, the negative log-likelihood function is convex, which guarantees the existence of a unique minimum (e.g., [1] and Chapter 8).

Two of the possible iterative minimization schemes to be used are

- Gradient descent (Section 5.2)

$$\boldsymbol{\theta}^{(i)} = \boldsymbol{\theta}^{(i-1)} - \mu_i X^T(\boldsymbol{s}^{(i-1)} - \boldsymbol{y}). \tag{7.44}$$

- Newton's scheme (Section 6.7)

$$\begin{aligned}\boldsymbol{\theta}^{(i)} &= \boldsymbol{\theta}^{(i-1)} - \mu_i \left(X^T R^{(i-1)} X\right)^{-1} X^T(\boldsymbol{s}^{(i-1)} - \boldsymbol{y}) \\ &= \left(X^T R^{(i-1)} X\right)^{-1} X^T R^{(i-1)} \boldsymbol{z}^{(i-1)},\end{aligned} \tag{7.45}$$

where

$$\boldsymbol{z}^{(i-1)} := X\boldsymbol{\theta}^{(i-1)} - \left(R^{(i-1)}\right)^{-1}(\boldsymbol{s}^{(i-1)} - \boldsymbol{y}). \tag{7.46}$$

Eq. (7.45) is a weighted version of the LS solution (Chapters 3 and 6); however, the involved quantities are iteration dependent and the resulting scheme is known as *iterative reweighted least-squares* scheme (IRLS) [36].

---

[2] Convexity is discussed in more detail in Chapter 8.

Maximizing the likelihood we may run into problems if the training data set is linearly separable. In this case, any point on a hyperplane, $\boldsymbol{\theta}^T \boldsymbol{x} = 0$, that solves the classification task and separates the samples from each class (note that there are infinitely many of such hyperplanes) results in $\sigma(\boldsymbol{x}) = 0.5$, and every training point from each class is assigned a posterior probability equal to one. Thus, ML forces the logistic sigmoid to become a step function in the feature space and equivalently $||\boldsymbol{\theta}|| \rightarrow \infty$. This can lead to overfitting and it is remedied by including a *regularization* term, e.g., $||\boldsymbol{\theta}||^2$, in the respective cost function.

The M-class case: For the more general M-class classification task, the logistic regression is defined for $m = 1, 2, \ldots, M$ as

$$\boxed{P(\omega_m|\boldsymbol{x}) = \frac{\exp(\boldsymbol{\theta}_m^T \boldsymbol{x})}{\sum_{j=1}^{M} \exp(\boldsymbol{\theta}_j^T \boldsymbol{x})}: \quad \text{multiclass logistic regression.}} \tag{7.47}$$

The previous definition is easily brought into the form of a linear model for the log ratio of the posteriors. Divide, for example, by $P(\omega_M|\boldsymbol{x})$ to obtain

$$\ln \frac{P(\omega_m|\boldsymbol{x})}{P(\omega_M|\boldsymbol{x})} = (\boldsymbol{\theta}_m - \boldsymbol{\theta}_M)^T \boldsymbol{x} = \hat{\boldsymbol{\theta}}_m^T \boldsymbol{x}.$$

Let us define, for notational convenience,

$$\phi_{nm} := P(\omega_m|\boldsymbol{x}_n), \quad n = 1, 2, \ldots, N, \ m = 1, 2, \ldots, M,$$

and

$$t_m := \boldsymbol{\theta}_m^T \boldsymbol{x}, \quad m = 1, 2, \ldots, M.$$

The likelihood function is now written as

$$P(\boldsymbol{y}; \boldsymbol{\theta}_1, \ldots, \boldsymbol{\theta}_M) = \prod_{n=1}^{N} \prod_{m=1}^{M} (\phi_{nm})^{y_{nm}}, \tag{7.48}$$

where $y_{nm} = 1$ if $\boldsymbol{x}_n \in \omega_m$ and zero otherwise. The respective negative log-likelihood function becomes

$$L(\boldsymbol{\theta}_1, \ldots, \boldsymbol{\theta}_M) = -\sum_{n=1}^{N} \sum_{m=1}^{M} y_{nm} \ln \phi_{nm}, \tag{7.49}$$

which is the generalization of the cross-entropy cost function for the case of M classes. Minimization with respect to $\boldsymbol{\theta}_m$, $m = 1, \ldots, M$, takes place iteratively. To this end, the following gradients are used (Problems 7.10–7.12):

$$\frac{\partial \phi_{nm}}{\partial t_j} = \phi_{nm}(\delta_{mj} - \phi_{nj}), \tag{7.50}$$

where δ_{mj} is one if $m = j$ and zero otherwise. Also,

$$\nabla_{\theta_j} L(\boldsymbol{\theta}_1, \ldots, \boldsymbol{\theta}_M) = \sum_{n=1}^{N} (\phi_{nj} - y_{nj}) \boldsymbol{x}_n. \tag{7.51}$$

The respective Hessian matrix is an $(lM) \times (lM)$ matrix, comprising $l \times l$ blocks. Its k, j block element is given by

$$\nabla_{\theta_k} \nabla_{\theta_j} L(\boldsymbol{\theta}_1, \ldots, \boldsymbol{\theta}_M) = \sum_{n=1}^{N} \phi_{nj} (\delta_{kj} - \phi_{nk}) \boldsymbol{x}_n \boldsymbol{x}_n^T. \tag{7.52}$$

The Hessian matrix is also positive definite, which guarantees uniqueness of the minimum as in the two-class case.

Remarks 7.4.

- *Probit regression*: Instead of using the logistic sigmoid function in (7.30) (for the two-class case), other functions can also be adopted. A popular function in the statistical community is the *probit* function, which is defined as

$$\begin{aligned}\Phi(t) &:= \int_{-\infty}^{t} \mathcal{N}(z|0, 1) dz \\ &= \frac{1}{2}\left(1 + \frac{1}{\sqrt{2}} \operatorname{erf}(t)\right),\end{aligned} \tag{7.53}$$

  where erf is the *error* function defined as

$$\operatorname{erf}(t) = \frac{2}{\sqrt{\pi}} \int_0^t \exp\left(-\frac{z^2}{2}\right) dz.$$

  In other words, $P(\omega_1|t)$ is modeled to be equal to the probability of a normalized Gaussian variable to lie in the interval $(-\infty, \ t]$. The graph of the probit function is very similar to that of the logistic one.

7.7 FISHER'S LINEAR DISCRIMINANT

We now turn our focus to designing linear classifiers. In other words, irrespective of the data distribution in each class, we decide to partition the space in terms of hyperplanes, so that

$$g(\boldsymbol{x}) = \boldsymbol{\theta}^T \boldsymbol{x} + \theta_0 = 0. \tag{7.54}$$

We have dealt with the task of designing linear classifiers in the framework of the LS method in Chapter 3. In this section, the unknown parameter vector will be estimated via a path that exploits a number of important notions relevant to classification. The method is known as *Fisher's discriminant* and it can be dressed up with different interpretations.

Thus, its significance lies not only in its practical use but also in its pedagogical value. Prior to presenting the method, let us first discuss some related issues concerning the selection of features that describe the input patterns and some associated measures that can quantify the "goodness" of a selected set of features.

7.7.1 SCATTER MATRICES

Two of the major phases in designing a pattern recognition system are the *feature generation* and *feature selection* phases. Selecting information-rich features is of paramount importance. If "bad" features are selected, whatever smart classifier one adopts, the performance is bound to be poor. Feature generaton/selection techniques are treated in detail in, e.g., [38,39], to which the interested reader may refer for further information. At this point, we only touch on a few notions that are relevant to our current design of a linear classifier. Let us first quantify what a "bad" and a "good" feature is. The main goal in selecting features, and, thus, in selecting the feature space in which one is going to work, can be summarized in the following way. Select the features to create a feature space in which the points, which represent the training patterns, are distributed so as to have

large between-class distance and small within-class variance.

Fig. 7.10 illustrates three different possible choices for the case of two-dimensional feature spaces. Each point corresponds to a different input pattern and each figure corresponds to a different choice of the pair of features; that is, each figure shows the distribution of the input patterns in the respective feature space. Common sense dictates that selecting as features to represent the input patterns those associated with Fig. 7.10C is the best one; the points in the three classes form groups that lie relatively far away from each other, and at the same time the data in each class are compactly clustered together. The worst of the three choices is that of Fig. 7.10B, where data in each class are spread around their mean value and the three groups are relatively close to each other. The goal in feature selection is to

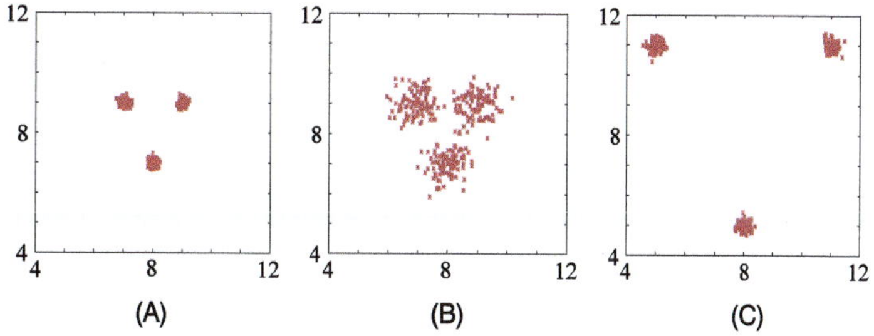

FIGURE 7.10

Three different choices of two-dimensional feature spaces: (A) small within-class variance and small between-class distance; (B) large within-class variance and small between-class distance; and (C) small within-class variance and large between-class distance. The last one is the best choice out of the three.

develop measures that quantify the "slogan" given in the box above. The notion of *scatter matrices* is of relevance to us here.

- *Within-class scatter matrix:*

$$\Sigma_w = \sum_{k=1}^{M} P(\omega_k)\Sigma_k, \tag{7.55}$$

where Σ_k is the covariance matrix of the points in the kth among M classes. In words, Σ_w is the average covariance matrix of the data in the specific l-dimensional feature space.

- *Between-class scatter matrix:*

$$\Sigma_b = \sum_{k=1}^{M} P(\omega_k)(\boldsymbol{\mu}_k - \boldsymbol{\mu}_0)(\boldsymbol{\mu}_k - \boldsymbol{\mu}_0)^T, \tag{7.56}$$

where $\boldsymbol{\mu}_0$ is the overall mean value defined by

$$\boldsymbol{\mu}_0 = \sum_{k=1}^{M} P(\omega_k)\boldsymbol{\mu}_k. \tag{7.57}$$

Another commonly used related matrix is the following.

- *Mixture scatter matrix:*

$$\Sigma_m = \Sigma_w + \Sigma_b. \tag{7.58}$$

A number of criteria that measure the "goodness" of the selected feature space are built around these scatter matrices; three typical examples are (e.g., [17,38])

$$J_1 := \frac{\text{trace}\{\Sigma_m\}}{\text{trace}\{\Sigma_w\}}, \quad J_2 = \frac{|\Sigma_m|}{|\Sigma_w|}, \quad J_3 = \text{trace}\{\Sigma_w^{-1}\Sigma_b\}, \tag{7.59}$$

where $|\cdot|$ denotes the determinant of a matrix.

The J_1 criterion is the easiest to understand. To simplify the arguments, let us focus on the two-dimensional (two features, x_1, x_2) case involving three classes. Recall from the definition of the covariance matrix of a random vector $\mathbf{x}$ in Eq. (2.31) of Chapter 2 that the elements across the main diagonal are the variances of the respective random variables. Thus, for each class, $k = 1, 2, 3$, the trace of the corresponding covariance matrix will be the sum of the variances for each one of the two features, i.e.,

$$\text{trace}\{\Sigma_k\} = \sigma_{k1}^2 + \sigma_{k2}^2.$$

Hence, the trace of Σ_w is the average total variance of the two features over all three classes,

$$\text{trace}\{\Sigma_w\} = \sum_{k=1}^{3} P(\omega_k)\left(\sigma_{k1}^2 + \sigma_{k2}^2\right) := s_w.$$

On the other hand, the trace of Σ_b is equal to the average, over all classes, of the total squared Euclidean distance of each individual feature mean value from the corresponding global one, i.e.,

$$\text{trace}\{\Sigma_b\} = \sum_{k=1}^{3} P(\omega_3)\left((\mu_{k1} - \mu_{01})^2 + (\mu_{k2} - \mu_{02})^2\right) \tag{7.60}$$

$$= \sum_{k=1}^{3} P(\omega_k)||\boldsymbol{\mu}_k - \boldsymbol{\mu}_0||^2 := s_b, \tag{7.61}$$

where $\boldsymbol{\mu}_k = [\mu_{k1}, \mu_{k2}]^T$ is the mean value for the kth class and $\boldsymbol{\mu}_0 = [\mu_{01}, \mu_{02}]^T$ is the global mean vector.[3] Thus, the J_1 criterion is equal to

$$J_1 = \frac{s_w + s_b}{s_w} = 1 + \frac{s_b}{s_w}.$$

In other words, the smaller the average total variance is and the larger the average squared Euclidean distance of the mean values from the global one gets, the larger the value of J_1 becomes. Similar arguments can be made for the other two criteria.

7.7.2 FISHER'S DISCRIMINANT: THE TWO-CLASS CASE

In Fisher's linear discriminant analysis, the emphasis in Eq. (7.54) is only on $\boldsymbol{\theta}$; the bias term θ_0 is left out of the discussion. The inner product $\boldsymbol{\theta}^T\boldsymbol{x}$ can be viewed as the projection of $\boldsymbol{x}$ along the vector $\boldsymbol{\theta}$. Strictly speaking, we know from geometry that the respective projection is also a vector, $\boldsymbol{y}$, given by (e.g., Section 5.6)

$$\boldsymbol{y} = \frac{\boldsymbol{\theta}^T\boldsymbol{x}}{||\boldsymbol{\theta}||}\frac{\boldsymbol{\theta}}{||\boldsymbol{\theta}||},$$

where $\frac{\boldsymbol{\theta}}{||\boldsymbol{\theta}||}$ is the unit norm vector in the direction of $\boldsymbol{\theta}$. From now on, we will focus on the scalar value of the projection, $y := \boldsymbol{\theta}^T\boldsymbol{x}$, and ignore the scaling factor in the denominator, because scaling all features by the same value has no effect on our discussion. The goal, now, is to select that direction $\boldsymbol{\theta}$ so that after projecting along this direction, (a) the data in the two classes are as far away as possible from each other, and (b) the respective variances of the points around their means, in each one of the classes, are as small as possible. A criterion that quantifies the aforementioned goal is *Fisher's discriminant ratio* (FDR), defined as

$$\boxed{\text{FDR} = \frac{(\mu_1 - \mu_2)^2}{\sigma_1^2 + \sigma_2^2}: \quad \text{Fisher's discriminant ratio,}} \tag{7.62}$$

where μ_1 and μ_2 are the (scalar) mean values of the two classes after the projection along $\boldsymbol{\theta}$, meaning

$$\mu_k = \boldsymbol{\theta}^T\boldsymbol{\mu}_k, \quad k = 1, 2.$$

[3] Note that the last equation can also be derived by using the property $\text{trace}\{A\} = \text{trace}\{A^T\}$, applied on Σ_b.

However, we have

$$(\mu_1 - \mu_2)^2 = \boldsymbol{\theta}^T(\boldsymbol{\mu}_1 - \boldsymbol{\mu}_2)(\boldsymbol{\mu}_1 - \boldsymbol{\mu}_2)^T\boldsymbol{\theta} = \boldsymbol{\theta}^T S_b \boldsymbol{\theta},$$
$$S_b := (\boldsymbol{\mu}_1 - \boldsymbol{\mu}_2)(\boldsymbol{\mu}_1 - \boldsymbol{\mu}_2)^T. \tag{7.63}$$

Note that if the classes are equiprobable, S_b is a scaled version of the between-class scatter matrix in (7.56) (this is easily checked, since under this assumption, $\boldsymbol{\mu}_0 = 1/2(\boldsymbol{\mu}_1 + \boldsymbol{\mu}_2)$), and we have

$$(\mu_1 - \mu_2)^2 \propto \boldsymbol{\theta}^T \Sigma_b \boldsymbol{\theta}. \tag{7.64}$$

Moreover,

$$\sigma_k^2 = \mathbb{E}\left[(\mathrm{y} - \mu_k)^2\right] = \mathbb{E}\left[\boldsymbol{\theta}^T(\mathbf{x} - \boldsymbol{\mu}_k)(\mathbf{x} - \boldsymbol{\mu}_k)^T\boldsymbol{\theta}\right] = \boldsymbol{\theta}^T \Sigma_k \boldsymbol{\theta}, \quad k = 1, 2, \tag{7.65}$$

which leads to

$$\sigma_1^2 + \sigma_2^2 = \boldsymbol{\theta}^T S_w \boldsymbol{\theta},$$

where $S_w = \Sigma_1 + \Sigma_2$. Note that if the classes are equiprobable, S_w becomes a scaled version of the within-class scatter matrix defined in (7.55), and we have

$$\sigma_1^2 + \sigma_2^2 \propto \boldsymbol{\theta}^T \Sigma_w \boldsymbol{\theta}. \tag{7.66}$$

Combining (7.62), (7.64), and (7.66) and neglecting the proportionality constants, we end up with

$$\boxed{\text{FDR} = \frac{\boldsymbol{\theta}^T \Sigma_b \boldsymbol{\theta}}{\boldsymbol{\theta}^T \Sigma_w \boldsymbol{\theta}}: \quad \text{generalized Rayleigh quotient.}} \tag{7.67}$$

Our goal now becomes that of maximizing the FDR with respect to $\boldsymbol{\theta}$. This is a case of the *generalized Rayleigh ratio*, and it is known from linear algebra that it is maximized if $\boldsymbol{\theta}$ satisfies

$$\Sigma_b \boldsymbol{\theta} = \lambda \Sigma_w \boldsymbol{\theta},$$

where λ is the maximum eigenvalue of the matrix $\Sigma_w^{-1}\Sigma_b$ (Problem 7.14). However, for our specific case[4] here, we can bypass the need for solving an eigenvalue–eigenvector problem. Taking into account that Σ_w is a scaled version of S_b in Eq. (7.63), the last equation can be rewritten as

$$\lambda \Sigma_w \boldsymbol{\theta} \propto (\boldsymbol{\mu}_1 - \boldsymbol{\mu}_2)(\boldsymbol{\mu}_1 - \boldsymbol{\mu}_2)^T\boldsymbol{\theta} \propto (\boldsymbol{\mu}_1 - \boldsymbol{\mu}_2),$$

since the inner product, $(\boldsymbol{\mu}_1 - \boldsymbol{\mu}_2)^T\boldsymbol{\theta}$, is a scalar. In other words, $\Sigma_w \boldsymbol{\theta}$ lies in the direction of $(\boldsymbol{\mu}_1 - \boldsymbol{\mu}_2)$, and because we are only interested in the direction, we can finally write

$$\boxed{\boldsymbol{\theta} = \Sigma_w^{-1}(\boldsymbol{\mu}_1 - \boldsymbol{\mu}_2),} \tag{7.68}$$

[4] Σ_b is a rank one matrix and, hence, there is only one nonzero eigenvalue; see also Problem 7.15.

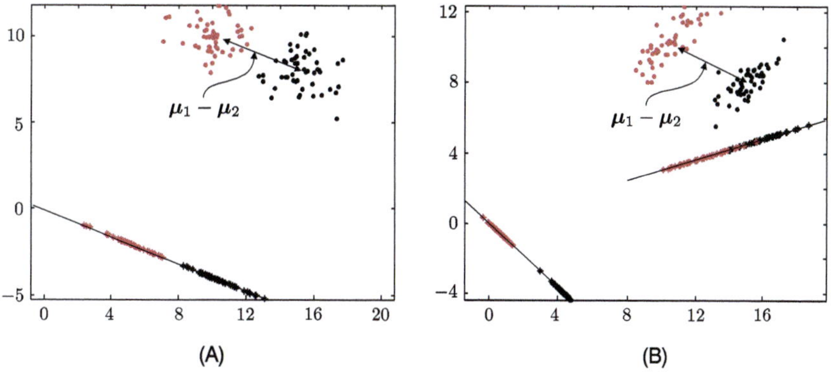

FIGURE 7.11

(A) The optimal direction resulting from Fisher's discriminant for two spherically distributed classes. The direction on which projection takes place is parallel to the segment joining the mean values of the data in the two classes. (B) The line on the bottom left of the figure corresponds to the direction that results from Fisher's discriminant; observe that it is no longer parallel to $\boldsymbol{\mu}_1 - \boldsymbol{\mu}_2$. For the sake of comparison, observe that projecting on the other line on the right results in class overlap.

assuming of course that Σ_w is invertible. In practice, Σ_w is obtained as the respective sample mean using the available observations.

Fig. 7.11A shows the resulting direction for two spherically distributed (isotropic) classes in the two-dimensional space. In this case, the direction for projecting the data is parallel to $(\boldsymbol{\mu}_1 - \boldsymbol{\mu}_2)$. In Fig. 7.11B, the distribution of the data in the two classes is not spherical, and the direction of projection (the line to the bottom left of the figure) is not parallel to the segment joining the two mean points. Observe that if the line to the right is selected, then after projection the classes do overlap.

In order for the Fisher discriminant method to be used as a classifier, a threshold θ_0 must be adopted, and decision in favor of a class is performed according to the rule

$$y = (\boldsymbol{\mu}_1 - \boldsymbol{\mu}_2)^T \Sigma_w^{-1} \boldsymbol{x} + \theta_0 \begin{cases} > 0, & \text{class } \omega_1, \\ < 0, & \text{class } \omega_2. \end{cases} \tag{7.69}$$

Compare, now, (7.69) with (7.20)–(7.22); the latter were obtained via the Bayes rule for the Gaussian case, when both classes share the same covariance matrix. Observe that for this case, the resulting hyperplanes for both methods are parallel and the only difference is in the threshold value. Note, however, that the Gaussian assumption was not needed for the Fisher discriminant. This justifies the use of (7.20)–(7.22), even when the data are not normally distributed. In practice, depending on the data, different threshold values may be used.

Finally, because the world is often small, it can be shown that Fisher's discriminant can also be seen as a special case of the LS solution if the target class labels, instead of ± 1, are chosen as $\frac{N_1}{N}$ and $\frac{-N_2}{N}$, respectively, where N is the total number of training samples, N_1 is the number of samples in class ω_1, and N_2 is the corresponding number in class ω_2, e.g., [41].

Another point of view for Fisher's discriminant method is that it performs dimensionality reduction by projecting the data from the original l-dimensional space to a lower one-dimensional space. This reduction in dimensionality is performed in a *supervised* way, by exploiting the class labels of the training data. As we will see in Chapter 19, there are other techniques in which the dimensionality reduction takes place in an unsupervised way. The obvious question now is whether it is possible to use Fisher's idea to reduce the dimensionality, not to one, but to another intermediate value between one and l, where l is the dimensionality of the feature space. It turns out that this is possible, but it also depends on the number of classes. More on dimensionality reduction techniques can be found in Chapter 19.

7.7.3 FISHER'S DISCRIMINANT: THE MULTICLASS CASE

Our starting point for generalizing to the multiclass case is the J_3 criterion defined in (7.59). It can be readily shown that the FDR criterion, used in the two-class case, is directly related to the J_3 one, once the latter is considered for the one-dimensional case and for equiprobable classes. For the more general multiclass formulation, the task becomes that of estimating an $l \times m,\ m < l$, matrix, A, such that the linear transformation from the original $\mathbb{R}^l$ to the new, lower-dimensional, $\mathbb{R}^m$ space, expressed as

$$\boldsymbol{y} = A^T \boldsymbol{x}, \tag{7.70}$$

retains as much classification-related information as possible. Note that in any dimensionality reduction technique, some of the original information is, in general, bound to be lost. Our goal is for the loss to be as small as possible. Because we chose to quantify classification-related information by the J_3 criterion, the goal is to compute A in order to maximize

$$J_3(A) = \text{trace}\{\Sigma_{wy}^{-1} \Sigma_{by}\}, \tag{7.71}$$

where Σ_{wy} and Σ_{by} are the within-class and between-class scatter matrices measured in the transformed lower-dimensional space. Maximization follows standard arguments of optimization with respect to matrices. The algebra gets a bit involved and we will state the final result. Details of the proof can be found in, e.g., [17,38]. Matrix A is given by the following equation:

$$(\Sigma_{wx}^{-1} \Sigma_{bx})A = A\Lambda. \tag{7.72}$$

Matrix Λ is a diagonal matrix having as elements m of the eigenvalues (Appendix A) of the $l \times l$ matrix, $\Sigma_{wx}^{-1} \Sigma_{bx}$, where Σ_{wx} and Σ_{bx} are the within-class and between-class scatter matrices, respectively, in the original $\mathbb{R}^l$ space. The matrix of interest, A, comprises columns that are the respective eigenvectors. The problem, now, becomes to select the m eigenvalues/eigenvectors. Note that by its definition, Σ_b, being the sum of M related (via $\boldsymbol{\mu}_0$) rank one matrices, is of rank $M - 1$ (Problem 7.15). Thus, the product $\Sigma_{wx}^{-1} \Sigma_{bx}$ has only $M - 1$ nonzero eigenvalues. This imposes a stringent constraint on the dimensionality reduction. The maximum dimension m that one can obtain is $m = M - 1$ (for the two-class task, $m = 1$), irrespective of the original dimension l. There are two cases that are worth focusing on:

- $m = M - 1$. In this case, it is shown that if A is formed having as columns all the eigenvectors corresponding to the nonzero eigenvalues, then

$$J_{3y} = J_{3x}.$$

In other words, there is no loss of information (as measured via the J_3 criterion) by reducing the dimension from l to $M-1$! Note that in this case, Fisher's method produces $m = M - 1$ discriminant (linear) functions. This complies with a general result in classification stating that the minimum number of discriminant functions needed for an M-classification problem is $M-1$ [38]. Recall that in Bayesian classification, we need M functions, $P(\omega_i|\boldsymbol{x}),\ i = 1, 2, \dots, M$; however, only $M-1$ of those are independent, because they must all add to one. Hence, Fisher's method provides the minimum number of linear discriminants required.

- $m < M - 1$. If A is built having as columns the eigenvectors corresponding to the maximum m eigenvalues, then

$$J_{3y} < J_{3x}.$$

However, the resulting value J_{3y} is the maximum possible one.

Remarks 7.5.

- If J_3 is used with other matrix combinations, as might be achieved by using Σ_m in place of Σ_b, the constraint of the rank being equal to $M-1$ is removed, and larger values for m can be obtained.
- In a number of practical cases, Σ_w may not be invertible. This is, for example, the case in the *small sample size* problems, where the dimensionality of the feature space, l, may be larger than the number of the training data, N. Such problems may be encountered in applications such as web document classification, gene expression profiling, and face recognition. There are different escape routes in this problem; see [38] for a discussion and related references.

7.8 CLASSIFICATION TREES

Classification trees are based on a simple yet powerful idea, and they are among the most popular techniques for classification. They are *multistage* systems, and classification of a pattern into a class is achieved *sequentially*. Through a series of tests, classes are *rejected* in a sequential fashion until a decision is finally reached in favor of one remaining class. Each one of the tests, whose outcome decides which classes are rejected, is of a *binary* "Yes" or "No" type and is applied to a *single* feature. Our goal is to present the main philosophy around a special type of trees known as *ordinary binary classification trees* (OBCTs). They belong to a more general class of methods that construct trees, both for classification and for regression, known as *classification and regression trees* (CARTs) [2,31]. Variants of the method have also been proposed [35].

The basic idea around OBCTs is to partition the feature space into *(hyper)rectangles*; that is, the space is partitioned via hyperplanes, which are parallel to the axes. This is illustrated in Fig. 7.12. The partition of the space in (hyper)rectangles is performed via a series of "questions" of this form: is the value of the feature $x_i < a$? This is also known as the *splitting criterion*. The sequence of questions can nicely be realized via the use of a tree. Fig. 7.13 shows the tree corresponding to the case illustrated in Fig. 7.12. Each node of the tree performs a test against an *individual* feature, and if it is not a leaf node, it is connected to two *descendant* nodes: one is associated with the answer "Yes" and the other with the answer "No."

Starting from the root node, a path of successive decisions is realized until a leaf node is reached. Each leaf node is associated with a *single* class. The assignment of a point to a class is done according to the label of the respective leaf node. This type of classification is conceptually simple and easily interpretable. For example, in a medical diagnosis system, one may start with the question, is the temperature high? If yes, a second question can be: is the nose runny? The process carries on until a final decision concerning the disease has been reached. Also, trees are useful in building up reasoning systems in artificial intelligence [37]. For example, the existence of specific objects, which is deduced via a series of related questions based on the values of certain (high-level) features, can lead to the recognition of a scene or of an object depicted in an image.

Once a tree has been developed, classification is straightforward. The major challenge lies in constructing the tree, by exploiting the information that resides in the training data set. The main questions one is confronted with while designing a tree are the following:

- Which splitting criterion should be adopted?
- When should one stop growing a tree and declare a node as final?
- How is a leaf node associated with a specific class?

Besides the above issues, there are more that will be discussed later on.

Splitting criterion: We have already stated that the questions asked at each node are of the following type:

$$\text{Is } x_i < a?$$

The goal is to select an appropriate value for the threshold value a. Assume that starting from the root node, the tree has grown up to the current node, t. Each node t is associated with a subset $X_t \subseteq X$ of the training data set X. This is the set of the training points that have survived to this node, after the tests that have taken place at the previous nodes in the tree. For example, in Fig. 7.13, a number of

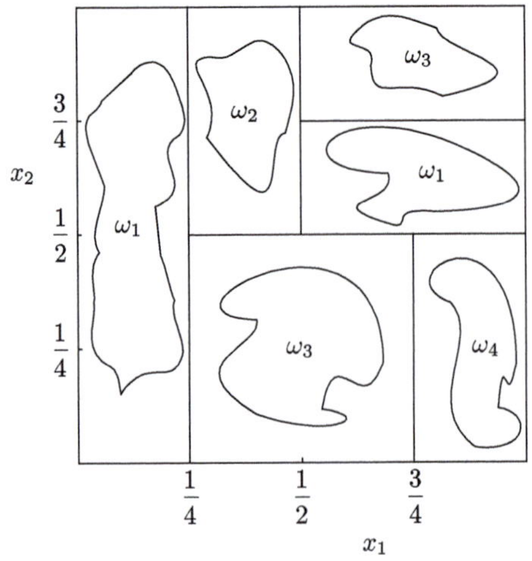

FIGURE 7.12

Partition of the two-dimensional features space, corresponding to three classes, via a classification (OBCT) tree.

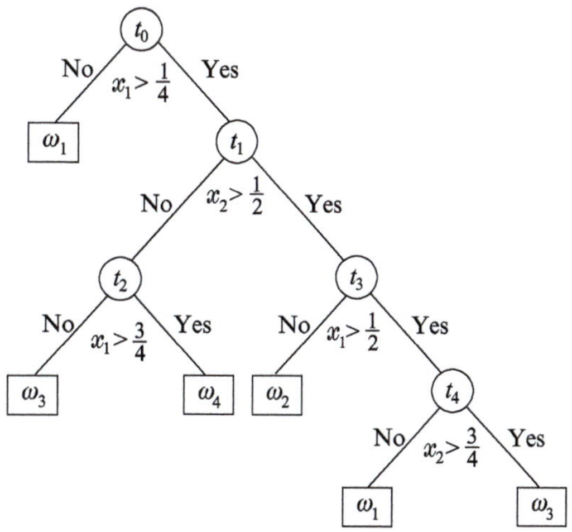

FIGURE 7.13

The classification tree that performs the space partitioning for the task indicated in Fig. 7.12.

points, which belong to, say, class ω_1, will not be involved in node t_1 because they have already been assigned in a previously labeled leaf node. The purpose of a splitting criterion is to split X_t into two *disjoint* subsets, namely, X_{tY} and X_{tN}, depending on the answer to the specific question at node t. For every split, the following is true:

$$\begin{aligned} X_{tY} \cap X_{tN} &= \emptyset, \\ X_{tY} \cup X_{tN} &= X_t. \end{aligned}$$

The goal in each node is to select which feature is to be tested and also what is the best value of the corresponding threshold value a. The adopted philosophy is to make the choice so that every split generates sets, X_{tY}, X_{tN}, which are more *class-homogeneous* compared to X_t. In other words, the data in each one of the two descendant sets must show a higher preference to specific classes, compared to the ancestor set. For example, assume that the data in X_t consist of points that belong to four classes, $\omega_1, \omega_2, \omega_3, \omega_4$. The idea is to perform the splitting so that most of the data in X_{tY} belong to, say, ω_1, ω_2, and most of the data in X_{tN} to ω_3, ω_4. In the adopted terminology, the sets X_{tY} and X_{tN} should be *purer* compared to X_t. Thus, we must first select a criterion that measures *impurity* and then compute the threshold value and choose the specific feature (to be tested) to maximize the decrease in node impurity. For example, a common measure to quantify impurity of node t is the *entropy*, defined as

$$I(t) = -\sum_{m=1}^{M} P(\omega_m|t) \log_2 P(\omega_m|t), \tag{7.73}$$

where $\log_2(\cdot)$ is the base-two logarithm. The maximum value of $I(t)$ occurs if all probabilities are equal (maximum impurity), and the smallest value, which is equal to zero, when only one of the probability values is one and the rest equal zero. Probabilities are approximated as

$$P(\omega_m|t) = \frac{N_t^m}{N_t}, \quad m = 1, 2, \ldots, M,$$

where N_t^m is the number of the points from class m in X_t and N_t is the total number of points in X_t. The decrease in node impurity, after splitting the data into two sets, is defined as

$$\Delta I(t) = I(t) - \frac{N_{tY}}{N_t} I(t_Y) - \frac{N_{tN}}{N_t} I(t_N), \tag{7.74}$$

where $I(t_Y)$ and $I(t_N)$ are the impurities associated with the two new sets, respectively. The goal now becomes to select the specific feature x_i and the threshold a_t so that $\Delta I(t)$ becomes maximum. This will now define two new descendant nodes of t, namely, t_N and t_Y; thus, the tree grows with two new nodes.

A way to search for different threshold values is the following. For each one of the features x_i, $i = 1, 2, \ldots, l$, rank the values x_{in}, $n = 1, 2, \ldots, N_t$, which this feature takes among the training points in X_t. Then define a sequence of corresponding threshold values, a_{in}, to be halfway between consecutive distinct values of x_{in}. Then test the impurity change that occurs for each one of these threshold values and keep the one that achieves the maximum decrease. Repeat the process for all features, and finally, keep the combination that results in the best maximum decrease.

Besides entropy, other impurity measuring indices can be used. A popular alternative, which results in a slightly sharper maximum compared to the entropy one, is the so-called *Gini index*, defined as

$$I(t) = \sum_{m=1}^{M} P(\omega_m|t)\big(1 - P(\omega_m|t)\big). \tag{7.75}$$

This index is also zero if one of the probability values is equal to 1 and the rest are zero, and it takes its maximum value when all classes are equiprobable.

Stop-splitting rule: The obvious question when growing a tree is when to stop growing it. One possible way is to adopt a threshold value T, and stop splitting a node once the maximum value $\Delta I(t)$, for all possible splits, is smaller than T. Another possibility is to stop when the cardinality of X_t becomes smaller than a certain number or if the node is pure, in the sense that all points in it belong to a single class.

Class assignment rule: Once a node t is declared to be a leaf node, it is assigned a class label; usually this is done on a majority voting rationale. That is, it is assigned the label of the class where most of the data in X_t belong.

Pruning a tree: Experience has shown that growing a tree and using a stopping rule does not always work well in practice; growing may either stop early or may result in trees of very large size. A common practice is to first grow a tree up to a large size and then adopt a pruning technique to eliminate nodes. Different pruning criteria can be used; a popular one is to combine an estimate of the error probability with a complexity measuring index (see [2,31]).

Remarks 7.6.

- Among the notable advantages of decision trees is the fact that they can naturally treat mixtures of numeric and categorical variables. Moreover, they scale well with large data sets. Also, they can treat missing variables in an effective way. In many domains, not all the values of the features are known for every pattern. The values may have gone unrecorded, or they may be too expensive to obtain. Finally, due to their structural simplicity, they are easily interpretable; in other words, it is possible for a human to understand the reason for the output of the learning algorithm. In some applications, such as in financial decisions, this is a legal requirement.

On the other hand, the prediction performance of the tree classifiers is not as good as other methods, such as support vector machines and neural networks, to be treated in Chapters 11 and 18, respectively.

- A major drawback associated with the tree classifiers is that they are *unstable*. That is, a small change in the training data set can result in a very different tree. The reason for this lies in the hierarchical nature of the tree classifiers. An error that occurs in a node at a high level of the tree propagates all the way down to the leaves below it.

  Bagging (bootstrap aggregating) [3] is a technique that can reduce the variance and improve the generalization error performance. The basic idea is to create B variants, $X_1, X_2, \dots, X_B$, of the training set X, using *bootstrap* techniques, by uniformly sampling from X with replacement. For each of the training set variants X_i, a tree T_i is constructed. The final decision for the classification of a given point is in favor of the class predicted by the majority of the subclassifiers T_i, $i = 1, 2, \dots, B$.

  Random forests use the idea of bagging in tandem with random feature selection [5]. The difference with bagging lies in the way the decision trees are constructed. The feature to split in each node is selected as the best among a set of F *randomly* chosen features, where F is a user-defined parameter. This extra introduced randomness is reported to have a substantial effect in performance improvement.

  Random forests often have very good predictive accuracy and have been used in a number of applications, including body pose recognition via Microsoft's popular Kinect sensor [34].

  Besides the previous methods, more recently, Bayesian techniques have also been suggested and used to stabilize the performance of trees (see [8,44]). Of course, the effect of using multiple trees is losing a main advantage of the trees, that is, their fairly easy interpretability.
- Besides the OBCT rationale, a more general partition of the feature space has also been proposed via hyperplanes that are not parallel to the axes. This is possible via questions of the following type: Is $\sum_{i=1}^{l} c_i x_i < a$? This can lead to a better partition of the space. However, the training now becomes more involved (see [35]).
- Decision trees have also been proposed for regression tasks, albeit with less success. The idea is to split the space into regions, and prediction is performed based on the average of the output values in the region where the observed input vector lies; such an averaging approach has as a consequence the lack of smoothness as one moves from one region to another, which is a major drawback of regression trees. The splitting into regions is performed based on the LS method [19].

7.9 COMBINING CLASSIFIERS

So far, we have discussed a number of classifiers, and more methods will be presented in Chapters 13, 11, and 18, concerning support vector machines, Bayesian methods, and neural/deep networks. The obvious question an inexperienced practitioner/researcher is confronted with is, which method then?

Unfortunately, there is no definitive answer. Furthermore, the choice of a method becomes a harder problem when the data size is rather small. The goal of this section is to discuss techniques that can benefit by combining different learners together.

NO FREE LUNCH THEOREM

The goal of the design of any classifier, and in general of any learning scheme, based on a training set of a finite size, is to provide a good generalization performance. However, there are no context independent or usage independent reasons to support one learning technique over another. Each learning task, represented by the available data set, will show a preference for a specific learning scheme that fits the specificities of the particular problem at hand. An algorithm that scores top in one problem can score low for another. This experimental finding is theoretically substantiated by the so-called *no free lunch theorem* for machine learning [43].

This important theorem states that, averaged over *all possible* data generating distributions, every classification algorithm results in the *same error rate* on data outside the training set. In other words, there is *no* learning algorithm that is universally optimal. However, note that these results hold only when one averages over all possible data generating distributions. If, on the other hand, when designing a learner, we exploit prior knowledge concerning the specificities of the particular data set, which is of interest to us, then we can design an algorithm that performs well on this data set.

In practice, one should try different learning methods from the available palette, each optimized to the specific task, and test its generalization performance against an independent data set different from the one used for training, using, for example, the leave-one-out method or any of its variants (Chapter 3). Then keep and use the method that scored best for the specific task.

To this end, there are a number of major efforts to compare different classifiers against different data sets and measure the "average" performance, via the use of different statistical indices, in order to quantify the overall performance of each classifier against the data sets.

SOME EXPERIMENTAL COMPARISONS

Experimental comparisons of methods always have a strong historical flavor, since, as time passes, new methods come into existence and, moreover, new and larger data sets can be obtained, which may change conclusions. In this subsection, we present samples of some major big projects that have been completed, whose goal was to compare different learners together. The terrain today may be different due to the advent of deep neural networks; yet, the knowledge that can be extracted from these previous projects is still useful and enlightening.

One of the very first efforts, to compare the performance of different classifiers, was the Statlog project [27]. Two subsequent efforts are summarized in [7,26]. In the former, 17 popular classifiers were tested against 21 data sets. In the latter, 10 classifiers and 11 data sets were employed. The results verify what has already been said: different classifiers perform better for different sets. However, it is reported that boosted trees (Section 7.11), random forests, bagged decision trees, and support vector machines were ranked among the top ones for most of the data sets.

The Neural Information Processing Systems Workshop (NIPS-2003) organized a classification competition based on five data sets. The results of the competition are summarized in [18]. The competition was focused on feature selection [28]. In a follow-up study [22], more classifiers were added. Among the considered classifiers, a Bayesian-type neural network scheme (Chapter 18) scored at the top, albeit at significantly higher run time requirements. The other classifiers considered were random forests and boosting, where trees and neural networks were used as base classifiers (Section 7.10). Random forests also performed well, at much lower computational times compared to the Bayesian-type classifier.

SCHEMES FOR COMBINING CLASSIFIERS

A trend to improve performance is to combine different classifiers and exploit their individual advantages. An observation that justifies such an approach is that during testing, there are patterns on which even the best classifier for a particular task fails to predict their true class. In contrast, the same patterns can be classified correctly by other classifiers, with an inferior overall performance. This shows that there may be some complementarity among different classifiers, and combination can lead to boosted performance compared to that obtained from the best (single) classifier. Recall that bagging, mentioned in Section 7.8, is a type of classifier combination.

The issue that arises now is to select a combination scheme. There are different schemes, and the results they provide can be different. Below, we summarize the more popular combination schemes.

- *Arithmetic averaging rule*: Assuming that we use L classifiers, where each one outputs a value of the posterior probability $P_j(\omega_i|\boldsymbol{x})$, $i = 1, 2, \ldots, M$, $j = 1, 2, \ldots, L$. A decision concerning the class assignment is based on the following rule:

$$\boxed{\text{Assign } \boldsymbol{x} \text{ to class } \omega_i = \arg\max_k \frac{1}{L} \sum_{j=1}^{L} P_j(\omega_k|\boldsymbol{x}), \; k = 1, 2, \ldots, M.} \tag{7.76}$$

  It can be shown that this rule is equivalent to computing the "final" posterior probability, $P(\omega_i|\boldsymbol{x})$, by minimizing the average Kullback–Leibler divergence (Problem 7.16),

$$D_{av} = \frac{1}{L} \sum_{j=1}^{L} D_j,$$

  where

$$D_j = \sum_{i=1}^{M} P_j(\omega_i|\boldsymbol{x}) \ln \frac{P_j(\omega_i|\boldsymbol{x})}{P(\omega_i|\boldsymbol{x})}.$$

- *Geometric averaging rule*: This rule is the outcome of minimizing the alternative formulation of the Kullback–Leibler divergence (note that KL divergence is not symmetric); in other words,

$$D_j = \sum_{i=1}^{M} P(\omega_i|\boldsymbol{x}) \ln \frac{P(\omega_i|\boldsymbol{x})}{P_j(\omega_i|\boldsymbol{x})},$$

  which results in (Problem 7.17)

$$\boxed{\text{Assign } \boldsymbol{x} \text{ to class } \omega_i = \arg\max_k \prod_{j=1}^{L} P_j(\omega_k|\boldsymbol{x}), \quad k = 1, 2, \ldots, M.} \tag{7.77}$$

- *Stacking*: An alternative way is to use a weighted average of the outputs of the individual classifiers, where the combination weights are obtained optimally using the training data. Assume that the output of each individual classifier, $f_j(\boldsymbol{x})$, is of a soft type, for example, an estimate of the posterior probability, as before. Then the combined output is given by

$$f(\boldsymbol{x}) = \sum_{j=1}^{L} w_j f_j(\boldsymbol{x}), \tag{7.78}$$

where the weights are estimated via the following optimization task:

$$\hat{\boldsymbol{w}} = \arg\min_{\boldsymbol{w}} \sum_{n=1}^{N} \mathcal{L}\big(y_n, f(\boldsymbol{x}_n)\big) = \arg\min_{\boldsymbol{w}} \sum_{n=1}^{N} \mathcal{L}\left(y_n, \sum_{j=1}^{L} w_j f_j(\boldsymbol{x}_n)\right), \tag{7.79}$$

where $\mathcal{L}(\cdot, \cdot)$ is a loss function, for example, the squared error one. However, adopting the previous optimization, based on the training data set, can lead to overfitting. According to stacking [42], a cross-validation rationale is adopted and instead of $f_j(\boldsymbol{x}_n)$, we employ the $f_j^{(-n)}(\boldsymbol{x}_n)$, where the latter is the output of the jth classifier trained on the data after *excluding* the pair $(y_n, \boldsymbol{x}_n)$. In other words, the weights are estimated by

$$\boxed{\hat{\boldsymbol{w}} = \arg\min_{\boldsymbol{w}} \sum_{n=1}^{N} \mathcal{L}\left(y_n, \sum_{j=1}^{L} w_j f_j^{(-n)}(\boldsymbol{x}_n)\right).} \tag{7.80}$$

Sometimes, the weights are constrained to be positive and add to one, giving rise to a constrained optimization task.
- *Majority voting rule*: The previous methods belong to the family of soft-type rules. A popular alternative is a hard-type rule, which is based on a voting scheme. One decides in favor of the class for which either there is a consensus or at least l_c of the classifiers agree on the class label, where

$$l_c = \begin{cases} \frac{L}{2} + 1, & L \text{ is even}, \\ \frac{L+1}{2}, & L \text{ is odd}. \end{cases}$$

Otherwise, the decision is rejection (i.e., no decision is taken).

In addition to the sum, product, and majority voting, other combinations rules have also been suggested, which are inspired by the inequalities [24]

$$\prod_{j=1}^{L} P_j(\omega_i|\boldsymbol{x}) \leq \min_{j=1}^{L} P_j(\omega_i|\boldsymbol{x}) \leq \frac{1}{L}\sum_{j=1}^{L} P_j(\omega_i|\boldsymbol{x}) \leq \max_{j=1}^{L} P_j(\omega_i|\boldsymbol{x}), \tag{7.81}$$

and classification is achieved by using the max or min bounds instead of the sum and product. When outliers are present, one can instead use the *median* value, i.e.,

$$\boxed{\text{Assign } \boldsymbol{x} \text{ to class } \omega_i = \arg\max_k \text{median}\left\{P_j(\omega_k|\boldsymbol{x})\right\}, \quad k = 1, 2, \ldots, M.} \tag{7.82}$$

It turns out that the no free lunch theorem is also valid for the combination rules; there is not a universally optimal rule. It all depends on the data at hand (see [21]).

There are a number of other issues related to the theory of combining classifiers; for example, how does one choose the classifiers to be combined? Should the classifiers be dependent or independent? Furthermore, combination does not necessarily imply improved performance; in some cases, one may experience a performance loss (higher error rate) compared to that of the best (single) classifier [20, 21]. Thus, combining has to take place with care. More on these issues can be found in [25,38] and the references therein.

7.10 THE BOOSTING APPROACH

The origins of the *boosting* method for designing learning machines are traced back to the work of Valiant and Kearns [23,40], who posed the question of whether a weak learning algorithm, meaning one that does slightly better than random guessing, can be *boosted* into a strong one with a good performance index. At the heart of such techniques lies the *base learner*, which is a weak one. Boosting consists of an iterative scheme, where at each step the base learner is optimally computed using a different training set; the set at the current iteration is generated either according to an iteratively obtained data distribution or, usually, via a weighting of the training samples, each time using a different set of weights. The latter are computed in order to take into account the achieved performance up to the current iteration step. The final learner is obtained via a *weighted average* of all the *hierarchically* designed base learners. Thus, boosting can also be considered a scheme for combining different learners.

It turns out that, given a sufficient number of iterations, one can significantly improve the (poor) performance of the weak learner. For example, in some cases in classification, the training error may tend to zero as the number of iterations increases. This is very interesting indeed. Training a weak classifier, by appropriate manipulation of the training data (as a matter of fact, the weighting mechanism identifies hard samples, the ones that keep failing, and places more emphasis on them) one can obtain a strong classifier. Of course, as we will discuss, the fact that the training error may tend to zero does *not* necessarily mean the test error goes to zero as well.

THE ADABOOST ALGORITHM

We now focus on the two-class classification task and assume that we are given a set of N training observations, $(y_n, \boldsymbol{x}_n)$, $n = 1, 2, \ldots, N$, with $y_n \in \{-1, 1\}$. Our goal is to design a binary classifier,

$$f(\boldsymbol{x}) = \text{sgn}\left\{F(\boldsymbol{x})\right\}, \tag{7.83}$$

where

$$\boxed{F(\boldsymbol{x}) := \sum_{k=1}^{K} a_k \phi(\boldsymbol{x}; \boldsymbol{\theta}_k),} \tag{7.84}$$

where $\phi(\boldsymbol{x}; \boldsymbol{\theta}_k) \in \{-1, 1\}$ is the base classifier at iteration k, defined in terms of a set of parameters, $\boldsymbol{\theta}_k,\ k = 1, 2, \ldots, K$, to be estimated. The base classifier is selected to be a binary one. The set of unknown parameters is obtained in a *step-wise* approach and in a *greedy* way; that is, at each iteration step i, we only optimize with respect to a single pair, $(a_i, \boldsymbol{\theta}_i)$, by keeping the parameters $a_k, \boldsymbol{\theta}_k,\ k = 1, 2, \ldots, i-1$, obtained from the previous steps, fixed. Note that ideally, one should optimize with respect to all the unknown parameters, $a_k, \boldsymbol{\theta}_k,\ k = 1, 2, \ldots, K$, simultaneously; however, this would lead to a very computationally demanding optimization task. Greedy algorithms are very popular, due to their computational simplicity, and lead to a very good performance in a wide range of learning tasks. Greedy algorithms will also be discussed in the context of sparsity-aware learning in Chapter 10.

Assume that we are currently at the ith iteration step; consider the partial sum of terms

$$F_i(\cdot) = \sum_{k=1}^{i} a_k \phi(\cdot; \boldsymbol{\theta}_k). \tag{7.85}$$

Then we can write the following recursion:

$$F_i(\cdot) = F_{i-1}(\cdot) + a_i \phi(\cdot; \boldsymbol{\theta}_i), \quad i = 1, 2, \ldots, K, \tag{7.86}$$

starting from an initial condition. According to the greedy rationale, $F_{i-1}(\cdot)$ is assumed to be known and the goal is to optimize with respect to the set of parameters $a_i, \boldsymbol{\theta}_i$. For optimization, a loss function has to be adopted. No doubt different options are available, giving different names to the derived algorithm. A popular loss function used for classification is the exponential loss, defined as

$$\boxed{\mathcal{L}\big(y, F(\boldsymbol{x})\big) = \exp\big(-yF(\boldsymbol{x})\big) : \quad \text{exponential loss function,}} \tag{7.87}$$

and it gives rise to the *adaptive boosting* (AdaBoost) algorithm. The exponential loss function is shown in Fig. 7.14, together with the 0-1 loss function. The former can be considered a (differentiable) upper bound of the (nondifferentiable) 0-1 loss function. Note that the exponential loss weighs misclassified ($yF(\boldsymbol{x}) < 0$) points more heavily compared to the correctly identified ones ($yF(\boldsymbol{x}) > 0$). Employing the exponential loss function, the set $a_i, \boldsymbol{\theta}_i$ is obtained via the respective empirical cost function, in the following manner:

$$(a_i, \boldsymbol{\theta}_i) = \arg\min_{a, \boldsymbol{\theta}} \sum_{n=1}^{N} \exp\Big(-y_n\big(F_{i-1}(\boldsymbol{x}_n) + a\phi(\boldsymbol{x}_n; \boldsymbol{\theta})\big)\Big). \tag{7.88}$$

This optimization is also performed in two steps. First, a is treated fixed and we optimize with respect to $\boldsymbol{\theta}$,

$$\boldsymbol{\theta}_i = \arg\min_{\boldsymbol{\theta}} \sum_{n=1}^{N} w_n^{(i)} \exp\big(-y_n a \phi(\boldsymbol{x}_n; \boldsymbol{\theta})\big), \tag{7.89}$$

where

$$w_n^{(i)} := \exp\big(-y_n F_{i-1}(\boldsymbol{x}_n)\big), \ n = 1, 2, \ldots, N. \tag{7.90}$$

Observe that $w_n^{(i)}$ depends neither on a nor on $\phi(\boldsymbol{x}_n; \boldsymbol{\theta})$; hence it can be considered a weight associated with sample n. Moreover, its value depends entirely on the results obtained from the previous recursions.

We now turn our focus on the cost in (7.89). The optimization depends on the specific form of the base classifier. Note, however, that the loss function is of an exponential form. Furthermore, the base classifier is a binary one, so that $\phi(\boldsymbol{x}; \boldsymbol{\theta}) \in \{-1, 1\}$. If we assume that $a > 0$ (we will come back to it soon) optimization of (7.89) is readily seen to be equivalent to optimizing the following cost:

$$\boldsymbol{\theta}_i = \arg\min_{\boldsymbol{\theta}} P_i, \tag{7.91}$$

where

$$P_i := \sum_{n=1}^{N} w_n^{(i)} \chi_{(-\infty,0]}\big(y_n \phi(\boldsymbol{x}_n; \boldsymbol{\theta})\big), \tag{7.92}$$

and $\chi_{[-\infty,0]}(\cdot)$ is the 0-1 loss function.[5] In other words, only misclassified points (i.e., those for which $y_n \phi(\boldsymbol{x}_n; \boldsymbol{\theta}) < 0$) contribute. Note that P_i is the weighted empirical classification error. Obviously, when the misclassification error is minimized, the cost in (7.89) is also minimized, because the exponential loss weighs the misclassified points heavier. To guarantee that P_i remains in the $[0, 1]$ interval, the weights are normalized to unity by dividing by the respective sum; note that this does not affect the optimization process. In other words, $\boldsymbol{\theta}_i$ can be computed in order to minimize the empirical misclas-

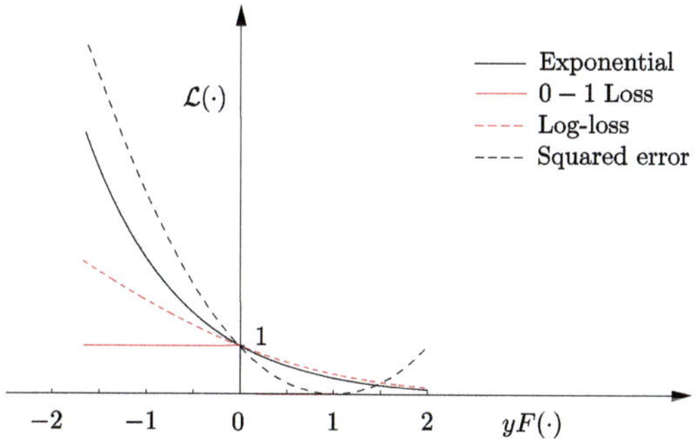

FIGURE 7.14

The 0-1, exponential, log-loss, and squared error loss functions. They have all been normalized to cross the point $(0, 1)$. The horizontal axis for the squared error corresponds to $y - F(\boldsymbol{x})$.

[5] The characteristic function $\chi_A(x)$ is equal to one if $x \in A$ and zero otherwise.

sification error committed by the base classifier. For base classifiers of very simple structure, such a minimization is computationally feasible.

Having computed the optimal $\boldsymbol{\theta}_i$, the following are easily established from the respective definitions,

$$\sum_{y_n \phi(\boldsymbol{x}_n; \boldsymbol{\theta}_i) < 0} w_n^{(i)} = P_i \tag{7.93}$$

and

$$\sum_{y_n \phi(\boldsymbol{x}_n; \boldsymbol{\theta}_i) > 0} w_n^{(i)} = 1 - P_i. \tag{7.94}$$

Combining (7.93) and (7.94) with (7.88) and (7.90), it is readily shown that

$$a_i = \arg\min_a \left\{ \exp(-a)(1 - P_i) + \exp(a) P_i \right\}. \tag{7.95}$$

Taking the derivative with respect to a and equating to zero results in

$$a_i = \frac{1}{2} \ln \frac{1 - P_i}{P_i}. \tag{7.96}$$

Note that if $P_i < 0.5$, then $a_i > 0$, which is expected to be the case in practice. Once a_i and $\boldsymbol{\theta}_i$ have been estimated, the weights for the next iteration are readily given by

$$w_n^{(i+1)} = \frac{\exp\left(-y_n F_i(\boldsymbol{x}_n)\right)}{Z_i} = \frac{w_n^{(i)} \exp\left(-y_n a_i \phi(\boldsymbol{x}_n; \boldsymbol{\theta}_i)\right)}{Z_i}, \tag{7.97}$$

where Z_i is the normalizing factor

$$Z_i := \sum_{n=1}^{N} w_n^{(i)} \exp\left(-y_n a_i \phi(\boldsymbol{x}_n; \boldsymbol{\theta}_i)\right). \tag{7.98}$$

Looking at the way the weights are formed, one can grasp one of the major secrets underlying the AdaBoost algorithm: The weight associated with a training sample $\boldsymbol{x}_n$ is increased (decreased) with respect to its value at the previous iteration, depending on whether the pattern has failed (succeeded) in being classified correctly. Moreover, the percentage of the decrease (increase) depends on the value of a_i, which controls the relative importance in the buildup of the final classifier. Hard samples, which keep failing over successive iterations, gain importance in their participation in the weighted empirical error value. For the case of the AdaBoost, it can be shown that the training error tends to zero exponentially fast (Problem 7.18). The scheme is summarized in Algorithm 7.1.

Algorithm 7.1 (The AdaBoost algorithm).

- Initialize: $w_n^{(1)} = \frac{1}{N}$, $i = 1, 2, \ldots, N$
- Initialize: $i = 1$
- Repeat

  - Compute the optimum $\boldsymbol{\theta}_i$ in $\phi(\cdot;\boldsymbol{\theta}_i)$ by minimizing P_i; (7.91)
  - Compute the optimum P_i; (7.92)
  - $a_i = \frac{1}{2}\ln\frac{1-P_i}{P_i}$
  - $Z_i = 0$
  - **For** $n = 1$ to N **Do**
    * $w_n^{(i+1)} = w_n^{(i)} \exp\left(-y_n a_i \phi(\boldsymbol{x}_n;\boldsymbol{\theta}_i)\right)$
    * $Z_i = Z_i + w_n^{(i+1)}$
  - **End For**
  - **For** $n = 1$ to N **Do**
    * $w_n^{(i+1)} = w_n^{(i+1)}/Z_i$
  - **End For**
  - $K = i$
  - $i = i + 1$
- Until a termination criterion is met.
- $f(\cdot) = \text{sgn}\left(\sum_{k=1}^{K} a_k \phi(\cdot,\boldsymbol{\theta}_k)\right)$

The AdaBoost was first derived in [14] in a different way. Our formulation follows that given in [15]. Yoav Freund and Robert Schapire received the prestigious Gödel award for this algorithm in 2003.

THE LOG-LOSS FUNCTION

In AdaBoost, the exponential loss function was employed. From a theoretical point of view, this can be justified by the following argument. Consider the mean value with respect to the binary label, y, of the exponential loss function

$$\mathbb{E}\left[\exp\left(-yF(\boldsymbol{x})\right)\right] = P(y=1)\exp\left(-F(\boldsymbol{x})\right) + P(y=-1)\exp\left(F(\boldsymbol{x})\right). \tag{7.99}$$

Taking the derivative with respect to $F(\boldsymbol{x})$ and equating to zero, we readily obtain that the minimum of (7.99) occurs at

$$F_*(\boldsymbol{x}) = \arg\min_f \mathbb{E}\left[\exp(-yf)\right] = \frac{1}{2}\ln\frac{P(y=1|\boldsymbol{x})}{P(y=-1|\boldsymbol{x})}. \tag{7.100}$$

The logarithm of the ratio on the right-hand side is known as the *log-odds* ratio. Hence, if one views the minimizing function in (7.88) as the empirical approximation of the mean value in (7.99), it fully justifies considering the sign of the function in (7.83) as the classification rule.

A major problem associated with the exponential loss function, as is readily seen in Fig. 7.14, is that it weights heavily wrongly classified samples, depending on the value of the respective margin, defined as

$$m_x := |yF(\boldsymbol{x})|. \tag{7.101}$$

Note that the farther the point is from the decision surface ($F(\boldsymbol{x}) = 0$), the larger the value of $|F(\boldsymbol{x})|$. Thus, points that are located at the wrong side of the decision surface ($yF(\boldsymbol{x}) < 0$) and far away are weighted with (exponentially) large values, and their influence in the optimization process is large compared to the other points. Thus, in the presence of outliers, the exponential loss is not the most appropriate one. As a matter of fact, in such environments, the performance of the AdaBoost can degrade dramatically.

An alternative loss function is the *log-loss* or *binomial deviance*, defined as

$$\boxed{\mathcal{L}\big(y, F(\boldsymbol{x})\big) := \ln\big(1+\exp\big(-yF(\boldsymbol{x})\big)\big): \quad \text{log-loss function,}} \tag{7.102}$$

which is also shown in Fig. 7.14. Observe that its increase is almost linear for large negative values. Such a function leads to a more balanced influence of the loss among all the points. We will return to the issue of *robust loss functions*, that is, loss functions that are more immune to the presence of outliers, in Chapter 11. Note that the function that minimizes the mean of the log-loss, with respect to y, is the same as the one given in (7.100) (try it). However, if one employs the log-loss instead of the exponential, the optimization task gets more involved, and one has to resort to gradient descent or Newton-type schemes for optimization (see [16]).

Remarks 7.7.

- For comparison reasons, in Fig. 7.14, the squared error loss is shown. The squared error loss depends on the value $(y - F(\boldsymbol{x}))$, which is the equivalent of the margin defined above, $yF(\boldsymbol{x})$. Observe that, besides the relatively large influence that large values of error have, the error is also penalized for patterns whose label has been predicted correctly. This is one more justification that the LS method is not, in general, a good choice for classification.
- Multiclass generalizations of the Boosting scheme have been proposed in [13,15]. In [10], regularized versions of the AdaBoost scheme have been derived in order to impose sparsity. Different regularization schemes are considered, including ℓ_1, ℓ_2, and ℓ_∞. The end result is a family of coordinate descent algorithms that integrate forward feature induction and back-pruning. In [33], a version is presented where a priori knowledge is brought into the scene. The so-called AdaBoost$^*_\upsilon$ is introduced in [29], where the margin is explicitly taken into account.
- Note that the boosting rationale can be applied equally well to regression tasks involving respective loss functions, such as the squared error one. A robust alternative to the squared error is the absolute error loss function [16].
- The boosting technique has attracted a lot of attention among researchers in the field in order to justify its good performance in practice and its relative immunity to overfitting. While the training error may become zero, this still does not necessarily imply overfitting. A first explanation was attempted in terms of bounds, concerning the respective generalization performance. The derived bounds are independent of the number of iterations, K, and they are expressed in terms of the margin [32]. However, these bounds tend to be very loose. Another explanation may lie in the fact that each time, optimization is carried out with only a single set of parameters. The interested reader may find the discussions following the papers [4,6,15] very enlightening on this issue.

Example 7.5. Consider a 20-dimensional two-class classification task. The data points from the first class (ω_1) stem from either of the two Gaussian distributions with means $\boldsymbol{\mu}_{11} = [0,\ 0, \dots, 0]^T$, $\boldsymbol{\mu}_{12} = [1,\ 1, \dots, 1]^T$, while the points of the second class (ω_2) stem from the Gaussian distribution with mean $\boldsymbol{\mu}_2 = [\overbrace{0, \dots, 0}^{10},\ \overbrace{1, \dots, 1}^{10}]^T$. The covariance matrices of all distributions are equal to the 20-dimensional identity matrix. Each one of the training and the test sets consists of 300 points, 200 from ω_1 (100 from each distribution) and 100 from ω_2.

For the AdaBoost, the base classifier was selected to be a *stump*. This is a very naive type of tree, consisting of a single node, and classification of a feature vector $\boldsymbol{x}$ is achieved on the basis of the value of only one of its features, say, x_i. Thus, if $x_i < a$, where a is an appropriate threshold, $\boldsymbol{x}$ is assigned to class ω_1. If $x_i > a$, it is assigned to class ω_2. The decision about the choice of the specific feature, x_i, to be used in the classifier was randomly made. Such a classifier results in a training error rate slightly better than 0.5. The AdaBoost algorithm was run on the training data for 2000 iteration steps. Fig. 7.15 verifies the fact that the training error rate converges to zero very fast. The test error rate keeps decreasing even after the training error rate becomes zero and then levels off at around 0.15.

7.11 BOOSTING TREES

In the discussion on experimental comparison of various methods in Section 7.9, it was stated the boosted trees are among the most powerful learning schemes for classification and data mining. Thus, it is worth spending some more time on this special type of boosting techniques.

Trees were introduced in Section 7.8. From the knowledge we have now acquired, it is not difficult to see that the output of a tree can be compactly written as

$$T(\boldsymbol{x}; \boldsymbol{\Theta}) = \sum_{j=1}^{J} \hat{y}_j \chi_{R_j}(\boldsymbol{x}), \tag{7.103}$$

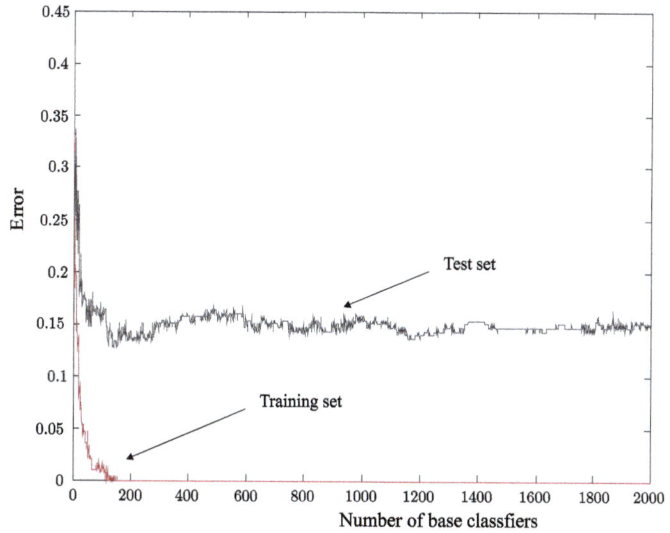

FIGURE 7.15

Training and test error rate curves as a function of the number of iterations, for the case of Example 7.5.

where J is the number of leaf nodes, R_j is the region associated with the jth leaf after the space partition imposed by the tree, $\hat{y}_j$ is the respective label associated with R_j (output/prediction value for regression), and χ is our familiar characteristic function. The set of parameters, $\boldsymbol{\Theta}$, consists of $(\hat{y}_j, R_j)$, $j = 1, 2, \ldots, J$, which are estimated during training. These can be obtained by selecting an appropriate cost function. Also, suboptimal techniques are usually employed, in order to build up a tree, as the ones discussed in Section 7.8.

In a boosted tree model, the base classifier comprises a tree. For example, the stump used in Example 7.5 is a very special case of a base tree classifier. In practice, one can employ trees of larger size. Of course, the size must not be very large, in order to be closer to a weak classifier. Usually, values of J between three and eight are advisable.

The boosted tree model can be written as

$$F(\boldsymbol{x}) = \sum_{k=1}^{K} T(\boldsymbol{x}; \boldsymbol{\Theta}_k), \tag{7.104}$$

where

$$T(\boldsymbol{x}; \boldsymbol{\Theta}_k) = \sum_{j=1}^{J} \hat{y}_{kj} \chi_{R_{kj}}(\boldsymbol{x}).$$

Eq. (7.104) is basically the same as (7.84), with the as being equal to one. We have assumed the size of all the trees to be the same, although this may not be necessarily the case. Adopting a loss function $\mathcal{L}$ and the greedy rationale used for the more general boosting approach, we arrive at the following recursive scheme of optimization:

$$\boldsymbol{\Theta}_i = \arg\min_{\boldsymbol{\Theta}} \sum_{n=1}^{N} \mathcal{L}\big(y_n, F_{i-1}(\boldsymbol{x}_n) + T(\boldsymbol{x}_n; \boldsymbol{\Theta})\big). \tag{7.105}$$

Optimization with respect to $\boldsymbol{\Theta}$ takes place in two steps: one with respect to $\hat{y}_{ij}$, $j = 1, 2, \ldots, J$, given R_{ij}, and then one with respect to the regions R_{ij}. The latter is a difficult task and only simplifies in very special cases. In practice, a number of approximations can be employed. Note that in the case of the exponential loss and the two-class classification task, the above is directly linked to the AdaBoost scheme.

For more general cases, numeric optimization schemes are mobilized (see [16]). The same rationale applies for regression trees, where now loss functions for regression, such as the squared error or the absolute error value, are used. Such schemes are also known as *multiple additive regression trees* (MARTs). A related implementation code for boosted trees is freely available in the R gbm package [30].

There are two critical factors concerning boosted trees. One is the size of the trees, J, and the other is the choice of K. Concerning the size of the trees, usually one tries different sizes, $4 \leq J \leq 8$, and selects the best one. Concerning the number of iterations, for large values, the training error may get close to zero, but the test error can increase due to overfitting. Thus, one has to stop early enough, usually by monitoring the performance.

Another way to cope with overfitting is to employ *shrinkage* methods, which tend to be equivalent to regularization. For example, in the stage-wise expansion of $F_i(\boldsymbol{x})$ used in the optimization step (7.105), one can instead adopt the following:

$$F_i(\cdot) = F_{i-1}(\cdot) + \nu T(\cdot; \boldsymbol{\Theta}_i).$$

The parameter ν takes small values and it can be considered as controlling the learning rate of the boosting procedure. Values smaller than $\nu < 0.1$ are advised. However, the smaller the value of ν, the larger the value K should be to guarantee good performance. For more on MARTs, the interested reader can peruse [19].

PROBLEMS

7.1 Show that the Bayesian classifier is optimal, in the sense that it minimizes the probability of error.
Hint: Consider a classification task of M classes and start with the probability of correct label prediction, $P(C)$. Then the probability of error will be $P(e) = 1 - P(C)$.

7.2 Show that if the data follow the Gaussian distribution in an M-class task, with equal covariance matrices in all classes, the regions formed by the Bayesian classifier are convex.

7.3 Derive the form of the Bayesian classifier for the case of two equiprobable classes, when the data follow the Gaussian distribution of the same covariance matrix. Furthermore, derive the equation that describes the LS linear classifier. Compare and comment on the results.

7.4 Show that the ML estimate of the covariance matrix of a Gaussian distribution, based on N i.i.d. observations, $\boldsymbol{x}_n,\ n = 1, 2, \ldots, N$, is given by

$$\hat{\Sigma}_{ML} = \frac{1}{N}\sum_{n=1}^{N}(\boldsymbol{x}_n - \hat{\boldsymbol{\mu}}_{ML})(\boldsymbol{x}_n - \hat{\boldsymbol{\mu}}_{ML})^T,$$

where

$$\hat{\boldsymbol{\mu}}_{ML} = \frac{1}{N}\sum_{n=1}^{N}\boldsymbol{x}_n.$$

7.5 Prove that the covariance estimate

$$\hat{\Sigma} = \frac{1}{N-1}\sum_{k=1}^{N}(\boldsymbol{x}_k - \hat{\boldsymbol{\mu}})(\boldsymbol{x}_k - \hat{\boldsymbol{\mu}})^T$$

defines an unbiased estimator, where

$$\hat{\boldsymbol{\mu}} = \frac{1}{N}\sum_{k=1}^{N}\boldsymbol{x}_k.$$

7.6 Show that the derivative of the logistic link function is given by

$$\frac{d\sigma(t)}{dt} = \sigma(t)(1-\sigma(t)).$$

7.7 Derive the gradient of the negative log-likelihood function associated with the two-class logistic regression.

7.8 Derive the Hessian matrix of the negative log-likelihood function associated with the two-class logistic regression.

7.9 Show that the Hessian matrix of the negative log-likelihood function of the two-class logistic regression is a positive definite matrix.

7.10 Show that if

$$\phi_m = \frac{\exp(t_m)}{\sum_{j=1}^{M}\exp(t_j)},$$

the derivative with respect to t_j, $j = 1, 2, \ldots, M$, is given by

$$\frac{\partial \phi_m}{\partial t_j} = \phi_m(\delta_{mj} - \phi_j).$$

7.11 Derive the gradient of the negative log-likelihood for the multiclass logistic regression case.

7.12 Derive the j, k block element of the Hessian matrix of the negative log-likelihood function for the multiclass logistic regression.

7.13 Consider the Rayleigh ratio,

$$R = \frac{\boldsymbol{\theta}^T A \boldsymbol{\theta}}{||\boldsymbol{\theta}||^2},$$

where A is a symmetric positive definite matrix. Show that R is maximized with respect to $\boldsymbol{\theta}$ if $\boldsymbol{\theta}$ is the eigenvector corresponding to the maximum eigenvalue of A.

7.14 Consider the generalized Rayleigh quotient,

$$R_g = \frac{\boldsymbol{\theta}^T B \boldsymbol{\theta}}{\boldsymbol{\theta}^T A \boldsymbol{\theta}},$$

where A and B are symmetric positive definite matrices. Show that R_g is maximized with respect to $\boldsymbol{\theta}$ if $\boldsymbol{\theta}$ is the eigenvector that corresponds to the maximum eigenvalue of $A^{-1}B$, assuming that the inversion is possible.

7.15 Show that the between-class scatter matrix Σ_b for an M-class problem is of rank $M-1$.

7.16 Derive the arithmetic rule for combination, by minimizing the average Kullback–Leibler divergence.

7.17 Derive the product rule via the minimization of the Kullback–Leibler divergence, as pointed out in the text.

7.18 Show that the error rate on the training set of the final classifier, obtained by boosting, tends to zero exponentially fast.

MATLAB® EXERCISES

7.19 Consider a two-dimensional class problem that involves two classes, ω_1 and ω_2, which are modeled by Gaussian distributions with means $\boldsymbol{\mu}_1 = [0,\ 0]^T$ and $\boldsymbol{\mu}_2 = [2,\ 2]^T$, respectively, and common covariance matrix $\Sigma = \begin{bmatrix} 1 & 0.25 \\ 0.25 & 1 \end{bmatrix}$.

(i) Form and plot a data set $\mathcal{X}$ consisting of 500 points from ω_1 and another 500 points from ω_2.

(ii) Assign each one of the points of $\mathcal{X}$ to either ω_1 or ω_2, according to the Bayes decision rule, and plot the points with different colors, depending on the class they are assigned to. Plot the corresponding classifier.

(iii) Based on (ii), estimate the error probability.

(iv) Let $L = \begin{bmatrix} 0 & 1 \\ 0.005 & 0 \end{bmatrix}$ be a loss matrix. Assign each one of the points of $\mathcal{X}$ to either ω_1 or ω_2, according to the average risk minimization rule (Eq. (7.13)), and plot the points with different colors, depending on the class they are assigned to.

(v) Based on (iv), estimate the average risk for the above loss matrix.

(vi) Comment on the results obtained by (ii)–(iii) and (iv)–(v) scenarios.

7.20 Consider a two-dimensional class problem that involves two classes, ω_1 and ω_2, which are modeled by Gaussian distributions with means $\boldsymbol{\mu}_1 = [0,\ 2]^T$ and $\boldsymbol{\mu}_2 = [0,\ 0]^T$ and covariance matrices $\Sigma_1 = \begin{bmatrix} 4 & 1.8 \\ 1.8 & 1 \end{bmatrix}$ and $\Sigma_2 = \begin{bmatrix} 4 & 1.2 \\ 1.2 & 1 \end{bmatrix}$, respectively.

(i) Form and plot a data set $\mathcal{X}$ consisting from 5000 points from ω_1 and another 5000 points from ω_2.

(ii) Assign each one of the points of $\mathcal{X}$ to either ω_1 or ω_2, according to the Bayes decision rule, and plot the points with different colors, according to the class they are assigned to.

(iii) Compute the classification error probability.

(iv) Assign each one of the points of $\mathcal{X}$ to either ω_1 or ω_2, according to the naive Bayes decision rule, and plot the points with different colors, according to the class they are assigned to.

(v) Compute the classification error probability for the naive Bayes classifier.

(vii) Repeat steps (i)–(v) for the case where $\Sigma_1 = \Sigma_2 = \begin{bmatrix} 4 & 0 \\ 0 & 1 \end{bmatrix}$.

(viii) Comment on the results.

Hint: Use the fact that the marginal distributions of $P(\omega_1|\boldsymbol{x})$, $P(\omega_1|x_1)$, and $P(\omega_1|x_2)$ are also Gaussians with means 0 and 2 and variances 4 and 1, respectively. Similarly, the marginal distributions of $P(\omega_2|\boldsymbol{x})$, $P(\omega_2|x_1)$, and $P(\omega_2|x_2)$ are also Gaussians with means 0 and 0 and variances 4 and 1, respectively.

7.21 Consider a two-class, two-dimensional classification problem, where the first class (ω_1) is modeled by a Gaussian distribution with mean $\boldsymbol{\mu}_1 = [0,\ 2]^T$ and covariance matrix $\Sigma_1 = \begin{bmatrix} 4 & 1.8 \\ 1.8 & 1 \end{bmatrix}$, while the second class (ω_2) is modeled by a Gaussian distribution with mean $\boldsymbol{\mu}_2 = [0,\ 0]^T$ and covariance matrix $\Sigma_2 = \begin{bmatrix} 4 & 1.8 \\ 1.8 & 1 \end{bmatrix}$.

(i) Generate and plot a training set $\mathcal{X}$ and a test set $\mathcal{X}_{test}$, each one consisting of 1500 points from each distribution.

(ii) Classify the data vectors of $\mathcal{X}_{test}$ using the Bayesian classification rule.

(iii) Perform logistic regression and use the data set $\mathcal{X}$ to estimate the involved parameter vector $\boldsymbol{\theta}$. Evaluate the classification error of the resulting classifier based on $\mathcal{X}_{test}$.

(iv) Comment on the results obtained by (ii) and (iii).

(v) Repeat steps (i)–(iv) for the case where $\Sigma_2 = \begin{bmatrix} 4 & -1.8 \\ -1.8 & 1 \end{bmatrix}$ and compare the obtained results with those produced by the previous setting. Draw your conclusions.

Hint: For the estimation of $\boldsymbol{\theta}$ in (iii), perform gradient descent (Eq. (7.44)) and set the learning parameter μ_i equal to 0.001.

7.22 Consider a two-dimensional classification problem involving three classes ω_1, ω_2, and ω_3. The data vectors from ω_1 stem from either of the two Gaussian with means $\boldsymbol{\mu}_{11} = [0,\ 3]^T$ and $\boldsymbol{\mu}_{12} = [11,\ -2]^T$ and covariance matrices $\Sigma_{11} = \begin{bmatrix} 0.2 & 0 \\ 0 & 2 \end{bmatrix}$ and $\Sigma_{12} = \begin{bmatrix} 3 & 0 \\ 0 & 0.5 \end{bmatrix}$, respectively. Similarly, the data vectors from ω_2 stem from either of the two Gaussian distributions with means $\boldsymbol{\mu}_{21} = [3,\ -2]^T$ and $\boldsymbol{\mu}_{22} = [7.5,\ 4]^T$ and covariance matrices $\Sigma_{21} = \begin{bmatrix} 5 & 0 \\ 0 & 0.5 \end{bmatrix}$ and $\Sigma_{22} = \begin{bmatrix} 7 & 0 \\ 0 & 0.5 \end{bmatrix}$, respectively. Finally, ω_3 is modeled by a single Gaussian distribution with mean $\boldsymbol{\mu}_3 = [7,\ 2]^T$ and covariance matrix $\Sigma_3 = \begin{bmatrix} 8 & 0 \\ 0 & 0.5 \end{bmatrix}$.

(i) Generate and plot a training data set $\mathcal{X}$ consisting of 1000 data points from ω_1 (500 from each distribution), 1000 data points from ω_2 (again 500 from each distribution), and 500 points from ω_3 (use 0 as the seed for the initialization of the Gaussian random number generator). In a similar manner, generate a test data set $\mathcal{X}_{test}$ (use 100 as the seed for the initialization of the Gaussian random number generator).

(ii) Generate and view a decision tree based on using $\mathcal{X}$ as the training set.

(iii) Compute the classification error on both the training and the test sets. Comment briefly on the results.

(iv) Prune the produced tree at levels 0 (no actual pruning), $1, \ldots, 11$ (in MATLAB®, trees are pruned based on an optimal pruning scheme that first prunes branches giving less improvement in error cost). For each pruned tree compute the classification error based on the test set.

(v) Plot the classification error versus the pruned levels and locate the pruned level that gives the minimum test classification error. What conclusions can be drawn by the inspection of this plot?

(vi) View the original decision tree as well as the best-pruned one.

Hint: The MATLAB® functions that generate a decision tree (DT), display a DT, prune a DT, and evaluate the performance of a DT on a given data set, are *classregtree*, *view*, *prune*, and *eval*, respectively.

7.23 Consider a two-class, two-dimensional classification problem where the classes are modeled as the first two classes in the previous exercise.

(i) Generate and plot a training set $\mathcal{X}$, consisting of 100 data points from each distribution of each class (that is, $\mathcal{X}$ contains 400 points in total, 200 points from each class). In a similar manner, generate a test set.

(ii) Use the training set to build a boosting classifier, utilizing as weak classifier a single-node decision tree. Perform 12,000 iterations.

(iii) Plot the training and the test error versus the number of iterations and comment on the results.

Hint:

- For (i) use $randn('seed', 0)$ and $randn('seed', 100)$ to initialize the random number generator for the training and the test set, respectively.
- For (ii) use $ens = fitensemble(X', y, 'AdaBoostM1', no\_of\_base\_classifiers, 'Tree')$, where X' has in its rows the data vectors, y is an ordinal vector containing the class where each row vector of X' belongs, $AdaBoostM1$ is the boosting method used, $no\_of\_base\_classifiers$ is the number of base classifiers that will be used, and $Tree$ denotes the weak classifier.
- For (iii) use $L = loss(ens, X', y, 'mode', 'cumulative')$, which for a given boosting classifier ens, returns the vector L of errors performed on X', such that $L(i)$ is the error committed when only the first i weak classifiers are taken into account.

REFERENCES

[1] S. Boyd, L. Vandenberghe, Convex Optimization, Cambridge University Press, 2004.
[2] L. Breiman, J. Friedman, R. Olshen, C. Stone, Classification and Regression Trees, Wadsworth, 1984.
[3] L. Breiman, Bagging predictors, Mach. Learn. 24 (1996) 123–140.
[4] L. Breiman, Arcing classifiers, Ann. Stat. 26 (3) (1998) 801–849.
[5] L. Breiman, Random forests, Mach. Learn. 45 (2001) 5–32.
[6] P. Bühlman, T. Hothorn, Boosting algorithms: regularization, prediction and model fitting (with discussion), Stat. Sci. 22 (4) (2007) 477–505.
[7] A. Caruana, A. Niculescu-Mizil, An empirical comparison of supervised learning algorithms, in: International Conference on Machine Learning, 2006.
[8] H. Chipman, E. George, R. McCulloch, BART: Bayesian additive regression trees, Ann. Appl. Stat. 4 (1) (2010) 266–298.
[9] L. Devroye, L. Gyorfi, G.A. Lugosi, A Probabilistic Theory of Pattern Recognition, Springer Verlag, New York, 1996.
[10] J. Duchi, Y. Singer, Boosting with structural sparsity, in: Proceedings of the 26th International Conference on Machine Learning, Montreal, Canada, 2009.
[11] R. Duda, P. Hart, D. Stork, Pattern Classification, second ed., Wiley, New York, 2000.
[12] B. Efron, The efficiency of logistic regression compared to normal discriminant analysis, J. Am. Stat. Assoc. 70 (1975) 892–898.
[13] G. Eibl, K.P. Pfeifer, Multiclass boosting for weak classifiers, J. Mach. Learn. Res. 6 (2006) 189–210.
[14] Y. Freund, R.E. Schapire, A decision theoretic generalization of on-line learning and an applications to boosting, J. Comput. Syst. Sci. 55 (1) (1997) 119–139.
[15] J. Friedman, T. Hastie, R. Tibshirani, Additive logistic regression: a statistical view of boosting, Ann. Stat. 28 (2) (2000) 337–407.
[16] J. Freidman, Greedy function approximation: a gradient boosting machine, Ann. Stat. 29 (5) (2001) 1189–1232.
[17] K. Fukunaga, Introduction to Statistical Pattern Recognition, second ed., Academic Press, 1990.
[18] I. Guyon, S. Gunn, M. Nikravesh, L. Zadeh (Eds.), Feature Extraction, Foundations and Applications, Springer Verlag, New York, 2006.

[19] T. Hastie, R. Tibshirani, J. Friedman, The Elements of Statistical Learning, second ed., Springer Verlag, 2009.
[20] R. Hu, R.I. Damper, A no panacea theorem for classifier combination, Pattern Recognit. 41 (2008) 2665–2673.
[21] A.K. Jain, P.W. Duin, J. Mao, Statistical pattern recognition: a review, IEEE Trans. Pattern Anal. Mach. Intell. 22 (1) (2000) 4–37.
[22] N. Johnson, A Study of the NIPS Feature Selection Challenge, Technical Report, Stanford University, 2009, http://statweb.stanford.edu/~tibs/ElemStatLearn/comp.pdf.
[23] M. Kearns, L.G. Valiant, Cryptographic limitations of learning Boolean formulae and finite automata, J. ACM 41 (1) (1994) 67–95.
[24] J. Kittler, M. Hatef, R. Duin, J. Matas, On combining classifiers, IEEE Trans. Pattern Anal. Mach. Intell. 20 (3) (1998) 228–234.
[25] I.L. Kuncheva, Pattern Classifiers: Methods and Algorithms, John Wiley, 2004.
[26] D. Meyer, F. Leisch, K. Hornik, The support vector machine under test, Neurocomputing 55 (2003) 169–186.
[27] D. Michie, D.J. Spiegelhalter, C.C. Taylor (Eds.), Machine Learning, Neural, and Statistical Classification, Ellis Horwood, London, 1994.
[28] R. Neal, J. Zhang, High dimensional classification with Bayesian neural networks and Dirichlet diffusion trees, in: I. Guyon, S. Gunn, M. Nikravesh, L. Zadeh (Eds.), Feature Extraction, Foundations and Applications, Springer Verlag, New York, 2006, pp. 265–296.
[29] G. Ratsch, M.K. Warmuth, Efficient margin maximizing with boosting, J. Mach. Learn. Res. 6 (2005) 2131–2152.
[30] G. Ridgeway, The state of boosting, Comput. Sci. Stat. 31 (1999) 172–181.
[31] B.D. Ripley, Pattern Recognition and Neural Networks, Cambridge University Press, 1996.
[32] R.E. Schapire, V. Freund, P. Bartlett, W.S. Lee, Boosting the margin: a new explanation for the effectiveness of voting methods, Ann. Stat. 26 (5) (1998) 1651–1686.
[33] R.E. Schapire, M. Rochery, M. Rahim, N. Gupta, Boosting with prior knowledge for call classification, IEEE Trans. Speech Audio Process. 13 (2) (2005) 174–181.
[34] J. Shotton, A. Fitzgibbon, M. Cook, T. Sharp, M. Finocchio, R. Moore, A. Kipman, A.A. Blake, Real-time human pose recognition in parts from single depth images, in: Proceedings of the Conference on Computer Vision and Pattern Recognition, CVPR, 2011.
[35] R. Quinlan, C4.5: Programs for Machine Learning, Morgan Kaufmann, San Mateo, 1993.
[36] D.B. Rubin, Iterative reweighted least squares, in: Encyclopedia of Statistical Sciences, vol. 4, John Wiley, New York, 1983, pp. 272–275.
[37] S. Russell, P. Norvig, Artificial Intelligence: A Modern Approach, third ed., Pearson, 2010.
[38] S. Theodoridis, K. Koutroumbas, Pattern Recognition, fourth ed., Academic Press, 2009.
[39] S. Theodoridis, A. Pikrakis, K. Koutroumbas, D. Cavouras, An Introduction to Pattern Recognition: A MATLAB Approach, Academic Press, 2010.
[40] L.G. Valiant, A theory of the learnable, Commun. ACM 27 (11) (1984) 1134–1142.
[41] A. Webb, Statistical Pattern Recognition, second ed., John Wiley, 2002.
[42] D. Wolpert, Stacked generalization, Neural Netw. 5 (1992) 241–259.
[43] D. Wolpert, The lack of a priori distinctions between learning algorithms, Neural Comput. 8 (7) (1996) 1341–1390.
[44] Y. Wu, H. Tjelmeland, M. West, Bayesian CART: prior structure and MCMC computations, J. Comput. Graph. Stat. 16 (1) (2007) 44–66.

CHAPTER

PARAMETER LEARNING: A CONVEX ANALYTIC PATH

8

CONTENTS

Machine Learning. https://doi.org/10.1016/B978-0-12-818803-3.00017-9

8.1 INTRODUCTION

The theory of convex sets and functions has a rich history and has been the focus of intense study for over a century in mathematics. In the terrain of applied sciences and engineering, the revival of interest in convex functions and optimization is traced back to the early 1980s. In addition to the increased processing power that became available via the use of computers, certain theoretical developments were catalytic in demonstrating the power of such techniques. The advent of the so-called interior point methods opened a new path in solving the classical linear programming task. Moreover, it was increasingly realized that, despite its advantages, the least-squares (LS) method also has a number of drawbacks, particularly in the presence of non-Gaussian noise and the presence of outliers. It has been demonstrated that the use of alternative cost functions, which may not even be differentiable, can alleviate a number of problems that are associated with the LS methods. Furthermore, the increased interest in robust learning methods brought into the scene the need for nontrivial constraints, which the optimized solution has to respect. In the machine learning community, the discovery of support vector machines, to be treated in Chapter 11, played an important role in popularizing convex optimization techniques.

The goal of this chapter is to present some basic notions and definitions related to convex analysis and optimization in the context of machine learning and signal processing. Convex optimization is a discipline in itself, and it cannot be summarized in one chapter. The emphasis here is on computationally light techniques with a focus on online versions, which are gaining in importance in the context of big data applications. A related discussion is also part of this chapter.

The material revolves around two families of algorithms. One goes back to the classical work of Von Neumann on projections on convex sets, which is reviewed together with its more recent online versions. The notions of projection and related properties are treated in some detail. The method of projections, in the context of constrained optimization, has been gaining in popularity recently.

The other family of algorithms that is considered builds around the notion of subgradient for optimizing nondifferentiable convex functions and generalizations of the gradient descent family, discussed in Chapter 5. Furthermore, a powerful tool for analyzing the performance of online algorithms for convex optimization, known as regret analysis, is discussed and a related case study is presented. Finally, some current trends in convex optimization, involving proximal and mirror descent methods are introduced.

8.2 CONVEX SETS AND FUNCTIONS

Although most of the algorithms we will discuss in this chapter refer to vector variables in Euclidean spaces, which is in line with what we have done so far in this book, the definitions and some of the fundamental theorems will be stated in the context of the more general case of Hilbert spaces.[1] This is because the current chapter will also serve the needs of subsequent chapters, whose setting is that of infinite-dimensional Hilbert spaces. For those readers who are not familiar with such spaces, all they need to know is that a Hilbert space is a generalization of the Euclidean one, allowing for infinite dimensions. To serve the needs of these readers, the differences between Euclidean and the more general Hilbert spaces in the theorems will be carefully pointed out whenever this is required.

8.2.1 CONVEX SETS

Definition 8.1. A nonempty subset C of a Hilbert space $\mathbb{H}$, $C \subseteq \mathbb{H}$, is called *convex* if $\forall\, \boldsymbol{x}_1, \boldsymbol{x}_2 \in C$ and $\forall \lambda \in [0, 1]$ the following holds true[2]:

$$\boldsymbol{x} := \lambda \boldsymbol{x}_1 + (1-\lambda)\boldsymbol{x}_2 \in C. \tag{8.1}$$

Note that if $\lambda = 1$, $\boldsymbol{x} = \boldsymbol{x}_1$, and if $\lambda = 0$, $\boldsymbol{x} = \boldsymbol{x}_2$. For any other value of λ in $[0, 1]$, $\boldsymbol{x}$ lies in the line segment joining $\boldsymbol{x}_1$ and $\boldsymbol{x}_2$. Indeed, from (8.1) we can write

$$\boldsymbol{x} - \boldsymbol{x}_2 = \lambda(\boldsymbol{x}_1 - \boldsymbol{x}_2), \quad 0 \le \lambda \le 1.$$

Fig. 8.1 shows two examples of convex sets, in the two-dimensional Euclidean space, $\mathbb{R}^2$. In Fig. 8.1A, the set comprises all points whose Euclidean (ℓ_2) norm is less than or equal to one,

$$C_2 = \left\{ \boldsymbol{x} : \sqrt{x_1^2 + x_2^2} \le 1 \right\}.$$

Sometimes we refer to C_2 as the *ℓ_2-ball* of radius equal to one. Note that the set includes all the points *on and inside the circle*. The set in Fig. 8.1B comprises all the points *on and inside the rhombus* defined by

$$C_1 = \left\{ \boldsymbol{x} : |x_1| + |x_2| \le 1 \right\}.$$

Because the sum of the absolute values of the components of a vector defines the ℓ_1 norm, that is, $\|\boldsymbol{x}\|_1 := |x_1| + |x_2|$, in analogy to C_2 we call the set C_1 as the ℓ_1 ball of radius equal to one. In contrast, the sets whose ℓ_2 and ℓ_1 norms are equal to one, or in other words,

$$\bar{C}_2 = \left\{ \boldsymbol{x} : x_1^2 + x_2^2 = 1 \right\}, \bar{C}_1 = \left\{ \boldsymbol{x} : |x_1| + |x_2| = 1 \right\},$$

are not convex (Problem 8.2). Fig. 8.2 shows two examples of nonconvex sets.

---

[1] The mathematical definition of a Hilbert space is provided in the appendix associated with this chapter and which can be downloaded from the book's website.

[2] In conformity with Euclidean vector spaces and for the sake of notational simplicity, we will keep the same notation and denote the elements of a Hilbert space with lower case bold letters.

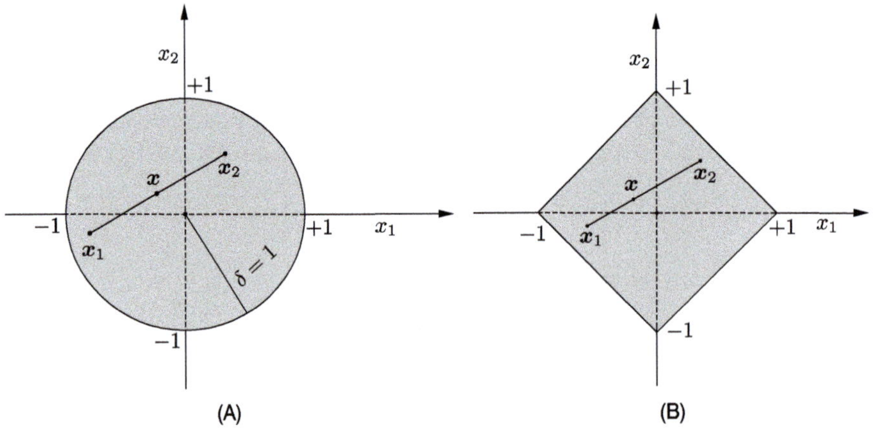

FIGURE 8.1

(A) The ℓ_2 ball of radius $\delta = 1$ comprises all points with Euclidean norm less than or equal to $\delta = 1$. (B) The ℓ_1 ball consists of all the points with ℓ_1 norm less than or equal to $\delta = 1$. Both are convex sets.

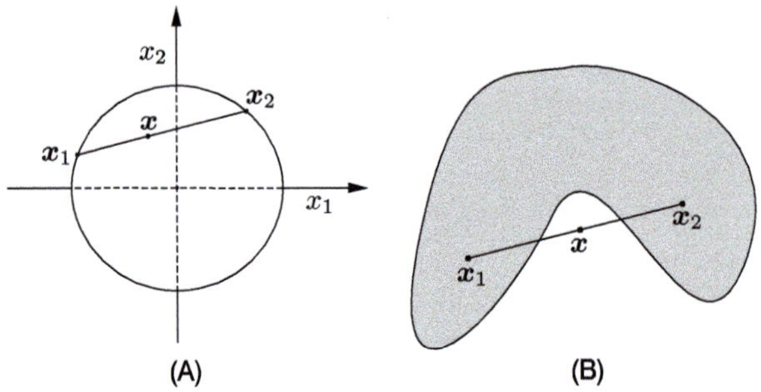

FIGURE 8.2

Examples of two nonconvex sets. In both cases, the point $\boldsymbol{x}$ does not lie on the same set in which $\boldsymbol{x}_1$ and $\boldsymbol{x}_2$ belong. In (A) the set comprises all the points whose Euclidean norm is equal to one.

8.2.2 CONVEX FUNCTIONS

Definition 8.2. A function

$$f : \mathcal{X} \subseteq \mathbb{R}^l \longmapsto \mathbb{R}$$

is called *convex* if $\mathcal{X}$ is convex and if $\forall\, \boldsymbol{x}_1, \boldsymbol{x}_2 \in \mathcal{X}$ the following holds true:

$$\boxed{f\big(\lambda \boldsymbol{x}_1 + (1-\lambda)\boldsymbol{x}_2\big) \leq \lambda f(\boldsymbol{x}_1) + (1-\lambda) f(\boldsymbol{x}_2),\ \lambda \in [0, 1].} \tag{8.2}$$

The function is called *strictly convex* if (8.2) holds true with strict inequality when $\lambda \in (0, 1)$, $\boldsymbol{x}_1 \neq \boldsymbol{x}_2$. The geometric interpretation of (8.2) is that the line segment joining the points $(\boldsymbol{x}_1, f(\boldsymbol{x}_1))$ and

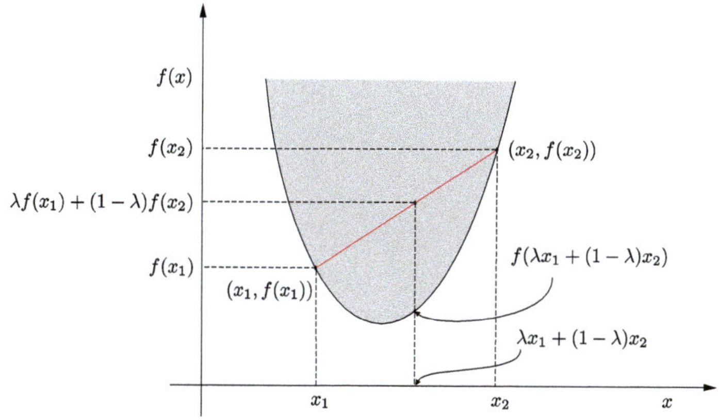

FIGURE 8.3

The line segment joining the points $(x_1, f(x_1))$ and $(x_2, f(x_2))$ lies above the graph of $f(x)$. The shaded region corresponds to the epigraph of the function.

$(\boldsymbol{x}_2, f(\boldsymbol{x}_2))$ lies above the graph of $f(\boldsymbol{x})$, as shown in Fig. 8.3. We say that a function is *concave* (*strictly concave*) if the negative, $-f$, is convex (strictly convex). Next, we state three important theorems.

Theorem 8.1 (First-order convexity condition). *Let $\mathcal{X} \subseteq \mathbb{R}^l$ be a convex set and let*

$$f : \mathcal{X} \longmapsto \mathbb{R}$$

be a differentiable function. Then f is convex if and only if $\forall\, \boldsymbol{x}, \boldsymbol{y} \in \mathcal{X}$,

$$\boxed{f(\boldsymbol{y}) \geq f(\boldsymbol{x}) + \nabla^T f(\boldsymbol{x})(\boldsymbol{y} - \boldsymbol{x}).} \tag{8.3}$$

The proof of the theorem is given in Problem 8.3. The theorem generalizes to nondifferentiable convex functions; it will be discussed in this context in Section 8.10.

Fig. 8.4 illustrates the geometric interpretation of this theorem. It means that the graph of the convex function is located above the graph of the affine function

$$g : \boldsymbol{y} \longmapsto \nabla^T f(\boldsymbol{x})(\boldsymbol{y} - \boldsymbol{x}) + f(\boldsymbol{x}),$$

which defines the tangent hyperplane of the graph at the point $(\boldsymbol{x}, f(\boldsymbol{x}))$.

Theorem 8.2 (Second-order convexity condition). *Let $\mathcal{X} \subseteq \mathbb{R}^l$ be a convex set. Then a twice differentiable function $f : \mathcal{X} \longmapsto \mathbb{R}$ is convex (strictly convex) if and only if the Hessian matrix is positive semidefinite (positive definite).*

The proof of the theorem is given in Problem 8.5. Recall that in previous chapters, when we dealt with the squared error loss function, we commented that it is a convex one. Now we are ready to justify

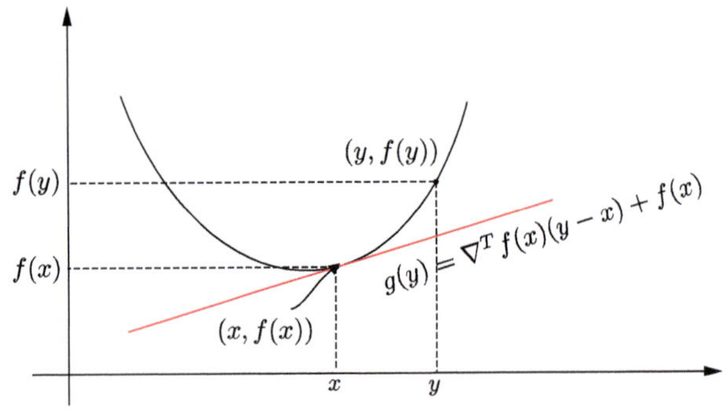

FIGURE 8.4

The graph of a convex function is above the tangent plane at any point of the respective graph.

this argument. Consider the quadratic function,

$$f(\boldsymbol{x}) := \frac{1}{2}\boldsymbol{x}^T Q\boldsymbol{x} + \boldsymbol{b}^T\boldsymbol{x} + c,$$

where Q is a positive definite matrix. Taking the gradient, we have

$$\nabla f(\boldsymbol{x}) = Q\boldsymbol{x} + \boldsymbol{b},$$

and the Hessian matrix is equal to Q, which by assumption is positive definite, hence f is a (strictly) convex function.

In the sequel, two very important notions in convex analysis and optimization are defined.

Definition 8.3. The *epigraph* of a function f is defined as the set of points

$$\boxed{\text{epi}(f) := \left\{(\boldsymbol{x}, r) \in \mathcal{X} \times \mathbb{R} : f(\boldsymbol{x}) \leq r\right\} : \text{ epigraph.}} \tag{8.4}$$

From a geometric point of view, the epigraph is the set of all points in $\mathbb{R}^l \times \mathbb{R}$ that lie on and above the graph of $f(\boldsymbol{x})$, as indicated by the gray shaded region in Fig. 8.3. It is important to note that a function is convex *if and only if* its epigraph is a convex set (Problem 8.6).

Definition 8.4. Given a real number ξ, the *lower level set* of function $f : \mathcal{X} \subseteq \mathbb{R}^l \longmapsto \mathbb{R}$, at *height* ξ, is defined as

$$\boxed{\text{lev}_{\leq\xi}(f) := \left\{\boldsymbol{x} \in \mathcal{X} : f(\boldsymbol{x}) \leq \xi\right\} : \quad \text{level set at } \xi.} \tag{8.5}$$

In words, it is the set of all points at which the function takes a value less than or equal to ξ. The geometric interpretation of the level set is shown in Fig. 8.5. It can easily be shown (Problem 8.7) that if a function f is convex, then its lower level set is convex for any $\xi \in \mathbb{R}$. The converse is not true. We can easily check that the function $f(x) = -\exp(x)$ is not convex (as a matter of fact it is concave) and all its lower level sets are convex.

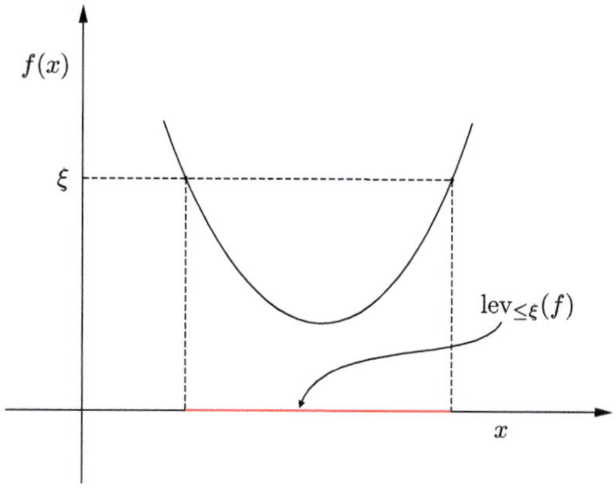

FIGURE 8.5

The level set at height ξ comprises all the points in the interval denoted as the "red" segment on the x-axis.

Theorem 8.3 (Local and global minimizers). *Let a convex function $f : \mathcal{X} \longmapsto \mathbb{R}$. Then if a point $\boldsymbol{x}_*$ is a local minimizer, it is also a global one, and the set of all minimizers is convex. Further, if the function is strictly convex, the minimizer is unique.*

Proof. Inasmuch as the function is convex we know that, $\forall\, \boldsymbol{x} \in \mathcal{X}$,

$$f(\boldsymbol{x}) \geq f(\boldsymbol{x}_*) + \nabla^T f(\boldsymbol{x}_*)(\boldsymbol{x} - \boldsymbol{x}_*),$$

and because at the minimizer the gradient is zero, we get

$$f(\boldsymbol{x}) \geq f(\boldsymbol{x}_*), \tag{8.6}$$

which proves the claim. Let us now denote

$$f_* = \min_{\boldsymbol{x}} f(\boldsymbol{x}). \tag{8.7}$$

Note that the set of all minimizers coincides with the level set at height f_*. Then, because the function is convex, we know that the level set $\text{lev}_{f_*}(f)$ is convex, which verifies the convexity of the set of minimizers. Finally, for strictly convex functions, the inequality in (8.6) is a strict one, which proves the uniqueness of the (global) minimizer. The theorem is also true, even if the function is not differentiable (Problem 8.10). □

8.3 PROJECTIONS ONTO CONVEX SETS

The projection onto a hyperplane in finite-dimensional Euclidean spaces was discussed and used in the context of the affine projection algorithm (APA) in Chapter 5. The notion of projection will now be gen-

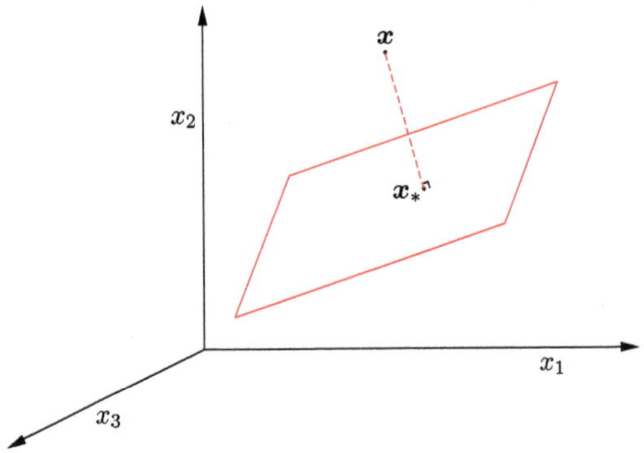

FIGURE 8.6

The projection $\boldsymbol{x}_*$ of $\boldsymbol{x}$ onto the plane minimizes the distance of $\boldsymbol{x}$ from all the points lying on the plane.

eralized to include any closed convex set and also in the framework of general (infinite-dimensional) Hilbert spaces.

The concept of projection is among the most fundamental concepts in mathematics, and everyone who has attended classes in basic geometry has used and studied it. What one may not have realized is that while performing a projection, for example, drawing a line segment from a point to a line or a plane, basically he or she solves an *optimization* task. The point $\boldsymbol{x}_*$ in Fig. 8.6, that is, the projection of $\boldsymbol{x}$ onto the plane H in the three-dimensional space, is that point, among all the points lying on the plane, whose (Euclidean) distance from $\boldsymbol{x} = [x_1, x_2, x_3]^T$ is minimum; in other words,

$$\boldsymbol{x}_* = \min_{\boldsymbol{y} \in H} \left((x_1 - y_1)^2 + (x_2 - y_2)^2 + (x_3 - y_3)^2 \right). \tag{8.8}$$

As a matter of fact, what we have learned to do in our early days at school is to solve a *constrained optimization* task. Indeed, (8.8) can equivalently be written as

$$\begin{aligned} \boldsymbol{x}_* &= \arg\min_{\boldsymbol{y}} ||\boldsymbol{x} - \boldsymbol{y}||^2, \\ \text{s.t.} \quad & \boldsymbol{\theta}^T \boldsymbol{y} + \theta_0 = 0, \end{aligned}$$

where the constraint is the equation describing the specific plane. Our goal herein focuses on generalizing the notion of projection, to employ it to attack more general and complex tasks.

Theorem 8.4. *Let C be a nonempty closed*[3] *convex set in a Hilbert space $\mathbb{H}$ and $\boldsymbol{x} \in \mathbb{H}$. Then there exists a unique point, denoted as $P_C(\boldsymbol{x}) \in C$, such that*

$$\boxed{\|\boldsymbol{x} - P_C(\boldsymbol{x})\| = \min_{\boldsymbol{y} \in C} \|\boldsymbol{x} - \boldsymbol{y}\| : \quad \textit{projection of } \boldsymbol{x} \textit{ on } C.}$$

[3] For the needs of this chapter, it suffices to say that a set C is closed if the limit point of any sequence of points in C lies in C.

The term $P_C(\boldsymbol{x})$ is called the (metric) projection of $\boldsymbol{x}$ onto C. Note that if $\boldsymbol{x} \in C$, then $P_C(\boldsymbol{x}) = \boldsymbol{x}$, since this makes the norm $\|\boldsymbol{x} - P_C(\boldsymbol{x})\| = 0$.

Proof. The proof comprises two paths. One is to establish *uniqueness* and the other to establish *existence*. Uniqueness is easier, and the proof will be given here. Existence is slightly more technical, and it is provided in Problem 8.11.

To show *uniqueness*, assume that there are two points, $\boldsymbol{x}_{*,1}$ and $\boldsymbol{x}_{*,2}$, $\boldsymbol{x}_{*,1} \neq \boldsymbol{x}_{*,2}$, such that

$$\|\boldsymbol{x} - \boldsymbol{x}_{*,1}\| = \|\boldsymbol{x} - \boldsymbol{x}_{*,2}\| = \min_{\boldsymbol{y} \in C} \|\boldsymbol{x} - \boldsymbol{y}\|. \tag{8.9}$$

(a) If $\boldsymbol{x} \in C$, then $P_C(\boldsymbol{x}) = \boldsymbol{x}$ is unique, since any other point in C would make $\|\boldsymbol{x} - P_C(\boldsymbol{x})\| > 0$.

(b) Let $\boldsymbol{x} \notin C$. Then, mobilizing the parallelogram law of the norm (Eq. (8.151) in the chapter's appendix, and Problem 8.8) we get

$$\|(\boldsymbol{x} - \boldsymbol{x}_{*,1}) + (\boldsymbol{x} - \boldsymbol{x}_{*,2})\|^2 + \|(\boldsymbol{x} - \boldsymbol{x}_{*,1}) - (\boldsymbol{x} - \boldsymbol{x}_{*,2})\|^2 = 2\left(\|\boldsymbol{x} - \boldsymbol{x}_{*,1}\|^2 + \|\boldsymbol{x} - \boldsymbol{x}_{*,2}\|^2\right),$$

or

$$\|2\boldsymbol{x} - (\boldsymbol{x}_{*,1} + \boldsymbol{x}_{*,2})\|^2 + \|\boldsymbol{x}_{*,1} - \boldsymbol{x}_{*,2}\|^2 = 2\left(\|\boldsymbol{x} - \boldsymbol{x}_{*,1}\|^2 + \|\boldsymbol{x} - \boldsymbol{x}_{*,2}\|^2\right),$$

and exploiting (8.9) and the fact that $\|\boldsymbol{x}_{*,1} - \boldsymbol{x}_{*,2}\| > 0$, we have

$$\left\|\boldsymbol{x} - \left(\frac{1}{2}\boldsymbol{x}_{*,1} + \frac{1}{2}\boldsymbol{x}_{*,2}\right)\right\|^2 < \|\boldsymbol{x} - \boldsymbol{x}_{*,1}\|^2. \tag{8.10}$$

However, due to the convexity of C, the point $\frac{1}{2}\boldsymbol{x}_{*,1} + \frac{1}{2}\boldsymbol{x}_{*,2}$ lies in C. Also, by the definition of projection, $\boldsymbol{x}_{*,1}$ is the point with the smallest distance, and hence (8.10) cannot be valid. □

For the existence, one has to use the property of closeness (every sequence in C has its limit in C) as well as the property of completeness of Hilbert spaces, which guarantees that every Cauchy sequence in $\mathbb{H}$ has a limit (see chapter's appendix). The proof is given in Problem 8.11.

Remarks 8.1.

- Note that if $\boldsymbol{x} \notin C \subseteq \mathbb{H}$, then its projection onto C lies on the *boundary* of C (Problem 8.12).

Example 8.1. Derive analytical expressions for the projections of a point $\boldsymbol{x} \in \mathbb{H}$, where $\mathbb{H}$ is a real Hilbert space, onto (a) a hyperplane, (b) a halfspace, and (c) the ℓ_2 ball of radius δ.

(a) A hyperplane H is defined as

$$H := \left\{\boldsymbol{y} : \langle \boldsymbol{\theta}, \boldsymbol{y} \rangle + \theta_0 = 0\right\},$$

for some $\boldsymbol{\theta} \in \mathbb{H}$, $\theta_0 \in \mathbb{R}$, and $\langle \cdot, \cdot \rangle$ denotes the respective inner product. If $\mathbb{H}$ breaks down to a Euclidean space, the projection is readily available by simple geometric arguments, and it is given by

$$\boxed{P_C(\boldsymbol{x}) = \boldsymbol{x} - \frac{\langle \boldsymbol{\theta}, \boldsymbol{x} \rangle + \theta_0}{\|\boldsymbol{\theta}\|^2}\boldsymbol{\theta} : \quad \text{projection onto a hyperplane,}} \tag{8.11}$$

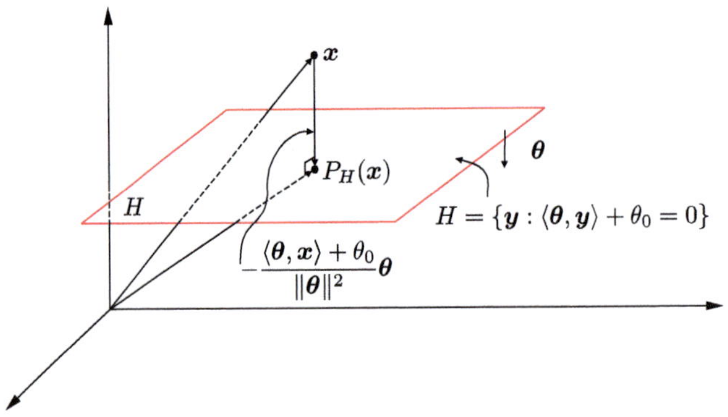

FIGURE 8.7

The projection onto a hyperplane in $\mathbb{R}^3$. The vector θ should be at the origin of the axes, but it is drawn to clearly show that it is perpendicular to H.

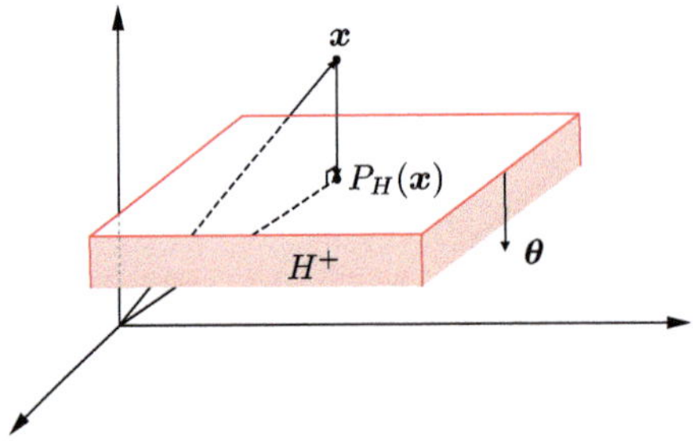

FIGURE 8.8

Projection onto a halfspace.

and it is shown in Fig. 8.7. For a general Hilbert space $\mathbb{H}$, the hyperplane is a closed convex subset of $\mathbb{H}$, and the projection is still given by the same formula (Problem 8.13).

(b) The definition of a halfspace, H^+, is given by

$$H^+ = \{y : \langle \theta, y\rangle + \theta_0 \geq 0\}, \tag{8.12}$$

and it is shown in Fig. 8.8 for the $\mathbb{R}^3$ case.

Because the projection lies on the boundary if $x \notin H^+$, its projection will lie on the hyperplane defined by θ and θ_0, and it will be equal to x if $x \in H^+$; thus, the projection is easily shown to be

$$\boxed{P_{H^+}(x) = x - \frac{\min\{0, \langle \theta, x\rangle + \theta_0\}}{\|\theta\|^2}\theta : \quad \text{projection onto a halfspace.}} \tag{8.13}$$

(c) The closed ball centered at $\mathbf{0}$ and of radius δ, denoted as $B[\mathbf{0},\delta]$, in a general Hilbert space, $\mathbb{H}$, is defined as

$$B[\mathbf{0},\delta]=\left\{\boldsymbol{y}:\|\boldsymbol{y}\|\leq\delta\right\}.$$

The projection of $\boldsymbol{x}\notin B[\mathbf{0},\delta]$ onto $B[\mathbf{0},\delta]$ is given by

$$\boxed{P_{B[\mathbf{0},\delta]}(\boldsymbol{x})=\begin{cases}\boldsymbol{x}, & \text{if } \|\boldsymbol{x}\|\leq\delta,\\ \delta\frac{\boldsymbol{x}}{\|\boldsymbol{x}\|}, & \text{if } \|\boldsymbol{x}\|>\delta,\end{cases}\quad : \quad \text{projection onto a closed ball,}} \tag{8.14}$$

and it is geometrically illustrated in Fig. 8.9, for the case of $\mathbb{R}^2$ (Problem 8.14).

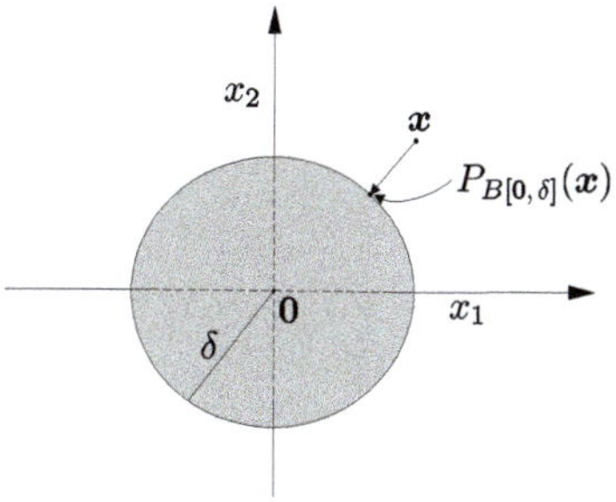

FIGURE 8.9

The projection onto a closed ball of radius δ centered at the origin of the axes.

Remarks 8.2.

- In the context of sparsity-aware learning, which is dealt with in Chapter 10, a key point is the projection of a point in $\mathbb{R}^l$ ($\mathbb{C}^l$) onto the ℓ_1 ball. There it is shown that, given the size of the ball, this projection corresponds to the so-called soft thresholding operation (see, also, Example 8.10 for a definition).

  It should be stressed that a linear space equipped with the ℓ_1 norm is no more Euclidean (Hilbert), inasmuch as this norm *is not induced* by an inner product operation; moreover, uniqueness of the projection with respect to this norm is not guaranteed (Problem 8.15).

8.3.1 PROPERTIES OF PROJECTIONS

In this section, we summarize some basic properties of the projections. These properties are used to prove a number of theorems and convergence results, associated with algorithms that are developed around the notion of projection.

Readers who are interested only in the algorithms can bypass this section.

Proposition 8.1. *Let $\mathbb{H}$ be a Hilbert space, $C \subseteq \mathbb{H}$ be a closed convex set, and $\boldsymbol{x} \in \mathbb{H}$. Then the projection $P_C(\boldsymbol{x})$ satisfies the following two properties*[4]*:*

$$\boxed{\text{Real}\{\langle \boldsymbol{x} - P_C(\boldsymbol{x}), \boldsymbol{y} - P_C(\boldsymbol{x})\rangle\} \leq 0, \ \forall \ \boldsymbol{y} \in C,} \tag{8.15}$$

and

$$\boxed{\|P_C(\boldsymbol{x}) - P_C(\boldsymbol{y})\|^2 \leq \text{Real}\{\langle \boldsymbol{x} - \boldsymbol{y}, P_C(\boldsymbol{x}) - P_C(\boldsymbol{y})\rangle\}, \ \forall \ \boldsymbol{x}, \boldsymbol{y} \in \mathbb{H}.} \tag{8.16}$$

The proof of the proposition is provided in Problem 8.16. The geometric interpretation of (8.15) for the case of real Hilbert space is shown in Fig. 8.10. Note that for a real Hilbert space, the first property becomes

$$\langle \boldsymbol{x} - P_C(\boldsymbol{x}), \boldsymbol{y} - P_C(\boldsymbol{x})\rangle \leq 0, \quad \forall \boldsymbol{y} \in C. \tag{8.17}$$

From the geometric point of view, (8.17) means that the angle formed by the two vectors $\boldsymbol{x} - P_C(\boldsymbol{x})$ and $\boldsymbol{y} - P_C(\boldsymbol{x})$ is *obtuse*. The hyperplane that crosses $P_C(\boldsymbol{x})$ and is orthogonal to $\boldsymbol{x} - P_C(\boldsymbol{x})$ is known as *supporting* hyperplane and it leaves all points in C on one side and $\boldsymbol{x}$ on the other. It can be shown that if C is closed and convex and $\boldsymbol{x} \notin C$, there is always such a hyperplane (see, for example, [30]).

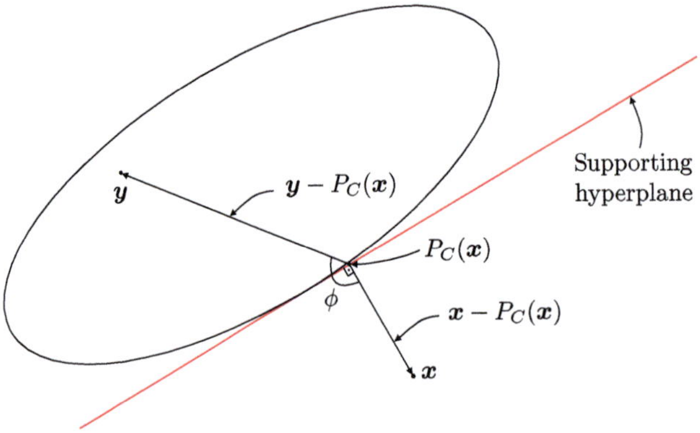

FIGURE 8.10

The vectors $\boldsymbol{y} - P_C(\boldsymbol{x})$ and $\boldsymbol{x} - P_C(\boldsymbol{x})$ form an obtuse angle, ϕ.

Lemma 8.1. *Let S be a closed subspace $S \subseteq \mathbb{H}$ in a Hilbert space $\mathbb{H}$. Then $\forall \ \boldsymbol{x}, \boldsymbol{y} \in \mathbb{H}$, the following properties hold true:*

$$\boxed{\langle \boldsymbol{x}, P_S(\boldsymbol{y})\rangle = \langle P_S(\boldsymbol{x}), \boldsymbol{y}\rangle = \langle P_S(\boldsymbol{x}), P_S(\boldsymbol{y})\rangle} \tag{8.18}$$

[4] The theorems are stated here for the general case of complex numbers.

and

$$\boxed{P_S(a\boldsymbol{x} + b\boldsymbol{y}) = aP_S(\boldsymbol{x}) + bP_S(\boldsymbol{y}),} \tag{8.19}$$

where a and b are arbitrary scalars. In other words, the projection operation on a closed subspace is a linear one (Problem 8.17). Recall that in a Euclidean space all subspaces are closed, and hence the linearity is always valid.

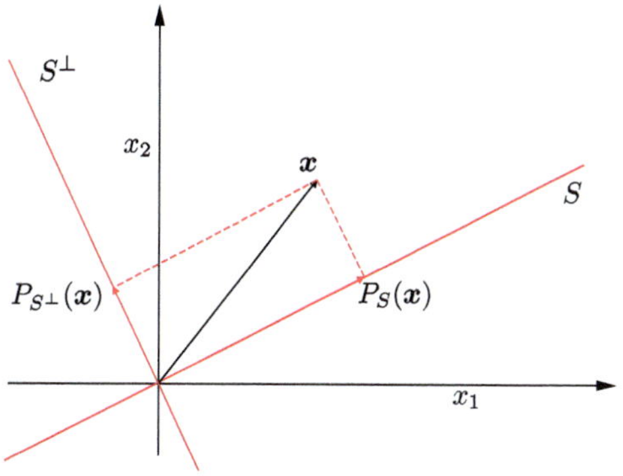

FIGURE 8.11

Every point in a Hilbert space $\mathbb{H}$ can be uniquely decomposed into the sum of its projections on any *closed* subspace S and its orthogonal complement $S^\perp$.

It can be shown (Problem 8.18) that if S is a closed subspace in a Hilbert space $\mathbb{H}$, its orthogonal complement, $S^\perp$, is also a closed subspace, such that $S \cap S^\perp = \{\mathbf{0}\}$; by definition, the orthogonal compliment $S^\perp$ is the set whose elements are orthogonal to each element of S. Moreover, $\mathbb{H} = S \oplus S^\perp$; that is, each element $\boldsymbol{x} \in \mathbb{H}$ can be *uniquely* decomposed as

$$\boxed{\boldsymbol{x} = P_S(\boldsymbol{x}) + P_{S^\perp}(\boldsymbol{x}), \;\; \boldsymbol{x} \in \mathbb{H}: \quad \text{for closed subspaces},} \tag{8.20}$$

as demonstrated in Fig. 8.11.

Definition 8.5. Let a mapping

$$T : \mathbb{H} \longmapsto \mathbb{H}.$$

T is called *nonexpansive* if $\forall\, \boldsymbol{x}, \boldsymbol{y} \in \mathbb{H}$

$$\boxed{\|T(\boldsymbol{x}) - T(\boldsymbol{y})\| \le \|\boldsymbol{x} - \boldsymbol{y}\| : \quad \text{nonexpansive mapping.}} \tag{8.21}$$

Proposition 8.2. *Let C be a closed convex set in a Hilbert space $\mathbb{H}$. Then the associated projection operator*

$$P_C : \mathbb{H} \longmapsto C$$

is nonexpansive.

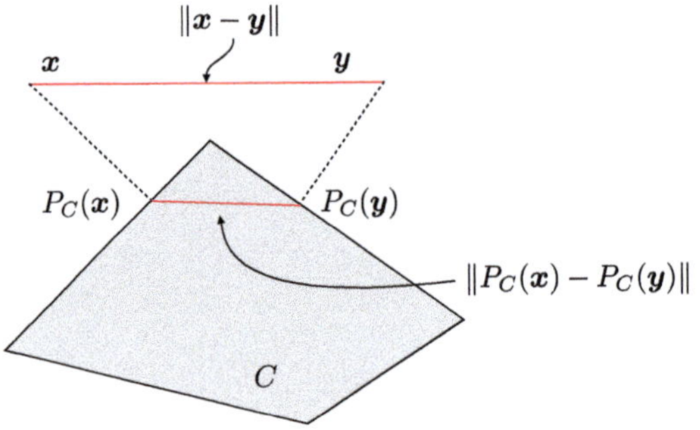

FIGURE 8.12

The nonexpansiveness property of the projection operator, $P_C(\cdot)$, guarantees that the distance between two points can never be smaller than the distance between their respective projections on a closed convex set.

Proof. Let $x, y \in \mathbb{H}$. Recall property (8.16), that is,

$$\|P_C(x) - P_C(y)\|^2 \leq \text{Real}\{\langle x - y, P_C(x) - P_C(y)\rangle\}. \tag{8.22}$$

Moreover, by employing the Schwarz inequality (Eq. (8.149) of chapter's appendix), we get

$$|\langle x - y, P_C(x) - P_C(y)\rangle| \leq \|x - y\| \|P_C(x) - P_C(y)\|. \tag{8.23}$$

Combining (8.22) and (8.23), we readily obtain

$$\|P_C(x) - P_C(y)\| \leq \|x - y\|. \tag{8.24}$$

□

Fig. 8.12 provides a geometric interpretation of (8.24). The property of nonexpansiveness, as well as a number of its variants (for example, [6,7,30,81,82]), is of paramount importance in convex set theory and learning. It is the property that guarantees the convergence of an algorithm, which comprises a sequence of successive projections (mappings), to the so-called *fixed point set*, that is, to the set whose elements are left unaffected by the respective mapping T, i.e.,

$$\boxed{\text{Fix}(T) = \{x \in \mathbb{H} : T(x) = x\} : \quad \text{fixed point set.}}$$

In the case of a projection operator on a closed convex set, we know that the respective fixed point set is the set C itself, since $P_C(x) = x, \; \forall x \in C$.

Definition 8.6. Let C be a closed convex set C in a Hilbert space. An operator

$$T_C : \mathbb{H} \longmapsto C$$

is called *relaxed projection* if

$$T_C := I + \mu(P_C - I), \quad \mu \in (0, 2),$$

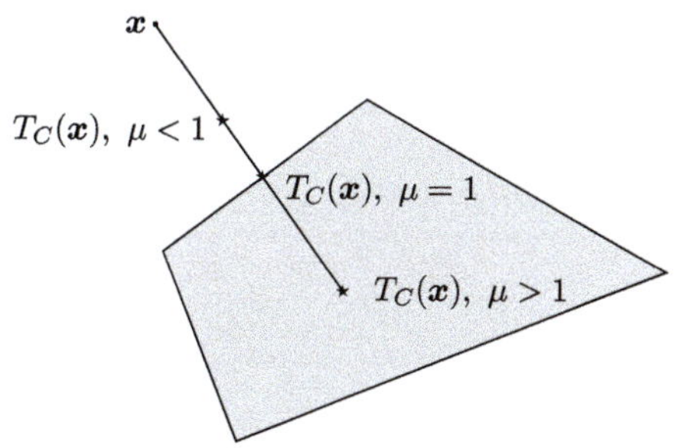

FIGURE 8.13

Geometric illustration of the relaxed projection operator.

or in other words, $\forall\, x \in \mathbb{H}$,

$$\boxed{T_C(x) = x + \mu\big(P_C(x) - x\big), \ \mu \in (0, 2): \quad \text{relaxed projection on } C.}$$

We readily see that for $\mu = 1$, $T_C(x) = P_C(x)$. Fig. 8.13 shows the geometric illustration of the relaxed projection. Observe that for different values of $\mu \in (0, 2)$, the relaxed projection traces all points in the line segment joining the points x and $x + 2(P_C(x) - x)$. Note that

$$T_C(x) = x, \ \forall x \in C,$$

that is, $\text{Fix}(T_C) = C$. Moreover, it can be shown that the relaxed projection operator is also nonexpansive, that is (Problem 8.19),

$$\|T_C(x) - T_C(y)\| \leq \|x - y\|, \ \forall \mu \in (0, 2).$$

A final property of the relaxed projection, which can also be easily shown (Problem 8.20), is the following:

$$\boxed{\forall\, y \in C, \ \|T_C(x) - y\|^2 \leq \|x - y\|^2 - \eta\|T_C(x) - x\|^2, \ \eta = \frac{2-\mu}{\mu}.} \tag{8.25}$$

Such mappings are known as *η-nonexpansive* or *strongly attracting* mappings; it is guaranteed that the distance $\|T_C(x) - y\|$ is *smaller* than $\|x - y\|$ at least by the positive quantity $\eta\|T_C(x) - x\|^2$; that is, the fixed point set $\text{Fix}(T_C) = C$ *strongly attracts* x. The geometric interpretation is given in Fig. 8.14.

8.4 FUNDAMENTAL THEOREM OF PROJECTIONS ONTO CONVEX SETS

In this section, one of the most celebrated theorems in the theory of convex sets is stated: the fundamental theorem of projections onto convex sets (POCS). This theorem is at the heart of a number of powerful algorithms and methods, some of which are described in this book. The origin of the theorem is traced back to Von Neumann [98], who proposed the theorem for the case of two subspaces.

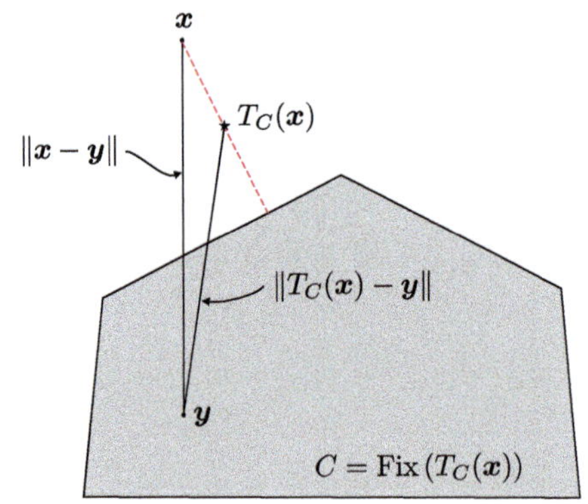

FIGURE 8.14

The relaxed projection is a strongly attracting mapping; $T_C(\boldsymbol{x})$ is closer to any point $\boldsymbol{y} \in C = \text{Fix}(T_C)$ than the point $\boldsymbol{x}$ is.

Von Neumann was a Hungarian-born American of Jewish descent. He was a child prodigy who earned his PhD at age 22. It is difficult to summarize his numerous significant contributions, which range from pure mathematics to economics (he is considered the founder of the game theory) and from quantum mechanics (he laid the foundations of the mathematical framework in quantum mechanics) to computer science (he was involved in the development of ENIAC, the first general purpose electronic computer). He was heavily involved in the Manhattan project for the development of the hydrogen bomb.

Let $C_k,\ k = 1, 2, \ldots, K$, be a *finite* number of *closed* convex sets in a Hilbert space $\mathbb{H}$, and assume that they share a *nonempty* intersection,

$$C = \bigcap_{k=1}^{K} C_k \neq \emptyset.$$

Let $T_{C_k},\ k = 1, 2, \ldots, K$, be the respective relaxed projection mappings

$$T_{C_k} = I + \mu_k (P_{C_k} - I),\ \mu_k \in (0, 2), \quad k = 1, 2, \ldots, K.$$

Form the concatenation of these relaxed projections,

$$T := T_{C_k} T_{C_{k-1}} \cdots T_{C_1},$$

where the specific order is not important. In words, T comprises a sequence of relaxed projections, starting from C_1. In the sequel, the obtained point is projected onto C_2, and so on.

Theorem 8.5. *Let $C_k, k = 1, 2, \ldots, K$, be closed convex sets in a Hilbert space $\mathbb{H}$, with nonempty intersection. Then for any $\boldsymbol{x}_0 \in \mathbb{H}$, the sequence $(T^n(\boldsymbol{x}_0)),\ n = 1, 2, \ldots,$ converges weakly to a point in $C = \bigcap_{k=1}^{K} C_k$.*

The theorem [17,42] states the notion of *weak convergence*. When $\mathbb{H}$ becomes a Euclidean (finite-dimensional) space, the notion of weak convergence coincides with the familiar "standard" definition of (strong) convergence. Weak convergence is a weaker version of strong convergence, and it is met in infinite-dimensional spaces. A sequence $\boldsymbol{x}_n \in \mathbb{H}$ is said to converge weakly to a point $\boldsymbol{x}_* \in \mathbb{H}$ if, $\forall\, \boldsymbol{y} \in \mathbb{H}$,

$$\langle \boldsymbol{x}_n, \boldsymbol{y} \rangle \xrightarrow[n\to\infty]{} \langle \boldsymbol{x}_*, \boldsymbol{y} \rangle,$$

and we write

$$\boldsymbol{x}_n \xrightarrow[n\to\infty]{w} \boldsymbol{x}_*.$$

As already said, in Euclidean spaces, weak convergence implies strong convergence. This is not necessarily true for general Hilbert spaces. On the other hand, strong convergence always implies weak convergence (for example, [87]) (Problem 8.21). Fig. 8.15 gives the geometric illustration of the theorem.

The proof of the theorem is a bit technical for the general case (for example, [87]). However, it can be simplified for the case where the involved convex sets are closed subspaces (Problem 8.23). At the heart of the proof lie (a) the nonexpansiveness property of $T_{C_k}, k = 1, 2, \ldots, K$, which is retained by T, and (b) the fact that the fixed point set of T is $\mathrm{Fix}(T) = \bigcap_{k=1}^{K} C_k$.

Remarks 8.3.

- In the special case where all $C_k, k = 1, 2, \ldots, K$, are closed subspaces, we have

$$T^n(\boldsymbol{x}_0) \longrightarrow P_C(\boldsymbol{x}_0).$$

In other words, the sequence of relaxed projections converges *strongly* to the projection of $\boldsymbol{x}_0$ on C. Recall that if each $C_k, k = 1, 2, \ldots, K$, is a closed subspace of $\mathbb{H}$, it can easily be shown that their

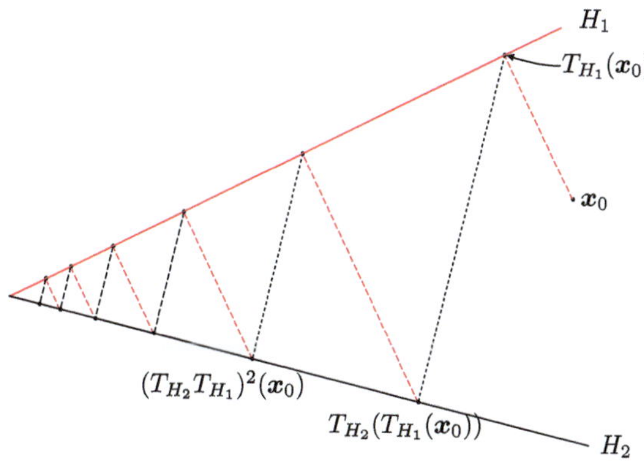

FIGURE 8.15

Geometric illustration of the fundamental theorem of projections onto convex sets (POCS), for $T_{C_i} = P_{C_i}$, $i = 1, 2$ ($\mu_{C_i} = 1$). The closed convex sets are the two straight lines in $\mathbb{R}^2$. Observe that the sequence of projections tends to the intersection of H_1, H_2.

intersection is also a closed subspace. As said before, in a Euclidean space $\mathbb{R}^l$, all subspaces are closed.

- The previous statement is also true for *linear varieties*. A linear variety is the translation of a subspace by a constant vector $\boldsymbol{a}$. That is, if S is a subspace and $\boldsymbol{a} \in \mathbb{H}$, then the set of points

$$S_a = \{\boldsymbol{y} : \boldsymbol{y} = \boldsymbol{a} + \boldsymbol{x},\ \boldsymbol{x} \in S\}$$

is a linear variety. Hyperplanes are linear varieties (see, for example, Fig. 8.16).

- The scheme resulting from the POCS theorem, employing the relaxed projection operator, is summarized in Algorithm 8.1,

Algorithm 8.1 (The POCS algorithm).

- Initialization.
  - Select $\boldsymbol{x}_0 \in \mathbb{H}$.
  - Select $\mu_k \in (0, 2),\ k = 1, 2, \ldots, K$
- **For** $n = 1, 2, \ldots,$ **Do**
  - $\hat{\boldsymbol{x}}_{0,n} = \boldsymbol{x}_{n-1}$
  - **For** $k = 1, 2, \ldots, K$, **Do**

$$\hat{\boldsymbol{x}}_{k,n} = \hat{\boldsymbol{x}}_{k-1,n} + \mu_k \left(P_{C_k}(\hat{\boldsymbol{x}}_{k-1,n}) - \hat{\boldsymbol{x}}_{k-1,n}\right) \tag{8.26}$$

  - **End For**
  - $\boldsymbol{x}_n = \hat{\boldsymbol{x}}_{K,n}$.
- **End For**

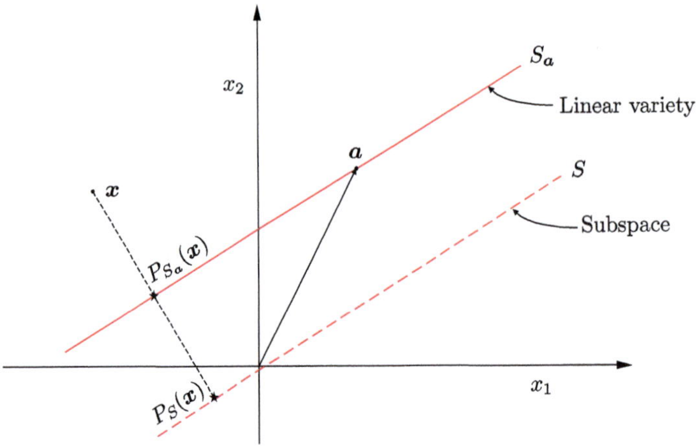

FIGURE 8.16

A hyperplane (not crossing the origin) is a linear variety; P_{S_a} and P_S are the projections of $\boldsymbol{x}$ onto S_a and S, respectively.

8.5 A PARALLEL VERSION OF POCS

In [68], a parallel version of the POCS algorithm was stated. In addition to its computational advantages (when parallel processing can be exploited), this scheme will be our vehicle for generalizations to the online processing, where one can cope with the case where the number of convex sets becomes infinite (or very large in practice). The proof for the parallel POCS is slightly technical and relies heavily on the results stated in the previous section. The concept behind the proof is to construct appropriate product spaces, and this is the reason that the algorithm is also referred to as POCS in *product spaces*. For the detailed proof, the interested reader may consult [68].

Theorem 8.6. *Let $C_k, k = 1, 2, \dots, K$, be closed convex sets in a Hilbert space $\mathbb{H}$. Then, for any $\boldsymbol{x}_0 \in \mathbb{H}$, the sequence $\boldsymbol{x}_n$, defined as*

$$\boldsymbol{x}_n = \boldsymbol{x}_{n-1} + \mu_n \left(\sum_{k=1}^{K} \omega_k P_{C_k}(\boldsymbol{x}_{n-1}) - \boldsymbol{x}_{n-1} \right), \tag{8.27}$$

weakly converges to a point in $\bigcap_{k=1}^{K} C_k$, if

$$0 < \mu_n \leq M_n,$$

and

$$M_n := \sum_{k=1}^{K} \frac{\omega_k \| P_{C_k}(\boldsymbol{x}_{n-1}) - \boldsymbol{x}_{n-1} \|^2}{\| \sum_{k=1}^{K} \omega_k P_{C_k}(\boldsymbol{x}_{n-1}) - \boldsymbol{x}_{n-1} \|^2}, \tag{8.28}$$

where $\omega_k > 0$, $k = 1, 2, \dots, K$, such that

$$\sum_{k=1}^{K} \omega_k = 1.$$

Update recursion (8.27) says that at each iteration, all projections on the convex sets take place *concurrently*, and then they are *convexly* combined. The extrapolation parameter μ_n is chosen in interval $(0, M_n]$, where M_n is recursively computed in (8.28), so that convergence is guaranteed. Fig. 8.17 illustrates the updating process.

8.6 FROM CONVEX SETS TO PARAMETER ESTIMATION AND MACHINE LEARNING

Let us now see how this elegant theory can be turned into a useful tool for parameter estimation in machine learning. We will demonstrate the procedure using two examples.

8.6.1 REGRESSION

Consider the regression model, relating input–output observation points,

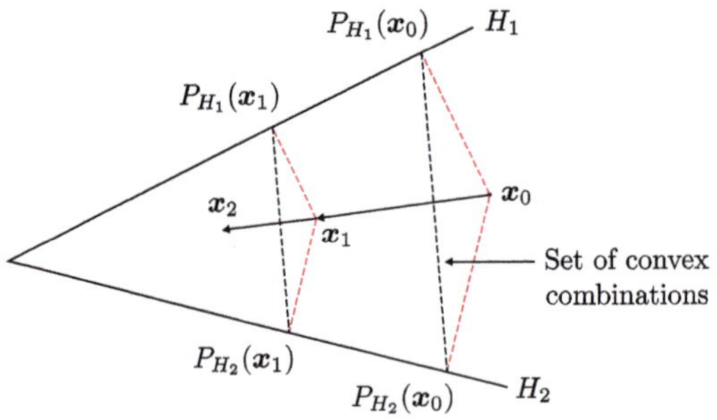

FIGURE 8.17

The parallel POCS algorithm for the case of two (hyperplanes) lines in $\mathbb{R}^2$. At each step, the projections on H_1 and H_2 are carried out in parallel and then they are convexly combined.

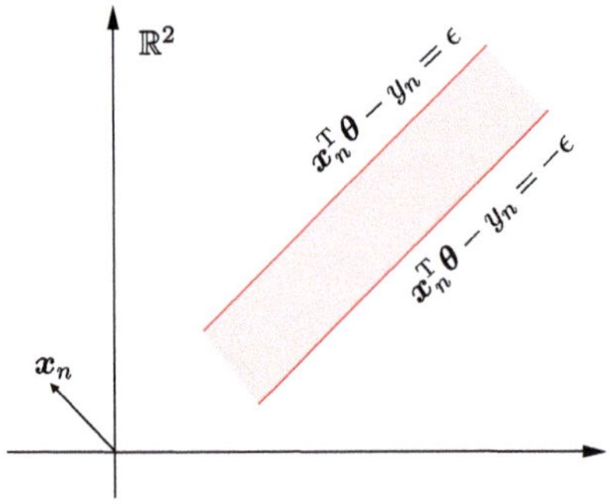

FIGURE 8.18

Each pair of training points, $(y_n, \boldsymbol{x}_n)$, defines a hyperslab in the parameters' space.

$$y_n = \boldsymbol{\theta}_o^T \boldsymbol{x}_n + \eta_n, \ (y_n, \boldsymbol{x}_n) \in \mathbb{R} \times \mathbb{R}^l, \quad n = 1, 2, \dots, N, \tag{8.29}$$

where $\boldsymbol{\theta}_o$ is the unknown parameter vector. Assume that η_n is a bounded noise sequence, that is,

$$|\eta_n| \leq \epsilon. \tag{8.30}$$

Then (8.29) and (8.30) guarantee that

$$|y_n - \boldsymbol{x}_n^T \boldsymbol{\theta}_o| \leq \epsilon. \tag{8.31}$$

Consider now the following set of points:

$$\boxed{S_\epsilon = \left\{\boldsymbol{\theta} : |y_n - \boldsymbol{x}_n^T \boldsymbol{\theta}| \leq \epsilon\right\} : \quad \text{hyperslab.}} \tag{8.32}$$

This set is known as a *hyperslab*, and it is geometrically illustrated in Fig. 8.18. The definition is generalized for any $\mathbb{H}$ by replacing the inner product notation as $\langle \boldsymbol{x}_n, \boldsymbol{\theta} \rangle$. The set comprises all the points that lie in the region formed by the two hyperplanes

$$\begin{aligned} \boldsymbol{x}_n^T\boldsymbol{\theta} - y_n &= \epsilon, \\ \boldsymbol{x}_n^T\boldsymbol{\theta} - y_n &= -\epsilon. \end{aligned}$$

This region is trivially shown to be a *closed convex* set. Note that *every pair* of training points, $(y_n, \boldsymbol{x}_n)$, $n = 1, 2, \ldots, N$, defines a hyperslab of different orientation (depending on $\boldsymbol{x}_n$) and position in space (determined by y_n). Moreover, (8.31) guarantees that the unknown, $\boldsymbol{\theta}_o$, lies within all these hyperslabs; hence, $\boldsymbol{\theta}_o$ lies in their *intersection*. All we need to do now is derive the projection operator onto hyperslabs (we will do it soon) and use one of the POCS schemes to find a point in the intersection. Assuming that enough training points are available and that the intersection is "small" enough, any point in this intersection will be "close" to $\boldsymbol{\theta}_o$. Note that such a procedure is not based on optimization arguments. Recall, however, that even in optimization techniques, iterative algorithms have to be used, and in practice, iterations have to stop after a finite number of steps. Thus, one can only approximately reach the optimal value. More on these issues and related convergence properties will be discussed later in this chapter.

The obvious question now is what happens if the noise is not bounded. There are two answers to this point. First, in any practical application where measurements are involved, the noise has to be bounded. Otherwise, the circuits will be burned out. So, at least conceptually, this assumption does not conflict with what happens in practice. It is a matter of selecting the right value for ϵ. The second answer is that one can choose ϵ to be a few times the standard deviation of the assumed noise model. Then $\boldsymbol{\theta}_o$ will lie in these hyperslabs with high probability. We will discuss strategies for selecting ϵ, but our goal in this section is to discuss the main rationale in using the theory in practical applications. Needless to say, there is nothing divine around hyperslabs. Other closed convex sets can also be used if the nature of the noise in a specific application suggests a different type of convex sets.

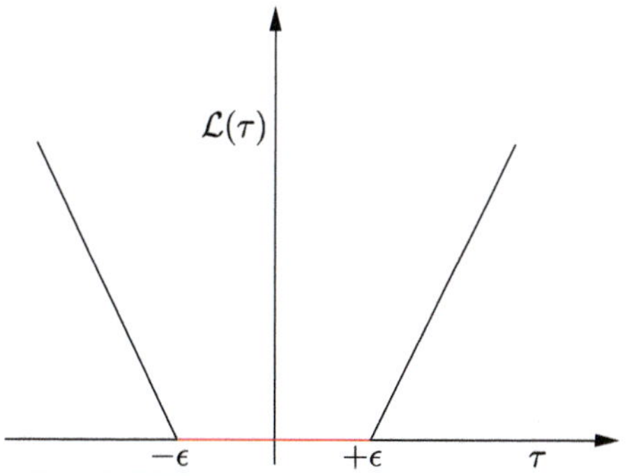

FIGURE 8.19

The linear ϵ-insensitive loss function $\tau = y - \boldsymbol{\theta}^T\boldsymbol{x}$. Its value is zero if $|\tau| < \epsilon$, and it increases linearly for $|\tau| \geq \epsilon$.

It is now interesting to look at the set where the solution lies, in this case at the hyperslab, from a different perspective. Consider the loss function

$$\boxed{\mathcal{L}(y, \boldsymbol{\theta}^T\boldsymbol{x}) = \max\left(0,\ |y - \boldsymbol{x}^T\boldsymbol{\theta}| - \epsilon\right): \quad \text{linear } \epsilon\text{-insensitive loss function,}} \tag{8.33}$$

which is illustrated in Fig. 8.19 for the case $\theta \in \mathbb{R}$. This is known as linear ϵ-insensitive loss function, and it has been popularized in the context of support vector regression (Chapter 11). For all $\boldsymbol{\theta}$s which lie within the hyperslab defined in (8.32), the loss function scores a zero. For points outside the hyperslab, there is a linear increase of its value. Thus, the hyperslab is the *zero level set* of the linear ϵ-insensitive loss function, defined *locally* according to the point $(y_n, \boldsymbol{x}_n)$. Thus, although no optimization concept is associated with POCS, the choice of the closed convex sets can be made to minimize "locally," at each point, a convex loss function by selecting its zero level set.

We conclude our discussion by providing the projection operator of a hyperslab, S_ϵ. It is trivially shown that, given $\boldsymbol{\theta}$, its projection onto S_ϵ (defined by $(y_n, \boldsymbol{x}_n, \epsilon)$) is given by

$$P_{S_\epsilon} = \boldsymbol{\theta} + \beta_{\boldsymbol{\theta}}(y_n, \boldsymbol{x}_n)\boldsymbol{x}_n, \tag{8.34}$$

where

$$\beta_{\boldsymbol{\theta}}(y_n, \boldsymbol{x}_n) = \begin{cases} \frac{y_n - \langle \boldsymbol{x}_n, \boldsymbol{\theta}\rangle - \epsilon}{\|\boldsymbol{x}_n\|^2}, & \text{if } \langle \boldsymbol{x}_n, \boldsymbol{\theta}\rangle - y_n < -\epsilon, \\ 0, & \text{if } |\langle \boldsymbol{x}, \boldsymbol{\theta}\rangle - y_n| \le \epsilon, \\ \frac{y_n - \langle \boldsymbol{x}_n, \boldsymbol{\theta}\rangle + \epsilon}{\|\boldsymbol{x}_n\|^2}, & \text{if } \langle \boldsymbol{x}_n, \boldsymbol{\theta}\rangle - y_n > \epsilon. \end{cases} \tag{8.35}$$

That is, if the point lies within the hyperslab, it coincides with its projection. Otherwise, the projection is on one of the two hyperplanes (depending on which side of the hyperslab the point lies on), which define S_ϵ. Recall that the projection of a point lies on the boundary of the corresponding closed convex set.

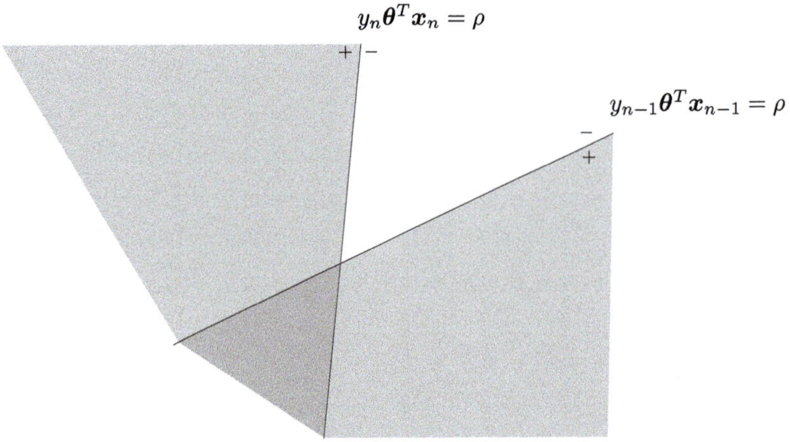

FIGURE 8.20

Each training point $(y_n, \boldsymbol{x}_n)$ defines a halfspace in the parameters $\boldsymbol{\theta}$-space, and the linear classifier will be searched in the intersection of all these halfspaces.

8.6.2 CLASSIFICATION

Let us consider the two-class classification task, and assume that we are given the set of training points $(y_n, \boldsymbol{x}_n), n = 1, 2, \ldots, N$.

Our goal now will be to design a linear classifier so as to score

$$\boldsymbol{\theta}^T \boldsymbol{x}_n \geq \rho, \text{ if } y_n = +1,$$

and

$$\boldsymbol{\theta}^T \boldsymbol{x}_n \leq -\rho, \text{ if } y_n = -1.$$

This requirement can be expressed as follows. Given $(y_n, \boldsymbol{x}_n) \in \{-1, 1\} \times \mathbb{R}^{l+1}$, design a linear classifier,[5] $\boldsymbol{\theta} \in \mathbb{R}^{l+1}$, such that

$$y_n \boldsymbol{\theta}^T \boldsymbol{x}_n \geq \rho > 0. \tag{8.36}$$

Note that, given $y_n, \boldsymbol{x}_n$, and ρ, (8.36) defines a *halfspace* (Example 8.1); this is the reason that we used "$\geq \rho$" rather than a strict inequality. In other words, all $\boldsymbol{\theta}$s which satisfy the desired inequality (8.36) lie in this halfspace. Since each pair $(y_n, \boldsymbol{x}_n)$, $n = 1, 2, \ldots, N$, defines a single halfspace, our goal now becomes that of trying to find a point at the intersection of all these halfspaces. This intersection is guaranteed to be nonempty if the classes are linearly separable. Fig. 8.20 illustrates the concept. The more realistic case of nonlinearly separable classes will be treated in Chapter 11, where a mapping in a high-dimensional (kernel) space makes the probability of two classes being linearly separable to tend to 1 as the dimensionality of the kernel space goes to infinity.

The halfspace associated with a training pair $(y_n, \boldsymbol{x}_n)$ can be seen as the level set of height zero of the so-called *hinge loss* function, defined as

$$\boxed{\mathcal{L}_\rho(y, \boldsymbol{\theta}^T \boldsymbol{x}) = \max\left(0, \rho - y\boldsymbol{\theta}^T \boldsymbol{x}\right): \quad \text{hinge loss function,}} \tag{8.37}$$

whose graph is shown in Fig. 8.21. Thus, choosing the halfspace as the closed convex set to represent $(y_n, \boldsymbol{x}_n)$ is equivalent to selecting the zero level set of the hinge loss, "adjusted" for the point $(y_n, \boldsymbol{x}_n)$.

Remarks 8.4.

- In addition to the two applications typical of the machine learning point of view, POCS has been applied in a number of other applications; see, for example, [7,24,87,89] for further reading.
- If the involved sets do not intersect, that is, $\bigcap_{k=1}^{K} C_k = \emptyset$, then it has been shown [25] that the parallel version of POCS in (8.27) converges to a point whose weighted squared distance from each one of the convex sets (defined as the distance of the point from its respective projection) is minimized.
- Attempts to generalize the theory to nonconvex sets have also been made (for example, [87] and more recently in the context of sparse modeling in [83]).

[5] Recall from Chapter 3, that this formulation covers the general case where a bias term is involved, by increasing the dimensionality of $\boldsymbol{x}_n$ and adding 1 as its last element.

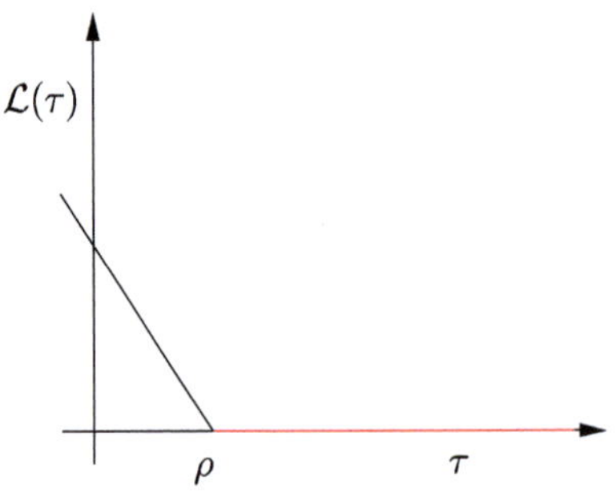

FIGURE 8.21

The hinge loss function. For the classification task $\tau = y\boldsymbol{\theta}^T\boldsymbol{x}$, its value is zero if $\tau \geq \rho$, and it increases linearly for $\tau < \rho$.

- When $C := \bigcap_{k=1}^{K} C_K \neq \emptyset$, we say that the problem is *feasible* and the intersection C is known as the *feasibility* set. The closed convex sets C_k, $k = 1, 2, \ldots, K$, are sometimes called the *property* sets, for obvious reasons. In both previous examples, namely, regression and classification, we commented that the involved property sets resulted as the 0-level sets of a loss function $\mathcal{L}$. Hence, assuming that the problem is feasible (the cases of bounded noise in regression and linearly separable classes in classification), any solution in the feasible set C will also be a *minimizer* of the respective loss functions in (8.33) and (8.37), respectively. Thus, although optimization did not enter into our discussion, there can be an optimizing flavor in the POCS method. Moreover, note that in this case, the loss functions need *not* be differentiable and the techniques we discussed in the previous chapters are not applicable. We will return to this issue in Section 8.10.

8.7 INFINITELY MANY CLOSED CONVEX SETS: THE ONLINE LEARNING CASE

In our discussion so far, we have assumed a finite number, K, of closed convex (property) sets. To land at their intersection (feasibility set) one has to cyclically project onto all of them or to perform the projections in parallel. Such a strategy is not appealing for the online processing scenario. At every time instant, a new pair of observations becomes available, which defines a new property set. Hence, in this case, the number of the available convex sets increases. Visiting all the available sets makes the complexity time dependent and after some time the required computational resources will become unmanageable.

An alternative viewpoint was suggested in [101–103], and later on extended in [76,104,105]. The main idea here is that at each time instant n, a pair of output–input training data is received and a (property) closed convex set, C_n, is constructed. The time index, n, is left to grow unbounded. However, at each time instant, q (a user-defined parameter) most recently constructed property sets are considered. In other words, the parameter q defines a *sliding window* in time. At each time instant, projections/relaxed projections are performed within this time window. The rationale is illustrated in

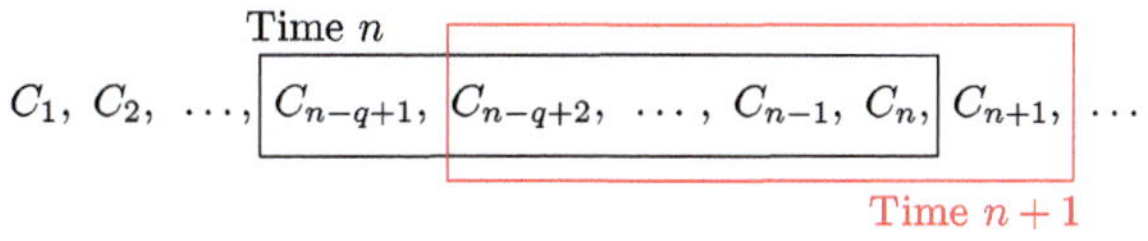

FIGURE 8.22

At time n, the property sets $C_{n-q+1}, \dots, C_n$ are used, while at time $n+1$, the sets $C_{n-q+2}, \dots, C_{n+1}$ are considered. Thus, the required number of projections does not grow with time.

Fig. 8.22. Thus, the number of sets onto which projections are performed does not grow with time, their number remains finite, and it is fixed by the user. The developed algorithm is an offspring of the parallel version of POCS and it is known as *adaptive projected subgradient method* (APSM). We will describe the algorithm in the context of regression. Following the discussion in Section 8.6, as each pair $(y_n, \boldsymbol{x}_n) \in \mathbb{R} \times \mathbb{R}^l$ becomes available, a hyperslab $S_{\epsilon,n}, n = 1, 2, \dots,$ is constructed and the goal is to find a $\boldsymbol{\theta} \in \mathbb{R}^l$ that lies in the intersection of all these property sets, starting from an arbitrary value, $\boldsymbol{\theta}_0 \in \mathbb{R}^l$.

Algorithm 8.2 (The APSM algorithm).

- Initialization
  - Choose $\boldsymbol{\theta}_0 \in \mathbb{R}^l$.
  - Choose q; The number of property sets to be processed at each time instant.
- **For** $n = 1, 2, \dots, q-1$, **Do**; Initial period, that is, $n < q$.
  - Choose $\omega_1, \dots, \omega_n: \quad \sum_{k=1}^{n} \omega_k = 1, \omega_k \geq 0$
  - Select μ_n

$$\boldsymbol{\theta}_n = \boldsymbol{\theta}_{n-1} + \mu_n \left(\sum_{k=1}^{n} \omega_k P_{S_{\epsilon,k}}(\boldsymbol{\theta}_{n-1}) - \boldsymbol{\theta}_{n-1} \right) \tag{8.38}$$

- **End For**
- **For** $n = q, q+1, \dots,$ **Do**
  - Choose $\omega_n, \dots, \omega_{n-q+1}$; usually $\omega_k = \frac{1}{q}, k = n-q+1, \dots, n$.
  - Select μ_n

$$\boldsymbol{\theta}_n = \boldsymbol{\theta}_{n-1} + \mu_n \left(\sum_{k=n-q+1}^{n} \omega_k P_{S_{\epsilon,k}}(\boldsymbol{\theta}_{n-1}) - \boldsymbol{\theta}_{n-1} \right) \tag{8.39}$$

- **End For**

The extrapolation parameter can now be chosen in the interval $(0, 2M_n)$ in order for convergence to be guaranteed. For the case of (8.39)

$$M_n = \sum_{k=n-q+1}^{n} \frac{\omega_k \| P_{S_{\epsilon,k}}(\boldsymbol{\theta}_{n-1}) - \boldsymbol{\theta}_{n-1} \|^2}{\left\| \sum_{k=n-q+1}^{n} \omega_k P_{S_{\epsilon,k}}(\boldsymbol{\theta}_{n-1}) - \boldsymbol{\theta}_{n-1} \right\|^2}. \tag{8.40}$$

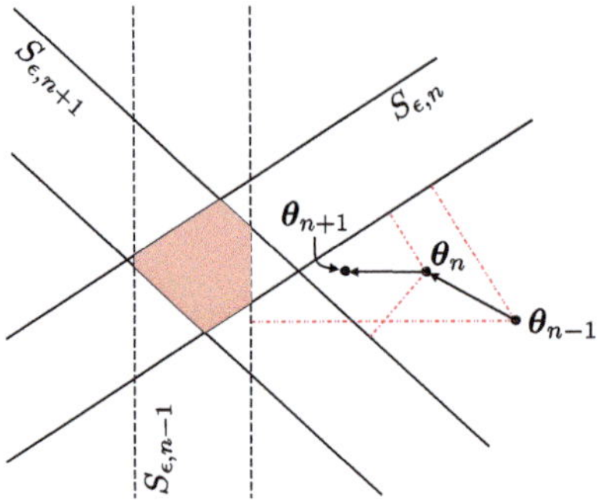

FIGURE 8.23

At time n, $q = 2$ hyperslabs are processed, namely, $S_{\epsilon,n}$, $S_{\epsilon,n-1}$; $\boldsymbol{\theta}_{n-1}$ is *concurrently* projected onto both of them and the projections are convexly combined. The new estimate is $\boldsymbol{\theta}_n$. Next, $S_{\epsilon,n+1}$ "arrives" and the process is repeated. Note that at every time, the estimate gets closer to the intersection; the latter will become smaller and smaller, as more hyperslabs arrive.

Note that this interval differs from that reported in the case of a finite number of sets in Eq. (8.27). For the first iteration steps associated with Eq. (8.38), the summations in the above formula starts from $k = 1$ instead of $k = n - q + 1$.

Recall that $P_{S_{\epsilon,k}}$ is the projection operation given in (8.34)–(8.35). Note that this is a generic scheme and can be applied with different property sets. All that is needed is to change the projection operator. For example, if classification is considered, all we have to do is to replace $S_{\epsilon,n}$ by the halfspace H_n^+, defined by the pair $(y_n, \boldsymbol{x}_n) \in \{-1, 1\} \times \mathbb{R}^{l+1}$, as explained in Section 8.6.2, and use the projection from (8.13) (see [78,79]). At this point, it must be emphasized that the original APSM form (e.g., [103, 104]) is more general and can cover a wide range of convex sets and functions.

Fig. 8.23 illustrates geometrically the APSM algorithm. We have assumed that the number of hyperslabs that are considered for projection at each time instant is $q = 2$. Each iteration comprises:

- q projections, which can be carried out in parallel,
- their convex combination, and
- the update step.

8.7.1 CONVERGENCE OF APSM

The proof of the convergence of the APSM is a bit technical and the interested reader can consult the related references. Here, we can be content with a geometric illustration that intuitively justifies the convergence, under certain assumptions. This geometric interpretation is at the heart of a stochastic approach to the APSM convergence, which was presented in [23].

Assume the noise to be bounded and that there is a true $\boldsymbol{\theta}_o$ that generates the data, that is,

$$y_n = \boldsymbol{x}_n^T \boldsymbol{\theta}_o + \eta_n. \tag{8.41}$$

By assumption,

$$|\eta_n| \le \epsilon.$$

Hence,

$$|\boldsymbol{x}_n^T \boldsymbol{\theta}_o - y_n| \le \epsilon.$$

Thus, $\boldsymbol{\theta}_o$ does lie in the intersection of all the hyperslabs of the form

$$|\boldsymbol{x}_n^T \boldsymbol{\theta} - y_n| \le \epsilon,$$

and in this case the problem is feasible. The question that is raised is how close one can go, even asymptotically as $n \longrightarrow \infty$, to $\boldsymbol{\theta}_o$. For example, if the volume of the intersection is large, even if the algorithm converges to a point in the boundary of this intersection this does not necessarily say much about how close the solution is to the true value $\boldsymbol{\theta}_o$. The proof in [23] establishes that the algorithm brings the estimate *arbitrarily close* to $\boldsymbol{\theta}_o$, under some general assumptions concerning the sequence of observations, and that the noise is bounded.

To understand what is behind the technicalities of the proof, recall that there are two main geometric issues concerning a hyperslab: (a) its orientation, which is determined by $\boldsymbol{x}_n$, and (b) its width. In finite-dimensional spaces, it is a matter of simple geometry to show that the width of a hyperslab is equal to

$$d = \frac{2\epsilon}{\|\boldsymbol{x}_n\|}. \tag{8.42}$$

This is a direct consequence of the fact that the distance[6] of a point, say, $\tilde{\boldsymbol{\theta}}$, from the hyperplane defined by the pair $(y, \boldsymbol{x})$, that is,

$$\boldsymbol{x}^T \boldsymbol{\theta} - y = 0,$$

is equal to

$$\frac{|\boldsymbol{x}^T \tilde{\boldsymbol{\theta}} - y|}{\|\boldsymbol{x}\|}.$$

Indeed, let $\bar{\boldsymbol{\theta}}$ be a point on one of the two boundary hyperplanes (e.g., $\boldsymbol{x}_n^T \boldsymbol{\theta} - y_n = \epsilon$) which define the hyperslab and consider its distance from the other one ($\boldsymbol{x}_n^T \boldsymbol{\theta} - y_n = -\epsilon$); then (8.42) is readily obtained.

Fig. 8.24 shows four hyperslabs in two different directions (one for the full lines and one for the dotted lines). The red hyperslabs are narrower than the black ones. Moreover, all four necessarily include $\boldsymbol{\theta}_o$. If $\boldsymbol{x}_n$ is left to vary randomly so that any orientation will occur, with high probability and for any orientation, the norm can also take small as well as arbitrarily large values, and then intuition says that the volume of the intersection around $\boldsymbol{\theta}_o$ will become arbitrarily small. Further results concerning the APSM algorithm can be found in, e.g., [82,89,103,104].

---

[6] For Euclidean spaces, this can be easily established by simple geometric arguments; see also Section 11.10.1.

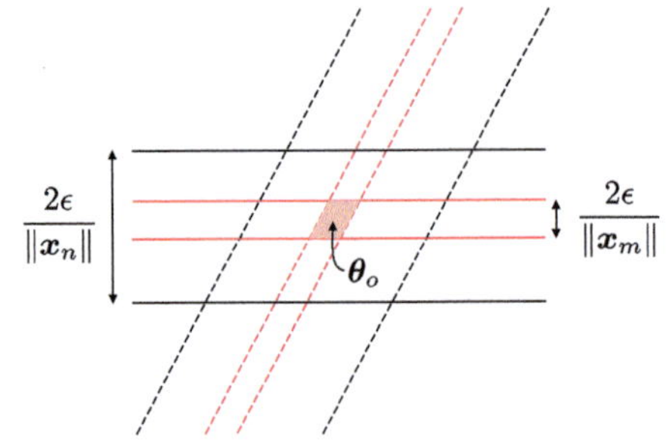

FIGURE 8.24

For each direction, the width of a hyperslab varies inversely proportional to $\|x_n\|$. In this figure, $||x_n|| < ||x_m||$ although both vectors point to the same direction. The intersection of hyperslabs of different directions and widths renders the volume of their intersection arbitrarily small around θ_o.

Some Practical Hints

The APSM algorithm needs the setting of three parameters, namely, ϵ, μ_n, and q. It turns out that the algorithm is not particularly sensitive in their choice:

- The choice of the parameter μ_n is similar in concept to the choice of the step size in the LMS algorithm. In particular, the larger μ_n, the faster the convergence speed, at the expense of a higher steady-state error floor. In practice, a step size approximately equal to $0.5M_n$ will lead to a low steady-state error, although the convergence speed will be relatively slow. On the contrary, if one chooses a larger step size, $1.5M_n$ approximately, then the algorithm enjoys a faster convergence speed, although the steady-state error after convergence is increased.
- Regarding the parameter ϵ. A typical choice is to set $\epsilon \approx \sqrt{2}\sigma$, where σ is the standard deviation of the noise. In practice (see, e.g., [47]), it has been shown that the algorithm is rather insensitive to this parameter. Hence, one needs only a *rough* estimate of the standard deviation.
- Concerning the choice of q, this is analogous to the q used for the APA in Chapter 5. The larger q is, the faster the convergence becomes; however, large values of q increase complexity as well as the error floor after convergence. In practice, relatively small values of q, for example, a small fraction of l, can significantly improve the convergence speed compared to the normalized least-mean-squares (NLMS) algorithm. Sometimes one can start with a relatively large value of q, and once the error decreases, q can be given smaller values to achieve lower error floors.

  It is important to note that the past data reuse within the sliding window of length q in the APA is implemented via the inversion of a $q \times q$ matrix. In the APSM, this is achieved via a sequence of q projections, leading to a complexity of linear dependence on q; moreover, these projections can be performed in parallel. Furthermore, the APA tends to be more sensitive to the presence of noise, since the projections are carried out on hyperplanes. In contrast, for the APSM case, projections are performed on hyperslabs, which implicitly care for the noise (for example, [102]).

Remarks 8.5.

- If the hyperslabs collapse to hyperplanes ($\epsilon = 0$) and $q = 1$, the algorithm becomes the NLMS. Indeed, for this case the projection in (8.39) becomes the projection on the hyperplane, H, defined by $(y_n, \boldsymbol{x}_n)$, that is,

$$\boldsymbol{x}_n^T \boldsymbol{\theta} = y_n,$$

and from (8.11), after making the appropriate notational adjustments, we have

$$P_H(\boldsymbol{\theta}_{n-1}) = \boldsymbol{\theta}_{n-1} - \frac{\boldsymbol{x}_n^T \boldsymbol{\theta}_{n-1} - y_n}{\|\boldsymbol{x}_n\|^2} \boldsymbol{x}_n. \tag{8.43}$$

Plugging (8.43) into (8.39), we get

$$\boldsymbol{\theta}_n = \boldsymbol{\theta}_{n-1} + \frac{\mu_n}{\|\boldsymbol{x}_n\|^2} e_n \boldsymbol{x}_n,$$

$$e_n = y_n - \boldsymbol{x}_n^T \boldsymbol{\theta}_{n-1},$$

which is the normalized LMS, introduced in Section 5.6.1.

- Closely related to the APSM algorithmic family are the *set-membership* algorithms (for example, [29,32–34,61]). This family can be seen as a special case of the APSM philosophy, where only special types of convex sets are used, for example, hyperslabs. Also, at each iteration step, a single projection is performed onto the set associated with the most recent pair of observations. For example, in [34,99] the update recursion of a set-membership APA is given by

$$\boldsymbol{\theta}_n = \begin{cases} \boldsymbol{\theta}_{n-1} + X_n (X_n^T X_n)^{-1} (\boldsymbol{e}_n - \boldsymbol{y}_n), & \text{if } |e_n| > \epsilon, \\ \boldsymbol{\theta}_{n-1}, & \text{otherwise}, \end{cases} \tag{8.44}$$

where $X_n = [\boldsymbol{x}_n, \boldsymbol{x}_{n-1}, \ldots, \boldsymbol{x}_{n-q+1}]$, $\boldsymbol{y}_n = [y_n, y_{n-1}, \ldots, y_{n-q+1}]^T$, and $\boldsymbol{e}_n = [e_n, e_{n-1}, \ldots, e_{n-q+1}]^T$, with $e_n = y_n - \boldsymbol{x}_n^T \boldsymbol{\theta}_{n-1}$. The stochastic analysis of the set-membership APA [34] establishes a mean-square error (MSE) performance, and the analysis is carried out by adopting energy conservation arguments (Chapter 5).

Example 8.2. The goal of this example is to demonstrate the comparative convergence performance of the NLMS, APA, APSM, and recursive least-squares RLS algorithms. The experiments were performed in two different noise settings, one for low and one for high noise levels, to demonstrate the sensitivity of the APA compared to the APSM. Data were generated according to our familiar model

$$y_n = \boldsymbol{\theta}_o^T \boldsymbol{x}_n + \eta_n.$$

The parameters $\boldsymbol{\theta}_o \in \mathbb{R}^{200}$ were randomly chosen from $\mathcal{N}(0, 1)$ and then fixed. The input vectors were formed by a white noise sequence with samples i.i.d. drawn from $\mathcal{N}(0, 1)$.

In the first experiment, the white noise sequence was chosen to have variance $\sigma^2 = 0.01$. The parameters for the three algorithms were chosen as $\mu = 1.2$ and $\delta = 0.001$ for the NLMS, $q = 30$, $\mu = 0.2$, and $\delta = 0.001$ for the APA, and $\epsilon = \sqrt{2}\sigma$, $q = 30$, and $\mu_n = 0.5 * M_n$ for the APSM. These parameters lead the algorithms to settle at the same error floor. Fig. 8.25 shows the obtained squared error, averaged over 100 realizations, in dBs ($10\log_{10}(e_n^2)$). For comparison, the RLS convergence curve is given for $\beta = 1$, which converges faster and at the same time settles at a lower error floor. If β is modified to a smaller value so that the RLS settles at the same error floor as the other algorithms, then its

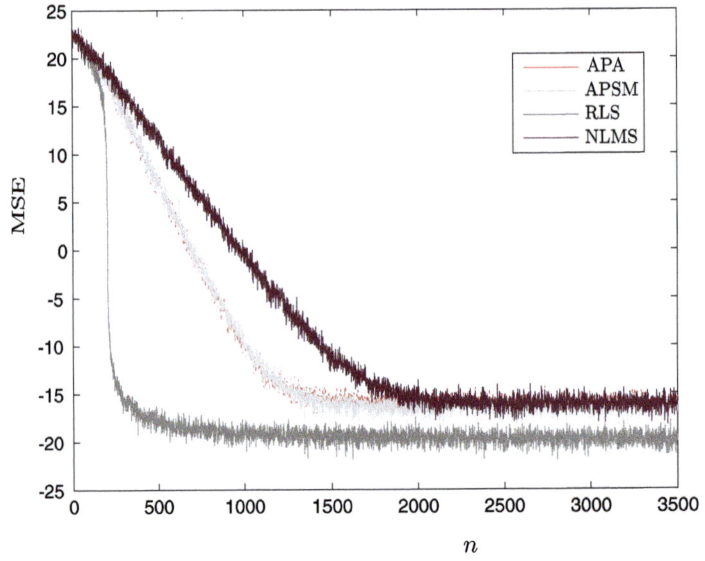

FIGURE 8.25

Mean-square error in dBs as a function of iterations. The data reuse ($q = 30$), associated with the APA and APSM, offers a significant improvement in convergence speed compared to the NLMS. The curves for the APA and APSM almost coincide in this low-noise scenario.

convergence gets even faster. However, this improved performance of the RLS is achieved at higher complexity, which becomes a problem for large values of l. Observe the faster convergence achieved by the APA and APSM, compared to the NLMS.

For the high-level noise, the corresponding variance was increased to 0.3. The obtained MSE curves are shown in Fig. 8.26. Observe that now, the APA shows an inferior performance compared to APSM in spite of its higher complexity, due to the involved matrix inversion.

8.8 CONSTRAINED LEARNING

Learning under a set of constraints is of significant importance in signal processing and machine learning, in general. We have already discussed a number of such learning tasks. Beamforming, discussed in Chapters 5 and 4, is a typical one. In Chapter 3, while introducing the concept of overfitting, we discussed the notion of regularization, which is another form of constraining the norm of the unknown parameter vector. In some other cases, we have available a priori information concerning the unknown parameters; this extra information can be given in the form of a set of constraints.

For example, if one is interested in obtaining estimates of the pixels in an image, then the values must be nonnegative. More recently, the unknown parameter vector may be known to be sparse; that is, only a few of its components are nonzero. In this case, constraining the respective ℓ_1 norm can significantly improve the accuracy as well as the convergence speed of an iterative scheme toward the

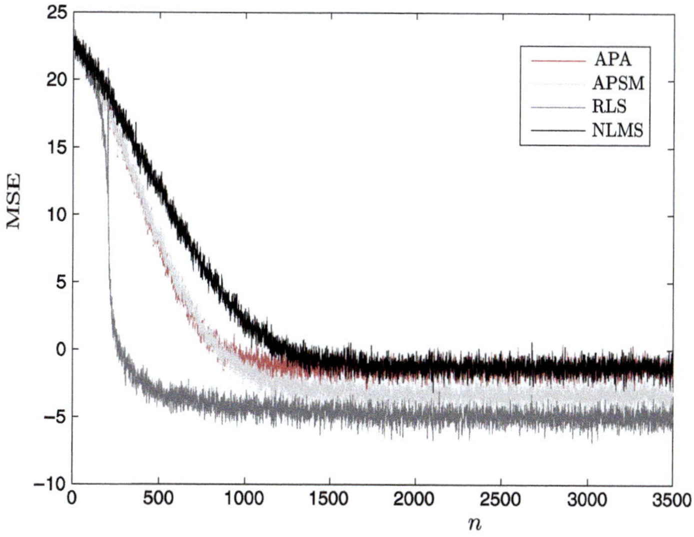

FIGURE 8.26

Mean-square error in dBs as a function of iterations for a high-noise scenario. Compared to Fig. 8.25, all curves settled at higher noise levels. Moreover, note that now the APA settles at higher error floor than the corresponding APSM algorithm, for the same convergence rate.

solution. Schemes that explicitly take into consideration the underlying sparsity are known as *sparsity promoting* algorithms. They will be considered in more detail in Chapter 10.

Algorithms that spring from the POCS theory are particularly suited to treat constraints in an elegant, robust, and rather straightforward way. Note that the goal of each constraint is to define a region in the solution space, where the required estimate is "forced" to lie. For the rest of this section, we will assume that the required estimate must satisfy M constraints, each one defining a *convex* set of points, $C_m, m = 1, 2, \dots, M$. Moreover,

$$\bigcap_{m=1}^{M} C_m \neq \emptyset,$$

which means that the constraints are consistent (there are also methods where this condition can be relaxed). Then it can be shown that the mapping T, defined as

$$T := P_{C_m} \dots P_{C_1},$$

is a *strongly attracting nonexpansive* mapping, (8.25) (for example, [6,7]). Note that the same holds true if instead of the concatenation of the projection operators, one could convexly combine them.

In the presence of a set of constraints, the only difference in the APSM in Algorithm 8.2 is that the update recursion (8.39) is now replaced by

$$\boldsymbol{\theta}_n = T\left(\boldsymbol{\theta}_{n-1} + \mu_n\left(\sum_{k=n-q+1}^{n} \omega_k P_{S_{\epsilon,k}}(\boldsymbol{\theta}_{n-1}) - \boldsymbol{\theta}_{n-1}\right)\right). \tag{8.45}$$

In other words, for M constraints, M extra projection operations have to be performed. The same applies to (8.38) with the difference being in the summation term in the brackets.

Remarks 8.6.

- The constrained form of the APSM has been successfully applied in the beamforming task, and in particular in treating nontrivial constraints, as required in the robust beamforming case [77,80, 103,104]. The constrained APSM has also been efficiently used for sparsity-aware learning (for example, [47,83]; see also Chapter 10). A more detailed review of related techniques is presented in [89].

8.9 THE DISTRIBUTED APSM

Distributed algorithms were discussed in Chapter 5. In Section 5.13.2, two versions of the diffusion LMS were introduced, namely, the adapt-then-combine and combine-then-adapt schemes. Diffusion versions of the APSM algorithm have also appeared in both configurations [20,22]. For the APSM case, both schemes result in very similar performance.

Following the discussion in Section 5.13.2, let the most recently received data pair at node $k = 1, 2, \ldots, K$ be $(y_k(n), \boldsymbol{x}_k(n)) \in \mathbb{R} \times \mathbb{R}^l$. For the regression task, a corresponding hyperslab is constructed, that is,

$$S_{\epsilon,n}^{(k)} = \left\{\boldsymbol{\theta} : |y_k(n) - \boldsymbol{x}_k^T(n)\boldsymbol{\theta}| \leq \epsilon_k\right\}.$$

The goal is the computation of a point that lies in the intersection of all these sets, for $n = 1, 2, \ldots$. Following similar arguments as those employed for the diffusion LMS, the combine-then-adapt version of the APSM, given in Algorithm 8.3, is obtained.

Algorithm 8.3 (**The combine-then-adapt diffusion APSM**).

- Initialization
  - **For** $k = 1, 2, \ldots, K$, **Do**
    - $\boldsymbol{\theta}_k(0) = \mathbf{0} \in \mathbb{R}^l$; or any other value.
  - **End For**
  - Select A: $A^T\mathbf{1} = \mathbf{1}$
  - Select q; The number of property sets to be processed at each time instant.
- **For** $n = 1, 2, \ldots, q-1$, **Do**; Initial period, that is, $n < q$.
  - **For** $k = 1, 2, \ldots, K$, **Do**
    - $\boldsymbol{\psi}_k(n-1) = \sum_{m \in \mathcal{N}_k} a_{mk}\boldsymbol{\theta}_m(n-1)$; $\mathcal{N}_k$ the neighborhood of node k.
  - **End For**
  - **For** $k = 1, 2, \ldots, K$, **Do**

 * Choose $\omega_1, \dots, \omega_n$: $\sum_{j=1}^{n} \omega_j = 1, \ \ \omega_j > 0$
 * Select $\mu_k(n) \in (0, \ 2M_k(n))$.
 · $\boldsymbol{\theta}_k(n) = \boldsymbol{\psi}_k(n-1) + \mu_k(n)\Big(\sum_{j=1}^{n} \omega_j P_{S_{\epsilon,j}^{(k)}} \big(\boldsymbol{\psi}_k(n-1)\big) - \boldsymbol{\psi}_k(n-1)\Big)$
 - **End For**
- **For** $n = q, q+1 \dots,$ **Do**
 - **For** $k = 1, 2, \dots, K,$ **Do**
 * $\boldsymbol{\psi}_k(n-1) = \sum_{m \in \mathcal{N}_k} a_{mk} \boldsymbol{\theta}_m(n-1)$
 - **End For**
 - **For** $k = 1, 2, \dots, K,$ **Do**
 * Choose $\omega_n, \dots, \omega_{n-q+1}$: $\sum_{j=n-q+1}^{n} \omega_j = 1, \ \ \omega_j > 0$.
 * Select $\mu_k(n) \in (0, \ 2M_k(n))$.
 · $\boldsymbol{\theta}_k(n) = \boldsymbol{\psi}_k(n-1) + \mu_k(n)\Big(\sum_{j=n-q+1}^{n} \omega_j P_{S_{\epsilon,j}^{(k)}} \big(\boldsymbol{\psi}_k(n-1)\big) - \boldsymbol{\psi}_k(n-1)\Big)$
 - **End For**
- **End For**

The interval $M_{k,n}$ is defined as

$$M_k(n) = \sum_{j=n-q+1}^{n} \frac{\omega_j \left\| P_{S_{\epsilon,j}^{(k)}} \big(\boldsymbol{\psi}_k(n-1)\big) - \boldsymbol{\psi}_k(n-1) \right\|^2}{\left\| \sum_{j=n-q+1}^{n} \omega_j P_{S_{\epsilon,j}^{(k)}} \big(\boldsymbol{\psi}_k(n-1)\big) - \boldsymbol{\psi}_k(n-1) \right\|^2},$$

and similarly for the initial period.

Remarks 8.7.

- An important theoretical property of the APSM-based diffusion algorithms is that they enjoy *asymptotic consensus*. In other words, the nodes converge asymptotically to the *same* estimate. This asymptotic consensus is not in the mean, as is the case with the diffusion LMS. This is interesting, since no explicit consensus constraints are employed.
- In [22], an extra projection step is used after the combination and prior to the adaptation step. The goal of this extra step is to "harmonize" the local information, which comprises the input/output observations, with the information coming from the neighborhood, that is, the estimates obtained from the neighboring nodes. This speeds up convergence, at the cost of only one extra projection.
- A scenario in which some of the nodes are damaged and the associated observations are very noisy is also treated in [22]. To deal with such a scenario, instead of the hyperslab, the APSM algorithm is rephrased around the Huber loss function, developed in the context of robust statistics, to deal with cases where outliers are present (see also Chapter 11).

Example 8.3. The goal of this example is to demonstrate the comparative performance of the diffusion LMS and APSM. A network of $K = 10$ nodes is considered, and there are 32 connections among the nodes. In each node, data are generated according to a regression model, using the same vector $\boldsymbol{\theta}_o \in \mathbb{R}^{60}$. The latter was randomly generated via a normal $\mathcal{N}(0, 1)$. The input vectors were i.i.d. generated according to the normal $\mathcal{N}(0, 1)$. The noise level at each node varied between 20 and

25 dBs. The parameters for the algorithms were chosen for optimized performance (after experimentation) and for similar convergence rate. For the LMS, $\mu = 0.035$, and for the APSM, $\epsilon = \sqrt{2}\sigma$, $q = 20$, $\mu_k(n) = 0.2 M_k(n)$. The combination weights were chosen according to the Metropolis rule and the data combination matrix was the identity one (no observations are exchanged). Fig. 8.27 shows the benefits of the data reuse offered by the APSM. The curves show the mean-square deviation (MSD $= \frac{1}{K}\sum_{k=1}^{K} ||\boldsymbol{\theta}_k(n) - \boldsymbol{\theta}_o||^2$) as a function of the number of iterations.

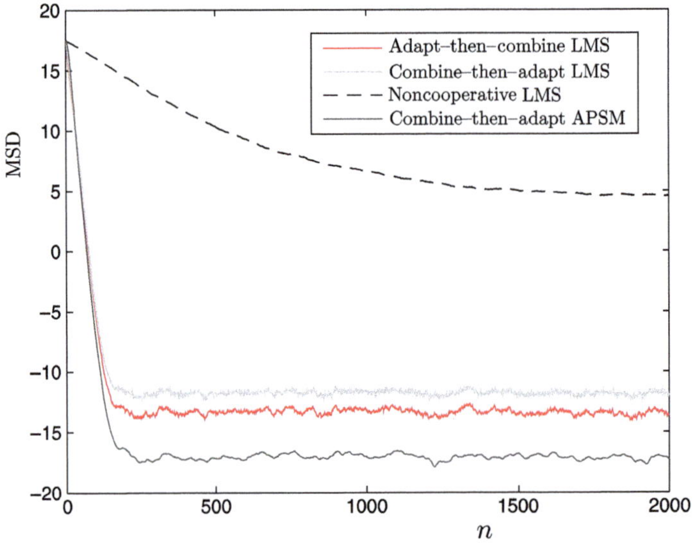

FIGURE 8.27

The MSD as a function of the number of iterations. The improved performance due to the data reuse offered by the diffusion ASPM is readily observed. Moreover, observe the significant performance improvement offered by all cooperation schemes, compared to the noncooperative LMS; for the latter, only one node is used.

8.10 OPTIMIZING NONSMOOTH CONVEX COST FUNCTIONS

Estimating parameters via the use of convex loss functions in the presence of a set of constraints is an established and well-researched field in optimization, with numerous applications in a wide range of disciplines. The mainstream methods follow either the Lagrange multiplier philosophy [11,15] or the rationale behind the so-called *interior point* methods [15,90]. In this section, we will focus on an alternative path and consider iterative schemes, which can be considered as the generalization of the gradient descent method, discussed in Chapter 5. The reason is that such techniques give rise to variants that scale well with the dimensionality and have inspired a number of algorithms, which have been suggested for online learning within the machine learning and signal processing communities. Later on, we will move to more advanced techniques that build on the so-called *operator/mapping* and *fixed point* theoretic framework.

Although the stage of our discussion will be that of Euclidean spaces, $\mathbb{R}^l$, everything that will be said can be generalized to infinite-dimensional Hilbert spaces; we will consider such cases in Chapter 11.

8.10.1 SUBGRADIENTS AND SUBDIFFERENTIALS

We have already met the first-order convexity condition in (8.3) and it was shown that this is a sufficient and necessary condition for convexity, provided, of course, that the gradient exists. The condition basically states that the graph of the convex function lies above the hyperplanes, which are tangent at any point $(\boldsymbol{x}, f(\boldsymbol{x}))$ that lies on this graph.

Let us now move a step forward and assume a function

$$f : \mathcal{X} \subseteq \mathbb{R}^l \longmapsto \mathbb{R}$$

to be convex and continuous, but *nonsmooth*. This means that there are points where the gradient is not defined. Our goal now becomes that of generalizing the notion of gradient, for the case of convex functions.

Definition 8.7. A vector $\boldsymbol{g} \in \mathbb{R}^l$ is said to be the *subgradient* of a convex function f at a point $\boldsymbol{x} \in \mathcal{X}$ if the following is true:

$$\boxed{f(\boldsymbol{y}) \geq f(\boldsymbol{x}) + \boldsymbol{g}^T(\boldsymbol{y} - \boldsymbol{x}), \quad \forall \boldsymbol{y} \in \mathcal{X}: \quad \text{subgradient.}} \tag{8.46}$$

It turns out that this vector is *not* unique. All the subgradients of a (convex) function at a point comprise a set.

Definition 8.8. The *subdifferential* of a convex function f at $\boldsymbol{x} \in \mathcal{X}$, denoted as $\partial f(\boldsymbol{x})$, is defined as the *set*

$$\boxed{\partial f(\boldsymbol{x}) := \left\{\boldsymbol{g} \in \mathbb{R}^l : f(\boldsymbol{y}) \geq f(\boldsymbol{x}) + \boldsymbol{g}^T(\boldsymbol{y} - \boldsymbol{x}), \ \forall \boldsymbol{y} \in \mathcal{X}\right\}: \quad \text{subdifferential.}} \tag{8.47}$$

If f is differentiable at a point $\boldsymbol{x}$, then $\partial f(\boldsymbol{x})$ becomes a singleton, that is,

$$\partial f(\boldsymbol{x}) = \{\nabla f(\boldsymbol{x})\}.$$

Note that if $f(\boldsymbol{x})$ is convex, then the set $\partial f(\boldsymbol{x})$ is *nonempty and convex*. Moreover, $f(\boldsymbol{x})$ is differentiable at a point $\boldsymbol{x}$ *if and only if* it has a unique subgradient [11]. From now on, we will denote a subgradient of f at the point $\boldsymbol{x}$ as $f'(\boldsymbol{x})$.

Fig. 8.28 gives a geometric interpretation of the notion of the subgradient. Each one of the subgradients at the point $\boldsymbol{x}_0$ defines a hyperplane that *supports* the graph of f. At $\boldsymbol{x}_0$, there is an infinity of subgradients, which comprise the subdifferential (set) at $\boldsymbol{x}_0$. At $\boldsymbol{x}_1$, the function is differentiable and there is a *unique* subgradient that coincides with the gradient at $\boldsymbol{x}_1$.

Example 8.4. Let $x \in \mathbb{R}$ and

$$f(x) = |x|.$$

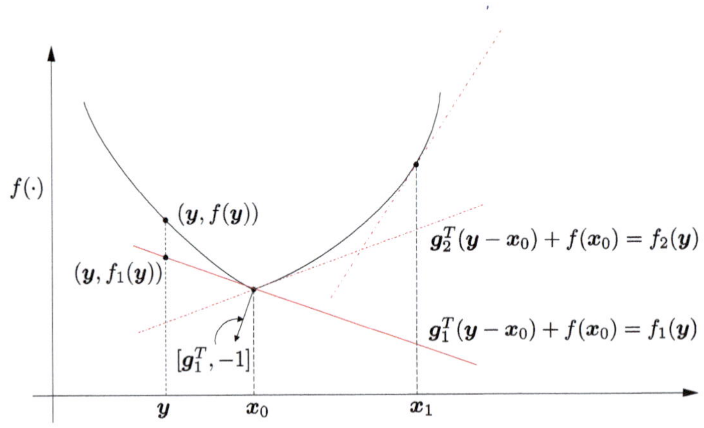

FIGURE 8.28

At $\boldsymbol{x}_0$, there is an infinity of subgradients, each one defining a hyperplane in the extended $(\boldsymbol{x}, f(\boldsymbol{x}))$ space. All these hyperplanes pass through the point $(\boldsymbol{x}_0, f(\boldsymbol{x}_0))$ and support the graph of $f(\cdot)$. At the point $\boldsymbol{x}_1$, there is a unique subgradient that coincides with the gradient and defines the unique tangent hyperplane at the respective point of the graph.

Show that

$$\partial f(x) = \begin{cases} \text{sgn}(x), & \text{if } x \neq 0, \\ g \in [-1, 1], & \text{if } x = 0, \end{cases}$$

where sgn$(\cdot)$ is the sign function, being equal to 1 if its argument is positive and -1 if the argument is negative.

Indeed, if $x > 0$, then

$$g = \frac{dx}{dx} = 1,$$

and similarly $g = -1$, if $x < 0$. For $x = 0$, any $g \in [-1, 1]$ satisfies

$$g(y - 0) + 0 = gy \leq |y|,$$

and it is a subgradient. This is illustrated in Fig. 8.29.

Lemma 8.2. *Given a convex function $f : \mathcal{X} \subseteq \mathbb{R}^l \longmapsto \mathbb{R}$, a point $\boldsymbol{x}_* \in \mathcal{X}$ is a minimizer of f if and only if the zero vector belongs to its subdifferential set, that is,*

$$\boxed{\mathbf{0} \in \partial f(\boldsymbol{x}_*): \quad \text{condition for a minimizer.}} \tag{8.48}$$

Proof. The proof is straightforward from the definition of a subgradient. Indeed, assume that $\mathbf{0} \in \partial f(\boldsymbol{x}_*)$. Then the following is valid:

$$f(\boldsymbol{y}) \geq f(\boldsymbol{x}_*) + \mathbf{0}^T(\boldsymbol{y} - \boldsymbol{x}_*), \ \forall \boldsymbol{y} \in \mathcal{X},$$

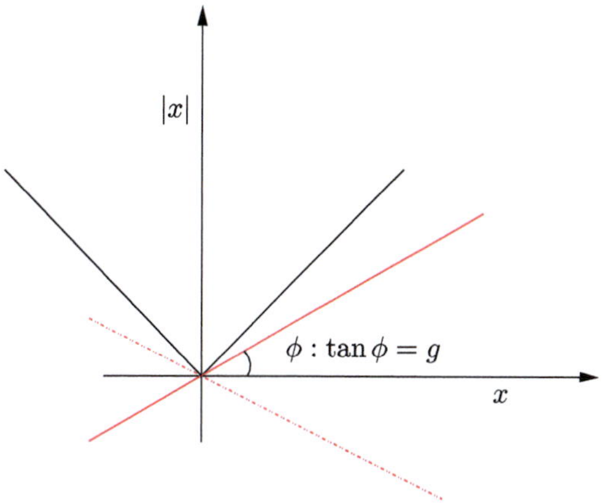

FIGURE 8.29

All lines with slope in $[-1, 1]$ comprise the subdifferential at $x = 0$.

and $\boldsymbol{x}_*$ is a minimizer. If now $\boldsymbol{x}_*$ is a minimizer, then we have

$$f(\boldsymbol{y}) \geq f(\boldsymbol{x}_*) = f(\boldsymbol{x}_*) + \boldsymbol{0}^T(\boldsymbol{y} - \boldsymbol{x}_*),$$

and hence $\boldsymbol{0} \in \partial f(\boldsymbol{x}_*)$. □

Example 8.5. Let the metric *distance* function

$$d_C(\boldsymbol{x}) := \min_{\boldsymbol{y} \in C} \|\boldsymbol{x} - \boldsymbol{y}\|.$$

This is the Euclidean distance of a point from its projection on a closed convex set C, as defined in Section 8.3. Then show (Problem 8.24) that the subdifferential is given by

$$\partial d_C(\boldsymbol{x}) = \begin{cases} \frac{\boldsymbol{x} - P_C(\boldsymbol{x})}{\|\boldsymbol{x} - P_C(\boldsymbol{x})\|}, & \boldsymbol{x} \notin C, \\ N_C(\boldsymbol{x}) \cap B[\boldsymbol{0}, 1], & \boldsymbol{x} \in C, \end{cases} \tag{8.49}$$

where

$$N_C(\boldsymbol{x}) := \left\{ \boldsymbol{g} \in \mathbb{R}^l : \boldsymbol{g}^T(\boldsymbol{y} - \boldsymbol{x}) \leq 0, \forall \boldsymbol{y} \in C \right\},$$

and

$$B[\boldsymbol{0}, 1] := \left\{ \boldsymbol{x} \in \mathbb{R}^l : \|\boldsymbol{x}\| \leq 1 \right\}.$$

Moreover, if $\boldsymbol{x}$ is an *interior* point of C, then

$$\partial d_C(\boldsymbol{x}) = \{\boldsymbol{0}\}.$$

Observe that for all points $\boldsymbol{x} \notin C$ as well as for all interior points of C, the subgradient is a singleton, which means that $d_C(\boldsymbol{x})$ is differentiable. Recall that the function $d_C(\cdot)$ is nonnegative, convex, and continuous [44]. Note that (8.49) is also generalized to infinite-dimensional Hilbert spaces.

8.10.2 MINIMIZING NONSMOOTH CONTINUOUS CONVEX LOSS FUNCTIONS: THE BATCH LEARNING CASE

Let J be a cost function[7]

$$J : \mathbb{R}^l \longmapsto [0, +\infty),$$

and let C be a closed convex set, $C \subseteq \mathbb{R}^l$. Our task is to compute a minimizer with respect to an unknown parameter vector, that is,

$$\boldsymbol{\theta}_* = \arg\min_{\boldsymbol{\theta}} J(\boldsymbol{\theta}),$$
$$\text{s.t.} \quad \boldsymbol{\theta} \in C, \tag{8.50}$$

and we will assume that the set of solutions is *nonempty*; J is assumed to be convex, continuous, but not necessarily differentiable at all points. We have already seen examples of such loss function, such as the ϵ-insensitive linear function in (8.33) and the hinge one (8.37). The ℓ_1 norm function is another example, and it will be treated in Chapters 9 and 10.

The Subgradient Method

Our starting point is the simplest of the cases, where $C = \mathbb{R}^l$; that is, the minimizing task is unconstrained. The first thought that comes into mind is to consider the generalization of the gradient descent method, which was introduced in Chapter 5, and replace the gradient by the subgradient operation. The resulting scheme is known as the *subgradient algorithm* [74,75].

Starting from an arbitrary estimate, $\boldsymbol{\theta}^{(0)} \in \mathbb{R}^l$, the update recursions become

$$\boxed{\boldsymbol{\theta}^{(i)} = \boldsymbol{\theta}^{(i-1)} - \mu_i J'\left(\boldsymbol{\theta}^{(i-1)}\right) : \quad \text{subgradient algorithm,}} \tag{8.51}$$

where J' denotes *any* subgradient of the cost function, and μ_i is a step size sequence judicially chosen so that convergence is guaranteed. In spite of the similarity in the appearance with our familiar gradient descent scheme, there are some major differences. The reader may have noticed that the new algorithm was not called subgradient "descent." This is because the update in (8.51) is not necessarily performed in the descent direction. Thus, during the operation of the algorithm, the value of the cost function may increase. Recall that in the gradient descent methods, the value of the cost function is guaranteed to decrease with each iteration step, which also led to a linear convergence rate, as we have pointed out in Chapter 5.

In contrast here, concerning the subgradient method, such comments cannot be stated. To establish convergence, a different route has to be adopted. To this end, let us define

$$J_*^{(i)} := \min\left\{J\left(\boldsymbol{\theta}^{(i)}\right), J\left(\boldsymbol{\theta}^{(i-1)}\right), \ldots, J\left(\boldsymbol{\theta}^{(0)}\right)\right\}, \tag{8.52}$$

[7] Recall that all the methods to be reported can be extended to general Hilbert spaces, $\mathbb{H}$.

which can also be recursively obtained by

$$J_*^{(i)} = \min\left\{J_*^{(i-1)}, J\left(\boldsymbol{\theta}^{(i)}\right)\right\}.$$

Then the following holds true.

Proposition 8.3. *Let J be a convex cost function. Assume that the subgradients at all points are bounded, that is,*

$$||J'(\boldsymbol{x})|| \leq G, \ \forall \boldsymbol{x} \in \mathbb{R}^l.$$

Let us also assume that the step size sequence be a diminishing one, such as

$$\sum_{i=1}^{\infty} \mu_i = \infty, \quad \sum_{i=1}^{\infty} \mu_i^2 < \infty.$$

Then

$$\lim_{i \longrightarrow \infty} J_*^{(i)} = J(\boldsymbol{\theta}_*),$$

where $\boldsymbol{\theta}_*$ is a minimizer, assuming that the set of minimizers is not empty.

Proof. We have

$$\begin{aligned} \left\|\boldsymbol{\theta}^{(i)} - \boldsymbol{\theta}_*\right\|^2 &= \left\|\boldsymbol{\theta}^{(i-1)} - \mu_i J'\left(\boldsymbol{\theta}^{(i-1)}\right) - \boldsymbol{\theta}_*\right\|^2 \\ &= \left\|\boldsymbol{\theta}^{(i-1)} - \boldsymbol{\theta}_*\right\|^2 - 2\mu_i J'^T\left(\boldsymbol{\theta}^{(i-1)}\right)\left(\boldsymbol{\theta}^{(i-1)} - \boldsymbol{\theta}_*\right) \\ &\quad + \mu_i^2 \left\|J'\left(\boldsymbol{\theta}^{(i-1)}\right)\right\|^2. \end{aligned} \tag{8.53}$$

By the definition of the subgradient, we have

$$J(\boldsymbol{\theta}_*) - J\left(\boldsymbol{\theta}^{(i-1)}\right) \geq J'^T\left(\boldsymbol{\theta}^{(i-1)}\right)\left(\boldsymbol{\theta}_* - \boldsymbol{\theta}^{(i-1)}\right). \tag{8.54}$$

Plugging (8.54) in (8.53) and after some algebraic manipulations, by applying the resulting inequality recursively (Problem 8.25), we finally obtain

$$J_*^{(i)} - J(\boldsymbol{\theta}_*) \leq \frac{\left\|\boldsymbol{\theta}^{(0)} - \boldsymbol{\theta}_*\right\|^2}{2\sum_{k=1}^{i} \mu_k} + \frac{\sum_{k=1}^{i} \mu_k^2}{2\sum_{k=1}^{i} \mu_k} G^2. \tag{8.55}$$

Leaving i to grow to infinity and taking into account the assumptions, the claim is proved. □

There are a number of variants of this proof. Other choices for the diminishing sequence can also guarantee convergence, such as $\mu_i = 1/\sqrt{i}$. Moreover, in certain cases, some of the assumptions may be relaxed. Note that the assumption of the subgradient being bounded is guaranteed if J is γ-Lipschitz continuous (Problem 8.26), that is, there is $\gamma > 0$ such that

$$|J(\boldsymbol{y}) - J(\boldsymbol{x})| \leq \gamma ||\boldsymbol{y} - \boldsymbol{x}||, \ \forall \boldsymbol{x}, \boldsymbol{y} \in \mathbb{R}^l.$$

Interpreting the proposition from a slightly different angle, we can say that the algorithm generates a *subsequence* of estimates, $\boldsymbol{\theta}_{i_*}$, which corresponds to the values of $J_*^{(i)}$, shown as $J(\boldsymbol{\theta}_{i_*}) \leq J(\boldsymbol{\theta}_i),\ i \leq i_*$, which converges to $\boldsymbol{\theta}_*$. The *best possible* convergence rate that may be achieved is of the order of $\mathcal{O}(\frac{1}{\sqrt{i}})$, if one optimizes the bound in (8.55), with respect to μ_k [64], which can be obtained if $\mu_i = \frac{c}{\sqrt{i}}$, where c is a constant. In any case, it is readily noticed that the convergence speed of such methods is rather slow. Yet due to their computational simplicity, they are still in use, especially in cases where the number of data samples is large. The interested reader can obtain more on the subgradient method from [10,75].

Example 8.6 (The perceptron algorithm). Recall the hinge loss function with $\rho = 0$, defined in (8.37),

$$\mathcal{L}(y, \boldsymbol{\theta}^T \boldsymbol{x}) = \max\left(0, -y\boldsymbol{\theta}^T \boldsymbol{x}\right).$$

In a two-class classification task, we are given a set of training samples, $(y_n, \boldsymbol{x}_n) \in \{-1, +1\} \times \mathbb{R}^{l+1},\ n = 1, 2, \ldots, N$, and the goal is to compute a linear classifier to minimize the empirical loss function

$$J(\boldsymbol{\theta}) = \sum_{n=1}^{N} \mathcal{L}(y_n, \boldsymbol{\theta}^T \boldsymbol{x}_n). \tag{8.56}$$

We will assume the classes to be *linearly separable*, which guarantees that there is a solution; that is, there exists a hyperplane that classifies correctly all data points. Obviously, such a hyperplane will score a zero for the cost function in (8.56). We have assumed that the dimension of our input data space has been increased by one, to account for the bias term for hyperplanes not crossing the origin.

The subdifferential of the hinge loss function is easily checked (e.g., use geometric arguments which relate a subgradient with a support hyperplane of the respective function graph) to be

$$\partial \mathcal{L}(y_n, \boldsymbol{\theta}^T \boldsymbol{x}_n) = \begin{cases} 0, & y_n \boldsymbol{\theta}^T \boldsymbol{x}_n > 0, \\ -y_n \boldsymbol{x}_n, & y_n \boldsymbol{\theta}^T \boldsymbol{x}_n < 0, \\ \boldsymbol{g} \in [-y_n \boldsymbol{x}_n, 0], & y_n \boldsymbol{\theta}^T \boldsymbol{x}_n = 0. \end{cases} \tag{8.57}$$

We choose to work with the following subgradient

$$\mathcal{L}'(y_n, \boldsymbol{\theta}^T \boldsymbol{x}_n) = -y_n \boldsymbol{x}_n \chi_{(-\infty, 0]}\left(y_n \boldsymbol{\theta}^T \boldsymbol{x}_n\right), \tag{8.58}$$

where $\chi_A(\tau)$ is the characteristic function, defined as

$$\chi_A(\tau) = \begin{cases} 1, & \tau \in A, \\ 0, & \tau \notin A. \end{cases} \tag{8.59}$$

The subgradient algorithm now becomes,

$$\boldsymbol{\theta}^{(i)} = \boldsymbol{\theta}^{(i-1)} + \mu_i \sum_{n=1}^{N} y_n \boldsymbol{x}_n \chi_{(-\infty, 0]}\left(y_n \boldsymbol{\theta}^{(i-1)T} \boldsymbol{x}_n\right). \tag{8.60}$$

This is the celebrated *perceptron algorithm*, which we are going to see in more detail in Chapter 18. Basically, what the algorithm in (8.60) says is the following. Starting from an arbitrary $\boldsymbol{\theta}^{(0)}$, test all training vectors with $\boldsymbol{\theta}^{(i-1)}$. Select all those vectors that fail to predict the correct class (for the correct class, $y_n\boldsymbol{\theta}^{(i-1)T}\boldsymbol{x}_n > 0$), and update the current estimate toward the direction of the weighted (by the corresponding label) average of the *misclassified* patterns. It turns out that the algorithm converges in a *finite* number of steps, even if the step size sequence is not a diminishing one. This is what was said before, i.e., in certain cases, convergence of the subgradient algorithm is guaranteed even if some of the assumptions in Proposition 8.3 do not hold.

The Generic Projected Subgradient Scheme

The generic scheme on which a number of variants draw their origin is summarized as follows. Select $\boldsymbol{\theta}^{(0)} \in \mathbb{R}^l$ arbitrarily. Then the iterative scheme

$$\boxed{\boldsymbol{\theta}^{(i)} = P_C\Big(\boldsymbol{\theta}^{(i-1)} - \mu_i J'\big(\boldsymbol{\theta}^{(i-1)}\big)\Big): \qquad \text{GPS scheme,}} \tag{8.61}$$

where J' denotes a respective subgradient and P_C is the projection operator onto C, converges (converges weakly in the more general case) to a solution of the constrained task in (8.50). The sequence of nonnegative real numbers, μ_i, is judicially selected. It is readily seen that this scheme is a generalization of the gradient descent scheme, discussed in Chapter 5, if we set $C = \mathbb{R}^l$ and J is differentiable.

The Projected Gradient Method (PGM)

This method is a special case of (8.61) if J is *smooth* and we set $\mu_i = \mu$. That is,

$$\boxed{\boldsymbol{\theta}^{(i)} = P_C\Big(\boldsymbol{\theta}^{(i-1)} - \mu \nabla J\big(\boldsymbol{\theta}^{(i-1)}\big)\Big): \qquad \text{PGM scheme.}} \tag{8.62}$$

It turns out that if the gradient is γ-Lipschitz, that is,

$$\|\nabla J(\boldsymbol{\theta}) - \nabla J(\boldsymbol{h})\| \le \gamma \|\boldsymbol{\theta} - \boldsymbol{h}\|, \ \gamma > 0, \ \forall \boldsymbol{\theta}, \ \boldsymbol{h} \in \mathbb{R}^l,$$

and

$$\mu \in \left(0, \frac{2}{\gamma}\right),$$

then starting from an arbitrary point $\boldsymbol{\theta}^{(0)}$, the sequence in (8.62) converges (weakly in a general Hilbert space) to a solution of (8.50) [41,52].

Example 8.7 (Projected Landweber method). Let our optimization task be

$$\begin{aligned} &\text{minimize} \ \frac{1}{2}\|\boldsymbol{y} - X\boldsymbol{\theta}\|^2, \\ &\text{subject to} \ \boldsymbol{\theta} \in C, \end{aligned}$$

where $X \in \mathbb{R}^{m\times l}$, $\boldsymbol{y} \in \mathbb{R}^m$. Expanding and taking the gradient, we get

$$J(\boldsymbol{\theta}) = \frac{1}{2}\boldsymbol{\theta}^T X^T X\boldsymbol{\theta} - \boldsymbol{y}^T X\boldsymbol{\theta} + \frac{1}{2}\boldsymbol{y}^T\boldsymbol{y},$$

$$\nabla J(\boldsymbol{\theta}) = X^T X\boldsymbol{\theta} - X^T \boldsymbol{y}.$$

First we check that $\nabla J(\boldsymbol{\theta})$ is γ-Lipschitz. To this end, we have

$$\|X^T X(\boldsymbol{\theta} - \boldsymbol{h})\| \leq \|X^T X\| \|\boldsymbol{\theta} - \boldsymbol{h}\| \leq \lambda_{\max} \|\boldsymbol{\theta} - \boldsymbol{h}\|,$$

where the spectral norm of a matrix has been used (Section 6.4) and $\lambda_{\max}$ denotes the maximum eigenvalue $X^T X$. Thus, if

$$\mu \in \left(0, \frac{2}{\lambda_{\max}}\right),$$

the corresponding iterations in (8.62) converge to a solution of (8.50). The scheme has been used in the context of compressed sensing where (as we will see in Chapter 10) the task of interest is

$$\begin{aligned} &\text{minimize} \quad \frac{1}{2}\|\boldsymbol{y} - X\boldsymbol{\theta}\|^2, \\ &\text{subject to} \quad \|\boldsymbol{\theta}\|_1 \leq \rho. \end{aligned}$$

Then it turns out that projecting on the ℓ_1 ball (corresponding to C) is equivalent to a soft thresholding operation[8] [35]. A variant occurs if a projection on a weighted ℓ_1 ball is used to speed up convergence (Chapter 10). Projection on a weighted ℓ_1 ball has been developed in [47], via fully geometric arguments, and it also results in soft thresholding operations.

Projected Subgradient Method

Starting from an arbitrary point $\boldsymbol{\theta}^{(0)}$, for the recursion [2,55]

$$\boxed{\boldsymbol{\theta}^{(i)} = P_C\left(\boldsymbol{\theta}^{(i-1)} - \frac{\mu_i}{\max\left\{1, \|J'(\boldsymbol{\theta}^{(i-1)})\|\right\}} J'(\boldsymbol{\theta}^{(i-1)})\right): \quad \text{PSMa},} \tag{8.63}$$

- either a solution of (8.50) is achieved in a finite number of steps,
- or the iterations converge (weakly in the general case) to a point in the set of solutions of (8.50),

provided that

$$\sum_{i=1}^{\infty} \mu_i = \infty, \quad \sum_{i=1}^{\infty} \mu_i^2 < \infty.$$

Another version of the projected subgradient algorithm was presented in [70]. Let $J_* = \min_{\boldsymbol{\theta}} J(\boldsymbol{\theta})$ be the minimum (strictly speaking the infimum) of a cost function, whose set of minimizers is assumed to

[8] See Chapter 10 and Example 8.10.

be nonempty. Then the iterative algorithm

$$\boxed{\theta^{(i)} = \begin{cases} P_C\left(\theta^{(i-1)} - \mu_i \frac{J\left(\theta^{(i-1)}\right) - J_*}{\left\|J'\left(\theta^{(i-1)}\right)\right\|^2} J'\left(\theta^{(i-1)}\right)\right), & \text{if } J'\left(\theta^{(i-1)}\right) \neq \mathbf{0}, \\ P_C\left(\theta^{(i-1)}\right), & \text{if } J'\left(\theta^{(i-1)}\right) = \mathbf{0}, \end{cases} \quad \text{PSMb}} \tag{8.64}$$

converges (weakly in infinite-dimensional spaces) for $\mu_i \in (0, 2)$, under some general conditions, and assuming that the subgradient is bounded. The proof is a bit technical and the interested reader can obtain it from, for example, [70,82].

Needless to say that, besides the previously reported major schemes discussed, there is a number of variants; for a related review see [82].

8.10.3 ONLINE LEARNING FOR CONVEX OPTIMIZATION

Online learning in the framework of the squared error loss function has been the focus in Chapters 5 and 6. One of the reasons that online learning was introduced was to give the potential to the algorithm to track time variations in the underlying statistics. Another reason was to cope with the unknown statistics when the cost function involved expectations, in the context of the stochastic approximation theory. Moreover, online algorithms are of particular interest when the number of the available training data and the dimensionality of the input space become very large, compared to the load that today's storage, processing, and networking devices can cope with. Exchanging information has now become cheap and databases have been populated with a massive number of data. This has rendered batch processing techniques for learning tasks with huge data sets impractical. Online algorithms that process one data point at a time have now become an indispensable algorithmic tool.

Recall from Section 3.14 that the ultimate goal of a machine learning task, given a loss function $\mathcal{L}$, is to minimize the expected loss/risk, which in the context of parametric modeling can be written as

$$\begin{aligned} J(\theta) &= \mathbb{E}\left[\mathcal{L}(\mathrm{y}, f_{\theta}(\mathbf{x}))\right] \\ &:= \mathbb{E}\left[\mathcal{L}(\theta, \mathrm{y}, \mathbf{x})\right]. \end{aligned} \tag{8.65}$$

Instead, the corresponding empirical risk function is minimized, given a set of N training points,

$$J_N(\theta) = \frac{1}{N} \sum_{n=1}^{N} \mathcal{L}(\theta, y_n, \boldsymbol{x}_n). \tag{8.66}$$

In this context, the subgradient scheme would take the form

$$\theta^{(i)} = \theta^{(i-1)} - \frac{\mu_i}{N} \sum_{n=1}^{N} \mathcal{L}_n'\left(\theta^{(i-1)}\right),$$

where for notational simplicity we used

$$\mathcal{L}_n(\theta) := \mathcal{L}(\theta, y_n, \boldsymbol{x}_n). \tag{8.67}$$

Thus, at each iteration, one has to compute N subgradient values, which for large values of N is computationally cumbersome. One way out is to adopt stochastic approximation arguments, as explained in Chapter 5, and come with a corresponding online version,

$$\boldsymbol{\theta}_n = \boldsymbol{\theta}_{n-1} - \mu_n \mathcal{L}_n'(\boldsymbol{\theta}_{n-1}), \tag{8.68}$$

where now the iteration index, i, coincides with the time index, n. There are two different ways to view (8.68). Either n takes values in the interval $[1, N]$ and one cycles periodically until convergence, or n is left to grow unbounded. The latter is very natural for very large values of N, and we focus on this scenario from now on. Moreover, such strategy can cope with slow time variations, if this is the case. Note that in the online formulation, at each time instant a *different* loss function is involved and our task becomes that of an *asymptotic* minimization. Furthermore, one has to study the asymptotic *convergence properties*, as well as the respected *convergence conditions*. Soon, we are going to introduce a relatively recent tool for analyzing the performance of online algorithms, namely, *regret analysis*.

It turns out that for each one of the optimization schemes discussed in Section 8.10.2, we can write its online version. Given the sequence of loss functions $\mathcal{L}_n$, $n = 1, 2 \ldots$, the online version of the generic projected subgradient scheme becomes

$$\boldsymbol{\theta}_n = P_C\left(\boldsymbol{\theta}_{n-1} - \mu_n \mathcal{L}_n'(\boldsymbol{\theta}_{n-1})\right), \quad n = 1, 2, 3, \ldots. \tag{8.69}$$

In a more general setting, the constraint-related convex sets can be left to be time-varying too; in other words, we can write C_n. For example, such schemes with time-varying constraints have been developed in the context of sparsity-aware learning, where in place of the ℓ_1 ball, a weighted ℓ_1 ball is used [47]. This has a really drastic effect in speeding up the convergence of the algorithm (see Chapter 10).

Another example is the so-called *adaptive gradient* (AdaGrad) algorithm [38]. The projection operator is defined in a more general context, in terms of the Mahalanobis distance, that is,

$$P_C^G(\boldsymbol{x}) = \min_{\boldsymbol{z} \in C} (\boldsymbol{x} - \boldsymbol{z})^T G (\boldsymbol{x} - \boldsymbol{z}), \quad \forall \boldsymbol{x} \in \mathbb{R}^l. \tag{8.70}$$

In place of G, the square root of the sum of the outer products of the computed subgradients is used, that is,

$$G_n = \sum_{k=1}^{n} \boldsymbol{g}_k \boldsymbol{g}_k^T,$$

where $\boldsymbol{g}_k = \mathcal{L}_k'(\boldsymbol{\theta}_{k-1})$ denotes the subgradient at time instant k. Also, the same matrix is used to weigh the gradient correction and the scheme has the form

$$\boldsymbol{\theta}_n = P_C^{G_n^{1/2}}\left(\boldsymbol{\theta}_{n-1} - \mu_n G_n^{-1/2} \boldsymbol{g}_n\right). \tag{8.71}$$

The use of the (time-varying) weighting matrix accounts for the geometry of the data observed in earlier iterations, which leads to a more informative gradient-based learning. For the sake of computational savings, the structure of G_n is taken to be diagonal. Different algorithmic settings are discussed in [38], alongside the study of the converging properties of the algorithm; see also Section 18.4.2.

Example 8.8 (The LMS algorithm). Let us assume that

$$\mathcal{L}_n(\boldsymbol{\theta}) = \frac{1}{2}\left(y_n - \boldsymbol{\theta}^T \boldsymbol{x}_n\right)^2,$$

and also set $C = \mathbb{R}^l$ and $\mu_n = \mu$. Then (8.69) becomes our familiar LMS recursion,

$$\boldsymbol{\theta}_n = \boldsymbol{\theta}_{n-1} + \mu\left(y_n - \boldsymbol{\theta}_{n-1}^T \boldsymbol{x}_n\right)\boldsymbol{x}_n,$$

whose convergence properties have been discussed in Chapter 5.

The PEGASOS Algorithm

The *primal estimated subgradient solver for SVM* (PEGASOS) algorithm is an online scheme built around the hinge loss function *regularized* by the squared Euclidean norm of the parameters vector [73]. From this point of view, it is an instance of the online version of the projected subgradient algorithm. This algorithm results if in (8.69) we set

$$\mathcal{L}_n(\boldsymbol{\theta}) = \max\left(0, 1 - y_n \boldsymbol{\theta}^T \boldsymbol{x}_n\right) + \frac{\lambda}{2}||\boldsymbol{\theta}||^2, \tag{8.72}$$

where in this case, ρ in the hinge loss function has been set equal to one. The associated empirical cost function is

$$J(\boldsymbol{\theta}) = \frac{1}{N}\sum_{n=1}^{N} \max\left(0, 1 - y_n \boldsymbol{\theta}^T \boldsymbol{x}_n\right) + \frac{\lambda}{2}||\boldsymbol{\theta}||^2, \tag{8.73}$$

whose minimization results in the celebrated *support vector machine*. Note that the only differences with the perceptron algorithm are the presence of the regularizer and the nonzero value of ρ. These seemingly minor differences have important implications in practice, and we are going to say more on this in Chapter 11, where nonlinear extensions treated in the more general context of Hilbert spaces will be considered.

The subgradient adopted by the PEGASOS is

$$\mathcal{L}_n'(\boldsymbol{\theta}) = \lambda\boldsymbol{\theta} - y_n \boldsymbol{x}_n \chi_{(-\infty,0]}\left(y_n \boldsymbol{\theta}^T \boldsymbol{x}_n - 1\right). \tag{8.74}$$

The step size is chosen as $\mu_n = \frac{1}{\lambda n}$. Furthermore, in its more general formulation, at each iteration step, an (optional) projection on the $\frac{1}{\sqrt{\lambda}}$ length ℓ_2 ball, $B[\mathbf{0}, \frac{1}{\sqrt{\lambda}}]$, is performed. The update recursion then becomes

$$\boldsymbol{\theta}_n = P_{B[\mathbf{0},\frac{1}{\sqrt{\lambda}}]}\left(\left(1 - \mu_n\lambda\right)\boldsymbol{\theta}_{n-1} + \mu_n y_n \boldsymbol{x}_n \chi_{(-\infty,0]}\left(y_n \boldsymbol{\theta}_{n-1}^T \boldsymbol{x}_n - 1\right)\right), \tag{8.75}$$

where $P_{B[\mathbf{0},\frac{1}{\sqrt{\lambda}}]}$ is the projection on the respective ℓ_2 ball given in (8.14). In (8.75), note that the effect of the regularization is to smooth out the contribution of $\boldsymbol{\theta}_{n-1}$. A variant of the algorithm for a fixed number of points, N, suggests to average out a number of m subgradient values in an index set, $A_n \subseteq \{1, 2, \ldots, N\}$, such that $k \in A_n$ if $y_k \boldsymbol{\theta}_{n-1}^T \boldsymbol{x}_k < 1$. Different scenarios for the choice of the m indices can be employed, with the random one being a possibility. The scheme is summarized in Algorithm 8.4.

Algorithm 8.4 (The PEGASOS algorithm).

- Initialization
  - Select $\boldsymbol{\theta}^{(0)}$; Usually set to zero.
  - Select λ.
  - Select m; Number of subgradient values to be averaged.
- **For** $n = 1, 2, \ldots, N$, **Do**
  - Select $A_n \subseteq \{1, 2, \ldots, N\}$: (cardinality) $|A_n| = m$, uniformly at random.
  - $\mu_n = \frac{1}{\lambda n}$
  - $\boldsymbol{\theta}_n = (1 - \mu_n \lambda)\boldsymbol{\theta}_{n-1} + \frac{\mu_n}{m} \sum_{k \in A_n} y_k \boldsymbol{x}_k$
  - $\boldsymbol{\theta}_n = \min\left(1, \frac{1}{\sqrt{\lambda}||\boldsymbol{\theta}_n||}\right)\boldsymbol{\theta}_n$; Optional.
- **End For**

Application of regret analysis arguments point out the required number of iterations for obtaining a solution of accuracy ϵ is $\mathcal{O}(1/\epsilon)$, when each iteration operates on a single training sample. The algorithm is very similar with the algorithms proposed in [45,112]. The difference lies in the choice of the step size. We will come back to these algorithms in Chapter 11. There, we are going to see that the online learning in infinite-dimensional spaces is more tricky. In [73], a number of comparative tests against well-established support vector machine algorithms have been performed using standard data sets. The main advantage of the algorithm is its computational simplicity, and it achieves comparable performance rates at lower computational costs.

8.11 REGRET ANALYSIS

A major effort when dealing with iterative learning algorithms is dedicated to the issue of convergence; where the algorithm converges, under which conditions it converges, and how fast it converges to its steady-state. A large part of Chapter 5 was focused on the convergence properties of the LMS. Furthermore, in the current chapter, when we discussed the various subgradient-based algorithms, convergence properties were also reported.

In general, analyzing the convergence properties of online algorithms tends to be quite a formidable task and classical approaches have to adopt a number of assumptions, sometimes rather strong. Typical assumptions refer to the statistical nature of the data (e.g., being i.i.d. or the noise being white). In addition, it may be assumed that the true model, which generates the data, to be known, and/or that the algorithm has reached a region in the parameter's space that is close to a minimizer.

More recently, an alternative methodology has been developed which bypasses the need for such assumptions. The methodology evolves around the concept of *cumulative loss*, which has already been introduced in Chapter 5, Section 5.5.2. The method is known as *regret analysis*, and its birth is due to developments in the interplay between game and learning theories (see, for example, [21]).

Let us assume that the training samples $(y_n, \boldsymbol{x}_n)$, $n = 1, 2, \ldots$, arrive sequentially and that an adopted online algorithm makes the corresponding predictions, $\hat{y}_n$. The quality of the prediction for each time instant is tested against a loss function, $\mathcal{L}(y_n, \hat{y}_n)$. The cumulative loss up to time N is given by

$$\mathcal{L}_{\text{cum}}(N) := \sum_{n=1}^{N} \mathcal{L}(y_n, \hat{y}_n). \tag{8.76}$$

Let f be a *fixed* predictor. Then the *regret* of the online algorithm relative to f, when running up to time instant N, is defined as

$$\boxed{\text{Regret}_N(f) := \sum_{n=1}^{N} \mathcal{L}(y_n, \hat{y}_n) - \sum_{n=1}^{N} \mathcal{L}(y_n, f(\boldsymbol{x}_n)) : \quad \text{regret relative to } f.} \tag{8.77}$$

The name regret is inherited from the game theory and it means how "sorry" the algorithm or the learner (in the machine learning jargon) is, in retrospect, not to have followed the prediction of the fixed predictor, f. The predictor f is also known as the *hypothesis*. Also, if f is chosen from a set of functions, $\mathcal{F}$, this set is called the *hypothesis class*.

The regret relative to the family of functions $\mathcal{F}$, when the algorithm runs over N time instants, is defined as

$$\text{Regret}_N(\mathcal{F}) := \max_{f \in \mathcal{F}} \text{Regret}_N(f). \tag{8.78}$$

In the context of regret analysis, the goal becomes that of designing an online learning rule so that the resulting regret with respect to an optimal fixed predictor is small; that is, the regret associated with the learner should grow *sublinearly* (slower than linearly) with the number of iterations, N. Sublinear growth guarantees that the difference between the *average* loss suffered by the learner and the average loss of the optimal predictor will tend to zero asymptotically.

For the linear class of functions, we have

$$\hat{y}_n = \boldsymbol{\theta}_{n-1}^T \boldsymbol{x}_n,$$

and the loss can be written as

$$\mathcal{L}(y_n, \hat{y}_n) = \mathcal{L}(y_n, \boldsymbol{\theta}_{n-1}^T \boldsymbol{x}_n) := \mathcal{L}_n(\boldsymbol{\theta}_{n-1}).$$

Adapting (8.77) to the previous notation, we can write

$$\text{Regret}_N(\boldsymbol{h}) = \sum_{n=1}^{N} \mathcal{L}_n(\boldsymbol{\theta}_{n-1}) - \sum_{n=1}^{N} \mathcal{L}_n(\boldsymbol{h}), \tag{8.79}$$

where $\boldsymbol{h} \in C \subseteq \mathbb{R}^l$ is a *fixed* parameter vector in the set C where solutions are sought.

Before proceeding, it is interesting to note that the cumulative loss is based on the loss suffered by the learner, against $y_n, \boldsymbol{x}_n$, using the estimate $\boldsymbol{\theta}_{n-1}$, which has been trained on data up to and including time instant $n-1$. The pair $(y_n, \boldsymbol{x}_n)$ is not involved in its training. From this point of view, the cumulative loss is in line with our desire to guard against overfitting.

In the framework of regret analysis, the path to follow is to derive an upper bound for the regret, exploiting the convexity of the employed loss function. We will demonstrate the technique via a case study, i.e., that of the online version of the simple subgradient algorithm.

REGRET ANALYSIS OF THE SUBGRADIENT ALGORITHM

The online version of (8.68), for minimizing the expected loss, $\mathbb{E}\left[\mathcal{L}(\boldsymbol{\theta}, \mathrm{y}, \mathbf{x})\right]$, is written as

$$\boldsymbol{\theta}_n = \boldsymbol{\theta}_{n-1} - \mu_n \boldsymbol{g}_n, \tag{8.80}$$

where for notational convenience the subgradient is denoted as

$$\boldsymbol{g}_n := \mathcal{L}_n'(\boldsymbol{\theta}_{n-1}).$$

Proposition 8.4. *Assume that the subgradients of the loss function are bounded, shown as*

$$||\boldsymbol{g}_n|| \leq G, \ \forall n. \tag{8.81}$$

Furthermore, assume that the set of solutions $\mathcal{S}$ is bounded; that is, $\forall\, \boldsymbol{\theta}, \boldsymbol{h} \in \mathcal{S}$, there exists a bound F such that

$$||\boldsymbol{\theta} - \boldsymbol{h}|| \leq F. \tag{8.82}$$

Let $\boldsymbol{\theta}_*$ be an optimal (desired) predictor. Then, if $\mu_n = \frac{1}{\sqrt{n}}$,

$$\boxed{\frac{1}{N}\sum_{n=1}^{N} \mathcal{L}_n(\boldsymbol{\theta}_{n-1}) \leq \frac{1}{N}\sum_{n=1}^{N} \mathcal{L}_n(\boldsymbol{\theta}_*) + \frac{F^2}{2\sqrt{N}} + \frac{G^2}{\sqrt{N}}.} \tag{8.83}$$

In words, as $N \longrightarrow \infty$ the average cumulative loss tends to the average loss of the optimal predictor.

Proof. Since the adopted loss function is assumed to be convex and by the definition of the subgradient, we have

$$\mathcal{L}_n(\boldsymbol{h}) \geq \mathcal{L}_n(\boldsymbol{\theta}_{n-1}) + \boldsymbol{g}_n^T(\boldsymbol{h} - \boldsymbol{\theta}_{n-1}), \quad \forall \boldsymbol{h} \in \mathbb{R}^l, \tag{8.84}$$

or

$$\mathcal{L}_n(\boldsymbol{\theta}_{n-1}) - \mathcal{L}_n(\boldsymbol{h}) \leq \boldsymbol{g}_n^T(\boldsymbol{\theta}_{n-1} - \boldsymbol{h}). \tag{8.85}$$

However, recalling (8.80), we can write

$$\boldsymbol{\theta}_n - \boldsymbol{h} = \boldsymbol{\theta}_{n-1} - \boldsymbol{h} - \mu_n \boldsymbol{g}_n, \tag{8.86}$$

which results in

$$\begin{aligned} ||\boldsymbol{\theta}_n - \boldsymbol{h}||^2 &= ||\boldsymbol{\theta}_{n-1} - \boldsymbol{h}||^2 + \mu_n^2 ||\boldsymbol{g}_n||^2 \\ &\quad -2\mu_n \boldsymbol{g}_n^T(\boldsymbol{\theta}_{n-1} - \boldsymbol{h}). \end{aligned} \tag{8.87}$$

Taking into account the bound of the subgradient, Eq. (8.87) leads to the inequality

$$\boldsymbol{g}_n^T(\boldsymbol{\theta}_{n-1} - \boldsymbol{h}) \leq \frac{1}{2\mu_n}\left(||\boldsymbol{\theta}_{n-1} - \boldsymbol{h}||^2 - ||\boldsymbol{\theta}_n - \boldsymbol{h}||^2\right) + \frac{\mu_n}{2} G^2. \tag{8.88}$$

Summing up both sides of (8.88), taking into account inequality (8.85) and after a bit of algebra (Problem 8.30), results in

$$\sum_{n=1}^{N}\mathcal{L}_n(\boldsymbol{\theta}_{n-1})-\sum_{n=1}^{N}\mathcal{L}_n(\boldsymbol{h})\leq\frac{1}{2\mu_N}F^2+\frac{G^2}{2}\sum_{n=1}^{N}\mu_n. \tag{8.89}$$

Setting $\mu_n=\frac{1}{\sqrt{n}}$, using the obvious bound

$$\sum_{n=1}^{N}\frac{1}{\sqrt{n}}\leq 1+\int_1^N\frac{1}{\sqrt{t}}\mathrm{d}t=2\sqrt{N}-1, \tag{8.90}$$

and dividing both sides of (8.89) by N, the proposition is proved for any $\boldsymbol{h}$. Hence, it will also be true for $\boldsymbol{\theta}_*$. □

The previous proof follows the one given in [113]; this was the first paper to adopt the notion of "regret" for the analysis of convex online algorithms. Proofs given later for more complex algorithms have borrowed, in one way or another, the arguments used there.

Remarks 8.8.

- Tighter regret bounds can be derived when the loss function is strongly convex [43]. A function $f:\mathcal{X}\subseteq\mathbb{R}^l\longmapsto\mathbb{R}$ is said to be σ-*strongly convex* if, $\forall\ \boldsymbol{y},\boldsymbol{x}\in\mathcal{X}$,

$$f(\boldsymbol{y})\geq f(\boldsymbol{x})+\boldsymbol{g}^T(\boldsymbol{y}-\boldsymbol{x})+\frac{\sigma}{2}||\boldsymbol{y}-\boldsymbol{x}||^2, \tag{8.91}$$

  for any subgradient $\boldsymbol{g}$ at $\boldsymbol{x}$. It also turns out that a function $f(\boldsymbol{x})$ is strongly convex if $f(\boldsymbol{x})-\frac{\sigma}{2}||\boldsymbol{x}||^2$ is convex (Problem 8.31).

  For σ-strongly convex loss functions, if the step size of the subgradient algorithm is diminishing at a rate $\mathcal{O}(\frac{1}{\sigma n})$, then the average cumulative loss is approaching the average loss of the optimal predictor at a rate $\mathcal{O}(\frac{\ln N}{N})$ (Problem 8.32). This is the case, for example, for the PEGASOS algorithm, discussed in Section 8.10.3.
- In [4,5], $\mathcal{O}(1/N)$ convergence rates are derived for a set of not strongly convex smooth loss functions (squared error and logistic regression) even for the case of constant step sizes. The analysis method follows statistical arguments.

8.12 ONLINE LEARNING AND BIG DATA APPLICATIONS: A DISCUSSION

Online learning algorithms have been treated in Chapters 4, 5, and 6. The purpose of this section is to first summarize some of the findings and at the same time present a discussion related to the performance of online schemes compared to their batch relatives.

Recall that the ultimate goal in obtaining a parametric predictor,

$$\hat{y}=f_{\boldsymbol{\theta}}(\boldsymbol{x}),$$

is to select $\boldsymbol{\theta}$ so as to optimize the *expected loss/risk* function (8.65). For practical reasons, the corresponding empirical formulation in (8.66) is most often adopted instead. From the learning theory's point of view, this is justified provided the respective class of functions is sufficiently restrictive [94]. The available literature is quite rich in obtaining performance bounds that measure how close the optimal value obtained via the expected loss is to that obtained via the empirical one, as a function of the number of points N. Note that as $N \longrightarrow \infty$, and recalling well-known arguments from probability theory and statistics, the empirical risk tends to the expected loss (under general assumptions). Thus, for very large training data sets, adopting the empirical risk may not be that different from using the expected loss. However, for data sets of shorter lengths, a number of issues occur. Besides the value of N, another critical factor enters the scene; this is the *complexity* of the family of the functions in which we search a solution. In other words, the generalization performance critically depends not only on N but also on how large or small this set of functions is. A related discussion for the specific case of the MSE was presented in Chapter 3, in the context of the bias–variance tradeoff. The roots of the more general theory go back to the pioneering work of Vapnik–Chervonenkis; see [31,95,96], and [88] for a less mathematical summary of the major points.

In the sequel, some results tailored to the needs of our current discussion will be discussed.

APPROXIMATION, ESTIMATION, AND OPTIMIZATION ERRORS

Recall that all we are given in a machine learning task is the available training set of examples. To set up the "game," the designer has to decide on the selection of (a) the loss function, $\mathcal{L}(\cdot,\cdot)$, which measures the deviation (error) between predicted and target values, and (b) the set $\mathcal{F}$ of (parametric) functions,

$$\mathcal{F} = \left\{ f_{\boldsymbol{\theta}} : \boldsymbol{\theta} \in \mathbb{R}^K \right\}.$$

Based on the choice of $\mathcal{L}(\cdot,\cdot)$, the *benchmark* function, denoted as f_*, is the one that minimizes the expected loss (see also Chapter 3), that is,

$$f_* = \arg\min_f \mathbb{E}\left[\mathcal{L}\big(\mathrm{y}, f(\mathbf{x})\big)\right],$$

or equivalently

$$f_*(\boldsymbol{x}) = \arg\min_{\hat{y}} \mathbb{E}\left[\mathcal{L}\left(y, \hat{y}\right) | \boldsymbol{x}\right]. \tag{8.92}$$

Let also $f_{\boldsymbol{\theta}_*}$ denote the optimal function that results by minimizing the expected loss constrained within the parametric family $\mathcal{F}$, that is,

$$\boxed{f_{\boldsymbol{\theta}_*} : \boldsymbol{\theta}_* = \arg\min_{\boldsymbol{\theta}} \mathbb{E}\left[\mathcal{L}\big(\mathrm{y}, f_{\boldsymbol{\theta}}(\mathbf{x})\big)\right].} \tag{8.93}$$

However, instead of $f_{\boldsymbol{\theta}_*}$, we obtain another function, denoted as f_N, by minimizing the empirical risk, $J_N(\boldsymbol{\theta})$,

$$\boxed{f_N(\boldsymbol{x}) := f_{\boldsymbol{\theta}_*(N)}(\boldsymbol{x}) : \ \boldsymbol{\theta}_*(N) = \arg\min_{\boldsymbol{\theta}} J_N(\boldsymbol{\theta}).} \tag{8.94}$$

Once f_N has been obtained, we are interested in evaluating its generalization performance; that is, to compute the value of the expected loss at f_N, $\mathbb{E}\left[\mathcal{L}(\mathrm{y}, f_N(\mathbf{x}))\right]$. The excess error with respect to the optimal value can then be decomposed as [13]

$$\mathcal{E} = \mathbb{E}\left[\mathcal{L}(\mathrm{y}, f_N(\mathbf{x}))\right] - \mathbb{E}\left[\mathcal{L}(\mathrm{y}, f_*(\mathbf{x}))\right] = \mathcal{E}_{\text{appr}} + \mathcal{E}_{\text{est}}, \tag{8.95}$$

where

$$\boxed{\mathcal{E}_{\text{appr}} := \mathbb{E}\left[\mathcal{L}(\mathrm{y}, f_{\boldsymbol{\theta}_*}(\mathbf{x}))\right] - \mathbb{E}\left[\mathcal{L}(\mathrm{y}, f_*(\mathbf{x}))\right]: \quad \text{approximation error,}}$$

and

$$\boxed{\mathcal{E}_{\text{est}} := \mathbb{E}\left[\mathcal{L}(\mathrm{y}, f_N(\mathbf{x}))\right] - \mathbb{E}\left[\mathcal{L}(\mathrm{y}, f_{\boldsymbol{\theta}_*}(\mathbf{x}))\right]: \quad \text{estimation error,}}$$

where $\mathcal{E}_{\text{appr}}$ is known as the *approximation* error and $\mathcal{E}_{\text{est}}$ is known as the *estimation* error. The former measures how well the chosen family of functions can perform compared to the optimal/benchmark value and the latter measures the performance loss within the family $\mathcal{F}$, due to the fact that optimization is performed via the empirical risk function. Large families of functions lead to low approximation error but higher estimation error, and vice versa. A way to improve upon the estimation error, while keeping the approximation error small, is to increase N. The size/complexity of the family $\mathcal{F}$ is measured by its *capacity*, which may depend on the number of parameters, but this is not always the whole story; see, for example, [88,95]. For example, the use of regularization, while minimizing the empirical risk, can have a decisive effect on the approximation–estimation error tradeoff.

In practice, while optimizing the (regularized) empirical risk, one has to adopt an iterative minimization or an online algorithm, which leads to an approximate solution, denoted as $\tilde{f}_N$. Then the excess error in (8.95) involves a third term [13,14],

$$\mathcal{E} = \mathcal{E}_{\text{appr}} + \mathcal{E}_{\text{est}} + \mathcal{E}_{\text{opt}}, \tag{8.96}$$

where

$$\boxed{\mathcal{E}_{\text{opt}} := \mathbb{E}\left[\mathcal{L}(\mathrm{y}, \tilde{f}_N(\mathbf{x}))\right] - \mathbb{E}\left[\mathcal{L}(\mathrm{y}, f_N(\mathbf{x}))\right]: \quad \text{optimization error.}}$$

The literature is rich in studies deriving bounds concerning the excess error. More detailed treatment is beyond the scope of this book. As a case study, we will follow the treatment given in [14].

Let the computation of $\tilde{f}_N$ be associated with a predefined accuracy,

$$\mathbb{E}\left[\mathcal{L}(\mathrm{y}, \tilde{f}_N(\mathbf{x}))\right] < \mathbb{E}\left[\mathcal{L}(\mathrm{y}, f_N(\mathbf{x}))\right] + \rho.$$

Then, for a class of functions that are often met in practice, for example, under strong convexity of the loss function [51] or under certain assumptions on the data distribution [92], the following equivalence relation can be established:

$$\mathcal{E}_{\text{appr}} + \mathcal{E}_{\text{est}} + \mathcal{E}_{\text{opt}} \sim \mathcal{E}_{\text{appr}} + \left(\frac{\ln N}{N}\right)^a + \rho, \quad a \in \left[\frac{1}{2}, 1\right], \tag{8.97}$$

which verifies the fact that as $N \longrightarrow \infty$ the estimation error decreases and provides a rule for the respective convergence rate. The excess error $\mathcal{E}$, besides the approximation component, on which we have no access to control (given the family of functions, $\mathcal{F}$), it depends on (a) the number of data and (b) on the accuracy, ρ, associated with the algorithm used. How one can control these parameters depends on the type of learning task at hand.

- *Small-scale tasks*: These types of tasks are constrained by the number of training points N. In this case, one can reduce the optimization error, since computational load is not a problem, and achieve the minimum possible estimation error, as this is allowed by the number of available training points. In this case, one achieves the approximation–estimation tradeoff.
- *Large-scale/big data tasks*: These types of tasks are constrained by the computational resources. Thus, a computationally cheap and less accurate algorithm may end up with lower excess error, since it has the luxury of exploiting more data, compared to a more accurate yet computationally more complex algorithm, given the maximum allowed computational load.

BATCH VERSUS ONLINE LEARNING

Our interest in this subsection lies in investigating whether there is a performance loss if in place of a batch algorithm an online one is used. There is a very subtle issue involved here, which turns out to be very important from a practical point of view. We will restrict our discussion to differentiable convex loss functions.

Two major factors associated with the performance of an algorithm (in a stationary environment) are its convergence rate and its accuracy after convergence. The general form of a batch algorithm in minimizing (8.66) is written as

$$\begin{aligned}\boldsymbol{\theta}^{(i)} &= \boldsymbol{\theta}^{(i-1)} - \mu_i \Phi_i \nabla J_N\left(\boldsymbol{\theta}^{(i-1)}\right) \\ &= \boldsymbol{\theta}^{(i-1)} - \frac{\mu_i}{N}\Phi_i \sum_{n=1}^{N} \mathcal{L}'\left(\boldsymbol{\theta}^{(i-1)}, y_n, \boldsymbol{x}_n\right). \end{aligned} \tag{8.98}$$

For gradient descent, $\Phi_i = I$, and for Newton-type recursions, Φ_i is the inverse Hessian matrix of the loss function (Chapter 6).

Note that these are not the only possible choices for matrix Φ. For example, in the Levenberg–Marquardt method, the square Jacobian is employed, that is,

$$\Phi_i = \left[\nabla J(\boldsymbol{\theta}^{(i-1)})\nabla^T J(\boldsymbol{\theta}^{(i-1)}) + \lambda I\right]^{-1},$$

where λ is a regularization parameter. In [3], the *natural gradient* is proposed, which is based on the Fisher information matrix associated with the noisy distribution implied by the adopted prediction model, $f_{\boldsymbol{\theta}}(\boldsymbol{x})$. In both cases, the involved matrices asymptotically behave like the Hessian, yet they may provide improved performance during the initial convergence phase. For a further discussion, the interested reader may consult [49,56].

As already mentioned in Chapters 5 and 6 (Section 6.7), the convergence rate to the respective optimal value of the simple gradient descent method is *linear*, that is,

$$\ln \frac{1}{\left\|\boldsymbol{\theta}^{(i)} - \boldsymbol{\theta}_*(N)\right\|^2} \propto i,$$

and the corresponding rate for a Newton-type algorithm is (approximately) quadratic, that is,

$$\ln \ln \frac{1}{\left\|\boldsymbol{\theta}^{(i)}-\boldsymbol{\theta}_*(N)\right\|^2} \propto i.$$

In contrast, the online version of (8.98), that is,

$$\boldsymbol{\theta}_n=\boldsymbol{\theta}_{n-1}-\mu_n \Phi_n \mathcal{L}'(\boldsymbol{\theta}_{n-1}, y_n, \boldsymbol{x}_n), \tag{8.99}$$

is based on a *noisy* estimation of the gradient, using the current sample point, $(y_n, \boldsymbol{x}_n)$, only. The effect of this is to slow down convergence, in particular when the algorithm gets close to the solution. Moreover, the estimate of the parameter vector fluctuates around the optimal value. We have extensively studied this phenomenon in the case of the LMS, when μ_n is assigned a constant value. This is the reason that in the stochastic gradient rationale, μ_n must be a decreasing sequence. However, it *must not* decrease very fast, which is guaranteed by the condition $\sum_n \mu_n \longrightarrow \infty$ (Section 5.4). Furthermore, recall from our discussion there that the rate of convergence toward $\boldsymbol{\theta}_*$ is, on average, $\mathcal{O}(1/n)$. This result also covers the more general case of online algorithms given in (8.99) (see, for example, [56]). Note, however, that all these results have been derived under a number of assumptions, for example, that the algorithm is close enough to a solution.

Our major interest now turns to comparing the rate at which a batch and a corresponding online algorithm converge to $\boldsymbol{\theta}_*$, that is, the value that minimizes the expected loss, which is the ultimate goal of our learning task. Since the aim is to compare performances, given the same number of training samples, let us use the same number, both for n for the online and for N for the batch. Following [12] and applying a second-order Taylor expansion on $J_n(\boldsymbol{\theta})$, it can be shown (Problem 8.28) that

$$\boldsymbol{\theta}_*(n)=\boldsymbol{\theta}_*(n-1)-\frac{1}{n}\Psi_n^{-1}\mathcal{L}'\big(\boldsymbol{\theta}_*(n-1), y_n, \boldsymbol{x}_n\big), \tag{8.100}$$

where

$$\Psi_n=\left(\frac{1}{n}\sum_{k=1}^{n}\nabla^2\mathcal{L}\big(\boldsymbol{\theta}_*(n-1), y_k, \boldsymbol{x}_k\big)\right).$$

Note that (8.100) is similar in structure to (8.99). Also, as $n \longrightarrow \infty$, Ψ_n converges to the Hessian matrix, H, of the expected loss function. Hence, for appropriate choices of the involved weighting matrices and setting $\mu_n = 1/n$, (8.99) and (8.100) can converge to $\boldsymbol{\theta}_*$ at similar rates; thus, in both cases, the critical factor that determines how close to the optimal $\boldsymbol{\theta}_*$ the resulting estimates are is the number of data points used. It can be shown [12,56,93] that

$$\boxed{\mathbb{E}\left[||\boldsymbol{\theta}_n-\boldsymbol{\theta}_*||^2\right]+\mathcal{O}\left(\frac{1}{n}\right)=\mathbb{E}\left[||\boldsymbol{\theta}_*(n)-\boldsymbol{\theta}_*||^2\right]+\mathcal{O}\left(\frac{1}{n}\right)=\frac{C}{n},}$$

where C is a constant depending on the specific form of the associated expected loss function used. *Thus, batch algorithms and their online versions can be made to converge at similar rates to* $\boldsymbol{\theta}_*$, *after appropriate fine-tuning of the involved parameters.* Once more, since the critical factor in big data applications is not data but computational resources, a cheap online algorithm can achieve enhanced performance (lower excess error) compared to a batch, yet being a computationally more thirsty scheme. This is because for a given computational load, the online algorithm can process more data points (Problem 8.33). More importantly, an online algorithm needs not to store the data, which can be processed on the fly as they arrive. For a more detailed treatment of the topic, the interested reader may consult [14].

In [14], two forms of batch linear support vector machines (Chapter 11) were tested against their online stochastic gradient counterparts. The tests were carried out on the RCV1 data basis [53], and the training set comprised 781.265 documents represented by (relatively) sparse feature vectors consisting of 47.152 feature values. The stochastic gradient online versions, appropriately tuned with a diminishing step size, achieved comparable error rates at substantially lower computational times (less than one tenth) compared to their batch processing relatives.

Remarks 8.9.

- Most of our discussion on the online versions has been focused on the simplest version, given in (8.99) for $\Phi_n = I$. However, the topic of stochastic gradient descent schemes, especially in the context of smooth loss functions, has a very rich history of over 60 years, and many algorithmic variants have been "born." In Chapter 5, a number of variations of the basic LMS scheme were discussed. Some more notable examples which are still popular are the following.
  Stochastic gradient descent with momentum: The basic iteration of this variant is

$$\boldsymbol{\theta}_n = \boldsymbol{\theta}_{n-1} - \mu_n \mathcal{L}'_n(\boldsymbol{\theta}_{n-1}) + \beta_n(\boldsymbol{\theta}_{n-1} - \boldsymbol{\theta}_{n-2}). \tag{8.101}$$

  Very often, $\beta_n = \beta$ is chosen to be a constant (see, for example, [91]).
  Gradient averaging: Another widely used version results if the place of the single gradient is taken by an average estimate, that is,

$$\boldsymbol{\theta}_n = \boldsymbol{\theta}_{n-1} - \frac{\mu_n}{n}\sum_{k=1}^{n} \mathcal{L}'_k(\boldsymbol{\theta}_{n-1}). \tag{8.102}$$

  Variants with different averaging scenarios (e.g., random selection instead of using all previously points) are also around. Such an averaging has a smoothing effect on the convergence of the algorithm. We have already seen this rationale in the context of the PEGASOS algorithm (Section 8.10.3). The general trend of all the variants of the basic stochastic gradient scheme is to improve with respect to the involved constants, but the convergence rate still remains $\mathcal{O}(1/n)$.

  In [50], the online learning rationale was used in the context of data sets of fixed size, N. Instead of using the gradient descent scheme in (8.98), the following version is proposed:

$$\boldsymbol{\theta}^{(i)} = \boldsymbol{\theta}^{(i-1)} - \frac{\mu_i}{N}\sum_{k=1}^{N} \boldsymbol{g}_k^{(i)}, \tag{8.103}$$

  where

$$\boldsymbol{g}_k^{(i)} = \begin{cases} \mathcal{L}'_k(\boldsymbol{\theta}^{(i-1)}), & \text{if } k = i_k, \\ \boldsymbol{g}_k^{(i-1)}, & \text{otherwise.} \end{cases} \tag{8.104}$$

The index i_k is randomly chosen every time from $\{1, 2, \ldots, N\}$. Thus, in each iteration only one gradient is computed and the rest are drawn from the memory. It turns out that for strongly convex smooth loss functions, the algorithm exhibits linear convergence to the solution of the empirical risk in (8.56). Of course, compared to the basic online schemes, an $\mathcal{O}(N)$ memory is required for keeping track of the gradient computations.

- The literature on deriving performance bounds concerning online algorithms is very rich, both in studies and in ideas. For example, another line of research involves bounds for *arbitrary* online algorithms (see [1,19,69] and references therein).

8.13 PROXIMAL OPERATORS

So far in the chapter, we have dealt with the notion of the projection operator. In this section, we go one step further and we will introduce an elegant generalization of the notion of projection. Note that when we refer to an operator, we mean a mapping from $\mathbb{R}^l \longmapsto \mathbb{R}^l$, in contrast to a function, which is a mapping $\mathbb{R}^l \longmapsto \mathbb{R}$.

Definition 8.9. Let

$$f : \mathbb{R}^l \longmapsto \mathbb{R}$$

be a convex function and $\lambda > 0$. The corresponding *proximal* or *proximity* operator of index λ [60,71],

$$\text{Prox}_{\lambda f} : \mathbb{R}^l \longmapsto \mathbb{R}^l, \tag{8.105}$$

is defined as

$$\boxed{\text{Prox}_{\lambda f}(\boldsymbol{x}) := \arg\min_{\boldsymbol{v} \in \mathbb{R}^l} \left\{ f(\boldsymbol{v}) + \frac{1}{2\lambda} ||\boldsymbol{x} - \boldsymbol{v}||^2 \right\} : \quad \text{proximal operator.}} \tag{8.106}$$

We stress that the proximal operator is a point in $\mathbb{R}^l$. The definition can also be extended to include functions defined as $f : \mathbb{R}^l \longmapsto \mathbb{R} \cup \{+\infty\}$. A closely related notion to the proximal operator is the following.

Definition 8.10. Let f be a convex function as before. We call the *Moreau* envelope the function

$$\boxed{e_{\lambda f}(\boldsymbol{x}) := \min_{\boldsymbol{v} \in \mathbb{R}^l} \left\{ f(\boldsymbol{v}) + \frac{1}{2\lambda} ||\boldsymbol{x} - \boldsymbol{v}||^2 \right\} : \quad \text{Moreau envelope.}} \tag{8.107}$$

Note that the Moreau envelope [59] is a *function* related to the proximal *operator* as

$$e_{\lambda f}(\boldsymbol{x}) = f\left(\text{Prox}_{\lambda f}(\boldsymbol{x})\right) + \frac{1}{2\lambda} ||\boldsymbol{x} - \text{Prox}_{\lambda f}(\boldsymbol{x})||^2. \tag{8.108}$$

The Moreau envelope can also be thought of as a regularized minimization, and it is also known as the *Moreau–Yosida regularization* [109]. Moreover, it can be shown that it is differentiable (e.g., [7]).

A first point to clarify is whether the minimum in (8.106) exists. Note that the two terms in the brackets are both convex; namely, $f(\boldsymbol{v})$ and the quadratic term $||\boldsymbol{x}-\boldsymbol{v}||^2$. Hence, as can easily be shown by recalling the definition of convexity, their sum is also convex. Moreover, the latter of the two terms is strictly convex, hence their sum is also strictly convex, which guarantees a *unique* minimum.

Example 8.9. Let us calculate $\text{Prox}_{\lambda\iota_C}$, where $\iota_C : \mathbb{R}^l \longmapsto \mathbb{R} \cup \{+\infty\}$ stands for the indicator function of a nonempty closed convex subset $C \subset \mathbb{R}^l$, defined as

$$\iota_C(\boldsymbol{x}) := \begin{cases} 0, & \text{if } \boldsymbol{x} \in C, \\ +\infty, & \text{if } \boldsymbol{x} \notin C. \end{cases}$$

It is not difficult to verify that

$$\begin{aligned} \text{Prox}_{\lambda\iota_C}(\boldsymbol{x}) &= \arg\min_{\boldsymbol{v}\in\mathbb{R}^l} \left\{ \iota_C(\boldsymbol{v}) + \frac{1}{2\lambda}\|\boldsymbol{x}-\boldsymbol{v}\|^2 \right\} \\ &= \arg\min_{\boldsymbol{v}\in C} \|\boldsymbol{x}-\boldsymbol{v}\|^2 = P_C(\boldsymbol{x}), \quad \forall \boldsymbol{x} \in \mathbb{R}^l,\ \forall\lambda > 0, \end{aligned}$$

where P_C is the (metric) projection mapping onto C.

Moreover,

$$\begin{aligned} e_{\lambda\iota_C}(\boldsymbol{x}) &= \min_{\boldsymbol{v}\in\mathbb{R}^l} \left\{ \iota_C(\boldsymbol{v}) + \frac{1}{2\lambda}\|\boldsymbol{x}-\boldsymbol{v}\|^2 \right\} \\ &= \min_{\boldsymbol{v}\in C} \frac{1}{2\lambda}\|\boldsymbol{x}-\boldsymbol{v}\|^2 = \frac{1}{2\lambda} d_C^2(\boldsymbol{x}), \end{aligned}$$

where d_C stands for the (metric) distance function to C (Example 8.5), defined as $d_C(\boldsymbol{x}) := \min_{\boldsymbol{v}\in C} \|\boldsymbol{x}-\boldsymbol{v}\|$.

Thus, as said in the beginning of this section, the proximal operator can be considered as the generalization of the projection one.

Example 8.10. In the case where f becomes the ℓ_1 norm of a vector, that is,

$$\|\boldsymbol{x}\|_1 = \sum_{i=1}^{l} |x_i|, \quad \forall \boldsymbol{x} \in \mathbb{R}^l,$$

it is easily determined that (8.106) decomposes into a set of l scalar minimization tasks, that is,

$$\text{Prox}_{\lambda\|\cdot\|_1}(\boldsymbol{x})|_i = \arg\min_{v_i\in\mathbb{R}} \left\{ |v_i| + \frac{1}{2\lambda}(x_i - v_i)^2 \right\}, \quad i = 1, 2, \ldots, l, \tag{8.109}$$

where $\text{Prox}_{\lambda\|\cdot\|_1}(\boldsymbol{x})|_i$ denotes the respective ith element. Minimizing (8.109) is equivalent to requiring the subgradient to be zero, which results in

$$\begin{aligned}
\text{Prox}_{\lambda\|\cdot\|_1}(\boldsymbol{x})|_i &= \begin{cases} x_i - \text{sgn}(x_i)\lambda, & \text{if } |x_i| > \lambda, \\ 0, & \text{if } |x_i| \leq \lambda \end{cases} \\
&= \text{sgn}(x_i)\max\{0, |x_i| - \lambda\}. \qquad (8.110)
\end{aligned}$$

For the time being, the proof is left as an exercise. The same task is treated in detail in Chapter 9, and the proof is provided in Section 9.3. The operation in (8.110) is also known as *soft thresholding*. In other words, it sets to zero all values with magnitude less than a threshold value (λ) and adds a constant bias (depending on the sign) to the rest. To provoke the unfamiliar reader a bit, this is a way to impose sparsity on a parameter vector.

Having calculated $\text{Prox}_{\lambda\|\cdot\|_1}(\boldsymbol{x})$, the Moreau envelope of $\|\cdot\|_1$ can be directly obtained by

$$\begin{aligned}
e_{\lambda\|\cdot\|_1}(\boldsymbol{x}) &= \sum_{i=1}^{l}\left(\frac{1}{2\lambda}\left(x_i - \text{Prox}_{\lambda\|\cdot\|_1}(\boldsymbol{x})|_i\right)^2 + \left|\text{Prox}_{\lambda\|\cdot\|_1}(\boldsymbol{x})|_i\right|\right) \\
&= \sum_{i=1}^{l}\left(\chi_{[0,\lambda]}(|x_i|)\frac{x_i^2}{2\lambda} + \chi_{(\lambda,+\infty)}(|x_i|)\left(|x_i - \text{sgn}(x_i)\lambda| + \frac{\lambda}{2}\right)\right) \\
&= \sum_{i=1}^{l}\left(\chi_{[0,\lambda]}(|x_i|)\frac{x_i^2}{2\lambda} + \chi_{(\lambda,+\infty)}(|x_i|)\left(|x_i| - \frac{\lambda}{2}\right)\right),
\end{aligned}$$

where $\chi_{\mathcal{A}}(\cdot)$ denotes the characteristic function of the set $\mathcal{A}$, defined in (8.59). For the one-dimensional case, $l = 1$, the previous Moreau envelope boils down to

$$e_{\lambda|\cdot|}(x) = \begin{cases} |x| - \frac{\lambda}{2}, & \text{if } |x| > \lambda, \\ \frac{x^2}{2\lambda}, & \text{if } |x| \leq \lambda. \end{cases}$$

This envelope and the original $|\cdot|$ functions are depicted in Fig. 8.30. It is worth noting here that $e_{\lambda|\cdot|}$ is a scaled version, more accurately $1/\lambda$ times, of the celebrated Huber function; a loss function vastly used against outliers in robust statistics, which will be discussed in more detail in Chapter 11. Note that the Moreau envelope is a "blown-up" smoothed version of the ℓ_1 norm function and although the original function is not differentiable, its Moreau envelope is continuously differentiable; moreover, *they both share the same minimum.* This is most interesting and we will come back to this very soon.

8.13.1 PROPERTIES OF THE PROXIMAL OPERATOR

We now focus on some basic properties of the proximal operator, which will soon be used to give birth to a new class of algorithms for the minimization of nonsmooth convex loss functions.

Proposition 8.5. *Consider a convex function*

$$f : \mathbb{R}^l \longmapsto \mathbb{R} \cup \{+\infty\},$$

and let $\text{Prox}_{\lambda f}(\cdot)$ *be its corresponding proximal operator of index* λ. *Then*

$$\boldsymbol{p} = \text{Prox}_{\lambda f}(\boldsymbol{x})$$

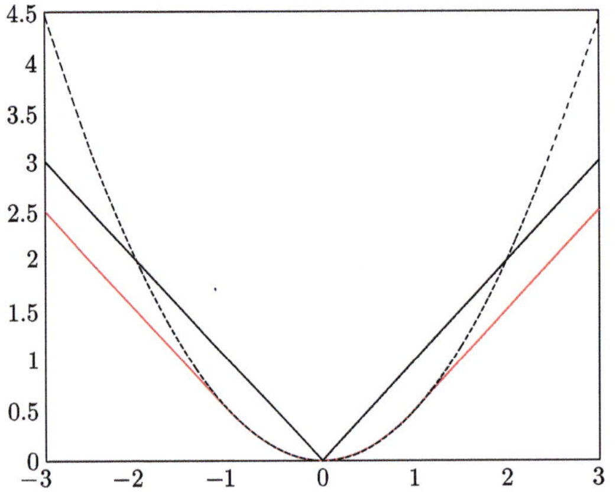

FIGURE 8.30

The $|x|$ function (black solid line), its Moreau envelope $e_{\lambda|\cdot|}(x)$ (red solid line), and $x^2/2$ (black dotted line), for $x \in \mathbb{R}$. Even if $|\cdot|$ is nondifferentiable at 0, $e_{\lambda|\cdot|}(x)$ is everywhere differentiable. Note also that although $x^2/2$ and $e_{\lambda|\cdot|}(x)$ behave exactly the same for small values of x, $e_{\lambda|\cdot|}(x)$ is more conservative than $x^2/2$ in penalizing large values of x; this is the reason for the extensive usability of the Huber function, a scaled-down version of $e_{\lambda|\cdot|}(x)$, as a robust tool against outliers in robust statistics.

if and only if

$$\boxed{\langle \boldsymbol{y} - \boldsymbol{p}, \boldsymbol{x} - \boldsymbol{p} \rangle \leq \lambda\big(f(\boldsymbol{y}) - f(\boldsymbol{p})\big), \quad \forall \boldsymbol{y} \in \mathbb{R}^l.} \tag{8.111}$$

Another necessary condition is

$$\boxed{\left\| \mathrm{Prox}_{\lambda f}(\boldsymbol{x}) - \mathrm{Prox}_{\lambda f}(\boldsymbol{y}) \right\|^2 \leq \left\langle \boldsymbol{x} - \boldsymbol{y}, \mathrm{Prox}_{\lambda f}(\boldsymbol{x}) - \mathrm{Prox}_{\lambda f}(\boldsymbol{y}) \right\rangle.} \tag{8.112}$$

The proofs of (8.111) and (8.112) are given in Problems 8.34 and 8.35, respectively. Note that (8.112) is of the same flavor as (8.16), which is inherited to the proximal operator from its more primitive ancestor. In the sequel, we will make use of these properties to touch upon the algorithmic front, where our main interest lies.

Lemma 8.3. *Consider the convex function*

$$f : \mathbb{R}^l \longmapsto \mathbb{R} \cup \{+\infty\}$$

and its proximal operator $\mathrm{Prox}_{\lambda f}$ *of index* λ. *Then the fixed point set of the proximal operator coincides with the set of minimizers of* f, *that is,*

$$\boxed{\mathrm{Fix}\big(\mathrm{Prox}_{\lambda f}\big) = \left\{ \boldsymbol{x} : \boldsymbol{x} = \arg\min_{\boldsymbol{y}} f(\boldsymbol{y}) \right\}.} \tag{8.113}$$

Proof. The definition of the fixed point set has been given in Section 8.3.1. We first assume that a point $\boldsymbol{x}$ belongs to the fixed point set, hence the action of the proximal operator leaves it unaffected, that is,

$$\boldsymbol{x} = \text{Prox}_{\lambda f}(\boldsymbol{x}),$$

and making use of (8.111), we get

$$\langle \boldsymbol{y} - \boldsymbol{x}, \boldsymbol{x} - \boldsymbol{x} \rangle \leq \lambda\big(f(\boldsymbol{y}) - f(\boldsymbol{x})\big), \quad \forall \boldsymbol{y} \in \mathbb{R}^l, \tag{8.114}$$

which results in

$$f(\boldsymbol{x}) \leq f(\boldsymbol{y}), \quad \forall \boldsymbol{y} \in \mathbb{R}^l. \tag{8.115}$$

That is, $\boldsymbol{x}$ is a minimizer of f. For the converse, we assume that $\boldsymbol{x}$ is a minimizer. Then (8.115) is valid, from which (8.114) is deduced, and since this is a necessary and sufficient condition for a point to be equal to the value of the proximal operator, we have proved the claim. □

Finally, note that f and $e_{\lambda f}$ share the same set of minimizers. Thus, one can minimize a nonsmooth convex function by dealing with an equivalent smooth one. From a practical point of view, the value of the method depends on how easy it is to obtain the proximal operator. For example, we have already seen that if the goal is to minimize the ℓ_1 norm, the proximal operator is a simple soft thresholding operation. Needless to say that life is not always that generous!

8.13.2 PROXIMAL MINIMIZATION

In this section, we will exploit our experience from Section 8.4 to develop iterative schemes which asymptotically land their estimates in the fixed point set of the respective operator. All that is required is for the operator to own a nonexpansiveness property.

Proposition 8.6. *The proximal operator associated with a convex function is nonexpansive, that is,*

$$\boxed{||\text{Prox}_{\lambda f}(\boldsymbol{x}) - \text{Prox}_{\lambda f}(\boldsymbol{y})|| \leq ||\boldsymbol{x} - \boldsymbol{y}||.} \tag{8.116}$$

Proof. The proof is readily obtained as a combination of the property in (8.112) with the Cauchy–Schwarz inequality. Moreover, it can be shown that the relaxed version of the proximal operator (also known as the *reflected* version),

$$R_{\lambda f}(\boldsymbol{x}) := 2\,\text{Prox}_{\lambda f}(\boldsymbol{x}) - I, \tag{8.117}$$

is also nonexpansive with the same fixed point set as that of the proximal operator (Problem 8.36). □

Proposition 8.7. *Let*

$$f : \mathbb{R}^l \longmapsto \mathbb{R} \cup \{+\infty\}$$

be a convex function, with $\text{Prox}_{\lambda f}$ *being the respective proximal operator of index* λ*. Then, starting from an arbitrary point,* $\boldsymbol{x}_0 \in \mathbb{R}^l$*, the iterative algorithm*

$$\boldsymbol{x}_k = \boldsymbol{x}_{k-1} + \mu_k \left(\text{Prox}_{\lambda f}(\boldsymbol{x}_{k-1}) - \boldsymbol{x}_{k-1}\right), \tag{8.118}$$

where $\mu_k \in (0,2)$ is such that

$$\sum_{k=1}^{\infty} \mu_k(2-\mu_k) = +\infty,$$

converges to an element of the fixed point set of the proximal operator; that is, it converges to a minimizer of f. Proximal minimization algorithms are traced back to the early 1970s [57,72].

The proof of the proposition is given in Problem 8.36 [84]. Observe that (8.118) is the counterpart of (8.26). A special case occurs if $\mu_k = 1$, which results in

$$\boldsymbol{x}_k = \text{Prox}_{\lambda f}(\boldsymbol{x}_{k-1}), \tag{8.119}$$

also known as the *proximal point* algorithm.

Example 8.11. Let us demonstrate the previous findings via the familiar optimization task of the quadratic function

$$f(\boldsymbol{x}) = \frac{1}{2}\boldsymbol{x}^T A\boldsymbol{x} - \boldsymbol{b}^T\boldsymbol{x}.$$

It does not take long to see that the minimizer occurs at the solution of the linear system of equations

$$A\boldsymbol{x}_* = \boldsymbol{b}.$$

From the definition in (8.106), taking the gradient of the quadratic function and equating to zero, we readily obtain

$$\text{Prox}_{\lambda f}(\boldsymbol{x}) = \left(A + \frac{1}{\lambda}I\right)^{-1}\left(\boldsymbol{b} + \frac{1}{\lambda}\boldsymbol{x}\right), \tag{8.120}$$

and setting $\epsilon = \frac{1}{\lambda}$, the recursion in (8.119) becomes

$$\boldsymbol{x}_k = (A + \epsilon I)^{-1}(\boldsymbol{b} + \epsilon\boldsymbol{x}_{k-1}). \tag{8.121}$$

After some simple algebraic manipulations (Problem 8.37), we finally obtain

$$\boldsymbol{x}_k = \boldsymbol{x}_{k-1} + (A + \epsilon I)^{-1}(\boldsymbol{b} - A\boldsymbol{x}_{k-1}). \tag{8.122}$$

This scheme is known from the numerical linear algebra as *iterative refinement* algorithm [58]. It is used when the matrix A is near-singular, so the regularization via ϵ helps the inversion. Note that at each iteration, $\boldsymbol{b} - A\boldsymbol{x}_{k-1}$ is the error committed by the current estimate. The algorithm belongs to a larger family of algorithms, known as stationary iterative or iterative relaxation schemes; we will meet such schemes in Chapter 10.

The interesting point here is that since the algorithm results as a special case of the proximal minimization algorithm, convergence to the solution is guaranteed even if ϵ is not small!

Resolvent of the Subdifferential Mapping

We will look at the proximal operator from a slightly different view, which will be useful to us soon, when it will be used for the solution of more general minimization tasks. We will follow a more descriptive and less mathematically formal path.

According to Lemma 8.2 and since the proximal operator is a minimizer of (8.106), it must be chosen so that

$$\mathbf{0} \in \partial f(\boldsymbol{v}) + \frac{1}{\lambda}\boldsymbol{v} - \frac{1}{\lambda}\boldsymbol{x}, \tag{8.123}$$

or

$$\mathbf{0} \in \lambda \partial f(\boldsymbol{v}) + \boldsymbol{v} - \boldsymbol{x}, \tag{8.124}$$

or

$$\boldsymbol{x} \in \lambda \partial f(\boldsymbol{v}) + \boldsymbol{v}. \tag{8.125}$$

Let us now define the mapping

$$(I + \lambda \partial f) : \mathbb{R}^l \longmapsto \mathbb{R}^l, \tag{8.126}$$

such that

$$(I + \lambda \partial f)(\boldsymbol{v}) = \boldsymbol{v} + \lambda \partial f(\boldsymbol{v}). \tag{8.127}$$

Note that this mapping is one-to-many, due to the definition of the subdifferential, which is a set.[9] However, its inverse mapping, denoted as

$$(I + \lambda \partial f)^{-1} : \mathbb{R}^l \longmapsto \mathbb{R}^l, \tag{8.128}$$

is single-valued, and as a matter of fact it coincides with the proximal operator; this is readily deduced from (8.125), which can equivalently be written as

$$\boldsymbol{x} \in \big(I + \lambda \partial f\big)(\boldsymbol{v}),$$

which implies that

$$(I + \lambda \partial f)^{-1}(\boldsymbol{x}) = \boldsymbol{v} = \text{Prox}_{\lambda f}(\boldsymbol{x}). \tag{8.129}$$

However, we know that the proximal operator is unique. The operator in (8.128) is known as the *resolvent of the subdifferential mapping* [72].

As an exercise, let us now apply (8.129) to the case of Example 8.11. For this case, the subdifferential set is a singleton comprising the gradient vector,

$$\text{Prox}_{\lambda f}(\boldsymbol{x}) = (I + \lambda \nabla f)^{-1}(\boldsymbol{x}) \Longrightarrow (I + \lambda \nabla f)\big(\text{Prox}_{\lambda f}(\boldsymbol{x})\big) = \boldsymbol{x},$$

or by definition of the mapping $(I + \lambda \nabla f)$ and taking the gradient of the quadratic function,

$$\text{Prox}_{\lambda f}(\boldsymbol{x}) + \lambda \nabla f\big(\text{Prox}_{\lambda f}(\boldsymbol{x})\big) = \text{Prox}_{\lambda f}(\boldsymbol{x}) + \lambda A\, \text{Prox}_{\lambda f}(\boldsymbol{x}) - \lambda \boldsymbol{b} = \boldsymbol{x},$$

[9] A point-to-set mapping is also called a relation on $\mathbb{R}^l$.

which finally results in

$$\text{Prox}_{\lambda f}(\boldsymbol{x}) = \left(A + \frac{1}{\lambda}I\right)^{-1}\left(\boldsymbol{b} + \frac{1}{\lambda}\boldsymbol{x}\right).$$

8.14 PROXIMAL SPLITTING METHODS FOR OPTIMIZATION

A number of optimization tasks often comes in the form of a summation of individual convex functions, some of them being differentiable and some of them nonsmooth. Sparsity-aware learning tasks are typical examples that have received a lot of attention recently, where the regularizing term is nonsmooth, for example, the ℓ_1 norm.

Our goal in this section is to solve the following minimization task:

$$\boldsymbol{x}_* = \arg\min_{\boldsymbol{x}} \left\{ f(\boldsymbol{x}) + g(\boldsymbol{x}) \right\}, \tag{8.130}$$

where both involved functions are convex,

$$f : \mathbb{R}^l \longmapsto \mathbb{R} \cup \{+\infty\}, \quad g : \mathbb{R}^l \longmapsto \mathbb{R},$$

and g is assumed to be differentiable while f is nonsmooth. It turns out that the iterative scheme

$$\boxed{\boldsymbol{x}_k = \underbrace{\text{Prox}_{\lambda_k f}}_{\text{backward step}} \underbrace{\left(\boldsymbol{x}_{k-1} - \lambda_k \nabla g(\boldsymbol{x}_{k-1})\right)}_{\text{forward step}}} \tag{8.131}$$

converges to a minimizer of the sum of the involved functions, that is,

$$\boldsymbol{x}_k \longrightarrow \arg\min_{\boldsymbol{x}} \left\{ f(\boldsymbol{x}) + g(\boldsymbol{x}) \right\}, \tag{8.132}$$

for a properly chosen sequence, λ_k, and provided that the gradient is continuous Lipschitz, that is,

$$||\nabla g(\boldsymbol{x}) - \nabla g(\boldsymbol{y})|| \leq \gamma ||\boldsymbol{x} - \boldsymbol{y}||, \tag{8.133}$$

for some $\gamma > 0$. It can be shown that if $\lambda_k \in \left(0, \frac{1}{\gamma}\right]$, then the algorithm converges to a minimizer at a sublinear rate, $\mathcal{O}(1/k)$ [8]. This family of algorithms is known as *proximal gradient* or *forward-backward splitting* algorithms. The term *splitting* is inherited from the split of the function into two (or more generally into more) parts. The term *proximal* indicates the presence of the proximal operator of f in the optimization scheme. The iteration involves an (explicit) forward gradient computation step performed on the smooth part and an (implicit) backward step via the use of the proximal operator of the nonsmooth part. The terms *forward-backward* are borrowed from numerical analysis methods involving discretization techniques [97]. Proximal gradient schemes are traced back to, for example, [18,54], but their spread in machine learning and signal processing matured later [26,36].

There are a number of variants of the previous basic scheme. A version that achieves an $\mathcal{O}(\frac{1}{k^2})$ rate of convergence is based on the classical Nesterov modification of the gradient algorithm [65], and it is summarized in Algorithm 8.5 [8]. In the algorithm, the update is split into two parts. In the proximal operator, one uses a smoother version of the obtained estimates, using an averaging that involves previous estimates.

Algorithm 8.5 (Fast proximal gradient splitting algorithm).

- Initialization
  - Select $\boldsymbol{x}_0$, $z_1 = \boldsymbol{x}_0$, $t_1 = 1$.
  - Select λ.
- **For** $k = 1, 2, \ldots,$ **Do**
  - $\boldsymbol{y}_k = z_k - \lambda \nabla g(z_k)$
  - $\boldsymbol{x}_k = \text{Prox}_{\lambda f}(\boldsymbol{y}_k)$
  - $t_{k+1} = \frac{1+\sqrt{4t_k^2+1}}{2}$
  - $\mu_k = 1 + \frac{t_k - 1}{t_{k+1}}$
  - $z_{k+1} = \boldsymbol{x}_k + \mu_k(\boldsymbol{x}_k - \boldsymbol{x}_{k-1})$
- **End For**

Note that the algorithm involves a step size μ_k. The computation of the variables t_k is done in such a way so that convergence speed is optimized. However, it has to be noted that convergence of the scheme is no more guaranteed, in general.

THE PROXIMAL FORWARD-BACKWARD SPLITTING OPERATOR

From a first look, the iterative update given in (8.131) seems to be a bit "magic." However, this is not the case and we can come to it by following simple arguments starting from the basic property of a minimizer. Indeed, let $\boldsymbol{x}_*$ be a minimizer of (8.130). Then we know that it has to satisfy

$$\begin{aligned} \mathbf{0} &\in \partial f(\boldsymbol{x}_*) + \nabla g(\boldsymbol{x}_*), \text{ or equivalently} \\ \mathbf{0} &\in \lambda \partial f(\boldsymbol{x}_*) + \lambda \nabla g(\boldsymbol{x}_*), \text{ or equivalently} \\ \mathbf{0} &\in \lambda \partial f(\boldsymbol{x}_*) + \boldsymbol{x}_* - \boldsymbol{x}_* + \lambda \nabla g(\boldsymbol{x}_*), \end{aligned}$$

or equivalently

$$\big(I - \lambda \nabla g\big)(\boldsymbol{x}_*) \in \big(I + \lambda \partial f\big)(\boldsymbol{x}_*),$$

or

$$\big(I + \lambda \partial f\big)^{-1}\big(I - \lambda \nabla g\big)(\boldsymbol{x}_*) = \boldsymbol{x}_*,$$

and finally

$$\boldsymbol{x}_* = \text{Prox}_{\lambda f}\big(I - \lambda \nabla g(\boldsymbol{x}_*)\big). \tag{8.134}$$

In other words, a minimizer of the task is a fixed point of the operator

$$(I + \lambda\partial f)^{-1}(I - \lambda\nabla g) : \mathbb{R}^l \longmapsto \mathbb{R}^l. \tag{8.135}$$

The latter is known as the *proximal forward-backward splitting operator* and it can be shown that if $\lambda \in \left(0, \frac{1}{\gamma}\right]$, where γ is the Lipschitz constant, then this operator is *nonexpansive* [108]. This short story justifies the reason that the iteration in (8.131) is attracted toward the set of minimizers.

Remarks 8.10.

- The proximal gradient splitting algorithm can be considered as a generalization of some previously considered algorithms. If we set $f(\boldsymbol{x}) = \iota_C(\boldsymbol{x})$, the proximal operator becomes the projection operator and the projected gradient algorithm of (8.62) results. If $f(\boldsymbol{x}) = 0$, we obtain the gradient algorithm and if $g(\boldsymbol{x}) = 0$ the proximal point algorithm comes up.
- Besides batch proximal splitting algorithms, online schemes have been proposed (see [36,48,106,107]), with an emphasis on the ℓ_1 regularization tasks.
- The application and development of novel versions of this family of algorithms in the fields of machine learning and signal processing is still an ongoing field of research and the interested reader can delve deeper into the field via [7,28,67,108].

ALTERNATING DIRECTION METHOD OF MULTIPLIERS (ADMM)

Extensions of the proximal splitting gradient algorithm for the case where both functions f and g are nonsmooth have also been developed, such as the *Douglas–Rachford* algorithm [27,54]. Here, we are going to focus on one of the most popular schemes, known as the *alternating direction method of multipliers* (ADMM) algorithm [40].

The ADMM algorithm is based on the notion of the *augmented Lagrangian* and at its very heart lies the Lagrangian duality concept (Appendix C).

The goal is to minimize the sum $f(\boldsymbol{x}) + g(\boldsymbol{x})$, where both f and g can be nonsmooth. This equivalently can be written as

$$\text{minimize with respect to } \boldsymbol{x}, \boldsymbol{y} \quad f(\boldsymbol{x}) + g(\boldsymbol{y}), \tag{8.136}$$

$$\text{subject to} \quad \boldsymbol{x} - \boldsymbol{y} = \boldsymbol{0}. \tag{8.137}$$

The augmented Lagrangian is defined as

$$L_\lambda(\boldsymbol{x}, \boldsymbol{y}, \boldsymbol{z}) := f(\boldsymbol{x}) + g(\boldsymbol{y}) + \frac{1}{\lambda}\boldsymbol{z}^T(\boldsymbol{x} - \boldsymbol{y}) + \frac{1}{2\lambda}||\boldsymbol{x} - \boldsymbol{y}||^2, \tag{8.138}$$

where we have denoted the corresponding Lagrange multipliers[10] by $\boldsymbol{z}$. The previous equation can be rewritten as

$$L_\lambda(\boldsymbol{x}, \boldsymbol{y}, \boldsymbol{z}) := f(\boldsymbol{x}) + g(\boldsymbol{y}) + \frac{1}{2\lambda}||\boldsymbol{x} - \boldsymbol{y} + \boldsymbol{z}||^2 - \frac{1}{2\lambda}||\boldsymbol{z}||^2. \tag{8.139}$$

The ADMM is given in Algorithm 8.6.

[10] In the book, we have used λ for the Lagrange multipliers. However, here, we have already reserved λ for the proximal operator.

Algorithm 8.6 (The ADMM algorithm).

- Initialization
  - Fix $\lambda > 0$.
  - Select $\boldsymbol{y}_0, \boldsymbol{z}_0$.
- **For** $k = 1, 2, \ldots,$ **Do**
  - $\boldsymbol{x}_k = \text{prox}_{\lambda f}(\boldsymbol{y}_{k-1} - \boldsymbol{z}_{k-1})$
  - $\boldsymbol{y}_k = \text{prox}_{\lambda g}(\boldsymbol{x}_k + \boldsymbol{z}_{k-1})$
  - $\boldsymbol{z}_k = \boldsymbol{z}_{k-1} + (\boldsymbol{x}_k - \boldsymbol{y}_k)$
- **End For**

Looking carefully at the algorithm and (8.139), observe that the first recursion corresponds to the minimization of the augmented Lagrangian with respect to $\boldsymbol{x}$, keeping $\boldsymbol{y}$ and $\boldsymbol{z}$ fixed from the previous iteration. The second recursion corresponds to the minimization with respect to $\boldsymbol{y}$, by keeping $\boldsymbol{x}$ and $\boldsymbol{z}$ frozen to their currently available estimates. The last iteration is an update of the *dual* variables (Lagrange multipliers) in the *ascent* direction; note that the difference in the parentheses is the gradient of the augmented Lagrangian with respect to $\boldsymbol{z}$. Recall from Appendix C that the saddle point is found as a max-min problem of the primal $(\boldsymbol{x}, \boldsymbol{y})$ and the dual variables. The convergence of the algorithm has been analyzed in [40]. For related tutorial papers the reader can look in, e.g., [16,46].

MIRROR DESCENT ALGORITHMS

A closely related algorithmic family to the forward-backward optimization algorithms is traced back to the work in [64]; it is known as *mirror descent algorithms* (MDAs). The method has undergone a number of evolutionary steps (for example, [9,66]). Our focus will be on adopting online schemes to minimize the regularized expected loss function

$$J(\boldsymbol{\theta}) = \mathbb{E}\left[\mathcal{L}(\boldsymbol{\theta}, \mathrm{y}, \mathbf{x})\right] + \phi(\boldsymbol{\theta}),$$

where the regularizing function, ϕ, is assumed to be convex, but not necessarily smooth. In a recent representative of this algorithmic class, also known as *regularized dual averaging* (ARD) algorithm [100], the main iterative equation is expressed as

$$\boldsymbol{\theta}_n = \min_{\boldsymbol{\theta}} \left\{ \langle \bar{\mathcal{L}}', \boldsymbol{\theta} \rangle + \phi(\boldsymbol{\theta}) + \mu_n \psi(\boldsymbol{\theta}) \right\}, \tag{8.140}$$

where ψ is a *strongly convex* auxiliary function. For example, one possibility is to choose $\phi(\boldsymbol{\theta}) = \lambda||\boldsymbol{\theta}||_1$ and $\psi(\boldsymbol{\theta}) = ||\boldsymbol{\theta}||_2^2$ [100]. The average subgradient of $\mathcal{L}$ up to and including time instant $n-1$ is denoted by $\bar{\mathcal{L}}'$, that is,

$$\bar{\mathcal{L}}' = \frac{1}{n-1} \sum_{j=1}^{n-1} \mathcal{L}_j'(\boldsymbol{\theta}_j),$$

where $\mathcal{L}_j(\boldsymbol{\theta}) := \mathcal{L}(\boldsymbol{\theta}, y_j, \boldsymbol{x}_j)$.

It can be shown that if the subgradients are bounded, and $\mu_n = \mathcal{O}\left(\frac{1}{\sqrt{n}}\right)$, then following regret analysis arguments an $\mathcal{O}\left(\frac{1}{\sqrt{n}}\right)$ convergence rate is achieved. If, on the other hand, the regularizing term is strongly convex and $\mu_n = \mathcal{O}\left(\frac{\ln n}{n}\right)$, then an $\mathcal{O}\left(\frac{\ln n}{n}\right)$ rate is obtained. In [100], different variants are proposed. One is based on Nesterov's arguments, as used in Algorithm 8.5, which achieves an $\mathcal{O}\left(\frac{1}{n^2}\right)$ convergence rate.

A closer observation of (8.140) reveals that it can be considered as a generalization of the recursion given in (8.131). Indeed, let us set in (8.140)

$$\psi(\boldsymbol{\theta}) = \frac{1}{2}||\boldsymbol{\theta} - \boldsymbol{\theta}_{n-1}||^2,$$

and in place of the average gradient consider the most recent value, $\mathcal{L}'_{n-1}$. Then (8.140) becomes equivalent to

$$\mathbf{0} \in \mathcal{L}'_{n-1} + \partial\phi(\boldsymbol{\theta}) + \mu_n(\boldsymbol{\theta} - \boldsymbol{\theta}_{n-1}). \tag{8.141}$$

This is the same relation that would result from (8.131) if we set

$$f \to \phi,\ \boldsymbol{x}_k \to \boldsymbol{\theta}_n,\ \boldsymbol{x}_{k-1} \to \boldsymbol{\theta}_{n-1},\ g \to \mathcal{L},\ \lambda_k \to \frac{1}{\mu_n}. \tag{8.142}$$

As a matter of fact, using these substitutions and setting $\phi(\cdot) = ||\cdot||_1$, the FOBOS algorithm results [36]. However, in the case of (8.140) one has the luxury of using other functions in place of the squared Euclidean distance from $\boldsymbol{\theta}_{n-1}$.

A popular auxiliary function that has been exploited is the *Bregman divergence*. The Bregman divergence with respect to a function, say, ψ, between two points $\boldsymbol{x}, \boldsymbol{y}$ is defined as

$$\boxed{B_\psi(\boldsymbol{x}, \boldsymbol{y}) = \psi(\boldsymbol{x}) - \psi(\boldsymbol{y}) - \langle \nabla\psi(\boldsymbol{y}), \boldsymbol{x} - \boldsymbol{y}\rangle : \quad \text{Bregman divergence.}} \tag{8.143}$$

It is left as a simple exercise to verify that the Euclidean distance results as the Bregman divergence if $\psi(\boldsymbol{x}) = ||\boldsymbol{x}||^2$.

Another algorithmic variant is the so-called *composite mirror descent*, which employs the currently available estimate of the subgradient, instead of the average, combined with the Bregman divergence; that is, $\bar{\mathcal{L}}$ is replaced by $\mathcal{L}'_{n-1}$ and $\psi(\boldsymbol{\theta})$ by $B_\psi(\boldsymbol{\theta}, \boldsymbol{\theta}_{n-1})$ for some function ψ [37]. In [38], a time-varying ψ_n is involved by using a weighted average of the Euclidean norm, as pointed out in Section 8.10.3. Note that in these modifications, although they may look simple, the analysis of the respective algorithms can be quite hard and substantial differences can be obtained in the performance.

At the time the current edition of the book is being compiled, this area is a hot topic of research, and it is still early to draw definite conclusions concerning the relative performance benefits among the various schemes. It may turn out, as is often the case, that different algorithms are better suited for different applications and data sets.

8.15 DISTRIBUTED OPTIMIZATION: SOME HIGHLIGHTS

Distributed optimization (e.g., [86]) has already been discussed in the context of the stochastic gradient descent with an emphasis on LMS and diffusion-type algorithms in Chapter 5 and also in Section 8.9. At the heart of distributed or decentralized optimization lies a cost function,

$$J(\boldsymbol{\theta}) = \sum_{k=1}^{K} J_k(\boldsymbol{\theta}),$$

which is defined over a connected network of K nodes (agents). Each J_k, $k = 1, 2, \ldots, K$, is fed with the kth node's data and quantifies the contribution of the corresponding agent to the overall cost. The goal is to compute a *common*, to all nodes, optimizer, $\boldsymbol{\theta}_*$. Moreover, each node communicates its updates, in the context of an iterative algorithm, *only* to its neighbors, instead of a common fusion center.

The ith step of the general iterative scheme of the so-called consensus algorithms (Section 5.13.4) is of the form

$$\boldsymbol{\theta}_k^{(i)} = \sum_{m \in \mathcal{N}_k} a_{mk} \boldsymbol{\theta}_m^{(i-1)} + \mu_i \nabla J_k \left(\boldsymbol{\theta}_k^{(i-1)} \right), \quad k = 1, 2, \ldots, K, \tag{8.144}$$

where $\mathcal{N}_k$ is the set of indices corresponding to the neighbors of the kth agent, that is, the nodes with which the kth node shares links in the corresponding graph (see Section 5.13). Matrix $A = [a_{mk}]$, $m, k = 1, 2, \ldots, K$, is a mixing matrix, which has to satisfy certain properties, e.g., to be left or doubly stochastic (Section 5.13.2).

In the above framework, a number of algorithms have been suggested based on a diminishing step size sequence that converge to a solution $\boldsymbol{\theta}_*$, i.e.,

$$\boldsymbol{\theta}_* = \arg\min_{\boldsymbol{\theta}} J(\boldsymbol{\theta}).$$

For example in [39], under the assumption of bounded gradients and a step size of $\mu_i = \frac{1}{\sqrt{i}}$, a convergence rate of $\mathcal{O}(\frac{\ln i}{i})$ is attained. Similar results hold for the so-called *push* method that is implemented on a dynamic graph [62]. Improved rates are obtained under the assumption of strong convexity [63].

In [85], a modification known as the EXTRA version of the basic iterative scheme, given in Eq. (8.144), is introduced, which adopts a constant step size, $\mu_i = \mu$, and it can achieve a convergence rate of $\mathcal{O}(\frac{1}{i})$ and a linear convergence rate under the strong convexity assumption for the cost function. In [110,111], a relation for diffusion-type algorithms is developed, which also relaxes some of the assumptions, concerning matrix A, compared to the original NEXT algorithm.

Distributed optimization is a topic with ongoing intense research at the time the second edition of book is compiled. Our goal in this section was not to present and summarize the related literature but more to make the reader alert of some key contributions and directions in the field.

PROBLEMS

8.1 Prove the Cauchy–Schwarz inequality in a general Hilbert space.

8.2 Show (a) that the set of points in a Hilbert space $\mathbb{H}$,

$$C = \{\boldsymbol{x} : \|\boldsymbol{x}\| \leq 1\},$$

is a convex set, and (b) that the set of points

$$C = \{\boldsymbol{x} : \|\boldsymbol{x}\| = 1\}$$

is a nonconvex one.

8.3 Show the first-order convexity condition.

8.4 Show that a function f is convex if the one-dimensional function

$$g(t) := f(\boldsymbol{x} + t\boldsymbol{y})$$

is convex, $\forall \boldsymbol{x}, \boldsymbol{y}$ in the domain of definition of f.

8.5 Show the second-order convexity condition.
Hint: Show the claim first for the one-dimensional case, and then use the result of the previous problem for the generalization.

8.6 Show that a function

$$f : \mathbb{R}^l \longmapsto \mathbb{R}$$

is convex if and only if its epigraph is convex.

8.7 Show that if a function is convex, then its lower level set is convex for any ξ.

8.8 Show that in a Hilbert space $\mathbb{H}$, the parallelogram rule

$$\|\boldsymbol{x} + \boldsymbol{y}\|^2 + \|\boldsymbol{x} - \boldsymbol{y}\|^2 = 2\left(\|\boldsymbol{x}\|^2 + \|\boldsymbol{y}\|^2\right), \quad \forall \boldsymbol{x}, \boldsymbol{y} \in \mathbb{H},$$

holds true.

8.9 Show that if $\boldsymbol{x}, \boldsymbol{y} \in \mathbb{H}$, where $\mathbb{H}$ is a Hilbert space, then the inner product-induced norm satisfies the triangle inequality, as required by any norm, that is,

$$\|\boldsymbol{x} + \boldsymbol{y}\| \leq \|\boldsymbol{x}\| + \|\boldsymbol{y}\|.$$

8.10 Show that if a point $\boldsymbol{x}_*$ is a local minimizer of a convex function, it is necessarily a global one. Moreover, it is the unique minimizer if the function is strictly convex.

8.11 Let C be a closed convex set in a Hilbert space $\mathbb{H}$. Then show that $\forall \boldsymbol{x} \in \mathbb{H}$, there exists a point, denoted as $P_C(\boldsymbol{x}) \in C$, such that

$$\|\boldsymbol{x} - P_C(\boldsymbol{x})\| = \min_{\boldsymbol{y} \in C} \|\boldsymbol{x} - \boldsymbol{y}\|.$$

8.12 Show that the projection of a point $\boldsymbol{x} \in \mathbb{H}$ onto a nonempty closed convex set, $C \subset \mathbb{H}$, lies on the boundary of C.

8.13 Derive the formula for the projection onto a hyperplane in a (real) Hilbert space $\mathbb{H}$.

8.14 Derive the formula for the projection onto a closed ball $B[\boldsymbol{0}, \delta]$.

8.15 Find an example of a point whose projection on the ℓ_1 ball is not unique.

8.16 Show that if $C \subset \mathbb{H}$ is a closed convex set in a Hilbert space, then $\forall \boldsymbol{x} \in \mathbb{H}$ and $\forall \boldsymbol{y} \in C$, the projection $P_C(\boldsymbol{x})$ satisfies the following properties:

- $\text{Real}\{\langle \boldsymbol{x} - P_C(\boldsymbol{x}),\ \boldsymbol{y} - P_C(\boldsymbol{x})\rangle\} \leq 0$,
- $\|P_C(\boldsymbol{x}) - P_C(\boldsymbol{y})\|^2 \leq \text{Real}\{\langle \boldsymbol{x} - \boldsymbol{y},\ P_C(\boldsymbol{x}) - P_C(\boldsymbol{y})\rangle\}$.

8.17 Prove that if S is a closed subspace $S \subset \mathbb{H}$ in a Hilbert space $\mathbb{H}$, then $\forall \boldsymbol{x}, \boldsymbol{y} \in \mathbb{H}$,

$$\langle \boldsymbol{x}, P_S(\boldsymbol{y})\rangle = \langle P_S(\boldsymbol{x}), \boldsymbol{y}\rangle = \langle P_S(\boldsymbol{x}), P_S(\boldsymbol{y})\rangle$$

and

$$P_S(a\boldsymbol{x} + b\boldsymbol{y}) = aP_S(\boldsymbol{x}) + bP_S(\boldsymbol{y}).$$

Hint: Use the result of Problem 8.18.

8.18 Let S be a closed convex subspace in a Hilbert space $\mathbb{H}$, $S \subset \mathbb{H}$. Let $S^{\perp}$ be the set of all elements $\boldsymbol{x} \in \mathbb{H}$ which are orthogonal to S. Then show that (a) $S^{\perp}$ is also a closed subspace, (b) $S \cap S^{\perp} = \{\boldsymbol{0}\}$, (c) $\mathbb{H} = S \oplus S^{\perp}$, that is, $\forall \boldsymbol{x} \in \mathbb{H}$, $\exists \boldsymbol{x}_1 \in S$ and $\boldsymbol{x}_2 \in S^{\perp}$: $\boldsymbol{x} = \boldsymbol{x}_1 + \boldsymbol{x}_2$, where $\boldsymbol{x}_1, \boldsymbol{x}_2$ are *unique*.

8.19 Show that the relaxed projection operator is a nonexpansive mapping.

8.20 Show that the relaxed projection operator is a strongly attractive mapping.

8.21 Give an example of a sequence in a Hilbert space $\mathbb{H}$ which converges weakly but not strongly.

8.22 Prove that if $C_1 \dots C_K$ are closed convex sets in a Hilbert space $\mathbb{H}$, then the operator

$$T = T_{C_K} \cdots T_{C_1}$$

is a *regular* one; that is,

$$\|T^{n-1}(\boldsymbol{x}) - T^n(\boldsymbol{x})\| \longrightarrow 0,\ n \longrightarrow \infty,$$

where $T^n := TT \dots T$ is the application of T n successive times.

8.23 Show the fundamental POCS theorem for the case of closed subspaces in a Hilbert space $\mathbb{H}$.

8.24 Derive the subdifferential of the metric distance function $d_C(\boldsymbol{x})$, where C is a closed convex set $C \subseteq \mathbb{R}^l$ and $\boldsymbol{x} \in \mathbb{R}^l$.

8.25 Derive the bound in (8.55).

8.26 Show that if a function is γ-Lipschitz, then any of its subgradients is bounded.

8.27 Show the convergence of the generic projected subgradient algorithm in (8.61).

8.28 Derive Eq. (8.100).

8.29 Consider the online version of PDMb in (8.64), that is,

$$\boldsymbol{\theta}_n = \begin{cases} P_C\left(\boldsymbol{\theta}_{n-1} - \mu_n \frac{J(\boldsymbol{\theta}_{n-1})}{\|J'(\boldsymbol{\theta}_{n-1})\|^2} J'(\boldsymbol{\theta}_{n-1})\right), & \text{if } J'(\boldsymbol{\theta}_{n-1}) \neq \boldsymbol{0}, \\ P_C(\boldsymbol{\theta}_{n-1}), & \text{if } J'(\boldsymbol{\theta}_{n-1}) = \boldsymbol{0}, \end{cases} \tag{8.145}$$

where we have assumed that $J_* = 0$. If this is not the case, a shift can accommodate for the difference. Thus, we assume that we know the minimum. For example, this is the case for a number of tasks, such as the hinge loss function, assuming linearly separable classes, or the linear ϵ-insensitive loss function, for bounded noise. Assume that

$$\mathcal{L}_n(\boldsymbol{\theta}) = \sum_{k=n-q+1}^{n} \frac{\omega_k d_{C_k}(\boldsymbol{\theta}_{n-1})}{\sum_{k=n-q+1}^{n} \omega_k d_{C_k}(\boldsymbol{\theta}_{n-1})} d_{C_k}(\boldsymbol{\theta}).$$

Then derive the APSM algorithm of (8.39).

8.30 Derive the regret bound for the subgradient algorithm in (8.83).

8.31 Show that a function $f(\boldsymbol{x})$ is σ-strongly convex if and only if the function $f(\boldsymbol{x}) - \frac{\sigma}{2}||\boldsymbol{x}||^2$ is convex.

8.32 Show that if the loss function is σ-strongly convex, then if $\mu_n = \frac{1}{\sigma n}$, the regret bound for the subgradient algorithm becomes

$$\frac{1}{N}\sum_{n=1}^{N} \mathcal{L}_n(\boldsymbol{\theta}_{n-1}) \leq \frac{1}{N}\sum_{n=1}^{N} \mathcal{L}_n(\boldsymbol{\theta}_*) + \frac{G^2(1+\ln N)}{2\sigma N}. \tag{8.146}$$

8.33 Consider a batch algorithm that computes the minimum of the empirical loss function, $\boldsymbol{\theta}_*(N)$, having a quadratic convergence rate, that is,

$$\ln \ln \frac{1}{||\boldsymbol{\theta}^{(i)} - \boldsymbol{\theta}_*(N)||^2} \sim i.$$

Show that an online algorithm, running for n time instants so as to spend the same computational processing resources as the batch one, achieves for large values of N better performance than the batch algorithm, shown as [12]

$$||\boldsymbol{\theta}_n - \boldsymbol{\theta}_*||^2 \sim \frac{1}{N \ln \ln N} << \frac{1}{N} \sim ||\boldsymbol{\theta}_*(N) - \boldsymbol{\theta}_*||^2.$$

Hint: Use the fact that

$$||\boldsymbol{\theta}_n - \boldsymbol{\theta}_*||^2 \sim \frac{1}{n} \quad \text{and} \quad ||\boldsymbol{\theta}_*(N) - \boldsymbol{\theta}_*||^2 \sim \frac{1}{N}.$$

8.34 Show property (8.111) for the proximal operator.

8.35 Show property (8.112) for the proximal operator.

8.36 Prove that the recursion in (8.118) converges to a minimizer of f.

8.37 Derive (8.122) from (8.121).

MATLAB® EXERCISES

8.38 Consider the regression model,

$$y_n = \boldsymbol{\theta}_o^T \boldsymbol{x}_n + \eta_n,$$

where $\boldsymbol{\theta}_o \in \mathbb{R}^{200}$ ($l = 200$) and the coefficients of the unknown vector are obtained randomly via the Gaussian distribution $\mathcal{N}(0, 1)$. The noise samples are also i.i.d., having zero mean and variance $\sigma_\eta^2 = 0.01$. The input sequence is a white noise one, i.i.d. generated via the Gaussian $\mathcal{N}(0, 1)$.

Using as training data the samples $(y_n, \boldsymbol{x}_n) \in \mathbb{R} \times \mathbb{R}^{200}$, $n = 1, 2, \ldots$, run the APA (Algorithm 5.2), the NLMS (Algorithm 5.3), the RLS (Algorithm 6.1), and the APSM (Algorithm 8.2) algorithms to estimate the unknown $\boldsymbol{\theta}_o$.

For the APA, choose $\mu = 0.2$, $\delta = 0.001$, and $q = 30$. For the APSM, $\mu = 0.5 \times M_n$, $\epsilon = \sqrt{2}\sigma$, and $q = 30$. Furthermore, in the NLMS set $\mu = 1.2$ and $\delta = 0.001$. Finally, for the RLS set the forgetting factor β equal to 1. Run 100 independent experiments and plot the average error per iteration in dBs, that is, $10\log_{10}(e_n^2)$, where $e_n^2 = (y_n - \boldsymbol{x}_n^T \boldsymbol{\theta}_{n-1})^2$. Compare the performance of the algorithms.

Keep the same parameters, but alter the noise variance so that it becomes 0.3. Plot the average error per iteration as in the previous experiment. What do you observe regarding the performance of the APA compared to the previous low-noise scenario?

Keep playing with different parameters and study the effect on the convergence speed and the error floor in which the algorithms converge.

8.39 Create an ad hoc network having 10 nodes and a total number of 32 connections. Generate at each node the data which adhere to the following model:

$$y_k(n) = \boldsymbol{\theta}_o^T \boldsymbol{x}_k(n) + \eta_k(n), \quad k = 1, \ldots, 10.$$

The unknown vector $\boldsymbol{\theta}_o \in \mathbb{R}^{60}$ and its coefficients are generated randomly via the Gaussian $\mathcal{N}(0, 1)$. The input vectors are i.i.d. and follow an $\mathcal{N}(0, 1)$. Moreover, the noise samples are i.i.d. generated from zero mean Gaussians with variances corresponding to different signal-to-noise levels, varying randomly from 20 to 25 dBs from node to node.

For the unknown vector estimation employ the combine-then-adapt diffusion APSM (Algorithm 8.3), adapt-then-combine LMS (Algorithm 5.7), combine-then-adapt LMS (Algorithm 5.8), and noncooperative LMS (Algorithm 5.1) algorithms. For the combine-then-adapt APSM, set $\mu_n = 0.5 \times M_n$, $\epsilon_k = \sqrt{2}\sigma_k$, and $q = 20$. For the adapt-then-combine, combine-then-adapt, and noncooperative LMS, set the step size equal to 0.03. Finally, choose the combination weights a_{mk} with respect to the Metropolis rule (Remarks 5.4).

Run 100 independent experiments and plot the average MSD per iteration in dBs, that is,

$$\mathrm{MSD}(n) = 10\log_{10}\left(\frac{1}{K}\sum_{k=1}^{K} \|\boldsymbol{\theta}_k(n) - \boldsymbol{\theta}_o\|^2\right).$$

Compare the performance of the combine-then-adapt APSM with the performance of the LMS-based algorithms.

Keep playing with different parameters for the involved algorithms and observe their influence on the obtained performance.

8.40 Download the *banknote authentication* data set.[11] Develop a MATLAB® program that implements the PEGASOS algorithm for classification (Algorithm 8.4). Keep 90% of the data as a

[11] https://archive.ics.uci.edu/ml/datasets/banknote+authentication.

training set and the remaining 10% as a test set. Set $\lambda = 0.1$ and $m = 1, 10, 30$. Once the training phase has been completed, freeze the parameters for the obtained classifier. Compute the classification error on the test set using the obtained classifier. Compare the classification error for the three choices of m.

REFERENCES

[1] A. Agarwal, P. Bartlett, P. Ravikumar, M.J. Wainwright, Information-theoretic lower bounds on the oracle complexity of convex optimization, IEEE Trans. Inf. Theory 58 (5) (2012) 3235–3249.

[2] A.E. Albert, L.A. Gardner, Stochastic Approximation and Nonlinear Regression, MIT Press, 1967.

[3] S. Amari, Natural gradient works efficiently in learning, Neural Comput. 10 (2) (1998) 251–276.

[4] F. Bach, E. Moulines, Non-strongly-convex smooth stochastic approximation with convergence rate $O(1/n)$, arXiv:1306.2119v1 [cs.LG], 2013.

[5] F. Bach, Adaptivity of averaged stochastic gradient descent to local strong convexity for logistic regression, J. Mach. Learn. Res. 15 (2014) 595–627.

[6] H.H. Bauschke, J.M. Borwein, On projection algorithms for solving convex feasibility problems, SIAM Rev. 38 (3) (1996) 367–426.

[7] H.H. Bauschke, P.L. Combettes, Convex Analysis and Monotone Operator Theory in Hilbert Spaces, Springer, 2011.

[8] A. Beck, M. Teboulle, Gradient-based algorithms with applications to signal recovery problems, in: D. Palomar, Y. Eldar (Eds.), Convex Optimization in Signal Processing and Communications, Cambridge University Press, 2010, pp. 42–88.

[9] A. Beck, M. Teboulle, Mirror descent and nonlinear projected subgradient methods for convex optimization, Oper. Res. Lett. 31 (2003) 167–175.

[10] D.P. Bertsekas, Nonlinear Programming, second ed., Athena Scientific, 1999.

[11] D.P. Bertsekas, A. Nedic, A.E. Ozdaglar, Convex Analysis and Optimization, Athena Scientific, 2003.

[12] L. Bottou, Y. Le Cun, Large scale online learning, in: Advances in Neural Information Processing Systems, NIPS, MIT Press, 2003, pp. 2004–2011.

[13] L. Bottou, O. Bousquet, The tradeoffs of large scale learning, Adv. Neural Inf. Process. Syst. 20 (2007) 161–168.

[14] L. Bottou, Large-scale machine learning with stochastic gradient descent, in: Y. Lechevallier, G. Saporta (Eds.), Proceedings 19th Intl. Conference on Computational Statistics, COMPSTAT 2010, Springer, Paris, France, 2010.

[15] S. Boyd, L. Vandenberghe, Convex Optimization, Cambridge University Press, 2004.

[16] S. Boyd, N. Parikh, E. Chu, P. Peleato, J. Eckstein, Distributed optimization and statistical learning via the alternating direction method of multipliers, Found. Trends Mach. Learn. 3 (1) (2011) 1122.

[17] L.M. Bregman, The method of successive projections for finding a common point of convex sets, Sov. Math. Dokl. 6 (1965) 688–692.

[18] R. Bruck, An iterative solution of a variational inequality for certain monotone operator in a Hilbert space, Bull. Am. Math. Soc. 81 (5) (1975) 890–892.

[19] N. Cesa-Bianchi, A. Conconi, C. Gentile, On the generalization ability of on-line learning algorithms, IEEE Trans. Inf. Theory 50 (9) (2004) 2050–2057.

[20] R.L.G. Cavalcante, I. Yamada, B. Mulgrew, An adaptive projected subgradient approach to learning in diffusion networks, IEEE Trans. Signal Process. 57 (7) (2009) 2762–2774.

[21] N. Cesa-Bianchi, G. Lugosi, Prediction, Learning, and Games, Cambridge University Press, 2006.

[22] S. Chouvardas, K. Slavakis, S. Theodoridis, Adaptive robust distributed learning in diffusion sensor networks, IEEE Trans. Signal Process. 59 (10) (2011) 4692–4707.

[23] S. Chouvardas, K. Slavakis, S. Theodoridis, I. Yamada, Stochastic analysis of hyperslab-based adaptive projected subgradient method under boundary noise, IEEE Signal Process. Lett. 20 (7) (2013) 729–732.

[24] P.L. Combettes, The foundations of set theoretic estimation, Proc. IEEE 81 (2) (1993) 182–208.

[25] P.L. Combettes, Inconsistent signal feasibility problems: least-squares solutions in a product space, IEEE Trans. Signal Process. 42 (11) (1994) 2955–2966.

[26] P.L. Combettes, V.R. Wajs, Signal recovery by proximal forward-backward splitting, Multiscale Model. Simul. 4 (2005) 1168–1200.

[27] P.L. Combettes, J.-C. Pesquet, A Douglas-Rachford splitting approach to nonsmooth convex variational signal recovery, IEEE J. Sel. Top. Signal Process. 1 (2007) 564–574.
[28] P.L. Combettes, J.-C. Pesquet, Proximal splitting methods in signal processing, in: H.H. Bauschke, R.S. Burachik, P.L. Combettes, V. Elser, D.R. Luke, H. Wolkowicz (Eds.), Fixed-Point Algorithms for Inverse Problems in Science and Engineering, Springer-Verlag, 2011, pp. 185–212.
[29] J.R. Deller, Set-membersip identification in digital signal processing, IEEE Signal Process. Mag. 6 (1989) 4–20.
[30] F. Deutsch, Best Approximation in Inner Product Spaces, CMS, Springer, 2000.
[31] L. Devroye, L. Györfi, G. Lugosi, A Probabilistic Theory of Pattern Recognition, Springer, 1991.
[32] P.S.R. Diniz, S. Werner, Set-membership binormalized data-reusing LMS algorithms, IEEE Trans. Signal Process. 52 (1) (2003) 124–134.
[33] P.S.R. Diniz, Adaptive Filtering: Algorithms and Practical Implementation, fourth ed., Springer Verlag, 2014.
[34] P.S.R. Diniz, Convergence performance of the simplified set-membership affine projection algorithm, J. Circuits Syst. Signal Process. 30 (2) (2011) 439–462.
[35] J. Duchi, S.S. Shwartz, Y. Singer, T. Chandra, Efficient projections onto the ℓ_1-ball for learning in high dimensions, in: Proceedings of the International Conference on Machine Leaning, ICML, 2008, pp. 272–279.
[36] J. Duchi, Y. Singer, Efficient online and batch learning using forward backward splitting, J. Mach. Learn. Res. 10 (2009) 2899–2934.
[37] J. Duchi, S. Shalev-Shwartz, Y. Singer, A. Tewari, Composite objective mirror descent, in: Proceedings of the 23rd Annual Conference on Computational Learning Theory, 2010.
[38] J. Duchi, E. Hazan, Y. Singer, Adaptive subgradient methods for online learning and stochastic optimization, J. Mach. Learn. Res. 12 (2011) 2121–2159.
[39] J. Duchi, A. Agrawal, M. Wainright, Dual averaging for distributed optimization: convergence analysis and network scaling, IEEE Trans. Autom. Control 57 (3) (2012) 592–606.
[40] M. Fortin, R. Glowinski, Augmented Lagrangian Methods: Applications to the Numerical Solution of Boundary-Value Problems, Elsevier Science/North-Holland, Amsterdam, 1983.
[41] A.A. Goldstein, Convex programming in Hilbert spaces, Bull. Am. Math. Soc. 70 (5) (1964) 709–710.
[42] L.G. Gubin, B.T. Polyak, E.V. Raik, The method of projections for finding the common point of convex sets, USSR Comput. Math. Phys. 7 (6) (1967) 1–24.
[43] E. Hazan, A. Agarwal, S. Kale, Logarithmic regret algorithms for online convex optimization, Mach. Learn. 69 (2–3) (2007) 169–192.
[44] J.B. Hiriart-Urruty, C. Lemarechal, Convex Analysis and Minimization Algorithms, Springer-Verlag, Berlin, 1993.
[45] J. Kivinen, A.J. Smola, R.C. Williamson, Online learning with kernels, IEEE Trans. Signal Process. 52 (8) (2004) 2165–2176.
[46] N. Komodakis, J.-C. Peusquet, Playing with duality: an overview of recent primal-dual approaches for solving large scale optimization problems, arXiv:1406.5429v1 [cs.NA], 20 June 2014.
[47] Y. Kopsinis, K. Slavakis, S. Theodoridis, Online sparse system identification and signal reconstruction using projections onto weighted ℓ_1-balls, IEEE Trans. Signal Process. 59 (3) (2011) 936–952.
[48] J. Langford, L. Li, T. Zhang, Sparse online learning via truncated gradient, J. Mach. Learn. Res. 10 (2009) 747–776.
[49] Y. LeCun, L. Bottou, G.B. Orr, K.R. Müller, Efficient BackProp, in: G.B. Orr, K.-R. Müller (Eds.), Neural Networks: Tricks of the Trade, Springer, 1998, pp. 9–50.
[50] N. Le Roux, M. Schmidt, F. Bach, A stochastic gradient method with an exponential convergence rate for finite training sets, arXiv:1202.6258v4 [math.OC], 2013.
[51] W.S. Lee, P.L. Bartlett, R.C. Williamson, The importance of convexity in learning with squared loss, IEEE Trans. Inf. Theory 44 (5) (1998) 1974–1980.
[52] B.S. Levitin, B.T. Polyak, Constrained minimization methods, Zh. Vychisl. Mat. Mat. Fiz. 6 (5) (1966) 787–823.
[53] D.D. Lewis, Y. Yang, T.G. Rose, F. Li, RCV1: a new benchmark collection for text categorization research, J. Mach. Learn. Res. 5 (2004) 361–397.
[54] P. Lions, B. Mercier, Splitting algorithms for the sum of two nonlinear operators, SIAM J. Numer. Anal. 16 (1979) 964–979.
[55] P.E. Maingé, Strong convergence of projected subgradient methods for nonsmooth and nonstrictly convex minimization, Set-Valued Anal. 16 (2008) 899–912.
[56] N. Murata, S. Amari, Statistical analysis of learning dynamics, Signal Process. 74 (1) (1999) 3–28.

[57] B. Martinet, Régularisation d' inéquations variationnelles par approximations successives, Rev. Fr. Inform. Rech. Oper. 4 (1970) 154–158.

[58] C. Moler, Iterative refinement in floating point, J. ACM 14 (2) (1967) 316–321.

[59] J.J. Moreau, Fonctions convexes duales et points proximaux dans un espace Hilbertien, Rep. Paris Acad. Sci. A 255 (1962) 2897–2899.

[60] J.J. Moreau, Proximité et dualité dans un espace hilbertien, Bull. Soc. Math. Fr. 93 (1965) 273–299.

[61] S. Nagaraj, S. Gollamudi, S. Kapoor, Y.F. Huang, BEACON: an adaptive set-membership filtering technique with sparse updates, IEEE Trans. Signal Process. 47 (11) (1999) 2928–2941.

[62] A. Nedic, A. Olshevsky, Distributed optimization over time varying directed graphs, in: Proceedings 52nd IEEE Conference on Decision and Control, 2013, pp. 6855–6860.

[63] A. Nedic, A. Olshevsky, Stochastic gradient-push for strongly convex functions on time varying directed graphs, IEEE Trans. Autom. Control 61 (12) (2016) 3936–3947.

[64] A. Nemirovsky, D. Yudin, Problem Complexity and Method Efficiency in Optimization, J. Wiley & Sons, New York, 1983.

[65] Y.E. Nesterov, A method of solving a convex programming problem with convergence rate $O(1/k^2)$, Sov. Math. Dokl. 27 (2) (1983) 372–376.

[66] Y. Nesterov, Primal-dual subgradient methods for convex problems, Math. Program. 120 (1) (2009) 221–259.

[67] N. Parish, S. Boyd, Proximal algorithms, Found. Trends Oprim. 1 (2) (2013) 123–231.

[68] G. Pierra, Decomposition through formalization in a product space, Math. Program. 8 (1984) 96–115.

[69] T. Poggio, S. Voinea, L. Rosasco, Online learning, stability and stochastic gradient descent, arXiv:1105.4701 [cs.LG], Sep 2011.

[70] B.T. Polyak, Minimization of unsmooth functionals, Zh. Vychisl. Mat. Mat. Fiz. 9 (3) (1969) 509–521.

[71] R.T. Rockafellar, Convex Analysis, Princeton University Press, Princeton, NJ, 1970.

[72] R.T. Rockafellar, Monotone operators and the proximal point algorithm, SIAM J. Control Optim. 14 (1976) 877–898.

[73] S. Shalev-Shwartz, Y. Singer, N. Srebro, A. Cotter, PEGASOS: primal estimated sub-gradient solver for SVM, Math. Program. B 127 (2011) 3–30.

[74] N.Z. Shor, On the Structure of Algorithms for the Numerical Solution of Optimal Planning and Design Problems, PhD Thesis, Cybernetics Institute, Academy of Sciences, Kiev, 1964.

[75] N.Z. Shor, Minimization Methods for Non-differentiable Functions, Springer Series in Computational Mathematics, Springer, 1985.

[76] K. Slavakis, I. Yamada, N. Ogura, The adaptive projected subgradient method over the fixed point set of strongly attracting nonexpansive mappings, Numer. Funct. Anal. Optim. 27 (7–8) (2006) 905–930.

[77] K. Slavakis, I. Yamada, Robust wideband beamforming by the hybrid steepest descent method, IEEE Trans. Signal Process. 55 (9) (2007) 4511–4522.

[78] K. Slavakis, S. Theodoridis, I. Yamada, Online classification using kernels and projection-based adaptive algorithms, IEEE Trans. Signal Process. 56 (7) (2008) 2781–2797.

[79] K. Slavakis, S. Theodoridis, Sliding window generalized kernel affine projection algorithm using projection mappings, EURASIP J. Adv. Signal Process. 2008 (2008) 830381, https://doi.org/10.1155/2008/830381.

[80] K. Slavakis, S. Theodoridis, I. Yamada, Adaptive constrained learning in reproducing kernel Hilbert spaces, IEEE Trans. Signal Process. 5 (12) (2009) 4744–4764.

[81] K. Slavakis, I. Yamada, The adaptive projected subgradient method constrained by families of quasi-nonexpansive mappings and its application to online learning, SIAM J. Optim. 23 (1) (2013) 126–152.

[82] K. Slavakis, A. Bouboulis, S. Theodoridis, Online learning in reproducing kernel Hilbert spaces, in: S. Theodoridis, R. Chellapa (Eds.), E-Reference for Signal Processing, Academic Press, 2013.

[83] K. Slavakis, Y. Kopsinis, S. Sheodoridis, S. McLaughlin, Generalized thresholding and online sparsity-aware learning in a union of subspaces, IEEE Trans. Signal Process. 61 (15) (2013) 3760–3773.

[84] K. Slavakis, Personal Communication, March 2014.

[85] W. Shi, Q. Ling, G. Wu, W. Yin, EXTRA: an exact first order algorithm for decentralized consensus optimization, SIAM J. Optim. 25 (2) (2015) 944–966.

[86] J. Tsitsiklis, Problems in Decentralized Decision Making and Computation, PhD Thesis, Department Electrical and Computer Engineering, MIT, 1984.

[87] H. Stark, Y. Yang, Vector Space Projections, John Wiley, 1998.

[88] S. Theodoridis, K. Koutroumbas, Pattern Recognition, fourth ed., Academic Press, 2009.

[89] S. Theodoridis, K. Slavakis, I. Yamada, Adaptive learning in a world of projections, IEEE Signal Process. Mag. 28 (1) (2011) 97–123.
[90] M. Todd, Semidefinite optimization, Acta Numer. 10 (2001) 515–560.
[91] O.P. Tseng, An incremental gradient (projection) method with momentum term and adaptive step size rule, SIAM J. Optim. 8 (2) (1998) 506–531.
[92] A.B. Tsybakov, Optimal aggregation of classifiers in statistical learning, Ann. Stat. 32 (1) (2004) 135–166.
[93] Y. Tsypkin, Foundation of the Theory of Learning Systems, Academic Press, 1973.
[94] V.N. Vapnik, Estimation of Dependences Based on Empirical Data, Springer Series in Statistics, Springer-Verlag, Berlin, 1982.
[95] V.N. Vapnik, Statistical Learning Theory, John Wiley & Sons, 1998.
[96] V.N. Vapnik, The Nature of Statistical Learning Theory, Springer, 2000.
[97] R.S. Varga, Matrix Iterative Analysis, second ed., Springer-Verlag, New York, 2000.
[98] J. von Neumann, Functional Operators, vol II. The Geometry of Orthogonal Spaces, Ann. Math. Stud., vol. 22, Princeton Univ. Press, NJ, 1950 (Reprint of lecture notes first distributed in 1933).
[99] S. Werner, P.S.R. Diniz, Set-membership affine projection algorithm, IEEE Signal Process. Lett. 8 (8) (2001) 231–235.
[100] L. Xiao, Dual averaging methods for regularized stochastic learning and online optimization, J. Mach. Learn. Res. 11 (2010) 2543–2596.
[101] I. Yamada, The hybrid steepest descent method for the variational inequality problem over the intersection of fixed point sets of nonexpansive mappings, Stud. Comput. Math. 8 (2001) 473–504.
[102] I. Yamada, K. Slavakis, K. Yamada, An efficient robust adaptive filtering algorithm based on parallel subgradient projection techniques, IEEE Trans. Signal Process. 50 (5) (2002) 1091–1101.
[103] I. Yamada, Adaptive projected subgradient method: a unified view of projection based adaptive algorithms, J. IEICE 86 (6) (2003) 654–658 (in Japanese).
[104] I. Yamada, N. Ogura, Adaptive projected subgradient method for asymptotic minimization of nonnegative convex functions, Numer. Funct. Anal. Optim. 25 (7–8) (2004) 593–617.
[105] I. Yamada, N. Ogura, Hybrid steepest descent method for variational inequality problem over the fixed point set of certain quasi-nonexpansive mappings, Numer. Funct. Anal. Optim. 25 (2004) 619–655.
[106] I. Yamada, S. Gandy, M. Yamagishi, Sparsity-aware adaptive filtering based on Douglas-Rachfold splitting, in: Proceedings, 19th European Signal Processing Conference, EUSIPCO, Barcelona, Spain, 2011.
[107] M. Yamagishi, M. Yukawa, I. Yamada, Acceleration of adaptive proximal forward-backward slitting method and its application in systems identification, in: Proceedings IEEE International Conference on Acoustics Speech and Signal Processing, ICASSP, Prague, 2011.
[108] I. Yamada, M. Yukawa, M. Yamagishi, Minimizing the Moreau envelope of non-smooth convex functions over the fixed point set of certain quasi-nonexpansive mappings, in: H.H. Bauschke, R.S. Burachik, P.L. Combettes, V. Elser, D.R. Luke, H. Wolkowicz (Eds.), Fixed-Point Algorithms for Inverse Problems in Science and Engineering, Springer, New York, 2011, pp. 345–390.
[109] K. Yosida, Functional Analysis, Springer, 1968.
[110] K. Yuan, B. Ying, X. Zhao, A. Sayed, Exact diffusion for distributed optimization and learning: Part I-Algorithm and development, arXiv:1702.05122v1 [math.OC], 16 Feb. 2017.
[111] K. Yuan, B. Ying, X. Zhao, A. Sayed, Exact diffusion for distributed optimization and learning: Part II-Convergence and analysis, arXiv:1702.05142v1 [math.OC], 16 Feb. 2017.
[112] T. Zhang, Solving large scale linear prediction problems using stochastic gradient descent algorithms, in: Proceedings of the 21st International Conference on Machine Learning, ICML, Banff, Alberta, Canada, 2004, pp. 919–926.
[113] M. Zinkevich, Online convex programming and generalized infinitesimal gradient ascent, in: Proceedings of the 20th International Conference on Machine Learning, ICML, Washington, DC, 2003.

CHAPTER

SPARSITY-AWARE LEARNING: CONCEPTS AND THEORETICAL FOUNDATIONS

9

CONTENTS

9.1 INTRODUCTION

In Chapter 3, the notion of regularization was introduced as a tool to address a number of problems that are usually encountered in machine learning. Improving the performance of an estimator by shrinking the norm of the minimum variance unbiased (MVU) estimator, guarding against overfitting, coping with ill-conditioning, and providing a solution to an underdetermined set of equations are some notable examples where regularization has provided successful answers. Some of the advantages were demon-

Machine Learning. https://doi.org/10.1016/B978-0-12-818803-3.00019-2

strated via the ridge regression concept, where the sum of squared errors cost function was combined, in a tradeoff rationale, with the squared Euclidean norm of the desired solution.

In this and the next chapter, our interest will be on alternatives in the Euclidean norm, and in particular the focus will revolve around the ℓ_1 norm; this is the sum of the absolute values of the components comprising a vector. Although seeking a solution to a problem via the ℓ_1 norm regularization of a cost function has been known and used since the 1970s, it is only recently that it has become the focus of attention of a massive volume of research in the context of compressed sensing. At the heart of this problem lies an underdetermined set of linear equations, which, in general, accepts an infinite number of solutions. However, in a number of cases, an extra piece of information is available: the true model, whose estimate we want to obtain, is sparse; that is, only a few of its coordinates are nonzero. It turns out that a large number of commonly used applications can be cast under such a scenario and can benefit by sparse modeling.

Besides its practical significance, sparsity-aware learning has offered to the scientific community novel theoretical tools and solutions to problems that only a few years ago seemed intractable. This is also a reason that this is an interdisciplinary field of research encompassing scientists from, for example, mathematics, statistics, machine learning, and signal processing. Moreover, it has already been applied in many areas, ranging from biomedicine to communications and astronomy. In this and the following chapters, I made an effort to present in a unifying way the basic notions and ideas that run across this field. The goal is to provide the reader with an overview of the major contributions that have taken place in the theoretical and algorithmic fronts and have been consolidated as a distinct scientific area.

In the current chapter, the focus is on presenting the main concepts and theoretical foundations related to sparsity-aware learning techniques. We start by reviewing various norms. Then we move on to establish conditions on the recovery of sparse vectors, or vectors that are sparse in a transform domain, using less observations than the dimension of the corresponding space. Geometry plays an important part in our approach. Finally, some theoretical advances that tie sparsity and sampling theory are presented. At the end of the chapter, a case study concerning image denoising is discussed.

9.2 SEARCHING FOR A NORM

Mathematicians have been very imaginative in proposing various norms in order to equip linear spaces. Among the most popular norms used in functional analysis are the ℓ_p norms. To tailor things to our needs, given a vector $\boldsymbol{\theta} \in \mathbb{R}^l$, its ℓ_p norm is defined as

$$\|\boldsymbol{\theta}\|_p := \left(\sum_{i=1}^{l} |\theta_i|^p\right)^{1/p}. \tag{9.1}$$

For $p = 2$, the Euclidean or ℓ_2 norm is obtained, and for $p = 1$, (9.1) results in the ℓ_1 norm, that is,

$$\|\boldsymbol{\theta}\|_1 = \sum_{i=1}^{l} |\theta_i|. \tag{9.2}$$

If we let $p \longrightarrow \infty$, then we get the ℓ_∞ norm; let $|\theta_{i_{\max}}| := \max\{|\theta_1|, |\theta_2|, \dots, |\theta_l|\}$, and note that

$$\|\boldsymbol{\theta}\|_\infty := \lim_{p\to\infty} \left(|\theta_{i_{\max}}|^p \sum_{i=1}^{l} \left(\frac{|\theta_i|}{|\theta_{i_{\max}}|} \right)^p \right)^{1/p} = |\theta_{i_{\max}}|, \tag{9.3}$$

that is, $\|\boldsymbol{\theta}\|_\infty$ is equal to the maximum of the absolute values of the coordinates of $\boldsymbol{\theta}$. One can show that all the ℓ_p norms are true norms for $p \geq 1$; that is, satisfy all four requirements that a function $\mathbb{R}^l \longmapsto [0, \infty)$ must respect in order to be called a norm, that is,

1. $\|\boldsymbol{\theta}\|_p \geq 0$,
2. $\|\boldsymbol{\theta}\|_p = 0 \Leftrightarrow \boldsymbol{\theta} = \mathbf{0}$,
3. $\|\alpha\boldsymbol{\theta}\|_p = |\alpha|\, \|\boldsymbol{\theta}\|_p$, $\forall \alpha \in \mathbb{R}$,
4. $\|\boldsymbol{\theta}_1 + \boldsymbol{\theta}_2\|_p \leq \|\boldsymbol{\theta}_1\|_p + \|\boldsymbol{\theta}_2\|_p$.

The third condition enforces the norm function to be (*positively*) *homogeneous* and the fourth one is the *triangle inequality*. These properties also guarantee that any function that is a norm is also a convex one (Problem 9.3). Though strictly speaking, if we allow $p > 0$ to take values less than one in (9.1), the resulting function is not a true norm (Problem 9.8), we may still call them norms, although we know that this is an abuse of the definition of a norm. An interesting case, which will be used extensively in this chapter, is the ℓ_0 norm, which can be obtained as the limit, for $p \longrightarrow 0$, of

$$\|\boldsymbol{\theta}\|_0 := \lim_{p\to 0} \|\boldsymbol{\theta}\|_p^p = \lim_{p\to 0} \sum_{i=1}^{l} |\theta_i|^p = \sum_{i=1}^{l} \chi_{(0,\infty)}(|\theta_i|), \tag{9.4}$$

where $\chi_{\mathcal{A}}(\cdot)$ is the characteristic function with respect to a set $\mathcal{A}$, defined as

$$\chi_{\mathcal{A}}(\tau) := \begin{cases} 1, & \text{if } \tau \in \mathcal{A}, \\ 0, & \text{if } \tau \notin \mathcal{A}. \end{cases}$$

That is, the ℓ_0 norm is equal to the number of nonzero components of the respective vector. It is very easy to check that this function is not a true norm. Indeed, this is not homogeneous, that is, $\|\alpha\boldsymbol{\theta}\|_0 \neq |\alpha|\, \|\boldsymbol{\theta}\|_0$, $\forall \alpha \neq 1$. Fig. 9.1 shows the isovalue curves, in the two-dimensional space, that correspond to $\|\boldsymbol{\theta}\|_p = 1$, for $p = 0, 0.5, 1, 2$, and ∞. Observe that for the Euclidean norm the isovalue curve has the shape of a circle and for the ℓ_1 norm the shape of a rhombus. We refer to them as the ℓ_2 and the ℓ_1 balls, respectively, by slightly "abusing" the meaning of a ball.[1] Observe that in the case of the ℓ_0 norm, the isovalue curve comprises both the horizontal and the vertical axes, excluding the $(0, 0)$ element. If we restrict the size of the ℓ_0 norm to be less than one, then the corresponding set of points becomes a singleton, that is, $(0, 0)$. Also, the set of all the two-dimensional points that have ℓ_0 norm less than or equal to two is the $\mathbb{R}^2$ space. This slightly "strange" behavior is a consequence of the discrete nature of this "norm."

[1] Strictly speaking, a ball must also contain all the points in the interior, that is, all concentric spheres of smaller radius (Chapter 8).

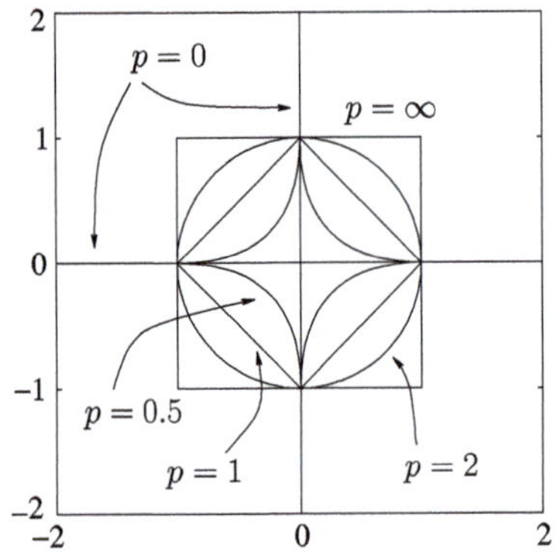

FIGURE 9.1

The isovalue curves for $\|\boldsymbol{\theta}\|_p = 1$ and for various values of p, in the two-dimensional space. Observe that for the ℓ_0 norm, the respective values cover the two axes with the exception of the point $(0, 0)$. For the ℓ_1 norm, the isovalue curve is a rhombus, and for the ℓ_2 (Euclidean) norm, it is a circle.

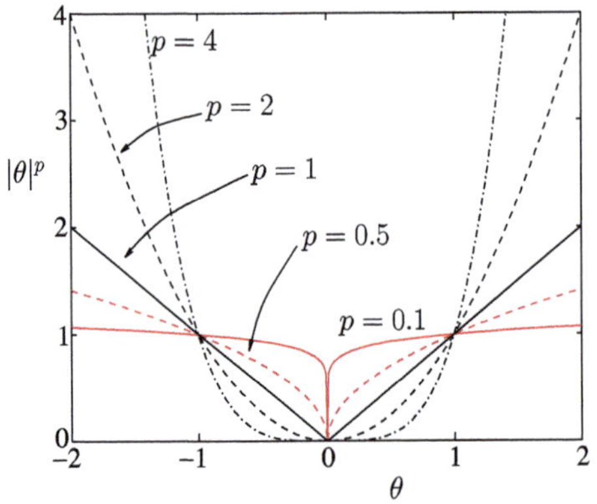

FIGURE 9.2

Observe that the epigraph, that is, the region above the graph, is nonconvex for values $p < 1$, indicating the nonconvexity of the respective $|\theta|^p$ function. The value $p = 1$ is the smallest one for which convexity is retained. Also note that, for large values of $p > 1$, the contribution of small values of $|\theta| < 1$ to the respective norm becomes insignificant.

Fig. 9.2 shows the graph of $|\theta|^p$, which is the individual contribution of each component of a vector to the ℓ_p norm, for different values of p. Observe that (a) for $p < 1$, the region that is formed above the graph (epigraph, see Chapter 8) is not a convex one, which verifies what we have already said, that is, the respective function is not a true norm; and (b) for values of the argument $|\theta| > 1$, the larger the value of $p \geq 1$ and the larger the value of $|\theta|$, the higher the contribution of the respective component to the norm. Hence, if ℓ_p norms, $p \geq 1$, are used in the regularization method, components with large values become the dominant ones and the optimization algorithm will concentrate on these by penalizing them to get smaller so that the overall cost can be reduced. The opposite is true for values $|\theta| < 1$; ℓ_p, $p > 1$, norms tend to push the contribution of such components to zero. The ℓ_1 norm is the only one (among $p \geq 1$) that retains relatively large values even for small values of $|\theta| < 1$ and, hence, components with small values can still have a say in the optimization process and can be penalized by being pushed to smaller values. Hence, if the ℓ_1 norm is used to replace the ℓ_2 one in Eq. (3.41), *only those components of the vector that are really significant in reducing the model misfit measuring term in the regularized cost function will be kept, and the rest will be forced to zero*. The same tendency, yet more aggressive, is true for $0 \leq p < 1$. The extreme case is when one considers the ℓ_0 norm. Even a small increase of a component from zero makes its contribution to the norm large, so the optimizing algorithm has to be very "cautious" in making an element nonzero.

In a nutshell, from all the true norms ($p \geq 1$), the ℓ_1 is the only one that shows respect to small values. The rest of the ℓ_p norms, $p > 1$, just squeeze them to make their values even smaller, and care mainly for the large values. We will return to this point very soon.

9.3 THE LEAST ABSOLUTE SHRINKAGE AND SELECTION OPERATOR (LASSO)

In Chapter 3, we discussed some of the benefits in adopting the regularization method for enhancing the performance of an estimator. In this chapter, we will see and study more reasons that justify the use of regularization. The first one refers to what is known as the *interpretation* power of an estimator. For example, in the regression task, we want to select those components θ_i of $\boldsymbol{\theta}$ that have the most important say in the formation of the output variable. This is very important if the number of parameters, l, is large and we want to concentrate on the most important of them. In a classification task, not all features are informative, hence one would like to keep the most informative of them and make the less informative ones equal to zero. Another related problem refers to those cases where we know, a priori, that a number of the components of a parameter vector are zero, but we do not know which ones. Now, the discussion at the end of the previous section becomes more meaningful. Can we use, while regularizing, an appropriate norm that can assist the optimization process (a) in unveiling such zeros or (b) to put more emphasis on the most significant of its components, those that play a decisive role in reducing the misfit measuring term in the regularized cost function, and set the rest of them equal to zero? Although the ℓ_p norms, with $p < 1$, seem to be the natural choice for such a regularization, the fact that they are not convex makes the optimization process hard. The ℓ_1 norm is the one that is "closest" to them, yet it retains the computationally attractive property of convexity.

The ℓ_1 norm has been used for such problems for a long time. In the 1970s, it was used in seismology [27,86], where the reflected signal that indicates changes in the various earth substrates is a sparse one, that is, very few values are relatively large and the rest are small and insignificant. Since then, it has been used to tackle similar problems in different applications (e.g., [40,80]). However, one can trace two papers that were catalytic in providing the spark for the current strong interest around the ℓ_1 norm. One came from statistics [89], which addressed the LASSO task (first formulated, to our knowledge, in [80]), to be discussed next, and the other came from the signal analysis community [26], which formulated the *basis pursuit*, to be discussed in a later section.

We first address our familiar regression task

$$y = X\theta + \eta, \quad y, \eta \in \mathbb{R}^N, \ \theta \in \mathbb{R}^l, \ N \geq l,$$

and obtain the estimate of the unknown parameter θ via the sum of squared error cost, regularized by the ℓ_1 norm, that is, for $\lambda \geq 0$,

$$\hat{\theta} := \arg\min_{\theta \in \mathbb{R}^l} L(\theta, \lambda) \tag{9.5}$$

$$:= \arg\min_{\theta \in \mathbb{R}^l} \left(\sum_{n=1}^{N} (y_n - x_n^T \theta)^2 + \lambda \|\theta\|_1 \right)$$

$$= \arg\min_{\theta \in \mathbb{R}^l} \left((y - X\theta)^T (y - X\theta) + \lambda \|\theta\|_1 \right). \tag{9.6}$$

Following the discussion with respect to the bias term given in Section 3.8 and in order to simplify the analysis, we will assume hereafter, without harming generality, that the data are of zero mean values. If this is not the case, the data can be centered by subtracting their respective sample means.

It turns out that the task in (9.6) can be equivalently written in the following two formulations:

$$\begin{aligned} \hat{\theta}: \quad & \min_{\theta \in \mathbb{R}^l} (y - X\theta)^T (y - X\theta), \\ \text{s.t.} \quad & \|\theta\|_1 \leq \rho, \end{aligned} \tag{9.7}$$

or

$$\begin{aligned} \hat{\theta}: \quad & \min_{\theta \in \mathbb{R}^l} \|\theta\|_1, \\ \text{s.t.} \quad & (y - X\theta)^T (y - X\theta) \leq \epsilon, \end{aligned} \tag{9.8}$$

given the user-defined parameters $\rho, \epsilon \geq 0$. The formulation in (9.7) is known as the LASSO and the one in (9.8) as the *basis pursuit denoising* (BPDN) (e.g., [15]). All three formulations are equivalent for specific choices of λ, ϵ, and ρ (see, e.g., [14]). Observe that the minimized cost function in (9.6) corresponds to the Lagrangian of the formulation in (9.7). However, this functional dependence among λ, ϵ, and ρ is hard to compute, unless the columns of X are mutually orthogonal. Moreover, this equivalence does not necessarily imply that all three formulations are equally easy or difficult to solve. As we will see later in this chapter, algorithms have been developed along each one of the previous formulations. From now on, we will refer to all three formulations as the LASSO task, in a slight abuse of the standard terminology, and the specific formulation will be apparent from the context, if not stated explicitly.

We know that ridge regression admits a closed form solution, that is,

$$\hat{\theta}_R = \left(X^T X + \lambda I \right)^{-1} X^T y.$$

In contrast, this is not the case for LASSO, and its solution requires iterative techniques. It is straightforward to see that LASSO can be formulated as a standard convex quadratic problem with linear inequalities. Indeed, we can rewrite (9.6) as

$$\min_{\{\theta_i,u_i\}_{i=1}^l} \quad (\boldsymbol{y}-X\boldsymbol{\theta})^T(\boldsymbol{y}-X\boldsymbol{\theta})+\lambda\sum_{i=1}^l u_i,$$

$$\text{s.t.} \quad \begin{cases} -u_i \le \theta_i \le u_i, \\ u_i \ge 0, \end{cases} \quad i=1,2,\ldots,l,$$

which can be solved by any standard convex optimization method (e.g., [14,101]). The reason that developing algorithms for the LASSO task has been at the center of an intense research activity is due to the emphasis on obtaining *efficient* algorithms by exploiting the specific nature of this task, especially for cases where l is very large, as is often the case in practice.

In order to get better insight into the nature of the solution that is obtained by LASSO, let us assume that the regressors are mutually orthogonal and of unit norm, hence $X^TX=I$. Orthogonality of the input matrix helps to decouple the coordinates and results to l one-dimensional problems that can be solved analytically. For this case, the LS estimate becomes

$$\hat{\boldsymbol{\theta}}_{\text{LS}}=(X^TX)^{-1}X^T\boldsymbol{y}=X^T\boldsymbol{y},$$

and the ridge regression gives

$$\hat{\boldsymbol{\theta}}_R=\frac{1}{1+\lambda}\hat{\boldsymbol{\theta}}_{\text{LS}}, \tag{9.9}$$

that is, every component of the LS estimate is simply shrunk by the *same* factor, $\frac{1}{1+\lambda}$; see, also, Section 6.5.

In the case of ℓ_1 regularization, the minimized Lagrangian function is no more differentiable, due to the presence of the absolute values in the ℓ_1 norm. So, in this case, we have to consider the notion of the subdifferential. It is known (Chapter 8) that if the zero vector belongs to the subdifferential set of a convex function at a point, this means that this point corresponds to a minimum of the function. Taking the subdifferential of the Lagrangian defined in (9.6) and recalling that the subdifferential set of a differentiable function includes as its *single* element the respective gradient, the estimate $\hat{\boldsymbol{\theta}}_1$, resulting from the ℓ_1 regularized task, must satisfy

$$\boldsymbol{0}\in-2X^T\boldsymbol{y}+2X^TX\boldsymbol{\theta}+\lambda\partial\left\|\boldsymbol{\theta}\right\|_1,$$

where ∂ stands for the subdifferential set (Chapter 8). If X has orthonormal columns, the previous equation can be written component-wise as follows:

$$0\in-\hat{\theta}_{\text{LS},i}+\hat{\theta}_{1,i}+\frac{\lambda}{2}\partial\left|\hat{\theta}_{1,i}\right|, \quad \forall i, \tag{9.10}$$

where the subdifferential of the function $|\cdot|$, derived in Example 8.4 (Chapter 8), is given as

$$\partial|\theta|=\begin{cases} \{1\}, & \text{if } \theta>0, \\ \{-1\}, & \text{if } \theta<0, \\ [-1,1], & \text{if } \theta=0. \end{cases}$$

Thus, we can now write for each component of the LASSO optimal estimate

$$\hat{\theta}_{1,i} = \begin{cases} \hat{\theta}_{\text{LS},i} - \frac{\lambda}{2}, & \text{if } \hat{\theta}_{1,i} > 0, \quad (9.11) \\ \hat{\theta}_{\text{LS},i} + \frac{\lambda}{2}, & \text{if } \hat{\theta}_{1,i} < 0. \quad (9.12) \end{cases}$$

Note that (9.11) can only be true if $\hat{\theta}_{\text{LS},i} > \frac{\lambda}{2}$, and (9.12) only if $\hat{\theta}_{\text{LS},i} < -\frac{\lambda}{2}$. Moreover, in the case where $\hat{\theta}_{1,i} = 0$, (9.10) and the subdifferential of $|\cdot|$ suggest that necessarily $\left|\hat{\theta}_{\text{LS},i}\right| \leq \frac{\lambda}{2}$. Concluding, we can write in a more compact way that

$$\boxed{\hat{\theta}_{1,i} = \text{sgn}(\hat{\theta}_{\text{LS},i}) \left(\left|\hat{\theta}_{\text{LS},i}\right| - \frac{\lambda}{2} \right)_+ \quad : \quad \text{soft thresholding operation,}} \quad (9.13)$$

where $(\cdot)_+$ denotes the "positive part" of the respective argument; it is equal to the argument if this is nonnegative, and zero otherwise. This is very interesting indeed. In contrast to the ridge regression which shrinks all coordinates of the unregularized LS solution by the same factor, LASSO forces all coordinates, whose absolute value is less than or equal to $\lambda/2$, to zero, and the rest of the coordinates are reduced, in absolute value, by the same amount $\lambda/2$. This is known as *soft thresholding*, to distinguish it from the *hard thresholding* operation; the latter is defined as $\theta \cdot \chi_{(0,\infty)} \left(|\theta| - \frac{\lambda}{2}\right)$, $\theta \in \mathbb{R}$, where $\chi_{(0,\infty)}(\cdot)$ stands for the characteristic function with respect to the set $(0, \infty)$. Fig. 9.3 shows the graphs illustrating the effect that the ridge regression, LASSO, and hard thresholding have on the unregularized LS solution, as a function of its value (horizontal axis). Note that our discussion here, simplified via the orthonormal input matrix case, has quantified what we said before about the tendency of the ℓ_1 norm to push small values to become *exactly zero*. This will be further strengthened, via a more rigorous mathematical formulation, in Section 9.5.

Example 9.1. Assume that the unregularized LS solution, for a given regression task, $\boldsymbol{y} = X\boldsymbol{\theta} + \boldsymbol{\eta}$, is given by

$$\hat{\boldsymbol{\theta}}_{\text{LS}} = [0.2, -0.7, 0.8, -0.1, 1.0]^T.$$

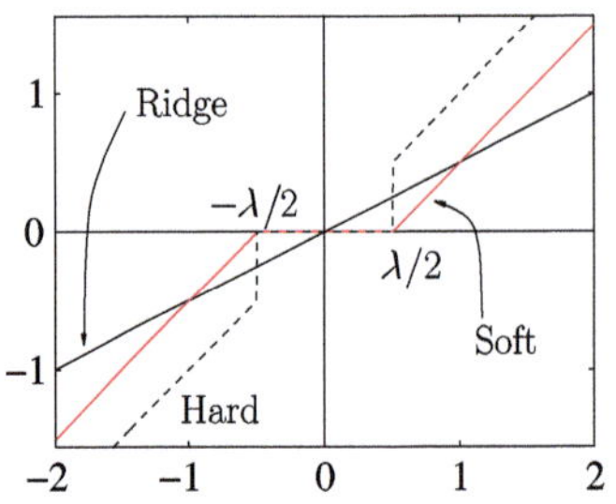

FIGURE 9.3

Curves relating the output (vertical) to the input (horizontal) for the hard thresholding and the soft thresholding operators shown together with the linear operator that is associated with the ridge regression, for the same value of $\lambda = 1$.

Derive the solutions for the corresponding ridge regression and ℓ_1 norm regularization tasks. Assume that the input matrix X has orthonormal columns and that the regularization parameter is $\lambda = 1$. Also, what is the result of hard thresholding the vector $\hat{\boldsymbol{\theta}}_{\text{LS}}$ with threshold equal to 0.5?

We know that the corresponding solution for the ridge regression is

$$\hat{\boldsymbol{\theta}}_R = \frac{1}{1+\lambda}\hat{\boldsymbol{\theta}}_{\text{LS}} = [0.1, -0.35, 0.4, -0.05, 0.5]^T.$$

The solution for the ℓ_1 norm regularization is given by soft thresholding, with threshold equal to $\lambda/2 = 0.5$, and hence the corresponding vector is

$$\hat{\boldsymbol{\theta}}_1 = [0, -0.2, 0.3, 0, 0.5]^T.$$

The result of the hard thresholding operation is the vector $[0, -0.7, 0.8, 0, 1.0]^T$.

Remarks 9.1.

- The hard and soft thresholding rules are only two possibilities out of a larger number of alternatives. Note that the hard thresholding operation is defined via a discontinuous function, and this makes this rule unstable in the sense of being very sensitive to small changes of the input. Moreover, this shrinking rule tends to exhibit large variance in the resulting estimates. The soft thresholding rule is a continuous function, but, as readily seen from the graph in Fig. 9.3, it introduces bias even for the large values of the input argument. In order to ameliorate such shortcomings, a number of alternative thresholding operators have been introduced and studied both theoretically and experimentally. Although these are not within the mainstream of our interest, we provide two popular examples for the sake of completeness—the *smoothly clipped absolute deviation* (SCAD) thresholding rule,

$$\hat{\theta}_{\text{SCAD}} = \begin{cases} \text{sgn}(\theta)\,(|\theta| - \lambda_{\text{SCAD}})_+, & |\theta| \leq 2\lambda_{\text{SCAD}}, \\ \dfrac{(\alpha-1)\theta - \alpha\lambda_{\text{SCAD}}\,\text{sgn}(\theta)}{\alpha - 2}, & 2\lambda_{\text{SCAD}} < |\theta| \leq \alpha\lambda_{\text{SCAD}}, \\ \theta, & |\theta| > \alpha\lambda_{\text{SCAD}}, \end{cases}$$

  and the *nonnegative garrote* thresholding rule,

$$\hat{\theta}_{\text{garr}} = \begin{cases} 0, & |\theta| \leq \lambda_{\text{garr}}, \\ \theta - \dfrac{\lambda_{\text{garr}}^2}{\theta}, & |\theta| > \lambda_{\text{garr}}. \end{cases}$$

  Fig. 9.4 shows the respective graphs. Observe that in both cases, an effort has been made to remove the discontinuity (associated with the hard thresholding) and to remove/reduce the bias for large values of the input argument. The parameter $\alpha > 2$ is a user-defined one. For a more detailed related discussion, the interested reader can refer, for example, to [2]. In [83], a generalized thresholding rule is suggested that encompasses all previously mentioned ones as special cases. Moreover, the proposed framework is general enough to provide means for designing novel thresholding rules and/or incorporating a priori information associated with the sparsity level, i.e., the number of nonzero components, of the sparse vector to be recovered.

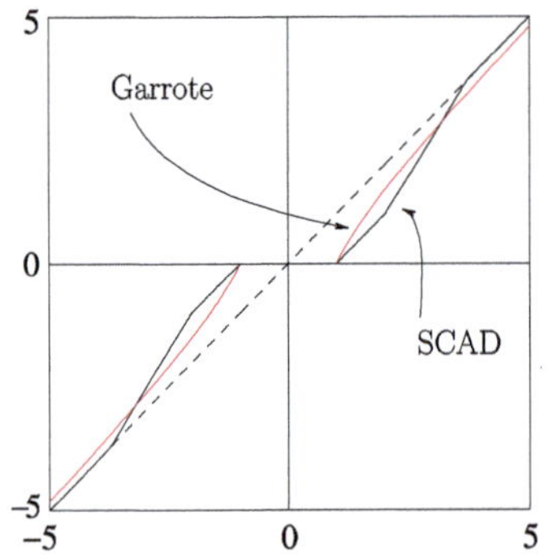

FIGURE 9.4

Output (vertical)–input (horizotal) graph for the SCAD and nonnegative garrote rules with parameters $\alpha = 3.7$ and $\lambda_{\text{SCAD}} = \lambda_{\text{garr}} = 1$. Observe that both rules smoothen out the discontinuity associated with the hard thresholding rule. Note, also, that the SCAD rule removes the bias associated with the soft thresholding rule for large values of the input variable. On the contrary, the garrote thresholding rule allows some bias for large input values, which diminishes as λ_{garr} gets smaller and smaller.

9.4 SPARSE SIGNAL REPRESENTATION

In the previous section, we brought into our discussion the need to take special care for zeros. Sparsity is an attribute that is met in a plethora of natural signals, because nature tends to be parsimonious. The notion of and need for parsimonious models was also discussed in Chapter 3, in the context of inverse problems in machine learning tasks. In this section, we will briefly present a number of application cases where the existence of zeros in a mathematical expansion is of paramount importance; hence, it justifies our search for and development of related analysis tools.

In Chapter 4, we discussed the task of echo cancelation. In a number of cases, the echo path, represented by a vector comprising the values of the impulse response samples, is a sparse one. This is the case, for example, in internet telephony and in acoustic and network environments (e.g., [3,10,73]). Fig. 9.5 shows the impulse response of such an echo path. The impulse response of the echo path is of short duration; however, the delay with which it appears is not known. So, in order to model it, one has to use a long impulse response, yet only a relatively small number of the coefficients will be significant and the rest will be close to zero. Of course, one could ask, why not use an LMS or an RLS, and eventually the significant coefficients will be identified? The answer is that this turns out not to be the most efficient way to tackle such problems, because the convergence of the algorithm can be very slow. In contrast, if one embeds, somehow, into the problem the a priori information concerning the existence of (almost) zero coefficients, then the convergence speed can be significantly increased and also better error floors can be attained.

A similar situation occurs in wireless communication systems, which involve multipath channels. A typical application is in high-definition television (HDTV) systems, where the involved communication channels consist of a *few* nonnegligible coefficients, some of which may have quite large time delays with respect to the main signal (see, e.g., [4,32,52,77]). If the information signal is transmitted at high symbol rates through such a dispersive channel, then the introduced intersymbol interference

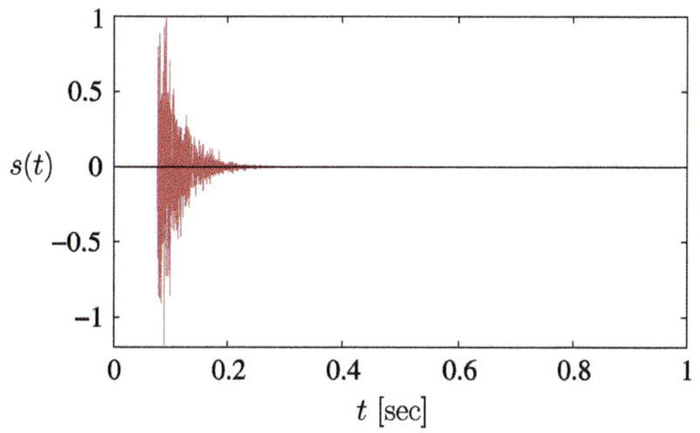

FIGURE 9.5

The impulse response function of an echo path in a telephone network. Observe that although it is of relatively short duration, it is not a priori known where exactly in time it will occur.

(ISI) has a span of several tens up to hundreds of symbol intervals. This in turn implies that quite long channel estimators are required at the receiver's end in order to reduce effectively the ISI component of the received signal, although only a small part of it has values substantially different from zero. The situation is even more demanding when the channel-frequency response exhibits deep nulls. More recently, sparsity has been exploited in channel estimation for multicarrier systems, both for single antenna as well as for multiple-input-multiple-output (MIMO) systems [46,47]. A thorough, in-depth treatment related to sparsity in multipath communication systems is provided in [5].

Another example, which might be more widely known, is that of signal compression. It turns out that if the signal modalities with which we communicate (e.g., speech) and also sense the world (e.g.,

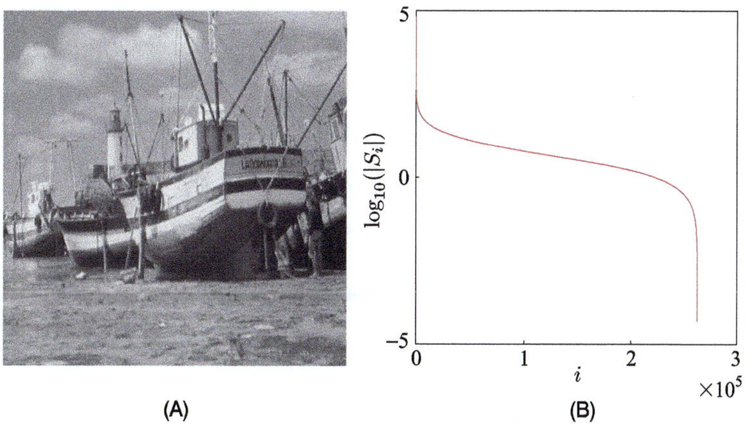

FIGURE 9.6

(A) A 512 × 512 pixel image and (B) the magnitude of its DCT components in descending order and logarithmic scale. Note that more than 95% of the total energy is contributed by only 5% of the largest components.

images, audio) are transformed into a suitably chosen domain, then they are sparsely represented; only a relatively small number of the signal components in this domain are large, and the rest are close to zero. As an example, Fig. 9.6A shows an image and Fig. 9.6B shows the plot of the magnitude of the obtained discrete cosine transform (DCT) components, which are computed by writing the corresponding image array as a vector in lexicographic order. Note that more than 95% of the total energy is contributed by only 5% of the largest components. This is at the heart of any compression technique. Only the large coefficients are chosen to be coded and the rest are considered to be zero. Hence, significant gains are obtained in memory/bandwidth requirements while storing/transmitting such signals, without much perceptual loss. Depending on the modality, different transforms are used. For example, in JPEG-2000, an image array, represented in terms of a vector that contains the intensity of the gray levels of the image pixels, is transformed via the discrete wavelet transform (DWT), resulting in a transformed vector that comprises only a few large components.

Let

$$\tilde{\boldsymbol{s}} = \Phi^H \boldsymbol{s}, \quad \boldsymbol{s}, \tilde{\boldsymbol{s}} \in \mathbb{C}^l, \tag{9.14}$$

where $\boldsymbol{s}$ is the vector of the "raw" signal samples, $\tilde{\boldsymbol{s}}$ is the (complex-valued) vector of the transformed ones, and Φ is the $l \times l$ transformation matrix. Often, this is an orthonormal/unitary matrix, $\Phi^H \Phi = I$. Basically, a transform is nothing more than a projection of a vector on a new set of coordinate axes, which comprise the columns of the transformation matrix Φ. Celebrated examples of such transforms are the wavelet, the discrete Fourier (DFT), and the discrete cosine (DCT) transforms (e.g., [87]). In such cases, where the transformation matrix is orthonormal, one can write

$$\boldsymbol{s} = \Psi \tilde{\boldsymbol{s}}, \tag{9.15}$$

where $\Psi = \Phi$. Eq. (9.14) is known as the *analysis* and (9.15) as the *synthesis* equation.

Compression via such transforms exploits the fact that many signals in nature, which are rich in context, can be *compactly* represented in an appropriately chosen basis, depending on the modality of the signal. Very often, the construction of such bases tries to "imitate" the sensory systems that the human brain has developed in order to sense these signals; and we know that nature (in contrast to modern humans) does not like to waste resources. A standard compression task comprises the following stages: (a) obtain the l components of $\tilde{\boldsymbol{s}}$ via the analysis step (9.14); (b) keep the k most significant of them; (c) code these values, as well as their respective locations in the transformed vector $\tilde{\boldsymbol{s}}$; and (d) obtain the (approximate) original signal $\boldsymbol{s}$ when needed (after storage or transmission), via the synthesis equation (Eq. (9.15)), where in place of $\tilde{\boldsymbol{s}}$ only its k most significant components are used, which are the ones that were coded, while the rest are set equal to zero. However, there is something unorthodox in this process of compression as it has been practiced until very recently. One processes (transforms) large signal vectors of l coordinates, where l in practice can be quite large, and then uses only a small percentage of the transformed coefficients, while the rest are simply ignored. Moreover, one has to store/transmit the location of the respective large coefficients that are finally coded.

A natural question that is raised is the following: Because $\tilde{\boldsymbol{s}}$ in the synthesis equation is (approximately) sparse, can one compute it via an alternative path to the analysis equation in (9.14)? The issue here is to investigate whether one could use a more informative way of sampling the available raw data so that fewer than l samples/observations are sufficient to recover all the necessary information. The ideal case would be to recover it via a set of k such samples, because this is the number of the significant free parameters. On the other hand, if this sounds a bit extreme, can one obtain N ($k < N \ll l$) such signal-related measurements, from which s can eventually be retrieved? It turns out that such an approach is possible and it leads to the solution of an *underdetermined* system of linear equations, under the constraint that the unknown target vector is a sparse one.

The importance of such techniques becomes even more apparent when, instead of an orthonormal basis, as discussed before, a more general type of expansion is adopted, in terms of what is known as *overcomplete dictionaries*. A dictionary [65] is a collection of parameterized waveforms, which are discrete-time signal samples, represented as vectors $\boldsymbol{\psi}_i \in \mathbb{C}^l$, $i \in \mathcal{I}$, where $\mathcal{I}$ is an integer index set. For example, the columns of a DFT or a discrete wavelet (DWT) matrix comprise a dictionary. These are two examples of what are known as *complete* dictionaries, which consist of l (orthonormal) vectors, that is, a number equal to the length of the signal vector. However, in many cases in practice, using such dictionaries is very restrictive. Let us take, for example, a segment of audio signal, from a news media or a video, that needs to be processed. This consists, in general, of different types of signals, namely, speech, music, and environmental sounds. For each type of these signals, different signal vectors may be more appropriate in the expansion for the analysis. For example, music signals are characterized by a strong harmonic content and the use of sinusoids seems to be best for compression, while for speech signals a Gabor-type signal expansion (sinusoids of various frequencies weighted by sufficiently narrow pulses at different locations in time [31,87]) may be a better choice. The same applies when one deals with an image. Different parts of an image, such as parts that are smooth or contain sharp edges, may demand a different expansion vector set for obtaining the best overall performance. The more recent tendency, in order to satisfy such needs, is to use *overcomplete* dictionaries. Such dictionaries can be obtained, for example, by concatenating different dictionaries together, for example, a DFT and a DWT matrix to result in a combined $l \times 2l$ transformation matrix. Alternatively, a dictionary can be "trained" in order to effectively represent a set of available signal exemplars, a task that is often referred to as dictionary learning [75,78,90,100]. While using such overcomplete dictionaries, the synthesis equation takes the form

$$s = \sum_{i \in \mathcal{I}} \theta_i \boldsymbol{\psi}_i. \tag{9.16}$$

Note that, now, the analysis is an ill-posed problem, because the elements $\{\boldsymbol{\psi}_i\}_{i \in \mathcal{I}}$ (usually called *atoms*) of the dictionary are not linearly independent, and there is not a unique set of parameters $\{\theta_i\}_{i \in \mathcal{I}}$ that generates s. Moreover, we expect most of these parameters to be (nearly) zero. Note that in such cases, the cardinality of $\mathcal{I}$ is larger than l. This necessarily leads to underdetermined systems of equations with infinitely many solutions. The question that is now raised is whether we can exploit the fact that most of these parameters are known to be zero, in order to come up with a unique solution. If yes, under which conditions is such a solution possible? We will return to the task of learning dictionaries in Chapter 19.

Besides the previous examples, there are a number of cases where an underdetermined system of equations is the result of our inability to obtain a sufficiently large number of measurements, due to physical and technical constraints. This is the case in MRI imaging, which will be presented in more detail in Section 10.3.

9.5 IN SEARCH OF THE SPARSEST SOLUTION

Inspired by the discussion in the previous section, we now turn our attention to the task of solving underdetermined systems of equations by imposing the sparsity constraint on the solution. We will develop the theoretical setup in the context of regression and we will adhere to the notation that has been adopted for this task. Moreover, we will focus on the real-valued data case in order to simplify the presentation. The theory can be readily extended to the more general complex-valued data case (see, e.g., [64,99]). We assume that we are given a set of observations/measurements, $\boldsymbol{y} := [y_1, y_2, \ldots, y_N]^T \in \mathbb{R}^N$, according to the linear model

$$\boldsymbol{y} = X\boldsymbol{\theta}, \quad \boldsymbol{y} \in \mathbb{R}^N, \ \boldsymbol{\theta} \in \mathbb{R}^l, \ l > N, \tag{9.17}$$

where X is the $N \times l$ input matrix, which is assumed to be of full row rank, that is, $\text{rank}(X) = N$. Our starting point is the noiseless case. The linear system of equations in (9.17) is an underdetermined one and accepts an infinite number of solutions. The set of possible solutions lies in the intersection of the N hyperplanes[2] in the l-dimensional space,

$$\left\{\boldsymbol{\theta} \in \mathbb{R}^l : y_n = \boldsymbol{x}_n^T \boldsymbol{\theta}\right\}, \quad n = 1, 2, \ldots, N.$$

We know from geometry that the intersection of N nonparallel hyperplanes (which in our case is guaranteed by the fact that X has been assumed to be full row rank, hence $\boldsymbol{x}_n$, $n = 1, 2, \ldots, N$, are linearly independent) is a plane of dimensionality $l - N$ (e.g., the intersection of two [nonparallel] [hyper]planes in the three-dimensional space is a straight line, that is, a plane of dimensionality equal to one). In a more formal way, the set of all possible solutions, to be denoted as Θ, is an *affine* set. An affine set is the translation of a linear subspace by a constant vector. Let us pursue this a bit further, because we will need it later on.

Let the null space of X be denoted as $\text{null}(X)$ (sometimes denoted as $\mathcal{N}(X)$), and it is defined as the linear subspace

$$\text{null}(X) = \left\{\boldsymbol{z} \in \mathbb{R}^l : X\boldsymbol{z} = \boldsymbol{0}\right\}.$$

Obviously, if $\boldsymbol{\theta}_0$ is a solution to (9.17), that is, $\boldsymbol{\theta}_0 \in \Theta$, then it is easy to verify that $\forall \boldsymbol{\theta} \in \Theta$, $X(\boldsymbol{\theta} - \boldsymbol{\theta}_0) = \boldsymbol{0}$, or $\boldsymbol{\theta} - \boldsymbol{\theta}_0 \in \text{null}(X)$. As a result,

$$\Theta = \boldsymbol{\theta}_0 + \text{null}(X),$$

and Θ is an affine set. We also know from linear algebra basics (and it is easy to show it; see Problem 9.9) that the null space of a full row rank matrix, $N \times l$, $l > N$, is a subspace of dimensionality $l - N$. Fig. 9.7 illustrates the case for one measurement sample in the two-dimensional space, $l = 2$ and $N = 1$. The set of solutions Θ is a straight line, which is the translation of the linear subspace crossing the origin (the $\text{null}(X)$). Therefore, if one wants to select a *single* point among all the points that lie in the affine set of solutions, Θ, then an extra constraint/a priori knowledge has to be imposed.

In the sequel, three such possibilities are examined.

[2] In $\mathbb{R}^l$, a hyperplane is of dimension $l - 1$. A plane has dimension lower than $l - 1$.

THE ℓ_2 NORM MINIMIZER

Our goal now becomes to pick a point in (the affine set) Θ that corresponds to the minimum ℓ_2 norm. This is equivalent to solving the following constrained task:

$$\min_{\boldsymbol{\theta}\in\mathbb{R}^l} \quad \|\boldsymbol{\theta}\|_2^2,$$
$$\text{s.t.} \quad \boldsymbol{x}_n^T\boldsymbol{\theta} = y_n, \quad n = 1, 2, \ldots, N. \tag{9.18}$$

We already know from Section 6.4 (and one can rederive it by employing Lagrange multipliers; see Problem 9.10) that the previous optimization task accepts a *unique* solution given in closed form as

$$\hat{\boldsymbol{\theta}} = X^T\left(XX^T\right)^{-1}\boldsymbol{y}. \tag{9.19}$$

The geometric interpretation of this solution is provided in Fig. 9.7A, for the case of $l = 2$ and $N = 1$. The radius of the Euclidean norm ball keeps increasing, until it touches the plane that contains the solutions. This point is the one with the minimum ℓ_2 norm or, equivalently, the point that lies closest to the origin. Equivalently, the point $\hat{\boldsymbol{\theta}}$ can be seen as the (metric) projection of $\mathbf{0}$ onto Θ.

Minimizing the ℓ_2 norm in order to solve a linear set of underdetermined equations has been used in various applications. The closest to us is in the context of determining the unknown parameters in an expansion using an overcomplete dictionary of functions (vectors) [35]. A main drawback of this method is that it is not sparsity preserving. There is no guarantee that the solution in (9.19) will give zeros even if the true model vector $\boldsymbol{\theta}$ has zeros. Moreover, the method is *resolution-limited* [26]. This means that even if there may be a sharp contribution of specific atoms in the dictionary, this is not portrayed in the obtained solution. This is a consequence of the fact that the information provided by XX^T

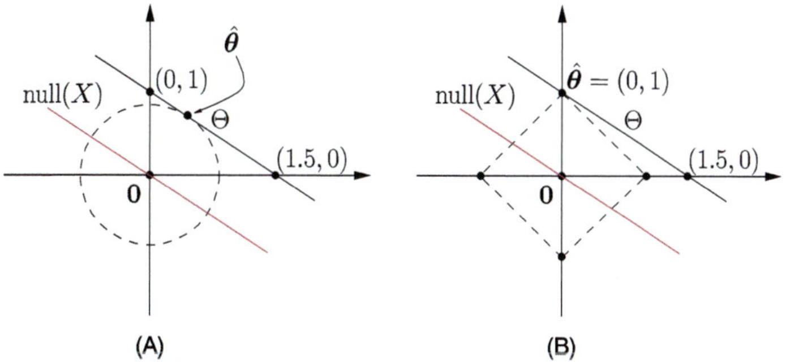

FIGURE 9.7

The set of solutions Θ is an affine set (gray line), which is a translation of the null(X) subspace (red line). (A) The ℓ_2 norm minimizer: The dotted circle corresponds to the smallest ℓ_2 ball that intersects the set Θ. As such, the intersection point, $\hat{\boldsymbol{\theta}}$, is the ℓ_2 norm minimizer of the task in (9.18). Note that the vector $\hat{\boldsymbol{\theta}}$ contains no zero component. (B) The ℓ_1 norm minimizer: The dotted rhombus corresponds to the smallest ℓ_1 ball that intersects Θ. Hence, the intersection point, $\hat{\boldsymbol{\theta}}$, is the solution of the constrained ℓ_1 minimization task of (9.21). Note that the obtained estimate $\hat{\boldsymbol{\theta}} = (0, 1)$ contains a zero.

is a global one, containing all atoms of the dictionary in an "averaging" fashion, and the final result tends to smoothen out the individual contributions, especially when the dictionary is overcomplete.

THE ℓ_0 NORM MINIMIZER

Now we turn our attention to the ℓ_0 norm (once more, it is pointed out that this is an abuse of the definition of the norm), and we make sparsity our new flag under which a solution will be obtained. The task now becomes

$$\begin{aligned} \min_{\boldsymbol{\theta}\in\mathbb{R}^l} \quad & \|\boldsymbol{\theta}\|_0, \\ \text{s.t.} \quad & \boldsymbol{x}_n^T\boldsymbol{\theta} = y_n, \; n = 1, 2, \ldots, N, \end{aligned} \tag{9.20}$$

that is, from all the points that lie on the plane of all possible solutions, find the *sparsest* one, that is, the one with the lowest number of nonzero elements. As a matter of fact, such an approach is within the spirit of *Occam's razor* rule—it corresponds to the smallest number of parameters that can explain the obtained observations. The points that are now raised are:

- Is a solution to this problem unique, and under which conditions?
- Can a solution be obtained with low enough complexity in realistic time?

We postpone the answer to the first question until later. As for the second one, the news is not good. Minimizing the ℓ_0 norm under a set of linear constraints is a task of combinatorial nature, and as a matter of fact, the problem is, in general, NP-hard [72]. The way to approach the problem is to consider all possible combinations of zeros in $\boldsymbol{\theta}$, removing the respective columns of X in (9.17), and check whether the system of equations is satisfied; keep as solutions the ones with the smallest number of nonzero elements. Such a searching technique exhibits complexity of an exponential dependence on l. Fig. 9.7A illustrates the two points ((1.5, 0) and (0, 1)) that comprise the solution set of minimizing the ℓ_0 norm for the single measurement (constraint) case.

THE ℓ_1 NORM MINIMIZER

The current task is now given by

$$\begin{aligned} \min_{\boldsymbol{\theta}\in\mathbb{R}^l} \quad & \|\boldsymbol{\theta}\|_1, \\ \text{s.t.} \quad & \boldsymbol{x}_n^T\boldsymbol{\theta} = y_n, \; n = 1, 2, \ldots, N. \end{aligned} \tag{9.21}$$

Fig. 9.7B illustrates the geometry. The ℓ_1 ball is increased until it touches the affine set of the possible solutions. For this specific geometry, the solution is the point (0, 1), which is a sparse solution. In our discussion in Section 9.2, we saw that the ℓ_1 norm is the one, out of all ℓ_p, $p \geq 1$ norms, that bears some similarity with the sparsity promoting (nonconvex) ℓ_p, $p < 1$ "norms." Also, we have commented that the ℓ_1 norm encourages zeros when the respective values are small. In the sequel, we will state one lemma that establishes this zero favoring property in a more formal way. The ℓ_1 norm minimization task is also known as *basis pursuit* and it was suggested for decomposing a vector signal in terms of the atoms of an overcomplete dictionary [26].

The ℓ_1 minimizer can be brought into the standard linear programming (LP) form and then can be solved by recalling any related method; the simplex method and the more recent interior point method are two possibilities (see, e.g., [14,33]). Indeed, consider the LP task

$$\begin{aligned} \min_{x} \quad & \boldsymbol{c}^T \boldsymbol{x}, \\ \text{s.t.} \quad & A\boldsymbol{x} = \boldsymbol{b}, \\ & \boldsymbol{x} \geq \boldsymbol{0}. \end{aligned}$$

To verify that our ℓ_1 minimizer can be cast in the previous form, note first that any l-dimensional vector $\boldsymbol{\theta}$ can be decomposed as

$$\boldsymbol{\theta} = \boldsymbol{u} - \boldsymbol{v}, \quad \boldsymbol{u} \geq \boldsymbol{0}, \boldsymbol{v} \geq \boldsymbol{0}.$$

Indeed, this holds true if, for example,

$$\boldsymbol{u} := \boldsymbol{\theta}_+, \quad \boldsymbol{v} := (-\boldsymbol{\theta})_+,$$

where $\boldsymbol{x}_+$ stands for the vector obtained after keeping the positive components of $\boldsymbol{x}$ and setting the rest equal to zero. Moreover, note that

$$\|\boldsymbol{\theta}\|_1 = [1, 1, \dots, 1] \begin{bmatrix} \boldsymbol{\theta}_+ \\ (-\boldsymbol{\theta})_+ \end{bmatrix} = [1, 1, \dots, 1] \begin{bmatrix} \boldsymbol{u} \\ \boldsymbol{v} \end{bmatrix}.$$

Hence, our ℓ_1 minimization task can be recast in the LP form if

$$\begin{aligned} \boldsymbol{c} &:= [1, 1, \dots, 1]^T, & \boldsymbol{x} &:= [\boldsymbol{u}^T, \boldsymbol{v}^T]^T, \\ A &:= [X, -X], & \boldsymbol{b} &:= \boldsymbol{y}. \end{aligned}$$

CHARACTERIZATION OF THE ℓ_1 NORM MINIMIZER

Lemma 9.1. *An element $\boldsymbol{\theta}$ in the affine set Θ of the solutions of the underdetermined linear system (9.17) has minimal ℓ_1 norm if and only if the following condition is satisfied:*

$$\left| \sum_{i:\, \theta_i \neq 0} \operatorname{sgn}(\theta_i) z_i \right| \leq \sum_{i:\, \theta_i = 0} |z_i|, \quad \forall z \in \operatorname{null}(X). \tag{9.22}$$

Moreover, the ℓ_1 minimizer is unique if and only if the inequality in (9.22) is a strict one for all $z \neq \boldsymbol{0}$ (see, e.g., [74] and Problem 9.11).

Remarks 9.2.

- The previous lemma has a very interesting and important consequence. If $\hat{\boldsymbol{\theta}}$ is the *unique* minimizer of (9.21), then

$$\operatorname{card}\{i : \hat{\theta}_i = 0\} \geq \dim(\operatorname{null}(X)), \tag{9.23}$$

where card{·} denotes the cardinality of a set. In words, the number of zero coordinates of the unique minimizer cannot be smaller than the dimension of the null space of X. Indeed, if this is not the case, then the unique minimizer could have fewer zeros than the dimensionality of null(X). This means that we can always find a $z \in \text{null}(X)$, which has zeros in the same locations where the coordinates of the unique minimizer are zero, and at the same time it is not identically zero, that is, $z \neq \mathbf{0}$ (Problem 9.12). However, this would violate (9.22), which in the case of uniqueness holds as a strict inequality.

Definition 9.1. A vector $\boldsymbol{\theta}$ is called k-sparse if it has *at most* k nonzero components.

Remarks 9.3.

- If the minimizer of (9.21) is *unique*, then it is a k-sparse vector with

$$k \leq N.$$

This is a direct consequence of Remarks 9.2, and the fact that for the matrix X,

$$\dim(\text{null}(X)) = l - \text{rank}(X) = l - N.$$

Hence, the number of the nonzero elements of the unique minimizer must be at most equal to N. If one resorts to geometry, all the previously stated results become crystal clear.

GEOMETRIC INTERPRETATION

Assume that our target solution resides in the three-dimensional space and that we are given one measurement,

$$y_1 = \boldsymbol{x}_1^T \boldsymbol{\theta} = x_{11}\theta_1 + x_{12}\theta_2 + x_{13}\theta_3.$$

Then the solution lies in the two-dimensional (hyper)plane, which is described by the previous equation. To get the minimal ℓ_1 solution we keep increasing the size of the ℓ_1 ball[3] (all the points that lie on the sides/faces of this ball have equal ℓ_1 norm) until it touches this plane. The only way that these two geometric objects have a single point in common (unique solution) is when they meet at a vertex of the diamond. This is shown in Fig. 9.8A. In other words, the resulting solution is 1-sparse, having two of its components equal to zero. This complies with the finding stated in Remarks 9.3, because now $N = 1$. For any other orientation of the plane, this will either cut across the ℓ_1 ball or will share with the diamond an edge or a side. In both cases, there will be infinite solutions.

Let us now assume that we are given an extra measurement,

$$y_2 = x_{21}\theta_1 + x_{22}\theta_2 + x_{23}\theta_3.$$

The solution now lies in the intersection of the two previous planes, which is a straight line. However, now, we have more alternatives for a unique solution. A line, for example, Θ_1, can either touch the ℓ_1 ball at a vertex (1-sparse solution) or, as shown in Fig. 9.8B, touch the ℓ_1 ball at one of its edges, for

[3] Observe that in the three-dimensional space the ℓ_1 ball looks like a diamond.

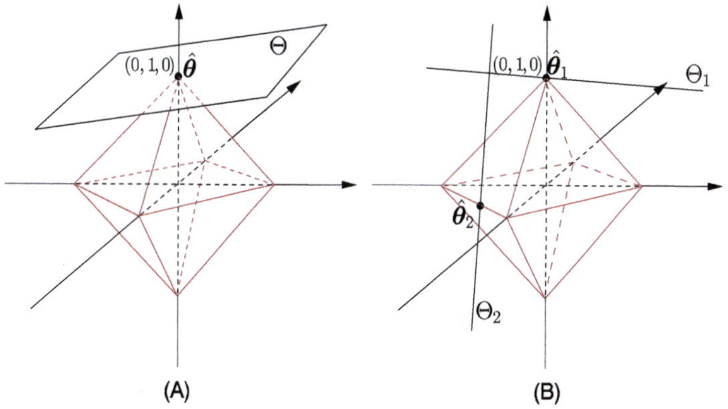

FIGURE 9.8

(A) The ℓ_1 ball intersecting with a plane. The only possible scenario, for the existence of a unique common intersecting point of the ℓ_1 ball with a plane in the Euclidean $\mathbb{R}^3$ space, is for the point to be located at one of the vertices of the ℓ_1 ball, that is, to be a 1-sparse vector. (B) The ℓ_1 ball intersecting with lines. In this case, the sparsity level of the unique intersecting point is relaxed; it could be a 1- or a 2-sparse vector.

example, Θ_2. The latter case corresponds to a solution that lies on a two-dimensional subspace; hence it will be a 2-sparse vector. This also complies with the findings stated in Remarks 9.3, because in this case we have $N = 2, l = 3$, and the sparsity level for a unique solution can be either 1 or 2.

Note that uniqueness is associated with the particular geometry and orientation of the affine set, which is the set of all possible solutions of the underdetermined system of equations. For the case of the squared ℓ_2 norm, the solution is always unique. This is a consequence of the (hyper)spherical shape formed by the Euclidean norm. From a mathematical point of view, the squared ℓ_2 norm is a strictly convex function. This is not the case for the ℓ_1 norm, which is convex, albeit not a strictly convex function (Problem 9.13).

Example 9.2. Consider a sparse vector parameter $[0, 1]^T$, which we assume to be unknown. We will use one measurement to *sense* it. Based on this single measurement, we will use the ℓ_1 minimizer of (9.21) to recover its true value. Let us see what happens.

We will consider three different values of the "sensing" (input) vector $\boldsymbol{x}$ in order to obtain the measurement $y = \boldsymbol{x}^T\boldsymbol{\theta}$: (a) $\boldsymbol{x} = [\frac{1}{2}, 1]^T$, (b) $\boldsymbol{x} = [1, 1]^T$, and (c) $\boldsymbol{x} = [2, 1]^T$. The resulting measurement, after sensing $\boldsymbol{\theta}$ by $\boldsymbol{x}$, is $y = 1$ for all three previous cases.

Case a: The solution will lie on the straight line

$$\Theta = \left\{ [\theta_1, \theta_2]^T \in \mathbb{R}^2 : \frac{1}{2}\theta_1 + \theta_2 = 1 \right\},$$

which is shown in Fig. 9.9A. For this setting, expanding the ℓ_1 ball, this will touch the straight line (our solution's affine set) at the vertex $[0, 1]^T$. This is a unique solution, hence it is sparse, and it coincides with the true value.

Case b: The solution lies on the straight line

$$\Theta = \left\{[\theta_1, \theta_2]^T \in \mathbb{R}^2 : \theta_1 + \theta_2 = 1\right\},$$

which is shown in Fig. 9.9B. For this setup, there is an infinite number of solutions, including two sparse ones.

Case c: The affine set of solutions is described by

$$\Theta = \left\{[\theta_1, \theta_2]^T \in \mathbb{R}^2 : 2\theta_1 + \theta_2 = 1\right\},$$

which is sketched in Fig. 9.9C. The solution in this case is sparse, but it is not the correct one.

This example is quite informative. *If we sense (measure) our unknown parameter vector with appropriate sensing (input) data, the use of the ℓ_1 norm can unveil the true value of the parameter vector, even if the system of equations is underdetermined, provided that the true parameter is sparse.* This becomes our new goal; to investigate whether what we have just said can be generalized, and under which conditions it holds true. In such a case, the choice of the regressors (which we called sensing vectors) and hence the input matrix (which we will refer to more and more frequently as the sensing matrix) acquire extra significance. It is not enough for the designer to care only for the rank of the matrix, that

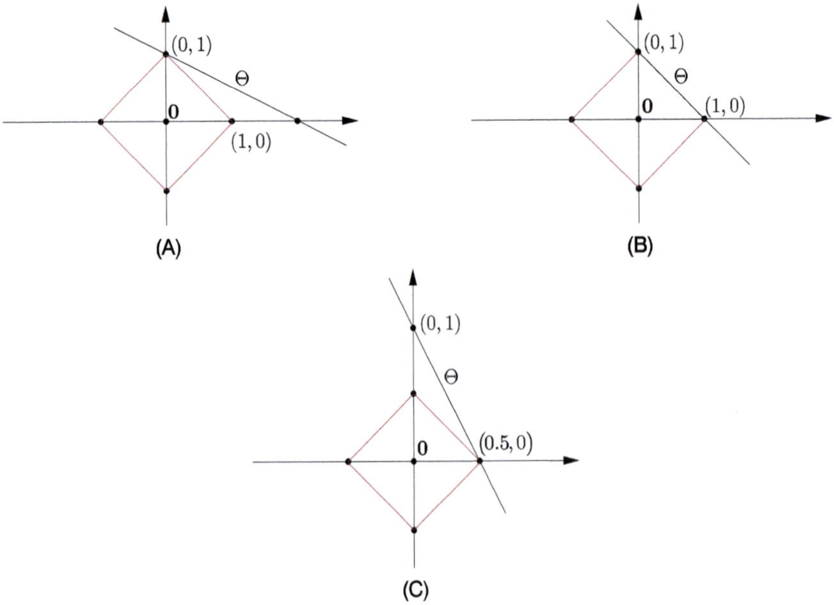

FIGURE 9.9

(A) Sensing with $x = [\frac{1}{2}, 1]^T$. (B) Sensing with $x = [1, 1]^T$. (C) Sensing with $x = [2, 1]^T$. The choice of the sensing vector x is crucial to unveiling the true sparse solution (0,1). Only the sensing vector $x = [\frac{1}{2}, 1]^T$ identifies uniquely the desired (0,1).

is, the linear independence of the sensing vectors. One has to make sure that the corresponding affine set of the solutions has such an orientation so that the touch with the ℓ_1 ball (as this increases from zero to meet this plane) is a "gentle" one; that is, they meet at a single point, and more importantly, at the correct one, which is the point that represents the true value of the sparse parameter vector.

Remarks 9.4.

- Often in practice, the columns of the input matrix, X, are normalized to unit ℓ_2 norm. Although ℓ_0 norm is insensitive to the values of the nonzero components of $\boldsymbol{\theta}$, this is not the case with the ℓ_1 and ℓ_2 norms. Hence, while trying to minimize the respective norms and at the same time fulfill the constraints, components that correspond to columns of X with high energy (norm) are favored over the rest. Hence, the latter become more popular candidates to be pushed to zero. In order to avoid such situations, the columns of X are normalized to unity by dividing each element of the column vector by the respective (Euclidean) norm.

9.6 UNIQUENESS OF THE ℓ_0 MINIMIZER

Our first goal is to derive *sufficient* conditions that guarantee uniqueness of the ℓ_0 minimizer, which has been defined in Section 9.5.

Definition 9.2. The *spark* of a full row rank $N \times l$ $(l \geq N)$ matrix, X, denoted as spark(X), is the *smallest* number of its linearly *dependent* columns.

According to the previous definition, *any* $m <$ spark(X) column of X is necessarily *linearly independent*. The spark of a square, $N \times N$, full rank matrix is equal to $N + 1$.

Remarks 9.5.

- In contrast to the rank of a matrix, which can be easily determined, its spark can only be obtained by resorting to a combinatorial search over all possible combinations of the columns of the respective matrix (see, e.g., [15,37]). The notion of the spark was used in the context of sparse representation, under the name *uniqueness representation property*, in [53]. The name "spark" was coined in [37]. An interesting discussion relating this matrix index with indices used in other disciplines is given in [15].
- Note that the notion of "spark" is related to the notion of the minimum Hamming weight of a linear code in coding theory (e.g., [60]).

Example 9.3. Consider the following matrix:

$$X = \begin{bmatrix} 1 & 0 & 0 & 0 & 1 & 0 \\ 0 & 1 & 0 & 0 & 1 & 1 \\ 0 & 0 & 1 & 0 & 0 & 1 \\ 0 & 0 & 0 & 1 & 0 & 0 \end{bmatrix}.$$

The matrix has rank equal to 4 and spark equal to 3. Indeed, any pair of columns is linearly independent. On the other hand, the first, second, and fifth columns are linearly dependent. The same is also true for the combination of the second, third, and sixth columns. Also, the maximum number of linearly independent columns is four.

Lemma 9.2. *If* null(X) *is the null space of* X, *then*

$$\|\boldsymbol{\theta}\|_0 \geq \text{spark}(X), \quad \forall \boldsymbol{\theta} \in \text{null}(X), \boldsymbol{\theta} \neq \mathbf{0}.$$

Proof. To derive a contradiction, assume that there exists a $\boldsymbol{\theta} \in \text{null}(X)$, $\boldsymbol{\theta} \neq \mathbf{0}$, such that $\|\boldsymbol{\theta}\|_0 < \text{spark}(X)$. Because by definition $X\boldsymbol{\theta} = \mathbf{0}$, there exists a number of $\|\boldsymbol{\theta}\|_0$ columns of X that are linearly dependent. However, this contradicts the minimality of spark(X), and the claim of Lemma 9.2 is established.

Lemma 9.3. *If a linear system of equations,* $X\boldsymbol{\theta} = \boldsymbol{y}$, *has a solution that satisfies*

$$\boxed{\|\boldsymbol{\theta}\|_0 < \frac{1}{2}\,\text{spark}(X),}$$

then this is the sparsest possible solution. In other words, this is necessarily the unique solution of the ℓ_0 *minimizer.*

Proof. Consider any other solution $\boldsymbol{h} \neq \boldsymbol{\theta}$. Then $\boldsymbol{\theta} - \boldsymbol{h} \in \text{null}(X)$, that is,

$$X(\boldsymbol{\theta} - \boldsymbol{h}) = \mathbf{0}.$$

Thus, according to Lemma 9.2,

$$\text{spark}(X) \leq \|\boldsymbol{\theta} - \boldsymbol{h}\|_0 \leq \|\boldsymbol{\theta}\|_0 + \|\boldsymbol{h}\|_0 . \tag{9.24}$$

Observe that although the ℓ_0 "norm" is not a true norm, it can be readily verified by simple inspection and reasoning that the triangular property is satisfied. Indeed, by adding two vectors together, the resulting number of nonzero elements will always be at most equal to the total number of nonzero elements of the two vectors. Therefore, if $\|\boldsymbol{\theta}\|_0 < \frac{1}{2}\,\text{spark}(X)$, then (9.24) suggests that

$$\|\boldsymbol{h}\|_0 > \frac{1}{2}\,\text{spark}(X) > \|\boldsymbol{\theta}\|_0 .$$

Remarks 9.6.

- Lemma 9.3 is a very interesting result. We have a sufficient condition to check whether a solution is the unique optimal in a generally NP-hard problem. Of course, although this is nice from a theoretical point of view, it is not of much use by itself, because the related bound (the spark) can only be obtained after a combinatorial search. In the next section, we will see that we can relax the bound by involving another index in place of the spark, which is easier to be computed.
- An obvious consequence of the previous lemma is that if the unknown parameter vector is a sparse one with k nonzero elements, then if matrix X is chosen in order to have spark$(X) > 2k$, the true parameter vector is necessarily the sparsest one that satisfies the set of equations, and the (unique) solution to the ℓ_0 minimizer.

- In practice, the goal is to sense the unknown parameter vector by a matrix that has a spark as high as possible, so that the previously stated sufficiency condition covers a wide range of cases. For example, if the spark of the input matrix is equal to three, then one can check for optimal sparse solutions up to a sparsity level of $k = 1$. From the respective definition, it is easily seen that the values of the spark are in the range $1 < \text{spark}(X) \leq N + 1$.
- Constructing an $N \times l$ matrix X in a random manner, by generating i.i.d. entries, guarantees with high probability that $\text{spark}(X) = N + 1$; that is, any N columns of the matrix are linearly independent.

9.6.1 MUTUAL COHERENCE

Because the spark of a matrix is a number that is difficult to compute, our interest shifts to another index, which can be derived more easily and at the same time offers a useful bound on the spark. The *mutual coherence* of an $N \times l$ matrix X [65], denoted as $\mu(X)$, is defined as

$$\boxed{\mu(X) := \max_{1 \leq i < j \leq l} \frac{\left|\boldsymbol{x}_i^{cT} \boldsymbol{x}_j^c\right|}{\left\|\boldsymbol{x}_i^c\right\| \left\|\boldsymbol{x}_j^c\right\|} : \quad \text{mutual coherence,}} \tag{9.25}$$

where $\boldsymbol{x}_i^c$, $i = 1, 2, \ldots, l$, denote the columns of X (note the difference in notation between a row $\boldsymbol{x}_i^T$ and a column $\boldsymbol{x}_i^c$ of the matrix X). This number reminds us of the correlation coefficient between two random variables. Mutual coherence is bounded as $0 \leq \mu(X) \leq 1$. For a square orthogonal matrix X, $\mu(X) = 0$. For general matrices with $l > N$, $\mu(X)$ satisfies

$$\sqrt{\frac{l - N}{N(l-1)}} \leq \mu(X) \leq 1,$$

which is known as the *Welch bound* [98] (Problem 9.15). For large values of l, the lower bound becomes approximately $\mu(X) \geq \frac{1}{\sqrt{N}}$. Common sense reasoning guides us to construct input (sensing) matrices of mutual coherence as small as possible. Indeed, the purpose of the sensing matrix is to "measure" the components of the unknown vector and "store" this information in the measurement vector $\boldsymbol{y}$. Thus, this should be done in such a way that $\boldsymbol{y}$ retains as much information about the components of $\boldsymbol{\theta}$ as possible. This can be achieved if the columns of the sensing matrix X are as "independent" as possible. Indeed, $\boldsymbol{y}$ is the result of a combination of the columns of X, each one weighted by a different component of $\boldsymbol{\theta}$. Thus, if the columns are as "independent" as possible, then the information regarding each component of $\boldsymbol{\theta}$ is contributed by a different direction, making its recovery easier. This is easier understood if X is a square orthogonal matrix. In the more general case of a nonsquare matrix, the columns should be made as "orthogonal" as possible.

Example 9.4. Assume that X is an $N \times 2N$ matrix, formed by concatenating two orthonormal bases together,

$$X = [I, W],$$

where I is the identity matrix, having as columns the vectors $\boldsymbol{e}_i$, $i = 1, 2, \ldots, N$, with elements equal to

$$\delta_{ir} = \begin{cases} 1, & \text{if } i = r, \\ 0, & \text{if } i \neq r, \end{cases}$$

for $r = 1, 2, \ldots, N$. The matrix W is the orthonormal DFT matrix, defined as

$$W = \frac{1}{\sqrt{N}} \begin{bmatrix} 1 & 1 & \ldots & 1 \\ 1 & W_N & \ldots & W_N^{N-1} \\ \vdots & \vdots & \ddots & \vdots \\ 1 & W_N^{N-1} & \ldots & W_N^{(N-1)(N-1)} \end{bmatrix},$$

where

$$W_N := \exp\left(-j\frac{2\pi}{N}\right).$$

For example, such an overcomplete dictionary could be used in the expansion (9.16) to represent signal vectors, which comprise the sum of sinusoids with spiky-like signal pulses. The inner products between any two columns of I and between any two columns of W are zero, due to orthogonality. On the other hand, it is easy to see that the inner product between any column of I and any column of W has absolute value equal to $\frac{1}{\sqrt{N}}$. Hence, the mutual coherence of this matrix is $\mu(X) = \frac{1}{\sqrt{N}}$. Moreover, observe that the spark of this matrix is $\text{spark}(X) = N + 1$.

Lemma 9.4. *For any $N \times l$ matrix X, the following inequality holds:*

$$\text{spark}(X) \geq 1 + \frac{1}{\mu(X)}. \tag{9.26}$$

The proof is given in [37] and it is based on arguments that stem from matrix theory applied on the Gram matrix, $X^T X$, of X (Problem 9.16). A "superficial" look at the previous bound is that for very small values of $\mu(X)$ the spark can be larger than $N + 1$! Looking at the proof, it is seen that in such cases the spark of the matrix attains its maximum value $N + 1$.

The result complies with common sense reasoning. The smaller the value of $\mu(X)$, the more independent the columns of X, hence the higher the value of its spark is expected to be. Based on this lemma, we can now state the following theorem, first given in [37]. Combining Lemma 9.3 and (9.26), we come to the following important theorem.

Theorem 9.1. *If the linear system of equations in (9.17) has a solution that satisfies the condition*

$$\boxed{\|\boldsymbol{\theta}\|_0 < \frac{1}{2}\left(1 + \frac{1}{\mu(X)}\right),} \tag{9.27}$$

then this solution is the sparsest one.

Remarks 9.7.

- The bound in (9.27) is "psychologically" important. It relates an easily computed bound to check whether the solution to an NP-hard task is the optimal one. However, it is not a particularly good bound and it restricts the range of values in which it can be applied. As it was discussed in Example 9.4, while the maximum possible value of the spark of a matrix was equal to $N+1$, the minimum possible value of the mutual coherence was $\frac{1}{\sqrt{N}}$. Therefore, the bound based on the mutual coherence restricts the range of sparsity, that is, $\|\boldsymbol{\theta}\|_0$, where one can check optimality, to around $\frac{1}{2}\sqrt{N}$. Moreover, as the previously stated Welch bound suggests, this $\mathcal{O}(\frac{1}{\sqrt{N}})$ dependence of the mutual coherence seems to be a more general trend and not only the case for Example 9.4 (see, e.g., [36]). On the other hand, as we have already stated in the Remarks 9.6, one can construct random matrices with spark equal to $N+1$; hence, using the bound based on the spark, one could expand the range of sparse vectors up to $\frac{1}{2}N$.

9.7 EQUIVALENCE OF ℓ_0 AND ℓ_1 MINIMIZERS: SUFFICIENCY CONDITIONS

We have now come to the crucial point where we will establish the conditions that guarantee the equivalence between the ℓ_1 and ℓ_0 minimizers. Hence, under such conditions, a problem that is in general an NP-hard one *can be solved via a tractable convex optimization task.* Under these conditions, the zero value encouraging nature of the ℓ_1 norm, which has already been discussed, obtains a much higher stature; it provides the sparsest solution.

9.7.1 CONDITION IMPLIED BY THE MUTUAL COHERENCE NUMBER

Theorem 9.2. *Consider the underdetermined system of equations*

$$y = X\boldsymbol{\theta},$$

where X is an $N \times l$ $(N < l)$ full row rank matrix. If a solution exists and satisfies the condition

$$\|\boldsymbol{\theta}\|_0 < \frac{1}{2}\left(1 + \frac{1}{\mu(X)}\right), \tag{9.28}$$

then this is the unique solution of both the ℓ_0 and ℓ_1 minimizers.

This is a very important theorem, and it was shown independently in [37,54]. Earlier versions of the theorem addressed the special case of a dictionary comprising two orthonormal bases [36,48]. A proof is also summarized in [15] (Problem 9.17). This theorem established, for the first time, what was until then empirically known: often, the ℓ_1 and ℓ_0 minimizers result in the same solution.

Remarks 9.8.

- The theory that we have presented so far is very satisfying, because it offers the theoretical framework and conditions that guarantee uniqueness of a sparse solution to an underdetermined system of equations. Now we know that under certain conditions, the solution which we obtain by solving the convex ℓ_1 minimization task, is the (unique) sparsest one. However, from a practical point of view, the theory, which is based on mutual coherence, does not tell the whole story and falls short in predicting what happens in practice. Experimental evidence suggests that the range of sparsity levels, for which the ℓ_0 and ℓ_1 tasks give the same solution, is much wider than the range guaranteed by the mutual coherence bound. Hence, there is a lot of theoretical happening in order to improve this bound. A detailed discussion is beyond the scope of this book. In the next section, we will present one of these bounds, because it is the one that currently dominates the scene. For more details and a related discussion, the interested reader may consult, for example, [39,49,50].

9.7.2 THE RESTRICTED ISOMETRY PROPERTY (RIP)

Definition 9.3. For each integer $k = 1, 2, \ldots,$ define the *isometry constant* δ_k of an $N \times l$ matrix X as the *smallest* number such that

$$\boxed{(1-\delta_k)\,\|\boldsymbol{\theta}\|_2^2 \leq \|X\boldsymbol{\theta}\|_2^2 \leq (1+\delta_k)\,\|\boldsymbol{\theta}\|_2^2 : \quad \text{the RIP condition}} \tag{9.29}$$

holds true for *all* k-sparse vectors $\boldsymbol{\theta}$.

This definition was introduced in [19]. We loosely say that matrix X obeys the RIP of order k if δ_k is not too close to one. When this property holds true, it implies that the Euclidean norm of $\boldsymbol{\theta}$ is approximately *preserved* after projecting it onto the rows of X. Obviously, if matrix X was orthonormal, we would have $\delta_k = 0$. Of course, because we are dealing with nonsquare matrices, this is not possible. However, the closer δ_k is to zero, the closer to orthonormal *all* subsets of k columns of X are. Another viewpoint of (9.29) is that X preserves Euclidean distances between k-sparse vectors. Let us consider two k-sparse vectors, $\boldsymbol{\theta}_1, \boldsymbol{\theta}_2$, and apply (9.29) to their difference, $\boldsymbol{\theta}_1 - \boldsymbol{\theta}_2$, which, in general, is a $2k$-sparse vector. Then we obtain

$$(1-\delta_{2k})\,\|\boldsymbol{\theta}_1-\boldsymbol{\theta}_2\|_2^2 \leq \|X(\boldsymbol{\theta}_1-\boldsymbol{\theta}_2)\|_2^2 \leq (1+\delta_{2k})\,\|\boldsymbol{\theta}_1-\boldsymbol{\theta}_2\|_2^2. \tag{9.30}$$

Thus, when δ_{2k} is small enough, the Euclidean distance is preserved after projection in the lower-dimensional observations' space. In words, if the RIP holds true, this means that searching for a sparse vector in the lower-dimensional subspace, $\mathbb{R}^N$, formed by the observations, and not in the original l-dimensional space, one can still recover the vector since distances are preserved and the target vector is not "confused" with others. After projection onto the rows of X, the discriminatory power of the method is retained. It is interesting to point out that the RIP is also related to the condition number of the Grammian matrix. In [6,19], it is pointed out that if X_r denotes the matrix that results by considering only r of the columns of X, then the RIP in (9.29) is equivalent to requiring the respective Grammian, $X_r^T X_r$, $r \leq k$, to have its eigenvalues within the interval $[1-\delta_k, 1+\delta_k]$. Hence, the more well-conditioned the matrix is, the better we dig out the information hidden in the lower-dimensional space.

Theorem 9.3. *Assume that for some k, $\delta_{2k} < \sqrt{2}-1$. Then the solution to the ℓ_1 minimizer of (9.21), denoted as $\boldsymbol{\theta}_*$, satisfies the following two conditions:*

$$\|\boldsymbol{\theta}-\boldsymbol{\theta}_*\|_1 \leq C_0\,\|\boldsymbol{\theta}-\boldsymbol{\theta}_k\|_1 \tag{9.31}$$

and

$$\|\boldsymbol{\theta} - \boldsymbol{\theta}_*\|_2 \leq C_0 k^{-\frac{1}{2}} \|\boldsymbol{\theta} - \boldsymbol{\theta}_k\|_1, \tag{9.32}$$

for some constant C_0. In the previously stated formulas, $\boldsymbol{\theta}$ is the true (target) vector that generates the observations in (9.21) and $\boldsymbol{\theta}_k$ is the vector that results from $\boldsymbol{\theta}$ if we keep its k largest components and set the rest equal to zero [18,19,22,23].

Hence, if the true vector is a sparse one, that is, $\boldsymbol{\theta} = \boldsymbol{\theta}_k$, then the ℓ_1 minimizer recovers the (unique) exact value. On the other hand, if the true vector is not a sparse one, then the minimizer results in a solution whose accuracy is dictated by a genie-aided procedure that knew in advance the locations of the k largest components of $\boldsymbol{\theta}$. This is a groundbreaking result. Moreover, it is deterministic; it is always true, not with high probability. Note that the isometry property of order $2k$ is used, because at the heart of the method lies our desire to preserve the norm of the differences between vectors.

Let us now focus on the case where there is a k-sparse vector that generates the observations, that is, $\boldsymbol{\theta} = \boldsymbol{\theta}_k$. Then it is shown in [18] that the condition $\delta_{2k} < 1$ guarantees that the ℓ_0 minimizer has a unique k-sparse solution. In other words, in order to get the equivalence between the ℓ_1 and ℓ_0 minimizers, the range of values for δ_{2k} has to be decreased to $\delta_{2k} < \sqrt{2} - 1$, according to Theorem 9.3. This sounds reasonable. If we relax the criterion and use ℓ_1 instead of ℓ_0, then the sensing matrix has to be more carefully constructed. Although we will not provide the proofs of these theorems here, because their formulation is well beyond the scope of this book, it is interesting to follow what happens if $\delta_{2k} = 1$. This will give us a flavor of the essence behind the proofs. If $\delta_{2k} = 1$, the left-hand side term in (9.30) becomes zero. In this case, there may exist two k-sparse vectors $\boldsymbol{\theta}_1, \boldsymbol{\theta}_2$ such that $X(\boldsymbol{\theta}_1 - \boldsymbol{\theta}_2) = \mathbf{0}$, or $X\boldsymbol{\theta}_1 = X\boldsymbol{\theta}_2$. Thus, it is not possible to recover all k-sparse vectors, after projecting them in the observations space, by any method.

The previous argument also establishes a connection between RIP and the spark of a matrix. Indeed, if $\delta_{2k} < 1$, this guarantees that any number of columns of X up to $2k$ are linearly independent, because for any $2k$-sparse $\boldsymbol{\theta}$, (9.29) guarantees that $\|X\boldsymbol{\theta}\|_2 > 0$. This implies that spark$(X) > 2k$. A connection between RIP and the coherence is established in [16], where it is shown that if X has coherence $\mu(X)$, and unit norm columns, then X satisfies the RIP of order k with δ_k, where $\delta_k \leq (k-1)\mu(X)$.

Constructing Matrices That Obey the RIP of Order k

It is apparent from our previous discussion that the higher the value of k, for which the RIP property of a matrix X holds true, the better, since a larger range of sparsity levels can be handled. Hence, a main goal toward this direction is to construct such matrices. It turns out that verifying the RIP for a matrix of a general structure is a difficult task. This reminds us of the spark of the matrix, which is also a difficult task to compute. However, it turns out that for a certain class of random matrices, the RIP can be established in an affordable way. Thus, constructing such sensing matrices has dominated the scene of related research. We will present a few examples of such matrices, which are also very popular in practice, without going into details of the proofs, because this is beyond the scope of this book. The interested reader may find this information in the related references.

Perhaps the most well-known example of a random matrix is the Gaussian one, where the entries $X(i, j)$ of the sensing matrix are i.i.d. realizations from a Gaussian PDF $\mathcal{N}(0, \frac{1}{N})$. Another popular example of such matrices is constructed by sampling i.i.d. entries from Bernoulli, or related, distributions

$$X(i,j) = \begin{cases} \frac{1}{\sqrt{N}}, & \text{with probability } \frac{1}{2}, \\ -\frac{1}{\sqrt{N}}, & \text{with probability } \frac{1}{2}, \end{cases}$$

or

$$X(i,j) = \begin{cases} +\sqrt{\frac{3}{N}}, & \text{with probability } \frac{1}{6}, \\ 0, & \text{with probability } \frac{2}{3}, \\ -\sqrt{\frac{3}{N}}, & \text{with probability } \frac{1}{6}. \end{cases}$$

Finally, one can adopt the uniform distribution and construct the columns of X by sampling uniformly at random on the unit sphere in $\mathbb{R}^N$. It turns out that such matrices obey the RIP of order k with overwhelming probability, provided that the number of observations, N, satisfies the inequality

$$N \geq Ck \ln(l/k), \tag{9.33}$$

where C is some constant which depends on the isometry constant δ_k. In words, having such a matrix at our disposal, one can recover a k-sparse vector from $N < l$ observations, where N is larger than the sparsity level by an amount controlled by inequality (9.33). More on these issues can be obtained from, for example, [6,67].

Besides random matrices, one can construct other matrices that obey the RIP. One such example includes the partial Fourier matrices, which are formed by selecting uniformly at random N rows drawn from the $l \times l$ DFT matrix. Although the required number of samples for the RIP to be satisfied may be larger than the bound in (9.33) (see [79]), Fourier-based sensing matrices offer certain computational advantages when it comes to storage ($\mathcal{O}(N \ln l)$) and matrix-vector products ($\mathcal{O}(l \ln l)$) [20]. In [56], the case of random Toeplitz sensing matrices containing statistical dependencies across rows is considered, and it is shown that they can also satisfy the RIP with high probability. This is of particular importance in signal processing and communications applications, where it is very common for a system to be excited in its input via a time series, and hence independence between successive input rows cannot be assumed. In [44,76], the case of separable matrices is considered where the sensing matrix is the result of a Kronecker product of matrices, which satisfy the RIP individually. Such matrices are of interest for multidimensional signals, in order to exploit the sparsity structure along each one of the involved dimensions. For example, such signals may occur while trying to "encode" information associated with an event whose activity spreads across the temporal, spectral, spatial, and other domains.

In spite of their theoretical elegance, the derived bounds that determine the number of the required observations for certain sparsity levels fall short of the experimental evidence (e.g., [39]). In practice, a rule of thumb is to use N of the order of $3k$–$5k$ [18]. For large values of l, compared to the sparsity level, the analysis in [38] suggests that we can recover most sparse signals when $N \approx 2k \ln(l/N)$. In an effort to overcome the shortcomings associated with the RIP, a number of other techniques have been proposed (e.g., [11,30,39,85]). Furthermore, in specific applications, the use of an empirical study may be a more appropriate path.

Note that, in principle, the minimum number of observations that are required to recover a k-sparse vector from $N < l$ observations is $N \geq 2k$. Indeed, in the spirit of the discussion after Theorem 9.3, the main requirement that a sensing matrix must fulfill is not to map two different k-sparse vectors to the same measurement vector $\boldsymbol{y}$. Otherwise, one can never recover both vectors from their (common) observations. If we have $2k$ observations and a sensing matrix that guarantees that any $2k$ columns are linearly independent, then the previously stated requirement is satisfied. However, the bounds on the number of observations set in order for the respective matrices to satisfy the RIP are larger. This is because RIP accounts also for the stability of the recovery process. We will come to this issue in Section 9.9, where we talk about *stable* embeddings.

9.8 ROBUST SPARSE SIGNAL RECOVERY FROM NOISY MEASUREMENTS

In the previous section, our focus was on recovering a sparse solution from an underdetermined system of equations. In the formulation of the problem, we assumed that there is no noise in the obtained observations. Having acquired some experience and insight from a simpler scenario, we now turn our attention to the more realistic task, where uncertainties come into the scene. One type of uncertainty may be due to the presence of noise, and our observations' model comes back to the standard regression form

$$\boldsymbol{y} = X\boldsymbol{\theta} + \boldsymbol{\eta}, \tag{9.34}$$

where X is our familiar nonsquare $N \times l$ matrix. A sparsity-aware formulation for recovering $\boldsymbol{\theta}$ from (9.34) can be cast as

$$\begin{aligned} \min_{\boldsymbol{\theta} \in \mathbb{R}^l} \quad & \|\boldsymbol{\theta}\|_1, \\ \text{s.t.} \quad & \|\boldsymbol{y} - X\boldsymbol{\theta}\|_2^2 \leq \epsilon, \end{aligned} \tag{9.35}$$

which coincides with the LASSO task given in (9.8). Such a formulation implicitly assumes that the noise is bounded and the respective range of values is controlled by ϵ. One can consider a number of different variants. For example, one possibility would be to minimize the $\|\cdot\|_0$ norm instead of the $\|\cdot\|_1$, albeit losing the computational elegance of the latter. An alternative route would be to replace the Euclidean norm in the constraints with another one.

Besides the presence of noise, one could see the previous formulation from a different perspective. The unknown parameter vector, $\boldsymbol{\theta}$, may not be exactly sparse, but it may consist of a few large components, while the rest are small and close to, yet not necessarily equal to, zero. Such a model misfit can be accommodated by allowing a deviation of $\boldsymbol{y}$ from $X\boldsymbol{\theta}$.

In this relaxed setting of a sparse solution recovery, the notions of uniqueness and equivalence concerning the ℓ_0 and ℓ_1 solutions no longer apply. Instead, the issue that now gains importance is that of *stability* of the solution. To this end, we focus on the computationally attractive ℓ_1 task. The counterpart of Theorem 9.3 is now expressed as follows.

Theorem 9.4. *Assume that the sensing matrix, X, obeys the RIP with $\delta_{2k} < \sqrt{2} - 1$, for some k. Then the solution $\boldsymbol{\theta}_*$ of (9.35) satisfies the following ([22,23]):*

$$\|\boldsymbol{\theta} - \boldsymbol{\theta}_*\|_2 \leq C_0 k^{-\frac{1}{2}} \|\boldsymbol{\theta} - \boldsymbol{\theta}_k\|_1 + C_1\sqrt{\epsilon}, \tag{9.36}$$

for some constants C_1, C_0, *and* $\boldsymbol{\theta}_k$ *as defined in Theorem 9.3.*

This is also an elegant result. If the model is exact and $\epsilon = 0$, we obtain (9.32). If not, the higher the uncertainty (noise) term in the model, the higher our ambiguity about the solution. Note, also, that the ambiguity about the solution depends on how far the true model is from $\boldsymbol{\theta}_k$. If the true model is k-sparse, the first term on the right-hand side of the inequality is zero. The values of C_1, C_0 depend on δ_{2k} but they are small, for example, close to five or six [23].

The important conclusion here is that *adopting the ℓ_1 norm and the associated LASSO optimization for solving inverse problems (which in general, as we noted in Chapter 3, tend to be ill-conditioned) is a stable one and the noise is not amplified excessively during the recovery process.*

9.9 COMPRESSED SENSING: THE GLORY OF RANDOMNESS

The way in which this chapter was deployed followed, more or less, the sequence of developments that took place during the evolution of the sparsity-aware parameter estimation field. We intentionally made an effort to follow such a path, because this is also indicative of how science evolves in most cases. The starting point had a rather strong mathematical flavor: to develop conditions for the solution of an underdetermined linear system of equations, under the sparsity constraint and in a mathematically tractable way, that is, using convex optimization. In the end, the accumulation of a sequence of individual contributions revealed that the solution can be (uniquely) recovered if the unknown quantity is sensed via randomly chosen data samples. This development has, in turn, given birth to a new field with strong theoretical interest as well as with an enormous impact on practical applications. This new emerged area is known as *compressed sensing* or *compressive sampling* (CS), and it has changed our view on how to sense and process signals efficiently.

COMPRESSED SENSING

In CS, the goal is to directly acquire as few samples as possible that encode the minimum information needed to obtain a compressed signal representation. In order to demonstrate this, let us return to the data compression example discussed in Section 9.4. There, it was commented that the "classical" approach to compression was rather unorthodox, in the sense that first all (i.e., a number of l) samples of the signal are used, and then they are processed to obtain l transformed values, from which only a small subset is used for coding. In the CS setting, the procedure changes to the following one.

Let X be an $N \times l$ sensing matrix, which is applied to the (unknown) signal vector, $\boldsymbol{s}$, in order to obtain the observations, $\boldsymbol{y}$, and let Ψ be the dictionary matrix that describes the domain where the signal $\boldsymbol{s}$ accepts a sparse representation, that is,

$$\begin{aligned} \boldsymbol{s} &= \Psi\boldsymbol{\theta}, \\ \boldsymbol{y} &= X\boldsymbol{s}. \end{aligned} \tag{9.37}$$

Assuming that at most k of the components of $\boldsymbol{\theta}$ are nonzero, this can be obtained by the following optimization task:

$$\begin{aligned} \min_{\boldsymbol{\theta}\in\mathbb{R}^l} & \quad \|\boldsymbol{\theta}\|_1, \\ \text{s.t.} & \quad \boldsymbol{y} = X\Psi\boldsymbol{\theta}, \end{aligned} \tag{9.38}$$

provided that the combined matrix $X\Psi$ *complies* with the RIP, and the number of observations, N, is large enough, as dictated by the bound in (9.33). Note that $\boldsymbol{s}$ needs not be stored and can be obtained any time, once $\boldsymbol{\theta}$ is known. Moreover, as we will soon discuss, there are techniques that allow observations, y_n, $n = 1, 2, \ldots, N$, to be acquired directly from an analog signal $s(t)$, prior to obtaining its sample (vector) version, $\boldsymbol{s}$! Thus, from such a perspective, CS fuses the data acquisition and the compression steps together.

There are different ways to obtain a sensing matrix, X, that lead to a product $X\Psi$, which satisfies the RIP. It can be shown (Problem 9.19) that if Ψ is orthonormal and X is a random matrix, which is constructed as discussed at the end of Section 9.7.2, then the product $X\Psi$ obeys the RIP, provided that (9.33) is satisfied. An alternative way to obtain a combined matrix that respects the RIP is to consider another orthonormal matrix Φ, whose columns have low coherence with the columns of Ψ (coherence between two matrices is defined in (9.25), where now the place of $\boldsymbol{x}_i^c$ is taken by a column of Φ and that of $\boldsymbol{x}_j^c$ by a column of Ψ). For example, Φ could be the DFT matrix and $\Psi = I$, or vice versa. Then choose N rows of Φ uniformly at random to form X in (9.37). In other words, for such a case, the sensing matrix can be written as $R\Phi$, where R is an $N \times l$ matrix that extracts N rows uniformly at random. The notion of incoherence (low coherence) between the sensing and the basis matrices is closely related to RIP. The more incoherent the two matrices, the lower the number of the required observations for the RIP to hold (e.g., [21,79]). Another way to view incoherence is that the rows of Φ cannot be sparsely represented in terms of the columns of Ψ. It turns out that if the sensing matrix X is a random one, formed as described in Section 9.7.2, then the RIP and the incoherence with any Ψ are satisfied with high probability.

It gets even better when we say that all the previously stated philosophy can be extended to the more general type of signals, which are not necessarily sparse or sparsely represented in terms of the atoms of a dictionary, and they are known as *compressible*. A signal vector is said to be compressible if its expansion in terms of a basis consists of just a few large parameters θ_i and the rest are small. In other words, the signal vector is *approximately* sparse in some basis. Obviously, this is the most interesting case in practice, where exact sparsity is scarcely (if ever) met. Reformulating the arguments used in Section 9.8, the CS task for this case can be cast as

$$\begin{aligned} \min_{\boldsymbol{\theta}\in\mathbb{R}^l} & \quad \|\boldsymbol{\theta}\|_1, \\ \text{s.t.} & \quad \|\boldsymbol{y} - X\Psi\boldsymbol{\theta}\|_2^2 \le \epsilon, \end{aligned} \tag{9.39}$$

and everything that has been said in Section 9.8 is also valid for this case, if in place of X we consider the product $X\Psi$.

Remarks 9.9.

- An important property in CS is that the sensing matrix, which provides the observations, may be chosen independently on the matrix Ψ, that is, the basis/dictionary in which the signal is sparsely represented. In other words, the sensing matrix can be "universal" and can be used to provide the observations for reconstructing any sparse or sparsely represented signal in any dictionary, provided the RIP is not violated.

- Each observation, y_n, is the result of an inner product of the signal vector with a row $\boldsymbol{x}_n^T$ of the sensing matrix X. Assuming that the signal vector $\boldsymbol{s}$ is the result of a sampling process on an analog signal, $s(t)$, y_n can be directly obtained, to a good approximation, by taking the inner product (integral) of $s(t)$ with a sensing waveform, $x_n(t)$, that corresponds to $\boldsymbol{x}_n$. For example, if X is formed by ± 1, as described in Section 9.7.2, then the configuration shown in Fig. 9.10 results in y_n. An important aspect of this approach, besides avoiding computing and storing the l components of $\boldsymbol{s}$, is that multiplying by ± 1 is a relatively easy operation. It is equivalent to changing the polarity of the signal and it can be implemented by employing inverters and mixers. It is a process that can be performed, in practice, at much higher rates than sampling. The sampling system shown in Fig. 9.10 is referred to as *random demodulator* [58,91]. It is one among the popular analog-to-digital (A/D) conversion architectures, which exploit the CS rationale in order to sample at rates much lower than those required for classical sampling. We will come back to this soon.

  One of the very first CS-based acquisition systems was an imaging system called the *one-pixel camera* [84], which followed an approach resembling the conventional digital CS. According to this, light of an image of interest is projected onto a random base generated by a micromirror device. A sequence of projected images is collected by a *single photodiode* and used for the reconstruction of the full image using conventional CS techniques. This was among the most catalytic examples that spread the rumor about the practical power of CS. CS is an example of common wisdom: "There is nothing more practical than a good theory!"

9.9.1 DIMENSIONALITY REDUCTION AND STABLE EMBEDDINGS

We will now shed light on what we have said so far in this chapter from a different point of view. In both cases, either when the unknown quantity was a k-sparse vector in a high-dimensional space, $\mathbb{R}^l$, or when the signal $\boldsymbol{s}$ was (approximately) sparsely represented in some dictionary ($\boldsymbol{s} = \Psi\boldsymbol{\theta}$), we chose to work in a lower-dimensional space ($\mathbb{R}^N$), that is, the space of the observations, $\boldsymbol{y}$. This is a typical task of dimensionality reduction (see Chapter 19). The main task in any (linear) dimensionality reduction technique is to choose the proper matrix X, which dictates the projection to the lower-dimensional space. In general, there is always a loss of information by projecting from $\mathbb{R}^l$ to $\mathbb{R}^N$, with $N < l$, in

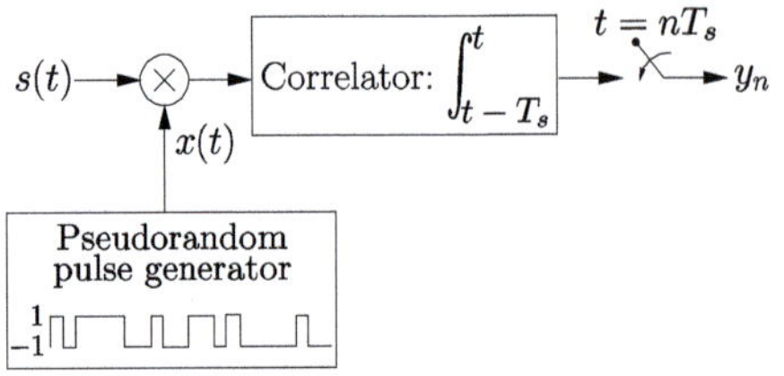

FIGURE 9.10

Sampling an analog signal $s(t)$ in order to generate the sample/observation y_n at time instant n. The sampling period T_s is much shorter than that required by the Nyquist sampling.

the sense that we cannot recover any vector $\boldsymbol{\theta}_l \in \mathbb{R}^l$ from its projection $\boldsymbol{\theta}_N \in \mathbb{R}^N$. Indeed, take any vector $\boldsymbol{\theta}_{l-N} \in \text{null}(X)$ that lies in the $(l - N)$-dimensional null space of the (full row rank) X (see Section 9.5). Then all vectors $\boldsymbol{\theta}_l + \boldsymbol{\theta}_{l-N} \in \mathbb{R}^l$ share the same projection in $\mathbb{R}^N$. However, what we have discovered in this chapter is that if the original vector is sparse, then we can recover it exactly. This is because all the k-sparse vectors do not lie anywhere in $\mathbb{R}^l$, but rather in a subset of it, that is, in a *union of subspaces*, each one having dimensionality k. If the signal $\boldsymbol{s}$ is sparse in some dictionary Ψ, then one has to search for it in the union of all possible k-dimensional subspaces of $\mathbb{R}^l$, which are spanned by k-column vectors from Ψ [8,62]. Of course, even in this case, where sparse vectors are involved, no projection can guarantee unique recovery. The guarantee is provided if the projection in the lower-dimensional space is a *stable embedding*. A stable embedding in a lower-dimensional space must guarantee that if $\boldsymbol{\theta}_1 \neq \boldsymbol{\theta}_2$, then their projections also remain different. Yet this is not enough. A stable embedding must guarantee that distances are (approximately) preserved; that is, vectors that lie far apart in the high-dimensional space have projections that also lie far apart. Such a property guarantees robustness to noise. The sufficient conditions, which have been derived and discussed throughout this chapter, and guarantee the recovery of a sparse vector lying in $\mathbb{R}^l$ from its projections in $\mathbb{R}^N$, are conditions that guarantee stable embeddings. The RIP and the associated bound on N provide a condition on X that leads to stable embeddings. We commented on this norm preserving property of RIP in the related section. The interesting fact that came from the theory is that we can achieve such stable embeddings via random projection matrices.

Random projections for dimensionality reduction are not new and have extensively been used in pattern recognition, clustering, and data mining (see, e.g., [1,13,34,82,87]). The advent of the big data era resparked the interest in random projection-aided data analysis algorithms (e.g., [55,81]) for two major reasons. The first is that data processing is computationally lighter in the lower-dimensional space, because it involves operations with matrices or vectors represented with fewer parameters. Moreover, the projection of the data to lower-dimensional spaces can be realized via well-structured matrices at a computational cost significantly lower compared to that implied by general matrix-vector multiplications [29,42]. The reduced computational power required by these methods renders them appealing when dealing with excessively large data volumes. The second reason is that there exist randomized algorithms which access the data matrix a (usually fixed) number of times that is much smaller than the number of accesses performed by ordinary methods [28,55]. This is very important whenever the full amount of data does not fit in fast memory and has to be accessed in parts from slow memory devices, such as hard discs. In such cases, the computational time is often dominated by the cost of memory access.

The spirit underlying CS has been exploited in the context of pattern recognition too. In this application, one need not return to the original high-dimensional space after the information digging activity in the low-dimensional subspace. Since the focus in pattern recognition is to identify the class of an object/pattern, this can be performed in the observations subspace, provided that there is no class-related information loss. In [17], it is shown, using compressed sensing arguments, that if the data are approximately linearly separable in the original high-dimensional space and the data have a sparse representation, even in an unknown basis, then projecting randomly in the observations subspace retains the structure of linear separability.

Manifold learning is another area where random projections have also applied. A manifold is, in general, a nonlinear k-dimensional surface, embedded in a higher-dimensional (ambient) space. For example, the surface of a sphere is a two-dimensional manifold in a three-dimensional space. In [7,96], the compressed sensing rationale is extended to signal vectors that live along a k-dimensional submanifold of the space $\mathbb{R}^l$. It is shown that if choosing a matrix X to project and a sufficient number N of observations, then the corresponding submanifold has a stable embedding in the observations subspace, under the projection matrix X; that is, pair-wise Euclidean and geodesic distances are approximately preserved after the projection mapping. More on these issues can be found in the given references and in, for example, [8]. We will come to the manifold learning task in Chapter 19.

9.9.2 SUB-NYQUIST SAMPLING: ANALOG-TO-INFORMATION CONVERSION

In our discussion in the remarks presented before, we touched on a very important issue—that of going from the analog domain to the discrete one. The topic of A/D conversion has been at the forefront of research and technology since the seminal works of Shannon, Nyquist, Whittaker, and Kotelnikof were published; see, for example, [92] for a thorough related review. We all know that if the highest frequency of an analog signal $s(t)$ is less than $F/2$, then Shannon's theorem suggests that no loss of information is achieved if the signal is sampled, at least, at the Nyquist rate of $F = 1/T$, where T is the corresponding sampling period, and the signal can be perfectly recovered by its samples

$$s(t) = \sum_n s(nT)\,\mathrm{sinc}(Ft - n),$$

where sinc is the sampling function

$$\mathrm{sinc}(t) = \frac{\sin(\pi t)}{\pi t}.$$

While this has been the driving force behind the development of signal acquisition devices, the increasing complexity of emerging applications demands increasingly higher sampling rates that cannot be accommodated by today's hardware technology. This is the case, for example, in wideband communications, where conversion speeds, as dictated by Shannon's bound, have become more and more difficult to obtain. Consequently, alternatives to high-rate sampling are attracting strong interest, with the goal of reducing the sampling rate by exploiting the *underlying structure* of the signals at hand. For example, in many applications, the signal comprises a few frequencies or bands; see Fig. 9.11 for an illustration. In such cases, sampling at the Nyquist rate is inefficient. This is an old problem investigated by a number of authors, leading to techniques that allow low-rate sampling whenever the locations of the nonzero bands in the frequency spectrum are known (see, e.g., [61,93,94]). CS theory has inspired research to study cases where the locations (carrier frequencies) of the bands are not known a priori. A typical application of this kind, of high practical interest, lies within the field of cognitive radio (e.g., [68,88,103]).

The process of sampling an analog signal with a rate lower than the Nyquist one is referred to as *analog-to-information* sampling or *sub-Nyquist* sampling. Let us focus on two among the most popular CS-based A/D converters. The first is the *random demodulator* (RD), which was first presented in [58] and later improved and theoretically developed in [91]. RD in its basic configuration is shown in Fig. 9.10, and it is designed for acquiring at sub-Nyquist rates sparse multitone signals, that is, signals having a sparse DFT. This implies that the signal comprises a few frequency components, but these components are constrained to correspond to integral frequencies. This limitation was pointed out in [91], and potential solutions have been sought according to the general framework proposed in [24]

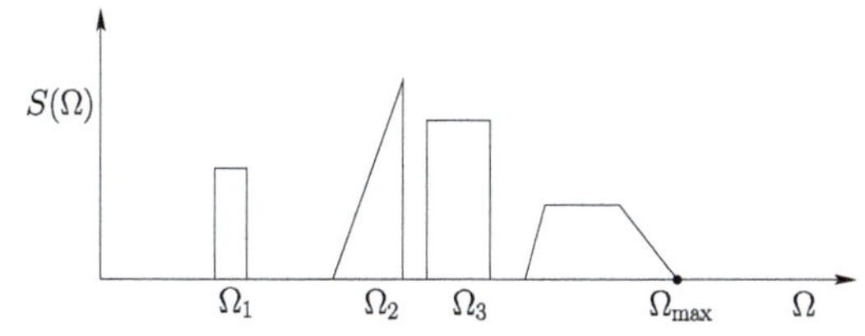

FIGURE 9.11

The Fourier transform of an analog signal, $s(t)$, which is sparse in the frequency domain; only a limited number of frequency bands contribute to its spectrum content $S(\Omega)$, where Ω stands for the angular frequency. Nyquist's theory guarantees that sampling at a frequency larger than or equal to twice the maximum Ω_{max} is sufficient to recover the original analog signal. However, this theory does not exploit information related to the sparse structure of the signal in the frequency domain.

and/or the heuristic approach described in [45]. Moreover, more elaborate RD designs, such as the *random-modulation preintegrator* (RMPI) [102], have the potential to deal with signals that are sparse in any domain.

Another CS-based sub-Nyquist sampling strategy that has received much attention is the *modulated wideband converter* (MWC) [68,69,71]. The MWC is very efficient in acquiring multiband signals such as the one depicted in Fig. 9.11. This concept has also been extended to accommodate signals with different characteristics, such as signals consisting of short pulses [66]. An in-depth investigation which sheds light on the similarities and differences between the RD and MWC sampling architectures can be found in [59].

Note that both RD and MWC sample the signal uniformly in time. In [97], a different approach is adopted, leading to much easier implementations. In particular, the preprocessing stage is avoided and nonuniformly spread in time samples are acquired directly from the raw signal. In total, fewer samples are obtained compared to Nyquist sampling. Then CS-based reconstruction is mobilized in order to recover the signal under consideration based on the values of the samples and the time information. Like in the basic RD case, the nonuniform sampling approach is suitable for signals sparse in the DFT basis. From a practical point of view, there are still a number of hardware implementation-related issues that more or less concern all the approaches above and need to be solved (see, e.g., [9,25,63]).

An alternative path to sub-Nyquist sampling embraces a different class of analog signals, known as *multipulse* signals, that is, signals that consist of a stream of short pulses. Sparsity now refers to the time domain, and such signals may not even be bandlimited. Signals of this type can be met in a number of applications, such as radar, ultrasound, bioimaging, and neuronal signal processing (see, e.g., [41]). An approach known as *finite rate of innovation sampling* passes an analog signal having k degrees of freedom per second through a linear time-invariant filter, and then samples at a rate of $2k$ samples per second. Reconstruction is performed via rooting a high-order polynomial (see, e.g., [12, 95] and the references therein). In [66], the task of sub-Nyquist sampling is treated using CS theory arguments and an expansion in terms of Gabor functions; the signal is assumed to consist of a sum of a few pulses of finite duration, yet of unknown shape and time positions. More on this topic can be obtained in [43,51,70] and the references therein.

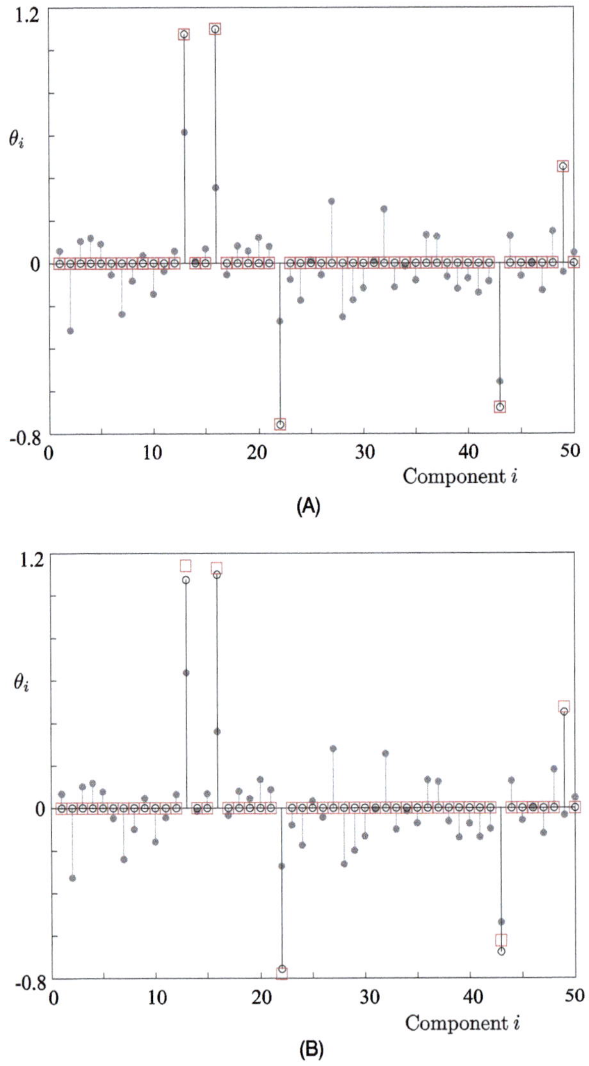

FIGURE 9.12

(A) Noiseless case. The values of the true vector, which generated the data for Example 9.5, are shown with stems topped with open circles. The recovered points are shown with squares. An exact recovery of the signal has been obtained. The stems topped with gray-filled circles correspond to the minimum Euclidean norm LS solution. (B) This figure corresponds to the noisy counterpart of that in (A). In the presence of noise, exact recovery is not possible and the higher the variance of the noise, the less accurate the results.

Example 9.5. We are given a set of $N = 20$ observations stacked in the $\boldsymbol{y} \in \mathbb{R}^N$ vector. These were taken by applying a sensing matrix X on an "unknown" vector in $\mathbb{R}^{50}$, which is known to be sparse with $k = 5$ nonzero components; the location of these nonzero components in the unknown vector is not known. The sensing matrix was a random matrix with elements drawn from a normal distribution $\mathcal{N}(0, 1)$, and then the columns were normalized to unit norm. There are two scenarios for the observations. In the first one, we are given their exact values while in the second one, white Gaussian noise of variance $\sigma^2 = 0.025$ was added.

In order to recover the unknown sparse vector, the CS matching pursuit (CoSaMP, Chapter 10) algorithm was used for both scenarios.

The results are shown in Fig. 9.12A and B for the noiseless and noisy scenarios, respectively. The values of the true unknown vector $\boldsymbol{\theta}$ are represented with black stems topped with open circles. Note that all but five of them are zero. In Fig. 9.12A, exact recovery of the unknown values is achieved; the estimated values of θ_i, $i = 1, 2 \ldots, 50$, are indicated with squares in red color. In the noisy case of Fig. 9.12B, the resulting estimates, which are denoted with squares, deviate from the correct values. Note that estimated values very close to zero ($|\theta| \leq 0.01$) have been omitted from the figure in order to facilitate visualization. In both figures, the stemmed gray-filled circles correspond to the minimum ℓ_2 norm LS solution. The advantages of adopting a sparsity promoting approach to recover the solution are obvious. The CoSaMP algorithm was provided with the exact number of sparsity. The reader is advised to reproduce the example and play with different values of the parameters and see how results are affected.

9.10 A CASE STUDY: IMAGE DENOISING

We have already discussed CS as a notable application of sparsity-aware learning. Although CS has acquired a lot of fame, a number of classical signal processing and machine learning tasks lend themselves to efficient modeling via sparsity-related arguments. Two typical examples are the following.

- *Denoising*: The problem in signal denoising is that instead of the actual signal samples, $\tilde{\boldsymbol{y}}$, a noisy version of the corresponding observations, $\boldsymbol{y}$, is available; that is, $\boldsymbol{y} = \tilde{\boldsymbol{y}} + \boldsymbol{\eta}$, where $\boldsymbol{\eta}$ is the vector of noise samples. Under the sparse modeling framework, the unknown signal $\tilde{\boldsymbol{y}}$ is modeled as a sparse representation in terms of a specific known dictionary Ψ, that is, $\tilde{\boldsymbol{y}} = \Psi\boldsymbol{\theta}$. Moreover, the dictionary is allowed to be redundant (overcomplete). Then the denoising procedure is realized in two steps.

  First, an estimate of the sparse representation vector, $\boldsymbol{\theta}$, is obtained via the ℓ_0 norm minimizer or via any LASSO formulation, for example,

$$\hat{\boldsymbol{\theta}} = \arg\min_{\boldsymbol{\theta}\in\mathbb{R}^l} \|\boldsymbol{\theta}\|_1, \tag{9.40}$$

$$\text{s.t.} \quad \|\boldsymbol{y} - \Psi\boldsymbol{\theta}\|_2^2 \leq \epsilon. \tag{9.41}$$

  Second, the estimate of the true signal is computed as $\hat{\boldsymbol{y}} = \Psi\hat{\boldsymbol{\theta}}$. In Chapter 19, we will study the case where the dictionary is not fixed and known, but is estimated from the data.
- *Linear inverse problems*: Such problems, which come under the more general umbrella of what is known as *signal restoration*, go one step beyond denoising. Now, the available observations are *distorted* as well as noisy versions of the true signal samples; that is, $\boldsymbol{y} = H\tilde{\boldsymbol{y}} + \boldsymbol{\eta}$, where H is a known linear operator. For example, H may correspond to the blurring point spread function of an

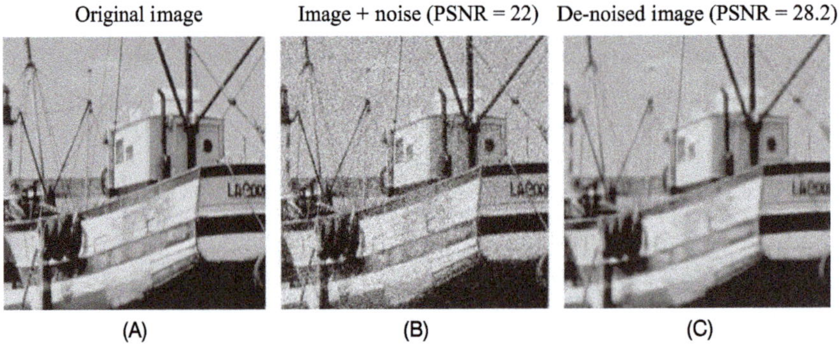

FIGURE 9.13

Denoising based on sparse and redundant representations.

image, as discussed in Chapter 4. Then, similar to the denoising example, assuming that the original signal samples can be efficiently represented in terms of an overcomplete dictionary, $\hat{\boldsymbol{\theta}}$ is estimated, via any sparsity promoting method, using $H\Psi$ in place of Ψ in (9.41), and the estimate of the true signal is obtained as $\hat{\boldsymbol{y}} = \Psi\hat{\boldsymbol{\theta}}$.

Besides deblurring, other applications that fall under this formulation include image inpainting, if H represents the corresponding sampling mask; inverse-Radon transform in tomography, if H comprises the set of parallel projections, and so on. See, for example, [49] for more details on this topic.

In this case study, the image denoising task, based on the sparse and redundant formulation as discussed above, is explored. Our starting point is the 256×256 image shown in Fig. 9.13A. In the sequel, the image is corrupted by zero mean Gaussian noise, leading to the noisy version of Fig. 9.13B, corresponding to a peak signal-to-noise ratio (PSNR) equal to 22 dB, which is defined as

$$\text{PSNR} = 20\log_{10}\left(\frac{m_I}{\sqrt{\text{MSE}}}\right), \tag{9.42}$$

where m_I is the maximum pixel value of the image and $\text{MSE} = \frac{1}{N_p}\|I - \tilde{I}\|_F^2$, where I and $\tilde{I}$ are the noisy and original image matrices, N_p is equal to the total number of pixels, and the Frobenius norm for matrices has been employed.

Denoising could be applied to the full image at once. However, a more efficient practice with respect to memory consumption is to split the image to patches of size much smaller than that of the image; for our case, we chose 12×12 patches. Then denoising is performed to each patch separately as follows: The ith patch image is reshaped in lexicographic order forming a one-dimensional vector, $\boldsymbol{y}_i \in \mathbb{R}^{144}$. We assume that each one of the patches can be reproduced in terms of an overcomplete dictionary, as discussed before; hence, the denoising task is equivalently formulated around (9.40)–(9.41). Denote by $\tilde{\boldsymbol{y}}_i$ the ith patch of the noise-free image. What is left is to choose a dictionary Ψ, which sparsely represents $\tilde{\boldsymbol{y}}_i$, and then solve for sparse $\boldsymbol{\theta}$ according to (9.40)–(9.41).

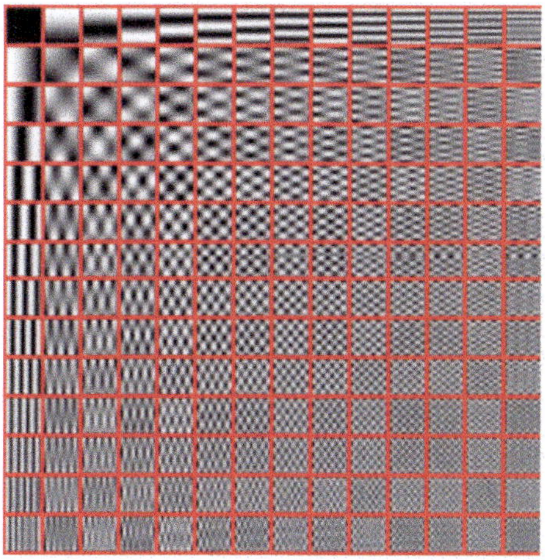

FIGURE 9.14

2D-DCT dictionary atoms, corresponding to 12 × 12 patch size.

It is known that images often exhibit sparse DCT transforms, so an appropriate choice for the dictionary is to fill the columns of Ψ with atoms of a redundant 2D-DCT reshaped in lexicographic order [49]. Here, 196 such atoms were used. There is a standard way to develop such a dictionary given the dimensionality of the image, described in Exercise 9.22. The *same* dictionary is used for *all* patches. The atoms of the dictionary, reshaped to form 12 × 12 blocks, are depicted in Fig. 9.14.

A question that naturally arises is how many patches to use. A straightforward approach is to tile the patches side by side in order to cover the whole extent of the image. This is feasible; however, it is likely to result in blocking effects at the edges of several patches. A better practice is to let the patches overlap. During the reconstruction phase ($\hat{\boldsymbol{y}} = \Psi\hat{\boldsymbol{\theta}}$), because each pixel is covered by more than one patch, the final value of each pixel is taken as the average of the corresponding predicted values from all the involved patches. The results of this method, for our case, are shown in Fig. 9.13C. The attained PSNR is higher than 28 dB.

PROBLEMS

9.1 If $x_i, y_i, \ i = 1, 2, \ldots, l$, are real numbers, then prove the Cauchy-Schwarz inequality,

$$\left(\sum_{i=1}^{l} x_i y_i\right)^2 \leq \left(\sum_{i=1}^{l} x_i^2\right)\left(\sum_{i=1}^{l} y_i^2\right).$$

9.2 Prove that the ℓ_2 (Euclidean) norm is a true norm, that is, it satisfies the four conditions that define a norm.
Hint: To prove the triangle inequality, use the Cauchy-Schwarz inequality.

9.3 Prove that any function that is a norm is also a convex function.

9.4 Show Young's inequality for nonnegative real numbers a and b,

$$ab \leq \frac{a^p}{p} + \frac{b^q}{q},$$

for $\infty > p > 1$ and $\infty > q > 1$ such that

$$\frac{1}{p} + \frac{1}{q} = 1.$$

9.5 Prove Hölder's inequality for ℓ_p norms,

$$||\boldsymbol{x}^T \boldsymbol{y}||_1 = \sum_{i=1}^{l} |x_i y_i| \leq ||\boldsymbol{x}||_p ||\boldsymbol{y}||_q = \left(\sum_{i=1}^{l} |x_i|^p\right)^{1/p} \left(\sum_{i=1}^{q} |y_i|^q\right)^{1/q},$$

for $p \geq 1$ and $q \geq 1$ such that

$$\frac{1}{p} + \frac{1}{q} = 1.$$

Hint: Use Young's inequality.

9.6 Prove Minkowski's inequality,

$$\left(\sum_{i=1}^{l} \left(|x_i| + |y_i|\right)^p\right)^{1/p} \leq \left(\sum_{i=1}^{l} |x_i|^p\right)^{1/p} + \left(\sum_{i=1}^{l} |y_i|^p\right)^{1/p},$$

for $p \geq 1$.
Hint: Use Hölder's inequality together with the identity

$$\left(|a| + |b|\right)^p = \left(|a| + |b|\right)^{p-1} |a| + \left(|a| + |b|\right)^{p-1} |b|.$$

9.7 Prove that for $p \geq 1$, the ℓ_p norm is a true norm.

9.8 Use a counterexample to show that any ℓ_p norm for $0 < p < 1$ is not a true norm and it violates the triangle inequality.

9.9 Show that the null space of a full row rank $N \times l$ matrix X is a subspace of dimensionality N, for $N < l$.

9.10 Show, using Lagrange multipliers, that the ℓ_2 minimizer in (9.18) accepts the closed form solution

$$\hat{\boldsymbol{\theta}} = X^T \left(XX^T\right)^{-1} \boldsymbol{y}.$$

9.11 Show that the necessary and sufficient condition for a $\boldsymbol{\theta}$ to be a minimizer of

$$\text{minimize} \quad ||\boldsymbol{\theta}||_1, \tag{9.43}$$

$$\text{subject to} \quad X\boldsymbol{\theta} = \boldsymbol{y} \tag{9.44}$$

is

$$\left|\sum_{i:\theta_i\neq 0} \text{sign}(\theta_i)z_i\right| \leq \sum_{i:\theta_i=0} |z_i|, \ \forall z \in \text{null}(X),$$

where null(X) is the null space of X. Moreover, if the minimizer is unique, the previous inequality becomes a strict one.

9.12 Prove that if the ℓ_1 norm minimizer is unique, then the number of its components, which are identically zero, must be at least as large as the dimensionality of the null space of the corresponding input matrix.

9.13 Show that the ℓ_1 norm is a convex function (as all norms), yet it is not strictly convex. In contrast, the squared Euclidean norm is a strictly convex function.

9.14 Construct in the five-dimensional space a matrix that has (a) rank equal to five and spark equal to four, (b) rank equal to five and spark equal to three, and (c) rank and spark equal to four.

9.15 Let X be a full row rank $N \times l$ matrix, with $l > N$. Derive the Welch bound for the mutual coherence $\mu(X)$,

$$\mu(X) \geq \sqrt{\frac{l-N}{N(l-1)}}. \tag{9.45}$$

9.16 Let X be an $N \times l$ matrix. Then prove that its spark is bounded as

$$\text{spark}(X) \geq 1 + \frac{1}{\mu(X)},$$

where $\mu(X)$ is the mutual coherence of the matrix.
Hint: Consider the Gram matrix $X^T X$ and the following theorem, concerning positive definite matrices: An $m \times m$ matrix A is positive definite if

$$|A(i,i)| > \sum_{j=1, j\neq i}^{m} |A(i,j)|, \ \forall i = 1, 2, \ldots, m$$

(see, for example, [57]).

9.17 Show that if the underdetermined system of equations $\boldsymbol{y} = X\boldsymbol{\theta}$ accepts a solution such that

$$||\boldsymbol{\theta}||_0 < \frac{1}{2}\left(1 + \frac{1}{\mu(X)}\right),$$

then the ℓ_1 minimizer is equivalent to the ℓ_0 one. Assume that the columns of X are normalized.

9.18 Prove that if the RIP of order k is valid for a matrix X and $\delta_k < 1$, then any $m < k$ columns of X are necessarily linearly independent.

9.19 Show that if X satisfies the RIP of order k and some isometry constant δ_k, so does the product $X\Psi$ if Ψ is an orthonormal matrix.

MATLAB® EXERCISES

9.20 Consider an unknown 2-sparse vector $\boldsymbol{\theta}_o$, which when measured with the sensing matrix

$$X = \begin{bmatrix} 0.5 & 2 & 1.5 \\ 2 & 2.3 & 3.5 \end{bmatrix},$$

that is, of $\boldsymbol{y} = X\boldsymbol{\theta}_o$, gives $\boldsymbol{y} = [1.25, \ 3.75]^T$. Perform the following tasks in MATLAB®. (a) Based on the pseudoinverse of X, compute $\hat{\boldsymbol{\theta}}_2$, which is the ℓ_2 norm minimized solution, (9.18). Next, check that this solution $\hat{\boldsymbol{\theta}}_2$ leads to zero estimation error (up to machine precision). Is $\hat{\boldsymbol{\theta}}_2$ a 2-sparse vector such as the true unknown vector $\boldsymbol{\theta}_o$, and if it is not, how is it possible to lead to zero estimation error? (b) Solve the ℓ_0 minimization task described in (9.20) (exhaustive search) for all possible 1- and 2-sparse solutions and get the best one, $\hat{\boldsymbol{\theta}}_o$. Does $\hat{\boldsymbol{\theta}}_o$ lead to zero estimation error (up to machine precision)? (c) Compute and compare the ℓ_2 norms of $\hat{\boldsymbol{\theta}}_2$ and $\hat{\boldsymbol{\theta}}_o$. Which is the smaller one? Was this result expected?

9.21 Generate in MATLAB® a sparse vector $\boldsymbol{\theta} \in \mathbb{R}^l$, $l = 100$, with its first five components taking random values drawn from a normal distribution $\mathcal{N}(0, 1)$ and the rest being equal to zero. Build, also, a sensing matrix X with $N = 30$ rows having samples normally distributed, $\mathcal{N}(0, \frac{1}{N})$, in order to get 30 observations based on the linear regression model $\boldsymbol{y} = X\boldsymbol{\theta}$. Then perform the following tasks. (a) Use the function "solvelasso.m",[4] or any other LASSO implementation you prefer, to reconstruct $\boldsymbol{\theta}$ from $\boldsymbol{y}$ and X. (b) Repeat the experiment with different realizations of X in order to compute the probability of correct reconstruction (assume the reconstruction is exact when $||\boldsymbol{y} - X\boldsymbol{\theta}||_2 < 10^{-8}$). (c) Construct another sensing matrix X having $N = 30$ rows taken uniformly at random from the $l \times l$ DCT matrix, which can be obtained via the built-in MATLAB® function "dctmtx.m". Compute the probability of reconstruction when this DCT-based sensing matrix is used and confirm that results similar to those in question (b) are obtained. (d) Repeat the same experiment with matrices of the form

$$X(i, j) = \begin{cases} +\sqrt{\frac{\sqrt{p}}{N}}, & \text{with probability } \frac{1}{2\sqrt{p}}, \\ 0, & \text{with probability } 1 - \frac{1}{\sqrt{p}}, \\ -\sqrt{\frac{\sqrt{p}}{N}}, & \text{with probability } \frac{1}{2\sqrt{p}}, \end{cases}$$

for p equal to 1, 9, 25, 36, 64 (make sure that each row and each column of X has at least a nonzero component). Give an explanation why the probability of reconstruction falls as p increases (observe that both the sensing matrix and the unknown vector are sparse).

9.22 This exercise reproduces the denoising results of the case study in Section 9.10, where the image depicting the boat can be downloaded from the book website. First, extract from the image all the possible sliding patches of size 12×12 using the im2col.m MATLAB® function. Confirm

[4] It can be found in the SparseLab MATLAB® toolbox, which is freely available from http://sparselab.stanford.edu/.

that $(256 - 12 + 1)^2 = 60{,}025$ patches in total are obtained. Next, a dictionary in which all the patches are sparsely represented needs to be designed.

Specifically, the dictionary atoms are going to be those corresponding to the two-dimensional redundant DCT transform, and are obtained as follows [49]:

a) Consider vectors $\boldsymbol{d}_i = [d_{i,1}, d_{i,2}, \ldots, d_{i,12}]^T$, $i = 0, \ldots, 13$, being the sampled sinusoids of the form

$$d_{i,t+1} = \cos\left(\frac{t\pi i}{14}\right), \quad t = 0, \ldots, 11.$$

Then make a (12×14) matrix $\bar{D}$, having as columns the vectors $\boldsymbol{d}_i$ normalized to unit norm; D resembles a redundant DCT matrix.

b) Construct the $(12^2 \times 14^2)$ dictionary Ψ according to $\Psi = D \bigotimes D$, where $\bigotimes$ denotes Kronecker product. Built in this way, the resulting atoms correspond to atoms related to the overcomplete 2D-DCT transform [49].

As a next step, denoise each image patch separately. In particular, assuming that $\boldsymbol{y}_i$ is the ith patch reshaped in column vector, use the function "solvelasso.m",[5] or any other suitable algorithm you prefer, to estimate a sparse vector $\boldsymbol{\theta}_i \in \mathbb{R}^{196}$ and obtain the corresponding denoised vector as $\hat{\boldsymbol{y}}_i = \Psi \boldsymbol{\theta}_i$. Finally, average the values of the overlapping patches in order to form the full denoised image.

REFERENCES

[1] D. Achlioptas, Database-friendly random projections, in: Proceedings of the Symposium on Principles of Database Systems, PODS, ACM Press, 2001, pp. 274–281.

[2] A. Antoniadis, Wavelet methods in statistics: some recent developments and their applications, Stat. Surv. 1 (2007) 16–55.

[3] J. Arenas-Garcia, A.R. Figueiras-Vidal, Adaptive combination of proportionate filters for sparse echo cancellation, IEEE Trans. Audio Speech Lang. Process. 17 (6) (2009) 1087–1098.

[4] S. Ariyavisitakul, N.R. Sollenberger, L.J. Greenstein, Tap-selectable decision feedback equalization, IEEE Trans. Commun. 45 (12) (1997) 1498–1500.

[5] W.U. Bajwa, J. Haupt, A.M. Sayeed, R. Nowak, Compressed channel sensing: a new approach to estimating sparse multipath channels, Proc. IEEE 98 (6) (2010) 1058–1076.

[6] R.G. Baraniuk, M. Davenport, R. DeVore, M.B. Wakin, A simple proof of the restricted isometry property for random matrices, Constr. Approx. 28 (2008) 253–263.

[7] R. Baraniuk, M. Wakin, Random projections of smooth manifolds, Found. Comput. Math. 9 (1) (2009) 51–77.

[8] R. Baraniuk, V. Cevher, M. Wakin, Low-dimensional models for dimensionality reduction and signal recovery: a geometric perspective, Proc. IEEE 98 (6) (2010) 959–971.

[9] S. Becker, Practical Compressed Sensing: Modern Data Acquisition and Signal Processing, PhD thesis, Caltech, 2011.

[10] J. Benesty, T. Gansler, D.R. Morgan, M.M. Sondhi, S.L. Gay, Advances in Network and Acoustic Echo Cancellation, Springer-Verlag, Berlin, 2001.

[11] P. Bickel, Y. Ritov, A. Tsybakov, Simultaneous analysis of LASSO and Dantzig selector, Ann. Stat. 37 (4) (2009) 1705–1732.

[12] T. Blu, P.L. Dragotti, M. Vetterli, P. Marziliano, L. Coulot, Sparse sampling of signal innovations, IEEE Signal Process. Mag. 25 (2) (2008) 31–40.

[5] It can be found in the SparseLab MATLAB® toolbox, which is freely available from http://sparselab.stanford.edu/.

[13] A. Blum, Random projection, margins, kernels and feature selection, in: Lecture Notes on Computer Science (LNCS), 2006, pp. 52–68.
[14] S. Boyd, L. Vandenberghe, Convex Optimization, Cambridge University Press, 2004.
[15] A.M. Bruckstein, D.L. Donoho, M. Elad, From sparse solutions of systems of equations to sparse modeling of signals and images, SIAM Rev. 51 (1) (2009) 34–81.
[16] T.T. Cai, G. Xu, J. Zhang, On recovery of sparse signals via ℓ_1 minimization, IEEE Trans. Inf. Theory 55 (7) (2009) 3388–3397.
[17] R. Calderbank, S. Jeafarpour, R. Schapire, Compressed Learning: Universal Sparse Dimensionality Reduction and Learning in the Measurement Domain, Tech. Rep., Rice University, 2009.
[18] E.J. Candès, J. Romberg, Practical signal recovery from random projections, in: Proceedings of the SPIE 17th Annual Symposium on Electronic Imaging, Bellingham, WA, 2005.
[19] E.J. Candès, T. Tao, Decoding by linear programming, IEEE Trans. Inf. Theory 51 (12) (2005) 4203–4215.
[20] E. Candès, J. Romberg, T. Tao, Robust uncertainty principles: exact signal reconstruction from highly incomplete Fourier information, IEEE Trans. Inf. Theory 52 (2) (2006) 489–509.
[21] E. Candès, T. Tao, Near optimal signal recovery from random projections: universal encoding strategies, IEEE Trans. Inf. Theory 52 (12) (2006) 5406–5425.
[22] E.J. Candès, J. Romberg, T. Tao, Stable recovery from incomplete and inaccurate measurements, Commun. Pure Appl. Math. 59 (8) (2006) 1207–1223.
[23] E.J. Candès, M.B. Wakin, An introduction to compressive sampling, IEEE Signal Process. Mag. 25 (2) (2008) 21–30.
[24] E.J. Candès, Y.C. Eldar, D. Needell, P. Randall, Compressed sensing with coherent and redundant dictionaries, Appl. Comput. Harmon. Anal. 31 (1) (2011) 59–73.
[25] F. Chen, A.P. Chandrakasan, V.M. Stojanovic, Design and analysis of hardware efficient compressed sensing architectures for compression in wireless sensors, IEEE Trans. Solid State Circuits 47 (3) (2012) 744–756.
[26] S. Chen, D.L. Donoho, M. Saunders, Atomic decomposition by basis pursuit, SIAM J. Sci. Comput. 20 (1) (1998) 33–61.
[27] J.F. Claerbout, F. Muir, Robust modeling with erratic data, Geophysics 38 (5) (1973) 826–844.
[28] K.L. Clarkson, D.P. Woodruff, Numerical linear algebra in the streaming model, in: Proceedings of the 41st Annual ACM Symposium on Theory of Computing, ACM, 2009, pp. 205–214.
[29] K.L. Clarkson, D.P. Woodruff, Low rank approximation and regression in input sparsity time, in: Proceedings of the 45th Annual ACM Symposium on Symposium on Theory of Computing, ACM, 2013, pp. 81–90.
[30] A. Cohen, W. Dahmen, R. DeVore, Compressed sensing and best k-term approximation, J. Am. Math. Soc. 22 (1) (2009) 211–231.
[31] R.R. Coifman, M.V. Wickerhauser, Entropy-based algorithms for best basis selection, IEEE Trans. Inf. Theory 38 (2) (1992) 713–718.
[32] S.F. Cotter, B.D. Rao, Matching pursuit based decision-feedback equalizers, in: IEEE Conference on Acoustics, Speech and Signal Processing, ICASSP, Istanbul, Turkey, 2000.
[33] G.B. Dantzig, Linear Programming and Extensions, Princeton University Press, Princeton, NJ, 1963.
[34] S. Dasgupta, Experiments with random projections, in: Proceedings of the 16th Conference on Uncertainty in Artificial Intelligence, Morgan-Kaufmann, San Francisco, CA, USA, 2000, pp. 143–151.
[35] I. Daubechies, Time-frequency localization operators: a geometric phase space approach, IEEE Trans. Inf. Theory 34 (4) (1988) 605–612.
[36] D.L. Donoho, X. Huo, Uncertainty principles and ideal atomic decomposition, IEEE Trans. Inf. Theory 47 (7) (2001) 2845–2862.
[37] D.L. Donoho, M. Elad, Optimally sparse representation in general (nonorthogonal) dictionaries via ℓ_1 minimization, in: Proceedings of National Academy of Sciences, 2003, pp. 2197–2202.
[38] D.L. Donoho, J. Tanner, Counting Faces of Randomly Projected Polytopes When the Projection Radically Lowers Dimension, Tech. Rep. 2006-11, Stanford University, 2006.
[39] D.L. Donoho, J. Tanner, Precise undersampling theorems, Proc. IEEE 98 (6) (2010) 913–924.
[40] D.L. Donoho, B.F. Logan, Signal recovery and the large sieve, SIAM J. Appl. Math. 52 (2) (1992) 577–591.
[41] P.L. Dragotti, M. Vetterli, T. Blu, Sampling moments and reconstructing signals of finite rate of innovation: Shannon meets Strang-Fix, IEEE Trans. Signal Process. 55 (5) (2007) 1741–1757.
[42] P. Drineas, M.W. Mahoney, S. Muthukrishnan, T. Sarlós, Faster least squares approximation, Numer. Math. 117 (2) (2011) 219–249.

[43] M.F. Duarte, Y. Eldar, Structured compressed sensing: from theory to applications, IEEE Trans. Signal Process. 59 (9) (2011) 4053–4085.
[44] M.F. Duarte, R.G. Baraniuk, Kronecker compressive sensing, IEEE Trans. Image Process. 21 (2) (2012) 494–504.
[45] M.F. Duarte, R.G. Baraniuk, Spectral compressive sensing, Appl. Comput. Harmon. Anal. 35 (1) (2013) 111–129.
[46] D. Eiwen, G. Taubock, F. Hlawatsch, H.G. Feichtinger, Group sparsity methods for compressive channel estimation in doubly dispersive multicarrier systems, in: Proceedings IEEE SPAWC, Marrakech, Morocco, June 2010.
[47] D. Eiwen, G. Taubock, F. Hlawatsch, H. Rauhut, N. Czink, Multichannel-compressive estimation of doubly selective channels in MIMO-OFDM systems: exploiting and enhancing joint sparsity, in: Proceedings International Conference on Acoustics, Speech and Signal Processing, ICASSP, Dallas, TX, 2010.
[48] M. Elad, A.M. Bruckstein, A generalized uncertainty principle and sparse representations in pairs of bases, IEEE Trans. Inf. Theory 48 (9) (2002) 2558–2567.
[49] M. Elad, Sparse and Redundant Representations: From Theory to Applications in Signal and Image Processing, Springer, 2010.
[50] Y.C. Eldar, G. Kutyniok, Compressed Sensing: Theory and Applications, Cambridge University Press, 2012.
[51] Y.C. Eldar, Sampling Theory: Beyond Bandlimited Systems, Cambridge University Press, 2014.
[52] M. Ghosh, Blind decision feedback equalization for terrestrial television receivers, Proc. IEEE 86 (10) (1998) 2070–2081.
[53] I.F. Gorodnitsky, B.D. Rao, Sparse signal reconstruction from limited data using FOCUSS: a re-weighted minimum norm algorithm, IEEE Trans. Signal Process. 45 (3) (1997) 600–614.
[54] R. Gribonval, M. Nielsen, Sparse decompositions in unions of bases, IEEE Trans. Inf. Theory 49 (12) (2003) 3320–3325.
[55] N. Halko, P.G. Martinsson, J.A. Tropp, Finding structure with randomness: probabilistic algorithms for constructing approximate matrix decompositions, SIAM Rev. 53 (2) (2011) 217–288.
[56] J. Haupt, W.U. Bajwa, G. Raz, R. Nowak, Toeplitz compressed sensing matrices with applications to sparse channel estimation, IEEE Trans. Inf. Theory 56 (11) (2010) 5862–5875.
[57] R.A. Horn, C.R. Johnson, Matrix Analysis, Cambridge University Press, New York, 1985.
[58] S. Kirolos, J.N. Laska, M.B. Wakin, M.F. Duarte, D. Baron, T. Ragheb, Y. Massoud, R.G. Baraniuk, Analog to information conversion via random demodulation, in: Proceedings of the IEEE Dallas/CAS Workshop on Design, Applications, Integration and Software, Dallas, USA, 2006, pp. 71–74.
[59] M. Lexa, M. Davies, J. Thompson, Reconciling compressive sampling systems for spectrally sparse continuous-time signals, IEEE Trans. Signal Process. 60 (1) (2012) 155–171.
[60] S. Lin, D.C. Constello Jr., Error Control Coding: Fundamentals and Applications, Prentice Hall, 1983.
[61] Y.-P. Lin, P.P. Vaidyanathan, Periodically nonuniform sampling of bandpass signals, IEEE Trans. Circuits Syst. II 45 (3) (1998) 340–351.
[62] Y.M. Lu, M.N. Do, Sampling signals from a union of subspaces, IEEE Signal Process. Mag. 25 (2) (2008) 41–47.
[63] P. Maechler, N. Felber, H. Kaeslin, A. Burg, Hardware-efficient random sampling of Fourier-sparse signals, in: Proceedings of the IEEE International Symposium on Circuits and Systems, ISCAS, 2012.
[64] A. Maleki, L. Anitori, Z. Yang, R. Baraniuk, Asymptotic analysis of complex LASSO via complex approximate message passing (CAMP), IEEE Trans. Inf. Theory 59 (7) (2013) 4290–4308.
[65] S. Mallat, S. Zhang, Matching pursuit in a time-frequency dictionary, IEEE Trans. Signal Process. 41 (1993) 3397–3415.
[66] E. Matusiak, Y.C. Eldar, Sub-Nyquist sampling of short pulses, IEEE Trans. Signal Process. 60 (3) (2012) 1134–1148.
[67] S. Mendelson, A. Pajor, N. Tomczak-Jaegermann, Uniform uncertainty principle for Bernoulli and subGaussian ensembles, Constr. Approx. 28 (2008) 277–289.
[68] M. Mishali, Y.C. Eldar, A. Elron, Xampling: analog data compression, in: Proceedings Data Compression Conference, Snowbird, Utah, USA, 2010.
[69] M. Mishali, Y. Eldar, From theory to practice: sub-Nyquist sampling of sparse wideband analog signals, IEEE J. Sel. Top. Signal Process. 4 (2) (2010) 375–391.
[70] M. Mishali, Y.C. Eldar, Sub-Nyquist sampling, IEEE Signal Process. Mag. 28 (6) (2011) 98–124.
[71] M. Mishali, Y.C. Eldar, A. Elron, Xampling: signal acquisition and processing in union of subspaces, IEEE Trans. Signal Process. 59 (10) (2011) 4719–4734.
[72] B.K. Natarajan, Sparse approximate solutions to linear systems, SIAM J. Comput. 24 (1995) 227–234.
[73] P.A. Naylor, J. Cui, M. Brookes, Adaptive algorithms for sparse echo cancellation, Signal Process. 86 (2004) 1182–1192.
[74] A.M. Pinkus, On ℓ_1-Approximation, Cambridge Tracts in Mathematics, vol. 93, Cambridge University Press, 1989.
[75] Q. Qiu, V.M. Patel, P. Turaga, R. Chellappa, Domain adaptive dictionary learning, in: Proceedings of the European Conference on Computer Vision, ECCV, Florence, Italy, 2012.

[76] Y. Rivenson, A. Stern, Compressed imaging with a separable sensing operator, IEEE Signal Process. Lett. 16 (6) (2009) 449–452.
[77] A. Rondogiannis, K. Berberidis, Efficient decision feedback equalization for sparse wireless channels, IEEE Trans. Wirel. Commun. 2 (3) (2003) 570–581.
[78] R. Rubinstein, A. Bruckstein, M. Elad, Dictionaries for sparse representation modeling, Proc. IEEE 98 (6) (2010) 1045–1057.
[79] M. Rudelson, R. Vershynin, On sparse reconstruction from Fourier and Gaussian measurements, Commun. Pure Appl. Math. 61 (8) (2008) 1025–1045.
[80] F. Santosa, W.W. Symes, Linear inversion of band limited reflection seismograms, SIAM J. Sci. Comput. 7 (4) (1986) 1307–1330.
[81] T. Sarlos, Improved approximation algorithms for large matrices via random projections, in: 47th Annual IEEE Symposium on Foundations of Computer Science, 2006, FOCS'06, IEEE, 2006, pp. 143–152.
[82] P. Saurabh, C. Boutsidis, M. Magdon-Ismail, P. Drineas, Random projections for support vector machines, in: Proceedings 16th International Conference on Artificial Intelligence and Statistics, AISTATS, Scottsdale, AZ, USA, 2013.
[83] K. Slavakis, Y. Kopsinis, S. Theodoridis, S. McLaughlin, Generalized thresholding and online sparsity-aware learning in a union of subspaces, IEEE Trans. Signal Process. 61 (12) (2013) 3760–3773.
[84] D. Takhar, V. Bansal, M. Wakin, M. Duarte, D. Baron, K.F. Kelly, R.G. Baraniuk, A compressed sensing camera: new theory and an implementation using digital micromirrors, in: Proceedings on Computational Imaging, SPIE, San Jose, CA, 2006.
[85] G. Tang, A. Nehorai, Performance analysis of sparse recovery based on constrained minimal singular values, IEEE Trans. Signal Process. 59 (12) (2011) 5734–5745.
[86] H.L. Taylor, S.C. Banks, J.F. McCoy, Deconvolution with the ℓ_1 norm, Geophysics 44 (1) (1979) 39–52.
[87] S. Theodoridis, K. Koutroumbas, Pattern Recognition, fourth ed., Academic Press, 2009.
[88] Z. Tian, G.B. Giannakis, Compressed sensing for wideband cognitive radios, in: Proceedings of the IEEE Conference on Acoustics, Speech and Signal Processing, ICASSP, 2007, pp. 1357–1360.
[89] R. Tibshirani, Regression shrinkage and selection via the LASSO, J. R. Stat. Soc. B 58 (1) (1996) 267–288.
[90] I. Tosić, P. Frossard, Dictionary learning, IEEE Signal Process. Mag. 28 (2) (2011) 27–38.
[91] J.A. Tropp, J.N. Laska, M.F. Duarte, J.K. Romberg, G. Baraniuk, Beyond Nyquist: efficient sampling of sparse bandlimited signals, IEEE Trans. Inf. Theory 56 (1) (2010) 520–544.
[92] M. Unser, Sampling: 50 years after Shannon, Proc. IEEE 88 (4) (2000) 569–587.
[93] R.G. Vaughan, N.L. Scott, D.R. White, The theory of bandpass sampling, IEEE Trans. Signal Process. 39 (9) (1991) 1973–1984.
[94] R. Venkataramani, Y. Bresler, Perfect reconstruction formulas and bounds on aliasing error in sub-Nyquist nonuniform sampling of multiband signals, IEEE Trans. Inf. Theory 46 (6) (2000) 2173–2183.
[95] M. Vetterli, P. Marzilliano, T. Blu, Sampling signals with finite rate of innovation, IEEE Trans. Signal Process. 50 (6) (2002) 1417–1428.
[96] M. Wakin, Manifold-based signal recovery and parameter estimation from compressive measurements, preprint, arXiv: 1002.1247, 2008.
[97] M. Wakin, S. Becker, E. Nakamura, M. Grant, E. Sovero, D. Ching, J. Yoo, J. Romberg, A. Emami-Neyestanak, E. Candes, A non-uniform sampler for wideband spectrally-sparse environments, IEEE Trans. Emerg. Sel. Top. Circuits Syst. 2 (3) (2012) 516–529.
[98] L.R. Welch, Lower bounds on the maximum cross correlation of signals, IEEE Trans. Inf. Theory 20 (3) (1974) 397–399.
[99] S. Wright, R. Nowak, M. Figueiredo, Sparse reconstruction by separable approximation, IEEE Trans. Signal Process. 57 (7) (2009) 2479–2493.
[100] M. Yaghoobi, L. Daudet, M. Davies, Parametric dictionary design for sparse coding, IEEE Trans. Signal Process. 57 (12) (2009) 4800–4810.
[101] Y. Ye, Interior Point Methods: Theory and Analysis, Wiley, New York, 1997.
[102] J. Yoo, S. Becker, M. Monge, M. Loh, E. Candès, A. Emami-Neyestanak, Design and implementation of a fully integrated compressed-sensing signal acquisition system, in: 2012 IEEE International Conference on Acoustics, Speech and Signal Processing, ICASSP, March 2012, pp. 5325–5328.
[103] Z. Yu, S. Hoyos, B.M. Sadler, Mixed-signal parallel compressed sensing and reception for cognitive radio, in: Proceedings IEEE Conference on Acoustics, Speech and Signal Processing, ICASSP, 2008, pp. 3861–3864.

CHAPTER

SPARSITY-AWARE LEARNING: ALGORITHMS AND APPLICATIONS

10

CONTENTS

10.1 INTRODUCTION

This chapter is the follow-up to the previous one concerning sparsity-aware learning. The emphasis now is on the algorithmic front. Following the theoretical advances concerning sparse modeling, a true scientific happening occurred in trying to derive algorithms tailored for the efficient solution of the related constrained optimization tasks. Our goal is to present the main directions that have been followed and to provide in a more explicit form some of the most popular algorithms. We will discuss batch as

Machine Learning. https://doi.org/10.1016/B978-0-12-818803-3.00020-9

well as online algorithms. This chapter can also be considered as a complement to Chapter 8, where some aspects of convex optimization were introduced; a number of algorithms discussed there are also appropriate for tasks involving sparsity-related constraints/regularization.

Besides describing various algorithmic families, some variants of the basic sparsity promoting ℓ_1 and ℓ_0 norms are discussed. Furthermore, some typical examples are considered and a case study concerning time-frequency analysis is presented. Finally, a discussion concernig the issue "synthesis versus analysis" models is provided.

10.2 SPARSITY PROMOTING ALGORITHMS

In the previous chapter, our emphasis was on highlighting some of the most important aspects underlying the theory of sparse signal/parameter vector recovery from an underdetermined set of linear equations. We now turn our attention to the algorithmic aspects of the problem (e.g., [52,54]). The issue now becomes that of discussing *efficient* algorithmic schemes, which can achieve the recovery of the unknown set of parameters. In Sections 9.3 and 9.5, we saw that the constrained ℓ_1 norm minimization (basis pursuit) can be solved via linear programming techniques and the LASSO task via convex optimization schemes. However, such general purpose techniques tend to be inefficient, because they often require many iterations to converge, and the respective computational resources can be excessive for practical applications, especially in high-dimensional spaces $\mathbb{R}^l$. As a consequence, a huge research effort has been invested for the goal of developing efficient algorithms that are tailored to these specific tasks. Our aim here is to provide the reader with some general trends and philosophies that characterize the related activity. We will focus on the most commonly used and cited algorithms, which at the same time are structurally simple, so the reader can follow them, without deeper knowledge of optimization. Moreover, these algorithms involve, in one way or another, arguments that are directly related to notions we have already used while presenting the theory; thus, they can also be exploited from a pedagogical point of view in order to strengthen the reader's understanding of the topic. We start our review with the class of batch algorithms, where all data are assumed to be available prior to the application of the algorithm, and then we will move on to online/time-adaptive schemes. Furthermore, our emphasis is on algorithms that are appropriate for any sensing matrix. This is stated in order to point out that in the literature, efficient algorithms have also been developed for specific forms of highly structured sensing matrices, and exploiting their particular structure can lead to reduced computational demands [61,93].

There are three rough types of families along which this algorithmic activity has grown: (a) greedy algorithms, (b) iterative shrinkage schemes, and (c) convex optimization techniques. We have used the word rough because in some cases, it may be difficult to assign an algorithm to a specific family.

10.2.1 GREEDY ALGORITHMS

Greedy algorithms have a long history; see, for example, [114] for a comprehensive list of references. In the context of dictionary learning, a greedy algorithm known as *matching pursuit* was introduced in [88]. A greedy algorithm is built upon a series of *locally* optimal *single-term* updates. In our context, the goals are (a) to unveil the "active" columns of the sensing matrix X, that is, those columns that correspond to the nonzero locations of the unknown parameters, and (b) to estimate the respective sparse parameter vector. The set of indices that correspond to the nonzero vector components is also known as the *support*. To this end, the set of active columns of X (and the support) is increased by one at each iteration step. In the sequel, an updated estimate of the unknown sparse vector is obtained. Let us assume that at the $(i-1)$th iteration step, the algorithm has selected the columns denoted as $\boldsymbol{x}^c_{j_1}, \boldsymbol{x}^c_{j_2}, \ldots, \boldsymbol{x}^c_{j_{i-1}}$, with $j_1, j_2, \ldots, j_{i-1} \in \{1, 2, \ldots, l\}$. These indices are the elements of the currently available support, $S^{(i-1)}$. Let $X^{(i-1)}$ be the $N \times (i-1)$ matrix, having $\boldsymbol{x}^c_{j_1}, \boldsymbol{x}^c_{j_2}, \ldots, \boldsymbol{x}^c_{j_{i-1}}$ as its columns. Let also the current estimate of the solution be $\boldsymbol{\theta}^{(i-1)}$, which is a $(i-1)$-sparse vector, with zeros at all locations with index outside the support. The *orthogonal matching pursuit* (OMP) scheme given in Algorithm 10.1 builds up recursively a sparse solution.

Algorithm 10.1 (The OMP algorithm).
The algorithm is initialized with $\boldsymbol{\theta}^{(0)} := \boldsymbol{0}$, $\boldsymbol{e}^{(0)} := \boldsymbol{y}$, and $S^{(0)} = \emptyset$. At iteration step i, the following computational steps are performed:

1. Select the column $\boldsymbol{x}^c_{j_i}$ of X, which is *maximally* correlated to (forms the smallest angle with) the respective error vector, $\boldsymbol{e}^{(i-1)} := \boldsymbol{y} - X\boldsymbol{\theta}^{(i-1)}$, that is,

$$\boldsymbol{x}^c_{j_i} : \; j_i := \arg\max_{j=1,2,\ldots,l} \frac{\left|\boldsymbol{x}^{c\,T}_j \boldsymbol{e}^{(i-1)}\right|}{\left\|\boldsymbol{x}^c_j\right\|_2}.$$

2. Update the support and the corresponding set of active columns, $S^{(i)} = S^{(i-1)} \cup \{j_i\}$ and $X^{(i)} = [X^{(i-1)}, \boldsymbol{x}^c_{j_i}]$.
3. Update the estimate of the parameter vector: Solve the least-squares (LS) problem that minimizes the norm of the error, using the active columns of X only, that is,

$$\tilde{\boldsymbol{\theta}} := \arg\min_{\boldsymbol{z} \in \mathbb{R}^i} \left\|\boldsymbol{y} - X^{(i)}\boldsymbol{z}\right\|_2^2.$$

   Obtain $\boldsymbol{\theta}^{(i)}$ by inserting the elements of $\tilde{\boldsymbol{\theta}}$ in the respective locations $(j_1, j_2, \ldots, j_i)$, which comprise the support (the rest of the elements of $\boldsymbol{\theta}^{(i)}$ retain their zero values).
4. Update the error vector

$$\boldsymbol{e}^{(i)} := \boldsymbol{y} - X\boldsymbol{\theta}^{(i)}.$$

The algorithm terminates if the norm of the error becomes less than a preselected user-defined constant, ϵ_0. The following observations are in order.

Remarks 10.1.

- Because $\boldsymbol{\theta}^{(i)}$, in Step 3, is the result of an LS task, we know from Chapter 6 that the error vector is orthogonal to the subspace spanned by the active columns involved, that is,

$$\boldsymbol{e}^{(i)} \perp \operatorname{span}\left\{\boldsymbol{x}^c_{j_1}, \ldots, \boldsymbol{x}^c_{j_i}\right\}.$$

  This guarantees that in the next step, taking the correlation of the columns of X with $\boldsymbol{e}^{(i)}$, none of the previously selected columns will be reselected; they result to zero correlation, being orthogonal to $\boldsymbol{e}^{(i)}$ (see Fig. 10.1).

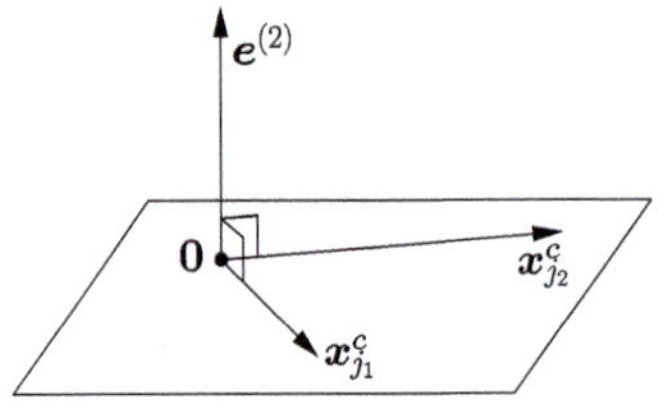

FIGURE 10.1

The error vector at the ith iteration is orthogonal to the subspace spanned by the currently available set of active columns. Here is an illustration for the case of the three-dimensional Euclidean space $\mathbb{R}^3$, and for $i = 2$.

- The column which has maximal correlation (maximum absolute value of the inner product) with the currently available error vector is the one that maximally reduces (compared to any other column) the ℓ_2 norm of the error, when $\boldsymbol{y}$ is approximated by linearly combining the currently available active columns. This is the point where the heart of the greedy strategy beats. This minimization is with respect to a *single term*, keeping the rest fixed, as they have been obtained from the previous iteration steps (Problem 10.1).
- Starting with all the components being zero, if the algorithm stops after k_0 iteration steps, the result will be a k_0-sparse solution.
- Note that there is no optimality in this searching strategy. The only guarantee is that the ℓ_2 norm of the error vector is decreased at every iteration step. In general, there is no guarantee that the algorithm can obtain a solution close to the true one (see, for example, [38]). However, under certain constraints on the structure of X, performance bounds can be obtained (see, for example, [37,115, 123]).
- The complexity of the algorithm amounts to $\mathcal{O}(k_0 l N)$ operations, which are contributed by the computations of the correlations, plus the demands raised by the solution of the LS task in step 3, whose complexity depends on the specific algorithm used. The k_0 is the sparsity level of the delivered solution and, hence, the total number of iteration steps that are performed.

Another more qualitative argument that justifies the selection of the columns based on their correlation with the error vector is the following. Assume that the matrix X is orthonormal. Let $\boldsymbol{y} = X\boldsymbol{\theta}$. Then $\boldsymbol{y}$ lies in the subspace spanned by the active columns of X, that is, those that correspond to the nonzero components of $\boldsymbol{\theta}$. Hence, the rest of the columns are orthogonal to $\boldsymbol{y}$, because X is assumed to be orthonormal. Taking the correlation of $\boldsymbol{y}$, at the first iteration step, with all the columns, it is certain that one among the active columns will be chosen. The inactive columns result in zero correlation. A similar argument holds true for all subsequent steps, because all activity takes place in a subspace that is orthogonal to all the inactive columns of X. In the more general case, where X is not orthonormal, we can still use the correlation as a measure that quantifies geometric similarity. The smaller the correlation/magnitude of the inner product is, the more orthogonal the two vectors are. This brings us back to the notion of mutual coherence, which is a measure of the maximum correlation (smallest angle) among the columns of X.

OMP Can Recover Optimal Sparse Solutions: Sufficiency Condition

We have already stated that, in general, there are no guarantees that OMP will recover optimal solutions. However, when the unknown vector is sufficiently sparse, with respect to the structure of the sensing matrix X, OMP can exactly solve the ℓ_0 minimization task in (9.20) and recover the solution in k_0 steps, where k_0 is the sparsest solution that satisfies the associated linear set of equations.

Theorem 10.1. *Let the mutual coherence (Section 9.6.1) of the sensing matrix X be $\mu(X)$. Assume, also, that the linear system $\boldsymbol{y} = X\boldsymbol{\theta}$ accepts a solution such as*

$$\|\boldsymbol{\theta}\|_0 < \frac{1}{2}\left(1 + \frac{1}{\mu(X)}\right). \tag{10.1}$$

Then OMP guarantees recovery of the sparsest solution in $k_0 = \|\boldsymbol{\theta}\|_0$ steps.

We know from Section 9.6.1 that under the previous condition, any other solution will be necessarily less sparse. Hence, there is a unique way to represent $\boldsymbol{y}$ in terms of k_0 columns of X. Without harming generality, let us assume that the true support corresponds to the first k_0 columns of X, that is,

$$\boldsymbol{y} = \sum_{j=1}^{k_0} \theta_j \boldsymbol{x}_j^c, \quad \theta_j \neq 0, \quad \forall j \in \{1, \ldots, k_0\}.$$

The theorem is a direct consequence of the following proposition.

Proposition 10.1. *If the condition* (10.1) *holds true, then the OMP algorithm will never select a column with index outside the true support (see, for example, [115] and Problem 10.2). In a more formal way, this is expressed as*

$$j_i = \arg\max_{j=1,2,\ldots,l} \frac{\left|\boldsymbol{x}_j^{cT} \boldsymbol{e}^{(i-1)}\right|}{\left\|\boldsymbol{x}_j^c\right\|_2} \in \{1, \ldots, k_0\}.$$

A geometric interpretation of this proposition is the following. If the angles formed between all the possible pairs among the columns of X are close to 90° (columns almost orthogonal) in the $\mathbb{R}^l$ space, which guarantees that $\mu(X)$ is small enough, then $\boldsymbol{y}$ will lean more (form a smaller angle) toward any one of the active columns that contribute to its formation, compared to the rest that are inactive and do not participate in the linear combination that generates $\boldsymbol{y}$. Fig. 10.2 illustrates the geometry, for the extreme case of mutually orthogonal vectors (Fig. 10.2A), and for the more general case where the vectors are not orthogonal, yet the angle between any pair of columns is close enough to 90° (Fig. 10.2B).

In a nutshell, the previous proposition guarantees that during the first iteration, a column corresponding to the true support will be selected. In a similar way, this is also true for all subsequent iterations. In the second step, another column, different from the previously selected one (as has already been stated) will be chosen. At step k_0, the last remaining active column corresponding to the true support is selected, and this necessarily results in zero error. To this end, it suffices to set ϵ_0 equal to zero.

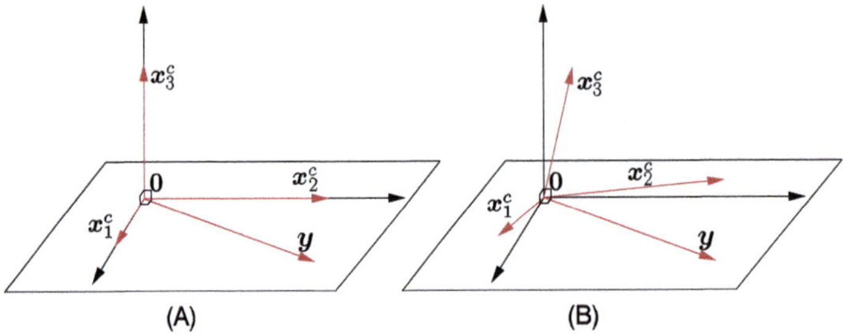

FIGURE 10.2

(A) In the case of an orthogonal matrix, the observations vector $\boldsymbol{y}$ will be orthogonal to any inactive column, here, $\boldsymbol{x}_3^c$. (B) In the more general case, it is expected to "lean" closer (form smaller angles) to the active than to the inactive columns.

The LARS Algorithm

The least-angle regression (LARS) algorithm [48] shares the first two steps with OMP. It selects j_i to be an index outside the currently available active set in order to maximize the correlation with the residual vector. However, instead of performing an LS fit to compute the nonzero components of $\boldsymbol{\theta}^{(i)}$, these are computed so that the residual will be equicorrelated with all the columns in the active set, that is,

$$\left|\boldsymbol{x}_j^{cT}\left(\boldsymbol{y} - X\boldsymbol{\theta}^{(i)}\right)\right| = \text{constant}, \quad \forall j \in S^{(i)},$$

where we have assumed that the columns of X are normalized, as is common in practice (recall, also, Remarks 9.4). In other words, in contrast to the OMP, where the error vector is forced to be orthogonal to the active columns, LARS demands this error to form equal angles with each one of them. Like OMP, it can be shown that, provided the target vector is sufficiently sparse and under incoherence of the columns of X, LARS can exactly recover the sparsest solution [116].

A further small modification leads to the LARS-LASSO algorithm. According to this version, a previously selected index in the active set can be removed at a later stage. This gives the algorithm the potential to "recover" from a previously bad decision. Hence, this modification departs from the strict rationale that defines the greedy algorithms. It turns out that this version solves the LASSO optimization task. This algorithm is the same as the one suggested in [99] and it is known as a *homotopy* algorithm. Homotopy methods are based on a continuous transformation from one optimization task to another. The solutions to this sequence of tasks lie along a continuous parameterized path. The idea is that while the optimization tasks may be difficult to solve by themselves, one can trace this path of solutions by slowly varying the parameters. For the LASSO task, it is the λ parameter that is varying (see, for example, [4,86,104]). Take as an example the LASSO task in its regularized version in (9.6). For $\lambda = 0$, the task minimizes the ℓ_2 error norm and for $\lambda \longrightarrow \infty$ the task minimizes the parameter vector's ℓ_1 norm, and for this case the solution tends to zero. It turns out that the solution path, as λ changes from large to small values, is polygonal. Vertices on this solution path correspond to vectors having nonzero elements only on a subset of entries. This subset remains unchanged until λ reaches the next critical value, which corresponds to a new vertex of the polygonal path and to a new subset of potential nonzero values. Thus, the solution is obtained via this sequence of steps along this polygonal path.

Compressed Sensing Matching Pursuit (CSMP) Algorithms

Strictly speaking, the algorithms to be discussed here are not greedy, yet as stated in [93], they are at heart greedy algorithms. Instead of performing a single term optimization per iteration step, in order to increase the support by one, as is the case with OMP, these algorithms attempt to obtain first an estimate of the support and then use this information to compute an LS estimate of the target vector, constrained on the respective active columns. The quintessence of the method lies in the near-orthogonal nature of the sensing matrix, assuming that this obeys the RIP condition.

Assume that X obeys the RIP for some small enough value δ_k and sparsity level k of the unknown vector. Let, also, the measurements be exact, that is, $\boldsymbol{y} = X\boldsymbol{\theta}$. Then $X^T\boldsymbol{y} = X^T X\boldsymbol{\theta} \approx \boldsymbol{\theta}$, due to the near-orthogonal nature of X. Therefore, intuition indicates that it is not unreasonable to select, in the first iteration step, the t (a user-defined parameter) largest in magnitude components of $X^T\boldsymbol{y}$ as indicative of the nonzero positions of the sparse target vector. This reasoning carries on for all subsequent steps, where, at the ith iteration, the place of $\boldsymbol{y}$ is taken by the residual $\boldsymbol{e}^{(i-1)} := \boldsymbol{y} - X\boldsymbol{\theta}^{(i-1)}$, where $\boldsymbol{\theta}^{(i-1)}$ indicates the estimate of the target vector at the $(i-1)$th iteration. Basically, this could be considered as a generalization of the OMP. However, as we will soon see, the difference between the two mechanisms is more substantial.

Algorithm 10.2 (The CSMP scheme).

1. Select the value of t.
2. Initialize the algorithm: $\boldsymbol{\theta}^{(0)} = \boldsymbol{0}$, $\boldsymbol{e}^{(0)} = \boldsymbol{y}$.
3. For $i = 1, 2, \ldots,$ execute the following;
   - **(a)** Obtain the current support:

$$S^{(i)} := \text{supp}\left(\boldsymbol{\theta}^{(i-1)}\right) \cup \left\{ \begin{array}{c} \text{indices of the } t \text{ largest in magnitude} \\ \text{components of } X^T\boldsymbol{e}^{(i-1)} \end{array} \right\}.$$

   - **(b)** Select the active columns: Construct $X^{(i)}$ to comprise the active columns of X in accordance to $S^{(i)}$. Obviously, $X^{(i)}$ is an $N \times r$ matrix, where r denotes the cardinality of the support set $S^{(i)}$.
   - **(c)** Update the estimate of the parameter vector: solve the LS task

$$\tilde{\boldsymbol{\theta}} := \arg\min_{\boldsymbol{z}\in\mathbb{R}^r} \left\| \boldsymbol{y} - X^{(i)}\boldsymbol{z} \right\|_2^2.$$

     Obtain $\hat{\boldsymbol{\theta}}^{(i)} \in \mathbb{R}^l$ having the r elements of $\tilde{\boldsymbol{\theta}}$ in the respective locations, as indicated by the support, and the rest of the elements being zero.
   - **(d)** $\boldsymbol{\theta}^{(i)} := H_k\left(\hat{\boldsymbol{\theta}}^{(i)}\right)$. The mapping H_k denotes the *hard thresholding* function; that is, it returns a vector with the k largest in magnitude components of the argument, and the rest are forced to zero.
   - **(e)** Update the error vector: $\boldsymbol{e}^{(i)} = \boldsymbol{y} - X\boldsymbol{\theta}^{(i)}$.

The algorithm requires as input the sparsity level k. Iterations carry on until a halting criterion is met. The value of t, which determines the largest in magnitude values in steps 1 and 3a, depends on the specific algorithm. In the compressive sampling matching pursuit (CoSaMP) [93], $t = 2k$ (Problem 10.3), and in the subspace pursuit (SP) [33], $t = k$.

Having stated the general scheme, a major difference with OMP becomes readily apparent. In OMP, only one column is selected per iteration step. Moreover, this remains in the active set for all subsequent steps. If, for some reason, this was not a good choice, the scheme cannot recover from such a bad decision. In contrast, the support and, hence, the active columns of X are continuously updated in CSMP, and the algorithm has the ability to correct a previously bad decision, as more information is accumulated and iterations progress. In [33], it is shown that if the measurements are exact ($\boldsymbol{y} = X\boldsymbol{\theta}$), then SP can recover the k-sparse true vector in a finite number of iteration steps, provided that X satisfies the RIP with $\delta_{3k} < 0.205$. If the measurements are noisy, performance bounds have been derived, which hold true for $\delta_{3k} < 0.083$. For the CoSaMP, performance bounds have been derived for $\delta_{4k} < 0.1$.

10.2.2 ITERATIVE SHRINKAGE/THRESHOLDING (IST) ALGORITHMS

This family of algorithms also have a long history (see, for example, [44,69,70,73]). However, in the "early" days, most of the developed algorithms had some sense of heuristic flavor, without establishing a clear bridge with optimizing a cost function. Later attempts were substantiated by sound theoretical arguments concerning issues such as convergence and convergence rate [31,34,50,56].

The general form of this algorithmic family has a striking resemblance to the classical linear algebra iterative schemes for approximating the solution of large linear systems of equations, known as *stationary iterative* or *iterative relaxation* methods. The classical Gauss–Seidel and Jacobi algorithms (e.g., [65]), in numerical analysis can be considered members of this family. Given a linear system of l equations with l unknowns, $\boldsymbol{z} = A\boldsymbol{x}$, the basic iteration at step i has the following form:

$$\begin{aligned}\boldsymbol{x}^{(i)} &= (I - QA)\boldsymbol{x}^{(i-1)} + Q\boldsymbol{z} \\ &= \boldsymbol{x}^{(i-1)} + Q\boldsymbol{e}^{(i-1)}, \qquad \boldsymbol{e}^{(i-1)} := \boldsymbol{z} - A\boldsymbol{x}^{(i-1)},\end{aligned}$$

which does not come as a surprise. It is of the same form as most of the iterative schemes for numerical solutions! The matrix Q is chosen in order to guarantee convergence, and different choices lead to different algorithms with their pros and cons. It turns out that this algorithmic form can also be applied to underdetermined systems of equations, $\boldsymbol{y} = X\boldsymbol{\theta}$, with a "minor" modification, which is imposed by the sparsity constraint of the target vector. This leads to the following general form of iterative computation:

$$\boldsymbol{\theta}^{(i)} = T_i\left(\boldsymbol{\theta}^{(i-1)} + Q\boldsymbol{e}^{(i-1)}\right), \qquad \boldsymbol{e}^{(i-1)} = \boldsymbol{y} - X\boldsymbol{\theta}^{(i-1)},$$

starting from an initial guess of $\boldsymbol{\theta}^{(0)}$ (usually $\boldsymbol{\theta}^{(0)} = \mathbf{0}$, $\boldsymbol{e}^{(0)} = \boldsymbol{y}$). In certain cases, Q can be made to be iteration dependent. The function T_i is a nonlinear thresholding that is applied *entry-wise*, that is, *component-wise*. Depending on the specific scheme, this can be either the hard thresholding function, denoted as H_k, or the soft thresholding function, denoted as S_α. Hard thresholding, as we already know, keeps the k largest components of a vector unaltered and sets the rest equal to zero. Soft thresholding was introduced in Section 9.3. All components with magnitude less than a threshold value α are forced to zero and the rest are reduced in magnitude by α; that is, the jth component of a vector $\boldsymbol{\theta}$, after soft thresholding, becomes

$$(S_\alpha(\boldsymbol{\theta}))_j = \text{sgn}(\theta_j)(|\theta_j| - \alpha)_+.$$

Depending on (a) the choice of T_i, (b) the specific value of the parameter k or α, and (c) the matrix Q, different instances occur. The most common choice for Q is μX^T, and the generic form of the main iteration becomes

$$\boxed{\boldsymbol{\theta}^{(i)} = T_i\left(\boldsymbol{\theta}^{(i-1)} + \mu X^T \boldsymbol{e}^{(i-1)}\right),} \tag{10.2}$$

where μ is a relaxation (user-defined) parameter, which can also be left to vary with each iteration step. The choice of X^T is intuitively justified, once more, by the near-orthogonal nature of X. For the first iteration step and for a linear system of the form $\boldsymbol{y} = X\boldsymbol{\theta}$, starting from a zero initial guess, we have $X^T\boldsymbol{y} = X^T X\boldsymbol{\theta} \approx \boldsymbol{\theta}$ and we are close to the solution.

Although intuition is most important in scientific research, it is not enough, by itself, to justify decisions and actions. The generic scheme in (10.2) has been reached from different paths, following different perspectives that lead to different choices of the involved parameters. Let us spend some more time on that, with the aim of making the reader more familiar with techniques that address optimization tasks of nondifferentiable loss functions. The term in parentheses in (10.2) coincides with the gradient descent iteration step if the cost function is the unregularized sum of squared errors cost (LS), that is,

$$J(\boldsymbol{\theta}) = \frac{1}{2}\|\boldsymbol{y} - X\boldsymbol{\theta}\|_2^2.$$

In this case, the gradient descent rationale leads to

$$\begin{aligned}\boldsymbol{\theta}^{(i-1)} - \mu\frac{\partial J(\boldsymbol{\theta}^{(i-1)})}{\partial\boldsymbol{\theta}} &= \boldsymbol{\theta}^{(i-1)} - \mu X^T\left(X\boldsymbol{\theta}^{(i-1)} - \boldsymbol{y}\right)\\ &= \boldsymbol{\theta}^{(i-1)} + \mu X^T\boldsymbol{e}^{(i-1)}.\end{aligned}$$

The gradient descent can alternatively be viewed as the result of minimizing a regularized version of the linearized cost function (verify it),

$$\begin{aligned}\boldsymbol{\theta}^{(i)} &= \arg\min_{\boldsymbol{\theta}\in\mathbb{R}^l}\Big\{J(\boldsymbol{\theta}^{(i-1)}) + (\boldsymbol{\theta} - \boldsymbol{\theta}^{(i-1)})^T\frac{\partial J(\boldsymbol{\theta}^{(i-1)})}{\partial\boldsymbol{\theta}}\\ &\quad + \frac{1}{2\mu}\|\boldsymbol{\theta} - \boldsymbol{\theta}^{(i-1)}\|_2^2\Big\}.\end{aligned} \tag{10.3}$$

One can adopt this view of the gradient descent philosophy as a kick-off point to minimize iteratively the following LASSO task:

$$\min_{\boldsymbol{\theta}\in\mathbb{R}^l}\left\{L(\boldsymbol{\theta}, \lambda) = \frac{1}{2}\|\boldsymbol{y} - X\boldsymbol{\theta}\|_2^2 + \lambda\|\boldsymbol{\theta}\|_1 = J(\boldsymbol{\theta}) + \lambda\|\boldsymbol{\theta}\|_1\right\}.$$

The difference now is that the loss function comprises two terms: one that is smooth (differentiable) and a nonsmooth one. Let the current estimate be $\boldsymbol{\theta}^{(i-1)}$. The updated estimate is obtained by

$$\boldsymbol{\theta}^{(i)} = \arg\min_{\boldsymbol{\theta}\in\mathbb{R}^l} \Big\{ J(\boldsymbol{\theta}^{(i-1)}) + (\boldsymbol{\theta}-\boldsymbol{\theta}^{(i-1)})^T \frac{\partial J(\boldsymbol{\theta}^{(i-1)})}{\partial \boldsymbol{\theta}} + \frac{1}{2\mu}\|\boldsymbol{\theta}-\boldsymbol{\theta}^{(i-1)}\|_2^2 + \lambda\|\boldsymbol{\theta}\|_1 \Big\},$$

which, after ignoring constants, is equivalently written as

$$\boldsymbol{\theta}^{(i)} = \arg\min_{\boldsymbol{\theta}\in\mathbb{R}^l} \Big\{ \frac{1}{2}\|\boldsymbol{\theta}-\tilde{\boldsymbol{\theta}}\|_2^2 + \lambda\mu\,\|\boldsymbol{\theta}\|_1 \Big\}, \tag{10.4}$$

where

$$\tilde{\boldsymbol{\theta}} := \boldsymbol{\theta}^{(i-1)} - \mu \frac{\partial J(\boldsymbol{\theta}^{(i-1)})}{\partial \boldsymbol{\theta}}. \tag{10.5}$$

Following exactly the same steps as those that led to the derivation of (9.13) from (9.6) (after replacing $\hat{\boldsymbol{\theta}}_{LS}$ with $\tilde{\boldsymbol{\theta}}$), we obtain

$$\boldsymbol{\theta}^{(i)} = S_{\lambda\mu}(\tilde{\boldsymbol{\theta}}) = S_{\lambda\mu}\Big(\boldsymbol{\theta}^{(i-1)} - \mu\frac{\partial J(\boldsymbol{\theta}^{(i-1)})}{\partial \boldsymbol{\theta}}\Big) \tag{10.6}$$

$$= S_{\lambda\mu}\big(\boldsymbol{\theta}^{(i-1)} + \mu X^T \boldsymbol{e}^{(i-1)}\big). \tag{10.7}$$

This is very interesting and practically useful. The only effect of the presence of the nonsmooth ℓ_1 norm in the loss function is an extra simple thresholding operation, which as we know is an operation performed *individually* on each component. It can be shown (e.g., [11,95]) that this algorithm converges to a minimizer $\boldsymbol{\theta}_*$ of the LASSO (9.6), provided that $\mu \in (0, 1/\lambda_{\max}(X^T X))$, where $\lambda_{\max}(\cdot)$ denotes the maximum eigenvalue of $X^T X$. The convergence rate is dictated by the rule

$$L(\boldsymbol{\theta}^{(i)}, \lambda) - L(\boldsymbol{\theta}_*, \lambda) \approx O(1/i),$$

which is known as *sublinear* global rate of convergence. Moreover, it can be shown that

$$L(\boldsymbol{\theta}^{(i)}, \lambda) - L(\boldsymbol{\theta}_*, \lambda) \le \frac{C\|\boldsymbol{\theta}^{(0)} - \boldsymbol{\theta}_*\|_2^2}{2i}.$$

The latter result indicates that if one wants to achieve an accuracy of ϵ, then this can be obtained by at most $\lfloor \frac{C\|\boldsymbol{\theta}^{(0)}-\boldsymbol{\theta}_*\|_2^2}{2\epsilon} \rfloor$ iterations, where $\lfloor\cdot\rfloor$ denotes the floor function.

In [34], (10.2) was obtained from a nearby corner, building upon arguments from the classical *proximal point* methods in optimization theory (e.g., [105]). The original LASSO regularized cost function is modified to the *surrogate objective*,

$$J(\boldsymbol{\theta}, \tilde{\boldsymbol{\theta}}) = \frac{1}{2}\|\boldsymbol{y} - X\boldsymbol{\theta}\|_2^2 + \lambda\,\|\boldsymbol{\theta}\|_1 + \frac{1}{2}d(\boldsymbol{\theta}, \tilde{\boldsymbol{\theta}}),$$

where

$$d(\boldsymbol{\theta}, \tilde{\boldsymbol{\theta}}) := c\|\boldsymbol{\theta} - \tilde{\boldsymbol{\theta}}\|_2^2 - \|X\boldsymbol{\theta} - X\tilde{\boldsymbol{\theta}}\|_2^2.$$

If c is appropriately chosen (larger than the largest eigenvalue of $X^T X$), the surrogate objective is guaranteed to be strictly convex. Then it can be shown (Problem 10.4) that the minimizer of the surrogate objective is given by

$$\hat{\boldsymbol{\theta}} = S_{\lambda/c}\Big(\tilde{\boldsymbol{\theta}} + \frac{1}{c}X^T(\boldsymbol{y} - X\tilde{\boldsymbol{\theta}})\Big). \tag{10.8}$$

In the iterative formulation, $\tilde{\boldsymbol{\theta}}$ is selected to be the previously obtained estimate; in this way, one tries to keep the new estimate close to the previous one. The procedure readily results in our generic scheme in (10.2), using soft thresholding with parameter λ/c. It can be shown that such a strategy converges to a minimizer of the original LASSO problem. The same algorithm was reached in [56], using *majorization-minimization* techniques from optimization theory. So, from this perspective, the IST family has strong ties with algorithms that belong to the convex optimization category.

In [118], the *sparse reconstruction by separable approximation* (SpaRSA) algorithm is proposed, which is a modification of the standard IST scheme. The starting point is (10.3); however, the multiplying factor, $\frac{1}{2\mu}$, instead of being constant, is now allowed to change from iteration to iteration according to a rule. This results in a speedup in the convergence of the algorithm. Moreover, inspired by the homotopy family of algorithms, where λ is allowed to vary, SpaRSA can be extended to solve a sequence of problems that are associated with a corresponding sequence of values of λ. Once a solution has been obtained for a particular value of λ, it can be used as a "warm start" for a nearby value. Solutions can therefore be computed for a range of values, at a small extra computational cost, compared to solving for a single value from a "cold start." This technique abides by the *continuation strategy*, which has been used in the context of other algorithms as well (e.g., [66]). Continuation has been shown to be a very successful tool to increase the speed of convergence.

An interesting variation of the basic IST scheme has been proposed in [11], which improves the convergence rate to $O(1/i^2)$, by only a simple modification with almost no extra computational burden. The scheme is known as *fast iterative shrinkage-thresholding algorithm* (FISTA). This scheme is an evolution of [96], which introduced the basic idea for the case of differentiable costs, and consists of the following steps:

$$\boldsymbol{\theta}^{(i)} = S_{\lambda\mu}\Big(\boldsymbol{z}^{(i)} + \mu X^T\big(\boldsymbol{y} - X\boldsymbol{z}^{(i)}\big)\Big),$$

$$\boldsymbol{z}^{(i+1)} := \boldsymbol{\theta}^{(i)} + \frac{t_i - 1}{t_{i+1}}\big(\boldsymbol{\theta}^{(i)} - \boldsymbol{\theta}^{(i-1)}\big),$$

where

$$t_{i+1} := \frac{1 + \sqrt{1 + 4t_i^2}}{2},$$

with initial points $t_1 = 1$ and $\boldsymbol{z}^{(1)} = \boldsymbol{\theta}^{(0)}$. In words, in the thresholding operation, $\boldsymbol{\theta}^{(i-1)}$ is replaced by $\boldsymbol{z}^{(i)}$, which is a specific linear combination of two successive updates of $\boldsymbol{\theta}$. Hence, at a marginal increase of the computational cost, a substantial increase in convergence speed is achieved.

In [17] the hard thresholding version has been used, with $\mu = 1$, and the thresholding function H_k uses the sparsity level k of the target solution, which is assumed to be known. In a later version, [19], the relaxation parameter is left to change so that, at each iteration step, the error is maximally reduced. It has been shown that the algorithm converges to a local minimum of the cost function $\|\boldsymbol{y} - X\boldsymbol{\theta}\|_2$, under the constraint that $\boldsymbol{\theta}$ is a k-sparse vector. Moreover, the latter version is a stable one and it results in a near-optimal solution if a form of RIP is fulfilled.

A modified version of the generic scheme given in (10.2), which evolves along the lines of [84], obtains the updates component-wise, one vector component at a time. Thus, a "full" iteration consists of l steps. The algorithm is known as *coordinate descent* and its basic iteration has the form (Problem 10.5)

$$\theta_j^{(i)} = S_{\lambda/\|\boldsymbol{x}_j^c\|_2^2}\left(\theta_j^{(i-1)} + \frac{\boldsymbol{x}_j^{cT}\boldsymbol{e}^{(i-1)}}{\|\boldsymbol{x}_j^c\|_2^2}\right), \quad j = 1, 2, \ldots, l. \tag{10.9}$$

This algorithm replaces the constant c in the previously reported soft thresholding algorithm with the norm of the respective column of X, if the columns of X are not normalized to unit norm. It has been shown that the parallel coordinate descent algorithm also converges to a LASSO minimizer of (9.6) [50]. Improvements of the algorithm, using line search techniques to determine the steepest descent direction for each iteration, have also been proposed (see [124]).

The main contribution to the complexity for the iterative shrinkage algorithmic family comes from the two matrix-vector products, which amounts to $\mathcal{O}(Nl)$, unless X has a special structure (e.g., DFT) that can be exploited to reduce the load.

In [85], the two-stage thresholding (TST) scheme is presented, which brings together arguments from the iterative shrinkage family and the OMP. This algorithmic scheme involves two stages of thresholding. The first step is exactly the same as in (10.2). However, this is now used only for determining "significant" nonzero locations, just as in CSMP algorithms, presented in the previous subsection. Then, an LS problem is solved to provide the updated estimate, under the constraint of the available support. This is followed by a second step of thresholding. The thresholding operations in the two stages can be different. In hard thresholding, H_k is used in both steps; this results in the algorithm proposed in [58]. For this latter scheme, convergence and performance bounds are derived if the RIP holds for $\delta_{3k} < 0.58$. In other words, the basic difference between the TST and CSMP approaches is that, in the latter case, the most significant nonzero coefficients are obtained by looking at the correlation term $X^T\boldsymbol{e}^{(i-1)}$, and in the TST family by looking at $\boldsymbol{\theta}^{(i-1)} + \mu X^T\boldsymbol{e}^{(i-1)}$. The differences between different approaches can be minor and the crossing lines between the different algorithmic categories are not necessarily crispy and clear. However, from a practical point of view, sometimes small differences may lead to substantially improved performance.

In [41], the IST algorithmic framework was treated as a *message passing* algorithm in the context of graphical models (Chapter 15), and the following modified recursion was obtained:

$$\boldsymbol{\theta}^{(i)} = T_i\left(\boldsymbol{\theta}^{(i-1)} + X^T\boldsymbol{z}^{(i-1)}\right), \tag{10.10}$$

$$\boldsymbol{z}^{(i-1)} = \boldsymbol{y} - X\boldsymbol{\theta}^{(i-1)} + \frac{1}{\alpha}\boldsymbol{z}^{(i-2)}\overline{T_i'\left(\boldsymbol{\theta}^{(i-2)} + X^T\boldsymbol{z}^{(i-2)}\right)}, \tag{10.11}$$

where $\alpha = \frac{N}{l}$, the overbar denotes the average over all the components of the corresponding vector, and T_i' denotes the respective derivative of the component-wise thresholding rule. The extra term on the right-hand side in (10.11) which now appears turns out to provide a performance improvement of the algorithm, compared to the IST family, with respect to the undersampling–sparsity tradeoff (Section 10.2.3). Note that T_i is iteration dependent and it is controlled via the definition of certain parameters. A parameterless version of it has been proposed in [91]. A detailed treatment on the message passing algorithms can be found in [2].

Remarks 10.2.

- The iteration in (10.6) bridges the IST algorithmic family with another powerful tool in convex optimization, which builds upon the notion of *proximal mapping* or *Moreau envelopes* (see Chapter 8 and, e.g., [32,105]). Given a convex function $h : \mathbb{R}^l \to \mathbb{R}$ and a $\mu > 0$, the proximal mapping, $\text{Prox}_{\mu h} : \mathbb{R}^l \longmapsto \mathbb{R}^l$, with respect to h and of index μ, is defined as the (unique) minimizer

$$\text{Prox}_{\mu h}(\boldsymbol{x}) := \arg\min_{\boldsymbol{v}\in\mathbb{R}^l} \left\{h(\boldsymbol{v}) + \frac{1}{2\mu}\|\boldsymbol{x}-\boldsymbol{v}\|_2^2\right\}, \quad \forall \boldsymbol{x} \in \mathbb{R}^l. \tag{10.12}$$

  Let us now assume that we want to minimize a convex function, which is given as the sum

$$f(\boldsymbol{\theta}) = J(\boldsymbol{\theta}) + h(\boldsymbol{\theta}),$$

  where J is convex and differentiable, and h is also convex, but not necessarily smooth. Then it can be shown (Section 8.14) that the following iterations converge to a minimizer of f,

$$\boldsymbol{\theta}^{(i)} = \text{Prox}_{\mu h}\left(\boldsymbol{\theta}^{(i-1)} - \mu\frac{\partial J(\boldsymbol{\theta}^{(i-1)})}{\partial \boldsymbol{\theta}}\right), \tag{10.13}$$

  where $\mu > 0$, and it can also be made iteration dependent, that is, $\mu_i > 0$. If we now use this scheme to minimize our familiar cost,

$$J(\boldsymbol{\theta}) + \lambda\|\boldsymbol{\theta}\|_1,$$

  we obtain (10.6); this is so because the proximal operator of $h(\boldsymbol{\theta}) := \lambda\|\boldsymbol{\theta}\|_1$ is shown (see [31,32] and Section 8.13) to be identical to the soft thresholding operator, that is,

$$\text{Prox}_h(\boldsymbol{\theta}) = S_\lambda(\boldsymbol{\theta}).$$

  In order to feel more comfortable with this operator, note that if $h(\boldsymbol{x}) \equiv 0$, its proximal operator is equal to $\boldsymbol{x}$, and in this case (10.13) becomes our familiar gradient descent algorithm.
- All the nongreedy algorithms that have been discussed so far have been developed to solve the task defined in the formulation (9.6). This is mainly because this is an easier task to solve; once λ has been fixed, it is an unconstrained optimization task. However, there are algorithms that have been developed to solve the alternative formulations.
  The NESTA algorithm has been proposed in [12] and solves the task in its (9.8) formulation. Adopting this path can have an advantage because ϵ may be given as an estimate of the uncertainty associated with the noise, which can readily be obtained in a number of practical applications. In contrast, selecting a priori the value for λ is more intricate. In [28], the value $\lambda = \sigma_\eta\sqrt{2\ln l}$, where σ_η is the noise standard deviation, is argued to have certain optimality properties; however, this

argument hinges on the assumption of the orthogonality of X. NESTA relies heavily on Nesterov's generic scheme [96], hence its name. The original Nesterov algorithm performs a constrained minimization of a smooth convex function $f(\boldsymbol{\theta})$, i.e.,

$$\min_{\boldsymbol{\theta}\in Q} f(\boldsymbol{\theta}),$$

where Q is a convex set, and in our case this is associated with the quadratic constraint in (9.8). The algorithm consists of three basic steps. The first one involves an auxiliary variable, and is similar to the step in (10.3), i.e.,

$$\boldsymbol{w}^{(i)} = \arg\min_{\boldsymbol{\theta}\in Q} \left\{ (\boldsymbol{\theta} - \boldsymbol{\theta}^{(i-1)})^T \frac{\partial f(\boldsymbol{\theta}^{(i-1)})}{\partial \boldsymbol{\theta}} + \frac{L}{2}\|\boldsymbol{\theta} - \boldsymbol{\theta}^{(i-1)}\|_2^2 \right\}, \tag{10.14}$$

where L is an upper bound on the Lipschitz coefficient, which the gradient of f has to satisfy. The difference with (10.3) is that the minimization is now a constrained one. However, Nesterov has also added a second step, involving another auxiliary variable, $\boldsymbol{z}^{(i)}$, which is computed in a similar way as $\boldsymbol{w}^{(i)}$, but the linearized term is now replaced by a weighted cumulative gradient,

$$\sum_{k=0}^{i-1} \alpha_k (\boldsymbol{\theta} - \boldsymbol{\theta}^{(k)})^T \frac{\partial f(\boldsymbol{\theta}^{(k)})}{\partial \boldsymbol{\theta}}.$$

The effect of this term is to smooth out the "zig-zagging" of the path toward the solution, whose effect is to increase the convergence speed significantly. The final step of the scheme involves an averaging of the previously obtained variables,

$$\boldsymbol{\theta}^{(i)} = t_i \boldsymbol{z}^{(i)} + (1 - t_i)\boldsymbol{w}^{(i)}.$$

The values of the parameters α_k, $k = 0, \dots, i-1$, and t_i result from the theory so that convergence is guaranteed. As was the case with its close relative FISTA, the algorithm enjoys an $O(1/i^2)$ convergence rate. In our case, where the function to be minimized, $\|\boldsymbol{\theta}\|_1$, is not smooth, NESTA uses a smoothed prox-function of it. Moreover, it turns out that closed form updates are obtained for $\boldsymbol{z}^{(i)}$ and $\boldsymbol{w}^{(i)}$. If X is chosen in order to have orthonormal rows, the complexity per iteration is $O(l)$ plus the computations needed for performing the product $X^T X$, which is the most computationally thirsty part. However, this complexity can substantially be reduced if the sensing matrix is chosen to be a submatrix of a unitary transform which admits fast matrix-vector product computations (e.g., a subsampled DFT matrix). For example, for the case of a subsampled DFT matrix, the complexity amounts to $O(l)$ plus the load to perform the two fast Fourier transforms (FFTs). Moreover, the continuation strategy can also be employed to accelerate convergence. In [12], it is demonstrated that NESTA exhibits good accuracy results, while retaining a complexity that is competitive with algorithms developed around the (9.6) formulation and scales in an affordable way for large-size problems. Furthermore, NESTA and, in general, Nesterov's scheme enjoy a generality that allows their use for other optimization tasks as well.

- The task in (9.7) has been considered in [14] and [99]. In the former, the algorithm comprises a projection on the ℓ_1 ball $\|\boldsymbol{\theta}\|_1 \leq \rho$ (see also Section 10.4.4) per iteration step. The most computationally dominant part of the algorithm consists of matrix-vector products. In [99], a homotopy algorithm is derived for the same task, where now the bound ρ becomes the homotopy parameter that is left to vary. This algorithm is also referred to as the LARS-LASSO, as has already been reported.

10.2.3 WHICH ALGORITHM? SOME PRACTICAL HINTS

We have already discussed a number of algorithmic alternatives to obtain solutions to the ℓ_0 or ℓ_1 norm minimization tasks. Our focus was on schemes whose computational demands are rather low and that scale well to very large problem sizes. We have not touched more expensive methods such as interior point methods for solving the ℓ_1 convex optimization task. A review of such methods is provided in [72]. Interior point methods evolve along the Newton-type recursion and their complexity per iteration step is at least of the order $\mathcal{O}(l^3)$. As is most often the case, there is a tradeoff. Schemes of higher complexity tend to result in enhanced performance. However, such schemes become impractical in problems of large size. Some examples of other algorithms that were not discussed can be found in [14,35,118,121]. Talking about complexity, it has to be pointed out that what really matters at the end is not so much the complexity per iteration step, but the overall required resources in computer time/memory for the algorithm to converge to a solution within a specified accuracy. For example, an algorithm may be of low complexity per iteration step, but it may need an excessive number of iterations to converge.

Computational load is only one among a number of indices that characterize the performance of an algorithm. Throughout the book so far, we have considered a number of other performance measures, such as convergence rate, tracking speed (for the adaptive algorithms), and stability with respect to the presence of noise and/or finite word length computations. No doubt, all these performance measures are of interest here as well. However, there is an additional aspect that is of particular importance when quantifying performance of sparsity promoting algorithms. This is related to the *undersampling–sparsity tradeoff* or the *phase transition curve*.

One of the major issues on which we focused in Chapter 9 was to derive and present the conditions that guarantee uniqueness of the ℓ_0 minimization and its equivalence to the ℓ_1 minimization task, under an underdetermined set of observations, $\boldsymbol{y} = X\boldsymbol{\theta}$, for the recovery of sparse enough signals/vectors. While discussing the various algorithms in this section, we reported a number of different RIP-related conditions that some of the algorithms have to satisfy in order to recover the target sparse vector. As a matter of fact, it has to be admitted that this was quite confusing, because each algorithm had to satisfy its own conditions. In addition, in practice, these conditions are not easy to be verified. Although such results are undoubtedly important to establish convergence, make us more confident, and help us better understand why and how an algorithm works, one needs further experimental evidence in order to establish good performance bounds for an algorithm. Moreover, all the conditions we have dealt with, including coherence and RIP, are sufficient conditions. In practice, it turns out that sparse signal recovery is possible with sparsity levels much higher than those predicted by the theory, for given N and l. Hence, proposing a new algorithm or selecting an algorithm from an available palette, one has to demonstrate experimentally the range of sparsity levels that can be recovered by the algorithm, as a percentage of the number of observations and the dimensionality. Thus, in order to select an algorithm, one should cast her/his vote for the algorithm that, for given l and N, has the potential to recover k-sparse vectors with k being as high as possible for most of the cases, that is, with *high probability*.

Fig. 10.3 illustrates the type of curve that is expected to result in practice. The vertical axis is the probability of exact recovery of a target k-sparse vector and the horizontal axis shows the ratio k/N, for a given number of observations, N, and the dimensionality of the ambient space, l. Three curves are shown. The red ones correspond to the same algorithm, for two different values of the dimensionality l, and the gray one corresponds to another algorithm. Curves of this shape are expected to result from experiments of the following setup. Assume that we are given a sparse vector $\boldsymbol{\theta}_o$, with k nonzero components in the l-dimensional space. Using a sensing matrix X, we generate N samples/observations $\boldsymbol{y} = X\boldsymbol{\theta}_o$. The experiment is repeated a number of M times, each time using a different realization of the sensing matrix and a different k-sparse vector. For each instance, the algorithm is run to recover the target sparse vector. This is not always possible. We count the number, m, of successful recoveries, and compute the corresponding percentage of successful recovery (probability), m/M, which is plotted on the vertical axis of Fig. 10.3. The procedure is repeated for a different value of k, $1 \leq k \leq N$. A number of issues now jump onto the stage: (a) how does one select the ensemble of sensing matrices, and (b) how does one select the ensemble of sparse vectors? There are different scenarios, and some typical examples are described next.

1. The $N \times l$ sensing matrices X are formed by:
   - **(a)** different i.i.d. realizations with elements drawn from a Gaussian $\mathcal{N}(0, 1/N)$,
   - **(b)** different i.i.d. realizations from the uniform distribution on the unit sphere in $\mathbb{R}^N$, which is also known as the uniform spherical ensemble,
   - **(c)** different i.i.d. realizations with elements drawn from Bernoulli-type distributions,
   - **(d)** different i.i.d. realizations of partial Fourier matrices, each time using a different set of N rows.
2. The k-sparse target vector $\boldsymbol{\theta}_o$ is formed by selecting the locations of (at most) k nonzero elements randomly, by "tossing a coin" with probability $p = k/l$, and filling the values of the nonzero elements according to a statistical distribution (e.g., Gaussian, uniform, double exponential, Cauchy).

Other scenarios are also possible. Some authors set all nonzero values to one [16], or to ± 1, with the randomness imposed on the choice of the sign. It must be stressed that the performance of an algorithm may vary significantly under different experimental scenarios, and this may be indicative of

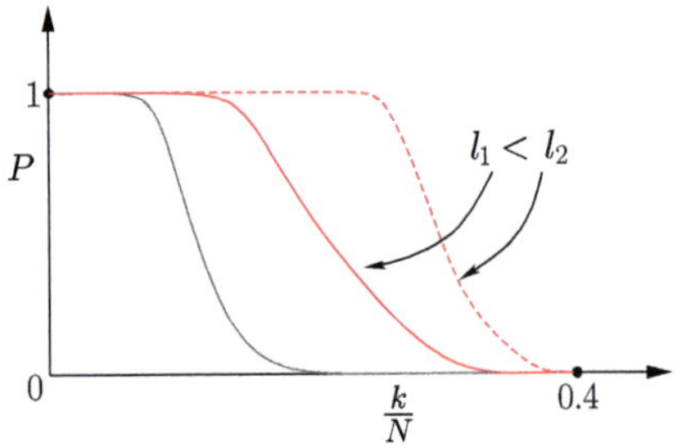

FIGURE 10.3

For any algorithm, the transition between the regions of 100% success and complete failure is very sharp. For the algorithm corresponding to the red curve, this transition occurs at higher sparsity values and, from this point of view, it is a better algorithm than the one associated with the gray curve. Also, given an algorithm, the higher the dimensionality, the higher the sparsity level where this transition occurs, as indicated by the two red curves.

the stability of an algorithm. In practice, a user may be interested in a specific scenario that is more representative of the available data.

Looking at Fig. 10.3, the following conclusions are in order. In all curves, there is a sharp transition between two levels, from the 100% success to the 0% success. Moreover, the higher the dimensionality, the sharper the transition. This has also been shown theoretically in [40]. For the algorithm corresponding to the red curves, this transition occurs at higher values of k, compared to the algorithm that generates the curve drawn in gray. Provided that the computational complexity of the "red" algorithm can be accommodated by the resources that are available for a specific application, this seems to be the more sensible choice between the two algorithms. However, if the resources are limited, concessions are unavoidable.

Another way to "interrogate" and demonstrate the performance of an algorithm, with respect to its robustness to the range of values of sparsity levels that can be successfully recovered, is via the *phase transition curve*. To this end, define:

- $\alpha := \frac{N}{l}$, which is a normalized measure of the problem indeterminacy,
- $\beta := \frac{k}{N}$, which is a normalized measure of sparsity.

In the sequel, plot a graph having $\alpha \in [0, 1]$ in the horizontal axis and $\beta \in [0, 1]$ in the vertical one. For each point (α, β) in the $[0, 1] \times [0, 1]$ region, compute the probability of the algorithm to recover a k-sparse target vector. In order to compute the probability, one has to adopt one of the previously stated scenarios. In practice, one has to form a grid of points that cover densely enough the region $[0, 1] \times [0, 1]$ in the graph. Use a varying intensity level scale to color the corresponding (α, β) point. Black corresponds to probability one and red to probability zero. Fig. 10.4 illustrates the type of graph that is expected to be recovered in practice for large values of l; that is, the transition from the region (phase) of "success" (black) to that of "fail" (red) is very sharp. As a matter of fact, there is a curve that separates the two regions. The theoretical aspects of this curve have been studied in the context of combinatorial geometry in [40] for the asymptotic case, $l \longrightarrow \infty$, and in [42] for finite values of l. Observe that the larger the value of α (larger percentage of observations), the larger the value of β at which the transition occurs. This is in line with what we have said so far in this chapter, and the problem gets increasingly difficult as one moves up and to the left in the graph. In practice, for smaller values of l, the transition region from red to black is smoother, and it gets narrower as l increases. In such cases, one can draw an approximate curve that separates the "success" and "fail" regions using regression techniques (see, e.g., [85]).

The reader may already be aware of the fact that, so far, we have avoided talking about the performance of individual algorithms. We have just discussed some "typical" behavior that algorithms tend to exhibit in practice. What the reader might have expected is a discussion of comparative performance tests and related conclusions. The reason is that at the time the current edition is being compiled there are not definite answers. Most authors compare their newly suggested algorithm with a few other algorithms, usually within a certain algorithmic family, and, more importantly, under some specific scenarios, where the advantages of the newly suggested algorithm are documented. However, the performance of an algorithm can change significantly by changing the experimental scenario under which the tests are carried out. The most comprehensive comparative performance study so far has been carried out in [85]. However, even in this work, the scenario of exact measurements has been considered and there are no experiments concerning the robustness of individual algorithms to the presence of noise. It is important to say that this study involved a huge effort of computation. We will comment on

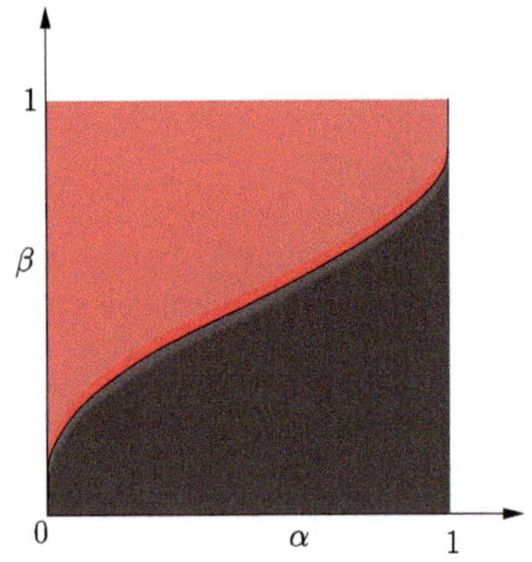

FIGURE 10.4

Typical phase transition behavior of a sparsity promoting algorithm. Black corresponds to 100% success of recovering the sparsest solution, and red to 0%. For high-dimensional spaces, the transition is very sharp, as is the case in the figure. For lower dimensionality values, the transition from black to red is smoother and involves a region of varying color intensity.

some of the findings from this study, which will also reveal to the reader that different experimental scenarios can significantly affect the performance of an algorithm.

Fig. 10.5A shows the obtained phase transition curves for (a) the iterative hard thresholding (IHT); (b) the iterative soft thresholding (IST) scheme of (10.2); (c) the two-stage thresholding (TST) scheme, as discussed earlier; (d) the LARS algorithm; and (e) the OMP algorithm, together with the theoretically obtained one using ℓ_1 minimization. All algorithms were tuned with the optimal values, with respect to the required user-defined parameters, after extensive experimentation. The results in the figure correspond to the uniform spherical scenario for the generation of the sensing matrices. Sparse vectors were generated according to the ± 1 scenario for the nonzero coefficients. The interesting observation is that, although the curves deviate from each other as they move to larger values of β, for smaller values, the differences in their performance become less and less. This is also true for computationally simple schemes such as the IHT one. The performance of LARS is close to the optimal one. However, this comes at the cost of computational increase. The required computational time for achieving the same accuracy, as reported in [85], favors the TST algorithm. In some cases, LARS required excessively longer time to reach the same accuracy, in particular when the sensing matrix was the partial Fourier one and fast schemes to perform matrix vector products can be exploited. For such matrices, the thresholding schemes (IHT, IST, TST) exhibited a performance that scales very well to large-size problems.

Fig. 10.5B indicates the phase transition curve for one of the algorithms (IST) as we change the scenarios for generating the sparse (target) vectors, using different distributions: (a) ± 1, with equiprobable selection of signs (constant amplitude random selection [CARS]); (b) double exponential (power); (c) Cauchy; and (d) uniform in $[-1, 1]$. This is indicative and typical for other algorithms as well, with some of them being more sensitive than others. Finally, Fig. 10.5C shows the transition curves for the IST algorithm by changing the sensing matrix generation scenario. Three curves are shown cor-

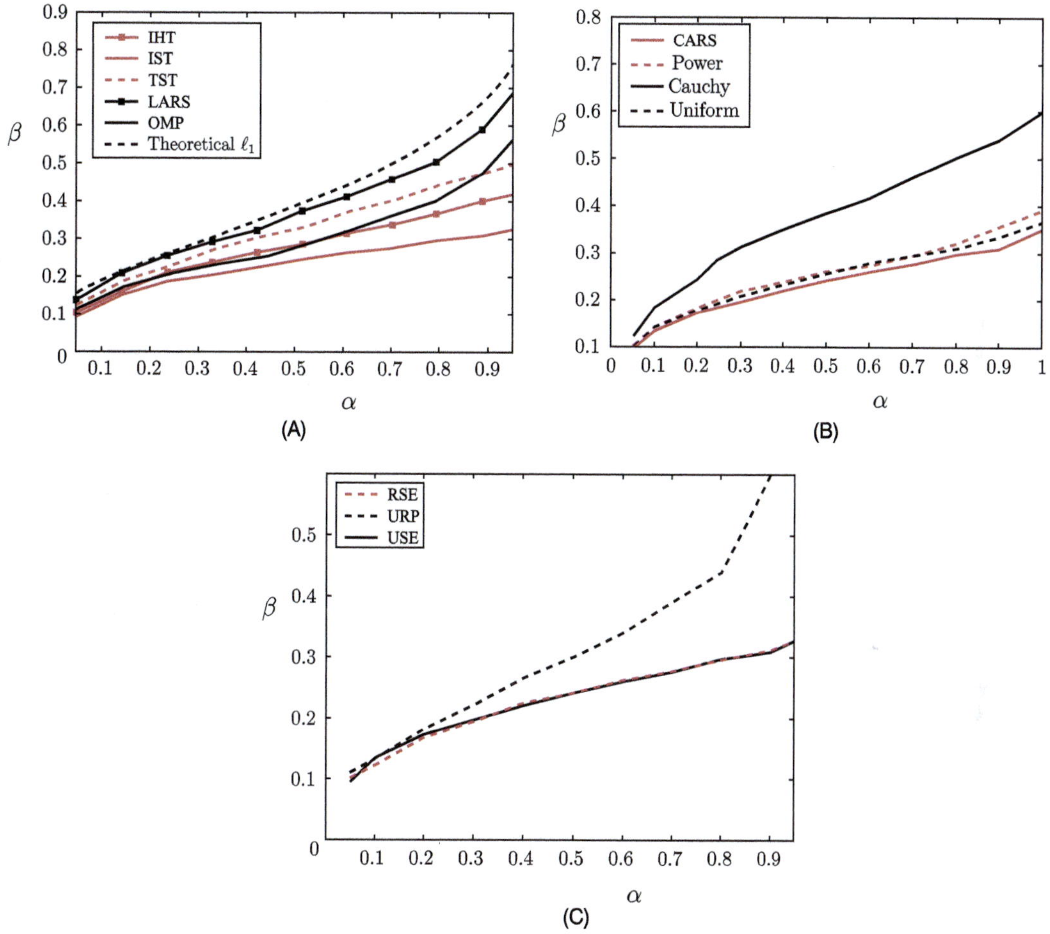

FIGURE 10.5

(A) The obtained phase transition curves for different algorithms under the same experimental scenario, together with the theoretical one. (B) Phase transition curve for the IST algorithm under different experimental scenarios for generating the target sparse vector. (C) The phase transition for the IST algorithms under different experimental scenarios for generating the sensing matrix X.

responding to (a) uniform spherical ensemble (USE); (b) random sign ensemble (RSE), where the elements are ± 1 with signs uniformly distributed; and (c) the uniform random projection (URP) ensemble. Once more, one can observe the possible variations that are expected due to the use of different matrix ensembles. Moreover, changing ensembles affects each algorithm in a different way.

Concluding this section, it must be emphasized that algorithmic development is still an ongoing research field, and it is early to come up with definite and concrete comparative performance conclusions. Moreover, besides the algorithmic front, existing theories often fall short in predicting what is observed in practice, with respect to their phase transition performance. For a related discussion, see, for example, [43].

10.3 VARIATIONS ON THE SPARSITY-AWARE THEME

In our tour so far, we have touched a number of aspects of sparsity-aware learning that come from mainstream theoretical developments. However, a number of variants have appeared and have been developed with the goal of addressing problems of a more special structure and/or proposing alternatives, which can be beneficial in boosting the performance in practice by serving the needs of specific applications. These variants focus on the regularization term in (9.6), and/or on the misfit measuring term. Once more, research activity in this direction has been dense, and our purpose is to simply highlight possible alternatives and make the reader aware of the various possibilities that spring forth from the basic theory.

In a number of tasks, it is a priori known that the nonzero parameters in the target signal/vector occur in groups and they are not randomly spread in all possible positions. A typical example is the echo path in internet telephony, where the nonzero coefficients of the impulse response tend to cluster together (see Fig. 9.5). Other examples of "structured" sparsity can be traced in DNA microarrays, MIMO channel equalization, source localization in sensor networks, magnetoencephalography, and neuroscience problems (e.g., [1,9,10,60,101]). As is always the case in machine learning, being able to incorporate a priori information into the optimization can only be of benefit for improving performance, because the estimation task is externally assisted in its effort to search for the target solution.

The *group LASSO* [8,59,97,98,117,122] addresses the task where it is a priori known that the nonzero components occur in groups. The unknown vector $\boldsymbol{\theta}$ is divided into L groups, that is,

$$\boldsymbol{\theta}^T = [\boldsymbol{\theta}_1^T, \dots, \boldsymbol{\theta}_L^T]^T,$$

each of them of a predetermined size, s_i, $i = 1, 2, \dots, L$, with $\sum_{i=1}^L s_i = l$. The regression model can then be written as

$$\boldsymbol{y} = X\boldsymbol{\theta} + \boldsymbol{\eta} = \sum_{i=1}^{L} X_i \boldsymbol{\theta}_i + \boldsymbol{\eta},$$

where each X_i is a submatrix of X comprising the corresponding s_i columns. The solution of the group LASSO is given by the following regularized task:

$$\hat{\boldsymbol{\theta}} = \arg\min_{\boldsymbol{\theta}\in\mathbb{R}^l} \left(\left\| \boldsymbol{y} - \sum_{i=1}^{L} X_i \boldsymbol{\theta}_i \right\|_2^2 + \lambda \sum_{i=1}^{L} \sqrt{s_i} \|\boldsymbol{\theta}_i\|_2 \right), \tag{10.15}$$

where $\|\boldsymbol{\theta}_i\|_2$ is the Euclidean norm (not the squared one) of $\boldsymbol{\theta}_i$, that is,

$$\|\boldsymbol{\theta}_i\|_2 = \sqrt{\sum_{j=1}^{s_i} |\theta_{i,j}|^2}.$$

In other words, the individual components of $\boldsymbol{\theta}$, which contribute to the formation of the ℓ_1 norm in the standard LASSO formulation, are now replaced by the square root of the energy of each individual block. In this setting, it is not the individual components but *blocks* of them that are forced to zero, when their contribution to the LS misfit measuring term is not significant. Sometimes, this type of regularization is coined as the ℓ_1/ℓ_2 regularization. An example of an ℓ_1/ℓ_2 ball for $\boldsymbol{\theta} \in \mathbb{R}^3$ can be seen in Fig. 10.6B in comparison with the corresponding ℓ_1 ball depicted in Fig. 10.6A.

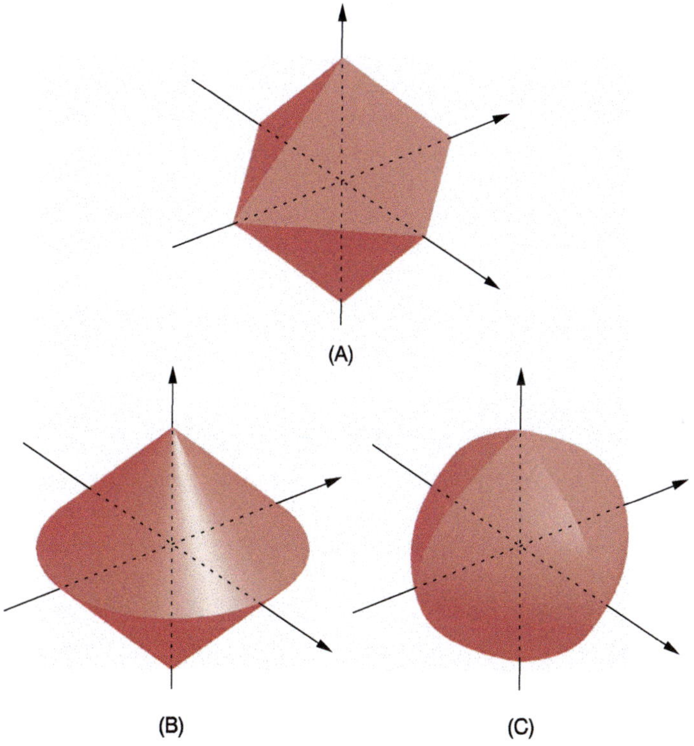

FIGURE 10.6

Representation of balls of $\boldsymbol{\theta} \in \mathbb{R}^3$ corresponding to: (A) the ℓ_1 norm; (B) an ℓ_1/ℓ_2 with nonoverlapping groups (one group comprises $\{\theta_1, \theta_2\}$ and the other one $\{\theta_3\}$); and (C) the ℓ_1/ℓ_2 with overlapping groups comprising $\{\theta_1, \theta_2, \theta_3\}$, $\{\theta_1\}$, and $\{\theta_3\}$ [6].

Beyond the conventional group LASSO, often referred to as *block sparsity*, research effort has been dedicated to the development of learning strategies incorporating more elaborate *structured sparse* models. There are two major reasons for such directions. First, in a number of applications the unknown set of parameters, $\boldsymbol{\theta}$, exhibit a structure that cannot be captured by the block sparse model. Second, even for cases where $\boldsymbol{\theta}$ is block sparse, standard grouped ℓ_1 norms require information about the partitioning of $\boldsymbol{\theta}$. This can be rather restrictive in practice. The adoption of overlapping groups has been proposed as a possible solution. Assuming that every parameter belongs to at least one group, such models lead to optimization tasks that, in many cases, are not hard to solve, for example, by resorting to proximal methods [6,7]. Moreover, by using properly defined overlapping groups [71], the allowed sparsity patterns can be constrained to form *hierarchical* structures, such as connected and rooted trees and subtrees that are met, for example, in multiscale (wavelet) decompositions. In Fig. 10.6C, an example of an ℓ_1/ℓ_2 ball for overlapping groups is shown.

Besides the previously stated directions, extensions of the compressed sensing principles to cope with structured sparsity led to the *model-based* compressed sensing [10,26]. The (k, C) model allows the significant parameters of a k-sparse signal to appear in at most C clusters, whose size is unknown. In Section 9.9, it was commented that searching for a k-sparse solution takes place in a union of subspaces, each one of dimensionality k. Imposing a certain structure on the target solution restricts the searching in a subset of these subspaces and leaves a number of these out of the game. This obviously facilitates the optimization task. In [27], structured sparsity is considered in terms of graphical models, and in [110] the C-HiLasso group sparsity model was introduced, which allows each block to have a sparse structure itself. Theoretical results that extend the RIP to the block RIP have been developed and reported (see, for example, [18,83]) and in the algorithmic front, proper modifications of greedy algorithms have been proposed in order to provide structured sparse solutions [53].

In [24], it is suggested to replace the ℓ_1 norm by a weighted version of it. To justify such a choice, let us recall Example 9.2 and the case where the "unknown" system was sensed using $\boldsymbol{x} = [2, 1]^T$. We have seen that by "blowing up" the ℓ_1 ball, the wrong sparse solution was obtained. Let us now replace the ℓ_1 norm in (9.21) with its weighted version,

$$\|\boldsymbol{\theta}\|_{1,w} := w_1|\theta_1| + w_2|\theta_2|, \quad w_1, w_2 > 0,$$

and set $w_1 = 4$ and $w_1 = 1$. Fig. 10.7A shows the isovalue curve $\|\boldsymbol{\theta}\|_{1,w} = 1$, together with that resulting from the standard ℓ_1 norm. The weighted one is sharply "pinched" around the vertical axis, and the larger the value of w_1, compared to that of w_2, the sharper the corresponding ball will be. Fig. 10.7B shows what happens when "blowing up" the weighted ℓ_1 ball. It will first touch the point (0, 1), which is the true solution. Basically, what we have done is "squeeze" the ℓ_1 ball to be aligned more to the axis that contains the (sparse) solution. For the case of our example, any weight $w_1 > 2$ would do the job.

Consider now the general case of a weighted norm,

$$\boxed{\|\boldsymbol{\theta}\|_{1,w} := \sum_{j=1}^{l} w_j|\theta_j|, \quad w_j > 0, : \quad \text{weighted } \ell_1 \text{ norm.}} \tag{10.16}$$

The ideal choice of the weights would be

$$w_j = \begin{cases} \frac{1}{|\theta_{o,j}|}, & \theta_{o,j} \neq 0, \\ \infty, & \theta_{o,j} = 0, \end{cases}$$

where $\boldsymbol{\theta}_o$ is the target true vector, and where we have tacitly assumed that $0 \cdot \infty = 0$. In other words, the smaller a parameter is, the larger the respective weight becomes. This is justified, because large weighting will force respective parameters toward zero during the minimization process. Of course, in practice the values of the true vector are not known, so it is suggested to use their estimates during each iteration of the minimization procedure. The resulting scheme is of the following form.

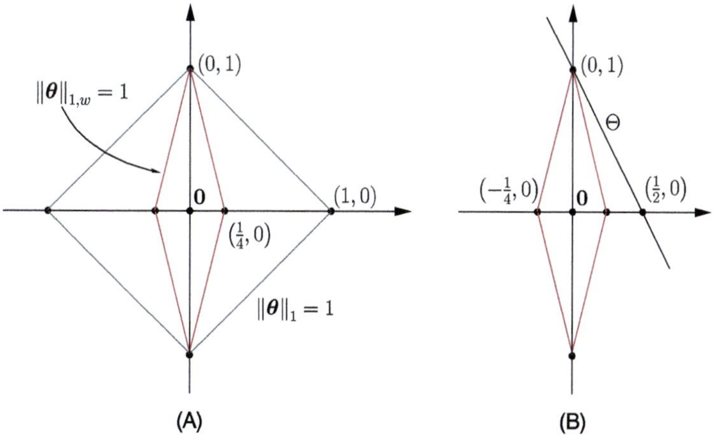

FIGURE 10.7

(A) The isovalue curves for the ℓ_1 and the weighted ℓ_1 norms for the same value. The weighted ℓ_1 is sharply pinched around one of the axes, depending on the weights. (B) Adopting to minimize the weighted ℓ_1 norm for the setup of Fig. 9.9C, the correct sparse solution is obtained.

Algorithm 10.3.

1. Initialize weights to unity, $w_j^{(0)} = 1$, $j = 1, 2, \ldots, l$.
2. Minimize the weighted ℓ_1 norm,

$$\begin{aligned} \boldsymbol{\theta}^{(i)} &= \arg\min_{\boldsymbol{\theta} \in \mathbb{R}^l} \|\boldsymbol{\theta}\|_{1,w} \\ \text{s.t.} \quad & \boldsymbol{y} = X\boldsymbol{\theta}. \end{aligned}$$

3. Update the weights

$$w_j^{(i+1)} = \frac{1}{\left|\theta_j^{(i)}\right| + \epsilon}, \quad j = 1, 2, \ldots, l.$$

4. Terminate when a stopping criterion is met, otherwise return to step 2.

The constant ϵ is a small user-defined parameter to guarantee stability when the estimates of the coefficients take very small values. Note that if the weights have constant preselected values, the task retains its convex nature; this is no longer true when the weights are changing. It is interesting to point out that this intuitively motivated weighting scheme can result if the ℓ_1 norm is replaced by $\sum_{j=1}^{l} \ln\left(|\theta_j| + \epsilon\right)$ as the regularizing term of (9.6). Fig. 10.8 shows the respective graph in the one-dimensional space together with that of the ℓ_1 norm. The graph of the logarithmic function reminds us of the ℓ_p, $p < 0 < 1$, "norms" and the comments made in Section 9.2. This is no longer a convex function, and the iterative scheme given before is the result of a majorization-minimization procedure in order to solve the resulting nonconvex task [24] (Problem 10.6).

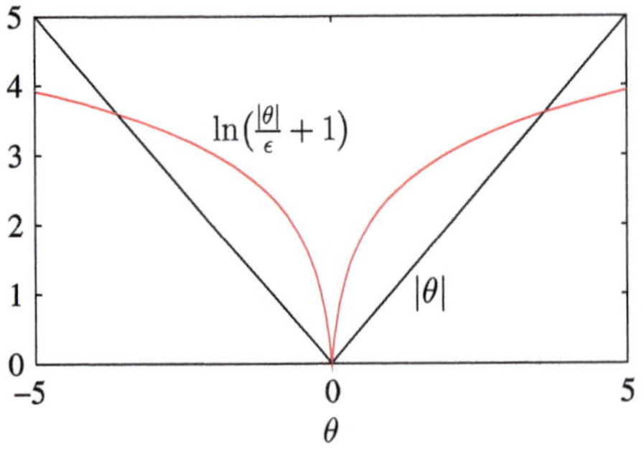

FIGURE 10.8

One-dimensional graphs of the ℓ_1 norm and the logarithmic regularizer $\ln\left(\frac{|\theta|}{\epsilon}+1\right)=\ln(|\theta|+\epsilon)-\ln\epsilon$, with $\epsilon=0.1$. The term $\ln\epsilon$ was subtracted for illustration purposes only and does not affect the optimization. Note the nonconvex nature of the logarithmic regularizer.

The concept of iterative weighting, as used before, has also been applied in the context of the *iterative reweighted LS algorithm*. Observe that the ℓ_1 norm can be written as

$$\|\boldsymbol{\theta}\|_1=\sum_{j=1}^{l}|\theta_j|=\boldsymbol{\theta}^T\mathcal{W}_\theta\boldsymbol{\theta},$$

where

$$\mathcal{W}_\theta=\begin{bmatrix}\frac{1}{|\theta_1|} & 0 & \cdots & 0\\ 0 & \frac{1}{|\theta_2|} & \cdots & 0\\ \vdots & \vdots & \ddots & \vdots\\ 0 & 0 & \cdots & \frac{1}{|\theta_l|}\end{bmatrix},$$

and where in the case of $\theta_i=0$, for some $i\in\{1,2,\ldots,l\}$, the respective coefficient of $\mathcal{W}_\theta$ is defined to be 1. If $\mathcal{W}_\theta$ were a constant weighting matrix, that is, $\mathcal{W}_\theta:=\mathcal{W}_{\tilde{\theta}}$, for some fixed $\tilde{\boldsymbol{\theta}}$, then obtaining the minimum

$$\hat{\boldsymbol{\theta}}=\arg\min_{\boldsymbol{\theta}\in\mathbb{R}^l}\|\boldsymbol{y}-X\boldsymbol{\theta}\|_2^2+\lambda\boldsymbol{\theta}^T\mathcal{W}_{\tilde{\theta}}\boldsymbol{\theta},$$

would be straightforward and similar to the ridge regression. In the iterative reweighted scheme, $\mathcal{W}_\theta$ is replaced by $\mathcal{W}_{\theta^{(i)}}$, formed by using the respected estimates of the parameters, which have been obtained from the previous iteration, that is, $\tilde{\boldsymbol{\theta}}:=\boldsymbol{\theta}^{(i)}$, as we did before. In the sequel, each iteration solves a weighted ridge regression task.

The *focal underdetermined system solver* (FOCUSS) algorithm [64] was the first one to use the concept of iterative-reweighted-least-squares (IRLS) to represent ℓ_p, $p\leq 1$, as a weighted ℓ_2 norm in order to find a sparse solution to an underdetermined system of equations. This algorithm is also of historical importance, because it is among the very first ones to emphasize the importance of sparsity; moreover, it provides comprehensive convergence analysis as well as a characterization of the stationary points of the algorithm. Variants of this basic iterative weighting scheme have also been proposed (see, e.g., [35] and the references therein).

In [126], the *elastic net* regularization penalty was introduced, which combines the ℓ_2 and ℓ_1 concepts together in a tradeoff fashion, that is,

$$\lambda \sum_{i=1}^{l} \left(\alpha \theta_i^2 + (1-\alpha)|\theta_i| \right),$$

where α is a user-defined parameter controlling the influence of each individual term. The idea behind the elastic net is to combine the advantages of the LASSO and the ridge regression. In problems where there is a group of variables in $\boldsymbol{x}$ that are highly correlated, LASSO tends to select one of the corresponding parameters in $\boldsymbol{\theta}$, and set the rest to zero in a rather arbitrary fashion. This can be understood by looking carefully at how the greedy algorithms work. When sparsity is used to select the most important of the variables in $\boldsymbol{x}$ (feature selection), it is better to select all the relevant components in the group. If one knew which of the variables are correlated, he/she could form a group and then use the group LASSO. However, if this is not known, involving the ridge regression offers a remedy to the problem. This is because the ℓ_2 penalty in ridge regression tends to shrink the coefficients associated with correlated variables toward each other (e.g., [68]). In such cases, it would be better to work with the elastic net rationale that involves LASSO and ridge regression in a combined fashion.

In [23], the LASSO task is modified by replacing the squared error term with one involving correlations, and the minimization task becomes

$$\begin{aligned} \hat{\boldsymbol{\theta}} : \quad & \min_{\boldsymbol{\theta} \in \mathbb{R}^l} \|\boldsymbol{\theta}\|_1, \\ \text{s.t.} \quad & \left\| X^T (\boldsymbol{y} - X\boldsymbol{\theta}) \right\|_\infty \leq \epsilon, \end{aligned}$$

where ϵ is related to l and the noise variance. This task is known as the *Dantzig selector*. That is, instead of constraining the energy of the error, the constraint now imposes an upper limit to the correlation of the error vector with any of the columns of X. In [5,15], it is shown that under certain conditions, the LASSO estimator and the Dantzig selector become identical.

Total variation (TV) [107] is a closely related to ℓ_1 sparsity promoting notion that has been widely used in image processing. Most of the grayscale image arrays, $I \in \mathbb{R}^{l \times l}$, consist of slowly varying pixel intensities except at the edges. As a consequence, the discrete gradient of an image array will be approximately sparse (compressible). The discrete directional derivatives of an image array are defined pixel-wise as

$$\nabla_x(I)(i,j) := I(i+1,j) - I(i,j), \quad \forall i \in \{1,2,\ldots,l-1\}, \tag{10.17}$$

$$\nabla_y(I)(i,j) := I(i,j+1) - I(i,j), \quad \forall j \in \{1,2,\ldots,l-1\}, \tag{10.18}$$

and

$$\nabla_x(I)(l,j) := \nabla_y(I)(i,l) := 0, \quad \forall i,j \in \{1,2,\ldots,l-1\}. \tag{10.19}$$

The discrete gradient transform

$$\nabla : \mathbb{R}^{l \times l} \to \mathbb{R}^{l \times 2l}$$

is defined in terms of a matrix form as

$$\nabla(I)(i,j) := [\nabla_x(i,j), \nabla_y(i,j)], \quad \forall i,j \in \{1,2,\ldots l\}. \tag{10.20}$$

The total variation of the image array is defined as the ℓ_1 norm of the *magnitudes* of the elements of the discrete gradient transform, that is,

$$\|I\|_{\text{TV}} := \sum_{i=1}^{l}\sum_{j=1}^{l} \|\nabla(I)(i,j)\|_2 = \sum_{i=1}^{l}\sum_{j=1}^{l} \sqrt{(\nabla_x(I)(i,j))^2 + (\nabla_y(I)(i,j))^2}. \tag{10.21}$$

Note that this is a mixture of ℓ_2 and ℓ_1 norms. The sparsity promoting optimization around the total variation is defined as

$$\begin{aligned} I_* \in \arg\min_I \|I\|_{TV}, \\ \text{s.t.} \quad \|\boldsymbol{y} - \mathcal{F}(I)\|_2 \le \epsilon, \end{aligned} \tag{10.22}$$

where $\boldsymbol{y} \in \mathbb{R}^N$ is the observations vector and $\mathcal{F}(I)$ denotes the result in vectorized form of the application of a linear operator on I. For example, this could be the result of the action of a partial two-dimensional DFT on the image. Subsampling of the DFT matrix as a means of forming sensing matrices has already been discussed in Section 9.7.2. The task in (10.22) retains its convex nature and it basically expresses our desire to reconstruct an image that is as smooth as possible, given the available observations. The NESTA algorithm can be used for solving the total variation minimization task; besides NESTA, other efficient algorithms for this task can be found in, for example, [63,120].

It has been shown in [22] for the exact measurements case ($\epsilon = 0$), and in [94] for the erroneous measurements case, that conditions and bounds that guarantee recovery of an image array from the task in (10.22) can be derived and are very similar to those we have discussed for the case of the ℓ_1 norm.

Example 10.1. (*Magnetic resonance imaging* (MRI)). In contrast to ordinary imaging systems, which directly acquire pixel samples, MRI scanners sense the image in an encoded form. Specifically, MRI scanners sample components in the spatial frequency domain, known as "k-space" in the MRI nomenclature. If all the components in this transform domain were available, one could apply the inverse 2D-DFT to recover the exact MR image in the pixel domain. Sampling in the k-space is realized along particular trajectories in a number of successive acquisitions. This process is time consuming, merely due to physical constraints. As a result, techniques for efficient image recovery from a *limited number of observations* is of high importance, because they can reduce the required acquisition time for performing the measurements. Long acquisition times are not only inconvenient but even impossible, because the patients have to stay still for long time intervals. Thus, MRI was among the very first applications where compressed sensing found its way to offering its elegant solutions.

Fig. 10.9A shows the "famous" Shepp–Logan phantom, and the goal is to recover it via a limited number of samples (measurements) in its frequency domain. The MRI measurements are taken across 17 radial lines in the spatial frequency domain, as shown in Fig. 10.9B. A "naive" approach to

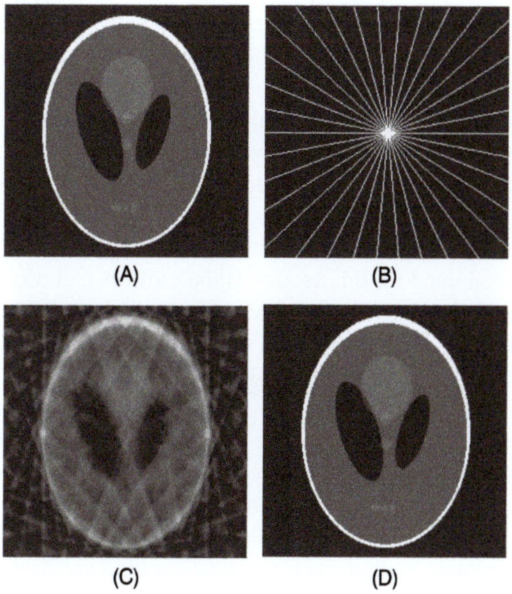

FIGURE 10.9

(A) The original Shepp–Logan image phantom. (B) The white lines indicate the directions across which the sampling in the spatial Fourier transform were obtained. (C) The recovered image after applying the inverse DFT, having first filled with zeros the missing values in the DFT transform. (D) The recovered image using the total variation minimization approach.

recovering the image from this limited number of samples would be to adopt a zero-filling rationale for the missing components. The recovered image according to this technique is shown in Fig. 10.9C. Fig. 10.9D shows the recovered image using the approach of minimizing the total variation, as explained before. Observe that the results for this case are astonishingly good. The original image is almost perfectly recovered. The constrained minimization was performed via the NESTA algorithm. Note that if the minimization of the ℓ_1 norm of the image array were used in place of the total variation, the results would not be as good; the phantom image is sparse in the discrete gradient domain, because it contains large sections that share constant intensities.

10.4 ONLINE SPARSITY PROMOTING ALGORITHMS

In this section, online schemes for sparsity-aware learning are presented. There are a number of reasons why one has to resort to such schemes. As has already been noted in previous chapters, in various signal processing tasks the data arrive sequentially. Under such a scenario, using batch processing techniques to obtain an estimate of an unknown target parameter vector would be highly inefficient, because the number of training points keeps increasing. Such an approach is prohibited for real-time applications. Moreover, time-recursive schemes can easily incorporate the notion of adaptivity, when the learning environment is not stationary but undergoes changes as time evolves. Besides signal processing applications, there are an increasing number of machine learning applications where online processing is of paramount importance, such as bioinformatics, hyperspectral imaging, and data mining. In such applications, the number of training points easily amounts to a few thousand up to hundreds of thousands of points. Concerning the dimensionality of the ambient (feature) space, one can claim numbers that lie in similar ranges. For example, in [82], the task is to search for sparse solutions in feature spaces with dimensionality as high as 10^9 having access to data sets as large as 10^7 points. Using batch techniques on a single computer is out of the question with today's technology.

The setting that we have adopted for this section is the same as that used in previous chapters (e.g., Chapters 5 and 6). We assume that there is an unknown parameter vector that generates data according to the standard regression model

$$y_n = \boldsymbol{x}_n^T \boldsymbol{\theta} + \eta_n, \quad \forall n,$$

and the training samples are received sequentially $(y_n, \boldsymbol{x}_n)$, $n = 1, 2, \dots$. In the case of a stationary environment, we would expect our algorithm to converge asymptotically, as $n \longrightarrow \infty$, to or "nearly to" the true parameter vector that gives birth to the observations, y_n, when it is sensed by $\boldsymbol{x}_n$. For time-varying environments, the algorithms should be able to track the underlying changes as time goes by. Before we proceed, a comment is important. Because the time index, n, is left to grow, all we have said in the previous sections with respect to underdetermined systems of equations loses its meaning. Sooner or later we are going to have more observations than the dimension of the space in which the data reside. Our major concern here becomes the issue of asymptotic convergence for the case of stationary environments. The obvious question that is now raised is, why not use a standard algorithm (e.g., LMS, RLS, or APSM), as we know that these algorithms converge to, or nearly enough in some sense, the solution (i.e., the algorithm will identify the zeros asymptotically)? The answer is that if such algorithms are modified to be aware of the underlying sparsity, convergence is significantly sped up; in real-life applications, one does not have the "luxury" of waiting a long time for the solution. In practice, a good algorithm should be able to provide a good enough solution, and in the case of sparse solutions to *obtain the support*, after a reasonably small number of iteration steps.

In Chapter 5, we commented on attempts to modify classical online schemes (for example, the proportionate LMS) in order to consider sparsity. However, these algorithms were of a rather ad hoc nature. In this section, the powerful theory around the ℓ_1 norm regularization will be used to obtain sparsity promoting time-adaptive schemes.

10.4.1 LASSO: ASYMPTOTIC PERFORMANCE

When presenting the basic principles of parameter estimation in Chapter 3, the notions of bias, variance, and consistency, which are related to the performance of an estimator, were introduced. In a number of cases, such performance measures were derived asymptotically. For example, we have seen that the maximum likelihood estimator is asymptotically unbiased and consistent. In Chapter 6, we saw that the LS estimator is also asymptotically consistent. Moreover, under the assumption that the noise samples are i.i.d., the LS estimator, $\hat{\boldsymbol{\theta}}_n$, based on n observations, is itself a random vector that satisfies the $\sqrt{n}$-estimation consistency, that is,

$$\sqrt{n}\left(\hat{\boldsymbol{\theta}}_n - \boldsymbol{\theta}_o\right) \longrightarrow \mathcal{N}\left(\mathbf{0}, \sigma_\eta^2 \Sigma^{-1}\right),$$

where $\boldsymbol{\theta}_o$ is the true vector that generates the observations, σ_η^2 denotes the variance of the noise source, and Σ is the covariance matrix $\mathbb{E}[\mathbf{x}\mathbf{x}^T]$ of the input sequence, which has been assumed to be zero mean and the limit denotes convergence in distribution.

The LASSO in its (9.6) formulation is the task of minimizing the ℓ_1 norm regularized version of the LS cost. However, nothing has been said so far about the statistical properties of this estimator. The only performance measure that we referred to was the error norm bound given in (9.36). However, this bound, although important in the context for which it was proposed, does not provide much statistical information. Since the introduction of the LASSO estimator, a number of papers have addressed problems related to its statistical performance (see, e.g., [45,55,74,127]).

When dealing with sparsity promoting estimators such as the LASSO, two crucial issues emerge: (a) whether the estimator, even asymptotically, can obtain the support, if the true vector parameter is a sparse one; and (b) to quantify the performance of the estimator with respect to the estimates of the nonzero coefficients, that is, those coefficients whose index belongs to the support. Especially for LASSO, the latter issue becomes to study whether LASSO behaves as well as the unregularized LS with respect to these nonzero components. This task was addressed for the first time and in a more general setting in [55]. Let the support of the true, yet unknown, k-sparse parameter vector $\boldsymbol{\theta}_o$ be denoted as S. Let also $\Sigma_{|S}$ be the $k \times k$ covariance matrix $\mathbb{E}[\mathbf{x}_{|S}\mathbf{x}_{|S}^T]$, where $\mathbf{x}_{|S} \in \mathbb{R}^k$ is the random vector that contains only the k components of $\mathbf{x}$, with indices in the support S. Then we say that an estimator satisfies asymptotically the *oracle properties* if:

- $\lim_{n\to\infty} \text{Prob}\left\{S_{\hat{\boldsymbol{\theta}}_n} = S\right\} = 1$; this is known as *support consistency*;
- $\sqrt{n}\left(\hat{\boldsymbol{\theta}}_{n|S} - \boldsymbol{\theta}_{o|S}\right) \longrightarrow \mathcal{N}\left(\mathbf{0}, \sigma_\eta^2 \Sigma_{|S}^{-1}\right)$; this is the *$\sqrt{n}$-estimation consistency*.

We denote as $\boldsymbol{\theta}_{o|S}$ and $\hat{\boldsymbol{\theta}}_{n|S}$ the k-dimensional vectors that result from $\boldsymbol{\theta}_o$, $\hat{\boldsymbol{\theta}}_n$, respectively, if we keep the components whose indices lie in the support S, and the limit is meant in distribution. In other words, according to the oracle properties, a good sparsity promoting estimator should be able to predict, asymptotically, the true support and its performance with respect to the nonzero components should be as good as that of a genie-aided LS estimator, which is informed in advance of the positions of the nonzero coefficients.

Unfortunately, the LASSO estimator *cannot* satisfy simultaneously both conditions. It has been shown [55,74,127] that:

- For support consistency, the regularization parameter $\lambda := \lambda_n$ should be time-varying, such that

  $$\lim_{n\to\infty} \frac{\lambda_n}{\sqrt{n}} = \infty, \qquad \lim_{n\to\infty} \frac{\lambda_n}{n} = 0.$$

  That is, λ_n must grow faster than $\sqrt{n}$, but slower than n.
- For $\sqrt{n}$-consistency, λ_n must grow as

  $$\lim_{n\to\infty} \frac{\lambda_n}{\sqrt{n}} = 0,$$

  that is, it grows slower than $\sqrt{n}$.

The previous two conditions are conflicting and the LASSO estimator cannot comply with the two oracle conditions simultaneously. The proofs of the previous two points are somewhat technical and are not given here. The interested reader can obtain them from the previously given references. However, before we proceed, it is instructive to see why the regularization parameter has to grow more slowly than n, in any case. Without being too rigorous mathematically, recall that the LASSO solution comes from Eq. (9.6). This can be written as

$$\mathbf{0} \in -\frac{2}{n}\sum_{i=1}^{n} \boldsymbol{x}_i y_i + \frac{2}{n}\left(\sum_{i=1}^{n} \boldsymbol{x}_i \boldsymbol{x}_i^T\right)\boldsymbol{\theta} + \frac{\lambda_n}{n}\partial \|\boldsymbol{\theta}\|_1, \tag{10.23}$$

where we have divided both sides by n. Taking the limit as $n \longrightarrow \infty$, if $\lambda_n/n \longrightarrow 0$, then we are left with the first two terms; this is exactly what we would have if the unregularized sum of squared errors had been chosen as the cost function. Recall from Chapter 6 that in this case, the solution asymptotically converges[1] (under some general assumptions, which are assumed to hold true here) to the true parameter vector; that is, we have strong consistency.

10.4.2 THE ADAPTIVE NORM-WEIGHTED LASSO

There are two ways to get out of the previously stated conflict. One is to replace the ℓ_1 norm with a nonconvex function, which can lead to an estimator that satisfies the oracle properties simultaneously [55]. The other is to modify the ℓ_1 norm by replacing it with a *weighted* version. Recall that the weighted ℓ_1 norm was discussed in Section 10.3 as a means to assist the optimization procedure to unveil the sparse solution. Here the notion of weighted ℓ_1 norm comes as a necessity imposed by our desire to satisfy the oracle properties. This gives rise to the *adaptive time-and-norm-weighted LASSO* (TNWL) cost estimate, defined as

$$\hat{\boldsymbol{\theta}} = \arg\min_{\boldsymbol{\theta}\in\mathbb{R}^l} \left\{ \sum_{j=1}^{n} \beta^{n-j}\left(y_j - \boldsymbol{x}_j^T\boldsymbol{\theta}\right)^2 + \lambda_n \sum_{i=1}^{l} w_{i,n}|\theta_i| \right\}, \tag{10.24}$$

where $\beta \leq 1$ is used as the forgetting factor to allow for tracking slow variations. The time-varying weighting sequences is denoted as $w_{i,n}$. There are different options. In [127] and under a stationary environment with $\beta = 1$, it is shown that if

$$w_{i,n} = \frac{1}{|\theta_i^{\text{est}}|^{\gamma}},$$

where θ_i^{est} is the estimate of the ith component obtained by *any* $\sqrt{n}$-consistent estimator, such as the unregularized LS, then for specific choices of λ_n and γ the corresponding estimator satisfies the oracle properties simultaneously. The main reasoning behind the weighted ℓ_1 norm is that as time goes by, and the $\sqrt{n}$-consistent estimator provides better and better estimates, the weights corresponding to indices outside the true support (zero values) are inflated and those corresponding to the true support converge

[1] Recall that this convergence is with probability 1.

to a finite value. This helps the algorithm simultaneously to locate the support and obtain unbiased (asymptotically) estimates of the large coefficients.

Another choice for the weighting sequence is related to the *smoothly clipped absolute deviation (SCAD)* [55,128]. This is defined as

$$w_{i,n} = \chi_{(0,\mu_n)}(|\theta_i^{\text{est}}|) + \frac{(\alpha\mu_n - |\theta_i^{\text{est}}|)_+}{(\alpha-1)\mu_n}\chi_{(\mu_n,\infty)}(|\theta_i^{\text{est}}|),$$

where $\chi(\cdot)$ stands for the characteristic function, $\mu_n = \lambda_n/n$, and $\alpha > 2$. Basically, this corresponds to a quadratic spline function. It turns out [128] that if λ_n is chosen to grow faster than $\sqrt{n}$ and slower than n, the adaptive LASSO with $\beta = 1$ satisfies both oracle conditions simultaneously.

A time-adaptive scheme for solving the TNWL LASSO was presented in [3]. The cost function of the adaptive LASSO in (10.24) can be written as

$$J(\boldsymbol{\theta}) = \boldsymbol{\theta}^T R_n \boldsymbol{\theta} - \boldsymbol{r}_n^T \boldsymbol{\theta} + \lambda_n \|\boldsymbol{\theta}\|_{1,w_n},$$

where

$$R_n := \sum_{j=1}^{n} \beta^{n-j} \boldsymbol{x}_j \boldsymbol{x}_j^T, \quad \boldsymbol{r}_n := \sum_{j=1}^{n} \beta^{n-j} y_j \boldsymbol{x}_j,$$

and $\|\boldsymbol{\theta}\|_{1,\boldsymbol{w}_n}$ is the weighted ℓ_1 norm. We know from Chapter 6, and it is straightforward to see, that

$$R_n = \beta R_{n-1} + \boldsymbol{x}_n \boldsymbol{x}_n^T, \quad \boldsymbol{r}_n = \beta \boldsymbol{r}_{n-1} + y_n \boldsymbol{x}_n.$$

The complexity for both of the previous updates, for matrices of a general structure, amounts to $\mathcal{O}(l^2)$ multiplication/addition operations. One alternative is to update R_n and $\boldsymbol{r}_n$ and then solve a convex optimization task for each time instant, n, using any standard algorithm. However, this is not appropriate for real-time applications, due to its excessive computational cost. In [3], a time-recursive version of a coordinate descent algorithm has been developed. As we saw in Section 10.2.2, coordinate descent algorithms update one component at each iteration step. In [3], iteration steps are associated with time updates, as is always the case with the online algorithms. As each new training pair $(y_n, \boldsymbol{x}_n)$ is received, a single component of the unknown vector is updated. Hence, at each time instant, a scalar optimization task has to be solved, and its solution is given in closed form, which results in a simple soft thresholding operation. One of the drawbacks of the coordinate techniques is that each coefficient is updated every l time instants, which, for large values of l, can slow down convergence. Variants of the basic scheme that cope with this drawback are also addressed in [3], referred to as online cyclic coordinate descent time-weighted LASSO (OCCD-TWL). The complexity of the scheme is of the order of $\mathcal{O}(l^2)$. Computational savings are possible if the input sequence is a time series and fast schemes for the updates of R_n and the RLS can then be exploited. However, if an RLS-type algorithm is used in parallel, the convergence of the overall scheme may be slowed down, because the RLS-type algorithm has to converge first in order to provide reliable estimates for the weights, as pointed out before.

10.4.3 ADAPTIVE CoSaMP ALGORITHM

In [90], an adaptive version of the CoSaMP algorithm, whose steps are summarized in Algorithm 10.2, was proposed. Iteration steps i now coincide with time updates n, and the LS solver in step 3c of the general CSMP scheme is replaced by an LMS one.

Let us focus first on the quantity $X^T \boldsymbol{e}^{(i-1)}$ in step 3a of the CSMP scheme, which is used to compute the support at iteration i. In the online setting and at (iteration) time n, this quantity is now "rephrased" as

$$X^T \boldsymbol{e}_{n-1} = \sum_{j=1}^{n-1} \boldsymbol{x}_j e_j.$$

In order to make the algorithm flexible to adapt to variations of the environment as the time index n increases, the previous correlation sum is modified to

$$\boldsymbol{p}_n := \sum_{j=1}^{n-1} \beta^{n-1-j} \boldsymbol{x}_j e_j = \beta \boldsymbol{p}_{n-1} + \boldsymbol{x}_{n-1} e_{n-1}.$$

The LS task, constrained on the active columns that correspond to the indices in the support S in step 3c, is performed in an online rationale by involving the basic LMS recursions, that is,[2]

$$\begin{aligned}\tilde{e}_n &:= y_n - \boldsymbol{x}_{n|S}^T \tilde{\boldsymbol{\theta}}_{|S}(n-1),\\ \tilde{\boldsymbol{\theta}}_{|S}(n) &:= \tilde{\boldsymbol{\theta}}_{|S}(n-1) + \mu \boldsymbol{x}_{n|S} \tilde{e}_n,\end{aligned}$$

where $\tilde{\boldsymbol{\theta}}_{|S}(\cdot)$ and $\boldsymbol{x}_{n|S}$ denote the respective subvectors corresponding to the indices in the support S. The resulting algorithm is summarized as follows.

Algorithm 10.4 (The AdCoSaMP scheme).

1. Select the value of $t = 2k$.
2. Initialize the algorithm: $\boldsymbol{\theta}(1) = \boldsymbol{0}$, $\tilde{\boldsymbol{\theta}}(1) = \boldsymbol{0}$, $\boldsymbol{p}_1 = \boldsymbol{0}$, $e_1 = y_1$.
3. Choose μ and β.
4. For $n = 2, 3, \ldots$, execute the following steps:
   - **(a)** $\boldsymbol{p}_n = \beta \boldsymbol{p}_{n-1} + \boldsymbol{x}_{n-1} e_{n-1}$.
   - **(b)** Obtain the current support:

$$S = \text{supp}\{\boldsymbol{\theta}(n-1)\} \cup \left\{ \begin{array}{c} \text{indices of the } t \text{ largest in} \\ \text{magnitude components of } \boldsymbol{p}_n \end{array} \right\}.$$

   - **(c)** Perform the LMS update:

$$\begin{aligned}\tilde{e}_n &= y_n - \boldsymbol{x}_{n|S}^T \tilde{\boldsymbol{\theta}}_{|S}(n-1),\\ \tilde{\boldsymbol{\theta}}_{|S}(n) &= \tilde{\boldsymbol{\theta}}_{|S}(n-1) + \mu \boldsymbol{x}_{n|S} \tilde{e}_n.\end{aligned}$$

---

[2] The time index for the parameter vector is given in parentheses, due to the presence of the other subscripts.

(d) Obtain the set S_k of the indices of the k largest components of $\tilde{\boldsymbol{\theta}}_{|S}(n)$.

(e) Obtain $\boldsymbol{\theta}(n)$ such that

$$\boldsymbol{\theta}_{|S_k}(n) = \tilde{\boldsymbol{\theta}}_{|S_k} \quad \text{and } \boldsymbol{\theta}_{|S_k^c}(n) = \mathbf{0},$$

where S_k^c is the complement set of S_k.

(f) Update the error: $e_n = y_n - \boldsymbol{x}_n^T \boldsymbol{\theta}(n)$.

In place of the standard LMS, its normalized version can alternatively be adopted. Note that step 4e is directly related to the hard thresholding operation.

In [90], it is shown that if the sensing matrix, which is now time dependent and keeps increasing in size, satisfies a condition similar to RIP for each time instant, called *exponentially weighted isometry property* (ERIP), which depends on β, then the algorithm asymptotically satisfies an error bound, which is similar to the one that has been derived for CoSaMP in [93], plus an extra term that is due to the excess mean-square error (see Chapter 5), which is the price paid for replacing the LS solver by the LMS.

10.4.4 SPARSE-ADAPTIVE PROJECTION SUBGRADIENT METHOD

The APSM family of algorithms was introduced in Chapter 8, as one among the most popular techniques for online/adaptive learning. As pointed out there, a major advantage of this algorithmic family is that one can readily incorporate convex constraints. In Chapter 8, APSM was used as an alternative to methods that build around the sum of squared errors cost function, such as the LMS and the RLS. The rationale behind APSM is that because our data are assumed to be generated by a regression model, the unknown vector could be estimated by finding a point in the intersection of a sequence of hyperslabs that are defined by the data points, that is, $S_n[\epsilon] := \left\{\boldsymbol{\theta} \in \mathbb{R}^l : \left|y_n - \boldsymbol{x}_n^T \boldsymbol{\theta}\right| \leq \epsilon\right\}$. Also, it was pointed out that such a model is very natural when the noise is bounded. When dealing with sparse vectors, there is an additional constraint that we want our solution to satisfy, that is, $\|\boldsymbol{\theta}\|_1 \leq \rho$ (see also the LASSO formulation (9.7)). This task fits nicely in the APSM rationale and the basic recursion can be readily written, without much thought or derivation, as follows. For any arbitrarily chosen initial point $\boldsymbol{\theta}_0$, define $\forall n$,

$$\boxed{\boldsymbol{\theta}_n = P_{B_{\ell_1}[\delta]}\left(\boldsymbol{\theta}_{n-1} + \mu_n \left(\sum_{i=n-q+1}^{n} \omega_i^{(n)} P_{S_i[\epsilon]}(\boldsymbol{\theta}_{n-1}) - \boldsymbol{\theta}_{n-1}\right)\right),} \tag{10.25}$$

where $q \geq 1$ is the number of hyperslabs that are considered each time and μ_n is a user-defined extrapolation parameter. In order for convergence to be guaranteed, theory dictates that it must lie in the interval $(0, 2\mathcal{M}_n)$, where

$$\mathcal{M}_n := \begin{cases} \dfrac{\sum_{i=n-q+1}^{n} \omega_i^{(n)} \left\| P_{S_i[\epsilon]}(\boldsymbol{\theta}_{n-1}) - \boldsymbol{\theta}_{n-1} \right\|^2}{\left\| \sum_{i=n-q+1}^{n} \omega_i^{(n)} P_{S_i[\epsilon]}(\boldsymbol{\theta}_{n-1}) - \boldsymbol{\theta}_{n-1} \right\|^2}, \\ \qquad \text{if } \left\| \sum_{i=n-q+1}^{n} \omega_i^{(n)} P_{S_i[\epsilon]}(\boldsymbol{\theta}_{n-1}) - \boldsymbol{\theta}_{n-1} \right\| \neq 0, \\ 1, \text{ otherwise}, \end{cases} \tag{10.26}$$

and $P_{B_{\ell_1}[\rho]}(\cdot)$ is the projection operator onto the ℓ_1 ball $B_{\ell_1}[\rho] := \left\{ \boldsymbol{\theta} \in \mathbb{R}^l : \|\boldsymbol{\theta}\|_1 \leq \rho \right\}$, because the solution is constrained to lie within this ball. Note that recursion (10.25) is analogous to the iterative soft thresholding shrinkage algorithm in the batch processing case (10.7). There, we saw that the only difference the sparsity imposes on an iteration, with respect to its unconstrained counterpart, is an extra soft thresholding. This is exactly the case here. The term in parentheses is the iteration for the unconstrained task. Moreover, as shown in [46], projection on the ℓ_1 ball is equivalent to a soft thresholding operation. Following the general arguments given in Chapter 8, the previous iteration converges arbitrarily close to a point in the intersection

$$B_{\ell_1}[\delta] \cap \bigcap_{n \geq n_0} S_n[\epsilon],$$

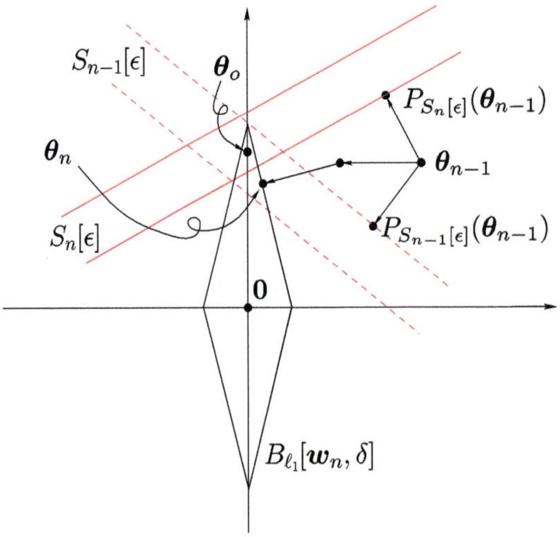

FIGURE 10.10

Geometric illustration of the update steps involved in the SpAPSM algorithm, for the case of $q = 2$. The update at time n is obtained by first convexly combining the projections onto the current and previously formed hyperslabs, $S_n[\epsilon]$, $S_{n-1}[\epsilon]$, and then projecting onto the weighted ℓ_1 ball. This brings the update closer to the target solution $\boldsymbol{\theta}_o$.

for some finite value of n_0. In [76,77], the weighted ℓ_1 ball (denoted here as $B_{\ell_1[\boldsymbol{w}_n,\rho]}$) has been used to improve convergence as well as the tracking speed of the algorithm, when the environment is time-varying. The weights were adopted in accordance with what was discussed in Section 10.3, that is,

$$w_{i,n} := \frac{1}{|\theta_{i,n-1}| + \acute{\epsilon}_n}, \quad \forall i \in \{1, 2, \ldots, l\},$$

where $(\acute{\epsilon}_n)_{n\geq 0}$ is a sequence (can be also constant) of small numbers to avoid division by zero. The basic time iteration becomes as follows. For any arbitrarily chosen initial point $\boldsymbol{\theta}_0$, define $\forall n$,

$$\boxed{\boldsymbol{\theta}_n = P_{B_{\ell_1}[\boldsymbol{w}_n,\rho]}\left(\boldsymbol{\theta}_{n-1} + \mu_n\left(\sum_{i=n-q+1}^{n} \omega_i^{(n)} P_{S_i[\epsilon]}(\boldsymbol{\theta}_{n-1}) - \boldsymbol{\theta}_{n-1}\right)\right),} \tag{10.27}$$

where $\mu_n \in (0, 2\mathcal{M}_n)$ and $\mathcal{M}_n$ is given in (10.26). Fig. 10.10 illustrates the associated geometry of the basic iteration in $\mathbb{R}^2$ and for the case of $q = 2$. It comprises two parallel projections on the hyperslabs, followed by one projection onto the weighted ℓ_1 ball. In [76], it is shown (Problem 10.7) that a good bound for the weighted ℓ_1 norm is the sparsity level k of the target vector, which is assumed to be known and is a user-defined parameter. In [76], it is shown that asymptotically, and under some general assumptions, this algorithmic scheme converges arbitrarily close to the intersection of the hyperslabs with the weighted ℓ_1 balls, that is,

$$\bigcap_{n\geq n_0}\left(P_{B_{\ell_1}[\boldsymbol{w}_n,\rho]} \cap S_n[\epsilon]\right),$$

for some nonnegative integer n_0. It has to be pointed out that in the case of weighted ℓ_1 norms, the constraint is *time-varying* and the convergence analysis is not covered by the standard analysis used for APSM, and had to be extended to this more general case.

The complexity of the algorithm amounts to $\mathcal{O}(ql)$. The larger q, the faster the convergence rate, at the expense of higher complexity. In [77], in order to reduce the dependence of the complexity on q, the notion of the *subdimensional* projection is introduced, where projections onto the q hyperslabs could be restricted along the directions of the most significant parameters of the currently available estimates. The dependence on q now becomes $\mathcal{O}(qk_n)$, where k_n is the sparsity level of the currently available estimate, which after a few steps of the algorithm gets much lower than l. The total complexity amounts to $\mathcal{O}(l) + \mathcal{O}(qk_n)$ per iteration step. This allows the use of large values of q, which (at only a small extra computational cost compared to $\mathcal{O}(l)$) drives the algorithm to a performance close to that of the adaptive weighted LASSO.

Projection Onto the Weighted ℓ_1 Ball

Projecting onto an ℓ_1 ball is equivalent to a soft thresholding operation. Projection onto the weighted ℓ_1 norm results in a slight variation of the soft thresholding, with different threshold values per component. In the sequel, we give the iteration steps for the more general case of the weighted ℓ_1 ball. The proof is a bit technical and lengthy and it will not be given here. It was derived, for the first time, via purely geometric arguments, and without the use of the classical Lagrange multipliers, in [76]. Lagrange multipliers have been used instead in [46], for the case of the ℓ_1 ball. The efficient computation of the projection on the ℓ_1 ball was treated earlier, in a more general context, in [100].

Recall from the definition of a projection, discussed in Chapter 8, that given a point outside the ball $\boldsymbol{\theta} \in \mathbb{R}^l \setminus B_{\ell_1}[\boldsymbol{w}, \rho]$, its projection onto the weighted ℓ_1 ball is the point $P_{B_{\ell_1}[\boldsymbol{w},\rho]}(\boldsymbol{\theta}) \in B_{\ell_1}[\boldsymbol{w}, \rho] := \{\boldsymbol{z} \in \mathbb{R}^l : \sum_{i=1}^{l} w_i |z_i| \leq \rho\}$ that lies closest to $\boldsymbol{\theta}$ in the Euclidean sense. If $\boldsymbol{\theta}$ lies within the ball, then it coincides with its projection. Given the weights and the value of ρ, the following iterations provide the projection.

Algorithm 10.5 (Projection onto the weighted ℓ_1 ball $B_{\ell_1}[\boldsymbol{w}, \rho]$).

1. Form the vector $[|\theta_1|/w_1, \ldots, |\theta_l|/w_l]^T \in \mathbb{R}^l$.
2. Sort the previous vector in a nonascending order, so that $|\theta_{\tau(1)}|/w_{\tau(1)} \geq \ldots \geq |\theta_{\tau(l)}|/w_{\tau(l)}$. The notation τ stands for the permutation, which is implicitly defined by the sorting operation. Keep in mind the inverse τ^{-1}, which is the index of the position of the element in the original vector.
3. $r_1 := l$.
4. Let $m = 1$. While $m \leq l$, do
   (a) $m_* := m$.
   (b) Find the maximum j_* among those $j \in \{1, 2, \ldots, r_m\}$ such that $\dfrac{|\theta_{\tau(j)}|}{w_{\tau(j)}} > \dfrac{\sum_{i=1}^{r_m} w_{\tau(i)} |\theta_{\tau(i)}| - \rho}{\sum_{i=1}^{r_m} w_{\tau(i)}^2}$.
   (c) If $j_* = r_m$ then break the loop.
   (d) Otherwise set $r_{m+1} := j_*$.
   (e) Increase m by 1 and go back to Step 4a.
5. Form the vector $\hat{\boldsymbol{p}} \in \mathbb{R}^{r_{m*}}$ whose jth component, $j = 1, \ldots, r_{m_*}$, is given by

$$\hat{p}_j := |\theta_{\tau(j)}| - \frac{\sum_{i=1}^{r_{m*}} w_{\tau(i)} |\theta_{\tau(i)}| - \rho}{\sum_{i=1}^{r_{m*}} w_{\tau(i)}^2} w_{\tau(j)}.$$

6. Use the inverse mapping τ^{-1} to insert the element $\hat{p}_j$ into the $\tau^{-1}(j)$ position of the l-dimensional vector $\boldsymbol{p}$, $\forall j \in \{1, 2, \ldots r_{m_*}\}$, and fill in the rest with zeros.
7. The desired projection is $P_{B_{\ell_1}[\boldsymbol{w},\rho]}(\boldsymbol{\theta}) = [\text{sgn}(\theta_1) p_1, \ldots, \text{sgn}(\theta_l) p_l]^T$.

Remarks 10.3.

- *Generalized thresholding rules*: Projections onto both ℓ_1 and weighted ℓ_1 balls impose convex sparsity inducing constraints via properly performed soft thresholding operations. More recent advances within the SpAPSM framework [78,109] allow the substitution of $P_{B_{\ell_1}[\rho]}$ and $P_{B_{\ell_1}[\boldsymbol{w},\rho]}$ with a *generalized thresholding*, built around the notions of SCAD, nonnegative garrote, and a number of thresholding functions corresponding to the *nonconvex* ℓ_p, $p < 1$, penalties. Moreover, it is shown that such generalized thresholding (GT) operators are nonlinear mappings with *their fixed point set being a union of subspaces*, that is, the nonconvex object that lies at the heart of any sparsity promoting technique. Such schemes are very useful for low values of q, where one can improve upon the performance obtained by the LMS-based AdCoSAMP, at comparable complexity levels.
- A comparative study of various online sparsity promoting low-complexity schemes, including the proportionate LMS, in the context of the echo cancelation task, is given in [80]. It turns out that the SpAPSM-based schemes outperform LMS-based sparsity promoting algorithms.

- More algorithms and methods that involve sparsity promoting regularization, in the context of more general convex loss functions, compared to the squared error, are discussed in Chapter 8, where related references are provided.
- *Distributed sparsity promoting algorithms*: Besides the algorithms reported so far, a number of algorithms in the context of distributed learning have also appeared in the literature. As pointed out in Chapter 5, algorithms complying to the consensus as well as the diffusion rationale have been proposed (see, e.g., [29,39,89,102]). A review of such algorithms appears in [30].

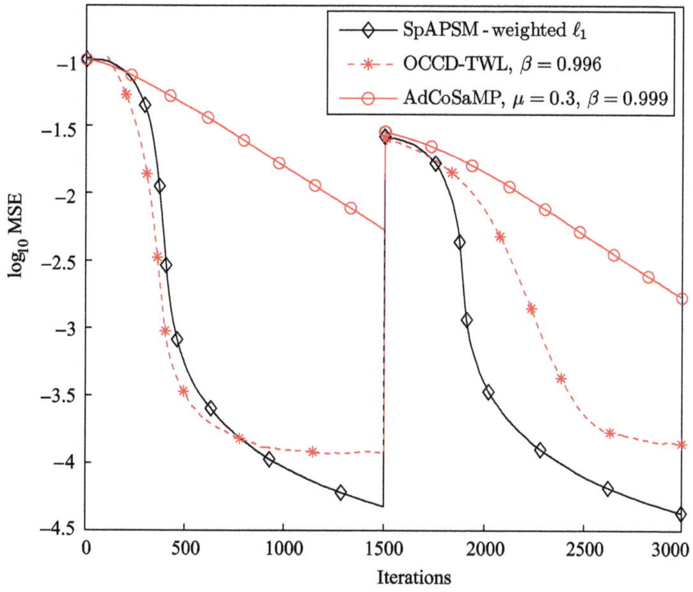

FIGURE 10.11

MSE learning curves for AdCoSAMP, SpAPSM, and OCCD-TWL for the simulation in Example 10.2. The vertical axis shows the $\log_{10}$ of the mean-square, that is, $\log_{10}\frac{1}{2}||s - \Psi\theta_n||^2$, and the horizontal axis shows the time index. At time $n = 1500$, the system undergoes a sudden change.

Example 10.2 (Time-varying signal). In this example, the performance curves of the most popular online algorithms, mentioned before, are studied in the context of a time-varying environment. A typical simulation setup, which is commonly adopted by the adaptive filtering community in order to study the tracking ability of an algorithm, is that of an unknown vector that undergoes an abrupt change after a number of observations. Here, we consider a signal, s, with a sparse wavelet representation, that is, $s = \Psi\theta$, where Ψ is the corresponding transformation matrix. In particular, we set $l = 1024$ with 100 nonzero wavelet coefficients. After 1500 observations, 10 arbitrarily picked wavelet coefficients change their values to new ones, selected uniformly at random from the interval $[-1, 1]$. Note that this may affect the sparsity level of the signal, and we can now end with up to 110 nonzero coefficients. A total of $N = 3000$ sensing vectors are used, which result from the wavelet transform of the input vectors $\boldsymbol{x}_n \in \mathbb{R}^l$, $n = 1, 2, \ldots, 3000$, having elements drawn from $\mathcal{N}(0, 1)$. In this way, the online algorithms do not estimate the signal itself, but its sparse wavelet representation, $\boldsymbol{\theta}$. The observations are corrupted by additive white Gaussian noise of variance $\sigma_n^2 = 0.1$. Regarding SpAPSM, the extrapolation parameter μ_n is set equal to $1.8 \times \mathcal{M}_n$, $\omega_i^{(n)}$ are all getting the same value $1/q$, the hyperslabs parameter ϵ was set equal to $1.3\sigma_n$, and $q = 390$. The parameters for all algorithms were selected in order to optimize their performance. Because the sparsity level of the signal may change (from $k = 100$ up to $k = 110$) and because in practice it is not possible to know in advance the exact value of k, we feed the algorithms with an overestimate, k, of the true sparsity value, and in particular we use $\hat{k} = 150$ (i.e., 50% overestimation up to the 1500th iteration).

The results are shown in Fig. 10.11. Note the enhanced performance obtained via the SpAPSM algorithm. However, it has to be pointed out that the complexity of the AdCoSAMP is much lower compared to the other two algorithms, for the choice of $q = 390$ for the SpAPSM. The interesting observation is that SpAPSM achieves a better performance compared to OCCD-TWL, albeit at significantly lower complexity. If on the other hand complexity is of major concern, use of SpAPSM offers the flexibility to use generalized thresholding operators, which lead to improved performance for small values of q, at complexity comparable to that of LMS-based sparsity promoting algorithms [79,80].

10.5 LEARNING SPARSE ANALYSIS MODELS

Our whole discussion so far has been spent in the terrain of signals that are either sparse themselves or that can be sparsely represented in terms of the atoms of a dictionary in a synthesis model, as introduced in (9.16), that is,

$$\boldsymbol{s} = \sum_{i \in \mathcal{I}} \theta_i \boldsymbol{\psi}_i.$$

As a matter of fact, most of the research activity has been focused on the synthesis model. This may be partly due to the fact that the synthesis modeling path may provide a more intuitively appealing structure to describe the generation of the signal in terms of the elements (atoms) of a dictionary. Recall from Section 9.9 that the sparsity assumption was imposed on $\boldsymbol{\theta}$ in the synthesis model and the corresponding optimization task was formulated in (9.38) and (9.39) for the exact and noisy cases, respectively.

However, this is not the only way to approach the task of sparse modeling. Very early in this chapter, in Section 9.4, we referred to the analysis model

$$\tilde{\boldsymbol{s}} = \Phi^H \boldsymbol{s}$$

and pointed out that in a number of real-life applications, the resulting transform $\tilde{\boldsymbol{s}}$ is sparse. To be fair, the most orthodox way to deal with the underlying model sparsity would be to consider $\left\|\Phi^H \boldsymbol{s}\right\|_0$. Thus, if one wants to estimate $\boldsymbol{s}$, a very natural way would be to cast the related optimization task as

$$\begin{aligned} &\min_{\boldsymbol{s}} \quad \left\|\Phi^H \boldsymbol{s}\right\|_0, \\ &\text{s.t.} \quad \boldsymbol{y} = X\boldsymbol{s}, \ \text{ or } \ \|\boldsymbol{y} - X\boldsymbol{s}\|_2^2 \leq \epsilon, \end{aligned} \tag{10.28}$$

depending on whether the measurements via a sensing matrix X are exact or noisy. Strictly speaking, the total variation minimization approach, which was used in Example 10.1, falls under this analysis model formulation umbrella, because what is minimized is the ℓ_1 norm of the gradient transform of the image.

The optimization tasks in either of the two formulations given in (10.28) build around the assumption that the signal of interest has *sparse analysis representation*. The obvious question that is now raised is whether the optimization tasks in (10.28) and their counterparts in (9.38) or (9.39) are any different. One of the first efforts to shed light on this problem was in [51]. There, it was pointed out that the two tasks, though related, are in general different. Moreover, their comparative performance depends on the specific problem at hand. Let us consider, for example, the case where the involved dictionary corresponds to an orthonormal transformation matrix (e.g., DFT). In this case, we already know that the analysis and synthesis matrices are related as

$$\Phi = \Psi = \Psi^{-H},$$

which leads to an equivalence between the two previously stated formulations. Indeed, for such a transform we have

$$\underbrace{\tilde{s} = \Phi^H s}_{\text{Analysis}} \quad \Leftrightarrow \quad \underbrace{s = \Phi \tilde{s}}_{\text{Synthesis}}.$$

Using the last formula in (10.28), the tasks in (9.38) or (9.39) are readily obtained by replacing $\boldsymbol{\theta}$ by s. However, this reasoning cannot be extended to the case of overcomplete dictionaries; in these cases, the two optimization tasks may lead to different solutions.

The previous discussion concerning the comparative performance between the synthesis or analysis-based sparse representations is not only of "philosophical" value. It turns out that often in practice, the nature of certain overcomplete dictionaries does not permit the use of the synthesis-based formulation. These are the cases where the columns of the overcomplete dictionary exhibit a high degree of dependence; that is, the coherence of the matrix, as defined in Section 9.6.1, has large values. Typical examples of such overcomplete dictionaries are the Gabor frames, the curvelet frames, and the oversampled DFT. The use of such dictionaries leads to enhanced performance in a number of applications (e.g., [111,112]). Take as an example the case of our familiar DFT transform. This transform provides a representation of our signal samples in terms of sampled exponential sinusoids, whose integral frequencies are multiples of $\frac{2\pi}{l}$, that is,

$$s := \begin{bmatrix} s_1 \\ s_2 \\ \vdots \\ s_{l-1} \end{bmatrix} = \sum_{i=0}^{l-1} \tilde{s}_i \boldsymbol{\psi}_i, \tag{10.29}$$

where, now, $\tilde{s}_i$ are the DFT coefficients and $\boldsymbol{\psi}_i$ is the sampled sinusoid with frequency equal to $\frac{2\pi}{l} i$, that is,

$$\boldsymbol{\psi}_i = \begin{bmatrix} 1 \\ \exp\left(-j\frac{2\pi}{l}i\right) \\ \vdots \\ \exp\left(-j\frac{2\pi}{l}i(l-1)\right) \end{bmatrix}. \tag{10.30}$$

However, this is not necessarily the most efficient representation. For example, it is highly unlikely that a signal comprises only integral frequencies and only such signals can result in a sparse representation using the DFT basis. Most probably, in general, there will be frequencies lying in between the frequency samples of the DFT basis that result in nonsparse representations. Using these extra frequencies, a much better representation of the frequency content of the signal can be obtained. However, in such a dictionary, the atoms are no longer linearly independent, and the coherence of the respective (dictionary) matrix increases.

Once a dictionary exhibits high coherence, then there is no way of finding a sensing matrix X so that $X\Psi$ obeys the RIP. Recall that at the heart of sparsity-aware learning lies the concept of stable embedding, which allows the recovery of a vector/signal after projecting it on a lower-dimensional space, which is what all the available conditions (e.g., RIP) guarantee. However, no stable embedding is possible with highly coherent dictionaries. Take as an extreme example the case where the first and second atoms are identical. Then no sensing matrix X can achieve a signal recovery that distinguishes the vector $[1, 0, \ldots, 0]^T$ from $[0, 1, 0, \ldots, 0]^T$. Can one then conclude that for highly coherent overcomplete dictionaries, compressed sensing techniques are not possible? Fortunately, the answer to this is negative. After all, our goal in compressed sensing has always been the recovery of the signal $\boldsymbol{s} = \Psi\boldsymbol{\theta}$ and not the identification of the sparse vector $\boldsymbol{\theta}$ in the synthesis model representation. The latter was just a means to an end. While the unique recovery of $\boldsymbol{\theta}$ cannot be guaranteed for highly coherent dictionaries, this does not necessarily cause any problems for the recovery of $\boldsymbol{s}$, using a small set of measurement samples. The escape route will come by considering the analysis model formulation.

10.5.1 COMPRESSED SENSING FOR SPARSE SIGNAL REPRESENTATION IN COHERENT DICTIONARIES

Our goal in this subsection is to establish conditions that guarantee recovery of a signal vector, which accepts a sparse representation in a redundant and coherent dictionary, using a small number of signal-related measurements. Let the dictionary at hand be a tight frame Ψ (see appendix of the chapter, that can be downloaded from the book's website). Then our signal vector is written as

$$\boldsymbol{s} = \Psi\boldsymbol{\theta}, \tag{10.31}$$

where $\boldsymbol{\theta}$ is assumed to be k-sparse. Recalling the properties of a tight frame, as they are summarized in the appendix, the coefficients in the expansion (10.31) can be written as $\langle\boldsymbol{\psi}_i, \boldsymbol{s}\rangle$, and the respective vector as

$$\boldsymbol{\theta} = \Psi^T\boldsymbol{s},$$

because a tight frame is self-dual. Then the analysis counterpart of the synthesis formulation in (9.39) can be cast as

$$\begin{aligned} \min_{s} \quad & \left\|\Psi^T s\right\|_1, \\ \text{s.t.} \quad & \|y - Xs\|_2^2 \leq \epsilon. \end{aligned} \tag{10.32}$$

The goal now is to investigate the accuracy of the recovered solution to this convex optimization task. It turns out that similar strong theorems are also valid for this problem, as with the case of the synthesis formulation, which was studied in Chapter 9.

Definition 10.1. Let Σ_k be the union of all subspaces spanned by all subsets of k columns of Ψ. A sensing matrix X obeys the restricted isometry property adapted to Ψ (Ψ-RIP) with δ_k, if

$$\boxed{(1-\delta_k)\,\|s\|_2^2 \leq \|Xs\|_2^2 \leq (1+\delta_k)\,\|s\|_2^2: \quad \Psi - \text{RIP condition},} \tag{10.33}$$

for all $s \in \Sigma_k$.

The union of subspaces, Σ_k, is the image under Ψ of all k-sparse vectors. This is the difference with the RIP definition given in Section 9.7.2. All the random matrices discussed earlier in this chapter can be shown to satisfy this form of RIP, with overwhelming probability, provided the number of observations, N, is at least of the order of $k \ln(l/k)$. We are now ready to establish the main theorem concerning our ℓ_1 minimization task.

Theorem 10.2. *Let Ψ be an arbitrary tight frame and X a sensing matrix that satisfies the Ψ-RIP with $\delta_{2k} \leq 0.08$, for some positive k. Then the solution, s_*, of the minimization task in (10.32) satisfies the property*

$$\|s - s_*\|_2 \leq C_0 k^{-\frac{1}{2}} \left\|\Psi^T s - (\Psi^T s)_k\right\|_1 + C_1\sqrt{\epsilon}, \tag{10.34}$$

where C_0, C_1 are constants depending on δ_{2k} and $(\Psi^T s)_k$ denotes the best k-sparse approximation of $\Psi^T s$, which results by setting all but the k largest in magnitude components of $\Psi^T s$ equal to zero.

The bound in (10.34) is the counterpart of that given in (9.36). In other words, the previous theorem states that if $\Psi^T s$ decays rapidly, then s can be reconstructed from just a few (compared to the signal length l) observations. The theorem was first given in [25] and it is the first time that such a theorem provides results for the sparse analysis model formulation in a general context.

10.5.2 COSPARSITY

In the sparse synthesis formulation, one searches for a solution in a union of subspaces that are formed by all possible combinations of k columns of the dictionary, Ψ. Our signal vector lies in one of these subspaces—the one that is spanned by the columns of Ψ whose indices lie in the support set (Section 10.2.1). In the sparse analysis approach, things get different. The kick-off point is the sparsity of the transform $\tilde{s} := \Phi^T s$, where Φ defines the transformation matrix or analysis operator. Because $\tilde{s}$ is assumed to be sparse, there exists an index set $\mathcal{I}$ such that $\forall i \in \mathcal{I}$, $\tilde{s}_i = 0$. In other words, $\forall i \in \mathcal{I}$, $\boldsymbol{\phi}_i^T s := \langle \boldsymbol{\phi}_i, s \rangle = 0$, where $\boldsymbol{\phi}_i$ stands for the ith column of Φ. Hence, the subspace in which s resides is the orthogonal complement of the subspace formed by those columns of Φ that correspond to a zero in the transform vector $\tilde{s}$. Assume, now, that card$(\mathcal{I}) = C_o$. The signal, s, can be identified by searching on the *orthogonal complements* of the subspaces formed by all possible combinations of C_o columns of Φ, that is,

$$\langle \boldsymbol{\phi}_i, \boldsymbol{s} \rangle = 0, \quad \forall i \in \mathcal{I}.$$

The difference between the synthesis and analysis problems is illustrated in Fig. 10.12. To facilitate the theoretical treatment of this new setting, the notion of *cosparsity* was introduced in [92].

Definition 10.2. The cosparsity of a signal $s \in \mathbb{R}^l$ with respect to a $p \times l$ matrix Φ^T is defined as

$$C_o := p - \|\Phi^T s\|_0. \tag{10.35}$$

In words, the cosparsity is the number of zeros in the obtained transform vector $\tilde{s} = \Phi^T s$; in contrast, the sparsity measures the number of the nonzero elements of the respective sparse vector. If one assumes that Φ has "full spark,"[3] that is, $l+1$, then any l of the columns of Φ, and thus any l rows of Φ^T, is guaranteed to be independent. This indicates that for such matrices, the maximum value that cosparsity can take is equal to $C_o = l - 1$. Otherwise, the existence of l zeros will necessarily correspond to a zero signal vector. Higher cosparsity levels are possible by relaxing the full spark requirement.

Let now the cosparsity of our signal with respect to a matrix Φ^T be C_o. Then, in order to dig out the signal from the subspace in which it is hidden, one must form all possible combinations of

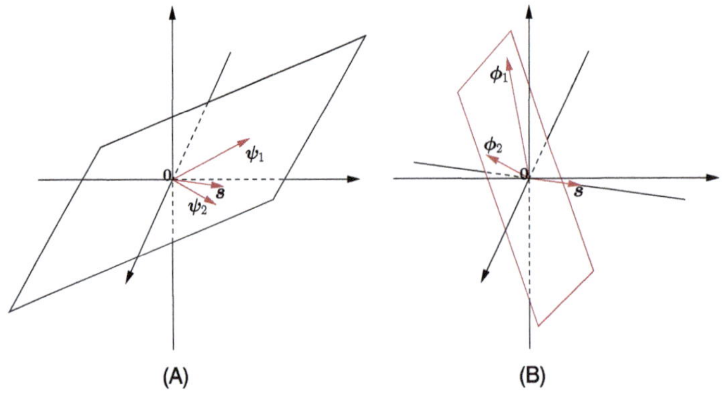

FIGURE 10.12

Searching for a sparse vector s. (A) In the synthesis model, the sparse vector lies in subspaces formed by combinations of k (in this case $k = 2$) columns of the dictionary Ψ. (B) In the analysis model, the sparse vector lies in the orthogonal complement of the subspace formed by C_o (in this case $C_o = 2$) columns of the transformation matrix Φ.

[3] Recall by Definition 9.2 that spark(Φ) is defined for an $l \times p$ matrix Φ with $p \geq l$ and of full rank.

C_o columns of Φ and search in their orthogonal complements. In the case that Φ is full rank, we have seen previously that $C_o < l$, and, hence, any set of C_o columns of Φ are linearly independent. In other words, the dimension of the span of those columns is C_o. As a result, the dimensionality of the orthogonal complement, into which we search for s, is $l - C_o$.

We have by now accumulated enough information to elaborate a bit more on the statement made before concerning the different nature of the synthesis and analysis tasks. Let us consider a synthesis task using an $l \times p$ dictionary, and let k be the sparsity level in the corresponding expansion of a signal in terms of this dictionary. The dimensionality of the subspaces in which the solution is sought is k (k is assumed to be less than the spark of the respective matrix). Let us keep the same dimensionality for the subspaces in which we are going to search for a solution in an analysis task. Hence, in this case $C_o = l - k$ (assuming a full spark matrix). Also, for the sake of comparison, assume that the analysis matrix is $p \times l$. Solving the synthesis task, one has to search $\binom{p}{k}$ subspaces, while solving the analysis task one has to search $\binom{p}{C_o=l-k}$ subspaces. These are two different numbers; assuming that $k \ll l$ and also that $l < p/2$, which are natural assumptions for overcomplete dictionaries, then the latter of the two numbers is much larger than the former (use your computer to play with some typical values). In other words, there are many more analysis than synthesis low-dimensional subspaces for which to search. The large number of low-dimensional subspaces makes the algorithmic recovery of a solution from the analysis model a tougher task [92]. However, it might reveal a much stronger descriptive power of the analysis model compared to the synthesis one.

Another interesting aspect that highlights the difference between the two approaches is the following. Assume that the synthesis and analysis matrices are related as $\Phi = \Psi$, as was the case for tight frames. Under this assumption, $\Phi^T s$ provides a set of coefficients for the synthesis expansion in terms of the atoms of $\Phi = \Psi$. Moreover, if $\|\Phi^T s\|_0 = k$, then $\Phi^T s$ is a possible k-sparse solution for the synthesis model. However, there is no guarantee that this is the sparsest one.

It is now time to investigate whether conditions that guarantee uniqueness of the solution for the sparse analysis formulation can be derived. The answer is in the affirmative, and it has been established in [92] for the case of exact measurements.

Lemma 10.1. *Let Φ be a transformation matrix of full spark. Then, for almost all $N \times l$ sensing matrices and for $N > 2(l - C_o)$, the equation*

$$y = Xs$$

has at most one solution with cosparsity at least C_o.

The above lemma guarantees the uniqueness of the solution, if one exists, of the optimization

$$\begin{aligned} \min_{s} \quad & \|\Phi^T s\|_0, \\ \text{s.t.} \quad & y = Xs. \end{aligned} \tag{10.36}$$

However, solving the previous ℓ_0 minimization task is a difficult one, and we know that its synthesis counterpart has been shown to be NP-hard, in general. Its relaxed convex relative is the ℓ_1 minimization

$$\begin{aligned} \min_{s} \quad & \|\Phi^T s\|_1, \\ \text{s.t.} \quad & y = Xs. \end{aligned} \tag{10.37}$$

In [92], conditions are derived that guarantee the equivalence of the ℓ_0 and ℓ_1 tasks, in (10.36) and (10.37), respectively; this is done in a way similar to that for the sparse synthesis modeling. Also, in [92], a greedy algorithm inspired by the orthogonal matching pursuit, discussed in Section 10.2.1, has been derived. A thorough study on greedy-like algorithms applicable to the cosparse model can be found in [62]. In [103] an iterative analysis thresholding is proposed and theoretically investigated. Other algorithms that solve the ℓ_1 optimization in the analysis modeling framework can be found in, for example, [21,49,108]. NESTA can also be used for the analysis formulation. Moreover, a critical aspect affecting the performance of algorithms obeying the cosparse analysis model is the choice of the analysis matrix Φ. It turns out that it is not always the best practice to use fixed and predefined matrices. As a promising alternative, problem-tailored analysis matrices can be learned using the available data (e.g., [106,119]).

10.6 A CASE STUDY: TIME-FREQUENCY ANALYSIS

The goal of this section is to demonstrate how all the previously stated theoretical findings can be exploited in the context of a real application. Sparse modeling has been applied to almost everything. So picking up a typical application would not be easy. We preferred to focus on a less "publicized" application, i.e., that of analyzing echolocation signals emitted by bats. However, the analysis will take place within the framework of time-frequency representation, which is one of the research areas that significantly inspired the evolution of compressed sensing theory. Time-frequency analysis of signals has been a field of intense research for a number of decades, and it is one of the most powerful signal processing tools. Typical applications include speech processing, sonar sounding, communications, biological signals, and EEG processing, to name but a few (see, e.g., [13,20,57]).

GABOR TRANSFORM AND FRAMES

It is not our intention to present the theory behind the Gabor transform. Our goal is to outline some basic related notions and use them as a vehicle for the less familiar reader to better understand how redundant dictionaries are used and to get better acquainted with their potential performance benefits.

The Gabor transform was introduced in the mid-1940s by Dennis Gabor (1900–1979), a Hungarian-British engineer. His most notable scientific achievement was the invention of holography, for which he won the Nobel Prize for Physics in 1971.

The discrete version of the Gabor transform can be seen as a special case of the short-time Fourier transform (STFT) (e.g., [57,87]). In the standard DFT transform, the full length of a time sequence, comprising l samples, is used all in "one go" in order to compute the corresponding frequency content. However, the latter can be time-varying, so the DFT will provide an average information, which cannot be of much use. The Gabor transform (and the STFT in general) introduces time localization via the use of a window function, which slides along the signal segment in time, and at each time instant focuses on a different part of the signal. This is a way that allows one to follow the slow time variations which take place in the frequency domain. The time localization in the context of the Gabor transform is achieved via a Gaussian window function, that is,

$$g(n) := \frac{1}{\sqrt{2\pi\sigma^2}} \exp\left(-\frac{n^2}{2\sigma^2}\right). \tag{10.38}$$

Fig. 10.13A shows the Gaussian window, $g(n-m)$, centered at time instant m. The choice of the window spreading factor, σ, will be discussed later on.

Let us now construct the atoms of the Gabor dictionary. Recall that in the case of the signal representation in terms of the DFT in (10.29), each frequency is represented only once, by the corresponding sampled sinusoid, (10.30). In the Gabor transform, each frequency appears l times; the corresponding sampled sinusoid is multiplied by the Gaussian window sequence, each time shifted by one sample. Thus, at the ith frequency bin, we have l atoms, $\mathbf{g}^{(m,i)}$, $m=0,1,\dots,l-1$, with elements given by

$$g^{(m,i)}(n)=g(n-m)\psi_i(n), \quad n,m,i=0,1,\dots,l-1, \tag{10.39}$$

where $\psi_i(n)$ is the nth element of the vector $\boldsymbol{\psi}_i$ in (10.30). This results in an overcomplete dictionary comprising l^2 atoms in the l-dimensional space. Fig. 10.13B illustrates the effect of multiplying different sinusoids with Gaussian pulses of different spread and at different time delays. Fig. 10.14 is a graphical interpretation of the atoms involved in the Gabor dictionary. Each node (m,i) in this time-frequency plot corresponds to an atom of frequency equal to $\frac{2\pi}{l}i$ and delay equal to m.

Note that the windowing of a signal of finite duration inevitably introduces boundary effects, especially when the delay m gets close to the time segment edges, 0 and $l-1$. A solution that facilitates the theoretical analysis is to use a modulo l arithmetic to wrap around at the edge points (this is equivalent to extending the signal periodically); see, for example, [113].

Once the atoms have been defined, they can be stacked one next to the other to form the columns of the $l \times l^2$ Gabor dictionary, G. It can be shown that the Gabor dictionary is a tight frame [125].

TIME-FREQUENCY RESOLUTION

By definition of the Gabor dictionary, it is readily understood that the choice of the window spread, as measured by σ, must be a critical factor, since it controls the localization in time. As known from our

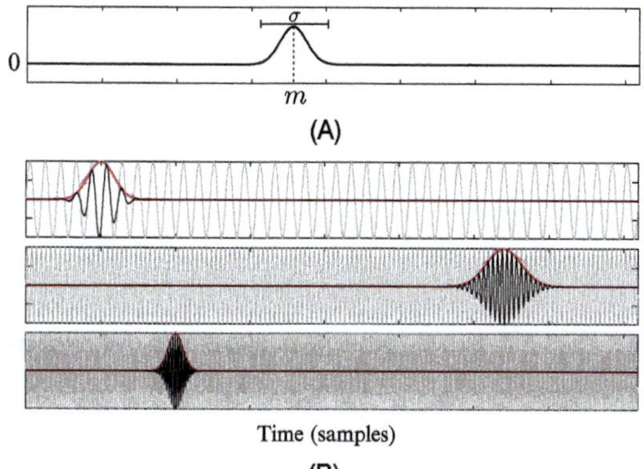

FIGURE 10.13

(A) The Gaussian window with spreading factor σ centered at time instant m. (B) Pulses obtained by windowing three different sinusoids with Gaussian windows of different spread and applied at different time instants.

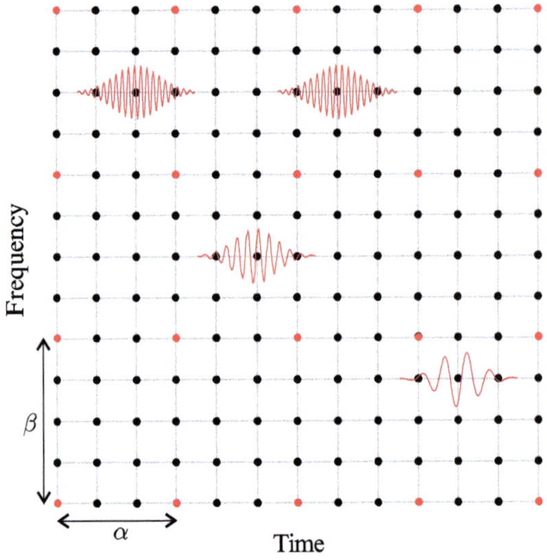

FIGURE 10.14

Each atom of the Gabor dictionary corresponds to a node in the time-frequency grid. That is, it is a sampled windowed sinusoid whose frequency and location in time are given by the coordinates of the respective node. In practice, this grid may be subsampled by factors α and β for the two axes, respectively, in order to reduce the number of the involved atoms.

Fourier transform basics, when the pulse becomes short, in order to increase the time resolution, its corresponding frequency content spreads out, and vice versa. From Heisenberg's principle, we know that we can never achieve high time and frequency resolution simultaneously; one is gained at the expense of the other. It is here where the Gaussian shape in the Gabor transform is justified. It can be shown that the Gaussian window gives the optimal tradeoff between time and frequency resolution [57, 87]. The time-frequency resolution tradeoff is demonstrated in Fig. 10.15, where three sinusoids are shown windowed with different pulse durations. The diagram shows the corresponding spread in the time-frequency plot. The value of σ_t indicates the time spread and σ_f the spread of the respective frequency content around the basic frequency of each sinusoid.

GABOR FRAMES

In practice, l^2 can take large values, and it is desirable to see whether one can reduce the number of the involved atoms without sacrificing the frame-related properties. This can be achieved by an appropriate subsampling, as illustrated in Fig. 10.14. We only keep the atoms that correspond to the red nodes. That is, we subsample by keeping every α nodes in time and every β nodes in frequency in order to form the dictionary, that is,

$$G_{(\alpha,\beta)} = \{\boldsymbol{g}^{(m\alpha,i\beta)}\}, \quad m = 0, 1, \ldots, \frac{l}{\alpha} - 1, \ i = 0, 1, \ldots, \frac{l}{\beta} - 1,$$

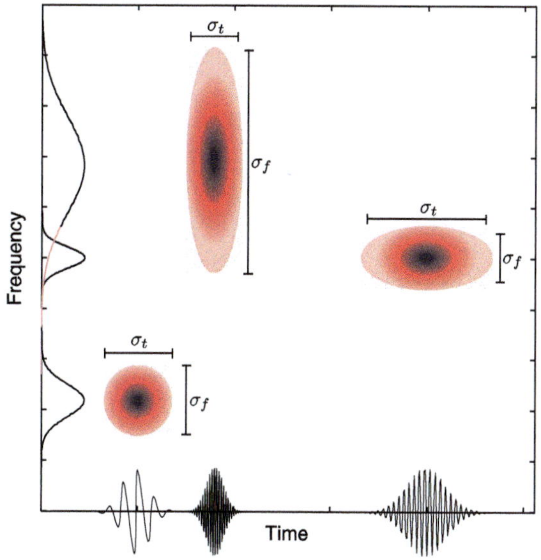

FIGURE 10.15

The shorter the width of the pulsed (windowed) sinusoid is in time, the wider the spread of its frequency content around the frequency of the sinusoid. The Gaussian-like curves along the frequency axis indicate the energy spread in frequency of the respective pulses. The values of σ_t and σ_f indicate the spread in time and frequency, respectively.

where α and β are divisors of l. Then it can be shown (e.g., [57]) that if $\alpha\beta < l$ the resulting dictionary retains its frame properties. Once $G_{(\alpha,\beta)}$ is obtained, the canonical dual frame is readily available via (10.47) (adjusted for complex data), from which the corresponding set of expansion coefficients, $\boldsymbol{\theta}$, results.

TIME-FREQUENCY ANALYSIS OF ECHOLOCATION SIGNALS EMITTED BY BATS

Bats use echolocation for navigation (flying around at night), for prey detection (small insects), and for prey approaching and catching; each bat adaptively changes the shape and frequency content of its calls in order to better serve the previous tasks. Echolocation is used in a similar way for sonars. Bats emit calls as they fly, and "listen" to the returning echoes in order to build up a sonic map of their surroundings. In this way, bats can infer the distance and the size of obstacles as well as of other flying creatures/insects. Moreover, all bats emit special types of calls, called social calls, which are used for socializing, flirting, and so on. The fundamental characteristics of the echolocation calls, for example, the frequency range and average time duration, differ from species to species because, thanks to evolution, bats have adapted their calls in order to become better suited to the environment in which a species operates.

Time-frequency analysis of echolocation calls provides information about the species (species identification) as well as the specific task and behavior of the bats in certain environments. Moreover, the bat biosonar system is studied in order for humans to learn more about nature and get inspired for subsequent advances in applications such as sonar navigation systems, radars, and medical ultrasonic devices.

Fig. 10.16 shows a case of a recorded echolocation signal from bats. Zooming at two different parts of the signal, we can observe that the frequency is changing with time. In Fig. 10.17, the DFT of the signal is shown, but there is not much information that can be drawn from it except that the signal is compressible in the frequency domain; most of the activity takes place within a short range of frequencies.

Our echolocation signal was a recording of total length $T = 21.845$ ms [75]. Samples were taken at the sampling frequency $f_s = 750$ kHz, which results in a total of $l = 16{,}384$ samples. Although the signal itself is not sparse in the time domain, we will take advantage of the fact that it is sparse in a transformed domain. We will assume that the signal is sparse in its expansion in terms of the Gabor dictionary.

Our goal in this example is to demonstrate that one does not really need all 16,384 samples to perform time-frequency analysis; all the processing can be carried out using a reduced number of observations, by exploiting the theory of compressed sensing. To form the observations vector, $\boldsymbol{y}$, the number of observations was chosen to be $N = 2048$. This amounts to a reduction of eight times with

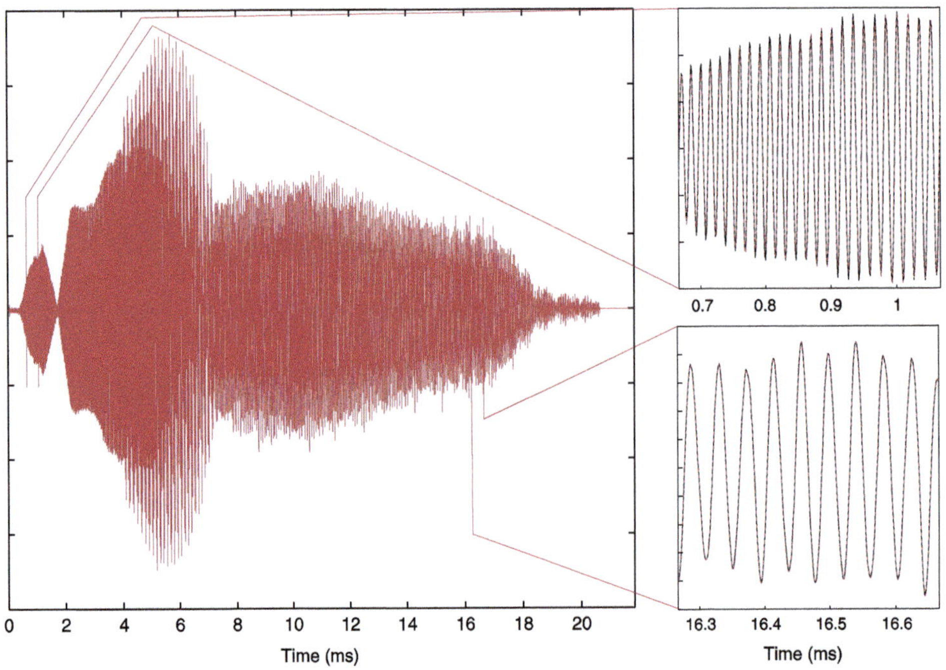

FIGURE 10.16

The recorded echolocation signal. The frequency of the signal is time-varying, which is indicated by focusing on two different parts of the signal.

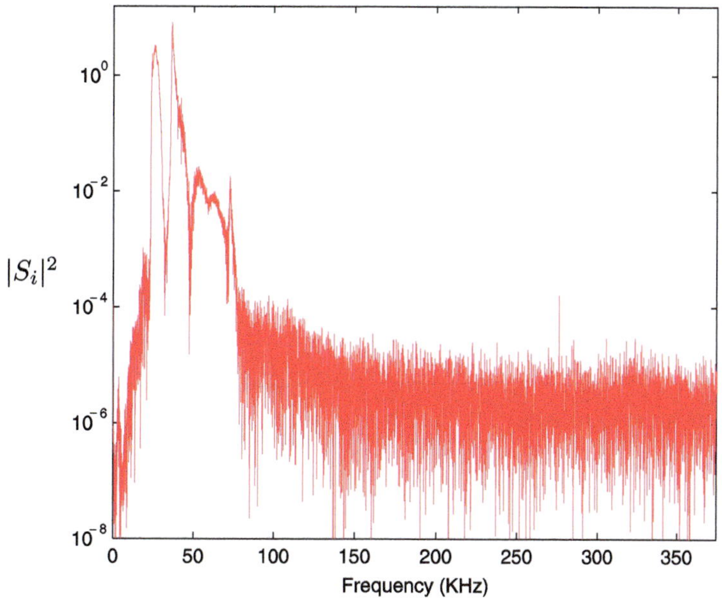

FIGURE 10.17

Plot of the energy of the DFT transform coefficients, S_i. Observe that most of the frequency activity takes place within a short frequency range.

respect to the number of available samples. The observations vector was formed as

$$y = Xs,$$

where X is an $N \times l$ sensing matrix comprising ± 1 generated in a random way. This means that once we obtain $\boldsymbol{y}$, we do not need to store the original samples anymore, leading to a savings in memory requirements. Ideally, one could have obtained the reduced number of observations by sampling directly the analog signal at sub-Nyquist rates, as has already been discussed at the end of Section 9.9. Another goal is to use both the analysis and synthesis models and demonstrate their difference.

Three different spectrograms were computed. Two of them, shown in Fig. 10.18B and C, correspond to the reconstructed signals obtained by the analysis (10.37) and the synthesis (9.37) formulations, respectively. In both cases, the NESTA algorithm was used and the $G_{(128,64)}$ frame was employed. Note that the latter dictionary is redundant by a factor of 2. The spectrograms are the result of plotting the time-frequency grid and coloring each node (t, i) according to the energy $|\theta|^2$ of the coefficient associated with the respective atom in the Gabor dictionary. The full Gabor transform was applied to the reconstructed signals to obtain the spectrograms, in order to get better coverage of the time-frequency grid. The scale is logarithmic and the darker areas correspond to larger values. The spectrogram of the original signal obtained via the full Gabor transform is shown in Fig. 10.18D. It is evident that the analysis model resulted in a more clear spectrogram, which resembles the original one better. When

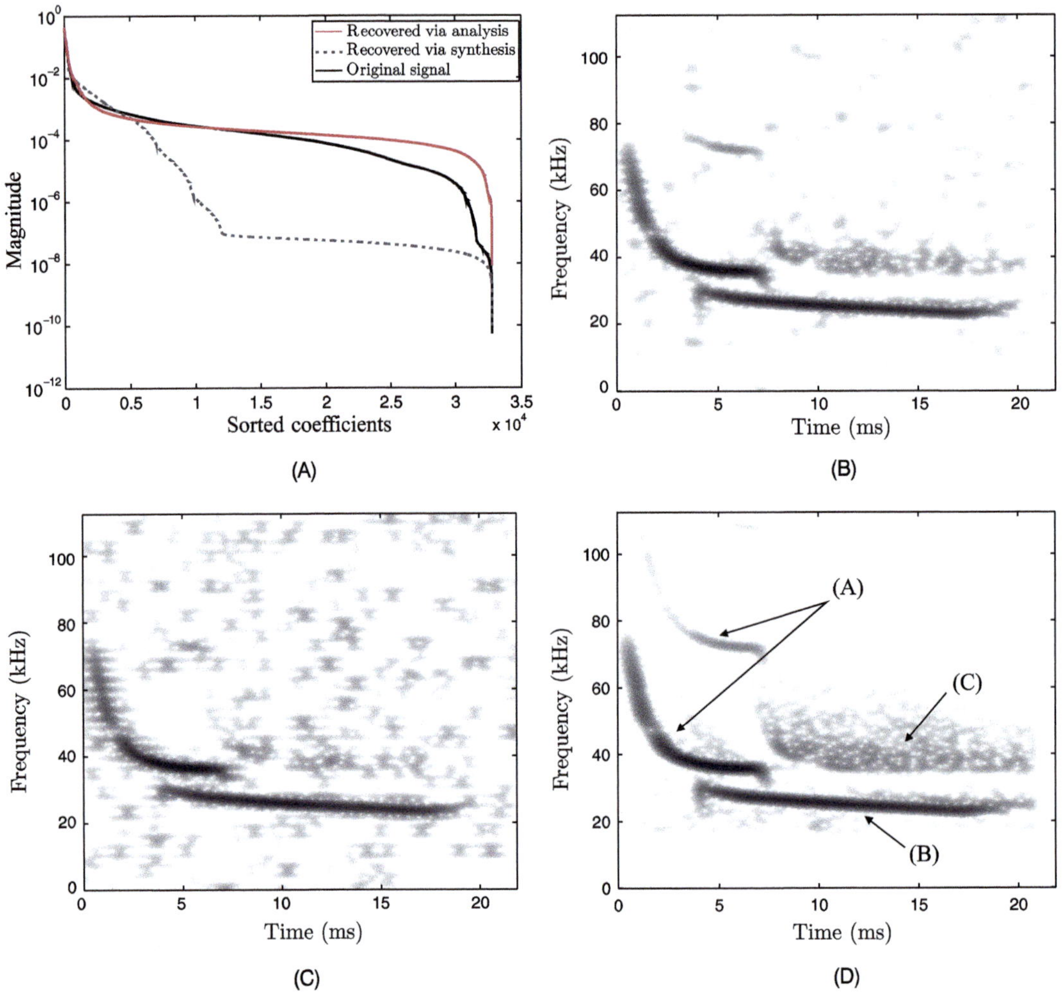

FIGURE 10.18

(A) Plot of the magnitude of the coefficients, sorted in decreasing order, in the expansion in terms of the $G_{(128,64)}$ Gabor frame. The results correspond to the analysis and synthesis model formulations. The third curve corresponds to the case of analyzing the original vector signal directly, by projecting it on the dual frame. (B) The spectrogram from the analysis and (C) the spectrogram from the synthesis formulations. (D) The spectrogram corresponding to the $G_{(64,32)}$ frame using the analysis formulation. For all cases, the number of observations used was one-eighth of the total number of signal samples. A, B, and C indicate different parts of the signal, as explained in the text.

the frame $G_{(64,32)}$ is employed, which is a highly redundant Gabor dictionary comprising $8l$ atoms, then the analysis model results in a recovered signal whose spectrogram is visually indistinguishable from the original one in Fig. 10.18D.

Fig. 10.18A is the plot of the magnitude of the corresponding Gabor transform coefficients, sorted in decreasing values. The synthesis model provides a sparser representation in the sense that the coefficients decrease much faster. The third curve is the one that results if we multiply the dual frame matrix $\tilde{G}_{(128,64)}$ directly with the vector of the original signal samples, and it is shown for comparison reasons.

To conclude, the curious reader may wonder what these curves in Fig. 10.18D mean after all. The call denoted by (A) belongs to a *Pipistrellus pipistrellus* (!) and the call denoted by (B) is either a social call or belongs to a different species. The signal (C) is the return echo from the signal (A). The large spread in time of (C) indicates a highly reflective environment [75].

PROBLEMS

10.1 Show that the step in a greedy algorithm that selects the column of the sensing matrix in order to maximize the correlation between the column and the currently available error vector $\boldsymbol{e}^{(i-1)}$ is equivalent to selecting the column that reduces the ℓ_2 norm of the error vector.
Hint: All the parameters obtained in previous steps are fixed, and the optimization is with respect to the new column as well as the corresponding weighting coefficient in the estimate of the parameter vector.

10.2 Prove the proposition that if there is a sparse solution to the linear system $\boldsymbol{y} = X\boldsymbol{\theta}$ such that

$$k_0 = ||\boldsymbol{\theta}||_0 < \frac{1}{2}\left(1 + \frac{1}{\mu(X)}\right),$$

where $\mu(X)$ is the mutual coherence of X, then the column selection procedure in a greedy algorithm will always select a column among the active columns of X, which correspond to the support of $\boldsymbol{\theta}$; that is, the columns that take part in the representation of $\boldsymbol{y}$ in terms of the columns of X.
Hint: Assume that

$$\boldsymbol{y} = \sum_{i=1}^{k_0} \theta_i \boldsymbol{x}_i^c.$$

10.3 Give an explanation to justify why in step 4 of the CoSaMP algorithm the value of t is taken to be equal to $2k$.

10.4 Show that if

$$J(\boldsymbol{\theta}, \tilde{\boldsymbol{\theta}}) = \frac{1}{2}||\boldsymbol{y} - X\boldsymbol{\theta}||_2^2 + \lambda||\boldsymbol{\theta}||_1 + \frac{1}{2}d(\boldsymbol{\theta}, \tilde{\boldsymbol{\theta}}),$$

where

$$d(\boldsymbol{\theta}, \tilde{\boldsymbol{\theta}}) := c||\boldsymbol{\theta} - \tilde{\boldsymbol{\theta}}||_2^2 - ||X\boldsymbol{\theta} - X\tilde{\boldsymbol{\theta}}||_2^2,$$

then minimization results in

$$\hat{\boldsymbol{\theta}} = S_{\lambda/c}\left(\frac{1}{c}X^T(\boldsymbol{y} - X\tilde{\boldsymbol{\theta}}) + \tilde{\boldsymbol{\theta}}\right).$$

10.5 Prove the basic recursion of the parallel coordinate descent algorithm.
Hint: Assume that at the ith iteration, it is the turn of the jth component to be updated, so that the following is minimized:

$$J(\theta_j) = \frac{1}{2}||\boldsymbol{y} - X\boldsymbol{\theta}^{(i-1)} + \theta_j^{(i-1)}\boldsymbol{x}_j - \theta_j \boldsymbol{x}_j||_2^2 + \lambda|\theta_j|.$$

10.6 Derive the iterative scheme to minimize the weighted ℓ_1 ball using a majorization-minimization procedure to minimize $\sum_{i=1}^{l} \ln(|\theta_i| + \epsilon)$, subject to the observations set $\boldsymbol{y} = X\boldsymbol{\theta}$.
Hint: Use the linearization of the logarithmic function to bound it from above, because it is a concave function and its graph is located below its tangent.

10.7 Show that the weighted ℓ_1 ball used in SpAPSM is upper bounded by the ℓ_0 norm of the target vector.

10.8 Show that the canonical dual frame minimizes the total ℓ_2 norm of the dual frame, that is,

$$\sum_{i \in \mathcal{I}} \|\tilde{\boldsymbol{\psi}}_i\|_2^2.$$

Note that this and the two subsequent problems are related to the appendix of this chapter.
Hint: Use the result of Problem 9.10.

10.9 Show that Parseval's tight frames are self-dual.

10.10 Prove that the bounds A, B of a frame coincide with the maximum and minimum eigenvalues of the matrix product $\Psi\Psi^T$.

MATLAB® EXERCISES

10.11 Construct a multitone signal having samples

$$\theta_n = \sum_{j=1}^{3} a_j \cos\left(\frac{\pi}{2N}(2m_j - 1)n\right), n = 0, \dots, l-1,$$

where $N = 30$, $l = 2^8$, $\boldsymbol{a} = [0.3, 1, 0.75]^T$, and $\boldsymbol{m} = [4, 10, 30]^T$. (a) Plot this signal in the time and in the frequency domain (use the "fft.m" MATLAB® function to compute the Fourier transform). (b) Build a sensing matrix 30×2^8 with entries drawn from a normal distribution, $\mathcal{N}(0, \frac{1}{N})$, and recover $\boldsymbol{\theta}$ based on these observations by ℓ_1 minimization using, for example, "solvelasso.m" (see MATLAB® exercise 9.21). (c) Build a sensing matrix 30×2^8, where each of its rows contains only a single nonzero component taking the value 1. Moreover, each column has at most one nonzero component. Observe that the multiplication of this sensing matrix with $\boldsymbol{\theta}$ just picks certain components of $\boldsymbol{\theta}$ (those that correspond to the position of the nonzero value of each row of the sampling matrix). Show by solving the corresponding ℓ_1 minimization task, as in question (b), that $\boldsymbol{\theta}$ can be recovered exactly using such a sparse sensing matrix (containing only 30 nonzero components!). Observe that the unknown $\boldsymbol{\theta}$ is sparse in the frequency domain and give an explanation why the recovery is successful with the specific sparse sensing matrix.

10.12 Implement the OMP algorithm (see Section 10.2.1) as well as the CSMP (see Section 10.2.1) with $t = 2k$. Assume a compressed sensing system using normally distributed sensing matrix.

(a) Compare the two algorithms in the case where $\alpha = \frac{N}{l} = 0.2$ for $\beta = \frac{k}{N}$ taking values in the set $\{0.1, 0.2, 0.3, \cdots, 1\}$ (choose yourself a signal and a sensing matrix in order to comply with the recommendations above). (b) Repeat the same test when $\alpha = 0.8$. Observe that this experiment, if it is performed for many different α values, $0 \leq \alpha \leq 1$, can be used for the estimation of phase transition diagrams such as the one depicted in Fig. 10.4. (c) Repeat (a) and (b) with the obtained measurements now contaminated with noise corresponding to 20 dB SNR.

10.13 Reproduce the MRI reconstruction experiment of Fig. 10.9 by running the MATLAB® script "MRIcs.m," which is available from the website of the book.

10.14 Reproduce the bat echolocation time-frequency analysis experiment of Fig. 10.18 by running the MATLAB® script "BATcs.m," which is available from the website of the book.

REFERENCES

[1] M.G. Amin (Ed.), Compressive Sensing for Urban Radar, CRC Press, 2014.

[2] M.R. Andersen, Sparse Inference Using Approximate Message Passing, MSc Thesis, Technical University of Denmark, Department of Applied Mathematics and Computing, 2014.

[3] D. Angelosante, J.A. Bazerque, G.B. Giannakis, Online adaptive estimation of sparse signals: where RLS meets the ℓ_1-norm, IEEE Trans. Signal Process. 58 (7) (2010) 3436–3447.

[4] M. Asif, J. Romberg, Dynamic updating for ℓ_1 minimization, IEEE J. Sel. Top. Signal Process. 4 (2) (2010) 421–434.

[5] M. Asif, J. Romberg, On the LASSO and Dantzig selector equivalence, in: Proceedings of the Conference on Information Sciences and Systems, CISS, Princeton, NJ, March 2010.

[6] F. Bach, Optimization with sparsity-inducing penalties, Found. Trends Mach. Learn. 4 (2012) 1–106.

[7] F. Bach, R. Jenatton, J. Mairal, G. Obozinski, Structured sparsity through convex optimization, Stat. Sci. 27 (4) (2012) 450–468.

[8] S. Bakin, Adaptive Regression and Model Selection in Data Mining Problems, PhD Thesis, Australian National University, 1999.

[9] R. Baraniuk, V. Cevher, M. Wakin, Low-dimensional models for dimensionality reduction and signal recovery: a geometric perspective, Proc. IEEE 98 (6) (2010) 959–971.

[10] R.G. Baraniuk, V. Cevher, M.F. Duarte, C. Hegde, Model-based compressive sensing, IEEE Trans. Inf. Theory 56 (4) (2010) 1982–2001.

[11] A. Beck, M. Teboulle, A fast iterative shrinkage algorithm for linear inverse problems, SIAM J. Imaging Sci. 2 (1) (2009) 183–202.

[12] S. Becker, J. Bobin, E.J. Candès, NESTA: a fast and accurate first-order method for sparse recovery, SIAM J. Imaging Sci. 4 (1) (2011) 1–39.

[13] A. Belouchrani, M.G. Amin, Blind source separation based on time-frequency signal representations, IEEE Trans. Signal Process. 46 (11) (1998) 2888–2897.

[14] E. van den Berg, M.P. Friedlander, Probing the Pareto frontier for the basis pursuit solutions, SIAM J. Sci. Comput. 31 (2) (2008) 890–912.

[15] P. Bickel, Y. Ritov, A. Tsybakov, Simultaneous analysis of LASSO and Dantzig selector, Ann. Stat. 37 (4) (2009) 1705–1732.

[16] A. Blum, Random projection, margins, kernels and feature selection, in: Lecture Notes on Computer Science (LNCS), 2006, pp. 52–68.

[17] T. Blumensath, M.E. Davies, Iterative hard thresholding for compressed sensing, Appl. Comput. Harmon. Anal. 27 (3) (2009) 265–274.

[18] T. Blumensath, M.E. Davies, Sampling theorems for signals from the union of finite-dimensional linear subspaces, IEEE Trans. Inf. Theory 55 (4) (2009) 1872–1882.

[19] T. Blumensath, M.E. Davies, Normalized iterative hard thresholding: guaranteed stability and performance, IEEE Sel. Top. Signal Process. 4 (2) (2010) 298–309.

[20] B. Boashash, Time Frequency Analysis, Elsevier, 2003.

[21] J.F. Cai, S. Osher, Z. Shen, Split Bregman methods and frame based image restoration, Multiscale Model. Simul. 8 (2) (2009) 337–369.
[22] E. Candès, J. Romberg, T. Tao, Robust uncertainty principles: exact signal reconstruction from highly incomplete Fourier information, IEEE Trans. Inf. Theory 52 (2) (2006) 489–509.
[23] E.J. Candès, T. Tao, The Dantzig selector: statistical estimation when p is much larger than n, Ann. Stat. 35 (6) (2007) 2313–2351.
[24] E.J. Candès, M.B. Wakin, S.P. Boyd, Enhancing sparsity by reweighted ℓ_1 minimization, J. Fourier Anal. Appl. 14 (5) (2008) 877–905.
[25] E.J. Candès, Y.C. Eldar, D. Needell, P. Randall, Compressed sensing with coherent and redundant dictionaries, Appl. Comput. Harmon. Anal. 31 (1) (2011) 59–73.
[26] V. Cevher, P. Indyk, C. Hegde, R.G. Baraniuk, Recovery of clustered sparse signals from compressive measurements, in: International Conference on Sampling Theory and Applications, SAMPTA, Marseille, France, 2009.
[27] V. Cevher, P. Indyk, L. Carin, R.G. Baraniuk, Sparse signal recovery and acquisition with graphical models, IEEE Signal Process. Mag. 27 (6) (2010) 92–103.
[28] S. Chen, D.L. Donoho, M. Saunders, Atomic decomposition by basis pursuit, SIAM J. Sci. Comput. 20 (1) (1998) 33–61.
[29] S. Chouvardas, K. Slavakis, Y. Kopsinis, S. Theodoridis, A sparsity promoting adaptive algorithm for distributed learning, IEEE Trans. Signal Process. 60 (10) (2012) 5412–5425.
[30] S. Chouvardas, Y. Kopsinis, S. Theodoridis, Sparsity-aware distributed learning, in: A. Hero, J. Moura, T. Luo, S. Cui (Eds.), Big Data Over Networks, Cambridge University Press, 2014.
[31] P.L. Combettes, V.R. Wajs, Signal recovery by proximal forward-backward splitting, SIAM J. Multiscale Model. Simul. 4 (4) (2005) 1168–1200.
[32] P.L. Combettes, J.-C. Pesquet, Proximal splitting methods in signal processing, in: Fixed-Point Algorithms for Inverse Problems in Science and Engineering, Springer-Verlag, 2011.
[33] W. Dai, O. Milenkovic, Subspace pursuit for compressive sensing signal reconstruction, IEEE Trans. Inf. Theory 55 (5) (2009) 2230–2249.
[34] I. Daubechies, M. Defrise, C. De-Mol, An iterative thresholding algorithm for linear inverse problems with a sparsity constraint, Commun. Pure Appl. Math. 57 (11) (2004) 1413–1457.
[35] I. Daubechies, R. DeVore, M. Fornasier, C.S. Güntürk, Iteratively reweighted least squares minimization for sparse recovery, Commun. Pure Appl. Math. 63 (1) (2010) 1–38.
[36] I. Daubechies, A. Grossman, Y. Meyer, Painless nonorthogonal expansions, J. Math. Phys. 27 (1986) 1271–1283.
[37] M.A. Davenport, M.B. Wakin, Analysis of orthogonal matching pursuit using the restricted isometry property, IEEE Trans. Inf. Theory 56 (9) (2010) 4395–4401.
[38] R.A. DeVore, V.N. Temlyakov, Some remarks on greedy algorithms, Adv. Comput. Math. 5 (1996) 173–187.
[39] P. Di Lorenzo, A.H. Sayed, Sparse distributed learning based on diffusion adaptation, IEEE Trans. Signal Process. 61 (6) (2013) 1419–1433.
[40] D.L. Donoho, J. Tanner, Neighborliness of randomly-projected simplifies in high dimensions, in: Proceedings on National Academy of Sciences, 2005, pp. 9446–9451.
[41] D.A. Donoho, A. Maleki, A. Montanari, Message-passing algorithms for compressed sensing, Proc. Natl. Acad. Sci. USA 106 (45) (2009) 18914–18919.
[42] D.L. Donoho, J. Tanner, Counting the faces of randomly projected hypercubes and orthants, with applications, Discrete Comput. Geom. 43 (3) (2010) 522–541.
[43] D.L. Donoho, J. Tanner, Precise undersampling theorems, Proc. IEEE 98 (6) (2010) 913–924.
[44] D.L. Donoho, I.M. Johnstone, Ideal spatial adaptation by wavelet shrinkage, Biometrika 81 (3) (1994) 425–455.
[45] D. Donoho, I. Johnstone, G. Kerkyacharian, D. Picard, Wavelet shrinkage: asymptopia? J. R. Stat. Soc. B 57 (1995) 301–337.
[46] J. Duchi, S.S. Shwartz, Y. Singer, T. Chandra, Efficient projections onto the ℓ_1-ball for learning in high dimensions, in: Proceedings of the International Conference on Machine Leaning, ICML, 2008, pp. 272–279.
[47] R.J. Duffin, A.C. Schaeffer, A class of nonharmonic Fourier series, Trans. Am. Math. Soc. 72 (1952) 341–366.
[48] B. Efron, T. Hastie, I.M. Johnstone, R. Tibshirani, Least angle regression, Ann. Stat. 32 (2004) 407–499.
[49] M. Elad, J.L. Starck, P. Querre, D.L. Donoho, Simultaneous cartoon and texture image inpainting using morphological component analysis (MCA), Appl. Comput. Harmon. Anal. 19 (2005) 340–358.
[50] M. Elad, B. Matalon, M. Zibulevsky, Coordinate and subspace optimization methods for linear least squares with non-quadratic regularization, Appl. Comput. Harmon. Anal. 23 (2007) 346–367.

[51] M. Elad, P. Milanfar, R. Rubinstein, Analysis versus synthesis in signal priors, Inverse Probl. 23 (2007) 947–968.
[52] M. Elad, Sparse and Redundant Representations: From Theory to Applications in Signal and Image Processing, Springer, 2010.
[53] Y.C. Eldar, P. Kuppinger, H. Bolcskei, Block-sparse signals: uncertainty relations and efficient recovery, IEEE Trans. Signal Process. 58 (6) (2010) 3042–3054.
[54] Y.C. Eldar, G. Kutyniok, Compressed Sensing: Theory and Applications, Cambridge University Press, 2012.
[55] J. Fan, R. Li, Variable selection via nonconcave penalized likelihood and its oracle properties, J. Am. Stat. Assoc. 96 (456) (2001) 1348–1360.
[56] M.A. Figueiredo, R.D. Nowak, An EM algorithm for wavelet-based image restoration, IEEE Trans. Image Process. 12 (8) (2003) 906–916.
[57] P. Flandrin, Time-Frequency/Time-Scale Analysis, Academic Press, 1999.
[58] S. Foucart, Hard thresholding pursuit: an algorithm for compressive sensing, SIAM J. Numer. Anal. 49 (6) (2011) 2543–2563.
[59] J. Friedman, T. Hastie, R. Tibshirani, A note on the group LASSO and a sparse group LASSO, arXiv:1001.0736v1 [math. ST], 2010.
[60] P.J. Garrigues, B. Olshausen, Learning horizontal connections in a sparse coding model of natural images, in: Advances in Neural Information Processing Systems, NIPS, 2008.
[61] A.C. Gilbert, S. Muthukrisnan, M.J. Strauss, Improved time bounds for near-optimal sparse Fourier representation via sampling, in: Proceedings of SPIE (Wavelets XI), San Diego, CA, 2005.
[62] R. Giryes, S. Nam, M. Elad, R. Gribonval, M. Davies, Greedy-like algorithms for the cosparse analysis model, Linear Algebra Appl. 441 (2014) 22–60.
[63] T. Goldstein, S. Osher, The split Bregman algorithm for ℓ_1 regularized problems, SIAM J. Imaging Sci. 2 (2) (2009) 323–343.
[64] I.F. Gorodnitsky, B.D. Rao, Sparse signal reconstruction from limited data using FOCUSS: a re-weighted minimum norm algorithm, IEEE Trans. Signal Process. 45 (3) (1997) 600–614.
[65] L. Hageman, D. Young, Applied Iterative Methods, Academic Press, New York, 1981.
[66] T. Hale, W. Yin, Y. Zhang, A Fixed-Point Continuation Method for l_1 Regularized Minimization With Applications to Compressed Sensing, Tech. Rep. TR07-07, Department of Computational and Applied Mathematics, Rice University, 2007.
[67] D. Han, D.R. Larson, Frames, Bases and Group Representations, American Mathematical Society, Providence, RI, 2000.
[68] T. Hastie, R. Tibshirani, J. Friedman, The Elements of Statistical Learning: Data Mining, Inference and Prediction, second ed., Springer, 2008.
[69] J.C. Hoch, A.S. Stern, D.L. Donoho, I.M. Johnstone, Maximum entropy reconstruction of complex (phase sensitive) spectra, J. Magn. Res. 86 (2) (1990) 236–246.
[70] P.A. Jansson, Deconvolution: Applications in Spectroscopy, Academic Press, New York, 1984.
[71] R. Jenatton, J.-Y. Audibert, F. Bach, Structured variable selection with sparsity-inducing norms, J. Mach. Learn. Res. 12 (2011) 2777–2824.
[72] S.-J. Kim, K. Koh, M. Lustig, S. Boyd, D. Gorinevsky, An interior-point method for large-scale ℓ_1-regularized least squares, IEEE J. Sel. Top. Signal Process. 1 (4) (2007) 606–617.
[73] N.G. Kingsbury, T.H. Reeves, Overcomplete image coding using iterative projection-based noise shaping, in: Proceedings IEEE International Conference on Image Processing, ICIP, 2002, pp. 597–600.
[74] K. Knight, W. Fu, Asymptotics for the LASSO-type estimators, Ann. Stat. 28 (5) (2000) 1356–1378.
[75] Y. Kopsinis, E. Aboutanios, D.E. Waters, S. McLaughlin, Time-frequency and advanced frequency estimation techniques for the investigation of bat echolocation calls, J. Acoust. Soc. Am. 127 (2) (2010) 1124–1134.
[76] Y. Kopsinis, K. Slavakis, S. Theodoridis, Online sparse system identification and signal reconstruction using projections onto weighted ℓ_1 balls, IEEE Trans. Signal Process. 59 (3) (2011) 936–952.
[77] Y. Kopsinis, K. Slavakis, S. Theodoridis, S. McLaughlin, Reduced complexity online sparse signal reconstruction using projections onto weighted ℓ_1 balls, in: 2011 17th International Conference on Digital Signal Processing, DSP, July 2011, pp. 1–8.
[78] Y. Kopsinis, K. Slavakis, S. Theodoridis, S. McLaughlin, Generalized thresholding sparsity-aware algorithm for low complexity online learning, in: Proceedings of the IEEE International Conference on Acoustics, Speech, and Signal Processing, ICASSP, Kyoto, Japan, March 2012, pp. 3277–3280.

[79] Y. Kopsinis, K. Slavakis, S. Theodoridis, S. McLaughlin, Thresholding-based online algorithms of complexity comparable to sparse LMS methods, in: 2013 IEEE International Symposium on Circuits and Systems, ISCAS, May 2013, pp. 513–516.
[80] Y. Kopsinis, S. Chouvardas, S. Theodoridis, Sparsity-aware learning in the context of echo cancelation: a set theoretic estimation approach, in: Proceedings of the European Signal Processing Conference, EUSIPCO, Lisbon, Portugal, September 2014.
[81] J. Kovacevic, A. Chebira, Life beyond bases: the advent of frames, IEEE Signal Process. Mag. 24 (4) (2007) 86–104.
[82] J. Langford, L. Li, T. Zhang, Sparse online learning via truncated gradient, J. Mach. Learn. Res. 10 (2009) 777–801.
[83] Y.M. Lu, M.N. Do, Sampling signals from a union of subspaces, IEEE Signal Process. Mag. 25 (2) (2008) 41–47.
[84] Z.Q. Luo, P. Tseng, On the convergence of the coordinate descent method for convex differentiable minimization, J. Optim. Theory Appl. 72 (1) (1992) 7–35.
[85] A. Maleki, D.L. Donoho, Optimally tuned iterative reconstruction algorithms for compressed sensing, IEEE J. Sel. Top. Signal Process. 4 (2) (2010) 330–341.
[86] D.M. Malioutov, M. Cetin, A.S. Willsky, Homotopy continuation for sparse signal representation, in: IEEE International Conference on Acoustics, Speech and Signal Processing, ICASSP, 2005, pp. 733–736.
[87] S. Mallat, A Wavelet Tour of Signal Processing: The Sparse Way, third ed., Academic Press, 2008.
[88] S. Mallat, S. Zhang, Matching pursuit in a time-frequency dictionary, IEEE Trans. Signal Process. 41 (1993) 3397–3415.
[89] G. Mateos, J. Bazerque, G. Giannakis, Distributed sparse linear regression, IEEE Trans. Signal Process. 58 (10) (2010) 5262–5276.
[90] G. Mileounis, B. Babadi, N. Kalouptsidis, V. Tarokh, An adaptive greedy algorithm with application to nonlinear communications, IEEE Trans. Signal Process. 58 (6) (2010) 2998–3007.
[91] A. Mousavi, A. Maleki, R.G. Baraniuk, Parameterless optimal approximate message passing, arXiv:1311.0035v1 [cs.IT], 2013.
[92] S. Nam, M. Davies, M. Elad, R. Gribonval, The cosparse analysis model and algorithms, Appl. Comput. Harmon. Anal. 34 (1) (2013) 30–56.
[93] D. Needell, J.A. Tropp, COSAMP: iterative signal recovery from incomplete and inaccurate samples, Appl. Comput. Harmon. Anal. 26 (3) (2009) 301–321.
[94] D. Needell, R. Ward, Stable image reconstruction using total variation minimization, SIAM J. Imaging Sci. 6 (2) (2013) 1035–1058.
[95] Y. Nesterov, Introductory Lectures on Convex Optimization: A Basic Course, Kluwer Academic Publishers, 2004.
[96] Y.E. Nesterov, A method for solving the convex programming problem with convergence rate $O(1/k^2)$, Dokl. Akad. Nauk SSSR 269 (1983) 543–547 (in Russian).
[97] G. Obozinski, B. Taskar, M. Jordan, Multi-Task Feature Selection, Tech. Rep., Department of Statistics, University of California, Berkeley, 2006.
[98] G. Obozinski, B. Taskar, M.I. Jordan, Joint covariate selection and joint subspace selection for multiple classification problems, Stat. Comput. 20 (2) (2010) 231–252.
[99] M.R. Osborne, B. Presnell, B.A. Turlach, A new approach to variable selection in least squares problems, IMA J. Numer. Anal. 20 (2000) 389–403.
[100] P.M. Pardalos, N. Kovoor, An algorithm for a singly constrained class of quadratic programs subject to upper and lower bounds, Math. Program. 46 (1990) 321–328.
[101] F. Parvaresh, H. Vikalo, S. Misra, B. Hassibi, Recovering sparse signals using sparse measurement matrices in compressed DNA microarrays, IEEE J. Sel. Top. Signal Process. 2 (3) (2008) 275–285.
[102] S. Patterson, Y.C. Eldar, I. Keidar, Distributed compressed sensing for static and time-varying networks, arXiv:1308.6086 [cs.IT], 2014.
[103] T. Peleg, M. Elad, Performance guarantees of the thresholding algorithm for the cosparse analysis model, IEEE Trans. Inf. Theory 59 (3) (2013) 1832–1845.
[104] M.D. Plumbley, Geometry and homotopy for ℓ_1 sparse representation, in: Proceedings of the International Workshop on Signal Processing With Adaptive Sparse Structured Representations, SPARS, Rennes, France, 2005.
[105] R.T. Rockafellar, Monotone operators and the proximal point algorithms, SIAM J. Control Optim. 14 (5) (1976) 877–898.
[106] R. Rubinstein, R. Peleg, M. Elad, Analysis KSVD: a dictionary-learning algorithm for the analysis sparse model, IEEE Trans. Signal Process. 61 (3) (2013) 661–677.
[107] L.I. Rudin, S. Osher, E. Fatemi, Nonlinear total variation based noise removal algorithms, Phys. D, Nonlinear Phenom. 60 (1–4) (1992) 259–268.

[108] I.W. Selesnick, M.A.T. Figueiredo, Signal restoration with overcomplete wavelet transforms: comparison of analysis and synthesis priors, in: Proceedings of SPIE, 2009.
[109] K. Slavakis, Y. Kopsinis, S. Theodoridis, S. McLaughlin, Generalized thresholding and online sparsity-aware learning in a union of subspaces, IEEE Trans. Signal Process. 61 (15) (2013) 3760–3773.
[110] P. Sprechmann, I. Ramirez, G. Sapiro, Y.C. Eldar, CHiLasso: a collaborative hierarchical sparse modeling framework, IEEE Trans. Signal Process. 59 (9) (2011) 4183–4198.
[111] J.L. Starck, E.J. Candès, D.L. Donoho, The curvelet transform for image denoising, IEEE Trans. Image Process. 11 (6) (2002) 670–684.
[112] J.L. Starck, J. Fadili, F. Murtagh, The undecimated wavelet decomposition and its reconstruction, IEEE Trans. Signal Process. 16 (2) (2007) 297–309.
[113] T. Strohmer, Numerical algorithms for discrete Gabor expansions, in: Gabor Analysis and Algorithms: Theory and Applications, Birkhauser, Boston, MA, 1998, pp. 267–294.
[114] V.N. Temlyakov, Nonlinear methods of approximation, Found. Comput. Math. 3 (1) (2003) 33–107.
[115] J.A. Tropp, Greed is good, IEEE Trans. Inf. Theory 50 (2004) 2231–2242.
[116] Y. Tsaig, Sparse Solution of Underdetermined Linear Systems: Algorithms and Applications, PhD Thesis, Stanford University, 2007.
[117] B.A. Turlach, W.N. Venables, S.J. Wright, Simultaneous variable selection, Technometrics 47 (3) (2005) 349–363.
[118] S. Wright, R. Nowak, M. Figueiredo, Sparse reconstruction by separable approximation, IEEE Trans. Signal Process. 57 (7) (2009) 2479–2493.
[119] M. Yaghoobi, S. Nam, R. Gribonval, M. Davies, Constrained overcomplete analysis operator learning for cosparse signal modelling, IEEE Trans. Signal Process. 61 (9) (2013) 2341–2355.
[120] J. Yang, Y. Zhang, W. Yin, A fast alternating direction method for TV ℓ_1 - ℓ_2 signal reconstruction from partial Fourier data, IEEE Trans. Sel. Top. Signal Process. 4 (2) (2010) 288–297.
[121] W. Yin, S. Osher, D. Goldfarb, J. Darbon, Bregman iterative algorithms for ℓ_1-minimization with applications to compressed sensing, SIAM J. Imaging Sci. 1 (1) (2008) 143–168.
[122] M. Yuan, Y. Lin, Model selection and estimation in regression with grouped variables, J. R. Stat. Soc. 68 (1) (2006) 49–67.
[123] T. Zhang, Sparse recovery with orthogonal matching pursuit under RIP, IEEE Trans. Inf. Theory 57 (9) (2011) 6215–6221.
[124] M. Zibulevsky, M. Elad, L1-L2 optimization in signal processing, IEEE Signal Process. Mag. 27 (3) (2010) 76–88.
[125] M. Zibulevsky, Y.Y. Zeevi, Frame analysis of the discrete Gabor scheme, IEEE Trans. Signal Process. 42 (4) (1994) 942–945.
[126] H. Zou, T. Hastie, Regularization and variable selection via the elastic net, J. R. Stat. Soc. B 67 (2) (2005) 301–320.
[127] H. Zou, The adaptive LASSO and its oracle properties, J. Am. Stat. Assoc. 101 (2006) 1418–1429.
[128] H. Zou, R. Li, One-step sparse estimates in nonconcave penalized likelihood models, Ann. Stat. 36 (4) (2008) 1509–1533.

CHAPTER

LEARNING IN REPRODUCING KERNEL HILBERT SPACES 11

CONTENTS

Machine Learning. https://doi.org/10.1016/B978-0-12-818803-3.00022-2

11.1 INTRODUCTION

Our emphasis in this chapter will be on learning nonlinear models. The necessity of adopting nonlinear models has already been discussed in Chapter 3, in the context of the classification as well as the regression tasks. For example, recall that given two jointly distributed random vectors $(\mathbf{y}, \mathbf{x}) \in \mathbb{R}^k \times \mathbb{R}^l$, we know that the optimal estimate of $\mathbf{y}$ given $\mathbf{x} = \boldsymbol{x}$ in the mean-square error (MSE) sense is the corresponding conditional mean, that is, $\mathbb{E}[\mathbf{y}|\boldsymbol{x}]$, which in general is a nonlinear function of $\boldsymbol{x}$.

There are different ways of dealing with nonlinear modeling tasks. Our emphasis in this chapter will be on a path through the so-called reproducing kernel Hilbert spaces (RKHS). The technique consists of mapping the input variables to a new space, such that the originally nonlinear task is transformed into a linear one. From a practical point of view, the beauty behind these spaces is that their rich structure allows us to perform inner product operations in a very efficient way, with complexity independent of the dimensionality of the respective RKHS. Moreover, note that the dimension of such spaces can even be infinite.

We start the chapter by reviewing some more "traditional" techniques concerning Volterra series expansions, and then we move slowly to exploring the RKHS concept. Cover's theorem, the basic properties of RKHSs, and their defining kernels are discussed. Kernel ridge regression and the support vector machine (SVM) framework is presented. Approximation techniques concerning the kernel function and the kernel matrix, such as random Fourier features (RFFs) and the Nyström method, and their implication to online and distributed learning in RKHS are discussed. Finally, some more advanced concepts related to sparsity and multikernel representations are presented. A case study in the context of text mining is presented at the end of the chapter.

11.2 GENERALIZED LINEAR MODELS

Given $(\mathrm{y}, \mathbf{x}) \in \mathbb{R} \times \mathbb{R}^l$, a generalized linear estimator $\hat{\mathrm{y}}$ of y has the form

$$\hat{\mathrm{y}} = f(\mathbf{x}) := \theta_0 + \sum_{k=1}^{K} \theta_k \phi_k(\mathbf{x}), \tag{11.1}$$

where $\phi_1, \ldots, \phi_K$ are *preselected* (nonlinear) functions. A popular family of functions is the polynomial one, for example,

$$\hat{\mathrm{y}} = \theta_0 + \sum_{i=1}^{l} \theta_i \mathrm{x}_i + \sum_{i=1}^{l-1} \sum_{m=i+1}^{l} \theta_{im} \mathrm{x}_i \mathrm{x}_m + \sum_{i=1}^{l} \theta_{ii} \mathrm{x}_i^2. \tag{11.2}$$

Assuming $l = 2$ ($\mathbf{x} = [\mathrm{x}_1, \mathrm{x}_2]^T$), (11.2) can be brought into the form of (11.1) by setting $K = 5$ and $\phi_1(\mathbf{x}) = \mathrm{x}_1$, $\phi_2(\mathbf{x}) = \mathrm{x}_2$, $\phi_3(\mathbf{x}) = \mathrm{x}_1\mathrm{x}_2$, $\phi_4(\mathbf{x}) = \mathrm{x}_1^2$, $\phi_5(\mathbf{x}) = \mathrm{x}_2^2$. The generalization of (11.2) to rth-order polynomials is readily obtained and it will contain products of the form $\mathrm{x}_1^{p_1}\mathrm{x}_2^{p_2}\cdots\mathrm{x}_l^{p_l}$, with $p_1 + p_2 + \cdots + p_l \le r$. It turns out that the number of free parameters, K, for an rth-order polynomial is equal to

$$K = \frac{(l+r)!}{r!l!}.$$

Just to get a feeling for $l = 10$ and $r = 3$, $K = 286$. The use of polynomial expansions is justified by the Weierstrass theorem, stating that every continuous function defined on a compact (closed and bounded) subset $S \subset \mathbb{R}^l$ can be uniformly approximated as closely as desired, with an arbitrarily small error, ϵ, by a polynomial function (for example, [95]). Of course, in order to achieve a good enough approximation, one may have to use a large value of r. Besides polynomial functions, other types of functions can also be used, such as splines and trigonometric functions.

A common characteristic of this type of models is that the basis functions in the expansion are *preselected* and they are fixed and independent of the data. The advantage of such a path is that the associated models are linear with respect to the unknown set of free parameters, and they can be estimated by following any one of the methods described for linear models, presented in Chapters 4–8. However, one has to pay a price for that. As shown in [7], for an expansion involving K fixed functions, the squared approximation error *cannot* be made smaller than order $\left(\frac{1}{K}\right)^{\frac{2}{l}}$. In other words, for high-dimensional spaces and in order to get a small enough error, one has to use large values of K. This is another face of the curse of dimensionality problem. In contrast, one can get rid of the dependence of the approximation error on the input space dimensionality, l, if the expansion involves data dependent functions, which are optimized with respect to the specific data set. This is, for example, the case for a class of neural networks, to be discussed in Chapter 18. In this case, the price one pays is that the dependence on the free parameters is now nonlinear, making the optimization with regard to the unknown parameters a harder task.

11.3 VOLTERRA, WIENER, AND HAMMERSTEIN MODELS

Let us start with the case of modeling nonlinear systems, where the involved input–output entities are time series/discrete-time signals denoted as (u_n, d_n), respectively. The counterpart of polynomial modeling in (11.2) is now known as the Volterra series expansion.

These types of models will not be pursued any more in this book, and they are briefly discussed here in order to put the nonlinear modeling task in a more general context as well as for historical reasons. *Thus, this section can be bypassed in a first reading.*

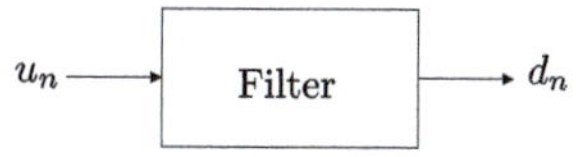

FIGURE 11.1

The nonlinear filter is excited by u_n and provides in its output d_n.

Volterra was an Italian mathematician (1860–1940) who made major contributions to mathematics as well as physics and biology. One of his landmark theories is the development of Volterra series, which is used to solve integral and integro-differential equations. He was one of the Italian professors who refused to take an oath of loyalty to the fascist regime of Mussolini and he was obliged to resign from his university post.

Fig. 11.1 shows an unknown nonlinear system/filter with the respective input–output signals. The output of a discrete-time Volterra model can be written as

$$d_n = \sum_{k=1}^{r} \sum_{i_1=0}^{M} \sum_{i_2=0}^{M} \cdots \sum_{i_k=0}^{M} w_k(i_1, i_2, \ldots, i_k) \prod_{j=1}^{k} u_{n-i_j}, \tag{11.3}$$

where $w_k(\cdot, \ldots, \cdot)$ denotes the kth-order *Volterra kernel*; in general, r can be infinite. For example, for $r = 2$ and $M = 1$, the input–output relation involves the linear combination of the terms

$$u_n, u_{n-1}, u_n^2, u_{n-1}^2, u_n u_{n-1}.$$

Special cases of the Volterra expansion are the *Wiener*, *Hammerstein*, and *Wiener–Hammerstein* models. These models are shown in Fig. 11.2. The systems $h(\cdot)$ and $g(\cdot)$ are linear systems with memory, that is,

$$s_n = \sum_{i=0}^{M_1} h_n u_{n-i}$$

and

$$d_n = \sum_{i=0}^{M_2} g_n x_{n-i}.$$

The central box corresponds to a memoryless nonlinear system, which can be approximated by a polynomial of degree r. Hence,

$$x_n = \sum_{k=1}^{r} c_k (s_n)^k.$$

In other words, a Wiener model is a linear time-invariant (LTI) system followed by the memoryless nonlinearity and the Hammerstein model is the combination of a memoryless nonlinearity followed by an LTI system. The Wiener–Hammerstein model is the combination of the two. Note that each one of these models is nonlinear with respect to the involved free parameters. In contrast, the equivalent Volterra model is linear with regard to the involved parameters; however, the number of the resulting free parameters is significantly increased with the order of the polynomial and the filter memory taps

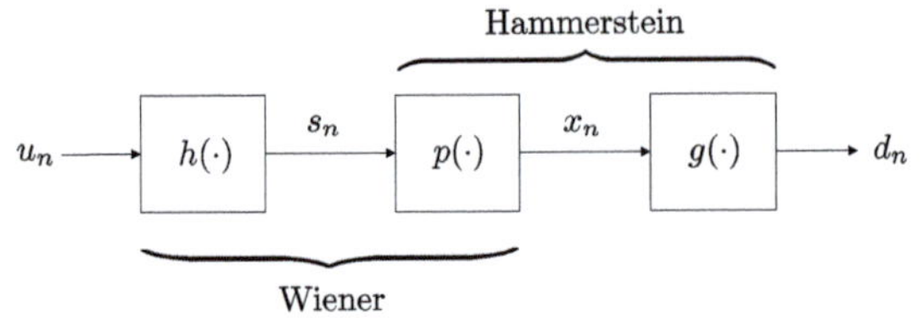

FIGURE 11.2

The Wiener model comprises a linear filter followed by a memoryless polynomial nonlinearity. The Hammerstein model consists of a memoryless nonlinearity followed by a linear filter. The Wiener–Hammerstein model is the combination of the two.

(M_1 and M_2). The interesting feature is that the equivalent to a Hammerstein model Volterra expansion consists only of the diagonal elements of the associated Volterra kernels. In other words, the output is expressed in terms of $u_n, u_{n-1}, u_{n-2}, \dots$ and their powers; there are no cross-product terms [60].

Remarks 11.1.

- The Volterra series expansion was first introduced as a generalization of the Taylor series expansion. Following [101], assume a memoryless nonlinear system. Then its input–output relationship is given by

$$d(t) = f(u(t)),$$

and adopting the Taylor expansion, for a particular time $t \in (-\infty, +\infty)$, we can write

$$d(t) = \sum_{n=0}^{+\infty} c_n (u(t))^n, \tag{11.4}$$

assuming that the series converges. The Volterra series is the extension of (11.4) to systems with memory, and we can write

$$\begin{aligned} d(t) &= w_0 + \int_{-\infty}^{+\infty} w_1(\tau_1) u(t-\tau_1) d\tau_1 \\ &+ \int_{-\infty}^{+\infty}\int_{-\infty}^{+\infty} w_2(\tau_1,\tau_2) u(t-\tau_1) u(t-\tau_2) d\tau_1 d\tau_2 \\ &+ \dots . \end{aligned} \tag{11.5}$$

In other words, the Volterra series is a power series with memory. The problem of convergence of the Volterra series is similar to that of the Taylor series. In analogy to the Weierstrass approximation theorem, it turns out that the output of a nonlinear system can be approximated arbitrarily close using a sufficient number of terms in the Volterra series expansion[1] [44].

A major difficulty in the Volterra series is the computation of the Volterra kernels. Wiener was the first to realize the potential of the Volterra series for nonlinear system modeling. In order to compute the involved Volterra kernels, he used the method of orthogonal functionals. The method resembles the method of using a set of orthogonal polynomials, when one tries to approximate a function via a polynomial expansion [130]. More on the Volterra modeling and related models can be obtained in, for example, [57,70,102]. Volterra models have been extensively used in a number of applications, including communications (e.g., [11]), biomedical engineering (e.g., [72]), and automatic control (e.g., [32]).

[1] The proof involves the theory of continuous functionals. A functional is a mapping of a function to the real axis. Observe that each integral is a functional, for a particular t and kernel.

11.4 COVER'S THEOREM: CAPACITY OF A SPACE IN LINEAR DICHOTOMIES

We have already justified the method of expanding an unknown nonlinear function in terms of a fixed set of nonlinear ones, by mobilizing arguments from the approximation theory. Although this framework fits perfectly to the regression task, where the output takes values in an interval in $\mathbb{R}$, such arguments are not well suited for the classification. In the latter case, the output value is of a discrete nature. For example, in a binary classification task, $y \in \{1, -1\}$, and as long as the sign of the predicted value, $\hat{y}$, is correct, we do not care how close y and $\hat{y}$ are. In this section, we will present an elegant and powerful theorem that justifies the expansion of a classifier f in the form of (11.1). It suffices to look at (11.1) from a different angle.

Let us consider N points, $\boldsymbol{x}_1, \boldsymbol{x}_2, \dots, \boldsymbol{x}_N \in \mathbb{R}^l$. We can say that these points are *in general position* if *there is no* subset of $l+1$ of them lying on an $(l-1)$-dimensional hyperplane. For example, in the two-dimensional space, any three of these points are not permitted to lie on a straight line.

Theorem 11.1 (Cover's theorem). *The number of groupings, denoted as $\mathcal{O}(N, l)$, that can be formed by $(l-1)$-dimensional hyperplanes to separate the N points in two classes, exploiting all possible combinations, is given by (see [31] and Problem 11.1)*

$$\mathcal{O}(N, l) = 2\sum_{i=0}^{l} \binom{N-1}{i},$$

where

$$\binom{N-1}{i} = \frac{(N-1)!}{(N-1-i)!i!}.$$

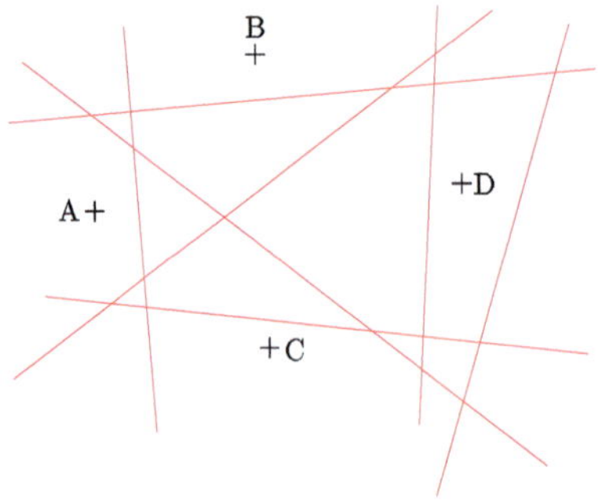

FIGURE 11.3

The possible number of linearly separable groupings of two, for four points in the two-dimensional space is $\mathcal{O}(4, 2) = 14 = 2 \times 7$.

Each one of these groupings in two classes is also known as a (linear) *dichotomy*. Fig. 11.3 illustrates the theorem for the case of $N = 4$ points in the two-dimensional space. Observe that the possible groupings are [(ABCD)], [A, (BCD)], [B, (ACD)], [C, (ABD)], [D, (ABC)], [(AB), (CD)], and

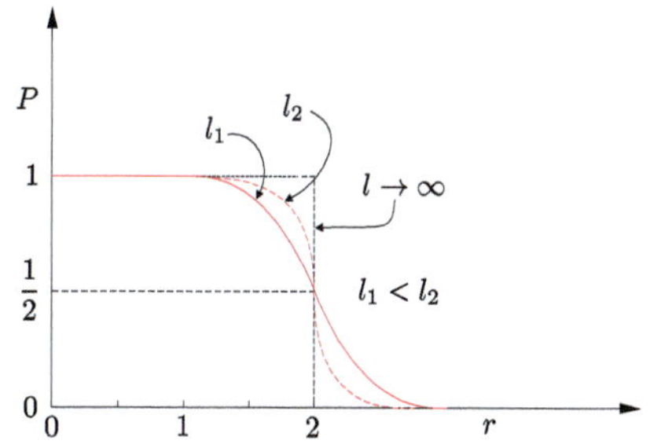

FIGURE 11.4

For $N > 2(l+1)$ the probability of linear separability becomes small. For large values of l, and provided $N < 2(l+1)$, the probability of any grouping of the data into two classes to be linearly separable tends to unity. Also, if $N \leq (l+1)$, all possible groupings in two classes are linearly separable.

[(AC), (BD)]. Each grouping is counted twice, as it can belong to either ω_1 or ω_2 class. Hence, the total number of groupings is 14, which is equal to $\mathcal{O}(4, 2)$. Note that the number of all possible combinations of N points in two groups is 2^N, which is 16 in our case. The grouping that is not counted in $\mathcal{O}(4, 2)$, as it cannot be linearly separated, is [(BC),(AD)]. Note that if $N \leq l+1$, then $\mathcal{O}(N, l) = 2^N$. That is, all possible combinations in groups of two are linearly separable; verify it for the case of $N = 3$ in the two-dimensional space.

Based on the previous theorem, given N points in the l-dimensional space, the probability of grouping these points in two *linearly* separable classes is

$$P_N^l = \frac{\mathcal{O}(N,l)}{2^N} = \begin{cases} \frac{1}{2^{N-1}} \sum_{i=0}^{l} \binom{N-1}{i}, & N > l+1, \\ 1, & N \leq l+1. \end{cases} \tag{11.6}$$

To visualize this finding, let us write $N = r(l+1)$, and express the probability P_N^l in terms of r, for a fixed value of l. The resulting graph is shown in Fig. 11.4. Observe that there are two distinct regions. One to the left of the point $r = 2$ and one to the right. At the point $r = 2$, that is, $N = 2(l+1)$, the probability is always $\frac{1}{2}$, because $\mathcal{O}(2l+2, l) = 2^{2l+1}$ (Problem 11.2). Note that the larger the value of l is, the sharper the transition from one region to the other becomes. Thus, for high-dimensional spaces and as long as $N < 2(l+1)$, the probability of any grouping of the points in two classes to be *linearly separable* tends to unity.

The way the previous theorem is exploited in practice is the following. Given N feature vectors $\boldsymbol{x}_n \in \mathbb{R}^l$, $n = 1, 2, \ldots, N$, a mapping

$$\boldsymbol{\phi} : \mathbb{R}^l \ni \boldsymbol{x}_n \longmapsto \boldsymbol{\phi}(\boldsymbol{x}_n) \in \mathbb{R}^K, \ K \gg l,$$

is performed. Then according to the theorem, the higher the value of K is, the higher the probability becomes for the images of the mapping, $\boldsymbol{\phi}(\boldsymbol{x}_n) \in \mathbb{R}^K$, $n = 1, 2, \ldots, N$, to be linearly separable in the space $\mathbb{R}^K$. Note that the expansion of a nonlinear classifier (that predicts the label in a binary classification task) is equivalent to using a linear one on the images of the original points after the mapping. Indeed,

$$f(\boldsymbol{x}) = \sum_{k=1}^{K} \theta_k \phi_k(\boldsymbol{x}) + \theta_0 = \boldsymbol{\theta}^T \begin{bmatrix} \boldsymbol{\phi}(\boldsymbol{x}) \\ 1 \end{bmatrix}, \tag{11.7}$$

with

$$\boldsymbol{\phi}(\boldsymbol{x}) := [\phi_1(\boldsymbol{x}), \phi_2(\boldsymbol{x}), \dots, \phi_K(\boldsymbol{x})]^T.$$

Provided that K is large enough, our task is *linearly separable* in the new space, $\mathbb{R}^K$, with high probability, which justifies the use of a linear classifier, $\boldsymbol{\theta}$, in (11.7). The procedure is illustrated in Fig. 11.5. The points in the two-dimensional space are not linearly separable. However, after the mapping in the three-dimensional space,

$$[x_1, x_2]^T \longmapsto \boldsymbol{\phi}(\boldsymbol{x}) = [x_1, x_2, f(x_1, x_2)]^T, \qquad f(x_1, x_2) = 4\exp\big(-(x_1^2 + x_2^2)/3\big) + 5,$$

the points in the two classes become linearly separable. Note, however, that after the mapping, the points lie on the surface of a paraboloid. This surface is fully described in terms of two free variables. Loosely speaking, we can think of the two-dimensional plane, on which the data lie originally, to be folded/transformed to form the surface of the paraboloid. This is basically the idea behind the more general problem. After the mapping from the original l-dimensional space to the new K-dimensional one, the images of the points $\boldsymbol{\phi}(\boldsymbol{x}_n)$, $n = 1, 2, \dots, N$, lie on an l-dimensional surface (*manifold*) in $\mathbb{R}^K$ [19]. We cannot fool nature. Because l variables were originally chosen to describe each pattern

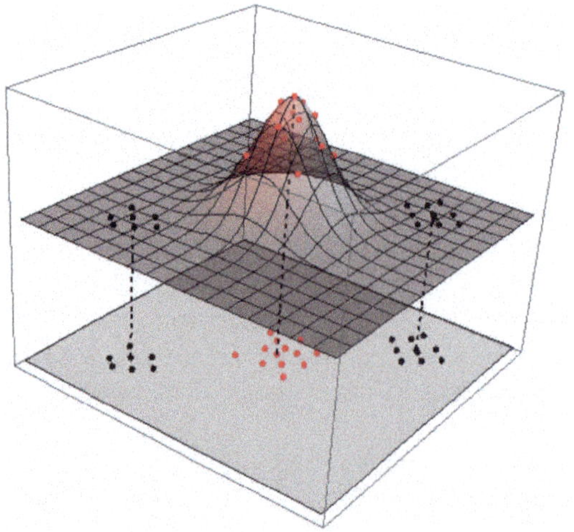

FIGURE 11.5

The points (red in one class and black for the other) that are not linearly separable in the original two-dimensional plane become linearly separable after the nonlinear mapping in the three-dimensional space; one can draw a plane that separates the "black" from the "red" points.

(dimensionality, number of free parameters) the same number of free parameters will be required to describe the same objects after the mapping in $\mathbb{R}^K$. In other words, after the mapping, we embed an l-dimensional manifold in a K-dimensional space in such a way that the data in the two classes become linearly separable.

We have by now fully justified the need for mapping the task from the original low-dimensional space to a higher-dimensional one, via a set of nonlinear functions. However, life is not easy to work in high-dimensional spaces. A large number of parameters are needed; this in turn poses computational complexity problems, and raises issues related to the generalization and overfitting performance of the designed predictors. In the sequel, we will address the former of the two problems by making a "careful" mapping to a higher-dimensional space of a specific structure. The latter problem will be addressed via regularization, as has already been discussed in various parts in previous chapters.

11.5 REPRODUCING KERNEL HILBERT SPACES

Consider a linear space $\mathbb{H}$ of real-valued functions defined on a set[2] $\mathcal{X} \subseteq \mathbb{R}^l$. Furthermore, suppose that $\mathbb{H}$ is a Hilbert space; that is, it is equipped with an inner product operation, $\langle \cdot, \cdot \rangle_{\mathbb{H}}$, that defines a corresponding norm $\| \cdot \|_{\mathbb{H}}$ and $\mathbb{H}$ is complete with respect to this norm.[3] From now on, and for notational simplicity, we omit the subscript $\mathbb{H}$ from the inner product and norm notations, and we are going to use them only if it is necessary to avoid confusion.

Definition 11.1. A Hilbert space $\mathbb{H}$ is called *reproducing kernel Hilbert space* (RKHS) if there exists a function

$$\kappa : \mathcal{X} \times \mathcal{X} \longmapsto \mathbb{R}$$

with the following properties:

- for every $\boldsymbol{x} \in \mathcal{X}$, $\kappa(\cdot, \boldsymbol{x})$ belongs to $\mathbb{H}$;
- $\kappa(\cdot, \cdot)$ has the so-called *reproducing property*, that is,

$$\boxed{f(\boldsymbol{x}) = \langle f, \kappa(\cdot, \boldsymbol{x}) \rangle, \ \forall f \in \mathbb{H}, \ \forall \boldsymbol{x} \in \mathcal{X} : \quad \text{reproducing property.}} \tag{11.8}$$

In words, the kernel function is a function of two arguments. Fixing the value of one of them to, say, $\boldsymbol{x} \in \mathcal{X}$, then the kernel becomes a function of single argument (associated with the $\cdot$) and this function belongs to $\mathbb{H}$. The reproducing property means that the value of any function $f \in \mathbb{H}$, at any $\boldsymbol{x} \in \mathcal{X}$, is equal to the respective inner product, performed in $\mathbb{H}$, between f and $\kappa(\cdot, \boldsymbol{x})$.

A direct consequence of the reproducing property, if we set $f(\cdot) = \kappa(\cdot, \boldsymbol{y})$, $\boldsymbol{y} \in \mathcal{X}$, is that

$$\langle \kappa(\cdot, \boldsymbol{y}), \kappa(\cdot, \boldsymbol{x}) \rangle = \kappa(\boldsymbol{x}, \boldsymbol{y}) = \kappa(\boldsymbol{y}, \boldsymbol{x}). \tag{11.9}$$

[2] Generalization to more general sets is also possible.

[3] For the unfamiliar reader, a Hilbert space is the generalization of Euclidean space allowing for infinite dimensions. More rigorous definitions and related properties are given in the appendix of Chapter 8.

Definition 11.2. Let $\mathbb{H}$ be an RKHS, associated with a kernel function $\kappa(\cdot,\cdot)$, and $\mathcal{X}$ a set of elements. Then the mapping

$$\boxed{\mathcal{X} \ni \boldsymbol{x} \longmapsto \phi(\boldsymbol{x}) := \kappa(\cdot, \boldsymbol{x}) \in \mathbb{H}: \quad \text{feature map}}$$

is known as *feature map* and the space $\mathbb{H}$ as the *feature space*.

In other words, if $\mathcal{X}$ is the set of our observation vectors, the feature mapping maps each vector to the RKHS $\mathbb{H}$. Note that, in general, $\mathbb{H}$ can be of infinite dimension and its elements are functions. That is, each training point is mapped to a function. In special cases, where $\mathbb{H}$ is of a finite dimension, K, the image can be represented as an equivalent vector $\boldsymbol{\phi}(\boldsymbol{x}) \in \mathbb{R}^K$. From now on, the general infinite-dimensional case will be treated and the images will be denoted as functions, $\phi(\cdot)$.

Let us now see what we have gained by choosing to perform the feature mapping from the original space to a high-dimensional RKHS one. Let $\boldsymbol{x}, \boldsymbol{y} \in \mathcal{X} \subseteq \mathbb{R}^l$. Then the inner product of the respective mapping images is written as

$$\langle \phi(\boldsymbol{x}), \phi(\boldsymbol{y}) \rangle = \langle \kappa(\cdot, \boldsymbol{x}), \kappa(\cdot, \boldsymbol{y}) \rangle,$$

or

$$\boxed{\langle \phi(\boldsymbol{x}), \phi(\boldsymbol{y}) \rangle = \kappa(\boldsymbol{x}, \boldsymbol{y}): \quad \text{kernel trick.}}$$

In other words, employing this type of mapping to our problem, we can perform inner product operations in $\mathbb{H}$ in a very efficient way; that is, via a function evaluation performed in the original low-dimensional space! This property is also known as the *kernel trick*, and it facilitates significantly the computations. As will become apparent soon, the way this property is exploited in practice involves the following steps:

1. Map (implicitly) the input training data to an RKHS

$$\boldsymbol{x}_n \longmapsto \phi(\boldsymbol{x}_n) \in \mathbb{H}, \quad n = 1, 2, \ldots, N.$$

2. Solve a *linear* estimation task in $\mathbb{H}$, involving the images $\phi(\boldsymbol{x}_n)$, $n = 1, 2, \ldots, N$.
3. Cast the algorithm that solves for the unknown parameters in terms of inner product operations in the form

$$\left\langle \phi(\boldsymbol{x}_i), \phi(\boldsymbol{x}_j) \right\rangle, \quad i, j = 1, 2, \ldots, N.$$

4. Replace each inner product by a kernel evaluation, that is,

$$\left\langle \phi(\boldsymbol{x}_i), \phi(\boldsymbol{x}_j) \right\rangle = \kappa(\boldsymbol{x}_i, \boldsymbol{x}_j).$$

It is apparent that one does not need to perform any explicit mapping of the data. All is needed is to perform the kernel operations at the final step. Note that the specific form of $\kappa(\cdot,\cdot)$ does not concern the analysis. Once the algorithm for the prediction, $\hat{y}$, has been derived, one can use different choices for $\kappa(\cdot,\cdot)$. As we will see, different choices for $\kappa(\cdot,\cdot)$ correspond to different types of nonlinearity. Fig. 11.6 illustrates the rationale behind the procedure. In practice, the four steps listed above are equivalent to (a) work in the original (low-dimensional Euclidean space) and express all operations in terms of inner products and (b) at the final step substitute the inner products with kernel evaluations.

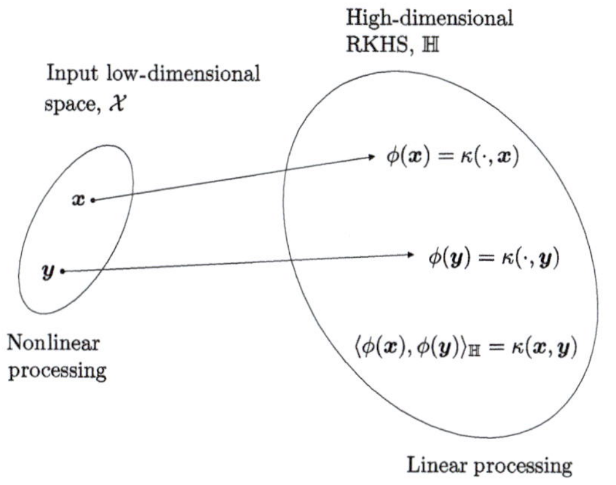

FIGURE 11.6

The nonlinear task in the original low-dimensional space is mapped to a linear one in the high-dimensional RKHS $\mathbb{H}$. Using feature mapping, inner product operations are efficiently performed via kernel evaluations in the original low-dimensional spaces.

Example 11.1. The goal of this example is to demonstrate that one can map the input space into another of higher dimension (finite-dimensional for this case[4]), where the corresponding inner product can be computed as a function performed in the lower-dimensional original one. Consider the case of the two-dimensional space and the mapping into a three-dimensional one, i.e.,

$$\mathbb{R}^2 \ni \boldsymbol{x} \longmapsto \boldsymbol{\phi}(\boldsymbol{x}) = [x_1^2, \ \sqrt{2}x_1x_2, \ x_2^2] \in \mathbb{R}^3.$$

Then, given two vectors $\boldsymbol{x} = [x_1, x_2]^T$ and $\boldsymbol{y} = [y_1, y_2]^T$, it is straightforward to see that

$$\boldsymbol{\phi}^T(\boldsymbol{x})\boldsymbol{\phi}(\boldsymbol{y}) = (\boldsymbol{x}^T\boldsymbol{y})^2.$$

That is, the inner product in the three-dimensional space, after the mapping, is given in terms of a function of the variables in the original space.

11.5.1 SOME PROPERTIES AND THEORETICAL HIGHLIGHTS

The reader who has no "mathematical anxieties" can bypass this subsection during a first reading.

Let $\mathcal{X}$ be a set of points. Typically $\mathcal{X}$ is a compact (closed and bounded) subset of $\mathbb{R}^l$. Consider a function

$$\kappa : \mathcal{X} \times \mathcal{X} \longmapsto \mathbb{R}.$$

[4] If a space of functions has finite dimension, then it is equivalent (isometrically isomorphic) to a finite Euclidean linear/vector space.

Definition 11.3. The function κ is called a *positive definite kernel* if

$$\boxed{\sum_{n=1}^{N}\sum_{m=1}^{N} a_n a_m \kappa(\boldsymbol{x}_n, \boldsymbol{x}_m) \geq 0: \quad \text{positive definite kernel,}} \tag{11.10}$$

for any real numbers, a_n, a_m, any points $\boldsymbol{x}_n, \boldsymbol{x}_m \in \mathcal{X}$, and any $N \in \mathbb{N}$.

Note that (11.10) can be written in an equivalent form. Define the so-called *kernel matrix* $\mathcal{K}$ of order N,

$$\mathcal{K} := \begin{bmatrix} \kappa(\boldsymbol{x}_1, \boldsymbol{x}_1) & \cdots & \kappa(\boldsymbol{x}_1, \boldsymbol{x}_N) \\ \vdots & \vdots & \vdots \\ \kappa(\boldsymbol{x}_N, \boldsymbol{x}_1) & \cdots & \kappa(\boldsymbol{x}_N, \boldsymbol{x}_N) \end{bmatrix}. \tag{11.11}$$

Then (11.10) is written as

$$\boldsymbol{a}^T \mathcal{K} \boldsymbol{a} \geq 0, \tag{11.12}$$

where

$$\boldsymbol{a} = [a_1, \ldots, a_N]^T.$$

Because (11.10) is true for any $\boldsymbol{a} \in \mathbb{R}^N$, (11.12) suggests that for a kernel to be positive definite, it suffices for the corresponding kernel matrix to be positive semidefinite.[5]

Lemma 11.1. *The reproducing kernel associated with an RKHS $\mathbb{H}$ is a positive definite kernel.*

The proof of the lemma is given in Problem 11.3. Note that the opposite is also true. It can be shown [82,106] that if $\kappa : \mathcal{X} \times \mathcal{X} \longmapsto \mathbb{R}$ is a positive definite kernel, then there exists an RKHS $\mathbb{H}$ of functions on $\mathcal{X}$ such that $\kappa(\cdot, \cdot)$ is a reproducing kernel of $\mathbb{H}$. This establishes the equivalence between reproducing and positive definite kernels. Historically, the theory of positive definite kernels was developed first in the context of integral equations by Mercer [76], and the connection to RKHS was developed later on (see, for example, [3]).

Lemma 11.2. *Let $\mathbb{H}$ be an RKHS on the set $\mathcal{X}$ with reproducing kernel $\kappa(\cdot, \cdot)$. Then the linear span of the function $\kappa(\cdot, \boldsymbol{x})$, $\boldsymbol{x} \in \mathcal{X}$, is dense in $\mathbb{H}$, that is,*

$$\mathbb{H} = \overline{span\{\kappa(\cdot, \boldsymbol{x}), \boldsymbol{x} \in \mathcal{X}\}}. \tag{11.13}$$

The proof of the lemma is given in Problem 11.4. The overbar denotes the closure of a set. In other words, $\mathbb{H}$ can be constructed by *all* possible linear combinations of the kernel function computed in $\mathcal{X}$, as well as the *limit points* of sequences of such combinations. Simply stated, $\mathbb{H}$ can be *fully generated from the knowledge* of $\kappa(\cdot, \cdot)$.

The interested reader can obtain more theoretical results concerning RKHSs from, for example, [64,83,87,103,106,108].

---

5 It may be slightly confusing that the definition of a positive definite kernel requires a positive semidefinite kernel matrix. However, this is what has been the accepted definition.

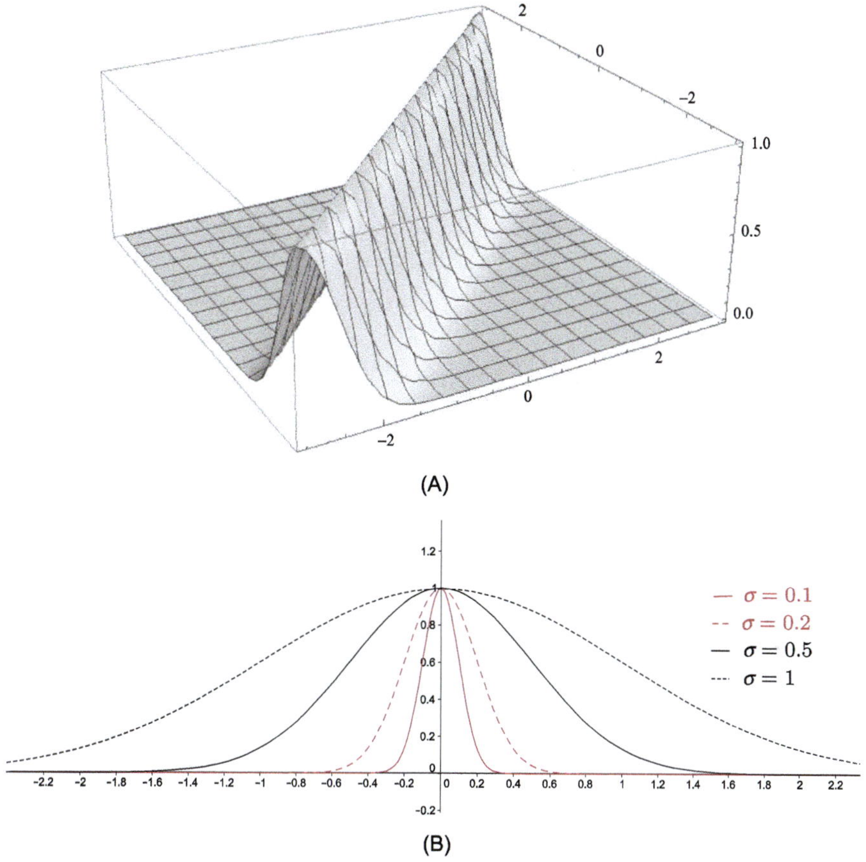

FIGURE 11.7

(A) The Gaussian kernel for $\mathcal{X} = \mathbb{R}$, $\sigma = 0.5$. (B) The function $\kappa(\cdot, 0)$ for different values of σ.

11.5.2 EXAMPLES OF KERNEL FUNCTIONS

In this subsection, we present some typical examples of kernel functions, which are commonly used in various applications.

- The *Gaussian* kernel is among the most popular ones, and it is given by our familiar form

$$\kappa(\boldsymbol{x}, \boldsymbol{y}) = \exp\left(-\frac{\|\boldsymbol{x} - \boldsymbol{y}\|^2}{2\sigma^2}\right),$$

with $\sigma > 0$ being a parameter. Fig. 11.7A shows the Gaussian kernel as a function of $x, y \in \mathcal{X} = \mathbb{R}$ and $\sigma = 0.5$. Fig. 11.7B shows the graph of resulting function if we set one of the kernel's arguments to 0, i.e., $\kappa(\cdot, 0)$, for various values of σ.

The dimension of the RKHS generated by the Gaussian kernel is *infinite*. A proof that the Gaussian kernel satisfies the required properties can be obtained, for example, from [108].

- The *homogeneous polynomial* kernel has the form

$$\kappa(\boldsymbol{x}, \boldsymbol{y}) = (\boldsymbol{x}^T \boldsymbol{y})^r,$$

where r is a parameter.

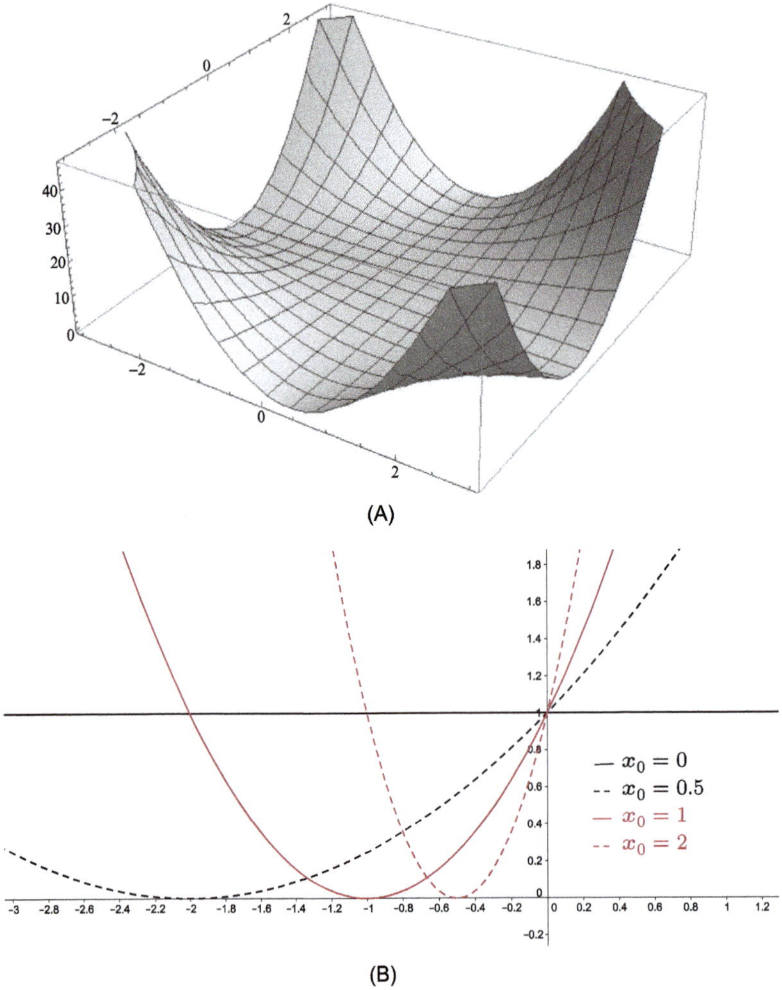

FIGURE 11.8

(A) The inhomogeneous polynomial kernel for $\mathcal{X} = \mathbb{R}$, $r = 2$. (B) The graph of $\kappa(\cdot, x_0)$ for different values of x_0.

- The *inhomogeneous* polynomial kernel is given by

$$\kappa(\boldsymbol{x}, \boldsymbol{y}) = (\boldsymbol{x}^T \boldsymbol{y} + c)^r,$$

where $c \geq 0$ and $r > 0$, $r \in \mathbb{N}$, are parameters. The graph of the kernel is given in Fig. 11.8A. In Fig. 11.8B the graphs of the function $\phi(\cdot, x_0)$ are shown for different values of x_0. The dimensionality of the RKHS associated with polynomial kernels is finite.

- The *Laplacian* kernel is given by

$$\kappa(\boldsymbol{x}, \boldsymbol{y}) = \exp\left(-t\|\boldsymbol{x} - \boldsymbol{y}\|\right),$$

where $t > 0$ is a parameter. The dimensionality of the RKHS associated with the Laplacian kernel is infinite.

- The *spline* kernels are defined as

$$\kappa(\boldsymbol{x}, \boldsymbol{y}) = B_{2p+1}(\|\boldsymbol{x} - \boldsymbol{y}\|^2),$$

where the B_n spline is defined via the $n+1$ convolutions of the unit interval $[-\frac{1}{2}, \frac{1}{2}]$, that is,

$$B_n(\cdot) := \bigotimes_{i=1}^{n+1} \chi_{[-\frac{1}{2},\frac{1}{2}]}(\cdot),$$

and $\chi_{[-\frac{1}{2},\frac{1}{2}]}(\cdot)$ is the characteristic function on the respective interval.[6]

- The *sampling function* or *sinc* kernel is of particular interest from a signal processing point of view. This kernel function is defined as

$$\operatorname{sinc}(x) = \frac{\sin(\pi x)}{\pi x}.$$

Recall that we have met this function in Chapter 9 while discussing sub-Nyquist sampling.
Let us now consider the set of all squared integrable functions, which are *bandlimited*, that is,

$$\mathcal{F}_B = \left\{ f : \int_{-\infty}^{+\infty} |f(x)|^2 dx < +\infty, \text{ and } |F(\omega)| = 0, \ |\omega| > \pi \right\},$$

where $F(\omega)$ is the respective Fourier transform

$$F(\omega) = \frac{1}{2\pi} \int_{-\infty}^{+\infty} f(x) e^{-j\omega x} dx.$$

It turns out that $\mathcal{F}_B$ is an RKHS whose reproducing kernel is the sinc function (e.g., [51]), that is,

$$\kappa(x, y) = \operatorname{sinc}(x - y).$$

[6] It is equal to one if the variable belongs to the interval and zero otherwise.

This takes us back to the classical sampling theorem through the RKHS route. Without going into details, a by-product of this view is the Shannon sampling theorem; any bandlimited function can be written as[7]

$$f(x) = \sum_n f(n)\,\text{sinc}(x-n). \tag{11.14}$$

Constructing Kernels

Besides the previous examples, one can construct more kernels by applying the following properties (Problem 11.6, [106]):

- If

$$\kappa_1(\boldsymbol{x}, \boldsymbol{y}) : \mathcal{X} \times \mathcal{X} \longmapsto \mathbb{R},$$
$$\kappa_2(\boldsymbol{x}, \boldsymbol{y}) : \mathcal{X} \times \mathcal{X} \longmapsto \mathbb{R}$$

  are kernels, then

$$\kappa(\boldsymbol{x}, \boldsymbol{y}) = \kappa_1(\boldsymbol{x}, \boldsymbol{y}) + \kappa_2(\boldsymbol{x}, \boldsymbol{y})$$

  and

$$\kappa(\boldsymbol{x}, \boldsymbol{y}) = \alpha\kappa_1(\boldsymbol{x}, \boldsymbol{y}),\ \alpha > 0,$$

  and

$$\kappa(\boldsymbol{x}, \boldsymbol{y}) = \kappa_1(\boldsymbol{x}, \boldsymbol{y})\kappa_2(\boldsymbol{x}, \boldsymbol{y})$$

  are also kernels.
- Let

$$f : \mathcal{X} \longmapsto \mathbb{R}.$$

  Then

$$\kappa(\boldsymbol{x}, \boldsymbol{y}) = f(\boldsymbol{x})f(\boldsymbol{y})$$

  is a kernel.
- Let a function

$$g : \mathcal{X} \longmapsto \mathbb{R}^l$$

  and a kernel function

$$\kappa_1(\cdot, \cdot) : \mathbb{R}^l \times \mathbb{R}^l \longmapsto \mathbb{R}.$$

  Then

$$\kappa(\boldsymbol{x}, \boldsymbol{y}) = \kappa_1\left(g(\boldsymbol{x}), g(\boldsymbol{y})\right)$$

[7] The key point behind the proof is that in $\mathcal{F}_B$, the kernel $\kappa(x, y)$ can be decomposed in terms of a set of orthogonal functions, that is, $\text{sinc}(x-n),\ n = 0, \pm 1, \pm 2, \ldots$.

is also a kernel.

- Let A be a positive definite $l \times l$ matrix. Then

$$\kappa(\boldsymbol{x}, \boldsymbol{y}) = \boldsymbol{x}^T A \boldsymbol{y}$$

is a kernel.

- If

$$\kappa_1(\boldsymbol{x}, \boldsymbol{y}) : \mathcal{X} \times \mathcal{X} \longmapsto \mathbb{R},$$

then

$$\kappa(\boldsymbol{x}, \boldsymbol{y}) = \exp\left(\kappa_1(\boldsymbol{x}, \boldsymbol{y})\right)$$

is also a kernel, and if $p(\cdot)$ is a polynomial with nonnegative coefficients,

$$\kappa(\boldsymbol{x}, \boldsymbol{y}) = p\left(\kappa_1(\boldsymbol{x}, \boldsymbol{y})\right)$$

is also a kernel.

The interested reader will find more information concerning kernels and their construction in, for example, [52,106,108].

String Kernels

So far, our discussion has been focused on input data that were vectors in a Euclidean space. However, as we have already pointed out, the input data need not necessarily be vectors, and they can be elements of more general sets.

Let us denote by $\mathcal{S}$ an alphabet set; that is, a set with a finite number of elements, which we call *symbols*. For example, this can be the set of all capital letters in the Latin alphabet. Bypassing the path of formal definitions, a *string* is a finite sequence of any length of symbols from $\mathcal{S}$. For example, two cases of strings are

$$T_1 = \text{“MYNAMEISSERGIOS”}, \quad T_2 = \text{“HERNAMEISDESPOINA”}.$$

In a number of applications, such as in text mining, spam filtering, text summarization, and bioinformatics, it is important to quantify how "similar" two strings are. However, kernels, by their definition, are similarity measures; they are constructed so as to express inner products in the high-dimensional feature space. An inner product is a similarity measure. Two vectors are most similar if they point to the same direction. Starting from this observation, there has been a lot of activity on defining kernels that measure similarity between strings. Without going into details, let us give such an example.

Let us denote by $\mathcal{S}^*$ the set of all possible strings that can be constructed using symbols from $\mathcal{S}$. Also, a string s is said to be a *substring* of x if $x = bsa$, where a and b are other strings (possibly empty) from the symbols of $\mathcal{S}$. Given two strings $x, y \in \mathcal{S}^*$, define

$$\kappa(x, y) := \sum_{s \in \mathcal{S}*} w_s \phi_s(x) \phi_s(y), \tag{11.15}$$

where $w_s \geq 0$, and $\phi_s(x)$ is the number of times substring s appears in x. It turns out that this is indeed a kernel, in the sense that it complies with (11.10); such kernels constructed from strings are known as *string kernels*.

Obviously, a number of different variants of this kernel are available. The so-called *k-spectrum* kernel considers common substrings only of length k. For example, for the two strings given before, the value of the 6-spectrum string kernel in (11.15) is equal to one (one common substring of length 6 is identified and appears once in each one of the two strings: "NAMEIS"). The interested reader can obtain more on this topic from, for example, [106]. We will use the notion of the string kernel in the case study in Section 11.15.

11.6 REPRESENTER THEOREM

The theorem to be stated in this section is of major importance from a practical point of view. It allows us to perform *empirical* loss function optimization, based on a finite set of training samples, in a very efficient way even if the function to be estimated belongs to a very high (even infinite)-dimensional RKHS $\mathbb{H}$.

Theorem 11.2. *Let*

$$\Omega : [0, +\infty) \longmapsto \mathbb{R}$$

be an arbitrary strictly monotonic increasing function. Let also

$$\mathcal{L} : \mathbb{R}^2 \longmapsto \mathbb{R} \cup \{\infty\}$$

be an arbitrary loss function. Then each minimizer $f \in \mathbb{H}$ of the regularized minimization task

$$\min_{f \in \mathbb{H}} J(f) := \sum_{n=1}^{N} \mathcal{L}\big(y_n, f(\boldsymbol{x}_n)\big) + \lambda\Omega\big(\|f\|^2\big) \tag{11.16}$$

admits a representation of the form[8]

$$\boxed{f(\cdot) = \sum_{n=1}^{N} \theta_n \kappa(\cdot, \boldsymbol{x}_n),} \tag{11.17}$$

where $\theta_n \in \mathbb{R}$, $n = 1, 2, \ldots, N$.

[8] The property holds also for regularization of the form $\Omega(\|f\|)$, since the quadratic function is strictly monotonic on $[0, \infty)$, and the proof follows a similar line.

Proof. The linear span, $A := \text{span}\{\kappa(\cdot, \boldsymbol{x}_1), \dots, \kappa(\cdot, \boldsymbol{x}_N)\}$, forms a closed subspace. Then each $f \in \mathbb{H}$ can be decomposed into two parts (see (8.20)), i.e.,

$$f(\cdot) = \sum_{n=1}^{N} \theta_n \kappa(\cdot, \boldsymbol{x}_n) + f_\perp,$$

where $f_\perp$ is the part of f that is orthogonal to A. From the reproducing property, we obtain

$$\begin{aligned} f(\boldsymbol{x}_m) = \langle f, \kappa(\cdot, \boldsymbol{x}_m) \rangle &= \left\langle \sum_{n=1}^{N} \theta_n \kappa(\cdot, \boldsymbol{x}_n), \kappa(\cdot, \boldsymbol{x}_m) \right\rangle \\ &= \sum_{n=1}^{N} \theta_n \kappa(\boldsymbol{x}_m, \boldsymbol{x}_n), \end{aligned}$$

where we used the fact that $\langle f_\perp, \kappa(\cdot, \boldsymbol{x}_n) \rangle = 0, \ n = 1, 2, \dots, N$. In other words, the expansion in (11.17) guarantees that at the training points, the value of f does not depend on $f_\perp$. Hence, the first term in (11.16), corresponding to the empirical loss, does *not* depend on $f_\perp$. Moreover, for all $f_\perp$ we have

$$\begin{aligned} \Omega(\|f\|^2) &= \Omega\left(\left\| \sum_{n=1}^{N} \theta_n \kappa(\cdot, \boldsymbol{x}_n) \right\|^2 + \|f_\perp\|^2 \right) \\ &\geq \Omega\left(\left\| \sum_{n=1}^{N} \theta_n \kappa(\cdot, \boldsymbol{x}_n) \right\|^2 \right). \end{aligned}$$

Thus, for *any* choice of $\theta_n, n = 1, 2, \dots, N$, the cost function in (11.16) is minimized for $f_\perp = 0$. Thus, the claim is proved. □

The theorem was first shown in [61]. In [2], the conditions under which the theorem exists were investigated and related sufficient and necessary conditions were derived. The importance of this theorem is that in order to optimize (11.16) with respect to f, one uses the expansion in (11.17) and minimization is carried out with respect to the *finite* set of parameters $\theta_n, n = 1, 2, \dots, N$.

Note that when working in high/infinite-dimensional spaces, the presence of a regularizer can hardly be avoided; otherwise, the obtained solution will suffer from overfitting, as only a finite number of data samples are used for training. The effect of regularization on the generalization performance and stability of the associated solution has been studied in a number of classical papers (for example, [18,39,79]).

Usually, a bias term is often added and it is assumed that the minimizing function admits the following representation:

$$\tilde{f} = f + b, \tag{11.18}$$

$$f(\cdot) = \sum_{n=1}^{N} \theta_n \kappa(\cdot, \boldsymbol{x}_n). \tag{11.19}$$

In practice, the use of a bias term (which does not enter in the regularization) turns out to improve performance. First, it enlarges the class of functions in which the solution is searched and potentially leads to better performance. Moreover, due to the penalization imposed by the regularizing term, $\Omega(\|f\|^2)$, the minimizer pushes the values which the function takes at the training points to smaller values. The existence of b tries to "absorb" some of this action (see, for example, [108]).

Remarks 11.2.

- We will use the expansion in (11.17) in a number of cases. However, it is interesting to apply this expansion to the RKHS of the bandlimited functions and see what comes out. Assume that the available samples from a function f are $f(n),\ n = 1, 2, \dots, N$ (assuming the case of normalized sampling period $x_s = 1$). Then according to the representer theorem, we can write the following approximation:

$$f(x) \approx \sum_{n=1}^{N} \theta_n \operatorname{sinc}(x - n). \tag{11.20}$$

  Taking into account the orthonormality of the $\operatorname{sinc}(\cdot - n)$ functions, we get $\theta_n = f(n),\ n = 1, 2, \dots, N$. However, note that in contrast to (11.14), which is exact, (11.20) is only an approximation. On the other hand, (11.20) can be used even if the obtained samples are contaminated by noise.

11.6.1 SEMIPARAMETRIC REPRESENTER THEOREM

The use of the bias term is also theoretically justified by the generalization of the representer theorem [103]. The essence of this theorem is to expand the solution into two parts, i.e., one that lies in an RKHS $\mathbb{H}$, and another that is given as a linear combination of a set of preselected functions.

Theorem 11.3. *Let us assume that in addition to the assumptions adopted in Theorem 11.2, we are given the set of real-valued functions*

$$\psi_m : \mathcal{X} \longmapsto \mathbb{R}, \quad m = 1, 2, \dots, M,$$

with the property that the $N \times M$ *matrix with elements* $\psi_m(\boldsymbol{x}_n),\ n = 1, 2, \dots, N,\ m = 1, 2, \dots, M$, *has rank* M. *Then any*

$$\tilde{f} = f + h, \ f \in \mathbb{H}, \quad h \in span\{\psi_m, m = 1, 2, \dots, M\},$$

solving the minimization task

$$\min_{\tilde{f}} J(\tilde{f}) := \sum_{n=1}^{N} \mathcal{L}\big(y_n, \tilde{f}(\boldsymbol{x}_n)\big) + \Omega(\|f\|^2) \tag{11.21}$$

admits the following representation:

$$\boxed{\tilde{f}(\cdot) = \sum_{n=1}^{N} \theta_n \kappa(\cdot, \boldsymbol{x}_n) + \sum_{m=1}^{M} b_m \psi_m(\cdot).} \tag{11.22}$$

Obviously, the use of a bias term is a special case of the expansion above. An example of successful application of this theorem was demonstrated in [13] in the context of image denoising. A set of nonlinear functions in place of ψ_m were used to account for the edges (nonsmooth jumps) in an image. The part of f lying in the RKHS accounted for the smooth parts in the image.

11.6.2 NONPARAMETRIC MODELING: A DISCUSSION

Note that searching a model function in an RKHS space is a typical task of *nonparametric* modeling. In contrast to the parametric modeling in Eq. (11.1), where the unknown function is *parameterized* in terms of a set of basis functions, the minimization in (11.16) or (11.21) is performed with regard to functions that are *constrained to belong in a specific space*. In the more general case, minimization could be performed with regard to any (continuous) function, for example,

$$\min_f \sum_{n=1}^{N} \mathcal{L}\big(y_n, f(\boldsymbol{x}_n)\big) + \lambda\phi(f),$$

where $\mathcal{L}(\cdot,\cdot)$ can be any loss function and ϕ an appropriately chosen regularizing functional. Note, however, that in this case, the presence of the regularization is crucial. If there is no regularization, then any function that *interpolates* the data is a solution; such techniques have also been used in interpolation theory (for example, [78,90]). The regularization term, $\phi(f)$, helps to smoothen out the function to be recovered. To this end, functions of derivatives have been employed. For example, if the minimization cost is chosen as

$$\sum_{n=1}^{N} (y_n - f(x_n))^2 + \lambda \int (f''(x))^2 \, dx,$$

then the solution is a cubic spline; that is, a piecewise cubic function with knots the points x_n, $n = 1, 2, \ldots, N$, and it is continuously differentiable to the second order. The choice of λ controls the degree of smoothness of the approximating function; the larger its value, the smoother the minimizer becomes.

If, on the other hand, f is constrained to lie in an RKHS and the minimizing task is as in (11.16), then the resulting function is of the form given in (11.17), where a kernel function is placed at each input training point. It must be pointed out that the parametric form that now results was not in our original intentions. It came out as a by-product of the theory. However, it should be stressed that, in contrast to the parametric methods, now the number of parameters to be estimated is not fixed but it depends on the number of the training points. Recall that this is an important difference and it was carefully pointed out when parametric methods were introduced and defined in Chapter 3.

11.7 KERNEL RIDGE REGRESSION

Ridge regression was introduced in Chapter 3 and it has also been treated in more detail in Chapter 6. Here, we will state the task in a general RKHS. The path to be followed is the typical one used to extend techniques which have been developed for linear models to the more general RKHS.

We assume that the generation mechanism of the data, represented by the training set $(y_n, \boldsymbol{x}_n) \in \mathbb{R} \times \mathbb{R}^l$, is modeled via a nonlinear regression task

$$y_n = g(\boldsymbol{x}_n) + \eta_n, \quad n = 1, 2, \ldots, N. \tag{11.23}$$

Let us denote by f the estimate of the unknown g. Sometimes, f is called the *hypothesis* and the space $\mathbb{H}$ in which f is searched is known as the *hypothesis space*. We will further assume that f lies in an RKHS, associated with a kernel

$$\kappa : \mathbb{R}^l \times \mathbb{R}^l \longmapsto \mathbb{R}.$$

Motivated by the representer theorem, we adopt the following expansion:

$$f(\boldsymbol{x}) = \sum_{n=1}^{N} \theta_n \kappa(\boldsymbol{x}, \boldsymbol{x}_n).$$

According to the kernel ridge regression approach, the unknown coefficients are estimated by the following task:

$$\begin{aligned} \hat{\boldsymbol{\theta}} &= \arg\min_{\boldsymbol{\theta}} J(\boldsymbol{\theta}), \\ J(\boldsymbol{\theta}) &:= \sum_{n=1}^{N} \left(y_n - \sum_{m=1}^{N} \theta_m \kappa(\boldsymbol{x}_n, \boldsymbol{x}_m) \right)^2 + C\langle f, f \rangle, \end{aligned} \tag{11.24}$$

where C is the regularization parameter.[9] Eq. (11.24) can be rewritten as (Problem 11.7)

$$J(\boldsymbol{\theta}) = (\boldsymbol{y} - \mathcal{K}\boldsymbol{\theta})^T (\boldsymbol{y} - \mathcal{K}\boldsymbol{\theta}) + C\boldsymbol{\theta}^T \mathcal{K}^T \boldsymbol{\theta}, \tag{11.25}$$

where

$$\boldsymbol{y} = [y_1, \ldots, y_N]^T, \quad \boldsymbol{\theta} = [\theta_1, \ldots, \theta_N]^T,$$

and $\mathcal{K}$ is the kernel matrix defined in (11.11); the latter is fully determined by the kernel function and the training points. Following our familiar-by-now arguments, minimization of $J(\boldsymbol{\theta})$ with regard to $\boldsymbol{\theta}$ leads to

$$(\mathcal{K}^T \mathcal{K} + C\mathcal{K}^T)\hat{\boldsymbol{\theta}} = \mathcal{K}^T \boldsymbol{y}$$

or

$$\boxed{(\mathcal{K} + CI)\hat{\boldsymbol{\theta}} = \boldsymbol{y} : \quad \text{kernel ridge regression},} \tag{11.26}$$

where $\mathcal{K}^T = \mathcal{K}$ has been assumed to be invertible.[10] Once $\hat{\boldsymbol{\theta}}$ has been obtained, given an unknown

[9] For the needs of this chapter, we denote the regularization constant as C, not to be confused with the Lagrange multipliers, to be introduced soon.

[10] This is true, for example, for the Gaussian kernel [103].

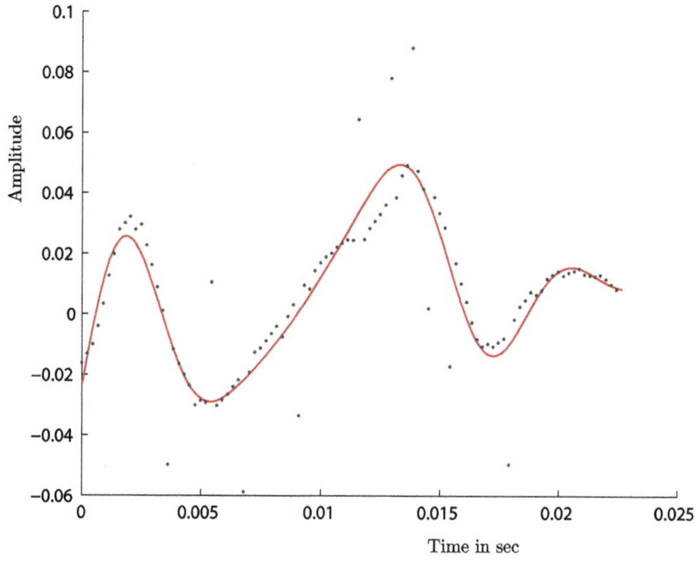

FIGURE 11.9

Plot of the data used for training together with the fitted (prediction) curve obtained via the kernel ridge regression, for Example 11.2. The Gaussian kernel was used.

vector $\boldsymbol{x} \in \mathbb{R}^l$, the corresponding prediction value of the dependent variable is given by

$$\hat{y} = \sum_{n=1}^{N} \hat{\theta}_n \kappa(\boldsymbol{x}, \boldsymbol{x}_n) = \hat{\boldsymbol{\theta}}^T \boldsymbol{\kappa}(\boldsymbol{x}),$$

where

$$\boldsymbol{\kappa}(\boldsymbol{x}) = [\kappa(\boldsymbol{x}, \boldsymbol{x}_1), \ldots, \kappa(\boldsymbol{x}, \boldsymbol{x}_N)]^T.$$

Employing (11.26), we obtain

$$\boxed{\hat{y}(\boldsymbol{x}) = \boldsymbol{y}^T (\mathcal{K} + CI)^{-1} \boldsymbol{\kappa}(\boldsymbol{x}).} \tag{11.27}$$

Example 11.2. In this example, the prediction power of the kernel ridge regression in the presence of noise and outliers will be tested. The original data were samples from a music recording of *Blade Runner* by Vangelis Papathanasiou. A white Gaussian noise was then added at a 15 dB level and a number of outliers were intentionally randomly introduced and "hit" some of the values (10% of them). The kernel ridge regression method was used, employing the Gaussian kernel with $\sigma = 0.004$. We allowed for a bias term to be present (see Problem 11.8). The prediction (fitted) curve, $\hat{y}(x)$, for various values of x is shown in Fig. 11.9 together with the (noisy) data used for training.

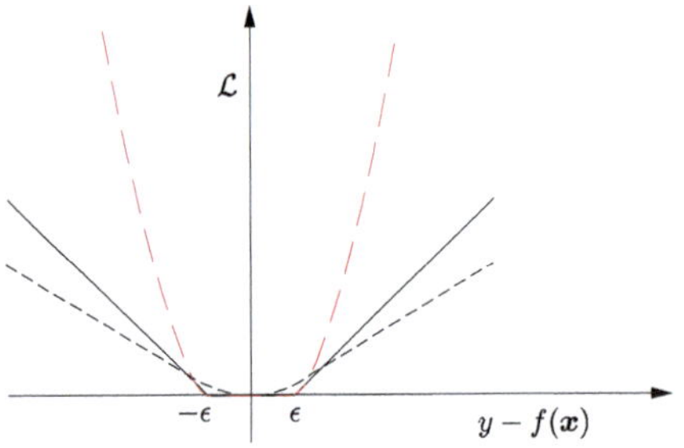

FIGURE 11.10

The Huber loss function (dotted-gray), the linear ϵ-insensitive (full-gray), and the quadratic ϵ-insensitive (red) loss functions, for $\epsilon = 0.7$.

11.8 SUPPORT VECTOR REGRESSION

The squared error loss function, associated with the LS method, is not always the best criterion for optimization, in spite of its merits. In the case of the presence of a non-Gaussian noise with long tails (i.e., high probability for relatively large values) and, hence, with an increased number of noise *outliers*, the square dependence on the error of the squared error criterion gets biased toward values associated with the presence of outliers. Recall from Chapter 3 that the LS method is equivalent to the maximum likelihood estimation under the assumption of white Gaussian noise. Moreover, under this assumption, the LS estimator achieves the Cramér–Rao bound and it becomes a minimum variance estimator. However, under other noise scenarios, one has to look for alternative criteria.

The task of optimization in the presence of outliers was studied by Huber [54], whose goal was to obtain a strategy for choosing the loss function that "matches" the noise model best. He proved that under the assumption that the noise has a symmetric probability density function (PDF), the optimal minimax strategy[11] for regression is obtained via the following loss function:

$$\mathcal{L}\big(y, f(\boldsymbol{x})\big) = |y - f(\boldsymbol{x})|,$$

which gives rise to the *least-modulus* method. Note from Section 5.8 that the stochastic gradient online version for this loss function leads to the sign-error LMS. Huber also showed that if the noise comprises two components, one corresponding to a Gaussian and another to an arbitrary PDF (which remains symmetric), then the best in the minimax sense loss function is given by

$$\mathcal{L}\big(y, f(\boldsymbol{x})\big) = \begin{cases} \epsilon|y - f(\boldsymbol{x})| - \frac{\epsilon^2}{2}, & \text{if } \ |y - f(\boldsymbol{x})| > \epsilon, \\ \frac{1}{2}|y - f(\boldsymbol{x})|^2, & \text{if } \ |y - f(\boldsymbol{x})| \leq \epsilon, \end{cases}$$

for some parameter ϵ. This is known as the Huber loss function, and it is shown in Fig. 11.10. A loss function that can approximate the Huber one and, as we will see, turns out to have some nice compu-

[11] The best L_2 approximation of the worst noise model. Note that this is a pessimistic scenario, because the only information that is used is the symmetry of the noise PDF.

tational properties, is the so-called *linear* ϵ-insensitive loss function, defined as (see also Chapter 8)

$$\mathcal{L}(y, f(\boldsymbol{x})) = \begin{cases} |y - f(\boldsymbol{x})| - \epsilon, & \text{if } |y - f(\boldsymbol{x})| > \epsilon, \\ 0, & \text{if } |y - f(\boldsymbol{x})| \leq \epsilon, \end{cases} \tag{11.28}$$

shown in Fig. 11.10. Note that for $\epsilon = 0$, it coincides with the absolute value loss function, and it is close to the Huber loss for small values of $\epsilon < 1$. Another version is the *quadratic* ϵ-insensitive, defined as

$$\mathcal{L}(y, f(\boldsymbol{x})) = \begin{cases} |y - f(\boldsymbol{x}))|^2 - \epsilon, & \text{if } |y - f(\boldsymbol{x})| > \epsilon, \\ 0, & \text{if } |y - f(\boldsymbol{x})| \leq \epsilon, \end{cases} \tag{11.29}$$

which coincides with the squared error loss for $\epsilon = 0$. The corresponding graph is given in Fig. 11.10. Observe that the two previously discussed ϵ-insensitive loss functions retain their convex nature; however, they are no more differentiable at all points.

11.8.1 THE LINEAR ϵ-INSENSITIVE OPTIMAL REGRESSION

Let us now adopt (11.28) as the loss function to quantify model misfit. We will treat the regression task in (11.23), employing a linear model for f, that is,

$$f(\boldsymbol{x}) = \boldsymbol{\theta}^T \boldsymbol{x} + \theta_0.$$

Once we obtain the solution expressed in inner product operations, the more general solution for the case where f lies in an RKHS will be obtained via the kernel trick; that is, inner products will be replaced by kernel evaluations.

Let us now introduce two sets of *auxiliary* variables. If

$$y_n - \boldsymbol{\theta}^T \boldsymbol{x}_n - \theta_0 \geq \epsilon,$$

define $\tilde{\xi}_n \geq 0$, such that

$$y_n - \boldsymbol{\theta}^T \boldsymbol{x}_n - \theta_0 \leq \epsilon + \tilde{\xi}_n.$$

Note that ideally, we would like to select $\boldsymbol{\theta}$, θ_0, so that $\tilde{\xi}_n = 0$, because this would make the contribution of the respective term in the loss function equal to zero. Also, if

$$y_n - \boldsymbol{\theta}^T \boldsymbol{x}_n - \theta_0 \leq -\epsilon,$$

define $\xi_n \geq 0$, such that

$$\boldsymbol{\theta}^T \boldsymbol{x}_n + \theta_0 - y_n \leq \epsilon + \xi_n.$$

Once more, we would like to select our unknown set of parameters so that ξ_n is zero.

We are now ready to formulate the minimizing task around the corresponding empirical cost, regularized by the norm of $\boldsymbol{\theta}$, which is cast in terms of the auxiliary variables as[12]

$$\text{minimize} \quad J(\boldsymbol{\theta}, \theta_0, \boldsymbol{\xi}, \tilde{\boldsymbol{\xi}}) = \frac{1}{2}\|\boldsymbol{\theta}\|^2 + C\left(\sum_{n=1}^{N} \xi_n + \sum_{n=1}^{N} \tilde{\xi}_n\right), \tag{11.30}$$

$$\text{subject to} \quad y_n - \boldsymbol{\theta}^T \boldsymbol{x}_n - \theta_0 \leq \epsilon + \tilde{\xi}_n, \; n = 1, 2, \ldots, N, \tag{11.31}$$

$$\boldsymbol{\theta}^T \boldsymbol{x}_n + \theta_0 - y_n \leq \epsilon + \xi_n, \; n = 1, 2, \ldots, N, \tag{11.32}$$

$$\tilde{\xi}_n \geq 0, \; \xi_n \geq 0, \; n = 1, 2, \ldots, N. \tag{11.33}$$

Before we proceed further some explanations are in order.

- The auxiliary variables $\tilde{\xi}_n$ and ξ_n, $n = 1, 2, \ldots, N$, which measure the excess error with regard to ϵ, are known as *slack variables*. Note that according to the ϵ-insensitive rationale, any contribution to the cost function of an error with absolute value less than or equal to ϵ is zero. The previous optimization task attempts to estimate $\boldsymbol{\theta}$, θ_0 so that the contribution of error values larger than ϵ and smaller than $-\epsilon$ is minimized. Thus, the optimization task in (11.30)–(11.33) is equivalent to minimizing the empirical loss function

$$\frac{1}{2}||\boldsymbol{\theta}||^2 + C\sum_{n=1}^{N} \mathcal{L}\left(y_n, \boldsymbol{\theta}^T \boldsymbol{x}_n + \theta_0\right),$$

  where the loss function is the linear ϵ-insensitive one. Note that any other method for minimizing (nondifferentiable) convex functions could be used (for example, Chapter 8). However, the constrained optimization involving the slack variables has a historical value and it was the path that paved the way in employing the kernel trick, as we will see soon.

The Solution

The solution of the optimization task is obtained by introducing Lagrange multipliers and forming the corresponding Lagrangian (see below for the detailed derivation). Having obtained the Lagrange multipliers, the solution turns out to be given in a simple and rather elegant form,

$$\hat{\boldsymbol{\theta}} = \sum_{n=1}^{N} (\tilde{\lambda}_n - \lambda_n)\boldsymbol{x}_n,$$

where $\tilde{\lambda}_n, \lambda_n$, $n = 1, 2, \ldots, N$, are the Lagrange multipliers associated with each one of the constraints in (11.31) and (11.32). It turns out that the Lagrange multipliers are nonzero *only* for those points $\boldsymbol{x}_n$ that correspond to error values *either equal to or larger than* ϵ. These are known as *support vectors*. Points that score error values *less* than ϵ correspond to zero Lagrange multipliers and do not participate

[12] It is common in the literature to formulate the regularized cost via the parameter C multiplying the loss term and not $||\boldsymbol{\theta}||^2$. In any case, they are both equivalent.

in the formation of the solution. The bias term can be obtained by anyone from the set of equations

$$y_n - \boldsymbol{\theta}^T \boldsymbol{x}_n - \theta_0 = \epsilon, \tag{11.34}$$

$$\boldsymbol{\theta}^T \boldsymbol{x}_n + \theta_0 - y_n = \epsilon, \tag{11.35}$$

where n above runs over the points that are associated with $\tilde{\lambda}_n > 0$ ($\lambda_n > 0$) *and* $\tilde{\xi}_n = 0$ ($\xi_n = 0$) (note that these points form a subset of the support vectors). In practice, $\hat{\theta}_0$ is obtained as the average from all the previous equations.

For the more general setting of solving a linear task in an RKHS, the solution is of the same form as the one obtained before All one needs to do is to replace $\boldsymbol{x}_n$ with the respective images, $\kappa(\cdot, \boldsymbol{x}_n)$, and the vector $\hat{\boldsymbol{\theta}}$ by a function, $\hat{\theta}$, i.e.,

$$\hat{\theta}(\cdot) = \sum_{n=1}^{N} (\tilde{\lambda}_n - \lambda_n) \kappa(\cdot, \boldsymbol{x}_n).$$

Once $\hat{\theta}, \hat{\theta}_0$ have been obtained, we are ready to perform prediction. Given a value $\boldsymbol{x}$, we first perform the (implicit) mapping using the feature map

$$\boldsymbol{x} \longmapsto \kappa(\cdot, \boldsymbol{x}),$$

and we get

$$\hat{y}(\boldsymbol{x}) = \left\langle \hat{\theta}, \kappa(\cdot, \boldsymbol{x}) \right\rangle + \hat{\theta}_0,$$

or

$$\boxed{\hat{y}(\boldsymbol{x}) = \sum_{n=1}^{N_s} (\tilde{\lambda}_n - \lambda_n) \kappa(\boldsymbol{x}, \boldsymbol{x}_n) + \hat{\theta}_0 : \quad \text{SVR prediction,}} \tag{11.36}$$

where $N_s \leq N$ is the number of nonzero Lagrange multipliers. Observe that (11.36) is an expansion in terms of nonlinear (kernel) functions. Moreover, as only a fraction of the points is involved (N_s), the use of the ϵ-insensitive loss function achieves a form of *sparsification* on the general expansion dictated by the representer theorem in (11.17).

Solving the Optimization Task

The reader who is not interested in proofs can bypass this part in a first reading.

The task in (11.30)–(11.33) is a *convex programming* minimization, with a set of linear inequality constraints. As discussed in Appendix C, a minimizer has to satisfy the following *Karush–Kuhn–Tucker* (KKT) conditions:

$$\frac{\partial L}{\partial \boldsymbol{\theta}} = \boldsymbol{0}, \quad \frac{\partial L}{\partial \theta_0} = 0, \quad \frac{\partial L}{\partial \tilde{\xi}_n} = 0, \quad \frac{\partial L}{\partial \xi_n} = 0, \tag{11.37}$$

$$\tilde{\lambda}_n (y_n - \boldsymbol{\theta}^T \boldsymbol{x}_n - \theta_0 - \epsilon - \tilde{\xi}_n) = 0, \; n = 1, 2, \ldots, N, \tag{11.38}$$

$$\lambda_n (\boldsymbol{\theta}^T \boldsymbol{x}_n + \theta_0 - y_n - \epsilon - \xi_n) = 0, \; n = 1, 2, \ldots, N, \tag{11.39}$$

$$\tilde{\mu}_n \tilde{\xi}_n = 0, \ \mu_n \xi_n = 0, \ n = 1, 2, \ldots, N, \tag{11.40}$$

$$\tilde{\lambda}_n \geq 0, \ \lambda_n \geq 0, \ \tilde{\mu}_n \geq 0, \ \mu_n \geq 0, \ n = 1, 2, \ldots, N, \tag{11.41}$$

where L is the respective Lagrangian,

$$\begin{aligned} L(\boldsymbol{\theta}, \theta_0, \tilde{\boldsymbol{\xi}}, \boldsymbol{\xi}, \boldsymbol{\lambda}, \boldsymbol{\mu}) = & \frac{1}{2}\|\boldsymbol{\theta}\|^2 + C\left(\sum_{n=1}^{N} \xi_n + \sum_{n=1}^{N} \tilde{\xi}_n\right) \\ & + \sum_{n=1}^{N} \tilde{\lambda}_n (y_n - \boldsymbol{\theta}^T \boldsymbol{x}_n - \theta_0 - \epsilon - \tilde{\xi}_n) \\ & + \sum_{n=1}^{N} \lambda_n (\boldsymbol{\theta}^T \boldsymbol{x}_n + \theta_0 - y_n - \epsilon - \xi_n) \\ & - \sum_{n=1}^{N} \tilde{\mu}_n \tilde{\xi}_n - \sum_{n=1}^{N} \mu_n \xi_n, \end{aligned} \tag{11.42}$$

where $\tilde{\lambda}_n, \lambda_n, \tilde{\mu}_n, \mu_n$ are the corresponding Lagrange multipliers. A close observation of (11.38) and (11.39) reveals that (why?)

$$\tilde{\xi}_n \xi_n = 0, \ \tilde{\lambda}_n \lambda_n = 0, \quad n = 1, 2, \ldots, N. \tag{11.43}$$

Taking the derivatives of the Lagrangian in (11.37) and equating to zero results in

$$\frac{\partial L}{\partial \boldsymbol{\theta}} = \mathbf{0} \longrightarrow \hat{\boldsymbol{\theta}} = \sum_{n=1}^{N} (\tilde{\lambda}_n - \lambda_n) \boldsymbol{x}_n, \tag{11.44}$$

$$\frac{\partial L}{\partial \theta_0} = 0 \longrightarrow \sum_{n=1}^{N} \tilde{\lambda}_n = \sum_{n=1}^{N} \lambda_n, \tag{11.45}$$

$$\frac{\partial L}{\partial \tilde{\xi}_n} = 0 \longrightarrow C - \tilde{\lambda}_n - \tilde{\mu}_n = 0, \tag{11.46}$$

$$\frac{\partial L}{\partial \xi_n} = 0 \longrightarrow C - \lambda_n - \mu_n = 0. \tag{11.47}$$

Note that all one needs in order to obtain $\hat{\boldsymbol{\theta}}$ are the values of the Lagrange multipliers. As discussed in Appendix C, these can be obtained by writing the problem in its dual representation form, that is,

$$\begin{aligned} \text{maximize with respect to } \boldsymbol{\lambda}, \tilde{\boldsymbol{\lambda}} \quad & \sum_{n=1}^{N} (\tilde{\lambda}_n - \lambda_n) y_n - \epsilon(\tilde{\lambda}_n + \lambda_n) \\ & -\frac{1}{2} \sum_{n=1}^{N} \sum_{m=1}^{N} (\tilde{\lambda}_n - \lambda_n)(\tilde{\lambda}_m - \lambda_m) \boldsymbol{x}_n^T \boldsymbol{x}_m, \end{aligned} \tag{11.48}$$

$$\text{subject to} \qquad 0 \le \tilde{\lambda}_n \le C, \ 0 \le \lambda_n \le C, \ n = 1, 2, \ldots, N, \tag{11.49}$$

$$\sum_{n=1}^{N} \tilde{\lambda}_n = \sum_{n=1}^{N} \lambda_n. \tag{11.50}$$

Concerning the maximization task in (11.48)–(11.50), the following comments are in order.

- Plugging into the Lagrangian the estimate obtained in (11.44) and following the steps as required by the dual representation form (Problem 11.10), (11.48) results.
- From (11.46) and (11.47), taking into account that $\mu_n \ge 0$, $\tilde{\mu}_n \ge 0$, (11.49) results.
- The beauty of the dual representation form is that it involves the observation vectors in the form of inner product operations. Thus, when the task is solved in an RKHS, (11.48) becomes

$$\begin{aligned} \text{maximize with respect to } \boldsymbol{\lambda}, \tilde{\boldsymbol{\lambda}} \qquad & \sum_{n=1}^{N} (\tilde{\lambda}_n - \lambda_n) y_n - \epsilon(\tilde{\lambda}_n + \lambda_n) \\ & -\frac{1}{2} \sum_{n=1}^{N} \sum_{m=1}^{N} (\tilde{\lambda}_n - \lambda_n)(\tilde{\lambda}_m - \lambda_m) \kappa(\boldsymbol{x}_n, \boldsymbol{x}_m). \end{aligned}$$

- The KKT conditions convey important information. The Lagrange multipliers, $\tilde{\lambda}_n, \lambda_n$, for points that score error less than ϵ, that is,

$$|\boldsymbol{\theta}^T \boldsymbol{x}_n + \theta_0 - y_n| < \epsilon,$$

  are zero. This is a direct consequence of (11.38) and (11.39) and the fact that $\tilde{\xi}_n, \xi_n \ge 0$. Thus, the Lagrange multipliers are *nonzero* only for points which score error either equal to ϵ ($\tilde{\xi}_n, \xi_n = 0$) or larger values ($\tilde{\xi}_n, \xi_n > 0$). In other words, only the points with nonzero Lagrange multipliers (support vectors) enter in (11.44), which leads to a *sparsification* of the expansion in (11.44).
- Due to (11.43), either $\tilde{\xi}_n$ or ξ_n can be nonzero, but not both of them. This also applies to the corresponding Lagrange multipliers.
- Note that if $\tilde{\xi}_n > 0$ (or $\xi_n > 0$), then from (11.40), (11.46), and (11.47) we obtain

$$\tilde{\lambda}_n = C \ \text{ or } \ \lambda_n = C.$$

  That is, the respective Lagrange multipliers get their maximum value. In other words, they have a "big say" in the expansion in (11.44). When $\tilde{\xi}_n$ and/or ξ_n are zero, then

$$0 \le \tilde{\lambda}_n \le C, \ 0 \le \lambda_n \le C.$$

- Recall what we have said before concerning the estimation of θ_0. Select any point corresponding to $0 < \tilde{\lambda}_n < C$ ($0 < \lambda_n < C$) which we know correspond to $\tilde{\xi}_n = 0$ ($\xi_n = 0$). Then $\hat{\theta}_0$ is computed from (11.38) and (11.39). In practice, one selects all such points and computes θ_0 as the respective mean.

- Fig. 11.11 illustrates $\hat{y}(\boldsymbol{x})$ for a choice of $\kappa(\cdot, \cdot)$. Observe that the value of ϵ forms a "tube" around the respective graph. Points lying outside the tube correspond to values of the slack variables larger than zero.

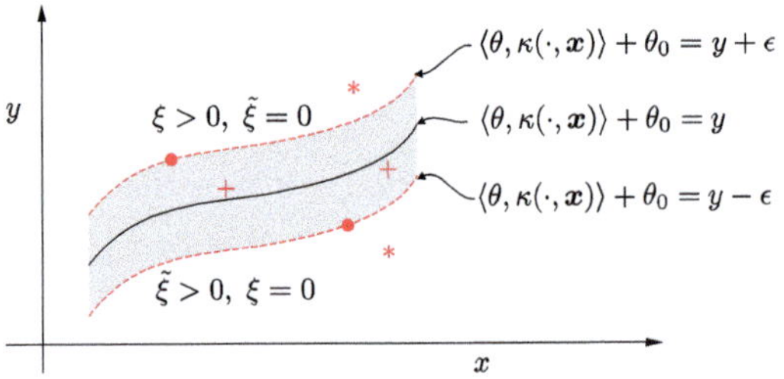

FIGURE 11.11

The tube around the nonlinear regression curve. Points outside the tube (denoted by stars) have either $\tilde{\xi} > 0$ and $\xi = 0$ or $\xi > 0$ and $\tilde{\xi} = 0$. The rest of the points have $\tilde{\xi} = \xi = 0$. Points that are inside the tube correspond to zero Lagrange multipliers.

Remarks 11.3.

- Besides the linear ϵ-insensitive loss, similar analysis is valid for the quadratic ϵ-insensitive and Huber loss functions (for example, [28]).It turns out that use of the Huber loss function results in a larger number of support vectors. Note that a large number of support vectors increases complexity, as more kernel evaluations are involved.
- *Sparsity and ϵ-insensitive loss function*: Note that (11.36) has exactly the same form as (11.18). However, in the former case, the expansion is a sparse one using $N_s < N$, and in practice often $N_s \ll N$. The obvious question that is now raised is whether there is a "hidden" connection between the ϵ-insensitive loss function and the sparsity promoting methods, discussed in Chapter 9. Interestingly enough, the answer is in the affirmative [47]. Assuming the unknown function, g, in (11.23) to reside in an RKHS, and exploiting the representer theorem, it is approximated by an expansion in an RKHS and the unknown parameters are estimated by minimizing

$$L(\boldsymbol{\theta}) = \frac{1}{2}\|y(\cdot) - \sum_{n=1}^{N} \theta_n \kappa(\cdot, \boldsymbol{x}_n)\|_{\mathbb{H}}^2 + \epsilon \sum_{n=1}^{N} |\theta_n|.$$

  This is similar to what we did for the kernel ridge regression with the notable exception that the ℓ_1 norm of the parameters is involved for regularization. The norm $\|\cdot\|_{\mathbb{H}}$ denotes the norm associated with the RKHS. Elaborating on the norm, it can be shown that for the noiseless case, the minimization task becomes identical with the SVR one.

Example 11.3. Consider the same time series used for the nonlinear prediction task in Example 11.2. This time, the SVR method was used optimized around the linear ϵ-insensitive loss function, with $\epsilon = 0.003$. The same Gaussian kernel with $\sigma = 0.004$ was employed as in the kernel ridge regression case. Fig. 11.12 shows the resulting prediction curve, $\hat{y}(x)$ as a function of x given in (11.36). The encircled points are the support vectors. Even without the use of any quantitative measure, the resulting curve fits the data samples much better compared to the kernel ridge regression, exhibiting the enhanced robustness of the SVR method relative to the kernel ridge regression, in the presence of outliers.

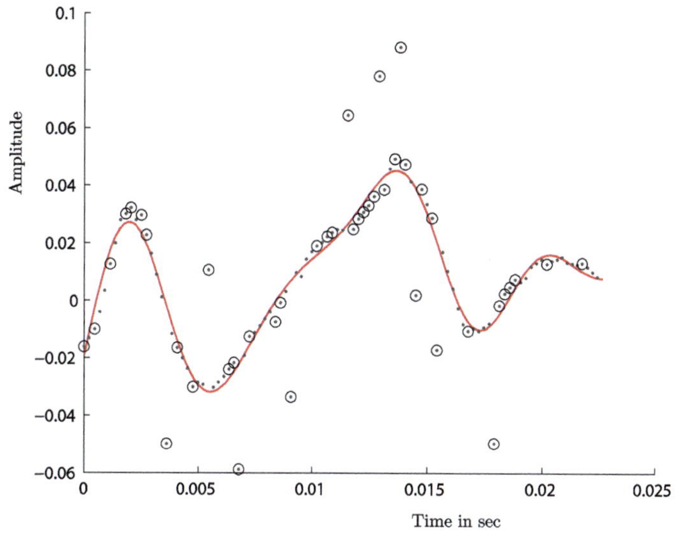

FIGURE 11.12

The resulting prediction curve for the same data points as those used for Example 11.2. The improved performance compared to the kernel ridge regression used for Fig. 11.9 is readily observed. The encircled points are the support vectors resulting from the optimization, using the ϵ-insensitive loss function.

Remarks 11.4.

- A more recent trend to deal with outliers is via their explicit modeling. The noise is split into two components, the inlier and the outlier. The outlier part has to be spare; otherwise, it would not be called outlier. Then, sparsity-related arguments are mobilized to solve an optimization task that estimates both the parameters as well as the outliers (see, for example, [15,71,80,85,86]).

11.9 KERNEL RIDGE REGRESSION REVISITED

The kernel ridge regression was introduced in Section 11.7. Here, it will be restated via its dual representation form. The ridge regression in its primal representation can be cast as

$$\begin{aligned} \text{minimize with respect to } \boldsymbol{\theta},\ \boldsymbol{\xi} \quad & J(\boldsymbol{\theta}, \boldsymbol{\xi}) = \sum_{n=1}^{N} \xi_n^2 + C\|\boldsymbol{\theta}\|^2, \\ \text{subject to} \quad & y_n - \boldsymbol{\theta}^T \boldsymbol{x}_n = \xi_n,\ n = 1, 2, \ldots, N, \end{aligned} \tag{11.51}$$

which leads to the following Lagrangian:

$$L(\boldsymbol{\theta}, \boldsymbol{\xi}, \boldsymbol{\lambda}) = \sum_{n=1}^{N} \xi_n^2 + C\|\boldsymbol{\theta}\|^2 + \sum_{n=1}^{N} \lambda_n (y_n - \boldsymbol{\theta}^T \boldsymbol{x}_n - \xi_n), \quad n = 1, 2, \ldots, N. \tag{11.52}$$

Differentiating with respect to $\boldsymbol{\theta}$ and ξ_n, $n = 1, 2, \ldots, N$, and equating to zero, we obtain

$$\boldsymbol{\theta} = \frac{1}{2C} \sum_{n=1}^{N} \lambda_n \boldsymbol{x}_n \tag{11.53}$$

and

$$\xi_n = \frac{\lambda_n}{2}, \quad n = 1, 2, \ldots, N. \tag{11.54}$$

To obtain the Lagrange multipliers, (11.53) and (11.54) are substituted in (11.52), which results in the dual formulation of the problem, that is,

$$\begin{aligned} \text{maximize with respect to } \boldsymbol{\lambda} \quad & \sum_{n=1}^{N} \lambda_n y_n - \frac{1}{4C} \sum_{n=1}^{N} \sum_{m=1}^{N} \lambda_n \lambda_m \kappa(\boldsymbol{x}_n, \boldsymbol{x}_m) \\ & - \frac{1}{4} \sum_{n=1}^{N} \lambda_n^2, \end{aligned} \tag{11.55}$$

where we have replaced $\boldsymbol{x}_n^T \boldsymbol{x}_m$ with the kernel operation according to the kernel trick. It is a matter of straightforward algebra to obtain (see [98] and Problem 11.9)

$$\boldsymbol{\lambda} = 2C(\mathcal{K} + CI)^{-1} \boldsymbol{y}, \tag{11.56}$$

which combined with (11.53) and involving the kernel trick we obtain the prediction rule for the kernel ridge regression, that is,

$$\hat{y}(\boldsymbol{x}) = \boldsymbol{y}^T (\mathcal{K} + CI)^{-1} \boldsymbol{\kappa}(\boldsymbol{x}), \tag{11.57}$$

which is the same as (11.27); however, via this path one needs not to assume invertibility of $\mathcal{K}$. An efficient scheme for solving the kernel ridge regression has been developed in [118,119].

11.10 OPTIMAL MARGIN CLASSIFICATION: SUPPORT VECTOR MACHINES

The optimal classifier, in the sense of minimizing the classification error, is the Bayesian classifier as discussed in Chapter 7. The method, being a member of the generative learning family, requires the knowledge of the underlying probability distributions. If these are not known, an alternative path is to resort to discriminative learning techniques and adopt a discriminant function f that realizes the corresponding classifier and try to optimize it so as to minimize the respective empirical loss, that is,

$$J(f) = \sum_{n=1}^{N} \mathcal{L}\big(y_n, f(\boldsymbol{x}_n)\big),$$

where

$$y_n = \begin{cases} +1, & \text{if } \boldsymbol{x}_n \in \omega_1, \\ -1, & \text{if } \boldsymbol{x}_n \in \omega_2. \end{cases}$$

For a binary classification task, the first loss function that comes to mind is

$$\mathcal{L}(y, f(\boldsymbol{x})) = \begin{cases} 1, & \text{if} \quad yf(\boldsymbol{x}) \leq 0, \\ 0, & \text{otherwise}, \end{cases} \tag{11.58}$$

which is also known as the (0, 1)-loss function. However, this is a discontinuous function and its optimization is a hard task. To this end, a number of alternative loss functions have been adopted in an effort to approximate the (0, 1)-loss function. Recall that the squared error loss can also be employed but, as already pointed out in Chapters 3 and 7, this is not well suited for classification tasks and bears little resemblance to the (0, 1)-loss function. In this section, we turn our attention to the so-called *hinge* loss function defined as (Chapter 8)

$$\mathcal{L}_\rho(y, f(\boldsymbol{x})) = \max\left\{0, \rho - yf(\boldsymbol{x})\right\}. \tag{11.59}$$

In other words, if the sign of the product between the true label (y) and that predicted by the discriminant function value ($f(\boldsymbol{x})$) is positive and larger than a threshold/margin (user-defined) value $\rho \geq 0$, the loss is zero. If not, the loss exhibits a linear increase. We say that a margin error is committed if $yf(\boldsymbol{x})$ cannot achieve a value of at least ρ. The hinge loss function is shown in Fig. 11.13, together with (0, 1) and squared error loss functions.

We will constrain ourselves to linear discriminant functions, residing in some RKHS, of the form

$$f(\boldsymbol{x}) = \theta_0 + \langle \theta, \phi(\boldsymbol{x}) \rangle,$$

where, by definition,

$$\phi(\boldsymbol{x}) := \kappa(\cdot, \boldsymbol{x})$$

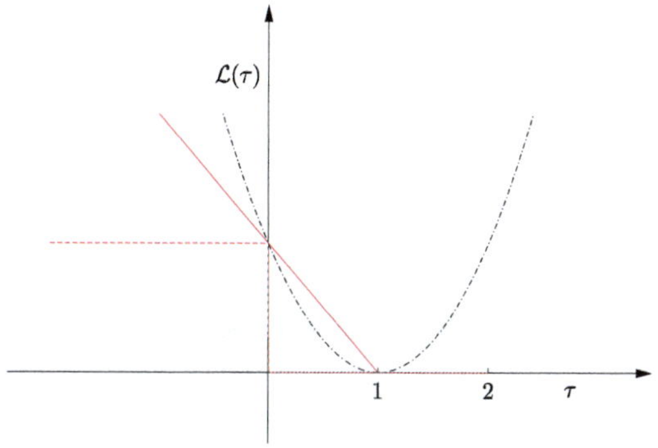

FIGURE 11.13

The (0, 1)-loss (dotted red), the hinge loss (full red), and the squared error (dotted black) functions tuned to pass through the (0, 1) point for comparison. For the hinge loss, $\rho = 1$. Also, for the hinge and (0, 1)-loss functions $\tau = yf(\boldsymbol{x})$, and for the squared error one, $\tau = y - f(\boldsymbol{x})$.

is the feature map. However, for the same reasons discussed in Section 11.8.1, we will cast the task as a linear one in the input space, $\mathbb{R}^l$, and at the final stage the kernel information will be "implanted" using the kernel trick.

The goal of designing a linear classifier now becomes equivalent to minimizing the cost

$$J(\boldsymbol{\theta}, \theta_0) = \frac{1}{2}\|\boldsymbol{\theta}\|^2 + C\sum_{n=1}^{N} \mathcal{L}_\rho(y_n, \boldsymbol{\theta}^T \boldsymbol{x}_n + \theta_0). \tag{11.60}$$

Alternatively, employing slack variables, and following a similar reasoning as in Section 11.8.1, minimizing (11.60) becomes equivalent to

$$\text{minimize with respect to } \boldsymbol{\theta},\ \theta_0,\ \boldsymbol{\xi} \qquad J(\boldsymbol{\theta}, \boldsymbol{\xi}) = \frac{1}{2}\|\boldsymbol{\theta}\|^2 + C\sum_{n=1}^{N} \xi_n, \tag{11.61}$$

$$\text{subject to} \qquad y_n(\boldsymbol{\theta}^T \boldsymbol{x}_n + \theta_0) \geq \rho - \xi_n, \tag{11.62}$$

$$\xi_n \geq 0,\ n = 1, 2, \ldots, N. \tag{11.63}$$

From now on, we will adopt the value $\rho = 1$, without harming generality. A margin error is committed if $y_n(\boldsymbol{\theta}^T \boldsymbol{x}_n + \theta_0) < 1$, corresponding to $\xi_n > 0$, otherwise the inequality in (11.62) is not satisfied. For the latter to be satisfied, in case of a margin error, it is necessary that $\xi_n > 0$ and this contributes to the cost in Eq. (11.61). On the other hand, if $\xi_n = 0$, then $y_n(\boldsymbol{\theta}^T \boldsymbol{x}_n + \theta_0) \geq 1$ and there is no contribution in the cost function. Observe that the smaller the value of $y_n(\boldsymbol{\theta}^T \boldsymbol{x}_n + \theta_0)$ is, with respect to the margin $\rho = 1$, the larger the corresponding ξ_n should be for the inequality to be satisfied and, hence, the larger the contribution in the cost in Eq. (11.61) is. Thus, the goal of the optimization task is to drive as many of the ξ_ns to zero as possible. The optimization task in (11.61)–(11.63) has an interesting and important geometric interpretation.

11.10.1 LINEARLY SEPARABLE CLASSES: MAXIMUM MARGIN CLASSIFIERS

Assuming linearly separable classes, there is an infinity of linear classifiers that solve the classification task exactly, without committing errors on the training set (see Fig. 11.14). It is easy to see, and it will become apparent very soon, that from this infinity of hyperplanes that solve the task, we can always identify a subset such as

$$y_n(\boldsymbol{\theta}^T \boldsymbol{x}_n + \theta_0) \geq 1, \quad n = 1, 2, \ldots, N,$$

which guarantees that $\xi_n = 0,\ n = 1, 2, \ldots, N$, in (11.61)–(11.63). Hence, for linearly separable classes, the previous optimization task is equivalent to

$$\text{minimize with respect to } \boldsymbol{\theta} \qquad \frac{1}{2}\|\boldsymbol{\theta}\|^2, \tag{11.64}$$

$$\text{subject to} \qquad y_n(\boldsymbol{\theta}^T \boldsymbol{x}_n + \theta_0) \geq 1,\ n = 1, 2, \ldots, N. \tag{11.65}$$

In other words, from this infinity of linear classifiers, which can solve the task and classify correctly all training patterns, our optimization task selects the one that has minimum norm. As will be explained next, the norm $\|\boldsymbol{\theta}\|$ is directly related to the margin formed by the respective classifier.

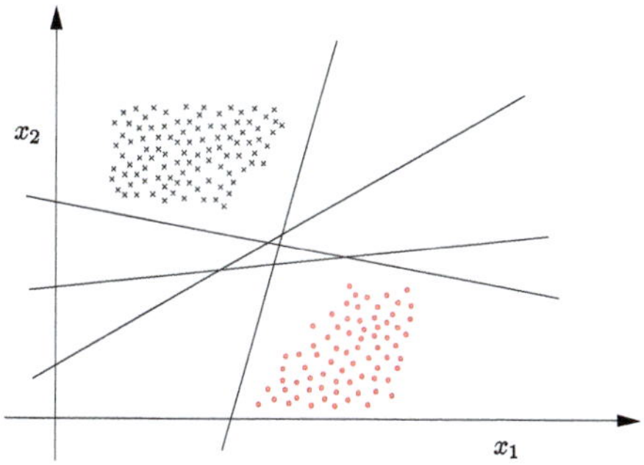

FIGURE 11.14

There is an infinite number of linear classifiers that can classify correctly all the patterns in a linearly separable class task.

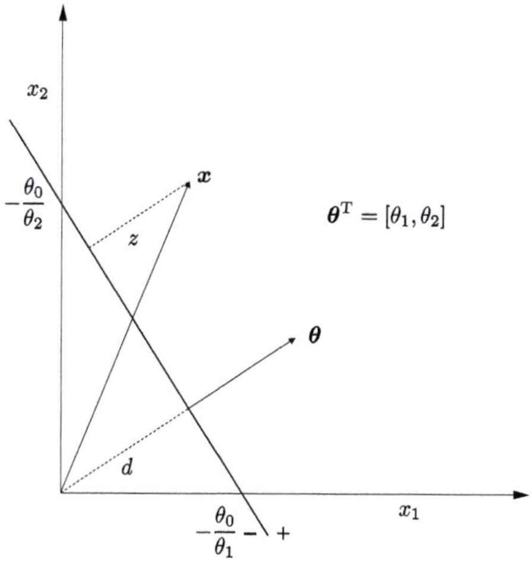

FIGURE 11.15

The direction of the hyperplane $\boldsymbol{\theta}^T \boldsymbol{x} + \theta_0 = 0$ is determined by $\boldsymbol{\theta}$ and its position in space by θ_0. Note that $\boldsymbol{\theta}$ has been shifted away from the origin, since here we are only interested to indicate its direction, which is perpendicular to the straight line.

Each hyperplane in space is described by the equation

$$f(\boldsymbol{x}) = \boldsymbol{\theta}^T \boldsymbol{x} + \theta_0 = 0. \tag{11.66}$$

From classical geometry (see also Problem 5.12), we know that its direction in space is controlled by $\boldsymbol{\theta}$ (which is perpendicular to the hyperplane) and its position is controlled by θ_0 (see Fig. 11.15). From the set of all hyperplanes that solve the task exactly and have certain direction (i.e., they share a common $\boldsymbol{\theta}$), we select θ_0 so as to place the hyperplane in between the two classes, such that its

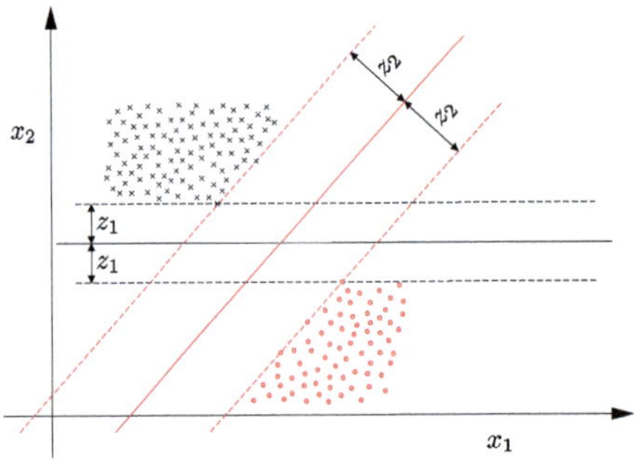

FIGURE 11.16

For each direction $\boldsymbol{\theta}$, "red" and "gray," the (linear) hyperplane classifier, $\boldsymbol{\theta}^T\boldsymbol{x}+\theta_0=0$ (full lines), is placed in between the two classes and normalized so that the nearest points from each class have a distance equal to one. The dotted lines, $\boldsymbol{\theta}^T\boldsymbol{x}+\theta_0=\pm 1$, which pass through the nearest points, are parallel to the respective classifier and define the margin. The width of the margin is determined by the direction of the corresponding classifier in space and is equal to $\frac{2}{\|\boldsymbol{\theta}\|}$.

distance from the nearest points from each one of the two classes is the same. Fig. 11.16 shows the linear classifiers (hyperplanes) in two different directions (full lines in gray and red). Both of them have been placed so as to have the same distance from the nearest points in both classes. Moreover, note that the distance z_1 associated with the "gray" classifier is smaller than the z_2 associated with the "red" one.

From basic geometry, we know that the distance of a point $\boldsymbol{x}$ from a hyperplane (see Fig. 11.15) is given by

$$z=\frac{|\boldsymbol{\theta}^T\boldsymbol{x}+\theta_0|}{\|\boldsymbol{\theta}\|},$$

which is obviously zero if the point lies on the hyperplane. Moreover, we can always scale by a constant factor, say, a, both $\boldsymbol{\theta}$ and θ_0 without affecting the geometry of the hyperplane, as described by Eq. (11.66). After an appropriate scaling, we can always make the distance of the nearest points from the two classes to the hyperplane equal to $z=\frac{1}{\|\boldsymbol{\theta}\|}$; equivalently, the scaling guarantees that $f(\boldsymbol{x})=\pm 1$ if $\boldsymbol{x}$ is a nearest to the hyperplane point and depending on whether the point belongs to ω_1 $(+1)$ or ω_2 (-1). The two hyperplanes, defined by $f(\boldsymbol{x})=\pm 1$, are shown in Fig. 11.16 as dotted lines, for both the "gray" and the "red" directions. The pair of these hyperplanes defines the corresponding margin, for each direction, whose width is equal to $\frac{2}{\|\boldsymbol{\theta}\|}$.

Thus, any classifier that is constructed as explained before, and which solves the task, has the following two properties:

- it has a *margin* of width equal to $\frac{1}{\|\boldsymbol{\theta}\|}+\frac{1}{\|\boldsymbol{\theta}\|}$;
- it satisfies the two sets of constraints,

$$\boldsymbol{\theta}^T\boldsymbol{x}_n+\theta_0\geq +1,\quad \boldsymbol{x}_n\in\omega_1,$$

$$\boldsymbol{\theta}^T \boldsymbol{x}_n + \theta_0 \leq -1, \quad \boldsymbol{x}_n \in \omega_2.$$

Hence, the optimization task in (11.64)–(11.65) computes the linear classifier which *maximizes the margin* subject to the constraints.

The margin interpretation of the regularizing term $\|\boldsymbol{\theta}\|^2$ ties nicely the task of designing classifiers, which maximize the margin, with the statistical learning theory and the pioneering work of Vapnik–Chervonenkis, which establishes elegant performance bounds on the generalization properties of such classifiers (see, for example, [28,124,128,129]).

The Solution

Following similar steps as for the support vector regression case, the solution is given as a linear combination of a subset of the training samples, that is,

$$\hat{\boldsymbol{\theta}} = \sum_{n=1}^{N_s} \lambda_n y_n \boldsymbol{x}_n, \tag{11.67}$$

where N_s is the number of the nonzero Lagrange multipliers. It turns out that only the Lagrange multipliers associated with the nearest-to-the-classifier points, that is, those points satisfying the constraints with equality ($y_n(\boldsymbol{\theta}^T \boldsymbol{x}_n + \theta_0) = 1$), are nonzero. These are known as the *support vectors*. The Lagrange multipliers corresponding to the points farther away ($y_n(\boldsymbol{\theta}^T \boldsymbol{x}_n + \theta_0) > 1$) are zero. The estimate of the bias term, $\hat{\theta}_0$, is obtained by selecting all constraints with $\lambda_n \neq 0$, corresponding to

$$y_n(\hat{\boldsymbol{\theta}}^T \boldsymbol{x}_n + \hat{\theta}_0) - 1 = 0, \quad n = 1, 2, \ldots, N_s,$$

solving for $\hat{\theta}_0$ and taking the average value.

For the more general RKHS case, the solution is a function given by

$$\hat{\theta}(\cdot) = \sum_{n=1}^{N_s} \lambda_n y_n \kappa(\cdot, \boldsymbol{x}_n),$$

which leads to the following prediction rule. Given an unknown $\boldsymbol{x}$, its class label is predicted according to the sign of

$$\hat{y}(\boldsymbol{x}) = \langle \hat{\theta}, \kappa(\cdot, \boldsymbol{x}) \rangle + \hat{\theta}_0,$$

or

$$\boxed{\hat{y}(\boldsymbol{x}) = \sum_{n=1}^{N_s} \lambda_n y_n \kappa(\boldsymbol{x}, \boldsymbol{x}_n) + \hat{\theta}_0 : \quad \text{support vector machine prediction.}} \tag{11.68}$$

Similarly to the linear case, we select all constraints associated with $\lambda_n \neq 0$, i.e.,

$$y_n \left(\sum_{m=1}^{N_s} \lambda_m y_m \kappa(\boldsymbol{x}_m, \boldsymbol{x}_n) + \hat{\theta}_0 \right) - 1 = 0, \; n = 1, 2, \ldots, N_s,$$

and $\hat{\theta}_0$ is computed as the average of the values obtained from each one of these constraints. Although the solution is unique, the corresponding Lagrange multipliers may not be unique (see, for example, [124]).

Finally, it must be stressed that the number of support vectors is related to the generalization performance of the classifier. The *smaller the number of support vectors, the better the generalization is expected to be* [28,124].

The Optimization Task

This part can also be bypassed in a first reading.

The task in (11.64)–(11.65) is a quadratic programming task and can be solved following similar steps to those adopted for the SVR task.

The associated Lagrangian is given by

$$L(\boldsymbol{\theta}, \theta_0, \boldsymbol{\lambda}) = \frac{1}{2}\|\boldsymbol{\theta}\|^2 - \sum_{n=1}^{N} \lambda_n \left(y_n \left(\boldsymbol{\theta}^T \boldsymbol{x}_n + \theta_0 \right) - 1 \right), \tag{11.69}$$

and the KKT conditions (Appendix C) become

$$\frac{\partial}{\partial \boldsymbol{\theta}} L(\boldsymbol{\theta}, \theta_0, \boldsymbol{\lambda}) = \mathbf{0} \longrightarrow \hat{\boldsymbol{\theta}} = \sum_{n=1}^{N} \lambda_n y_n \boldsymbol{x}_n, \tag{11.70}$$

$$\frac{\partial}{\partial \theta_0} L(\boldsymbol{\theta}, \theta_0, \boldsymbol{\lambda}) = 0 \longrightarrow \sum_{n=1}^{N} \lambda_n y_n = 0, \tag{11.71}$$

$$\lambda_n \left(y_n (\boldsymbol{\theta}^T \boldsymbol{x}_n + \theta_0) - 1 \right) = 0, \ n = 1, 2, \ldots, N, \tag{11.72}$$

$$\lambda_n \geq 0, \ n = 1, 2, \ldots, N. \tag{11.73}$$

The Lagrange multipliers are obtained via the dual representation form after plugging (11.70) into the Lagrangian (Problem 11.11), that is,

$$\text{maximize with respect to } \boldsymbol{\lambda} \quad \sum_{n=1}^{N} \lambda_n - \frac{1}{2} \sum_{n=1}^{N} \sum_{m=1}^{N} \lambda_n \lambda_m y_n y_m \boldsymbol{x}_n^T \boldsymbol{x}_m, \tag{11.74}$$

$$\text{subject to} \quad \lambda_n \geq 0, \tag{11.75}$$

$$\sum_{n=1}^{N} \lambda_n y_n = 0. \tag{11.76}$$

For the case where the original task has been mapped to an RKHS, the cost function becomes

$$\sum_{n=1}^{N} \lambda_n - \frac{1}{2} \sum_{n=1}^{N} \sum_{m=1}^{N} \lambda_n \lambda_m y_n y_m \kappa(\boldsymbol{x}_n, \boldsymbol{x}_m). \tag{11.77}$$

- According to (11.72), if $\lambda_n \neq 0$, then necessarily

$$y_n (\boldsymbol{\theta}^T \boldsymbol{x}_n + \theta_0) = 1.$$

That is, the respective points are the closest points, from each class, to the classifier (distance $\frac{1}{\|\boldsymbol{\theta}\|}$). They lie on either of the two hyperplanes forming the border of the margin. These points are the support vectors and the respective constraints are known as the *active constraints*. The rest of the points, associated with

$$y_n(\boldsymbol{\theta}^T \boldsymbol{x}_n + \theta_0) > 1, \tag{11.78}$$

which lie outside the margin, correspond to $\lambda_n = 0$ (*inactive constraints*).

- The cost function in (11.64) is strictly convex and, hence, the solution of the optimization task is *unique* (Appendix C).

11.10.2 NONSEPARABLE CLASSES

We now turn our attention to the more realistic case of overlapping classes and the corresponding geometric interpretation of the task in (11.61)–(11.63). In this case, there is no (linear) classifier that can classify correctly all the points, and some errors are bound to occur. Fig. 11.17 shows the respective geometry for a linear classifier. Note that although the classes are not separable, we still define the margin as the area between the two hyperplanes $f(\boldsymbol{x}) = \pm 1$. There are three types of points.

- Points that lie on the border or outside the margin and in the correct side of the classifier, that is,

$$y_n f(\boldsymbol{x}_n) \geq 1.$$

These points commit no (margin) error, and hence the inequality in (11.62) is satisfied for

$$\xi_n = 0.$$

- Points which lie on the correct side of the classifier, but lie inside the margin (circled points), that is,

$$0 < y_n f(\boldsymbol{x}_n) < 1.$$

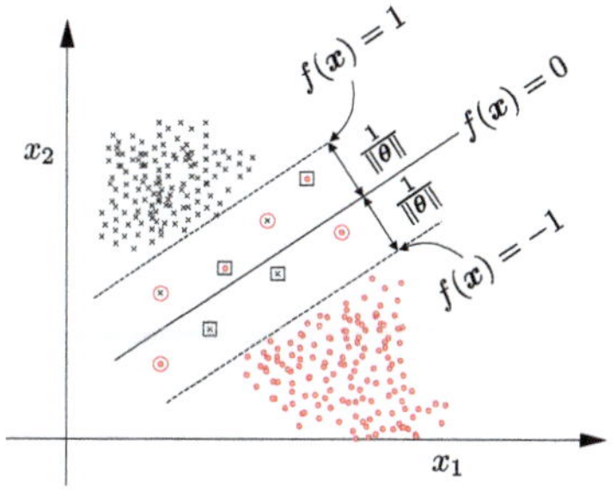

FIGURE 11.17

When classes are overlapping, there are three types of points: (a) points that lie outside or on the borders of the margin and are classified correctly ($\xi_n = 0$); (b) points inside the margin and classified correctly ($0 < \xi_n < 1$), denoted by circles; and (c) misclassified points, denoted by a square ($\xi_n \geq 1$).

These points commit a margin error, and the inequality in (11.62) is satisfied for

$$0 < \xi_n < 1.$$

- Points that lie on the wrong side of the classifier (points in squares), that is,

$$y_n f(\boldsymbol{x}_n) \leq 0.$$

These points commit an error and the inequality in (11.62) is satisfied for

$$1 \leq \xi_n.$$

Our desire would be to estimate a hyperplane classifier, so as to *maximize the margin and at the same time keep the number of errors (including margin errors) as small as possible*. This goal could be expressed via the optimization task in (11.61)–(11.62), if in place of ξ_n we had the indicator function, $I(\xi_n)$, where

$$I(\xi) = \begin{cases} 1, & \text{if } \xi > 0, \\ 0, & \text{if } \xi = 0. \end{cases}$$

However, in such a case the task becomes a combinatorial one. So, we relax the task and use ξ_n in place of the indicator function, leading to (11.61)–(11.62). Note that optimization is achieved in a tradeoff rationale; the user-defined parameter C controls the influence of each of the two contributions to the minimization task. If C is large, the resulting margin (the distance between the two hyperplanes defined by $f(\boldsymbol{x}) = \pm 1$) will be small, in order to commit a smaller number of margin errors. If C is small, the opposite is true. As we will see from the simulation examples, the choice of C is critical.

The Solution

Once more, the solution is given as a linear combination of a subset of the training points,

$$\hat{\boldsymbol{\theta}} = \sum_{n=1}^{N_s} \lambda_n y_n \boldsymbol{x}_n, \tag{11.79}$$

where $\lambda_n, \; n = 1, 2, \ldots, N_s$, are the nonzero Lagrange multipliers associated with the support vectors. In this case, support vectors are all points that lie either (a) on the pair of the hyperplanes that define the margin or (b) inside the margin or (c) outside the margin but on the wrong side of the classifier. That is, correctly classified points that lie outside the margin do no contribute to the solution, because the corresponding Lagrange multipliers are zero. For the RKHS case, the class prediction rule is the same as in (11.68), where $\hat{\theta}_0$ is computed from the constraints corresponding to $\lambda_n \neq 0$ and $\xi_n = 0$; these correspond to the points that lie on the hyperplanes defining the margin and on the correct side of the classifier.

The Optimization Task

As before, this part can be bypassed in a first reading.

The Lagrangian associated with (11.61)–(11.63) is given by

$$L(\boldsymbol{\theta}, \theta_0, \boldsymbol{\xi}, \boldsymbol{\lambda}) = \frac{1}{2}\|\boldsymbol{\theta}\|^2 + C\sum_{n=1}^{N} \xi_n - \sum_{n=1}^{N} \mu_n \xi_n - \sum_{n=1}^{N} \lambda_n \left(y_n \left(\boldsymbol{\theta}^T \boldsymbol{x}_n + \theta_0 \right) - 1 + \xi_n \right),$$

leading to the following KKT conditions:

$$\frac{\partial L}{\partial \boldsymbol{\theta}} = \mathbf{0} \longrightarrow \hat{\boldsymbol{\theta}} = \sum_{n=1}^{N} \lambda_n y_n \boldsymbol{x}_n, \tag{11.80}$$

$$\frac{\partial L}{\partial \theta_0} = 0 \longrightarrow \sum_{n=1}^{N} \lambda_n y_n = 0, \tag{11.81}$$

$$\frac{\partial L}{\partial \xi_n} = 0 \longrightarrow C - \mu_n - \lambda_n = 0, \tag{11.82}$$

$$\lambda_n \left(y_n (\boldsymbol{\theta}^T \boldsymbol{x}_n + \theta_0) - 1 + \xi_n \right) = 0, \; n = 1, 2, \dots, N, \tag{11.83}$$

$$\mu_n \xi_n = 0, \; n = 1, 2, \dots, N, \tag{11.84}$$

$$\mu_n \geq 0, \; \lambda_n \geq 0, \; n = 1, 2, \dots, N, \tag{11.85}$$

and in our by-now-familiar procedure, the dual problem is cast as

$$\text{maximize with respect to } \boldsymbol{\lambda} \quad \sum_{n=1}^{N} \lambda_n - \frac{1}{2} \sum_{n=1}^{N} \sum_{m=1}^{N} \lambda_n \lambda_m y_n y_m \boldsymbol{x}_n^T \boldsymbol{x}_m \tag{11.86}$$

$$\text{subject to} \quad 0 \leq \lambda_n \leq C, \; n = 1, 2, \dots, N, \tag{11.87}$$

$$\sum_{n=1}^{N} \lambda_n y_n = 0. \tag{11.88}$$

When working in an RKHS, the cost function becomes

$$\sum_{n=1}^{N} \lambda_n - \frac{1}{2} \sum_{n=1}^{N} \sum_{m=1}^{N} \lambda_n \lambda_m y_n y_m \kappa(\boldsymbol{x}_n, \boldsymbol{x}_m).$$

Observe that the only difference compared to its linearly class-separable counterpart in (11.74)–(11.76) is the existence of C in the inequality constraints for λ_n. The following comments are in order:

- From (11.83), we conclude that for all the points outside the margin, and on the correct side of the classifier, which correspond to $\xi_n = 0$, we have

$$y_n (\boldsymbol{\theta}^T \boldsymbol{x}_n + \theta_0) > 1,$$

hence, $\lambda_n = 0$. That is, these points do not participate in the formation of the solution in (11.80).

- We have $\lambda_n \neq 0$ only for the points that live either on the border hyperplanes or inside the margin or outside the margin but on the wrong side of the classifier. These comprise the support vectors.
- For points lying inside the margin or outside but on the wrong side, $\xi_n > 0$; hence, from (11.84), $\mu_n = 0$ and from (11.82) we get

$$\lambda_n = C.$$

- Support vectors which lie on the margin border hyperplanes satisfy $\xi_n = 0$ and therefore μ_n can be nonzero, which leads to

$$0 \leq \lambda_n \leq C.$$

Remarks 11.5.

- *ν-SVM*: An alternative formulation for the SVM classification has been given in [104], where the margin is defined by the pair of hyperplanes,

$$\boldsymbol{\theta}^T \boldsymbol{x} + \theta_0 = \pm\rho,$$

  and $\rho \geq 0$ is left as a free variable, giving rise to the ν-SVM; ν controls the importance of ρ in the associated cost function. It has been shown [23] that the ν-SVM and the formulation discussed above, which is sometimes referred to as the C-SVM, lead to the same solution for appropriate choices of C and ν. However, the advantage of ν-SVM lies in the fact that ν can be directly related to bounds concerning the number of support vectors and the corresponding error rate (see also [124]).
- *Reduced convex hull interpretation*: In [58], it has been shown that for linearly separable classes, the SVM formulation is equivalent to finding the nearest points between the convex hulls formed by the data in the two classes. This result was generalized for overlapping classes in [33]; it is shown that in this case the ν-SVM task is equivalent to searching for the nearest points between the *reduced convex hulls* (RCHs) associated with the training data. Searching for the RCH is a computationally hard task of combinatorial nature. The problem was efficiently solved in [73–75,122], who came up with efficient iterative schemes to solve the SVM task, via nearest point searching algorithms. More on these issues can be obtained from [66,123,124].
- *ℓ_1 Regularized versions*: The regularization term, which has been used in the optimization tasks discussed so far, has been based on the ℓ_2 norm. A lot of research effort has been focused on using ℓ_1 norm regularization for tasks treating the linear case. To this end, a number of different loss functions have been used in addition to the squared error, hinge loss, and ϵ-insensitive versions, as for example the logistic loss. The solution of such tasks comes under the general framework discussed in Chapter 8. As a matter of fact, some of these methods have been discussed there. A related concise review is provided in [132].
- *Multitask learning*: In multitask learning, two or more related tasks, for example, classifiers, are jointly optimized. Such problems are of interest in, for example, econometrics and bioinformatics. In [40], it is shown that the problem of estimating many task functions with regularization can be cast as a single task learning problem if a family of appropriately defined multitask kernel functions is used.

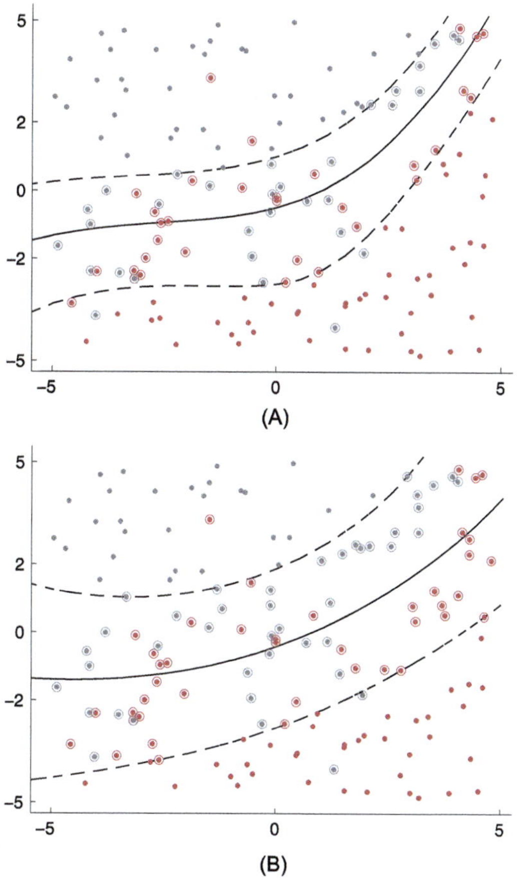

FIGURE 11.18

(A) The training data points for the two classes (red and gray, respectively) of Example 11.4. The full line is the graph of the obtained SVM classifier and the dotted lines indicate the margin, for $C = 20$. (B) The result for $C = 1$. For both cases, the Gaussian kernel with $\sigma = 20$ was used.

Example 11.4. In this example, the performance of the SVM is tested in the context of a two-class two-dimensional classification task. The data set comprises $N = 150$ points uniformly distributed in the region $[-5, 5] \times [-5, 5]$. For each point $\boldsymbol{x}_n = [x_{n,1},\ x_{n,2}]^T$, $n = 1, 2, \ldots, N$, we compute

$$y_n = 0.05x_{n,1}^3 + 0.05x_{n,1}^2 + 0.05x_{n,1} + 0.05 + \eta,$$

where η stands for zero mean Gaussian noise of variance $\sigma_\eta^2 = 4$. The point is assigned to either of the two classes, depending on the value of the noise as well as its position with respect to the graph of the function

$$f(x) = 0.05x^3 + 0.05x^2 + 0.05x + 0.05$$

in the two-dimensional space. That is, if $x_{n,2} \geq y_n$, the point is assigned to class ω_1; otherwise, it is assigned to class ω_2.

The Gaussian kernel was used with $\sigma = 10$, as this resulted in the best performance. Fig. 11.18A shows the obtained classifier for $C = 20$ and Fig. 11.18B that for $C = 1$. Observe how the obtained classifier, and hence the performance, depends heavily on the choice of C. In the former case, the number of the support vectors was equal to 64, and for the latter it was equal to 84.

11.10.3 PERFORMANCE OF SVMs AND APPLICATIONS

A notable characteristic of the SVMs is that the complexity is independent of the dimensionality of the respective RKHS. This has an influence on the generalization performance; SVMs exhibit very good generalization performance in practice. Theoretically, such a claim is substantiated by their maximum margin interpretation, in the framework of the elegant *structural risk minimization theory* [28,124,128].

An extensive comparative study concerning the performance of SVMs against 16 other popular classifiers has been reported in [77]. The results verify that the SVM ranks at the very top among these classifiers, although there are cases for which other methods score better performance. Another comparative performance study is reported in [25].

It is hard to find a discipline related to machine learning/pattern recognition where SVMs and the concept of working in kernel spaces have not been applied. Early applications included data mining, spam categorization, object recognition, medical diagnosis, optical character recognition (OCR), and bioinformatics (see, for example, [28] for a review). More recent applications include cognitive radio (for example, [35]), spectrum cartography and network flow prediction [9], and image denoising [13].

In [117], the notion of *kernel embedding* of conditional probabilities is reviewed, as a means to address challenging problems in graphical models. The notion of kernelization has also been extended in the context of *tensor-based* models [50,107,133]. Kernel-based hypothesis testing is reviewed in [49]. In [121], the use of kernels in manifold learning is discussed in the framework of diffusion maps. The task of analyzing the performance of kernel techniques with regard to dimensionality, signal-to-noise ratio, and local error bars is reviewed in [81]. In [120], a collection of articles related to kernel-based methods and applications is provided.

11.10.4 CHOICE OF HYPERPARAMETERS

One of the main issues associated with SVM/SVRs is the choice of the parameter C, which controls the relative influence of the loss and the regularizing parameter in the cost function. Although some efforts have been made in developing theoretical tools for the respective optimization, the path that has survived in practice is that of cross-valuation techniques against a test data set. Different values of C are used to train the model, and the value that results in the best performance over the test set is selected.

The other main issue is the choice of the kernel function. Different kernels lead to different performance. Let us look carefully at the expansion in (11.17). One can think of $\kappa(\boldsymbol{x}, \boldsymbol{x}_n)$ as a function that measures the *similarity* between $\boldsymbol{x}$ and $\boldsymbol{x}_n$; in other words, $\kappa(\boldsymbol{x}, \boldsymbol{x}_n)$ *matches* $\boldsymbol{x}$ to the training sample $\boldsymbol{x}_n$. A kernel is local if $\kappa(\boldsymbol{x}, \boldsymbol{x}_n)$ takes relatively large values in a small region around $\boldsymbol{x}_n$. For example, when the Gaussian kernel is used, the contribution of $\kappa(\boldsymbol{x}, \boldsymbol{x}_n)$ away from $\boldsymbol{x}_n$ decays exponentially fast, depending of the value of σ^2. Thus, the choice of σ^2 is very crucial. If the function to be approximated is smooth, then large values of σ^2 should be employed. On the contrary, if the function is highly varying in input space, the Gaussian kernel may not be the best choice. As a matter of fact, if for such cases the Gaussian kernel is employed, one must have access to a large number of training data, in order to fill in the input space densely enough, so as to be able to obtain a good enough approximation of such a function. This brings into the scene another critical factor in machine learning, related to the size of the training set. The latter is not only dictated by the dimensionality of the input space (curse of dimensionality) but it also depends on the type of variation that the unknown function undergoes (see, for example, [12]). In practice, in order to choose the right kernel function, one uses different kernels and after cross-validation selects the "best" one for the specific problem.

A line of research is to design kernels that match the data at hand, based either on some prior knowledge or via some optimization path (see, for example, [29,65]). In Section 11.13 we will discuss techniques which use multiple kernels in an effort to optimally combine their individual characteristics.

11.10.5 MULTICLASS GENERALIZATIONS

The SVM classification task has been introduced in the context of a two-class classification task. The more general M-class case can be treated in various ways:

- *One-against-all*: One solves M two-class problems. Each time, one of the classes is classified against all the others using a different SVM. Thus, M classifiers are estimated, that is,

$$f_m(\boldsymbol{x}) = 0, \; m = 1, 2, \dots, M,$$

  which are trained so that $f_m(\boldsymbol{x}) > 0$ for $\boldsymbol{x} \in \omega_m$ and $f_m(\boldsymbol{x}) < 0$ if $\boldsymbol{x}$ otherwise. Classification is achieved via the rule

$$\text{assign } \boldsymbol{x} \text{ in } \omega_k: \; \text{if } k = \arg\max_m f_m(\boldsymbol{x}).$$

  According to this method, there may be regions in space where more than one of the discriminant functions score a positive value [124]. Moreover, another disadvantage of this approach is the so-called *class imbalance* problem; this is caused by the fact that the number of training points in one of the classes (which comprises the data from $M - 1$ classes) can be much larger than the points in the other. Issues related to the class imbalance problem are discussed in [124].
- *One-against-one*. According to this method, one solves $\frac{M(M-1)}{2}$ binary classification tasks by considering all classes in pairs. The final decision is taken on the basis of the majority rule.
- In [129], the SVM rationale is extended in estimating simultaneously M hyperplanes. However, this technique ends up with a large number of parameters, equal to $N(M - 1)$, which have to be estimated via a single minimization task; this turns out to be rather prohibitive for most practical problems.
- In [34], the multiclass task is treated in the context of error correcting codes. Each class is associated with a binary code word. If the code words are properly chosen, an error resilience is "embedded" into the process (see also [124]). For a comparative study of multiclass classification schemes, see, for example, [41] and the references therein.
- *Division and Clifford algebras*: The SVM framework has also been extended to treat complex and hypercomplex data, both for the regression and the classification cases, using either division algebras or Clifford algebras [8]. In [125], the case of quaternion RKH spaces is considered.

A more general method for the case of complex-valued data, which exploits the notion of *widely linear estimation* as well as *pure complex kernels*, has been presented in [14]. In this paper, it is shown that any complex SVM/SVR task is equivalent to solving two real SVM/SVR tasks exploiting a specific real kernel, which is generated by the chosen complex one. Moreover, in the classification case, it is shown that the proposed framework inherently splits the complex space into four parts. This leads naturally to solving the four-class task (quaternary classification), instead of the standard two-class scenario of the real SVM. This rationale can be used in a multiclass problem as a split-class scenario.

11.11 COMPUTATIONAL CONSIDERATIONS

Solving a quadratic programming task, in general, requires $\mathcal{O}(N^3)$ operations and $\mathcal{O}(N^2)$ memory operations. To cope with such demands a number of decomposition techniques have been devised (for example, [22,55]), which "break" the task into a sequence of smaller ones. In [59,88,89], the *sequential minimal optimization* (SMO) algorithm breaks the task into a sequence of problems comprising two points, which can be solved analytically. Efficient implementation of such schemes leads to an empirical training time that scales between $\mathcal{O}(N)$ and $\mathcal{O}(N^{2.3})$.

The schemes derived in [74,75] treat the task as a minimum distance points search between reduced convex hulls and end up with an iterative scheme which projects the training points on hyperplanes. The scheme leads to even more efficient implementations compared to [59,88]; moreover, the minimum distance search algorithm has a built-in enhanced parallelism.

The issue of parallel implementation is also discussed in [21]. Issues concerning complexity and accuracy are reported in [53]. In the latter, polynomial time algorithms are derived that produce approximate solutions with a guaranteed accuracy for a class of QP problems including SVM classifiers.

Incremental versions for solving the SVM task which deal with sequentially arriving data have also appeared (for example, [26,37,100]). In the latter, at each iteration a new point is considered and the previously selected set of support vectors (active set) is updated accordingly by adding/removing samples.

Online versions that apply in the primal problem formulation have also been proposed. In [84], an iterative reweighted LS approach is followed that alternates weight optimization with cost constraint forcing. A structurally and computationally simple scheme, named *Primal Estimated sub-Gradient SOlver for SVM* (PEGASOS), has been proposed in [105]. The algorithm is of an iterative subgradient form applied on the regularized empirical hinge loss function in (11.60) (see also Chapter 8). The algorithm, for the case of linear kernels, exhibits very good convergence properties and it finds an ϵ-accurate solution in $\mathcal{O}\left(\frac{C}{\epsilon}\right)$ iterations.

In [43,56] the classical technique of *cutting planes* for solving convex tasks has been employed in the context of SVMs in the primal domain. The resulting algorithm is very efficient and, in particular for the linear SVM case, the complexity becomes of order $O(N)$.

In [126] the concept of the *core vector machine* (CVM) is introduced to lead to efficient solutions. The main concept behind this method is that of the minimum enclosing ball (MEB) in computational geometry. Then a subset of points is used, known as the *core set*, to achieve an approximation to the MEB, which is employed during the optimization.

In [27], a comparative study between solving the SVM tasks in the primal and dual domains is carried out. The findings of the paper point out that both paths are equally efficient, for the linear as well as for the nonlinear cases. Moreover, when the goal is to resort to approximate solutions, opting for the primal task can offer certain benefits. In addition, working in the primal also offers the advantage of tuning the hyperparameters by resorting to joint optimization. One of the main advantages of resorting to the dual domain was the luxury of casting the task in terms of inner products. However, this is also possible in the primal, by appropriate exploitation of the representer theorem. We will see such examples in Section 11.12.1.

More recently, a version of SVM for distributed processing has been presented in [42]. In [99], the SVM task is solved in a subspace using the method of random projections.

11.12 RANDOM FOURIER FEATURES

In Section 11.11, a number of techniques were discussed in order to address and cope with the computational load and the way this scales with the number of training data points, N, in the context of SVMs. Furthermore, at the heart of a number of kernel-based methods, such as the kernel ridge regression (Section 11.7) and the Gaussian processes (to be discussed in Section 13.12) lies the inversion of the kernel matrix, which has been defined in Section 11.5.1. The latter is of dimension $N \times N$ and it is formed by all possible kernel evaluation pairs in the set of the training examples, i.e., $\kappa(\boldsymbol{x}_i, \boldsymbol{x}_j)$, $i, j = 1, 2, \ldots, N$. Inverting an $N \times N$ matrix amounts, in general, to $\mathcal{O}(N^3)$ algebraic operations, which gets out of hand as N increases. In this vein, a number of techniques have been suggested in an effort to reduce related computational costs. Some examples of such methods are given in [1,45], where random projections are employed to "throw away" individual elements or entire rows of the matrix in an effort to form related low rank or sparse matrix approximations.

An alternative "look" at the computational load reduction task was presented in [91]. The essence of this approach is to bypass the implicit mapping to a higher (possibly infinite)-dimensional space via the kernel function, i.e., $\phi(\boldsymbol{x}) = \kappa(\cdot, \boldsymbol{x})$, which comprises the spine of any kernel-based method (Section 11.5). Instead, an explicit mapping to a *finite-dimensional* space, $\mathbb{R}^D$, is performed, i.e.,

$$\boldsymbol{x} \in \mathcal{X} \subseteq \mathbb{R}^l \longmapsto \boldsymbol{z}(\boldsymbol{x}) \in \mathbb{R}^D.$$

However, the mapping is done in a "thoughtful" way so that the inner product between two vectors in this D-dimensional space is approximately equal to the respective value of the kernel function, i.e.,

$$\kappa(\boldsymbol{x}, \boldsymbol{y}) = \langle \phi(\boldsymbol{x}), \phi(\boldsymbol{y}) \rangle \approx \boldsymbol{z}(\boldsymbol{x})^T \boldsymbol{z}(\boldsymbol{y}), \ \forall \boldsymbol{x}, \boldsymbol{y} \in \mathcal{X}.$$

To establish such an approximation, Bochner's theorem from harmonic analysis was mobilized (e.g., [96]). To grasp the main rationale behind this theorem, without resorting to mathematical details, let us recall the notion of power spectral density (PSD) from Section 2.4.3. The PSD of a stationary process is defined as the Fourier transform of its autocorrelation sequence and there it was shown to be a real and nonnegative function. Furthermore, its integral is finite (Eq. (2.125)). Although the proof given there followed a rather "practical" and conceptual path, a more mathematically rigorous proof reveals that this result is a direct by-product of the positive definite property of the autocorrelation sequence (Eq. (2.116)).

In this line of thought, everything that was said above for the PSD is also valid for the Fourier transform of a shift-invariant kernel function. As is well known, a kernel function is a positive definite one (Section 11.5.1). Also, the shift-invariant property, i.e., $\kappa(\boldsymbol{x}, \boldsymbol{y}) = \kappa(\boldsymbol{x} - \boldsymbol{y})$, could be seen as the equivalent of the stationarity. Then one can show that the Fourier transform of a shift-invariant kernel is a nonnegative real function and that its integral is finite. After some normalization, we can always make the integral equal to one. Thus, the Fourier transform of a shift-invariant kernel function can be treated and interpreted as a PDF, that is, a real nonnegative function that integrates to one.

Let $\kappa(\boldsymbol{x} - \boldsymbol{y})$ be a shift-invariant kernel and let $p(\boldsymbol{\omega})$ be its (multidimensional) Fourier transform, i.e.,

$$p(\boldsymbol{\omega}) = \frac{1}{(2\pi)^l} \int_{\mathbb{R}^l} \kappa(\boldsymbol{r}) e^{-j\boldsymbol{\omega}^T \boldsymbol{r}} d\boldsymbol{r}.$$

Then, via the inverse Fourier transform, we have

$$\kappa(\boldsymbol{x} - \boldsymbol{y}) = \int_{\mathbb{R}^l} p(\boldsymbol{\omega}) e^{j\boldsymbol{\omega}^T (\boldsymbol{x}-\boldsymbol{y})} d\boldsymbol{\omega}. \tag{11.89}$$

Having interpreted $p(\boldsymbol{\omega})$ as a PDF, the right-hand side in Eq. (11.89) is the respected expectation, i.e.,

$$\kappa(\boldsymbol{x} - \boldsymbol{y}) = \mathbb{E}_{\boldsymbol{\omega}} \left[e^{j\boldsymbol{\omega}^T \boldsymbol{x}} e^{-j\boldsymbol{\omega}^T \boldsymbol{y}} \right].$$

One can get rid of the complex exponentials by mobilizing Euler's formula[13] and the fact that the kernel is a real function. To this end, we define

$$z_{\boldsymbol{\omega},\mathrm{b}}(\boldsymbol{x}) := \sqrt{2} \cos(\boldsymbol{\omega}^T \boldsymbol{x} + \mathrm{b}), \tag{11.90}$$

where $\boldsymbol{\omega}$ is a random vector that follows $p(\boldsymbol{\omega})$ and b is a uniformly distributed random variable in $[0, 2\pi]$. Then it is easy to show that (Problem 11.12)

$$\kappa(\boldsymbol{x} - \boldsymbol{y}) = \mathbb{E}_{\boldsymbol{\omega},\mathrm{b}} \left[z_{\boldsymbol{\omega},\mathrm{b}}(\boldsymbol{x}) z_{\boldsymbol{\omega},\mathrm{b}}(\boldsymbol{y}) \right], \tag{11.91}$$

where the expectation is taken with respect to $p(\boldsymbol{\omega})$ and $p(b)$. The previous expectation can be approximated by

$$\kappa(\boldsymbol{x} - \boldsymbol{y}) \approx \frac{1}{D} \sum_{i=1}^{D} z_{\boldsymbol{\omega}_i, b_i}(\boldsymbol{x}) z_{\boldsymbol{\omega}_i, b_i}(\boldsymbol{y}),$$

where $(\boldsymbol{\omega}_i, b_i)$, $i = 1, 2, \ldots, D$, are i.i.d. generated samples from the respective distributions.

We now have all the ingredients to define an appropriate mapping that builds upon the previous findings.

- *Step 1*: Generate D i.i.d. samples $\boldsymbol{\omega}_i \sim p(\boldsymbol{\omega})$ and $b_i \sim \mathcal{U}(0, 2\pi)$, $i = 1, 2, \ldots, D$.
- *Step 2*: Perform the following mapping to $\mathbb{R}^D$:

[13] Euler's formula: $e^{j\phi} = \cos\phi + j \sin\phi$.

$$x \in \mathcal{X} \longmapsto z_\Omega(x) = \sqrt{\frac{2}{D}}\left[\cos(\omega_1^T x + b_1), \dots, \cos(\omega_D^T x + b_D)\right]^T.$$

The above mapping defines an inner product approximation of the original kernel, i.e.,

$$\kappa(x - y) \approx z_\Omega(x)^T z_\Omega(y),$$

where, by definition,

$$\Omega := \begin{bmatrix} \omega_1 & \omega_2 & \dots & \omega_D \\ b_1 & b_2 & \dots & b_D \end{bmatrix}.$$

Once this mapping has been performed, one can mobilize any linear modeling method to work in a finite-dimensional space. For example, instead of solving the kernel ridge regression problem, the ridge regression task in Section 3.8 can be used. However, now, the involved matrix to be inverted is of dimensions $D \times D$ instead of $N \times N$, which can lead to significant gains for large enough values of N.

The choice of D is user-defined and problem dependent. In [91], related accuracy bounds have been derived. For within ϵ accuracy approximation to a shift-invariant kernel, one needs only $D = \mathcal{O}(d\epsilon^{-2} \ln \frac{1}{\epsilon^2})$ dimensions. In practice, values of D in the range of a few tens to a few thousands, depending on the data set, seem to suffice to obtain significant computational gains compared to some of the more classical methods.

Remarks 11.6.

- *Nyström approximation*: Another popular and well-known technique to obtain a low rank approximation to the kernel matrix is known as the Nyström approximation.
  Let $\mathcal{K}$ be an $N \times N$ kernel matrix. Then select randomly $q < N$ points out of N. Various sampling scenarios can be used. It turns out that there exists an approximation matrix $\tilde{\mathcal{K}}$ of rank q, such that

$$\tilde{\mathcal{K}} = \mathcal{K}_{nq} \mathcal{K}_q^{-1} \mathcal{K}_{nq}^T,$$

  where $\mathcal{K}_q$ is the respective invertible kernel matrix associated with the subset of the q points and $\mathcal{K}_{nq}$ comprises the columns of $\mathcal{K}$ associated with the previous q points [131]. Generalizations and related theoretical results can be found in [36]. If the approximate matrix, $\tilde{\mathcal{K}}$, is used in place of the original kernel matrix, $\mathcal{K}$, the memory requirements are reduced to $\mathcal{O}(Nq)$ from N^2, and operations of the order of $\mathcal{O}(N^3)$ can be reduced to $\mathcal{O}(Nq^2)$.

11.12.1 ONLINE AND DISTRIBUTED LEARNING IN RKHS

One of the major drawbacks that one encounters when dealing with online algorithms in an RKHS is the growing memory problem. This is a direct consequence of the representer theorem (Section 11.6), where the number of terms (kernels) in the expansion of the estimated function grows with N. Thus, in most of the online versions that have been proposed, the emphasis is on devising ad hoc techniques to reduce the number of terms in the expansion. Usually, these techniques build around the concept of a dictionary, where a subset of the points is judicially selected, according to a criterion, and subsequently used in the related expansion. In this vein, various kernel versions of the LMS ([68,94]), RLS ([38,

127]), APSM and APA ([108–113]), and PEGASOS (Section 8.10.3, [62,105]) algorithms have been suggested. A more extended treatment of this type of algorithms can be downloaded from the book's website, under the Additional Material part that is associated with the current chapter.

In contrast, one can bypass the growing memory problem if the RFF rationale is employed. The RFF framework offers a theoretically pleasing approach, without having to resort to techniques to limit the growing memory, which, in one way or another, have an ad hoc flavor. In the RFF approach, all one has to do is to generate randomly the D points for $\boldsymbol{\omega}$ and b and then use the standard LMS, RLS, etc., algorithms. The power of the RFF framework becomes more evident when the online learning takes place in a distributed environment. In such a setting, all the dictionary-based methods that have been developed so far break down; this is because the dictionaries have to be exchanged among the nodes and, so far, no practically feasible method has been proposed.

The use of the RFF method in the context of online learning and in particular in a distributed environment has been presented in [16,17], where, also, related theoretical convergence issues are discussed. Simulation comparative results reported there reveal the power and performance gains that one can achieve. Values of D in the range of a few hundred seem to suffice and the performance is reported to be rather insensitive on the choice of its value, provided that this is not too small.

11.13 MULTIPLE KERNEL LEARNING

A major issue in all kernel-based algorithmic procedures is the selection of a suitable kernel as well as the computation of its defining parameters. Usually, this is carried out via cross-validation, where a number of different kernels are used on a validation set separate from the training data (see Chapter 3 for different methods concerning validation), and the one with the best performance is selected. It is obvious that this is not a universal approach; it is time consuming and definitely not theoretically appealing. The ideal would be to have a set of different kernels (this also includes the case of the same kernel with different parameters) and let the optimization procedure decide how to choose the proper kernel, or the proper combination of kernels. This is the scope of a research activity, which is usually called *multiple kernel learning* (MKL). To this end, a variety of MKL methods have been proposed to treat several kernel-based algorithmic schemes. A complete survey of the field is outside the scope of this book. Here, we will provide a brief overview of some of the major directions in MKL methods that relate to the content of this chapter. The interested reader is referred to [48] for a comparative study of various techniques.

One of the first attempts to develop an efficient MKL scheme is the one presented in [65], where the authors considered a linear combination of kernel matrices, that is, $K = \sum_{m=1}^{M} a_m K_m$. Because we require the new kernel matrix to be positive definite, it is reasonable to impose in the optimization task some additional constraints. For example, one may adopt the general constraint $K \geq 0$ (the inequality indicating semidefinite positiveness), or a more strict one, for example, $a_m > 0$, for all $m = 1, \dots, M$. Furthermore, one needs to bound the norm of the final kernel matrix. Hence, the general MKL SVM task can be cast as

$$\begin{aligned} \text{minimize with respect to } K \quad & \omega_C(K), \\ \text{subject to} \quad & K \geq 0, \\ & \text{trace}\{K\} \leq c, \end{aligned} \tag{11.92}$$

where $\omega_C(K)$ is the solution of the dual SVM task, given in (11.75)–(11.77), which can be written in a more compact form as

$$\omega_C(K) = \max_{\boldsymbol{\lambda}} \left\{ \boldsymbol{\lambda}^T \mathbf{1} - \frac{1}{2}\boldsymbol{\lambda}^T G(K)\boldsymbol{\lambda} : 0 \leq \lambda_i \leq C, \ \boldsymbol{\lambda}^T \boldsymbol{y} = 0 \right\},$$

with each element of $G(K)$ given as $[G(K)]_{i,j} = [K]_{i,j} y_i y_j$; $\boldsymbol{\lambda}$ denotes the vector of the Lagrange multipliers and $\mathbf{1}$ is the vector having all its elements equal to one. In [65], it is shown how (11.92) can be transformed to a semidefinite programming optimization (SDP) task and solved accordingly.

Another path that has been exploited by many authors is to assume that the modeling nonlinear function is given as a summation,

$$\begin{aligned} f(\boldsymbol{x}) &= \sum_{m=1}^{M} a_m f_m(\boldsymbol{x}) = \sum_{m=1}^{M} a_m \langle f_m, \kappa_m(\cdot, \boldsymbol{x}) \rangle_{\mathbb{H}_m} + b \\ &= \sum_{m=1}^{M} \sum_{n=1}^{N} \theta_{m,n} a_m \kappa_m(\boldsymbol{x}, \boldsymbol{x}_n) + b, \end{aligned}$$

where each one of the functions, f_m, $m = 1, 2, \ldots M$, lives in a different RKHS, $\mathbb{H}_m$. The respective composite kernel matrix, associated with a set of training data, is given by $K = \sum_{m=1}^{M} a_m^2 K_m$, where $K_1, \ldots, K_M$ are the kernel matrices of the individual RKHSs. Hence, assuming a data set $\{(y_n, \boldsymbol{x}_n),\ n = 1, \ldots N\}$, the MKL learning task can be formulated as follows:

$$\min_f \sum_{n=1}^{N} \mathcal{L}(y_n, f(\boldsymbol{x}_n)) + \lambda \Omega(f), \tag{11.93}$$

where $\mathcal{L}$ represents a loss function and $\Omega(f)$ the regularization term. There have been two major trends following this rationale. The first one gives priority toward a sparse solution, while the second aims at improving performance.

In the context of the first trend, the solution is constrained to be sparse, so that the kernel matrix is computed fast and the strong similarities between the data set are highlighted. Moreover, this rationale can be applied to the case where the type of kernel has been selected beforehand and the goal is to compute the optimal kernel parameters. One way (e.g., [4,116]) is to employ a regularization term of the form $\Omega(f) = \left(\sum_{m=1}^{M} a_m \| f_m \|_{\mathbb{H}_m} \right)^2$, which has been shown to promote sparsity among the set $\{a_1, \ldots a_M\}$, as it is associated to the group LASSO, when the squared error loss is employed in place of $\mathcal{L}$ (see Chapter 10).

In contrast to the sparsity promoting criteria, another trend (e.g., [30,63,92]) revolves around the argument that, in some cases, the sparse MKL variants may not exhibit improved performance compared to the original learning task. Moreover, some data sets contain multiple similarities between individual data pairs that cannot be highlighted by a single type of kernel, but require a number of different kernels to improve learning. In this context, a regularization term of the form $\Omega(f) = \sum_{m=1}^{M} a_m \| f_m \|_{\mathbb{H}_m}^2$ is preferred. There are several variants of these methods that either employ additional constraints to the task (e.g., $\sum_{m=1}^{M} a_m = 1$) or define the summation of the spaces a little differently (e.g., $f(\cdot) = \sum_{m=1}^{M} \frac{1}{a_m} f_m(\cdot)$). For example, in [92] the authors reformulate (11.93) as follows:

$$\begin{aligned} &\text{minimize with respect to } \boldsymbol{a} && J(\boldsymbol{a}), \\ &\text{subject to} && \sum_{m=1}^{M} a_m = 1, \\ & && a_m \geq 0, \end{aligned} \tag{11.94}$$

where

$$J(\boldsymbol{a}) = \min_{f_{1:M}, \xi, b} \left\{ \begin{array}{c} \frac{1}{2}\sum_{m=1}^{M} \frac{1}{a_m} \|f_m\|_{\mathbb{H}_m}^2 + C \sum_{n=1}^{N} \xi_n, \\ y_i \left(\sum_{m=1}^{M} f_m(\boldsymbol{x}_n) + b \right) \geq 1 - \xi_n, \\ \xi_n \geq 0. \end{array} \right\}$$

The optimization is cast in RKHSs; however, the problem is always formulated in such a way so that the kernel trick can be mobilized.

11.14 NONPARAMETRIC SPARSITY-AWARE LEARNING: ADDITIVE MODELS

We have already pointed out that the representer theorem, as summarized by (11.17), provides an approximation of a function, residing in an RKHS, in terms of the respective kernel centered at the points $\boldsymbol{x}_1, \ldots, \boldsymbol{x}_N$. However, we know that the accuracy of any interpolation/approximation method depends on the number of points, N. Moreover, as discussed in Chapter 3, how large or small N needs to be depends heavily on the dimensionality of the space, exhibiting an exponential dependence on it (curse of dimensionality); basically, one has to fill in the input space with "enough" data in order to be able to "learn" with good enough accuracy the associated function.[14] In Chapter 7, the naive Bayes classifier was discussed; the essence behind this method is to consider each dimension of the input random vectors, $\mathbf{x} \in \mathbb{R}^l$, *individually*. Such a path breaks the problem into a number l of *one-dimensional* tasks. The same idea runs across the so-called *additive* models approach.

According to the additive models rationale, the unknown function is constrained within the family of *separable* functions, that is,

$$\boxed{f(\boldsymbol{x}) = \sum_{i=1}^{l} h_i(x_i): \quad \text{additive model},} \tag{11.95}$$

where $\boldsymbol{x} = [x_1, \ldots, x_l]^T$. Recall that a special case of such expansions is the linear regression, where $f(\boldsymbol{x}) = \boldsymbol{\theta}^T \boldsymbol{x}$.

We will further assume that each one of the functions, $h_i(\cdot)$, belongs to an RKHS, $\mathbb{H}_i$, defined by a respective kernel, $\kappa_i(\cdot,\cdot)$,

$$\kappa_i(\cdot,\cdot): \mathbb{R} \times \mathbb{R} \longmapsto \mathbb{R}.$$

[14] Recall from the discussion in Section 11.10.4 that the other factor that ties accuracy and N together is the rate of variation of the function.

Let the corresponding norm be denoted as $\|\cdot\|_i$. For the regularized total squared error cost [93], the optimization task is now cast as

$$\text{minimize with respect to } f \quad \frac{1}{2}\sum_{n=1}^{N}\big(y_n - f(\boldsymbol{x}_n)\big)^2 + \lambda\sum_{i=1}^{l}\|h_i\|_i, \tag{11.96}$$

$$\text{s.t.} \quad f(\boldsymbol{x}) = \sum_{i=1}^{l} h_i(x_i). \tag{11.97}$$

If one plugs (11.97) into (11.96) and following arguments similar to those used in Section 11.6, it is readily obtained that we can write

$$\hat{h}_i(\cdot) = \sum_{n=1}^{N}\theta_{i,n}\kappa_i(\cdot, x_{i,n}), \tag{11.98}$$

where $x_{i,n}$ is the ith component of $\boldsymbol{x}_n$. Moving along the same path as that adopted in Section 11.7, the optimization can be rewritten in terms of

$$\boldsymbol{\theta}_i = [\theta_{i,1}, \ldots, \theta_{i,N}]^T, \quad i = 1, 2, \ldots, l,$$

as

$$\{\hat{\boldsymbol{\theta}}_i\}_{i=1}^{l} = \arg\min_{\{\boldsymbol{\theta}_i\}_{i=1}^{l}} J(\boldsymbol{\theta}_1, \ldots, \boldsymbol{\theta}_l),$$

where

$$J(\boldsymbol{\theta}_1, \ldots, \boldsymbol{\theta}_l) := \frac{1}{2}\left\|\boldsymbol{y} - \sum_{i=1}^{l}\mathcal{K}_i\boldsymbol{\theta}_i\right\|^2 + \lambda\sum_{i=1}^{l}\sqrt{\boldsymbol{\theta}_i^T\mathcal{K}_i\boldsymbol{\theta}_i}, \tag{11.99}$$

and $\mathcal{K}_i$, $i = 1, 2, \ldots, l$, are the respective $N \times N$ kernel matrices

$$\mathcal{K}_i := \begin{bmatrix} \kappa_i(x_{i,1}, x_{i,1}) & \cdots & \kappa_i(x_{i,1}, x_{i,N}) \\ \vdots & \ddots & \vdots \\ \kappa_i(x_{i,N}, x_{i,1}) & \cdots & \kappa_i(x_{i,N}, x_{i,N}) \end{bmatrix}.$$

Observe that (11.99) is a (weighted) version of the *group* LASSO, defined in Section 10.3. Thus, the optimization task enforces sparsity by pushing some of the vectors $\boldsymbol{\theta}_i$ to zero values. Any algorithm developed for the group LASSO can be employed here as well (see, for example, [10,93]).

Besides the squared error loss, other loss functions can also be employed. For example, in [93] the *logistic regression* model is also discussed. Moreover, if the separable model in (11.95) cannot adequately capture the whole structure of f, models involving combination of components can be considered, such as the ANOVA model (e.g., [67]).

Analysis of variance (ANOVA) is a method in statistics to analyze interactions among variables. According to this technique, a function $f(\boldsymbol{x})$, $\boldsymbol{x} \in \mathbb{R}^l$, $l > 1$, is decomposed into a number of terms; each term is given as a sum of functions involving a *subset* of the components of $\boldsymbol{x}$. From this point of view, separable functions of the form in (11.95) are a special case of an ANOVA decomposition. A more general decomposition would be

$$f(\boldsymbol{x}) = \theta_0 + \sum_{i=1}^{l} h_i(x_i) + \sum_{i<j=1}^{l} h_{ij}(x_i, x_j) + \sum_{i<j<k=1}^{l} h_{ijk}(x_i, x_j, x_k), \tag{11.100}$$

and this can be generalized to involve larger number of components. Comparing (11.100) with (11.2), ANOVA decomposition can be considered as a generalization of the polynomial expansion. The idea of ANOVA decomposition has already been used to decompose kernels, $\kappa(\boldsymbol{x}, \boldsymbol{y})$, in terms of other kernels that are functions of a subset of the involved components of $\boldsymbol{x}$ and $\boldsymbol{y}$, giving rise to the so-called ANOVA kernels [115].

Techniques involving sparse additive kernel-based models have been used in a number of applications, such as matrix and tensor completion, completion of gene expression, kernel-based dictionary learning, network flow, and spectrum cartography (see [10] for a related review).

11.15 A CASE STUDY: AUTHORSHIP IDENTIFICATION

Text mining is the part of data mining that analyzes textual data to detect, extract, represent, and evaluate patterns that appear in texts and can be transformed into real-world knowledge. A few examples of text mining applications include spam e-mail detection, topic-based classification of texts, sentiment analysis in texts, text authorship identification, text indexing, and text summarization. In other words, the focus can be on detecting basic morphology idiosyncracies in a text that identify its author (authorship identification), on identifying complexly expressed emotions (sentiment analysis), or on condensing redundant information from multiple texts to a concise summary (summarization).

In this section, our focus will be on the case of authorship identification (for example, [114]), which is a special case of text classification. The task is to determine the author of a given text, given a set of training texts labeled with their corresponding author. To fulfill this task, one needs to represent and act upon textual data. Thus, the first decision related to text mining is how one represents the data.

It is very common to represent texts in a vector space, following the vector space model (VSM) [97]. According to the VSM, a text document, T, is represented by a set of terms w_i, $0 \leq i \leq k$, $k \in \mathbb{N}$, each mapped to a dimension of the vector $\mathbf{t} = [t_1, t_2, \ldots, t_k]^T \in \mathbb{R}^k$. The elements, t_i, in each dimension indicate the importance of the corresponding term w_i when describing the document. For example, each w_i can be one word in the vocabulary of a language. Thus, in practical applications, k can be very large, as large as the number of words that are considered sufficient for the specific task of interest. Typical examples of k are of the order of 10^5.

Widely used approaches to assign values to t_i are the following:

- *Bag-of-words, frequency approach*: The bag-of-words approach to text relies on the assumption that a text T is nothing more than a histogram of the words it contains. Thus, in the bag-of-words assumption, we do not care about the order of the words in the original text. What we care about is how many times term w_i appears in the document. Thus, t_i is assigned the number of occurrences of w_i in T.

- *Bag-of-words, Boolean approach*: The Boolean approach to the bag-of-words VSM model restricts the t_i values, such that $t_i \in \{0, 1\}$; t_i is assigned a value of 1 if w_i is found at least once in T, otherwise $t_i = 0$.

The bag-of-words approaches have proven efficient, for example, in early spam filtering applications, where individual words were clear indicators of whether an e-mail is spam. However, the spammers soon adapted and changed the texts they sent out by breaking words using spaces: the word "pills" would be replaced by "p i l l s." The bag-of-words approach then needed to apply preprocessing to deal with such cases, or even to deal with changes in different word forms (via word stemming or lemmatization) or spelling mistakes (by using dictionaries to correct the errors).

An alternative representation, with high resilience to noise, is that of the *character n-grams*. This representation is based on character subsequences of a text of length n (also called the order of an n-gram). To represent a text using n-grams, we split the text T in (usually contiguous) groups of characters of length n, which are then mapped to dimensions in the vector space (bag-of-n-grams). Below we give an example on how n-grams can be extracted from a given text.

Example 11.5. Given the input text T =*A_fine_day_today*, the character and word 2-grams ($n = 2$) would be as follows:

- *Unique* character 2-grams: "*A_*", "*_f*", "*fi*", "*in*", "*ne*", "*e_*", "*_d*", "*da*", "*ay*", "*y_*", "*_t*", "*to*", "*od*".
- Unique word 2-grams: "*A_fine*", "*fine_day*", "*day_today*".

Observe that "ay" and "da" appear twice in the phrase, yet only once in the sequence, since each n-gram is uniquely represented. Even though this approach has proven very useful [20], the sequence of the n-grams is still not exploited.

In the following, we describe a more complex representation, which takes into account the sequence of n-grams, and at the same time allows for noise. The representation is termed *n-gram graph* [46]. An n-gram graph represents the way on how n-grams *cooccur* in a text.

Definition 11.4. If $S = \{S_1, S_2, \ldots\}$, $S_k \neq S_l$, for $k \neq l$, $k, l \in \mathrm{N}$ is the set of *distinct* n-grams extracted from a text T, and S_i is the ith extracted n-gram, then $G = \{V, E, W\}$ is a graph where $V = S$ is the set of vertices v, E is the set of edges e of the form $e = \{v_1, v_2\}$, and $W : E \to \mathbb{R}$ is a function assigning a weight w_e to every edge.

To generate an n-gram graph, we first extract the unique n-grams from a text, creating one vertex for each. Then, given a user-defined parameter distance D, we consider the n-grams that are found *within* a distance D of each other in the text; these are considered *neighbors* and their respective vertices are connected with an edge. For each edge, a weight is assigned. The edge weighting function that is most commonly used in n-gram graphs is the number of cooccurrences of the linked (in the graph) n-grams in the text.

Two examples of 2-gram graphs, with directed neighborhood edges, with $D = 2$ are presented in Figs. 11.19 and 11.20.

Between two n-gram graphs, G^i and G^j, there exists a symmetric, normalized similarity function called *value similarity* (VS). This measure quantifies the ratio of common edges between two graphs, taking into account the ratio of weights of common edges. In this measure, each matching edge e, having weights w_e^i and w_e^j, in graphs G^i and G^i, respectively, contributes to VS the amount $\frac{\mathrm{VR}(e)}{\max(|G^i|,|G^j|)}$,

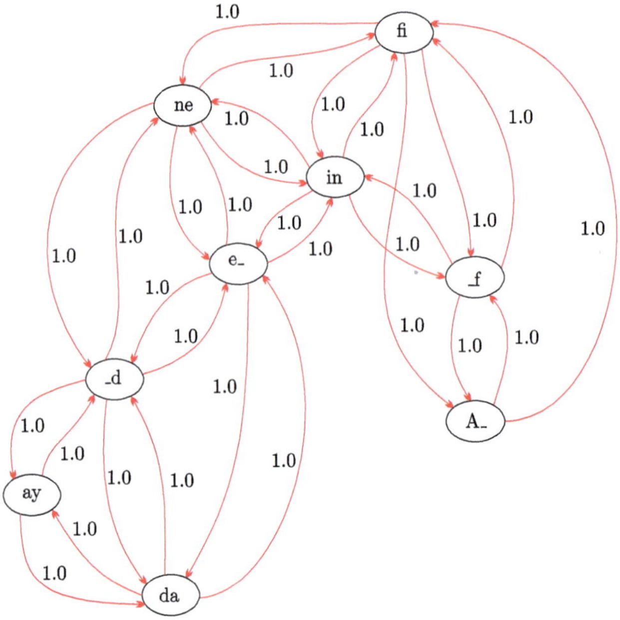

FIGURE 11.19

The 2-gram graph ($n = 2$), with $D = 2$ of the string "A fine day."

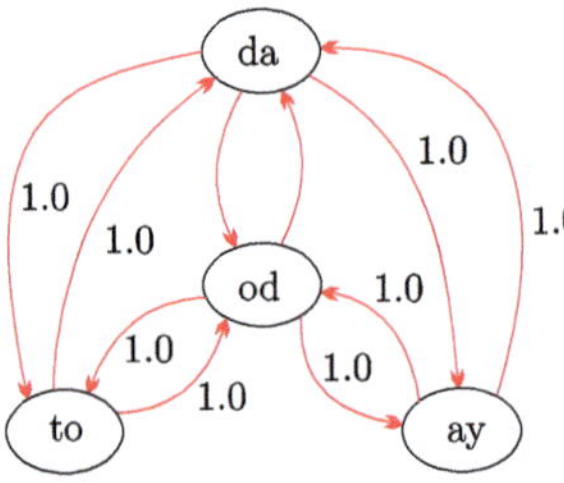

FIGURE 11.20

The 2-gram graph ($n = 2$), with $D = 2$ of the string "today."

where $|\cdot|$ indicates cardinality with respect to the number of edges and

$$\mathrm{VR}(e) := \frac{\min(w_e^i, w_e^j)}{\max(w_e^i, w_e^j)}. \tag{11.101}$$

Not matching edges have no contribution (consider that for an edge $e \notin G^i(G^j)$ we define $w_e^i = 0$ ($w_e^j = 0$)). The equation indicates that the *Value ratio* (VR) takes values in [0, 1], and is symmetric. Thus, the full equation for the VS is

$$\mathrm{VS}(G^i, G^j) = \frac{\sum_{e \in G^i} \frac{\min(w_e^i, w_e^j)}{\max(w_e^i, w_e^j)}}{\max(|G^i|, |G^j|)}. \tag{11.102}$$

The VS takes a value close to 1 for graphs that share many edges with similar weights. A value of VS = 1 indicates a perfect match between the compared graphs. For the two graphs shown in Figs. 11.19 and 11.20, the calculation of the VS returns a value of VS = 0.067 (verify it).

For our case study, the VS will be used to build an SVM kernel classifier, in order to assign (classify) texts to their authors. We should stress here that the VS function is a similarity function, but not a kernel. Fortunately, the VS on the specific data set is an (ϵ, γ)-good similarity function as defined in [6, Theorem 3], and hence this allows its use as a kernel function, in line with what we have already said about string kernels in Section 11.5.2.

It is interesting to note that given the VSM representation of a text, one can also employ any vector-based kernel on strings. However, there have been several works that use *string kernels* [69] to avoid any loss of information in the transformation between the original string and its VSM equivalent. In this example, we demonstrate how one can combine even a complex representation with the SVM via an appropriate kernel function.

The data set we used to examine the effectiveness of this combination is a subset of the Reuter_50_50 Data Set from the UCI repository [5]. This data set contains 2500 training and 2500 test texts from 50 different authors (50 training and 50 test texts per author).

- First, we used the ngg2svm command line tool[15] to represent the strings as n-gram graphs and generate a precomputed kernel matrix file. The tool extracts are—by default—3-gram graphs ($n = 3$, $D = 3$) from training texts. It then uses the VS to estimate a kernel and calculates the kernel values over all pairs of training instances into a kernel matrix. The matrix is the output of this phase.
- Given the kernel matrix, we can use the LibSVM software [24] to design a classifier, without caring about the original texts, but only relying on precomputer kernel values.

Then, a 10-fold cross-validation over the data was performed, using the "svmtrain" program in LibSVM. The achieved accuracy of the cross-validation in our example was 94%.

PROBLEMS

11.1 Derive the formula for the number of groupings $\mathcal{O}(N, l)$ in Cover's theorem.
Hint: Show first the following recursion:

$$\mathcal{O}(N+1, l) = \mathcal{O}(N, l) + \mathcal{O}(N, l-1).$$

[15] You can download the tool from https://github.com/ggianna/ngg2svm.

To this end, start with N points and add an extra one. Show that the extra number of linear dichotomies is solely due to those, for the N data point case, which could be drawn via the new point.

11.2 Show that if $N = 2(l+1)$, the number of linear dichotomies in Cover's theorem is equal to 2^{2l+1}.

Hint: Use the identity

$$\sum_{i=1}^{j} \binom{j}{i} = 2^j$$

and recall that

$$\binom{2n+1}{n-i+1} = \binom{2n+1}{n+i}.$$

11.3 Show that the reproducing kernel is a positive definite one.

11.4 Show that if $\kappa(\cdot,\cdot)$ is the reproducing kernel in an RKHS $\mathbb{H}$, then

$$\mathbb{H} = \overline{\text{span}\{\kappa(\cdot, \boldsymbol{x}),\ \boldsymbol{x} \in \mathcal{X}\}}.$$

11.5 Show the Cauchy–Schwarz inequality for kernels, that is,

$$\|\kappa(\boldsymbol{x}, \boldsymbol{y})\|^2 \leq \kappa(\boldsymbol{x}, \boldsymbol{x})\kappa(\boldsymbol{y}, \boldsymbol{y}).$$

11.6 Show that if

$$\kappa_i(\cdot,\cdot) : \mathcal{X} \times \mathcal{X} \longmapsto \mathbb{R},\ i = 1, 2,$$

are kernels, then

- $\kappa(\boldsymbol{x}, \boldsymbol{y}) = \kappa_1(\boldsymbol{x}, \boldsymbol{y}) + \kappa_2(\boldsymbol{x}, \boldsymbol{y})$ is also a kernel;
- $a\kappa(\boldsymbol{x}, \boldsymbol{y}),\ a > 0$, is also a kernel;
- $\kappa(\boldsymbol{x}, \boldsymbol{y}) = \kappa_1(\boldsymbol{x}, \boldsymbol{y})\kappa_2(\boldsymbol{x}, \boldsymbol{y})$ is also a kernel.

11.7 Derive Eq. (11.25).

11.8 Show that the solution for the parameters $\hat{\boldsymbol{\theta}}$ for the kernel ridge regression, if a bias term b is present, is given by

$$\begin{bmatrix} \mathcal{K} + CI & \mathbf{1} \\ \mathbf{1}^T\mathcal{K} & N \end{bmatrix} \begin{bmatrix} \boldsymbol{\theta} \\ b \end{bmatrix} = \begin{bmatrix} \boldsymbol{y} \\ \boldsymbol{y}^T\mathbf{1} \end{bmatrix},$$

where $\mathbf{1}$ is the vector with all its elements being equal to one. Invertibility of the kernel matrix has been assumed.

11.9 Derive Eq. (11.56).

11.10 Derive the dual cost function associated with the linear ϵ-insensitive loss function.

11.11 Derive the dual cost function for the separable class SVM formulation.

11.12 Derive the kernel approximation in Eq. (11.91).

11.13 Derive the subgradient for the Huber loss function.

MATLAB® EXERCISES

11.14 Consider the regression problem described in Example 11.2. Read an audio file using MATLAB®'s wavread function and take 100 data samples (use *Blade Runner*, if possible, and take 100 samples starting from the 100,000th sample). Then add white Gaussian noise at a 15 dB level and randomly "hit" 10 of the data samples with outliers (set the outlier values to 80% of the maximum value of the data samples).

(a) Find the reconstructed data samples using the unbiased kernel ridge regression method, that is, Eq. (11.27). Employ the Gaussian kernel with $\sigma = 0.004$ and set $C = 0.0001$. Plot the fitted curve of the reconstructed samples together with the data used for training.

(b) Find the reconstructed data samples using the biased kernel ridge regression method (employ the same parameters as for the unbiased case). Plot the fitted curve of the reconstructed samples together with the data used for training.

(c) Repeat steps (a) and (b) using $C = 10^{-6},\ 10^{-5}, 0.0005,\ 0.001,\ 0.01,$ and 0.05.

(d) Repeat steps (a) and (b) using $\sigma = 0.001,\ 0.003,\ 0.008,\ 0.01,$ and 0.05.

(e) Comment on the results.

11.15 Consider the regression problem described in Example 11.2. Read the same audio file as in Problem 11.14. Then add white Gaussian noise at a 15 dB level and randomly "hit" 10 of the data samples with outliers (set the outlier values to 80% of the maximum value of the data samples).

(a) Find the reconstructed data samples obtained by the support vector regression (you can use libsvm[16] for training). Employ the Gaussian kernel with $\sigma = 0.004$ and set $\epsilon = 0.003$ and $C = 1$. Plot the fitted curve of the reconstructed samples together with the data used for training.

(b) Repeat step (a) using $C = 0.05,\ 0.1,\ 0.5,\ 5,\ 10,$ and 100.

(c) Repeat step (a) using $\epsilon = 0.0005,\ 0.001,\ 0.01,\ 0.05,$ and 0.1.

(d) Repeat step (a) using $\sigma = 0.001,\ 0.002,\ 0.01,\ 0.05,$ and 0.1.

(e) Comment on the results.

11.16 Consider the two-class two-dimensional classification task described in the SVM example in the book and slides. The data set comprises $N = 150$ points, $\boldsymbol{x}_n = [x_{n,1},\ x_{n,2}]^T$, $n = 1, 2, \ldots, N$, uniformly distributed in $[-5, 5] \times [-5, 5]$. For each point, compute

$$y_n = 0.05x_{n,1}^3 + 0.05x_{n,1}^2 + 0.05x_{n,1} + 0.05 + \eta,$$

where η denotes zero mean Gaussian noise of variance $\sigma_\eta^2 = 4$. Then, if $x_{n,2} \geq y_n$, assign it to ω_1, and if $x_{n,2} < y_n$, assign it to class ω_2.

(a) Plot the points $[x_{n,1}, x_{n,2}]$ using different colors for each class.

(b) Obtain the SVM classifier using libsvm or any other related MATLAB® package. Use the Gaussian kernel with $\sigma = 20$ and set $C = 1$. Plot the classifier and the margin (for the latter you can employ MATLAB®'s contour function). Moreover, find the support vectors (i.e., the points with nonzero Lagrange multipliers that contribute to the expansion of the classifier) and plot them as circled points.

[16] http://www.csie.ntu.edu.tw/cjlin/libsvm/.

(c) Repeat step 11.16b using $C = 0.5,\ 0.1$, and 0.05.
(d) Repeat step 11.16b using $C = 5,\ 10,\ 50$, and 100.
(e) Comment on the results.

11.17 Consider the authorship identification problem described in Section 11.15.
(a) Using the training texts of the authors whose names start with "T" in the data set, perform 10-fold cross-validation using the n-gram graph representation of the texts and the VS as a kernel function.
(b) Using the same subset of the data set, create the vector space model representation of the texts and apply classification using a Gaussian kernel. Compare the results with the results of (i).

REFERENCES

[1] D. Achlioptas, F. McSherry, B. Schölkopf, Sampling techniques for kernel methods, in: Advances in Neural Information Processing Systems, NIPS, 2001.
[2] A. Argyriou, C.A. Micchelli, M. Pontil, When is there a representer theorem? Vector versus matrix regularizers, J. Mach. Learn. Res. 10 (2009) 2507–2529.
[3] N. Aronszajn, Theory of reproducing kernels, Trans. Am. Math. Soc. 68 (3) (1950) 337–404.
[4] F.R. Bach, Consistency of the group LASSO and multiple kernel learning, J. Mach. Learn. Res. 9 (2008) 1179–1225.
[5] K. Bache, M. Lichman, UCI Machine Learning Repository, University of California, Irvine, School of Information and Computer Sciences, http://archive.ics.uci.edu/ml, 2013.
[6] M.F. Balcan, A. Blum, N. Srebro, A theory of learning with similarity functions, Mach. Learn. 72 (1–2) (2008) 89–112.
[7] A.R. Barron, Universal approximation bounds for superposition of a sigmoid function, Trans. Inf. Theory 39 (3) (1993) 930–945.
[8] E.J. Bayro-Corrochano, N. Arana-Daniel, Clifford support vector machines for classification, regression and recurrence, IEEE Trans. Neural Netw. 21 (11) (2010) 1731–1746.
[9] J.A. Bazerque, G.B. Giannakis, Nonparametric basis pursuit via sparse kernel-based learning, IEEE Signal Process. Mag. 30 (4) (2013) 112–125.
[10] J.A. Bazerque, G. Mateos, G.B. Giannakis, Group-LASSO on splines for spectrum cartography, Trans. Signal Process. 59 (10) (2011) 4648–4663.
[11] S. Beneteto, E. Bigleiri, Principles of Digital Transmission, Springer, 1999.
[12] Y. Bengio, O. Delalleau, N. Le Roux, The curse of highly variable functions for local kernel machines, in: Y. Weiss, B. Scholkopf, J. Platt (Eds.), Advances in Neural Information Processing Systems, NIPS, MIT Press, 2006, pp. 107–114.
[13] P. Bouboulis, K. Slavakis, S. Theodoridis, Adaptive kernel-based image denoising employing semi-parametric regularization, IEEE Trans. Image Process. 19 (6) (2010) 1465–1479.
[14] P. Bouboulis, S. Theodoridis, C. Mavroforakis, L. Dalla, Complex support vector machines for regression and quaternary classification, IEEE Trans. Neural Netw. Learn. Syst. 26 (6) (2015) 1260–1274.
[15] P. Bouboulis, S. Theodoridis, Kernel methods for image denoising, in: J.A.K. Suykens, M. Signoretto, A. Argyriou (Eds.), Regularization, Optimization, Kernels, and Support Vector Machines, Chapman and Hall/CRC, Boca Raton, FL, 2014.
[16] P. Bouboulis, S. Chouvardas, S. Theodoridis, Online distributed learning over networks in RKH spaces using random Fourier features, IEEE Trans. SP 66 (7) (2018) 1920–1932.
[17] P. Bouboulis, S. Theodoridis, S. Chouvardas, A random Fourier features perspective of KAFs with application to distributed learning over networks, in: D. Comminielo, J. Principe (Eds.), Adaptive Learning Methods in Nonlinear Modelling, Butterworth-Heinemann, 2018.
[18] O. Bousquet, A. Elisseeff, Stability and generalization, J. Mach. Learn. Res. 2 (2002) 499–526.
[19] C.J.C. Burges, Geometry and invariance in kernel based methods, in: B. Scholkopf, C.J. Burges, A.J. Smola (Eds.), Advances in Kernel Methods, MIT Press, 1999, pp. 90–116.
[20] W.B. Cavnar, J.M. Trenkle, et al., N-gram-based text categorization, in: Symposium on Document Analysis and Information Retrieval, University of Nevada, Las Vegas, 1994, pp. 161–176.

[21] L.J. Cao, S.S. Keerthi, C.J. Ong, J.Q. Zhang, U. Periyathamby, X.J. Fu, H.P. Lee, Parallel sequential minimal optimization for the training of support vector machines, IEEE Trans. Neural Netw. 17 (4) (2006) 1039–1049.
[22] C.C. Chang, C.W. Hsu, C.J. Lin, The analysis of decomposition methods for SVM, IEEE Trans. Neural Netw. 11 (4) (2000) 1003–1008.
[23] C.C. Chang, C.J. Lin, Training ν-support vector classifiers: theory and algorithms, Neural Comput. 13 (9) (2001) 2119–2147.
[24] C.-C. Chang, C.J. Lin, LIBSVM: a library for support vector machines, ACM Trans. Intell. Syst. Technol. 2 (3) (2011) 27.
[25] R. Caruana, A. Niculescu-Mizil, An empirical comparison of supervised learning algorithms, in: International Conference on Machine Learning, 2006.
[26] G. Cauwenberghs, T. Poggio, Incremental and decremental support vector machine learning, in: Advances in Neural Information Processing Systems, NIPS, MIT Press, 2001, pp. 409–415.
[27] O. Chapelle, Training a support vector machine in the primal, Neural Comput. 19 (5) (2007) 1155–1178.
[28] N. Cristianini, J. Shawe-Taylor, An Introduction to Support Vector Machines, Cambridge University Press, 2000.
[29] C. Cortes, P. Haffner, M. Mohri, Rational kernels: theory and algorithms, J. Mach. Learn. Res. 5 (2004) 1035–1062.
[30] C. Cortes, M. Mohri, A. Rostamizadeh, L_2 regularization for learning kernels, in: Proc. of the 25th Conference on Uncertainty in Artificial Intelligence, 2009, pp. 187–196.
[31] T.M. Cover, Geometrical and statistical properties of systems of linear inequalities with applications in pattern recognition, IEEE Trans. Electron. Comput. 14 (1965) 326.
[32] P. Crama, J. Schoukens, Hammerstein-Wiener system estimator initialization, Automatica 40 (9) (2004) 1543–1550.
[33] D.J. Crisp, C.J.C. Burges, A geometric interpretation of ν-SVM classifiers, in: Proceedings of Neural Information Processing, NIPS, vol. 12, MIT Press, 1999.
[34] T.G. Dietterich, G. Bakiri, Solving multiclass learning problems via error-correcting output codes, J. Artif. Intell. Res. 2 (1995) 263–286.
[35] G. Ding, Q. Wu, Y.-D. Yao, J. Wang, Y. Chen, Kernel-based learning for statistical signal processing in cognitive radio networks, IEEE Signal Process. Mag. 30 (4) (2013) 126–136.
[36] P. Drineas, M.W. Mahoney, On the Nyström method for approximating a Gram matrix for improved kernel-based learning, J. Mach. Learn. Res. (JMLR) 6 (2005) 2153–2175.
[37] C. Domeniconi, D. Gunopulos, Incremental support vector machine construction, in: IEEE International Conference on Data Mining, San Jose, USA, 2001, pp. 589–592.
[38] Y. Engel, S. Mannor, R. Meir, The kernel recursive least-squares algorithm, IEEE Trans. Signal Process. 52 (8) (2004) 2275–2285.
[39] T. Evgeniou, M. Pontil, T. Poggio, Regularization networks and support vector machines, Adv. Comput. Math. 13 (2000) 1–50.
[40] T. Evgeniou, C.A. Michelli, M. Pontil, Learning multiple tasks with kernel methods, J. Mach. Learn. Res. 6 (2005) 615–637.
[41] B. Fei, J. Liu, Binary tree of SVM: a new fast multiclass training and classification algorithm, IEEE Trans. Neural Netw. 17 (5) (2006) 696–704.
[42] P.A. Forero, A. Cano, G.B. Giannakis, Consensus-based distributed support vector machines, J. Mach. Learn. Res. 11 (2010) 1663–1707.
[43] V. Franc, S. Sonnenburg, Optimized cutting plane algorithm for large-scale risk minimization, J. Mach. Learn. Res. 10 (2009) 2157–2192.
[44] M. Frechet, Sur les fonctionnelles continues, Ann. Sci. Éc. Supér. 27 (1910) 193–216.
[45] A. Frieze, R. Kannan, S. Vempala, Fast Monte-Carlo algorithms for finding low-rank approximations, in: Foundations of Computer Science, FOCS, 1998, pp. 378–390.
[46] G. Giannakopoulos, V. Karkaletsis, G. Vouros, P. Stamatopoulos, Summarization system evaluation revisited: N-gram graphs, ACM Trans. Speech Lang. Process. 5 (3) (2008) 1–9.
[47] F. Girosi, An equivalence between sparse approximation and support vector machines, Neural Comput. 10 (6) (1998) 1455–1480.
[48] M. Gonen, E. Alpaydin, Multiple kernel learning algorithms, J. Mach. Learn. Res. 12 (2011) 2211–2268.
[49] Z. Harchaoui, F. Bach, O. Cappe, E. Moulines, Kernel-based methods for hypothesis testing, IEEE Signal Process. Mag. 30 (4) (2013) 87–97.
[50] D. Hardoon, J. Shawe-Taylor, Decomposing the tensor kernel support vector machine for neuroscience data with structured labels, Neural Netw. 24 (8) (2010) 861–874.

[51] J.R. Higgins, Sampling Theory in Fourier and Signal Analysis Foundations, Oxford Science Publications, 1996.
[52] T. Hofmann, B. Scholkopf, A. Smola, Kernel methods in machine learning, Ann. Stat. 36 (3) (2008) 1171–1220.
[53] D. Hush, P. Kelly, C. Scovel, I. Steinwart, QP algorithms with guaranteed accuracy and run time for support vector machines, J. Mach. Learn. Res. 7 (2006) 733–769.
[54] P. Huber, Robust estimation of location parameters, Ann. Math. Stat. 35 (1) (1964) 73–101.
[55] T. Joachims, Making large scale support vector machine learning practical, in: B. Scholkopf, C.J. Burges, A.J. Smola (Eds.), Advances in Kernel Methods: Support Vector Learning, MIT Press, 1999.
[56] T. Joachims, T. Finley, C.-N. Yu, Cutting-plane training of structural SVMs, Mach. Learn. 77 (1) (2009) 27–59.
[57] N. Kalouptsidis, Signal Processing Systems: Theory and Design, John Wiley, 1997.
[58] S.S. Keerthi, S.K. Shevade, C. Bhattacharyya, K.R.K. Murthy, A fast iterative nearest point algorithm for support vector machine classifier design, IEEE Trans. Neural Netw. 11 (1) (2000) 124–136.
[59] S.S. Keerthi, S.K. Shevade, C. Bhattacharyya, K.R.K. Murthy, Improvements to Platt's SMO algorithm for SVM classifier design, Neural Comput. 13 (2001) 637–649.
[60] A.Y. Kibangou, G. Favier, Wiener-Hammerstein systems modeling using diagonal Volterra kernels coefficients, Signal Process. Lett. 13 (6) (2006) 381–384.
[61] G. Kimeldorf, G. Wahba, Some results on Tchebycheffian spline functions, J. Math. Anal. Appl. 33 (1971) 82–95.
[62] J. Kivinen, A.J. Smola, R.C. Williamson, Online learning with kernels, IEEE Trans. Signal Process. 52 (8) (2004) 2165–2176.
[63] M. Kloft, U. Brefeld, S. Sonnenburg, P. Laskov, K.R. Müller, A. Zien, Efficient and accurate lp-norm multiple kernel learning, Adv. Neural Inf. Process. Syst. 22 (2009) 997–1005.
[64] S.Y. Kung, Kernel Methods and Machine Learning, Cambridge University Press, 2014.
[65] G. Lanckriet, N. Cristianini, P. Bartlett, L. El Ghaoui, M. Jordan, Learning the kernel matrix with semidefinite programming, in: C. Summu, A.G. Hoffman (Eds.), Proceedings of the 19th International Conference on Machine Learning, ICML, Morgan Kaufmann, 2002, pp. 323–330.
[66] J.L. Lázaro, J. Dorronsoro, Simple proof of convergence of the SMO algorithm for different SVM variants, IEEE Trans. Neural Netw. Learn. Syst. 23 (7) (2012) 1142–1147.
[67] Y. Lin, H.H. Zhang, Component selection and smoothing in multivariate nonparametric regression, Ann. Stat. 34 (5) (2006) 2272–2297.
[68] W. Liu, J.C. Principe, S. Haykin, Kernel Adaptive Filtering: A Comprehensive Introduction, John Wiley, 2010.
[69] H. Lodhi, C. Saunders, J. Shawe-Taylor, N. Cristianini, C. Watkins, Text classification using string kernels, J. Mach. Learn. Res. 2 (2002) 419–444.
[70] V.J. Mathews, G.L. Sicuranza, Polynomial Signal Processing, John Wiley, New York, 2000.
[71] G. Mateos, G.B. Giannakis, Robust nonparametric regression via sparsity control with application to load curve data cleansing, IEEE Trans. Signal Process. 60 (4) (2012) 1571–1584.
[72] P.Z. Marmarelis, V.Z. Marmarelis, Analysis of Physiological Systems – The White Noise Approach, Plenum, New York, 1978.
[73] M. Mavroforakis, S. Theodoridis, Support vector machine classification through geometry, in: Proceedings XII European Signal Processing Conference, EUSIPCO, Anatlya, Turkey, 2005.
[74] M. Mavroforakis, S. Theodoridis, A geometric approach to support vector machine classification, IEEE Trans. Neural Netw. 17 (3) (2006) 671–682.
[75] M. Mavroforakis, M. Sdralis, S. Theodoridis, A geometric nearest point algorithm for the efficient solution of the SVM classification task, IEEE Trans. Neural Netw. 18 (5) (2007) 1545–1549.
[76] J. Mercer, Functions of positive and negative type and their connection with the theory of integral equations, Phil. Trans. R. Soc. Lond. 209 (1909) 415–446.
[77] D. Meyer, F. Leisch, K. Hornik, The support vector machine under test, Neurocomputing 55 (2003) 169–186.
[78] C.A. Micceli, Interpolation of scattered data: distance matrices and conditionally positive definite functions, Constr. Approx. 2 (1986) 11–22.
[79] C.A. Miccheli, M. Pontil, Learning the kernel function via regularization, J. Mach. Learn. Res. 6 (2005) 1099–1125.
[80] K. Mitra, A. Veeraraghavan, R. Chellappa, Analysis of sparse regularization based robust regression approaches, IEEE Trans. Signal Process. 61 (2013) 1249–1257.
[81] G. Montavon, M.L. Braun, T. Krueger, K.R. Müller, Analysing local structure in kernel-based learning, IEEE Signal Process. Mag. 30 (4) (2013) 62–74.
[82] E.H. Moore, On properly positive Hermitian matrices, Bull. Am. Math. Soc. 23 (1916) 59.

[83] K. Muller, S. Mika, G. Ratsch, K. Tsuda, B. Scholkopf, An introduction to kernel-based learning algorithms, IEEE Trans. Neural Netw. 12 (2) (2001) 181–201.
[84] A. Navia-Vázquez, F. Perez-Cruz, A. Artes-Rodriguez, A. Figueiras-Vidal, Weighted least squares training of support vector classifiers leading to compact and adaptive schemes, IEEE Trans. Neural Netw. 15 (5) (2001) 1047–1059.
[85] G. Papageorgiou, P. Bouboulis, S. Theodoridis, K. Themelis, Robust linear regression analysis - a greedy approach, arXiv: 1409.4279 [cs.IT], 2014.
[86] G. Papageorgiou, P. Bouboulis, S. Theodoridis, Robust linear regression analysis - a greedy approach, IEEE Trans. Signal Process. (2015).
[87] V.I. Paulsen, An Introduction to Theory of Reproducing Kernel Hilbert Spaces, Notes, 2009.
[88] J. Platt, Sequential Minimal Optimization: A Fast Algorithm for Training Support Vector Machines, Technical Report, Microsoft Research, MSR-TR-98-14, April 21, 1998.
[89] J.C. Platt, Using analytic QP and sparseness to speed training of support vector machines, in: Proceedings Neural Information Processing Systems, NIPS, 1999.
[90] M.J.D. Powell, Radial basis functions for multivariate interpolation: a review, in: J.C. Mason, M.G. Cox (Eds.), Algorithms for Approximation, Clarendon Press, Oxford, 1987, pp. 143–167.
[91] A. Rahimi, B. Recht, Random features for large-scale kernel machines, in: Neural Information Processing Systems, NIPS, vol. 20, 2007.
[92] A. Rakotomamonjy, F.R. Bach, S. Canu, Y. Grandvalet, SimpleMKL, J. Mach. Learn. Res. 9 (2008) 2491–2521.
[93] P. Ravikumar, J. Lafferty, H. Liu, L. Wasserman, Sparse additive models, J. R. Stat. Soc. B 71 (5) (2009) 1009–1030.
[94] C. Richard, J. Bermudez, P. Honeine, Online prediction of time series data with kernels, IEEE Trans. Signal Process. 57 (3) (2009) 1058–1067.
[95] W. Rudin, Principles of Mathematical Analysis, third ed., McGraw-Hill, 1976.
[96] W. Rudin, Fourier Analysis on Groups, Wiley Classics Library, Wiley-Interscience, New York, 1994.
[97] G. Salton, A. Wong, C.-S. Yang, A vector space model for automatic indexing, Commun. ACM 18 (11) (1975) 613–620.
[98] C. Saunders, A. Gammerman, V. Vovk, Ridge regression learning algorithm in dual variables, in: J. Shavlik (Ed.), Proceedings 15th International Conference on Machine Learning, ICMM'98, Morgan Kaufman, 1998.
[99] P. Saurabh, C. Boutsidis, M. Magdon-Ismail, P. Drineas, Random projections for support vector machines, in: Proceedings of the 16th International Conference on Artificial Intelligence and Statistics, AISTATS, Scottsdale, AZ, USA, 2013.
[100] A. Shilton, M. Palaniswami, D. Ralph, A.C. Tsoi, Incremental training of support vector machines, IEEE Trans. Neural Netw. 16 (1) (2005) 114–131.
[101] M. Schetzen, Nonlinear system modeling based on the Wiener theory, Proc. IEEE 69 (12) (1981) 1557–1573.
[102] M. Schetzen, The Volterra and Wiener Theories of Nonlinear Systems, John Wiley, New York, 1980.
[103] B. Schölkopf, A.J. Smola, Learning With Kernels, MIT Press, Cambridge, 2001.
[104] B. Schölkopf, A.J. Smola, R.C. Williamson, P.L. Bartlett, New support vector algorithms, Neural Comput. 12 (2000) 1207–1245.
[105] S. Shalev-Shwartz, Y. Singer, N. Srebro, A. Cotter, PEGASOS: primal estimated sub-gradient solver for SVM, Math. Program. 127 (1) (2011) 3–30.
[106] J. Shawe-Taylor, N. Cristianini, Kernel Methods for Pattern Analysis, Cambridge University Press, 2004.
[107] M. Signoretto, L. De Lathauwer, J. Suykens, A kernel-based framework to tensorial data analysis, Neural Netw. 24 (8) (2011) 861–874.
[108] K. Slavakis, P. Bouboulis, S. Theodoridis, Online learning in reproducing kernel Hilbert spaces, in: Academic Press Library in Signal Processing, Signal Process. Theory Machine Learn., vol. 1, 2013, pp. 883–987.
[109] K. Slavakis, S. Theodoridis, I. Yamada, Online kernel-based classification using adaptive projections algorithms, IEEE Trans. Signal Process. 56 (7) (2008) 2781–2796.
[110] K. Slavakis, S. Theodoridis, Sliding window generalized kernel affine projection algorithm using projection mappings, EURASIP J. Adv. Signal Process. (2008) 735351, https://doi.org/10.1155/2008/735351.
[111] K. Slavakis, S. Theodoridis, I. Yamada, Adaptive constrained learning in reproducing kernel Hilbert spaces, IEEE Trans. Signal Process. 5 (12) (2009) 4744–4764.
[112] K. Slavakis, A. Bouboulis, S. Theodoridis, Adaptive multiregression in reproducing kernel Hilbert spaces, IEEE Trans. Neural Netw. Learn. Syst. 23 (2) (2012) 260–276.
[113] K. Slavakis, A. Bouboulis, S. Theodoridis, Online learning in reproducing kernel Hilbert spaces, in: S. Theodoridis, R. Chellapa (Eds.), E-reference for Signal Processing, Academic Press, 2013.
[114] E. Stamatatos, A survey of modern authorship attribution methods, J. Am. Soc. Inf. Sci. Technol. 60 (3) (2009) 538–556.

[115] M.O. Stitson, A. Gammerman, V. Vapnik, V. Vovk, C. Watkins, J. Weston, Support vector regression with ANOVA decomposition kernels, in: B. Scholkopf, C.J. Burges, A.J. Smola (Eds.), Advances in Kernel Methods: Support Vector Learning, MIT Press, 1999.
[116] S. Sonnenburg, G. Rätsch, C. Schaffer, B. Scholkopf, Large scale multiple kernel learning, J. Mach. Learn. Res. 7 (2006) 1531–1565.
[117] L. Song, K. Fukumizu, A. Gretton, Kernel embeddings of conditional distributions, IEEE Signal Process. Mag. 30 (4) (2013) 98–111.
[118] J.A.K. Suykens, J. Vandewalle, Least squares support vector machine classifiers, Neural Process. Lett. 9 (1999) 293–300.
[119] J.A.K. Suykens, T. van Gestel, J. de Brabanter, B. de Moor, J. Vandewalle, Least Squares Support Vector Machines, World Scientific, Singapore, 2002.
[120] J.A.K. Suykens, M. Signoretto, A. Argyriou (Eds.), Regularization, Optimization, Kernels, and Support Vector Machines, Chapman and Hall/CRC, Boca Raton, FL, 2014.
[121] R. Talmon, I. Cohen, S. Gannot, R. Coifman, Diffusion maps for signal processing, IEEE Signal Process. Mag. 30 (4) (2013) 75–86.
[122] Q. Tao, G.-W. Wu, J. Wang, A generalized S-K algorithm for learning ν-SVM classifiers, Pattern Recognit. Lett. 25 (10) (2004) 1165–1171.
[123] S. Theodoridis, M. Mavroforakis, Reduced convex hulls: a geometric approach to support vector machines, Signal Process. Mag. 24 (3) (2007) 119–122.
[124] S. Theodoridis, K. Koutroumbas, Pattern Recognition, fourth ed., Academic Press, 2009.
[125] F.A. Tobar, D.P. Mandic, Quaternion reproducing kernel Hilbert spaces: existence and uniqueness conditions, IEEE Trans. Inf. Theory 60 (9) (2014) 5736–5749.
[126] I.W. Tsang, J.T. Kwok, P.M. Cheung, Core vector machines: fast SVM training on very large data sets, J. Mach. Learn. Res. (JMLR) 6 (2005) 363–392.
[127] S. Van Vaerenbergh, M. Lazaro-Gredilla, I. Santamaria, Kernel recursive least-squares tracker for time-varying regression, IEEE Trans. Neural Netw. Learn. Syst. 23 (8) (2012) 1313–1326.
[128] V.N. Vapnik, The Nature of Statistical Learning, Springer, 2000.
[129] V.N. Vapnik, Statistical Learning Theory, Wiley, 1998.
[130] N. Wiener, Nonlinear Problems in Random Theory, Technology Press, MIT and Wiley, 1958.
[131] C.K.I. Williams, M. Seeger, Using the Nyström method to speed up kernel machines, in: Advances in Neural Information Processing Systems, NIPS, vol. 13, 2001.
[132] G.-X. Yuan, K.W. Chang, C.-J. Hsieh, C.J. Lin, A comparison of optimization methods and software for large-scale ℓ_1-regularized linear classification, J. Mach. Learn. Res. 11 (2010) 3183–3234.
[133] Q. Zhao, G. Zhou, T. Adali, L. Zhang, A. Cichocki, Kernelization of tensor-based models for multiway data analysis, IEEE Signal Process. Mag. 30 (4) (2013) 137–148.

CHAPTER

BAYESIAN LEARNING: INFERENCE AND THE EM ALGORITHM

12

CONTENTS

12.1 INTRODUCTION

The Bayesian approach to parameter inference was introduced in Chapter 3. Compared to other methods for parameter estimation we have covered, the Bayesian method adopts a radically different viewpoint. The unknown set of parameters are treated as random variables instead of as a set of fixed (yet unknown) values. This was a revolutionary idea at the time it was introduced by Bayes and later on by Laplace, as pointed out in Chapter 3. Even now, after more than two centuries, it may seem strange to assume that a physical phenomenon/mechanism is controlled by a set of random parameters. However, there is a subtle point here. Treating the underlying set of parameters as random variables, $\boldsymbol{\theta}$, we do not really imply a random nature for them. The associated randomness, in terms of the prior distribution

Machine Learning. https://doi.org/10.1016/B978-0-12-818803-3.00023-4

$p(\boldsymbol{\theta})$, encapsulates our *uncertainty* about their values, prior to receiving any measurements/observations. Stated differently, the prior distribution represents our *belief* about the different possible values, although only one of them is actually true. From this perspective, probabilities are viewed in a more open-minded way, that is, as measures of uncertainty, as discussed in the beginning of Chapter 2.

Recall that parameter learning from data is an inverse problem. Basically, all we do is deduce the "causes" (parameters) from the "effects" (observations). The Bayes theorem can be seen as an inversion procedure expressed in a probabilistic context. Indeed, given the set of observations, say, $\mathcal{X}$, which are controlled by the unknown set of parameters, we write

$$p(\boldsymbol{\theta}|\mathcal{X}) = \frac{p(\mathcal{X}|\boldsymbol{\theta})p(\boldsymbol{\theta})}{p(\mathcal{X})}.$$

All that is needed for the above inversion is to have a guess about $p(\boldsymbol{\theta})$. This term has brought a lot of controversy in the statistical community for a number of years. However, once a reasonable guess of the prior is available, a number of advantages associated with the Bayesian approach emerge, compared to the alternative route. The latter embraces methods that view the parameters deterministically as constants of unknown values, and they are also referred to as *frequentist* techniques. The term comes from the more classical view of probabilities as frequencies of occurrence of repeatable events. A typical example of this family of methods is the maximum likelihood approach, which estimates the values of the parameters by maximizing $p(\mathcal{X}|\boldsymbol{\theta})$; the value of the latter conditional probability density function (PDF) is solely controlled by the obtained observations in a sequence of experiments.

This is the first of two chapters dedicated to Bayesian learning. We present the main concepts and philosophy behind Bayesian inference. We introduce the expectation-maximization (EM) algorithm and apply it in some typical machine learning parametric modeling tasks, such as regression, mixture modeling, and mixture of experts. Finally, the exponential family of distributions is introduced and the notion of conjugate priors is discussed.

12.2 REGRESSION: A BAYESIAN PERSPECTIVE

The Bayesian inference treatment of the linear regression task was introduced in Chapter 3. In the current chapter, we go beyond the basic definitions and reveal and exploit various possibilities that the Bayesian philosophy offers to the study of this important machine learning task. Let us first summarize the findings of Chapter 3 and then start building upon them.[1]

Recall the (generalized) linear regression task, as it was introduced in previous chapters, that is,

$$\mathrm{y} = \boldsymbol{\theta}^T \boldsymbol{\phi}(\mathbf{x}) + \eta = \theta_0 + \sum_{k=1}^{K-1} \theta_k \phi_k(\mathbf{x}) + \eta, \qquad (12.1)$$

where $\mathrm{y} \in \mathbb{R}$ is the output random variable, $\mathbf{x} \in \mathbb{R}^l$ is the input random vector, $\eta \in \mathbb{R}$ is the noise disturbance, $\boldsymbol{\theta} \in \mathbb{R}^K$ is the unknown parameter vector, and

[1] Recall our adopted notation: random variables and vectors are denoted with roman and their respected measured values/observations with Times Roman fonts.

$$\boldsymbol{\phi}(\mathbf{x}) := [\phi_1(\mathbf{x}), \ldots, \phi_{K-1}(\mathbf{x}), 1]^T,$$

where ϕ_k, $k = 1, \ldots, K-1$, are some (fixed) basis functions. As we already know, typical examples of such functions can be the Gaussian function, splines, monomials, and others. We are given a set of N output–input training points, $(y_n, \boldsymbol{x}_n)$, $n = 1, 2, \ldots, N$. In our current setting, we assume that the respective (unobserved) noise values η_n, $n = 1, 2, \ldots, N$, are samples of jointly Gaussian distributed random variables with covariance matrix Σ_η, that is,

$$p(\boldsymbol{\eta}) = \frac{1}{(2\pi)^{N/2}|\Sigma_\eta|^{1/2}} \exp\left(-\frac{1}{2}\boldsymbol{\eta}^T \Sigma_\eta^{-1} \boldsymbol{\eta}\right), \tag{12.2}$$

where $\boldsymbol{\eta} = [\eta_1, \eta_2, \ldots, \eta_N]^T$.

12.2.1 THE MAXIMUM LIKELIHOOD ESTIMATOR

The maximum likelihood (ML) method was introduced in Chapter 3. According to the method, the unknown set of parameters are treated as a deterministic vector variable, $\boldsymbol{\theta}$, which parameterizes the PDF that describes the output vector of observations

$$\boldsymbol{y} = \Phi\boldsymbol{\theta} + \boldsymbol{\eta}, \tag{12.3}$$

where

$$\Phi = \begin{bmatrix} \boldsymbol{\phi}^T(\boldsymbol{x}_1) \\ \boldsymbol{\phi}^T(\boldsymbol{x}_2) \\ \vdots \\ \boldsymbol{\phi}^T(\boldsymbol{x}_N) \end{bmatrix} \tag{12.4}$$

and

$$\boldsymbol{y} = [y_1, y_2, \ldots, y_N]^T.$$

A simple replacement of X with Φ in (3.61) changes the ML estimate to

$$\hat{\boldsymbol{\theta}}_{\mathrm{ML}} = \left(\Phi^T \Sigma_\eta^{-1} \Phi\right)^{-1} \Phi^T \Sigma_\eta^{-1} \boldsymbol{y}. \tag{12.5}$$

For the simple case of uncorrelated noise samples of equal variance σ_η^2 ($\Sigma_\eta = \sigma_\eta^2 I$), Eq. (12.5) becomes identical to the least-squares (LS) solution

$$\hat{\boldsymbol{\theta}}_{\mathrm{ML}} = \left(\Phi^T \Phi\right)^{-1} \Phi^T \boldsymbol{y} = \hat{\boldsymbol{\theta}}_{\mathrm{LS}}. \tag{12.6}$$

A major drawback of the ML approach is that it is vulnerable to overfitting, because no care is taken of complex models that try to "learn" the specificities of the particular training set, as already discussed in Chapter 3.

12.2.2 THE MAP ESTIMATOR

According to the maximum a posteriori probability (MAP) method, the unknown set of parameters is treated as a random vector $\boldsymbol{\theta}$ and its posterior, for a given set of output observations, $\boldsymbol{y}$, is expressed as

$$p(\boldsymbol{\theta}|\boldsymbol{y}) = \frac{p(\boldsymbol{y}|\boldsymbol{\theta})p(\boldsymbol{\theta})}{p(\boldsymbol{y})}, \tag{12.7}$$

where $p(\boldsymbol{\theta})$ is the associated prior PDF. We have eliminated from the notation the dependence on $\mathcal{X}$, to make it look simpler. We emphasize that the input set, $\mathcal{X} = \{\boldsymbol{x}_1, \ldots, \boldsymbol{x}_N\}$, is considered fixed, so all the randomness associated with $\boldsymbol{y}$ is due to the noise source. Assuming both the prior and the conditional PDFs to be Gaussians,[2] that is,

$$p(\boldsymbol{\theta}) = \mathcal{N}(\boldsymbol{\theta}|\boldsymbol{\theta}_0, \Sigma_\theta) \tag{12.8}$$

and

$$p(\boldsymbol{y}|\boldsymbol{\theta}) = \mathcal{N}(\boldsymbol{y}|\Phi\boldsymbol{\theta}, \Sigma_\eta), \tag{12.9}$$

where (12.2) and (12.3) have been used, the posterior $p(\boldsymbol{\theta}|\boldsymbol{y})$ turns out also to be Gaussian with mean vector

$$\boldsymbol{\mu}_{\theta|y} := \mathbb{E}[\boldsymbol{\theta}|\boldsymbol{y}] = \boldsymbol{\theta}_0 + \left(\Sigma_\theta^{-1} + \Phi^T \Sigma_\eta^{-1} \Phi\right)^{-1} \Phi^T \Sigma_\eta^{-1} (\boldsymbol{y} - \Phi\boldsymbol{\theta}_0). \tag{12.10}$$

Because the maximum of a Gaussian coincides with its mean, we have

$$\hat{\boldsymbol{\theta}}_{\text{MAP}} = \mathbb{E}[\boldsymbol{\theta}|\boldsymbol{y}]. \tag{12.11}$$

In the present chapter's appendix, an analytical proof of (12.10) is provided.[3] It suffices to replace in (12.143) $\boldsymbol{t} \rightarrow \boldsymbol{y}$, $\boldsymbol{z} \rightarrow \boldsymbol{\theta}$, $A \rightarrow \Phi$, $\Sigma_{t|z} \rightarrow \Sigma_\eta$, and $\Sigma_z \rightarrow \Sigma_\theta$. Note that the MAP estimate is a regularized version of $\hat{\boldsymbol{\theta}}_{\text{ML}}$. Regularization is achieved via $\boldsymbol{\theta}_0$ and Σ_θ, which are imposed by the prior $p(\boldsymbol{\theta})$. If one assumes $\Sigma_\theta = \sigma_\theta^2 I$, $\Sigma_\eta = \sigma_\eta^2 I$, and $\boldsymbol{\theta}_0 = \boldsymbol{0}$, then (12.10) coincides with the solution of the regularized LS (ridge) regression,[4]

$$\hat{\boldsymbol{\theta}}_{\text{MAP}} = \left(\lambda I + \Phi^T \Phi\right)^{-1} \Phi^T \boldsymbol{y}, \tag{12.12}$$

where we have set $\lambda := \frac{\sigma_\eta^2}{\sigma_\theta^2}$. We already know from Chapter 3 that the value of λ is critical to the performance of the estimator with respect to the mean-square error (MSE) performance. The main issue now becomes how to choose a good value for λ, or equivalently for Σ_θ, Σ_η in the more general case. In practice, the cross-validation method (Chapter 3) is employed; different values of λ are tested and

---

2 Because in this chapter many random variables will be involved, we explicitly state the name of the variable to which we refer in $\mathcal{N}(\cdot|\cdot,\cdot)$.

3 Because the appendix serves the needs of various parts of the book, each time involving different variables, one has to make the necessary notational substitutions. Note that the appendix can be downloaded from the book's site.

4 Recall from Section 3.8 that this is valid if either the data have been centered or the intercept (bias) is involved in the regularizing norm term.

the one that leads to the best MSE (or some other criterion) is selected. However, this is a computationally costly procedure, especially for complex models, where a large number of parameters is involved. Moreover, such a procedure forces us to use only a fraction of the available data for training, to reserve the rest for testing. The reader may wonder why we do not use the training data to optimize with respect to both the unknown parameter vector $\boldsymbol{\theta}$ and the regularization parameter. Let us consider as an example the simpler case of ridge regression, for centered data ($\theta_0 = 0$). The cost function comprises two terms, one that is data dependent and measures the misfit, and one that depends only on the unknown parameters,

$$J(\boldsymbol{\theta}, \lambda) = \|\boldsymbol{y} - \Phi\boldsymbol{\theta}\|^2 + \lambda\|\boldsymbol{\theta}\|^2. \tag{12.13}$$

It is obvious that the only value of λ that leads to the minimum squared error fit *over the training data set* (empirical loss) is $\lambda = 0$. Any other value of λ would result in an estimate of $\boldsymbol{\theta}$ which scores larger values of the squared error term; this is natural, because for $\lambda \neq 0$ the optimization has to take care of the extra regularizing term, too. It is only when *test data sets* are employed, where values of $\lambda \neq 0$ lead to an overall decrease of the MSE (not the empirical one).

12.2.3 THE BAYESIAN APPROACH

The Bayesian approach to regression attempts to overcome the previously reported drawbacks, which are associated with the overfitting. All the involved parameters can be estimated on the training set. In this vein, the parameters will be treated as random variables. At the same time, because the main task now becomes that of inferring the PDF that describes the unknown set of parameters, instead of obtaining a single vector estimate, one has more information at her/his disposal. Having said that, it does not mean that Bayesian techniques are necessarily free from cross-validation; this will be needed to assess their overall performance. We will comment further on this in the Remarks at the end of Section 12.3.

As we know, the starting point is the same as that for MAP, and in particular (12.7). However, instead of taking just the maximum of the numerator in (12.7), we will make use of $p(\boldsymbol{\theta}|\boldsymbol{y})$ as a whole. Most of the secrets here lie in the denominator $p(\boldsymbol{y})$, which is basically the normalizing constant,

$$\boxed{p(\boldsymbol{y}) = \int p(\boldsymbol{y}|\boldsymbol{\theta})p(\boldsymbol{\theta})\,d\boldsymbol{\theta}.} \tag{12.14}$$

As we will soon see, there is much more information hidden in $p(\boldsymbol{y})$ that goes beyond the need of just computing $p(\boldsymbol{\theta}|\boldsymbol{y})$. The difficulty with (12.14) is that, in general, the evaluation of the integral cannot be performed analytically. In such cases, one has to resort to approximate techniques to obtain the required information. To this end, a number of approaches are available, and a large part of this book is dedicated to their study. More specifically, the following methods have been proposed and will be considered:

- the Laplacian approximation method, presented in Section 12.3;
- the variational approximation method, presented in Section 13.2;
- the variational bound approximation method, presented in Section 13.8;
- Monte Carlo techniques for the evaluation of the integral, which are discussed in Chapter 14;
- message passing algorithms, to be discussed in Chapter 15.

For the case under study in this section, where $p(\mathbf{y}|\boldsymbol{\theta})$ and $p(\boldsymbol{\theta})$ are both assumed to be Gaussians, $p(\mathbf{y})$ can be evaluated analytically; it turns out that the joint distribution $p(\mathbf{y}, \boldsymbol{\theta})$ is also Gaussian and hence the marginal $p(\mathbf{y})$ is Gaussian as well. All these are shown in detail in the appendix of this chapter. Indeed, if we set in (12.146) and (12.151) of the appendix $\mathbf{z} \to \boldsymbol{\theta}$, $\mathbf{t} \to \mathbf{y}$, and $A \to \Phi$, it turns out that for the regression model of (12.3) and the prior PDF in (12.8) as well as the noise model of (12.2), we obtain

$$p(\mathbf{y}) = \mathcal{N}\left(\mathbf{y}|\Phi\boldsymbol{\theta}_0, \Sigma_\eta + \Phi\Sigma_\theta\Phi^T\right). \tag{12.15}$$

Moreover, the posterior $p(\boldsymbol{\theta}|\mathbf{y})$ is also Gaussian,

$$p(\boldsymbol{\theta}|\mathbf{y}) = \mathcal{N}\left(\boldsymbol{\theta}|\boldsymbol{\mu}_{\theta|y}, \Sigma_{\theta|y}\right), \tag{12.16}$$

where $\boldsymbol{\mu}_{\theta|y}$ is given by (12.10) and the covariance matrix results from (12.147), after the appropriate notational substitutions, that is,

$$\Sigma_{\theta|y} = \left(\Sigma_\theta^{-1} + \Phi^T\Sigma_\eta^{-1}\Phi\right)^{-1}. \tag{12.17}$$

The posterior PDF in (12.16) encapsulates our knowledge about $\boldsymbol{\theta}$, after the observations $\mathbf{y}$ have been obtained. Hence, our uncertainty about $\boldsymbol{\theta}$ has been reduced, which is the main reason that (12.16) is different from the prior PDF in (12.8); the latter represents only our initial guess. The covariance matrix in (12.17) provides the information about our uncertainty with respect to $\boldsymbol{\theta}$. If the Gaussian in (12.16) is very broad around its mean $\boldsymbol{\mu}_{\theta|y}$, it indicates that in spite of the reception of the observations still much uncertainty about $\boldsymbol{\theta}$ remains. This can be due (a) to the nature of the problem, for example, high noise variance, as this is conveyed by Σ_η, (b) to the number of observations, N, which may not be enough, and/or (c) to modeling inaccuracies, as this is conveyed by Φ in (12.17). The opposite comments are in order if the posterior PDF is sharply peaked around its mean.

As we have already stated in Chapter 3, the Bayesian philosophy provides the means for a direct inference of the output variable, which in many applications is the quantity of interest; given the input vector, the task is to predict the output. In such cases, estimating a value for the unknown $\boldsymbol{\theta}$ is only the means to an end. To formulate the prediction task directly, without involving $\boldsymbol{\theta}$, one has to integrate the contribution of $\boldsymbol{\theta}$. Having learned the posterior $p(\boldsymbol{\theta}|\mathbf{y})$, given a new input vector $\mathbf{x}$, for the regression model in (12.1), the conditional PDF of the output variable, y, given the set of observations, is written as

$$p(y|\mathbf{x}, \mathbf{y}) = \int p(y|\mathbf{x}, \boldsymbol{\theta})p(\boldsymbol{\theta}|\mathbf{y})\, d\boldsymbol{\theta}. \tag{12.18}$$

Note that we have used $p(y|\mathbf{x}, \mathbf{y}, \boldsymbol{\theta}) = p(y|\mathbf{x}, \boldsymbol{\theta})$ because y is conditionally independent of $\mathbf{y}$ given the value of $\boldsymbol{\theta}$. As already stated, strictly speaking, the posterior should have been denoted as $p(\boldsymbol{\theta}|\mathbf{y}; \mathcal{X})$ to indicate the dependence on the input training samples. However, the dependence on $\mathcal{X}$ has been suppressed to unclutter notation.

In the sequel, and in order to simplify algebra and focus on the concepts, we assume that the noise model in (12.2) is such that $\Sigma_\eta = \sigma_\eta^2 I$ and also $\Sigma_\theta = \sigma_\theta^2 I$ for the prior PDF in (12.8). Then we have

$$p(y|\mathbf{x}, \boldsymbol{\theta}) = \mathcal{N}(y|\boldsymbol{\theta}^T\boldsymbol{\phi}(\mathbf{x}), \sigma_\eta^2),$$

and (12.17) and (12.10) for the posterior covariance matrix and mean, respectively, become

$$\Sigma_{\theta|y} = \left(\frac{1}{\sigma_\theta^2} I + \frac{1}{\sigma_\eta^2} \Phi^T \Phi \right)^{-1}, \tag{12.19}$$

$$\boldsymbol{\mu}_{\theta|y} = \boldsymbol{\theta}_0 + \frac{1}{\sigma_\eta^2} \left(\frac{1}{\sigma_\theta^2} I + \frac{1}{\sigma_\eta^2} \Phi^T \Phi \right)^{-1} \Phi^T (\mathbf{y} - \Phi \boldsymbol{\theta}_0). \tag{12.20}$$

The integration in (12.18) can now be carried out analytically as in (12.136) and (12.137) and using (12.150) and (12.151) in the appendix, with $\boldsymbol{z} \rightarrow \boldsymbol{\theta}$, $\boldsymbol{t} \rightarrow \mathbf{y}$, $A \rightarrow \boldsymbol{\phi}^T$, $\boldsymbol{\mu}_z \rightarrow \boldsymbol{\mu}_{\theta|y}$, $\Sigma_z \rightarrow \Sigma_{\theta|y}$, $\Sigma_{t|z} \rightarrow \sigma_\eta^2$, and we obtain

$$\boxed{p(y|\boldsymbol{x}, \mathbf{y}) = \mathcal{N}\left(y|\mu_y, \sigma_y^2\right): \quad \text{predictive distribution,}} \tag{12.21}$$

where

$$\mu_y = \boldsymbol{\phi}^T(\boldsymbol{x}) \boldsymbol{\mu}_{\theta|y}, \tag{12.22}$$

$$\begin{aligned} \sigma_y^2 &= \sigma_\eta^2 + \boldsymbol{\phi}^T(\boldsymbol{x}) \Sigma_{\theta|y} \boldsymbol{\phi}(\boldsymbol{x}) \\ &= \sigma_\eta^2 + \boldsymbol{\phi}^T(\boldsymbol{x}) \left(\frac{1}{\sigma_\theta^2} I + \frac{1}{\sigma_\eta^2} \Phi^T \Phi \right)^{-1} \boldsymbol{\phi}(\boldsymbol{x}) \\ &= \sigma_\eta^2 + \sigma_\eta^2 \sigma_\theta^2 \boldsymbol{\phi}^T(\boldsymbol{x}) \left(\sigma_\eta^2 I + \sigma_\theta^2 \Phi^T \Phi \right)^{-1} \boldsymbol{\phi}(\boldsymbol{x}). \end{aligned} \tag{12.23}$$

Hence, given $\boldsymbol{x}$ one can predict the respective value of y using the most probable value, that is, μ_y in (12.22). Note that the same prediction value would result via the MAP estimate in (12.10) (or (12.12), if $\boldsymbol{\theta}_0 = \mathbf{0}$, also obtained via the ridge regression task). Have we then gained anything extra by adopting the Bayesian approach? The answer is in the affirmative. *More information concerning the predicted value is now available, because* (12.23) *quantifies the associated uncertainty.*

To investigate (12.23) further, let us simplify it by adopting the following approximation:

$$R_\phi := \mathbb{E}[\boldsymbol{\phi}(\mathbf{x})\boldsymbol{\phi}^T(\mathbf{x})] \simeq \frac{1}{N} \sum_{n=1}^{N} \boldsymbol{\phi}(\boldsymbol{x}_n)\boldsymbol{\phi}^T(\boldsymbol{x}_n) = \frac{1}{N} \Phi^T \Phi,$$

or

$$\Phi^T \Phi \simeq N R_\phi, \tag{12.24}$$

where R_ϕ is the autocorrelation matrix of the random vector $\boldsymbol{\phi}(\mathbf{x})$. Employing (12.24) into (12.23) leads to

$$\sigma_y^2 \simeq \sigma_\eta^2 \left(1 + \sigma_\theta^2 \boldsymbol{\phi}^T(\boldsymbol{x}) \left(\sigma_\eta^2 I + N \sigma_\theta^2 R_\phi \right)^{-1} \boldsymbol{\phi}(\boldsymbol{x}) \right), \tag{12.25}$$

which for large enough N becomes

$$\sigma_y^2 \simeq \sigma_\eta^2 \left(1 + \frac{1}{N}\boldsymbol{\phi}^T(\boldsymbol{x})R_\phi^{-1}\boldsymbol{\phi}(\boldsymbol{x})\right).$$

Thus, for a large number of observations, $\sigma_y^2 \longrightarrow \sigma_\eta^2$, and our uncertainty is contributed by the noise source, which cannot be reduced anymore. For smaller values of N, there is extra uncertainty associated with the parameter $\boldsymbol{\theta}$, measured by σ_θ^2 in (12.25).

So far in this section, we dealt with Gaussians, which led to tractable and analytically computed integrals. Are there ways to attack more general cases? Moreover, even in the case of Gaussian PDFs, we have assumed the covariance matrices Σ_θ, Σ_η to be known. In practice, they are not. Even if one assumes that Σ_η can be experimentally measured, there still remains Σ_θ. Can one select the related parameters via an optimization process? If the answer is yes, can this optimization be carried out on the training set, or one would necessarily run into problems similar to the ones we faced with the regularization approach? We will indulge in all these challenges in the sections to follow.

Remarks 12.1.

- The MAP estimator is sometimes referred to as *Type I* estimator, to be distinguished from the Type II estimation method, which will be discussed in Remarks 12.2, in the next section.
- The posterior mean in (12.10) can be met in different variants, which are obtained via the application of the matrix inversion lemmas given in Appendix A.1. In the chapter's appendix, it is shown that (Eq. (12.152))

$$\boldsymbol{\mu}_{\theta|y} = \left(\Sigma_\theta^{-1} + \Phi^T \Sigma_\eta^{-1}\Phi\right)^{-1}\left(\Phi^T \Sigma_\eta^{-1}\boldsymbol{y} + \Sigma_\theta^{-1}\boldsymbol{\theta}_0\right) \tag{12.26}$$

or (Eq. (12.148))

$$\boldsymbol{\mu}_{\theta|y} = \boldsymbol{\theta}_0 + \Sigma_\theta\Phi^T\left(\Sigma_\eta + \Phi\Sigma_\theta\Phi^T\right)^{-1}(\boldsymbol{y} - \Phi\boldsymbol{\theta}_0). \tag{12.27}$$

Also, using Woodbury's identity from Appendix A.1, we can readily see that

$$\Sigma_{\theta|y} = \Sigma_\theta - \Sigma_\theta\Phi^T\left(\Sigma_\eta + \Phi\Sigma_\theta\Phi^T\right)^{-1}\Phi\Sigma_\theta. \tag{12.28}$$

In practice, one uses the most computationally convenient form, depending on the dimensionality of the involved matrices to invert the one of lower dimension.

Example 12.1. This example demonstrates the prediction task summarized in (12.22) and (12.23). Data are generated based on the following nonlinear model:

$$y_n = \theta_0 + \theta_1 x_n + \theta_2 x_n^2 + \theta_3 x_n^3 + \theta_5 x_n^5 + \eta_n, \; n = 1, 2, \ldots, N,$$

where η_n are i.i.d. noise samples drawn form a zero mean Gaussian with variance σ_η^2. Samples x_n are equidistant points in the interval [0, 2]. The goal of the task is to predict the value y given a measured value x, using (12.22). The parameter values used to generate the data were equal to

$$\theta_0 = 0.2, \; \theta_1 = -1, \; \theta_2 = 0.9, \; \theta_3 = 0.7, \; \theta_5 = -0.2.$$

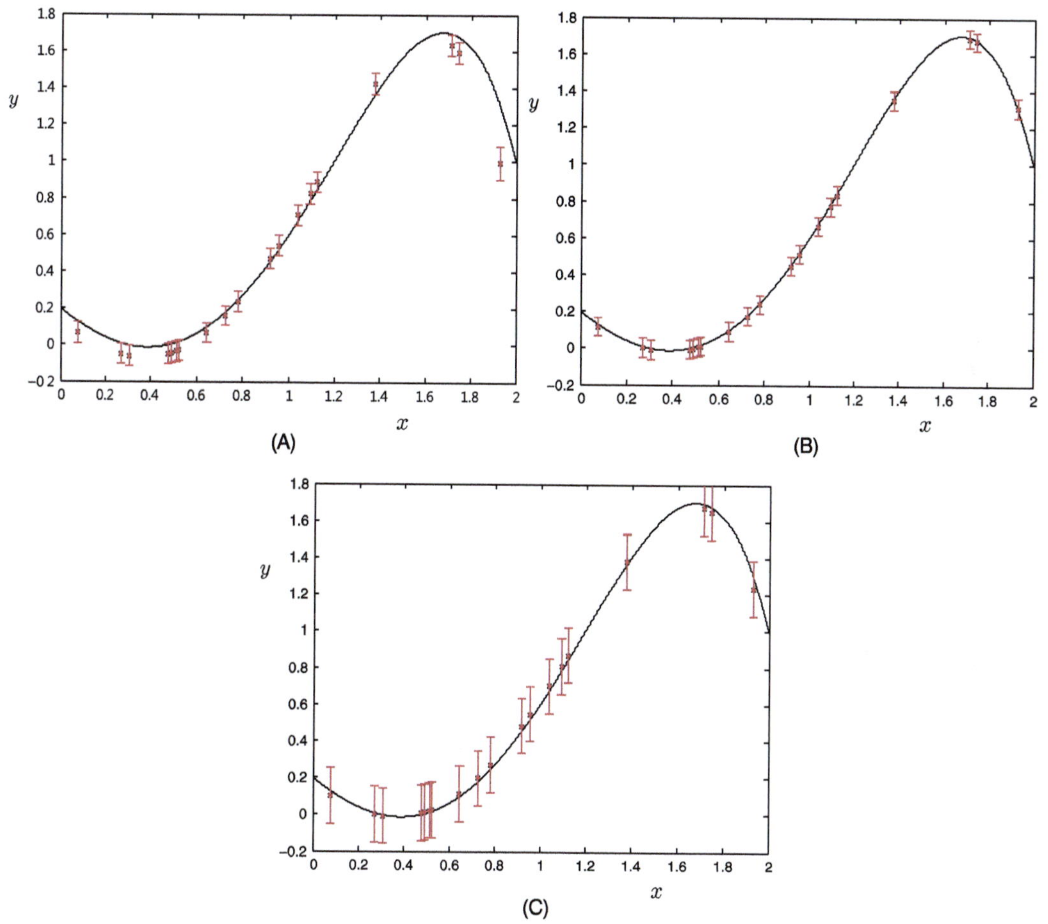

FIGURE 12.1

Each one of the red points (y, x) indicates the prediction (y) corresponding to the input value (x). The error bars are dictated by the computed variance, σ_y^2. The mean values used in the Gaussian prior are equal to the true values of the unknown model. (A) $\sigma_\eta^2 = 0.05$, $N = 20$, $\sigma_\theta^2 = 0.1$. (B) $\sigma_\eta^2 = 0.05$, $N = 500$, $\sigma_\theta^2 = 0.1$. (C) $\sigma_\eta^2 = 0.15$, $N = 500$, $\sigma_\theta^2 = 0.1$. Observe that the larger the data set is, the better the predictions are and the larger the noise variance is, the larger the error bars become.

(a) In the first set of experiments, a Gaussian prior for the unknown $\boldsymbol{\theta}$ was used with mean $\boldsymbol{\theta}_0$ equal to the previous true set of parameters and $\Sigma_\theta = 0.1I$. Also, the true model structure was used to construct the matrix Φ. Fig. 12.1A shows the points (y, x) in red together with the error bars, as measured by the computed σ_y^2, for the case of $N = 20$ training points and $\sigma_\eta^2 = 0.05$. Fig. 12.1B demonstrates the obtained improvement when the training points are increased to $N = 500$, while keeping the values of the other two parameters unchanged. Fig. 12.1C corresponds to the latter case, where the noise variance is increased to $\sigma_\eta^2 = 0.15$.

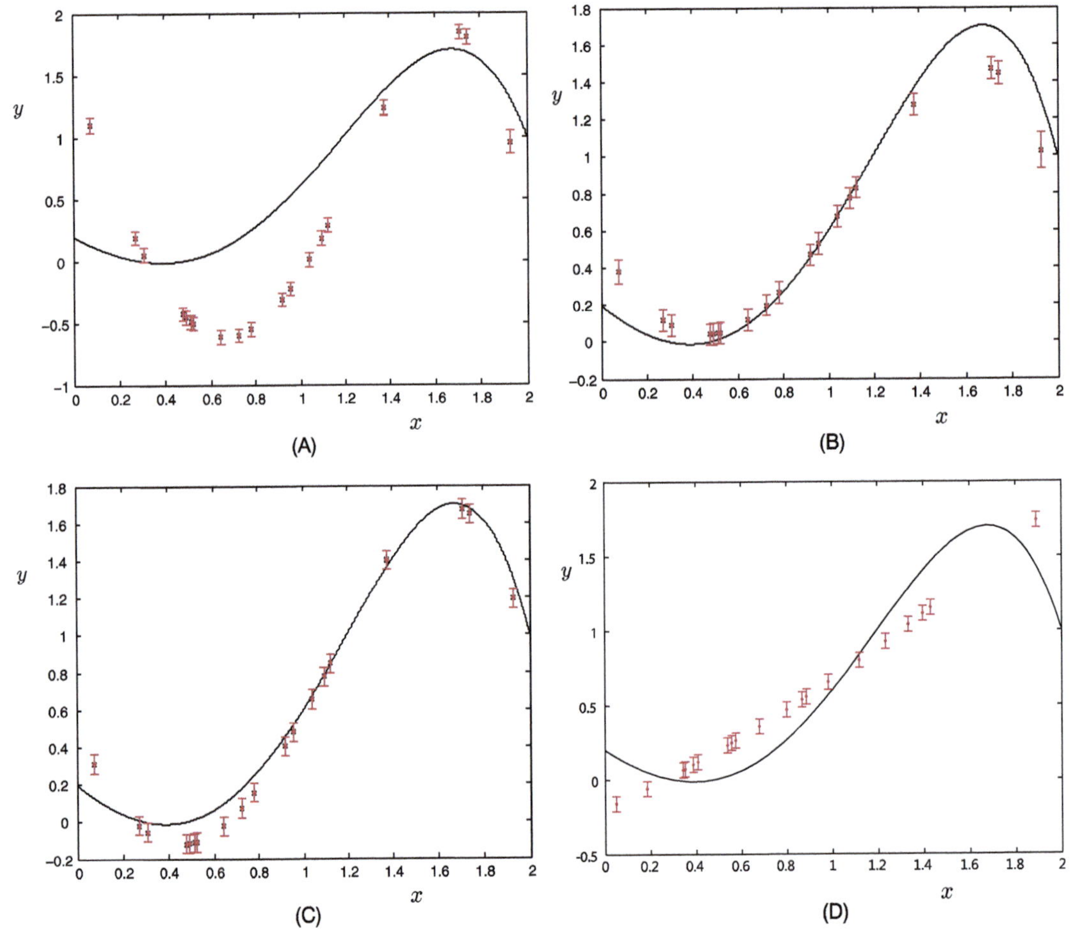

FIGURE 12.2

In this set of figures the mean values of the prior are different from that of the true model. (A) $\sigma_\eta^2 = 0.05$, $N = 20$, $\sigma_\theta^2 = 0.1$. (B) $\sigma_\eta^2 = 0.05$, $N = 20$, $\sigma_\theta^2 = 2$; observe the effect of using larger variance for the prior. (C) $\sigma_\eta^2 = 0.05$, $N = 500$, $\sigma_\theta^2 = 0.1$; observe the effect of the larger training data set. (D) The points correspond to a wrong model.

(b) In the second set of experiments, we kept the correct model, but the mean of the prior was given a different value to that of the true model, namely,

$$\boldsymbol{\theta}_0 = [-10.54, 0.465, 0.0087, -0.093, -0.004]^T.$$

Fig. 12.2A corresponds to the case of $\sigma_\eta^2 = 0.05$, $N = 20$, and $\sigma_\theta^2 = 0.1$. Note the improvement that is obtained when increasing $\sigma_\theta^2 = 2$, shown in Fig. 12.2B, while N and σ_η^2 remain the same as before; this is because the model takes into consideration our uncertainty about the prior mean being away from the true value. Fig. 12.2C corresponds to $\sigma_\eta^2 = 0.05$, $N = 500$, and $\sigma_\theta^2 = 0.1$ and shows the advantage of using a large number of training points.

(c) Fig. 12.2D corresponds to the case where the adopted model for prediction is the wrong one, that is,

$$y = \theta_0 + \theta_1 x + \theta_2 x^2 + \eta.$$

The used values were $\sigma_\eta^2 = 0.05$, $N = 500$, and $\sigma_\theta^2 = 2$. Observe that once a wrong model has been adopted, one must not have "high expectations" for good prediction performance.

12.3 THE EVIDENCE FUNCTION AND OCCAM'S RAZOR RULE

In the previous section, we made a comment about the importance of the marginal PDF $p(\mathbf{y})$. This section is fully dedicated to this quantity. In the notation used in (12.14), we did tacitly suppress the dependence on the adopted model. For example, the Gaussian assumption for the prior in (12.8) and for the conditional in (12.9) should have been reflected in the marginal as $p(\mathbf{y}; \boldsymbol{\phi}, \Sigma_\eta, \Sigma_\theta)$, because different Gaussians, different basis functions, and different orders K of the model can be used. Furthermore, non-Gaussian PDFs can also be adopted. In a more general setting, let us make the dependence on the model explicit as $p(\mathbf{y}|\mathcal{M}_i)$. Assuming the choice of a model to be random, mobilizing the Bayes theorem once more, we have

$$P(\mathcal{M}_i|\mathbf{y}) = \frac{P(\mathcal{M}_i)p(\mathbf{y}|\mathcal{M}_i)}{p(\mathbf{y})}, \tag{12.29}$$

where

$$p(\mathbf{y}) = \sum_i P(\mathcal{M}_i)p(\mathbf{y}|\mathcal{M}_i), \tag{12.30}$$

and $P(\mathcal{M}_i)$ is the prior probability of $\mathcal{M}_i$; $P(\mathcal{M}_i)$ provides a measure of the subjective prior over all possible models, which expresses our guess on how plausible a model is with respect to alternative ones, prior to the data arrival. Because the denominator in (12.29) is independent of the model, one can obtain the most probable model, after observing $\mathbf{y}$, by maximizing the numerator. If one assigns to all possible models equal probabilities, then detecting the most probable model under the given set of observations becomes a task of maximizing $p(\mathbf{y}|\mathcal{M}_i)$ with respect to the model, $\mathcal{M}_i$. This is the reason that this PDF is known as the *evidence function* for the model or simply as the *evidence*. In practice, we content ourselves with using the most probable model, although an orthodox Bayesian would suggest to average all obtained quantities over all possible models, as in (12.30). In an ideal Bayesian setting, one does not choose among models; predictions are performed by summing over all possible models, each one weighted by the respective probability. However, in many practical problems we may have reasons to suggest that the evidence function is strongly peaked around a specific model; after all, such an assumption may simplify the task considerably.

From a mathematical formulation's point of view, each model is expressed in terms of a set of nonrandom (deterministic) parameters. To get the model that maximizes the evidence is equivalent to maximizing with respect to these parameters. For example, let us assume that in the regression task we adopt a Gaussian distribution for the noise and a Gaussian prior for the regression task random

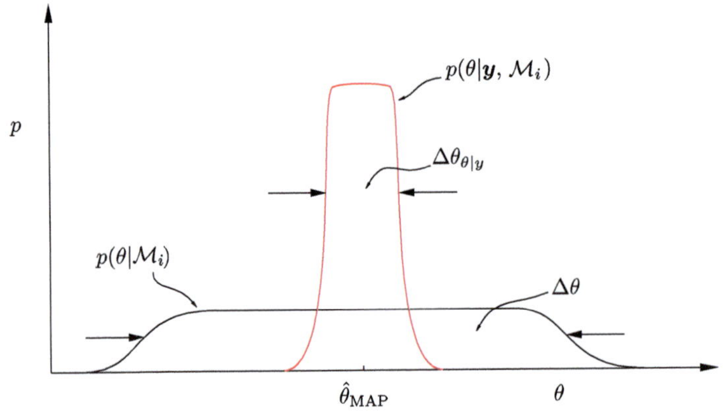

FIGURE 12.3

The posterior peaks around the value $\hat{\theta}_{\text{MAP}}$ and the posterior PDF can be approximated by $p(\hat{\theta}_{\text{MAP}}|\mathbf{y}; \mathcal{M}_i)$ over an interval of values equal to $\Delta\theta_{\theta|y}$.

parameters, $\boldsymbol{\theta}$. Then this pair of Gaussians comprises our model and the parameters that describe it are the respective covariance matrices for the two Gaussians as well as the mean value of the prior. To distinguish from the random parameters, $\boldsymbol{\theta}$, which describe the original learning task (e.g., regression), these additional parameters are often called *hyperparameters*. Later on, in Chapter 13, we will see that the hyperparameters can also be treated as random variables.

We now turn our attention to what is hidden behind the optimization of $p(\mathbf{y}|\mathcal{M}_i)$ with respect to different models. Before we proceed, it is worth making a comment. A superficial first look may lead one to think whether this is any different from maximizing the likelihood $p(\mathbf{y}; \boldsymbol{\theta})$, as done in Section 12.2.1. As a matter of fact, the two cases belong to two different worlds. ML maximizes with respect to a single (vector) parameter within an adopted model, and this is the weak point that makes ML vulnerable to overfitting. Maximizing the evidence is an optimization task with respect to the model itself, a wise alternative that guards us against overfitting, as we explain next.

From (12.14) we have

$$\boxed{p(\mathbf{y}|\mathcal{M}_i) = \int p(\mathbf{y}|\mathcal{M}_i, \boldsymbol{\theta})p(\boldsymbol{\theta}|\mathcal{M}_i)d\boldsymbol{\theta} : \quad \text{evidence function.}} \tag{12.31}$$

Let us assume for simplicity that θ is a scalar, $\theta \in \mathbb{R}$, and that the integrand in (12.31), which according to the Bayes theorem is analogous to the posterior $p(\theta|\mathbf{y}, \mathcal{M}_i)$, peaks around a value; this is obviously the value that would result as the MAP estimate, $\hat{\theta}_{\text{MAP}}$. Fig. 12.3 illustrates the respective graphs. Thus, (12.31) can be approximated by

$$p(\mathbf{y}|\mathcal{M}_i) \simeq p(\mathbf{y}|\mathcal{M}_i, \hat{\theta}_{\text{MAP}})p(\hat{\theta}_{\text{MAP}}|\mathcal{M}_i)\Delta\theta_{\theta|y}. \tag{12.32}$$

To get a better feeling for each one of the factors involved in (12.32), let us also assume that the prior PDF is (almost) uniform with a width equal to $\Delta\theta$. Then (12.32) is rewritten as

$$p(y|\mathcal{M}_i) \simeq p(y|\mathcal{M}_i, \hat{\theta}_{\text{MAP}})\frac{\Delta\theta_{\theta|y}}{\Delta\theta}. \tag{12.33}$$

The first factor in the product on the right-hand side in (12.33) coincides with the likelihood function at its optimal value, because for this case of uniform prior, $\hat{\theta}_{\text{MAP}} = \hat{\theta}_{\text{ML}}$. In other words, this factor provides us with the best fit that model $\mathcal{M}_i$ can achieve on the given set of observations. However, now, in contrast to the ML method, the evidence function also depends on the second factor, $\frac{\Delta\theta_{\theta|y}}{\Delta\theta}$. As it has been pointed out in the insightful papers [13,27,28], this term accounts for the complexity of the model, and it is named the Occam factor for obvious reasons. Let us elaborate on this a bit more by following the reasoning given in [28].

The Occam factor penalizes those models which are finely tuned to the received observations. As an example, if two different models $\mathcal{M}_i$ and $\mathcal{M}_j$ have a similar range of values for their prior PDFs, then if, say, $\Delta\theta_{\theta|y}(\mathcal{M}_i) \ll \Delta\theta_{\theta|y}(\mathcal{M}_j)$, then $\mathcal{M}_i$ will be penalized more; only a small range of values for θ survive (i.e., correspond to high probability values) after the reception of $\boldsymbol{y}$. So, if this fine-tuned (to the data) model $\mathcal{M}_i$ resulted in a large value of the ML term, it would not be certain that the evidence would be maximized for it, because the Occam factor would be small. Which model, between the two, finally wins it depends on the product of the two involved terms. Soon we will see that the Occam term is also related to the number of parameters; that is, to the complexity of the adopted model.

LAPLACIAN APPROXIMATION AND THE EVIDENCE FUNCTION

To investigate the evidence function for the general multiparameter case, we will employ the method of Laplacian approximation of a PDF. This is a general methodology that approximates any PDF *locally* in terms of a Gaussian one. To this end, define[5]

$$g(\boldsymbol{\theta}) = \ln\big(p(\boldsymbol{y}|\mathcal{M}_i, \boldsymbol{\theta})p(\boldsymbol{\theta}|\mathcal{M}_i)\big). \tag{12.34}$$

Use Taylor's expansion around $\hat{\boldsymbol{\theta}}_{\text{MAP}}$ and keep terms up to the second order,

$$\begin{aligned} g(\boldsymbol{\theta}) &= g(\hat{\boldsymbol{\theta}}_{\text{MAP}}) + (\boldsymbol{\theta} - \hat{\boldsymbol{\theta}}_{\text{MAP}})^T \frac{\partial g(\boldsymbol{\theta})}{\partial \boldsymbol{\theta}}\Big|_{\boldsymbol{\theta}=\hat{\boldsymbol{\theta}}_{\text{MAP}}} \\ &\quad + \frac{1}{2}(\boldsymbol{\theta} - \hat{\boldsymbol{\theta}}_{\text{MAP}})^T \frac{\partial^2 g(\boldsymbol{\theta})}{\partial \boldsymbol{\theta}^2}\Big|_{\boldsymbol{\theta}=\hat{\boldsymbol{\theta}}_{\text{MAP}}} (\boldsymbol{\theta} - \hat{\boldsymbol{\theta}}_{\text{MAP}}) \\ &= g(\hat{\boldsymbol{\theta}}_{\text{MAP}}) - \frac{1}{2}(\boldsymbol{\theta} - \hat{\boldsymbol{\theta}}_{\text{MAP}})^T \Sigma^{-1} (\boldsymbol{\theta} - \hat{\boldsymbol{\theta}}_{\text{MAP}}), \end{aligned} \tag{12.35}$$

where

$$\Sigma^{-1} := -\frac{\partial^2 g(\boldsymbol{\theta})}{\partial \boldsymbol{\theta}^2}\Big|_{\boldsymbol{\theta}=\hat{\boldsymbol{\theta}}_{\text{MAP}}},$$

[5] Similarly, to obtain the Laplacian approximation of a general PDF, $p(\boldsymbol{x})$, we set $g(\boldsymbol{x}) = \ln p(\boldsymbol{x})$.

which leads to the approximation

$$p(\boldsymbol{y}|\mathcal{M}_i,\boldsymbol{\theta})p(\boldsymbol{\theta}|\mathcal{M}_i) \simeq p(\boldsymbol{y}|\mathcal{M}_i,\hat{\boldsymbol{\theta}}_{\text{MAP}})p(\hat{\boldsymbol{\theta}}_{\text{MAP}}|\mathcal{M}_i)\times \exp\left(-\frac{1}{2}(\boldsymbol{\theta}-\hat{\boldsymbol{\theta}}_{\text{MAP}})^T\Sigma^{-1}(\boldsymbol{\theta}-\hat{\boldsymbol{\theta}}_{\text{MAP}})\right). \tag{12.36}$$

Plugging (12.36) into the integral of (12.31) we obtain

$$p(\boldsymbol{y}|\mathcal{M}_i) = p(\boldsymbol{y}|\mathcal{M}_i,\hat{\boldsymbol{\theta}}_{\text{MAP}})p(\hat{\boldsymbol{\theta}}_{\text{MAP}}|\mathcal{M}_i)(2\pi)^{\frac{K}{2}}|\Sigma|^{1/2}, \tag{12.37}$$

and taking the logarithms we have

$$\underbrace{\ln p(\boldsymbol{y}|\mathcal{M}_i)}_{\text{Evidence}} = \underbrace{\ln p(\boldsymbol{y}|\mathcal{M}_i,\hat{\boldsymbol{\theta}}_{\text{MAP}})}_{\text{Best likelihood fit}} + \underbrace{\ln p(\hat{\boldsymbol{\theta}}_{\text{MAP}}|\mathcal{M}_i) + \frac{K}{2}\ln(2\pi) + \frac{1}{2}\ln|\Sigma|}_{\text{Occam factor}}. \tag{12.38}$$

The direct dependence of the Occam term on the complexity (number of basis functions) of the adopted model is now readily spotted. Moreover, the complexity-related Occam term depends on the prior PDF and the second derivatives (via Σ) of the posterior PDF, too; that is, it depends on how "sharp" the shape of the latter is in the K-dimensional space. In other words, the covariance term provides the "error bar" information. Moreover, its determinant does depend on K, too. That is, the dependence on the complexity term K is more involved than what a naive look at Eq. (12.38) suggests (see Remarks 12.2, concerning the BIC criterion). Hence, in a single equation, besides the number of parameters and the associated best-fit term, the evidence also takes into account information related to the associated variance; maximizing the evidence leads to the best tradeoff. Fig. 12.4 illustrates the essence behind

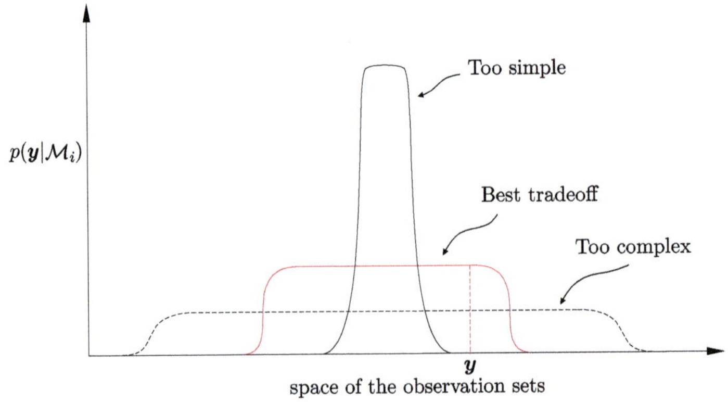

FIGURE 12.4

Too simple models can explain well a very small range of data. On the other hand, too complex models can explain a wide range of data; however, they do not provide any confidence because they assign low probability to all data sets. For the observation set, $\boldsymbol{y}$, the evidence is maximized for the model with intermediate complexity.

the evidence maximization for model selection. If the model is too complex, it can fit well a wide range of data sets, and because $p(\mathbf{y}|\mathcal{M}_i)$ has to integrate to one, its value for any value of $\mathbf{y}$ is expected to be low. The opposite is true for models that are too simple; such models can model well some data sets but not a wide range of them, and consequently, the evidence function peaks sharply around a value in the space of observation sets. Thus, selecting a data set at random it is rather unlikely that this has been generated by such a model. Having said that, it is important to emphasize, once more, the Occam term does not depend *solely* on the number of parameters; hence, complexity here should be interpreted in a more "open-minded" way. This robustness against overfitting, which is intrinsic in the Bayesian inference approach, is the consequence of integrating the parameters for any specific model in (12.31); this integration penalizes models of high complexity because such models can model a large range of data.

Historically, the Occam's razor rule in its Bayesian interpretation was first demonstrated in [13] and later on in [27,43], although the foundations go back to the pioneering work of Sir Herald Jeffrey in the 1930s [23]. Two insightful reviews on the Bayesian inference approach that are well worth reading are given in [22,26].

Returning to (12.14) and assuming for simplicity that

$$p(\mathbf{y}|\boldsymbol{\theta}) = \mathcal{N}\left(\mathbf{y}|\Phi\boldsymbol{\theta}, \sigma_\eta^2 I\right) \quad \text{and} \quad p(\boldsymbol{\theta}) = \mathcal{N}\left(\boldsymbol{\theta}|\boldsymbol{\theta}_0, \sigma_\theta^2 I\right), \tag{12.39}$$

we can express the evidence as $p(\mathbf{y}; \sigma_\eta^2, \sigma_\theta^2)$, which, for this case, turns out to be Gaussian (appendix of this chapter); thus, it is available in closed form. Hence for this specific case, the model space is described via the hyperparameters $\sigma_\eta^2, \sigma_\theta^2$ and maximization of the evidence with respect to these (unknown) model parameters can take place iteratively, for the given set of observations stacked in $\mathbf{y}$ (see, for example, [27]). However, in general we do not have the ability to express the evidence in closed form. The EM algorithm, which is described in Section 12.4, is a popular way and a powerful tool to this end. One could also resort to the Laplacian approximation to approximate the involved PDFs as Gaussians, but this approximation turns out not always to be a good choice; furthermore, in high-dimensional parameter spaces, the computations of the second-order derivatives and the determinant can become burdensome [3].

Finally, let us make a final comment concerning the Laplacian approximation. In the discussion above, our goal was to get an approximation of the integral (normalizing constant/evidence) of $p(\mathbf{y}|\mathcal{M}_i, \boldsymbol{\theta})p(\boldsymbol{\theta}|\mathcal{M}_i)$. However, if our interest were to approximate the PDF itself, we should be careful in selecting the normalizing constant, which by the nature of the Gaussian function leads to

$$p(\mathbf{y}|\mathcal{M}_i, \boldsymbol{\theta})p(\boldsymbol{\theta}|\mathcal{M}_i) \simeq \frac{1}{(2\pi)^{K/2}|\Sigma|^{1/2}} \exp\left(-\frac{1}{2}(\boldsymbol{\theta} - \hat{\boldsymbol{\theta}}_{\text{MAP}})^T \Sigma^{-1}(\boldsymbol{\theta} - \hat{\boldsymbol{\theta}}_{\text{MAP}})\right).$$

This is also the case for the Laplacian approximation for any PDF $p(\cdot)$.

Remarks 12.2.

- In the Bayesian approach, one makes all the modeling assumptions explicit, and it is then left to the rules of probability theory to provide the answers. One does not have to "worry" about the choice of an optimizing criterion, where different criteria lead to different estimators, and there is not an objective, systematic way to decide which criterion is best. On the other hand, in the Bayesian approach one has to make sure to select the prior that explains the data in the best possible way.

- The choice of the prior PDF is very critical in the performance of Bayesian methods and must be carried out in such a way to encapsulate prior knowledge as fully as possible. In practice, different alternatives can be adopted [3].
  - *Subjective priors.* According to this path, we choose the prior $p(\boldsymbol{\theta})$ to make the manipulation of the integration tractable, by employing conjugate priors (Section 3.11.1) within the exponential family of PDFs. This family of PDFs will be presented in Section 12.8.
  - *Hierarchical priors.* Each one of the components of $\theta_k,\ k=0,2,\ldots,K-1$, of $\boldsymbol{\theta}$, is controlled by a different parameter; for example, all θ_ks may be assumed to be independent Gaussian variables, each one with a different variance. In turn, variances are considered to be random variables that follow a statistical distribution controlled by another set of deterministic (not random) hyperparameters; thus, a hierarchy of priors is adopted. As we will see later on, hierarchical priors are often designed using conjugate pairs of PDFs.
  - *Noninformative or objective priors.* The choice of the prior is done in such a way to embed as little extra information as possible and to exploit knowledge that is conveyed only by the available data. One way to construct such priors is to resort to information theoretic arguments. For example, one can estimate $p(\boldsymbol{\theta})$ by minimizing its Kullback–Leibler (KL) divergence from $p(\boldsymbol{\theta}|\mathbf{y})$.
- The fact that the Bayesian approach allows the recovery of all the desired information from a single data set does not suggest that the method is "cross-validation-free". Maximizing the evidence, which at the same time guards against overfitting, *does not* necessarily mean that the performance of the designed estimator is optimized. This is more true in practice where, as we are going to see very soon, most often a bound of the evidence is optimized instead, to bypass computational obstacles. As is always the case in life, the proof of the cake is in the eating. Thus, the final verdict should only come from the generalization ability of the designed estimator, that is, its ability to make reliable predictions using before unseen data. Moreover, there is no reason to suggest that the evidence may be a reliable predictor of the generalization performance. This has been known and explicitly stated since the method's infancy stage (see, for example, [27]). The generalization performance depends heavily on whether the adopted prior matches the "true" distribution of the unknown parameters. This is nicely demonstrated with a toy example in [17]. It is shown that the Bayesian average is optimal only if the adopted prior coincides with the true one. The situation is less clear when this is not the case. A more theoretical treatment of the topic, when there is a mismatch between the true and the selected prior, can be found in [18]. Thus, to be able to assess the generalization performance of a model learned via Bayesian inference, cross-validation is required, unless an independent test set can be afforded (for example, [44]).

  To avoid the need for cross-validation, an alternative way has been adopted by a number of authors. The cost function, to be minimized in (12.13), is built to quantify the generalization performance of an estimator; optimization then takes place concurrently for the unknown weights as well as the regularization parameter (see, for example, [15,32]). In general, this leads to a nonconvex optimization task and such techniques have not (yet) been widely embraced by the machine learning community.
- The Laplacian approximation to the evidence function is closely related to the *Bayesian information criterion* (BIC) [39] for model selection, which is expressed as

$$\ln p(\mathbf{y}|\mathcal{M}_i) \approx \ln p(\mathbf{y}|\mathcal{M}_i, \hat{\boldsymbol{\theta}}_{\text{MAP}}) - \frac{1}{2}K \ln N.$$

The BIC is obtained as a large N approximation to (12.38), assuming a broad enough Gaussian prior, and manipulating a bit the determinant involved in the last term. For a discussion including other related criteria, see [3,41].

- The Bayesian framework is also closely related to the minimum description length (MDL) methods. The log-evidence is associated with the number of bits in the shortest message that encodes the data via model $\mathcal{M}_i$ (for example, [45]).
- *Type II maximum likelihood*: Note that the evidence is the marginal likelihood function after integrating out the parameters $\boldsymbol{\theta}$. To distinguish it from the MAP method, when the evidence function is maximized with respect to a set of unknown parameters, it is usually referred to as *generalized maximum likelihood* or *Type II maximum likelihood* and sometimes as *empirical Bayes*. Recall from Remarks 12.1 that the MAP is named Type I estimator.

12.4 LATENT VARIABLES AND THE EM ALGORITHM

At the end of Section 12.3, it was pointed out that if we assume that $p(\boldsymbol{y}|\boldsymbol{\theta})$ and $p(\boldsymbol{\theta})$ are Gaussians of the form given in (12.39), then the evidence function associated with the regression task in Eq. (12.3) is also Gaussian parameterized via the (hyper)parameters $\sigma_\eta^2, \sigma_\theta^2$. Let us denote this set of unknown nonrandom parameters as $\boldsymbol{\xi} = [\sigma_\eta^2, \sigma_\theta^2]^T$, and we can write $p(\boldsymbol{y}; \boldsymbol{\xi})$. Maximizing the evidence with respect to $\boldsymbol{\xi}$ becomes a typical maximum likelihood one. However, in general, such closed form expressions for the evidence function are not possible, and the integration in (12.14) is intractable. The main source of difficulty is the fact that our regression model is described via two sets of random variables, that is, $\mathbf{y}$ and $\boldsymbol{\theta}$, yet only one of them, $\mathbf{y}$, can be directly observed. The other one, $\boldsymbol{\theta}$, cannot be observed, and this is the reason that the Bayesian philosophy tries to integrate it out of the joint PDF $p(\boldsymbol{y}, \boldsymbol{\theta})$. If $\boldsymbol{\theta}$ could be observed, then the unknown set of parameters $\boldsymbol{\xi}$ could be obtained by maximizing the likelihood $p(\boldsymbol{y}, \boldsymbol{\theta}; \boldsymbol{\xi})$, given a set of (joint) observations of $(\boldsymbol{y}, \boldsymbol{\theta})$. Because they cannot be observed, the random variables in the vector $\boldsymbol{\theta}$ are known as *hidden* variables.

Although we introduced the notion of hidden variables via our familiar regression task, unobserved variables (besides the noise) occur very often in a number of problems in probability and statistics. In a number of cases, from a larger set of jointly distributed random variables, only some can be observed and the rest remain hidden. Moreover, it is often useful to *build* hidden variables into a model by design. These variables are meant to represent latent causes that influence the observed variables and their introduction may facilitate the analysis. Often, such models associate one extra variable for *each one* of the observations. We will refer to such unobserved variables as *latent* variables. Their difference with the hidden ones is that their number is equal to that of the observations and grows accordingly as more observations get available. In contrast, unobserved random variables that are associated with the model and not with each one of the observations, individually, will be referred to as hidden variables.

12.4.1 THE EXPECTATION-MAXIMIZATION ALGORITHM

The *expectation-maximization* (EM) algorithm is an elegant algorithmic tool to maximize the likelihood (evidence) function for problems with latent/hidden variables. We will state the problem in a general formulation, and then we will apply it to different tasks, including regression.

Let $\mathbf{x}$ be a random vector and let $\mathcal{X}$ be the respective set of observations. Let $\mathcal{X}^l := \{\boldsymbol{x}_1^l, \ldots, \boldsymbol{x}_N^l\}$ be the corresponding set of latent variables; these can be either of a discrete or of a continuous nature. Each observation in $\mathcal{X}$ is associated with a latent vector $\boldsymbol{x}^l$ in $\mathcal{X}^l$. These latent variables are also known as *local* ones and each one expresses the hidden structure associated with the corresponding observation. We will refer to the set $\left\{\mathcal{X}, \mathcal{X}^l\right\}$ as the *complete* data set and to the set of observations, $\mathcal{X}$, as the *incomplete* one. Hidden random parameters, $\boldsymbol{\theta}$, can also be dealt with as latent variables; however, their number remains fixed (independent of N) and they are also known as *global* variables. In such cases, the complete data set is $\{\mathcal{X}, \mathcal{X}^l, \boldsymbol{\theta}\}$. To unclutter notation, we will focus on the set $\mathcal{X}^l$; yet, everything to be said also applies to hidden/global as well as to a combination of local/latent and global/hidden variables. Furthermore, let the corresponding joint distribution be parameterized in terms of a set of unknown nonrandom (hyper)parameters, $\boldsymbol{\xi}$. We further assume that, although $\mathcal{X}^l$ cannot be observed, the posterior distribution $p(\mathcal{X}^l|\mathcal{X}; \boldsymbol{\xi})$ ($P(\mathcal{X}^l|\mathcal{X}; \boldsymbol{\xi})$ for the discrete case) is fully specified, given the values in $\boldsymbol{\xi}$ and the observations in $\mathcal{X}$. This is a critical assumption for the EM algorithm. If the posterior PDF is not known, then one has to resort to variants of the EM which attempt to approximate it. We will come to such schemes in Section 13.2.

If the complete log-likelihood $p(\mathcal{X}, \mathcal{X}^l; \boldsymbol{\xi})$ were available, then the problem would be a typical maximum likelihood one. However, because no observations for the latent variables are available, the EM algorithm considers the *expectation* of the complete log-likelihood with respect to the latent variables associated with $\mathcal{X}^l$; this operation is possible, because the posterior distribution $p(\mathcal{X}^l|\mathcal{X}; \boldsymbol{\xi})$ is assumed to be known, provided that $\boldsymbol{\xi}$ is known. It can be shown that maximizing this expectation is equivalent to maximizing the corresponding evidence function $p(\mathcal{X}; \boldsymbol{\xi})$ (see Problem 12.3 and Section 12.7). To this end, the EM algorithm builds on an iterative philosophy, initialized by an arbitrary value $\boldsymbol{\xi}^{(0)}$. Then it proceeds along the following steps.

The EM algorithm

1. Expectation E-step: At the $(j+1)$th iteration, compute $p(\mathcal{X}^l|\mathcal{X}, \boldsymbol{\xi}^{(j)})$ and

$$\boxed{\mathcal{Q}(\boldsymbol{\xi}, \boldsymbol{\xi}^{(j)}) = \mathbb{E}\left[\ln p(\mathcal{X}, \mathcal{X}^l; \boldsymbol{\xi})\right],} \tag{12.40}$$

   where the expectation is taken with respect to $p(\mathcal{X}^l|\mathcal{X}; \boldsymbol{\xi}^{(j)})$.
2. Maximization M-step: Determine $\boldsymbol{\xi}^{(j+1)}$ so that

$$\boxed{\boldsymbol{\xi}^{(j+1)} = \arg\max_{\boldsymbol{\xi}} \mathcal{Q}(\boldsymbol{\xi}, \boldsymbol{\xi}^{(j)}).} \tag{12.41}$$

3. Check for convergence according to a criterion. If it is not satisfied go back to step 1.

A possible convergence criterion is to check whether $\|\boldsymbol{\xi}^{(j+1)} - \boldsymbol{\xi}^{(j)}\| < \epsilon$, for some user-defined constant ϵ. The use of the EM algorithm presupposes that working with the joint PDF $p(\mathcal{X}, \mathcal{X}^l; \boldsymbol{\xi})$ is computationally tractable. This is, for example, the case when working within the exponential family of PDFs, where the E-step may require only the computation of a few statistics of the latent variables. The exponential family of distributions is a computationally convenient one and it is treated in more detail in Section 12.8.

Remarks 12.3.

- The EM algorithm was proposed and given its name in the seminal 1977 paper by Arthur Dempster, Nan Laird, and Donald Rubin [12]. The paper generalized previously published results, as for example [2,38], and had a significant impact as a powerful tool in statistics. The complete convergence proof was given in [47]. See, for example, [29] for a related discussion.
- It can be shown that the EM algorithm converges to an (in general, local) maximum of $p(\mathcal{X}; \boldsymbol{\xi})$, which was our original goal. The likelihood never decreases. The convergence is slower than the quadratic convergence of Newton-type searching techniques, although near an optimal point a speedup may be possible. However, the convergence of the algorithm is smooth and its complexity more attractive to Newton-type schemes, with no matrix inversions involved. The keen reader may obtain more information in, for example, [14,30,33,43].
- The EM algorithm can be modified to obtain the MAP estimate. To this end, the M-step is changed to (Problem 12.4)

$$\boldsymbol{\xi}^{(j+1)} = \arg\max_{\boldsymbol{\xi}} \left\{ \mathcal{Q}(\boldsymbol{\xi}, \boldsymbol{\xi}^{(j)}) + \ln p(\boldsymbol{\xi}) \right\}, \tag{12.42}$$

  where $p(\boldsymbol{\xi})$ is the prior PDF associated with $\boldsymbol{\xi}$, if it is considered to be a random vector.
- The EM algorithm can be sensitive to the choice of the initial point $\boldsymbol{\xi}^{(0)}$. In practice, one can run the algorithm a number of times, starting from different initial points, and keep the best of the results. Other initialization procedures have also been used, depending on the application.
- *Missing data*: The EM algorithm can also be used to cope with cases where some of the values from the observed training data are missing. Missing values can be treated as hidden variables and maximization of the likelihood can be done by marginalizing over them. Such a procedure makes sense only if data are *missing at random*; that is, the cause of missing data is a random event and does not depend on the values of the unobserved samples.

12.5 LINEAR REGRESSION AND THE EM ALGORITHM

The Bayesian viewpoint to the regression task was considered in Section 12.2.3 via the Gaussian model assumption for $p(\boldsymbol{y}|\boldsymbol{\theta})$ and $p(\boldsymbol{\theta})$, given in (12.9) and (12.8), which subsequently led to a Gaussian posterior for $p(\boldsymbol{\theta}|\boldsymbol{y})$, given in (12.16). In the current section, and for the sake of presentation simplicity, we will adopt the special case of diagonal covariance matrices, that is, $\Sigma_\eta = \sigma_\eta^2 I$, $\Sigma_\theta = \sigma_\theta^2 I$, and $\boldsymbol{\theta}_0 = \boldsymbol{0}$.

Our goal now becomes to consider σ_η^2 and σ_θ^2 as (nonrandom) parameters and to obtain their values by maximizing the corresponding evidence function in (12.15). To this end, we will use the EM algorithm. Following the notation that we have adopted so far for the regression task, the observed variables are the outputs, $\boldsymbol{y}$, and the unobserved ones comprise the random parameters, $\boldsymbol{\theta}$, that define the regression model. Hence, in the current context, $\boldsymbol{y}$ will replace $\mathcal{X}$ and $\boldsymbol{\theta}$ will take the place of $\mathcal{X}^l$ in the general formulation of the EM algorithm in Section 12.4.1.

A prerequisite in order to apply the EM procedure is the knowledge of the posterior, which for this case is known, given the values of the parameters. We will work with the precision variables, and the parameter vector becomes

$$\boldsymbol{\xi} = [\alpha, \beta]^T, \ \alpha = \frac{1}{\sigma_\theta^2} \quad \text{and} \quad \beta = \frac{1}{\sigma_\eta^2}.$$

The EM algorithm is initialized with some arbitrary positive values $\alpha^{(0)}$ and $\beta^{(0)}$. Then the algorithm at the $(j+1)$th iteration step, where $\alpha^{(j)}$ and $\beta^{(j)}$ are assumed known, proceeds as follows:

- E-Step: Compute the posterior $p(\boldsymbol{\theta}|\boldsymbol{y};\boldsymbol{\xi}^{(j)})$, which according to (12.16) and for $\boldsymbol{\theta}_0 = \boldsymbol{0}$ is fully specified if we compute its mean and covariance matrix, using (12.19) and (12.20), that is,

$$\Sigma_{\theta|y}^{(j)} = \left(\alpha^{(j)} I + \beta^{(j)} \Phi^T \Phi\right)^{-1}, \tag{12.43}$$

$$\boldsymbol{\mu}_{\theta|y}^{(j)} = \beta^{(j)} \Sigma_{\theta|y}^{(j)} \Phi^T \boldsymbol{y}. \tag{12.44}$$

Compute the expected value of the log-likelihood associated with the complete data set; this is given by

$$\ln p(\boldsymbol{y},\boldsymbol{\theta};\boldsymbol{\xi}) := \ln p(\boldsymbol{y},\boldsymbol{\theta};\alpha,\beta) = \ln\left(p(\boldsymbol{y}|\boldsymbol{\theta};\beta)p(\boldsymbol{\theta};\alpha)\right),$$

or

$$\ln p(\boldsymbol{y},\boldsymbol{\theta};\alpha,\beta) = \frac{N}{2}\ln\beta + \frac{K}{2}\ln\alpha - \frac{\beta}{2}\|\boldsymbol{y} - \Phi\boldsymbol{\theta}\|^2 - \frac{\alpha}{2}\boldsymbol{\theta}^T\boldsymbol{\theta} - \left(\frac{N}{2} + \frac{K}{2}\right)\ln(2\pi). \tag{12.45}$$

Treating the hidden parameters as random variables, the expected value of (12.45), with respect to $\boldsymbol{\theta}$, is carried out via the Gaussian posterior defined by (12.43) and (12.44). To this end, the following steps are adopted.

1. To compute $\mathbb{E}\left[\boldsymbol{\theta}^T\boldsymbol{\theta}\right]$, recall the definition of the respective covariance matrix,

$$\Sigma_{\theta|y}^{(j)} = \mathbb{E}\left[(\boldsymbol{\theta} - \boldsymbol{\mu}_{\theta|y}^{(j)})(\boldsymbol{\theta} - \boldsymbol{\mu}_{\theta|y}^{(j)})^T\right] \tag{12.46}$$

or

$$\mathbb{E}\left[\boldsymbol{\theta}\boldsymbol{\theta}^T\right] = \Sigma_{\theta|y}^{(j)} + \boldsymbol{\mu}_{\theta|y}^{(j)}\boldsymbol{\mu}_{\theta|y}^{(j)T}, \tag{12.47}$$

which results in

$$\begin{aligned} A := \mathbb{E}\left[\boldsymbol{\theta}^T\boldsymbol{\theta}\right] &= \mathbb{E}\left[\text{trace}\left\{\boldsymbol{\theta}\boldsymbol{\theta}^T\right\}\right] \\ &= \text{trace}\left\{\boldsymbol{\mu}_{\theta|y}^{(j)}\boldsymbol{\mu}_{\theta|y}^{(j)T} + \Sigma_{\theta|y}^{(j)}\right\} \\ &= \left\|\boldsymbol{\mu}_{\theta|y}^{(j)}\right\|^2 + \text{trace}\left\{\Sigma_{\theta|y}^{(j)}\right\}. \end{aligned} \tag{12.48}$$

2. To compute $\mathbb{E}\left[\|\boldsymbol{y}-\Phi\boldsymbol{\theta}\|^2\right]$, define $\boldsymbol{\psi} := \boldsymbol{y}-\Phi\boldsymbol{\theta}$, and use the previous rationale to compute $\mathbb{E}\left[\boldsymbol{\psi}^T\boldsymbol{\psi}\right]$, which leads to (Problem 12.5)

$$B := \mathbb{E}\left[\|\boldsymbol{y}-\Phi\boldsymbol{\theta}\|^2\right] = \left\|\boldsymbol{y}-\Phi\boldsymbol{\mu}_{\theta|y}^{(j)}\right\|^2 + \text{trace}\left\{\Phi\Sigma_{\theta|y}^{(j)}\Phi^T\right\}. \tag{12.49}$$

Hence,

$$\mathcal{Q}\left(\alpha,\beta;\alpha^{(j)},\beta^{(j)}\right) = \frac{N}{2}\ln\beta + \frac{K}{2}\ln\alpha - \frac{\beta}{2}B - \frac{\alpha}{2}A - \left(\frac{N}{2}+\frac{K}{2}\right)\ln(2\pi). \tag{12.50}$$

- M-Step: Compute

$$\alpha^{(j+1)} : \frac{\partial}{\partial\alpha}\mathcal{Q}\left(\alpha,\beta;\alpha^{(j)},\beta^{(j)}\right) = 0,$$
$$\beta^{(j+1)} : \frac{\partial}{\partial\beta}\mathcal{Q}\left(\alpha,\beta;\alpha^{(j)},\beta^{(j)}\right) = 0,$$

which trivially lead to

$$\alpha^{(j+1)} = \frac{K}{\left\|\boldsymbol{\mu}_{\theta|y}^{(j)}\right\|^2 + \text{trace}\left\{\Sigma_{\theta|y}^{(j)}\right\}}, \tag{12.51}$$

$$\beta^{(j+1)} = \frac{N}{\left\|\boldsymbol{y}-\Phi\boldsymbol{\mu}_{\theta|y}^{(j)}\right\|^2 + \text{trace}\left\{\Phi\Sigma_{\theta|y}^{(j)}\Phi^T\right\}}. \tag{12.52}$$

Once the algorithm converges, the resulting values for α and β are used to completely specify the involved PDFs, which can be used either to obtain an estimate of $\hat{\boldsymbol{\theta}}$, for example, $\hat{\boldsymbol{\theta}} = \mathbb{E}[\boldsymbol{\theta}|\boldsymbol{y}]$, or make predictions via (12.21).

Example 12.2. In this example, the generalized linear regression model of Example 12.1 is reconsidered. The goal is to use the EM algorithm of Section 12.5, as summarized by the recursions (12.43), (12.44), (12.51), and (12.52). The variance of the Gaussian noise used in the model to generate the data was set equal to $\sigma_\eta^2 = 0.05$. The number of training points was $N = 500$. For the EM algorithm, both α and β were initialized to one. The correct dimensionality for the unknown parameter vector was used. The recovered values after the convergence of the EM were $\alpha = 1.32$ corresponding to $\sigma_\theta^2 = 0.756$ and $\beta = 19.96$ corresponding to $\sigma_\eta^2 = 0.0501$. Note that the latter is very close to the true variance of the noise. Then, predictions of the output variable y were performed at 20 points, using (12.22) and the value of $\boldsymbol{\mu}_{\theta|y}$ recovered by the EM algorithm, via (12.44).

Fig. 12.5A shows the predictions together with the associated error bars, computed from (12.23) using the values of σ_η^2 and σ_θ^2 obtained via the EM algorithm. Fig. 12.5B shows the convergence curve for σ_η^2 as a function of the number of iterations of the EM algorithm.

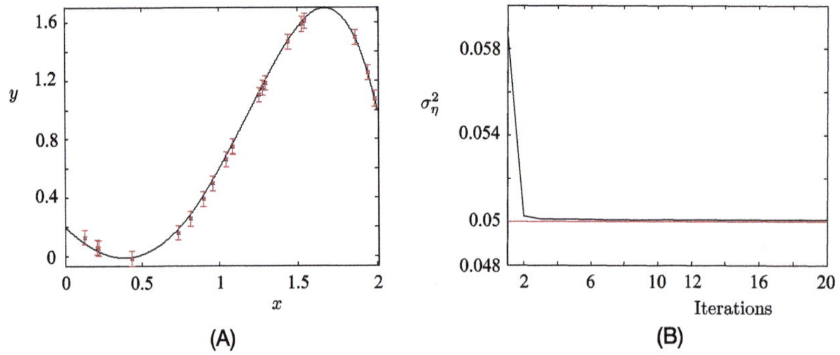

FIGURE 12.5

(A) The original graph from which the training points were sampled. In red, the respective predictions $\hat{y}$ and associated error bars for 20 randomly chosen points are shown. (B) The convergence curve for σ_η^2 as a function of the iterations of the EM algorithm. The red line corresponds to the true value.

12.6 GAUSSIAN MIXTURE MODELS

So far, we have seen a number of PDFs that can be used to model the distribution of an unknown random vector $\mathbf{x} \in \mathbb{R}^l$. However, all these models restrict the PDF to a specific functional term. Mixture modeling provides the freedom to model the unknown PDF, $p(\boldsymbol{x})$, as a linear combination of different distributions, that is,

$$p(\boldsymbol{x}) = \sum_{k=1}^{K} P_k p(\boldsymbol{x}|k), \tag{12.53}$$

where P_k is the parameter weighting the specific contributing PDF, $p(\boldsymbol{x}|k)$. To guarantee that $p(\boldsymbol{x})$ is a PDF, the weighting parameters must be nonnegative and add to one ($\sum_{k=1}^{K} P_k = 1$). The physical interpretation of (12.53) is that we are given a set of K distributions, $p(\boldsymbol{x}|k)$, $k = 1, 2, \ldots, K$. Each observation $\boldsymbol{x}_n$, $n = 1, 2, \ldots, N$, is drawn from one of these K distributions, but we are not told from which one. All we know is a set of parameters, P_k, $1, 2, \ldots, K$, each one providing the probability that a sample has been drawn from the corresponding PDF, $p(\boldsymbol{x}|k)$. It can be shown that for a large enough number of *mixtures*, K, and appropriate choice of the involved parameters, one can approximate arbitrarily close any continuous PDF.

Mixture modeling is a typical task involving latent variables; that is, the labels k of the PDF from which an obtained observation has originated. In practice, each $p(\boldsymbol{x}|k)$ is chosen from a known PDF family, parameterized via a set of parameters, and (12.53) can be rewritten as

$$p(\boldsymbol{x}) = \sum_{k=1}^{K} P_k p(\boldsymbol{x}|k; \boldsymbol{\xi}_k), \tag{12.54}$$

and the task is to estimate $(P_k, \boldsymbol{\xi}_k)$, $k = 1, 2, \ldots, K$, based on a set of observations $\boldsymbol{x}_n$, $n = 1, 2, \ldots, N$. The set of observations $\mathcal{X} = \{\boldsymbol{x}_n, \; n = 1, \ldots, N\}$ forms the incomplete set while the complete set $\{\mathcal{X}, \mathcal{K}\}$ comprises the sample pairs $(\boldsymbol{x}_n, k_n)$, $n = 1, \ldots, N$, with k_n being the label of the distribution (PDF) from which $\boldsymbol{x}_n$ was drawn. Parameter estimation for such a problem naturally lends itself to be treated via the EM algorithm. We will demonstrate the procedure via the use of Gaussian mixtures.

Let

$$p(\boldsymbol{x}|k;\boldsymbol{\xi}_k) = p(\boldsymbol{x}|k;\boldsymbol{\mu}_k,\Sigma_k) = \mathcal{N}\left(\boldsymbol{x}|\boldsymbol{\mu}_k,\Sigma_k\right),$$

where for simplicity we will assume that $\Sigma_k = \sigma_k^2 I,\ k = 1,\dots,K$. We will further assume the observations to be i.i.d. For such a modeling, the following hold true:

- The log-likelihood of the *complete* data set is given by

$$\ln p(\mathcal{X},\mathcal{K};\Xi,\boldsymbol{P}) = \sum_{n=1}^{N} \ln p(\boldsymbol{x}_n,k_n;\boldsymbol{\xi}_{k_n}) = \sum_{n=1}^{N} \ln\left(p(\boldsymbol{x}_n|k_n;\boldsymbol{\xi}_{k_n})P_{k_n}\right). \tag{12.55}$$

  We have used the notation

$$\Xi = [\boldsymbol{\xi}_1^T,\dots,\boldsymbol{\xi}_K^T]^T,\quad \boldsymbol{P} = [P_1,P_2,\dots,P_K]^T,\quad \text{and}\quad \boldsymbol{\xi}_k = [\boldsymbol{\mu}_k^T,\sigma_k^2]^T.$$

  In other words, the deterministic parameters that have to be estimated via the EM algorithm are the mean values and variances of all the Gaussian mixtures, as well as the respective mixing probabilities.
- The posterior probabilities of the latent discrete variables are given by

$$P(k|\boldsymbol{x};\Xi,\boldsymbol{P}) = \frac{p(\boldsymbol{x}|k;\boldsymbol{\xi}_k)P_k}{p(\boldsymbol{x};\Xi,\boldsymbol{P})}, \tag{12.56}$$

  where

$$p(\boldsymbol{x};\Xi,\boldsymbol{P}) = \sum_{k=1}^{K} P_k p(\boldsymbol{x}|k;\boldsymbol{\xi}_k). \tag{12.57}$$

  We have now all the ingredients required by the EM algorithm. Starting from $\Xi^{(0)}$ and $\boldsymbol{P}^{(0)}$, the $(j+1)$th iteration comprises the following steps:
- E-step: Using (12.56) and (12.57), compute

$$\boxed{P\left(k|\boldsymbol{x}_n;\Xi^{(j)},\boldsymbol{P}^{(j)}\right) = \frac{p\left(\boldsymbol{x}_n|k;\boldsymbol{\xi}_k^{(j)}\right)P_k^{(j)}}{\sum_{k=1}^{K} P_k^{(j)} p\left(\boldsymbol{x}_n|k;\boldsymbol{\xi}_k^{(j)}\right)},\quad n = 1,2,\dots,N,} \tag{12.58}$$

  which in turn defines

$$\begin{aligned}\mathcal{Q}\left(\Xi, \boldsymbol{P}; \Xi^{(j)}, \boldsymbol{P}^{(j)}\right) &= \sum_{n=1}^{N} \mathbb{E}\left[\ln\left(p(\boldsymbol{x}_n|\mathrm{k}_n; \boldsymbol{\xi}_{k_n})P_{k_n}\right)\right] \\ &:= \sum_{n=1}^{N}\sum_{k=1}^{K} P\left(k|\boldsymbol{x}_n; \Xi^{(j)}, \boldsymbol{P}^{(j)}\right)\left(\ln P_k - \frac{l}{2}\ln\sigma_k^2\right. \\ &\left. - \frac{1}{2\sigma_k^2}\|\boldsymbol{x}_n - \boldsymbol{\mu}_k\|^2\right) + C, \end{aligned} \tag{12.59}$$

where C includes all the terms corresponding to the normalization constant. Note that we have finally relaxed the notation from k_n to k, because we sum up over all k, which does not depend on n.

- M-step: Maximization of $\mathcal{Q}(\Xi, \boldsymbol{P}; \Xi^{(j)}, \boldsymbol{P}^{(j)})$ with respect to all the involved parameters results in the following set of recursions (Problem 12.6):
  Set, for notational convenience,

$$\gamma_{kn} := P(k|\boldsymbol{x}_n; \Xi^{(j)}, \boldsymbol{P}^{(j)}).$$

Then

$$\boldsymbol{\mu}_k^{(j+1)} = \frac{\sum_{n=1}^{N}\gamma_{kn}\boldsymbol{x}_n}{\sum_{n=1}^{N}\gamma_{kn}}, \tag{12.60}$$

$$\sigma_k^{2(j+1)} = \frac{\sum_{n=1}^{N}\gamma_{kn}\left\|\boldsymbol{x}_n - \boldsymbol{\mu}_k^{(j+1)}\right\|^2}{l\sum_{n=1}^{N}\gamma_{kn}}, \tag{12.61}$$

$$P_k^{(j+1)} = \frac{1}{N}\sum_{n=1}^{N}\gamma_{kn}. \tag{12.62}$$

Iterations continue until a convergence criterion is met. The extension to the case of a general covariance matrix is straightforward by replacing (12.61) by

$$\Sigma_k^{(j+1)} = \frac{\sum_{n=1}^{N}\gamma_{kn}\left(\boldsymbol{x}_n - \boldsymbol{\mu}_k^{(j+1)}\right)\left(\boldsymbol{x}_n - \boldsymbol{\mu}_k^{(j+1)}\right)^T}{\sum_{n=1}^{N}\gamma_{kn}}.$$

Remarks 12.4.

- To get good initialization for the EM algorithm, sometimes a simpler clustering algorithm, for example, the k-means (Section 12.6.1 and [41]), is run to provide an initial estimate of the means and shapes of clusters (covariance matrices), by associating each mixture with a cluster in the input space. Another simpler way is to select K points randomly from the data set. A more elaborate technique, which is commonly used, is to select them randomly but in such a way to make sure that the whole data set is represented in a balanced way (see, for example, [1]).
- The number of mixtures, K, is usually determined by cross-validation (Chapter 3; see also [16]).
- The mixing parameters P_k, $k = 1, \dots, K$, should be initialized by keeping in mind that they are probabilities and they have to add to one.

- One of the problems that may be encountered in practice in the Gaussian mixture task is when one of the mixture components is centered at (or very close to) one of the data points, for example, $\boldsymbol{\mu}_k^{(j+1)} = \boldsymbol{x}_n$, for some values of k and n. In such a case, the exponent term of the respective Gaussian becomes one and the contribution of this particular component in the log-likelihood is equal to $(2\pi\sigma_k^2)^{-l/2}$. If, in addition, σ_k is very small, this will lead the likelihood to a large value, although this is not indicative that the true model has been learned. Soon, we will see that the use of priors can alleviate such problems.
- *Identifiability*: A further issue associated with the EM algorithm, in the context of distribution mixtures, is that the obtained solution in the parameter space is not unique. For the case of K mixtures, for each solution (point in the parameter space) there are $K! - 1$ other points which give rise to the same distribution. For example, let us fit a model of two Gaussians in the one-dimensional space, which will result in estimates for the respective mean values, $\hat{\mu}_1$ and $\hat{\mu}_2$. However, in the corresponding parameter space, there is an uncertainty on whether these values define the point $\boldsymbol{\mu}_a = [\hat{\mu}_1, \hat{\mu}_2]^T$ or the point $\boldsymbol{\mu}_b = [\hat{\mu}_2, \hat{\mu}_1]$. Both of these points give rise to the same distribution. We say that the parameters in our model are not identifiable. A parameter (vector) which defines a family of distributions $p(\boldsymbol{x}; \boldsymbol{\theta})$ is said to be *identifiable* if $p(\boldsymbol{x}; \boldsymbol{\theta}_1) \neq p(\boldsymbol{x}; \boldsymbol{\theta}_2)$ for $\boldsymbol{\theta}_1 \neq \boldsymbol{\theta}_2$ (see, e.g., [7]). Although in our context, where our interest is in computing $p(\boldsymbol{x})$, unidentifiability does not cause any problems, this can be an issue in cases where the focus of interest lies on the parameters (see, for example, [36]).
- *Mixtures of Student's t distributions*: A significant shortcoming of mixtures based on normal distributions is their vulnerability to outliers. The replacement of normal distributions with the heavier-tailed Student's t distributions (see Section 13.5) has been proposed as a way to mitigate these shortcomings and a related treatment of the resulting model under an EM algorithmic framework has been conducted. Although the steps get a bit more involved, the ideas explored so far transfer nicely in this case too (see, for example, [9,10,35,37]).

Example 12.3. The goal of this example is to demonstrate the application of the EM algorithm in the context of the Gaussian mixture modeling. The data are generated according to three Gaussians in the two-dimensional space, with parameters

$$\boldsymbol{\mu}_1 = [10, 3]^T, \quad \boldsymbol{\mu}_2 = [1, 1]^T, \quad \boldsymbol{\mu}_3 = [5, 4]^T$$

and covariance matrices

$$\Sigma_1 = \begin{bmatrix} 1 & 0 \\ 0 & 1 \end{bmatrix}, \quad \Sigma_2 = \begin{bmatrix} 1.5 & 0 \\ 0 & 1.5 \end{bmatrix}, \quad \Sigma_3 = \begin{bmatrix} 2 & 0 \\ 0 & 2 \end{bmatrix},$$

respectively. The number of the generated points is 300, with 100 points per mixture. The points are shown in Fig. 12.6, together with the gray circles indicating the 80% probability regions, for each one of the clusters. The EM algorithm comprising the steps (12.58) and (12.60)–(12.62) was run with the following initial values:

$$\boldsymbol{\mu}_1^{(0)} = [3, 5]^T, \quad \boldsymbol{\mu}_2^{(0)} = [2, 0.4]^T, \quad \boldsymbol{\mu}_3^{(0)} = [4, 3]^T,$$

and

$$\Sigma_1^{(0)} = \Sigma_2^{(0)} = \Sigma_3^{(0)} = \begin{bmatrix} 1 & 0 \\ 0 & 1 \end{bmatrix}.$$

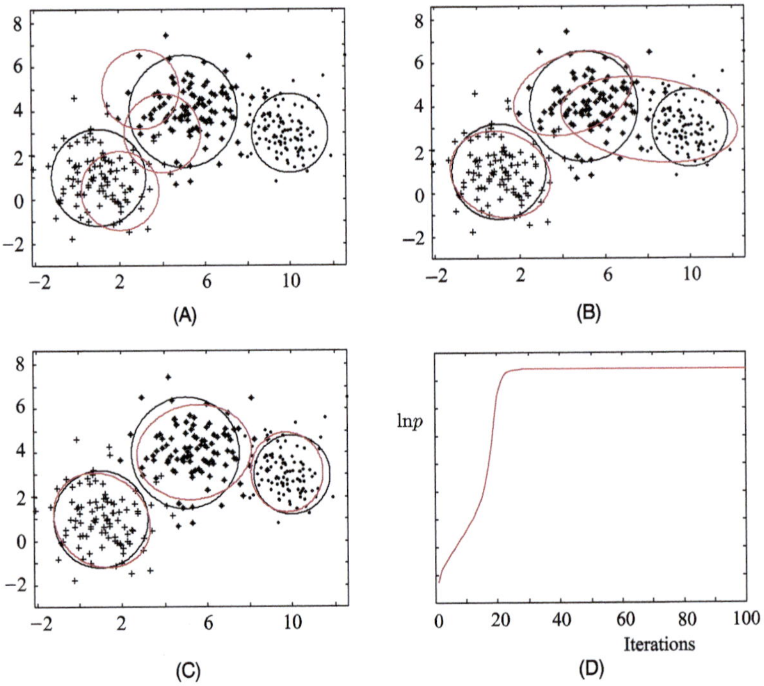

FIGURE 12.6

The curves (ellipses) indicate the 80% probability regions. The gray curves correspond to the true Gaussian clusters of Example 12.3. The red curves correspond to (A) the initial values for the mean and the covariance matrices, (B) the mixtures recovered by the EM algorithm after five iterations, and (C) after 30 iterations. (D) The log-likelihood as a function of the number of iterations.

The probabilities were initialized to their true values $P_1^{(0)} = P_2^{(0)} = P_3^{(0)} = 1/3$. The red curves in Fig. 12.6 correspond to the mixtures recovered by the EM algorithm at (A) the initial estimates, (B) after five iterations, and (C) after convergence. Fig. 12.6D shows the log-likelihood as a function of the number of iterations.

Fig. 12.7 corresponds to a different setup. This time, the mean values were initialized at points very far from the true ones, that is,

$$\boldsymbol{\mu}_1^{(0)} = [10, 13]^T, \ \boldsymbol{\mu}_2^{(0)} = [11, 12]^T, \ \boldsymbol{\mu}_3^{(0)} = [13, 11]^T,$$

while the covariances and probabilities were initialized as before. Observe that in this case, the EM algorithm fails to capture the true nature of the problem, having been trapped in a local minimum.

12.6.1 GAUSSIAN MIXTURE MODELING AND CLUSTERING

Clustering or unsupervised learning is an important part of machine learning, which is not treated in this book. Extensive coverage of clustering is given in, for example, [41]. However, the mixture modeling

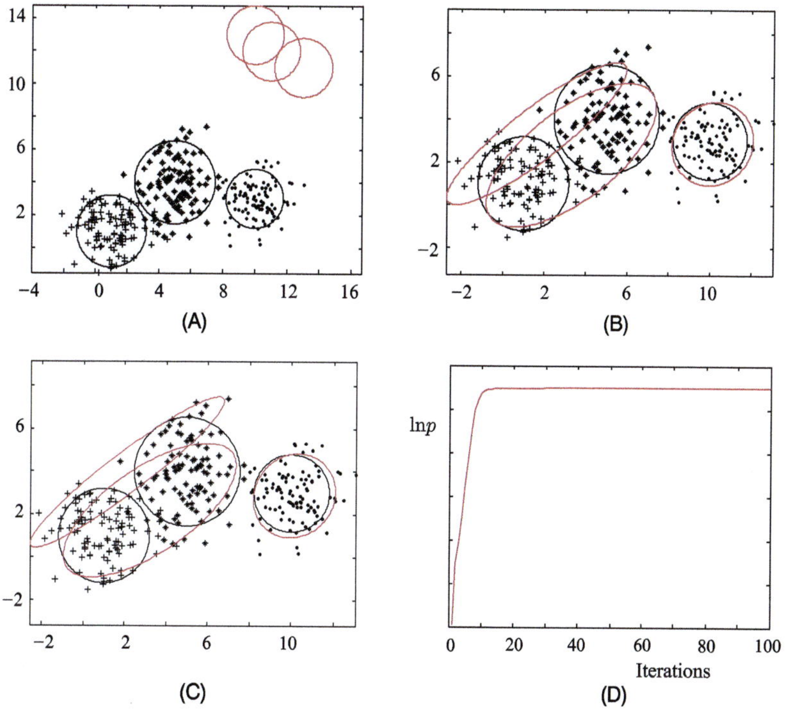

FIGURE 12.7

This is the counterpart of Fig. 12.6, where now the initial values for the means are very far from the true ones. In this case, the EM fails to recover the true nature of the mixtures and has been trapped in a local minimum.

task via the EM offers us a good excuse to say a few words. Without going into formal definitions, the task of clustering is to assign a number of points, $\boldsymbol{x}_1, \dots, \boldsymbol{x}_N$, into K groups or clusters. Points that are assigned to the same cluster must be more "similar" than points which are assigned to different clusters. Some clustering algorithms need the number of clusters, K, to be provided by the user as an input variable. Other schemes treat it as a free parameter to be recovered from the data by the algorithm. The other major issue in clustering is to quantify "similarity." Different definitions end up with different clusterings. A clustering is a specific allocation of the points to clusters. In general, assigning points to clusters according to an optimality criterion is an NP-hard task (see, for example, [41]). Thus, in general, any clustering algorithm provides a suboptimal solution.

Gaussian mixture modeling is among the popular clustering algorithms. The main assumption is that the points which belong to the same cluster are distributed according to the same Gaussian distribution (this is how similarity is defined in this case), of unknown mean and covariance matrix. Each mixture component defines a different cluster. Thus, the goal is to run the EM algorithm over the available data points to provide, after convergence, the posterior probabilities $P(k|\boldsymbol{x}_n)$, $k = 1, 2, \dots, K$, $n = 1, 2, \dots, N$, where each k corresponds to a cluster. Then, each point is assigned to cluster k according to the rule

$$\text{assign } \boldsymbol{x}_n \text{ to cluster } k = \arg\max_i P(i|\boldsymbol{x}_n), \; i = 1, 2, \dots, K.$$

The EM algorithm for clustering can be considered to be a refined version of a more primitive scheme, known as the *k-means* or *isodata* algorithm. In the EM algorithm, the posterior probability of each point $\boldsymbol{x}_n$, with respect to each one of the clusters k, is computed recursively. Moreover, the mean value $\boldsymbol{\mu}_k$ of the points associated with cluster k is computed as a weighted average of *all* the training points (12.60). In contrast, in the k-means algorithm, at each iteration the posterior probability gets a binary value in $\{1, 0\}$; for each point $\boldsymbol{x}_n$, the Euclidean distance from all the currently available estimates of the mean values is computed, and the posterior probability is estimated according to the following rule:

$$P(k|\boldsymbol{x}_n) = \begin{cases} 1, & \text{if } ||\boldsymbol{x}_n - \boldsymbol{\mu}_k||^2 < ||\boldsymbol{x}_n - \boldsymbol{\mu}_j||^2, \; j \neq k, \\ 0, & \text{otherwise.} \end{cases}$$

The k-means algorithm is not concerned about covariance matrices. Despite its simplicity, it is not an exaggeration to say that it is the most well-known clustering algorithm, and a number of theoretical papers and improved versions have been proposed over the years (see, for example, [41]). Due to its popularity, we will take the liberty to state it in Algorithm 12.1.

Algorithm 12.1 (The k-means or isodata clustering algorithm).

- Initialize
  - Select the number of clusters K.
  - Set $\boldsymbol{\mu}_k, \; k = 1, 2, \dots, K$, to arbitrary values.
- **For** $n = 1, 2, \dots, N$, **Do**
  - Determine the closest cluster mean, say, $\boldsymbol{\mu}_k$, to $\boldsymbol{x}_n$.
  - Set $b(n) = k$.
- **End For**
- **For** $k = 1, 2, \dots, K$, **Do**
  - Update $\boldsymbol{\mu}_k, \; k = 1, 2, \dots, K$, as the mean of all the points with $b(n) = k, \; n = 1, 2, \dots, N$.
- **End For**
- Until no change in $\boldsymbol{\mu}_k, \; k = 1, 2, \dots, K$, occurs between two successive iterations.

The k-means algorithm can also be derived as a limiting case of the EM scheme (for example, [34]). Note that both the EM algorithm and the k-means one can only recover *compact clusters*. In other words, if the points are distributed in ring-shaped clusters, then this type of clustering algorithms is not appropriate.

Fig. 12.8A shows the data points generated by two Gaussians; 200 points from each one. The points are shown by red and gray colors, depending on the Gaussian that generated them. Of course in clustering, the data points are given to the algorithm without the "color" (labeling). It is up to the algorithm to make the partition in clusters. For both the EM and the k-means algorithm, the correct number of clusters ($K = 2$) was given. The k-means was initialized with zero mean values. Fig. 12.8B shows the clusters formed by the k-means and Fig. 12.8D shows the clusters formed by the EM algorithm. Fig. 12.8C shows the Gaussians that were used for the initialization of the EM.

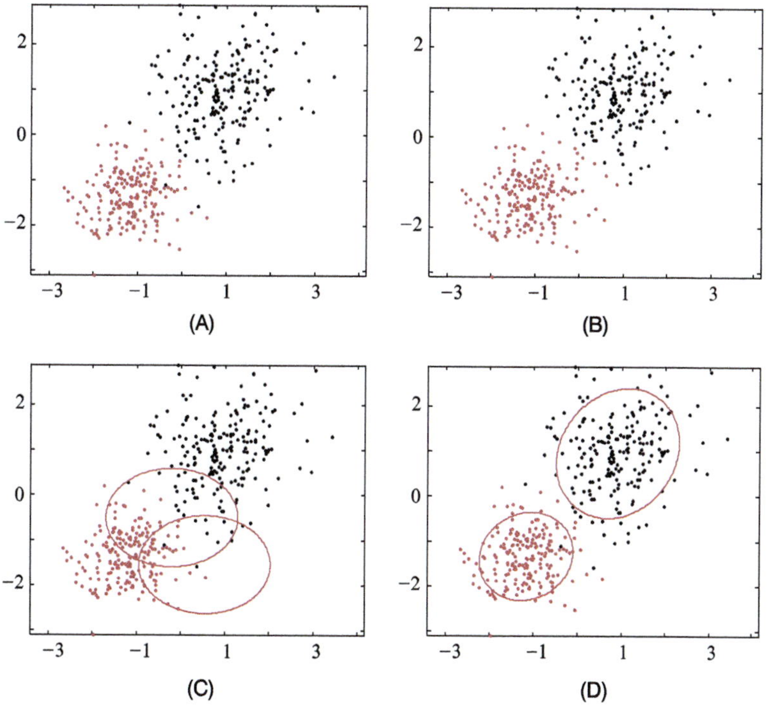

FIGURE 12.8

(A) The data points generated by two Gaussians (red and gray). (B) The recovered clusters by the k-means (red and gray). (C) The 80% probability curves for the initialization of the EM algorithm. (D) The final Gaussians obtained by the EM algorithm with the respective clusters.

Fig. 12.9 shows the respective sequence of figures, which corresponds to points obtained by the same Gaussians; however, now, there is an imbalance in the number of the points, as only 20 points spring forth from the first one and 200 points from the second. Observe that the k-means has a problem in recovering the true clustering structure; it attempts to make the two clusters more equally sized. A number of techniques and versions of the basic k-means scheme have been proposed to overcome its drawbacks (see [41]). Finally, it must be stressed that both the EM and the k-means algorithm will always recover as many clusters as the user-defined input variable K dictates. In the case of the EM algorithm, this drawback is overcome when the variational EM algorithm is used, as will be discussed in Section 13.4.

12.7 THE EM ALGORITHM: A LOWER BOUND MAXIMIZATION VIEW

In Section 12.4, the EM algorithm was introduced and it was stated that maximizing the expectation of the complete log-likelihood with respect to the set of parameters $\boldsymbol{\xi}$ is equivalent to maximizing the

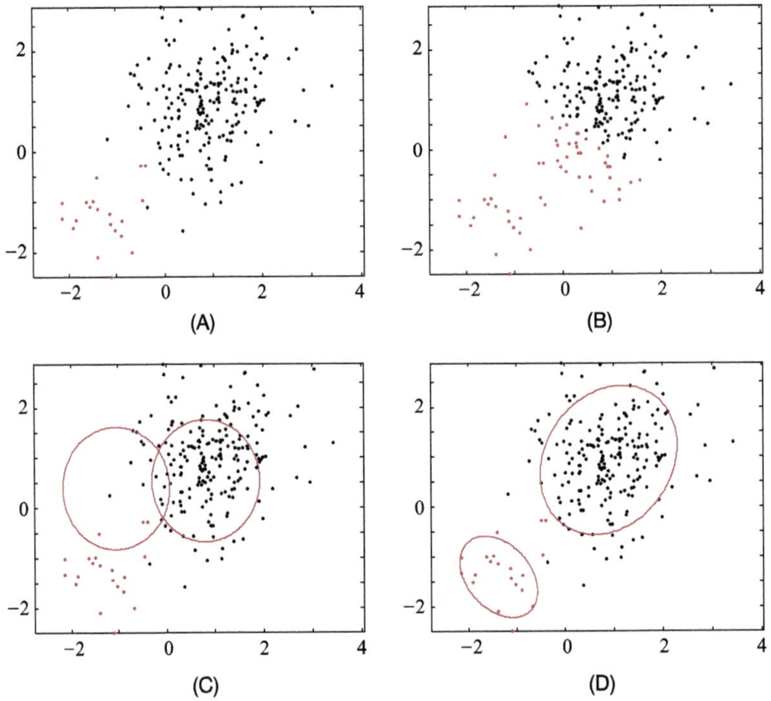

FIGURE 12.9

(A) The data points generated by two Gaussians (red and gray). One of the clusters consists of only 20 points and the other one of 200 points. (B) The recovered clusters by the k-means (red and gray). Observe that the algorithm has not identified the correct clusters, by assigning more points to the "smaller" one. (C) The 80% probability curves for the initialization of the EM algorithm. (D) The final Gaussians obtained by the EM algorithm, with the respective clusters.

corresponding evidence function, i.e., $p(\mathcal{X}; \boldsymbol{\xi})$. However, this was not explicitly shown. In this subsection, this connection will become clear. It will be shown that the EM algorithm basically maximizes a tight *lower bound* of the evidence. Furthermore, this interpretation of the EM will allow for generalizations to extend it to cases where the computations involving the posterior, $p(\mathcal{X}^l|\mathcal{X}; \boldsymbol{\xi})$, are not computationally tractable.

Let us consider the functional[6]

$$\boxed{\mathcal{F}(q, \boldsymbol{\xi}) := \int q(\mathcal{X}^l) \ln \frac{p(\mathcal{X}, \mathcal{X}^l; \boldsymbol{\xi})}{q(\mathcal{X}^l)} \, d\mathcal{X}^l,} \tag{12.63}$$

[6] A functional is an operator that takes as input a function and returns a real value. It is a generalization of our familiar functions, where now the inputs are also functions.

where $q(\mathcal{X}^l)$ is any nonnegative function that integrates to one; that is, it is a PDF defined over the latent variables. The functional $\mathcal{F}$ depends on $\boldsymbol{\xi}$ and on q, and its definition bears a strong similarity to the notion of free energy, used in statistical physics. Indeed, (12.63) can be written as

$$\mathcal{F}(q, \boldsymbol{\xi}) = \int q(\mathcal{X}^l) \ln p(\mathcal{X}, \mathcal{X}^l; \boldsymbol{\xi})\, d\mathcal{X}^l + H, \tag{12.64}$$

where

$$H = -\int q(\mathcal{X}^l) \ln q(\mathcal{X}^l)\, d\mathcal{X}^l$$

is the entropy associated with $q(\mathcal{X}^l)$. If one defines $-\ln p(\mathcal{X}, \mathcal{X}^l; \boldsymbol{\xi})$ as the *energy* of the system, $(\mathcal{X}, \mathcal{X}^l)$, then $\mathcal{F}(q, \boldsymbol{\xi})$ represents the negative of the so-called *free energy* [34]. Focussing on the first term on the right-hand side in Eq. (12.64), this can be written as

$$\boxed{\mathcal{F}(q, \boldsymbol{\xi}) = \mathbb{E}_q \Big[\ln p(\mathcal{X}, \mathcal{X}^l; \boldsymbol{\xi}) \Big] + H.} \tag{12.65}$$

In other words, the first term on the right-hand side is very similar to the $\mathcal{Q}$ term in Eq. (12.40). The only difference is that the expectation here is taken with respect to q.

Taking into account Bayes' theorem, Eq. (12.63) becomes

$$\begin{aligned} \mathcal{F}(q, \boldsymbol{\xi}) &= \int q(\mathcal{X}^l) \ln \frac{p(\mathcal{X}^l|\mathcal{X}; \boldsymbol{\xi}) p(\mathcal{X}; \boldsymbol{\xi})}{q(\mathcal{X}^l)} d\mathcal{X}^l, \\ &= \int q(\mathcal{X}^l) \ln \frac{p(\mathcal{X}^l|\mathcal{X}; \boldsymbol{\xi})}{q(\mathcal{X}^l)} d\mathcal{X}^l + \ln p(\mathcal{X}; \boldsymbol{\xi}), \end{aligned} \tag{12.66}$$

where the latter results because $\ln p(\mathcal{X}; \boldsymbol{\xi})$ does not depend on $q(\mathcal{X}^l)$ and the latter integrates to one, being a probability distribution. The first term on the right-hand side is the negative of the so-called KL divergence (Eq. (2.161)) between $q(\mathcal{X}^l)$ and $p(\mathcal{X}^l|\mathcal{X}; \boldsymbol{\xi})$, which we will denote as $\mathrm{KL}(q \parallel p)$. Recall that the KL divergence measures how different two distributions are. If the two involved distributions are equal, then their KL divergence becomes zero. Thus, finally, we get

$$\ln p(\mathcal{X}; \boldsymbol{\xi}) = \mathcal{F}(q, \boldsymbol{\xi}) + \mathrm{KL}(q \parallel p). \tag{12.67}$$

Because the KL divergence is a nonnegative quantity, i.e., $\mathrm{KL}(q \parallel p) \geq 0$ (Problem 12.7), it turns out that

$$\boxed{\ln p(\mathcal{X}; \boldsymbol{\xi}) \geq \mathcal{F}(q, \boldsymbol{\xi}).} \tag{12.68}$$

In other words, the functional $\mathcal{F}(q, \boldsymbol{\xi})$ is a lower bound of the log-likelihood function. Also, the bound becomes tight if $\mathrm{KL}(q \parallel p) = 0$, which is true *if and only if* $q(\mathcal{X}^l) = p(\mathcal{X}^l|\mathcal{X}; \boldsymbol{\xi})$. Moreover, the previous bound is valid for *all* distributions q and *all* parameters $\boldsymbol{\xi}$.

The previous findings pave the way of maximizing $\ln p(\mathcal{X}; \boldsymbol{\xi})$ by trying to maximize its lower bound. This is in line with a more general class of optimization algorithms, known as *minorize-maximization* (or mazorize-minimization) (MM) methods [20]. A surrogate function that minorizes the cost function (lower bound), which is easier to be maximized, is employed. Then maximizing this lower bound iteratively pushes the cost function to a local maximum. Note that in our case maximization of $\mathcal{F}$ involves two terms, namely, q and $\boldsymbol{\xi}$. We will adopt a widely used technique known as *alternating optimization*. Such an approach imposes an iterative procedure. Starting from some initial conditions, one "freezes" the value of one of the involved terms and maximizes with respect to the other. Then the value of the latter term is frozen and optimization is carried out with respect to the former. These alternating steps carry on until convergence. In our current context, starting from an arbitrary $\boldsymbol{\xi}^{(0)}$, the $(j+1)$th iteration comprises the following steps:

- Step 1: Keeping $\boldsymbol{\xi}^{(j)}$ fixed, optimize with respect to q. This step tightens the lower bound in (12.68). This is achieved if $\mathrm{KL}(q \parallel p) = 0$, and it can only happen if

$$q^{(j+1)}(\mathcal{X}^l) = p(\mathcal{X}^l|\mathcal{X};\boldsymbol{\xi}^{(j)}), \tag{12.69}$$

  that is, if we set $q(\mathcal{X}^l)$ equal to the posterior given $\mathcal{X}$ and $\boldsymbol{\xi}^{(j)}$; as (12.67) suggests, this makes the bound tight, that is,

$$\ln p(\mathcal{X};\boldsymbol{\xi}^{(j)}) = \mathcal{F}\left(p(\mathcal{X}^l|\mathcal{X};\boldsymbol{\xi}^{(j)}), \boldsymbol{\xi}^{(j)}\right). \tag{12.70}$$

- Step 2: Fixing $q^{(j+1)}$, insert it in the place of q in (12.68), and because the bound holds for any q, maximize with respect to $\boldsymbol{\xi}$, that is,

$$\boldsymbol{\xi}^{(j+1)} = \arg\max_{\boldsymbol{\xi}} \mathcal{F}\left(p\left(\mathcal{X}^l|\mathcal{X};\boldsymbol{\xi}^{(j)}\right), \boldsymbol{\xi}\right).$$

Hence, we have now obtained the following inequalities:

$$\ln p\left(\mathcal{X};\boldsymbol{\xi}^{j+1}\right) \geq \mathcal{F}\left(p\left(\mathcal{X}^l|\mathcal{X};\boldsymbol{\xi}^{(j)}\right), \boldsymbol{\xi}^{j+1}\right) \geq \mathcal{F}\left(p\left(\mathcal{X}^l|\mathcal{X};\boldsymbol{\xi}^{(j)}\right), \boldsymbol{\xi}^{j}\right),$$

and taking into account that the last term on the right-hand side is equal to $p\left(\mathcal{X};\boldsymbol{\xi}^{j}\right)$, we have

$$\boxed{\ln p\left(\mathcal{X};\boldsymbol{\xi}^{j+1}\right) \geq \ln p\left(\mathcal{X};\boldsymbol{\xi}^{j}\right).}$$

It is now readily seen that we have rederived the EM algorithm. Indeed, from the definition of $\mathcal{F}(\cdot,\cdot)$ in (12.63) we obtain

$$\mathcal{F}\left(p(\mathcal{X}^l|\mathcal{X};\boldsymbol{\xi}^{(j)}), \boldsymbol{\xi}\right) = \mathcal{Q}(\boldsymbol{\xi}, \boldsymbol{\xi}^{(j)}) - \int p(\mathcal{X}^l|\mathcal{X};\boldsymbol{\xi}^{(j)}) \ln p(\mathcal{X}^l|\mathcal{X};\boldsymbol{\xi}^{(j)}) d\mathcal{X}^l, \tag{12.71}$$

where $\mathcal{Q}(\boldsymbol{\xi}, \boldsymbol{\xi}^{(j)})$ is the same as in (12.40) and the second term on the right-hand side is independent of $\boldsymbol{\xi}$; this latter term is equal to the entropy associated with $q^{(j+1)}(\mathcal{X}^l)$. The rederivation of the EM via this path makes it clear that the quantity that is maximized is the log-likelihood, $\ln p(\mathcal{X};\boldsymbol{\xi})$, and that its value is guaranteed not to decrease after each combined iteration step. Fig. 12.10 illustrates schematically the two EM steps comprising the $(j+1)$th iteration.

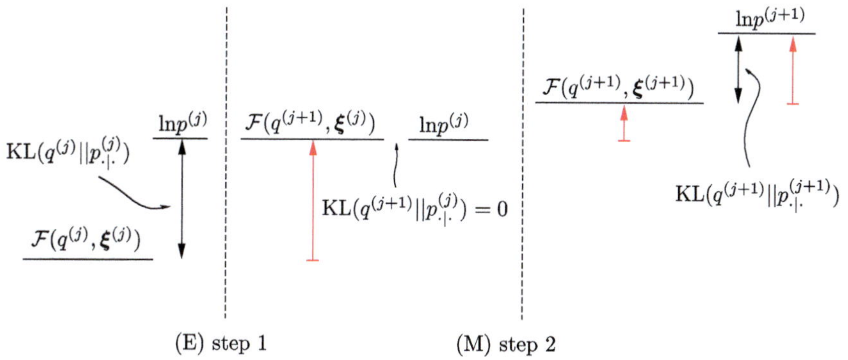

FIGURE 12.10

The E-step adjusts $q^{(j)} := q^{(j)}(\mathcal{X}^l)$ so that its KL divergence from $p^{(j)}_{\cdot|\cdot} := p(\mathcal{X}^l|\mathcal{X}; \boldsymbol{\xi}^{(j)})$ becomes zero. The M-step maximizes with respect to $\boldsymbol{\xi}$. We used p^j to denote the evidence function.

Needless to say that the EM algorithm is not a panacea. We will soon seek variants to deal with cases where the posterior cannot be given in an analytic form. Moreover, there are still cases where the M-step can be computationally intractable. To this end, several variants have been proposed (see, for example, [31,34]).

Remarks 12.5.

- *Online versions of EM*: We have already pointed out that in many cases of large data applications, online versions are the preferable choice in practice. The EM algorithm is no exception, and a number of related versions have been proposed. In [34], an online EM algorithm is proposed based on the lower bound interpretation. In [6], stochastic approximation arguments have been employed. In [25], a comparative study of different techniques is reported.
- Often in practice, carrying out the expectation step may be intractable. Later on in the next chapter, we will see variational methods as a way to overcome this obstacle. An alternative path is to employ Monte Carlo sampling techniques (Chapter 14) to generate samples from the involved distributions and approximate the expectation with the computation of the respective sample mean (see, for example, [8,11]).

12.8 EXPONENTIAL FAMILY OF PROBABILITY DISTRIBUTIONS

It must be clear by now that the Bayesian setting starts by adopting a specific functional form for the conditional distribution, which "explains" the generation of the observations given the parameters, and a prior distribution that describes the randomness of the associated parameters. The latter is equivalent to regularizing the corresponding learning task and expresses our uncertainty about the values of the parameters, prior to receiving any observations. The goal in Bayesian learning is to obtain the posterior distribution of the parameters given the values of the observed variables. The computation of the posterior can be greatly facilitated if the conditional and the prior distributions are carefully chosen. Selecting these distributions from the *exponential family* makes the computation of the posterior a rather trivial task. Distributions of the exponential family will be used in the next chapter, where approximate Bayesian inference methods are discussed.

We will treat the topic of the exponential family of probability distributions in a general setting. Let $\mathbf{x} \in \mathbb{R}^l$ be a random vector and $\boldsymbol{\theta} \in \mathbb{R}^K$ a random (parameter) vector. We say that the parameterized PDF $p(\boldsymbol{x}|\boldsymbol{\theta})$ is of the exponential form if

$$p(\boldsymbol{x}|\boldsymbol{\theta}) = g(\boldsymbol{\theta}) f(\boldsymbol{x}) \exp\left(\boldsymbol{\phi}^T(\boldsymbol{\theta})\boldsymbol{u}(\boldsymbol{x})\right), \tag{12.72}$$

where

$$g(\boldsymbol{\theta}) = \frac{1}{\int f(\boldsymbol{x}) \exp\left(\boldsymbol{\phi}^T(\boldsymbol{\theta})\boldsymbol{u}(\boldsymbol{x})\right) d\boldsymbol{x}} \tag{12.73}$$

is the normalizing constant of the PDF. A similar definition holds if $\mathbf{x}$ is a discrete random variable and the respective function represents the probability mass function $P(\boldsymbol{x}|\boldsymbol{\theta})$; in this case, the integration in (12.73) becomes a summation. The vector $\boldsymbol{\phi}(\boldsymbol{\theta})$ comprises the set of the so-called *natural parameters*, and $f, \boldsymbol{u}$ are functions defining the distribution. It is readily seen from the factorization theorem in Section 3.7 that $\boldsymbol{u}(\boldsymbol{x})$ is a *sufficient statistic* for the parameter $\boldsymbol{\theta}$. Note that an attribute of the exponential family is that the number of sufficient statistics, that is, the dimensionality of $\boldsymbol{u}$, is finite and remains independent of the number of observations. If $\boldsymbol{\phi}(\boldsymbol{\theta}) = \boldsymbol{\theta}$, then the exponential family is said to be in *canonical* form. A number of widely used distributions belongs to the exponential family, for example, the normal, exponential, gamma, chi-squared, beta, Dirichlet, Bernoulli, binomial, and multinomial distributions. Examples of distributions that do not belong in this family are the uniform with unknown bounds, Student's t, and most mixture distributions (for example, [40,46]).

An advantage of the exponential family is that one can find conjugate priors for $\boldsymbol{\theta}$; that is, priors that lead to posteriors, $p(\boldsymbol{\theta}|\mathcal{X})$, of the same functional form as $p(\boldsymbol{\theta})$ (Section 3.11.1, Remarks 3.4).

Given (12.72) its conjugate prior is given by

$$p(\boldsymbol{\theta}; \lambda, \boldsymbol{v}) = h(\lambda, \boldsymbol{v})\big(g(\boldsymbol{\theta})\big)^{\lambda} \exp\left(\boldsymbol{\phi}^T(\boldsymbol{\theta})\boldsymbol{v}\right), \tag{12.74}$$

where $\lambda > 0$ and $\boldsymbol{v}$ are known as *hyperparameters*; that is, parameters that control other parameters. The factor $h(\lambda, \boldsymbol{v})$ is an appropriate normalizing constant. It is easy to see that defining the prior as in (12.74) and the likelihood function as in (12.72), the posterior $p(\boldsymbol{\theta}|\boldsymbol{x})$ is of the same form as in (12.74).

Before we give some examples, let us investigate a bit more the role played by λ and $\boldsymbol{v}$, as well as the presence of $g(\boldsymbol{\theta})$ and $\boldsymbol{\phi}(\boldsymbol{\theta})$, in both (12.74) and (12.72). Assume that $\mathbf{x}$ and $\boldsymbol{\theta}$ obey (12.72)–(12.74) and let $\mathcal{X} = \{\boldsymbol{x}_1, \ldots, \boldsymbol{x}_N\}$ be a set of i.i.d. observations. Then

$$p(\mathcal{X}|\boldsymbol{\theta}) = (g(\boldsymbol{\theta}))^N \prod_{n=1}^{N} f(\boldsymbol{x}_n) \exp\left(\boldsymbol{\phi}^T(\boldsymbol{\theta}) \sum_{i=1}^{N} \boldsymbol{u}(\boldsymbol{x}_i)\right) \tag{12.75}$$

and

$$p(\boldsymbol{\theta}|\mathcal{X}) \propto p(\mathcal{X}|\boldsymbol{\theta})p(\boldsymbol{\theta}) \propto (g(\boldsymbol{\theta}))^{\lambda+N} \exp\left(\boldsymbol{\phi}^T(\boldsymbol{\theta})\left(\boldsymbol{v} + \sum_{n=1}^{N} \boldsymbol{u}(\boldsymbol{x}_n)\right)\right). \tag{12.76}$$

In other words, the posterior has hyperparameters equal to

$$\tilde{\lambda} = \lambda + N, \quad \tilde{\boldsymbol{v}} = \boldsymbol{v} + \sum_{n=1}^{N} \boldsymbol{u}(\boldsymbol{x}_n). \tag{12.77}$$

Interpreting (12.77), one can view λ as being the effective number of observations that, implicitly, the prior information contributes to the Bayesian learning process and $\boldsymbol{v}$ is the total amount of information that these (implicit) λ observations contribute to the sufficient statistic. Their exact values, basically, quantify the amount of prior knowledge that the designer wants to embed into the problem.

Example 12.4. *The Gaussian–gamma pair*: Let our random variable x be a scalar and assume that

$$p(x|\sigma^2) = \mathcal{N}(x|\mu, \sigma^2), \tag{12.78}$$

where μ is known and σ^2 will be treated as an unknown random parameter. We will show that:

1. $p(x|\sigma^2)$ belongs to the exponential family.

   It is algebraically more convenient to work with the precision $\beta = \frac{1}{\sigma^2}$. Thus,

$$p(x|\beta) = \frac{\beta^{1/2}}{\sqrt{2\pi}} \exp\left(-\frac{1}{2}\beta(x-\mu)^2\right). \tag{12.79}$$

   Thus, $p(x|\beta)$ belongs to the exponential family with

$$f(x) = \frac{1}{\sqrt{2\pi}}, \quad \phi(\beta) = -\beta, \quad u(x) = \frac{1}{2}(x-\mu)^2$$

   and

$$g(\beta) = \frac{1}{\int_{-\infty}^{+\infty} \frac{1}{\sqrt{2\pi}} \exp\left(-\frac{1}{2}\beta(x-\mu)^2\right)dx} = \beta^{1/2}.$$

2. The conjugate prior of (12.78) follows the gamma distribution.

   The respective conjugate prior from (12.74) becomes

$$p(\beta; \lambda, v) = h(\lambda, v)\beta^{\frac{\lambda}{2}} \exp(-\beta v). \tag{12.80}$$

   This has the form of

$$\text{Gamma}(\beta|a, b) = \frac{1}{\Gamma(a)} b^a \beta^{a-1} \exp(-b\beta), \tag{12.81}$$

   with parameters (Chapter 2) $a = \frac{\lambda}{2} + 1$ and $b = v$ and the normalizing constant $h(\lambda, v)$ being necessarily equal to $b^a / \Gamma(a)$. The function $\Gamma(a)$ is defined as

$$\Gamma(a) = \int_0^{\infty} x^{a-1} e^{-x} dx.$$

If we are given multiple observations x_n, $n = 1, 2, \ldots, N$, then the resulting posterior according to (12.76) and (12.77) will be a gamma distribution with

$$\tilde{b} = b + \frac{1}{2}\sum_{n=1}^{N}(x_n - \mu)^2 = b + \frac{N}{2}\hat{\sigma}^2_{\mathrm{ML}},$$

where $\hat{\sigma}^2_{\mathrm{ML}}$ denotes the maximum likelihood estimate of the variance (Problem 3.22). Hence, the physical meaning of b is that it quantifies our prior estimate about the unknown variance. This also ties nicely with what we have said in Section 3.7; $\hat{\sigma}^2_{\mathrm{ML}}$ is a sufficient statistic for the variance if this is the unknown parameter in a Gaussian. It can easily be shown that the conjugate prior with respect to μ, if σ^2 is known, is a Gaussian (Problem 12.10).

In the case of a multivariate Gaussian of known mean $\boldsymbol{\mu}$ and unknown covariance matrix Σ (precision matrix $Q = \Sigma^{-1}$), it can also be shown that it is of the exponential form and its conjugate prior is given by the Wishart distribution,

$$\mathcal{W}(Q|W, \nu) = h|Q|^{\frac{\nu-l-1}{2}} \exp\left(-\frac{1}{2}\mathrm{trace}\left\{W^{-1}Q\right\}\right), \tag{12.82}$$

where h is the normalizing constant (Problem 12.11) and W is an $l \times l$ matrix. The normalizing constant is given by

$$h = |W|^{-\frac{\nu}{2}}\left(2^{\frac{\nu l}{2}}\pi^{\frac{l(l-1)}{4}}\prod_{i=1}^{l}\Gamma\left(\frac{\nu+1-i}{2}\right)\right)^{-1}, \tag{12.83}$$

which admittedly is quite intimidating; however, in Bayesian learning we have the luxury of bypassing the computation of the normalizing factor in (12.82). Once we express a PDF in terms of Q as in (12.82), then the normalizing constant has to be given by (12.83). The Wishart distribution is a multivariate analogue of the gamma distribution.

Example 12.5. *The Gaussian Gaussian–gamma pair*: We will now treat both μ and the precision β as unknown random parameters. We will show that:

1. $p(x|\mu, \sigma^2) = \mathcal{N}(x|\mu, \sigma^2)$ is also of an exponential form. Indeed, for this case we have

$$p(x|\mu, \sigma^2) = p(x|\mu, \beta^{-1}) = \frac{\beta^{1/2}\exp\left(-\beta\frac{\mu^2}{2}\right)}{\sqrt{2\pi}}\exp\left(\left[-\frac{\beta}{2}, \beta\mu\right]\begin{bmatrix} x^2 \\ x \end{bmatrix}\right).$$

Hence,

$$\boldsymbol{\theta} = [\beta, \mu]^T, \ \boldsymbol{\phi}(\boldsymbol{\theta}) = \begin{bmatrix} -\frac{\beta}{2} \\ \beta\mu \end{bmatrix}, \ \boldsymbol{u}(\boldsymbol{x}) = \begin{bmatrix} x^2 \\ x \end{bmatrix},$$

and performing the respective integration, we obtain

$$f(\boldsymbol{x}) = \frac{1}{\sqrt{2\pi}}, \ g(\boldsymbol{\theta}) = \beta^{1/2}\exp\left(-\frac{\beta\mu^2}{2}\right),$$

which proves the claim.

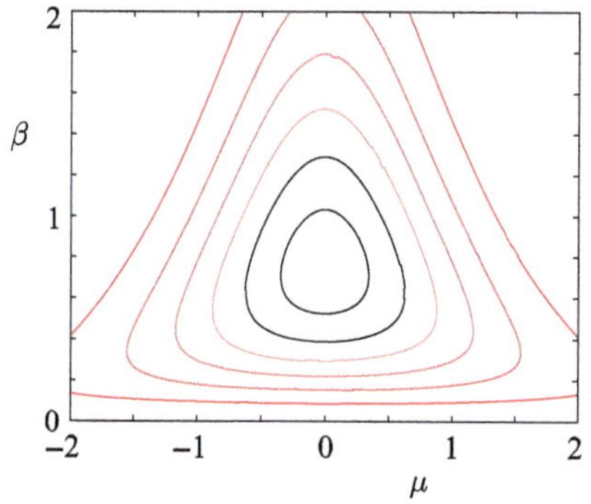

FIGURE 12.11

Contour plots of the Gaussian–gamma distribution with parameter values $\lambda = 2,\ v_1 = 4,\ v_2 = 0$.

2. The conjugate prior of $p(x|\mu, \sigma^2)$ is of a Gaussian–gamma form.

   We have

$$p(\mu, \beta; \lambda, \boldsymbol{v}) = h(\lambda, \boldsymbol{v})\beta^{\frac{\lambda}{2}} \exp\left(-\frac{\lambda\beta\mu^2}{2}\right) \exp\left(\left[-\frac{\beta}{2}, \beta\mu\right]\begin{bmatrix} v_1 \\ v_2 \end{bmatrix}\right),$$

   which after some trivial algebra (Problem 12.12) gives

$$p(\mu, \beta; \lambda, \boldsymbol{v}) = \mathcal{N}\left(\mu\left|\frac{v_2}{\lambda}, (\lambda\beta)^{-1}\right.\right) \text{Gamma}\left(\beta\left|\frac{\lambda+1}{2}, \frac{v_1}{2} - \frac{v_2^2}{2\lambda}\right.\right), \tag{12.84}$$

   which is known as the Gaussian–gamma distribution with the Gaussian having mean value $\mu_0 = \frac{v_2}{\lambda}$ and variance $\sigma_\mu^2 = (\lambda\beta)^{-1}$ and the defining parameters of the gamma PDF are $a = \frac{\lambda+1}{2}$ and $b = \frac{v_1}{2} - \frac{v_2^2}{2\lambda}$. Fig. 12.11 shows the contour plot of the Gaussian–gamma distribution of (12.84).

   For the more general case of a multivariate Gaussian, $\mathcal{N}(\boldsymbol{x}|\boldsymbol{\mu}, \Sigma)$, it turns out that it is also of an exponential form and its conjugate prior is of the Gaussian–Wishart form (Problem 12.13), that is,

$$p\left(\boldsymbol{\mu}, Q; \boldsymbol{\mu}_0, \lambda, W, \nu\right) = \mathcal{N}\left(\boldsymbol{\mu}\left|\boldsymbol{\mu}_0, (\lambda Q)^{-1}\right.\right) \mathcal{W}(Q|W, \nu),$$

   where $Q = \Sigma^{-1}$.

Example 12.6. We now turn our attention to discrete variables, and we will show that the multinomial distribution is of an exponential form and that its conjugate prior is given by the Dirichlet distribution.

1. Let $z_1, z_2, \ldots, z_K$ be K mutually exclusive and exhaustive events. Let $P_1, P_2, \ldots, P_K$ be the respective probabilities, hence $\sum_{k=1}^{K} P_k = 1$. Let the experiment be repeated N times. Then the probability of the joint event, z_1 occurred x_1 times, z_2 occurred x_2 times, and so on, is given by the multinomial

distribution

$$P(x_1, x_2, \dots, x_K) = \begin{pmatrix} N \\ x_1 \dots x_K \end{pmatrix} \prod_{k=1}^{K} P_k^{x_k}, \tag{12.85}$$

where

$$\begin{pmatrix} N \\ x_1 \dots x_K \end{pmatrix} = \frac{N!}{x_1! x_2!, \dots, x_K!}.$$

Defining $\boldsymbol{P} = [P_1, \dots, P_K]^T$, Eq. (12.85) can be rewritten as

$$\begin{aligned} P(x_1, \dots, x_K | \boldsymbol{P}) &= \begin{pmatrix} N \\ x_1 \dots x_K \end{pmatrix} \prod_{k=1}^{K} \exp(x_k \ln P_k) \\ &= \begin{pmatrix} N \\ x_1, \dots, x_K \end{pmatrix} \exp\left(\sum_{k=1}^{K} x_k \ln P_k \right). \end{aligned} \tag{12.86}$$

Thus, the multinomial is of an exponential form with

$$\boldsymbol{\phi}(\boldsymbol{P}) = [\ln P_1, \ln P_2, \dots, \ln P_K]^T,$$
$$\boldsymbol{u}(\boldsymbol{x}) = [x_1, x_2, \dots, x_K]^T,$$

and because probabilities sum to one, we obtain

$$g(\boldsymbol{P}) = 1, \quad f(\boldsymbol{x}) = \begin{pmatrix} N \\ x_1 \dots x_K \end{pmatrix}.$$

2. The conjugate prior of (12.86) can then be written as

$$\begin{aligned} p(\boldsymbol{P}; \lambda, \boldsymbol{v}) &= h(\lambda, \boldsymbol{v}) \exp\left(\sum_{k=1}^{K} v_k \ln P_k \right) \\ &\propto \prod_{k=1}^{K} P_k^{v_k}, \end{aligned} \tag{12.87}$$

which is a Dirichlet PDF. If we let $v_k := a_k - 1$, we bring (12.87) in the more standard formulation

$$p(\boldsymbol{P}; \boldsymbol{a}) = \frac{\Gamma(\overline{a})}{\Gamma(a_1) \dots \Gamma(a_K)} \prod_{k=1}^{K} P_k^{a_k - 1}, \quad \sum_{k=1}^{K} P_k = 1, \tag{12.88}$$

where the normalization constant (Chapter 2) has been plugged in, with

$$\overline{a} := \sum_{k=1}^{K} a_k.$$

12.8.1 THE EXPONENTIAL FAMILY AND THE MAXIMUM ENTROPY METHOD

Besides the computational advantages associated with the exponential family, there is another reason that justifies its high popularity. Assume that we are given a set of observations, $x_n \in \mathcal{A}_x \subseteq \mathbb{R}$, $n = 1, 2, \ldots, N$, drawn from a distribution whose functional form is unknown. Our goal is to estimate the unknown PDF; however, we require that it will respect certain empirical expectations, which are computed from the available observations, that is,

$$\hat{\mu}_i := \frac{1}{N}\sum_{n=1}^{N} u_i(x_n), \; i \in \mathcal{I}, \tag{12.89}$$

where $\mathcal{I}$ is an index set and $u_i : \mathcal{A}_x \longmapsto \mathbb{R}$, $i \in \mathcal{I}$ are specific functions. For example, if $u_i(x) = x$, then $\hat{\mu}_i$ is the sample mean. In such cases, it is not sensible to adopt a parametric functional form for the PDF and try to optimize with respect to the unknown parameters, for example, via the maximum likelihood method; in general, we cannot know if an adopted functional form can comply with the available empirical expectations.

The maximum entropy (ME) method (sometimes called principle) offers a possible way to estimate the unknown PDF, subject to the set of the available constraints [21]. According to this method, the cost function to be maximized is the *entropy* (Section 2.5.2) associated with the PDF, that is,

$$H := -\int_{\mathcal{A}_x} p(x) \ln p(x)\, dx. \tag{12.90}$$

It is well known from Shannon's information theory that the entropy is a measure of uncertainty or randomness. Maximization of the entropy with respect to $p(x)$ results in the most random PDF, subject to the available constraints. Seen from another point of view, such a procedure guarantees that the estimation of an unknown PDF is carried out by adopting the lowest number of assumptions, that is, only the available set of constraints. For our case, the maximum entropy estimation method is cast as follows:

$$\begin{aligned} &\text{maximize with respect to } p(x) && -\int_{\mathcal{A}_x} p(x) \ln p(x) dx, \\ &\text{subject to} && \mathbb{E}[u_i(\mathrm{x})] = \int_{\mathcal{A}_x} p(x) u_i(x) dx, \; i \in \mathcal{I}. \end{aligned} \tag{12.91}$$

In addition to the previous set of constraints, one has to consider the obvious one that guarantees that $p(x)$ integrates to one, that is,

$$\int_{\mathcal{A}_x} p(x)\, dx = 1.$$

In the case of discrete variables, the involved integrations are replaced by summations. Solving the optimization task in (12.91), it turns out that (Problem 12.15)

$$\hat{p}(x) = C \exp\left(\sum_{i \in \mathcal{I}} \theta_i u_i(x)\right), \tag{12.92}$$

that is, the ME estimate is of an exponential form. The parameters $\theta_i,\ i \in \mathcal{I}$, are the Lagrange multipliers used in the optimization task and their values are determined via the constraints and are given in terms of the available empirical expectations, $\hat{\mu}_i,\ i \in \mathcal{I}$. If no constraint is used other than the obvious (normalizing) one and $\mathcal{A}_x = [a, b] \subset \mathbb{R}$, then the resulting PDF is the uniform distribution, $p(x) = C$; indeed, this is the most random one, because it shows no preference for any specific interval of values. If two constraints are used such that $u_1(x) = x$ and $u_2(x) = x^2$ the resulting PDF is the Gaussian one, because the exponent is of a quadratic form (chapter's appendix). In other words, the Gaussian is the most random PDF, subject to two constraints related to the mean and the variance. Note that although we focused on real-valued random variables, everything is trivially extended to vector-valued ones. An interesting discussion concerning the ME method and alternative views of the problem is provided in [42].

12.9 COMBINING LEARNING MODELS: A PROBABILISTIC POINT OF VIEW

The idea of combining different learners to boost the overall performance by exploiting their individual characteristics was introduced in Section 7.9. We now return to this task via probabilistic arguments. The section is also useful from a pedagogical point of view, to familiarize the reader further with the use of the EM algorithm.

Our starting consideration is that the data are distributed in different regions of the input space. Thus, it seems reasonable to fit different learning models, one for each region. This idea reminds us of the decision trees treated in Chapter 7. There, axis-aligned (linear) splits of the input space were performed. Here, the input space will be split via hyperplanes (generalizations to more general hypersurfaces are also possible) in a general position. Moreover, the main difference lies in the fact that in CARTS, the splits were of the hard-type decision rule. In the current setting, we adopt a more relaxed attitude and we are going to consider soft-type probabilistic splits, at the expense of some loss in interpretability.

The basic concept of the combining scheme of this section is illustrated in Fig. 12.12. It is common to refer to each one of the K learners as an *expert*. At the heart of our modeling approach lie the so-called *gating* functions, $g_k(\boldsymbol{x}),\ k = 1, 2, \ldots, K$, which control the importance of each expert towards the final decision. These are optimally tuned during the training phase, together with the set of parameters, $\boldsymbol{\theta}_k,\ k = 1, 2, \ldots, K$, which parameterize the experts, respectively. In the general case, the gating functions are functions of the input variables. We refer to this type of modeling as *mixture of experts*. In contrast, the special type of combination, where these are parameters and not functions, that is, $g_k(\boldsymbol{x}) = g_k$, will be referred to as *mixing of learners*. We will focus on the latter case and present the method in the context of the regression and classification tasks, using linear models.

12.9.1 MIXING LINEAR REGRESSION MODELS

Our starting point is that each model is a linear regression model, $\boldsymbol{\theta}_k,\ k = 1, 2, \ldots, K$, where the dimensionality of the input space has been assumed to increase by one to account for the intercepts and the output variables are related to the input according to our familiar equation,

$$\mathrm{y}_k = \boldsymbol{\theta}_k^T \boldsymbol{x} + \eta, \tag{12.93}$$

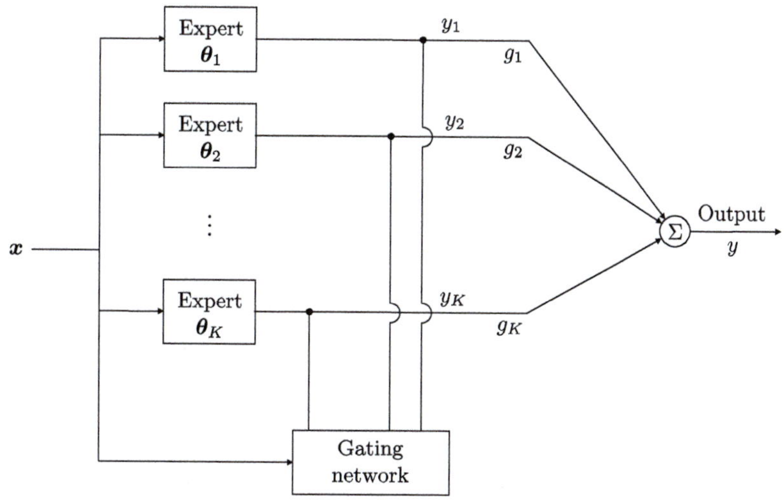

FIGURE 12.12

A block diagram of a mixture of experts. The output of each expert is weighted according to the outputs of the gating network. In the general case, these weights are considered functions of the input.

where η is a white Gaussian noise source with variance σ_η^2 and it is assumed to be common for all models; extension to more general cases can be obtained in a straightforward way. For generalized linear models, $\boldsymbol{x}$ can simply be replaced by the nonlinear mapping $\boldsymbol{\phi}(\boldsymbol{x})$. We assume that the gating parameters are interpreted as probabilities, and they will be denoted as $g_k = P_k$.

Under the previous assumptions, the following mixing model is adopted:

$$p(y; \Xi, \boldsymbol{P}) = \sum_{k=1}^{K} P_k \mathcal{N}(y|\boldsymbol{\theta}_k^T \boldsymbol{x}, \sigma_\eta^2), \tag{12.94}$$

where

$$\Xi := [\boldsymbol{\theta}_1^T, \ldots, \boldsymbol{\theta}_K^T, \sigma_\eta^2]^T, \quad \boldsymbol{P} := [P_1, \ldots, P_K]^T \tag{12.95}$$

are the vectors of the unknown parameters, to be estimated during the training phase using the set of training points, $(y_n, \boldsymbol{x}_n)$, $n = 1, 2, \ldots, N$. Because each model is designed to be "in charge" of one region in space, the corresponding parameters should be trained using input samples that originate from the respective region; note, however, that the regions are not known and have to be learned during training as well. This is in analogy with the task of Gaussian mixture modeling; recall that during training, each observation was associated with a specific mixture component via the use of a hidden variable. In the current setting, each input sample will be associated with a specific learner. Thus, our current task is a close relative of the one treated in Section 12.6 and we could follow similar steps to derive our results. However, for the sake of variety, a slightly different route will be taken. This will also prove useful later on and at the same time fits slightly better with the jargon used for the current formulation. Instead of the indices k_n, used in Gaussian mixture modeling, we will introduce a new set of *hidden* variables, $z_{nk} \in \{0, 1\}$, $k = 1, 2, \ldots, K$, $n = 1, 2, \ldots, N$. If $z_{nk} = 1$, then sample $\boldsymbol{x}_n$ is processed by expert k. At the same time, for each n, z_{nk} becomes equal to one only for a single value of k and zero for the rest. We are now ready to write down the likelihood for the *complete* training data set,[7] (y_n, z_{nk}), i.e.,

$$p(\boldsymbol{y}, Z; \Xi, \boldsymbol{P}) = \prod_{n=1}^{N}\prod_{k=1}^{K}\left(P_k \mathcal{N}\left(y_n | \boldsymbol{\theta}_k^T \boldsymbol{x}_n, \sigma_\eta^2\right)\right)^{z_{nk}}, \tag{12.96}$$

where $\boldsymbol{y}$ is the vector of the output observations and Z the matrix of the respective hidden variables. The log-likelihood is readily obtained as

$$\ln p(\boldsymbol{y}, Z; \Xi, \boldsymbol{P}) = \sum_{n=1}^{N}\sum_{k=1}^{K} z_{nk} \ln\left(P_k \mathcal{N}\left(y_n | \boldsymbol{\theta}_k^T \boldsymbol{x}_n, \sigma_\eta^2\right)\right). \tag{12.97}$$

We can now state the steps for the EM algorithm. Starting from some initial conditions, $\Xi^{(0)}$, $\boldsymbol{P}^{(0)}$, the $(j+1)$th iteration is given by:

- E-Step:

$$\begin{aligned} Q(\Xi, \boldsymbol{P}; \Xi^{(j)}, \boldsymbol{P}^{(j)}) &= \mathbb{E}_Z\left[\ln p(\boldsymbol{y}, Z; \Xi, \boldsymbol{P})\right] \\ &= \sum_{n=1}^{N}\sum_{k=1}^{K} \mathbb{E}[z_{nk}] \ln\left(P_k \mathcal{N}\left(y_n | \boldsymbol{\theta}_k^T \boldsymbol{x}_n, \sigma_\eta^2\right)\right). \end{aligned}$$

  However,

$$\begin{aligned} \mathbb{E}[z_{nk}] &= P(k | y_n; \Xi^{(j)}, \boldsymbol{P}^{(j)}) \\ &= \frac{P_k^{(j)} \mathcal{N}\left(y_n | \boldsymbol{\theta}_k^{(j)T} \boldsymbol{x}_n, \sigma_\eta^2\right)}{\sum_{i=1}^{K} P_i^{(j)} \mathcal{N}\left(y_n | \boldsymbol{\theta}_i^{(j)T} \boldsymbol{x}_n, \sigma_\eta^2\right)}, \end{aligned} \tag{12.98}$$

  or

$$\begin{aligned} Q(\Xi, \boldsymbol{P}; \Xi^{(j)}, \boldsymbol{P}^{(j)}) = \sum_{n=1}^{N}\sum_{k=1}^{K} \gamma_{nk}\Big(\ln P_k - \frac{1}{2}\ln \sigma_\eta^2 - \\ \frac{1}{2\sigma_\eta^2}\left(y_n - \boldsymbol{\theta}_k^T \boldsymbol{x}_n\right)^2\Big) + C, \end{aligned} \tag{12.99}$$

  where C is a constant not affecting the optimization, and

$$\gamma_{nk} := P(k | y_n; \Xi^{(j)}, \boldsymbol{P}^{(j)}).$$

---

[7] Strictly speaking, the data set depends also on $\boldsymbol{x}_n$; to simplify notation, we only give y_n, because this is the one that is treated as a random variable, with the input variables being fixed.

As expected, (12.99) looks like (12.59).

- M-Step: This step comprises the computation of the unknown parameters via three different optimization problems.
  Gating parameters: Following similar steps as for (12.62), we obtain

$$\boxed{P_k^{(j+1)} = \frac{1}{N}\sum_{n=1}^{N} \gamma_{nk}.} \tag{12.100}$$

Learners' parameters: For each $k = 1, 2, \ldots, K$, we have

$$Q(\Xi, \boldsymbol{P}; \Xi^{(j)}, \boldsymbol{P}^{(j)}) = -\sum_{n=1}^{N} \frac{\gamma_{nk}}{2\sigma_\eta^2}\left(y_n - \boldsymbol{\theta}_k^T \boldsymbol{x}_n\right)^2 + C_1, \tag{12.101}$$

where C_1 includes all terms that do not depend on $\boldsymbol{\theta}_k$. Taking the gradient and equating to zero, we readily obtain

$$\sum_{n=1}^{N} \gamma_{nk}\boldsymbol{x}_n\left(y_n - \boldsymbol{x}_n^T \boldsymbol{\theta}_k\right) = \boldsymbol{0},$$

or, employing the input data matrix $X^T := [\boldsymbol{x}_1, \ldots, \boldsymbol{x}_N]$,

$$X^T \Gamma_k (\boldsymbol{y} - X\boldsymbol{\theta}_k) = \boldsymbol{0},$$

with

$$\Gamma_k := \text{diag}\{\gamma_{1k}, \ldots, \gamma_{Nk}\},$$

and finally

$$\boxed{\boldsymbol{\theta}_k^{(j+1)} = \left(X^T \Gamma_k X\right)^{-1} X^T \Gamma_k \boldsymbol{y},\ k = 1, 2, \ldots, K.} \tag{12.102}$$

Eq. (12.102) is the solution to a weighted LS problem, similar in form as the one met in Section 7.6, while dealing with the logistic regression. Note that the weighting matrix involves the posterior probabilities associated with the kth expert.

Noise variance: We have

$$Q(\Xi, \boldsymbol{P}; \Xi^{(j)}, \boldsymbol{P}^{(j)}) = \sum_{n=1}^{N}\sum_{k=1}^{K} \gamma_{nk}\Big(-\frac{1}{2}\ln\sigma_\eta^2 - \frac{1}{2\sigma_\eta^2}\left(y_n - \boldsymbol{\theta}_k^{T(j+1)}\boldsymbol{x}_n\right)^2\Big) + C_2, \tag{12.103}$$

whose optimization with respect to σ_η^2 leads to

$$\boxed{\sigma_\eta^{2(j+1)} = \frac{1}{N}\sum_{n=1}^{N}\sum_{k=1}^{K}\gamma_{nk}\left(y_n - \boldsymbol{\theta}_k^{T(j+1)}\boldsymbol{x}_n\right)^2.} \tag{12.104}$$

Mixture of Experts

In mixture of experts [24], the gating parameters are expressed in a parametric form, as functions of the input variables $\boldsymbol{x}$. A common choice is to assume that

$$g_k(\boldsymbol{x}) := P_k(\boldsymbol{x}) = \frac{\exp\left(\boldsymbol{w}_k^T\boldsymbol{x}\right)}{\sum_{i=1}^{K}\exp\left(\boldsymbol{w}_i^T\boldsymbol{x}\right)}. \tag{12.105}$$

Referring to Fig. 12.12, the gating weights are the outputs of the gating network, which is also excited by the same inputs as the experts. In the neural networks context, as we will see in Chapter 18, we can consider the gating network as a neural network, with activation function given by (12.105), which is known as the *softmax* activation [5]. Note that (12.105) is of exactly the same form as (7.47), used in the multiclass logistic regression. Under such a setting, P_k in (12.99) is replaced by $P_k(\boldsymbol{x})$, and the respective M-step becomes equivalent to optimizing with respect to $\boldsymbol{w}_k,\ k = 1, 2, \ldots, K$. We have

$$Q(\Xi, \boldsymbol{P}; \Xi^{(j)}, \boldsymbol{P}^{(j)}) = \sum_{n=1}^{N}\sum_{k=1}^{K}\gamma_{nk}\ln P_k(\boldsymbol{x}) + C_3. \tag{12.106}$$

Observe that (12.106) is of the same form as (7.49), used for the multiclass logistic regression, and optimization follows similar steps (see also, for example, [19]).

A mixture of experts has been used in a number of applications with a typical one being that of inverse problems, where from the output, one has to deduce the input. However, in many cases, this is a one-to-many task and the mixture of experts is useful to model the choice among these "many" options. For example, in [4], mixture of experts is used for tracking people in video recordings, where the mapping from the image to pose is not unique, due to occlusion.

Hierarchical Mixture of Experts

A direct generalization of the mixture of experts concept is to add more levels of gating functions in a hierarchical fashion, giving rise to what is known as a *hierarchial mixture of experts* (HME). The idea is illustrated in the block diagram of Fig. 12.13. This architecture resembles that of trees, having the experts as leaves, the gating networks as nonterminal nodes, and the output (summing) node as the root one. A hierarchical mixture of experts divides the space into a nested set of regions, with the information combined among the experts under the control of the hierarchically placed gating networks. This hierarchy conforms with the more general idea of *conquer and divide* strategies.

Compared to decision trees, an HME evolves around soft decision rules, in contrast to the hard ones that are employed in CARTs. A hard decision, usually, leads to a loss of information. Once a decision is taken, it cannot change later on. In contrast, soft decision rules provide the luxury to the network to preserve information until a final decision is taken. For example, according to a hard decision rule, if a sample is located close to a decision surface, it will be labeled according to the label on which side it

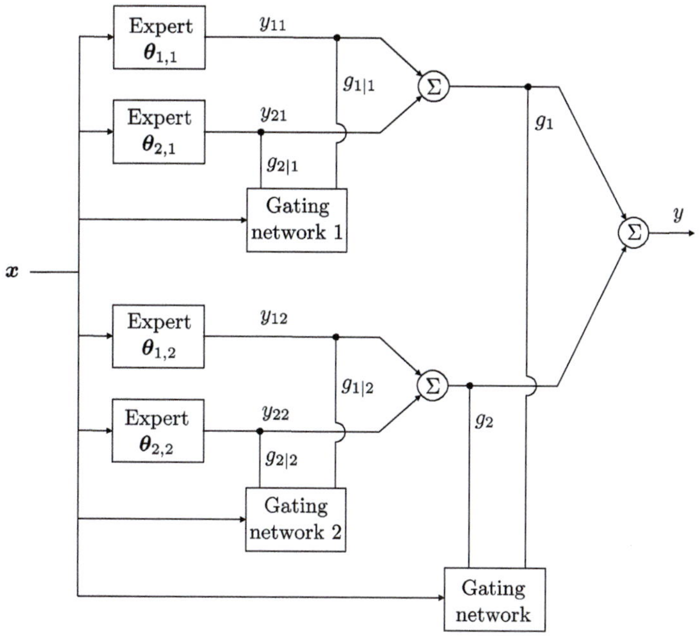

FIGURE 12.13

A block diagram of a hierarchical mixture of experts with two levels of hierarchy.

lies. However, in a soft decision rule, the information, related to the position of the point with respect to the decision surface, will be retained until the stage at which the final decision must be made, by taking into consideration more information that becomes available as the processing develops.

Note that training a mixture of experts can also be carried out via a different path, by optimizing a cost function without it being necessary to employ probabilistic arguments (see, for example, [19]).

12.9.2 MIXING LOGISTIC REGRESSION MODELS

Following Section 12.9.1, the combination rationale can also be applied to classification tasks. To this end, we employ the two-class logistic regression model for each one of the experts, and the combination rule, given the input value $\boldsymbol{x}$, is now written as

$$P(y; \Xi, \boldsymbol{P}) = \sum_{k=1}^{K} P_k s_k^y (1 - s_k)^{1-y}, \tag{12.107}$$

where the definition of logistic regression from Section 7.6 has been used, the values of the label $y \in \{0, 1\}$ correspond to the two classes ω_1 and ω_2, respectively, and

$$s_k := \sigma(\boldsymbol{\theta}_k^T \boldsymbol{x}) \tag{12.108}$$

denotes the output of the kth expert. As in Section 12.9.1, Ξ is the set of the unknown parameters and $\boldsymbol{P}$ the corresponding set of the gating network. Mobilizing similar arguments as for the case of linear regression, we can easily state that the likelihood of the complete data set is given by

$$P(\boldsymbol{y}, Z; \Xi, \boldsymbol{P}) = \prod_{n=1}^{N}\prod_{k=1}^{K}\left(P_k s_{nk}^{y_n}(1-s_{nk})^{1-y_n}\right)^{z_{nk}}, \tag{12.109}$$

where $s_{nk} := \sigma(\boldsymbol{\theta}_k^T \boldsymbol{x}_n)$ and $\boldsymbol{y}$ is the set of labels, y_n, $n = 1, 2, \ldots, N$, of the training samples. Following the standard arguments of the EM algorithm applied on the respective log-likelihood function, it is readily shown that the E-step at the jth iteration is given by

$$Q(\Xi, \boldsymbol{P}; \Xi^{(j)}, \boldsymbol{P}^{(j)}) = \sum_{n=1}^{N}\sum_{k=1}^{K}\gamma_{nk}\Big(\ln P_k + y_n \ln s_{nk} + (1-y_n)\ln(1-s_{nk})\Big), \tag{12.110}$$

where

$$\gamma_{nk} = \mathbb{E}[z_{nk}] = P(k|y_n, \Xi^{(j)}, \boldsymbol{P}^{(j)}) = \frac{P_k^{(j)} s_{nk}^{y_n}(1-s_{nk})^{1-y_n}}{\sum_{i=1}^{K} P_i^{(j)} s_{ni}^{y_n}(1-s_{ni})^{1-y_n}}. \tag{12.111}$$

Note that in (12.111), the notation $s_{nk}^{(j)}, s_{ni}^{(j)}$ should have been used, but we tried to unclutter it slightly.

In the M-step, minimization with respect to P_k is of the same form as it was for the regression task, and it leads to

$$P_k^{(j+1)} = \frac{1}{N}\sum_{n=1}^{N}\gamma_{nk}. \tag{12.112}$$

To obtain the parameters for the experts, one has to resort to an iterative scheme. Observe that the only differences of (12.110) with (7.38) are (a) the presence of the term involving P_k, (b) the summation over k, and (c) the existence of the multiplicative factors γ_{nk}. The first two make no difference in the optimization with respect to a single $\boldsymbol{\theta}_k$ and the latter is just a constant. Hence the optimization is similar to the one used for the two-class logistic regression in Section 7.6, with the gradient and the Hessian matrices being the same except for the multiplicative factors (and the sign, because there, the negative log-likelihood was considered). The extension to the multiclass case is straightforward and follows similar steps.

Example 12.7. This example demonstrates the application of a mixture of two linear regression models to a synthetic data set. The input and the output are scalars, x_n and y_n. Fig. 12.14A shows the setup. The data reside in different parts of the input space and in each region the input–output relation is of a different form. The goal is to estimate the two linear functions, $\theta_{1,k}x + \theta_{0,k}$, $k \in \{1, 2\}$. The EM algorithm of Section 12.9.1 was initialized with the true value of the noise variance σ_η^2.

Fig. 12.14A–C shows the resulting linear models after the first, the seventh, and finally the 15th iteration. Fig. 12.14D–F shows the resulting posteriors $P(k|y_n, x_n)$ (measured by the length of the bar) associated with each learner as a function of x_n. After convergence, they are of a bimodal nature, depending on where each sample resides in the input space. In this way, significant probability mass

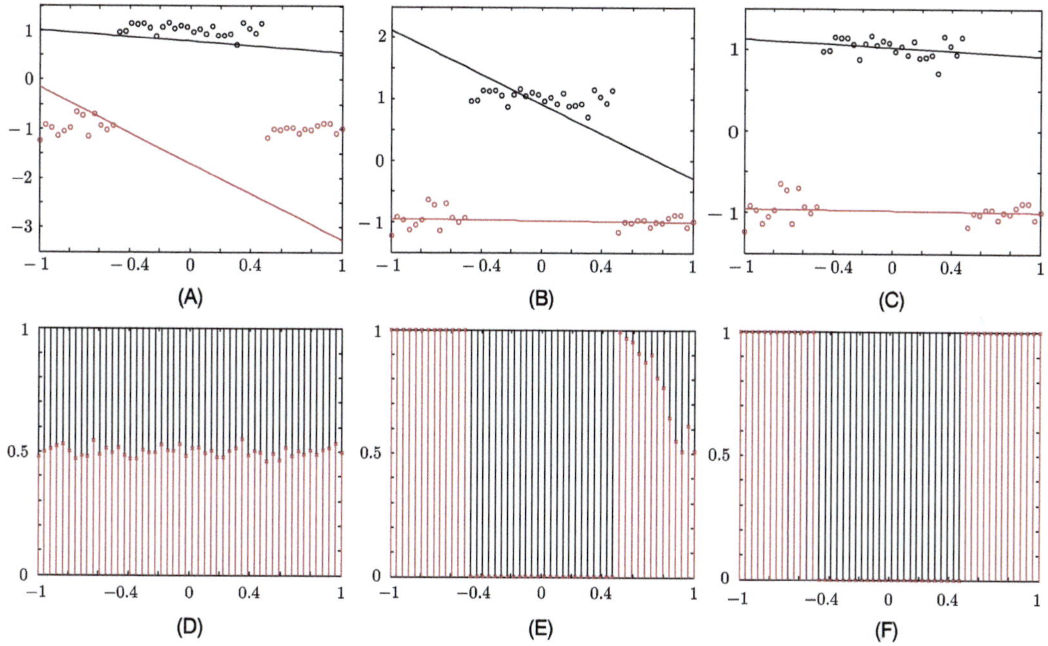

FIGURE 12.14

The two fitted lines as estimated by the EM algorithm after (A) the first, (B) the seventh, and (C) the 15th iterations. Figures (D)–(F) show the corresponding posterior probabilities, for each one of the training points x_n. The length of each segment is equal to the value of the respective probability.

is assigned even to regions where data points do not exist. A smoother and more accurate, from a generalization point of view, estimate results if we let the gating parameters be functions of the input variables themselves.

PROBLEMS

12.1 Show that if

$$p(z) = \mathcal{N}(z|\boldsymbol{\mu}_z, \Sigma_z)$$

and

$$p(\boldsymbol{t}|z) = \mathcal{N}(\boldsymbol{t}|Az, \Sigma_{t|z}),$$

then

$$\mathbb{E}[\mathbf{z}|\boldsymbol{t}] = (\Sigma_z^{-1} + A^T \Sigma_{t|z}^{-1} A)^{-1}(A^T \Sigma_{t|z}^{-1} \boldsymbol{t} + \Sigma_z^{-1} \boldsymbol{\mu}_z).$$

12.2 Let $\mathbf{x} \in \mathbb{R}^l$ be a random vector following the normal $\mathcal{N}(\boldsymbol{x}|\boldsymbol{\mu}, \Sigma)$. Consider $\boldsymbol{x}_n$, $n = 1, 2, \dots, N$, to be i.i.d. observations. If the prior for $\boldsymbol{\mu}$ follows $\mathcal{N}(\boldsymbol{\mu}|\boldsymbol{\mu}_0, \Sigma_0)$, show that the posterior $p(\boldsymbol{\mu}|\boldsymbol{x}_1, \dots, \boldsymbol{x}_N)$ is normal $\mathcal{N}(\boldsymbol{\mu}|\tilde{\boldsymbol{\mu}}, \tilde{\Sigma})$ with

$$\tilde{\Sigma}^{-1} = \Sigma_0^{-1} + N\Sigma^{-1}$$

and

$$\tilde{\boldsymbol{\mu}} = \tilde{\Sigma}(\Sigma_0^{-1}\boldsymbol{\mu}_0 + N\Sigma^{-1}\bar{\boldsymbol{x}}),$$

where $\bar{\boldsymbol{x}} = \frac{1}{N}\sum_{n=1}^{N} \boldsymbol{x}_n$.

12.3 If $\mathcal{X}$ is the set of observed variables and $\mathcal{X}^l$ the set of the corresponding latent ones, show that

$$\frac{\partial \ln p(\mathcal{X}; \boldsymbol{\xi})}{\partial \boldsymbol{\xi}} = \mathbb{E}\left[\frac{\partial \ln p(\mathcal{X}, \mathcal{X}^l; \boldsymbol{\xi})}{\partial \boldsymbol{\xi}}\right],$$

where $\mathbb{E}[\cdot]$ is with respect to $p(\mathcal{X}^l|\mathcal{X}; \boldsymbol{\xi})$ and $\boldsymbol{\xi}$ is an unknown vector parameter. Note that if one fixes the value of $\boldsymbol{\xi}$ in $p(\mathcal{X}^l|\mathcal{X}; \boldsymbol{\xi})$, then one has obtained the M-step of the EM algorithm.

12.4 Show Eq. (12.42).

12.5 Let $\boldsymbol{y} \in \mathbb{R}^N$, let $\boldsymbol{\theta} \in \mathbb{R}^l$, and let Φ be a matrix of appropriate dimensions. Derive the expected value of $\|\boldsymbol{y} - \Phi\boldsymbol{\theta}\|^2$ with respect to $\boldsymbol{\theta}$, given $\mathbb{E}[\boldsymbol{\theta}]$ and the corresponding covariance matrix Σ_θ.

12.6 Derive recursions (12.60)–(12.62).

12.7 Show that the Kullback–Leibler divergence, $\mathrm{KL}(p \parallel q)$, is a nonnegative quantity.
Hint: Recall that $\ln(\cdot)$ is a concave function and use Jensen's inequality, i.e.,

$$f\left(\int g(\boldsymbol{x})p(\boldsymbol{x})d\boldsymbol{x}\right) \leq \int f(g(\boldsymbol{x}))p(\boldsymbol{x})d\boldsymbol{x},$$

where $p(\boldsymbol{x})$ is a PDF and f is a convex function.

12.8 Prove that the binomial and beta distributions are conjugate pairs with respect to the mean value.

12.9 Show that the normalizing constant C in the Dirichlet PDF

$$\mathrm{Dir}(\boldsymbol{x}|\boldsymbol{a}) = C\prod_{k=1}^{K} x_k^{a_k-1}, \quad \sum_{k=1}^{K} x_k = 1$$

is given by

$$C = \frac{\Gamma(a_1 + a_2 + \dots + a_K)}{\Gamma(a_1)\Gamma(a_2)\dots\Gamma(a_K)}.$$

Hint: Use the property $\Gamma(a+1) = a\Gamma(a)$.

(a) Use induction. Because the proposition is true for $k = 2$ (beta distribution), assume that it is true for $k = K - 1$, and prove that it will be true for $k = K$.

(b) Note that due to the constraint $\sum_{k=1}^{K} x_k = 1$, only $K-1$ of the variables are independent. So, basically, the Dirichlet PDF implies that

$$p(x_1, x_2, \ldots, x_{K-1}) = C \prod_{k=1}^{K-1} x_k^{a_k - 1} \left(1 - \sum_{k=1}^{K-1} x_k\right)^{a_K - 1}.$$

12.10 Show that $\mathcal{N}(\boldsymbol{x}|\boldsymbol{\mu}, \Sigma)$ for known Σ is of an exponential form and that its conjugate prior is also Gaussian.

12.11 Show that the conjugate prior of the multivariate Gaussian with respect to the precision matrix Q is a Wishart distribution.

12.12 Show that the conjugate prior of the univariate Gaussian $\mathcal{N}(x|\mu, \sigma^2)$ with respect to the mean and the precision $\beta = \frac{1}{\sigma^2}$, is the Gaussian–gamma product

$$p(\mu, \beta; \lambda, \boldsymbol{v}) = \mathcal{N}\left(\mu \middle| \frac{v_2}{\lambda}, (\lambda\beta)^{-1}\right) \text{Gamma}\left(\beta \middle| \frac{\lambda+1}{2}, \frac{v_1}{2} - \frac{v_2^2}{2\lambda}\right),$$

where $\boldsymbol{v} := [v_1, v_2]^T$.

12.13 Show that the multivariate Gaussian $\mathcal{N}(\boldsymbol{x}|\boldsymbol{\mu}, Q^{-1})$ has as a conjugate prior, with respect to the mean and the precision matrix Q, the Gaussian–Wishart product.

12.14 Show that the distribution

$$P(x|\mu) = \mu^x (1-\mu)^{1-x}, \ x \in \{0, 1\},$$

is of an exponential form and derive its conjugate prior with respect to μ.

12.15 Show that estimating an unknown PDF by maximizing the respective entropy, subject to a set of empirical expectations, results in a PDF that belongs to the exponential family.

MATLAB® EXERCISES

12.16 Sample $N = 20$ equally spaced points x_n in the interval $[0, 2]$. Create the output samples y_n according to the nonlinear model of Example 12.1, where the noise variance is set equal to $\sigma_\eta^2 = 0.05$.

(a) Let the parameters of the Gaussian prior be $\boldsymbol{\theta}_0 = [0.2, -1, 0.9, 0.7, -0.2]^T$ and $\Sigma_\theta = 0.1I$. Compute the covariance matrix and the mean of the posterior Gaussian distribution using Eq. (12.19) and Eq. (12.20), respectively. Then, select randomly $K = 20$ points x_k in the interval $[0, 2]$. Compute the predictions for the mean values μ_y and the associated variances σ_y^2, utilizing Eq. (12.22) and Eq. (12.23), respectively. Plot the graph of the true function together with the predicted mean values μ_y, and use MATLAB®'s "errorbar" function to show the confidence intervals on these predictions.
Repeat the experiment using $N = 500$ points, and try different values of σ_η^2 to notice the change in the estimated confidence intervals.

(b) Repeat the previous experiment using a randomly chosen value for $\boldsymbol{\theta}_0$ and different values on the parameters, for example, $\sigma_\eta^2 = 0.05$, $\sigma_\eta^2 = 0.15$, $\sigma_\theta^2 = 0.1$, or $\sigma_\theta^2 = 2$, and $N = 500$ or $N = 20$.

(c) Repeat the experiment once more using a wrong order for the model, e.g., second- or third-order polynomial. Use different values for the parameters than the corresponding correct ones for the initialization. See also Example 12.1.

12.17 Consider Example 12.1 as before. Sample $N = 500$ equally spaced points x_n in the interval $[0, 2]$. Create the output samples y_n according to the nonlinear model of the example, where the noise variance is set equal to $\sigma_\eta^2 = 0.05$. Implement the linear regression EM algorithm of Section 12.5. Assume the correct number of parameters. Then repeat Example 12.2. After the convergence of the EM, sample ten points, x_k, randomly, in the same interval as before and compute the predictive means μ_y and variances σ_y^2. Plot the true signal curve, the predictive means μ_y, and the respective confidence intervals using MATLAB®'s "errorbar" function. Repeat the EM run, using different initial values and an incorrect number of parameters. Comment on the results.

12.18 Generate 100 data points from each of the three two-dimensional Gaussian distributions of Example 12.3. Plot the data points along with the confidence ellipsoids for each Gaussian with coverage probability 80%. Implement the Gaussian mixture model via the EM algorithm, whose steps are described in Eqs. (12.58)–(12.62). Moreover, compute the log-likelihood function in every iteration of the EM algorithm using Eq. (12.55).

(a) In separate figures (always containing the data), plot the ellipsoids of the Gaussian distributions estimated by the EM algorithm during iterations $j = 1$, $j = 5$, and $j = 30$, and the log-likelihood function versus the number of iterations.

(b) Repeat the experiment after bringing the cluster means closer together. Compare the results.

12.19 Generate 100 data points from each of the two-dimensional Gaussian distributions with parameters

$$\boldsymbol{\mu}_1^T = [0.9, 1.02]^T, \quad \boldsymbol{\mu}_2^T = [-1.2, -1.3]^T$$

and

$$\Sigma_1 = \begin{bmatrix} 0.5 & 0.081 \\ 0.081 & 0.7 \end{bmatrix}, \quad \Sigma_2 = \begin{bmatrix} 0.4 & 0.02 \\ 0.02 & 0.3 \end{bmatrix}.$$

Plot the data points using different colors for the two Gaussian distributions. Implement the k-means algorithm presented in Algorithm 12.1.

(a) Run the k-means algorithm for $K = 2$ and plot the results. Run, also, the Gaussian mixtures EM of the previous exercise and plot the 80% probability confidence ellipsoids to compare the results.

(b) Now, sample $N_1 = 100$ and $N_2 = 20$ points from each distribution and repeat the experiment to reproduce the results of Fig. 12.9.

(c) Try different configurations and play with K different than the true number of clusters. Comment on the results.

(d) Play with different initialization points and also try points which are too far from the true mean values of the clusters. Comment on the results.

12.20 Generate 50 equidistant input data points in the interval $[-1, 1]$. Assume two linear regression models, the first with scale 0.005 and intercept -1 and the second with scale 0.018 and intercept 1. Generate observations from these two models by using the first model for the input points in the interval $[-0.5, 0.5]$, and the second model for the inputs in the interval $[-1, -0.5] \cup [0.5, 1]$. Also, add Gaussian noise of zero mean and variance 0.01. Next, implement the EM algorithm developed in Section 12.9.1. Initialize the noise precision β to its true value. For iterations 1, 5, and 30, plot the data points and the estimated linear functions $\theta_{1,k}x + \theta_{0,k}$, $k \in \{1, 2\}$, of the models, to reproduce the results of Fig. 12.14.

REFERENCES

[1] D. Arthur, S. Vassilvitskii, k-means++: the advantages of careful seeding, in: Proceedings 18th ACM-SIAM Symposium on Discrete Algorithms, SODA, 2007, pp. 1027–1035.

[2] L.E. Baum, T. Petrie, G. Soules, N. Weiss, A maximization technique occurring in the statistical analysis of probabilistic functions of Markov chains, Ann. Math. Stat. 41 (1970) 164–171.

[3] M.J. Beal, Variational Algorithms for Approximate Bayesian Inference, PhD Thesis, University College London, 2003.

[4] L. Bo, C. Sminchisescu, A. Kanaujia, D. Metaxas, Fast algorithms for large scale conditional 3D prediction, in: Proceedings International Conference to Computer Vision and Pattern Recognition, CVPR, Anchorage, AK, 2008.

[5] J.S. Bridle, Probabilistic interpretation of feedforward classification network outputs with relationship to statistical pattern recognition, in: F. Fougelman-Soulie, J. Heurault (Eds.), Neuro-Computing: Algorithms, Architectures and Applications, Springer Verlag, 1990.

[6] O. Cappe, E. Mouline, Online EM algorithm for latent data models, J. R. Stat. Soc. B 71 (3) (2009) 593–613.

[7] G. Casella, R.L. Berger, Statistical Inference, second ed., Duxbury Press, 2002.

[8] G. Celeux, J. Diebolt, The SEM algorithm: a probabilistic teacher derived from the EM algorithm for the mixture problem, Comput. Stat. Q. 2 (1985) 73–82.

[9] S.P. Chatzis, D.I. Kosmopoulos, T.A. Varvarigou, Signal modeling and classification using a robust latent space model based on t-distributions, IEEE Trans. Signal Process. 56 (3) (2008) 949–963.

[10] S.P. Chatzis, D. Kosmopoulos, T.A. Varvarigou, Robust sequential data modeling using an outlier tolerant hidden Markov model, IEEE Trans. Pattern Anal. Mach. Intell. 31 (9) (2009) 1657–1669.

[11] B. Delyon, M. Lavielle, E. Moulines, Convergence of a stochastic approximation version of the EM algorithm, Ann. Stat. 27 (1) (1999) 94–128.

[12] A.P. Dempster, N.M. Laird, D.B. Rubin, Maximum likelihood from incomplete data via the EM algorithm, J. R. Stat. Soc. B 39 (1) (1977) 1–38.

[13] S.F. Gull, Bayesian inductive inference and maximum entropy, in: G.J. Erickson, C.R. Smith (Eds.), Maximum Entropy and Bayesian Methods in Science and Engineering, Kluwer, 1988.

[14] M.R. Gupta, Y. Chen, Theory and use of the EM algorithm, Found. Trends Signal Process. 4 (3) (2010) 223–299.

[15] L.K. Hansen, C.E. Rasmussen, Pruning from adaptive regularization, Neural Comput. 6 (1993) 1223–1232.

[16] L.K. Hansen, J. Larsen, Unsupervised learning and generalization, in: IEEE International Conference on Neural Networks, 1996, pp. 25–30.

[17] L.K. Hansen, Bayesian averaging is well-temperated, in: S. Solla (Ed.), Proceedings Neural Information Processing, NIPS, MIT Press, 2000, pp. 265–271.

[18] D. Haussler, M. Kearns, R. Schapire, Bounds on the sample complexity of Bayesian learning using information theory and the VC dimension, Mach. Learn. 14 (1994) 83–113.

[19] S. Haykin, Neural Networks: A Comprehensive Foundation, Prentice Hall, 1999.

[20] D.R. Hunter, K. Lange, A tutorial on MM algorithms, Am. Stat. 58 (2004) 30–37.

[21] E.T. Jaynes, On the rationale of the maximum entropy methods, Proc. IEEE 70 (9) (1982) 939–952.

[22] E.T. Jaynes, Bayesian methods-an introductory tutorial, in: J.H. Justice (Ed.), Maximum Entropy and Bayesian Methods in Science and Engineering, Cambridge University Press, 1986.

[23] H. Jeffreys, Theory of Probability, Oxford University Press, 1992.

[24] M.I. Jordan, R.A. Jacobs, Hierarchical mixture of experts and the EM algorithm, Neural Comput. 6 (1994) 181–214.

[25] P. Liang, D. Klein, Online EM for unsupervised models, in: Proceeding of Human Language Technologies: The 2009 Annual Conference of the North American Chapter of the Association for Computational Linguistics, NAACL, 2009, pp. 611–619.

[26] T.J. Loredo, From Laplace to supernova SN 1987A: Bayesian inference in astrophysics, in: P. Fougere (Ed.), Maximum Entropy and Bayesian Methods, Kluwer, 1990, pp. 81–143.
[27] D.J.C. McKay, Bayesian interpolation, Neural Comput. 4 (3) (1992) 417–447.
[28] D.J.C. McKay, Probable networks and plausible predictions – a review of practical Bayesian methods for supervised neural networks, Netw. Comput. Neural Syst. 6 (1995) 469–505.
[29] X.L. Meng, D. Van Dyk, The EM algorithm—an old folk-song sung to a fast new tune, J. R. Stat. Soc. B 59 (3) (1997) 511–567.
[30] G.J. McLachlan, K.E. Basford, Mixture Models. Inference and Applications to Clustering, Marcel Dekker, 1988.
[31] X.L. Meng, D.B. Rubin, Maximum likelihood estimation via the ECM algorithm: a generalization framework, Biometrika 80 (1993) 267–278.
[32] J.E. Moody, Note on generalization, regularization, and architecture selection in nonlinear learning systems, in: Proceedings, IEEE Workshop on Neural Networks for Signal Processing, Princeton, NJ, USA, 1991, pp. 1–10.
[33] T. Moon, The expectation maximization algorithm, Signal Process. Mag. 13 (6) (1996) 47–60.
[34] R.M. Neal, G.E. Hinton, A new view of the EM algorithm that justifies incremental, sparse and other variants, in: M.J. Jordan (Ed.), Learning in Graphical Models, Kluwer Academic Publishers, 1998, pp. 355–369.
[35] S. Shoham, Robust clustering by deterministic agglomeration EM of mixtures of multivariate t distributions, Pattern Recognit. 35 (5) (2002) 1127–1142.
[36] M. Stephens, Dealing with label-switching in mixture models, J. R. Stat. Soc. B 62 (2000) 795–809.
[37] M. Svensen, C.M. Bishop, Robust Bayesian mixture modeling, Neurocomputing 64 (2005) 235–252.
[38] R. Sundberg, Maximum likelihood theory for incomplete data from an exponential family, Scand. J. Stat. 1 (2) (1974) 49–58.
[39] G. Schwarz, Estimating the dimension of a model, Ann. Stat. 6 (1978) 461–464.
[40] J. Shao, Mathematical Statistics: Exercises and Solutions, Springer, 2005.
[41] S. Theodoridis, K. Koutroumbas, Pattern Recognition, fourth ed., Academic Press, 2009.
[42] Y. Tikochinsky, N.Z. Tishby, R.D. Levin, Alternative approach to maximum-entropy inference, Phys. Rev. A 30 (5) (1985) 2638–2644.
[43] D.M. Titterington, A.F.M. Smith, U.E. Makov, Statistical Analysis of Finite Mixture Distributions, John Wiley & Sons, 1985.
[44] A. Vehtari, J. Lampinen, Bayesian model assessment and comparison using cross-validation predictive densities, Neural Comput. 14 (10) (2002) 2439–2468.
[45] C.S. Wallace, P.R. Freeman, Estimation and inference by compact coding, J. R. Stat. Soc. B 493 (1987) 240–265.
[46] R.L. Wolpert, Exponential Families, Technical Report, University of Duke, 2011, www.stat.duke.edu/courses/Spring11/sta114/lec/expofam.pdf.
[47] C. Wu, On the convergence properties of the EM algorithm, Ann. Stat. 11 (1) (1983) 95–103.

CHAPTER

BAYESIAN LEARNING: APPROXIMATE INFERENCE AND NONPARAMETRIC MODELS

13

CONTENTS

Machine Learning. https://doi.org/10.1016/B978-0-12-818803-3.00025-8

13.1 INTRODUCTION

This chapter is the second one dedicated to Bayesian learning. The emphasis here, compared to Chapter 12, is on more advanced topics, dealing with approximate inference methods. Such methods are employed when the involved integrations are no longer computationally tractable. Two paths for approximate inference, known as variational techniques, are discussed. One is based on the mean field approximation and the lower bound interpretation of the EM, and the other on convex duality and variational bounds. Regression and mixture modeling are discussed in this framework. Emphasis is given to sparse Bayesian modeling techniques and hierarchical Bayesian models. The relevance vector machine framework is presented. Expectation propagation is also discussed as an alternative to variational methods for approximate inference. At the end of the chapter, Bayesian learning in the context of nonparametric models is presented, including Dirichlet processes (DPs), the Chinese restaurant process (CRP), the Indian buffet process (IBP), and Gaussian processes. Finally, a case study concerning hyperspectral imaging is presented.

13.2 VARIATIONAL APPROXIMATION IN BAYESIAN LEARNING

Recall that in order to apply the EM algorithm, the functional form of the posterior of the latent/hidden variables, given the observations, must be known. However, analytic computations related to the posterior are not always tractable. In such cases, the EM algorithm, in its standard form as discussed in the previous chapter, is not applicable. In this section, we will describe an alternative path that builds upon the EM interpretation given in Section 12.7.

Once more, we will adopt a general notation, which can then be adapted to the needs of specific problems. Let $\mathcal{X} = \{\boldsymbol{x}_1, \ldots, \boldsymbol{x}_N\}$ be the set of observed variables and $\mathcal{X}^l = \{\boldsymbol{x}_1^l, \ldots, \boldsymbol{x}_N^l\}$ the set of the N corresponding (local) latent ones, as they have been defined in Section 12.4. Furthermore, in the current section, in addition to the latent variables, we will explicitly bring into the game a set of K (global) hidden random parameters, $\boldsymbol{\theta} \in \mathbb{R}^K$, whose number is fixed and which will be accompanied by a prior PDF. The complete likelihood function is now written $p(\mathcal{X}, \mathcal{X}^l, \boldsymbol{\theta}; \boldsymbol{\xi})$, where $\boldsymbol{\xi}$ is the set of unknown nonrandom (hyper)parameters that have to be estimated. As is always the case in Bayesian learning, the goal is to infer the posterior probability distributions that describe the latent as well as the hidden random variables.

The functional in Eq. (12.63) is now redefined as

$$\mathcal{F}(q,\boldsymbol{\xi}) = \int q(\mathcal{X}^l,\boldsymbol{\theta}) \ln \frac{p(\mathcal{X},\mathcal{X}^l,\boldsymbol{\theta};\boldsymbol{\xi})}{q(\mathcal{X}^l,\boldsymbol{\theta})} d\mathcal{X}^l\, d\boldsymbol{\theta}. \tag{13.1}$$

The counterpart of Eq. (12.66) becomes

$$\mathcal{F}(q,\boldsymbol{\xi}) = \ln p(\mathcal{X};\boldsymbol{\xi}) + \int q(\mathcal{X}^l,\boldsymbol{\theta}) \ln \frac{p(\mathcal{X}^l,\boldsymbol{\theta}|\mathcal{X};\boldsymbol{\xi})}{q(\mathcal{X}^l,\boldsymbol{\theta})} d\mathcal{X}^l\, d\boldsymbol{\theta}, \tag{13.2}$$

which leads to

$$\ln p(\mathcal{X};\boldsymbol{\xi}) = \mathcal{F}(q,\boldsymbol{\xi}) + \mathrm{KL}\left(q(\mathcal{X}^l,\boldsymbol{\theta})||p(\mathcal{X}^l,\boldsymbol{\theta}|\mathcal{X};\boldsymbol{\xi})\right).$$

The difference with Eq. (12.67) lies in the fact that in our current setting the posterior, $p(\mathcal{X}^l,\boldsymbol{\theta}|\mathcal{X};\boldsymbol{\xi})$, is *not* assumed to be known; thus, setting the Kullback–Leibler (KL) divergence, $\mathrm{KL}\big(q(\mathcal{X}^l,\boldsymbol{\theta})||p(\mathcal{X}^l,\boldsymbol{\theta}|\mathcal{X};\boldsymbol{\xi})\big)$, to zero, in order to maximize the value of the functional, is no longer possible. The only alternative is to resort to computationally tractable approximations concerning the functional form of q, which will allow the optimization of the functional with respect to q (and with respect to the nonrandom parameters, $\boldsymbol{\xi}$).

Optimizing a functional with respect to a function is known in mathematics as *calculus of variations*. The simplest example of this problem is to compute the geodesic that connects two points on a surface. Two names whose contributions are considered significant breakthroughs that consolidated this field are the Swiss German mathematician Leonhard Euler (1707–1783) and the Italian-born mathematician and astronomer Joseph-Louis Lagrange (1736–1813). It is interesting to note that Lagrange succeeded Euler as director of mathematics in the Prussian Academy of Sciences in Berlin.

In order to deal with the current problem, we will *constrain* $q(\mathcal{X}^l,\boldsymbol{\theta})$ to lie within a specific family of functions. Note that in this case, if the unknown $p(\mathcal{X}^l,\boldsymbol{\theta}|\mathcal{X};\boldsymbol{\xi})$ does not belong to the selected family of functions, the KL divergence *cannot* become zero, and the lower bound, $\mathcal{F}(q,\boldsymbol{\xi})$, of the marginal log-likelihood *cannot* be made *tight*. This is the reason the method is called a variational approximation.

THE MEAN FIELD APPROXIMATION

This type of approximation results by constraining $q(\mathcal{X}^l,\boldsymbol{\theta})$ to be factorized, that is,

$$q(\mathcal{X}^l,\boldsymbol{\theta}) = q_{\mathcal{X}^l}(\mathcal{X}^l) q_{\boldsymbol{\theta}}(\boldsymbol{\theta}). \tag{13.3}$$

This factorization can be, and usually is, extended to

$$\boxed{q(\mathcal{X}^l,\boldsymbol{\theta}) = q_{\mathbf{x}_1^l}(\boldsymbol{x}_1^l)\dots q_{\mathbf{x}_N^l}(\boldsymbol{x}_N^l) q_{\boldsymbol{\theta}}(\boldsymbol{\theta}): \quad \text{mean field approximation.}} \tag{13.4}$$

Furthermore, the hidden variables can be further factorized, i.e., $q_{\boldsymbol{\theta}}(\boldsymbol{\theta}) = \prod_i^K q_{\theta_i}(\theta_i)$. This type of factorized function approximation has been inspired from the field of statistical physics and is known as *mean field approximation* (e.g., [17,46,60]). No doubt, a number of combinations that group different variables together can also be used. To simplify the notation, without sacrificing generality, we will work with Eq. (13.3), which involves only two factors.

Having adopted a factorized model as in Eq. (13.3), Eq. (13.1) becomes (Problem 13.1)

$$\mathcal{F}(q_{\mathcal{X}^l}, q_{\boldsymbol{\theta}}, \boldsymbol{\xi}) = \int q_{\mathcal{X}^l}(\mathcal{X}^l)\left(\int q_{\boldsymbol{\theta}}(\boldsymbol{\theta}) \ln p(\mathcal{X}, \mathcal{X}^l, \boldsymbol{\theta}; \boldsymbol{\xi})\, d\boldsymbol{\theta}\right) d\mathcal{X}^l \\ - \int q_{\mathcal{X}^l}(\mathcal{X}^l) \ln q_{\mathcal{X}^l}(\mathcal{X}^l)\, d\mathcal{X}^l - \int q_{\boldsymbol{\theta}}(\boldsymbol{\theta})\, d\boldsymbol{\theta}, \tag{13.5}$$

which can alternatively be written as

$$\mathcal{F}(q_{\mathcal{X}^l}, q_{\boldsymbol{\theta}}, \boldsymbol{\xi}) = \mathbb{E}_{q_{\mathcal{X}^l}} \mathbb{E}_{q_{\boldsymbol{\theta}}} \left[\ln p(\mathcal{X}, \mathcal{X}^l, \boldsymbol{\theta}; \boldsymbol{\xi})\right] + H_{q_{\mathcal{X}^l}} + H_{q_{\boldsymbol{\theta}}}. \tag{13.6}$$

Taking into account that the order of integration in (13.5) can be interchanged, we can also write

$$\mathcal{F}(q_{\mathcal{X}^l}, q_{\boldsymbol{\theta}}, \boldsymbol{\xi}) = \mathbb{E}_{q_{\boldsymbol{\theta}}} \mathbb{E}_{q_{\mathcal{X}^l}} \left[\ln p(\mathcal{X}, \mathcal{X}^l, \boldsymbol{\theta}; \boldsymbol{\xi})\right] + H_{q_{\mathcal{X}^l}} + H_{q_{\boldsymbol{\theta}}}, \tag{13.7}$$

where the last two terms on the right-hand side are the entropies associated with the two distribution terms, $q_{\mathcal{X}^l}$ and $q_{\boldsymbol{\theta}}$. Note that Eqs. (13.6) and (13.7) are direct generalizations of Eq. (12.65) in Chapter 12.

The lower bound functional $\mathcal{F}(q_{\mathcal{X}^l}, q_{\boldsymbol{\theta}}, \boldsymbol{\xi})$ depends on three terms. Following the alternating optimization rationale, in every iteration step we will freeze two of them and maximization will be performed with respect to the remaining one, in an alternating fashion. Optimizing the lower bound functional with respect to a distribution will take place by setting to zero a corresponding and appropriately defined KL divergence. To this end, for example, let us rewrite Eq. (13.6) in a slightly different and more convenient form. Define the quantity $\tilde{p}$ as

$$\mathbb{E}_{q_{\boldsymbol{\theta}}} \left[\ln p(\mathcal{X}, \mathcal{X}^l, \boldsymbol{\theta}; \boldsymbol{\xi})\right] := \ln \tilde{p}(\mathcal{X}, \mathcal{X}^l; \boldsymbol{\xi}). \tag{13.8}$$

Then Eq. (13.6) (or, equivalently, (13.5)) is written as

$$\mathcal{F}(q_{\mathcal{X}^l}, q_{\boldsymbol{\theta}}; \boldsymbol{\xi}) = \int q_{\mathcal{X}^l}(\mathcal{X}^l) \ln \frac{\tilde{p}(\mathcal{X}, \mathcal{X}^l; \boldsymbol{\xi})}{q_{\mathcal{X}^l}(\mathcal{X}^l)} d\mathcal{X}^l + H_{q_{\boldsymbol{\theta}}}. \tag{13.9}$$

Observe that the first term on the right-hand side has the form of a KL divergence, where the hidden variables, $\boldsymbol{\theta}$, have been averaged out. Note, however, that by its definition, $\tilde{p}$ is not necessarily a distribution. As we will soon see, in order to become a distribution, a normalizing constant is required.

We have by now all the ingredients to go ahead with the maximization of $\mathcal{F}(q_{\mathcal{X}^l}, q_{\boldsymbol{\theta}}, \boldsymbol{\xi})$. The optimization will take place by maximizing first with respect to $q_{\mathcal{X}^l}$, in the sequel with respect to $q_{\boldsymbol{\theta}}$, and finally with respect to $\boldsymbol{\xi}$. The algorithm is initialized to some arbitrary values for $\boldsymbol{\xi}^{(0)}$ as well as for $q_{\boldsymbol{\theta}}^{(0)}$. The latter is achieved by initializing parameters (statistics) related to $q_{\boldsymbol{\theta}}^{(0)}$ (this will become more clear while dealing with the examples). The $(j+1)$th iteration comprises the following steps:

E-Step 1a: Holding $\boldsymbol{\xi}^{(j)}$ and $q_{\boldsymbol{\theta}}^{(j)}$ fixed, optimize Eq. (13.9) with respect to $q_{\mathcal{X}^l}$; that is,

$$\begin{aligned} q_{\mathcal{X}^l}^{(j+1)}(\mathcal{X}^l) &= \max_{q_{\mathcal{X}^l}} \mathcal{F}\left(q_{\mathcal{X}^l}, q_{\boldsymbol{\theta}}^{(j)}, \boldsymbol{\xi}^{(j)}\right) \\ &= \max_{q_{\mathcal{X}^l}} \int q_{\mathcal{X}^l}(\mathcal{X}^l) \ln \frac{\tilde{p}\left(\mathcal{X}, \mathcal{X}^l; \boldsymbol{\xi}^{(j)}\right)}{q_{\mathcal{X}^l}(\mathcal{X}^l)} d\mathcal{X}^l + \text{constant}, \end{aligned} \tag{13.10}$$

where "constant" contains all the terms that do not depend on $\mathcal{X}^l$. The negative KL divergence in Eq. (13.10) is maximized if we set

$$q_{\mathcal{X}^l}^{(j+1)}(\mathcal{X}^l) \propto \tilde{p}(\mathcal{X}, \mathcal{X}^l; \boldsymbol{\xi}^{(j)}), \tag{13.11}$$

where $\propto$ denotes proportionality. Combining Eqs. (13.11) and (13.8), we can now write

$$\boxed{q_{\mathcal{X}^l}^{(j+1)}(\mathcal{X}^l) = \frac{\exp\left(\mathbb{E}_{q_{\boldsymbol{\theta}}^{(j)}}\left[\ln p(\mathcal{X}, \mathcal{X}^l|\boldsymbol{\theta}; \boldsymbol{\xi}^{(j)})\right]\right)}{\int \exp\left(\mathbb{E}_{q_{\boldsymbol{\theta}}^{(j)}}\left[\ln p(\mathcal{X}, \mathcal{X}^l|\boldsymbol{\theta}; \boldsymbol{\xi}^{(j)})\right]\right) d\mathcal{X}^l},} \tag{13.12}$$

where the proportionality constant has, necessarily, been replaced by the normalizing factor to guarantee that $q_{\mathcal{X}^l}$ is a distribution. For the specific form in (13.12), the Bayes theorem, i.e., $p(\mathcal{X}, \mathcal{X}^l, \boldsymbol{\theta}; \boldsymbol{\xi}^{(j)}) = p(\mathcal{X}, \mathcal{X}^l|\boldsymbol{\theta}; \boldsymbol{\xi}^{(j)})p(\boldsymbol{\theta}; \boldsymbol{\xi}^{(j)})$, has been employed in both the numerator and the denominator.

E-Step 1b: In this step, $\boldsymbol{\xi}^{(j)}$ and $q_{\mathcal{X}^l}^{(j+1)}$ are frozen. Following similar steps as before (repeat the steps as an exercise), starting from the formulation in Eq. (13.7) and maximizing with respect to $q_{\boldsymbol{\theta}}$, we obtain

$$\boxed{q_{\boldsymbol{\theta}}^{(j+1)}(\boldsymbol{\theta}) = \frac{p(\boldsymbol{\theta}; \boldsymbol{\xi}^{(j)}) \exp\left(\mathbb{E}_{q_{\mathcal{X}^l}^{(j+1)}}\left[\ln p(\mathcal{X}, \mathcal{X}^l|\boldsymbol{\theta}; \boldsymbol{\xi}^{(j)})\right]\right)}{\int p(\boldsymbol{\theta}; \boldsymbol{\xi}^{(j)}) \exp\left(\mathbb{E}_{q_{\mathcal{X}^l}^{(j+1)}}\left[\ln p(\mathcal{X}, \mathcal{X}^l|\boldsymbol{\theta}; \boldsymbol{\xi}^{(j)})\right]\right) d\boldsymbol{\theta}}.} \tag{13.13}$$

Steps 1a and 1b comprise the E-step of the variational Bayesian EM.

M-Step 2: Freezing $q_{\boldsymbol{\theta}}^{(j+1)}$ and $q_{\mathcal{X}^l}^{(j+1)}$, maximize the lower bound with respect to $\boldsymbol{\xi}$, that is,

$$\boxed{\boldsymbol{\xi}^{(j+1)} = \arg\max_{\boldsymbol{\xi}} \mathcal{F}\left(q_{\mathcal{X}^l}^{(j+1)}, q_{\boldsymbol{\theta}}^{(j+1)}, \boldsymbol{\xi}\right).} \tag{13.14}$$

The counterpart of the EM illustration of Fig. 12.10 is given in Fig. 13.1. There are two observations to be made. Step 1 is now split into two parts, and more importantly, the KL divergence does *not* (in general) go to zero; hence the bound does not become tight. This comprises the E-step of the variational Bayesian EM.

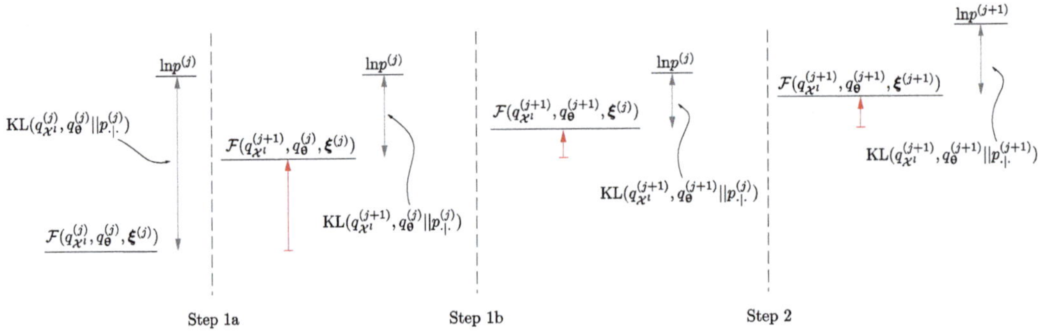

FIGURE 13.1

Illustration of the stepwise increase of $\ln p^{(j)}$ at the $(j+1)$th iteration of the variational Bayesian EM algorithm. Observe that $\ln p^{(j+1)} > \ln p^{(j)}$, where we have used the notation $p^{(j)} = p(\mathcal{X}; \boldsymbol{\xi}^{(j)})$ and $p^{(j)}_{\cdot|\cdot} := p(\mathcal{X}^l, \boldsymbol{\theta}|\mathcal{X}; \boldsymbol{\xi}^{(j)})$.

If there are more than two factors in $q(\mathcal{X}^l, \boldsymbol{\theta})$, as in Eq. (13.4), then there are more than two substeps in step 1, and each time we estimate one of the factors by averaging $\ln p(\mathcal{X}, \mathcal{X}^l, \boldsymbol{\theta}; \boldsymbol{\xi})$ with respect to the rest. Let q be factorized in M factors,

$$q(\mathcal{X}^l) = q_1(\mathcal{X}^l_1) \dots q_M(\mathcal{X}^l_M),$$

where for notational uniformity we have not differentiated between parameters and latent variables and the dependence on $\boldsymbol{\xi}$ has been suppressed. Then the general form of update becomes

$$\boxed{\ln q_m(\mathcal{X}^l_m) = \mathbb{E}\Big[\ln p(\mathcal{X}, \mathcal{X}^l_1, \dots, \mathcal{X}^l_M)\Big] + \text{constant},} \tag{13.15}$$

where the expectation is with respect to $\prod_{r=1_{r\neq m}}^M q_r(\mathcal{X}^l_r)$.

Remarks 13.1.

- Note that $q(\mathcal{X}^l, \boldsymbol{\theta})$ is an estimate of the posterior $p(\mathcal{X}^l, \boldsymbol{\theta}|\mathcal{X})$ and each one of the factors is the respective posterior estimate given the observations $\mathcal{X}$, for example, $q_{\boldsymbol{\theta}}(\boldsymbol{\theta}) \simeq p(\boldsymbol{\theta}|\mathcal{X})$.
- Once $q(\mathcal{X}^l, \boldsymbol{\theta})$ is factorized, *no additional assumptions on the functional form* of $q_{\mathcal{X}^l}$ and $q_{\boldsymbol{\theta}}$ are made.
- Note that factorization of a PDF implies independence. Thus, if this is not the case for the data at hand, the recovered approximations may not be faithful representations of the underlying data structure. Hence, choosing a specific factorization has to be carried out with care. In practice, one may have to use a number of alternatives and keep the best one. However, computational complexity is the other face of the coin, which one must consider in a tradeoff game. In general, the factorized variational approach tends to provide approximations to the posterior PDF that are more compact than the true ones (e.g., [46]).
- Recall our discussion in Section 12.3 related to model selection and Occam's rule. This was a kick-off point for our efforts to maximize the evidence with respect to different models, in order

to achieve the complexity–accuracy tradeoff. However, having resorted to approximate solutions (even if we forget convergence to local maxima, which somehow can be bypassed by using different initializations) we do not maximize the evidence but a lower bound of it; the latter is not, in general, tight. How tight it is depends on the KL divergence, which, unfortunately, cannot be trivially computed. Hence, if the lower bound is used for model selection, it has to be treated with caution [7].

- The variational approximation to Bayesian inference was first proposed in [25] and later on used in a number of areas, ranging from machine learning to decoding (e.g., [7,26,31–33,35]).
- *Online versions*: An online version of the variational Bayes algorithm was first proposed in [71]. There, the exponential family has been employed to show that parameter updating via the variational Bayes philosophy is equivalent to a natural gradient descent method (see Section 8.12 for the natural gradient) with step size equal to one. This equivalence is further discussed in [28], where, similarly, a stochastic approximation algorithm is proposed in order to process chunks of data in parallel. An online variational Bayes algorithm for parameter estimation in the context of sparse linear regression modeling has also been proposed in [83].

13.2.1 THE CASE OF THE EXPONENTIAL FAMILY OF PROBABILITY DISTRIBUTIONS

Looking carefully at Eqs. (13.12) and (13.13), it becomes clear that the practical application of the variational Bayesian EM depends on the computational tractability of the expected values of $\ln p(\mathcal{X}, \mathcal{X}^l|\boldsymbol{\theta}; \boldsymbol{\xi})$. Let us now see the form that the iterative steps take when one adopts the PDF models from the exponential family.

Let us assume that the samples $(\boldsymbol{x}_n, \boldsymbol{x}_n^l)$, $n = 1, 2, \ldots, N$, in the complete data set are i.i.d. Then

$$p(\mathcal{X}, \mathcal{X}^l|\boldsymbol{\theta}) = \prod_{n=1}^{N} p(\boldsymbol{x}_n, \boldsymbol{x}_n^l|\boldsymbol{\theta}). \tag{13.16}$$

We further assume $p(\boldsymbol{x}_n, \boldsymbol{x}_n^l|\boldsymbol{\theta})$ to lie within the exponential family (Section 12.8), that is,

$$p(\boldsymbol{x}_n, \boldsymbol{x}_n^l|\boldsymbol{\theta}) = g(\boldsymbol{\theta}) f(\boldsymbol{x}_n, \boldsymbol{x}_n^l) \exp\left(\boldsymbol{\phi}^T(\boldsymbol{\theta})\boldsymbol{u}(\boldsymbol{x}_n, \boldsymbol{x}_n^l)\right). \tag{13.17}$$

We also adopt a prior for $\boldsymbol{\theta}$ to be of the respective conjugate form, that is,

$$p(\boldsymbol{\theta}|\lambda, \boldsymbol{v}) = h(\lambda, \boldsymbol{v})(g(\boldsymbol{\theta}))^{\lambda} \exp\left(\boldsymbol{\phi}^T(\boldsymbol{\theta})\boldsymbol{v}\right). \tag{13.18}$$

The parameters $\lambda, \boldsymbol{v}$ constitute $\boldsymbol{\xi}$, which will be considered fixed, because our current emphasis is to follow up the specific functional forms that $q_{\mathcal{X}^l}$ and $q_{\boldsymbol{\theta}}$ get as iterations progress. So we relax the notational dependence on these parameters.

E-Step 1a: We have from Eq. (13.12)

$$\begin{aligned} q_{\mathcal{X}^l}^{(j+1)}(\mathcal{X}^l) &\propto \exp\left(\mathbb{E}_{q_{\boldsymbol{\theta}}^{(j)}}\left[\ln p(\mathcal{X}, \mathcal{X}^l|\boldsymbol{\theta})\right]\right) \\ &= \exp\left(\mathbb{E}_{q_{\boldsymbol{\theta}}^{(j)}}\left[\sum_{n=1}^{N} \ln p(\boldsymbol{x}_n, \boldsymbol{x}_n^l|\boldsymbol{\theta})\right]\right) \end{aligned}$$

$$= \prod_{n=1}^{N} \exp\left(\mathbb{E}_{q_{\theta}^{(j)}}\left[\ln p(\boldsymbol{x}_n, \boldsymbol{x}_n^l|\boldsymbol{\theta})\right]\right),$$

which then suggests that

$$q_{\mathbf{x}_n^l}^{(j+1)}(\boldsymbol{x}_n^l) \propto \exp\left(\mathbb{E}_{q_{\theta}^{(j)}}\left[\ln p(\boldsymbol{x}_n, \boldsymbol{x}_n^l|\boldsymbol{\theta})\right]\right),$$

and combined with Eq. (13.17) results in

$$q_{\mathbf{x}_n^l}^{(j+1)}(\boldsymbol{x}_n^l) = \tilde{g} f(\boldsymbol{x}_n, \boldsymbol{x}_n^l) \exp\left(\tilde{\boldsymbol{\phi}}^T \boldsymbol{u}(\boldsymbol{x}_n, \boldsymbol{x}_n^l)\right),$$

where $\tilde{g}$ is the respective normalization constant and

$$\tilde{\boldsymbol{\phi}}^T = \mathbb{E}_{q_{\theta}^{(j)}}\left[\boldsymbol{\phi}^T(\boldsymbol{\theta})\right]. \tag{13.19}$$

This is very interesting indeed. Although no functional form was assumed for $q_{\mathcal{X}^l}$, it turns out to be a member of the exponential family!

E-Step 1b: In a similar way, from Eqs. (13.13), (13.16), and (13.17), we obtain

$$q_{\theta}^{(j+1)}(\boldsymbol{\theta}) \propto p(\boldsymbol{\theta}) \exp\left(N \ln g(\boldsymbol{\theta}) + \sum_{n=1}^{N} \mathbb{E}_{q_{\mathbf{x}_n^l}^{(j+1)}}\left[\ln\left(f(\boldsymbol{x}_n, \mathbf{x}_n^l)\right)\right]\right.$$
$$\left. + \boldsymbol{\phi}^T(\boldsymbol{\theta}) \sum_{n=1}^{N} \mathbb{E}_{q_{\mathbf{x}_n^l}^{(j+1)}}\left[\boldsymbol{u}(\boldsymbol{x}_n, \mathbf{x}_n^l)\right]\right),$$

which combined with Eq. (13.18) results in

$$q_{\theta}^{(j+1)}(\boldsymbol{\theta}) \propto (g(\boldsymbol{\theta}))^{\lambda+N} \exp\left(\boldsymbol{\phi}^T(\boldsymbol{\theta})\left(\boldsymbol{v} + \sum_{n=1}^{N} \mathbb{E}_{q_{\mathbf{x}_n^l}^{(j+1)}}\left[\boldsymbol{u}(\boldsymbol{x}_n, \mathbf{x}_n^l)\right]\right)\right). \tag{13.20}$$

Thus, the approximation $q_{\theta}^{(j+1)}(\boldsymbol{\theta})$ of the posterior $p(\boldsymbol{\theta}|\mathcal{X})$ is of the same form as the conjugate prior with

$$\tilde{\lambda} = \lambda + N, \quad \tilde{v} = v + \sum_{n=1}^{N} \mathbb{E}_{q_{\mathbf{x}_n^l}^{(j+1)}}\left[\boldsymbol{u}(\boldsymbol{x}_n, \mathbf{x}_n^l)\right]. \tag{13.21}$$

Note that Eq. (13.21) is of the same form as (12.77). We only have to average out the hidden variables. This is a very elegant result, because nothing has been assumed about the functional form of $q_{\boldsymbol{\theta}}$. In other words, once we adopt the functional form for the PDFs of the complete set as well as that of the prior of the parameters to be of the exponential type, then subsequent iterations become a "family business."

13.3 A VARIATIONAL BAYESIAN APPROACH TO LINEAR REGRESSION

Once more, let us consider our familiar regression task

$$\mathbf{y} = \Phi\boldsymbol{\theta} + \boldsymbol{\eta}, \quad \mathbf{y} \in \mathbb{R}^N, \ \boldsymbol{\theta} \in \mathbb{R}^K.$$

In Section 12.5, we treated the case where $\boldsymbol{\eta}$ was Gaussian and the prior $p(\boldsymbol{\theta})$ was also Gaussian. We used the EM in order to optimize the evidence $p(\boldsymbol{y})$ with respect to the parameters that define the two adopted Gaussian PDFs; note that for this case, one could bypass the EM and resort to analytical computations in order to obtain the evidence and subsequently use an optimization technique to estimate the unknown parameters.

In this section, we will adopt assumptions that do not allow for tractable analytic computations of the posterior, $p(\boldsymbol{\theta}|\boldsymbol{y})$, which is a prerequisite both for the standard EM and for the analytic computations of the evidence $p(\boldsymbol{y})$. This approach is far from a pedagogic toy and has strong practical flavor. We will develop the task in some detail, and the reader is advised to go through the computations, because they are typical of what will be encountered in practice, once the variational Bayesian approach is chosen for addressing a task.

Assume that

$$p(\boldsymbol{y}|\boldsymbol{\theta}, \beta) = \mathcal{N}(\Phi\boldsymbol{\theta}, \beta^{-1}I). \tag{13.22}$$

That is, the noise is Gaussian and for simplicity we have considered it to be white, $\Sigma_\eta = \sigma_\eta^2 I$, and $\beta = \frac{1}{\sigma_\eta^2}$. In contrast to what we did in Section 12.5, now we will be more democratic and give the freedom to each one of the parameter components, θ_k, to have a different variance, $\sigma_k^2 := \frac{1}{\alpha_k}$, $k = 0, 1, \ldots, K-1$. Moreover, we go one step further. The values of β and α_k, $k = 0, \ldots, K-1$, will not be treated as deterministic variables. We will also treat them as random ones, (β, α_k), which will be assigned prior PDFs; these prior PDFs are in turn controlled by another set of hyperparameters. More specifically, our model, in addition to Eq. (13.22), comprises [8]

$$p(\boldsymbol{\theta}|\boldsymbol{\alpha}) = \prod_{k=0}^{K-1} \mathcal{N}(\theta_k|0, \alpha_k^{-1}), \tag{13.23}$$

$$p(\boldsymbol{\alpha}) = \prod_{k=0}^{K-1} \text{Gamma}(\alpha_k|a, b), \tag{13.24}$$

and

$$p(\beta) = \text{Gamma}(\beta|c, d). \tag{13.25}$$

Note that the previous choice of the priors indicates our will to "play" the game within the exponential family terrain. The prior $p(\boldsymbol{\alpha})$ is the conjugate pair of Eq. (13.23) (see Chapter 12). Also, Eq. (13.25) would be the conjugate of Eq. (13.22), if we had considered $\boldsymbol{\theta}$ fixed. Fig. 13.2 provides a graphical representation of the dependencies among the various variables involved in our model. Arrows indicate conditional dependencies. Graphical models will be considered in a formal way in Chapter 15. Note that such a model forms various levels of *hierarchy* in the dependency among the involved parameters.

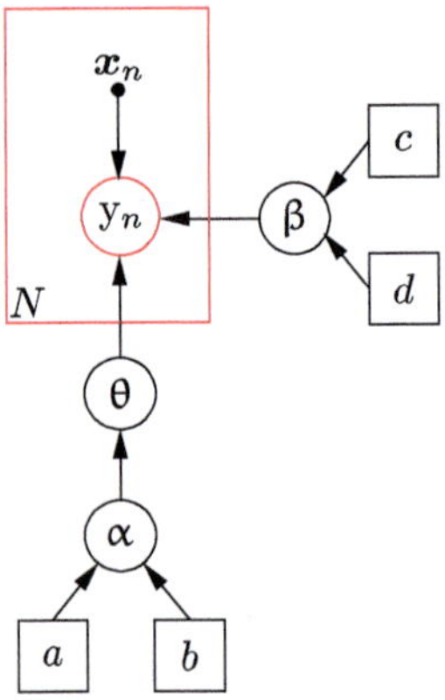

FIGURE 13.2

A graphical illustration of the dependencies among the various variables involved in the model of linear regression. The red circle indicates the random variable that is observed, gray circles indicate hidden random variables, and squares correspond to deterministic parameters. The direction of each arrow indicates the direction of the dependence between the connected variables. The red box indicates that the above dependencies hold for all N time instants.

This concept of hierarchy is at the heart of what we call *hierarchical Bayesian modeling*. Each one of the involved PDFs is expressed in terms of certain parameters. Because the values of these parameters are unknown, they are also treated as random variables whose priors are expressed in terms of a new set of hyperparameters. Each one of them is in turn treated as a random variable associated with a new prior, known as *hyperprior*. This rationale can be extended in order to construct different levels of hierarchy. Often, at the higher level of hierarchy, the corresponding (unknown) hyperparameters are assigned values by the user, based on experience; for example, the overall model can be relatively insensitive to their specific values, which makes the corresponding choice a fairly easy job.

Our current task comprises hidden variables in the form of parameters grouped in $\boldsymbol{\theta}$, $\boldsymbol{\alpha}$, and β and it involves no other unobserved variables. The set of observations is now given by $\boldsymbol{y}$. Also, observe that the posterior $p(\boldsymbol{\theta}, \boldsymbol{\alpha}, \beta|\boldsymbol{y})$ is not analytically tractable. We will resort to the variational Bayesian EM to obtain an estimate of the previous posterior PDF.

Using the mean field approximation, we assume that the approximation to the posterior (the dependence on $\boldsymbol{y}$ has been suppressed for notational convenience) factorizes as

$$q(\boldsymbol{\theta}, \boldsymbol{\alpha}, \beta) = q_{\theta}(\boldsymbol{\theta}) q_{\alpha}(\boldsymbol{\alpha}) q_{\beta}(\beta), \tag{13.26}$$

where we have relaxed our notation, for simplicity, from the explicit dependence on a, b, c, and d. We will bring them back into the game whenever needed. The variational EM consists of three substeps, one for each factor in Eq. (13.26). Starting from some initial guesses, for $\mathbb{E}[\beta]$, $\mathbb{E}[\alpha_k]$, $k = 0, \ldots, K-1$, (it will become clear soon why we need to start with those[1]) we get:

[1] If a, b, c, d were not fixed, then one would need initialization for these parameters too.

E-Step 1a: From the general update form of Eq. (13.15), we have

$$\ln q_{\theta}^{(j+1)}(\boldsymbol{\theta}) = \mathbb{E}_{q_{\alpha}^{(j)} q_{\beta}^{(j)}}\Big[\ln p(\boldsymbol{y}, \boldsymbol{\theta}, \boldsymbol{\alpha}, \beta)\Big] + \text{constant}, \tag{13.27}$$

where now

$$\begin{aligned}\ln p(\boldsymbol{y}, \boldsymbol{\theta}, \boldsymbol{\alpha}, \beta) &= \ln\big(p(\boldsymbol{y}|\boldsymbol{\theta}, \boldsymbol{\alpha}, \beta) p(\boldsymbol{\theta}, \boldsymbol{\alpha}, \beta)\big) \\ &= \ln\big(p(\boldsymbol{y}|\boldsymbol{\theta}, \beta) p(\boldsymbol{\theta}|\boldsymbol{\alpha}) p(\boldsymbol{\alpha}) p(\beta)\big),\end{aligned} \tag{13.28}$$

where the independence of $\mathbf{y}$ on $\boldsymbol{\alpha}$, given the values $\boldsymbol{\theta}$, has been taken into account. Using Eqs. (13.22) and (13.25) and some trivial algebra we get

$$\begin{aligned}\ln p(\boldsymbol{y}, \boldsymbol{\theta}, \boldsymbol{\alpha}, \beta) = \ln \frac{\beta^{N/2}}{(2\pi)^{N/2}} - \frac{\beta}{2}\|\boldsymbol{y} - \Phi\boldsymbol{\theta}\|^2 - \frac{1}{2}\sum_{k=0}^{K-1} a_k \theta_k^2 \\ + \sum_{k=0}^{K-1} \ln\sqrt{\frac{\alpha_k}{2\pi}} + \ln p(\boldsymbol{\alpha}) + \ln p(\beta),\end{aligned}$$

or

$$\ln p(\boldsymbol{y}, \boldsymbol{\theta}, \boldsymbol{\alpha}, \beta) = -\frac{\beta}{2}\|\boldsymbol{y} - \Phi\boldsymbol{\theta}\|^2 - \frac{1}{2}\sum_{k=0}^{K-1} a_k \theta_k^2 + \text{constant}, \tag{13.29}$$

where "constant" includes all terms that do not depend on $\boldsymbol{\theta}$, because in this step our goal is to estimate a function of $\boldsymbol{\theta}$. Expanding Eq. (13.29) and taking expectations with respect to β and $\boldsymbol{\alpha}$, considering $q_{\beta}^{(j)}(\beta)$ and $q_{\alpha}^{(j)}(\boldsymbol{\alpha})$ known, we get

$$\begin{aligned}\ln q_{\theta}^{(j+1)}(\boldsymbol{\theta}) = \mathbb{E}_{q_{\beta}^{(j)} q_{\alpha}^{(j)}}\big[\ln p(\boldsymbol{y}, \boldsymbol{\theta}, \boldsymbol{\alpha}, \beta)\big] + \text{constant} = -\frac{1}{2}\mathbb{E}[\beta]\boldsymbol{\theta}^T\Phi^T\Phi\boldsymbol{\theta} \\ -\frac{1}{2}\mathbb{E}[\beta]\boldsymbol{y}^T\boldsymbol{y} + \mathbb{E}[\beta]\boldsymbol{\theta}^T\Phi^T\boldsymbol{y} - \frac{1}{2}\boldsymbol{\theta}^T A\boldsymbol{\theta} + \text{constant},\end{aligned} \tag{13.30}$$

where by definition

$$A := \text{diag}\big\{\mathbb{E}[\alpha_0], \ldots, \mathbb{E}[\alpha_{K-1}]\big\},$$

and we have used for notational simplifications

$$\mathbb{E}[\beta] := \mathbb{E}_{q_{\beta}^{(j)}}[\beta] \quad \text{and} \quad \mathbb{E}[\alpha_k] := \mathbb{E}_{q_{\alpha}^{(j)}}[\alpha_k], \quad k = 0, 1, 2, \ldots, K-1. \tag{13.31}$$

It is readily noticed that the right-hand side of Eq. (13.30) is of a quadratic form with respect to $\boldsymbol{\theta}$, and hence $q_{\theta}^{(j+1)}(\boldsymbol{\theta})$ is Gaussian; in order to completely specify it, it suffices to compute the respective mean and covariance (precision) matrix.

Reshuffling the terms in Eq. (13.30), we get

$$\ln q_{\theta}^{(j+1)}(\boldsymbol{\theta}) = -\frac{1}{2}\boldsymbol{\theta}^T(A + \mathbb{E}[\beta]\Phi^T\Phi)\boldsymbol{\theta} + \mathbb{E}[\beta]\boldsymbol{\theta}^T\Phi^T\boldsymbol{y} + \text{constant},$$

which according to Eqs. (12.114), (12.116), and (12.117) of the appendix of the previous chapter (available via the book's site) results in

$$q_{\theta}^{(j+1)}(\boldsymbol{\theta}) = \mathcal{N}\left(\boldsymbol{\theta}\big|\boldsymbol{\mu}_{\theta}^{(j+1)}, \Sigma_{\theta}^{(j+1)}\right), \tag{13.32}$$

$$\Sigma_{\theta}^{(j+1)} = \left(A + \mathbb{E}[\beta]\Phi^T\Phi\right)^{-1}, \tag{13.33}$$

and

$$\boldsymbol{\mu}_{\theta}^{(j+1)} = \mathbb{E}[\beta]\Sigma_{\theta}^{(j+1)}\Phi^T \boldsymbol{y}. \tag{13.34}$$

During the first iteration step, $\mathbb{E}[\beta]$ and $\mathbb{E}[\alpha_k]$ are provided by their initial values. For the subsequent iterations, they have to be obtained together with $q_{\beta}^{(j)}$ and $q_{\alpha}^{(j)}$. Note that the approximation to the posterior $p(\boldsymbol{\theta}|\boldsymbol{y})$ (Eq. (13.32)) turns out to be Gaussian, although we did not assume it to be so. This is a consequence of the particular form of the adopted PDFs, which spring forth from the exponential family.

E-Step 1b: We have

$$\ln q_{\alpha}^{(j+1)}(\boldsymbol{\alpha}) = \mathbb{E}_{q_{\theta}^{(j+1)}q_{\beta}^{(j)}}\left[\ln p(\boldsymbol{y}, \boldsymbol{\theta}, \boldsymbol{\alpha}, \beta)\right] + \text{constant} \tag{13.35}$$

$$= \mathbb{E}_{q_{\theta}^{(j+1)}q_{\beta}^{(j)}}\left[\ln p(\boldsymbol{\theta}|\boldsymbol{\alpha}) + \ln p(\boldsymbol{\alpha})\right] + \text{constant}, \tag{13.36}$$

where the constant contains all terms that do not depend on $\boldsymbol{\alpha}$. Because no term in the bracket in the right-hand side of Eq. (13.36) depends on β, we have

$$\ln q_{\alpha}^{(j+1)}(\boldsymbol{\alpha}) = \mathbb{E}_{q_{\theta}^{(j+1)}}\left[\frac{1}{2}\sum_{k=0}^{K-1}\ln\alpha_k - \frac{1}{2}\sum_{k=0}^{K-1}\alpha_k\theta_k^2\right] + \ln p(\boldsymbol{\alpha}) + \text{constant}. \tag{13.37}$$

Taking into account Eq. (13.24) and after some algebra (Problem 13.2), we obtain

$$q_{\alpha}^{(j+1)}(\boldsymbol{\alpha}) = \prod_{k=0}^{K-1}\text{Gamma}(\alpha_k|\tilde{a}, \tilde{b}_k), \tag{13.38}$$

where

$$\tilde{a} = a + \frac{1}{2}, \tag{13.39}$$

$$\tilde{b}_k = b + \frac{1}{2}\mathbb{E}_{q_{\theta}^{(j+1)}}[\theta_k^2], \quad k = 0, \ldots, K-1. \tag{13.40}$$

In order to compute $\mathbb{E}[\theta_k^2]$, recall Eqs. (12.46) and (12.47) and apply them into our setting to give

$$\mathbb{E}_{q_{\theta}^{(j+1)}}[\boldsymbol{\theta}\boldsymbol{\theta}^T] = \Sigma_{\theta}^{(j+1)} + \boldsymbol{\mu}_{\theta}^{(j+1)}\boldsymbol{\mu}_{\theta}^{(j+1)T},$$

or

$$\mathbb{E}[\theta_k^2] = \left[\mathbb{E}_{q_\theta^{(j+1)}}[\boldsymbol{\theta}\boldsymbol{\theta}^T]\right]_{kk} = \left[\Sigma_\theta^{(j+1)} + \boldsymbol{\mu}_\theta^{(j+1)}\boldsymbol{\mu}_\theta^{(j+1)^T}\right]_{kk}, \quad k = 0, 1, \ldots, K-1, \tag{13.41}$$

where $[A]_{kk}$ denotes the (k,k) element of a matrix A. To complete the computations, we have to compute $\mathbb{E}[\alpha_k]$, $k = 0, 1, \ldots, K-1$, to be used during the next iteration in Eq. (13.33). However, because each α_k follows a gamma distribution, we know that (Section 2.3.2)

$$\mathbb{E}_{q_\alpha^{(j+1)}}[\alpha_k] = \frac{\tilde{a}}{\tilde{b}_k}. \tag{13.42}$$

E-Step 1c: We have

$$\begin{aligned}\ln q_\beta^{(j+1)}(\beta) &= \mathbb{E}_{q_\theta^{(j+1)} q_\alpha^{(j+1)}}\left[\ln p(\boldsymbol{y}, \boldsymbol{\theta}, \boldsymbol{\alpha}, \beta)\right] + \text{constant} \\ &= \mathbb{E}_{q_\theta^{(j+1)} q_\alpha^{(j+1)}}\left[\ln p(\boldsymbol{y}|\boldsymbol{\theta}, \beta) + \ln p(\beta)\right] + \text{constant}.\end{aligned}$$

This is of the same form as Eq. (13.36), and following similar steps (Problem 13.3), it can be shown that

$$q_\beta^{(j+1)}(\beta) = \text{Gamma}(\beta|\tilde{c}, \tilde{d}), \tag{13.43}$$

$$\tilde{c} = c + \frac{N}{2}, \tag{13.44}$$

$$\tilde{d} = d + \frac{1}{2}\mathbb{E}_{q_\theta^{(j+1)}}[\|\boldsymbol{y} - \Phi\boldsymbol{\theta}\|^2]. \tag{13.45}$$

To compute the expectation in Eq. (13.45), recall Eq. (12.49), which for our needs becomes

$$\mathbb{E}_{q_\theta^{(j+1)}}[\|\boldsymbol{y} - \Phi\boldsymbol{\theta}\|^2] = \|\boldsymbol{y} - \Phi\boldsymbol{\mu}_\theta^{(j+1)}\|^2 + \text{trace}\left\{\Phi\Sigma_\theta^{(j+1)}\Phi^T\right\}. \tag{13.46}$$

Finally, we have

$$\mathbb{E}_{q_\beta^{(j+1)}}[\beta] = \frac{\tilde{c}}{\tilde{d}}, \tag{13.47}$$

which completes all the computations associated with the E-step of the variational EM. Note that $q_\alpha^{(j+1)}(\boldsymbol{\alpha}) \simeq p(\boldsymbol{\alpha}|\boldsymbol{y})$ and $q_\beta^{(j+1)}(\beta) \simeq p(\beta|\boldsymbol{y})$ retain the gamma functional form of the corresponding priors that were originally adopted, without forcing them to.

In principle, one can add an extra M-step in the algorithm to maximize the bound with respect to the unknown parameters a, b, c, and d. However, in practice, for computational simplicity these parameters are fixed to very small values, that is, $a = b = c = d = 10^{-6}$, which correspond to *uninformative* gamma prior distributions, in the sense of giving no preference to any specific range of values. Note that for such small values, the gamma distribution falls as $\frac{1}{x}$. Indeed, for $a, b \simeq 0$

$$\text{Gamma}(x|a, b) \simeq \frac{1}{x}, \quad x > 0.$$

Because every positive x can be expressed as

$$x = \exp(z), \quad z = \ln x, \quad z \in \mathbb{R},$$

it can be easily checked (Problem 13.4) that the PDF that describes z is uniform. This is a typical procedure in practice; that is, one allows enough levels of hierarchy and fixes the hyperparameters in the highest level to define uninformative hyperpriors.

In summary, the variational Bayesian EM steps are given in Algorithm 13.1.

Algorithm 13.1 (Variational EM for linear regression).

- Initialization
  - Select initial values for $\mathbb{E}[\beta]$, $\mathbb{E}_{q_\alpha}[\alpha_k]$, $\quad k = 0, 1, \ldots, K-1$.
- **For**, $j = 1, 2, \ldots$, **Do**
  - $A = \text{diag}\{\mathbb{E}_{q_\alpha}[\alpha_0,], \mathbb{E}_{q_\alpha}[\alpha_1], \ldots, \mathbb{E}_{q_\alpha}[\alpha_{K-1}]\}$.
  - Compute Σ_θ from Eq. (13.33) and $\boldsymbol{\mu}_\theta$ from Eq. (13.34).
  - Compute $\tilde{a}$ from Eq. (13.39).
  - Compute $\tilde{b}_k$, $k = 0, 1, \ldots, K-1$, from Eqs. (13.40) and (13.41).
  - Compute $\mathbb{E}_{q_\alpha}[\alpha_k]$, $k = 0, 1, \ldots, K-1$, from Eq. (13.42).
  - Compute $\tilde{c}$ from Eq. (13.44) and $\tilde{d}$ from Eqs. (13.45) and (13.46).
  - Compute $\mathbb{E}_{q_\beta}[\beta]$ from Eq. (13.47).
  - If convergence criterion is met, Stop.
- **End For**

Once the algorithm has converged, predictions can be made on the basis of the predictive distribution given in Eqs. (12.21)–(12.23), by replacing $\Sigma_{\theta|y}$, $\boldsymbol{\mu}_{\theta|y}$, and σ_η^2 by the converged values of Σ_θ, $\boldsymbol{\mu}_\theta$, and $\mathbb{E}[\beta]$, respectively. Note, however, that this is only an approximation, because the Gaussian form for the posterior of the parameters is a result of the mean field approximation and also we have used the mean value, $\mathbb{E}[\beta]$, in place of the noise variance. The latter can be justified because as the number of training samples increases, the distribution of β sharply peaks around its mean value [8].

COMPUTATION OF THE LOWER BOUND

Once the algorithm has converged, the quantities $q_\theta(\boldsymbol{\theta}), q_\alpha(\boldsymbol{\alpha}), q_\beta(\beta)$ are available and the lower bound $\mathcal{F}(q_\theta, q_\alpha, q_\beta)$ can be computed. The computation of this lower bound can also be done at every iteration to check how much it changes from iteration to iteration, and then this can be used as a convergence criterion. Let $\tilde{q}_\theta, \tilde{q}_\alpha, \tilde{q}_\beta$ be the approximate posteriors after convergence, defined by the parameters $\tilde{\Sigma}_\theta, \tilde{\boldsymbol{\mu}}_\theta, \tilde{a}, \tilde{b}_k, k = 0, 1, 2, \ldots, K-1, \tilde{c}$, and $\tilde{d}$. The lower bound is then given as

$$\begin{aligned} \mathcal{F}(\tilde{q}_\theta, \tilde{q}_\alpha, \tilde{q}_\beta) &= \mathbb{E}_{\tilde{q}_\theta \tilde{q}_\alpha \tilde{q}_\beta}\left[\ln p(\boldsymbol{y}, \boldsymbol{\theta}, \boldsymbol{\alpha}, \beta)\right] - \mathbb{E}_{\tilde{q}_\theta}\left[\ln \tilde{q}_\theta(\boldsymbol{\theta})\right] \\ &\quad - \mathbb{E}_{\tilde{q}_\alpha}\left[\ln \tilde{q}_\alpha(\boldsymbol{\alpha})\right] - \mathbb{E}_{\tilde{q}_\beta}\left[\ln \tilde{q}_\beta(\beta)\right]. \end{aligned} \tag{13.48}$$

Performing the expectations can be a bit tedious, but it is straightforward (Problem 13.5).

13.4 A VARIATIONAL BAYESIAN APPROACH TO GAUSSIAN MIXTURE MODELING

Dealing with the Gaussian mixture modeling in Section 12.6, it was pointed out in the remarks that the standard EM approach may lead to singularities. One way to bypass this drawback is to enforce priors on the involved parameters and resort to a variational Bayesian philosophy to estimate the quantities of interest. The task was first treated in [4] and later in [13]. We will present the latter approach and comment on the underlying differences between the two later on.

Given a set of observations $\mathcal{X} = \{\boldsymbol{x}_1, \ldots, \boldsymbol{x}_N\}$, the respective PDF model is

$$p(\boldsymbol{x}) = \sum_{k=1}^{K} P_k \mathcal{N}(\boldsymbol{x}|\boldsymbol{\mu}_k, Q_k^{-1}), \quad \boldsymbol{x} \in \mathbb{R}^l.$$

The task is to estimate the unknown parameters $(P_k, \boldsymbol{\mu}_k, Q_k)$, $k = 1, 2, \ldots, K$. We already know that this is a typical task with latent variables, and the complete set comprises $(\boldsymbol{x}_n, k_n)$, $n = 1, 2, \ldots, N$, with k_n being the index of the respective mixture, $k_n = 1, 2, \ldots, K$. In Section 12.6, the information about each of the latent variables, k_n, entered into the problem via the posterior $P(k_n|\boldsymbol{x}_n)$ for every time instant n, the summation over all possible values of k_n was performed, and hence one could drop out the time index. However, in the current context, a different path has to be followed and we have to consider the latent variables together with their corresponding time index. To this end, and following [13], an auxiliary latent random vector is introduced, $\mathbf{z}_n \in \mathbb{R}^K$, for each observation, $n = 1, 2, \ldots, N$. Its components take binary values, such as

$$z_{n_k} \in \{0, 1\} \quad \text{and} \quad \sum_{k=1}^{K} z_{n_k} = 1, \tag{13.49}$$

and they are used as indicators of the respective mixture from which the observation at time n, $\boldsymbol{x}_n$, was drawn; that is, if $z_{n_k} = 1$ it indicates that $\boldsymbol{x}_n$ was drawn from the kth distribution. Obviously,

$$P(z_{n_k} = 1) = P_k,$$

and for any $\boldsymbol{z}_n \in \mathbb{R}^K$ that satisfies Eq. (13.49),

$$P(\boldsymbol{z}_n) = \prod_{k=1}^{K} P_k^{z_{n_k}}. \tag{13.50}$$

Hence, the probability of occurrence of the set $\mathcal{Z} = \{\boldsymbol{z}_1, \ldots, \boldsymbol{z}_N\}$ is

$$P(\mathcal{Z}) = \prod_{n=1}^{N} \prod_{k=1}^{K} P_k^{z_{n_k}}, \tag{13.51}$$

and in this way, we have described the random nature of the N latent variables using a multinomial probability distribution.

In the sequel, we will treat the mean values as well as the precision matrices as random quantities adopting the following prior PDFs:

$$p(\boldsymbol{\mu}_k) = \mathcal{N}\left(\boldsymbol{\mu}_k|\mathbf{0}, \beta^{-1}I\right)$$

and

$$p(Q_k) = \mathcal{W}(Q_k|W_0, \nu_0),$$

for fixed ν_0, W_0, and β. That is, the adopted priors are Gaussians for the mean values and Wishart PDFs for the precision matrices. We will treat $\boldsymbol{P} = [P_1, \ldots, P_k]^T$ as deterministic parameters whose optimized values are obtained in the M-step.

Following the philosophy of the variational Bayesian EM, we adopt

$$q(\mathcal{Z}, \boldsymbol{\mu}_{1:K}, Q_{1:K}) = q_{\mathcal{Z}}(\mathcal{Z})q_{\mu}(\boldsymbol{\mu}_{1:K})q_{Q}(Q_{1:K}),$$

where $\boldsymbol{\mu}_{1:K}$ and $Q_{1:K}$ indicate the collections $\{\boldsymbol{\mu}_1, \ldots, \boldsymbol{\mu}_K\}$ and $\{Q_1, \ldots, Q_K\}$, respectively.

Furthermore, observe that the conditional PDF of the observations can now be written as

$$p(\mathcal{X}|\mathcal{Z}, \boldsymbol{\mu}_{1:K}, Q_{1:K}) = \prod_{n=1}^{N}\prod_{k=1}^{K}\left(\mathcal{N}(\boldsymbol{x}_n|\boldsymbol{\mu}_k, Q_k^{-1})\right)^{z_{nk}}.$$

Fig. 13.3 shows the corresponding graphical model.

Computational steps of the variational EM for Gaussian mixture modeling

Initialization: (a) $\boldsymbol{P}^{(0)}$, (b) $\mathbb{E}_{q_Q^{(0)}}[\mathrm{Q}_k]$, (c) $\mathbb{E}_{q_Q^{(0)}}[\ln|\mathrm{Q}_k|]$, (d) $\mathbb{E}_{q_\mu^{(0)}}[\boldsymbol{\mu}_k] := \tilde{\boldsymbol{\mu}}_k^{(0)}$, and (e) $\mathbb{E}_{q_\mu^{(0)}}[\boldsymbol{\mu}_k\boldsymbol{\mu}_k^T] := \tilde{\Sigma}_k^{(0)} + \tilde{\boldsymbol{\mu}}_k^{(0)}\tilde{\boldsymbol{\mu}}_k^{(0)T}$, $k = 1, 2, \ldots, K$, where $|\cdot|$ denotes the corresponding determinant.

The $(j+1)$th iteration consists of the following computations (Problem 13.6):

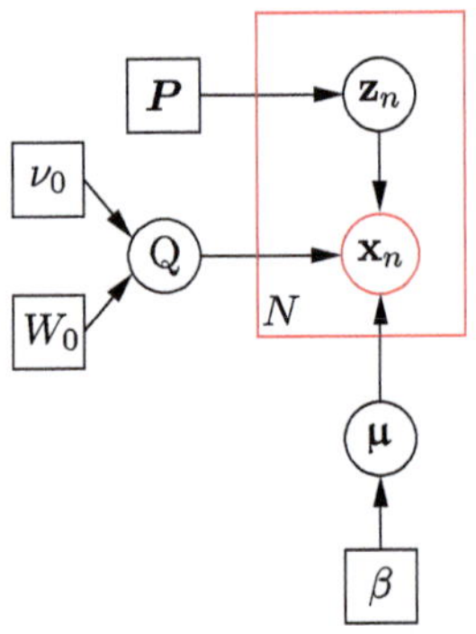

FIGURE 13.3

The graphical model associated with the Gaussian mixture modeling of Section 13.4.

E-Step 1a:

$$\begin{aligned}
\pi_{n_k} &= P_k^{(j)} \exp\Big(\frac{1}{2}\,\mathbb{E}_{q_Q^{(j)}}[\ln|Q_k|] - \frac{1}{2}\text{trace}\big\{\,\mathbb{E}_{q_Q^{(j)}}[Q_k]\big(\boldsymbol{x}_n\boldsymbol{x}_n^T \\
&\quad -\boldsymbol{x}_n\,\mathbb{E}_{q_\mu^{(j)}}[\boldsymbol{\mu}_k^T] - \mathbb{E}_{q_\mu^{(j)}}[\boldsymbol{\mu}_k]\boldsymbol{x}_n^T + \mathbb{E}_{q_\mu^{(j)}}[\boldsymbol{\mu}_k\boldsymbol{\mu}_k^T]\big)\big\}\Big), \\
\rho_{n_k} &= \frac{\pi_{n_k}}{\sum_{k=1}^K \pi_{n_k}}, \\
q_{\mathcal{Z}}^{(j+1)}(\mathcal{Z}) &= \prod_{n=1}^N\prod_{k=1}^K \rho_{n_k}^{z_{n_k}}.
\end{aligned}$$

E-Step 1b:

$$\tilde{Q}_k = \beta I + \mathbb{E}_{q_Q^{(j)}}[Q_k]\sum_{n=1}^N \rho_{n_k},$$

$$\tilde{\boldsymbol{\mu}}_k = \tilde{Q}_k^{-1}\,\mathbb{E}_{q_Q^{(j)}}[Q_k]\sum_{n=1}^N \rho_{n_k}\boldsymbol{x}_n,$$

$$q_\mu^{(j+1)}(\boldsymbol{\mu}_{1:K}) = \prod_{k=1}^K \left(\boldsymbol{\mu}_k|\tilde{\boldsymbol{\mu}}_k, \tilde{Q}_k^{-1}\right),$$

and adopting Eq. (12.47) to the current needs,

$$\mathbb{E}_{q_\mu^{(j+1)}}[\boldsymbol{\mu}_k\boldsymbol{\mu}_k^T] = \tilde{\Sigma}_k + \tilde{\boldsymbol{\mu}}_k\tilde{\boldsymbol{\mu}}_k^T = \tilde{Q}_k^{-1} + \tilde{\boldsymbol{\mu}}_k\tilde{\boldsymbol{\mu}}_k^T.$$

E-Step 1c:

$$\begin{aligned}
\tilde{\nu}_k &= \nu + \sum_{n=1}^N \rho_{n_k}, \\
\tilde{W}_k^{-1} &= \tilde{W}_0^{-1} + \sum_{n=1}^N \rho_{n_k}\left(\boldsymbol{x}_n\boldsymbol{x}_n^T - \tilde{\boldsymbol{\mu}}_k\boldsymbol{x}_n^T - \boldsymbol{x}_n\tilde{\boldsymbol{\mu}}_k^T + \mathbb{E}_{q_\mu^{(j+1)}}[\boldsymbol{\mu}_k\boldsymbol{\mu}_k^T]\right), \\
q_Q^{(j+1)}(Q_{1:K}) &= \prod_{k=1}^K \mathcal{W}(Q_k|\tilde{\nu}_k, \tilde{W}_k), \\
\mathbb{E}_{q_Q^{(j+1)}}[Q_k] &= \tilde{\nu}_k\tilde{W}_k, \\
\mathbb{E}_{q_Q^{(j+1)}}\Big[\ln|Q_k|\Big] &= \sum_{i=1}^l \psi\left(\frac{\tilde{\nu}_{k+1-i}}{2}\right) + l\ln 2 + \ln|\tilde{W}_k|,
\end{aligned}$$

where $\psi(\cdot)$ is the *digamma* function, defined as

$$\psi(a) := \frac{d \ln \Gamma(a)}{da},$$

and the gamma function has been defined in Eq. (2.91).

M-Step 2: We have

$$P_k^{(j+1)} = \frac{1}{N} \sum_{n=1}^{N} \rho_{nk}.$$

The previous steps have concluded the algorithm. Observe that the iterations retain the functional form of the PDFs that were adopted for the respective priors; this is a consequence of their exponential family origin. In [13], it is suggested that this procedure can also be used to determine the number of mixtures, instead of adopting a cross-validation technique, as pointed out in the remarks of Section 12.6. By adopting a large enough value for K, the probabilities P_k associated with the irrelevant components will be driven to zero during the M-step. Note that such a modeling is possible in the Bayesian framework, because it automatically achieves a tradeoff between model complexity and data fitting. In [4], the probabilities P_k, $k = 1, 2, \ldots, K$, were considered as random variables and a Dirichlet prior was also imposed on them (Problem 13.7). However, such priors need to be selected with some care; otherwise it may affect the sparsification potential of the algorithm (e.g., [9]).

Example 13.1. The purpose of this example is to demonstrate the power of the variational Bayesian method for mixture modeling compared to the more classical EM algorithm, which was discussed in Section 12.6. Five clusters of data were generated using a corresponding number of Gaussians, as shown in Fig. 13.4. The parameters used for each one of these Gaussians were

$$\boldsymbol{\mu}_1 = [-2.5, 2.5]^T, \quad \boldsymbol{\mu}_2 = [-4.0, -2.0]^T, \quad \boldsymbol{\mu}_3 = [2.0, -1.0]^T,$$
$$\boldsymbol{\mu}_4 = [0.1, 0.2]^T, \quad \boldsymbol{\mu}_5 = [3.0, 3.0]^T$$

and

$$\Sigma_1 = \begin{bmatrix} 0.5 & 0.081 \\ 0.081 & 0.7 \end{bmatrix}, \quad \Sigma_2 = \begin{bmatrix} 0.4 & 0.02 \\ 0.002 & 0.3 \end{bmatrix}, \quad \Sigma_3 = \begin{bmatrix} 0.6 & 0.531 \\ 0.531 & 0.9 \end{bmatrix},$$
$$\Sigma_4 = \begin{bmatrix} 0.5 & 0.22 \\ 0.22 & 0.8 \end{bmatrix}, \quad \Sigma_5 = \begin{bmatrix} 0.88 & 0.2 \\ 0.2 & 0.22 \end{bmatrix}.$$

Prior to running the algorithms, we assumed that we do not know the exact number of mixtures, so a number of $K = 25$ clusters was used, that is, a much larger number than the true one.

For the EM algorithm, the initial mean values were generated randomly, using a Gaussian $\mathcal{N}(\boldsymbol{\mu} | \boldsymbol{0}, I)$ and the respective initial covariance matrices, $\Sigma_k^{(0)}$, $k = 1, 2, \ldots, 25$, with random elements, making sure that it is positive definite. One way to achieve this is to generate a matrix Φ with random elements from $\mathcal{N}(0, 1)$ and then form $\Phi^T \Phi$. Another possibility is to start with a diagonal matrix, for example, the identity one I.

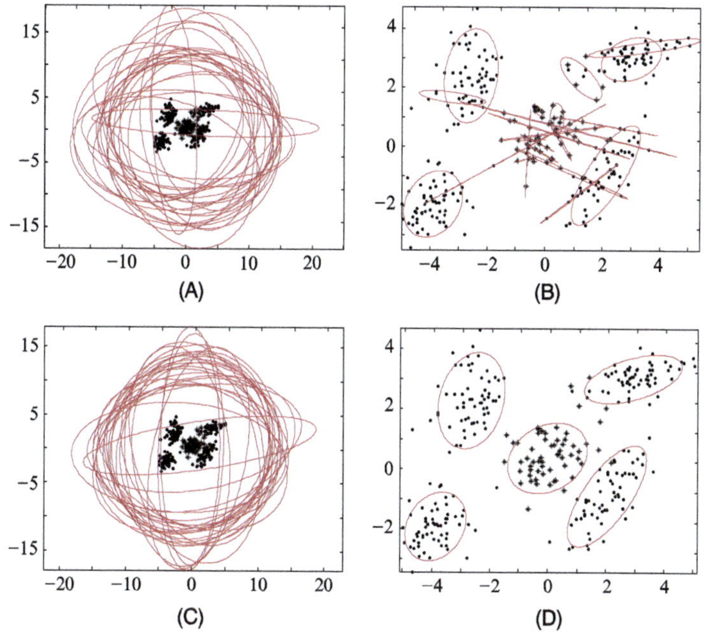

FIGURE 13.4

Figure for Example 13.1. (A) The initial (25) Gaussians for the EM algorithm. (B) The final clusters obtained after convergence by the EM algorithm. (C) The initial (25) Gaussians for the variational EM. (D) The final Gaussians obtained by the variational EM, after convergence. All the curves correspond to the 80% probability regions. Observe that the variational EM identifies the five clusters associated with the data; the rest of the mixtures correspond to zero probability weights.

For the variational EM algorithm, the following initial values were used: the mean values $\tilde{\boldsymbol{\mu}}_k^{(0)}$ and the initial covariance matrices $\tilde{\Sigma}_k^{(0)}$, $k = 1, 2, \ldots, 25$, were generated as before. Also, $\mathbb{E}_{q_Q^{(0)}}[Q_k] = I$, $\mathbb{E}_{q_Q^{(0)}}[\ln |Q_k|] = 1$. In both cases, the initial probabilities were set to be equal.

Observe that the variational EM identifies the five clusters associated with the data; the rest of the mixtures correspond to zero probability weights. In contrast, the EM algorithm tries to identify all 25 mixtures and the result is not satisfactory.

13.5 WHEN BAYESIAN INFERENCE MEETS SPARSITY

The Bayesian approach to sparsity-aware learning will soon become our major concern. However, we will use this subsection to "warm" us up. The close relationship between the use of a prior PDF and the regularization of a cost function has already been discussed in Section 12.2.2. There, the adoption of a Gaussian prior together with a Gaussian noise for the regression task led to the equivalence of MAP with the ridge regression. It will not take a minute to show that the use of a Gaussian model for the noise together with a Laplacian prior for each one of the weights, that is,

$$p(\theta_k) = \frac{\lambda}{2} \exp\big(- \lambda |\theta_k| \big),$$

renders MAP equivalent to the ℓ_1 norm regularization of the LS cost. For a Bayesian, however, who is not interested in cost functions, the secret that lies within the Laplacian prior is hidden in the heavy tails of this distribution. This is in contrast to a Gaussian PDF, which has very light tails. In other words, the probability that an observation of a Gaussian random variable can take values far from its mean decreases very fast. For example, the probability of observing variables that deviate from the mean by more than 2σ, 3σ, 4σ, and 5σ are 0.046, 0.003, 6×10^{-5}, and 6×10^{-7}, respectively. That is, if we provide a Gaussian prior, we basically inform the learning process to look for values "around" the mean; values away from the mean are heavily penalized. However, in sparsity-aware learning this would be the wrong information to pass over to our learning mechanism. Assuming the mean of the prior to be zero, although we expect most of the components of our parameters to be zero, still we want a few of them to be large. Hence, our prior information should be selected so as to assign small (but not too small) probabilities to large values. Hence, to a Bayesian, sparsity-aware learning becomes synonymous with imposing heavy-tail priors. Let us now turn back to our current task, and see how this brief introduction is related to our model. Our prior PDF, $p(\boldsymbol{\theta})$, according to the model of Eqs. (13.23) and (13.24), is obtained by marginalizing out the hyperparameters $\boldsymbol{\alpha}$ (Problem 13.10), that is,

$$\begin{aligned} p(\boldsymbol{\theta}; a, b) &= \int p(\boldsymbol{\theta}|\boldsymbol{\alpha}) p(\boldsymbol{\alpha})\, d\boldsymbol{\alpha} \\ &= \int \prod_{k=0}^{K-1} \mathcal{N}(\theta_k|0, \alpha_k^{-1}) \text{Gamma}(\alpha_k|a, b)\, d\boldsymbol{\alpha} \\ &= \prod_{k=0}^{K-1} \text{st}\left(\theta_k|0, \frac{a}{b}, 2a\right), \end{aligned} \tag{13.52}$$

where $\text{st}(x|\mu, \lambda, \nu)$ is the Student's t PDF, defined by

$$\text{st}(x|\mu, \lambda, \nu) = \frac{\Gamma(\frac{\nu+1}{2})}{\Gamma(\frac{\nu}{2})} \left(\frac{\lambda}{\pi\nu}\right)^{1/2} \frac{1}{\left(1 + \frac{\lambda(x-\mu)^2}{\nu}\right)^{\frac{\nu+1}{2}}}. \tag{13.53}$$

The parameter ν is known as the number of degrees of freedom. Fig. 13.5 shows the graph of Student's t PDFs for different values of ν. For $\nu \longrightarrow \infty$, the Student's t distribution tends to a Gaussian of the same mean and precision λ. Observe the heavy-tail feature of Student's t PDF, especially for low values of ν. Recall that in our case, where we have used uninformative hyperpriors, the hyperparameter, a, was given a small value. Thus, our treatment in this section favors sparse solutions for the regression model. It will push as many of the coefficients θ_k as possible toward zero. That is, it prunes the less relevant basis functions $\phi_k(\boldsymbol{x})$ by setting the corresponding coefficients to zero. This is also the reason for using different hyperparameters α_k for each one of the parameters θ_k, $k = 0, 2, \ldots, K-1$, which provide more freedom to the learning procedure to adjust each one of the parameters individually. In

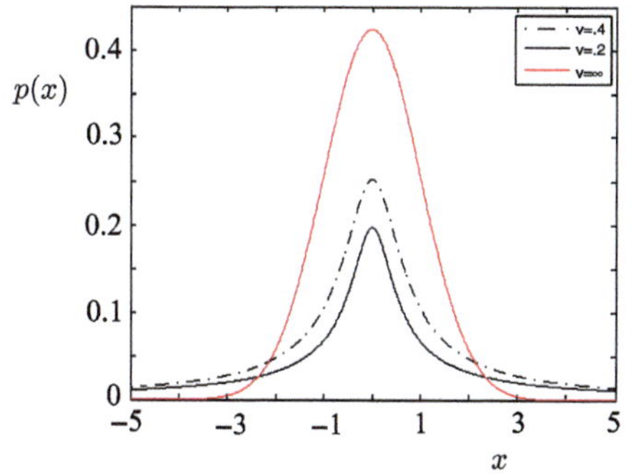

FIGURE 13.5

Observe that for low values of the degrees of freedom, ν, Student's t PDF has very high tails. In contrast, the Gaussian PDF is a low-tailed PDF.

the earlier days, this approach was coined *automatic relevance determination* (ARD) [47,52,54]. An interesting discussion relating adaptive regularization and pruning is provided in [24].

Fig. 13.6A provides a clear demonstration of the sparsity imposing properties of the Student's t distribution. In the two-dimensional space, and as we move away from zero, probability mass is skewed toward the coordinate axes; that is, the PDF peaks around sparse solutions and *sparsity is now enforced probabilistically*. In contrast, the Gaussian does not give much chance to large values (see Fig. 13.6B).

13.6 SPARSE BAYESIAN LEARNING (SBL)

In Section 13.3, the prior for each one of the unknown parameters θ_k, $k = 0, 1, \ldots, K-1$, was given the liberty to have their own variances, $\sigma_k^2 := \frac{1}{\alpha_k}$. In turn, these variances were treated as hidden random variables and a prior was assigned to each of them in terms of a number of hyperparameters.

In [85,92], the model was slightly modified. The concept of using different variances for the priors was retained, but the variances were treated as deterministic parameters and not as random ones.[2] In this context, the task becomes a generalization of the one treated in Section 12.5, and it is built upon the following assumptions:

$$p(\boldsymbol{y}|\boldsymbol{\theta};\beta) = \mathcal{N}(\boldsymbol{y}|\Phi\boldsymbol{\theta}, \beta^{-1}I), \tag{13.54}$$

$$p(\boldsymbol{\theta};\boldsymbol{\alpha}) = \mathcal{N}(\boldsymbol{\theta}|\boldsymbol{0}, A^{-1}), \tag{13.55}$$

where

$$A := \operatorname{diag}\{\alpha_0, \ldots, \alpha_{K-1}\}. \tag{13.56}$$

[2] A slightly different yet equivalent view, employing uniform priors and using the respective modes instead of marginalizing out the variances, is followed in [85].

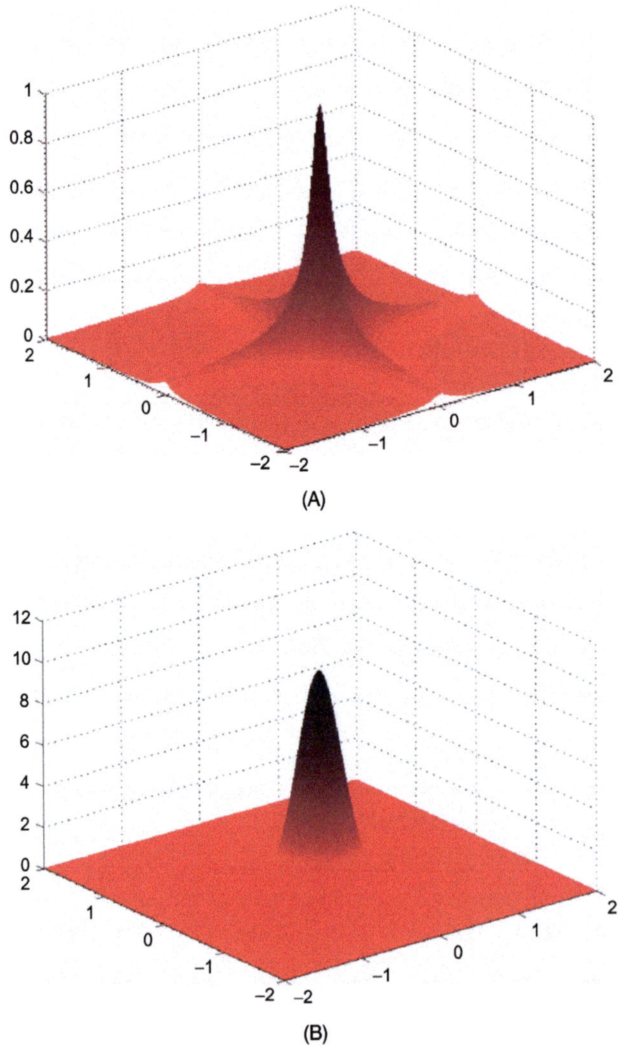

FIGURE 13.6

(A) The Student's t peaks sharply around zero and falls slowly along the axes; hence, sparse solutions are favored. (B) The Gaussian peaks around zero and decays very fast, along all directions.

The precision β is also treated as an unknown deterministic parameter. Our goal is (a) to obtain estimates for β and α_k, $k = 0, 1, \ldots, K - 1$, and (b) to compute the predictive distribution, $p(y|\boldsymbol{x}, \boldsymbol{y})$, where $\boldsymbol{y}$ is the vector of observations.[3] To this end, one could adopt the EM algorithm and follow

[3] Notational dependence on the input training data, $\mathcal{X}$, has been suppressed.

similar steps as in Section 12.5; the only difference is that there, a common variance was shared by all the involved prior PDFs. The method is usually referred to as *sparse Bayesian learning* (SBL) and complies with the ARD rationale discussed in Section 13.5.

In this section, we will adopt a different path, exploiting the Gaussian nature of the involved PDFs. A Type II maximum likelihood method will be employed, which was introduced in Remarks 12.2. Type II likelihood is defined as the marginal one, after integrating out the parameters $\boldsymbol{\theta}$. Following the discussion in Section 12.2 and for our current needs, Eq. (12.15) is written as

$$p(\boldsymbol{y};\boldsymbol{\alpha},\beta)=\mathcal{N}(\boldsymbol{y}|\boldsymbol{0},\beta^{-1}I+\Phi A^{-1}\Phi^T). \tag{13.57}$$

Also, for the sake of completeness, Eqs. (12.16), (12.17), and (12.10) take the form

$$p(\boldsymbol{\theta}|\boldsymbol{y};\boldsymbol{\alpha},\beta)=\mathcal{N}(\boldsymbol{\theta}|\boldsymbol{\mu},\Sigma;\boldsymbol{\alpha},\beta), \tag{13.58}$$

with

$$\boldsymbol{\mu}=\beta\Sigma\Phi^T\boldsymbol{y},\quad \Sigma=\left(A+\beta\Phi^T\Phi\right)^{-1}. \tag{13.59}$$

The objective now becomes to maximize with respect to $\alpha_k,\ k=0,\dots,K-1$, and β the cost function

$$\begin{aligned} L(\boldsymbol{\alpha},\beta) &:= \ln p(\boldsymbol{y};\boldsymbol{\alpha},\beta) \\ &= -\frac{N}{2}\ln(2\pi)-\frac{1}{2}\ln|\beta^{-1}I+\Phi A^{-1}\Phi^T| \\ &\quad -\frac{1}{2}\boldsymbol{y}^T\left(\beta^{-1}I+\Phi A^{-1}\Phi^T\right)^{-1}\boldsymbol{y}. \end{aligned} \tag{13.60}$$

Maximizing the above cost cannot be carried out analytically, and the following iterative scheme is derived (Problem 13.11, the proof is a bit tedious):

$$\gamma_k = 1-\alpha_k^{(\text{old})}\Sigma_{kk}^{(\text{old})}, \tag{13.61}$$

$$\alpha_k^{(\text{new})} = \frac{\gamma_k}{(\mu_k^{(\text{old})})^2},\quad k=0,1,\dots,K-1, \tag{13.62}$$

$$\beta^{(\text{new})} = \frac{N-\sum_{k=0}^{K-1}\gamma_k}{||\boldsymbol{y}-\Phi\boldsymbol{\mu}^{(\text{new})}||^2}. \tag{13.63}$$

The iterative scheme is initialized by an arbitrary set of values and it is repeated until a convergence criterion is met; Σ_{kk} is the respective diagonal element of the matrix Σ. Note that both Σ and $\boldsymbol{\mu}$ depend on the values of β and α_k. The main complexity per iteration step is due to the matrix inversion involved in the respective definition in (13.59), which amounts to $O(K^3)$ operations. Moreover, because a matrix inversion is involved, one must take care of near singularities, due to numerical errors. This can be the case in practice, because some of the values of α_k may become very large. Thus, care must be taken so that once such values occur, one removes the corresponding columns in Φ and sets the respective values of θ_k to zero. As a matter of fact, this is how sparsity is enforced by the method. Parameters with mean value equal to zero and a variance that becomes very small (precision very large) are set to zero. This behavior has empirically been observed in practice.

The alternative path to deal with the method is via the EM algorithm. This leads to an equivalent set of recursions [85], but practical experience has shown that the previously given set of updates converge faster.

Extensions of the SBL framework in the context of block sparsity and for the case of multiple measurement vectors (MMVs), when elements in each nonzero row of the solution matrix are temporally correlated, are reported in [96,97]. Moreover, in the latter one, a theoretical analysis is provided, which shows that the SBL cost function has the very desirable property that its global minimum coincides with the sparsest solution to the MMV problem.

Example 13.2. The goal of this example is to demonstrate the comparative performance, via a simulation example, of (a) the variational Bayesian method, (b) the maximum likelihood/LS (12.6), and (c) the EM algorithm of Section 12.5 in the context of linear regression and in particular in the sparse modeling framework. The SBL method gave results very similar to the variational approach, and it is not discussed any further. To this end, we generated the training data according to the following scenario.

The interval in the real axis $[-10,\ 10]$ was sampled at $N = 100$ equidistant points x_n, $n = 1, 2, \ldots,$ 100. The training data comprise the pairs (y_n, x_n), $n = 1, 2, \ldots, N$, where

$$y_n = \exp\left(-\frac{1}{2}\frac{(x_n + 5.8)^2}{0.1}\right) + \exp\left(-\frac{1}{2}\frac{(x_n - 2.6)^2}{0.1}\right) + \eta_n,$$

where η_n are i.i.d. zero mean Gaussian noise samples of variance $\sigma_\eta^2 = 0.015$. To fit the data, the following model was adopted:

$$y = \sum_{k=1}^{N} \theta_k \exp\left(-\frac{1}{2}\frac{(x - x_k)^2}{0.1}\right).$$

Thus, the matrix Φ has the following elements:

$$[\Phi]_{nk} = \exp\left(-\frac{1}{2}\frac{(x_n - x_k)^2}{0.1}\right), \quad n = 1, 2, \ldots, N,\ k = 1, 2, \ldots, N.$$

Note that we have as many parameters as the number of training points. This is in line with the relevance vector machine rationale, which will be discussed in Section 13.7. Fig. 13.7 illustrates the results. The red full-line curve corresponds to the true function that generates the data. The gray full curve corresponds to the model, having plugged in as estimated values $\hat{\theta}_k$ the respective posterior mean values from Eq. (13.34). The dotted red curve corresponds to the ML solution and the dotted gray curve to the EM, where the estimates correspond to the mean of the respective posterior (Eq. (12.44)). The performance advantages of the variational approach are obvious, which almost coincide with the true one. Observe how the variational Bayesian approach managed to cope with the overfitting and pushed most of the parameters to zero values.

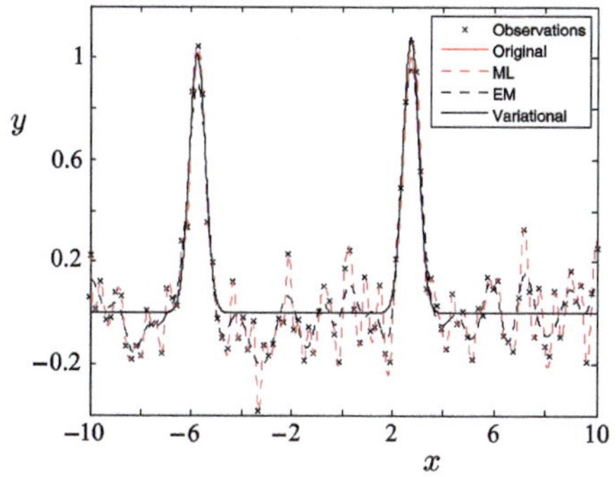

FIGURE 13.7

The figure corresponds to the setup of Example 13.2. Observe that the fitting curve obtained via the variational method is almost identical to the true one.

13.6.1 THE SPIKE AND SLAB METHOD

This is an old technique for imposing sparsity [42,50]. Let us consider our familiar regression model,

$$\mathrm{y} = \boldsymbol{\theta}^T \boldsymbol{\phi}(\boldsymbol{x}) + \eta = \sum_{k=0}^{K-1} \theta_k \phi_k(\boldsymbol{x}) + \eta. \tag{13.64}$$

A new set of auxiliary binary *indicator* variables are introduced, $\mathrm{s}_k \in \{0, 1\}$, $k = 0, 1, \dots, K-1$. Let also the prior imposed on $\boldsymbol{\theta}$ be a Gaussian, $p(\boldsymbol{\theta}) = \mathcal{N}(\boldsymbol{\theta}|\mathbf{0}, \sigma^2 I)$. As the name suggests, the indicator variables control the presence or absence of a parameter in the summation in Eq. (13.64). For example, if $s_k = 1$, then the corresponding parameter θ_k is present, and if $s_k = 0$, then θ_k is removed; this is the way sparsity is imposed onto the model. To this end, a joint Bernoulli prior distribution (Chapter 2) is adopted for the indicator variables, i.e.,

$$P(\boldsymbol{s}) = \prod_{k=0}^{K-1} p^{s_k}(1-p)^{1-s_k}, \tag{13.65}$$

where the parameter $0 \le p \le 1$ specifies a prior level of sparsity. This turns out to be equivalent to adopting the following prior on the parameters:

$$p(\boldsymbol{\theta}) = \prod_{k=0}^{K-1} \left(s_k \mathcal{N}(\theta_k|0, \sigma^2) + (1 - s_k)\delta(\theta_k) \right) : \quad \text{spike and slab prior.} \tag{13.66}$$

The latter is known as the *spike and slab* prior. The name comes from the fact that if $s_k = 0$, then a "spike" is imposed at the zero and the values $s_k = 1$ impose a "slab," because a Gaussian is a broad one (for large enough σ^2). The corresponding posterior is not Gaussian and its computation can be done by mobilizing approximate inference techniques, such as variational or Monte Carlo (see, e.g., [29] and the references therein).

Variants of the basic spike and slab scheme do also exist (see, e.g., [78]). In the latter reference, it is shown that one can obtain the classical ℓ_0-based sparsity enforcing constraint on the LS criterion (Chapter 9) as a limiting case of one of these variants. Such a path provides another connection between probabilistic and optimization-based techniques for sparsity. Another connection will be discussed in Section 13.9.

13.7 THE RELEVANCE VECTOR MACHINE FRAMEWORK

An important aspect of the work in [85] was the introduction of *relevance vector machines* for regression as well as for classification. Inspired by the support vector regression (SVR), which was discussed in Chapter 11, a specific regression model was considered, that is,

$$y(\mathbf{x}) = \theta_0 + \sum_{k=1}^{N} \theta_k \kappa(\mathbf{x}, \boldsymbol{x}_k) + \eta. \tag{13.67}$$

In other words, the general regression model of Eq. (12.1) is considered for $K = N + 1$, where N is the number of observations and

$$\phi_k(\mathbf{x}) = \kappa(\mathbf{x}, \boldsymbol{x}_k),$$

where $\kappa(\cdot, \cdot)$ is a kernel function, as defined in Chapter 11, centered at the input observation points, $\boldsymbol{x}_k$, $k = 1, 2, \ldots, N$. Thus, the number of parameters becomes equal (plus one) to the number of training points.

The task can be treated either via the SBL philosophy or via the employment of the variational approximation rationale, in order to impose sparsity. In [85], it is pointed out that the variational Bayesian approach is computationally more intensive and in practice it results in mean values for the hyperparameters, which are identical to the values obtained by using the SBL approach.

Inspired by the definition of the support vectors in the SVR, the surviving data points that contribute to Eq. (13.67) are called *relevance vectors*. Also, the kernels to be used in the RVM framework need not be symmetric positive definite functions, because the modeling is not necessarily associated to a reproducing kernel Hilbert space (RKHS).

13.7.1 ADOPTING THE LOGISTIC REGRESSION MODEL FOR CLASSIFICATION

Besides the relevance vector regression, the relevance vector classification was also introduced in [85]. Recall that in the support vector machine (SVM) classification, a linear (in an RKHS) classifier was designed. The same model is also adopted for the RVM. Given the value of a feature vector, $\boldsymbol{x}$, classification is performed according to the sign of the discriminant function, namely,

$$f(\boldsymbol{x}) := \boldsymbol{\theta}^T \boldsymbol{\phi}(\boldsymbol{x}) := \theta_0 + \sum_{k=1}^{N} \theta_k \phi_k(\boldsymbol{x}).$$

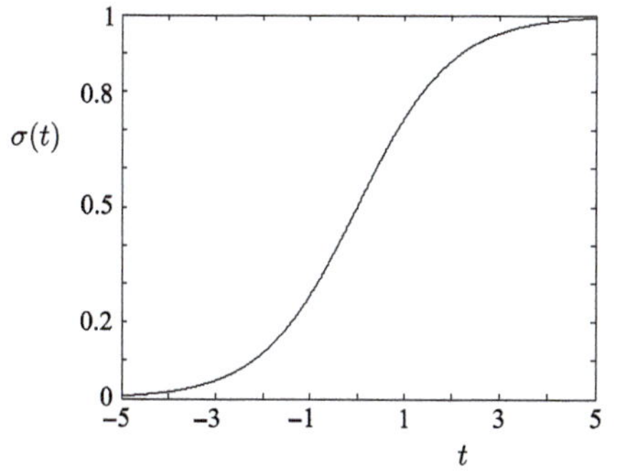

FIGURE 13.8

The logistic sigmoid function.

The goal is to obtain an estimate of the parameters $\boldsymbol{\theta}$ in the Bayesian framework; thus, somehow, we have to "embed" $\boldsymbol{\theta}$ into a PDF that relates the input–output data. In this vein, a well-known and widely used technique is the *logistic regression* model, which was introduced in Section 7.6.

According to this model and for a two-class (ω_1, ω_2) classification task, the posterior probabilities, as required by the Bayesian classifier, are modeled as

$$\boxed{P(\omega_1|\boldsymbol{x}) = \frac{1}{1+\exp\left(-\boldsymbol{\theta}^T\boldsymbol{\phi}(\boldsymbol{x})\right)}: \quad \text{logistic regression model,}} \tag{13.68}$$

and

$$P(\omega_2|\boldsymbol{x}) = 1 - P(\omega_1|\boldsymbol{x}). \tag{13.69}$$

There is more than one reason that justifies such a choice (see, e.g., [46] and Problem 13.12). Multiclass generalizations are also possible (e.g., [84] and Chapter 7).

For the sake of the less familiar reader, let us look at Eq. (13.68) more closely. The graph of the function

$$\sigma(t) = \frac{1}{1+\exp(-t)}, \tag{13.70}$$

known as the *logistic sigmoid* function, is shown in Fig. 13.8. For $t > 0$ $(\boldsymbol{\theta}^T\boldsymbol{\phi}(\boldsymbol{x}) > 0)$, $P(\omega_1|\boldsymbol{x}) > \frac{1}{2}$ and the decision is in favor of ω_1. The opposite holds true for $t < 0$ $(\boldsymbol{\theta}^T\boldsymbol{\phi}(\boldsymbol{x})) < 0)$. Considering the training set $(y_n, \boldsymbol{x}_n), \boldsymbol{x}_n \in \mathbb{R}^l$ and $y_n \in \{0, 1\}$, and adopting a Bernoulli distribution for $P(y|\boldsymbol{x})$, the respective likelihood function can be defined as

$$P(\boldsymbol{y}|\boldsymbol{\theta}) = \prod_{n=1}^{N}\left(\sigma\left(\boldsymbol{\theta}^T\boldsymbol{\phi}(\boldsymbol{x}_n)\right)\right)^{y_n}\left(1-\sigma\left(\boldsymbol{\theta}^T\boldsymbol{\phi}(\boldsymbol{x}_n)\right)\right)^{1-y_n}, \tag{13.71}$$

which is the counterpart of Eq. (13.22) for the regression case. We also adopt a Gaussian prior for $\boldsymbol{\theta}$, as in Eqs. (13.55) and (13.56). As in the SBL approach, our goal is to maximize the Type II log-likelihood with respect to the unknown parameters, $\boldsymbol{\alpha}$. However, $p(\boldsymbol{y}|\boldsymbol{\theta})$ is no longer Gaussian, and marginalizing out $\boldsymbol{\theta}$ cannot be carried out analytically. In [85], the Laplacian approximation is employed and the following stepwise procedure is adopted:

1. Assuming $\boldsymbol{\alpha}$ to be currently available, maximize with respect to $\boldsymbol{\theta}$ the posterior, which by simple arguments is easily shown to be

$$p(\boldsymbol{\theta}|\boldsymbol{y},\boldsymbol{\alpha}) = \frac{P(\boldsymbol{y}|\boldsymbol{\theta})p(\boldsymbol{\theta}|\boldsymbol{\alpha})}{P(\boldsymbol{y}|\boldsymbol{\alpha})},$$

or equivalently,

$$\begin{aligned}
\hat{\boldsymbol{\theta}}_{\text{MAP}} &= \arg\max_{\boldsymbol{\theta}} \ln\left(P(\boldsymbol{y}|\boldsymbol{\theta})p(\boldsymbol{\theta}|\boldsymbol{\alpha})\right) \\
&= \arg\max_{\boldsymbol{\theta}} \left\{\sum_{n=1}^{N}\left[y_n \ln\sigma\left(\boldsymbol{\theta}^T\boldsymbol{\phi}(\boldsymbol{x}_n)\right) + \right.\right. && (13.72) \\
&\quad \left.(1-y_n)\ln\left(1-\sigma\left(\boldsymbol{\theta}^T\boldsymbol{\phi}(\boldsymbol{x}_n)\right)\right)\right] - \\
&\quad \left.\frac{1}{2}\boldsymbol{\theta}^T A\boldsymbol{\theta} + \text{constant}\right\}, && (13.73)
\end{aligned}$$

where $A := \text{diag}\{\alpha_0, \alpha_2, \ldots, \alpha_N\}$. Maximizing Eq. (13.73) with respect to $\boldsymbol{\theta}$ results in (Problem 13.13)

$$\boxed{\hat{\boldsymbol{\theta}}_{\text{MAP}} = A^{-1}\Phi^T(\boldsymbol{y}-\boldsymbol{s}),} \tag{13.74}$$

where $\boldsymbol{s} := [s_1, \ldots, s_N]^T$ and $s_n := \sigma\left(\boldsymbol{\theta}^T\boldsymbol{\phi}(\boldsymbol{x}_n)\right)$, $n = 1, 2, \ldots, N$.

2. Use $\hat{\boldsymbol{\theta}}_{\text{MAP}}$ and the Laplace approximation method (Section 12.3) to approximate $p(\boldsymbol{\theta}|\boldsymbol{y},\boldsymbol{\alpha})$ by a Gaussian centered at $\hat{\boldsymbol{\theta}}_{\text{MAP}}$ [45]. Recall from Section 12.3 that the covariance matrix of the approximate Gaussian is given by

$$\Sigma^{-1} = -\left.\frac{\partial^2 \ln\left(P(\boldsymbol{y}|\boldsymbol{\theta})p(\boldsymbol{\theta}|\boldsymbol{\alpha})\right)}{\partial\boldsymbol{\theta}^2}\right|_{\boldsymbol{\theta}=\hat{\boldsymbol{\theta}}_{\text{MAP}}},$$

or (Problem 13.14)

$$\boxed{\Sigma^{-1} = (\Phi^T T\Phi + A),} \tag{13.75}$$

where $T = \text{diag}\{t_1, t_2, \ldots, t_N\}$ and

$$t_n = \sigma\left(\boldsymbol{\theta}^T\boldsymbol{\phi}(\boldsymbol{x}_n)\right)\left(1-\sigma\left(\boldsymbol{\theta}^T\boldsymbol{\phi}(\boldsymbol{x}_n)\right)\right)\Big|_{\boldsymbol{\theta}=\hat{\boldsymbol{\theta}}_{\text{MAP}}}.$$

3. Having obtained $\hat{\boldsymbol{\theta}}_{\text{MAP}}$ and computed Σ, adapting Eq. (12.37) to our current notation we obtain

$$P(\boldsymbol{y}|\boldsymbol{\alpha}) = P(\boldsymbol{y}|\hat{\boldsymbol{\theta}}_{\text{MAP}})p(\hat{\boldsymbol{\theta}}_{\text{MAP}}|\boldsymbol{\alpha})(2\pi)^{\frac{N}{2}}|\Sigma|^{1/2}. \tag{13.76}$$

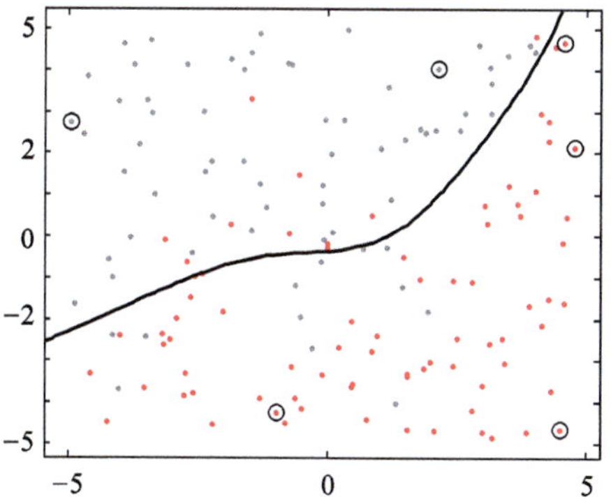

FIGURE 13.9

The decision curve that separates the two classes (red versus gray), which is obtained by the RVM classifier, corresponds to posterior probability values $P(\omega_1|\boldsymbol{x}) = 0.5$. The Gaussian kernel was used with $\sigma^2 = 3$. Only six relevance vectors survive—the ones that have been circled.

Next, maximization of Eq. (13.76) with respect to $\boldsymbol{\alpha}$ provides the updated iteration estimate. Note that the first term of the product on the right-hand side is independent of $\boldsymbol{\alpha}$. Taking the logarithm and maximizing easily results in (Problem 13.15)

$$-\frac{1}{2}\theta_{\mathrm{MAP},k}^2 + \frac{1}{2\alpha_k} - \frac{1}{2}\Sigma_{kk} = 0. \tag{13.77}$$

Because Σ_{kk} and $\theta_{\mathrm{MAP},k}$ depend on $\boldsymbol{\alpha}$, the equation is solved iteratively and results in exactly the same scheme as in Eq. (13.62), that is,

$$\alpha_k^{(\mathrm{new})} = \frac{1 - \alpha_k^{(\mathrm{old})}\Sigma_{kk}^{(\mathrm{old})}}{\left(\theta_{\mathrm{MAP},k}^{(\mathrm{old})}\right)^2}.$$

The procedure continues until a convergence criterion is met [44,85].

As pointed out in [85], although in general the Laplacian local approximation to a Gaussian may not be a good one, in the case of the current classification task, due to the specific nature of the adopted models, the approximation is expected to provide good accuracy.

Fig. 13.9 shows the decision curve that results from the RVM method[4] and classifies the points of the red/gray classes. The data set is the same as the one used in Example 11.4 of Chapter 11 when dealing with the SVM classifier. Six points, which have been circled, are the surviving relevance vectors. The Gaussian kernel was used with $\sigma^2 = 3$, which was found to give the best results. Observe that the number of support vectors surviving is significantly less compared to the case of SVM of Chapter 11.

[4] The software used was that from http://www.miketipping.com/sparsebayes.htm#software.

Remarks 13.2.

- Compared to SVM (SVR), the RVM machinery presents advantages and disadvantages. The SVM approach has the mathematically elegant property of, theoretically, giving a single minimum due to the convexity of the associated cost functions. This is not the case for the RVM framework, where the involved optimization steps refer to a nonconvex cost. It must be kept in mind that solving a nonconvex task, one may have to run the optimization algorithm a number of times, starting each time from different initial conditions, because a nonconvex problem can be trapped in a local minimum.

  Concerning complexity, the algorithmic steps for the RVM involve the inversion of the Hessian matrix, which amounts to $O(N^3)$ complexity. As discussed in Section 11.11, the complexity range of the efficient schemes for solving the SVM scales from linear to (approximately) quadratic. Also, the memory for the RVM exhibits an $O(N^2)$ dependence on the size of the training set, as opposed to a linear dependence in the SVM case. Besides complexity, inverting (big) matrices must be done with care in order to avoid numerical instabilities due to possible (near) singularity. Also, in general, RVMs need longer training times to converge, compared to SVMs, for similar error rates.

  A fast RVM algorithm has been developed in [86] by analyzing the properties of the marginal likelihood. This enables a sequential addition and deletion of candidate basis functions (columns of Φ) to monotonically maximize the marginal likelihood. This iterative algorithm operates in a constructive manner, until all relevant basis functions (for which the associated weights are nonzero) have been included. If M denotes the number of relevant terms, the complexity amounts to $O(M^3)$, which for $M << N$ is more efficient than the original RVM.

  The main advantage of RVMs is that, in general, they result in sparser solutions compared to SVMs for similar levels of generalization errors. This makes the prediction step, after the training has been completed, more efficient compared to the prediction model resulting from SVM. Moreover, SVMs suffer from their dependence on the user dependent hyperparameter, C (ϵ for regression), and they are generally found by cross-validation, which involves multiple training for different values.
- In [8], a different algorithmic approach has been adopted based on the *variational bound approximation* method, to be described next.

13.8 CONVEX DUALITY AND VARIATIONAL BOUNDS

In the previous chapter, the Laplacian technique for the approximation of a general PDF by a Gaussian was introduced. The driving force behind such an approximation was to benefit from the computationally friendly nature of the Gaussian PDF. In this section, we will approach this task from a different perspective, involving maximization of a lower bound of the PDF at hand with respect to an extra parameter, which is introduced into the problem and on which the lower bound depends. Our theoretical framework is that of convex duality, a well-known and powerful tool in convex analysis.

Let a function $f : \mathbb{R}^l \longmapsto \mathbb{R}$. The function

$$f^* : \mathbb{R}^l \longmapsto \mathbb{R},$$

defined as[5]

$$f^*(\boldsymbol{\xi}) = \max_{\boldsymbol{x}} \left\{ \boldsymbol{\xi}^T \boldsymbol{x} - f(\boldsymbol{x}) \right\}, \tag{13.78}$$

is called the *conjugate* of f. The domain of the conjugate function consists of all $\boldsymbol{\xi} \in \mathbb{R}^l$ for which the maximum is finite. A notable property of the conjugate function is its *convexity*; this is true whether or not f is convex. The convexity is the outcome of the point-wise maximization of a family of (convex with respect to $\boldsymbol{\xi}$) affine functions [12].

Maximizing Eq. (13.78) with respect to $\boldsymbol{x}$ results in a value of $\boldsymbol{x}_*$, such that

$$\boldsymbol{x}_* : \nabla f(\boldsymbol{x}_*) = \boldsymbol{\xi}, \tag{13.79}$$

which leads to the value

$$f^*(\boldsymbol{\xi}) = \boldsymbol{\xi}^T \boldsymbol{x}_* - f(\boldsymbol{x}_*). \tag{13.80}$$

Eqs. (13.79) and (13.80) provide the geometric interpretation of the conjugate function. The graph of the linear function $\boldsymbol{\xi}^T \boldsymbol{x}$ defines a hyperplane whose direction is controlled by $\boldsymbol{\xi}$; the latter is now equal to $\nabla f(\boldsymbol{x}_*)$, which defines the direction of the tangent hyperplane of the graph of $f(\boldsymbol{x})$ at $\boldsymbol{x}_*$. This tangent hyperplane is described by

$$g(\boldsymbol{x}) = f(\boldsymbol{x}_*) + (\boldsymbol{x} - \boldsymbol{x}_*)^T \nabla f(\boldsymbol{x}_*),$$

or using Eq. (13.80),

$$g(\boldsymbol{x}) = \boldsymbol{\xi}^T \boldsymbol{x} - f^*(\boldsymbol{\xi}). \tag{13.81}$$

For $\boldsymbol{x} = 0$, Eq. (13.81) becomes $g(\boldsymbol{0}) = -f^*(\boldsymbol{\xi})$. This is illustrated in Fig. 13.10. Thus, $f^*(\boldsymbol{\xi})$ corresponds to the displacement that the graph of $\boldsymbol{\xi}^T \boldsymbol{x}$ has to undergo in order to "touch" that of $f(\boldsymbol{x})$.

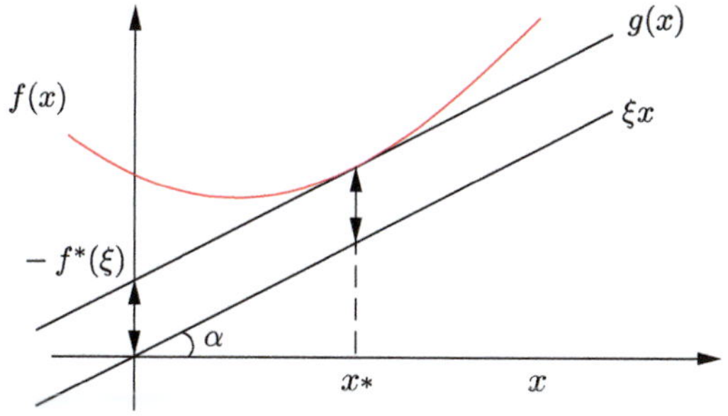

FIGURE 13.10

The direction of the line $y = \xi x$ that crosses the origin is controlled by ξ ($\xi = \tan\alpha$). The (negative) value of the conjugate function f^* at ξ defines the point where the line $y = g(x)$ cuts the vertical axis; $g(x)$ is formed by translating ξx until it becomes tangent to f at x_*.

[5] Strictly speaking, one should use the sup instead of max and inf instead of min, throughout.

A by-product of all this turns out to be very useful for us. It can be shown (Problem 13.16) that if f is a convex function, then $(f^*)^* = f$ and in this case we can write

$$f(\boldsymbol{x}) = \max_{\boldsymbol{\xi}} \left\{ \boldsymbol{x}^T \boldsymbol{\xi} - f^*(\boldsymbol{\xi}) \right\}. \tag{13.82}$$

Thus, once f^* is computed, a lower bound for f becomes readily available, that is,

$$f(\boldsymbol{x}) \geq \boldsymbol{x}^T \boldsymbol{\xi} - f^*(\boldsymbol{\xi}), \tag{13.83}$$

where now $\boldsymbol{\xi}$ is interpreted as a parameter. To investigate this bound a bit further, we plug Eqs. (13.79) and (13.80) into Eq. (13.83) to obtain[6]

$$f(\boldsymbol{x}) \geq f(\boldsymbol{x}_*) + (\boldsymbol{x} - \boldsymbol{x}_*)^T \nabla f(\boldsymbol{x}_*),$$

where the right-hand side is the linear function $g(\boldsymbol{x})$ describing the hyperplane tangent to $f(\boldsymbol{x})$ at $\boldsymbol{x}_*$. The bound becomes tight at $\boldsymbol{x} = \boldsymbol{x}_*$ (see Fig. 13.10). We will soon see how we can make this linear function bound a nonlinear one; it suffices to transform the argument of the function.

All that has been said for convex functions applies to concave ones if we replace the max operation in Eqs. (13.78) and (13.82) with min operations. Note that following this definition, the conjugate function is *concave*, being the result of point-wise minimization of a set of concave functions (an affine function can be considered either convex or concave). Furthermore, if the involved function is neither convex nor concave, one can search for *invertible* transformations that render it convex or concave.

In our context, the purpose of resorting to the notion of the conjugate function is our expectation that such a function, which bounds a (PDF) function as in Eq. (13.83), may lead to a functional form that lends itself to tractable computations of the involved integrations.

Example 13.3. Compute the conjugate of the logarithmic function $f(x) = \ln x, x > 0$.

The logarithmic function is known to be concave. Hence

$$f^*(\xi) = \min_{x>0} \{\xi x - \ln x\},$$

or

$$x_* : \frac{1}{x} = \xi \Rightarrow x_* = \frac{1}{\xi}.$$

Hence,

$$f^*(\xi) = 1 + \ln \xi.$$

Therefore,

$$\ln x = \min_{\xi>0} \{\xi x - 1 - \ln \xi\}.$$

Fig. 13.11 shows the respective graphs of the logarithmic function as well as the resulting *linear* function bound for different values of ξ.

[6] Note that this is a necessary and sufficient condition for convexity.

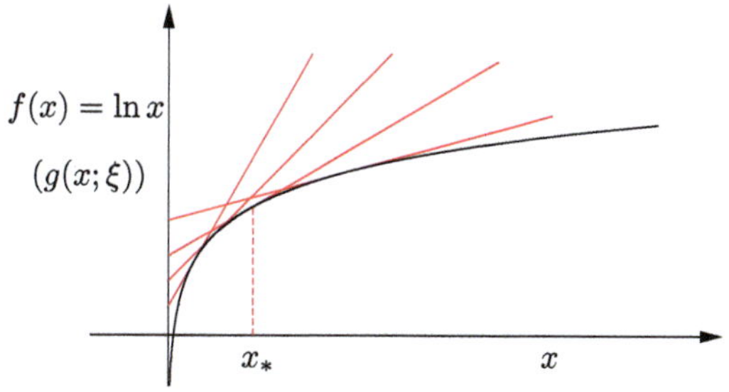

FIGURE 13.11

The linear functions $g(x;\xi) = \xi x - 1 - \ln \xi$ provide upper bounds to $f(x) = \ln x$. Each one of the lines is tangent to $f(x) = \ln x$ at the point $x_* = \frac{1}{\xi}$.

Example 13.4. Consider the univariate Laplacian PDF, which we have already seen in Section 13.5,

$$p(\theta) = \frac{\lambda}{2}\exp(-\lambda|\theta|), \quad \theta \in \mathbb{R}. \tag{13.84}$$

Our goal is to derive a lower bound in terms of its conjugate function. From Eq. (13.84), we get

$$\ln p(\theta) = \ln \frac{\lambda}{2} - \lambda|\theta|.$$

Define

$$f(x) = \ln \frac{\lambda}{2} - \lambda\sqrt{x}, \quad x > 0. \tag{13.85}$$

Then

$$\ln p(\theta) = f(\theta^2). \tag{13.86}$$

Note that $f(x)$ is a convex function with respect to x (Problem 13.16). The conjugate of $f(x)$ is obtained as[7]

$$f^*(\xi) = \max_x \left\{-\frac{\xi}{2}x - f(x)\right\}, \ \xi > 0. \tag{13.87}$$

ξ is constrained to positive values, because for $\xi \leq 0$ the maximum with respect to x becomes infinite, which violates the definition of the conjugate functions. Recalling Eq. (13.85), maximization leads to

$$x_* : \lambda x^{-\frac{1}{2}} = \xi \Rightarrow x_* = \lambda^2\xi^{-2}. \tag{13.88}$$

---

[7] Because maximization takes place for all ξ, we use $-\frac{\xi}{2}$. This is only for notational convenience in order to obtain the result in a convenient form.

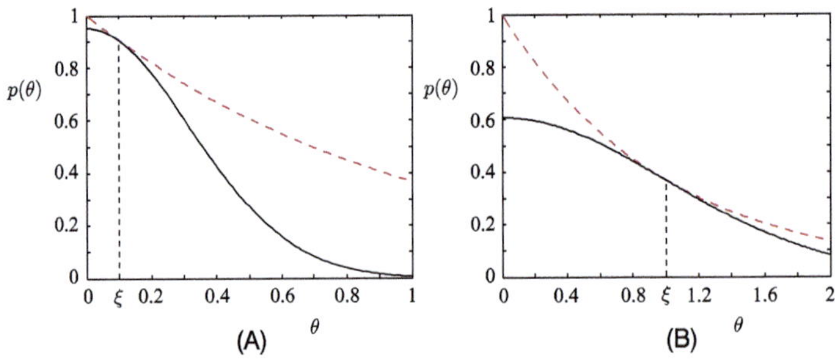

FIGURE 13.12

The Laplacian (red curves) and the approximating Gaussians for two different values of ξ.

Combining Eqs. (13.87) and (13.88) gives

$$f^*(\xi) = \frac{\lambda^2}{2}\xi^{-1} - \ln\frac{\lambda}{2}. \tag{13.89}$$

Hence, we obtain the bound

$$f(x) \geq -\frac{\xi}{2}x - \frac{\lambda^2}{2}\xi^{-1} + \ln\frac{\lambda}{2},$$

or

$$\ln p(\theta) \geq -\frac{\xi}{2}\theta^2 - \frac{\lambda^2}{2}\xi^{-1} + \ln\frac{\lambda}{2}, \quad \xi > 0.$$

Because this is true $\forall \xi > 0$, we can replace ξ with ξ^{-1}, for notational convenience, which results in

$$p(\theta) \geq \frac{\lambda}{2}\exp\left(-\frac{\xi^{-1}}{2}\theta^2\right)\exp\left(-\frac{\lambda^2}{2}\xi\right), \tag{13.90}$$

which, after mobilizing the Gaussian notation and its integration property, can be rewritten as

$$p(\theta) \geq \mathcal{N}(\theta|0, \xi)\phi(\xi), \quad \xi > 0, \tag{13.91}$$

with

$$\phi(\xi) = \frac{\lambda}{2}\sqrt{2\pi\xi}\exp\left(-\frac{\lambda^2}{2}\xi\right), \quad \xi > 0.$$

This is very interesting indeed. The obtained lower bound has a functional dependence on θ, which is of a Gaussian nature; the Gaussian term is centered at zero with variance ξ. Maximizing with respect to ξ, we will obtain the required approximation. Fig. 13.12 shows the obtained approximation for different values of ξ. Observe that introducing transformations in the involved variables can render the function in the obtained bound a *nonlinear* one.

For a multivariate Laplacian and assuming a parameter vector with independent components, it can be trivially shown that

$$p(\boldsymbol{\theta}) = \prod_{k=0}^{K-1} p(\theta_k) \geq \mathcal{N}(\boldsymbol{\theta}|\mathbf{0}, \Xi) \prod_{k=0}^{K-1} \phi(\xi_k) := \hat{p}(\boldsymbol{\theta}; \boldsymbol{\xi}), \tag{13.92}$$

where $\boldsymbol{\xi} = [\xi_0, \ldots, \xi_{K-1}]^T$ and

$$\Xi := \text{diag}\{\xi_0, \xi_1, \ldots, \xi_{K-1}\}.$$

Remarks 13.3.

- The method of representing a convex function via the optimization of the lower bound in terms of its conjugate (dual form) is known as the *variational method*, and the associated parameters ξ_k as *variational* parameters [67]. Its use in the context of machine learning was first reported in [31] (see also [35]), and its use subsequently proliferated and was adopted in different scenarios.
- The method has been used to obtain variational approximations for a number of PDFs that are suitable for sparsity-aware learning, for example, Jeffreys', Student's t, generalized Gaussians (see, e.g., [59]), and the logistic regression model [33] (Problem 13.18). Compared to the Laplacian approximation method, the variational approach provides the extra flexibility of optimizing with respect to the corresponding variational parameters (see [33] for a related discussion). The reader, however, has to keep in mind that both approximations need not always be good ones. This is readily observed from Fig. 13.12. One may obtain a good approximation locally, but not everywhere. However, it turns out that in practice, in the context of Bayesian learning, a poor approximation of the prior (for which such approximations are used) does not necessarily lead to a poor approximation of the posterior. There is no guarantee of it, and it is only the performance in practice that has the final verdict. This can be considered a drawback of the Bayesian technique compared to the deterministic methods based on optimization criteria. The latter ones, by adopting convex cost functions, can lead to solutions that are well characterized. In contrast, Bayesian inference techniques suffer from their nonconvex nature and the fact that very often the imposed approximations may not necessarily be good ones. However, this is always the case in life. There is no free lunch. At the time this book was written, both of these paths to machine learning still comprised viable and powerful techniques, with their pros and cons.

13.9 SPARSITY-AWARE REGRESSION: A VARIATIONAL BOUND BAYESIAN PATH

The goal of this section is to demonstrate the use of convex duality and the respective variational bounds in order to approximate the computation of the evidence function in cases where the corresponding integral is intractable. We have chosen to describe the method in the framework of sparsity-aware learning; this can also help us establish bridges with Chapter 9, and some of the results will be used in Section 13.9 for this purpose.

To comply with the assumptions made in Chapter 9, and without loss of generality, let us assume that the involved data have zero mean values. If not, the training data can be centered by subtracting

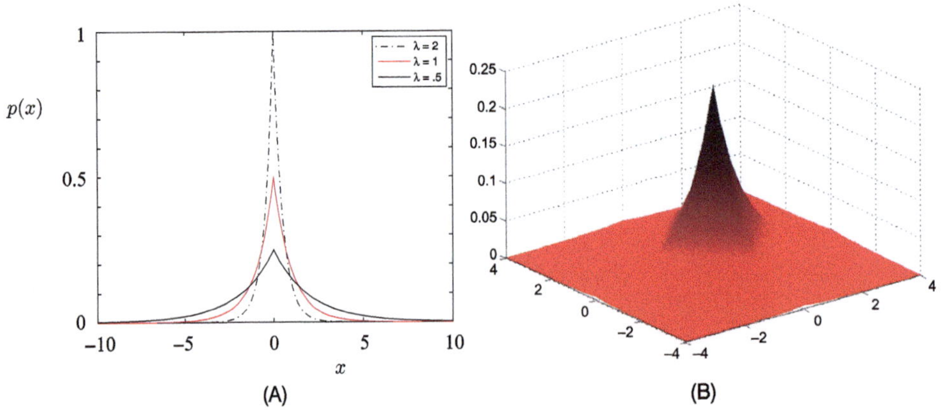

FIGURE 13.13

(A) The one-dimensional Laplacian for different values of λ. (B) A plot of a two-dimensional Laplacian PDF.

their respective sample means; thus, we set $\theta_0 = 0$ and assume that the number of parameters is K. Then our regression model becomes

$$\mathbf{y} = \Phi\boldsymbol{\theta} + \boldsymbol{\eta}, \quad \boldsymbol{\eta}, \mathbf{y} \in \mathbb{R}^N, \ \boldsymbol{\theta} \in \mathbb{R}^K, \ K > N,$$

where we are informed about the "secret" that most of the components of $\boldsymbol{\theta}$ are (almost) zero. In Section 13.5, we commented on the inadequacy of a Gaussian prior to provide a reasonable statistical description of a sparse random vector. A heavy-tailed distribution that enjoys popularity for sparse modeling is the Laplacian one. After all, adopting a Laplacian prior to $\boldsymbol{\theta}$, that is,

$$p(\boldsymbol{\theta}) = \prod_{k=1}^{K} p(\theta_k) = \prod_{k=1}^{K} \frac{\lambda}{2} \exp(-\lambda|\theta_k|),$$

and a Gaussian conditional PDF, $p(\mathbf{y}|\boldsymbol{\theta})$, for the observations $\mathbf{y}$ as in Eq. (13.22), makes the MAP estimation identical to our familiar LASSO task, discussed in Chapter 9. In this vein, we will build this section around the Laplacian PDF. Fig. 13.13A shows the Laplacian for different values of λ. In 13.13B the two-dimensional plot is provided, from which the respect that this PDF shows to sparse solutions is readily observed.

The problem with the Laplacian PDF is that its presence in Eq. (12.14) makes the computation of the integral computationally intractable. Also, recall from Section 12.8 that the Laplacian PDF does not belong to the computationally attractive exponential family. To facilitate its treatment, we will employ the variational bound approximation method in order to approximate the Laplacian by a Gaussian, following Eqs. (13.90) and (13.92). The variational parameters will be determined by maximizing the respective evidence via the EM algorithm. This method in recovering sparse solutions was first introduced in [21] in the context of dictionary learning.

For simplicity, let the noise sequence in our regression model be white with variance $\sigma_\eta^2 := \frac{1}{\beta}$. Then

$$p(\mathbf{y}|\boldsymbol{\theta};\beta) = \mathcal{N}(\mathbf{y}|\Phi\boldsymbol{\theta}, \beta^{-1}I),$$

and using Eq. (13.92), we can write

$$\begin{aligned} p(\mathbf{y};\beta) &= \int \mathcal{N}(\mathbf{y}|\Phi\boldsymbol{\theta}, \beta^{-1}I)p(\boldsymbol{\theta})\,d\boldsymbol{\theta} \\ &\geq \left(\int \mathcal{N}(\mathbf{y}|\Phi\boldsymbol{\theta}, \beta^{-1}I)\mathcal{N}(\boldsymbol{\theta}|\mathbf{0}, \Xi)\,d\boldsymbol{\theta}\right)\prod_{k=1}^{K}\phi(\xi_k). \end{aligned}$$

We know by now that the integral results in a new Gaussian function (recall Eq. (12.15)), and hence

$$p(\mathbf{y};\beta) \geq \mathcal{N}(\mathbf{y}|\mathbf{0}, \beta^{-1}I + \Phi\Xi\Phi^T)\prod_{k=1}^{K}\phi(\xi_k) := \hat{p}(\mathbf{y};\beta,\Xi). \tag{13.93}$$

The unknown values of β and Ξ could be obtained by direct maximization of the previous bound. However, here we will adopt an EM algorithm approach, in a similar way as in Section 12.5; the difference here lies in the existence of the multiplicative terms $\phi(\xi_k)$ that differentiates the M-step. In order to employ the EM, we need to know the posterior $p(\boldsymbol{\theta}|\mathbf{y};\beta)$. We will accept the following,

$$p(\boldsymbol{\theta}|\mathbf{y};\beta) \simeq \hat{p}(\boldsymbol{\theta}|\mathbf{y};\beta,\Xi) := \frac{\mathcal{N}(\mathbf{y}|\Phi\boldsymbol{\theta}, \beta^{-1}I)\mathcal{N}(\boldsymbol{\theta}|\mathbf{0}, \Xi)}{\int \mathcal{N}(\mathbf{y}|\Phi\boldsymbol{\theta}, \beta^{-1}I)\mathcal{N}(\boldsymbol{\theta}|\mathbf{0}, \Xi)\,d\boldsymbol{\theta}}, \tag{13.94}$$

where in place of $p(\boldsymbol{\theta})$ we have used its respective bound, $\hat{p}(\boldsymbol{\theta};\boldsymbol{\xi})$, from Eq. (13.92). Note that irrespective of which method one adopts to optimize with respect to the unknown parameters, the approximate posterior given in Eq. (13.94) is the quantity of interest in regression; this is used either to predict $\boldsymbol{\theta}$ or to perform predictions of the output value (Eq. (12.18)).

It must be stressed, however, that Eq. (13.94) is not a bound of $p(\boldsymbol{\theta}|\mathbf{y};\beta)$ anymore, because normalization has taken place and division does not necessarily respect bounds. Recalling what we said in Section 12.2.2 (Eqs. (12.27) and (12.28)) we obtain

$$\hat{p}(\boldsymbol{\theta}|\mathbf{y};\beta,\Xi) = \mathcal{N}(\boldsymbol{\theta}|\boldsymbol{\mu}_{\theta|y}, \Sigma_{\theta|y}), \tag{13.95}$$

where

$$\boldsymbol{\mu}_{\theta|y} = \Xi\Phi^T\left(\frac{1}{\beta}I + \Phi\Xi\Phi^T\right)^{-1}\mathbf{y}, \tag{13.96}$$

$$\Sigma_{\theta|y} = \Xi - \Xi\Phi^T\left(\frac{1}{\beta}I + \Phi\Xi\Phi^T\right)^{-1}\Phi\Xi. \tag{13.97}$$

We are now ready to give the algorithmic steps. Recall that in EM, the goal is to maximize the expected value of the complete log-likelihood with respect to the unknown set of deterministic parameters. In our case, our goal will be to maximize the corresponding bound with respect to β and $\boldsymbol{\xi}$,

$$\mathbb{E}\left[\ln p(\mathbf{y}, \boldsymbol{\theta}; \beta)\right] = \mathbb{E}\left[\ln\left(p(\mathbf{y}|\boldsymbol{\theta}; \beta)p(\boldsymbol{\theta})\right)\right] \geq \mathbb{E}\left[\ln\left(p(\mathbf{y}|\boldsymbol{\theta}; \beta)\hat{p}(\boldsymbol{\theta}; \boldsymbol{\xi})\right)\right].$$

Assuming $\boldsymbol{\xi}^{(0)}$, $\beta^{(0)}$ are known, the $(j+1)$th iteration comprises the following computations:

- E-step: From Eq. (13.96) and Eq. (13.97) compute

$$\boldsymbol{\mu}_{\theta|y}^{(j)} \quad \text{and} \quad \Sigma_{\theta|y}^{(j)}.$$

Following similar steps as in Section 12.5, we readily obtain

$$\begin{aligned} Q(\boldsymbol{\xi}, \beta; \boldsymbol{\xi}^{(j)}, \beta^{(j)}) = & \frac{N}{2}\ln\beta - \frac{N}{2}\ln(2\pi) - \frac{\beta}{2}\mathbb{E}_{\theta|y}\left[\|\mathbf{y} - \Phi\boldsymbol{\theta}\|^2\right] \\ & + \sum_{k=1}^{K}\ln\phi(\xi_k) - \frac{K}{2}\ln(2\pi) - \frac{1}{2}\ln|\Xi| \\ & - \frac{1}{2}\sum_{k=1}^{K}\frac{\mathbb{E}_{\theta|y}\left[\theta_k^2\right]}{\xi_k}, \end{aligned} \tag{13.98}$$

where (recall Eq. (12.49))

$$\mathbb{E}_{\theta|y}\left[\|\mathbf{y} - \Phi\boldsymbol{\theta}\|^2\right] = \|\mathbf{y} - \Phi\boldsymbol{\mu}_{\theta|y}^{(j)}\|^2 + \text{trace}\left\{\Phi\Sigma_{\theta|y}^{(j)}\Phi^T\right\},$$

and (recall Eq. (12.47))

$$\mathbb{E}_{\theta|y}\left[\theta_k^2\right] = \left[\boldsymbol{\mu}_{\theta|y}^{(j)}\boldsymbol{\mu}_{\theta|y}^{(j)T} + \Sigma_{\theta|y}^{(j)}\right]_{kk}.$$

- M-Step: Taking the derivative of $Q(\boldsymbol{\xi}, \beta; \boldsymbol{\xi}^{(j)}, \beta^{(j)})$ with respect to β and equating to 0 we get

$$\boxed{\beta^{(j+1)} = \frac{N}{\|\mathbf{y} - \Phi\boldsymbol{\mu}_{\theta|y}^{(j)}\|^2 + \text{trace}\left\{\Phi\Sigma_{\theta|y}^{(j)}\Phi^T\right\}}}. \tag{13.99}$$

The derivation with respect to ξ_k, $k = 1, 2, \ldots, K$, results in (Problem 13.19)

$$\boxed{\xi_k^{(j+1)} = \sqrt{\frac{\mathbb{E}_{\theta|y}\left[\theta_k^2\right]}{\lambda^2}}}, \tag{13.100}$$

which completes the loop. Iterations continue until a termination criterion is met.

An alternative viewpoint that justifies the maximization of the bound of the evidence with respect to the variational parameters is the following (see also [95] for a related discussion). At each iteration step, the EM algorithm maximizes $\mathbb{E}\left[p(\mathbf{y}|\boldsymbol{\theta};\beta)\hat{p}(\boldsymbol{\theta};\boldsymbol{\xi})\right]$ due to the monotonicity of the logarithmic function. Equivalently, this can be seen as the following minimization task:

$$\boldsymbol{\xi} = \arg\min_{\boldsymbol{\xi}} \mathbb{E}\left[p(\mathbf{y}|\boldsymbol{\theta};\beta)|p(\boldsymbol{\theta}) - \hat{p}(\boldsymbol{\theta};\boldsymbol{\xi})|\right], \tag{13.101}$$

where the lower bound property (Eq. (13.92)) has been used in order to involve the absolute value. Looking at Eq. (13.101), one may think of a reason that justifies what in practice is commonly observed; that is, the method results in good performance although the overall approximation of the prior may not be a good one. The important issue is to have a good approximation in values of $\boldsymbol{\theta}$ that correspond to relatively large values of $p(\mathbf{y}|\boldsymbol{\theta})$. The approximation in ranges of $\boldsymbol{\theta}$ where $p(\mathbf{y}|\boldsymbol{\theta}) \approx 0$ does not affect the main goal of the task. Moreover, Eq. (13.101) could also provide a justification of the relative advantage of the variational approximation method compared to the Laplacian method; in the latter, there is no room left to leverage any extra parameters in order to improve the final goal.

Remarks 13.4.

- As we have already commented in Chapter 9, sparsity-aware learning has been a field of intense research. Undoubtedly this is also the case for the Bayesian approach to sparsity promoting models. So far, we presented a hierarchical approach in Section 13.5, where sparsity was indirectly imposed by associating a gamma PDF prior on each one of the precision variables individually; this led to an equivalent high-tail Student's t PDF description of the involved parameters $\boldsymbol{\theta}$. In the current section, a Laplacian prior was imposed on $\boldsymbol{\theta}$ in order to promote sparseness. These are not the only possibilities. We focused on them in order to demonstrate two possible paths to treat the evidence maximization whenever the resulting integral is computationally "awkward."
- In [18], sparsity in the Bayesian framework is attacked by imposing a Gaussian prior on the parameters, treating variance as latent variables with an exponential prior. Such modeling is equivalent to a Laplacian PDF, once variances are integrated out. The EM procedure is then used to compute the required estimates. Also in this paper, the use of Jeffreys' prior ($p(x) \simeq \frac{1}{x}$) is proposed as an alternative to the Laplacian one.
- In [5], sparsity is imposed in a similar way as before, but in the hierarchical model, the parameter controlling the exponential prior of the precisions is also treated as a latent variable with a Jeffreys' prior. In [6], sparsity on the unknown parameters was imposed via a generalized Gaussian PDF, that is,

$$p(\boldsymbol{\theta}|\alpha) \propto \exp\left(-\lambda \sum_{k=1}^{K} |\theta_k|^p\right). \tag{13.102}$$

  Combining this prior with a Gaussian PDF for the conditional, $p(\mathbf{y}|\boldsymbol{\theta})$ in Eq. (13.22), would result in a MAP that corresponds to the LS regularized by a nonconvex ℓ_p, $p<1$, norm; we know that such norms are more aggressive, compared to the ℓ_1 norm, in recovering sparse solutions. For $p=1$, Eq. (13.102) becomes the Laplacian prior. In [6], gamma priors are used in association with the hyperparameter α and the noise variance, and the variational Bayesian approach is used to obtain the solution.
- In [34], the RVM framework is exploited to obtain sparse solutions and the information related to the variance of the obtained estimates is used to determine the number of measurements, which is sufficient for recovering the solution in the framework of compressed sensing.

SPARSITY-AWARE LEARNING: SOME CONCLUDING REMARKS

The task of sparsity-aware learning has been treated in a number of parts in this book. In Chapter 9, it was treated as an optimization problem of a regularized cost function. In Section 13.3 the automatic relevance determination (ARD) concept in Bayesian learning was discussed, and in Section 13.9, the variational bound technique was exploited to overcome the computational obstacle associated with the Laplacian prior.

In this vein, there is a number of questions that are naturally posed. The first one concerns the relationship between the Bayesian and the regularized cost function optimization approaches. How different are they? Are there any paths that establish connections among them? The second question addresses theoretical issues associated with the performance of the Bayesian techniques compared to their counterparts that were discussed in Chapter 9. A first systematic attempt to address both questions was made in [73,74,91,93,94]. Furthermore, another important theoretical question addresses the task of *identifiability* concerning the SBL models [64].

A summary of results concerning the previous tasks can be downloaded from the book' s website under the Additional Material part that is associated with the current chapter.

13.10 EXPECTATION PROPAGATION

Expectation propagation is an alternative to the variational techniques for approximating posterior PDFs. The task of interest is the same as the one treated in the beginning of this chapter, in Section 13.2. Assume that we are given a set of observations $\mathcal{X}$, which are distributed according to $p(\mathcal{X}|\boldsymbol{\theta})$, and a prior, $p(\boldsymbol{\theta})$, corresponding to the set of the unknown parameters[8] $\boldsymbol{\theta}$. The goal is to obtain an estimate of the posterior, $p(\boldsymbol{\theta}|\mathcal{X})$, assuming that its computation is intractable.

Let us denote by $q(\boldsymbol{\theta})$ the estimate of the posterior. The starting point is to compute q by minimizing the KL divergence,

$$\mathrm{KL}(p||q) = \int p(\boldsymbol{\theta}|\mathcal{X}) \ln \frac{p(\boldsymbol{\theta}|\mathcal{X})}{q(\boldsymbol{\theta})} \, d\boldsymbol{\theta}. \tag{13.103}$$

Note that $\mathrm{KL}(p||q)$ is different from the $\mathrm{KL}(q||p)$ divergence, which is involved in the bound in Eq. (13.2). Because the KL divergence is not symmetric, the two methods minimize a different cost. Before proceeding any further, it is important to highlight some implications associated with the two forms of the KL divergence.

- *I-Projection*: The $\mathrm{KL}(q||p)$ divergence is given by

$$\mathrm{KL}(q||p) = \int q(\boldsymbol{\theta}) \ln \frac{q(\boldsymbol{\theta})}{p(\boldsymbol{\theta}|\mathcal{X})} \, d\boldsymbol{\theta}. \tag{13.104}$$

  This is sometimes known as *I-projection* or *information projection*. Looking carefully at it, note that in regions of the parameter space where $p(\boldsymbol{\theta}|\mathcal{X})$ assumes small values, $\mathrm{KL}(q||p)$ gets large values and minimization pushes $q(\boldsymbol{\theta})$ to small values as well. Consider now the case that $p(\boldsymbol{\theta}|\mathcal{X})$ is

[8] If other hidden variables are also involved, we consider them as part of $\boldsymbol{\theta}$.

bimodal, while $q(\boldsymbol{\theta})$ is constrained to be unimodal. Then minimizing $\text{KL}(q||p)$ will force q to be placed close to either of the two peaks of p in order to get small values in the regions where p takes small values too.

- *M-Projection*: We now turn our focus to $\text{KL}(p||q)$ divergence, defined in Eq. (13.103). This is also known as *M-projection* or *moment projection*. For the case discussed before, in the regions where p assumes large values, $\text{KL}(p||q)$ gets large values and minimization estimates q in order to have large values in these regions too. Thus, the estimate q is placed in order for its mode to lie somewhere between the two modes of p, as a compromise between the two. Obviously, this is not a good result, because the estimate puts high-probability mass in regions where p assumes small values. This discussion points out some limitations on the performance that the expectation propagation method is expected to exhibit in practice, because it is based on $\text{KL}(p||q)$ minimization.

We now assume that $p(\mathcal{X}, \boldsymbol{\theta})$ can be factorized, that is,

$$p(\mathcal{X}, \boldsymbol{\theta}) = \prod_j f_j(\boldsymbol{\theta}). \tag{13.105}$$

For example, such a product can cover the case where

$$p(\mathcal{X}, \boldsymbol{\theta}) = \prod_n p(\boldsymbol{x}_n|\boldsymbol{\theta})p(\boldsymbol{\theta}),$$

where $p(\boldsymbol{\theta})$ is the corresponding prior. The more general formulation of the factorization used in Eq. (13.105) can serve the needs for more general tasks, as for example graphical models to be treated in Chapter 15. Thus, we can now write

$$\boxed{p(\boldsymbol{\theta}|\mathcal{X}) = \frac{1}{p(\mathcal{X})} \prod_j f_j(\boldsymbol{\theta}),} \tag{13.106}$$

where $p(\mathcal{X})$ is the evidence of the model. The estimate q will be chosen to be given in a factorized form, as in the variational approach in Section 13.2, that is,

$$\boxed{q(\boldsymbol{\theta}) = \frac{1}{Z} \prod_j \hat{f}_j(\boldsymbol{\theta}),} \tag{13.107}$$

where $\hat{f}_j(\boldsymbol{\theta})$ corresponds to $f_j(\boldsymbol{\theta})$ and Z is the normalizing constant. The next assumption is that $q(\boldsymbol{\theta})$ is constrained to lie within the exponential family of PDFs (Section 12.8) and for our current needs it can be written as

$$q(\boldsymbol{\theta}) := g(\boldsymbol{a})h(\boldsymbol{\theta}) \exp\left(\boldsymbol{a}^T \boldsymbol{u}(\boldsymbol{\theta})\right). \tag{13.108}$$

where $\boldsymbol{a}$ is the associated set of parameters.

MINIMIZING THE KL DIVERGENCE

Plugging into Eq. (13.103) the definition in Eq. (13.108) and collecting all terms that are independent of $\boldsymbol{a}$ in a constant, we readily obtain

$$\text{KL}(p||q) = -\ln g(\boldsymbol{a}) - \int p(\boldsymbol{\theta}|\mathcal{X}) \left(\boldsymbol{a}^T \boldsymbol{u}(\boldsymbol{\theta})\right) d\boldsymbol{\theta} + \text{constants}. \tag{13.109}$$

Taking the gradient with respect to $\boldsymbol{a}$ and equating to zero we get

$$-\frac{1}{g(\boldsymbol{a})} \nabla g(\boldsymbol{a}) = \mathbb{E}_p [\boldsymbol{u}(\boldsymbol{\theta})]. \tag{13.110}$$

However, from Eq. (13.108) we have

$$g(\boldsymbol{a}) \int h(\boldsymbol{\theta}) \exp\left(\boldsymbol{a}^T \boldsymbol{u}(\boldsymbol{\theta})\right) d\boldsymbol{\theta} = 1,$$

and taking the gradient with respect to $\boldsymbol{a}$ results in

$$\begin{aligned} \boldsymbol{0} = & \nabla g(\boldsymbol{a}) \int h(\boldsymbol{\theta}) \exp\left(\boldsymbol{a}^T \boldsymbol{u}(\boldsymbol{\theta})\right) d\boldsymbol{\theta} \\ & + g(\boldsymbol{a}) \int h(\boldsymbol{\theta}) \exp\left(\boldsymbol{a}^T \boldsymbol{u}(\boldsymbol{\theta})\right) \boldsymbol{u}(\boldsymbol{\theta})\, d\boldsymbol{\theta} \end{aligned}$$

or

$$-\frac{1}{g(\boldsymbol{a})} \nabla g(\boldsymbol{a}) = \mathbb{E}_q [\boldsymbol{u}(\boldsymbol{\theta})],$$

which combined with Eq. (13.110) finally results in

$$\boxed{\mathbb{E}_q [\boldsymbol{u}(\boldsymbol{\theta})] = \mathbb{E}_p [\boldsymbol{u}(\boldsymbol{\theta})] : \quad \text{moment matching.}} \tag{13.111}$$

The latter is an elegant equation known as *moment matching*. It basically states that at the optimum, $q(\boldsymbol{\theta})$, the expectations of its sufficient statistics are equal to the expectations associated with the PDF to be learned. For example, if q is chosen to be a Gaussian, the sufficient statistics involves the mean and the covariance matrix. Thus, all one has to do is compute the mean and covariance with respect to $p(\boldsymbol{\theta})$ (assuming that they can be obtained) and use them to define the respective Gaussian.

THE EXPECTATION PROPAGATION ALGORITHM

We will now make use of the moment matching result to obtain the factors, $\hat{f}_j(\boldsymbol{\theta})$, one at a time. The algorithm starts from some initial estimates, $\hat{f}_j^{(0)}$. Let us assume that we are currently seeking to update factor $\hat{f}_k(\boldsymbol{\theta})$. Let $q^{(i)}(\boldsymbol{\theta})$ be the currently available estimate of $q(\boldsymbol{\theta})$, at the ith iteration.

Step 1: Remove $\hat{f}_k^{(i)}(\boldsymbol{\theta})$ from $q^{(i)}(\boldsymbol{\theta})$, and define

$$q_{/k}^{(i)}(\boldsymbol{\theta}) := \frac{q^{(i)}(\boldsymbol{\theta})}{\hat{f}_k^{(i)}(\boldsymbol{\theta})}. \tag{13.112}$$

Step 2: Define the PDF

$$\frac{1}{Z_k} f_k(\boldsymbol{\theta}) q_{/k}^{(i)}(\boldsymbol{\theta}). \tag{13.113}$$

In other words, in the current estimate $q^{(i)}(\boldsymbol{\theta})$, $\hat{f}_k^{(i)}(\boldsymbol{\theta})$ is replaced by $f_k(\boldsymbol{\theta})$, and Z_k is the corresponding normalizing constant.

Step 3: Compute the normalizing constant,

$$Z_k = \int f_k(\boldsymbol{\theta}) q_{/k}^{(i)}(\boldsymbol{\theta}) d\boldsymbol{\theta}. \tag{13.114}$$

Step 4: In this step, the optimization is performed by minimizing the KL divergence,

$$\text{KL}\left(\frac{1}{Z_k} f_k(\boldsymbol{\theta}) q_{/k}^{(i)}(\boldsymbol{\theta}) || q^{(i+1)}(\boldsymbol{\theta})\right).$$

This is achieved by moment matching, and the new $q^{(i+1)}$ is defined so that the expectations of the respective sufficient statistics are matched to those of $\frac{1}{Z_k} f_k(\boldsymbol{\theta}) q_{/k}^{(i)}(\boldsymbol{\theta})$, and this operation is assumed to be computationally tractable.

Step 5: Compute $\hat{f}_k^{(i+1)}$ such that

$$\hat{f}_k^{(i+1)}(\boldsymbol{\theta}) := K \frac{q^{(i+1)}(\boldsymbol{\theta})}{q_{/k}^{(i)}(\boldsymbol{\theta})}, \tag{13.115}$$

where the proportionality constant is computed so that

$$\int \hat{f}_k^{(i+1)}(\boldsymbol{\theta}) q_{/k}^{(i)}(\boldsymbol{\theta})\, d\boldsymbol{\theta} = \int f_k(\boldsymbol{\theta}) q_{/k}^{(i)}(\boldsymbol{\theta})\, d\boldsymbol{\theta}, \tag{13.116}$$

which results in $K = Z_k$.

The procedure is then applied for the estimation of $\hat{f}_{k+1}^{(i+1)}$. For convergence, more than one passes have to be performed. The evidence can be approximated as

$$p(\mathcal{X}) \approx \int \prod_j \hat{f}_j(\boldsymbol{\theta})\, d\boldsymbol{\theta}. \tag{13.117}$$

A detailed application of the algorithm in the context of a simple-to-follow example is given in [48].

Remarks 13.5.

- In general, there is no guarantee that the algorithm will converge, which is a major disadvantage of the method. However, it can be shown that if the iterations do converge, the solution is a stationary point of a particular energy function [48]. Recall that in the variational Bayes approach, there is guarantee of convergence to a local optimum point. Of course, one could optimize the KL divergence for the expectation propagation method directly, which guarantees convergence, but in this case the algorithm is more complex and slow.

- Taking into account our discussion concerning the two forms of KL divergence, the expectation propagation method results in poor performance when the true posterior is multimodal. However, for other scenarios, such as logistic-type models, the expectation propagation method can offer competitive and sometimes better performance compared to the variational methods or methods built around the Laplacian approximation (see, e.g., [39,48]).
- The expectation propagation algorithm was first proposed in [48], and it is a modification of what was known before as *assumed density filtering* (ADF) or *moment matching* (e.g., [57] and the references therein).
- Looking at the factors $f_j(\boldsymbol{\theta})$ in a more general view, it turns out that the expectation propagation offers the vehicle for obtaining a range of message passing algorithms in the context of probabilistic graphical models (Chapter 15) [49].
- *α-Divergence*: Having spent time discussing in some detail the two forms of KL divergence, it is interesting to point out that both formulations can be obtained as special cases of a more general family known as the α family of divergences, defined as

$$\boxed{D_\alpha(p||q) := \frac{4}{1-\alpha^2}\left(1 - \int p(x)^{(1+\alpha)/2} q(x)^{(1-\alpha)/2}\, dx\right) : \ \alpha\text{-divergence},} \tag{13.118}$$

where $\alpha \in \mathbb{R}$ is a parameter. Note that $\mathrm{KL}(p||q)$ is obtained at the limit $\alpha \to 1$ and $\mathrm{KL}(q||p)$ is obtained for $\alpha \to -1$; $D_\alpha(p||q)$ is nonnegative and it becomes zero if $p = q$ (see, e.g., [2]).

13.11 NONPARAMETRIC BAYESIAN MODELING

The Bayesian approach to parametric modeling has been the focus of our attention in the current and previous chapters. The underlying assumption was that the number of the unknown parameters was fixed and finite. We now turn our attention to a more general task. We will assume that the hidden structure of our model is not fixed but is allowed to grow with the data. In other words, its complexity is not specified a priori but is left to be determined from the data. This is the reason that such models are called *nonparametric*; recall from Chapter 3 that a model is called parametric if the number of free parameters is fixed and independent of the size of the data set.

We will avoid treating nonparametric Bayesian models in a mathematically rigorous sense. Such a path would take us a bit far from the purpose of this book and also from the mathematical skills of the average reader. Thus, we will be content with presenting the main concepts in a mathematically "humble" way. Once the basics have been grasped, the keen reader can delve deeper into the topic by referring to more specialized literature (see, e.g., [27]). To demonstrate the idea behind nonparametric Bayesian models, we will start with our familiar mixture modeling task. In the sequel, the so-called matrix factorization problem will be introduced and it will be treated in the nonparametric context.

In Section 13.4, K mixtures (clusters) were assumed. Each mixture was modeled via a Gaussian PDF with unknown mean and precision matrix, $(\boldsymbol{\mu}_k, Q_k)$, $k = 1, 2, \ldots, K$. These, in turn, were considered as random entities and were dressed up with a prior—a Gaussian PDF for the mean and a Wishart one for the precision matrix. The probabilities P_k, $k = 1, 2, \ldots, K$, for each mixture were treated either as constants, which were optimized during the M-step, or they were considered random variables and a Dirichlet PDF prior was associated with them (Problem 13.7). The goal was to obtain an estimate of the *posterior probabilities* of the labels associated with each observation point, $P(k_n|\mathcal{X})$. There, we resorted to variational techniques via the mean field approximation and the posterior was approximated by the function $q_z(\mathcal{Z}) = q_z(z_1, \dots, z_N)$, where z_n, $n = 1, 2, \dots, N$, were 0-1 coding vectors, with a one placed at the location corresponding to a specific mixture (k) and the rest of the elements being zero. We refer to such vectors as *one-hot* vectors. In this way, an equivalent clustering of the observation points was achieved in K Gaussian-distributed clusters.

The nonparametric counterpart of the previous task is expressed in almost the same way, albeit with a single important difference. The number of mixtures, K, is not fixed to a finite value; as a matter of fact, the number of mixtures is allowed to be *countably infinite*. There are two questions that now pop up: (a) how can one deal with an infinite number of clusters, and (b) how can one deal with a prior distribution related to infinitely many probability values?

To give an indication of how to deal with infinity, recall that in nonparametric modeling the number of data points is still finite and equal to N. Thus, whatever model one adopts, there is no way of having more than N mixtures (clusters); the latter corresponds to the worst-case scenario, where each one of the points belongs to a different cluster. Hence, although in theory one can have infinitely many clusters, only a finite subset of them is *nonempty*. Thus, all we need to do is to obtain an explicit representation of the nonempty mixture components.

Concerning the second question, it can be shown that a *prior* distribution over an infinite number of groupings, $P(\mathcal{Z})$, that favors assigning data to a small number of groups is the *Chinese restaurant process* (CRP); this is a distribution over infinite partitions of the integers [1,22,63].

We will first state the algorithm that generates a draw (realization) from such a process and then we will provide more details for those who are interested in further theoretical aspects that are associated with such processes.

13.11.1 THE CHINESE RESTAURANT PROCESS

The name draws from the seemingly infinite number of tables in some very large Chinese restaurants in California. Each table is associated with one cluster/mixture and each customer with one observation. The first customer sits at the first table. The second customer sits at the first table with probability $\frac{1}{1+\alpha}$ and at a new table with probability $\frac{\alpha}{1+\alpha}$. The nth customer sits at one of the previously occupied tables with a probability proportional to the number of people who already sit at it, and he/she sits at a new table with a probability proportional to α. The parameter α is known as the *concentration parameter*. The larger its value, the more tables are occupied and the fewer the customers who sit at a single table. In a more formal way, let k_n denote the table for the nth customer. Then we can write

$$P(k_n = k|k_{1:n-1}) = \begin{cases} \dfrac{n_k}{n-1+\alpha}, & \text{if } k \leq K_{n-1}, \\ \dfrac{a}{n-1+\alpha}, & \text{otherwise}, \end{cases} \tag{13.119}$$

where K_{n-1} is the number of tables occupied by the previously $n-1$ arrived customers and n_k the number of customers already sitting at table k. It can be shown that the expected number of occupied tables grows as $\alpha \ln n$; that is, the expected number of clusters grows with the number of data (e.g., [19]). The rule in Eq. (13.119) provides the sampling philosophy for assigning data to clusters as new data arrive sequentially. It can be shown (see, e.g., [19]) that the resulting probability $P(k_1, \dots, k_n)$ is

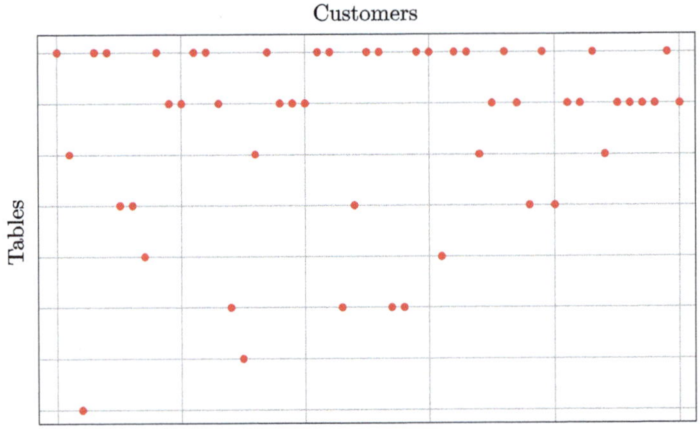

FIGURE 13.14

Every customer sits at one of the previously occupied tables or selects to sit at a new one. Observe that the probability of a customer to select to sit at a new table decreases fast. Dots indicate the tables where customers sit. The first table is occupied by many customers. As we move downwards to the second, third, etc., table, the number of customers that select them is fast decreasing.

independent (up to label changes) of the sequence in which data arrive; this is an important invariance property.

Fig. 13.14 illustrates a draw from a CRP process. Each customer sits at one of the previously occupied tables with some probability or prefers to sit at a new one. Observe that the probability of a customer to select a new table decreases fast. In contrast, the probability of a customer sitting at a table already occupied increases with the popularity of the table (rich gets richer).

13.11.2 DIRICHLET PROCESSES

This section provides a brief discussion of the more general mathematical framework in which CRPs belong. This section can be bypassed in a first reading.

The notion of a stochastic process was introduced in Section 2.4. A *Dirichlet process* (DP), G, which was first introduced in [16], is a *distribution over distributions* and it is defined in terms of (a) the *concentration parameter*, α, and (b) the so-called *base distribution*, G_0, over a space Θ, and we write $G \sim \text{DP}(\alpha, G_0)$. We say that $G \sim \text{DP}(\alpha, G_0)$ is a DP if for *any* partition[9] T_k, $k = 1, 2, \ldots, K$, of Θ, i.e., $\Theta = \cup_{k=1}^{K} T_k$, the following holds true:

$$\big(G(T_1), \ldots, G(T_K)\big) \sim \text{Dir}\big(\alpha G_0(T_1), \ldots, \alpha G_0(T_K)\big), \tag{13.120}$$

[9] Strictly speaking, we should say measurable partition; a partition is measurable if it is closed under complementation and countable union.

where $G_0(T_k)$ is the probability (according to G_0) corresponding to the occurrence of T_k, and $G(T_k)$ is similarly defined. In other words, $G(T_k),\ k = 1, 2, \ldots, K$, are jointly distributed according to a Dirichlet distribution. To establish the connection with Chapter 2, where the Dirichlet probability distribution was defined, $\alpha G_0(T_k)$ correspond to the associated parameters of the distribution in Eq. (2.95), i.e., $a_k,\ k = 1, 2, \ldots, K$, and $G(T_k)$ to the values of the respective random variables, x_k. Recall that by the definition of a Dirichlet distribution, $x_k,\ k = 1, 2, \ldots, K$, lie in the interval [0, 1] and they add to one, and hence they can be interpreted as *probabilities*.

Also, from the definition of a random process in Section 2.4, we know that every realization that results from an experiment comprises an infinite number of samples (countable or uncountable), depending on whether it is a discrete or a continuous one. It has been shown in [16] that a DP is of a *discrete* nature[10]; moreover, every realization can be interpreted as a probability distribution. In other words, the realizations of the process comprise distinct samples and each one corresponds to a probability value. Obviously, their sum is equal to one. Moreover, they comply with the defining property in Eq. (13.120). Mathematically, the above can be formulated as

$$\boxed{G = \sum_{i=1}^{\infty} P_i \delta_{\boldsymbol{\theta}_i}(\boldsymbol{\theta}),} \tag{13.121}$$

where $\delta_{\boldsymbol{\theta}_i}(\boldsymbol{\theta})$ is equal to 1 if $\boldsymbol{\theta} = \boldsymbol{\theta}_i$ and zero otherwise. In other words, a draw (realization) $G \sim \mathrm{DP}(\alpha, G_0)$ puts a probability mass, P_i, on a specific countably infinite set of samples $\boldsymbol{\theta}_i,\ i = 1, 2, \ldots$. The points $\boldsymbol{\theta}_i$ are known as *atoms* and they are i.i.d. drawn from the *base distribution*, G_0, i.e., $\boldsymbol{\theta}_i \sim G_0$. Let us now elaborate on the result in Eq. (13.121) and its connection with Eq. (13.120) a bit more.

Take *any* subset[11] $T_k \subset \Theta$ and collect all indices in Eq. (13.121) $I_k := \{i : \boldsymbol{\theta}_i \in T_k\}$. Then $G(T_k) = \sum_{i\in I_k} P_i$. Moreover, it is guaranteed that $G(\Theta) = 1$, since G is a probability distribution and $\Theta = \cup_{k=1}^{K} T_k$. A DP *guarantees* that the probability values $G(T_k),\ k = 1, 2, \ldots, K$, follow a Dirichlet distribution as in (13.120), for any finite number K. Note that G_0 may not be discrete. For example, it can be Gaussian or another probability density function. Note that G is random in two ways; both the probability values P_i and the locations $\boldsymbol{\theta}_i$ are randomly obtained.

Mean and variance: For any subset $T_k \subset \Theta$, the mean value is

$$\mathbb{E}\big[\mathrm{G}(T_k)\big] = G_0(T_k),$$

and the variance is given by

$$\mathrm{var}\big[\mathrm{G}(T_k)\big] = \frac{G_0(T_k)\big(1 - G_0(T_k)\big)}{\alpha + 1}.$$

The proofs are given in Problem 13.20. The above indicates that the draws from a DP distribution stay "around" the base one and the variance is inversely proportional to the concentration parameter.

Posterior from prior distributions: We know that every draw $G \sim \mathrm{DP}(\alpha, G_0)$ is a distribution and it is used to draw samples from Θ. Let $\boldsymbol{\theta}_i \in \Theta \sim G$ be a sequence of such i.i.d. drawn samples. Our

[10] Strictly speaking, with probability one.
[11] Strictly speaking, measurable.

goal now is to derive the posterior probabilities $G(T_k)$, $k = 1, 2, \ldots, K$, given a set of n observations, $\boldsymbol{\theta}_1, \ldots, \boldsymbol{\theta}_n$.

To this end, recall from Example 12.6 that the Dirichlet distribution is the conjugate of the multinomial one, given in Eq. (2.58), i.e.,

$$P(n_1, n_2, \ldots, n_K) = \binom{n}{n_1, \ldots, n_K} \prod_{k=1}^{K} G(T_k)^{n_k}, \tag{13.122}$$

where we used n_k in place to x_k to be in line with a more standard notation used in DPs. Recall that n_k is the number of times the kth variable associated with probability $P = G(T_k)$ occurs, after n successive experiments. In our setting, n_k is the cardinality of the previously defined set I_k, i.e.,

$$n_k = \#\{i : \boldsymbol{\theta}_i \in T_k\}.$$

In words, n_k is the number of draws from Θ that lie in T_k.

Having observed $\boldsymbol{\theta}_1, \ldots, \boldsymbol{\theta}_n$, we see they are equivalent to specific occurrence numbers, $n_1, \ldots, n_K$. Thus, taking into account the property of the conjugate pairs, as expressed in Eq. (12.77) in the context of Example 12.6, it can be readily seen that

$$\big(G(T_1), \ldots, G(T_K)|\boldsymbol{\theta}_1, \ldots, \boldsymbol{\theta}_n\big) = \text{Dir}\big(\alpha G_0(T_1) + n_1, \ldots, \alpha G_0(T_K) + n_K\big). \tag{13.123}$$

It is not difficult to see (e.g., Problem 13.21) that the above can be equivalently written as

$$\big(G(T_1), \ldots, G(T_K)|\boldsymbol{\theta}_1, \ldots, \boldsymbol{\theta}_n\big) = \text{Dir}\big(\alpha' G_0'(T_1), \ldots, \alpha' G_0'(T_K)\big),$$

where the new concentration parameter and base distribution are given by

$$\alpha' = \alpha + n \text{ and } G_0' = \frac{1}{\alpha + n}\left(\alpha G_0 + \sum_{i=1}^{n} \delta_{\boldsymbol{\theta}_i}(\boldsymbol{\theta})\right).$$

Hence, we can compactly write that

$$\boxed{G|\boldsymbol{\theta}_1, \ldots, \boldsymbol{\theta}_n \sim \text{DP}\left(\alpha + n, \; \frac{1}{\alpha + n}\left(\alpha G_0 + \sum_{i=1}^{n} \delta_{\boldsymbol{\theta}_i}(\boldsymbol{\theta})\right)\right).} \tag{13.124}$$

The above is an elegant result. The concentration parameter increases by the number of observations that we have at our disposal. Recall that the variance around the mean of the associated probability values is inversely proportional to the concentration parameter. Hence, the obtained result is in line with common sense. The more observations we get, the lower our related uncertainty becomes. Also, the base distribution associated with the posterior DP comprises two components, i.e., the original one and a distribution which is of a *discrete* nature. As a matter of fact, it imposes probability masses on the observed values.

Predictive Distribution and the Pólya Urn Model

Let us now build upon the posterior DP formulation in Eq. (13.124). To this end, the reasoning provided in [80] will be followed. Having observed the samples $\boldsymbol{\theta}_1, \ldots, \boldsymbol{\theta}_n$, we focus on the next draw, $\boldsymbol{\theta}_{n+1}$, and we will compute the probability of the sample to lie in an interval T. To simplify notation, let us use G' to denote the posterior $G|\boldsymbol{\theta}_1, \ldots, \boldsymbol{\theta}_n$. Then we write

$$P(\boldsymbol{\theta}_{n+1} \in T|\boldsymbol{\theta}_1, \ldots, \boldsymbol{\theta}_n) = G'(T).$$

However, we know that G' is itself a random draw from the posterior DP. Marginalizing out with respect to G', i.e., taking the expectation, and recalling the property of the mean value stated before, we get

$$P(\boldsymbol{\theta}_{n+1} \in T|\boldsymbol{\theta}_1, \ldots, \boldsymbol{\theta}_n) = \frac{1}{\alpha + n}\left(\alpha G_0(T) + \sum_{i=1}^{n} \delta_{\boldsymbol{\theta}_i}(\boldsymbol{\theta} \in T)\right).$$

The above is true for any $T \subset \Theta$, and it can be reformulated as

$$\boxed{\boldsymbol{\theta}_{n+1}|\boldsymbol{\theta}_1, \ldots, \boldsymbol{\theta}_n \sim \frac{\alpha}{\alpha + n} G_0 + \frac{n}{\alpha + n}\left(\frac{1}{n}\sum_{i=1}^{n} \delta_{\boldsymbol{\theta}_i}(\boldsymbol{\theta})\right).} \qquad (13.125)$$

In the above formulation, the term in parentheses is written in a way to remind us of its discrete probability nature that adds to one.

The last formula leads to the following physical interpretation. Consider an empty urn and assume that each value in $\boldsymbol{\theta}$ corresponds to a *unique* color. Also, we are given a countably infinite number of balls. Pick at random a color from the basic distribution, i.e., $\boldsymbol{\theta}_i \sim G_0$, and paint the ball with the corresponding color. After the ball has been painted, it is placed in the urn, which now contains one ball. In the second step, we pick up a new ball and we either (a) pick a new color, $\boldsymbol{\theta}_2 \sim G_0$, with probability $\frac{\alpha}{\alpha+1}$, paint it accordingly, and place the painted ball in the urn, or (b) with probability $\frac{1}{\alpha+1}$, paint the ball with the same color as that of the ball which is already in the urn and then place it in the urn. Now the urn contains two balls. The steps are repeated. So, at the $(n+1)$th step, the urn contains n balls. We pick a new ball and we either (a) pick a new color $\boldsymbol{\theta}_{n+1} \sim G_0$ with probability $\frac{\alpha}{\alpha+n}$, paint the new ball, and place it in the urn, or (b) with probability $\frac{n}{\alpha+n}$, we randomly pick one of the n balls in the urn, and choose its color to paint the new one; after painting, it is also placed in the urn, which will now contain $n+1$ balls. This interpretation has been used in [10] to show the existence of DP processes.

There are two important properties. Note that as n increases, the probability $\frac{n}{\alpha+n}$ of selecting a color from the urn keeps increasing, relative to $\frac{\alpha}{\alpha+n}$ of selecting a color from G_0. Thus, for long sequences, we are going to select more and more previously used colors, which will be repeated. This justifies the discrete nature of the process.

The other important property is that it turns out that the probability of generating a sequence of colors, i.e., $\boldsymbol{\theta}_1, \ldots, \boldsymbol{\theta}_n$, is equal to the probability of generating these colors in *any* sequence. That is,

$$P(\boldsymbol{\theta}_1, \ldots, \boldsymbol{\theta}_n) = P(\boldsymbol{\theta}_{\pi(1)}, \ldots, \boldsymbol{\theta}_{\pi(n)}),$$

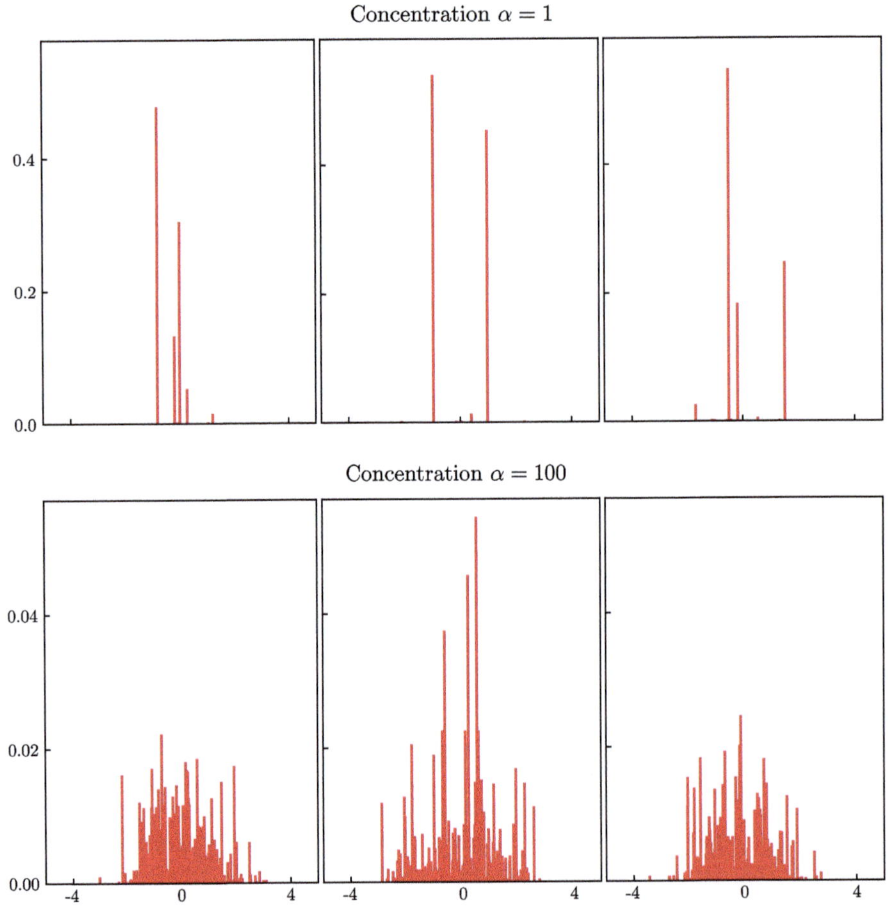

FIGURE 13.15

Three different realizations of a DP for two different values of the concentration parameter, α. For $\alpha = 1$, only a few (discrete) probability mass values survive, in all three corresponding draws (top row). On the contrary, for $\alpha = 100$, a much larger number of probability masses are obtained. Note that, in all cases, the generated masses stay around the base distribution, i.e., the standard normal one for this case.

where $\pi(\cdot)$ denotes *any* permutation of the numbers in $\{1, 2, \ldots, n\}$. Such sequences are known as *exchangeable* (see, e.g., [36] for an insightful discussion).

Fig. 13.15 shows different realizations for two different values of the concentration parameter α. The base distribution is the standard Gaussian with zero mean and unit variance. Two different values of the concentration parameter are used, and for each one three different draws are performed. For the smaller value, $\alpha = 1$, only a small number of discrete probability masses survive. This is natural, because in Eq. (13.125), when the value of α is small, the significance of the first of the two terms on the right-hand side is small. Thus, the probability of selecting previously generated values is large. The opposite is true when α gets large values. In the latter case, the significance of the first term becomes small, only when n becomes relatively large. Thus, a larger number of probability masses survive.

Chinese Restaurant Process Revisited

We have already commented on the discrete nature of the DPs. Furthermore, the rule in Eq. (13.125) implies a *clustering* structure of the resulting values from the draws. As said, successive draws (colors) repeatedly pick up previously drawn values. Let $\boldsymbol{\theta}_1^*, \ldots, \boldsymbol{\theta}_{K_n}^*$ be the K_n colors that have survived after n successive draws. Let $n_1, \ldots, n_{K_n}$ be the corresponding numbers of occurrence for each one of the colors. Then Eq. (13.125) can be rewritten as

$$\boldsymbol{\theta}_{n+1}|\boldsymbol{\theta}_1, \ldots, \boldsymbol{\theta}_n \sim \frac{\alpha}{\alpha+n} G_0 + \frac{n}{\alpha+n}\left(\frac{1}{n}\sum_{i=1}^{K_n} n_i \delta_{\boldsymbol{\theta}_i^*}(\boldsymbol{\theta})\right).$$

Observe that the above formulation leads directly to Eq. (13.119) (if we apply it for the nth instead of the $n+1$th customer). Indeed, we either select a new color (table in the case of CPR) or the ith previously selected one, with probability $\frac{n_i}{\alpha+n}$, $i \leq K_n$. Recall from Section 13.11.1 that according to this rule, the expected number of clusters grows as $\alpha \ln n$.

An alternative way to arrive at the CRP is as a limiting case ($K \longrightarrow \infty$) of the finite mixture modeling, by adopting a Dirichlet prior with parameter α/K (see, e.g., [23,72]).

The cluster assignments are exchangeable: An important aspect of the CRP is that the probability of a specific clustering does not depend on the order customers have arrived and it is unchanged after reshuffling the customers, up to a permutation of the table labels (see Problem 13.23). In words, this means that in Fig. 13.14, what is the important information is that, for example, customers #1, #4, #5 sit at the *same* table and customer, for example, #3 sits on his/her own. The table numbering is of no importance.

13.11.3 THE STICK BREAKING CONSTRUCTION OF A DP

We have already discussed the Pólya urn model and the related CRP. The *stick breaking representation* is an alternative representation of a DP and it was developed in [75]. The method builds upon Eq. (13.121) and proposes the steps to generate P_i, as well as the respective values of $\boldsymbol{\theta}_i$, $i = 1, 2, \ldots$.

Consider a stick of unit length (Fig. 13.16). The stick is divided into a sequence of infinitely many segments of length P_i, $i = 1, 2, \ldots$, according to the following algorithm. First, make a draw of a beta (Chapter 2) distributed random variable, $\beta_1 \sim \text{Beta}(\beta|1, \alpha)$, and break off a segment of the stick of length equal to β_1; we set $P_1 = \beta_1$. The length of the remaining segment is $1 - \beta_1$. Then make another draw, $\beta_2 \sim \text{Beta}(\beta|1, \alpha)$, and break off another segment in proportion to β_2. Thus, the length of the discarded segment is equal to $\beta_2(1 - \beta_1)$ and we set P_2 equal to its length. The length of the remaining segment is, obviously, $(1 - \beta_1) - \beta_2(1 - \beta_1) = (1 - \beta_1)(1 - \beta_2)$. Following this rationale recursively, by breaking off pieces from the remaining segments, and starting at $P_1 = \beta_1 \sim \text{Beta}(\beta|1, \alpha)$, the ith step of the algorithm is

$$\beta_i \sim \text{Beta}(\beta|1, \alpha), \tag{13.126}$$

$$P_i = \beta_i \prod_{j=1}^{i-1}(1 - \beta_j), \quad i = 2, 3, \ldots. \tag{13.127}$$

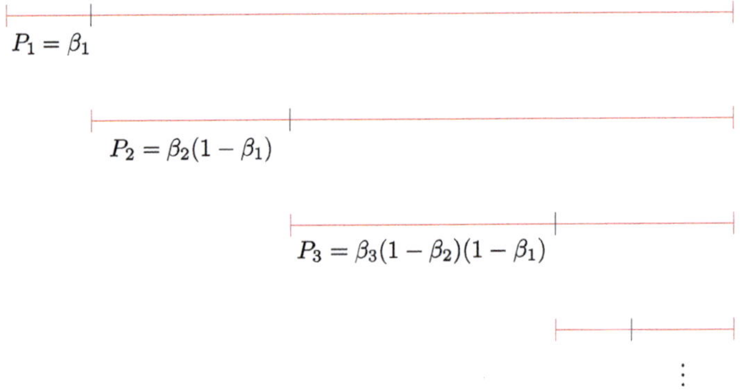

FIGURE 13.16

The stick breaking construction of a DP. At each iteration, we sample from a beta distribution $\text{Beta}(\beta|1, \alpha)$ and we break off a piece from the remaining segment in proportion to the value of the sample.

Using the resulting sequence of probability values, P_i, a random distribution is formed according to Eq. (13.121), with $\boldsymbol{\theta}_i$ drawn i.i.d. from G_0. It can be shown that the resulting distribution is a DP, i.e., $G \sim \text{DP}(\alpha, G_0)$.

Note that as i increases, the probability values decrease, because smaller and smaller fractions of the stick remain. Thus, out of the infinite possible terms in Eq. (13.121) only a relatively small number survive that have significant contribution. Moreover, the probability values associated with these surviving terms are the first ones to be generated, which, in practice, is very important when one implements a DP to generate a prior (see, e.g., [58]).

A concise tutorial concerning DPs can also be found in [19]. There, a number of sites with publicly available software tools are also provided.

13.11.4 DIRICHLET PROCESS MIXTURE MODELING

Having established the discrete and clustering properties of a DP, let us now see how it can be used as a prior to Bayesian learning of a mixture distribution, e.g., Gaussian mixture modeling. Our framework will be that of a nonparametric setting; that is, the number of mixture components is *not* a priori fixed. This forces us to replace priors over a fixed number of parameters, with *priors over distributions*. The Bayesian learning rationale offers the tools for updating the previous *prior* information to that related to *posterior* distributions, once observations have been obtained.

Let us assume that we are given a set of observations, $\boldsymbol{x}_n \in \mathbb{R}^l$, $n = 1, 2, \ldots, N$. We assume that each one of them is emitted by a corresponding PDF (distribution) parameterized in terms of a respective set of hidden variables, $\boldsymbol{\theta}_1, \boldsymbol{\theta}_2, \ldots, \boldsymbol{\theta}_N \in \Theta$. That is, $\boldsymbol{x}_n \sim p(\boldsymbol{x}|\boldsymbol{\theta}_n)$. Observe that the set of parameters grows with N, in contrast to the fixed number of parameters used in the mixture modeling of Section 13.4. In turn, we assume that the parameters are i.i.d. drawn from a distribution, G, which is itself a draw from a DP (see, e.g., [3]). In summary, the data generation according to this mixture model is described as

$$G|(\alpha, G_0) \sim \text{DP}(\alpha, G_0), \tag{13.128}$$

$$\boldsymbol{\theta}_n|G \sim G_0, \tag{13.129}$$

$$\boldsymbol{x}_n|\boldsymbol{\theta}_n \sim p(\boldsymbol{x}|\boldsymbol{\theta}_n). \tag{13.130}$$

Note that since G is discrete and bears a strong clustering structure, a number of the observations samples, $\boldsymbol{x}_n$, will share the same hidden parameters, and this is how the mixture modeling is imposed by the DP prior.

If the stick breaking construction is employed, the above three basic steps are "rephrased," and the data generation mechanism is described as

1. Draw the beta variables, $\beta_i \sim \text{Beta}(\beta|1, \alpha),\ i = 1, 2, \ldots$.
2. Generate the probability values, $P_i = \beta_i \prod_{j=1}^{i-1}(1 - \beta_i), i = 2, 3, \ldots,$ with $P_1 = \beta_1$.
3. Draw the corresponding parameters from the base distribution, $\boldsymbol{\theta}_i \sim G_0,\ i = 1, 2, \ldots$.
4. For all the observations, $\boldsymbol{x}_n,\ n = 1, 2, \ldots, N$, Do:

- $k_n \sim \text{Cat}(P_1, P_2, \ldots)$.
- $\boldsymbol{x}_n|k_n \sim p(\boldsymbol{x}_n|\boldsymbol{\theta}_{k_n})$.

The labels k_n comprise the latent variables that indicate the corresponding mixture component (cluster) for $\boldsymbol{x}_n$ and they are drawn from a *categorical* distribution, denoted as Cat. The categorical distribution coincides with the multinomial when only one experiment in involved, i.e., $n = 1$ in Eq. (13.122). A categorical distribution involves a number of random variables, each one associated with a probability value. The draw consists of picking up one of the variables, according to the given probability distribution (see, also, Section 2.3). The categorical is sometimes referred to as *multinulli* distribution. Concerning the generation of probabilities via the beta distribution, in practice, one selects a value T, for which we assume that $P_i = 0,\ i > T$.

INFERENCE

The task of inference consists of obtaining the posteriors, given the observations and the functional forms of the priors that are associated with the latent variables as well as the involved (hidden) random (nondeterministic) parameters. In Section 13.4, the chosen priors were the Gaussian for the mean values and the Wishart PDFs for the covariance matrices. As also commented there, at the end of the section, the probabilities could also be treated as random variables and adopt a Dirichlet distribution as the respective prior.

In the current setting, because the number of mixtures, K, is not preselected, we will adopt a DP mixture process model as a prior. Under this modeling, the following latent variables are involved:

- The variables β_i, which are associated with the computation of the probabilities $P_i, i = 1, 2, \ldots$.
- The random vectors $\boldsymbol{\theta}_i$, which are used to "place" the preselected form of the PDF, which "emits" the observations $p(\boldsymbol{x}|\boldsymbol{\theta})$ in the input space. Often, the functional form of this PDF is chosen so as to form a conjugate pair with the base distribution G_0.
- The cluster assignment latent variables $k_n, n = 1, 2, \ldots, N$, which are also known as indicator variables. As already stated before, sometimes, the indicator variables are written as one-hot vectors, $\boldsymbol{k}_n$, with all components being 0 except the one which corresponds to the position that indicates the number of the mixture that the nth variable is associated with, which is set equal to 1.

During the learning phase of the posteriors, one has to estimate (a) the parameter α of the beta distribution and (b) the parameters that define G_0 and $p(\cdot|\cdot)$. As already said, in practice, one assumes that $P_i = 0$, $i > T$, for a preselected value of T. This is known as the *truncated stick breaking* representation. After all, the maximum number of mixtures one can have is N, i.e., the number of points. In the worst case, each point belongs to a different cluster. Moreover, we know that the average number of clusters grows as $O(\alpha \ln N)$.

As said, it is common to stay within the exponential family. In words this means that we select the following forms for the PDFs involved in the DP model.

- For the observations' emitting PDF,

$$p(\boldsymbol{x}|\boldsymbol{\theta}) = g(\boldsymbol{\theta}) f(\boldsymbol{x}) \exp\left(\boldsymbol{\theta}^T \boldsymbol{x}\right),$$

  where we have used the canonical form for the exponential family, as defined in Section 12.8.
- The base distribution of the DP is chosen as the respective conjugate function,

$$G_0(\boldsymbol{\theta}; \lambda, \boldsymbol{v}) = h(\lambda, \boldsymbol{v})(g(\boldsymbol{\theta}))^{\lambda} \exp\left(\boldsymbol{\theta}^T \boldsymbol{v}\right).$$

  With such a choice, the parameters that have to be learned, besides α, during learning are λ and $\boldsymbol{v}$.

We are now ready to formulate the task. Collecting all the random variables together, we form the corresponding matrix,

$$W := [\boldsymbol{\beta}, \Theta, \boldsymbol{k}],$$

where $\boldsymbol{\beta}$ is the vector with all stick lengths, Θ is the matrix comprising all the vectors, $\boldsymbol{\theta}_i$, $i = 1, 2, \ldots, T$, because we use the truncated stick breaking representation, and $\boldsymbol{k}$ is the N-dimensional vector of the indicator variables.

The goal of the inference task is to estimate the joint posterior, $p(W|\mathcal{X})$, where $\mathcal{X}$ is the set of the observations, $\mathcal{X} = \{\boldsymbol{x}_1, \boldsymbol{x}_2, \ldots, \boldsymbol{x}_N\}$. However, computing $p(W|\mathcal{X})$ turns out to be intractable. One way to overcome the associated difficulties is to resort to the mean field approximation (Section 13.2). To this end, the following *factorized* family for the variational approximation of the posterior distribution is adopted:

$$q(W) := q(\boldsymbol{\beta}, \Theta, \boldsymbol{k}) = \prod_{i=1}^{T-1} q_{\gamma_i}(\beta_i) \prod_{i=1}^{T} q_{\lambda_i}(\boldsymbol{\theta}_i) \prod_{n=1}^{N} q_{\phi_n}(k_n). \tag{13.131}$$

Note that the estimates of the posterior PDFs of the $\boldsymbol{\theta}_i$ are governed by variational parameters $\boldsymbol{\lambda}_i$ and are members of the exponential family, as discussed before. The estimates of the posteriors for the indicator variables are chosen to be multinomials (categorical) with variational parameters $\boldsymbol{\phi}_n$. The estimates of the posteriors for the beta variables are chosen to be beta distributions with variational parameters $\boldsymbol{\gamma}_i$. Note that the first product involves $T - 1$ factors. This is because of the truncation and the fact that probabilities add to one. This necessarily makes $\beta_T = 1$ and, hence, $q(\beta_T) = 1$ (Problem 13.22).

The next step is to learn the above posteriors as well as the involved parameters by maximizing the lower bound in inequality (12.68), which for our case becomes

$$\ln p(\mathcal{X};\boldsymbol{\xi}) \geq \mathcal{F}(q;\boldsymbol{\xi}) = \mathbb{E}_q\left[\ln p(W,\mathcal{X};\boldsymbol{\xi})\right] - \mathbb{E}_q\left[\ln q(W)\right],$$

and $\boldsymbol{\xi} := [\alpha, \lambda, \boldsymbol{v}^T]^T$. The algorithm for maximizing the lower bound follows exactly the rationale and the steps established in Section 13.2. Note that maximization with respect to q takes place via the respective variational parameters, as they are defined in Eq. (13.131). The algorithmic details can be found in [11]; see also [38]. The parameters $\boldsymbol{\xi}$ are computed during the M-step of the algorithm. The other alternative is to assume that the parameters are in turn random entities, and in this case specific priors have to be adopted. Once the learning phase has been terminated, assignment to mixtures is performed according to the posteriors $q(k_n)$ that have resulted from the training.

The alternative path to learn the posteriors is via Monte Carlo sampling techniques, to be discussed in Chapter 14. For such cases, the CPR model is particularly convenient for Gibbs sampling (Section 14.9); see, for example, [55].

Example 13.5. This example illustrates the computational evolution of the variational inference method for a two-dimensional Gaussian mixture model. The data are generated according to five separate Gaussian distributions, with parameters

$$\boldsymbol{\mu}_1 = [-12.5, 2.5]^T, \ \boldsymbol{\mu}_2 = [-4, -0.1]^T, \ \boldsymbol{\mu}_3 = [2, -3.5]^T,$$
$$\boldsymbol{\mu}_4 = [10, 8]^T, \ \boldsymbol{\mu}_5 = [3, 3]^T$$

and

$$\Sigma_1 = \begin{bmatrix} 1.4 & 0.81 \\ 0.81 & 1.3 \end{bmatrix}, \quad \Sigma_2 = \begin{bmatrix} 1.5 & 0.2 \\ 0.2 & 2.1 \end{bmatrix}, \quad \Sigma_3 = \begin{bmatrix} 1.6 & 1 \\ 1 & 2.9 \end{bmatrix},$$
$$\Sigma_4 = \begin{bmatrix} 0.5 & 0.22 \\ 0.22 & 0.8 \end{bmatrix}, \quad \Sigma_5 = \begin{bmatrix} 1.5 & 1.4 \\ 1.4 & 2.4 \end{bmatrix},$$

for the means and covariance matrices, respectively. One hundred data points were generated from the Gaussian mixture, where each Gaussian was assigned an arbitrary number of points. Fig. 13.17 depicts the data points as red circles. Variational inference on the model was performed based on the MATLAB® implementation of the method[12] in [11]. A data set comprising equidistant and closely spaced test points in the area $[-20, 15] \times [-8, 12]$ was used to compute the approximate predictive distribution estimated by the variational inference method. The contours of the predictive distribution computed during the first, second, and fifth iterations of the algorithm are plotted in Fig. 13.17. The algorithm has clearly identified the clusters of the data.

13.11.5 THE INDIAN BUFFET PROCESS

We have discussed the clustering promoting nature of a DP, which was exploited in the context of the mixture modeling task. The CRP and the stick breaking construction were two paths to represent and implement a DP as a prior in practice.

[12] http://sites.google.com/site/kenichikurihara/academic-software.

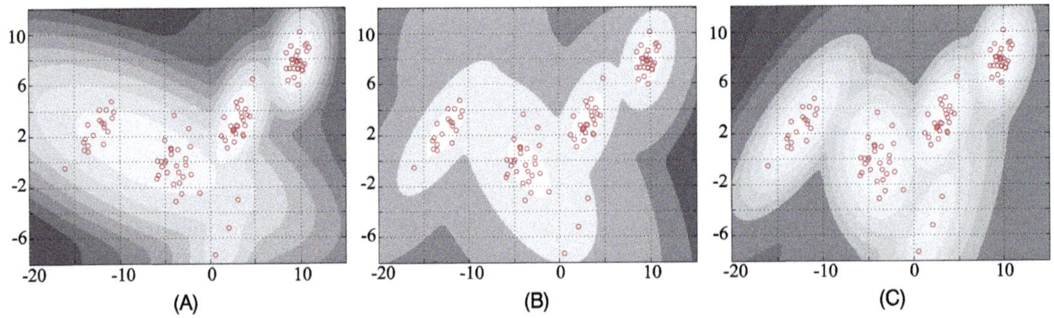

FIGURE 13.17

Contours of predictive distribution for Example 13.5, after (A) the first, (B) the second, and (C) the fifth iteration. Observe that finally five clusters have survived.

We now turn our attention to a different task, which bears a close affinity to mixture modeling. Recall that mixture modeling *imposes a structure* on the observations, by assigning groups of them to the same mixture component. In the current section, a different type of structure will be imposed.

At the heart of the current task lies the assumption that the observations are *controlled* via a set of unobserved *latent* variables that are stacked together in vectors, known as the feature vectors. The basic assumption of the model is that the available observations are *linear combinations* of the set of the unknown latent variables, i.e.,

$$\boldsymbol{x}_n = A\boldsymbol{z}_n, \ \boldsymbol{x}_n \in \mathbb{R}^l, \ \boldsymbol{z}_n \in \mathbb{R}^K, \ n = 1, 2, \ldots, N.$$

Depending on the relative size of l and K, we refer to the task with different names. For example, if $K < l$ we refer to it as *dimensionality reduction*, since less variables are needed to describe the l-dimensional observations.

Collecting all observations together, the previous equation is written as

$$X = AZ, \ X \in \mathbb{R}^{l\times N}, \ Z \in \mathbb{R}^{K\times N}, \tag{13.132}$$

where $X = [\boldsymbol{x}_1, \ldots, \boldsymbol{x}_N]$, $Z = [\boldsymbol{z}_1, \ldots, \boldsymbol{z}_N]$. The task comprises the computation of A and Z, given the matrix X. In general, there are infinitely many solutions to this problem. In practice, one has to mobilize further assumptions in order to restrict the possible set of solutions. A number of such external constraints are treated in Chapter 19. Depending on the imposed conditions, the task is given a different name. In this subsection, we treat the task via nonparametric Bayesian arguments and we are going to impose a prior that will play the role of this extra "something" that will make the task well defined.

The added difficulty is that the number of latent variables, K, is not known a priori.[13] In the mixture modeling, when the number, K, of mixtures was not known, we left it going to infinity, under a Dirichlet prior over the K mixture probabilities; we have already stated that this limit leads to the CRP. In a

[13] We use K to denote the number of latent variables to stress the fact that latent variables here play the role of the mixture components in the mixture modeling task.

similar rationale, we will allow the number of latent variables $K \longrightarrow \infty$, in order to come up with an appropriate prior. Moreover, the prior that will result should have some sparsifying properties, in a similar way that while the number of mixtures K was left to grow to infinity, the resulting prior promoted the formation of a small number of clusters/mixtures.

Since sparsity becomes crucial in our discussion, we will assume that Z comprises zeros and ones. Before we go any further, let us elaborate a bit more on the meaning of the zeros. Take as an example the case of $N = 5$ and $K = 3$ and make some of the matrix elements zeros. Then the matrix factorization in Eq. (13.132) takes the form

$$[\boldsymbol{x}_1, \boldsymbol{x}_2, \dots, \boldsymbol{x}_5] = [\boldsymbol{a}_1, \boldsymbol{a}_2, \boldsymbol{a}_3] \begin{bmatrix} z_{11} & z_{12} & z_{13} & 0 & z_{15} \\ z_{21} & 0 & 0 & z_{24} & z_{25} \\ 0 & z_{32} & z_{33} & z_{34} & 0 \end{bmatrix}.$$

The above implies that

$$\boldsymbol{x}_1 = z_{11}\boldsymbol{a}_1 + z_{21}\boldsymbol{a}_2, \ \boldsymbol{x}_5 = z_{15}\boldsymbol{a}_1 + z_{25}\boldsymbol{a}_2, \tag{13.133}$$

$$\boldsymbol{x}_2 = z_{12}\boldsymbol{a}_1 + z_{32}\boldsymbol{a}_3, \ \boldsymbol{x}_3 = z_{13}\boldsymbol{a}_1 + z_{33}\boldsymbol{a}_3, \tag{13.134}$$

$$\boldsymbol{x}_4 = z_{24}\boldsymbol{a}_2 + z_{34}\boldsymbol{a}_3. \tag{13.135}$$

It does not need much of a thought to realize that the existence of zeros in Z imposes a structure on the observations. Indeed, $\boldsymbol{x}_1$ and $\boldsymbol{x}_5$ lie in the same subspace, spanned by $\boldsymbol{a}_1$ and $\boldsymbol{a}_2$. Observations $\boldsymbol{x}_2$ and $\boldsymbol{x}_3$ lie in the subspace that is spanned by $\boldsymbol{a}_1$ and $\boldsymbol{a}_3$, which is different from the previous one. That is, the existence of zeros imposes a *clustering* structure on the input data.

Thus, our starting point is to select a prior that promotes zeros on Z. Furthermore, we will assume that the elements of Z are either 1 or 0, i.e., $z_{kn} \in \{0, 1\}$, $k = 1, 2, \dots, K$, $n = 1, 2, \dots, N$. Such a simplified treatment will reveal the secrets behind the method. Later on, one can generalize to the more practical terrain, where the nonzero values can take real values. For example, one can write $Z' = Z \circ B$, where $\circ$ denotes element-wise multiplication, and then impose an extra prior on the values of B.

Searching for a Prior on Infinite Binary Matrices

The magic words behind the prior we are looking for is that "it should impose a clustering structure on the data." Although this was the goal in the mixture modeling, the problem here is distinctly different.

- *Mixture modeling*: Each observation belongs to (is emitted from) a *single* mixture component. The underlying structure results from grouping observations together in a number of different mixture components. Observations that are assigned in the same mixture are more "similar" than observations that are assigned to different clusters.
- *Matrix factorization*: Every observation is expressed as a linear combination of a number of columns of matrix A. That is, every observation vector is associated with a number of columns of A. Similarity between observations is established by collecting together observations that are associated with the *same* columns of A.

  In a slightly different jargon, the columns of A are known as *features* and if two observed vectors are given as a combination of the same columns, we say that they *share* the same features. If $z_{kn} = 1$, we say that the kth feature is present in the nth observation.

For a fixed value of K, let P_k, $k = 1, 2, \ldots, K$, be the probability that $z_{kn} = 1$, for any value of n. Then we can write

$$P(Z|\boldsymbol{P}) = \prod_{k=1}^{K}\prod_{n=1}^{N} P(z_{kn}|\boldsymbol{P}) \tag{13.136}$$

$$= \prod_{k=1}^{K} P_k^{m_k}(1 - P_k)^{N-m_k}, \tag{13.137}$$

where $\boldsymbol{P} = [P_1, \ldots, P_K]^T$ and $m_k := \sum_{n=1}^{N} z_{kn}$. That is, m_k is the number of nonzero elements in the kth row of Z; its physical interpretation is that it counts the number of observations that share the kth feature. Obviously, for Eq. (13.137) to hold true, independence among the involved variables has been assumed.

Select a prior for the probabilities: As a prior for P_k, $k = 1, 2 \ldots, K$, we adopt the beta distribution, i.e.,

$$P_k \sim \text{Beta}(P|a, b).$$

Recall that according to the beta distribution, $P \in [0, 1]$. Note that these K probabilities need *not* add to one. In contrast, as we pointed out in the case of CRP, the prior over the K mixture probabilities was taken to be the Dirichlet distribution. This is because they should add to one. Every point was *necessarily* emitted by any one among the K mixtures. In contrast, in our current setting, the kth feature is either shared by an observation, with probability P_k, or not, with probability $1 - P_k$. In order to get the limiting case, $K \longrightarrow \infty$, the parameters of the beta distribution are set equal to $a = \frac{\alpha}{K}$ and $b = 1$. Such a choice makes the normalizing constant (Chapter 2) equal to

$$B(\frac{\alpha}{K}, 1) = \frac{\Gamma(\frac{\alpha}{K})\Gamma(1)}{\Gamma(\frac{\alpha}{K} + 1)} = \frac{K}{\alpha},$$

where the recursive property of the gamma function, $\Gamma(x + 1) = x\Gamma(x)$, has been taken into account.

Thus, adopting the previous prior, the elements of Z are generated according to the following model:

$$P_k \sim \text{Beta}\left(P|\frac{\alpha}{K}, 1\right), \tag{13.138}$$

$$z_{kn} \sim \text{Bern}(z|P_k), \tag{13.139}$$

where the latter distribution is the Bernoulli distribution (Chapter 2). By the way, recall from Problem 12.8 that the beta and Bernoulli distributions form a conjugate pair.

Taking the limit $K \longrightarrow \infty$: The proof is a bit technical and can be found in [23] and Problem 13.24. However, we are going to comment on some of the steps involved in the proof, since these reveal some interesting properties. The first step in deriving a prior by taking the limit of the above model is to compute the probability $P(Z)$. This is done by marginalizing (integrating) P_k out in Eq. (13.137), taking into consideration that they are beta distributed. This results in a dependence of $P(Z)$ that is inversely proportional to K that tends to zero as K tends to infinity. This is natural. If $K \longrightarrow \infty$, the probability of any binary matrix, Z, to occur tends to zero. However, here comes a crucial point. What we are interested in is not $P(Z)$ but something else!

Equivalence classes of binary matrices: The probability $P(Z)$ does not provide the representation we are looking for. The reason is that two matrices may be different, yet they can convey the same information. For example, assume $N = 8$ and consider the kth row of Z to be

$$\tilde{z}_k^T = [1, 0, 0, 1, 1, 0, 0, 1].$$

Also by definition,

$$X = AZ = [\boldsymbol{a}_1, \ldots, \boldsymbol{a}_k, \ldots, \boldsymbol{a}_K] \begin{bmatrix} \tilde{z}_1^T \\ \vdots \\ \tilde{z}_k^T \\ \vdots \\ \tilde{z}_K^T \end{bmatrix},$$

or focusing only on the contribution of the kth feature, we get

$$X = [\boldsymbol{x}_1, \boldsymbol{x}_2, \ldots, \boldsymbol{x}_8] = \ldots + \boldsymbol{a}_k \tilde{z}_k^T + \ldots.$$

Taking into account the specific values of the rows $\tilde{z}_k^T$, given before, it is readily seen from the above that feature $\boldsymbol{a}_k$ is shared only by the observations $\boldsymbol{x}_1$, $\boldsymbol{x}_4$, $\boldsymbol{x}_5$, and $\boldsymbol{x}_8$. However, the important attribute of the above is that the latter four vectors *share* the *same* feature. It is of no importance to us whether this feature is called k or 1 or 3 or whatever. This is in analogy to the mixture modeling, where one does not care if two observations are emitted by the, say, first or second mixture. What is important is that they are emitted by the *same* mixture component. Labeling features is of no importance. The crucial information is that a specific group of observations share a number of *common* features. In a more formal way, the structure of X does *not* depend on the order in which the columns of A and the corresponding rows of Z appear. The information related to the clustering structure of X is *invariant* to *permutations* of A and Z. We are now close to defining the equivalence class concept.

Consider a specific matrix Z for some fixed value of K. Let us now generate *all* possible permutations of the rows of Z. Then, as far as the clustering structure of the corresponding matrix X is concerned (feature sharing), all these permuted matrices are equivalent. We say that they form an *equivalence class*. Take, as an example, the following permuted matrices for $K = 3$ and $N = 5$:

$$\begin{bmatrix} 1 & 1 & 0 & 1 & 0 \\ 1 & 1 & 1 & 0 & 1 \\ 1 & 0 & 1 & 0 & 0 \end{bmatrix}, \quad \begin{bmatrix} 1 & 1 & 1 & 0 & 1 \\ 1 & 1 & 0 & 1 & 0 \\ 1 & 0 & 1 & 0 & 0 \end{bmatrix}.$$

Observe that both matrices unveil the same structure for the input matrix. The only difference is that what we call feature #2 in the matrix on the left, we call #1 in the other.

Let us now denote by $[Z]$ the set of all matrices equivalent to Z. What we are really interested in is to compute the probability $P([Z])$, for all possible equivalent classes. It turns out that as $K \longrightarrow \infty$, $P([Z])$ *remains finite*. The secret behind the difference between $P(Z)$ and $P([Z])$ is that as K keeps increasing to infinity, the number of possible permutations, and as a consequence the cardinality of each equivalence class, also increases; this leads to a finite probability for each equivalent class. Note that each row comprises N zeros and ones. Hence, the maximum number of possible nonzero rows is $2^N - 1$. Thus, the nonzero rows of Z are formed by a random repetition of these binary numbers. What characterizes each equivalence class is that all its members are formed by the same binary numbers. For example, both matrices above consist of the same binary numbers, namely, 11010, 11101, 10100. The sequence in which these numbers occur is of no importance. We have touched upon all the required ingredients one needs to derive $P([Z])$. The proof is given in Problem 13.24.

It turns out that

$$\lim_{K \longrightarrow \infty} P([Z]) \propto \alpha^{K_+} \exp\left(-\alpha H_N\right), \tag{13.140}$$

where K_+ is the number of nonzero rows in the class and $H_N = \sum_{j=1}^{N} \frac{1}{j}$ is a constant. The exact form of the involved constants in the above proportionality relation is of no importance to us (see, e.g., [23] for details).

Having derived the above prior probability over the class of binary matrices, the issue now is how one can sample from or implement such a distribution in practice. In the same vein as with the DPs in Section 13.11.2, there are two alternatives.

Restaurant Construction

In analogy to the CRP, it turns out that the following metaphor is a way to sample from the prior in Eq. (13.140). Imagine that the columns of Z correspond to customers and that the rows correspond to dishes in an infinitely long buffet, which is inspired by the large variety of dishes in a typical Indian restaurant! This is the reason that the method is known as the *Indian buffet process* (IBP).

1. The first customer, $n = 1$, takes $K^{(1)}$ dishes, according to a Poisson distribution with parameter α, i.e.,

$$P\left(K^{(1)}; \alpha\right) = \frac{\alpha^{K(1)} \exp(-\alpha)}{K^{(1)}!}.$$

2. The nth customer takes dishes that have been previously sampled with probability $\frac{m_k}{n}$, where m_k is the number of customers who have sampled the kth dish (feature). He/she also takes $K^{(n)}$ new dishes, according to the Poisson distribution with parameter α/n, i.e.,

$$P\left(K^{(n)}; \frac{\alpha}{n}\right) = \frac{\left(\frac{\alpha}{n}\right)^{K(n)} \exp\left(-\frac{\alpha}{n}\right)}{K^{(n)}!}.$$

Thus, after N customers have been served, a matrix $Z \in \mathbb{R}^{K_+} \times N$ is formed, with K_+ being the total number of dishes that have been sampled. It can be shown that following such a sampling procedure, as $K \longrightarrow \infty$, the probability of any equivalence class $P([Z])$ is equal to that in Eq. (13.140) (e.g., [23]).

FIGURE 13.18

The distribution of dishes per customer. Each new customer selects some of the previously selected dishes and some new ones. Observe that the probability of selecting a new one decreases fast.

It turns out that:

- The number K_+ of nonzero rows of Z is also distributed according to a Poisson with parameter αH_N, i.e.,

$$P(K_+; \alpha H_N) = \frac{(\alpha H_N)^{K_+} \exp(-\alpha H_N)}{K_+!}. \tag{13.141}$$

- As $K \longrightarrow \infty$, matrix Z remains *sparse*. As a matter of fact, the number of nonzero elements follows a Poisson distribution with parameter αN and its mean value is αN. Furthermore, the probability of nonzero values higher than the mean decreases *exponentially*. Also, taking into account that the mean value of a Poisson is equal to its parameter, the average number of dishes selected (features used) is equal to αH_N (e.g., [23]). That is, for a fixed N, the larger the concentration parameter is, the more dishes are selected.

In analogy to Fig. 13.14, Fig. 13.18 shows the distribution of dishes per customer for a specific choice of the α parameter. In CRP, each customer is associated with a *single* table. In CRP, the crucial point is how many customers sit at the same table. In IBP, each customer is associated with *multiple* dishes and the crucial point is how many customers select the same dishes. Note that the number of new dishes keeps decreasing fast. Dishes associated with the first (top) rows are shared by many customers. The dishes associated with rows to the bottom of the figure are shared by fewer and fewer customers.

Fig. 13.19 shows the evolution of the number of dishes per customer, for two different concentration parameters. To avoid confusion, note that, for practical reasons for saving space in presenting the figure, customers correspond to rows and dishes to columns (in contrast to Fig. 13.18). The larger the concentration parameter is, the more dishes (features) are selected, which is in line with what has been said before in relation to Eq. (13.141).

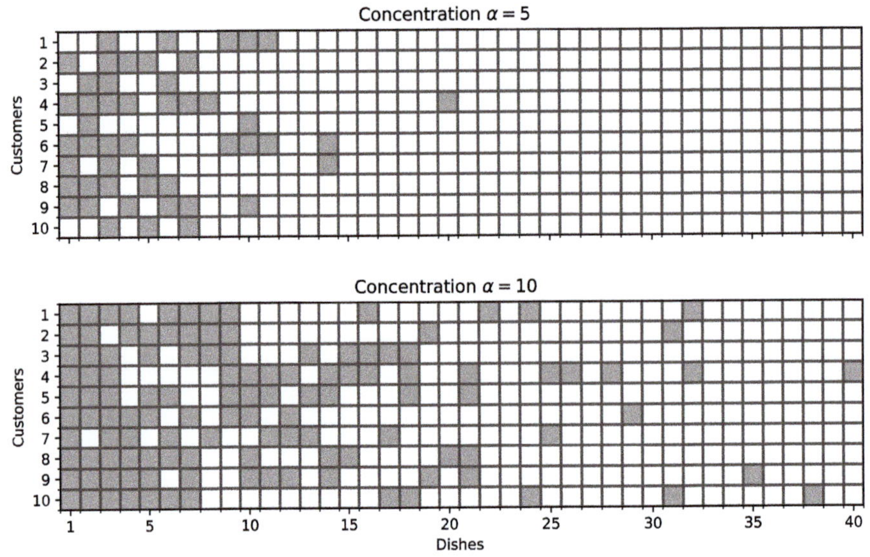

FIGURE 13.19

In this figure, customers are shown in rows and dishes in columns, for graphical convenience. Gray squares correspond to selected dishes. Note that as the concentration parameter increases, more dishes are selected.

Stick Breaking Construction

As was the case with the CRP, the restaurant construction of an IBP fits nicely when inference is done via Monte Carlo sampling, e.g., Gibbs sampling (Chapter 14). However, when variational approximation methods are to be used, a stick breaking construction is more appropriate.

Choose a beta distribution, $\text{Beta}(\beta|\alpha, 1)$. Sample $\beta_1 \sim \text{Beta}(\beta|\alpha, 1)$ and set $P_1 = \beta_1$. Then the kth step of the algorithm is given by

$$\beta_k \sim \text{Beta}(\beta|\alpha, 1), \tag{13.142}$$

$$P_k = \prod_{j=1}^{k} \beta_j, \ k = 1, 2, \ldots. \tag{13.143}$$

Fig. 13.20 illustrates the process. Starting with a stick of unit length, we first break a piece of length β_1, which we keep and we set $P_1 = \beta_1$. The remaining part of the stick of length π_1 is thrown away. In the sequel, we break a piece of length $\beta_2\beta_1$, and we set $P_2 = \beta_1\beta_2$. We keep the piece and the other part of length π_2 is discarded, and so on. Associating each P_k with the probability of the kth feature to occur, it can be shown that this construction of the sequence of probabilities is equivalent to an IBP process [79].

It is interesting to point out that the sequence π_k of the discarded segments implements a sequence related to the DP stick breaking construction given in Eq. (13.127) (Problem 13.25). This parallelism

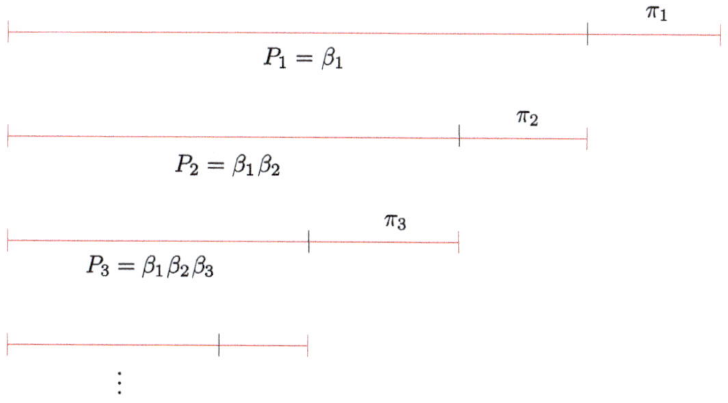

FIGURE 13.20

The stick breaking construction of an IBP. At each iteration, we sample from a Beta(α, 1) and we keep part of the segment in proportion to the sample (left part), while we discard the remaining (right) part. We set the probabilities equal to the lengths of the segments that we keep. The lengths π_k of the discarded parts correspond to the probabilities associated with a DP process.

reveals nicely the difference between an IBP and a DP. In an IBP, the sequence of probabilities is a decreasing one and the respective values do not add to one. In contrast, the sequence of probabilities in a DP is not necessarily decreasing, and the respective values do add to one.

Inference

For the inference, the data generation model should be first explicitly written. To this end, adopting the stick breaking construction of an IBP, let us assume that the data follow a Gaussian distribution with a binary latent feature model; this leads to the following sequence of steps:

1. Generate the beta variables, $\beta_k \sim \text{Beta}(\beta|\alpha, 1)$, $k = 1, 2, \ldots$.
2. Generate the probability values, $P_k = \prod_{j=1}^{k} \beta_j$, $k = 1, 2, 3, \ldots$.
3. Populate matrix Z with elements, $z_{kn} \sim \text{Bern}(z|P_k)$, $n = 1, 2, \ldots, N$ and $k = 1, 2, \ldots$.
4. Generate the features in matrix A, $\boldsymbol{a}_k \sim \mathcal{N}(\boldsymbol{0}, \sigma_A^2 I)$, $k = 1, 2, \ldots$.
5. Generate the observations, $\boldsymbol{x}_n \sim \mathcal{N}(AZ, \sigma_\eta^2 I)$, $n = 1, 2, \ldots, N$.

The hidden/latent variables are (a) the stick breaking lengths, β_k, (b) the elements of Z, and (c) the elements of matrix A. In practice, a truncated stick breaking process is considered, where we assume that $P_k = 0$, $k > K$. Thus, the matrix of the hidden/latent variables is $W = [\boldsymbol{\beta}, A, Z]$. Given the set of observations, $\mathcal{X} = \{\boldsymbol{x}_1, \ldots, \boldsymbol{x}_N\}$, the posterior to be maximized with respect to the unknown parameters, α, σ_A^2 and σ_η^2 is given by

$$p\left(\boldsymbol{\beta}, A, Z|\mathcal{X}; \alpha, \sigma_A^2, \sigma_\eta^2\right).$$

However, it turns out that this is not in a tractable form and the mean field approximation technique can be mobilized (e.g., [15]). In analogy to Eq. (13.131), the factorized family for the variational approximation of the posterior is given by

$$q(W) = \prod_{k=1}^{K} q_{\gamma_k}(\beta_k) \prod_{k=1}^{K} q_{\Phi_k}(\boldsymbol{a}_k) \prod_{n=1}^{N} q_{\nu_{nk}}(z_{nk}),$$

where $q_{\gamma_k}(\beta_k)$ are beta distributions with variational parameters $\boldsymbol{\gamma}_k$, the variational posterior estimates for the columns of A are Gaussians with parameters Φ_k (i.e., means and covariance matrices), and $q_{\nu_{nk}}(z_{nk})$ are Bernoulli with parameters (probabilities) ν_{nk}. Inference is carried out by maximizing the bound in Eq. (12.68). Note that if the sparse matrix is not binary, then it is replaced by $Z \circ B$ and the elements of B are also latent variables; different priors can be used depending on the task; for example, it can be Gaussian or Laplacian (see, e.g., [15,23]). In the latter reference, the alternative path to variational inference via Gibbs sampling techniques is discussed.

Remarks 13.6.

- In analogy to CPR, which draws samples according to a DP, it can be shown that the IBP draws samples according to the so-called beta process [79,88].
- By replacing the parameters $(\alpha, 1)$ in the beta distribution with the more general case Beta(a, b), and changing their values, different distributions result. One such example is the so-called Pitman–Yor IBP. For such processes, the resulting probabilities decay in expectation following a power law; in contrast, in the IBP presented before, the decrease is exponentially fast (see, e.g., [62]).

13.12 GAUSSIAN PROCESSES

In Section 13.11, the way to impose priors onto the model was similar in spirit with that used for parametric modeling techniques; that is, priors were imposed on the set of unknown parameters. In this section, a different rationale will be adopted. The prior will be placed directly over the space of nonlinear functions, rather than specifying a parametric family of nonlinear functions and placing priors over their parameters.

Let us recall the nonlinear regression task given in Eq. (12.1), that is,

$$\mathrm{y} = \theta_0 + \sum_{k=1}^{K-1} \theta_k \phi_k(\boldsymbol{x}) + \eta = \boldsymbol{\theta}^T \boldsymbol{\phi}(\boldsymbol{x}) + \eta, \tag{13.144}$$

where the parameters $\boldsymbol{\theta}$ are treated as a random vector. Let us define

$$\mathrm{f}(\boldsymbol{x}) = \boldsymbol{\theta}^T \boldsymbol{\phi}(\boldsymbol{x}),$$

where $\mathrm{f}(\boldsymbol{x})$ is a *random process*. From Chapter 2, we know that a random process is a random entity whose realization (the outcome of an experiment) is a function, $f(\boldsymbol{x})$, instead of a single value. The idea that spans this section is to work directly on $\mathrm{f}(\boldsymbol{x})$ instead of the indirect approach of modeling it via the set of parameters, $\boldsymbol{\theta}$. This is not the first time we have adopted such a path. We silently did it in Chapter 11 while searching for functions in RKHSs. As a matter of fact, this section can be considered a bridge between the current chapter and Chapter 11.

Recall from Chapter 11 that instead of expanding an unknown function in parameterized form in terms of a number of *preselected* basis functions as in Eq. (13.144), we preferred to search directly for functions that reside in an RKHS; the optimization was carried out with respect to the function itself (not with respect to a set of parameters). In the context of the squared error loss function, the optimization was cast as

$$\min_{f \in \mathbb{H}} \sum_{n=1}^{N} \left(y_n - f(\boldsymbol{x}_n)\right)^2 + C\|f\|^2,$$

where $\|\cdot\|$ denotes the norm in $\mathbb{H}$. The goal in this section is to state the "Bayesian counterpart" to this approach. To this end, we will focus on a specific family of processes, known as Gaussian processes, proposed in [56].

Definition 13.1. A random process, $\mathrm{f}(\boldsymbol{x})$, is called a *Gaussian process* if and only if for *any* finite number of points, $\boldsymbol{x}_{(1)}, \ldots, \boldsymbol{x}_{(N)}$, the respective joint probability density function, $p\left(f(\boldsymbol{x}_{(1)}), \ldots, f(\boldsymbol{x}_{(N)})\right)$, is Gaussian.

We know that a set of jointly Gaussian distributed random variables is fully described by the respective mean value and the covariance matrix. In a similar spirit, a Gaussian process is fully determined by its mean value and its *covariance function*, that is,

$$\mu_x = \mathbb{E}\left[\mathrm{f}(\boldsymbol{x})\right], \quad \mathrm{cov}_f(\boldsymbol{x}, \boldsymbol{x}') = \mathbb{E}\left[(\mathrm{f}(\boldsymbol{x}) - \mu_x)(\mathrm{f}(\boldsymbol{x}') - \mu_{x'})\right].$$

A Gaussian process is said to be *stationary* if $\mu_x = \mu$ and its covariance function is of the form (see also Chapter 2)

$$\mathrm{cov}_f(\boldsymbol{x}, \boldsymbol{x}') = \mathrm{cov}_f(\boldsymbol{x} - \boldsymbol{x}').$$

In addition, if $\mathrm{cov}_f(\cdot, \cdot)$ depends on the *magnitude* of the distance between $\boldsymbol{x}$ and $\boldsymbol{x}'$ (i.e., $\|\boldsymbol{x} - \boldsymbol{x}'\|$), the Gaussian process is called *homogeneous*. From now on, we will assume $\mu_x = 0$. Before we proceed further, let us establish another connection with Chapter 11.

13.12.1 COVARIANCE FUNCTIONS AND KERNELS

For any N and *any* collection of N points, $\boldsymbol{x}_{(1)}, \ldots, \boldsymbol{x}_{(N)}$, the respective covariance matrix is defined by

$$\Sigma = \mathbb{E}[\mathbf{f}\mathbf{f}^T],$$

where

$$\mathbf{f} := [\mathrm{f}(\boldsymbol{x}_{(1)}), \ldots, \mathrm{f}(\boldsymbol{x}_{(N)})]^T, \tag{13.145}$$

with elements given by

$$[\Sigma]_{ij} = \mathrm{cov}_f(\boldsymbol{x}_{(i)}, \boldsymbol{x}_{(j)}), \quad i, j = 1, 2, \ldots, N.$$

Because Σ is a positive semidefinite matrix, this guarantees that the covariance function is a *kernel* function (Section 11.5.1). To stress this, from now on we will use the notation

$$\mathrm{cov}_f(\boldsymbol{x}, \boldsymbol{x}') = \kappa(\boldsymbol{x}, \boldsymbol{x}'),$$

and the covariance matrix becomes the corresponding *kernel matrix* denoted as $\mathcal{K}$ (Chapter 11). This change of notation will make the connections with RKHSs readily spotted. Some typical examples of kernel functions used for Gaussian processes are:

- *Linear kernel*:
$$\kappa(\boldsymbol{x}, \boldsymbol{x}') = \boldsymbol{x}^T \boldsymbol{x}'.$$
Note that this kernel does not correspond to a stationary process.
- *Squared exponential or Gaussian kernel*:
$$\kappa(\boldsymbol{x}, \boldsymbol{x}') = \exp\left(-\frac{\|\boldsymbol{x} - \boldsymbol{x}'\|^2}{2h^2}\right),$$
where h is a parameter determining the *length scale* of the process. The larger the value of h, the larger the "statistical" similarity (stronger correlation) of two points having a distance $d = ||\boldsymbol{x} - \boldsymbol{x}'||$ apart.
- *Ornstein–Uhlenbeck kernel*:
$$\kappa(\boldsymbol{x}, \boldsymbol{x}') = \exp\left(-\frac{\|\boldsymbol{x} - \boldsymbol{x}'\|}{h}\right).$$
- *Rational quadratic kernel*:
$$\kappa(\boldsymbol{x}, \boldsymbol{x}') = \left(1 + \|\boldsymbol{x} - \boldsymbol{x}'\|^2\right)^{-\alpha}, \quad \alpha \geq 0.$$

Recall from Chapter 2, where random processes were first presented, that a stationary covariance function/kernel has as its Fourier transform the power spectrum of the respective random process; by definition, the power spectrum of a process is a nonnegative function in the frequency domain. This suggests a way of constructing kernels for random processes; that is, take the inverse Fourier transform of a positive function in the frequency domain. Moreover, in principle, all the rules for constructing kernels, which are discussed in Section 11.5.2, can also be applied to construct covariance functions. For example, a popular choice of a kernel for a Gaussian process is

$$\kappa(\boldsymbol{x}, \boldsymbol{x}'; \boldsymbol{\theta}) = \theta_1 \exp\left(-\sum_{m=1}^{M} \frac{(x_i - x_i')^2}{2h_i^2}\right) + \theta_2,$$

where θ_1, θ_2 are hyperparameters, which define the process.

Fig. 13.21A shows examples of different realizations of a stationary Gaussian process using the Gaussian covariance kernel with $h = 2$, and Fig. 13.21B for $h = 0.2$.

13.12.2 REGRESSION

Let us assume that we are given a set $\mathcal{X}$ of input observations, $\mathcal{X} = \{\boldsymbol{x}_1, \ldots, \boldsymbol{x}_N\}$. Recall from Section 12.2 that the main goal in a Bayesian regression task is to obtain the two PDFs,

$$p(\boldsymbol{y}|\mathcal{X}) \quad \text{and} \quad p(y|\boldsymbol{x}, \boldsymbol{y}, \mathcal{X}),$$

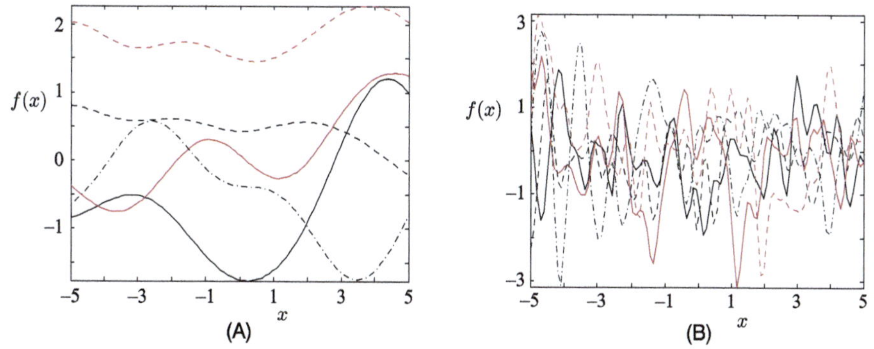

FIGURE 13.21

Different realizations of a Gaussian process. Gaussian covariance kernel with (A) $h = 2$ and (B) $h = 0.2$. Note that when the correlation function fades away fast, the graph of the respective realizations shows a fast variation as a function of the free variable (x).

where

$$\mathbf{y} = \mathbf{f} + \boldsymbol{\eta}, \quad \mathbf{y} := [\mathrm{y}_1, \ldots, \mathrm{y}_N]^T, \tag{13.146}$$

and

$$\mathrm{y} = \mathrm{f}(\boldsymbol{x}) + \eta,$$

and $\mathbf{f}$ is defined in Eq. (13.145). The first of the two PDFs is the joint probability density of the output variables, which are generated by input points in $\mathcal{X}$; the associated randomness is due to f as well as to the noise η. The second PDF refers to the prediction of the value of y, given the value of $\boldsymbol{x}$ and the training data $(y_n, \boldsymbol{x}_n)$, $n = 1, 2, \ldots, N$. We will omit $\mathcal{X}$ to unclutter notation, as we did in Section 12.2.

Assuming f(·) to be a zero mean Gaussian process, $\mathbf{f}$ is jointly Gaussian with zero mean and covariance matrix $\mathcal{K}$, dictated by the covariance function/kernel $\kappa(\cdot, \cdot)$, that is,

$$p(\boldsymbol{f}) = \mathcal{N}(\boldsymbol{f}|\mathbf{0}, \mathcal{K}).$$

Also, let $\boldsymbol{\eta}$ be of zero mean with covariance matrix Σ_η and independent of f(·); without harming generality, let $\Sigma_\eta = \sigma_\eta^2 I$. Thus,

$$p(\boldsymbol{y}|\boldsymbol{f}) = \mathcal{N}(\boldsymbol{y}|\boldsymbol{f}, \sigma_\eta^2 I).$$

Then, following exactly the same arguments as in Section 12.2, we obtain

$$p(\boldsymbol{y}) = \mathcal{N}(\boldsymbol{y}|\mathbf{0}, \mathcal{K} + \sigma_\eta^2 I). \tag{13.147}$$

This is also obvious from the fact that the sum of two independent Gaussian variables is also Gaussian and the mean and covariance matrix can directly be obtained from Eq. (13.146).

To obtain $p(y|\boldsymbol{x}, \mathbf{y})$ we can use (13.147) and apply it recursively. It will also be useful here to bring into the notation the number of available observations, N, explicitly and write

$$\mathbf{y}_{N+1} = \begin{bmatrix} y \\ \mathbf{y}_N \end{bmatrix}, \quad \mathbf{y}_N := [y_1, \ldots, y_N]^T.$$

From Eq. (13.147), $\mathbf{y}_{N+1}$ follows a Gaussian distribution

$$p(\mathbf{y}_{N+1}|\mathbf{0}, \Sigma_{N+1}),$$

with

$$\Sigma_{N+1} := \mathcal{K}_{N+1} + \sigma_\eta^2 I_{N+1}.$$

Then, from the Bayes theorem, we have

$$p(y|\mathbf{y}_N) = \frac{p(\mathbf{y}_{N+1})}{p(\mathbf{y}_N)}. \tag{13.148}$$

However, because the joint PDF is Gaussian, the conditional in Eq. (13.148) is also Gaussian. The respective mean and variance are computed by partitioning the matrix Σ_{N+1} (see Appendix of Chapter 12, Eqs. (12.134) and (12.133))

$$\Sigma_{N+1} = \begin{bmatrix} \kappa(\boldsymbol{x}, \boldsymbol{x}) + \sigma_\eta^2, & \boldsymbol{\kappa}^T(\boldsymbol{x}) \\ \boldsymbol{\kappa}(\boldsymbol{x}), & \Sigma_N \end{bmatrix}, \quad \boldsymbol{\kappa}(\boldsymbol{x}) := [\kappa(\boldsymbol{x}, \boldsymbol{x}_1), \ldots, \kappa(\boldsymbol{x}, \boldsymbol{x}_N)]^T,$$

i.e.,

$$\boxed{\begin{aligned} \mu_y(\boldsymbol{x}) &= \boldsymbol{\kappa}^T(\boldsymbol{x})\Sigma_N^{-1}\mathbf{y}, \\ \sigma_y^2(\boldsymbol{x}) &= \sigma_\eta^2 + \kappa(\boldsymbol{x}, \boldsymbol{x}) - \boldsymbol{\kappa}^T(\boldsymbol{x})\Sigma_N^{-1}\boldsymbol{\kappa}(\boldsymbol{x}). \end{aligned}} \tag{13.149}$$

Compare Eq. (13.149) with Eq. (11.27). Taking into account that $\Sigma_N = \mathcal{K}_N + \sigma_\eta^2 I$, $\mu_y(\boldsymbol{x})$ is identical to $\hat{y}$ obtained by the kernel ridge regression, for appropriate choices of C and σ_η^2. However, now we have also obtained information concerning the respective variance of the resulting estimate.

At this point, it is interesting to look back at the Bayesian regression task for parametric modeling in Section 12.2.3, and to remember that the obtained mean value in Eq. (12.20) was the same (for a zero mean prior $p(\boldsymbol{\theta})$) as that provided by the ridge regression, for an appropriate choice of λ.

Remarks 13.7.

- From the previous discussion it is apparent that solving the regression task by resorting to Gaussian processes is the Bayesian answer to solving a regression task in an RKHS. Both approaches share a common advantage. Although the underlying mapping to an RKHS (implied by the adopted kernel) may live in a high-dimensional space, the complexity for solving the task depends on the number of training points, N. The source of complexity associated with the Gaussian processes is the inversion of the matrix, which amounts to $\mathcal{O}(N^3)$ operations.

- Both equations in (13.149) can be obtained from the corresponding equation derived for the linear case of Bayesian learning, covered in Section 12.2.3. Indeed, setting $\boldsymbol{\theta}_0 = \mathbf{0}$ in Eq. (12.27) and combining it with Eq. (12.22), we obtain

$$\mu_y(\boldsymbol{x}) = \sigma_\theta^2 \boldsymbol{x}^T X^T \left(\sigma_\eta^2 I + \sigma_\theta^2 X X^T\right)^{-1} \boldsymbol{y}, \tag{13.150}$$

where X has replaced Φ, because the linear case is treated. Applying now the kernel trick, as discussed in Chapter 11, to replace $\sigma_\theta^2 \boldsymbol{x}_i^T \boldsymbol{x}_j$ with a kernel $\kappa(\boldsymbol{x}_i, \boldsymbol{x}_j)$ operation, one readily obtains the corresponding equation in Eq. (13.149).

In a similar way, one can obtain $\sigma_y^2(\boldsymbol{x})$ in Eq. (13.149) from Eq. (12.23) by using Woodbury's formula for matrix inversion from Appendix A.1 to reformulate Eq. (12.23) according to Eq. (12.28) (try it).

Dealing With Hyperparameters

As we have already stated, the kernel function can be given in terms of some parameters, say, $\boldsymbol{\theta}$, which in turn have to be estimated from the data. There are various ways to deal with this task. The first that comes to mind is to optimize the resulting parameterized log-likelihood, $\ln p(\boldsymbol{y}; \boldsymbol{\theta})$, with respect to $\boldsymbol{\theta}$, by taking the gradient and equating to zero. Another way is to assume a prior on the parameters and use Bayesian arguments to integrate them out. The integration is usually intractable and approximate techniques must be used, for example, Monte Carlo methods (Chapter 14). Needless to say, both techniques have their drawbacks. Optimizing the log-likelihood is a nonconvex task that cannot guarantee, in general, a global maximum. On the other hand, Monte Carlo techniques tend to be computationally intensive, requiring many iterations to converge. More on these issues can be found in [66].

Computational Considerations

In order to reduce the $\mathcal{O}(N^3)$ computational load associated with the inversion of Σ_N, a number of approximate techniques have been proposed. A possible path is the *sparse Gaussian processes*; in these methods, the full Gaussian process model is approximated by using an expansion in terms of a finite set of basis functions. For example, it is common to use as bases the set $\kappa(\boldsymbol{x}, \boldsymbol{u}_m)$, where $\boldsymbol{u}_m$, $m = 1, 2, \ldots, M \ll N$, is a subset of the input samples known as *active set*. Such techniques can lead to a reduced cost of the order of $\mathcal{O}(M^2 N)$ (e.g., [65]). Other alternatives that do not require the active set to be a subset of the training samples have also been proposed (e.g., [40,77]). In [87], a variational sparse method is proposed that attempts to alleviate problems encountered when one increases the size of the active set.

A variation of the Gaussian processes approach is to equip it with the ability to forget past samples for time-varying environments; this method has been proposed in [61,89] as an alternative to the kernel RLS algorithm discussed in Chapter 11. Other variants use transformations of the output variables to make Gaussian models applicable to a wider range of problems [41,76].

In [70], the connection between Gaussian processes and Kalman filtering is exploited and the solution is obtained via the involvement of stochastic differential equations, which makes the dependence of the complexity on time to be linear.

Finally, an extended review of related techniques is provided in [43].

13.12.3 CLASSIFICATION

In contrast to the regression task, under the Gaussian assumptions for the noise and the involved random process, the classification task gets more involved. In Section 13.7, the logistic regression in its parametric form, given in Eq. (13.68), was employed. In the context of the Gaussian processes, the model becomes

$$P(\omega_1|\boldsymbol{x}) = \frac{1}{1+\exp\big(-f(\boldsymbol{x})\big)} = \sigma\big(f(\boldsymbol{x})\big),$$

where now $f(\boldsymbol{x})$ will be treated in terms of a Gaussian random process, associated with a kernel function $\kappa(\cdot,\cdot)$. Given a set of training samples, $(y_n, \boldsymbol{x}_n)$, $n = 1, 2, \ldots, N$, $y_n \in \{0, 1\}$, and following the same arguments as in Section 13.7, we can now write

$$P(\boldsymbol{y}|\boldsymbol{f}) = \prod_{n=1}^{N} \sigma(f_n)^{y_n}(1-\sigma(f_n))^{1-y_n},$$

where $f_n := f(x_n)$, and

$$p(\boldsymbol{f}) = \mathcal{N}(\boldsymbol{f}|\boldsymbol{0}, \mathcal{K}).$$

Note that $P(\boldsymbol{y}|\boldsymbol{f})$ is no longer Gaussian and the involved integrations needed to obtain $P(\boldsymbol{y})$ and/or $P(y|\boldsymbol{x}, \boldsymbol{y})$ cannot be performed analytically. There are various ways to perform approximations. One path is to resort to the Laplacian approximation of $p(f(\boldsymbol{x})|\boldsymbol{y})$ (see Section 12.3) [90]. Another is to use Monte Carlo techniques [53]. In [20], a variational approach has been used to obtain bounds on the logistic sigmoid and approximate the respective product with a product of Gaussians. The expectation propagation method has been used in [57].

For further reading on Gaussian processes, the interested reader may consult the classical reference [66].

Example 13.6. The goal of this example is to demonstrate the usage of Gaussian processes in regression. To this end, $N = 20$ points were randomly sampled from a realization of a Gaussian process, with zero mean and covariance function based on the Gaussian kernel with length scale $h = 0.5$. The corresponding input points were drawn according to a normal distribution of zero mean and unit variance. In the sequel, Gaussian noise was added to these Gaussian process points, with variance 0.01, to form the set of observed data (shown as "+" in Fig. 13.22). Using these as the training data, predictions of the output variables, corresponding to $D = 1000$ equidistant input points in the interval $[-3, 4]$, were performed; for the prediction, the expressions for the posterior Gaussian process mean and variance in Eq. (13.149) were used. The mean of the posterior Gaussian process is illustrated in Fig. 13.22 as a solid red line. The shaded area surrounding the curve of the posterior mean corresponds to the error bars $\mu_y \pm 2\sigma_y$ of the posterior prediction. Note the increase of the posterior prediction variance in regions where observed data points are scarce.

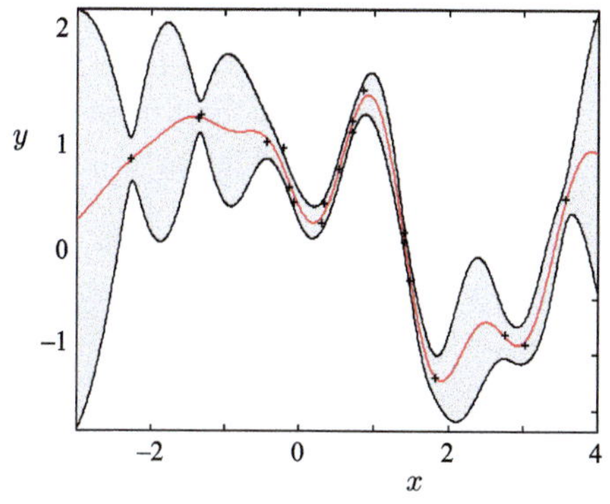

FIGURE 13.22

The red line corresponds to the mean of the posterior Gaussian process. The shaded area corresponds to ± twice the standard deviation.

13.13 A CASE STUDY: HYPERSPECTRAL IMAGE UNMIXING

Hyperspectral image unmixing (HSI) is a typical application of sparse regression modeling under a set of constraints. It is a good "excuse" for us to demonstrate the application of the hierarchical Bayesian modeling approach via a task of great practical importance.

In *hyperspectral remote sensing*, the electromagnetic solar energy emanating from the earth's surface is measured by sensitive scanners located aboard a satellite, an aircraft, or a space station. The scanners are sensitive to a number of wavelength bands of the electromagnetic radiation. Different properties of the earth's surface contribute to the reflection of the energy in the different bands. For example, in the visible–infrared range, properties such as the mineral and moisture contents of soils, the sedimentation of water, and the moisture content of vegetation are the main contributors to the reflected energy. In contrast, at the thermal end of the infrared, it is the thermal capacity and thermal properties of the surface that contribute to the reflection. Thus, each band measures different properties of the same patch of the earth's surface. In this way, images of the earth's surface corresponding to the spatial distribution of the reflected energy in each band can be created. The task now is to exploit this information in order to identify the various ground cover types, that is, built-up land, agricultural land, forest, fire burn, water, diseased crop, and so on.

Fig. 13.23 illustrates the process of generating a pixel's spectral signature out of a hyperspectral image data cube (the cube consists of two spatial and one spectral dimension). Each image corresponds to a single wavelength (band) and each pixel to a specific patch of the earth's surface. The *spectral signature* of a pixel is simply a vector containing radiance values measured in the various spectral bands. Technological advances in recent years have allowed the implementation of imaging spectrometers, which have the ability to collect data in hundreds of adjacent spectral bands. The highly increased volume of data conveys spatial/spectral information that can be properly exploited to accurately determine the type and nature of the objects being imaged.

An intimate limitation of hyperspectral remote sensing is that a single pixel often records a mixed spectral signature of different distinct materials, due to the low spatial resolution of the remote sensor.

This raises the need for spectral unmixing (SU) [37], which is a very important step in hyperspectral image processing that has recently attracted strong scientific interest. SU is the procedure of decomposing the measured spectrum of an observed pixel into a collection of constituent spectral signatures (or *endmembers*) and their corresponding proportions (or *abundances*). A widely used model to perform SU is the linear mixing model.

Assume a remotely sensed hyperspectral image consisting of M spectral bands, and let $\boldsymbol{y} \in \mathbb{R}^M$ be the vector containing the measured spectral signature (i.e., the radiance values in all spectral bands) of a single pixel (specific earth patch). Also let $X = [\boldsymbol{x}_1, \boldsymbol{x}_2, \ldots, \boldsymbol{x}_l]$ stand for the $M \times l$ endmember signature matrix, where $\boldsymbol{x}_i \in \mathbb{R}^M$, $i = 1, 2, \ldots, l$, comprises the spectral signatures of the ith endmember, and l is the total number of (possible) distinct endmembers (earth surface/material types) present in the scene. Finally, let $\boldsymbol{\theta} = [\theta_1, \theta_2, \ldots, \theta_l]^T$ be the *abundance vector* associated with $\boldsymbol{y}$, where θ_i denotes the abundance fraction of $\boldsymbol{x}_i$ in $\boldsymbol{y}$. The linear mixing model assumes that there is a linear relationship between the spectra of the measured pixel and the endmembers, expressed as

$$\boldsymbol{y} = X\boldsymbol{\theta} + \boldsymbol{\eta}, \tag{13.151}$$

where $\boldsymbol{\eta}$ stands for the additive noise values, which are assumed to be samples of a zero mean Gaussian distributed random vector, with (i.i.d.) elements, that is, $\boldsymbol{\eta} \sim \mathcal{N}(\boldsymbol{\eta}|\boldsymbol{0}, \beta^{-1} I_M)$, where β denotes the inverse of the noise variance (precision), and I_M is the $M \times M$ identity matrix. Note that the model in Eq. (13.151) is a typical regression model in its multivariate formulation, because now the output for each measurement is a vector and not a scalar (see also Section 4.9 of Chapter 4). The output

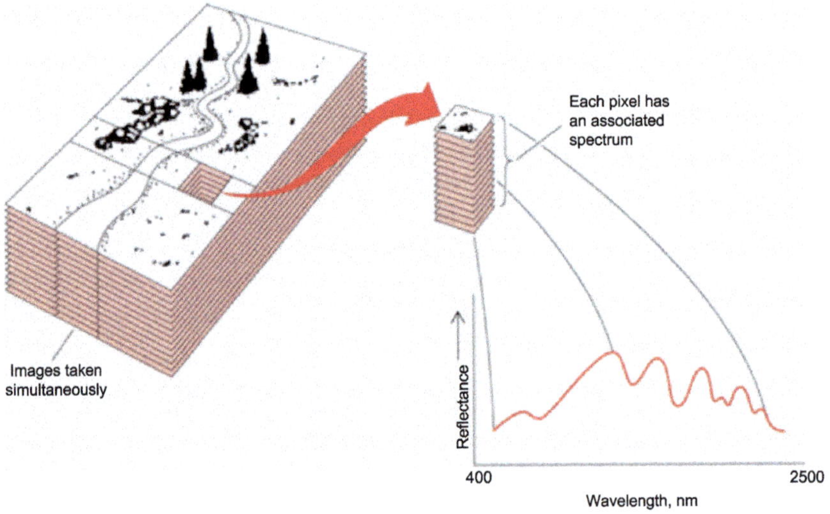

FIGURE 13.23

Each image corresponds to a specific wavelength band and each pixel to a particular patch of the earth's surface. The signature of a pixel is a vector whose coefficients measure the radiance of the respective patch of the earth in the different bands (modified image taken from [69]).

variables are measured and the matrix X is assumed known, and indeed there are methods to estimate its elements.

Treating such a model to recover the abundance coefficients would be a straightforward application of what has been said so far in the current and previous chapters of this book. However, there is a physical constraint that has to be considered and that makes the task more interesting. The abundance coefficients are nonnegative, that is,

$$\theta_i \geq 0, \quad i = 1, 2, \ldots, l. \tag{13.152}$$

Additionally, a valid assumption is that only a few of the endmembers present in the image will contribute to the spectrum of a single pixel $\boldsymbol{y}$. In other words, the abundance vector $\boldsymbol{\theta}$ accepts a *sparse* representation in X.

Thus, our goal is to estimate $\boldsymbol{\theta}$ subject to the nonnegativity as well as the sparsity constraints, given the spectral measurements, $\boldsymbol{y}$, and the endmember matrix, X. Obviously, there are different paths to achieve this goal. Because we are currently exploring the Bayesian world, we will employ the Bayesian framework. To this end, an appropriate prior model that expresses our prior belief on the parameters of interest will first be adopted, and we will then perform Bayesian inference using the variational Bayes methodology, as has been previously discussed.

13.13.1 HIERARCHICAL BAYESIAN MODELING

The presence of Gaussian noise in Eq. (13.151) dictates that

$$\begin{aligned} p(\boldsymbol{y}|\boldsymbol{\theta}, \beta) &= \mathcal{N}(\boldsymbol{y}|X\boldsymbol{\theta}, \beta^{-1} I_M) \\ &= (2\pi)^{-\frac{M}{2}} \beta^{\frac{M}{2}} \exp\left(-\frac{\beta}{2}\|\boldsymbol{y} - X\boldsymbol{\theta}\|^2\right). \end{aligned} \tag{13.153}$$

We now turn our attention to selecting suitable priors for the model parameters, which are treated as random variables, $\boldsymbol{\theta}, \beta$. As a prior for the nonnegative noise precision β we adopt a Gamma distribution (Section 13.3, Eq. (13.25)), expressed as

$$p(\beta) = \text{Gamma}(\beta|c, d) = \frac{d^c}{\Gamma(c)} \beta^{c-1} \exp(-d\beta), \tag{13.154}$$

where c and d are the respective parameters (set equal to 10^{-6} in the experiments).

For the abundance vector $\boldsymbol{\theta}$, we define a two-level hierarchical prior that is expressed in a conjugate form and imposes sparsity as well as nonnegativity on the abundance coefficients. Inspired by [68], a nonnegatively truncated Gaussian prior is selected, i.e.,

$$p(\boldsymbol{\theta}|\boldsymbol{\alpha}) = \mathcal{N}_{\mathbb{R}_+^l}\left(\boldsymbol{\theta}|\boldsymbol{0}, A^{-1}\right), \tag{13.155}$$

where $\boldsymbol{\alpha} := [\alpha_1, \alpha_2, \ldots, \alpha_l]^T$ is the precision parameter vector, $A = \text{diag}\{\alpha_1, \ldots, \alpha_l\}$ is the corresponding diagonal matrix, and $\mathcal{N}_{\mathbb{R}_+^l}$ signifies the l-variate normal distribution truncated at the nonnegative orthant of $\mathbb{R}^l$, denoted by $\mathbb{R}_+^l$ [81]. In the second level of hierarchy, the precision parameters are also considered random variables, $\alpha_i, i = 1, 2, \ldots, l$, that follow an inverse Gamma distribution, that is,

$$p(\alpha_i) = \text{IGamma}\left(\alpha_i|1, \frac{b_i}{2}\right) = \frac{b_i}{2}\alpha_i^{-2}\exp\left(-\frac{b_i}{2}\frac{1}{\alpha_i}\right), \tag{13.156}$$

where $b_i, i = 1, 2, \ldots, N$, are scale hyperparameters. These two levels of hierarchy form a nonnegatively truncated multivariate Laplace prior over the abundance vector $\boldsymbol{\theta}$, which can be established by integrating out the precision $\boldsymbol{\alpha}$ [81], that is,

$$p(\boldsymbol{\theta}|\mathbf{b}, \beta) = \prod_{i=1}^{l}\sqrt{\beta b_i}\exp\left(-\sqrt{\beta b_i}|\theta_i|\right) I_{\mathbb{R}_+^l}(\boldsymbol{\theta}), \tag{13.157}$$

where $I_{\mathbb{R}_+^l}(\boldsymbol{\theta})$ is the indicator function, with $I_{\mathbb{R}_+^l}(\boldsymbol{\theta}) = 1$ (resp. 0) if $\boldsymbol{\theta} \in \mathbb{R}_+^l$ (resp. $\boldsymbol{\theta} \notin \mathbb{R}_+^l$). In our formulation, the sparsity promoting scale hyperparameters in Eq. (13.156) are also assumed to be random and are inferred from the data, by assuming the following Gamma prior distribution for each $b_i, \ i = 1, 2, \ldots, l$:

$$p(b_i) = \text{Gamma}(b_i|\kappa, \nu) = \frac{\nu^\kappa}{\Gamma(\kappa)}b_i^{\kappa-1}\exp(-\nu b_i). \tag{13.158}$$

Hyperparameters κ and ν in Eq. (13.158) are also set to small values (10^{-6} in the experiments).

Having adopted the hierarchical Bayesian model, the variational EM algorithm discussed in Section 13.3 is applied with the goal of obtaining estimates, $q(\theta_i), \ i = 1, 2, \ldots, l$, of the posteriors of the abundance parameters given the observations. In the experiments, the respective mean values of $q(\theta_i)$ will be used as estimates of the unknown parameter values. Details on the derivation can be obtained from [82]. The alternative path to the variational EM algorithm is to employ Monte Carlo techniques (see, e.g., [14]).

13.13.2 EXPERIMENTAL RESULTS

The previously described model was applied to a real hyperspectral image, collected by the Airborne Visible/Infrared Imaging Spectrometer (AVIRIS) over a Cuprite mining district in Nevada in the summer of 1997.[14] The Cuprite data set has been extensively used to evaluate remote sensing technologies and spectral unmixing algorithms (e.g., [30,51,81]). It comprises 224 spectral bands in the range from 400 to 2500 nanometers. A subimage of the Cuprite data set with size 250 × 191 pixels is used in our experiments. Fig. 13.24 displays a composite of our image, where bands 183, 193, and 203 have been used.

After removing some low signal-to-noise ratio (SNR) bands and water vapor absorption bands, $M = 188$ spectral bands remain available for processing. As a preprocessing step, the VCA algorithm[15] has been used to extract 14 endmembers from our hyperspectral image, as in [51]. The vertex component analysis (VCA) algorithm identifies the signatures of the "pure" pixels in the image and considers

[14] The data are publicly available at http://aviris.jpl.nasa.gov/data/free_data.html.

[15] The VCA code is available at http://www.lx.it.pt/~bioucas/code.htm.

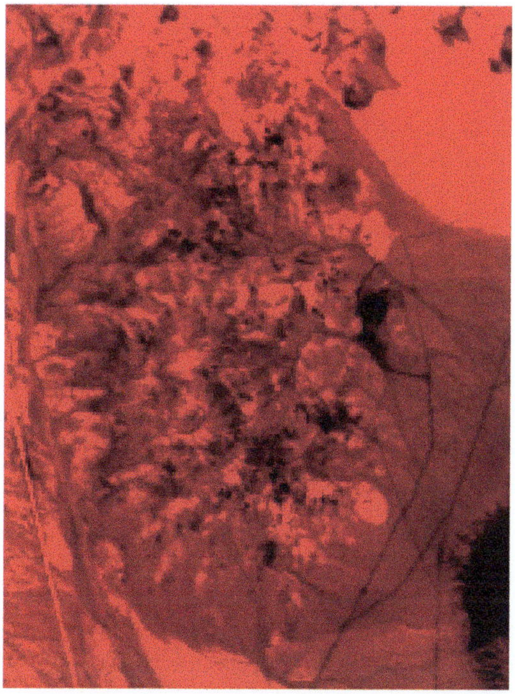

FIGURE 13.24

Composite of the AVIRIS Cuprite subimage using bands 183, 193, and 203 (from [69]). The full RGB color image is available from the site of this book.

them pure material signatures. A plot of the spectral signatures of the extracted endmembers versus the wavelength is displayed in Fig. 13.25.

Fig. 13.26 shows the resulting abundance maps for six different endmembers, using the variational Bayes method. A dark (resp. light) pixel reveals a low (resp. high) proportional percentage for the respective endmember in that pixel. In other words, each image shows the distribution of values of a specific abundance coefficient, θ_i, over the sensed earth surface.

More important, we are able to identify the presented endmembers in Fig. 13.26 as muscovite, alunite, buddingtonite, montmorillonite, kaolinite 1, and kaolinite 2.

PROBLEMS

13.1 Show Eq. (13.5).

13.2 Show Eq. (13.38).

13.3 Show Eqs. (13.43)–(13.45).

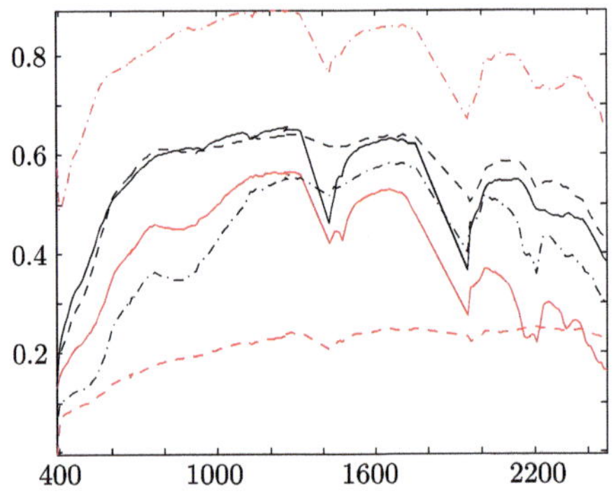

FIGURE 13.25

Spectral signatures of six out of the 14 endmembers extracted from the Cuprite image using the VCA algorithm [51]. A figure showing all 14 signatures can be downloaded from the site of this book.

13.4 Show that if

$$p(x) \propto \frac{1}{x},$$

then the random variable $z := \ln x$ follows a uniform distribution.

13.5 Derive the lower bound after convergence of the variational Bayesian EM for the linear regression task, which is modeled in Section 13.3.

13.6 Consider the Gaussian mixture model

$$p(\boldsymbol{x}) = \sum_{k=1}^{K} P_k \mathcal{N}(\boldsymbol{x}|\boldsymbol{\mu}_k, Q_k^{-1}),$$

with priors

$$p(\boldsymbol{\mu}_k) = \mathcal{N}(\boldsymbol{\mu}_k|0, \beta^{-1}I) \tag{13.159}$$

and

$$p(Q_k) = \mathcal{W}(Q_k|\nu_0, W_0).$$

Given the set of observations $\mathcal{X} = \{\boldsymbol{x}_1, \dots, \boldsymbol{x}_N\}, \boldsymbol{x} \in \mathbb{R}^l$, derive the respective variational Bayesian EM algorithm, using the mean field approximation for the involved posterior PDFs. Consider $P_k,\ k = 1, 2, \dots, K$, as deterministic parameters and optimize the respective lower bound of the evidence with respect to the P_ks.

13.7 Consider the Gaussian mixture model of Problem 13.6, with the following priors imposed on μ, Q, and **P**:

$$\begin{aligned} p(\boldsymbol{\mu}, Q) &= p(\boldsymbol{\mu}|Q)p(Q) \\ &= \prod_{k=1}^{K} \mathcal{N}\left(\boldsymbol{\mu}_k|\mathbf{0}, (\lambda Q_k)^{-1}\right) \mathcal{W}(Q_k|\nu_0, W_o), \end{aligned}$$

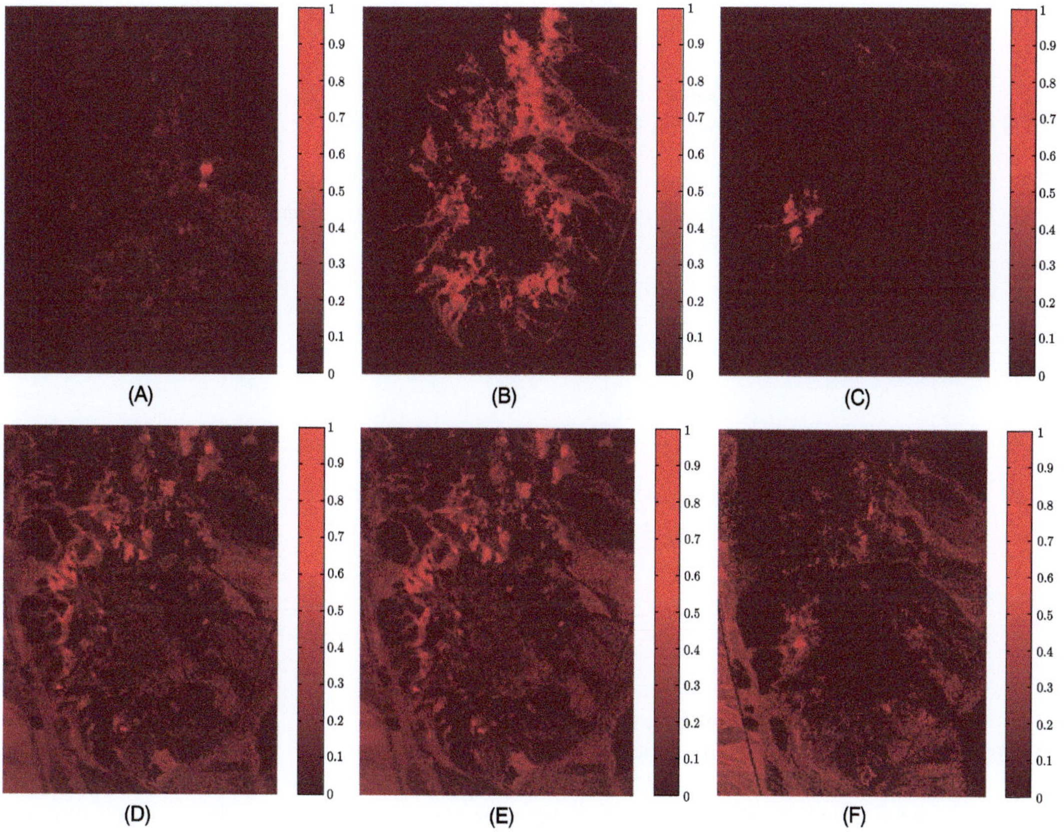

FIGURE 13.26

Estimated abundance maps for the materials (A) muscovite, (B) alunite, (C) buddingtonite, (D) montmorillonite, (E) kaolinite 1, and (F) kaolinite 2. The full-color image is available from the site of this book.

that is, a Gaussian–Wishart product, and

$$p(\boldsymbol{P}) = \mathrm{Dir}(\boldsymbol{P}|a) \propto \prod_{k=1}^{K} P_k^{a-1},$$

that is, a Dirichlet prior. Thus, **P** is treated as a random vector. Derive the E algorithmic steps of the variational Bayesian approximation, adopting the mean field approximation for the involved posterior PDFs. We have adopted the notation $\boldsymbol{\mu}$ in place of $\boldsymbol{\mu}_{1:K}$ and Q in place of $Q_{1:K}$ for notational simplicity.

13.8 If μ and Q are distributed according to a Gaussian–Wishart product,

$$p(\boldsymbol{\mu}, Q) = \mathcal{N}(\boldsymbol{\mu}|\hat{\boldsymbol{\mu}}, (\lambda Q)^{-1})\mathcal{W}(Q|\nu, W),$$

compute the expectation

$$\mathbb{E}[\boldsymbol{\mu}^T Q \boldsymbol{\mu}].$$

13.9 Derive the Hessian matrix with respect to $\boldsymbol{\theta}$ of the cost function

$$J(\boldsymbol{\theta}) = \sum_{n=1}^{N} \left[y_n \ln \sigma\left(\boldsymbol{\phi}^T(x_n)\boldsymbol{\theta}\right) + (1-y_n) \ln\left(1 - \sigma\left(\boldsymbol{\phi}^T(x_n)\boldsymbol{\theta}\right)\right)\right] - \frac{1}{2}\boldsymbol{\theta}^T A \boldsymbol{\theta},$$

where

$$\sigma(z) = \frac{1}{1+\exp(-z)}.$$

13.10 Show that the marginal of a Gaussian PDF with a gamma prior on the variance, after integrating out the variance, is the Student's t PDF, given by

$$\mathrm{st}(x|\mu, \lambda, \nu) = \frac{\Gamma(\frac{\nu+1}{2})}{\Gamma(\frac{\nu}{2})} \left(\frac{\lambda}{\pi \nu}\right)^{1/2} \frac{1}{\left(1 + \frac{\lambda(x-\mu)^2}{\nu}\right)^{\frac{\nu+1}{2}}}. \tag{13.160}$$

13.11 Derive the pair of recursions Eqs. (13.62)–(13.63).

13.12 Consider a two-class classification task and assume that the feature vectors in each one of the two classes, ω_1, ω_2, are distributed according to the Gaussian PDF. Both classes share the same covariance matrix Σ, and the mean values are $\boldsymbol{\mu}_1$ and $\boldsymbol{\mu}_2$, respectively. Prove that, given an observed feature vector, $\boldsymbol{x} \in \mathbb{R}^l$, the posterior probabilities for deciding in favor of one of the classes is given by the logistic function, i.e.,

$$P(\omega_2|\boldsymbol{x}) = \frac{1}{1 + \exp\left(-\boldsymbol{\theta}^T \boldsymbol{x} + \theta_0\right)},$$

where

$$\boldsymbol{\theta} := \Sigma^{-1}(\boldsymbol{\mu}_2 - \boldsymbol{\mu}_1)$$

and

$$\theta_0 = \frac{1}{2}(\boldsymbol{\mu}_2 - \boldsymbol{\mu}_1)^T \Sigma^{-1} (\boldsymbol{\mu}_2 + \boldsymbol{\mu}_1) + \ln \frac{P(\omega_1)}{P(\omega_2)}.$$

13.13 Derive Eq. (13.74).

13.14 Show Eq. (13.75).

13.15 Derive the recursion Eq. (13.77).

13.16 Show that if f is a convex function, $f : \mathbb{R}^l \longrightarrow \mathbb{R}$, then it is equal to the conjugate of its conjugate, i.e., $(f^*)^* = f$.

13.17 Prove that

$$f(x) = \ln \frac{\lambda}{2} - \lambda\sqrt{x}, \quad x \geq 0$$

is a convex function.

13.18 Derive variational bounds for the logistic regression function

$$\sigma(x) = \frac{1}{1+e^{-x}},$$

one of them in terms of a Gaussian function. For the latter case, use the transformation $t = \sqrt{x}$.

13.19 Prove Eq. (13.100).

13.20 Derive the mean and variance of $G(T_k)$ for a DP process.

13.21 Show that the posterior DP, after having obtained n observations from the set Θ, is given by

$$G|\boldsymbol{\theta}_1, \ldots, \boldsymbol{\theta}_n \sim \mathrm{DP}\left(\alpha + n, \frac{1}{\alpha+n}\left(\alpha G_0 + \sum_{i=1}^{n} \delta_{\boldsymbol{\theta}_i}(\boldsymbol{\theta})\right)\right).$$

13.22 The stick breaking construction of a DP is built around the following rule: $P_1 = \beta_1 \sim \mathrm{Beta}(\beta|1,\alpha)$ and

$$\beta_i \sim \mathrm{Beta}(\beta|1,\alpha), \tag{13.161}$$

$$P_i = \beta_i \prod_{j=1}^{i-1}(1-\beta_j),\ i \geq 2. \tag{13.162}$$

Show that if the number of steps is finite, i.e., we assume that $P_i = 0,\ i > T$, for some T, then $\beta_T = 1$.

13.23 Show that in CRP, the cluster assignments are exchangeable and do not depend on the sequence that customers arrive, up to a permutation of the labels of the tables.

13.24 Show that in an IBP, the probabilities for $P(Z)$ and the equivalence classes $P([Z])$, are given by the formulas

$$P(Z) = \prod_{k=1}^{K} \frac{\alpha}{K} \frac{\Gamma\left(m_k + \frac{\alpha}{K}\right)\Gamma(N - m_k + 1)}{\Gamma\left(N + 1 + \frac{\alpha}{K}\right)}$$

and

$$P([Z]) = \frac{K!}{\prod_{h=0}^{2^N-1} K_h!} \prod_{k=1}^{K} \frac{\alpha}{K} \frac{\Gamma\left(m_k + \frac{\alpha}{K}\right)\Gamma(N - m_k + 1)}{\Gamma\left(N + 1 + \frac{\alpha}{K}\right)},$$

respectively. Note that $K_h,\ h = 1, 2, \ldots, 2^N - 1$, is the number of times the row vector associated with the hth nonzero binary number appears in Z.

13.25 Show that the discarded pieces, π_k, in the stick breaking construction of an IBP are equal to the sequence of probabilities produced in a DP stick breaking construction.

MATLAB® EXERCISES

13.26 Generate $N = 60$ data points from each of the five Gaussian distributions given in Example 13.1. Implement the EM algorithm to obtain estimates of the parameters of the Gaussian mixture model (Exercise 12.18). Run the EM algorithm on our generated data, assuming $K = 25$ clusters, using randomly chosen values for the initial mean values and the covariance matrices. Next, implement the variational Bayes algorithm that treats the same problem, according to the steps reported in Section 13.4. Plot the initial and final estimates of the EM and the variational Bayes algorithm to reproduce the results of Fig. 13.4. Play with different values of the parameters.

13.27 Generate a vector comprising $N = 100$ equidistant sampling points x_n in the interval $[-10, 10]$. Compute N basis functions, each one located at a sampling point x_n, of the form $\phi_n(x) = \exp\left(-(x - x_n)^2/2\sigma_\phi^2\right)$, where $\sigma_\phi^2 = 0.1$. Select two of the basis functions randomly to compute the output samples, y_n, according to the regression model of Example 13.2. The additive noise power should correspond to an SNR level of 6 dB. Implement the EM algorithm expressed in Eqs. (12.43), (12.44), (12.51), and (12.52), in order to fit a (generalized) linear regression model comprising the N basis functions to the generated data y_n. Also, implement the variational Bayes EM, summarized in Algorithm 13.1. Plot the reconstructed signals and compare the results.

13.28 Generate $N = 150$ two-dimensional data points $\boldsymbol{x}_n$, uniformly distributed in the region $[-5, 5] \times [-5, 5]$. Assign a binary label to each $\boldsymbol{x}_n$, depending a) on which side of the graph of the function

$$f(x) = 0.05x^3 + 0.05x^2 + 0.05x + 0.05$$

the point lies and b) on the value of a noise variable. To generate the training data, for each sample $\boldsymbol{x}_n = [x_{n1}, x_{n2}]^T$, compute

$$y_n = 0.05x_{n1}^3 + 0.05x_{n1}^2 + 0.05x_{n1} + 0.05 + \eta,$$

where η stands for zero-mean Gaussian noise of variance $\sigma_\eta^2 = 4$. If $x_{n2} \geq y_n$ assign $\boldsymbol{x}_n$ to ω_1, otherwise assign it to class ω_2. Download and run the MATLAB code of the RVM classifier[16] for the generated data set. Use the Gaussian kernel with $\sigma^2 = 3$. Repeat the experiments with different values of σ^2. Plot the points $\boldsymbol{x}_n$ using different colors for each class. Plot the obtained decision curves (classifier) and discuss the results.

13.29 Download the MATLAB® code for the CRP mixture model from *http://sites.google.com/site/kenichikurihara/academic-software*. Generate two-dimensional data from the Gaussian mixture model of Example 13.5 and reproduce the results in Fig. 13.17.

13.30 Consider a one-dimensional Gaussian process with zero mean and Gaussian (kernel) covariance function, with length scale $h = 0.5$

(a) Sample $D = 100$ equidistant input points in the interval $[-2, 2]$. Use these as input points to compute the covariance function of the Gaussian process and form the respective 100×100 covariance matrix. Use the corresponding multivariate Gaussian to generate

[16] The RVM software can be found at http://www.miketipping.com/sparsebayes.htm.

samples for five different realizations and plot the results, as in Fig. 13.21. Repeat the same experiment with different values for the parameter h.

(b) Now, sample $N = 20$ input points from a zero mean, unit-variance normal distribution. Based on these input points, evaluate the covariance function and the respective 20×20 covariance matrix, as before. Then, generate noisy Gaussian process data, by first sampling N points from our Gaussian process, and then adding zero mean Gaussian noise with variance 0.1. Next, sample $D = 100$ points in the interval $[-3, 4]$. Compute the corresponding mean and the variance of the predictive Gaussian process, as given in Eq. (13.149). In a single figure, plot the observed data, the posterior mean, and the error bars of the predictive mean, as in Fig. 13.22.

13.31 Reproduce the hyperspectral unmixing results of Fig. 13.26 by running the script "HSIvB.m," which is available at the website of the book.

REFERENCES

[1] D. Aldous, Exchangeability and related topics, in: École d'Été de Probabilités de Saint-Flour XIII-1983, in: Lecture Notes in Mathematics, Springer, New York, 1985, pp. 1–198.

[2] S. Amari, Differential Geometrical Methods in Statistics, Springer, New York, 1985.

[3] C. Antoniak, Mixtures of Dirichlet processes with applications to Bayesian nonparametric problems, Ann. Stat. 2 (6) (1974) 1152–1174.

[4] H. Attias, Inferring parameters and structure of latent variable models by variational Bayes, in: K.B. Laskey, H. Prade (Eds.), Proceedings of the 15th Conference on Uncertainty in Artificial Intelligence, Morgan-Kaufmann, San Mateo, 1999, pp. 21–30.

[5] S. Babacan, R. Molina, A. Katsaggelos, Fast Bayesian compressive sensing using Laplace priors, in: Proceedings International Conference on Acoustics, Speech and Signal Processing, ICASSP, Taipei, Taiwan, 2009.

[6] S.D. Babacan, L. Maniera, R. Molina, A. Katsaggelos, Non-convex priors in Bayesian compressive sensing, in: Proceedings, 17th European Signal Processing Conference, EURASIP, Glasgow, Scotland, 2009.

[7] M.J. Beal, Variational Algorithms for Approximate Bayesian Inference, PhD Thesis, University College London, 2003.

[8] C. Bishop, M. Tipping, Variational relevance vector machines, in: Proceedings of the 16th Conference on Uncertainty in Artificial Intelligence, 2000, pp. 46–53.

[9] C.M. Bishop, Pattern Recognition and Machine Learning, Springer, New York, 2006.

[10] D. Blackwell, J.B. MacQueen, Ferguson distributions via Pólya urn schemes, Ann. Stat. 1 (2) (1973) 353–355.

[11] D. Blei, M. Jordan, Variational inference for Dirichlet process mixtures, Bayesian Anal. 1 (1) (2006) 121–144.

[12] S. Boyd, L. Vandenberghe, Convex Optimization, Cambridge University Press, Cambridge, 2004.

[13] A. Ben-Israel, T.N.E. Greville, Variational Bayesian model selection for mixture distribution, in: T. Jaakula, T. Richardshon (Eds.), Artificial Intelligence and Statistics, Morgan-Kaufmann, San Mateo, 2001, pp. 27–34.

[14] N. Dobigeon, J.-Y. Tourneret, C.-I. Chang, Semi-supervised linear spectral unmixing using a hierarchical Bayesian model for hyperspectral imagery, IEEE Trans. Signal Process. 56 (7) (2008) 2684–2695.

[15] F. Doshi-Velez, K.T. Miller, J. Van Gael, Y.W. Teh, Variational inference for the Indian buffet process, in: 12th International Conference on Artificial Intelligence and Statistics, AISTATS, 2009.

[16] T. Ferguson, A Bayesian analysis of some nonparametric problems, Ann. Stat. 1 (2) (1973) 209–230.

[17] R.P. Feyman, A Set of Lectures, Perseus, Reading, MA, 1972.

[18] M.A.P. Figuerido, Adaptive sparseness for supervised learning, IEEE Trans. Pattern Anal. Mach. Learn. 25 (9) (2003) 1150–1159.

[19] J. Gershman, D.M. Blei, A tutorial on Bayesian nonparametric models, J. Math. Psychol. 56 (2012) 1–12.

[20] M.N. Gibs, D.J.C. MacKay, Variational Gaussian process classifiers, IEEE Trans. Neural Netw. 11 (6) (2000) 1458–1464.

[21] M. Girolami, A variational method for learning sparse and overcomplete representations, Neural Comput. 13 (2001) 2517–2532.

[22] P. Green, S. Richardson, Modeling heterogeneity with and without the Dirichlet process, Scand. J. Stat. 28 (2) (2001) 355–375.
[23] T.L. Griffiths, Z. Ghahramani, The Indian buffet process: an introduction and review, J. Mach. Learn. Res. (JMLR) 12 (2011) 1185–1224.
[24] L.K. Hansen, C.E. Rasmussen, Pruning from adaptive regularization, Neural Comput. 6 (1993) 1223–1232.
[25] G.E. Hinton, D. Van Camp, Keeping neural networks simple by minimizing the description length of weight, in: Proceedings 6th ACM Conference on Computing Learning, Santa Cruz, 1993.
[26] G.E. Hinton, D.S. Zemel, Autoencoders, minimum description length and Helmholtz free energy, in: J.D. Conan, G. Tesauro, J. Alspector (Eds.), Advances in Neural Information Processing System, vol. 6, Morgan-Kaufmann, San Mateo, 1999.
[27] N. Hjort, C. Holmes, P. Muller, S. Walker, Bayesian Nonparametrics, Cambridge University Press, Cambridge, 2010.
[28] M.D. Hoffman, M.D. Blei, C. Wang, J. Paisley, Stochastic variational inference, J. Mach. Learn. Res. 14 (2013) 1303–1347.
[29] H. Ishwaran, J.S. Rao, Spike and slab variable selection: frequentist and Bayesian strategies, Ann. Stat. 33 (2) (2005) 730–773.
[30] M.D. Iordache, J.M. Bioucas-Dias, A. Plaza, Collaborative sparse regression for hyperspectral unmixing, IEEE Trans. Geosci. Remote Sens. 52 (1) (2014) 341–354.
[31] T.J. Jaakola, Variational Methods for Inference and Estimation in Graphical Models, PhD Thesis, Department of Brain and Cognitive Sciences, MIT, Cambridge, USA, 1997.
[32] T.J. Jaakola, M.I. Jordan, Improving the mean field approximation via the use of mixture distributions, in: M.I. Jordan (Ed.), Learning in Graphical Models, Kluwer, Dordrecht, 1998, pp. 163–173.
[33] T.J. Jaakola, M.I. Jordan, Bayesian logistic regression: a variational approach, Stat. Comput. 10 (2000) 25–37.
[34] S. Ji, Y. Xue, L. Carin, Bayesian compressive sensing, IEEE Trans. Signal Process. 56 (6) (2008) 2346–2356.
[35] M.I. Jordan, Z. Ghahramaniz, T.J. Jaakola, L.K. Saul, An introduction to variational methods in graphical models, Mach. Learn. 37 (1999) 183–233.
[36] M. Jordan, Bayesian nonparametric learning: expressive priors for intelligent systems, in: R. Dechter, H. Geffner, J.Y. Halpern (Eds.), Heuristics, Probability and Causality, College Publications, 2010.
[37] N. Keshava, A survey of spectral unmixing algorithms, Linc. Lab. J. 14 (1) (2003) 55–78.
[38] K. Kurihara, M. Welling, Y. Teh, Collapsed variational Dirichlet process mixture models, in: Proceedings of the International Joint Conference on Artificial Intelligence, vol. 20, 2007, pp. 2796–2801.
[39] M. Kuss, C. Rasmussen, Assessing approximations for Gaussian classification, in: Advances in Neural Information Processing Systems, vol. 18, MIT Press, Cambridge, MA, 2006.
[40] M. Lazaro-Gredilla, A. Figueiras-Vidal, Inter-domain Gaussian processes for sparse inference using inducing features, in: Advances in Neural Information Processing Systems, vol. 22, MIT Press, Cambridge, MA, 2010.
[41] M. Lazaro-Gredilla, Bayesian warped Gaussian processes, in: Advances in Neural Information Processing Systems, vol. 25, MIT Press, Cambridge, MA, 2013.
[42] F.B. Lempers, Posterior Probabilities of Alternative Linear Models, Rotterdam University Press, Rotterdam, 1971.
[43] H. Liu, Y.S. Ong, X. Shen, J. Cai, When Gaussian process meets big data: a review of scalable GPs, arXiv:1807.01065v2 [stat.ML], 9 Apr 2019.
[44] D.J.C. McKay, Bayesian interpolation, Neural Comput. 4 (3) (1992) 417–447.
[45] D.J.C. MacKay, The evidence framework applied to classification networks, Neural Comput. 4 (1992) 720–736.
[46] D.J.C. MacKay, Information Theory, Inference and Learning Algorithms, Cambridge University Press, Cambridge, 2003.
[47] D.J.C. MacKay, Bayesian nonlinear modeling for the energy prediction competition, ASHRAE Trans. 100 (2) (1994) 1053–1062.
[48] T. Minka, Expectation propagation for approximate Bayesian inference, in: J. Breese, D. Koller (Eds.), Proceedings 17th Conference on Uncertainty in Artificial Intelligence, 2001, pp. 362–369.
[49] T. Minka, Divergence Measures and Message Passing, Technical Report, Microsoft Research Laboratory, Cambridge, UK, 2005.
[50] T. Mitchell, J. Beauchamp, Bayesian variable selection in linear regression, J. Am. Stat. Assoc. 83 (1988) 1023–1036.
[51] J.M.P. Nascimento, J.M. Bioucas-Dias, Vertex component analysis: a fast algorithm to unmix hyperspectral data, IEEE Trans. Geosci. Remote Sens. 43 (4) (2005) 898–910.
[52] R.M. Neal, Bayesian Learning for Neural Networks, Lecture Notes in Statistics, vol. 118, Springer-Verlag, New York, 1996.
[53] R.M. Neal, Monte Carlo Implementation for Gaussian Process Models for Bayesian Regression and Classification, Technical Report CRG-TR-97-2, Department of Computer Science, University of Toronto, 1997.

[54] R.M. Neal, Assessing relevance determination methods using DELVE, in: C. Bishop (Ed.), Neural Networks and Machine Learning, Springer-Verlag, New York, 1998, pp. 97–120.

[55] R. Neal, Markov chain sampling methods for Dirichlet process mixture models, J. Comput. Graph. Stat. 9 (2) (2000) 249–265.

[56] A. O'Hagan, J.F. Kingman, Curve fitting and optimal design for prediction, J. R. Stat. Soc. B 40 (1) (1978) 1783–1816.

[57] M. Opper, O. Winther, A Bayesian approach to on-line learning, in: D. Saad (Ed.), On-Line Learning in Neural Networks, Cambridge University Press, Cambridge, 1999, pp. 363–378.

[58] J. Paisley, A. Zaas, C.W. Woods, G.S. Ginsburg, L. Carin, A stick-breaking construction of the beta process, in: Proceedings of the 27th International Conference on Machine Learning, 2010.

[59] J. Palmer, D. Wipf, K. Krentz-Delgade, B. Rao, Variational EM algorithms for non-Gaussian latent variable models, in: Advances in Neural Information Systems, vol. 18, 2006, pp. 1059–1066.

[60] G. Parisi, Statistical Field Theory, Addison Wesley, New York, 1988.

[61] F. Perez-Cruz, S. Van Vaerenbergh, J.J. Murillo-Fuentes, M. Lazaro-Gredilla, I. Santamaria, Gaussian processes for non-linear signal processing, IEEE Signal Process. Mag. 30 (4) (2013) 40–50.

[62] J. Pittman, M. Yor, The two-parameter Poisson-Dirichlet distribution derived from a stable subordinator, Ann. Stat. 25 (1997) 855–900.

[63] J. Pitman, Combinatorial Stochastic Processes, Technical report 621, Notes for Saint Flour Summer School, Department of Statistics, UC, Berkeley, 2002.

[64] P. Pal, P.P. Vaidyanathan, Parameter identifiability in sparse Bayesian learning, in: Proceedings International Conference on Acoustics, Speech and Signal Processing, ICASSP, Florence, Italy, 2014.

[65] J. Quionero-Candela, C.E. Rasmussen, A unifying view of sparse approximate Gaussian process regression, Mach. Learn. Res. 6 (2005) 1939–1959.

[66] C.E. Rasmussen, C.K.I. Williams, Gaussian Processes for Machine Learning, MIT Press, Cambridge, MA, 2006.

[67] R. Rockaffelar, Convex Analysis, Princeton University Press, Princeton, NJ, 1970.

[68] G.A. Rodriguez-Yam, R.A. Davis, L.L. Scharf, A Bayesian model and Gibbs sampler for hyperspectral imaging, in: Proceedings, IEEE Sensor Array and Multichannel Signal Processing Workshop, 2002, pp. 105–109.

[69] S. Ryan, M. Lewis, Mapping soils using high resolution airborne imagery, Barossa Valley, SA, in: Proceedings of the Inaugural Australian Geospatial Information and Agriculture Conference Incorporating Precision Agriculture in Australasia 5th Annual Symposium, 2001, pp. 17–19.

[70] S. Sarkka, A. Solin, J. Hartikainen, Spatiotemporal learning via infinite-dimensional Bayesian filtering and smoothing, IEEE Signal Process. Mag. 30 (4) (2013) 51–61.

[71] M.A. Sato, Online model selection based on the variational Bayes, Neural Comput. 13 (7) (2001) 1649–1681.

[72] M.N. Schmidt, M. Morup, Nonparametric Bayesian modeling of complex networks, IEEE Signal Process. Mag. 30 (3) (2013) 110–128.

[73] M.W. Seeger, H. Nickish, Large Scale Variational Inference and Experimental Design for Sparse Generalized Linear Models, Technical report, # TR-175, Max Plank Institute für Biologische Kybernetic, 2008.

[74] M.W. Seeger, D.P. Wipf, Variational Bayesian inference techniques, IEEE Signal Process. Mag. 27 (1) (2010) 81–91.

[75] J. Sethuraman, A constructive definition of Dirichlet priors, Stat. Sin. 4 (2) (1994) 639–650.

[76] E. Snelson, C.E. Rasmussen, Z. Ghahramani, Warped Gaussian processes, in: Advances in Neural Information Processing Systems, vol. 16, MIT Press, Cambridge, MA, 2003.

[77] E. Snelson, Z. Ghahramani, Sparse Gaussian processes using pseudo-inputs, in: Advances in Neural Information Processing Systems, vol. 18, MIT Press, Cambridge, MA, 2006, pp. 1259–1266.

[78] C. Soussen, J. Idier, D. Brie, J. Duan, From Bernoulli-Gaussian deconvolution to sparse signal restoration, IEEE Trans. Signal Process. 59 (10) (2011) 4572–4584.

[79] Y.W. Teh, D. Görür, Z. Gahramani, Stick-breaking construction for the Indian buffet process, in: Proceedings 11th Conference on Artificial Intelligence and Statistics, AISTATS, 2007.

[80] Y.W. Teh, Dirichlet Process, Technical Report, University of London, 2010, http://www.gatsby.ucl.ac.uk/~ywteh/research/npbayes/Teh2010a.pdf.

[81] K.E. Themelis, A.A. Rontogiannis, K.D. Koutroumbas, A novel hierarchical Bayesian approach for sparse semisupervised hyperspectral unmixing, IEEE Trans. Signal Process. 60 (2) (2012) 585–599.

[82] K.E. Themelis, A.A. Rontogiannis, K.D. Koutroumbas, Semisupervised hyperspectral image unmixing using a variational Bayes algorithm, arXiv:1406.4705, 2014.

[83] K. Themelis, A. Rontogiannis, K. Koutroumbas, A variational Bayes framework for sparse adaptive estimation, IEEE Trans. Signal Process. 62 (18) (2014) 4723–4736.
[84] S. Theodoridis, K. Koutroumbas, Pattern Recognition, fourth ed., Academic Press, Boston, 2009.
[85] M.E. Tipping, Sparse Bayesian learning and the relevance vector machine, J. Mach. Learn. Res. 1 (2001) 211–244.
[86] M.E. Tipping, A.C. Faul, Fast marginal likelihood maximisation for sparse Bayesian models, in: C.M. Bishop, B.J. Frey (Eds.), Proceedings of the Ninth International Workshop on Artificial Intelligence and Statistics, Key West, FL, 2003.
[87] M.K. Titsias, Variational learning of inducing variables in sparse Gaussian processes, in: Proceedings 12th International Workshop on Artificial Intelligence and Statistics, 2009, pp. 567–574.
[88] R. Thibaux, M.I. Jordan, Hierarchical beta processes and the Indian buffet process, in: Proceedings 11th Conference on Artificial Intelligence and Statistics, AISTATS, 2007.
[89] S. Van Vaerenbergh, M. Lazaro-Gredilla, I. Santamaria, Kernel recursive least-squares tracker for time-varying regression, IEEE Trans. Neural Netw. Learn. Syst. 23 (8) (2012) 1313–1326.
[90] C.K.I. Williams, D. Barber, Bayesian classification with Gaussian processes, IEEE Trans. Pattern Anal. Mach. Intell. 20 (1998) 1342–1351.
[91] D. Wipf, Bayesian Methods for Finding Sparse Representations, PhD Thesis, University of California, San Diego, 2006.
[92] D.P. Wipf, B.D. Rao, An empirical Bayesian strategy for solving the simultaneous sparse approximation problem, IEEE Trans. Signal Process. 55 (7) (2007) 3704–3716.
[93] D. Wipf, S. Nagarajan, A new view of automatic relevance determination, in: Advances in Neural Information Systems, NIPS, vol. 20, 2008.
[94] D. Wipf, B. Rao, S. Nagarajan, Latent variable models for promoting sparsity, IEEE Trans. Inf. Theory 57 (9) (2011) 6236–6255.
[95] D.P. Wipf, B.D. Rao, S. Nagarajan, Latent variable Bayesian methods for promoting sparsity, IEEE Trans. Inf. Theory 57 (9) (2011) 6236–6255.
[96] X. Zhang, B.D. Rao, Sparse signal recovery with temporally correlated source vectors using sparse Bayesian learning, IEEE Trans. Sel. Areas Signal Process. 5 (5) (2011) 912–926.
[97] X. Zhang, B.D. Rao, Extension of SBL algorithms for the recovery of block sparse signals with intra-block correlation, IEEE Trans. Signal Process. 61 (8) (2013) 2009–2015.

CHAPTER

MONTE CARLO METHODS 14

CONTENTS

14.1 INTRODUCTION

In Chapters 12 and 13, the Bayesian inference task was considered. A large part of the latter chapter was dedicated to dealing with approximation techniques, which offered escape routes when the involved PDFs were complex enough to render integral computations intractable. All these techniques were of a deterministic nature; that is, the goal was to approximate the mathematical expression of the corresponding PDF by another one that could ease the associated calculations. Such methods include the Laplacian approximation, as well as the variational methods based on the mean field theory or the convex duality concept. Deterministic approximation methods will also be used for approximate inference in Chapter 15, to deal with graphical models.

In this chapter, we turn our attention to approximation methods with a much stronger statistical flavor, which are based on randomly generating samples using numerical techniques; these samples are typical of an underlying distribution, which may be of either continuous or discrete nature. This is an old field, with origins tracing back to the late 1940s and early 1950s in the pioneering work of Stanislav Ulam, John Von-Neumann, and Nicholas Metropolis in Los Alamos, when the term *Monte Carlo* was coined as an umbrella name of such techniques, inspired by the famous casino in Monaco

Machine Learning. https://doi.org/10.1016/B978-0-12-818803-3.00026-X

(see, e.g., [28] for a historical note). The first application of such techniques, which coincided with the development of the first computers, was in the context of the Manhattan project for developing the hydrogen bomb; soon after, Monte Carlo methods were embraced by almost every scientific area where statistical computations are involved.

As is often the case with pioneering ideas, when they are looked at a posteriori, that is, once they have been stated, the basic idea seems simple. Our current task of interest is the computation of an integral, which involves a PDF; this can alternatively be interpreted as the computation of an "expectation." Such a view provides the permit to approximate the integral as the sample mean of the involved quantities, given a sufficient number of samples and exploiting the law of large numbers.

To condense a field with a history of a number of decades in a single chapter is obviously impossible. Our goal is to present the basic concepts, definitions, and directions, with the aim of serving the needs associated with typical machine learning tasks rather than looking at it as an entity on its own.

We start with the more classical methods using transformations and then move on to the rejection and importance sampling techniques. In the sequel, the more powerful methods based on arguments from the theory of Markov chains are reviewed. The Metropolis–Hastings and Gibbs sampling methods are presented and discussed. Finally, a case study concerning the change-point detection task is considered.

14.2 MONTE CARLO METHODS: THE MAIN CONCEPT

Our starting point is the evaluation of integrals of the form

$$\mathbb{E}\left[f(\mathbf{x})\right] := \int_{-\infty}^{\infty} f(\boldsymbol{x})p(\boldsymbol{x})\,d\boldsymbol{x}, \tag{14.1}$$

where $\mathbf{x} \in \mathbb{R}^l$ is a random vector and $p(\boldsymbol{x})$ is the corresponding distribution.[1] Our interest lies in cases where the forms of $f(\boldsymbol{x})$ and/or $p(\boldsymbol{x})$ are such that the evaluation of such integrals is intractable. For example, such integrations occur in the evaluation of the evidence function (Eq. (12.14)), in the prediction task (Eq. (12.18)), and in the E-step of the EM algorithm (Eq. (12.40)). In Eq. (12.14), the random variable is the parameter vector $\boldsymbol{\theta}$ and $f(\boldsymbol{\theta}) = p(\boldsymbol{y}|\boldsymbol{\theta})$.

Coming back to Eq. (14.1), assume that one has at her/his disposal a number of i.i.d. samples, $\boldsymbol{x}_1, \ldots, \boldsymbol{x}_N$, drawn from $p(\boldsymbol{x})$. Then the approximation

$$\mathbb{E}\left[f(\mathbf{x})\right] \simeq \frac{1}{N}\sum_{i=1}^{N} f(\boldsymbol{x}_i) := \bar{\mathbb{E}}_{f,N} \tag{14.2}$$

is justified by (a) the *law of large numbers* and (b) the *central limit theorem* [32]. Let us denote $\mathbb{E}[f(\mathbf{x})] = \mu_f$ and the respective variance as $\text{var}[f(\mathbf{x})] := \mathbb{E}\left[(f(\mathbf{x}) - \mathbb{E}[f(\mathbf{x})])^2\right] = \sigma_f^2$. Then the pre-

[1] In the case of discrete variables, $p(\boldsymbol{x})$ becomes the probability mass function $P(\boldsymbol{x})$, and integrations are replaced by summations.

viously referred two theorems guarantee that

$$\lim_{N\to\infty} \bar{\mathbb{E}}_{f,N} = \mu_f, \tag{14.3}$$

and

$$p(\bar{\mathbb{E}}_{f,N}) \simeq \mathcal{N}\left(\bar{\mathbb{E}}_{f,N} | \mu_f, \frac{\sigma_f^2}{N}\right). \tag{14.4}$$

The limit in Eq. (14.3) refers to the notion of *almost sure convergence*, that is,

$$\text{Prob}\left\{\lim_{N\to\infty} \left|\mu_f - \bar{\mathbb{E}}_{f,N}\right| = 0\right\} = 1.$$

The approximate Gaussian distribution in Eq. (14.4) guarantees that the variance (as one changes the set of N samples) of the obtained estimate $\bar{\mathbb{E}}_{f,N}$ around the true value μ_f decreases with N.

Thus, if one generates the samples $\boldsymbol{x}_n$, $n = 1, 2, \ldots, N$, from the distribution $p(\boldsymbol{x})$, the use of Monte Carlo techniques offers the means for an approximation of the integral in Eq. (14.1) with the following nice properties: (a) the approximation error is decreasing as $\frac{1}{\sqrt{N}}$; (b) the obtained estimate using N samples is an unbiased estimate of the true value; and (c) the convergence rate is *independent* of the dimensionality l. The latter property is in contrast to methods based on the deterministic numerical integration, which, in general, have a rate of convergence that slows down as the dimensionality increases. In Monte Carlo techniques, if one is not satisfied with the obtained accuracy, all he/she has to do is generate more samples.

The crucial point now becomes that of developing techniques to generate i.i.d. samples from $p(\boldsymbol{x})$. This is not an easy task, especially for high-dimensional spaces. Note that achieving a certain accuracy for the estimator in Eq. (14.2) is independent of the dimensionality, once i.i.d. samples drawn from $p(\boldsymbol{x})$ are available. On the other hand, drawing i.i.d. samples typical of $p(\boldsymbol{x})$ becomes harder as the dimensionality increases. We will return to this point soon. In the sequel, we will focus on some basic directions to achieve the aforementioned goal.

14.2.1 RANDOM NUMBER GENERATION

Random number generation can be achieved either as the result of an experiment or via the use of computers. For example, the tosses of a fair coin can generate a random sequence of 0s (heads) or 1s (tails). Another example is the sequence of numbers corresponding to the distance between radioactive emissions; such an experiment generates a sequence of exponentially distributed samples. However, such approaches are not of much practical value and the emphasis has been on techniques that generate samples via a computer, using a *pseudorandom number generator*. At the heart of such methods lie algorithms that guarantee the generation of a sequence of *integers*, z_i, which approximately follow a *uniform distribution* in an interval in the real axis. In the sequel, the generation of random numbers/vectors, which follow an arbitrary distribution, is obtained *indirectly* via a variety of methods, each with its pros and cons. The path for generating integers in an interval $(0, M)$ follows the general recursion,

$$z_i = g(z_{i-1}, \ldots, z_{i-m}) \bmod M,$$

where g is a function depending on the m previously generated samples and mod denotes the modulus operation; that is, z_i is the remainder of the division of $g(z_{i-1}, \dots, z_{i-m})$ by M. The simpler form is the linear version,

$$z_i = \alpha z_{i-1} \bmod M, \quad z_0 = 1, \ i \geq 1, \tag{14.5}$$

where M is a large prime number and α is an integer. Recursion (14.5) generates a sequence of numbers between 1 and $M-1$. The method is known as *linear congruential generator* or Lehmer's algorithm [20].

If α is properly chosen, then the resulting sequence of numbers turns out to be periodic with period $M-1$. This is the reason we call these generators pseudorandom, because a periodic sequence can never be claimed to be random. However, for large values of M, the obtained sequence can be sufficiently random with uniform distribution, provided, of course, that $N \leq M-1$. For example, a value of M of the order of 10^9 is sufficient for most applications. Note that not all possible choices of the parameter α guarantee a good generator. In practice, a sequence is accepted as being random only if it meets a number of related tests of randomness and is subsequently used successfully in a variety of applications (see, e.g., [32]). A common choice of parameters that leads to a reasonably good uniformly distributed random sequence is $\alpha = 7^5$ and $M = 2^{31} - 1$ (see, e.g., [34]). More on this topic can be found in Knuth's classical text and the references therein [18]. Once a sequence of integers is available, a sequence of uniformly distributed real random numbers is obtained as the ratio $x_i = \frac{z_i}{M} \in (0, 1)$ (as a matter of fact, this is the sequence on which the randomness tests are applied).

Remarks 14.1.

- Note that even the generation of a sequence of (pseudo)random numbers with uniform distribution in (0, 1) is not an easy task, in spite of the fact that the uniform distribution is an "easy" one; that is, all values are equally probable. Moreover, often in practice, a PDF is known up to its normalizing constant, i.e.,

  $$p(\boldsymbol{x}) = \frac{\phi(\boldsymbol{x})}{Z},$$

  where

  $$Z = \int_{-\infty}^{+\infty} \phi(\boldsymbol{x})\, d\boldsymbol{x}.$$

  However, if $\phi(\boldsymbol{x})$ has a complicated form, the previous integration may be intractable. This is often met when computing posterior PDFs. The previous points make the process of sampling from a general $p(\boldsymbol{x})$ much harder than for the case of a uniform one. The task becomes even harder in high dimensions, even if Z is available. The required number of points, in order to cover sufficiently a region in a high-dimensional space, exhibits an exponential dependence on the respective dimensionality (curse of dimensionality). Thus, a huge number of points is needed in order to get a good representation of $p(\boldsymbol{x})$ in high-dimensional spaces. In practice, one would be more content to generate samples from the regions where $p(\boldsymbol{x})$ gets relatively high values. However, the higher the dimensionality, the more difficult the task of locating the high-probability regions. Similar arguments hold for random variables of a discrete nature, where the number of states that the variable can take is very large. Ideally, in order to have a representative sequence of samples, all states have to be visited.

14.3 RANDOM SAMPLING BASED ON FUNCTION TRANSFORMATION

In this section, we deal with some of the most basic techniques for drawing samples from a PDF, $p(\boldsymbol{x})$.

Function inversion. Let x be a real random variable with a PDF, $p(x)$, and a corresponding cumulative distribution function

$$F_{\mathrm{x}}(x) = \int_{-\infty}^{x} p(\tau)\, d\tau.$$

It is known from probability theory that the random variable, u, defined as

$$\mathrm{u} := F_{\mathrm{x}}(\mathrm{x}), \tag{14.6}$$

is *uniformly* distributed in the interval $0 \leq u \leq 1$ *irrespective of the nature* of $p(x)$ (see [32] and Problem 14.1). If, in addition, we assume that the function F_{x} has an inverse, F_{x}^{-1}, then we can write

$$\mathrm{x} = F_{\mathrm{x}}^{-1}(\mathrm{u}). \tag{14.7}$$

Thus, following the reverse arguments, samples from $p(x)$ can be generated by first generating samples from the uniform distribution, $\mathcal{U}(u|0,1)$, and then applying on them the inverse function, F_{x}^{-1} (Problem 14.2).

This method works well provided that F_{x} has an inverse that can be easily computed. However, only a few PDFs can be "proud" of having inverses that can be expressed in an analytical form.

Example 14.1. Generate samples, x_n, that follow the exponential distribution,

$$p(x) = \lambda \exp(-\lambda x), \quad x \geq 0,\ \lambda > 0, \tag{14.8}$$

using a pseudorandom generator that generates samples, u_n, from the uniform distribution $\mathcal{U}(u|0,1)$.

We have

$$F_{\mathrm{x}}(x) = \int_{0}^{x} \lambda \exp(-\lambda \tau)\, d\tau = 1 - \exp(-\lambda x).$$

By letting

$$u := F_{\mathrm{x}}(x)$$

and solving for x, we get

$$x = -\frac{1}{\lambda} \ln(1-u) := F_{\mathrm{x}}^{-1}(u).$$

Hence, if u_n are samples drawn from a uniform distribution, the sequence

$$x_n = -\frac{1}{\lambda} \ln(1-u_n), \quad n = 1, 2, \ldots, N,$$

are samples drawn from the exponential PDF in Eq. (14.8). Fig. 14.1 shows the histogram of the generated samples for $N = 1000$, alongside $p(x)$ for $\lambda = 1$.

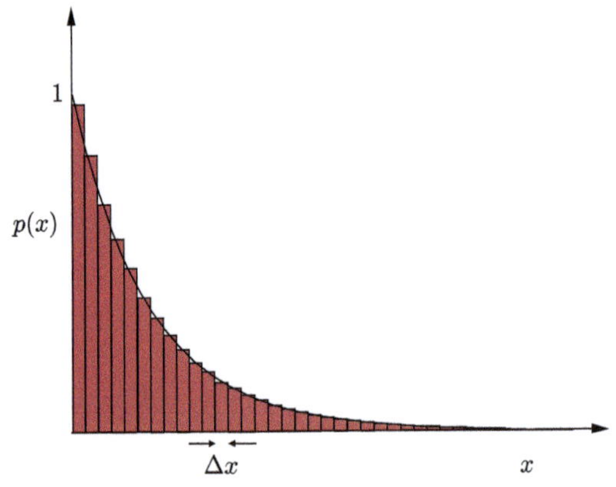

FIGURE 14.1

The histogram of the samples generated from the uniform distribution, and using the inverse of F_{x}, which describes the exponential PDF, whose curve is shown in black. The length of the bin interval, Δx, was chosen equal to 0.02.

Example 14.2. *Generating samples from discrete distributions.* Here, an intuitive method for generating samples from discrete distributions is presented. We will use such distributions in Section 17.2.

Let $\mathrm{x}_1, \mathrm{x}_2, \ldots, \mathrm{x}_K$ denote discrete random events occurring with probabilities $P_1, P_2, \ldots, P_K$, respectively, such that $\sum_{k=1}^{K} P_k = 1$. Then the following simple algorithm draws samples from this distribution.

Algorithm 14.1 (Sampling discrete distributions).

- Define $a_k = \sum_{i=1}^{k-1} P_i$, $b_k = \sum_{i=1}^{k} P_i$, $k = 1, 2, \ldots, K$, $a_1 = 0$.
- **For** $i = 1, 2, \ldots,$ **Do**
  - $u \sim \mathcal{U}(0, 1)$
  - Select
    - x_k if $u \in [a_k, b_k)$, $\quad k = 1, 2, \ldots, K$
- **End For**

Fig. 14.2 provides an illustration of the algorithm. Note that the probability jumps at the beginning of each interval and the corresponding cumulative distribution function (CDF) is constructed; the algorithm basically computes the inverse of u, according to this CDF (see, e.g., [4]).

Function transformation. We will demonstrate the method via an example involving the transformation of two random variables, say, r and ϕ, to two new ones, x and y. Let

$$\mathrm{x} = g_{\mathrm{x}}(\mathrm{r}, \phi)$$

and

$$\mathrm{y} = g_{\mathrm{y}}(\mathrm{r}, \phi).$$

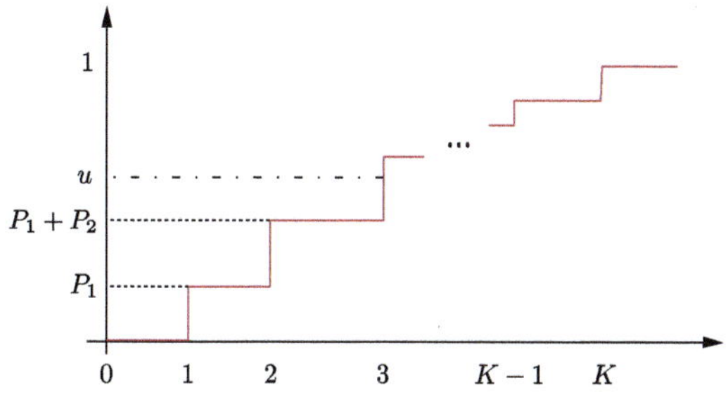

FIGURE 14.2

The CDF for a discrete distribution of K discrete random events. If $P_1 + P_2 \leq u < P_1 + P_2 + P_3$, the event x_3 is drawn. Note that the higher the probability of an event is, the larger the corresponding interval jump in the CDF is, hence the higher the probability of this event being drawn.

Let us now assume that there is a unique solution for the inverses and that they can be expressed in an analytic form (which is not the case in general), that is,

$$\mathrm{r} = g_{\mathrm{r}}(\mathrm{x}, \mathrm{y}),$$
$$\phi = g_{\phi}(\mathrm{x}, \mathrm{y}).$$

We know from Section 2.2.5 that if $p_{\mathrm{r},\phi}(r, \phi)$ is the joint distribution of r and ϕ, then the joint distribution of x and y is given by

$$\begin{aligned} p_{\mathrm{x,y}}(x, y) &= \frac{p_{\mathrm{r},\phi}\big(g_{\mathrm{r}}(x, y), g_{\phi}(x, y)\big)}{\big|\det\big(J(\mathrm{x}, \mathrm{y}; \mathrm{r}, \phi)\big)\big|} \\ &= p_{\mathrm{r},\phi}\big(g_{\mathrm{r}}(x, y), g_{\phi}(x, y)\big)\big|\det\big(J(\mathrm{r}, \phi; \mathrm{x}, \mathrm{y})\big)\big|, \end{aligned} \tag{14.9}$$

where $\big|\det\big(J(\mathrm{x}, \mathrm{y}; \mathrm{r}, \phi)\big)\big|$ is the absolute value of the determinant of the Jacobian matrix,

$$J(\mathrm{x}, \mathrm{y}; \mathrm{r}, \phi) = \begin{bmatrix} \frac{\partial g_{\mathrm{x}}}{\partial r} & \frac{\partial g_{\mathrm{x}}}{\partial \phi} \\ \frac{\partial g_{\mathrm{y}}}{\partial r} & \frac{\partial g_{\mathrm{y}}}{\partial \phi} \end{bmatrix}, \tag{14.10}$$

$J(\mathrm{r}, \phi; \mathrm{x}, \mathrm{y})$ is analogously defined, and we have assumed, for simplicity, that to each value of (r, ϕ) there corresponds one value of (x, y). Let us now see how one can generate samples from a Gaussian $p(x) = \mathcal{N}(x|0, 1)$ by using samples drawn from a uniform and an exponential distribution, respectively, for ϕ and r; recall that in Example 14.1, we described a technique for generating samples from an exponential distribution.

The Box–Muller method. Let r be distributed according to an exponential distribution,

$$p_{\mathrm{r}}(r) = \frac{1}{2}\exp\left(-\frac{r}{2}\right), \quad r \geq 0, \tag{14.11}$$

and ϕ to a uniform distribution, $\mathcal{U}(\phi|0, 1)$,

$$p_{\phi}(\phi) = \begin{cases} \frac{1}{2\pi}, & 0 \le \phi \le 2\pi, \\ 0, & \text{otherwise}, \end{cases} \tag{14.12}$$

and also assume that they are independent, that is,

$$p_{r,\phi}(r, \phi) = p_r(r) p_{\phi}(\phi). \tag{14.13}$$

Generate two new random variables as

$$x = \sqrt{r} \cos \phi, \tag{14.14}$$

$$y = \sqrt{r} \sin \phi. \tag{14.15}$$

The physical interpretation of the previous transformation is that x, y correspond to the Cartesian coordinates of a point and r, ϕ are its polar ones (Fig. 14.3). From Eqs. (14.14) and (14.15), we can write

$$r = x^2 + y^2, \tag{14.16}$$

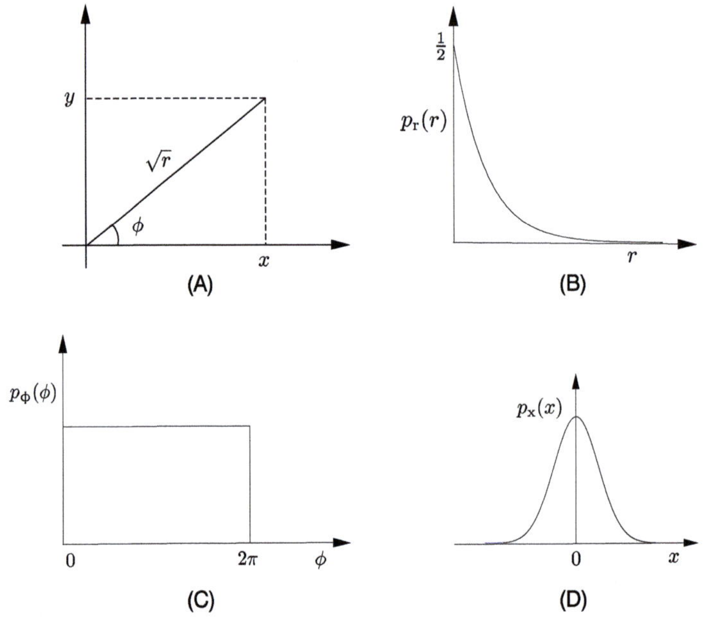

FIGURE 14.3

(A) Relation of the Cartesian (x, y) to the polar coordinates (r, ϕ). (B, C) If r and ϕ are random variables following an exponential and a uniform in $[0, 2\pi]$ distribution, respectively, then x and y are independent and they both follow a normalized Gaussian, as shown in (D) for the x-variable.

$$\phi = \arctan\left(\frac{\mathrm{y}}{\mathrm{x}}\right). \tag{14.17}$$

Adjusting Eq. (14.9) to our current needs, using Eqs. (14.11)–(14.13), we obtain

$$\begin{aligned} p_{\mathrm{x,y}}(x, y) &= \frac{1}{2\pi}\frac{1}{2}\exp\left(-\frac{x^2+y^2}{2}\right)2 \\ &= \frac{1}{\sqrt{2\pi}}\exp\left(-\frac{x^2}{2}\right)\frac{1}{\sqrt{2\pi}}\exp\left(-\frac{y^2}{2}\right), \end{aligned} \tag{14.18}$$

where we have used that (Problem 14.3)

$$|J(\mathrm{x}, \mathrm{y}; \mathrm{r}, \phi)| = \frac{1}{2}.$$

Thus, we have shown that using the transformation given in Eqs. (14.14) and (14.15), we can generate samples from normalized Gaussians.

Once samples from a normalized Gaussian, $\mathcal{N}(x|0, 1)$, are available, samples from a general Gaussian, $\mathcal{N}(y|\mu, \sigma^2)$, are obtained via the obvious transformation

$$\mathrm{y} = \sigma \mathrm{x} + \mu. \tag{14.19}$$

The previous approach is also generalized to random vectors in $\mathbb{R}^l$. One can first draw samples from $\mathbf{x} \sim \mathcal{N}(\boldsymbol{x}|\mathbf{0}, I)$ by stacking together l i.i.d. samples drawn from a normalized Gaussian, $\mathcal{N}(x|0, 1)$, and then apply the transformation

$$\mathbf{y} = L\mathbf{x} + \boldsymbol{\mu},$$

which is equivalent to drawing samples from

$$\mathbf{y} \sim \mathcal{N}(\boldsymbol{y}|\boldsymbol{\mu}, \Sigma), \tag{14.20}$$

where $\Sigma = LL^T$ (Cholesky factorization, Problem 14.4).

Example 14.3. Generate $N = 100$ samples, r_n, $n = 1, 2, \ldots, 100$, from the exponential distribution in Eq. (14.11) (following Example 14.1) and $N = 100$ samples, ϕ_n, $n = 1, 2, \ldots, 100$, from the uniform in Eq. (14.12). Then use the transformations in Eq. (14.14), (14.15), and (14.19) to obtain samples, x_n, $n = 1, 2, \ldots, 100$, from $p(x) = \mathcal{N}(x|1, 0.5)$. The histogram of the obtained samples is shown in Fig. 14.4.

14.4 REJECTION SAMPLING

Applying the previously reported transformation techniques relies on having the involved transform functions available in a convenient (analytic) form, which in general is the exception instead of the rule. From now on, we turn our attention to alternative methods.

Rejection sampling (e.g., [7,37]) is conceptually a simple technique; in order to generate independent samples from a desired PDF, $p(\boldsymbol{x})$, one draws samples from another one, say, $q(\boldsymbol{x})$, that is easier

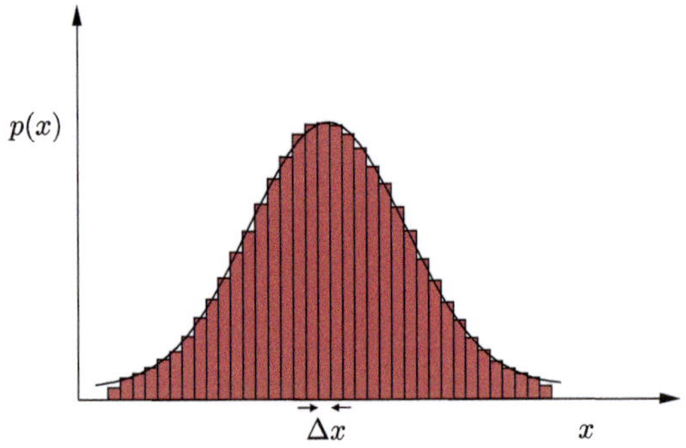

FIGURE 14.4

The histogram of $N = 100$ points generated in Example 14.3 together with the graph of $p(x) = \mathcal{N}(x|1, 0.5)$. The bin length was chosen to be equal to $\Delta x = 0.05$.

to handle, and then, instead of applying a transformation, some of the points are *rejected* according to an appropriate criterion.

Given two random variables, x and u, recall that the marginal, $p(x)$, is obtained by integrating the joint PDF, $p_{x,u}(x, u)$, that is,

$$p(x) = \int_{-\infty}^{+\infty} p_{x,u}(x, u)\, du. \tag{14.21}$$

Let us now consider the following identity:

$$p(x) \equiv \int_{0}^{p(x)} 1\, dx = \int_{-\infty}^{+\infty} \chi_{[0,p(x)]}(u)\, du, \tag{14.22}$$

where $\chi_{[0,p(x)]}(\cdot)$ is our familiar characteristic function in the interval $[0, p(x)]$, that is,

$$\chi_{[0,p(x)]}(u) = \begin{cases} 1, & 0 \leq u \leq p(x), \\ 0, & \text{otherwise.} \end{cases}$$

Comparing Eqs. (14.21) and (14.22), it turns out that $\chi_{[0,p(x)]}(u)$ can be interpreted as the joint PDF of the pair (x, u) defined over the set

$$\mathcal{A} = \big\{(x, u) : \, x \in \mathbb{R}, \, 0 \leq u \leq p(x)\big\}. \tag{14.23}$$

Looking more carefully at $p_{x,u}(x, u) = \chi_{[0,p(x)]}(u)$, it does not take long to realize that this is the *uniform density* under the area of the graph $u = p(x)$, as seen in Fig. 14.5A. In other words, if one fills in the shaded area in Fig. 14.5A uniformly at random with points (x, u) and then neglects the u dimension, then the obtained points are samples drawn from $p(x)$. We can now go one step further and assume that $p(x)$ is not exactly known, that is,

$$p(x) = \frac{1}{Z}\phi(x),$$

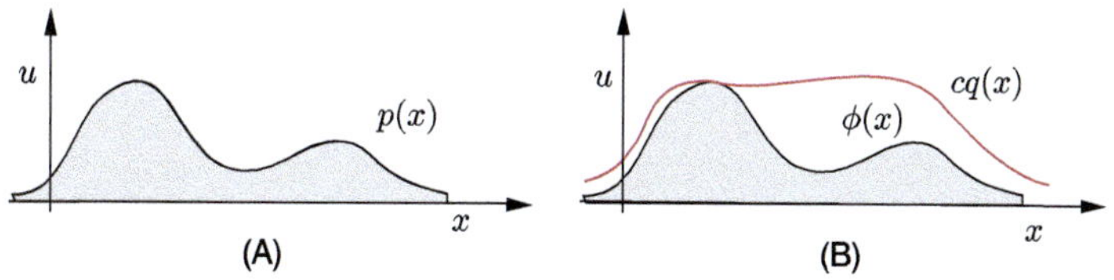

FIGURE 14.5

(A) Filling in the shaded area uniformly at random with points (x_n, u_n), after neglecting the coordinate u_n, is equivalent to drawing points x_n from $p(x)$. (B) The proposal distribution, $cq(x)$, is everywhere larger than or equal to $\phi(x)$.

and that the normalizing constant is not available (as we know, often, the computation of the normalizing constant is not easy). Then we have

$$p(x) = \frac{1}{Z}\phi(x) = \frac{1}{Z}\int_0^{\phi(x)} du = \frac{1}{Z}\int_{-\infty}^{+\infty} \chi_{[0,\phi(x)]}(u)\, du,$$

and Z is given by

$$Z = \int_{-\infty}^{+\infty}\int_{-\infty}^{+\infty} \chi_{[0,\phi(x)]}(u)\, du\, dx.$$

Hence,

$$p(x) = \frac{\int_{-\infty}^{+\infty} \chi_{[0,\phi(x)]}(u)\, du}{\int_{-\infty}^{+\infty}\int_{-\infty}^{+\infty} \chi_{[0,\phi(x)]}(u)\, du\, dx}. \tag{14.24}$$

In other words, even if $p(x)$ is not exactly known, $p(x)$ can still be obtained, this time in terms of the uniform distribution, $\chi_{[0,\phi(x)]}(x, u)$, normalized appropriately. However, rescaling the uniform does not affect the marginal. It suffices to sample uniformly at random the region $\mathcal{A}$, which now should be defined in terms of $\phi(x)$ instead of $p(x)$. What we have said so far applies also to random vectors, $\mathbf{x} \in \mathbb{R}^l$, by considering the extended space $(\boldsymbol{x}, u)$, and we talk about the volume under the surface $\phi(\boldsymbol{x})$ (or $p(\boldsymbol{x})$).

We now turn our attention to see how one can fill in the volume under the surface formed by $u = \phi(\boldsymbol{x})$ (or $u = p(\boldsymbol{x})$ if it is fully available), with points uniformly at random. Let $q(\boldsymbol{x})$ be a distribution from which we know how to draw samples; we refer to it as the *proposal distribution*. We select a constant c such that[2]

$$\phi(\boldsymbol{x}) \le cq(\boldsymbol{x}), \ \forall \boldsymbol{x} \in \mathbb{R}^l.$$

The respective geometry is shown in Fig. 14.5B. The goal is to draw points in the interval $[0, cq(\boldsymbol{x})]$ and then keep only those that lie in the region under the surface $u = \phi(\boldsymbol{x})$. The following algorithm does the job.

[2] If $q(\boldsymbol{x}) = 0$ in an interval, then $\phi(\boldsymbol{x})$ should be zero there.

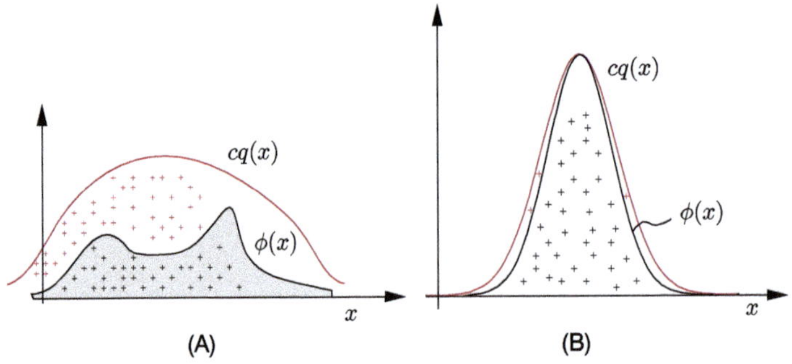

FIGURE 14.6

(A) If $cq(x)$ is much larger than $\phi(x)$ most of the samples are rejected (red points), which is inefficient. (B) If $cq(x)$ and $\phi(x)$ have a good match, most of the samples are retained.

Algorithm 14.2 (Rejection sampling).

- **For** $i = 1, 2, \dots, N$, **Do**
  - Draw $x_i \sim q(\boldsymbol{x})$
  - Draw $u_i \sim \mathcal{U}(0, cq(x_i))$
  - Retain the sample if
    * $u_i \le \phi(x_i)$
- **End For**

The probability of accepting a point, $\boldsymbol{x}$, is given by

$$\text{Prob}\{\text{u} \le \phi(\boldsymbol{x})\} = \frac{1}{cq(\boldsymbol{x})}\phi(\boldsymbol{x}),$$

and the total probability, over all the possible values of $\boldsymbol{x}$, for accepting samples is equal to

$$\text{Prob}\{\text{acceptance}\} = \frac{1}{c}\int \frac{\phi(\boldsymbol{x})}{q(\boldsymbol{x})} q(\boldsymbol{x})\, d\boldsymbol{x} = \frac{1}{c}\int \phi(\boldsymbol{x}).$$

Hence, if c has a large value, only a small percentage of points is finally retained. In order to have a practical algorithm, $cq(\boldsymbol{x})$ must be chosen in order to be a good fit of $\phi(\boldsymbol{x})$. Fig. 14.6A is an example of a bad choice, while Fig. 14.6B corresponds to a good example. This is a reason that rejection sampling *does not scale* well with *dimensionality*. In high dimensions, guaranteeing that $cq(\boldsymbol{x}) \ge \phi(\boldsymbol{x})$ may oblige us to select a c with an excessively large value (Problem 14.5).

Besides the basic rejection scheme, a number of variants have also been proposed, in order to overcome the difficulty of selecting a proposal distribution that "looks like" the desired one. *Adaptive rejection sampling* is such a technique (see, e.g., [11] and the references therein). According to this method, the proposal distribution is adaptively constructed, based on the derivatives of $\ln p(x)$. For log-concave functions, this is a nondecreasing function and can be used to construct an envelope function of $p(x)$.

Although rejection sampling is not appropriate for difficult tasks, it has still been used, sometimes in its more refined forms, to generate samples from a number of standard distributions, such as the Gaussian, gamma, and Student's t (see, e.g., [22]).

14.5 IMPORTANCE SAMPLING

Importance sampling (IS) is a method for estimating expectations. Let $f(\mathbf{x})$ be a known function of a random vector variable, $\mathbf{x}$, which is distributed according to $p(\boldsymbol{x})$. If one could draw samples from $p(\boldsymbol{x})$, then the expectation in Eq. (14.1) could be approximated as in Eq. (14.2). We will now assume that we are not able to draw samples from $p(\boldsymbol{x})$, and to go one step further, assume that $p(\boldsymbol{x})$ is only known up to a normalizing constant, that is,

$$p(\boldsymbol{x}) = \frac{1}{Z}\phi(\boldsymbol{x}).$$

Let $q(\boldsymbol{x})$ be another distribution from which samples can be drawn. Then we can write

$$\begin{aligned}\mathbb{E}\left[f(\mathbf{x})\right] &= \frac{1}{Z}\int_{-\infty}^{\infty} f(\boldsymbol{x})\phi(\boldsymbol{x})\,d\boldsymbol{x} = \frac{1}{Z}\int_{-\infty}^{\infty} f(\boldsymbol{x})\frac{\phi(\boldsymbol{x})}{q(\boldsymbol{x})}q(\boldsymbol{x})\,d\boldsymbol{x} \\ &\simeq \frac{1}{NZ}\sum_{i=1}^{N} f(\boldsymbol{x}_i)w(\boldsymbol{x}_i), \end{aligned} \tag{14.25}$$

where $\boldsymbol{x}_i$, $i = 1, 2, \ldots, N$, are samples drawn from $q(\boldsymbol{x})$ and

$$w(\boldsymbol{x}) := \frac{\phi(\boldsymbol{x})}{q(\boldsymbol{x})}. \tag{14.26}$$

The normalizing constant can readily be obtained as

$$Z = \int_{-\infty}^{\infty} \phi(\boldsymbol{x})\,d\boldsymbol{x} = \int_{-\infty}^{\infty}\left(\frac{\phi(\boldsymbol{x})}{q(\boldsymbol{x})}\right)q(\boldsymbol{x})\,d\boldsymbol{x} \simeq \frac{1}{N}\sum_{i=1}^{N} w(\boldsymbol{x}_i). \tag{14.27}$$

Combining Eqs. (14.25) and (14.27), we finally obtain

$$\mathbb{E}\left[f(\mathbf{x})\right] \simeq \frac{\sum_{i=1}^{N} w(\boldsymbol{x}_i)f(\boldsymbol{x}_i)}{\sum_{i=1}^{N} w(\boldsymbol{x}_i)}, \tag{14.28}$$

or

$$\boxed{\mathbb{E}\left[f(\mathbf{x})\right] \simeq \sum_{i=1}^{N} W(\boldsymbol{x}_i)f(\boldsymbol{x}_i): \quad \text{importance sampling approximation,}}$$

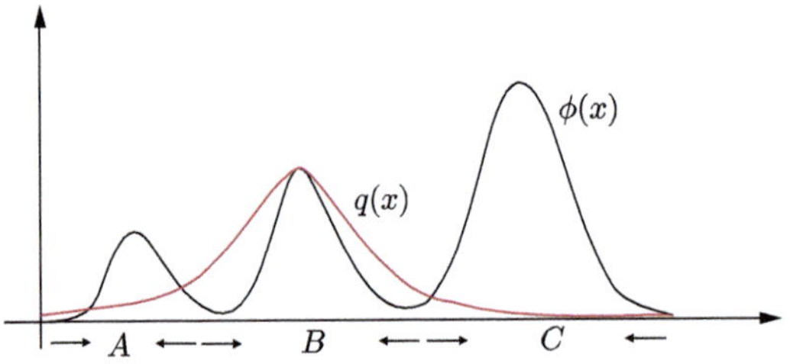

FIGURE 14.7

If $q(\boldsymbol{x})$ is not a good match of $\phi(\boldsymbol{x})$, a number of undesired effects appear. Samples from region A will give rise to weights of much larger values compared to those in region B. Due to the extremely low values of $q(\boldsymbol{x})$ in C, it is highly likely that given the finite size of the number of the samples, N, no samples will be drawn from this region, in spite of the fact that this is the most dominant region for $\phi(\boldsymbol{x})(p(\boldsymbol{x}))$.

where $W(\boldsymbol{x}_i) = \frac{w(\boldsymbol{x}_i)}{\sum_{i=1}^N w(\boldsymbol{x}_i)}$ are the normalized weights. It is not difficult to show (Problem 14.6) that the estimate

$$\hat{Z} = \frac{1}{N}\sum_{i=1}^{N} w(\boldsymbol{x}_i) \tag{14.29}$$

corresponds to an unbiased estimator of the normalizing constant. This is very interesting, because computing the normalizing constant is particularly useful information in a number of tasks. Recall that the evidence function, discussed in Chapter 12, is a normalizing constant; see also [26] for related comments.

In contrast, the estimator associated with Eq. (14.28), being the result of a ratio, is unbiased only asymptotically and it is a *biased* one for finite values of N (Problem 14.6). Hence, if one would have the luxury of a very large number N of samples, Eq. (14.28) would be a good enough estimate. However, in practice, N cannot be made arbitrarily large and the resulting estimate may not be satisfactory.

If $q(\boldsymbol{x}) \simeq p(\boldsymbol{x})$, or at least $q(\boldsymbol{x})$ is a fairly good approximation of $\phi(\boldsymbol{x})$, then Eq. (14.28) would approximate Eq. (14.2). However, for most practical cases, this is not easy to obtain, especially in high-dimensional spaces. If $q(\boldsymbol{x})$ is not a good match to $\phi(\boldsymbol{x})$, it is very likely that there will be regions where $\phi(\boldsymbol{x})$ is large while $q(\boldsymbol{x})$ is much smaller. The corresponding weights will have large values, relative to those from other regions, and they will be the dominant ones in the summation (Eq. (14.28)).

The effect of it is equivalent to reducing the number, N, of samples. Moreover, it is also possible that $q(\boldsymbol{x})$ takes very small values in some regions, which makes it very likely that samples from such regions are completely absent in Eq. (14.28) (see Fig. 14.7). In such cases, not only may the resulting estimate be wrong, but we will not be aware of it, and the variance of the weights, $w(\boldsymbol{x}_n)$ and $w(\boldsymbol{x}_n)f(\boldsymbol{x}_n)$, may exhibit low values. *These phenomena are accentuated in high-dimensional spaces* (see, e.g., [26] and Problem 14.7).

To alleviate the previous shortcomings, a number of variants have been proposed to search for high-probability regions and make local approximations around the modes and use them in order to generate samples (see, e.g., [30] and the references therein).

14.6 MONTE CARLO METHODS AND THE EM ALGORITHM

In Section 12.4.1, the EM algorithm was introduced for maximizing the log-likelihood function when some of the variables are hidden or missing. During the E-step (Eq. (12.40)), the function $\mathcal{Q}$ is computed, which at the $(j+1)$th iteration step of the algorithms is written as

$$\begin{aligned} \mathcal{Q}(\boldsymbol{\xi}, \boldsymbol{\xi}^{(j)}) &= \mathbb{E}\left[\ln p(\mathcal{X}^l, \mathcal{X}; \boldsymbol{\xi})\right] \\ &= \int p(\mathcal{X}^l | \mathcal{X}; \boldsymbol{\xi}^{(j)}) \ln p(\mathcal{X}^l, \mathcal{X}; \boldsymbol{\xi}) d\mathcal{X}^l, \end{aligned} \tag{14.30}$$

where $\mathcal{X}^l$ is the set of hidden variables, $\mathcal{X}$ the set of observed values, and $\boldsymbol{\xi}$ the unknown set of parameters. In case the computation of the integral is not tractable, Monte Carlo techniques can be mobilized to generate L samples for the hidden variables, $\mathcal{X}_1^l, \ldots, \mathcal{X}_L^l$, from the posterior $p(\mathcal{X}^l | \mathcal{X}; \boldsymbol{\xi}^{(j)})$ and obtain an approximation

$$\hat{\mathcal{Q}}(\boldsymbol{\xi}, \boldsymbol{\xi}^{(j)}) \approx \frac{1}{L} \sum_{i=1}^{L} \ln p(\mathcal{X}_i^l, \mathcal{X}; \boldsymbol{\xi}). \tag{14.31}$$

Maximization with respect to $\boldsymbol{\xi}$ is now carried out via $\hat{\mathcal{Q}}$.

A specific form of Monte Carlo EM results in the context of mixture modeling; this is known as *stochastic EM*. The idea is to generate a *single* sample from the posterior (which now refers to the labels of the mixtures) and assign corresponding observations in the respective mixtures. That is, a hard assignment takes place. The M-step is then applied based on this approximation [3].

14.7 MARKOV CHAIN MONTE CARLO METHODS

As we have already discussed, a major drawback associated with rejection as well as importance sampling methods is that they cannot tackle tasks in high-dimensional spaces very well.

In this section, we will deal with methods that scale well with the dimensionality of the sample space. Such techniques build upon arguments that come from the theory of Markov chains; we start by presenting some definitions and basics related to this important theory. Hidden Markov models, treated in Chapter 16.5, are instances of Markov chains. Here we will shed more light on such models from a different perspective.

Markov chains/processes are named after the Russian mathematician Andrey Andreyevish Markov (1856–1922), who contributed seminal papers in the field of stochastic processes. As a professor at Saint Petersburg University during the students' riots in 1908, he refused the government's order to monitor and spy on his students, and he retired from the university.

Definition 14.1. A *Markov chain* is a sequence of random (vector) variables, $\mathbf{x}_0, \mathbf{x}_1, \mathbf{x}_2 \ldots$, with conditional distributions that obey the rule

$$p(\boldsymbol{x}_n | \boldsymbol{x}_{n-1}, \{\boldsymbol{x}_t : t \in \mathcal{I}\}) = p(\boldsymbol{x}_n | \boldsymbol{x}_{n-1}), \tag{14.32}$$

where $\mathcal{I} = \{0, 1, \ldots, n-2\}$. The index n is usually interpreted as time.

In words, Eq. (14.32) says that $\mathbf{x}_n$ is independent of the variables with indices in $\mathcal{I}$, given the values of the variables in $\mathbf{x}_{n-1}$. The distribution p can be either a density function or a probability distribution corresponding to discrete variables, taking values in a discrete set, known as *states*. We will assume that all variables share a common range, known as *state space*. Most of our discussion will evolve along finite state spaces, where states take values in a finite discrete set, say, $\{1, 2, \ldots, K\}$. A Markov chain is specified in terms of (a) the distribution (vector of probabilities), $\boldsymbol{p}_0$, associated with the first vector in the sequence, $\mathbf{x}_0$, and (b) the $K \times K$ matrices of the *transition probabilities*, that is,

$$P_n(\boldsymbol{x}_n|\boldsymbol{x}_{n-1}) = [P_n(i|j)],$$

where

$$P_n(i|j) := P_n(\boldsymbol{x}_n = i|\boldsymbol{x}_{n-1} = j), \quad i, j = 1, 2, \ldots, K,$$

denotes the probability of the variable at time $n-1$ to be at state j *and* the variable at time n to be at state i.[3] Given the matrix of the transition probabilities, we write

$$\boldsymbol{p}_n = P_n(\boldsymbol{x}_n|\boldsymbol{x}_{n-1})\boldsymbol{p}_{n-1}, \tag{14.33}$$

where

$$\boldsymbol{p}_n := [P(\boldsymbol{x}_n = 1), P(\boldsymbol{x}_n = 2), \ldots, P(\boldsymbol{x}_n = K)]^T \tag{14.34}$$

$$:= [P_n(1), \ldots, P_n(K)]^T \tag{14.35}$$

is the vector of the respective probabilities at time n. The Markov chain is said to be *homogeneous* or *stationary* if the transition matrix is independent on time, that is,

$$P_n(\boldsymbol{x}_n = i|\boldsymbol{x}_{n-1} = j) = P(i|j) := P_{ij}, \quad i, j = 1, 2, \ldots, K,$$

and

$$P_n(\boldsymbol{x}_n|\boldsymbol{x}_{n-1}) = P = \left[P_{ij}\right].$$

In this case, we can write

$$\boldsymbol{p}_n = P\boldsymbol{p}_{n-1} = P^2\boldsymbol{p}_{n-2} = \cdots = P^n\boldsymbol{p}_0, \tag{14.36}$$

or equivalently,

$$P_n(i) = \sum_{j=1}^{K} P_{ij} P_{n-1}(j). \tag{14.37}$$

In the sequel, we will focus on stationary Markov chains.

Properties of the transition probabilities matrix. The transition matrix has a special structure leading to certain properties, which will be used later on.

[3] Note that equality here, $\boldsymbol{x}_n = i$, means that the (vector) variable $\mathbf{x}_n$ is at state i.

- The matrix P is a *stochastic* matrix. That is, all its entries are nonnegative, and the entries across each column add to one, that is,

$$\sum_{i=1}^{K} P_{ij} = 1,$$

which is a direct consequence of the definition of probabilities.
- The value $\lambda = 1$ is always an eigenvalue of P (Problem 14.8). Moreover, there is no eigenvalue with magnitude larger than one (Problem 14.9).
- The eigenvectors corresponding to eigenvalues $\lambda \neq 1$ comprise components that add to zero (Problem 14.10).
- The left eigenvector corresponding to $\lambda = 1$,

$$\boldsymbol{b}_1^T P = \boldsymbol{b}_1^T, \tag{14.38}$$

has all its elements equal. This is easily verified by plugging in $\boldsymbol{b}_1 = [1, 1, \ldots, 1]^T$ and checking that this is indeed an eigenvector.
- *Invariant distribution.* A distribution is said to be invariant over the states of a Markov chain if

$$\boldsymbol{p} = P\boldsymbol{p}.$$

Note that $\boldsymbol{p}$ is necessarily an eigenvector corresponding to the eigenvalue $\lambda = 1$. Moreover, because $\boldsymbol{p}$ consists of probabilities, its elements must add to one. Depending on the multiplicity of $\lambda = 1$, there may be more than one invariant distribution. For example, if $P = I$ any probability distribution is an invariant for the respective Markov chain. It turns out that any Markov chain with a finite number of states has at least one invariant distribution. However, if the elements of P are strictly positive, then there is a unique invariant distribution that coincides with the unique eigenvector corresponding to the maximum eigenvalue, $\lambda = 1$, which in this case has a multiplicity of one. Furthermore, the eigenvector corresponding to $\lambda = 1$ comprises positive elements, which after scaling can always be made to add to one. This is a by-product of the celebrated *Perron–Frobenius theorem*, which is ensured if P has strictly positive entries[4] (see, e.g., [32]). We will cover invariant distributions in more detail soon.
- *Detailed balanced condition.* Let P be the transition probability matrix of a stationary Markov chain. Let also $\boldsymbol{p} = [P_1, \ldots, P_K]^T$ be the set of probabilities describing a discrete distribution. We say that the *detailed balanced condition* is satisfied if

$$P(i|j)P_j = P(j|i)P_i. \tag{14.39}$$

That is, there exists a type of *symmetry*. If this condition holds, then the respective distribution is invariant for the Markov chain. Indeed,

[4] This is also true for a class of matrices with nonnegative elements, known as *primitive* matrices. That is, there exists an n such that P^n has positive elements.

$$\sum_{j=1}^{K} P(i|j)P_j = \sum_{j=1}^{K} P(j|i)P_i = P_i, \tag{14.40}$$

or

$$\boldsymbol{p} = P\boldsymbol{p}. \tag{14.41}$$

Although this is not a necessary condition for distribution invariance, it is very useful in practice; it helps us construct Markov chains with a desired invariant distribution. As we will soon see, this will be the type of distributions from which we want to draw samples.

14.7.1 ERGODIC MARKOV CHAINS

We now turn our attention to a specific type of Markov chains, which are known as *ergodic*. Such chains have a *unique* invariant distribution, which can be obtained as the limit

$$\lim_{n\to\infty} \boldsymbol{p}_n = \lim_{n\to\infty} P^n \boldsymbol{p}_0,$$

which is *independent* of the choice of the initial values in $\boldsymbol{p}_0$. We will now focus on a class of ergodic processes and elaborate on their convergence.

Let us consider a stationary Markov chain, with a transition matrix P, with eigenvalues $1 = \lambda_1 > |\lambda_2| \geq \cdots \geq |\lambda_K|$. That is, only one eigenvalue has the maximum value and the rest have magnitude strictly less than one. Moreover, we assume that one can find a complete set of *linearly independent* eigenvectors. Such assumptions are not restrictive and hold true for a wide class of stochastic matrices. Then we can write

$$P = A\Lambda A^{-1}, \tag{14.42}$$

where Λ is the diagonal matrix $\Lambda = \text{diag}\{1, \lambda_2, \ldots, \lambda_K\}$ and A has as columns the respective eigenvectors. Hence, from Eq. (14.36) we get

$$\boldsymbol{p}_n = A\Lambda^n A^{-1}\boldsymbol{p}_0 = A \begin{bmatrix} 1 & & & O \\ & \lambda_2^n & & \\ O & & \ddots & \\ & & & \lambda_K^n \end{bmatrix} A^{-1}\boldsymbol{p}_0,$$

with $\lambda_k^n \longrightarrow 0, k = 2, \ldots, K$.

Hence,

$$\boldsymbol{p}_\infty := \lim_{n\to\infty} \boldsymbol{p}_n = P_\infty \boldsymbol{p}_0, \tag{14.43}$$

where

$$P_\infty = A \begin{bmatrix} 1 & & & \\ & 0 & & O \\ & O & \ddots & \\ & & & 0 \end{bmatrix} A^{-1} = \boldsymbol{a}_1 \boldsymbol{b}_1^T, \tag{14.44}$$

with $\boldsymbol{a}_1$ being the first eigenvector (first column of A), corresponding to $\lambda = 1$, and $\boldsymbol{b}_1^T$ the first row of A^{-1}. In other words, P_∞ is a rank one matrix. However, it is straightforward to see from Eq. (14.42) $\left(A^{-1}P = \Lambda A^{-1}\right)$ that $\boldsymbol{b}_1^T$ is a left eigenvector of P, that is,

$$\boldsymbol{b}_1^T P = \boldsymbol{b}_1^T,$$

and recalling the properties (Eq. (14.38) and the comments just after it) of P, $\boldsymbol{b}_1^T = [1, 1, \ldots, 1]$ (within a proportionality constant c). Thus,

$$P_\infty = [\boldsymbol{a}_1, \ldots, \boldsymbol{a}_1],$$

and from Eq. (14.43), because the components of $\boldsymbol{p}_0$ add to one, we finally obtain

$$\boldsymbol{p}_\infty = \boldsymbol{a}_1.$$

That is, the limiting distribution is equal (after scaling) to the unique eigenvector of P corresponding to $\lambda_1 = 1$; moreover, this is true irrespective of the values of $\boldsymbol{p}_0$. In other words, the limiting distribution is the invariant distribution of P, that is,

$$P\boldsymbol{p} = \boldsymbol{p}. \tag{14.45}$$

Note that the convergence rate is controlled by the magnitude of $|\lambda_2|$. Other, more theoretically refined, convergence results and bounds can be found in, for example, [24,39,40].

Remarks 14.2.

- Needless to say, not all Markov chains are ergodic. For example, if the eigenvalue $\lambda_1 = 1$ of the transition matrix has multiplicity higher than one, then the limiting distribution depends on the values of the initial choice in $\boldsymbol{p}_0$. On the other hand, if the transition matrix has more than one eigenvalue with magnitude equal to one (e.g., $\lambda_1 = 1, \lambda_2 = -1$) then, again, it does not have a limiting distribution but instead exhibits a *periodic* limit cycle (e.g., [32]).
- *Building Markov chains.* In practice, one can construct transition probability matrices for ergodic chains using a set of simpler transition matrices, $B_1, B_2, \ldots, B_M$, which are known as *base transition* matrices. Each one of them may not be ergodic, but it is required to accept the desired distribution as its invariant. Then the transition matrix is built as

$$P = \sum_{m=1}^{M} \alpha_m B_m, \quad \alpha_m > 0, \; \sum_{m=1}^{M} \alpha_m = 1.$$

If a distribution is invariant with respect to each B_m, $m = 1, 2, \ldots, M$, it will be invariant for P. The same applies with the detailed balance condition.
Another way is to combine individual transition matrices sequentially, that is,

$$P = B_1 B_2 \cdots B_M.$$

For example, each B_m, $m = 1, 2, \ldots, M$, may act and change a subset of the random entries comprising the random vector $\mathbf{x}$. We will see that this is the case with the Gibbs sampling method, to be reviewed soon. It is easy to see that if $\boldsymbol{p}$ is invariant for each individual B_m, $m = 1, 2, \ldots, M$, it will also be invariant for P.

- In this section, we focused our discussion on Markov chains with finite state space. Everything we have said can be generalized to Markov chains with countably infinite or continuous state spaces. In the latter case, the place of the probability transition matrix is taken by the transition density or kernel, $p(\boldsymbol{x}_n|\boldsymbol{x}_{n-1})$, and the probability density of $\mathbf{x}_n$, at time n, is given by

$$p_n(\boldsymbol{x}) = \int p(\boldsymbol{x}|\boldsymbol{y}) p_{n-1}(\boldsymbol{y})\, d\boldsymbol{y}. \tag{14.46}$$

The analysis in this case is more difficult and care has to be taken because not all results obtained for the finite discrete case are readily valid for the continuous one. The reason we focused on the discrete finite state space is that one can get the feeling of the theory of Markov chains by spending less "budget" on the required mathematical effort.

Example 14.4. Consider the Markov chain with the transition probability matrix

$$P = \begin{bmatrix} 0.2 & 0.4 & 0.6 \\ 0.5 & 0.1 & 0.3 \\ 0.3 & 0.5 & 0.1 \end{bmatrix}.$$

Its eigenvalues are $\lambda_1 = 1$, $\lambda_2 = -0.3 + 0.1732j$, and $\lambda_3 = -0.3 - 0.1732j$, and the respective eigenvectors are

$$\begin{aligned} \boldsymbol{a}_1 &= [0.6608,\ 0.5406,\ 0.5206]^T, \\ \boldsymbol{a}_2 &= [0.5774,\ \ -0.2887 - 0.5j,\ \ -0.2887 + 0.5j]^T, \\ \boldsymbol{a}_3 &= [0.5774,\ \ -0.2887 + 0.5j,\ \ -0.2887 - 0.5j]^T. \end{aligned}$$

Observe that all the elements of the eigenvector corresponding to $\lambda = 1$ are positive. Also, the elements of the other two eigenvectors add to zero.

We can now write

$$\begin{aligned} P = &\begin{bmatrix} 0.6608 & 0.5774 & 0.5774 \\ 0.5406 & -0.2887 - 0.5j & -0.2887 + 0.5j \\ 0.5206 & -0.2887 + 0.5j & -0.2887 - 0.5j \end{bmatrix} \\ &\times \begin{bmatrix} 1 & 0 & 0 \\ 0 & -0.3 + 0.1732j & 0 \\ 0 & 0 & -0.3 - 0.1732j \end{bmatrix} \end{aligned}$$

$$\times \begin{bmatrix} 0.5807 & 0.5807 & 0.5807 \\ 0.5337 - 0.0058i & -0.3323 + 0.4942j & -0.3323 - 0.5058j \\ 0.5337 + 0.0058i & -0.3323 - 0.4942j & -0.3323 + 0.5058j \end{bmatrix}.$$

Observe that the first row of the last matrix (A^{-1}) has all its elements equal. Having written P in a product form, the following is easy to obtain:

$$P^2 = \begin{bmatrix} 0.42 & 0.42 & 0.30 \\ 0.24 & 0.36 & 0.36 \\ 0.34 & 0.22 & 0.34 \end{bmatrix},$$

$$P^{10} = \begin{bmatrix} 0.3837 & 0.3837 & 0.3837 \\ 0.3140 & 0.3140 & 0.3139 \\ 0.3023 & 0.3023 & 0.3029 \end{bmatrix}.$$

The sequence has converged for $n = 10$. Note that after convergence, P^n has all its column vectors equal, and the elements add to one. Moreover, observe that

$$P_\infty \propto [\boldsymbol{a}_1, \boldsymbol{a}_1, \boldsymbol{a}_1].$$

Example 14.5. *Random walk with finite states.* Random walks are popular models that can model faithfully a number of real-world phenomena, such as thermal noise, the motion of gas molecules, and stock value variations. Moreover, such chains can help us understand the behavior of more complex Markov chains, to be discussed soon. There are various random walk models, depending on the choice of the transition probabilities (see, e.g., [32]). Here, we assume the variables to be discrete and take integer values in a finite set, [0, N]. Hence, the total number of states is $N + 1$. At every time instant, the value of the variable can either increase or decrease by one with probability p, respectively, or stay unchanged, with probability q, provided that the current state is in the interval $[1, N - 1]$. That is, if $0 < x_{n-1} < N$,

$$P(x_n = x_{n-1} + 1) = P(x_n = x_{n-1} - 1) = p,$$
$$P(x_n = x_{n-1}) = q.$$

If $x_{n-1} = 0$, then x_n can either stay in the same state with probability q_e or increase by one with probability p. If $x_{n-1} = N$, then x_n can either stay in the same state with probability q_e or decrease by one with probability p. Obviously,

$$2p + q = 1, \quad p + q_e = 1.$$

The transition probability matrix, for the case of $N = 4$, $p = \frac{1}{4}$, $q = \frac{1}{2}$, and $q_e = \frac{3}{4}$, is

$$P = \begin{bmatrix} 3/4 & 1/4 & 0 & 0 & 0 \\ 1/4 & 1/2 & 1/4 & 0 & 0 \\ 0 & 1/4 & 1/2 & 1/4 & 0 \\ 0 & 0 & 1/4 & 1/2 & 1/4 \\ 0 & 0 & 0 & 1/4 & 3/4 \end{bmatrix}.$$

The respective eigenvalues are $\lambda_1 = 1$, $\lambda_2 = 0.904$, $\lambda_3 = 0.654$, $\lambda_4 = 0.345$, and $\lambda_5 = 0.095$. Observe that all eigenvalues, except $\lambda_1 = 1$, have magnitude less than one. The corresponding eigenvectors are

$$\begin{aligned} \boldsymbol{a}_1 &= [0.447,\ 0.447,\ 0.447,\ 0.447,\ 0.447]^T, \\ \boldsymbol{a}_2 &= [-0.601,\ -0.371,\ 0,\ 0.371,\ 0.601]^T, \\ \boldsymbol{a}_3 &= [-0.511,\ 0.195,\ 0.632,\ 0.195,\ -0.511]^T, \\ \boldsymbol{a}_4 &= [-0.371,\ 0.6015,\ 0,\ -0.601,\ 0.371]^T, \\ \boldsymbol{a}_5 &= [0.195,\ -0.511,\ 0.632,\ -0.511,\ 0.195]^T. \end{aligned}$$

The eigenvector corresponding to λ_1 has all its components equal and positive. Hence, the invariant distribution ($\boldsymbol{p}$: $P\boldsymbol{p} = \boldsymbol{p}$), after the required scaling, becomes the uniform one, $\boldsymbol{p} = [1/5,\ 1/5,\ 1/5,\ 1/5,\ 1/5]^T$. Similar arguments apply for any value of N. Observe that the components of all the other eigenvectors add to zero.

Fig. 14.8 shows the probability distribution $\boldsymbol{p}_n$ for the case of $N = 4$, at times $n = 10,\ 50$, and 100. The components of $\boldsymbol{p}_0$ were randomly chosen. Fig. 14.9 corresponds to the case of $N = 9$. Observe that the larger the value of N, the slower the convergence.

Example 14.6. In this example, we consider a random walk with (countable) infinitely many states; at every time instant, the value of the random variable can either increase or decrease by one, with probability p, or stay in the same state with probability q, that is,

$$\begin{aligned} P(x_n = x_{n-1} + 1) &= P(x_n = x_{n-1} - 1) = p, \\ P(x_n = x_{n-1}) &= q, \end{aligned}$$

and

$$2p + q = 1.$$

The difference with the previous example is that now there are no "barrier" points and the random variable can take any integer value. Our goal is to compute the mean and variance as functions of time n, when the starting point is deterministically chosen to be $x_0 = 0$.

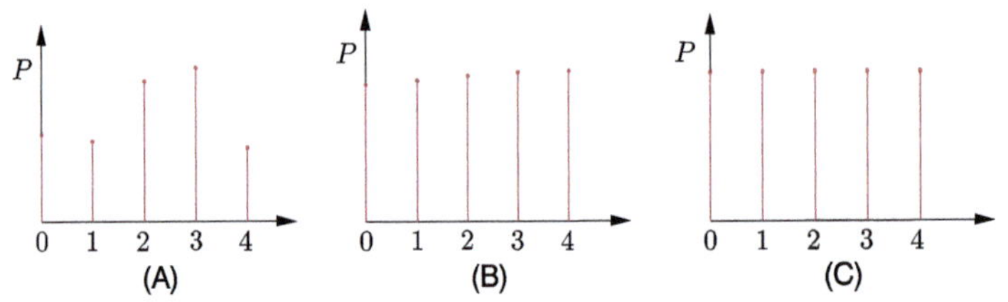

FIGURE 14.8

The probability distribution for the random walk chain of Example 14.5 for $N = 4$ and at time instants (A) $n = 10$, (B) $n = 50$, and (C) $n = 100$.

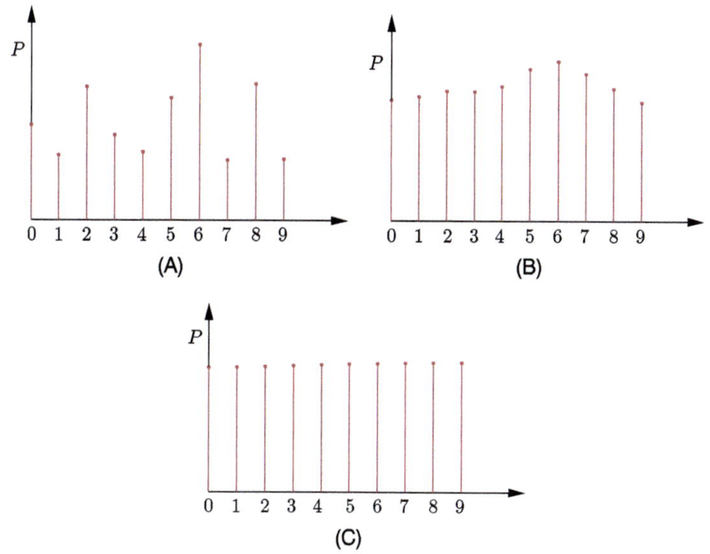

FIGURE 14.9

The probability distribution for the random walk chain of Example 14.5 for $N = 9$ and at time instants (A) $n = 10$, (B) $n = 50$, and (C) $n = 100$. Compared to Fig. 14.8, observe that the higher the number N, the slower the convergence.

It is readily seen that $\mathbb{E}[\mathrm{x}_n] = 0$, because the variable is equally likely to increase or decrease and hence it is equally likely to assume any positive or negative value.

For the variance, we obtain (Problem 14.11)

$$\begin{aligned}\mathbb{E}[\mathrm{x}_n^2] &= \mathbb{E}[\mathrm{x}_{n-1}^2] + 2p \\ &= 2pn + \mathbb{E}[\mathrm{x}_0^2] = 2pn,\end{aligned} \tag{14.47}$$

because $\mathbb{E}[\mathrm{x}_0^2] = 0$. Note that the variance tends to infinity with time, hence the infinite state-space random walk *does not* have a limiting distribution. This verifies what we said before; results that hold true for finite state spaces do not necessarily carry on to the case where the number of states becomes infinite.

Looking at Eq. (14.47) more carefully reveals that, on average, after n time instants, x_n would be within $\pm\sqrt{2pn}$. If x_n denotes the distance of a point from the origin from where it starts and moves backward or forward, then the distance it travels is proportional only to the square root of the time it has spent traveling. Although this result has been derived for the infinite state space case, still it can shed light on the slow convergence to the invariant distribution that we saw in the previous example, for the case of finite state space. As stated in [30], convergence to the invariant distribution can be achieved once all points in the state space have been visited, and this has a square root dependence on time. In order to get a good enough approximation of the limited distribution, one must be patient enough to compute $\mathcal{O}(N^2)$ iterations.

14.8 THE METROPOLIS METHOD

The Metropolis method or algorithm, as it is sometimes called, builds upon a surprisingly simple idea, and it is the first method that exploited the Markov chain theory for sampling. It appeared in the classical paper [27] and it may be the most popular and widely known sampling technique, which has inspired a wealth of variants. In contrast to rejection and importance sampling, the proposal distribution is now time-varying, following the evolution of a Markov chain; the latter is constructed such that its transition probability matrix (density kernel) has the desired distribution $p(\boldsymbol{x})$ as its invariant. Moreover, in contrast to the rejection and importance sampling techniques, it is not required that the proposal distribution "looks like" the desired one in order for the method to be useful in practice. The proposal distribution depends on the value of the previous state, $\boldsymbol{x}_{n-1}$, that is, $q(\cdot|\boldsymbol{x}_{n-1})$. In words, drawing a new sample (generating a new state) depends on the value of the previous one. In its original version, the proposal distribution was chosen to be symmetric, that is,

$$q(\boldsymbol{x}|\boldsymbol{y}) = q(\boldsymbol{y}|\boldsymbol{x}).$$

Later on, it was generalized by Hastings [14] to include nonsymmetric ones. The general scheme is known as the Metropolis–Hastings algorithm, which is summarized next.

Algorithm 14.3 (Metropolis–Hastings algorithm).

- Let the desired distribution be $p(\cdot) = \frac{1}{Z}\phi(\cdot)$.
- Choose the proposal distribution to be $q(\cdot|\cdot)$.
- Choose the value of the initial state $\boldsymbol{x}_0$.
- **For** $n = 1, 2, \dots, N$, **Do**
  - Draw $\boldsymbol{x} \sim q(\cdot|\boldsymbol{x}_{n-1})$
  - Compute the acceptance ratio
    - $\alpha(\boldsymbol{x}|\boldsymbol{x}_{n-1}) = \min\left\{1, \frac{q(\boldsymbol{x}_{n-1}|\boldsymbol{x})\phi(\boldsymbol{x})}{q(\boldsymbol{x}|\boldsymbol{x}_{n-1})\phi(\boldsymbol{x}_{n-1})}\right\}$
  - Draw
    - $u \sim \mathcal{U}(0, 1)$
  - **If** $u \leq \alpha(\boldsymbol{x}|\boldsymbol{x}_{n-1})$
    - $\boldsymbol{x}_n = \boldsymbol{x}$
  - **Else**
    - $\boldsymbol{x}_n = \boldsymbol{x}_{n-1}$
- **End For**

The following points are readily deduced from the algorithm:

- The algorithm does not need the exact form of p. It suffices to know it up to its normalizing constant Z. This is due to the fact that p enters into the algorithm only in the ratio for computing the acceptance ratio.

- If the proposal distribution is symmetric, the acceptance ratio becomes

$$\alpha(\boldsymbol{x}|\boldsymbol{x}_{n-1}) = \min\left\{1, \frac{\phi(\boldsymbol{x})}{\phi(\boldsymbol{x}_{n-1})}\right\}, \tag{14.48}$$

  and in this case, we sometimes refer to it as the Metropolis algorithm.
- Note that if a sample is not accepted, we retain the value of the previous state.
- Observe that a sample is accepted or rejected depending on the value of $\alpha(\boldsymbol{x}|\boldsymbol{x}_{n-1})$. This is easier understood by looking at the original form of the algorithm based on Eq. (14.48). If the probability $p(\boldsymbol{x})$ is larger than $p(\boldsymbol{x}_{n-1})$, then the new sample is accepted. If not, it is accepted/rejected based on its relative value.
- Successive samples are *not* independent.

There are variants of the previous basic scheme, concerning the choice of the function for the acceptance ratio. In [33], an argument in support of the rationale behind the Metropolis–Hastings scheme is based on an optimality proof concerning the variance of the obtained approximations.

Let us now turn our focus to understanding how the previously stated algorithm relates to the Markov chain theory. We will work with the more general continuous state-space models, and we define

$$p(\boldsymbol{x}|\boldsymbol{y}) = q(\boldsymbol{x}|\boldsymbol{y})\alpha(\boldsymbol{x}|\boldsymbol{y}) + \delta(\boldsymbol{x} - \boldsymbol{y})r(\boldsymbol{x}), \tag{14.49}$$

where $r(\boldsymbol{x})$ is the rejection probability,

$$r(\boldsymbol{x}) = \int (1 - \alpha(\boldsymbol{x}|\boldsymbol{y}))q(\boldsymbol{x}|\boldsymbol{y})\,d\boldsymbol{y}, \tag{14.50}$$

and $\delta(\cdot)$ is Dirac's delta function. A little thought reveals that $p(\cdot|\cdot)$, as defined above, is the transition density kernel (transition matrix for finite discrete spaces), $p(\boldsymbol{x}_n|\boldsymbol{x}_{n-1})$, for an equivalent Markov chain. Moreover, this Markov chain has the desired distribution $p(\boldsymbol{x})$ as its invariant, that is,

$$p(\boldsymbol{x}) = \int p(\boldsymbol{x}|\boldsymbol{y})p(\boldsymbol{y})\,d\boldsymbol{y},$$

which, as already pointed out in Section 14.7, is a direct outcome of the fact that the following detailed balance condition is satisfied (Problem 14.12):

$$p(\boldsymbol{x}|\boldsymbol{y})p(\boldsymbol{y}) = p(\boldsymbol{y}|\boldsymbol{x})p(\boldsymbol{x}).$$

It turns out that the equivalent Markov chain is ergodic, and hence converging to the invariant (desired) distribution, provided that $p(\boldsymbol{x}|\boldsymbol{y})$ and $p(\boldsymbol{x})$ are *strictly positive*. This guarantees that any state has a nonzero probability to be reached starting from any state.

Hence, the Metropolis–Hastings algorithm equivalently draws samples from the Markov chain defined by the transition density given in Eq. (14.49), although the samples are drawn from the chosen (easily sampled) proposal distribution. Typical distributions used to play the role of the proposal distribution are the Gaussian and Cauchy distributions. The latter, due to its heavy-tail property, allows large changes to occur from time to time. Sometimes, the uniform distribution is also used. For the discrete case, the uniform distribution seems to be a popular choice.

Burn-in phase: After convergence, the process becomes equivalent to drawing samples from the desired $p(\boldsymbol{x})$! However, nothing is perfect in this world; a major weakness of the Markov chain Monte Carlo techniques is that it is difficult to assess whether the Markov chain *has converged*, and hence to be sure that the samples one generates are indeed effectively independent and truly representative of $p(\boldsymbol{x})$. Samples generated before the chain has converged are not representative of the desired distribution and have to be rejected; this is known as the *burn-in phase*. The interval that a Markov chain takes to converge is known as the *mixing time* (e.g., [21]).

To this end, a number of diagnostics have been proposed, though none of them can be considered a panacea (see, e.g., [6,8,35]) for a discussion. A theoretical justification concerning the difficulty of accessing convergence of such techniques is provided in [2], where it is shown that this is a computationally intractable task.

In practice, after the rejection of the samples during the burn-in phase, one runs a long chain and discards one out of, say, M samples. For large enough values of M, one expects to obtain independent samples. This process is also known as *thinning*. An alternative path is to run a few, say, three to four, different (starting from different initial points) chains of medium size (e.g., 100,000) and take samples from each of them, having discarded the samples in the respective burn-in phases (e.g., the first half of them).

14.8.1 CONVERGENCE ISSUES

When dealing with the rejection and importance sampling, we discussed that these methods do not scale well with dimensionality. In contrast, the Metropolis approach shows much better behavior, and it is an algorithm that lends itself to applications in large spaces. Having said that, the method is not without shortcomings. To elaborate, we will use our experience gained from the random walk examples and employ similar arguments as those given in [30].

Consider a two-dimensional task and adopt as the proposal distribution, $q(\boldsymbol{x}|\boldsymbol{x}_{n-1})$, the Gaussian one with covariance matrix $\sigma^2 I$ and each time-centered at $\boldsymbol{x}_{n-1}$. The desired distribution, from which samples are to be drawn, is another elongated Gaussian $\mathcal{N}(\boldsymbol{x}|\boldsymbol{0}, \Sigma)$, as shown in Fig. 14.10A. The values $\sigma_{\max}, \sigma_{\min}$ denote the scales (standard deviations) associated with the two axes of the ellipse (recall from Chapter 2 that this is defined by the eigenstructure of Σ), which corresponds to the one standard deviation contour (the exponent in the Gaussian is equal to $-1/2$) of $p(\boldsymbol{x})$.

Every time a sample is drawn from $\mathcal{N}(\boldsymbol{x}|\boldsymbol{x}_{n-1}, \sigma^2 I)$, the new sample will be within the circle of radius σ around $\boldsymbol{x}_{n-1}$, with high probability. In order for the new sample to have a large chance to lie within the high-probability elliptical region, σ must be of the order of $\sigma_{\min}$ or smaller. If σ is chosen to have a large value, there is a high probability for the sample to end up outside the ellipse and be rejected. Hence, once sampling starts inside the ellipse, small values of σ guarantee, with high probability, that samples remain within the ellipse, and hence are accepted. On the other hand, if σ is small, a large number of iterations will be required to sufficiently cover the interior of the ellipse with points. If one looks at the process of sampling as a random walk, with approximate step size σ, then the number of iterations needed to cover a scale of the order of $\sigma_{\max}$ will be $(\frac{\sigma_{\max}}{\sigma})^2$; if $\sigma \simeq \sigma_{\min}$, this becomes $\left(\frac{\sigma_{\max}}{\sigma_{\min}}\right)^2$. In high dimensions, where there is high probability for one of the dimensions to be of relatively small scale compared to the maximum one, this square dependence rule of thumb can slow convergence substantially.

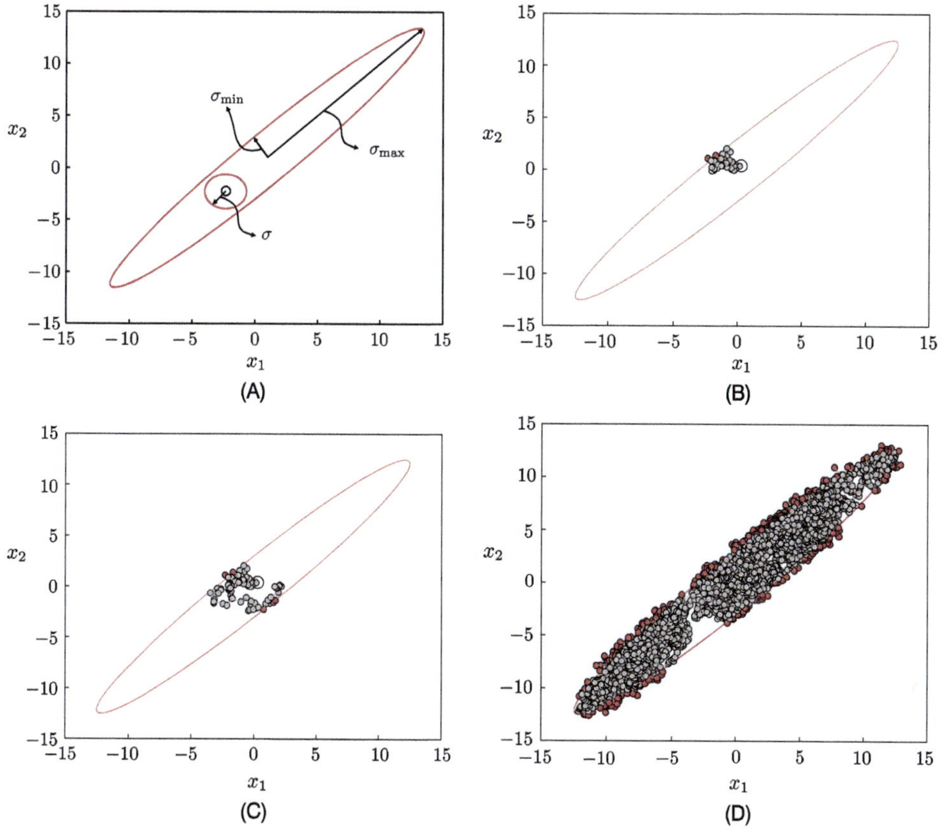

FIGURE 14.10

(A) The region of significant probability mass is enclosed by the ellipse, with scales σ_{max} and σ_{min}, respectively, as defined by the major and minor axes. The region of significant probability mass of the proposal distribution is spherical of scale equal to σ, which is of the same order as σ_{min}. (B) Starting from the point shown as a circle, 50 generated points are shown using a proposal distribution of covariance equal to $\sigma^2 = 0.1I$; rejected points are shown in red. (C) The snapshot with 100 points and (D) with 3000 points. Observe that even in the latter case, still there are parts in the high-probability region of the desired distribution that have not been covered.

Figs. 14.10B–D show the case where the desired two-dimensional Gaussian has zero mean and covariance matrix given by

$$\Sigma = \begin{bmatrix} 1.00 & 0.99 \\ 0.99 & 1.00 \end{bmatrix}.$$

The proposal distribution is a Gaussian with covariance matrix $0.1I$. The figures show three snapshots in the sequence of point generation, corresponding to 50, 100, and 3000 points. The rejected ones are denoted in red.

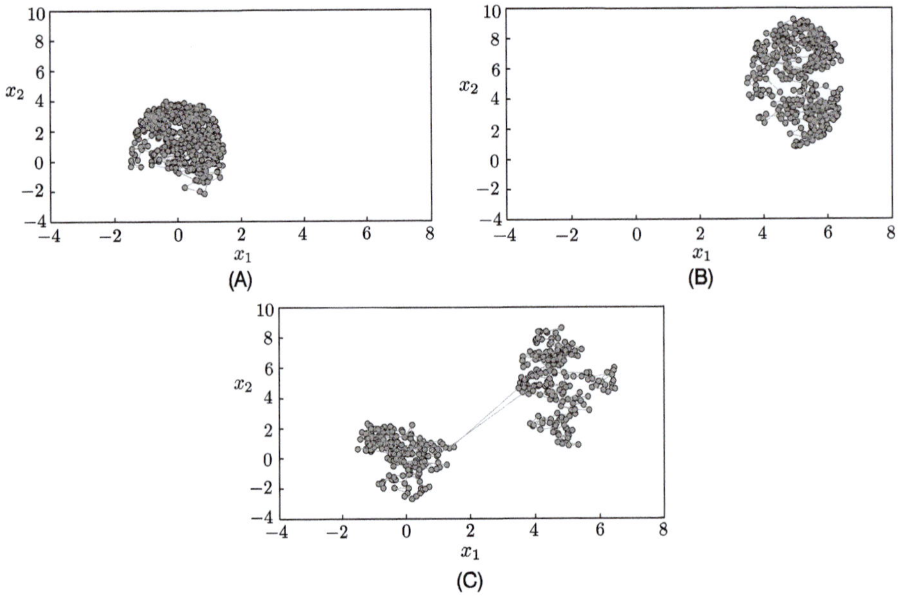

FIGURE 14.11

The desired distribution comprises the mixture of two Gaussians. In each figure, a different initialization point is selected. In all cases, 400 points have been generated. Note that in two of the cases, the process seems to be trapped in either one of the two Gaussians.

Another problem that may arise with the Metropolis method is that of *local trapping*. This may occur when the desired distribution is multimodal, which is common in high-dimensional complex problems. We will demonstrate the case via a simple example in the two-dimensional space. Let the desired distribution comprise a mixture of two Gaussians,

$$p(\boldsymbol{x}) = \frac{1}{2}\mathcal{N}(\boldsymbol{x}|\boldsymbol{\mu}_1, \Sigma_1) + \frac{1}{2}\mathcal{N}(\boldsymbol{x}|\boldsymbol{\mu}_2, \Sigma_2),$$

where $\boldsymbol{\mu}_1 = [0, 0]^T$, $\boldsymbol{\mu}_2 = [5, 5]^T$, $\Sigma_1 = \Sigma_2 = \text{diag}\{0.25, 2\}$, and the proposal distribution is $\mathcal{N}(\boldsymbol{x}|\boldsymbol{\mu}, I)$, where $\boldsymbol{\mu} = [2.5, 2.5]^T$. Fig. 14.11 shows the paths traveled by the drawn and accepted points over three different runs. In Figs. 14.11A and B after 400 iterations the points drawn cover only one of the two mixtures. Both mixtures are visited in the run corresponding to Fig. 14.11C.

14.9 GIBBS SAMPLING

Gibbs sampling is among the most popular and widely used sampling methods. It is also known as the *heat bath* algorithm. Although Gibbs sampling was already known and used in statistical physics, two papers [9,10] were catalytic for its widespread use in the Bayesian and machine learning communities.

Josiah Willard Gibbs (1839–1903) was an American scientist whose work on thermodynamics laid the foundations of physical chemistry. Together with James Clark Maxwell (1831–1879) and Ludwig Eduard Boltzmann (1844–1906), Gibbs pioneered the field of statistical mechanics. He is also the father, together with Oliver Heaviside (1850–1925), of what is today known as vector calculus.

Gibbs sampling is appropriate for drawing samples from multidimensional distributions, and it can be considered a special instance of the more general Metropolis method.

Let the random vector in the Markov chain at time n be given as

$$\mathbf{x}_n = [\mathrm{x}_n(1), \ldots, \mathrm{x}_n(l)]^T.$$

The basic assumption underlying Gibbs sampling is that the *conditional* distributions of each one of the variables, $x_n(d)$, $d = 1, 2, \ldots, l$, given the rest, that is,

$$p\big(x_n(d)|\{x_n(i) : i \neq d\}\big), \tag{14.51}$$

are known and they can be easily sampled. At each iteration, a sample is drawn for only one of the variables, based on Eq. (14.51), by freezing the values of the rest to those already available from the previous iteration. The scheme is summarized next.

Algorithm 14.4 (Gibbs sampler).

- Initialize, $x_0(1), \ldots, x_0(l)$, arbitrarily.
- **For** $n = 1, 2, \ldots, N$, **Do**
  - **For** $d = 1, 2, \ldots, l$, **Do**
    - Draw
      - $x_n(d) \sim p\big(x \mid \{x_n(i), i < d \neq 1\}, \{x_{n-1}(i),\ i > d \neq l\}\big)$
  - **End For**
- **End For**

Note that in the previous scheme, all dimensions are visited in sequence. Another version is to visit them in random order.

The Gibbs scheme can be viewed as a realization of a Markov chain, where the transition matrix/PDF is sequentially constructed from l base transitions, that is,

$$T = B_1 \cdots B_l,$$

where each individual base transition acts on the corresponding dimension, that is, coordinate-wise. For continuous variables, it can be readily checked that

$$B_d(\boldsymbol{x}|\boldsymbol{y}) = p\big(x(d)|\{y(i)\} : i \neq d\big) \prod_{i \neq d} \delta\big(y(i) - x(i)\big), \quad d = 1, 2, \ldots, l.$$

In words, only the component $x(d)$ changes while the rest are left unchanged. It is not difficult to see that the desired joint distribution $p(\boldsymbol{x}) = p(x(1), \ldots, x(l))$ is invariant with respect to each one of B_d, $d = 1, 2, \ldots, l$ (Problem 14.13). Hence, it will be invariant under their product

$$T = B_1 \cdots B_l.$$

Ergodicity is ensured by requiring that all conditional probabilities are *strictly* positive, which guarantees the convergence of the chain to the desired $p(\boldsymbol{x})$.

Remarks 14.3.

- Gibbs sampling, being an instance of the Metropolis method, inherits its random walk-like convergence performance.
- Gibbs sampling is appropriate for many graphical models (Chapter 16) that are described in terms of conditional distributions. Often, these distributions can be sampled in an easy way, using techniques such as rejection sampling and its variants, as discussed in Section 14.4.
- Note that in Gibbs sampling, no samples are rejected. This can also be shown if Gibbs sampling is considered as an instance of the Metropolis method, via the specific choice of Eq. (14.51) as the proposal distribution (Problem 14.14).
- *Blocking Gibbs sampling*: Gibbs sampling samples one variable at a time. This can make the algorithm move very slowly through the state space in case the variables are highly correlated. In such cases, it is preferable to sample *groups* of variables, not necessarily disjoint, and sample from the variables in the block, conditioned on the remaining. This is known as blocking Gibbs sampling [16], and it improves performance by achieving much bigger moves through the state space.
- *Collapsed Gibbs sampling*: In collapsed Gibbs sampling, one integrates out (marginalizes over) one or more variables and samples from the remaining ones. For example, in the case of three variables, Gibbs sampling samples from $p(x_1|x_2, x_3)$, then $p(x_2|x_1, x_3)$, and finally $p(x_3|x_1, x_2)$ to complete the iteration step. In collapsed Gibbs sampling, we can integrate out, for example, x_3 (which is *collapsed*), and sample sequentially from $p(x_1|x_2)$ and $p(x_2|x_1)$. Then sampling is performed in a lower-dimensional space, and hence it is more efficient. Collapsing one variable is tractable if it is a conjugate prior of another involved variable; for example, they are both members of the exponential family. Thus, x_3 does not participate in Gibbs sampling. In the sequel, we can sample $p(x_3|x_1, x_2)$. This can be justified by the *Rao–Blackwell* theorem, which states that the variance of the estimate created by analytically integrating out x_3 will always be lower than (or equal to) the variance of direct Gibbs sampling [23].

14.10 IN SEARCH OF MORE EFFICIENT METHODS: A DISCUSSION

In order to sidestep the drawbacks associated with the described basic Markov chain-based schemes, namely, the slow random walk-like convergence and the local trap problem, a number of more advanced methods have been suggested. It is beyond the scope of this chapter to present such schemes in more detail and the interested reader may consult more specialized books and articles, e.g., [5,22,24,30,38]. Below, we provide a short discussion on some of the most popular directions that have been proposed.

A family of algorithms known as *auxiliary variable Markov chain Monte Carlo* methods is a popular one. Such methods augment with auxiliary variables either the desired or the proposal distribution in the Metropolis–Hastings algorithm. The presence of the auxiliary variable is intended to either help the algorithm to escape from possible local traps, or to cancel out the normalizing constant, if this is intractable. Such methods include algorithms like the *simulating annealing* [17], the *simulated tempering* [25], and the *slice sampler* [15]. The rationale behind the slice sampling techniques builds around our discussion in Section 14.4; recall that sampling from $p(\boldsymbol{x})$ is equivalent to sampling uniformly from the region in

$$\mathcal{A} = \left\{ (\boldsymbol{x}, u) : \boldsymbol{x} \in \mathbb{R}^l, \ 0 \le u \le p(\boldsymbol{x}) \right\}.$$

In [31], a Gibbs-type implementation of the slice sampler is suggested; each component of $\boldsymbol{x}$ is updated sequentially using a single-variable slicing sampling strategy. It turns out that the slice sampler improves upon the convergence speed of the standard Metropolis–Hastings algorithm.

In [29], an auxiliary variable is used so that the computation of the normalizing constant is bypassed. This is important in cases where its computation is intractable.

Another sampling philosophy spans the *population-based* methods. In order to overcome the local trap problem, a number (population) of Markov chains are run in parallel under an information exchange strategy, which improves convergence. Typical examples of such techniques include *adaptive direction sampling* [12] and the *evolutionary* Monte Carlo method, which builds upon arguments used in genetic algorithms [22].

Another direction is that of the *Hamiltonian* Monte Carlo methods, which exploit arguments from classical mechanics around the elegant Hamiltonian equations [26,30]. For PDFs of the form

$$p(\boldsymbol{x}) = \frac{1}{Z_E} \exp\big(- E(\boldsymbol{x})\big),$$

$E(\boldsymbol{x})$ can be given the interpretation of the system's potential energy (Section 15.4.2). Once such a bridge has been established, an auxiliary random vector, $\boldsymbol{q}$, is introduced and interpreted as the momentum of the system; hence, the corresponding kinetic energy is expressed as

$$K(\boldsymbol{q}) = \frac{1}{2} \sum_{i=1}^{l} q_i^2.$$

The Hamiltonian function is then given by

$$H(\boldsymbol{x}, \boldsymbol{q}) = E(\boldsymbol{x}) + K(\boldsymbol{q}),$$

and it defines the distribution

$$\begin{aligned} p(\boldsymbol{x}, \boldsymbol{q}) &= \frac{1}{Z_H} \exp\big(- H(\boldsymbol{x}, \boldsymbol{q})\big) \\ &= \frac{1}{Z_E} \exp(-E) \frac{1}{Z_K} \exp\big(- K(\boldsymbol{q})\big) \\ &:= p(\boldsymbol{x}) p(\boldsymbol{q}), \end{aligned}$$

where Z_K is the normalizing constant of the respective Gaussian term associated with the kinetic energy. The desired distribution, $p(\boldsymbol{x})$, is obtained as the marginal of $p(\boldsymbol{x}, \boldsymbol{q})$. Hence, if sampling from $p(\boldsymbol{x}, \boldsymbol{q})$ is possible, then discarding $\boldsymbol{q}$ results in samples drawn from the desired distribution. The evolution of the variables in time is given by the associated Hamiltonian dynamics of the equivalent system.

Such methods may lead to a substantial improvement in convergence speed; the reason is that via the Hamiltonian interpretation, information hidden in the derivatives of $E(\boldsymbol{x})$ (i.e., $\dot{\boldsymbol{q}} = -\frac{\partial E(\boldsymbol{x})}{\partial \boldsymbol{x}}$) is exploited in order for the system to detect directions toward high-probability mass.

In the *reversible jump Markov chain Monte Carlo* algorithms, the Metropolis–Hastings algorithm is extended to account for state spaces of varying dimensionality [13]. Such methods are appropriate in cases where multiple parameter models of varying dimensionality are involved. Thus, the Markov chain is given the liberty to jump between models of different dimensionality.

VARIATIONAL INFERENCE OR MONTE CARLO METHODS

In the beginning of this chapter we mentioned that the variational inference techniques, which were considered in Chapter 13, are the deterministic alternatives of the Monte Carlo methods. We will now attempt to sketch in a few lines the pros and cons of each of the two approaches. The main advantages concerning the former path to Bayesian learning are as follows:

- They are computationally more efficient for small- and medium-scale tasks.
- It is fairly easy to determine when to stop iterations and to know when convergence has been achieved.
- One can compute lower bounds for the likelihood function.

The advantages concerning Monte Carlo methods are as follows:

- They can be applied to more general cases, for example, models without computationally convenient priors, or to models whose structure is changing.
- They do not rely on approximations such as the mean field approximation.
- They can be more efficient for large-scale tasks.

14.11 A CASE STUDY: CHANGE-POINT DETECTION

The task of change-point detection is of major importance in a number of scientific disciplines, ranging from engineering and sociology to economics and environmental studies. The accumulated literature is vast; see, for example, [1,19,36] and the references therein. The aim of the change-point identification task is to detect partitions in a sequence of observations, in order for the data in each block to be statistically "similar," in other words, to be distributed according to a common probability distribution. The hidden Markov models and the dynamical Bayesian methods, which are discussed in Chapter 17, come under this more general umbrella of problems. Our goal in this example is to demonstrate the use of Gibbs sampling in the context of the change-point detection task (see, e.g., [4]).

Let x_n be a discrete random variable that corresponds to the count of an event, for example, the number of requests for telephone calls within an interval of time, requests for individual documents on a web server, particle emissions in radioactive materials, or number of accidents in a working environment. We adopt the Poisson process to model the distribution of x_n, that is,

$$P(x;\lambda) = \frac{(\lambda\tau)^x}{x!} e^{-\lambda\tau}. \tag{14.52}$$

Poisson processes have been widely used to model the number of events that take place in a time interval, τ. For our example, we have chosen $\tau = 1$. The parameter λ is known as the *intensity* of the process (see, e.g., [32]).

We assume that our observations, $x_n,\ n = 1, 2, \dots, N$, have been generated by two different Poisson processes, $P(x; \lambda_1)$ and $P(x; \lambda_2)$. Also, the change of the model has taken place suddenly at an unknown time instant, n_0. Our goal is to estimate the posterior,

$$P(n_0|\lambda_1, \lambda_2, \boldsymbol{x}_{1:N}).$$

Moreover, the exact values of λ_1 and λ_2 are not known. The only available information is that the Poisson process intensities, $\lambda_i,\ i = 1, 2$, are distributed according to a (prior) gamma distribution, that is,

$$p(\lambda) = \text{Gamma}(\lambda|a, b) = \frac{1}{\Gamma(a)} b^a \lambda^{a-1} \exp(-b\lambda),$$

for some known positive values a, b. We will finally assume that we have no prior information on when the time of change occurred; thus, the prior is chosen to be the uniform distribution $P(n_0) = \frac{1}{N}$. Based on the previous assumptions, the corresponding joint distribution is given by

$$p(n_0, \lambda_1, \lambda_2, \boldsymbol{x}_{1:N}) = p(\boldsymbol{x}_{1:N}|\lambda_1, \lambda_2, n_0) p(\lambda_1) p(\lambda_2) P(n_0),$$

or

$$p(n_0, \lambda_1, \lambda_2, \boldsymbol{x}_{1:N}) = \prod_{n=1}^{n_0} P(x_n|\lambda_1) \prod_{n=n_0+1}^{N} P(x_n|\lambda_2) p(\lambda_1) p(\lambda_2) P(n_0).$$

Taking the logarithm in order to get rid of the products, and integrating out respective variables, the following conditionals needed in Gibbs sampling are obtained (Problem 14.15):

$$p(\lambda_1|n_0, \lambda_2, \boldsymbol{x}_{1:N}) = \text{Gamma}(\lambda_1|a_1, b_1), \tag{14.53}$$

with

$$a_1 = a + \sum_{n=1}^{n_0} x_n, \quad b_1 = b + n_0,$$

$$p(\lambda_2|n_0, \lambda_1, \boldsymbol{x}_{1:N}) = \text{Gamma}(\lambda_2|a_2, b_2), \tag{14.54}$$

$$a_2 = a + \sum_{n=n_0+1}^{N} x_n, \quad b_2 = b + (N - n_0),$$

and

$$\begin{aligned} P(n_0|\lambda_1, \lambda_2, \boldsymbol{x}_{1:N}) &= \ln \lambda_1 \sum_{n=1}^{n_0} x_n - n_0 \lambda_1 + \ln \lambda_2 \sum_{n=n_0+1}^{N} x_n \\ &- (N - n_0)\lambda_2, \quad n_0 = 1, 2, \dots, N. \end{aligned} \tag{14.55}$$

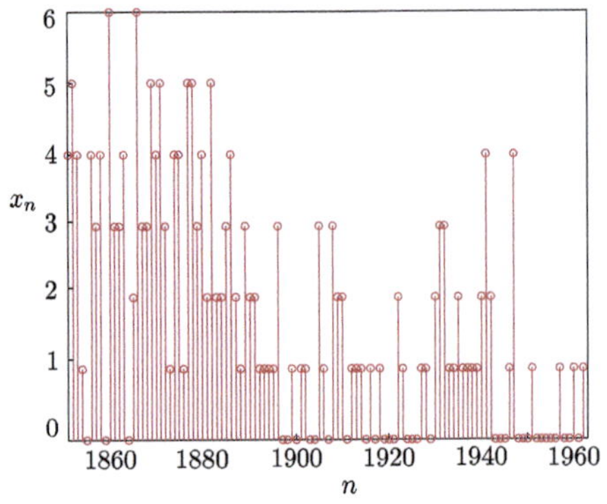

FIGURE 14.12

Number of deadly accidents per year in the coal mines in England over the period 1851–1962.

Note that the first two conditionals are gamma distributed and, as we said at the end of Section 14.4, a number of different approaches are available for generating samples from it. The last distribution is a discrete one, and samples can be drawn as discussed in Algorithm 14.1. We are now ready to apply Gibbs sampling.

Algorithm 14.5 (Gibbs sampling for change-point detection).

- Having obtained $\boldsymbol{x}_{1:N} := \{x_1, \ldots, x_N\}$, select a and b.
- Initialize $n_0^{(0)}$
- **For** $i = 1, 2, \ldots,$ **Do**
  - $\lambda_1^{(i)} \sim \text{Gamma}\big(\lambda|a + \sum_{n=1}^{n_0^{(i-1)}} x_n,\ b + n_0^{(i-1)}\big)$
  - $\lambda_2^{(i)} \sim \text{Gamma}\big(\lambda|a + \sum_{n=n_0^{(i-1)}+1}^{N} x_n,\ b + (N - n_0^{(i-1)})\big)$
  - $n_0^{(i)} \sim P(n_0|\lambda_1^{(i)}, \lambda_2^{(i)}, x_{1:N})$
- **End For**

Fig. 14.12 shows the number of deadly accidents per year in the coal mines in England spanning the years 1851–1962. Looking at the graph, it is readily observed that the "front" part of the graph looks different from its "back" end, with a change around 1890–1900. As a matter of fact, in 1890, new health and safety regulations were introduced, following pressure from the coal miners' unions. We will use the model explained before and draw samples according to Algorithm 14.5 in order to determine the point, n_0, where a change in the statistical distributions describing the data occurred [4]. The values of a and b were chosen equal to $a = 2$ and $b = 1$, although the obtained results are not sensitive in their choice. The burn-in phase was 200 samples. Fig. 14.13 shows the obtained histogram of the values of n_0 drawn by the algorithm, which clearly indicates a peak at the year 1890. Fig. 14.14 shows the plot of the points drawn for λ_1 and λ_2. The plot clearly indicates that the intensity of the Poisson process dropped from $\lambda_1 = 3$ to $\lambda_2 = 1$ after the introduction of the safety regulations.

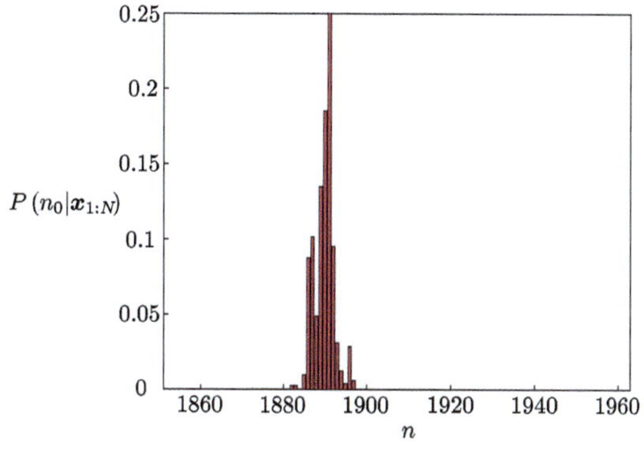

FIGURE 14.13

The histogram obtained from the values of n_0 generated by the algorithm, which approximates the posterior for n_0, for the case study of Example 14.11. Observe that the histogram peaks at 1890, the year when the new regulations were introduced.

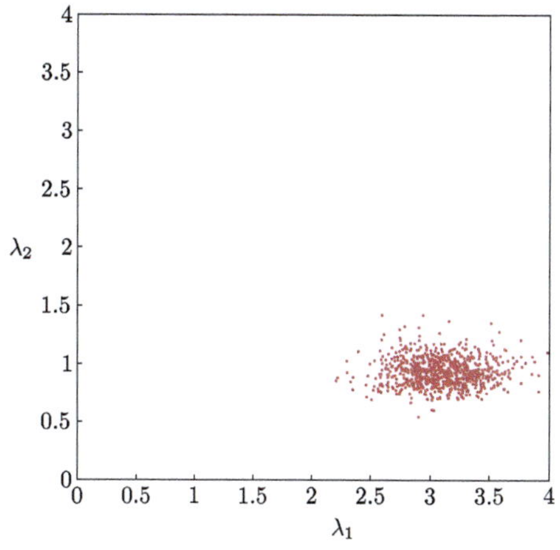

FIGURE 14.14

Case study of Example 14.11: The cluster formed by the obtained values of λ_1 and λ_2.

PROBLEMS

14.1 Show that if $F_{\mathrm{x}}(x)$ is the cumulative distribution function of a random variable x, then the random variable $\mathrm{u} = F_{\mathrm{x}}(\mathrm{x})$ follows the uniform distribution in [0, 1].

14.2 Show that if u follows the uniform distribution and

$$\mathrm{x} = F_{\mathrm{x}}^{-1}(\mathrm{u}) := g(\mathrm{u}), \tag{14.56}$$

then indeed x is distributed according to $F_{\mathrm{x}}(x) = \int_{-\infty}^{x} p(x)\,dx$.

14.3 Consider the random variables r and ϕ with exponential and uniform distributions

$$p_{\mathrm{r}}(r) = \frac{1}{2}\exp\left(-\frac{r}{2}\right), \quad r \geq 0,$$

and

$$p_{\phi}(\phi) = \begin{cases} \frac{1}{2\pi}, & 0 \leq \phi \leq 2\pi, \\ 0, & \text{otherwise,} \end{cases}$$

respectively. Show that the transformation

$$\mathrm{x} = \sqrt{\mathrm{r}}\cos\phi = g_{\mathrm{x}}(\mathrm{r}, \phi),$$

$$\mathrm{y} = \sqrt{\mathrm{r}}\sin\phi = g_{\mathrm{y}}(\mathrm{r}, \phi)$$

renders both x and y to follow the normalized Gaussian $\mathcal{N}(0, 1)$.

14.4 Show that if

$$p_{\mathrm{x}}(\boldsymbol{x}) = \mathcal{N}(\boldsymbol{x}|0, I),$$

then $\mathbf{y}$ given by the transformation

$$\mathbf{y} = L\mathbf{x} + \boldsymbol{\mu}$$

is distributed according to

$$p_{\mathrm{y}}(\boldsymbol{y}) = \mathcal{N}(\boldsymbol{y}|\boldsymbol{\mu}, \Sigma),$$

where $\Sigma = LL^T$.

14.5 Consider two Gaussians

$$p(\boldsymbol{x}) = \mathcal{N}(\boldsymbol{x}|0, \sigma_p^2 I), \quad \sigma_p^2 = 0.1$$

and

$$q(\boldsymbol{x}) = \mathcal{N}(\boldsymbol{x}|0, \sigma_q^2 I), \quad \sigma_q^2 = 0.11,$$

with $\boldsymbol{x} \in \mathbb{R}^l$. In order to use $q(\boldsymbol{x})$ for drawing samples from $p(\boldsymbol{x})$ via the rejection sampling method, a constant c has to be computed so that

$$cq(\boldsymbol{x}) \geq p(\boldsymbol{x}).$$

Show that

$$c \geq \left(\frac{\sigma_q}{\sigma_p}\right)^l,$$

and compute the probability of accepting samples.

14.6 Show that using importance sampling leads to an unbiased estimator for the normalizing constant of the desired distribution,

$$p(\boldsymbol{x}) = \frac{1}{Z}\phi(\boldsymbol{x}).$$

However, the estimator of $\mathbb{E}[f(\mathbf{x})]$ for a function f is a biased one.

14.7 Let $p(\boldsymbol{x}) = \mathcal{N}(\boldsymbol{x}|\mathbf{0}, \sigma_1^2 I)$. Choose the proposal distribution for importance sampling as

$$q(\boldsymbol{x}) = \mathcal{N}(\boldsymbol{x}|\mathbf{0}, \sigma_2^2 I).$$

The weights are computed as

$$w(\boldsymbol{x}) = \frac{p(\boldsymbol{x})}{q(\boldsymbol{x})}.$$

If $w(\mathbf{0})$ is the weight at $\boldsymbol{x} = \mathbf{0}$, then the ratio $\frac{w(\boldsymbol{x})}{w(\mathbf{0})}$ is given by

$$\frac{w(\boldsymbol{x})}{w(\mathbf{0})} = \exp\frac{1}{2}\left(\frac{\sigma_1^2 - \sigma_2^2}{\sigma_1^2\sigma_2^2}\sum_{i=1}^{l} x_i^2\right).$$

Observe that even for a very good match between $q(\boldsymbol{x})$ and $p(\boldsymbol{x})$ ($\sigma_1^2 \simeq \sigma_2^2$), for large values of l, the values of the weights can change significantly due to the exponential dependence.

14.8 Show that a stochastic matrix P always has the value $\lambda = 1$ as its eigenvalue.

14.9 Show that if the eigenvalue of a transition matrix is not equal to one, its magnitude cannot be larger than one, that is, $|\lambda| \le 1$.

14.10 Prove that if P is a stochastic matrix and $\lambda \neq 1$, then the elements of the corresponding eigenvector add to zero.

14.11 Prove the square root dependence of the distance traveled by a random walk with infinitely many integer states on the time n.

14.12 Prove, using the detailed balance condition, that the invariant distribution associated with the Markov chain implied by the Metropolis–Hastings algorithm is the desired distribution, $p(\boldsymbol{x})$.

14.13 Show that in Gibbs sampling, the desired joint distribution is invariant with respect to each one of the base transition PDFs.

14.14 Show that the acceptance rate for the Gibbs sampling is equal to one.

14.15 Derive the formulas for the conditional distributions of Example 14.11.

MATLAB® EXERCISE

14.16 Develop a MATLAB® code for the Gibbs sampler. Then, use it to draw samples from the two-dimensional Gaussian distribution, with mean value and covariance matrix equal to

$$\boldsymbol{\mu} = [0, 0]^T, \quad \Sigma = \begin{bmatrix} 1 & 0.5 \\ 0.5 & 1 \end{bmatrix}.$$

Derive the conditional PDF of each one of the variables with respect to the other (use the appendix associated with Chapter 12, which can be downloaded from the book's website). Then use the conditional PDFs to implement the Gibbs sampler. Plot the generated points in the two-dimensional space after 20, 50, 100, 300, and 1000 iterations. What do you observe concerning convergence?

REFERENCES

[1] D. Barry, J.A. Hartigan, A Bayesian analysis for change point problems, J. Am. Stat. Assoc. 88 (1993) 309–319.

[2] N. Bhatnagar, A. Bogdanov, E. Mossel, The computational complexity of estimating convergence time, arXiv:1007.0089v1 [cs.DS], 2010.

[3] G. Celeux, J. Diebolt, The SEM algorithm: a probabilistic teacher derive from the EM algorithm for the mixture problem, Comput. Stat. Q. 2 (1985) 73–82.

[4] A.T. Cemgil, A tutorial introduction to Monte Carlo methods, Markov chain Monte Carlo and particle filtering, in: R. Chellappa, S. Theodoridis (Eds.), Academic Press Library in Signal Processing, vol. 1, Academic Press, San Diego, CA, 2014, pp. 1065–1113.

[5] M.H. Chen, Q.M. Shao, J.G. Ibrahim (Eds.), Monte Carlo Methods in Bayesian Computation, Springer, New York, 2001.

[6] M.K. Cowles, B.P. Carlin, Markov chain Monte Carlo convergence diagnostics: a comparative review, J. Am. Stat. Assoc. 91 (1996) 883–904.

[7] L. Devroye, Non-Uniform Random Variate Generation, Springer-Verlag, New York, 1986.

[8] A. Gelman, D.B. Rubin, Inference from iterative simulation using multiple sequences, Stat. Sci. 7 (1992) 457–511.

[9] A.E. Gelfand, A.F.M. Smith, Sampling based approaches to calculating marginal densities, J. Am. Stat. Assoc. 85 (1990) 398–409.

[10] S. Geman, D. Geman, Stochastic relaxation Gibbs distributions and the Bayesian restoration of images, IEEE Trans. Pattern Anal. Mach. Intell. 6 (1984) 721–741.

[11] W.R. Gilks, P. Wild, Adaptive rejection sampling for Gibbs sampling, Appl. Stat. 41 (1992) 337–348.

[12] W.R. Gilks, G.O. Roberts, E.I. George, Adaptive direction sampling, Statistician 43 (1994) 179–189.

[13] P.J. Green, Reversible jump Markov chain Monte Carlo computation and Bayesian model determination, Biometrika 82 (1995) 711–732.

[14] W.K. Hastings, Monte Carlo sampling methods using Markov chains and their applications, Biometrika 57 (1970) 97–109.

[15] D.M. Higdon, Auxiliary variable methods for Markov chain Monte Carlo with applications, J. Am. Stat. Assoc. 93 (1994) 179–189.

[16] C.A. Jensen, A. Kong, U. Kjaeruff, Blocking Gibbs sampling in very large probabilistic expert systems, Int. J. Hum.-Comput. Stud. 42 (1995) 647–666.

[17] S. Kirkpatrick, C.D. Gelatt, M.P. Vecchi, Optimization by simulated annealing, Science 220 (1983) 671–680.

[18] D.E. Knuth, The Art of Computer Programming, second ed., Addison Wesley, Reading, MA, 1981.

[19] T.L. Lai, Sequential change point detection in quality control and dynamical systems, J. R. Stat. Soc. B 57 (1995) 613–658.

[20] D.H. Lehmer, Mathematical methods in large scale computing units, Ann. Comput. Lab. Harvard Univ. 26 (1951).

[21] D.A. Levin, Y. Peres, E.L. Wilmer, Markov Chains and Mixing Times, American Mathematical Society, Providence, RI, 2008.

[22] F. Liang, C. Liu, R.J. Caroll, Advanced Markov Chain Monte Carlo Methods: Learning From Past Samples, John Wiley, New York, 2010.

[23] J.S. Liu, The collapsed Gibbs sampler in Bayesian computations with applications to a gene regulation problem, J. Am. Stat. Assoc. 89 (427) (1994) 958–966.

[24] J.S. Liu, Monte Carlo Strategies in Scientific Computing, Springer, New York, 2001.

[25] E. Marinari, G. Parisi, Simulated tempering: a new Monte Carlo scheme, Europhys. Lett. 19 (6) (1992) 451–458.

[26] D.J.C. MacKay, Information Theory, Inference, and Learning Algorithms, Cambridge University Press, Cambridge, 2003.

[27] N. Metropolis, A.W. Rosenbluth, M.N. Rosenbluth, A.H. Teller, E. Teller, Equation of state calculations by fast computing machines, J. Chem. Phys. 21 (1953) 1087–1091.

[28] N. Metropolis, The beginning of Monte Carlo methods, Los Alamos Sci. (1987) 125–130.

[29] J. Moller, A.N. Pettitt, R. Reeves, K.K. Berthelsen, An efficient Markov chain Monte Carlo method for distributions with intractable normalising constants, Biometrica 93 (2006) 451–458.

[30] R.M. Neal, Probabilistic Inference Using Markov Chain Monte Carlo Methods, Technical Report (GR-TR-93-1), Department of Computer Science, University of Toronto, Canada, 1993.

[31] R.M. Neal, Slice sampling, Ann. Stat. 31 (2003) 705–767.

[32] A. Papoulis, S.U. Pillai, Probability, Random Variables and Stochastic Processes, fourth ed., McGraw-Hill, New York, 2002.

[33] P.H. Peskun, Optimum Monte Carlo sampling using Markov chains, Biometrika 60 (1973) 607–612.

[34] S.K. Park, K.W. Miller, Random number generations: good ones are hard to find, Commun. ACM 31 (10) (1988) 1192–1201.
[35] M. Plummer, N. Best, K. Cowles, CODA: output analysis and diagnostics for Markov chain Monte Carlo simulations, http://cran.r-project.org, 2006.
[36] J. Reeves, J. Chen, X.L. Wang, R. Lund, Q.Q. Lu, A review and comparison of changepoint detection techniques for climate data, J. Appl. Meteorol. Climatol. 46 (2007) 900–915.
[37] B. Ripley, Stochastic Simulation, John Wiley, New York, 1987.
[38] C.P. Robert, G. Casella, Monte Carlo Statistical Methods, second ed., Springer, New York, 2004.
[39] A. Sinclair, Algorithms for Random Generation and Counting: A Markov Chain Approach, Birkhäuser, Boston, 1993.
[40] L. Tierney, Markov chains for exploring posterior distribution, Ann. Stat. 22 (1994) 1701–1762.

CHAPTER

PROBABILISTIC GRAPHICAL MODELS: PART I

15

CONTENTS

15.1 INTRODUCTION

In Fig. 13.2, we used a graphical description to indicate conditional dependencies among various parameters that control the "fusion" of the prior and conditional PDFs in a hierarchical manner. Our purpose there was more of a pedagogical nature; we could live without it. In this chapter, graphical models emerge out of necessity. In many everyday machine learning applications involving multivariate statistical modeling, even simple inference tasks can easily become computationally intractable. Typical applications involve bioinformatics, speech recognition, machine vision, and text mining, to name but a few.

Graph theory has proved a powerful and elegant tool that has extensively been used in optimization and computational theory. A graph encodes dependencies among interacting variables and can be used

Machine Learning. https://doi.org/10.1016/B978-0-12-818803-3.00027-1

to formalize the probabilistic structure that underlies our modeling assumptions. This can then be used to facilitate computations in a number of inference tasks, such as the calculation of marginals, modes, and conditional probabilities. Moreover, graphical models can be used as a vehicle to impose approximations onto the models when computational needs go beyond the available resources.

Early celebrated examples of the use of such models in learning tasks are the hidden Markov models, Kalman filtering, and error correcting coding, which have been popular since the early 1960s.

This is the first of two chapters dedicated to probabilistic graphical models. This chapter focuses on the basic definitions and concepts, and most of its material is a must for a first reading on the topic. A number of basic graphical models are discussed, such as Bayesian networks (BNs) and Markov random fields (MRFs). Exact inference is presented, and the elegant message passing algorithm for inference on chains and trees is introduced.

15.2 THE NEED FOR GRAPHICAL MODELS

Let us consider a simplified example of a learning system in the context of a medical application. Such a system comprises a set of m diseases that correspond to hidden variables and a set of n symptoms (findings). The diseases are treated as random variables, $\mathrm{d}_1, \mathrm{d}_2, \ldots, \mathrm{d}_m$, and each of them can be absent or present and thus can be encoded by a zero or a one, that is, $\mathrm{d}_j \in \{0, 1\},\ j = 1, 2, \ldots, m$. The same applies to the symptoms f_i, which can either be absent or present; hence, $\mathrm{f}_i \in \{0, 1\},\ i = 1, 2, \ldots, n$. The symptoms comprise the observed variables.[1]

The goal of the system is to predict a disease hypothesis, that is, the presence of a number of diseases, given the presence of a set of symptoms which have been observed. During the training, which is based on experts' assessments, the system learns the prior probabilities $P(d_j)$ and the conditional probabilities $P_{ij} = P(f_i = 1|d_j = 1),\ i = 1, 2, \ldots, n,\ j = 1, 2, \ldots, m$. The latter comprise a table of nm entries. For a realistic system, these numbers can be very large. For example, in [41], m is of the order of 500–600 and n of the order of 4000. Let $\boldsymbol{f}$ be the vector that corresponds to a specific set of observations for the findings, indicating the presence or absence of the respective symptoms. Assuming that symptoms are *conditionally independent*, given any disease hypothesis, $\boldsymbol{d}$, we can write

$$P(\boldsymbol{f}|\boldsymbol{d}) = \prod_{i=1}^{n} P(f_i|\boldsymbol{d}). \tag{15.1}$$

Ideally, one should be able to obtain the conditional probabilities $P(f_i|\boldsymbol{d})$ for each disease hypothesis. However, for all possible 2^m combinations of $\boldsymbol{d}$, this should require a huge amount of training data, which is impossible to collect for any practical system. This is bypassed by adopting the following model:

$$P(f_i = 0|\boldsymbol{d}) = \prod_{j=1}^{m} (1 - P_{ij})^{d_j}, \tag{15.2}$$

where the exponent is set to zero, $d_j = 0$, when the disease is not related to the symptom. This is known as the *noisy-OR* model. That is, it is assumed that for a negative finding the individual causes

[1] In a more realistic system, some of the findings may not be available, that is, they may be unobservable.

are independent [37]. Obviously,

$$P(f_i = 1|\boldsymbol{d}) = 1 - P(f_i = 0|\boldsymbol{d}).$$

Let us now assume that we observe a set of findings, $\boldsymbol{f}$, and we want to infer $P(d_j|\boldsymbol{f})$ for some j. Then

$$\begin{aligned} P(d_j = 1|\boldsymbol{f}) &= \frac{P(\boldsymbol{f}|d_j = 1)P(d_j = 1)}{P(\boldsymbol{f})} \\ &= \frac{\sum_{\boldsymbol{d}:d_j=1} P(\boldsymbol{f}|\boldsymbol{d})P(\boldsymbol{d})}{\sum_{\boldsymbol{d}} P(\boldsymbol{f}|\boldsymbol{d})P(\boldsymbol{d})}. \end{aligned} \tag{15.3}$$

The summation in the denominator involves 2^m terms. For $m \sim 500$, this is a formidable task that simply cannot be carried out in a realistic time.

The previous example indicates that once one gets involved with complex systems, even innocent looking tasks turn out to be computationally intractable. Thus, one has either to be more clever in exploiting possible independencies in the data, which can reduce the required number of computations, or make certain assumptions/approximations. In this chapter, we will study both alternatives.

Before we proceed further, it is interesting to point out another source of computational obstacles besides the calculation of Eq. (15.3). In practice, it may be more convenient to perform addition instead of implementing multiplication; multiplying a large number of variables of small values such as probabilities may cause arithmetic accuracy problems. One way to bypass products is either via logarithmic or exponential operations, which transform products into summations. For example, Eq. (15.2) can be rewritten as

$$P(f_i = 0|\boldsymbol{d}) = \exp\left(-\sum_{j=1}^{m} \theta_{ij} d_j\right), \tag{15.4}$$

where $\theta_{ij} := -\ln(1 - P_{ij})$ and

$$P(f_i = 1|\boldsymbol{d}) = 1 - \exp\left(-\sum_{j=1}^{m} \theta_{ij} d_j\right). \tag{15.5}$$

Observe that the presence in Eq. (15.1) of terms corresponding to negative findings contributes linearly to the complexity (product of exponentials correspond to summations). However, this is not the case with the terms associated with positive findings. Take, for example, the extreme case where all findings are negative. Then

$$\begin{aligned} P(\boldsymbol{f} = \boldsymbol{0}|\boldsymbol{d}) &= \prod_{i=1}^{n} \exp\left(-\sum_{j=1}^{m} \theta_{ij} d_j\right) \\ &= \exp\left(-\sum_{i=1}^{n}\left(\sum_{j=1}^{m} \theta_{ij} d_j\right)\right). \end{aligned} \tag{15.6}$$

Consider now $f_1 = 1$ and the rest to be $f_i = 0$, $i = 2, \ldots, n$. Then

$$P(f|d) = \left(1 - \exp\left(-\sum_{j=1}^{m} \theta_{1j} d_j\right)\right) \exp\left(-\sum_{i=2}^{n}\left(\sum_{j=1}^{m} \theta_{ij} d_j\right)\right), \tag{15.7}$$

where the number of exponents to be computed is two. It can easily be shown that the cross-product terms lead to an exponential computational growth [20] (Problem 15.1).

The path we will follow in order to derive efficient exact inference algorithms as well as to derive efficient approximation rules, when exact inference is not possible, will be via the use of graphical models.

15.3 BAYESIAN NETWORKS AND THE MARKOV CONDITION

Before we move on to definitions, let us first see how the existence of some structure in a joint distribution can simplify the task of marginalization. We will demonstrate it using discrete probabilities, where the use of counting can make things simpler.

Let us consider l discrete jointly distributed random variables. Applying the product rule of probability, we obtain

$$P(x_1, x_2, \ldots, x_l) = P(x_l|x_{l-1}, x_{l-2}, \ldots, x_1)\, P(x_{l-1}|x_{l-2}, \ldots, x_1) \ldots P(x_1). \tag{15.8}$$

Assume that each one of these variables takes values in the discrete set $\{1, 2, \ldots, k\}$. In the general case, if we want to marginalize with respect to one of the variables, say, x_1, we must sum over the others, that is,

$$P(x_1) = \sum_{x_2} \cdots \sum_{x_l} P(x_1, x_2, \ldots, x_l),$$

where each one of the summations is over k possible values, which is equivalent to $\mathcal{O}(k^l)$ summations; for large values of k and/or l, this is a formidable and sometimes impossible task. Let us consider now one extreme case, where all the involved variables are mutually independent. Then the product rule becomes

$$P(x_1, x_2, \ldots, x_l) = \prod_{i=1}^{l} P(x_i),$$

and marginalization turns out to be the trivial identity

$$P(x_1) = \Big(\sum_{x_l} P(x_l) \sum_{x_{l-1}} P(x_{l-1}) \cdots \sum_{x_2} P(x_2)\Big) P(x_1), \tag{15.9}$$

because each summation is carried out independently, and of course results to one. In other words, exploiting the product rule and the statistical independence can bypass the obstacle of the exponential growth of the computational load. As a matter of fact, the previous full-independence assumption gives birth to the naive Bayes classifier (Chapter 7).

In this chapter, we are going to study cases that lie between the previous two extremes. The general idea is to be able to express the joint probability distribution (probability density/mass function) in terms of *products* of factors, where each one of them depends on a *subset* of the involved variables. This can be expressed by writing the joint distribution as

$$\boxed{p(x_1, x_2, \ldots, x_l) = \prod_{i=1}^{l} p(x_i|\mathrm{Pa}_i),} \tag{15.10}$$

where Pa_i denotes the subset of variables associated with the random variable x_i. Take the following example:

$$p(x_1, x_2, x_3, x_4, x_5, x_6) = p(x_6|x_4)\, p(x_5|x_3, x_4)\, p(x_4|x_1, x_2)\, p(x_3|x_1)\, p(x_2)\, p(x_1). \tag{15.11}$$

Then $\mathrm{Pa}_6 = \{\mathrm{x}_4\}$, $\mathrm{Pa}_5 = \{\mathrm{x}_3, \mathrm{x}_4\}$, $\mathrm{Pa}_4 = \{\mathrm{x}_1, \mathrm{x}_2\}$, $\mathrm{Pa}_3 = \{\mathrm{x}_1\}$, $\mathrm{Pa}_2 = \emptyset$, $\mathrm{Pa}_1 = \emptyset$. The variables in the set, Pa_i, are defined as the *parents* of the respective x_i, and from a statistical point of view this means that x_i is statistically independent of *all* the variables *given* the values of its parents. Every $p(x_i|\mathrm{Pa}_i)$ expresses a *conditional independence* relationship and it imposes a *probabilistic structure* that underlies our multivariate set. It is such types of independencies that we will exploit in order to perform inference tasks at a lower computational cost.

15.3.1 GRAPHS: BASIC DEFINITIONS

A graph $G = \{V, E\}$ is a collection of nodes/vertices $V = \{x_1, \ldots, x_l\}$ and a collection of edges (arcs) $E \subset V \times V$. Each edge connects two vertices and it is denoted as a pair, $(x_i, x_j) \in E$. An edge can be either *directed*—then we write $(x_i \rightarrow x_j)$ to indicate the direction—or *undirected*—then we simply write (x_i, x_j). Suppose we have a set of nodes $x_1, x_2, \ldots, x_k$, $k \geq 2$, and a corresponding set of edges $(x_{i-1}, x_i) \in E$ or $(x_{i-1} \rightarrow x_i) \in E$, $2 \leq i \leq k$; that is, the edges connect pairs of nodes in *sequence* and they can be either directed or not. This sequence of edges is called a *path* from x_1 to x_k. If there is at least one directed edge, the path is called *directed*. A *cycle* is a path from a node to itself. A *chain* or a *trail* is a path that can be "run" either from x_1 to x_k or from x_k to x_1; that is, all directed edges are replaced by undirected ones.

A *directed* graph comprises directed edges only, and it is called a *directed acyclic graph* (DAG) if it contains *no* cycles. Given a DAG, a node in it, x_i, is called *parent* of x_j if there is a (directed) edge from x_i to x_j, and we call x_j the *child* of x_i. A node x_j is called a *descendant* of x_i and x_i an *ancestor* of x_j if there is a path from x_i to x_j. A node x_j is called *nondescendant* of x_i if it is not a descendant of x_i. A graph is said to be *fully connected* or *complete* if there is an edge between every pair of nodes. Fig. 15.1 illustrates the previous definitions.

Definition 15.1. A *Bayesian network structure* is a DAG whose nodes represent random variables, $\mathrm{x}_1, \ldots, \mathrm{x}_l$, and every variable (node), x_i, is *conditionally independent* of the set of *all* its nondescendants, given the set of all its *parents*. Sometimes this is also known as the *Markov condition*.

If we denote the set of the nondescendants of a node x_i as ND_i, the Markov condition can be written as [12] $\mathrm{x}_i \perp ND_i|\mathrm{Pa}_i$, $\forall\, i = 1, 2, \ldots, l$. Sometimes, the conditional independencies are also known as *local independencies*. Stated differently, a BN graphical structure is a convenient way to encode

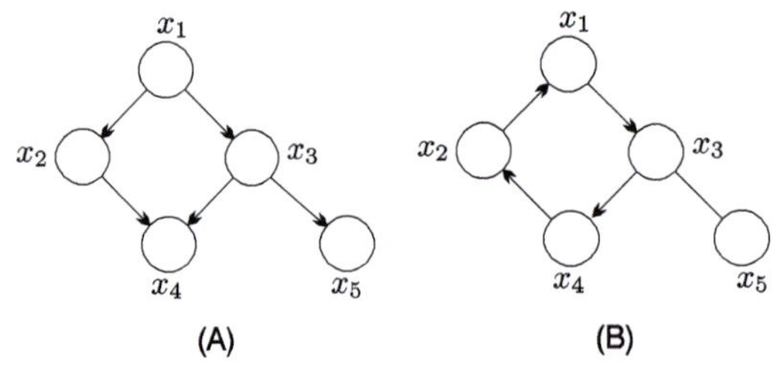

FIGURE 15.1

(A) This is a DAG because there are no cycles; x_1 is a parent of both x_2 and x_3; x_4 and x_5 are children of x_3; x_1, x_2, and x_3 are ancestors of x_4, while x_4 and x_5 are descendants of x_1; x_5 is a nondescendant of x_2 and x_4. (B) This is not a DAG and the sequence $(x_2, x_1, x_3, x_4, x_2)$ comprises a cycle. The edge (x_3, x_5) is undirected. The sequence of nodes (x_1, x_3, x_5) forms a directed path, and the sequence (x_1, x_2, x_4, x_3) forms a chain, once directed edges are replaced by undirected ones.

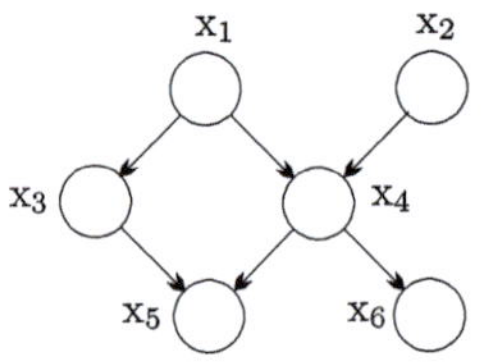

FIGURE 15.2

The BN structure corresponding to the PDF in Eq. (15.11). Observe that x_5 is conditionally independent of x_1 and x_2 given the values of x_3 and x_4. Note that nodes in a BN structure correspond to random variables.

conditional independencies. Fig. 15.2 shows the DAG that expresses the conditional independencies used in Eq. (15.11), in order to express the joint distribution as a product of factors. Conditional independence among random variables, that is, $\mathrm{x} \perp \mathrm{y} | \mathrm{z}$, or, equivalently, $p(x|y, z) = p(x|z)$, means that once we know the value of z, observing the value of y gives no additional information about x (note that the previous makes sense only if $p(y, z) > 0$). For example, the probability of children having a good education depends on whether they grow up in a poor or a rich (low or high gross national product [GNP]) country. The probability of someone getting a high-paying job depends on her/his level of education. The probability of someone getting a high-paying job is independent of the country in which he or she was born and raised, given the level of her/his education.

Theorem 15.1. *Let G be a BN structure and let p be the joint probability distribution of the random variables associated with the graph. Then p is equal to the product of the conditional distributions of all the nodes given the values of their parents, and we say that p factorizes over G.*

The proof of the theorem is done by induction (Problem 15.2). Moreover, the reverse of this theorem is also true. The previous theorem assumed a distribution and built the BN based on the underlying conditional independencies. The next theorem deals with the reverse procedure. One builds a graph based on a set of conditional distributions—one for each node of the network.

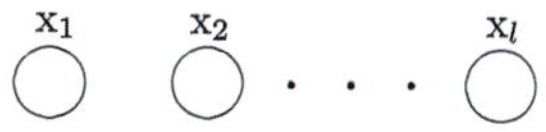

FIGURE 15.3

Bayesian network structure for independent variables. No edges are present because every variable is independent of all the others and no parents can be identified.

Theorem 15.2. *Let G be a DAG and associate a conditional probability for each node, given the values of its parents. Then the product of these conditional probabilities yields a joint probability of the variables. Moreover, the Markov condition is satisfied.*

The proof of this theorem is given in Problem 15.4. Note that in this theorem, we used the term probability and not distribution. The reason is that the theorem is not true for every form of conditional densities (PDFs) [14]. However, it holds true for a number of widely used PDFs, such as the Gaussians. This theorem is very useful because, often in practice, this is the way we construct a probabilistic graphical model—building it hierarchically, using reasoning on the corresponding physical process that we want to model, and encoding conditional independencies in the graph.

Fig. 15.3 shows the BN structure describing a set of mutually independent variables (naive Bayes assumption).

Definition 15.2. A *Bayesian network* (BN) is a pair (G, p), where the distribution p factorizes over the DAG G, in terms of a set of conditional probability distributions, associated with the nodes of G.

In other words, a BN is associated with a specific distribution. In contrast, a BN structure refers to any distribution that satisfies the Markov condition as expressed by the network structure.

Example 15.1. Consider the following simplified study relating the GNP of a country to the level of education and the type of a job an adult gets later in her/his professional life. Variable x_1 is binary with two values, HGP and LGP, corresponding to countries with high and low GNP, respectively. Variable x_2 gets three values, NE, LE, and HE, corresponding to no education, low-level, and high-level education, respectively. Finally, variable x_3 gets also three possible values, UN, LP, HP, corresponding to unemployed, low-paying, and high-paying jobs, respectively. Using a large enough sample of data, the following probabilities are learned:

1. Marginal probabilities:

$$P(x_1 = LGP) = 0.8, \; P(x_1 = HGP) = 0.2.$$

2. Conditional probabilities:

$$\begin{aligned}
&P(x_2 = NE|x_1 = LGP) = 0.1, \; P(x_2 = LE|x_1 = LGP) = 0.7,\\
&P(x_2 = HE|x_1 = LGP) = 0.2,\\
&P(x_2 = NE|x_1 = HGP) = 0.05, \; P(x_2 = LE|x_1 = HGP) = 0.2,\\
&P(x_2 = HE|x_1 = HGP) = 0.75,\\
&P(x_3 = UN|x_2 = NE) = 0.15, \; P(x_3 = LP|x_2 = NE) = 0.8,
\end{aligned}$$

$$P(x_3 = HP|x_2 = NE) = 0.05.$$
$$P(x_3 = UN|x_2 = LE) = 0.10,\ P(x_3 = LP|x_2 = LE) = 0.85,$$
$$P(x_3 = HP|x_2 = LE) = 0.05,$$
$$P(x_3 = UN|x_2 = HE) = 0.05,\ P(x_3 = LP|x_2 = HE) = 0.15,$$
$$P(x_3 = HP|x_2 = HE) = 0.8.$$

Note that these values are not the result of a specific experiment. However, they are in line with the general trend provided by more professional studies, which involve many more random variables. However, for pedagogical reasons we keep the example simple.

The first observation is that even for this simplistic example involving only three variables, one has to obtain 17 probability values. This verifies the high computational load that may be required for such tasks.

Fig. 15.4 shows the BN that captures the previously stated conditional probabilities. Note that the Markov condition renders x_3 independent of x_1, given the value of x_2. Indeed, the job that one finds is independent of the GNP of the country, given her/his education level. We will verify that by playing with the laws of probability for the previously defined values.

According to Theorem 15.2, the joint probability of an event is given by the product

$$P(x_1, x_2, x_3) = P(x_3|x_2)P(x_2|x_1)P(x_1). \tag{15.12}$$

In other words, the probability of someone coming from a rich country, having a good education, and getting a high-paying job will be equal to $(0.8)(0.75)(0.2) = 0.12$; similarly, the probability of somebody coming from a poor country, having a low-level education, and getting a low-paying job is 0.476.

As a next step, we will verify the Markov condition, implied by the BN structure, using the probability values given before. That is, we will verify that using conditional probabilities to build the network, these probabilities basically encode *conditional independencies*, as Theorem 15.2 suggests. Let us consider

$$P(x_3 = HP|x_2 = HE, x_1 = HGP) = \frac{P(x_3 = HP, x_2 = HE, x_1 = HGP)}{P(x_2 = HE, x_1 = HGP)}$$
$$= \frac{0.12}{P(x_2 = HE, x_1 = HGP)}.$$

Also,

$$P(x_2 = HE, x_1 = HGP) = P(x_2 = HE|x_1 = HGP)P(x_1 = HGP)$$
$$= 0.75 \times 0.2 = 0.15,$$

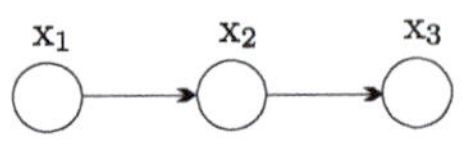

FIGURE 15.4

BN for Example 15.1. Note that $x_3 \perp x_1 | x_2$.

which finally results to

$$P(x_3 = HP|x_2 = HE, x_1 = HGP) = 0.8$$
$$= P(x_3 = HP|x_2 = HE),$$

which verifies the claim. The reader can check that this is true for all possible combinations of values.

15.3.2 SOME HINTS ON CAUSALITY

The existence of directed links in a Bayesian network *does not* necessarily reflect a cause–effect relationship from a parent to a child node.[2] It is a well-known fact in statistics that correlation between two variables does not always establish a causal relationship between them. For example, their correlation may be due to the fact that they both relate to a latent (unknown) variable. A typical example is the discussion related to whether smoking causes cancer or they are both due to an unobserved genotype that causes cancer and at the same time a craving for nicotine; for many years, this argument was used as a defense line of the tobacco companies.

Let us return to Example 15.1. Although GNP and quality of education are correlated, one cannot say that GNP is a cause of the educational system. No doubt there is a multiplicity of reasons, such as the political system, the social structure, the economic system, historical reasons, and tradition, all of which need to be taken into consideration. As a matter of fact, the structure of the graph relating the three variables in the example could be reversed. We could collect data the other way around; obtain the probabilities $P(x_3 = UN)$, $P(x_3 = LP)$, $P(x_3 = HP)$, and then the conditional probabilities $P(x_2|x_3)$ (e.g., $P(x_2 = HE|x_3 = UN)$) and finally $P(x_1|x_2)$ (e.g., $P(x_1 = HGP|x_2 = HE)$). In principle, such data can also be collected from a sample of people. In such a case, the resulting BN would comprise again three nodes as in Fig. 15.4, but with the direction of the arrows reversed. This is also reasonable because the probability of someone coming from a rich or a poor country is independent of her/his job, given the level of education. Moreover, both models should result in the same joint probability distribution for any joint event. Thus, if the direction of the arrows were to indicate causality, then this time, it would be that the educational system has a cause–effect relationship on the GNP. This, for the same reasons stated before, cannot be justified. Having said all that, it does not necessarily mean that cause–effect relationships are either absent in a BN or it is not important to know them. On the contrary, in many cases, there is good reason to strive to unveil the underlying cause–effect relationships while building a BN.

Let us elaborate a bit more on this and see why exploiting any underlying cause–effect relationships can be to our benefit. Take, for example, the BN in Fig. 15.5 relating the presence or absence of a disease with the findings from two medical tests. Let x_1 indicate the presence or absence of a disease and x_2, x_3 the discrete outcomes that can result from the two tests.

The BN in Fig. 15.5A complies with our common sense reasoning that x_1 (disease) causes x_2 and x_3 (tests). However, this is not possible to deduce by *simply* looking at the available probabilities. This is because the probability laws are *symmetric*. Even if x_1 is the cause, we can still compute $P(x_1|x_2)$

[2] This topic will not be pursued any further; its purpose is to make the reader aware of the issue. It can be bypassed in a first reading.

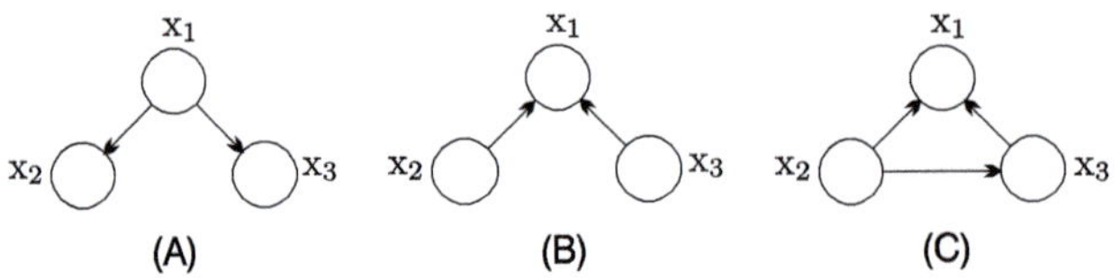

FIGURE 15.5

Three possible graphs relating a disease, x_1, to the results of two tests, x_2, x_3. (A) The dependencies in this graph comply with common sense. (B) This graph renders x_2, x_3 statistically independent, which is not reasonable. (C) Training this graph needs an extra probability value compared to that in (A).

once $P(x_1, x_2, x_3)$ and $P(x_2)$ are available, that is,

$$P(x_1|x_2) = \frac{P(x_1, x_2)}{P(x_2)} = \frac{\sum_{x_3} P(x_1, x_2, x_3)}{P(x_2)}.$$

Previously, in order to say that x_1 causes x_2 and x_3, we used some *extra* information/knowledge, which we called common sense reasoning. Note that in this case, training requires the knowledge of the values of three probabilities, namely, $P(x_1)$, $P(x_2|x_1)$, and $P(x_3|x_1)$. Let us now assume that we choose the graph model in Fig. 15.5B. This time, ignoring the cause–effect relationship has resulted in the wrong model. This model renders x_2 and x_3 independent, which, obviously, cannot be the case. These should only be *conditionally* independent given x_1. The only sensible way to keep x_2 and x_3 as parents of x_1 is to add an extra link, as shown in Fig. 15.5C, which establishes a relation between the two. However, to train such a network, besides the values of the three probabilities $P(x_2)$, $P(x_3)$, and $P(x_1|x_2, x_3)$, one needs to know the values for an extra one, $P(x_3|x_2)$. Thus, when building a BN, it is always good to know any underlying cause–effect directions. Moreover, there are other reasons, too. For example, this may be related to the *interventions*, which are actions that change the state of a variable in order to study the respective impact on other variables; because a change propagates in the causal direction, such a study is only possible if the network has been structured in a cause–effect hierarchy. For example, in biology, there is a strong interest in understanding which genes affect activation levels of other genes, and in predicting the effects of turning certain genes on or off.

The notion of causality is not an easy one, and philosophers have been arguing about it for centuries. Although our intention here is by no means to touch this issue, it is interesting to quote two well-known philosophers.

According to David Hume, causality is not a property of the real world but a concept of the mind that helps us explain our perception of the world. Hume (1711–1776) was a Scottish philosopher best known for his philosophical empiricism and skepticism. His most well-known work is the "Treatise of Human Nature," and in contrast to the rationalistic philosophy school, he advocated that human nature is mainly governed by desire and not reason.

According to Bertrand Russell, the law of causality has nothing to do with the laws of physics, which are symmetrical (recall our statement before concerning conditional probabilities) and indicate no cause–effect relationship. For example, Newton's gravity law can be expressed in any of the following forms:

$$B = mg \text{ or } g = \frac{B}{m} \text{ or } m = \frac{B}{g},$$

and looking only at them, no cause–effect relationship can be deduced. Bertrand Russell (1872–1970) was a British philosopher, mathematician, and logician. He is considered one of the founders of analytic philosophy. In *Principia Mathematica*, coauthored with A.N. Whitehead, they made an attempt to ground mathematics on mathematical logic. He was also an antiwar activist and a liberal.

The previously stated provocative arguments have been inspired by Judea Pearl's book [38], and we provided them in order to persuade the reader to read this book; he or she can only become wiser. Pearl has made a number of significant contributions to the field and was the recipient of the Turing award in 2011.

Although one cannot deduce causality by looking only at the laws of physics or probabilities, ways of identifying it have been developed. One way is to carry out *controlled* experiments; one can change the values of the variable and study the effect of the change on another. However, this has to be done in a controlled way in order to guarantee that the caused effects are not due to other related factors.

Besides experimentation, there has been a major effort to discover causal relationships from nonexperimental evidence. In modern applications, such as microarray measurements for gene expressions or fMRI brain imaging, the number of the involved variables can easily reach the order of a few thousand. Performing experiments for such tasks is out of the question. In [38], the notion of causality is related to that of the minimality in the structure of the obtained possible DAGs. Such a view ties causality with Occam's razor. More recently, inferring causality was attempted by comparing the conditional distributions of variables given their direct causes, for all hypothetical causal directions, and choosing the most plausible. The method builds upon some smoothness arguments that underlie the conditional distributions of the effect given the causes, compared to the marginal distributions of the effect/cause [43]. In [24], an interesting alternative for inferring causality is built upon arguments from Kolmogorov's complexity theory; causality is verified by comparing shortest description lengths of strings associated with the involved distributions. For further information, the interested reader may consult, for example, [42] and the references therein.

15.3.3 *d*-SEPARATION

Dependencies and independencies among a set of random variables play a key role in understanding their statistical behavior. Moreover, as we have already commented, they can be exploited to substantially reduce the computational load for solving inference tasks.

By the definition and the properties of a BN structure, G, we know that certain independencies hold and are readily observed via the parent–child links. The question that is now raised is whether there are additional independencies that the structure of the graph imposes on any joint probability distribution that factorizes over G. Unveiling extra independencies offers the designer more freedom to deal with computational complexity issues more aggressively.

We will attack the task of searching for conditional independencies across a network by observing whether probabilistic evidence that becomes available at a node x can propagate and influence our certainty about another node, y.

Serial or head-to-tail connection. This type of node connection is shown in Fig. 15.6A. Evidence on x will influence the certainty about y, which in turn will influence that of z. This is also true for the reverse direction, starting from z and propagating to x. However, if the state of y is known, then x and

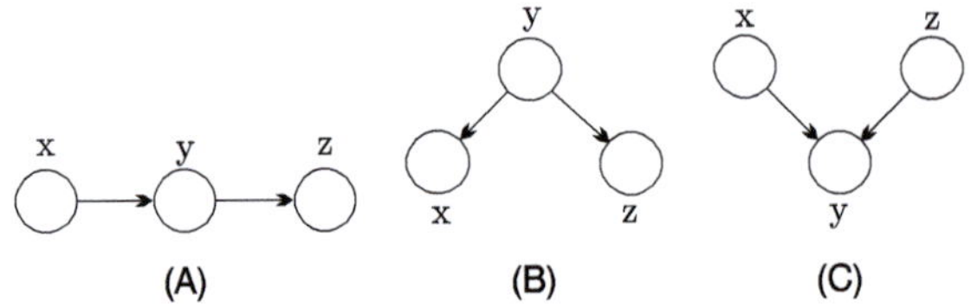

FIGURE 15.6

Three different types of connections: (A) serial, (B) diverging, and (C) converging.

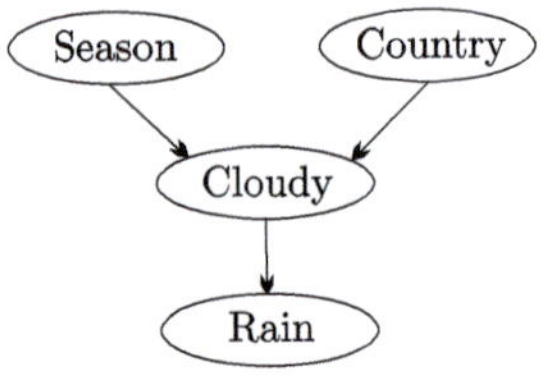

FIGURE 15.7

Having some evidence about the weather being either cloudy or rainy establishes the path for information flow between the nodes "season" and "country."

z become (conditionally) independent. In this case, we say that y *blocks* the path from x to z, and vice versa. When the state at a node is fixed/known, we say that the node is *instantiated.*

Diverging or tail-to-tail connection. In this type of connection, shown in Fig. 15.6B, evidence can propagate from y to x and from y to z, and also from x to z and from z to x via y, unless y is instantiated. In the latter case, y *blocks* the path from x to z, and vice versa. That is, x and z become independent given the value of y. For example, if y represents "flu," x "runny nose," and z "sneezing," then if we do not know whether someone has the flu, a runny nose is evidence that can change our certainty about her/him having the flu; this in turn changes our belief about sneezing. However, if we know that someone has the flu, seeing the nose running gives no extra information about sneezing.

Converging or head-to-head connection or v-structure. This type of connection is slightly more subtle than the previous two cases, and it is shown in Fig. 15.6C. Evidence from x does not propagate to z and thus cannot change our certainty about it. Knowing something about x tells us nothing about z. For example, let z denote either of two countries (e.g., England and Greece), x "season," and y "cloudy weather." Obviously, knowing the season says nothing about a country. However, having some evidence about cloudy weather y, knowing that it is summer provides information that can change our certainty about the country. This is in accordance with our intuition. Knowing that it is summer and that the weather is cloudy *explains away* that the country is Greece. This is the reason we sometimes refer to this type of reasoning as *explaining away.* Explaining away is an instance of a general reasoning pattern called *intercausal reasoning,* where different causes of the same effect can interact; this is a very common pattern of reasoning in humans.

For this particular type of connection, explaining away is also achieved by evidence that is provided by *any one* of the descendants of y. Fig. 15.7 illustrates the case via an example. Having evidence about the rain will also establish a path so that evidence about the season (country), x (z), changes our certainty about the country (season), z (x).

To recapitulate, let us stress the delicate point here. For the first two cases, head-to-tail and tail-to-tail, the path is blocked if node y is instantiated, that is, when its state is disclosed to us. However, in the head-to-head connection, the path between x and z "*opens*" when probabilistic evidence becomes available, either at y or *at any one* of its descendants.

Definition 15.3. Let G be a BN structure, and let $\mathrm{x}_1, \ldots, \mathrm{x}_k$ comprise a chain of nodes. Let Z be a subset of *observed* variables. The chain $\mathrm{x}_1, \ldots, \mathrm{x}_k$ is said to be *active given* the set Z, if

- whenever a converging connection, $\mathrm{x}_{i-1} \rightarrow \mathrm{x}_i \leftarrow \mathrm{x}_{i+1}$, is present in the chain, then either x_i or one of its descendants is in Z;
- no other node in the chain is in Z.

In other words, in an active chain, probabilistic evidence can flow from x_1 to x_k, and vice versa, because no nodes (links) which can block this information flow are present.

Definition 15.4. Let G be a BN structure and let X, Y, Z be three mutually disjoint sets of nodes in G. We say that X and Y are *d-separated* given Z if there is *no* active chain between *any* node $\mathrm{x} \in X$ and $\mathrm{y} \in Y$ given Z. If these are not d-separated, we say that they are *d-connected.*

In other words, if two variables x and y are d-separated by a third one, z, then observing the state of z blocks any evidence propagation from x to y and vice versa. That is, d-separation implies *conditional independence*. Moreover, the following very important theorem holds.

Theorem 15.3. *Let the pair* (G, p) *be a BN. For every three mutually disjoint subsets of nodes* X, Y, Z, *whenever* X *and* Y *are d-separated, given* Z, *for every pair* $(\mathrm{x}, \mathrm{y}) \in X \times Y$, x *and* y *are conditionally independent in* p *given* Z.

The proof of the theorem was given in [45]. In other words, this theorem guarantees that d-separation implies conditional independence on any probability distribution that factorizes over G. Note that, unfortunately, the opposite is not true. There may be conditional independencies that cannot be identified by d-separation (e.g., Problem 15.5). However, for most practical applications, the reverse is also true. The number of distributions that do not comply with the reverse statement of the theorem is infinitesimally small (see, e.g., [25]). Identification of all d-separations in a graph can be carried out via a number of efficient algorithms (e.g., [25,32]).

Example 15.2. Consider the DAG G of Fig. 15.8, connecting two nodes x, y. It is obvious that these nodes are not d-separated and comprise an active chain. Consider the following probability distribution, which factorizes over G:

$$P(y=0|x=0)=0.2, \quad P(y=1|x=0)=0.8,$$
$$P(y=0|x=1)=0.2, \quad P(y=1|x=1)=0.8.$$

It can easily be checked that $P(y|x)=P(y)$ (independent of the values of $P(x=1)$ and $P(x=0)$) and the variables x and y are independent; this cannot be predicted by observing the d-separations. Note, however, that if we slightly perturb the values of the conditional probabilities, then the resulting

FIGURE 15.8

This DAG involves no nodes that are d-separated.

distribution has as many independencies as those predicted by the d-separations, that is, in this case, none. As a matter of fact, this is a more general result. If we have a distribution which factorizes over a graph that has independencies that are not predicted by the respective d-separations, a small perturbation will almost always eliminate them (e.g., [25]).

Example 15.3. Consider the DAG shown in Fig. 15.9. The red nodes indicate that the respective random variables have been observed; that is, these nodes have been instantiated. Node x_5 is d-connected to x_1, x_2, x_6. In contrast, node x_9 is d-separated from all the rest. Indeed, evidence starting from x_1 is blocked by x_3. However, it propagates via x_4 (instantiated and converging connection) to x_2, x_6, and then to x_5 (x_7 is instantiated and a converging connection). In contrast, any flow of evidence toward x_9 is blocked by the instantiation of x_7. It is interesting to note that although all neighbors of x_5 have been instantiated, it still remains d-connected with other nodes.

Definition 15.5. The *Markov blanket* of a node is the set of nodes comprising (a) its parents, (b) its children, and (c) the nodes sharing a child with this node. Once all the nodes in the blanket of a node are instantiated, the node becomes d-separated from the rest of the network (Problem 15.7).

For example, in Fig. 15.9, the Markov blanket of x_5 comprises the nodes x_3, x_4, x_8, x_7, x_6. Note that if all these nodes are instantiated, then x_5 becomes d-separated from the rest of the nodes.

In the sequel, we give some examples of machine learning tasks, which can be cast in terms of a Bayesian graphical representation. As we will discuss, for many practical cases, the involved conditional probability distributions are expressed in terms of a set of parameters.

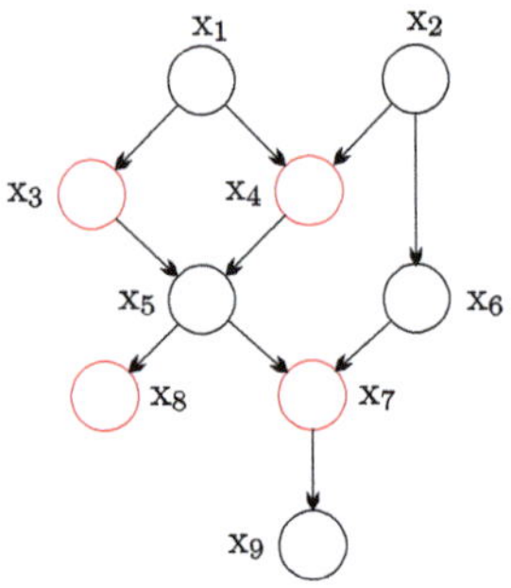

FIGURE 15.9

Red nodes are instantiated. Node x_5 is d-connected to x_1, x_2, x_6 and node x_9 is d-separated from all the nonobserved variables.

15.3.4 SIGMOIDAL BAYESIAN NETWORKS

We have already seen that when the involved random variables are discrete, the conditional probabilities $P(x_i|\text{Pa}_i), i = 1, \dots, l$, associated with the nodes of a Bayesian graph structure have to be learned from the training data. If the number of possible states and/or the number of the variables in Pa_i is large enough, this amounts to a large number of probabilities that have to be learned; thus, a large number of training points is required in order to obtain good estimates. This can be alleviated by expressing the conditional probabilities in a parametric form, that is,

$$P(x_i|\text{Pa}_i) = P(x_i|\text{Pa}_i; \boldsymbol{\theta}_i), \quad i = 1, 2, \dots, l. \tag{15.13}$$

In the case of binary-valued variables, a common functional form is to view P as a logistic regression model; we used this model in the context of relevance vector machines in Chapter 13. Adopting this model, we have

$$P(x_i = 1|\text{Pa}_i; \boldsymbol{\theta}_i) = \sigma(t_i) = \frac{1}{1 + \exp(-t_i)}, \tag{15.14}$$

$$t_i := \theta_{i0} + \sum_{k: x_k \in \text{Pa}_i} \theta_{ik} x_k. \tag{15.15}$$

This reduces the number of parameter vectors to be learned to $O(l)$. The exact number of parameters depends on the size of the parent sets. Assuming the maximum number of parents for a node to be K, the unknown number of parameters to be learned from the training data is less than or equal to lK. Taking into account the binary nature of the variables, we can write

$$P(x_i|\text{Pa}_i; \boldsymbol{\theta}_i) = x_i \sigma(t_i) + (1 - x_i)(1 - \sigma(t_i)), \tag{15.16}$$

where t_i is given in Eq. (15.15).

Such models are also known as *sigmoidal Bayesian networks*, and they have been proposed as one type of neural network (Chapter 18) (e.g., [33]). Fig. 15.10 presents the graphical structure of such a network. The network can be treated as a BN structure by associating a binary variable at each node and interpreting nodes' activations as probabilities, as dictated by Eq. (15.16). Performing inference and training of the parameters in such networks is not an easy task. We have to resort to approximations. We will come to this in Section 16.3.

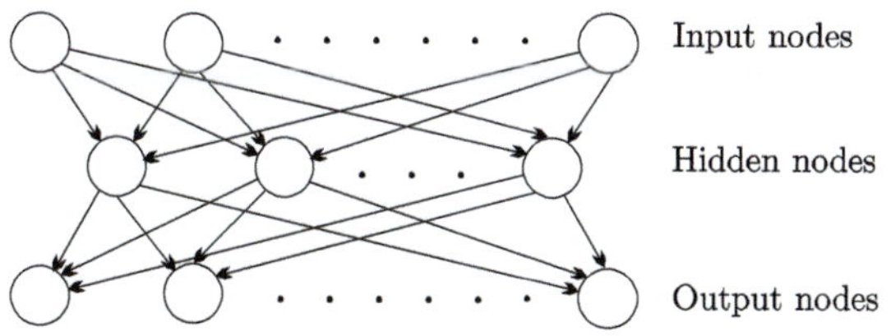

FIGURE 15.10

A sigmoidal Bayesian network.

15.3.5 LINEAR GAUSSIAN MODELS

The computational advantages of the Gaussian PDF have recurrently been exploited in this book. We will now see the advantage gains in the framework of graphical models when the conditional PDF at every node, given the values of its parents, is expressed in a Gaussian form. Let

$$\boxed{p(x_i|\text{Pa}_i) = \mathcal{N}\left(x_i \,\Bigg|\, \sum_{k:x_k \in \text{Pa}_i} \theta_{ik} x_k + \theta_{i0}, \sigma_i^2\right),} \tag{15.17}$$

where σ_i^2 is the respective variance and θ_{i0} is the bias term. From the properties of a BN, the joint PDF will be given by the product of the conditional probabilities (Theorem 15.2, which is valid for Gaussians), and the respective logarithm is given by

$$\ln p(\boldsymbol{x}) = \sum_{i=1}^{l} \ln p(x_i|\text{Pa}_i) = -\sum_{i=1}^{l} \frac{1}{2\sigma_i^2}\left(x_i - \sum_{k:x_k \in \text{Pa}_i} \theta_{ik} x_k - \theta_{i0}\right)^2 + \text{constant}. \tag{15.18}$$

This is of a quadratic form, and hence it is also of a Gaussian nature. The mean values and the covariance matrices for each one of the variables can be computed *recursively* in a straightforward way (Problem 15.8).

Note the computational elegance of such a BN. In order to obtain the joint PDF, one has only to sum up all the exponents, that is, an operation of linear complexity. Moreover, concerning training, one could readily think of a way to learn the unknown parameters; adopting the maximum likelihood method (although it may not be necessarily the best method), optimization with respect to the unknown parameters is a straightforward task. In contrast, one cannot make similar comments for the training of the sigmoidal BN. Unfortunately, products of sigmoid functions do not lead to an easy computational procedure. In such cases, one has to resort to approximations. For example, one way is to employ the variational bound approximation, as discussed in Chapter 13, in order to enforce, locally, a Gaussian functional form. We will discuss this technique in Section 16.3.1.

15.3.6 MULTIPLE-CAUSE NETWORKS

In the beginning of this chapter, we started with an example from the field of medical informatics. We were given a set of diseases and a set of symptoms/findings. The conditional probabilities for each symptom being absent, given the presence of a disease (Eq. (15.2)), were assumed known. We can consider the diseases as hidden causes (h) and the symptoms as observed variables (y) in a learning task. This can be represented in terms of a BN structure as in Fig. 15.11. For the previous medical example, the variables h correspond to d (diseases), and the observed variables y to the findings f.

However, the Bayesian structure given in Fig. 15.11 can serve the needs of a number of inference and pattern recognition tasks, and sometimes it is referred to as a *multiple-cause network*, for obvious reasons. For example, in a machine vision application, the hidden causes, $\text{h}_1, \text{h}_2, \ldots, \text{h}_k$, may refer to the presence or absence of an object, and y_n, $n = 1, \ldots, N$, may correspond to the values of the observed pixels in an image [15]. The hidden variables can be binary (presence or absence of the respective object) and the conditional PDF can be formulated into a parameterized form, that is,

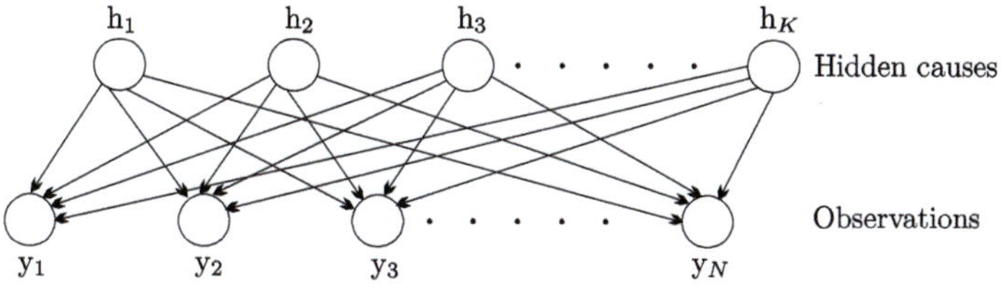

FIGURE 15.11

The general structure of a multiple-cause Bayesian network. The top-level nodes correspond to the hidden causes and the bottom ones to the observations.

$p(y_n|\boldsymbol{h};\boldsymbol{\theta})$. The specific form of the PDF captures the way objects interact as well as the effects of the noise. Note that in this case, the BN has a mixed set of variables, the observations are continuous, and the hidden causes are binary. We will return to this type of Bayesian structure when discussing approximate inference methods in Section 16.3.

15.3.7 I-MAPS, SOUNDNESS, FAITHFULNESS, AND COMPLETENESS

We have seen a number of definitions and theorems referring to the notion of conditional independence in graphs and probability distributions. Before we proceed further, it will be instructive to summarize what has been said and provide some definitions that will dress up our findings in a more formal language. This will prove useful for subsequent generalizations.

We have seen that a BN is a DAG that encodes a number of conditional independencies. Some of them are local ones, defined by the parent–child links, and some of them are of a more global nature and are the result of d-separations. Given a DAG, G, we denote as $I(G)$ the set of all independencies that correspond to d-separations. Also, let p be a probability distribution over a set of random variables, $\mathrm{x}_1, \ldots, \mathrm{x}_l$. We denote as $I(p)$ the set of all independence assertions of the type $\mathrm{x}_i \perp \mathrm{x}_j|Z$ that hold true for the distribution p.

Let G be a DAG and p a distribution that factorizes over G; in other words, it satisfies the local independencies as suggested by G. Then we have seen (Theorem 15.3) that

$$I(G) \subseteq I(p). \tag{15.19}$$

We say that G is an *I-map* (independence map) for p. This property is sometimes referred to as *soundness*.

Definition 15.6. A distribution p is *faithful* to a graph G if *any* independence in p is reflected in the d-separation properties of the graph.

In other words, the graph can represent all (and only) the conditional independence properties of the distribution. In such a case we write $I(p) = I(G)$. If equality is valid, we say that the graph G is a *perfect* map for p. Unfortunately, this is *not* valid for any distribution, p, that factorizes over G. However, for most practical purposes, $I(G) = I(P)$, which is true for *almost all* distributions that factorize over G.

Although $I(p) = I(G)$ is not valid for all distributions that factorize over G, the following two properties are always valid for any BN structure (e.g., [25]):

- If $x \perp y|Z$ for *all* distributions p that factorize over G, then x and y are d-separated given Z.
- If x and y are d-connected given Z, then there will be some distribution that factorizes over G where x and y are *dependent*.

A final definition concerns minimality.

Definition 15.7. A graph G is said to be a *minimal* I-map for a set of independencies if the removal of any of its edges renders it *not* to be an I-map.

Note that a minimal I-map is not necessarily a perfect map. In the same way that there exist algorithms to find the set of d-separations, there exist algorithms to find perfect and minimal I-maps for a distribution (e.g., [25]).

15.4 UNDIRECTED GRAPHICAL MODELS

Bayesian structures and networks are not the only way to encode independencies in distributions. As a matter of fact, the directionality assigned to the edges of a DAG, while being advantageous and useful in some cases, becomes a disadvantage in others. A typical example is that of four variables, x_1, x_2, x_3, x_4. There is no directed graph that can encode the following conditional independencies simultaneously: $x_1 \perp x_4|\{x_2, x_3\}$ *and* $x_2 \perp x_3|\{x_1, x_4\}$. Fig. 15.12 shows the possible DAGs; note that both fail to capture the desired independencies.

In 15.12A, $x_1 \perp x_4|\{x_2, x_3\}$ because both paths which connect x_1 and x_4 are blocked. However, x_2 and x_3 are d-connected given x_1 and x_4 (why?). In Fig. 15.12B, $x_2 \perp x_3|\{x_1, x_4\}$ because the diverging links are blocked. However, we have violation of the other independence (why?).

Such situations can be overcome by resorting to undirected graphs. We will also see that this type of graphical modeling leads to a simplification concerning our search for conditional independencies.

Undirected graphical models or *Markov networks* or *Markov random fields* (MRFs) have their roots in statistical physics. As was the case with the Bayesian models, each node of the graph is associated

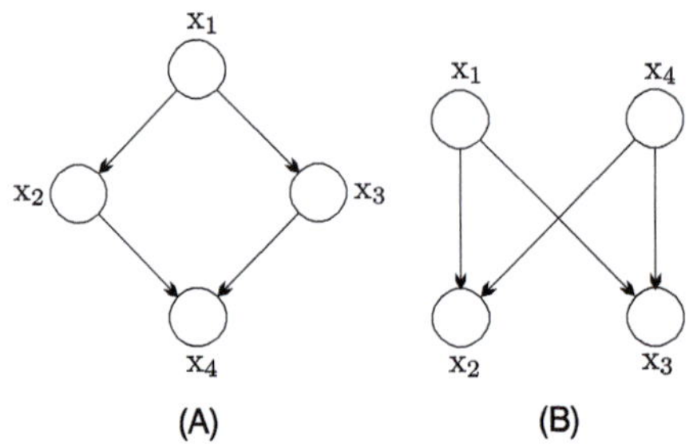

FIGURE 15.12

None of these DAGs can capture the two independencies $x_1 \perp x_4|\{x_2, x_3\}$ and $x_2 \perp x_3|\{x_1, x_4\}$.

with a random variable. Edges connecting nodes are undirected, giving no preference to either of the two directions. Local interactions among connected nodes are expressed via functions of the involved variables, but they do not necessarily express probabilities. One can view these local functional interactions as a way to encode information related to the affinity/similarity among the involved variables. These local functions are known as *potential functions* or *compatibility functions* or *factors*, and they are *nonnegative*, usually positive, functions of their arguments. Moreover, as we will soon see, the global description of such a model is the result of the product of these local potential functions; this is in analogy to what holds true for the BNs.

Following a path similar to that used for the directed graphs, we will begin with the factorization properties of a distribution over an MRF and then move on to study conditional independencies. Let $\mathrm{x}_1, \dots, \mathrm{x}_l$ be a set of random variables that are grouped in K groups, $\mathbf{x}_1, \dots, \mathbf{x}_K$; each random vector $\mathbf{x}_k,\ k = 1, 2, \dots, K$, involves a subset of the random variables, $\mathrm{x}_i,\ i = 1, 2, \dots, l$.

Definition 15.8. A distribution is called a *Gibbs* distribution if it can be factorized in terms of a set of potential functions $\psi_1, \dots, \psi_K$, such that

$$p(x_1, \dots, x_l) = \frac{1}{Z} \prod_{k=1}^{K} \psi_k(\boldsymbol{x}_k). \tag{15.20}$$

The constant Z is known as the *partition* function and it is the normalizing constant to guarantee that $p(x_1, \dots, x_l)$ is a probability distribution. Hence,

$$Z = \int \cdots \int \prod_{k=1}^{K} \psi_k(\boldsymbol{x}_k)\, dx_1, \dots, dx_l, \tag{15.21}$$

which becomes a summation for the case of probabilities.

Note that nobody can prohibit us from assigning conditional probability distributions as potential functions and making (15.20) identical to Eq. (15.10); in this case, normalization is not explicitly required because each one of the conditional distributions is normalized. However, MRFs can deal with more general cases.

Definition 15.9. We say that a Gibbs distribution p factorizes over an MRF H if *each* group of the variables $\mathbf{x}_k,\ k = 1, 2, \dots, K$, involved in the K factors of the distribution p forms a *complete subgraph* of H. Every complete subgraph of an MRF is known as a *clique*, and the corresponding factors of the Gibbs distribution are known as *clique potentials*.

Fig. 15.13A shows an MRF and two cliques. Note that the set of nodes $\{\mathrm{x}_1, \mathrm{x}_3, \mathrm{x}_4\}$ does not comprise a clique because the respective subgraph is not fully connected. The same applies to the set $\{\mathrm{x}_1, \mathrm{x}_2, \mathrm{x}_3, \mathrm{x}_4\}$. In contrast, the sets $\{\mathrm{x}_1, \mathrm{x}_2, \mathrm{x}_3\}$ and $\{\mathrm{x}_3, \mathrm{x}_4\}$ form cliques. The fact that all variables in a group $\mathbf{x}_k$ that are involved in the respective factor $\psi_k(\boldsymbol{x}_k)$ form a clique means that *all* these variables mutually interact, and the factor is a measure of such an interaction/dependence.

A clique is called *maximal* if we cannot include any other node from the graph in the set without it ceasing to be a clique. For example, both cliques in Fig. 15.13A are maximal cliques. On the other hand, the clique in Fig. 15.13B formed by $\{\mathrm{x}_1, \mathrm{x}_2, \mathrm{x}_3\}$ is not maximal because bringing x_4 into the new set $\{\mathrm{x}_1, \mathrm{x}_2, \mathrm{x}_3, \mathrm{x}_4\}$ also gives a clique. The same holds true for the clique formed by $\{\mathrm{x}_2, \mathrm{x}_3, \mathrm{x}_4\}$.

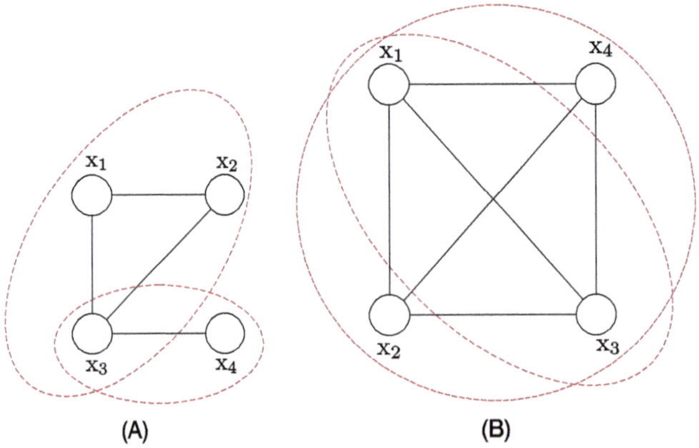

FIGURE 15.13

(A) There are two cliques, encircled by red lines. (B) There are as many possible cliques as the combinations of the points in pairs, in triples, and so on. Considering all points together also forms a clique, and this is a maximal clique.

15.4.1 INDEPENDENCIES AND I-MAPS IN MARKOV RANDOM FIELDS

We will now state the equivalent theorem of d-separation, which was established for BN structures (recall the respective definition in Section 15.3.3), via the notion of an active chain.

Definition 15.10. Let H be an MRF and let $x_1, x_2, \dots, x_k$ comprise a path.[3] If Z is a set of observed variables/nodes, the path is said to be *active given* Z if none of $x_1, x_2, \dots, x_k$ is in Z.

Given three disjoint sets, X, Y, Z, we say that the nodes of X are *separated* by the nodes of Y, given Z, if there is no active path between X and Y, given Z. Note that the previous definition is much simpler compared to the respective definition given for the BN structures. According to the current definition, for a set X to be separated from a set Y given a third set Z, it suffices that *all* possible paths from X to Y pass via Z. Fig. 15.14 illustrates the geometry. In 15.14A, there is no active path connecting the nodes in X from the nodes in Y given the nodes in Z. In 15.14B, there exist active paths connecting X and Y given Z.

Let us now denote by $I(H)$ the set of all possible statements of the type "X separated by Y given Z." This is in analogy to the set of all possible d-separations associated with a BN structure. The following theorem (soundness) holds true (Problem 15.10).

Theorem 15.4. *Let p be a Gibbs distribution that factorizes over an MRF H. Then this is an I-map for p, that is,*

$$I(H) \subseteq I(p). \tag{15.22}$$

This is the counterpart of Theorem 15.3 in its "I-map formulation" as introduced in Section 15.3.7. Moreover, given that an MRF H is an I-map for a distribution p, p factorizes over H. Note that this

[3] Because edges are undirected, the notions of "chain" and "path" become identical.

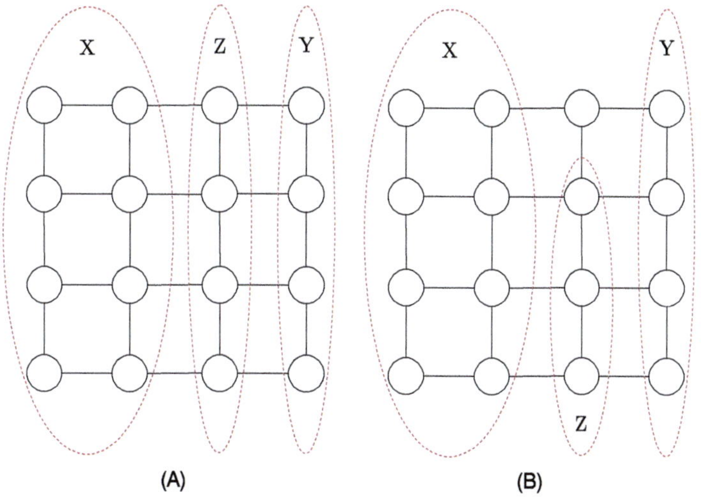

FIGURE 15.14

(A) The nodes of X and Y are separated by the nodes of Z. (B) There exist active paths that connect the nodes of X with the nodes of Y, given Z.

holds true for BN structures; indeed, if $I(G) \subseteq I(p)$, then p factorizes over G (Problem 15.12). However, for MRFs, it is only true for strictly positive Gibbs distributions and it is given by the following *Hammersley–Clifford theorem*.

Theorem 15.5. *Let H be an MRF over a set of random variables, $\mathrm{x}_1, \ldots, \mathrm{x}_l$, described by a probability distribution, $p > 0$. If H is an I-map for p, then p is a Gibbs distribution that factorizes over H.*

For a proof of this theorem, the interested reader is referred to the original paper [22] and also [5].

Our final touch on independencies in the context of MRFs concerns the notion of completeness. As was the case with the BNs, if p factorizes over an MRF, this does not necessarily establish completeness, although it is true for almost all practical cases. However, the weaker version holds. That is, if x and y are two nodes in an MRF, which are *not* separated given a set Z, then there exists a Gibbs distribution p which factorizes over H and according to which x and y are dependent, given the variables in Z (see, e.g., [25]).

15.4.2 THE ISING MODEL AND ITS VARIANTS

The origin of the theory on MRFs is traced back to the discipline of statistical physics, and since then it has extensively been used in a number of different disciplines, including machine learning. In particular, in image processing and computer vision, MRFs have been established as a major tool in tasks such as denoising, image segmentation, and stereo reconstruction (see, e.g., [29]). The goal of this section is to state a basic and rather primitive model, which, however, demonstrates the way information is captured and subsequently processed by such models.

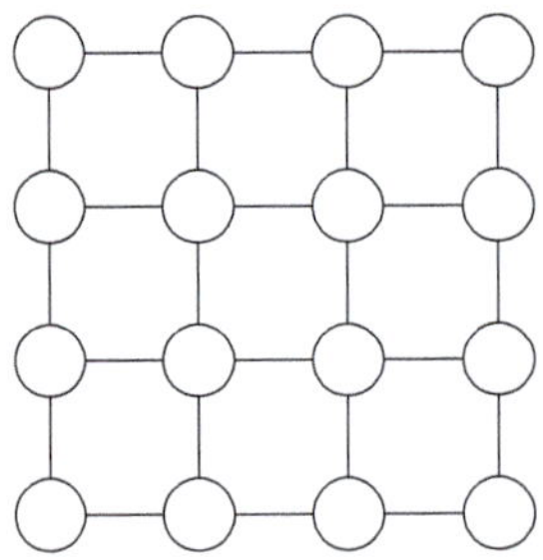

FIGURE 15.15

The graph of an MRF with pair-wise dependencies among the nodes.

Assume that each random variable takes binary values in $\{-1, 1\}$ and that the joint probability distribution is given by the following model:

$$p(x_1, \ldots, x_l) := p(\boldsymbol{x}) = \frac{1}{Z} \exp\left(-\sum_i \left(\sum_{j>i} \theta_{ij} x_i x_j + \theta_{i0} x_i\right)\right), \tag{15.23}$$

where $\theta_{ij} = 0$ if the respective nodes are not connected. It is readily seen that this model is the result of the product of potential functions (factors), each one of an exponential form, defined on cliques of size two. Also, $\theta_{ij} = \theta_{ji}$ and we sum such that $i < j$ in order to avoid duplication. This model was originally used by Ising in 1924 in his doctoral thesis to model phase transition phenomena in magnetic materials. The ± 1 of each node in the lattice models the two possible spin directions of the respective atoms. If $\theta_{ij} > 0$, interacting atoms tend to align spins in the same direction in order to decrease energy (ferromagnetism). The opposite is true if $\theta_{ij} < 0$. The corresponding graph is given in Fig. 15.15.

This basic model has been exploited in computer vision and image processing for tasks such as image denoising, image segmentation, and scene analysis. Let us take, as an example, a binarized image and let x_i denote the noiseless pixel values (± 1). Let y_i be the observed noisy pixels whose values have been corrupted by noise and have changed polarity; see Fig. 15.16 for the respective graph. The task is to obtain the noiseless pixel values. One can rephrase the model in Eq. (15.23) to the needs of this task and rewrite it as [5,21]

$$P(\boldsymbol{x}|\boldsymbol{y}) = \frac{1}{Z} \exp\left(\sum_i \left(\alpha \sum_{j>i} x_i x_j + \beta\, x_i y_i\right)\right), \tag{15.24}$$

where we have used only two parameters, α and β. Moreover, the summation $\sum_{j>i}$ involves only *neighboring* pixels. The goal now becomes that of estimating the pixel values, x_i, by maximizing the conditional (on the observations) probability. The adopted model is justified by the following two facts: (a) for low enough noise levels, most of the pixels will have the same polarity as the respective observations; this is encouraged by the presence of the product $x_i y_i$, where similar signs contribute to

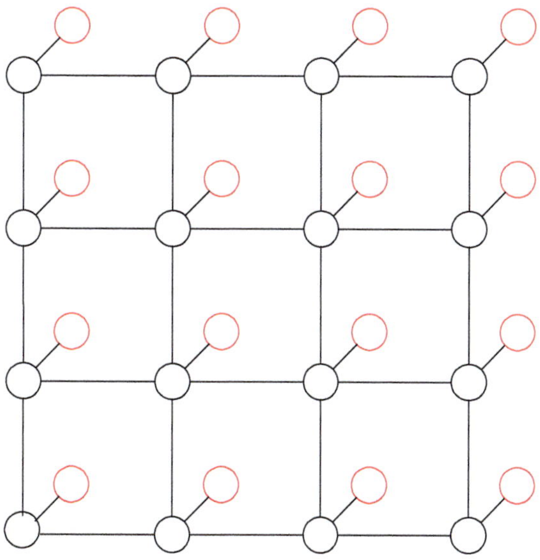

FIGURE 15.16

A pair-wise MRF as in Fig. 15.15, but now the observed values associated with each node are separately denoted as red nodes. For the image denoising task, black nodes correspond to the noiseless pixel values (hidden variables) and red nodes to the observed pixel values.

higher probability values; and (b) neighboring pixels are encouraged to have the same polarity because we know that real-world images tend to be smooth, except at the points that lie close to the edges in the image. Sometimes, a term cx_i, for an appropriately chosen value of c, is also present if we want to penalize either of the two polarities. The max-product or max-sum algorithms, to be discussed later in this chapter, are possible algorithmic alternatives for the maximization of the joint probability given in Eq. (15.24). However, these are not the only algorithmic possibilities to perform the optimization task. A number of alternative schemes that deal with inference in MRFs have been developed and studied. Some of them are suboptimal, yet they enjoy computational efficiency. Some classical references on the use of MRFs in image processing are [7–9,47]. A number of variants of the basic Ising model result if one writes it as

$$P(\boldsymbol{x}) = \frac{1}{Z} \exp\left(-\sum_i \left(\sum_{j>i} f_{ij}(x_i x_j) + f_i(x_i) \right) \right) \tag{15.25}$$

and uses different functional forms for $f_{ij}(\cdot,\cdot)$ and $f_i(\cdot)$, and also by allowing the variables to take more than two values. This is sometimes known as the *Potts* model. In general, MRF models of the general form of Eq. (15.23) are also known as *pair-wise MRFs undirected graphs* because the dependence among nodes is expressed in terms of products of pairs of variables. Further information on the applications of MRFs in image processing can be found in, for example, [29,40].

Another name for Eq. (15.23) is Boltzmann distribution, where, usually, the variables take values in $\{0, 1\}$. Such a distribution has been used in *Boltzmann machines* [23]. Boltzmann machines can be seen as the stochastic counterpart of Hopfield networks; the latter have been proposed to act as *associative memories* as well as a way to attack combinatoric optimization problems (e.g., [31]). The interest in Boltzmann machines has been revived in the context of deep learning, and we will discuss them in more detail in Chapter 18.

15.4.3 CONDITIONAL RANDOM FIELDS (CRFS)

All the graphical models (directed and undirected) that have been discussed so far evolve around the joint distribution of the involved random variables and its factorization on a corresponding graph. More recently, there is a trend to focus on the conditional distribution of some of the variables given the rest. The focus on the joint PDF originates from our interest in developing generative learning models. However, this may not always be the most efficient way to deal with learning tasks, and we have already discussed in Chapters 3 and 7 about the discriminative learning alternative. Let us assume that from the set of the jointly distributed variables, some correspond to output target variables, whose variables are to be inferred when the rest are observed. For example, the target variables may correspond to the labels in a classification task and the rest to the (input) features.

Let us denote the former set by the vector $\mathbf{y}$ and the latter by $\mathbf{x}$. Instead of focusing on the joint distribution $p(\boldsymbol{x}, \boldsymbol{y})$, it may be more sensible to focus on $p(\boldsymbol{y}|\boldsymbol{x})$. In [27], graphical models were adopted to encode the conditional distribution, $p(\boldsymbol{y}|\boldsymbol{x})$.

A *conditional random Markov field* is an undirected graph H whose nodes correspond to the joint set of random variables $(\mathbf{x}, \mathbf{y})$, but we now assume that the conditional distribution is factorized, that is,

$$p(\boldsymbol{y}|\boldsymbol{x}) = \frac{1}{Z(\boldsymbol{x})} \prod_{k=1}^{K} \psi_k(\boldsymbol{x}_k, \boldsymbol{y}_k), \tag{15.26}$$

where $\{\boldsymbol{x}_k, \boldsymbol{y}_k\} \subseteq \{\boldsymbol{x}, \boldsymbol{y}\}$, $k = 1, 2, \dots, K$, and

$$Z(\boldsymbol{x}) = \int p(\boldsymbol{y}|\boldsymbol{x}) d\boldsymbol{y}, \tag{15.27}$$

where for discrete distributions the integral becomes summation. To stress the difference with Eq. (15.20), note that there it is the joint distribution of all the involved variables that is factorized. As a result, and observing Eqs. (15.20) and (15.26), it turns out that the normalization constant is now a function of $\boldsymbol{x}$. This seemingly minor difference can offer a number of advantages in practice. Avoiding the explicit modeling of $p(\boldsymbol{x})$, we have the benefit of using as inputs variables with complex dependencies, because we do not care to model them. This has led CRFs to be applied in a number of applications, such as text mining, bioinformatics, and computer vision. Although we are not going to get involved with CRFs from now on, it suffices to say that the efficient inference techniques, which will be discussed in subsequent sections, can also be adapted, with only minor modifications, to the case of CRFs. For a tutorial on CRFs, including a number of variants and techniques concerning inference and learning, the interested reader is referred to [44].

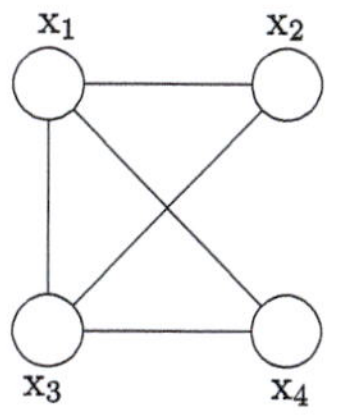

FIGURE 15.17

The Gibbs distribution can be written as a product of factors involving the cliques (x_1, x_2), (x_1, x_3), (x_3, x_4), (x_3, x_2), (x_1, x_4) or of (x_1, x_2, x_3), (x_1, x_3, x_4).

15.5 FACTOR GRAPHS

In contrast to a BN, an MRF does not necessarily indicate the specific form of factorization of the corresponding Gibbs distribution. Looking at a BN, the factorization evolves along the conditional distributions allocated in each node. Let us look at the MRF of Fig. 15.17. The corresponding Gibbs distribution could be written as

$$p(x_1, x_2, x_3, x_4) = \frac{1}{Z}\psi_1(x_1, x_2)\psi_2(x_1, x_3)\psi_3(x_3, x_2)\psi_4(x_3, x_4)\psi_5(x_1, x_4), \tag{15.28}$$

or

$$p(x_1, x_2, x_3, x_4) = \frac{1}{Z}\psi_1(x_1, x_2, x_3)\psi_2(x_1, x_3, x_4). \tag{15.29}$$

As an extreme case, if all the points of an MRF form a maximal clique, as is the case in Fig. 15.13B, we could include only a single product term. Note that aiming at maximal cliques reduces the number of factors, but at the same time the complexity is increased; for example, this can amount to an exponential explosion in the number of terms that have to be learned in the case of discrete variables. At the same time, using large cliques hides modeling details. On the other hand, smaller cliques allow us to be more explicit and detailed in our description.

Factor graphs provide us with the means of making the decomposition of a probability distribution into a product of factors more explicit. A factor graph is an undirected *bipartite* graph involving two types of nodes (thus the term bipartite): one that corresponds to the random variables, denoted by circles, and one that corresponds to the potential functions, denoted by squares. Edges exist only between two different types of nodes, that is, between "potential function" nodes and "variable" nodes [15,16,26].

Fig. 15.18A is an MRF for four variables. The respective factor graph in Fig. 15.18B corresponds to the product

$$p(x_1, x_2, x_3, x_4) = \frac{1}{Z}\psi_{c_1}(x_1, x_2, x_3)\psi_{c_2}(x_3, x_4), \tag{15.30}$$

and the one in Fig. 15.18C to

$$p(x_1, x_2, x_3, x_4) = \frac{1}{Z}\psi_{c_1}(x_1)\psi_{c_2}(x_1, x_2)\psi_{c_3}(x_1, x_2, x_3)\psi_{c_4}(x_3, x_4). \tag{15.31}$$

As an example, if the potential functions were chosen to express "interactions" among variables using probabilistic information, the involved functions in Fig. 15.18 may be chosen as

$$\psi_{c_1}(x_1, x_2, x_3) = p(x_3|x_1, x_2)p(x_2|x_1)p(x_1) \tag{15.32}$$

and

$$\psi_{c_2}(x_3, x_4) = p(x_4|x_3). \tag{15.33}$$

For the case of Fig. 15.18C,

$$\psi_{c_1}(x_1) = p(x_1),\ \psi_{c_2}(x_1, x_2) = p(x_2|x_1)$$
$$\psi_{c_3}(x_1, x_2, x_3) = p(x_3|x_1, x_2),\ \psi_{c_4}(x_3, x_4) = p(x_4|x_3).$$

For such an example, in both cases, it is readily seen that $Z = 1$. We will soon see that factor graphs turn out to be very useful for inference computations.

Remarks 15.1.

- A variant of the factor graphs, known as *normal factor graphs* (NFG), has been more recently introduced. In an NFG, edges represent variables and vertices represent factors. Moreover, latent and observable variables (internal and external) are distinguished by being represented by edges of degree 2 and degree 1, respectively. Such models can lead to simplified learning algorithms and can nicely unify a number of previously proposed models (see, e.g., [2,3,18,19,30,34,35]).

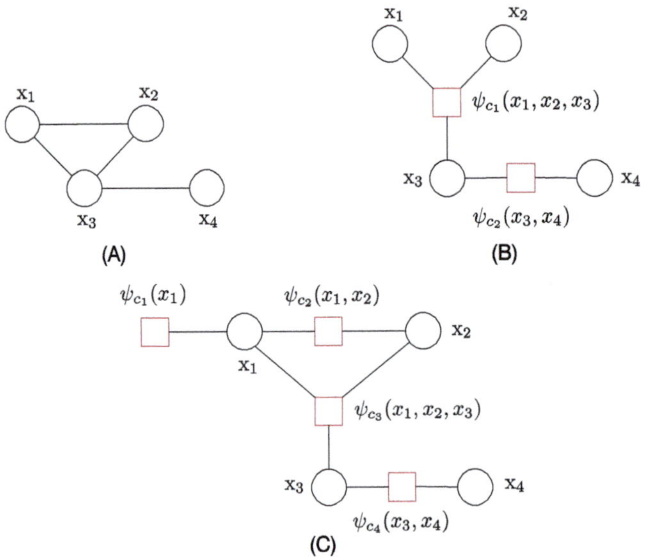

FIGURE 15.18

(A) An MRF and (B, C) possible equivalent factor graphs at different fine-grained factorization in terms of product factors.

15.5.1 GRAPHICAL MODELS FOR ERROR CORRECTING CODES

Graphical models are extensively used for representing a class of error correcting codes. In the *block parity-check* codes (e.g., [31]), one sends k information bits (0, 1 for a binary code) in a block of N bits, $N > k$; thus, redundancy is introduced into the system to cope with the effects of noise in the transmission channel. The extra bits are known as parity-check bits. For each code, a parity-check matrix, H, is defined; in order to be a valid one, for each code word, $\boldsymbol{x}$, it must satisfy the parity-check constraint (modulo-2 operations), $H\boldsymbol{x} = \boldsymbol{0}$. Take as an example the case of $k = 3$ and $N = 6$. The code comprises 2^3 (2^k in general) code words, each of them of length $N = 6$ bits. For the parity-check matrix,

$$H = \begin{bmatrix} 1 & 1 & 0 & 1 & 0 & 0 \\ 1 & 0 & 1 & 0 & 1 & 0 \\ 0 & 1 & 1 & 0 & 0 & 1 \end{bmatrix},$$

the eight code words that satisfy the parity-check constraint are 000000, 001011, 010101, 011110, 100110, 101101, 110011, and 111000. In each one of the eight words, the first three bits are the information bits and the remaining ones the parity-check bits, which are uniquely determined in order to satisfy the parity-check constraint. Each one of the three parity-check constraints can be expressed via a function, that is,

$$\psi_1(x_1, x_2, x_4) = \delta(x_1 \oplus x_2 \oplus x_4),$$
$$\psi_2(x_1, x_3, x_5) = \delta(x_1 \oplus x_3 \oplus x_5),$$
$$\psi_3(x_2, x_3, x_6) = \delta(x_2 \oplus x_3 \oplus x_6),$$

where $\delta(\cdot)$ is equal to one or zero, depending on whether its argument is one or zero, respectively, and $\oplus$ denotes the modulo-2 addition. The code words are transmitted to a noisy memoryless binary symmetric channel, where each transmitted bit, x_i, may be flipped over and be received as y_i, according to the following rule:

$$P(y = 0|x = 1) = p, \quad P(y = 1|x = 1) = 1 - p,$$
$$P(y = 1|x = 0) = p, \quad P(y = 0|x = 0) = 1 - p.$$

Upon reception of the observation sequence, y_i, $i = 1, 2, \ldots, N$, one has to decide the value x_i that was transmitted. Because the channel has been assumed memoryless, every bit is affected by the noise independently of the other bits, and the overall posterior probability of each code word is proportional to

$$\prod_{i=1}^{N} P(x_i|y_i).$$

In order to guarantee that only valid code words are considered, and assuming *equiprobable* information bits, we write the joint probability as

$$P(\boldsymbol{x}, \boldsymbol{y}) = \frac{1}{Z}\psi_1(x_1, x_2, x_4)\psi_2(x_1, x_3, x_5)\psi_3(x_2, x_3, x_6)\prod_{i=1}^{N} P(y_i|x_i),$$

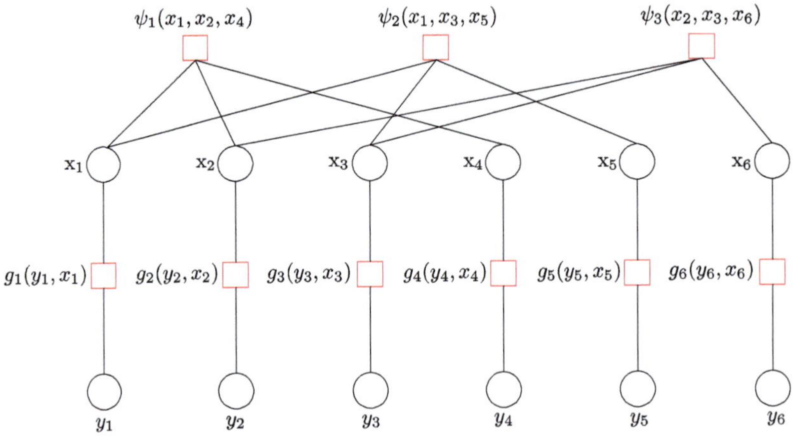

FIGURE 15.19

Factor graph for a (3,3) parity-check code.

where the parity-check constraints have been taken into account. The respective factor model is shown in Fig. 15.19, where

$$g_i(y_i, x_i) = P(y_i|x_i).$$

The task of decoding is to derive an efficient inference scheme to compute the posteriors, and based on that to decide in favor of 1 or 0.

15.6 MORALIZATION OF DIRECTED GRAPHS

At a number of points we have already made bridges between BNs and MRFs. In this section, we will formalize the bridge and see how one can convert a BN to an MRF and discuss the subsequent effects of such a conversion on the implied conditional independencies.

We can trust common sense to drive us to construct such a conversion. Because the conditional distributions will play the role of the potential functions (factors), one has to make sure that edges do exist among all the involved variables in each one of these factors. Because edges from the parents to children exist, we have to (a) retain these edges and make them undirected and (b) add edges between nodes that are parents of a common child. This is shown in Fig. 15.20. In Fig. 15.20A, a DAG is shown, which is converted to the MRF of Fig. 15.20B by adding undirected edges between x_1, x_2 (parents of x_3) and x_3, x_6 (parents of x_5). The procedure is known as *moralization* and the resulting undirected graph as a *moral graph*. The terminology stems from the fact that "parents are forced to be married." This conversion will be very useful soon, when an inference algorithm will be stated that covers both BNs and MRFs in a unifying framework.

The obvious question that is now raised is how moralization affects independencies. It turns out that if H is the resulting moral graph, then $I(H) \subseteq I(G)$ (Problem 15.11). In other words, the moral

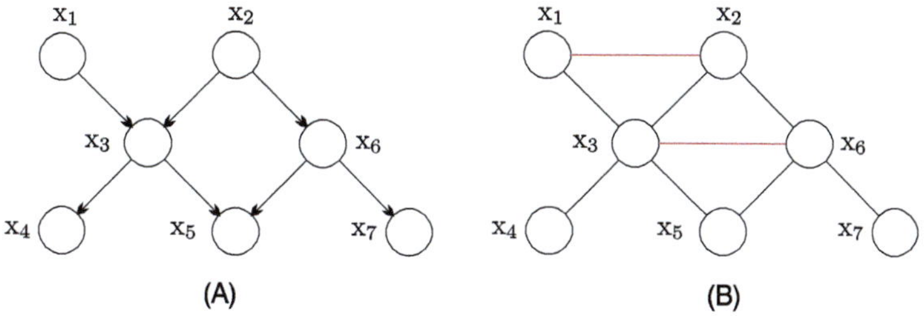

FIGURE 15.20

(A) A DAG and (B) the resulting MRF after applying moralization on the DAG. Directed edges become undirected and new edges, shown in red, are added to "marry" parents with common child nodes.

graph can guarantee a smaller number of independencies compared to the original BN via its set of d-separations. This is natural because one adds extra links. For example, in Fig. 15.20A, x_1, x_2 in the converging node $x_1 \rightarrow x_3 \leftarrow x_2$ are marginally independent, not given x_3. However, in the resulting moral graph in 15.20B, this independence is lost. It can be shown that moralization adds the fewest extra links and hence retains the maximum number of independence (see, for example, [25]).

15.7 EXACT INFERENCE METHODS: MESSAGE PASSING ALGORITHMS

This section deals with efficient techniques for inference on undirected graphical models. So, even if our starting point was a BN, we assume that it was converted to an undirected one prior to the inference task. We will begin with the simplest case of graphical models—graphs comprising a chain of nodes. This will help the reader to grasp the basic notions behind exact inference schemes. It is interesting to note that, in general, the inference task in graphical models is an NP-hard one [10]. Moreover, it has been shown that for general BNs, approximate inference to a desired number of digits of precision is also an NP-hard task [11]; that is, the time required has an exponential dependence on the number of digits of accuracy. However, as we will see in a number of cases that are commonly encountered in practice, the exponential growth in computational time can be bypassed by exploiting the underlying independencies and factorization properties of the associated distributions.

The inference tasks of interest are (a) computing the likelihood, (b) computing the marginals of the involved variables, (c) computing conditional probability distributions, and (d) finding modes of distributions.

15.7.1 EXACT INFERENCE IN CHAINS

Let us consider the chain graph of Fig. 15.21 and focus our interest on computing marginals. The naive approach, which overlooks factorization and independencies, would be to work directly on the joint distribution. Let us concentrate on discrete variables and assume that each one of them, l in total, has K states. Then, in order to compute the marginal of, say, x_j, we have to obtain the sum

FIGURE 15.21

An undirected chain graph with l nodes. There are $l-1$ cliques consisting of pairs of nodes.

$$\begin{aligned} P(x_j) := & \sum_{x_1} \cdots \sum_{x_{j-1}} \sum_{x_{j+1}} \cdots \sum_{x_l} P(x_1, \ldots, x_l) \\ = & \sum_{x_i : i \neq j} P(x_1, \ldots, x_l). \end{aligned} \tag{15.34}$$

Each summation is over K values; hence, the number of the required computations amounts to $\mathcal{O}(K^l)$. Let us now bring factorization into the game and concentrate on computing $P(x_1)$. Assume that the joint probability factorizes over the graph. Hence, we can write

$$\boxed{P(\boldsymbol{x}) := P(x_1, x_2, \ldots, x_l) = \frac{1}{Z} \prod_{i=1}^{l-1} \psi_{i,i+1}(x_i, x_{i+1}),} \tag{15.35}$$

and

$$P(x_1) = \frac{1}{Z} \sum_{x_i : i \neq 1} \prod_{i=1}^{l-1} \psi_{i,i+1}(x_i, x_{i+1}). \tag{15.36}$$

Note that the only term that depends on x_l is $\psi_{l-1,l}(x_{l-1}, x_l)$. Let us start by summing with respect to this last term, which leaves unaffected all the preceding factors in the sequence of products in Eq. (15.36), i.e.,

$$P(x_1) = \frac{1}{Z} \sum_{x_i : i \neq 1, l} \prod_{i=1}^{l-2} \psi_{i,i+1}(x_i, x_{i+1}) \sum_{x_l} \psi_{l-1,l}(x_{l-1}, x_l), \tag{15.37}$$

where we exploited the basic property of arithmetic

$$\sum_i \alpha \beta_i = \alpha \sum_i \beta_i. \tag{15.38}$$

- Define

$$\sum_{x_l} \psi_{l-1,l}(x_{l-1}, x_l) := \mu_b(x_{l-1}).$$

  Because the possible values of the pair (x_{l-1}, x_l) comprise a table with K^2 elements, the summation involves K^2 terms and $\mu_b(x_{l-1})$ consists of K possible values.

- After marginalizing out x_l, the only factor in the product that depends on x_{l-1} is

$$\psi_{l-2,l-1}(x_{l-2}, x_{l-1})\mu_b(x_{l-1}).$$

Then, in a similar way as before, we obtain

$$P(x_1) = \frac{1}{Z} \sum_{x_i: i \neq 1, l-1, l} \prod_{i=1}^{l-3} \psi_{i,i+1}(x_i, x_{i+1}) \sum_{x_{l-1}} \psi_{l-2,l-1}(x_{l-2}, x_{l-1})\mu_b(x_{l-1}),$$

where this summation also involves K^2 terms.

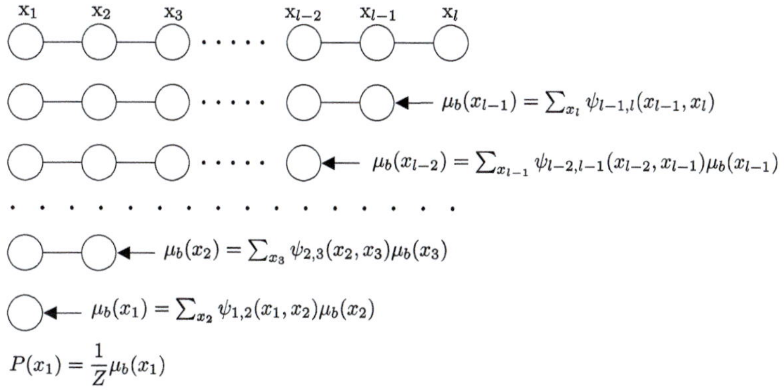

FIGURE 15.22

To compute $P(x_1)$, starting from the last node, x_l, each node (a) receives a message and (b) processes it locally via sum and product operations, which produces a new message; and (c) the latter is passed backward, to the node on its left. We have assumed that $\mu_b(x_l) = 1$.

We are now ready to define the general recursion as

$$\boxed{\begin{aligned} &\mu_b(x_i) := \sum_{x_{i+1}} \psi(x_i, x_{i+1})\mu_b(x_{i+1}), \ i = 1, 2, \ldots, l-1, \\ &\mu_b(x_l) = 1, \end{aligned}} \tag{15.39}$$

whose repeated application leads to

$$\mu_b(x_1) = \sum_{x_1} \psi_{1,2}(x_1, x_2)\mu_b(x_2),$$

and finally

$$P(x_1) = \frac{1}{Z}\mu_b(x_1). \tag{15.40}$$

The series of recursions is illustrated in Fig. 15.22.

We can think that every node, x_i, (a) receives a message from its right, $\mu_b(x_i)$, which for our case comprises K values, (b) performs locally sum-multiply operations and a new message $\mu_b(x_{i-1})$ is computed, which (c) is passed to its left, to node x_{i-1}. The subscript "b" in μ_b denotes "backward" to remind us of the flow of the message passing activity from right to left.

If we wanted to compute $P(x_l)$, we would adopt the same reasoning but start summation from x_1. In this case, message passing takes place forward (from left to right) and messages are defined as

$$\boxed{\begin{aligned} &\mu_f(x_{i+1}) := \sum_{x_i} \psi_{i,i+1}(x_i, x_{i+1})\mu_f(x_i), \quad i = 1, \ldots, l-1, \\ &\mu_f(x_1) = 1, \end{aligned}} \tag{15.41}$$

where "f" has been used to denote "forward" flow. The procedure is shown in Fig. 15.23.

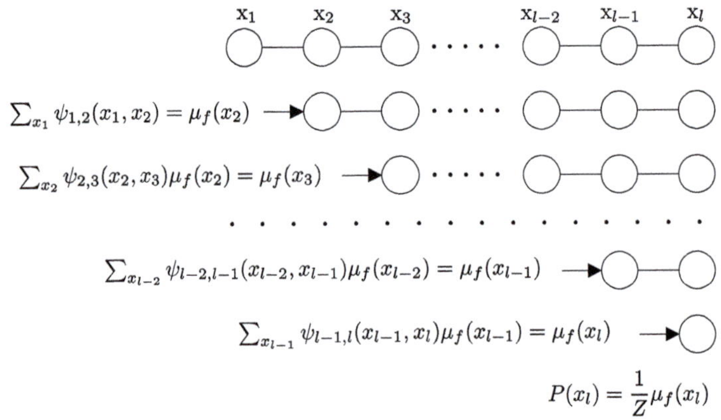

FIGURE 15.23

To compute $P(x_l)$, message passing takes place in the forward direction, from left to right. As opposed to Fig. 15.22, messages are denoted as μ_f, to remind us of the forward flow.

The term $\mu_b(x_j)$ is the result of summing the products over $x_{j+1}, x_{j+2}, \ldots, x_l$, and the term $\mu_f(x_j)$ over $x_1, x_2, \ldots, x_{j-1}$. At each iteration step, one variable is *eliminated* by summing up over all its possible values. It can be easily shown, following similar arguments, that the marginal at any point, x_j, $2 \leq j \leq l-1$, is obtained by (Problem 15.13)

$$P(x_j) = \frac{1}{Z}\mu_f(x_j)\mu_b(x_j), \quad j = 2, 3, \ldots, l-1. \tag{15.42}$$

The idea is to perform one forward and one backward message passing operation, store the values, and then compute any one of the marginals of interest. The total cost will be $\mathcal{O}(2K^2 l)$, instead of K^l of the naive approach.

We still have to compute the normalizing constant Z. This is readily done by summing up both sides of Eq. (15.42), which requires $\mathcal{O}(K)$ operations,

$$Z = \sum_{x_j=1}^{K} \mu_f(x_j)\mu_b(x_j). \tag{15.43}$$

So far, we have considered the computation of marginal probabilities. Let us now turn our attention to their conditional counterparts. We start with the simplest case, for example, to compute $P(x_j|\mathrm{x}_k = \hat{x}_k)$, $k \neq j$. That is, we assume that variable x_k has been observed and its value is $\hat{x}_k$. The first step in computing the conditional is to recover the joint $P(x_j, \mathrm{x}_k = \hat{x}_k)$. This is a normalized version of the respective conditional, which can then be obtained as

$$P(x_j|\mathrm{x}_k = \hat{x}_k) = \frac{P(x_j, \mathrm{x}_k = \hat{x}_k)}{P(\hat{x}_k)}. \tag{15.44}$$

The only difference in computing $P(x_j, \mathrm{x}_k = \hat{x}_k)$ compared to the previous computations of the marginals is that now, in order to obtain the messages, we *do not sum* with respect to x_k. We just clump the respected potential function to its value $\hat{x}_k$. That is, the computations

$$\mu_b(x_{k-1}) = \sum_{x_k} \psi_{k-1,k}(x_{k-1}, x_k)\mu_b(x_k),$$
$$\mu_f(x_{k+1}) = \sum_{x_k} \psi_{k,k+1}(x_k, x_{k+1})\mu_f(x_k)$$

are replaced by

$$\mu_b(x_{k-1}) = \psi_{k-1,k}(x_{k-1}, \hat{x}_k)\mu_b(\hat{x}_k),$$
$$\mu_f(x_{k+1}) = \psi_{k,k+1}(\hat{x}_k, x_{k+1})\mu_f(\hat{x}_k).$$

In other words, x_k is considered a delta function at the instantiated value. Once $P(x_j, \mathrm{x}_k = \hat{x}_k)$ has been obtained, normalization is straightforward and is locally performed at the jth node. The procedure can be generalized when more than one variable is observed.

15.7.2 EXACT INFERENCE IN TREES

Having gained experience and learned the basic secrets in developing efficient inference algorithms for chains, we turn our attention to the more general case involving *tree-structured* undirected graphical models.

A tree is a graph in which there is a single path between any two nodes of the graph; thus, there are no cycles in the graph. Figs. 15.24A and B are two examples of trees. Note that in a directed tree, any node has only a single parent. A tree can be directed or undirected. Furthermore, because in a directed one there are no children with two parents, the moralization step, which converts a directed graph to an undirected one, adds no extra links. The only change consists of making the edges undirected.

There is an important property of the trees, which will prove very important for our current needs. Let us denote a tree graph as T, which is a collection of vertices/nodes V and edges E, which link the nodes, that is, $T = \{V, E\}$. Consider any node $\mathrm{x} \in V$ and consider the set of all its neighbors, that is, all nodes that share an edge with x. Denote this set as

$$\mathcal{N}(\mathrm{x}) = \left\{\mathrm{y} \in V : (\mathrm{x}, \mathrm{y}) \in E\right\}.$$

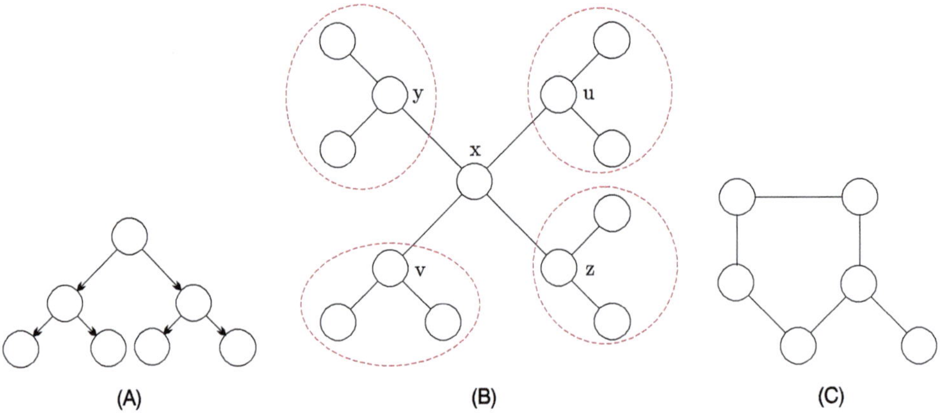

FIGURE 15.24

Examples of (A) directed and (B) undirected trees. Note that in the directed one, any node has a single parent. In both cases, there is only a single chain that connects any two nodes. (C) The graph is not a tree because there is a cycle.

Looking at Fig. 15.24B, we have $\mathcal{N}(\mathrm{x}) = \{\mathrm{y, u, z, v}\}$. Then, for *each* element $\mathrm{r} \in \mathcal{N}(\mathrm{x})$, define the subgraph $T_r = \{V_r, E_r\}$ such that *any* node in this subgraph can be reached from r via paths that *do not pass* through x. In Fig. 15.24B, the respective subgraphs, each associated with one element in (y, u, z, v), are encircled by dotted lines. By the definition of a tree, it can easily be deduced that each one of these subgraphs is also a tree. Moreover, these subgraphs are *disjoint*. In other words, *each one* of the neighboring nodes of a node, for example, x, can be viewed as a *root* of a subtree, and these subtrees are mutually disjoint, having no common nodes. This property will allow us to break a large problem into a number of smaller ones. Moreover, each one of the smaller problems can be further divided in the same way, being itself a tree. We now have all the basic ingredients to derive an efficient scheme for inference on trees (recall that such a breaking of a large problem into a sequence of smaller ones was at the heart of the message passing algorithm for chains). However, let us first bring the notion of factor graphs into the scene. The reason is that using factor graphs allows us to deal with some more general graph structures, such as *polytrees*.

A directed polytree is a graph in which, although there are no cycles, a child may have more than one parent. Fig. 15.25A shows an example of a polytree. The unpleasant situation results after the moralization step because marrying the parents results in cycles, and we *cannot* derive *exact* inference algorithms in graphs with cycles, Fig. 15.25B. However, if one converts the original directed polytree into a factor graph, the resulting bipartite entity has a tree structure, with no cycles involved, Fig. 15.25C. Thus, everything we said before about tree structures applies to these factor graphs.

15.7.3 THE SUM-PRODUCT ALGORITHM

We will develop the algorithm in a "bottom-up" approach, via the use of an example. Once the rationale is understood, the generalization can readily be obtained. Let us consider the factor tree of Fig. 15.26.

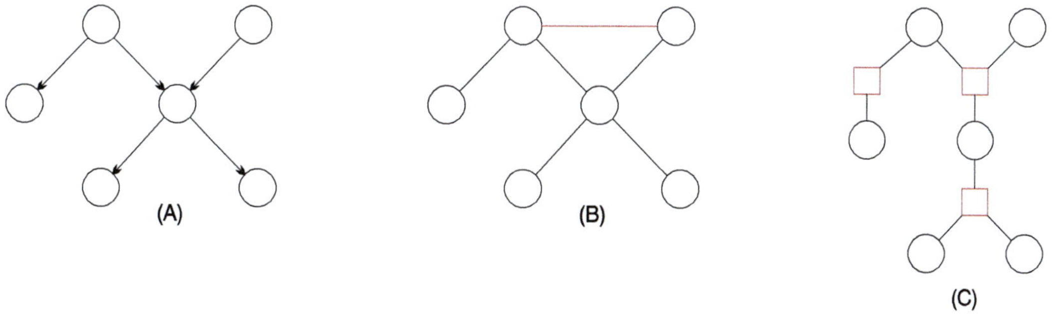

FIGURE 15.25

(A) Although there are no cycles, one of the nodes has two parents, and hence the graph is a polytree. (B) The resulting structure after moralization has a cycle. (C) A factor graph for the polytree in (A).

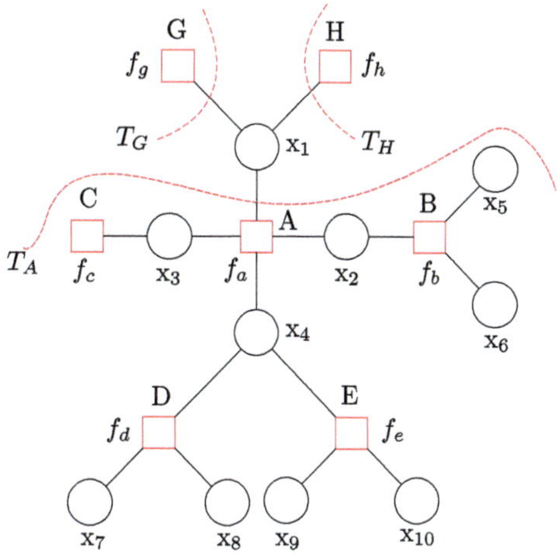

FIGURE 15.26

The tree is subdivided into three subtrees, each one having as its root one of the factor nodes connected to x_1. This is the node whose marginal is computed in the text. Messages are initiated from the leaf nodes toward x_1. Once messages arrive at x_1, a new propagation of messages starts, this time from x_1 to the leaves.

The factor nodes are denoted by capital letters and squares, and each one is associated with a potential function. The rest are variable nodes, denoted by circles. Assume that we want to compute the marginal $P(x_1)$. Node x_1, being a variable node, is connected to factor nodes only. We split the graph in as many (tree) subgraphs as the factor nodes connected to x_1 (three in our case). In the figure, each one of these subgraphs is encircled, having as roots the nodes A, H, and G, respectively. Recall that the joint $P(\boldsymbol{x})$ is given as the product of all the potential functions, each one associated with one factor node, divided by the normalizing constant, Z. Focusing on the node of interest, x_1, this product can be written as

$$P(\boldsymbol{x}) = \frac{1}{Z}\psi_A(x_1, \boldsymbol{x}_A)\psi_H(x_1, \boldsymbol{x}_H)\psi_G(x_1, \boldsymbol{x}_G), \tag{15.45}$$

where $\boldsymbol{x}_A$ denotes the vector corresponding to all the variables in T_A; the vectors $\boldsymbol{x}_H$ and $\boldsymbol{x}_G$ are similarly defined. The function $\psi_A(x_1, \boldsymbol{x}_A)$ is the product of all the potential functions associated with the factor nodes in T_A, and $\psi_H(x_1, \boldsymbol{x}_H)$ and $\psi_G(x_1, \boldsymbol{x}_G)$ are defined in an analogous way. Then the marginal of interest is given by

$$P(x_1) = \frac{1}{Z}\sum_{\boldsymbol{x}_A \in V_A}\sum_{\boldsymbol{x}_H \in V_H}\sum_{\boldsymbol{x}_G \in V_G}\psi_A(x_1, \boldsymbol{x}_A)\psi_H(x_1, \boldsymbol{x}_H)\psi_G(x_1, \boldsymbol{x}_G). \tag{15.46}$$

We will concentrate on the subtree with root A, denoted as $T_A := \{V_A, E_A\}$, where V_A stands for the nodes in T_A and E_A for the respective set of edges. Because the three subtrees are disjoint, we can split the previous expression in Eq. (15.46) into

$$P(x_1) = \frac{1}{Z}\sum_{\boldsymbol{x}_A \in V_A}\psi_A(x_1, \boldsymbol{x}_A)\sum_{\boldsymbol{x}_H \in V_H}\psi_H(x_1, \boldsymbol{x}_H)\sum_{\boldsymbol{x}_G \in V_G}\psi_G(x_1, \boldsymbol{x}_G). \tag{15.47}$$

Note that $x_1 \notin V_A \cup V_H \cup V_G$. Having reserved the symbol $\psi_A(\cdot, \cdot)$ to denote the product of all the potentials in the subtree T_A (and similarly for T_H and T_G), let us denote the individual potential functions, for each one of the factor nodes, via the symbol f, as shown in Fig. 15.26. Thus, we can now write

$$\begin{aligned}
\sum_{\boldsymbol{x}_A \in V_A}\psi_A(x_1, \boldsymbol{x}_A) &= \sum_{\boldsymbol{x}_A \in V_A} f_a(x_1, x_2, x_3, x_4) f_c(x_3) f_b(x_2, x_5, x_6) \\
&\quad \times f_d(x_4, x_7, x_8) f_e(x_4, x_9, x_{10}) \\
&= \sum_{x_2}\sum_{x_3}\sum_{x_4} f_a(x_1, x_2, x_3, x_4) f_c(x_3) \sum_{x_7}\sum_{x_8} f_d(x_4, x_7, x_8) \\
&\quad \times \sum_{x_9}\sum_{x_{10}} f_e(x_4, x_9, x_{10}) \sum_{x_6}\sum_{x_5} f_b(x_2, x_5, x_6).
\end{aligned} \tag{15.48}$$

Recall from our treatment of the chain graph that messages were nothing but locally computed summations over products. Having this experience, let us define

$$\mu_{f_b \to x_2}(x_2) = \sum_{x_6}\sum_{x_5} f_b(x_2, x_5, x_6),$$

$$\mu_{f_e \to x_4}(x_4) = \sum_{x_9}\sum_{x_{10}} f_e(x_4, x_9, x_{10}),$$

$$\mu_{f_d \to x_4}(x_4) = \sum_{x_7}\sum_{x_8} f_d(x_4, x_7, x_8),$$

$$\mu_{f_c \to x_3}(x_3) = f_c(x_3),$$

$$\mu_{x_4 \to f_a}(x_4) = \mu_{f_d \to x_4}(x_4)\mu_{f_e \to x_4}(x_4),$$
$$\mu_{x_2 \to f_a}(x_2) = \mu_{f_b \to x_2}(x_2),$$
$$\mu_{x_3 \to f_a}(x_3) = \mu_{f_c \to x_3}(x_3),$$

and

$$\mu_{f_a \to x_1}(x_1) = \sum_{x_2}\sum_{x_3}\sum_{x_4} f_a(x_1, x_2, x_3, x_4)\mu_{x_2 \to f_a}(x_2)\mu_{x_3 \to f_a}(x_3)\mu_{x_4 \to f_a}(x_4).$$

Observe that we were led to define two types of messages; one type is passed from variable nodes to factor nodes and the other type is passed from factor nodes to variable nodes.

- Variable node to factor node messages (Fig. 15.27A):

$$\boxed{\mu_{x \to f}(x) = \prod_{s: f_s \in \mathcal{N}(\mathrm{x}) \setminus f} \mu_{f_s \to x}(x).} \tag{15.49}$$

  We use $\mathcal{N}(\mathrm{x})$ to denote the set of the nodes with which a variable node x is connected; $\mathcal{N}(\mathrm{x}) \setminus f$ refers to all nodes *excluding* the factor node f; note that all these nodes are factor nodes. In other words, the action of a variable node, as far as message passing is concerned, is to multiply incoming messages. Obviously, if it is connected only to one factor node (except f), then such a variable node passes what it receives, without any computation. This is, for example, the case with $\mu_{x_2 \to f_a}$, as previously defined.
- Factor node to variable node messages (Fig. 15.27B):

$$\boxed{\mu_{f \to x}(x) = \sum_{x_i \in \mathcal{N}(f) \setminus \mathrm{x}} f(\boldsymbol{x}^f) \prod_{i: x_i \in \mathcal{N}(f) \setminus \mathrm{x}} \mu_{x_i \to f}(x_i),} \tag{15.50}$$

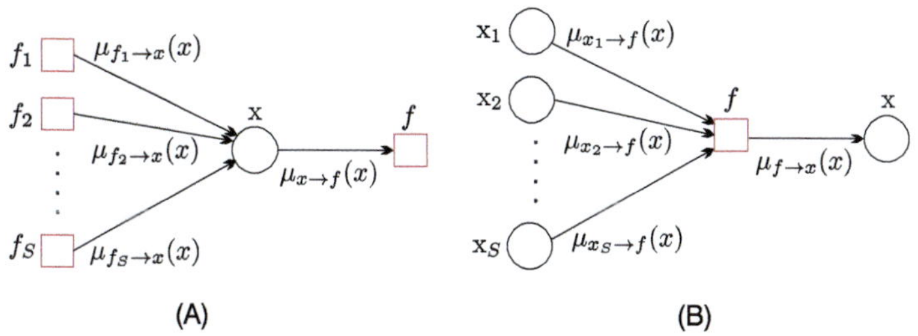

FIGURE 15.27

(A) Variable x is connected to S factor nodes, besides f; that is, $\mathcal{N}(x) \setminus f = \{f_1, f_2, \ldots, f_S\}$. The output message from x to f is the product of the incoming messages. The arrows indicate directions of flow of the message propagation. (B) The factor node f is connected to S node variables, besides x; that is, $\mathcal{N}(f) \setminus \mathrm{x} = \{\mathrm{x}_1, \mathrm{x}_2, \ldots, \mathrm{x}_S\}$.

where $\mathcal{N}(f)$ denotes the set of the (variable) nodes connected to f and $\mathcal{N}(f) \setminus \mathrm{x}$ the corresponding set if we exclude node x. The vector $\boldsymbol{x}^f$ comprises all the variables involved as arguments in f, that is, all the variables/nodes in $\mathcal{N}(f)$.

If a node is a leaf, we adopt the following convention. If it is a variable node, x, connected to a factor node, f, then

$$\mu_{x \to f}(x) = 1. \tag{15.51}$$

If it is a factor node, f, connected to variable node, x, then

$$\mu_{f \to x}(x) = f(x). \tag{15.52}$$

Adopting the previously stated definitions, Eq. (15.48) is now written as

$$\begin{aligned} &\sum_{\boldsymbol{x}_A \in V_A} \psi_A(x_1, \boldsymbol{x}_A) \\ &= \sum_{x_2} \sum_{x_3} \sum_{x_4} f_a(x_1, x_2, x_3, x_4) \mu_{x_2 \to f_a}(x_2) \mu_{x_3 \to f_a}(x_3) \mu_{x_4 \to f_a}(x_4) \\ &= \mu_{f_a \to x_1}(x_1). \end{aligned} \tag{15.53}$$

Working similarly for the other two subtrees, T_G and T_H, we finally obtain

$$P(x_1) = \frac{1}{Z} \mu_{f_a \to x_1}(x_1) \mu_{f_g \to x_1}(x_1) \mu_{f_h \to x_1}(x_1). \tag{15.54}$$

Note that each summation can be viewed as a step that "removes" a variable and produces a message. This is the reason that sometimes this procedure is called *variable elimination*. We are now ready to summarize the steps of the algorithm.

The algorithmic steps

1. Pick the variable x, whose marginal, $P(x)$, will be computed.
2. Divide the tree into as many subtrees as the number of factor nodes to which the variable node x is connected.
3. For each one of these subtrees, identify the leaf nodes.
4. Start message passing, toward x, by initializing the leaf nodes according to Eqs. (15.51) and (15.52), by utilizing Eqs. (15.49) and (15.50).
5. Compute the marginal according to Eq. (15.54), or in general

$$P(x) = \frac{1}{Z} \prod_{s: f_s \in \mathcal{N}(\mathrm{x})} \mu_{f_s \to x}(x). \tag{15.55}$$

The normalizing constant, Z, can be obtained by adding both sides of Eq. (15.55) over all possible values of x. As was the case with the chain graphs, if a variable is observed, then we replace the summation over this variable, in the places where this is required, by a single term evaluated on the observed value.

Although so far we have considered discrete variables, everything that has been said also applies to continuous variables by substituting summations by integrals. For such integrations, Gaussian models turn out to be very convenient.

Remarks 15.2.

- Thus far, we have concentrated on the computation of the marginal for a single variable, x. If the marginal of another variable is required, the obvious way would be for the whole process to be repeated. However, as we already commented in Section 15.7.1, such an approach is computationally wasteful because many of the computations are common and can be shared for the evaluation of the marginals of the various nodes. Note that in order to compute the marginal at any variable node, one needs all the messages from the factor nodes, to which the specific variable node is connected, to be available (Eq. (15.55)). Assume now that we pick one node, say, x_1, and compute the marginal, once all the required messages have "arrived." Then this node initiates a new message propagation phase, this time toward the leaves. It is not difficult to see (Problem 15.15) that this process, once completed, will make available to every node all the required messages for the computation of the respective marginals. In other words, this two-stage message passing procedure, in two opposite-flow directions, suffices to provide all the necessary information for the computation of the marginals at every node. The total number of messages passed is just twice the number of edges in the graph. Similar to the case of chain graphs, in order to compute conditional probabilities, say, $P(x_i|\mathrm{x}_k = \hat{x}_k)$, node x_k has to be instantiated. Running the sum-product algorithm will provide the joint probability $P(x_i, \mathrm{x}_k = \hat{x}_k)$, from which the respective conditional is obtained after normalization; this is performed locally at the respective variable node.
- The (joint) marginal probability of all the variables $\mathrm{x}_1, \mathrm{x}_2, \ldots, \mathrm{x}_S$, associated with a factor node f, is given by (Problem 15.16)

$$P(x_1, \ldots, x_S) = \frac{1}{Z} f(x_1, \ldots, x_S) \prod_{s=1}^{S} \mu_{x_s \to f}(x_s). \tag{15.56}$$

- Earlier versions of the sum-product algorithm, known as *belief propagation*, were independently developed in the context of singly connected graphs in [28,36,37]. However, the problem of variable elimination has an older history and has been discovered in different communities (e.g., [4,6,39]). Sometimes, the general sum-product algorithm, as described before, is also called *the generalized forward-backward algorithm*.

15.7.4 THE MAX-PRODUCT AND MAX-SUM ALGORITHMS

Let us now turn our attention from marginals to modes of distributions. That is, given a distribution $P(\boldsymbol{x})$ that factorizes over a tree (factor) graph, the task is to compute efficiently the quantity

$$\max_{\boldsymbol{x}} P(\boldsymbol{x}).$$

We will focus on discrete variables. Following similar arguments as before, one can readily write the counterpart of Eq. (15.46), for the case of Fig. 15.26, as

$$\max_{\boldsymbol{x}} P(\boldsymbol{x}) = \frac{1}{Z} \max_{x_1} \max_{\boldsymbol{x}_A \in V_A} \max_{\boldsymbol{x}_H \in V_H} \max_{\boldsymbol{x}_G \in V_G} \psi_A(x_1, \boldsymbol{x}_A)\psi_H(x_1, \boldsymbol{x}_H)\psi_G(x_1, \boldsymbol{x}_G). \tag{15.57}$$

Exploiting the property of the max operator, that is,

$$\max_{b,c}(ab, ac) = a \max_{b,c}(b, c)\ , \quad a \geq 0,$$

we can rewrite Eq. (15.57) as

$$\max_{\boldsymbol{x}} P(\boldsymbol{x}) = \frac{1}{Z} \max_{x_1} \max_{\boldsymbol{x}_A \in V_A} \psi_A(x_1, \boldsymbol{x}_A) \max_{\boldsymbol{x}_H \in V_H} \psi_H(x_1, \boldsymbol{x}_H) \max_{\boldsymbol{x}_G \in V_G} \psi_G(x_1, \boldsymbol{x}_G).$$

Following similar arguments as for the sum-product rule, we arrive at the counterpart of Eq. (15.48), that is,

$$\begin{aligned} \max_{x_1} \max_{\boldsymbol{x}_A \in V_A} \psi_A(x_1, \boldsymbol{x}_A) = \max_{x_1} \max_{x_2, x_3, x_4} & f_a(x_1, x_2, x_3, x_4) f_c(x_3) \max_{x_7, x_8} f_d(x_4, x_7, x_8) \\ & \times \max_{x_9, x_{10}} f_e(x_4, x_9, x_{10}) \max_{x_5, x_6} f_b(x_2, x_6, x_5). \end{aligned} \tag{15.58}$$

Eq. (15.58) suggests that everything that was said before for the sum-product message passing algorithm holds true here, provided we replace summations with the max operations, and the definitions of the messages passed between nodes change to

$$\mu_{x \to f}(x) = \prod_{s: f_s \in \mathcal{N}(\mathrm{x}) \setminus f} \mu_{f_s \to x}(x), \tag{15.59}$$

and

$$\boxed{\mu_{f \to x}(x) = \max_{x_i : x_i \in \mathcal{N}(f) \setminus \mathrm{x}} f(\boldsymbol{x}^f) \prod_{i: x_i \in \mathcal{N}(f) \setminus \mathrm{x}} \mu_{x_i \to f}(x_i),} \tag{15.60}$$

with the same definition of symbols as for Eq. (15.50). Then the mode of $P(\boldsymbol{x})$ is given by

$$\max_{\boldsymbol{x}} P(\boldsymbol{x}) = \frac{1}{Z} \max_{x_1} \mu_{f_a \to x_1}(x_1) \mu_{f_g \to x_1}(x_1) \mu_{f_h \to x_1}(x_1), \tag{15.61}$$

or in general

$$\max_{\boldsymbol{x}} P(\boldsymbol{x}) = \frac{1}{Z} \max_{x} \prod_{s: f_s \in \mathcal{N}(\mathrm{x})} \mu_{f_s \to x}(x), \tag{15.62}$$

where x is the node chosen to play the role of the root, toward which the flow of the messages is directed, starting from the leaves. The resulting scheme is known as the *max-product algorithm.*

In practice, an alternative formulation of the previously stated max-product algorithm is usually adopted. Often, the involved potential functions are probabilities (by absorbing the normalization constant) and their magnitude is less than one; however, if a large number of product terms is involved it may lead to arithmetic inaccuracies. A way to bypass this is to involve the logarithmic function, which transforms products into summations. This is justified by the fact that the logarithmic function is monotonic and increasing; hence it does not affect the point $\boldsymbol{x}$ at which a maximum occurs, that is,

$$\boldsymbol{x}_* := \arg\max_{\boldsymbol{x}} P(\boldsymbol{x}) = \arg\max_{\boldsymbol{x}} \ \ln P(\boldsymbol{x}). \tag{15.63}$$

Under this formulation, the following *max-sum* version of the algorithm results. It is straightforward to see that Eqs. (15.59) and (15.60) now take the form of

$$\mu_{x\to f}(x) = \sum_{s: f_s \in \mathcal{N}(\mathrm{x})\setminus f} \mu_{f_s \to x}(x), \tag{15.64}$$

$$\mu_{f\to x}(x) = \max_{x_i : x_i \in \mathcal{N}(f)\setminus \mathrm{x}} \left\{ \ln f(\boldsymbol{x}^f) + \sum_{i: x_i \in \mathcal{N}(f)\setminus \mathrm{x}} \mu_{x_i \to f}(x_i) \right\}. \tag{15.65}$$

In place of Eqs. (15.51) and (15.52) for the initial messages, sent by the leaf nodes, we now define

$$\mu_{x\to f}(x) = 0 \text{ and } \mu_{f\to x}(x) = \ln f(x). \tag{15.66}$$

Note that after one pass of the message flow, the maximum value of $P(\boldsymbol{x})$ has been obtained. However, one is also interested in knowing the corresponding value $\boldsymbol{x}_*$, for which the maximum occurs, that is,

$$\boldsymbol{x}_* = \arg\max_{\boldsymbol{x}} P(\boldsymbol{x}).$$

This is achieved by a reverse message passing process, which is slightly different from what we have discussed so far, and it is known as back-tracking.

Back-tracking: Assume that x_1 is the chosen node to play the role of the root, where the flow of messages "converges." From Eq. (15.61), we get

$$x_{1*} = \arg\max_{x_1} \mu_{f_a\to x_1}(x_1)\mu_{f_g\to x_1}(x_1)\mu_{f_h\to x_1}(x_1). \tag{15.67}$$

A new message passing flow now starts, and the root node, x_1, passes the obtained optimal value to the factor nodes to which it is connected. Let us follow this message passing flow within the nodes of the subtree T_A.

- Node A: It receives x_{1*} from node x_1.
  - Selection of the optimal values: Recall that

$$\mu_{f_a\to x_1}(x_1) = \max_{x_2,x_3,x_4} f_a(x_1,x_2,x_3,x_4)\mu_{x_4\to f_a}(x_4) \times \mu_{x_3\to f_a}(x_3)\mu_{x_2\to f_a}(x_2).$$

  Thus, for different values of x_1, different optimal values for (x_2, x_3, x_4) will result. For example, assume that in our discrete variable setting, each variable can take one out of four possible values, that is, $\mathrm{x} \in \{1,2,3,4\}$. Then, if $x_{1*} = 2$, say that the resulting optimal values are $(x_{2*}, x_{3*}, x_{4*}) = (1, 1, 3)$. On the other hand, if $x_{1*} = 4$, then maximization may result to, let us say, $(x_{2*}, x_{3*}, x_{4*}) = (2, 3, 4)$. However, having obtained a *specific* value for x_{1*} via the maximization at node x_1, we choose the triplet (x_{2*}, x_{3*}, x_{4*}) such that

$$\begin{aligned}(x_{2*}, x_{3*}, x_{4*}) &= \arg\max_{x_2,x_3,x_4} f_a(x_{1*}, x_2, x_3, x_4)\mu_{x_4 \to f_a}(x_4) \\ &\times \mu_{x_3 \to f_a}(x_3)\mu_{x_2 \to f_a}(x_2).\end{aligned} \tag{15.68}$$

Hence, during the first pass, the obtained optimal values have to be stored to be used during the second (backward) pass.

  - Message passing: Node A passes x_{4*} to node x_4, x_{2*} to node x_2, and x_{3*} to node x_3.
- Node x_4 passes x_{4*} to nodes D and E.
- Node D
  - Selection of the optimal values: Select (x_{7*}, x_{8*}) such that

$$(x_{7*}, x_{8*}) = \arg\max_{x_7,x_8} f_d(x_{4*}, x_7, x_8)\mu_{x_7 \to f_d}(x_7)\mu_{x_8 \to f_d}(x_8).$$

  - Message Passing: Node D passes (x_{7*}, x_{8*}) to nodes x_7 and x_8, respectively.

This type of flow spreads toward all the leaves, and finally

$$\boldsymbol{x}_* = \arg\max_{\boldsymbol{x}} P(\boldsymbol{x}) \tag{15.69}$$

is obtained. One may wonder why not use a similar two-stage message passing as we did with the sum-product rule and recover x_{i*} for each node i. This would be possible if there were a guarantee for a unique optimum, $\boldsymbol{x}_*$. If this is not the case and we have two optimal values, say, $\boldsymbol{x}_*^1$ and $\boldsymbol{x}_*^2$, which result from Eq. (15.69), then we run the danger of failing to obtain them. To see this, let us take an example of four variables, $\mathrm{x}_1, \mathrm{x}_2, \mathrm{x}_3, \mathrm{x}_4$, each taking values in the discrete set $\{1, 2, 3, 4\}$. Assume that $P(\boldsymbol{x})$ does not have a unique maximum and the two combinations for optimality are

$$(x_{1*}, x_{2*}, x_{3*}, x_{4*}) = (1, 1, 2, 3) \tag{15.70}$$

and

$$(x_{1*}, x_{2*}, x_{3*}, x_{4*}) = (1, 2, 2, 4). \tag{15.71}$$

Both of them are acceptable because they correspond to max $P(x_1, x_1, x_3, x_4)$. The back-tracking procedure guarantees to give either of the two. In contrast, using two-stage message passing may result to a combination of values, for example,

$$(x_{1*}, x_{2*}, x_{3*}, x_{4*}) = (1, 1, 2, 4), \tag{15.72}$$

which does not correspond to the maximum. Note that this result is correct in its own rationale. It provides, for every node, a value for which an optimum may result. Indeed, searching for a maximum of $P(\boldsymbol{x})$, node x_2 can take either the value of 1 or 2. However, what we want to find is the correct *combination* for all nodes. This is guaranteed by the back-tracking procedure.

Remarks 15.3.

- The max-product (max-sum) algorithm is a generalization of the celebrated Viterbi algorithm [46], which has extensively been used in communications [17] and speech recognition [39]. The algorithm has been generalized to arbitrary commutative semirings on tree-structured graphs (e.g., [1, 13]).

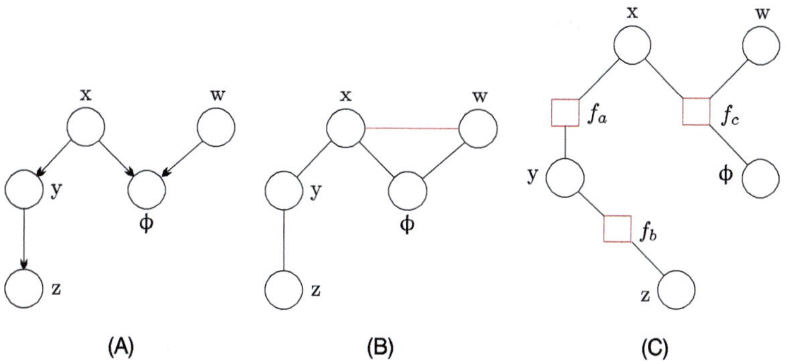

FIGURE 15.28

(A) The Bayesian network of Example 15.4; (B) its moralized version, where the two parents of ϕ have been connected; and (C) a possible factor graph.

Example 15.4. Consider the BN of Fig. 15.28A. The involved variables are binary, {0, 1}, and the respective probabilities are

$$\begin{aligned}
P(\mathrm{x}=1)=0.7,\ & P(\mathrm{x}=0)=0.3,\\
P(\mathrm{w}=1)=0.8,\ & P(\mathrm{w}=0)=0.2,\\
P(\mathrm{y}=1|\mathrm{x}=0)=0.8,\ & P(\mathrm{y}=0|\mathrm{x}=0)=0.2,\\
P(\mathrm{y}=1|\mathrm{x}=1)=0.6,\ & P(\mathrm{y}=0|\mathrm{x}=1)=0.4,\\
P(\mathrm{z}=1|\mathrm{y}=0)=0.7,\ & P(\mathrm{z}=0|\mathrm{y}=0)=0.3,\\
P(\mathrm{z}=1|\mathrm{y}=1)=0.9,\ & P(\mathrm{z}=0|\mathrm{y}=1)=0.1,\\
P(\phi=1|\mathrm{x}=0,\mathrm{w}=0)=0.25,\ & P(\phi=0|\mathrm{x}=0,\mathrm{w}=0)=0.75,\\
P(\phi=1|\mathrm{x}=1,\mathrm{w}=0)=0.3,\ & P(\phi=0|\mathrm{x}=1,\mathrm{w}=0)=0.7,\\
P(\phi=1|\mathrm{x}=0,\mathrm{w}=1)=0.2,\ & P(\phi=0|\mathrm{x}=0,\mathrm{w}=1)=0.8,\\
P(\phi=1|\mathrm{x}=1,\mathrm{w}=1)=0.4,\ & P(\phi=0|\mathrm{x}=1,\mathrm{w}=1)=0.6.
\end{aligned}$$

Compute the combination $x_*, y_*, z_*, \phi_*, w_*$, which results in the maximum of the joint probability

$$\begin{aligned}
P(x,y,z,\phi,w) &= P(z|y,x,\phi,w)P(y|x,\phi,w)P(\phi|x,w)P(x|w)P(w)\\
&= P(z|y)P(y|x)P(\phi|x,w)P(x)P(w),
\end{aligned}$$

which is the factorization imposed by the BN.

In order to apply the max-product rule, we first moralize the graph and then form a factor graph version, as shown in Figs. 15.28B and C, respectively. The factor nodes realize the following potential (factor) functions:

$$f_a(x, y) = P(y|x)P(x),$$
$$f_b(y, z) = P(z|y),$$
$$f_c(\phi, x, w) = P(\phi|x, w)P(w),$$

and obviously

$$P(x, y, z, \phi, w) = f_a(x, y) f_b(y, z) f_c(\phi, x, w).$$

Note that in this case, the normalizing constant $Z = 1$. Thus, the values these factor functions take, according to their previous definitions, are

$$f_a(x, y) : \begin{cases} f_a(1, 1) = 0.42, \\ f_a(1, 0) = 0.28, \\ f_a(0, 1) = 0.24, \\ f_a(0, 0) = 0.06, \end{cases} \quad f_b(y, z) : \begin{cases} f_b(1, 1) = 0.9, \\ f_b(1, 0) = 0.1, \\ f_b(0, 1) = 0.7, \\ f_b(0, 0) = 0.3, \end{cases}$$

$$f_c(\phi, x, w) : \begin{cases} f_c(1, 1, 1) = 0.32, \\ f_c(1, 1, 0) = 0.06, \\ f_c(1, 0, 1) = 0.48, \\ f_c(1, 0, 0) = 0.14, \\ f_c(0, 1, 1) = 0.16, \\ f_c(0, 1, 0) = 0.05, \\ f_c(0, 0, 1) = 0.64, \\ f_c(0, 0, 0) = 0.15. \end{cases}$$

Note that the number of possible values of a factor explodes by increasing the number of the involved variables.

Application of the max-product algorithm: Choose as root the node x. Then the nodes z, ϕ, and w become the leaves.

- Initialization:

$$\mu_{z \to f_b}(z) = 1, \quad \mu_{\phi \to f_c}(\phi) = 1, \ \mu_{w \to f_c}(w) = 1.$$

- Begin message passing:
  - $f_b \to$ y:

$$\mu_{f_b \to y}(y) = \max_z f_b(y, z) \mu_{z \to f_b}(z),$$

  or

$$\mu_{f_b \to y}(1) = 0.9, \quad \mu_{f_b \to y}(0) = 0.7,$$

  where the first one occurs at $z = 1$ and the second one at $z = 1$.

- $y \rightarrow f_a$:

$$\mu_{y\rightarrow f_a}(y) = \mu_{f_b\rightarrow y}(y),$$

or

$$\mu_{y\rightarrow f_a}(1) = 0.9, \quad \mu_{y\rightarrow f_a}(0) = 0.7.$$

- $f_a \rightarrow x$:

$$\mu_{f_a\rightarrow x}(x) = \max_y f_a(x, y)\mu_{y\rightarrow f_a}(y),$$

or

$$\mu_{f_a\rightarrow x}(1) = 0.42 \cdot 0.9 = 0.378,$$

which occurs for $y = 1$. Note that for $y = 0$, the value for $\mu_{f_a\rightarrow x}(1)$ would be $0.7 \cdot 0.28 = 0.196$, which is smaller than 0.378. Also,

$$\mu_{f_a\rightarrow x}(0) = 0.24 \cdot 0.9 = 0.216,$$

which also occurs for $y = 1$.

- $f_c \rightarrow x$:

$$\mu_{f_c\rightarrow x}(x) = \max_{w,\phi} f_c(\phi, x, w)\mu_{w\rightarrow f_c}(w)\mu_{\phi\rightarrow f_c}(\phi),$$

or

$$\mu_{f_c\rightarrow x}(1) = 0.48,$$

which occurs for $\phi = 0$ and $w = 1$, and

$$\mu_{f_c\rightarrow x}(0) = 0.64,$$

which occurs for $\phi = 0$ and $w = 1$.

- Obtain the optimal value:

$$x_* = \arg\max \mu_{f_a\rightarrow x}(x)\mu_{f_c\rightarrow x}(x),$$

or

$$x_* = 1,$$

and the corresponding maximum value is

$$\max P(x, y, z, w, \phi) = 0.378 \cdot 0.48 = 0.1814.$$

- Back-tracking:
  - Node f_c:

$$\max_{w,\phi} f_c(1, \phi, w)\mu_{w\rightarrow f_c}(w)\mu_{\phi\rightarrow f_c}(\phi),$$

which has occurred for

$$\phi_* = 0 \text{ and } w_* = 1.$$

- Node f_a:

$$\max_y f_a(1, y)\mu_{y \to f_a}(y),$$

which has occurred for

$$y_* = 1.$$

- Node f_b:

$$\max_z f_b(1, z)\mu_{z \to f_b}(z),$$

which has occurred for

$$z_* = 1.$$

Thus, the optimizing combination is

$$(x_*, y_*, z_*, \phi_*, w_*) = (1, 1, 1, 0, 1).$$

PROBLEMS

15.1 Show that in the product

$$\prod_{i=1}^{n}(1 - x_i),$$

the number of cross-product terms, $x_1, x_2, \ldots, x_k$, $1 \le k \le n$, for all possible combinations of $x_1, \ldots, x_n$ is equal to $2^n - n - 1$.

15.2 Prove that if a probability distribution p satisfies the Markov condition, as implied by a BN, then p is given as the product of the conditional distributions given the values of the parents.

15.3 Show that if a probability distribution factorizes according to a BN structure, then it satisfies the Markov condition.

15.4 Consider a DAG and associate each node with a random variable. Define for each node the conditional probability of the respective variable given the values of its parents. Show that the product of the conditional probabilities yields a valid joint probability and that the Markov condition is satisfied.

15.5 Consider the graph in Fig. 15.29. Random variable x has two possible outcomes, with probabilities $P(x_1) = 0.3$ and $P(x_2) = 0.7$. Variable y has three possible outcomes, with conditional probabilities

$$P(y_1|x_1) = 0.3, \; P(y_2|x_1) = 0.2, \; P(y_3|x_1) = 0.5,$$
$$P(y_1|x_2) = 0.1, \; P(y_2|x_2) = 0.4, \; P(y_3|x_2) = 0.5.$$

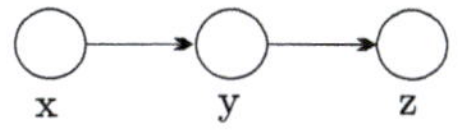

FIGURE 15.29

Graphical model for Problem 15.5.

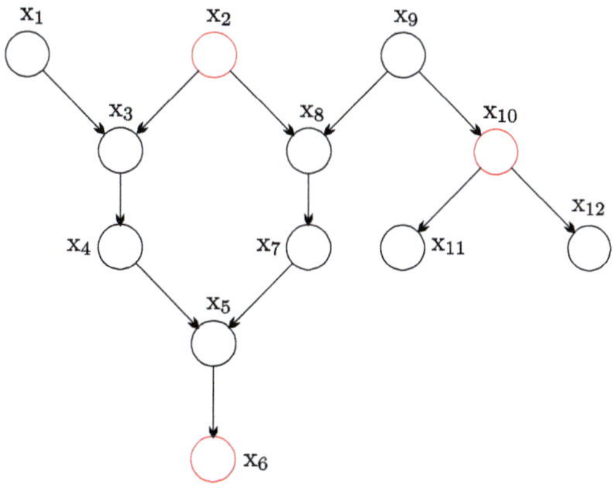

FIGURE 15.30

DAG for Problem 15.6. Nodes in red have been instantiated.

Finally, the conditional probabilities for z are

$$P(z_1|y_1) = 0.2, \quad P(z_2|y_1) = 0.8,$$
$$P(z_1|y_2) = 0.2, \quad P(z_2|y_2) = 0.8,$$
$$P(z_1|y_3) = 0.4, \quad P(z_2|y_3) = 0.6.$$

Show that this probability distribution, which factorizes over the graph, renders x and z independent. However, x and z in the graph are not d-separated because y is not instantiated.

15.6 Consider the DAG in Fig. 15.30. Detect the d-separations and d-connections in the graph.

15.7 Consider the DAG of Fig. 15.31. Detect the blanket of node x_5 and verify that if all the nodes in the blanket are instantiated, then the node becomes d-separated from the rest of the nodes in the graph.

15.8 In a linear Gaussian BN model, derive the mean values and the respective covariance matrices for each one of the variables in a recursive manner.

15.9 Assuming the variables associated with the nodes of the Bayesian structure of Fig. 15.32 to be Gaussian, find the respective mean values and covariances.

15.10 Prove that if p is a Gibbs distribution that factorizes over an MRF H, then H is an I-map for p.

15.11 Show that if H is the moral graph that results from moralization of a BN structure, then

$$I(H) \subseteq I(G).$$

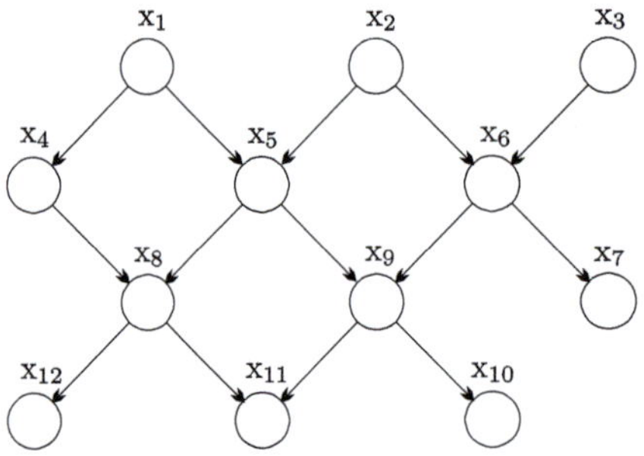

FIGURE 15.31

The graph structure for Problem 15.7.

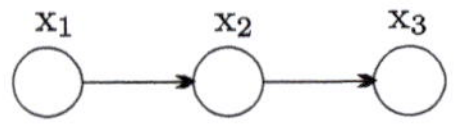

FIGURE 15.32

Network for Problem 15.9.

15.12 Consider a BN structure and a probability distribution p. Show that if $I(G) \subseteq I(p)$, then p factorizes over G.

15.13 Show that in an undirected chain graphical model, the marginal probability $P(x_j)$ of a node x_j is given by

$$P(x_j) = \frac{1}{Z}\mu_f(x_j)\mu_b(x_j),$$

where $\mu_f(x_j)$ and $\mu_b(x_j)$ are the forward and backward messages received by the node.

15.14 Show that the joint distribution of two neighboring nodes in an undirected chain graphical model is given by

$$P(x_j, x_{j+1}) = \frac{1}{Z}\mu_f(x_j)\psi_{j,j+1}(x_j, x_{j+1})\mu_b(x_{j+1}).$$

15.15 Using Fig. 15.26, prove that if there is a second message passing, starting from x, toward the leaves, then any node will have the available information for the computation of the respective marginals.

15.16 Consider the tree graph of Fig. 15.26. Compute the marginal probability $P(x_1, x_2, x_3, x_4)$.

15.17 Repeat the message passing procedure to find the optimal combination of variables for Example 15.4 using the logarithmic version and the max-sum algorithm.

REFERENCES

[1] S.M. Aji, R.J. McEliece, The generalized distributive law, IEEE Trans. Inf. Theory 46 (2000) 325–343.

[2] A. Al-Bashabsheh, Y. Mao, Normal factor graphs and holographic transformations, IEEE Trans. Inf. Theory 57 (4 February, 2011) 752–763.
[3] A. Al-Bashabsheh, Y. Mao, Normal factor graphs as probabilistic models, arXiv:1209.3300v1 [cs.IT], 14 September 2012.
[4] U. Bertele, F. Brioschi, Nonserial Dynamic Programming, Academic Press, Boston, 1972.
[5] J. Besag, Spatial interaction and the statistical analysis of lattice systems, J. R. Stat. Soc. B 36 (2) (1974) 192–236.
[6] C.E. Cannings, A. Thompson, M.H. Skolnick, The recursive derivation of likelihoods on complex pedigrees, Adv. Appl. Probab. 8 (4) (1976) 622–625.
[7] R. Chellappa, R.L. Kashyap, Digital image restoration using spatial interaction models, IEEE Trans. Acoust. Speech Signal Process. 30 (1982) 461–472.
[8] R. Chellappa, S. Chatterjee, Classification of textures using Gaussian Markov random field models, IEEE Trans. Acoust. Speech Signal Process. 33 (1985) 959–963.
[9] R. Chellappa, A.K. Jain (Eds.), Markov Random Fields: Theory and Applications, Academic Press, Boston, 1993.
[10] G.F. Cooper, The computational complexity of probabilistic inference using Bayesian belief networks, Artif. Intell. 42 (1990) 393–405.
[11] P. Dagum, M. Luby, Approximating probabilistic inference in Bayesian belief networks is NP-hard, Artif. Intell. 60 (1993) 141–153.
[12] A.P. Dawid, Conditional independence in statistical theory, J. R. Stat. Soc. B 41 (1978) 1–31.
[13] A.P. Dawid, Applications of a general propagation algorithm for probabilistic expert systems, Stat. Comput. 2 (1992) 25–36.
[14] A.P. Dawid, M. Studeny, Conditional products: an alternative approach to conditional independence, in: D. Heckerman, J. Whittaker (Eds.), Artificial Intelligence and Statistics, Morgan-Kaufmann, San Mateo, 1999.
[15] B.J. Frey, Graphical Models for Machine Learning and Digital Communications, MIT Press, Cambridge, MA, 1998.
[16] B.J. Frey, F.R. Kschishany, H.A. Loeliger, N. Wiberg, Factor graphs and algorithms, in: Proceedings of the 35th Alerton Conference on Communication, Control and Computing, 1999.
[17] G.D. Forney Jr., The Viterbi algorithm, Proc. IEEE 61 (1973) 268–277.
[18] G.D. Forney Jr., Codes on graphs: normal realizations, IEEE Trans. Inf. Theory 47 (2001) 520–548.
[19] G.D. Forney Jr., Codes on graphs: duality and MacWilliams identities, IEEE Trans. Inf. Theory 57 (3) (2011) 1382–1397.
[20] D. Geiger, T. Verma, J. Pearl, d-Separation: from theorems to algorithms, in: M. Henrion, R.D. Shachter, L.N. Kanal, J.F. Lemmer (Eds.), Proceedings 5th Annual Conference on Uncertainty in Artificial Intelligence, 1990.
[21] S. Geman, D. Geman, Stochastic relaxation, Gibbs distributions and the Bayesian restoration of images, IEEE Trans. Pattern Anal. Mach. Intell. 6 (1) (1984) 721–741.
[22] J.M. Hammersley, P. Clifford, Markov fields on finite graphs and lattices, unpublished manuscript available the web, 1971.
[23] G.E. Hinton, T. Sejnowski, Learning and relearning in Boltzmann machines, in: D.E. Rumelhart, J.L. McClelland (Eds.), Parallel Distributed Processing, vol. 1, MIT Press, Cambridge, MA, 1986.
[24] D. Janzing, B. Schölkopf, Causal inference using the algorithmic Markov condition, IEEE Trans. Inf. Theory 56 (2010) 5168–5194.
[25] D. Koller, N. Friedman, Probabilistic Graphical Models: Principles and Techniques, MIT Press, Cambridge, MA, 2009.
[26] F.R. Kschischang, B.J. Frey, H.A. Loeliger, Factor graphs and the sum-product algorithm, IEEE Trans. Inf. Theory 47 (2) (2001) 498–519.
[27] J. Lafferty, A. McCallum, F. Pereira, Conditional random fields: probabilistic models for segmenting and labeling sequence data, in: International Conference on Machine Learning, 2001, pp. 282–289.
[28] S.L. Lauritzen, D.J. Spiegelhalter, Local computations with probabilities on graphical structures and their application to expert systems, J. R. Stat. Soc. B 50 (1988) 157–224.
[29] S.Z. Li, Markov Random Field Modeling in Image Analysis, Springer-Verlag, New York, 2009.
[30] H.A. Loeliger, J. Dauwels, J. Hu, S. Korl, L. Ping, F.R. Kschischang, The factor graph approach to model-based signal processing, Proc. IEEE 95 (6) (2007) 1295–1322.
[31] D.J.C. MacKay, Information Inference and Learning Algorithms, Cambridge University Press, Cambridge, 2003.
[32] R.E. Neapolitan, Learning Bayesian Networks, Prentice Hall, Upper Saddle River, NJ, 2004.
[33] R.M. Neal, Connectionist learning of belief networks, Artif. Intell. 56 (1992) 71–113.
[34] F.A.N. Palmieri, Learning nonlinear functions with factor graphs, IEEE Trans. Signal Process. 61 (12) (2013) 4360–4371.
[35] F.A.N. Palmieri, A comparison of algorithms for learning hidden variables in normal graphs, arXiv:1308.5576v1 [stat.ML], 26 August 2013.
[36] J. Pearl, Fusion, propagation, and structuring in belief networks, Artif. Intell. 29 (1986) 241–288.

[37] J. Pearl, Probabilistic Reasoning in Intelligent Systems: Networks of Plausible Inference, Morgan-Kaufmann, San Mateo, 1988.
[38] J. Pearl, Causality, Reasoning and Inference, second ed., Cambridge University Press, Cambridge, 2012.
[39] L. Rabiner, A tutorial on hidden Markov models and selected applications in speech processing, Proc. IEEE 77 (1989) 257–286.
[40] U. Schmidt, Learning and Evaluating Markov Random Fields for Natural Images, Master's Thesis, Department of Computer Science, Technishe Universität Darmstadt, Germany, 2010.
[41] M.A. Shwe, G.F. Cooper, An empirical analysis of likelihood-weighting simulation on a large, multiply connected medical belief network, Comput. Biomed. Res. 24 (1991) 453–475.
[42] P. Spirtes, Introduction to causal inference, J. Mach. Learn. Res. 11 (2010) 1643–1662.
[43] X. Sun, D. Janzing, B. Schölkopf, Causal inference by choosing graphs with most plausible Markov kernels, in: Proceedings, 9th International Symposium on Artificial Intelligence and Mathematics, Fort Lauderdale, 2006, pp. 1–11.
[44] C. Sutton, A. McCallum, An introduction to conditional random fields, arXiv:1011.4088v1 [stat.ML], 17 November 2010.
[45] T. Verma, J. Pearl, Causal networks: semantics and expressiveness, in: R.D. Schachter, T.S. Levitt, L.N. Kanal, J.F. Lemmer (Eds.), Proceedings of the 4th Conference on Uncertainty in Artificial Intelligence, North-Holland, 1990.
[46] A.J. Viterbi, Error bounds for convolutional codes and an asymptotically optimum decoding algorithm, IEEE Trans. Inf. Theory IT-13 (1967) 260–269.
[47] J.W. Woods, Two-dimensional discrete Markovian fields, IEEE Trans. Inf. Theory 18 (2) (1972) 232–240.

CHAPTER

PROBABILISTIC GRAPHICAL MODELS: PART II

16

CONTENTS

16.1 INTRODUCTION

This is the follow-up to Chapter 15 and it builds upon the notions and models introduced there. The emphasis of this chapter is on more advanced topics for probabilistic graphical models. It wraps up the topic of exact inference in the context of junction trees and then moves on to introduce approximate inference techniques. This establishes a bridge with Chapter 13. Then, dynamic Bayesian networks are introduced, with an emphasis on hidden Markov models (HMMs). Inference and training of HMMs is seen as a special case of the message passing algorithm that was introduced in Chapter 15 and the EM scheme discussed in Chapter 12. Finally, the more general concept of training graphical models is briefly discussed.

Machine Learning. https://doi.org/10.1016/B978-0-12-818803-3.00028-3

16.2 TRIANGULATED GRAPHS AND JUNCTION TREES

In Chapter 15, we discussed three efficient schemes for exact inference in graphical entities of a tree structure. Our focus in this section is on presenting a methodology that can transform an arbitrary graph into an equivalent one having a tree structure. Thus, in principle, such a procedure offers the means for exact inference in arbitrary graphs. This transformation of an arbitrary graph to a tree involves a number of stages. Our goal is to present these stages and explain the procedure more via examples and less via formal mathematical proofs. A more detailed treatment can be obtained from more specialized sources (for example, [32,45]).

We assume that our graph is undirected. Thus, if the original graph was a directed one, then it is assumed that the moralization step has previously been applied.

Definition 16.1. An undirected graph is said to be *triangulated* if and only if for *every* cycle of length greater than three the graph possesses a *chord*. A chord is an edge joining two *nonconsecutive* nodes in the cycle.

In other words, in a triangulated graph, the largest "minimal cycle" is a triangle. Fig. 16.1A shows a graph with a cycle of length $n = 4$ and Figs. 16.1B and C show two triangulated versions; note that the process of triangulation does not lead to unique answers. Fig. 16.2A is an example of a graph with a cycle of $n = 5$ nodes. Fig. 16.2B, although it has an extra edge joining two nonconsecutive nodes, is not triangulated. This is because there still remains a chordless cycle of four nodes ($\mathrm{x}_2 - \mathrm{x}_3 - \mathrm{x}_4 - \mathrm{x}_5$). Fig. 16.2C is a triangulated version. There are no cycles of length $n > 3$ without a chord. Note that by joining nonconsecutive edges in order to triangulate a graph, we divide it into cliques (Section 15.4); we will appreciate this very soon. Figs. 16.1B and C comprise two (three-node) cliques and Fig. 16.2C comprises three cliques. This is not the case with Fig. 16.2B, where the subgraph ($\mathrm{x}_2, \mathrm{x}_3, \mathrm{x}_4, \mathrm{x}_5$) is not a clique.

Let us now see how the previous definition relates to the task of variable elimination, which underlies the message passing philosophy. In our discussion on such algorithmic schemes, we started with a node and marginalized out the respective variable (e.g., in the sum-product algorithm); as a matter of fact, this is not quite true. The message passing was initialized at the leaves of the tree graphs; this was done on purpose, although not explicitly stated there. We will soon realize why.

Consider Fig. 16.3 and let

$$P(\boldsymbol{x}) := \psi_1(x_1)\psi_2(x_1, x_2)\psi_3(x_1, x_3)\psi_4(x_2, x_4)\psi_5(x_2, x_3, x_5)\psi_6(x_3, x_6), \tag{16.1}$$

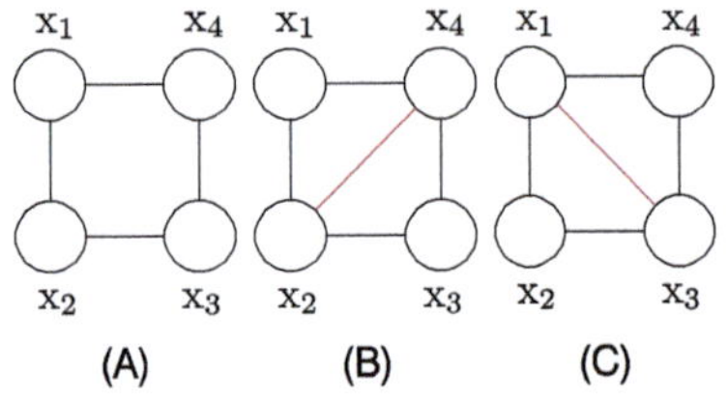

FIGURE 16.1

(A) A graph with a cycle of length $n = 4$. (B, C) Two possible triangulated versions.

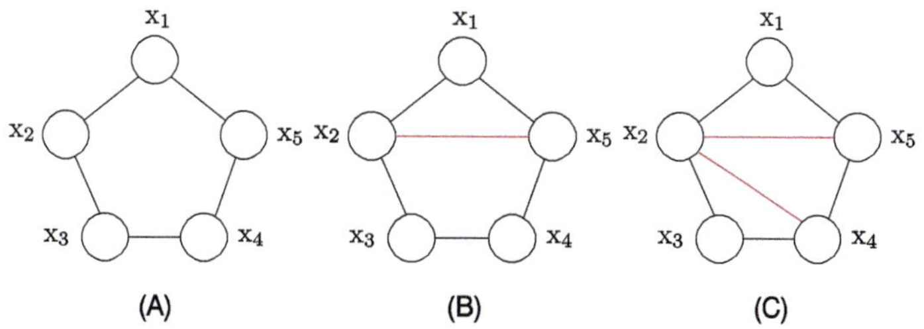

FIGURE 16.2

(A) A graph of cycle of length $n = 5$. (B) Adding one edge still leaves a cycle of length $n = 4$ chordless. (C) A triangulated version; there are no cycles of length $n > 3$ without a chord.

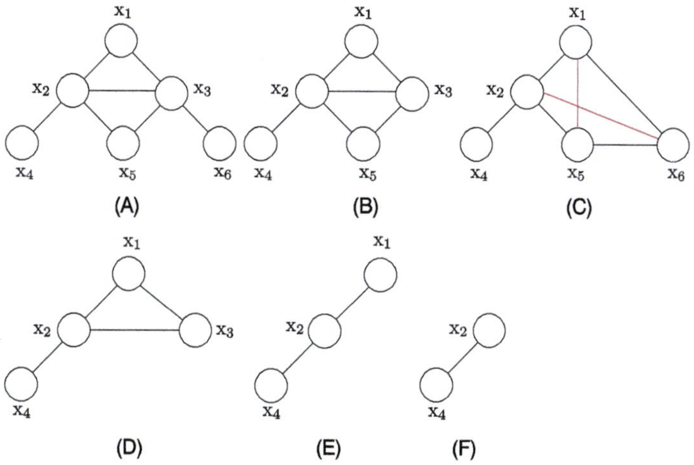

FIGURE 16.3

(A) An undirected graph with potential (factor) functions $\psi_1(x_1)$, $\psi_2(x_1, x_2)$, $\psi_3(x_1, x_3)$, $\psi_4(x_2, x_4)$, $\psi_5(x_2, x_3, x_5)$, and $\psi_6(x_3, x_6)$. (B) The graph resulting after the elimination of x_6. (C) The graph that would have resulted if the first node to be eliminated were x_3. Observe the fill-in edges denoted by red. (D–F) are the graphs that would result if the elimination process had continued from the topology shown in (B) and sequentially removing the nodes: x_5, x_3, and finally x_1.

assuming that $Z = 1$.

Let us eliminate x_6 first, that is,

$$\sum_{x_6} P(\boldsymbol{x}) = \psi^{(1)}(x_1, x_2, x_3, x_4, x_5) \sum_{x_6} \psi_6(x_3, x_6)$$
$$= \psi^{(1)}(x_1, x_2, x_3, x_4, x_5)\psi^{(3)}(x_3), \tag{16.2}$$

where the definitions of $\psi^{(1)}$ and $\psi^{(3)}$ are self-explained, by comparing Eqs. (16.1) and (16.2). The result of elimination is equivalent to a new graph, shown in Fig. 16.3B with $P(\boldsymbol{x})$ given as the product of the same potential functions as before with the exception of ψ_3, which is now replaced by the product $\psi_3(x_1, x_3)\psi^{(3)}(x_3)$. Basically, $\psi^{(3)}(\cdot)$ is the message passed to x_3.

In contrast, let us now start by eliminating x_3 first. Then we have

$$\sum_{x_3} P(\boldsymbol{x}) = \psi^{(2)}(x_1, x_2, x_4) \sum_{x_3} \psi_3(x_1, x_3)\psi_5(x_2, x_3, x_5)\psi_6(x_3, x_6)$$
$$= \psi^{(2)}(x_1, x_2, x_4)\tilde{\psi}^{(3)}(x_1, x_2, x_5, x_6).$$

Note that this summation is more difficult to perform. It involves four variables $(\mathrm{x}_1, \mathrm{x}_2, \mathrm{x}_5, \mathrm{x}_6)$ besides x_3, which requires many more combination terms be computed than before. Fig. 16.3C shows the resulting equivalent graph, after eliminating x_3. Due to the resulting factor $\tilde{\psi}^{(3)}(x_1, x_2, x_5, x_6)$ *new* connections implicitly appear, known as *fill-in edges*. This is not a desired situation, as it introduces factors depending on new combinations of variables. Moreover, the new factor depends on four variables, and we know that the larger the number of variables, or the *domain* of the factor as we say, the larger the number of terms involved in the summations, which increases the computational load.

Thus, the choice of the sequence of elimination is very important and far from innocent. For example, for the case of Fig. 16.3A an elimination sequence that does not introduce fill-ins is the following: $\mathrm{x}_6, \mathrm{x}_5, \mathrm{x}_3, \mathrm{x}_1, \mathrm{x}_2, \mathrm{x}_4$. For such an elimination sequence, every time a variable is eliminated, the new graph results from the previous one by just removing one node. This is shown by the sequence of graphs in Figs. 16.3A, B, and D–F, for the case of the previously given elimination sequence. An elimination sequence that *does not* introduce fill-ins is known as a *perfect elimination sequence*.

Proposition 16.1. *An undirected graph is triangulated if and only if it has a perfect elimination sequence (for example, [32]).*

Definition 16.2. A tree T is said to be a *join tree* if (a) its nodes correspond to the cliques of an (undirected) graph G *and* (b) the intersection of *any* two nodes, $U \cap V$, is contained in every node in the *unique* path between U and V. The latter property is also known as the *running intersection property*.

Moreover, if a probability distribution p factorizes over G so that each of the product factors (potential functions) is attached to a clique (i.e., depends only on variables associated with the nodes in the clique), then the join tree is said to be a *junction tree* for p [7].

Example 16.1. Consider the triangulated graph of Fig. 16.2C. It comprises three cliques, namely, $(\mathrm{x}_1, \mathrm{x}_2, \mathrm{x}_5)$, $(\mathrm{x}_2, \mathrm{x}_3, \mathrm{x}_4)$, and $(\mathrm{x}_2, \mathrm{x}_4, \mathrm{x}_5)$. Associating each clique with a node of a tree, Fig. 16.4 presents three possibilities. The trees in Fig. 16.4A and B are not join trees. Indeed, the intersection $\{\mathrm{x}_1, \mathrm{x}_2, \mathrm{x}_5\} \cap \{\mathrm{x}_2, \mathrm{x}_4, \mathrm{x}_5\} = \{\mathrm{x}_2, \mathrm{x}_5\}$ does not appear in node $(\mathrm{x}_2, \mathrm{x}_3, \mathrm{x}_4)$. Similar arguments hold true for the case of Fig. 16.4B. In contrast, the tree in Fig. 16.4C is a join tree, because the intersection $\{\mathrm{x}_1, \mathrm{x}_2, \mathrm{x}_5\} \cap \{\mathrm{x}_2, \mathrm{x}_3, \mathrm{x}_4\} = \{\mathrm{x}_2\}$ is contained in $(\mathrm{x}_2, \mathrm{x}_4, \mathrm{x}_5)$. If, now, we have a distribution such as

$$p(\boldsymbol{x}) = \psi_1(x_1, x_2, x_5)\psi_2(x_2, x_3, x_4)\psi_3(x_2, x_4, x_5),$$

the graph in Fig. 16.4C is a junction tree for $p(\boldsymbol{x})$.

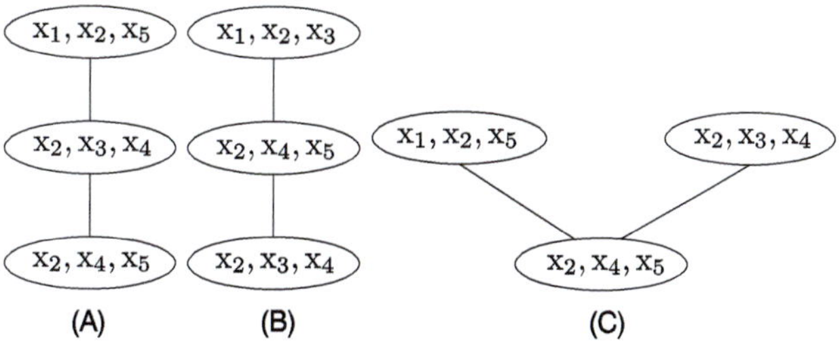

FIGURE 16.4

The graphs resulting from Fig. 16.2C and shown in (A) and (B) are not join trees. (C) This is a join tree, because the node in the path from (x_1, x_2, x_5) to (x_2, x_3, x_4) contains their intersection, x_2.

We are now ready to state the basic theorem of this section; the one that will allow us to transform an arbitrary graph into a graph of a tree structure.

Theorem 16.1. *An undirected graph is triangulated if and only if its cliques can be organized into a join tree (Problem 16.1).*

Once a triangulated graph, which is associated with a factorized probability distribution $p(\boldsymbol{x})$, has been transformed into a junction tree, then any of the message passing algorithms, described in Chapter 15, can be adopted to perform exact inference.

16.2.1 CONSTRUCTING A JOIN TREE

Starting from a triangulated graph, the following algorithmic steps construct a join tree ([32]):

- Select a node in a maximal clique of the triangulated graph, which is *not* shared by other cliques. Eliminate this node and keep removing nodes from the clique, as long as they are *not* shared by other cliques. Denote the set of the remaining nodes of this clique as S_i, where i is the number of the nodes eliminated so far. This set is called a *separator*. Use V_i to denote the set of all the nodes in the clique, prior to the elimination process.
- Select another maximal clique and repeat the process with the index counting the node elimination starting from i.
- Continue the process until all cliques have been eliminated. Once the previous peeling off procedure has been completed, join together the parts that have resulted, so that each separator, S_i, is joined to V_i on one of its sides and to a clique node (set) V_j, $(j > i)$, such that $S_i \subset V_j$. This is in line with the running intersection property. It can be shown that the resulting graph is a join tree (part of the proof in Problem 16.1).

An alternative algorithmic path to construct a join tree, once the cliques have been formed, is the following. Build an undirected graph having as nodes the maximal cliques of the triangulated graph. For each pair of linked nodes, V_i, V_j, assign a weight w_{ij} on the respective edge equal to the cardinality of $V_i \cap V_j$. Then run the *maximal spanning tree* algorithm (e.g., [43]) to identify a tree in this graph such that the sum of weights is maximal [41]. It turns out that such a procedure guarantees the running intersection property.

Example 16.2. Consider the graph of Fig. 16.5, which is described in the seminal paper [44]. Smoking can cause lung cancer or bronchitis. A recent visit to Asia increases the probability of tuberculosis. Both tuberculosis and cancer can result in a positive X-ray finding. Also, all three diseases can cause difficulty in breathing (dyspnea). In the context of the current example, we are not interested in the values of the respective probability table, and our goal is to construct a join tree, following the previous algorithm. Fig. 16.6 shows a triangulated graph that corresponds to Fig. 16.5.

The elimination sequence of the nodes in the triangulated graph is graphically illustrated in Fig. 16.7. First, node A is eliminated from the clique (A, T) and the respective separator set comprises T. Because only one node can be eliminated ($i = 1$), we indicate the separator as S_1. Next, node T is eliminated from the clique (T, L, E) and the S_2 ($i = i + 1$) separator comprises L, E. The process continues until clique (B, D, E) is the only remaining one. It is denoted as V_8, as all three nodes can

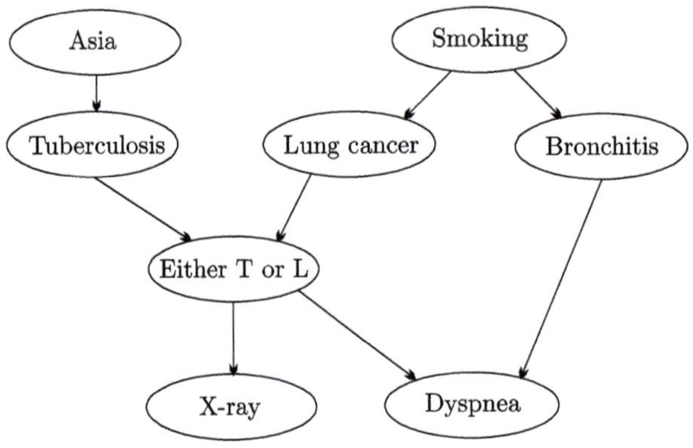

FIGURE 16.5

The Bayesian network structure of the example given in [44].

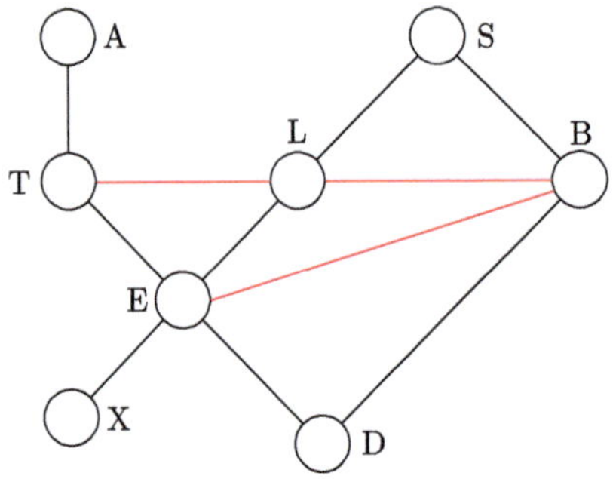

FIGURE 16.6

The graph resulting from the Bayesian network structure of Fig. 16.5, after having been moralized and triangulated. The inserted edges are drawn in red.

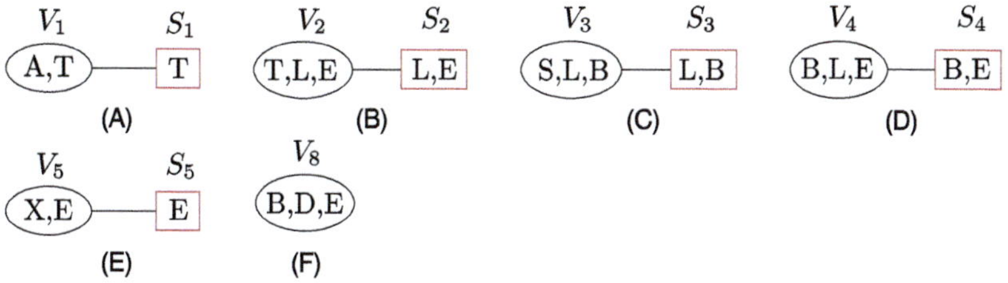

FIGURE 16.7

The sequence of elimination of nodes from the respective cliques of Fig. 16.6 and the resulting separators.

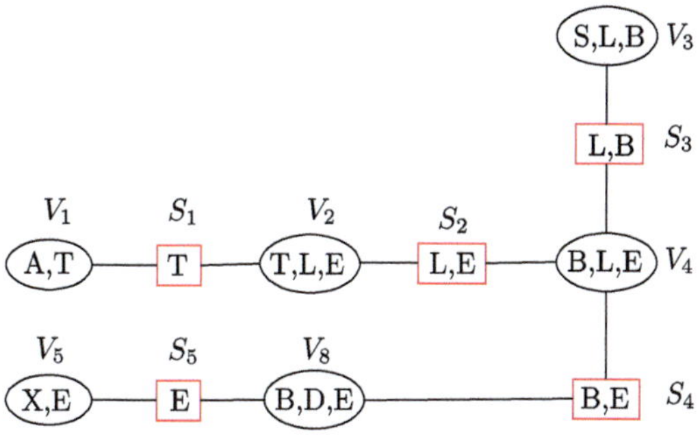

FIGURE 16.8

The resulting join tree from the graph of Fig. 16.6. A separator S_i is linked to a clique V_j $(j > i)$ so that $S_i \subset V_j$.

be eliminated sequentially (hence, $8 = 5 + 3$); there is no other neighboring clique. Fig. 16.8 shows the resulting junction tree. Verify the running intersection property.

16.2.2 MESSAGE PASSING IN JUNCTION TREES

By its definition, a junction tree is a join tree where we have associated a factor, say, ψ_c, of a probability distribution, p, with each one of the cliques. Each factor can be considered as the product of all potential functions, which are defined in terms of the variables associated with the nodes of the corresponding clique; hence, the domain of each one of these potential functions is a subset of the variables-nodes comprising the clique. Then, focusing on the discrete probability case, we can write

$$P(\boldsymbol{x}) = \frac{1}{Z} \prod_{c} \psi_c(\boldsymbol{x}_c), \tag{16.3}$$

where c runs over the cliques and $\boldsymbol{x}_c$ denotes the variables comprising the respective clique. Because a junction tree is a graph with a tree structure, exact inference can take place in the same way as we have already discussed in Section 15.7, via a message passing rationale. A two-way message passing is also required here. There are, however, some small differences. In the case of the factor graphs, which we have considered previously in Chapter 15, the exchanged messages were functions of one variable. This is not necessarily the case here. Moreover, after the bidirectional flow of the messages has been completed, what is recovered from each node of the junction tree is the *joint probability of the variables associated with the clique,* $P(\boldsymbol{x}_c)$. The computation of the marginal probabilities for individual variables requires extra summations in order to marginalize with respect to the rest.

Note that in the message passing, the following take place:

- A separator receives messages and passes their product to one of its connected cliques, depending on the direction of the message passing flow, that is,

$$\boxed{\mu_{S\to V}(\boldsymbol{x}_S) = \prod_{v\in\mathcal{N}(S)\setminus V} \mu_{v\to S}(\boldsymbol{x}_S),} \tag{16.4}$$

  where $\mathcal{N}(S)$ is the index set of the clique nodes connected to S and $\mathcal{N}(S)\setminus V$ is this set excluding the index for clique node V. Note that the message is a function of the variables comprising the separator.
- Each clique node performs marginalization and passes the message to each one of its connected separators, depending on the direction of the flow. Let V be a clique node and $\boldsymbol{x}_V$ the vector of the involved variables in it, and let S be a separator node connected to it. The message passed to S is given by

$$\boxed{\mu_{V\to S}(\boldsymbol{x}_S) = \sum_{\boldsymbol{x}_V\setminus \boldsymbol{x}_S} \psi_V(\boldsymbol{x}_V) \prod_{s\in\mathcal{N}(V)\setminus S} \mu_{s\to V}(\boldsymbol{x}_s).} \tag{16.5}$$

By $\boldsymbol{x}_S$ we denote the variables in the separator S. Obviously, $\boldsymbol{x}_S \subset \boldsymbol{x}_V$ and $\boldsymbol{x}_V\setminus\boldsymbol{x}_S$ denotes all variables in $\boldsymbol{x}_V$ excluding those in $\boldsymbol{x}_S$; $\mathcal{N}(V)$ is the index set of all separators connected to V and $\boldsymbol{x}_s$, the set of variables in the respective separator ($\boldsymbol{x}_s \subset \boldsymbol{x}_V$, $s\in\mathcal{N}(V)$); $\mathcal{N}(V)\setminus S$ denotes the index set of all separators connected to V excluding S. This is basically the counterpart of Eq. (15.50). Fig. 16.9 shows the respective configuration.

Once the two-way message passing has been completed, marginals in the clique as well as the separator nodes are computed as (Problem 16.3):

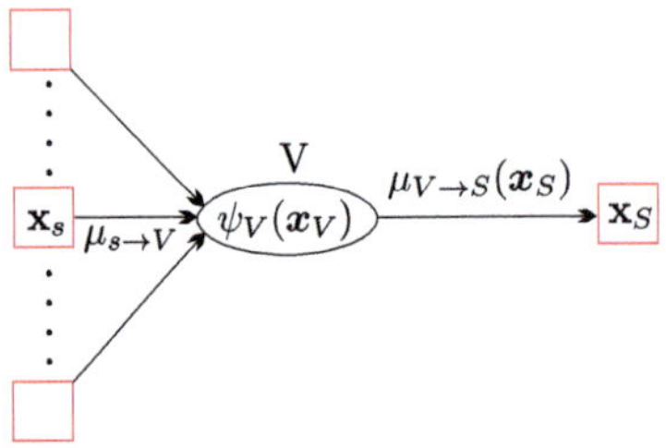

FIGURE 16.9

Clique node V "collects" all incoming messages from the separators it is connected with (except S); then it outputs a message to S, after the marginalization performed on the product of $\psi_V(\boldsymbol{x}_V)$ with the incoming messages.

- Clique nodes:

$$P(\boldsymbol{x}_V) = \frac{1}{Z}\psi_V(\boldsymbol{x}_V) \prod_{s \in \mathcal{N}(V)} \mu_{s \to V}(\boldsymbol{x}_s). \tag{16.6}$$

- Separator nodes: Each separator is connected only to clique nodes. After the two-way message passing, every separator has received messages from both flow directions. Then it is shown that

$$P(\boldsymbol{x}_S) = \frac{1}{Z} \prod_{v \in \mathcal{N}(S)} \mu_{v \to S}(\boldsymbol{x}_S). \tag{16.7}$$

An important by-product of the previous message passing algorithm in junction trees concerns the joint distribution of all the involved variables, which turns out to be independent of Z, and it is given by (Problem 16.4)

$$\boxed{P(\boldsymbol{x}) = \frac{\prod_v P_v(\boldsymbol{x}_v)}{\prod_s [P_s(\boldsymbol{x}_s)]^{d_s - 1}},} \tag{16.8}$$

where $\prod_v$ and $\prod_s$ run over the sets of clique nodes and separators, respectively, and d_s is the number of the cliques separator S is connected to.

Example 16.3. Let us consider the junction tree of Fig. 16.8. Assume that $\psi_1(A,T)$, $\psi_2(T,L,E)$, $\psi_3(S,L,B)$, $\psi_4(B,L,E)$, $\psi_5(X,E)$, and $\psi_6(B,D,E)$ are known. For example,

$$\psi_1(A,T) = P(T|A)P(A)$$

and

$$\psi_3(S,L,B) = P(L|S)P(B|S)P(S).$$

The message passing can start from the leaves, (A,T) and (X,E), toward (S,L,B); once this message flow has been completed, message passing takes place in the reverse direction. Some examples of message computations are given below.

The message received by node (T,L,E) is equal to

$$\mu_{S_1 \to V_2}(T) = \sum_A \psi_1(A,T).$$

Also,

$$\mu_{V_2 \to S_2}(L,E) = \sum_T \psi_2(T,L,E)\mu_{S_1 \to V_2}(T) = \mu_{S_2 \to V_4}(L,E),$$

and

$$\mu_{V_4 \to S_3}(L,B) = \sum_E \psi_4(B,L,E)\mu_{S_2 \to V_4}(L,E)\mu_{S_4 \to V_4}(B,E).$$

The rest of the messages are computed in a similar way.

For the marginal probability $P(T, L, E)$ of the variables in clique node V_2, we get

$$P(T, L, E) = \psi_{V_2}(T, L, E)\mu_{S_2 \to V_2}(L, E)\mu_{S_1 \to V_2}(T).$$

Observe that in this product, all other variables, besides T, L, and E, have been marginalized out. Also,

$$P(L, E) = \mu_{V_4 \to S_2}(L, E)\mu_{V_2 \to S_2}(L, E).$$

Remarks 16.1.

- Note that a variable is part of more than one node in the tree. Hence, if one is interested in obtaining the marginal probability of an individual variable, this can be obtained by marginalizing over different variables in different nodes. The properties of the junction tree guarantee that all of them give the same result (Problem 16.5).
- We have already commented that there is not a unique way to triangulate a graph. A natural question is now raised: Are all the triangulated versions equivalent from a computational point of view? Unfortunately, the answer is no. Let us consider the simple case where all the variables have the same number of possible states, k. Then the number of probability values for each clique node depends on the number of variables involved in it, and we know that this dependence is of an exponential form. Thus, our goal while triangulating a graph should be to implement it in such a way that the resulting cliques are as small as possible with respect to the number of nodes-variables involved. Let us define the size of a clique, V_i, as $s_i = k^{n_i}$, where n_i denotes the number of nodes comprising the clique. Ideally, we should aim at obtaining a triangulated version (or equivalently an elimination sequence) so that the total size of the triangulated graph, $\sum_i s_i$, where i runs over all cliques, is minimum. Unfortunately, this is an NP-hard task [1]. One of the earliest algorithms proposed to obtain low-size triangulated graphs is given in [71]. A survey of related algorithms is provided in [39].

16.3 APPROXIMATE INFERENCE METHODS

So far, our focus has been on presenting efficient algorithms for exact inference in graphical models. Although such schemes form the basis of inference and have been applied in a number of applications, often one encounters tasks where exact inference is not practically possible. At the end of the previous section, we discussed the importance of small-sized cliques. However, in a number of cases, the graphical model may be so densely connected that it renders the task of obtaining cliques of a small size impossible. We will soon consider some examples.

In such cases, resorting to methods for tractable approximate inference is the only viable alternative. Obviously, there are various paths to approach this problem and a number of techniques have been proposed. Our goal in this section is to discuss the main directions that are currently popular. Our approach will be more on the descriptive side than that of rigorous mathematical proofs and theorems. The reader who is interested in delving deeper into this topic can refer to more specialized references, which are given in the text below.

16.3.1 VARIATIONAL METHODS: LOCAL APPROXIMATION

The current and the next subsections draw a lot of their theoretical basis on the variational approximation methods, which were introduced in Chapter 13 and in particular in Sections 13.2 and 13.8.

The main goal in variational approximation methods is to replace probability distributions with computationally attractive bounds. The effect of such *deterministic* approximation methods is that it simplifies the computations; as we will soon see, this is equivalent to simplifying the graphical structure. Yet, these simplifications are carried out in the context of an associated optimization process. The functional form of these bounds is very much problem dependent, so we will demonstrate the methodology via some selected examples.

Two main directions are followed: the *sequential* one and the *block* one [34]. The former will be treated in this subsection and the latter in the next one.

In the sequential methods, the approximation is imposed on individual nodes in order to modify the functional form of the local probability distribution functions. This is the reason we called them *local methods*. One can impose the approximation on some of the nodes or to all of them. Usually, some of the nodes are selected, whose number is sufficient so that exact inference can take place with the remaining ones, within practically acceptable computational time and memory size. An alternative viewpoint is to look at the method as a sparsification procedure that removes nodes so as to transform the original graph to a "computationally" manageable one. There are different scenarios on how to select nodes. One way is to introduce approximation to one node at a time until a sufficiently simplified structure occurs. The other way is to introduce approximation to all the nodes and then reinstate the exact distributions one node at a time. The latter of the two has the advantage that the network is computationally tractable all the way (see, for example, [30]). Local approximations are inspired by the method of bounding convex/concave functions in terms of their conjugate ones, as discussed in Section 13.8. Let us now unveil the secrets behind the method.

Multiple-Cause Networks and the Noisy-OR Model

In the beginning of Chapter 15 (Section 15.2) we presented a simplified case from the medical diagnosis field, concerning a set of diseases and findings. Adopting the so-called noisy-OR model, we arrived at Eqs. (15.4) and (15.5), which are repeated here for convenience. We have

$$P(f_i = 0|\boldsymbol{d}) = \exp\left(-\sum_{j\in \text{Pa}_i} \theta_{ij} d_j\right), \tag{16.9}$$

$$P(f_i = 1|\boldsymbol{d}) = 1 - \exp\left(-\sum_{j\in \text{Pa}_i} \theta_{ij} d_j\right), \tag{16.10}$$

where we have exploited the experience we have gained so far and we have introduced in the notation the set of the parents Pa_i of the ith finding. The respective graphical model belongs to the family of multiple-cause networks (Section 15.3.6) and it is shown in Fig. 16.10A. We will now pick a specific node, say the ith node, assume that it corresponds to a positive finding ($f_i = 1$), and demonstrate how the variational approximation method can offer a way out from the "curse" of the exponential dependence of the joint probability on the number of the involved terms; recall that this is caused by the form of Eq. (16.10).

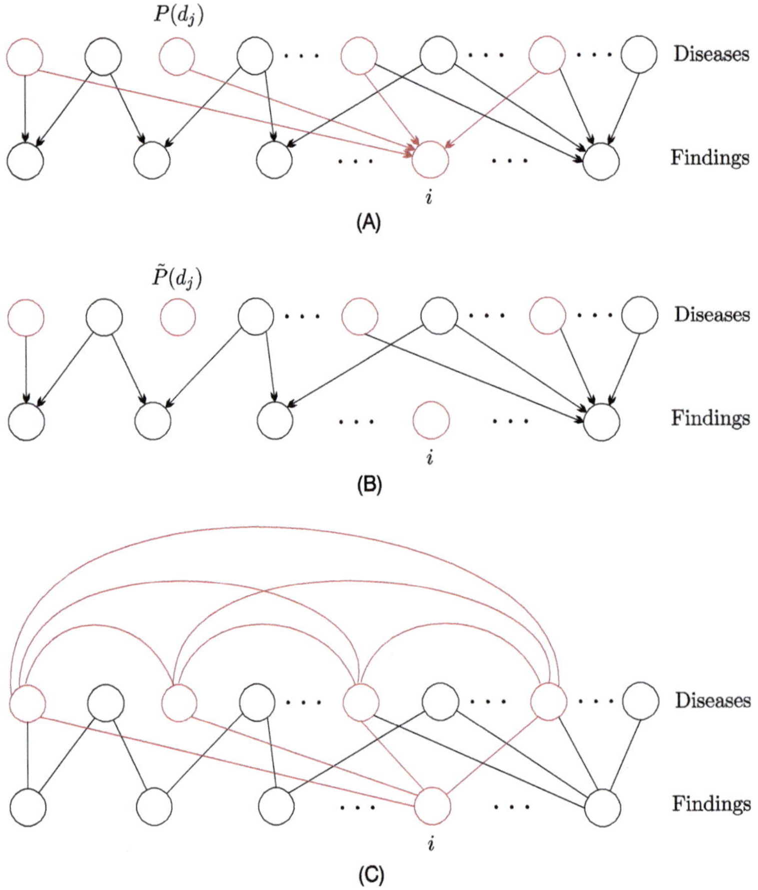

FIGURE 16.10

(A) A Bayesian network for a set of findings and diseases. To simplify the figure only a few connections are shown. The node associated with the ith finding, together with its parents and respective edges, are shown in red; it is the node on which variational approximation is introduced. (B) After the variational approximation is performed for node i, the edges joining it with its parents are removed. At the same time, the prior probabilities of the respective parent nodes change values. This is shown in the figure for the jth disease. (C) The graph that would have resulted after the moralization step, focusing on node i.

Derivation of the variational bound: The function $1 - \exp(-x)$ belongs to the so-called *log-concave* family of functions, meaning that

$$f(x) = \ln(1 - \exp(-x)), \quad x > 0,$$

is concave (Problem 16.9). Being a concave function, we know from Section 13.8 that it is upper bounded by

$$f(x) \leq \xi x - f^*(\xi),$$

where $f^*(\xi)$ is its conjugate function. Tailoring it to the needs of Eq. (16.10) and using ξ_i in place of ξ, to explicitly indicate the dependence on node i, we obtain

$$P(f_i = 1|\boldsymbol{d}) \leq \exp\left(\xi_i \left(\sum_{j \in \text{Pa}_i} \theta_{ij} d_j\right) - f^*(\xi_i)\right), \tag{16.11}$$

or

$$P(f_i = 1|\boldsymbol{d}) \leq \exp\left(-f^*(\xi_i)\right) \prod_{j \in \text{Pa}_i} \left(\exp(\xi_i \theta_{ij})\right)^{d_j}, \tag{16.12}$$

where (Problem 16.10)

$$f^*(\xi_i) = -\xi_i \ln(\xi_i) + (\xi_i + 1)\ln(\xi_i + 1), \quad \xi_i > 0.$$

Note that usually, a constant θ_{i0} is also present in the linear terms ($\sum_{j \in \text{Pa}_i} \theta_{ij} d_j + \theta_{i0}$) and in this case the first exponent in the upper bound becomes $\exp(-f^*(\xi_i) + \xi_i \theta_{i0})$.

Let us now observe Eq. (16.12). The first factor on the right-hand side is a constant, once ξ_i is determined. Moreover, each one of the factors, $\exp(\xi_i \theta_{ij})$, is also a constant raised in d_j. Thus, substituting Eq. (16.12) in the products in Eq. (15.1), in order to compute, for example, Eq. (15.3), each one of these constants can be absorbed by the respective $P(d_j)$, that is,

$$\tilde{P}(d_j) \propto P(d_j) \exp(\xi_i \theta_{ij} d_j), \quad j \in \text{Pa}_i.$$

Basically, from a graphical point of view, we can equivalently consider that the ith node is delinked and its influence on any subsequent processing is via the modified factors associated with its parent nodes (see Fig. 16.10B). In other words, the variational approximation *decouples* the parent nodes. In contrast, for exact inference, during the moralization stage, all parents of node i are connected. This is the source of computational explosion (see Fig. 16.10C). The idea is to remove a sufficient number of nodes, so that the remaining network can be handled using exact inference methods.

There is still a main point to be addressed: how the various ξ_is are obtained. These are computed to make the bound as tight as possible, and any standard optimization technique can be used. Note that this minimization corresponds to a convex cost function (Problem 16.11). Besides the upper bound, a lower bound can also be derived [27]. Experiments performed in [30] verify that reasonably good accuracies can be obtained in affordable computational times. The method was first proposed in [27].

The Boltzmann Machine

The Boltzmann machine, which was introduced in Section 15.4.2, is another example where any attempt for exact inference is confronted with cliques of sizes that make the task computationally intractable [28].

We will demonstrate the use of the variational approximation in the context of the computation of the normalizing constant Z. Recall from Eq. (15.23) that

$$Z = \sum_{x} \exp\left(-\sum_{i}\left(\sum_{j>i}\theta_{ij}x_i x_j + \theta_{i0}x_i\right)\right)$$
$$= \sum_{x\setminus x_k}\sum_{x_k=0}^{1} \exp\left(-\sum_{i}\left(\sum_{j>i}\theta_{ij}x_i x_j + \theta_{i0}x_i\right)\right), \tag{16.13}$$

where we chose node x_k to impose variational approximation. We split the summation into two, one with regard to x_k and one with regard to the rest of the variables; $x\setminus x_k$ denotes summation over all variables excluding x_k. Performing the inner sum in Eq. (16.13) (terms different to x_k and $x_k = 0$, $x_k = 1$), we get

$$Z = \sum_{x\setminus x_k} \exp\left(-\sum_{i\neq k}\left(\sum_{i<j\neq k}\theta_{ij}x_i x_j + \theta_{i0}x_i\right)\right)\left(1 + \exp\left(-\sum_{i\neq k}\theta_{ki}x_i - \theta_{k0}\right)\right),$$

where $i < j \neq k$ indicates that both i and j are different from k.

Derivation of the variational bound: The function $1 + \exp(-x)$, $x \in \mathbb{R}$, is *log-convex* (Problem 16.12); thus, following similar arguments to those adopted for the derivation of the bound in Eq. (16.12), but for convex instead of concave functions, we obtain

$$Z \geq \sum_{x\setminus x_k} \exp\left(-\sum_{i\neq k}\left(\sum_{i<j\neq k}\theta_{ij}x_i x_j + \theta_{i0}x_i\right)\right)$$
$$\times \exp\left(\xi_k\left(\sum_{i\neq k}\theta_{ki}x_i + \theta_{k0}\right) - f^*(\xi_k)\right), \tag{16.14}$$

where $f^*(\xi)$ is the respective conjugate function (Problem 16.12). Note that the second exponential in the bound can be combined with the first one and we can write

$$Z \geq \exp\left(-f^*(\xi_k) + \xi_k\theta_{k0}\right) \sum_{x\setminus x_k} \exp\left(-\sum_{i\neq k}\left(\sum_{i<j\neq k}\tilde{\theta}_{ij}x_i x_j + \tilde{\theta}_{i0}x_i\right)\right),$$

where

$$\tilde{\theta}_{ij} = \theta_{ij}, \quad i \neq k,\ j \neq k,$$

and

$$\tilde{\theta}_{i0} = \theta_{i0} - \xi_k\theta_{ki}, \quad i \neq k.$$

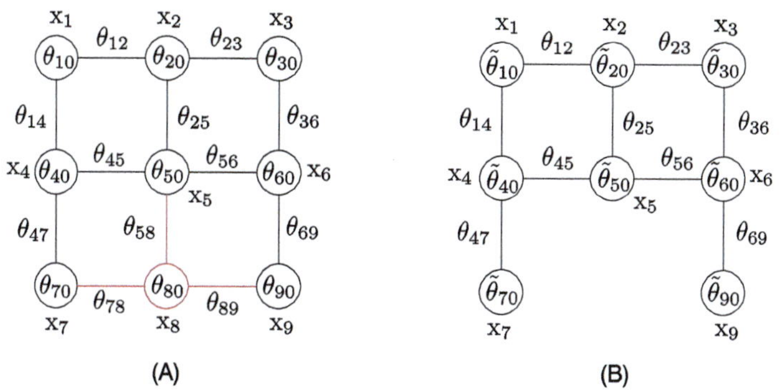

FIGURE 16.11

(A) An MRF corresponding to a Boltzmann machine. (B) The resulting MRF after removing x_8 via variational approximation. Note that if x_5 is removed next, then the remaining graphical model is a chain.

In other words, if from now on we replace Z with the bound, it is as if node x_k has been removed and the remaining network is a Boltzmann machine with one node less and the respective parameters modified, compared to the original ones. The value of ξ_k can be obtained via optimization so as to make the bound as tight as possible.

Fig. 16.11 illustrates the effect of applying variational approximation to node x_8. For the case of this figure, note that if x_5 is removed next (after x_8) the remaining graphical structure becomes a chain and exact inference can be carried out.

Following exactly similar arguments and employing the conjugate of the sigmoid function (Problem 13.18), one can apply the technique to the sigmoidal Bayesian networks, discussed in Section 15.3.4 (see also [27]).

16.3.2 BLOCK METHODS FOR VARIATIONAL APPROXIMATION

In contrast to the previous approach, where the approximation is introduced on selected nodes individually, here the approximation is imposed on a set of nodes. Once more, a derived bound of the involved probability distribution is optimized. In principle, the method is equivalent to imposing a specific graphical structure on the nodes, which can then be addressed by tractable exact inference techniques. In the sequel, the family of distributions, which can be factorized over this simplified substructure, is optimized with respect to a set of variational parameters. The method builds upon the same arguments as those used in Section 13.2. We will retain the same notation and we will provide the related formulas for the discrete variable case, to be used later on in our selected examples.

Let $\mathcal{X}$ be the set of observed and $\mathcal{X}^l$ the set of the latent random variables associated with the nodes of a graphical structure; in the graphical model terminology we can refer to them as evidence and hidden nodes, respectively. Define

$$\mathcal{F}(Q) = \sum_{x \in \mathcal{X}^l} Q(\mathcal{X}^l) \ln \frac{P(\mathcal{X}, \mathcal{X}^l)}{Q(\mathcal{X}^l)}, \tag{16.15}$$

where Q is any probability function. Then, from Eq. (16.15), we readily obtain

$$\mathcal{F}(Q) = \ln P(\mathcal{X}) + \sum_{x \in \mathcal{X}^l} Q(\mathcal{X}^l) \ln \frac{P(\mathcal{X}^l|\mathcal{X})}{Q(\mathcal{X}^l)},$$

or

$$\ln P(\mathcal{X}) = \mathcal{F}(Q) + \sum_{x \in \mathcal{X}^l} Q(\mathcal{X}^l) \ln \frac{Q(\mathcal{X}^l)}{P(\mathcal{X}^l|\mathcal{X})}. \tag{16.16}$$

Note that the second term in Eq. (16.16) is the Kullback–Leibler (KL) divergence between $P(\mathcal{X}^l|\mathcal{X})$ and $Q(\mathcal{X}^l)$. Because KL divergence is always nonnegative (Problem 12.7), we can write

$$\ln P(\mathcal{X}) \geq \mathcal{F}(Q),$$

and the lower bound is maximized if we minimize the KL divergence.

Let us now return to our goal, i.e., given the evidence, to perform inference on the graph associated with $P(\mathcal{X}^l|\mathcal{X})$. If this cannot be performed in a tractable way, the method adopts an approximation, $Q(\mathcal{X}^l)$, of $P(\mathcal{X}^l|\mathcal{X})$ and at the same time imposes a specific factorization on $Q(\mathcal{X}^l)$ (which equivalently induces a specific graphical structure) so that exact inference techniques can be employed. From the adopted family of distributions, we choose the one that minimizes the KL divergence between $P(\mathcal{X}^l|\mathcal{X})$ and $Q(\mathcal{X}^l)$. Such a choice guarantees the maximum lower bound for the log-evidence function. Among the different ways of factorization, the so-called *mean field* factorization is the simplest and, possibly, the most popular. The imposed structure on the graph has no edges, which leads to a complete factorization of $Q(\mathcal{X}^l)$, that is,

$$\boxed{Q(\mathcal{X}^l) = \prod_{i:x_i \in \mathcal{X}^l} Q_i(x_i): \quad \text{mean field factorization.}} \tag{16.17}$$

The Mean Field Approximation and the Boltzmann Machine

As we already know, the joint probability for the Boltzmann machine is given by

$$P(\mathcal{X}, \mathcal{X}^l) = \frac{1}{Z} \exp\left(-\sum_i \left(\sum_{j>i} \theta_{ij} x_i x_j + \theta_{i0} x_i\right)\right),$$

where some of x_i (x_j) belong to $\mathcal{X}$ and some to $\mathcal{X}^l$. Our first goal is to compute $P(\mathcal{X}^l|\mathcal{X})$ so as to use it in the KL divergence. Note that if both $x_i, x_j \in \mathcal{X}$, their contribution results to a constant, which is finally absorbed by the normalizing factor Z. If one is observed and the other one is latent, then the product contribution becomes linear with regard to the hidden variable and it is absorbed by the respective linear term. Then we can write

$$P(\mathcal{X}^l|\mathcal{X}) = \frac{1}{\tilde{Z}} \exp\left(-\sum_{i:x_i \in \mathcal{X}^l} \left(\sum_{x_j \in \mathcal{X}^l: j>i} \theta_{ij} x_i x_j + \tilde{\theta}_{i0} x_i\right)\right). \tag{16.18}$$

We now turn our attention to the form of Q. Due to the (assumed) binary nature of the variables, a sensible completely factorized form of Q is [34]

$$Q(\mathcal{X}^l;\boldsymbol{\mu}) = \prod_{i:x_i\in\mathcal{X}^l} \mu_i^{x_i}(1-\mu_i)^{(1-x_i)}, \tag{16.19}$$

where the dependence on the variational parameters, $\boldsymbol{\mu}$, is explicitly shown. Also, due to the adopted Bernoulli distribution for each variable, $\mathbb{E}[\mathrm{x}_i] = \mu_i$ (Chapter 2). The goal now is to optimize the KL divergence with respect to the variational parameters. Plugging Eqs. (16.18) and (16.19) into

$$KL(Q||P) = \sum_{x_i\in\mathcal{X}^l} Q(\mathcal{X}^l;\boldsymbol{\mu}) \ln \frac{Q(\mathcal{X}^l;\boldsymbol{\mu})}{P(\mathcal{X}^l|\mathcal{X})}, \tag{16.20}$$

we obtain (Problem 16.13, [34])

$$KL(Q||P) = \sum_i \left(\mu_i \ln \mu_i + (1-\mu_i)\ln(1-\mu_i) + \sum_{j>i} \theta_{ij}\mu_i\mu_j + \tilde{\theta}_{i0}\mu_i \right) + \ln \tilde{Z},$$

whose minimization with regard to μ_i finally results in (Problem 16.13)

$$\boxed{\mu_i = \sigma\left(-\left(\sum_{j\neq i} \theta_{ij}\mu_j + \tilde{\theta}_{i0}\right)\right) : \quad \text{mean field equations,}} \tag{16.21}$$

where $\sigma(\cdot)$ is the sigmoid link function; recall from the definition of the Ising model that $\theta_{ij} = \theta_{ji} \neq 0$ if x_i and x_j are connected and zero otherwise. Plugging the values μ_i into Eq. (16.19), an approximation of $P(\mathcal{X}^l|\mathcal{X})$ in terms of $Q(\mathcal{X}^l;\boldsymbol{\mu})$ has been obtained.

Eq. (16.21) is equivalent to a set of coupled equations known as *mean field equations* and they are used in a recursive manner to compute a solution fixed point set, assuming that one exists. Eq. (16.21) is quite interesting. Although we assumed independence among hidden nodes, imposing minimization of the KL divergence, information related to the (true) mutually dependent nature of the variables (as this is conveyed by $P(\mathcal{X}^l|\mathcal{X})$) is "embedded" into the mean values with respect to $Q(\mathcal{X}^l|\boldsymbol{\mu})$; the mean values of the respective variables are *interrelated*. Eq. (16.21) can also be viewed as a message passing algorithm (see Fig. 16.12). Fig. 16.13 shows the graph associated with a Boltzmann machine prior to and after the application of the mean field approximation.

Note that what we have said before is nothing but an instance of the variational EM algorithm, presented in Section 13.2; as a matter of fact, Eq. (16.21) is the outcome of the E-step for each one of the factors of Q_i, assuming the rest are fixed.

Thus far in the chapter, we have not mentioned the important task of how to obtain estimates of the parameters describing a graphical structure; in our current context, these are the parameters θ_{ij} and θ_{i0}, comprising the set $\boldsymbol{\theta}$. Although the parameter estimation task is discussed at the end of this chapter, there is no harm in saying a few words at this point. Let us give the dependence of $\boldsymbol{\theta}$ explicitly and denote the involved probabilities as $Q(\mathcal{X}^l;\boldsymbol{\mu},\boldsymbol{\theta})$, $P(\mathcal{X},\mathcal{X}^l;\boldsymbol{\theta})$, and $P(\mathcal{X}^l|\mathcal{X};\boldsymbol{\theta})$. Treating $\boldsymbol{\theta}$ as an unknown parameter vector, we know from the variational EM that this can be iteratively estimated by adding the M-step in the algorithm and optimizing the lower bound, $\mathcal{F}(Q)$, with regard to $\boldsymbol{\theta}$, fixing the rest of the involved parameters (see, for example, [29,69]).

Remarks 16.2.

- The mean field approximation method has also been applied in the case of sigmoidal neural networks, defined in Section 15.3.4 (see, e.g., [69]).
- The mean field approximation involving the completely factorized form of Q is the simplest and crudest approximation. More sophisticated attempts have also been suggested, where Q is allowed to have a richer structure while retaining its computational tractability (see, e.g., [17,31,80]).
- In [82], the mean field approximation has been applied to a general Bayesian network, where, as we know, the joint probability distribution is given by the product of the conditionals across the nodes,

$$p(\boldsymbol{x}) = \prod_i p(x_i|\mathrm{Pa}_i).$$

Unless the conditionals are given in a structurally simple form, exact message passing can become computationally tough. For such cases, the mean field approximation can be introduced in the hidden variables, i.e.,

$$Q(\mathcal{X}^l) = \prod_{i:x_i \in \mathcal{X}^l} Q_i(x_i),$$

which are then estimated so as to maximize the lower bound $\mathcal{F}(Q)$ in (16.15). Following the arguments that were introduced in Section 13.2, this is achieved iteratively starting from some initial estimates and at each iteration step optimization takes place with respect to a single factor, holding the rest fixed. At the $(j+1)$th step, the mth factor is obtained as (Eq. (13.15))

$$\begin{aligned}\ln Q_m^{(j+1)}(x_m^l) &= \mathbb{E}\left[\ln \prod_i p(x_i|\mathrm{Pa}_i)\right] + \text{constant}\\ &= \mathbb{E}\left[\sum_i \ln p(x_i|\mathrm{Pa}_i)\right] + \text{constant},\end{aligned}$$

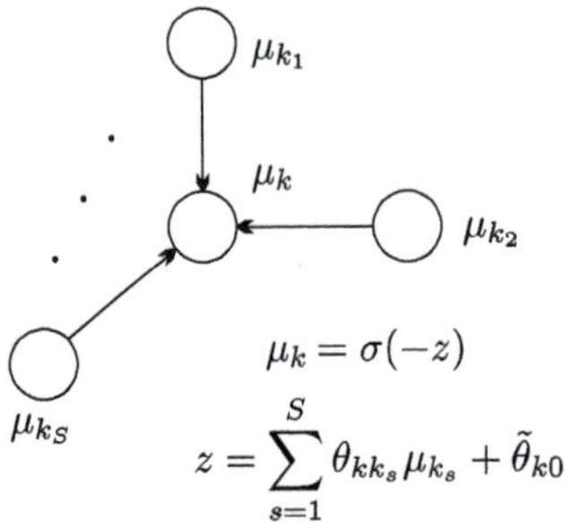

FIGURE 16.12

Node k is connected to S nodes and receives messages from its neighbors, and then passes messages to its neighbors.

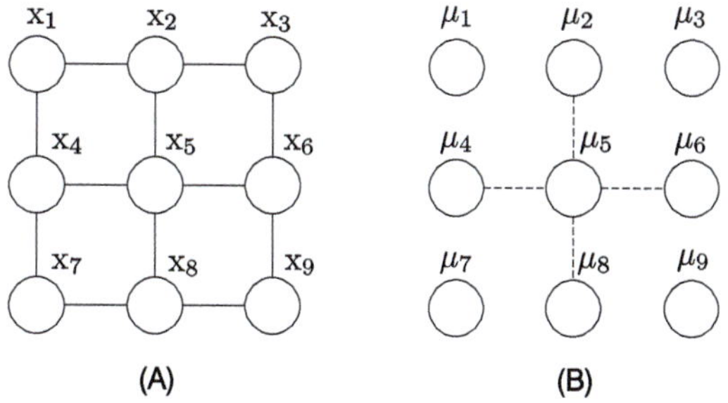

FIGURE 16.13

(A) The nodes of the graph representing a Boltzmann machine. (B) The mean field approximation results in a graph without edges. The dotted lines indicate the deterministic relation that is imposed among nodes, which were linked prior to the approximation with node x_5.

where the expectation is with respect to the currently available estimates of the factors, *excluding* Q_m, and x_m^l is the respective hidden variable. When restricting the conditionals within the conjugate-exponential family, the computations of expectations of the logarithms become tractable. The resulting scheme is equivalent to a message passing algorithm, known as *variational message passing*, and it comprises passing moments and parameters associated with the exponential distributions. An implementation example of the variational message passing scheme in the context of MIMO-OFDM communication systems is given in [6,23,38].

16.3.3 LOOPY BELIEF PROPAGATION

The message passing algorithms, which were considered previously for exact inference in graphs with a tree structure, can also be used for approximate inference in general graphs with cycles (loops). Such schemes are known as *loopy belief propagation* algorithms.

The idea of using the message passing (sum-product) algorithm with graphs with cycles goes back to Pearl [57]. Note that algorithmically, there is nothing to prevent us from applying the algorithm in such general structures. On the other hand, if we do it, there is no guarantee that the algorithm will converge in two passes and, more importantly, that it will recover the true values for the marginals. As a matter of fact, there is no guarantee that such a message propagation will ever converge. Thus, without any clear theoretical understanding, the idea of using the algorithm in general graphs was rather forgotten. Interestingly enough, the spark for its comeback was ignited by a breakthrough in coding theory, under the name *turbo codes* [5]. It was empirically verified that the scheme can achieve performance very close to the theoretical Shannon limit.

Although, in the beginning, such coding schemes seemed to be unrelated to belief propagation, it was subsequently shown [49] that the turbo decoding is just an instance of the sum-product algorithm, when applied to a graphical structure that represents the turbo code. As an example of the use of the loopy belief propagation for decoding, consider the case of Fig. 15.19. This is a graph with cycles. Applying belief propagation on this graph, we can obtain after convergence the conditional probabilities $P(x_i|y_i)$ and, hence, decide on the received sequence of bits. This finding revived interest in loopy belief propagation; after all, it "may be useful" in practice. Moreover, it initiated activity in theoretical research in order to understand its performance as well as its more general convergence properties.

In [83], it is shown, on the basis of pair-wise connected MRFs (undirected graphical models with potential functions involving at most pairs of variables, e.g., trees), that whenever the sum-product algorithm converges on loopy graphs, the fixed points of the message passing algorithm are actually stationary points of the so-called *Bethe free energy* cost. This is directly related to the KL divergence between the true and an approximating distribution (Section 12.7). Recall from Eq. (16.20) that one of the terms in KL divergence is the negative entropy associated with Q. In the mean field approximation, this entropy term can be easily computed. However, this is not the case for more general structures with cycles and one has to be content to settle for an approximation. The so-called Bethe entropy approximation is employed, which in turn gives rise to the Bethe free energy cost function. To obtain the Bethe entropy approximation, one "embeds" into the approximating distribution, Q, a structure that is in line with (16.8), which holds true for trees. Indeed, it can be checked (try it) that for (singly connected) trees, the product in the numerator runs over all pairs of connected nodes in the tree; let us denote it as $\prod_{(i,j)} P_{ij}(x_i, x_j)$. Also, d_s is equal to the number of nodes that node s is connected with. Thus, we can write the joint probability as

$$P(\boldsymbol{x}) = \frac{\prod_{(i,j)} P_{ij}(x_i, x_j)}{\prod_s [P_s(x_s)]^{d_s - 1}}.$$

Note that nodes that are connected to only one node have no contribution in the denominator. Then the entropy of the tree, that is,

$$E = -\mathbb{E}[\ln P(\boldsymbol{x})],$$

can be written as

$$\boxed{E = -\sum_{(i,j)} \sum_{x_i} \sum_{x_j} P_{ij}(x_i, x_j) \ln P_{ij}(x_i, x_j) + \sum_s (d_s - 1) \sum_{x_s} P_s(x_s) \ln P_s(x_s).} \tag{16.22}$$

Thus, this expression for the entropy is exact for trees. However, for more general graphs with cycles, this can only hold approximately true, and it is known as the *Bethe approximation of the entropy*. The closer to a tree a graph is, the better the approximation becomes; see [83] for a concise, related introduction.

It turns out that in the case of trees, the sum-product algorithm leads to the true marginal values because no approximation is involved and minimizing the free energy is equivalent to minimizing the KL divergence. Thus, from this perspective, the sum-product algorithm gets an optimization flavor. In a number of practical cases, the Bethe approximation is accurate enough, which justifies the good performance that is often achieved in practice by the loopy belief algorithm (see, e.g., [52]). The loopy belief propagation algorithm is not guaranteed to converge in graphs with cycles, so one may choose to minimize the Bethe energy cost directly; although such schemes are slower compared to message passing, they are guaranteed to converge (e.g., [85]).

An alternative interpretation of the sum-product algorithm as an optimization algorithm of an appropriately selected cost function is given in [75,77]. A unifying framework for exact, as well as approximate, inference is provided in the context of the exponential family of distributions. Both the mean field approximation and the loopy belief propagation algorithm are considered and viewed as different ways to approximate a convex set of realizable mean parameters, which are associated with the corresponding distribution. Although we will not proceed in a detailed presentation, we will provide a few "brush strokes," which are indicative of the main points around which this theory develops. At the same time, this is a good excuse for us to be exposed to an interesting interplay among the notions of convex duality, entropy, cumulant generating function, and mean parameters, in the context of the exponential family.

The general form of a probability distribution in the exponential family is given by (Section 12.8.1)

$$\begin{aligned} p(\boldsymbol{x};\boldsymbol{\theta}) &= C\exp\left(\sum_{i\in I}\theta_i u_i(\boldsymbol{x})\right) \\ &= \exp\left(\boldsymbol{\theta}^T\boldsymbol{u}(\boldsymbol{x}) - A(\boldsymbol{\theta})\right), \end{aligned}$$

with

$$A(\boldsymbol{\theta}) = -\ln C = \ln\int \exp\left(\boldsymbol{\theta}^T\boldsymbol{u}(\boldsymbol{x})\right)\,\mathrm{d}\boldsymbol{x},$$

where the integral becomes summation for discrete variables; $A(\boldsymbol{\theta})$ is a convex function and it is known as the *log-partition* or *cumulant generating function* (Problem 16.14, [75,77]). It turns out that the conjugate function of $A(\boldsymbol{\theta})$, denoted as $A^*(\boldsymbol{\mu})$, is the *negative entropy* function of $p(\boldsymbol{x};\boldsymbol{\theta}(\boldsymbol{\mu}))$, where $\boldsymbol{\theta}(\boldsymbol{\mu})$ is the value of $\boldsymbol{\theta}$ where the maximum (in the definition of the conjugate function) occurs given the value of $\boldsymbol{\mu}$; we say that $\boldsymbol{\theta}(\boldsymbol{\mu})$ and $\boldsymbol{\mu}$ are dually coupled (Problem 16.15). Moreover,

$$\mathbb{E}[\boldsymbol{u}(\mathbf{x})] = \boldsymbol{\mu},$$

where the expectation is with respect to $p(\boldsymbol{x};\boldsymbol{\theta}(\boldsymbol{\mu}))$. This is an interesting interpretation of $\boldsymbol{\mu}$ as a mean parameter vector; recall from Section 12.8.1 that these mean parameters define the respective exponential distribution. Then

$$A(\boldsymbol{\theta}) = \max_{\boldsymbol{\mu}\in\mathcal{M}}\left(\boldsymbol{\theta}^T\boldsymbol{\mu} - A^*(\boldsymbol{\mu})\right), \tag{16.23}$$

where $\mathcal{M}$ is the set that guarantees that $A^*(\boldsymbol{\mu})$ is finite, according to the definition of the conjugate function in Eq. (13.78). It turns out that in graphs of a tree structure, the sum-product algorithm is an iterative scheme of solving a Lagrangian dual formulation of Eq. (16.23) [75,77]. Moreover, in this case, the set $\mathcal{M}$, which can be shown to be a convex one, is possible to be characterized explicitly in a straightforward way and the negative entropy $A^*(\boldsymbol{\mu})$ has an explicit form. These properties are no more valid in graphs with cycles. The mean field approximation involves an inner approximation of the set $\mathcal{M}$; hence, it restricts optimization to a limited class of distributions, for which the entropy can be recovered exactly. On the other hand, the loopy belief algorithm provides an outer approximation and, hence, enlarges the class of distributions; entropy can only approximately be recovered, which for the case of pair-wise MRFs can take the form of the Bethe approximation.

The previously summarized theoretical findings have been generalized to the case of junction trees, where the potential functions involve more than two variables. Such methods involve the so-called *Kikuchi* energy, which is a generalization of the Bethe approximation [77,84]. Such arguments have their origins in statistical physics [37].

Remarks 16.3.

- Following the success of loopy belief propagation in turbo decoding, further research verified its performance potential in a number of tasks, such as low-density parity-check codes [15,47], network diagnostics [48], sensor network applications [24], and multiuser communications [70]. Furthermore, a number of modified versions of the basic scheme have been proposed. In [74], the so-called tree-reweighted belief propagation is proposed. In [26], arguments from information geometry are employed and in [78], projection arguments in the context of information geometry are used. More recently, the belief propagation algorithm and the mean field approximation were proposed to be optimally combined to exploit their respective advantages [65]. A related review can be found in [76]. In a nutshell, this old scheme is still alive and kicking!
- In Section 13.10, the expectation propagation algorithm was discussed in the context of parameter inference. The scheme can also be adopted in the more general framework of graphical models, if the place of parameters is taken by the hidden variables. Graphical models are particularly tailored for this approach because the joint PDF is factorized. It turns out that if the approximate PDF is completely factorized, corresponding to a partially disconnected network, the expectation propagation algorithm turns out to be the loopy belief propagation algorithm [50]. In [51], it is shown that a new family of message passing algorithms can be obtained by utilizing a generalization of the KL divergence as the optimizing cost. This family encompasses a number of previously developed schemes.
- Besides the approximation techniques that were previously presented, another popular pool of methods is the Markov chain Monte Carlo (MCMC) framework. Such techniques were discussed in Chapter 14 (see, for example, [25] and the references therein).

16.4 DYNAMIC GRAPHICAL MODELS

All the graphical models that have been discussed so far were developed to serve the needs of random variables whose statistical properties remained fixed over time. However, this is not always the case. As a matter of fact, the terms *time adaptivity* and *time variation* are central for most parts of this book. Our focus in this section is to deal with random variables whose statistical properties are not fixed but are allowed to undergo changes. A number of time series as well as sequentially obtained data fall under this setting with applications ranging from signal processing and robotics to finance and bioinformatics.

A key difference here, compared to what we have discussed in the previous sections of this chapter, is that now observations are sensed sequentially and the *specific sequence* in which they occur carries important information, which has to be respected and exploited in any subsequent inference task. For example, in speech recognition, the sequence in which the feature vectors result is very important. In a typical speech recognition task, the raw speech data are *sequentially* segmented in short (usually overlapping) time windows and from each window a feature vector is obtained (e.g., DFT of the samples in the respective time slot). This is illustrated in Fig. 16.14. These feature vectors constitute the observation sequence. Besides the information that resides in the specific values of these observation vectors, the sequence in which the observations appear discloses important information about the word that is spoken; our language and spoken words are highly structured human activities. Similar argu-

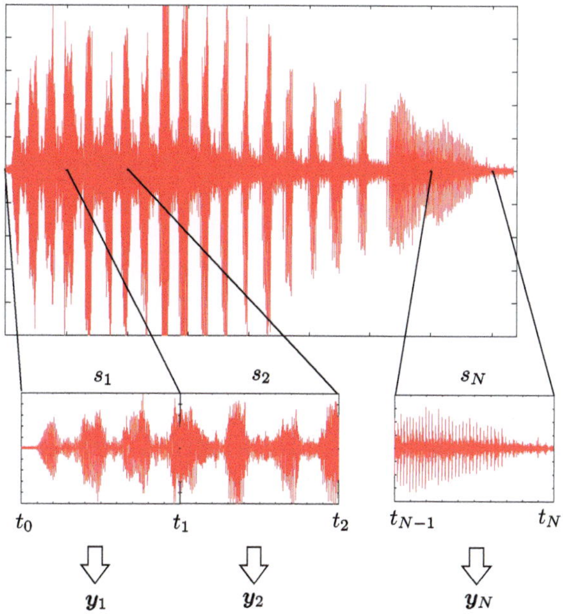

FIGURE 16.14

A speech segment and N time windows, each one of length equal to 500 ms. They correspond to time intervals $[0, 500]$, $[500, 1000]$, and $[3500, 4000]$, respectively. From each one of them, a feature vector, $\boldsymbol{y}$, is generated. In practice, an overlap between successive windows is allowed.

ments hold true for applications such as learning and reasoning concerning biological molecules, for example, DNA and proteins.

Although any type of graphical model has its dynamic counterpart, we will focus on the family of *dynamic Bayesian networks* and, in particular, a specific type known as *hidden Markov models*.

A very popular and effective framework to model sequential data is via the so-called *state-observation* or *state-space* models. Each set of random variables, $\mathbf{y}_n \in \mathbb{R}^l$, which are observed at time n, is associated with a corresponding hidden/latent random vector $\mathbf{x}_n$ (not necessarily of the same dimensionality as that of the observations). The system dynamics are modeled via the latent variables and observations are considered to be the output of a measuring *noisy* sensing device. The so-called *latent Markov models* are built around the following two independence assumptions:

$$(1)\ \mathbf{x}_{n+1} \perp (\mathbf{x}_1, \ldots, \mathbf{x}_{n-1}) | \mathbf{x}_n, \tag{16.24}$$

$$(2)\ \mathbf{y}_n \perp (\mathbf{x}_1, \ldots, \mathbf{x}_{n-1}, \mathbf{x}_{n+1}, \ldots, \mathbf{x}_N) | \mathbf{x}_n, \tag{16.25}$$

where N is the total number of observations. The first condition defines the system dynamics via the transition model

$$p(\boldsymbol{x}_{n+1}|\boldsymbol{x}_1, \ldots, \boldsymbol{x}_n) = p(\boldsymbol{x}_{n+1}|\boldsymbol{x}_n), \tag{16.26}$$

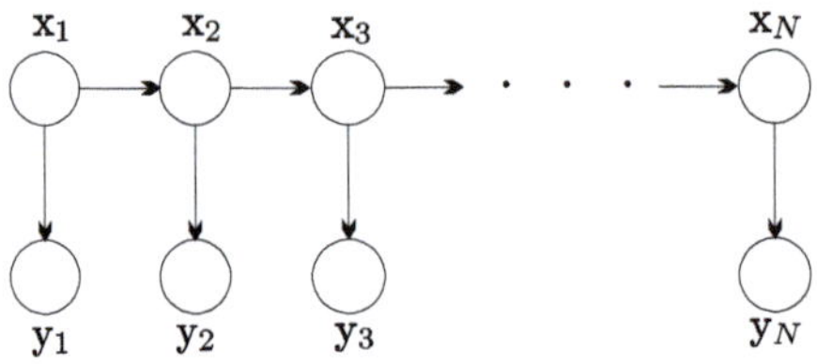

FIGURE 16.15

The Bayesian network corresponding to a latent Markov model. If latent variables are of a discrete nature, this corresponds to an HMM. If both observed and latent variables are continuous and follow a Gaussian distribution, this corresponds to a linear dynamic system (LDS). Note that the observed variables comprise the leaves of the graph.

and the second one via the observation model

$$p(\boldsymbol{y}_n|\boldsymbol{x}_1, \ldots, \boldsymbol{x}_N) = p(\boldsymbol{y}_n|\boldsymbol{x}_n). \tag{16.27}$$

In words, *the future is independent of the past given the present, and the observations are independent of the future and past given the present.*

The previously stated independencies are graphically represented via the graph of Fig. 16.15. If the hidden variables are of a discrete nature, the resulting model is known as a hidden Markov model. If, on the other hand, both hidden and observation variables are of a continuous nature, the resulting model gets rather involved to deal with. However, analytically tractable tools can be and have been developed for some special cases. In the so-called *linear dynamic systems* (LDSs), the system dynamics and the generation of the observations are modeled as

$$\mathbf{x}_n = F_n \mathbf{x}_{n-1} + \boldsymbol{\eta}_n, \tag{16.28}$$

$$\mathbf{y}_n = H_n \mathbf{x}_n + \boldsymbol{v}_n, \tag{16.29}$$

where $\boldsymbol{\eta}_n$ and $\boldsymbol{v}_n$ are zero mean, mutually independent noise disturbances modeled by Gaussian distributions. This is the celebrated Kalman filter, which we have already discussed in Chapter 4 and it will also be considered, from a probabilistic perspective, in Chapter 17. The probabilistic counterparts of Eqs. (16.28) and (16.29) are

$$p(\boldsymbol{x}_n|\boldsymbol{x}_{n-1}) = \mathcal{N}(\boldsymbol{x}_n|F_n\boldsymbol{x}_{n-1}, Q_n), \tag{16.30}$$

$$p(\boldsymbol{y}_n|\boldsymbol{x}_n) = \mathcal{N}(\boldsymbol{y}_n|H_n\boldsymbol{x}_n, R_n), \tag{16.31}$$

where Q_n and R_n are the covariance matrices of $\boldsymbol{\eta}_n$ and $\boldsymbol{v}_n$, respectively.

16.5 HIDDEN MARKOV MODELS

Hidden Markov models are represented by the graphical model in Fig. 16.15 and Eqs. (16.26) and (16.27). The latent variables are discrete; hence, we write the *transition* probability as $P(\boldsymbol{x}_n|\boldsymbol{x}_{n-1})$ and this corresponds to a table of probabilities. Observation variables can either be discrete or continuous.

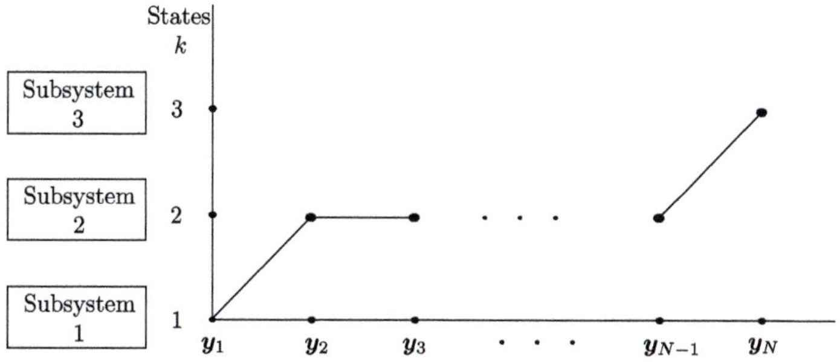

FIGURE 16.16

The unfolding in time of a trajectory that associates observations with states.

Basically, an HMM is used to model a *quasistationary* process that undergoes *sudden* changes among a number of, say, K subprocesses. Each one of these subprocesses is described by different statistical properties. One could alternatively view it as a combined system comprising a number of subsystems; each one of these subsystems generates data/observations according to a different statistical model; for example, one may follow a Gaussian and the other one a Student's t distribution. Observations are emitted by these subsystems; however, once an observation is received, we do not know which subsystem this was emitted from. This reminds us of the mixture modeling task of a PDF; however, in mixture modeling, we did not care about the sequence in which observations occur.

For modeling purposes, we associate with each observation $\boldsymbol{y}_n$ a hidden variable, $\mathrm{k}_n = 1, 2, \ldots, K$, which is the (random) index indicating the subsystem/subprocess that generated the respective observation vector. We will call it the *state*. Each k_n corresponds to $\mathbf{x}_n$ of the general model. The sequence of the complete observation set $(\boldsymbol{y}_n, k_n)$, $n = 1, 2, \ldots, N$, forms a trajectory in a two-dimensional grid, having the states on one axis and the observations on the other. This is shown in Fig. 16.16 for $K = 3$. Such a path reveals the origin of each observation; $\boldsymbol{y}_1$ was emitted from state $k_1 = 1$, $\boldsymbol{y}_2$ from $k_2 = 2$, $\boldsymbol{y}_3$ from $k_3 = 2$, and $\boldsymbol{y}_N$ from $k_N = 3$. Note that each trajectory is associated with a probability distribution, that is, the joint distribution of the complete set. Indeed, the probability that the trajectory of Fig. 16.16 will occur depends on the value of $P\left((\boldsymbol{y}_1, k_1 = 1), (\boldsymbol{y}_2, k_2 = 2), (\boldsymbol{y}_3, k_3 = 2), \ldots, (\boldsymbol{y}_N, k_N = 3)\right)$. We will soon see that some of the possible trajectories that can be drawn in the grid are not allowed in practice; this may be due to physical constraints concerning the data generation mechanism that underlies the corresponding system/process.

Transition probabilities. As already said, the dynamics of a latent Markov model are described in terms of the distribution $p(\boldsymbol{x}_n|\boldsymbol{x}_{n-1})$, which for an HMM becomes the set of probabilities

$$P(k_n|k_{n-1}), \quad k_n, k_{n-1} = 1, 2, \ldots, K,$$

indicating the probability of the system to "jump" at time n to state k_n from state k_{n-1}, where it was at time $n - 1$. In general, this table of probabilities may be time-varying. In the standard form of HMMs, this is considered to be independent of time and we say that our model is *homogeneous*. Thus, we can write

$$P(k_n|k_{n-1}) = P(i|j) := P_{ij}, \quad i, j = 1, 2, \ldots, K.$$

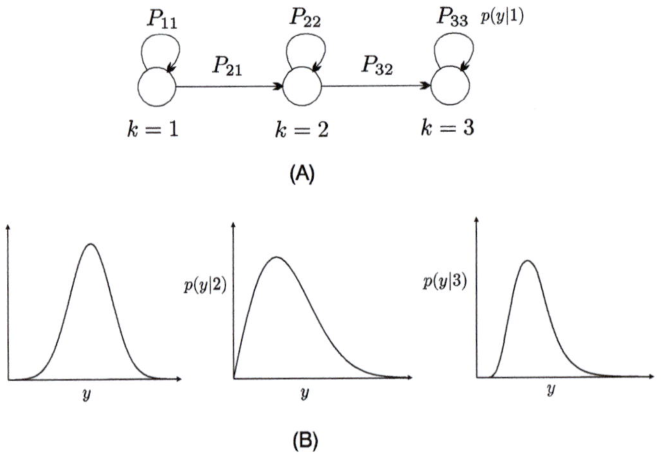

FIGURE 16.17

(A) A three-state left-to-right HMM model. (B) Each state is characterized by different statistical properties.

Note that some of these transition probabilities can be zero, depending on the modeling assumptions. Fig. 16.17A shows an example of a three-state system. The model is of the so-called *left-to-right* type, where two types of transitions are allowed: (a) self transitions and (b) transitions from a state of a lower index to a state of a higher index. The system, once it jumps into a state k, emits data according to a probability distribution $p(\mathbf{y}|k)$, as illustrated in Fig. 16.17B. Besides the left-to-right models, other alternatives have also been proposed [8,63]. The states correspond to certain physical characteristics of the corresponding system. For example, in speech recognition, the number of states that are chosen to model a spoken word depends on the expected number of sound phenomena (phonemes) within the

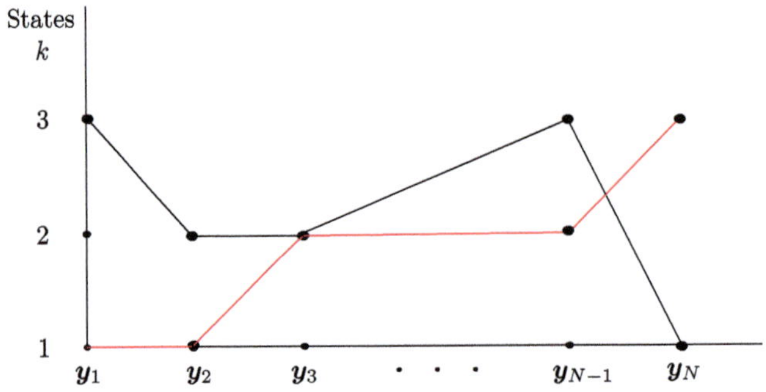

FIGURE 16.18

The black trajectory is not allowed to occur under the HMM model of Fig. 16.17. Transitions from state $k = 3$ to state $k = 2$ and from $k = 3$ to $k = 1$ are not permitted. In contrast, the state unfolding in the red curve is in agreement with the model.

word. Typically, three to four states are used per phoneme. Another modeling path uses the average number of observations resulting from various versions of a spoken word as an indication of the number of states. Seen from the transition probabilities perspective, an HMM is basically a stochastic finite state automaton that generates an observation string. Note that the semantics of Fig. 16.17 is different from and must not be confused with the graphical structure given in Fig. 16.15. Fig. 16.17 is a graphical interpretation of the transition probabilities among the states; it says nothing about independencies among the involved random variables. Once a state transition model has been adopted, some trajectories in the trellis diagram of Fig. 16.16 will not be allowed. In Fig. 16.18, the red trajectory is not in line with the model of Fig. 16.17.

16.5.1 INFERENCE

As in any graphical modeling task, the ultimate goal is inference. Two types of inference are of particular interest in the context of classification/recognition. Let us discuss it in the framework of speech recognition; similar arguments hold true for other applications. We are given a set of (output variables) observations, $\boldsymbol{y}_1, \dots, \boldsymbol{y}_N$, and we have to decide to which spoken word these correspond. In the database, each spoken word is represented by an HMM model, which is the result of extensive training. An HMM model is fully described by the following set of parameters:

HMM model parameters

1. Number of states K.
2. The probabilities for the initial state at $n = 1$ to be at state k, that is, $P_k,\ k = 1, 2, \dots, K$.
3. The set of transition probabilities $P_{ij},\ i, j = 1, 2, \dots, K$.
4. The state emission distributions $p(\boldsymbol{y}|k), k = 1, 2, \dots, K$, which can be either discrete or continuous. Often, these probability distributions may be parameterized, $p(\boldsymbol{y}|k; \boldsymbol{\theta}_k), k = 1, 2, \dots, K$.

Prior to inference, all the involved parameters are assumed to be known. Learning of the HMM parameters takes place in the training phase; we will come to it shortly.

For the recognition, a number of scores can be used. Here we will discuss two alternatives that come as a direct consequence of our graphical modeling approach. For a more detailed discussion, see, for example, [72].

In the first one, the joint distribution for the observed sequence is computed, after marginalizing out all hidden variables; this is done for each one of the models/words. Then the word that scores the larger value is selected. This method corresponds to the sum-product rule. The other path is to compute, for each model/word, the optimal trajectory in the trellis diagram; that is, the trajectory that scores the highest joint probability. In the sequel, we decide in favor of the model/word that corresponds to the largest optimal value. This method is an implementation of the max-sum rule.

The Sum-Product Algorithm: the HMM Case

The first step is to transform the directed graph of Fig. 16.15 to an undirected one; a factor graph or a junction tree graph. Note that this is trivial for this case, as the graph is already a tree. Let us work with the junction tree formulation. Also, in order to use the message passing formulas of (16.4) and (16.5) as well as (16.6) and (16.7) for computing the distribution values, we will first adopt a more compact way of representing the conditional probabilities. We will employ the technique that was used in Section 13.4 for the mixture modeling case. Let us denote each latent variable as a K-dimensional vector, $\boldsymbol{x}_n \in \mathbb{R}^K$, $n = 1, 2, \ldots, N$, whose elements are all zero except at the kth location, where k is the index of the (unknown) state from which $\boldsymbol{y}_n$ has been emitted, that is,

$$\boldsymbol{x}_n^T = [x_{n,1}, x_{n,2}, \ldots, x_{n,K}] : \begin{cases} x_{n,i} = 0, & i \neq k \\ x_{n,k} = 1. \end{cases}$$

Then we can compactly write

$$P(\boldsymbol{x}_1) = \prod_{k=1}^{K} P_k^{x_{1,k}}, \tag{16.32}$$

and

$$P(\boldsymbol{x}_n|\boldsymbol{x}_{n-1}) = \prod_{i=1}^{K}\prod_{j=1}^{K} P_{ij}^{x_{n-1,j}x_{n,i}}. \tag{16.33}$$

Indeed, if the jump is from a specific state j at time $n-1$ to a specific state i at time n, then the only term that survives in the previous product is the corresponding factor, P_{ij}. The joint probability distribution of the complete set, as a direct consequence of the Bayesian network model of Fig. 16.15, is written as

$$p(Y, X) = P(\boldsymbol{x}_1)p(\boldsymbol{y}_1|\boldsymbol{x}_1)\prod_{n=2}^{N} P(\boldsymbol{x}_n|\boldsymbol{x}_{n-1})p(\boldsymbol{y}_n|\boldsymbol{x}_n), \tag{16.34}$$

where

$$p(\boldsymbol{y}_n|\boldsymbol{x}_n) = \prod_{k=1}^{K} \left(p(\boldsymbol{y}_n|k; \boldsymbol{\theta}_k)\right)^{x_{n,k}}. \tag{16.35}$$

The corresponding junction tree is trivially obtained from the graph in Fig. 16.15. Replacing directed links with undirected ones and considering cliques of size two, the graph in Fig. 16.19A results. However, as all the $\boldsymbol{y}_n$ variables are observed (*instantiated*) and no marginalization is required, their

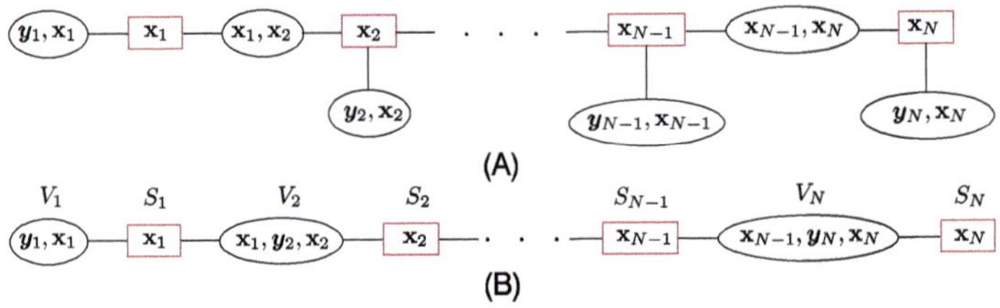

FIGURE 16.19

(A) The junction tree that results from the graph of Fig. 16.15. (B) Because $\boldsymbol{y}_n$ are observed, their effect is only of a multiplicative nature (no marginalization is involved) and its contribution can be trivially absorbed by the potential functions (distributions) associated with the latent variables.

multiplicative contribution can be absorbed by the respective conditional probabilities, which leads to the graph of Fig. 16.19B. Alternatively, this junction tree can be obtained if one considers the nodes $(\mathbf{x}_1, \boldsymbol{y}_1)$ and $(\mathbf{x}_{n-1}, \mathbf{x}_n, \boldsymbol{y}_n)$, $n = 2, 3, \ldots, N$, to form cliques associated with the potential functions

$$\psi_1(\boldsymbol{x}_1, \boldsymbol{y}_1) = P(\boldsymbol{x}_1)p(\boldsymbol{y}_1|\boldsymbol{x}_1) \tag{16.36}$$

and

$$\psi_n(\boldsymbol{x}_{n-1}, \boldsymbol{y}_n, \boldsymbol{x}_n) = P(\boldsymbol{x}_n|\boldsymbol{x}_{n-1})p(\boldsymbol{y}_n|\boldsymbol{x}_n), \quad n = 2, \ldots, N. \tag{16.37}$$

The junction tree of Fig. 16.19B results by eliminating nodes from the cliques starting from $\mathbf{x}_1$. Note that the normalizing constant is equal to one, $Z = 1$.

To apply the sum-product rule for junction trees, Eq. (16.5) now becomes

$$\begin{aligned}\mu_{V_n \to S_n}(\boldsymbol{x}_n) &= \sum_{\boldsymbol{x}_{n-1}} \psi_n(\boldsymbol{x}_{n-1}, \boldsymbol{y}_n, \boldsymbol{x}_n)\mu_{S_{n-1} \to V_n}(\boldsymbol{x}_{n-1}) \\ &= \sum_{\boldsymbol{x}_{n-1}} \mu_{S_{n-1} \to V_n}(\boldsymbol{x}_{n-1})P(\boldsymbol{x}_n|\boldsymbol{x}_{n-1})p(\boldsymbol{y}_n|\boldsymbol{x}_n).\end{aligned}$$

Also,

$$\mu_{S_{n-1} \to V_n}(\boldsymbol{x}_{n-1}) = \mu_{V_{n-1} \to S_{n-1}}(\boldsymbol{x}_{n-1}). \tag{16.38}$$

Thus,

$$\mu_{V_n \to S_n}(\boldsymbol{x}_n) = \sum_{\boldsymbol{x}_{n-1}} \mu_{V_{n-1} \to S_{n-1}}(\boldsymbol{x}_{n-1})P(\boldsymbol{x}_n|\boldsymbol{x}_{n-1})p(\boldsymbol{y}_n|\boldsymbol{x}_n), \tag{16.39}$$

with

$$\mu_{V_1 \to S_1}(\boldsymbol{x}_1) = P(\boldsymbol{x}_1)p(\boldsymbol{y}_1|\boldsymbol{x}_1). \tag{16.40}$$

In the HMM literature, it is common to use the "alpha" symbol for the exchanged messages, that is,

$$\alpha(\boldsymbol{x}_n) := \mu_{V_n \to S_n}(\boldsymbol{x}_n). \tag{16.41}$$

If one considers that the message passing terminates at a node V_n, then based on (16.7), and taking into account that the variables $\boldsymbol{y}_1, \ldots, \boldsymbol{y}_n$ are clumped to the observed values (recall the related comment following Eq. (15.44)), it is readily seen that

$$\alpha(\boldsymbol{x}_n) = p(\boldsymbol{y}_1, \boldsymbol{y}_2, \ldots, \boldsymbol{y}_n, \boldsymbol{x}_n), \tag{16.42}$$

which can also be deduced by the respective definitions in (16.39) and (16.40); all hidden variables, except $\mathbf{x}_n$, have been marginalized out. This is a set of K probability values (one for each value of $\boldsymbol{x}_n$). For example, for $\boldsymbol{x}_n : \boldsymbol{x}_{n,k} = 1$, $\alpha(\boldsymbol{x}_n)$ is the probability of the trajectory to be at time n at state k *and* having obtained the specific observations up to and including time n. From (16.42), one can readily obtain the joint probability distribution (evidence) over the observation sequence, comprising N time instants, that is,

$$\boxed{p(Y) = \sum_{x_N} p(y_1, y_2, \ldots, y_N, x_N) = \sum_{x_N} \alpha(x_N): \text{ evidence of observations,}}$$

which, as said in the beginning of the section, is a quantity used for classification/recognition.

In the signal processing "jargon," the computation of $\alpha(x_n)$ is referred to as the *filtering* recursion. By the definition of $\alpha(x_n)$, we have [2]

$$\boxed{\alpha(x_n) = \underbrace{p(y_n|x_n)}_{\text{corrector}} \cdot \underbrace{\sum_{x_{n-1}} \alpha(x_{n-1}) P(x_n|x_{n-1})}_{\text{predictor}}: \quad \text{filtering recursion.}} \tag{16.43}$$

As is the case with the Kalman filter, to be treated in Chapter 17, the only difference there will be that the summation is replaced by integration. Having adopted Gaussian distributions, these integrations translate into updates of the respective mean values and covariance matrices. The physical meaning of (16.43) is that the predictor provides a prediction on the state using all the past information prior to n. Then this information is corrected based on the observation y_n, which is received at time n. Thus, the updated information, based on the entire observation sequence up to and including the current time n, is readily available by

$$P(x_n|Y_{[1:n]}) = \frac{\alpha(x_n)}{p(Y_{[1:n]})},$$

where the denominator is given by $\sum_{x_n} \alpha(x_n)$, and $Y_{[1:n]} := (y_1, \ldots, y_n)$.

Let us now carry on with the second message passing phase, in the opposite direction than before, in order to obtain

$$\mu_{V_{n+1} \to S_n}(x_n) = \sum_{x_{n+1}} \mu_{S_{n+1} \to V_{n+1}}(x_{n+1}) P(x_{n+1}|x_n) p(y_{n+1}|x_{n+1}),$$

with

$$\mu_{S_{n+1} \to V_{n+1}}(x_{n+1}) = \mu_{V_{n+2} \to S_{n+1}}(x_{n+1}).$$

Hence,

$$\mu_{V_{n+1} \to S_n}(x_n) = \sum_{x_{n+1}} \mu_{V_{n+2} \to S_{n+1}}(x_{n+1}) P(x_{n+1}|x_n) p(y_{n+1}|x_{n+1}), \tag{16.44}$$

with

$$\mu_{V_{N+1} \to S_N}(x_N) = 1. \tag{16.45}$$

Note that $\mu_{V_{n+1} \to S_n}(x_n)$ involves K values and for the computation of each one of them K summations are performed. So, the complexity scales as $\mathcal{O}(K^2)$ per time instant. In the HMM literature, the symbol "beta" is used,

$$\beta(x_n) = \mu_{V_{n+1} \to S_n}(x_n). \tag{16.46}$$

From the recursive definition in (16.44) and (16.45), where $\boldsymbol{x}_{n+1}, \boldsymbol{x}_{n+2}, \dots, \boldsymbol{x}_N$ have been marginalized out, we can equivalently write

$$\beta(\boldsymbol{x}_n) = p(\boldsymbol{y}_{n+1}, \boldsymbol{y}_{n+2}, \dots, \boldsymbol{y}_N|\boldsymbol{x}_n). \tag{16.47}$$

That is, conditioned on the values of $\boldsymbol{x}_n$, for example, $\boldsymbol{x}_n : x_{nk} = 1$, $\beta(\boldsymbol{x}_n)$ is the value of the joint distribution for the observed values, $\boldsymbol{y}_{n+1}, \dots, \boldsymbol{y}_N$, to be emitted when the system is at state k at time n.

We have now all the "ingredients" in order to compute marginals. From (16.7), we obtain (explain it based on the independence properties that underlie an HMM)

$$\begin{aligned} p(\boldsymbol{x}_n, \boldsymbol{y}_1, \boldsymbol{y}_2, \dots, \boldsymbol{y}_N) &= \mu_{V_{n-1}\to S_n}(\boldsymbol{x}_n)\mu_{V_{n+1}\to S_n}(\boldsymbol{x}_n) \\ &= \alpha(\boldsymbol{x}_n)\beta(\boldsymbol{x}_n), \end{aligned} \tag{16.48}$$

which in turns leads to

$$\boxed{\gamma(\boldsymbol{x}_n) := P(\boldsymbol{x}_n|Y) = \frac{\alpha(\boldsymbol{x}_n)\beta(\boldsymbol{x}_n)}{p(Y)} : \quad \text{smoothing recursion.}} \tag{16.49}$$

This part of the recursion is known as the *smoothing* recursion. Note that in this computation, both past (via $\alpha(\boldsymbol{x}_n)$) and future (via $\beta(\boldsymbol{x}_n)$) data are involved.

An alternative way to obtain $\gamma(\boldsymbol{x}_n)$ is via its own recursion together with $\alpha(\boldsymbol{x}_n)$, by avoiding $\beta(\boldsymbol{x}_n)$ (Problem 16.16). In such a scenario, both passing messages are related to densities with regard to $\boldsymbol{x}_n$, which has certain advantages for the case of linear dynamic systems.

Finally, from (16.6) and recalling (16.38), (16.41), and (16.46), we obtain

$$\begin{aligned} p(\boldsymbol{x}_{n-1}, \boldsymbol{x}_n, Y) &= P(\boldsymbol{x}_n|\boldsymbol{x}_{n-1})p(\boldsymbol{y}_n|\boldsymbol{x}_n)\mu_{S_n\to V_n}(\boldsymbol{x}_n)\mu_{S_{n-1}\to V_n}(\boldsymbol{x}_{n-1}) \\ &= \alpha(\boldsymbol{x}_{n-1})P(\boldsymbol{x}_n|\boldsymbol{x}_{n-1})p(\boldsymbol{y}_n|\boldsymbol{x}_n)\beta(\boldsymbol{x}_n), \end{aligned} \tag{16.50}$$

or

$$\begin{aligned} p(\boldsymbol{x}_{n-1}, \boldsymbol{x}_n|Y) &= \frac{\alpha(\boldsymbol{x}_{n-1})P(\boldsymbol{x}_n|\boldsymbol{x}_{n-1})p(\boldsymbol{y}_n|\boldsymbol{x}_n)\beta(\boldsymbol{x}_n)}{p(Y)} \\ &:= \xi(\boldsymbol{x}_{n-1}, \boldsymbol{x}_n). \end{aligned} \tag{16.51}$$

Thus, $\xi(\cdot,\cdot)$ is a table of K^2 probability values. Let $\xi(x_{n-1,j}, x_{n,i})$ correspond to $x_{n-1,j} = x_{n,i} = 1$. Then $\xi(x_{n-1,j}, x_{n,i})$ is the probability of the system being at states j and i at times $n-1$ and n, respectively, conditioned on the transmitted sequence of observations.

In Section 15.7.4 a message passing scheme was proposed for the efficient computation of the maximum of the joint distribution. This can also be applied in the junction tree associated with an HMM. The resulting algorithm is known as the *Viterbi* algorithm. The Viterbi algorithm results in a straightforward way from the general max-sum algorithm. The algorithm is similar to the one derived before; all one has to do is to replace summations with the maximum operations. As we have already commented, while discussing the max-product rule, computing the sequence of the complete set $(\boldsymbol{y}_n, \boldsymbol{x}_n)$,

$n = 1, 2, \ldots, N$, that maximizes the joint probability, using back-tracking, equivalently defines the optimal trajectory in the two-dimensional grid.

Another inference task that is of interest in practice, besides recognition, is prediction, that is, given an HMM and the observation sequence $\boldsymbol{y}_n$, $n = 1, 2, \ldots, N$, to optimally predict the value $\boldsymbol{y}_{n+1}$. This can also be performed efficiently by appropriate marginalization (Problem 16.17).

16.5.2 LEARNING THE PARAMETERS IN AN HMM

This is the second time we refer to the learning of graphical models. The first time was at the end of Section 16.3.2. The most natural way to obtain the unknown parameters is to maximize the likelihood/evidence of the joint probability distribution. Because our task involves both observed and latent variables, the EM algorithm is the first one that comes to mind. However, the underlying independencies in an HMM will be employed in order to come up with an efficient learning scheme. The set of the unknown parameters, Θ, involves (a) the initial state probabilities, P_k, $k = 1, \ldots, K$, (b) the transition probabilities, P_{ij}, $i, j = 1, 2, \ldots, K$, and (c) the parameters in the probability distributions associated with the observations, $\boldsymbol{\theta}_k$, $k = 1, 2, \ldots, K$.

Expectation step: From the general scheme presented in Section 12.4.1 (with Y in place of $\mathcal{X}$, X in place of $\mathcal{X}^l$, and Θ in place of $\boldsymbol{\xi}$) at the $(t+1)$th iteration, we have to compute

$$Q(\Theta, \Theta^{(t)}) = \mathbb{E}\left[\ln p(Y, X; \Theta)\right],$$

where $\mathbb{E}[\cdot]$ is the expectation with respect to $P(X|Y; \Theta^{(t)})$. From (16.32)–(16.35) we obtain

$$\begin{aligned}
\ln p(Y, X; \Theta) &= \sum_{k=1}^{K} \left(x_{1,k} \ln P_k + \ln p(\boldsymbol{y}_1|k; \boldsymbol{\theta}_k) \right) \\
&\quad + \sum_{n=2}^{N} \sum_{i=1}^{K} \sum_{j=1}^{K} (x_{n-1,j} x_{n,i}) \ln P_{ij} \\
&\quad + \sum_{n=2}^{N} \sum_{k=1}^{K} x_{n,k} \ln p(\boldsymbol{y}_n|k; \boldsymbol{\theta}_k),
\end{aligned}$$

thus,

$$\begin{aligned}
Q(\Theta, \Theta^{(t)}) &= \sum_{k=1}^{K} \mathbb{E}[\mathrm{x}_{1,k}] \ln P_k + \sum_{n=2}^{N} \sum_{i=1}^{K} \sum_{j=1}^{K} \mathbb{E}[\mathrm{x}_{n-1,j} \mathrm{x}_{n,i}] \ln P_{ij} \\
&\quad + \sum_{n=1}^{N} \sum_{k=1}^{K} \mathbb{E}[\mathrm{x}_{n,k}] \ln p(\boldsymbol{y}_n|k; \boldsymbol{\theta}_k).
\end{aligned} \tag{16.52}$$

Let us now recall (16.49) to obtain

$$\mathbb{E}[\mathrm{x}_{n,k}] = \sum_{\boldsymbol{x}_n} P(\boldsymbol{x}_n|Y; \Theta^{(t)}) x_{n,k} = \sum_{\boldsymbol{x}_n} \gamma(\boldsymbol{x}_n; \Theta^{(t)}) x_{n,k}.$$

Note that $x_{n,k}$ can either be zero or one; hence, its mean value will be equal to the probability that $\boldsymbol{x}_n$ has the kth element $x_{n,k}=1$ and we denote it as

$$\mathbb{E}[\mathrm{x}_{n,k}] = \gamma(x_{n,k}=1;\Theta^{(t)}). \tag{16.53}$$

Recall that given $\Theta^{(t)}$, $\gamma(\cdot;\Theta^{(t)})$ can be efficiently computed via the sum-product algorithm described before. In a similar spirit and mobilizing the definition in (16.51), we can write

$$\begin{aligned}\mathbb{E}[\mathrm{x}_{n-1,j}\mathrm{x}_{n,i}] &= \sum_{\boldsymbol{x}_n}\sum_{\boldsymbol{x}_{n-1}} P(\boldsymbol{x}_n,\boldsymbol{x}_{n-1}|Y;\Theta^{(t)})x_{n-1,j}x_{n,i} \\ &= \sum_{\boldsymbol{x}_n}\sum_{\boldsymbol{x}_{n-1}} \xi(\boldsymbol{x}_n,\boldsymbol{x}_{n-1};\Theta^{(t)})x_{n-1,j}x_{n,i} \\ &= \xi(x_{n-1,j}=1,x_{n,i}=1;\Theta^{(t)}).\end{aligned} \tag{16.54}$$

Note that $\xi(\cdot,\cdot;\Theta^{(t)})$ can also be efficiently computed as a by-product of the sum-product algorithm, given $\Theta^{(t)}$. Thus, we can summarize the E-step as

$$\begin{aligned}\mathcal{Q}(\Theta,\Theta^{(t)}) = &\sum_{k=1}^{K}\gamma(x_{1,k}=1;\Theta^{(t)})\ln P_k \\ &+\sum_{n=2}^{N}\sum_{i=1}^{K}\sum_{j=1}^{K}\xi(x_{n-1,j}=1,x_{n,i}=1;\Theta^{(t)})\ln P_{ij} \\ &+\sum_{n=1}^{N}\sum_{k=1}^{K}\gamma(x_{n,k}=1;\Theta^{(t)})\ln p(\boldsymbol{y}_n|k;\boldsymbol{\theta}_k).\end{aligned} \tag{16.55}$$

Maximization step: In this step, it suffices to obtain the derivatives/gradients with regard to P_k, P_{ij}, and $\boldsymbol{\theta}_k$ and equate them to zero in order to obtain the new estimates, which will comprise $\Theta^{(t+1)}$. Note that P_k and P_{ij} are probabilities; hence, their maximization should be constrained so that

$$\sum_{k=1}^{K} P_k = 1 \quad \text{and} \quad \sum_{i=1}^{K} P_{ij}=1, \quad j=1,2,\dots K.$$

The resulting reestimation formulas are (Problem 16.18)

$$P_k^{(t+1)} = \frac{\gamma(x_{1,k}=1;\Theta^{(t)})}{\sum_{i=1}^{K}\gamma(x_{1,i}=1;\Theta^{(t)})}, \tag{16.56}$$

$$P_{ij}^{(t+1)} = \frac{\sum_{n=2}^{N}\xi(x_{n-1,j}=1,x_{n,i}=1;\Theta^{(t)})}{\sum_{n=2}^{N}\sum_{k=1}^{K}\xi(x_{n-1,j}=1,x_{n,k}=1;\Theta^{(t)})}. \tag{16.57}$$

The reestimation of $\boldsymbol{\theta}_k$ depends on the form of the corresponding distribution $p(\boldsymbol{y}_n|k;\boldsymbol{\theta}_k)$. For example, in the Gaussian scenario, the parameters are the mean values and the elements of the covariance matrix.

In this case, we obtain exactly the same iterations as those resulting for the problem of Gaussian mixtures (see Eqs. (12.60) and (12.61)), if in place of the posterior we use γ.

In summary, training an HMM comprises the following steps:

1. Initialize the parameters in Θ.
2. Run the sum-product algorithm to obtain $\gamma(\cdot)$ and $\xi(\cdot,\cdot)$, using the current set of parameter estimates.
3. Update the parameters as in (16.56) and (16.57).

Iterations in steps 2 and 3 continue until a convergence criterion is met, such as in EM. This iterative scheme is also known as the *Baum–Welch* or *forward-backward* algorithm. Besides the forward-backward algorithm for training HMMs, the literature is rich in a number of alternatives with the goal of either simplifying computations or improving performance. For example, a simpler training algorithm can be derived tailored to the Viterbi scheme for computing the optimum path (e.g., [63,72]). Also, to further simplify the training algorithm, we can assume that our state observation variables, $\mathbf{y}_n$, are discretized (quantized) and can take values from a finite set of L possible ones, $\{1, 2, \dots, L\}$. This is often the case in practice. Furthermore, assume that the first state is also known. This is, for example, the case for left-to-right models like the one shown in Fig. 16.17. In such a case, we need not compute estimates of the initial probabilities. Thus, the unknown parameters to be estimated are the transition probabilities and the probabilities $P_y(r|i),\ r = 1, 2, \dots, L,\ i = 1, 2, \dots, K$, that is, the probability of emitting symbol r from state i.

Viterbi reestimation: The goal of the algorithm is to obtain the best path and compute the associated cost, say, D, along the path. In the speech literature, the algorithm is also known as the *segmental k-means training* algorithm [63].

Definitions:

- $n_{i|j}$:= number of transitions from state j to state i.
- $n_{\cdot|j}$:= number of transitions originated from state j.
- $n_{i|\cdot}$:= number of transitions terminated at state i.
- $n(r|i)$:= number of times observation $r \in \{1, 2, \dots, L\}$ occurs jointly with state i.

Iterations:

- Initial conditions: Assume the initial estimates of the unknown parameters.
- Step 1: From the available best path, reestimate the new model parameters as

$$P^{(\text{new})}(i|j) = \frac{n_{i|j}}{n_{\cdot|j}},$$

$$P_x^{(\text{new})}(r|i) = \frac{n(r|i)}{n_{i|\cdot}}.$$

- Step 2: For the new model parameters, obtain the best path and compute the corresponding overall cost $D^{(\text{new})}$. Compare it with the cost D of the previous iteration. If $D^{(\text{new})} - D > \epsilon$, set $D = D^{(\text{new})}$ and go to step 1. Otherwise stop.

The Viterbi reestimation algorithm can be shown to converge to a proper characterization of the underlying observations [14].

Remarks 16.4.

- *Scaling*: The probabilities α and β, being less than one, as iterations progress can take very small values. In practice, the dynamic range of their computed values may exceed that of the computer. This phenomenon can be efficiently dealt with within an appropriate scaling. If this is done properly on both α and β, then the effect of scaling cancels out [63].
- *Insufficient training data set*: Generally, a large amount of training data is necessary to learn the HMM parameters. The observation sequence must be sufficiently long with respect to the number of states of the HMM model. This will guarantee that all state transitions will appear a sufficient number of times, so that the reestimation algorithm learns their respective parameters. If this is not the case, a number of techniques have been devised to cope with the issue. For a more detailed treatment, the reader may consult [8,63] and the references therein.

16.5.3 DISCRIMINATIVE LEARNING

Discriminative learning is another path that has attracted a lot of attention. Note that the EM algorithm optimizes the likelihood with respect to the unknown parameters of a *single* HMM in "isolation"; that is, without considering the rest of the HMMs, which model the other words (in the case of speech recognition) or other templates/prototypes that are stored in the database. Such an approach is in line with what we defined as generative learning in Chapter 3. In contrast, the essence of discriminative learning is to optimize the set of parameters so that the models become optimally discriminated over the training sets (e.g., in terms of the error probability criterion). In other words, the parameters describing the different statistical models (HMMs) are optimized in a *combined* way, not individually. The goal is to make the different HMM models as distinct as possible, according to a criterion. This has been an intense line of research and a number of techniques have been developed around criteria that lead to either convex or nonconvex optimization methods (see, for example, [33] and the references therein).

Remarks 16.5.

- Besides the basic HMM scheme, which was described in this section, a number of variants have been proposed in order to overcome some of its shortcomings. For example, alternative modeling paths concern the first-order Markov property and propose models to extend correlations to longer times.
  In the *autoregressive HMM* [11], links are added among the observation nodes of the basic HMM scheme in Fig. 16.15; for example, $\mathbf{y}_n$ is not only linked to $\mathbf{x}_n$, but it shares direct links with, for example, $\mathbf{y}_{n-2}, \mathbf{y}_{n-1}, \mathbf{y}_{n+1}$, and $\mathbf{y}_{n+2}$, if the model extends correlations up to two time instants away. A different concept has been introduced in [56] in the context of *segment modeling*. According to this model, each state is allowed to emit, say, d successive observations, which comprise a segment. The length of the segment, d, is itself a random variable and it is associated with a probability $P(d|k)$, $k = 1, 2, \ldots, K$. In this way, correlation is introduced via the joint distribution of the samples comprising the segment.
- *Variable duration HMM*: A serious shortcoming of the HMMs, which is often observed in practice, is associated with the self-transition probabilities, $P(k|k)$, which are among the model parameters associated with an HMM. Note that the probability of the model being at state k for d successive

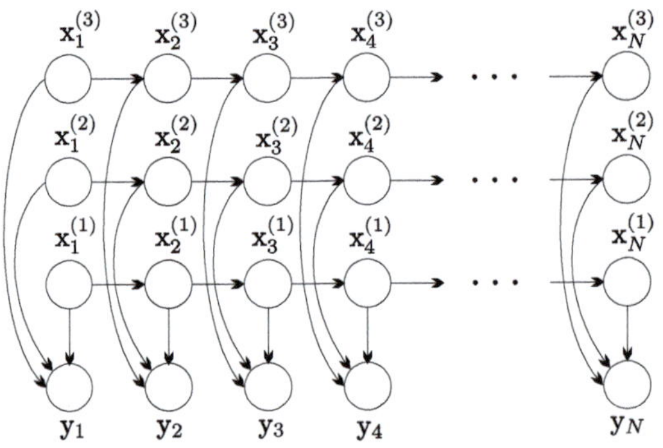

FIGURE 16.20

A factorial HMM with three chains of hidden variables.

instants (initial transition to the state and $d-1$ self-transitions) is given by

$$P_k(d) = (P(k|k))^{d-1}(1 - P(k|k)),$$

where $1 - P(k|k)$ is the probability of leaving the state. For many cases, this exponential state duration dependence is not realistic. In variable-duration HMMs, $P_k(d)$ is explicitly modeled. Different models for $P_k(d)$ can be employed (see, e.g., [46,68,72]).

- Hidden Markov modeling is among the most powerful tools in machine learning and has been widely used in a large number of applications besides speech recognition. Some sampled references are [9] in bioinformatics, [16,36] in communications, [4,73] in optical character recognition (OCR), and [40,61,62] in music analysis/recognition, to name but a few. For a further discussion on HMMs, see, for example, [8,64,72].

16.6 BEYOND HMMS: A DISCUSSION

In this section, some notable extensions of the hidden Markov models, which were previously discussed, are considered in order to meet requirements of applications where either the number of states is large or the homogeneity assumption is no more justified.

16.6.1 FACTORIAL HIDDEN MARKOV MODELS

In the HMMs considered before, the system dynamics is described via the hidden variables, whose graphical representation is a chain. However, such a model may turn out to be too simple for certain applications. A variant of the HMM involves M chains, instead of one chain, where each chain of hidden variables unfolds in time independently of the others. Thus at time n, M hidden variables are involved, denoted as $\mathbf{x}_n^{(m)}$, $m = 1, 2, \ldots, M$ [17,34,81]. The observations occur as a combined emission where all hidden variables are involved. The respective graphical structure is shown in Fig. 16.20 for $M = 3$. Each one of the chains develops on its own, as the graphical model suggests. Such models are known as *factorial HMMs* (FHMMs). One obvious question is, why not use a single chain of hidden variables by increasing the number of possible states? It turns out that such a naive approach would blow up complexity. Take as an example the case of $M = 3$, where for each one of the hidden variables, the number of states is equal to 10. The table of transition probabilities for each chain requires 10^2 entries, which amounts to a total number of 300, i.e., $P_{ij}^{(m)}$, $i, j = 1, 2, \ldots, 10$, $m = 1, 2, 3$. Moreover, the total number of state combinations which can be realized is $10^3 = 1000$. To implement the same number of states via a single chain one would need a table of transition probabilities equal to $(10^3)^2 = 10^6$!

Let $\mathbb{X}_n$ be the M-tuple $(\boldsymbol{x}_n^{(1)}, \ldots, \boldsymbol{x}_n^{(M)})$, where each $\boldsymbol{x}_n^{(m)}$ has only one of its elements equal to 1 (indicating a state) and the rest are zero. Then

$$P(\mathbb{X}_n|\mathbb{X}_{n-1}) = \prod_{m=1}^{M} P^{(m)}\left(\boldsymbol{x}_n^{(m)}|\boldsymbol{x}_{n-1}^{(m)}\right).$$

In [17], the Gaussian distribution was employed for the observations, that is,

$$p(\boldsymbol{y}_n|\mathbb{X}_n) = \mathcal{N}\left(\boldsymbol{y}_n \middle| \sum_{m=1}^{M} \mathcal{M}^{(m)}\boldsymbol{x}_n^{(m)}, \Sigma\right), \tag{16.58}$$

where

$$\mathcal{M}^{(m)} = \left[\boldsymbol{\mu}_1^{(m)}, \ldots, \boldsymbol{\mu}_K^{(m)}\right], \quad m = 1, 2, \ldots, M, \tag{16.59}$$

are the matrices comprising the mean vectors associated with each state and the covariance matrix is assumed to be known and the same for all. The joint probability distribution is given by

$$p(\mathbb{X}_1 \ldots \mathbb{X}_N, Y) = \prod_{m=1}^{M}\left(P^{(m)}\left(\boldsymbol{x}_1^{(m)}\right)\prod_{n=2}^{N} P^{(m)}\left(\boldsymbol{x}_n^{(m)}|\boldsymbol{x}_{n-1}^{(m)}\right)\right) \times \prod_{n=1}^{N} p(\boldsymbol{y}_n|\mathbb{X}_n). \tag{16.60}$$

The challenging task in factional HMMs is complexity. This is illustrated in Fig. 16.21, where the explosion in the size of cliques after performing the moralization and triangulation steps is readily deduced.

In [17], the variational approximation method is adopted to simplify the structure. However, in contrast to the complete factorization scheme, which was adopted for the approximating distribution, Q, in Section 16.3.2 for the Boltzmann machine (corresponding to the removal of all edges in the graph), here the approximating graph will have a more complex structure. Only the edges connected to the output nodes are removed; this results in the graphical structure of Fig. 16.22, for $M = 3$. Because this structure is tractable, there is no need for further simplifications. The approximate conditional distribution, Q, of the simplified structure is parameterized in terms of a set of variational parameters,

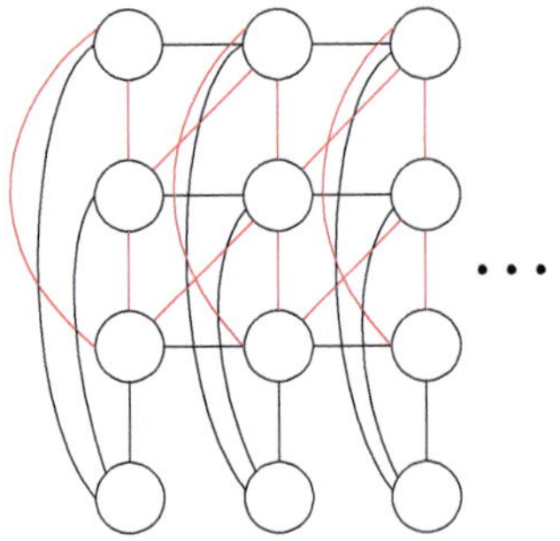

FIGURE 16.21

The graph resulting from a factorial HMM with three chains of hidden variables, after the moralization (linking variables in the same time instant) and triangulation (linking variables between neighboring time instants) steps.

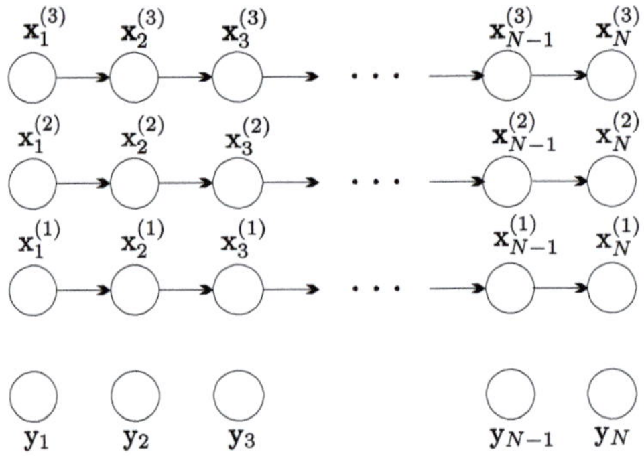

FIGURE 16.22

The simplified graphical structure of an FHMM comprising three chains used in the framework of variational approximation. The nodes associated with the observed variables are delinked.

$\lambda_n^{(m)}$ (one for each delinked node), and it is written as

$$Q(\mathbb{X}_1 \dots \mathbb{X}_N | Y; \boldsymbol{\lambda}) = \prod_{m=1}^{M} \left(\tilde{P}^{(m)} \left(\boldsymbol{x}_1^{(m)} \right) \prod_{n=2}^{N} \tilde{P}^{(m)} \left(\boldsymbol{x}_n^{(m)} | \boldsymbol{x}_{n-1}^{(m)} \right) \right), \tag{16.61}$$

where,

$$\tilde{P}^{(m)} \left(\boldsymbol{x}_n^{(m)} | \boldsymbol{x}_{n-1}^{(m)} \right) = P^{(m)} \left(\boldsymbol{x}_n^{(m)} | (\boldsymbol{x}_{n-1}^{(m)} \right) \lambda_n^{(m)}, \quad m = 2, \dots, M, \; n = 1, 2, \dots, N,$$

and

$$\tilde{P}^{(1)} (\boldsymbol{x}_1) = P^{(1)} (\boldsymbol{x}_1) \lambda_1^{(m)}.$$

The variational parameters are estimated by minimizing the KL distance between Q and the conditional distribution associated with (16.60). This compensates for some of the information loss caused by the removal of the observation nodes. The optimization process renders the variational parameters interdependent; this (deterministic) interdependence can be viewed as an approximation to the probabilistic dependence imposed by the exact structure prior to the approximation.

16.6.2 TIME-VARYING DYNAMIC BAYESIAN NETWORKS

Hidden Markov as well as factorial hidden Markov models are homogeneous; hence, both the structure and the parameters are fixed throughout time. However, such an assumption is not satisfying for a number of applications where the underlying relationships as well as the structural pattern of a system undergoes changes as time evolves. For example, the gene interactions do not remain the same throughout life; the appearance of an object across multiple cameras is continuously changing. For systems that are described by parameters whose values are varying slowly in an interval, we have already discussed a number of alternatives in previous chapters. The theory of graphical models provides the tools to study systems with a mixed set of parameters (discrete and continuous); also, graphical models lend themselves to modeling of nonstationary environments, where step changes are also involved.

One path toward time-varying modeling is to consider graphical models of fixed structure but with time-varying parameters, known as *switching linear dynamic systems* (SLDSs). Such models serve the needs of systems in which a linear dynamic model jumps from one parameter setting to another; hence, the latent variables are both of discrete and of continuous nature. At time instant n, a switch discrete variable, $\mathrm{s}_n \in \{1, 2, \dots, M\}$, selects a single LDS from an available set of M (sub)systems. The dynamics of s_n is also modeled to comply with the Markovian philosophy, and transitions from one LDS to another are governed by $P(\mathrm{s}_n|\mathrm{s}_{n-1})$. This problem has a long history and its origins can be traced back to the time just after the publication of the seminal paper by Kalman [35]; see, for example, [18] and [2,3] for a more recent review of related techniques concerning the approximate inference task in such networks.

Another path is to consider that both the structure as well as the parameters change over time. One route is to adopt a quasistationary rationale, and assume that the data sequence is piece-wise stationary in time (for example, [12,54,66]). Nonstationarity is conceived as a cascade of stationary models, which have previously been learned by presegmented subintervals. The other route assumes that the structure and parameters are continuously changing (for example, [42,79]). An example for the latter case is a Bayesian network where the parents of each node and the parameters, which define the conditional distributions, are time-varying. A separate variable is employed that defines the structure at each time instant; that is, the set of linking directed edges. The concept is illustrated in Fig. 16.23. The method has been applied to the task of active camera tracking [79].

16.7 LEARNING GRAPHICAL MODELS

Learning a graphical model consists of two parts. Given a number of observations, one has to specify both the graphical structure and the associated parameters.

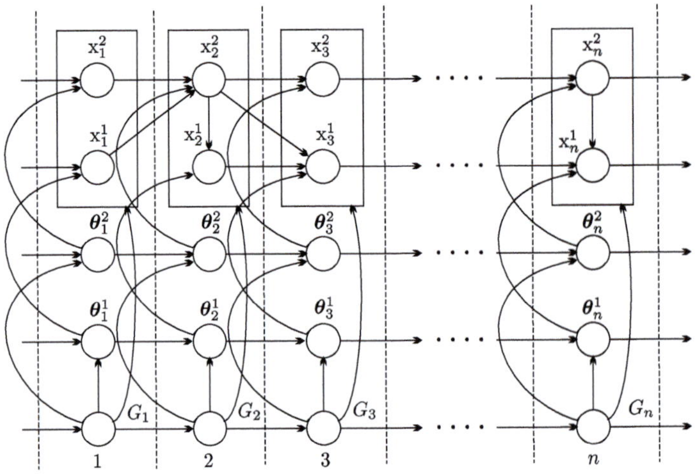

FIGURE 16.23

The figure corresponds to a time-varying dynamic Bayesian network with two variables x_n^1 and x_n^2, $n = 1, 2, \ldots$. The parameters controlling the conditional distributions are considered as separate nodes, $\boldsymbol{\theta}_n^1$ and $\boldsymbol{\theta}_n^2$, respectively, for the two variables. The structure variable, G_n, controls the values of the parameters as well as the structure of the network, which is continuously changing.

16.7.1 PARAMETER ESTIMATION

Once a graphical model has been adopted, one has to estimate the unknown parameters. For example, in a Bayesian network involving discrete variables one has to estimate the values of the conditional probabilities. In Section 16.5, the case of learning the unknown parameters in the context of an HMM was presented. The key point was to maximize the joint PDF over the observed output variables. This is among the most popular criteria used for parameter estimation in different graphical structures. In the HMM case, some of the variables were latent, and hence the EM algorithm was mobilized. If all the variables of the graph can be observed, then the task of parameter learning becomes a typical maximum likelihood one. More specifically, consider a network with l nodes representing the variables $\mathrm{x}_1, \ldots, \mathrm{x}_l$, which are compactly written as a random vector $\mathbf{x}$. Let also $\boldsymbol{x}_1, \boldsymbol{x}_2, \ldots, \boldsymbol{x}_N$ be a set of observations; then

$$\hat{\boldsymbol{\theta}} = \arg\max_{\boldsymbol{\theta}} p(\boldsymbol{x}_1, \boldsymbol{x}_2, \ldots, \boldsymbol{x}_N; \boldsymbol{\theta}),$$

where $\boldsymbol{\theta}$ comprises all the parameters in the graph. If latent variables are involved, then one has to marginalize them out. Any of the parameter estimation techniques that were discussed in Chapters 12 and 13 can be used. Moreover, one can take advantage of the special structure of the graph (i.e., the underlying independencies) to simplify computations. In the HMM case, its Bayesian network structure was exploited by bringing the sum-product algorithm into the game. Besides maximum likelihood, one can adopt any other method related to parameter estimation/inference. For example, one can impose a prior $p(\boldsymbol{\theta})$ on the unknown parameters and resort to a MAP estimation. Moreover, the full Bayesian scenario can also be employed, and we assume the parameters to be random variables. Such a line presupposes that the unknown parameters have been included as extra nodes to the network, linked appropriately to those of the variables that they affect. As a matter of fact, this is what we did in Fig. 13.2, although there, we had not talked about graphical models yet (see also Fig. 16.24). Note that in this case, in order to perform any inference on the variables of the network one should marginalize out the parameters. For example, assume that our l variables correspond to the nodes of a Bayesian network, where the local conditional distributions,

$$p(x_i|\text{Pa}_i; \boldsymbol{\theta}_i), \quad i = 1, 2, \ldots, l,$$

depend on the parameters $\boldsymbol{\theta}_i$. Also, assume that the (random) parameters $\boldsymbol{\theta}_i$, $i = 1, 2, \ldots, l$, are mutually independent. Then the joint distribution over the variables is given by

$$p(x_1, x_2, \ldots, x_l) = \prod_{i=1}^{l} \int_{\boldsymbol{\theta}_i} p(x_i|\text{Pa}_i; \boldsymbol{\theta}_i) p(\boldsymbol{\theta}_i) \, d\boldsymbol{\theta}_i.$$

Using convenient priors, that is, conjugate priors, computations can be significantly facilitated; we have demonstrated such examples in Chapters 12 and 13.

Besides the previous techniques, which are offspring of the generative modeling of the underlying processes, discriminative techniques have also been developed.

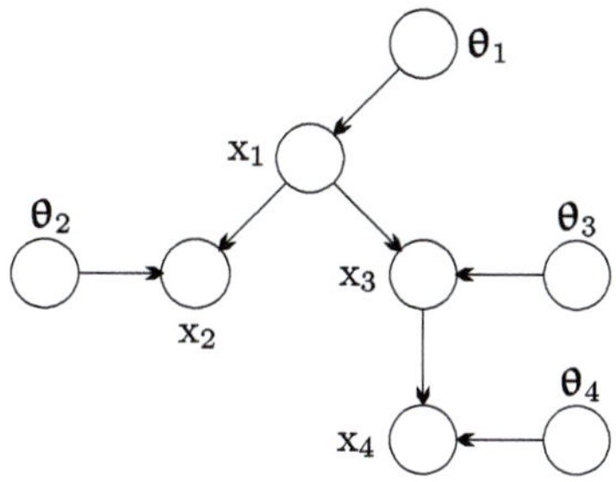

FIGURE 16.24

An example of a Bayesian network, where new nodes associated with the parameters have been included in order to treat parameters as random variables, as required by the Bayesian parameter learning approach.

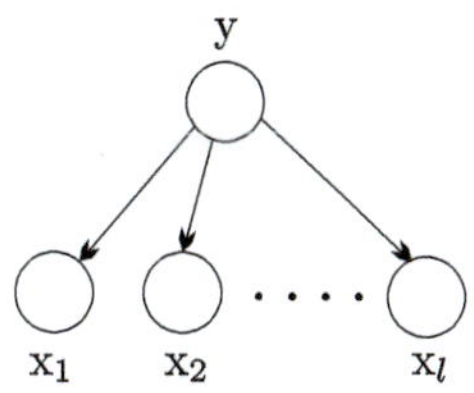

FIGURE 16.25

The Bayesian network associated with the naive Bayes classifier. The joint PDF factorizes as $p(y, x_1, \ldots, x_l) = p(y) \prod_{i=1}^{l} p(x_i|y)$.

In a general setting, let us consider a pattern recognition task where the (output) label variable y and the (input) feature variables $\mathrm{x}_1, \ldots, \mathrm{x}_l$ are jointly distributed according to a distribution that can be factorized over a graph, which is parameterized in terms of a vector parameter $\boldsymbol{\theta}$, i.e., $p(y, x_1, x_2, \ldots, x_l; \boldsymbol{\theta}) := p(y, \boldsymbol{x}; \boldsymbol{\theta})$ [13]. A typical example of such modeling is the naive Bayes classifier, which was discussed in Chapter 7, whose graphical representation is given in Fig. 16.25. For a given set of training data, $(y_n, \boldsymbol{x}_n)$, $n = 1, 2, \ldots, N$, the log-likelihood function becomes

$$L(Y, X; \boldsymbol{\theta}) = \sum_{n=1}^{N} \ln p(y_n, \boldsymbol{x}_n; \boldsymbol{\theta}). \tag{16.62}$$

Estimating $\boldsymbol{\theta}$ by maximizing $L(\cdot, \cdot; \boldsymbol{\theta})$, one would obtain an estimate that guarantees the best (according to the maximum likelihood criterion) fit of the corresponding distribution to the available training set. However, our ultimate goal is not to model the generation "mechanism" of the data. Our ultimate goal is to classify them correctly. Let us rewrite (16.62) as

$$L(Y, X; \boldsymbol{\theta}) = \sum_{n=1}^{N} \ln P(y_n | \boldsymbol{x}_n; \boldsymbol{\theta}) + \sum_{n=1}^{N} \ln p(\boldsymbol{x}_n; \boldsymbol{\theta}).$$

Getting biased toward the classification task, it is more sensible to obtain $\boldsymbol{\theta}$ by maximizing the first of the two terms only, that is,

$$\begin{aligned} L_c(Y, X; \boldsymbol{\theta}) &= \sum_{n=1}^{N} \ln P(y_n | \boldsymbol{x}_n; \boldsymbol{\theta}) \\ &= \sum_{n=1}^{N} \left(\ln p(y_n, \boldsymbol{x}_n; \boldsymbol{\theta}) - \ln \sum_{y_n} p(y_n, \boldsymbol{x}_n; \boldsymbol{\theta}) \right), \end{aligned} \tag{16.63}$$

where the summation over y_n is over all possible values of y_n (classes). This is known as the *conditional log-likelihood* (see, for example [19,20,67]). The resulting estimate, $\hat{\boldsymbol{\theta}}$, guarantees that overall the posterior class probabilities, given the feature values, are maximized over the training data set; after all, Bayesian classification is based on selecting the class of $\boldsymbol{x}$ according to the maximum of the posterior probability. However, one has to be careful. The price one pays for such approaches is that the conditional log-likelihood is not decomposable, and more sophisticated optimization schemes have to be mobilized. Maximizing the conditional log-likelihood does not guarantee that the error probability is also minimized. This can only be guaranteed if one estimates $\boldsymbol{\theta}$ so as to minimize the empirical error probability. However, such a criterion is hard to deal with, as it is not differentiable; attempts to deal with it by using approximate smoothing functions or hill climbing greedy techniques have been proposed (for example, [58] and the references therein). Note that the rationale behind the conditional log-likelihood is closely related to that behind conditional random fields, discussed in Section 15.4.3.

Another route in discriminative learning is to obtain the estimate of $\boldsymbol{\theta}$ by *maximizing the margin*. The probabilistic class margin (for example, [21,59]) is defined as

$$d_n = \min_{y \neq y_n} \frac{P(y_n|\boldsymbol{x}_n;\boldsymbol{\theta})}{P(y|\boldsymbol{x}_n;\boldsymbol{\theta})} = \frac{P(y_n|\boldsymbol{x}_n;\boldsymbol{\theta})}{\max_{y\neq y_n} P(y|\boldsymbol{x}_n;\boldsymbol{\theta})}$$
$$= \frac{p(y_n,\boldsymbol{x}_n;\boldsymbol{\theta})}{\max_{y\neq y_n} p(y,\boldsymbol{x}_n;\boldsymbol{\theta})}.$$

The idea is to estimate $\boldsymbol{\theta}$ so as to maximize the minimum margin over all training data, that is,

$$\hat{\boldsymbol{\theta}} = \arg\max_{\boldsymbol{\theta}} \min(d_1, d_2, \ldots, d_N).$$

The interested reader may also consult [10,55,60] for related reviews and methodologies.

Example 16.4. The goal of this example is to obtain the values in the conditional probability table in a general Bayesian network, which consists of l discrete random nodes/variables, $\mathrm{x}_1, \mathrm{x}_2, \ldots, \mathrm{x}_l$. We assume that all the involved variables can be observed and we have a training set of N observations. The maximum likelihood method will be employed. Let $x_i(n)$, $n = 1, 2, \ldots, N$, denote the nth observation of the ith variable.

The joint PDF under the Bayesian network assumption is given by

$$P(x_1, \ldots, x_l) = \prod_{i=1}^{l} P(x_i|\mathrm{Pa}_i;\boldsymbol{\theta}_i),$$

and the respective log-likelihood is

$$L(X;\boldsymbol{\theta}) = \sum_{n=1}^{N}\sum_{i=1}^{l} \ln P\big(x_i(n)|\mathrm{Pa}_i(n);\boldsymbol{\theta}_i\big).$$

Assuming $\boldsymbol{\theta}_i$ to be disjoint with $\boldsymbol{\theta}_j$, $i \neq j$, optimization over each $\boldsymbol{\theta}_i$, $i = 1, 2, \ldots, l$, can take place separately. This property is referred to as the *global decomposition* of the likelihood function. Thus, it suffices to perform the optimization locally on each node, that is,

$$l(\boldsymbol{\theta}_i) = \sum_{n=1}^{N} \ln P\big(x_i(n)|\mathrm{Pa}_i(n);\boldsymbol{\theta}_i\big), \quad i = 1, 2, \ldots, l. \tag{16.64}$$

Let us now focus on the case where all the involved variables are discrete, and the unknown quantities at any node i are the values of the conditional probabilities in the respective conditional probability table. For notational convenience, denote as $\boldsymbol{h}_i$ the vector comprising the state indices of the parent variables of x_i. Then the respective (unknown) probabilities are denoted as $P_{x_i|\boldsymbol{h}_i}(x_i, \boldsymbol{h}_i)$, for all possible combinations of values of x_i and $\boldsymbol{h}_i$. For example, if all the involved variables are binary and x_i has two parent nodes, then $P_{x_i|\boldsymbol{h}_i}(x_i, \boldsymbol{h}_i)$ can take a total of eight values that have to be estimated. Eq. (16.64) can now be rewritten as

$$l(\boldsymbol{\theta}_i) = \sum_{\boldsymbol{h}_i}\sum_{x_i} s(x_i, \boldsymbol{h}_i) \ln P_{x_i|\boldsymbol{h}_i}(x_i, \boldsymbol{h}_i), \tag{16.65}$$

where $s(x_i, \boldsymbol{h}_i)$ is the number of times the specific combination of $(x_i, \boldsymbol{h}_i)$ appeared in the N samples of the training set. We assume that N is large enough so that all possible combinations occurred at least once, that is, $s(x_i, \boldsymbol{h}_i) \neq 0, \ \forall (x_i, \boldsymbol{h}_i)$. All one has to do now is to maximize (16.65) with respect to $P_{x_i|\boldsymbol{h}_i}(\cdot, \cdot)$, taking into account that

$$\sum_{x_i} P_{x_i|\boldsymbol{h}_i}(x_i, \boldsymbol{h}_i) = 1.$$

Note that $P_{x_i|\boldsymbol{h}_i}$ are independent for different values of $\boldsymbol{h}_i$. Thus, maximization of (16.65) can take place separately for each $\boldsymbol{h}_i$, and it is straightforward to see that

$$\hat{P}_{x_i|\boldsymbol{h}_i} = \frac{s(x_i, \boldsymbol{h}_i)}{\sum_{x_i} s(x_i, \boldsymbol{h}_i)}. \tag{16.66}$$

In words, the maximum likelihood estimate of the unknown conditional probabilities complies with our common sense; given a specific combination of the parent values, $\boldsymbol{h}_i$, $P_{x_i|\boldsymbol{h}_i}$ is approximated by the fraction of times the specific combination $(x_i, \boldsymbol{h}_i)$ appeared in the data set, of the total number of times $\boldsymbol{h}_i$ occurred (relate (16.66) to the Viterbi algorithm in Section 16.5.2). One can now see that in order to obtain good estimates, the number of training points, N, should be large enough so that each combination occurs a sufficiently large number of times. If the average number of parent nodes is large and/or the number of states is large, this poses heavy demands on the size of the training set. This is where parameterization of the conditional probabilities can prove to be very helpful.

16.7.2 LEARNING THE STRUCTURE

In the previous subsection, we considered the structure of the graph to be known and our task was to estimate the unknown parameters. We now turn our attention to learning the structure. In general, this is a much harder task. We only intend to provide a sketch of some general directions.

One path, known as *constrained-based*, is to try to build up a network that satisfies the data independencies, which are "measured" using different statistical tests on the training data set. The method relies a lot on intuition and such methods are not particularly popular in practice.

The other path comes under the name of *s-based* methods. This path treats the task as a typical model selection problem. The score that is chosen to be maximized provides a tradeoff between model complexity and accuracy of the fit to the data. Classical model fitting criteria such as Bayesian information criterion (BIC) and minimum description length (MDL) have been used, among others. The main difficulty with all these criteria is that their optimization is an NP-hard task and the issue is to find appropriate approximate optimization schemes.

The third main path draws its existence from the Bayesian philosophy. Instead of a single structure, an ensemble of structures is employed by embedding appropriate priors into the problem. The readers who are interested in a further and deeper study are referred to more specialized books and papers (for example, [22,41,53]).

PROBLEMS

16.1 Prove that an undirected graph is triangulated if and only if its cliques can be organized into a join tree.

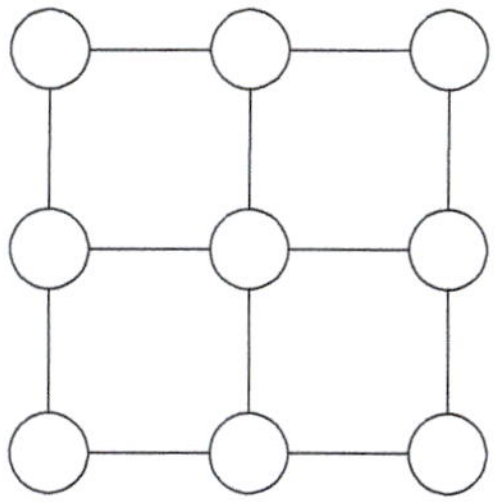

FIGURE 16.26

The graph for Problem 16.6.

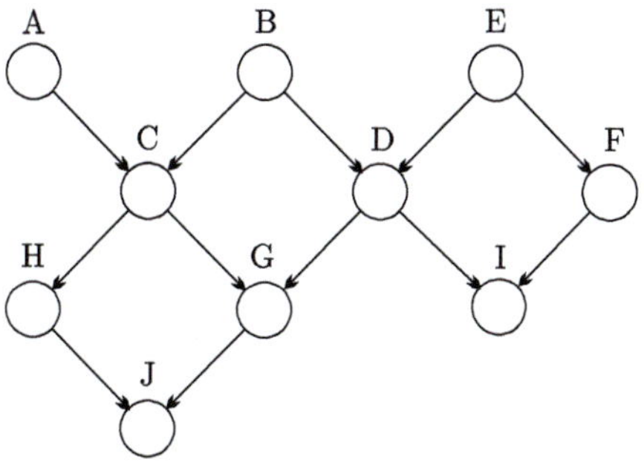

FIGURE 16.27

The Bayesian network structure for Problem 16.7.

16.2 For the graph of Fig. 16.3A, give all possible perfect elimination sequences and draw the resulting sequence of graphs.

16.3 Derive the formulas for the marginal probabilities of the variables in (a) a clique node and (b) in a separator node in a junction tree.

16.4 Prove that in a junction tree the joint PDF of the variables is given by Eq. (16.8).

16.5 Show that obtaining the marginal over a single variable is independent of which one from the clique/separator nodes that contain the variable the marginalization is performed.
Hint: Prove it for the case of two neighboring clique nodes in the junction tree.

16.6 Consider the graph in Fig. 16.26. Obtain a triangulated version of it.

16.7 Consider the Bayesian network structure given in Fig. 16.27. Obtain an equivalent join tree.

16.8 Consider the random variables A, B, C, D, E, F, G, H, I, J and assume that the joint distribution is given by the product of the following potential functions:

$$p = \frac{1}{Z}\psi_1(A, B, C, D)\psi_2(B, E, D)\psi_3(E, D, F, I)\psi_4(C, D, G)\psi_5(C, H, G, I).$$

Construct an undirected graphical model on which the previous joint probability factorizes and in the sequence derive an equivalent junction tree.

16.9 Prove that the function

$$g(x) = 1 - \exp(-x), \quad x > 0,$$

is log-concave.

16.10 Derive the conjugate function of

$$f(x) = \ln(1 - \exp(-x)).$$

16.11 Show that minimizing the bound in (16.12) is a convex optimization task.

16.12 Show that the function $1 + \exp(-x)$, $x \in \mathbb{R}$, is log-convex and derive the respective conjugate one.

16.13 Derive the KL divergence between $P(\mathcal{X}^l|\mathcal{X})$ and $Q(\mathcal{X}^l)$ for the mean field Boltzmann machine and obtain the respective l variational parameters.

16.14 Given a distribution in the exponential family

$$p(\boldsymbol{x}) = \exp\left(\boldsymbol{\theta}^T \boldsymbol{u}(\boldsymbol{x}) - A(\boldsymbol{\theta})\right),$$

show that $A(\boldsymbol{\theta})$ generates the respective mean parameters that define the exponential family

$$\frac{\partial A(\boldsymbol{\theta})}{\partial \theta_i} = \mathbb{E}\left[u_i(\mathbf{x})\right] = \mu_i.$$

Also, show that $A(\boldsymbol{\theta})$ is a convex function.

16.15 Show that the conjugate function of $A(\boldsymbol{\theta})$, associated with an exponential distribution such as that in Problem 16.14, is the corresponding negative entropy function. Moreover, if $\boldsymbol{\mu}$ and $\boldsymbol{\theta}(\boldsymbol{\mu})$ are doubly coupled, then

$$\boldsymbol{\mu} = \mathbb{E}\left[\boldsymbol{u}(\mathbf{x})\right],$$

where $\mathbb{E}[\cdot]$ is with respect to $p(\boldsymbol{x}; \boldsymbol{\theta}(\boldsymbol{\mu}))$.

16.16 Derive a recursion for updating $\gamma(\boldsymbol{x}_n)$ in HMMs independent of $\beta(\boldsymbol{x}_n)$.

16.17 Derive an efficient scheme for prediction in HMM models, that is, to obtain $p(\boldsymbol{y}_{N+1}|Y)$, where $Y = \{\boldsymbol{y}_1, \boldsymbol{y}_2, \ldots, \boldsymbol{y}_N\}$.

16.18 Prove the estimation formulas for the probabilities P_k, $k = 1, 2, \ldots, K$, and P_{ij}, $i, j = 1, 2, \ldots, K$, in the context of the forward-backward algorithm for training HMMs.

16.19 Consider the Gaussian Bayesian network of Section 15.3.5 defined by the local conditional PDFs

$$p(x_i|\text{Pa}_i) = \mathcal{N}\left(x_i \middle| \sum_{k: x_k \in \text{Pa}_i} \theta_{ik} x_k + \theta_{i0}, \sigma^2\right), \quad i = 1, 2, \ldots, l.$$

Assume a set of N observations, $x_i(n)$, $n = 1, 2, \ldots, N$, $i = 1, 2, \ldots, l$, and derive a maximum likelihood estimate of the parameters $\boldsymbol{\theta}$; assume the common variance σ^2 to be known.

REFERENCES

[1] S. Arnborg, D. Cornell, A. Proskurowski, Complexity of finding embeddings in a k-tree, SIAM J. Algebraic Discrete Methods 8 (2) (1987) 277–284.

[2] D. Barber, A.T. Cemgil, Graphical models for time series, IEEE Signal Process. Mag. 27 (2010) 18–28.

[3] D. Barber, Bayesian Reasoning and Machine Learning, Cambridge University Press, Cambridge, 2013.

[4] R. Bartolami, H. Bunke, Hidden Markov model-based ensemble methods for off-line handwritten text line recognition, Pattern Recognit. 41 (11) (2008) 3452–3460.

[5] C.A. Berrou, A. Glavieux, P. Thitimajshima, Near Shannon limit error-correcting coding and decoding: turbo-codes, in: Proceedings IEEE International Conference on Communications, Geneva, Switzerland, 1993.

[6] L. Christensen, J. Zarsen, On data and parameter estimation using the variational Bayesian EM algorithm for block-fading frequency-selective MIMO channels, in: International Conference on Acoustics Speech and Signal Processing, ICASSP, vol. 4, 2006, pp. 465–468.

[7] R.G. Cowell, A.P. Dawid, S.L. Lauritzen, D.J. Spiegehalter, Probabilistic Networks and Expert Systems, Springer-Verlag, New York, 1999.

[8] J. Deller, J. Proakis, J.H.L. Hansen, Discrete-Time Processing of Speech Signals, Macmillan, New York, 1993.

[9] R. Durbin, S. Eddy, A. Krogh, G. Mitchison, Biological Sequence Analysis: Probabilistic Models of Proteins and Nuclear Acids, Cambridge University Press, Cambridge, 1998.

[10] J. Domke, Learning graphical parameters with approximate marginal inference, arXiv:1301.3193v1 [cs,LG], 15 January 2013.

[11] Y. Ephraim, D. Malah, B.H. Juang, On the application of hidden Markov models for enhancing noisy speech, IEEE Trans. Acoust. Speech Signal Process. 37 (12) (1989) 1846–1856.

[12] P. Fearhead, Exact and efficient Bayesian inference for multiple problems, Stat. Comput. 16 (2) (2006) 203–213.

[13] N. Friedman, D. Geiger, M. Goldszmidt, Bayesian network classifiers, Mach. Learn. 29 (1997) 131–163.

[14] K.S. Fu, Syntactic Pattern Recognition and Applications, Prentice Hall, Upper Saddle River, NJ, 1982.

[15] R.G. Gallager, Low density parity-check codes, IEEE Trans. Inf. Theory 2 (1968) 21–28.

[16] C. Georgoulakis, S. Theodoridis, Blind and semi-blind equalization using hidden Markov models and clustering techniques, Signal Process. 80 (9) (2000) 1795–1805.

[17] Z. Ghahramani, M.I. Jordan, Factorial hidden Markov models, Mach. Learn. 29 (1997) 245–273.

[18] Z. Ghahramani, G.E. Hinton, Variational learning for switching state space models, Neural Comput. 12 (4) (1998) 963–996.

[19] R. Greiner, W. Zhou, Structural extension to logistic regression: discriminative parameter learning of belief net classifiers, in: Proceedings 18th International Conference on Artificial Intelligence, 2002, pp. 167–173.

[20] D. Grossman, P. Domingos, Learning Bayesian network classifiers by maximizing conditional likelihood, in: Proceedings 21st International Conference on Machine Learning, Bauff, Canada, 2004.

[21] Y. Guo, D. Wilkinson, D. Schuurmans, Maximum margin Bayesian networks, in: Proceedings, International Conference on Uncertainty in Artificial Intelligence, 2005.

[22] D. Heckerman, D. Geiger, M. Chickering, Learning Bayesian networks: the combination of knowledge and statistical data, Mach. Learn. 20 (1995) 197–243.

[23] B. Hu, I. Land, L. Rasmussen, R. Piton, B. Fleury, A divergence minimization approach to joint multiuser decoding for coded CDMA, IEEE J. Sel. Areas Commun. 26 (3) (2008) 432–445.

[24] A. Ihler, J.W. Fisher, P.L. Moses, A.S. Willsky, Nonparametric belief propagation for self-localization of sensor networks, J. Sel. Areas Commun. 23 (4) (2005) 809–819.

[25] A. Ihler, D. McAllester, Particle belief propagation, in: International Conference on Artificial Intelligence and Statistics, 2009, pp. 256–263.

[26] S. Ikeda, T. Tanaka, S.I. Amari, Information geometry of turbo and low-density parity-check codes, IEEE Trans. Inf. Theory 50 (6) (2004) 1097–1114.

[27] T.S. Jaakola, Variational Methods for Inference and Estimation in Graphical Models, PhD Thesis, Department of Brain and Cognitive Sciences, M.I.T., 1997.

[28] T.S. Jaakola, M.I. Jordan, Recursive algorithms for approximating probabilities in graphical models, in: M.C. Mozer, M.I. Jordan, T. Petsche (Eds.), Proceedings in Advances in Neural Information Processing Systems, NIPS, MIT Press, Cambridge, MA, 1997.

[29] T.S. Jaakola, M.I. Jordan, Improving the mean field approximation via the use of mixture distributions, in: M.I. Jordan (Ed.), Learning in Graphical Models, MIT Press, Cambridge, MA, 1999.

[30] T.S. Jaakola, M.I. Jordan, Variational methods and the QMR-DT database, J. Artif. Intell. Res. 10 (1999) 291–322.
[31] T.S. Jaakola, Tutorial on variational approximation methods, in: M. Opper, D. Saad (Eds.), Advanced Mean Field Methods: Theory and Practice, MIT Press, Cambridge, MA, 2001, pp. 129–160.
[32] F.V. Jensen, Bayesian Networks and Decision Graphs, Springer, New York, 2001.
[33] H. Jiang, X. Li, Parameter estimation of statistical models using convex optimization, IEEE Signal Process. Mag. 27 (3) (2010) 115–127.
[34] M.I. Jordan, Z. Ghahramani, T.S. Jaakola, L.K. Saul, An introduction to variational methods for graphical models, Mach. Learn. 37 (1999) 183–233.
[35] R.E. Kalman, A new approach to linear filtering and prediction problems, Trans. ASME J. Basic Eng. 82 (1960) 34–45.
[36] G.K. Kaleh, R. Vallet, Joint parameter estimation and symbol detection for linear and nonlinear channels, IEEE Trans. Commun. 42 (7) (1994) 2406–2414.
[37] R. Kikuchi, The theory of cooperative phenomena, Phys. Rev. 81 (1951) 988–1003.
[38] G.E. Kirkelund, C.N. Manchon, L.P.B. Christensen, E. Riegler, Variational message-passing for joint channel estimation and decoding in MIMO-OFDM, in: Proceedings, IEEE Globecom, 2010.
[39] U. Kjærulff, Triangulation of Graphs: Algorithms Giving Small Total State Space, Technical Report, R90-09, Aalborg University, Denmark, 1990.
[40] A.P. Klapuri, A.J. Eronen, J.T. Astola, Analysis of the meter of acoustic musical signals, IEEE Trans. Audio Speech Lang. Process. 14 (1) (2006) 342–355.
[41] D. Koller, N. Friedman, Probabilistic Graphical Models: Principles and Techniques, MIT Press, Cambridge, MA, 2009.
[42] M. Kolar, L. Song, A. Ahmed, E.P. Xing, Estimating time-varying networks, Ann. Appl. Stat. 4 (2010) 94–123.
[43] J.B. Kruskal, On the shortest spanning subtree and the travelling salesman problem, Proc. Am. Math. Soc. 7 (1956) 48–50.
[44] S.L. Lauritzen, D.J. Spiegelhalter, Local computations with probabilities on graphical structures and their application to expert systems, J. R. Stat. Soc. B 50 (1988) 157–224.
[45] S.L. Lauritzen, Graphical Models, Oxford University Press, Oxford, 1996.
[46] S.E. Levinson, Continuously variable duration HMMs for automatic speech recognition, Comput. Speech Lang. 1 (1986) 29–45.
[47] D.J.C. MacKay, Good error-correcting codes based on very sparse matrices, IEEE Trans. Inf. Theory 45 (2) (1999) 399–431.
[48] Y. Mao, F.R. Kschischang, B. Li, S. Pasupathy, A factor graph approach to link loss monitoring in wireless sensor networks, J. Sel. Areas Commun. 23 (4) (2005) 820–829.
[49] R.J. McEliece, D.J.C. MacKay, J.F. Cheng, Turbo decoding as an instance of Pearl's belief propagation algorithm, IEEE J. Sel. Areas Commun. 16 (2) (1998) 140–152.
[50] T.P. Minka, Expectation propagation for approximate inference, in: Proceedings 17th Conference on Uncertainty in Artificial Intelligence, Morgan-Kaufmann, San Mateo, 2001, pp. 362–369.
[51] T.P. Minka, Divergence Measures and Message Passing, Technical Report MSR-TR-2005-173, Microsoft Research Cambridge, 2005.
[52] K.P. Murphy, T. Weiss, M.J. Jordan, Loopy belief propagation for approximate inference: an empirical study, in: Proceedings 15th Conference on Uncertainties on Artificial Intelligence, 1999.
[53] K.P. Murphy, Machine Learning: A Probabilistic Perspective, MIT Press, Cambridge, MA, 2012.
[54] S.H. Nielsen, T.D. Nielsen, Adapting Bayesian network structures to non-stationary domains, Int. J. Approx. Reason. 49 (2) (2008) 379–397.
[55] S. Nowozin, C.H. Lampert, Structured learning and prediction in computer vision, Found. Trends Comput. Graph. Vis. 6 (2011) 185–365.
[56] M. Ostendorf, V. Digalakis, O. Kimball, From HMM's to segment models: a unified view of stochastic modeling for speech, IEEE Trans. Audio Speech Process. 4 (5) (1996) 360–378.
[57] J. Pearl, Probabilistic Reasoning in Intelligent Systems: Networks of Plausible Inference, Morgan-Kaufmann, San Mateo, 1988.
[58] F. Pernkopf, J. Bilmes, Efficient heuristics for discriminative structure learning of Bayesian network classifiers, J. Mach. Learn. Res. 11 (2010) 2323–2360.
[59] F. Pernkopf, M. Wohlmayr, S. Tschiatschek, Maximum margin Bayesian network classifiers, IEEE Trans. Pattern Anal. Mach. Intell. 34 (3) (2012) 521–532.
[60] F. Pernkopf, R. Peharz, S. Tschiatschek, Introduction to probabilistic graphical models, in: R. Chellappa, S. Theodoridis (Eds.), E-Reference in Signal Processing, vol. 1, 2013.

[61] A. Pikrakis, S. Theodoridis, D. Kamarotos, Recognition of musical patterns using hidden Markov models, IEEE Trans. Audio Speech Lang. Process. 14 (5) (2006) 1795–1807.
[62] Y. Qi, J.W. Paisley, L. Carin, Music analysis using hidden Markov mixture models, IEEE Trans. Signal Process. 55 (11) (2007) 5209–5224.
[63] L. Rabiner, A tutorial on hidden Markov models and selected applications in speech processing, Proc. IEEE 77 (1989) 257–286.
[64] L. Rabiner, B.H. Juang, Fundamentals of Speech Recognition, Prentice Hall, Upper Saddle River, NJ, 1993.
[65] E. Riegler, G.E. Kirkeland, C.N. Manchon, M.A. Bodin, B.H. Fleury, Merging belief propagation and the mean field approximation: a free energy approach, arXiv:1112.0467v2 [cs.IT], 2012.
[66] J.W. Robinson, A.J. Hartemink, Learning nonstationary dynamic Bayesian networks, J. Mach. Learn. Res. 11 (2010) 3647–3680.
[67] T. Roos, H. Wertig, P. Grunvald, P. Myllmaki, H. Tirvi, On discriminative Bayesian network classifiers and logistic regression, Mach. Learn. 59 (2005) 267–296.
[68] M.J. Russell, R.K. Moore, Explicit modeling of state occupancy in HMMs for automatic speech recognition, in: Proceedings of the Intranational Conference on Acoustics, Speech and Signal Processing, ICASSP, vol. 1, 1985, pp. 5–8.
[69] L.K. Saul, M.I. Jordan, A mean field learning algorithm for unsupervised neural networks, in: M.I. Jordan (Ed.), Learning in Graphical Models, MIT Press, Cambridge, MA, 1999.
[70] Z. Shi, C. Schlegel, Iterative multiuser detection and error control coding in random CDMA, IEEE Trans. Inf. Theory 54 (5) (2006) 1886–1895.
[71] R. Tarjan, M. Yanakakis, Simple linear-time algorithms to test chordality of graphs, test acyclicity of hypergraphs, and selectively reduce acyclic hypergraphs, SIAM J. Comput. 13 (3) (1984) 566–579.
[72] S. Theodoridis, K. Koutroumbas, Pattern Recognition, fourth ed., Academic Press, Boston, 2009.
[73] J.A. Vlontzos, S.Y. Kung, Hidden Markov models for character recognition, IEEE Trans. Image Process. 14 (4) (1992) 539–543.
[74] M.J. Wainwright, T.S. Jaakola, A.S. Willsky, A new class of upper bounds on the log partition function, IEEE Trans. Inf. Theory 51 (7) (2005) 2313–2335.
[75] M.J. Wainwright, M.I. Jordan, A variational principle for graphical models, in: S. Haykin, J. Principe, T. Sejnowski, J. Mcwhirter (Eds.), New Directions in Statistical Signal Processing, MIT Press, Cambridge, MA, 2005.
[76] M.J. Wainwright, Sparse graph codes for side information and binning, IEEE Signal Process. Mag. 24 (5) (2007) 47–57.
[77] M.J. Wainwright, M.I. Jordan, Graphical models, exponential families, and variational inference, Found. Trends Mach. Learn. 1 (1–2) (2008) 1–305.
[78] J.M. Walsh, P.A. Regalia, Belief propagation, Dykstra's algorithm, and iterated information projections, IEEE Trans. Inf. Theory 56 (8) (2010) 4114–4128.
[79] Z. Wang, E.E. Kuruoglu, X. Yang, T. Xu, T.S. Huang, Time varying dynamic Bayesian network for nonstationary events modeling and online inference, IEEE Trans. Signal Process. 59 (2011) 1553–1568.
[80] W. Wiegerinck, Variational approximations between mean field theory and the junction tree algorithm, in: Proceedings 16th Conference on Uncertainty in Artificial Intelligence, 2000.
[81] C.K.I. Williams, G.E. Hinton, Mean field networks that learn to discriminate temporally distorted strings, in: D.S. Touretzky, J.L. Elman, T.J. Sejnowski, G.E. Hinton (Eds.), Proceedings of 1990 Connectionist Models Summer School, Morgan-Kauffman, San Mateo, CA, 1991.
[82] J. Win, C.M. Bishop, Variational massage passing, J. Mach. Learn. Res. 6 (2005) 661–694.
[83] J. Yedidia, W.T. Freeman, T. Weiss, Generalized belief propagation, in: Advances on Neural Information Processing System, NIPS, MIT Press, Cambridge, MA, 2001, pp. 689–695.
[84] J. Yedidia, W.T. Freeman, T. Weiss, Understanding Belief Propagation and Its Generalization, Technical Report TR-2001-22, Mitsubishi Electric Research Laboratories, 2001.
[85] A.L. Yuille, CCCP algorithms to minimize the Bethe and Kikuchi free energies: convergent alternatives to belief propagation, Neural Comput. 14 (7) (2002) 1691–1722.

CHAPTER

PARTICLE FILTERING

17

CONTENTS

17.1 INTRODUCTION

This chapter is a follow-up to Chapter 14, whose focus was on Monte Carlo methods. Our interest now turns to a special type of sampling techniques known as sequential-sampling methods. In contrast to the Monte Carlo methods, considered in Chapter 14, here we will assume that distributions from which we want to sample are time-varying, and that sampling will take place in a sequential fashion. The main emphasis of this chapter is on particle filtering techniques for inference in state-space dynamic models. In contrast to the classical form of Kalman filtering, here the model is allowed to be nonlinear and/or the distributions associated with the involved variables non-Gaussians.

17.2 SEQUENTIAL IMPORTANCE SAMPLING

Our interest in this section shifts toward tasks where data are sequentially arriving, and our goal becomes that of sampling from their joint distribution. In other words, we are receiving observations $\boldsymbol{x}_n \in \mathbb{R}^l$ of random vectors $\mathbf{x}_n$. At some time n, let $\boldsymbol{x}_{1:n} = \{\boldsymbol{x}_1, \ldots, \boldsymbol{x}_n\}$ denote the set of the available samples and let the respective joint distribution be denoted as $p_n(\boldsymbol{x}_{1:n})$. No doubt, this new task is to be treated with special care. Not only is the dimensionality of the task (number of random variables, i.e., $\mathbf{x}_{1:n}$) now time-varying, but also, after some time has elapsed, the dimensionality will be very large and,

Machine Learning. https://doi.org/10.1016/B978-0-12-818803-3.00029-5

in general, we expect the corresponding distribution, $p_n(\boldsymbol{x}_{1:n})$, to be of a rather complex form. Moreover, at time instant n, even if we knew how to sample from $p_n(\boldsymbol{x}_{1:n})$, the required time for sampling would be at least of the order of n. Hence, as n increases, even such a case could not be computationally feasible for large values of n. Sequential sampling is of particular interest in dynamic systems in the context of particle filtering, and we will come to deal with such systems very soon. Our discussion will develop around the importance sampling method, which was introduced in Section 14.5.

17.2.1 IMPORTANCE SAMPLING REVISITED

Recall from Eq. (14.28) that given (a) a function $f(\boldsymbol{x})$, (b) a desired distribution $p(\boldsymbol{x}) = \frac{1}{Z}\phi(\boldsymbol{x})$, and (c) a proposal distribution $q(\boldsymbol{x})$, we have

$$\mathbb{E}\left[f(\mathbf{x})\right] := \mu_f \simeq \sum_{i=1}^{N} W(\boldsymbol{x}_i) f(\boldsymbol{x}_i) := \hat{\mu}, \tag{17.1}$$

where $\boldsymbol{x}_i$ are samples drawn from $q(\boldsymbol{x})$. Recall, also, that the estimate

$$\hat{Z} = \frac{1}{N}\sum_{i=1}^{N} w(\boldsymbol{x}_i) \tag{17.2}$$

defines an unbiased estimator of the true normalizing constant Z, where $w(\boldsymbol{x}_i)$ are the nonnormalized weights, $w(\boldsymbol{x}_i) = \frac{\phi(\boldsymbol{x}_i)}{q(\boldsymbol{x}_i)}$. Note that the approximation in Eq. (17.1) equivalently implies the following approximation of the desired distribution:

$$\boxed{p(\boldsymbol{x}) \simeq \sum_{i=1}^{N} W(\boldsymbol{x}_i)\delta(\boldsymbol{x} - \boldsymbol{x}_i): \quad \text{discrete random measure approximation.}} \tag{17.3}$$

In other words, even a continuous PDF is approximated by a set of discrete points and weights assigned to them. We say that the distribution is approximated by a *discrete random measure* defined by the *particles* $\boldsymbol{x}_i$, $i = 1, 2, \ldots, N$, with respective normalized weights $W(\boldsymbol{x}_i) := W^{(i)}$. The approximating random measure is denoted as $\chi = \{\boldsymbol{x}_i, W^{(i)}\}_{i=1}^{N}$.

Also, we have already commented in Section 14.5 that a major drawback of importance sampling is the large variance of the weights, which becomes more severe in high-dimensional spaces, where our interest will be from now on. Let us elaborate on this variance problem a bit more and seek ways to bypass/reduce this undesired behavior.

It can be shown (e.g., [33] and Problem 17.1) that the variance of the corresponding estimator, $\hat{\mu}$, in Eq. (17.1) is given by

$$\text{var}[\hat{\mu}] = \frac{1}{N}\left(\int \frac{f^2(\boldsymbol{x})p^2(\boldsymbol{x})}{q(\boldsymbol{x})}d\boldsymbol{x} - \mu_f^2\right). \tag{17.4}$$

Observe that if the numerator $f^2(\boldsymbol{x})p^2(\boldsymbol{x})$ tends to zero slower than $q(\boldsymbol{x})$ does, then for fixed N, the variance $\text{var}[\hat{\mu}] \longrightarrow \infty$. This demonstrates the significance of selecting q very carefully. It is not difficult to see, by minimizing Eq. (17.4), that the optimal choice for $q(\boldsymbol{x})$, leading to the minimum (zero) variance, is proportional to the product $f(\boldsymbol{x})p(\boldsymbol{x})$. We will make use of this result later on. Note, of course, that the proportionality constant is $1/\mu_f$, which is not known. Thus, this result can only be considered as a benchmark.

Concerning the variance issue, let us turn our attention to the unbiased estimator $\hat{Z}$ of Z in Eq. (17.2). It can be shown (Problem 17.2) that

$$\mathrm{var}[\hat{Z}] = \frac{Z^2}{N}\left(\int \frac{p^2(\boldsymbol{x})}{q(\boldsymbol{x})}d\boldsymbol{x} - 1\right). \tag{17.5}$$

By its definition, the variance of $\hat{Z}$ is directly related to the variance of the weights. It turns out that, in practice, the variance in Eq. (17.5) exhibits an exponential dependence on the dimensionality (e.g., [11, 15] and Problem 17.5). In such cases, the number of samples, N, has to be excessively large in order to keep the variance relatively small. One way to *partially* cope with the variance-related problem is the *resampling* technique.

17.2.2 RESAMPLING

Resampling is a very intuitive approach where one attempts a *randomized* pruning of the available samples (particles), drawn from q, by (most likely) discarding those associated with low weights and replacing them with samples whose weights have larger values. This is achieved by drawing samples from the approximation of $p(\boldsymbol{x})$, denoted as $\hat{p}(\boldsymbol{x})$, which is based on the discrete random measure $\{\boldsymbol{x}_i, W^{(i)}\}_{i=1}^N$, in Eq. (17.3). In importance sampling, the involved particles are drawn from $q(\boldsymbol{x})$ and the weights are appropriately computed in order to "match" the desired distribution. Adding the extra step of resampling, a *new* set of *unweighted* samples is drawn from the discrete approximation $\hat{p}$ of p. Using the resampling step, we still obtain samples that are approximately distributed as p; moreover, particles of low weight have been removed with high probability and thereby, for the next time instant, the probability of exploring regions with larger probability masses is increased. There are different ways of sampling from a discrete distribution.

- *Multinomial resampling.* This method is equivalent to the one presented in Example 14.2. Each particle $\boldsymbol{x}_i$ is associated with a probability $P_i = W^{(i)}$. Redrawing N (new) particles will generate from each particle, $\boldsymbol{x}_i$, $N^{(i)}$ "offsprings" ($\sum_{i=1}^N N^{(i)} = N$), depending on their respective probability P_i. Hence, $N^{(1)}, \ldots, N^{(N)}$ will follow a multinomial distribution (Section 2.3), that is,

$$P(N^{(1)}, \ldots, N^{(N)}) = \binom{N}{N^{(1)} \ldots N^{(N)}} \prod_{i=1}^N P_i^{N^{(i)}}.$$

  In this way, the higher the probability (weight $W^{(i)}$) of an originally drawn particle, the higher the number of times, $N^{(i)}$, this particle will be redrawn.

  The new discrete estimate of the desired distribution will now be given by

$$\bar{p}(\boldsymbol{x}) = \sum_{i=1}^N \frac{N^{(i)}}{N}\delta(\boldsymbol{x} - \boldsymbol{x}_i).$$

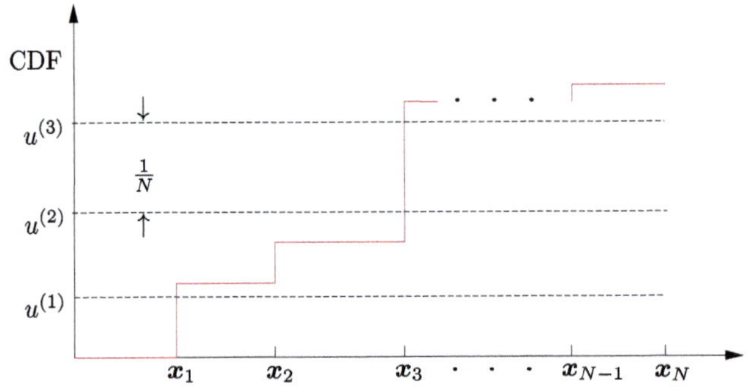

FIGURE 17.1

The sample $u^{(1)}$ drawn from $\mathcal{U}\left(0, \frac{1}{N}\right)$ determines the first point that defines the set of N equidistant lines, which are drawn and cut across the cumulative distribution function (CDF). The respective intersections determine the number of times, $N^{(i)}$, the corresponding particle, $\boldsymbol{x}_i$, will be represented in the set. For the case of the figure, $\boldsymbol{x}_1$ will appear once, $\boldsymbol{x}_2$ is missed, and $\boldsymbol{x}_3$ appears two times.

From the properties of the multinomial distribution, we have $\mathbb{E}[\mathrm{N}^{(i)}] = NP_i = NW^{(i)}$, and hence $\bar{p}(\boldsymbol{x})$ is an unbiased approximation of $\hat{p}(\boldsymbol{x})$.

- *Systematic resampling*. Systematic resampling is a variant of the multinomial approach. Recall from Example 14.2 that every time a particle is to be (re)drawn, a new sample is generated from the uniform distribution $\mathcal{U}(0, 1)$. In contrast, in systematic resampling, the process is not entirely random. To generate N particles, we only select randomly one sample, $u^{(1)} \sim \mathcal{U}(0, \frac{1}{N})$. Then define

$$u^{(j)} = u^{(1)} + \frac{j-1}{N}, \quad j = 2, 3, \ldots, N,$$

and set

$$N^{(i)} = \text{card}\left\{\text{All } j : \sum_{k=1}^{i-1} W^{(k)} \le u^{(j)} < \sum_{k=1}^{i} W^{(k)}\right\},$$

where card$\{\cdot\}$ denotes the cardinality of the respective set. Fig. 17.1 illustrates the method. The resampling algorithm is summarized next.

Algorithm 17.1 (Resampling).

- Initialization
  - Input the samples $\boldsymbol{x}_i$ and respective weights $W^{(i)}$, $i = 1, 2, \ldots, N$.
  - $c_0 = 0$, $N^{(i)}$=0, $i = 1, 2, \ldots, N$.
- **For** $i = 1, 2, \ldots, N$, **Do**
  - $c_i = c_{i-1} + W^{(i)}$; Construct CDF.
- **End For**
- Draw $u^{(1)} \sim \mathcal{U}(0, \frac{1}{N})$

- $i = 1$
- **For** $j = 1, 2, \dots, N$, **Do**
  - $u^{(j)} = u^{(1)} + \frac{j-1}{N}$
  - **While** $u^{(j)} > c_i$
    - $i = i + 1$
  - **End While**
  - $\bar{\boldsymbol{x}}_j = \boldsymbol{x}_i$; Assign sample.
  - $N^{(i)} = N^{(i)} + 1$
- **End For**

The output comprises the new samples $\bar{\boldsymbol{x}}_j$, $j = 1, 2, \dots, N$, and all the weights are set equal to $\frac{1}{N}$. The sample $\boldsymbol{x}_i$ will now appear, after resampling, $N^{(i)}$ times.

The two previously stated resampling methods are not the only possibilities (see, e.g., [12]). However, the systematic resampling is the one that is usually adopted due mainly to its easy implementation. Systematic resampling was introduced in [25].

Resampling schemes result in estimates that converge to their true values, as long as the number of particles tends to infinity (Problem 17.3).

17.2.3 SEQUENTIAL SAMPLING

Let us now apply the experience we have gained in importance sampling to the case of sequentially arriving particles. The first examples of such techniques date back to the 1950s (e.g., [19,36]). At time n, our interest is to draw samples from the joint distribution

$$p_n(\boldsymbol{x}_{1:n}) = \frac{\phi_n(\boldsymbol{x}_{1:n})}{Z_n}, \tag{17.6}$$

based on a proposal distribution $q_n(\boldsymbol{x}_{1:n})$, where Z_n is the normalizing constant at time n. However, we are going to set the same goal that we have adopted for any time-recursive setting throughout this book, that is, to keep computational complexity *fixed, independent* of the time instant, n. Such a rationale dictates a *time-recursive computation* of the involved quantities. To this end, we select a proposal distribution of the form

$$q_n(\boldsymbol{x}_{1:n}) = q_{n-1}(\boldsymbol{x}_{1:n-1}) q_n(\boldsymbol{x}_n | \boldsymbol{x}_{1:n-1}). \tag{17.7}$$

From Eq. (17.7), it is readily seen that

$$q_n(\boldsymbol{x}_{1:n}) = q_1(\boldsymbol{x}_1) \prod_{k=2}^{n} q_k(\boldsymbol{x}_k | \boldsymbol{x}_{1:k-1}). \tag{17.8}$$

This means that one has only to choose $q_k(\boldsymbol{x}_k | \boldsymbol{x}_{1:k-1})$, $k = 2, 3, \dots, n$, together with the initial (prior) $q_1(\boldsymbol{x}_1)$. Note that the dimensionality of the involved random vector in $q_k(\cdot|\cdot)$, given the past, remains *fixed* for all time instants. Eq. (17.8), viewed from another angle, reveals that in order to draw a single (multivariate) sample that spans the time interval up to time n, that is, $\boldsymbol{x}_{1:n}^{(i)} = \{\boldsymbol{x}_1^{(i)}, \boldsymbol{x}_2^{(i)}, \dots, \boldsymbol{x}_n^{(i)}\}$, we build it up *recursively*; we first draw $\boldsymbol{x}_1^{(i)} \sim q_1(\boldsymbol{x})$ and then draw $\boldsymbol{x}_k^{(i)} \sim q_k(\boldsymbol{x}|\boldsymbol{x}_{1:k-1}^{(i)})$, $k = 2, 3, \ldots, n$. The corresponding nonnormalized weights are also computed recursively [15]. Indeed,

$$\begin{aligned} w_n(\boldsymbol{x}_{1:n}) &:= \frac{\phi_n(\boldsymbol{x}_{1:n})}{q_n(\boldsymbol{x}_{1:n})} = \frac{\phi_{n-1}(\boldsymbol{x}_{1:n-1})}{q_n(\boldsymbol{x}_{1:n})} \frac{\phi_n(\boldsymbol{x}_{1:n})}{\phi_{n-1}(\boldsymbol{x}_{1:n-1})} \\ &= \frac{\phi_{n-1}(\boldsymbol{x}_{1:n-1})}{q_{n-1}(\boldsymbol{x}_{1:n-1})} \frac{\phi_n(\boldsymbol{x}_{1:n})}{\phi_{n-1}(\boldsymbol{x}_{1:n-1}) q_n(\boldsymbol{x}_n|\boldsymbol{x}_{1:n-1})} \\ &= w_{n-1}(\boldsymbol{x}_{1:n-1}) a_n(\boldsymbol{x}_{1:n}) \\ &= w_1(\boldsymbol{x}_1) \prod_{k=2}^{n} a_k(\boldsymbol{x}_{1:k}), \end{aligned} \tag{17.9}$$

where

$$a_k(\boldsymbol{x}_{1:k}) := \frac{\phi_k(\boldsymbol{x}_{1:k})}{\phi_{k-1}(\boldsymbol{x}_{1:k-1}) q_k(\boldsymbol{x}_k|\boldsymbol{x}_{1:k-1})}, \quad k = 2, 3, \ldots, n. \tag{17.10}$$

The question that is now raised is how to choose $q_n(\boldsymbol{x}_n|\boldsymbol{x}_{1:n-1})$, $n = 2, 3, \ldots$. A sensible strategy is to select it in order to minimize the variance of the weight $w_n(\boldsymbol{x}_{1:n})$, given the samples $\boldsymbol{x}_{1:n-1}$. It turns out that the optimal value, which actually makes the variance zero (Problem 17.4), is given by

$$\boxed{q_n^{opt}(\boldsymbol{x}_n|\boldsymbol{x}_{1:n-1}) = p_n(\boldsymbol{x}_n|\boldsymbol{x}_{1:n-1}) : \text{ optimal proposal distribution.}} \tag{17.11}$$

However, most often in practice, $p_n(\boldsymbol{x}_n|\boldsymbol{x}_{1:n-1})$ is not easy to sample and one has to be content with adopting some approximation of it. We are now ready to state our first algorithm for sequential importance sampling (SIS).

Algorithm 17.2 (Sequential importance sampling).

- Select $q_1(\cdot)$, $q_n(\cdot|\cdot)$, $n = 2, 3, \ldots$
- Select number of particles, N.
- **For** $i = 1, 2, \ldots, N$, **Do**; Initialize N different realizations/streams.
  - Draw $\boldsymbol{x}_1^{(i)} \sim q_1(\boldsymbol{x})$
  - Compute the weights $w_1(\boldsymbol{x}_1^{(i)}) = \frac{\phi_1(\boldsymbol{x}_1^{(i)})}{q_1(\boldsymbol{x}_1^{(i)})}$
- **End For**
- **For** $i = 1, 2, \ldots, N$, **Do**
  - Compute the normalized weights $W_1^{(i)}$.
- **End For**
- **For** $n = 2, 3, \ldots$, **Do**
  - **For** $i = 1, 2, \ldots, N$, **Do**
    * Draw $\boldsymbol{x}_n^{(i)} \sim q_n(\boldsymbol{x}|\boldsymbol{x}_{1:n-1}^{(i)})$
    * Compute the weights
      · $w_n(\boldsymbol{x}_{1:n}^{(i)}) = w_{n-1}(\boldsymbol{x}_{1:n-1}^{(i)}) a_n(\boldsymbol{x}_{1:n}^{(i)})$; from Eq. (17.9).

  - **End For**
  - **For** $i = 1, 2, \ldots, N$, **Do**
    * $W_n^{(i)} \propto w_n(\boldsymbol{x}_{1:n}^{(i)})$
  - **End For**
- **End For**

Once the algorithm has been completed, we can write

$$\hat{p}_n(\boldsymbol{x}_{1:n}) = \sum_{i=1}^{N} W_n^{(i)} \delta(\boldsymbol{x}_{1:n} - \boldsymbol{x}_{1:n}^{(i)}).$$

However, as we have already said, the variance of the weights has the tendency to increase with n (see Problem 17.5). Thus, the resampling version of the sequential importance sampling is usually employed.

Algorithm 17.3 (SIS with resampling).

- Select $q_1(\cdot)$, $q_n(\cdot|\cdot)$, $n = 1, 2, \ldots$
- Select number of particles N.
- **For** $i = 1, 2, \ldots, N$, **Do**
  - Draw $\boldsymbol{x}_1^{(i)} \sim q_1(\boldsymbol{x})$
  - Compute the weights $w_1(\boldsymbol{x}_1^{(i)}) = \frac{\phi_1(\boldsymbol{x}_1^{(i)})}{q_1(\boldsymbol{x}_1^{(i)})}$.
- **End For**
- **For** $i = 1, \ldots, N$, **Do**
  - Compute the normalized weights $W_1^{(i)}$
- **End For**
- Resample $\{\boldsymbol{x}_1^{(i)}, W_1^{(i)}\}_{i=1}^N$ to obtain $\{\bar{\boldsymbol{x}}_1^{(i)}, \frac{1}{N}\}_{i=1}^N$, using Algorithm 17.1
- **For** $n = 2, 3, \ldots,$ **Do**
  - **For** $i = 1, 2, \ldots, N$, **Do**
    * Draw $\boldsymbol{x}_n^{(i)} \sim q_n(\boldsymbol{x}|\bar{\boldsymbol{x}}_{1:n-1}^{(i)})$
    * Set $\boldsymbol{x}_{1:n}^{(i)} = \{\boldsymbol{x}_n^{(i)}, \bar{\boldsymbol{x}}_{1:n-1}^{(i)}\}$
    * Compute $w_n(\boldsymbol{x}_{1:n}^{(i)}) = \frac{1}{N}\alpha_n(\boldsymbol{x}_{1:n}^{(i)})$; (Eq. (17.9)).
  - **End For**
  - **For** $i = 1, 2, \ldots, N$, **Do**
    * Compute $W_n^{(i)}$
  - **End Do**
  - Resample $\{\boldsymbol{x}_{1:n}^{(i)}, W_n^{(i)}\}_{i=1}^N$ to obtain $\{\bar{\boldsymbol{x}}_{1:n}^{(i)}, \frac{1}{N}\}_{i=1}^N$
- **End For**

Remarks 17.1.

- Convergence results concerning sequential importance sampling can be found in, for example, [5–7]. It turns out that, in practice, the use of resampling leads to substantially smaller variances.
- From a practical point of view, sequential importance methods with resampling are expected to work reasonably well if the desired successive distributions at different time instants do not differ much and the choice of $q_n(\boldsymbol{x}_n|\boldsymbol{x}_{1:n-1})$ is *close to the optimal* one (see, e.g., [15]).

17.3 KALMAN AND PARTICLE FILTERING

Particle filtering is an instance of the sequential Monte Carlo methods. Particle filtering is a technique born in the 1990s and it was first introduced in [18] as an attempt to solve estimation tasks in the context of state-space modeling for the more general nonlinear and non-Gaussian scenarios. The term "particle filtering" was coined in [3], although the term "particle" had been used in [25].

Hidden Markov models (HMMs), which are treated in Section 16.4, and Kalman filters, treated in Chapter 4, are special types of state-space (state-observation) modeling. The former address the case of discrete state (latent) variables and the latter the continuous case, albeit in the very special case of linear and Gaussian scenario. In particle filtering, the interest shifts to models of the following form:

$$\mathbf{x}_n = \boldsymbol{f}_n(\mathbf{x}_{n-1}, \boldsymbol{\eta}_n): \quad \text{state equation,} \tag{17.12}$$

$$\mathbf{y}_n = \boldsymbol{h}_n(\mathbf{x}_n, \mathbf{v}_n): \quad \text{observations equation,} \tag{17.13}$$

where $\boldsymbol{f}_n$ and $\boldsymbol{h}_n$ are nonlinear, in general, (vector) functions, $\boldsymbol{\eta}_n$ and $\mathbf{v}_n$ are noise sequences, and the dimensions of $\mathbf{x}_n$ and $\mathbf{y}_n$ can be different. The random vector $\mathbf{x}_n$ is the (latent) state vector and $\mathbf{y}_n$ corresponds to the observations. There are two inference tasks that are of interest in practice.

Filtering: Given the set of observations, $\boldsymbol{y}_{1:n}$, in the time interval $[1, n]$, compute

$$p(\boldsymbol{x}_n|\boldsymbol{y}_{1:n}).$$

Smoothing: Given the set of observations $\boldsymbol{y}_{1:N}$ in a time interval $[1, N]$, compute

$$p(\boldsymbol{x}_n|\boldsymbol{y}_{1:N}), \quad 1 \le n \le N.$$

Before we proceed to our main goal, let us review the simpler case, that of Kalman filters, this time from a Bayesian viewpoint.

17.3.1 KALMAN FILTERING: A BAYESIAN POINT OF VIEW

Kalman filtering was discussed in Section 4.10 in the context of linear estimation methods and the mean-square error criterion. In the current section, the Kalman filtering algorithm will be rederived following concepts from the theory of graphical models and Bayesian networks, which are treated in Chapters 15 and 16. This probabilistic view will then be used for the subsequent nonlinear generalizations in the framework of particle filtering. For the linear case model, Eqs. (17.12) and (17.13)

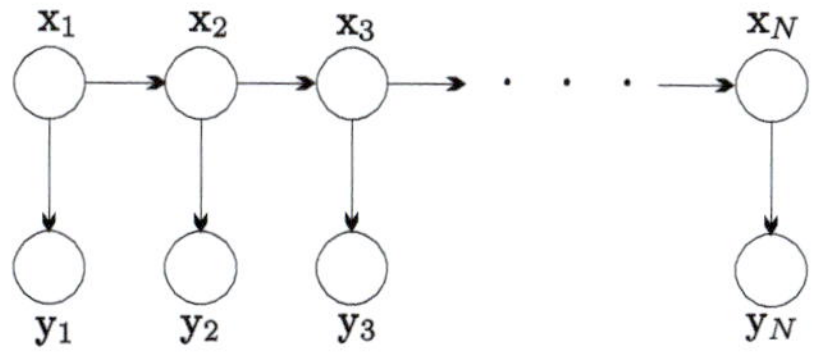

FIGURE 17.2

Graphical model corresponding to the state-space modeling for Kalman and particle filters.

become

$$\mathbf{x}_n = F_n \mathbf{x}_{n-1} + \boldsymbol{\eta}_n, \tag{17.14}$$

$$\mathbf{y}_n = H_n \mathbf{x}_n + \mathbf{v}_n, \tag{17.15}$$

where F_n and H_n are matrices of appropriate dimensions. We further assume that the two noise sequences are statistically independent and of a Gaussian nature, that is,

$$p(\boldsymbol{\eta}_n) = \mathcal{N}(\boldsymbol{\eta}_n|\mathbf{0}, Q_n), \tag{17.16}$$

$$p(\boldsymbol{v}_n) = \mathcal{N}(\boldsymbol{v}_n|\mathbf{0}, R_n). \tag{17.17}$$

The kick-off point for deriving the associated recursions is the Bayes rule,

$$\begin{aligned} p(\boldsymbol{x}_n|\boldsymbol{y}_{1:n}) &= \frac{p(\boldsymbol{y}_n|\boldsymbol{x}_n, \boldsymbol{y}_{1:n-1})p(\boldsymbol{x}_n|\boldsymbol{y}_{1:n-1})}{Z_n} \\ &= \frac{p(\boldsymbol{y}_n|\boldsymbol{x}_n)p(\boldsymbol{x}_n|\boldsymbol{y}_{1:n-1})}{Z_n}, \end{aligned} \tag{17.18}$$

where

$$\begin{aligned} Z_n &= \int p(\boldsymbol{y}_n|\boldsymbol{x}_n)p(\boldsymbol{x}_n|\boldsymbol{y}_{1:n-1})\,d\boldsymbol{x}_n \\ &= p(\boldsymbol{y}_n|\boldsymbol{y}_{1:n-1}), \end{aligned} \tag{17.19}$$

and we have used the fact that $p(\boldsymbol{y}_n|\boldsymbol{x}_n, \boldsymbol{y}_{1:n-1}) = p(\boldsymbol{y}_n|\boldsymbol{x}_n)$, which is a consequence of Eq. (17.15). For those who have already read Chapter 15, recall that Kalman filtering is a special case of a Bayesian network and corresponds to the graphical model given in Fig. 17.2. Hence, due to the Markov property, $\mathbf{y}_n$ is independent of the past given the values in $\boldsymbol{x}_n$. Moreover, note that

$$\begin{aligned} p(\boldsymbol{x}_n|\boldsymbol{y}_{1:n-1}) &= \int p(\boldsymbol{x}_n|\boldsymbol{x}_{n-1}, \boldsymbol{y}_{1:n-1})p(\boldsymbol{x}_{n-1}|\boldsymbol{y}_{1:n-1})\,d\boldsymbol{x}_{n-1} \\ &= \int p(\boldsymbol{x}_n|\boldsymbol{x}_{n-1})p(\boldsymbol{x}_{n-1}|\boldsymbol{y}_{1:n-1})\,d\boldsymbol{x}_{n-1}, \end{aligned} \tag{17.20}$$

where, once more, the Markov property (i.e., Eq. (17.14)) has been used.

Eqs. (17.18)–(17.20) comprise the set of recursions which lead to the update

$$p(\boldsymbol{x}_{n-1}|\boldsymbol{y}_{1:n-1}) \longrightarrow p(\boldsymbol{x}_n|\boldsymbol{y}_{1:n}),$$

starting from the initial (prior) $p(\boldsymbol{x}_0|\boldsymbol{y}_0) := p(\boldsymbol{x}_0)$. If $p(\boldsymbol{x}_0)$ is chosen to be Gaussian, then all the involved PDFs turn out to be Gaussian due to Eqs. (17.16) and (17.17) and the linearity of Eqs. (17.14) and (17.15); this makes the computation of the integrals a trivial task following the recipe rules in the Appendix of Chapter 12.

Before we proceed further, note that the recursions in Eqs. (17.18) and (17.20) are an instance of the sum-product algorithm for graphical models. Indeed, to put our current discussion in this context, let us compactly write the previous recursions as

$$p(\boldsymbol{x}_n|\boldsymbol{y}_{1:n}) = \underbrace{\frac{p(\boldsymbol{y}_n|\boldsymbol{x}_n)}{Z_n}}_{\text{corrector}} \underbrace{\int p(\boldsymbol{x}_{n-1}|\boldsymbol{y}_{1:n-1})p(\boldsymbol{x}_n|\boldsymbol{x}_{n-1})\,d\boldsymbol{x}_{n-1}}_{\text{predictor}} : \quad \text{filtering.} \tag{17.21}$$

Note that this is of exactly the same form, within the normalizing factor, as Eq. (16.43) of Chapter 16; just replace summation with integration. One can rederive Eq. (17.21) using the sum-product rule, following similar steps as for Eq. (16.43). The only difference is that the normalizing constant has to be involved in all respective definitions and we replace summations with integrations. Because all the involved PDFs are Gaussians, the computation of the involved normalizing constants is trivially done; moreover, it suffices to derive recursions only for the respective mean values and covariances.

In Eq. (17.20), we have

$$p(\boldsymbol{x}_n|\boldsymbol{x}_{n-1}) = \mathcal{N}(\boldsymbol{x}_n|F_n\boldsymbol{x}_{n-1}, Q_n).$$

Let, also, $p(\boldsymbol{x}_{n-1}|\boldsymbol{y}_{1:n-1})$ be Gaussian with mean and covariance matrix

$$\boldsymbol{\mu}_{n-1|n-1}, \quad P_{n-1|n-1},$$

respectively, where the notation is chosen for the derived recursions to comply with the algorithm given in Section 4.10. Then, according to the Appendix of Chapter 12, $p(\boldsymbol{x}_n|\boldsymbol{y}_{1:n-1})$ is a Gaussian marginal PDF with mean and covariance given by (see Eqs. (12.150) and (12.151))

$$\boldsymbol{\mu}_{n|n-1} = F_n\boldsymbol{\mu}_{n-1|n-1}, \tag{17.22}$$

$$P_{n|n-1} = Q_n + F_n P_{n-1|n-1} F_n^T. \tag{17.23}$$

Also, in Eq. (17.18) we have

$$p(\boldsymbol{y}_n|\boldsymbol{x}_n) = \mathcal{N}(\boldsymbol{y}_n|H_n\boldsymbol{x}_n, R_n).$$

From the Appendix of Chapter 12, and taking into account Eqs. (17.22) and (17.23), we find that $p(\boldsymbol{x}_n|\boldsymbol{y}_{1:n})$ is the posterior (Gaussian) with mean and covariance given by (see Eqs. (12.148) and (12.149))

$$\boldsymbol{\mu}_{n|n} = \boldsymbol{\mu}_{n|n-1} + K_n(\boldsymbol{y}_n - H_n\boldsymbol{\mu}_{n|n-1}), \tag{17.24}$$

$$P_{n|n} = P_{n|n-1} - K_n H_n P_{n|n-1}, \tag{17.25}$$

where

$$K_n = P_{n|n-1} H_n^T S_n^{-1}, \tag{17.26}$$

and

$$S_n = R_n + H_n P_{n|n-1} H_n^T. \tag{17.27}$$

Note that these are exactly the same recursions that were derived in Section 4.10 for the state estimation; recall that under the Gaussian assumption, the posterior mean coincides with the least-squares estimate.

Here we have assumed that matrices F_n, H_n as well as the covariance matrices are known. This is most often the case. If not, these can be learned using similar arguments as those used in learning the HMM parameters, which are discussed in Section 16.5.2 (see, e.g., [2]).

17.4 PARTICLE FILTERING

In Section 4.10, extended Kalman filtering (EKF) was discussed as one possibility to generalize Kalman filtering to nonlinear models. Particle filtering, to be discussed next, is a powerful alternative technique to EKF. The involved PDFs are approximated by *discrete random measures*. The underlying theory is that of sequential importance sampling (SIS); as a matter of fact, particle filtering is an instance of SIS.

Let us now consider the state-space model of the general form in Eqs. (17.12) and (17.13). From the specific form of these equations (and by the Bayesian network nature of such models, for the more familiar reader) we can write

$$p(\boldsymbol{x}_n|\boldsymbol{x}_{1:n-1}, \boldsymbol{y}_{1:n-1}) = p(\boldsymbol{x}_n|\boldsymbol{x}_{n-1}) \tag{17.28}$$

and

$$p(\boldsymbol{y}_n|\boldsymbol{x}_{1:n}, \boldsymbol{y}_{1:n-1}) = p(\boldsymbol{y}_n|\boldsymbol{x}_n). \tag{17.29}$$

Our starting point is the *sequential* estimation of $p(\boldsymbol{x}_{1:n}|\boldsymbol{y}_{1:n})$; the estimation of $p(\boldsymbol{x}_n|\boldsymbol{y}_{1:n})$, which comprises our main goal, will be obtained as a by-product. Note that [15]

$$\begin{aligned} p(\boldsymbol{x}_{1:n}, \boldsymbol{y}_{1:n}) &= p(\boldsymbol{x}_n, \boldsymbol{x}_{1:n-1}, \boldsymbol{y}_n, \boldsymbol{y}_{1:n-1}) \\ &= p(\boldsymbol{x}_n, \boldsymbol{y}_n|\boldsymbol{x}_{1:n-1}, \boldsymbol{y}_{1:n-1}) p(\boldsymbol{x}_{1:n-1}, \boldsymbol{y}_{1:n-1}) \\ &= p(\boldsymbol{y}_n|\boldsymbol{x}_n) p(\boldsymbol{x}_n|\boldsymbol{x}_{n-1}) p(\boldsymbol{x}_{1:n-1}, \boldsymbol{y}_{1:n-1}), \end{aligned} \tag{17.30}$$

where Eqs. (17.28) and (17.29) have been employed.

Our goal is to obtain an approximation, via the generation of particles, of the conditional PDF,

$$p(\boldsymbol{x}_{1:n}|\boldsymbol{y}_{1:n}) = \frac{p(\boldsymbol{x}_{1:n}, \boldsymbol{y}_{1:n})}{\int p(\boldsymbol{x}_{1:n}, \boldsymbol{y}_{1:n}) d\boldsymbol{x}_{1:n}} = \frac{p(\boldsymbol{x}_{1:n}, \boldsymbol{y}_{1:n})}{Z_n}, \tag{17.31}$$

where

$$Z_n := \int p(\boldsymbol{x}_{1:n}, \boldsymbol{y}_{1:n})\, d\boldsymbol{x}_{1:n}.$$

To put the current discussion in the general framework of SIS, compare Eq. (17.31) with Eq. (17.6), which leads to the definition

$$\phi_n(\boldsymbol{x}_{1:n}) := p(\boldsymbol{x}_{1:n}, \boldsymbol{y}_{1:n}). \tag{17.32}$$

Then Eq. (17.9) becomes

$$w_n(\boldsymbol{x}_{1:n}) = w_{n-1}(\boldsymbol{x}_{1:n-1})\alpha_n(\boldsymbol{x}_{1:n}), \tag{17.33}$$

where now

$$\alpha_n(\boldsymbol{x}_{1:n}) = \frac{p(\boldsymbol{x}_{1:n}, \boldsymbol{y}_{1:n})}{p(\boldsymbol{x}_{1:n-1}, \boldsymbol{y}_{1:n-1})q_n(\boldsymbol{x}_n|\boldsymbol{x}_{1:n-1}, \boldsymbol{y}_{1:n})},$$

which from Eq. (17.30) becomes

$$\alpha_n(\boldsymbol{x}_{1:n}) = \frac{p(\boldsymbol{y}_n|\boldsymbol{x}_n)p(\boldsymbol{x}_n|\boldsymbol{x}_{n-1})}{q_n(\boldsymbol{x}_n|\boldsymbol{x}_{1:n-1}, \boldsymbol{y}_{1:n})}. \tag{17.34}$$

The final step is to select the proposal distribution. From Section 17.2, recall that the optimal proposal distribution is given from Eq. (17.11), which for our case takes the form

$$\begin{aligned} q_n^{opt}(\boldsymbol{x}_n|\boldsymbol{x}_{1:n-1}, \boldsymbol{y}_{1:n}) &= p(\boldsymbol{x}_n|\boldsymbol{x}_{1:n-1}, \boldsymbol{y}_{1:n}) \\ &= p(\boldsymbol{x}_n|\boldsymbol{x}_{n-1}, \boldsymbol{x}_{1:n-2}, \boldsymbol{y}_n, \boldsymbol{y}_{1:n-1}), \end{aligned}$$

and exploiting the underlying independencies, as they are imposed by the Bayesian network structure of the state-space model, we finally get

$$\boxed{q^{opt}(\boldsymbol{x}_n|\boldsymbol{x}_{1:n-1}, \boldsymbol{y}_{1:n}) = p(\boldsymbol{x}_n|\boldsymbol{x}_{n-1}, \boldsymbol{y}_n): \quad \text{optimal proposal distribution.}} \tag{17.35}$$

The use of the optimal proposal distribution leads to the following weight update recursion (Problem 17.6):

$$\boxed{w_n(\boldsymbol{x}_{1:n}) = w_{n-1}(\boldsymbol{x}_{1:n-1})p(\boldsymbol{y}_n|\boldsymbol{x}_{n-1}): \quad \text{optimal weights.}} \tag{17.36}$$

However, as is most often the case in practice, optimality is not always easy to obtain. Note that Eq. (17.36) requires the following integration:

$$p(\boldsymbol{y}_n|\boldsymbol{x}_{n-1}) = \int p(\boldsymbol{y}_n|\boldsymbol{x}_n)p(\boldsymbol{x}_n|\boldsymbol{x}_{n-1})\, d\boldsymbol{x}_n,$$

which may not be tractable. Moreover, even if the integral can be computed, sampling from $p(\boldsymbol{y}_n|\boldsymbol{x}_{n-1})$ directly may not be feasible. In any case, even if the optimal proposal distribution cannot be used, we can still select the proposal distribution to be of the form

$$q_n(\boldsymbol{x}_n|\boldsymbol{x}_{1:n-1}, \boldsymbol{y}_{1:n}) = q(\boldsymbol{x}_n|\boldsymbol{x}_{n-1}, \boldsymbol{y}_n). \tag{17.37}$$

Note that such a choice is particularly convenient, because sampling at time n only depends on $\boldsymbol{x}_{n-1}$ and $\boldsymbol{y}_n$, and *not* on the entire history. If, in addition, the goal is to obtain estimates of $p(\boldsymbol{x}_n|\boldsymbol{y}_{1:n})$, then one need not keep in memory all previously generated samples, but only the most recent one, $\boldsymbol{x}_n$.

We are now ready to write the first particle filtering algorithm.

Algorithm 17.4 (SIS particle filtering).

- Select a prior distribution, p, to generate the initial state $\boldsymbol{x}_0$.
- Select the number of particle streams, N.
- **For** $i = 1, 2, \ldots, N$, **Do**
  - Draw $\boldsymbol{x}_0^{(i)} \sim p(\boldsymbol{x})$; Initialize the N streams of particles.
  - Set $w_0^{(i)} = \frac{1}{N}$; Set all initial weights equal.
- **End For**
- **For** $n = 1, 2, \ldots,$ **Do**
  - **For** $i = 1, 2, \ldots, N$, **Do**
    * Draw $\boldsymbol{x}_n^{(i)} \sim q(\boldsymbol{x}|\boldsymbol{x}_{n-1}^{(i)}, \boldsymbol{y}_n)$
    * $w_n^{(i)} = w_{n-1}^{(i)} \frac{p(\boldsymbol{x}_n^{(i)}|\boldsymbol{x}_{n-1}^{(i)})p(\boldsymbol{y}_n|\boldsymbol{x}_n^{(i)})}{q(\boldsymbol{x}_n^{(i)}|\boldsymbol{x}_{n-1}^{(i)}, \boldsymbol{y}_n)}$; formulas (17.33), (17.34), and (17.37).
  - **End For**
  - **For** $i = 1, 2, \ldots, N$, **Do**
    * Compute the normalized weights $W_n^{(i)}$
  - **End For**
- **End For**

Note that the generation of the N streams of particles can take place *concurrently*, by exploiting parallel processing capabilities, if they are available in the processor.

The particles generated along the ith stream $\boldsymbol{x}_n^{(i)}$, $n = 1, 2, \ldots,$ represent a *path/trajectory* through the state space. Once the particles have been drawn and the normalized weights computed, we obtain the estimate

$$\hat{p}(\boldsymbol{x}_{1:n}|\boldsymbol{y}_{1:n}) = \sum_{i=1}^{N} W_n^{(i)} \delta(\boldsymbol{x}_{1:n} - \boldsymbol{x}_{1:n}^{(i)}).$$

If, as commented earlier, our interest lies in keeping the terminal sample, $\boldsymbol{x}_n^{(i)}$, only, then discarding the path history, $\boldsymbol{x}_{1:n-1}^{(i)}$, we can write

$$\hat{p}(\boldsymbol{x}_n|\boldsymbol{y}_{1:n}) = \sum_{i=1}^{N} W_n^{(i)} \delta(\boldsymbol{x}_n - \boldsymbol{x}_n^{(i)}).$$

Note that as the number of particles, N, tends to infinity, the previous approximations tend to the true posterior densities. Fig. 17.3 provides a graphical interpretation of the SIS Algorithm 17.4.

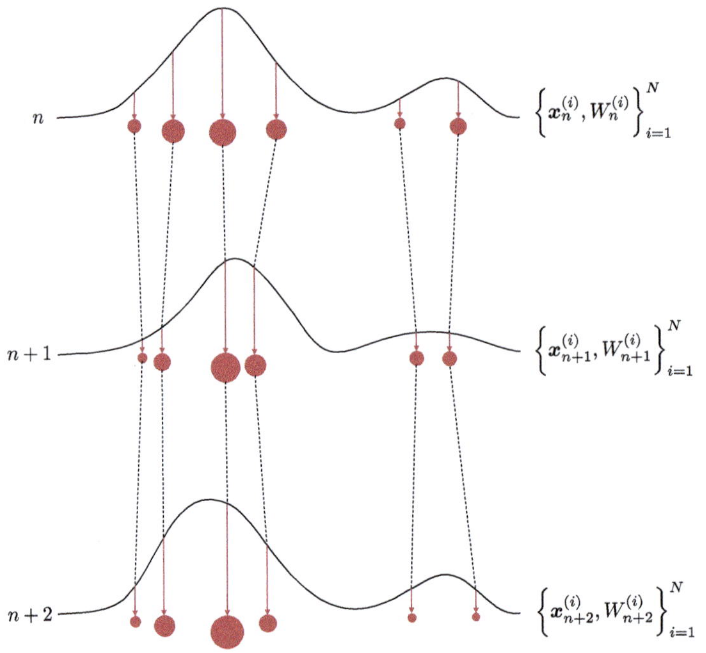

FIGURE 17.3

Three consecutive recursions, for the particle filtering scheme given in Algorithm 17.4, with $N = 7$ streams of particles. The area of the circles corresponds to the size of the normalized weights of the respective particles drawn from the proposal distribution.

Example 17.1. Consider the one-dimensional random walk model written as

$$\mathrm{x}_n = \mathrm{x}_{n-1} + \eta_n, \tag{17.38}$$

$$\mathrm{y}_n = \mathrm{x}_n + \mathrm{v}_n, \tag{17.39}$$

where $\eta_n \sim \mathcal{N}(\eta_n|0, \sigma_\eta^2)$, $\mathrm{v}_n \sim \mathcal{N}(v_n|0, \sigma_u^2)$, with $\sigma_\eta^2 = 1$, $\sigma_u^2 = 1$. Although this is a typical task for (linear) Kalman filtering, we will attack it here via the particle filtering rationale in order to demonstrate some of the previously reported performance-related issues. The proposal distribution is selected to be

$$q(x_n|x_{n-1}, y_n) = p(x_n|x_{n-1}) = \mathcal{N}(x_n|x_{n-1}, \sigma_\eta^2).$$

1. Generate $T = 100$ observations, y_n, $n = 1, 2, \ldots, T$, to be used by Algorithm 17.4. To this end, start with an arbitrary state value, for example, $x_0 = 0$, and generate a realization of the random walk, drawing samples from the Gaussians (we know how to generate Gaussian samples) $\mathcal{N}(\cdot|0, \sigma_n^2)$ and $\mathcal{N}(\cdot|0, \sigma_u^2)$, according to Eqs. (17.38) and (17.39). Fig. 17.4 shows a realization for the output variable. Our goal is to use the sequence of the resulting observations to generate particles and demonstrate the increase of the variance of the associated weights as time goes by.

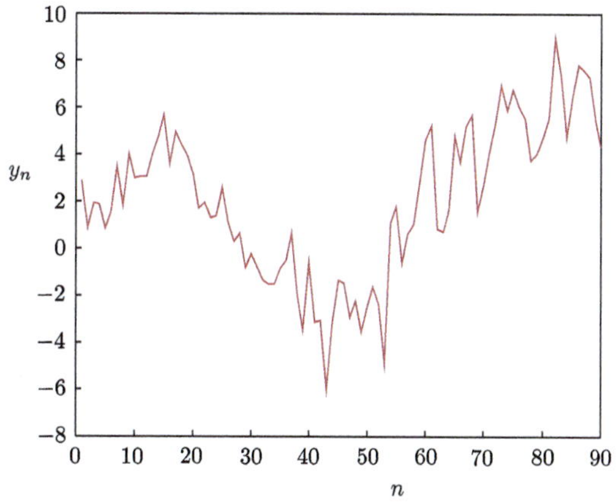

FIGURE 17.4

The observation sequence for Example 17.1.

2. Use $\mathcal{N}(\cdot|0, 1)$ to initialize $N = 200$ particle streams, $x^{(i)}$, $i = 1, 2, \dots, N$, and initialize the normalized weights to equal values, $W_0^{(i)} = \frac{1}{N}$, $i = 1, 2, \dots, N$. Fig. 17.5 provides the corresponding plot.
3. Use Algorithm 17.4 and plot the resulting particles together with the respective weights at time instants $n = 0$, $n = 1$, $n = 3$, and $n = 30$. Observe how the variance of the weights increases with time. At time $n = 30$ only a few particles have nonzero weights.
4. Repeat the experiment with $N = 1000$. Fig. 17.6 is the counterpart of Fig. 17.5 for the snapshots of $n = 3$ and $n = 30$. Observe that increasing the number of particles improves the performance with respect to the variance of weights. This is one path to obtain more particles with significant weight values. The other path is via resampling techniques.

17.4.1 DEGENERACY

Particle filtering is a special case of sequential importance sampling; hence, everything that has been said in Section 17.2 concerning the respective performance is also applied here.

A major problem is the *degeneracy* phenomenon. The variance of the importance weights increases in time, and after a few iterations only very few (or even only one) of the particles are assigned non-negligible weights, and the discrete random measure degenerates quickly. There are two methods for reducing degeneracy: one is selecting a good proposal distribution and the other is resampling.

We know the optimal choice for the proposal distribution is

$$q(\cdot|\boldsymbol{x}_{n-1}^{(i)}, \boldsymbol{y}_n) = p(\cdot|\boldsymbol{x}_{n-1}^{(i)}, \boldsymbol{y}_n).$$

There are cases where this is available in analytic form. For example, this happens if the noise sources are Gaussian and the observation equation is linear (e.g., [11]). If analytic forms are not available and direct sampling is not possible, approximations of $p(\cdot|\boldsymbol{x}_{n-1}^{(i)}, \boldsymbol{y}_n)$ are mobilized. Our familiar (from

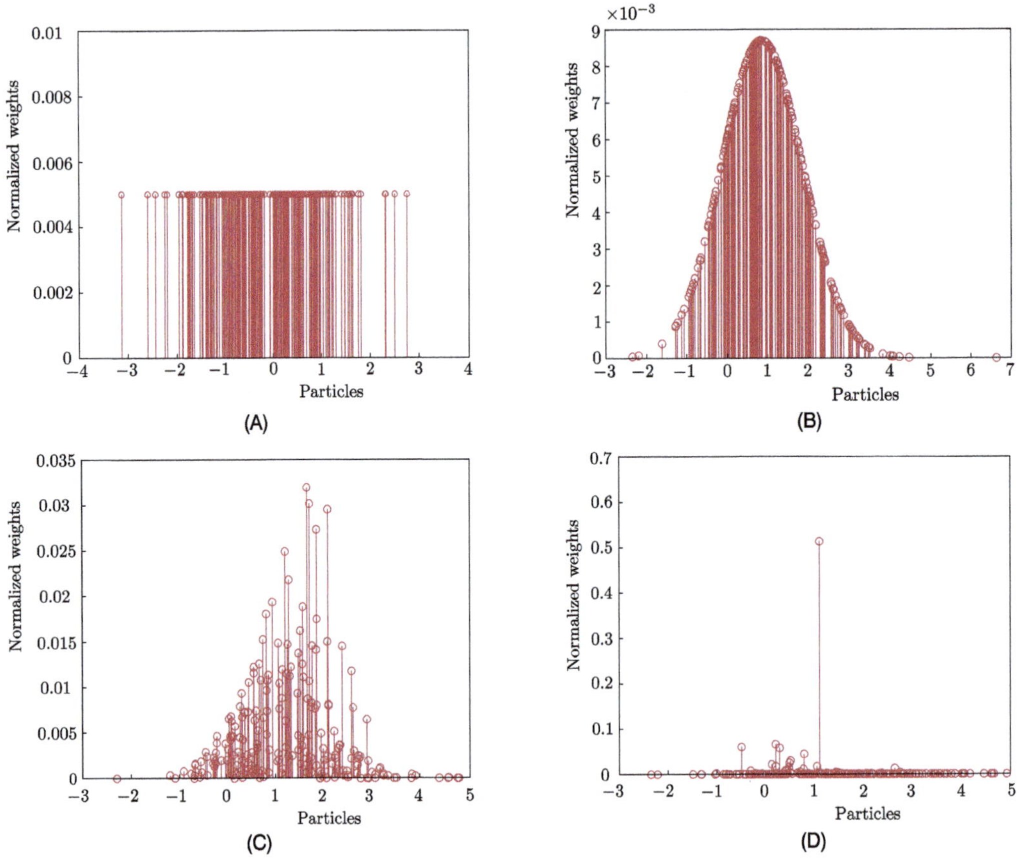

FIGURE 17.5

Plot of $N = 200$ generated particles with the corresponding (normalized) weights, for Example 17.1, at time instants (A) $n = 0$, (B) $n = 1$, (C) $n = 3$, and (D) $n = 30$. Observe that as time goes by, the variance of the weights increases. At time $n = 30$, only very few particles have a nonzero weight value.

Chapter 12) Gaussian approximation via local linearization of $\ln p(\cdot|\boldsymbol{x}_{n-1}^{(i)}, \boldsymbol{y}_n)$ is a possibility [11]. The use of suboptimal filtering techniques such as the extended/unscented Kalman filter have also been advocated [37]. In general, it must be kept in mind that the choice of the proposal distribution plays a *crucial* role in the performance of particle filtering. Resampling is the other path that has been discussed in Section 17.2.2. The counterpart of Algorithm 17.1 can also be adopted for the case of particle filtering. However, we are going to give a slightly modified version of it.

17.4.2 GENERIC PARTICLE FILTERING

Resampling has a number of advantages. It discards, with high probability, particles of low weights; that is, only particles corresponding to regions of high-probability mass are propagated. Of course,

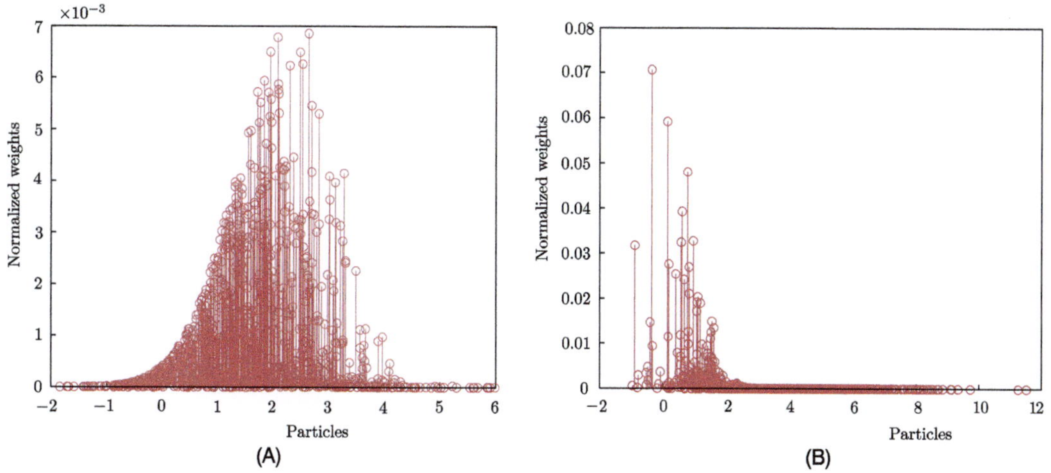

FIGURE 17.6

Plot of $N = 1000$ generated particles with the corresponding (normalized) weights, for Example 17.1, at time instants (A) $n = 3$ and (B) $n = 30$. As expected, compared to Fig. 17.5, more particles with significant weights survive.

resampling has its own limitations. For example, a particle of low weight at time n will not necessarily have a low weight at later time instants. In such a case, resampling is rather wasteful. Moreover, resampling limits the potential of parallelizing the computational process, because particles along the different streams have to be "combined" at each time instant. However, some efforts for enhancing parallelism have been reported (see, e.g., [21]). Also, particles corresponding to high values of weights are drawn many times and lead to a set of samples of low diversity; this phenomenon is also known as *sample impoverishment*. The effects of this phenomenon become more severe in cases of low state/process noise, $\boldsymbol{\eta}_n$, in Eq. (17.12), where the set of the sampling points may end up comprising a single point (e.g., [1]).

Hence, avoiding resampling can be beneficial. In practice, resampling is performed only if a related metric of the variance of the weights is below a threshold. In [28,29], the *effective number* of samples is approximated by

$$N_{eff} \approx \frac{1}{\sum_{i=1}^{N} \left(W_n^{(i)}\right)^2}. \tag{17.40}$$

The value of this index ranges from 1 to N. Resampling is performed if $N_{eff} \leq N_T$, typically with $N_T = \frac{N}{2}$.

Algorithm 17.5 (Generic particle filtering).

- Select a prior distribution, p, to generate particles for the initial state $\mathbf{x}_0$.
- Select the number of particle streams, N.

- **For** $i = 1, 2, \ldots, N$, **Do**
  - Draw $\boldsymbol{x}_0^{(i)} \sim p(\boldsymbol{x})$; Initialize N streams.
  - set $W_0^{(i)} = \frac{1}{N}$; All initial normalized weights are equal.
- **End For**
- **For** $n = 1, 2, 3, \ldots,$ **Do**
  - **For** $i = 1, 2, \ldots, N$, **Do**
    - $*$ Draw $\boldsymbol{x}_n^{(i)} \sim q(\boldsymbol{x}|\boldsymbol{x}_{n-1}^{(i)}, \boldsymbol{y}_n)$
    - $*$ $w_n^{(i)} = w_{n-1}^{(i)} \frac{p(\boldsymbol{x}_n^{(i)}|\boldsymbol{x}_{n-1}^{(i)})p(\boldsymbol{y}_n|\boldsymbol{x}_n^{(i)})}{q(\boldsymbol{x}_n^{(i)}|\boldsymbol{x}_{n-1}^{(i)}, \boldsymbol{y}_n)}$
  - **End For**
  - **For** $i = 1, 2, \ldots, N$, **Do**
    - $*$ Compute the normalized $W_n^{(i)}$.
  - **End For**
  - Compute N_{eff}; Eq. (17.40).
  - **If** $N_{eff} \le N_T$; preselected value N_T.
    - $*$ Resample $\{\boldsymbol{x}_n^{(i)}, W_n^{(i)}\}_{i=1}^N$ to obtain $\{\bar{\boldsymbol{x}}_n^{(i)}, \frac{1}{N}\}_{i=1}^N$
    - $*$ $\boldsymbol{x}_n^{(i)} = \bar{\boldsymbol{x}}_n^{(i)}$, $w_n^{(i)} = \frac{1}{N}$
  - **End If**
- **End For**

Fig. 17.7 presents a graphical illustration of the time evolution of the algorithm.

Remarks 17.2.

- A popular choice for the proposal distribution is the prior

$$q(\boldsymbol{x}|\boldsymbol{x}_{n-1}^{(i)}, \boldsymbol{y}_n) = p(\boldsymbol{x}_n|\boldsymbol{x}_{n-1}^{(i)}),$$

which yields the following weights' update recursion,

$$w_n^{(i)} = w_{n-1}^{(i)} p(\boldsymbol{y}_n|\boldsymbol{x}_n^{(i)}).$$

The resulting algorithm is known as *sampling-importance-resampling* (SIR). The great advantage of such a choice is its simplicity. However, the generation mechanism of particles ignores important information that resides in the observation sequence; the proposal distribution is independent of the observations. This may lead to poor results. A remedy can be offered by the use of auxiliary particle filtering, to be reviewed next. Another possibility is discussed in [22], via a combination of the prior and the optimal proposal distributions.

Example 17.2. Repeat Example 17.1, using $N = 200$ particles, for Algorithm 17.5. Use the threshold value $N_T = 100$. Observe in Fig. 17.8 that, for the corresponding time instants, more particles with significant weights are generated compared to Fig. 17.5.

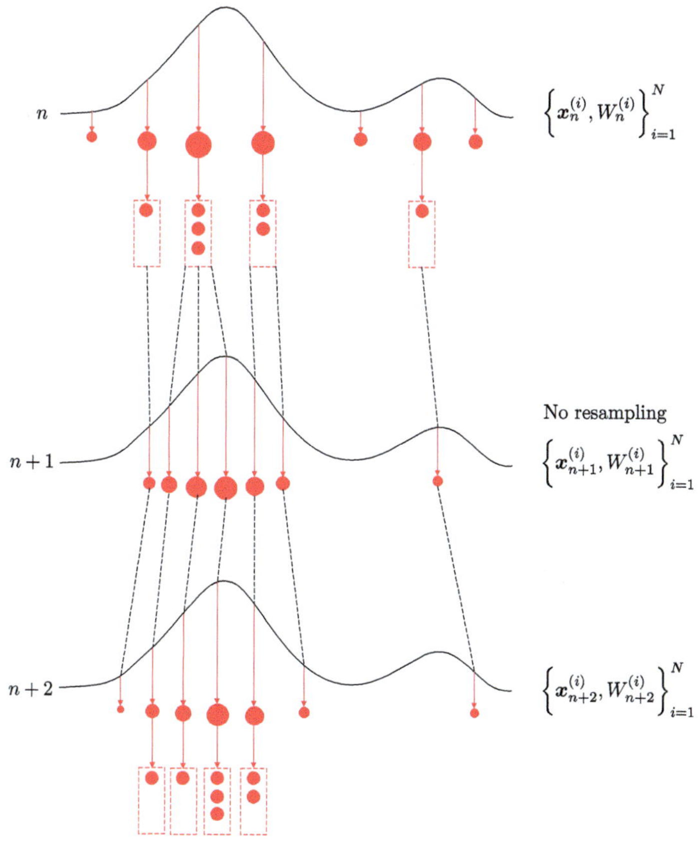

FIGURE 17.7

Three successive time iterations for $N = 7$ streams of particles corresponding to Algorithm 17.5. At steps n and $n+2$ resampling is performed. At step $n+1$ no resampling is needed.

17.4.3 AUXILIARY PARTICLE FILTERING

Auxiliary particle filters were introduced in [34] in order to improve performance when dealing with heavy-tailed distributions. The method introduces an auxiliary variable; this is the index of a particle at the previous time instant. We allow for a particle in the ith stream at time n to be drawn using a particle from a different stream at time $n-1$. Let the ith particle at time n be $\boldsymbol{x}_n^{(i)}$ and let the index of its "parent" particle at time $n-1$ be i_{n-1}. The idea is to sample for the pair $(\boldsymbol{x}_n^{(i)}, i_{n-1})$, $i = 1, 2, \ldots, N$. Employing the Bayes rule, we obtain

$$\begin{aligned} p(\boldsymbol{x}_n, i|\boldsymbol{y}_{1:n}) &\propto p(\boldsymbol{y}_n|\boldsymbol{x}_n)p(\boldsymbol{x}_n, i|\boldsymbol{y}_{1:n-1}) \\ &= p(\boldsymbol{y}_n|\boldsymbol{x}_n)p(\boldsymbol{x}_n|i, \boldsymbol{y}_{1:n-1})P(i|\boldsymbol{y}_{1:n-1}), \end{aligned} \tag{17.41}$$

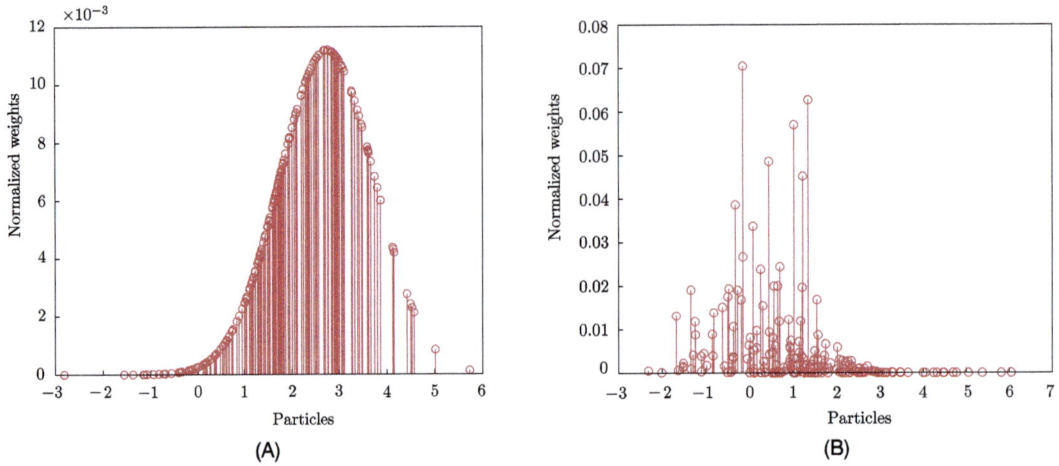

FIGURE 17.8

Plot of $N = 200$ generated particles with the corresponding (normalized) weights, for Example 17.2 using resampling, at time instants (A) $n = 3$ and (B) $n = 30$. Compared to Figs. 17.5C and D, more particles with significant weights survive.

where the conditional independencies underlying the state-space model have been used. To unclutter notation, we have used $\boldsymbol{x}_n$ in place of $\boldsymbol{x}_n^{(i)}$, and the subscript $n-1$ has been omitted from i_{n-1} and we use i instead. Note that by the definition of the index i_{n-1}, we have

$$p(\boldsymbol{x}_n|i, \boldsymbol{y}_{1:n-1}) = p(\boldsymbol{x}_n|\boldsymbol{x}_{n-1}^{(i)}, \boldsymbol{y}_{1:n-1}) = p(\boldsymbol{x}_n|\boldsymbol{x}_{n-1}^{(i)}), \tag{17.42}$$

and also

$$P(i|\boldsymbol{y}_{1:n-1}) = W_{n-1}^{(i)}. \tag{17.43}$$

Thus, we can write

$$p(\boldsymbol{x}_n, i|\boldsymbol{y}_{1:n}) \propto p(\boldsymbol{y}_n|\boldsymbol{x}_n)p(\boldsymbol{x}_n|\boldsymbol{x}_{n-1}^{(i)})W_{n-1}^{(i)}. \tag{17.44}$$

The proposal distribution is chosen as

$$q(\boldsymbol{x}_n, i|\boldsymbol{y}_{1:n}) \propto p(\boldsymbol{y}_n|\boldsymbol{\mu}_n^{(i)})p(\boldsymbol{x}_n|\boldsymbol{x}_{n-1}^{(i)})W_{n-1}^{(i)}. \tag{17.45}$$

Note that we have used $\boldsymbol{\mu}_n^{(i)}$ in place of $\boldsymbol{x}_n$ in $p(\boldsymbol{y}_n|\boldsymbol{x}_n)$, because $\boldsymbol{x}_n$ is still to be drawn. The estimate $\boldsymbol{\mu}_n^{(i)}$ is chosen in order to be easily computed and at the same time to be a good representative of $\boldsymbol{x}_n$. Typically, $\boldsymbol{\mu}_n^{(i)}$ can be the mean, the mode, a draw, or another value associated with the distribution $p(\boldsymbol{x}_n|\boldsymbol{x}_{n-1}^{(i)})$; for example, $\boldsymbol{\mu}_n^{(i)} \sim p(\boldsymbol{x}_n|\boldsymbol{x}_{n-1}^{(i)})$. Also, if the state equation is $\boldsymbol{x}_n = \boldsymbol{f}(\boldsymbol{x}_{n-1}) + \boldsymbol{\eta}_n$, a good choice would be $\boldsymbol{\mu}_n^{(i)} = \boldsymbol{f}(\boldsymbol{x}_{n-1}^{(i)})$.

Applying the Bayes rule in Eq. (17.45) and adopting

$$q(\boldsymbol{x}_n|i, \boldsymbol{y}_{1:n}) = p(\boldsymbol{x}_n|\boldsymbol{x}_{n-1}^{(i)}),$$

we obtain

$$q(i|\boldsymbol{y}_{1:n}) \propto p(\boldsymbol{y}_n|\boldsymbol{\mu}_n^{(i)})W_{n-1}^{(i)}. \tag{17.46}$$

Hence, we draw the value of the index i_{n-1} from a multinomial distribution, i.e.,

$$i_{n-1} \sim q(i|\boldsymbol{y}_{1:n}) \propto p(\boldsymbol{y}_n|\boldsymbol{\mu}_n^{(i)})W_{n-1}^{(i)}, \quad i = 1, 2, \ldots, N. \tag{17.47}$$

The index i_{n-1} identifies the distribution from which $\boldsymbol{x}_n^{(i)}$ will be drawn, i.e.,

$$\boldsymbol{x}_n^{(i)} \sim p(\boldsymbol{x}_n|\boldsymbol{x}_{n-1}^{(i_{n-1})}), \quad i = 1, 2, \ldots, N. \tag{17.48}$$

Note that Eq. (17.47) actually performs a resampling. However, now, the resampling at time $n-1$ takes into consideration information that becomes available at time n, via the observation $\boldsymbol{y}_n$. This information is exploited in order to determine which particles are to survive, after resampling at a given time instant, so that their "offsprings" are likely to land in regions of high-probability mass. Once sample $\boldsymbol{x}_n^{(i)}$ has been drawn, the index i_{n-1} is discarded, which is equivalent to marginalizing $p(\boldsymbol{x}_n, i|\boldsymbol{y}_{1:n})$ to obtain $p(\boldsymbol{x}_n|\boldsymbol{y}_{1:n})$. Each sample $\boldsymbol{x}_n^{(i)}$ is finally assigned a weight according to

$$w_n^{(i)} \propto \frac{p(\boldsymbol{x}_n^{(i)}, i_{n-1}|\boldsymbol{y}_{1:n})}{q(\boldsymbol{x}_n^{(i)}, i_{n-1}|\boldsymbol{y}_{1:n})} = \frac{p(\boldsymbol{y}_n|\boldsymbol{x}_n^{(i)})}{p(\boldsymbol{y}_n|\boldsymbol{\mu}_n^{(i)})},$$

which results by dividing the right-hand sides of Eqs. (17.44) and (17.45). Note that the weight accounts for the mismatch between the likelihood $p(\boldsymbol{y}_n|\cdot)$ at the actual sample and at the predicted point, $\boldsymbol{\mu}_n^{(i)}$. The resulting algorithm is summarized next.

Algorithm 17.6 (Auxiliary particle filtering).

- Initialization: Select a prior distribution, p, to generate the initial state $\boldsymbol{x}_0$.
- Select N.
- **For** $i = 1, 2, \ldots, N$, **Do**
  - Draw $\boldsymbol{x}_0^{(i)} \sim p(\boldsymbol{x})$; Initialize N streams of particles.
  - Set $W_0^{(i)} = \frac{1}{N}$; Set all normalized weights to equal values.
- **End For**
- **For** $n = 1, 2, \ldots,$ **Do**
  - **For** $i = 1, 2, \ldots, N$, **Do**
    - Draw/compute $\boldsymbol{\mu}_n^{(i)}$
    - $Q_i = p(\boldsymbol{y}_n|\boldsymbol{\mu}_n^{(i)})W_{n-1}^{(i)}$; This corresponds to $q(i|\boldsymbol{y}_{1:n})$ in Eq. (17.46).
  - **End For**

- **For** $i = 1, 2, \ldots, N$, **Do**
  - $*$ Compute normalized Q_i
- **End For**
- **For** $i = 1, 2, \ldots, N$, **Do**
  - $*$ $i_{n-1} \sim Q_i$; Eq. (17.47).
  - $*$ Draw $\boldsymbol{x}_n^{(i)} \sim p(\boldsymbol{x}|\boldsymbol{x}_{n-1}^{(i_{n-1})})$
  - $*$ Compute $w_n^{(i)} = \frac{p(y_n|\boldsymbol{x}_n^{(i)})}{p(y_n|\boldsymbol{\mu}_n^{(i)})}$
- **End For**
- **For** $i = 1, 2, \ldots, N$, **Do**
  - $*$ Compute normalized $W_n^{(i)}$
- **End For**

• **End For**

Fig. 17.9 shows $N = 200$ particles and their respective normalized weights, generated by Algorithm 17.6 for the observation sequence of Example 17.1 and using the same proposal distribution. Observe that compared to the corresponding Figs. 17.5 and 17.8, a substantially larger number of particles with significant weights survive.

The previous algorithm is sometimes called the *single-stage* auxiliary particle, filter as opposed to the *two-stage* one, which was originally proposed in [34]. The latter involved an extra resampling step to obtain samples with equal weights. It has been experimentally verified that the single-stage version leads to enhanced performance, and it is the one that is more widely used. It has been reported that

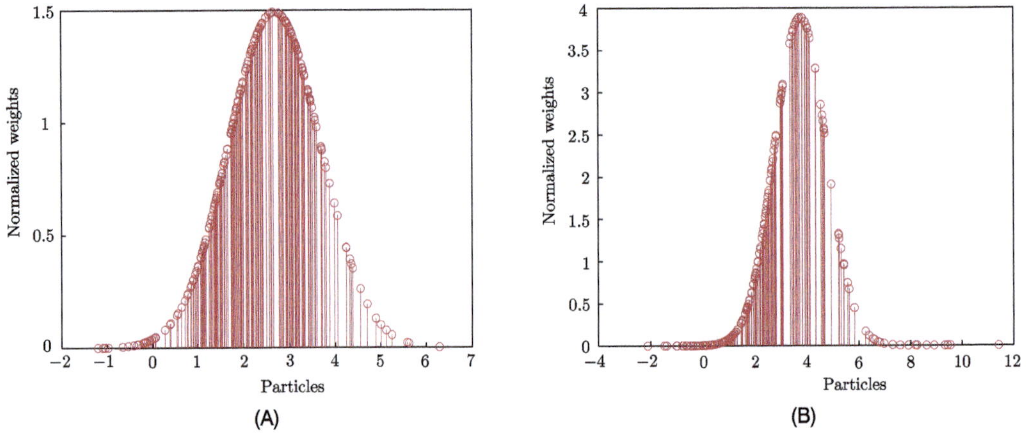

FIGURE 17.9

Plot of $N = 200$ generated particles with the corresponding (normalized) weights, for the same observation sequence as that in Example 17.1, using the auxiliary particle filtering algorithm, at time instants (A) $n = 3$ and (B) $n = 30$. Compared to Figs. 17.5C and D and Fig. 17.8, more particles with significant weights survive.

the auxiliary particle filter may lead to enhanced performance compared to Algorithm 17.5, for high signal-to-noise ratios. However, for high-noise terrains its performance degrades (see, e.g., [1]). More results concerning the performance and analysis of the auxiliary filter can be found in, for example, [13,23,35].

Remarks 17.3.

- Besides the algorithms presented earlier, a number of variants have been proposed over the years in order to overcome the main limitations of particle filters, associated with the increasing variance and the sample impoverishment problem. In *resample-move* [17] and *block sampling* [14], instead of just sampling for $\boldsymbol{x}_n^{(i)}$ at time instant n, one also tries to modify past values, over a window $[n-1, n-L+1]$ of fixed size L, in light of the newly arrived observation $\boldsymbol{y}_n$. In the *regularized particle filter* [32], in the resampling stage of Algorithm 17.5, instead of sampling from a discrete distribution, samples are drawn from a smooth approximation,

$$p(\boldsymbol{x}_n|\boldsymbol{y}_{1:n}) \simeq \sum_{i=1}^{N} W_n^{(i)} K(\boldsymbol{x}_n - \boldsymbol{x}_n^{(i)}),$$

  where K is a smooth kernel density function. In [26,27], the posteriors are approximated by Gaussians; as opposed to the more classical extended Kalman filters, the updating and filtering is accomplished via the propagation of particles.
  The interested reader may find more information concerning particle filtering in the tutorial papers [1,9,15].
- *Rao–Blackwellization* is a technique used to reduce the variance of estimates that are obtained via Monte Carlo sampling methods (e.g., [4]). To this end, this technique has also been employed in particle filtering of dynamic systems. It turns out that often in practice, some of the states are conditionally linear given the nonlinear ones. The main idea consists of treating the linear states differently by viewing them as nuisance parameters and *marginalizing* them out of the estimation process. The particles of the nonlinear states are propagated randomly, and then the task is treated linearly via the use of a Kalman filter (see, e.g., [10,12,24]).
- *Smoothing* is closely related to filtering processing. In filtering, the goal lies in obtaining estimates of $\boldsymbol{x}_{1:n}$ $(\boldsymbol{x}_n)$ based on observations taken in the interval $[1, n]$, that is, on $\boldsymbol{y}_{1:n}$. In smoothing, one obtains estimates of $\boldsymbol{x}_n$ based on an observation set $\boldsymbol{y}_{1:n+k}$, $k > 0$. There are two paths to smoothing. One is known as *fixed lag* smoothing, where k is a fixed lag. The other is known as *fixed interval*, where one is interested in obtaining estimates based on observations taken over an interval $[1, T]$, that is, based on a fixed set of measurements $\boldsymbol{y}_{1:T}$.
  There are different algorithmic approaches to smoothing. The naive one is to run the particle filtering up to time k or T and use the obtained weights for weighting the particles at time n, in order to form the random measure, i.e.,

$$p(\boldsymbol{x}_n|\boldsymbol{y}_{1:n+k}) \simeq \sum_{i=1}^{N} W_{n+k}^{(i)} \delta(\boldsymbol{x}_n - \boldsymbol{x}_n^{(i)}).$$

  This can be a reasonable approximation for small values of k (or $T-n$). Other, more refined, techniques adopt a two-pass rationale. First, a particle filtering is run, and then a backward set of recursions is used to modify the weights (see, e.g., [10]).

- A summary concerning convergence results related to particle filtering can be found in, for example, [6].
- A survey on applications of particle filtering in signal processing-related tasks is given in [8,9].
- Following the general trend for developing algorithms for distributed learning, a major research effort has been dedicated in this direction in the context of particle filtering. For a review on such schemes, see, for example, [20].
- One of the main difficulties of the particle filtering methods is that the number of particles required to approximate the underlying distributions increases exponentially with the state dimension. To overcome this problem, several methods have been proposed. In [30], the authors propose to partition the state and estimate each partition independently. In [16], the annealed particle filter is proposed, which implements a coarse-to-fine strategy by using a series of smoothed weighting functions. The unscented particle filter [37] proposes to use the unscented transform for each particle to avoid wasting resources in low likelihood regions. In [31], a hierarchical search strategy is proposed that uses auxiliary lower dimension models to guide the search in the higher-dimensional one.

Example 17.3. *Stochastic volatility model.* Consider the following state-space model for generating the observations:

$$\mathrm{x}_n = \alpha \mathrm{x}_{n-1} + \eta_n,$$
$$\mathrm{y}_n = \beta \mathrm{v}_n \exp\left(\frac{\mathrm{x}_n}{2}\right).$$

This model belongs to a more general class known as *stochastic volatility* models, where the variance of a process is itself randomly distributed. Such models are used in financial mathematics to model derivative securities, such as options. The state variable is known as the *log-volatility*. We assume the two noise sequences to be i.i.d. and mutually independent Gaussians with zero mean and variances σ_η^2 and σ_v^2, respectively. The model parameters α and β are known as the *persistence in volatility shocks* and *modal volatility*, respectively. The adopted values for the parameters are $\sigma_\eta^2 = 0.178$, $\sigma_v^2 = 1$, $\alpha = 0.97$, and $\beta = 0.69$.

The goal of the example is to generate a sequence of observations and then, based on these measurements, to predict the state, which is assumed to be unknown. To this end, we generate a sequence of $N = 2000$ particles, and the state variable at each time instant is estimated as the weighted average of the generated particles, that is,

$$\hat{x}_n = \sum_{i=1}^{N} W_n^{(i)} x_n^{(i)}.$$

Both the SIR Algorithm 17.5 and the auxiliary filter method of Algorithm 17.6 were used. The proposal distribution was

$$q(x_n|x_{n-1}) = \mathcal{N}(x_n|\alpha x_{n-1}, \sigma_\eta^2).$$

Fig. 17.10 shows the observation sequence together with the obtained estimate. For comparison reasons, the corresponding true state value is also shown. Both methods for generating particles gave almost identical results, and we only show one of them. Observe how closely the estimates follow the true values.

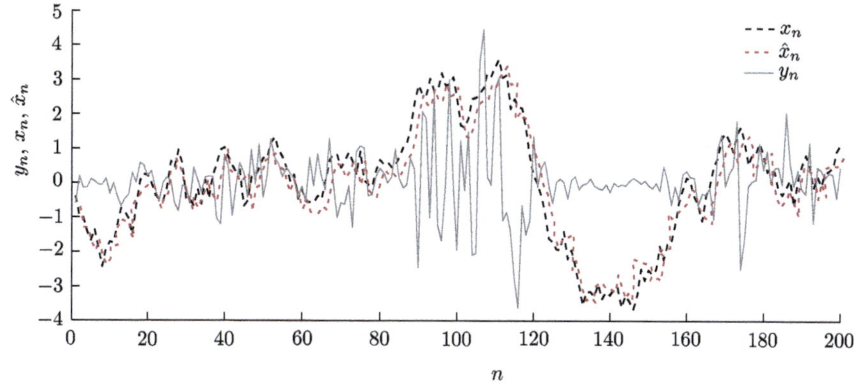

FIGURE 17.10

The observation sequence generated by the volatility model together with the true and estimated values of the state variable.

Example 17.4. *Visual tracking*. Consider the problem of visual tracking of a circle, which has a constant and known radius. We seek to track its position, that is, the coordinates of its center, $\mathbf{x} = [\mathrm{x}_1, \mathrm{x}_2]^T$. This vector will comprise the state variable. The model for generating the observations is given by

$$\begin{aligned} \mathbf{x}_n &= \mathbf{x}_{n-1} + \boldsymbol{\eta}_n, \\ \mathbf{y}_n &= \mathbf{x}_n + \mathbf{v}_n, \end{aligned} \tag{17.49}$$

where $\boldsymbol{\eta}_n$ is a uniform noise in the interval $[-10, 10]$ pixels, for each dimension. Note that due to the uniform nature of the noise, Kalman filtering, in its standard formulation, is no longer the optimal choice, in spite of the linearity of the model. The noise $\mathbf{v}_n$ follows a Gaussian PDF $\mathcal{N}(\mathbf{0}, \Sigma_{\mathrm{v}})$, where

$$\Sigma_{\mathrm{v}} = \begin{bmatrix} 2 & 0.5 \\ 0.5 & 2 \end{bmatrix}.$$

Initially, the target circle is located in the image center. The particle filter employs N=50 particles and the SIS sampling method was used (see, also, MATLAB® Exercise 17.12).

Fig. 17.11 shows the circle and the generated particles, which attempt to track the center of the circle from the noisy observations, for different time instants. Observe how closely the particles track the center of the circle as it moves around. A related video is available from the companion site of this book.

PROBLEMS

17.1 Let

$$\mu := \mathbb{E}[f(\mathbf{x})] = \int f(\boldsymbol{x}) p(\boldsymbol{x})\, d\boldsymbol{x}$$

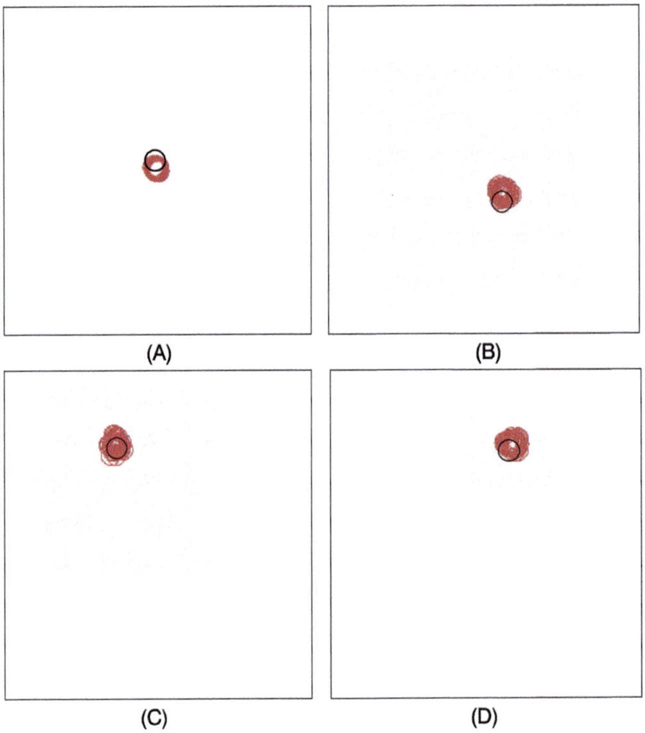

FIGURE 17.11

The circle (in gray) and the generated particles (in red) for time instants $n = 1$, $n = 30$, $n = 60$, and $n = 120$.

and let $q(\boldsymbol{x})$ be the proposal distribution. Show that if

$$w(\boldsymbol{x}) := \frac{p(\boldsymbol{x})}{q(\boldsymbol{x})}$$

and

$$\hat{\mu} = \frac{1}{N}\sum_{i=1}^{N} w(\boldsymbol{x}_i) f(\boldsymbol{x}_i),$$

then the variance

$$\sigma_f^2 = \mathbb{E}\left[\left(\hat{\mu} - \mathbb{E}\left[\hat{\mu}\right]\right)^2\right] = \frac{1}{N}\left(\int \frac{f^2(\boldsymbol{x})p^2(\boldsymbol{x})}{q(\boldsymbol{x})} d\boldsymbol{x} - \mu^2\right).$$

Observe that if $f^2(\boldsymbol{x})p^2(\boldsymbol{x})$ goes to zero slower than $q(\boldsymbol{x})$, then for fixed N, $\sigma_f^2 \longrightarrow \infty$.

17.2 In importance sampling, with weights defined as

$$w(\boldsymbol{x}) = \frac{\phi(\boldsymbol{x})}{q(\boldsymbol{x})},$$

where

$$p(\boldsymbol{x}) = \frac{1}{Z}\phi(\boldsymbol{x}),$$

we know from Problem 14.6 that the estimate

$$\hat{Z} = \frac{1}{N}\sum_{i=1}^{N} w(\boldsymbol{x}_i)$$

defines an unbiased estimator of the normalizing constant, Z. Show that the respective variance is given by

$$\text{var}[\hat{Z}] = \frac{Z^2}{N}\left(\int \frac{p^2(\boldsymbol{x})}{q(\boldsymbol{x})} d\boldsymbol{x} - 1\right).$$

17.3 Show that using resampling in importance sampling, as the number of particles tends to infinity, the approximating, by the respective discrete random measure, distribution, $\bar{p}$, tends to the true (desired) one, p.
Hint: Consider the one-dimensional case.

17.4 Show that in sequential importance sampling, the proposal distribution that minimizes the variance of the weight at time n, conditioned on $\boldsymbol{x}_{1:n-1}$, is given by

$$q_n^{opt}(\boldsymbol{x}_n|\boldsymbol{x}_{1:n-1}) = p_n(\boldsymbol{x}_n|\boldsymbol{x}_{1:n-1}).$$

17.5 In a sequential importance sampling task, let

$$p_n(\boldsymbol{x}_{1:n}) = \prod_{k=1}^{n} \mathcal{N}(x_k|0, 1),$$

$$\phi_n(\boldsymbol{x}_{1:n}) = \prod_{k=1}^{n} \exp\left(-\frac{x_k^2}{2}\right),$$

and let the proposal distribution be

$$q_n(\boldsymbol{x}_{1:n}) = \prod_{k=1}^{n} \mathcal{N}(x_k|0, \sigma^2).$$

Let the estimate of $Z_n = (2\pi)^{\frac{n}{2}}$ be

$$\hat{Z}_n = \frac{1}{N}\sum_{i=1}^{N} w(\boldsymbol{x}_{1:n}^{(i)}).$$

Show that the variance of the estimator is given by

$$\text{var}[\hat{Z}_n] = \frac{Z_n^2}{N}\left(\left(\frac{\sigma^4}{2\sigma^2 - 1}\right)^{\frac{n}{2}} - 1\right).$$

Observe that for $\sigma^2 > 1/2$, which is the range of values for the above formula to make sense and guarantees a finite value for the variance, the variance exhibits an exponential increase with respect to n. To keep the variance small, one has to make N very large, that is, generate a very large number of particles [15].

17.6 Prove that the use of the optimal proposal distribution in particle filtering leads to

$$w_n(\boldsymbol{x}_{1:n}) = w_{n-1}(\boldsymbol{x}_{1:n-1})p(\boldsymbol{y}_n|\boldsymbol{x}_{n-1}).$$

MATLAB® EXERCISES

17.7 For the state-space model of Example 17.1, implement the generic particle filtering algorithm for different numbers of particle streams N and different thresholds of effective particle sizes N_{eff}.
Hint: Start by selecting a distribution (the normal should be a good start) and initialize. Then update the particles in each step according to the algorithm. Finally, check whether N_{eff} is lower than the threshold, and if it is, continue with the resampling process.

17.8 For the same example as before, implement the SIS particle filtering algorithm and plot the resulting particles together with the normalized weights for various time instances n. Observe the degeneracy phenomenon of the weights as time evolves.

17.9 For Example 17.1, implement the SIR particle filtering algorithm for different numbers of particle streams N and for various time instances n. Use $N_T = N/2$. Compare the performance of SIR and SIS algorithms.

17.10 Repeat the previous exercise, implement the auxiliary particle filtering (APF) algorithm, and compare the particle-weight histogram with the ones obtained from SIS and SIR algorithms. Observe that the number of particles with significant weights that survive is substantially larger.

17.11 Reproduce Fig. 17.10 for the stochastic volatility model of Example 17.3 and observe how the estimated sequence $\hat{x}_n$ follows the true sequence x_n based on the observations y_n.

17.12 Develop the MATLAB® code to reproduce the visual tracking of the circle of Example 17.4. Because, at each time instant, we are only interested in $\boldsymbol{x}_n$ and not in the whole sequence, modify the SIS sampling in Algorithm 17.4 to care for this case.

Specifically, given Eqs. (17.28) and (17.29), in order to estimate $\mathbf{x}_n$ instead of $\mathbf{x}_{1:n}$, Eq. (17.31) is simplified to

$$p(\boldsymbol{x}_n|\boldsymbol{y}_{1:n}) = \frac{p(\boldsymbol{y}_n|\boldsymbol{x}_n)p(\boldsymbol{x}_n|\boldsymbol{y}_{1:n-1})}{p(\boldsymbol{y}_n|\boldsymbol{y}_{1:n-1})}, \tag{17.50}$$

where

$$p(\boldsymbol{x}_n|\boldsymbol{y}_{1:n-1}) = \int_{\boldsymbol{x}_{n-1}} p(\boldsymbol{x}_n|\boldsymbol{x}_{n-1})p(\boldsymbol{x}_{n-1}|\boldsymbol{y}_{1:n-1})\,d\boldsymbol{x}_{n-1}. \tag{17.51}$$

The samples are now weighted as

$$w_n^{(i)} = \frac{p(\boldsymbol{x}_n^{(i)}|\boldsymbol{y}_{1:n})}{q(\boldsymbol{x}_n^{(i)}|\boldsymbol{y}_{1:n})}, \tag{17.52}$$

and a popular selection for the proposal distribution is

$$q(\boldsymbol{x}_n|\boldsymbol{y}_{1:n}) \equiv p(\boldsymbol{x}_n|\boldsymbol{y}_{1:n-1}). \tag{17.53}$$

Substituting Eqs. (17.53) and (17.50) into Eq. (17.52), we get the following rule for the weights:

$$w_n^{(i)} \propto p(\boldsymbol{y}_n|\boldsymbol{x}_n^{(i)}). \tag{17.54}$$

REFERENCES

[1] M.S. Arulampalam, S. Maskell, N. Gordon, T. Clapp, A tutorial on particle filters for online nonlinear/non-Gaussian Bayesian tracking, IEEE Trans. Signal Process. 50 (2) (2002) 174–188.

[2] C.M. Bishop, Pattern Recognition and Machine Learning, Springer, New York, 2006.

[3] J. Carpenter, P. Clifford, P. Fearnhead, Improved particle filter for nonlinear problems, in: Proceedings IEE, Radar, Sonar and Navigation, vol. 146, 1999, pp. 2–7.

[4] G. Casella, C.P. Robert, Rao-Blackwellisation of sampling schemes, Biometrika 83 (1) (1996) 81–94.

[5] N. Chopin, Central limit theorem for sequential Monte Carlo methods and its application to Bayesian inference, Ann. Stat. 32 (2004).

[6] D. Crisan, A. Doucet, A survey of convergence results on particle filtering methods for practitioners, IEEE Trans. Signal Process. 50 (3) (2002) 736–746.

[7] P. Del Moral, Feynman-Kac Formulae: Genealogical and Interacting Particle Systems With Applications, Springer-Verlag, New York, 2004.

[8] P.M. Djuric, Y. Huang, T. Ghirmai, Perfect sampling: a review and applications to signal processing, IEEE Trans. Signal Process. 50 (2002) 345–356.

[9] P.M. Djuric, J.H. Kotecha, J. Zhang, Y. Huang, T. Ghirmai, M.F. Bugallo, J. Miguez, Particle filtering, IEEE Signal Process. Mag. 20 (2003) 19–38.

[10] P.M. Djuric, M. Bugallo, Particle filtering, in: T. Adali, S. Haykin (Eds.), Adaptive Signal Processing: Next Generation Solutions, John Wiley & Sons, Inc., New York, 2010.

[11] A. Doucet, S. Godsill, C. Andrieu, On sequential Monte Carlo sampling methods for Bayesian filtering, Stat. Comput. 10 (2000) 197–208.

[12] R. Douc, O. Cappe, E. Moulines, Comparison of resampling schemes for particle filtering, in: 4th International Symposium on Image and Signal Processing and Analysis, ISPA, 2005.

[13] R. Douc, E. Moulines, J. Olsson, On the auxiliary particle filter, arXiv:0709.3448v1 [math.ST], 2010.

[14] A. Doucet, M. Briers, S. Sénécal, Efficient block sampling strategies for sequential Monte Carlo methods, J. Comput. Graph. Stat. 15 (2006) 693–711.

[15] A. Doucet, A.M. Johansen, A tutorial on particle filtering and smoothing: fifteen years later, in: Handbook of Nonlinear Filtering, Oxford University Press, Oxford, 2011.

[16] J. Deutscher, A. Blake, I. Reid, Articulated body motion capture by annealed particle filtering, in: Proceedings of the IEEE Conference on Computer Vision and Pattern Recognition, vol. 2, 2000, pp. 126–133.

[17] W.R. Gilks, C. Berzuini, Following a moving target—Monte Carlo inference for dynamic Bayesian models, J. R. Stat. Soc. B 63 (2001) 127–146.

[18] N.J. Gordon, D.J. Salmond, A.F.M. Smith, Novel approach to nonlinear/non-Gaussian Bayesian state estimation, Proc. IEEE F 140 (2) (1993) 107–113.

[19] J.M. Hammersley, K.W. Morton, Poor man's Monte Carlo, J. R. Stat. Soc. B 16 (1) (1954) 23–38.

[20] O. Hinka, F. Hlawatz, P.M. Djuric, Distributed particle filtering in agent networks, IEEE Signal Process. Mag. 30 (1) (2013) 61–81.

[21] S. Hong, S.S. Chin, P.M. Djurić, M. Bolić, Design and implementation of flexible resampling mechanism for high-speed parallel particle filters, J. VLSI Signal Process. 44 (1–2) (2006) 47–62.

[22] Y. Huang, P.M. Djurić, A blind particle filtering detector of signals transmitted over flat fading channels, IEEE Trans. Signal Process. 52 (7) (2004) 1891–1900.

[23] A.M. Johansen, A. Doucet, A note on auxiliary particle filters, Stat. Probab. Lett. 78 (12) (2008) 1498–1504.

[24] R. Karlsson, F. Gustafsson, Complexity analysis of the marginalized particle filter, IEEE Trans. Signal Process. 53 (11) (2005) 4408–4411.

[25] G. Kitagawa, Monte Carlo filter and smoother for non-Gaussian nonlinear state space models, J. Comput. Graph. Stat. 5 (1996) 1–25.

[26] J.H. Kotecha, P.M. Djurić, Gaussian particle filtering, IEEE Trans. Signal Process. 51 (2003) 2592–2601.

[27] J.H. Kotecha, P.M. Djurić, Gaussian sum particle filtering, IEEE Trans. Signal Process. 51 (2003) 2602–2612.

[28] J.S. Liu, R. Chen, Sequential Monte Carlo methods for dynamical systems, J. Am. Stat. Assoc. 93 (1998) 1032–1044.

[29] J.S. Liu, Monte Carlo Strategies in Scientific Computing, Springer, New York, 2001.

[30] J. MacCormick, M. Isard, Partitioned sampling, articulated objects, and interface-quality hand tracking, in: Proceedings of the 6th European Conference on Computer Vision, Part II, ECCV, Springer-Verlag, London, UK, 2000, pp. 3–19.

[31] A. Makris, D. Kosmopoulos, S. Perantonis, S. Theodoridis, A hierarchical feature fusion framework for adaptive visual tracking, Image Vis. Comput. 29 (9) (2011) 594–606.

[32] C. Musso, N. Oudjane, F. Le Gland, Improving regularised particle filters, in: A. Doucet, N. de Freitas, N.J. Gordon (Eds.), Sequential Monte Carlo Methods in Practice, Springer-Verlag, New York, 2001.

[33] A. Owen, Y. Zhou, Safe and effective importance sampling, J. Am. Stat. Assoc. 95 (2000) 135–143.

[34] M.K. Pitt, N. Shephard, Filtering via simulation: auxiliary particle filters, J. Am. Stat. Assoc. 94 (1999) 590–599.

[35] M.K. Pitt, R.S. Silva, P. Giordani, R. Kohn, Auxiliary particle filtering within adaptive Metropolis-Hastings sampling, arXiv:1006.1914 [stat.Me], 2010.

[36] M.N. Rosenbluth, A.W. Rosenbluth, Monte Carlo calculation of the average extension of molecular chains, J. Chem. Phys. 23 (2) (1956) 356–359.

[37] R. van der Merwe, N. de Freitas, A. Doucet, E. Wan, The unscented particle filter, in: Proceedings Advances in Neural Information Processing Systems, NIPS, 2000.

CHAPTER

NEURAL NETWORKS AND DEEP LEARNING

18

CONTENTS

Machine Learning. https://doi.org/10.1016/B978-0-12-818803-3.00030-1

18.1 INTRODUCTION

Neural networks have a long history that goes back to the first attempts to understand how the human (and more generally, the mammal) brain works and how what we call intelligence is formed.

From a physiological point of view, one can trace the beginning of the field back to the work of Santiago Ramon y Cajal [187], who discovered that the basic building element of the brain is the *neuron*. The brain comprises approximately 60 to 100 billions neurons; that is, a number of the same order as the number of stars in our galaxy! Each neuron is connected with other neurons via elementary structural and functional units/links, known as *synapses*. It is estimated that there are 50 to 100 trillions of synapses. These links mediate information between connected neurons. The most common type of synapses are the chemical ones, which convert electric pulses, produced by a neuron, to a chemical signal and then back to an electrical one. Via these links, each neuron is connected to other neurons and this happens in a hierarchically structured way, in a layer-wise fashion.

Santiago Ramon y Cajal (1852–1934) was a Spanish pathologist, histologist, neuroscientist, and Nobel laureate. His many pioneering investigations of the microscopic structure of the brain have established him as the father of modern neuroscience.

A milestone from the learning theory's point of view occurred in 1943, when Warren McCulloch and Walter Pitts [158] developed a computational model for the basic neuron. Moreover, they provided results that tie neurophysiology with mathematical logic. They showed that given a sufficient number of neurons and adjusting appropriately the synaptic links, each one associated with a weight, one can compute, in principle, any computable function. As a matter of fact, it is generally accepted that this is the paper that gave birth to the fields of neural networks and artificial intelligence.

Warren McCulloch (1898–1969) was an American psychiatrist and neuroanatomist who spent many years studying the representation of an event in the neural system. Walter Pitts (1923–1969) was an American logician who worked in the field of cognitive psychology. He was a mathematical prodigy and he taught himself logic and mathematics. At the age of 12, he read *Principia Mathematica* by Alfred North Whitehead and Bertrand Russell and he wrote a letter to Russell commenting on certain parts of the book. He worked with a number of great mathematicians and logicians, including Wiener, Householder, and Carnap. When he met McCulloch at the University of Chicago, he was familiar with the work of Leibnitz on computing, which inspired them to study whether the nervous system could be considered to be a type of universal computing device. This gave birth to their 1943 paper, mentioned in the reference before.

Frank Rosenblatt [195,196] borrowed the idea of a neuron model, as suggested by McCulloch and Pitts, to build a true learning machine which learns from a set of training data. In the most basic version of operation, he used a single neuron and adopted a rule that can learn to separate data, which belong to two linearly separable classes. That is, he built a pattern recognition system. He called the basic neuron a *perceptron* and developed a rule/algorithm, the *perceptron algorithm*, for the respective training. The perceptron will be the kick-off point for our tour in this chapter.

Frank Rosenblatt (1928–1971) was educated at Cornell, where he obtained his PhD in 1956. In 1959, he took over as director of Cornell's Cognitive Systems Research Program and also as a lecturer in the psychology department. He used an IBM 704 computer to simulate his perceptron and later built a special purpose hardware which implemented the perceptron learning rule.

Neural networks are learning machines, comprising a large number of neurons, which are connected in a layered fashion. Learning is achieved by adjusting the synaptic weights to minimize a preselected cost function. It took almost 25 years, after the pioneering work of Rosenblatt, for neural networks to find their widespread use in machine learning. This is the time period needed for the basic McCulloch–Pitts model of a neuron to be generalized and lead to an algorithm for training such networks. A breakthrough came under the name *backpropagation algorithm*, which was developed for training neural networks based on a set of input–output training samples. Backpropagation is also treated in detail in this chapter.

It is interesting to note that neural networks dominated the field of machine learning for almost a decade, from 1986 until the middle of the 1990s. Then, they were superseded, to a large extent, by the support vector machines, which established their reign until 2010 or so. After that, neural networks with many layers, known as *deep networks*, have taken over the "kingdom" of machine learning. Earlier works, associated with what is known as convolutional networks [129] and recurrent neural networks

[94] have inspired the field that is now flourishing. This comeback, however, would not have been possible without the availability of computational power, due to advances in computer architectures, as well as the buildup of big data sets that are needed for training such networks.

Interestingly enough, there is one name that is associated with the revival of interest on neural networks, both in the mid-1980s and in the first decade of the 21st century; this is the name of Geoffrey Hinton [86,199]. Geoffrey Hinton, together with Yoshua Bengio and Yann Lecun, received the 2019 Turing award for their contributions to the field of neural networks.

18.2 THE PERCEPTRON

Our starting point is the simple problem of a *linearly separable* two-class (ω_1, ω_2) classification task. In other words, we are given a set of training samples, $(y_n, \boldsymbol{x}_n)$, $n = 1, 2, \ldots, N$, with $y_n \in \{-1, +1\}$, $\boldsymbol{x}_n \in \mathbb{R}^l$ and it is assumed that there is a hyperplane,

$$\boldsymbol{\theta}_*^T \boldsymbol{x} = 0,$$

such that

$$\boldsymbol{\theta}_*^T \boldsymbol{x} > 0, \quad \text{if } \boldsymbol{x} \in \omega_1,$$
$$\boldsymbol{\theta}_*^T \boldsymbol{x} < 0, \quad \text{if } \boldsymbol{x} \in \omega_2.$$

In other words, such a hyperplane classifies correctly *all* the points in the training set. For notational simplification, the bias term of the hyperplane has been absorbed in $\boldsymbol{\theta}_*$ after extending the dimensionality of the input space by one, as has been explained in Chapter 3 and used in various parts of this book.

The goal now becomes that of developing an algorithm that iteratively computes a hyperplane that classifies correctly all the patterns from both classes. To this end, a cost function is adopted.

The perceptron cost: Let the available vector estimate at the current iteration step of the unknown parameters be $\boldsymbol{\theta}$. Then there are two possibilities. The first one is that all points are classified correctly; this means that a solution has been obtained. The other alternative is that $\boldsymbol{\theta}$ classifies correctly some of the points and the rest are misclassified. Let $\mathcal{Y}$ be the set of all misclassified samples. The perceptron cost is defined as

$$\boxed{J(\boldsymbol{\theta}) = -\sum_{n: \boldsymbol{x}_n \in \mathcal{Y}} y_n \boldsymbol{\theta}^T \boldsymbol{x}_n : \quad \text{perceptron cost},} \tag{18.1}$$

where

$$y_n = \begin{cases} +1, & \text{if } \boldsymbol{x} \in \omega_1, \\ -1, & \text{if } \boldsymbol{x} \in \omega_2. \end{cases} \tag{18.2}$$

Observe that the cost function is nonnegative. Indeed, because the sum is over the misclassified points, if $\boldsymbol{x}_n \in \omega_1$ (ω_2), then $\boldsymbol{\theta}^T \boldsymbol{x}_n \leq (\geq)\ 0$, rendering the product $-y_n \boldsymbol{\theta}^T \boldsymbol{x}_n \geq 0$. A solution is achieved if there are no misclassified points, that is, $\mathcal{Y} = \emptyset$. By convention, we can say that in this case $J(\boldsymbol{\theta}) = 0$.

The perceptron cost function is not differentiable at all points. It is a *continuous piece-wise linear* function. Indeed, let us write it in a slightly different way,

$$J(\boldsymbol{\theta}) = \left(- \sum_{n:\boldsymbol{x}_n \in \mathcal{Y}} y_n \boldsymbol{x}_n^T \right) \boldsymbol{\theta}.$$

This is a linear function with respect to $\boldsymbol{\theta}$, as long as the number of misclassified points remains the same. However, as one slowly changes the value of $\boldsymbol{\theta}$, which corresponds to a change of the position of the respective hyperplane, there will be a point where the number of misclassified samples in $\mathcal{Y}$ suddenly changes; this is the time where a sample in the training set changes its relative position with respect to the (moving) hyperplane and as a consequence the set $\mathcal{Y}$ is modified. After this change, $J(\boldsymbol{\theta})$ will correspond to a new linear function.

The perceptron algorithm: It can be shown (for example, [167,196]) that, starting from an arbitrary point, $\boldsymbol{\theta}^{(0)}$, the following iterative update,

$$\boxed{\boldsymbol{\theta}^{(i)} = \boldsymbol{\theta}^{(i-1)} + \mu_i \sum_{n:\boldsymbol{x}_n \in \mathcal{Y}} y_n \boldsymbol{x}_n : \quad \text{the perceptron rule,}} \tag{18.3}$$

converges after a *finite number of steps*. The parameter sequence μ_i is judicially chosen to guarantee convergence. Note that this is the same algorithm as the one derived in Section 8.10.2, for minimizing the hinge loss function via the notion of subgradient.

Besides the previous scheme, another version of the algorithm considers one sample per iteration in a cyclic fashion, until the algorithm converges. Let us denote by $y_{(i)}, \boldsymbol{x}_{(i)}, \ (i) \in \{1, 2, \ldots, N\}$, the training pair that is presented in the algorithm at the ith iteration step.[1] Then the update iteration becomes

$$\boxed{\boldsymbol{\theta}^{(i)} = \begin{cases} \boldsymbol{\theta}^{(i-1)} + \mu_i y_{(i)} \boldsymbol{x}_{(i)}, & \text{if } \boldsymbol{x}_{(i)} \text{ is misclassified by } \boldsymbol{\theta}^{(i-1)}, \\ \boldsymbol{\theta}^{(i-1)}, & \text{otherwise.} \end{cases}} \tag{18.4}$$

In other words, starting from an initial estimate, e.g., randomly initializing $\boldsymbol{\theta}^{(0)}$ with some small values, we test each one of the samples, $\boldsymbol{x}_n, \ n = 1, 2, \ldots, N$. Every time a sample is misclassified, action is taken for a correction. Otherwise no action is required. Once all samples have been considered, we say that one *epoch* has been completed. If no convergence has been attained, all samples are reconsidered in a second epoch, and so on. This algorithmic version is known as a *pattern-by-pattern* scheme. Sometimes it is also referred to as *online* algorithm. However, note that the term "online" has been used in previous chapters in a different context, when data were received in a streaming/sequential fashion and their number was unbounded. In contrast, in the current context, the total number of data

---

[1] The symbol (i) has been adopted to denote the index of the samples, instead of i, because we do not know which point will be presented to the algorithm at the ith iteration. Recall that each training point is considered many times, until convergence is achieved.

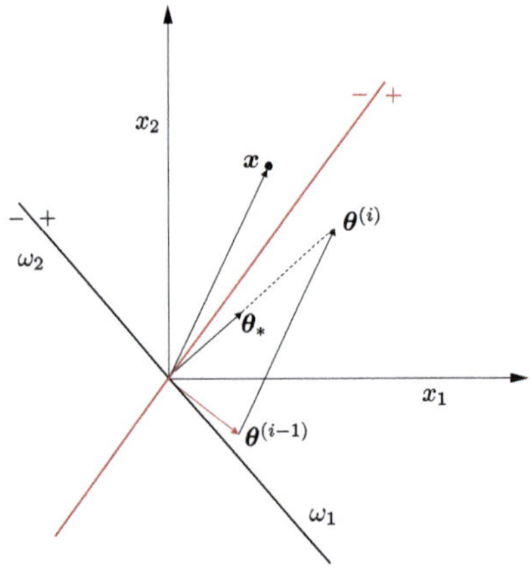

FIGURE 18.1

Pattern $\boldsymbol{x}$ is misclassified by the red line. The action of the perceptron rule is to turn the hyperplane toward the point $\boldsymbol{x}$, in an attempt to include it in the correct side of the new hyperplane and classify it correctly. The new hyperplane is defined by $\boldsymbol{\theta}^{(i)}$ and it is shown by the black line.

samples is fixed and the algorithm considers them in a *cyclic fashion*, epoch after epoch. To avoid confusion the former pattern-by-pattern term will be adopted.

After a successive *finite* number of epochs, the algorithm is guaranteed to converge. Note that for convergence, the sequence μ_i must be appropriately chosen. This is pretty familiar to us by now. However, for the case of the perceptron algorithm, convergence is still guaranteed even if μ_i is a positive constant, $\mu_i = \mu > 0$, usually taken to be equal to one (Problem 18.1).

The formulation in (18.4) brings the perceptron algorithm under the umbrella of the so-called *reward-punishment* philosophy of learning. If the current estimate succeeds in predicting the class of the respective pattern, no action is taken (reward). Otherwise, the algorithm is punished to perform an update.

Fig. 18.1 provides a geometric interpretation of the perceptron rule. Assume that sample $\boldsymbol{x}$ is misclassified by the hyperplane, $\boldsymbol{\theta}^{(i-1)}$. As we know from geometry, $\boldsymbol{\theta}^{(i-1)}$ corresponds to a vector that is perpendicular to the hyperplane which is defined by this vector (see also Fig. 11.15 in Section 11.10.1). Because $\boldsymbol{x}$ lies in the $(-)$ side of the hyperplane and it is misclassified, it belongs to class ω_1. Hence, assuming $\mu = 1$, the applied correction by the algorithm is

$$\boldsymbol{\theta}^{(i)} = \boldsymbol{\theta}^{(i-1)} + \boldsymbol{x},$$

and its effect is to turn the hyperplane to the direction toward $\boldsymbol{x}$ to place it in the $(+)$ side of the new hyperplane, which is defined by the updated estimate $\boldsymbol{\theta}^{(i)}$.

The perceptron algorithm in its pattern-by-pattern mode of operation is summarized in Algorithm 18.1.

Algorithm 18.1 (The pattern-by-pattern perceptron algorithm).

- Initialization

  - Initialize $\boldsymbol{\theta}^{(0)}$; usually, randomly, to some small values.
  - Select μ; usually it is set equal to one.
  - $i = 1$.
- **Repeat**; Each iteration corresponds to one epoch.
  - counter = 0; Counts the number of updates per epoch.
  - **For** $n = 1, 2, \ldots, N$, **Do**; For each epoch, all samples are presented once.
    * **If** $(y_n \boldsymbol{x}_n^T \boldsymbol{\theta}^{(i-1)} \leq 0)$ **Then**

      $\boldsymbol{\theta}^{(i)} = \boldsymbol{\theta}^{(i-1)} + \mu y_n \boldsymbol{x}_n$

      $i = i + 1$

      counter=counter +1
  - **End For**
- **Until** counter=0

Once the perceptron algorithm has run and converged, we have the weights, θ_i, $i = 1, 2, \ldots, l$, of the synapses of the associated neuron/perceptron as well as the bias term θ_0. These can now be used to classify unknown patterns. Fig. 18.2A shows the corresponding architecture of the basic neuron element. The features x_i, $i = 1, 2, \ldots, l$, are applied to the input nodes. In turn, each feature is multiplied by the respective synapse (weight), and then the bias term is added on their linear combination. The outcome of this operation then goes through a nonlinear function, f, known as the *activation* function. Depending on the form of the nonlinearity, different types of neurons occur. In the more classical one, known as the McCulloch–Pitts neuron, the activation function is the Heaviside one, that is,

$$f(z) = \begin{cases} 1, & \text{if } z > 0, \\ 0, & \text{if } z \leq 0. \end{cases} \tag{18.5}$$

Usually, the summation operation and the nonlinearity are merged to form a *node* in the respective graph and the architecture in Fig. 18.2B occurs. In the sequel, both terms, neuron and node, will be used interchangeably, the latter indicating a neuron within a larger network.

Thus, the basic model of an (artificial) neuron comprises the concatenation of (a) a linear combiner, (b) a threshold value (bias), and (c) a nonlinearity. Note that because of the existence of the nonlinearity,

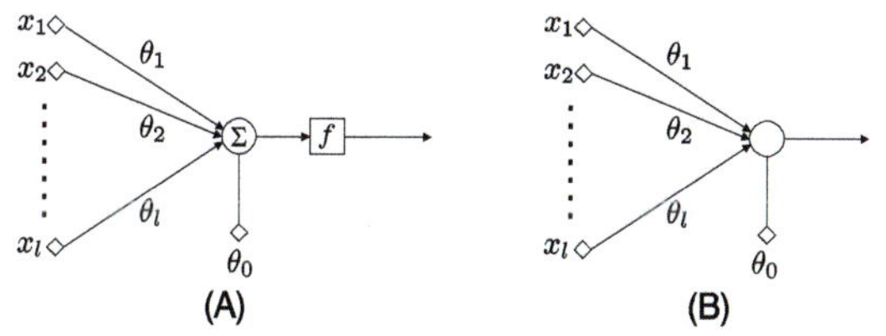

FIGURE 18.2

(A) In the basic neuron/perceptron architecture the input features are applied to the input nodes and are weighted by the respective weights that define the synapses. The bias term is then added on their linear combination and the result is pushed through the nonlinearity. In the McCulloch–Pitts neuron, the output fires a 1 for patterns in class ω_1 or a zero for class ω_2. (B) The summation and nonlinear operation are merged together for graphical simplicity.

the output of the neuron indicates the class from which the input pattern originates. For the Heaviside nonlinearity, the output is 1 for patterns from class ω_1 and 0 for those from class ω_2.

Remarks 18.1.

- *ADALINE*: Soon after Rosenblatt proposed the perceptron, Widrow and Hopf proposed the *adaptive line element* (ADALINE), which is a linear version of the perceptron [252]. That is, during training the nonlinearity of the activation function is not involved. The resulting scheme is the LMS algorithm, treated in detail in Chapter 5. It is interesting to note that the LMS was readily adopted and widely used for online learning within the signal processing and communications communities.
- A kernelized version of the perceptron algorithm has also been derived and the interested reader can obtain it via the book's site under the Additional Material part that is associated with the current chapter.

18.3 FEED-FORWARD MULTILAYER NEURAL NETWORKS

A single neuron is associated with a hyperplane

$$H: \theta_1 x_1 + \theta_2 x_2 + \ldots + \theta_l x_l + \theta_0 = 0,$$

in the input (feature) space. Moreover, classification is performed via the nonlinearity, which fires a one or stays at zero, depending on which side of H a point lies. We will now show how to combine neurons, in a layer-wise fashion, to construct nonlinear classifiers. We will follow a simple constructive proof, which will unveil certain aspects of neural networks. These will be useful later on, when dealing with deep architectures.

As a starting point, we consider the case where the classes in the feature space are formed by unions of polyhedral regions. This is shown in Fig. 18.3, for the case of the two-dimensional feature space. Polyhedral regions are formed as intersections of halfspaces, each one associated with a hyperplane. In Fig. 18.3, there are three hyperplanes (straight lines in $\mathbb{R}^2$), indicated as H_1, H_2, H_3, giving rise to seven polyhedral regions. For each hyperplane, the $(+)$ and $(-)$ sides (halfspaces) are indicated. In the sequel, each one of the regions is labeled using a triplet of binary numbers, depending on which side it is located with respect to H_1, H_2, H_3. For example, the region labeled as (101) lies in the (+) side of H_1, the (–) side of H_2, and the (+) side of H_3.

Fig. 18.4A shows three neurons, realizing the three hyperplanes, H_1, H_2, and H_3, of Fig. 18.3, respectively. The associated outputs, denoted as y_1, y_2, and y_3, form the label of the region in which the corresponding input pattern lies. Indeed, if the weights of the synapses have been appropriately set, then if a pattern originates from the region, say, (010), then the first neuron on the left will fire a zero ($y_1 = 0$), the middle a one ($y_2 = 1$), and the right-most a zero ($y_3 = 0$). In other words, combining the outputs of the three neurons together, we have achieved a *mapping* of the input feature space into the three-dimensional space. More specifically, the mapping is performed on the vertices of the *unit* cube in $\mathbb{R}^3$, as shown in Fig. 18.5. Each region of the input space uniquely corresponds to one vertex of the cube. In the more general case, where p neurons are employed, the mapping will be on the vertices of the unit hypercube in $\mathbb{R}^p$. This layer of neurons comprises the first *hidden layer* of the network, which we are developing.

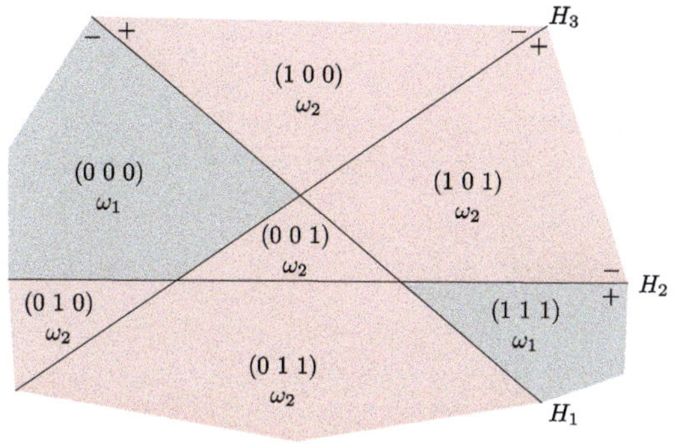

FIGURE 18.3

Classes are formed by unions of polyhedral regions. Regions are labeled according to the side on which they lie, with respect to the three lines, H_1, H_2, and H_3. The number 1 indicates the (+) side and the 0 the (−) side. Class ω_1 consists of the union of the (000) and (111) regions.

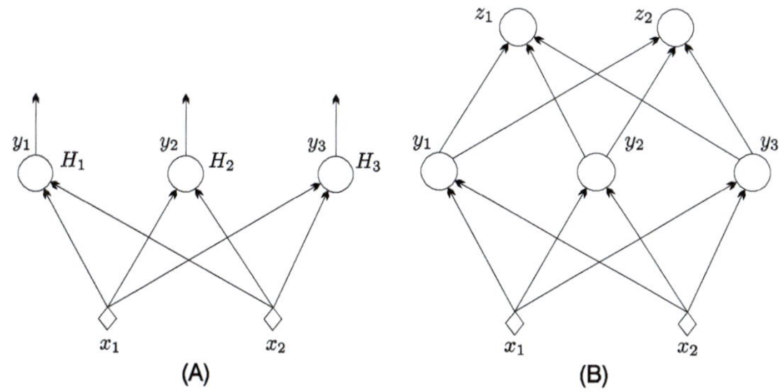

FIGURE 18.4

(A) The neurons of the first hidden layer are excited by the feature values applied at the input nodes and form the polyhedral regions. (B) The neurons of the second layer have as inputs the outputs of the first layer, and they thereby form the classes. To simplify the figure, the bias terms for each neuron are not shown.

An alternative way to view this mapping is as a new *representation* of the input patterns in terms of code words. For three neurons/hyperplanes we can form 2^3 binary code words, each corresponding to a vertex of the unit cube, which can represent $2^3 - 1 = 7$ regions (there is one remaining vertex, i.e., (110), which does not correspond to any region). Note, however, that this mapping encodes information concerning some *structure* of the input data; that is, information relating to how the input patterns are grouped together in the feature space in different regions.

We will now use this new representation, as it is provided by the outputs of the neurons of the first hidden layer, as input which feeds the neurons of a second hidden layer, which is constructed as follows. We choose all regions that belong to one class. For the sake of our example in Fig. 18.3, we

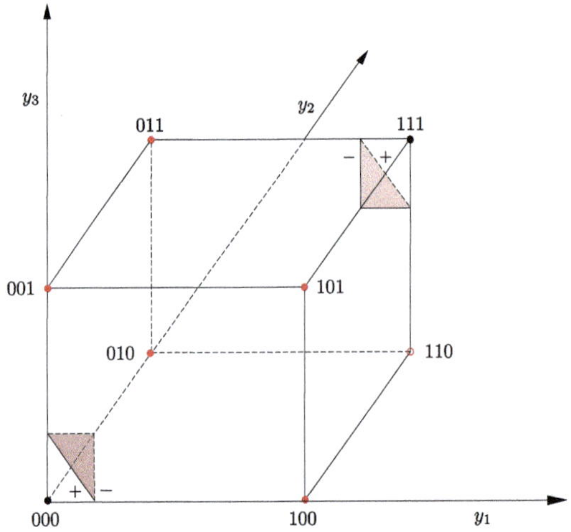

FIGURE 18.5

The neurons of the first hidden layer perform a mapping from the input feature space to the vertices of a unit hypercube. Each region is mapped to a vertex. Each vertex of the hypercube is now linearly separable from all the rest and it can be separated by a hyperplane realized by a neuron. The vertex 110, denoted as an unshaded circle, does not correspond to any region.

select the two regions that correspond to class ω_1, that is, (000) and (111). Recall that all the points from these regions are mapped to the respective vertices of the unit cube in the $\mathbb{R}^3$. However, in this new transformed space, each one of the vertices is *linearly separable* from the rest. This means that we can use a neuron/perceptron in the transformed space, which will place a single vertex in the (+) side and the rest in the (−) one of the associated hyperplane. This is shown in Fig. 18.5, where two such planes are shown, which separate the respective vertices from the rest. Each of these planes is realized by a neuron, operating in $\mathbb{R}^3$, as shown in Fig. 18.4B, where a second layer of *hidden* neurons has been added.

Note that the output z_1 of the left neuron will fire a 1 only if the input pattern originates from the region 000 and it will be at 0 for all other patterns. For the neuron on the right, the output z_2 will be 1 for all the patterns coming from region (111) and zero for all the rest. Note that this second layer of neurons has performed a second mapping, this time to the vertices of the unit rectangle in $\mathbb{R}^2$. This mapping provides a new representation of the input patterns, and this representation encodes information related to the classes of the regions. Fig. 18.6 shows the mapping to the vertices of the unit rectangle in the (z_1, z_2) space.

Note that all the points originating from class ω_2 are mapped to (00) and the points from class ω_1 are mapped either to (10) or to (01). This is very interesting; by successive mappings, we have transformed our originally nonlinearly separable task to one that is linearly separable. Indeed, the point (00) can be linearly separated from (01) and (10) and this can be realized by an extra neuron operating in the (z_1, z_2) space; it is known as the *output neuron*, because it provides the final classification decision. The final resulting network is shown is Fig. 18.7. We call this network *feed-forward*, because information

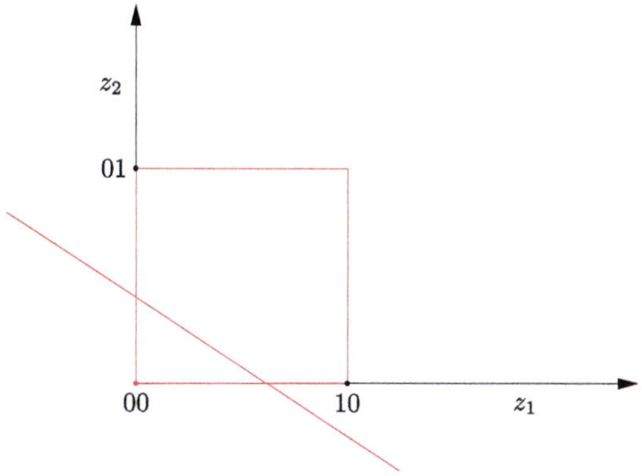

FIGURE 18.6

The patterns from class ω_1 are mapped either to (01) or to (10) and patterns from class ω_2 are all mapped to (00). Thus the classes have now become linearly separable and can be separated via a straight line realized by a neuron.

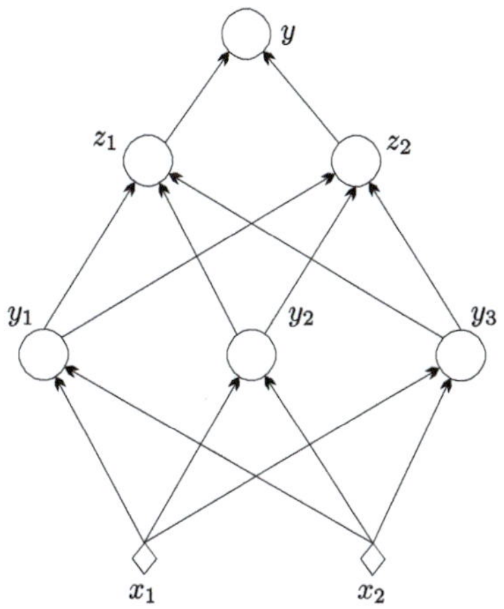

FIGURE 18.7

A three-layer feed-forward neural network. It comprises the input (nonprocessing) layer, two hidden layers, and one output layer of neurons. Such a three-layer neural network can solve *any* classification task, where classes are formed by unions of polyhedral regions.

flows forward from the input to the output layer. It comprises the input layer, which is a nonprocessing one, two hidden layers (the term "hidden" is self-explained), and one output layer. We call such a neural network a three-layer network, without counting the input layer of nonprocessing nodes.

We have constructively shown that a three-layer feed-forward neural network can, in principle, solve *any* classification task whose classes are formed by unions of polyhedral regions. Although we focused on the two-class case, the generalization to multiclass cases is straightforward, by employing more output neurons depending on the number of classes. Note that in some cases, one hidden layer of nodes may be sufficient. This depends on whether the vertices on which the regions are mapped are assigned to classes so that linear separability is possible. For example, this would be the case if class ω_1 was the union of (000) and (100) regions. Then these two vertices could be separated from the rest via a single plane and a second hidden layer of neurons would not be required (check why). In any case, we will not take our discussion any further. The reason is that such a construction is important to demonstrate the power of building a multilayer neural network, in analogy to what is happening in our brain. However, from a practical point of view, such a construction has not much to offer. In practice, when the data live in high-dimensional spaces, there is no chance of determining the parameters that define the neurons analytically to realize the hyperplanes, which form the polyhedral regions. Furthermore, in real life, classes are not necessarily formed by the union of polyhedral regions and more important classes do overlap. Hence, one needs to devise a training procedure based on a cost function and a set of training data.

All that we have to keep from our previous discussion is the structure of the multilayer network; our focus will turn on seeking ways for estimating the unknown weights of the synapses and biases of the neurons. However, from a conceptual point of view, we have to remember that each layer performs a mapping into a new space, and each mapping provides a different, hopefully more informative, representation of the input data, until the last layer, where the task has been transformed into one that it is easy to solve.

18.3.1 FULLY CONNECTED NETWORKS

The feed-forward networks that have been introduced before are also known as *fully connected networks*. This name is to stress out that each one of the neurons/nodes in any layer is directly connected to *every* node of the previous layer. The nodes of the first hidden layer are *fully* connected to those of the input layer. In other words, *each* neuron is associated with a *vector* of parameters, whose dimension is equal to the number of nodes of the previous (input) layer. The algebraic operations which are performed are *inner products*.

To summarize in a more formal way the type of operations that take place in a fully connected network, let us focus on, say, the rth layer of a multilayer neural network and assume that it comprises k_r neurons. The input vector to this layer consists of the outputs at the nodes of the previous layer, denoted as $\boldsymbol{y}^{r-1}$. Let $\boldsymbol{\theta}_j^r$ be the vector of the synaptic weights, including the bias term, associated with the jth neuron of the rth layer, where $j = 1, 2, \ldots k_r$. The respective dimension is $k_{r-1} + 1$, where k_{r-1} is the number of the neurons of the previous, $r - 1$, layer and the increase by 1 accounts for the bias term. Then the performed operations, prior to the nonlinearity, are the inner products

$$z_j^r = \boldsymbol{\theta}_j^{rT} \boldsymbol{y}^{r-1}, \; j = 1, 2, \ldots, k_r.$$

Collecting all the output values into a vector, $\boldsymbol{z}^r = [z_1^r, z_2^r, \ldots, z_{k_r}^r]^T$, and stacking all the synaptic vectors as rows, one under the other, in a matrix Θ, we can write collectively

$$\boldsymbol{z}^r = \Theta \boldsymbol{y}^{r-1}, \text{ where } \Theta := [\boldsymbol{\theta}_1^r, \boldsymbol{\theta}_2^r, \ldots, \boldsymbol{\theta}_{k_r}^r]^T.$$

The vector of the outputs of the rth hidden layer, after pushing each z_i^r through the nonlinearity f, is finally given by

$$\boldsymbol{y}^r = \begin{bmatrix} 1 \\ f(\boldsymbol{z}^r) \end{bmatrix},$$

where the notation above means that f acts on each one of the respective vector components, individually, and the extension of the vector by one is to account for the bias terms in the standard practice.

For large networks, with many layers and many nodes per layer, this type of connectivity turns out to be very costly in terms of the number of parameters (weights), which is of the order of $k_r k_{r-1}$. For example, if $k_{r-1} = 1000$ and $k_r = 1000$, this amounts to an order of 1 million parameters. Note that this number is the contribution from the parameters of only one of the layers. However, a large number of parameters makes a network vulnerable to overfitting, when training is involved, as has already been discussed in Section 3.8.

In contrast, one can employ the so-called *weight sharing* techniques (e.g., [181,231]), where a set of parameters are *shared* among a number of connections, via appropriately built-in constraints. The convolutional networks, to be discussed in Section 18.12, belong to this family of weight sharing networks. As we will see, in a convolutional network, convolutions replace the inner product operations, which allows for a significant weight sharing that leads to a substantial reduction in the required number of parameters.

Remarks 18.2.

- *Shallow and deep networks*: From now on, we will refer to the number of layers in a network as the *depth* of the network. Networks with up to three (two hidden) layers are known as *shallow*, whereas those with more than three layers are called *deep* networks.

18.4 THE BACKPROPAGATION ALGORITHM

A feed-forward neural network consists of a number of layers of neurons, and each neuron is determined by the corresponding set of synaptic weights and its bias term. From this point of view, a neural network realizes a nonlinear parametric function, $\hat{y} = f_{\boldsymbol{\theta}}(\boldsymbol{x})$, where $\boldsymbol{\theta}$ stands for all the weights/biases present in the network. Thus, training a neural network seems not to be any different from training any other parametric prediction model. All that is needed is (a) a set of training samples, (b) a loss function, $\mathcal{L}(y, \hat{y})$, and (c) an iterative scheme, for example, gradient descent, to perform the optimization of the associated cost function (empirical loss; see Section 3.14, Chapter 3),

$$J(\boldsymbol{\theta}) = \sum_{n=1}^{N} \mathcal{L}\big(y_n, f_{\boldsymbol{\theta}}(\boldsymbol{x}_n)\big). \tag{18.6}$$

The difficulty with training neural networks lies in their multilayer structure that complicates the computation of the gradients, which are involved in the optimization. Moreover, the McCulloch–Pitts neuron is based on the discontinuous Heaviside activation function, which is not differentiable. A first step in developing a practical algorithm for training a neural network is to replace the Heaviside activation function with a differentiable approximation of it.

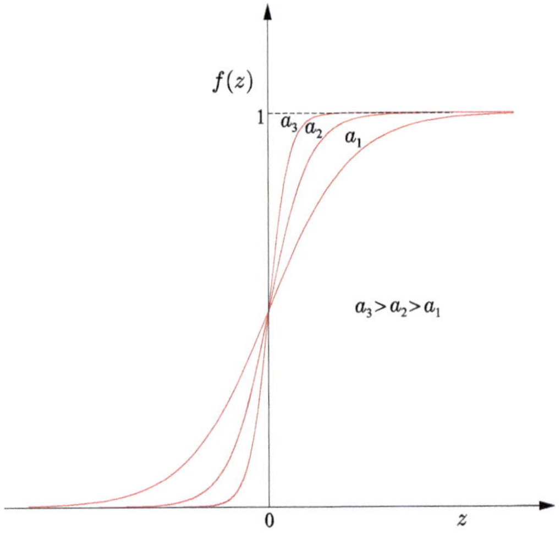

FIGURE 18.8

The logistic sigmoid function for different values of the parameter a.

The logistic sigmoid neuron: One possibility is to adopt the logistic sigmoid function, that is,

$$f(z) = \sigma(z) := \frac{1}{1 + \exp(-az)}. \tag{18.7}$$

The graph of the function is shown in Fig. 18.8. Note that the larger the value of the parameter a is, the closer the corresponding graph gets to that of the Heaviside function. Another possibility would be to use

$$f(z) = a \tanh\left(\frac{cz}{2}\right), \tag{18.8}$$

where c and a are controlling parameters. The graph of this function is shown in Fig. 18.9. Note that in contrast to the logistic sigmoid one, this is an antisymmetric function, that is, $f(-z) = -f(z)$. Both are also known as *squashing* functions, because they limit the output to a finite range of values.

NONCONVEXITY OF THE COST FUNCTION

Optimization of cost functions has been a recurrent theme that runs across this book. In Chapter 5, the gradient descent optimization algorithm was introduced and some of its convergence properties were outlined. In the sequel, the stochastic approximation framework for optimizing expectations of loss functions in an online, sample after sample, mode of operation was discussed. The gradient descent scheme and the online optimization rationale comprise the spine for a number of optimization algorithms that have been proposed and used for training feed-forward neural networks. The reason that we need to discuss these algorithms specifically in the context of neural networks is that the *multilayer* structure of these networks poses difficulties that have to be understood, prior to applying an off-the-shelf optimization scheme.

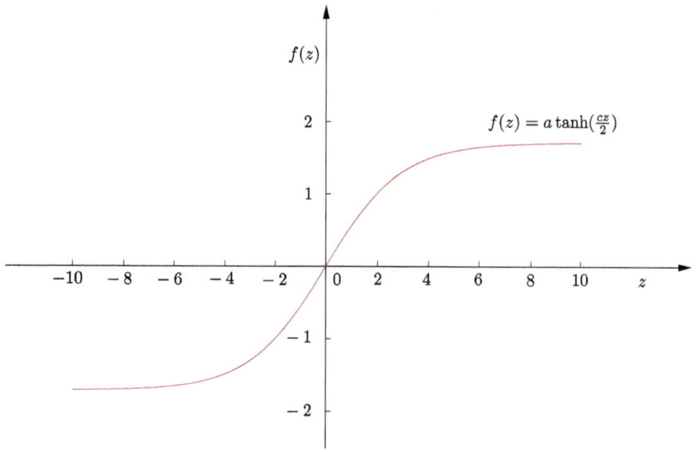

FIGURE 18.9

The hyperbolic tangent squashing function for $a = 1.7$ and $c = 4/3$.

A major difficulty when dealing with the minimization of a cost function, such as in Eq. (18.6), in the framework of neural networks, is its *nonconvexity*. Convex functions were discussed in Chapter 8. There, it was pointed out that for the case of a convex function, every local minimum is also a global one. Recall, for example, a typical graph of a convex function in Fig. 8.3. By the definition of a minimum of a function, we know that at this point the gradient (derivative in the simplest single-parameter space) becomes zero. Points where the gradient of a function becomes zero are known as *critical* or *stationary* points.

However, if the function is not convex, a stationary point can belong to one of the following three categories: to be (a) a local minimum (maximum), (b) a global minimum (maximum), or (c) a saddle point; see Fig. 18.10 for the case of the one-dimensional space. All these stationary points are of importance in neural networks. A local minimum is the point where the value of the cost, $J(\theta_l)$, becomes minimum within a region around θ_l. The global minimum, θ_g, is the point where $J(\theta_g) \leq J(\theta)$, $\forall \theta \in \mathbb{R}$. A saddle point, θ_s, is neither a minimum nor a maximum, yet the derivative (gradient in general) is equal to zero.

When adopting a gradient descent scheme to minimize a nonconvex cost function, the algorithm can converge to either a local or to a global minimum. Take, for example, the case of Fig. 18.10. Recall from Chapter 5 that the update rule of the gradient descent algorithm, in its one-dimensional version, becomes

$$\theta(\text{new}) = \theta(\text{old}) - \mu \frac{dJ}{d\theta}\bigg|_{\theta(\text{old})},$$

and the iterations start from an arbitrary initial point, $\theta^{(0)}$. If at the current iteration the algorithm is, say, at the point $\theta(\text{old}) = \theta_1$, then it will move towards the local minimum, θ_l. This is because the derivative of the cost at θ_1 is equal to the tangent of ϕ_1, which is negative (the angle is obtuse) and the update, $\theta(\text{new})$, will move to the right towards the local minimum, θ_l. In contrast, if the algorithm had

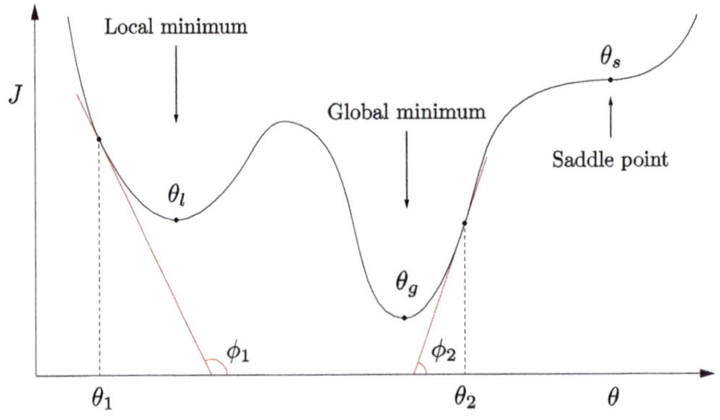

FIGURE 18.10

A nonconvex function, besides the global, usually comprises a number of local minima and saddle points. To which minimum, out of the many, the algorithm will converge, depends on the point from which the algorithm was initialized. However, if the value of the cost at a local minimum, e.g., $J(\theta_l)$, is not much larger than that of the global minimum, i.e., $J(\theta_g)$, then θ_l can be a satisfactory solution in practice.

been initialized from a different point and the algorithm is currently at, say, $\theta(\text{old}) = \theta_2$, the update will move towards the global minimum, θ_g, since the derivative is now equal to the tangent of ϕ_2, which is positive (the angle is acute). As we know from Chapter 5, the choice of the step size, μ, is critical for the convergence of the algorithm.

In real problems in multidimensional spaces, the number of local minima can be large, so the algorithm can converge to a local one. However, this is not necessarily very bad news. If this local minimum is deep enough, that is, if the value of the cost function at this point, e.g., $J(\theta_l)$, is not much larger than that achieved at the global minimum, i.e., $J(\theta_g)$, convergence to such a local minimum can correspond to a good solution. In practice, one has to be careful with how to initialize an algorithm when dealing with nonconvex cost functions. We are going to discuss this issue later on. Also, the interplay between local, global, and saddle points, when the dimension of the parameter space becomes large, which is the case for deep networks, is discussed in Section 18.11.1.

18.4.1 THE GRADIENT DESCENT BACKPROPAGATION SCHEME

Having adopted a differentiable activation function, we are ready to proceed with developing the gradient descent iterative scheme for the minimization of the cost function. We will formulate the task in a general framework.

Let $(\boldsymbol{y}_n, \boldsymbol{x}_n)$, $n = 1, 2, \ldots, N$, be the set of training samples. Note that we have assumed multiple output variables, assembled as a vector. We assume that the network comprises L layers; $L-1$ hidden layers and one output layer. Each layer consists of k_r, $r = 1, 2, \ldots, L$, neurons. Thus, the (target/desired) output vectors are

$$\boldsymbol{y}_n = [y_{n1}, y_{n2}, \ldots, y_{nk_L}]^T \in \mathbb{R}^{k_L}, \quad n = 1, 2, \ldots, N.$$

For the sake of the mathematical derivations, we also denote the number of input nodes as k_0; that is, $k_0 = l$, where l is the dimensionality of the input feature space.

Let $\boldsymbol{\theta}_j^r$ denote the vector of the synaptic weights associated with the jth neuron in the rth layer, with $j = 1, 2, \ldots, k_r$ and $r = 1, 2, \ldots, L$, where the bias term is included in $\boldsymbol{\theta}_j^r$, that is,

$$\boldsymbol{\theta}_j^r := [\theta_{j0}^r, \theta_{j1}^r, \ldots, \theta_{jk_{r-1}}^r]^T. \tag{18.9}$$

The synaptic weights link the respective neuron to all neurons in layer k_{r-1} (see Fig. 18.11). The basic iterative step for the gradient descent scheme is written as

$$\boldsymbol{\theta}_j^r(\text{new}) = \boldsymbol{\theta}_j^r(\text{old}) + \Delta\boldsymbol{\theta}_j^r, \tag{18.10}$$

$$\Delta\boldsymbol{\theta}_j^r := -\mu \left.\frac{\partial J}{\partial \boldsymbol{\theta}_j^r}\right|_{\boldsymbol{\theta}_j^r(\text{old})}. \tag{18.11}$$

The parameter μ is the user-defined step size (it can also be iteration dependent) and J denotes the cost function.

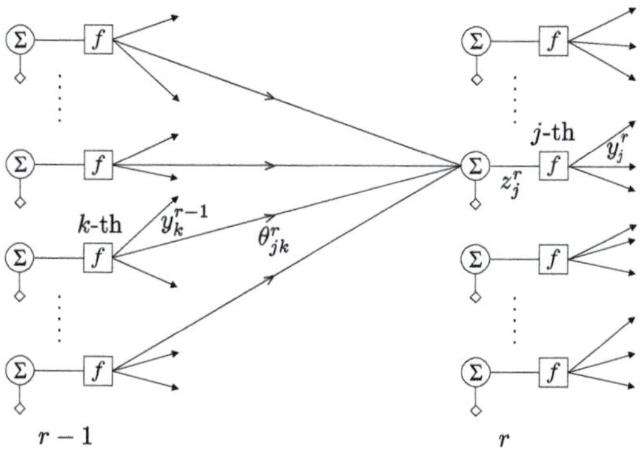

FIGURE 18.11

The links and the associated variables of the jth neuron at the rth layer; y_k^{r-1} is the output of the kth neuron at the $(r-1)$th layer and θ_{jk}^r is the respective weight connecting these two neurons. The dependence on index n has been suppressed for notational convenience.

Update Eqs. (18.10) and (18.11) comprise the pair of the gradient descent scheme for optimization (e.g., Chapter 5). As previously stated, the difficulty in feed-forward neural networks arises from their multilayer structure. In order to compute the gradients in Eq. (18.11), for *all* neurons in *all* layers, one has to follow two steps of computations.

- *Forward computations*: For a given input vector $\boldsymbol{x}_n$, $n = 1, 2, \ldots, N$, use the current estimates of the parameters (synaptic weights) ($\boldsymbol{\theta}_j^r(\text{old})$) and compute all the outputs of all the neurons in all layers, denoted as y_{nj}^r; in Fig. 18.11, the index n has been suppressed to unclutter notation.

- *Backward computations*: Using the above computed neuronal outputs together with the known target values, y_{nk}, of the output layer, compute the gradients of the cost function. This involves L steps, that is, as many as the number of layers. The sequence of the algorithmic steps is given below:
  - Compute the gradient of the cost function with respect to the parameters of the neurons of the *last layer*, i.e., $\frac{\partial J}{\partial \boldsymbol{\theta}_j^L}$, $j = 1, 2, \ldots, k_L$.
  - **For** $r = L - 1$ to 1, **Do**
    * Compute the gradients with respect to the parameters associated with the neurons of the rth layer, i.e., $\frac{\partial J}{\partial \boldsymbol{\theta}_k^r}$, $k = 1, 2, \ldots, k_r$, based on all the gradients $\frac{\partial J}{\partial \boldsymbol{\theta}_j^{r+1}}$, $j = 1, 2, \ldots, k_{r+1}$, with respect to the parameters of the layer $r + 1$ that have been computed in the previous step.
  - **End For**

The backward computations scheme is a direct application of the chain rule for derivatives, and it starts with the initial step of computing the derivatives associated with the last (output) layer, which turns out to be straightforward. Then the algorithm "flows" backwards in the hierarchy of layers. This is because of the nature of the multilayer network, where the outputs, layer after layer, are formed as functions of functions. Indeed, let us focus on the output y_k^r of the kth neuron at layer r. Then we have

$$y_k^r = f\left(\boldsymbol{\theta}_k^{r^T} \boldsymbol{y}^{r-1}\right), \quad k = 1, 2, \ldots, k_r,$$

where $\boldsymbol{y}^{r-1}$ is the (extended) vector comprising all the outputs at the previous layer, $r - 1$, and f denotes the nonlinearity. Based on the above, the output of the jth neuron at the next layer is given by

$$y_j^{r+1} = f\left(\boldsymbol{\theta}_j^{r+1^T} \boldsymbol{y}^r\right) = f\left(\boldsymbol{\theta}_j^{r+1^T} \begin{bmatrix} 1 \\ f\left(\Theta^r \boldsymbol{y}^{r-1}\right) \end{bmatrix}\right),$$

where $\Theta^r := [\boldsymbol{\theta}_1^r, \boldsymbol{\theta}_2^r, \ldots, \boldsymbol{\theta}_{k_r}]^T$ denotes the matrix having as rows the weight vectors at layer r. One can easily spot what we called before as "a function of a function." Obviously, this goes on as we move on in the hierarchy. This function-over-function-over-function structure is the by-product of the multilayer nature of the neural networks, and it is a highly nonlinear operation that gives rise to the complication for computing the gradients, in contrast to other learners that we have studied in previous chapters.

However, one can easily spot that computing the gradients with respect to the parameters defining the output layer does not pose any difficulties. Indeed, the output of the jth neuron of the last layer (which is actually the respective current output estimate) is written as

$$\hat{y}_j := y_j^L = f(\boldsymbol{\theta}_j^{L^T} \boldsymbol{y}^{L-1}).$$

Since $\boldsymbol{y}^{L-1}$ is known, after the computations during the forward pass, taking the derivative with respect to $\boldsymbol{\theta}_j^L$ is straightforward; no function-over-function operation is involved here. This is why we start from the top layer and then move backwards.

Due to its historical importance, the full derivation of the backpropagation algorithm will be given. Those of the readers who are not interested in the details can bypass this part in a first reading.

For the detailed derivation of the backpropagation algorithm, the squared error loss function is adopted as an example, i.e.,

$$J(\boldsymbol{\theta}) = \sum_{n=1}^{N} J_n(\boldsymbol{\theta}), \tag{18.12}$$

and

$$J_n(\boldsymbol{\theta}) = \frac{1}{2}\sum_{k=1}^{k_L}\left(\hat{y}_{nk} - y_{nk}\right)^2, \tag{18.13}$$

where $\hat{y}_{nk},\ k = 1, 2, \ldots, k_L$, are the estimates provided at the corresponding output nodes of the network. We will consider them as the elements of a corresponding vector, $\hat{\boldsymbol{y}}_n$.

Computation of the gradients: Let z_{nj}^r denote the output of the linear combiner of the jth neuron in the rth layer at time instant n, when the pattern $\boldsymbol{x}_n$ is applied at the input nodes (see Fig. 18.11). Then we can write

$$z_{nj}^r = \sum_{m=1}^{k_{r-1}} \theta_{jm}^r y_{nm}^{r-1} + \theta_{j0}^r = \sum_{m=0}^{k_{r-1}} \theta_{jm}^r y_{nm}^{r-1} = \boldsymbol{\theta}_j^{r^T} \boldsymbol{y}_n^{r-1}, \tag{18.14}$$

where by definition

$$\boldsymbol{y}_n^{r-1} := [1, y_{n1}^{r-1}, \ldots, y_{nk_{r-1}}^{r-1}]^T, \tag{18.15}$$

and $y_{n0}^r \equiv 1,\ \forall\, r, n$, and $\boldsymbol{\theta}_j^r$ has been defined in Eq. (18.9). For the neurons at the output layer, $r = L$, $y_{nm}^L = \hat{y}_{nm},\ m = 1, 2, \ldots, k_L$, and for $r = 1$, we have $y_{nm}^0 = x_{nm},\ m = 1, 2, \ldots, k_0$; that is, y_{nm}^0 are set equal to the input feature values.

Hence, we can now write

$$\frac{\partial J_n}{\partial \boldsymbol{\theta}_j^r} = \frac{\partial J_n}{\partial z_{nj}^r}\frac{\partial z_{nj}^r}{\partial \boldsymbol{\theta}_j^r} = \frac{\partial J_n}{\partial z_{nj}^r}\boldsymbol{y}_n^{r-1}. \tag{18.16}$$

Let us now define

$$\delta_{nj}^r := \frac{\partial J_n}{\partial z_{nj}^r}. \tag{18.17}$$

Then Eq. (18.11) becomes

$$\boxed{\Delta\boldsymbol{\theta}_j^r = -\mu\sum_{n=1}^{N}\delta_{nj}^r \boldsymbol{y}_n^{r-1}, \quad r = 1, 2, \ldots, L.} \tag{18.18}$$

Computation of δ_{nj}^r: Here is where the heart of the backpropagation algorithm beats. For the computation of the gradients, δ_{nj}^r, one starts at the last layer, $r = L$, and proceeds *backwards* toward $r = 1$; this "philosophy" justifies the name given to the algorithm.

1. $r = L$: We have

$$\delta_{nj}^{L} = \frac{\partial J_n}{\partial z_{nj}^{L}}. \tag{18.19}$$

For the squared error loss function,

$$J_n = \frac{1}{2}\sum_{k=1}^{k_L}\left(f(z_{nk}^{L}) - y_{nk}\right)^2. \tag{18.20}$$

Hence,

$$\begin{aligned}\delta_{nj}^{L} &= (\hat{y}_{nj} - y_{nj})f'(z_{nj}^{L}) \\ &= e_{nj}f'(z_{nj}^{L}), \quad j = 1, 2, \dots, k_L,\end{aligned} \tag{18.21}$$

where f' denotes the derivative of f and e_{nj} is the error associated with the jth output variable at time n. Note that for the last layer, the computation of the gradient, δ_{nj}^{L}, is straightforward.

2. $r < L$: Due to the successive dependence between the layers, the value of z_{nj}^{r-1} influences all the values z_{nk}^{r}, $k = 1, 2, \dots, k_r$, of the next layer. Employing the chain rule for differentiation, we get

$$\delta_{nj}^{r-1} = \frac{\partial J_n}{\partial z_{nj}^{r-1}} = \sum_{k=1}^{k_r}\frac{\partial J_n}{\partial z_{nk}^{r}}\frac{\partial z_{nk}^{r}}{\partial z_{nj}^{r-1}}, \tag{18.22}$$

or

$$\delta_{nj}^{r-1} = \sum_{k=1}^{k_r}\delta_{nk}^{r}\frac{\partial z_{nk}^{r}}{\partial z_{nj}^{r-1}}. \tag{18.23}$$

However,

$$\frac{\partial z_{nk}^{r}}{\partial z_{nj}^{r-1}} = \frac{\partial\left(\sum_{m=0}^{k_{r-1}}\theta_{km}^{r}y_{nm}^{r-1}\right)}{\partial z_{nj}^{r-1}}, \tag{18.24}$$

where

$$y_{nm}^{r-1} = f(z_{nm}^{r-1}), \tag{18.25}$$

which leads to

$$\frac{\partial z_{nk}^{r}}{\partial z_{nj}^{r-1}} = \theta_{kj}^{r}f'(z_{nj}^{r-1}), \tag{18.26}$$

and combining with Eqs. (18.22) and (18.23), we obtain the recursive rule

$$\boxed{\delta_{nj}^{r-1} = \left(\sum_{k=1}^{k_r}\delta_{nk}^{r}\theta_{kj}^{r}\right)f'(z_{nj}^{r-1}), \quad j = 1, 2, \dots, k_{r-1}.} \tag{18.27}$$

For uniformity with (18.21), define

$$e_{nj}^{r-1} := \sum_{k=1}^{k_r} \delta_{nk}^r \theta_{kj}^r, \tag{18.28}$$

and we finally get

$$\delta_{nj}^{r-1} = e_{nj}^{r-1} f'(z_{nj}^{r-1}). \tag{18.29}$$

The only remaining computation is the derivative of f. For the case of the logistic sigmoid function, it is easily shown to be equal to (Problem 18.2)

$$f'(z) = af(z)\big(1 - f(z)\big). \tag{18.30}$$

The derivation has been completed and the backpropagation scheme is summarized in Algorithm 18.2.

Algorithm 18.2 (The gradient descent backpropagation algorithm).

- Initialization
  - Initialize all synaptic weights and biases randomly with small, but not very small, values.
  - Select step size μ.
  - Set $y_{nj}^0 = x_{nj}, \ j = 1, 2, \ldots, k_0 := l, \ n = 1, 2, \ldots, N$.
- **Repeat**; Each repetition completes one epoch.
  - **For** $n = 1, 2, \ldots, N$, **Do**
    - **For** $r = 1, 2, \ldots, L$, **Do**; Forward computations.
      - **For** $j = 1, 2, \ldots, k_r$, **Do**

        Compute z_{nj}^r from (18.14)

        Compute $y_{nj}^r = f(z_{nj}^r)$
      - **End For**
    - **End For**
    - **For** $j = 1, 2, \ldots, k_L$, **Do**; Backward computations (output layer).

      Compute δ_{nj}^L from (18.21)
    - **End For**
    - **For** $r = L, L-1, \ldots, 2$, **Do**; Backward computations (hidden layers).
      - **For** $j = 1, 2, \ldots, k_r$, **Do**

        Compute δ_{nj}^{r-1} from (18.29).
      - **End For**
    - **End For**
  - **End For**
  - **For** $r = 1, 2, \ldots, L$, **Do**; Update the weights.

      * **For** $j = 1, 2, \ldots, k_r$, **Do**
        Compute $\Delta\boldsymbol{\theta}_j^r$ from (18.18)
        $\boldsymbol{\theta}_j^r = \boldsymbol{\theta}_j^r + \Delta\boldsymbol{\theta}_j^r$
      * **End For**
    - **End For**
- **Until** a stopping criterion is met.

The backpropagation algorithm can claim a number of fathers. The popularization of the algorithm is associated with the classical paper [199], where the derivation of the algorithm is provided. However, the algorithm had been derived much earlier in [250]. The idea of backpropagation also appears in [26] in the context of optimal control.

Remarks 18.3.

- A number of criteria have been suggested for terminating the backpropagation algorithm. One possibility is to track the value of the cost function, and stop the algorithm when this gets smaller than a preselected threshold. An alternative path is to check for the gradient values and stop when these become small; this means that the values of the weights do not change much from iteration to iteration (see, for example, [124]).
- As is the case with all gradient descent schemes, the choice of the step size, μ, is very critical; it has to be small enough to guarantee convergence, but not too small; otherwise the convergence speed slows down. The choice depends a lot on the specific problem at hand. Adaptive values of μ, whose value depends on the iteration, are also popular alternatives and they will be discussed soon.
- Due to the highly nonlinear nature of the neural network problem, the cost function in the parameter space has, in general, a complicated landscape and there are a number of local minima, where the algorithm can be trapped. If such a local minimum is deep enough, the associated solution can be acceptable. However, this may not be the case and the algorithm can be trapped in a shallow minimum resulting in a bad solution. Ideally, one should reinitialize randomly the weights a number of times and keep the best solution. Initialization has to be performed with care; we discuss this issue later on. A discussion concerning local and global minima in the context of deep networks is discussed in Section 18.11.1.

Pattern-by-Pattern/Online Scheme

The scheme discussed in Algorithm 18.2 is of the *batch* type, where the weights are updated once per epoch, that is, after all N training samples have been presented to the algorithm. An alternative route is the *pattern-by-pattern* version; for this case, the weights are updated every time a new training sample appears in the input. Sometimes, such schemes are known as *online*. Yet, we will avoid to use this term, because it usually refers to the case where streaming data are considered and observations are continuously coming in, instead of having a fixed sized training set where data points are considered repeatedly, epoch after epoch. Online algorithms have been considered in Chapters 5 and 8. Another name that is more recently used to describe such algorithms is *stochastic* optimization. This is a reminiscent of the stochastic approximation method, introduced in Chapter 5, where in order to minimize the expected loss, individual observations are sequentially used to update the estimates of the unknown parameters/weights. Of course, keep in mind that, in the true stochastic gradient descent, there is *no* data reuse (i.e., epoch after epoch) and data are assumed to arrive in a streaming way and convergence is achieved *asymptotically*.

Pattern-by-pattern versions exploit better the training set, when redundancies in the data are present or training samples are very similar. Averaging, as is done in the batch mode, wastes resources, because averaging the contribution to the gradient of similar patterns does not add much information. In contrast, in the pattern-by-pattern implementations, all examples are equally exploited, inasmuch as an update takes place for each one of the training samples.

The pattern-by-pattern mode leads to a less smooth convergence trajectory; however, such a randomness may have the advantage of helping the algorithm to escape from local minima.

Minibatch Schemes

An intermediate way, where the parameter updates are performed every $K < N$ samples, has also been considered, which is referred to as a *minibatch* or *stochastic minibatch*. However, more and more and in a rather abuse of terminology, such schemes are referred to as simply stochastic. The choice of the minibatch size, K, is influenced by a number of factors and it depends on the application and the specific data as well as the available computing power.

For example, multicore architectures are underutilized if the batch size is very small. Also, when GPUs are used, employing minibatch sizes of power of 2 can offer better run times. For such cases, some typical values are ranging from 16 to 256, depending on the size of the training set. On the other hand, the use of small minibatch sizes can have a regularizing effect, due to the noise that is implicitly added in the computation of the gradients. However, in such cases, one has to use smaller values for the step size, for stability reasons, and this can increase the overall training time.

Batch and minibatch schemes have an averaging effect on the computation of the gradients. In [213], it is advised to add a small white noise sequence to the training data, which may have a beneficial effect for the algorithm to escape from poor local minima.

To exploit randomness even further in the pattern-by-pattern/minibatch schemes, it is advisable that prior to the pass of a new epoch, the sequence in which data are presented to the algorithm is randomized (see, for example, [79]). This has no meaning in the batch mode, because updates take place once all data have been considered. This random shuffling of the data "breaks" possible correlations that underlie successive samples, making successive gradient computations independent. Such correlations may be due to specific experimental protocols that have been followed for generating the data. Although it is advisable to shuffle the data for every epoch, this may not be practical when large data sets are used, e.g., of the order of a billion. In such cases, even a single data shuffling in the beginning is beneficial.

These days, in the context of deep learning with large data training sets, minibatch schemes seem to be the ones that are favored for most applications.

A major problem in all gradient descent schemes and their stochastic variants is to select the step size or *learning rate*, μ. As we know from Chapter 5, if its value is small the learning curves are smooth, but convergence can be slow. In contrast, if its value is relatively large, learning curves tend to be oscillatory but convergence can get faster, provided that the value does not become large enough for the algorithm to diverge. In practice, time-varying step sizes are more appropriate and they are in line with the stochastic gradient rationale. One can employ various strategies, yet these strategies are more an engineering "art" than the result of a theoretical analysis. It is advisable for a user to first look at what previous researchers/practitioners have done in similar situations and start with this experience.

A possible strategy is to start training with a linearly decreasing step size and then switch to a fixed one after a number of iterations. For example, one can proceed with a rule such as

$$\mu_i = (1 - a_i)\mu_0 + a_i \mu_I,$$

where $a_i = i/I$ for some fixed iteration I. After iteration I, the step size is fixed. The issue now is to set the parameters μ_0, μ_I, and I. In practice, one has to monitor the learning curve for a number of iterations first and adjust the values of the parameters accordingly as a tradeoff between how fast or how slow the initial iterations move towards convergence and at the same time being cautious about instabilities.

18.4.2 VARIANTS OF THE BASIC GRADIENT DESCENT SCHEME

The basic gradient descent scheme inherits all the advantages (low computational demands per iteration step) and all the disadvantages (slow convergence rate) of the gradient descent algorithmic family, as it was first presented in this book in Chapter 5. To speed up the convergence rate, a lot of research effort has been invested and a large number of variants of the basic gradient descent backpropagation scheme have been proposed. In this section, we provide some directions, which at the time the current edition is being compiled seem to be the trend in practice.

For simplicity, and having derived the backpropagation descent scheme, let us get rid of the indices r and j referring to the layer and the specific neuron in the respective layer. Instead, the parameter/weight vector $\boldsymbol{\theta}$ will be used that comprises all the parameters of the network. Also, the corresponding estimate at the ith iteration of a learning algorithm will be denoted by $\boldsymbol{\theta}^{(i)}$.

Let $\mathcal{L}(y, \boldsymbol{x}, \boldsymbol{\theta})$ be the adopted loss function. The latter measures the deviation between the true output, y, and the corresponding predicted value, $\hat{y}$, which depends on the respective input, $\boldsymbol{x}$, as well as the parameters that define the network, $\boldsymbol{\theta}$. The cost function, over all the training points, is given by

$$J(\boldsymbol{\theta}) = \sum_{n=1}^{N} \mathcal{L}(y_n, \boldsymbol{x}_n, \boldsymbol{\theta}).$$

If batch processing is employed, then at each iteration the gradient of the above cost function is used for the update of the parameters, i.e., $\frac{\partial J(\boldsymbol{\theta})}{\partial \boldsymbol{\theta}}$, which involves *all* the training points. That is, for every epoch, only one parameter vector update is performed. On the other extreme, if a pattern-by-pattern scheme is adopted, at each iteration the updates are estimated on the basis of the gradient of the loss function computed at the current input–output pair, i.e., $\frac{\partial \mathcal{L}(y_n, \boldsymbol{x}_n, \boldsymbol{\theta})}{\partial \boldsymbol{\theta}}$. That is, in this case, N updates for the parameters are obtained for every epoch.

The minibatch operation needs a bit more of elaboration. Let us assume that the data set is split in M minibatches, each of size K. Obviously, N should be a multiple of K. Assuming that the data have been randomly shuffled, then at the current iteration epoch, the N input–output data pairs will appear in the algorithm in the following sequence:

$$\underbrace{(y_{(1)}, \boldsymbol{x}_{(1)}), \ldots, (y_{(K)}, \boldsymbol{x}_{(K)})}_{\text{1st minibatch}} \ldots \underbrace{(y_{((M-1)K+1)}, \boldsymbol{x}_{((M-1)K+1)}), \ldots, (y_{(N)}, \boldsymbol{x}_{(N)})}_{M\text{th minibatch}},$$

where, as we have already pointed out in Section 18.2, the notation $(i) \in \{1, 2, \ldots, N\}$ is used, due to the data shuffling. Thus, during one epoch, M updates will be performed, each one involving the gradient of a different cost; this is because a different minibatch will be considered each time, i.e.,

$$\frac{\partial J^{(m)}(\boldsymbol{\theta})}{\partial \boldsymbol{\theta}}, \quad m = 1, 2, \ldots, M, \tag{18.31}$$

$$J^{(m)}(\boldsymbol{\theta}) := \sum_{k=1}^{K} \mathcal{L}\Big(y_{((m-1)K+k)}, \boldsymbol{x}_{((m-1)K+k)}, \boldsymbol{\theta}\Big). \tag{18.32}$$

In the sequel, to simplify notations, we will use a common symbol, $\frac{\partial J}{\partial \boldsymbol{\theta}}$, to denote the gradient operation for all three possible cases discussed before. The true value of the function in the place of J will depend on (a) the specific scheme (i.e., batch, minibatch, pattern-by-pattern) and (b) the current iteration step, as different samples are involved in each step (e.g., in the stochastic versions).

Gradient Descent With a Momentum Term

One way to improve the convergence rate, while remaining within the gradient descent rationale, is to employ the so-called *momentum term* [72,247]. The correction term as well as the update recursion are now modified as

$$\begin{aligned} \Delta\boldsymbol{\theta}^{(i)} &= a\Delta\boldsymbol{\theta}^{(i-1)} - \mu \frac{\partial J}{\partial \boldsymbol{\theta}}\bigg|_{\boldsymbol{\theta}^{(i-1)}}, \\ \boldsymbol{\theta}^{(i)} &= \boldsymbol{\theta}^{(i-1)} + \Delta\boldsymbol{\theta}^{(i)}, \end{aligned} \tag{18.33}$$

starting from an initial value for $\Delta\boldsymbol{\theta}^{(0)}$, e.g., $\mathbf{0}$. The gradient is computed via the backpropagation and a is the so-called *momentum factor*. In other words, the algorithm takes into account the correction used in the previous iteration step as well as the current gradient computations. Its effect is to increase the step size in regions where the cost function exhibits low curvature. The step size implicitly increases when a number of successive gradients point in exactly the same direction. Assuming that the gradient is approximately constant over, say, I successive iterations, it can be shown (Problem 18.3) that using the momentum term, the corresponding correction becomes

$$\Delta\boldsymbol{\theta} \approx -\frac{\mu}{1-a}\boldsymbol{g}, \tag{18.34}$$

where $\boldsymbol{g}$ is the gradient value over the I successive iteration steps. That is, the step size has been increased by the factor $\frac{1}{1-a}$. Typical values of a are in the range of 0.5 to 0.9. In essence, the use of a momentum term helps to dampen the zig-zags of the convergence trajectory, as discussed in Chapter 5 (see Fig. 5.9). It has been reported that the use of a momentum term can speed up the convergence rate with a factor of up to two [214]. Experience seems to suggest that the use of a momentum factor helps the batch mode of operation more compared to its stochastic counterparts.

Iteration dependent step size: A heuristic variant of the previous version results if the step size is left to vary as iterations progress. A rule is to change its value according to whether the cost function in the current iteration step is larger or smaller than the previous one. Let $J^{(i)}$ be the computed cost value at the current iteration. Then if $J^{(i)} < J^{(i-1)}$, the learning rate is increased by a factor of r_i. If, on the other hand, the new value is larger than the previous one by a factor larger than c, then the learning rate is reduced by a factor of r_d. Otherwise, the learning rate remains unaltered. Typical values for the involved parameters are $r_i = 1.05$, $r_d = 0.7$, and $c = 1.04$. For iteration steps where the value of the cost increases, it is advisable to set the momentum factor, a, equal to zero [240].

Such techniques, also known as *adaptive momentum*, are more appropriate for batch processing, because for online versions the values of the cost tend to oscillate from iteration to iteration.

Using different step size for each weight: It is beneficial for improving the convergence rate, to employ a *different step size* for each individual weight; this gives the freedom to the algorithm to exploit better the dependence of the cost function on each direction in the parameter space. In [109], it is suggested to increase the learning rate, associated with a weight, if the respective gradient value has the same sign for two successive iterations. Conversely, the learning rate is decreased if the sign changes, because this is indicative of possible oscillation.

In the sequel, more recent schemes that use iteration dependent as well as different for each weight step sizes will be discussed.

Example 18.1. In this toy example, the capability of a multilayer perceptron to classify nonlinearly separable classes is demonstrated in the two-dimensional space, where visualization is possible. The

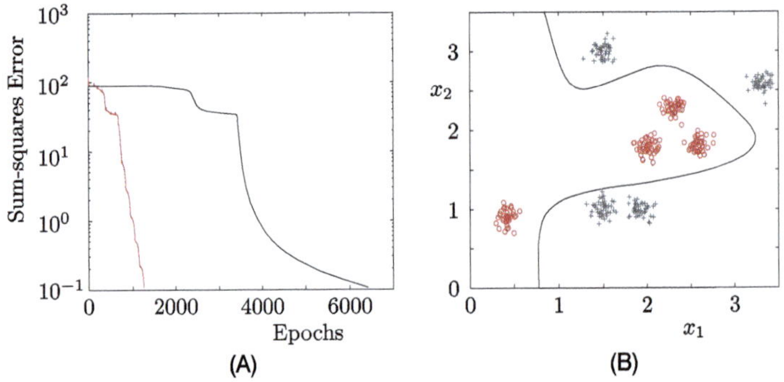

FIGURE 18.12

(A) Error convergence curves for the adaptive momentum (red line) and the momentum algorithms, for Example 18.1. Note that the adaptive momentum leads to faster convergence. (B) The decision line implemented by the classifier that results from the adaptive momentum algorithm.

classification task consists of two classes, each being modeled by a four-component Gaussian mixture model, where the covariance matrix for each component is $\sigma^2 I$, with $\sigma^2 = 0.08$. The mean values are different for each of the Gaussians. Specifically, the samples of the class denoted by a red $\circ$ (see Fig. 18.12) are spread around the mean vectors

$$[0.4, 0.9]^T, \ [2.0, 1.8]^T, \ [2.3, 2.3]^T, \ [2.6, 1.8]^T,$$

and those of the class denoted by a black $+$ around the values

$$[1.5, 1.0]^T, \ [1.9, 1.0]^T, \ [1.5, 3.0]^T, \ [3.3, 2.6]^T.$$

A total of 400 training vectors were generated, 50 from each distribution. A multilayer perceptron with three neurons in the first and two neurons in the second hidden layer were used, with a single output neuron. The activation function was the logistic one with $a = 1$ and the desired outputs 1 and 0, respectively, for the two classes. Two different algorithms were used for the training, namely, the momentum and the adaptive momentum. After some experimentation, the parameters employed were (a) for the momentum $\mu = 0.05$, $\alpha = 0.85$ and (b) for the adaptive momentum $\mu = 0.01$, $\alpha = 0.85$, $r_i = 1.05$, $c = 1.05$, $r_d = 0.7$. The weights were initialized by a uniform pseudorandom distribution between 0 and 1. Fig. 18.12A shows the respective output error convergence curves for the two algorithms as a function of the number of epochs. The respective curves can be considered typical and the adaptive momentum algorithm leads to faster convergence. Both curves correspond to the batch mode of operation. Fig. 18.12B shows the graph of the decision line of the resulting classifier, using the parameters/weights that are estimated via the adaptive momentum training.

Nesterov's Momentum Algorithm

A variant of the celebrated Nesterov algorithm [169] was suggested in [223]. The update rule for the correction term becomes

$$\Delta\boldsymbol{\theta}^{(i)} = a\Delta\boldsymbol{\theta}^{(i-1)} - \mu \frac{\partial J\left(\boldsymbol{\theta} + a\Delta\boldsymbol{\theta}^{(i-1)}\right)}{\partial\boldsymbol{\theta}}\bigg|_{\boldsymbol{\theta}^{(i-1)}}.$$

The difference with the previous momentum update rule lies in that the gradient of the cost function is computed after pushing $\boldsymbol{\theta}$ in the direction of the current correction term, $\Delta\boldsymbol{\theta}^{(i-1)}$. In words, the gradient is not computed at the currently available estimates of the parameters but at their approximate future values. As has been reported in [223], this can lead to substantial gains in convergence speed. The values for a are in the same range as in the case of the momentum algorithm.

The AdaGrad Algorithm

The AdaGrad algorithm was introduced in its online form in the convex cost functions setting in Chapter 8, Eq. (8.71). Borrowing the main rationale behind the algorithm, in our current framework it can be rephrased as

$$\boldsymbol{\theta}^{(i)} = \boldsymbol{\theta}^{(i-1)} - \mu G_{i-1}^{-1/2}\boldsymbol{g}^{(i-1)},$$

where by definition

$$\boldsymbol{g}^{(i)} := \frac{\partial J}{\partial\boldsymbol{\theta}}\bigg|_{\boldsymbol{\theta}^{(i)}}$$

is the gradient computed at the respective estimate, $\boldsymbol{\theta}^{(i)}$, and the matrix G_i is the of sum of the outer products of all the gradient vectors up to and including iteration i, that is,

$$G_i := \sum_{t=0}^{i} \boldsymbol{g}^{(t)}\boldsymbol{g}^{(t)^T}.$$

In practice, matrix G_i is taken to be diagonal; hence the inverse of its root becomes computationally attractive. In such a case, if D is the total number of parameters of the network, the $D \times D$ matrix G_i

involves on its diagonal the squares of the respective gradients, i.e.,

$$G_i(d,d) = \sum_{t=0}^{i} \left(g_d^{(t)}\right)^2, \quad d = 1, 2, \ldots, D,$$

where g_d denotes the dth component of the respective gradient at iteration (t), i.e., $g_d^{(t)} = \frac{\partial J}{\partial \theta_d^{(t)}}$, and θ_d is the dth element of $\boldsymbol{\theta} \in \mathbb{R}^D$. Usually, a small constant ϵ is used to avoid possible division by zero, and the corresponding updates for each one of the parameter vector components take the form

$$\theta_d^{(i)} = \theta_d^{(i-1)} - \frac{\mu}{\sqrt{G_{i-1}(d,d) + \epsilon}} g_d^{(i-1)}, \quad d = 1, 2, \ldots, D. \tag{18.35}$$

The value of ϵ can be of the order of 10^{-8} or so. In words, the step size in the AdaGrad is iteration dependent and also different for each component. The value of μ is usually set close to 0.01. The correction for each individual parameter is inversely proportional to the values of its gradients. Larger values lead to smaller step sizes. Thus, as iterations evolve, each dimension evens out over time. This can be beneficial when training deep networks, because the scale of the gradients can vary a lot in magnitude in different layers.

However, AdaGrad's main drawback is the accumulation of the squared gradients in the denominator, which can freeze the updates after a number of iterations.

The RMSProp With Nesterov Momentum

In the RMSProp, the sum of the squares of the gradients used in AdaGrad are replaced by a recursively defined decaying average of all past squared gradients. In such a way, the remote past samples are exponentially discarded. At the same time, a Nesterov-type momentum rationale is employed. The main recursions are summarized as follows:

$$\begin{aligned}
\boldsymbol{g} &= \left.\frac{\partial J(\boldsymbol{\theta} + a\Delta\boldsymbol{\theta}^{(i-1)})}{\partial \boldsymbol{\theta}}\right|_{\boldsymbol{\theta}^{(i-1)}}, \\
\boldsymbol{v}^{(i)} &= \beta \boldsymbol{v}^{(i-1)} + (1-\beta)\boldsymbol{g} \circ \boldsymbol{g}, \\
\Delta\theta_d^{(i)} &= a\Delta\theta_d^{(i-1)} - \frac{\mu}{\sqrt{v_d^{(i)}} + \epsilon} g_d, \quad d = 1, 2, \ldots, D, \\
\boldsymbol{\theta}^{(i)} &= \boldsymbol{\theta}^{(i-1)} + \Delta\boldsymbol{\theta}^{(i)},
\end{aligned}$$

where β is a user-defined parameter. The first of the recursions computes the gradient after the Nesterov-type correction. The second one updates the squares of the gradient values in the decaying average rationale. The "$\circ$" operation denotes element-wise multiplication. The third recursion computes the correction term in the AdaGrad rationale and the fourth one provides the updates. The algorithm starts with initial conditions on $\boldsymbol{v}^{(0)} = \boldsymbol{0}$, $\Delta\boldsymbol{\theta}^{(0)}$ and $\boldsymbol{\theta}^{(0)}$. Some typical values for the hyperparameters are $\epsilon = 10^{-8}$, $\beta = 0.9$, and $\mu = 0.001$.

The RMSProp was proposed by Hinton et al. in a lecture.[2] A simpler version, the RMSProp, does not use the Nesterov step and it coincides with the so-called AdaDelta rule, which was independently proposed in [260]. The scheme is similar to the AdaGrad with the difference that the decaying average for the square gradients is employed.

The Adaptive Moment Estimation Algorithm (Adam)

The Adam algorithm proposed in [121] borrows the concept of forgetting past values of the squared gradients, yet it introduces and propagates the values of the gradients as well, in a similar way as the momentum algorithm does. Furthermore, it introduces some important normalization that takes care of bias that may be introduced. The major recursions around which the Adam evolves are

$$\begin{aligned} \boldsymbol{g} &= \left.\frac{\partial J}{\partial \boldsymbol{\theta}}\right|_{\boldsymbol{\theta}^{(i-1)}}, \\ \boldsymbol{m}^{(i)} &= \beta_1 \boldsymbol{m}^{(i-1)} + (1-\beta_1)\boldsymbol{g}, \\ \boldsymbol{v}^{(i)} &= \beta_2 \boldsymbol{v}^{(i-1)} + (1-\beta_2)\boldsymbol{g} \circ \boldsymbol{g}, \end{aligned}$$

where $\circ$ denotes element-wise multiplication and β_1 and β_2 are user-defined parameters. In the sequel, the obtained values of the moments are normalized, i.e.,

$$\hat{\boldsymbol{m}}^{(i)} = \frac{\boldsymbol{m}^{(i)}}{1-\beta_1^i}, \quad \hat{\boldsymbol{v}}^{(i)} = \frac{\boldsymbol{v}^{(i)}}{1-\beta_2^i}.$$

These normalizations account for the tendency of the two gradient moments to be biased towards zero, especially during the early iterations, due to their zero value initialization. As iterations proceed, the normalizing coefficients tend to one. The update for the dth component of the parameter vector is computed as

$$\theta_d^{(i)} = \theta_d^{(i-1)} - \frac{\mu}{\sqrt{\hat{v}_d^{(i)}} + \epsilon} \hat{m}_d^{(i)}, \; d = 1, 2, \ldots, D.$$

The Adam algorithm is given in Algorithm 18.3.

Algorithm 18.3 (The Adam algorithm).

- Initialization
  - Initialize $\boldsymbol{\theta}^{(0)}$.
  - Initialize, $\boldsymbol{v}^{(0)} = \boldsymbol{0}$, $\boldsymbol{m}^{(0)} = \boldsymbol{0}$.
  - Select step size μ; Typical value 0.001.
  - Select β_1 and β_2; Typical values 0.9 and 0.99, respectively.
  - Select ϵ; Typical value 10^{-8}.
  - Set $i = 0$.

[2] www.cs.toronto.edu/~tijmen/csc321/slides/lecture_slides_lec6.pdf.

- **While** a stopping criterion not met **Do**
  - Select a minibatch of K samples, e.g.,

    $(y_{((m-1)K+1)}, \boldsymbol{x}_{((m-1)K+1)}), \dots, (y_{(mK)}, \boldsymbol{x}_{(mK)})$
  - Compute the gradient

    $$\boldsymbol{g} = \left.\frac{\partial J^{(m)}}{\partial \boldsymbol{\theta}}\right|_{\boldsymbol{\theta}^{(i-1)}} \text{ (see Eq. (18.32)).}$$
  - $i = i + 1$
  - $\boldsymbol{m}^{(i)} = \beta_1 \boldsymbol{m}^{(i-1)} + (1 - \beta_1)\boldsymbol{g}$
  - $\boldsymbol{v}^{(i)} = \beta_2 \boldsymbol{v}^{(i-1)} + (1 - \beta_2)\boldsymbol{g} \circ \boldsymbol{g}$; element-wise multiplication.
  - $\hat{\boldsymbol{m}}^{(i)} = \frac{\boldsymbol{m}^{(i)}}{1-\beta_1^i}$
  - $\hat{\boldsymbol{v}}^{(i)} = \frac{\boldsymbol{v}^{(i)}}{1-\beta_2^i}$
  - **For** $d = 1, 2, \dots, D$, **Do**

    $$\theta_d^{(i)} = \theta_d^{(i-1)} - \frac{\mu}{\sqrt{\hat{v}_d^{(i)}} + \epsilon} \hat{m}_d^{(i)}$$
  - **End Do**
- **End While**

Remarks 18.4.

- At the time this edition of the book is compiled, the most popular algorithms for training deep neural networks seem to be the Adam, the simple stochastic gradient with or without momentum, and the RMSProp with or without Nesterov's momentum term. It all depends on the application and also the familiarity of the user. Furthermore, the use of minibatches seems to be the trend.
- Besides the standard forms of the previous schemes, various techniques have been proposed for their more efficient running and utilization. For example, one can employ warm restarts during training (e.g., [141]) or use ensemble combining techniques (Section 7.9 and, e.g., [101]). The idea in the latter is to train a single network, converging to several local minima along its optimization path, and saving the model parameters. These are then combined in an ensemble rationale known as *snapshot ensembling*.

Some Practical Hints

Training a neural network still has a lot of practical engineering flavor compared to mathematical rigorousness. In this section, some practical hints are presented that experience has shown to be useful in improving the performance of the backpropagation algorithm (see, for example, [131] for a more detailed discussion).

Preprocessing the input features/variables: It is advisable to preprocess the input variables so they have (approximately) zero mean over the training set. Also, one should scale them so they all have similar variances, assuming that all variables are equally important. In the case that the nonlinearity used is of a squashing type, it is advisable that their variance should also match the range of values of the activation (squashing) function. Moreover, it is beneficial for the convergence of the algorithm if the input variables are uncorrelated. A way to achieve this is via an appropriate transformation, for example, PCA.

Selecting symmetric activation functions: In the cases where squashing types of activation functions are employed, it is desirable that the outputs of the neurons assume equally likely positive and negative values. After all, the outputs of one layer become inputs to the next. To this end, the hyperbolic activation function in Eq. (18.8) can be used. Recommended values are $a = 1.7159$ and $c = 4/3$. These values guarantee that if the inputs are preprocessed as suggested before, that is, to be normalized to variances equal to one, then the variance at the output of the activation function is also equal to one and the respective mean value equal to zero. However, as we will soon see in Section 18.6.1, currently, such activation functions seem to be less popular.

Target values: The target values should be carefully chosen to be in line with the activation function used. The values should be selected to offset by some small amount the limiting value of the squashing function. Otherwise, the algorithm tends to push the weights to large values and this slows down the convergence; the activation function is driven to saturation, making the derivative of the activation function very small, which in turn renders small gradient values. For the hyperbolic tangent function, using the parameters discussed before, the choice of ± 1 for the target class labels seems to be the right one. Note that in this case, the saturation values are $a = \pm 1.7158$. In Section 18.5, we will see that the choice of the output activation function should be dictated by the adopted optimality criterion.

Initialization: Initialization of the weight parameters for any optimization scheme is of major importance when training neural networks. Starting from a "wrong" set of values can have a number of unwanted effects on training. Initialization can affect how fast or slow an algorithm converges. Also, the initial values play a critical part in whether the algorithm converges to a point of low or high cost (good or bad local minimum). Furthermore, the generalization performance can be affected by the initialization. However, it must be pointed out that the theoretical underpinnings related to such issues are not yet clear and a number of notions are not well understood and currently constitute ongoing research areas (see, also, the discussion in Section 18.11.1).

Weights are randomly initialized to some *small values*. To this end, a number of different scenarios have been developed, and this is clear by the number of options one is given when using standard related software packages.

For example, if squashing-type activation functions are used and large initial values are assigned, then all activation functions will operate in their saturation point. This drives the gradients to very small values, which, in turn, slows down convergence. The effect on the gradients is the same when the weights are initialized to very small values. Initialization must be done so that the operation in each neuron takes place in the (approximate) linear region of the graph of the activation function and not in the saturated one. It can be shown ([131], Problem 18.4) that if the input variables are preprocessed to zero mean and unit variance and the tangent hyperbolic function is used with parameter values as discussed before, then the best choice for initializing the weights is to assign values drawn from a distribution with zero mean and standard deviation equal to

$$\sigma = m_{in}^{-1/2},$$

where m_{in} is the number of inputs (synaptic connections) in the corresponding neuron.

A critical issue during initialization is the so-called *symmetry breaking* between neurons. It can easily be shown that if two hidden neurons use the same activation function (which is usually the case) and are connected to the same input, then if they are initialized with the same values, their gradients will be equal and after the descent iteration their updated values will remain to be equal. No doubt, this is an undesirable effect that leads to redundancies. This is the reason that setting all values initially to zero is not a good practice. Random initialization helps to avoid such scenarios. On the other hand, trying to avoid large values is in line with trying to avoid what is known as exploding gradient phenomenon, which is intrinsic to backpropagation (see Section 18.6). Thus, it is advisable to start with small random values.

To implement the above, possible scenarios are the uniform or the Gaussian distributions. For example, in [58], it is suggested to initialize the weights via the uniform distribution

$$\theta_{ij}^r \sim \mathcal{U}\left(-\sqrt{\frac{6}{m_{in}+m_{out}}},\ \sqrt{\frac{6}{m_{in}+m_{out}}}\right),$$

where m_{in} and m_{out} are the number of inputs and outputs at the respective layer r. The goal behind this heuristic is to initialize all layers to have the same activation variance and the same gradient variance. However, the derivation is based on the assumption of linear units. A number of variants of the above are also possible, e.g., with the draw from $\mathcal{U}(-\frac{1}{\sqrt{m_{in}}}, \frac{1}{\sqrt{m_{in}}})$. Also, instead of a uniform, the zero mean Gaussian of the same variance can be used or sometimes the truncated Gaussian.

In contrast to the weights, biases can be set to zero, and this seems to be the most often used scenario.

However, all the recipes should always be treated with some care and before using them it is a good idea to see what other researchers have used before in similar situations. Furthermore, the type of the adopted nonlinearity should also be considered (see, e.g., [77,113]). In [77], the weights are initialized by taking into account the size of the previous layer only. More specifically, it is suggested to draw samples from a zero mean (truncated) Gaussian with variance $2/m_{in}$. It is reported that the latter is more appropriate when rectified linear units (Section 18.6.1) are employed. For hyperbolic tangent nonlinearities, the uniform distributions mentioned before seems to be a more popular choice.

Batch Normalization

A major difference while training multilayer structures compared to other single-layer models is that the distribution of the inputs for each layer changes during training, as the parameters of the previous layers are iteratively updated. This slows down the training process by requiring the use of lower learning rates/step sizes. Furthermore, it makes the training sensitive to the parameter initialization, especially in the presence of saturating nonlinearities.

One way to address and cope with such phenomena is to employ the so-called *batch normalization* [104]. As has already been discussed, during preprocessing, it is advisable to scale the input variables to unit variance around a zero mean value, in order to avoid large differences in the dynamic range of the values of the various inputs. Batch normalization is inspired by this fact and tries to impose such a scaling on the activation values that are produced by the network in all the layers. There are different variants of the basic idea. For example, some apply the normalization prior to the nonlinearity and some after. We will follow the latter option to demonstrate the method.

Referring to Fig. 18.11, let us consider the output y_j of the jth neuron, where for notational simplicity the superscript indicating the respective layer, r, has been omitted. The activation y_j comprises an input to all the neurons of the next layer, $r+1$. Furthermore, assume that minibatches of size K are used. Prior to updating the parameters to the descent direction, while the, say, mth minibatch is being considered, all the values of the respective activations are tracked and their sample mean and corresponding variance are computed as

$$\mu_j^{(m)} = \frac{1}{K}\sum_{k=1}^{K} y_j^{(k)}, \quad \sigma_j^{(m)} = \sqrt{\frac{1}{K}\sum_{k=1}^{K}\left(y_j^{(k)} - \mu_k^{(m)}\right)^2},$$

where $y_j^{(k)}$ is the response of the jth neuron when it is excited by the kth sample of the respective (e.g., mth) minibatch. Then the corresponding normalized activations are computed as

$$\hat{y}_j^{(k)} = \frac{y_j^{(k)} - \mu_j^{(m)}}{\sigma_j^{(m)}}. \tag{18.36}$$

In words, every activation is normalized to unit variance around a zero mean, prior to passing it to the layer above.

However, batch normalization goes one step ahead. Instead of using the normalized variables in their previous primitive form, it imposes a further linear transformation, i.e.,

$$\bar{y}_j^{(k)} = \gamma_j \hat{y}_j^{(k)} + \beta_j, \tag{18.37}$$

where γ_j and β_j are learned during the training as extra parameters in the backpropagation framework. In such a way, we make up for the equal treatment that normalization implicitly applies to each neuron, and we offer the freedom to the network to adjust the expressive power of each neuron separately.

It turns out that batch normalization allows for higher learning rates and speeds up convergence. Furthermore, it is reported to be less sensitive to initialization and can, also, act as a regularizer [104]. Further theoretical findings related to batch normalization can be found in, e.g., [123].

Once the network has been trained, one wonders how to employ the transformation given in Eq. (18.37), since Eq. (18.36) requires the mean and variance of a specific minibatch. In practice, once the parameters γ and β have been learned, the mean and standard deviation are replaced by averages over the various minibatches, i.e.,

$$\mu_j = \frac{1}{M}\sum_{m=1}^{M} \mu_j^{(m)}, \quad \sigma_j^2 = \frac{1}{M}\sum_{m=1}^{M} (\sigma_j^{(m)})^2,$$

and

$$\bar{y}_j = \gamma_j \frac{y_j - \mu_j}{\sigma_j + \epsilon} + \beta_j,$$

where a small constant ϵ has been used to avoid a possible division by zero and M is the total number of minibatches. Often, instead of σ_j^2, its corrected unbiased version (see, e.g., Problem 7.5) is used, i.e., $\frac{K}{K-1}\sigma_j^2$.

In practice, however, one uses running averages that are collected during training. For example, one can employ a momentum-based running average, and update the statistics after each minibatch has been processed, i.e.,

$$\mu_j(\text{new}) = a\mu_j(\text{old}) + (1-a)\mu_j^{(m)}, \quad m = 1, 2, \ldots, M,$$

where a is a user-defined momentum parameter. A similar recursion applies to the variances.

18.4.3 BEYOND THE GRADIENT DESCENT RATIONALE

The other path to follow to improve upon the convergence rate of the gradient descent-based backpropagation algorithm, at the expense of increased complexity, is to resort to schemes that involve, in one way or another, information related to the second-order derivatives. We have already discussed such families in this book, for example, the Newton family introduced in Chapter 6. For each one of the available families, a backpropagation version can be derived to serve the needs of the neural network training. We will not delve into details, because the concept remains the same as that discussed for the gradient descent. The difference is that now second-order derivatives have to be propagated backwards. The interested reader can look at the respective references and also in [19,33,75,131,263] for more details.

In [11,112,124], schemes based on the conjugate gradient philosophy have been developed, and members of the Newton family have been proposed in, for example, [13,191,244]. In all these schemes, the computation of the elements of the Hessian matrix, that is,

$$\frac{\partial^2 J}{\partial \theta_{jk}^r \partial \theta_{j'k'}^{r'}},$$

is required, where j and k run over all the parameters in the rth layer and j' and k' over all the parameters associated with the r'th layer, for all values of r and r'. To this end, various simplifying assumptions are employed in the different papers (see, also, Problems 18.5 and 18.6).

An algorithm which is loosely based on Newton's scheme has been proposed in [50], known as the *quickprop* algorithm. It is a heuristic method that treats the synaptic weights as if they were quasi-independent. It then approximates the error surface, as a function of each weight, via a quadratic polynomial. If this has its minimum at a sensible value, it is used as the updated value in the iterations; otherwise, a number of heuristics are mobilized. A common formulation for the resulting updating rule is given by

$$\Delta\theta_{ij}^r(\text{new}) = \begin{cases} a_{ij}^r(\text{new})\Delta\theta_{ij}^r(\text{old}), & \text{if } \Delta\theta_{ij}^r(\text{old}) \neq 0, \\ -\mu \dfrac{\partial J}{\partial \theta_{ij}^r}, & \text{if } \Delta\theta_{ij}^r(\text{old}) = 0, \end{cases} \tag{18.38}$$

where

$$a_{ij}^r(\text{new}) = \min\left\{ \frac{\dfrac{\partial J(\text{new})}{\partial \theta_{ij}^r}}{\dfrac{\partial J(\text{old})}{\partial \theta_{ij}^r} - \dfrac{\partial J(\text{new})}{\partial \theta_{ij}^r}}, a_{\max}^r \right\}, \tag{18.39}$$

with typical values of the parameters used being $0.01 \le \mu \le 0.6$, and $a_{\max} \approx 1.75$. An algorithm in similar spirit with the quickprop has been proposed in [192].

In practice, when large networks and data sets are involved, simpler methods, such as carefully tuned gradient descent schemes and their versions, as for example those previously discussed in Section 18.4.2, seem to work better and are currently the trend. The more complex second-order techniques can offer improvements in smaller networks, especially in the context of regression tasks.

18.5 SELECTING A COST FUNCTION

As we have already commented, feed-forward neural networks belong to the more general class of parametric models; thus, in principle, any loss function we have met so far in this book can be employed. Over the years, certain loss functions have gained in popularity in the context of regression and classification tasks. However, in the context of feed-forward multilayer networks, the choice of the loss function is tightly coupled to the type of the *output* nonlinearity that is used. A "wrong" *combination* can severely affect the speed at which the network learns during training. To understand this claim, let us consider the following combination.

Adopt as loss function the squared error one and as output nonlinearity the logistic sigmoid function, $\sigma(z)$, given in Eq. (18.7). For the sake of simplicity, let us assume that we only have a single output node and also suppress the indices n, which relates to observations, and r, which relates to layers. Then, if y is the target value and $\hat{y}$ the estimated one, the contribution to the cost function will be

$$J = \frac{1}{2}(y - \hat{y})^2 = \frac{1}{2}(y - \sigma(z))^2, \; z = \boldsymbol{\theta}^T \boldsymbol{y}, \tag{18.40}$$

where $\boldsymbol{y}$ is the (extended) vector of the outputs of the last hidden layer and $\boldsymbol{\theta}$ is the vector of synaptic weights connecting the nodes of the last hidden layer to the output one. The bias term has been absorbed in $\boldsymbol{\theta}$, according to standard practice. Then the gradient of J with respect to $\boldsymbol{\theta}$, using the chain rule for derivatives, is trivially seen to be

$$\frac{\partial J}{\partial \boldsymbol{\theta}} = \frac{\partial J}{\partial z}\frac{\partial z}{\partial \boldsymbol{\theta}} = \delta \boldsymbol{y}, \; \delta := \frac{\partial J}{\partial z},$$

where, after taking into account Eq. (18.40), we get

$$\delta = \frac{\partial J}{\partial \hat{y}}\frac{\partial \hat{y}}{\partial z} = (\hat{y} - y)\,\sigma'(z).$$

Thus,

$$\frac{\partial J}{\partial \boldsymbol{\theta}} = (\hat{y} - y)\sigma'(z)\boldsymbol{y}.$$

In words, the above means that the gradient of the cost, with respect to the parameters leading to the output node, depends on the error, $(\hat{y} - y)$, and also on the derivative of the logistic sigmoid function. The dependence on the error is "healthy." Indeed, this is what we want. The larger the error, the larger the gradient (in absolute value); this will make the correction term in Eq. (18.10) large to account for a large error or keep the update close to the current estimate when the error is small. However, the dependence on the derivative of the nonlinear function is not good news. Looking at the graph in Fig. 18.8, for values of the argument that are not close to zero, the function quickly saturates and its derivative becomes very small. Hence, although the error can be large, the respective correction will be small. Also, recall that in the backpropagation, the gradient of the cost with respect to the parameters of the last (output) layer, say, L, is passed to the previous one, layer $L-1$, and is used to compute the respective gradients, and so on. If the values of the gradients of the last layer are small, this affects all the gradients in all layers and as a result convergence can significantly slow down.

In contrast, one can readily check that if the output node is a linear one, instead of being sigmoid, then the combination of the squared error loss function with a linear output unit is a good combination. The resulting gradient becomes proportional to the error, since now the derivative becomes a constant.

We will now focus on an alternative loss function that bypasses the previous drawback. Let us adopt, for a classification task, as targets the 0, 1 values. Then the true and predicted values, y_{nm}, $\hat{y}_{nm}$, $n = 1, 2, \ldots, N$, $m = 1, 2, \ldots, k_L$, for k_L output nodes, can be interpreted as probabilities and a commonly used cost function in this setting is the *cross-entropy*, which is defined as

$$\boxed{J = -\sum_{n=1}^{N}\sum_{k=1}^{k_L} y_{nk} \ln \hat{y}_{nk}: \quad \text{cross-entropy cost.}} \tag{18.41}$$

Note that this is exactly the same cost used in Eq. (7.49) for the multiclass logistic regression treated in Chapter 7. For the classification task, when the number of classes are equal to the output nodes, i.e., we have k_L classes, then the cross-entropy is the negative log-likelihood of the observations. Indeed, for the 0, 1-class coding scheme for the observed label variables, y_{nk}, the corresponding target output vector, $\boldsymbol{y}_n \in \mathbb{R}^{k_L}$, will have a 1 at the position corresponding to the true class and the rest of the elements will be zero (one-hot vector). If we interpret $\hat{y}_{nk}$ as the respective posterior probability, i.e., $P(\omega_k|\boldsymbol{x}_n;\boldsymbol{\theta})$, where $\boldsymbol{\theta}$ denotes all the parameters that define the corresponding network, then the likelihood of the observations is given by

$$P(\boldsymbol{y}_1, \ldots, \boldsymbol{y}_N|\mathcal{X};\boldsymbol{\theta}) = \prod_{n=1}^{N}\prod_{k=1}^{k_L} (\hat{y}_{nk})^{y_{nk}}. \tag{18.42}$$

Taking the logarithm and changing the sign, Eq. (18.41) is obtained.

A different formulation of the cross-entropy loss function, known as *relative entropy*, is sometimes used, i.e.,

$$J = -\sum_{n=1}^{N}\sum_{k=1}^{k_L} y_{nk} \ln \frac{\hat{y}_{nk}}{y_{nk}}. \tag{18.43}$$

This formulation brings up the affinity with the Kullback–Leibler (KL) divergence (Section 2.5.2) that measures deviation between probabilities (distributions). In our context, it measures how different y_{nk} is from $\hat{y}_{nk}$.

Besides the previously defined cross-entropy cost function, a variant of it is also used and it is given by

$$J = -\sum_{n=1}^{N}\sum_{k=1}^{k_L}\left(y_{nk}\ln \hat{y}_{nk} + (1-y_{nk})\ln(1-\hat{y}_{nk})\right), \tag{18.44}$$

whose minimum occurs when $y_{nk} = \hat{y}_{nk}$, which for binary target values is equal to zero. Comparing the above cost in Eq. (18.44) with the cross-entropy in Eq. (18.41), it is easily seen that the former, while trying to "push" $\hat{y}_{nk}$ towards 1 for the correct class ($y_{nk} = 1$), at the same time it tries to "push" the corresponding values for the wrong classes towards 0, in order to minimize J. Note that the cost in Eq. (18.44) can be seen as a generalization of the cross-entropy for the two-class case, using a single output neuron (see also Eq. (7.38)). Adopting the 0, 1-class coding scheme and assuming, this time, that the classes are not mutually exclusive and are independent, the counterpart of Eq. (18.42) becomes

$$P(\boldsymbol{y}_n|\mathcal{X};\boldsymbol{\theta}) = \prod_{n=1}^{N}\prod_{k=1}^{k_L}(\hat{y}_{nk})^{y_{nk}}(1-\hat{y}_{nk})^{1-y_{nk}},$$

which leads to Eq. (18.44). Note, however, that this interpretation does not hold true if classes are mutually exclusive, which is the case in most classification tasks.

One can easily see that combining the above versions of the cross-entropy function with the logistic sigmoid nonlinearity "frees" the gradient of the cost, with respect to the synaptic weights of the kth output neuron, of the dependence on the derivative of the respective activation function. It can easily be shown that (Problem 18.7)

$$\frac{\partial J}{\partial \boldsymbol{\theta}_j^L} = \sum_{n=1}^{N}\delta_{nj}^L \boldsymbol{y}_n^{L-1}, \quad j = 1, 2, \ldots, k_L,$$

where $\delta_{nj}^L = y_{nk}(\hat{y}_{nk} - 1)$ for the cross-entropy loss in Eq. (18.41) and $\delta_{nj}^L = \hat{y}_{nk} - y_{nk}$ for the case of Eq. (18.44). Note that $\boldsymbol{y}_n^{L-1}$ is the vector of the outputs of the last hidden layer, $L-1$.

It can be shown (Problem 18.9) that the cost functions in (18.41) and (18.44) depend on the relative errors and not on the absolute errors, as is the case in the squared error loss; thus, small and large error values are equally weighted during the optimization. Furthermore, it can be shown that the cross-entropy belongs to the so-called well-formed loss functions, in the sense that if there is a solution that classifies correctly all the training data, the gradient descent scheme will find it [2]. In [215], it is pointed out that the cross-entropy loss function may lead to improved generalization and faster training for classification, compared to the squared error loss.

Having said all that, recall that we have interpreted the outputs of the network as probabilities; yet, there is no guarantee that these add to one. This can be enforced by selecting the activation function in the last layer of nodes to be

$$\hat{y}_{nk} = \frac{\exp(z_{nk}^L)}{\sum_{m=1}^{k_L}\exp(z_{nm}^L)}: \quad \text{softmax activation function,} \tag{18.45}$$

which is known as the *softmax activation* function [21]. Recall that $z_{nk}^L := \boldsymbol{\theta}_k^{L^T} \mathbf{y}_n^{L-1}$ is the value of the linear combiner (prior to the nonlinearity) associated with the kth output neuron (layer $r = L$, Fig. 18.11) that corresponds to the nth input training sample. For those familiar with logistic regression, discussed in Chapter 7, compare (18.45) with Eq. (7.47); after all, the world is small!

As was the case with the sigmoid activation function, it is easy to show that combining the softmax activation with the cross-entropy loss function leads to gradients that are *independent* of derivatives of the nonlinearity (Problem 18.10).

18.6 VANISHING AND EXPLODING GRADIENTS

We have already discussed the importance of trying to avoid to involve nonlinearities in the output that promote small values of gradients, when they are combined with some loss functions. Let us now look at what is happening in the hidden layers. The secret lies in Eq. (18.27) of the backpropagation algorithm. Let us investigate it a bit more. As we have seen in the previous subsection, at the heart of the computation of the gradients lie the quantities

$$\delta_j^r := \frac{\partial J}{\partial z_j^r}, \quad j = 1, 2, \ldots, k_r, \quad r = 1, 2, \ldots, L,$$

where the index n has been suppressed to unclutter notation, r is the index indicating the corresponding layer in an L-layer network, and j is the index of a neuron at the rth layer that consists of k_r neurons. By definition, $z_j^r := \boldsymbol{\theta}_j^{r^T} \mathbf{y}^{r-1}$ is the linear output, prior to the nonlinearity f, of the jth neuron (associated with the parameter vector, $\boldsymbol{\theta}_j^r$) and $\mathbf{y}^{r-1}$ is the vector of the output values of the $(r-1)$th layer (Fig. 18.11). Then the propagation of the various δ-derivatives follows the recursive rule (Eq. (18.27)),

$$\delta_j^{r-1} = \left(\sum_{k=1}^{k_r} \delta_k^r \theta_{kj}^r \right) f'(z_j^{r-1}), \quad j = 1, 2, \ldots, k_{r-1}.$$

In words, the δ-derivatives for layer $r-1$ depend on the respective δ-derivatives of the layer above, r, and (a) on the respective weights and (b) on the *derivative* of the nonlinearity, both in a *multiplicative way*. For the sake of clarity, let us write the above formula for two successive steps,

$$\delta_j^{r-1} = \left(\sum_{k=1}^{k_r} \left(\sum_{i=1}^{k_{r+1}} \delta_i^{r+1} \theta_{ik}^{r+1} \right) f'(z_k^r) \theta_{kj}^r \right) f'(z_j^{r-1}), \quad j = 1, 2, \ldots, k_{r-1}.$$

All one has to keep in mind from the above is not its exact formulation, but the notion that as the backpropagation algorithm flows backwards, the *derivatives of the nonlinearities* as well as *the weights* are *multiplied* and the number of the involved products *grows*. The lower (closer to the input layer) in hierarchy δ_j^r is, the more products are involved for its computation. There are two extremes that may, and often do, occur in practice.

Taking into account that the derivatives of the activation function can be smaller than one (and in the case of the sigmoid-type nonlinearities can take very small values) if the estimated values of the parameters are not very large, the gradients of the cost, with respect to the parameters of the lower layers, can take vanishingly small values and this can make the training extremely slow (see, e.g., [95,225] for further discussion and insight). This phenomenon becomes more prominent when deep networks with many layers are involved.

On the other extreme, if the values of the parameters happen to get very large values, this can lead to an explosion of the gradient estimates. As a result, this can affect the learning by pushing the estimates of the parameters to the wrong region in the parameter space.

Another related problem is that the gradients along the various layers may take values of different scale. This means that some layers can learn faster than others. This is basically a form of instability.

All the above are some of the difficulties one encounters while training multilayer neural networks, especially when many layers are involved.

To cope with such problems, a number of modifications and tricks have been developed. Playing with different cost functions is one, playing with different optimization variants of the backpropagation algorithm is another (see, e.g., [162] for a discussion). In the subsection to follow, we will see how to cope with nonlinearities to be used in the hidden layers, which are not of the saturating type; hence, they can "guard" us from the small derivative values tendency.

18.6.1 THE RECTIFIED LINEAR UNIT

Besides the activation functions that we have considered so far, the *rectified linear unit* (ReLU), defined as

$$\boxed{f(z) = \max\{0, z\}: \quad \text{ReLU},} \tag{18.46}$$

has been proposed more recently (see Fig. 18.13).

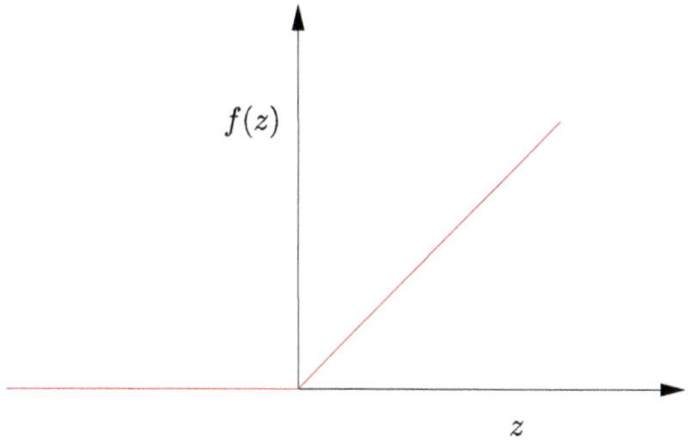

FIGURE 18.13

The graph of ReLU. Note that for $z > 0$ the derivative is equal to one and for $z < 0$ it is equal to zero.

It has been reported that in the context of deep networks, the use of the ReLU nonlinearity in the hidden layers can significantly speed up the training time [126]. Such an activation function does not suffer from saturation and its derivative is equal to one when the neuron operates in its active region ($z > 0$). Thus, it is desirable to set the biases of the neurons, during initialization, to some small positive value, e.g., $\theta_0 = 0.1$, in order to increase the probability that the input to the activation function is positive. For negative values, the derivative is zero. At $z = 0$ the derivative is not defined and if this happens during training, one can either choose 1 or 0; for those familiar with the notion of the subgradient (Chapter 8) such a choice is fully justified.

Thus, selecting the ReLU as the activation function, one bypasses problems related to the slowing down when derivatives get small values. Recall that in the backpropagation algorithm, when gradients with respect to the parameters of hidden layers are computed, the derivative of the activation function enters in a multiplicative fashion.

A drawback of the ReLU is that training freezes when $z < 0$. To overcome this problem, variants of the ReLU have been proposed. For example, one can employ

$$f(z) = \max\{0, z\} + \alpha \min\{0, z\}.$$

Depending on the value of α, different variants result. For $\alpha = -1$, the so-called *absolute value rectification* is obtained [108]. If α is assigned a fixed small value, e.g., 0.01, the resulting nonlinearity is known as the *leaky* ReLU [150]. In [77], α is left as a parameter to be learned during training. A further modification, known as the *max output* unit, a number of, say, k different ReLUs are employed, whose parameters are learned during the training, and, each time, the one that results in the maximum value is selected to activate the corresponding neuron [60]. The ReLU was first introduced in the context of dynamical networks in [71] and it was motivated by biological arguments.

The question that is now raised concerns the choice of the specific nonlinearity to be used in practice. There is no definite answer to that and the choice depends on the specific application. At the time this edition is compiled, it seems that the use of ReLU, in any one of its versions, is the more popular choice for the hidden layers. For classification tasks and for the output layer, the softmax nonlinearity, combined with the cross-entropy cost function, is most commonly used.

18.7 REGULARIZING THE NETWORK

A crucial factor in training neural networks is to decide the size of the network. The size is directly related to the number of parameters to be estimated. Concerning feed-forward neural networks, two issues are involved. The first concerns the number of layers and the other the number of neurons per layer. As we will discuss in Section 18.11, a number of reasons support the use of more than two hidden layers. Such networks comprise a large number of parameters and can be vulnerable to overfitting, which affects the generalization performance of the learner. One path to cope with overfitting is via regularization. Over the years, various regularization approaches have been proposed. A brief presentation and some guidelines are given below. We will return to overfitting and generalization issues in the context of deep networks and from a different perspective in Section 18.11.1.

Weight decay: This path refers to a typical cost function regularization via the square Euclidean norm of the weights. Instead of minimizing a cost function, $J(\boldsymbol{\theta})$, its regularized version is used, such that (e.g., [84])

$$J'(\boldsymbol{\theta}) = J(\boldsymbol{\theta}) + \lambda ||\boldsymbol{\theta}||^2. \tag{18.47}$$

We have already discussed in Chapter 3 in the context of ridge regression that involving the bias terms in the regularizing norm is not a good practice, because it affects the translation-invariant property of the estimator. A more sensible way to regularize is to remove the bias terms from the norm. Although this simple type of regularization helps in improving the generalization performance of the network, and it can be sufficient for some cases, in general it is not the most appropriate way to go.

Besides the square Euclidean norm, other norms can be used in Eq. (18.47). For example, an alternative would be to employ the ℓ_1 norm. From Chapter 9, we know that the latter norm promotes sparsity. This is a welcome property when training a large network so that one can push less informative parameters to zero.

Over the years, various scenarios in applying the regularized cost function have been proposed. One scenario is to group all the parameters (excluding biases) for each layer together and apply Eq. (18.47), involving different regularizing constants for each group. Another scenario is to deal with the parameters of each neuron separately. Another path is to look at regularization as being equivalent to constraining the norm to be less than a preselected value (Chapter 3). This turns out to be equivalent to a projection step onto the respective ball associated with the norm, e.g., the ℓ_2 ball (see Chapter 8), which leads to a normalization operation (see [91]).

Weight elimination: Instead of employing the norm of the weights, another approach involves more general functions for the regularization term, that is,

$$J'(\boldsymbol{\theta}) = J(\boldsymbol{\theta}) + \lambda h(\boldsymbol{\theta}). \tag{18.48}$$

For example, in [246] the following is used:

$$h(\boldsymbol{\theta}) = \sum_{k=1}^{K} \frac{\theta_k^2}{\theta_h^2 + \theta_k^2}, \tag{18.49}$$

where K is the total number of the parameters involved and θ_h is a preselected threshold value. A careful look at this function reveals that if $\theta_k < \theta_h$, the penalty term goes to zero very fast. In contrast, for values $\theta_k > \theta_h$, the penalty term tends to unity. In this way, less significant weights are pushed toward zero. A number of variants of this method have also appeared (see, for example, [201]).

Methods based on sensitivity analysis: In [130], the so-called *optimal brain damage* technique is proposed. A perturbation analysis of the cost function in terms of the weights is performed, via the second-order Taylor expansion, that is,

$$\delta J = \sum_{i=1}^{K} g_i \delta\theta_i + \frac{1}{2}\sum_{i=1}^{K} h_{ii}\delta\theta_i^2 + \frac{1}{2}\sum_{i=1}^{K}\sum_{j=1, j\neq i}^{K} h_{ij}\delta\theta_i\delta\theta_j, \tag{18.50}$$

where

$$g_i := \frac{\partial J}{\partial \theta_i}, \quad h_{ij} := \frac{\partial^2 J}{\partial \theta_i \partial \theta_j}.$$

Then, assuming the Hessian matrix to be diagonal and if the algorithm operates near the optimum (zero gradient), we can approximately set

$$\delta J \approx \frac{1}{2} \sum_{i=1}^{K} h_{ii} \delta \theta_i^2. \tag{18.51}$$

The method works as follows:

- The network is trained using the backpropagation algorithm. After a few iteration steps, the training is frozen.
- The so-called *saliencies*, defined as

$$s_i = \frac{h_{ii}\theta_i^2}{2},$$

  are computed for each weight, and weights with a small saliency are removed. Basically, the saliency measures the effect on the cost function if one removes (sets equal to zero) the respective weight.
- Training is continued and the process is repeated every few iterations, until a stopping criterion is satisfied.

In [74], the full Hessian matrix is computed, giving rise to the *optimal brain surgeon* method. Note that regularization techniques that remove connections are also known as *pruning* techniques.

Early stopping: An alternative, primitive in concept yet particularly useful in practice, technique to avoid overfitting is the so-called *early stopping*. The idea is to stop the training when the test error starts increasing. Training the network over many epochs can lead the training error to converge to small values. However, this is an indication of overfitting rather than indicative of a good solution. According to the early stopping method, training is performed for some iterations and then it is frozen. The network, using the currently available estimates of the weights/biases, is evaluated against a validation/test data set and the value of the cost function is computed. Then training is resumed, and after some iterations the previous process is repeated. When the value of the cost function, computed on the test set, starts increasing, then training is stopped.

Early stopping is also used in combination with other regularization strategies. Even when using regularization strategies that modify the objective function, the decision when to stop training cannot be based solely on focusing on the training error.

Regularization via noise injection: The regularization effect of the presence of noise during training has been well known since the early 1990s. There are various ways of adding extra noise to the input samples (e.g., [19]). This is equivalent to modifying the cost function by adding an extra term which acts as a regularizer.

The other alternative is to add noise to the parameters as they are being estimated during training (see, e.g., [166]). Looking at the noise as a small perturbation on the parameters and employing a first-order Taylor expansion of the output, one can show that this is equivalent to adding an extra regularizing term to the cost function. Large gradient values are penalized and this pushes the algorithm to converge to solutions in the parameter space, where they are relatively insensitive to small variations.

Besides the two previous possibilities, there are strategies that attempt to cope with noisy output labels. This is known as *label smoothing* (see, e.g., [228]).

Regularizing via the artificial expansion of the data set: As has already been discussed in various parts of the book (e.g., Chapter 3), the cause of overfitting is the large size (capacity) of the number of unknown parameters of the model, with respect to the size of the training set. Keeping the model fixed, while increasing the number of the training data, acts beneficially on the overfitting problem. However, labeled data may not be always practical to have. In such cases and in certain applications, a way out is to generate artificially "fake" data and use them as part of the training set. For example, in tasks such as object recognition and OCR, one can generate many replicas of the objects or digits and letters, by applying linear transformations, such as rotations, translations and scaling, on existing images. In [110], data augmentation has also been applied to speech recognition. Later on, in Section 18.15, we are going to discuss more recent techniques on how to artificially generate data.

Example 18.2. The goal of this example is to show the effect that regularization has on overfitting. Fig. 18.14 shows the resulting decision lines that separate the samples of the two classes, denoted by

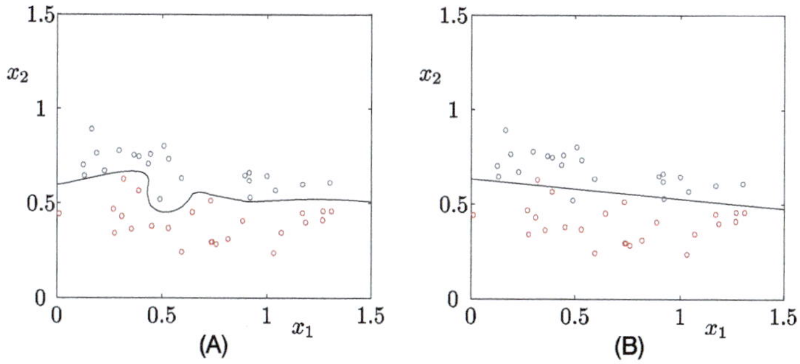

FIGURE 18.14

Decision curve (A) before pruning and (B) after pruning.

black and red o, respectively. Fig. 18.14A corresponds to a multilayer perceptron with two hidden layers and 20 neurons in each of them, amounting to a total of 480 weights. Training was performed via the batch gradient descent backpropagation algorithm. The overfitting nature of the resulting curve is readily observed. Fig. 18.14B corresponds to the same multilayer perceptron; however, this time a pruning algorithm was employed. Specifically, the method based on parameter sensitivity was used, testing the saliency values of the weights every 100 epochs and removing weights with saliency value below a chosen threshold. Finally, only 25 of the 480 weights survived and the curve is simplified to a straight line.

DROPOUT

This technique for regularizing deep networks follows a different concept from what we have already discussed. Its origin borrows ideas from the concept of combining learners, as has already been discussed in Sections 7.8 and 7.9 and in particular to the bagging rationale. However, dropout cleverly modifies the basic idea of bagging to make it more efficient and suitable to large networks.

The term "dropout" refers to dropping out units/nodes (in the hidden and input layers) in a neural network. At each iteration of the training algorithm, a number of nodes are removed (along with their incoming and outgoing associated connections). The parameters of the remaining nodes are updated according to the updating rule. In other words, at each iteration step only a *subset* of the parameters are updated while the rest (the ones associated with the removed nodes) are frozen to their currently available estimates from the previous iteration. The subset consisting of the remaining nodes defines a *subnetwork* of the original larger one. The procedure is illustrated in Fig. 18.15. In Fig. 18.15A, the

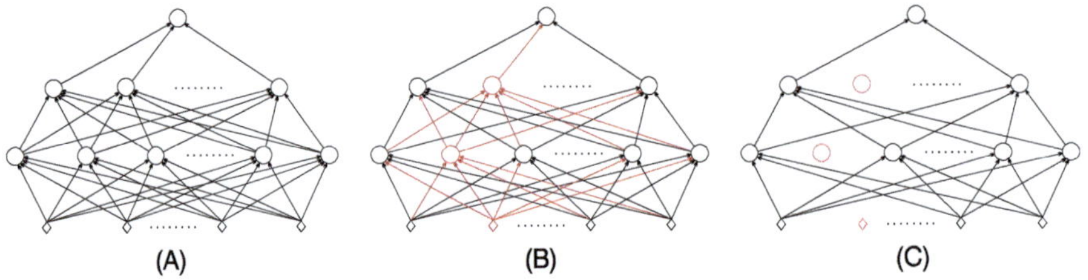

FIGURE 18.15

(A) The full network. (B) The nodes to be removed are shown in red, together with all incoming and outgoing connections. (C) The red ones have been removed. The estimates of their associated parameters are frozen to the values obtained in the previous iteration and are not updated in the current iteration step.

full network is shown. In Fig. 18.15B, the nodes to be removed are shown in red, together with all their incoming and outgoing connections. In Fig. 18.15C, the subnetwork to be updated in the current iteration involves only the gray nodes. The red ones have been removed and the estimates of their associated parameters are frozen to the values that have already been obtained in the previous iteration and which are *not updated* in the current step. In the figure, one of the removed nodes is an input one and the rest lie in hidden layers.

The removal of nodes is carried out probabilistically. That is, each node is retained with probability P. Usually, the value of P is equal to 0.5 for the hidden layers and is set equal to 0.8 for the input layer nodes. For a network with, say, K nodes, the total number of possible subnetworks is 2^K. This is indeed a large number for large values of K, which is the case in practice. However, keep in mind that there is a high degree of parameter *sharing* among the various subnetworks and the total number of parameters to be learned is equal to the number of parameters that constitute the original network.

Once the training phase has been completed and the backpropagation algorithm has converged, the obtained estimates are multiplied by the respective probability, P. This corresponds to an averaging operation over all possible subnetworks that have been trained [91]. It can be shown that (e.g., [64,91]) for a network with one hidden layer and K nodes with a softmax output unit for computing the probabilities of the class labels, using the mean network is equivalent to taking the geometric mean of the probability distributions over labels predicted by all 2^K possible networks. In [245], it is shown that applying the dropout rationale to a linear regression task is equivalent to an ℓ_2 regularization with different regularizing weights per model parameter. However, this equivalence does not carry to general deep networks.

A heuristic explanation on why dropout works is provided in [91]. Dropout reduces coadaptation of neurons, because in every iteration the parameters of different sets of neurons are updated. Thus the network is "forced" to learn more robust features. In other words, the network learns while, each time, parts of it are missing. A more theoretically pleasing explanation is given in [255], where it is shown that the dropout technique is equivalent to an approximate variational inference with specific priors in a deep Gaussian process.

At the time the current edition is compiled, the dropout technique seems to be the most popular method for regularization and has been applied to a wide range of applications (see, e.g., [210]). Also, using the dropout does not prevent to combine it with other types of regularization [91]. Needless to say that the method is not a "panacea" and has its drawbacks. A major shortcoming is that typically one has to train much larger networks to account for the loss of capacity imposed by the regularization. Moreover, one needs more iterations for the training. A dropout network typically takes 2–3 times longer to train than a standard neural network of the same architecture. To this end, besides the previously reported scheme, computational efficient alternatives have also been proposed (see, e.g., [241,243]).

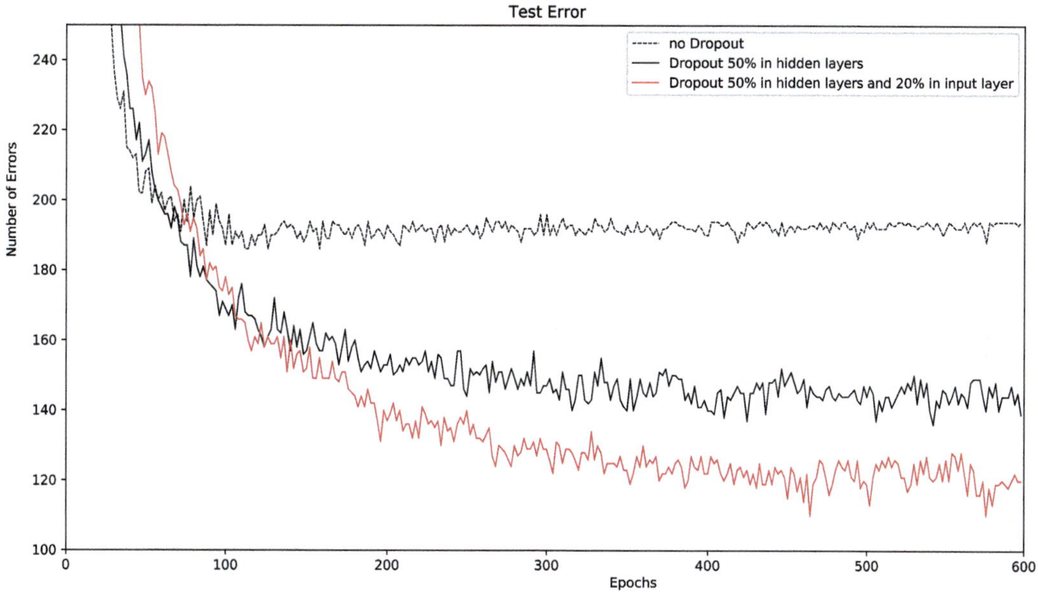

FIGURE 18.16

The use of dropout during training has a significant effect on the generalization performance of the network. This is verified by the test error measured by the number of errors committed in the test set. The probabilities for retaining nodes in the hidden layers was $P = 0.5$ and for the input nodes $P = 0.8$.

Example 18.3. Fig. 18.16 illustrates the effect of the dropout regularization on the test error, in the context of an optical character recognition (OCR) classification task. A fully connected feed-forward network is used with 784 nodes in the input layer, 2000 nodes in each one of the two hidden layers, and 10 output nodes, one per class. The inputs to the network come from the MNIST database for digit recognition. The input images are of 28×28 (784) and the pixel values were normalized in the range [0,1]. For the training, 55000 images were used and 10000 were kept for the testing phase. The nonlinearity used in the hidden layers was the ReLU and for the output nodes the softmax one. The loss function was the cross-entropy one. The standard gradient descent algorithm was used with step size

$\mu = 0.01$. The network was trained for 600 epochs and the minibatch size was equal to 100. To obtain the curves, the error over the test set is computed using the estimates obtained during the training phase, every time an epoch iteration has been completed. Observe that the use of dropout has a significant effect on the obtained test error. Also, a combination of dropping out hidden as well as input nodes has a beneficial effect. One should notice that although the use of dropout is beneficial, even without regularization, the network is doing quite well. We will come back to that in Section 18.11.1.

18.8 DESIGNING DEEP NEURAL NETWORKS: A SUMMARY

So far in our discussion, we have touched upon a number of challenges that one has to address while building up a deep neural network for a specific application. No doubt, a practitioner has to make a number of critical decisions prior to running the code, using, usually, one of the off-the-self software packages.

Adopting the neural network architecture and selecting the related hyperparameters is still a task with a strong engineering flavor. It may not be extreme to compare this task to the task of building up a complex circuit in the times before the development of related sophisticated software packages. The engineer had to employ a number of tricks, which were learned rather from experience than from theory. The goal of this section is to summarize and put together the major challenges and some general tips that have to be followed.

- Think of the problem at hand carefully, try to understand its specificities and the nature and the statistical properties of the data and the goals, prior to building up the network. Avoid to start playing with algorithms before making sure that the problem at hand has been understood.
- Look at the data and make sure that they are well collected and highly representative of the problem. Make sure that there are no "biases" that favor certain decisions or classes. Biases of special concern, which are also the most difficult ones to identify, are those that take place at a subconscious level and are driven by social stereotypes, e.g., issues related to gender, race, religion, and social class.
- Make sure to use the appropriate type of nonlinearities for the hidden as well as the output units. The right choice is also problem dependent, although we have already made some suggestions concerning the "typical" case. Yet, "typical" does not necessarily mean "must."
- Make sure that the right loss function is selected to optimize the data. The related issues have already been discussed before. Yet, one may be more imaginative and need to use other loss functions, which are more appropriate for the problem at hand. One should not be, necessarily, biased to the few examples that are given in the book. In the so many decades of machine learning history, a number of alternatives have been suggested and used in the various scientific communities, where application-specific loss functions may be more appropriate.
- Selecting the size of the network is the first challenge, that is, to decide the number of layers and the number of nodes per layer. If the network is too big, even the use of regularization may not be enough to cope with the overfitting issue. If it is too small, the performance may be poor. In some cases, one may have to reconsider and obtain more data, if possible. Before one starts developing the architecture, searching what others have done in similar situations and settings is advisable. Searching for the "best" size, it is a common practice to split the available data set into three parts, namely, the *training* set, the *validation*, set and the *test* set.

The validation set is used during training, for evaluating the comparative performance of different models, which are trained on the training set. This is a "hybrid" set that is used as an independent set to assist the designer, during the training phase, to select the model. Evaluating the final performance is entirely done via the test set, which has not participated in any way during training. The validation set should not be confused with the cross-validation method discussed in Chapter 3. Such techniques cannot be employed when large data sets are involved, due to computational timing constraints. Moreover, when large data sets are available, one can afford the split of the data set into parts to serve different purposes.

- For each algorithm that will be used, one has to carefully select the hyperparameters of the employed stochastic gradient optimizer. For example, one has to follow the evolution of the algorithm's convergence and may have to reconsider the choice that has been made. The choice of the minibatch size should also be carefully made, to serve the needs of the optimizing algorithm and at the same time to exploit the computational aspects of the specific architecture that is used to run the algorithm.
- Initialize carefully the parameters and normalize the values obtained by the network appropriately, e.g., by batch normalization. Guidelines have already been given before. Yet, guidelines are to help and one may have to be more careful and imaginative.
- Employ regularization. To this end, one has to experiment with the probabilities of removing nodes for the dropout or the involved regularizing parameters, if another method is involved.
- Modern neural network architectures comprise millions of parameters; yet, their entailed computations are highly parallelizable. GPUs are relatively cheap and ubiquitous hardware devices that can afford massive parallelization of the computation, as they comprise thousands of cores. Thus, acquiring multiple GPUs is a most needed investment. As the memory capacity of a GPU increases (or multiple GPUs are used), one can increase the minibatch size, which leads to faster training (due to parallelization) and often better convergence (due to large batches).
- Exploit the experience gained by others in similar setups, when implementing and running an algorithm. Furthermore, keep in mind that the field is still being developed and new results and techniques may appear any time. Thus, follow closely the advances as they happen.
- Never run the algorithm in a *black-box* rationale. Try to understand what each one of the hyperparameters means and how it can affect the performance.

A final tip to the practitioner that I keep repeating to my students is: *understand before you run.*

18.9 UNIVERSAL APPROXIMATION PROPERTY OF FEED-FORWARD NEURAL NETWORKS

In Section 18.3, the classification power of a three-layer feed-forward neural network, built around the McCulloch–Pitts neuron, was discussed. Then, we moved on to employ smooth versions of the activation function, for the sake of differentiability. The issue now is whether we can say something more concerning the prediction power of such networks. It turns out that some strong theoretical results have been produced, which provide support for the use of neural networks in practice (see, for example, [36,54,96,105]).

Let us consider a *two-layer* network, with one hidden layer, involving sigmoidal nonlinearities and with a single output *linear* node. The output of the network is then written as

$$\hat{g}(\boldsymbol{x}) = \sum_{k=1}^{K} \theta_k^o f(\boldsymbol{\theta}_k^{h^T} \boldsymbol{x}) + \theta_0^o, \tag{18.52}$$

where the vector $\boldsymbol{\theta}_k^h$ consists of the synaptic weights and the bias term that define the kth hidden neuron and the superscript "o" refers to the output neuron. Then the following theorem holds true.

Theorem 18.1. *Let $g(\boldsymbol{x})$ be a continuous function defined in a compact*[3] *subset $S \subset \mathbb{R}^l$. Then, for any $\epsilon > 0$, there is a two-layer network with $K(\epsilon)$ hidden nodes of the form in Eq. (18.52), so that*

$$|g(\boldsymbol{x}) - \hat{g}(\boldsymbol{x})| < \epsilon, \quad \forall \boldsymbol{x} \in S. \tag{18.53}$$

In [12], it is shown that the approximation error decreases according to an $\mathcal{O}(1/K)$ rule. In other words, the input dimensionality does not enter into the scene and the error depends on the number of neurons used. The theorem states that a two-layer neural network is sufficient to approximate any continuous function; that is, it can be used to realize any nonlinear discriminant surface in a classification task or any nonlinear function for prediction in a general regression problem. This is a strong theorem indeed. Related universal approximation theorems have been proved for a more general class of activation functions, including the ReLU (see, e.g., [138,218]).

However, what the theorem does not say is how big such a network can be in terms of the required number of neurons in the single layer. It may be that a very large number of neurons are needed to obtain a good enough approximation. This is where the use of more layers can be advantageous. Using more layers, the overall number of neurons needed to achieve certain approximation may be much smaller. We will come to this issue soon, when discussing deep architectures.

Remarks 18.5.

- *Extreme learning machines* (ELMs): These are single-layered feed-forward networks (SLFNs) with output of the form [97]

$$g_K(\boldsymbol{x}) = \sum_{i=1}^{K} \theta_i^o f(\boldsymbol{\theta}_i^{h^T} \boldsymbol{x} + b_i), \tag{18.54}$$

  where f is the respective activation function and K is the number of hidden nodes. The main difference with standard single-layer feed-forward networks is that the parameters associated with each node (i.e., $\boldsymbol{\theta}_i^h$ and b_i) are generated *randomly*, whereas the weights of the output function (i.e., θ_i^o) are selected so that the squared error over the training points is minimized. This implies solving

$$\min_{\boldsymbol{\theta}^o} \sum_{n=1}^{N} \left(y_n - g_K(\boldsymbol{x}_n)\right)^2. \tag{18.55}$$

[3] Closed and bounded.

Hence, according to the ELM rationale, we do not need to compute the values of the parameters for the hidden layer. It turns out that such a training philosophy has a solid theoretical foundation, as convergence to a unique solution is guaranteed. It is interesting to note that although the node parameters are randomly generated, for infinitely differentiable activation functions, the training error can become arbitrarily small if K approaches N (it becomes zero if $K = N$). Furthermore, the universal approximation theorem ensures that for sufficiently large values of K and N, g_K can approximate any nonconstant piece-wise continuous function [98]. A number of variations and generalizations of this simple idea can be found in the respective literature. The interested reader is referred to, for example, [99,186] for related reviews.

18.10 NEURAL NETWORKS: A BAYESIAN FLAVOR

In Chapter 12, the (generalized) linear regression and the classification tasks were treated in the framework of Bayesian learning. Because a feed-forward neural network realizes a parametric input–output mapping, $f_{\boldsymbol{\theta}}(\boldsymbol{x})$, there is nothing to prevent us from looking at the problem from a fully statistical point of view. Let us focus on the regression task and assume that the noise variable is a zero mean Gaussian one. Then the output variable, given the value of $f_{\boldsymbol{\theta}}(\boldsymbol{x})$, is described in terms of a Gaussian distribution,

$$p(y|\boldsymbol{\theta};\beta) = \mathcal{N}\left(y|f_{\boldsymbol{\theta}}(\boldsymbol{x}), \beta^{-1}\right), \tag{18.56}$$

where β is the noise precision variable. Assuming successive training samples, $(y_n, \boldsymbol{x}_n)$, $n = 1, 2, \dots, N$, to be independent, we can write

$$p(\boldsymbol{y}|\boldsymbol{\theta};\beta) = \prod_{n=1}^{N} \mathcal{N}\left(y_n|f_{\boldsymbol{\theta}}(\boldsymbol{x}_n), \beta^{-1}\right). \tag{18.57}$$

Adopting a Gaussian prior for $\boldsymbol{\theta}$, that is,

$$p(\boldsymbol{\theta};\alpha) = \mathcal{N}(\boldsymbol{\theta}|\boldsymbol{0}, \alpha^{-1}I), \tag{18.58}$$

the posterior distribution, given the output values $\boldsymbol{y}$, can be written as

$$p(\boldsymbol{\theta}|\boldsymbol{y}) \propto p(\boldsymbol{\theta};\alpha)p(\boldsymbol{y}|\boldsymbol{\theta};\beta). \tag{18.59}$$

However, in contrast to Eq. (12.16), the posterior is not a Gaussian one, owing to the nonlinearity of the dependence on $\boldsymbol{\theta}$. Here is where complications arise and one has to employ a series of approximations to deal with it.

Laplacian approximation: The Laplacian approximation method, introduced in Chapter 12, is adopted to approximate $p(\boldsymbol{\theta}|\boldsymbol{y})$ to a Gaussian one. To this end, the maximum, $\boldsymbol{\theta}_{\text{MAP}}$, has to be computed, which is carried out via an iterative optimization scheme. Once this is found, the posterior can be replaced by a Gaussian approximation, denoted as $q(\boldsymbol{\theta}|\boldsymbol{y})$.

Taylor expansion of the neural network mapping: The final goal is to compute the predictive distribution,

$$p(y|\boldsymbol{x}, \boldsymbol{y}) = \int p(y|f_{\boldsymbol{\theta}}(\boldsymbol{x}))q(\boldsymbol{\theta}|\boldsymbol{y})\,d\boldsymbol{\theta}. \tag{18.60}$$

However, although the involved PDFs are Gaussians, the integration is intractable, because of the nonlinear nature of $f_{\boldsymbol{\theta}}$. In order to carry this out, a first-order Taylor expansion is performed,

$$f_{\boldsymbol{\theta}}(\boldsymbol{x}) \approx f_{\boldsymbol{\theta}_{\text{MAP}}}(\boldsymbol{x}) + \boldsymbol{g}^T(\boldsymbol{\theta} - \boldsymbol{\theta}_{\text{MAP}}), \tag{18.61}$$

where $\boldsymbol{g}$ is the respective gradient computed at $\boldsymbol{\theta}_{\text{MAP}}$, which can be computed using backpropagation arguments. After this linearization, the involved PDFs become linear with respect to $\boldsymbol{\theta}$ and the integration leads to an approximate Gaussian predictive distribution as in Eq. (12.21). For the classification, instead of the Gaussian PDF, the logistic regression model as in Section 13.7.1 of Chapter 13 is adopted and similar approximations as before are employed. More on this more classical view on the Bayesian approach to neural networks can be obtained in [151,152].

Bayesian inference methods: More recently, an interest in the Bayesian view of deep networks has been revived in an effort to enforce regularization and pruning in a more efficient way. In this vein, all parameters are treated as random variables that are described in terms of conditional and prior distributions. The latter act as regularizers. In the sequel, variational Bayesian techniques are mobilized to infer the posteriors. For example, in [142], hierarchical priors are introduced to prune nodes instead of individual weights. Also, the posterior uncertainties are exploited to determine optimal fixed point precision to encode the weights.

In [174], nonparametric Bayesian arguments via the Indian buffer process (IBP) (Chapter 13) are mobilized to prune weights, nodes, or whole kernels in the case of convolutional networks (see Section 18.12). Also, the nonlinearities are substituted by fully probabilistic units involving competing local winner-take-all (LTWA) arguments. It is reported that the approach leads to very efficient structures in terms of number of units as well as in terms of bit precision requirements.

18.11 SHALLOW VERSUS DEEP ARCHITECTURES

In our tour so far in this chapter, we have discussed various aspects of learning feed-forward networks involving a number of layers of nodes. The backpropagation gradient descent scheme, in its various formulations, was introduced as a popular algorithmic framework for training multilayer architectures. We also established some very important properties of the multilayer neural networks that concern their universal approximation property and also their power to solve any classification task comprising classes formed by the union of polyhedra regions in the input space. Two or three layers were, theoretically, enough to perform such tasks. Thus, it seems that everything has been said. Unfortunately (or maybe fortunately) this is far from the truth.

Following the mid-1980s, feed-forward neural networks, after almost one decade of intense research, lost their initial glory and were superseded, to a large extent, by other techniques, such as kernel-based schemes, boosting and boosted trees, and Bayesian approaches. A major reason for this loss of popularity was that their training can become difficult and often backpropagation-related algorithms exhibit a slow convergence speed, or they can converge to a "bad" local minimum, which was the general belief at the time. Although various "tricks" and techniques were proposed in order to improve convergence or find a better minimum, after training multiple times via different random initializations, still their generalization performance was not competitive compared to other methods. This drawback appeared to be more severe if more than two hidden layers were used. As a matter of fact, efforts to use more than two hidden layers were soon abandoned.

In this section, we are going to discuss whether there is any need for deep networks that involve more than two hidden layers. Furthermore, we are going to offer arguments that challenge the view that the existence of poor local minima comprises a major obstacle while training deep neural networks.

18.11.1 THE POWER OF DEEP ARCHITECTURES

In Section 18.3, we discussed how each layer of a neural network provides a different representation of the input patterns. The input layer described each pattern as a point in the feature space. The first hidden layer of nodes formed a partition of the input space and placed the input point in one of the regions, using a coding scheme of zeros and ones (for the Heaviside activation) at the outputs of the respective neurons. This can be considered as a more abstract representation of the input patterns. The second hidden layer of nodes, based on the information provided by the previous layer, encoded information related to the classes; this is a further representation abstraction, which carries some type of "semantic meaning." For example, it provides the information of whether a tumor is malignant or benign, in a related medical application.

The previously reported hierarchical type of representation of the input patterns mimics the way that a mammal's brain follows in order to "sense" and "perceive" the world around us. The brain of the mammals is organized in a number of layers of neurons, and each layer provides a different representation of the input percept. In this way, different levels of abstraction are formed, via a hierarchy of transformations. For example, in the primate visual system, this hierarchy involves detection of edges and primitive shapes, and, as we move to higher hierarchy levels, more complex visual shapes are formed, until finally a semantics concept is established; for example, a car moving in a video scene, a person sitting in an image. The cortex of our brain can be seen as a multilayer architecture with 5–10 layers dedicated only to our visual system [212].

On the Representation Properties of Deep Networks

Following the previous discussion, an issue that is now raised is whether one can obtain an input–output representation via a relatively simple functional formulation, such as the one implied by the support vector machines or via networks with less than three layers of neurons/processing elements, that is equivalent in performance, maybe at the expense of more elements per layer.

The answer to the first point is yes, as long as the input–output dependence relation is simple enough. However, for more complex tasks, where more complex dependencies have to be learned, for example, recognition of a scene in a video recording or in language and speech recognition, the underlying functional dependence is of a very complex nature so that we are unable to express it analytically in a simple way.

The answer to the second point, concerning shallow networks consisting of only a few layers, lies in what is known as *compactness* of representation. We say that a network, realizing an input–output functional dependence, is compact if it consists of relatively few free parameters (few computational elements) to be learned/tuned during the training phase. Thus, for a given number of training points, we expect compact representations to result in better generalization performance.

It turns out that using networks with more layers, one can obtain more compact representations of the input–output relation. Although there are no theoretical findings for general learning tasks to prove such a claim, theoretical results from the theory of circuits of Boolean functions suggest that a function which can compactly be realized by, say, k layers of logic elements may need an exponentially large number of elements if it is realized via $k-1$ layers. Some of these results have been generalized and are valid for learning algorithms in some special cases. For example, the parity function with l inputs requires $\mathcal{O}(2^l)$ training samples and parameters to be represented by a Gaussian support vector machine, $\mathcal{O}(l^2)$ parameters for a neural network with one hidden layer, and $\mathcal{O}(l)$ parameters and nodes for a multilayer network with $\mathcal{O}(\log_2 l)$ layers (see, for example, [16,17,172]).

In [160,183], it is shown that for a special class of deep networks and target outputs, one needs a substantially smaller number of nodes to achieve a predefined accuracy compared to a shallow network. In [163], for networks employing ReLU, it has been shown that the composition of layers identifies linear regions in the input space, whose number has an exponential dependence on the depth of the network. In [46], it is shown that there is a simple function in $\mathbb{R}^l$, which is expressive by a small three-layer feed-forward neural network, while it cannot be adequately approximated by *any* two-layer network, unless the number of nodes is exponentially large with respect to the dimension. This result holds true for virtually all known activation functions, including the ReLU. Formally, these results demonstrate that depth—even if increased by one—can be exponentially more valuable than width (number of nodes per layer) for standard feed-forward neural networks. Results similar in spirit have been derived in [230]. In [34], the focus is on convolutional networks; employing arguments from tensor algebra, it is shown that besides a negligible (zero measure) set, all functions that can be realized by a deep network of polynomial size require exponential size in order to be realized, or even approximated, by a shallow network.

In [146], the interplay between width (number of nodes in a layer) and depth is considered. The question posed is whether there are wide networks that *cannot be realized by narrow networks whose size is not substantially larger*. This is on the antipodal of the so-called "existence" path, that is, to find functions that are efficiently realizable with a certain depth but cannot be efficiently realized with shallower depths. It is shown that there exists a family of ReLU networks that cannot be approximated by narrower networks whose depth increase is no more than polynomial. The theoretical and the experimental evidence in the paper points out that depth may be more effective than width for the expressiveness of ReLU networks. Moreover, the paper raises a number of open problems. At the time the current edition is compiled, the topic constitutes an area of ongoing research.

Such arguments as the one before may seem a bit confusing to a newcomer in the field, because we have already stated that networks with two layers of nodes are universal approximators for a certain class of functions. However, this theorem does not say how one can achieve this in practice. For example, any continuous function can be approximated arbitrarily close by a sum of monomials. Nevertheless, a huge number of monomials may be required, which is not practically feasible. In any learning task, we have to be concerned with what is feasibly "learnable" in a given representation. The interested reader may refer to, for example, [237] for a discussion on the benefits one is expected to get when using many-layer architectures.

Let us now elaborate a bit more on the aforementioned issues and also make bridges to some of the techniques that have been discussed in previous chapters. Recall from Chapter 11 that nonparametric techniques, modeling the input–output relation in RKHSs, establish a functional dependence of the form

$$f(\boldsymbol{x}) = \sum_{n=1}^{N} \theta_n \kappa(\boldsymbol{x}, \boldsymbol{x}_n) + \theta_0. \tag{18.62}$$

This can be seen as a network with one hidden layer, whose processing nodes perform kernel computations and the output node performs a linear combination. As already commented in Section 11.10.4, the kernel function $\kappa(\boldsymbol{x}, \boldsymbol{x}_n)$ can be thought of as a measure of similarity between $\boldsymbol{x}$ and the respective training sample, $\boldsymbol{x}_n$. For kernels such as the Gaussian one, the action of the kernel function is of a *local* nature, in the sense that the contribution of $\kappa(\boldsymbol{x}, \boldsymbol{x}_n)$ in the summation tends to zero as the distance of $\boldsymbol{x}$ from $\boldsymbol{x}_n$ increases (the rate of decreasing influence depends on the variance σ^2 of the Gaussian). Thus, if the true input–output functional dependence undergoes fast variations, then a large number of such local kernels will be needed to model sufficiently well the input–output relation. This is natural, as one attempts to approximate a fast-changing function in terms of smooth bases of a local extent. Similar arguments hold true for the Gaussian processes discussed in Chapter 13. Besides the kernel methods, other widely used learning schemes are also of a local nature, as is the case for the decision trees, discussed in Chapter 7. This is because the input space is partitioned into regions via rules that are of a local nature.

In contrast, assuming that the above stated input–output dependence variations are not random in nature but that there exist underlying (unknown) regularities, resorting to models with a more compact representation, such as networks with many layers, one expects to learn the regularities and exploit them to improve the performance. As stated in [194], exploiting the regularities that are hidden in the training data is likely to aid in the design of an excellent predictor for future events. The interested reader may explore more on these issues from the insightful tutorial [18].

Distributed Representations

A notable characteristic of multilayer neural networks is that they offer what is known in machine learning as *distributed representation* of the input patterns. Take, as an example, the simple case where the neuron outputs are either 1 or 0, as discussed in Section 18.3. Interpreting the output of each node as a feature, the vector comprising these feature values, in a layer, provides information with respect to the input patterns; this is a distributed representation that is spread among all the features in a layer. Moreover, these are *not* mutually exclusive. It turns out that such a distributed representation is sparse, because only a few of the neurons are active each time. This is in line with what we believe happens in the human brain, where at each time less than 5% of the neurons, in each layer, fire, and the rest remain inactive. In the antipodal of such distributive representations would be to have a single neuron firing each time.

In the case of neural networks with more general (compared to 0 and 1) activation functions, the features that are generated as outputs of the neurons in each one of the layers are *shared* by all patterns in all of the classes. For example, the same neurons are used and shared when learning to discriminate between, say, "airplanes" from "cars." This makes a lot of sense, since many attributes are shared and are common to both classes. For example, the metal structure and the existence of wheels are common. In contrast, decision trees (Chapter 7) are not based on distributed representation. If a pattern from the "airplane" class is presented to the input, only the corresponding class leaf and the nodes on the path from the root to this leaf are activated. What underlies distributed representations is that they build a similarity space in which semantically close input patterns remain close in some "distance" sense.

At the other extreme of representation is the one offered by local methods, where a different model is attached to each region in space and parameters are optimized locally. However, it turns out that distributed representations can be exponentially more compact, compared to local representations. Take as an example the representation of integers in the interval $[1, 2, \ldots, N]$. One way is to use a vector of length N and to set for each integer the respective position equal to 1. However, a more efficient way in terms of the number of bits would be to employ a distributed representation, that is, use a vector of size $\log_2 N$ and encode each integer via ones and zeros positioned to express the number as a sum of powers of two. An early discussion on the benefits of distributed representation in learning tasks can be found in [82]. A more detailed treatment of these issues is provided in [18].

On the Optimization of Deep Networks: Some Theoretical Highlights

In the beginning of this section, it was stated that dealing with deep networks in the 1980s and 1990s was abandoned due to the difficulty in their training. The general belief at the time was that because the number of the parameters becomes very large, the cost function in the parameter space becomes complicated and the probability of getting stuck in a local minimum significantly increases.

The above belief has been seriously challenged after 2010. At that time, it was discovered that one can train large networks, provided that enough training data were used. It was around this time that large data sets were built and could be used for training in parallel with the advances in computer technology that offered the necessary computational power. As a matter of fact, this is the crucial factor for the comeback of the multilayer feed-forward neural networks, that is, computer *technology* together with the availability of *large data sets*. Of course, a number of techniques that in the mean time have been developed, such as the use of ReLU nonlinearity and the dropout method for regularization, combined with experience gained for the associated training, also have their share in the popularity and success of neural networks; however, these advances had a rather secondary contribution.

As a consequence, the success of training big networks raised questions on issues related to local minima. Even if one has huge amounts of data, if the terrain of the cost function in the parameter space is "full" of local minima, then once the algorithm has been trapped in one of those, keeping training with more data would not make much sense.

The previous setting ignited a related research happening and many interesting findings have challenged the previous belief. At the time this edition is compiled, this is still a hot topic of research. Our goal here is not to provide an extensive related discussion, but to give some basic directions and insights and make the reader alert of the issue. For example, in [38], it is argued that a more profound difficulty, especially in high-dimensional problems, originates from the proliferation of *saddle points*. The existence of such points can slow down the convergence of the training algorithm dramatically. In [62], it is argued that the effect of the large number of saddle points on the gradient descent is unclear, since the value of the gradient can be very small and slow down the convergence rate, yet it seems that the algorithm can escape such critical points. In [35], it is claimed that in large-size networks, most of the local minima yield low cost values and result in similar performance on a test set. Moreover, the probability of finding a poor (high cost value) local minimum decreases fast as the network size increases. Both papers borrow results from statistical physics on Gaussian random fields. A result similar in essence, yet via a different mathematical path, is derived in [116]. It is shown that for the square loss function and deep linear neural networks, every local minimum is also a global one. Also, every critical point that is not a global minimum is necessarily a saddle point. The results are extended to nonlinear networks under certain independence assumptions. In a more recent paper [45], the issue is answered via the convergence of the gradient descent algorithm. For the case of the squared error loss function, it is shown that gradient descent finds a global minimum in training deep neural networks; this is in spite of the fact that the cost function is a nonconvex one with respect to the involved parameters. It is shown that the gradient descent achieves zero training loss in polynomial time. The paper deals with three different types of overparameterized neural networks. A fully connected, a convolutional, and a residual deep network, which will be discussed in Section 18.12. In [122], the convergence of the stochastic gradient descent is studied and it is argued that the algorithm will not get stuck at local minima with small diameters, as long as the neighborhoods of these regions contain enough gradient information. The neighborhood size is controlled by the step size and the gradient noise. The case of saddle points and how one can escape from them is treated in, e.g., [111,171].

On the Generalization Power of Deep Networks

In the previous paragraph, some issues related to the convergence of the training algorithm and the nonconvexity of the optimizing task were briefly presented. The focus of the discussion now turns to issues related to the generalization performance of deep networks. This is still an open problem, and at the time this edition is compiled the topic is a very active area of research.

The generalization performance of any learner, including of course deep neural networks, is quantified by the difference between the training error and the test error. Good learners are those where the test error and the training error have close values. Also, as we have discussed in Chapter 3, if the number of parameters is large enough with respect to the size of the training data, overfitting occurs and we expect the test error to deviate from the training one. However, in the case of deep networks we are faced with a "paradox."

Training very large overparameterized networks, where the number of parameters is larger than the size of the training set, even *without regularization*, often (not always) the resulting network exhibits good generalization performance! In [256], it is shown that deep neural networks easily fit *random labels*. In other words, the effective capacity of a neural network is large enough for a brute-force *memorization* of the entire data set. Furthermore, it is pointed out that in contrast with classical convex empirical risk minimization, where explicit regularization is necessary to rule out trivial solutions, in the world of deep neural networks regularization seems to help improve the final test error of a model, yet its absence does not necessarily lead to poor generalization performance.

An explanation for this phenomenon has been attempted via the implicit regularization imposed by the gradient descent algorithm. Take for example the least-squares and a linear model. We know (e.g., Sections 6.4 and 9.5) that when a system is underdetermined the least-squares solution is the minimum norm one. Similar arguments hold for other cost functions and models, when the gradient descent algorithm is used (e.g., [73]) or if its stochastic gradient version is employed (see, e.g., [184]). As it is stated in [219], there are many global minima of the training objective, most of which will not generalize well, but the optimization algorithm (e.g., gradient descent) biases the solution toward a particular minimum that does generalize well. The effect of the batch size on the generalization ability is studied experimentally in [119], where it is pointed out that small batch sizes have a beneficial effect. Some of the arguments posed in this paper are challenged in [44].

In [14], the generalization power of over-parameterized networks, where the number of the unknown parameters is larger than that of the training data samples, is discussed in the context of function smoothness and small norm solutions. The classical U-shaped graph of the test error versus the model's complexity, as discussed in Section 3.13, is reconsidered and it is pointed out that for over-parameterized networks the test error can be made to decrease even if the training error tends to zero.

As said in the beginning, this is a new field of research, and a number of questions and issues are still open and different points of view come to contribute to the discussion (see, e.g., [117]).

18.12 CONVOLUTIONAL NEURAL NETWORKS

In our discussion so far, we have assumed that neural networks are fed to their input layer with *feature vectors*. This is in line with any other classifier/predictor that has been discussed in previous chapters. Feature vectors are generated from the raw data in an effort to compact information that is relevant to the machine learning task at hand. To this end, for many years, a large pool of techniques has been developed to fit the nature of the data in different applications (see, e.g., [231]).

An alternative breakthrough came in the late 1980s, when the feature generation phase was integrated as part of the training of a neural network. The idea was to *learn* the features from the data *together* with the parameters of the neural network and *not* independently. Such networks were called *convolutional neural networks* (CNNs) and their success was first demonstrated in the OCR task in recognizing digits of numbers [129]. The name "convolutional" comes from the fact that the first layers in a neural network perform convolutions instead of inner products, which are the basic operations in the fully connected networks presented in Section 18.3.1

18.12.1 THE NEED FOR CONVOLUTIONS

Let us first make sure that we understand the reason of why we cannot, in practice, feed the input of a neural network directly with raw data, e.g., an image array or the samples of a digitized version of a speech segment, and why *preprocessing* is necessary in order to generate features. This is also true for any predictor/learner and not for neural networks only. We have commented on this issue in the introductory chapter, but there it was too early to grasp the true need. In many applications, working directly with raw data makes the task simply unmanageable.

Let us take, as an example, the case of a 256×256 image array. Vectorizing it results in an input vector $\boldsymbol{x} \in \mathbb{R}^l$, where $l \approx 65000$. Assume that the number of nodes of the first layer is $k_1 = 1000$. Then the number of the involved parameters, θ_{jk}, $j = 1, 2, \ldots, 65000$, $k = 1, 2, \ldots 1000$, connecting all the input nodes to all the nodes of the first layer in a fully connected network, would be of the order of 65 million. This number explodes further if the input image is a high-resolution one with pixels of the order of 1000×1000. Furthermore, this number increases if we deal with, for example, color images where the dimensionality of the input is multiplied by three for an RGB (red-green-blue) color representation scheme. Moreover, as one adds more hidden layers, the number of parameters keeps increasing. Besides the associated computational load issues, we know that training networks with a large number of parameters seriously challenges its generalization performance. Such networks would require a huge amount of training data in order to cope with the overfitting tendency.

Besides the explosion in the number of parameters, vectorizing an image array leads to a loss of information; this is because we throw away important information with respect to how the pixels are interrelated within an area in an image. As a matter of fact, the goal of the various feature generation techniques that have been developed over the years is exactly that. That is, to extract information that quantifies *correlations* or other statistical dependencies that *relate* pixel values within the image. In this way, one can efficiently "encode" learning-related information that resides in the raw data.

By employing convolutions, one can simultaneously tackle both issues, i.e., that of the parameter explosion as well as the extraction of useful statistical information. The basic steps involved in any convolutional network are:

- the convolution step,
- the nonlinearity step,
- the pooling step.

The Convolution Step

One way to reduce the number of parameters is via *weight sharing*, as has been briefly discussed at the end of Section 18.3.1. We will now borrow this idea of weight sharing and use it in a more sophisticated fashion. To this end, let us focus on the case where the input to the network comprises images. The input image array is denoted as I. For the case of Fig. 18.17A, this is a 3×3 array; note that the input is

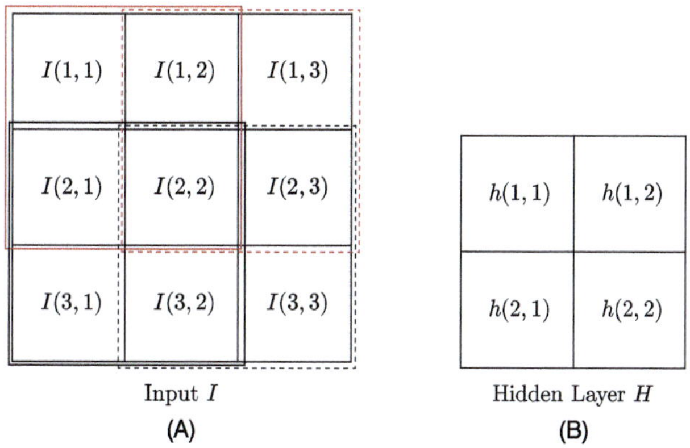

FIGURE 18.17

(A) A 3×3 input image array. (B) The "nodes" of the hidden layer can be thought of as elements of a two-dimensional array. Each node corresponds to a single parameter that is associated with the corresponding element of the array, H. In this case, the corresponding array is of size 2×2. To perform convolutions, one slides matrix H over matrix I. For the specific setup of the figure, four different positions are possible, indicated in (A) by different colors and/or types of lines.

not vectorized. To stress out that we will depart from the rationale of the multiply-add (inner product) operations of the fully connected feed-forward networks, we will use a different symbol, h, instead of θ, to denote the associated parameters. Let us now introduce the concept of weight sharing. Recall that in a fully connected network, each node is associated with a *vector* of parameters, $\boldsymbol{\theta}_i$, for the ith node, whose dimensionality is equal to the number of nodes of the previous layer. In contrast, now, each node will be associated with a *single* parameter. To this end, arrange the nodes in the form of a two-dimensional array, as shown in Fig. 18.17B. For the case of the figure, we have assumed four nodes arranged in a 2×2 array, H. The first node is characterized by $h(1,1)$, the second by $h(1,2)$, and so on. In other words, any connection that ends at the first node will be multiplied by the same weight, $h(1,1)$, and a similar argument holds true for the rest of the nodes. This rationale reduces the number of parameters dramatically. However, in order for this to make sense, we have to move away from the inner product operations rationale of the fully connected networks. Indeed to understand why, let us assume that we use a single parameter per node in a fully connected network. Then the output of the linear combiner associated with the first node would be $O(1,1) = h(1,1)a$, where a is the sum of all the inputs to the node received from the previous layer. The respective output of the second node would be $O(1,2) = h(1,2)a$, and so on. Hence, all nodes would, basically, provide the same information with respect to the input values; the only difference would be the different weights acting upon the *same* input information from the previous layer.

Let us now introduce a different concept, where we keep a single parameter per node, yet each one of the outputs of a hidden layer conveys *different* information with respect to the various inputs that are received from the previous layer. To this end, we will introduce convolutions. In this context, the nodes of the hidden layer are interpreted as elements of an array H, and we convolve H with the input array I. The first output value of the hidden layer will be

$$O(1,1) = h(1,1)I(1,1) + h(1,2)I(1,2) + h(2,1)I(2,1) + h(2,2)I(2,2).$$

The above result is obtained if we place the 2×2 matrix H on top of I, starting at the left top corner (the full red square in Fig. 18.17A indicates the position of the H matrix). Then, we multiply corresponding elements in the overlapping parts of the two matrices and add them together. From a physical point of view, the resulting $O(1,1)$ value is a weighted average over a *local* area within the I array. In the previous operation, the corresponding area of the image consists of the pixels within the 2×2 top left part of the array I. To obtain the second output value, $O(1,2)$, we slide H one pixel to the right, as indicated by the dotted red square box, and repeat the operations, i.e.,

$$O(1,2) = h(1,1)I(1,2) + h(1,2)I(1,3) + h(2,1)I(2,2) + h(2,2)I(2,3).$$

Following the same rationale, we slide H so as to "scan" the whole image array; thus, two more output values are obtained, i.e., $O(2,1)$ and $O(2,2)$. The four possible positions of H on top of I are indicated in Fig. 18.17A by the full-red, dotted red, dark gray, and dotted gray square boxes. For each position, one output value is obtained. Hence, under the previous described scenario, the outputs of the first hidden layer form a 2×2 array, O. Each one of the elements of the output array encodes information from a different area of the input image.

In the more general setting, the convolution operation between two matrices, $H \in \mathbb{R}^{m \times m}$ and $I \in \mathbb{R}^{l \times l}$, is another matrix, defined as

$$O(i,j) = \sum_{t=1}^{m} \sum_{r=1}^{m} h(t,r) I(i+t-1, j+r-1), \tag{18.63}$$

where for our case, $m < l$. In words, $O(i,j)$ contains information in a window area of the input array. According to the definition in Eq. (18.63), the element $I(i,j)$ is the top left element in this window area. The size of the window depends on the value of m. The size of the output matrix depends on the assumptions that one adopts on how to deal with the elements/pixels at the borders of I. We will come back to that soon. Strictly speaking, in the signal processing jargon, Eq. (18.63) is known as the *cross-correlation* operation. For the convolution operation, as has already been defined in Eq. (4.49), a flipping of the indices has first to take place.[4] However, this is the name that has "survived" in the machine learning community and we well adhere to it. After all, both operations perform a weighted averaging over the pixels within a window area of an image.

The previous discussion "forces" us to seize thinking of a hidden layer as a collection of nodes one next to the other. Instead, in a CNN, each hidden layer corresponds to a (or more than one, as we will soon see) matrix H. Furthermore, H is used to perform convolutions. From a signal processing point of view, this matrix is a *filter*, which acts upon the input to provide the output. In the machine learning jargon, it is also called the *kernel* matrix instead of a filter. The output matrix is usually referred to as the *feature map* array.

In summary, by performing convolutions, instead of inner product operations, we have achieved what was our original goal: (a) the parameters comprising the hidden layer are *shared* by all the input pixels and we do not have a dedicated set of parameters per input element (pixel), and (b) the outputs of the hidden layer encode local neighborhood correlation information from the various areas within the input image. Also, since the output of the hidden layer is also an image array, one can consider it as the input to a second hidden layer and in this way build a network with many layers, each one performing convolutions.

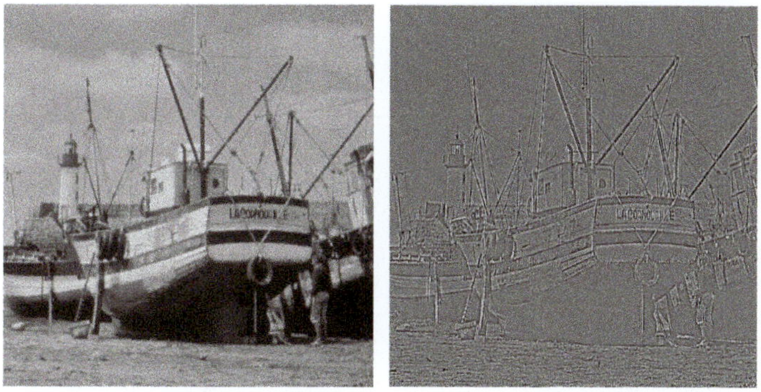
(A) (B)

FIGURE 18.18

(A) The original image and (B) the image edges extracted after filtering the original image array with the filter matrix, H, in Eq. (18.64).

As a matter of fact, such filtering operations have traditionally been used in order to generate features from images. The difference was that the elements of the filter matrix were *preselected*. Take, as an example, the following matrix:

[4] Also, here, we subtract 1, since we start counting the indices from 1 and not from 0.

$$H = \begin{bmatrix} -1 & -1 & -1 \\ -1 & 8 & -1 \\ -1 & -1 & -1 \end{bmatrix}. \tag{18.64}$$

The above filter is known as *edge detector*. Convolving an image array, I, with the previous matrix, H, detects the edges in an image. Fig. 18.18A shows the boat image and Fig. 18.18B shows the output after filtering the image on the left with the filter matrix H above. Detecting edges is of major importance in image understanding. Moreover, by changing appropriately the values in H, one can detect edges in different orientations, e.g., diagonal, vertical, horizontal; in words, changing the values of H one can generate different types of features. We have come closer to the idea behind CNNs.

- Instead of using a *fixed* filter/kernel matrix, as in the edge detector example, leave the computation of the values of the filter matrix, H, to the training phase. In other words, we make H *data-adaptive* and not preselected.
- Instead of using a single filter matrix, employ more than one. Each of them will generate a different type of features. For example, one may generate diagonal edges, the other one horizontal, etc. Hence, each hidden layer will comprise more than one filter matrix. The values of the elements of each one of the filter matrices will be computed during the training phase, by optimizing some criterion. In other words, each hidden layer of a CNN generates a set of features *optimally*.

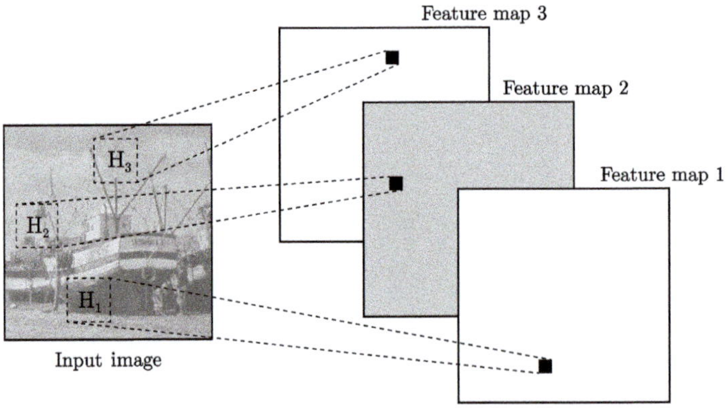

FIGURE 18.19

Each pixel in a feature map corresponds to a specific area in the input image, which is known as the receptive field of the corresponding pixel. In this figure, three filters/kernels are used. The number of filters is known as depth or sometimes we refer to it as the number of channels. Hence, in this case, the depth of the hidden layer is three.

Fig. 18.19 illustrates the input and the first hidden layer of a CNN. The input comprises an image array. The hidden layer consists of three filter matrices, namely, H_1, H_2, H_3. Observe that each feature map array is the result of sliding (convolving) a different filter matrix over the input image. The more filters are employed, the more feature maps are extracted and, in principle, the better the network should perform. However, the more filters we use, the more parameters have to be learned, raising computational as well as overfitting issues. Note that each pixel in an output feature map array encodes information within the window area that is defined by the corresponding position of the respective filter matrix.

An important characteristic of a CNN is that *translation invariance* is naturally built into the network and it is a by-product of the involved convolutions. Indeed, the latter are performed by sliding the *same* filter matrix over the entire image. Thus, if an object, which is present in an image, is placed in another position, the only difference would be that the contribution of this object to the output will also move the same amount in the number of pixels.

It is interesting to note that there is strong evidence from the visual neuroscience field that similar computations are performed in the human brain (e.g., [102,212]). The notion of convolutions was first used in [53] in the context of unsupervised learning.

Below, we provide some jargon terms used in conjunction with CNNs.

- *Depth*: The depth of a layer is the number of filter matrices that are employed in this layer. This is not to be confused with the depth of the network, which corresponds to the total number of hidden layers used. Sometimes, we refer to the number of filters as the number of *channels*.
- *Receptive field*: Each pixel in an output feature map array results as a weighted average of the pixels within a specific area of the input (or of the output of the previous layer) image array. The specific area that corresponds to a pixel is known as its receptive field (see Fig. 18.19).

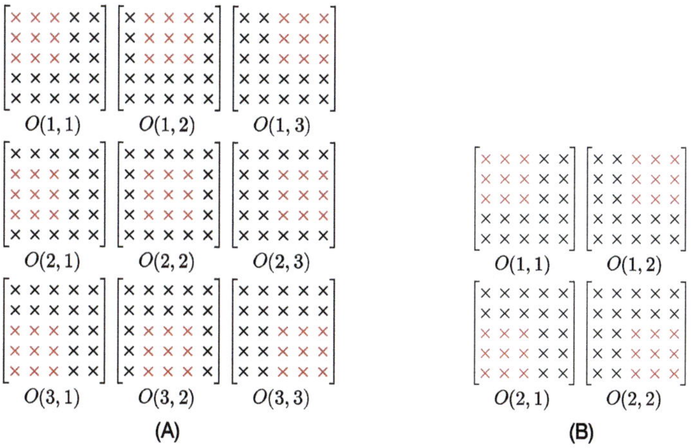

FIGURE 18.20

The figure presents the case of an input matrix of size 5×5 and of a filter matrix 3×3. In (A), the stride is equal to $s = 1$ and in (B) it is equal to $s = 2$. In (A) the output is a matrix of size 3×3 and in (B) of size 2×2.

- *Stride*: In practice, instead of sliding the filter matrix one pixel at a time, one can slide it by, say, s pixels. This value is known as the stride. For values of $s > 1$, feature map arrays that are smaller in size result. This is illustrated in Figs. 18.20A and B.
- *Zero padding*: Sometimes, zeros are used to pad the input matrix around the border pixels. In this way, the dimension of the matrix increases. If the original matrix has dimensions $l \times l$, after expanding it with p columns and rows, the new dimensions become $(l + 2p) \times (l + 2p)$. This is shown in Fig. 18.21.

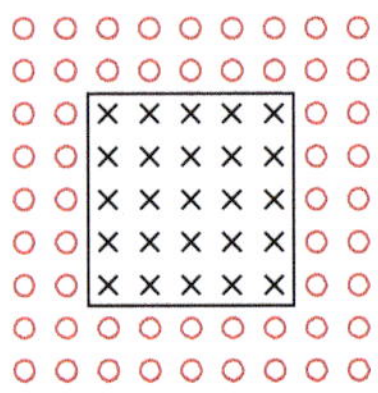

FIGURE 18.21

An example where the original matrix is 5×5 and after padding with $p = 2$ rows and columns, its size becomes 9×9.

- *Bias term*: After each convolution operation that generates a feature map pixel, a bias term, b, is added. The value of this term is also computed during training. Note that a common bias term is used for all the pixels in the same feature map. This is in line with the weight sharing rationale; in the same way that all parameters of a filter matrix are shared by all the input image array pixels, the same bias term is used for all pixel locations.

One can adjust the size of an output feature map array by adjusting the value of the stride, s, and the number of extra zero columns and rows in the padding. In general, it can easily be checked that if $I \in \mathbb{R}^{l \times l}$, $H \in \mathbb{R}^{m \times m}$, s is the stride, and p is the number of extra rows and columns for padding, then the feature map has dimensions $k \times k$, where

$$k = \left\lfloor \frac{l + 2p - m}{s} + 1 \right\rfloor, \tag{18.65}$$

and $\lfloor \cdot \rfloor$ denotes the floor operation, i.e., $\lfloor 3.7 \rfloor = 3$. For example, if $l = 5$, $m = 3$, $p = 0$, and $s = 1$, then $k = 3$. On the other hand, if $l = 5$, $m = 3$, $p = 0$, and $s = 2$, then $k = 2$ (see Figs. 18.20A and B).

Note that if the values of l, m, p, and s are such that the filter matrix, as it slides over I, falls outside I, such operations are not performed. We only perform operations as long as the filter matrix is *contained* within I.

One may wonder why one has to pad with zeros. Note that by the way convolutions are performed, the size of the feature map is smaller than that of the input array. As we will soon see, in a deep network, the output feature map is used as input to the next layer. Thus, the size of the arrays would be decreasing as we move on deeper into the network. In contrast, after padding with zeros, as Eq. (18.65) suggests, we can control the size of the involved arrays. For example, if $l = 5$, $m = 3$, $s = 1$, and $p = 1$, then the output will have the same size $k = 5$ as the input array. As a matter of fact, if $p = (m - 1)/2$, for an odd value of m, $k = l$. In such cases, we call the operation *same convolution*. The other reason that padding may be used is that the border pixels of the input image contribute less to the output, compared to the pixels that are located in the interior of the image. Take as an example Fig. 18.17. Pixel $I(1, 1)$ contributes only to $O(1, 1)$. In contrast, pixel $I(2, 2)$ contributes to all output elements, since it is contained within the window area, in all four positions. Using padding with zeros, we give the chance to the border elements to have a more equal "say" to the output values, when compared with the interior pixels.

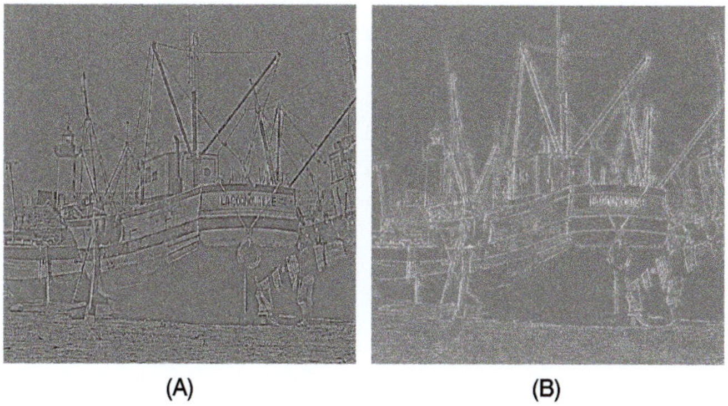

FIGURE 18.22

(A) The image where the edges have been extracted and (B) the resulting image after applying the ReLU on each one of the pixels. Note that after filtering, the image array in (A) may involve negative values, which are set to zero after the ReLU activation.

The Nonlinearity Step

Once convolutions have been performed and the bias term has been added to all feature map values, the next step is to apply a nonlinearity (activation function) to each one of the pixels of every feature map array. Any one of the nonlinearities that have been previously discussed can be employed. Currently, the rectified linear activation function, ReLU, seems to be the most popular one.

Fig. 18.22A shows the image obtained after filtering the original boat image with the edge detector filter in Eq. (18.64) and Fig. 18.22B shows the result that is obtained after the application of the ReLU nonlinearity on each individual pixel.

The Pooling Step

The purpose of this step is to reduce the dimensionality of each feature map array. Sometimes, the step is also referred to as spatial pooling. To this end, one defines a window and slides it over the corresponding matrix. The sliding can be done by adopting a value for the respective stride parameter, s.

The pooling operation consists of choosing a *single* value to represent all the pixels that lie within the window. The most commonly used operation is the *max pooling*; that is, among all the pixels that lie within the window, the one with the maximum value is selected. Another possibility is the *average pooling*, where the average value of all the pixels is selected; sometimes, this is known as sum pooling. The pooling operation is illustrated in Fig. 18.23. The original image array is 6×6 and the window is of size 2×2. We have chosen the stride to be equal to $s = 2$. That is, every time the window slides two pixels to the right or to the bottom. Different colors have been used to indicate the various positions of the window. Within each window, the maximum value is selected. The resulting matrix is of size 3×3. The same formula as in Eq. (18.65) can be used to compute the size of the resulting matrix in the general case. Thus, the effect of pooling is to reduce (via downsampling) the dimensionality and make the size of the arrays smaller. This is important because the output of each layer is presented as the input to the next one. Hence, controlling the size of the arrays is of paramount importance, in order

2	3	**7**	1	4	5
4	**5**	0	6	**7**	1
6	2	1	3	2	3
4	5	6	**8**	4	**5**
1	3	2	1	2	4
4	2	1	**8**	**6**	3

(A)

5	7	7
6	8	5
4	8	6

(B)

FIGURE 18.23

(A) The original matrix is of size 6 × 6. For pooling by a 2 × 2 window and stride $s = 2$, there are nine possible locations of the window. These locations are indicated by using different colors to show the elements that are grouped together in each one of the window locations. The maximum value per window location is indicated in bold. (B) The resulting 3 × 3 matrix after max-pooling.

to control the number of the involved parameters. Of course, the reduction in size should be done in such a way so that the loss of information is as small as possible.

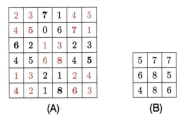

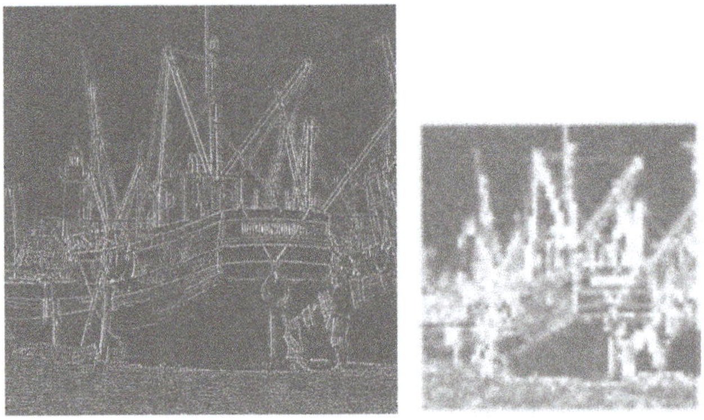

(A) (B)

FIGURE 18.24

(A) The edges of the boat image after the application of the ReLU and (B) the resulting image after applying max-pooling using an 8 × 8 window. Although the resolution is lower, the basic edge information has not been lost.

Fig. 18.24 shows the effect of applying pooling to the image shown on the left. No doubt, the edges become coarser, yet the information related to edges can still be extracted. Note that after pooling, the size of the image array is reduced.

Looking at pooling from a different view, it can be said that it summarizes the statistics within the pooling area. Pooling can be considered as a special type of filtering, where instead of convolution, the maximum (or average) value is selected. It turns out that pooling helps the representation to become approximately invariant to small translations of the input. This can be understood by the following simple argument. If a small translation does not bring in the window a new largest element and also does not remove the largest element by taking it outside the pooling window, then the maximum does not change.

18.12.2 CONVOLUTION OVER VOLUMES

In Fig. 18.19, the output of the first hidden layer comprises three image arrays. These will constitute the input to the next layer. Such an input setting that consists of multiple images is also the case when the input image is in color and its representation is given in terms of an RGB representation; that is, the input consists of three arrays, one per color. Another example is that of hyperspectral imaging, where the number of images is equal to that of the spectral bands (Section 13.13). Thus, in general, the inputs to the various layers are not two-dimensional arrays but sets of two-dimensional arrays. In mathematics, such entities are known as *multilinear arrays*, or *three-dimensional arrays*, or *three-dimensional tensors*, or *volumes*. We will adhere to the latter term, because it is reminiscent of the associated geometry, in the same way that we think of an image as a two-dimensional square. The question now is to see how one can perform convolutions when volumes are involved.

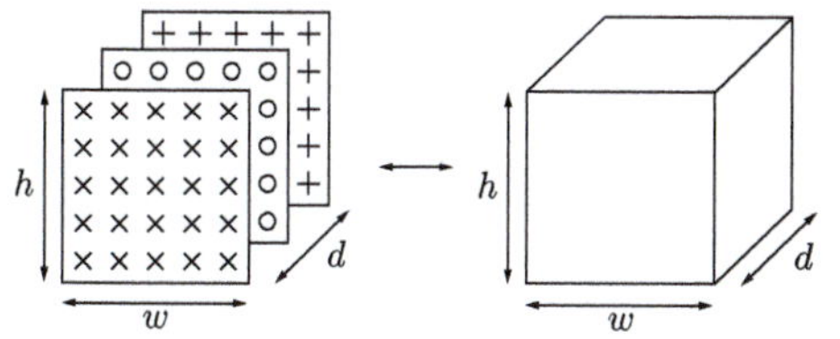

FIGURE 18.25

A number of d matrices each of size $h \times w$ are stacked together to form a volume of size $h \times w \times d$. In this case, $h = w = 5$ and $d = 3$.

By convention, the three dimensions of a volume will be represented as h for the height, w for the width, and d for the depth. Note that the depth d corresponds to the number of images involved. So, if we have three 256×256 images, then $h = w = 256$ and $d = 3$ and we will say that the volume is of size (dimension) $256 \times 256 \times 3$. Fig. 18.25 illustrates the geometry associated with the respective definitions.

Let the input to a layer be an $h \times w \times d$ volume. When volumes are involved, hidden layers consist of filter/kernel volumes, too. However, there is a crucial point here. The filter volume associated with the hidden layer *must* be of the same depth as the input volume. The height and width dimensions can be (and in practice usually are) different. We are going to use bold capital letters to denote volumes. Assume that the input is the $l \times l \times d$ volume $\boldsymbol{I}$. Obviously, this comprises d images, say, $I_r,\ r = 1, 2, \ldots, d$, each one of dimensions $l \times l$. Let the filter be the $m \times m \times d$ volume $\boldsymbol{H}$. The latter comprises the set of d images, $H_r,\ r = 1, 2, \ldots, d$, each one of dimensions $m \times m$. Then the operation of convolution is defined via the following steps:

1) Convolve corresponding two-dimensional image arrays to generate d two-dimensional output arrays, i.e., $O_r = I_r * H_r,\ r = 1, 2, \ldots, d$.

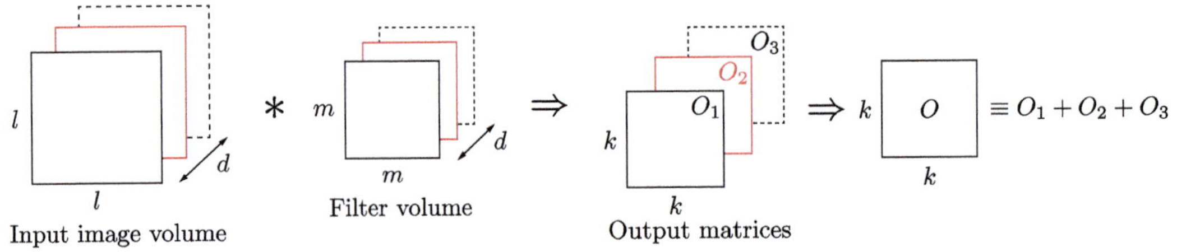

FIGURE 18.26

The figure illustrates the case of depth $d = 3$. Convolving the $l \times l \times d$ input volume ($\boldsymbol{I}$) with a filter volume ($\boldsymbol{H}$) with dimensions $m \times m \times d$ is equivalent to convolving the d ($l \times l$) matrices which comprise the input volume with an equal number of d ($m \times m$) filters. This operation results in d output (feature map) matrices, which are subsequently added together. The value of k is determined by the specific values of l, m, and the stride s. No padding is involved in the case of this figure.

2) The convolution of the two volumes, $\boldsymbol{I}$ and $\boldsymbol{H}$, is defined as

$$O = \sum_{r=1}^{d} O_r.$$

In words, the convolution (denoted by $*$) of two volumes is a *two-dimensional array*, i.e.,

$$\boxed{\text{3D volume} * \text{3D volume} = \text{2D array.}}$$

The operation is illustrated in Fig. 18.26. Corresponding arrays (shown by different colors and types of lines) are convolved. The three ($d = 3$) output arrays are subsequently added together to form the convolution of the two volumes. The dimension k of the output depends on the values of l and m, the stride s, and the padding p, if used, according to Eq. (18.65).

In practice, each layer of a convolutional network comprises a number of such filter volumes. For example, if the input to a layer is an $l \times l \times d$ volume, and there are, say, c kernel volumes, each one of dimensions $m \times m \times d$, the output of the layer will be a $k \times k \times c$ volume, where k is determined as explained before.

Network in Network and 1 × 1 Convolution

The 1×1 convolution [139] does not make sense when two-dimensional arrays are involved. Indeed, a 1×1 filter matrix is a scalar. Convolving an $l \times l$ matrix I with a scalar a is equivalent to sliding the scalar value over all the pixels and multiplying each one of them with a. The result is the trivial multiplication aI. However, when volumes are involved, the 1×1 convolution makes sense. In this case, the corresponding filter, $\boldsymbol{H}$, is a volume of size $1 \times 1 \times d$. Geometrically, this is a "tube," with $h = w = 1$ and d elements in depth, $h(1, 1, r)$, $r = 1, 2, \ldots, d$. Hence, the output of convolving an $l \times l \times d$ volume $\boldsymbol{I}$ with a $1 \times 1 \times d$ volume $\boldsymbol{H}$ is the weighted average,

$$O = \boldsymbol{I} * \boldsymbol{H} = \sum_{r=1}^{d} h(1,1,r)I_r,$$

where I_r, $r = 1, 2, \ldots, d$, are the d arrays, each of dimensions $l \times l$, that comprise $\boldsymbol{I}$. Now, one may wonder why we need such an operation in practice. The answer is related to the size of the involved volumes; via the use of 1×1 convolutions, one can control and change their sizes to fit the needs of the network.

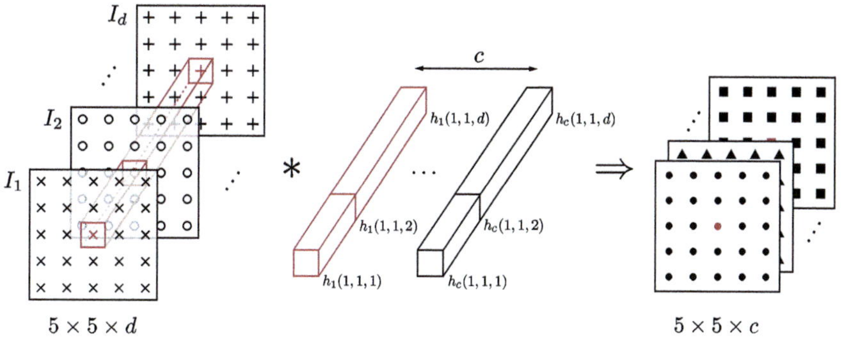

FIGURE 18.27

Illustration of the 1×1 convolution. There are c $1 \times 1 \times d$ tubes, where d is the depth of the input volume. The convolution of each one of the tubes with the input volume results in a two-dimensional array of the same dimensions as the input ones (5×5). The elements of the output arrays are weighted averages of the corresponding elements of the input arrays. For example, the red elements, at position (3, 3) of the 5×5 output arrays, are the weighted averages of the red points at the (3, 3) position of the input arrays. The weights used in the weighted average are the respective elements that define the corresponding tube. Since we have c tubes, the output volume will be of depth c.

Let us assume that at one stage/layer of a deep network we have obtained a volume $\boldsymbol{I}$ of dimensions $k \times k \times d$. To change the depth from d to c, while retaining the same size k, for the height and the width, we employ c volumes, $\boldsymbol{H}_t$, $t = 1, 2, \ldots, c$, each of dimensions $1 \times 1 \times d$. Performing the c convolutions, we obtain

$$O_t = \boldsymbol{I} * \boldsymbol{H}_t = \sum_{r=1}^{d} h_t(1,1,r)I_r, \quad t = 1, 2, \ldots, c. \tag{18.66}$$

Stacking O_t, $t = 1, 2 \ldots, c$, together, we obtain the volume $\boldsymbol{O}$ of dimension $k \times k \times c$. The operation is illustrated in Fig. 18.27.

For example, we can employ 1×1 convolution to reduce the size, by selecting $c < d$. In this way, although we reduce the depth, the elements of the new volume are weighted averages of the original $\boldsymbol{I}$. So, the original information is still retained within the new volume, in an averaging fashion. Often, once the new volume $\boldsymbol{O}$ is obtained, its elements are "pushed" through a nonlinearity, e.g., ReLU. Sometimes, the 1×1 convolution followed by the nonlinearity is referred to as *network in network* operation and its purpose is to add an extra nonlinearity stage in the flow of operations through the network. Thus, in this context, if $c < d$, the network in network operation can be thought of as a nonlinear dimensionality reduction technique.

Looking at the 1×1 convolution from a different viewpoint, it is nothing but a layer of a fully connected neural network, with c nodes. The weights connecting the tth node with the d input values are $h_t(1, 1, r)$, $r = 1, 2, \ldots, d$. This is the reason that we also call the operation network in network, in the sense that we embed a fully connected neural network between two successive convolution layers in the network. In this way, one can add an extra nonlinearity in the flow of operations through the network. In practice, the parameters of the respective $\boldsymbol{H}_t$ tubes are computed during the training phase of the overall network.

Example 18.4. Let us consider the input to a layer to be a $28 \times 28 \times 192$ volume, $\boldsymbol{I}$. The goal is to produce at the output of the layer a volume, $\boldsymbol{O}$, which is of dimension $28 \times 28 \times 32$. To this end, employ 5×5 same convolutions. Compute the number of required operations.

a) *The direct way*: Since same convolutions are required, we should first pad all the arrays that are stacked in $\boldsymbol{I}$ with p zero columns and rows, where $p = (5 - 1)/2 = 2$ (as has been explained in the text before, when "same convolutions" were defined). After padding, the height and width of each image become $h = w = 32$. Using $s = 1$ for the stride, the total number of possible locations, as we slide the 5×5 window over a 32×32 image array, is $28^2 = 784$. For each location of the window, we need to perform $5^2 = 25$ multiplications and additions (MADS), or $784 \times 25 = 19600$ MADS operations per image. Since the volume involves 192 images, the total number of MADS operations needed is $19600 \times 192 \approx 3.7 \times 10^6$. These operations are needed for each channel. Since the depth of the output should be 32, we need 32 such filter volumes (channels) and the total number of required MADS operations will be $3.7 \times 10^6 \times 32 \approx 120$ million.

b) *Via the use of* 1×1 *convolutions*: We will now produce a $28 \times 28 \times 32$ output volume using a substantially lower number of operations. Our path will involve two stages. We will first employ 1×1 convolutions to generate an intermediate volume, $\boldsymbol{O}'$, of dimensions $28 \times 28 \times 16$. To this end, we use 16 $1 \times 1 \times 192$ filter volumes and perform the respective convolutions, according to Eq. (18.66). The corresponding number of MADS is $28^2 \times 192 \times 16 \approx 2.4$ million. In the sequel, we pad each one of the arrays contained in volume $\boldsymbol{O}'$ with $p = 2$ extra zero columns and rows (as before); then, we perform same convolution with 32 filter volumes $\boldsymbol{H}_t$, $t = 1, 2, \ldots 32$, each one of dimensions $5 \times 5 \times 16$, to obtain the output volume, $\boldsymbol{O}$, of dimensions $28 \times 28 \times 32$. The respective number of operations is $28^2 \times 25 \times 16 \times 32 \approx 10$ million. Thus, now, the total number of operations, for both stages, amounts to approximately 12.4 million, which is substantially lower than the 120 million that were needed via the previous path (a).

Often, the intermediate volume, $\boldsymbol{O}'$, is known as the *bottleneck* layer; its role is to "shrink" the size of the input volume first, prior to obtaining the final output one. The overall layout is shown in Fig. 18.28.

18.12.3 THE FULL CNN ARCHITECTURE

The typical form of a full convolutional network consists of a sequence of convolutional layers, each comprising the three basic steps, namely, convolution, nonlinearity, and pooling, as described in the beginning of Section 18.12.1. Depending on the application, one can stack as many layers as required,

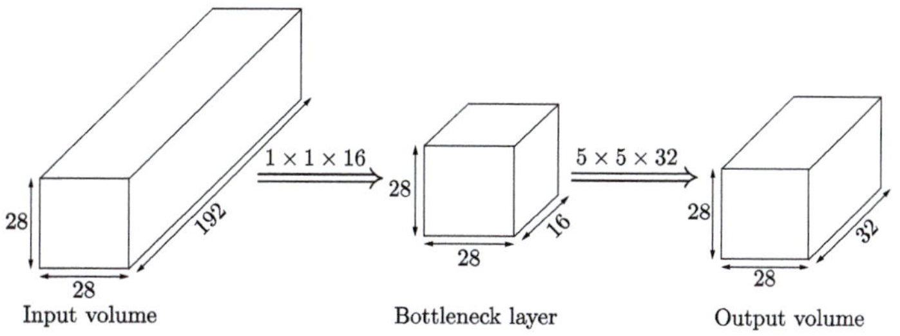

FIGURE 18.28

The bottleneck layer. In the first step, $c = 16$ 1×1 convolutions are performed and the output is a $28 \times 28 \times 16$ volume. Then, we apply a same $5 \times 5 \times 32$ convolution to obtain the final $28 \times 28 \times 32$ volume.

where the output of one layer becomes the input to the next one. Inputs and outputs to each layer are volumes, as described before. The general architecture is illustrated in Fig. 18.29. In the first layer, a number of filter volumes (channels) is employed to perform convolutions followed by the ReLU (usually) nonlinear operation. Then, the pooling stage takes over to reduce the height and width of each output volume, which is then used as input to the second layer, and so on. Finally, the output volume of the last layer is vectorized. Sometimes, this is also referred to as *flattening* operation. In words, all the elements of the output volume are stacked one under the other to form a vector. Vectorization can take place via various strategies. As a matter of fact, the obtained vector forms the *feature vector* that has finally been generated via the various *transformations* that the convolutions implement layer after layer. This feature vector will then be used as input to a learner, for example, to a fully connected neural network (lower part of the figure) or to any other predictor, such as a kernel machine.

The general strategy is to keep reducing the height and width while increasing the depth of the volumes. Larger depth corresponds to more filters per stage, which translates to more features. Both the number of convolutional layers and the number of layers of the fully connected network depend heavily on the application and, up to now, there is not a formal method to determine automatically the number of layers as well as the number of filters or nodes per layer. The choice is a matter of "engineering" and evaluation of different combinations to select the best. A good practice is to select an existing architecture that has been used before in the related application and start from there. A more recent research path towards developing more systematic ways of learning the number of nodes/filters per layer is via Bayesian learning arguments (see, e.g., [174]).

Training of a convolutional network follows a similar rationale as that of backpropagation, which has been discussed in Section 18.4. However, certain modifications should be involved in order to take care of the constraints imposed by the weight sharing (see, e.g., [59] for details).

As a final remark, recall that the crucial point for training deep networks is, besides the available computational power, the existence of large data sets (e.g., [200,262]), which have made the training of such big networks possible.

What Deep Neural Networks Learn

It is by now well established that convolutional networks do work well and at the time this edition is compiled, they constitute the state of the art in a large number of diverse applications. However, a

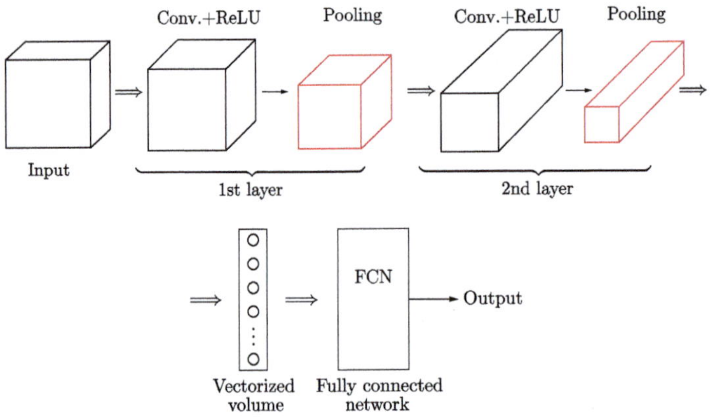

FIGURE 18.29

The full CNN comprises a series of layers. In the figure, two such layers are shown. Each layer consists of the convolution step, followed by the application of the nonlinearity, and then the pooling step, which reduces the height and width of the images that comprise the respective volume. The depth remains the same. As we move to higher layers, the tendency is to reduce the height and width and increase the depth of the volumes. The final output volume is vectorized and presented as input to a learner, usually a fully connected network.

critical question is: what type of features does a CNN learn? In other words, what is the information that is propagated from one layer to the other? This is crucial to the understanding of why they work so well. The answer to this question could facilitate their *interpretability*, which is of paramount importance in specific applications, such as in medical and financial fields. Also, such understanding could help to develop improved models.

To this end, visualization techniques have been mobilized to reveal the specific input stimuli that excite individual feature maps at any layer (e.g., [261]). In this paper, an experimental study has been conducted in the context of computer vision. The findings reveal a hierarchical nature of the features produced, as one moves from the input to the final output. For example, layer #2 seems to respond to the corners and other edge/color conjunctions associated with the objects present in the input image. Layer #3 has more complex invariances, capturing similar textures. Higher layers reveal more class-specific information, e.g., dog faces, bird lengths, etc. Such findings are in line with the discussion in Section 18.11.1.

Another interesting finding is that convergence in the lower layers, closer to the input, is rather fast. In contrast, the upper layers, closer to the output, develop after a considerable number of epochs. Furthermore, concerning feature invariance, with respect to translation and scaling, it seems that small transformations may have a dramatic effect in the first layers but a smaller impact on the top layers.

With the growing success of deep neural networks, the need of being able to *explain* the predictions is of paramount importance in building up confidence for their deployment in real-world applications. To this end, there are still a number of open questions and the topic is currently an ongoing research area. A more detailed coverage is beyond the scope of this chapter and the interested reader may obtain a good feeling of the current trends from, e.g., [27] and the references therein.

More recently (e.g., [56] and the references therein), the importance of *texture* in recognizing objects in images has been outlined. It is postulated that local textures can provide sufficient information about object classes and object recognition could, in principle, be achieved through texture recognition alone. This seems to be in line with the results obtained in [22], where it is demonstrated that using small local patches, rather than integrating object parts for shape recognition, can achieve surprisingly high accuracies. Such findings may also facilitate the interpretability issue that has been previously discussed, since one can follow easier on how the evidence from smaller image patches is integrated to reach the final decision.

In a different direction, research has been focused on revealing various aspects associated with the multilayer structure of such networks in an effort to shed light from different viewpoints, which can help in their understanding. For example, in [24,156] deep architectures are implemented via a cascade of wavelet transform convolutions, combined with a nonlinear operation followed by an averaging operation, so as to build translation-invariant representations. Furthermore, such networks preserve high-frequency information related to classification. In [25], it is shown that the pooling step in deep CNNs results in shift invariance. In [57], it is shown that deep neural networks with random Gaussian weights perform a distance preserving embedding of the data. In this analysis, tools from the compressed sensing and dictionary learning tasks are employed, which establishes a bridge of deep learning with the topics treated in Chapters 9 and 19. A closely related dictionary learning approach is also followed in [185]. A multilayer convolutional sparse coding scheme is adopted and the similarity with deep convolutional networks is established. The ReLU nonlinearity is seen as a special type of a soft thresholding operation (Chapter 9, [48]).

In [204,233], information theoretic arguments are mobilized and deep networks are viewed as a succession of intermediate representations in a Markov chain; this is closely related to the successive refinement of information in rate distortion theory. Each layer in the network can now be characterized by the amount of information it retains from the input variables, on the target output variables, and on the predicted outputs of the network. In [10], a bridge between deep networks and approximation theory via spline functions is established. It is shown that a large class of deep networks can be written as a composition of max-affine spline operators.

Finally, another path is the one that establishes bridges between Gaussian processes and deep networks. This field is not new, and its origins can be traced back to the early 1990s [168]. Since then, it is well known that a single-layer fully connected neural network with an i.i.d. prior over its parameters is equivalent to a Gaussian process, in the limit of infinite network width. In more recent years, generalizations to more layers have been established and the topic seems to regain popularity (see, e.g., [136], [7], [30] and the references therein).

18.12.4 CNNS: THE EPILOGUE

What we have described in the last subsection are the basic steps that are used to design a CNN. There are a number of variants around the architecture given in Fig. 18.29. Also, there are different tricks and algorithms that can be used to perform computations, e.g., for the efficient computation of the involved convolutions. Undoubtedly, there is a lot of "engineering" involved to make such big networks to learn the parameters and run efficiently in practical applications. Below we provide a brief description of some classical convolutional networks. The reader who wants to become familiar and get a deeper understanding of CNNs is advised to read the related papers. Although some of the implementation tricks adopted there may not be in use today, still these papers can help the reader to get further insight and understanding of CNNs.

LeNet-5: This is a typical example of the first generation of CNNs and it was built to recognize digits of numbers (see, e.g., [132]). For historical reasons, let us comment a bit on its architecture. The input of the network consists of grayscale images of size $32 \times 32 \times 1$. The network employs two convolution layers. In the first layer, the output volume has size $28 \times 28 \times 6$, which after pooling becomes $14 \times 14 \times 6$. The dimensions of the volume in the second layer were $10 \times 10 \times 16$ and after pooling $5 \times 5 \times 16$. The nonlinearity used at the time was of the sigmoid type. Observe that the height and the width of the volumes decreases and the depth increases, as pointed out before. The number of elements of the last volume is equal to 400. These elements are stacked into a vector and feed the corresponding input nodes of a fully connected network. The latter consists of two hidden layers with 120 nodes in the first and 84 nodes in the second. There are 10 output nodes, one per digit, using a softmax nonlinearity. The total number of the involved parameters is of the order of 60 thousand.

AlexNet: This network is also a historical one since it demonstrated that the crucial point for making big networks to work is the availability of large training sets [126]. The related paper is the one that really brought CNNs back into the scene and acted as a catalyst for their adoption much beyond the digit recognition task. The Alexnet is a development of LeNet-5, yet it is much bigger and involves approximately 60 million parameters. The inputs to the network are RGB images of size $227 \times 227 \times 3$. It comprises five hidden layers and the final volume consists of 9216 elements that feed a fully connected network with two hidden layers of 4096 units each. The output consists of 1000 softmax nodes (one per class) to recognize images from the ImageNet data set for object recognition [200]. The ReLU has been used as the nonlinearity in the hidden layers.

VGG-16: This network [216] is much larger than AlexNet. It involves a total of approximately 140 million parameters. The main characteristic of this network is its regularity. It involves 3×3 filters to perform same convolutions using padding and stride $s = 1$ and 2×2 windows for maxpooling with stride $s = 2$. Every time pooling is involved, the height and the width of the volumes are halved and every time the depth is increased by two. Starting with $224 \times 224 \times 3$ input images and after 13 layers, the final volume has size $7 \times 7 \times 512$, a total of 7168 elements, which after its vectorization is fed to a fully connected network with 2 hidden layers, each comprising 4096 nodes. The 1000 output nodes are built around the softmax nonlinearity and ReLU has been used for the hidden units throughout the network.

GoogleNet and the Inception network: The architecture used in this network deviates from the "archetypal" one given in Fig. 18.29. At the heart of this network lies the so-called inception module [226]. An inception module consists of filters of different sizes and depths as well as a different pooling path. A typical architecture of an inception model that provides its rationale is shown in Fig. 18.30. Note that the output volume of the previous layer becomes the input to different paths. One involves a 1×1 convolution that acts on the depth of the input volume. Another path performs pooling and then feeds a 5×5 convolution. Two paths feed separately two different convolution stages, one based on a 3×3 filter and the other on a 5×5. Prior to the convolutions, a bottleneck layer, via 1×1 convolution, is employed to reduce the respective computational load (Example 18.4). The output volumes of all these

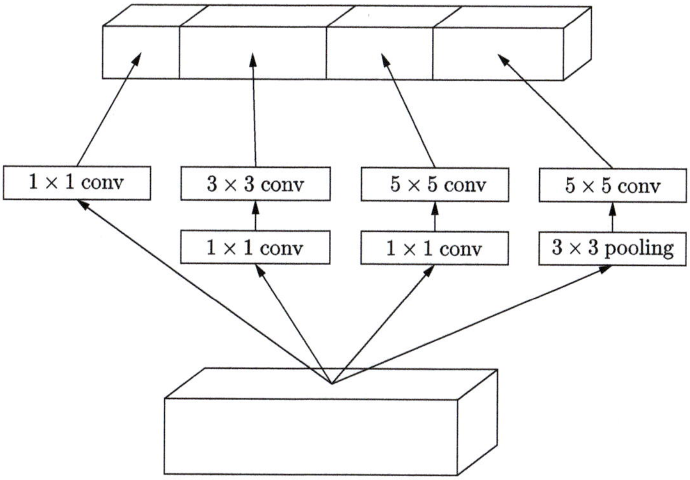

FIGURE 18.30

The inception module concept. Each layer comprises different paths of convolutions. The intermediate outputs from all the paths are concatenated together, across the channel dimension, to build up the final output volume of the corresponding layer.

paths are then concatenated to form the final output of this stage. The idea behind the inception module is to leave the network, during the training phase, to "decide" what operations fit best for the different layers and inputs. For example, as we have already commented, features of higher abstraction are captured by layers closer to the output of the network. Hence, the spatial concentration of the respective information is expected to decrease. This suggests that the ratio of 3×3 and 5×5 convolutions should increase as we move to higher layers. The number of layers of the network was 22 and the total number of parameters reported in the paper was of the order of 6 million.

Residual networks (ResNets): The benefits of designing deep networks have already been discussed. We also addressed ways on how to cope with the problem of vanishing/exploding gradients by a combination of methods and tricks that enable the backpropagation algorithm to converge sufficiently fast. However, once we start building very deep networks (of the order of tens or even hundreds of layers) we are confronted with the following "unorthodox" behavior.

One would expect that, adding more and more layers, the *training* error improves or at least does not increase. However, what one observes in practice is that beyond a certain number of layers, the training error starts increasing. This is graphically illustrated in Fig. 18.31. This phenomenon has nothing to do with overfitting. After all, we are talking about the training error and not the generalization one. It seems that this may be due to the optimization task that becomes harder and harder as more and more layers are added. Mathematically, any layer, say, the rth layer, can be seen as a mapping that maps the corresponding input, e.g., $\boldsymbol{y}^{r-1}$, to the output, e.g., $\boldsymbol{y}^r$. Let us denote the respective mapping as

$$\boldsymbol{y}^r = H(\boldsymbol{y}^{r-1}).$$

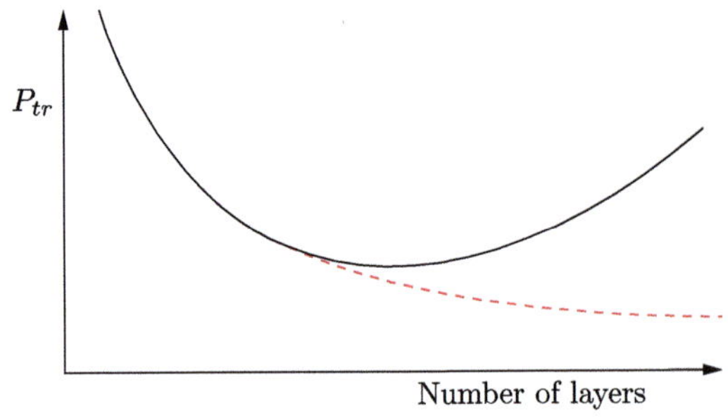

FIGURE 18.31

When a network is very deep, it turns out that the training error starts increasing when the number of layers exceeds some number, instead of being decreased (red curve) as expected from theory.

Taking this view, it is easy to see why, when adding more layers, the training error should not be increased. In the worst case, where all information has been extracted up to a layer r, we expect that adding an extra layer this should implement the identity mapping, i.e., $\boldsymbol{y}^r = H(\boldsymbol{y}^{r-1}) = \boldsymbol{y}^r$. That is, the extra layer adds no information and simply copies the input to the output. However, it seems that once the network starts becoming very deep, accuracy gets "saturated" and the optimization tools have a difficulty to come up with an accurate enough solution to this identity mapping, at least within a feasible time.

One way to bypass this difficulty is proposed in [76]. The idea is to fit an alternative equivalent mapping, i.e.,

$$F(\boldsymbol{y}^r) = H(\boldsymbol{y}^{r-1}) - \boldsymbol{y}^{r-1}.$$

Then the original mapping, $H(\boldsymbol{y}^{r-1})$, becomes equal to $F(\boldsymbol{y}^{r-1}) + \boldsymbol{y}^{r-1}$. In practice, it turns out that optimizing with respect to the residual mapping, F, is easier than optimizing with respect to the original one, H. In the extreme case, when an identity mapping should be realized, it seems that it is easier to push the residual to zero than fitting the identity one.

The use of residual representation is not new and has been used before in the context of vector quantization. The essence of the residual learning is to introduce the so-called *residual* building block, shown in Fig. 18.32. In this way, a number of layers, say, two, as in the case of the figure, are stacked together, and we *explicitly* let these layers fit the residual mapping, via the so-called shortcut or skip connections. Each weight layer performs a transformation over its input, e.g., convolutions. If $\boldsymbol{y}^r$ and $\boldsymbol{y}^{r-1}$ are of different dimension, then the identity mapping shortcut is modified to $W\boldsymbol{y}^{r-1}$, where W is a matrix of appropriate dimensions.

Fig. 18.33A shows a schematic example of a so-called plain network (no residuals involved) that comprises a sequence of convolutional layers. Its residual counterpart is shown in Fig. 18.33B. Note that if the identity shortcut is used, then no extra parameters are involved and training follows the standard backpropagation rationale. A variant of the residual network, known as *highway* network, has been proposed (e.g., [69,211]), where extra data dependent gating functions that control the flow in the shortcuts have been introduced.

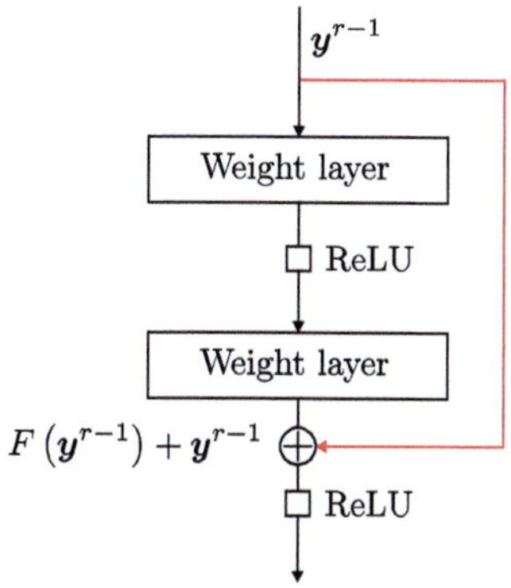

FIGURE 18.32

Two successive layers in a network have been combined and their implied combined transformation is performed via the identity mapping, implemented by the red line shortcut/skip connection.

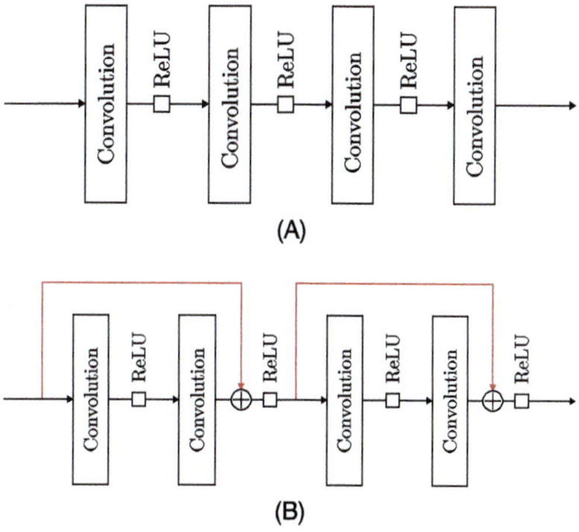

FIGURE 18.33

(A) The layout of a plain network and (B) the corresponding layout with the use of shortcuts.

In [76], networks as deep as 50 to 152 layers have been constructed. In spite of such sizes, it is reported that the overall number of arithmetic operations remains substantially lower than that required by VGG-16, at an improved performance in error rates. In [229], the concept of residual networks is combined with that of the inception networks. It is reported that training with residual connections can accelerate the training of inception networks significantly.

DenseNet: As explained before, ResNets consist of a sequence of connected residual blocks. In this way, one layer accepts input not only from its direct predecessor layer, but also by a previous one, via the shortcut (skip) connection and these two are added together. In contrast, in the DensNet [100], the building blocks are the so-called dense blocks, which combine a number of layers together, and every layer within the block receives inputs from *all* the previous ones within the block. Furthermore, these inputs are not added but are concatenated together. The reported results indicate that in this way one can reduce the number of operations and parameters involved without sacrificing performance. Numbers of layers as high as 250 were tried.

18.13 RECURRENT NEURAL NETWORKS

Recall from the previous section that at the heart of convolutional networks lies the concept of weight sharing. That is, the same filter matrix is sliding over an image array instead of dedicating a specific weight for each image pixel. In this way, a neural network can easily scale to images of different dimensions.

Our interest in the current section turns on the case of *sequential* data. That is, the input vectors are not independent but occur in sequence. Moreover, the specific order in which they occur encapsulates important information. For example, such sequences occur in speech recognition and in language processing, e.g., machine translation. Undoubtedly, the sequence in which words occur is of paramount importance. Dynamic graphical models, such as Kalman filtering and hidden Markov models (HMMs), which have been treated in Chapters 16 and 17, are models that deal with sequential data.

Weight sharing via convolutions could also be and have been used for such cases (see, e.g., [128]). Such networks are known as *time-delay* neural networks. However, sliding a filter across time to perform convolutions is an operation of a local nature. The output is a function of the input samples within a time window spanned by the length of the respective impulse response of the filter, which for practical reasons cannot be very long.

To bypass the aforementioned drawback of limited memory, in the current section we focus on networks that build upon the concept of the *state*. As was the case with the HMMs and Kalman filtering, the state vector "encodes" the past history *up to* the current time n. The idea behind *recurrent neural networks* (RNNs) is to apply the *same* type of operations (weight sharing) at each time instant (which justifies the term recurrent) by involving the *current state* (previous history) as well as the value of the current *input*. In this way, a network can scale well to sequences of different lengths; this is because one does not assign specific weights at the different instants and the same weights are shared across the whole time axis.

The variables that are involved in an RNN are:

- the state vector at time n, denoted as $\boldsymbol{h}_n$. The symbol reminds us that $\boldsymbol{h}$ is a vector of hidden variables (hidden layer in the neural network jargon); the state vector constitutes the *memory* of the system,
- the input vector at time n, denoted as $\boldsymbol{x}_n$,
- the output vector at time n, $\hat{\boldsymbol{y}}_n$, and the target output vector, $\boldsymbol{y}_n$.

The model is described via a set of unknown parameter matrices and vectors, namely, U, W, V, $\boldsymbol{b}$, and $\boldsymbol{c}$, which have to be learned during training, in analogy to the unknown parameters in an HMM, which are also learned during training.

The equations that describe an RNN model are

$$\boldsymbol{h}_n = f\left(U\boldsymbol{x}_n + W\boldsymbol{h}_{n-1} + \boldsymbol{b}\right), \tag{18.67}$$

$$\hat{\boldsymbol{y}}_n = g\left(V\boldsymbol{h}_n + \boldsymbol{c}\right), \tag{18.68}$$

where the nonlinear functions f and g act element-wise and are applied individually on each element of their vector arguments. In words, once a new input vector has been observed, the state vector is updated. Its new value depends on the most recent information, which is conveyed by the input $\boldsymbol{x}_n$ as well as the past history, as this has been accumulated in $\boldsymbol{h}_{n-1}$. The output depends on the updated state vector, $\boldsymbol{h}_n$. That is, it depends on the "history" up to the current time instant n, as this is expressed by $\boldsymbol{h}_n$.

Note that Eqs. (18.67) and (18.68) are very similar to the extended Kalman filter defined in Chapter 4. Note, however, that in the present case the involved matrices and vectors, U, W, V, $\boldsymbol{b}$, and $\boldsymbol{c}$ are unknown and have to be learned. Such a view of the RNNs, in the context of extended Kalman filtering, has been adopted in [32]. Typical choices for f are the hyperbolic tangent, tanh, or the ReLU nonlinearities. The initial value $\boldsymbol{h}_0$ is typically set equal to the zero vector. The output nonlinearity, g, is often chosen to be the softmax function, introduced in Eq. (18.45).

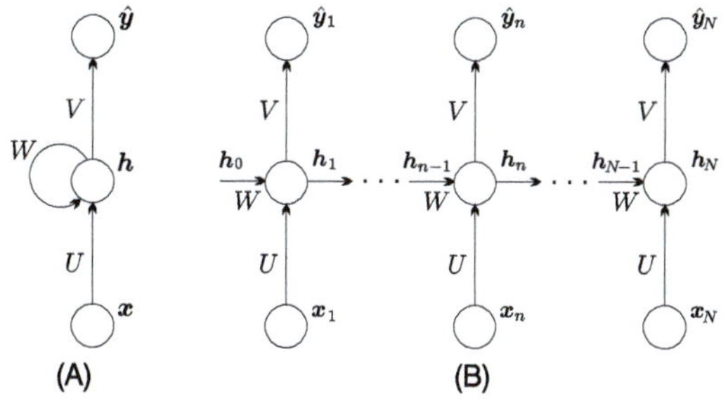

FIGURE 18.34

(A) The input, $\boldsymbol{x}$, "feeds" the (hidden) state, $\boldsymbol{h}$, which is updated using also its previous value (self-loop in the graph). In turn, it generates the output, $\hat{\boldsymbol{y}}$. (B) The operations involved in an RNN as time evolves are shown, starting from an initial value $\boldsymbol{h}_0$ of $\boldsymbol{h}$. As input vectors are sequentially observed, the corresponding output vectors are produced and the updated state vectors are passed to the next stage (time instant). The process goes on, until the final output vector is computed, for an input sequence of length N.

From the above equations, it is clear that the parameter matrices and vectors are *shared* across all the time instants. During training, they are initialized via random numbers. The graphical model associated with the pair of Eqs. (18.67) and (18.68) is given in Fig. 18.34A. In Fig. 18.34B, the graph is *unfolded* over the various time instants for which observations are available. For example, if the sequence of interest is a sentence of 10 words, then N is set equal to 10, while $\boldsymbol{x}_n$ is the vector that codes the respective input words.

18.13.1 BACKPROPAGATION THROUGH TIME

Training RNNs follows a similar rationale as that of the backpropagation algorithm for training feed-forward neural networks, as has already been discussed in Section 18.4.1. After all, an RNN can be seen as a feed-forward network with N layers. The top layer is that at time instant N and the first layer corresponds to time $n = 1$. A difference lies in that the hidden layers in RNN also produce outputs, i.e., $\hat{\boldsymbol{y}}_n$, and are fed directly with inputs. However, as far as the training is concerned, these differences do not affect the main rationale.

Learning the unknown parameter matrices and vectors is achieved via a gradient descent scheme, in line with Eqs. (18.10) and (18.11). It turns out that the required *gradients* of the cost function, with respect to the unknown parameters, take place *recursively*, by starting at the *latest* time instant, N, and going *backwards in time*, $n = N - 1, N - 2, \ldots, 1$. This is the reason that the algorithm is known as *backpropagation through time* (BPTT).

The cost function is the sum over time, n, of the corresponding loss function contributions, which depend on the respective values of $\boldsymbol{h}_n$, $\boldsymbol{x}_n$, i.e.,

$$J(U, W, V, \boldsymbol{b}, \boldsymbol{c}) = \sum_{n=1}^{N} J_n(U, W, V, \boldsymbol{b}, \boldsymbol{c}).$$

For example, for the cross-entropy loss function case,

$$J_n(U, W, V, \boldsymbol{b}, \boldsymbol{c}) := -\sum_k y_{nk} \ln \hat{y}_{nk},$$

where the summation is over the dimensionality of $\boldsymbol{y}$, and

$$\hat{\boldsymbol{y}}_n = g(\boldsymbol{h}_n, V, \boldsymbol{c}) \text{ and } \boldsymbol{h}_n = f(\boldsymbol{x}_n, \boldsymbol{h}_{n-1}, U, W, \boldsymbol{b}).$$

It turns out that at the heart of the computation of the *gradients* of the cost function with respect to the various parameter matrices and vectors lies the computation of the gradients of J with respect to the *state vectors*, $\boldsymbol{h}_n$. Once the latter have been computed, the rest of the gradients, with respect to the unknown parameter matrices and vectors, is straightforward. To this end, note that each $\boldsymbol{h}_n$, $n = 1, 2, \ldots, N-1$, affects J in two ways:

- directly, through J_n,
- indirectly, via the *chain* that is imposed by the RNN structure, i.e.,

$$\boldsymbol{h}_n \to \boldsymbol{h}_{n+1} \to \ldots \to \boldsymbol{h}_N.$$

That is, $\boldsymbol{h}_n$, besides J_n, also affects all the subsequent cost values, $J_{n+1}, \ldots, J_N$.

Employing the chain rule for derivatives, the above dependencies lead to the following recursive computation:

$$\boxed{\frac{\partial J}{\partial \boldsymbol{h}_n} = \underbrace{\left(\frac{\partial \boldsymbol{h}_{n+1}}{\partial \boldsymbol{h}_n}\right)^T \frac{\partial J}{\partial \boldsymbol{h}_{n+1}}}_{\text{indirect recursive part}} + \underbrace{\left(\frac{\partial \hat{\boldsymbol{y}}_n}{\partial \boldsymbol{h}_n}\right)^T \frac{\partial J}{\partial \hat{\boldsymbol{y}}_n}}_{\text{direct part}},} \tag{18.69}$$

where, by definition, the derivative of a vector, say, $\boldsymbol{y}$, with respect to another vector, say, $\boldsymbol{x}$, is defined as the matrix $\left[\frac{\partial \boldsymbol{y}}{\partial \boldsymbol{x}}\right]_{ij} := \frac{\partial y_i}{\partial x_j}$. Note that the gradient of the cost function, with respect to the hidden parameters (state vector) at layer "n," is given as a function of the respective gradient in the layer above, i.e., with respect to the state vector at time $n+1$. The full proof of the backpropagation in time is given in Problem 18.12.

The two passes required by the backpropagation through time are summarized below.

- *Forward pass*:
  - Starting at $n = 1$ and using the current estimates of the involved parameter matrices and vectors, compute in sequence,

$$(\boldsymbol{h}_1, \hat{\boldsymbol{y}}_1) \rightarrow (\boldsymbol{h}_2\ \hat{\boldsymbol{y}}_2) \rightarrow \ \dots \ \rightarrow (\boldsymbol{h}_N, \hat{\boldsymbol{y}}_N).$$

- *Backward pass*:
  - Starting at $n = N$, compute in sequence,

$$\frac{\partial J}{\partial \boldsymbol{h}_N} \rightarrow \frac{\partial J}{\partial \boldsymbol{h}_{N-1}} \rightarrow \ \dots \ \rightarrow \frac{\partial J}{\partial \boldsymbol{h}_1}.$$

Note that the computation of the gradient $\frac{\partial J}{\partial \boldsymbol{h}_N}$ is straightforward and it only involves the direct part in Eq. (18.69).

For the implementation of the BPTT, one proceeds by (a) randomly initializing the involved unknown matrices and vectors, (b) computing all the required gradients, following the previously stated two passes, and (c) performing the updates according to the gradient descent scheme. Steps (b) and (c) are performed in an iterative manner until a convergence criterion is met, in analogy to the standard backpropagation Algorithm 18.2.

Vanishing and Exploding Gradients

The task of vanishing and exploding gradients has been introduced and discussed in Section 18.6, in the context of the backpropagation algorithm. The same problems are present in the BPTT algorithm. After all, the latter is a specific form of the backpropagation concept, and, as said, an RNN can be seen as a multilayer network, where each time instant corresponds to a different layer. As a matter of fact, in RNNs, the vanishing/exploding gradient phenomenon appears in a rather "aggressive" way, taking into account that N can get large values.

The multiplicative nature of the propagation of gradients can be readily spotted in Eq. (18.69). To help the reader grasp the main concept, let us simplify the setting and assume that only one state variable is involved. Then the state vectors become scalars, h_n, and the matrix W a scalar w. Furthermore, assume the outputs to be scalars, too. Then the recursion in Eq. (18.69) is simplified as

$$\frac{\partial J}{\partial h_n} = \frac{\partial h_{n+1}}{\partial h_n}\frac{\partial J}{\partial h_{n+1}} + \frac{\partial \hat{y}_n}{\partial h_n}\frac{\partial J}{\partial \hat{y}_n}. \tag{18.70}$$

Assuming in Eq. (18.67) f to be the standard tanh function and using its respective derivative from known tables,[5] it is readily seen that

$$\frac{\partial h_{n+1}}{\partial h_n} = w(1 - h_{n+1}^2),$$

where, by the definition of the tanh function, the magnitude of h_{n+1} is smaller than 1. Writing the recursion for two successive steps, we get

$$\frac{\partial J}{\partial h_n} = w^2(1 - h_{n+1}^2)(1 - h_{n+2}^2)\frac{\partial J}{\partial h_{n+2}} + \text{other terms}.$$

It is not difficult to see that the multiplication of the terms smaller than one can lead to vanishing values, especially if we take into account that, in practice, sequences can be quite large, e.g., $N = 100$. Hence, for time instants close to $n = 1$, the contribution to the gradient of the first term on the right-hand side in Eq. (18.70) will involve a large number of products of numbers less than one in magnitude. On the other hand, the value of w will be contributing in w^n power. So, if its value is larger than one, it can lead to exploding values of the respective gradients (see, e.g., [179]).

In a number of cases, one can *truncate* the backpropagation algorithm to a few time steps. Another way is to replace the tanh nonlinearity with the ReLU one. For the exploding value case, one can introduce a *clipping* technique that clips the values to a predetermined threshold, once values become larger than that.

However, another technique that is usually employed in practice is to replace the previously described standard RNN formulation with an alternative structure, which can cope better with such phenomena that are caused by the long-term dependencies.

Remarks 18.6.

- *Deep RNNs*: Besides the basic RNN network that comprises a single layer of states, extensions have been proposed that involve multiple layers of states, one above the other (see, e.g., [180]).
- *Bidirectional RNNs*: As the name suggests, in the bidirectional RNNs, there are two state variables, i.e., one denoted as $\overrightarrow{\boldsymbol{h}}$, which propagates forward, and another one, $\overleftarrow{\boldsymbol{h}}$, which propagates backwards. In this way, the outputs are left to depend on both the past and the future (see, e.g., [65]).

The Long Short-Term Memory (LSTM) Network

The key idea behind the LSTM network, proposed in the seminal paper [94], is the so-called *cell state*, which helps to overcome the problems associated with the vanishing/exploding phenomena that are caused by the long-term dependencies within the network.

The LSTM networks have the built-in ability to *control* the information flow into and out of the system's *memory* via nonlinear elements known as *gates*. These gates are implemented via the logistic sigmoid nonlinearity and a multiplier. From an algorithmic point of view, the gates are equivalent to applying a *weighting* on the related information flow. The weights lie in the [0,1] range and depend on

[5] Recall, $\frac{d\tanh(x)}{dx} = 1 - \tanh^2(x)$.

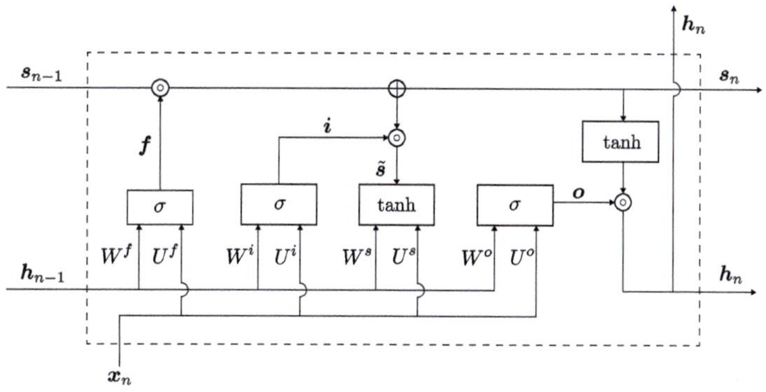

FIGURE 18.35

The LSTM unit. Note that in this case, there are two types of memory-related variables that are propagated, i.e., the cell state vector, $\boldsymbol{s}$, and the hidden variables vector, $\boldsymbol{h}$. The involved bias vectors are not shown to unclutter notation.

the values of the involved variables that activate the sigmoid nonlinearity. In other words, the weighting (control) of information takes place in *context*. According to such a rationale, the network has the agility to forget information that has already been used and is no more needed. The basic LSTM cell/unit is shown in Fig. 18.35. It is built around two sets of variables, stacked in the vector $\boldsymbol{s}$, which is known as the cell or unit state, and the vector $\boldsymbol{h}$, which is known as the hidden variables vector. An LSTM network is built upon the successive *concatenation* of this basic unit. The unit corresponding to time n, besides the input vector, $\boldsymbol{x}_n$, receives $\boldsymbol{s}_{n-1}$ and $\boldsymbol{h}_{n-1}$ from the previous stage and passes $\boldsymbol{s}_n$ and $\boldsymbol{h}_n$ to the next one.

The associated updating equations are summarized below,

$$
\begin{aligned}
\boldsymbol{f} &= \sigma\left(U^f \boldsymbol{x}_n + W^f \boldsymbol{h}_{n-1} + \boldsymbol{b}^f\right), \\
\boldsymbol{i} &= \sigma\left(U^i \boldsymbol{x}_n + W^i \boldsymbol{h}_{n-1} + \boldsymbol{b}^i\right), \\
\tilde{\boldsymbol{s}} &= \tanh\left(U^s \boldsymbol{x}_n + W^s \boldsymbol{h}_{n-1} + \boldsymbol{b}^s\right), \\
\boldsymbol{o} &= \sigma\left(U^o \boldsymbol{x}_n + W^o \boldsymbol{h}_{n-1} + \boldsymbol{b}^o\right), \\
\boldsymbol{s}_n &= \boldsymbol{s}_{n-1} \circ \boldsymbol{f} + \boldsymbol{i} \circ \tilde{\boldsymbol{s}}, \\
\boldsymbol{h}_n &= \boldsymbol{o} \circ \tanh(\boldsymbol{s}_n),
\end{aligned}
$$

where $\circ$ denotes element-wise product between vectors or matrices (Hadamard product), that is, $(\boldsymbol{s} \circ \boldsymbol{f})_i = s_i f_i$, and σ denotes the logistic sigmoid function.

Observe that the cell state, $\boldsymbol{s}$, passes direct information from the previous instant to the next one. This information is first controlled by the first gate, according to the elements in $\boldsymbol{f}$, which take values in the range [0,1], depending on the current input and the hidden variables that are received from the previous stage. This is what we said before, i.e., that the weighting is adjusted in "context." Next, new information, i.e., $\tilde{\boldsymbol{s}}$, is added to $\boldsymbol{s}_{n-1}$, which is also controlled by the second sigmoid gate network (i.e., $\boldsymbol{i}$). Thus, there is a guarantee that information from the past is forwarded to the future in a direct way, which helps the network to memorize information. It turns out that this type of memory exploits better the long-range dependencies in the data, compared to the basic RNN structure. The hidden variables vector $\boldsymbol{h}$ is controlled by both the cell state and the current values of the input and the previous state variables. All the involved matrices and vectors are learned via the training phase. Note that there are two lines associated with $\boldsymbol{h}_n$. The one to the right leads to the next stage and the one on the top is used to provide the output, $\hat{\boldsymbol{y}}_n$, at time n, via, say, the softmax nonlinearity, as in the standard RNNs in Eq. (18.68).

Besides the previously discussed LSTM structure, a number of variants have been proposed. An extensive comparative study among different LSTM and RNN architectures can be found in, e.g., [68,113].

RNNs and LSTMs have been successfully used in a wide range of applications, such as language modeling (e.g., [222]), machine translation (e.g., [147]), speech recognition (e.g., [66]), machine vision for the generation of image descriptors (e.g., [115]), and in fMRI data analysis in order to grasp the time dynamics in the associated brain networks (e.g., [207]). For example, in language processing, the input is typically a sequence of words, which are encoded as numbers (these are pointers to the available dictionary). The output is the sequence of words to be predicted. During training, one sets $\boldsymbol{y}_n = \boldsymbol{x}_{n+1}$. That is, the network is trained as a nonlinear predictor.

18.13.2 ATTENTION AND MEMORY

The use of *attention* schemes in neural networks has a rather long history (see, e.g., [41]). As the name suggests, the concept of "attention" draws on the idea of the attention mechanism found in humans. For example, our vision system provides us with the ability to focus more on the most important information that resides in a scene; this information is in *context*; that is, it depends on what we have in mind to look at. In machine learning, a number of different models have been proposed on how one can implement the attention concept.

One of the most popular paths is to apply a type of weighting (transformation) on various variables on which the output depends. These weights are *learned* during training. Let us take as an example an RNN. In its basic form, as has been discussed before, the output, $\hat{\boldsymbol{y}}_n$, depends on the corresponding state vector, $\boldsymbol{h}_n$. However, although the state vector encodes/summarizes the system's memory up to the most recent time instant, n, this may not necessarily be the most important information that is needed in certain tasks. As a matter of fact, it is rather unreasonable to assume that in a long input sequence in a machine language translation system, for example, the most recent state vector is the most representative information to get a reliable output. A typical case to demonstrate the previous statement is when one translates from Japanese to English, where the last word of a Japanese sentence could be highly predictive of the first word in its English translation. Similarly in life, what action we decide to take at a specific response heavily depends on our total previous experience. Yet, some specific experiences in the past may have a much stronger influence than the most recent ones.

To deal with such cases, one can employ an attention mechanism so that the output, at time n, depends on a *weighted combination* of all the previous, to time n, state vectors and leave to the system to learn the values of the weights during the learning phase. So, during training, it will be decided what is the most important piece of information which the output should be based on. That is, the system learns to "attend" to the most important contextual information.

For example, the output vector may be modified to depend on all the previously computed state vectors, i.e.,

$$\hat{y}_n = f\left(\sum_{i=1}^{n} \alpha_{ni} \boldsymbol{h}_i + \boldsymbol{c}\right),$$

where a_{ni} are the corresponding weights at time n. The above idea of combining all previous state vectors has been employed, in a somewhat different formulation, in the machine translation system that is described in [8] (see also Section 18.18).

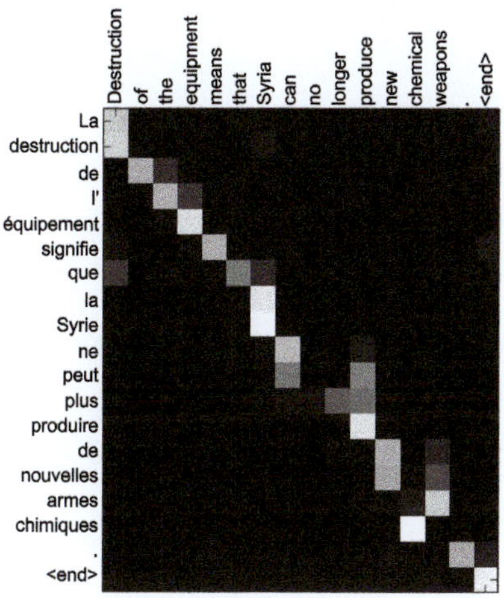

FIGURE 18.36

The values of the attention weights in grayscale, showing the degree of dependence of the words in the output sequence (English) to those of the input sequence (French) (from [8]). Note, for example, that in order to produce the word "Syria," the network "attends" the words "La Syrie."

Fig. 18.36, taken from the previous reference, illustrates the rationale of employing a weighting mechanism. The French words present the input and the English words the corresponding output sequences. The values of the corresponding attention weights are visualized as pixels; the larger the weight, the whiter the pixel. Note, for example, that the output word "produce" is the result of weighting information from three successive time instants, associated with the words "peut plus produire," and the word "destruction" with two words, i.e., "la destruction."

In [253], a network with attention mechanism is used for the automatic generation of image description. In the described system, a variant of a CNN network is employed to generate a set of feature vectors. Each feature vector corresponds to a *portion* of the original image. This step is equivalent to an *encoding* phase of the original image. These vectors are in turn used to form the input sequence to an LSTM network, where attention weights have been employed. The network is trained to compute the output word probability given the LSTM state. Fig. 18.37, taken from [253], shows the part of the image that the model "attends" while generating a word.

An interesting aspect of integrating an attention mechanism within the model is that one can follow what the model does and how the output information is formed; this can be useful when the issue of

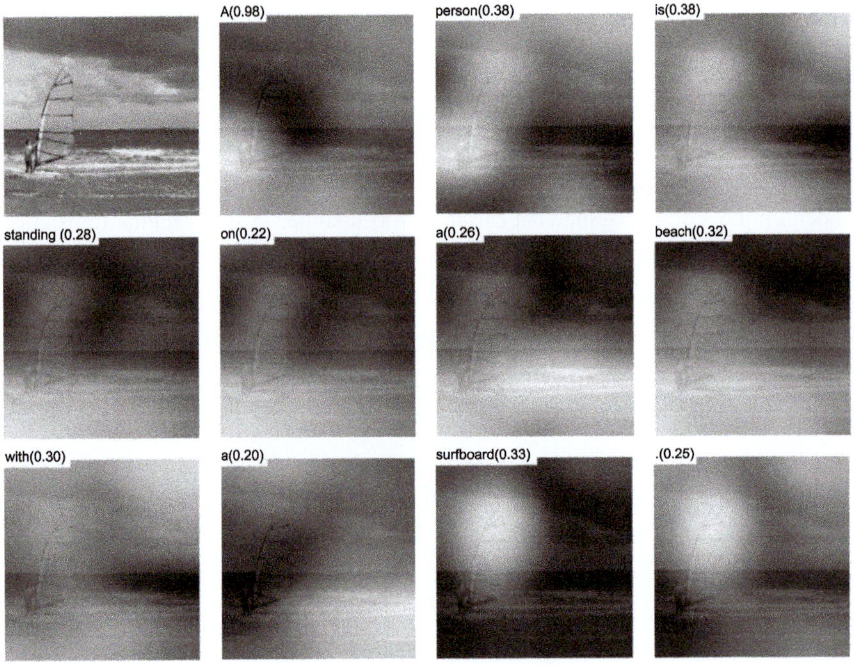

FIGURE 18.37

The original image is shown at the top left. The resulting image description is "A person is standing on the beach with a surfboard." The use of attention weights highlights the corresponding part of the image, where the output word more heavily depends on. As an example, observe that while the output produces the word "surfboard," the attention weights that are associated with the pixels within the corresponding region in the image get the largest values (from [253]).

interpretability of the network becomes important. That is, to understand "why" and "how" the network makes a decision. For a related discussion, see, e.g., [206].

Different paths to attention are also possible. In [67], the so-called *neural Turing machine* (NTM) is proposed, where a memory module is applied in parallel with the neural network (feed-forward or LSTM). A *learnable* attention mechanism is used to read and write to the memory selectively. An extension of the NTM that involves reinforcement learning techniques is given in [259]. In [221,251], a memory generated by the input data is allowed to be read multiple times before producing an output. This is in analogy to making multiple reasoning steps based on the content of the memory; that is, based on the input "story."

Remarks 18.7.

- *Reservoir computing*: The term reservoir computing refers mainly to two closely related families of recurrent networks that have been independently proposed, i.e., the *echo state networks* (ESNs) [106] and the *liquid state machines* [149]. The latter implements spiking neurons instead of continuous-valued neurons.

The main idea behind the original ESN is to train only the parameters associated with the output neurons. The weights associated with the input and the state vectors are generated *randomly*, following certain rules. The untrained part is called the *reservoir* and the resulting states *echoes*. The rationale springs forth from the idea that if a random RNN possesses certain properties, training the output parameters suffices. The main property that the reservoir should possess is the so-called *echo state property*. This is basically a stability condition related to the dynamics of the network [144]. A nonparametric Bayesian formulation is discussed in [29]; a prior distribution is imposed over the output weights, which are in turn marginalized out in the context of prediction generation, given the training data.

18.14 ADVERSARIAL EXAMPLES

At the moment this edition is compiled, deep neural networks are state of the art in achieving performance and accuracies that are often comparable to, and sometimes better than, those achieved by humans. However, it seems that we are not yet in a position to claim that these models truly "understand" the task they have "learned" to perform; this is in spite of the fact that, for example, they can predict correct labels in classification tasks with very high probability. In [227], it is demonstrated that one can construct *adversarial* examples that *consistently* fool machine learning models. The term "adversarial" means that one can intentionally impose *small worst-case* perturbations on patterns in the input set, which will result in wrong label prediction with high probability. The most interesting issue is that adding this small noise perturbation is hardly perceptible to the human eye, in case of images (e.g., [227]), and to the human ear, in case of music (e.g., [118,220]). Fig. 18.38 (taken from [227]) shows nine images in total. A neural network (AlexNet) has been trained to recognize the content of images. All three images on the left, taken from the respective test set, were recognized correctly. The images in the middle are noise images that are added to the corresponding ones on the left. The resulting images are shown on the right. No human has any difficulty to predict the correct label. Yet, AlexNet classified all three images to the class "ostrich, struthio camelus"!

There are various ways to generate adversarial examples. In [227], an optimization task is employed that finds the minimum perturbation that can lead to a change of a label. In [63], the perturbation is performed in the direction of the sign of the gradient of the cost function with respect to the input pattern. In [164], a method is proposed to construct a *universal* small perturbation that can cause all images in a data set to be misclassified with high probability.

It seems that at the heart of this "strange" behavior lies the high dimensionality of the input space. In general, one expects that in a learning task, the *smoothness* assumption is valid. That is, for small enough positive ϵ and an input pattern $\boldsymbol{x}$, we would expect that for any $\boldsymbol{v} : ||\boldsymbol{v}|| \leq \epsilon$, the pattern $\boldsymbol{x}' := \boldsymbol{x} + \boldsymbol{v}$ is assigned in the same class as $\boldsymbol{x}$, with high probability. The effect of the high dimensionality on the smoothness condition can easily be seen for the case of a linear classifier (see, e.g., [63]). Let the trained classifier be described in terms of its parameters, $\boldsymbol{\theta}$. Given an input pattern, $\boldsymbol{x}$, the label is computed according to the sign of the inner product, $\boldsymbol{\theta}^T\boldsymbol{x}$. For the case of $\boldsymbol{x}'$, the inner product is given by $\boldsymbol{\theta}^T\boldsymbol{x}' = \boldsymbol{\theta}^T\boldsymbol{x} + \boldsymbol{\theta}^T\boldsymbol{v}$. Let us now intentionally set $\boldsymbol{v} = \pm\epsilon\,\text{sgn}(\boldsymbol{\theta})$, where the sign operation acts in an element-wise fashion. Then it turns out that

FIGURE 18.38

The images on the left have been classified correctly. All the images on the right have been classified as "ostrich, Struthio camelus"! The images in the center show the noise (after some magnification) added to obtain the images on the right (example taken from [227]).

$$\boldsymbol{\theta}^T \boldsymbol{x}' = \boldsymbol{\theta}^T \boldsymbol{x} + \boldsymbol{\theta}^T \boldsymbol{v} = \boldsymbol{\theta}^T \boldsymbol{x} \pm \epsilon \sum_{i=1}^{l} |\theta_i|.$$

Hence, if the input dimensionality l is large, large deviations are expected between the values of the respective inner products and this can lead to different predicted labels for $\boldsymbol{x}$ and $\boldsymbol{x}'$. In words, the combination of linearity with high dimensionality violates the smoothness assumption.

The above explanation can be extended to the deep networks, when, for example, ReLU is employed or when the involved nonlinearities operate in their linear region. An alternative geometric viewpoint in explaining the adversarial phenomenon is provided in [49,164]. There, it is pointed out that at the heart of the adversarial examples lie some distinct geometric properties that are associated with the imposed perturbation in relation to geometric correlations between different parts of the decision boundary. Their analysis demonstrates the difference that exists between a general type of random noise and a worst-case adversarial type of perturbations.

The question that comes in mind is whether a more careful sampling of the input space would result in a richer representation, where adversarial examples could be included in the training set and hence the network can learn them. As claimed in [227], the set of adversarial negatives is of extremely low probability, and thus is never (or rarely) observed in the data set, yet it is dense (much like the rational numbers); so, it can be found near to virtually every test case. It should be noted, however, that such a statement is not substantiated by a theoretical proof.

ADVERSARIAL TRAINING

Having constructed adversarial examples and making some attempts to understand their existence, the next front is focused more on the practical aspects of the phenomenon. Although it seems that adversarial examples are far from common in the input (training and test sets) data, it is a rather disturbing phenomenon. Moreover, it can always be used to fool a network *intentionally*. To this end, a number of techniques have already appeared in an attempt to "robustify" the networks against adversaries.

In [227], adversarial examples were generated and fed back to the training set. This is a type of regularization via data generation. However, in [118], it is argued that in the case of music data the method did not really improve performance.

In [63], the loss function J is modified appropriately as

$$J'(\boldsymbol{\theta}, \boldsymbol{x}, \boldsymbol{y}) = \alpha J(\boldsymbol{\theta}, \boldsymbol{x}, \boldsymbol{y}) + (1-\alpha) J(\boldsymbol{\theta}, \boldsymbol{x} + \Delta\boldsymbol{x}, \boldsymbol{y}), \ 0 < \alpha < 1,$$

where

$$\Delta\boldsymbol{x} = \epsilon \operatorname{sgn}\left(\frac{\partial}{\partial \boldsymbol{x}} J(\boldsymbol{\theta}, \boldsymbol{x}, \boldsymbol{y})\right), \quad \epsilon > 0,$$

is shown to be a direction for adversarial perturbation.

In [161], a regularizer is used to promote smoothness of the model distributions with respect to the input, around every input data point. In [209], a robust optimization method is proposed, which is built around a minmax formulation, where the cost function is optimized with respect to a worst-case realization of a perturbation. In [175], the *distillation* technique is proposed as a way to cope with adversarial examples. Distillation is a training procedure initially designed to train a deep neural network in the context of transfer learning (see Section 18.17), e.g., [92]. In the adversarial training framework, it is claimed that distillation can reduce the gradients that lead to adversarial sample creation by many orders of magnitude. Moreover, distillation can significantly increase the average minimum number of features that need to be modified to create adversarial samples.

As a final touch, it must be said that at the time the current edition is being compiled, adversarial examples constitute an ongoing hot topic of research. Adversarial examples have the potential to be dangerous. Consider, for example, an attacker who targets autonomous vehicles by using stickers or paint to create an adversarial "stop" sign that the vehicle would interpret as a "yield" or another sign (see, e.g., [177]). In [127], it is demonstrated that feeding adversarial images obtained from a cell phone camera to an ImageNet inception classifier, a large fraction of adversarial examples were misclassified, even when perceived through the camera. In [176], a threat model is proposed where attacks and defenses within an adversarial framework are categorized. It is shown that there are (possibly unavoidable) tensions between model complexity, accuracy, and resilience that must be calibrated for the environments in which they will be used.

In [208], it is discussed that in spite of a wide range of defenses that have been proposed to robustify neural networks against adversarial attacks, it seems that such defenses are quickly broken. Moreover, the theoretical analysis in the paper indicates that for certain classes of problems, adversarial examples are inescapable.

18.15 DEEP GENERATIVE MODELS

So far, our emphasis was on supervised learning via neural network architectures. The other main direction in machine learning is that of unsupervised learning, where one has to "learn" and unveil the underlying dependencies and structure that are hidden in the input data. Clustering, as introduced in Chapter 12 (see also [231] for an extensive coverage) is a major path of learning via unlabeled data. Another one is to learn explicit probabilistic dependencies with the final goal being that of estimating probability distributions, as was the focus in Chapters 15 and 16.

Another direction, where the use of unlabeled samples is of importance, is that of learning an efficient representation of the input data. As a matter of fact, feature learning is a facet of data representation. The convolutional layers in a CNN, which precede the fully connected network part, are dedicated to this goal; that is, to obtain an efficient and information-rich representation of the input data. However, in the framework of CNNs, this takes place in the context of a specific learning task; that is, the output labels are also considered during training to reach an optimal representation that serves the needs of the task at hand.

In this section, the discussion setting is to learn representations that are independent of a specific target task; the goal is to extract such information using the input data only. The reason for such a focus is twofold. First, learning a model representation of the input data can be used subsequently in different tasks in order to facilitate the training. Sometimes, this is also known as *pretraining*, where parameters learned using unlabeled data can be used as initial estimates of the parameters for another supervised learning. This can be useful when the number of labeled examples is not large enough (see, e.g., [148] for a discussion). It is worth pointing out that such a pretraining rationale is of a historical importance, because it led to the revival of neural networks, as will be discussed soon [86].

Another path, which is currently of high interest, is to exploit such learned representations in order to *generate* new data; recall the discussion in Section 18.7. Such techniques may *not* necessarily learn the underlying probability distribution *explicitly*, yet they acquire the necessary knowledge needed to be able to draw samples according to the distribution that "explains" the data.

18.15.1 RESTRICTED BOLTZMANN MACHINES

A *restricted Boltzmann machine* (RBM) is a special type of the more general class of Boltzmann machines, which were introduced in Chapter 15 [1,217]. Fig. 18.39 shows the probabilistic graphical model corresponding to an RBM. There are no connections among nodes of the same layer. Moreover, the upper level comprises nodes corresponding to hidden variables and the lower level consists of visible ones. That is, observations are applied to the nodes of the lower layer only. Deep RBMs can be constructed by stacking one on top of the other.

Following the general definition of a Boltzmann machine, the joint distribution of the involved random variables is of the form

$$P(v_1, \ldots, v_J, h_1, \ldots, h_I) = \frac{1}{Z} \exp\big(-E(\boldsymbol{v}, \boldsymbol{h})\big), \tag{18.71}$$

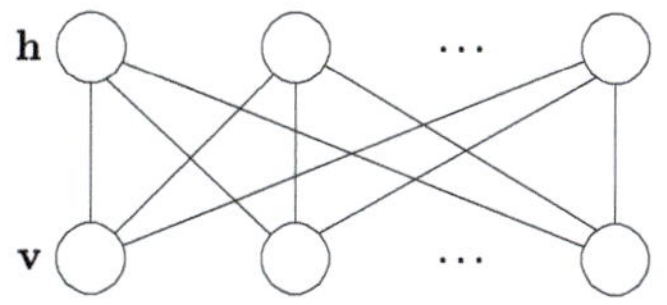

FIGURE 18.39

An RBM is an undirected graphical model with no connections among nodes of the same layer. The lower level comprises visible nodes and the upper layer consists of hidden nodes only.

where we have used different symbols for the J visible ($\mathrm{v}_j,\ j=1,2,\dots J$) and the I hidden variables ($\mathrm{h}_i,\ i=1,2,\dots,I$). The *energy* is defined in terms of a set of unknown parameters,[6] that is,

$$E(\boldsymbol{v},\boldsymbol{h}) = -\sum_{i=1}^{I}\sum_{j=1}^{J}\theta_{ij}h_i v_j - \sum_{i=1}^{I} b_i h_i - \sum_{j=1}^{J} c_j v_j. \tag{18.72}$$

The normalizing constant is obtained as

$$Z = \sum_{\boldsymbol{v}}\sum_{\boldsymbol{h}} \exp\big(-E(\boldsymbol{v},\boldsymbol{h})\big). \tag{18.73}$$

We will focus on discrete variables; hence the involved distributions are probabilities. More specifically, we will focus on variables of a binary nature, that is, $\mathrm{v}_j,\ \mathrm{h}_i \in \{0,1\},\ j=1,\dots,J,\ i=1,\dots,I$. Observe from Eq. (18.72) that, in contrast to a general Boltzmann machine, only products between hidden and visible variables are present in the energy term.

The goal in training an RBM is to learn the set of unknown parameters, $\theta_{ij},\ b_i,\ c_j$, which will be collectively denoted as Θ, $\boldsymbol{b}$, and $\boldsymbol{c}$, respectively. A major path to this end is to maximize the log-likelihood, using N observations of the visible variables, denoted as $\boldsymbol{v}_n,\ n=1,2,\dots,N$, where

$$\boldsymbol{v}_n := [v_{1n},\dots,v_{Jn}]^T$$

is the vector of the corresponding observations at time n. We will say that the visible nodes are *clamped* on the respective observations. The corresponding (average) log-likelihood is given by

$$\begin{aligned} L(\Theta,\boldsymbol{b},\boldsymbol{c}) &= \frac{1}{N}\sum_{n=1}^{N}\ln P(\boldsymbol{v}_n;\Theta,\boldsymbol{b},\boldsymbol{c}) \\ &= \frac{1}{N}\sum_{n=1}^{N}\ln\Big(\frac{1}{Z}\sum_{\boldsymbol{h}}\exp\big(-E(\boldsymbol{v}_n,\boldsymbol{h};\Theta,\boldsymbol{b},\boldsymbol{c})\big)\Big) \end{aligned}$$

[6] Compared to the notation used in Section 15.4.2 we use a negative sign. This is only to suit better the needs of the section, and it is obviously of no importance for the derivations.

$$
\begin{aligned}
&= \frac{1}{N}\sum_{n=1}^{N} \ln\Big(\sum_{h} \exp\big(-E(\boldsymbol{v}_n,\boldsymbol{h};\Theta,\boldsymbol{b},\boldsymbol{c})\big)\Big) \\
&\quad - \ln \sum_{\boldsymbol{v}}\sum_{\boldsymbol{h}} \exp\big(-E(\boldsymbol{v},\boldsymbol{h})\big),
\end{aligned}
$$

where the index n in the energy refers to the respective observations onto which the visible nodes have been clamped, and Θ has explicitly been brought into the notation.

Taking the derivative of $L(\Theta, \boldsymbol{b}, \boldsymbol{c})$ with respect to θ_{ij} (similar is the case of the derivatives with respect to b_i and c_j) and applying standard properties of derivatives, it is not difficult to show (Problem 18.13) that

$$
\boxed{\frac{\partial L(\Theta,\boldsymbol{b},\boldsymbol{c})}{\partial \theta_{ij}} = \frac{1}{N}\sum_{n=1}^{N}\left(\sum_{\boldsymbol{h}} P(\boldsymbol{h}|\boldsymbol{v}_n) h_i v_{jn}\right) - \sum_{\boldsymbol{v}}\sum_{\boldsymbol{h}} P(\boldsymbol{v},\boldsymbol{h}) h_i v_j,} \tag{18.74}
$$

where the following has been used:

$$
P(\boldsymbol{h}|\boldsymbol{v}) = \frac{P(\boldsymbol{v},\boldsymbol{h})}{\sum_{\boldsymbol{h}'} P(\boldsymbol{v},\boldsymbol{h}')}.
$$

The gradient in (18.74) involves two terms. The first one can be computed once $P(\boldsymbol{h}|\boldsymbol{v})$ is available. Basically, this term is the mean *firing rate* or *correlation* when the RBM is operating in its clamped phase; often, we refer to it as the *positive* phase, and the term is denoted as $< \mathrm{h}_i \mathrm{v}_j >^+$. The second term is the corresponding correlation when the RBM is working in its so-called *free running* or *negative* phase, and it is denoted as $< \mathrm{h}_i \mathrm{v}_j >^-$. Thus, a gradient ascent scheme for maximizing the log-likelihood will be of the form

$$
\theta_{ij}(\text{new}) = \theta_{ij}(\text{old}) + \mu\left(< \mathrm{h}_i \mathrm{v}_j >^+ - < \mathrm{h}_i \mathrm{v}_j >^-\right).
$$

Let us take a minute to justify why we have named the two phases of operation as positive and negative, respectively. These terms appear in the seminal papers on Boltzmann machines by Hinton and Senjowski [81,83]. The first one, corresponding to the clamped condition, can be thought of as a form of a *Hebbian* learning rule. Hebb was a neurobiologist and stated the first ever (to the best of my knowledge) learning rule [78]: "If two neurons on either side of a synapse are activated simultaneously, the strength of this synapse is selectively increased." Note that this is exactly the effect of the positive phase correlation in the parameter's update recursion. On the contrary, the effect of the negative phase correlation term is the opposite. Thus, the latter term can be thought of as a *forgetting* or *unlearning* contribution; it can be considered as a control condition of a purely "internal" nature (note that it does not depend on the observations), compared to the "external" information received from the environment (observations).

Details concerning the optimization of the log-likelihood as well as the finally derived algorithm can be obtained from the site of the book under the Additional Material part that is associated with the current chapter.

18.15.2 PRETRAINING DEEP FEED-FORWARD NETWORKS

This subsection presents the rationale of employing RMBs to pretrain a deep forward neural network. This concept was first presented in [86]. Although pretraining is no more widely used, it can still offer advantages in cases where labeled data are not enough to train a big neural network.

Fig. 18.40 presents a block diagram of a deep neural network with three hidden layers. The vector of the input random variables is denoted as $\mathbf{x}$ and those associated with the hidden ones as $\mathbf{h}^i$, $i = 1, 2, 3$. The vector of the output nodes is denoted as $\mathbf{y}$. Pretraining evolves in a sequential fashion, starting from the weights connecting the input nodes to the nodes of the first hidden layer. This can be achieved by maximizing the likelihood of the observed samples of the input observations, $\boldsymbol{x}$, and treating the variables associated with the first layer as hidden ones. Once the weights corresponding to the first layer have been computed, the respective nodes are allowed to fire an output value and a vector of values, $\boldsymbol{h}^1$, is formed. This is the reason that a generative model for the unsupervised pretraining is adopted (such as the RBM), to be able to generate *in a probabilistic way* outputs at the hidden nodes. These values are in turn used as observations for the pretraining of the next hidden layer, and so on.

Once pretraining has been completed, a supervised learning rule, such as backpropagation, is then employed to obtain the values of weights leading to the output nodes, as well as to fine-tune the weights associated with the hidden layers, using as initial weight values those obtained during the pretraining phase.

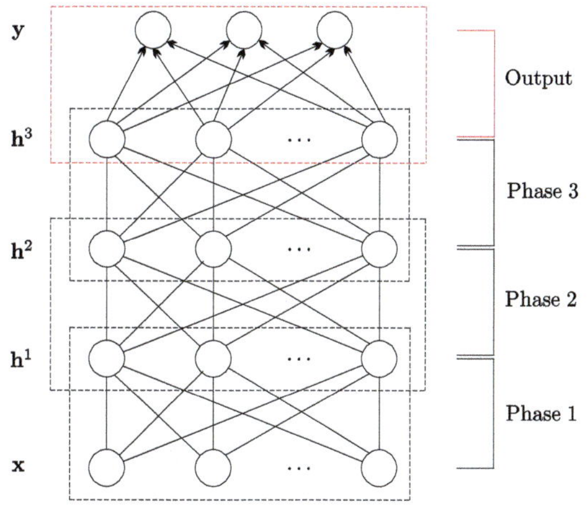

FIGURE 18.40

Block diagram of a deep neural network architecture, with three hidden layers and one output layer. The vector of the input random variables at the input layer is denoted as $\mathbf{x}$. The vector of the variables associated with the nodes of the ith hidden layer is denoted as $\mathbf{h}^i$, $i = 1, 2, 3$. The output variables are denoted as $\mathbf{y}$. At each phase of the pretraining, the weights associated with one hidden layer are computed, one at a time. For the network of the figure, comprising three hidden layers, pretraining consists of three stages of unsupervised learning. Once pretraining of the hidden units has been completed, the weights associated with the output nodes are pretrained via a supervised learning algorithm. During the final fine-tuning, all the parameters are estimated via a supervised learning rule, such as the backpropagation scheme, using as initial values those obtained during pretraining.

As already said before, unsupervised learning is a way to discover and unveil information hidden in the data, by learning the underlying regularities and the statistical structure of the data. In this way, pretraining can be thought of as a data dependent *regularizer* that pushes the unknown parameters to regions where good solutions exist, by exploiting the extra information acquired by the unsupervised learning (see, for example, [47]).

Fig. 18.40 illustrates a multilayer perceptron with three hidden layers. As is always the case with any supervised learning task, the kick-off point is a set of training examples, $(y_n, \boldsymbol{x}_n),\ n = 1, 2, \ldots, N$. Training a deep multilayer perceptron, employing what we have said before, involves two major phases: (a) pretraining and (b) supervised fine-tuning. Pretraining the weights associated with hidden nodes involves unsupervised learning via the RBM rationale. Assuming K hidden layers, $\mathbf{h}^k,\ k = 1, 2, \ldots, K$, we look at them in pairs, that is, $(\mathbf{h}^{k-1}, \boldsymbol{h}^k),\ k = 1, 2, \ldots, K$, with $\mathbf{h}^0 := \mathbf{x}$ being the input layer. Each pair will be treated as an RBM, in a hierarchical manner, with the outputs of the previous one becoming the inputs to the next. It can be shown (for example, [86]) that adding a new layer each time increases a variational lower bound on the log-probability of the training data.

Pretraining of the weights leading to the output nodes is performed via a supervised learning algorithm. The last hidden layer together with the output layer is not treated as an RBM, but as a one-layer feed-forward network. In other words, the input to this *supervised* learning task are the features formed in the last hidden layer.

Finally, fine-tuning involves retraining in a typical backpropagation algorithm rationale, using the values obtained during pretraining for initialization. This is very important for getting a better feeling and understanding of how deep learning works. The label information is used in the hidden layers *only* at the fine-tuning stage. During pretraining, the feature values in each layer grasp information related to the input distribution and the underlying regularities. The label information does not participate in the process of discovering the features. Most of this part is left to the unsupervised phase, during pretraining. Note that this type of learning can also work even if some of the data are unlabeled. Unlabeled information is useful, because it provides valuable extra information concerning the input data. As a matter of fact, this is at the heart of semisupervised learning (see, e.g., [231]).

More details on RBM-based pretraining can be found on the site of the book under the Additional Material part associated with the current chapter (see also [88]).

18.15.3 DEEP BELIEF NETWORKS

The emphasis given in this chapter, so far, was on discussing feed-forward architectures. Our focus was on the information flow in the feed-forward or bottom-up direction. However, this is only part of the whole story. The other part concerns training *generative* models. The goal of such learning tasks is to "teach" the model to generate data. One way to achieve this is via learning probabilistic models that relate a set of variables, which can be observed, with another set of hidden ones. RBMs are just an instance of such models. Moreover, it has to be emphasized that RBMs can represent any discrete distribution if enough hidden units are used [52,137].

In our discussion up to now in this section, we viewed a deep network as a mechanism forming layer-by-layer features of features, that is, more and more higher-level representations of the input data. The issue now becomes whether one can start from the last layer, corresponding to the highest level representation, and follow a *top-down* path with the new goal of generating data. Besides the need in some practical applications, there is an additional reason to look at this reverse direction of information flow.

Some studies suggest that such top-down connections exist in our visual system to generate lower-level features of images starting from higher-level representations. Such a mechanism can explain the creation of vivid imagery during dreaming, as well as the disambiguating effect on the interpretation of local image regions by providing contextual prior information from previous frames (for example, [134,135,165]).

A popular way to represent statistical generative models is via the use of probabilistic graphical models, which were treated in Chapters 15 and 16. A typical example of a generative model is that of sigmoidal networks, introduced in Section 15.3.4, which belong to the family of parametric Bayesian (belief) networks. A sigmoidal network is illustrated in Fig. 18.41A, where a directed acyclic graph (Bayesian) is shown. Following the theory developed in Chapter 15, the joint probability of the observed ($\boldsymbol{x}$) and hidden variables, distributed in K layers, is given by

$$P(\boldsymbol{x}, \boldsymbol{h}^1, \ldots, \boldsymbol{h}^K) = P(\boldsymbol{x}|\boldsymbol{h}^1) \left(\prod_{k=1}^{K-1} P\left(\boldsymbol{h}^k|\boldsymbol{h}^{k+1}\right) \right) P(\boldsymbol{h}^K),$$

where the conditionals for each one of the I_k nodes of the kth layer are defined as

$$P(h_i^k|\boldsymbol{h}^{k+1}) = \sigma\left(\sum_{j=1}^{I_{k+1}} \theta_{ij}^{k+1} h_j^{k+1} \right), \quad k = 1, 2, \ldots, K-1, \; i = 1, 2, \ldots, I_k.$$

A variant of the sigmoidal network was proposed in [86], which has become known as *deep belief network*. The difference with a sigmoidal one is that the top two layers comprise an RBM. Thus, it is a mixed type of network consisting of both directed and undirected edges. The corresponding graphical model is shown in Fig. 18.41B. The respective joint probability of all the involved variables is given

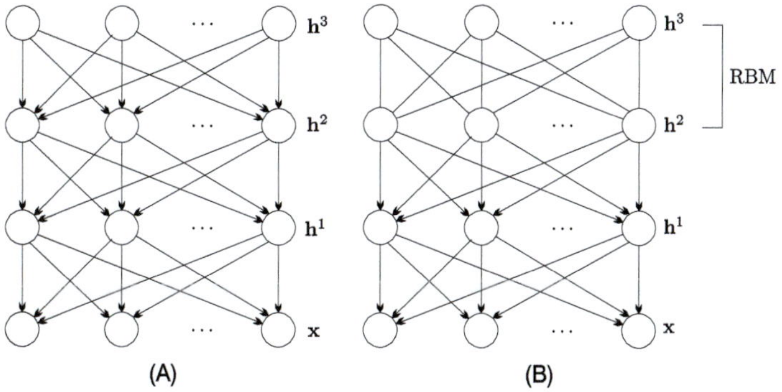

FIGURE 18.41

(A) A graphical model corresponding to a sigmoidal belief (Bayesian) network. (B) A graphical model corresponding to a deep belief network. It is a mixture of directed and undirected edges connecting nodes. The top layer involves undirected connections and it corresponds to an RBM.

by

$$P(\boldsymbol{x}, \boldsymbol{h}^1, \ldots, \boldsymbol{h}^K) = P(\boldsymbol{x}|\boldsymbol{h}^1) \left(\prod_{k=1}^{K-2} P\left(\boldsymbol{h}^k|\boldsymbol{h}^{k+1}\right) \right) P\left(\boldsymbol{h}^{K-1}, \boldsymbol{h}^K\right). \tag{18.75}$$

It is known that learning Bayesian networks of relatively large size is intractable, because of the presence of converging edges (see Section 15.3.3). To this end, variational approximation methods can be mobilized to bypass this obstacle (see Section 16.3).

In [86], an alternative path was proposed following a scheme similar to that used before for pretraining neural networks. In other words, all hidden layers, starting from the input one, are treated as RBMs, and a greedy layer-by-layer pretraining bottom-up philosophy is adopted. It should be emphasized that the conditionals, which are recovered by such a scheme, can only be thought of as approximations of the true ones. After all, the original graph is a directed one and not undirected, as the RBM assumption imposes. The only exception lies at the top level, where the RBM assumption is a valid one.

Once the bottom-up pass has been completed, the estimated values of the unknown parameters are used for initializing another fine-tuning training algorithm; such a scheme has been developed in [85] for training sigmoidal networks and is known as *wake-sleep* algorithm. The objective behind the wake-sleep scheme is to adjust the weights during the top-down pass, so as to maximize the probability of the network to generate the observed data. The scheme has a variational approximation flavor, and if initialized randomly takes a long time to converge. However, using the values obtained from the pretraining for initialization, the process can significantly be sped up [89].

Once training of the weights has been completed, data generation is achieved by the scheme summarized in Algorithm 18.4.

Algorithm 18.4 (Generating samples via a DBN).

- Obtain samples $\boldsymbol{h}^{K-1}$, for the nodes at level $K-1$. This can be done via running a Gibbs chain, by alternating samples, $\boldsymbol{h}^K \sim P(\boldsymbol{h}|\boldsymbol{h}^{K-1})$ and $\boldsymbol{h}^{K-1} \sim P(\boldsymbol{h}|\boldsymbol{h}^K)$. This can be carried out in analogy to the technique used to train RBMs (see the Additional Material for the current chapter on the book's site), as the top two layers comprise an RBM. The convergence of the Gibbs chain can be sped up by initializing the chain with a feature vector formed at the $(K-1)$th layer by one of the input patterns; this can be done by following a bottom-up pass to generate features in the hidden layers, as the one used during pretraining.
- **For** $k = K-2, \ldots, 1$, **Do**; Top-down pass.
  - **For** $i = 1, 2, \ldots, I_k$, **Do**
    - $h_i^{k-1} \sim P\left(h_i|\boldsymbol{h}^k\right)$; Sample for each one of the nodes.
  - **End For**
- **End For**
- $\boldsymbol{x} = \boldsymbol{h}^0$; Generated pattern.

18.15.4 AUTOENCODERS

Autoencoders have been proposed in [9,199] as methods for dimensionality reduction. An autoencoder consists of two parts, the *encoder* and the *decoder*. The output of the encoder is the reduced representation of the input pattern, and it is defined in terms of a vector function,

$$f : \boldsymbol{x} \in \mathbb{R}^l \longmapsto \boldsymbol{h} \in \mathbb{R}^m, \tag{18.76}$$

where

$$h_i := f_i(\boldsymbol{x}) = \phi_e(\boldsymbol{\theta}_i^T \boldsymbol{x} + b_i), \quad i = 1, 2, \ldots, m, \tag{18.77}$$

with ϕ_e being the activation function; the latter is usually taken to be the logistic sigmoid function, $\phi_e(\cdot) = \sigma(\cdot)$. In other words, the encoder is a single hidden layer feed-forward neural network.

The decoder is another function $\boldsymbol{g}$,

$$\boldsymbol{g} : \boldsymbol{h} \in \mathbb{R}^m \longmapsto \hat{\boldsymbol{x}} \in \mathbb{R}^l, \tag{18.78}$$

where

$$\hat{x}_j = g_j(\boldsymbol{h}) = \phi_d(\boldsymbol{\theta}'^T_j \boldsymbol{h} + b'_j), \; j = 1, 2, \ldots, l. \tag{18.79}$$

The activation ϕ_d is, usually, taken to be either the identity (linear reconstruction) or the logistic sigmoid one. The task of training is to estimate the parameters

$$\Theta := [\boldsymbol{\theta}_1, \ldots, \boldsymbol{\theta}_m,], \; \boldsymbol{b}, \; \Theta' := [\boldsymbol{\theta}'_1, \ldots, \boldsymbol{\theta}'_l], \; \boldsymbol{b}'.$$

It is common to assume that $\Theta' = \Theta^T$. The parameters are estimated so the reconstruction error, $\boldsymbol{e} = \boldsymbol{x} - \hat{\boldsymbol{x}}$, over the available input samples is minimum in some sense. Usually, the least-squares cost is employed, but other choices are also possible. Regularized versions, involving a norm of the parameters, is also a possibility (for example, [193]). If the activation ϕ_e is chosen to be the identity (linear representation) and $m < l$ (to avoid triviality), the autoencoder is equivalent to the PCA technique [9]. PCA is treated in more detail in Chapter 19.

Another version of autoencoders results if during training one adds noise to the input [238,239]. This is a stochastic counterpart, known as the *denoising autoencoder*. For reconstruction, the uncorrupted input is employed. The idea behind this version is that by trying to undo the effect of noise, one captures statistical dependencies between inputs. More specifically, in [238], the corruption process randomly sets some of the inputs (as many as half of them) to zero. Hence, the denoising autoencoder is forced to predict the missing values from the nonmissing ones, for randomly selected subsets of missing patterns.

Autoencoders have also been used for pretraining deep networks, in place of the RBMs discussed before (see, e.g., [87]). In the latter, autoencoders with many layers, instead of a single one, have been employed.

18.15.5 GENERATIVE ADVERSARIAL NETWORKS

In Sections 18.15.1 and 18.15.3, we considered generative probabilistic graphical models for probability distribution estimation as well as for data generation. The major drawback of such models comes from the computational intractability when maximizing the associated likelihood function and related costs in order to compute the unknown parameters that define the graphical model.

An alternative path for designing generative models was first described in [61]. The breakthrough concept was to abandon the idea of modeling the probability distribution *directly*; instead, a game theoretic scenario was adopted, where the generator network was left to *compete* against an *adversary*.[7] The essence behind the method is to "make" the generator to produce examples that are *indistinguishable* from the available observations, which are used for training. Fig. 18.42 illustrates the main idea of such networks, known as *generative adversarial networks* (GANs). In the figure, there are two networks, each of them being a deep neural network, i.e., the *generator* (G) and the *discriminator* (D) network.

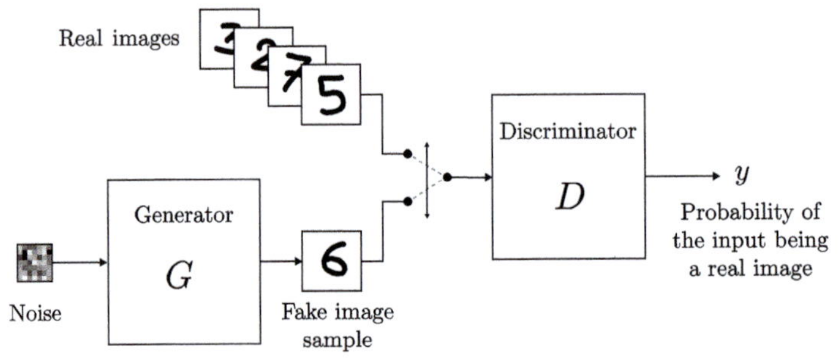

FIGURE 18.42

One class comprises the real images and the other the fake ones. The latter are produced by the generator, which is excited by noise. The output of the discriminator represents the probability of an input pattern to originate from the real data class. If the classes were perfectly separable, the discriminator's output, y, would be equal to 1 for real images and 0 for the fake ones. The goal of training a GAN is to "confuse" the discriminator so that its output becomes $1/2$ both for the real and for the fake ones.

The generator is fed to its input with noise samples, according to a probability distribution $p_{\mathbf{z}}(z)$, which is typically a uniform or a Gaussian PDF. The generator transforms the input noise vector, z, into a sample $\boldsymbol{x} = G(z; \boldsymbol{\theta}_g)$ of the same dimension as that of the available observations, $\boldsymbol{x}_n, n = 1, 2, \ldots, N$. The parameter vector $\boldsymbol{\theta}_g$ comprises all the parameters that define the generator neural network and these parameters have to be estimated during the training phase. The goal of the training is to compute the parameters so that the generated patterns, $\boldsymbol{x}$, are statistically indistinguishable from the available observations. From now on, we will refer to the observations as *real* and those that are produced by the generator as *fake* data.

The discriminator is a *binary* classifier that is implemented as a deep neural network. Its input is fed with samples $\boldsymbol{x}$ and it outputs a corresponding probability value, $y = D(\boldsymbol{x}; \boldsymbol{\theta}_d)$. The parameter vector $\boldsymbol{\theta}_d$ comprises all the parameters that define the associated neural network and are learned during training. The output value, y, represents the probability that the corresponding input pattern, $\boldsymbol{x}$, originates from the observations rather than from the generator. In other words, $D(\boldsymbol{x}; \boldsymbol{\theta}_d)$ is the probability of $\boldsymbol{x}$

[7] Note that the notion of adversary here is used in a different context than that of the adversarial examples, discussed in Section 18.14.

being real, and $1 - D(\boldsymbol{x}; \boldsymbol{\theta}_d)$ is the probability of $\boldsymbol{x}$ being fake. If the discriminator were designed to make perfect decisions, then the output should be 1 for the real data and 0 for the fake ones.

During training, the parameter vectors are optimally estimated so that the discriminator *confuses* the real with the fake data points. To this end in [61], the following two-player minmax game value (cost) is adopted:

$$\min_{\boldsymbol{\theta}_g} \max_{\boldsymbol{\theta}_d} J(\boldsymbol{\theta}_g, \boldsymbol{\theta}_d), \tag{18.80}$$

where

$$\boxed{J(\boldsymbol{\theta}_g, \boldsymbol{\theta}_d) = \mathbb{E}_{\mathbf{x} \sim p_r(\boldsymbol{x})} \left[\ln D(\mathbf{x}; \boldsymbol{\theta}_d) \right] + \mathbb{E}_{\mathbf{z} \sim p_{\mathbf{z}}(\boldsymbol{z})} \left[\ln \left(1 - D(G(\mathbf{z}; \boldsymbol{\theta}_g); \boldsymbol{\theta}_d) \right) \right],} \tag{18.81}$$

and $p_r(\boldsymbol{x})$ denotes the probability distribution associated with the real data (subscript r reminds us of *real*).

In words, if $\boldsymbol{\theta}_g$ is kept fixed, then the discriminator is trained, via $\boldsymbol{\theta}_d$, so that its output is maximized both for real $(D(\boldsymbol{x}; \boldsymbol{\theta}_d))$ and for fake $(1 - D(G(\boldsymbol{z}; \boldsymbol{\theta}_g); \boldsymbol{\theta}_d))$ examples. Furthermore, keeping $\boldsymbol{\theta}_d$ fixed, G is *simultaneously* trained through $\boldsymbol{\theta}_g$ to minimize $1 - D(G(\boldsymbol{z}; \boldsymbol{\theta}_g); \boldsymbol{\theta}_d)$, in order to confuse the discriminator.

Algorithm 18.5 summarizes the main concept, adopting a minibatch (of size K) optimization rationale.

Algorithm 18.5 (The GAN algorithm).

- Initialize $\boldsymbol{\theta}_d^{(0)}$ and $\boldsymbol{\theta}_g^{(0)}$.
- Set the minibatch size equal to K.
- Set the number of iterations, m, for the discriminator; the simplest case is $m = 1$, and it is used in the original paper.
- **While** $\boldsymbol{\theta}_d$ and $\boldsymbol{\theta}_g$ have not converged, **Do**
  - **For** $t = 1, 2, \ldots, m$, **Do**
    - Sample $\boldsymbol{z}^{(i)}$, $i = 1, 2, \ldots, K$, from $p_{\mathbf{z}}(\boldsymbol{z})$
    - Sample $\boldsymbol{x}^{(i)}$, $i = 1, 2, \ldots, K$, from $p_r(\boldsymbol{x})$
    - Compute the gradient

$$\nabla_{\boldsymbol{\theta}_d} \left\{ \frac{1}{K} \sum_{i=1}^{K} \left(\ln D(\boldsymbol{x}^{(i)}; \boldsymbol{\theta}_d) + \ln \left(1 - D(G(\boldsymbol{z}^{(i)}; \boldsymbol{\theta}_g); \boldsymbol{\theta}_d) \right) \right) \right\}$$

    - Update $\boldsymbol{\theta}_d$ via a gradient *ascent* scheme

$$\boldsymbol{\theta}_d \leftarrow \boldsymbol{\theta}_d$$

  - **End For**
- Sample $\boldsymbol{z}^{(i)}$, $i = 1, 2, \ldots, K$, from $p_{\mathbf{z}}(\boldsymbol{z})$

- Compute the gradient

$$\nabla_{\boldsymbol{\theta}_g} \left\{ \frac{1}{K} \sum_{i=1}^{K} \ln\left(1 - D(G(\boldsymbol{z}^{(i)}; \boldsymbol{\theta}_g); \boldsymbol{\theta}_d)\right) \right\}$$

- Update $\boldsymbol{\theta}_g$ via a gradient *descent* scheme

$$\boldsymbol{\theta}_g \leftarrow \boldsymbol{\theta}_g$$

- **End While**

Remarks 18.8.

- In [61], it is pointed out that rather than training G to minimize $\ln(1 - D(G(z)))$, it is better to train G to maximize $\ln(D(G(z))$. This turned out to deal better with the involved gradients. Further theoretical justification for this modification, together with comparative experimental results, can be found in [51].

On the Optimality of the Solution

The first question that is now raised concerns the optimal solution associated with the minmax optimization task in (18.81). Note that the generator G *implicitly* defines a probability distribution, $p_g(\boldsymbol{x})$, of the samples $\boldsymbol{x} = G(z; \boldsymbol{\theta}_g)$ that are generated if $\mathbf{z} \sim p_{\mathbf{z}}(z)$.

For our analysis, we will free ourselves from the parametric modeling of the involved functions via $\boldsymbol{\theta}_g$ and $\boldsymbol{\theta}_d$, and we will study the optimal solution in terms of functions $D(\boldsymbol{x})$ and $G(z)$, considered in a *general* nonparametric formulation. Under such a scenario, Eq. (18.81) is rephrased as

$$\min_G \max_D J(G, D) := \mathbb{E}_{\mathbf{x} \sim p_r(x)} \left[\ln D(\mathbf{x}) \right] + \mathbb{E}_{\mathbf{z} \sim p_{\mathbf{z}}(z)} \left[\ln\left(1 - D(G(\mathbf{z}))\right) \right]. \tag{18.82}$$

By the definition of the expectation we can write

$$\begin{aligned} J(G, D) &= \int_x p_r(\boldsymbol{x}) \ln(D(\boldsymbol{x}))d\boldsymbol{x} + \int_z p_{\mathbf{z}}(z) \ln\left(1 - D(G(z))\right)dz \\ &= \int_x \left(p_r(\boldsymbol{x}) \ln(D(\boldsymbol{x})) + p_g(\boldsymbol{x}) \ln\left(1 - D(\boldsymbol{x})\right) \right) d\boldsymbol{x}, \end{aligned}$$

where the subscript g stands for the distribution associated with the *generator* output. Fixing function G (equivalently, fixing the probability distribution function p_g), the optimal value of the discriminator's output probability function D is easily derived by taking the derivative of the integrand with respect to the function D and setting it equal to zero (Problem 18.15). It is easily shown that

$$D^* = \frac{p_r}{p_r + p_g}. \tag{18.83}$$

Thus, plugging the above optimal value in (18.82), the cost with respect to G of the minimax game is reformulated as

$$C(G) := \mathbb{E}_{\mathbf{x} \sim p_r(x)} \left[\ln \frac{p_r(\mathbf{x})}{p_r(\mathbf{x}) + p_g(\mathbf{x})} \right] + \mathbb{E}_{\mathbf{x} \sim p_g(\mathbf{x})} \left[\ln \frac{p_g(\mathbf{x})}{p_r(\mathbf{x}) + p_g(\mathbf{x})} \right], \tag{18.84}$$

where the above is a function of p_g. Recalling the definition of the KL divergence (Sections 2.5.2 and 12.7), Eq. (18.84) is compactly rewritten as

$$C(G) = \mathrm{KL}(p_r||p_r + p_g) + \mathrm{KL}(p_g||p_r + p_g),$$

where, as we know, the KL divergence is *not* symmetric with respect to its arguments. Normalizing by 2, and taking into account the definition of the KL divergence, the above is easily rewritten as

$$C(G) = -\ln 4 + \mathrm{KL}\left(p_r||\frac{p_r + p_g}{2}\right) + \mathrm{KL}\left(p_g||\frac{p_r + p_g}{2}\right),$$

or

$$C(G) = -\ln 4 + 2\mathrm{JS}(p_r||p_g), \tag{18.85}$$

where JS denotes the *Jensen–Shannon divergence* between p_r and p_g, defined as

$$\boxed{\mathrm{JS}(p_r||p_g) := \frac{1}{2}\mathrm{KL}\left(p_r||\frac{p_r + p_g}{2}\right) + \frac{1}{2}\mathrm{KL}\left(p_g||\frac{p_r + p_g}{2}\right).}$$

Thus, as Eq. (18.85) suggests, the cost $C(G)$ associated with the generator depends on the JS divergence between the p_r of the real data and the p_g that the generator implements. It can easily be checked that the JS divergence, in contrast to KL divergence, is a symmetric one. As conjectured in [232], it is exactly the use of this divergence, compared to the KL one, that gave GANs the advantage over more traditional maximum likelihood-based approaches.

It can easily be seen that the JS divergence is nonnegative and the minimum value is achieved if and only if $p_r = p_g$. Hence, the solution of the minmax task defined in (18.82) is obtained when D is given by (18.83) and when G minimizes (18.85), leading to

$$\boxed{p_g = p_r, \ D^* = \frac{1}{2}, \ \text{and } C^* = -\ln 4.}$$

In other words, optimality of the game is achieved when the generator learns the distribution of the true data and the discriminator D is "confused" and becomes unable to discriminate between the true and the fake data.

Problems in Training GANs

In the previous section, the optimality of the GANs has been established. Yet, the task is far from solved. The main reason is that the optimality analysis was carried out in the probability function space. However, in practice, we use parametric models to achieve an approximation to the above, and this is where problems arise. As a matter of fact, as Algorithm 18.5 suggests, in practice, optimization is carried out via parameter optimization and the involved gradients are computed via the backpropagation rationale and updates follow one of the available gradient-based optimization schemes. However, updating the parameters of the generator and the discriminator concurrently, to solve a nonconvex optimization task, does not necessarily mean that the scheme converges to the optimal game solution.

Looking at Eq. (18.85) and following the arguments on how we reached it, one would expect that training the discriminator close to its optimality (so that the cost function on $\boldsymbol{\theta}_g$ is a good approximation of Eq. (18.85)) and then applying the gradient steps on $\boldsymbol{\theta}_g$ is the way to proceed. However, in practice, this is not the case. As the discriminator improves, the updates of the generator get worse. An explanation of this phenomenon is provided in [5]. At the heart of this "strange" behavior lies the fact that *parameterized* approximations of p_r and p_g lie in low-dimensional manifolds and this makes it very difficult to share common supports (regions in the input space where both functions share nonzero values). This is illustrated in Fig. 18.43. As a consequence, the trained discriminator, instead of achiev-

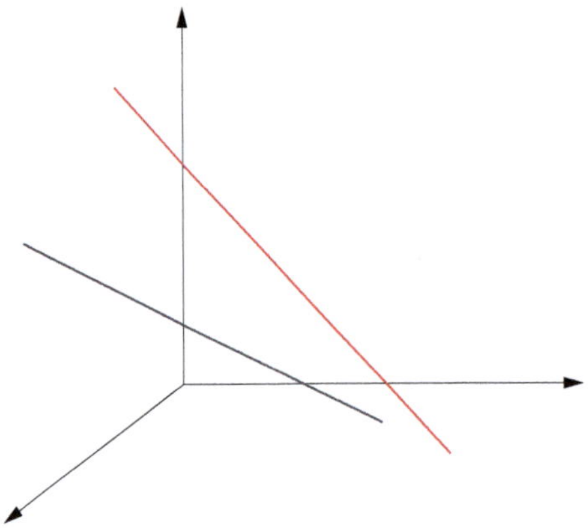

FIGURE 18.43

In the three-dimensional space, the two straight lines are low-dimensional (linear) manifolds. It is highly unlikely that they share a common intersection. So if, say, p_r is confined to one and p_g to the other of the two lines, it is highly improbable for the two distributions to share a nonnegligible support.

ing a cost according to Eq. (18.85), achieves a zero error, meaning that it can easily recognize perfectly the fake from the true data. This leads the cost to zero and the gradients with respect to $\boldsymbol{\theta}_g$ to vanishingly small values that makes the training of GANs, via the JS cost, difficult. While training, one has to decide how much to train the discriminator. In other words, there is a tradeoff between inaccuracy and vanishing gradients for training.

Another problem associated with training GANs is the so-called *mode collapse*. This means that although the generator may "fool" the discriminator, the generator produces "same" outputs. It seems that the generator gets stuck in a "small" region, which leads to outputs of low variability. This may be due to the fact that the generator cannot learn sufficiently well the data distribution, and it focuses on some parts of it, while ignoring other parts.

In order to bypass the aforementioned drawbacks, a number of tricks and techniques have been proposed (see, e.g., [203,258]). In [188], convolutional generative networks are introduced. In [154], least-squares arguments are adopted for the design of GANs. In [140], the concept of combining models is employed based on a competitive training procedure that splits the data distribution into components, which are approximated well by independent generative models.

The Wasserstein GAN

In [6], the task of matching p_r and p_g in low-dimensional manifolds is systematically tackled via the so-called *Wasserstein distance* between distributions. It is first shown that when learning distributions in low-dimensional manifolds, the divergence that measures the difference between distributions must have certain properties. Both JS and KL divergence do not have these properties. In contrast, the Wasserstein distance is a proper one in this context.

The Wasserstein distance, or *earth moving distance* (EMD), between two distributions p_r and p_g, is defined as

$$\boxed{W(p_r, p_g) = \inf_{\gamma \in \Pi(p_r, p_g)} \mathbb{E}_{(\mathbf{x},\mathbf{y})\sim\gamma}\left[||\mathbf{x}-\mathbf{y}||\right], \text{ Wasserstein distance.}} \tag{18.86}$$

Although the previous definition may look a bit complicated to the unfamiliar reader, it makes a lot of sense; $\Pi(p_r, p_g)$ is the set of *all* joint distributions between the random vectors, whose marginals are equal to p_r and p_g, respectively. Intuitively, the expectation indicates how much probability mass has to be transferred from $\mathbf{x}$ to $\mathbf{y}$ in order to transform p_r to p_g. Their distance corresponds to the "minimum" value among all possible joint distributions.[8]

In practice, Eq. (18.86) cannot easily be implemented. An alternative way is via its dual formulation, known as the Kantorovich–Rubinstein duality (e.g., [236]), given by

$$W(p_r, p_g) = \sup_{||f||_L<1} \left\{\mathbb{E}_{\mathbf{x}\sim p_r}\left[f(\mathbf{x})\right] - \mathbb{E}_{\mathbf{x}\sim p_g}\left[f(\mathbf{x})\right]\right\}, \tag{18.87}$$

where $||f||_L < 1$ denotes all the 1-Lipschitz functions (see also Section 8.10.2), $f : \mathcal{X} \mapsto \mathbb{R}$, i.e.,

$$|f(\boldsymbol{x}_1) - f(\boldsymbol{x}_2)| < ||\boldsymbol{x}_1 - \boldsymbol{x}_2||, \ \forall \boldsymbol{x}_1, \boldsymbol{x}_2 \in \mathcal{X}.$$

Thus, the Wasserstein probability distance between two distributions, in the form of Eq. (18.87), is given by the "maximum" in the difference of the mean values, over all possible 1-Lipschitz functions, with respect to the two distributions.

Designing a GAN according to Eq. (18.87) makes the setting different from the one that evolved around the cost in Eq. (18.81). The goal is no more to design a binary classifier (discriminator) to distinguish real from fake data, but a Lipschitz function, f. The latter is parameterized via a deep neural network, in terms of a set of parameters, $\boldsymbol{\theta}_f$, i.e., $f(\boldsymbol{x}; \boldsymbol{\theta}_f)$. The respective parameters are estimated by solving the task

$$\max_{\boldsymbol{\theta}_f} \left\{\mathbb{E}_{\mathbf{x}\sim p_r}\left[f(\mathbf{x}; \boldsymbol{\theta}_f)\right] - \mathbb{E}_{\mathbf{z}\sim p_{\mathbf{z}}}\left[f(G(z; \boldsymbol{\theta}_g); \boldsymbol{\theta}_f)\right]\right\}.$$

Fixing the parameters $\boldsymbol{\theta}_f$, the parameters $\boldsymbol{\theta}_g$ that define the generator are estimated by minimizing the difference in the bracket above. Note that this gradient depends only on the second term, the first being independent of $\boldsymbol{\theta}_g$. As the loss function decreases during the training, the Wasserstein distance gets smaller and the output of the generator gets closer to the real data distribution. The Wasserstein GAN scheme is outlined in Algorithm 18.6.

8 For a nice discussion see, e.g., https://vincentherrmann.github.io/blog/wasserstein/.

Algorithm 18.6 (**The Wasserstein GAN algorithm**).

- Initialize $\boldsymbol{\theta}_f^{(0)}$ and $\boldsymbol{\theta}_g^{(0)}$.
- Set the minibatch size equal to K.
- Set the number of iterations, m, for the function parameters $\boldsymbol{\theta}_f$.
- Set the clipping parameter c; this parameter takes care of the Lipschitz condition. Typically, $c = 0.001$.
- **While** $\boldsymbol{\theta}_f$ and $\boldsymbol{\theta}_g$ have not converged, **Do**
  - **For** $t = 1, 2, \ldots, m$, **Do**
    * Sample $\boldsymbol{z}^{(i)}$, $i = 1, 2, \ldots, K$, from $p_{\boldsymbol{z}}(\boldsymbol{z})$
    * Sample $\boldsymbol{x}^{(i)}$, $i = 1, 2, \ldots, K$, from $p_r(\boldsymbol{x})$
    * Compute the gradient

$$\nabla_{\boldsymbol{\theta}_f}\left\{\frac{1}{K}\sum_{i=1}^{K}\Big(f(\boldsymbol{x}^{(i)};\boldsymbol{\theta}_f) - f\big(G(\boldsymbol{z}^{(i)};\boldsymbol{\theta}_g);\boldsymbol{\theta}_f\big)\Big)\right\}$$

    * Update $\boldsymbol{\theta}_f$ via a gradient *ascent* scheme

$$\boldsymbol{\theta}_f \leftarrow \boldsymbol{\theta}_f$$

    * Clip the values in the interval $[-c, c]$
  - **End For**
- Sample $\boldsymbol{z}^{(i)}$, $i = 1, 2, \ldots, K$, from $p_{\boldsymbol{z}}(\boldsymbol{z})$
- Compute the gradient

$$\nabla_{\boldsymbol{\theta}_g}\left\{-\frac{1}{K}\sum_{i=1}^{K} f\big(G(\boldsymbol{z}^{(i)};\boldsymbol{\theta}_g);\boldsymbol{\theta}_f\big)\right\}$$

- Update $\boldsymbol{\theta}_g$ via a gradient *descent* scheme

$$\boldsymbol{\theta}_g \leftarrow \boldsymbol{\theta}_g$$

- **End While**

Remarks 18.9.

- Note that since Algorithm 18.6 does not train a discriminator, the better the estimate of the function f is, the higher the quality of the gradient with respect to $\boldsymbol{\theta}_g$ is expected to be. Hence, m can be given a fairly large value, without having to worry about balancing the discriminator and generator, as was the case with Algorithm 18.5.
- In [6], it is reported that the Wasserstein GAN leads to improved stability and robustness compared to previously developed schemes.
- However, as everything in life, nothing is perfect. For example, clipping is a rather "crude" way to enforce the Lipschitz condition. Improvements of the basic scheme have been proposed in, e.g., [248] and [70].

Which Algorithm Then

So far, in our discussion on GANs, we focused on two possible algorithms and we also provided references to a number of other alternatives. The Wasserstein GAN helped us to outline some important issues that underlie the task of learning distributions (implicitly or explicitly) where, most often, the learning process takes place in lower-dimensional manifolds. The reader and any practitioner may wonder which algorithm is "best" or more suitable to be used in a practical application.

One of the major problems associated with data generation is to be able to test the performance of the generator by assessing the quality of the generated patterns. The main issue in evaluating the performance stems from the fact that one cannot explicitly compute the distribution $p_g(\boldsymbol{x})$. Thus, classical costs such as the likelihood function cannot be employed. Attempts to approximate it in high-dimensional spaces seem to be problematic in the current context (e.g., [103]). To this end, related metrics have been suggested. For example, the so-called *inception score* (IS) builds upon the concept that a good model should generate samples for which, when evaluated by the discriminator, the class distribution has low entropy. At the same time, the generated samples should exhibit large variation [203]. On the other hand, the *Fréchet inception distance* focuses on a metric that quantifies the difference between true and fake data samples, after a specific embedding in a feature space via an inception network [80]. Assuming that the embedded data follow a multivariate Gaussian distribution, the distance between the two distributions is quantified by the Fréchet distance between the corresponding Gaussians.

A comparative experimental study on various GANs has been presented in [143]. The findings there, using a number of metrics, including the previously stated quality measuring ones, do not seem to favor any one of the proposed GANs against the others. It is reported that most models can achieve similar scores after careful hyperparameter tuning in optimization, higher computational budget, and random restarts.

However, it should be stated that at the time this edition is compiled, designing GANs is an ongoing research area and it is rather early to come up with definite statements.

Example 18.5. The purpose of this example is to describe a basic experiment on GANs to generate images of hand-written characters. To this end, the generator as well as the discriminator networks have to be designed.

- The generator network has 100 input nodes that are excited by i.i.d. Gaussian noise samples of zero mean and unit variance. The network comprises three hidden layers, with 256, 512, and 1024 neurons, respectively. The nonlinearity used for the neurons of the hidden layers was the leaky ReLU (Section 18.6.1), with $\alpha = 0.2$. There are 784 output units and the output nonlinearity was the tanh function. The number of output units is chosen to be equal to the size of the MNIST image data set,[9] which is used for training.
- The discriminator has 784 input nodes, to match the 1×784 size vector associated with the 28×28 MNIST images. The discriminator consists of three hidden layers, of 1024, 512, and 256 neurons, respectively. The leaky ReLU unit, with $\alpha = 0.2$, was used as the activation function for the hidden layer neurons. The output is a single binary sigmoidal node and the cost function was the two-class

[9] http://yann.lecun.com/exdb/mnist/.

cross-entropy function (Eq. (18.44)). For the training, the dropout method was used with the probability of discarding hidden neurons being equal to 0.3. The Adam optimizer was employed for training (Problem 18.21). The input images were normalized in the range $[-1, 1]$. The number of images used for training was 60000 and the size of the minibatches equal to 100.

Fig. 18.44 shows examples of generated images by the generator after 1, 20, and 400 epochs of training. Observe how the quality of the fake images improves as the training algorithm converges.

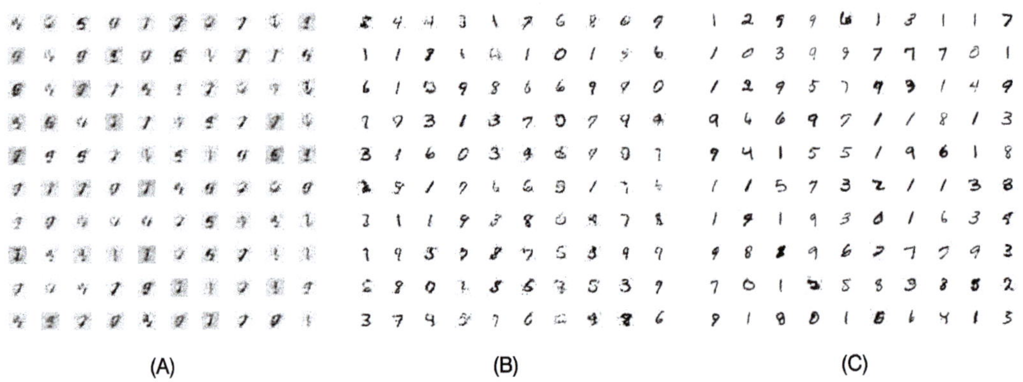

FIGURE 18.44

Fake images produced at the output of the generator network after training for (A) 1, (B) 20, and (C) 400 epochs, respectively.

18.15.6 VARIATIONAL AUTOENCODERS

At the heart of GANs lies the concept of generating random samples, $\mathbf{x}$, by exciting the generator network with noise random samples, $\mathbf{z}$. *Variational autoencoders* (VAEs) follow a similar concept; yet the generator, known as decoder in this case, is excited by random variables whose PDF has been learned from the data. The training builds upon Bayesian learning arguments (see Chapters 12 and 13). Variational autoencoders were proposed in [120,190].

The basic assumption behind VAEs is that we are given a set of observations, $\boldsymbol{x}_n, n = 1, 2, \ldots, N$, which are samples of a random vector, $\mathbf{x} \in \mathbb{R}^l$. Moreover, the latter is the result of another process that involves a set of continuous *latent* random variables, $\mathbf{z} \in \mathbb{R}^m$, where usually $m \ll l$. That is, the underlying generation mechanism evolves along the following steps:

- A sample z_n is generated according to a prior PDF, $p(z; \boldsymbol{\theta})$, where $\boldsymbol{\theta}$ is a set of unknown parameters.
- A sample $\boldsymbol{x}_n$ is generated from a conditional PDF, $p(\boldsymbol{x}|z_n; \boldsymbol{\theta})$.

Both $z_n, n = 1, 2, \ldots, N$, and $\boldsymbol{\theta}$ are unknown and have to be learned from the available observations $\boldsymbol{x}_n, n = 1, 2, \ldots, N$. As we know from the Bayesian-related chapters, a way to estimate a set of parameters, $\boldsymbol{\theta}$, in the presence of latent (unobserved) variables is via the EM algorithm. A prerequisite for its application is to know the posterior, $p(z|\boldsymbol{x})$; this is needed for averaging out the (unknown) latent variables. In general, this posterior is not available (or the required by the EM computations may not be computationally tractable) and one has to resort to approximations. In Chapter 13, a factorized form was adopted in the framework of the mean field approximation. In contrast, according to the VAEs, the approximation $q(z|\boldsymbol{x};\boldsymbol{\phi})$ of the posterior is implemented via the use of a neural network; the vector $\boldsymbol{\phi}$ denotes the set of the unknown parameters that have to be learned together with $\boldsymbol{\theta}$.

In the VAE jargon, $p(z|\boldsymbol{x};\boldsymbol{\phi})$ is known as the probabilistic *encoder*; indeed, given an observation $\boldsymbol{x}$, it produces a distribution (e.g., a Gaussian) over the possible values of $\mathbf{z}$, which is also known as the *code*. On the other hand, $p(\boldsymbol{x}|z;\boldsymbol{\theta})$ is referred to as the probabilistic *decoder*; given a code z, it produces a distribution over the possible corresponding values of $\mathbf{x}$. The major difference with the autoencoders considered in Section 18.15.4 is that in the variational autoencoder context, the encoder is not designed to output a *single value* for each latent variable; in contrast, it is designed to provide a *probability distribution* for each latent attribute.

The concept behind VAEs should be clear by now. Once we learn the parameters associated with the approximate models of the posterior and the conditional, $\boldsymbol{\phi}$ and $\boldsymbol{\theta}$, then we can generate new samples. Generate z_n according to the posterior and then use the generated sample, together with the conditional, to produce the corresponding $\boldsymbol{x}_n$. All that remains for us now is (a) to adopt the explicit parametric models for the conditional $p(\boldsymbol{x}|z;\boldsymbol{\theta})$ and for the approximation to the posterior $q(z|\boldsymbol{x};\boldsymbol{\phi})$ and (b) to establish the method for estimating the unknown parameters.

The parametric models: In [120], the following model is proposed:

- For the prior, the standard multivariate Gaussian is adopted, i.e., $p(z)=\mathcal{N}(z;\mathbf{0},I)$. Thus, according to such a choice, the prior does not depend on any parameters.
- For the posterior approximation $q(z|\boldsymbol{x};\boldsymbol{\phi})$, the multivariate Gaussian is also selected, with diagonal covariance matrix ($\Sigma_n=\text{diag}\{\sigma_n^2(i)\}_{i=1}^m$), i.e.,

$$q(z|\boldsymbol{x}_n;\boldsymbol{\phi})=\mathcal{N}(z;\boldsymbol{\mu}_n,\Sigma_n).$$

  The crucial point here is that the above mean values and variances are provided as outputs of the encoding network; in the simplest case, a network of a single hidden layer is employed (deeper networks and other variants can also be considered). More specifically,

$$\begin{aligned}\boldsymbol{h}_n &= \tanh\left(W_1\boldsymbol{x}_n+\boldsymbol{b}_1\right),\\ \boldsymbol{\mu}_n &= W_2\boldsymbol{h}_n+\boldsymbol{b}_2,\\ \ln\boldsymbol{\sigma}_n^2 &= W_3\boldsymbol{h}_n+\boldsymbol{b}_3,\end{aligned}$$

  where $\ln\boldsymbol{\sigma}_n^2$ in the last equation denotes an element-wise action on the vector comprising the variances. Thus, the parameters $\boldsymbol{\phi}$ comprise the set of all the parameters that define the network as well as the linear combiners, i.e., W_1, W_2, W_3, $\boldsymbol{b}_1$, $\boldsymbol{b}_2$, $\boldsymbol{b}_3$, which are matrices and bias vectors of appropriate dimensions. Note that in order to guarantee nonnegative values for the variance, we model its logarithm. Then, its value can be obtained via the exponent operation.
- For the conditional PDF, if it is also Gaussian, a similar model as the one above is used, where the roles of z and $\boldsymbol{x}$ are swapped and $\boldsymbol{\theta}$ comprises the parameters of the associated decoder network. If, on the other hand, the observations are of a discrete nature, a different model should be used. For example, for binary images, a Bernoulli distribution can be employed to implement the decoder network (e.g., [120]).

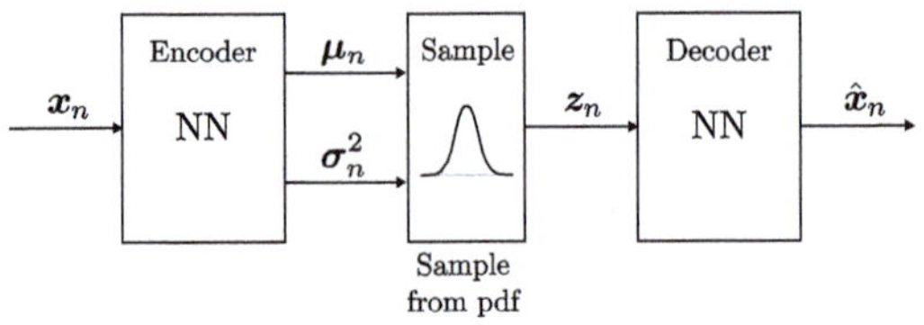

FIGURE 18.45

The encoder network outputs the mean values and variances of the multivariate Gaussian posterior that describes the latent variables, which correspond to the current input. The decoder network, is excited by samples of the latent variables, drawn according to their posterior estimate, to generate samples similar to the original input.

Fig. 18.45 illustrates the VAEs architecture. For each observation $\boldsymbol{x}_n$, the encoder's network outputs the corresponding mean values and variances, which define the specific posterior multivariate Gaussian PDF of the respective latent variables, z_n. Sampling from the posterior PDF, one can generate samples, according to the conditional, at the output of the decoder network that match the original input.

The cost function: Following the arguments in Chapter 13, ideally one should maximize the likelihood of the observations with respect to the unknown parameters. However, the computation of the likelihood is not tractable for this case. Instead, we will maximize a related *lower bound* of it, i.e., Eqs. (13.1) and (13.2). Assuming the observations to be i.i.d. and adjusting to the current notational context, this lower bound, $\mathcal{F}$, becomes

$$\ln p(\boldsymbol{x}_1, \boldsymbol{x}_2, \dots, \boldsymbol{x}_n; \boldsymbol{\theta}) = \sum_{n=1}^{N} \ln p(\boldsymbol{x}_n; \boldsymbol{\theta}) \geq \sum_{n=1}^{N} \mathcal{F}(\boldsymbol{\theta}, \boldsymbol{\phi}; \boldsymbol{x}_n),$$

where

$$\mathcal{F}(\boldsymbol{\theta}, \boldsymbol{\phi}; \boldsymbol{x}_n) := \mathbb{E}_q \left[\ln p(\boldsymbol{x}_n, \mathbf{z}; \boldsymbol{\theta}) - \ln q(\mathbf{z}|\boldsymbol{x}_n; \boldsymbol{\phi}) \right].$$

After applying the Bayes theorem on the joint distribution term and using the definition of the KL divergence, we get

$$\mathcal{F}(\boldsymbol{\theta}, \boldsymbol{\phi}; \boldsymbol{x}_n) = -\mathrm{KL}\big(q(z|\boldsymbol{x}_n; \boldsymbol{\phi}) || p(z; \boldsymbol{\theta})\big) + \mathbb{E}_q \left[\ln p(\boldsymbol{x}_n|\mathbf{z}; \boldsymbol{\theta}) \right].$$

Once the parametric forms of the involved distributions have been adopted, all one has to do is to maximize $\mathcal{F}$ with respect to the unknown parameters. Taking into account that the adopted models are feed-forward neural networks, the necessary gradients for the optimization task are computed in the backpropagation rationale. To this end, several "tricks" are in order, and the details can be found in, e.g., [120].

The lower bound involves two terms. The second term on the right-hand side is the expected value over the latent variables. To maximize the bound, this term should be maximized for the respective observations, $\boldsymbol{x}_n$; this encourages the decoder to learn to reconstruct the data. The first term is negative (the KL divergence is a nonnegative quantity). To maximize the bound, the KL should be minimized; thus, the corresponding posterior will remain close to the prior; the latter has been chosen to be the standard multivariate Gaussian, $\mathcal{N}(\mathbf{0}, I)$. Hence, the prior acts as a regularizer that "pushes" the network's latent variables to match the standard Gaussian distribution as closely as possible. The existence of this term is very important. The regularizer "forces" the PDF of the latent variables to be broad enough. We do not want, for example, two different images from the same class to be mapped to two different regions of the latent space; i.e., we do not want their corresponding PDFs to be very narrow, with small variance, placed at mean values that are far away. Our desire is to map/code every image of the same class via latent variables that lie "close" enough in the same region of space.

The reparameterization trick: Before, it was mentioned that in order to apply the backpropagation algorithm some "tricks" are required. A major problem in the current setting, when dealing with the backpropagation for gradient computations, is that the decoder is excited by random samples. However, the dependence of the random samples on the unknown parameters is not explicit, e.g., via a function dependence. Their dependence is implicit via the respective PDF. In other words, if we draw samples from, say, a Gaussian, the sample values are not explicitly expressed in terms of the respective mean value and variance, although, of course, they depend on them. Undoubtedly, it is counterintuitive to try to compute derivatives of a random variable with respect to parameters that define the associated distribution.

The previous difficulty is bypassed by the so-called reparameterization trick. The idea is to view the latent variables as being *deterministic* and express them as a function in terms of the parameters, which define the respective distribution, and of an independent auxiliary random variable that takes care of the randomness. For example, in the case of Gaussians, these parameters are the mean values and variances (covariance matrix). The auxiliary random variable can follow a simple PDF (e.g., Gaussian or uniform). It suffices for the function to be differentiable. In our context, the implied transformation is a linear one, and each component of the latent variables is expressed as

$$z_i = \mu_i + \sigma\epsilon, \ i = 1, 2, \ldots, m,$$

where ϵ is the auxiliary random variable, with zero mean and unit variance. This transformation makes it possible to compute derivatives with respect to the parameters of the distribution, while still maintaining the ability to randomly sample from that distribution.

Although VAEs are easier to train compared to GANs, experimental evidence indicates that when VAEs are used for generating images, the obtained images tend to be more blurred compared to the ones that are produced via GANs. In [234], the Wasserstein distance is used to replace the KL divergence in the cost function for training VAEs and it is reported that this has a beneficial effect on the obtained performance. Another line of research that is currently pursued is along the convergence of GANs and VAEs, in an effort to combine the benefits of both (see, e.g., [155,159]).

At the time the current edition of the book is being compiled, this is an ongoing active research field.

18.16 CAPSULE NETWORKS

The convolutional neural networks, treated in Section 18.12, and their many variants are currently state of the art in a diverse range of applications. Yet, in spite of their successes, CNNs are not beyond shortcomings. The need for large data sets for their training is a major one. One of the reasons for the requirement of large training sets is that it is not easy for CNNs to learn and extrapolate geometric relationships to new viewpoints. For example, at the end of Section 18.12.1, it was pointed out that pooling introduces invariance to small translations. However, in order to deal with more general invariances, such as scaling and rotation, one has to train the network on different viewpoints of the same object, which necessarily leads to an increase of the required size of the data set. Furthermore, the subsampling associated with the pooling "throws away" information with respect to the position of an entity within an area in the input image; this can affect information concerning the precise spatial relationships that are critical for recognition of objects. For example, the spatial relationships among the eyes, the mouth, and the nose are important in recognizing a face. To this end, a number of variants have been suggested in order to embed into a network affine transformation stages, which are learned during training, in an attempt to robustify it and boost its transformation invariance properties (see, e.g., [107] and the references therein).

In [202], an alternative concept was introduced as a way to overcome the previous drawbacks. The proposed architecture builds upon the notion of convolutions as in CNNs, but it further embeds a new type of layers based on the concept of *capsules*. The term "capsule" refers to a group of scalar activations that are collectively combined to form an *activity* vector. Fig. 18.46, inspired by [202], is an example of such an architecture. To grasp the main rationale behind the capsule networks, let us follow

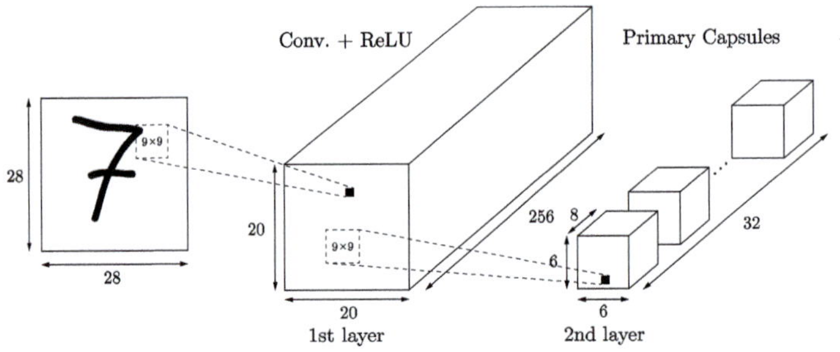

FIGURE 18.46

The input is a 28×28 image. Using 9×9 convolutions with stride $s = 1$ and 256 channels, the produced volume in the first layer is of size $20 \times 20 \times 256$. The convolutions performed for the second layer are 9×9 with stride $s = 2$. They are combined together in groups of 8, to form 32 $6 \times 6 \times 8$ volumes.

step-by-step the various blocks in the figure.

- *Input*: The inputs to the network are 28×28 images from the MNIST database.
- *First convolutional layer*: This is a standard convolutional layer. The filter matrix (kernel) is 9×9, the stride is $s = 1$, and the depth (number of channels) is $d = 256$. Thus, the size of the output volume of the first hidden layer is $k \times k \times 256$, where $k = \lfloor \frac{28-9}{1} + 1 \rfloor = 20$.
- *Primary capsules*: This is also a convolutional layer. The difference with the previous one is that in this layer a *grouping* of activities takes place. The filter matrix for the convolutions is also 9×9, the stride is $s = 2$, and the depth is $d = 256$. However, the output channels are clustered together in groups of 8. Thus, 32 volumes are formed, each of depth 8 (total 256) (see Fig. 18.46). Each one of the volumes has size $k \times k \times 8$, where $k = \lfloor \frac{20-9}{2} + 1 \rfloor = 6$.

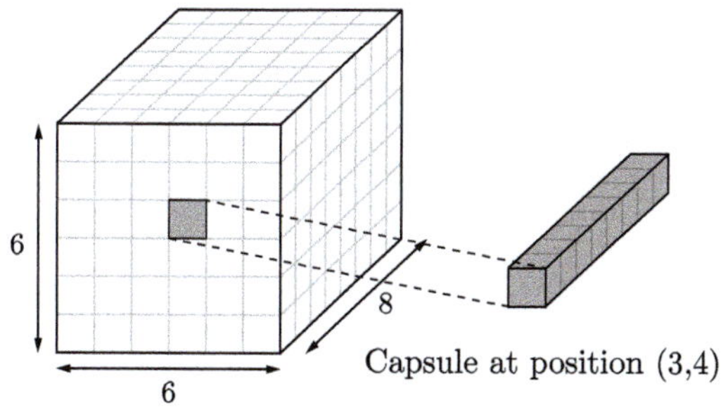

FIGURE 18.47

For each position/pixel (i, j) of the 6×6 frontal image grid, a capsule is formed of dimension 8. It comprises the eight elements of the depth of the volume at the respective grid position, i.e., (i, j, k), $k = 1, 2 \ldots, 8$.

Fig. 18.47 illustrates how the capsules are formed. For the example of the figure, each capsule is of dimension $D = 8$. For each one of the 32 volumes, 36 (6×6) eight-dimensional tubes/vectors are formed. In other words, each capsule corresponds to a specific location in the feature map arrays. Hence, each one of the eight dimensions in a capsule *encapsulates* activity within a corresponding receptive field in the volume of the previous layer (see Fig. 18.46). Thus, a total of $36 \times 32 = 1152$ capsules are obtained, each one of dimension equal to 8. These vectors comprise the *outputs* of the second layer. Let us denote each one of the produced capsules as $\boldsymbol{u}_i$, $i = 1, 2, \ldots, I$, where $I = 1152$.

- *Digit capsules*: This layer comprises three stages, which are illustrated in Fig. 18.48.
  - *Stage 1*: In this stage, affine transformations are performed on each one of the input (to this layer) capsules, i.e.,

$$\hat{\boldsymbol{u}}_{j|i} = W_{ij}\boldsymbol{u}_i, \quad i = 1, 2, \ldots, I, \quad j = 0, 2, \ldots, 9,$$

and $W_{ij} \in \mathbb{R}^{16 \times 8}$. In words, each one of the input capsules is transformed to *ten* 16-dimensional vectors. The number "ten" corresponds to the number of the classes, one per digit. Hence, the total number of performed transformations is $10I$. The W_{ij} matrices are learned during training and their role is to establish spatial and other types of relationships between the lower-level features (as they are encoded in the input capsules) and the higher-level ones. Simply stated, the input information is "optimally" transformed to "match" the higher-level information, which in our case are the ten classes. The W_{ij} matrices could be seen as the model to establish a "part-to-whole" relationship; the classes can be viewed as the "whole" and the input capsules as the "parts," each encoding individual features, e.g., type of a stroke, its thickness, and its width, which form the input images; however, these images may have been, for example, scaled and/or rotated and the role of these transformations is to take care of such variants.

  - *Stage 2*: The obtained transformed capsules, for each one of the classes, are then combined together via a weighted sum using *coupling coefficients* to form ten (one per class) vectors/cap-

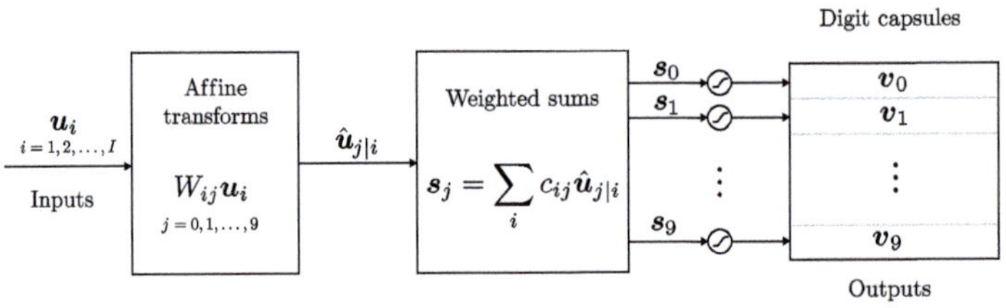

FIGURE 18.48

This layer consists of three stages. In the first one, ten (one per class) affine transformations are performed on each one of the input capsules, which were formed in the previous layer. In the next stage, a weighting sum is computed, over the transformed capsules for each one of the ten classes. At the final stage, the obtained vectors are "pushed" through a squashing nonlinearity to generate the ten output capsules. The input capsules are of dimension 8 and the output ones of dimension 16, using transform matrices, W_{ij}, of appropriate dimensions.

sules,

$$s_j = \sum_{i=1}^{I} c_{ij}\hat{u}_{j|i}, \quad j = 0, 1, \dots, 9. \tag{18.88}$$

As we will see in stage 3, it is these capsules that go through the nonlinearity to produce the outputs that finally define the winning class. Looking at Eq. (18.88) more carefully, the coupling coefficients c_{ij} weigh the importance of each one of the involved transformed capsules while computing the corresponding s_j. The nonnegative coupling coefficients have a probability interpretation and they add to one summing over j, for each fixed value of i. As will become apparent soon, their computation comprises a critical part of the capsule networks.

- *Stage 3*: Once the s_js have been computed, they are "pushed" through the nonlinearity to provide the output capsules, v_j, $j = 0, 2, \dots, 9$. However, the nonlinearity has been carefully chosen, so that the lengths (Euclidean norms) of the output capsules are indicative of the class of the input image. For example, if the input digit is, say, "4," the network is trained so that the length of the respective capsule, v_3, is the largest one compared to the other nine competitors. The idea is that each dimension (16 in total) of the output capsules represents different activities that are present in the input images. Hence, for the capsule that corresponds to the correct class, it is expected that all dimensions "exhibit" high activity.
  The adopted nonlinearity is

$$v_j = \frac{||s_j||^2}{1+||s_j||^2}\frac{s_j}{||s_j||}, \quad j = 0, 1, 2, \dots, 9. \tag{18.89}$$

Observe that the nonlinearity affects the length but *not* the direction of the vector. For small values of $||s_j||$,

$$v_j \approx ||s_j|| s_j \Rightarrow ||v_j|| \approx ||s_j||^2.$$

For large values of $||\boldsymbol{s}_j||$,

$$\boldsymbol{v}_j \approx \frac{\boldsymbol{s}_j}{||\boldsymbol{s}_j||} \Rightarrow ||\boldsymbol{v}_j|| \approx 1.$$

Note that here lies a major difference of the capsule networks with the more "traditional" neural networks. Both the input and the output of the nonlinearity are *vectors*, instead of scalars. Activations are coded not only in terms of magnitude but also in terms of directions.

TRAINING

The training phase of the capsule networks evolves around the following interrelated concepts.

The loss function bears a close similarity to the hinge loss function that has been used in the context of support vector machines (Section 11.10), and it is defined as

$$\mathcal{L} = \sum_{j=0}^{9} \left\{ T_j \max\{0, m^+ - ||\boldsymbol{v}_j||^2\} + \lambda(1 - T_j) \max\{0, ||\boldsymbol{v}_j||^2 - m^-\} \right\}. \tag{18.90}$$

The loss function comprises two terms. During training, if the input digit corresponds to, say, the kth capsule, then $T_k = 1$ and $T_{j \neq k} = 0$. Thus, for minimum loss value, the length of the correct class capsule k will be adjusted towards values larger than a threshold, m^+, whose value in practice is set equal to 0.9. In contrast, the lengths of the rest of the capsules should be "pushed" towards lengths smaller than m^-, which in practice is set equal to 0.1. The parameter λ is used to downweigh the contribution of the second term and in [202] it is suggested to be set equal to 1/2. The loss value is used to adjust the values of the W_{ij} matrices and the coefficients of the filters used in the convolutions.

The iterative dynamics routing algorithm refers to the update of the coupling coefficients. These coefficients have been "dressed" up with a probability interpretation, which is guaranteed by the following softmax operation:

$$c_{ij} = \frac{\exp(b_{ij})}{\sum_{k=0}^{9} \exp(b_{ik})}, \quad i = 1, 2, \ldots, I, \; j = 0, 1, \ldots, 9.$$

In words, c_{ij} represents the probability that the ith capsule of the *previous* layer is *coupled* to the jth output (class) capsule of the *higher* layer. The computation of the coupling coefficients takes place *iteratively*, according to Algorithm 18.7.

Algorithm 18.7 (Dynamic routing algorithm).

- Initialize $b_{ij} = 0, i = 1, 2, \ldots, I, \; j = 0, 1, 2, \ldots, 9.$
- **For** $r = 1, 2, \ldots, R$, **Do**; In practice, $R = 3$ iterations suffices.
  - **For** $j = 0, 1, 2, \ldots, 9$, **Do**
    - **For** $i = 1, 2, \ldots, I$, **Do**

      $$c_{ij} = \frac{\exp(b_{ij})}{\sum_{k=0}^{9} \exp(b_{ik})}$$

    - **End For**

    * $s_j = \sum_{i=1}^{I} c_{ij} \hat{u}_{j|i}$
    * $v_j = \frac{||s_j||^2}{1+||s_j||^2} \frac{s_j}{||s_j||}$
    * **For** $i = 1, 2, \ldots, I$, **Do**

      $b_{ij} \leftarrow b_{ij} + v_j^T \hat{u}_{j|i}$

    * **End For**
  - **End For**
- **End For**
- **Return** v_j, $j = 0, 1, \ldots, 9$.

A major feature in the algorithm is that the output capsules, v_j, are produced by iteratively adjusting their *similarity* or *agreement* with the transformed outputs/capsules of the previous level. This is achieved via the updates of the b_{ij}s, which are done in a way that "strengthens" or "weakens" their values according to the similarity (inner product) between v_j and $\hat{u}_{j|i}$. Note that every time iterations start by setting $b_{ij} = 0$, i.e., they start from equiprobable values for the coupling coefficients. That is, each i has equal probability with respect to all classes. Then, their values are updated according to the matching between the computed outputs and the respective transformed input capsules. In words, the coupling coefficients *route* the information from the lower layer to the higher one. A large c_{ij} value means a stronger connection of the ith capsule of the lower level to the jth capsule of the layer above. It is important to emphasize that the values of the coupling coefficients are *not* adjusted by the backpropagation that tries to minimize the cost function. The cost function, via the backpropagation of the gradients, is used to adjust the elements of the W_{ij} matrices as well as the filters of the convolutions. The dynamic routing algorithm is the algorithm that produces the outputs. This is done both during *training* and during *testing*, given the values of the matrices W_{ij}. As indicated in Algorithm 18.7, it seems that three iterations suffice.

In [202], a *regularizer via reconstruction* is also used combined with the loss in Eq. (18.90), to form the cost over the training data. During training, the output capsule v_k, associated with the correct class k, is provided as input to a fully connected network, which acts as a decoder to reconstruct the corresponding input. The loss function to train the fully connected network is the sum of squared differences between the input image pixels and the outputs, using sigmoid output nonlinearities. That is, the loss function that is used for training the capsule network becomes

$$\text{loss} = \mathcal{L} + a \times \text{reconstruction error},$$

where $\mathcal{L}$ is given in Eq. (18.90) and $a = 0.005$.

Experiments in [202] with the MNIST database for digit recognition verify that different dimensions (16 in total) of the output capsules are associated with different features of the input digits, e.g., thickness, skew, and width. In an extension of the original work in [93], EM-type arguments are employed for the dynamic routing algorithm.

It should be emphasized that capsule networks are still an evolving methodology and it is early to discuss clear statements concerning their performance. Although for small data sets, such as the MNIST, capsule networks have provided state-of-the art results, their performance on larger data sets is still to be shown.

18.17 DEEP NEURAL NETWORKS: SOME FINAL REMARKS

The topic of deep architectures has been one of intense research and of continuously increasing interest since the early years of this millennium. It would be impossible to cover in a single chapter all techniques, all algorithmic variants, and the related literature that have appeared, and in particular in the framework of specific applications. In the previous sections, major directions and algorithms were reviewed that constitute the spine around which a plethora of variants have been developed. Also, it is the author's view that the previously discussed techniques and models constitute the basic knowledge that one has to grasp in a first reading to be able to proceed further. Some related tutorial papers are [40,90,133,205].

Below, some further concepts are highlighted, which are of high interest at the time the current edition of this book is compiled, and seem to offer ways to improve upon the performance of the basic notions and architectures that have been discussed before.

TRANSFER LEARNING

Transfer learning is not a new idea; it has been around for many years. There are various approaches and concepts that have been developed and applied in different applications (see, for example, [173,242,249]). Our aim here is to present the main concept and not to delve into a more detailed presentation of the various techniques that have been proposed over the years.

As has already been stated in a number of points throughout this chapter, deep neural networks currently represent the state of the art in machine learning. Deep networks have been applied in almost any scientific discipline and in many cases offer human or even superhuman performances (see, e.g., [114,197,229] for such example applications).

However, training of such networks requires huge amounts of labeled data. Collecting and annotating data to build up big data sets is a rather painstaking task. Sometimes, large data sets do exist and are open and public, such as the ImageNet one [39], which contains 1.2 million images with 1000 categories. However, often, they are proprietary or expensive to obtain. For example, this is the case for a number of speech-related data sets. Also, in a number of applications, obtaining big data sets is not possible, as for example in a number of medical-related applications, e.g., X-ray imaging and fMRI, one reason being that it is not easy to obtain large numbers of images corresponding to various types of a disease, e.g., malignant (cancerous) tumors, because a minority of people suffer from it.

In contrast to this demand for huge training data sets, the human brain does not need to be trained from scratch every time it is faced with a new situation. Humans have an inherent mechanism to transfer knowledge and experience acquired from one situation (task) to another one that is sufficiently similar. Such mechanisms mainly operate at a subconscious level. For example, having learned how to bike, the related knowledge and experience can be transferred to learn how to ride a motorbike or a car.

Transfer learning in machine learning is the methodology of bypassing the rationale of the isolated-task learning paradigm. Instead, the goal is to develop techniques so as to utilize knowledge learned from one task to another related one.

To demonstrate the concept, let us focus on an example in the image recognition context. Assume that one is interested in training a network for automatic recognition and characterization of tumors in X-ray imaging. As said before, it is difficult to acquire a large amount of data for training. However, images share a number of features irrespective of their specific type. In other words, an image with a dog, or a car, or a tumor is formed by a combination of edges, primitive shapes, changes in light intensity, etc. As a matter of fact, as we have already commented at the end of Section 18.12.3, the purpose of the various layers in a CNN is to learn representations related to this type of information and code it via the features that are formed hierarchically, layer after layer. Recalling what we have said in our discussion of CNNs, the convolutional layers learn the features. The features of the last convolutional layer are presented as input to the fully connected part of the network, whose output provides the associated prediction.

In the context of our example, the idea of transfer learning is to "borrow" the architecture and the associated parameters of the *convolutional layers*, which have been previously learned by training via a large image database, e.g., ImagNet. Then, fix these parameters and train the network *only* for the parameters comprising the fully connected part of the network (see Fig. 18.29), using the available X-ray images. In other words, we use the same features produced by training the network with images of a different kind. Then, these features are used to train the fully connected network using a task-specific database. Hence, we transfer knowledge previously acquired in an image-related task to facilitate the learning of a similar task. The first task, on which the network is trained, is usually referred to as the *source* task, and the second one, to which knowledge is transferred, as the *target* task.

Another scenario is to use the previously learned parameters via the source task as initial values, instead of fixing them, and retrain the whole network using the X-ray images of the target task. Various scenarios and combinations can be devised, depending on the application and the data size of the target task.

In summary, the main idea is this. Train a (or use a pretrained) deep network employing data acquired for a source task, for which the available data set is large enough for the respective training. Fix the parameters of the lower layers (or use them as initial values) and train for the parameters of the higher layers using the smaller-size data set of the target task (see, e.g., [189]). If the data size for the specific target application task is small, a simple single output layer can be used, e.g., a softmax or a linear SVM.

MULTITASK LEARNING

The information sharing between tasks in transfer learning takes place *sequentially*. One, first, learns the parameters via training on the source task; then, the obtained set of parameters is used for learning the target task, via its associated set of data.

In the so-called *multitask* learning, a number of different tasks are learned *simultaneously*. Multitask learning can be considered as a type of inductive learning, where knowledge is transferred by embedding bias into the learning task. In a way, this is similar in concept with using a regularizer, whose purpose is to bias the solution towards the constraint that is imposed by the regularizer. In the context of multitask learning, the model for each one of the involved tasks is trained so as to *bias* its solution towards the rest of the tasks. It turns out that training networks in the multitask framework robustifies the network model against overfitting, when compared to model training on each task separately.

As it was the case with transfer learning, the idea of multitask learning is not new; it has been around for a number of years in the context of different learners (see, e.g., [4,28]). Among the large number of techniques that have been proposed, two paths will be highlighted here that are more popular in the framework of deep networks (see e.g., [198] and [257] for more recent reviews).

In the so-called *hard parameter sharing* approach, a number of hidden layers are *shared* by all tasks. At the same time, a number of (output) layers are allowed to be task-specific. This is schematically shown in Fig. 18.49. The parameters of all the layers, both shared and task-specific, are learned

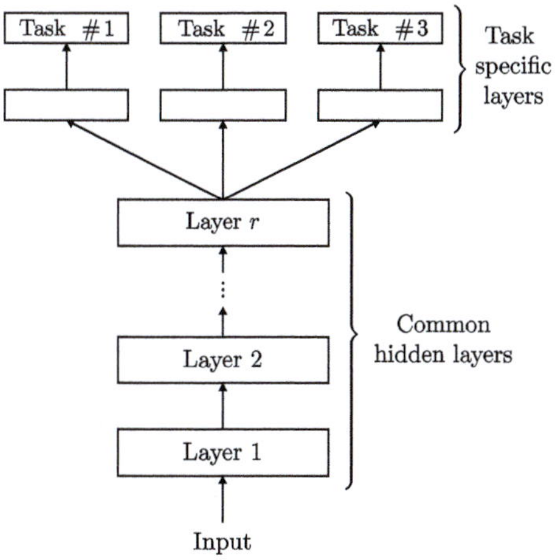

FIGURE 18.49

In the hard parameter sharing multitask learning, a number of hidden layers are shared among the various tasks, while some higher-level layers (e.g., the output layer) are task-specific.

simultaneously. In such a training concept, the overfitting tendency is reduced. This is natural, because learning more than one tasks simultaneously makes the obtained representation able to capture information form all tasks, and hence the tendency for overfitting to any specific one is reduced. An application of this type of multitasking is, for example, in computer vision, where the goal of each task could be to predict the label of a different object. The setting is also very convenient when more than one object is present in an image and the task is to detect the objects' class and their bounding boxes.

Another path is the so-called *soft parameter sharing* multitask approach, which is illustrated in Fig. 18.50. According to this rationale, the models do not share parameters, yet their parameters are constrained to be similar. Similarity is measured according to some distance norm. For example, parameters are constrained so that their distance according to the Euclidean ℓ_2 norm is smaller than a threshold. Other norms can be and have been used, too.

GEOMETRIC DEEP LEARNING

The setting of our discussion in this chapter referred to data that reside in Euclidean spaces. Indeed, all the examples we referred to were images, videos, and time series signals/sequential data. Also, the basic operation in CNNs was convolution, which was defined as an operation on images and time series. However, in many scientific fields the underlying structure is non-Euclidean. Such examples include information that resides in social networks, sensor networks in communications, and brain networks in fMRI, to name but a few. Dealing with such types of data, we no longer have the regularity that underlies sequential data or images. The definition of convolution has to be modified and extended to include information which is related to the geometry of the non-Euclidean domains, such as graphs

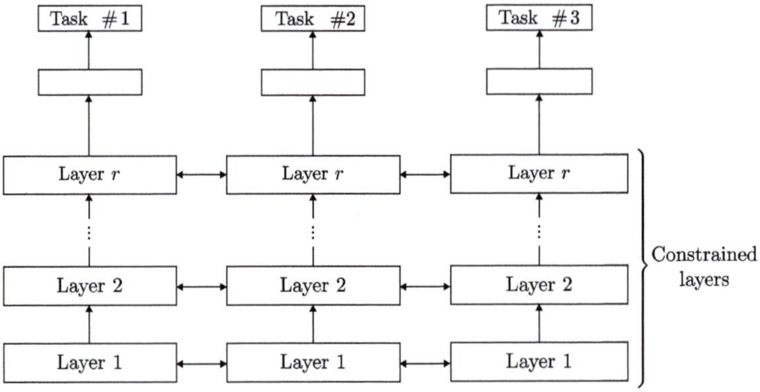

FIGURE 18.50

In the soft parameter sharing multitask learning, the constrained hidden layers exchange parameters; during training, care is taken so that the parameters of the layers for the different tasks, at the same level, are similar in some sense; for example, their ℓ_2 distance is small.

and manifolds. The treatment of such data goes beyond the scope of this chapter. The interested reader can consult, for example, the tutorial given in [23] and the references therein.

OPEN PROBLEMS

Undoubtedly, deep neural networks as a technology have revolutionized the discipline of machine learning. Yet, as we have commented in various parts of this chapter, they are not free of drawbacks. The reader should be aware and alert of these for two reasons. The first is that drawbacks, necessarily, define topics of research with a number of challenging and interesting problems. So, drawbacks have also their positive side; they constitute opportunities. Science and technology are evolutionary processes and no method or theory solves all the problems in one go. The second reason is that we should learn not to "worship" this new tool and consider it as a panacea that provides solutions to all problems.

As has already been stated in other parts of this chapter, deep neural networks have achieved accuracy performance that matches that obtained by humans and in some cases superhuman performance has been demonstrated; yet, these reported results come from carefully curated data sets. At the same time, deep neural networks can be fooled by adversarial examples, while no judicious human would. Furthermore, such impressive performances may not hold true when such models are applied in the "wild" real world and the model is challenged to deal with conditions quite different from those represented in the training set. Also, the reported accuracies are obtained at significantly high computational cost, using energy consuming devices, such as GPUs, and huge amounts of data are required for the training. Such data sets are, sometimes, open but in a number of cases are private property.

Deep neural networks suffer from what we call interpretability of the obtained results. Although advances in this direction have been reported (see, for example, [170] and the references therein), the problem is far from solved. Any machine leaning system should be designed in a way that the user can follow the rationale of each prediction/decision, that is, to be able to answer questions of the type, "why is this person diagnosed with cancer?"

Deep neural networks suffer from what is known as *catastrophic forgetting* [157]. When our human brain learns, say, task A, it can generalize and learn a second one, B, without forgetting A. Existing deep networks tend to forget the previous task A when they learn the new one, B. This is a major drawback for tasks where sequential or incremental learning is of importance, for example, in applications such as gesture recognition, network traffic analysis, or face and object recognition in mobile robots, where the network has to make updates on site and in time. This is an ongoing research topic (see, for example, [182] and the references therein).

Deep neural networks suffer from significant parameter redundancy (e.g., [42]). It has been reported that, in some cases, more than 95% of the parameters can be removed without sacrificing performance. This is the basis behind the pruning techniques that have been reported in this chapter. This is also a topic of ongoing research (see, e.g., [142,174] and the references therein).

Deep neural networks do not scale well for low-power devices, such as those needed in mobile phones, robots, and autonomous cars. Having to make billions of predictions per day has substantial energy costs. As commented before, GPUs are energy consuming devices. Real-time predictions are often orders of magnitude away from what deep neural networks can deliver. Thus, compression and efficiency comprise an important topic of interest in the deep learning research (see, e.g., [125,142,174] and the references therein). *Federated* learning, in the context of what is known as "AI at the edge", is another path in this direction, see, e.g, [20,178].

In a nutshell, although one should be careful in making predictions for the future, I can predict that, in a few years, a new edition will be needed to cover advances related to the ongoing research "happening" in this exciting field.

18.18 A CASE STUDY: NEURAL MACHINE TRANSLATION

Neural machine translation (NMT) is a subtopic of the *natural language processing* (NLP) discipline. The latter is broadly defined as the scientific area whose goal is the automatic manipulation (process, analyze, and synthesize) of natural languages, such as speech and text, via the use of computers. In Section 11.15, a case study on the closely related topic of authorship identification was discussed. There, a number of terms such as bag-of-words and n-grams were defined and used in the context of SVM classification.

In this concluding section of the current chapter, our goal is not to introduce the reader in the general and broad NLP field. After all, such a treatment could justify a dedicated book. Our aim is more humble and we are only going to focus on some basic neural network models that can efficiently be employed in the framework of automatic translation of phrases from one language to another. The stage of our discussion is that of the RNNs, which have been introduced in Section 18.13. Like speech, the more general language processing/modeling lends itself naturally to be treated via RNNs, since any language is sequential in nature. Moreover, very often, the meaning of a word depends on the part in a phrase this is "located." More importantly, the respective meaning may be in "context"; that is, it often depends on the other words that are present and comprise a specific phrase.

In the sequel, we are only going to focus on some basic models. Once the basics have been grasped, one can search for more detailed and advanced techniques in the related literature. The source of "inspiration" lies in the results summarized in some key papers in the area, such as [8,31,145,224] and the references therein.

The starting point refers to the representation of each word in a computer. We will assume that each word corresponds to a single vector. For example, in the so-called *one-hot representation*, each word corresponds to a specific one-hot vector. The dimension l of these vectors is equal to the number of the words in the vocabulary that is used for the translation purposes. For example, 10000 words seems to be a reasonable size that suffices for many purposes and, of course, for everyday communication. By convention, each one of the, say, 10000 words is assigned a specific number/index between 1 and 10000 and the corresponding one-hot vector, which represents it, has a 1 as its element in the respective position, while the rest of the elements are set equal to 0. There are other, more "fancy," representations but this is out of our scope (see, e.g., [15]). It must be noted that the number of existing words in a language vocabulary can be much larger than the, say, 10000 that are used for practical purposes. For example, in the Greek language, being an ancient one, it is estimated that there are approximately 100000 words and 300000 meanings.

To formulate mathematically the automatic translation task, let us consider, as an example, the French phrase "Je suis étudiant," the English translation of which is "I am a student." The phrase in French consists of three words, each one represented by a one-hot vector, i.e., $\boldsymbol{x}_1, \boldsymbol{x}_2, \boldsymbol{x}_3$, respectively. The breaking of a phrase into individual words is often referred to as *tokenization* and the individual words/vectors as *tokens*. Usually, the end of a phrase is indicated by a special symbol, $\langle\text{EOS}\rangle$, which means "end of sequence," and a special token can be reserved for it. Also, often, the beginning of a phrase is denoted as $\langle\text{SOS}\rangle$, i.e., start of sequence, which is usually a vector of zeros. Finally, since a word may not exist in the selected vocabulary, a token corresponding to an unknown word, e.g., $\langle\text{UNK}\rangle$, may also be used. Thus, the input sequence is

$$\langle\text{SOS}\rangle, \boldsymbol{x}_1, \boldsymbol{x}_2, \boldsymbol{x}_3, \langle\text{EOS}\rangle.$$

For the one-hot representation, each one of the vectors is of the following form:

$$\boldsymbol{x}_n = \overbrace{[0, \dots, 0, \underbrace{1}_{l_n}, 0, \dots, 0]}^{l}{}^T \in \mathbb{R}^l, \;\; n = 1, 2, 3,$$

with the 1 being at the l_n position, which is the corresponding index of the word in the respective vocabulary, assumed to be of size l. In a similar way, the output sequence, which is the translated phrase into English, is written as

$$\langle\text{SOS}\rangle, \boldsymbol{y}_1, \boldsymbol{y}_2, \boldsymbol{y}_3, \boldsymbol{y}_4, \langle\text{EOS}\rangle,$$

where the one-hot vectors of the output sequence indicate the corresponding index of the respective word in the English vocabulary.

In practice, most often, the so-called *embeddings* are employed, instead of the one-hot-vectors. That is, each one-hot vector is projected into a lower-dimensional space via its multiplication with an *embedding matrix*, i.e., $\tilde{\boldsymbol{x}}_n = E\boldsymbol{x}_n$, where E is of appropriate dimensions and its elements are learned during training. Besides the reduction in the respective dimensions, which leads to a substantial reduction of the associated parameters in the involved RNNs, such an embedding can also exploit further semantic relations that exist among different words (see, e.g., [15]). To keep our discussion simple, let us stick to the one-hot representation.

The first observation one can readily make is that the input and output sequences are of a different size. The input sequence is of length $N = 3$ and the output one of length $N' = 4$. Thus, using a standard RNN formulation, as illustrated in Fig. 18.34, is not appropriate, since, there, the input and output sequences are assumed to be of the same length. Hence, a modification to the standard RNN formulation, as this was treated in Section 18.13, is required. Besides the appropriate model architecture, the second major goal is to adopt a suitable loss function that quantifies the fit and the "similarity" between the input and the output sequences. In the current setting, the first criterion that comes to mind is the conditional probability,

$$P(\boldsymbol{y}_1, \ldots, \boldsymbol{y}_{N'} | \boldsymbol{x}_1, \ldots, \boldsymbol{x}_N). \tag{18.91}$$

That is, from *all* possible word combinations of size N' in the English (for our case) dictionary, the *target sequence* should be the one that maximizes the *conditional* probability, given the input (French) sequence. We will now provide answers on how to deal with the previous issues and come up with methods concerning (a) the design of appropriate architectures, (b) the related formulation for the optimizing criterion, and (c) an inference algorithm.

Encoder-decoder design: Since the input and output sequences are of different size, a way out is to use two different RNNs. One will be dedicated to the encoding of the input sequence and the other to

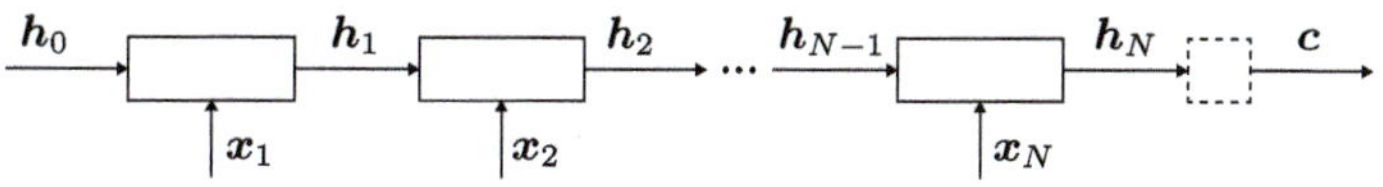

FIGURE 18.51

The RNN encoder receives the input words sequentially and outputs the state vector at time instant N, after all the words have been presented to it. The initial state vector, $\boldsymbol{h}_0$, is set equal to zero. The summary vector, $\boldsymbol{c}$, "summarizes" the information that resides in the input sequence and it is subsequently used to excite the decoder.

the generation (decoding) of the output one. Recall that at the heart of an RNN lies the state vector, which, as stated in Section 18.13, encodes the history of the input sequence. The RNN encoder is shown in Fig. 18.51. It is basically the same as that in Fig. 18.34B, with one exception. There is no output sequence. Furthermore, we have used boxes instead of circles. This is to stress out that, often in practice, the LSTM version is used as the basic architectural unit. So, starting from a, usually, zero input state vector, $\boldsymbol{h}_0$, the output of the encoder is the so-called *representation* or *summary* or *context* vector, $\boldsymbol{c}$, associated with the input sequence. In the simplest model, $\boldsymbol{c}$ is set equal to the state vector at time instant N, i.e., $\boldsymbol{c} = \boldsymbol{h}_N$. Later on, we are going to discuss possible modifications on such a choice. Also, in practice, each one of the boxes in Fig. 18.51 may correspond to a multilayer LSTM, as discussed in Remarks 18.6.

The RNN decoder is shown in Fig. 18.52. Looking at this figure, the following comments are in order. First, the context vector $\boldsymbol{c}$ is directly fed to each one of the RNN stages. There are variants to this. For example, in [224], $\boldsymbol{c}$ only feeds the initial state $\boldsymbol{h}'_0$ and is not utilized in the subsequent stages. That is, in this context, $\boldsymbol{c}$ is only used to "excite" the RNN decoder, instead of the zero value that is

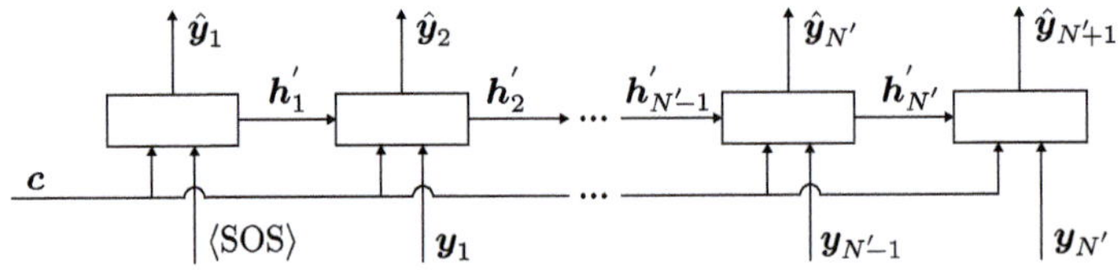

FIGURE 18.52

The RNN decoder uses the summary vector, c, which is provided by the encoder, and the respective inputs are the (delayed) words/vectors of the target output sequence. The decoder is trained to act as a predictor of the target output sequence. At each stage, it outputs a vector of softmax probability estimates for each one of the output vocabulary words, conditioned on the input sequence as well as the previous target output vectors/words (or previous decisions when it is operating in the inference phase). In the figure, we have used an output at time $N' + 1$, and the respective target symbol during training is the ⟨EOS⟩ token.

used to initialize h_0 in the encoder part. Second, the output target values of the previous stage are used as an input to the next one. This is natural. The goal of the decoder is to act as a *predictor*. Having decided on the output word at stage $n-1$, the goal is to *predict* the next word, at stage n, in the output sequence. The update of the state equation for each stage can be written as

$$h'_n = f(y_{n-1}, h'_{n-1}, c),$$

for some nonlinear function f, and the output is of the form

$$\hat{y}_n = g(h'_n). \tag{18.92}$$

We have refrained the above equations from the dependence on the involved parameters (e.g., multiplying matrices, related bias vectors; Section 18.13) for notational simplification and we only focus on the recursive dependence of the involved variables. Recall that the action of the nonlinear functions is element-wise upon their vector arguments. In some of the previously referenced papers, an explicit dependence on c and y_{n-1} is also given in Eq. (18.92).

The nonlinear function, f, can be a simple element-wise sigmoid one, as in simple RNNs, or of a more complex nature, when the LSTM module is involved (see Section 18.13). The nonlinearity, g, is of the softmax type, involving l possible values. That is, $\hat{y}_n$ is an l-dimensional vector that contains the predicted probabilities of selecting each one of the words in the available dictionary. Eventually, selection of the words is performed by making use of these predicted probabilities and the procedure is explained later on in the inference algorithm part. To make sure that there is no confusion with the notation, $\hat{y}_n$ is not the one-hot vector associated with the predicted word. It is the output vector of probabilities at the nth stage of the RNN decoder, in line with the notation reserved for the RNN outputs in Section 18.13.

Thus, we can write

$$P(y_n|y_{n-1}, \ldots, y_1, \mathcal{X}_{1:N}) \approx g(h'_n), \tag{18.93}$$

where we used the symbol $\mathcal{X}_{1:N}$ in place of the sequence $x_1, \ldots, x_N$ for notational convenience. For the sake of clarity and avoiding possible confusion, note that the previous approximation in Eq. (18.93) notationally means that on the right-hand side (vector), only the element corresponding to the index of the word $\boldsymbol{y}_n$ is used. Recall that the conditional probability that is of interest to us is the joint conditional distribution in Eq. (18.91), where, after employing the product rule of probabilities (Section 2.2.2), it is written as

$$P(\boldsymbol{y}_1, \ldots, \boldsymbol{y}_{N'} | \mathcal{X}_{1:N}) = P(\boldsymbol{y}_1 | \mathcal{X}_{1:N}) \prod_{n=2}^{N'} P(\boldsymbol{y}_n | \boldsymbol{y}_{n-1}, \ldots, \boldsymbol{y}_1, \mathcal{X}_{1:N}). \tag{18.94}$$

Note that one can insert for each factor in Eq. (18.94) the corresponding terms as Eq. (18.93) suggests.

The optimizing criterion: The goal is to maximize the conditional joint probability in Eq. (18.91). Considering the corresponding log-likelihood and taking into account Eqs. (18.93) and (18.94), this is equivalent to minimizing the respective cross-entropy over the whole length, N' of the output sequence, i.e.,

$$J(\boldsymbol{\theta}) = -\sum_{n=1}^{N'} \sum_{i=1}^{l} y_{ni} \ln \hat{y}_{ni}, \tag{18.95}$$

where $\boldsymbol{\theta}$ includes *all* the unknown parameters that are involved both in the encoder as well as the decoder part. Indeed, minimizing Eq. (18.95), one maximizes the likelihood at the target sequence, since $y_{ni} = 0$ for any word with index other than the target output one and the probabilities are estimated via the softmax nonlinearity.

In practice, the optimization takes place over all available phrases in the corpus and the cost can be written as

$$J(\boldsymbol{\theta}) = - \sum_{(\mathcal{Y}_{N'}, \mathcal{X}_N) \in \mathcal{D}} \sum_{n=1}^{N'} \sum_{i=1}^{l} y_{ni} \ln \hat{y}_{ni},$$

where $\mathcal{D}$ denotes the corpus and $\mathcal{Y}_{N'}$, $\mathcal{X}_N$ all the possible input–output sequences.

Inference: Having learned the model, given an input phrase one has to decide on the output one. In theory, one should try in the decoder all possible combinations of words and sequence lengths and come up with the one that results in the highest joint conditional probability. Undoubtedly, this is impossible for vocabularies of the size that have been stated. In practice, suboptimal searching techniques are used. On the other extreme to the previously mentioned exhaustive search lie the greedy algorithms that decide on the optimal choice on a stage-by-stage fashion, starting at $n = 1$. Greedy algorithms have been used and commented on in the context of Adaboost (Section 7.10) and in the context of sparsity promoting optimization algorithms (Section 10.2.1).

In a greedy algorithm setting, the optimization becomes computationally feasible, since each time optimization takes place with respect to a single word. For example, assume that at stage $n - 1$ a decision has been reached. Then, the one-hot vector of this word is used in the place of $\boldsymbol{y}_{n-1}$ in the decoder, e.g., via hard thresholding of the respective vector of probabilities at stage $n - 1$ (in the case, of course, that one-hot representation is employed). By searching over all words in the dictionary, the winner at stage n is the one that scores the highest probability. The procedure is repeated until the final word has been predicted; for example, an ⟨EOS⟩ token is detected. The major drawback of such suboptimal algorithms is that they do not consider word combinations, while optimizing stage-by-stage.

Extensions of the previous basic greedy algorithm are usually used, as for example the *beam search* algorithm (e.g., [224]). According to this algorithm, more than one (conditional) probability values, e.g., the highest K, with $K = 5$ being a typical value, are selected at each stage and propagated through time. Although such an algorithm is still suboptimal, in this way, the final decision is based on information that relies on successive symbol combinations and not on each stage individually.

RNN with attention mechanism: The rationale underlying the attention mechanism has been briefly introduced in Section 18.13.2. An interesting variant of the previously discussed RNN encoder-decoder basic model is to combine it with an attention mechanism technique (e.g., [8]). The corresponding architecture is illustrated in Fig. 18.53. In this variant, a bidirectional RNN has been employed for the encoder (see Remarks 18.6). There are two state vectors that are propagated, i.e., one in the forward, $\overrightarrow{\boldsymbol{h}}_n$, and one in the backward direction, $\overleftarrow{\boldsymbol{h}}_n$.

In this case, the context vector becomes time dependent, i.e., $\boldsymbol{c}_n, n = 1, 2, \ldots, N'$, and it is defined as

$$\boldsymbol{c}_n = \sum_{t=1}^{N} \alpha_{nt} \boldsymbol{h}_t,$$

where the weighting coefficients are given by

$$\alpha_{nt} = \frac{\exp(e_{nt})}{\sum_{k=1}^{N} \exp(e_{nk})}, \text{ with } e_{nt} = \phi(\boldsymbol{h}_t^T \boldsymbol{h}'_{n-1}),$$

where ϕ is a nonlinearity that is implemented via a neural network, which is "learned" during training, and $\boldsymbol{h}_n$ is formed as the concatenation of the respective forward ($\overrightarrow{\boldsymbol{h}}_n$) and backward ($\overleftarrow{\boldsymbol{h}}_n$) state vectors. The variable e_{nt} is known as the *alignment* model and quantifies how well the state vectors of the encoder, at time t, and the decoder, at time n, match. Related simulation results have been presented and commented on following Fig. 18.36.

Remarks 18.10.

- A similar modeling rationale has been used in the so-called *language modeling*. In this context, an RNN is trained as a predictor of the next word, given the previous one(s). The RNN learns the respective conditional probabilities. Once the model is trained, one can draw samples from the learned distributions and generate phrases starting from an initial word or a few words.
- The rationale that was previously presented has also been used for image annotations. For example, one could feed the decoder with the vectorized output of a convolutional network that has been trained with images. In other words, the summary vector, $\boldsymbol{c}$, is replaced by the output of the CNN; the latter, instead of being passed to a fully connected network for classification, as Fig. 18.29 suggests, is used to excite the state $\boldsymbol{h}'_0$ of the decoder. The decoder is then trained to annotate the images accordingly (see, e.g., [153]).
- Besides RNNs, other paths are also used for NMT. For example, CNNs have been employed in [55]. In [3,37,235] neither RNNs nor CNNs are used and the neural model is solely based on attention mechanism concepts, and it is claimed that much longer word dependencies can be exploited

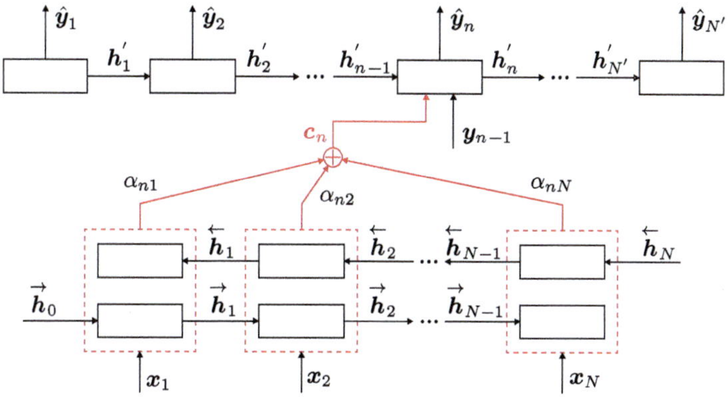

FIGURE 18.53

A bidirectional RNN is used as an encoder with forward ($\overrightarrow{\boldsymbol{h}}_n$) and backward ($\overleftarrow{\boldsymbol{h}}_n$) state vectors propagation. When an attention mechanism is used, at each stage, the decoder is fed with a different context vector, $\boldsymbol{c}_n$. The latter is a linear combination, over all time instants/stages of the encoder, of the combined state vectors (concatenation of $\overrightarrow{\boldsymbol{h}}_n$ and $\overleftarrow{\boldsymbol{h}}_n$). The coefficients α_{ni}, $i = 1, \ldots, N$, are learned during training. The figure focuses on the nth stage of the decoder to unclutter the illustration and assumes N' stages in total.

compared to LSTMs. In [43] and [254] further enhancements are reported via the use of pretraining techniques. At the time this edition is compiled, NMT is still a field of ongoing intense research.

18.19 PROBLEMS

18.1 Prove that the perceptron algorithm, in its pattern-by-pattern mode of operation, converges in a finite number of iteration steps. Assume that $\boldsymbol{\theta}^{(0)} = \mathbf{0}$.
Hint: Note that because classes are assumed to be linearly separable, there is a normalized hyperplane, $\boldsymbol{\theta}_*$, and $\gamma > 0$, so that

$$\gamma \leq y_n \boldsymbol{\theta}_*^T \boldsymbol{x}_n, \quad n = 1, 2, \ldots, N,$$

where, y_n is the respective label, being $+1$ for ω_1 and -1 for ω_2. By the term *normalized hyperplane*, we mean that

$$\boldsymbol{\theta}_*^T = [\hat{\boldsymbol{\theta}}_*, \theta_{0*}]^T, \text{ with } ||\hat{\boldsymbol{\theta}}_*|| = 1.$$

In this case, $y_n \boldsymbol{\theta}_*^T \boldsymbol{x}_n$ is the distance of $\boldsymbol{x}_n$ from the hyperplane $\boldsymbol{\theta}_*$ [167].

18.2 The derivative of the sigmoid functions has been computed in Problem 7.6. Compute the derivative of the hyperbolic tangent activation function (Eq. (18.8)), and show that it is equal to

$$f'(z) = ac\big(1 - f^2(z)\big).$$

18.3 Show that the effect of the momentum term in the gradient descent backpropagation scheme is to effectively increase the learning convergence rate of the algorithm.
Hint: Assume that the gradient is approximately constant over I successive iterations.

18.4 Show that if (a) the activation function is the hyperbolic tangent and (b) the input variables are normalized to zero mean and unit variance, then to guarantee that all the outputs of the neurons have zero mean and unit variance, the weights must be drawn from a distribution of zero mean and standard deviation equal to

$$\sigma = m^{-1/2},$$

where m is the number of synaptic weights associated with the corresponding neuron.
Hint: For simplicity, consider the bias to be zero, and assume that the inputs to each neuron are mutually uncorrelated.

18.5 Consider the sum of the squared errors cost function

$$J = \frac{1}{2}\sum_{n=1}^{N}\sum_{m=1}^{k_L}(\hat{y}_{nm} - y_{nm})^2. \tag{18.96}$$

Compute the elements of the Hessian matrix

$$\frac{\partial^2 J}{\partial \theta_{kj}^r \partial \theta_{k'j'}^{r'}}. \tag{18.97}$$

Near the optimum, show that the second-order derivatives can be approximated by

$$\frac{\partial^2 J}{\partial \theta_{kj}^r \partial \theta_{k'j'}^{r'}} = \sum_{n=1}^{N}\sum_{m=1}^{k_L} \frac{\partial \hat{y}_{nm}}{\partial \theta_{kj}^r}\frac{\partial \hat{y}_{nm}}{\partial \theta_{k'j'}^{r'}}. \tag{18.98}$$

In other words, the second-order derivatives can be approximated as products of the first-order derivatives. The derivatives can be computed by following similar arguments as the gradient descent backpropagation scheme [74].

18.6 It is common when computing the Hessian matrix to assume that it is diagonal. Show that under this assumption, the quantities

$$\frac{\partial^2 E}{\partial (\theta_{kj})^2},$$

where

$$E = \sum_{m=1}^{k_L}\left(f(z_m^L) - y_m\right)^2,$$

propagate backward according to the following:

- $$\frac{\partial^2 E}{\partial (\theta_{kj}^r)^2} = \frac{\partial^2 E}{\partial (z_j^r)^2}(y_k^{r-1})^2,$$

- $\dfrac{\partial^2 E}{\partial (z_L^r)^2} = f''(z_j^L)e_j + (f'(z_j^L))^2,$
- $\dfrac{\partial^2 E}{\partial (z_k^{r-1})^2} = (f'(z_j^{r-1}))^2 \sum_k \dfrac{\partial^2 E}{\partial (z_k^r)^2}(\theta_{kj}^r)^2 + f''(z_j^{r-1}) \sum_{k=1}^{k_r} \theta_{kl}^r \delta_k^r.$

18.7 Show that if the activation function is the logistic sigmoid and the loss function in (18.44) is used, then δ_{nj}^L in (18.21) becomes

$$\delta_{nj}^L = a(\hat{y}_{nj} - y_{nj}).$$

If the cross-entropy in Eq. (18.41) is used, then it is equal to

$$\delta_{nj}^L = a y_{nj}(\hat{y}_{nj} - 1).$$

18.8 Show that if the activation function in the output layer nodes is the logistic sigmoid and the relative entropy cost function is used, then δ_{nj}^L in Eq. (18.21) becomes

$$\delta_{nj}^L = a y_{nj}(\hat{y}_{nj} - 1).$$

18.9 Show that the cross-entropy loss function depends on the relative output errors.

18.10 Show that if the activation function is the softmax and the loss function is the cross-entropy (or the loss in (18.44)), then δ_{nj}^L in (18.21) does not depend on the derivatives of the nonlinearity.

18.11 As in the previous problem, use the relative entropy as the cost function and the softmax activation function. Then show that

$$\delta_{nj}^L = \hat{y}_{nj} - y_{nj}.$$

18.12 Derive the backpropagation through time algorithm for training RNNs.

18.13 Derive the gradient of the log-likelihood in Eq. (18.74).

18.14 Prove that for the case of RBMs, the conditional probabilities are given by the following factorized form:

$$P(\boldsymbol{h}|\boldsymbol{v}) = \prod_{i=1}^{I} \frac{\exp\left(\sum_{j=1}^{J} \theta_{ij} v_j + b_i\right) h_i}{\sum_{h_i'} \left[\exp\left(\sum_{j=1}^{J} \theta_{ij} v_j + b_i\right) h_i'\right]}.$$

18.15 Derive Eq. (18.83).

COMPUTER EXERCISES

18.16 Consider a two-dimensional class problem that involves two classes ω_1 (+1) and ω_2 (−1). Each one of them is modeled by a mixture of equiprobable Gaussian distributions. Specifically, the means of the Gaussians associated with ω_1 are $[-5,\ 5]^T$ and $[5,\ -5]^T$, while the means of the Gaussians associated with ω_2 are $[-5,\ -5]^T$, $[0,\ 0]^T$, and $[5,\ 5]^T$. The covariances of all Gaussians are $\sigma^2 I$, where $\sigma^2 = 1$.

(a) Generate and plot a data set X_1 (training set) containing 100 points from ω_1 (50 points from each associated Gaussian) and 150 points from ω_2 (again 50 points from each associated Gaussian). In the same way, generate an additional set X_2 (test set).

(b) Based on X_1, train a two-layer neural network with two nodes in the hidden layer, each one having the hyperbolic tangent as activation function and a single output node with linear activation function,[10], using the standard backpropagation algorithm for 9000 iterations and step size equal to 0.01. Compute the training and test errors, based on X_1 and X_2, respectively. Also, plot the test points as well as the decision lines formed by the network. Finally, plot the training error versus the number of iterations.

(c) Repeat step 18.16b for step size equal to 0.0001 and comment on the results.

(d) Repeat step 18.16b for $k = 1, 4, 20$ hidden layer nodes and comment on the results.

Hint: Use different seeds in the $rand$ MATLAB® function for the train and the test sets. To train the neural networks, use the $newff$ MATLAB® function. To plot the decision region performed by a neural network, first determine the boundaries of the region where the data live (for each dimension determine the minimum and the maximum values of the data points), then apply a rectangular grid on this region, and for each point in the grid compute the output of the network. Then draw this point with different colors according to the class it is assigned to (use, e.g., the "magenta" and "cyan" colors).

18.17 Consider the classification problem of the previous exercise, as well as the same data sets X_1 and X_2.

Consider a two-layer feed-forward neural network as the one in (b) and train it using the adaptive backpropagation algorithm with initial step size equal to 0.0001 and $r_i = 1.05$, $r_d = 0.7$, $c = 1.04$, for 6000 iterations. Compute the training and test errors, based on X_1 and X_2, respectively, and plot the error during training against the number of iterations. Compare the results with those obtained from the previous exercise (b).

18.18 Repeat the previous exercise for the case where the covariance matrix for the Gaussians is $6I$, for 2, 20, and 50 hidden layer nodes, compute the training and the test errors in each case, and draw the corresponding decision regions. Draw your conclusions.

18.19 This is a classification-related exercise to gain experience by playing with the dropout regularization in the context of the MNIST[11] database.

Train a feed-forward neural network using the TensorFlow[12] machine learning framework. You are encouraged to use the Keras[13] high-level API that is already implemented in TensorFlow for simplicity.

(a) Load the data set where the 55000 images are used for training and the 10000 for testing. Normalize image values so that all values are in the [0, 1] interval. In Keras, you may use the functions provided by the *tensorflow.keras.datasets.mnist* module.

(b) Create a feed-forward neural network that consists of an input layer of 784 neurons (number of pixels of the 28×28 input images), two hidden layers of 2000 neurons each, and

---

10 The number of input nodes is equal to the dimensionality of the feature space.

11 http://yann.lecun.com/exdb/mnist/.

12 https://www.tensorflow.org.

13 https://www.tensorflow.org/guide/keras.

an output layer of 10 neurons, where 10 is the number of all possible digit classes. Use the ReLU activation function for all layers except for the output layer, where softmax activation is used. Train this network for 600 epochs using the gradient descent optimizer with constant learning rate 0.01. Employ the cross-entropy cost function. Use a minibatch size equal to 100 or any other value that fits your RAM. In Keras, you may define and train your network using the *Sequential* module.

(c) Repeat step (b) using 50% dropout in the hidden layers.

(d) Repeat step (b) using 50% dropout in the hidden layers and 20% dropout in the input layer.

(e) For the models in (b)–(d), plot the number of classification errors in the testing set for each epoch. Comment on the results.

18.20 The goal of this exercise is to gain experience in using different optimizers and sizes of convolution kernels. Consider the CIFAR-10 common benchmark classification problem using TensorFlow.[14] The task is to classify 32×32 pixel RGB images in 10 categories: *airplane*, *automobile*, *bird*, *cat*, *deer*, *dog*, *frog*, *horse*, *ship*, and *truck*.

(a) Try to reproduce the results for the model described in the tutorial, which is described in the previous link given in the footnote.

(b) Play with the hyperparameters of the model. For example, as a set of parameters (*optimizer*, *initial learning rate*, *convolutional kernel size*), use some of the following combinations:

- optimizer: Adam and stochastic gradient descent.
- initial learning rate: 0.1, 0.01, 0.001, 0.0001, 0.00001.
- convolutional kernel size: 3, 5, or 7.

(c) Comment on the results using the TensorBoard[15] visualization tool provided by TensorFlow.

Hint: You may use and edit the source code provided by the tutorial. Alternatively, you may use the Keras high-level API in order to define, train, and evaluate your model.

18.21 This exercise focuses on the original GAN, which is trained on the MNIST database to "learn" to generate fake hand-written characters.

To implement and train the network, use the TensorFlow framework. Moreover, as in the previous exercise, you are encouraged to use the Keras high-level API for simplicity.

(a) Load the data set of 60000 to be used for training. Normalize image values so that all values are in the $[-1, 1]$ interval. In Keras, you may use the functions provided by the *tensorflow.keras.datasets.mnist* module.

(b) The generator is fed to its input with a noise vector and outputs an image. The dimension of the input noise vector is 100, and its elements are i.i.d. sampled from a normal distribution of zero mean and unit variance. The generator consists of three fully connected layers with 256, 512, and 1024 neurons, respectively. The activation function used for the neurons in the hidden layers is the leaky ReLU with parameter $\alpha = 0.2$. The output layer comprises 784 nodes and the tanh is employed as the respective activation function.

[14] https://www.tensorflow.org/tutorials/images/deep_cnn.

[15] https://www.tensorflow.org/guide/summaries_and_tensorboard.

(c) The discriminator takes as input a 1×784 vector, corresponding to the vectorized form of the 28×28 MNIST images. The discriminator consists of three fully connected layers with 1024, 512, and 256 neurons, respectively. The leaky ReLU activation function is also employed, with $\alpha = 0.2$. During training, use the dropout regularization method, with probability of discarding nodes equal to 0.3. The output layer consist of a single node with a sigmoid activation function.

(d) To train the implemented network, use the two-class (binary) cross-entropy loss function. Adopt the Adam minimizer with step size (learning rate) equal to 2×10^{-3}, $\beta_1 = 0.5$, and $\beta_2 = 0.999$ as parameters for the optimizer. The recommended batch size is 100. Train the network for 400 epochs as follows. For each training loop, (a) generate a random set of input noise and images, (b) generate fake images via the generator, (c) train only the discriminator, and (d) then train only the generator, according to Algorithm 18.5. Play with the number of iterations, associated with the discriminator training.

(e) During training and every 20 epochs, visualize the generated images created by the generator and comment on the evolution of the learning process.

(f) Play with all the various parameters that have been suggested above and see the effect on the training.

18.22 This exercise focuses on the NMT task. For the purposes of this exercise, use the NMT tutorial provided by Tensorflow.[16]

(a) Install the tutorial and the instructions available at the provided link. Subsequently, train the default configuration of the model, which uses LSTMs as both the encoder and the decoder of the model with 128-dimensional hidden states and embeddings, and performs regularization via dropout. Embeddings are obtained through a simple matrix multiplication of the one-hot vectors with a trainable weight matrix. The available data set pertains to a standard German-to-English translation benchmark.

(b) Run the inference algorithm, so as to obtain translations for the available test set. Select the greedy search algorithm, which simply picks at each stage the word corresponding to the highest predictive probability, paying no attention to the underlying temporal dynamics.

(c) How does the performance change if you repeat the previous experiment using beam search with a beam width of K=5. What factor do you think contributes to the observed inferential improvement?

(d) How does the performance change if you increase the beam size $K = 10$? What is your take-home lesson?

(e) Increase the latent representation space dimensionality. Let us postulate hidden layers (as well as embedding layers) of two and three times the initially considered one. Hoes do training and decoding time change? What about the obtained accuracy?

(f) The model comprises a lot of parameters. Therefore, one would expect it to overfit if no countermeasure is adopted. Here, the model uses simple dropout. What happens if you remove this regularization layer across the network?

---

[16] https://github.com/tensorflow/nmt/blob/master/nmt/scripts/wmt16_en_de.sh.

REFERENCES

[1] D. Ackle, G.E. Hinton, T. Sejnowski, A learning algorithm for Boltzmann machines, Cogn. Sci. 9 (1985) 147–169.

[2] T. Adali, X. Liu, K. Sonmez, Conditional distribution learning with neural networks and its application to channel equalization, IEEE Trans. Signal Process. 45 (4) (1997) 1051–1064.

[3] R. Al-Rfou, D. Choe, N. Constant, M. Guo, L. Jones, Character-level language modeling with deeper self-attention, arXiv:1808.04444v2 [cs.CL], 10 December 2018.

[4] A. Argyriou, M. Pontil, Multi-task feature learning, in: Proceedings Advances in Neural Information Processing Systems, NIPS, 2007.

[5] M. Arjovsky, L. Bottou, Towards principled methods for training generative adversarial networks, arXiv:1701.04862v1 [stat.ML], 17 January 2017.

[6] M. Arjovsky, S. Chintala, L. Bottou, Wasserstein GAN, arXiv:1701.07875v3 [stat.ML], 6 December 2017.

[7] S. Arora, S.S. Du, W. Hu, Z. Li, R. Salakhutdinov, R. Wang, On exact computation with an infinitely wide neural net, arXiv:1904.11955v1 [cs.LG], 26 Apr 2019.

[8] D. Bahdanau, K.-H. Cho, Y. Bengio, Neural machine translation by jointly learning to align and translate, arXiv:1409.0473v7 [cs.CL], 19 May 2016.

[9] P. Baldi, K. Hornik, Neural networks and principal component analysis: learning from examples, without local minima, Neural Netw. 2 (1989) 53–58.

[10] R. Balestriero, R. Baraniuk, A spline theory of deep networks, in: Proceedings of the 35th International Conference on Machine Learning, ICML, 2018.

[11] E. Barnard, Optimization for training neural networks, IEEE Trans. Neural Netw. 3 (2) (1992) 232–240.

[12] R.A. Barron, Universal approximation bounds for superposition of a sigmoidal function, IEEE Trans. Inf. Theory 39 (3) (1993) 930–945.

[13] R. Battiti, First and second order methods for learning: between steepest descent and Newton's methods, Neural Comput. 4 (1992) 141–166.

[14] M. Belkin, D. Hsu, S. Ma, S. Mandala, Reconciling modern machine learning practice and the bias-variance trade-off, arXiv:1812.11118v2 [stat.ML], 10 Sep 2019.

[15] Y. Bengio, R. Ducharme, P. Vincent, C. Jauvin, A neural probabilistic language model, J. Mach. Learn. Res. 3 (2003) 1137–1155.

[16] Y. Bengio, O. Delalleau, N. Le Roux, The curse of highly variable functions for local kernel machines, in: Y. Weiss, B. Schölkopf, J. Platt (Eds.), Advances in Neural Information Processing Systems, NIPS, vol. 18, MIT Press, Cambridge, MA, 2006, pp. 107–114.

[17] Y. Bengio, P. Lamblin, D. Popovici, H. Larochelle, Greedy layer-wise training of deep networks, in: B. Schölkopf, J. Platt, T. Hofmann (Eds.), Advances in Neural Information Processing Systems, NIPS, vol. 19, MIT Press, Cambridge, MA, 2007, pp. 153–161.

[18] Y. Bengio, Learning deep architectures for AI, Found. Trends Mach. Learn. 2 (1) (2009) 1–127, https://doi.org/10.1561/2200000006.

[19] C.M. Bishop, Neural Networks for Pattern Recognition, Oxford University Press, Oxford, 1995.

[20] K. Bonawitz, et al., Towards federated learning at scale: system design, arXiv:1902.01046v2 [cs.LG], 22 March 2019.

[21] J.S. Bridle, Training stochastic model recognition algorithms as networks can lead to maximum information estimation parameters, in: D.S. Touretzky, et al. (Eds.), Neural Information Processing Systems, NIPS, vol. 2, Morgan Kaufmann, San Francisco, CA, 1990, pp. 211–217.

[22] W. Brendel, M. Bethge, Approximating CNNs with bag-of-words local features models works surprisingly well on ImageNet, in: Proceedings International Conference on Learning Representations, ICLR, 2019.

[23] M.M. Bronstein, J. Bruna, Y. LeCun, A. Szlam, P. Vandergheynst, Geometric deep learning: going beyond Euclidean data, IEEE Signal Process. Mag. 34 (4) (2017) 18–42.

[24] J. Bruna, S. Mallat, Invariant scattering convolutional networks, IEEE Trans. Pattern Anal. Mach. Intell. (PAMI) 35 (8) (2013) 1872–1886.

[25] J. Bruna, Y. Le Cun, A. Szlam, Learning stable invariant representations with convolutional networks, arXiv:1301.3537v1 [cs.AI], 10 January 2013.

[26] A. Bryson, W. Denham, S. Dreyfus, Optimal programming problems with inequality constraints I: necessary conditions for extremal solutions, J. Am. Inst. Aeronaut. Astronaut. 1 (1963) 25–44.

[27] S. Carter, Z. Armstrong, L. Schubert, I. Johnson, C. Olah, Exploring neural networks with activation atlases, https://distill.pub/2019/activation-atlas/, 2019.

[28] R. Caruana, Multitask learning, Auton. Agents Multi-Agent Syst. 27 (1) (2009) 95–133.

[29] S. Chatzis, Y. Demiris, Echo state Gaussian process, IEEE Trans. Neural Netw. 22 (9) (2011) 1435–1445.

[30] Y. Cho, L.K. Saul, Kernel methods for deep learning, in: Advances in Neural Information Processing Systems (NIPS), 2009, pp. 342–350.

[31] K. Cho, B. van Merriënboer, C. Gulcehre, D. Bahdanau, F. Bougares, H. Schwenk, Y. Bengio, Learning phrase representations using RNN encoder–decoder for statistical machine translation, in: Proceedings of the Conference on Empirical Methods in Natural Language Processing, EMNLP, 2014, pp. 1724–1734.

[32] J. Choi, A.C. Lima, S. Haykin, Kalman filter-trained recurrent neural network equalizers for time varying channels, IEEE Trans. Commun. 53 (3) (2005) 472–480.

[33] A. Cichoki, R. Unbenhauen, Neural Networks for Optimization and Signal Processing, John Wiley, New York, 1993.

[34] N. Cohen, O. Sharir, A. Shashua, On the expressive power of deep learning: a tensor analysis, arXiv:1509.05009v3 [cs.NE], 27 May 2016.

[35] A. Choromanska, M. Henaff, M. Mathieu, G.B. Arous, Y. Le Cun, The loss surfaces of multilayer networks, in: Proceedings of the 18th International Conference on Artificial Intelligence and Statistics, AISTATS, 2015.

[36] G. Cybenko, Approximation by superpositions of a sigmoidal function, Math. Control Signals Syst. 2 (1989) 304–314.

[37] Z. Dai, Z. Yang, Y. Yang, J. Carbonell, Q.V. Le, R. Salakhutdinov, Transformer-XL: attentive language models beyond a fixed-length context, arXiv:1901.02860v3 [cs.LG], 2 June 2019.

[38] Y. Dauphin, R. Pascanu, C. Gulcehre, K. Cho, S. Ganguli, Y. Bengio, Identifying and attacking the saddle point problem in high-dimensional nonconvex optimization, arXiv:1406.2572v1 [cs.LG], 10 June 2014.

[39] J. Deng, W. Dong, R. Socher, L. Li, K. Li, L. Fei-fei, ImageNet: a large-scale hierarchical image database, in: IEEE Conference on Computer Vision and Pattern Recognition, ICCVPR, 2009.

[40] L. Deng, Y. Dong, Deep Learning: Methods and Applications, vol. 7(3–4), Now Publishers, 2014.

[41] M. Denil, L. Bazzani, H. Larochelle, N. de Freitas, Learning where to attend with deep architectures for image tracking, arXiv:1109.3737v1 [cs.AI], 16 September 2011.

[42] M. Denil, B. Shakibi, L. Dinh, N. de Freitas, et al., Predicting parameters in deep learning, in: Proceedings Advances in Neural Information Processing Systems, NIPS, 2013, pp. 2148–2156.

[43] J. Devlin, M.W. Chang, K. Lee, K. Toutanova, BERT: pre-training of deep bidirectional transformers for language understanding, arXiv:1810.04805v2 [cs.CL], 24 May 2019.

[44] L. Dinh, R. Pascanu, S. Bengio, Y. Bengio, Sharp minima can generalize for deep nets, arXiv:1703.04933v2 [cs.LG], 15 May 2017.

[45] S. Du, J. Lee, H. Li, L. Wang, X. Zhai, Gradient descent finds global minima of deep neural networks, arXiv:1811.03804v1 [cs.LG], 9 November 2018.

[46] R. Eldan, O. Shamir, The power of depth for feed-forward neural networks, arXiv:1512.03965v4 [cs.LG], 9 May 2016.

[47] D. Erhan, P.A. Manzagol, Y. Bengio, S. Bengio, P. Vincent, The difficulty of training deep architectures and the effect of unsupervised pretraining, in: Proceedings of the Twelfth International Conference on Artificial Intelligence and Statistics, AISTATS09, 2009, pp. 153–160.

[48] A. Fawzi, M. Davies, P. Frossad, Dictionary learning for fast classification based on soft thresholding, Int. J. Comput. Vis. 114 (2–3) (2015) 306–321.

[49] A. Fawzi, S.M. Moosavi-Dezfooli, P. Frossad, The robustness of deep networks: a geometrical perspective, IEEE Signal Process. Mag. 34 (6) (2017) 50–62.

[50] S.E. Fahlman, Faster learning variations on back-propagation: an empirical study, in: Proceedings Connectionist Models Summer School, Morgan Kaufmann, San Francisco, CA, 1988, pp. 38–51.

[51] W. Fedus, M. Rosca, B. Lakshminarayanan, A.M. Dai, S. Mohamed, I.M. Goodfellow, Many paths to equilibrium: GANs do not need to decrease a divergence at every step, in: Proceedings in International Conference on Learning Representations, ICLR, 2018.

[52] Y. Freund, D. Haussler, Unsupervised Learning of Distributions of Binary Vectors Using Two Layer Networks, Technical Report, UCSC-CRL-94-25, 1994.

[53] K. Fukushima, Neocognitron: a self-organizing neural network model for a mechanism of pattern recognition unaffected by shift in position, Biol. Cybern. 36 (1980) 193–202.

[54] K. Funashashi, On the approximation realization of continuous mappings by neural networks, Neural Netw. 2 (3) (1989) 183–192.

[55] J. Gehring, M. Auli, D. Grangier, D. Yarats, Y.N. Dauphin, Convolutional sequence to sequence learning, arXiv:1705.03122v3 [cs.CL], 25 July 2017.
[56] R. Geirhos, P. Rubisch, C. Michaelis, M. Bethge, F.A. Wichmann, W. Brendel, ImageNet-trained CNNs are biased towards texture; increasing shape bias improves accuracy and robustness, in: Proceedings International Conference on Learning Representations, ICLR, 2019.
[57] R. Girges, G. Sapiro, A.M. Bronstein, Deep neural networks with random Gaussian weights: a universal classification strategy, IEEE Trans. Signal Process. 64 (13) (2015) 3444–3457.
[58] X. Glorot, Y. Bengio, Understanding the difficulty of training deep feedforward neural networks, in: Proceedings of the 13th International Conference on Artificial Intelligence and Statistics, AISTATS, 2010.
[59] I.J. Goodfellow, Multidimensional Downsampled Convolution for Autoencoders, Technical report, University of Montreal, 2010.
[60] I.J. Goodfellow, D. Warde-Farley, M. Mirza, A. Courville, Y. Bengio, Maxout networks, in: Proceedings 30th Proceedings International Conference on Learning Representations, ICLR, 2013.
[61] I.J. Goodfellow, J. Pouget-Abadie, M. Mirza, B. Xu, D. Warde-Farlay, S. Ozair, A. Courville, Y. Bengio, Generative adversarial nets, in: Proceedings Advances in Neural Information Processing Systems, NIPS, 2014, pp. 2672–2680.
[62] I.J. Goodfellow, O. Vinyals, A.M. Saxe, Qualitatively characterizing neural network optimization problems, in: Proceedings International Conference on Learning Representations, ICLR, 2015.
[63] I.J. Goodfellow, J. Slens, C. Szegedy, Explaining and harnessing adversarial examples, in: Proceedings International Conference on Learning Representations, ICLR, 2015.
[64] I.J. Goodfellow, Y. Bengio, A. Courville, Deep Learning, MIT Press, 2016.
[65] A. Graves, A. Mohamed, G.E. Hinton, Speech recognition with deep recurrent neural networks, in: Proceedings International Conference on Acoustics, Speech and Signal Processing, ICASSP, 2013, pp. 6645–6649.
[66] A. Graves, N. Jaitly, Towards end-to-end speech recognition with recurrent neural networks, in: 31st International Conference on Learning Representations, ICLR, 2014.
[67] A. Graves, G. Wagne, I. Danihelka, Neural Turing machines, arXiv:1410.5401 [cs.NE], 10 December 2014.
[68] K. Greff, R.K. Shrivastava, J. Koutnik, B.R. Steunebrick, J. Schmidhuber, LSTM: a search space odyssey, arXiv:1503.04069v1 [cs.NE], 13 March 2015.
[69] K. Greff, R.K. Shrivastava, J. Schmidhuber, Highway and residual networks learn unrolled iterative estimation, arXiv:1612.07771v3 [cs.NE], 14 March 2017.
[70] I. Gulrajani, F. Ahmed, M. Arjovsky, V. Dumoulin, A. Courville, Improved training of Wasserstein GANs, arXiv:1704.00028v3 [cs.LG], 25 December 2017.
[71] R. Hahnloser, R. Sarpeshkar, M.A. Mahowald, R.J. Douglas, H.S. Seung, Digital selection and analogue amplification coexist in a cortex-inspired silicon circuit, Nature 405 (2000) 947–951.
[72] M. Hagiwara, Theoretical derivation of momentum term in backpropagation, in: International Joint Conference on Neural Networks, Baltimore, vol. I, 1991, pp. 682–686.
[73] M. Hardt, B. Recht, Y. Singer, Train fast, generalize better: stability of stochastic gradient descent, arXiv:1509.01240 [cs.LG], 7 February 2016.
[74] B. Hassibi, D.G. Stork, G.J. Wolff, Optimal brain surgeon and general network pruning, in: Proceedings IEEE Conference on Neural Networks, vol. 1, 1993, pp. 293–299.
[75] S. Haykin, Neural Networks, second ed., Prentice Hall, Upper Saddle River, NJ, 1999.
[76] K. He, X. Zhang, S. Ren, J. Sun, Deep residual learning for image recognition, arXiv:1512.03385v1 [cs.CV], 10 December 2015.
[77] K. He, X. Zhang, S. Ren, J. Sun, Delving deep into rectifiers: surpassing human-level performance on image net classification, arXiv:1502.01852v1 [cs.CV], 6 February 2015.
[78] D.O. Hebb, The Organization of Behavior: A Neuropsychological Theory, Wiley, New York, 1949.
[79] J. Hertz, A. Krogh, R.G. Palmer, Introduction to the Theory of Neural Computation, Addison-Wesley, Reading, MA, 1991.
[80] M. Heusel, H. Ramsauer, T. Unterthiner, B. Nessler, S. Hochreiter, GANs trained by a two time-scale update rule converge to a local Nash equilibrium, in: Proceedings in Advances in Neural Information Processing Systems, NIPS, 2017.
[81] G.E. Hinton, T.J. Sejnowski, Optimal perceptual inference, in: Proceedings of the IEEE Conference on Computer Vision and Pattern Recognition, Washington, DC, June, 1983.
[82] G.E. Hinton, Learning distributed representations of concepts, in: Proceedings of the Eighth Annual Conference of the Cognitive Science Society, Amherst, Lawrence Erlbaum, Hillsdale, 1986, pp. 1–12.

[83] G.E. Hinton, T.J. Sejnowski, Learning and relearning in Boltzmann machines, in: D.E. Rumelhart, J.L. McClelland (Eds.), Parallel Distributed Processing: Explorations in the Microstructure of Cognition, vol. 1, MIT Press, Cambridge, MA, 1986, pp. 282–317.

[84] G.E. Hinton, Learning translation invariant recognition in massively parallel networks, in: Proceedings Intl. Conference on Parallel Architectures and Languages, PARLE, 1987, pp. 1–13.

[85] G.E. Hinton, P. Dayan, B.J. Frey, R.M. Neal, The wake-sleep algorithm for unsupervised neural networks, Science 268 (1995) 1158–1161.

[86] G.E. Hinton, S. Osindero, Y. Teh, A fast learning algorithm for deep belief nets, Neural Comput. 18 (2006) 1527–1554.

[87] G.E. Hinton, R. Salakhutdinov, Reducing the dimensionality of data with neural networks, Science 313 (2006) 504–507.

[88] G. Hinton, A Practical Guide to Training Restricted Boltzmann Machines, Technical Report, UTML TR 2010-003, University of Toronto, 2010, http://learning.cs.toronto.edu.

[89] G.E. Hinton, Learning multiple layers of representation, Trends Cogn. Sci. 11 (10) (2010) 428–434.

[90] G. Hinton, L. Deng, D. Yu, G.E. Dahl, A.R. Mohamed, N. Jaitly, A. Senior, V. Vanhoucke, P. Nguyen, T.N. Sainath, B. Kinsbury, Deep neural networks for acoustic modeling in speech recognition: the shared views of four research groups, IEEE Signal Process. Mag. 29 (6) (2012) 82–97.

[91] G.E. Hinton, N. Srivastava, A. Krizhevsky, I. Sutskever, R.R. Salakhutdinov, Improving neural networks by preventing co-adaptation of feature detectors, arXiv:1207.0580v1 [cs.NE], 3 July 2012.

[92] G.E. Hinton, O. Vinyals, J. Dean, Distilling the knowledge in a neural network, arXiv:1503.02531 [stat.ML], 9 March 2014.

[93] G.E. Hinton, S. Sabour, N. Frosst, Matrix capsules with EM routing, in: Proceedings International Conference on Learning Representations, ICLR, 2018.

[94] S. Hochreiter, J. Schmidhuber, Long short-term memory, Neural Netw. 9 (8) (1997) 1735–1780.

[95] S. Hochreiter, Y. Bengio, P. Frascani, J. Schmidhuber, Gradient flow in recurrent nets: the difficulty of learning long-term dependencies, in: J. Kolen, S. Kremer (Eds.), A Field Guide to Dynamical Recurrent Networks, IEEE-Wiley Press, 2001.

[96] K. Hornik, M. Stinchcombe, H. White, Multilayer feedforward networks are universal approximators, Neural Netw. 2 (5) (1989) 359–366.

[97] G.B. Huang, Q.-Y. Zhu, C.-K. Siew, Extreme learning machine: theory and applications, Neurocomputing 70 (2006) 489–501.

[98] G.B. Huang, L. Chen, C.K. Siew, Universal approximation using incremental constructive feedforward networks with random hidden nodes, IEEE Trans. Neural Netw. 17 (4) (2006) 879–892.

[99] G.B. Huang, D.H. Wang, Y. Lan, Extreme learning machines: a survey, Int. J. Mach. Learn. Cybern. 2 (2011) 107–122.

[100] G. Huang, Z. Liu, L. Maaten, K.Q. Weinberger, Densely connected convolutional networks, arXiv:1608.06993v3 [cs.CV], 28 June 2018.

[101] G. Huang, Y. Li, G. Pleiss, Z. Liu, J.E. Hopcroft, K.Q. Weinberger, Snapshot ensembles: train 1, get M for free, in: Proceedings International Conference on Learning Representations, ICLR, 2017.

[102] D.H. Hubel, T.N. Wiesel, Receptive fields, binocular interaction, and functional architecture in the cats visual cortex, J. Physiol. 160 (1962) 106–154.

[103] F. Huszár, How (not) to train your generative model: scheduled sampling, likelihood, adversary?, arXiv:1511.05101v1 [stat.ML], 16 November 2015.

[104] S. Ioffe, C. Szegedy, Batch normalization: accelerating deep network training by reducing internal covariate shift, arXiv: 1502.03167v3 [cs.LG], 2 March 2015.

[105] Y. Ito, Representation of functions by superpositions of a step or sigmoid function and their application to neural networks theory, Neural Netw. 4 (3) (1991) 385–394.

[106] H. Jaeger, The "Echo State" Approach to Analysing and Training Recurrent Neural Networks, Technical Report GMD 148, German National Research Center for Information Technology, 2001.

[107] M. Jaderberg, K. Simonyan, A. Zisserman, K. Kavukcuoglu, Spatial transformer networks, arXiv:1506.02025v3 [cs.CV], 4 February 2016.

[108] K. Jarrett, K. Kavukcuoglu, M. Ranzato, Y. LeCun, What is the best multi-stage architecture for object recognition?, in: IEEE 12th International Conference on Computer Vision, 2009.

[109] R.A. Jacobs, Increased rates of convergence through learning rate adaptation, Neural Netw. 2 (1988) 359–366.

[110] N. Jaitly, G.E. Hinton, Learning a better representation of speech sound waves using restricted Boltzmann machines, in: Proceedings International Conference on Acoustics, Speech and Signal Processing, ICASSP, 2011, pp. 5884–5887.

[111] C. Jin, R. Ge, P. Netrapalli, S.M. Kakade, M.I. Jordan, How to escape saddle points efficiently, in: Proceedings of the 34th International Conference on Machine Learning, ICML, 2017.

[112] E.M. Johanson, F.U. Dowla, D.M. Goodman, Backpropagation learning for multilayer feedforward neural networks using conjugate gradient method, Int. J. Neural Syst. 2 (4) (1992) 291–301.

[113] R. Jozefowitcz, W. Zaremba, I. Sutskever, An empirical exploration of recurrent network architectures, in: Proceedings 32nd International Conference on Machine Learning, ICML, Lille, France, 2015.

[114] A. Kannan, et al., Smart reply: automated response suggestion for email, in: Proceedings of the 22nd ACM SIGKDD International Conference on Knowledge Discovery and Data Mining, 2016.

[115] A. Karpathy, L. Fei-Fei, Deep visual semantic alignments for generating image descriptors, in: IEEE Conference on Computer Vision, CPVR, 2015, pp. 3128–3137.

[116] K. Kawaguchi, Deep learning without poor local minima, arXiv:1605.07110v3 [stat.ML], 27 December 2016.

[117] K. Kawaguchi, L. Kaelbling, Y. Bengio, Generalization in deep networks, arXiv:1710.05468v3 [stat.ML], 22 February 2018.

[118] C. Kerelick, B.L. Sturm, J. Larsen, Deep learning and music adversaries, IEEE Trans. Multimed. 17 (11) (2015) 2059–2071.

[119] N.S. Keskar, D. Mudigere, J. Nocedal, M. Smelyanskiy, P.T.P. Tank, On large batch training for deep learning: generalization gap and sharp minima, arXiv:1609.04836v2 [cs.LG], 9 February 2017.

[120] D.P. Kingma, M. Welling, Auto-encoding variational Bayes, in: Proceedings of the International Conference on Learning Representations, ICLR, 2014.

[121] D.P. Kingma, J.L. Ba, Adam: a method for stochastic optimization, in: Proceedings International Conference on Learning Representations, ICLR, 2015.

[122] R. Kleinberg, Y. Li, An alternative view: when does SGD escape local minima?, arXiv:1802.06175v2 [cs.LG], 16 August 2018.

[123] J. Kohler, H. Daneshmand, A. Lucchi, T. Hofmann, M. Zhou, K. Neymeyr, Exponential convergence rates for batch normalization: the power of length-direction decoupling in non-convex optimization, in: Proceedings of the 22nd International Conference on Artificial Intelligence and Statistics, AISTATS, Naha, Okinawa, Japan, 2019.

[124] A.H. Kramer, A. Sangiovanni-Vincentelli, Efficient parallel learning algorithms for neural networks, in: D.S. Touretzky (Ed.), Advances in Neural Information Processing Systems 1, NIPS, Morgan Kaufmann, San Francisco, CA, 1989, pp. 40–48.

[125] R. Krishnamoorthi, Quantizing deep convolutional networks for efficient inference: a white paper, arXiv:1806.08342v1 [cs.LG], 21 June 2018.

[126] A. Krizhevsky, I. Sutskever, G.E. Hinton, Imagenet classification with deep convolutional networks, in: Advances in Neural Information Processing Systems, NIPS, vol. 25, 2012, pp. 1097–1105.

[127] A. Kurakin, I.J. Goodfellow, S. Bengio, Adversarial examples in the physical world, arXiv:1607.02533v4 [cs.CV], 11 February 2017.

[128] K.J. Lang, A.H. Waibel, G.E. Hinton, A time delay neural network architecture for isolated word recognition, Neural Netw. 3 (1) (1990) 23–43.

[129] Y. LeCun, B. Boser, J.S. Denker, D. Henderson, R.E. Howard, W. Hubbard, L.D. Jackel, Backpropagation applied to handwritten zip code recognition, Neural Comput. 1 (4) (1989) 541–551.

[130] Y. LeCun, J.S. Denker, S.A. Solla, Optimal brain damage, in: D.S. Touretzky (Ed.), Advances in Neural Information Systems, vol. 2, Morgan Kaufmann, San Francisco, CA, 1990, pp. 598–605.

[131] Y. LeCun, L. Bottou, G.B. Orr, K.R. Müller, Efficient BackProp, in: G.B. Orr, K.-R. Müller (Eds.), Neural Networks: Tricks of the Trade, Springer, New York, 1998, pp. 9–50.

[132] Y. LeCun, L. Bottou, Y. Bengio, P. Haffner, Gradient-based learning applied to document recognition, Proc. IEEE 86 (11) (1998) 2278–2324.

[133] J. LeCun, Y. Bengio, G.E. Hinton, Deep learning, Nature 521 (2015) 436–444.

[134] T.S. Lee, D.B. Mumford, R. Romero, V.A.F. Lamme, The role of the primary visual cortex in higher level vision, Vis. Res. 38 (1998) 2429–2454.

[135] T.S. Lee, D. Mumford, Hierarchical Bayesian inference in the visual cortex, J. Opt. Soc. Am. A 20 (7) (2003) 1434–1448.

[136] J. Lee, Y. Bahri, R. Novak, S.S. Schoenholz, J. Pennington, J. Sohl-Dickstein, Deep neural networks as Gaussian processes, in: Proceedings, International Conference on Machine Learning, ICML, 2018.

[137] N. Le Roux, Y. Bengio, Representational power of restricted Boltzmann machines and deep belief networks, Neural Comput. 20 (6) (2008) 1631–1649.

[138] M. Lesno, V.Y. Lin, A. Pinkus, S. Schocken, Multilayer feed-forward networks with a polynomial activation function can approximate any function, Neural Netw. 6 (1993) 861–867.
[139] M. Lin, Q. Chen, S. Yan, Network-in-network, arXiv:1312.4400v3 [CS.NE], 4 March 2014.
[140] F. Locatello, D. Vincent, I. Tolstikhin, G. Rätsch, S. Gelly, B. Schölkopf, Competitive training of mixtures of independent deep generative models, arXiv:1804.11130v4 [cs.LG], 3 March 2019.
[141] I. Loshchilov, F. Hutter, SGDR: stochastic gradient descent with warm restarts, in: Proceedings International Conference on Learning Representations, ICLR, 2017.
[142] C. Louizos, K. Ullrich, M. Welling, Bayesian compression for deep learning, in: 31st Conference on Neural Information Processing Systems, NIPS, 2017.
[143] M. Lucic, K. Kurach, M. Michalski, O. Bousquet, S. Gelly, Are GANs created equal? A large-scale study, arXiv:1711.10337v4 [stat.ML], 29 October 2018.
[144] M. Lukoševičius, H. Jaeger, Reservoir computing approaches to recurrent neural network training, Comput. Sci. Rev. 3 (3) (2009) 127–149.
[145] M.T. Luong, H. Pham, C.D. Manning, Effective approaches to attention-based neural machine translation, arXiv:1508.04025v5 [cs.CL], 20 September 2015.
[146] Z. Lu, H. Pu, F. Wang, Z. Hu, L. Wang, The expressive power of neural networks: a view from the width, in: Advances in Neural Information Processing Systems, NIPS, 2017.
[147] S. Liu, N. Yang, M. Li, M. Zhou, A recursive recurrent neural network for statistical machine translation, in: Proceedings 52nd Annual Meeting of the Association of Computational Linguistics, 2014.
[148] J. Ma, R.P. Sheridan, A. Liaw, G.E. Dahl, V. Svetnik, Deep neural nets as a method for quantitative structure – activity relationships, J. Chem. Inf. Model. 55 (2) (2015) 263–274.
[149] W. Maass, T. Natschläger, H. Markram, Real time computing without stable states: a new framework for neural computation based on perturbations, Neural Comput. 14 (11) (2002) 2531–2560.
[150] A.L. Maas, A.Y. Hannun, A.Y. Ng, Rectifier nonlinearities improve neural network acoustic models, in: Proceedings 30th Proceedings International Conference on Learning Representations, ICLR, 2013.
[151] D.J.C. MacKay, A practical Bayesian framework for back-propagation networks, Neural Comput. 4 (3) (1992) 448–472.
[152] D.J.C. MacKay, The evidence framework applied to classification networks, Neural Comput. 4 (5) (1992) 720–736.
[153] J. Mao, W. Xu, Y. Yang, J. Wang, Z. Huang, A. Yuille, Deep captioning with multimodal recurrent neural networks (m-RNN), in: Proceedings International Conference on Machine Learning, ICML, 2015.
[154] X. Mao, Q. Li, H. Xie, R. Lau, Z. Wang, S.P. Smalley, Least squares generative adversarial networks, in: International Conference on Computer Vision, ICCV, 2017.
[155] A. Makhzani, J. Shlens, N. Jaitly, I.J. Goodfellow, B. Frey, Adversarial autoencoders, arXiv:1511.05644v2 [cs.LG], 25 May 2016.
[156] S. Mallat, Group invariant scattering, Commun. Pure Appl. Math. 65 (10) (2012) 1331–1398.
[157] M. McCloskey, N.J. Cohen, Catastrophic interference in connectionist networks: the sequential learning problem, Psychol. Learn. Motiv. 24 (1989) 109–165.
[158] W. McCulloch, W. Pitts, A logical calculus of ideas immanent in nervous activity, Bull. Math. Biophys. 5 (1943) 115–133.
[159] L. Mescheder, S. Nowozin, A. Geiger, Adversarial variational Bayes: unifying variational autoencoders and generative adversarial networks, arXiv:1701.04722v4 [cs.LG], 11 June 2018.
[160] H. Mhaskar, T. Poggio, Deep vs shallow networks: an approximation theory perspective, Anal. Appl. 14 (6) (2016) 829–848.
[161] T. Miyato, S.I. Maeda, M. Koyama, K. Nakae, S. Ishii, Distributional smoothing with virtual adversarial training, in: Proceedings International Conference of Learning Representations, ICLR, 2016.
[162] G. Montoron, G. Orr, K.R. Müller (Eds.), Neural Networks: Tricks of the Trade, Lecture Notes in Computer Science, 2nd ed., 2012.
[163] G. Montufar, R. Pascanu, K. Cho, Y. Bengio, On the number of linear regions of deep networks, arXiv:1402.1869v2 [stat.ML], 7 June 2014.
[164] S.M. Moosavi-Dezfooli, A. Fawzi, O. Fawzi, P. Frossad, Universal adversarial perturbations, arXiv:1610.08401v1 [cs.CV], 26 October 2016.
[165] D.B. Mumford, On the computational architecture of the neocortex. II. The role of cortico-cortical loops, Biol. Cybern. 66 (1992) 241–251.
[166] A.F. Murray, P.J. Edwards, Enhanced MLP performance and fault tolerance resulting from synaptic weight noise during training, IEEE Trans. Neural Netw. 5 (5) (1994) 792–802.

[167] A.B. Navikoff, On convergence proofs on perceptrons, in: Symposium on the Mathematical Theory of Automata, vol. 12, Polytechnic Institute of Brooklyn, 1962, pp. 615–622.
[168] R.M. Neal, Bayesian Learning for Neural Networks, PhD Thesis, University of Toronto, Dept. of Computer Science, 1994.
[169] Y. Nesterov, A method for unconstrained convex minimization problem with the rate of convergence $\mathcal{O}(1/k^2)$, Sov. Math. Dokl. 27 (2) (1983) 372–376 (translated from Russian Doklady ANSSSR).
[170] C. Olah, A. Satyanarayan, I. Johnson, S. Carter, L. Schubert, K. Ye, A. Mordvintsev, The building blocks of interpretability, https://distill.pub/2018/building-blocks/.
[171] M. O Neill, S.J. Wright, Behavior of accelerated gradient methods near critical points of nonconvex functions, Math. Program. 176 (2019) 403–427.
[172] P. Orponen, Computational complexity of neural networks: a survey, Nord. J. Comput. 1 (1) (1994) 94–110.
[173] S.J. Pan, Q. Yang, A survey on transfer learning, IEEE Trans. Knowl. Data Eng. 22 (10) (2010) 1345–1359.
[174] K. Panousis, S. Chatzis, S. Theodoridis, Nonparametric Bayesian deep networks with local competition, in: Proceedings, International Conference on Machine Learning, ICML, 2019.
[175] N. Papernot, P. McDaniel, X. Wux, S. Jhax, A. Swami, Distillation as a defence to adversarial perturbations against deep neural networks, in: Proceedings 37th IEEE Symposium on Security & Privacy, 2016.
[176] N. Papernot, P. McDaniel, A. Sinhay, M. Wellmany, SoK: towards the science of security and privacy in machine learning, arXiv:1611.03814v1 [cs.CR], 11 November 2016.
[177] N. Papernot, P. McDaniel, I. Goodfellow, S. Jha, Z.B. Celik, A. Swami, Practical black-box attacks against machine learning, arXiv:1602.02697v4 [cs.CR], 19 March 2017.
[178] J. Park, S. Samarakoon, M. Bennis, M. Debbah, Wireless network intelligence at the edge, arXiv:1812.02858v1 [cs.IT], 2 December 2019.
[179] R. Pascanu, T. Mikolov, Y. Bengio, On the difficulty of training recurrent neural networks, in: Proceedings on the 30th International Conference on Machine Learning, Atlanta, Georgia, 2013.
[180] R. Pascanu, C. Gülcehre, K. Cho, Y. Bengio, How to construct deep recurrent neural networks, arXiv:1312.6026v5 [cs. NE], 24 April 2014.
[181] S.J. Perantonis, P.J.G. Lisboa, Translation, rotation, and scale invariant pattern recognition by high-order neural networks and moment classifiers, IEEE Trans. Neural Netw. 3 (2) (1992) 241–251.
[182] B. Pfülb, A. Gepperth, A comprehensive, application-oriented study of catastrophic forgetting in DNNs, in: Proceedings 36th International Conference on Machine Learning, ICML, 2019.
[183] T. Poggio, H. Mhaskar, L. Rosasco, B. Miranda, Q. Liao, Why and when can deep but not shallow networks avoid the curse of dimensionality: a review, Int. J. Autom. Comput. 14 (5) (2017) 503–519.
[184] T. Poggio, Q. Liao, B. Miranda, A. Banburski, X. Boix, J. Hidary, Theory IIIb: generalization in deep networks, arXiv: 1806.11379v1 [cs.LG], 29 June 2018.
[185] V. Popyan, Y. Romano, M. Elad, Convolutional neural networks analysed via convolutional sparse coding, J. Mach. Learn. Res. (JMLR) 18 (2017) 1–52.
[186] R. Rajesh, J.S. Prakash, Extreme learning machines—a review and state-of-the-art, Int. J. Wisdom Based Comput. 1 (1) (2011) 35–49.
[187] S. Ramon y Cajal, Histologia du Systéms Nerveux de l' Homme et des Vertebes, vols. I, II, Maloine, Paris, 1911.
[188] A. Radford, L. Metz, S. Chintala, Unsupervised representation learning with deep convolutional generative adversarial networks, in: Proceedings International Conference on Learning Representations, ICLR, 2015.
[189] A.S. Razavian, H. Azizpour, J. Sullivan, S. Carlsson, CNN features off-the-shelf: an astounding baseline for recognition, arXiv:1403.6382v3 [cs.CV], 12 May 2014.
[190] D.J. Rezende, S. Mohamed, D. Wierstra, Stochastic backpropagation and approximate inference in deep generative models, in: Proceedings 31st International Conference on Machine Learning, PMLR, vol. 32 (2), 2014, pp. 1278–1286.
[191] L.P. Ricotti, S. Ragazzini, G. Martinelli, Learning the word stress in a suboptimal second order backpropagation neural network, in: Proceedings IEEE International Conference on Neural Networks, San Diego, vol. 1, 1988, pp. 355–361.
[192] M. Riedmiller, H. Brau, A direct adaptive method for faster backpropagation learning: the prop algorithm, in: Proceedings of the IEEE Conference on Neural Networks, San Francisco, 1993.
[193] S. Rifai, P. Vincent, X. Muller, X. Gloro, Y. Bengio, Contractive auto-encoders: explicit invariance during feature extraction, in: Proceedings of the 28th International Conference on Machine Learning, ICML, Bellevue, WA, USA, 2011.
[194] J. Rissanen, G.G. Langdon, Arithmetic coding, IBM J. Res. Dev. 23 (1979) 149–162.
[195] F. Rosenblatt, The perceptron: a probabilistic model for information storage and organization in the brain, Psychol. Rev. 65 (1958) 386–408.

[196] F. Rosenlatt, Principles of Neurodynamics: Perceptrons and the Theory of Brain Mechanisms, Spartan, Washington, DC, 1962.
[197] S. Ruan, J.O. Wobbrock, K. Liou, A. Ng, J. Landay, Speech is 3x faster than typing for English and Mandarin text entry on mobile devices, arXiv:submit/1646347 [cs.HC], 23 August 2016.
[198] S. Ruder, An overview of multi-task learning in deep neural networks, arXiv:1706.05098v1 [cs.LG], 15 June 2017.
[199] D.E. Rumelhart, G.E. Hinton, R.J. Williams, Learning representations by backpropagating errors, Nature 323 (1986) 533–536.
[200] O. Russakovsky, et al., Imagenet large scale visual recognition challenge, Int. J. Comput. Vis. 115 (3) (2015) 211–252.
[201] R. Russel, Pruning algorithms: a survey, IEEE Trans. Neural Netw. 4 (5) (1993) 740–747.
[202] S. Sabour, N. Frosst, G.E. Hinton, Dynamic routing between capsules, in: Proceedings 31st Conference on Neural Information Processing Systems, NIPS, 2017.
[203] T. Saliman, I.J. Goodfellow, W. Zaremba, V. Cheung, A. Radford, X. Chen, Improved techniques for training GANs, in: Proceedings, 30th International Conference on Neural Information Processing, NIPS, 2016.
[204] A.M. Saxe, et al., On the information bottleneck theory of deep learning, in: Proceedings International Conference on Learning Representations, 2018.
[205] J. Schmidhuber, Deep learning in neural networks: an overview, arXiv:1404.7828v4 [cs.NE], 8 October 2014.
[206] R.R. Selvaraju, M. Cogswell, A. Das, R. Vedantam, Grad-CAM: visual explanations from deep networks via gradient-based localization, arXiv:1610.02391v3 [cs.CV], 21 March 2017.
[207] Y. Seo, M. Morante, Y. Kopsinis, S. Theodoridis, Unsupervised pre-training of the brain connectivity dynamic using residual D-net, in: Proceedings of the 26th International Conference on Neural Information Processing of the Asian-Pacific Neural Network, 2019.
[208] A. Shafahi, W.R. Huang, C. Studer, S. Feizi, T. Goldstein, Are adversarial examples inevitable?, in: Proceedings of International Conference on Learning Representations, ICLR, 2019.
[209] U. Shaham, Y. Yamada, S. Negahban, Understanding adversarial training: increasing local stability of neural networks through robustness, Neurocomputing 307 (2018) 195–204.
[210] N. Srivastava, G. Hinton, A. Krizhevsky, I. Sutskever, R. Salakhutdinov, Dropout: a simple way to prevent neural networks form overfitting, J. Mach. Learn. Res. (JMLR) 15 (2014) 1929–1958.
[211] R.K. Shrivastava, K. Greff, J. Schmidhuber, Highway networks, arXiv:1505.00387v2 [cs.LG], 3 November 2015.
[212] T. Serre, G. Kreiman, M. Kouh, C. Cadieu, U. Knoblich, T. Poggio, A quantitative theory of immediate visual recognition, in: Progress in Brain Research, Computational Neuroscience: Theoretical Insights Into Brain Function, vol. 165, 2007, pp. 33–56.
[213] J. Sietsma, R.J.F. Dow, Creating artificial neural networks that generalize, Neural Netw. 4 (1991) 67–79.
[214] E.M. Silva, L.B. Almeida, Acceleration techniques for the backpropagation algorithm, in: L.B. Almeida, et al. (Eds.), Proceedings on the EURASIP Workshop on Neural Networks, Portugal, 1990, pp. 110–119.
[215] P.Y. Simard, D. Steinkraus, J. Platt, Best practice for convolutional neural networks applied to visual document analysis, in: Proceedings International Conference on Document Analysis and Recognition, ICDAR, 2003, pp. 958–962.
[216] K. Simonyan, A. Zisserman, Very deep convolutional networks for large scale image recognition, arXiv:1409.1556v6 [cs.CV], 12 April 2015.
[217] P. Smolensky, Information processing in dynamical systems: foundations of harmony theory, in: Parallel Distributed Processing: Explorations in the Microstructure of Cognition, vol. 1, 1986, pp. 194–281.
[218] S. Sonoda, N. Murata, Neural network with unbounded activation functions is universal approximator, Appl. Comput. Harmon. Anal. 42 (2) (2017) 233–268.
[219] D. Soudry, E. Hoffer, M.S. Nason, N. Srebro, The implicit bias of gradient descent on separable data, in: Proceedings International Conference on Learning Representations, ICLR, 2018.
[220] B.L. Sturm, A simple method to determine if a music information retrieval system is a horse, IEEE Trans. Multimed. 16 (6) (2014) 1636–1644.
[221] S. Sukhbaart, A. Szlam, J. Weston, R. Fergus, End-to-end memory networks, arXiv:1503.08895v5 [cs.NE], 24 November 2015.
[222] I. Sutskever, J. Martens, G.E. Hinton, Generating text with recurrent neural networks, in: International Conference on Machine Learning, ICML, 2011.
[223] I. Sutskever, Training Recurrent Neural Networks, PhD Thesis, Department of Computer Science, University of Toronto, 2013.

[224] I. Sutskever, O. Vinyals, Q.V. Le, Sequence to sequence learning with neural networks, arXiv:1409.3215v3 [cs.CL], 14 December 2014.
[225] D. Sussilo, L.F. Abbott, Random walk initialization for training very deep feedforward networks, arXiv:1412.6558v3 [cs.NE], 27 February 2015.
[226] C. Szegedy, et al., Going deeper with convolutions, arXiv:1409.4842v1 [cs.CV], 17 September 2014.
[227] C. Szegedy, W. Zaremba, I. Sutskever, J. Bruma, D. Erhan, I. Goodfellow, R. Fergus, Intriguing properties of neural networks, in: Proceedings International Conference on Learning Representations, ICLR, 2014.
[228] C. Szegedy, V. Vanhoucke, S. Ioffe, J. Shlens, Z. Wojna, Rethinking the inception architecture for computer vision, arXiv: 1512.00567v3 [cs.CC], 11 December 2015.
[229] C. Szegedy, S. Ioffe, V. Vanhoucke, A. Alemi, Inception-v4, inception-ResNet and the impact of residual connections on learning, arXiv:1602.07261v2 [cs.CV], 23 August 2016.
[230] M. Telgarsky, Benefits of depth in neural networks, arXiv:1602.04485 [cs.LG], 27 May 2016.
[231] S. Theodoridis, K. Koutroumbas, Pattern Recognition, fourth ed., Academic Press, Boston, 2009.
[232] L. Theis, A. van den Oord, M. Bethge, A note on the evaluation of generative models, in: Proceedings International Conference on Learning Representation, ICML, 2016.
[233] N. Tishby, N. Zavlansky, Deep learning and the information bottleneck principle, arXiv:1503.02406v1 [cs.LG], 9 March 2015.
[234] I. Tolstikhin, O. Bousquet, S. Gelly, Bernhard Schölkopf, Wasserstein auto-encoders, arXiv:1711.01558v3 [stat.ML], 12 March 2018.
[235] A. Vaswani, et al., Attention is all you need, arXiv:1706.03762v5 [cs.CL], 6 December 2017.
[236] C. Villani, Optimal Transport: Old and New, Springer, Berlin, 2009.
[237] P.E. Utgoff, D.J. Stracuzzi, Many-layered learning, Neural Comput. 14 (2002) 2497–2539.
[238] P. Vincent, H. Larochelle, I. Lajoie, Y. Bengio, P.A. Manzagol, Stacked denoising autoencoders: learning useful representations in a deep network with a local denoising criterion, J. Mach. Learn. Res. 11 (2010) 3371–3408.
[239] P. Vincent, A connection between score matching and denoising autoencoders, Neural Comput. 23 (7) (2011) 1661–1674.
[240] T.P. Vogl, J.K. Mangis, A.K. Rigler, W.T. Zink, D.L. Allcon, Accelerating the convergence of the backpropagation method, Biol. Cybern. 59 (1988) 257–263.
[241] S. Wang, C. Manning, Fast dropout training, in: Proceedings 30th International Conference on Learning Representations, ICLR, 2013.
[242] M. Wang, W. Deng, Deep visual domain adaptation: a survey, Neurocomputing 312 (2018) 135–153.
[243] D. Warde-Farley, I.J. Goodfellow, A. Courville, Y. Bengio, An empirical analysis of dropout in piecewise linear networks, arXiv:1312.6197v2 [stat.ML], 2 January 2014.
[244] R.L. Watrous, Learning algorithms for connectionist networks: applied gradient methods of nonlinear optimization, in: Proceedings on the IEEE International Conference on Neural Networks, vol. 2, 1988, pp. 619–627.
[245] S. Wager, S. Wang, P. Liang, Dropout training as adaptive regularization, in: Proceedings in Advances on Neural Information Processing Systems, NIPS, vol. 26, 2013, pp. 351–359.
[246] A.S. Weigend, D.E. Rumerlhart, B.A. Huberman, Backpropagation, weight elimination and time series prediction, in: D. Touretzky, J. Elman, T. Sejnowski, G. Hinton (Eds.), Proceedings, Connectionist Models Summer School, 1990, pp. 105–116.
[247] W. Wiegerinck, A. Komoda, T. Heskes, Stochastic dynamics on learning with momentum in neural networks, J. Phys. A 25 (1994) 4425–4437.
[248] X. Wei, B. Gong, Z. Liu, W. Lu, L. Wang, Improving the improved training of Wasserstein GANs: a consistency term and its dual effect, in: Proceedings International Conference on Learning Representation, ICLR, 2018.
[249] K. Weiss, T.M. Khoshgoftaar, D. Wang, A survey of transfer learning, J. Big Data 3 (2016) 9, https://doi.org/10.1186/s40537-016-0043-6.
[250] P.J. Werbos, Beyond Regression: New Tools for Prediction and Analysis in the Behavioral Sciences, PhD Thesis, Harvard University, Cambridge, MA, 1974.
[251] S. Weston, S. Chopra, A. Bordes, Memory networks, arXiv:1410.3916v11 [cs.AI], 29 November 2015.
[252] B. Widrow, M.E. Hoff Jr., Adaptive switching networks, in: IRE WESCON Convention Record, 1960, pp. 96–104.
[253] K. Xu, J. Lei Ba, R. Kiros, K. Cho, A. Courville, R. Salakhutdinov, R.S. Zemel, Y. Bengio, Show, attend and tell: neural image caption generation with visual attention, arXiv:1502.03044v3 [cs.LG], 19 April 2016.
[254] Z. Yang, Z. Dai, Y. Yang, J. Carbonell, R. Salakhutdinov, Q.V. Le, XLNet: generalized autoregressive pre-training for language understanding, arXiv:1906.08237v1 [cs.CL], 19 June 2019.

[255] G. Yarin, Z. Ghahramani, Dropout as a Bayesian approximation: representing model uncertainty in deep learning, in: Proceedings International Conference on Learning Representations, ICLR, 2016.
[256] C. Zhang, S. Bengio, M. Hardt, B. Recht, O. Vinyals, Understanding deep learning requires rethinking generalization, arXiv:1611.03530v2 [cs.LG], 26 February 2017.
[257] Y. Zhang, Q. Yang, A survey on multi-task learning, arXiv:1707.08114v2 [cs.LG], 27 July 2018.
[258] J. Zhao, M. Mathiew, Y. Le Cun, Energy-based generative adversarial networks, in: Proceedings, International Conference on Learning Representations, ICLR, 2017.
[259] W. Zarembe, I. Sutskever, Reinforcement learning neural Turing machines, arXiv:1505.00521v3 [cs.LG], 12 January 2016.
[260] M.D. Zeiler, AdaDelta: an adaptive learning rate method, arXiv:1212.5701v1 [cs.LG], 22 December 2012.
[261] M. Zeiler, R. Fergus, Visualizing and understanding convolutional networks, arXiv:1311.2901v3 [cs.CV], 28 November 2013.
[262] B. Zhou, A. Lapedvisa, J. Xiao, A. Torrabla, A. Oliva, Learning deep features for scene recognition using places database, in: Advances in Neural Information Processing Systems, NIPS, vol. 27, 2014, pp. 487–495.
[263] J. Zourada, Introduction to Artificial Neural Networks, West Publishing Company, St. Paul, MN, 1992.

CHAPTER

DIMENSIONALITY REDUCTION AND LATENT VARIABLE MODELING

19

CONTENTS

Machine Learning. https://doi.org/10.1016/B978-0-12-818803-3.00031-3

19.1 INTRODUCTION

In many practical applications, although the data reside in a high-dimensional space, the true dimensionality, known as *intrinsic dimensionality*, can be of a much lower value. We have met such cases in the context of sparse modeling in Chapter 9. There, although the data lay in a high-dimensional space, a number of the components were known to be zero. The task was to learn the locations of the zeros; this is equivalent to learning the specific subspace, which is determined by the locations of the nonzero components. In this chapter, the goal is to treat the task in a more general setting and assume that the data can lie in any possible subspace (not only the ones formed by the removal of coordinate axes) or manifold. For example, in a three-dimensional space, the data may cluster around a straight line, or around the circumference of a circle or the graph of a parabola, arbitrarily placed in $\mathbb{R}^3$. In all previous cases, the intrinsic dimensionality of the data is equal to one, as any of these curves can equivalently be described in terms of a single parameter. Fig. 19.1 illustrates the three cases. Learning the lower-dimensional structure associated with a given set of data is gaining in importance in the context of *big data* processing and analysis. Some typical examples are the disciplines of computer vision, robotics, medical imaging, and computational neuroscience.

The goal of this chapter is to introduce the reader to the main directions, which are followed in this topic, starting from more classical techniques, such as principal component analysis (PCA) and factor analysis, both in their standard as well as in their probabilistic formulations. Canonical correlation analysis (CCA), independent component analysis (ICA), nonnegative matrix factorization (NMF), and

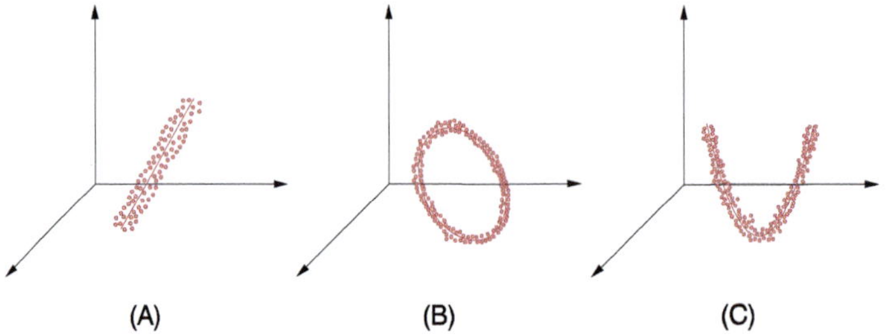

FIGURE 19.1

The data reside close to (A) a straight line, (B) the circumference of a circle, and (C) the graph of a parabola in the three-dimensional space. In all three cases, the intrinsic dimensionality of the data is equal to one. In (A) the data are clustered around a (translated/affine) linear subspace and in (B) and (C) around one-dimensional manifolds.

dictionary learning techniques are also discussed; in the latter case, data are represented via an expansion in terms of overcomplete dictionaries, and sparsity-related arguments are mobilized to detect the most relevant atoms in the dictionary. Finally, nonlinear techniques for learning (nonlinear) manifolds are presented such as the kernel PCA, the local linear embedding (LLE), and the isometric mapping (ISOMAP) techniques. At the end of the chapter, a case study in the context of fMRI data analysis is presented.

19.2 INTRINSIC DIMENSIONALITY

A data set $\mathcal{X} \subset \mathbb{R}^l$ is said to have *intrinsic dimensionality* $m \leq l$ if $\mathcal{X}$ can be (approximately) described in terms of m free parameters. Take as an example the case where the vectors in $\mathcal{X}$ are generated as functions in terms of m random variables, that is, $\mathbf{x} = \mathbf{g}(\mathrm{u}_1, \ldots, \mathrm{u}_m)$, $\mathrm{u}_i \in \mathbb{R}$, $i = 1, \ldots, m$. The corresponding geometric interpretation is that the respective observation vectors will lie along a manifold, whose form depends on the vector-valued function $\boldsymbol{g} : \mathbb{R}^m \longmapsto \mathbb{R}^l$. Let us consider the case where

$$\mathbf{x} = [r \cos\theta, r \sin\theta]^T,$$

where r is a constant and the random variable $\theta \in [0, 2\pi]$. The data lie along the circumference of a circle of radius r and a single free parameter suffices to describe the data. If now a small amount of noise is added, then the data will be clustered close to the circumference, as for example in Fig. 19.1B, and the intrinsic dimensionality is equal to one. From a statistical point of view, it means that the components of the random vectors are highly correlated. Sometimes, we say that the "effective" dimensionality is lower than the apparent one of the "ambient" space, in which the lower-dimensional manifold lies.

In a more general setting, the data may lie in groups of manifolds or even in groups of clusters or they may follow a special spatial or temporal structure. For example, in the wavelet domain most of the coefficients of an image are close to zero and can be neglected, yet the larger (nonzero) ones have a particular structure that is characteristic of natural images. Such a structured sparsity has been exploited in the JPEG2000 coding scheme. Structured sparsity representations are often met in many big data applications (see, for example, [42]). In this chapter, we will only focus on identifying manifold structures, linear (subspaces/affine subspaces) in the beginning and nonlinear ones later on.

Learning the manifold in which a data set resides can be used to provide a compact low-dimensional encoding of a high-dimensional data set, which can subsequently be exploited for performing processing and learning tasks in a much more efficient way. Also, dimensionality reduction can be used for data visualization.

19.3 PRINCIPAL COMPONENT ANALYSIS

Principal component analysis (PCA) or *Karhunen–Loève transform* is among the oldest and most widely used methods for dimensionality reduction [105]. The assumption underlying PCA, as well as any dimensionality reduction technique, is that the observed data are generated by a system or process that is driven by a (relatively) small number of *latent* (not directly observed) variables. The goal is to learn this latent structure.

Given a set of observation vectors, $\boldsymbol{x}_n \in \mathbb{R}^l$, $n = 1, 2, \ldots, N$, of a random vector $\mathbf{x}$, which will be assumed to be of zero mean (otherwise the mean/sample mean is subtracted), PCA determines a *subspace* of dimension $m \leq l$, such that after projection on this subspace, the statistical variation of the data is optimally retained. This subspace is defined in terms of m *mutually orthogonal axes*, known as *principal axes* or *principal directions*, which are computed so that the variance of the data, after projection on the subspace, is maximized [95].

We will derive the principal axes in a step-wise fashion. First, assume that $m = 1$ and the goal is to find a single direction in $\mathbb{R}^l$ so that the variance of the corresponding projections of the data points is maximized. Let $\boldsymbol{u}_1$ denote the principal axis. The variance of the projections (having assumed centered data) is given by

$$\begin{aligned} J(\boldsymbol{u}_1) &= \frac{1}{N}\sum_{n=1}^{N}(\boldsymbol{u}_1^T\boldsymbol{x}_n)^2 = \frac{1}{N}\sum_{n=1}^{N}(\boldsymbol{u}_1^T\boldsymbol{x}_n)(\boldsymbol{x}_n^T\boldsymbol{u}_1) \\ &= \boldsymbol{u}_1^T\hat{\Sigma}\boldsymbol{u}_1, \end{aligned}$$

where

$$\hat{\Sigma} := \frac{1}{N}\sum_{n=1}^{N}\boldsymbol{x}_n\boldsymbol{x}_n^T \tag{19.1}$$

is the sample covariance matrix of the data. For large values of N or if the statistics can be computed, the covariance (instead of the sample covariance) matrix can be used. The task now becomes that of maximizing the variance. However, because we are only interested in directions, the principal axis will be represented by the respective unit norm vector. Thus, the optimization task is cast as

$$\boldsymbol{u}_1 = \arg\max_{\boldsymbol{u}} \boldsymbol{u}^T\hat{\Sigma}\boldsymbol{u}, \tag{19.2}$$

$$\text{s.t.} \quad \boldsymbol{u}^T\boldsymbol{u} = 1. \tag{19.3}$$

This is a constrained optimization problem and the corresponding Lagrangian is given by

$$L(\boldsymbol{u}, \lambda) = \boldsymbol{u}^T\hat{\Sigma}\boldsymbol{u} - \lambda(\boldsymbol{u}^T\boldsymbol{u} - 1). \tag{19.4}$$

Taking the gradient and setting it equal to zero we get

$$\hat{\Sigma}\boldsymbol{u} = \lambda\boldsymbol{u}. \tag{19.5}$$

In other words, the principal direction is an eigenvector of the sample covariance matrix. Plugging Eq. (19.5) into Eq. (19.2) and taking into account (19.3), we obtain

$$\boldsymbol{u}^T\hat{\Sigma}\boldsymbol{u} = \lambda. \tag{19.6}$$

Hence, the variance is maximized if $\boldsymbol{u}_1$ is the eigenvector that corresponds to the maximum eigenvalue, λ_1. Recall that because the (sample) covariance matrix is symmetric and positive semidefinite, all the eigenvalues are real and nonnegative. Assuming $\hat{\Sigma}$ to be invertible (hence, necessarily, $N > l$), the eigenvalues are all positive, that is, $\lambda_1 > \lambda_2 > \ldots \lambda_l > 0$, and we also assume they are distinct, in order to simplify the discussion.

The second principal component is selected so that it (a) is orthogonal to $\boldsymbol{u}_1$ and (b) maximizes the variance after projecting the data onto this direction. Following similar arguments as before, a similar optimization task results with an extra constraint, $\boldsymbol{u}^T\boldsymbol{u}_1 = 0$. It can easily be shown (Problem 19.1) that the second principal axis is the eigenvector corresponding to the second-largest eigenvalue, λ_2. The process continues until m principal axes have been obtained; they are the eigenvectors corresponding to the m largest eigenvalues.

PCA, SVD, AND LOW RANK MATRIX FACTORIZATION

The SVD decomposition of a matrix was discussed in Section 6.4. Given a matrix $X \in \mathbb{R}^{l\times N}$, we can write

$$X = UDV^T. \tag{19.7}$$

For a rank r matrix X, U is the $l \times r$ matrix having as columns the eigenvectors corresponding to the r nonzero eigenvalues of XX^T, and V is the $N \times r$ matrix having as columns the respective eigenvectors of X^TX; D is a square $r \times r$ diagonal matrix comprising the singular values[1] $\sigma_i := \sqrt{\lambda_i}$, $i = 1, 2, \ldots, r$. If we construct X to have as columns the data vectors $\boldsymbol{x}_n$, $n = 1, 2, \ldots, N$, then XX^T is a scaled version of the corresponding sample covariance matrix, $\hat{\Sigma}$; hence, the respective eigenvectors coincide and the corresponding eigenvalues are equal within a scaling factor (N). Without harming generality, we can assume XX^T to be full rank ($r = l < N$), and Eq. (19.7) becomes

$$X = \underbrace{[\boldsymbol{u}_1, \ldots, \boldsymbol{u}_l]}_{l\times l} \underbrace{\begin{bmatrix} \sqrt{\lambda_1}\boldsymbol{v}_1^T \\ \vdots \\ \sqrt{\lambda_l}\boldsymbol{v}_l^T \end{bmatrix}}_{l\times N} = [\boldsymbol{u}_1, \ldots, \boldsymbol{u}_l] \begin{bmatrix} \sqrt{\lambda_1}v_{11} & \ldots & \sqrt{\lambda_1}v_{1n} & \ldots & \sqrt{\lambda_1}v_{1N} \\ \vdots & \vdots & \vdots & \vdots & \vdots \\ \sqrt{\lambda_l}v_{l1} & \ldots & \sqrt{\lambda_l}v_{ln} & \ldots & \sqrt{\lambda_l}v_{lN} \end{bmatrix}. \tag{19.8}$$

Thus, the columns of X can be written in terms of the following expansion[2]:

$$\boldsymbol{x}_n = \sum_{i=1}^{l} z_{ni}\boldsymbol{u}_i = \sum_{i=1}^{m} z_{ni}\boldsymbol{u}_i + \sum_{i=m+1}^{l} z_{ni}\boldsymbol{u}_i, \tag{19.9}$$

where $\boldsymbol{z}_n^T := [z_{n1}, \ldots, z_{nl}]$ is the nth column of the $l \times N$ matrix on the right-hand side in Eq. (19.8). That is, by definition, $z_{n1} = \sqrt{\lambda_1}v_{1n}$ and so on. The sum in Eq. (19.9) has been split into two terms,

[1] Because in some places we are going to involve the variance σ^2, we will carry on working with the square root of the eigenvalues, to avoid possible confusion.

[2] Note that what we have defined in previous chapters as the data matrix is the transpose of X. This is because, for dimensionality reduction tasks, it is more common to work with the current notational convention. If the transpose of X is used, the expansion of the data vectors is in terms of the columns of V and the analysis carries on in a similar way.

where m can be any value $1 \le m \le l$. Note that, due to the orthonormality of the $\boldsymbol{u}_i$s,

$$z_{ni} = \boldsymbol{u}_i^T \boldsymbol{x}_n, \; i = 1, 2, \ldots, l, \; n = 1, 2, \ldots, N.$$

From Section 6.4, we know that the best, in the Frobenius sense, m rank matrix approximation of X is given by

$$\hat{X} = \underbrace{[\boldsymbol{u}_1, \ldots, \boldsymbol{u}_m]}_{l \times m} \underbrace{\begin{bmatrix} \sqrt{\lambda_1}\boldsymbol{v}_1^T \\ \vdots \\ \sqrt{\lambda_m}\boldsymbol{v}_m^T \end{bmatrix}}_{m \times N}, \tag{19.10}$$

$$= \sum_{i=1}^{m} \sqrt{\lambda_i}\boldsymbol{u}_i \boldsymbol{v}_i^T. \tag{19.11}$$

Recalling the previous definition of z_{ni}, the nth column vector of $\hat{X}$ can now be written as

$$\hat{\boldsymbol{x}}_n = \sum_{i=1}^{m} z_{ni}\boldsymbol{u}_i. \tag{19.12}$$

Comparing Eqs. (19.9) and (19.12) and taking into account the orthonormality of $\boldsymbol{u}_i$, $i = 1, 2 \ldots, l$, we readily see that $\hat{\boldsymbol{x}}_n$ is the projection of the original observation vectors, $\boldsymbol{x}_n$, $n = 1, 2, \ldots, N$, onto the subspace span$\{\boldsymbol{u}_1, \ldots, \boldsymbol{u}_m\}$ generated by the m principal axes of XX^T ($\hat{\Sigma}$) (Fig. 19.2).

The previous arguments establish a bridge between PCA and SVD. In other words, the principal axes can be obtained via the SVD decomposition of X. Moreover, the columns of the best m rank matrix approximation, $\hat{X}$, of X are the projections of the observation vectors $\boldsymbol{x}_n$ on the (optimally) reduced in dimension subspace, spanned by the principal axes.

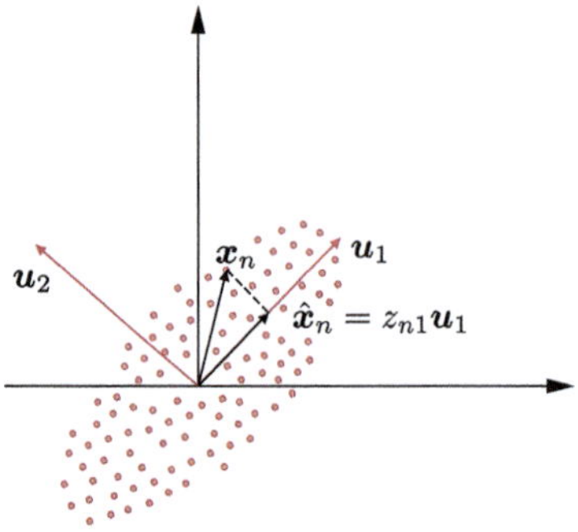

FIGURE 19.2

The projection of $\boldsymbol{x}_n$ on the principal axis $\boldsymbol{u}_1$ is given by $\hat{\boldsymbol{x}}_n = z_{n1}\boldsymbol{u}_1$, where $z_{n1} = \boldsymbol{u}_1^T \boldsymbol{x}_n$.

Looking at Eq. (19.10), PCA can also be seen as a *low rank matrix factorization* method. Matrix factorization will be a recurrent theme in this chapter. Given a matrix X, there is not a unique way to factorize it, in terms of two matrices. PCA provides an m rank matrix factorization of X, by imposing *orthogonality* on the structure of the involved factors. Later on, we are going to discuss other approaches.

Finally, it is important to emphasize that the bridge between PCA and SVD establishes a connection between the low rank factorization of a matrix X and the intrinsic dimensionality of the subspace in which its column vectors reside, since this is the subspace where maximum variance of the data is guaranteed.

MINIMUM ERROR INTERPRETATION

Having established the bridge between PCA and SVD, another interpretation of the PCA method becomes readily available. Because $\hat{X}$ is the best m rank matrix approximation of X in the Frobenius sense, the quantity

$$||\hat{X} - X||_F^2 := \sum_i \sum_j |\hat{X}(i,j) - X(i,j)|^2 = \sum_{n=1}^{N} ||\hat{\boldsymbol{x}}_n - \boldsymbol{x}_n||^2$$

is minimum; that is, obtaining any other m-dimensional approximation (say, $\tilde{\boldsymbol{x}}_n$) of $\boldsymbol{x}_n$, by choosing to project onto another m-dimensional subspace, would result in higher squared error norm approximation, compared to that resulting from PCA. This is also a strong result that establishes a notable merit of the PCA method as a dimensionality reduction technique. This interpretation goes back to Pearson [146].

PCA AND INFORMATION RETRIEVAL

The previous minimum error interpretation paves the way to build around PCA an efficient searching procedure in identifying similar patterns in large databases. Assume that a number N of prototypes are represented in terms of l features, giving rise to feature vectors, $\boldsymbol{x}_n \in \mathbb{R}^l$, $n = 1, 2, \ldots, N$, which are stored in a database. Given an unknown object, which is represented by a feature vector $\boldsymbol{x}$, the task is to identify to which one among the prototypes this pattern is most similar. Similarity is measured in terms of the Euclidean distance $||\boldsymbol{x} - \boldsymbol{x}_n||^2$. If N and l are large, searching for the minimum Euclidean distance can be computationally very expensive. The idea is to keep in the database the components $\boldsymbol{z}_n^{(m)} := [z_{n1}, \ldots, z_{nm}]^T$ (see Eq. (19.12)) that describe the projections of the N prototypes in span$\{\boldsymbol{u}_1, \ldots, \boldsymbol{u}_m\}$, instead of the original l-dimensional feature vectors. Assuming that m is large enough to capture most of the variability of the original data (i.e., the intrinsic dimensionality of the data is m to a good approximation), then $\boldsymbol{z}_n^{(m)}$ is a good feature vector description because we know that in this case $\hat{\boldsymbol{x}}_n \approx \boldsymbol{x}_n$. Given now an unknown pattern, $\boldsymbol{x}$, we first project it onto span$\{\boldsymbol{u}_1, \ldots, \boldsymbol{u}_m\}$, resulting in

$$\hat{\boldsymbol{x}} = \sum_{i=1}^{m} (\boldsymbol{u}_i^T \boldsymbol{x})\boldsymbol{u}_i := \sum_{i=1}^{m} z_i \boldsymbol{u}_i. \tag{19.13}$$

Then we have

$$||\boldsymbol{x}_n - \boldsymbol{x}||^2 \approx ||\hat{\boldsymbol{x}}_n - \hat{\boldsymbol{x}}||^2 = \left\| \sum_{i=1}^{m} z_{ni}\boldsymbol{u}_i - \sum_{i=1}^{m} z_i \boldsymbol{u}_i \right\|^2$$
$$= ||\boldsymbol{z}_n^{(m)} - \boldsymbol{z}||^2,$$

where $\boldsymbol{z} := [z_1, \dots, z_m]^T$. In other words, Euclidean distances are computed in the lower-dimensional subspace, which leads to substantial computational gains (see, for example, [22,63,160] and the references therein). This method is also known as *latent semantics indexing*.

ORTHOGONALIZING PROPERTIES OF PCA AND FEATURE GENERATION

We will now shed light on PCA from a different angle. We have just discussed, in the context of the information retrieval application, that PCA can also be seen as a feature generation method that generates a set of new feature vectors, $\boldsymbol{z}$, whose components describe a pattern in terms of the principal axes. Let us now assume (to make life easier) that N is large enough and the sample covariance matrix is a good approximation of the (full rank) covariance matrix $\Sigma = \mathbb{E}[\mathbf{x}\mathbf{x}^T]$. We know that any vector $\boldsymbol{x} \in \mathbb{R}^l$ can be described in terms of $\boldsymbol{u}_1, \dots, \boldsymbol{u}_l$, that is,

$$\boldsymbol{x} = \sum_{i=1}^{l} z_i \boldsymbol{u}_i = \sum_{i=1}^{l} (\boldsymbol{u}_i^T \boldsymbol{x})\boldsymbol{u}_i.$$

Our focus now turns to the covariance matrix of the random vectors, $\mathbf{z}$, as $\mathbf{x}$ changes randomly. Taking into account that

$$\mathrm{z}_i = \boldsymbol{u}_i^T \mathbf{x}, \tag{19.14}$$

and the definition of U in Eqs. (19.7) and (19.8), we can write $\mathbf{z} = U^T\mathbf{x}$, and hence

$$\mathbb{E}[\mathbf{z}\mathbf{z}^T] = \mathbb{E}\left[U^T\mathbf{x}\mathbf{x}^T U\right] = U^T \Sigma U.$$

However, we know from linear algebra (Appendix A.2) that U is the matrix that diagonalizes Σ; hence,

$$\mathbb{E}[\mathbf{z}\mathbf{z}^T] = \mathrm{diag}\{\lambda_1, \dots, \lambda_l\}. \tag{19.15}$$

In other words, the new features are *uncorrelated*, that is,

$$\boxed{\mathbb{E}[\mathrm{z}_i \mathrm{z}_j] = 0, \quad i \neq j,\ i, j = 1, 2, \dots, l.} \tag{19.16}$$

Furthermore, note that the variances of z_i are equal to the eigenvalues λ_i, $i = 1, 2, \dots, l$, respectively. Hence, by selecting as features the ones that correspond to the dominant eigenvalues, one has maximally retained the total variance associated with the original features, x_i; indeed, the corresponding total variance is given by the trace of the covariance matrix, which in turn is equal to the sum of the eigenvalues, as we know from linear algebra. In other words, the new set of features, z_i, $i = 1, 2, \dots, m$, represent the patterns in a more compact way, as they are *mutually uncorrelated* and most of the variance is retained. It is common in practice, when the goal is that of feature generation, for each one of the z_is to be normalized to unit variance.

Later on, we will see that a more recent method, known as ICA, imposes the constraint that after a linear transformation (a projection is a linear transformation, after all) the obtained latent variables (components) are statistically independent, which is a much stronger condition than being uncorrelated.

LATENT VARIABLES

The random components z_i, $i = 1, 2, \ldots, m$, are known as *principal components*. Sometimes, their observed values, z_i, are known as *principal scores*. As a matter of fact, the principal components comprise the *latent* variables, which we mentioned at the beginning of this section.

According to the general (linear) latent variable modeling approach, we assume that our l variables comprising $\mathbf{x}$ are modeled as

$$\mathbf{x} \approx A\mathbf{z}, \tag{19.17}$$

where A is an $l \times m$ matrix and $\mathbf{z} \in \mathbb{R}^m$ is the corresponding set of latent variables. Adopting the PCA model, we have shown that

$$A = [\boldsymbol{u}_1, \ldots, \boldsymbol{u}_m] := U_m,$$

and the model implies that each one of the l components of $\mathbf{x}$ is (approximately) generated in terms of these mutually uncorrelated m latent random variables, that is,

$$\mathrm{x}_i \approx u_{i1}\mathrm{z}_1 + \ldots + u_{im}\mathrm{z}_m. \tag{19.18}$$

Alternatively, in linear latent variable modeling, we can assume that the latent variables can also be recovered by a linear model from the original random variables, as for example,

$$\mathbf{z} = W\mathbf{x}. \tag{19.19}$$

In the case of the PCA approach, we have already seen that

$$W = U_m^T.$$

Eqs. (19.17) and (19.19) constitute the backbone of this chapter, and different methods provide different solutions for computing A or W.

Let us now collect all the principal score vectors, z_n, $n = 1, 2, \ldots, N$, as the columns of the $m \times N$ score matrix Z, that is,

$$Z := [z_1, \ldots, z_N]. \tag{19.20}$$

Then (19.10) can be rewritten in terms of the score matrix

$$\boxed{X \approx U_m Z.} \tag{19.21}$$

Moreover, taking into account the definition of the principal components in Eq. (19.14), we can also write

$$\boxed{Z = U_m^T X.} \tag{19.22}$$

Remarks 19.1.

- A major issue in practice is to select the m dominant eigenvalues. One way is to rank them in descending order, and determine m so that the gap between λ_m and λ_{m+1} is "large." The interested reader can obtain more on this issue in [55,104].
- The treatment so far involved centered quantities. In case we want to approximate the original observation vectors by taking into consideration the respective mean value of the data set, Eq. (19.13) is rephrased as

$$\hat{\boldsymbol{x}} = \bar{\boldsymbol{x}} + \sum_{i=1}^{m} \boldsymbol{u}_i^T (\boldsymbol{x} - \bar{\boldsymbol{x}}) \boldsymbol{u}_i, \tag{19.23}$$

  where $\bar{\boldsymbol{x}}$ is the sample mean (mean if it is known)

$$\bar{\boldsymbol{x}} = \frac{1}{N} \sum_{n=1}^{N} \boldsymbol{x}_n$$

  and $\boldsymbol{x}$ denotes the original (not centered) vector.
- PCA builds upon *global* information spread over *all* the data observations in the set $\mathcal{X}$. Indeed, the main source of information is the sample covariance matrix (XX^T). Thus, PCA is effective if the covariance matrix provides a sufficiently rich description of the data at hand. For example, this is the case for Gaussian-like distributions. In [41], modifications of the standard approach are suggested in order to deal with data having a clustered nature. Soon, we are going to discuss techniques alternative to PCA in order to overcome this drawback.
- Computing the SVD of large matrices can be computationally costly and a number of efficient techniques have been proposed (see, e.g., [1,83,194]). In a number of cases in practice, it turns out that $l > N$. Of course, in this case, the sample covariance is not invertible and some of the eigenvalues are zero. In such scenarios, it is preferable to work with an $X^T X$ ($N \times N$) instead of an XX^T ($l \times l$) matrix. To this end, the relationships given in Section 6.4, in order to obtain $\boldsymbol{u}_i$ from $\boldsymbol{v}_i$, can be employed.
- The treatment of PCA bears a similarity with the Fisher linear discriminant (FLD) method (Chapter 7). They both rely on the eigenstructure of matrices that, in one way or another, encode (co)variance information. However, note that PCA is an *unsupervised* method, in contrast to FLD, which is a *supervised* one. As a consequence, PCA performs dimensionality reduction so as to preserve data variability (variance) while FLD class separability. Fig. 19.3 demonstrates the difference in the resulting (hyper)planes.
- *Multidimensional scaling* (MDS) is another linear technique used to project in a lower-dimensional space, while respecting certain constraints. Given the set $\mathcal{X} \subset \mathbb{R}^l$, the goal is to project onto a lower-dimensional space, so that inner products are optimally preserved; that is, the cost

$$E = \sum_i \sum_j \left(\boldsymbol{x}_i^T \boldsymbol{x}_j - \boldsymbol{z}_i^T \boldsymbol{z}_j \right)^2$$

  is minimized, where $\boldsymbol{z}_i$ is the image of $\boldsymbol{x}_i$ and the sum runs over all the training points in $\mathcal{X}$. The problem is similar to PCA and it can be shown that the solution is given by the eigendecomposition

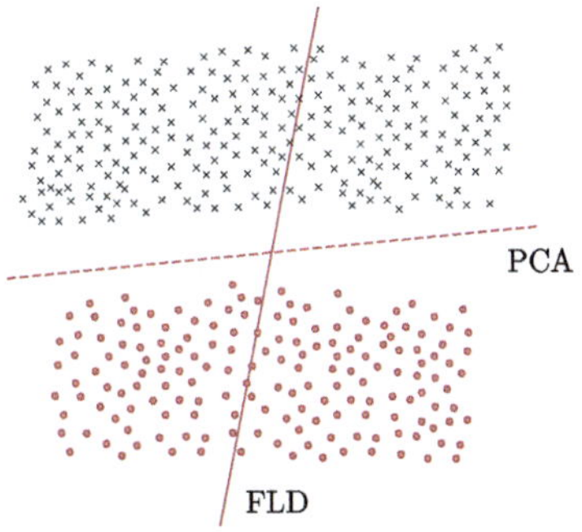

FIGURE 19.3

The case of a two-class task in the two-dimensional space. PCA computes the direction along which the variance is maximally retained after the projections of the data on it. In contrast, FLD computes the line so that the class separability is maximized.

of the Gram matrix,[3] $\mathcal{K} := X^T X$. Another side of the same coin is to require the Euclidean distances, instead of the inner products, to be optimally preserved. A Gram matrix, consistent with the squared Euclidean distances, can then be formed, leading to the same solution as before. It turns out that the solutions obtained by PCA and MDS are equivalent. This can readily be understood as $X^T X$ and XX^T share the same (nonzero) eigenvalues. The corresponding eigenvectors are different, yet they are related, as we have seen while introducing SVD in Section 6.4.

More on these issues can be found in [29,60]. As we will soon see in Section 19.9, the main idea behind MDS of preserving the distances is used, in one way or another, in a number of more recently developed nonlinear dimensionality reduction techniques.

- In a variant of the basic PCA, known as *supervised PCA* [16,195], the output variables in regression or in classification (depending on the problem at hand) are used together with the input ones, in order to determine the principal directions.

Example 19.1. This example demonstrates the power of PCA as a method to represent data in a lower-dimensional space. Each pattern in a database, described in terms of a feature vector, $\boldsymbol{x}_n \in \mathbb{R}^l$, will be represented by a corresponding vector of a reduced dimensionality, $\boldsymbol{z}_n^{(m)} \in \mathbb{R}^m$, $n = 1, 2, \ldots, N$. In this example, each feature vector comprises the pixels of a 168×168 face image. These face images are members of the software-based aligned version [191] of the *Labeled Faces in the Wild* (LFW) database [102]. In particular, among the over 13,000 face images of this database, $N = 1924$ have been selected with criteria such as the quality of the image and the face angle (portraits were of preference). Moreover, the images are zoomed in order to omit most of the background. Examples of the face images used are depicted in Fig. 19.4 and the full collection of all the 1924 images can be found in the companion site of this book.

3 In order to avoid confusion, recall that here X has been defined as the transpose of what we called a data matrix in previous chapters.

FIGURE 19.4

Indicative examples of the face images used.

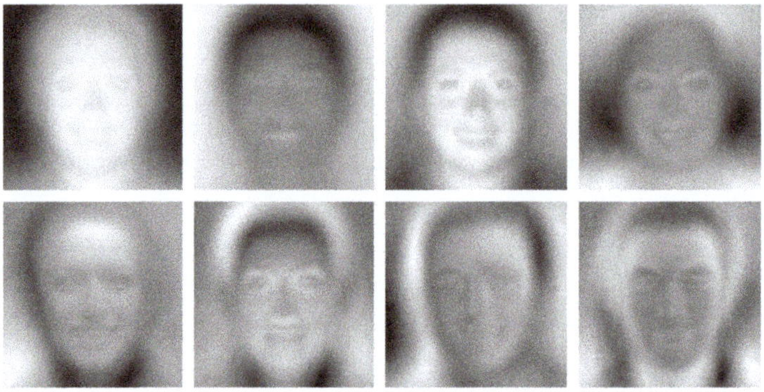

FIGURE 19.5

Examples of eigenfaces.

The images are first vectorized (in $\mathbb{R}^l$, $l = 168 \times 168 = 28,224$) and in the sequel are concatenated in the columns of the $28,224 \times 1924$ matrix X. Moreover, the mean value across each one of the rows is computed and then subtracted from the corresponding element of each column.

In this case, where $l > N$, it is convenient to compute the eigenvectors of $X^T X$, denoted by $\boldsymbol{v}_i$, $i = 1, \dots, N$, and then the principal axis directions, that is, the eigenvectors of XX^T are computed by $\boldsymbol{u}_i \propto X\boldsymbol{v}_i$ (Chapter 6, Eq. (6.18)). These eigenvectors can be rearranged in a matrix form to give 168×168 images, known as *eigenimages*, which in the particular case of face images are referred to as *eigenfaces*. Fig. 19.5 shows examples of eigenfaces resulting from the PCA of matrix X and specifically those corresponding, from top left to bottom right, to the 1st, 2nd, 6th, 7th, 8th, 10th, 11th, and 17th largest eigenvalues.

Next, the quality of reconstruction of an original image, in terms of its lower-dimensional representation, is examined according to Eq. (19.13) for different values of m. As an example, the images depicting Marilyn Monroe and Andy Warhol, shown in Fig. 19.4, are chosen. The results are illustrated in Fig. 19.6. It is observed that for $m = 100$, or even better, for $m = 600$, the resulting approximation is very close to the original images. Note that exact reconstruction will be achieved when the full set of the 1924 eigenfaces is used.

To put our previous findings in an information retrieval context, assume that one has available an image and wants to know what person is depicted in it. Assuming that the image of this person is in the database, the procedure would be (a) to vectorize the image, (b) to project it onto the subspace

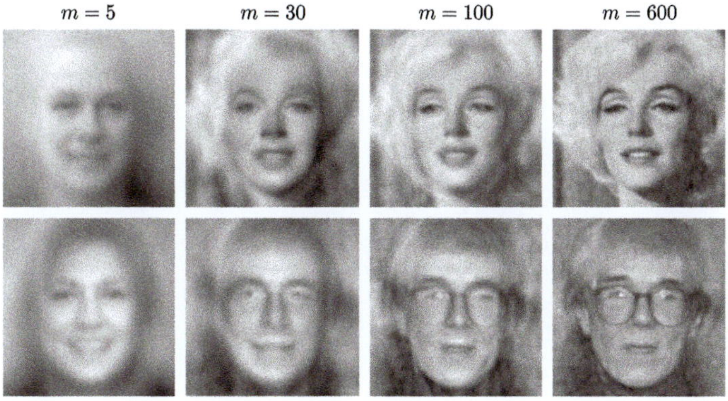

FIGURE 19.6

Image compression and reconstruction based on the first m eigenvectors.

spanned by the, say, $m = 100$ eigenfaces, and (c) to search in this lower-dimensional space to identify the vectorized image in the database that is closest in the Euclidean norm sense. Usually, it is preferable to identify the, say, five or ten most similar images and rank them according to the Euclidean distance (or any other distance) similarity. Then, through the database, he/she can have the name and all the associated information that is kept in the database.

In information retrieval, each one of the images in the database could be stored in terms of the corresponding vector of the principal scores.

Example 19.2. In this example, the use of PCA for image compression is demonstrated. In the previous example, PCA was performed across the different images of a database. Here, the focus will be on a single image.

The pixel values of the image are stored in an $l \times N$ matrix X and the columns of this matrix are considered to be the observation vectors $\boldsymbol{x}_n \in \mathbb{R}^l$, $n = 1, 2, \ldots, N$. Note that X needs to be zero mean along the rows so the mean vector, $\bar{\boldsymbol{x}}$, is computed and subtracted from each column. Then the eigenvectors corresponding to the m, $1 \leq m < l$, largest eigenvalues are obtained either via the sample covariance matrix or directly through SVD. Exploiting the matrix factorization formulation of PCA in Eq. (19.22) a compressed representation of X, comprising m instead of l rows, is given by

$$Z^{(m)} = \underbrace{[\boldsymbol{u}_1, \ldots, \boldsymbol{u}_m]^T}_{m \times l} X, \tag{19.24}$$

where the dimensionality m has been explicitly brought into the notation. Thus, only $Z^{(m)}$ and $\boldsymbol{u}_1, \ldots, \boldsymbol{u}_m$ are needed to get an estimate of the, mean-subtracted, X via Eq. (19.21). Finally, in order to reconstruct the image, the mean vector $\bar{\boldsymbol{x}}$ needs to be added back to each column (see Eq. (19.23)).

The effectiveness of the PCA-based image compression will be demonstrated with the aid of the top-left image depicted in Fig. 19.7. This image is square, having $l = N = 400$. For any m chosen, the compression ratio is easily computed considering that instead of 400×400 values of the original image, after compression the storage of $2 \times m \times 400$ values for the matrix $Z^{(m)}$ and the eigenvectors,

FIGURE 19.7

PCA-based image compression. The image is from the Greek island Andros.

$\boldsymbol{u}_1, \dots, \boldsymbol{u}_m$, plus 400 values for the mean vector $\bar{\boldsymbol{x}}$ is needed. This amounts to a compression ratio of $400 : (2m + 1)$. The reconstructed images together with the corresponding MSE between the original and the reconstructed image for different compression rates are shown in Fig. 19.7.

Remarks 19.2.

- *Subspace tracking*: Online subspace tracking is another old area with a revived interest recently.

  A well-known algorithm of relative low complexity, for tracking the signal subspace, is the so-called projection approximation subspace tracking (PAST), proposed in [197]. In PAST, the recursive least-squares (RLS) technique is employed for subspace estimation. Alternative algorithms in this line of philosophy have been presented in, for example, [69,115,170,179].

  More recently, the work in [25,50,137] tackles the problem of subspace tracking with missing/unobserved data. The methodology presented in [25] is based on gradient descent iterations on the Grassmannian manifold. Furthermore, the algorithms of [50,137] attempt to estimate the unknown subspace by minimizing properly constructed loss functions.

  Finally, [51,52,92,132,162] attack the subspace tracking problem in environments where observations are contaminated by outlier noise.

19.4 CANONICAL CORRELATION ANALYSIS

PCA is a dimensionality reduction technique focusing on a *single* data set. However, in a number of cases, one has to deal with multiple data sets which, although they may originate from different sources, are closely related. For example, many problems in medical imaging fall under this umbrella. A typical case occurs in the study of brain activity where one can use different modalities, for example, electroencephalogram (EEG), functional magnetic resonance imaging (fMRI), or structural MRI. Each one of these modalities can grasp a different type of information and it is beneficial to exploit all of them in a complementary fashion. Thus, the respective experimental data can appropriately be fused in order to get a better description concerning the brain activity that gives birth to the data. Another scenario where multiple data sets are of interest is when a single modality is used but different data are available measured on different subjects; thus, jointly analyzing the results can be beneficial for the finally reached conclusions (see, e.g., [56]).

Canonical correlation analysis (CCA) is an old technique developed in [96] to process two data sets jointly. Our starting point is the fact that when two sets of random variables (two random vectors) are involved, the value of their correlation *does depend* on the coordinate system in which the random vectors are represented. The goal behind CCA is to seek a pair of linear transformations, one for each set of variables, such that after the transformation, the resulting transformed variables are *maximally correlated.*

Let us assume that we are given two sets of random variables comprising the components of two random vectors, $\mathbf{x} \in \mathbb{R}^p$ and $\mathbf{y} \in \mathbb{R}^q$, and let the corresponding sets of observations be $\boldsymbol{x}_n, \boldsymbol{y}_n$, $n = 1, 2, \ldots, N$, respectively. Following a step-wise procedure, as we did for PCA, we will first compute a single pair of directions, namely, $\boldsymbol{u}_{x,1}, \boldsymbol{u}_{y,1}$, so that the correlation between the projections onto these directions is maximized. Let $\mathrm{z}_{x,1} := \boldsymbol{u}_{x,1}^T \mathbf{x}$ and $\mathrm{z}_{y,1} := \boldsymbol{u}_{y,1}^T \mathbf{y}$ be the (zero mean) random variables after the linear transformation (projection). Note that these variables are the counterparts of what we called principal components in PCA. The corresponding correlation coefficient (normalized covariance) is defined as

$$\rho := \frac{\mathbb{E}[\mathrm{z}_{x,1}\mathrm{z}_{y,1}]}{\sqrt{\mathbb{E}\left[\mathrm{z}_{x,1}^2\right]\mathbb{E}\left[\mathrm{z}_{y,1}^2\right]}} = \frac{\mathbb{E}\left[(\boldsymbol{u}_{x,1}^T\mathbf{x})(\mathbf{y}^T\boldsymbol{u}_{y,1})\right]}{\sqrt{\mathbb{E}\left[(\boldsymbol{u}_{x,1}^T\mathbf{x})^2\right]\mathbb{E}\left[(\boldsymbol{u}_{y,1}^T\mathbf{y})^2\right]}}$$

or

$$\rho := \frac{\boldsymbol{u}_{x,1}^T \Sigma_{xy} \boldsymbol{u}_{y,1}}{\sqrt{(\boldsymbol{u}_{x,1}^T \Sigma_{xx} \boldsymbol{u}_{x,1})(\boldsymbol{u}_{y,1}^T \Sigma_{yy} \boldsymbol{u}_{y,1})}}, \tag{19.25}$$

where

$$\mathbb{E}\left[\begin{bmatrix}\mathbf{x}\\ \mathbf{y}\end{bmatrix}[\mathbf{x}^T, \mathbf{y}^T]\right] := \begin{bmatrix}\Sigma_{xx} & \Sigma_{xy}\\ \Sigma_{yx} & \Sigma_{yy}\end{bmatrix}. \tag{19.26}$$

Note that, by the respective definition, we have $\Sigma_{xy} = \Sigma_{yx}^T$. When expectations are not available, covariances are replaced by the corresponding sample covariance values. This is the most common case in practice, so we will adhere to it and use the notation with the "hat." Furthermore, it can easily be checked that the correlation coefficient is invariant to scaling (changing, e.g., $\mathbf{x} \to b\mathbf{x}$). Thus, maximizing it with respect to the directions $\boldsymbol{u}_{x,1}$ and $\boldsymbol{u}_{y,1}$ can equivalently be cast as the following constrained optimization task:

$$\max_{\boldsymbol{u}_x,\boldsymbol{u}_y} \quad \boldsymbol{u}_x^T \hat{\Sigma}_{xy} \boldsymbol{u}_y, \tag{19.27}$$

$$\text{s.t.} \quad \boldsymbol{u}_x^T \hat{\Sigma}_{xx} \boldsymbol{u}_x = 1, \tag{19.28}$$

$$\boldsymbol{u}_y^T \hat{\Sigma}_{yy} \boldsymbol{u}_y = 1. \tag{19.29}$$

Compare Eqs. (19.27)–(19.29) with the optimization task defining PCA in Eqs. (19.2) and (19.3). For CCA, two directions have to be computed and the constraints involve the weighted Σ norm instead of the Euclidean one. Moreover, in PCA the variance is maximized, while CCA cares for the correlation between the projections of the two involved vectors onto the new axes.

Employing Lagrange multipliers, the corresponding Lagrangian of Eqs. (19.27)–(19.29) is given by

$$L(\boldsymbol{u}_x, \boldsymbol{u}_y, \lambda_x, \lambda_y) = \boldsymbol{u}_x^T \hat{\Sigma}_{xy} \boldsymbol{u}_y - \frac{\lambda_x}{2}\left(\boldsymbol{u}_x^T \hat{\Sigma}_{xx} \boldsymbol{u}_x - 1\right) - \frac{\lambda_y}{2}\left(\boldsymbol{u}_y^T \hat{\Sigma}_{yy} \boldsymbol{u}_y - 1\right).$$

Taking the gradients with respect to $\boldsymbol{u}_x$ and $\boldsymbol{u}_y$ and equating to zero, we obtain (Problem 19.2)

$$\lambda_x = \lambda_y := \lambda$$

and

$$\hat{\Sigma}_{xy} \boldsymbol{u}_y = \lambda \hat{\Sigma}_{xx} \boldsymbol{u}_x, \tag{19.30}$$

$$\hat{\Sigma}_{yx} \boldsymbol{u}_x = \lambda \hat{\Sigma}_{yy} \boldsymbol{u}_y. \tag{19.31}$$

Solving the latter of the two with respect to $\boldsymbol{u}_y$ and substituting into the first one, we finally get

$$\hat{\Sigma}_{xy} \hat{\Sigma}_{yy}^{-1} \hat{\Sigma}_{yx} \boldsymbol{u}_x = \lambda^2 \hat{\Sigma}_{xx} \boldsymbol{u}_x \tag{19.32}$$

and

$$\boldsymbol{u}_y = \frac{1}{\lambda} \hat{\Sigma}_{yy}^{-1} \hat{\Sigma}_{yx} \boldsymbol{u}_x, \tag{19.33}$$

assuming, of course, invertibility of $\hat{\Sigma}_{yy}$. Furthermore, assuming invertibility of $\hat{\Sigma}_{xx}$, too, we end up with the following eigenvalue-eigenvector problem:

$$\left(\hat{\Sigma}_{xx}^{-1} \hat{\Sigma}_{xy} \hat{\Sigma}_{yy}^{-1} \hat{\Sigma}_{yx}\right) \boldsymbol{u}_x = \lambda^2 \boldsymbol{u}_x. \tag{19.34}$$

Thus, the axis $\boldsymbol{u}_{x,1}$ is obtained as an eigenvector of the product of matrices in the parentheses in Eq. (19.34). Taking into account Eq. (19.30) and the constraints, it turns out that the corresponding optimal value of the correlation, ρ, is equal to

$$\rho = \boldsymbol{u}_{x,1}^T \hat{\Sigma}_{xy} \boldsymbol{u}_{y,1} = \lambda \boldsymbol{u}_{x,1}^T \hat{\Sigma}_{xx} \boldsymbol{u}_{x,1} = \lambda.$$

Hence, selecting $\boldsymbol{u}_{x,1}$ to be the eigenvector corresponding to the maximum eigenvalue, λ^2, results in maximum correlation.

The eigenvectors $\boldsymbol{u}_{x,1}, \boldsymbol{u}_{y,1}$ are known as the *normalized canonical correlation basis vectors*, the eigenvalue λ^2 as the squared *canonical correlation*, and the projections $\mathrm{z}_{x,1}, \mathrm{z}_{y,1}$ as the *canonical variates*.

The previous idea can now be taken further, and we can compute a pair of subspaces, span$\{\boldsymbol{u}_{x,1}, \ldots, \boldsymbol{u}_{x,m}\}$ and span$\{\boldsymbol{u}_{y,1}, \ldots, \boldsymbol{u}_{y,m}\}$, where $m \leq \min(p, q)$. One way to achieve this goal is in a step-wise fashion, as was done for PCA. Assuming that k pairs of basis vectors have already been computed, the $k+1$ is obtained by solving the following constrained optimization task:

$$\max_{\boldsymbol{u}_x, \boldsymbol{u}_y} \quad \boldsymbol{u}_x^T \hat{\Sigma}_{xy} \boldsymbol{u}_y, \tag{19.35}$$

$$\text{s.t.} \quad \boldsymbol{u}_x^T \hat{\Sigma}_{xx} \boldsymbol{u}_x = 1, \quad \boldsymbol{u}_y^T \hat{\Sigma}_{yy} \boldsymbol{u}_y = 1, \tag{19.36}$$

$$\boldsymbol{u}_x^T \hat{\Sigma}_{xx} \boldsymbol{u}_{x,i} = 0, \quad \boldsymbol{u}_y^T \hat{\Sigma}_{yy} \boldsymbol{u}_{y,i} = 0, \quad i = 1, 2, \ldots, k, \tag{19.37}$$

$$\boldsymbol{u}_x^T \hat{\Sigma}_{xy} \boldsymbol{u}_{y,i} = 0, \quad \boldsymbol{u}_y^T \hat{\Sigma}_{yx} \boldsymbol{u}_{x,i} = 0, \quad i = 1, 2, \ldots, k. \tag{19.38}$$

In other words, every new pair of vectors is computed so as to be normalized (Eq. (19.36)) and at the same time, each one is orthogonal (in the generalized sense) to those obtained in the previous iteration steps (Eqs. (19.37) and (19.38)). Note that this guarantees that the derived canonical variates are uncorrelated to all previously derived ones. This reminds us of the uncorrelatedness property of the principal components in PCA. The only nonzero correlation in CCA, which is maximized at every iteration step, is the one between $\mathrm{z}_{x,k} = \boldsymbol{u}_{x,k}^T \mathbf{x}$ and $\mathrm{z}_{y,k} = \boldsymbol{u}_{y,k}^T \mathbf{y}$, $k = 1, 2, \ldots, m$.

More on CCA can be found in [6,26]. Extensions of CCA in reproducing kernel Hilbert spaces have also been developed and used (see, for example, [9,89,117] and the references therein). In [89], the kernel CCA is used for content-based image retrieval. The aim is to allow retrieval of images from a text query but without reference to any labeling associated with the image. The task is treated as a cross-modal problem. A probabilistic Bayesian formulation of CCA has been given in [15,113]. A regularized CCA version, using sparsity-based arguments, has been derived in [90]. In [65], a variant of CCA is proposed, named *correlated component analysis*; instead of two directions (subspaces), a common direction is derived for both data sets. The idea behind this method is that the two data sets may not be much different, so a single direction is enough. In this way, the task has fewer free parameters to estimate. Moreover, the constraint on orthogonality is dropped, which in some cases may not be physically justifiable. A Bayesian extension of the method is provided in [147].

Example 19.3. Let $\mathrm{x} \in \mathbb{R}^2$ be a normally distributed random vector, $\mathcal{N}(\mathbf{0}, I)$. The pair of random variables $(\mathrm{y}_1, \mathrm{y}_2)$ is related to $(\mathrm{x}_1, \mathrm{x}_2)$ as

$$\mathbf{y} = \begin{bmatrix} 0.7 & 0.3 \\ 0.3 & 0.7 \end{bmatrix} \mathbf{x}.$$

Note the strong correlation that exists between the involved variables, because

$$\mathrm{y}_1 + \mathrm{y}_2 = \mathrm{x}_1 + \mathrm{x}_2.$$

However, the cross-covariance matrix Σ_{yx},

$$\Sigma_{yx} = AI = \begin{bmatrix} 0.7 & 0.3 \\ 0.3 & 0.7 \end{bmatrix},$$

indicates a rather low correlation. After performing CCA, the resulting directions are

$$\boldsymbol{u}_{x,1} = \boldsymbol{u}_{y,1} = -\frac{1}{\sqrt{2}}[1, 1]^T,$$

which actually is the direction where the linear equality of the involved variables lies. The maximum correlation coefficient value is equal to 1, indicating strong correlation indeed.

19.4.1 RELATIVES OF CCA

CCA is not the only multivariate technique to process and deal with different data sets jointly. Various techniques have been developed, using different optimizing criteria/constraints, each one serving different needs and goals.

The aim of this subsection is to briefly discuss some of these methods under a common framework. Recall that the eigenvalue-eigenvector problem for computing the pair of canonical basis vectors results from the pair of equations in Eq. (19.30) and (19.31). These can be combined into a single one [26], namely,

$$C\boldsymbol{u} = \lambda B\boldsymbol{u}, \tag{19.39}$$

where

$$\boldsymbol{u} := [\boldsymbol{u}_x^T, \boldsymbol{u}_y^T]^T$$

and

$$C := \begin{bmatrix} O & \hat{\Sigma}_{xy} \\ \hat{\Sigma}_{yx} & O \end{bmatrix}, \quad B := \begin{bmatrix} \hat{\Sigma}_{xx} & O \\ O & \hat{\Sigma}_{yy} \end{bmatrix}.$$

Changing the structure of the two matrices, C and B, different methods result. For example, if we set $C = \hat{\Sigma}_{xx}$ and $B = I$, we get the eigenvalue-eigenvector task of PCA.

In [189], algorithmic procedures for the solution of the related equations, in a numerically robust way, are discussed.

Partial Least-Squares

The *partial least-squares* (PLS) method was first introduced in [186], and it has been used extensively in a number of applications, such as chemometrics, bioinformatics, food research, medicine, pharmacology, social sciences, and physiology, to name but a few. The corresponding eigenanalysis problem results if we set in Eq. (19.39)

$$B = \begin{bmatrix} I & O \\ O & I \end{bmatrix}$$

and keep C the same as for CCA. This eigenvalue-eigenvector problem arises (try it) if instead of maximizing the correlation coefficient ρ in Eq. (19.25), one maximizes the covariance, that is,

$$\text{cov}\left(z_{x,1}, z_{y,1}\right) = \mathbb{E}[z_{x,1} z_{y,1}]. \tag{19.40}$$

This means that while trying to reduce the dimensionality, not only our concern focuses on the correlation but *at the same* we want to identify directions that *also* care for maximum variance for both sets of variables. The optimizing task for identifying the first pair of axes, $\boldsymbol{u}_{x,1}, \boldsymbol{u}_{y,1}$, now becomes

$$\text{maximize} \quad \boldsymbol{u}_x^T \hat{\Sigma}_{xy} \boldsymbol{u}_y, \tag{19.41}$$

$$\text{s.t.} \quad \boldsymbol{u}_x^T \boldsymbol{u}_x = 1, \tag{19.42}$$

$$\boldsymbol{u}_y^T \boldsymbol{u}_y = 1. \tag{19.43}$$

PLS has been used both for classification and for regression tasks. For example, in Chapter 6, we used PCA for regression in order to reduce the dimensionality of the space and the least-squares solution was expressed in this lower-dimensional space. However, the principal axes were determined only on the basis of the input data so as to retain maximum variance. In contrast, PLS can be employed by considering the output observations as the second set of variables, and one can select the axes so as to maximize the variances as well as the correlation between the two data sets. The latter can be understood from the fact that maximizing the covariance (PLS) is equivalent to maximizing the product of the correlation coefficient (used for CCA) times the two variance terms.

The literature on PLS is extensive and the method has been studied both algorithmically and from its performance point of view. The interested reader can obtain more on PLS from [153]. In all the techniques we have discussed so far, a major focus is on computing the eigenvalues-eigenvectors. To this end, although one can use general packages and algorithms, a number of more efficient alternatives have been derived. A common approach is to solve the task in a two-step iterative procedure. In the first step, the largest eigenvalue (eigenvector) is computed, for which there exist efficient algorithms, such as the power method (e.g., [83]). Then, a procedure known as *deflation* is adopted; this consists of removing from the covariance matrices the variance that has been explained with the features extracted from the first step (see, for example, [135]). Kernelized versions of PLS have also been proposed (for example, [9,152]).

Remarks 19.3.

- Another dimensionality reduction method results if we set in Eq. (19.39)

$$B = \begin{bmatrix} \hat{\Sigma}_{xx} & O \\ O & I \end{bmatrix}.$$

  The resulting method is known as *multivariate linear regression* (MLR). This is the task of finding a set of basis vectors and corresponding regressors such that the MSE in a regression problem is minimized [26].
- CCA is *invariant* with respect to affine transformations. This is an important advantage with respect to the ordinary correlation analysis (for example, [6]).
- Extensions of CCA and PLS to more than two data sets have also been proposed (see, for example, [56,110,185]).

19.5 INDEPENDENT COMPONENT ANALYSIS

The latent variable interpretation of PCA was summarized in Eqs. (19.17)–(19.19), where each one of the observed random variables, x_i, is (approximately) written as a linear combination of the latent variables (principal components in this case), z_i, which are in turn obtained via Eq. (19.19), imposing the uncorrelatedness constraint.

The kick-off point for ICA is to assume that the following latent model is true:

$$\mathbf{x} = A\mathbf{s}, \tag{19.44}$$

where the (unknown) latent variables of $\mathbf{s}$ are assumed to be mutually statistically *independent* and we refer to them as the *independent components* (ICs). The task then comprises obtaining estimates of both the matrix A and the independent components. We will focus on the case where A is an $l \times l$ square matrix. Extensions to fat and tall matrices, corresponding to scenarios where the number of latent variables, m, is smaller or larger than the number of the observed random variables, l, have also been considered and developed (see, e.g., [100]).

Matrix A is known as the *mixing matrix* and its elements, a_{ij}, as the *mixing coefficients*. The resulting estimates of the latent variables will be denoted as z_i, $i = 1, 2 \ldots, l$, and we will also refer to them as independent components. The observed random variables, x_i, $i = 1, 2, \ldots, l$, are sometimes called the *mixture variables* or simply mixtures.

To obtain the estimates of the latent variables, we adopt the model

$$\hat{\mathbf{s}} := \mathbf{z} = W\mathbf{x}, \tag{19.45}$$

where W is also known as the *unmixing* or *separating* matrix. Note that

$$\mathbf{z} = WA\mathbf{s},$$

and we have to estimate the unknown parameters, so that $\mathbf{z}$ is as close to $\mathbf{s}$ as possible. For square matrices, $A = W^{-1}$, assuming invertibility.

19.5.1 ICA AND GAUSSIANITY

Although in general in statistics adopting the Gaussian assumption for a PDF seems to be rather a "blessing," in the case of ICA this is not true anymore. This can easily be understood if we look at the consequences of adopting the Gaussian assumption. If the independent components follow Gaussian distributions, their joint PDF is given by

$$p(\boldsymbol{s}) = \frac{1}{(2\pi)^{l/2}} \exp\left(-\frac{||\boldsymbol{s}||^2}{2}\right), \tag{19.46}$$

where for simplicity we have assumed that all the variables are normalized to unit variance. Let the mixing matrix, A, be an orthogonal one, that is, $A^{-1} = A^T$. Then the joint PDF of the mixtures is readily obtained as (see Eq. (2.45))

$$p(\boldsymbol{x}) = \frac{1}{(2\pi)^{l/2}} \exp\left(-\frac{||A^T\boldsymbol{x}||^2}{2}\right) |\det(A^T)|. \tag{19.47}$$

However due to the orthogonality of A, we have $||A^T \boldsymbol{x}||^2 = ||\boldsymbol{x}||^2$ and $|\det(A^T)| = 1$, which makes $p(\boldsymbol{s})$ indistinguishable from $p(\boldsymbol{x})$. That is, no conclusion about A can be drawn by observing $\mathbf{x}$, as all related information has been lost. Seen from another point of view, the mixtures x_i are mutually uncorrelated, as $\Sigma_x = I$, and ICA can provide no further information. This is a direct consequence of the fact that uncorrelatedness for jointly Gaussian variables is equivalent to independence (see Section 2.3.2). In other words, if the latent variables are Gaussians, ICA cannot take us any further than PCA, because the latter provides uncorrelated components. That is, the mixing matrix, A, *is not identifiable* for Gaussian independent components. In a more general setting, in a case where some of the components are Gaussians and some are not, ICA can identify the non-Gaussian ones. Thus, for a matrix A to be identifiable, *at most one* of the independent components can be Gaussian.

From a mathematical point of view, the ICA task is ill-posed for Gaussian variables. Indeed, assume that a set of independent Gaussian components, $\mathbf{z}$, have been obtained; then any linear transformation on $\mathbf{z}$ by a unitary matrix will also be a solution (as shown previously). Note that this problem is bypassed in PCA, because the latter imposes a specific structure on the transformation matrix.

In order to deal with independence one has to involve, in one way or another, higher-order statistical information. Second-order statistical information suffices for imposing uncorrelatedness, as is the case with PCA, but it is not enough for ICA. To this end, a large number of techniques and algorithms have been developed over the years and reviewing all these techniques is far beyond the limits imposed on a book section. The goal here is to provide the reader with the essence behind these techniques and emphasize the need to bring higher-order statistics into the game. The interested reader can delve deeper in this field from [55,58,81,100,120].

19.5.2 ICA AND HIGHER-ORDER CUMULANTS

Imposing the constraint on the components of $\mathbf{z}$ to be independent is equivalent to demanding all higher-order *cross-cumulants* (Appendix B.3) to be zero. One possibility to achieve this is to restrict ourselves up to the fourth-order cumulants [57]. As stated in Appendix B.3, the first three cumulants for zero mean variables are equal to the corresponding moments, that is,

$$\kappa_1(\mathrm{z}_i) = \mathbb{E}[\mathrm{z}_i] = 0,$$
$$\kappa_2(\mathrm{z}_i, \mathrm{z}_j) = \mathbb{E}[\mathrm{z}_i \mathrm{z}_j],$$
$$\kappa_3(\mathrm{z}_i, \mathrm{z}_j, \mathrm{z}_k) = \mathbb{E}[\mathrm{z}_i \mathrm{z}_j \mathrm{z}_k],$$

and the fourth-order cumulants are given by

$$\begin{aligned}\kappa_4(\mathrm{z}_i, \mathrm{z}_j, \mathrm{z}_k, \mathrm{z}_r) = &\ \mathbb{E}[\mathrm{z}_i \mathrm{z}_j \mathrm{z}_k \mathrm{z}_r] - \mathbb{E}[\mathrm{z}_i \mathrm{z}_j]\mathbb{E}[\mathrm{z}_k \mathrm{z}_r] \\ &- \mathbb{E}[\mathrm{z}_i \mathrm{z}_k]\mathbb{E}[\mathrm{z}_j \mathrm{z}_r] - \mathbb{E}[\mathrm{z}_i \mathrm{z}_r]\mathbb{E}[\mathrm{z}_j \mathrm{z}_k].\end{aligned}$$

An assumption that is employed is that the involved PDFs are symmetric, which renders odd-order cumulants to zero. Thus, we are left only with the second- and fourth-order cumulants. Under the previous assumptions, our goal is to estimate the unmixing matrix, W, so that (a) the second-order and (b) the fourth-order cumulants become zero. This is achieved in two steps.

Step 1: Compute

$$\hat{\mathbf{z}} = U^T \mathbf{x}, \tag{19.48}$$

where U is the unitary $l \times l$ matrix associated with PCA. This transformation guarantees that the components of $\hat{\mathbf{z}}$ are uncorrelated, that is,

$$\mathbb{E}[\hat{z}_i \hat{z}_j] = 0, \quad i \neq j, \; i, j = 1, 2, \ldots, l.$$

Step 2: Compute an orthogonal matrix, $\hat{U}$, such that the fourth-order cross-cumulants of the components of the transformed random vector

$$\mathbf{z} = \hat{U}^T \hat{\mathbf{z}} \tag{19.49}$$

are zero. In order to achieve this, the following maximization task is solved:

$$\max_{\hat{U}\hat{U}^T = I} \sum_{i=1}^{l} \kappa_4^2(\mathrm{z}_i). \tag{19.50}$$

Step 2 is justified as follows. It can be shown [57] that the sum of the squares of the fourth-order cumulants is invariant under a linear transformation by an orthogonal matrix. Therefore, as the sum of the squares of the fourth-order cumulants is fixed for $\mathbf{z}$, maximizing the sum of the squares of the autocumulants of $\mathbf{z}$ will force the corresponding cross-cumulants to zero. Observe that this is basically a diagonalization problem of the fourth-order cumulant multidimensional array. In practice, this can be achieved by generalizing the method of Givens rotations, used for matrix diagonalization [57]. Note that the sum that is maximized is a function of (a) the elements of the unknown matrix $\hat{U}$, (b) the elements of the known (for this step) matrix U, and (c) the cumulants of the random components of the mixtures $\mathbf{x}$, which have to be estimated prior to the application of the method. In practice, it usually turns out that setting the cross-cumulants to zero is only approximately achieved. This is because the model in Eq. (19.44) may not be exact, for example, due to the existence of noise. Also, the cumulants of the mixtures are only approximately known, because they are estimated by the available observations.

Once U and $\hat{U}$ have been computed, the unmixing matrix is readily available and we can write

$$\mathbf{z} = W\mathbf{x} = (U\hat{U})^T \mathbf{x},$$

and the mixing matrix is given as $A = W^{-1}$.

A number of algorithms have been developed around the idea of higher-order cumulants, which are also known as *tensorial methods*. Tensors are generalizations of matrices and cumulant tensors are generalizations of the covariance matrix. Moreover, note that as the eigenanalysis of the covariance matrix leads to uncorrelated (principal) components, the eigenanalysis of the cumulant tensor leads to independent components. The interested reader can obtain a more detailed account of such techniques from [39,57,119].

ICA Ambiguities

Any ICA method can (approximately) recover the independent components within the following two indeterminacies.

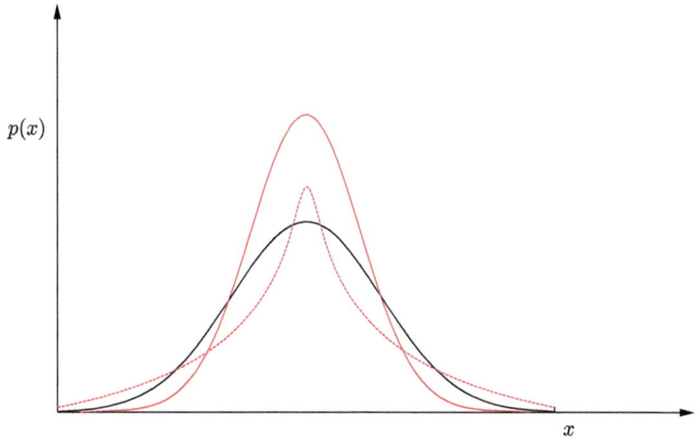

FIGURE 19.8

A Gaussian (full-gray line) a super-Gaussian (dotted red line) and a sub-Gaussian (full-red line).

- Independent components (ICs) are recovered to within a constant factor. Indeed, if A and $\mathbf{z}$ are the recovered quantities by an ICA algorithm, then $(1/a)A$ and $a\mathbf{z}$ is also a solution, as is readily seen from Eq. (19.44). Thus, usually the recovered latent variables (ICs) are normalized to unit variance.
- We cannot determine the order of the ICs. Indeed, if A and $\mathbf{z}$ have been recovered and P is a permutation matrix, then AP^{-1} and $P\mathbf{z}$ is also a solution, because the components of $P\mathbf{z}$ are the same as those of $\mathbf{z}$ in a different order (with the same statistical properties).

19.5.3 NON-GAUSSIANITY AND INDEPENDENT COMPONENTS

The fourth-order (auto)cumulant, of a random variable, z,

$$\kappa_4(\mathrm{z}) = \mathbb{E}[\mathrm{z}^4] - 3\left(\mathbb{E}[\mathrm{z}^2]\right)^2,$$

is known as the *kurtosis* of the variable and it is a measure of *non-Gaussianity*. Variables following the Gaussian distribution have zero kurtosis. Sub-Gaussian variables (variables whose PDF falls at a slower rate than the Gaussian for the same variance) have negative kurtosis. Super-Gaussian variables (corresponding to PDFs that fall at a faster rate than the Gaussian) have positive kurtosis. Thus, if we keep the variance fixed (e.g., for variables normalized to unit variance), maximizing the sum of squared kurtosis, it results in maximizing the non-Gaussianity of the recovered ICs. Usually, the absolute value of the kurtosis of the recovered ICs is used as a measure of ranking them. This is important if ICA is used as a feature generation technique. Fig. 19.8 shows some typical examples of a sub-Gaussian and a super-Gaussian together with the corresponding Gaussian distribution. Also, another typical example of a sub-Gaussian distribution is the uniform one.

Recall from Chapter 12 (Section 12.8.1) that the Gaussian distribution is the one that maximizes the entropy under the variance and mean constraints. In other words, it is the most random one, under these constraints, and from this point of view the least informative with respect to the underlying structure of the data. In contrast, distributions that have the least resemblance to the Gaussian are more interesting as they are able to better unveil the structure associated with the data. This observation is at the heart of *projection pursuit*, which is closely related to the ICA family of techniques. The essence of these techniques is to search for directions in the feature space where the data projections are described in terms of non-Gaussian distributions [97,106].

19.5.4 ICA BASED ON MUTUAL INFORMATION

The approach based on zeroing the second- and fourth-order cross-cumulants is not the only one. An alternative path is to estimate W by minimizing the *mutual information* among the latent variables. The notion of mutual information was introduced in Section 2.5. Elaborating a bit on Eq. (2.158) and performing the integrations on the right-hand side (for the case of more than two variables), it is readily shown that

$$I(\mathbf{z}) = -H(\mathbf{z}) + \sum_{i=1}^{l} H(z_i), \tag{19.51}$$

where $H(z_i)$ is the associated entropy of z_i, defined in Eq. (2.157). In Section 2.5 it has been shown that $I(\mathbf{z})$ is equal to the Kullback–Leibler (KL) divergence between the joint PDF $p(z)$ and the product of the respective marginal probability densities, $\prod_{i=1}^{l} p_i(z_i)$. The KL divergence (and, hence, the associated mutual information $I(\mathbf{z})$) is a nonnegative quantity and it becomes zero if the components z_i are statistically independent. This is because only in this case the joint PDF becomes equal to the product of the corresponding marginal PDFs, leading the KL divergence to zero. Hence, the idea now becomes to compute W so as to force $I(\mathbf{z})$ to be minimum, as this will make the components of $\mathbf{z}$ *as independent as possible*. Plugging Eq. (19.45) into Eq. (19.51) and taking into account the formula that relates the two PDFs associated with $\mathbf{x}$ and $\mathbf{z}$ (Eq. (2.45)), we end up with

$$I(\mathbf{z}) = -H(\mathbf{x}) - \ln|\det(W)| - \sum_{i=1}^{l} \int p_i(z_i) \ln p_i(z_i)\, dz_i. \tag{19.52}$$

The elements of the unknown matrix, W, are also hidden in the marginal PDFs of the latent variables, z_i. However, it is not easy to express this dependence explicitly. One possibility is to expand each one of the marginal densities around the Gaussian PDF, denoted here as $g(z)$, following Edgeworth's expansion (Appendix B), and truncate the series to a reasonable approximation. For example, keeping the first two terms in the Edgeworth expansion we have

$$p_i(z_i) = g(z_i)\left(1 + \frac{1}{3!}\kappa_3(z_i)H_3(z_i) + \frac{1}{4!}\kappa_4(z_i)H_4(z_i)\right), \tag{19.53}$$

where $H_k(z_i)$ is the Hermite polynomial of order k (Appendix B). To obtain an approximate expression for $I(\mathbf{z})$, in terms of cumulants of z_i and W, we can (a) insert in Eq. (19.52) the PDF approximation in Eq. (19.53), (b) adopt the approximation $\ln(1+y) \simeq y - y^2$, and (c) perform the integrations. This is no doubt a rather painful task! For the case of Eq. (19.53) and constraining W to be orthogonal, the following is obtained (e.g., [100]):

$$I(\mathbf{z}) \approx C - \sum_{i=1}^{l} \left(\frac{1}{12}\kappa_3^2(z_i) + \frac{1}{48}\kappa_4^2(z_i) + \frac{7}{48}\kappa_4^4(z_i) - \frac{1}{8}\kappa_3^2(z_i)\kappa_4(z_i) \right), \tag{19.54}$$

where C is a quantity independent of W. Under the assumption that the PDFs are symmetric (thus, third-order cumulants are zero), it can be shown that minimizing the approximate expression of the mutual information in Eq. (19.54) is equivalent to maximizing the sum of the squares of the fourth-order cumulants. Note that the orthogonal W constraint is not necessary, and if it is not adopted other approximate expressions for $I(\mathbf{z})$ result (for example, [91]).

Minimization of $I(\mathbf{z})$ in Eq. (19.54) can be carried out by a gradient descent technique (Chapter 5), where the involved expectations (associated with the cumulants) are replaced by the respective instantaneous values. Although we will not treat the derivation of algorithmic schemes in detail, in order to get a flavor of the involved tricks, let us go back to Eq. (19.52), before we apply the approximations. Because $H(\mathbf{x})$ does not depend on W, minimizing $I(\mathbf{z})$ is equivalent to the maximization of

$$J(W) = \ln|\det(W)| + \mathbb{E}\left[\sum_{i=1}^{l} \ln p_i(z_i)\right]. \tag{19.55}$$

Taking the gradient of the cost function with respect to W results in

$$\frac{\partial J(W)}{\partial W} = W^{-T} - \mathbb{E}[\boldsymbol{\phi}(\mathbf{z})\mathbf{x}^T], \tag{19.56}$$

where

$$\boldsymbol{\phi}(\mathbf{z}) := \left[-\frac{p_1'(z_1)}{p_1(z_1)}, \ldots, -\frac{p_l'(z_l)}{p_l(z_l)} \right]^T, \tag{19.57}$$

and

$$p_i'(z_i) := \frac{dp_i(z_i)}{dz_i}, \tag{19.58}$$

and we used the formula

$$\frac{\partial \det(W)}{\partial W} = W^{-T}\det(W).$$

Obviously, the derivatives of the marginal probability densities depend on the type of approximation adopted in each case. The general gradient ascent scheme at the ith iteration step can now be written as

$$W^{(i)} = W^{(i-1)} + \mu_i \left((W^{(i-1)})^{-T} - \mathbb{E}\left[\boldsymbol{\phi}(\mathbf{z})\mathbf{x}^T\right] \right),$$

or

$$W^{(i)} = W^{(i-1)} + \mu_i \left(I - \mathbb{E}\left[\boldsymbol{\phi}(\mathbf{z})\mathbf{z}^T\right] \right) (W^{(i-1)})^{-T}. \tag{19.59}$$

In practice, the expectation operator is neglected and random variables are replaced by respective observations, in the spirit of the stochastic approximation rationale (Chapter 5).

The update equation in Eq. (19.59) involves the inversion of the transpose of the current estimate of W. Besides the computational complexity issues, there is no guarantee of the invertibility in the process of adaptation. The use of the so-called *natural gradient* [68], instead of the gradient in Eq. (19.56), results in

$$W^{(i)} = W^{(i-1)} + \mu_i \left(I - \mathbb{E}[\boldsymbol{\phi}(\mathbf{z})\mathbf{z}^T] \right) W^{(i-1)}, \tag{19.60}$$

which does not involve matrix inversion and at the same time improves convergence. A more detailed treatment of this issue is beyond the scope of this book. Just to give an incentive to the mathematically inclined reader for indulging more deeply this field, it suffices to say that our familiar gradient, that is, Eq. (19.56), points to the steepest ascent direction if the space is Euclidean. However, in our case the parameter space consists of all the nonsingular $l \times l$ matrices, which is a multiplicative group. The space is Riemannian and it turns out that the natural gradient, pointing to the steepest ascent direction, results if we multiply the gradient in Eq. (19.56) by $W^T W$, which is the corresponding Riemannian metric tensor [68].

Remarks 19.4.

- From the gradient in Eq. (19.56), it is easy to see that at a stationary point the following is true:

$$\frac{\partial J(W)}{\partial W} W^T = \mathbb{E}[I - \boldsymbol{\phi}(\mathbf{z})\mathbf{z}^T] = O. \tag{19.61}$$

  In other words, what we achieve with ICA is a *nonlinear generalization* of PCA. Recall that for the latter, the uncorrelatedness condition can be written as

$$\mathbb{E}[I - \mathbf{z}\mathbf{z}^T] = O. \tag{19.62}$$

  The presence of the *nonlinear* function $\boldsymbol{\phi}$ takes us beyond simple uncorrelatedness, and brings the cumulants into the scene. As a matter of fact, Eq. (19.61) was the one that inspired the early pioneering work on ICA, as a direct nonlinear generalization of PCA [93,107].
- The origins of ICA are traced back to the seminal paper [93]. For a number of years, it remained an activity pretty much within the French signal processing and statistics communities. Two papers were catalytic for its widespread use and popularity, namely, [18] in the mid-1990s and the development of the FastICA[4] [99], which allowed for efficient implementations (see [108] for a related review).
- In machine learning, the use of ICA as a feature generation technique is justified by the following argument. In [17], it is suggested that the outcome of the early processing performed by the visual cortical feature detectors might be the result of a *redundancy reduction* process. Thus, searching for independent features, conditioned on the input data, is in line with such a claim (see, for example, [75,114] and the references therein).

[4] http://research.ics.aalto.fi/ica/fastica/index.shtml.

- Although we have focused on the noiseless case, extensions of ICA to noisy tasks have also been proposed (see, e.g., [100]). For an extension of ICA in the complex-valued case, see [2]. Nonlinear extensions have also been considered, including kernelized ICA versions (for example, [13]).
- In [3], the treatment of ICA also involves random processes and a wider class of signals, including Gaussians, can be identified.
- In [7], the multiset ICA framework of *independent vector analysis* (IVA) is discussed. It is shown that it generalizes the multiset CCA if higher-order, besides second-order, statistics are taken into account.

19.5.5 ALTERNATIVE PATHS TO ICA

Besides the previously discussed two paths to ICA, a number of alternatives have been suggested, shedding light on different aspects of the problem. Some notable directions are the following.

- *Infomax principle*: This method assumes that the latent variables are the outputs of a nonlinear system (neural network, Chapter 18) of the form

$$z_i = \phi_i(\boldsymbol{w}_i^T \mathbf{x}) + \eta, \quad i = 1, 2, \ldots, l,$$

  where ϕ_i are nonlinear functions and η is additive Gaussian noise. The weight vectors $\boldsymbol{w}_i$ are computed so as to maximize the entropy of the outputs; the reasoning is based on some information theoretic arguments concerning the information flow in the network [18].
- *Maximum likelihood*: Starting from Eq. (19.44), the PDF of the observed variables is expressed in terms of the PDFs of the independent components

$$p(\boldsymbol{x}) = |\det(W)| \prod_{i=1}^{l} p_i(\boldsymbol{w}^T x_i),$$

  where we used

$$W := A^{-1}.$$

  Assuming that we have N observations, $\boldsymbol{x}_1, \boldsymbol{x}_2, \ldots, \boldsymbol{x}_N$, and taking the logarithm of the joint $p(\boldsymbol{x}_1, \ldots, \boldsymbol{x}_N)$, one can maximize the log-likelihood with respect to W. It is straightforward to derive the log-likelihood function and to observe that it is very similar to $J(W)$ given in Eq. (19.55). The p_is are chosen so as to belong to families of non-Gaussians (for example, [100]). A connection between the infomax approach and the maximum likelihood one has been established in [37,38].
- *Negentropy*: According to this method, the starting point is to maximize the non-Gaussianity, which is now measured in terms of the *negentropy*, defined as

$$J(\mathbf{z}) := H(\mathbf{z}_{\text{Gauss}}) - H(\mathbf{z}),$$

  where $\mathbf{z}_{\text{Gauss}}$ corresponds to Gaussian distributed variables of the same covariance matrix, which we know corresponds to the maximum entropy, H. Thus, maximizing the negentropy, which is a nonnegative function, is equivalent to making the latent variables as less Gaussian as possible. Usually, approximations of the negentropy are employed, which are expressed in terms of higher-order cumulants, or by matching the nonlinearity to source distribution [100,141].

- If the unmixing matrix is constrained to be orthogonal, the negentropy and the maximum likelihood approaches become equivalent [2].

THE COCKTAIL PARTY PROBLEM

A classical application that demonstrates the power of the ICA is the so-called *cocktail party problem*. In a party, there are various people speaking; in our case, we are going to consider music as well. Let us say that there are people (a female and a male) and there is also monophonic music, making three sources of sound in total. Then, three microphones (as many as the sources) are placed in different places in the room and the mixed speech signals are recorded. We denote the inputs to the three microphones as $x_1(t)$, $x_2(t)$, and $x_3(t)$, respectively. In the simplest of the models, the three recorded signals can be considered as linear combinations of the individual source signals. Delays are not considered. The goal is to use ICA and recover the original speech and music from the recorded mixed signals.

To this end and in order to bring the task in the formulation we have previously adopted, we consider the values of the three signals at different time instants as different observations of the corresponding random variables, x_1, x_2, and x_3, which are put together to form the random vector $\mathbf{x}$. We further adopt the very reasonable assumption that the original source signals, denoted as $s_1(t)$, $s_2(t)$, and $s_3(t)$, are independent and (similarly as before) the values at different time instants correspond to the values of three latent variables, denoted together as a random vector $\mathbf{s}$.

We are ready now to apply ICA to compute the unmixing matrix W, from which we can obtain the estimates of the ICs corresponding to the observations received by the three microphones,

$$z(t) = [z_1(t), z_2(t), z_3(t)]^T = W[x_1(t), x_2(t), x_3(t)]^T.$$

Fig. 19.9A shows the three different signals, which are linearly combined (by a set of mixing coefficients defining a mixing matrix A) to form the three "microphone signals." Fig. 19.9B shows the resulting signals, which are then used as described before for the ICA analysis. Fig. 19.9C shows the recovered original signals, as the corresponding ICs. The FastICA algorithm was employed.[5] Fig. 19.10 is the result when PCA is used and the original signals are obtained via the (three) principal components.

One can observe that ICA manages to separate the signals with very good accuracy, whereas PCA fails. The reader can also listen to the signals by downloading the corresponding ".wav" files from the site of this book.

Note that the cocktail party problem is representative of a large class of tasks where a number of recorder signals result as linear combinations of other independent signals; the goal is the recovery of the latter. A notable application of this kind is found in electroencephalography (EEG). EEG data consist of electrical potentials recorded at different locations on the scalp (or more recently, in the ear [112]), which are generated by the combination of different underlying components of brain and muscle activity. The task is to use ICA to recover the components, which in turn can unveil useful information about the brain activity (for example, [158]).

The cocktail party problem is a typical example of a more general class of tasks known as *blind source separation* (BSS). The goal in these tasks is to estimate the "causes" (sources, original signals)

[5] http://research.ics.aalto.fi/ica/fastica/.

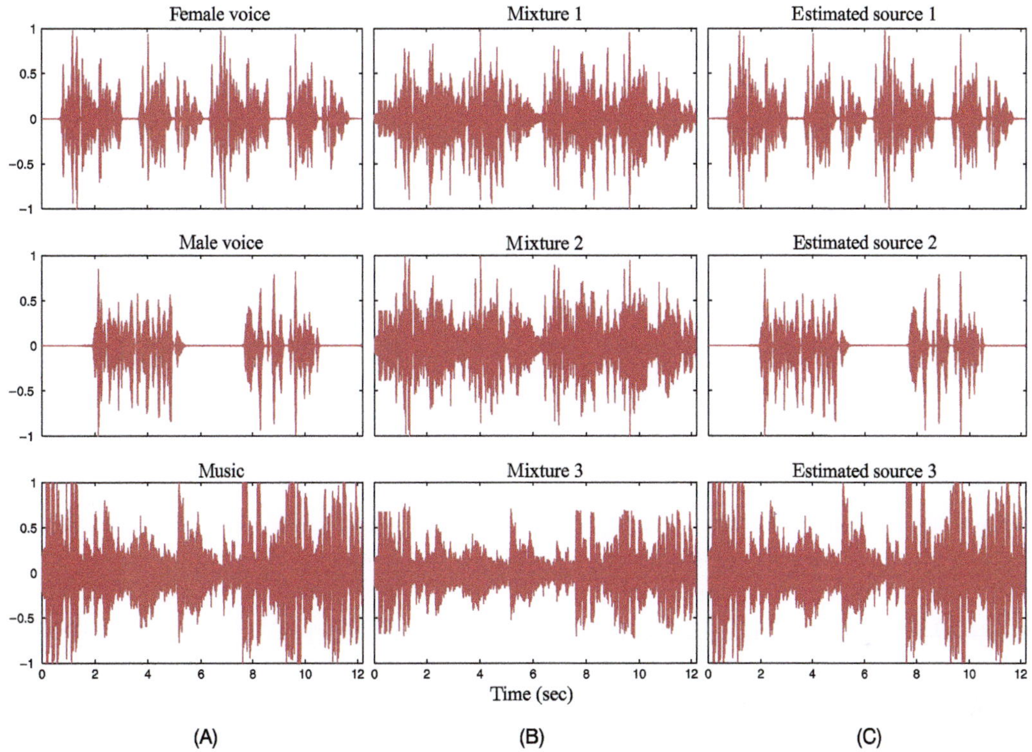

FIGURE 19.9

ICA source separation in the cocktail party setting.

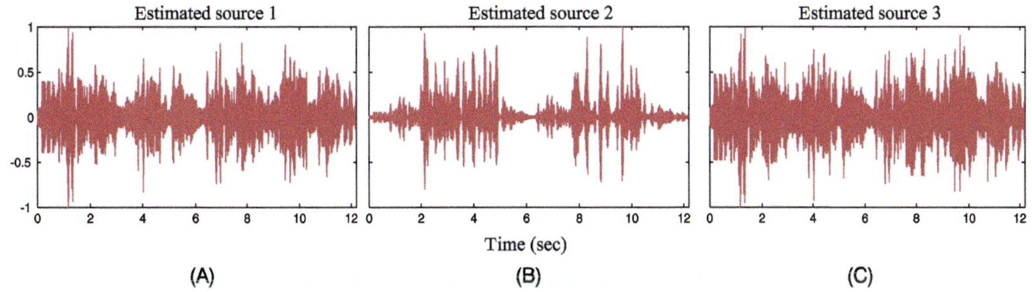

FIGURE 19.10

PCA source separation in the cocktail party setting.

based only on information residing in the observations, without any other extra information, and this is the reason that the word "blind" is used. Viewed in another way, BSS is an example of unsupervised learning. ICA is one among the most widely used techniques for such problems.

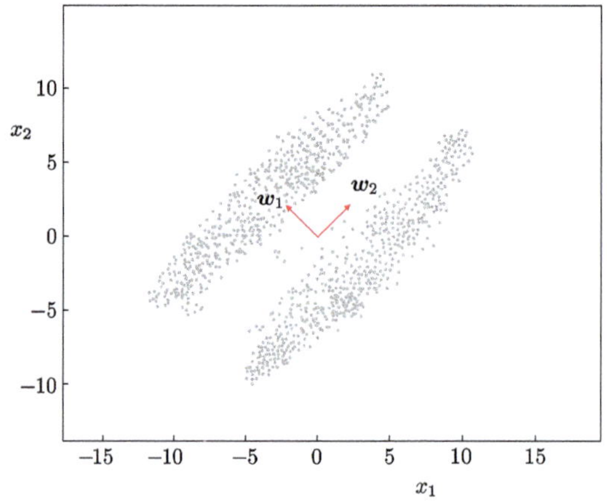

FIGURE 19.11

The setup for the ICA simulation example. The two vectors point to the projection directions resulting from the analysis. The optimal direction for projection, resulting from the ICA analysis, is that of w_2.

Example 19.4. The goal of this example is to demonstrate the power of ICA as a feature generation technique, where the most informative of the generated features are to be kept.

The example is a realization of the case shown in Fig. 19.11. A number of 1024 samples of a two-dimensional normal distribution were generated.

The mean and covariance matrix of the normal PDF were

$$\boldsymbol{\mu} = [-2.6042, 2.5]^T, \quad \Sigma = \begin{bmatrix} 10.5246 & 9.6313 \\ 9.6313 & 11.3203 \end{bmatrix}.$$

Similarly, 1024 samples from a second normal PDF were generated with the same covariance matrix and mean $-\boldsymbol{\mu}$. For the ICA, the method based on the second- and fourth-order cumulants, presented in this section, was used. The resulting transformation matrix W is

$$W = \begin{bmatrix} -0.7088 & 0.7054 \\ 0.7054 & 0.7088 \end{bmatrix} := \begin{bmatrix} \boldsymbol{w}_1^T \\ \boldsymbol{w}_2^T \end{bmatrix}.$$

The vectors $\boldsymbol{w}_2$ and $\boldsymbol{w}_1$ point in the principal and minor axis directions, respectively, obtained from the PCA analysis. According to PCA, the most informative direction is along the principal axis $\boldsymbol{w}_2$, which is the one with maximum variance. However, the most interesting direction for projection, according to the ICA analysis, is that of $\boldsymbol{w}_1$. Indeed, the kurtosis of the obtained ICs $\mathrm{z}_1, \mathrm{z}_2$ along these directions are

$$\kappa_4(\mathrm{z}_1) = -1.7,$$
$$\kappa_4(\mathrm{z}_2) = 0.1,$$

respectively. Thus, projection in the principal (PCA) axis direction results in a variable with a PDF close to a Gaussian. The projection on the minor axis direction results in a variable with a PDF that deviates from the Gaussian (it is bimodal) and it is the more interesting one from the classification point of view. This can be easily verified by looking at the figure; projecting on the direction $\boldsymbol{w}_2$ leads to class overlapping.

19.6 DICTIONARY LEARNING: THE k-SVD ALGORITHM

The concept of *overcomplete dictionaries* and their importance in modeling real-world signals have been introduced in Chapter 9. We return to this topic, this time in a more general setting. There, the dictionary was assumed known with preselected atoms. In this section, the blind version of this task is considered; that is, the atoms of the dictionary are unknown and have to be estimated from the observed data. Recall that this was the case with ICA; however, instead of the independence concept, used for ICA, sparsity arguments will be mobilized here. Giving the freedom to the dictionary to adapt to the needs of the specific, each time, input can lead to enhanced performance compared to dictionaries with preselected atoms.

Our starting point is that the observed l random variables are expressed in terms of $m > l$ latent ones according to the linear model

$$\mathbf{x} = A\mathbf{z}, \ \mathbf{x} \in \mathbb{R}^l, \quad \mathbf{z} \in \mathbb{R}^m, \tag{19.63}$$

and A is an unknown $l \times m$ matrix. Usually, $m \gg l$. Even if A were known and fixed, it does not need special mathematical skills to see that this task has not a single solution and one has to embed constraints into the problem. To this end, we are going to adopt sparsity promoting constraints, as we have already discussed in various parts in this book.

Let $\boldsymbol{x}_n, \ n = 1, 2, \ldots, N$, be the observations that will constitute the only available information. The task is to obtain the atoms (columns of A) of the dictionary as well as the latent variables that are assumed to be sparse; that is, we are going to establish a sparse representation of our input observations (vectors). No doubt, there are different paths to achieve the goal. We are going to focus on one of the most widely known and used methods, known as *k-SVD*, proposed in [4].

Let $X := [\boldsymbol{x}_1, \ldots, \boldsymbol{x}_N]$, $A := \{\boldsymbol{a}_1, \ldots, \boldsymbol{a}_m\}$, and $Z := [z_1, \ldots, z_N]$, where z_n is the latent vector corresponding to the input $\boldsymbol{x}_n, \ n = 1, 2, \ldots, N$. The dictionary learning task is cast as the following optimization problem

$$\text{minimize with respect to } A, Z \quad ||X - AZ||_F^2, \tag{19.64}$$

$$\text{subject to} \quad ||z_n||_0 \leq T_0, \ n = 1, 2, \ldots, N, \tag{19.65}$$

where T_0 is a threshold value and $\|\cdot\|_0$ denotes the ℓ_0 norm, as discussed in Chapter 9. This is a nonconvex optimization task, and it is performed iteratively; each iteration comprises two stages. In the first one, A is assumed to be fixed and optimization is carried out with respect to $z_n, \ n = 1, 2, \ldots, N$. In the second stage, the latent vectors are assumed fixed and optimization is carried out with respect to the columns of A.

In k-SVD, a slightly different rationale is adopted. While optimizing with respect to the columns of A, one at a time, an update of some of the elements of Z is also performed. This is a crucial difference of k-SVD with the more standard optimization techniques; it appears to lead to improved performance in practice.

Stage 1: Assume A to be known and fixed to the value obtained from the previous iteration. Then the associated optimization task becomes

$$\min_{Z} \quad ||X - AZ||_F^2,$$
$$\text{s.t.} \quad ||z_n||_0 \leq T_0, \quad n = 1, 2, \ldots, N,$$

which, due to the definition of the Frobenius norm, is equivalent to solving N distinct optimization tasks,

$$\min_{z_n} \quad ||\boldsymbol{x}_n - Az_n||^2, \tag{19.66}$$
$$\text{s.t.} \quad ||z_n||_0 \leq T_0, \quad n = 1, 2, \ldots, N. \tag{19.67}$$

A similar objective is met if the following optimization tasks are considered instead:

$$\min_{z_n} \quad ||z_n||_0,$$
$$\text{s.t.} \quad ||\boldsymbol{x}_n - Az_n||^2 < \epsilon, \quad n = 1, 2, \ldots, N,$$

where ϵ is a constant acting as an upper bound of the error.

The task in Eqs. (19.66) and (19.67) can be solved by any one of the ℓ_0 minimization solvers, which have been considered in Chapter 10, for example, the OMP. This stage is known as *sparse coding*.

Stage 2: This stage is known as the *codebook update*. Having obtained z_n, $n = 1, 2, \ldots, N$ (for fixed A), from stage 1, the goal now is to optimize with respect to the columns of A. This is achieved on a *column-by-column* basis. Assume that we currently consider the update of $\boldsymbol{a}_k$; this is carried out so as to minimize the (squared) Frobenius norm, $||X - AZ||_F^2$. To this end, we can write the product AZ as a sum of rank one matrices, that is,

$$AZ = [\boldsymbol{a}_1, \ldots, \boldsymbol{a}_m][\boldsymbol{z}_1^r, \ldots, \boldsymbol{z}_m^r]^T = \sum_{i=1}^{m} \boldsymbol{a}_i \boldsymbol{z}_i^{rT}, \tag{19.68}$$

where $\boldsymbol{z}_i^{rT}$, $i = 1, 2, \ldots, m$, are the *rows* of Z. Note that in the above sum, the vectors for indices, $i = 1, 2, \ldots, k - 1$, are fixed to their recently updated values during this second stage of the current iteration step, while and vectors corresponding to $i = k + 1, \ldots, m$, are fixed to the values that are available from the previous iteration step. This strategy allows for the use of the most recent updated information. We will now minimize with respect to the rank one outer product matrix, $\boldsymbol{a}_k \boldsymbol{z}_k^{rT}$. Observe that this product, besides the kth column of A, also involves the kth row of Z; both of them will be updated. The rank one matrix is estimated so as to minimize

$$||E_k - \boldsymbol{a}_k \boldsymbol{z}_k^{rT}||_F^2, \tag{19.69}$$

where

$$E_k := X - \sum_{i=1, i \neq k}^{m} \boldsymbol{a}_i \boldsymbol{z}_i^{rT}.$$

In other words, we seek to find the best, in the Frobenius sense, rank one approximation of E_k. Recall from Chapter 6 (Section 6.4) that the solution is given via the SVD of E_k. However, if we do that, there is no guarantee that whatever sparse structure has been embedded in z_k^r, from the update in stage 1, will be retained. According to the k-SVD, this is bypassed by focusing on the active set, that is, involving only the nonzero of its coefficients. Thus, we first search for the locations of the nonzero coefficients in z_k^r and let

$$\omega_k := \left\{ j_k, 1 \leq j_k \leq N : z_k^r(j_k) \neq 0 \right\}.$$

Then, we form the reduced vector $\tilde{z}_k^r \in \mathbb{R}^{|\omega_k|}$, where $|\omega_k|$ denotes the cardinality of ω_k, which contains only the nonzero elements of z_k^r. A little thought reveals that when writing $X = AZ$, the column of current interest, $\boldsymbol{a}_k$, contributes (as part of the corresponding linear combination) only to the columns $\boldsymbol{x}_{j_k}$, $j_k \in \omega_k$, of X. We then collect the corresponding columns of E_k to construct a reduced-order matrix, $\tilde{E}_k$, which comprises the columns that are associated with the locations of the nonzero elements of z_k^r, and select $\boldsymbol{a}_k \tilde{z}_k^{rT}$ so as to minimize

$$||\tilde{E}_k - \boldsymbol{a}_k \tilde{z}_k^{rT}||_F^2. \tag{19.70}$$

Performing SVD, $\tilde{E}_k = UDV^T$, $\boldsymbol{a}_k$ is set equal to $\boldsymbol{u}_1$ corresponding to the largest of the singular values and $\tilde{z}_k^r = D(1,1)\boldsymbol{v}_1$. Thus, the atoms of the dictionary are obtained in *normalized* form (recall from the theory of SVD that $||\boldsymbol{u}_1|| = 1$). In the sequel, the updated values obtained for $\tilde{z}_k^r$ are placed in the corresponding locations in z_k^r. The latter now has at least as many zeros as it had before, as some of the elements in $\boldsymbol{v}_1$ may be zeros. Simple arguments (Problem 19.3) show that at each iteration the error decreases and the algorithm converges to a local minimum. The success of the algorithm depends on the ability of the greedy algorithm to provide a sparse solution during the first stage. As we know from Chapter 10, greedy algorithms work well for sparsity levels, T_0, small enough compared to l.

In summary, each iteration step of the k-SVD algorithm comprises the following computation steps.

- Initialize $A^{(0)}$ with columns normalized to unit ℓ_2 norm.
- Set $i = 1$.
- *Stage 1*: Solve the optimization task in Eqs. (19.66) and (19.67) to obtain the sparse coding representation vectors, z_n, $n = 1, 2, \ldots, N$; use any algorithm developed for this task.
- *Stage 2*: For any column, $k = 1, 2, \ldots, m$, in $A^{(i-1)}$, update it according to the following:
  - Identify the locations of the nonzero elements in the kth row of the computed, from stage 1, matrix Z.
  - Select the columns in E_k, which correspond to the locations of the nonzero elements of the kth row of Z and form a reduced-order error matrix, $\tilde{E}_k$.
  - Perform SVD on $\tilde{E}_k$: $\tilde{E}_k = UDV^T$.
  - Update the kth column of $A^{(i)}$ to be the eigenvector corresponding to the largest singular value, $\boldsymbol{a}_k^{(i)} = \boldsymbol{u}_1$.
  - Update Z, by embedding in the nonzero locations of its kth row the values $D(1,1)\boldsymbol{v}_1^T$.
- Stop if a convergence criterion is met.
- If not, $i = i + 1$, and continue.

WHY THE NAME k-SVD?

The SVD part of the name is pretty obvious. However, the reader may wonder about the presence of "k" in front. As stated in [4], the algorithm can be considered a generalization of the k-means algorithm, introduced in Chapter 12 (Algorithm 12.1). There, we can consider the mean values, which represent each cluster, as the code words (atoms) of a dictionary. During the first stage of the k-means learning, given the representatives of each cluster, a sparse coding scheme is performed; that is, each input vector is assigned to a single cluster. Thus, we can think of the k-means clustering as a sparse coding scheme that associates a latent vector with each one the observations. Note that each of the latent vectors has only one nonzero element, pointing to the cluster where the respective input vector is assigned, according to the smallest Euclidean distance from all cluster representatives. This is a major difference with the k-SVD dictionary learning, during which each observation vector can be associated with more than one atom; hence, the sparsity level of the corresponding latent vector can be larger than one. Furthermore, based on the assignment of the input vectors to the clusters, in the second stage of the k-means algorithm, an update of the cluster representatives is performed, and for each representative only the input vectors assigned to it are used. This is also similar in spirit to what happens in the second stage of k-SVD. The difference is that each input observation may be associated with more than one atom. As pointed out in [4], if one sets $T_0 = 1$, the k-means algorithm can result from k-SVD.

DICTIONARY LEARNING AND DICTIONARY IDENTIFIABILITY

Dictionary learning was introduced via the ℓ_0 norm in (19.64)–(19.65). In a more general setting, the dictionary learning task is cast as follows:

$$\text{minimize with respect to } A \in \mathcal{A} \text{ and } Z \quad \left\{ ||X - AZ||_F^2 + \lambda g(Z) \right\}, \tag{19.71}$$

where $g(Z)$ is a sparsity promoting function of the elements of matrix Z. For example,

$$g(Z) = \sum_{i=1}^{m} \sum_{j=1}^{N} |Z(i, j)|, \tag{19.72}$$

and λ is a regularization parameter. The set $\mathcal{A} \subseteq \mathbb{R}^{l \times m}$ is a compact constraint set. For example, often we assume that the columns of A have unit norm. This guarantees *scale invariance*. Indeed, assuming A, Z is a solution, if we do not use any constraint, the set $A' = (cA),\ Z' = (\frac{1}{c} Z)$ would also be a solution, for any value of c. Other constraints can also be used to account for extra a priori knowledge concerning the available data (see, e.g., [85]).

In practice, the solution involves two stages, as with k-SVD. Assuming that $A^{(i)}$ is known at the iteration step i, minimization with respect to the code vectors first takes place, i.e.,

Stage 1:

$$Z^{(i)} = \min_{Z} \left\{ ||X - A^{(i)} Z||_F^2 + \lambda g(Z) \right\}. \tag{19.73}$$

Then, fixing $Z^{(i)}$, the code book update results as

Stage 2:

$$A^{(i+1)} = \min_{A \in \mathcal{A}} ||X - AZ^{(i)}||_F^2. \quad (19.74)$$

In the last equation, the sparsity promoting term is not involved, since it is independent of A. Assuming $g(Z)$ is a convex function, at each stage a convex optimization task is solved. Such problems are known as biconvex, in the sense that by fixing one of the arguments in the cost function the optimization problem becomes a convex one with respect to the other. The solution of the optimization task can be obtained via different paths; for example, via majorize-minimize (MM) techniques or the ADMM scheme (Section 8.14); see, e.g., [80,140,196] for some application cases under various constraints.

Solving the dictionary learning task is a nonconvex optimization problem. An important related issue is that of the characterization of the associated local minima. This task is another face of what we call *identifiability*. In other words, assuming that there exists a dictionary that generates the data, can this be recovered? To know the answer to this question is of major importance in a number of applications. For example, in a source localization task, the atoms of the dictionary relate to the directions of the arrival of the involved signals. Also, in fMRI (see Section 19.11) the atoms of the dictionary provide information related to the time responses of the neurons associated with the activated regions in the brain. In [86], it is shown that, under certain assumptions related to sparsity, the minimized cost function in the dictionary learning task has a *guaranteed* local minimum around the dictionary that generates the data, with high probability. Moreover, the bound on the number of samples that are sufficient for the existence of such a minimum is also derived.

Remarks 19.5.

- Alternative paths to k-SVD to dictionary learning have also been suggested. For example, in [73] a dictionary learning technique referred to as method of optimal directions (MOD) was proposed, which differs from k-SVD in the dictionary update step. In particular, the full dictionary is updated via direct minimization of the Frobenius norm. In [126,143] probabilistic arguments are employed, using a Laplacian prior to enforce sparsity. We know from Chapter 13 (Section 13.5) that in this case, the involved integrations are not analytically tractable and the different methods differ in the different approximations used to bypass this obstacle. In the former, the maximum value of the integrand is used and in the latter a Gaussian approximation of the posterior is adopted in order to handle the integration. In [82], variational bound techniques are mobilized (see Section 13.9).
- The method proposed in [125] bears some similarities to k-SVD, because it also revolves around SVD, but the dictionary is constrained to be a union of orthonormal bases. This can lead to some computational advantages; on the other hand, k-SVD puts no constraints on the atoms of the dictionary, which gives more freedom in modeling the input. Another difference lies in the column-by-column update introduced in k-SVD.
- A more detailed comparative study of k-SVD with other methods is given in [4].
- Dictionary learning is essentially a matrix factorization problem where a certain type of constraint is imposed on the right matrix factor. This approach can be considered to be just a manifestation of a wider class of constrained matrix factorization methods that allow several types of constraints to hold. Such techniques include the regularized PCA, where functional and/or sparsity constraints are imposed to the left and to the right factors [12,190,200], as well as the structured sparse matrix factorization in [14].

- Besides the previously reported algorithms, a number of *online* and *distributed* dictionary learning schemes have been proposed. In [127], a distributed version of the k-SVD algorithm is developed. A cloud-based version of k-SVD is proposed in [149] with an emphasis on big data applications. A dictionary learning algorithm employing arguments inspired by the EXTRA optimization scheme, which is discussed in Section 8.15, is described in [182]. In [49], the various agents/nodes learn different parts of the dictionary. Such an approach is also suited for big data applications. An online algorithm for dictionary learning has been presented in [139] and an online version for a distributed setting, with provable convergence to a stationary point, has been proposed in [53]. In [62], the decentralized case over time-varying digraphs is considered that involves a general set of constraints, where a number of matrix factorization schemes result as special cases.

Example 19.5. The goal of this example is to show the performance of the dictionary learning technique in the context of the image denoising task. In the case study of Section 9.10, image denoising, based on a predetermined and fixed DCT dictionary, was considered. Here, k-SVD will be employed in order to learn the dictionary using information of the image *itself.* The two (256×256) images, without and with noise corresponding to PSNR = 22 are shown in Figs. 19.13A and B, respectively. The noisy image is divided in overlapping patches of size 12×12 (144), resulting in $(256 - 12 + 1)^2 = 60,025$ patches in total; these will constitute the training data set used for the learning of the dictionary. Specifically, the patches are sequentially extracted from the noisy image, vectorized in lexicographic order, and used as columns, one after the other, to define the $(144 \times 60,025)$ matrix X. Then, k-SVD is mobilized to train an overcomplete dictionary of size 144×196. The resulting atoms, reshaped in order to form 12×12 pixel patches, are shown in Fig. 19.12. Compare the atoms of this dictionary with atoms of the fixed DCT dictionary of Fig. 9.14.

Next, we follow the same procedure as in Section 9.10, by replacing the DCT dictionary with the one obtained by the k-SVD method. The resulting denoised image is shown in Fig. 19.13C. Note that although the dictionary was trained based on *the noisy data,* it led to about 2 dB PSNR improvement over the fixed-dictionary case. As a matter of fact, because the number of patches is large and each one of them carries a different noise realization, the noise, during the dictionary learning stage, is averaged out leading to nearly noise-free dictionary atoms. More advanced use of dictionary learning techniques to further improve performance in tasks such as denoising and inpainting can be found in [71,72,138].

19.7 NONNEGATIVE MATRIX FACTORIZATION

The strong connection between dimensionality reduction and low rank matrix factorization has already been stressed while discussing PCA. ICA can also be considered as a low rank matrix factorization, if a smaller number, compared to the l observed random variables, of independent components is retained (e.g., selecting the $m < l$ least Gaussian ones).

An alternative to the previously discussed low rank matrix factorization schemes was suggested in [144,145], which guarantees the *nonnegativity* of the elements of the resulting matrix factors. Such a constraint is enforced in certain applications because negative elements contradict physical reality. For example, in image analysis, the intensity values of the pixels cannot be negative. Also, probability values cannot be negative. The resulting factorization is known as *nonnegative matrix factorization*

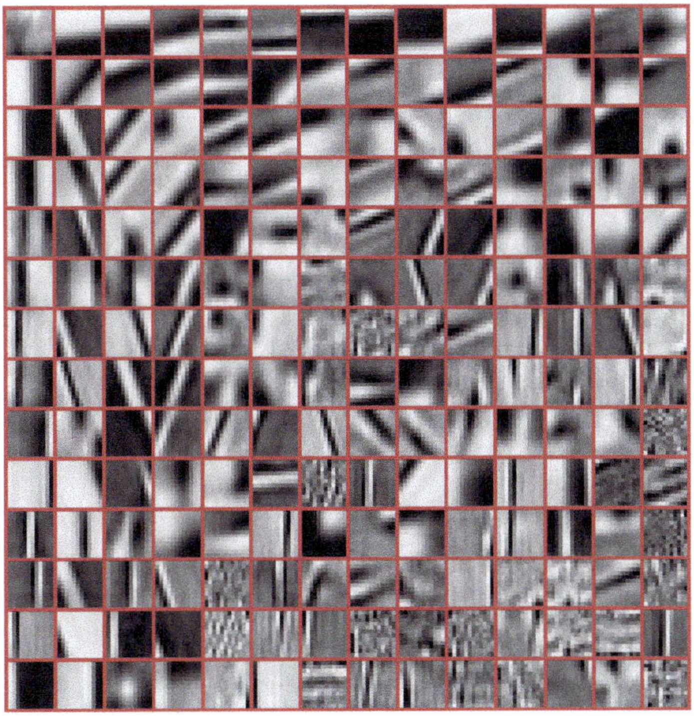

FIGURE 19.12

Dictionary resulting from k-SVD.

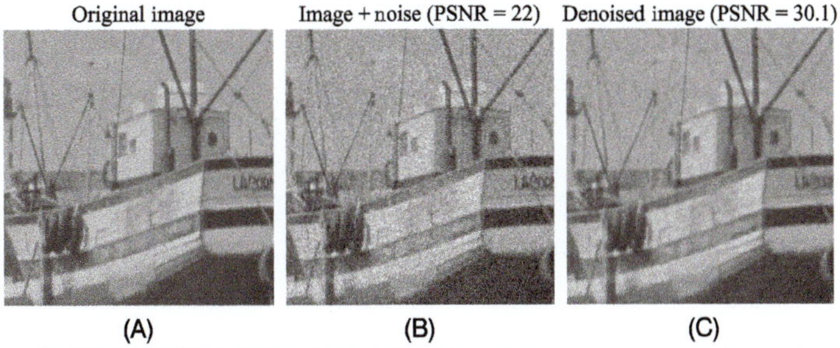

FIGURE 19.13

Image denoising based on dictionary learning.

(NMF) and it has been used successfully in a number of applications, including document clustering [192], molecular pattern discovery [28], image analysis [122], clustering [171], music transcription, music instrument classification [20,166], and face verification [198].

Given an $l \times N$ matrix X, the task of NMF consists of finding an approximate factorization of X, that is,

$$X \approx AZ, \tag{19.75}$$

where A and Z are $l \times m$ and $m \times N$ matrices, respectively, $m \leq \min(N, l)$, and all the matrix elements are nonnegative, that is, $A(i,k) \geq 0,\ Z(k,j) \geq 0,\ i = 1, 2, \ldots, l,\ k = 1, 2, \ldots, m,\ j = 1, 2, \ldots,\ N$. Clearly, if matrices A and Z are of low rank, their product is also a low rank, at most m, approximation of X. The significance of the above is that every column vector in X is represented by the expansion

$$\boldsymbol{x}_i \approx \sum_{k=1}^{m} Z(k,i)\boldsymbol{a}_k,\ i = 1, 2, \ldots, N,$$

where $\boldsymbol{a}_k,\ k = 1, 2, \ldots, m$, are the column vectors of A and constitute the basis of the expansion. The number of vectors in the basis is less than the dimensionality of the vector itself. Hence, NMF can also be seen as a method for *dimensionality reduction*.

To get a good approximation in Eq. (19.75) one can adopt different costs. The most common cost is the Frobenius norm of the error matrix. In such a setting, the NMF task is cast as follows:

$$\min_{A,Z} \quad ||X - AZ||_F^2 := \sum_{i=1}^{l}\sum_{j=1}^{N}\left(X(i,j) - [AZ](i,j)\right)^2, \tag{19.76}$$

$$\text{s.t.} \quad A(i,k) \geq 0,\ Z(k,j) \geq 0, \tag{19.77}$$

where $[AZ](i,j)$ is the (i,j) element of matrix AZ, and i, j, k run over all possible values. Besides the Frobenius norm, other costs have also been suggested (see, e.g., [168]).

Once the problem has been formulated, the major issue rests at the solution of the optimization task. To this end, a number of algorithms have been proposed, for example, Newton-type or gradient descent-type algorithms. Such algorithmic issues, as well as a number of related theoretic ones, are beyond the scope of this book, and the interested reader may consult, for example, [54,67,178]. More recently, regularized versions, including sparsity promoting regularizers, have been proposed (see, for example, [55] for a more recent review on the topic).

19.8 LEARNING LOW-DIMENSIONAL MODELS: A PROBABILISTIC PERSPECTIVE

In this section, the emphasis is on looking at the dimensionality reduction task from a Bayesian perspective. Our focus will be more on presenting the main ideas and less on algorithmic procedures; the latter depend on the specific model and can be dug out from the palette of algorithms that have already been presented in Chapters 12 and 13. Our path to low-dimensional modeling traces its origin to the so-called *factor analysis*.

19.8.1 FACTOR ANALYSIS

Factor analysis was originally proposed in the work of Charles Spearman [169]. Charles Spearman (1863–1945) was an English psychologist who has made important contributions to statistics. Spearman was interested in human intelligence and developed the method in 1904, for analyzing multiple measures of cognitive performance. He argued that there exists a general intelligence factor (the so-called g-factor) that can be extracted by applying the factor analysis method on intelligence test data. However, this notion has been strongly disputed, as intelligence comprises a multiplicity of components (see, e.g., [84]).

Let $\mathbf{x} \in \mathbb{R}^l$. The factor analysis model assumes that there are $m < l$ underlying (latent) zero mean variables or *factors* $\mathbf{z} \in \mathbb{R}^m$ so that

$$\mathrm{x}_i - \mu_i = \sum_{j=1}^{m} a_{ij}\mathrm{z}_j + \epsilon_i, \quad i = 1, 2, \dots, l, \tag{19.78}$$

or

$$\mathbf{x} - \boldsymbol{\mu} = A\mathbf{z} + \boldsymbol{\epsilon}, \tag{19.79}$$

where $\boldsymbol{\mu}$ is the mean of $\mathbf{x}$ and $A \in \mathbb{R}^{l \times m}$ is formed by the weights a_{ij} known as *factor loadings*. The variables z_j, $j = 1, 2, \dots, m$, are sometimes called *common factors*, because they contribute to all observed variables, x_i, and ϵ_i are the *unique* or *specific factors*. As we have already done so far and without loss of generality, we will assume our data are centered, that is, $\boldsymbol{\mu} = \mathbf{0}$. In factor analysis, we assume ϵ_i to be of zero mean and mutually uncorrelated, that is, $\Sigma_\epsilon = \mathbb{E}[\boldsymbol{\epsilon}\boldsymbol{\epsilon}^T] := \text{diag}\{\sigma_1^2, \sigma_2^2, \dots, \sigma_l^2\}$. We also assume that $\mathbf{z}$ and $\boldsymbol{\epsilon}$ are independent. The m $(< l)$ columns of A form a lower-dimensional subspace, and $\boldsymbol{\epsilon}$ is that part of $\mathbf{x}$ not contained in this subspace. The first question that is now raised is whether the model in Eq. (19.79) is any different from our familiar regression task. The answer is in the affirmative. Note that here the matrix A is not known. All that we are given is the set of observations, $\boldsymbol{x}_n$, $n = 1, 2, \dots, N$, and we have to obtain the subspace described by A. It is basically the same linear model that we have considered so far in this chapter, with the difference that now we have introduced the noise term. Once A is known, z_n can be obtained for each $\boldsymbol{x}_n$.

From Eq. (19.79), it is readily seen that

$$\Sigma_x = \mathbb{E}[\mathbf{x}\mathbf{x}^T] = A\,\mathbb{E}[\mathbf{z}\mathbf{z}^T]A^T + \Sigma_\epsilon.$$

We will further assume that $\mathbb{E}[\mathbf{z}\mathbf{z}^T] = I$; hence, we can write

$$\Sigma_x = AA^T + \Sigma_\epsilon. \tag{19.80}$$

Hence, A results as a factor of $(\Sigma_x - \Sigma_\epsilon)$. However, such a factorization, if it exists, is not unique. This can be easily checked if we consider $\bar{A} = AU$, where U is an orthonormal matrix. Then $\bar{A}\bar{A}^T = AA^T$. This has brought a lot of controversy around the factor analysis method when it comes to interpreting individual factors (see, for example, [44] for a discussion). To remedy this drawback, a number of authors have suggested methods and criteria that deal with the rotation (orthogonal or oblique) in order to gain improved interpretation of the factors [157]. However, from our perspective, where the goal is to express our problem in a lower-dimensional space, this is not a problem. Any orthonormal matrix imposes a rotation within the subspace spanned by the columns of A, but we do not care about the exact choice of the coordinates, that is, the common factors.

There are different methods to obtain A (see, e.g., [64]). A popular one is to assume $p(\boldsymbol{x})$ to be Gaussian and employ the maximum likelihood method to optimize with respect to the unknown parameters that define Σ_x in Eq. (19.80). Once A becomes available, one way to estimate the factors is to further assume that these can be expressed as linear combinations of the observations, that is,

$$\mathbf{z} = W\mathbf{x}.$$

Postmultiplying by $\mathbf{x}$, taking expectations, and recalling Eq. (19.79) and that $\mathbb{E}[\mathbf{z}\mathbf{z}^T] = I$, we get

$$\mathbb{E}[\mathbf{z}\mathbf{x}^T] = \mathbb{E}[\mathbf{z}\mathbf{z}^T A^T] + \mathbb{E}[\mathbf{z}\boldsymbol{\epsilon}^T] = A^T. \tag{19.81}$$

Also,

$$\mathbb{E}[\mathbf{z}\mathbf{x}^T] = W\,\mathbb{E}[\mathbf{x}\mathbf{x}^T] = W\Sigma_x. \tag{19.82}$$

Hence,

$$W = A^T \Sigma_x^{-1}.$$

Thus, given a value $\boldsymbol{x}$, the values of the corresponding latent variables are obtained by

$$z = A^T \Sigma_x^{-1}\boldsymbol{x}. \tag{19.83}$$

19.8.2 PROBABILISTIC PCA

New light on this old problem was shed via the Bayesian rationale in the late 1990s [154,175,176]; the task was treated for the special case $\Sigma_\epsilon = \sigma^2 I$ and it was named *probabilistic PCA* (PPCA). The latent variables, $\mathbf{z}$, are dressed with a Gaussian prior,

$$p(z) = \mathcal{N}(z|\mathbf{0}, I),$$

which is in agreement with the earlier assumption $\mathbb{E}[\mathbf{z}\mathbf{z}^T] = I$, and the conditional PDF is chosen as

$$p(\boldsymbol{x}|z) = \mathcal{N}\left(\boldsymbol{x}|Az, \sigma^2 I\right),$$

where, for simplicity, we assume $\boldsymbol{\mu} = \mathbf{0}$ (otherwise the mean would be $Az + \boldsymbol{\mu}$). We are by now pretty familiar with writing down

$$p(z|\boldsymbol{x}) = \mathcal{N}\left(z|\boldsymbol{\mu}_{z|x}, \Sigma_{z|x}\right), \tag{19.84}$$

and

$$p(\boldsymbol{x}) = \mathcal{N}(\boldsymbol{x}|\mathbf{0}, \Sigma_x), \tag{19.85}$$

where (see Eqs. (12.17), (12.10), and (12.15), Chapter 12)

$$\Sigma_{z|x} = \left(I + \frac{1}{\sigma^2}A^T A\right)^{-1}, \tag{19.86}$$

$$\mu_{z|x} = \frac{1}{\sigma^2}\Sigma_{z|x}A^T\boldsymbol{x}, \tag{19.87}$$

$$\Sigma_x = \sigma^2 I + AA^T. \tag{19.88}$$

Note that using the Bayesian framework, the computation of the latent variables corresponding to a given set of observations, $\boldsymbol{x}$, can naturally be obtained via the posterior $p(\boldsymbol{z}|\boldsymbol{x})$ in Eq. (19.84). For example, one can pick the respective mean value

$$\boldsymbol{z} = \frac{1}{\sigma^2}\Sigma_{z|x}A^T\boldsymbol{x}. \tag{19.89}$$

Using the matrix inversion lemma (Problem 19.4), it turns out that Eqs. (19.83) and (19.89) are exactly the same; however, now, it comes as a natural consequence of our Bayesian assumptions.

One way to compute A is to apply the maximum likelihood method on $\prod_{n=1}^N p(\boldsymbol{x}_n)$ and maximize with regard to A and σ^2 (and $\boldsymbol{\mu}$, if $\boldsymbol{\mu} \neq \boldsymbol{0}$). It turns out that the maximum likelihood solution for A is given by [175]

$$A_{\mathrm{ML}} = U_m \mathrm{diag}\{\lambda_1 - \sigma^2, \dots, \lambda_m - \sigma^2\} R,$$

where U_m is the $l \times m$ matrix with columns the eigenvectors corresponding to the m largest eigenvalues, $\lambda_i,\ i = 1, 2, \dots, m$, of the sample covariance matrix of $\mathbf{x}$, and R is an arbitrary orthogonal matrix ($RR^T = I$). Setting $R = I$, the columns of A are the (scaled) principal directions as computed by the classical PCA, discussed in Section 19.3. In any case, the columns of A span the principal subspace of the standard PCA. Note that as $\sigma^2 \to 0$, PPCA tends to PCA (Problem 19.5). Also, it turns out that

$$\sigma^2_{\mathrm{ML}} = \frac{1}{l-m}\sum_{i=m+1}^{l}\lambda_i. \tag{19.90}$$

The previously established connection with PCA does not come as a surprise. It has been well known for a long time (e.g., [5]) that if in the factor analysis model one assumes $\Sigma_\epsilon = \sigma^2 I$, then at stationary points of the likelihood function the columns of A are scaled eigenvectors of the sample covariance matrix. Furthermore, σ^2 is the average of the discarded eigenvalues, as suggested in Eq. (19.90).

Another way to estimate A and σ^2 is via the EM algorithm [154,175]. This is possible because we have $p(\boldsymbol{z}|\boldsymbol{x})$ in an analytic form. Given the set $(\boldsymbol{x}_n, \boldsymbol{z}_n),\ n = 1, 2, \dots, N$, of the observed and latent variables, the complete log-likelihood function is given by

$$\begin{aligned}
\ln p(\mathcal{X}, \mathcal{Z}; A, \sigma^2) &= \sum_{n=1}^{N}\Big(\ln p(\boldsymbol{x}_n|\boldsymbol{z}_n; A, \sigma^2) + \ln p(\boldsymbol{z}_n)\Big)\\
&= -\sum_{n=1}^{N}\Big(\frac{l}{2}\ln(2\pi) - \frac{l}{2}\ln\beta + \frac{\beta}{2}\|\boldsymbol{x}_n - A\boldsymbol{z}_n\|^2\\
&+ \frac{m}{2}\ln(2\pi) + \frac{1}{2}\boldsymbol{z}_n^T\boldsymbol{z}_n\Big),
\end{aligned}$$

which is of the same form as the one given in Eq. (12.45). We have used $\beta = \frac{1}{\sigma^2}$. Thus, following similar steps as for Eq. (12.45) and rephrasing Eqs. (12.46)–(12.50) to our current notation, the E-step becomes

- E-step:

$$\mathcal{Q}(A, \beta; A^{(j)}, \beta^{(j)}) = -\sum_{n=1}^{N} \Big(-\frac{l}{2}\ln\beta + \frac{1}{2}\left\|\boldsymbol{\mu}_{z|x}^{(j)}(n)\right\|^2 + \frac{1}{2}\text{trace}\left\{\Sigma_{z|x}^{(j)}\right\} + \frac{\beta}{2}\left\|\boldsymbol{x}_n - A\boldsymbol{\mu}_{z|x}^{(j)}(n)\right\|^2 + \frac{\beta}{2}\text{trace}\left\{A\Sigma_{z|x}^{(j)}A^T\right\}\Big) + C,$$

where C is a constant and

$$\boxed{\boldsymbol{\mu}_{z|x}^{(j)}(n) = \beta^{(j)}\Sigma_{z|x}^{(j)}A^{(j)T}\boldsymbol{x}_n, \quad \Sigma_{z|x}^{(j)} = \left(I + \beta^{(j)}A^{(j)T}A^{(j)}\right)^{-1}.}$$

- M-step: Taking the derivatives with regard to β and A and equating to zero (Problem 19.6), we obtain

$$\boxed{A^{(j+1)} = \left(\sum_{n=1}^{N}\boldsymbol{x}_n\boldsymbol{\mu}_{z|x}^{(j)T}(n)\right)\left(N\Sigma_{z|x}^{(j)} + \sum_{n=1}^{N}\boldsymbol{\mu}_{z|x}^{(j)}(n)\boldsymbol{\mu}_{z|x}^{(j)T}(n)\right)^{-1},} \tag{19.91}$$

and

$$\boxed{\beta^{(j+1)} = \frac{Nl}{\sum_{n=1}^{N}\left(\left\|\boldsymbol{x}_n - A^{(j+1)}\boldsymbol{\mu}_{z|x}^{(j)}(n)\right\|^2 + \text{trace}\left\{A^{(j+1)}\Sigma_{z|x}^{(j)}A^{(j+1)T}\right\}\right)}.} \tag{19.92}$$

Observe that having adopted the EM algorithm, one does not need to compute the eigenvalues/eigenvectors of Σ_x. Even retrieving only the m principal components, the lowest cost one has to pay is $\mathcal{O}(ml^2)$ operations. Beyond that, $\mathcal{O}(Nl^2)$ are needed to compute Σ_x. For the EM approach, the covariance matrix need not be computed and the most demanding part comprises the matrix vector products, which amount to $\mathcal{O}(Nml)$. Hence for $m \ll l$, computational savings are expected compared to the classical PCA. Keep in mind, though, that the two methods optimize different criteria. PCA guarantees minimum least-squares error reconstruction, PPCA via the EM optimizes the likelihood. Thus, for applications where the error reconstruction is important, such as in compression, one has to be aware of this fact; see [175] for a related discussion.

The other alternative route to solve PPCA is by considering A and σ^2 as random variables with appropriate priors and applying the variational EM algorithm (see [24]). This has the added advantage that if one uses as the prior

$$p(A|\boldsymbol{\alpha}) = \prod_{k=1}^{m}\left(\frac{\alpha_k}{2\pi}\right)^{l/2}\exp\left(-\frac{\alpha_k}{2}\boldsymbol{a}_k^T\boldsymbol{a}_k\right),$$

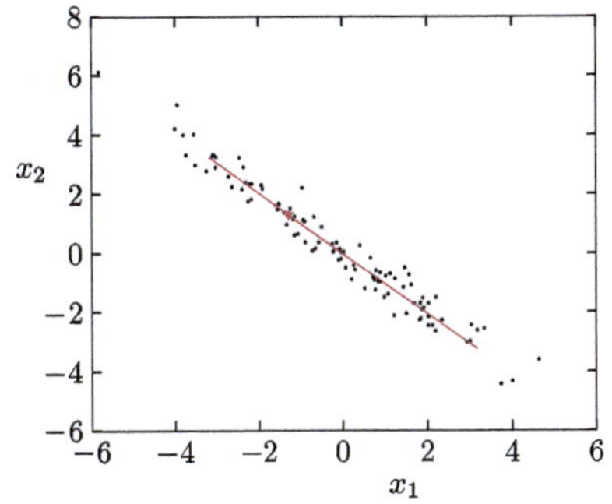

FIGURE 19.14

Data points are distributed around a straight line (one-dimensional subspace) in $\mathbb{R}^2$. The subspace is fully recovered by the PPCA, running the EM algorithm.

with different precisions, α_k, $k = 1, 2, \ldots, m$, per column, then using large enough m one can achieve pruning of the unnecessary components; this was discussed in Section 13.5. Hence, such an approach could provide the means of automatic determination of m. The interested reader, besides the references given before, can dig out useful related information from [47].

Example 19.6. Fig. 19.14 shows a set of data which have been generated via a two-dimensional Gaussian, with zero mean value and covariance matrix equal to

$$\Sigma = \begin{bmatrix} 5.05 & -4.95 \\ -4.95 & 5.05 \end{bmatrix}.$$

The corresponding eigenvalues/eigenvectors are computed as

$$\lambda_1 = 0.05, \quad \boldsymbol{a}_1 = [1, 1]^T,$$
$$\lambda_2 = 5.00, \quad \boldsymbol{a}_2 = [-1, 1]^T.$$

Observe that the data are distributed mainly around a straight line. The EM PPCA algorithm was run on this set of data for $m = 1$. The resulting matrix A, which now becomes a vector, is

$$\boldsymbol{a} = [-1.71, 1.71]^T$$

and $\beta = 0.24$. Note that the obtained vector $\boldsymbol{a}$ points in the direction of the line (subspace) around which the data are distributed.

Remarks 19.6.

- In PPCA, a special diagonal structure was assumed for Σ_ϵ. An EM algorithm for the more general case has also been derived in the early 1980s [156]. Moreover, if the Gaussian prior imposed on the latent variables is replaced by another one, different algorithms result. For example, if non-Gaussian priors are used, then ICA versions are obtained. As a matter of fact, employing different priors, probabilistic versions of the canonical correlation analysis (CCA) and the partial least-squares (PLS) methods result; related references have already been given in the respective sections. Sparsity promoting priors have also been used, resulting in what is known as *sparse factor analysis* (for example, [8,24]). Once the priors have been adopted, one uses standard arguments, more or less, to solve the task, like those discussed in Chapters 12 and 13.

- Besides real-valued variables, extensions to categorical variables have also been considered (for example, [111]). A unifying view of various probabilistic dimensionality reduction techniques is provided in [142].

19.8.3 MIXTURE OF FACTORS ANALYZERS: A BAYESIAN VIEW TO COMPRESSED SENSING

Let us go back to our original model in Eq. (19.79) and rephrase it into a more "trendy" fashion. Matrix A had dimensions $l \times m$ with $m < l$, and $\mathbf{z} \in \mathbb{R}^m$. Let us now make $m > l$. For example, the columns of A may comprise vectors of an overcomplete dictionary. Thus, this section can be considered as the probabilistic counterpart of Section 19.6. The required low dimensionality of the modeling is expressed by imposing sparsity on $\mathbf{z}$; we can rewrite the model in terms of the respective observations as [47]

$$\boldsymbol{x}_n = A(z_n \circ \boldsymbol{b}) + \boldsymbol{\epsilon}_n, \quad n = 1, 2, \ldots, N,$$

where N is the number of our training points and the vector $\boldsymbol{b} \in \mathbb{R}^m$ has elements $b_i \in \{0, 1\}$, $i = 1, 2, \ldots, m$. The product $z_n \circ \boldsymbol{b}$ is the point-wise vector product, that is,

$$z_n \circ \boldsymbol{b} = [z_n(1)b_1, z_n(2)b_2, \ldots, z_n(m)b_m]^T. \tag{19.93}$$

If $\|\boldsymbol{b}\|_0 \ll l$, then $\boldsymbol{x}_n$ is sparsely represented in terms of the columns of A and its intrinsic dimensionality is equal to $\|\boldsymbol{b}\|_0$. Adopting the same assumptions as before,

$$p(\boldsymbol{\epsilon}) = \mathcal{N}(\boldsymbol{\epsilon}|\mathbf{0}, \beta^{-1}I_l), \quad p(z) = \mathcal{N}(z|\mathbf{0}, \alpha^{-1}I_m),$$

where now we have explicitly brought l and m into the notation in order to remind us of the associated dimensions. Also, for the sake of generality, we have assumed that the elements of $\mathbf{z}$ correspond to precision values different than one. Following our familiar standard arguments (as for Eq. (12.15)), it is readily shown that the observations $\boldsymbol{x}_n$, $n = 1, 2, \ldots, N$, are drawn from

$$\mathbf{x} \sim \mathcal{N}(\boldsymbol{x}|\mathbf{0}, \Sigma_x), \tag{19.94}$$

$$\Sigma_x = \alpha^{-1} A \Lambda A^T + \beta^{-1} I_l, \tag{19.95}$$

where

$$\Lambda = \text{diag}\{b_1, \ldots, b_m\}, \tag{19.96}$$

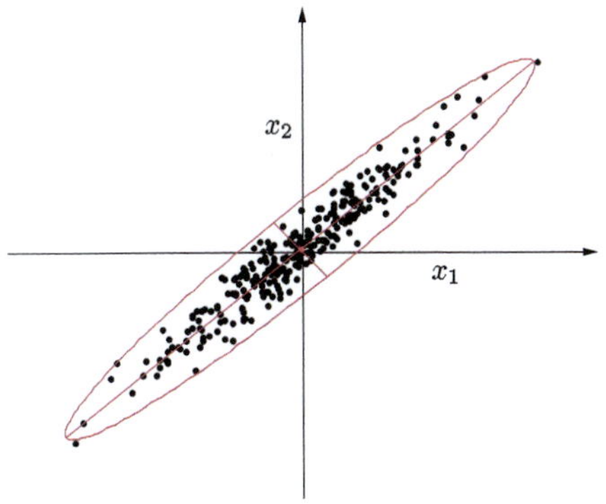

FIGURE 19.15

Data points that lie close to a hyperplane can be sufficiently modeled by a Gaussian PDF whose high probability region corresponds to a sufficiently flat (hyper)ellipsoid.

which guarantees that in Eq. (19.95) only the columns of A, which correspond to nonzero values of $\boldsymbol{b}$, contribute to the formation of Σ_x. We can rewrite the matrix product in the following form:

$$A\Lambda A^T = \sum_{i=1}^{m} b_i \boldsymbol{a}_i \boldsymbol{a}_i^T,$$

and because only $\|\boldsymbol{b}\|_0 := k \ll l$ nonzero terms contribute to the summation, this corresponds to a rank $k < l$ matrix, provided that the respective columns of A are linearly independent. Furthermore, assuming that β^{-1} is small, Σ_x turns out to have a rank approximately equal to k.

Our goal now becomes the learning of the involved parameters; that is, A, β, α, and Λ. This can be done in a standard Bayesian setting by imposing priors on α, β (typically gamma PDFs) and for the columns of A,

$$p(\boldsymbol{a}_i) = \mathcal{N}\left(\boldsymbol{a}_i | 0, \frac{1}{l} I_l\right), \quad i = 1, 2, \ldots, m,$$

which guarantees unit expected norm for each column. The prior for the elements of $\boldsymbol{b}$ are chosen to follow a Bernoulli distribution (see [47] for more details).

Before generalizing the model, let us see the underlying geometric interpretation of the adopted model. Recall from our statistics basics (see also Section 2.3.2, Chapter 2) that most of the activity of a set of jointly Gaussian variables takes place within an (hyper)ellipsoid whose principal axes are determined by the eigenstructure of the covariance matrix. Thus, assuming that the values of **x** lie close to a subspace/(hyper)plane, the resulting Gaussian model Eqs. (19.94) and (19.95) can sufficiently model it by adjusting the elements of Σ_x (after training) so that the corresponding high probability region forms a sufficiently flat ellipsoid (see Fig. 19.15 for an illustration).

Once we have established the geometric interpretation of our factor model, let us leave our imagination free to act. Can this viewpoint be extended for modeling data that originate from a union of subspaces? A reasonable response to this challenge would be to resort to a mixture of factors; one for

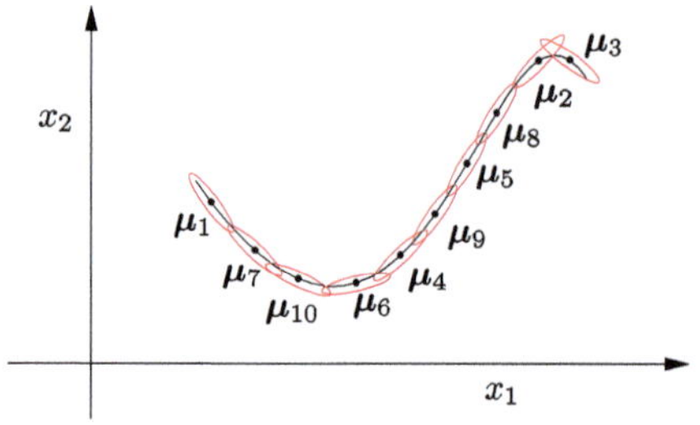

FIGURE 19.16

The curve (manifold) is covered by a number of sufficiently flat ellipsoids centered at the respective mean values.

each subspace. However, there is more to it than that. It has been shown (for example, [27]) that a compact manifold can be covered by a finite number of topological disks, whose dimensionality is equal to the dimensionality of the manifold. Associating topological disks with the principal hyperplanes that define sufficiently flat hyperellipsoids, one can model the data activity, which takes place along a manifold, by a sufficient number of factors, one per ellipsoid [47].

A *mixture of factor analyzers* (MSA) is defined as

$$p(\boldsymbol{x}) = \sum_{j=1}^{J} P_j \mathcal{N}\left(\boldsymbol{x}|\boldsymbol{\mu}_j, \alpha_j^{-1} A_j \Lambda_j A_j^T + \beta^{-1} I_l\right), \tag{19.97}$$

where $\sum_{j=1}^{J} P_j = 1$, $\Lambda_j = \text{diag}\left\{b_{j1}, \dots, b_{jm}\right\}$, $b_{ji} \in \{0, 1\}$, $i = 1, 2, \dots, m$. The expansion in Eq. (19.97) for fixed J and preselected Λ_j, for the jth factor, has been known for some time; in this context, learning of the unknown parameters is achieved in the Bayesian framework, by imposing appropriate priors and mobilizing techniques such as the variational EM (e.g., [79]), EM [175], and maximum likelihood [180]. In a more recent treatment of the problem, the dimensionality of each Λ_j, $j = 1, 2, \dots, J$, as well as the number of factors, J, can be learned by the learning scheme. To this end, *nonparametric priors* are mobilized (see, for example, [40,47,94] and Section 13.11). The model parameters are then computed via Gibbs sampling (Chapter 14) or variational Bayesian techniques. Note that in general, different factors may turn out to have different dimensionality.

For *nonlinear manifold learning*, the geometric interpretation of Eq. (19.97) is illustrated in Fig. 19.16. The number J is the number of flat ellipsoids used to cover the manifold, $\boldsymbol{\mu}_j$ are the sampled points on the manifold, the columns $A_j \Lambda_j$ (approximately) span the local k-dimensional tangent subspace, the noise variance β^{-1} depends on the manifold curvature, and the weights P_j reflect the respective density of the points across the manifold. The method has also been used for matrix completion (see also Section 19.10.1) via a low rank matrix approximation of the involved matrices [40].

Once the model has been learned, using Bayesian inference on a set of training data $\boldsymbol{x}_n$, $n = 1, 2, \dots, N$, it can subsequently be used for compressed sensing, that is, to be able to obtain any $\boldsymbol{x}$, which belongs in the ambient space $\mathbb{R}^l$ but "lives" in the learned k-dimensional manifold, modeled by Eq. (19.97), using $K \ll l$ measurements. To this end, in complete analogy with what has been said in

Chapter 9, one has to determine a sensing matrix, which is denoted here as $\Phi \in \mathbb{R}^{K \times l}$, so as to be able to recover $\boldsymbol{x}$ from the measured (projection) vector

$$\boldsymbol{y} = \Phi \boldsymbol{x} + \boldsymbol{\eta},$$

where $\boldsymbol{\eta}$ denotes the vector of the (unobserved) samples of the measurement noise; all that is now needed is to compute the posterior $p(\boldsymbol{x}|\boldsymbol{y})$. Assuming the noise samples follow a Gaussian and because $p(\boldsymbol{x})$ is a sum of Gaussians, it can be readily seen that the posterior is also a sum of Gaussians determined by the parameters in Eq. (19.97) and the covariance matrix of the noise vector. Hence, $\boldsymbol{x}$ can be recovered by a substantially smaller number of measurements, K, compared to l. In [47], a theoretical analysis is carried out that relates the dimensionality of the manifold, k, the dimensionality of the ambient space, l, and Gaussian/sub-Gaussian types of sensing matrices Φ; this is an analogy to the RIP so that a stable embedding is guaranteed (Section 9.9).

One has to point out a major difference between the techniques developed in Chapter 9 and the current section. There, the model that generates the data was assumed to be known; the signal, denoted there by $\boldsymbol{s}$, was written as

$$\boldsymbol{s} = \Psi \boldsymbol{\theta},$$

where Ψ was the matrix of the dictionary and $\boldsymbol{\theta}$ the sparse vector. That is, the signal was assumed to reside in a subspace, which is spanned by some of the columns of Ψ; in order to recover the signal vector, one had to search for it in a union of subspaces. In contrast, in the current section, we had to "learn" the manifold in which the signal, denoted here by $\boldsymbol{x}$, lies.

19.9 NONLINEAR DIMENSIONALITY REDUCTION

All the techniques that have been considered so far build around linear models, which relate the observed and the latent variables. In this section, we turn our attention to their nonlinear relatives. Our aim is to discuss the main directions that are currently popular and we will not delve into many details. The interested reader can get a deeper understanding and related implementation details from the references provided in the text.[6]

19.9.1 KERNEL PCA

As its name suggests, this is a kernelized version of the classical PCA, and it was first introduced in [161]. As we have seen in Chapter 11, the idea behind any kernelized version of a linear method is to map the variables that originally lie in a low-dimensional space, $\mathbb{R}^l$, into a high (possibly infinite)-dimensional reproducing kernel Hilbert space (RKHS). This is achieved by adopting an implicit mapping,

$$\boldsymbol{x} \in \mathbb{R}^l \longmapsto \boldsymbol{\phi}(\boldsymbol{x}) \in \mathbb{H}. \tag{19.98}$$

[6] Much of this section is based on [174].

Let $\boldsymbol{x}_n,\ n=1,2,\ldots,N$, be the available training examples. The sample covariance matrix of the images, after mapping into $\mathbb{H}$ and assuming centered data, is given by[7]

$$\hat{\Sigma} = \frac{1}{N}\sum_{n=1}^{N} \boldsymbol{\phi}(\boldsymbol{x}_n)\boldsymbol{\phi}(\boldsymbol{x}_n)^T. \tag{19.99}$$

The goal is to perform the eigendecomposition of $\hat{\Sigma}$, that is,

$$\hat{\Sigma}\boldsymbol{u} = \lambda\boldsymbol{u}. \tag{19.100}$$

By the definition of $\hat{\Sigma}$, it can be shown that $\boldsymbol{u}$ lies in the span$\{\boldsymbol{\phi}(\boldsymbol{x}_1), \boldsymbol{\phi}(\boldsymbol{x}_2), \ldots, \boldsymbol{\phi}(\boldsymbol{x}_N)\}$. Indeed,

$$\lambda\boldsymbol{u} = \left(\frac{1}{N}\sum_{n=1}^{N}\boldsymbol{\phi}(\boldsymbol{x}_n)\boldsymbol{\phi}^T(\boldsymbol{x}_n)\right)\boldsymbol{u} = \frac{1}{N}\sum_{n=1}^{N}\left(\boldsymbol{\phi}^T(\boldsymbol{x}_n)\boldsymbol{u}\right)\boldsymbol{\phi}(\boldsymbol{x}_n),$$

and for $\lambda \neq 0$ we can write

$$\boldsymbol{u} = \sum_{n=1}^{N} a_n \boldsymbol{\phi}(\boldsymbol{x}_n). \tag{19.101}$$

Combining Eqs. (19.100) and (19.101), it turns out (Problem 19.7) that the problem is equivalent to performing an eigendecomposition of the corresponding kernel matrix (Chapter 11)

$$\mathcal{K}\boldsymbol{a} = N\lambda\boldsymbol{a}, \tag{19.102}$$

where

$$\boldsymbol{a} := [a_1, a_2, \ldots, a_N]^T. \tag{19.103}$$

As we already know (Section 11.5.1), the elements of the kernel matrix are $\mathcal{K}(i,j) = \kappa(\boldsymbol{x}_i, \boldsymbol{x}_j)$ with $\kappa(\cdot,\cdot)$ being the adopted kernel function. Thus, the kth eigenvector of $\hat{\Sigma}$, corresponding to the kth (nonzero) eigenvalue of $\mathcal{K}$ in Eq. (19.102), is expressed as

$$\boldsymbol{u}_k = \sum_{n=1}^{N} a_{kn}\boldsymbol{\phi}(\boldsymbol{x}_n), \quad k = 1, 2, \ldots, p, \tag{19.104}$$

where $\lambda_1 \geq \lambda_2 \geq \ldots \geq \lambda_p$ denote the respective eigenvalues in descending order and λ_p is the smallest nonzero one and $\boldsymbol{a}_k^T := [a_{k1}, \ldots, a_{kN}]$ is the kth eigenvector of the kernel matrix. The latter is assumed to be normalized so that $\langle \boldsymbol{u}_k, \boldsymbol{u}_k \rangle = 1,\ k = 1, 2, \ldots, p$, where $\langle \cdot, \cdot \rangle$ is the inner product in the Hilbert space $\mathbb{H}$. This imposes an equivalent normalization on the respective $\boldsymbol{a}_k$s, resulting from

$$1 = \langle \boldsymbol{u}_k, \boldsymbol{u}_k \rangle = \left\langle \sum_{i=1}^{N} a_{ki}\boldsymbol{\phi}(\boldsymbol{x}_i), \sum_{j=1}^{N} a_{kj}\boldsymbol{\phi}(\boldsymbol{x}_j) \right\rangle$$

[7] If the dimension of $\mathbb{H}$ is infinite, the definition of the covariance matrix needs a special interpretation, but we will not bother with it here.

$$= \sum_{i=1}^{N} \sum_{j=1}^{N} a_{ki} a_{kj} \mathcal{K}(i, j)$$

$$= \boldsymbol{a}_k^T \mathcal{K} \boldsymbol{a}_k = N\lambda_k \boldsymbol{a}_k^T \boldsymbol{a}_k, \quad k = 1, 2, \ldots, p. \tag{19.105}$$

We are now ready to summarize the basic steps for performing a kernel PCA; that is, to compute the corresponding latent variables (kernel principal components). Given $\boldsymbol{x}_n \in \mathbb{R}^l$, $n = 1, 2, \ldots, N$, and a kernel function $\kappa(\cdot, \cdot)$:

- Compute the $N \times N$ kernel matrix, with elements $\mathcal{K}(i, j) = \kappa(\boldsymbol{x}_i, \boldsymbol{x}_j)$.
- Compute the m dominant eigenvalues/eigenvectors λ_k, $\boldsymbol{a}_k$, $k = 1, 2, \ldots, m$, of $\mathcal{K}$ (Eq. (19.102)).
- Perform the required normalization (Eq. (19.105)).
- Given a feature vector $\boldsymbol{x} \in \mathbb{R}^l$, obtain its low-dimensional representation by computing the m projections onto each one of the dominant eigenvectors,

$$z_k := \langle \boldsymbol{\phi}(\boldsymbol{x}), \boldsymbol{u}_k \rangle = \sum_{n=1}^{N} a_{kn} \kappa(\boldsymbol{x}, \boldsymbol{x}_n), \; k = 1, 2, \ldots, m. \tag{19.106}$$

The operations given in Eq. (19.106) correspond to a *nonlinear mapping* in the input space. Note that, in contrast to the linear PCA, the dominant eigenvectors $\boldsymbol{u}_k$, $k = 1, 2, \ldots, m$, are not computed explicitly. All we know are the respective (nonlinear) projections z_k along them. However, after all, this is what we are finally interested in.

Remarks 19.7.

- Kernel PCA is equivalent to performing a standard PCA in the RKHS $\mathbb{H}$. It can be shown that all the properties associated with the dominant eigenvectors, as discussed for PCA, are still valid for the kernel PCA. That is, (a) the dominant eigenvector directions optimally retain most of the variance; (b) the MSE in approximating a vector (function) in $\mathbb{H}$ in terms of the m dominant eigenvectors is minimal with respect to any other m directions; and (c) projections onto the eigenvectors are uncorrelated [161].
- Recall from Remarks 19.1 that the eigendecomposition of the Gram matrix was required for the metric multidimensional scaling (MDS) method. Because the kernel matrix is the Gram matrix in the respective RKHS, kernel PCA can be considered as a kernelized version of MDS, where inner products in the input space have been replaced by kernel operations in the Gram matrix.
- Note that the kernel PCA method does not consider an explicit underlying structure of the manifold on which the data reside.
- A variant of kernel PCA, known as the kernel entropy component analysis (ECA), has been developed in [103], where the dominant directions are selected so as to maximize the Renyi entropy.

19.9.2 GRAPH-BASED METHODS

Laplacian Eigenmaps

The starting point of this method is the assumption that the points in the data set, $\mathcal{X}$, lie on a smooth manifold $\mathcal{M} \supset \mathcal{X}$, whose intrinsic dimension is equal to $m < l$ and it is embedded in $\mathbb{R}^l$, that is, $\mathcal{M} \subset$

$\mathbb{R}^l$. The dimension m is given as a parameter by the user. In contrast, this is not required in the kernel PCA, where m is the number of dominant components, which, in practice, is determined so that the gap between λ_m and λ_{m+1} has a "large" value.

The main philosophy behind the method is to compute the low-dimensional representation of the data so that *local neighborhood information* in $\mathcal{X} \subset \mathcal{M}$ is optimally preserved. In this way, one attempts to get a solution that reflects the geometric structure of the manifold. To achieve this, the following steps are in order.

Step 1: Construct a graph $G = (V, E)$, where $V = \{v_n,\ n = 1, 2, \ldots, N\}$ is a set of vertices and $E = \{e_{ij}\}$ is the corresponding set of edges connecting vertices $(v_i, v_j),\ i, j = 1, 2, \ldots, N$ (see also Chapter 15). Each node v_n of the graph corresponds to a point $\boldsymbol{x}_n$ in the data set $\mathcal{X}$. We connect v_i, v_j, that is, insert the edge e_{ij} between the respective nodes, if points $\boldsymbol{x}_i, \boldsymbol{x}_j$ are "close" to each other. According to the method, there are two ways of quantifying "closeness." Vertices v_i, v_j are connected with an edge if:

1. $||\boldsymbol{x}_i - \boldsymbol{x}_j||^2 < \epsilon$, for some user-defined parameter ϵ, where $||\cdot||$ is the Euclidean norm in $\mathbb{R}^l$, or
2. $\boldsymbol{x}_j$ is among the k-nearest neighbors of $\boldsymbol{x}_i$ or $\boldsymbol{x}_i$ is among the k-nearest neighbors of $\boldsymbol{x}_j$, where k is a user-defined parameter and neighbors are chosen according to the Euclidean distance in $\mathbb{R}^l$. The use of the Euclidean distance is justified by the smoothness of the manifold that allows to approximate, locally, manifold geodesics by Euclidean distances in the space where the manifold is embedded. The latter is a known result from differential geometry.

   For those who are unfamiliar with such concepts, think of a sphere embedded in the three-dimensional space. If somebody is constrained to live on the surface of the sphere, the shortest path to go from one point to another is the geodesic between these two points. Obviously this is not a straight line but an arc across the surface of the sphere. However, if these points are close enough, their geodesic distance can be approximated by their Euclidean distance, computed in the three-dimensional space.

Step 2: Each edge, e_{ij}, is associated with a weight, $W(i, j)$. For nodes that are not connected, the respective weights are zero. Each weight, $W(i, j)$, is a measure of the "closeness" of the respective neighbors, $\boldsymbol{x}_i, \boldsymbol{x}_j$. A typical choice is

$$W(i, j) = \begin{cases} \exp\left(-\frac{||\boldsymbol{x}_i - \boldsymbol{x}_j||^2}{\sigma^2}\right), & \text{if } v_i, v_j \text{ correspond to neighbors,} \\ 0, & \text{otherwise,} \end{cases}$$

where σ^2 is a user-defined parameter. We form the $N \times N$ weight matrix W having as elements the weights $W(i, j)$. Note that W is symmetric and it is *sparse* because, in practice, many of its elements turn out to be zero.

Step 3: Define the diagonal matrix D with elements $D_{ii} = \sum_j W(i, j),\ i = 1, 2, \ldots, N$, and also the matrix $L := D - W$. The latter is known as the *Laplacian matrix of the graph*, $G(V, E)$. Perform the generalized eigendecomposition

$$L\boldsymbol{u} = \lambda D\boldsymbol{u}.$$

Let $0 = \lambda_0 \leq \lambda_1 \leq \lambda_2 \leq \ldots \leq \lambda_m$ be the smallest $m + 1$ eigenvalues.[8] Ignore the $\boldsymbol{u}_o$ eigenvector corresponding to $\lambda_0 = 0$ and choose the next m eigenvectors $\boldsymbol{u}_1, \boldsymbol{u}_2, \ldots, \boldsymbol{u}_m$. Then map

$$\boldsymbol{x}_n \in \mathbb{R}^l \longmapsto \boldsymbol{z}_n \in \mathbb{R}^m, \quad n = 1, 2, \ldots, N,$$

where

$$\boldsymbol{z}_n^T = [u_{1n}, u_{2n}, \ldots, u_{mn}], \quad n = 1, 2, \ldots, N. \tag{19.107}$$

That is, $\boldsymbol{z}_n$ comprises the nth components of the m previous eigenvectors. The computational complexity of a general eigendecomposition solver amounts to $O(N^3)$ operations. However, for sparse matrices, such as the Laplacian matrix, L, efficient schemes can be employed to reduce complexity to be subquadratic in N, e.g., the Lanczos algorithm [83].

The proof concerning the statement of step 3 will be given for the case of $m = 1$. For this case, the low-dimensional space is the real axis. Our path evolves along the lines adopted in [19]. The goal is to compute $z_n \in \mathbb{R}$, $n = 1, 2, \ldots, N$, so that connected points (in the graph, i.e., neighbors) stay as close as possible after the mapping onto the one-dimensional subspace. The criterion used to satisfy the closeness after the mapping is

$$E_L = \sum_{i=1}^{N}\sum_{j=1}^{N}(z_i - z_j)^2 W(i, j) \tag{19.108}$$

to become minimum. Observe that if $W(i, j)$ has a large value (i.e., $\boldsymbol{x}_i, \boldsymbol{x}_j$ are close in $\mathbb{R}^l$), then if the respective z_i, z_j are far apart in $\mathbb{R}$ it incurs a heavy penalty in the cost function. Also, points that are not neighbors do not affect the minimization as the respective weights are zero. For the more general case, where $1 < m < l$, the cost function becomes

$$E_L = \sum_{i=1}^{N}\sum_{j=1}^{N}||\boldsymbol{z}_i - \boldsymbol{z}_j||^2 W(i, j).$$

Let us now reformulate Eq. (19.108). After some trivial algebra, we obtain

$$\begin{aligned} E_L &= \sum_i z_i^2 \sum_j W(i, j) + \sum_j z_j^2 \sum_i W(i, j) - 2\sum_i\sum_j z_i z_j W(i, j) \\ &= \sum_i z_i^2 D_{ii} + \sum_j z_j^2 D_{jj} - 2\sum_i\sum_j z_i z_j W(I, j) \\ &= 2\boldsymbol{z}^T L \boldsymbol{z}, \end{aligned} \tag{19.109}$$

where

$$\boxed{L := D - W: \quad \text{Laplacian matrix of the graph,}} \tag{19.110}$$

[8] In contrast to the notation used for PCA, the eigenvalues here are marked in ascending order. This is because, in this subsection, we are interested in determining the smallest values and such a choice is notationally more convenient.

and $z^T = [z_1, z_2, \ldots, z_N]$. The Laplacian matrix, L, is symmetric and positive semidefinite. The latter is readily seen from the definition in Eq. (19.109), where E_L is always a nonnegative scalar. Note that the larger the value of D_{ii} is, the more "important" is the sample $\boldsymbol{x}_i$. This is because it implies large values for $W(i, j)$, $j = 1, 2, \ldots, N$, and plays a dominant role in the minimization process. Obviously, the minimum of E_L is achieved by the trivial solution $z_i = 0$, $i = 1, 2, \ldots, N$. To avoid this, as is common in such cases, we constrain the solution to a prespecified norm. Hence, our problem now becomes

$$\min_{z} \quad z^T L z,$$

$$\text{s.t.} \quad z^T D z = 1.$$

Although we can work directly on the previous task, we will slightly reshape it in order to use tools that are more familiar to us. Define

$$y = D^{1/2} z, \tag{19.111}$$

and

$$\tilde{L} = D^{-1/2} L D^{-1/2}, \tag{19.112}$$

which is known as the *normalized graph Laplacian* matrix. It is now readily seen that our optimization problem becomes

$$\min_{y} \quad y^T \tilde{L} y, \tag{19.113}$$

$$\text{s.t.} \quad y^T y = 1. \tag{19.114}$$

Using Lagrange multipliers and equating the gradient of the Lagrangian to zero, it turns out that the solution is given by

$$\tilde{L} y = \lambda y. \tag{19.115}$$

In other words, computing the solution becomes equivalent to solving an eigenvalue-eigenvector problem. Substituting Eq. (19.115) into the cost function in (19.113) and taking into account the constraint (19.114), it turns out that the value of the cost associated with the optimal y is equal to λ. Hence, the solution is the eigenvector corresponding to the minimum eigenvalue. However, the minimum eigenvalue of $\tilde{L}$ is zero and the corresponding eigenvector corresponds to a trivial solution. Indeed, observe that

$$\tilde{L} D^{1/2} \mathbf{1} = D^{-1/2} L D^{-1/2} D^{1/2} \mathbf{1} = D^{-1/2} (D - W) \mathbf{1} = \mathbf{0},$$

where $\mathbf{1}$ is the vector having all its elements equal to 1. In words, $y = D^{1/2}\mathbf{1}$ is an eigenvector corresponding to the zero eigenvalue and it results in the trivial solution, $z_i = 1$, $i = 1, 2, \ldots, N$. That is, all the points are mapped onto the same point in the real line. To exclude this undesired solution, recall that $\tilde{L}$ is a positive semidefinite matrix and, hence, 0 is its smallest eigenvalue. In addition, if the graph is assumed to be connected, that is, there is at least one path (see Chapter 15) that connects any pair of vertices, $D^{1/2}\mathbf{1}$ is the only eigenvector associated with the zero eigenvalue, λ_0 [19]. Also, as $\tilde{L}$ is a symmetric matrix, we know (Appendix A.2) that its eigenvectors are orthogonal to each other. In the sequel, we impose an extra constraint and we now require the solution to be *orthogonal* to $D^{1/2}\mathbf{1}$. Constraining the solution to be orthogonal to the eigenvector corresponding to the smallest (zero) eigenvalue drives the solution to the next eigenvector corresponding to the next smallest (nonzero) eigenvalue λ_1. Note that the eigendecomposition of $\tilde{L}$ is equivalent to what we called generalized eigendecomposition of L in step 3 before.

For the more general case of $m > 1$, we have to compute the m eigenvectors associated with $\lambda_1 \leq \ldots \leq \lambda_m$. As a matter of fact, for this case, the constraints prevent us from mapping into a subspace of dimension less than the desired m. For example, we do not want to project in a three-dimensional space and the points to lie on a two-dimensional plane or on a one-dimensional line. For more details, the interested reader is referred to the insightful paper [19].

Local Linear Embedding (LLE)

As was the case with the Laplacian eigenmap method, *local linear embedding* (LLE) assumes that the data points rest on a smooth enough manifold of dimension m, which is embedded in the $\mathbb{R}^l$ space, with $m < l$ [155]. The smoothness assumption allows us to further assume that, provided there are sufficient data and the manifold is "well" sampled, nearby points lie on (or close to) a "locally" *linear* patch of the manifold (see, also, related comments in Section 19.8.3). The algorithm in its simplest form is summarized in the following three steps:

Step 1: For each point, $\boldsymbol{x}_n$, $n = 1, 2, \ldots, N$, search for its nearest neighbors.

Step 2: Compute the weights $W(n, j)$, $j = 1, 2, \ldots, N$, that best reconstruct each point, $\boldsymbol{x}_n$, from its nearest neighbors, so as to minimize the cost

$$\arg\min_{W} \; E_W = \sum_{n=1}^{N} \|\boldsymbol{x}_n - \sum_{j=1}^{N} W(n, j)\boldsymbol{x}_{n_j}\|^2, \tag{19.116}$$

where $\boldsymbol{x}_{n_j}$ denotes the jth neighbor of the nth point. (a) The weights are constrained to be zero for points which are not neighbors and (b) the rows of the weight matrix add to one, that is,

$$\sum_{j=1}^{N} W(n, j) = 1, \; n = 1, 2, \ldots, N. \tag{19.117}$$

That is, the sum of the weights, over all neighbors, must be equal to one.

Step 3: Once the weights have been computed from the previous step, use them to obtain the corresponding points $\boldsymbol{z}_n \in \mathbb{R}^m$, $n = 1, 2, \ldots, N$, so as to minimize the cost with respect to the unknown set of points $\mathcal{Z} = \{\boldsymbol{z}_n, \; n = 1, 2, \ldots, N\}$,

$$\arg\min_{\boldsymbol{z}_n:\, n=1,\ldots,N} \; E_{\mathcal{Z}} = \sum_{n=1}^{N} \|\boldsymbol{z}_n - \sum_{j=1}^{N} W(n, j)\boldsymbol{z}_j\|^2. \tag{19.118}$$

The above minimization takes place subject to two constraints, to avoid degenerate solutions: (a) the outputs are centered, $\sum_n \boldsymbol{z}_n = \mathbf{0}$, and (b) the outputs have unit covariance matrix [159]. Nearest points, in step 1, are searched in the same way as for the Laplacian eigenmap method. Once again, the use of the Euclidean distance is justified by the smoothness of the manifold, as long as the search is limited "locally" among neighboring points. For the second step, the method exploits the local linearity of a smooth manifold and tries to predict *linearly* each point by its neighbors using the least-squares method. Minimizing the cost subject to the constraint given in Eq. (19.117) results in a solution that satisfies the following three properties:

1. rotation invariance,
2. scale invariance,
3. translation invariance.

The first two can easily be verified by the form of the cost function and the third one is the consequence of the imposed constraints. The implication of this is that the computed weights encode information about the intrinsic characteristics of each neighborhood and they do not depend on the particular point.

The resulting weights, $W(i, j)$, reflect the intrinsic properties of the local geometry underlying the data, and because our goal is to retain the local information after the mapping, these weights are used to reconstruct each point in the $\mathbb{R}^m$ subspace by its neighbors. As is nicely stated in [159], it is as if we take a pair of scissors to cut small linear patches of the manifold and place them in the low-dimensional subspace.

It turns out that solving (19.118) for the unknown points, z_n, $n = 1, 2, \ldots, N$, is equivalent to:

- performing an eigendecomposition of the matrix $(I - W)^T(I - W)$,
- discarding the eigenvector that corresponds to the smallest eigenvalue,
- taking the eigenvectors that correspond to the next (smaller) eigenvalues. These yield the low-dimensional latent variable scores, z_n, $n = 1, 2, \ldots, N$.

Once again, the involved matrix W is sparse and if this is taken into account the eigenvalue problem scales relatively well to large data sets with complexity subquadratic in N. The complexity for step 2 scales as $O(Nk^3)$ and it is contributed by the solver of the linear set of equations with k unknowns for each point. The method needs two parameters to be provided by the user, the number of nearest neighbors, k (or ϵ) and the dimensionality m. The interested reader can find more on the LLE method in [159].

Isometric Mapping (ISOMAP)

In contrast to the two previous methods, which unravel the geometry of the manifold on a local basis, the ISOMAP algorithm adopts the view that only the geodesic distances between all pairs of the data points can reflect the true structure of the manifold. Euclidean distances between points in a manifold cannot represent it properly, because points that lie far apart, as measured by their geodesic distance, may be close when measured in terms of their Euclidean distance (see Fig. 19.17). ISOMAP is basically a variant of the multidimensional scaling (MDS) algorithm, in which the Euclidean distances are substituted by the respective geodesic distances along the manifold. The essence of the method is to estimate geodesic distances between points that lie far apart. To this end, a two-step procedure is adopted.

Step 1: For each point, $\boldsymbol{x}_n$, $n = 1, 2, \ldots, N$, compute the nearest neighbors and construct a graph $G(V, E)$ whose vertices represent the data points and the edges connect nearest neighbors. (Nearest neighbors are computed with either of the two alternatives used for the Laplacian eigenmap method. The parameters k or ϵ are user-defined parameters.) The edges are assigned weights based on the

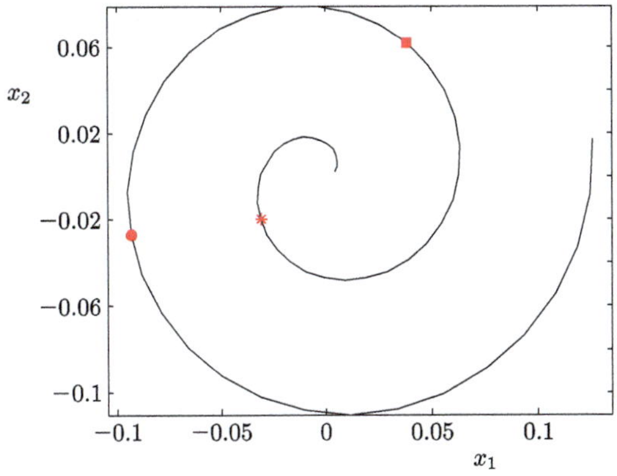

FIGURE 19.17

The point denoted by a "star" is deceptively closer to the point denoted by a "dot" than to the point denoted by a "box" if distance is measured in terms of the Euclidean distance. However, if one is constrained to travel along the spiral, the geodesic distance is the one that determines closeness and it is the "box" point that is closer to the "star."

respective Euclidean distance (for nearest neighbors, this is a good approximation of the respective geodesic distance).

Step 2: Compute the pair-wise geodesic distances among all pairs (i, j), $i, j = 1, 2, \ldots, N$, along shortest paths through the graph. The key assumption is that the geodesic between any two points on the manifold can be approximated by the *shortest path* connecting the two points along the graph $G(V, E)$. To this end, efficient algorithms can be used to achieve it with complexity $\mathcal{O}(N^2 \ln N + N^2 k)$ (e.g., Djikstar's algorithm, [59]). This cost can be prohibitive for large values of N.

Having estimated the geodesics between all pairs of point, the MDS method is mobilized. Thus, the problem becomes equivalent to performing the eigendecomposition of the respective Gram matrix and selecting the m most dominant eigenvectors to represent the low-dimensional space. After the mapping, Euclidean distances between points in the low-dimensional subspace match the respective geodesic distances on the manifold in the original high-dimensional space. As is the case in PCA and MDS, m is estimated by the number of significant eigenvalues. It can be shown that ISOMAP is guaranteed asymptotically ($N \longrightarrow \infty$) to recover the true dimensionality of a class of nonlinear manifolds [66,173].

All three graph-based methods share a common step for computing nearest neighbors in a graph. This is a problem of complexity $\mathcal{O}(N^2)$ but more efficient search techniques can be used by employing a special type of data structures (for example, [23]). A notable difference between the ISOMAP on the one side and the Laplacian eigenmap and LLE methods on the other is that the latter two approaches rely on the eigendecomposition of sparse matrices as opposed to the ISOMAP that relies on the eigendecomposition of the dense Gram matrix. This gives a computational advantage to the Laplacian eigenmap and LLE techniques. Moreover, the calculation of the shortest paths in the ISOMAP is another computationally demanding task. Finally, it is of interest to note that the three graph-based techniques perform the task of dimensionality reduction while trying to unravel, in one way or another, the geometric properties of the manifold on which the data (approximately) lie. In contrast, this is not the case with the kernel PCA, which shows no interest in any manifold learning. However, as the world is very small, in [88] it is pointed out that the graph-based techniques can be seen as special cases of kernel PCA! This becomes possible if data dependent kernels, derived from graphs encoding neighborhood information, are used in place of predefined kernel functions.

The goal of this section was to present some of the most basic directions that have been suggested for nonlinear dimensionality reduction. Besides the previous basic schemes, a number of variants have been proposed in the literature (e.g., [21,70,163]). In [150] and [118] (*diffusion maps*), the low-dimensional embedding is achieved so as to preserve certain measures that reflect the connectivity of the graph $G(V, E)$. In [30,101], the idea of preserving the local information in the manifold has been carried out to define linear transforms of the form $\boldsymbol{z} = A^T\boldsymbol{x}$, and the optimization is now carried out with respect to the elements of A. The task of incremental manifold learning for dimensionality reduction was more recently considered in [121]. In [172,184], the *maximum variance unfolding* method is introduced. The variance of the outputs is maximized under the constraint that (local) distances and angles are preserved among neighbors in the graph. Like the ISOMAP, it turns out that the top eigenvectors of a Gram matrix have to be computed, albeit avoiding the computationally demanding step of estimating geodesic distances, as is required by the ISOMAP. In [164], a general framework, called *graph embedding*, is presented that offers a unified view for understanding and explaining a number of known (including PCA and nonlinear PCA) dimensionality reduction techniques and it also offers a platform for developing new ones. For a more detailed and insightful treatment of the topic, the interested reader is referred to [29]. A review of nonlinear dimensionality reduction techniques can be found in, for example, [32,123].

Example 19.7. Let a data set consisting of 30 points be in the two-dimensional space. The points result from sampling the spiral of Archimedes (see Fig. 19.18A), described by

$$x_1 = a\theta\cos\theta, \quad x_2 = a\theta\sin\theta.$$

The points of the data set correspond to the values $\theta = 0.5\pi,\ 0.7\pi,\ 0.9\pi, \ldots, 2.05\pi$ (θ is expressed in radians), and $a = 0.1$. For illustration purposes and in order to keep track of the "neighboring" information, we have used a sequence of six symbols, "x," "+," "★," "□," "◇," and "o" with black color, followed by the same sequence of symbols in red color, repeatedly.

To study the performance of PCA for this case, where data lie on a nonlinear manifold, we first performed the eigendecomposition of the covariance matrix, estimated from the data set. The resulting eigenvalues are

$$\lambda_2 = 0.089 \quad \text{and} \quad \lambda_1 = 0.049.$$

Observe that the eigenvalues are comparable in size. Thus, if one would trust the "verdict" coming from PCA, the answer concerning the dimensionality of the data would be that it is equal to 2. Moreover, after projecting along the direction of the principal component (the straight line in Fig. 19.18B), corresponding to λ_2, neighboring information is lost because points from different locations are mixed together.

In the sequel, the Laplacian eigenmap technique for dimensionality reduction is employed, with $\epsilon = 0.2$ and $\sigma = \sqrt{0.5}$. The obtained results are shown in Fig. 19.18C. Looking from right to left, we see that the Laplacian method nicely "unfolds" the spiral in a one-dimensional straight line. Furthermore, neighboring information is retained in this one-dimensional representation of the data. Black and red

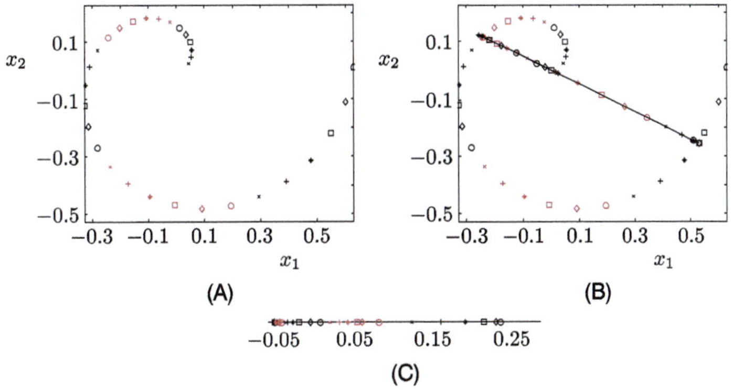

FIGURE 19.18

(A) A spiral of Archimedes in the two-dimensional space. (B) The previous spiral together with the projections of the sampled points on the direction of the first principal component, resulting from PCA. It is readily seen that neighboring information is lost after the projection and points corresponding to different parts of the spiral overlap. (C) The one-dimensional map of the spiral using the Laplacian method. In this case, the neighboring information is retained after the nonlinear projection and the spiral nicely unfolds to a one-dimensional line.

areas are succeeding each other in the right order, and also, observing the symbols, one can see that neighbors are mapped to neighbors.

Example 19.8. Fig. 19.19 shows samples from a three-dimensional spiral, parameterized as $x_1 = a\theta\cos\theta,\ \ x_2 = a\theta\sin\theta$, and sampled at $\theta = 0.5\pi,\ 0.7\pi\ ,0.9\pi, \ldots, 2.05\pi$ (θ is expressed in radians), $a = 0.1$, and $x_3 = -1,\ -0.8,\ -0.6, \ldots, 1$.

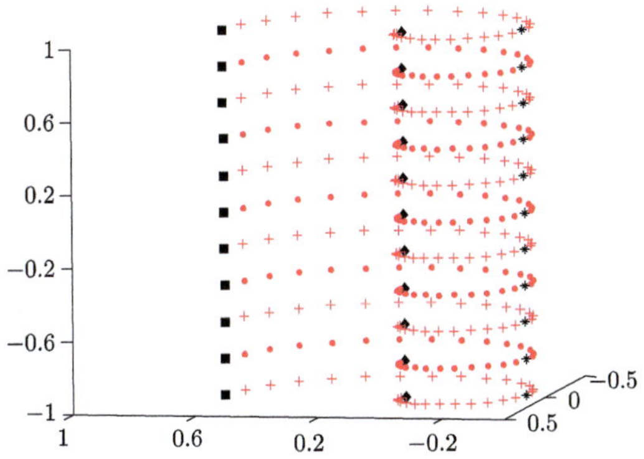

FIGURE 19.19

Samples from a three-dimensional spiral. One can think of it as a number of two-dimensional spirals one above the other. Different symbols have been used in order to track neighboring information.

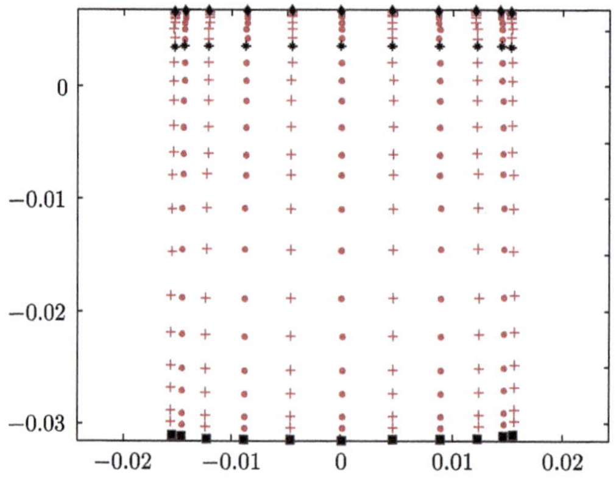

FIGURE 19.20

Two-dimensional mapping of the spiral of Fig. 19.19 using the Laplacian eigenmap method. The three-dimensional structure is unfolded to the two-dimensional space by retaining the neighboring information.

For illustration purposes and in order to keep track of the "identity" of each point, we have used red crosses and dots interchangeably, as we move upward in the x_3 dimension. Also, the first, the middle, and the last points for each level of x_3 are denoted by black "$\diamond$," black "$\star$," and black "$\square$," respectively. Basically, all points at the same level lie on a two-dimensional spiral.

Fig. 19.20 shows the two-dimensional mapping of the three-dimensional spiral using the Laplacian method for dimensionality reduction, with parameter values $\epsilon = 0.35$ and $\sigma = \sqrt{0.5}$. Comparing Figs. 19.19 and 19.20, we see that all points corresponding to the same level x_3 are mapped across the same line, with the first point being mapped to the first one, and so on. That is, as was the case of Example 19.7, the Laplacian method unfolds the three-dimensional spiral into a two-dimensional surface, while retaining neighboring information.

19.10 LOW RANK MATRIX FACTORIZATION: A SPARSE MODELING PATH

The low rank matrix factorization task has already been discussed from different perspectives. In this section, the task will be considered in a specific context; that of missing entries and/or in the presence of outliers. Such a focus is dictated by a number of more recent applications, especially in the framework of big data problems. To this end, sparsity promoting arguments will be mobilized to offer a fresh look at this old problem. We are not going to delve into many details and our purpose is to highlight the main directions and methods which have been considered.

19.10.1 MATRIX COMPLETION

To recapitulate some of the main findings in Chapters 9 and 10, let us consider a signal vector $\boldsymbol{s} \in \mathbb{R}^l$, where only N of its components are observed and the rest are unknown. This is equivalent to sensing $\boldsymbol{s}$ via a sensing matrix having its N rows picked uniformly at random from the standard (canonical) basis

$\Phi = I$, where I is the $l \times l$ identity matrix. The question that was posed there was whether it is possible to recover $\boldsymbol{s}$ exactly based on these N components. From the theory presented in Chapter 9, we know that one can recover all the components of $\boldsymbol{s}$, provided that $\boldsymbol{s}$ is sparse in some basis or dictionary, Ψ, which exhibits low mutual coherence with $\Phi = I$, and N is large enough, as has been pointed out in Section 9.9.

Inspired by the theoretical advances in compressed sensing, a question similar in flavor and with a prominent impact regarding practical applications was posed in [33]. Given an $l_1 \times l_2$ matrix M, assume that only $N \ll l_1 l_2$ among its entries are known. Concerning notation, we refer to a general matrix M, irrespective of how this matrix was formed. For example, it may correspond to an image array. The question now is whether one is able to recover the exact full matrix. This problem is widely known as *matrix completion* [33]. The answer, although it might come as a surprise, is "yes" with high probability, provided that (a) the matrix is *well structured* and complies with certain assumptions, (b) it has a *low rank*, $r \ll l$, where $l = \min(l_1, l_2)$, and (c) N is large enough. Intuitively, this is plausible because a low rank matrix is fully described in terms of a number of parameters (degrees of freedom), which is much smaller than its total number of entries. These parameters are revealed via its SVD

$$M = \sum_{i=1}^{r} \sigma_i \boldsymbol{u}_i \boldsymbol{v}_i^T = U \begin{bmatrix} \sigma_1 & & O \\ & \ddots & \\ O & & \sigma_r \end{bmatrix} V^T, \tag{19.119}$$

where r is the rank of the matrix, $\boldsymbol{u}_i \in \mathbb{R}^{l_1}$ and $\boldsymbol{v}_i \in \mathbb{R}^{l_2}$, $i = 1, 2, \ldots, r$, are the left and right orthonormal singular vectors, spanning the column and row spaces of M, respectively, σ_i, $i = 1, 2, \ldots, r$, are the corresponding singular values, and $U = [\boldsymbol{u}_1, \boldsymbol{u}_2, \cdots, \boldsymbol{u}_r]$, $V = [\boldsymbol{v}_1, \boldsymbol{v}_2, \cdots, \boldsymbol{v}_r]$.

Let $\boldsymbol{\sigma}_M$ denote the vector containing all the singular values of M, that is, $\boldsymbol{\sigma}_M = [\sigma_1, \sigma_2, \cdots, \sigma_l]^T$. Then $\text{rank}(M) := \|\boldsymbol{\sigma}_M\|_0$. Counting the parameters associated with the singular values and vectors in Eq. (19.119), it turns out that the number of degrees of freedom of a rank r matrix is equal to $d_M = r(l_1 + l_2) - r^2$ (Problem 19.8). When r is small, d_M is much smaller than l.

Let us denote by Ω the set of N pairs of indices, (i, j), $i = 1, 2, \ldots, l_1$, $j = 1, 2, \ldots, l_2$, of the locations of the known entries of M, which have been sampled uniformly at random. Adopting a similar rationale to the one running across the backbone of sparsity-aware learning, one would attempt to recover M based on the following rank minimization problem:

$$\begin{aligned} \min_{\hat{M} \in \mathbb{R}^{l_1 \times l_2}} \quad & \|\boldsymbol{\sigma}_{\hat{M}}\|_0, \\ \text{s.t.} \quad & \hat{M}(i, j) = M(i, j), \quad (i, j) \in \Omega. \end{aligned} \tag{19.120}$$

It turns out that assuming there exists a unique low rank matrix having as elements the specific known entries, the task in (19.120) leads to the exact solution [33]. However, compared to the case of sparse vectors, in the matrix completion problem the uniqueness issue gets much more involved. The following issues play a crucial part concerning the uniqueness of the task in (19.120).

1. If the number of known entries is lower than the number of degrees of freedom, that is, $N < d_M$, then there is no way to recover the missing entries whatsoever, because there is an infinite number of low rank matrices consistent with the N observed entries.

2. Even if $N \geq d_M$, uniqueness is not guaranteed. It is required that the N elements with indices in Ω are such that at least one entry per column and one entry per row are observed. Otherwise, even a rank one matrix $M = \sigma_1 \boldsymbol{u}_1 \boldsymbol{v}_1^T$ cannot be recovered. This becomes clear with a simple example. Assume that M is a rank one matrix and that no entry in the first column as well as in the last row is observed. Then, because for this case $M(i, j) = \sigma_1 u_{1i} v_{1j}$, it is clear that no information concerning the first component of $\boldsymbol{v}_1$ as well as the last component of $\boldsymbol{u}_1$ is available; hence, it is impossible to recover these singular vector components, regardless of which method is used. As a consequence, the matrix cannot be completed. On the other hand, if the elements of Ω are picked at random and N is large enough, one can only hope that Ω is such so as to comply with the requirement above, i.e., at least one entry per row and column to be observed, with high probability. It turns out that this problem resembles the famous theorem in probability theory known as the *coupon collector's* problem. According to this, at least $N = C_0 l \ln l$ entries are needed, where C_0 is a constant [134]. This is the information theoretic limit for exact matrix completion [35] of any low rank matrix.

3. Even if points (1) and (2) above are fulfilled, uniqueness is still not guaranteed. In fact, not every low rank matrix is liable to exact completion, regardless of the number and the positions of the observed entries. Let us demonstrate this via an example. Let one of the singular vectors be sparse. Assume, without loss of generality, that the third left singular vector, $\boldsymbol{u}_3$, is sparse with sparsity level $k = 1$ and also that its nonzero component is the first one, that is, $u_{31} \neq 0$. The rest of $\boldsymbol{u}_i$ and all $\boldsymbol{v}_i$ are assumed to be dense. Let us return to SVD for a while in Eq. (19.119). Observe that the matrix M is written as the sum of r $l_1 \times l_2$ matrices $\sigma_i \boldsymbol{u}_i \boldsymbol{v}_i^T$, $i = 1, \ldots, r$. Thus, in this specific case where $\boldsymbol{u}_3$ is $k = 1$ sparse, the matrix $\sigma_3 \boldsymbol{u}_3 \boldsymbol{v}_3^T$ has zeros everywhere except for its first row. In other words, the information that $\sigma_3 \boldsymbol{u}_3 \boldsymbol{v}_3^T$ brings to the formation of M is concentrated in its first row only. This argument can also be viewed from another perspective; the entries of M obtained from any row except the first one do not provide any useful information with respect to the values of the free parameters σ_3, $\boldsymbol{u}_3$, $\boldsymbol{v}_3$. As a result, in this case, unless one incorporates extra information about the sparse nature of the singular vector, the entries from the first row that are missed are not recoverable, because the number of parameters concerning this row is larger than the number of the available data.

Intuitively, when a matrix has dense singular vectors it is better rendered for exact completion as each one among the observed entries carries information associated with all the d_M parameters that fully describe it. To this end, a number of conditions which evaluate the suitability of the singular vectors have been established. The simplest one is the following [33]:

$$\|\boldsymbol{u}_i\|_\infty \leq \sqrt{\frac{\mu_B}{l_1}}, \quad \|\boldsymbol{v}_i\|_\infty \leq \sqrt{\frac{\mu_B}{l_2}}, \; i = 1, \ldots, r, \tag{19.121}$$

where μ_B is a bound parameter. In fact, μ_B is a measure of the coherence of matrix U (and similarly of V)[9] (vis-à-vis the standard basis), defined as follows:

$$\mu(U) := \frac{l_1}{r} \max_{1 \leq i \leq l_1} \|P_U \boldsymbol{e}_i\|^2, \tag{19.122}$$

[9] This is a quantity different than the mutual coherence already discussed in Section 9.6.1.

where P_U defines the orthogonal projection to subspace U and $\boldsymbol{e}_i$ is the ith vector of the canonical basis. Note that when U results from SVD, then $\|P_U \boldsymbol{e}_i\|^2 = \left\|U^T \boldsymbol{e}_i\right\|^2$. In essence, coherence is an index quantifying the extent to which the singular vectors are correlated with the standard basis $\boldsymbol{e}_i$, $i = 1, 2, \ldots, l$. The smaller μ_B, the less "spiky" the singular vectors are likely to be, and the corresponding matrix is better suited for exact completion. Indeed, assuming for simplicity a square matrix M, that is, $l_1 = l_2 = l$, if *any one* among the singular vectors is sparse having a single nonzero component only, then, taking into account that $\boldsymbol{u}_i^T \boldsymbol{u}_i = \boldsymbol{v}_i^T \boldsymbol{v}_i = 1$, this value will have magnitude equal to one and the bound parameter will take its largest value possible, that is, $\mu_B = l$. On the other hand, the smallest value that μ_B can get is 1, something that occurs when the components of *all* the singular vectors assume the same value (in magnitude). Note that in this case, due to the normalization, this common component value has magnitude $\frac{1}{l}$. Tighter bounds to a matrix coherence result from the more elaborate incoherence property [33,151] and the strong incoherence property [35]. In all cases, the larger the bound parameter is, the larger the number of known entries becomes, which is required to guarantee uniqueness.

In Section 19.10.3, the aspects of uniqueness will be discussed in the context of a real-life application.

The task formulated in (19.120) is of limited practical interest because it is an NP-hard task. Thus, borrowing the arguments used in Chapter 9, the ℓ_0 (pseudo)norm is replaced by a *convexly* relaxed counterpart of it, that is,

$$\begin{aligned} \min_{\hat{M} \in \mathbb{R}^{l_1 \times l_2}} \quad & \left\|\boldsymbol{\sigma}_{\hat{M}}\right\|_1, \\ \text{s.t.} \quad & \hat{M}(i,j) = M(i,j), \quad (i,j) \in \Omega, \end{aligned} \tag{19.123}$$

where $\left\|\boldsymbol{\sigma}_{\hat{M}}\right\|_1$, that is, the sum of the singular values, is referred to as the *nuclear norm* of the matrix $\hat{M}$, often denoted as $\left\|\hat{M}\right\|_*$. The nuclear norm minimization was proposed in [74] as a convex approximation of rank minimization, which can be cast as a semidefinite programming task.

Theorem 19.1. *Let M be an $l_1 \times l_2$ matrix of rank r, which is a constant much smaller than $l = \min(l_1, l_2)$, obeying (19.121). Suppose that we observe N entries of M with locations sampled uniformly at random. Then there is a positive constant C such that if*

$$N \geq C \mu_B^4 l \ln^2 l, \tag{19.124}$$

then M is the unique solution to the task in (19.123) with probability at least $1 - l^{-3}$.

There might be an ambiguity on how small the rank should be in order for the corresponding matrix to be characterized as "low rank." More rigorously, a matrix is said to be of low rank if $r = \mathcal{O}(1)$, which means that r is a constant with no dependence (not even logarithmic) on l. Matrix completion is also possible for more general rank cases where, instead of the mild coherence property of (19.121), the incoherence and the strong incoherence properties [33,35,87,151] are mobilized in order to get similar theoretical guarantees. The detailed exposition of these alternatives is beyond the scope of this book. In fact, Theorem 19.1 embodies the essence of the matrix completion task: with high probability, nuclear norm minimization recovers all the entries of a low rank matrix M with no error. More importantly, the number of entries, N, that the convexly relaxed problem needs is only by a logarithmic factor larger than the information theoretic limit, which, as was mentioned before, equates to $C_0 l \ln l$. Moreover, similar to compressed sensing, robust matrix completion in the presence of noise is also possible as long as the request $\hat{M}(i, j) = M(i, j)$ in Eqs. (19.120) and (19.123) is replaced by $\|\hat{M}(i, j) - M(i, j)\|_2 \leq \epsilon$ [34]. Furthermore, the notion of matrix completion has also been extended to tensors (for example, [77, 165]).

19.10.2 ROBUST PCA

The developments on matrix completion theory led, more recently, to the formulation and solution of another problem of high significance. To this end, the notation $\|M\|_1$, that is, the ℓ_1 norm of a matrix, is introduced and defined as the sum of the absolute values of its entries, that is, $\|M\|_1 = \sum_{i=1}^{l_1} \sum_{j=1}^{l_2} |M(i, j)|$. In other words, it acts on the matrix as if this were a long vector.

Assume now that M is expressed as the sum of a low rank matrix L and a sparse matrix S, that is, $M = L + S$. Consider the following convex minimization problem task, [36,43,187,193], which is usually referred to as *principal component pursuit* (PCP):

$$\min_{\hat{L}, \hat{S}} \quad \|\sigma_M\|_1 + \lambda\|\hat{S}\|_1, \tag{19.125}$$

$$\text{s.t.} \quad \hat{L} + \hat{S} = M, \tag{19.126}$$

where $\hat{L}$ and $\hat{S}$ are both $l_1 \times l_2$ matrices. It can be shown that solving the task in (19.125)–(19.126) recovers *both* L and S according to the following theorem [36].

Theorem 19.2. *The PCP recovers both L and S with probability at least $1 - cl_1^{-10}$, where c is a constant, provided that:*

1. *The support set Ω of S is uniformly distributed among all sets of cardinality N.*
2. *The number k of nonzero entries of S is relatively small, that is, $k \leq \rho l_1 l_2$, where ρ is a sufficiently small positive constant.*
3. *L obeys the incoherence property.*
4. *The regularization parameter λ is constant with value $\lambda = \frac{1}{\sqrt{l_2}}$.*
5. *We have* $\text{rank}(L) \leq C\frac{l_2}{\ln^2 l_1}$, *with C being a constant.*

In other words, based on *all* the entries of a matrix M, which is known to be the sum of two unknown matrices L and S, with the first one being of low rank and the second being sparse, PCP recovers exactly, with probability almost 1, both L and S, irrespective of how large the magnitude of the entries of S are, provided *that both r and k are sufficiently small.*

The applicability of the previous task is very broad. For example, PCP can be employed in order to find a low rank approximation of M. In contrast to the standard PCA (SVD) approach, PCP is robust and insensitive in the presence of outliers, as these are naturally modeled, via the presence of S. Note that outliers are sparse by their nature. For this reason, the above task is widely known as *robust PCA via nuclear norm minimization.* (More classical PCA techniques are known to be sensitive to outliers and a number of alternative approaches have in the past been proposed toward its robustification [for example, [98,109]].)

When PCP serves as a robust PCA approach, the matrix of interest is L and S accounts for the outliers. However, PCP estimates both L and S. As will be discussed soon, another class of applications are well accommodated when the focus of interest is turned to the sparse matrix S itself.

Remarks 19.8.

- Just as ℓ_1 minimization is the tightest convex relaxation of the combinatorial ℓ_0 minimization problem in sparse modeling, the nuclear norm minimization is the tightest convex relaxation of the NP-hard rank minimization task. Besides the nuclear norm, other heuristics have also been proposed, such as the log-determinant heuristic [74] and the max norm [76].
- The nuclear norm as a rank minimization approach is the generalization of the trace-related cost, which is often used in the control community for the rank minimization of positive semidefinite matrices [133]. Indeed, when the matrix is symmetric and positive semidefinite, the nuclear norm of M is the sum of the eigenvalues and, thus, it is equal to the trace of M. Such problems arise when, for example, the rank minimization task refers to covariance matrices and positive semidefinite Toeplitz or Hankel matrices (see, e.g., [74]).
- Both matrix completion (19.123) and PCP (19.126) can be formulated as semidefinite programs and are solved based on interior point methods. However, whenever the size of a matrix becomes large (e.g., 100×100), these methods are deemed to fail in practice due to excessive computational load and memory requirements. As a result, there is an increasing interest, which has propelled intensive research efforts, for the development of efficient methods to solve both optimization tasks, or related approximations, which scale well with large matrices. Many of these methods revolve around the philosophy of the iterative soft and hard thresholding techniques, as discussed in Chapter 9. However, in the current low rank approximation setting, it is the singular values of the estimated matrix that are thresholded. As a result, in each iteration, the estimated matrix, after thresholding its singular values, tends to be of lower rank. Either the thresholding of the singular values is imposed, such as in the case of the singular value thresholding (SVT) algorithm [31], or it results as a solution of regularized versions of (19.123) and (19.126) (see, e.g., [48,177]). Moreover, algorithms inspired by greedy methods such as CoSaMP have also been proposed (e.g., [124,183]).
- Improved versions of PCP that allow for exact recovery even if some of the constraints of Theorem 19.2 are relaxed have also been developed (see, e.g., [78]). Fusions of PCP with matrix completion and compressed sensing are possible, in the sense that only a subset of the entries of M is available and/or linear measurements of the matrix in a compressed sensing fashion can be used instead of matrix entries (for example, [183,188]). Moreover, stable versions of PCP dealing with noise have also been investigated (for example, [199]).

19.10.3 APPLICATIONS OF MATRIX COMPLETION AND ROBUST PCA

The number of applications in which these techniques are involved is ever increasing and their extensive presentation is beyond the scope of this book. Next, some key applications are selectively discussed in order to reveal the potential of these methods and at the same time to assist the reader in better understanding the underlying notions.

Matrix Completion

A typical application where the matrix completion problem arises is in the *collaborative filtering* task (e.g., [167]), which is essential for building up successful recommender systems. Let us consider that a group of individuals provide their ratings concerning products that they have enjoyed. Then, a matrix with ratings can be filled, where each row indexes a different individual and the columns index the products. As a popular example, take the case where the products are different movies. Inevitably, the associated matrix will be partially filled because it is not common that all customers have watched all the movies and submitted ratings for all of them. Matrix completion comes to provide an answer, potentially in the affirmative, to the following question: Can we predict the ratings that the users would give to films that they have not seen yet? This is the task of a recommender system in order to encourage users to watch movies, which are likely to be of their preference. The exact objective of competition for the famous Netflix Prize[10] was the development of such a recommender system.

The aforementioned problem provides a good opportunity to build up our intuition about the matrix completion task. First, an individual's preferences or taste in movies are typically governed by a small number of factors, such as gender, the actors that appear in it, and the continent of origin. As a result, a matrix fully filled with ratings is expected to be low rank. Moreover, it is clear that each user needs to have at least one movie rated in order to have any hope of filling out her/his ratings across all movies. The same is true for each movie. This requirement complies with the second assumption in Section 19.10.1, concerning uniqueness; that is, one needs to know at least one entry per row and column. Finally, imagine a single user who rates movies with criteria that are completely different from those used by the rest of the users. One could, for example, provide ratings at random or depending on, let us say, the first letter of the movie title. The ratings of this particular user cannot be described in terms of the singular vectors that model the ratings of the rest of the users. Accordingly, for such a case, the rank of the matrix increases by one and the user's preferences will be described by an extra set of left and right singular vectors. However, the corresponding left singular vector will comprise a single nonzero component, at the place corresponding to the row dedicated to this user, and the right singular vector will comprise her/his ratings normalized to unit norm. Such a scenario complies with the third point concerning uniqueness in the matrix completion problem, as previously discussed. Unless all the ratings of the specific user are known, the matrix cannot be fully completed.

Other applications of matrix completion include system identification [128], recovering structure from motion [46], multitask learning [10], and sensor network localization [136].

Robust PCA/PCP

In the collaborative filtering task, robust PCA offers an extra attribute compared to matrix completion, which can be proved very crucial in practice. The users are allowed to even tamper with some of the ratings without affecting the estimation of the low rank matrix. This seems to be the case whenever the rating process involves many individuals in an environment, which is not strictly controlled, because some of them occasionally are expected to provide ratings in an ad hoc, or even malicious manner.

One of the first applications of PCP was in video surveillance systems (e.g., [36]) and the main idea behind it appeared to be popular and extendable to a number of computer vision applications. Take the example of a camera recording a sequence of frames consisting of a merely static background and a foreground with a few moving objects, for example, vehicles and/or individuals. A common task in surveillance recording is to extract from the background the foreground, in order, for example, to detect any activity or to proceed with further processing, such as face recognition. Suppose the

[10] http://www.netflixprize.com/.

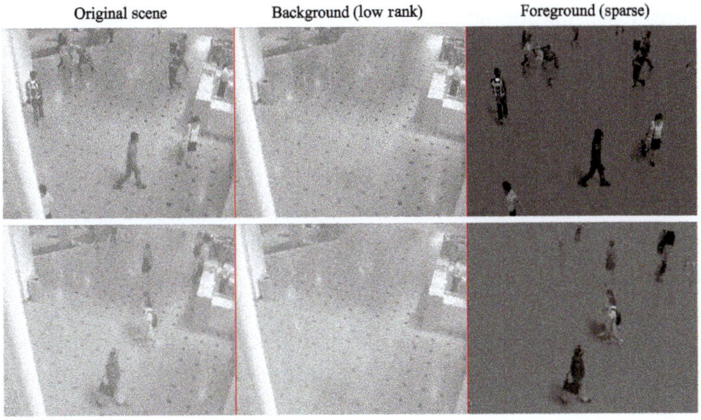

FIGURE 19.21

Background-foreground separation via PCP.

successive frames are converted to vectors in lexicographic order and are then placed as columns in a matrix M. Due to the background, even though this may slightly vary due, for example, to changes in illumination, successive columns are expected to be highly correlated. As a result, the background contribution to the matrix M can be modeled as an approximately low rank matrix L. On the other hand, the objects in the foreground appear as "anomalies" and correspond to only a fraction of pixels in each frame; that is, to a limited number of entries in each column of M. Moreover, due to the motion of the foreground objects, the positions of these anomalies are likely to change from one column of M to the next. Therefore, they can be modeled as a sparse matrix S.

Next, the above discussed philosophy is applied to a video acquired from a shopping mall surveillance camera [129], with the corresponding PCP task being solved with a dedicated accelerated proximal gradient algorithm [130]. The results are shown in Fig. 19.21. In particular, two randomly selected frames are depicted together with the corresponding columns of the matrices L and S reshaped back to pictures.

19.11 A CASE STUDY: FMRI DATA ANALYSIS

In the brain, tasks involving action, perception, cognition, and so forth, are performed via the simultaneous activation of a number of so-called *functional brain networks* (FBNs), which are engaged in proper interactions in order to effectively execute the task. Such networks are usually related to low-level brain functions and they are defined as a number of *segregated* specialized small brain regions, potentially distributed over the whole brain. For each FBN, the involved segregated brain regions define the *spatial map*, which characterizes the specific FBN. Moreover, these brain regions, irrespective of their anatomical proximity or remoteness, exhibit *strong* functional connectivity, which is expressed as strong coherence in the activation time patterns of these regions. Examples of such functional brain networks are the visual, sensorimotor, auditory, default-mode, dorsal attention, and executive control networks [148].

Functional magnetic resonance imaging (fMRI) [131] is a powerful noninvasive tool for detecting brain activity along time. Most commonly, it is based on *blood oxygenation level dependent* (BOLD) contrast, which translates to detecting localized changes in the hemodynamic flow of oxygenated blood in activated brain areas. This is achieved by exploiting the different magnetic properties of oxygen-saturated versus oxygen-desaturated hemoglobin. The detected fMRI signal is recorded in both the spatial (three-dimensional) and the temporal (one-dimensional) domain. The spatial domain is segmented with a three-dimensional grid to elementary cubes of edge size 3–5 mm, which are named *voxels*. Indicatively, a complete volume scan typically consists of $64 \times 64 \times 48$ voxels and it is acquired in one or two seconds [131]. Relying on adequate postprocessing, which effectively compensates for possible time lags and other artifacts [131], it is fairly accurate to assume that each acquisition is performed instantly. The, say, l in total voxel values, corresponding to a single scan, are collected in a flattened (row) one-dimensional vector, $\boldsymbol{x}_n \in \mathbb{R}^l$. Considering $n = 1, 2, \ldots, N$ successive acquisitions, the full amount of data is collected in a data matrix $X \in \mathbb{R}^{N \times l}$. Thus, each column, $i = 1, 2, \ldots, l$, of X represents the evolution in time of the values of the corresponding ith voxel. Each row, $n = 1, 2, \ldots, N$, corresponds to the activation pattern, at the corresponding time n, over all l voxels.

The recorded voxel values result from the cumulative contribution of several FBNs, where each one of them is activated following certain time patterns, depending on the tasks that the brain is performing. The above can be mathematically modeled according to the following factorization of the data matrix:

$$X = \sum_{j=1}^{m} \boldsymbol{a}_j \boldsymbol{z}_j^T := AZ, \tag{19.127}$$

where $\boldsymbol{z}_j \in \mathbb{R}^l$ is a sparse vector of latent variables, representing the spatial map of the jth FBN having nonzero values only in positions that correspond to brain regions associated with the specific FBN, and $\boldsymbol{a}_j \in \mathbb{R}^N$ represents the activation *time course* of the respective FBN. The model assumes that m FBNs have been activated. In order to understand better the previous model, take as an example the extreme case where only one set of brain regions (one FBN) is activated. Then matrix X is written as

$$X = \boldsymbol{a}_1 \boldsymbol{z}_1^T := \begin{bmatrix} a_1(1) \\ a_1(2) \\ \vdots \\ a_1(N) \end{bmatrix} \underbrace{[\ldots, *, \ldots, *, \ldots, *, \ldots]}_{l \text{ (voxels)}},$$

where $*$ denotes a nonzero element (active voxel in the FBN) and the dots zero ones. Observe that according to this model, all nonzero elements in the nth row of X result from the nonzero elements of z_1 multiplied by the *same* number, $a_1(n)$, $n = 1, 2, \ldots, N$. If now two FBNs are active, the model for the data matrix becomes

$$X = \boldsymbol{a}_1 \boldsymbol{z}_1^T + \boldsymbol{a}_2 \boldsymbol{z}_2^T = [\boldsymbol{a}_1, \boldsymbol{a}_2] \begin{bmatrix} \boldsymbol{z}_1^T \\ \boldsymbol{z}_2^T \end{bmatrix}.$$

Obviously, for m FBNs, Eq. (19.127) results.

One of the major goals of fMRI analysis is to detect, study, and characterize the different FBNs and to relate them to particular mental and physical activities. In order to achieve this, the subject (person) subjected to fMRI is presented with carefully designed experimental procedures, so that the activation of the FBNs will be as controlled as possible.

ICA has been successfully employed for fMRI unmixing, that is, for estimating matrices A and Z above. If we consider each column of X to be a realization of a random vector $\mathbf{x}$, the fMRI data generation mechanism can be modeled to follow the classical ICA latent model, that is, $\mathbf{x} = A\mathbf{s}$, where the components of $\mathbf{s}$ are statistically independent and A is an unknown mixing matrix. The goal of ICA is to recover the unmixing matrix, W and Z. Matrix A is then obtained from W. The use of ICA in the fMRI task could be justified by the following argument. Nonzero elements of Z in the same column contribute to the formation of a single element of X, for each time instant n, and correspond to different FBNs. Thus, they are assumed to correspond to two statistically independent sources.

As a result of the application of ICA on X, one hopes that each row of the obtained matrix Z could be associated with an FBN, that is, to a spatial activity map. Furthermore, the corresponding column of A could represent the respective time activation pattern.

This approach will be applied next for the case of the following experimental procedure [61]. A visual pattern was presented to the subject, in which an 8-Hz reversing black and white checkerboard was shown intermittently in the left and right visual ends for 30 s at a time. This is a typical block design paradigm in fMRI, consisting of three different conditions to which the subject is exposed, i.e., checkerboard on the left (red block), checkerboard on the right (black block), and no visual stimulus (white block). The subject was instructed to focus on the cross at the center during the full time of the experiment (Fig. 19.22). More details about the scanning procedure and the preprocessing of the data can be found in [61]. The Group ICA of fMRI Toolbox (GIFT)[11] simulation tool was used.

When ICA is performed on the obtained data set,[12] the aforementioned matrices Z and A are computed. Ideally, at least some of the rows of Z should constitute spatial maps of the true FBNs, and

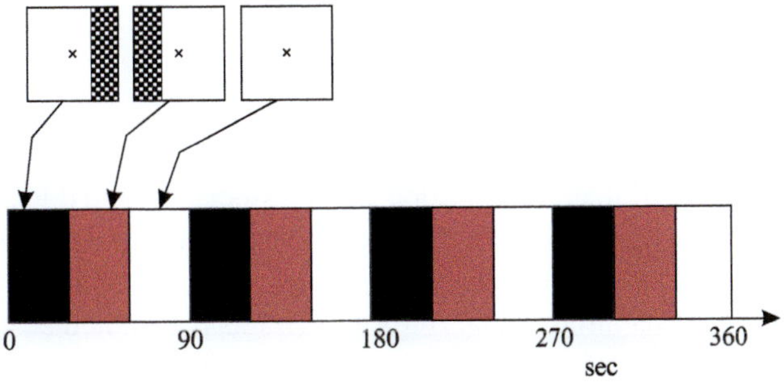

FIGURE 19.22

The fMRI experimental procedure used.

[11] It can be obtained from http://mialab.mrn.org/software/gift/.

[12] The specific data set is provided as test data with GIFT.

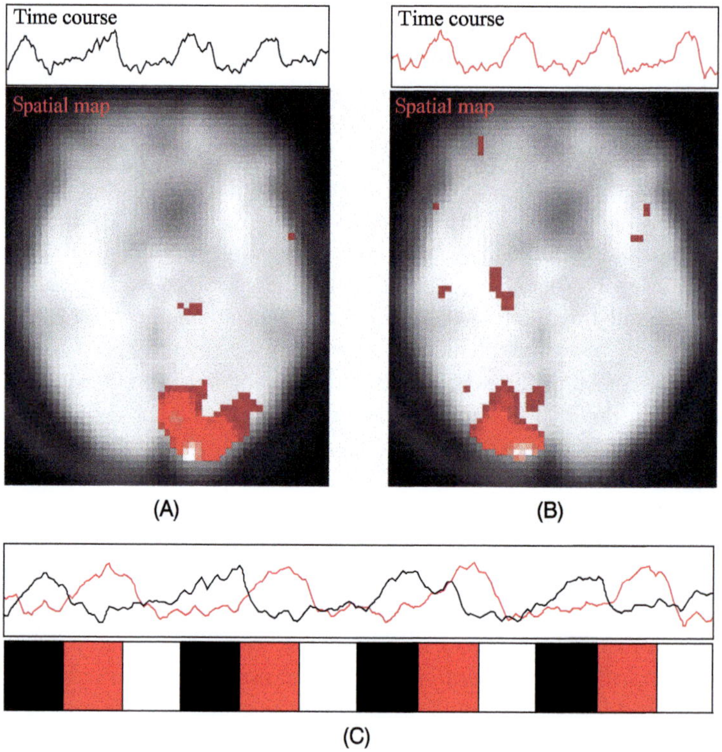

FIGURE 19.23

The two time courses follow well the fMRI experimental setup used.

the corresponding columns of A should represent the activation patterns of the respective FBNs, which correspond to the specific experimentation procedure.

The news is good, as shown in Fig. 19.23. In particular, Figs. 19.23A and B show two time courses (columns of A). For each one of them, one section of the associated spatial map (corresponding row of Z) is considered, which represents voxels of a slice in the brain. The areas that are activated (red) are those corresponding to the (a) left and (b) right visual cortex, which are the regions of the brain responsible for processing visual information. Activation of this part should be expected according to the characteristics of the specific experimental procedure. More interestingly, as seen in Fig. 19.23C, the two activation patterns, as represented by the two time courses, follow closely the two different conditions, namely, the checkerboard to be placed on the left or on the right from the point the subject is focusing on.

Besides ICA, alternative methods, discussed in this chapter, can also be used, which can better exploit the low rank nature of X. Dictionary learning is a promising candidate leading to notably good results (see, for example, [11,116,140,181]). In contrast to ICA, dictionary learning techniques build upon the sparse nature of the spatial maps, as already pointed out before. Besides the matrix factorization techniques, tensor-based methods have also been used, via related low rank tensor factorizations, that can exploit the 3D-structure of the brain, see, e.g., [45] and the reference therein.

PROBLEMS

19.1 Show that the second principal component in PCA is given as the eigenvector corresponding to the second largest eigenvalue.

19.2 Show that the pair of directions, associated with CCA, which maximize the respective correlation coefficient, satisfy the following pair of relations:

$$\Sigma_{xy}\boldsymbol{u}_y = \lambda\Sigma_{xx}\boldsymbol{u}_x,$$
$$\Sigma_{yx}\boldsymbol{u}_x = \lambda\Sigma_{yy}\boldsymbol{u}_y.$$

19.3 Establish the arguments that verify the convergence of the k-SVD.

19.4 Prove that Eqs. (19.83) and (19.89) are the same.

19.5 Show that the ML PPCA tends to PCA as $\sigma^2 \to 0$.

19.6 Show Eqs. (19.91) and (19.92).

19.7 Show Eq. (19.102).

19.8 Show that the number of degrees of freedom of a rank r matrix is equal to $r(l_1 + l_2) - r^2$.

MATLAB® EXERCISES

19.9 This exercise reproduces the results of Example 19.1. Download the faces from this book's website and read them one by one using the imread.m MATLAB® function. Store them as columns in a matrix X. Then, compute and subtract the mean in order for the rows to become zero mean.

A direct way to compute the eigenvectors (eigenfaces) would be to use the svd.m MATLAB® function in order to perform an SVD to the matrix X, that is, $X = UDV^T$. In this way, the eigenfaces are the columns of U. However, this needs a lot of computational effort because X has too many rows. Alternatively, you can proceed as follows. First compute the product $A = X^T X$ and then the SVD of A (using the SVD.m MATLAB® function) in order to compute the right singular vectors of X via $A = VD^2V^T$. Then calculate each eigenface according to $\boldsymbol{u}_i = \frac{1}{\sigma_i}X\boldsymbol{v}_i$, where σ_i is the ith singular value of the SVD. In the sequel, select one face at random in order to reconstruct it using the first 5, 30, 100, and 600 eigenvectors.

19.10 Recompute the eigenfaces as in Exercise 19.9, using all the face images apart from one, which you choose. Then reconstruct that face that did not take part in the computation of the eigenfaces, using the first 300 and 1000 eigenvectors. Is the reconstructed face anywhere close to the true one?

19.11 Download the fast ICA MATLAB® software package[13] to reproduce the results of the cocktail party example described in Section 19.5.5. The two voice and the music signals can be downloaded from this book's website and read using the wavread.m MATLAB® function. Generate

[13] http://research.ics.aalto.fi/ica/fastica/.

a random mixing matrix A (3×3) and produce with it the three mixture signals. Each one of them simulates the signal received in each microphone. Then, apply FastICA in order to estimate the source signals. Use the MATLAB® function wavplay.m to listen to the original signals, the mixtures, and the recovered ones. Repeat the previous steps performing PCA instead of ICA and compare the results.

19.12 This exercise reproduces the dictionary learning-based denoising Example 19.5. The image depicting the boat can be obtained from this book's website. Moreover, k-SVD either needs to be implemented according to Section 19.6 or an implementation available on the web can be downloaded.[14]

Then the next steps are to be followed. First, extract from the image all the possible sliding patches of size 12×12 using the im2col.m MATLAB® function and store them as columns in a matrix X. Using this matrix, train an overcomplete dictionary, Ψ, of size (144×196), for 100 k-SVD iterations with $T_0 = 5$. For the first iteration, the initial dictionary atoms are drawn from a zero mean Gaussian distribution and then normalized to unit norm. As a next step, denoise each image patch separately. In particular, assuming that $\boldsymbol{y}_i$ is the ith patch reshaped in column vector use the OMP (Section 10.2.1) in order to estimate a sparse vector $\boldsymbol{\theta}_i \in \mathbb{R}^{196}$ with $\|\boldsymbol{\theta}_i\|_0 = 5$, such that $\left\|\boldsymbol{y}_i - A\boldsymbol{\theta}_i\right\|$ is small. Then $\hat{\boldsymbol{y}}_i = \Psi\boldsymbol{\theta}_i$ is the ith denoised patch. Finally, average the values of the overlapped patches to form the full denoised image.

19.13 Download one of the videos (they are provided in the form of a sequence of bitmap images) from http://perception.i2r.a-star.edu.sg/bk_model/bk_index.html.

In Section 19.10.3, the "shopping center" bitmap image sequence has been used. Read one by one the bitmap images using the imread.m MATLAB® function, convert them from color to grayscale using rgb2gray.m, and finally store them as columns in a matrix X. Download one of the MATLAB® implementations of an algorithm performing the robust PCA task from http://perception.csl.illinois.edu/matrix-rank/sample_code.html.

The "Accelerated Proximal Gradient" method and the accompanied proximal_gradient_rpca.m MATLAB® function are a good and easy to use choice. Set $\lambda = 0.01$. Note, however, that depending on the video used, this regularization parameter might need to be fine-tuned.

REFERENCES

[1] D. Achlioptas, F. McSherry, Fast computation of low rank approximations, in: Proceedings of the ACM STOC Conference, 2001, pp. 611–618.

[2] T. Adali, H. Li, M. Novey, J.F. Cardoso, Complex ICA using nonlinear functions, IEEE Trans. Signal Process. 56 (9) (2008) 4536–4544.

[3] T. Adali, M. Anderson, G.S. Fu, Diversity in independent component and vector analyses: identifiability, algorithms, and applications in medical imaging, IEEE Signal Process. Mag. 31 (3) (2014) 18–33.

[4] M. Aharon, M. Elad, A. Bruckstein, k-SVD: an algorithm for designing overcomplete dictionaries for sparse representation, IEEE Trans. Signal Process. 54 (11) (2006) 4311–4322.

[5] T.W. Anderson, Asymptotic theory for principal component analysis, Ann. Math. Stat. 34 (1963) 122–148.

[6] T.W. Anderson, An Introduction to Multivariate Analysis, second ed., John Wiley, New York, 1984.

[14] For example, http://www.cs.technion.ac.il/~elad/software/.

[7] M. Anderson, X.L. Li, T. Adali, Joint blind source separation with multivariate Gaussian model: algorithms and performance analysis, IEEE Trans. Signal Process. 60 (4) (2012) 2049–2055.
[8] C. Archambeau, F. Bach, Sparse probabilistic projections, in: D. Koller, D. Schuurmans, Y. Bengio, L. Bottou (Eds.), Neural Information Processing Systems, NIPS, Vancouver, Canada, 2008.
[9] J. Arenas-García, K.B. Petersen, G. Camps-Valls, L.K. Hansen, Kernel multivariate analysis framework for supervised subspace learning, IEEE Signal Process. Mag. 30 (4) (2013) 16–29.
[10] A. Argyriou, T. Evgeniou, M. Pontil, Multi-task feature learning, in: Advances in Neural Information Processing Systems, vol. 19, MIT Press, Cambridge, MA, 2007.
[11] V. Abolghasemi, S. Ferdowsi, S. Sanei, Fast and incoherent dictionary learning algorithms with application to fMRI, Signal Image Video Process. (2013), https://doi.org/10.1007/s11760-013-0429-2.
[12] G.I. Allen, Sparse and functional principal components analysis, arXiv preprint, arXiv:1309.2895, 2013.
[13] F.R. Bach, M.I. Jordan, Kernel independent component analysis, J. Mach. Learn. Res. 3 (2002) 1–48.
[14] F. Bach, R. Jenatton, J. Mairal, G. Obozinski, Structured sparsity through convex optimization, Stat. Sci. 27 (4) (2012) 450–468.
[15] F. Bach, M. Jordan, A Probabilistic Interpretation of Canonical Correlation Analysis, Technical Report 688, University of Berkeley, 2005.
[16] E. Barshan, A. Ghodsi, Z. Azimifar, M.Z. Jahromi, Supervised principal component analysis: visualization, classification and regression on subspaces and submanifolds, Pattern Recognit. 44 (2011) 1357–1371.
[17] H.B. Barlow, Unsupervised learning, Neural Comput. 1 (1989) 295–311.
[18] A.J. Bell, T.J. Sejnowski, An information maximization approach to blind separation and blind deconvolution, Neural Comput. 7 (1995) 1129–1159.
[19] M. Belkin, P. Niyogi, Laplacian eigenmaps for dimensionality reduction and data representation, Neural Comput. 15 (6) (2003) 1373–1396.
[20] E. Benetos, M. Kotti, C. Kotropoulos, Applying supervised classifiers based on non-negative matrix factorization to musical instrument classification, in: Proceedings IEEE International Conference on Multimedia and Expo, Toronto, Canada, 2006, pp. 2105–2108.
[21] Y. Bengio, J.-F. Paiement, P. Vincent, O. Delalleau, N. Le Roux, M. Quimet, Out of sample extensions for LLE, Isomap, MDS, eigenmaps and spectral clustering, in: S. Thrun, L. Saul, B. Schölkopf (Eds.), Advances in Neural Information Processing Systems Conference, MIT Press, Cambridge, MA, 2004.
[22] M. Berry, S. Dumais, G. O'Brie, Using linear algebra for intelligent information retrieval, SIAM Rev. 37 (1995) 573–595.
[23] A. Beygelzimer, S. Kakade, J. Langford, Cover trees for nearest neighbor, in: Proceedings of the 23rd International Conference on Machine Learning, Pittsburgh, PA, 2006.
[24] C.M. Bishop, Variational principal components, in: Proceedings 9th International Conference on Artificial Neural Networks, ICANN, vol. 1, 1999, pp. 509–514.
[25] L. Bolzano, R. Nowak, B. Recht, Online identification and tracking of subspaces from highly incomplete information, arXiv:1006.4046v2 [cs.IT], 12 July 2011.
[26] M. Borga, Canonical Correlation Analysis: a Tutorial, Technical Report, 2001, www.imt.liu.se/~magnus/cca/tutorial/tutorial.pdf.
[27] M. Brand, Charting a manifold, in: Advances in Neural Information Processing Systems, vol. 15, MIT Press, Cambridge, MA, 2003, pp. 985–992.
[28] J.-P. Brunet, P. Tamayo, T.R. Golub, J.P. Mesirov, Meta-genes and molecular pattern discovery using matrix factorization, Proc. Natl. Acad. Sci. 101 (2) (2004) 4164–4169.
[29] C.J.C. Burges, Geometric Methods for Feature Extraction and Dimensional Reduction: A Guided Tour, Technical Report MSR-TR-2004-55, Microsoft Research, 2004.
[30] D. Cai, X. He, Orthogonal locally preserving indexing, in: Proceedings 28th Annual International Conference on Research and Development in Information Retrieval, 2005.
[31] J.-F. Cai, E.J. Candès, Z. Shen, A singular value thresholding algorithm for matrix completion, SIAM J. Optim. 20 (4) (2010) 1956–1982.
[32] F. Camastra, Data dimensionality estimation methods: a survey, Pattern Recognit. 36 (2003) 2945–2954.
[33] E.J. Candès, B. Recht, Exact matrix completion via convex optimization, Found. Comput. Math. 9 (6) (2009) 717–772.
[34] E.J. Candès, P. Yaniv, Matrix completion with noise, Proc. IEEE 98 (6) (2010) 925–936.
[35] E.J. Candès, T. Tao, The power of convex relaxation: near-optimal matrix completion, IEEE Trans. Inf. Theory 56 (3) (2010) 2053–2080.

[36] E.J. Candès, X. Li, Y. Ma, J. Wright, Robust principal component analysis, J. ACM 58 (3) (2011) 1–37.
[37] J.F. Cardoso, Infomax and maximum likelihood for blind source separation, IEEE Signal Process. Lett. 4 (1997) 112–114.
[38] J.-F. Cardoso, Blind signal separation: statistical principles, Proc. IEEE 9 (10) (1998) 2009–2025.
[39] J.-F. Cardoso, High-order contrasts for independent component analysis, Neural Comput. 11 (1) (1999) 157–192.
[40] L. Carin, R.G. Baraniuk, V. Cevher, D. Dunson, M.I. Jordan, G. Sapiro, M.B. Wakin, Learning low-dimensional signal models, IEEE Signal Process. Mag. 34 (2) (2011) 39–51.
[41] V. Casteli, A. Thomasian, C.-S. Li, CSVD: clustering and singular value decomposition for approximate similarity searches in high-dimensional space, IEEE Trans. Knowl. Data Eng. 15 (3) (2003) 671–685.
[42] V. Cevher, P. Indyk, L. Carin, R.G. Baraniuk, Sparse signal recovery and acquisition with graphical models, IEEE Signal Process. Mag. 27 (6) (2010) 92–103.
[43] V. Chandrasekaran, S. Sanghavi, P.A. Parrilo, A.S. Willsky, Rank-sparsity incoherence for matrix decomposition, SIAM J. Optim. 21 (2) (2011) 572–596.
[44] C. Chatfield, A.J. Collins, Introduction to Multivariate Analysis, Chapman Hall, London, 1980.
[45] C. Chatzichristos, E. Kofidis, M. Morante, S. Theodoridis, Blind fMRI source unmixing via higher-order tensor decompositions, J. Neurosci. Methods 315 (8) (2019) 17–47.
[46] P. Chen, D. Suter, Recovering the missing components in a large noisy low-rank matrix: application to SFM, IEEE Trans. Pattern Anal. Mach. Intell. 26 (8) (2004) 1051–1063.
[47] M. Chen, J. Silva, J. Paisley, C. Wang, D. Dunson, L. Carin, Compressive sensing on manifolds using nonparametric mixture of factor analysers: algorithms and performance bounds, IEEE Trans. Signal Process. 58 (12) (2010) 6140–6155.
[48] C. Chen, B. He, X. Yuan, Matrix completion via an alternating direction method, IMA J. Numer. Anal. 32 (2012) 227–245.
[49] J. Chen, Z.J. Towfic, A. Sayed, Dictionary learning over distributed models, IEEE Trans. Signal Process. 63 (4) (2015) 1001–1016.
[50] Y. Chi, Y.C. Eldar, R. Calderbank, PETRELS: subspace estimation and tracking from partial observations, in: IEEE International Conference on Acoustics, Speech and Signal Processing, ICASSP, 2012, pp. 3301–3304.
[51] S. Chouvardas, Y. Kopsinis, S. Theodoridis, An adaptive projected subgradient based algorithm for robust subspace tracking, in: Proc. International Conference on Acoustics Speech and Signal Processing, ICASSP, Florence, Italy, May 4–9, 2014.
[52] S. Chouvardas, Y. Kopsinis, S. Theodoridis, Robust subspace tracking with missing entries: the set-theoretic approach, IEEE Trans. Signal Process. 63 (19) (2015) 5060–5070.
[53] S. Chouvardas, Y. Kopsinis, S. Theodoridis, An online algorithm for distributed dictionary learning, in: Proceedings International Conference on Acoustics, Speech and Signal Processing, ICASSP, 2015, pp. 3292–3296.
[54] M. Chu, F. Diele, R. Plemmons, S. Ragni, Optimality, computation and interpretation of the nonnegative matrix factorization, available at http://www.wfu.edu/~plemmons, 2004.
[55] A. Cichoki, Unsupervised learning algorithms and latent variable models: PCA/SVD, CCA, ICA, NMF, in: R. Chelappa, S. Theodoridis (Eds.), E-Reference for Signal Processing, Academic Press, Boston, 2014.
[56] N.M. Correa, T. Adal, Y.-Q. Li, V.D. Calhoun, Canonical correlation analysis for group fusion and data inferences, IEEE Signal Process. Mag. 27 (4) (2010) 39–50.
[57] P. Comon, Independent component analysis: a new concept, Signal Process. 36 (1994) 287–314.
[58] P. Comon, C. Jutten, Handbook of Blind Source Separation: Independent Component Analysis and Applications, Academic Press, 2010.
[59] T.H. Cormen, C.E. Leiserson, R.L. Rivest, C. Stein, Introduction to Algorithms, second ed., MIT Press/McGraw-Hill, Cambridge, MA, 2001.
[60] T. Cox, M. Cox, Multidimensional Scaling, Chapman & Hall, London, 1994.
[61] V. Calhoun, T. Adali, G. Pearlson, J. Pekar, A method for making group inferences from functional MRI data using independent component analysis, Hum. Brain Mapp. 14 (3) (2001) 140–151.
[62] A. Daneshmand, Y. Sun, G. Scutari, F. Fracchinei, B.M. Sadler, Decentralized dictionary learning over time-varying digraphs, arXiv:1808.05933v1 [math.OC], 17 August 2018.
[63] S. Deerwester, S. Dumais, G. Furnas, T. Landauer, R. Harshman, Indexing by latent semantic analysis, J. Soc. Inf. Sci. 41 (1990) 391–407.
[64] W.R. Dillon, M. Goldstein, Multivariable Analysis Methods and Applications, John Wiley, New York, 1984.
[65] J.P. Dmochowski, P. Sajda, J. Dias, L.C. Parra, Correlated components of ongoing EEG point to emotionally laden attention—a possible marker of engagement? Front. Human Neurosci. 6 (2012), https://doi.org/10.3389/fnhum.2012.00112.

[66] D.L. Donoho, C.E. Grimes, When Does ISOMAP Recover the Natural Parameterization of Families of Articulated Images? Technical Report 2002-27, Department of Statistics, Stanford University, 2002.
[67] D. Donoho, V. Stodden, When does nonnegative matrix factorization give a correct decomposition into parts? in: S. Thrun, L. Saul, B. Schölkopf (Eds.), Advances in Neural Information Processing Systems, MIT Press, Cambridge, MA, 2004.
[68] S.C. Douglas, S. Amari, Natural gradient adaptation, in: S. Haykin (Ed.), Unsupervised Adaptive Filtering, Part I: Blind Source Separation, John Wiley & Sons, New York, 2000, pp. 13–61.
[69] X. Doukopoulos, G.V. Moustakides, Fast and stable subspace tracking, IEEE Trans. Signal Process. 56 (4) (2008) 1452–1465.
[70] V. De Silva, J.B. Tenenbaum, Global versus local methods in nonlinear dimensionality reduction, in: S. Becker, S. Thrun, K. Obermayer (Eds.), Advances in Neural Information Processing Systems, vol. 15, MIT Press, Cambridge, MA, 2003, pp. 721–728.
[71] M. Elad, M. Aharon, Image denoising via sparse and redundant representations over learned dictionaries, IEEE Trans. Image Process. 15 (12) (2006) 3736–3745.
[72] M. Elad, Sparse and Redundant Representations: From Theory to Applications in Signal and Image Processing, Springer, New York, 2010.
[73] K. Engan, S.O. Aase, J.H.A. Husy, Multi-frame compression: theory and design, Signal Process. 80 (10) (2000) 2121–2140.
[74] M. Fazel, H. Hindi, S. Boyd, Rank minimization and applications in system theory, in: Proceedings American Control Conference, vol. 4, 2004, pp. 3273–3278.
[75] D.J. Field, What is the goal of sensory coding? Neural Comput. 6 (1994) 559–601.
[76] R. Foygel, N. Srebro, Concentration-based guarantees for low-rank matrix reconstruction, in: Proceedings, 24th Annual Conference on Learning Theory, COLT, 2011.
[77] S. Gandy, B. Recht, I. Yamada, Tensor completion and low-n-rank tensor recovery via convex optimization, Inverse Probl. 27 (2) (2011) 1–19.
[78] A. Ganesh, J. Wright, X. Li, E.J. Candès, Y. Ma, Dense error correction for low-rank matrices via principal component pursuit, in: Proceedings IEEE International Symposium on Information Theory, 2010, pp. 1513–1517.
[79] Z. Ghahramani, M. Beal, Variational inference for Bayesian mixture of factor analysers, in: Advances in Neural Information Processing Systems, vol. 12, MIT Press, Cambridge, MA, 2000, pp. 449–455.
[80] P. Giampouras, K. Themelis, A. Rontogiannis, K. Koutroumbas, Simultaneously sparse and low-rank abundance matrix estimation for hyperspectral image unmixing, IEEE Trans. Geosci. Remote Sens. 54 (8) (2016) 4775–4789.
[81] M. Girolami, Self-Organizing Neural Networks, Independent Component Analysis and Blind Source Separation, Springer-Verlag, New York, 1999.
[82] M. Girolami, A variational method for learning sparse and overcomplete representations, Neural Comput. 13 (2001) 2517–2532.
[83] G.H. Golub, C.F. Van Loan, Matrix Computations, Johns Hopkins Press, Baltimore, 1989.
[84] S. Gould, The Mismeasure of Man, second ed., Norton, New York, 1981.
[85] R. Gribonval, R. Jenatton, F. Bach, M. Kleinsteuber, M. Seibert, Sample complexity of dictionary learning and other matrix factorizations, IEEE Trans. Inf. Theory 61 (6) (2015) 3469–3486.
[86] R. Gribonval, R. Jenatton, F. Bach, Sparse and spurious: dictionary learning with noise and outliers, IEEE Trans. Inf. Theory 61 (11) (2015) 6298–6319.
[87] D. Gross, Recovering low-rank matrices from few coefficients in any basis, IEEE Trans. Inf. Theory 57 (3) (2011) 1548–1566.
[88] J. Ham, D.D. Lee, S. Mika, B. Schölkopf, A kernel view of the dimensionality reduction of manifolds, in: Proceedings of the 21st International Conference on Machine Learning, Banff, Canada, 2004, pp. 369–376.
[89] D.R. Hardoon, S. Szedmak, J. Shawe-Taylor, Canonical correlation analysis: an overview with application to learning methods, Neural Comput. 16 (2004) 2639–2664.
[90] D.R. Hardoon, J. Shawe-Taylor, Sparse canonical correlation analysis, Mach. Learn. 83 (3) (2011) 331–353.
[91] S. Haykin, Neural Networks: A Comprehensive Foundation, second ed., Prentice Hall, Upper Saddle River, NJ, 1999.
[92] J. He, L. Balzano, J. Lui, Online robust subspace tracking from partial information, arXiv preprint, arXiv:1109.3827, 2011.
[93] J. Hérault, C. Jouten, B. Ans, Détection de grandeurs primitive dans un message composite par une architecture de calcul neuroimimétique en apprentissage non supervisé, in: Actes du Xème colloque GRETSI, Nice, France, 1985, pp. 1017–1022.

[94] N. Hjort, C. Holmes, P. Muller, S. Walker, Bayesian Nonparametrics, Cambridge University Press, Cambridge, 2010.
[95] H. Hotelling, Analysis of a complex of statistical variables into principal components, J. Educ. Psychol. 24 (1933) 417–441.
[96] H. Hotelling, Relations between two sets of variates, Biometrika 28 (34) (1936) 321–377.
[97] P.J. Huber, Projection pursuit, Ann. Stat. 13 (2) (1985) 435–475.
[98] M. Hubert, P.J. Rousseeuw, K. Vanden Branden, ROBPCA: a new approach to robust principal component analysis, Technometrics 47 (1) (2005) 64–79.
[99] A. Hyvärinen, Fast and robust fixed-point algorithms for independent component analysis, IEEE Trans. Neural Netw. 10 (3) (1999) 626–634.
[100] A. Hyvärinen, J. Karhunen, E. Oja, Independent Component Analysis, John Wiley, New York, 2001.
[101] X. He, P. Niyogi, Locally preserving projections, in: Proceedings Advances in Neural Information Processing Systems Conference, 2003.
[102] B.G. Huang, M. Ramesh, T. Berg, Labeled Faces in the Wild: A Database for Studying Face Recognition in Unconstrained Environments, Technical Report, No. 07-49, University of Massachusetts, Amherst, 2007.
[103] R. Jenssen, Kernel entropy component analysis, IEEE Trans. Pattern Anal. Mach. Intell. 32 (5) (2010) 847–860.
[104] J.E. Jackson, A User's Guide to Principal Components, John Wiley, New York, 1991.
[105] I. Jolliffe, Principal Component Analysis, Springer-Verlag, New York, 1986.
[106] M.C. Jones, R. Sibson, What is projection pursuit? J. R. Stat. Soc. A 150 (1987) 1–36.
[107] C. Jutten, J. Herault, Blind separation of sources, Part I: an adaptive algorithm based on neuromimetic architecture, Signal Process. 24 (1991) 1–10.
[108] C. Jutten, Source separation: from dusk till dawn, in: Proceedings 2nd International Workshop on Independent Component Analysis and Blind Source Separation, ICA'2000, Helsinki, Finland, 2000, pp. 15–26.
[109] J. Karhunen, J. Joutsensalo, Generalizations of principal component analysis, optimization problems, and neural networks, Neural Netw. 8 (4) (1995) 549–562.
[110] J. Kettenring, Canonical analysis of several sets of variables, Biometrika 58 (3) (1971) 433–451.
[111] M.E. Khan, M. Marlin, G. Bouchard, K.P. Murphy, Variational bounds for mixed-data factor analysis, in: J.D. Lafferty, C.K.I. Williams, J. Shawe-Taylor, R.S. Zemel, A. Culotta (Eds.), Neural Information Processing Systems, NIPS, Vancouver, Canada, 2010.
[112] P. Kidmose, D. Looney, M. Ungstrup, M.L. Rank, D.P. Mandic, A study of evoked potentials from ear-EEG, IEEE Trans. Biomed. Eng. 60 (10) (2013) 2824–2830.
[113] A. Klami, S. Virtanen, S. Kaski, Bayesian canonical correlation analysis, J. Mach. Learn. Res. 14 (2013) 965–1003.
[114] O.W. Kwon, T.W. Lee, Phoneme recognition using the ICA-based feature extraction and transformation, Signal Process. 84 (6) (2004) 1005–1021.
[115] S.-Y. Kung, K.I. Diamantaras, J.-S. Taur, Adaptive principal component extraction (APEX) and applications, IEEE Trans. Signal Process. 42 (5) (1994) 1202–1217.
[116] Y. Kopsinis, H. Georgiou, S. Theodoridis, fMRI unmixing via properly adjusted dictionary learning, in: Proceedings of the 20th European Signal Processing Conference, EUSIPCO, 2012, pp. 61–65.
[117] P.L. Lai, C. Fyfe, Kernel and nonlinear canonical correlation analysis, Int. J. Neural Syst. 10 (5) (2000) 365–377.
[118] S. Lafon, A.B. Lee, Diffusion maps and coarse-graining: a unified framework for dimensionality reduction, graph partitioning and data set parameterization, IEEE Trans. Pattern Anal. Mach. Intell. 28 (9) (2006) 1393–1403.
[119] L.D. Lathauer, Signal Processing by Multilinear Algebra, PhD Thesis, Faculty of Engineering, K.U. Leuven, Belgium, 1997.
[120] T.W. Lee, Independent Component Analysis: Theory and Applications, Kluwer, Boston, MA, 1998.
[121] M.H.C. Law, A.K. Jain, Incremental nonlinear dimensionality reduction by manifold learning, IEEE Trans. Pattern Anal. Mach. Intell. 28 (3) (2006) 377–391.
[122] D.D. Lee, S. Seung, Learning the parts of objects by nonnegative matrix factorization, Nature 401 (1999) 788–791.
[123] J.A. Lee, M. Verleysen, Nonlinear Dimensionality Reduction, Springer, New York, 2007.
[124] K. Lee, Y. Bresler, ADMiRA: atomic decomposition for minimum rank approximation, IEEE Trans. Inf. Theory 56 (9) (2010) 4402–4416.
[125] S. Lesage, R. Gribonval, F. Bimbot, L. Benaroya, Learning unions of orthonormal bases with thresholded singular value decomposition, in: IEEE International Conference on Acoustics, Speech and Signal Processing, 2005.
[126] M.S. Lewicki, T.J. Sejnowski, Learning overcomplete representations, Neural Comput. 12 (2000) 337–365.

[127] J. Liang, M. Zhang, X. Zeng, G. Yu, Distributed dictionary learning for sparse representation in sensor networks, IEEE Trans. Image Process. 23 (6) (2014) 2528–2541.
[128] Z. Liu, L. Vandenberghe, Interior-point method for nuclear norm approximation with application to system identification, SIAM J. Matrix Anal. Appl. 31 (3) (2010) 1235–1256.
[129] L. Li, W. Huang, I.-H. Gu, Q. Tian, Statistical modeling of complex backgrounds for foreground object detection, IEEE Trans. Image Process. 13 (11) (2004) 1459–1472.
[130] Z. Lin, A. Ganesh, J. Wright, L. Wu, M. Chen, Y. Ma, Fast convex optimization algorithms for exact recovery of a corrupted low-rank matrix, in: Intl. Workshop on Comp. Adv. in Multi-Sensor Adapt. Processing, Aruba, Dutch Antilles, 2009.
[131] M.A. Lindquist, The statistical analysis of fMRI data, Stat. Sci. 23 (4) (2008) 439–464.
[132] G. Mateos, G.B. Giannakis, Robust PCA as bilinear decomposition with outlier-sparsity regularization, IEEE Trans. Signal Process. 60 (2012) 5176–5190.
[133] M. Mesbahi, G.P. Papavassilopoulos, On the rank minimization problem over a positive semidefinite linear matrix inequality, IEEE Trans. Autom. Control 42 (2) (1997) 239–243.
[134] R. Motwani, P. Raghavan, Randomized Algorithms, Cambridge University Press, Cambridge, 1995.
[135] L. Mackey, Deflation methods for sparse PCA, in: D. Koller, D. Schuurmans, Y. Bengio, L. Bottou (Eds.), Advances in Neural Information Processing Systems, vol. 21, 2009, pp. 1017–1024.
[136] G. Mao, B. Fidan, B.D.O. Anderson, Wireless sensor network localization techniques, Comput. Netw. 51 (10) (2007) 2529–2553.
[137] M. Mardani, G. Mateos, G.B. Giannakis, Subspace learning and imputation for streaming big data matrices and tensors, arXiv preprint, arXiv:1404.4667, 2014.
[138] J. Mairal, M. Elad, G. Sapiro, Sparse representation for color image restoration, IEEE Trans. Image Process. 17 (1) (2008) 53–69.
[139] J. Mairal, F. Bach, J. Ponce, G. Sapiro, Online learning for matrix factorization and sparse coding, J. Mach. Learn. Res. 11 (2010).
[140] M. Morante, Y. Kopsinis, S. Theodoridis, Information assisted dictionary learning for fMRI data analysis, arXiv:1802.01334v2 [stat.ML], 11 May 2018.
[141] M. Novey, T. Adali, Complex ICA by negentropy maximization, IEEE Trans. Neural Netw. 19 (4) (2008) 596–609.
[142] M.A. Nicolaou, S. Zafeiriou, M. Pantic, A unified framework for probabilistic component analysis, in: European Conference Machine Learning and Principles and Practice of Knowledge Discovery in Databases, ECML/PKDD'14, Nancy, France, 2014.
[143] B.A. Olshausen, B.J. Field, Sparse coding with an overcomplete basis set: a strategy employed by v1, Vis. Res. 37 (1997) 3311–3325.
[144] P. Paatero, U. Tapper, R. Aalto, M. Kulmala, Matrix factorization methods for analysis diffusion battery data, J. Aerosol Sci. 22 (Supplement 1) (1991) 273–276.
[145] P. Paatero, U. Tapper, Positive matrix factor model with optimal utilization of error, Environmetrics 5 (1994) 111–126.
[146] K. Pearson, On lines and planes of closest fit to systems of points in space, Lond. Edinb. Dubl. Philos. Mag. J. Sci., Sixth Ser. 2 (1901) 559–572.
[147] A.T. Poulsen, S. Kamronn, L.C. Parra, L.K. Hansen, Bayesian correlated component analysis for inference of joint EEG activation, in: 4th International Workshop on Pattern Recognition in Neuroimaging, 2014.
[148] V. Perlbarg, G. Marrelec, Contribution of exploratory methods to the investigation of extended large-scale brain networks in functional MRI: methodologies, results, and challenges, Int. J. Biomed. Imaging 2008 (2008) 1–14.
[149] H. Raja, W. Bajwa, Cloud k-SVD: a collaborative dictionary learning algorithm for big distributed data, IEEE Trans. Signal Process. 64 (1) (2016) 173–188.
[150] H. Qui, E.R. Hancock, Clustering and embedding using commute times, IEEE Trans. Pattern Anal. Mach. Intell. 29 (11) (2007) 1873–1890.
[151] B. Recht, A simpler approach to matrix completion, J. Mach. Learn. Res. 12 (2011) 3413–3430.
[152] R. Rosipal, L.J. Trejo, Kernel partial least squares regression in reproducing kernel Hilbert spaces, J. Mach. Learn. Res. 2 (2001) 97–123.
[153] R. Rosipal, N. Krämer, Overview and recent advances in partial least squares, in: C. Saunders, M. Grobelnik, S. Gunn, J. Shawe-Taylor (Eds.), Subspace, Latent Structure and Feature Selection, Springer, New York, 2006.
[154] S. Roweis, EM algorithms for PCA and SPCA, in: M.I. Jordan, M.J. Kearns, S.A. Solla (Eds.), Advances in Neural Information Processing Systems, vol. 10, MIT Press, Cambridge, MA, 1998, pp. 626–632.

[155] S.T. Roweis, L.K. Saul, Nonlinear dimensionality reduction by locally linear embedding, Science 290 (2000) 2323–2326.
[156] D.B. Rubin, D.T. Thayer, EM algorithm for ML factor analysis, Psychometrika 47 (1) (1982) 69–76.
[157] C.A. Rencher, Multivariate Statistical Inference and Applications, John Wiley & Sons, New York, 2008.
[158] S. Sanei, Adaptive Processing of Brain Signals, John Wiley, New York, 2013.
[159] L.K. Saul, S.T. Roweis, An introduction to locally linear embedding, http://www.cs.toronto.edu/~roweis/lle/papers/lleintro.pdf.
[160] N. Sebro, T. Jaakola, Weighted low-rank approximations, in: Proceedings of the ICML Conference, 2003, pp. 720–727.
[161] B. Schölkopf, A. Smola, K.R. Muller, Nonlinear component analysis as a kernel eigenvalue problem, Neural Comput. 10 (1998) 1299–1319.
[162] F. Seidel, C. Hage, M. Kleinsteuber, pROST: a smoothed ℓ_p-norm robust online subspace tracking method for background subtraction in video, in: Machine Vision and Applications, 2013, pp. 1–14.
[163] F. Sha, L.K. Saul, Analysis and extension of spectral methods for nonlinear dimensionality reduction, in: Proceedings of the 22nd International Conference on Machine Learning, Bonn, Germany, 2005.
[164] Y. Shuicheng, D. Xu, B. Zhang, H.-J. Zhang, Q. Yang, S. Lin, Graph embedding and extensions: a general framework for dimensionality reduction, IEEE Trans. Pattern Anal. Mach. Intell. 29 (1) (2007) 40–51.
[165] M. Signoretto, R. Van de Plas, B. De Moor, J.A.K. Suykens, Tensor versus matrix completion: a comparison with application to spectral data, IEEE Signal Process. Lett. 18 (7) (2011) 403–406.
[166] P. Smaragdis, J.C. Brown, Nonnegative matrix factorization for polyphonic music transcription, in: Proceedings IEEE Workshop on Applications of Signal Processing to Audio and Acoustics, 2003.
[167] X. Su, T.M. Khoshgoftaar, A survey of collaborative filtering techniques, Adv. Artif. Intell. 2009 (2009) 1–19.
[168] S. Sra, I.S. Dhillon, Non-negative Matrix Approximation: Algorithms and Applications, Technical Report TR-06-27, University of Texas at Austin, 2006.
[169] C. Spearman, The proof and measurement of association between two things, Am. J. Psychol. 100 (3–4) (1987) 441–471 (republished).
[170] G.W. Stewart, An updating algorithm for subspace tracking, IEEE Trans. Signal Process. 40 (6) (1992) 1535–1541.
[171] A. Szymkowiak-Have, M.A. Girolami, J. Larsen, Clustering via kernel decomposition, IEEE Trans. Neural Netw. 17 (1) (2006) 256–264.
[172] J. Sun, S. Boyd, L. Xiao, P. Diaconis, The fastest mixing Markov process on a graph and a connection to a maximum variance unfolding problem, SIAM Rev. 48 (4) (2006) 681–699.
[173] J.B. Tenenbaum, V. De Silva, J.C. Langford, A global geometric framework for dimensionality reduction, Science 290 (2000) 2319–2323.
[174] S. Theodoridis, K. Koutroumbas, Pattern Recognition, fourth ed., Academic Press, Boston, 2009.
[175] M.E. Tipping, C.M. Bishop, Probabilistic principal component analysis, J. R. Stat. Soc. B 21 (3) (1999) 611–622.
[176] M.E. Tipping, C.M. Bishop, Mixtures probabilistic principal component analysis, Neural Comput. 11 (2) (1999) 443–482.
[177] K.C. Toh, S. Yun, An accelerated proximal gradient algorithm for nuclear norm regularized linear least squares problems, Pac. J. Optim. 6 (2010) 615–640.
[178] J.A. Tropp, Literature survey: nonnegative matrix factorization, unpublished note, http://www.personal.umich.edu/~jtropp/, 2003.
[179] C.G. Tsinos, A.S. Lalos, K. Berberidis, Sparse subspace tracking techniques for adaptive blind channel identification in OFDM systems, in: IEEE International Conference on Acoustics, Speech and Signal Processing, ICASSP, 2012, pp. 3185–3188.
[180] N. Ueda, R. Nakano, Z. Ghahramani, G.E. Hinton, SMEM algorithm for mixture models, Neural Comput. 12 (9) (2000) 2109–2128.
[181] G. Varoquaux, A. Gramfort, F. Pedregosa, V. Michel, B. Thirion, Multi-subject dictionary learning to segment an atlas of brain spontaneous activity, in: Information Processing in Medical Imaging, Springer, Berlin/Heidelberg, 2011.
[182] H.T. Wai, T.H. Chang, A. Scaglione, A consensus-based decentralized algorithm for non-convex optimization with application to DL, in: Proceedings International Conference on Acoustics, Speech and Signal Processing, ICASSP, 2015, pp. 3546–3550.
[183] A.E. Waters, A.C. Sankaranarayanan, R.G. Baraniuk, SpaRCS: recovering low-rank and sparse matrices from compressive measurements, in: Advances in Neural Information Processing Systems, NIPS, Granada, Spain, 2011.
[184] K.Q. Weinberger, L.K. Saul, Unsupervised learning of image manifolds by semidefinite programming, in: Proceedings of the IEEE Conference on Computer Vision and Pattern Recognition, vol. 2, Washington, DC, USA, 2004, pp. 988–995.

[185] J. Westerhuis, T. Kourti, J. MacGregor, Analysis of multiblock and hierarchical PCA and PLS models, J. Chemom. 12 (1998) 301–321.

[186] H. Wold, Nonlinear estimation by iterative least squares procedures, in: F. David (Ed.), Research Topics in Statistics, John Wiley, New York, 1966, pp. 411–444.

[187] J. Wright, Y. Peng, Y. Ma, A. Ganesh, S. Rao, Robust principal component analysis: exact recovery of corrupted low-rank matrices by convex optimization, in: Neural Information Processing Systems, NIPS, 2009.

[188] J. Wright, A. Ganesh, K. Min, Y. Ma, Compressive principal component pursuit, arXiv:1202.4596, 2012.

[189] D. Weenink, Canonical correlation analysis, in: Institute of Phonetic Sciences, University of Amsterdam, Proceedings, vol. 25, 2003, pp. 81–99.

[190] D.M. Witten, R. Tibshirani, T. Hastie, A penalized matrix decomposition, with applications to sparse principal components and canonical correlation analysis, Biostatistics 10 (3) (2009) 515–534.

[191] L. Wolf, T. Hassner, Y. Taigman, Effective unconstrained face recognition by combining multiple descriptors and learned background statistics, IEEE Trans. Pattern Anal. Mach. Intell. 33 (10) (2011) 1978–1990.

[192] W. Xu, X. Liu, Y. Gong, Document clustering based on nonnegative matrix factorization, in: Proceedings 26th Annual International ACM SIGIR Conference, ACM Press, New York, 2003, pp. 263–273.

[193] H. Xu, C. Caramanis, S. Sanghavi, Robust PCA via outlier pursuit, IEEE Trans. Inf. Theory 58 (5) (2012) 3047–3064.

[194] J. Ye, Generalized low rank approximation of matrices, in: Proceedings of the 21st International Conference on Machine Learning, Banff, Alberta, Canada, 2004, pp. 887–894.

[195] S.K. Yu, V. Yu, K.H.-P. Tresp, M. Wu, Supervised probabilistic principal component analysis, in: Proceedings International Conference on Knowledge Discovery and Data Mining, 2006.

[196] M. Yaghoobi, T. Blumensath, M.E. Davies, Dictionary learning for sparse approximations with the majorization method, IEEE Trans. Signal Process. 57 (6) (2009) 2178–2191.

[197] B. Yang, Projection approximation subspace tracking, IEEE Trans. Signal Process. 43 (1) (1995) 95–107.

[198] S. Zafeiriou, A. Tefas, I. Buciu, I. Pitas, Exploiting discriminant information in non-negative matrix factorization with application to frontal face verification, IEEE Trans. Neural Netw. 17 (3) (2006) 683–695.

[199] Z. Zhou, X. Li, J. Wright, E.J. Candès, Y. Ma, Stable principal component pursuit, in: Proceedings, IEEE International Symposium on Information Theory, 2010, pp. 1518–1522.

[200] H. Zou, T. Hastie, R. Tibshirani, Sparse principal component analysis, J. Comput. Graph. Stat. 15 (2) (2006) 265–286.

Index

D

E

M

S

V

W

Y